R. Nieuwenhuys J. Voogd C. van Huijzen

The Human Central Nervous System

R. Nieuwenhuys J. Voogd
C. van Huijzen

The Human Central Nervous System

Fourth Edition

With 391 Figures

Springer

Rudolf Nieuwenhuys M.D., Ph.D.
Professor emeritus of Neuroanatomy
The Netherlands Institute
for Neuroscience
Meibergdreef 47
1105 BA Amsterdam
The Netherlands
s.bakker@nki.nl

(home)
Papehof 25
1391 BD Abcoude
The Netherlands

Jan Voogd M.D., Ph.D.
Professor emeritus of Anatomy
Department of Neuroscience
Erasmus University Rotterdam
P.O. Box 2040
3000 CA Rotterdam
The Netherlands
janvoogd@bart.nl

(home)
Rhijngeesterstraatweg 1
2342 AN Oegstgeest
The Netherlands

Christiaan van Huijzen F.M.A.A.
Medical Artist (retired)

(home)
Willem Degenstraat 2
6525 BW Nijmegen
The Netherlands

ISBN 978-3-662-52682-8 Springer-Verlag Berlin Heidelberg New York
DOI 10.1007/978-3-540-34686-9

ISBN 3-662-52682-8 3. Auflage Springer-Verlag Berlin Heidelberg New York

Springer is a part of Springer Science+Business Media

springer.com

Softcover re-print of the Hardcover 4th edition 2008

Production Manager: K. Schwind
Medical Editor: Dr. Maria Magdalene Nabbe
Typesetting: K+V Fotosatz GmbH, Beerfelden
Cover Design: WMX Design GmbH, Heidelberg

SPIN 11768005 80/7231 – 5 4 3 2 1 0 Printed on acid-free paper

Preface to the Fourth Edition

The present edition of *The Human Central Nervous System* differs considerably from its predecessors. In previous editions, the text was essentially confined to a section dealing with the various functional systems of the brain. This section, which has been rewritten and updated, is now preceded by 15 newly written chapters, which introduce the pictorial material of the gross anatomy, the blood vessels and meninges and the microstructure of its various parts and deal with the development, topography and functional anatomy of the spinal cord, the brain stem and the cerebellum, the diencephalon and the telencephalon. Great pains have been taken to cover the most recent concepts and data. As suggested by the front cover, there is a focus on the evolutionary development of the human brain. Throughout the text numerous correlations with neuropathology and clinical neurology have been made.

After much thought, we decided to replace the full Latin terminology, cherished in all previous editions, with English and Anglicized Latin terms. It has been an emotional farewell from beautiful terms such as *decussatio hipposideriformis Wernekinkii* and *pontes grisei caudatolenticulares*. Not only the text, but also the pictorial material has been extended and brought into harmony with the present state of knowledge. More than 230 new illustrations have been added and many others have been revised. The number of macroscopical sections through the brain has been extended considerably. Together, these illustrations now comprise a complete and convenient atlas for interpreting neuroimaging studies.

The bibliography increased from 272 references in the first and second editions, via 1553 references in the third edition, to almost 4000 references in the present one. Therefore, the book has grown again considerably; however, its aim – to provide a straightforward, clear and reliable guide to the structural and functional organization of the human central nervous system for both apprentices and specialists in the neurological sciences – has not changed.

During the preparation of this and previous editions we received advice and help from many persons; hence, our gratitude is manifold. Professor Luis Puelles and Dr. Iain H.M. Smart were so kind as to critically read the chapter on the development of the central nervous system. Professor Jean Büttner-Ennever read and offered valuable advice for changes in the chapter on the visual system. The many consultations we had with Professor Harry B.M. Uylings on the structure and subdivision of the cerebral cortex were also of considerable help to us. Professor Jaak Duysens gave valuable suggestions for the section on Locomotion of Chapter 21, Motor systems.

We would also like to acknowledge the artists: Mr. Wil P.M. Maas, who prepared numerous line drawings, and Mr. Gerben van der Meulen, who gave many figures their final shape and also aided in labelling them. Mrs. Jenneke Kruisbrink, librarian of the Netherlands Institute of Neuroscience assisted in retrieving the literature. The secretarial assistance afforded by Mrs. Inge Eijkhout and the reference management of Mrs. Irene S. Benne are especially acknowledged.

In concluding this preamble, the invaluable moral and practical support of Mrs. Suzanne Bakker M.Sc. in all phases of this project is gratefully acknowledged and, finally, we extend our most sincere thanks to our publisher, Springer-Verlag, and their staff - especially Dr. Maria Magdalene Nabbe, Mrs. Sherryl Sundell, and Mr. Klemens Schwind, for their kind help during the preparation of this book.

October 2007

Rudolf Nieuwenhuys
Jan Voogd
Chris van Huijzen

Contents

Section I Orientation, Development, Gross Anatomy, Blood Supply and Meninges

1 Orientation

The human central nervous system or neuraxis consists of the brain (encephalon) and the spinal cord (medulla spinalis). The brain is encased by the skull; the spinal cord lies within the spinal canal, extending from the foramen magnum to the level of the second lumbar vertebra. Globally, the brain can be subdivided into the cerebrum, the cerebellum and the truncus cerebri or brain stem (Fig. 1.1). On the basis of its ontogeny the brain can be divided into three main parts, the rhombencephalon or hindbrain, the mesencephalon or midbrain and the prosencephalon or forebrain (Fig. 1.2). The rhombencephalon classically includes the medulla oblongata or myelencephalon and the metencephalon. The medulla oblongata represents an expanded rostral extension of the spinal cord. The metencephalon forms a large ventral protrusion, known as the pons, whereas its much expanded dorsal portion is differentiated into the cerebellum. The isthmus is the rostralmost portion of the hindbrain [1], bordering on the midbrain. The mesencephalon is a short wedge-shaped segment of the brain between the rhombencephalon and the prosencephalon. Together, the medulla oblongata, pons and mesencephalon form the brain stem.

The prosencephalon comprises the diencephalon, or betweenbrain and the telencephalon, or endbrain. The diencephalon contains two important nuclear complexes, the large thalamus and the much smaller, somewhat funnel-shaped hypothalamus. The telencephalon is mainly composed of the two cerebral hemispheres, but also comprises the small telencephalon impar, which is caudally directly continuous with the diencephalon (Fig. 1.2). The cerebral hemispheres form by far the largest parts of the human brain: They surround and cover both the diencephalon and the midbrain. The surface of the cerebral hemispheres is thrown into a number of gyri or convolutions, separated by sulci or fissures.

The central nervous system develops from a tube-like anlage (Chap. 2), in which the primordial neural walls surround a continuous, fluid-filled ventricular cavity. This cavity persists as a system of communicating ventricles in the adult brain (Fig. 1.3). It comprises the diamond-shaped, rhombencephalic, fourth ventricle; the narrow, slit-like diencephalic third ventricle; and the large, crescent-shaped, lateral ventricles in the cerebral hemispheres. On both sides the lateral ventricle communicates with the third ventricle via the interventricular foramen. The third and fourth ventricles are interconnected by the narrow, mesencephalic cerebral aqueduct. At the spinal level, the ventricular system is reduced to the thin central canal.

The cerebral hemispheres are subdivided into five main parts or lobes: lobus frontalis, lobus parietalis, lobus temporalis, lobus occipitalis and lobus limbicus (Fig. 1.4). Most, though not all, of the boundaries between these lobes coincide with sulci or fissures.

The weight of the brain in normal young adults ranges from 1050 g to 1800 g [2]. It is positively correlated with the body weight. The mean weight of the male brain is 1350 g and that of the female is 1250 g. The spinal cord weighs about 35 g.

References

1. His W (1893) Vorschläge zur Eintheilung des Gehirns. Arch Anat Physiol Anat Abt 172–180
2. Roland PE, Zilles K (1994) Brain atlases: a new research tool. Trends Neurosci 17:458–467

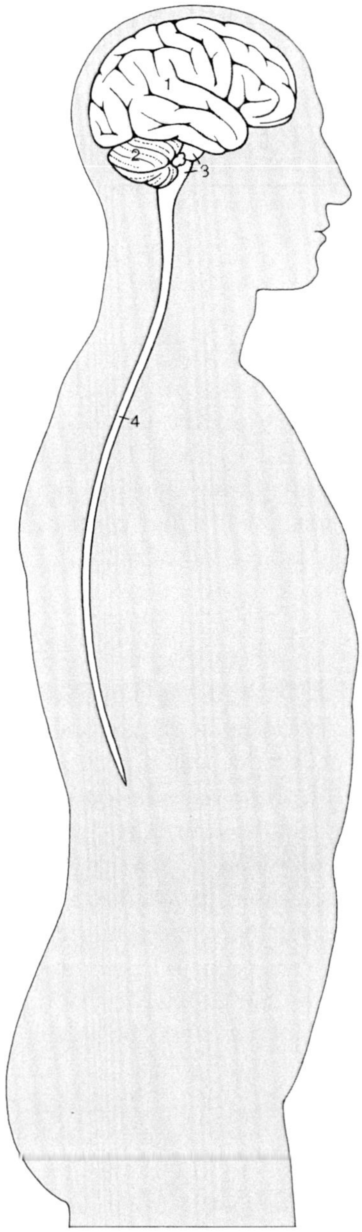

1 Cerebrum
2 Cerebellum
3 Truncus cerebri
4 Medulla spinalis

Fig. 1.1. The central nervous system in situ (1/6×)

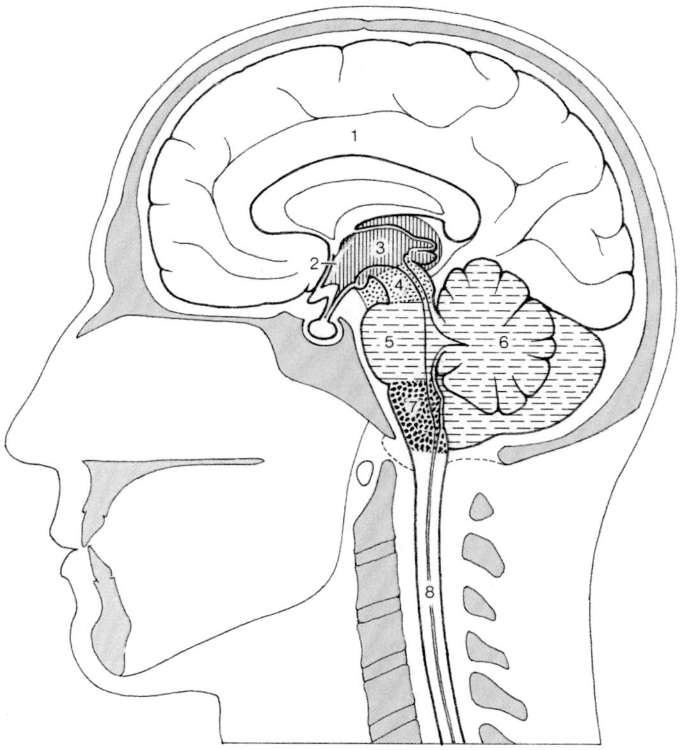

1 Telecephalon (Cerebrum)	Prosencephalon		Encephalon	Neuraxis (systema nervosum centrale)
2 Telencephalon impar				
3 Diencephalon				
4 Mesencephalon		Truncus cerebri		
5 Pons, 6 Cerebellum } Metencephalon	Rhombencephalon			
7 Myelencephalon (Medulla oblongata)				
8 Medulla spinalis				

Fig. 1.2. Medial surface of the right half of the brain in the bisected head indicating the position of its major subdivisions (2/5×)

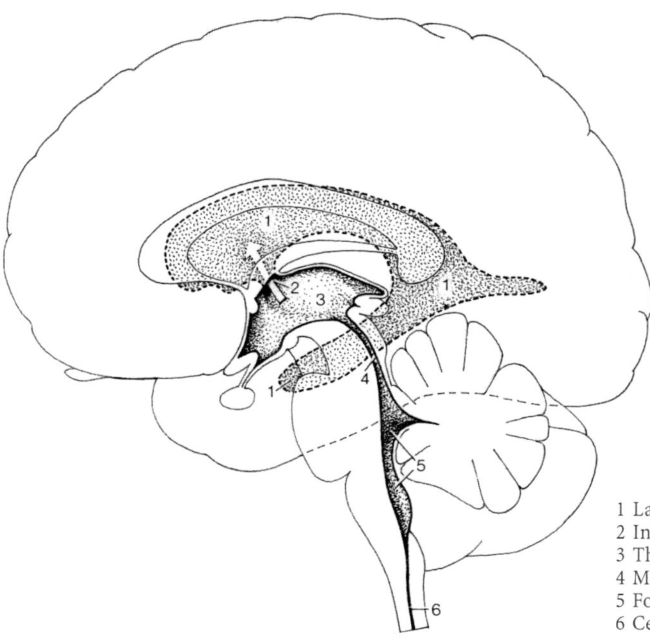

1 Lateral ventricle
2 Interventricular foramen
3 Third ventricle
4 Mesencephalic aqueduct
5 Fourth ventricle
6 Central canal

Fig. 1.3. The ventricular system of the brain. The *arrow* passes through the interventricular foramen from the third ventricle to the lateral ventricle (3/5×)

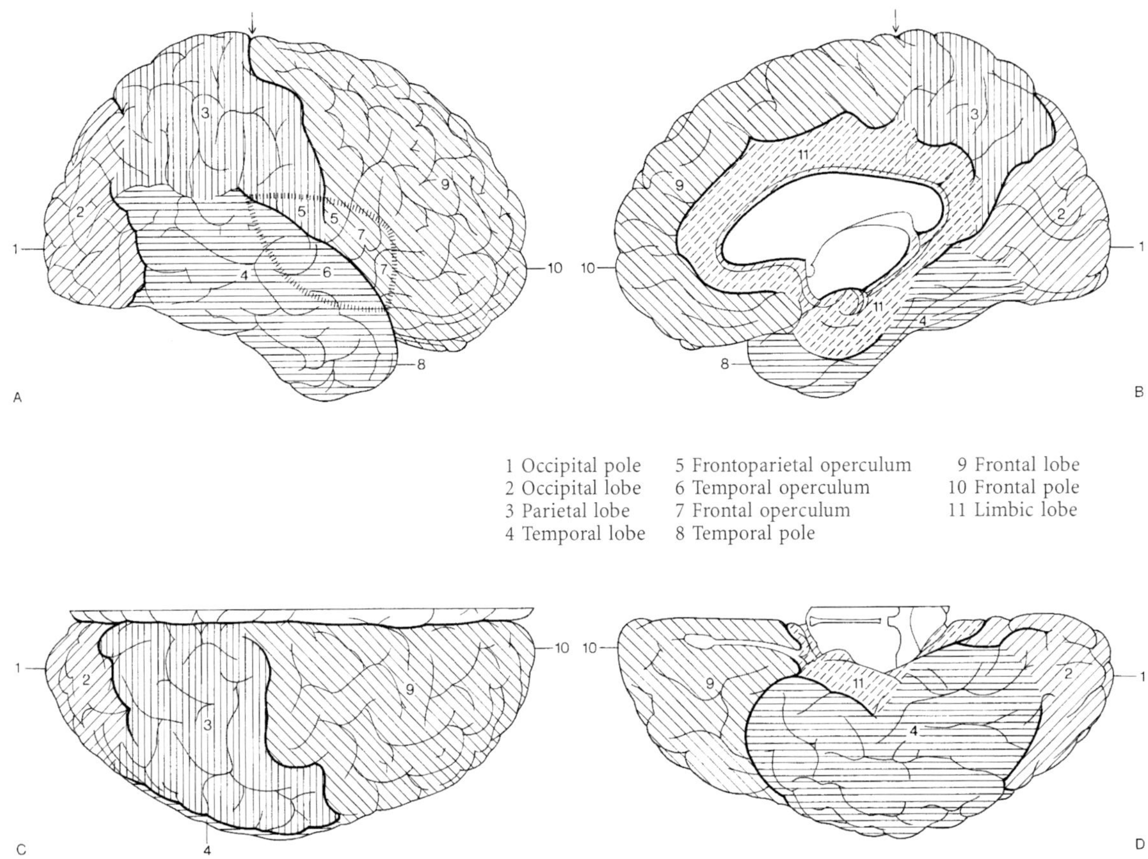

Fig. 1.4 A–D. Subdivisions of the right cerebral hemisphere into lobes. **A** Lateral view; **B** medial view; **C** superior view; **D** inferior view (1/2×)

2 Development

Early Morphogenesis

The central nervous system in humans and other vertebrates is derived from the neural plate, a thickened and elongated paramedian zone of the external germ layer or ectoderm (Fig. 2.1 A, E). The ectoderm along the lateral edges of the neural plate forms bilaterally a band-like strip, the primordial neural crest, which separates the primordial neural ectoderm from the primordial general body ectoderm. As the neural plate grows, its lateral edges rise to form the neural folds, whereas its midline region is depressed to form the neural groove (Fig. 2.1 B, F). During further development the neural groove deepens and the neural folds meet dorsally and eventually fuse to form the neural tube. It should be noted that, as the edges of the neural groove approach each other, they carry with them the adjoining primordial general body ectoderm, and that, when the process of fusion occurs, not only the neural ectoderm but also the body ectoderm fuses in the median plane. As a result of this double fusion, the neural ectoderm detaches completely from the body ectoderm to form a submerged neural tube. With the closure of the neural tube, the cells of the bilateral primordia of the neural crest separate off and move into the space between the dorsal part of the neural tube and the overlying ectoderm (Fig. 2.1 C, G). A proportion of these neural crest cells aggregates to form the sensory ganglia of cranial nerves and spinal dorsal roots.

The closure of the neural groove is initiated at multiple sites at the levels of the primordia of the forebrain, midbrain and cervical spinal cord, and then proceeds rostrally and caudally [137]. The temporary openings at either end of the closing neural tube are known as the rostral and caudal neuropores (Fig. 2.1 D). The rostral neuropore closes $\approx$ 30 days after fertilization; the caudal neuropore closes 1 day later [143].

The transformation of the neural plate into the neural tube is called primary neurulation. It is known that the most caudal part of the spinal cord develops from a solid cord of cells, which is transformed into a hollow tube by cavitation [134, 184]. The process by which a hollow neural structure is formed without direct involvement of the ectoderm and without the intermediate phase of a neural plate is known as secondary neurulation [143].

The early neural tube can be divided into a *floor plate*, a *roof plate* and *lateral plates* on both sides, which together enclose a slit-like, fluid-filled ventricular cavity (Fig. 2.1 H). The floor plate and roof plate are thin and consist of a single layer of epithelial cells, but the lat-

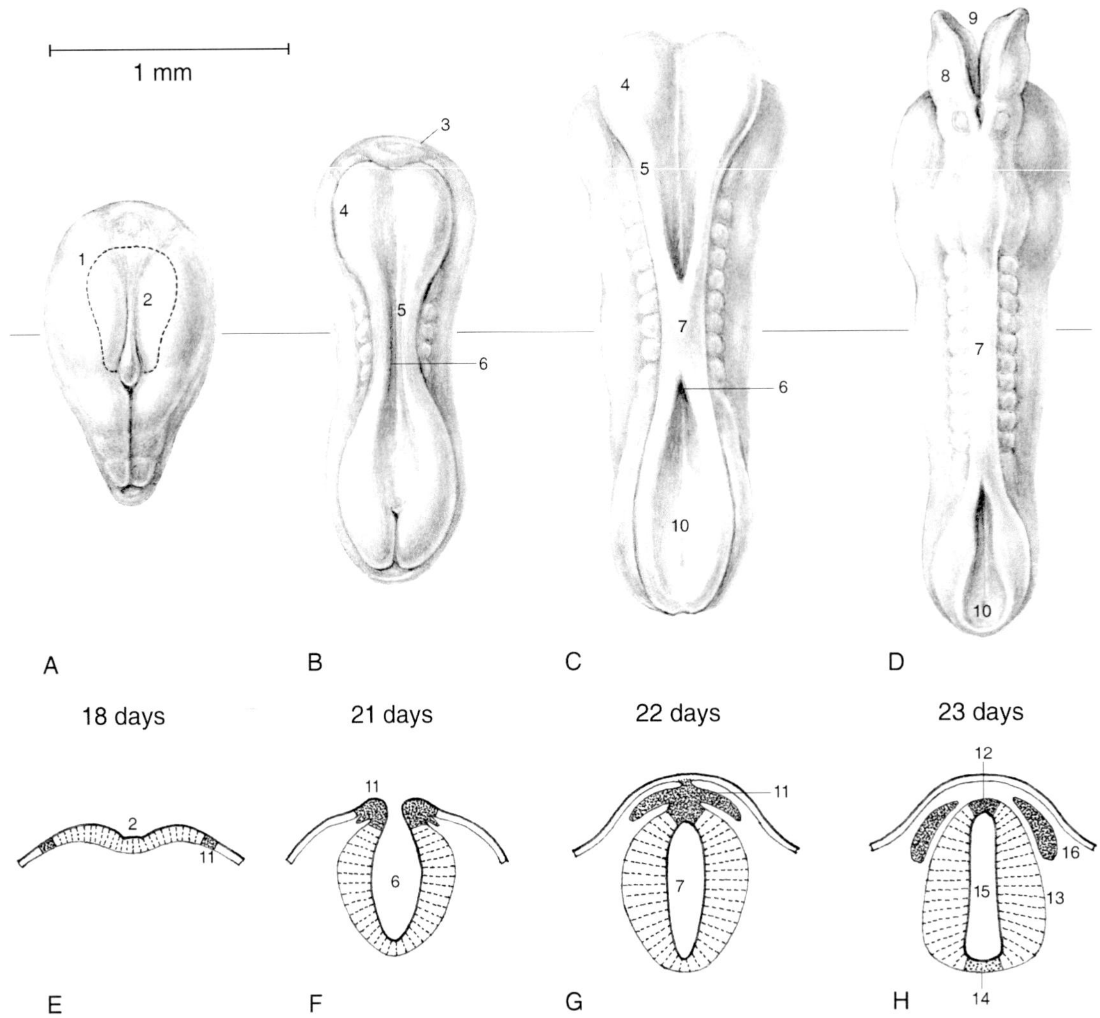

1 Ectoderm
2 Neural plate
3 Cut edge of amnion
4 Brain plate
5 Neural fold
6 Neural groove
7 Neural tube
8 Brain
9 Rostral neuropore
10 Caudal neuropore
11 Neural crest
12 Roof plate
13 Lateral plate
14 Floor plate
15 Ventricular cavity
16 Spinal ganglion

Fig. 2.1 A–D. Dorsal aspects of reconstructions of human embryos. **A** Presomite stage; the approximate extension of the neural plate is marked by a *dashed line;* **B** early somite and neural groove stage; **C** seven-somite embryo; **D** ten-somite embryo. Adapted from Noback and Demarest [139]. **E–H** Schematic transverse sections through the anlagen of the CNS of the embryos depicted in **A–D**. The *horizontal line* indicates the level of these sections

eral plates soon thicken, and from these structures the neurally differentiated parts of brain and spinal cord arise. A longitudinal ventricular groove develops along most of the extent of the neural tube. This groove, the *sulcus limitans of His* [66, 67], divides the lateral plates on each side into a ventral *basal plate* and a dorsal *alar plate*. This separation indicates a fundamental functional difference because the primary sensory centres will develop in the alar plate, but the primary motor centres in the basal plate. Incidentally, in the *Nomina Embryologica* associated with the *Nomina Anatomica* (fourth edition) by the International Anatomical Nomenclature Committee (Tokyo 1977), the terms floor plate, basal plate, alar plate and roof plate have been changed to ventral lamina, ventrolateral lamina, dorsolateral lamina and dorsal lamina, respectively. However, we do not regard these modifications as improvements and will continue to use the old terms.

Formation of the Brain Regions

According to the classical description of von Kupffer [215], which has since been paraphrased in numerous textbooks, the following processes lead to the differentiation of the major divisions of the brain:

1. Early in development the rostral part of the *neural plate* becomes wider than the caudal part. The former gives rise to the brain whereas the spinal cord develops from the latter.

2. Prior to closure of the anterior neuropore, the rostral part of the primordial brain shows a transversely oriented fold, the *plica encephali ventralis*. This fold, situated at the level of the rostral end of the notochord, is considered to mark the boundary between two principal regions, the *archencephalon* and the *deuterencephalon*. The position of the archencephalon is typically prechordal, whereas the deuterencephalon and its caudal continuation, the spinal cord, arise dorsal to the notochord.

3. Soon after formation of the neural tube, its rostral, primordial brain part shows three rostrocaudally arranged dilatations. These so-called *primary brain vesicles* are known as the primary forebrain or *prosencephalon*, the midbrain or *mesencephalon* and the primary hindbrain or *rhombencephalon*. The prosencephalon develops from the archencephalon, whereas the mesencephalon and the rhombencephalon are derivatives of the deuterencephalon. The boundaries between the three primary brain vesicles are marked not only by constrictions, but also by early-developing, transversely oriented fibre bundles, some of them commissures. Thus, the *posterior commissure* develops dorsally in the pretectal part of the caudal diencephalon and abuts the mesencephalon caudally. The narrowing between the mesencephalon and the rhombencephalon correlates with the dorsal crossing and origin of the trochlear nerve; this area is known as the isthmus, or *isthmus rhombencephali*. Most authors use these terms to denote just the constriction, but others, such as His [66], or, more recently, Puelles [157], regard the area surrounding this constriction as a brain segment in its own right. Before the neural tube is entirely closed up, small bulges, the optic vesicles, extend laterally from each side of the developing forebrain (Figs. 2.2A, 2.3A).

4. Ultimately, the prosencephalon and the rhombencephalon each differentiate into two parts. The rhombencephalon becomes subdivided into a rostral *metencephalon* and a caudal *myelencephalon*, the latter being continuous with the spinal cord. The prosencephalon becomes divided into the rostral *telencephalon* or endbrain and the caudal *diencephalon* or betweenbrain. These subdivisions and the midbrain and the spinal cord, which remain undivided, make up the six major regions of the human CNS. The boundary between the telencephalon and the diencephalon is usually defined as a plane passing through either the anterior commissure or the decussation of optic fibres and the velum transversum. The latter is a ventrally directed fold of the membranous prosencephalic roof, which is present in all vertebrates (Fig. 2.5C) [215]. Telencephalon,

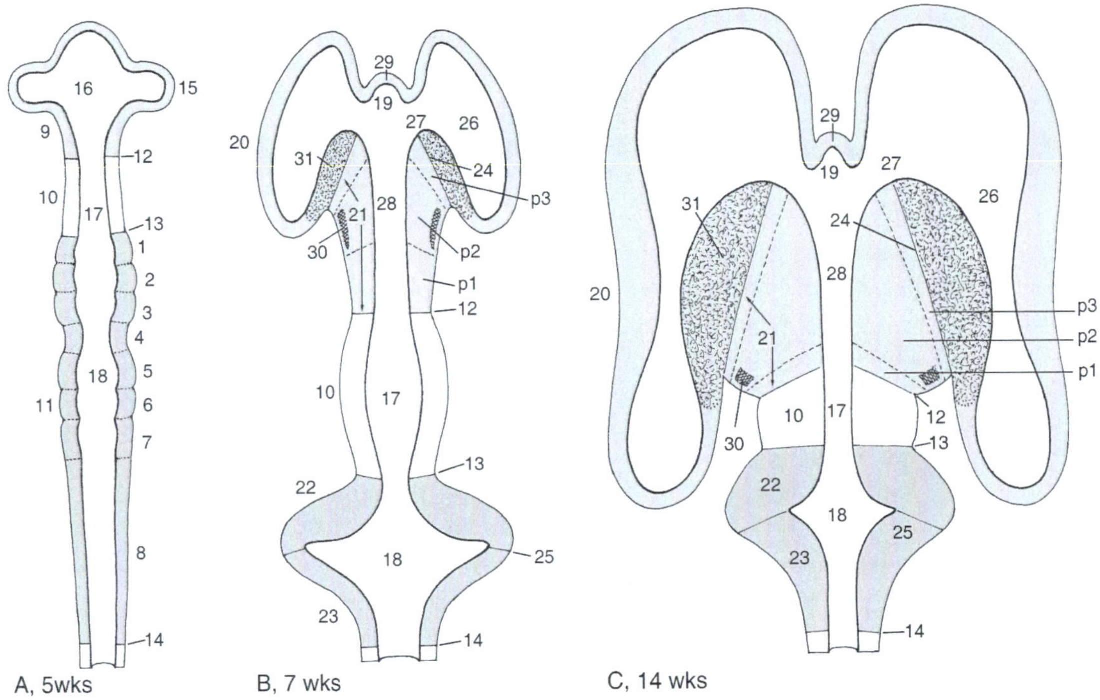

1–8 Rhombomeres 1–8
9 Prosencephalon (in *red*)
10 Mesencephalon
11 Rhombencephalon (in grey)
12 Prosencephalomesencephalic border
13 Mesencephalorhombencephalic border
14 Rhombencephalospinal border
15 Optic vesicle
16 Prosencephalic ventricle
17 Mesencephalic ventricle (mesencephalic aqueduct)
18 Rhombencephalic ventricle
19 Telencephalon impar
20 Cerebral hemisphere
21 Diencephalon
22 Metencephalon
23 Myelencephalon
24 Telencephalodiencephalic border
25 Metencephalomyelencephalic border
26 Lateral ventricle
27 Interventricular foramen (Monro)
28 Third ventricle
29 Lamina terminalis
30 Primordium of dorsal part of lateral geniculate nucleus (*hatched*)
31 Primordia of basal ganglia and amygdaloid complex (*dotted*)
p1–3 Prosomeres

Fig. 2.2 A–C. Diagrammatic horizontal sections through the developing human brain at 5 weeks (**A**), 7 weeks (**B**) and 14 weeks (**C**). Partly based on [83, 188]

diencephalon, mesencephalon, metencephalon and myelencephalon are commonly denoted as the *secondary brain vesicles*. Thus, the classical view on the development of the brain in humans (and other vertebrates) can be epitomized as follows: neural plate → neural groove → neural tube → formation of three primary brain vesicles → formation of five secondary brain vesicles. It should be emphasized that this classical scheme, insofar as the development of the human brain is concerned, is incorrect at several points.

1. Careful reconstructions based on serial sections of early human embryos have revealed that the brain begins to differentiate well before closure of the neural tube and manifests itself in the as yet completely unfused neural folds, as a series of rostrocaudally arranged segment-like structures. Thus, Bartelmez [14] and Bartelmez and Evans [15] observed that in two-somite embryos (approximate age: 26 days), there are five enlargements of the neural folds: forebrain, midbrain and three hindbrain segments. During further development these primary segments appeared to subdivide to form the definitive neuromeres. Just before closure of the rostral neuropore 11 neuromeres were counted. These observations have been largely confirmed by O'Rahilly and Müller [143]. In the folds of the completely open neural groove of a 26-day embryo they distinguished six primary neuromeres: prosencephalon, mesencephalon and four rhombomeres A, B, C and D. In the following days the number of neuromeres increased steadily, there being 8 at 29 days, 12 at 30 days, 13 at 31 and 15 at 32 days, until day 33, i.e. about 3 days after the closure of the rostral neuropore, when the full complement of 16 (secondary) neuromeres is attained (Table 2.1). During further development the neuromeres, or rather the interneuromeric constrictions, fade rapidly. In 35-day embryos, the diencephalic neuromeres can only be vaguely distinguished and in 40-day embryos the rhombomeres are no longer recognizable as separate entities. (As will be discussed in detail, the neuromeres, in spite of their rela-

Table 2.1. Development of neuromeres in the human embryo, according to O'Rahilly and Müller [143]

Approximate age (days)					
26	**29**	**30**	**31**	**32**	**33**
	T	T	T	T	T
	D1	D1	D1	D1	D1
P					Par r
				Par	
	D2	D2	D2		Par c
				Syn	Syn
			M1	M1	M1
M	M	M	M2	M2	M2
				Isth	Isth
		Rh 1	Rh 1	Rh 1	Rh 1
Rh A	Rh A	Rh 2	Rh 2	Rh 2	Rh 2
		Rh 3	Rh 3	Rh 3	Rh 3
Rh B	Rh B	Rh 4	Rh 4	Rh 4	Rh 4
		Rh 5	Rh 5	Rh 5	Rh 5
Rh C	Rh C	Rh 6	Rh 6	Rh 6	Rh 6
		Rh 7	Rh 7	Rh 7	Rh 7
Rh D	Rh D	Rh 8	Rh 8	Rh 8	Rh 8

D1, D2, diencephalic neuromeres; Isth, isthmus neuromere; M, mesencephalon; M1, M2, mesencephalic neuromeres; P, prosencephalon; Par, parencephalon; Par c, r, parencephalon caudalis, rostralis; Rh A to D, primary rhombomeres; Rh 1 to 8, secondary rhombomeres; Syn, synencephalon

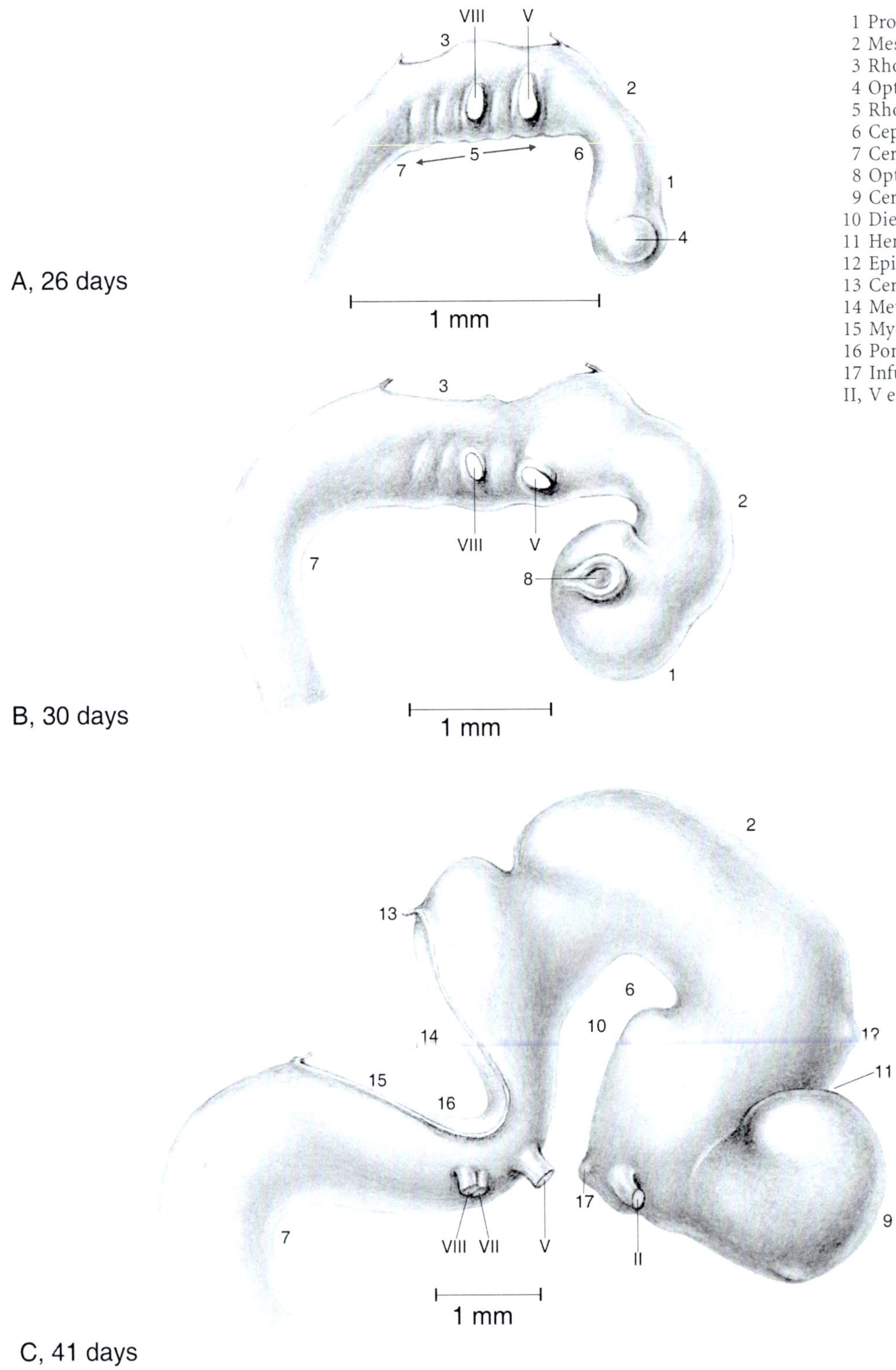

1 Prosencephalon
2 Mesencephalon
3 Rhombencephalon
4 Optic vesicle
5 Rhombomeres
6 Cephalic flexure
7 Cervical flexure
8 Optic cup
9 Cerebral hemisphere
10 Diencephalon
11 Hemispheric sulcus
12 Epiphysis
13 Cerebellum
14 Metencephalon
15 Myelencephalon
16 Pontine flexure
17 Infundibulum
II, V etc. Cranial nerves

Fig. 2.3 A–C. Lateral views of the brains of human embryos at 26 days (**A**), 30 days (**B**) and 41 days (**C**). Based on [69]

1 Olfactory bulb
2 Cerebral hemisphere
3 Epiphysis
4 Mesencephalon
5 Cerebellum
6 Metencephalon
7 Roof of fourth ventricle (cut edge)
8 Pontine flexure
9 Myelencephalon
10 Hypothalamus
11 Adenohypophysis
12 Neurohypophysis
13 External cerebellar swelling
14 Internal cerebellar swelling
15 Isthmus
16 Mamillary pouch
II, V etc. Cranial nerves

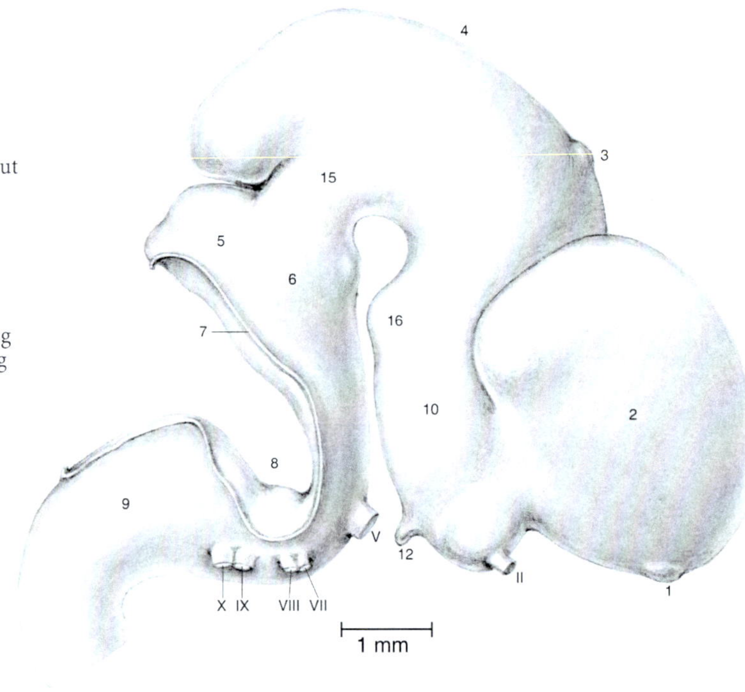

A, 6 wks

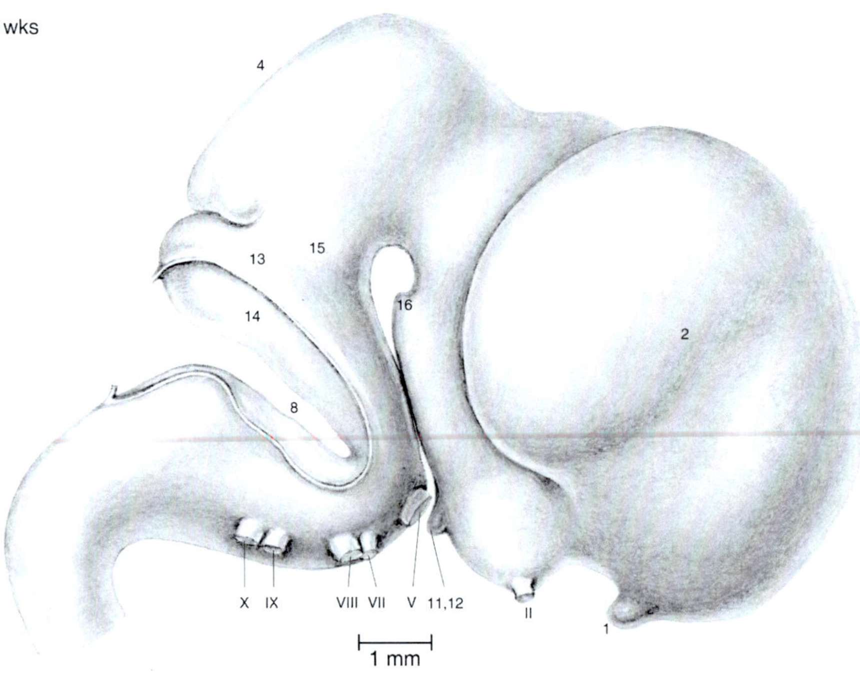

B, 8 wks

Fig. 2.4 A, B. Lateral views of the brains of human embryos at 6 weeks (A) and 8 weeks (B). Based on [69]

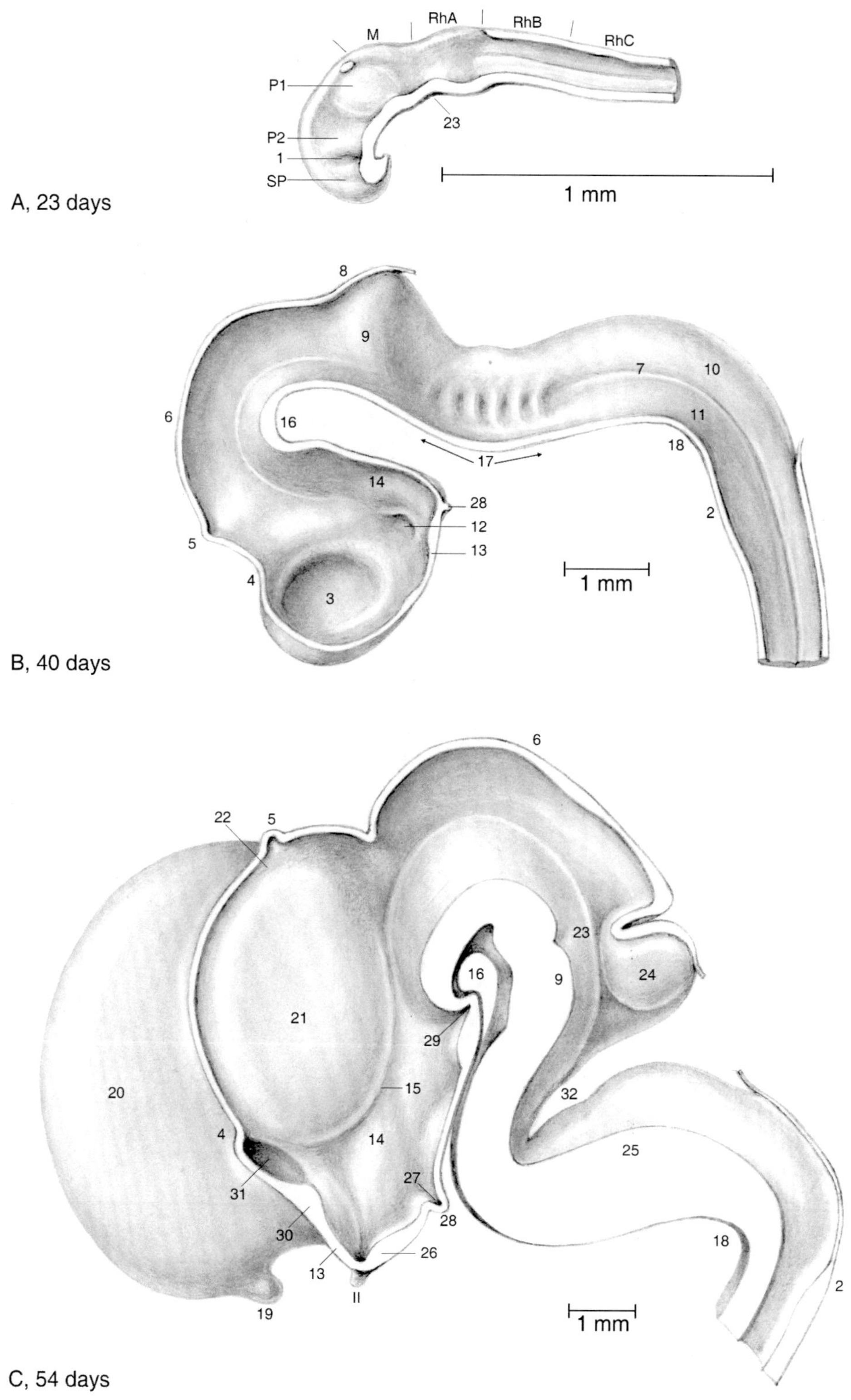

Fig. 2.5. For legends see opposite page

M Mesencephalic neuromere
P1,2 Prosomeres
Rh A, B etc. Rhombencephalic neuromeres
SP Secondary prosencephalon
1 Optic evagination
2 Spinomedullary junction
3 Lateral ventricle
4 Velum transversum
5 Epiphysis
6 Mesencephalon
7 Sulcus limitans
8 Cerebellum
9 Metencephalon
10 Alar lamina
11 Basal lamina
12 Lumen of optic stalk
13 Lamina terminalis
14 Hypothalamus
15 Hypothalamic sulcus
16 Cephalic flexure
17 Rhombomeres
18 Cervical flexure
19 Olfactory bulb
20 Cerebral hemisphere
21 Thalamus
22 Epithalamus
23 Isthmus
24 Internal cerebellar swelling
25 Myelencephalon
26 Optic chiasm
27 Infundibulum
28 Neurohypophysis
29 Mamillary recess
30 Anterior commissure
31 Interventricular foramen
32 Pontine flexure

Fig. 2.5 A–C. Medial views of the right halves of the bisected brains of human embryos at 23 days (**A**), 40 days (**B**) and 54 days (**C**). **A** is modified from [143]; **B** and **C** are based on [65]

tively brief presence as overt structures, are nevertheless considered fundamental morphological units.)

2. After closure of the rostral neuropore, the prosencephalon, mesencephalon and rhombencephalon do not present themselves as three rostrocaudally arranged brain vesicles. External views of the early neural tube (Fig. 2.3 A, B) show at most a swelling of the prosencephalon that is connected to that of the rhombencephalon by the angulated and relatively narrow mesencephalon [142].

3. The *five brain vesicles* were first described by von Baer [214] in the developing avian brain. His [66] considered this subdivision applicable to the human brain, with the reservation already mentioned that he interpreted the isthmus rhombencephali, situated between the mesencephalon and the metencephalon, as a separate entity. It should be emphasized, however, that of these five secondary "brain vesicles", only the telencephalon manifests itself as a pair of true vesicles (Figs. 2.2 B, C, 2.3 C, 2.4).

As a result of unequal growth of its different regions, three flexures appear in the developing brain (Figs. 2.3–2.5). The *cephalic flexure*, which is associated with the formation of the plica encephali ventralis, becomes manifest before the closure of the neural tube. The *cervical flexure*, which like the cephalic flexure is concave ventrally, appears at the junction of the hindbrain and spinal cord. The *pontine flexure*, which differs from the other two in that its convexity is directed ventrally, manifests itself in the middle of the rhombencephalon. It eventually attains such a depth that the morphologically dorsal sides of the rhombencephalic parts situated in front of and behind the flexure approach each other (Fig. 2.4). At these developmental stages the pontine flexure may be considered as approximately marking the boundary between the metencephalon and the myelencephalon. The *cerebellum* develops from the rostrodorsal part of the metencephalon (isthmus and rhombomere 1), whereas the *pons* is formed in its ventral part (rhombomeres 2–3). During the fetal period the pontine and cervical flexures gradually decrease in depth (Fig. 2.6).

Histogenesis

The neural plate and early neural tube are formed by a single layer of columnar cells, the neuroepithelium (Fig. 2.7: 1). As this layer thickens it gradually acquires the configuration of a pseudostratified epithelium; that is to say, its nuclei become arranged in more and more layers, but all elements remain in contact with the external and internal surface. Mitotic figures are found exclusively along the ventricular surface (Fig. 2.7: 2–7).

The wall of the early neural tube consists of a single type of epithelial cell in various stages of the mitotic cycle. The resting cells reside in the superficial part of the wall. The nuclei of the elements that are going to divide are translocated toward the ventricular surface. At the end of this migration phase, the peripheral process of the cell loses contact with the external surface and retracts; the cell then rounds up and divides into two daughter cells. Each daughter cell produces a new peripheral process and their nuclei move away from the ventricle (Fig. 2.8).

At a certain developmental stage the nuclei of the elongated neuroepithelial cells withdraw from the most superficial zone of the neural tube and the wall divides into an outer anuclear margin and an inner nuclear zone (Fig. 2.7: 3–4). The outer zone, or *marginal layer*, consists exclusively of the external cytoplasmic processes of the neuroepithelial cells for a certain period, but it is soon invaded by the axonal processes of maturing neuroblasts. The inner zone is termed the *matrix layer* or ventricular zone [45, 46, 80]. It contains the densely crowded nuclei of a morphologically homogeneous cell population, all elements of which participate in the proliferation process. The matrix cells are the precursors of all neuronal and macroglial elements of the CNS.

The matrix layer can be subdivided into three zones, the M or mitotic, the I or intermediate and the S or synthetic zone (Fig. 2.8). Fujita [45, 46] characterized the translocation of the nuclei of the matrix cells during a generation cycle as an elevator movement. He

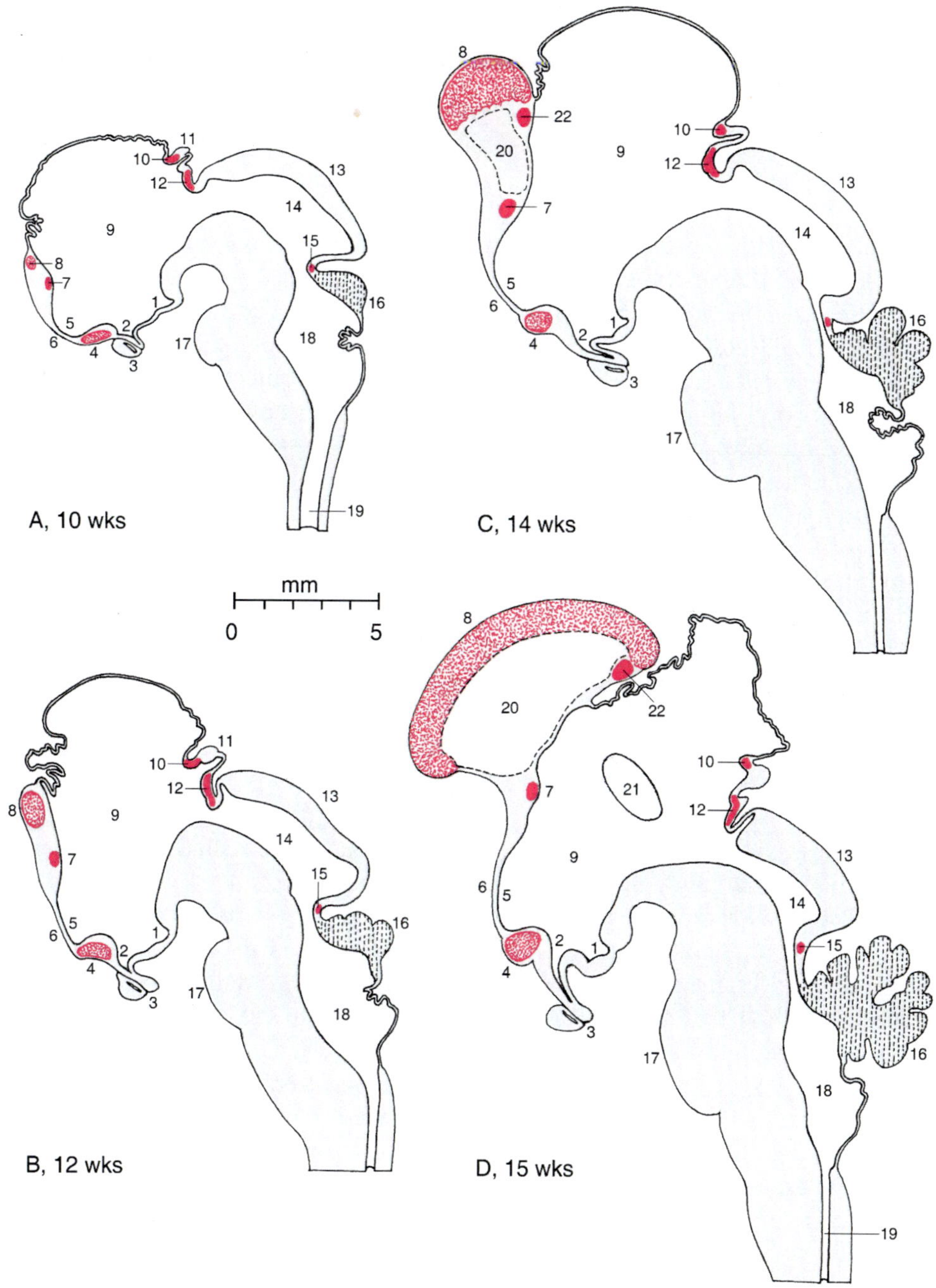

1 Mamillary recess
2 Infundibular recess
3 Hypophysis
4 Optic chiasm
5 Optic recess
6 Lamina terminalis
7 Anterior commissure
8 Corpus callosum
9 Third ventricle
10 Habenular commissure
11 Epiphysis
12 Posterior commissure
13 Mesencephalic tectum
14 Mesencephalic aqueduct
15 Trochlear decussation
16 Cerebellum
17 Pons
18 Fourth ventricle
19 Spinal canal
20 Cavum of septum pellucidum
21 Interthalamic adhesion
22 Commissure of fornix

Fig. 2.6 A–D. Semidiagrammatic median sections through the brains of human fetuses at 10 weeks (**A**), 12 weeks (**B**), 14 weeks (**C**) and 15 weeks (**D**). Based on Hochstetter [69]

showed that at the time of DNA synthesis (t_s) the nuclei of the matrix cells are located in the superficial half of the matrix layer (S-zone). When the nuclei have finished DNA synthesis, they descend during a postsynthetic or premitotic period, t_2, through the I zone to enter the M zone. The matrix cells divide there and, after mitotic time, t_m, both nuclei of the daughter cells pass to the I zone, where they spend the postmitotic and presynthetic period, t_1. Finally, they enter the S zone once again, where a new generation cycle begins.

During a certain period the matrix layer represents a purely proliferative compartment. Mother cells produce more mother cells, with the surface area as well as the thickness of the tube increasing steadily (Fig. 2.7: 2–4). This period of symmetrical division of germinal cells is followed by a period of asymmetrical division, in which one of the daughter cells resulting from each mitosis differs from both its mother cell and its sister cell in that it withdraws from the mitotic cycle and migrates out from the matrix layer (Fig. 2.8: n). These postmitotic elements or neuroblasts then form a third compartment, the *mantle layer*, which is situated between the matrix and marginal layers (Fig. 2.7: 5). The elements giving rise to one postmitotic and one proliferative daughter cell are denoted as stem cells. In a later stage, both daughter cells resulting from a mitosis differ from their mother cell and become postmitotic elements.

When stem cells appear in the matrix layer, the period of pure proliferation has come to an end. As more and more dividing neuroepithelial cells switch to the presumptive stem cell mode and begin to generate postmitotic daughter cells, the mantle layer increases rapidly in thickness. During this phase, proliferation and stem cells coexist in the matrix layer. At some later point in time matrix cells start to produce two postmitotic cells, and gradually more and more matrix elements switch to this developmental mode. Within the mantle layer, which further expands *pari passu* with the progressive depletion of the matrix layer (Fig. 2.7: 5–8), the grisea of the CNS originate by migration and aggregation of neuroblasts.

The three classical layers, matrix, mantle layer and marginal layer, can be clearly observed in all parts of the lateral plates of the neural tube. An international group of neuroembryologists [23] has proposed replacing the names of the three layers just mentioned by the "geographical" terms ventricular zone, intermediate zone and marginal zone. We do not consider these changes improvements and suggest that the old terms be maintained.

Apart from the ubiquitous ventricular matrix, two other proliferative compartments, i.e. the subventricular zone and the external germinal layer, have been observed in the developing brain.

The *subventricular zone*, sometimes also denoted as the subependymal layer or cell plate, has only been described in the lateral and basal walls of the mammalian telencephalon. This layer develops at the junction of the matrix zone and the mantle layer. Most authors agree that this layer appears relatively late, at the moment when neurogenesis is largely or entirely completed (Fig. 2.7: 9). However, according to Rakic [170], the subventricular zone can be recognized in the pallial wall of the rhesus monkey at as early as 45 days of gestation (the total duration of gestation is about 165 days in this animal). It persists after birth and, in vestigial manner, into adult life and even senescence. Contrary to the ventricular matrix cells, the subventricular elements divide in situ and do not exhibit interkinetic nuclear migration.

The subventricular layer gives rise to special classes of neurons and to all types of macroglial elements, with the possible exception of ependymal cells. A particularly strongly developed subventricular zone can be observed in the telencephalic ganglionic eminences, i.e. the mammalian striatal primordia.

The third proliferative compartment, i.e. the *external germinal layer*, is confined to the cerebellum. This layer, also referred to as the external granular layer, develops from the ventricular matrix in the rostral part of the rhombic lip. The *rhombic lip* is a thickened germinal zone in the rhombencephalic alar plate, situated directly adjacent to the attachment of the roof of the fourth ventricle (Fig. 2.9). From the

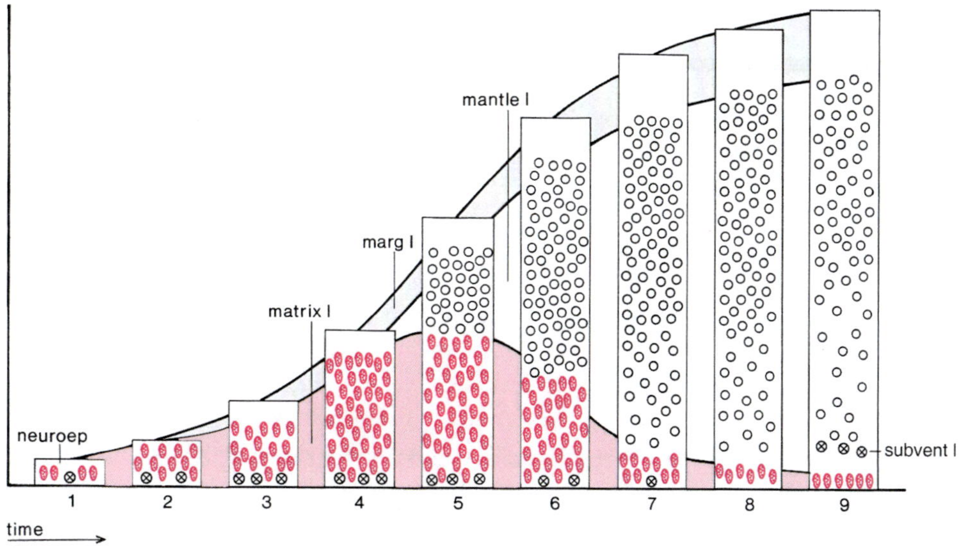

Fig. 2.7. Ontogeny of the CNS. The histogenesis of the wall of the neural tube is subdivided into nine phases. The following developmental events are indicated: Transformation of monolayered neuroepithelium into a pseudostratified epithelium (*1* → *4*); increase (*2* → *4*), culmination (*5*), decrease (*5* → *7*) and depletion (*8*) of matrix layer; appearance (*3*) and development (*3* → *9*) of marginal layer; appearance (*5*) and expansion (*5* → *9*) of mantle layer; appearance of subventricular layer (*9*). *mantle l*, mantle layer; *marg l*, marginal layer; *matrix l*, matrix layer; *neuroep*, monolayered neuroepithelium; *subvent l*, subventricular layer (modified from [89] Fig. 33 a)

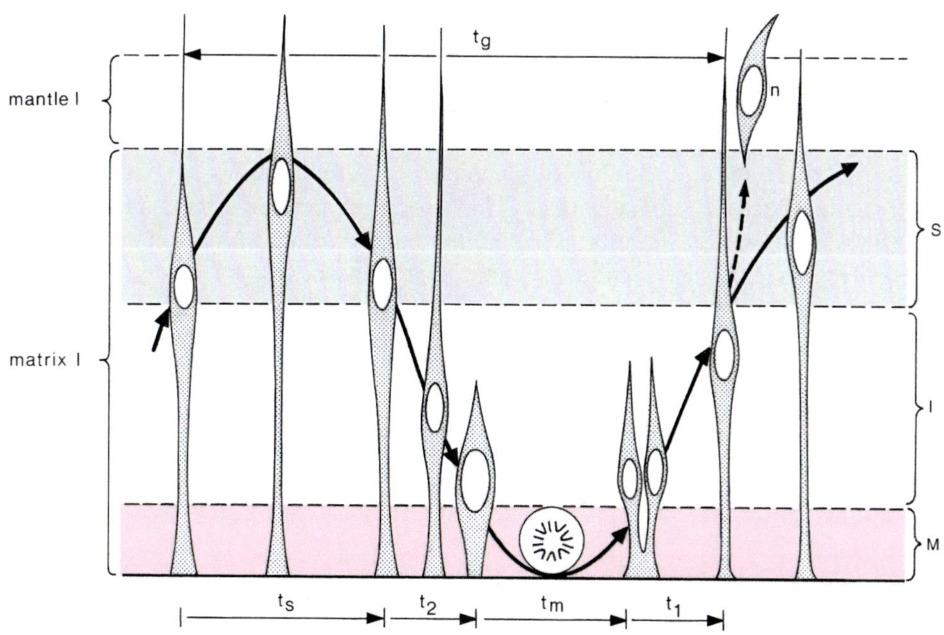

Fig. 2.8. Representation of an "elevator movement" of the matrix cell. *I*, Intermediate zone; *M*, mitotic zone; *mantle l*, mantle layer; *matrix l*, matrix layer; *n*, neuroblast just differentiated from the matrix cells; S,s-zone, i.e. zone of DNA synthesis; *tg*, generation time of the matrix cell; *tm*, mitotic time; *ts*, DNA synthetic time; t_1, postmitotic resting time; t_2, premitotic resting time (modified from [46])

rostral segment of this zone, the layer spreads by tangential migration of its elements over the entire external surface of the cerebellar anlage; these superficial cells continue to proliferate, giving rise to enormous numbers of postmitotic cells. The elements born within this transitory germinal zone ultimately migrate radially inward to form the deepest stratum, i.e. the granular layer of the cerebellar cortex.

After formation, young neuroblasts migrate away from their origins in the proliferative zones to their final positions in cell assemblies. In many places in the human CNS there is a considerable distance between the site where a given neuron was generated and the position which it occupies in the mature brain. The intervening migration process involves active movement of the neuroblasts through the surrounding cellular environment to their final destination, before they settle to establish their specific synaptic connections. Two distinct modes of migration have been identified: radial and tangential migration. During a *radial migration* the neuroblasts simply move from the ventricular to the pial surface. The radially oriented peripheral processes of early differentiating ependymal neuroglial cells provide contact guidance paths for these migrating neuroblasts [168, 169, 171, 190]. During a *tangential migration*, neuroblasts move parallel to the pial surface. Elements involved in this mode of displacement may form discrete and compact streams, which are either located subpially or take a deeper, intramural course. Streams of the latter type usually pass along the external surface of the matrix layer. The caudal segment of the rhombic lip has long been known to be the source of a number of conspicuous and remarkably long tangential migratory streams [2, 3 4, 5, 41, 42] (Fig. 2.9). However, substantial tangential migrations of neuroblasts have also been observed in the spinal cord [109, 154], the midbrain [34, 164, 213] and the telencephalon. Recently, it has been shown that in the latter brain part numerous neuroblasts, originating from the basal or subpallial region, migrate to the dorsal, pallial region [12, 135, 144, 155, 224] and participate in the formation of the cerebral cortex.

Although the neuroblasts destined to form a given cell mass may travel over great distances before attaining their ultimate position, many grisea in the human brain result from relatively short, straight, or radial migrations from the matrix layer into the local mantle layer.

The mode of differentiation of the cell masses in the human CNS differs from place to place. In the spinal cord and in some parts of the brain, the radially migrating neuroblasts remain in direct contact with the matrix zone and transform into neurons there. After the matrix disappears, these cells form a zone of periventricular grey, whereas the marginal layer is transformed into a peripheral fibre zone (Fig. 2.10A,B). Local aggregation of neurons within the grey zone may lead to the formation of periventricular nuclei (Fig. 2.10C). In many parts of the brain, migrating cells detach themselves from the ventricular matrix zone to form one or several layers of migrated cells. These migration layers may split up into two, or sometimes three sublayers. Each of the layers or sublayers within a given area may form one or more separate cell masses [21]. A simple example is presented in Fig. 2.10D–F. Within a given area two migration layers, m1 and m2, are formed. The former fractionates into two separate nuclei, whereas the latter give rise to a single cell mass.

In the largest centre of the human brain, the neocortex, histogenesis is particularly complex. The neocortex occupies most of the dorsal or pallial part of the cerebral hemisphere. This structure consists of six layers, usually numbered I–VI, and is separated from the ventricular surface by a wide fibre zone (Fig. 2.10K). The early development of the pallium does not differ from that observed in other parts of the neuraxis. Matrix, mantle and marginal layers are formed in the usual way (Fig. 2.10G,H). The marginal layer contains some scattered, early differentiating neurons and also a plexus of long, monoaminergic fibres, originating from the brain stem. During the sixth week of gestation radially migrating neuroblasts start to settle in the marginal layer, where together they form a compact *cortical plate* (Fig. 2.10J). This cortical plate divides the marginal layer

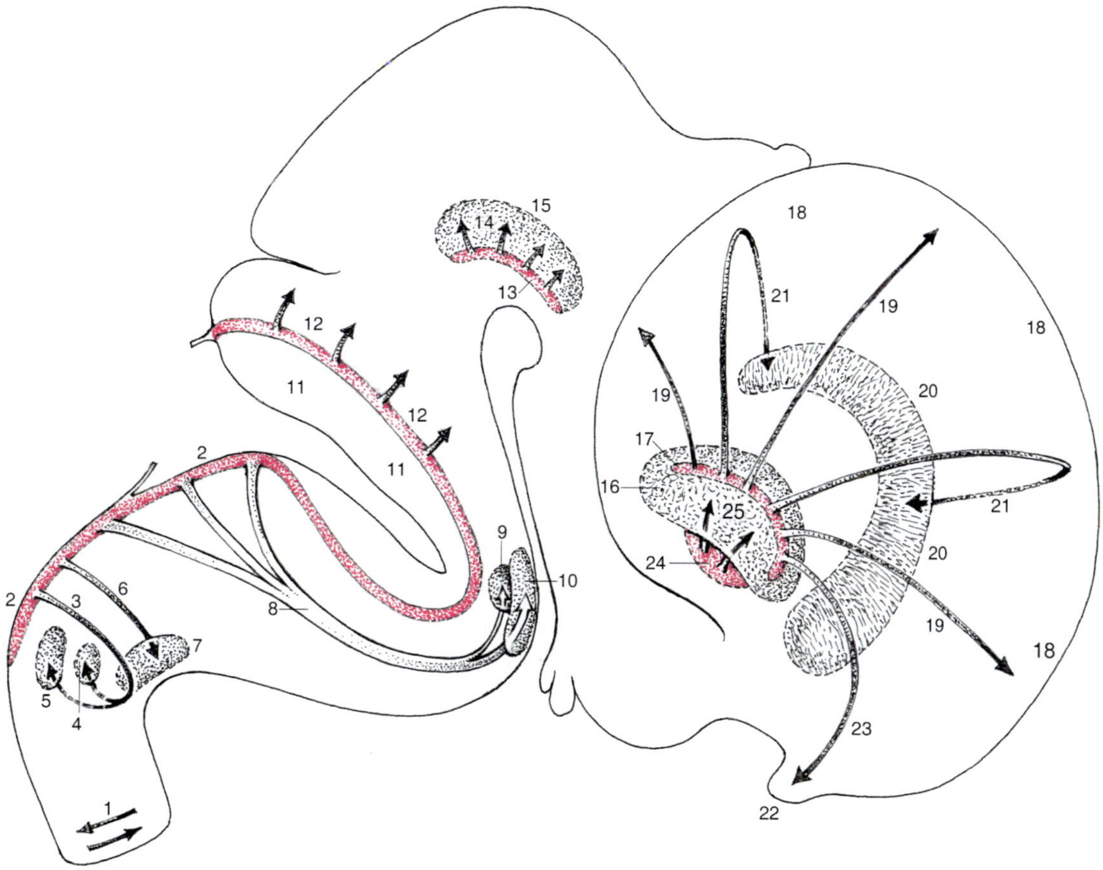

1 Tangential migrations in the spinal cord
2 Caudal rhombic lip
3 Posterior extramural migratory stream
4 Nucleus of the lateral funiculus
5 Lateral cuneate nucleus
6 Intramural migratory stream
7 Inferior olivary complex
8 Anterior extramural migratory stream
9 Nucleus reticularis tegmenti pontis
10 Basal pontine nuclei
11 Rostral rhombic lip
12 Tangential migration leading to the formation of the cerebellar external germinal and granular layers
13 Mesencephalic (and diencephalic) midventral proliferation zone
14 Mesencephalic marginal migratory stream
15 Compact part of the substantia nigra
16 Medial and lateral ganglionic eminence
17 Subventricular germinal zone of 16
18 Anlage of neocortex
19 Precursors of GABAergic interneurons migrating from 17 to 18
20 Anlage of hippocampal formation
21 Precursors of GABAergic interneurons migrating from 17 to 20
22 Anlage of olfactory bulb
23 Compact rostral migratory stream along which precursors of GABAergic interneurons pass from the rostrolateral part of 17 to 22
24 Entopeduncular area
25 Precursors of cholinergic interneurons migrating from 24 to the lateral ganglionic eminence, i.e. the anlage of the striatum

Fig. 2.9. Diagrammatic representation of the principal tangential migrations of neuronal precursors in the human CNS, using a lateral view of an embryo of about 8 weeks as a topographical framework

into an outer *zonal layer* and an inner *subplate* [95, 126, 127]. The subpial zonal layer is transformed into the first cortical layer. The subplate is, just like the cortical plate, a transient structure. Its neurons play an important role in the guidance of thalamic afferents into the developing cortex [189]. The neuroblasts that first arrive in the cortical plate form layer VI, followed sequentially by those destined for the more superficial layers V–II. The latter migrate through the subplate and the earlier formed layers to settle more peripherally in an "inside-out" fashion.

The massive migration of neuroblasts to the developing cortex lasts until about the 26th week of gestation [128]. After that period, the pallial mantle layer is gradually transformed into a fibre zone, consisting of cortical afferents and efferents. Some subplate neurons remain in the most superficial part of this fibre zone (Fig. 2.10 K).

Recent studies [123, 135, 136, 155] have conclusively shown that, apart from the pallial matrix, the subpallial matrix of the primordial basal ganglia contributes substantially to the formation of the cortex. Whereas the elements produced in the pallial matrix, after a radial migration, differentiate into glutamatergic pyramidal neurons (large arrow in Fig. 2.10 J), the subpallial matrix gives rise to GABA-producing local circuit neurons. The latter migrate tangentially to the cortex (Fig. 2.9) through the subplate/mantle layer complex (small arrows in Fig. 2.10 J). Interestingly, a certain proportion of these elements actively migrate into the ventricular matrix. On reaching the pallium, they pause there for an extended period of time and then resume migration radially to take up their position in the cortical plate (dashed arrows in Fig. 2.10 J) [68, 98, 136]. Tangentially migrating cells also enter the most superficial zone of the developing cortex, where they form a transient subpial granular layer [48]. The fate of these elements is uncertain [210].

Several authors [169, 194, 196] have reported the presence of a subventricular matrix layer in the developing neocortex. As already mentioned, such a layer is characterized by undifferentiated elements that divide in situ and do not exhibit the interkinetic nuclear migration shown in Fig. 2.8. Smart et al. [198] recently studied the ontogenesis of the primary visual cortex (area 17) in the macaque monkey. They found that in this species a prominent subventricular matrix layer is present which, in terms of size and duration of activity, clearly exceeds the relatively early declining ventricular matrix. The authors concluded that this subventricular matrix layer is the principal source of cortical neurons in the monkey. Smart et al. [198] also reported that the subventricular matrix layer in the macaque can be subdivided into a narrow inner subventricular layer, consisting of randomly organized cells, and a much wider outer subventricular layer, showing a highly radial organization (Fig. 2.10 L). They presume that the latter is unique to the primate. Thus, it appears that no less than three separate proliferative compartments participate in the formation of the primate cerebral cortex. It would be interesting to know whether the subventricular matrix is involved in the production of the enormous numbers of small excitatory interneurons – spiny stellate cells – present in the primate visual and other primary sensory cortices.

We conclude this section on the histogenesis of the CNS with some general remarks.

1. The waxing and waning of the ventricular matrix layer (Fig. 2.7), with regard to both amplitude and duration, shows profound differences among the various parts of the neuraxis. In general, it can be said that in the basal plate the matrix declines much earlier than in the alar plate and that in the pallial region the matrix persists for a remarkably long time [80].

2. Until quite recently it was generally thought that neurogenesis in the mammalian brain ends in the period just after birth. However, it has now been shown that three regions, namely, the olfactory bulb, the hippocampus and the substantia nigra, continue to produce new nerve cells throughout life. The precursors of the olfactory bulb cells proliferate in the rostral part of the telencephalic subventricular zone and migrate along a restricted migratory stream to the olfactory bulb, where they differentiate into two different types of interneurons:

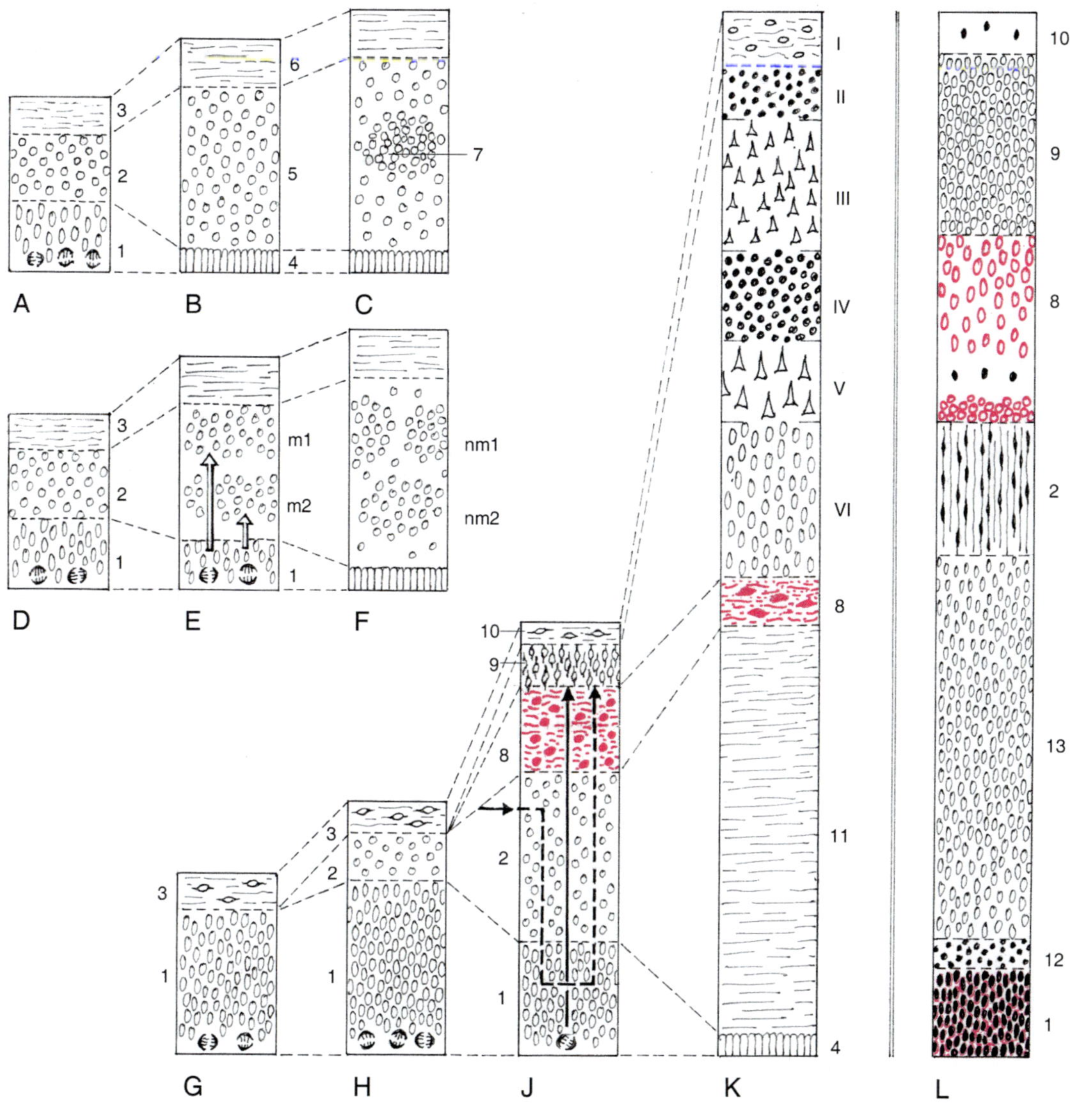

1 Ventricular matrix layer
2 Mantle layer
3 Marginal layer
4 Ependymal layer
5 Periventricular grey
6 Peripheral fibre zone
7 Periventricular nucleus
8 Subplate
9 Cortical plate

10 Zonal layer
11 Subcortical white matter
12 Inner subventricular matrix layer
13 Outer subventricular matrix layer
I, II etc. Layers of the mature neocortex
m1,2 Layers of migrated cells
nm1 Nuclei derived from m1
nm2 Nucleus derived from m2

Fig. 2.10 A–L. Differentiation of grisea in the human CNS. **A–C** Development of periventricular formations. **D–F** Development of nuclei from layers of migrated cells. **G–J** Histogenesis of the neocortex. **K** Cytoarchitecture of the mature neocortex. **L** Structure of the putative primary visual cortex (area 17) of a 72-day-old macaque embryo. G and H are based on [127, 143]; **J** is reproduced from [198]. Elements belonging to the cortical subplate are shown in *red*. The *arrows* indicate migration routes, which are explained in the text

granular and periglomerular cells [16, 92, 116–118]. (It is important to note that this rostrally directed migratory stream, which has been demonstrated in several mammals, including non-human primates, according to a recent report [185], is lacking in the adult human brain.) The precursors of the hippocampal cells are produced in the subgranular zone of that structure and migrate into the granular layer, where they differentiate into hippocampal granule cells [40, 49, 88, 151]. In both the subventricular zone and the hippocampus astrocytes have been shown to function as neuronal precursors [9]. Samai et al. [185] also recently reported the presence of a band of astrocytes in the subventricular zone lining the lateral ventricles of the adult human brain that proliferate *in vivo* and behave as multipotent progenitor cells *in vitro*. Remarkably, such an astroglial band has not been observed in non-human primates, or in any other mammal studied so far. The substantia nigra cells, generated during adult life, could be identified as dopaminergic projection neurons. They appeared to derive from stem cells lining the most ventral part of the mesencephalic ventricle [230].

3. Recent studies, mainly on mice, have drawn attention to the remarkable versatility of radial glial cells. It has been known for some time already that these elements (a) guide the radial migration of newborn neurons [58, 171], (b) form canal-like compartments, which determine growth and orientation of outgrowing axons [91, 141, 191] and (c) may transform into astrocytes [31, 114, 186]. Now, it has become clear that radial glial cells, in addition, (d) are mitotically active throughout the development of the CNS [57, 131] and, most remarkably, (e) are able to produce neurons [13, 29, 52, 120, 140, 219].

4. In the course of the past few years numerous laboratories have been involved in identifying and isolating neural stem cells with the goal of using these elements for neuronal replacement therapies in Parkinson's, Alzheimer's and other neurodegenerative diseases [90, 115, 150]. In light of the data reviewed above under items 3 and 4, it seems likely that the stem cells in the CNS are contained within the neuroepithelial → radial glia → astrocyte lineage [8]. It has recently been shown that neural stem cells can give rise to radial glial cells and that radial glial cell-guided neuronal migration can be recapitulated in the adult CNS [53].

5. In addition to proliferation, migration and differentiation, cell death is also an important event in CNS development. This normal loss of cells, which is genetically regulated, is known as programmed cell death or apoptosis. It plays an important role in the development of the spinal cord and of several parts of the brain [149]. To give an example: many of the neurons in the pallial subplate (Fig. 2.10J) disappear after fulfilling their role in guiding thalamic afferents to the developing neocortex [210].

Fundamental Morphological Pattern

The human brain shares a basic and fundamental morphological pattern with all other vertebrates. The following synopsis of the main features of this pattern is primarily based on the classical embryological studies of the Swedish investigators Bergquist and Källén and their associates [17–19, 75, 76, 84, 85], and on the recent work of Puelles and Rubenstein and their collaborators [25, 153, 156, 159–161, 164, 180, 182, 213].

1. The neural walls of the brain can be divided into a number of basic units or fields formed by the intersection of transversely oriented neuromeres and longitudinally arranged zones. The boundaries between these units are indicated in Fig. 2.11 by dashed curves.

2. Two longitudinal domains, the basal and alar plates, extend along the length of the neuraxis. The border zone of these two domains, i.e. the *zona limitans longitudinalis* or alar-basal boundary, ends rostrally just behind the optic chiasm. This border roughly approximates the sulcus limitans concept of His [66, 68], which basically separated the precociously differentiated basal plate neurons from the

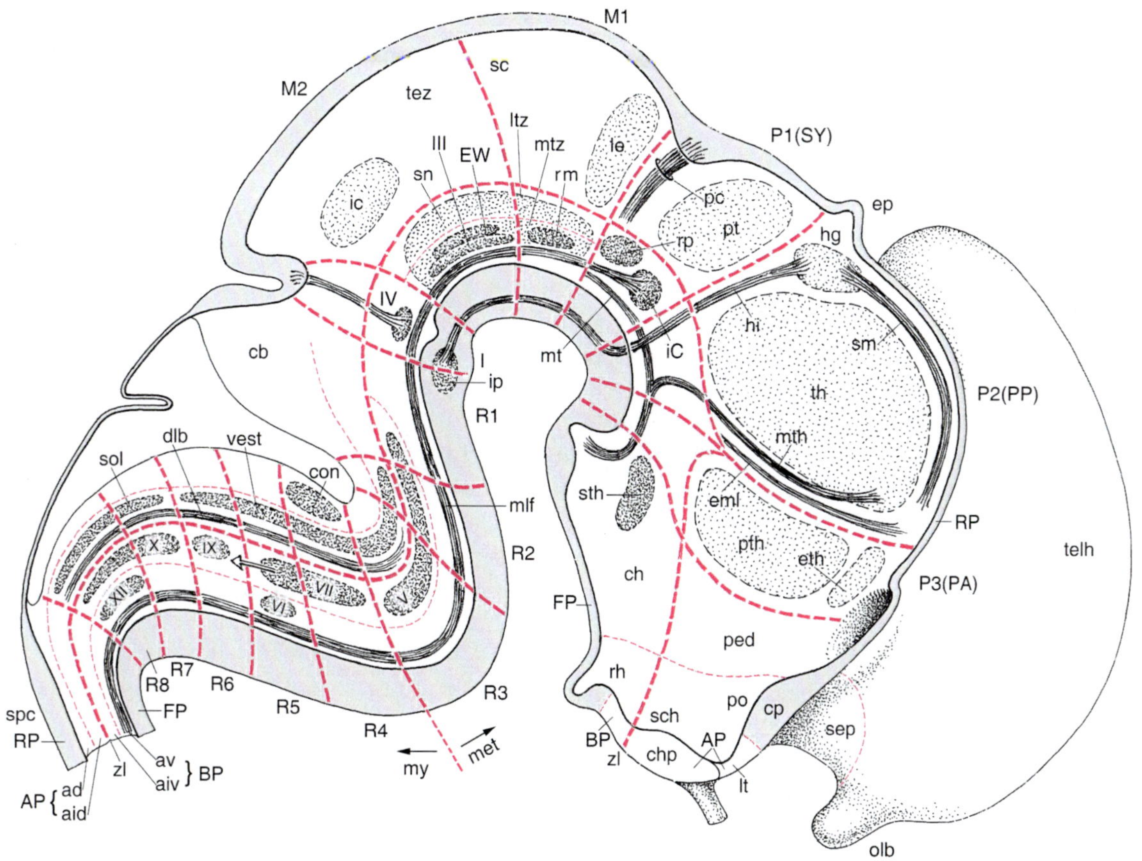

ad Area dorsalis
aid Area intermediodorsalis
aiv Area intermedioventralis
AP Alar plate
av Area ventralis
BP Basal plate
cb Cerebellum
ch Caudal hypothalamic area
chp Chiasmatic plate
con Cochlear nuclei
cp Commissural plate
dlb Dorsolateral bundle
eml External medullary lamina
ep Epiphysis
eth Eminentia thalami
EW Edinger-Westphal nucleus
FP Floor plate
hg Habenular ganglion
hi Habenulo-interpeduncular tract
I Isthmic neuromere
iC Interstitial nucleus of Cajal
ic Inferior colliculus
ip Interpeduncular nucleus
le Lentiform or posterior pretectal nucleus
lt Lamina terminalis
ltz Lateral tegmental zone
M1, M2 Mesomeres
met Metencephalon
mlf Medial longitudinal fasciculus
mt Mamillotegmental tract
mth Mamillothalamic tract
mtz Medial tegmental zone
my Myencephalon
olb Olfactory bulb
P1, P2 etc. Prosomeres
PA Parencephalon anterius
pc Posterior commissure
ped Peduncular area
po Preoptic area
PP Parencephalon posterius
pt Pretectum
pth Prethalamus
R1, R2 etc. Rhombomeres
rh Rostral hypothalamic area
rm Magnocellular red nucleus
rp Parvocellular red nucleus
RP Roof plate
sc Superior colliculus
sch Suprachiasmatic nucleus
sep Septum
sm Stria medullaris
sn Substantia nigra
sol Solitary nucleus
spc Spinal cord
sth Subthalamic nucleus
SY Synencephalon
telh Telencephalic hemisphere
tez Tectal zone
th Thalamus
vest Vestibular nuclei
zl Zona limitans
III, IV etc. Motor cranial nerve nuclei

Fig. 2.11. Basic organization plan of the human brain. Morphological units and early-developing fibre systems mapped in a medial view of an embryo of about 8 weeks. The ventricular surface of the rhombencephalon has been exaggerated and turned towards the beholder to show the zonal and neuromeric organization of this brain part. The *arrow* indicates that the facial nucleus shifts caudally during ontogeny. The interpretation of the rostral parts of the brain is mainly based on [18, 157, 163, 164]. The entire region in front of P3 can be designated as secondary prosencephalon

more tardive alar plate elements. This boundary can be traced in all vertebrates by use of molecular markers [156], but only rarely relates to a distinct sulcus in the ventricular surface.

3. At the rhombencephalic level, the alar and basal plates can both be subdivided into two longitudinal zones, as proposed in classical schemes. These zones are termed here: area dorsalis, area intermediodorsalis, area intermedioventralis and area ventralis (Fig. 2.11).

4. Genes involved in regulating CNS development are often expressed in regionally restricted patterns. The boundaries of their expression domains often coincide with those of the morphological units. In recent years, considerable progress has been made in defining the genetic mechanisms involved in the transverse and dorsoventral patterning of the CNS. A discussion of the literature on this subject is beyond the scope of the present synopsis. (For reviews, see [110, 156, 180, 182].) However, to illustrate this, let us mention that, throughout the midbrain and forebrain, the gene *Nkx-2.2* is expressed in a thin strip of tissue abutting the zona limitans longitudinalis or alar-basal boundary.

5. The basic morphological units represent tridimensional radial complexes, stretching from the ventricular to the meningeal surface. Each of these units forms a centre of proliferation, producing radially migrating postmitotic cells which, after attaining their ultimate position, participate in the formation of different kinds of grisea. The nuclear and laminar modes of differentiation, discussed in the previous section, come about essentially within the confines of the radial units. However, in the formation of certain nuclei (such as the motor nuclei of V, VII and X; Fig. 2.11) two or three adjacent units are involved, and some neuronal populations (such as those forming the catecholaminergic cell groups) constitute highly patterned plurisegmental and plurizonal complexes [159, 164, 213]. The relationship between a number of nuclei *casu quo* nuclear complexes and the fundamental morphological units is indicated in Fig. 2.11. It should, however, be emphasized that the origin of a large number of cell masses in the human brain is still unknown.

6. Tangential migrations may lead to an admixture of cells derived from different radial units and in some cases to the "colonization" of foreign territory. As for admixtures of cells, a remarkable neuronal exchange takes place between the lateral and medial ganglionic eminences during development [148], and it has already been discussed that many of the neocortical interneurons are derived from the ganglionic eminences. As for "colonization" of foreign territory, we have seen that, in the rhombencephalon, massive streams of neuroblasts, originating from the most dorsal parts of the alar plate, invade the basal plate to form the inferior olivary and pontine nuclei (Fig. 2.9).

7. In early embryonic brains most fibres pass close to the border zones of the fundamental morphological units, together forming a scaffold of transverse and longitudinal bundles. The relationship between this primordial pattern and the fibre organization observed in adult brains has not been sufficiently investigated. However, it is known that several adult fibre systems (among them are those shown in Fig. 2.11) derive directly from the early embryonic axonal scaffold.

Spinal Cord

Of all parts of the CNS, the spinal cord preserves the early embryonic tube like shape most clearly. Until the third month of intrauterine life the cord occupies the entire length of the vertebral canal. However, from that time onwards, the vertebral column grows more rapidly than the cord itself, which leads to a gradual rostral shift of the caudal end of the cord. This upward retreat of the spinal cord is known as *ascensus medullae spinalis*. At birth, the lower end of the cord is opposite the body of the third lumbar vertebra. In the adult it usually lies at the level of the lower border of the first lumbar vertebra. During the ascensus, the most caudal portion of the spinal cord, which is devoid of developing neurons, is drawn out into a thin glial thread, the *filum terminale*.

During development, the initially slit-like ventricular cavity of the spinal cord (Fig. 2.12) is gradually reduced to a narrow central canal. According to Smart [192] this decrease in ventricular surface area results from a continued loss of cells from the ependymal layer and not from fusion of the opposing sides, as is commonly believed.

By the sixth week, the three embryonic layers, matrix, mantle and marginal layer, can be clearly distinguished (Fig. 2.12). In the spinal cord, histogenesis proceeds much faster in the basal plate than in the alar plate. Hence, in the stage shown in Fig. 2.12, the matrix zone is wide and the mantle zone is still narrow in the alar plate, whereas in the basal plate these relations are reversed.

The radial glia is well developed in the embryonic human cord [31] and the majority of the spinal neuroblasts migrate from the matrix into the mantle layer along the peripheral processes of these elements. Apart from this prevailing radial migration, tangential migration also occurs in the spinal cord. The axons of early-differentiating alar plate neurons pass ventrally and decussate in the floor plate. These decussating axons are collectively designated as the *arcuate fibres* or the *circumferential pathway* [71, 72]. It has been shown that neuroblasts migrate both dorsally [154] and ventrally [109] along the fibres of this pathway (Fig. 2.12).

Neuronal differentiation occurs in the developing cord entirely within the confines of the mantle layer. By the 14th week all the major cell groups that characterize the mature cord can be recognized [179]. Rexed [175] subdivided the grey matter in the spinal cord of the cat into nine dorsoventrally arranged layers (I through IX), and this subdivision has also been applied to the human spinal grey matter [187] (Fig. 6.2). Although Rexed's laminar scheme provides a convenient framework for the description of structural features of the spinal grey matter, it should be emphasized that only the dorsal part of the spinal dorsal horn (layers I–III) shows a true laminar organization.

Somatic motor neurons differentiate early in the outermost zone of the basal plate mantle layer. Their axons penetrate the marginal layer and emerge from the spinal cord as ventral root fibres. The cells in the spinal ganglia, which are derived from the neural crest (Fig. 2.1 E–H), take a bipolar shape and develop two processes, one central and one peripheral. The central processes enter the marginal zone of the alar plate as dorsal root fibres and bifurcate there into ascending and descending branches. The peripheral processes of the spinal ganglion cells unite with the ventral root fibres in the region of the intervertebral foramina to form the mixed (i.e. afferent and efferent) spinal nerves. Peripherally, a spinal nerve is distributed to a segment of the body, including a myotome and a dermatome. Centrally, the regular arrangement of the sites of emergence of the dorsal and ventral root fibres from the surface of the spinal cord allow a subdivision of this organ into segments. (A spinal segment is usually defined as the region of the spinal cord associated with the emergence of a pair of spinal nerves.) There are usually 31 pairs of spinal nerves, grouped as eight pairs of cervical (C), 12 pairs of thoracic (T), five pairs of lumber (L), five pairs of sacral (S) and one pair of coccygeal (Co) spinal nerves (Fig. 3.14). The corresponding spinal segments are designated accordingly as C1–C8, T1–T12, etc.

As a result of the ascent of the spinal cord, all the spinal nerves below the upper cervical region pass out laterally with increasing degrees of obliquity in a rostrocaudal direction to reach their corresponding intervertebral foramina. Below the level of the intervertebral disk between the first and second lumbar vertebra, where the spinal cord tip comes to lie, the vertebral canal is filled with a large number of longitudinal nerve roots. This collection of dorsal and ventral roots, which surrounds the filum terminale, is known as the *cauda equina* (Fig. 3.14).

Two fusiform spinal enlargements, cervical and lumbosacral, develop simultaneously with the anlagen of the upper and lower extremities. The *cervical enlargement* encompasses the segments C4–T1, whereas the *lumbosacral enlargement* extends between L1 and S2 (Fig. 3.15).

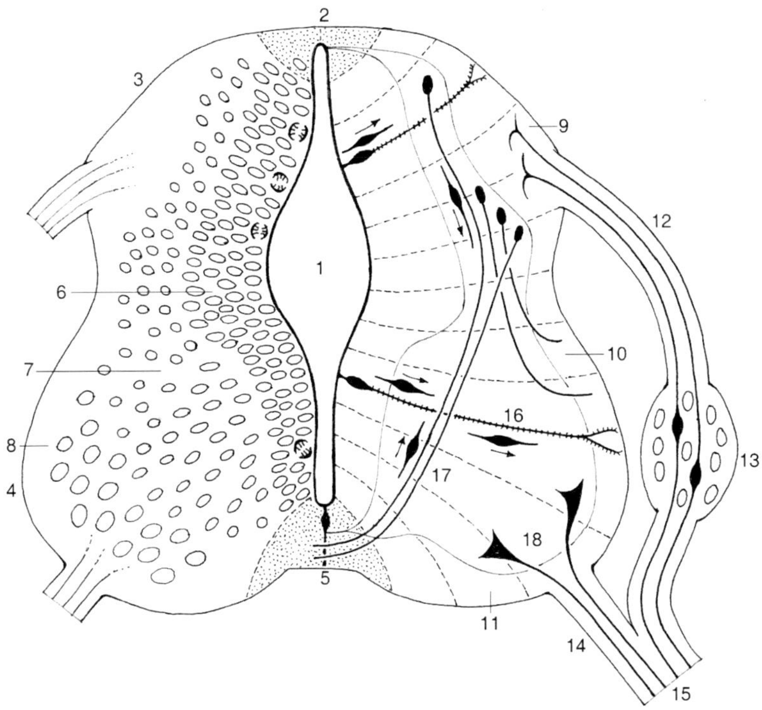

1 Spinal canal
2 Roof plate
3 Alar plate
4 Basal plate
5 Floor plate
6 Matrix layer
7 Mantle layer
8 Marginal layer
9 Posterior funiculus
10 Lateral funiculus
11 Anterior funiculus
12 Dorsal root
13 Dorsal root ganglion
14 Ventral root
15 Spinal nerve
16 Ependymal cell, guiding radially migrating neuroblasts
17 Arcuate fibres, guiding tangentially migrating neuroblasts
18 Motoneurons

Fig. 2.12. Diagrammatic transverse section through the spinal cord of a 6-week-old human embryo, showing the general relations and the migratory paths

As development proceeds, cells in the mantle layer send their axons to lower and higher levels of the spinal cord, and the axons of others ascend to supraspinal levels. Conversely, axons of neurons in the brain start to descend to the cord. All of these ascending and descending axons, as well as the primary afferent fibres already discussed, assemble in the marginal zone which, as a result, becomes greatly thickened. As during later development the mantle layer differentiates into the *dorsal* and *ventral horns,* this thickened fibre zone becomes divided into *dorsal, lateral* and *ventral funiculi.* Because many of the ascending and descending spinal axons acquire a myelin sheath, the adult spinal cord shows on cross section the characteristic pattern of an H- or butterfly-shaped central zone of grey matter, surrounded by a peripheral zone of white matter (Figs. 6.6–6.10).

The spinal floor and roof plates are formed by ependymal cells. Their initially very narrow marginal zones become pathways for decussating fibres (Fig. 2.12).

Hindbrain

The dominant morphogenetic events in the hindbrain are represented by a gradual divergence of the lateral plates, with consequent thinning and stretching of the roof plate (Fig. 2.13 A) and transformation of the initially narrow, ventricular cavity into the characteristic, rhomboid-shaped fourth ventricle (Figs. 2.2 B, C, 2.14 A). During the formation of the pontine flexure, which attains its maximum at the end of the second month of development (Fig. 2.4 B), the ventricular surfaces of the caudal (myelencephalic) and the rostral (metencephalic) portions of the hindbrain approach each other, and the intervening part of the fourth ventricle becomes very narrow. At about the same time the lateral angles of the ventricular cavity extend to form the lateral recesses of the fourth ventricle (Fig. 2.14 B).

Neuromeres can be clearly distinguished in the hindbrain of about 1-month-old human embryos (Figs. 2.2 A, 2.3 A). Recent studies in chick and mouse embryos have shown that these rhombencephalic segments or rhombomeres represent rostrocaudal domains of differential gene expression, with characteristic fates. Each rhombomere is characterized by a unique combination of *Hox* genes [35, 99, 223]. Moreover, it has been demonstrated that rhombomeres act as true embryonic compartments [44, 62, 209]. In the human embryo eight rhombomeres, R1–R8, can be distinguished. An "independent" isthmic neuromere, I, forms the transition from the hindbrain to the midbrain (Fig. 2.11, Table 2.1) [143], but is as *isthmus rhombencephali* usually included in the hindbrain. The *metencephalon* encompasses the segments I and R1-3, whereas the remaining rhombomeres form the *myelencephalon* or *medulla oblongata.*

Shortly after the appearance of the neuromeres, the rhombencephalic mantle layer roughly differentiates into four longitudinal zones, dorsal, intermediodorsal, intermedioventral and ventral. Intersection of the neuromeres and the longitudinal zones yields the fundamental histogenetic units of the rhombencephalon (Fig. 2.11) [17, 21, 75, 76]. A complicating factor is, however, that all motoneurons are produced in the ventral zone and only the somatomotor neurons remain there, while the branchiomotor and parasympathetic preganglionic elements migrate tangentially into the intermedioventral zone and thus are not born there [143, 144, 225]. In general, the four histogenetic zones correspond to the functional zones, somatosensory, viscerosensory, visceromotor and somatomotor, described by Johnston [78] and Herrick [64] at the beginning of the last century. Thus, the somatomotor nuclei of IV, VI and XII are born and develop within the ventral zone; the visceromotor (branchiomotor) and preganglionic parasympathetic nuclei of V, VII, IX, X and XI move from their ventral origin into the intermedioventral zone. The viscerosensory gustatory and solitary nuclei develop in the intermediodorsal zone and the somatosensory cochlear nuclei develop from the dorsal zone. This correspondence is, however, not complete. For example, it has been demonstrated that the somatosensory vestibular nuclei develop from the intermediodorsal and not from the dorsal zone [75, 76]. In the midbrain, the somatomotor and visceromotor nuclei of the oculomotor complex also develop from the ventral zone, but do not separate later on [17]. Part of the oculomotor motoneurons migrate across the mesencephalic midline after having grown their axons peripherally, thus stretching them into the oculomotor decussation.

As a result of studies on the central relations of the functional components of the various cranial nerves three of the four functional zones mentioned above have been subdivided into separate general and special moieties. The resultant, very useful, "seven-column scheme" (Fig. 6.12) will be discussed in a later section. It should, however, be mentioned that the nuclei belonging to these functional columns do not develop from seven separate embryonic cell cords, as is suggested in many textbooks.

Two large somatosensory relay centres, the *gracile* and *medial cuneate nuclei* (collectively designated as the dorsal column nuclei), originate from the most caudal part of the rhombencephalic alar plates. These nuclei produce elongated tubercles on the external surface of the developing rhombencephalon (Fig. 2.14 C). They are in receipt of primary afferent fibres that ascend in the spinal dorsal funiculi. Their efferents decussate as internal arcuate fibres to the contralateral side and then pass as *medial lemniscus* to the thalamus.

The origin of the cells forming the human (and in general: mammalian) *rhombencephalic reticular formation* is unknown. In the avian rhombencephalon, however, it has been experimentally established that the matrix zones of the basal plates and alar plates both contribute to the medial magnocellular and to the lateral parvocellular parts of the reticular formation [206].

In the hindbrain many of the neurons produced remain close to their site of generation and aggregate in periventricular nuclei. This holds true for the somatomotor nuclei of IV, VI and XII, for the viscerosensory solitary nucleus, and for the special somatosensory cochlear and vestibular nuclei. Other cells move away from the ventricular surface to attain their permanent residence after a shorter or longer radial or tangential migration. The branchiomotor nuclei of VII and of IX + X + XI (ambiguus nucleus) are the product of radial migrations (Fig. 2.13 B), and it has been shown that the cells destined for the facial nucleus move outward in close association with radial glial fibres [204]. It should be noted, however, that the facial motoneurons, before migrating peripherally, show a highly remarkable caudal shift from R4, where they are born, into R6, past the anlage of the abducens nucleus in R5 (Fig. 2.11) [164]. A "gliophilic" radial migration has also been demonstrated for the cells of the principal sensory trigeminal nucleus [1], but the exact site of origin of these cells, or of those forming the spinal trigeminal nucleus, is not actually known. The same holds true for many other rhombencephalic cell masses. It is appropriate to quote Puelles [156, p. 333] at this point: "Morphological analysis should aim to reduce the large numbers of nuclei that 'float' unexplained in the usual no-man's-land of conventional atlases into developmentally integrated complexes."

Tangential migrations underlie the formation of a considerable number of cell masses in the hindbrain, including the *lateral cuneate nucleus*, the *nucleus of the lateral funiculus*, the *inferior olivary complex*, the *nucleus reticularis tegmenti pontis* and the *pontine nuclei*. Because all of these nuclei project to the cerebellum, they are collectively designated as precerebellar nuclei. The neuroblasts involved in the formation of these nuclei arise from the *rhombic lip*, a proliferative zone along the line of attachment of the roof of the fourth ventricle (Fig. 2.14 A, B). After leaving the rhombic lip, these neuroblasts assemble in three migratory streams, the anterior and posterior extramural streams and the intramural migratory stream (Figs. 2.9, 2.13 B) [2–5]. The cells constituting the *posterior extramural migratory stream* are destined to form the external cuneate nucleus and the nucleus of the lateral funiculus. They move circumferentially over the wall of the caudal myelencephalon and cross the median

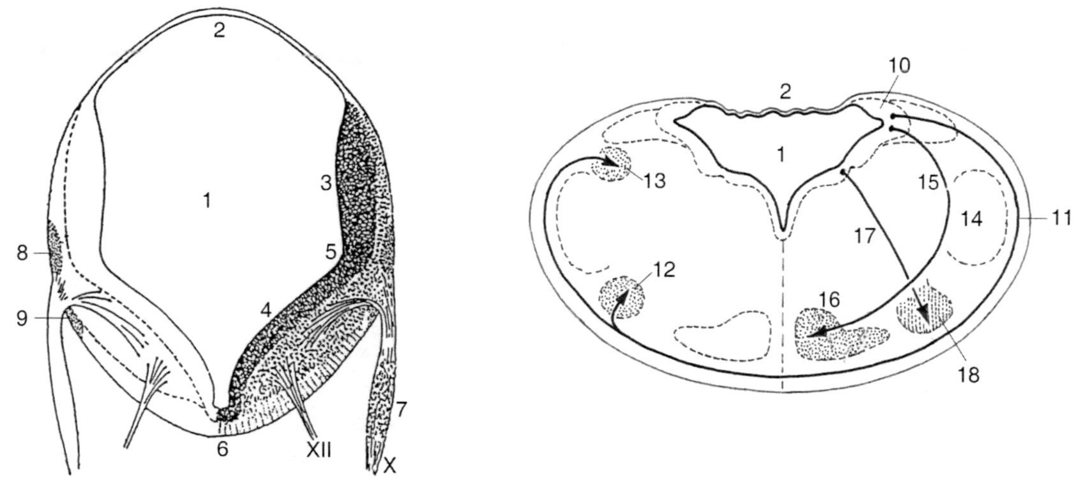

A, 30 days

1 Fourth ventricle
2 Roof plate
3 Alar plate
4 Basal plate
5 Sulcus limitans
6 Floor plate
7 Nodose ganglion
8 Solitary tract
9 Spinal trigeminal tract
10 Caudal rhombic lip

B

11 Posterior extramural migratory stream
12 Nucleus of the lateral funiculus
13 Lateral cuneate nucleus
14 Nucleus of the spinal trigeminal tract
15 Intramural migratory stream
16 Inferior olivary complex
17 Migration route of neuroblasts to 18
18 Facial nucleus
X, XII etc. Cranial nerves

Fig. 2.13 A, B. Development of rhombencephalon. **A** Transverse section through the rhombencephalon of a human embryo, reproduced from [201]. **B** Diagrammatic transverse section through the caudal rhombencephalon of an older rat embryo, showing the tangential migratory paths from the caudal rhombic lip and the radial path of migration leading to the formation of the motor nucleus of VII. Based on data from [4, 5, 204]

plane to continue dorsolaterally on the contralateral side (Fig. 2.13 B). The cells in the *intramural migratory stream* pass ventromedially through the superficial part of the myelencephalic mantle layer. They remain on the ipsilateral side and contribute to the developing inferior olivary complex (Fig. 2.13 B). The cells in the large *anterior extramural migratory stream*, finally, move downward and forward to attain the basal plate region of the first rhombomere, where they settle to form the reticular tegmental and basal pontine nuclei (Fig. 2.9). The exact segmental origin of the various precerebellar nuclei in the mammalian brain is unknown, but it has been experimentally established that in the avian brain the pontine and inferior olivary nuclei derive from R1–6 and R7–8, respectively [10, 28, 122, 227]. The cells of the inferior olivary complex provide climbing fibres to the Purkinje cells whereas the remaining precerebellar nuclei project mossy fibres to granule cells. Interestingly, it has recently been established that, in the mouse, the cells that project mossy fibres and those that produce climbing fibres stem from two precursor populations with different gene expression patterns [178].

Initially, most longitudinally running rhombencephalic fibres occupy a superficial position in the marginal zone (Fig. 2.13 A). However, due to the massive peripheral migration of branchiomotor, precerebellar and other nuclei, the simple "cells inside – fibres outside" pattern, which characterizes the embryonic (Fig. 2.12) and adult spinal cord (Figs. 6.6–6.10), becomes disturbed at rhombencephalic levels.

The rhombencephalic floor plate, which connects to the basal plates, is still thin in early embryonic stages (Fig. 2.5 B). During further development it increases in thickness and its constituent ependymal cells become stretched to form a median glial septum or *raphe*. Numerous fibres, among which internal arcuate fibres from the dorsal column nuclei and the sensory trigeminal nuclei, olivocerebellar fibres, pontocerebellar fibres and secondary auditory fibres decussate through the raphe to the contralateral side. Cells originating from the medial zones of both basal plates migrate ventromedially to form the median *raphe nuclei*.

The rhombencephalic roof plate initially forms a thin ependymal tela. At about the sixth week transversely oriented folds of this tela start to project inward to form the *choroid plexus of the IVth ventricle* (Fig. 2.6).

Cerebellum

The cerebellum develops from the rostral parts of the rhombencephalic alar plates. Its bilateral anlagen, which are first discernable at the beginning of the second month (Fig. 2.14 A), are mainly derived from the first rhombomere [183, 227], but the isthmus neuromere also contributes to the formation of the cerebellum [143] (Fig. 2.11). Initially, the bilateral cerebellar anlagen form an angle of about 90° to each other (Fig. 2.14 A), but due to the accentuation of the pontine flexure, this angle gradually increases. At the end of the second month, the cerebellar anlagen together form a transversely oriented ridge, stretched across the rostral part of the fourth ventricle (Fig. 2.14 B). During the same period, the cerebellar anlagen thicken by the formation of elongated intraventricular and extraventricular swellings (Figs. 2.4 B, 2.14 B). The intraventricular swellings rapidly increase in size and their medial surfaces, which are initially separated by a deep furrow, contact each other and fuse [70, 86]. By this fusion the bilateral cerebellar anlagen are transformed into a single cerebellar primordium. This cerebellar primordium is rostrally connected with the tectal region of the midbrain by the *anterior medullary velum*, formed by the roof plate of the isthmus neuromere (Figs. 2.6, 2.15). Caudally, the cerebellar primordium thins out into the epithelial roof of the IVth ventricle, and laterally it is directly continuous with the remainder of the rhombencephalon. Here the three peduncles, inferior, middle and superior, develop, through which the cerebellar afferent and efferent fibres pass.

The adult human cerebellum contains a superficially situated cortex, whereas its periventricular zone is occupied by a series of

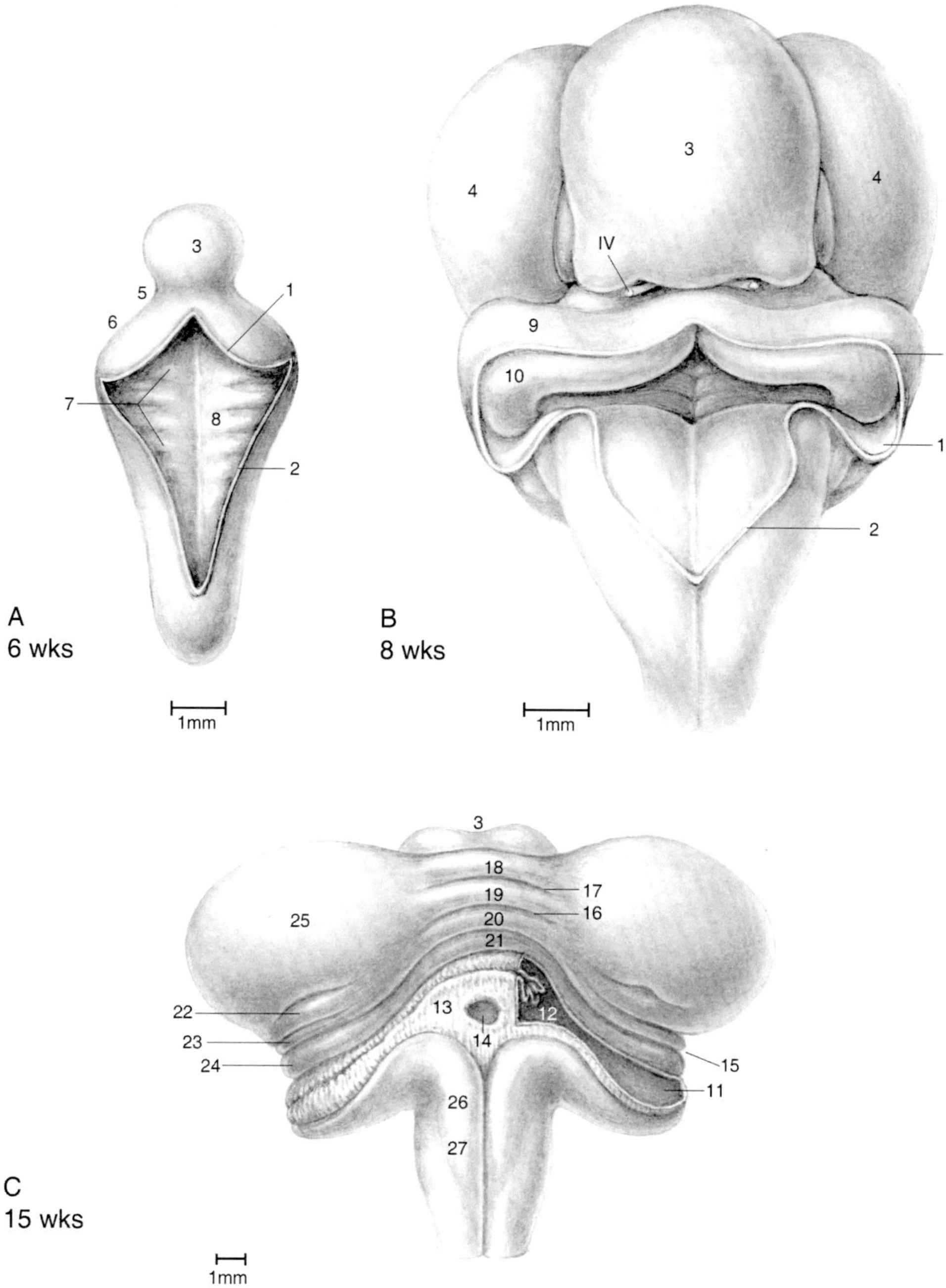

1 Rostral rhombic lip
2 Caudal rhombic lip
3 Mesencephalon
4 Telencephalic vesicles
5 Isthmus
6 Cerebellar anlage
7 Rhombomeres
8 Rhomboid fossa
9 Extraventricular cerebellar swelling
10 Intraventricular cerebellar swelling
11 Lateral recess of fourth ventricle
12 Choroid plexus
13 Ependymal roof of IVth ventricle
14 Median aperture
15 Posterolateral fissure
16 Fissura secunda
17 Prepyramidal fissure
18 Declive + folium vermis + tuber vermis
19 Pyramis
20 Uvula
21 Nodulus
22 Tonsilla
23 Paraflocculus
24 Flocculus
25 Cerebellar hemisphere
26 Gracile tubercle
27 Cuneate tubercle
IV Trochlear nerve

Fig. 2.14 A–C. Dorsal views of the developing human brain stem and cerebellum at 6 weeks (A), 8 weeks (B) and 15 weeks (C). A and B are based on [70]; C is modified from [55]

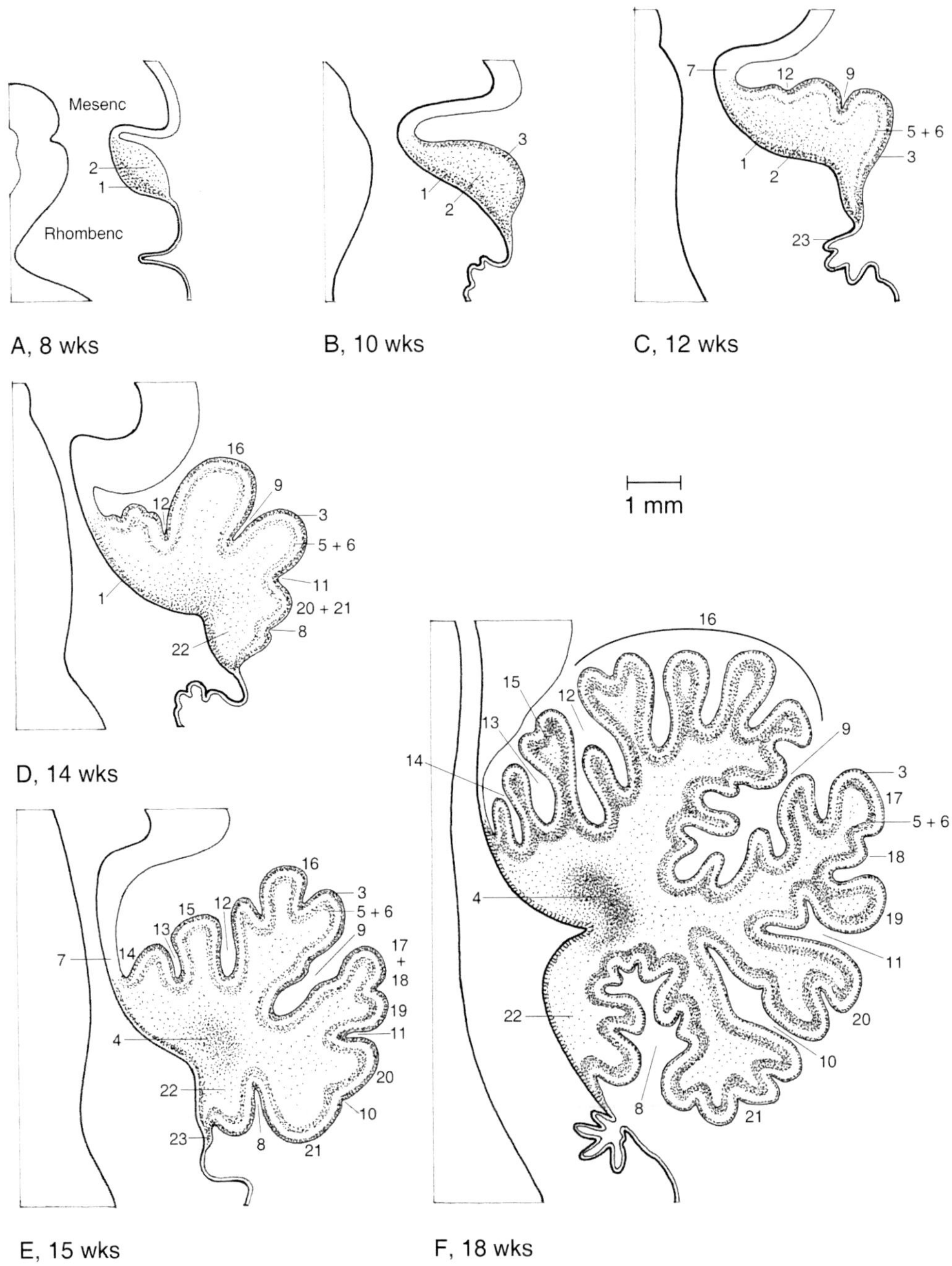

1 Ventricular matrix
2 Mantle layer
3 External granular layer
4 Deep cerebellar nuclei
5 Purkinje layer
6 Internal granular layer
7 Superior medullary velum
8 Posterolateral fissure
9 Fissura prima
10 Fissura secunda
11 Prepyramidal fissure
12 Preculminate fissure
13 Precentral fissure
14 Lingula
15 Central lobe
16 Culmen
17 Declive
18 Folium vermis
19 Tuber vermis
20 Pyramis
21 Uvula
22 Nodulus
23 Inferior medullary velum

Fig. 2.15 A–F. Sagittal sections of the developing cerebellum at 8 weeks (**A**), 10 weeks (**B**), 12 weeks (**C**), 14 weeks (**D**), 15 weeks (**E**) and 18 weeks (**F**). Mainly based on [105]

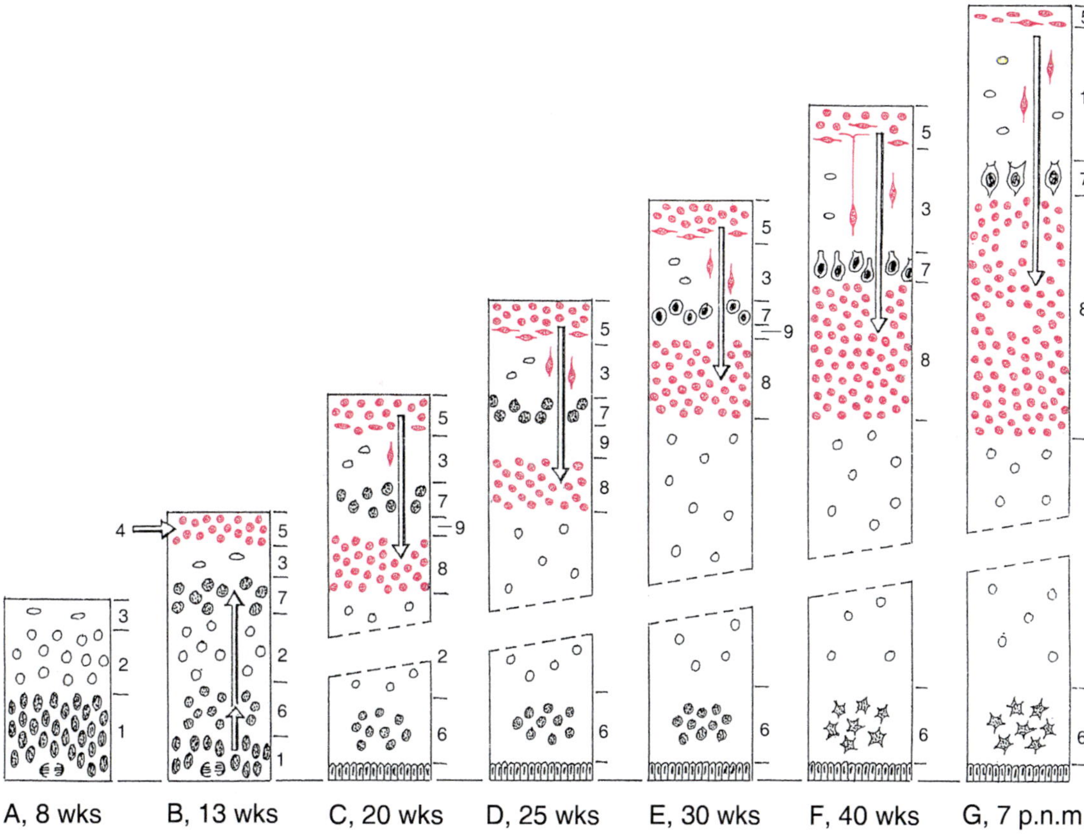

1 Ventricular matrix
2 Mantle layer
3 Marginal layer
4 Rhombic lip
5 External granular layer
6 Deep cerebellar nuclei
7 Purkinje layer
8 Internal granular layer
9 Lamina dissecans
10 Molecular layer

Fig. 2.16 A–G. Semidiagrammatic summary of the histogenesis of the human cerebellum at 8 weeks (**A**), 13 weeks (**B**), 20 weeks (**C**), 25 weeks (**D**), 30 weeks (**E**), 40 weeks (**F**) and 7 months after birth (**G**). The *arrows* indicate the main directions of cell migration. *p.n.m.*, postnatal month; the external granular layer and its derivatives are shown in *red*. Based on [47]

mediolaterally arranged nuclei (Fig. 2.16 G). The cortex and the nuclei are separated by a layer of fibres (Figs. 5.22, 5.23). The latter extends as *"arbor vitae"* into the cerebellar lobules and folia.

The *cerebellar cortex*, which shows a uniform structure throughout, consists of an external, cell-poor molecular layer, an intermediate layer formed by the large piriform somata of the Purkinje cells and a deep layer made up by densely packed, small granule elements (Fig. 2.16 G).

The Purkinje cells, which form the only output elements of the cerebellar cortex, send their axons to the deep cerebellar nuclei. Collectively, they are designated as the *corticonuclear projection*. The flattened, sagittally oriented dendritic trees of the Purkinje cells extend into the molecular layer.

Apart from the Purkinje and the granule cells, the cerebellar cortex contains three types of interneurons, the stellate and basket cells, which are situated in the molecular layer and the Golgi cells, which are found in the superficial zone of the granular layer. All cerebellar neurons, except for the granule cells, are inhibitory in nature.

The cerebellar cortex receives two types of afferents, climbing fibres, which arise from the inferior olivary complex, and mossy fibres, which originate from the remaining precerebellar nuclei. The climbing fibres synapse with the proximal parts of the Purkinje dendritic trees. The mossy fibres ramify in the granular layer, where they contact the dendrites of the granule cells. The axons of the latter ascend to the molecular layer where they bifurcate into transversely oriented parallel fibres. Each of these parallel fibres makes synaptic contact with many Purkinje cells.

Two germinal zones, the ventricular matrix and the external granular layer, are involved in the histogenesis of the cerebellum [27, 58, 61, 77, 171, 190, 226, 227] (Figs. 2.15, 2.16).

The cerebellar ventricular matrix is mainly active during the second and third month. It produces cells of the deep cerebellar nuclei and the Purkinje cells, in that order. The primordia of the nuclei arise throughout the cerebellar plate and aggregate rostrolaterally via a tangential migration. They are first discernable as distinct masses at the tenth week. The Purkinje cells start to be produced as the nuclear neurons move out; they first accumulate periventricularly and later migrate radially into the overlying cerebellar cortex. At about the tenth week, the Purkinje layer manifests itself as a concentration of neuroblasts in the outer zone of the mantle layer (Fig. 2.16 B). During further development, the Purkinje elements slowly mature and the Purkinje layer, which is initially several cells thick, gradually thins out via cell intercalation to the monolayer that characterizes the mature cerebellum (Fig. 2.16 B–G). It is important (see below) that during early fetal development the primordial Purkinje layer shows clear interruptions and, hence, has a clustered appearance. Apart from Purkinje and cerebellar nuclear cells, the cortical interneurons, stellate, basket and Golgi, all arise subsequently from the ventricular matrix [59] or from free precursors displaced into the white matter.

The second cerebellar germinal zone originates from the rostral, metencephalic portion of the rhombic lip (Figs. 2.9, 2.14 A, B). From the tenth week onwards rhombic lip cells migrate rostrally through the outermost part of the marginal zone to form the external granular layer. Whereas the cerebellar ventricular matrix declines at the end of the third month, the external granular layer remains mitotically active until well after birth. During this protracted proliferative period, the steadily expanding external granular layer produces enormous numbers of neuroblasts destined to form the internal granular layer. (The cerebellar granular cells are by far the most numerous neuron in the human central nervous system; according to some estimates [101] 80–85% of all human neurons are cerebellar granule cells.) After their final mitosis, the precursors of the granule cells become bipolar and send out two processes, oriented parallel to the long axis of the folia. Then, another process develops and grows out centrally. The nucleus and surrounding cytoplasm move into this process and the cell soma now assumes a radially oriented spindle shape. Finally, the radially oriented neuroblasts migrate inward to form the internal (or definitive) granular layer (Fig. 2.16C-G). During their passage through the marginal layer, the spindle-shaped somata take the lead, "trailing" their lengthening, radial process behind them. Throughout their inward migration, the granule precursors are guided by the long radial processes of Bergmann cells [133, 168, 190]. During most of the fetal period, the Purkinje and deep granular layers are separated by a cell-poor *lamina dissecans* (Fig. 2.16C–E).

The most noticeable morphogenetic processes during the postembryonic development of the cerebellum are the extraventricular expansion and folding of its external surface (Figs. 2.6, 2.15B–F), which occur concomitantly with the formation of the definite internal granular layer from the external granular layer. During the extraventricular growth, the lateral parts of the cerebellar primordium (the future *cerebellar hemispheres*) expand rapidly, whereas the intervening median portion (the future *cerebellar vermis*) lags somewhat behind. Due to this differential growth, the cerebellar primordium shows a typical dumb-bell shape at the end of the fourth month (Fig. 2.14C).

The folding of the external cerebellar surface involves the development of lobes, lobules and folia, as well as the formation of numerous intervening fissures (Figs. 2.6, 2.15C–F, 3.8). These processes will not be described in detail here. Let it suffice to mention that the fissures in the anterior part of the cerebellum, including the *fissura prima* (i.e. the first fissure to appear), first develop in the region of the vermis and later extend laterally. In the posterior regions of the cerebellum most fissures develop independently in vermis and hemispheres, and sometimes become confluent at later stages. This mode of development is also followed by the *posterolateral fissure*, which forms the approximate border between the caudal vestibulocerebellum and the somesthetic corpus cerebelli [22, 104, 106, 107]. The folding and fissuration results in a vast increase in surface area of the cerebellum. This surface area is about 150 mm^2 in a 4-month-old fetus, but it amounts to 66,000 mm^2 in the adult [212]. Because of the transverse orientation of the cerebellar folds and fissures the cerebellar surface area extends mainly in a rostrocaudal direction. The flattened and extended surface of the adult human cerebellum has a maximal rostrocaudal extension of 120 cm and a maximal transverse width of 17 cm [24].

Although transverse folding is by far the most prominent external feature of the human cerebellum (Fig. 3.11), hodological, physiological and chemoarchitectural studies have conclusively shown that this organ is fundamentally composed of a number of longitudinal compartments or modules. (For reviews, see [217, 218].) Each of these compartments contains a strip of Purkinje cells and a corresponding sector of the mediolaterally arranged deep cerebellar nuclei. The Purkinje cells project to "their own" sector of the deep nuclei and the climbing fibres, originating from the inferior olivary complex, project in a highly orderly fashion to both the cortical and the nuclear parts of the cerebellar modules.

The longitudinal zonal organization manifests itself clearly during the development of the cerebellum. Thus, Korneliussen [93, 94] observed that during early development the precursors of the Purkinje and the deep nuclear cells form four mediolaterally arranged condensations in the homogeneous cerebellar mantle layer. Kappel [86] studied the regional aggregation of future Purkinje cells in embryos and fetuses of the rhesus monkey. Her reconstructions revealed that the Purkinje elements are arranged in five longitudinally arranged principal zones (designated as I, II, III, "X" and IV) and that two of these zones (II and IV) can be subdivided further (Fig. 2.17A). Moreover, she found that there is a close topical relationship between the developing Purkinje zones and (parts of) the deep cerebellar nuclei. Similar observations have been reported for human fetuses [216] (Fig. 2.17B). Most probably, some of the longitudinal zones in rhesus monkey and human fetuses may correspond to the arrays of roundish elevations or "foci" observed by Langelaan [103] in the vermal region of the cerebellum of 3- to 5-month-old human fetuses (Fig. 2.17C).

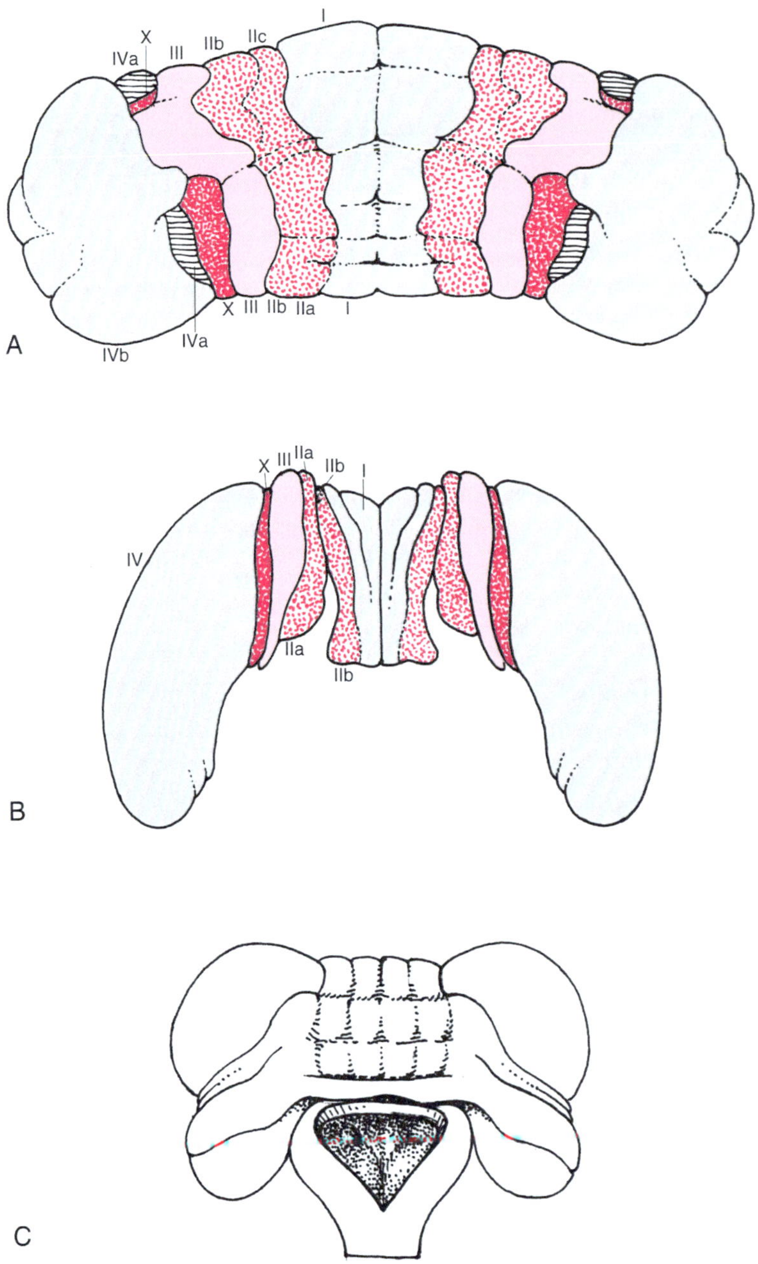

Fig. 2.17 A–C. Features of a longitudinal zonal organization in the developing cerebellum. **A** Graphical reconstruction of a dorsal view of the aggregations of future Purkinje cells in a 70-day-old rhesus monkey fetus. **B** Graphical reconstruction of an anterior view of the clusters of putative Purkinje cells in a 65-mm (crown-rump) human fetus. **C** Dorsal view of a human cerebellum at the end of the third month or beginning of the fourth. **A**, **B** and **C** are reproduced from [86], [216] and [103], respectively. For explanation of symbols, see text

Recent molecular biological analyses (for review, see [60, 146]) have also revealed that a variety of molecular markers are expressed in specific mediolaterally arranged clusters in the cerebellum.

Midbrain

The midbrain or mesencephalon is the only major brain part that develops directly from a primary brain vesicle. During most of the embryonic period, the thin-walled mesencephalic vesicle is strongly curved and encloses a ventricular cavity of some size (Figs. 2.2B, 2.3B,C, 2.4), but during further development its walls gradually thicken and the ventricle becomes reduced to form the *cerebral aqueduct* (Figs. 2.2C, 2.6).

Two neural segments, the mesomeres M1 and M2, have been postulated to participate in the formation of the midbrain [21, 54, 143, 152] (Fig. 2.11, Table 2.1), although in recent studies, some authors, among them Puelles [156, 157], hold that there is no such subdivision. Within the lateral walls, three longitudinal zones, medial tegmental, lateral tegmental and tectal, can be delimited [17] (Fig. 2.18). Together the medial and lateral tegmental zones form the mesencephalic part of the basal plate, whereas the tectal zone represents the alar plate. The relationship between the individual mesencephalic and rhombencephalic zones is not entirely clear. Bengmark et al. [17] and Bergquist and Källén [21] reported that the rhombencephalic ventral, intermedioventral and intermediodorsal zones correspond to the medial tegmental, lateral tegmental and tectal zones, respectively. According to these authors, the rhombencephalic dorsal zone has no mesencephalic equivalent. However, on comparative grounds we consider it likely that the intermedioventral zone is confined to the rhombencephalon [138]. Gene markers, e.g. *Nkx-6.1*, habitually show a distinct interruption at the rhombomesencephalic boundary.

The *oculomotor nuclei* and the *Edinger-Westphal nucleus* develop from the caudal part of the medial tegmental zone, whereas the rostral part of that zone gives rise to the *nucleus of Darkschewitsch* and the *magnocellular part of the red nucleus*. In contrast to conventional accounts, the *trochlear nucleus* clearly arises in the isthmic neuromere (Fig. 2.11). According to Verney et al. [213], the large *parvocellular part of the red nucleus* is of diencephalic origin. The *mesencephalic reticular formation*, the *paralemniscal nucleus* and the *reticular part of the substantia nigra* (see below) develop from the lateral tegmental zone. The tectal zone forms two pairs of elevations, the *inferior and superior colliculi*. The inferior colliculi are relay centres in the auditory projection, whereas the superior colliculi serve as visual reflex centres. In the primordial superior colliculus a zone of migrated cells is formed, which later splits up into the cellular layers typical of the adult structure (Fig. 2.18). After the production of the cells destined for the superior and inferior colliculi, the matrix layer of the tectal zone is gradually transformed into the *mesencephalic central grey*. As indicated in Fig. 2.11, the *posterior pretectal nucleus* develops from the rostral part of the tectal zone.

The ventral part of the embryonic midbrain contains a conspicuous midventral proliferation zone (Fig. 2.18A). Neuroblasts originating from this area spread ventrally almost to the meningeal surface and then migrate laterally through the marginal zone of the mesencephalic basal plate (arrows in Fig. 2.18) [34, 56, 87, 89, 159, 213]. The crescent of cells that develops from this remarkable tangential migration represents the dopaminergic *compact part of the substantia nigra* [213]. The GABAergic cells, which form the *reticular part of the substantia nigra*, stem from quite another source. These elements are generated in the ventricular matrix of the lateral tegmentum and migrate radially towards the brain surface [213].

At the end of the fourth month, fibres descending from the developing cerebral cortex reach the midbrain. Initially, these corticofugal fibres bilaterally form a narrow band on the superficial aspect of the nigral crescent. However, due to the steady arrival of new axons, these thin sheets rapidly increase in thickness to become the massive and prominent *cerebral peduncles* (Figs. 5.16, 6.37).

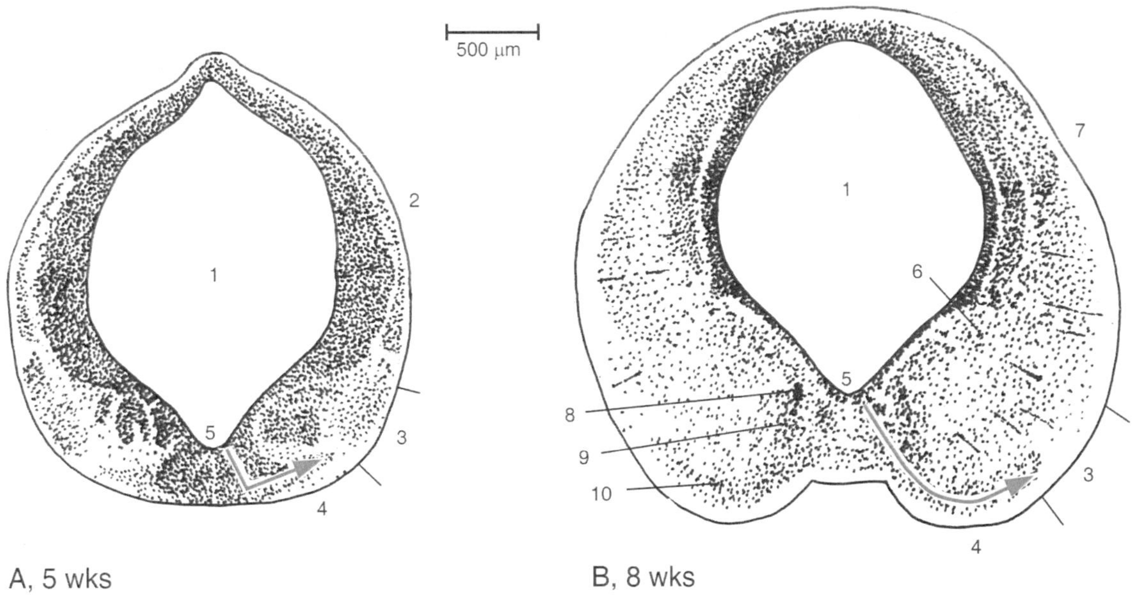

1 Mesocoel
2 Tectal zone
3 Lateral tegmental zone
4 Medial tegmental zone
5 Midventral proliferation zone
6 Habenulo-interpeduncular tract
7 Superior colliculus
8 Oculomotor nucleus
9 Red nucleus
10 Substantia nigra

Fig. 2.18 A, B. Transverse sections of the developing midbrain at 5 weeks (**A**) and 8 weeks (**B**). The *red arrows* indicate the direction of the marginal migration stream, leading to the formation of the compact part of the substantia nigra. Based on photomicrographs in [34] and [80]

Forebrain

At the end of the fourth week the prosencephalon appears as the slightly dilated rostral end of the neural tube (Fig. 2.3 A). It is generally agreed that the prosencephalon, just like the other parts of the brain, comprise lateral plates on both sides connected by thin floor and roof plates and that the lateral plates can be divided into a ventral basal plate and a dorsal alar plate. However, regarding the spatial relations of these various components in the most rostral part of the neuraxis, widely divergent opinions have been expressed. (For reviews, see [138, 165]). Recent experimental and gene expression data [156, 157, 161, 165, 213] have shown that the rostral, prechordal part of the floor plate ends directly in front of the infundibular recess, and that the roof plate extends for some distance into the commissural plate (Fig. 2.11). It has already been mentioned that the boundary between the alar and basal plates is marked throughout the prosencephalon by a narrow strip-like zone where the gene *Nkx-2.2* is expressed. This strip, which makes a marked, unexplained dorsal deflection in the thalamic region, crosses the median plane directly behind the optic chiasm (Fig. 2.11). It follows that in the area between the rostral ends of the floor and roof plates, the alar and basal plates of both sides are directly continuous with each other across the median plane (Fig. 2.11).

In the prosencephalon, three subsequent evagination processes occur. The laterally directed optic evaginations start to develop very early, before the closure of the anterior neuropore, and rapidly enlarge to form the optic vesicles, from which the eyes and the optic nerves develop (Figs. 2.2 A, 2.3). Somewhat later, a bilateral evagination of the dorsolateral part of the prosencephalon, caudal to the eye vesicles, leads to the formation of the telencephalic hemispheres (Figs. 2.2 B,C, 2.3 C, 2.4, 2.5 B,C) and, by a secondary evagination of both hemispheric walls, the olfactory bulbs are formed (Figs. 2.4, 2.5 B).

As a result of the hemispheric evagination, the prosencephalic ventricular cavity forms two lateral dilatations, the future *lateral ventricles*. The rest of the prosencephalon remains unevaginated. The diencephalon develops from the caudal portion. The unevaginated, most rostral portion later forms part of the telencephalon and is called the *telencephalon impar*. The preoptic nuclei develop within its walls. After the cerebral hemispheres are formed, the cavity of the diencephalon together with that of the telencephalon impar is called the *third ventricle*. Figure 2.11 shows that the rostral closure of the third ventricle is a heterogeneous structure. It includes a basal plate component (the *retrochiasmatic area*), two alar plate components (the *chiasmatic plate* and the thin *lamina terminalis*) and the most rostral part of the roof plate. The latter thickens and forms the commissural plate, within which the commissures between the two hemispheres will be formed (see below). The lateral ventricles communicate with the unpaired third ventricle via a wide interventricular foramen. During later development the interventricular foramina become narrowed by the expansion of the medial ventricular eminences (see below).

The olfactory vesicles, which develop on the basal side of the cerebral hemispheres (Fig. 2.4), initially communicate widely with the lateral ventricles. During further development these vesicles lengthen in a rostrocaudal direction and become stalked (see Fig. 2.23 E). The rostral portion thickens and dilates to form the olfactory bulb, while the proximal portion or stalk remains thinner and is transformed into an elongated olfactory tract. The ventricular cavity in the olfactory bulb and tract is gradually obliterated. Secondary olfactory fibres, originating from the olfactory bulb, pass to the olfactory or piriform cortex via the olfactory tract. During early development the primordium of this cortical region occupies a superficial position in the ventrolateral hemisphere wall (Figs. 2.19 A, 2.20 A), but due to the expansion of the neocortex it turns ventrally and ultimately comes to lie on the dorsomedial surface of the temporal lobe (Fig. 6.41).

A small median evagination of the diencephalic roof plate leads to the formation of the *epiphysis* (Figs. 2.5, 2.6). Similar local evagina-

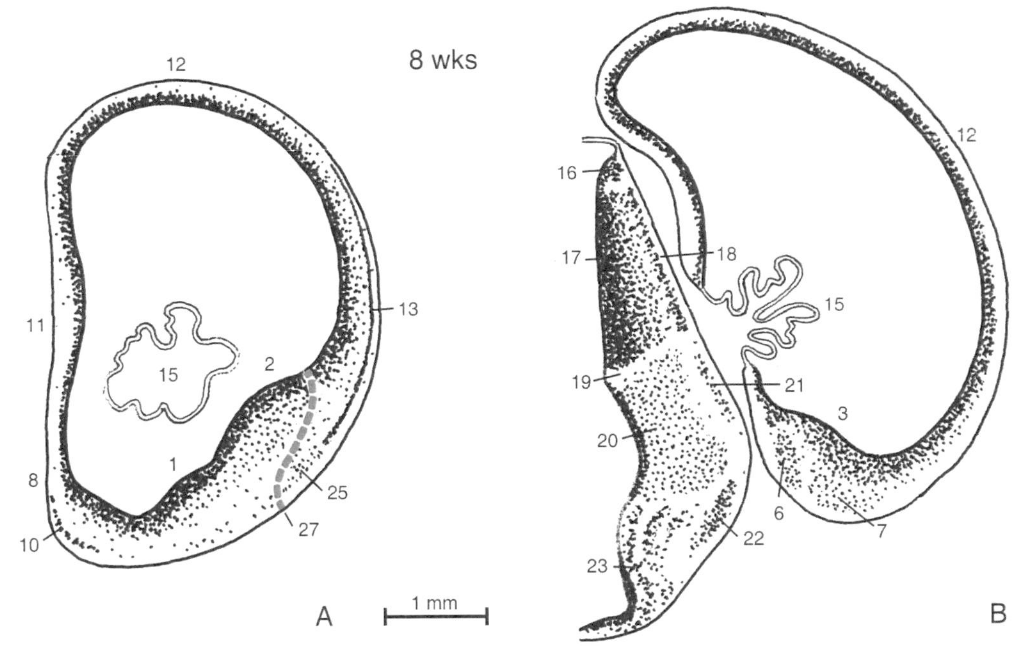

Fig. 2.19. Transverse sections through the rostral telencephalon (A) and the caudal telencephalon and diencephalon (B) of a human embryo of about 25 mm. Based on [69, 82, 100, 143, 174, 176]. For abbreviations see Fig. 2.20

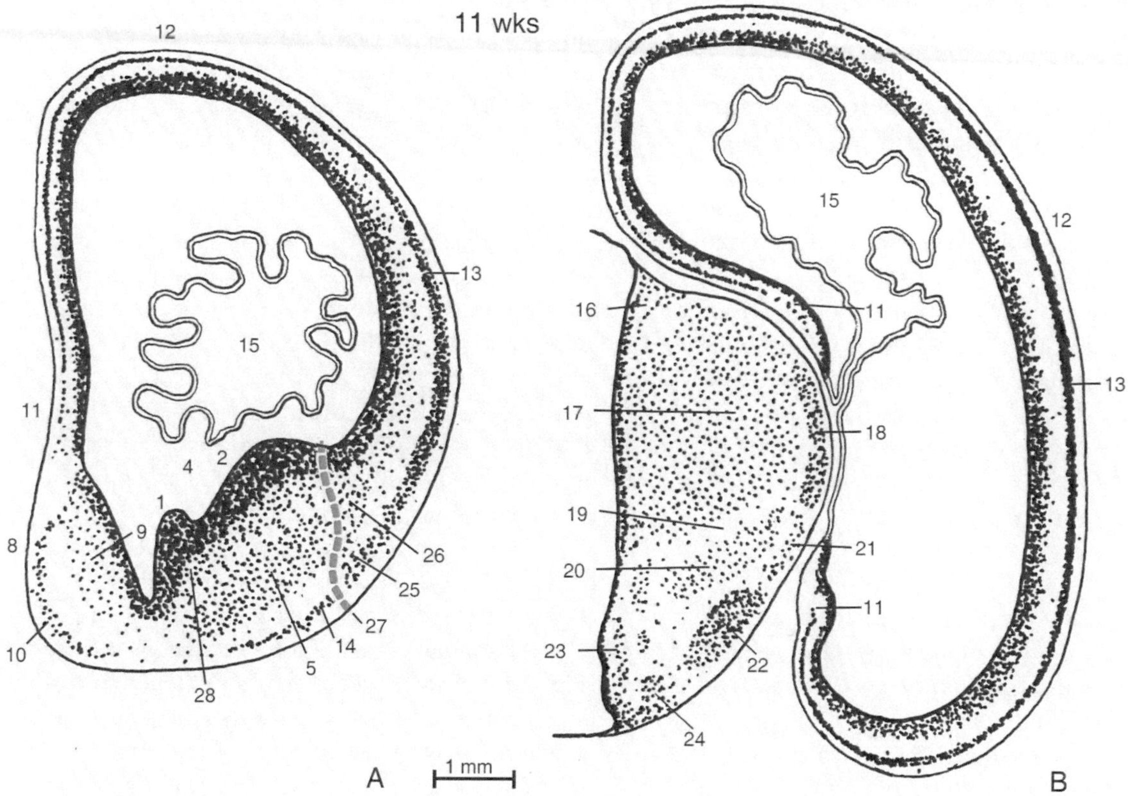

1 Medial ganglionic eminence
2 Lateral ganglionic eminence
3 Caudal part of ganglionic eminence
4 Internal capsule
5 Putamen
6 Medial amygdaloid nucleus
7 Basal amygdaloid nucleus
8 Septum
9 Accumbens nucleus
10 Nucleus of diagonal band
11 Hippocampal formation
12 Neopallium
13 Cortical plate
14 Olfactory tubercle
15 Choroid plexus
16 Habenular ganglion
17 Dorsal thalamus
18 Dorsal lateral geniculate nucleus
19 Zona limitans intrathalamica
20 Ventral thalamus (zona incerta)
21 Ventral lateral geniculate nucleus
22 Subthalamic nucleus
23 Hypothalamus
24 Mamillary body
25 Piriform cortex
26 Ventral pallium (claustral stream)
27 Pallio-subpallial boundary
28 Pallidum

Fig. 2.20. Transverse sections through the rostral telencephalon (**A**) and the caudal telencephalon and diencephalon (**B**) of a human embryo of about 48 mm. Based on [69, 82, 100, 143, 174, 176]

tions of the rostral, prechordal part of the floor plate give origin to two niches of the third ventricle, the *mamillary* and *infundibular recesses* (Fig. 2.6). The wall of the latter participates in the formation of the *neurohypophysis*.

During the second month, a small, transversely oriented fold appears in the rostral part of the roof plate, just dorsal to the commissural plate (Fig. 2.5 B, C). According to Johnston [79] this *velum transversum*, which disappears during the fetal period, marks the dorsal end of the telodiencephalic boundary.

Diencephalon

Traditionally, the human diencephalon is subdivided into four dorsoventrally arranged zones: *epithalamus, dorsal thalamus, ventral thalamus* and *hypothalamus* (Figs. 2.19 B, 2.20 B). This subdivision, which originated from the comparative anatomical studies of Herrick [63], soon became widely accepted and was also applied to the embryonic human diencephalon [50, 96, 100, 174]. In their ontogenetic studies of the human diencephalon, Kahle [81] and Richter [176, 177] distinguished five rather than four longitudinal zones. According to these authors, the extra - subthalamic - zone is intercalated between the ventral thalamus and the hypothalamus.

The *epithalamic zone* gives origin to the *habenular nuclei* and the *pineal gland* (*epiphysis*). The habenular nuclei receive their main afferents from limbic structures (septum, hypothalamus), via the *stria medullaris*. This fibre system passes along *the taenia thalami*, i.e. the seam of attachment of the diencephalic roof plate (Fig. 2.11). The principal efferent pathway of the habenular nuclei is the *habenulo-interpeduncular tract* (also known as *fasciculus retroflexus* or *bundle of Meynert*), which descends steeply to the base of the midbrain, and then terminates mainly in the interpeduncular nucleus (Fig. 2.11). The habenular nuclei, the stria medullaris and the habenulo-interpeduncular tract develop early. All of these structures can be clearly distinguished before the end of the second month [96, 228].

The *dorsal thalamus* occupies initially only a narrow strip of the diencephalic wall [174]. However, during further development this inconspicuous primordium grows out and differentiates into an enormous nuclear complex, which is designated as the *thalamus*. The thalamus is the main relay station for subcortical information being sent to the telencephalon. It entertains massive and reciprocal connections with the cerebral cortex and the striatum. Due to the expansion of the thalamic primordium this structure as well as all adjacent structures become considerably deformed, as is apparent from the following enumeration.

1. The habenular nuclei are displaced caudally. Consequently, the stria medullaris becomes extraordinarily long and the habenulo-interpeduncular tract shows a sharp flexure (Fig. 7.1). In fact, the latter tract is truly "retroflexed" only in humans and other primates, i.e. in mammals with a greatly expanded thalamus.
2. The expansion of the thalamus itself and of the thalamotelencephalic connections leads to a widening and a change in the orientation of the boundary plane between diencephalon and telencephalon. In early embryonic stages, this plane is more or less transversely oriented, but gradually it assumes an almost longitudinal orientation (Fig. 2.2). Afferent and efferent fibres connecting the thalamus with the telencephalon accumulate in the ventral part of the enlarged telodiencephalic contact zone to form the so-called hemispheric stalk. Rostrolaterally, this fibre mass is directly continuous with the *internal capsule* (see below) and caudally with the *pes pedunculi*.
3. The early embryonic free lateral surface of the thalamus rotates to become its caudal surface (Figs. 2.2, 2.21) [69, 188]. This apparent vanishing of the free lateral wall of the thalamus is accompanied by a remarkable caudal displacement of the anlage of the *dorsal lateral geniculate nucleus* (Fig. 2.2). Late in development, the enormous outgrowth of the *pulvinar*, i.e. the most caudal portion of the thalamus, leads to a further ventral displacement of the dorsal lateral geniculate nucleus [172].

4. The dorsal surface of the expanding thalamus ends dorsomedially at the level of the thinned *taenia thalami*, where the choroidal roof of the third ventricle is attached. The latter is continuous rostrally at the interventricular foramen with a thin, epithelial part of the medial wall of the caudally extending telencephalic hemispheres. This epithelial structure, which gives rise to the *choroid plexus of the lateral ventricles*, represents a stretched and much deformed part of the roof plate, which during the evagination process is drawn into the telencephalic hemispheres. In the adult, it almost reaches the tip of the temporal lobe. As will be detailed below, the thalamus proper, or dorsal thalamus, is separated from the prosencephalic region, in which the evagination of the telencephalic hemispheres occurs, by the *ventral thalamus* or *prethalamus*. The dorsal part of this intervening structure is known as the *eminentia thalami* (Fig. 2.11). The choroidal roof of the third ventricle is attached to the eminentia thalami, its taenia being directly continuous with that of the thalamus proper. The morphologically caudal portion of the eminentia thalami forms a protrusion at the back side of the interventricular foramen, whereas its morphologically rostral portion is deflected caudolaterally and drawn into the developing hemispheres. This remarkable morphogenetic event explains why a traditionally diencephalic part of the forebrain borders directly on the traditionally purely telencephalic lateral ventricle. In the literature these relations are generally "explained" by postulating that a thin, membranous part of the telencephalic wall adheres to the dorsal surface of the adjoining thalamus during ontogenesis, thus forming the so-called *lamina affixa*. However, it now appears that this structure does not exist. The territory in question, which is clearly visible in Figs. 3.10, 3.12, 3.17 and 5.29, actually represents the surface area of the prethalamic territory, intercalated between the thalamus proper and the cerebral hemispheres [158].

5. As will be discussed later in the chapter, the shape of the ventral thalamic and subthalamic zones is also strongly influenced by the outgrowth of the dorsal thalamus.

The differentiation of the dorsal thalamus begins late and progresses slowly in comparison with the development of the ventral thalamus and hypothalamus. During the second half of the second month and the first half of the third month, the thalamic anlage is characterized by the presence of a wide and compact ventricular zone, a cell-rich mantle layer and a narrow marginal zone (Fig. 2.19B). The most superficial part of the mantle zone is occupied by a sheet of densely packed cells, the thalamic plate [96]. The *dorsal lateral geniculate nucleus*, which differentiates much earlier than the remaining thalamic nuclei, develops from this plate [172].

The anlage of the dorsal thalamus shows a remarkable pattern of proliferation. Mitotic figures crowd along its ventricular surface, but many of these are also observed away from the ventricular surface, throughout the compact ventricular zone. Such non-surface-mitoses are particularly numerous in the dorsocaudal part of the thalamus, which is the area giving rise to the largest and the most populous of the thalamic nuclei [193]. Thus, the thalamic ventricular zone combines features of a classical ventricular matrix (interkinetic nuclear migration) with those of a subventricular matrix (proliferation in situ). Only by this combination is the thalamic anlage able to produce the enormous number of cells required for the huge thalamic nuclear complex [193]. A similar developmental mechanism operates in other areas of the CNS, in which particularly large numbers of neurons are produced, such as the spinal dorsal horn, the neopallium and the striatal primordia [192, 194, 195, 197, 198].

During the second half of the third month, the thalamic ventricular zone rapidly declines in width, and by the end of that month, its proliferative activity comes to an end (Fig. 2.20B) [80, 172]. During the same period, the thalamic anlage grows out to become a large, but as yet entirely undifferentiated mass of cells (Fig. 2.20B).

The diencephalic ventricular zone represents the principal, but by no means only source of thalamic neurons. The investigations of Rakic et al. [111, 112, 172, 190] have shown that the human thalamus, after the diencephalic mitotic activity has ceased, receives a massive influx of neuroblasts from the ganglionic eminence, i.e. a proliferative zone in the basal telencephalon. These elements form a compact, medially directed stream, which is designated as the *corpus gangliothalamicum*. Its constituent cells, which are all GABAergic, contribute to the expansion of those thalamic nuclei that project selectively to the association cortex, i.e. the *medial* and *pulvinar nuclei*. This remarkable additional complement of neurons is imported to the thalamus during a protracted period of fetal development, extending from the 15th to the 34th week [112]. A comparable telo-diencephalic migratory stream has not been identified in the developing macaque monkey brain, nor in any other mammalian species that has been investigated so far [112]. An interesting and as yet entirely open question is: how do these migrating cells traverse the intervening ventral thalamic territory on their way to the dorsal thalamus?

The nuclear differentiation of the dorsal thalamus begins at the end of the third month and 2 months later most of the adult thalamic nuclei can be readily recognized. The *central median nucleus* appears shortly after the *dorsal lateral geniculate nucleus* and the expansion of the *pulvinar* occurs relatively late [96, 172].

During early development, the *ventral thalamus* or *prethalamus* occupies a considerable sector of the wall of the diencephalon (Fig. 2.19 A) [50, 174]. Its matrix zone, which never attains the width of that of the dorsal thalamus, has been completely depleted by the end of the tenth week [80]. The dorsal and ventral thalami are separated by a cell-free band, the *zona limitans intrathalamica* (Fig. 2.19 A). Within this zone, the *lamina medullaris externa* develops. A similar, but less conspicuous limiting zone separates the ventral thalamus from the subthalamus and becomes the *fasciculus thalamicus*. The column of grey matter between these two limiting zones represents the ventral thalamic primordium (Fig. 2.19B), from which the *ventral lateral geniculate nucleus*, the *zona incerta* and the *reticular nucleus* develop, nuclei which share GABA as a transmitter of (most of) their neurons (Fig. 2.20 B). These nuclei differentiate early; they are already clearly distinguishable in the second half of the third month, during which time the dorsal thalamus is still represented by an undifferentiated primordium (Fig. 2.20 B). Due to the expansion of the thalamic nuclear complex, the reticular nucleus is transformed into a thin shell of neurons covering this complex rostrally, laterally and ventrally (Fig. 6.37).

The *subthalamus*, which is characterized by the very early depletion of its ventricular matrix, gives rise to the subthalamic nucleus and, according to Kahle [81] and Richter [176, 177], to the internal and external parts of the globus pallidus as well. These authors reported that the anlagen of these three cell masses initially show a caudorostral arrangement (Fig. 2.21 A, B), but that due to the expansion of the thalamus this longitudinal orientation gradually changes into an almost transverse one (Fig. 2.21 C–E). During later development the two pallidal primordia were observed to unite with the primordial putamen to form the lentiform nucleus. Recent studies of gene expression patterns (in particular that of *Nkx2.1*) in mice have conclusively shown that the globus pallidus derives from an entirely telencephalic primordium [123].

In the rat, a germinative zone lying caudally to the mamillary recess is responsible for formation of the neurons of the subthalamic nucleus. From this site of origin, the neurons migrate radially and then tangentially and dorsally along the marginal zone of the ventral diencephalon [121], deep to the developing peduncle.

The interpretation of the subthalamus as a separate diencephalic entity is not generally accepted. Kuhlenbeck [100] incorporated the entire subthalamic zone in the hypothalamus and, according to several recent authors who used molecular genetic techniques [157, 166], the globus pallidus is of telencephalic rather than of diencephalic origin.

The *hypothalamus* represents the ventralmost part of the diencephalon, where it forms the floor and contributes to development of the lateral walls of the third ventricle. Its upper

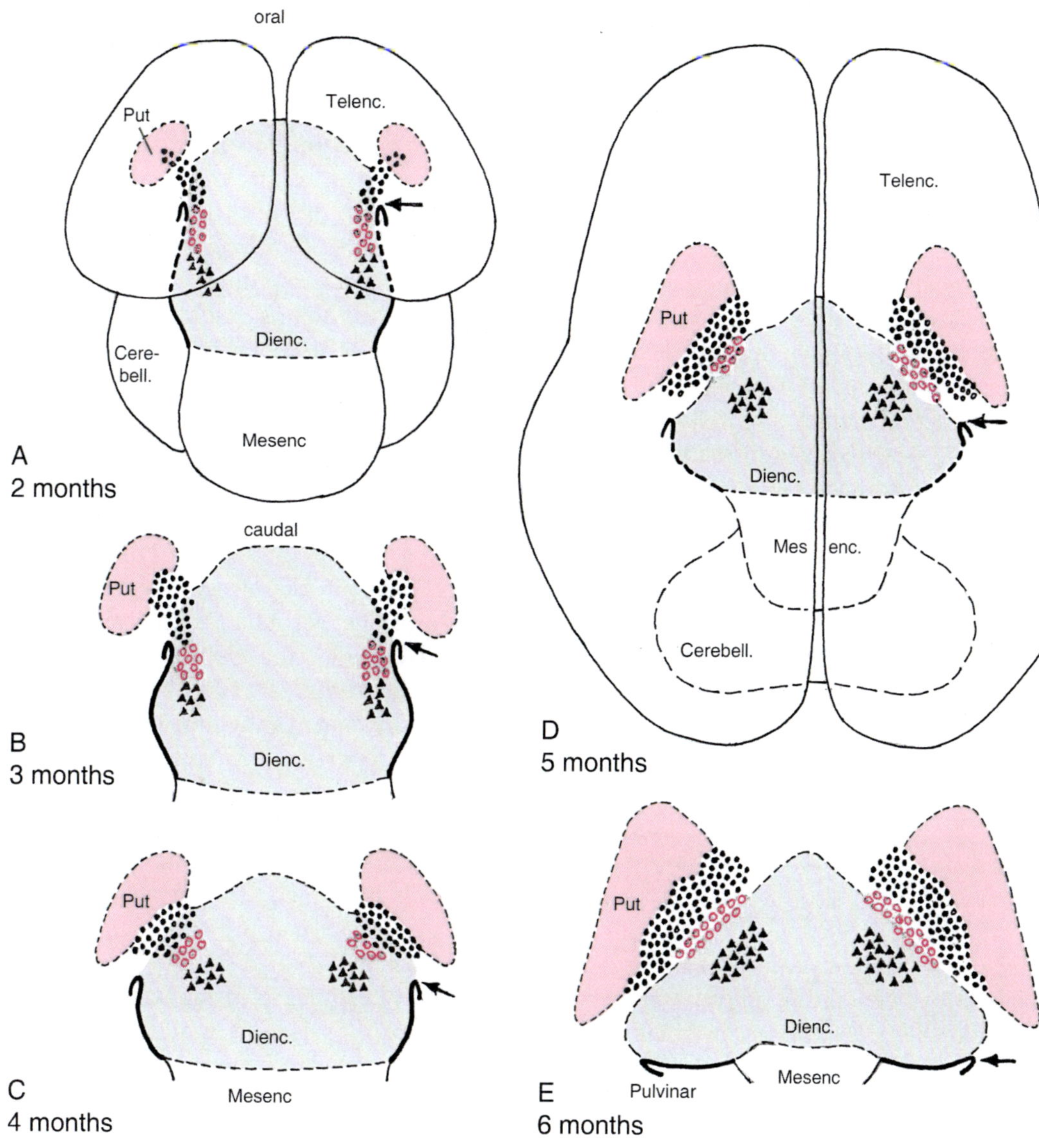

Fig. 2.21 A–E. Development of the diencephalon and adjacent parts of the telencephalon, as shown by a series of human embryos ranging from **A**, the second, to **E** the sixth month. The structures depicted are projected upon a horizontal plane. The anlage of the diencephalon is tinted *light grey* and its external surface is indicated by *heavy lines. Arrows* indicate the transition of the diencephalic surface to the medial hemispheric wall. *Black triangles* indicate the subthalamic nucleus; *open circles* in *red*, the internal part of the globus pallidus; *closed circles*, the external part of the globus pallidus. The putamen is indicated by a *light red colour.* Modified from [176]

boundary is marked on the ventricular side by a shallow groove, the *hypothalamic sulcus* (Fig. 3.8). Gilbert [50] distinguished four caudorostrally arranged regions in the hypothalamus of a 9.5-mm embryo, the mamillary region, the premamillary region, a dense band of cells which she designated as the hypothalamic cord and the preoptic region. During later development the hypothalamic cord differentiated into several cell masses, including the anterior, ventromedial and lateral hypothalamic nuclei. For a recent, detailed analysis of the ontogenesis of the human hypothalamus, we refer to Koutcherov et al. [97]. Let it suffice to mention that, according to these authors, the primordia of most hypothalamic nuclei can be distinguished by the end of the third month, the *lateral* and *posterior hypothalamic areas* as well as the *supraoptic* and *paraventricular nuclei* [96] differentiate early (9 weeks), and the *suprachiasmatic nucleus* cannot be clearly distinguished before the 23rd week [97].

Telencephalon

The telencephalic hemispheres can be subdivided into a dorsal pallium and a ventral subpallium. The subpallium comprises a small medial or septal zone and a much larger lateral zone. During the second month, the walls of the lateral subpallium start to thicken and form an intraventricular protrusion on both sides. A ventricular groove divides the rostral part of this elongated protrusion into separate *medial* and *lateral ganglionic eminences* (Figs. 2.19, 2.20). The lateral ganglionic eminence gives rise to the *caudate nucleus* and the *putamen*, whereas the *globus pallidus* develops from the medial ganglionic eminence. The *amygdaloid complex* arises largely from the caudal, undivided part of the intraventricular swelling. The cells produced in the lateral ganglionic eminence initially form a single striatal primordium. However, during further development, increasing numbers of corticofugal and corticopetal fibres pass through this primordium. These fibres, which together form a compact sheet of white matter known as the *internal capsule*, separate the periventricular *caudate nucleus* from the peripheral *putamen* (Fig. 2.20 A). Cells destined to form the *globus pallidus* migrate in a ventrolateral direction from the matrix zone of the medial ganglionic eminence and come to lie directly medially to those of the primordial putamen. In the adult brain putamen and globus pallidus together form a discrete, macroscopically recognizable complex, the *nucleus lentiformis* (Figs. 5.7, 6.38). The caudate and lentiform nuclei together are designated as the *basal ganglia*.

At the level of the intraventricular foramen, the medial surface of the medial ganglionic eminence forms part of the wall of the third ventricle (Fig. 2.11). It has already been mentioned that on that account several authors [100, 176, 177] consider the globus pallidus to be of diencephalic origin.

It is noteworthy that throughout the ganglionic eminences a particularly strongly developed subventricular zone can be observed [195, 197]. In these eminences, which, as discussed, give rise to the basal ganglia, a zone of proliferative cells is formed at the external side of the ventricular matrix at the end of the second month. During the third month, this subventricular zone becomes a prominent feature of the striatal primordia and gradually takes over from the ventricular matrix as the major site of mitotic activity. After the fourth month, this zone gradually declines. The thin subventricular layer observed in the late fetal and early postnatal stages, which is concerned mainly with the production of macroglial elements, is considered to be a remnant of this conspicuous, neuron-producing compartment.

Whereas many developing neurons settle in the periventricular zone in the subpallium, in the pallium, the neuroblasts migrate away from this zone and become incorporated into superficially situated cortical formations. Three such formations, the medial *hippocampal cortex*, the dorsolateral *neocortex* and the ventrolateral *piriform cortex*, can be distinguished (Figs. 2.19, 2.20). The pallial neuroblasts migrate peripherally along the processes of ependymal gliocytes [169, 171]. In the early cerebral hemispheres, these fibres show a radial orientation. However, later in development the ependymoglial fibres in the striatal region lose contact with the meningeal surface and those emanating from the adjacent lateral part of the pallial ventricular surface deflect ventrally in association with the appearance of the primordial cortex. Thus, the ependymal processes pursue curving courses, which circumscribe the striatal anlage. Given the close association between the ependymal fibres and the migration of pallial neuroblasts, these glial processes not only reveal the formal changes in the lateral hemisphere wall, but also indicate the remarkable route along which the neuroblasts destined for the piriform cortex reach their destination [132, 198] (see Fig. 2.25).

In all parts of the pallium, zones of migrated cells are formed, but a compact cortical plate becomes manifests only in the neopallium, i.e. the region within which the *neocortex* develops (Figs. 2.10 J, 2.19, 2.20). The adult *hippocampal formation* contains three layers, an intermediate cellular layer, sandwiched between an internal and an external plexiform layer (Figs. 6.35-6.37). Likewise, the *piriform cortex* comprises three layers: a deep layer displaying a moderate cell density, an intermediate layer consisting of densely packed neuronal somata, and a superficial plexiform or zonal layer (Fig. 6.41). By contrast, the *neocortex* consists of six layer and, as already mentioned, all of these layers, except for the most superficial one, originate from the cortical plate (Fig. 2.10 J, K).

The expansion of the evaginated telencephalic vesicles, which leads to the formation of the huge cerebral hemispheres, begins in the fifth week (Fig. 2.3 C) and proceeds steadily throughout development. The development, outgrowth and differentiation of the neocortex is doubtless the driving force behind this expansion. The telencephalic vesicles enlarge rostrally, dorsally and caudally. Beyond the level of the lamina terminalis and the roof plate the rostral and dorsal outgrowth brings the medial wall of each vesicle into apposition (Fig. 2.5 C). The telencephalic vesicles expanding backwards, or *cerebral hemispheres*, successively overlap the diencephalon (Fig. 2.4), the midbrain (Fig. 2.2 C), and ultimately even the cerebellum (Fig. 3.7).

The outgrowth of the neopallium not only leads to an overall telencephalic expansion, but also involves a rotation of the rostral and caudal portions of the cerebral hemispheres around a transverse axis. The rostral rotation causes a caudal displacement of the site of attachment of the olfactory tracts. The caudal rotation leads to the formation of the temporal lobes. Figure 2.22 illustrates how the expansion and the resulting curvature of the neocortex determines the shape of the lateral ventricle. Due to the caudal rotation, the early fetal posterior pole of the cerebral hemisphere moves downwards and then forwards to become the definitive temporal pole. During fetal life, a new caudal expansion results in the formation of the definitive occipital pole. This development entails the formation of a caudal extension of the ventricular cavity, known as the posterior horn (Fig. 2.22).

The further expansion of the neopallium is coupled with convolution of its external surface. At the end of the third month, a depression, the *lateral cerebral fossa*, appears on the lateral surface of the cerebral hemispheres. The bottom of this depression is known as the *insula*. During further development, the depressed area is gradually overgrown by the adjacent parts of the hemispheres (Fig. 2.23). The overlapping, neocortical areas are termed the *frontal*, *frontoparietal* and *temporal opercula*. Finally, the insula becomes completely covered, and the superficial part of the lateral cerebral fossa is transformed into a cleft, the *sulcus lateralis* (Figs. 3.2, 3.3). The folding of the external surface of the neopallium results in the formation of a stereotypical pattern of sulci that demarcate the different convolutions or gyri. During the fifth month the first convolutions make their appearance and by the end of the seventh month most of the important sulci and gyri can be recognized (Fig. 2.23).

It is important to note that the morphogenesis of the neopallium is closely correlated with its histogenesis and differentiation. During the late embryonic and early fetal phase the walls

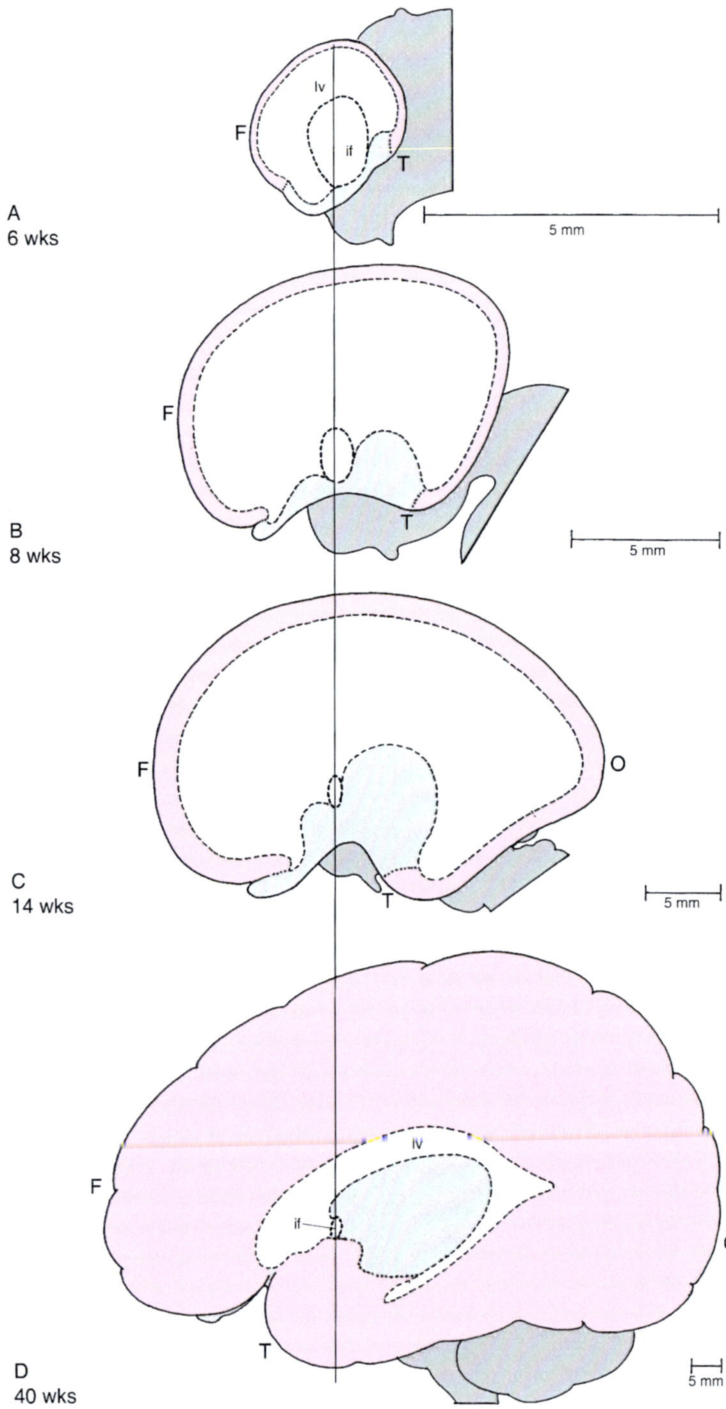

Fig. 2.22 A–D. Four stages in the development of the human brain at 6 weeks (**A**), 8 weeks (**B**), 14 weeks (**C**) and 40 weeks (**D**) to show changes in form and proportion of the cerebral hemispheres. The latter are depicted as if transparent. Neopallium in *red*; hippocampus and central grey masses in *light grey*; brain stem, cerebellum and diencephalon in *dark grey*. *F*, frontal pole; *if*, interventricular foramen; *lv*, lateral ventricle; *O*, occipital pole; *T*, temporal pole

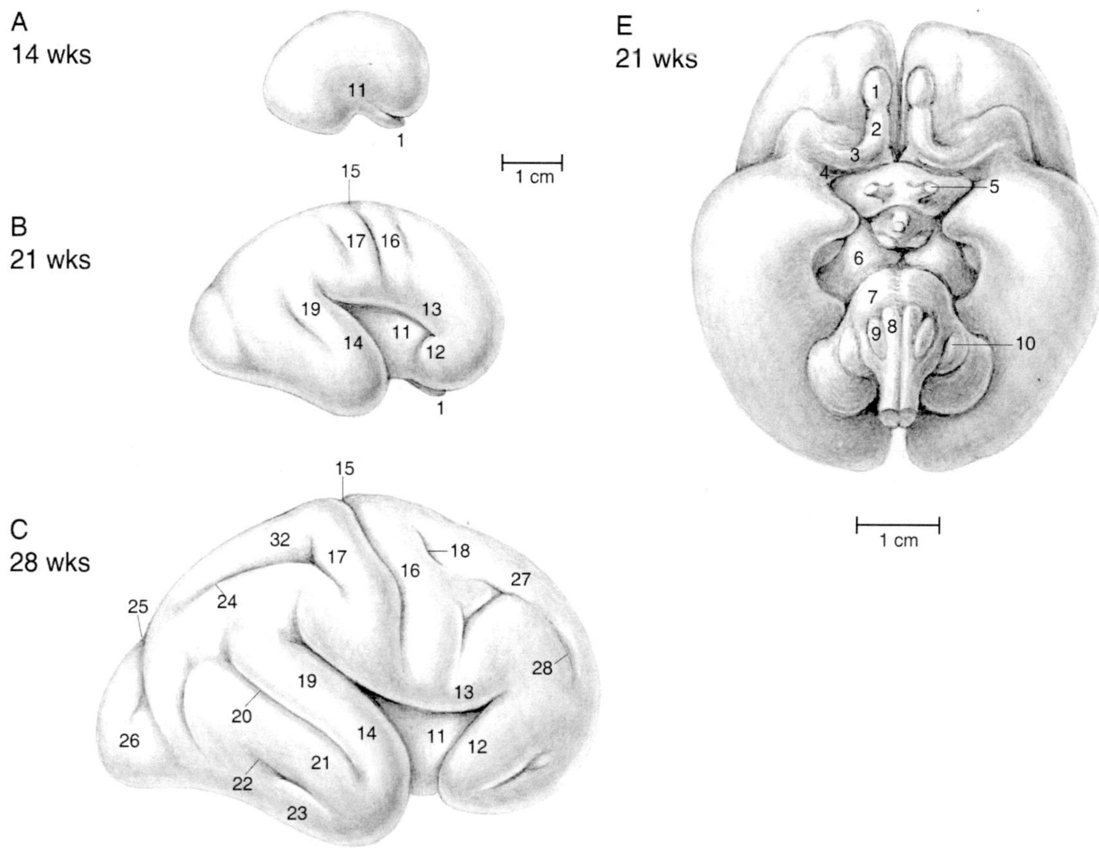

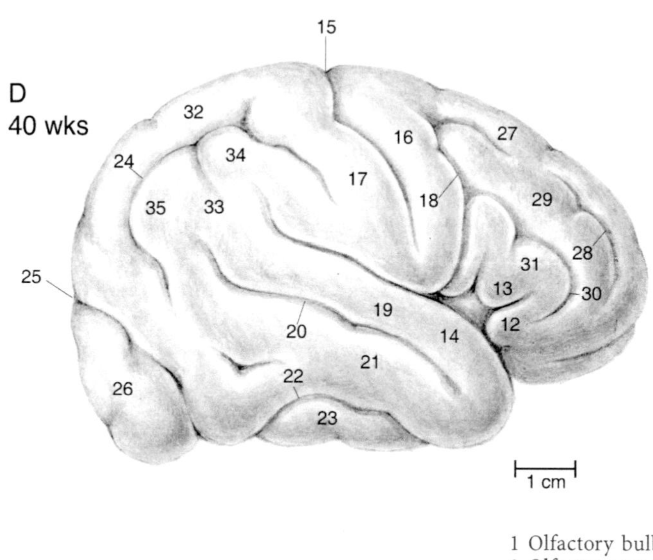

1 Olfactory bulb
2 Olfactory tract
3 Lateral olfactory tract
4 Piriform cortex
5 Optic nerve
6 Cerebral peduncle
7 Pons
8 Pyramis
9 Oliva
10 Flocculus
11 Lateral cerebral fossa; insula
12 Frontal operculum
13 Frontoparietal operculum
14 Temporal operculum
15 Central sulcus
16 Precentral gyrus
17 Postcentral gyrus
18 Precentral sulcus
19 Superior temporal gyrus
20 Superior temporal sulcus
21 Middle temporal gyrus
22 Inferior temporal sulcus
23 Inferior temporal gyrus
24 Intraparietal sulcus
25 Parieto-occipital sulcus
26 Occipital lobe
27 Superior frontal gyrus
28 Superior frontal sulcus
29 Middle frontal gyrus
30 Inferior frontal sulcus
31 Inferior frontal gyrus
32 Superior parietal lobule
33 Inferior parietal lobule
34 Supramarginal gyrus
35 Angular gyrus

Fig. 2.23. Lateral views of cerebral hemispheres (**A–D**) and basal view of the brain (**E**) at the ages indicated. E is based on Fig. 88 in [102]

of the rapidly expanding cerebral vesicles remain thin (Figs. 2.19, 2.20). This "inflation" greatly enlarges the ventricular surface area, and by this active enlargement, the capacity required to produce enormous numbers of neocortical neurons is created in the adjacent germinal zone. Later, when most of the neuroblasts produced have crossed to the cortical plate, the accumulation of corticopetal and corticofugal projection fibres in the deep neopallial zone leads to a reduction of the lateral ventricles. The convolution of the external surface of the neopallium, finally, is a direct consequence of the transformation of the compact cortical plate into the definitive neocortex.

Several commissures are formed in the prosencephalic region (Fig. 2.6). Two of these, the *habenular* and the *posterior commissures*, develop in the diencephalic portion of the roof plate. These two commissures are separated by the epiphysial evagination.

The median strip in front of the infundibulum is known as the *lamina terminalis*. It is dorsally continuous with the thin prosencephalic roof plate. Decussating fibres of the optic nerve invade the most caudal portion of the lamina terminalis to form the *optic chiasm*. After the appearance of telencephalic vesicles, the rostral part of the lamina terminalis become thickened to form the *commissural plate*, i.e. the route by which fibres can pass from one cerebral hemisphere to the other. Three distinct commissural systems, the *anterior commissure*, the *corpus callosum* and the *hippocampal commissure*, develop in the commissural plate (Figs. 2.5 C, 2.6). The *anterior commissure*, which is the first to appear, interconnects the olfactory regions of the cerebral hemispheres. In a later phase of development, fibres originating from the temporal neocortex join this commissure. Soon after the appearance of the anterior commissure, fibres connecting the hippocampal regions of the two hemispheres assemble in the most dorsal portion of the commissural plate to form the *hippocampal* or *fornical commissure*. The largest and most important commissure to cross in the commissural plate is the *corpus callosum*, which connects the neopallial regions of the two hemispheres. The first fibres of this commissure appear in the ninth week as a small bundle directly on top of the hippocampal commissure. Parallel with the growth and differentiation of the neopallium, the corpus callosum increases rapidly in size, extending both rostrally and caudally (Fig. 2.6). By this expansion, the underlying part of the commissural plate stretches out to form the thin *septum pellucidum* (Fig. 2.6 E, F). As the corpus callosum grows backwards, it arches over the thin roof of the diencephalon.

Before the corpus callosum develops, the primordium of the hippocampal formation forms part of the medial wall of the telencephalic vesicles, and it has already been mentioned that, due to the curvature of these vesicles, the caudal part of the hippocampal formation is carried into the temporal lobes [65]. The caudally extending corpus callosum splits the rostral part of the hippocampal formation into a dorsal and a ventral part. The dorsal part can be divided into a *precommissural* and a *supracommissural hippocampus*. The precommissural hippocampus is formed by the *taenia tecta*, a small area in the rostral part of the medial hemisphere wall (Fig. 3.18). The supracommissural hippocampus is represented by a tiny band of grey matter, known as the *indusium griseum*, which bilaterally follows the dorsal aspect of the corpus callosum (Figs. 3.18, 5.7). The *subcommissural part of the hippocampus* is mainly formed by the *fornix*, a large fibre system that connects the hippocampal formation with the septum and the hypothalamus. Its course is strongly influenced by the curvature of the hemispheres (Fig. 3.18). In the adult human brain, the precommissural and supracallosal parts of the hippocampus are small, vestigial structures; only the *retrocommissural hippocampus*, which is carried into the temporal lobe, is well-developed and differentiated into the three layers already mentioned (Fig. 3.18). At the end of the third month, the hippocampal formation starts to roll in on itself along a longitudinal groove, the *hippocampal sulcus* or *fissure*. Due to this infolding, the mature retrocommissural hippocampus protrudes into the inferior horn of the lateral ventricle (Fig. 6.37). It is remarkable that, due to the enlargement, curvature and infolding of the telencephalic vesicles, morphogenetic processes that are all driven by the expansion of the neopallium, two limbic structures, i.e. the *amygdaloid complex* and the *retrocommissural hippocampus*, are displaced towards the temporal lobe (Figs. 12.4).

Parts of the diencephalic roof plate and adjacent medial walls of the developing telencephalic hemispheres become transformed into a bilateral *choroid plexus*. The development of this plexus is similar to that of the fourth ventricle. An ependymal fold starts to invaginate into the ventricular cavity, together with a richly vascularized mesenchyme. During further development, this fold produces numerous villous projections, but its site of origin remains narrow and forms a choroid fissure.

The formation of the prosencephalic plexus begins in the posterior wall of the interventricular foramina during the sixth week of development. Subsequently, the plexus extends caudally in the roof of the third ventricle, as well as in the medial wall of the lateral ventricle. The latter, telencephalic part of the plexus follows the curvature of the expanding hemispheres, extending caudally, ventrally and ultimately rostrally into the inferior horns of the lateral ventricles (Figs. 2.19 B, 2.20 B, 3.17).

During the third and fourth months, the choroid plexuses in the lateral ventricles are relatively large (Fig. 2.20). During that period they probably provide nutritive materials to the metabolically very active, but entirely avascular telencephalic ventricular matrix zone.

Finally, it may be mentioned that during later development, coarctations, i.e. fusions of the ventricular surface of the ganglionic eminences with those of the adjacent surfaces of the septal and pallial portions of the cerebral hemispheres, reduce the lateral ventricles somewhat [202, 220, 222].

New Insights into the Development of Forebrain

In the preceding pages, the classical and universally used subdivision of the forebrain, summarized in Table 2.2, has been employed. During the last decade, a combination of factors have led to new insights into the morphological pattern of the prosencephalon and into the composition of its constituent parts. Of these, I would like to mention:

(a) A reappraisal of the classical work of the Swedish school of comparative neuroembryology [138]
(b) The slowly penetrating notion that neuromorphological research should focus on topological rather than on topographical relations [157, 160]
(c) The results of the morphological and molecular genetic studies of Puelles and Rubenstein and their associates [25, 156, 157, 160, 161, 180, 182, 213]
(d) The results of analyses of the migratory paths of neuroblasts [112, 123, 124, 135, 155, 224].

In the following, these new insights will be summarized and some comments will be provided.

1. The prosencephalon is, just as the remaining parts of the neuraxis, composed of a number of rostrocaudally arranged neuromeres, designated as *prosomeres* [21, 54, 89, 156, 161, 181]. Different numbers of such units have been proposed. There is strong evidence in support of three prosomeres in the caudal forebrain (diencephalon proper, or caudal diencephalon). The rostral forebrain may be designated as the secondary prosencephalon (after the segregation of the diencephalon proper). It consists of the hypothalamus (rostral diencephalon) and the overlying telencephalon. This complex area poses special difficulties for tracing prosomeres, whose existence and number in this domain are still under discussion. Possibly, the entire secondary prosencephalon represents a single, peculiarly patterned, prosomere [163].

Table 2.2. Classical subdivision of prosencephalon

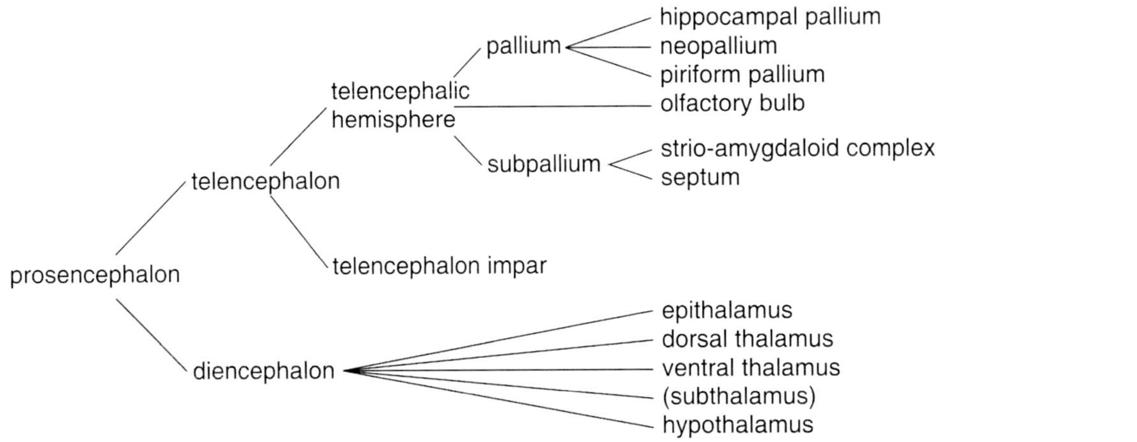

2. The four classical subdivisions of the diencephalon, epithalamus, dorsal thalamus, ventral thalamus and hypothalamus are, contrary to the belief of Herrick [63], Kuhlenbeck [100] and many others [50, 81, 96, 174, 176, 177], not oriented parallel to the longitudinal axis of the brain. The true, i.e. topological relations of these subdivisions become instantly clear if the strong curvature of the neural tube at the junction of the midbrain and the forebrain is taken into account (Fig. 2.11) [21, 25, 43, 54, 89, 129, 156, 157, 161].

3. The diencephalon proper, that is, the classical diencephalon minus the hypothalamus, is composed of three neuromeres, *synencephalon, parencephalon posterius* and *parencephalon anterius* (Fig. 2.11).

The boundaries of the diencephalic neuromeres are well-documented. Radial glial cells, specified by the putative cell adhesion molecule CD15, are concentrated in the interneuromeric border zones [119]. The boundary between the parencephalon posterius and the parencephalon anterius is, as already mentioned, marked by the cell-free *zona limitans intrathalamica*. During later development this zone is transformed into a fibre sheet, the *lamina medullaris externa*. Some early-developing, transversely oriented fibre segments, which pass close to the interneuromeric borders, represent permanent landmarks for recognizing the diencephalic segmentation pattern. Thus, the *posterior commissure* is situated close to the caudal border of the synencephalon. The *habenulo-interpeduncular tract* occupies a similar position in the parencephalon posterius, whereas the *mamillothalamic tract* passes through the most rostral zone of the same neuromere (Fig. 2.11).

The *synencephalon* generates the *pretectal region* and a tegmental area within which the *interstitial nucleus of Cajal* and the *parvocellular part of the red nucleus* differentiate. The *parencephalon posterius* gives rise to the *epithalamus* (*habenular nuclei* plus *epiphysis*) and the *dorsal thalamus*, whereas the *ventral thalamus* or *prethalamus* develops within the *parencephalon anterius* (Fig. 2.11) [18, 19, 21]. According to Puelles [157], the third diencephalic neuromere, P3, includes the dorsally placed *eminentia thalami*. This eminence abuts, via the interventricular foramen, the primordia of the *amygdaloid complex*.

4. The remainder of the prosencephalon, which includes the hypothalamus, the telencephalon impar and the telencephalic hemispheres, may be designated as the *secondary prosencephalon* [157].

5. The boundary between the classical entities diencephalon and telencephalon does not coincide with an interneuromeric border and is, hence, artificial [20, 21, 84].

6. The interpretation of the parts of the secondary prosencephalon is fraught with difficulties. Two evaginations, the early optic and the somewhat later hemispheric, disturb the primary pattern of this region. The classical subdivision of the telencephalic hemispheres, as represented in Table 2.2, actually considers these formations as cylindrical structures, and does not account for the relations in their morphologically most rostral and caudal parts. In particular, the morphological continuities in the transitory zone of the evaginated and unevaginated portions of the secondary prosencephalon are hard to judge. For an adequate study of these continuities, carefully prepared three-dimensional reconstructions of closely graded embryonic stages or complex topological transformation, such as the deevagination procedure proposed by Alvarez-Bolado and Swanson [6] are required. In the latter procedure, the rostral, dorsal and caudal rims of the interventricular foramen are stretched and moved peripherally to simulate the original, unevaginated condition displayed by the early embryonic prosencephalon (Fig. 2.24).

There is no consensus as regards the number of neuromeres participating in the formation of the secondary prosencephalon [54, 138]. Bergquist and Källén [18–21, 84, 85] carefully analysed the ontogenesis of this region in a large number of vertebrates, including humans. They concluded that two neuromeres, a caudal one and a rostral one, both extending from the hypothalamic area into the telencephalic hemispheres, are involved in its formation. The caudal neuromere encompasses the bulk of the hypothalamus and passes, via a transitional area, over into the telencephalic pallium, whereas the rostral neuromere gives rise to the rostral hypothalamus, the optic and preoptic regions and the subpallium.

Recently, Puelles and Rubenstein et al. [32, 33, 157, 166, 173, 203, 229] re-analysed the morphological pattern of the structural constituents of the telencephalon in mouse and chicken embryos by means of molecular genetic, immunohistochemical and grafting techniques. Their main results, which are here visualized provisionally in the forebrain of a human embryo (Fig. 2.24), may be summarized as follows:

(a) The molecularly defined pallio-subpallial boundary corresponds to the morphological boundary established by Holmgren [73] and Källén [84, 85]. It runs all the way across the telencephalic hemispheres, extending from the septum rostromedially to the amygdaloid complex caudally.

(b) The cerebral hemispheres are composed of seven longitudinally arranged divisions, each of which represents a histogenetic unit that radially extends from the ventricular to the pial surface.

(c) The subpallium can be divided into three parallel zones, designated, from ventral to dorsal, as the *entopeduncular area, pallidum* and *striatum*. The entopeduncular area borders ventrally on the *preoptic region.*

(d) The pallium comprises four parallel subdivisions, which are referred to as the *ventral, lateral, dorsal* and *medial pallial regions*. Three of these, the lateral, dorsal and medial, correspond to the piriform pallium, neopallium and hippocampal pallium, respectively, of classical descriptions. The ventral pallium represents a novel domain, intercalated between the striatum and the lateral piriform pallium. It is characterized by the absence of *Emx-1*, a gene that is expressed throughout the remainder of the pallium. The rostral part of the ventral pallium gives rise to the olfactory bulb and to the anterior olfactory nucleus. The *lateral olfactory tract* relates precisely to its subpial zone.

(e) All of the subpallial zones as well as the ventral and lateral pallial zones extend rostrally into the septum and participate caudally in the formation of the amygdaloid complex.

(f) The nature of the pallio-subpallial boundary is uncertain. Puelles [157] tested the hypothesis of Bergquist and Källén [21], according to which this boundary represents the dorsal part of the border zone between two telencephalo-hypothalamic neuromeres. However, so far he failed to find any morphological or molecular evidence for a continuity of the pallio-subpallial boundary with any border zone in the developing hypothalamus.

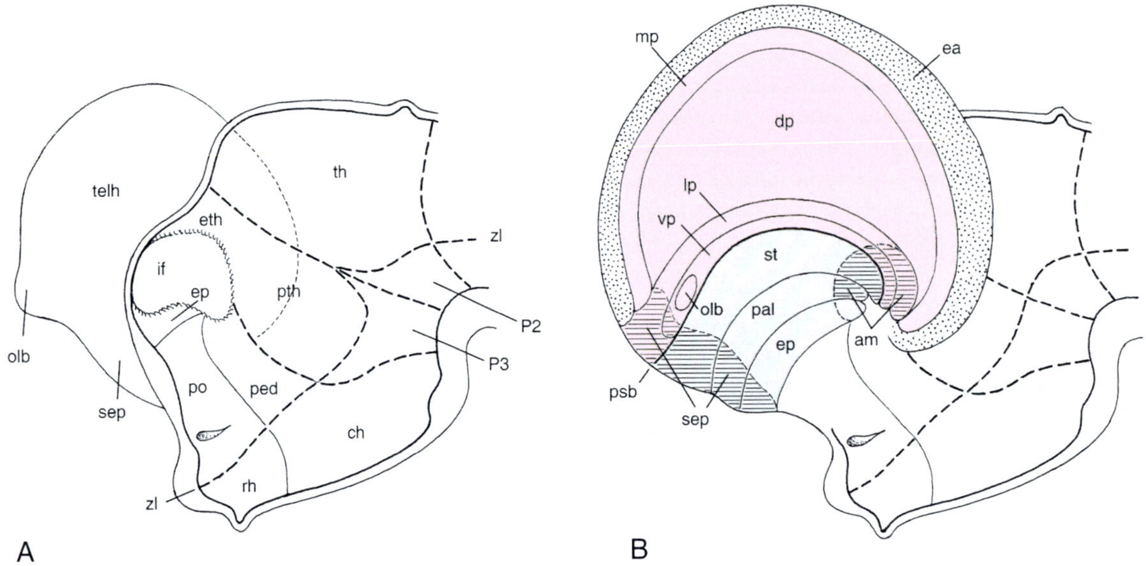

am	Amygdaloid complex
ch	Caudal hypothalamic area
dp	Dorsal pallium (neopallium)
ea	Epithelial (choroidal) area [65]
ep	Entopeduncular area
eth	Eminentia thalami
if	Interventricular foramen
lp	Lateral pallium (piriform pallium)
mp	Medial pallium (hippocampal pallium)
olb	Olfactory bulb
P2, P3	Prosomeres
pal	Pallidum
ped	Peduncular area
po	Preoptic area
psb	Palliosubpallial boundary
pth	Prethalamus
rh	Rostral hypothalamic area
sep	Septum
st	Striatum
telh	Telencephalic hemispheres
th	Thalamus
vp	Ventral pallium
zl	Zona limitans

Fig. 2.24 A, B. An attempt to show the topological relations of the telencephalic longitudinal zones, distinguished by Puelles et al. [130, 157, 162, 166] in the embryonic human brain. **A** Medial view of the rostral part of the brain of a 19-mm embryo. **B** The same brain; the ventricular wall of the telencephalic hemisphere is flattened by deevagination according to [6]. For this preliminary transformation, the figures in [6, 7, 18, 65, 69, 143, 166] were used. *Thick interrupted lines*, zona limitans and interneuromeric boundaries; *thin dotted line*, outline of cerebral hemisphere. Pallial zones are shown in *red*, subpallial zones in *grey*. Note that both pallial and subpallial zones participate in the formation of the septum and the amygdaloid complex. The primordia of both these structures are *hatched*. *This figure is dedicated to Larry Swanson*

Recent studies on the origin and migratory pathways of neuronal precursors have given new insights into the composition of the various telencephalic grisea [123, 124]. The precursors of projection neurons generally migrate radially and remain within the confines of their respective histogenetic units, whereas developing local circuit neurons often migrate tangentially and, in doing so, cross the boundaries of one or more histogenetic units. The various telencephalic migratory routes to be discussed are diagrammatically indicated in Fig. 2.25 (cf. also Fig. 2.9). The letters in the right half of that figure correspond to those used in the text. As for projection neurons, it is known that (a) the glutamatergic pallial projection neurons (i.e. the pyramidal cells) migrate radially to their destination [38, 51, 169, 170], and that the same holds true for the GABA-containing striatal (b) and pallidal (c) projection neurons [37, 125, 148], as well as for the cholinergic entopeduncular elements (d), which will form the basal magnocellular complex [125, 224]. As for local circuit neurons (interneurons), it has been established that GABA-containing neurons, originating from the lateral ganglionic eminence, migrate tangentially to (e) the neocortex and (f) the hippocampal formation [12, 36, 38, 135, 144, 205], and that GABA-containing elements with the same destinations also originate from (g) the medial ganglionic eminence [38, 108, 224]. The question arises as to whether *all* cortical GABA interneurons derive from the subpallial telencephalon. There is experimental evidence indicating that, in the mouse, most – if not all – of these neurons are indeed of subpallial origin [11, 30, 74], but it has been recently established [113] that, in humans, only about one third of the cortical GABAergic interneurons derive from the ganglionic eminences. Other tangential migrations originating from the basal telencephalon include the precursors of (h) cholinergic striatal interneurons arising from the entopeduncular area [224], (j) GABA-containing granule and periglomerular cells in the olfactory bulb, stemming from the rostrolateral part of the lateral ganglionic eminence [92, 102, 162], and (k) GABA-containing interneurons in thalamic association nuclei, produced in the medial ganglionic eminence [112, 172]. The neuroblasts destined for the olfactory bulb form a compact *rostral migratory stream* (Fig. 2.9). Remarkably, the production and migration of these elements proceeds continuously into adulthood [92]. Conceivably, the olfactory bulb neurons in question are not produced in the lateral ganglionic eminence, but rather in the germinative zone of the adjacent, recently identified ventral pallium. It is important to note that in their recent publication Sanai et al. [185] found no evidence of a compact rostral migratory stream in the adult human brain. The precursors of the GABA-containing thalamic interneurons mentioned make their amazing telo-diencephalic journey also jointly, via another compact migratory stream, the *corpus gangliothalamicum*. It has already been mentioned that this migratory stream has been identified only in human fetuses so far. Long tangential migrations are not reserved to the precursors of local circuit neurons. It has recently been established that the entopeduncular area gives origin to large numbers of oligodendrocyte progenitors, which migrate tangentially into more dorsal telencephalic regions, including the cerebral cortex [147, 167, 200, 207]. It is not clear yet whether telencephalic oligodendrogenesis is entirely confined to the entopeduncular area.

If we survey these data on the ontogenesis of the telencephalon, the following tentative conclusions seem to be warranted.

The developing telencephalon can be subdivided into seven, radially organized, histogenetic units. Each of these units is characterized by a particular pattern of expression of a number of developmental regulatory genes. These genes specify, *inter alia*, precursor populations that produce neurons that synthesize different neurotransmitters within the proliferative compartments of the various units. The proliferative compartments of the various units produce two kinds of neurons, projection neurons that migrate radially to their final locations, guided by the processes of radial glial cells, and local circuit neurons that migrate tangentially to populate adjacent units. It has been es-

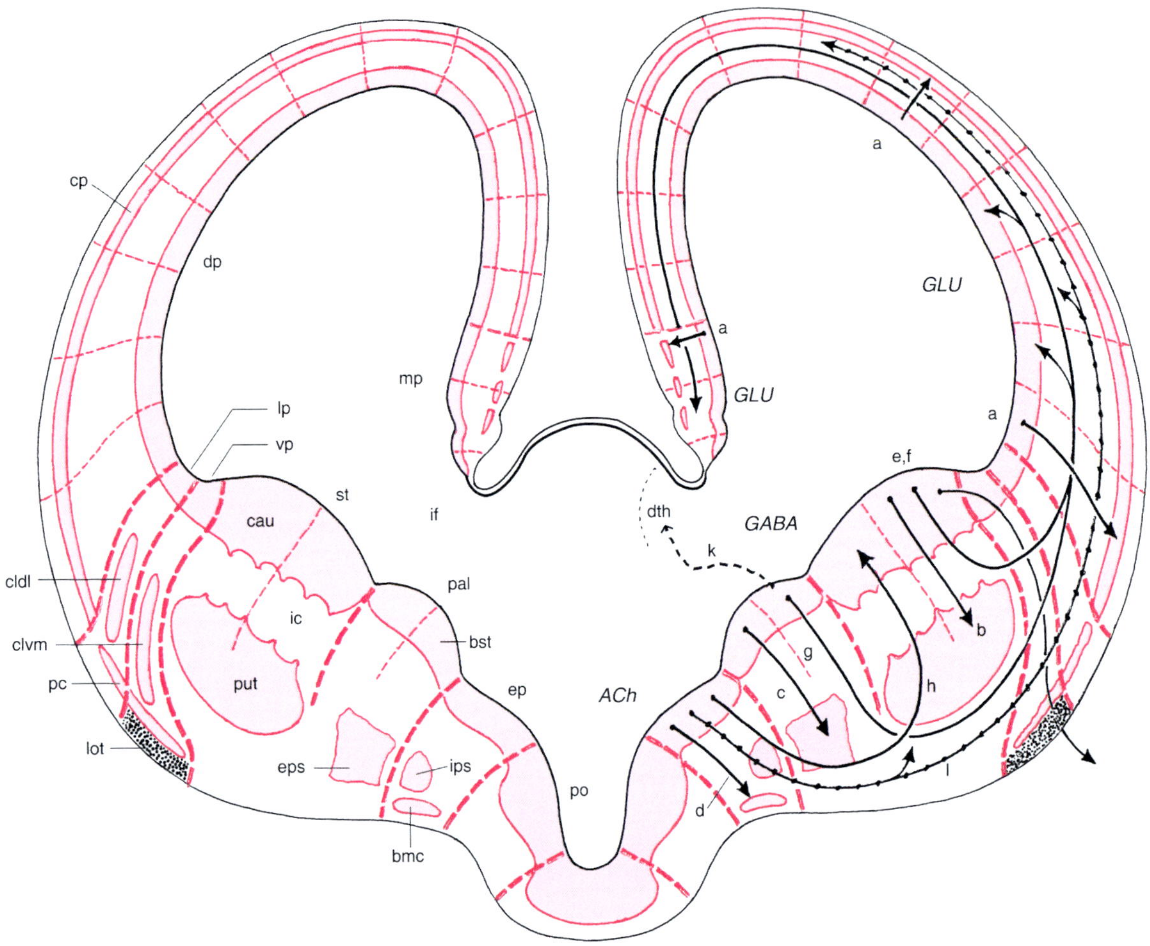

a, b etc.	Migratory routes described in the text
bmc	Basal magnocellular complex
bst	Bed nucleus of stria terminalis
cau	Caudate nucleus
cldl	Claustrum, dorsolateral part
clvm	Claustrum, ventromedial part
cp	Cortical plate
dp	Dorsal pallium (neopallium)
dth	Dorsal thalamus
ep	Entopeduncular area
eps	External pallidal segment
ic	Internal capsule
if	Interventricular foramen
ips	Internal pallidal segment
lot	Lateral olfactory tract
lp	Lateral pallium (piriform pallium)
mp	Medial pallium (hippocampal pallium)
pal	Pallidum
pc	Piriform or olfactory cortex
po	Preoptic area
put	Putamen
st	Striatum
vp	Ventral pallium

Fig. 2.25. Diagrammatic transverse section through the telencephalon at the level of the interventricular foramina of a human fetus of about 11 weeks. The histogenetic units, distinguished by Puelles et al. [157, 162, 166] are shown on the *left*. Areas producing neurons that synthesize a particular neurotransmitter and migratory routes are indicated on the *right*. *Coarse interrupted lines* indicate the boundaries of the histogenetic units. These *curves* and the intervening *fine dotted curves* indicate the course of radial glial fibres, along which the precursors of projection neurons migrate towards their destination

tablished that at least three telencephalic units, the striatum, the neocortex, and the hippocampal formation, obtain their neurons from two different proliferative compartments. It might be added that a similar conclusion can be drawn for a number of non-telencephalic structures such as the dorsal thalamus, the substantia nigra and the cerebellar cortex. Most, if not all, telencephalic oligodendrocytes originate from the ventralmost zone of the telencephalon. A similar focal generation of oligodendrocytes from the ventral neuroepithelium has also been observed in more caudal regions of the neuraxis [39, 167, 199, 208]. However, it has been recently shown that restricted sectors of the dorsal spinal cord and rhombencephalon are also able to produce oligodendrocytes [26, 211].

References

1. Al-Ghoul WM, Miller MW (1993) Orderly migration of neurons to the principal sensory nucleus of the trigeminal nerve of the rat. J Comp Neurol 330:464–475
2. Altman J, Bayer SA (1987) Development of the precerebellar nuclei in the rat. I. The precerebellar neuroepithelium of the rhombencephalon. J Comp Neurol 257:477–489
3. Altman J, Bayer SA (1987) Development of the precerebellar nuclei in the rat. II. The intramural olivary migratory stream and the neurogenetic organization of the inferior olive. J Comp Neurol 257:490–512
4. Altman J, Bayer SA (1987) Development of the precerebellar nuclei in the rat. III. The posterior precerebellar extramural migratory stream and the lateral reticular and external cuneate nuclei. J Comp Neurol 257:513–528
5. Altman J, Bayer SA (1987) Development of the precerebellar nuclei in the rat. IV. The anterior precerebellar extramural migratory stream and the nucleus reticularis tegmenti pontis and the basal pontine gray. J Comp Neurol 257:529–552
6. Alvarez-Bolado G, Swanson LW (1995) Appendix: on mapping patterns in the embryonic forebrain. J Comp Neurol 355:287–295
7. Alvarez-Bolado G, Rosenfeld MG, Swanson LW (1995) Model of forebrain regionalization based on spatiotemporal patterns of POU-III homeobox gene expression, birthdates, and morphological features. J Comp Neurol 355:237–295
8. Alvarez-Buylla A, García-Verdugo JM, Tramontin AD (2001) A unified hypothesis on the lineage of neural stem cells. Nat Rev Neurosci 2:287–293
9. Alvarez-Buylla A, Seri B, Doetsch F (2002) Identification of neural stem cells in the adult vertebrate brain. Brain Res 57:751–758
10. Ambrosiani J, Armengol JA, Martínez S, Puelles L (1996) The avian inferior olive derives from the alar neuroepithelium of the rhombomeres 7 and 8: an analysis by using chick-quail chimeric embryos. Neuroreport 7:1285–1288
11. Anderson S, Mione M, Yun K, Rubenstein JLR (1999) Differential origins of neocortical projection and local circuit neurons: role of *Dlx* genes in neocortical interneuronogenesis. Cereb Cortex 9:646–654
12. Anderson SA, Eisenstat DD, Shi L, Rubenstein JLR (1997) Interneuron migration from basal forebrain to neocortex: dependence on Dlx genes. Science 278:474–476
13. Anthony TE, Klein C, Fishell G, Heintz N (2004) Radial glia serve as neuronal progenitors in all regions of the central nervous system. Neuron 41:881–890
14. Bartelmez GW (1923) The subdivisions of the neural folds in man. J Comp Neurol 35:231–295
15. Bartelmez GW, Evans HM (1926) Development of the human embryo during the period of somite formation. Contrib Embryol Carnegie Inst 17:1–67
16. Bédard A, Lévesque M, Bernier PJ, Parent A (2002) The rostral migratory stream in adult squirrel monkeys: contribution of new neurons to the olfactory tubercle and involvement of the antiapoptotic protein Bcl-2. Eur J Neurosci 16:1917–1924
17. Bengmark S, Hugosson R, Källén B (1953) Studien über Kernanlagen im Mesencephalon sowie im Rostralteil des Rhombencephalon von Mus musculus. Z Anat Entwicklungsgesch 117:73–91
18. Bergquist H (1954) Morphogenesis of diencephalic nuclei in homo. Kgl Fysiogr Sällsk Lund Handl N F 64:1–47
19. Bergquist H (1954) Ontogenesis of diencephalic nuclei in vertebrates. Kgl Fysiogr Sällsk Lund Handl N F 65:1–34
20. Bergquist H (1964) Die Entwicklung des Diencephalons im Lichte neuer Forschung. Progr Brain Res 5:223–229
21. Bergquist H, Källén B (1954) Notes on the early histogenesis and morphogenesis of the central nervous system in vertebrates. J Comp Neurol 100: 627–659
22. Bolk L (1906) Das Cerebellum der Säugetiere. Fisher, Haarlem
23. Boulder Committee (1969) Embryonic vertebrate central nervous system: revised terminology. Anat Rec 166:257–262
24. Braitenberg V, Atwood RP (1958) Morphological observations on the cerebellar cortex. J Comp Neurol 109:1–34
25. Bulfone A, Puelles L, Porteus MH et al (1993) Spatially restricted expression of Dix-1, Dix-2 (Tes-1),

Gbx-2, and Wnt-3 in the embryonic day 12.5 mouse forebrain defines potential transverse and longitudinal boundaries. J Neurosci 13:3155–3172
26. Cai J, Qi Y, Hu X et al (2005) Generation of oligodendrocyte precursor cells from mouse dorsal spinal cord independent of Nkx6 regulation and SAhh signaling. Neuron 45:41–53
27. Cajal SR (1972) Histologie du système nerveus de l'homme et des vertébrés. Consejo Superior de Investigaciones Cientificas. Instituto Ramon y Cajal, Madrid
28. Cambronero F, Puelles L (2000) Rostrocaudal nuclear relationships in the avian medulla oblongata: a fate map with quail chick chimeras. J Comp Neurol 427:522–545
29. Campbell K, Götz M (2002) Radial glia: multi-purpose cells for vertebrate brain development. Trends Neurosci 25:235–238
30. Casarosa S, Fode C, Guillemot F (1999) Mash 1 regulates neurogenesis in the ventral telencephalon. Development 126:525–534
31. Choi BH (1981) Radial glia of developing human fetal spinal cord: Golgi, immunohistochemical and electron microscopic study. Brain Res Dev Brain Res 1:249–267
32. Cobos I, Puelles L, Martínez S (2001) The avian telencephalic subpallium originates inhibitory neurons that invade tangentially the pallium (dorsal ventricular ridge and cortical areas). Dev Biol 239: 30–45
33. Cobos I, Shimamura K, Rubenstein JLR, Martínez S, Puelles L (2001) Fate map of the avian anterior forebrain at the four-somite stage, based on the analysis of quail-chick chimeras. Dev Biol 239:46–67
34. Cooper ERA (1946) The development of the substantia nigra. Brain 69:22–33
35. Cordes SP (2001) Molecular genetics of cranial nerve development in mouse. Nat Rev Neurosci 2:611–623
36. De Carlos JA, López-Mascaraque L, Valverde F (1996) Dynamics of cell migration from the lateral ganglionic eminence in the rat. J Neurosci 16: 6146–6156
37. Deacon TW, Pakzaban P, Isacson O (1994) The lateral ganglionic eminence is the origin of cells committed to striatal phenotypes: neural transplantation and developmental evidence. Brain Res 668:211–219
38. Denaxa M, Chan CH, Schachner CH, Parnavelas JG, Karagogeos D (2001) The adhesion molecule TAG-1 mediates the migration of cortical interneurons from the ganglionic eminence along the corticofugal fiber system. Development 128:4635–4644
39. Dickinson PJ, Fanarraga ML, Griffiths IR et al (1996) Oligodendrocyte progenitors in the embryonic spinal cord express DM-20. Neuropathol Appl Neurobiol 22:188–198
40. Eriksson PS, Perfilieva E, Bjork-Eriksson T et al (1998) Neurogenesis in the adult human hippocampus. Nat Med 4:1313–1317
41. Essick CR (1907) The corpus ponto-bulbare – a hitherto undescribed nuclear mass in the human hindbrain. Am J Anat 7:119–135
42. Essick CR (1912) The development of the nuclei pontis and the nucleus arcuatus in man. Am J Anat 13:25–54
43. Figdor MC, Stern CD (1993) Segmental organization of the embryonic diencephalon. Nature 363: 630–634
44. Fraser S, Keynes R, Lumsden A (1990) Segmentation in the chick embryo hindbrain is defined by cell lineage restrictions. Nature 344:431–435
45. Fujita S (1963) The matrix cell and cytogenesis in the developing central nervous system. J Comp Neurol 120:37–42
46. Fujita S (1966) Application of light and electron microscopic autoradiography to the study of cytogenesis of the forebrain. In: Hassler R, Stephan H (eds) Evolution of the forebrain. Thieme, Stuttgart, pp 180–196
47. Fujita S (1969) Autoradiographic studies on histogenesis of the cerebellar cortex. In: Llinás R (ed) Neurobiology of cerebellar evolution and development. AMA, Chicago, pp 743–748
48. Gadisseux J-F, Goffinet AM, Lyon G, Evrard P (1992) The human transient subpial granular layer: an optical, immunohistochemical, and ultrastructural analysis. J Comp Neurol 324:94–114
49. Gage FH, Kempermann G, Palmer T, Peterson DA, Ray J (1998) Multipotent progenitor cells in the adult dentate gyrus. J Neurobiol 36:249–266
50. Gilbert MS (1935) The early development of the human diencephalon. J Comp Neurol 62:1–115
51. Gorski JA, Talley T, Qiu M et al (2002) Cortical excitatory neurons and glia, but not GABAergic neurons, are produced in the Emx1-expressing lineage. J Neurosci 22:6309–6314
52. Götz M, Hartfuss E, Malatesta P (2002) Radial glial cells as neuronal precursors: a new perspective on the correlation of morphology and lineage restriction in the developing cerebral cortex of mice. Brain Res Bull 57:777–780
53. Gregg C, Weiss S (2003) Generation of functional radial glial cells by embryonic and adult forebrain neural stem cells. J Neurosci 23:11587–11601
54. Gribnau AAM, Geijsberts LGM (1985) Morphogenesis of the brain in staged *Rhesus* monkey embryos. Adv Anat Embryol Cell Biol 91:1–69
55. Hamilton WJ, Boyd JD, Mossman A (1947) Human embryology. Heffer, Cambridge
56. Hanaway J, McConnell JA, Netsky MG (1971) Histogenesis of the substantia nigra, ventral tegmental area of tsai and interpeduncular nucleus: an autoradiographic study of the mesencephalon in the rat. J Comp Neurol 142:59–74
57. Hartfuss E, Galli R, Heins N, Götz M (2001) Characterization of CNS precursor subtypes and radial glia. Dev Biol 229:15–30
58. Hatten ME (1999) Central nervous system neuronal migration. Annu Rev Neurosci 22:511–539

59. Hatten ME, Heintz N (1995) Mechanisms of neural patterning and specification in the developing cerebellum. Annu Rev Neurosci 18:385–408
60. Hawkes R (1997) An anatomical model of cerebellar modules. Prog Brain Res 114:39–52
61. Hayashi M (1924) Einige wichtige Tatsachen aus der ontogenetischen Entwicklung des menschlichen Kleinhirns. Dtsch Z Nervenheilkd 81:74–82
62. Hemond SG, Glover JC (1993) Clonal patterns of cell proliferation, migration, and dispersal in the brainstem of the chicken embryo. J Neurosci 13: 1387–1402
63. Herrick CJ (1910) The morphology of the forebrain in amphibia and reptilia. J Comp Neurol 20:413–547
64. Herrick CJ (1913) Anatomy of the brain. In: The reference handbook of the medical sciences, vol 2, 3rd edn. Wood, New York, pp 274–342
65. Hines M (1922) Studies in the growth and differentiation of the telencephalon in man. The Fissura hippocampi. J Comp Neurol 34:79–171
66. His W (1893) Vorschläge zur Eintheilung des Gehirns. Arch Anat Physiol Anat Abt 172–180
67. His W (1893) Über das frontale Ende des Gehirnrohres. Arch Anat Physiol Anat Abt 157–172
68. His W (1904) Die Entwicklung des menschlichen Gehirns während der ersten Monate. Hirzel, Leipzig
69. Hochstetter F (1919) Beiträge zur Entwicklungsgeschichte des menschlichen Gehirns. Deuticke, Vienna
70. Hochstetter F (1929) Beiträge zur Entwicklungsgeschichte des menschlichen Gehirns. II. Die Entwicklung des Mittel- und Rautenhirns. Deuticke, Vienna
71. Holley JA (1982) Early development of the circumferential axonal pathway in mouse and chick spinal cord. J Comp Neurol 205:371–382
72. Holley JA, Nornes H, Morita M (1982) Guidance of neuritic growth in the transverse plane of embryonic mouse spinal cord. J Comp Neurol 205: 360–370
73. Holmgren N (1925) Points of view concerning forebrain morphology in higher vertebrates. Acta Zool 6:413–477
74. Horton S, Meredith A, Richardson JA, Johnson JE (1999) Correct coordination of neuronal differentiation events in ventral forebrain requires the bHLH factor MASH1. Mol Cell Neurosci 14:355–369
75. Hugosson R (1955) Studien über die Entwicklung der longitudinalen Zellsäulen und der Anlagen der Gehirnnervenkerne in der Medulla oblongata bei verschiedenen Vertebraten. Z Anat Entwicklungsgesch 118:543–566
76. Hugosson R (1957) Morphologic and experimental studies on the development and significance of the rhombencephalic longitudinal cell columns. Thesis, University of Lund
77. Jakob A (1928) Das Kleinhirn. In: Von Möllendorff's Handbuch der Mikroskopischen Anatomie des Menschen, IV/I, Nervensystem. Springer, Berlin
78. Johnston JB (1902) An attempt to define the primitive functional divisions of the central nervous system. J Comp Neurol 12:87–106
79. Johnston JB (1909) The morphology of the forebrain vesicle in vertebrates. J Comp Neurol 19:457–539
80. Kahle W (1951) Studien über die Matrixphasen und die örtlichen Reifungsunterschiede im embryonalen menschlichen Gehirn. Dtsch Z Nervenheilkd 166:272–302
81. Kahle W (1956) Zur Entwicklung des menschlichen Zwischenhirns. Dtsch Z Nervenheilkd 175: 259–318
82. Kahle W (1969) Die Entwicklung der menschlichen Grosshirnhemisphäre: Mit 55 Abbildungen. Springer, Berlin Heidelberg New York
83. Kahle W (1986) Nervous system and sensory organs. Thieme, New York
84. Källén B (1951) Embryological studies on the nuclei and their homologization in the vertebrate forebrain. Kgl Fysiogr Sällsk Lund Handl N F 62:1–36
85. Källén B (1951) The nuclear development in the mammalian forebrain with special regard to the subpallium. Kgl Fysiogr Sällsk Lund Handl N F 61:1–43
86. Kappel RM (1981) The development of the cerebellum in *Macaca mulatta*. Thesis, University of Leiden
87. Kawano H, Ohyama K, Kawamura K, Nagatsu I (1995) Migration of dopaminergic neurons in the embryonic mesencephalon of mice. Brain Res Dev Brain Res 86:101–113
88. Kempermann G, Jessberger S, Steiner B, Kronenberg G (2004) Milestones of neuronal development in the adult hippocampus. Trends Neurosci 27:447–452
89. Keyser A (1972) The development of the diencephalon of the chinese hamster: an investigation of the validity of the criteria of subdivision of the brain. Acta Anat (Basel) 59:1–178
90. Kim J-H, Auerbach JM, Rodriguez-Gómez JA et al (2002) Dopamine neurons derived from embryonic stem cells function in an animal model of Parkinson's disease. Nature 418:50–56
91. Knyihar-Csillik E, Csillik B, Rakic P (1995) Structure of the embryonic primate spinal cord. Anat Embryol (Berl) 191:319–540
92. Kornack DR, Rakic P (2001) The generation, migration, and differentiation of olfactory neurons in the adult primate brain. Proc Natl Acad Sci USA 98:4752–4757
93. Korneliussen HK (1968) Comments on the cerebellum and its division. Brain Res 8:229–235
94. Korneliussen HK (1969) Cerebellar organization in the light of cerebellar nuclear morphology and cerebellar corticogenesis. In: Llinás R (ed) Neurobiology of cerebellar evolution and development. AMA, Chicago, pp 515–523

95. Kostovic I (1990) Structural and histochemical reorganization of the human prefrontal cortex during perinatal and postnatal life. Prog Brain Res 85:223–240
96. Kostovic I (1990) Zentralnervensystem. In: Hinrichsen KV (ed) Humane Embryologie. Springer, Berlin Heidelberg New York, pp 381–448
97. Koutcherov Y, Mai JK, Ashwell KWS, Paxinos G (2002) Organization of human hypothalamus in fetal development. J Comp Neurol 446:301–324
98. Kriegstein AR, Noctor SC (2004) Patterns of neuronal migration in the embryonic cortex. Trends Neurosci 27:392–399
99. Krumlauf R, Marshall H, Studer M et al (1993) Hox Homeobox genes and regionalisation of the nervous system. J Neurobiol 24:1328–1340
100. Kuhlenbeck H (1954) The human diencephalon: a summary of development, structure, function and pathology. Confin Neurol 14:1–230
101. Lange W (1975) Cell number and cell density in the cerebellar cortex of man and some other mammals. Cell Tissue Res 157:115–124
102. Langelaan JW (1910) Voordrachten over den bouw van het centrale zenuwstelsel. Versluys, Amsterdam
103. Langelaan JW (1919) On the development of the external form of the human cerebellum. Brain 42:130–170
104. Larsell O (1934) Morphogenesis and evolution of the cerebellum. Arch Neurol 31:373–395
105. Larsell O (1947) The development of the cerebellum in man in relations to its comparative anatomy. J Comp Neurol 87:85–129
106. Larsell O (1970) The comparative anatomy and histology of the cerebellum from monotremes through apes. University of Minneapolis Press, Minneapolis
107. Larsell O, Jansen J (1972) The comparative anatomy and histology of the cerebellum. III. The human cerebellum, cerebellar connections, and cerebellar cortex. University of Minneapolis Press, Minneapolis
108. Lavdas AA, Grigoriou M, Pachnis V, Parnavelas JG (1999) The medial ganglionic eminence gives rise to a population of early neurons in the developing cerebral cortex. J Neurosci 19:7881–7888
109. Leber SM, Sanes JR (1995) Migratory paths of neurons and glia in the embryonic chick spinal cord. J Neurosci 15:1236–1248
110. Lee KJ, Jessell TM (1999) The specification of dorsal cell fates in the vertebrate central nervous system. Annu Rev Neurosci 22:261–294
111. Letinic K, Kostovic I (1997) Transient fetal structure, the gangliothalamic body, connects telencephalic germinal zone with all thalamic regions in the developing human brain. J Comp Neurol 384: 373–395
112. Letinic K, Rakic P (2002) Telencephalic origin of human thalamic GABAergic neurons. Nat Neurosci 4:931–936
113. Letinic K, Zoncu R, Rakic P (2002) Origin of GABAergic neurons in the human neocortex. Nature 417:645–649
114. Levitt P, Rakic P (1980) Immunoperoxidase localization of glial fibrillary acidic protein in radial glial cells and astrocytes of the developing rhesus monkey brain. J Comp Neurol 193:815–840
115. Lim DA, Flames N, Collado L, Herrera DG (2002) Investigating the use of primary adult subventricular zone neural precursor cells for neuronal replacement therapies. Brain Res Bull 57:759–764
116. Lois C, Alvarez-Buylla A (1993) Proliferating subventricular zone cells in the adult mammalian forebrain can differentiate into neurons and glia. Proc Natl Acad Sci USA 90:2074–2077
117. Lois C, Alvarez-Buylla A (1994) A long-distance neuronal migration in the adult mammalian brain. Science 264:1145–1148
118. Luskin MB (1993) Restricted proliferation and migration of postnatally generated neurons derived from the forebrain subventricular zone. Neuron 11:173–189
119. Mai JK, Andressen C, Ashwell KWS (1998) Demarcation of prosencephalic regions by CD15-positive radial glia. Eur J Neurosci 10:746–751
120. Malatesta P, Hartfuss E, Götz M (2000) Isolation of radial glial cells by fluorescent-activated cell sorting reveals a neuronal lineage. Development 127:5253–5263
121. Marchand R (1987) Histogenesis of the subthalamic nucleus. Neuroscience 21:183–195
122. Marin F, Puelles L (1995) Morphological fate of rhombomeres in quail/chick chimeras: a segmental analysis of hindbrain nuclei. Eur J Neurosci 7:1714–1738
123. Marín O, Rubenstein JLR (2001) A long, remarkable journey: tangential migration in the telencephalon. Nat Rev Neurosci 2:780–790
124. Marín O, Rubenstein JLR (2003) Cell migration in the forebrain. Annu Rev Neurosci 26:441–483
125. Marín O, Anderson SA, Rubenstein JLR (2000) Origin and molecular specification of striatal interneurons. J Neurosci 20:6063–6076
126. Marin-Padilla M (1978) Dual origin of the mammalian neocortex and evolution of the cortical plate. Anat Embryol 152:109–126
127. Marin-Padilla M (1988) Early ontogenesis of the human cerebral cortex. In: Jones EG, Peters A (eds) Development and maturation of cerebral cortex. Plenum, New York, pp 1–34 (Cerebral cortex, vol 7)
128. Marin-Padilla M (1992) Ontogenesis of the pyramidal cell of the mammalian neocortex and developmental cytoarchitecture: a unifying theory. J Comp Neurol 321:223–240
129. Martínez S, Puelles L (2000) Neurogenetic compartments of the mouse diencephalon and some characteristic gene expression patterns. In: Goffinet AM, Rakic P (eds) Mouse brain development. Springer, Heidelberg, pp 91–106 (Results and problems in cell differentiation, vol 30)

130. Medina L, Legaz I, Gonzalez G et al (2004) Expression of Dbx1, Neurogenin 2, Semaphorin 5A, Cadherin 8, and Emx1 distinguish ventral and lateral pallial histogenetic divisions in the developing mouse claustroamygdaloid complex. J Comp Neurol 474:504–523
131. Misson J-P, Edwards MA, Yamamoto M, Caviness VS Jr (1988) Mitotic cycling of radial glial cells of the fetal murine cerebral wall: a combined autoradiographic and immunohistochemical study. Brain Res Dev Brain Res 38:183–190
132. Misson J-P, Austin CP, Takahashi T, Cepko CL, Caviness VS Jr (1991) The alignment of migrating neural cells in relation to the murine neopallial radial glial fiber system. Cereb Cortex 1:221–229
133. Mugnaini E, Forströnen PF (1967) Ultrastructural studies on the cerebellar histogenesis. I. Differentiation of granule cells and development of glomeruli in the chick embryo. Z Zellforsch 77:115–143
134. Müller F, O'Rahilly R (1987) The development of the human brain, the closure of the caudal neuropore, and the beginning of secondary neurulation at stage 12. Anat Embryol 176:413–430
135. Nadarajah B, Parnavelas JG (2002) Modes of neuronal migration in the developing cerebral cortex. Nat Rev Neurosci 3:423–432
136. Nadarajah B, Alifragis P, Wong R, Parnavelas JG (2002) Ventricle-directed migration in the developing cerebral cortex. Nat Neurosci 5:218–224
137. Nakatsu T, Uwabe C, Shiota K (2000) Neural tube closure initiates at multiple sites: evidence from human embryos and implications for the pathogenesis of neural tube defects. Anat Embryol 201:455–466
138. Nieuwenhuys R (1998) Morphogenesis and general structure. In: Nieuwenhuys R, Ten Donkelaar HJ, Nicholson C (eds) The central nervous system of vertebrates, vol 1. Springer, Berlin Heidelberg New York, pp 159–228
139. Noback CR, Demarest RJ (1975) The human nervous system. McGraw-Hill, New York
140. Noctor SC, Flint AC, Weissman TA, Dammerman RS, Kriegstein AR (2001) Neurons derived from radial glial cells establish radial units in neocortex. Nature 409:714–720
141. Nordlander RH (1987) Axonal growth cones in the developing amphibian spinal cord. J Comp Neurol 263:485–496
142. O'Rahilly R, Gardner E (1979) The initial development of the human brain. Acta Anat (Basel) 104:123–133
143. O'Rahilly R, Müller F (1999) The embryonic human brain. An atlas of developmental stages, 2nd edn. Wiley, New York
144. O'Rahilly R, Müller F, Hutchins GM, Moore GW (1984) Computer ranking of the sequence of appearance of 100 features of the brain and related structures in staged human embryos during the first 5 weeks of development. Am J Anat 171: 243–257
145. O'Rourke NA, Sullivan DP, Kaznowski CE, Jacobs AA, McConnell SK (1995) Tangential migration of neurons in the developing cerebral cortex. Development 121:2166–2176
146. Oberdick J, Baader SL, Schilling K (1998) From zebra stripes to postal zones: deciphering patterns of gene expression in the cerebellum. Trends Neurosci 21:383–390
147. Olivier C, Cobos I, Perez-Villegas E et al (2001) Monofocal origin of telencephalic oligodendrocytes in the anterior entopeduncular area of the chick embryo. Development 128:1757–1769
148. Olsson M, Björklund A, Campbell K (1998) Early specification of striatal projection neurons and interneuronal subtypes in the lateral and medial ganglionic eminence. Neuroscience 84:867–876
149. Oppenheim RW (1991) Cell death during development of the nervous system. Annu Rev Neurosci 14:453–501
150. Ourednik V, Ourednik J, Flax JD et al (2001) Segregation of human neural stem cells in the developing primate forebrain. Science 293:1820–1824
151. Palmer TD, Willhoite AR, Gage FH (2000) Vascular niche for adult hippocampal neurogenesis. J Comp Neurol 425:479–494
152. Palmgren A (1921) Embryological and morphological studies on the midbrain and cerebellum of vertebrates. Acta Zool (Stockh) 2:1–94
153. Parmantier E, Braun C, Thomas J-L et al (1997) *PMP-22* expression in the central nervous system of the embryonic mouse defines potential transverse segments and longitudinal columns. J Comp Neurol 378:159–172
154. Phelps PE, Vaughn JE (1995) Commissural fibers may guide cholinergic neuronal migration in developing rat cervical spinal cord. J Comp Neurol 355:38–50
155. Pleasure S, Anderson S, Hevner R et al (2000) Cell migration from the ganglionic eminences is required for the development of hippocampal GABAergic interneurons. Neuron 28:727–740
156. Puelles L (1995) A segmental morphological paradigm for understanding vertebrate forebrains. Brain Behav Evol 46:319–337
157. Puelles L (2001) Brain segmentation and forebrain development in amniotes. Brain Res Bull 55:695–710
158. Puelles L (2002) Morphogenetic deformation at the thalamotelencephalic boundary and the lamina affixa myth. In: The human brain 2002. Abstracts of an IRCCS meeting, Rome, Oct 5–10, 2002, p 41
159. Puelles L, Medina L (1994) Development of neurons expressing tyrosine hydroxylase and dopamine in the chicken brain: a comparative segmental analysis. In: Smeets WJAJ, Reiner A (eds) Phylogeny and development of catecholamine systems in the CNS of vertebrates. Cambridge University Press, Cambridge, pp 381–404
160. Puelles L, Medina L (2002) Field homology as a way to reconcile genetic and developmental

variability with adult homology. Brain Res 57: 243–255
161. Puelles L, Rubenstein JL (1993) Expression patterns of homeobox and other putative regulatory genes in the embryonic mouse forebrain suggest a neuromeric organization. Trends Neurosci 16:472–479
162. Puelles L, Rubenstein JLR (2002) Forebrain. In: Ramachandran VS (ed) Encyclopedia of the human brain, Vol 2. Academic Press, Amsterdam, pp 299–315
163. Puelles L, Rubenstein JLR (2003) Forebrain gene expression domains and the evolving prosomeric model. Trends Neurosci 469–476
164. Puelles L, Verney C (1998) Early neuromeric distribution of tyrosine-hydroxylase-immunoreactive neurons in human embryos. J Comp Neurol 394:283–308
165. Puelles L, Domenech-Ratto G, Martinez-de-la-Torre M (1987) Location of the rostral end of the longitudinal brain axis: review of an old topic in the light of marking experiments on the closing rostral neuropore. J Morphol 194:163–171
166. Puelles L, Kuwana E, Puelles E et al (2000) Pallial and subpallial derivatives in the embryonic chick and mouse telencephalon, traced by the expression of the genes Dlx-2, Emx-1, Nkx-2.1, Pax-6, and Tbr-1. J Comp Neurol 424:409–438
167. Qi Y, Stapp D, Qiu M (2002) Origin and molecular specification of oligodendrocytes in the telencephalon. Trends Neurosci 25:223–225
168. Rakic P (1971) Neuron-glia relationship during granule cell migration in developing cerebellar cortex. A Golgi and electronmicroscopic study in *Macacus rhesus*. J Comp Neurol 141:283–312
169. Rakic P (1972) Mode of cell migration to the superficial layers of fetal monkey neocortex. J Comp Neurol 145:61–84
170. Rakic P (1974) Neurons in rhesus monkey visual cortex: systematic relation between time of origin and eventual disposition. Science 183:425–427
171. Rakic P (1990) Principles of neural cell migration. Experientia 46:883–891
172. Rakic P, Sidman RL (1969) Telencephalic origin of pulvinar neurons in the fetal human brain. Z Anat Entwicklungsgesch 129:53–82
173. Redies C, Medina L, Puelles L (2001) Cadherin expression by embryonic divisions and derived gray matter structures in the telencephalon of the chicken. J Comp Neurol 438:253–285
174. Reinoso-Suarez F (1966) Development of the human diencephalon. In: Hassler R, Stephan H (eds) Evolution of the forebrain. Thieme, Stuttgart, pp 296–304
175. Rexed B (1954) A cytoarchitectonic atlas of the spinal cord in the cat. J Comp Neurol 100:297–351
176. Richter C (1966) Über die Entwicklung des Globus pallidus und des Corpus subthalamicum beim Menschen. In: Hassler R, Stephan H (eds) Evolution of the forebrain. Thieme, Stuttgart, pp 285–295
177. Richter E (1965) Die Entwicklung des Globus Pallidus und des Corpus Subthalamicum. Springer, Berlin Heidelberg New York
178. Rodriguez CI, Dymecki SM (2000) Origin of the precerebellar system. Neuron 27:475–486
179. Romanes GJ (1941) Cell columns in the spinal cord of a human foetus of fourteen weeks. J Anat 75:145–152
180. Rubenstein JLR, Beachy PA (1998) Patterning of the embryonic forebrain. Curr Opin Neurobiol 8:18–26
181. Rubenstein JLR, Martínez S, Shimamura K, Puelles L (1994) The embryonic vertebrate forebrain: the prosomeric model. Science 266:578–580
182. Rubenstein JLR, Shimamura K, Martínez S, Puelles L (1998) Regionalization of the prosencephalic neural plate. Annu Rev Neurosci 21:445–477
183. Rüdeberg SI (1961) Morphogenetic studies on the cerebellar nuclei and their homologization in different vertebrates including man. Thesis, Lund
184. Saitsu H, Yamada S, Uwabe C, Ishibashi M, Shiota K (2004) Development of the posterior neural tube in human embryos. Anat Embryol 209:107–117
185. Sanai N, Tramontin AD, Quiñones-Hinojosa A et al (2004) Unique astrocyte ribbon in adult human brain contains neural stem cells but lacks chain migration. Nature 427:740–744
186. Schmechel DE, Rakic P (1979) A Golgi study of radial glial cells in developing monkey telencephalon: morphogenesis and transformation into astrocytes. Anat Embryol (Berl) 156:115–152
187. Schoenen J, Faull RLM (1990) Spinal cord: cytoarchitectural, dendroarchitectural and myeloarchitectural organization. In: Paxinos G (ed) The human nervous system. Academic Press, San Diego, pp 19–53
188. Schwalbe G (1880) Beiträge zur Entwicklungsgeschichte des Zwischenhirns. Sitz Ber Jen Ges Med Naturwiss 20:2–7
189. Shatz CJ (1992) How are specific connections formed between thalamus and cortex? Curr Opin Neurobiol 2:79–82
190. Sidman RL, Rakic P (1973) Neuronal migration, with special reference to developing human brain: a review. Brain Res 62:1–35
191. Singer M, Nordlander RH, Egar M (1979) Axonal guidance during embryogenesis and regeneration in the spinal cord of the newt: the blueprint hypothesis of neuronal pathway patterning. J Comp Neurol 185:1–22
192. Smart IHM (1972) Proliferative characteristics of the ependymal layer during the early development of the spinal cord in the mouse. J Anat 111:365–380
193. Smart IHM (1972) Proliferative characteristics of the ependymal layer during the early development of the mouse diencephalon, as revealed by recording the number, location, and plane of cleavage of mitotic figures. J Anat 113:109–129

194. Smart IHM (1973) Proliferative characteristics of the ependymal layer during the early development of the mouse neocortex: a pilot study based on recording the number, location and plane of cleavage of mitotic figures. J Anat 116:67–91
195. Smart IHM (1976) A pilot study of cell production by the ganglionic eminences of the developing mouse brain. J Anat 121:71–84
196. Smart IHM, McSherry GM (1982) Growth patterns in the lateral wall of the mouse telencephalon: II. Histological changes during and subsequent to the period of isocortical neuron production. J Anat 134:415–442
197. Smart IHM, Sturrock RR (1979) Ontogeny of the neostriatum. In: Divac I, Oberg RGE (eds) The neostriatum. Pergamon, Oxford, pp 127–146
198. Smart IHM, Dehay C, Giroud P, Berland M, Kennedy H (2002) Unique morphological features of the proliferative zones and postmitotic compartments of the neural epithelium giving rise to striate and extrastriate cortex in the monkey. Cereb Cortex 12:37–53
199. Spassky N, Goujet-Zalc C, Parmantier E et al (1998) Multiple restricted origins of oligodendrocytes. J Neurosci 18:8331–8343
200. Spassky N, Olivier C, Perez-Villegas E et al (2000) Single or multiple oligodendroglial lineages: a controversy. Glia 29:143–148
201. Streeter GL (1911) Die Entwicklung des centralen Nervensystems. In: Keibel F, Mall FP (eds) Handbuch der Entwicklungsgeschichte des Menschen, vol 2. Hirzel, Leipzig, pp 28–125
202. Sturrock RR (1979) A comparison of the processes of ventricular coarctation and choroid and ependymal fusion in the mouse brain. J Anat 129:235–242
203. Sussel L, Marin O, Kimura S, Rubenstein JLR (1999) Loss of *Nkx2.1* homeobox gene function results in a ventral to dorsal molecular respecification within the basal telencephalon: evidence for a transformation of the pallidum into the striatum. Development 126:3359–3370
204. Swanson JJ, Kuehl-Kovarik MC, Elmquist JK, Sakaguchi DS, Jacobson CD (1999) Development of the facial and hypoglossal motor nuclei in the neonatal Brazilian opossum brain. Dev Brain Res 112:159–172
205. Tamamaki N, Fujimori KE, Takauji R (1997) Origin and route of tangentially migrating neurons in the developing neocortical intermediate zone. J Neurosci 17:8313–8323
206. Tan K, Le Douarin M (1991) Development of the nuclei and cell migration in the medulla oblongata: application of the quail-chick chimera system. Anat Embryol 183:321–343
207. Tekki-Kessaris N, Woodruff R, Hall AC et al (2001) Hedgehog-dependent oligodendrocyte lineage specification in the telencephalon. Development 128:2545–2554
208. Timsit S, Martinez S, Allinquant B et al (1995) Oligodendrocytes originate from a restricted zone of the embryonic ventral neural tube defined by DM-20 mRNA. J Neurosci 15:1012–1024
209. Trainor PA, Krumlauf R (2000) Patterning the cranial neural crest: hindbrain segmentation and Hox gene plasticity. Nat Rev Neurosci 1:116–124
210. Uylings HBM (2001) The human cerebral cortex in development. In: Kalverboer AF, Gramsbergen A (eds) Handbook of brain and behaviour in human development. Kluwer, Dordrecht, pp 63–80
211. Vallstedt A, Klos JM, Ericson J (2005) Multiple dorsoventral origins of oligodendrocyte generation in the spinal cord and hindbrain. Neuron 45:55–67
212. Verbitskaya LB (1969) Some aspects of the ontophylogenesis of the cerebellum. In: Llinás R (ed) Neurobiology of cerebellum evolution and development. AMA, Chicago, pp 859–879
213. Verney C, Zecevic N, Puelles L (2001) Structure of longitudinal brain zones that provide the origin for the substantia nigra and ventral tegmental area in human embryos, as revealed by cytoarchitecture and tyrosine hydroxylase, calretinin, calbindin, and GABA immunoreactions. J Comp Neurol 429:22–44
214. Von Baer KE (1828) Über die Entwicklungsgeschichte der Thiere. Bornträger, Königsberg
215. Von Kupffer C (1906) Die Morphogenie des Zentralnervensystems. In: Hertwig O (ed) Handbuch der Vergleichenden und Experimentellen Entwicklungslehre der Wirbeltiere, vol 2, part 3. Fischer, Jena, pp 1–272
216. Voogd J (1992) The morphology of the cerebellum the last 25 years. Eur J Morphol 30:81–96
217. Voogd J, Feirabend HKP, Schoen JHR (1990) Cerebellum and precerebellar nuclei. In: Paxinos G (ed) The human nervous system. Academic Press, San Diego, pp 321–386
218. Voogd J, Jaarsma D, Marani E (1996) The cerebellum, chemoarchitecture and anatomy. In: Swanson LW, Björklund A, Hökfelt T (eds) Integrated systems of the CNS, part III. Cerebellum, basal ganglia, olfactory system. Elsevier, Amsterdam, pp 1–369 (Handbook of chemical neuroanatomy, vol 12)
219. Weissman T, Noctor SC, Clinton BK, Honig LS, Kriegstein AR (2003) Neurogenic radial glial cells in reptile, rodent and human: from mitosis to migration. Cereb Cortex 13:550–559
220. Westergaard E (1969) The cerebral ventricles of the golden hamster during growth. Acta Anat (Basel) 72:533–548
221. Westergaard E (1969) The cerebral ventricles of the rat during growth. Acta Anat (Basel) 74:405–423
222. Westergaard E (1971) The lateral cerebral ventricles of human foetuses with a crown-rump length of 26–178. Acta Anat (Basel) 79:409–422
223. Wilkinson DG, Krumlauf R (1990) Molecular approaches to the segmentation of the hindbrain. Trends Neurosci 13:335–339

224. Wilson SW, Rubenstein JLR (2000) Induction and dorsoventral patterning of the telencephalon. Neuron 28:641–651
225. Windle WF (1970) Development of neural elements in human embryos of four to seven weeks gestation. Exp Neurol 28 (Suppl 5):44–83
226. Wingate RJT (2001) The rhombic lip and early cerebellar development. Curr Opin Neurobiol 11:82–88
227. Wingate RJ, Hatten ME (1999) The role of the rhombic lip in avian cerebellum development. Development 126:4395–4404
228. Yamadori T (1965) Die Entwicklung des Thalamuskerns mit ihren ersten Fasersystemen bei menschlichen Embryonen. J Hirnforsch 7:393–413
229. Yun K, Potter S, Rubenstein JLR (2001) *Gsh2* and *Pax6* play complementary roles in dorsoventral patterning of the mammalian telencephalon. Development 128:193–205
230. Zhao M, Momma S, Delfani K et al (2003) Evidence for neurogenesis in the adult mammalian substantia nigra. Proc Natl Acad Sci USA 100:7925–7930

3 Gross Anatomy

The nomenclature used for the gross anatomy of the human brain is the latinized version (or its Anglo-American derivative) of the mainly German nomenclature introduced by Burdach and Reil in the late eighteenth and early nineteenth centuries, with some terms thrown in by Italian and French anatomists such as Malacarne, Rolando and Gratiolet [10]. This nomenclature has been repeatedly revised, adapted, cleaned of most of its eponyms and codified by the Federation Committee on Anatomical Terminology of the International Organization of Anatomists [10].

The fundamental subdivision of the brain into the cerebrum, the cerebellum and the brain stem is immediately apparent. The *cerebral hemispheres* appear to be symmetrical, although left/right hand differences already are apparent upon casual inspection of Figs. 3.1 and 3.4. Systematic differences in the configuration of the convolutions of the cerebral hemisphere related to the lateralization of functions are considered in Chap. 15. The term "*convolutions*" is reminiscent of Vesalius' [19] description of the surface of the cerebral hemispheres as resembling the convolutions of the gut. The Dutch anatomist de la Boe Sylvius [1, 2, 16] drew attention to the horseshoe shape of the hemisphere and gave his name to the lateral sulcus in its centre, which, when opened, reveals the insula (Fig. 3.3). It took a long time, and the discerning eyes of Rolando [9] and Gratiolet [14], to discover a common pattern in the gyri and sulci of the brain.

The *central sulcus* separates the frontal and parietal lobes on the external surface. The *precentral gyrus* and the three frontal gyri, with their subdivisions, occupy the external surface of the frontal lobe; its orbital surface consists of the *orbital gyri* and the *gyrus rectus* (Figs. 3.4, 3.5). The three subdivisions of the *inferior frontal gyrus* derived their names from their location in the roof of the orbita, their triangular shape and their position as part of the coverage or "operculum" (gill-cover) of the insula. The opercular and triangular subdivisions are known as *Broca's speech area* [4–6, 15]. Other parts of the operculum are provided by the inferior portion of the precentral gyrus and the parietal and temporal lobes (Fig. 3.3).

At the convexity of the hemisphere, the border between the parietal and occipital lobes is not distinct, but an obvious *parieto-occipital sulcus* is present on the medial surface (Fig. 3.6). The *parietal lobe* can be subdivided into the *postcentral gyrus* and the *superior* and *inferior parietal lobules*. The latter includes the *supramarginal gyrus*, which folds over the posterior branch of the sulcus lateralis, and the *angular gyrus* (Fig. 3.2). The supramarginal gyrus is also known as the Wernicke speech area [11, 20]. The mesial surface of the superior parietal lobule includes part of the *paracentral lobule* and the *precuneus*.

The border of the occipital and temporal lobes is indicated by the *incisura preoccipitalis*, a notch at the ventral surface. On its medial aspect, the deep *calcarine sulcus* –, named after the *calcar avis* (avian spur), a prominence in the medial wall of the posterior horn of the lateral ventricle caused by this sulcus, – separates the cuneus from the occipital part of the *medial occipitotemporal gyrus* (also known as the *lingual gyrus*). The collateral sulcus separates the medial from the *lateral occipitotemporal gyrus* (*fusiform gyrus*). The *striate* (primary visual) *cortex*, characterized by the presence of a myelinated streak (*line of Gennari* [12]; Figs. 5.13, 5.14, 5.21, 5.30, 5.31) is located in the banks of the calcarine sulcus.

The convexity of the *temporal lobe* can be subdivided into *superior*, *middle* and *inferior gyri*. The superior temporal gyrus forms the *temporal operculum*. The ventral and medial aspect of the temporal lobe is dominated by the occipitotemporal gyri and by structures belonging to the inferior portion of the limbic lobe (Figs. 3.5, 3.6). The temporal operculum, formed by the superior temporal gyrus, includes regions related to hearing and speech that display a clear left/right hand difference (Fig. 18.1).

The *insula* becomes visible when the opercula have been removed (Fig. 3.3). Its surface is subdivided into short and long gyri, separated by a central sulcus. At its base it opens at the *limen insulae*, which borders on the fibers of the olfactory tract contained in the lateral olfactory stria (Figs. 13.5, 13.6).

Limbic lobe is the term (introduced by Broca [1, 2]) used to describe the gyri located at the inneredge (or limbus) of the hemisphere. It consists of the *cingulate gyrus*, which concentrically surrounds the corpus callosum, and the *parahippocampal* and *hippocampal gyri*, which are located medially on the ventral side of the temporal lobe (Figs. 3.6, 3.18). On the medial aspect of the hemisphere, the only portion of the hippocampus that is visible from the outside is the *dentate gyrus* (Figs. 3.6, 23.3). The choroid plexus of the inferior horn of the lateral ventricle is attached to the fornix, where it emerges as the *fimbria fornicis* from the hippocampus at the medial side of the dentate gyrus. This attachment is known as the *taenia fimbriae* (Figs. 3.17, 12.4). At the temporal pole, the parahippocampal gyrus turns back on itself to form the *uncus* (a hook). The anatomy of the limbic lobe and the hippocampal formation is considered in more detail in Chap. 12.

The *corpus callosum* is the great commissure of the cerebral cortex (Fig. 3.7). In the unfixed brain, it feels much harder (callous) than the soft grey matter covering the hemispheres. Ventrally, the corpus callosum tapers into the *rostrum*; its anterior portion is formed by the *genu* (knee), and its caudal extremity widens into the *splenium*. Its central portion is known as the *truncus* of the corpus callosum. The *septum pellucidum*, the thin medial wall of the anterior horn of the lateral ventricle, stretches between the inner surface of the corpus callosum and the fornix (Figs. 3.6, 3.7). The area subcallosa can be found at the medial side of the frontal lobe located beneath the genu and rostrum of the corpus callosum (Figs. 3.6, 6.34, 6.41, 23.3).

When attempting to trace the boundaries of the brain stem's subdivisions in the median section shown in Figs. 3.7 and 3.8, one should realize that, although their traditional surface markings are quite distinct, their location within the continuum of the brain stem often is arbitrary. The transition of the medulla oblongata into the spinal cord is marked by the *decussation of the pyramidal tract*, which is visible as an interruption of the anterior median fissure (Fig. 3.12) but is not obvious in the median section. The prominence of the *pons* marks the caudal and rostral borders of the metencephalon. This entire segment of the brain stem is usually indicated as the pons. At the dorsal surface of the brain stem the border between the medulla oblongata is located at the level of the greatest width of the rhomboid fourth ventricle (Fig. 3.10). At the midline, this border is marked by the *fastigium*, the extension of the fourth ventricle into the base of the cerebellum (Figs. 3.8, 3.13). The caudal wall of the fastigium (tent) consists of the choroid plexus, which covers the caudal half of the fourth ventricle. The rostral wall of the fastigium is formed by the *superior medullary velum*, an elongated portion of the cerebellum located in the roof of the rostral half of the fourth ventricle, which contains the rostral-most lobule of the cerebellum, the *lingula* (Fig. 3.8). The border between the metencephalon and the mesencephalon is dorsally marked by the decussation of the trochlear nerve in the superior medullary velum (Fig. 3.10).

The caudal border of the mesencephalon with the metencephalon is strongly curved and follows the outline of the decussation of the superior cerebellar peduncles, visible as a patch of white matter in the median section (Fig. 5.21). The border between the mesencephalon and the diencephalon is located at the *posterior commissure* dorsally, and caudal to the *mamillary bodies* ventrally. The roof of the mesencephalon is elevated into the four colliculi. The *brachium of the inferior colliculus*, which contains the efferent pathway of this structure, is visible as a ridge on the lateral surface of the mesencephalon. It borders the *superior colliculus* on its lateral side (Fig. 3.10).

The midsagittal section through the diencephalon (Fig. 3.8) passes through the third ventricle, which is closed off by the *lamina terminalis*. The lamina terminalis belongs to the telencephalon impar. It extends between the optic chiasm and the *anterior commissure*. The *optic chiasm* contains the partial decussation on the optic nerves. The crossed and uncrossed fibres of this nerve continue as the *optic tract* on their way to the lateral geniculate body. The optic chiasm and the optic tract demarcate the border of the diencephalon and the telencephalon (Fig. 3.4). *Talairach's plane* passes immediately dorsal to the anterior and ventral to the posterior commissures [18]. This plane is often used in imaging studies of the brain (Figs. 4.10, 5.27).

The lamina terminalis dorsally continues into the rostrum of the corpus callosum and the *commissura of the fornix*. This commissure is located at the base of the septum pellucidum, where the left and right fornix meet, and continues caudally as a triangular sheet, extending from fornix to fornix under cover of the corpus callosum. It allows the exchange of fibres between these bundles (Fig. 3.17).

The *thalamus* and the *hypothalamus*, separated by the *sulcus hypothalamicus*, are visible in the lateral wall of the third ventricle. The *massa intermedia* interconnects the thalami of both sides through the third ventricle. The floor of the hypothalamus protrudes as the *infundibulum*, into the stalk of the hypophysis. The *interventricular foramen* is located in front of the thalamus. The choroid plexus in the roof of the diencephalon is attached to a dorsal ridge, the *taenia thalami* (Figs. 3.6, 3.10, 3.12, 3.17). The plexus continues via the interventricular foramen into the medial wall of the cerebral hemisphere. Caudally, the third ventricle extends as the *pineal recess* into the *pineal gland*. The pineal recess is bordered by the *commissure of the habenula* rostrally, and by the *posterior commissure* on its caudal side (Fig. 3.8).

The ventral aspect of the brain stem is dominated by the large descending fibre systems of the cerebral cortex (Fig. 3.12). They emerge from their coverage by the optic tract as the *cerebral peduncles* on the ventral aspect of the diencephalon and the mesencephalon. The central bundle of the cerebral peduncle, which takes its origin mainly from the pericentral cortex, passes through the pons and emerges on the ventral side of the medulla oblongata as the *pyramid*. The medial and lateral components of the peduncle, which take their origin from more frontal and postcentral cortices, respectively, terminate in the pons. The pyramidal tracts decussate at the transition of the medulla oblongata and the spinal cord. Crossed fibres descend in the lateral funiculi of the cord to lumbar levels; a smaller uncrossed component of the pyramidal tract descends in the ventral funiculus to high thoracic levels (Fig. 21.8). The pontine nuclei are the main relay between the cerebral cortex and the cerebellum. Crossing pontocerebellar fibres are responsible for the transverse relief of the pons. These crossed fibres reach the cerebellum through the middle cerebellar peduncle (brachium pontis) to connect with the contralateral cerebellum.

The twelve pairs of cranial nerves are visible on the ventral aspect of the brain, with the exception of the fourth (trochlear) nerve, which originates from the dorsal side of the brain stem. The first, or olfactory, "nerve" consists of fascicles of unmyelinated *fila olfactoria*, the axons of the receptor cells in the olfactory mucosa. They pass through the lamina cribrosa of the ethmoid and terminate in the *olfactory bulb*. The olfactory tract, located in the *olfactory sulcus*, connects the bulbus with the region of the *anterior perforated substance* (Fig. 13.2). In front of the anterior perforated area, the tract continues into the *medial* and *lateral olfactory striae* (Fig. 13.2). The central connections of the olfactory system are discussed in Chaps. 12 and 22. The decussation of the second, or optic, nerve in the optic chiasm and the course of the optic tract have already been mentioned. Central connections of the optic nerve are considered in Chap. 19.

The cranial nerves, which take their origin from the brain stem, emerge from its ventral and lateral sides, with the exception of the fourth, or *trochlear nerve*, which decussates in, and emerges from, the superior medullary velum at the border of the metencephalon and the mesencephalon (Fig. 3.10). Ventral nerves include the third, sixth and twelfth cranial nerves. The third, or *oculomotor nerve*, appears at the ventral aspect of the mesencephalon in the depression (*interpeduncular fossa*) between the cerebral peduncles (Fig. 3.8). The sixth, or *abducens nerve*, emerges at the caudal border of the pons. The twelfth, or *hypoglossal nerve*, originates beween the pyramid and the *inferior olive*. The latter forms an oval elevation on the ventral surface of the medulla oblongata. The roots of the hypoglossal nerve continue uninterruptedly into the fila radicularia of the ventral root exit zone of the spinal cord (Fig. 3.12).

The fifth and the seventh to the eleventh cranial nerves originate from the lateral aspect of the brain stem (Figs. 3.9, 3.10, 3.12). The fifth, or *trigeminal nerve*, emerges from the lateral aspect of the pons. It consists of a small, rostral *portio minor* and a much larger, caudal *portio major*. The seventh, or *facial nerve*, and the eighth, or *vestibulo-cochlear nerves*, are located laterally along the caudal border of the pons. A furrow separates the large cochlear nerve from the more rostrally located, smaller vestibular nerve. The thin "*intermediate*" *nerve*, which emerges between the seventh and eighth nerves, carries gustatory fibres from the chorda tympani and forms part of the facial nerve. The ninth (*glossopharyngeal*), tenth (*vagal*) and eleventh (*accessory*) cranial nerves take their origin from the region lateral to the inferior olive. The "bulbar" root of the accessory nerve, which takes its origin from the medulla oblongata, is joined by a bundle of ascending root fibres that originate from the spinal cord (the spinal root of the accessory nerve (Figs. 3.9, 3.10, 3.14). The spinal roots of the accessory nerve emerge from the lateral aspect of the upper three cervical segments. They are situated ventral to the dorsal root entrance zone of the cord.

The *cerebellum* is attached to the brain stem in the dorsal region of the metencephalon, rostral to the greatest width of the fourth ventricle (Fig. 3.10). Caudally, the cerebellum bulges over the tela choroidea in the roof of the fourth ventricle (Figs. 3.7, 3.8). Transverse fissures of different depth subdivide the cerebellar surface in lobes, lobules and folia (leaflets). The deepest fissure on the superior surface of the cerebellum is the *primary fissure*, the border between the anterior and posterior lobes (Figs. 3.8, 3.11). Two longitudinal paramedian fissures demarcate the vermis from the cerebellar hemispheres. The *paramedian fissures* are shallow on the superior surface, but deep on the inferior aspect of the cerebellum (Fig. 3.11). The subdivision of the cerebellum will be considered in Chap. 20.

Three cerebellar peduncles connect the cerebellum with the brain stem. Their position relative to the fourth ventricle and the superior medullary velum can be appreciated in Figs. 3.10 and 3.13, where the connection of the cerebellum with the brain stem has been transected, and in a lateral dissection (Fig. 20.1). The *middle cerebellar peduncle*, or *brachium pontis*, connects the pons with the cerebellum. The *inferior cerebellar peduncle*, or *restiform body*, carries connections from the spinal cord and the medulla oblongata to the cerebellum, whereas the *superior cerebellar peduncle*, or *brachium conjunctivum*, represents the main efferent pathway from the cerebellum.

A mould of the ventricular system is illustrated in Fig. 3.16. The position of the *choroid plexus* of the third and the lateral ventricles is shown in Figs. 3.17, 5.7–5.9. The plexus continues from the roof of the third ventricle, as the roof of the interventricular foramina, into the plexus of the lateral ventricles. In the lateral ventricles the choroid plexus is found in the *central part* and the *inferior horn* of the ventricle. The attachment of the choroid plexus to the fornix medially (*taenia fornicis*) and to the *lamina affixa* of the thalamus laterally is shown in Fig. 3.17. The position of the choroid plexus of the fourth ventricle and its extension in the lateral recess of the fourth ventricle is illustrated in Fig. 3.10. The communication of the fourth ventricle with the subarachnoid space in the lateral recess (*lateral aperture of the fourth ventricle*) is marked by the emergence of a small tuft of choroid plexus, visible on the ventral aspect of the brain, known as *Bochdalek's flower basket* [3] (Figs. 3.4, 3.10, 3.12, 4.22). The *median aperture* of the fourth ventricle is located caudally in its ependymal roof.

The internal structures of the prosencephalon and their relations to the lateral ventricle have been illustrated in Figs. 3.21–3.24. The thalamus is located in the lateral wall of the third ventricle. Its position in the floor of the pars centralis of the lateral ventricle, as illustrated in Fig. 3.23, is an indirect one. It is caused by the fusion of part of the original medial wall of the hemisphere, the lamina affixa, with the thalamus (see also Chap. 2 and Fig. 3.17). The subdivision of the thalamus in its nuclei (Figs. 3.19, 3.20) will be discussed in Chap. 8.

The *corpus striatum* is located in the lateral wall of the lateral ventricle and in the region underneath the insula. The *caudate nucleus*, one of the subdivisions of the corpus striatum, is located in the lateral wall of the lateral ventricle (Figs. 3.21–3.24). The *head* (caput) *of the caudate* is located in the lateral wall of the anterior horn of the lateral ventricle. It tapers off into the *tail* (cauda) *of the caudate nucleus*, which is located in the lateral wall of the pars centralis and the cornu inferius. The striatal division located beneath the cortex of the insula is known as the *putamen*. The putamen and caudate nucleus are separated by the *internal capsule*, the great fibre system connecting the cerebral cortex with the thalamus and all lower levels of the central nervous system. Fibres of the internal capsule radiate into the hemisphere as the *corona radiata* (Fig. 3.23). Small extensions of striatal tissue (*ponticuli* or *pontes grisei*) connect the caudate nucleus with the putamen through the internal capsule (Figs. 3.22–3.24). A broad connection between the two nuclei exists ventrally and rostrally. The *globus pallidus* is wedged between the putamen and the internal capsule (Fig. 3.22). Its position can be appreciated from the brain sections illustrated in Figs. 5.5–5.8, 5.23, 5.24, 5.31–5.33. The putamen and the globus pallidus, collectively, are known as the *lentiform nucleus*. The globus pallidus derives its name from its pallid appearence, which is due to its high content of myelinated fibres and which distinguishes it from the dark, almost translucent appearence of the corpus striatum. The striatum derived its name from the striated appearence of the region, which is caused by the fibre bundles of the internal capsule that traverse it [21]. The *amygdaloid nucleus* is located ventromedial to the head of the caudate (Fig. 3.22), underneath the uncus (Fig. 3.6). The term "*basal ganglia*" is often used to denote the striatum and the globus pallidus; sometimes it is extended to include the thalamus.

The gross anatomy of the spinal cord is illustrated in Figs. 3.14 and 3.15. Its salient features will be reviewed in Chap. 6.

References

1. Bartholin C (1641) Institutiones anatomicae, novis rectiorum opiniobus. Observationibus quarum inumerae hactenus editae non sunt, figurisque aucta ab auctoris filio Thoma Bartholino
2. Beukers H (2000) The Sylvian fissure. In: Koehler PJ, Bruyn GW, Pearce JMS (eds) Neurological eponyms. Oxford University Press, Oxford, pp 51–55
3. Bochdalek VA (1833) Anleitung zur praktischen Zergliederung des menschlichen Gehirns nebst einer anatomischen Beschreibung desselben mit besonderer Rücksicht auf das kleine Gehirn. Hause Söhne, Prague
4. Broca P (1861) Sur la principe des localizations cérébrales. Bull Soc Anthropol 2:190–204
5. Broca P (1861) Perte de parole, ramollissement chronique et destruction partielle lobe antérieur gauche. Sur le siège de la faculté du langage. Bull Soc Anthropol 2:235–238
6. Broca P (1861) Remarques sur la siège de la faculté du langage articulé, suivies d'un observation d'aphémie. Bull Soc Anat 36:330–357
7. Broca P (1877) Sur la circonvolution limbique et la scissure limbique. Bull Soc Anthropol 12:646–657
8. Broca P (1878) Le grand lobe limbique et la scissure limbique dans les series des mammalifères. Rev Anthropol 2:385–498
9. Caputi F, Spaziante R, de Divitiis E, Nashold BS Jr (1995) Luigi Rolando and his pioneering efforts to relate structure to function in the nervous system. J Neurosurg 83:933–937

10. Clarke E, O'Malley CD (1968) The human brain and spinal cord: a historical study illustrated by writings from antiquiry to the twentieth century. University of California Press, Berkeley
11. Dalman AJE, Eling E (2000) Wernicke's aphasia. In: Koehler PJ, Bruyn GW, Pearce JMS (eds) Neurological eponyms. Oxford University Press, Oxford, pp 244–249
12. Gennari F (1782) Francisci Gennari Parmensis Medicinae Doctoris Collefiati de Peculiari Structura Cerebri Nonnullisque Eius Morbis. Regio Typographeo, Parma
13. Giacomini C (1883) Fascia dentata del grande hippocampo nel cervello umano. Giorg d R Acad di Medic, Torino
14. Gratiolet P (1854) Mémoire sur les plis cérébraux de l'homme et des primatés., Paris
15. Moffie D, Schiller F (2000) Broca's aphasia. In: Koehler PJ, Bruyn GW, Pearce JMS (eds) Neurological eponyms. Oxford University Press, Oxford, pp 194–199
16. Pearce JMS (2000) The fissure of Sylvius (1614–72). J Neurol Neurosurg Psychiatry 69:463
17. Retzius G (1896) Das Menschenhirn. Studien in der Makroskopischen Morphologie. Norstedt & Sohne, Stockholm
18. Talairach J, Toumoux P (1988) Co-planar stereotaxic atlas of the human brain. Thieme, New York
19. Vesalius A (1952) Vesalius on the human brain. English translation of the text from the latin. Oxford University Press, London
20. Wernicke C (1874) Der aphasische Symptomenkomplex: Eine psychologische Studie auf anatomischer Basis. Max Cohn & Weigert, Breslau
21. Willis T (1664 (1965)) Of the anatomy of the brain. Englished by Samual Pordage, Esq. London. Printed for Dring, Leigh and Martin. Fascimile edition, McGill University Press, Montreal

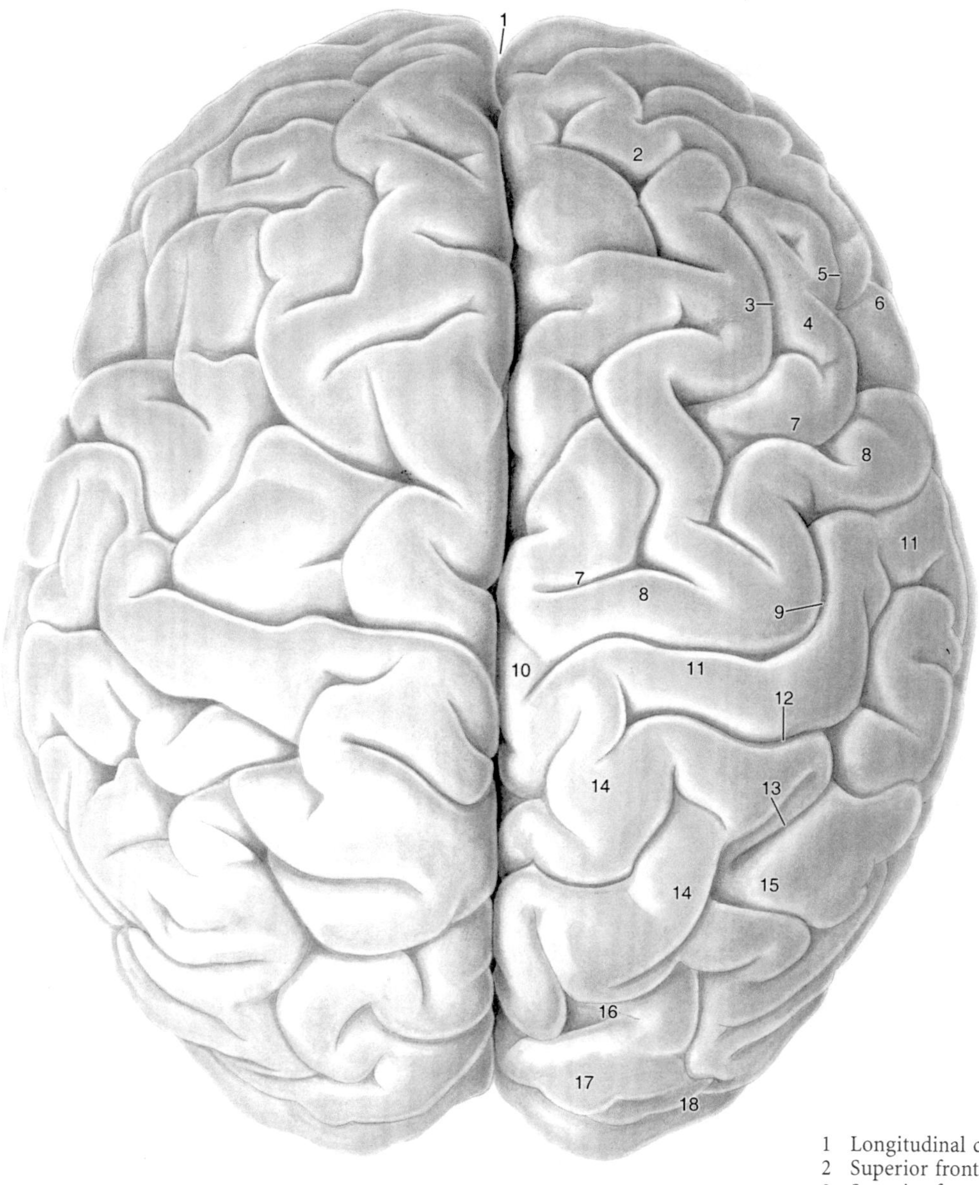

1 Longitudinal cerebral fissure
2 Superior frontal gyrus
3 Superior frontal sulcus
4 Middle frontal gyrus
5 Inferior frontal sulcus
6 Inferior frontal gyrus
7 Precentral sulcus
8 Precentral gyrus
9 Central sulcus
10 Paracentral lobule
11 Postcentral gyrus
12 Postcentral sulcus
13 Intraparietal sulcus
14 Superior parietal lobule
15 Inferior parietal lobule
16 Parieto-occipital sulcus
17 Occipital gyri
18 Occipital sulci

Fig. 3.1. The brain seen from above (1/1×)

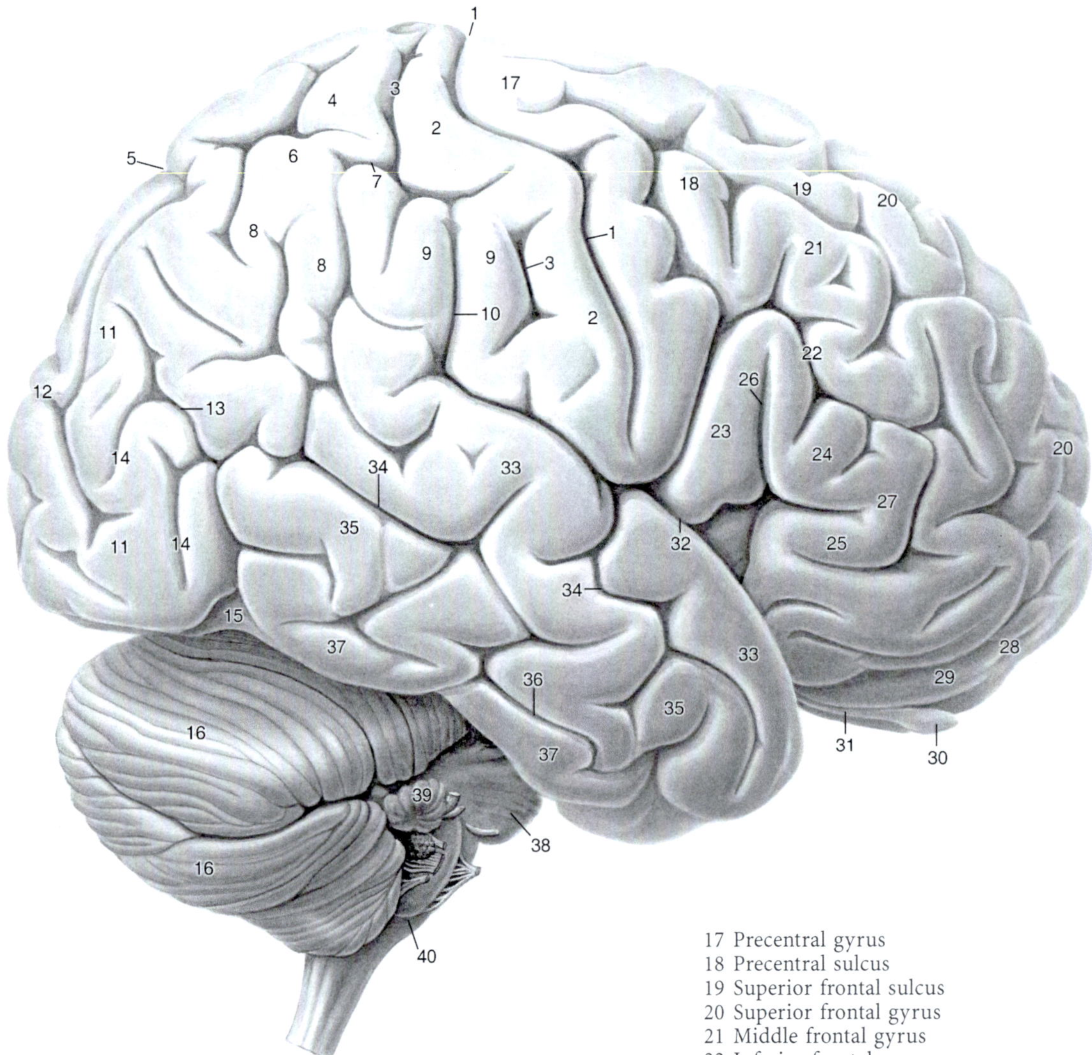

1 Central sulcus
2 Postcentral gyrus
3 Postcentral sulcus
4 Superior parietal lobule
5 Parieto-occipital sulcus
6 Inferior parietal lobule
7 Intraparietal sulcus
8 Angular gyrus
9 Supramarginal gyrus
10 Lateral sulcus, posterior branch
11 Occipital gyri
12 Lunate sulcus
13 Anterior occipital sulcus
14 Occipital sulci
15 Preoccipital notch
16 Cerebellar hemisphere
17 Precentral gyrus
18 Precentral sulcus
19 Superior frontal sulcus
20 Superior frontal gyrus
21 Middle frontal gyrus
22 Inferior frontal gyrus
23 Opercular part of inferior frontal gyrus
24 Triangular part of inferior frontal gyrus
25 Orbital part of inferior frontal gyrus
26 Lateral sulcus, ascending branch
27 Lateral sulcus, anterior branch
28 Orbital sulci
29 Orbital gyri
30 Olfactory bulb
31 Olfactory tract
32 Lateral sulcus
33 Superior temporal gyrus
34 Superior temporal sulcus
35 Middle temporal gyrus
36 Inferior temporal sulcus
37 Inferior temporal gyrus
38 Pons
39 Flocculus
40 Medulla oblongata

Fig. 3.2. Lateral view of the brain (1/1×)

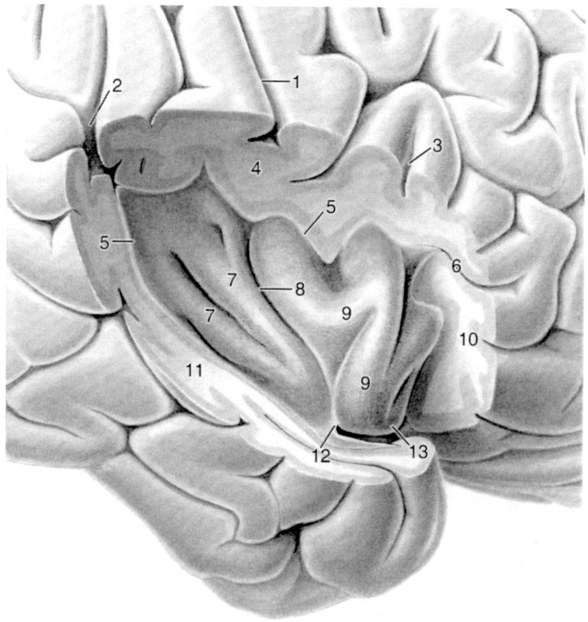

1 Central sulcus
2 Lateral sulcus, posterior branch
3 Lateral sulcus, ascending branch
4 Frontoparietal operculum
5 Circular sulcus of the insula
6 Lateral sulcus, anterior branch
7 Long gyrus of the insula
8 Central sulcus of the insula
9 Short gyri of the insula
10 Frontal operculum
11 Temporal operculum
12 Limen insulae
13 Anterior pole of the insula

Fig. 3.3. Dissection of the right cerebral hemisphere to display the insula (1/1×)

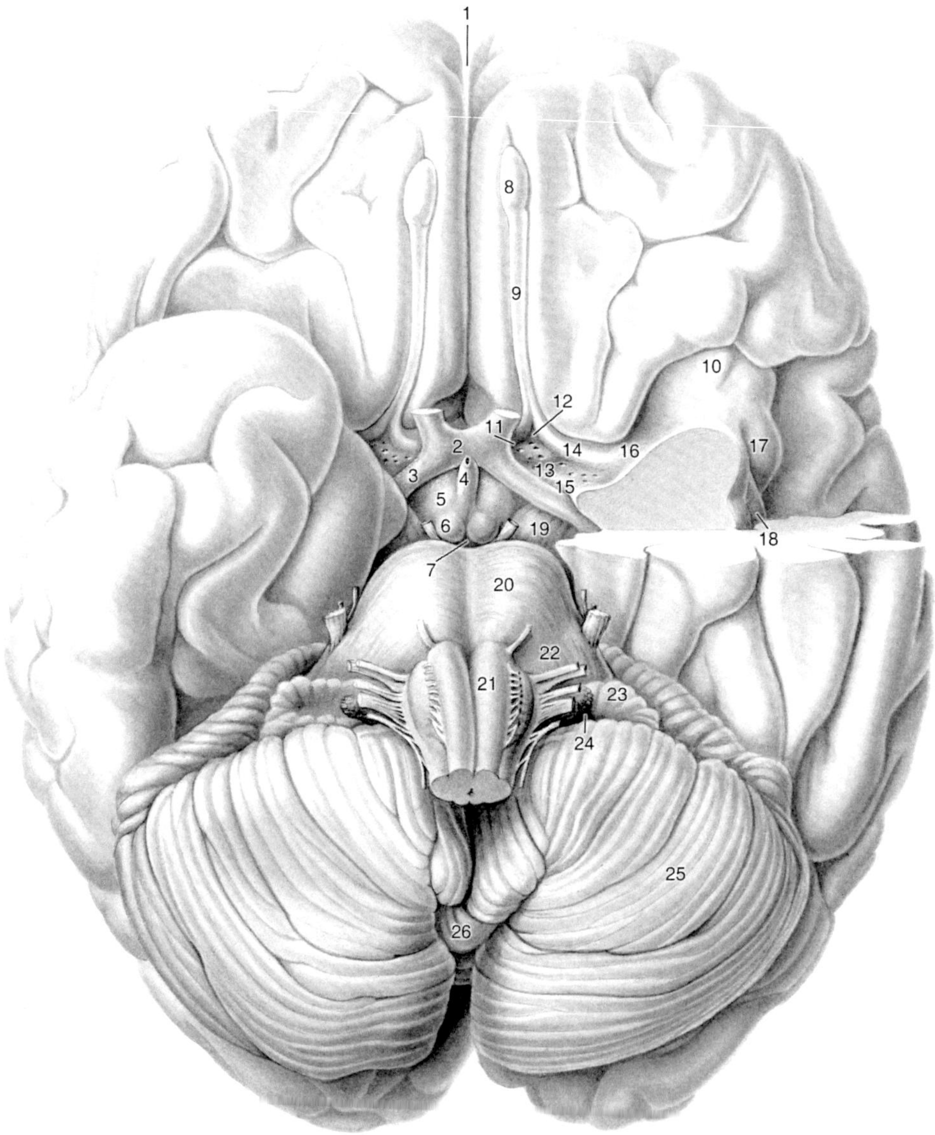

1 Longitudinal cerebral fissure
2 Optic chiasm
3 Optic tract
4 Infundibulum
5 Tuber cinereum
6 Mamillary body
7 Interpeduncular fossa
8 Olfactory bulb
9 Olfactory tract
10 Anterior pole of the insula
11 Medial olfactory stria
12 Olfactory triangle
13 Anterior perforated substance
14 Lateral olfactory stria
15 Diagonal gyrus
16 Limen insulae
17 Short gyri of the insula
18 Long gyrus of the insula
19 Cerebral peduncle
20 Pons
21 Pyramid
22 Inferior olive
23 Flocculus
24 Choroid plexus of the fourth ventricle
25 Cerebellar hemisphere
26 Cerebellar vermis

Fig. 3.4. Basal view of the brain. The frontal portion of the left temporal lobe has been removed to expose the underlying structures (1/1×)

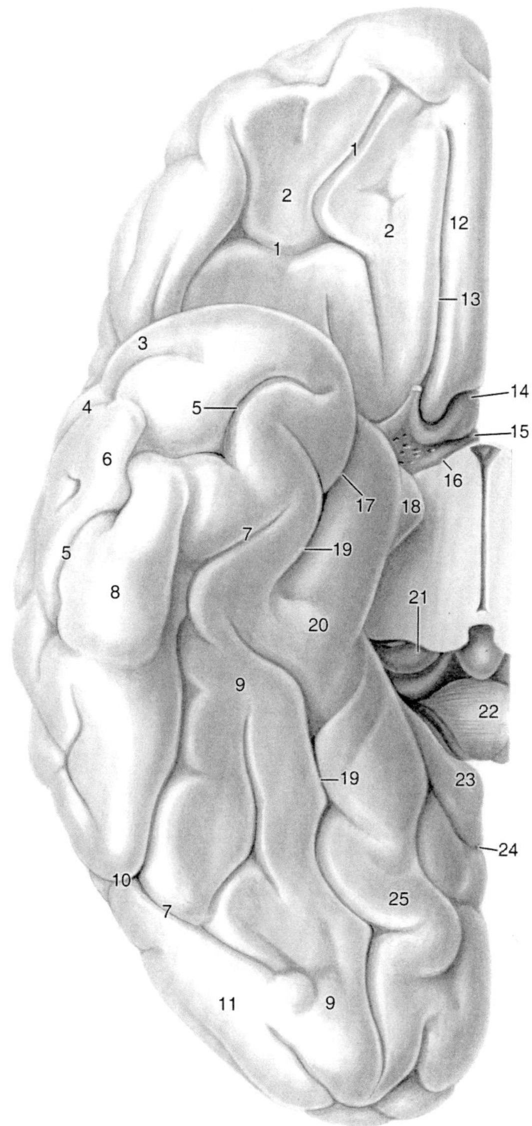

1 Orbital sulci
2 Orbital gyri
3 Superior temporal gyrus
4 Superior temporal sulcus
5 Inferior temporal sulcus
6 Middle temporal gyrus
7 Occipitotemporal sulcus
8 Inferior temporal gyrus
9 Lateral occipitotemporal (fusiform) gyrus
10 Preoccipital notch
11 Occipital gyri
12 Straight gyrus (gyrus rectus)
13 Olfactory sulcus
14 Subcallosal area
15 Paraterminal gyrus
16 Diagonal gyrus
17 Rhinal sulcus
18 Gyrus ambiens
19 Collateral sulcus
20 Parahippocampal gyrus
21 Pulvinar of the thalamus
22 Splenium of the corpus callosum
23 Isthmus of the cingular gyrus
24 Calcarine sulcus
25 Medial occipitotemporal (lingual) gyrus

Fig. 3.5. Basal view of the right hemisphere. The olfactory tract has been sectioned (1/1×)

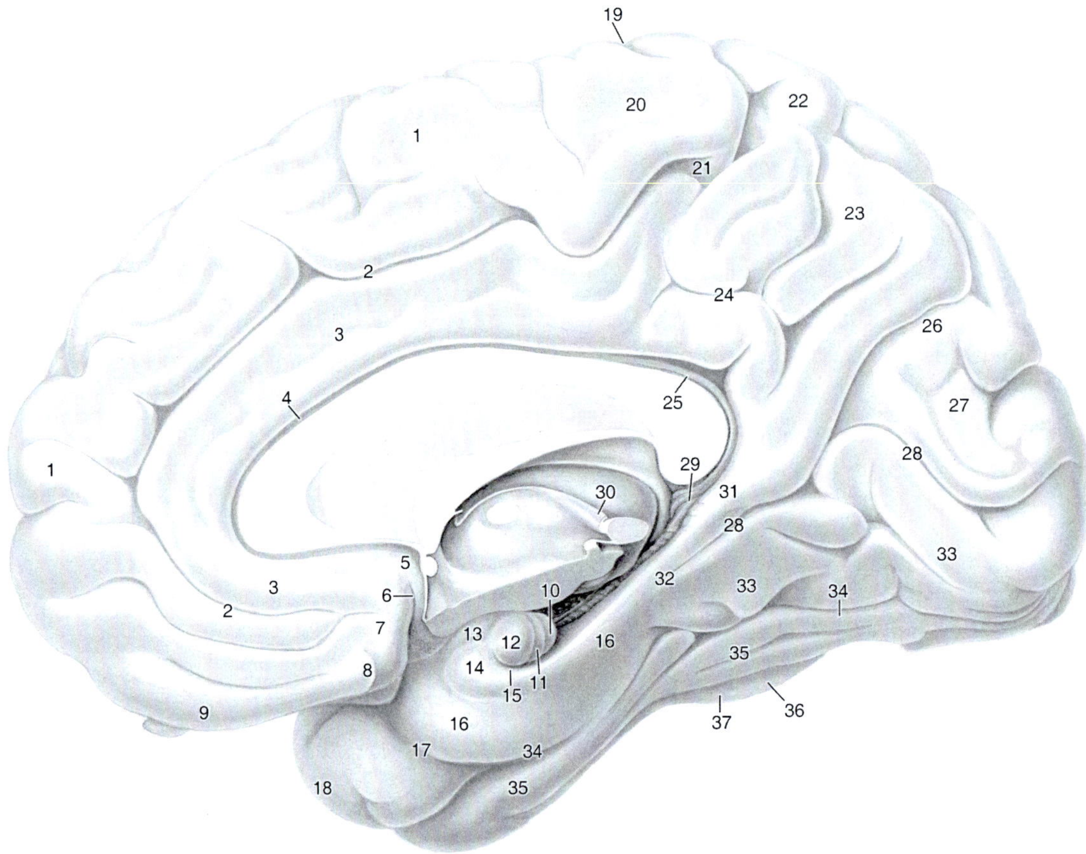

1 Superior frontal gyrus
2 Cingulate sulcus
3 Cingulate gyrus
4 Sulcus of the corpus callosum
5 Paraterminal gyrus
6 Posterior parolfactory sulcus
7 Subcallosal area
8 Anterior parolfactory sulcus
9 Straight gyrus (gyrus rectus)
10 Intralimbic gyrus of the uncus
11 Band of Giacomini of the uncus [13]
12 Uncinate gyrus of the uncus
13 Semilunar gyrus
14 Gyrus ambiens
15 Notch of the uncus
16 Parahippocampal gyrus
17 Rhinal sulcus
18 Superior temporal gyrus
19 Central sulcus
20 Paracentral lobule
21 Marginal branch of the cingulate sulcus
22 Superior parietal lobule
23 Precuneus
24 Subparietal sulcus
25 Indusium griseum
26 Parieto-occipital sulcus
27 Cuneus
28 Calcarine sulcus
29 Fasciolar gyrus
30 Taenia thalami
31 Isthmus of the cingulate gyrus
32 Dentate gyrus
33 Medial occipitotemporal (lingual) gyrus
34 Collateral sulcus
35 Lateral occipitotemporal (fusiform) gyrus
36 Occipitotemporal sulcus
37 Inferior temporal gyrus

Fig. 3.6. Medial aspect of the right cerebral hemisphere (1/1×)

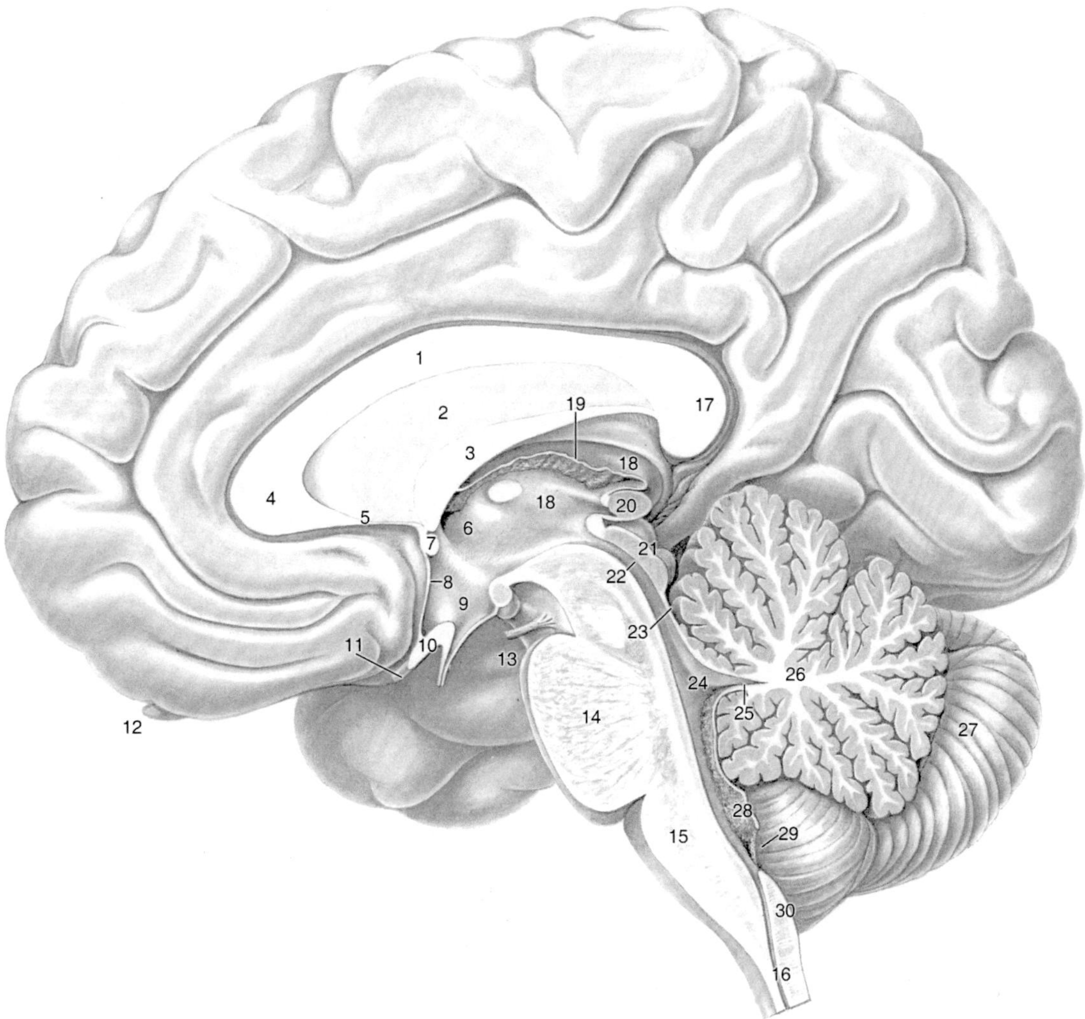

1 Trunk of the corpus callosum
2 Septum pellucidum
3 Fornix
4 Genu of the corpus callosum
5 Rostrum of the corpus callosum
6 Interventricular foramen
7 Anterior commissure
8 Lamina terminalis
9 Hypothalamus
10 Optic chiasm
11 Optic nerve
12 Olfactory bulb
13 Oculomotor nerve
14 Pons
15 Medulla oblongata
16 Spinal cord
17 Splenium of the corpus callosum
18 Thalamus
19 Choroid plexus of the third ventricle
20 Pineal body
21 Lamina quadrigemina
22 Cerebral aquaduct
23 Superior medullary velum
24 Fourth ventricle
25 Inferior medullary velum
26 Cerebellar vermis
27 Cerebellar hemisphere
28 Choroid plexus of the fourth ventricle
29 Median aperture of the fourth ventricle
30 Central canal

Fig. 3.7. Medial view of the right half of the bisected brain (1/1 ×)

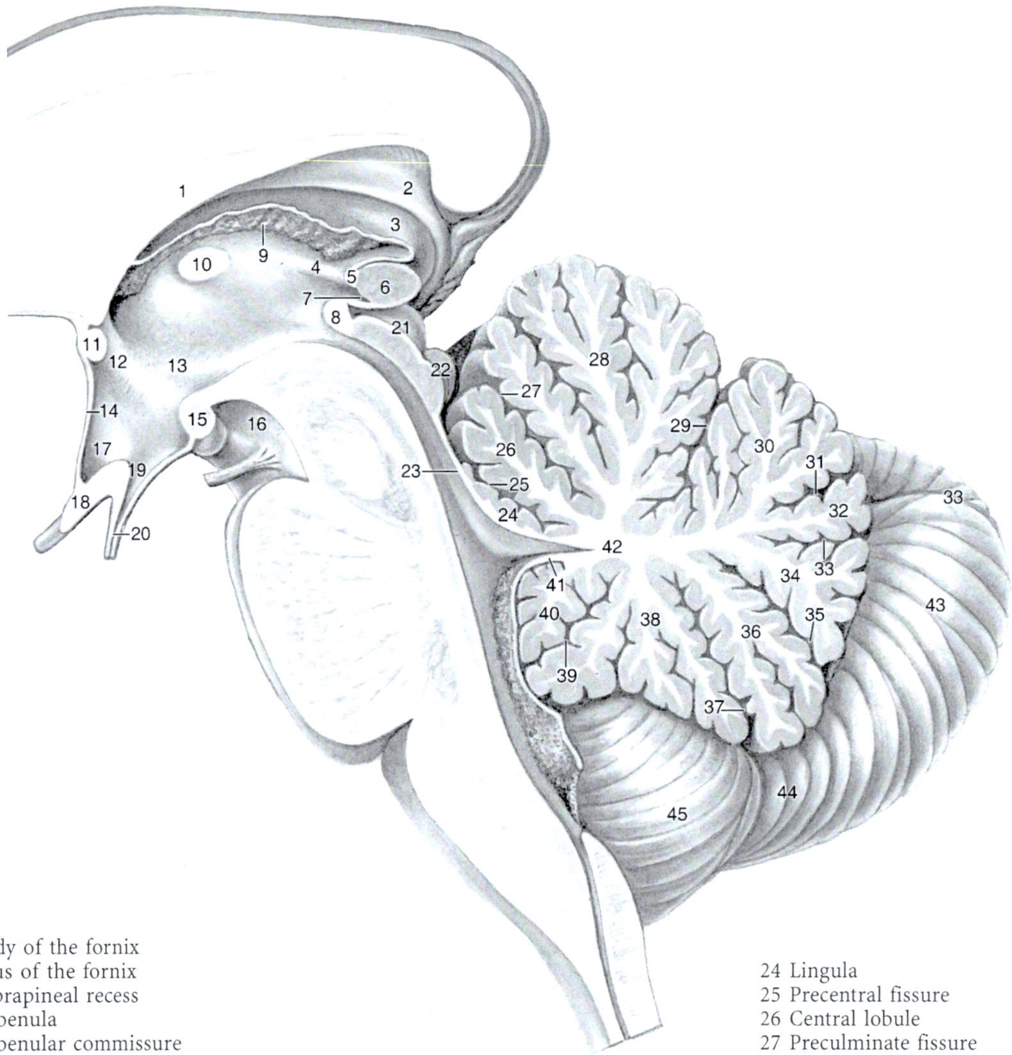

1 Body of the fornix
2 Crus of the fornix
3 Suprapineal recess
4 Habenula
5 Habenular commissure
6 Pineal body
7 Pineal recess
8 Posterior commissure
9 Choroid plexus of the third ventricle
10 Massa intermedia
11 Anterior commissure
12 Column of the fornix
13 Hypothalamic sulcus
14 Lamina terminalis
15 Mamillary body
16 Interpeduncular fossa
17 Optic recess
18 Optic chiasm
19 Infundibular recess
20 Infundibulum
21 Superior colliculus
22 Inferior colliculus
23 Superior medullary velum
24 Lingula
25 Precentral fissure
26 Central lobule
27 Preculminate fissure
28 Culmen
29 Primary fissure
30 Declive
31 Posterior superior fissure
32 Folium vermis
33 Horizontal fissure
34 Tuber vermis
35 Prepyramidal fissure
36 Pyramis
37 Secondary fissure
38 Uvula
39 Posterolateral fissure
40 Nodulus
41 Inferior medullary velum
42 Fastigium
43 Inferior semilunar lobule
44 Biventral lobule
45 Tonsil

Fig. 3.8. Medial view of the bisected brain stem and cerebellum (3/2×)

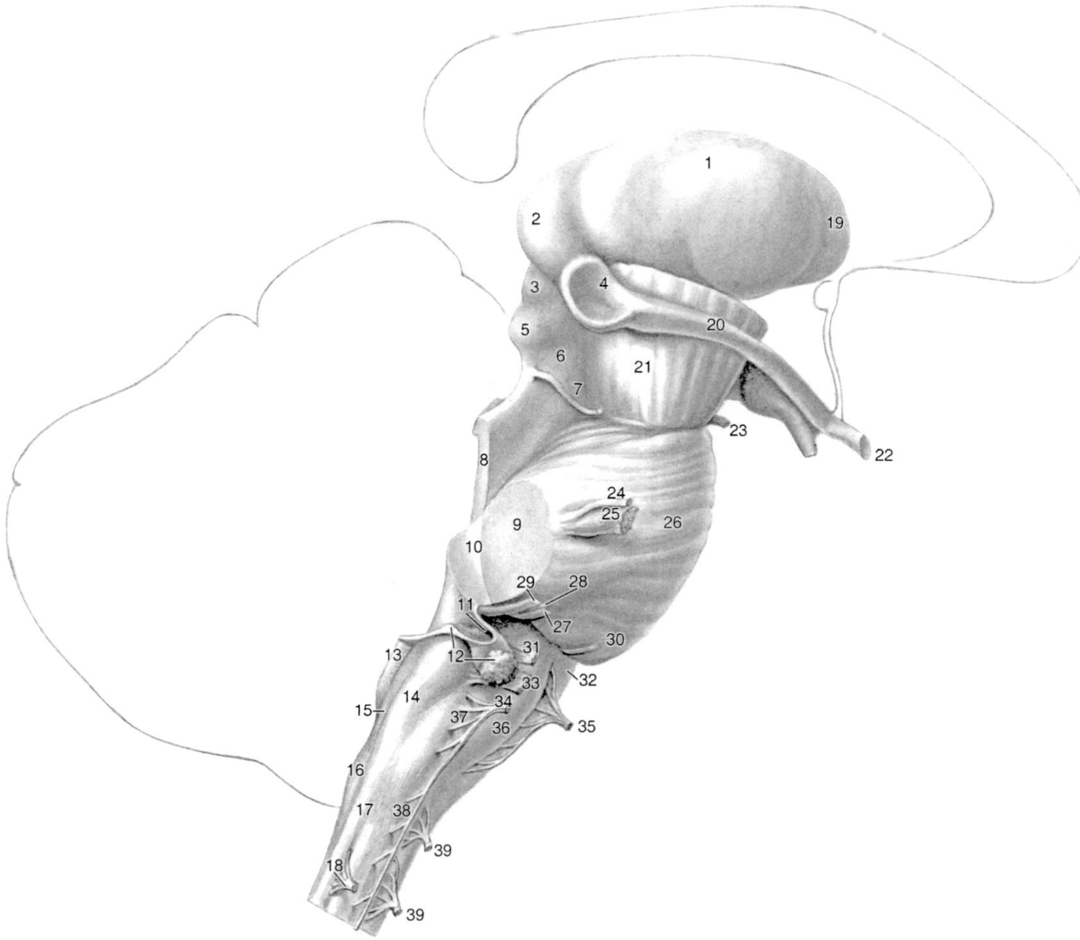

1 Thalamus
2 Pulvinar
3 Superior colliculus
4 Lateral geniculate body
5 Inferior colliculus
6 Lemniscal triangle
7 Trochlear nerve
8 Superior cerebellar peduncle (brachium conjunctivum)
9 Middle cerebellar peduncle (brachium pontis)
10 Inferior cerebellar peduncle (restiform body)
11 Lateral recess of the fourth ventricle
12 Choroid plexus of the fourth ventricle
13 Attachment of the choroid plexus of the fourth ventricle
14 Cuneate tubercle
15 Obex
16 Gracile tubercle
17 Lateral posterior sulcus
18 Dorsal root
19 Anterior tubercle of the thalamus
20 Optic tract
21 Cerebral peduncle
22 Optic nerve
23 Oculomotor nerve
24 Trigeminal nerve, portio minor
25 Trigeminal nerve, portio major
26 Pons
27 Facial nerve
28 Intermediate nerve
29 Vestibulocochlear nerve
30 Abducens nerve
31 Glossopharyngeal nerve
32 Pyramis
33 Vagal nerve
34 Accessory nerve
35 Hypoglossal nerve
36 Inferior olive
37 Bulbar roots of the accessory nerve
38 Spinal roots of the accessory nerve
39 Ventral roots

Fig 3.9. Lateral view of the brain stem and the diencephalon after removal of the structures surrounding the thalamus (3/2×)

1 Lateral ventricle
2 Third ventricle
3 Pineal body
4 Brachium of the superior colliculus
5 Superior colliculus
6 Brachium of the inferior colliculus
7 Inferior colliculus
8 Cerebral peduncle

25 Taenia choroidea: attachment of the choroid plexus to the lamina affixa
26 Lamina affixa
27 Stria terminalis
28 Stria medullaris of the thalamus
29 Taenia thalami: attachment of the choroid plexus of the third ventricle to the thalamus
30 Habenular triangle
31 Pulvinar of the thalamus
32 Medial geniculate body
33 Lateral geniculate body

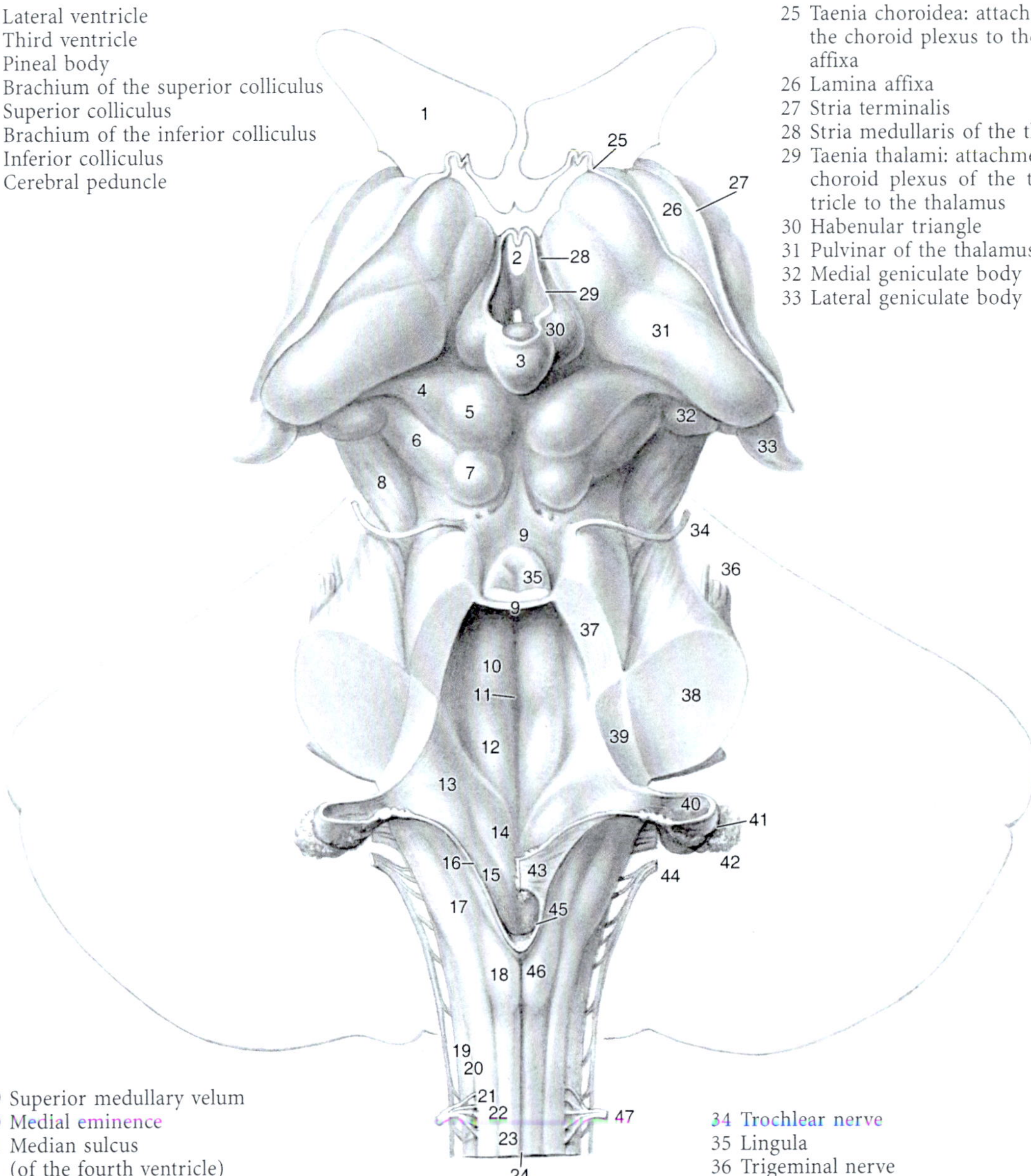

9 Superior medullary velum
10 Medial eminence
11 Median sulcus (of the fourth ventricle)
12 Facial collicle
13 Vestibular area
14 Hypoglossal triangle
15 Vagal triangle
16 Attachment of the roof of the fourth ventricle
17 Cuneate tubercle
18 Gracile tubercle
19 Lateral funiculus
20 Lateral posterior sulcus
21 Cuneate fascicle
22 Posterior intermediate sulcus
23 Gracile fascicle
24 Posterior median sulcus

34 Trochlear nerve
35 Lingula
36 Trigeminal nerve
37 Superior cerebellar peduncle (brachium conjunctivum)
38 Middle cerebellar peduncle (brachium pontis)
39 Inferior cerebellar peduncle (restiform body)
40 Lateral recess of the fourth ventricle
41 Lateral aperture of the fourth ventricle
42 Bochdalek's flowerbasket [3]
43 Ependymal roof of the fourth ventricle
44 Accessory nerve
45 Median aperture of the fourth ventricle
46 Obex
47 Dorsal root

Fig. 3.10. Dorsal view of the brain stem and the diencephalon after removal of the structures surrounding the thalamus. The contour of the cerebellum is indicated (3/2×)

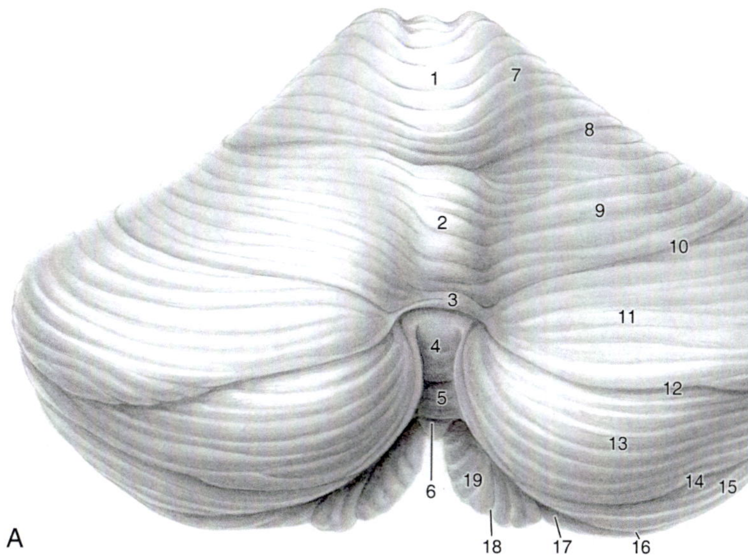

1 Culmen
2 Declive
3 Folium vermis
4 Tuber vermis
5 Pyramis
6 Uvula
7 Quadrangular lobule
8 Primary fissure
9 Lobulus simplex
10 Posterior superior fissure
11 Superior semilunar lobule
12 Horizontal fissure
13 Inferior semilunar lobule
14 Pregracile fissure (variable)
15 Gracile lobule
16 Prebiventral fissure
17 Biventral lobule
18 Secondary fissure
19 Tonsil

Fig. 3.11 A. Dorsal view of the cerebellum (6/5×)

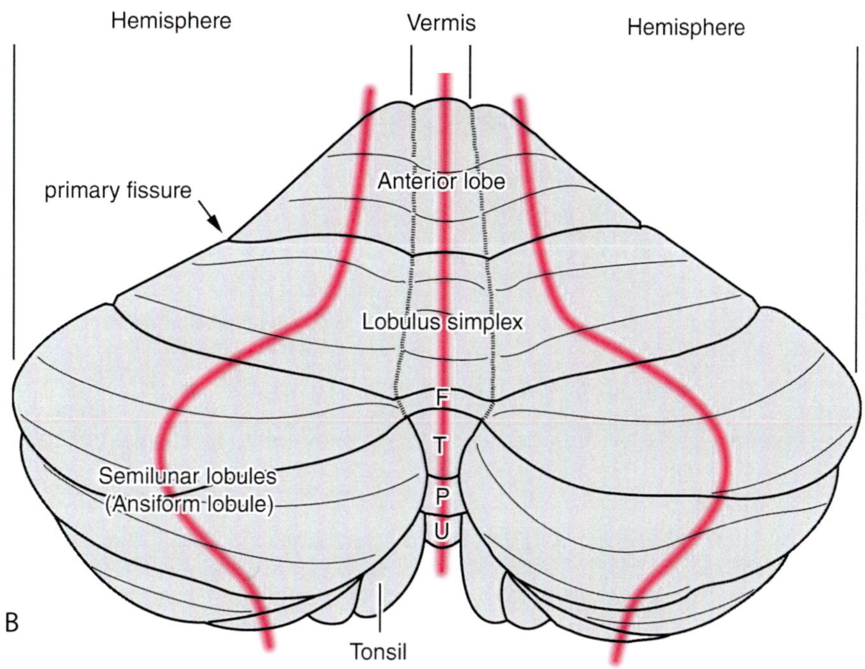

Fig. 3.11 B. Diagram of a dorsal view of the cerebellum. The direction of the folial chains of vermis and hemispheres is indicated by *red lines*. Note folial loop of the semilunar lobules. (The ansiform lobule of the comparative anatomical nomenclature, see also Fig. 20.2.) *F*, folium ; *P*, pyramis; *T*, tuber ; *U*, uvula

1 Taenia choroidea: attachment of the choroid plexus to the lamina affixa
2 Lamina affixa
3 Taenia thalami: attachment of the choroid plexus of the third ventricle to the thalamus
4 Anterior tubercle of the thalamus
5 Thalamus
6 Massa intermedia
7 Optic chiasm
8 Optic nerve
9 Optic tract
10 Lateral geniculate body
11 Oculomotor nerve
12 Trochlear nerve

24 Lateral ventricle
25 Third ventricle
26 Infundibulum
27 Mamillary bodies
28 Cerebral peduncle
29 Posterior perforated substance
30 Interpeduncular fossa
31 Pons
32 Basilar sulcus of the pons

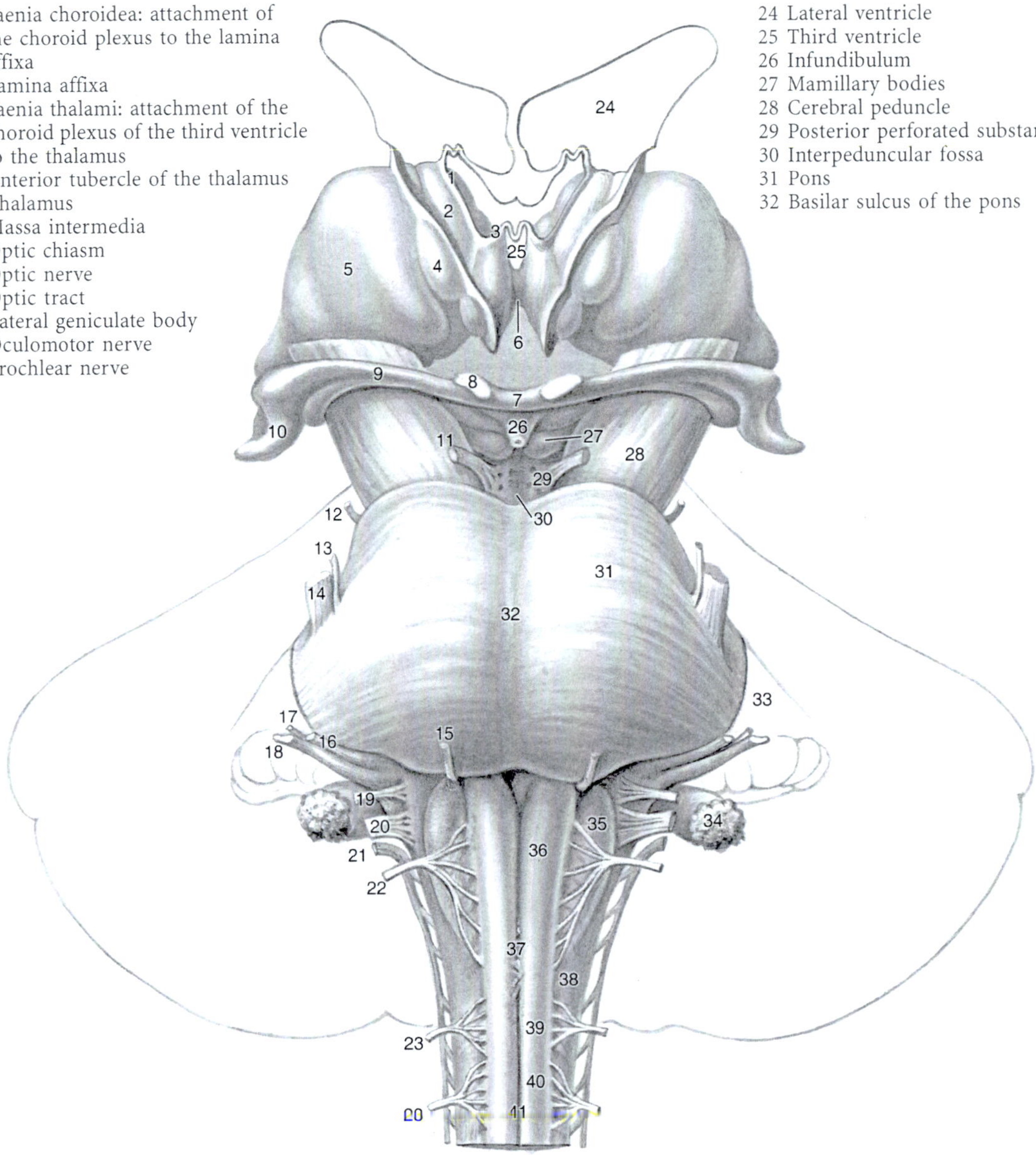

13 Trigeminal nerve, portio minor
14 Trigeminal nerve, portio major
15 Abducens nerve
16 Facial nerve
17 Intermediate nerve
18 Vestibulocochlear nerve
19 Glossopharyngeal nerve
20 Vagal nerve
21 Accessory nerve
22 Hypoglossal nerve
23 Ventral spinal roots

33 Middle cerebellar peduncle (brachium pontis)
34 Choroid plexus of the fourth ventricle
35 Inferior olive
36 Pyramid
37 Pyramidal decussation
38 Lateral funiculus
39 Anterior lateral sulcus
40 Anterior funiculus
41 Anterior median fissure

Fig. 3.12. Ventral view of the brain stem and the diencephalon. The structures surrounding the thalamus have been removed (3/2×)

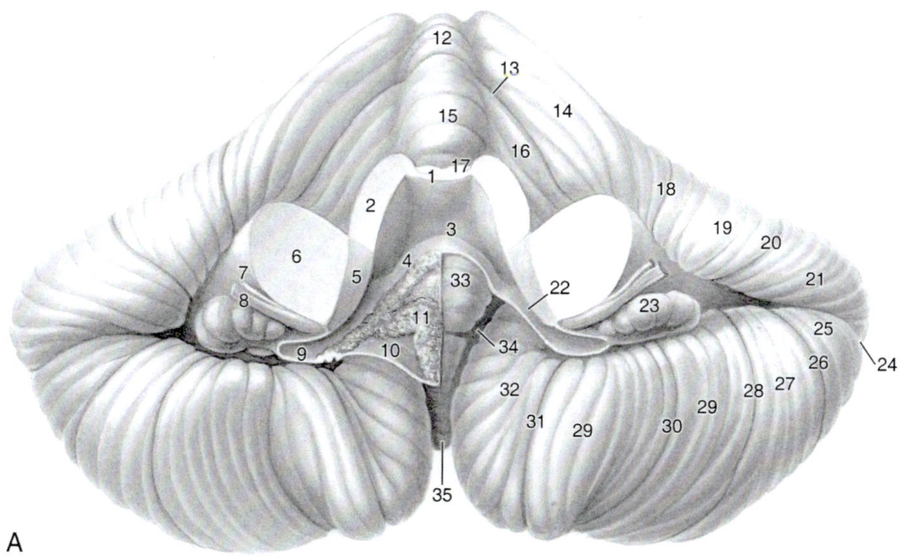

1 Superior medullary velum
2 Superior cerebellar peduncle (brachium conjunctivum)
3 Fastigium
4 Inferior medullary velum
5 Inferior cerebellar peduncle (restiform body)
6 Middle cerebellar peduncle (brachium pontis)
7 Intermediate nerve
8 Vestibulocochlear nerve
9 Lateral recess of the fourth ventricle
10 Roofplate of the fourth ventricle
11 Choroid plexus of the fourth ventricle
12 Culmen
13 Preculminate fissure
14 Anterior quadrangular lobule
15 Central lobule
16 Ala (wing) of the central lobule
17 Lingula
18 Primary fissure
19 Posterior quadrangular lobule (lobulus simplex)
20 Superior posterior fissure
21 Superior semilunar lobule
22 Floccular peduncle
23 Flocculus
24 Horizontal fissure
25 Inferior semilunar lobule
26 Ansoparamedian fissure
27 Gracile lobule
28 Prebiventral fissure
29 Biventral lobule
30 Intrabiventral fissure
31 Secondary fissure
32 Tonsil
33 Nodulus
34 Posterolateral fissure
35 Uvula

Fig. 3.13 A. Ventral view of the cerebellum (6/5×)

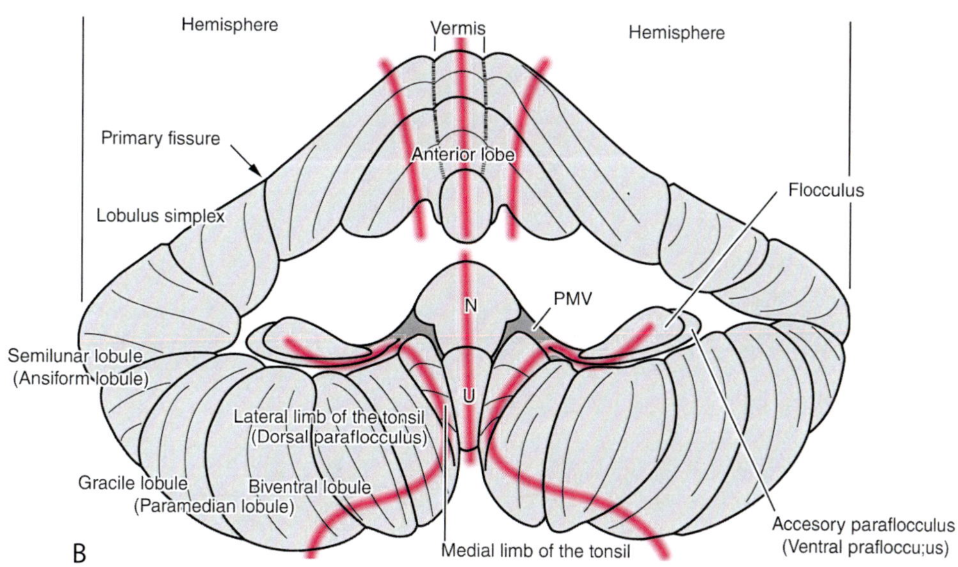

Fig. 3.13 B. Diagram of a ventral view of the cerebellum. The direction of the folial chains of vermis and hemispheres is indicated by *red lines*. Note medially directed folial loop of the tonsil (the dorsal paraflocculus of the comparative anatomical nomenclature, see also Fig. 20.2). The medial limb of the folial loop of the tonsil, and the accessory paraflocculus are hidden from view in **A.** *N*, nodule; *PMV*, inferior medullary velum; *U*, uvula

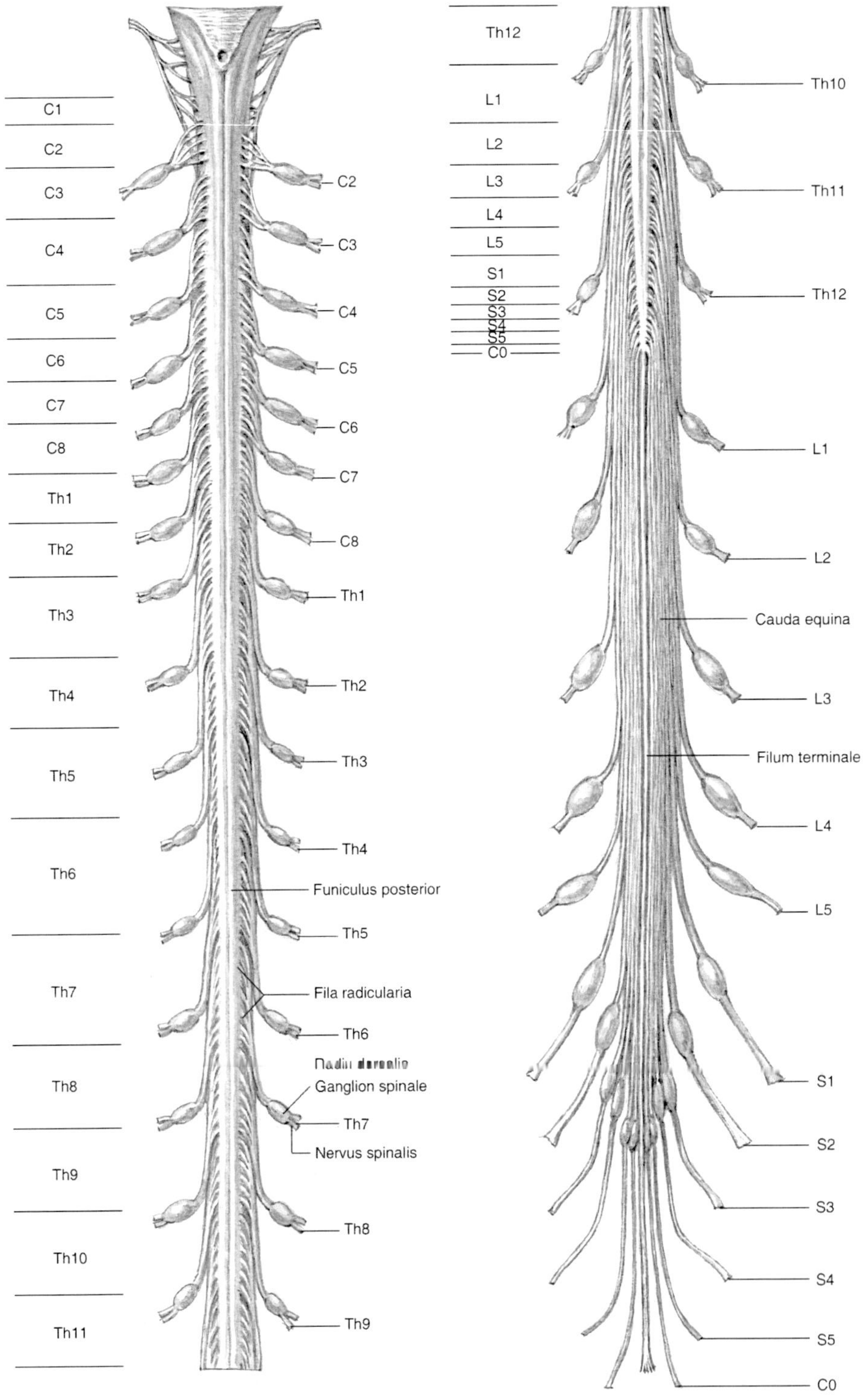

Fig. 3.14. Dorsal view of the spinal cord showing attached dorsal root filaments and spinal ganglia. The cervical (*C*), thoracic (*T*), lumbar (*L*), sacral (*S*) and coccygeal (*Co*) spinal nerves have been transected at their site of exit from the intervertebral foramina. The position of the spinal segments is indicated on the *left* side of the cord (2/3×)

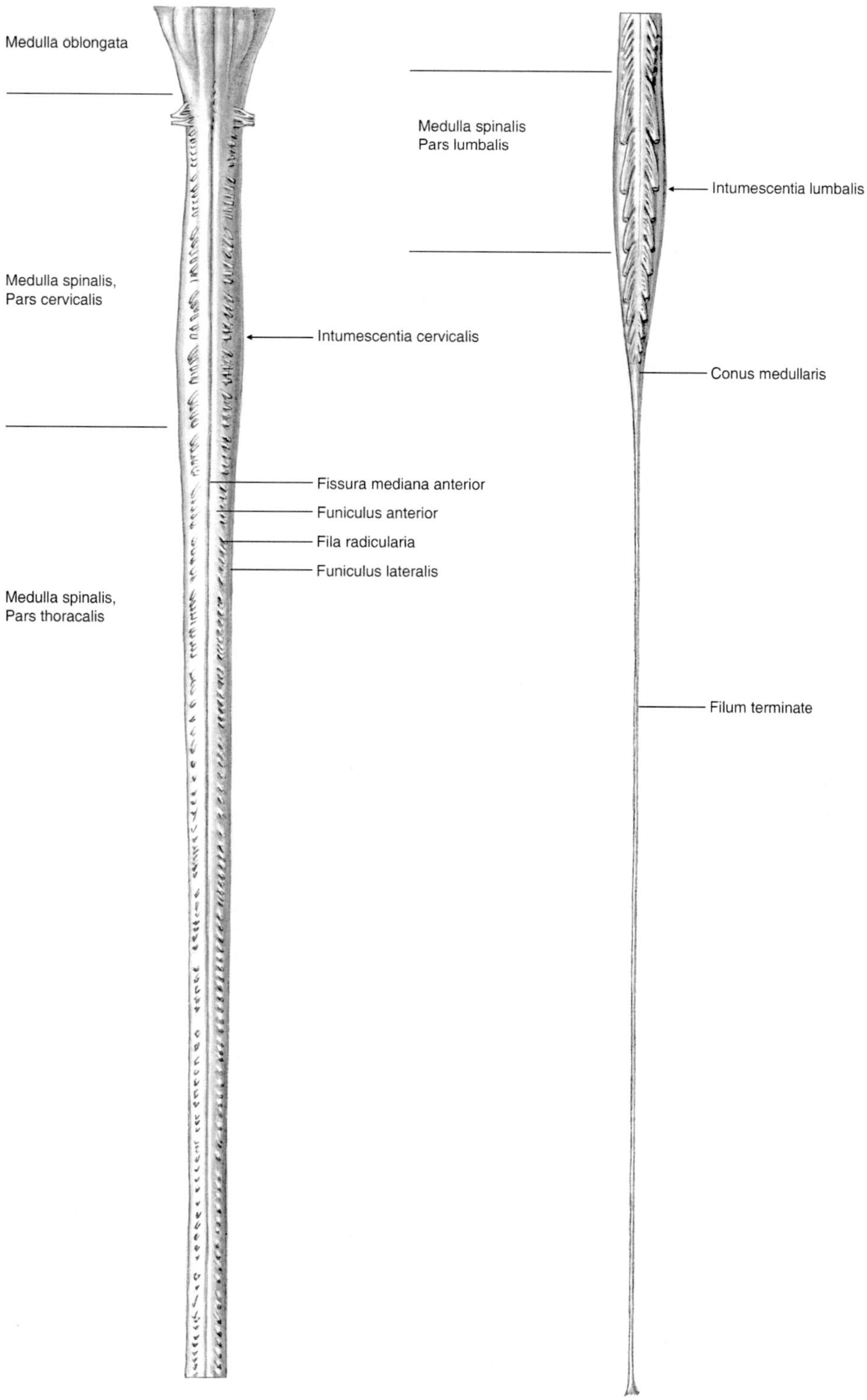

Fig. 3.15. Ventral view of the spinal cord; the ventral root filaments have been transected (2/3×)

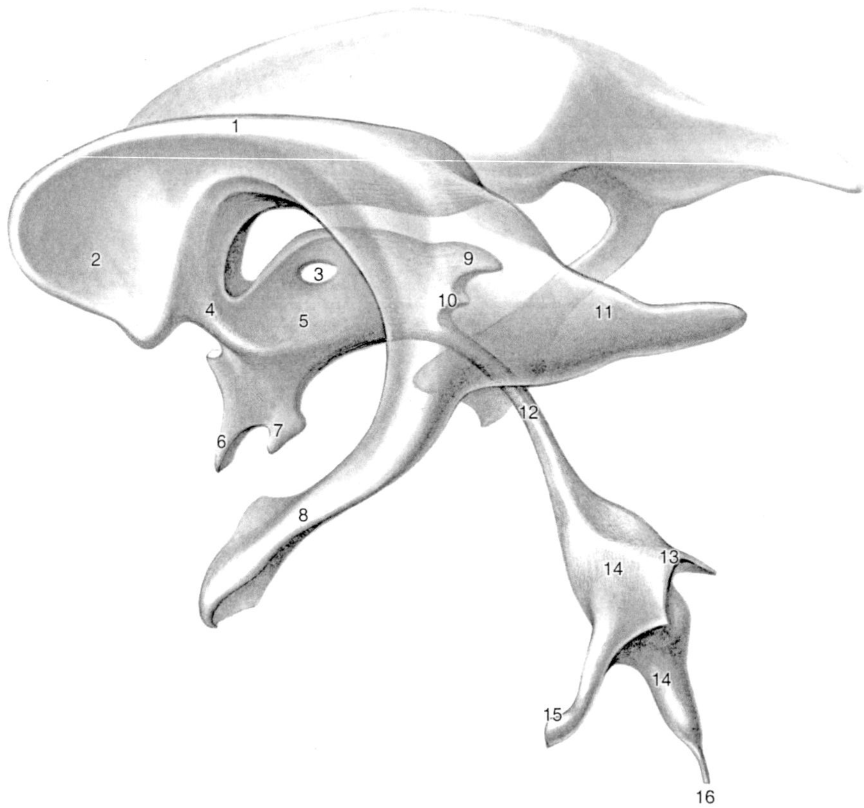

1 Lateral ventricle, central part
2 Lateral ventricle, anterior horn
3 Massa intermedia
4 Interventricular foramen
5 Third ventricle
6 Optic recess
7 Infundibular recess
8 Lateral ventricle, inferior horn
9 Suprapineal recess
10 Pineal recess
11 Lateral ventricle, posterior horn
12 Cerebral aqueduct
13 Fastigium
14 Fourth ventricle
15 Lateral recess of the fourth ventricle
16 Central canal

Fig. 3.16. The ventricles of the brain; oblique view from behind and above (6/5×)

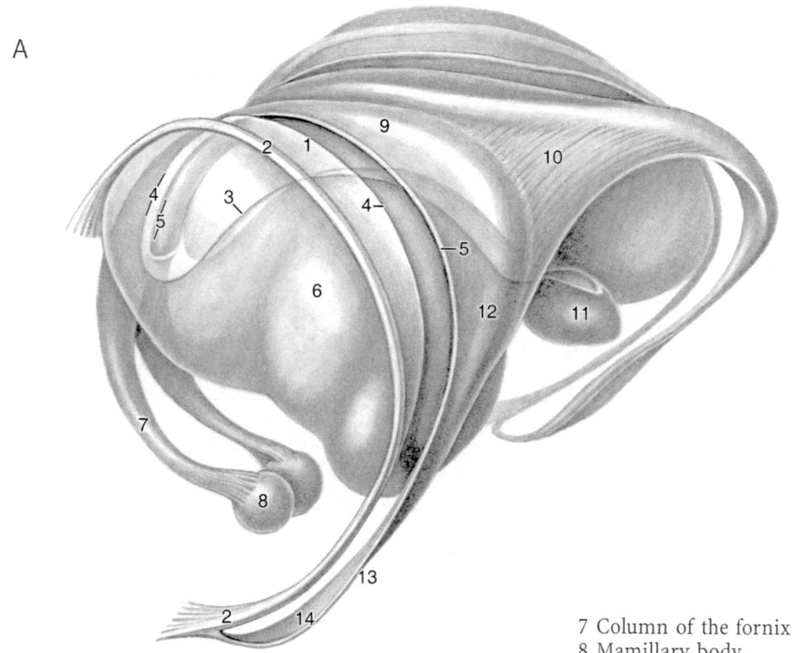

1 Lamina affixa
2 Stria terminalis
3 Taenia thalami: attachment of choroid plexus of the third ventricle to the thalamus
4 Taenia choroidea: attachment of the choroid plexus to the lamina affixa
5 Taenia fornicis: attachment of the choroid plexus to the fornix
6 Thalamus
7 Column of the fornix
8 Mamillary body
9 Corpus of the fornix
10 Commissure of the fornix
11 Pineal body
12 Crus of the fornix
13 Taenia fimbriae: attachment of the choroid plexus to the fimbria
14 Fimbria

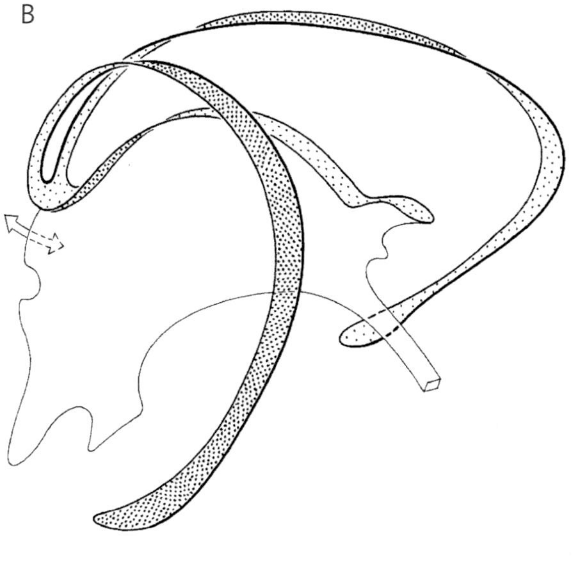

Fig. 3.17 A, B. Topography of diencephalic and telencephalic attachments (taeniae) of the choroid plexus; oblique view from behind and above (**A**). In the complementary diagram (**B**), the choroid walls of the lateral and third ventricles are shown. *Dense stippling*: ventricular surface; *light stippling*: meningeal surface; *double arrow*: interventricular foramen (2/1×; diagram 5/3×)

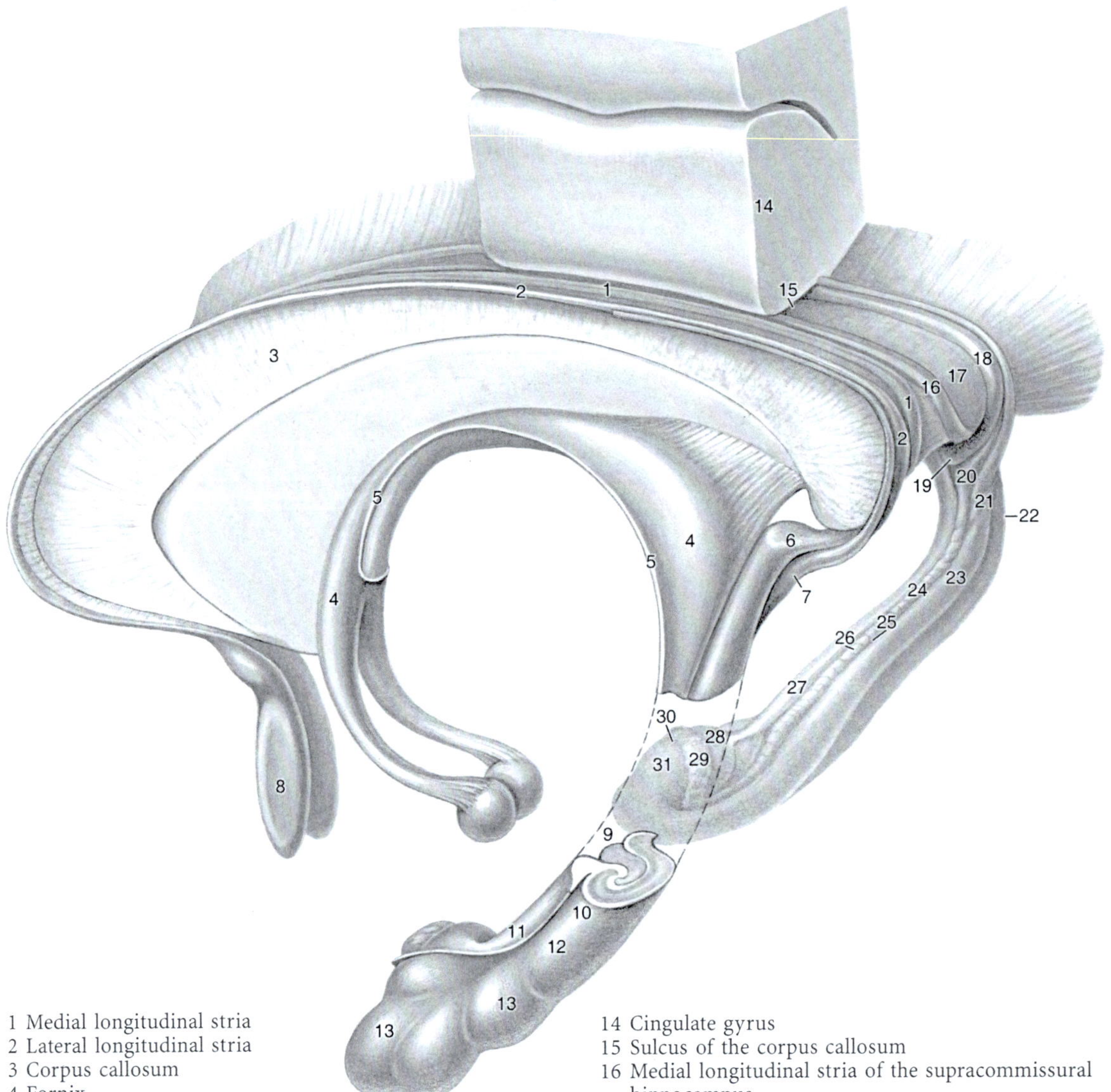

1 Medial longitudinal stria
2 Lateral longitudinal stria
3 Corpus callosum
4 Fornix
5 Taenia fornicis: attachment of the choroid plexus to the fornix
6 Fasciolar gyrus
7 Gyrus of Anders Retzius [17]
8 Precommissural hippocampus
9 Dentate gyrus
10 Alveus
11 Fimbria
12 Ammon's horn
13 Digitations of the hippocampus
14 Cingulate gyrus
15 Sulcus of the corpus callosum
16 Medial longitudinal stria of the supracommissural hippocampus
17 Indusium griseum of the supracommissural hippocampus
18 Lateral longitudinal stria of the supracommissural hippocampus
19 Fasciolar gyrus
20 Fasciola cinerea
21 Gyrus of Anders Retzius [17]
22 Ammon's horn of the retrocommissural hippocampus
23 Subiculum of the retrocommissural hippocampus
24 Dentate gyrus of the retrocommissural hippocampus
25 Hippocampal sulcus
26 Fimbriodentate sulcus
27 Fimbria
28 Intralimbic gyrus
29 Limbus of Giacomini [13]
30 Hippocampal sulcus
31 Uncinate gyrus

Fig. 3.18. Dissection showing the hippocampus and some related structures in oblique view from behind and above (2/1×)

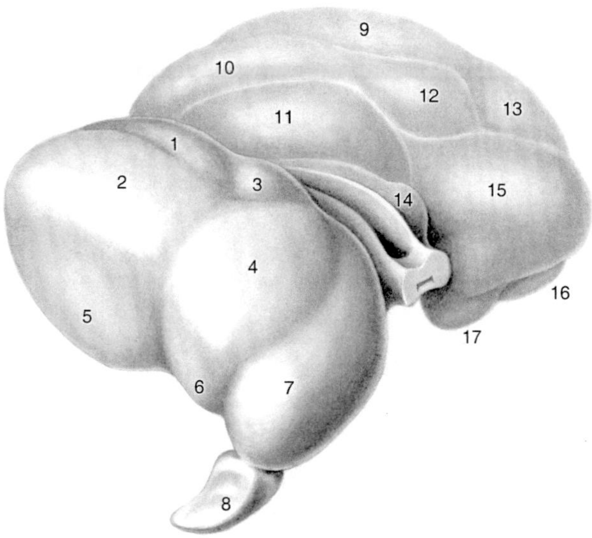

1 Anterior nucleus
2 Ventral lateral nucleus
3 Lateral dorsal nucleus
4 Lateral posterior nucleus
5 Ventral anterior nucleus
6 Ventral posterolateral nucleus
7 Pulvinar
8 Lateral geniculate body
9 Ventral lateral nucleus

10 Anterior nucleus
11 Medial nucleus
12 Lateral dorsal nucleus
13 Lateral posterior nucleus
14 Nuclei of the habenula
15 Pulvinar
16 Lateral geniculate body
17 Medial geniculate body

Fig. 3.19. A model of both thalami: oblique view from behind and above. The reticular nucleus and the midline nuclei have been omitted (2/1×)

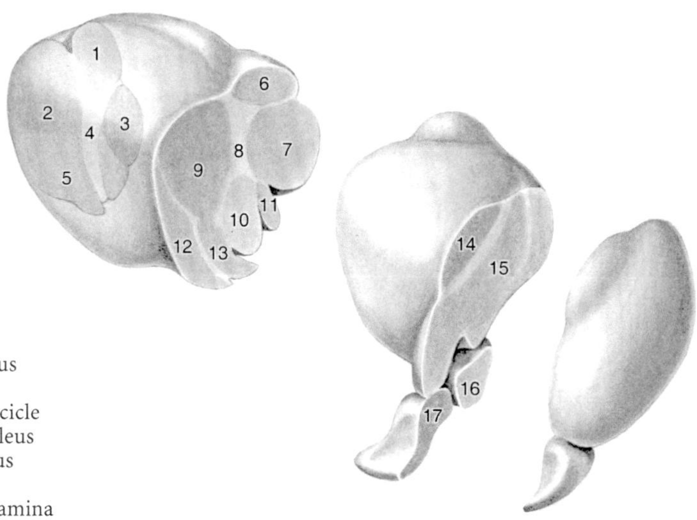

1 Anterior nucleus
2 Ventral lateral nucleus
3 Medial nucleus
4 Mamillothalamic fascicle
5 Ventral anterior nucleus
6 Lateral dorsal nucleus
7 Medial nucleus
8 Internal medullary lamina
9 Ventral lateral nucleus
10 Nucleus of the centre médian
11 Parafascicular nucleus
12 Ventral posterolateral nucleus
13 Ventral posteromedial nucleus

14 Lateral posterior nucleus
15 Pulvinar
16 Medial geniculate body
17 Lateral geniculate body

Fig. 3.20. Left half of the same model as shown in Fig. 3.19. The position of the major thalamic nuclei is indicated on three frontal sections (2/1×)

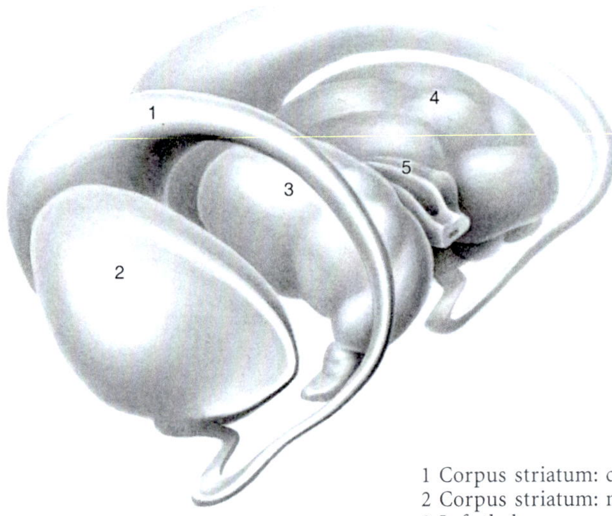

1 Corpus striatum: caudate nucleus
2 Corpus striatum: nucleus lentiformis
3 Left thalamus
4 Right thalamus
5 Stria medullaris

Fig. 3.21. The corpus striatum and the thalamus of both sides in oblique view from behind and above (6/5×)

1 Isolated fibre bundles of the corona radiata
2 Tail of the nucleus caudatus
3 Strands of grey matter (pontes grisei) connecting the caudate nucleus with the putamen
4 Corpus of the caudate nucleus
5 Putamen
6 Outline of the thalamus
7 Internal capsule, one isolated fibre bundle indicated
8 Globus pallidus, external segment
9 Globus pallidus, internal segment

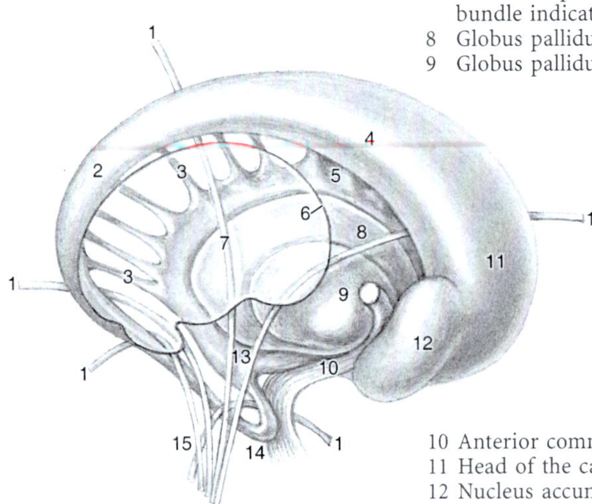

10 Anterior commissure
11 Head of the caudate nucleus
12 Nucleus accumbens
13 Peduncle of the lentiform nucleus
14 Junction of the tail of the caudate nucleus with the peduncle of the lentiform nucleus
15 Cerebral peduncle

Fig. 3.22. The basal ganglia in medial view (6/5×)

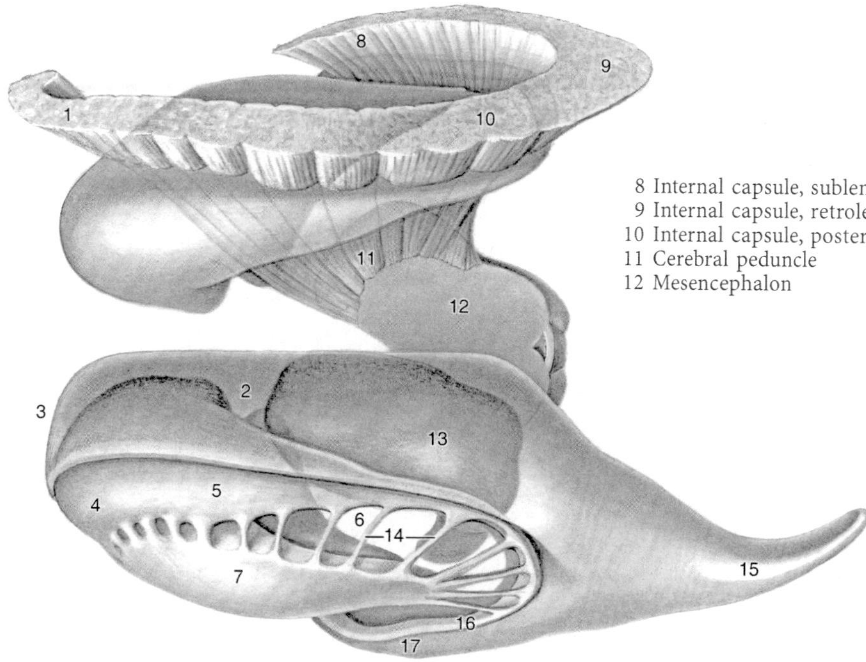

8 Internal capsule, sublenticular part
9 Internal capsule, retrolenticular part
10 Internal capsule, posterior limb
11 Cerebral peduncle
12 Mesencephalon

1 Internal capsule, anterior limb
2 Lateral ventricle, central portion
3 Lateral ventricle, anterior horn
4 Head of the caudate nucleus
5 Corpus of the caudate nucleus
6 Globus pallidus
7 Putamen

13 Thalamus
14 Strands of grey matter (pontes grisei) connecting the caudate nucleus with the putamen
15 Lateral ventricle, posterior horn
16 Tail of the caudate nucleus
17 Lateral ventricle, inferior horn

Fig. 3.23. The corpus striatum of both sides, viewed from above. The thalamus and the extent of the lateral ventricle can be seen on the *left side*. The internal capsule and its convergence upon the cerebral peduncle are indicated on the *right side* (6/5×)

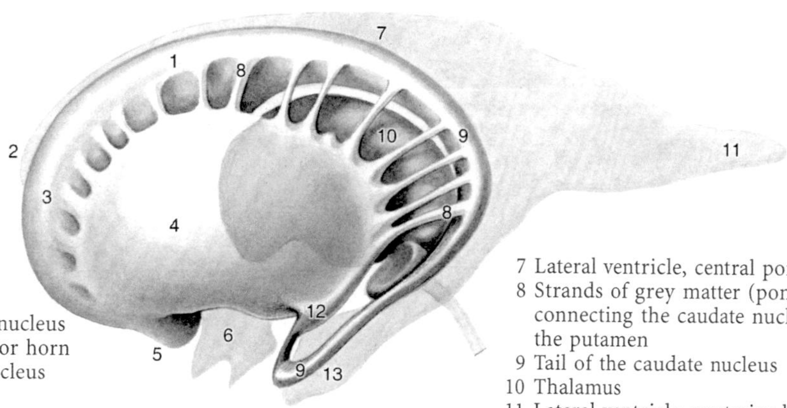

1 Corpus of the caudate nucleus
2 Lateral ventricle, anterior horn
3 Head of the caudate nucleus
4 Putamen
5 Nucleus accumbens
6 Third ventricle

7 Lateral ventricle, central portion
8 Strands of grey matter (pontes grisei) connecting the caudate nucleus with the putamen
9 Tail of the caudate nucleus
10 Thalamus
11 Lateral ventricle, posterior horn
12 Peduncle of the lentiform nucleus
13 Lateral ventricle, inferior horn

Fig. 3.24. The corpus striatum and the thalamus in lateral view. The contours of some parts of the ventricular system are indicated (6/5×)

4 Blood Supply, Meninges and Cerebrospinal Fluid Circulation

Introduction

The vascularization and the circulation of the cerebrospinal fluid (liquor cerebrospinalis, CSF) of the brain and the spinal cord are of great clinical importance. The main vascular syndromes are summarized in Table 4.1. In this chapter, the anatomy of blood vessels, meninges and circumventricular organs will be discussed. The central nervous system, which is of ectodermal origin (Chap. 2), is surrounded by mesodermal structures. A system of three connective tissue layers, the meninges, and a fluid compartment containing CSF are located between the bony skull and vertebral column and the nervous tissue of the brain and the spinal cord. Blood vessels, themselves of mesodermal origin, are surrounded by derivatives of the meninges over their full extent, until the interface between the capillary wall and the glial basal membrane makes exchange of substances possible. CSF is produced by the choroid plexus of the ventricles. It circulates from the interstitial spaces of the nervous tissue and the choroid plexus, through the ventricles and their apertures in the roof of the fourth ventricle, to the CSF compartment of the subarachnoid space and its exit through the arachnoid villi to the venous system. The nervous tissue of the central nervous system and the CSF spaces remain segregated from the rest of the body by barrier layers in the meninges (the barrier layer of the arachnoid), the choroid plexus (the blood-CSF barrier) and the capillaries (the blood-brain barrier). The circulation of the CSF plays an important role in maintaining the environment of the nervous tissue; moreover, the subarachnoidal space forms a bed that absorbs external shocks.

Arteries of the Brain

The arteries of the brain originate from two of the greater vessels in the neck: the paired internal carotid and vertebral arteries (Fig. 4.1). The *internal carotid artery* enters the skull through the petrosal bone in the carotid canal. It loops through the cavernous sinus (carotis syphon), where it emits the *ophthalmic artery*. Immediately beyond the origin of the posterior communicating artery it splits into the middle and anterior cerebral arteries. The vertebral arteries enter the skull through the foramen magnum. After their passage through the dura, the arteries become located within connective tissue derived from the pia mater and the arachnoid.

The middle cerebral artery supplies the convexity of the hemisphere (Figs. 4.3, 4.4, 4.5) and the anterior cerebral artery vascularizes approximately the anterior and upper half of the medial aspect of the hemisphere, up to the precuneus (Fig. 4.2). The vertebral arteries unite into the basilar artery at the ventral aspect of the medulla oblongata. Its terminal branches are the left and right posterior cerebral arteries, which supply the posterior, medial and basal aspects of the cerebral hemisphere. The vertebro-basilar arteries also supply the brain stem and the cerebellum. It gives rise to the inferior, middle and superior cerebellar arteries (Fig. 4.11). Frontal and lateral projections of the arterial system are shown in Figs. 4.9 and 4.10.

A system of communicating arteries, known as the *circle of Willis* [18, 36], interconnects the anterior and middle cerebral arteries of both sides with the vertebro-basilar system (Figs. 4.6 and 4.14). It is located at the base of the brain and surrounds the infundibulum and the optic chiasm. It includes the *anterior communicating artery*, which interconnects the anterior cerebral arteries, immediately in front of the optic chiasm, and the two *posterior communicating arteries*, which form an anastomosis between the most distal part of the internal carotic and the posterior cerebral artery near their origin from the basilar artery.

The initial segments of the middle and anterior cerebral arteries give rise to *central arteries* (Figs. 4.6, 4.8, 4.12), which enter the brain in the anterior perforated substance (Fig. 3.4). Together with branches from the posterior communicating artery, they supply the basal ganglia, the internal capsule and the thalamus. The *middle cerebral artery* enters the sulcus lateralis. Just before this point it emits the anterior choroidal artery, which also supplies a branch to the globus pallidus (Fig. 4.8). At the surface of the insula the middle cerebral artery branches into anterior, middle and posterior trunks. The branches of the middle cerebral artery loop over the opercula and ramify over the surface of the cerebral hemisphere to supply the cerebral cortex and the adjacent white matter (Figs. 4.3, 4.4). The *anterior cerebral artery* enters the longitudinal fissure to branch on the medial aspect of the hemisphere. The anterior communicating artery, which connects the two anterior cerebral arteries, is located immediately rostral to the optic chiasm (Fig. 4.2).

The *vertebral arteries* enter the skull through the foramen magnum. They give rise to the *anterior spinal artery*, which descends in the anterior median fissure of the cord, and emit the *posterior inferior cerebellar arteries*. The vertebral arteries unite into the *basilar artery* at the ventral aspect of the brain stem. The basilar artery gives origin to the *anterior inferior* and *superior cerebellar arteries* and splits into the posterior cerebral arteries. The oculomotor nerve emerges between the superior cerebellar and the posterior cerebral arteries and thus marks the bifurcation of the basilar artery (Fig. 4.11). The posterior cerebellar artery makes a characteristic, caudally directed curve before it reaches the cerebellum. Both the posterior inferior and superior cerebellar arteries contribute branches to the dorsolateral brain stem.

The *posterior cerebral artery* supplies the medial aspect of the temporal and occipital lobes. The border region of the vascularization territories of the posterior and middle cerebral arteries include the temporal and occipital poles. The latter contains the posterior portion of the primary visual (striate) cortex with the representation of the fovea. Occlusion of the posterior cerebral artery thus leads to loss of peripheral vision, with maintained central vision ("tunnel vision") (Table 4.1). The borders of the arterial territories of the cerebral hemisphere do not correspond to the borders of the the four major lobes distinguished in the gross anatomy of the cerebral hemipheres (Fig. 4.7). Asymmetries of the brain's arterial system are frequently observed, most often in the calibre of the vertebral, the posterior cerebral and the posterior communicating arteries.

The vertebral, basilar and posterior cerebral arteries also give rise to smaller branches, which enter the brain stem in the median sulcus and more laterally (Fig. 4.11). Branches from the basilar and posterior cerebral arteries (Fig. 4.18) enter the mesencephalon in the posterior perforated substance, located in the floor of the interpeduncular fossa (Fig. 3.12). The vascularization territories of these arteries have been thoroughly studied by Duvernoy [12]. These territories are illustrated in a number of transverse sections in which both arterial supply and venous drainage are visualized (Figs. 4.18–4.20). These figures also document the important contributions of the cerebellar arteries to the vascularization of the brain stem.

The existence of a collateral circulation is of great significance for the vascularization of the CNS. There are different types and different sites of anastomoses; moreover, the diameter of these anastomoses may differ considerably. Anastomoses between arteries can be found in relation to three arterial systems, i.e. between the two main arterial systems of the carotid and vertebral arteries and between the arterial systems of the brain and the external carotid artery.

Apart from the main arterial anastomosis between the systems of the internal carotid and the vertebral-basilar arterial system in the arterial circle of Willis, cerebro-cerebral anastomoses are present between the branches of the middle cerebral artery (Figs. 4.2, 4.14). Anastomoses between the cerebellar arteries are documented in Figs. 4.9 and 4.11.

Anastomoses with the external carotid artery occur both with meningeal and extracranial branches of this artery. Four types of anastomoses with branches of the external carotid artery are indicated with asterisks in Fig. 4.1. Orbital anastomoses with branches of the ophthalmic artery are enumerated as two special categories.

Meninges, Cisterns and Cerebrospinal Fluid Circulation

The brain is completely enclosed by three connective tissue layers: the meninges. These are, starting from the brain's surface, the *pia mater*, the *arachnoid* and the *dura mater*. The dura is also known as the pachymeninx, due to its strength and thickness, which is imparted by multiple layers of collagen tissue. The thin and loose tissue of the pia mater and the arachnoid is collectively known as the leptomeninx.

The cranial dura is merged with the periosteum of the inner table of the skull. As a consequence, the dura is firmly attached to the skull, especially at the sites of the sutures. Dural septa are located between the main divisions of the brain. In the midline, the *falx cerebri* is located between the cerebral hemispheres and the tentorium cerebelli extends between the occipital and temporal lobes of the hemisphere and the cerebellum. Venous sinuses occupy the inner and outer margins of the falx (superior and inferior sagittal sinus), the junction of the falx and the tentorium (straight sinus, or sinus rectus), and the attachment of the tentorium to the skull transverse sinus and superior petrosal sinus (Figs. 4.14–4.16). The *pia mater* closely covers the surface of the brain and intrudes into its sulci and depressions. The *arachnoid* covers the brain at a variable distance, thus creating a subarachnoidal space between the pia and the arachnoid. This space is bridged by many trabeculae. It contains the CSF. Widenings of the subarachnoid space are known as the cisterns.

For the understanding of the production, circulation and drainage of CSF, the fine structure of the interface of the CSF compartments, the nervous tissue and the mesenchymal tissue of the meninges is important. The central nervous system is isolated from the rest of the body by a series of cellular barriers, which limit the flux of hydrophilic molecules between these cells. These barriers generally consist of extensive tight junctions between the cells, where the outer leaflets of the plasma membranes of two opposing cells are fused. These barriers are found in the epithelium of the choroid plexus (blood-CSF barrier), the outer (barrier) layer of the arachnoid and in the endothelium of capillaries located within arachnoid and the pia mater and nervous tissue (blood-brain barrier).

The CSF is produced by the *choroid plexus* of the lateral, third and fourth ventricles. The wall of the capillaries of the stroma of the choroid plexus is fenestrated by local fusion of the inner leaflets of the plasma membranes of the endothelial cells, and the tight junction between the endothelial cells of these capillaries is less well developed. These fenestrated capillaries, therefore, allow passage of fluid and lyophilic molecules. CSF is secreted by the cells of the choroid epithelium; passage between these epithelial cells is prevented by extensive tight junctions.

Interstitial fluids and residues produced by the nervous tissue may leak into the ventricles, where the covering layer of ependymal cells lacks such an impediment. Passage of CSF from the ventricular system into the subarachnoid space is made possible by three openings in the roof of the fourth ventricle: the *median aperture* in the caudal midline, and the two *lateral apertures* in the lateral recesses (Figs. 3.8, 3.10).

The dura is composed of fibroblasts and extracellular collagen (Fig. 4.21). The innermost border layer of the dura consists of flattened cells with sinuous processes connected by occasional desmosomes and extracellular spaces filled with an amorphous substance. A real *subdural space* does not exist; the border layers of the dura and the arachnoid are interconnected by occasional cell processes and desmosomes. The arachnoid can be subdivided into two layers. The outer layer of the arachnoid, located next to the dura, is known as the *barrier layer of the arachnoid*. It is impassable for lyophilic molecules, due to the presence of a tight junction between its constituting cells. The inner layers of the arachnoid and the pia mater actually form a single loose structure containing the smaller or larger interconnected spaces for the CSF. Where these spaces coalesce into continuous fluid compartments, we speak of the *subarachnoid space* and its cisterns. Near the surface of the brain the cells of the pia form a membrane, which remains separated from the basal membrane of the glia limitans of the brain by a narrow, subpial space. Cells of the pia mater are joined by desmosomes and gap junctions, but tight junctions, which would prevent the passage of fluids, are absent here. The ventricular system and the subarachnoidal space thus constitute a compartment, which is closed off from the rest of the body by the tight junctions of the choroid epithelium and the barrier layer of the arachnoid, but which is accessible for fluids from the side of the nervous tissue [14, 15, 21] (Fig. 4.21).

CSF can be drained from the subarachnoidal space by protrusions of the arachnoid into the venous sinuses, known as the *arachnoid villi*, and along the cerebral and spinal nerves into lymph vessels. Drainage along the arachnoid villi is pressure-dependent. Arachnoid villi are especially abundant in the superior sagittal sinus and its lateral extensions, the lacunae (Figs. 4.15, 4.24, 4.25).

The structure, molecular biology and clinical implications of the *blood-brain barrier* have recently been reviewed by Ballabh [2]. Blood vessels that pass through the dura into the subarachnoidal space become surrounded by a sheath of pial cells. When entering the central nervous system, this sheath only becomes lost at the level of the capillaries. Capillaries in the central nervous system are distinguished by the absence of fenestrations and the presence of extensive tight junctions between the endothelial cells and a basement membrane surrounding the capillary. Astrocyte endfeet and pericytes embedded in the basement membrane ensheath the capillaries. Passage between the endothelial cells is blocked by tight junctions. Nutrients and large molecules enter the brain via transporters and receptor-mediated endocytosis of endothelial cells. Fenestrated capillaries occur in the circumventricular organs; this is discussed in the next section of this chapter.

Arachnoid cisterns are local widenings of the subarachnoidal space. In Figs. 4.26 and 4.27 the cisterns are presented as moulds. The most important cisterns surround the cerebellum and the brain stem. The *cerebellomedullary cistern*, also known as the *cisterna magna*, is located between the cerebellum and the medulla oblongata (Fig. 4.24). It is accessible for puncture through the foramen magnum. The *pontine cistern* surrounds the pons. The *medullary cistern* surrounds the medulla oblongata and continues as the subarachnoidal space surrounding the spinal cord.

The vascularization of the meninges is derived from the *external carotid artery* as the *middle meningeal artery*, which enters the skull through the foramen spinosum (Fig. 4.1). Smaller meningeal arteries detach from the ethmoidal, internal carotid, ascending pharyngeal and occipital arteries.

The choroid plexus of the lateral ventricle is supplied by the *anterior choroidal artery*, a branch from the middle cerebral artery, and the *posterior choroidal arteries* branches from the posterior cerebral arteries. The choroid plexus of the fourth ventricle receives its arterial supply from the inferior cerebellar artery (Fig. 4.23).

Circumventricular Organs

Circumventricular organs (CVOs) are small loci of specialized tissue that, as their name implies, are situated in the immediate vicinity of the ventricular system [10]. Most of these loci contain typical neurons, and all of them lie in regions of the central nervous system devoid of a blood-brain barrier. The capillaries in these regions are fenestrated. The CVOs function primarily as chemosensitive neurohaemal organs, which monitor the changing levels of circulating hormones and other substances. However, they may also be influenced by substances carried by the CSF and by neural afferents entering the organs from other parts of the central nervous system or from the periphery. Apart from the subfornical organ, the vascular organ of the lamina terminalis and the area postrema (to be considered here) the CVOs include the posterior pituitary, the median eminence and the pineal gland.

The *subfornical organ* (SFO) and the *vascular organ of the lamina terminalis* (OVLT) are both unpaired structures situated in the wall of the small non-evaginated portion of the telencephalon, i.e. the telencephalon impar (Fig. 1.2) – the former rostrodorsally, just between the two interventricular foramina; the latter rostroventrally, immediately dorsal to the optic chiasm.

The SFO contains many neurons, a dense fenestrated capillary plexus and a high density of receptors for the peptide hormone angiotensin II (AII). It is known that hypovolemic thirst is accompanied by increased circulating levels of this hormone, which elicits, by way of the SFO neurons and their targets, a spectrum of responses aimed at maintaining blood pressure and replacing lost body water [26].

The OVLT and the immediately adjacent median preoptic nucleus can be designated together as the anteroventral third ventricular area (AV3V). Like the SFO, the OVLT is rich in AII receptors. The AV3V is known to be especially sensitive to hyperosmolarity of the blood, which is another stimulus of thirst [8].

As its name implies, the *area postrema* is, contrary to the SFO and the OVLT, located in the caudal-most part of the brain. It is a small, bilateral, spongy structure that protrudes into the fourth ventricle on either side of the midline, immediately rostral to the obex. It has been suggested that the area postrema is a chemoreceptive trigger zone forming part of the neurocircuitry underlying water and energy balance, as well as cardiovascular regulation. However, the only well-established function of this organ seems to be that it elicits nausea and vomiting in response to circulating emetic substances [7].

Veins of the Brain

Blood is drained from the brain by different pathways, mainly converging upon the internal jugular vein, with minor outlets through the external jugular vein and the intravertebral venous plexus (Fig. 4.13). Valves are lacking in the venous system of the brain. Venous sinuses are venous channels, enclosed in duplicatures of the dura mater. Blood from the convexity of the hemisphere is collected by the *superior cerebral veins*, which drain into the *superior sagittal sinus*. This sinus is located in the superior margin of the falx cerebri (Figs. 4.14, 4.15). It joins the *straight sinus* (sinus rectus) located at the junction of the falx and the tentorium cerebelli, together with the great cerebral vein, which carries blood from deep structures of the cerebral hemisphere and the choroid plexus of the lateral ventricles. The straight sinus and the superior sagittal sinus drain in the left and right *transverse sinus*, located in the posterior margin of the tentorium (Figs. 4.12, 4.13). The meeting point of the sinus rectus, the superior sagittal and the transverse sinuses is known as the *confluens sinuum*. It also receives the *occipital sinus*, located in the falx cerebelli, a dural protrusion located over the midline of the cerebellum (Figs. 4.14–4.17).

Where the transverse sinus leaves the tentorium, it continues as the *sigmoid sinus*, located in a groove on the mastoid temporal bone and the jugular process of the occipital bone. The sigmoid sinus drains into the jugular bulb and the *internal jugular vein*, which leaves the skull through the jugular foramen.

The basal sinuses and the subdural venous plexus at the clivus are illustrated in Figs. 4.13 and 4.14. The *cavernous sinuses* are paired, reticulated structures located at the lateral surface of the body of the sphenoid. They give passage to the internal carotid artery and several cranial nerves. The cavernous sinuses receive blood from the cerebral hemisphere through the superficial middle cerebral vein, from the meninges, from the orbita through the inferior and superior orbital veins, from the central retinal vein of the eyeball and from deep facial structures through the plexus pterygoideus. They are interconnected through the *intercavernous sinus*. The superior and inferior *petrosal sinuses* connect them with the sigmoid sinus and the jugular vein, respectively.

Blood from the choroid plexus of the lateral and third ventricles, and from deep structures of the brain, is collected by the great cerebral vein, which discharges into the sinus rectus. The *great cerebral vein* (of Galen [31]) is formed by fusion of the internal cerebral and *basal* veins. The latter collects blood from the upper brain stem, the insula and the choroid plexus of the inferior horn of the lateral ventricle. The *internal cerebral vein* drains blood from the choroid plexus of the corpus of the lateral ventricle and from the *thalamostriate vein*, which runs from the level of the interventricular foramen caudally all the way along the caudate-thalamic border (Figs. 4.13, 4.16, 4.17).

Venous anastomoses are more frequent than arterial ones. Formation of venous plexuses is a widespread phenomenon. Venous plexuses around the brain and the spinal cord play an important role in the collateral circulation: the cavernous plexus, the basilar plexus, the (extracranial) pterygoid plexus and the internal venous plexus of the cord (see the next section of this chapter on vessels and meninges of the spinal cord). *Emissary veins* traverse cranial apertures and make connections between venous sinuses and extracranial veins or diploical veins within the diploe of the calvaria (Figs. 4.13, 4.24). Some are constant, others sometimes absent. Two venous anastomoses should be mentioned separately. The *superior anastomotic vein of Trolard* [30] connects the middle cerebral vein with the superior sagittal sinus, whereas the *inferior anastomotic vein of Labbé* [4] links this vein to the sinus transversus (Fig. 4.13).

Vessels and Meninges of the Spinal Cord

All segments of the cord are supplied by three longitudinal spinal arteries (Fig. 4.28 B). The *anterior spinal artery* is situated ventrally, at the anterior median fissure; the *posterior spinal arteries* run at some distance from the midline, along the dorsal root entry zone. The spinal arteries receive their blood through segmentally organized anterior and posterior radicular arteries, which take their origin from the vertebral, intercostal and lumbar arteries (Figs. 4.28 A, 4.30). These segmental arteries are only present at about half the number of spinal segments. Usually there is only one prominent radicular artery present as the *arteria radicularis magna of Adamkiewicz* [20], an intersegmental branch from the aorta at the lower thoracic or upper lumbar level (Fig. 4.28 A). From the longitudinal spinal arteries, tiny branches enter the cord, where they supply rather circumscript territories.

The veins that carry blood from the spinal cord through the intervertebral, intercostal and lumbar veins to the superior vena cava show a pattern comparable to that of the arterial supply (Fig. 4.29). However, only one spinal vein is present on the dorsal side: the *posterior spinal vein*, situated in the posterior median sulcus. Moreover, there are fewer segmental veins than corresponding segmental arteries, and there is no single, prominent vein among them. The segmental veins also communicate with the *internal spinal venous plexus* (located in the epidural space), the vertebra and the external spinal venous plexus (Figs. 4.28–31). Like the cerebral veins, the spinal veins and the veins of the spinal plexuses do not possess valves.

The meninges of the spinal cord are a continuation of those of the brain, with the exception of the endocranial periosteum, the outer table of the cranial dura. In the spinal canal the periosteum remains attached to the bony skeleton, and the epidural space separates it from the spinal dura (Figs. 4.30–4.32). The epidural internal venous plexus is contained in the epidural space. Caudal to the spinal cord (i.e., caudal to the first lumbar vertebra) the dura extends in the spinal canal to low sacral levels and encloses the *dural sac* (a compartment of the subarachnoid space), which is traversed by the anterior and posterior roots of the *cauda equina*, where cerebrospinal fluid can be tapped by lumbar puncture. Laterally, the cord is attached to the dura by the triangular *denticular ligaments*. The proportions of the vertebral canal at four levels of the vertebral column, the meningeal spaces, the denticular ligaments and the neural elements are illustrated in Fig. 4.32.

References

1. Adamkiewicz A (1882) Die Blutgefässe des menschlichen Rückenmarkes, II: Die Gefässe der Rückenmarks-Oberfläche. Sitzb Akad Wiss 85:101–130
2. Ballabh P, Braun A, Nedergaard M (2004) The blood-brain barrier: an overview: structure, regulation, and clinical implications. Neurobiol Dis 16:1–13
3. Barnett HJM, Meldrum H (2000) Wallenberg's syndrome. In: Koehler PJ, Bruyn GW, Pearce JMS (eds) Neurological eponyms. Oxford University Press, Oxford, pp 258
4. Bartels RH, van Overbeeke JJ (1997) Charles Labbé (1851–1889). J Neurosurg 87:477–480
5. Benedikt M (1889) Tremblement avec paralyse croisée du moteur oculaire commun. Le Bulletin Médicale (Paris) 3: 547–548
6. Bernard C (1862) Des phénomènes oculo-pupillaires produit par section du nerf sympatique cervical. Comptes rendues de l'Académie des Sciences (Paris) 55:381–388
7. Borison HL (1989) Area postrema: chemoreceptor circumventricular organ of the medulla oblongata. Prog Neurobiol 32: 351–390
8. Bourque CW, Oliet SH (1997) Osmoreceptors in the central nervous system. Annu Rev Physiol 59: 601–619
9. Bruyn GW, Gooddy W (2000) Horner's syndrome. In: Koehler PJ, Bruyn GW, Pearce JMS (eds) Neurological eponyms. Oxford University Press, Oxford, pp 227–233
10. Cottrell GT, Ferguson AV (2004) Sensory circumventricular organs: central roles in integrated autonomic regulation. Regul Pept 117:11–23
11. Djindjian R (1970) L'angiographie de la moelle épinière. Masson, Paris
12. Duvernoy HM (1978) Human brainstem vessels. Springer, Berlin
13. Gillilan LA (1974) Potential collateral circulation to the human cerebral cortex. Neurology 24: 941-948
14. Haines DE (1991) On the question of a subdural space. Anat Rec 230:3–21
15. Haines DE, Harkey HL, al-Mefty O (1993) The "subdural" space: a new look at an outdated concept. Neurosurgery 32:111–120
16. Heubner JOL (1872) Topographie der Ernährungsgebiete der einzelnen Hirnarterien. Zentralbl Med Wiss 10: 817–821
17. Horner JF (1869) Über eine Form von Ptosis. Klinische Monatsblätter für Augenheilkunde (Stuttgart) 7:193–198
18. Isler HE (2000) The circle of Willis. In: Koehler PJ, Bruyn GW, Pearce JMS (eds) Neurological eponyms. Oxford University Press, Oxford, pp 56–62
19. Koppejan EH (2000) Parinaud's syndrome. In: Koehler PJ, Bruyn GW, Pearce JMS (eds) Neurological eponyms. Oxford University Press, Oxford, pp 239–243
20. Milen MT, Bloom DA, Culligan J, Murasko K (1999) Albert Adamkiewicz (1850–1921) – his artery and its significance for the retroperitoneal surgeon. World J Urol 17:168–170
21. Nabeshima S, Reese TS, Landis DM, Brightman MW (1975) Junctions in the meninges and marginal glia. J Comp Neurol 164: 127–169
22. Pacchioni A (1705) Dissertatio epistolaris ad Lucam Schroeckium de glandulis conglobatis durae meningis humanae. Francesco Buagni, Rome
23. Parinaud H (1883) Paralyse des mouvements associés des yeux. Archives de Neurologie 5:145–172
24. Renard M, Masson JP, Larde (1976) Les veines épidurales lombaires. Les veines du canal sacré. Bull Assoc Anat 60:787–803
25. Rickenbacher J (1982) Rücken. In: Lang J (ed) Praktische Anatomie, vol II/7. Springer, Berlin, pp 460
26. Swanson LW (1989) The neural basis of motivated behavior. Acta Morphol Neerl-Scand 26:165–176
27. Talairach J, Toumoux P (1988) Co-planar stereotaxic atlas of the human brain. Thieme, New York
28. Théron J, Moret J (1978) Spinal phlebography. Springer, Berlin
29. Thijssen HOM (1971) Over angiografie bij trauma capitis. Thesis, Nijmegen, pp 248
30. Trolard P (1842–1910) French anatomist

31. Ustun C (2004) Galen and his anatomic eponym: vein of Galen. Clin Anat 17:454–457
32. Voogd J (1983) Morphology and clinical neurology. The history and evolution of Wallenberg's syndrome. Acta Morph Neerl-Scand 21:27–47
33. Wallenberg A (1901) Acute Bulbäraffection (Embolie der Art. cerebellar. post. inf. sinistr.?). Arch Psychiatr 27:504–540
34. Wallenberg A (1915) Anatomischer Befund in einem als „acute Bulbäraffection" (Embolie der Art. cerebelli post. sinistra) beschriebenen Falle. Arch Psych Nervenkrankh 34:923–959
35. Weber HD (1863) A contribution to the pathology of the crura cerebri. Medico-Chirurgical Transactions (London) 46:121–139
36. Willis T (1664) Of the anatomy of the brain. Englished by Samual Pordage, Esq. Dring Harper Leigh & Martyn, London. Fascimile edition, McGill University Press, Montreal 1965
37 Wolters EC, Groenewegen HJ (1996) Neurologie. Structuur, functie en dysfunctie van het zenuwstelsel. Bohn Stafleu Van Loghum, Houten

Table 4.1. The main vascular syndromes [37]

- **Occlusion of the anterior cerebral artery** (Fig. 4.7) causes a contralateral paresis and sensory disturbances in the contralateral lower extremity, sometimes combined with an imperative urge to micturate.
- **Occlusion of the middle cerebral artery** or its branches (Fig. 4.7) occurs most frequently. Ischemia of the anterior branches always leads to a contralateral hemiplegia, which is expressed most in the distal upper extremity and accompanied by sensory disturbances and an expressive aphasia when the dominant hemisphere is affected. Involvement of the posterior branches of the middle cerebral artery causes loss of spatial orientation (constructive apraxia when the posterior parietal cortex is onvolved, homonymous hemianopia when the infarct includes optic radiation and as visual inattentiveness when the representation of the fovea is affected. A sensory aphasia is present when the infarct is located in the dominant hemisphere.
- **Occlusion of the posterior cerebral artery** (Fig. 4.7) generally causes a cortical homonymous hemianopia for the contralateral side, with sparing of central vision (tunnel vision). When the infarct affects the dominant hemisphere there is a chance alexia or visual agnosia may develop. Bilateral occlusion of the posterior cerebral arteries, cortical causes blindness. Loss of memory and prosopagnosia (inability to recognize faces) are due to involvement of structures in the medial temporal lobe.
- **Medial syndromes of the caudal brain stem** are caused by occlusion of caudal, paramedian branches of the vertebral and basilar arteries (Fig. 4.20). Interruption of the pyramidal tract and the medial lemniscus are responsible for, and are characterized by, a contralateral hemiplegia and an ipsilateral loss of sensibility. Depending on the level of the infarct, different cranial nerves may be involved.
- **Wallenberg's syndrome** [3, 32–34] is caused by an infarct of the dorsolateral medulla oblongata on occlusion of the posterior inferior cerebellar artery and/or its side branches supplying this region of the brain stem (Fig. 4.20). Interruption of the spinothalamic tract in the anterolateral fascicle, and of the spinal tract of the trigeminal nerve with its nucleus, causes an alternating loss of sensibility in the ipsilateral face and the contralateral extremities. These symptoms can be combined with an ipsilateral ataxia due to the interruption of the restiform body and involvement of the cerebellum, as well as signs of a lesion to the nucleus ambiguus (dysphagia, dysarthria, dysphonia), the respiratory centres in the reticular formation (hiccouph) and the vestibular nerve and its nuclei. **Horner's syndrome** [6, 9, 17] may be present when the descending connections to the sympathetic centres in the spinal cord are interrupted.
- **A dorsolateral infarct of the pons** is caused by occlusion of the superior cerebellar artery and/or the branches of the basilar artery supplying this part of the brain stem (Fig. 4.19). The motor nucleus of the facial nerve and the cochlear nucleus are affected, with an ipsilateral facial paralysis and ipsilateral deafness as a result. More medial lesions may affect the motor nucleus of the trigeminal nerve, with ipsilateral deviation of the jaw and problems with mastication on the ipsilateral side, and/or affection of the horizontal gaze centre in the paramedian pontine reticular formation, with a paralysis of horizontal gaze to the side of the lesion.
- **Occlusion of the superior cerebellar artery** affects the rostral, dorsolateral pons, interrupting the superior cerebellar peduncle. An ipsilateral cerebellar syndrome may be combined with signs of a lesion to the vestibular nuclei and its efferent tracts, or a contralateral loss of sensibility when the medial lemniscus is involved.

Three vascular, mesencephalic syndromes are distinguished (Fig. 4.18):

- **The syndrome of Weber** [35] is caused by ventral infarcts involving the cerebral peduncle and the oculomotor nerve after occlusion of peduncular branches of the posterior cerebral, posterior communicating and the posterior choroidal arteries. It is characterized by a contralateral spastic hemiplegia, an ipsilateral supranuclear palsy of the facial nerve, an oculomotor palsy and a fixed dilated pupil.
- **The syndrome of Benedikt** [5] follows occlusion of branches of the posterior cerebral artery supplying the central part of the mesencephalic tegmentum containing the oculomotor nucleus and the red nucleus with the crossed ascending branch of the superior cerebellar peduncle. In this case the oculomotor palsy is combined with a contralateral cerebellar syndrome.
- **The syndrome of Perinaud** [19, 23] consists of a vertical gaze paralysis combined with fixed, dilated pupils and loss of convergence. It is due to damage to the tectum, the pretectum and the underlying vertical gaze centre. It may be caused by occlusion of tectal branches of the posterior choroidal and posterior cerebral arteries but, more frequently, is caused by compression by a tumour of the epiphysis.

1 Cerebromeningeal anastomosis*
2 Calvaria (outer and inner surfaces)
3 Cerebrum (outer surface)
4 Callosomarginal artery
5 Pericallosal artery
6 Corpus callosum
7 Anterior cerebral artery
8 Supratrochlear artery
9 Dorsal nasal artery***
10 Frontal foramen

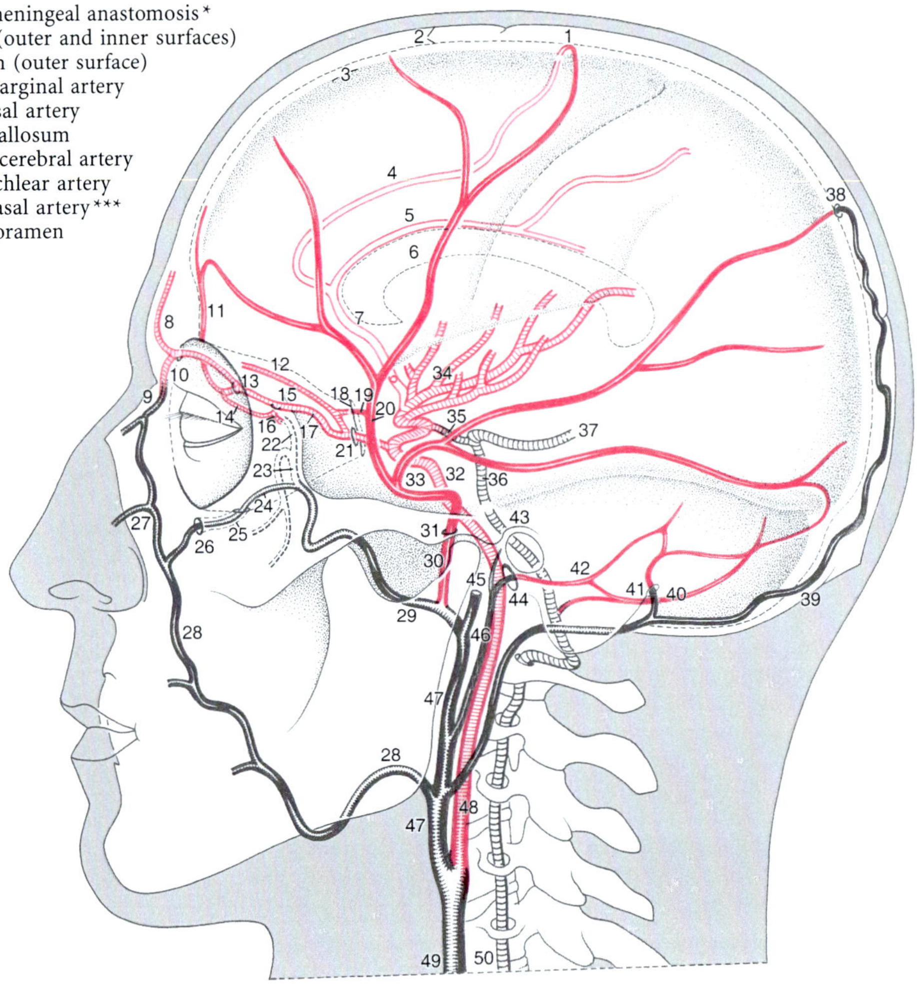

11 Anterior meningeal artery
12 Lacrimal artery
13 Anterior ethmoidal foramen
14 Anterior ethmoidal artery***
15 Posterior ethmoidal foramen
16 Posterior ethmoidal artery
17 Ophthalmic artery
18 Superior orbital fissure
19 Middle meningeal artery, anastomotic branch****
20 Middle meningeal artery, frontal branch
21 Optic canal
22 Superior conchal artery (anastomosis)***
23 Sphenopalatine artery
24 Infraorbital artery
25 Infraorbital canal
26 Infraorbital foramen***
27 Angular artery
28 Facial artery
29 Maxillary artery
30 Middle meningeal artery
31 Foramen spinosum
32 Internal carotid artery, petrous part
33 Middle meningeal artery, parietal branch
34 Middle cerebral artery, insular part
35 Posterior communicating artery
36 Basilar artery
37 Posterior cerebral artery
38 Parietal foramen**
39 Occipital artery
40 Occipital artery, mastoid branch**
41 Mastoid foramen
42 Posterior meningeal artery
43 Junction of the vertebral arteries
44 Jugular foramen
45 Superficial temporal artery
46 Ascending pharyngeal artery
47 External carotid artery
48 Internal carotid artery
49 Common carotid artery
50 Vertebral artery

Anastomoses

1	Cerebromeningeal	*
38+40	Extracraniomeningeal	**
9+22+26	Extracranial-orbital	***
14+19	Orbitomeningeal	****

Fig. 4.1. Collateral circulation in the arterial system of the head; semidiagrammatic lateral view (2/3x). *Black*: external carotid artery with extracranial branches; *black hatched*: system of the vertebral artery (main trunk); *solid red*: meningeal arteries; *red hatched*: internal carotid artery with orbital and lateral cortical branches; *open red*: medial cortical branches of internal carotid artery

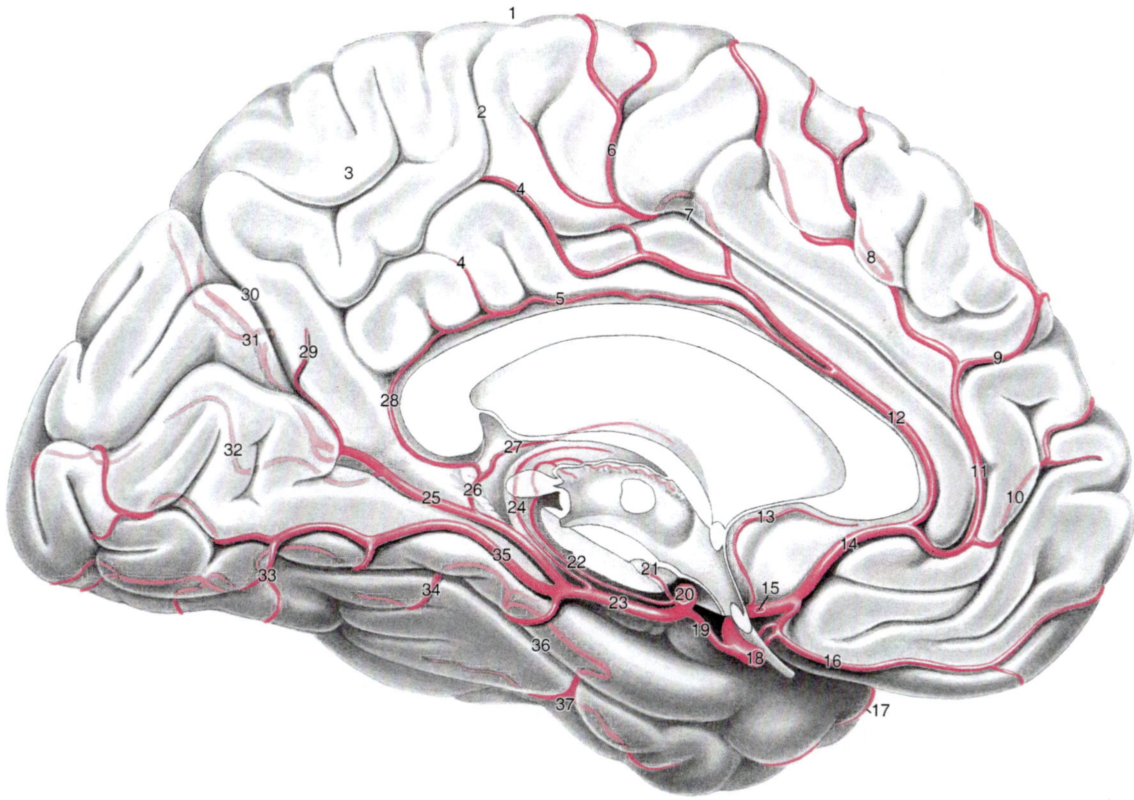

1 Central sulcus
2 Marginal branch of the cingulate sulcus
3 Precuneus
4 Artery of the precuneus
5 Pericallosal artery, posterior branch (anastomosis with 28)
6 Paracentral artery
7 Cingulate sulcus
8 Posteromedial frontal artery
9 Intermediomedial frontal artery
10 Anteromedial frontal artery
11 Callosomarginal artery
12 Pericallosal artery
13 Median artery of the corpus callosum
14 Anterior cerebral artery, postcommunicating part
15 Anterior communicating artery
16 Medial frontobasal artery
17 Temporopolar artery
18 Internal carotid artery
19 Posterior communicating artery
20 Posterior cerebral artery, precommunicating part
21 Posteromedial central arteries
22 Posterior medial choroidal branch
23 Posterior cerebral artery, postcommunicating branch
24 Posterior thalamic branches
25 Medial occipital artery
26 Cingulatethalamic artery
27 Superior thalamic branch
28 Dorsal branch of the corpus callosum (anastomosis with 5)
29 Parietal branch
30 Parieto-occipital sulcus
31 Parieto-occipital branch
32 Calcarine branch (in calcarine sulcus)
33 Posterior temporal branches
34 Medial intermediate temporal branch
35 Lateral occipital artery
36 Collateral sulcus
37 Anterior temporal branches

Fig. 4.2. The arteries of the medial hemisphere; the anterior and posterior cerebral arteries (1/1×). Some central branches of the posterior cerebral artery are also shown. End branches of the anterior cerebral artery that reach the lateral side of the superior frontal gyrus are illustrated in Fig. 4.3. Figures 4.2–4.6 are all derived from the same specimen

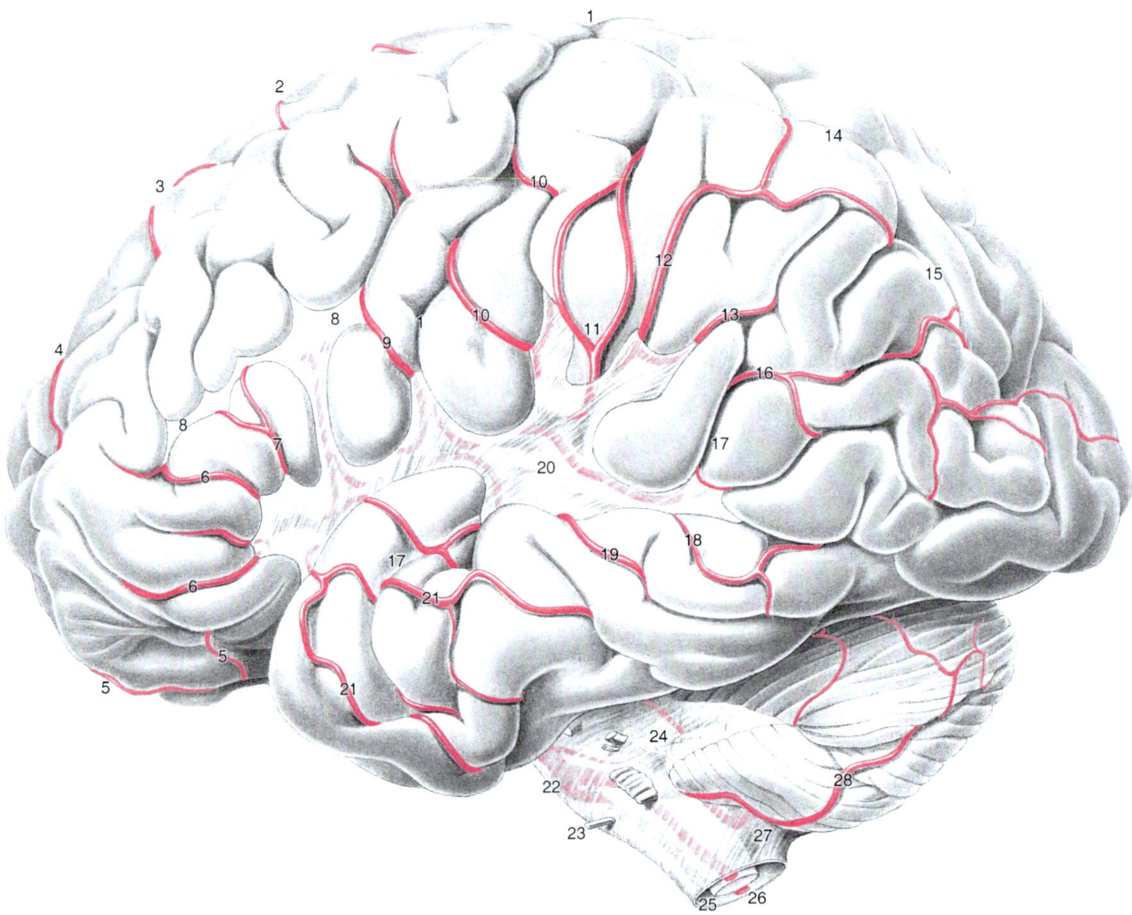

1 Central sulcus
2 Posteromedial frontal branch
3 Intermediomedial frontal branch
4 Anteromedial frontal branch
5 Medial frontobasal artery
6 Lateral frontobasal artery
7 Prefrontal artery
8 Inferior frontal sulcus
9 Artery of the precentral sulcus
10 Artery of the central sulcus
11 Artery of the postcentral sulcus (anterior parietal artery)
12 Posterior parietal artery
13 Artery of the angular gyrus
14 Intraparietal sulcus
15 Transverse occipital sulcus
16 Temporo-occipital artery
17 Superior temporal sulcus
18 Posterior temporal artery
19 Middle temporal artery
20 Cistern of the lateral cerebral fossa
21 Anterior temporal artery
22 Pontine cistern
23 Abducens nerve
24 Pontocerebellar cistern
25 Medullary cistern
26 Vertebral artery
27 Cerebellomedullary cistern (cisterna magna)
28 Posterior inferior cerebellar artery, lateral branch

Fig. 4.3. The arteries of the lateral cerebral cortex: the middle cerebral artery (1/1×). In this figure the lateral and medullary cisterns are left intact. On the lateral surface of the cerebellum one inferior and two superior cerebellar branches are illustrated (see Fig. 4.11). On the superior frontal gyrus some end branches of the anterior cerebral artery can be seen

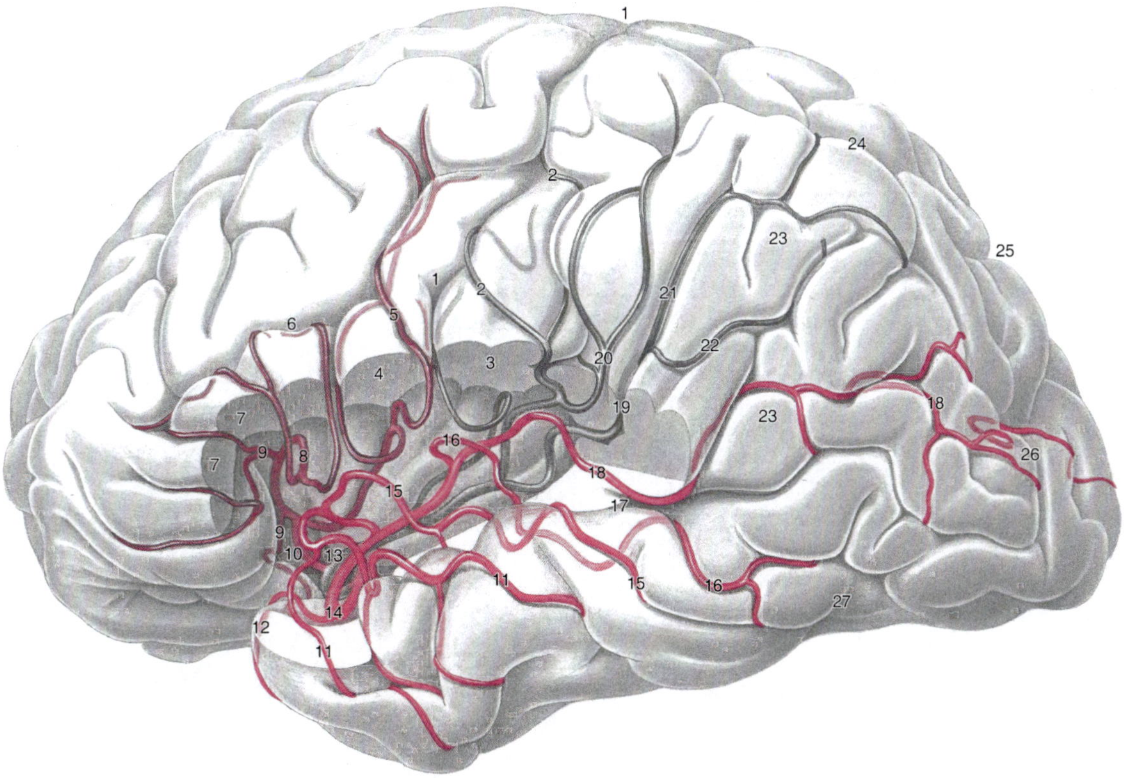

1 Central sulcus
2 Artery of the central sulcus (branches)
3 Postcentral gyrus
4 Precentral gyrus
5 Artery of the precentral sulcus
6 Inferior frontal sulcus
7 Inferior frontal gyrus, triangular part
8 Prefrontal artery (candelabrum artery)
9 Lateral frontobasal artery (branched)
10 Anterior trunk of the middle cerebral artery (ascending frontal artery)
11 Anterior temporal artery (branches)
12 Temporopolar artery
13 Middle trunk of the middle cerebral artery
14 Posterior trunk of the middle cerebral artery
15 Middle temporal artery
16 Posterior temporal artery
17 Superior temporal sulcus
18 Temporo-occipital artery
19 Lateral sulcus, posterior branch
20 Artery of the postcentral sulcus (anterior parietal artery)
21 Posterior parietal artery
22 Artery of the angular gyrus
23 Angular gyrus
24 Intraparietal sulcus
25 Parieto-occipital sulcus
26 Lunate sulcus
27 Anterior occipital sulcus

Alternative subdivision
11+15 Anterior temporal artery
16 Middle temporal artery
18 Posterior temporal artery
20 Parietal artery
21+22 Artery of the angular gyrus

Fig. 4.4. The branches of the middle cerebral artery seen at their full extent: lateral view (1/1×). In this specimen, as in most cases, a trifurcation can be seen of the artery. The branches of the anterior (frontal) trunk are shown in *black* and *red*; the branches of the middle (parietal) trunk are in *black* only; the branches of the posterior (temporal) trunk are in *red*. The candelabrum-like branching, especially of the anterior trunk, is a common phenomenon

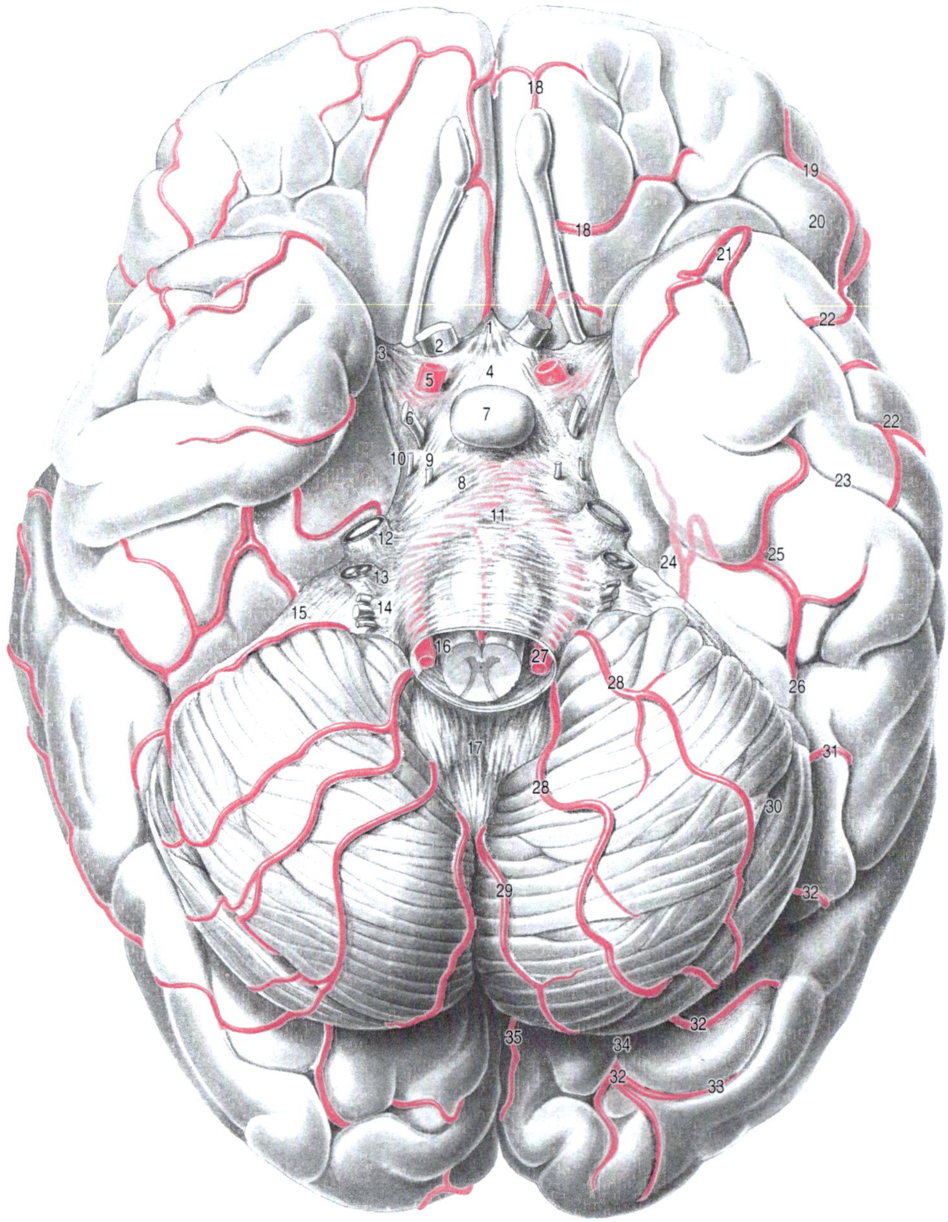

1 Cistern of the lamina terminalis
2 Optic nerve
3 Cistern of the vallecula cerebri
4 Cistern of the chiasm
5 Internal carotid artery (cerebral part)
6 Oculomotor nerve
7 Hypophysis
8 Interpeduncular cistern
9 Abducens nerve
10 Trochlear nerve
11 Pontine cistern
12 Trigeminal cistern
13 Cistern of the internal acoustic meatus with facial and vestibulocochlear nerves
14 Glossopharyngeal, vagal and accessory nerves
15 Pontocerebellar cistern
16 Medullary cistern
17 Cerebellomedullary cistern
18 Medial frontobasal artery (branch of the anterior cerebral artery)
19 Lateral frontobasal artery (branch of the middle cerebral artery)
20 Inferior frontal gyrus, orbital part
21 Temporopolar artery
22 Anterior temporal artery
23 Inferior temporal sulcus
24 Collateral sulcus with lateral occipital artery
25 Anterior temporal branches
26 Occipitotemporal sulcus
27 Vertebral artery
28 Posterior inferior cerebellar artery, lateral branches
29 Posterior inferior cerebellar artery, medial branches
30 Horizontal fissure of the cerebellum
31 Medial middle temporal branch
32 Posterior temporal branches
33 Occipitotemporal sulcus
34 Collateral sulcus
35 Lateral occipital artery

Fig. 4.5. The arteries of the brain viewed from the basal side (1/1×). In this figure the basal, cerebellar and medullary cisterns are left intact

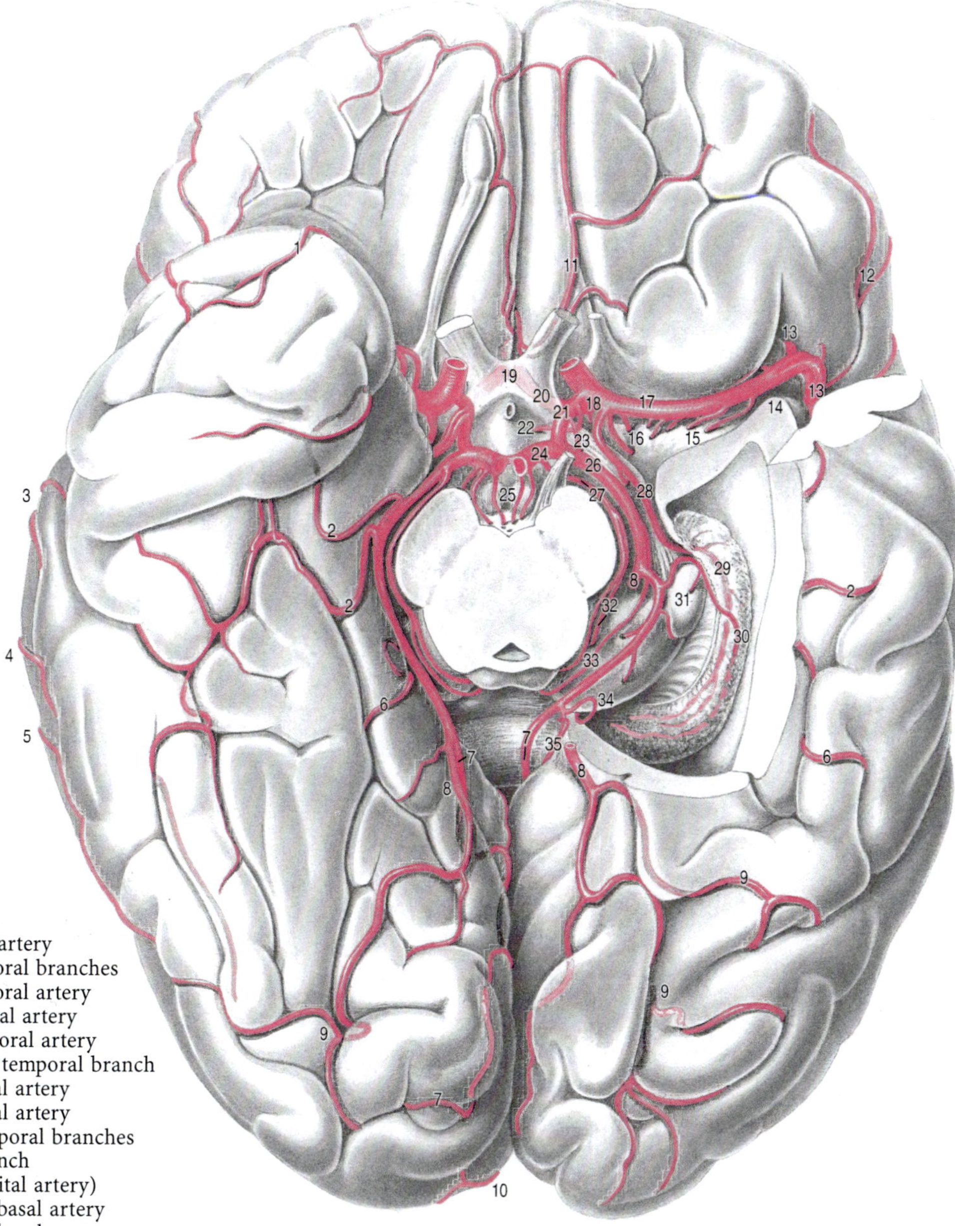

1 Temporopolar artery
2 Anterior temporal branches
3 Anterior temporal artery
4 Middle temporal artery
5 Posterior temporal artery
6 Medial middle temporal branch
7 Medial occipital artery
8 Lateral occipital artery
9 Posterior temporal branches
10 Calcarine branch (medial occipital artery)
11 Medial frontobasal artery
12 Lateral frontobasal artery
13 Middle cerebral artery, insular part
14 Limen insulae
15 Anterolateral central arteries, lateral branches
16 Anterolateral central arteries, medial branches
17 Middle cerebral artery, sphenoid part
18 Anteromedial central arteries
19 Anterior communicating artery
20 Anterior cerebral artery, precommunicating part
21 Posterior communicating artery
22 Hypothalamic artery
23 Thalamic branch (anteroinferior)
24 Posterior cerebral artery, precommunicating part
25 Posteromedial central arteries
26 Posterior cerebral artery, postcommunicating part
27 Medial posterior choroidal branch
28 Anterior choroidal artery
29 Choroidal branches of the anterior choroidal artery
30 Lateral posterior choroidal branch
31 Lateral geniculate body
32 Thalamic branch (inferior)
33 Thalamic branch (posterior)
34 Thalamic branch (superior)
35 Dorsal branch of the corpus callosum

19 + 20 + 21 + 24 Arterial circle (left half)

Fig. 4.6. The cerebral arteries viewed from the basal side (1/1×). Part of the left temporal lobe has been removed to show the sphenoid part of the middle cerebral artery and the arterial supply of the choroid plexus of the lateral ventricle. The lateral occipital artery has been interrupted to gain a clear view of the diencephalic, mesencephalic and retrosplenial branches of the posterior cerebral artery

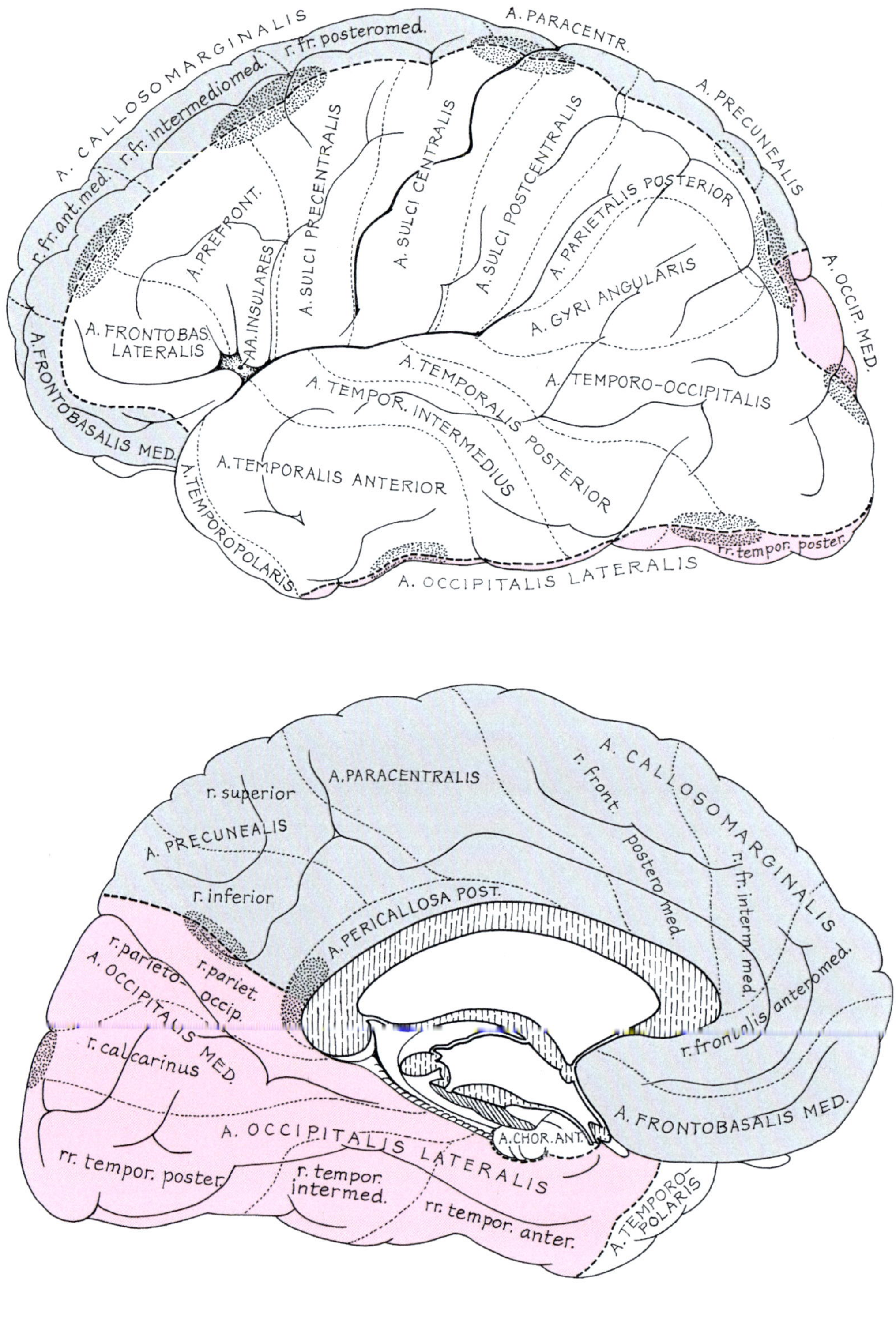

Fig. 4.7. Cortical territories of the three cerebral arteries; semidiagrammatic lateral and medial views of the left cerebral hemisphere (2/3×). The territories correspond to the vascularization pattern illustrated in Figs. 4.2–4.4. *Stippled areas*: sites of possible cerebrocerebral arterial anastomoses, mostly according to Gillilan [13]

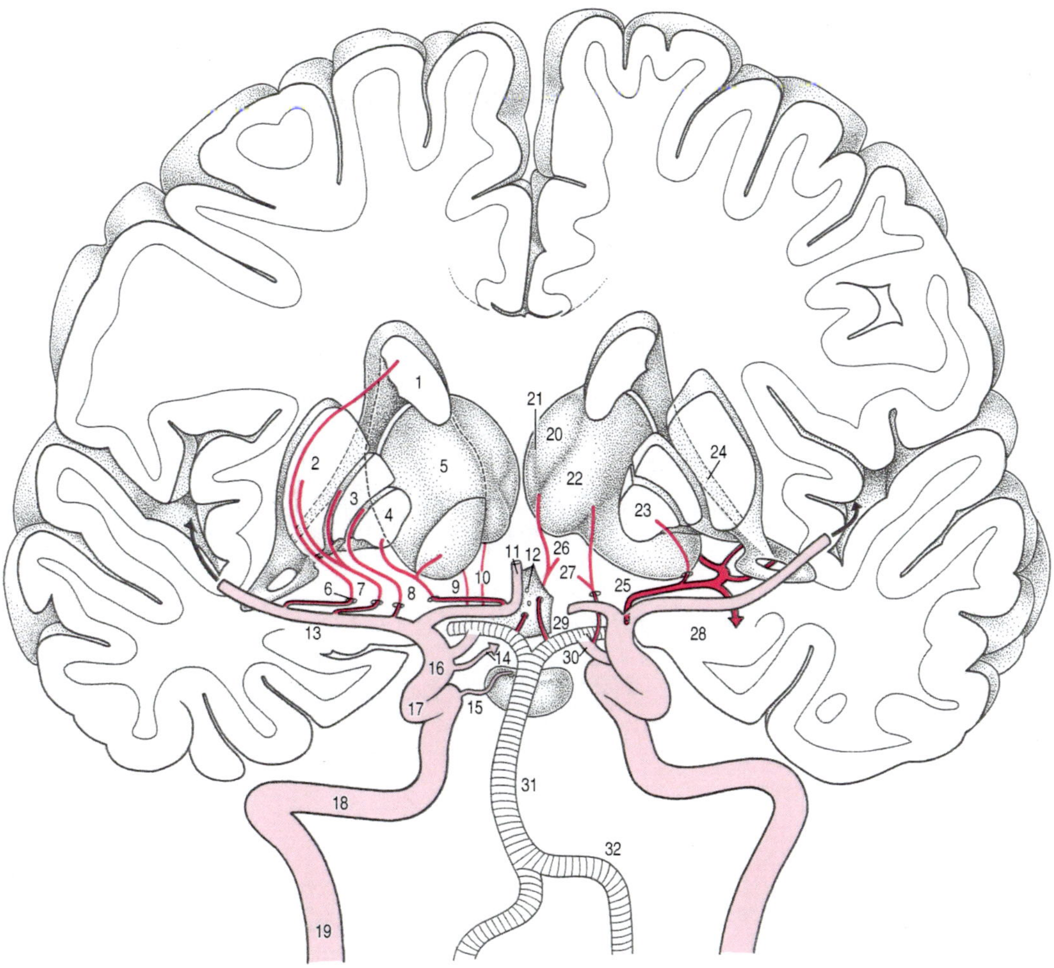

1 Caudate nucleus
2 Putamen
3 Globus pallidus, external segment
4 Globus pallidus, internal segment
5 Thalamus
6 Anterior perforated substance
7 Anterolateral central arteries, lateral branches
8 Anterolateral central arteries, medial branches
9 Long central artery (Heubner [16])
10 Anteromedial central arteries
11 Anterior cerebral artery
12 Posterior perforated substance
13 Middle cerebral artery, sphenoid part
14 Superior hypophyseal artery
15 Inferior hypophyseal artery
16 Internal carotid artery, cerebral part
17 Internal carotid artery, cavernous part
18 Internal carotid artery, petrous part
19 Internal carotid artery, cervical part
20 Medial nucleus of the thalamus
21 Midline nuclei of the thalamus
22 Anterior nucleus of the thalamus
23 Globus pallidus, internal segment
24 Tail of the caudate nucleus
25 Anterior choroidal artery
26 Subthalamus with posteromedial central arteries
27 Hypothalamus with hypothalamic branch
28 Amygdaloid nucleus
29 Posterior cerebral artery
30 Posterior communicating artery
31 Basilar artery
32 Vertebral artery

Fig. 4.8. The central arteries from the carotid and vertebral system in a frontal view (1/1×). Substrate based on a reconstruction. The frontal section is perpendicular to the horizontal plane of Frankfurt, passing through the centre of the insula. The central arteries have been derived from different sources

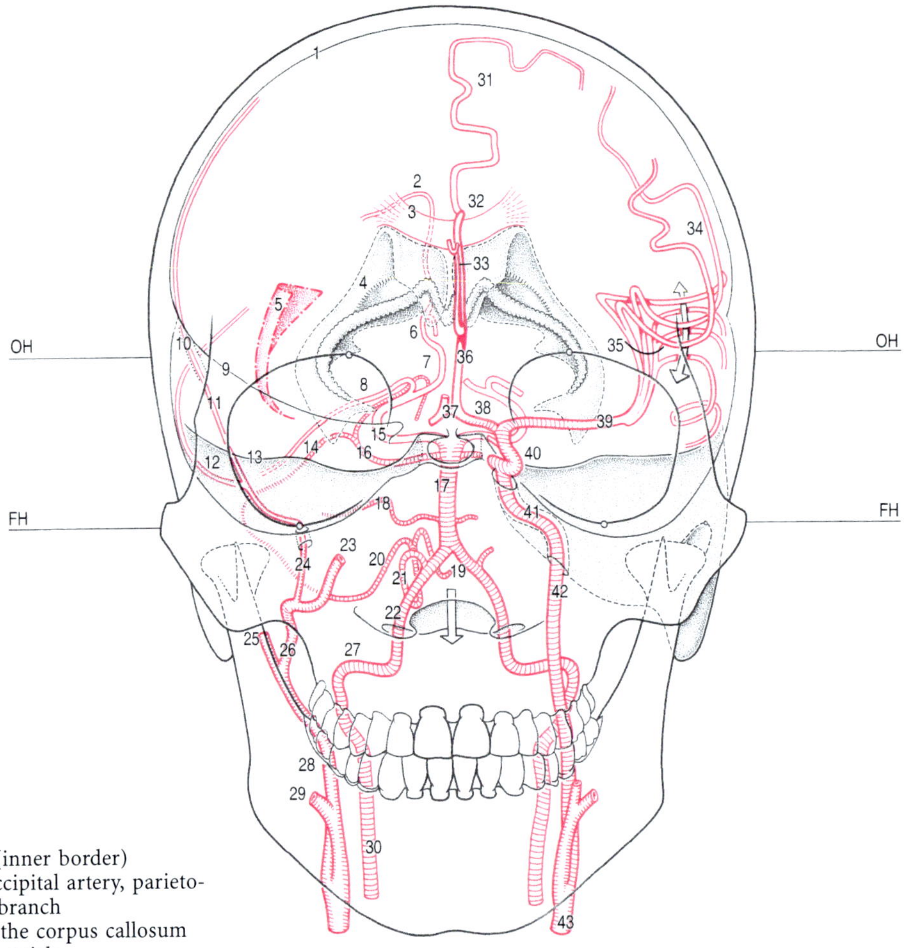

1 Calvaria (inner border)
2 Medial occipital artery, parieto-occipital branch
3 Trunk of the corpus callosum
4 Lateral ventricle
5 Insula
6 Medial occipital artery
7 Superior cerebellar artery, medial branch
8 Lateral occipital artery
9 Free margin of the lesser wing of the sphenoid bone
10 Middle meningeal artery, intraosseous part (inconstant)
11 Middle meningeal artery, frontal branch
12 Middle meningeal artery, parietal branch
13 Superior margin of petrous part of the temporal bone
14 Superior cerebellar artery, lateral branch
15 Posterior cerebral artery
16 Superior cerebellar artery
17 Basilar artery
18 Anterior inferior cerebellar artery
19 Posterior inferior cerebellar artery, medial branch
20 Posterior inferior cerebellar artery, lateral branch
21 Posterior inferior cerebellar artery
22 Vertebral artery, intracranial part
23 Maxillary artery, pterygoid part
24 Middle meningeal artery
25 Superficial temporal artery
26 Maxillary artery, mandibular part
27 Vertebral artery, atlantal part
28 External carotid artery
29 Facial artery
30 Vertebral artery, cervical part
31 Paracentral artery
32 Pericallosal artery
33 Callosomarginal artery
34 Middle cerebral artery, terminal part
35 Middle cerebral artery, insular part
36 Anterior cerebral artery, postcommunicating part
37 Anterior communicating artery
38 Anterior cerebral artery, precommunicating part
39 Middle cerebral artery, sphenoid part
40 Internal carotid artery, cavernous part
41 Internal carotid artery, petrous part
42 Internal carotid artery, cervical part
43 Common carotid artery

Fig. 4.9. Orthogonal frontal projection of the cerebral and cerebellar arteries in situ, together with some bony landmarks and the lateral ventricles (2/3×). The projection was made parallel to the horizontal plane of Frankfurt by using a graphical reconstruction from the frontal slices of one specimen, and by cross-reference with Fig. 4.10. In this figure and the next, ample use has been made of indications by Thijssen [29]. Most vessels are illustrated only in one half of the skull; the vertebral artery is shown bilaterally. *OH*, Upper horizontal plane (Krönlein): tangential to supraorbital margin; *FH*, Horizontal plane of Frankfurt (Reid): tangential to infraorbital margin; *double arrow,* sulcus lateralis; *single arrow:* foramen magnum

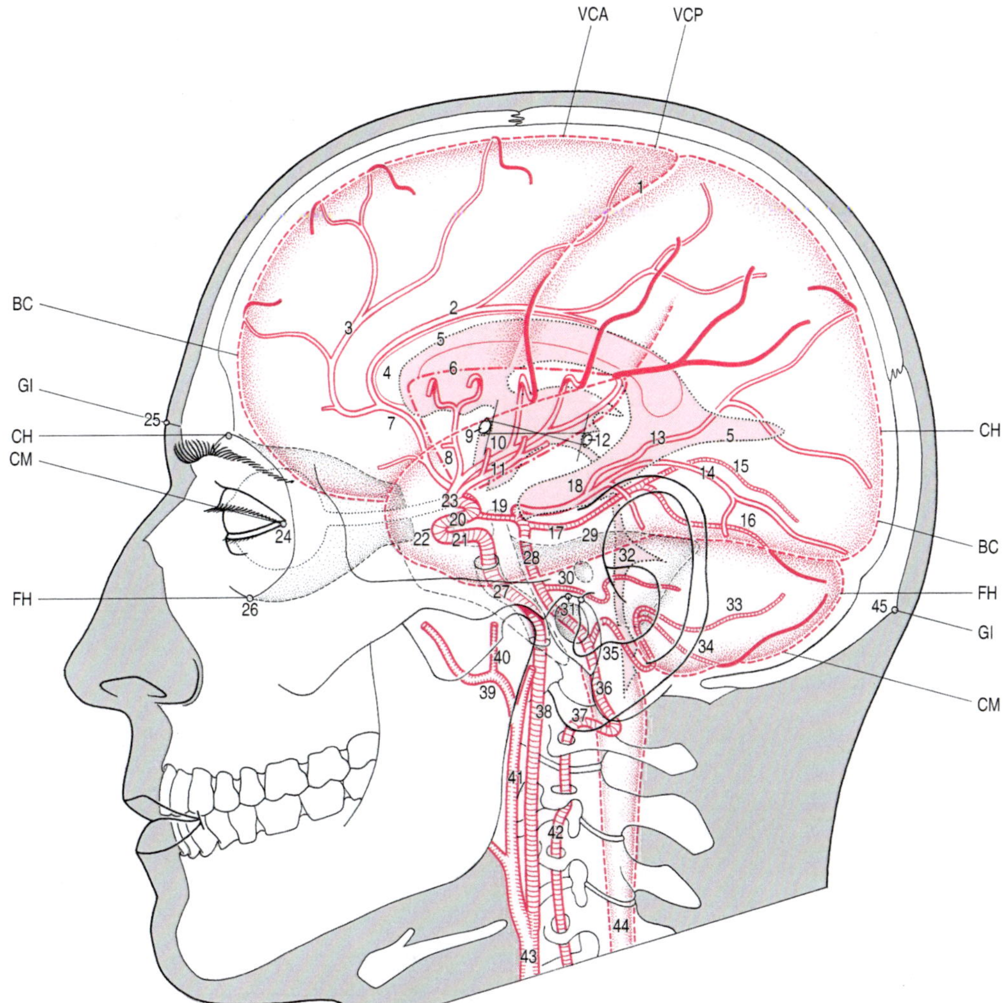

1 Central sulcus
2 Pericallosal artery
3 Callosomarginal artery
4 Corpus callosum
5 Outline of ventricles
6 Outline of insula
7 Anterior cerebral artery
8 Middle cerebral artery, frontal trunk
9 Anterior commissure
10 Middle cerebral artery, parietal trunk
11 Middle cerebral artery, temporal trunk
12 Posterior commissure
13 Medial occipital artery
14 Lateral occipital artery
15 Superior cerebellar artery, medial branch
16 Superior cerebellar artery, lateral branch
17 Superior cerebellar artery
18 Posterior cerebral artery
19 Posterior communicating artery
20 Internal carotid artery, cerebral part
21 Internal carotid artery, cavernous part
22 Siphon point
23 Middle cerebral artery, sphenoid part
24 Ektocanthion (Canthus externus)
25 Glabella
26 Orbital (on infraorbital margin)
27 Internal carotid artery, petrous part
28 Basilar artery
29 Superior margin of petrous part of the temporal bone
30 Anterior inferior cerebellar artery
31 Porion (on suprameatal margin)
32 Fourth ventricle
33 Posterior inferior cerebellar artery, medial branch
34 Posterior inferior cerebellar artery, lateral branch
35 Posterior inferior cerebellar artery
36 Vertebral artery, intracranial part
37 Vertebral artery, atlantal part
38 Internal carotid artery, cervical part
39 Maxillary artery
40 Middle meningeal artery
41 External carotid artery
42 Vertebral artery, cervical part
43 Common carotid artery
44 Spinal cord
45 Inion (external occipital protuberance)

Fig. 4.10. Orthogonal lateral projection of the cerebral and cerebellar arteries, together with external and bony landmarks, in a schematized composition of data from different specimens and publications (2/3×). Some neural structures are also illustrated in their outlines: the left hemisphere, cerebellum, left insula, corpus callosum and ventricular system. Within the outlines of the orbita the bulbus oculi and the optic nerve are indicated. On the outer side of the figure a number of reference lines are added. In the centre, two lines tangential to the anterior (AC) and posterior (PC) commissures can be seen: the one passing above the AC and beneath the PC is part of the bicommissural line of Talairach [27] (*BC*); the other tangent is part of the upper horizontal line of Krönlein (*OH*); *CM*, canthus-meatus line; *FH*, horizontal line or plane of Frankfurt (Reid); *GI*, glabella–inion line; *VCA*, vertical tangential to anterior commissure; *VCP*, vertical tangential to posterior commissure

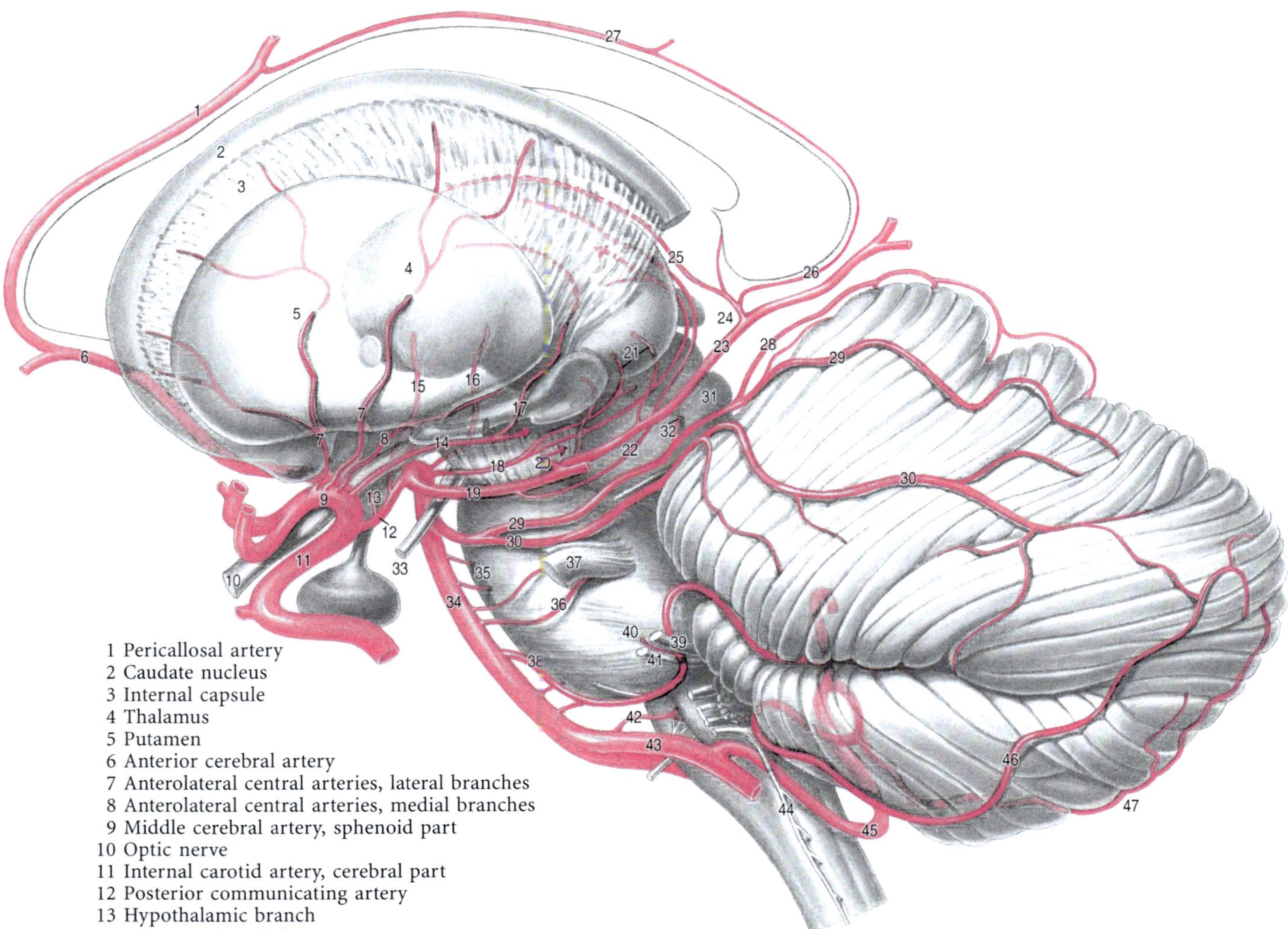

1 Pericallosal artery
2 Caudate nucleus
3 Internal capsule
4 Thalamus
5 Putamen
6 Anterior cerebral artery
7 Anterolateral central arteries, lateral branches
8 Anterolateral central arteries, medial branches
9 Middle cerebral artery, sphenoid part
10 Optic nerve
11 Internal carotid artery, cerebral part
12 Posterior communicating artery
13 Hypothalamic branch
14 Anterior choroidal artery
15 Anteroinferior thalamic branch
16 Posteromedial central arteries
17 Branch of the internal capsule (lateroinferior thalamic branch)
18 Posteromedial choroidal branch
19 Posterior cerebral artery, postcommunicating branch
20 Lateral posterior choroidal branch
21 Posteroinferior thalamic branches
22 Posterior thalamic branch
23 Medial occipital artery
24 Cingulothalamic artery
25 Superior thalamic branch
26 Dorsal branch of the corpus callosum (anastomosis with 27)
27 Pericallosal artery: posterior branch
28 Superior vermian artery
29 Medial branch of the superior cerebellar artery
30 Lateral branch of the superior cerebellar artery
31 Inferior colliculus
32 Mesencephalic branch
33 Oculomotor nerve
34 Basilar artery
35 Medial pontine arteries
36 Lateral pontine arteries
37 Trigeminal nerve
38 Anterior inferior cerebellar artery
39 Vestibulo-cochlear nerve
40 Labyrinthine artery
41 Facial nerve
42 Medullary branches
43 Vertebral artery
44 Spinal root of the accessory nerve
45 Posterior inferior cerebellar artery
46 Posterior inferior cerebellar artery, lateral branch
47 Posterior inferior cerebellar artery, medial branch

Fig. 4.11. The arteries of cerebellum, brain stem, thalamus and the corpus striatum in a lateral view (3/2×). Some arteries are slightly simplified in order to show their course and relations more clearly. The three *arrow points* indicate the choroidal branches of the three choroidal arteries. The same specimen as in Figs. 4.2–4.6, with some slight simplifications

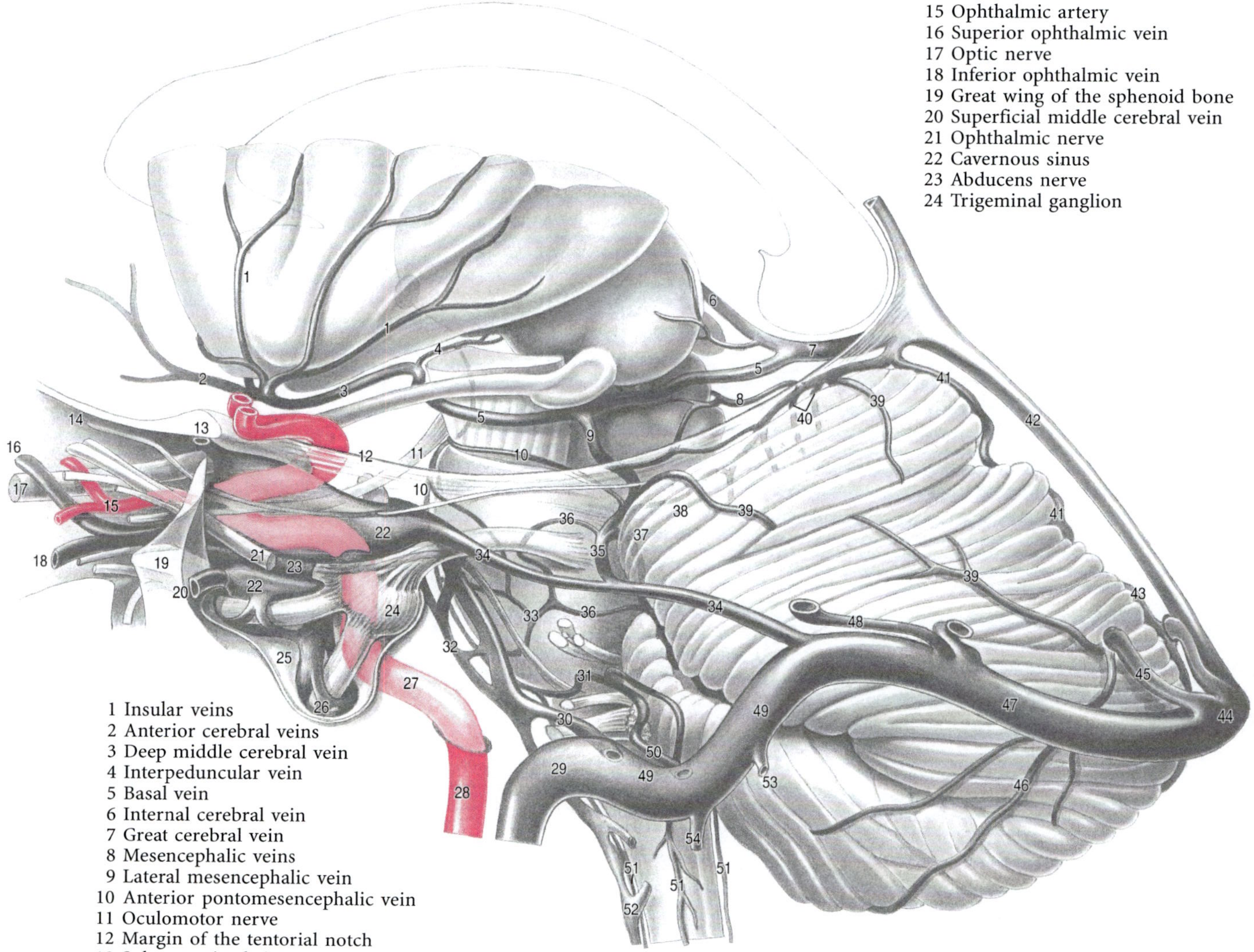

1 Insular veins
2 Anterior cerebral veins
3 Deep middle cerebral vein
4 Interpeduncular vein
5 Basal vein
6 Internal cerebral vein
7 Great cerebral vein
8 Mesencephalic veins
9 Lateral mesencephalic vein
10 Anterior pontomesencephalic vein
11 Oculomotor nerve
12 Margin of the tentorial notch
13 Sphenoparietal sinus
14 Trochlear nerve
15 Ophthalmic artery
16 Superior ophthalmic vein
17 Optic nerve
18 Inferior ophthalmic vein
19 Great wing of the sphenoid bone
20 Superficial middle cerebral vein
21 Ophthalmic nerve
22 Cavernous sinus
23 Abducens nerve
24 Trigeminal ganglion
25 Trigeminal space
26 Venous plexus of the foramen ovale
27 Internal carotid artery, petrous part
28 Internal carotid artery, cervical part
29 Bulbus of the superior jugular vein
30 Inferior petrosal sinus
31 Vein of the pontomedullary sulcus
32 Basilar plexus
33 Pontine veins
34 Superior petrosal sinus
35 Petrosal vein
36 Superior and inferior transverse pontine veins
37 Lateral pontine vein
38 Vein of the superior cerebellar peduncle
39 Superior veins of the cerebellar hemisphere
40 Precentral cerebellar vein
41 Superior vein of the cerebellar vermis
42 Straight sinus
43 Inferior vein of the cerebellar vermis
44 Confluence of sinuses
45 Sinus of the tentorium (collecting infratentorial veins)
46 Inferior veins of the cerebellar hemisphere
47 Transverse sinus
48 Sinus of the tentorium (collecting supratentorial veins)
49 Sigmoid sinus
50 Inferior petrosal vein (inconstant)
51 Anterior, lateral and posterior spinal veins
52 Anterior internal vertebral venous plexus
53 Mastoid emissary vein
54 Condylar emissary vein

Fig. 4.12. Sinuses and veins of the diencephalon, brain stem and cerebellum in a lateral view (3/2×). Composite drawing from two specimens with additions from other sources. The cortical origins of the basal vein have been added, i.e. the insular veins, the deep middle cerebral vein and the anterior cerebral veins. The tentorium has been made fully transparent and the cavernous sinus has been deprived of its lateral dural wall. The inner lateral wall of the trigeminal space has also been removed. The orbit has been opened by a sagittal cut through its centre

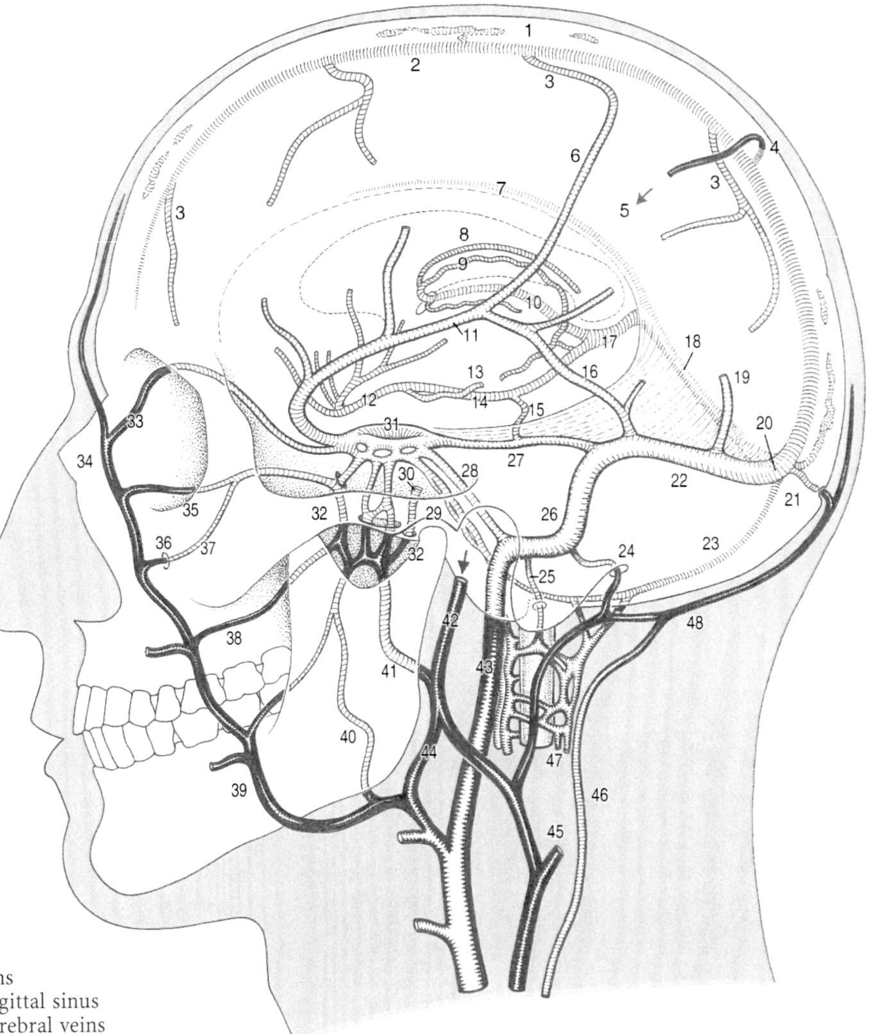

1 Diploic veins
2 Superior sagittal sinus
3 Superior cerebral veins
4 Parietal emissary vein
5 Superficial temporal veins (parietal branch)
6 Superior anastomotic vein (Trolard [30])
7 Inferior sagittal sinus
8 Superior thalamostriate vein
9 Superior choroidal vein
10 Internal cerebral vein
11 Superficial middle cerebral vein
12 Deep middle cerebral vein
13 Inferior choroidal vein
14 Basal vein
15 Lateral mesencephalic vein and petrosal vein
16 Inferior anastomotic vein (Labbé [4])
17 Great cerebral vein
18 Straight sinus (sinus rectus)
19 Inferior cerebral veins
20 Confluens of the sinuses
21 Occipital emissary vein
22 Transverse sinus
23 Occipital sinus
24 Mastoid emissary vein
25 Condylar emissary vein
26 Sigmoid sinus
27 Superior petrosal sinus
28 Inferior petrosal sinus
29 Basilar plexus
30 Middle meningeal veins
31 Cavernous sinus
32 Pterygoid plexus
33 Superior ophthalmic vein
34 Angular vein
35 Inferior ophthalmic vein
36 Infraorbital foramen
37 Infraorbital vein
38 Deep facial vein
39 Facial vein
40 Palatine vein
41 Maxillary veins
42 Superficial temporal veins (see no. 5)
43 Internal jugular vein
44 Retromandibular vein
45 External jugular vein
46 Deep cervical vein
47 Internal vertebral venous plexus
48 Occipital vein

Fig. 4.13. Collateral circulation in the venous system of the head; semidiagrammatic lateral view (2/3×). Unpaired sinuses in the median plane are drawn *without outlines;* the extracranial veins draining into the internal and external jugular veins are in *black;* between the intravertebral venous plexuses a fragment of the spinal medulla can be seen. The *arrows* indicate the continuity of the superficial temporal veins

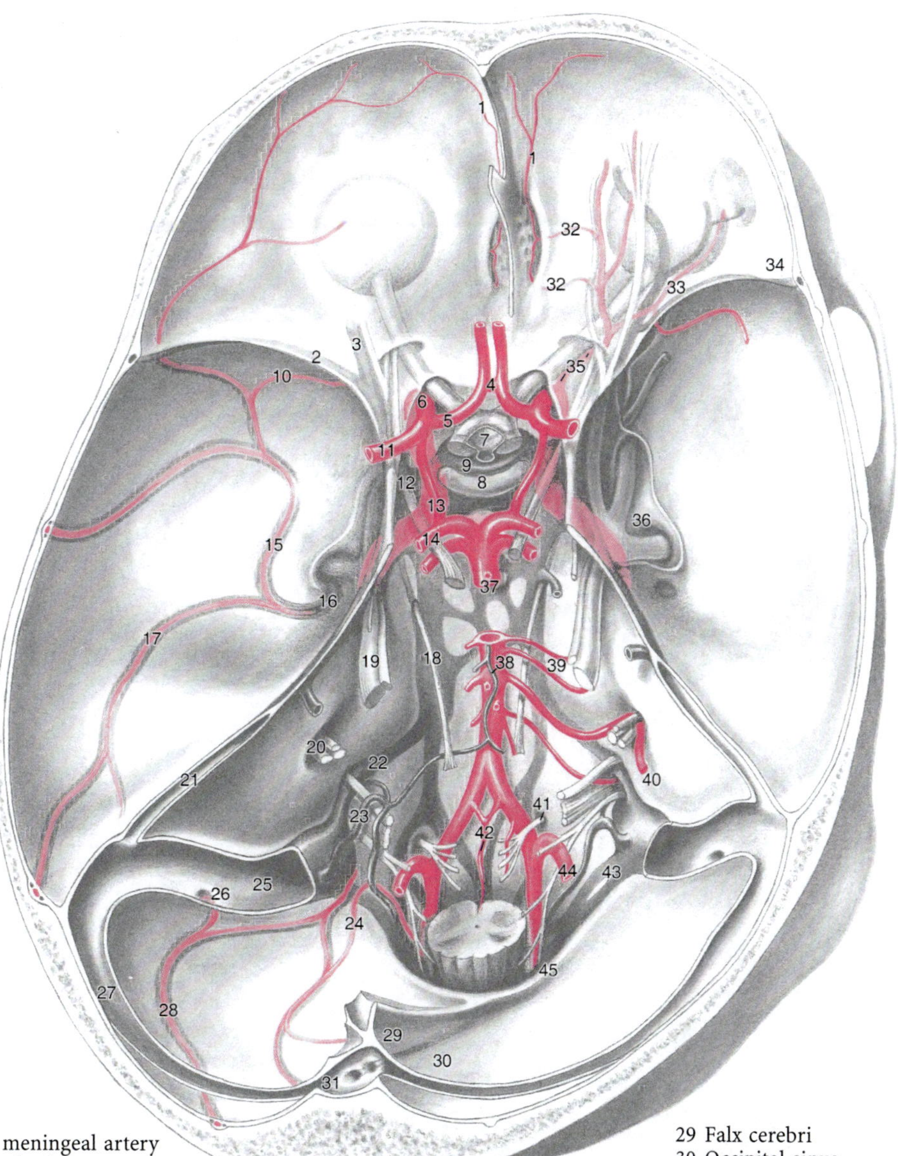

1 Anterior meningeal artery
2 Sphenoparietal sinus
3 Superior ophthalmic vein
4 Anterior communicating artery
5 Anterior cerebral artery
6 Internal carotid artery, cerebral part
7 Optic recess (overlying optic chiasm)
8 Dorsum sellae
9 Intercavernous sinus (posterior part)
10 Middle meningeal artery, anastomosing branch
11 Middle cerebral artery
12 Oculomotor nerve
13 Posterior communicating artery
14 Posterior cerebral artery
15 Middle meningeal artery, frontal branch
16 Middle meningeal veins
17 Middle meningeal artery, parietal branch
18 Abducens nerve (piercing the dura mater)
19 Trigeminal nerve
20 Internal acoustic meatus with facial and vestibulo-cochlear nerves
21 Superior petrosal sinus
22 Inferior petrosal sinus
23 Inferior petrosal vein (inconstant)
24 Posterior meningeal branch of the vertebral artery and vertebral arteries
25 Sigmoid sinus
26 Mastoid emissary vein and meningeal branch of the occipital artery
27 Transverse sinus
28 Posterior meningeal artery
29 Falx cerebri
30 Occipital sinus
31 Confluence of the sinuses
32 Anterior and posterior ethmoidal arteries
33 Lacrimal artery
34 Anterior temporal diploic vein, draining into sphenoparietal sinus
35 Ophthalmic artery
36 Trigeminal space
37 Basilar artery and basilar plexus
38 Anterior pontomesencephalic vein
39 Pontine arteries, lateral branches
40 Anterior inferior cerebellar artery
41 Hypoglossal nerve in hypoglossal canal
42 Anterior spinal artery
43 Marginal sinus
44 Posterior inferior cerebellar artery
45 Spinal root of accessory nerve

4 + 5 + 11 + 13 + 14 Circulus arteriosus (left half)

Fig. 4.14. The skull base in an oblique view from the posterior right side (1/1 ×). The dura has been left in place with the exception of the roof of the greater part of the sinuses and the inner, lateral sheet of the wall of the right trigeminal space. The arterial circle of Willis [36] is left intact; the basilar artery has been interrupted to show the underlying subdural venous plexus of the clivus. On the *left* the meningeal vessels are illustrated; on the *right* the arteries to the cranial nerves can be seen, as can the vessels to the orbita shining through the orbital roof

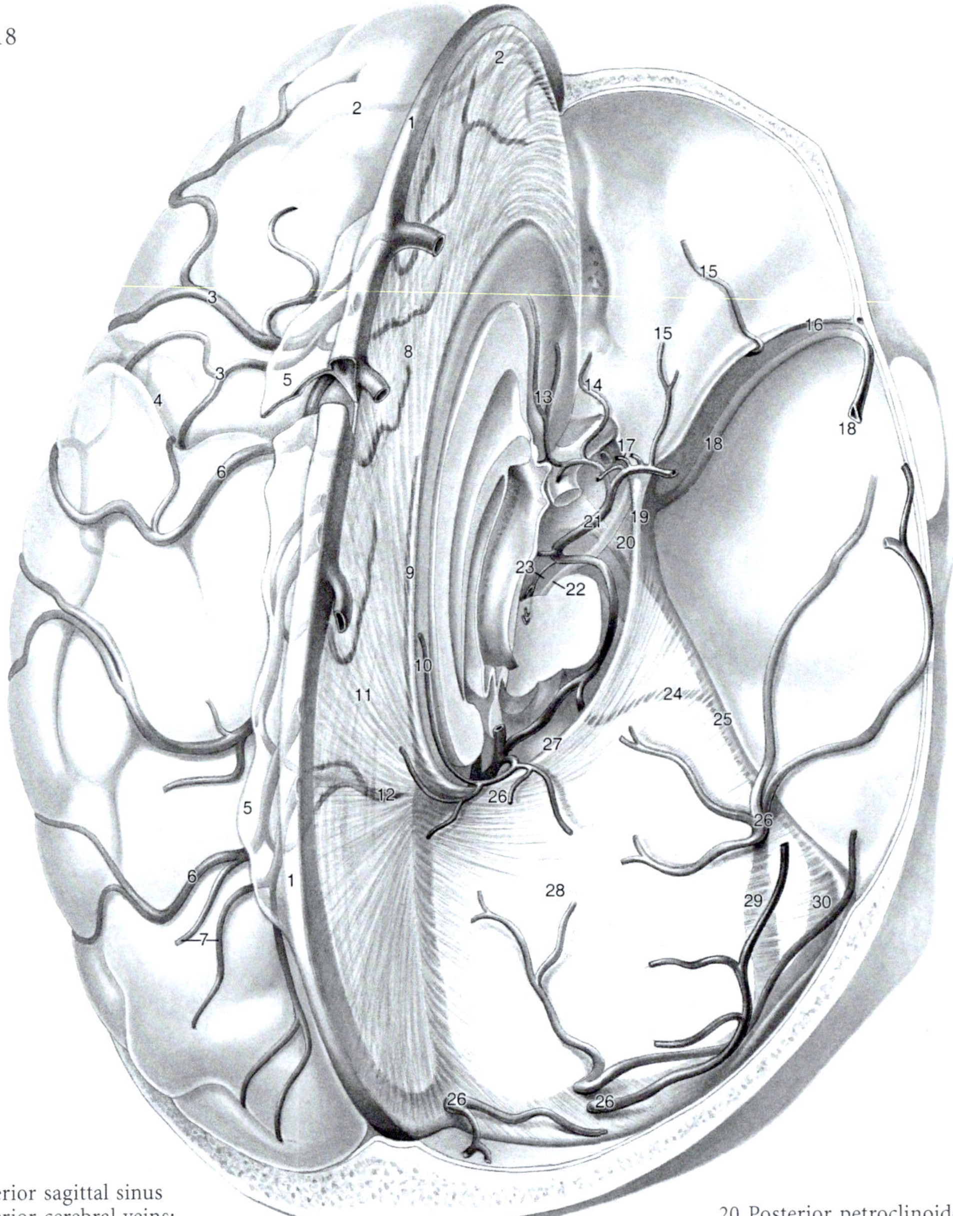

1 Superior sagittal sinus
2 Superior cerebral veins: prefrontal veins
3 Superior cerebral veins: frontal veins
4 Central sulcus
5 Lateral lacunae of the superior sagittal sinus
6 Superior cerebral veins: parietal veins
7 Superior cerebral veins: occipital veins
8 Cingulate sulcus (behind falx)
9 Inferior sagittal sinus
10 Falx cerebri
11 Dorsal vein of the corpus callosum
12 Parieto-occipital sulcus (behind falx)
13 Anterior cerebral veins
14 Vein of the olfactory gyri
15 Vein of the orbital gyri
16 Sphenoparietal sinus
17 Inferior thalamostriate veins
18 Superficial middle cerebral vein
19 Anterior petroclinoideal fold
20 Posterior petroclinoideal fold
21 Basal vein
22 Oculomotor nerve
23 Interpeduncular vein
24 Petrosal vein
25 Superior petrosal sinus (attachment of the tentorium)
26 Inferior cerebral veins
27 Notch of the tentorium
28 Tentorium cerebelli
29 Sinus of the tentorium (collecting supratentorial veins)
30 Transverse sinus

Fig. 4.15. The veins of the cerebral cortex; same view and same specimen as Fig. 4.14 (1/1×). The falx cerebri has been made transparent. In the drawing the most common drainage of the superficial medial cerebral vein (into the cavernous sinus) and the basal vein (into the great cerebral vein of Galen [31]) has been illustrated, contrary to the actual situation in the specimen: drainage into the transverse sinus and via the petrosal vein only into the superior petrosal sinus, respectively. Variability of venous patterns does not occur only interindividually, but also bilaterally: on the *left*, seven individual superior cerebral veins can be seen, among them one coming only from the medial side and one frontopolar vein draining into the anterior tip of the superior sagittal sinus. On the *right* all superior veins drain into the sinus via three stems all together. For insular veins see Fig. 4.16

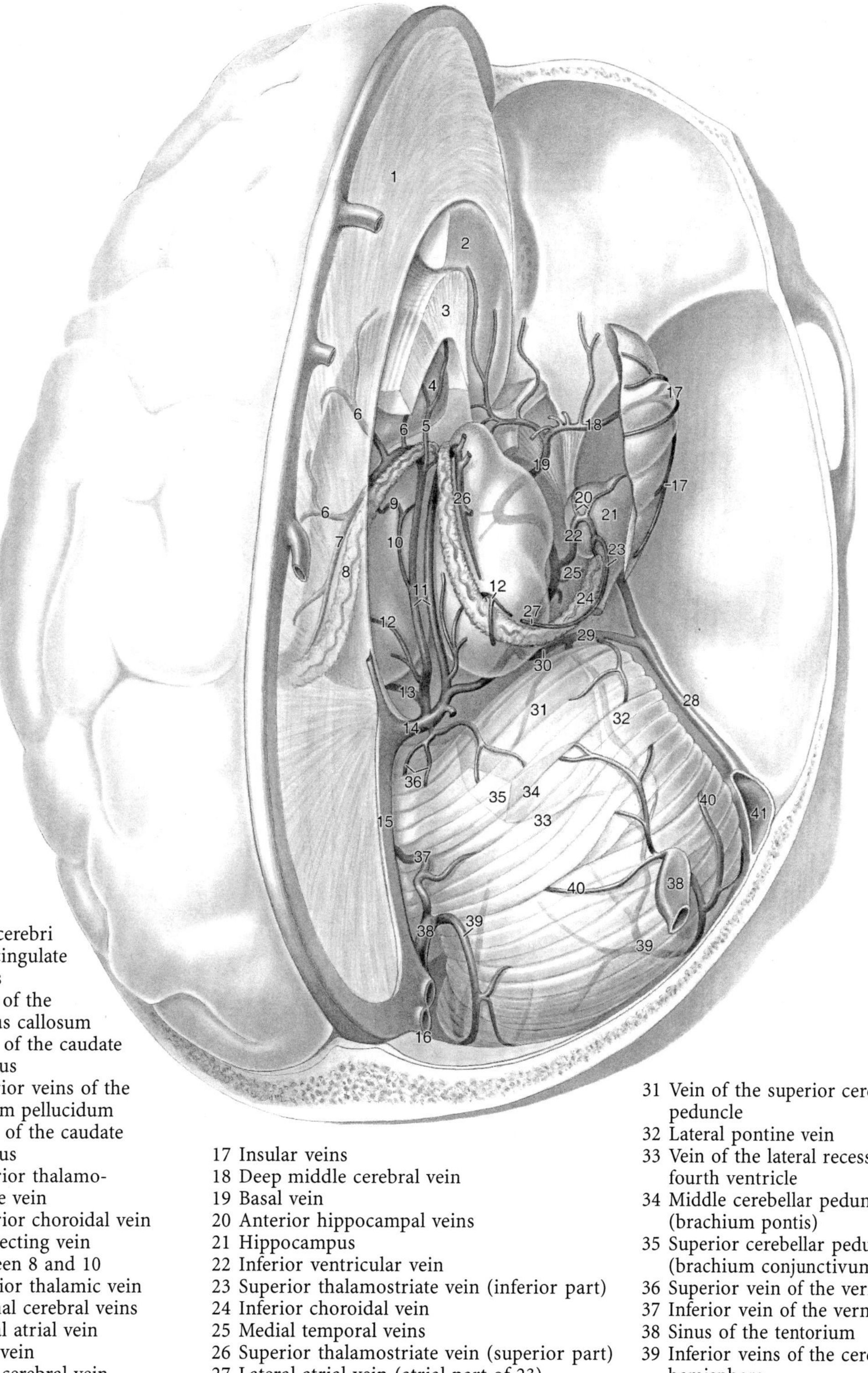

1 Falx cerebri
2 Left cingulate gyrus
3 Knee of the corpus callosum
4 Head of the caudate nucleus
5 Anterior veins of the septum pellucidum
6 Veins of the caudate nucleus
7 Superior thalamo-striate vein
8 Superior choroidal vein
9 Connecting vein between 8 and 10
10 Superior thalamic vein
11 Internal cerebral veins
12 Medial atrial vein
13 Basal vein
14 Great cerebral vein
15 Straight sinus (sinus rectus)
16 Transverse sinus (medial extension)
17 Insular veins
18 Deep middle cerebral vein
19 Basal vein
20 Anterior hippocampal veins
21 Hippocampus
22 Inferior ventricular vein
23 Superior thalamostriate vein (inferior part)
24 Inferior choroidal vein
25 Medial temporal veins
26 Superior thalamostriate vein (superior part)
27 Lateral atrial vein (atrial part of 23)
28 Superior petrosal sinus
29 Petrosal vein
30 Lateral mesencephalic vein
31 Vein of the superior cerebellar peduncle
32 Lateral pontine vein
33 Vein of the lateral recess of the fourth ventricle
34 Middle cerebellar peduncle (brachium pontis)
35 Superior cerebellar peduncle (brachium conjunctivum)
36 Superior vein of the vermis
37 Inferior vein of the vermis
38 Sinus of the tentorium
39 Inferior veins of the cerebellar hemisphere
40 Superior veins of the cerebellar hemisphere
41 Transverse sinus (lateral extremity)

Fig. 4.16. The deep cerebral and the cerebellar veins; same view and same specimen as Figs. 4.14 and 4.15 (1/1×). The falx cerebri has been made transparent in order to show the left periventricular veins. The veins of the medial and posterior thalamus can be seen on both sides; the inferior cerebellar veins are shining through the cerebellum. The insula has been added to present the insular origins of the deep middle cerebral vein. In the inferior part of the specimen the choroid plexus ends by a cut at the site where the hippocampus appears, the posterosuperior part of it being cut away

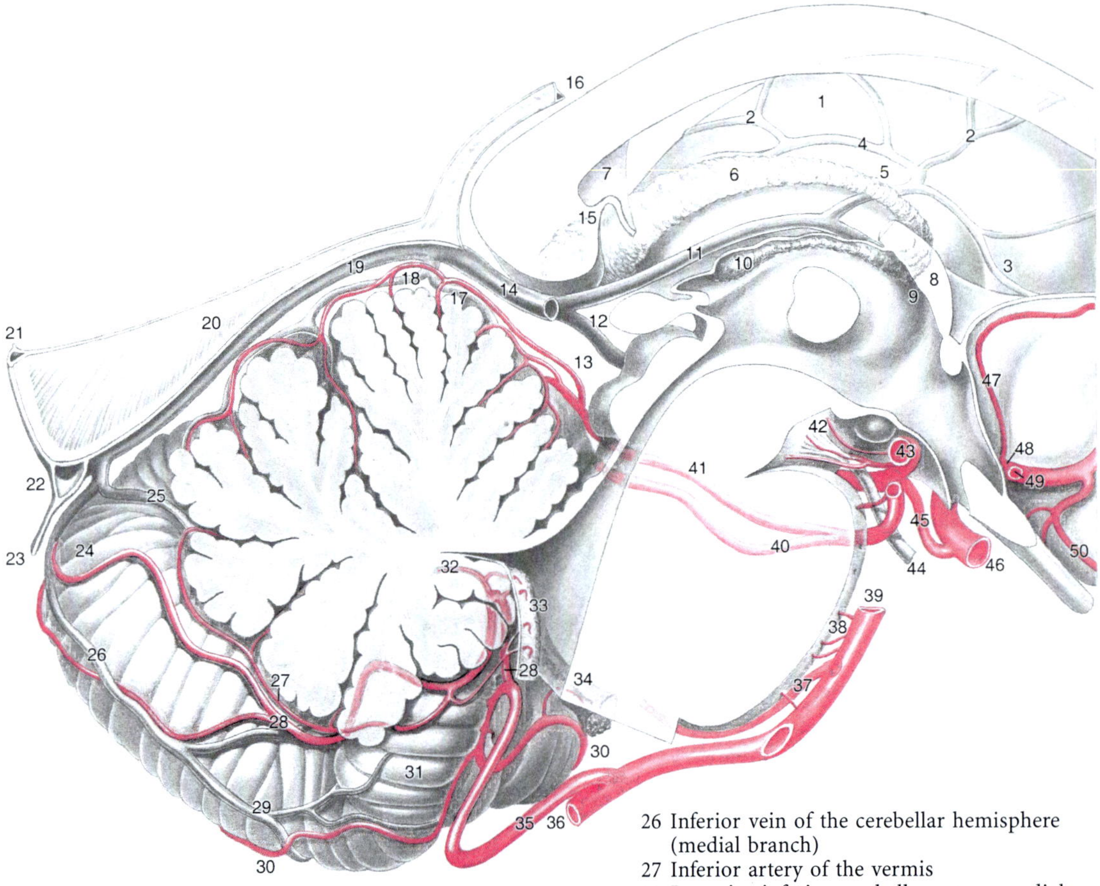

1 Caudate nucleus, ventricular surface
2 Veins of the caudate nucleus
3 Anterior vein of the septum pellucidum
4 Superior thalamostriate vein
5 Lamina affixa
6 Choroid plexus of the lateral ventricle
7 Septum pellucidum
8 Fornix
9 Interventricular foramen
10 Choroid plexus of the third ventricle
11 Internal cerebral vein
12 Basal vein
13 Superior vein of vermis
14 Great cerebral vein
15 Commissure of the fornix
16 Inferior sagittal sinus
17 Precentral cerebellar vein
18 Superior vein of the vermis
19 Straight sinus (sinus rectus)
20 Falx cerebri
21 Superior sagittal sinus
22 Confluence of the sinuses
23 Occipital sinus
24 Horizontal fissure
25 Inferior vein of the vermis
26 Inferior vein of the cerebellar hemisphere (medial branch)
27 Inferior artery of the vermis
28 Posterior inferior cerebellar artery, medial branch
29 Retrotonsillar veins
30 Posterior inferior cerebellar artery, lateral branches
31 Tonsil
32 Tonsillar branch of 35
33 Branches to the choroid plexus of the fourth ventricle (from 35)
34 Branch to the lateral recess of the fourth ventricle (from 35)
35 Posterior inferior cerebellar artery
36 Vertebral artery
37 Anterior inferior cerebellar artery
38 Pontine arteries, medial branches
39 Basilar artery
40 Superior cerebellar artery, lateral branch
41 Superior cerebellar artery, medial branch
42 Posteromedial central arteries
43 Posterior cerebral artery, precommunicating part
44 Oculomotor nerve
45 Posterior communicating artery
46 Internal carotid artery
47 Median callosal artery
48 Anterior cerebral artery, precommunicating part
49 Anterior communicating artery
50 Medial frontobasal artery

Fig. 4.17. Medial arteries and veins of the cerebellum; medial view of the left half of a bisected brain (3/2×). Most of the medial perforating arteries to the pons and the mesencephalon are also illustrated. Superiorly, the major subependymal veins on the caudate nucleus and the internal cerebral vein can be seen

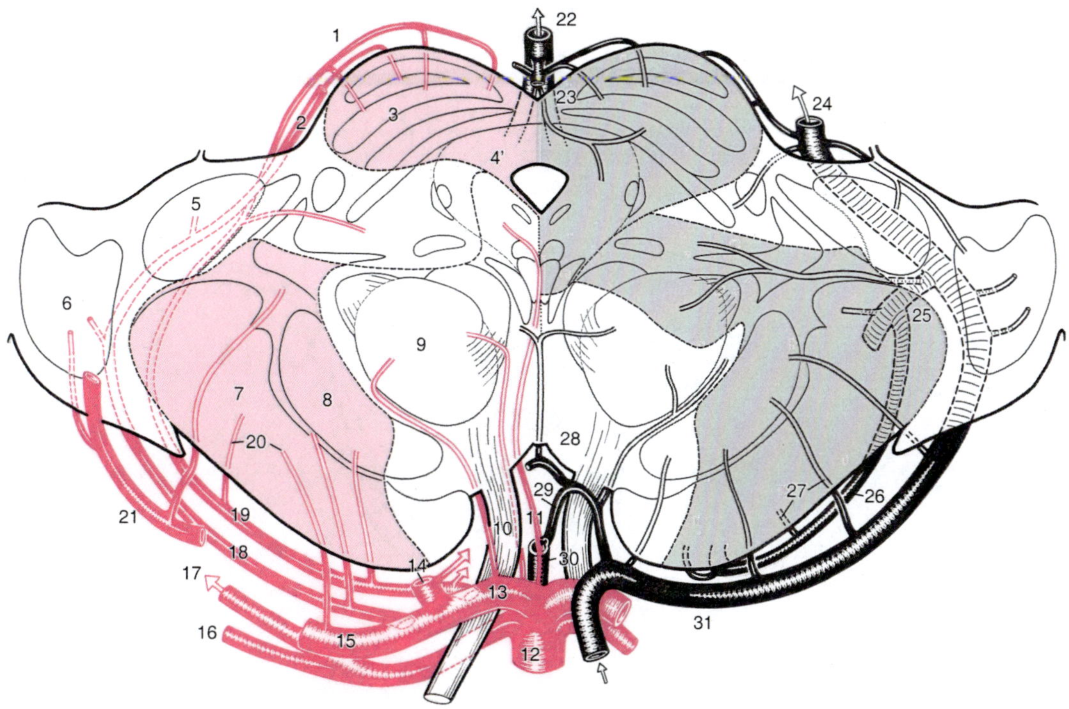

1 Quadrigeminal artery
2 Medial posterior choroidal branch
3 Superior colliculus
4 Periaqueductal grey
5 Medial geniculate body
6 Lateral geniculate body
7 Cerebral peduncle
8 Substantia nigra
9 Red nucleus
10 Oculomotor nerve
11 Posteromedial central arteries
12 Basilar artery
13 Posterior cerebral artery, precommunicating part
14 Posterior communicating artery (with mesencephalic branch)
15 Posterior cerebral artery, postcommunicating part
16 Superior cerebellar artery, lateral branch
17 Superior cerebellar artery, medial branch; the mesencephalic branch (arrow) is located more distally
18 Medial posterior choroidal branch
19 Quadrigeminal artery
20 Peduncular branches
21 Anterior choroidal artery
22 Precentral cerebellar vein
23 Mesencephalic veins (superior median collicular vein)
24 Basal vein (posterior segment)
25 Lateral mesencephalic vein
26 Anterior pontomesencephalic vein (peripeduncular branch)
27 Peduncular veins
28 Posterior communicating vein
29 Interpeduncular vein
30 Anterior pontomesencephalic vein (median branch)
31 Basal vein (anterior segment)

Fig. 4.18. Vascularization of the mesencephalon; transverse section in a superior view, with the vessels of about that level (4/1×). The plane of section is that of Fig. 6.18. Territories according to Duvernoy [12], arteries and arterial territories in the *right half*, veins and their territories in the *left half*. *Red arrows* indicate mesencephalic branches of 14 and 19

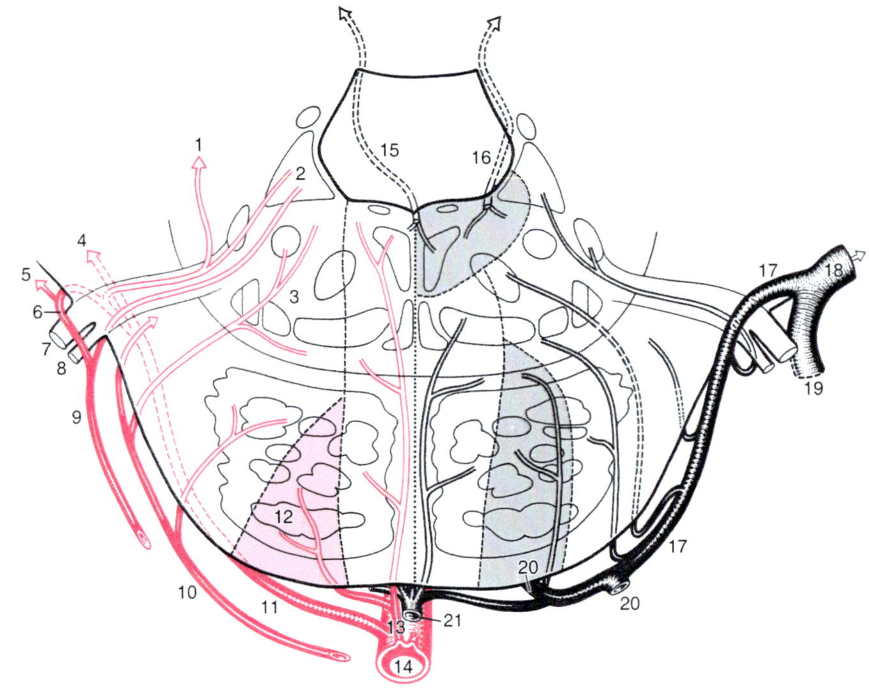

1 Branch to inferior cerebellar peduncle
2 Vestibular nuclei
3 Pontine tegmentum
4 Branch to horizontal fissure
5 Branch to middle cerebellar peduncle
6 Anastomosis around trigeminal nerve (inconstant)
7 Trigeminal nerve, portio major
8 Trigeminal nerve, portio minor
9 Pontine arteries, lateral branches (superolateral pontine arteries)
10 Pontine arteries, lateral branches (inferolateral pontine arteries)
11 Anterior inferior cerebellar artery
12 Pons
13 Pontine arteries, medial branches
14 Basilar artery
15 Fourth ventricle
16 Posterior tegmental veins, ascending to the lateral mesencephalic vein
17 Transverse pontine veins
18 Petrosal vein
19 Lateral pontine vein (anterior cerebellar vein)
20 Pontine veins
21 Anterior pontomesencephalic vein

Fig. 4.19. Vascularization of the metencephalon; transverse section in a superior view, with the vessels of about that level (4/1×). The plane of sectioning is that of Fig. 6.16. Source and execution as in Fig. 4.18

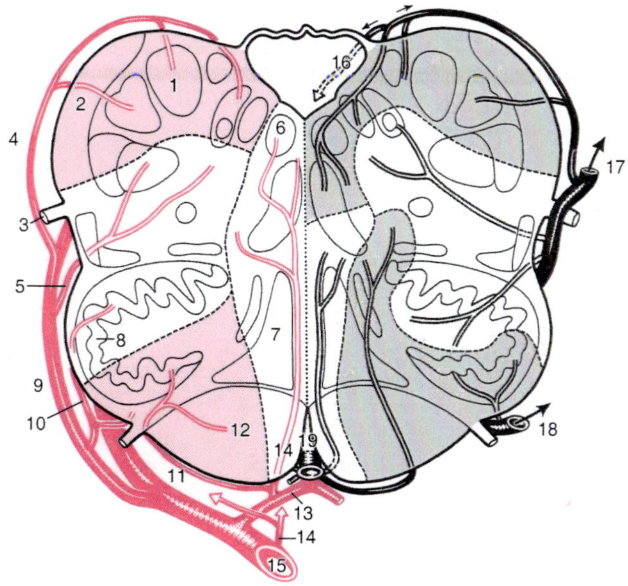

1 Medial cuneate nucleus
2 Inferior cerebellar peduncle
3 Vagal nerve
4 Posterior medullary branches
5 Lateral medullary branches
6 Nucleus of the hypoglossal nerve
7 Medial lemniscus
8 Inferior olive
9 Posterior inferior cerebellar artery
10 Medial medullary branches
11 Anterolateral medullary branches
12 Pyramidal tract
13 Anterior spinal artery
14 Anteromedial medullary branches
15 Vertebral artery

16 Vein descending to posteromedian medullary vein
17 Lateral vein of the medulla oblongata
18 Anterolateral vein of the medulla oblongata
19 Median vein of the medulla oblongata
17–19 ascend to the plexus of corresponding pontine veins

Fig. 4.20. Vascularization of the medulla oblongata; transverse section in a superior view, with the vessels of about that level (4/1×). The plane of section is that of Fig. 6.14. Source and execution as in Fig. 4.18

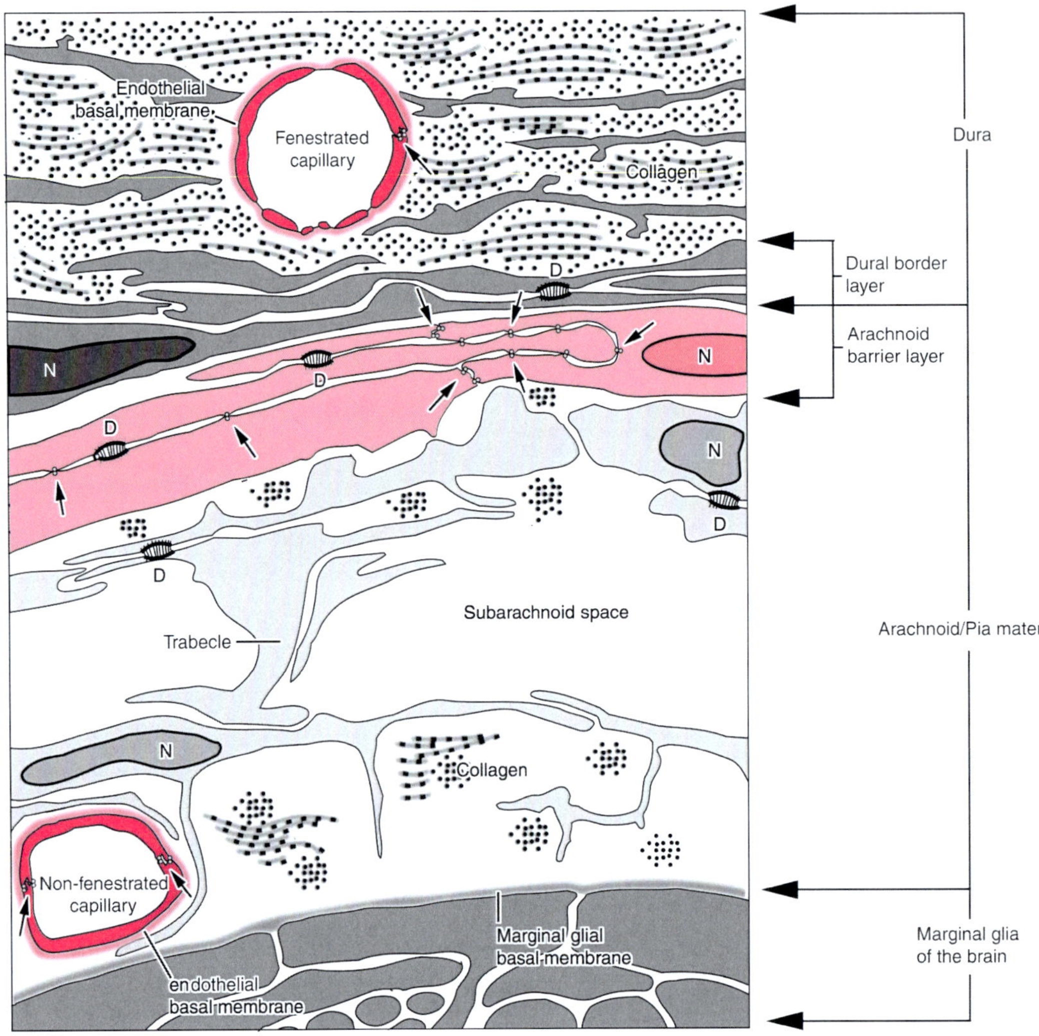

Fig. 4.21. Diagram illustrating the ultrastructure of the cranial meninges. The dura consists of an outer layer of fibroblast and collagen fibres and a layer of dural border cells. A subdural space does not exist. In the case of a subdural haematoma, arterial pressure splits the dural border layer and a subdural space is formed. The arachnoid consists of an outer barrier layer, where the cells are connected by tight junctions (*small arrows*) and an inner layer that fuses with the pia mater. The subarachnoid space is formed from the coalition of intracellular spaces of the arachnoid/pia mater. The subarachnoid space is traversed by trabecula interconnecting, the cell layers bordering it on its dural and neural sides. Capillaries outside the barrier layer of the arachnoid are fenestrated. Capillaries located inside the barrier layer and in the brain lack these fenestrations. Based on a diagram published by Nabeshima et al. [21]. D, desmosome; N, nucleus

A

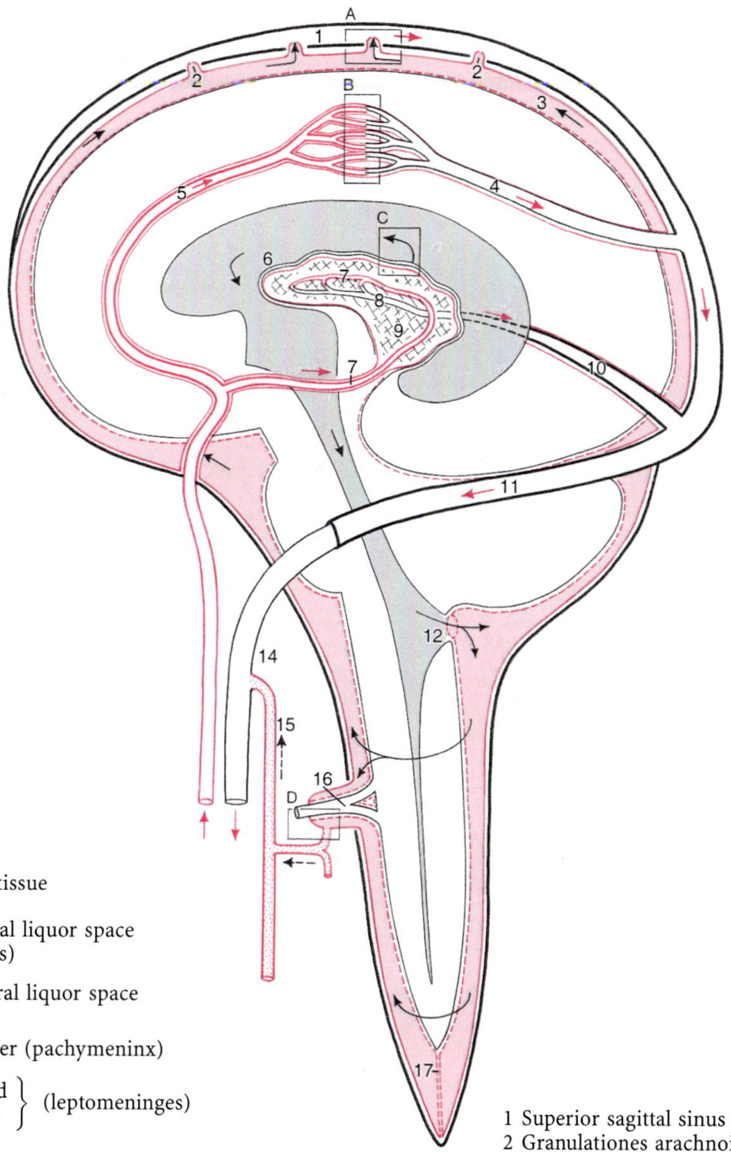

B

- Nervous tissue
- Intraneural liquor space (ventricles)
- Extraneural liquor space (cisterns)
- Dura mater (pachymeninx)
- Arachnoid } (leptomeninges)
- Pia mater } (leptomeninges)
- Arteries
- Veins
- Venous sinuses
- Lymphe vessels
- Blood
- Liquor cerebrospinalis
- Lymphe

A Liquor-bloods (valve-like) passage
B Bloods-brain barrier of non-fenestrated endothelial wall + arachnoidal barrier layer
C Blood-liquor barrier of non-fenestrated epithelial wall + arachnoidal barrier layer
D Liquor-blood passage via lymphatic system

1 Superior sagittal sinus with lateral lacunae
2 Granulationes arachnoidales
3 Subarachnoidal space
4 Cortical and deep cerebral veins
5 Cortical and deep cerebral arteries
6 Choroid plexus
7 Anterior and posterior choroidal arteries
8 Superior and inferior choroidal veins
9 Connective tissue
10 Straight sinus
11 Transverse and sigmoid sinuses
12 Apertures of the fourth ventricle
13 Arterial supply: internal carotid and vertebral artery
14 Venous drainage by the internal jugular vein
15 Thoracic duct
16 Spinal nerves
17 Filum terminale

Fig. 4.22 A, B. Diagram of the blood and CSF circulation in the brain. For reasons of simplification one channel represents the arterial supply instead of the two systems that exist in reality (see Fig. 4.23)

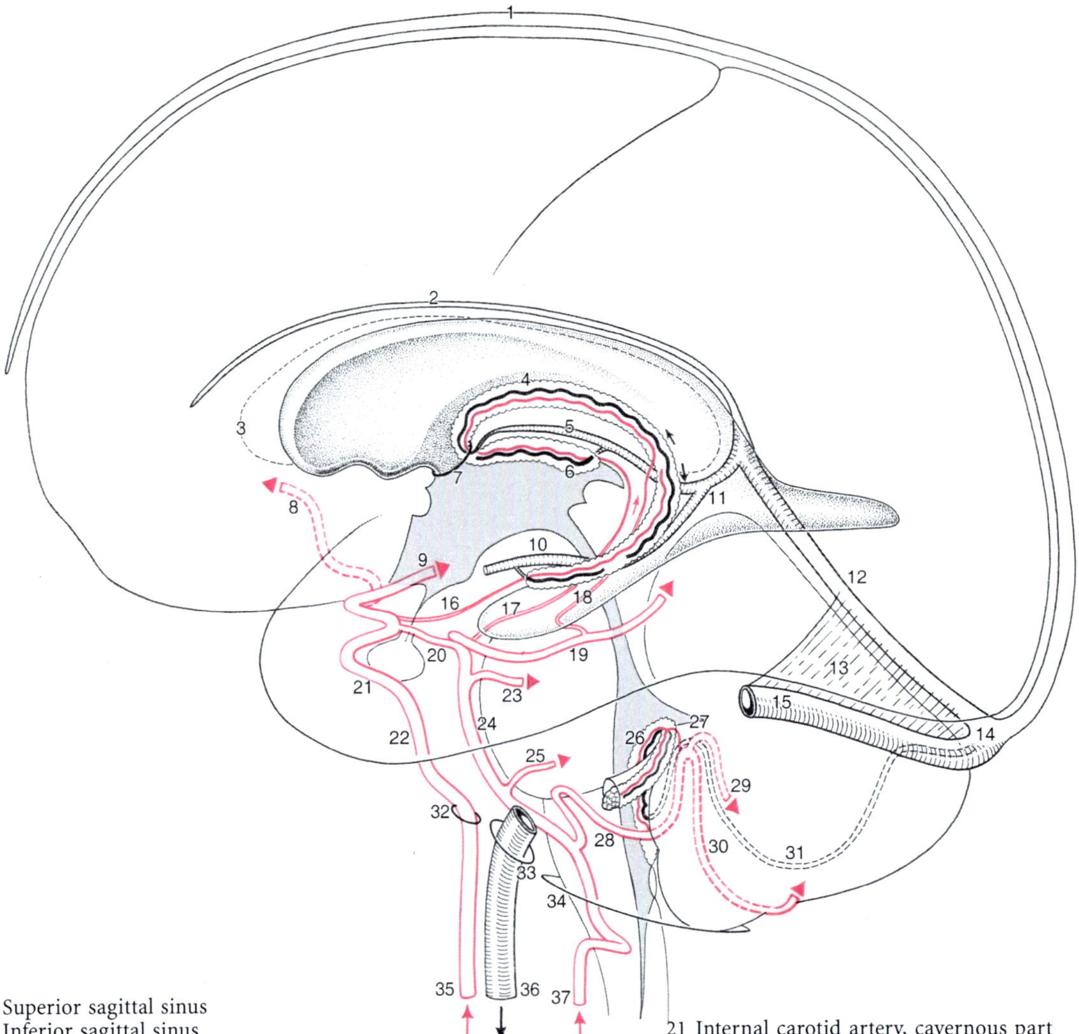

1 Superior sagittal sinus
2 Inferior sagittal sinus
3 Corpus callosum
4 Choroid plexus of the lateral ventricle
5 Internal cerebral vein
6 Choroid plexus of the fourth ventricle
7 Interventricular foramen
8 Anterior cerebral artery
9 Middle cerebral artery
10 Basal vein
11 Great cerebral vein
12 Straight sinus (sinus rectus)
13 Tentorium cerebelli
14 Confluence of the sinuses
15 Transverse sinus
16 Anterior choroidal artery
17 Posteromedial choroidal branches
18 Posterolateral choroidal branches
19 Posterior cerebral artery (displaced inferiorly)
20 Posterior communicating artery
21 Internal carotid artery, cavernous part
22 Internal carotid artery, petrosal part
23 Superior cerebellar artery
24 Basilar artery
25 Anterior inferior cerebellar artery
26 Choroid plexus of the fourth ventricle
27 Choroidal branches of 28
28 Posterior inferior cerebellar artery
29 Posterior inferior cerebellar artery, medial branch
30 Posterior inferior cerebellar artery, lateral branch
31 Inferior vein of the vermis
32 Carotid canal (inferior opening)
33 Foramen jugulare
34 Foramen magnum
35 Internal carotid artery, cervical part
36 Internal jugular vein
37 Vertebral artery

Fig. 4.23. Vascular supply of the choroid plexus, as seen in a semidiagrammatic lateral view of the left side (1/1×). All vessels are paired, with the exception of the basilar artery (*24*), the superior and inferior sagittal sinuses (*1, 2*), the great cerebral vein (*11*), the straight sinus (sinus rectus) (*12*) and the inferior vermian vein (*31*)

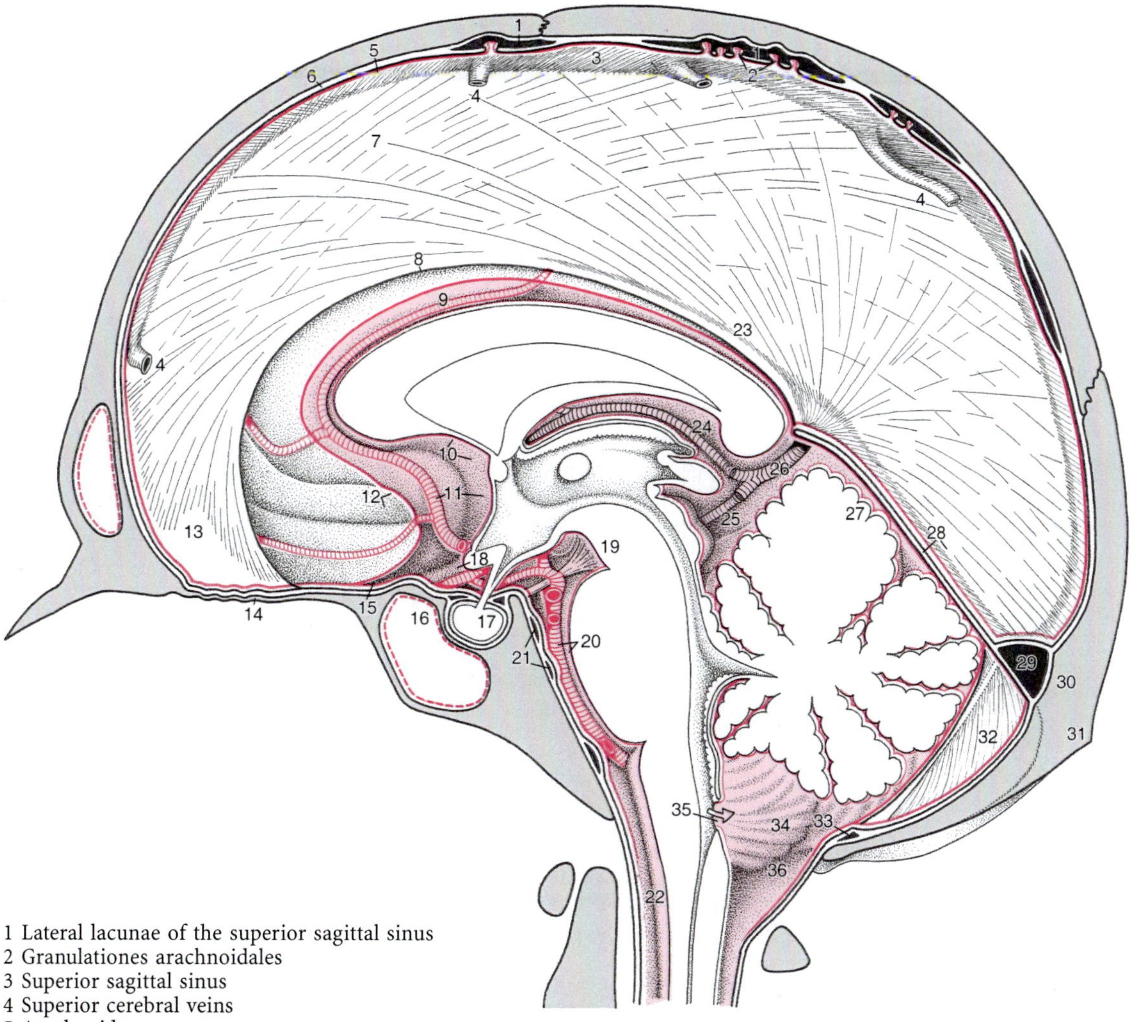

1 Lateral lacunae of the superior sagittal sinus
2 Granulationes arachnoidales
3 Superior sagittal sinus
4 Superior cerebral veins
5 Arachnoid
6 Dura mater
7 Falx cerebri
8 Free margin of the falx
9 Pericallosal cistern with artery
10 Pia mater (surface and cut edge)
11 Cistern of the lamina terminalis with anterior cerebral artery
12 Arachnoid (surface and cut edge)
13 Crista galli
14 Lamina cribrosa
15 Connection with cistern of the olfactory gyri
16 Anterior intercavernosal sinus in diaphragma sellae
17 Posterior part of intercavernosal sinus
18 Cistern of the optic chiasm
19 Interpeduncular cistern
20 Pontine cistern with basilar artery
21 Basilar plexus
22 Medullary cistern
23 Inferior sagittal sinus
24 Cistern of the transverse fissure with internal cerebral vein
25 Cistern of the tectum with basal vein
26 Cistern of the great cerebral vein
27 Superior cerebellar cistern
28 Straight sinus (sinus rectus) with tentorium cerebelli (cut)
29 Confluence of the sinuses
30 Internal occipital protuberance
31 External occipital protuberance
32 Falx cerebelli
33 Marginal sinus
34 Tonsil (pial surface)
35 Median aperture of the fourth ventricle
36 Cerebellomedullary cistern (cisterna magna)

Fig. 4.24. Meninges and cisterns from the medial side (1/1×). The sagittal section passes just left of the falx cerebri and the sinuses situated in the median plane, thus cutting through the lateral lacunae of the superior sagittal sinus. The pachymeninx (dura mater) is *outlined in black*, the leptomeninges (arachnoidea and pia mater) are in *red*. The cisterns are indicated with *red*

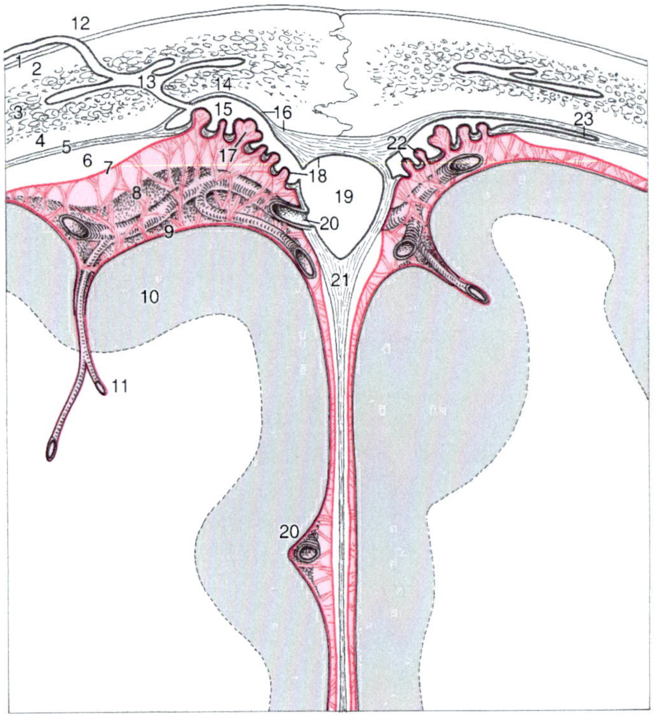

1 Cranial periost
2 Tabula externa
3 Diploe
4 Tabula interna
5 Dura mater
6 Subdural space (artefactually produced)
7 Arachnoid
8 Subarachnoid space
9 Pia mater
10 Cerebral cortex
11 Small venous branch
12 Emissary vein
13 Diploic vein
14 Foveola granularis (Pacchioni [22])
15 Lateral lacunae of the superior sagittal sinus
16 Endosteal layer of cranial dura
17 Granulationes arachnoidales (Pacchioni [22])
18 Endothelium
19 Superior sagittal sinus
20 Superior cerebral veins
21 Falx cerebri
22 Arachnoid villi
23 Meningeal vein

Fig. 4.25. Diagrammatic frontal section through the superior sagittal sinus and surrounding structures, showing the different compartments of blood and CSF and the layers involved (about 3/1×)

A

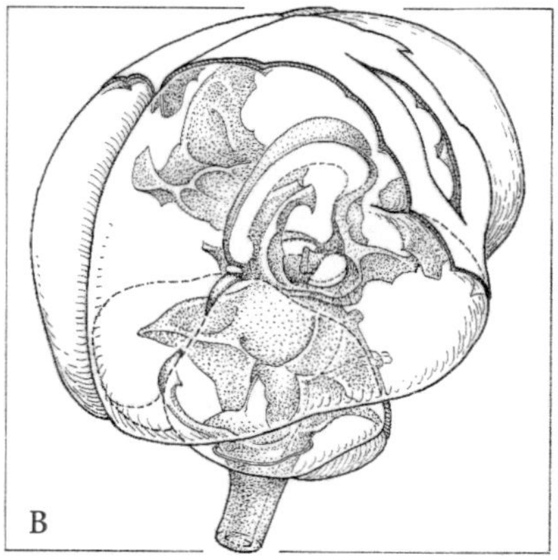

Subarachnoid cisterns

- a Granulationes arachnoideales
- b Cistern of the central sulcus
- c Pericallosal cistern
- d Cistern of the lamina terminalis
- e Cistern of the fossa lateralis
- f Cistern of the vallecula cerebri
- g Cistern of the optic chiasm
- h Interpeduncular cistern
- i Cistern of the transverse fissure
- j Cistern of the great cerebral vein
- k Superior cerebellar cistern
- l Cisterna ambiens
- m Pontine cistern
- n Trigeminal cistern
- o Cistern of the internal acoustic meatus
- p Pontocerebellar cistern
- q Cerebellomedullary cistern (cisterna magna)
- r Medullary cistern

Impressions (structures in contact with cisterns)

1 Precentral gyrus
2 Insula
3 Knee of the corpus callosum
4 Superior temporal gyrus
5 Parahippocampal gyrus (temporal pole)
6 Optic tract
7 Fornix
8 Splenium of the corpus callosum
9 Great cerebral vein
10 Pulvinar
11 Lateral geniculate body
12 Subiculum
13 Notch of the tentorium (*arrow*)
14 Basilar sulcus
15 Middle cerebellar peduncle
16 Glossopharyngeal, vagal and accessory nerves
17 Horizontal fissure
18 Biventral lobule
19 Vallecula cerebelli
20 Foramen magnum

Fig. 4.26 A, B. Model of the cisterns after a spatial reconstruction in an oblique view from behind (1/1×). In the *inset*, the site within the brain is indicated. The places of contact with different neural structures can be recognized and are labelled

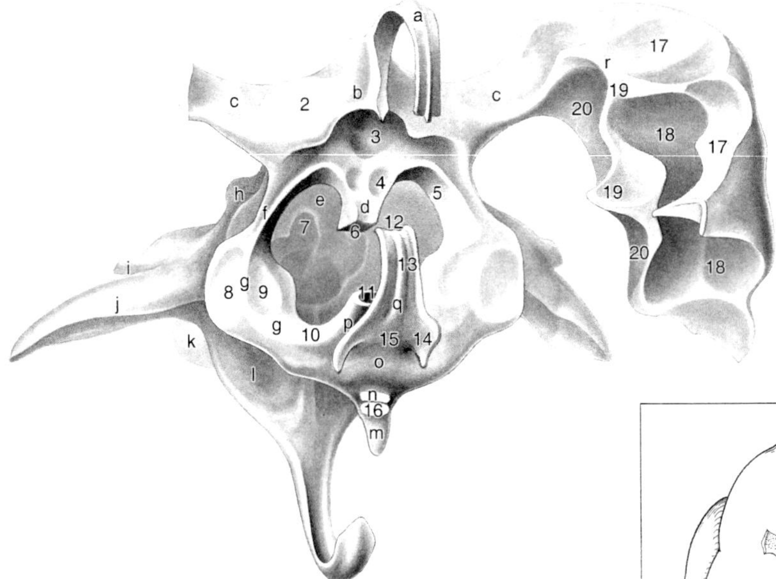

A

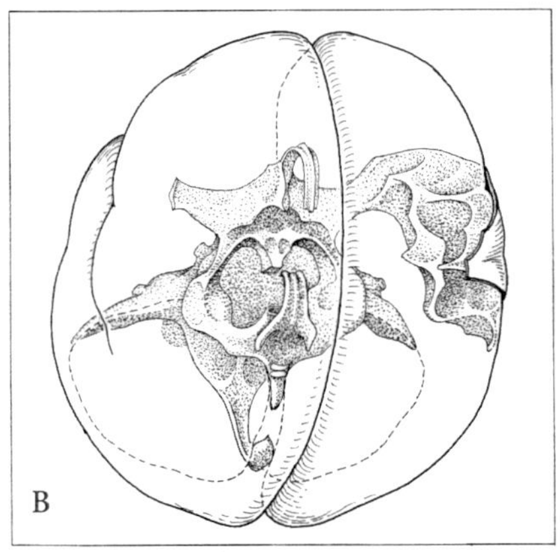

B

Subarachnoid cisterns
a Pericallosal cistern
b Cistern of the lamina terminalis
c Cistern of the vallecula cerebri
d Interpeduncular cistern
e Pontine cistern
f Cisterna ambiens: peripeduncular cistern
g Cisterna ambiens: wing
h Trigeminal cistern
i Cistern of the internal acoustic meatus
j Pontocerebellar cistern
k Medullary cistern
l Cerebellomedullary cistern (cisterna magna)
m Superior cerebellar cistern
n Posterior pericallosal cistern
o Cistern of the great cerebral vein
p Cistern of the tectum
q Cistern of the transverse fissure
r Cistern of the lateral cerebral sulcus

Impressions (structures in contact with cisterns)
1 Knee of the corpus callosum
2 Orbital gyri
3 Tuber cinereum
4 Mamillary body
5 Cerebral peduncle
6 Interpeduncular fossa
7 Pyramid
8 Lateral geniculate body
9 Medial geniculate body
10 Pulvinar
11 Suprapineal recess
12 Column of the fornix
13 Corpus of the fornix
14 Crus of the fornix
15 Splenium of the corpus callosum
16 Great cerebral vein
17 Precentral sulcus
18 Frontoparietal operculum
19 Circular sulcus of the insula
20 Insula

Fig. 4.27 A, B. The same model as in Fig. 4.26, in an oblique view from above (1/1×). To gain a clear view of the central and lower cisterns, the cortical and sylvian cisterns of the left side, and part of the cortical cisterns of the right side, were removed. Conventions as in Fig. 4.26

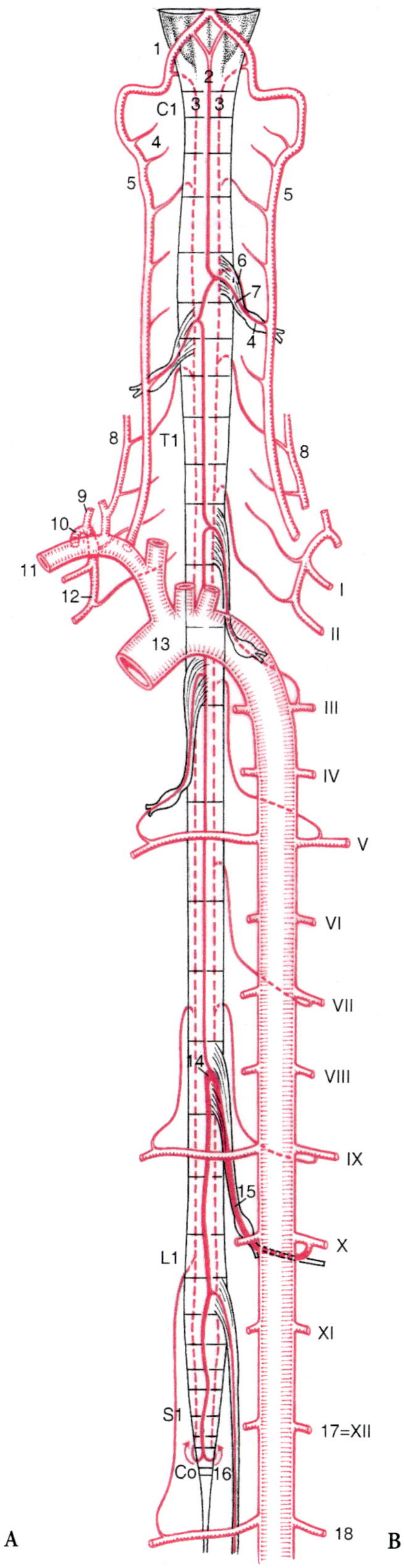

◀ 1 Vertebral artery, intracranial part
2 Anterior spinal artery
3 Posterior spinal artery
4 Spinal branches
5 Vertebral artery, transverse part
6 Posterior radicular artery
7 Anterior radicular artery
8 Ascending cervical artery
9 Deep cervical artery (may be source of radicular arteries)
10 Costocervical trunk
11 Subclavian artery
12 Supreme intercostal artery
13 Aorta
14 Hairpin junction of radicular artery to spinal artery
15 Arteria radicularis magna (Adamkiewicz [1]
16 Ansa anastomotica
17 Subcostal artery
18 Lumbar artery

I–XII, posterior intercostal arteries
I–XII

C Cervical segment of the spinal cord
Co Coccygeal segment of the spinal cord
L Lumbar segment of the spinal cord
S Sacral segment of the spinal cord
T Thoracic segment of the spinal cord

---

1 Posterior spinal artery (left side)
2 Posterior spinal vein
3 Posterior spinal artery (right side)
4 Posterior radicular artery (C5, right side)
5 Anterior radicular vein (C5, left side)
6 Anterior radicular vein (C5, right side)
7 Anterior radicular artery (C5, right side)
8 Anterior radicular artery (C6, right side)
9 Anterior spinal vein
10 Anterior spinal artery
▼ 11 C6 ventral root

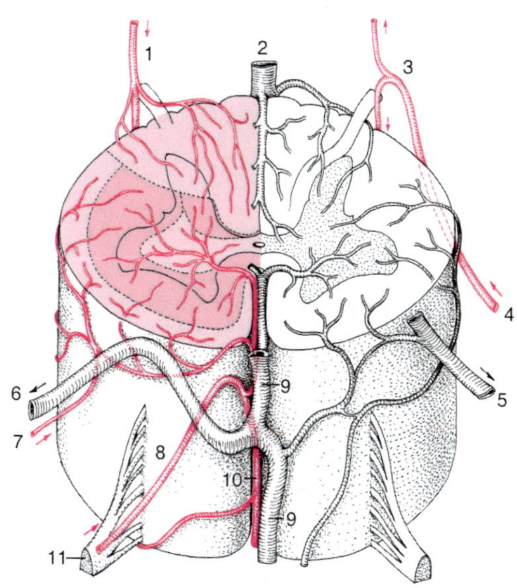

Fig. 4.28 A. Arterial supply of the spinal cord in a semidiagrammatic ventral view (slightly less than 1/2×). Spinal and spinal root arteries are in *solid red;* dorsal arteries behind the medulla shown as a *broken line.* The aorta and other arteries of origin are *outlined in red.* Of the spinal nerve roots, only those that belong to the same segment as the illustrated spinal root arteries are illustrated. The levels of these arteries are the most frequent according to different, through not fully concordant, sources (Rickenbacher [25] and Djindjian [11])
Fig. 4.28 B. Vascularization areas in the spinal cord (C5/6). *Left* arterial supply; *right* venous drainage

1 Anterior spinal vein
2 Posterior spinal vein
3 Vertebral vein
4 Deep cervical vein
5 Posterior radicular vein
6 Anterior radicular vein
7 Spinal branch of the posterior intercostal vein
8 Supreme intercostal vein
9 Subclavian vein
10 Inferior bulb of internal jugular vein
11 Right brachiocephalic vein
12 Left brachiocephalic vein
13 Superior vena cava
14 Right superior intercostal vein
15 Left superior intercostal vein
16 Accessory vena azygos
17 Vena azygos
18 Vena hemiazygos
19 Subcostal vein
20 Ascending lumbar vein
21 Lumbar vein

I–XII. Posterior intercostal veins I–XII

C Cervical segment of the spinal cord
Co Coccygeal segment of the spinal cord
L Lumbar segment of the spinal cord
S Sacral segment of the spinal cord
T Thoracic segment of the spinal cord

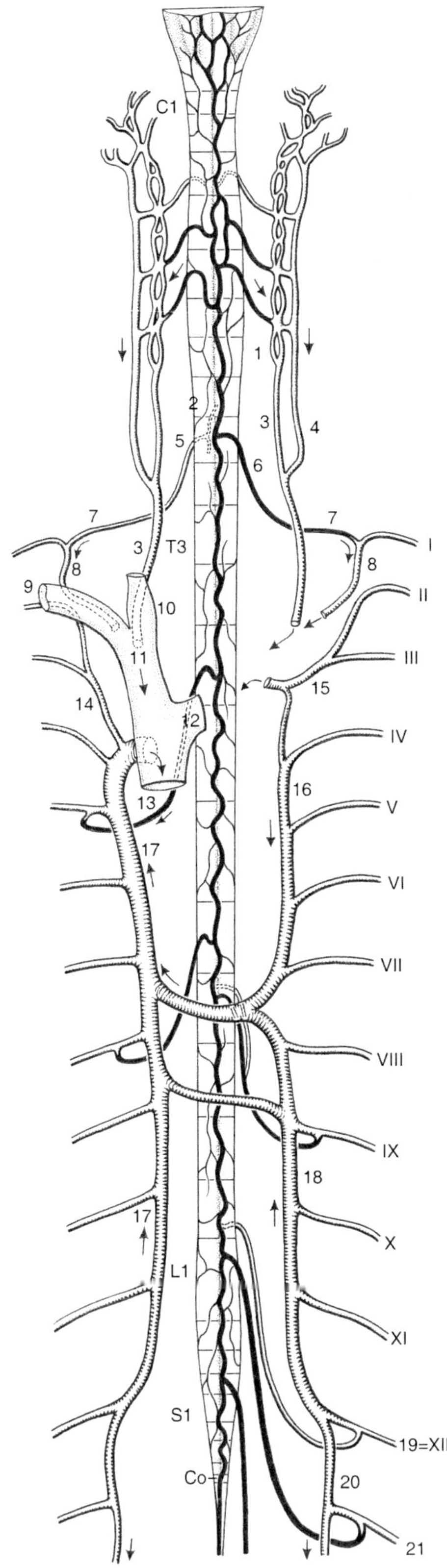

Fig. 4.29. Venous drainage of the spinal cord in a semidiagrammatic ventral view (slightly less than 1/2×). Veins of the ventral roots are in *solid black*, dorsal ones are indicated by *open outlines*. Collecting veins are *half-hatched*; the final path (the main branches of the superior vena cava) is indicated by *stippling*. The *lowest arrows* indicate the drainage via the ascending lumbar veins into the common iliac veins. Levels of the spinal root veins are mostly according to Djindjian [11]

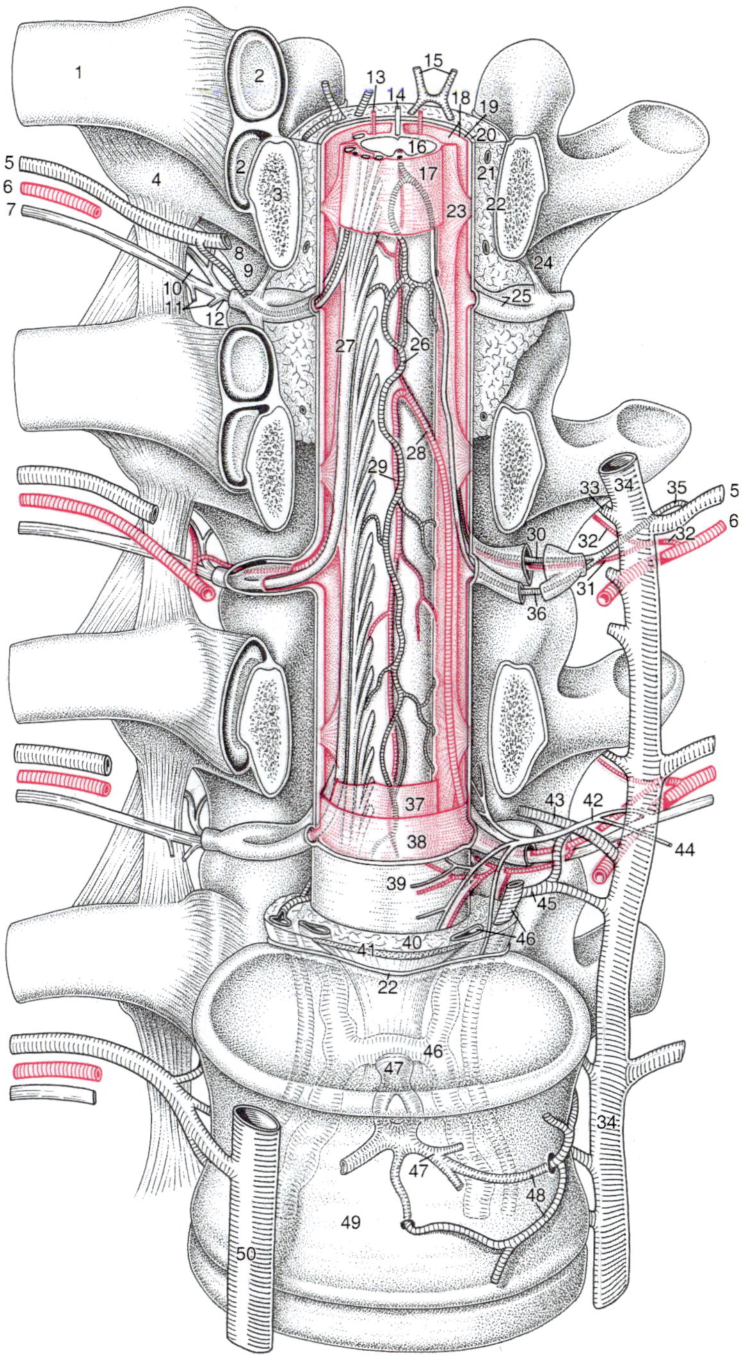

1 Eighth rib (right side)
2 Articular facet of a rib
3 Pedicle
4 Transverse process
5 Posterior intercostal vein
6 Posterior intercostal artery
7 Intercostal nerve (ventral branch of a thoracic nerve)
8 Spinal branch of intercostal vein
9 Dorsal branch of eighth thoracic nerve
10 Costotransverse foramen
11 Communicating branches
12 Trunk of the eighth thoracic nerve
13 Right posterior spinal artery
14 Posterior spinal vein
15 Posterior part of internal vertebral venous plexus
16 Spinal cord
17 Pia mater
18 Arachnoid and subarachnoidal space
19 Subdural space (exists only in cases of subdural haematoma)
20 Dura mater
21 Epidural space
22 Periosteum
23 Denticulate ligament
24 Fibrous closure of intervertebral foramen
25 Dural sac of spinal ganglion and roots
26 Anterior spinal veins
27 Ventral root of ninth thoracic nerve
28 Arteria radicularis magna
29 Anterior spinal artery
30 Posterior radicular vein
31 Spinal branch of intercostal artery
32 Dorsal branch of intercostal artery
33 Dorsal branch of vena hemiazygos
34 Vena hemiazygos
35 Spinal branch of posterior intercostal vein
36 Anterior radicular vein
37 Pia mater
38 Arachnoid
39 Dura mater
40 Epidural space
41 Posterior longitudinal ligament
42 Meningeal branch of 10th thoracic nerve
43 Superior intervertebral vein
44 Communicating branch of 42
45 Inferior intervertebral vein
46 Anterior part of internal vertebral venous plexus
47 Basivertebral veins
48 Anterior part of external vertebral venous plexus
49 Ninth thoracic vertebra
50 Vena azygos

Fig. 4.30. Topography of the contents of the vertebral canal at the level of the 8th to 11th rib; ventral view, slightly from above (3/2×). Semidiagrammatic representation. Vascular configurations are derived from Figs. 4.28, 4.29 and 4.31. Dura in *black outlines*; leptomeninges in *red*. The dural sac around the spinal ganglion and the ventral root has been cut in different ways. At the left side the spinal roots are left out, with the exception of part of the ventral root of T10

1 Dura
2 Conus medullaris
3 Anterior part of internal vertebral venous plexus: medial epidural vein
4 Anterior part of internal vertebral venous plexus: lateral epidural vein
5 Vena azygos
6 Medial arcuate ligament
7 Ascending lumbar vein
8 Pedicle of first lumbar vertebra
9 Posterior part of internal vertebral venous plexus
10 Spinal roots within dural sheath (between epidural veins)
11 Superior intervertebral vein
12 Inferior intervertebral vein
13 Segmental communicating vein
14 Dorsal branch of ascending lumbar vein
15 Dorsal branch of external vertebral vein
16 Posterior part of internal vertebral venous plexus
17 Posterior longitudinal ligament
18 Retrocorporeal venous anastomosis
19 Basovertebral vein
20 Anterior part of internal vertebral venous plexus
21 Dorsal branch of lateral sacral vein (ascending sacral vein)
22 Lateral sacral vein
23 Common iliac vein
24 External iliac vein
25 Internal iliac vein
26 Pedicle of third sacral vertebra
27 Pelvic sacral foramina
28 Dorsal sacral foramina

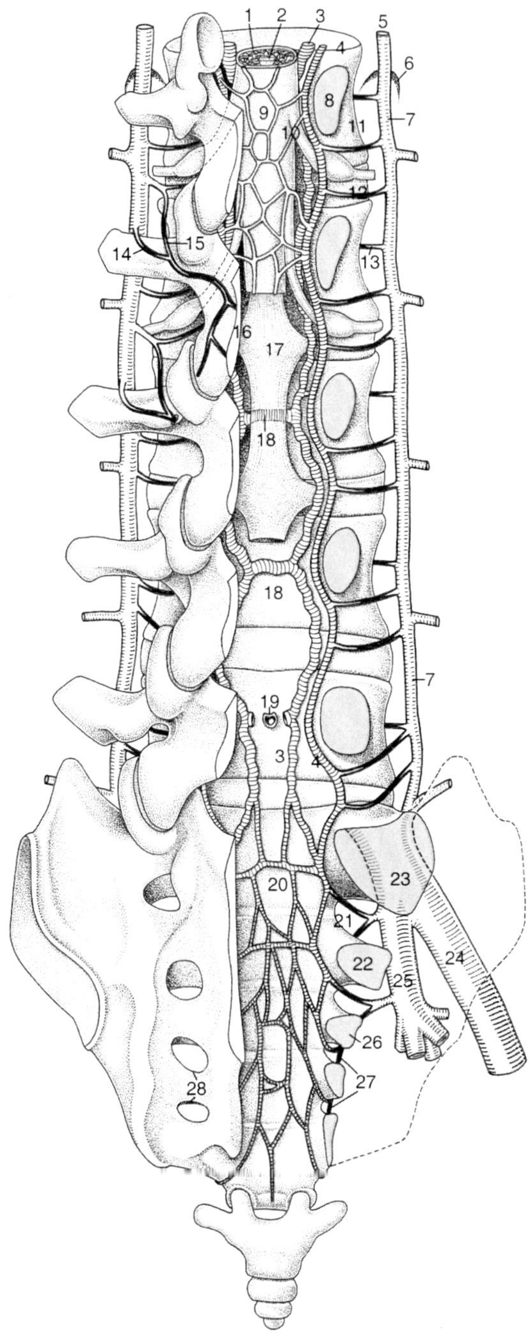

Fig. 4.31. The epidural veins of the lumbar and sacral vertebral canal in a dorsal view (2/3×). At the dorsal and right sides the wall of the bony canal has been removed unto the middle of the pedicles of the vertebral arch. At the right side the lateral outline of the sacrum is indicated in a *broken line*. At the level of the fifth lumbar vertebra the transverse vein has been interrupted to show the sagittal vein piercing the body of the vertebra (basivertebral vein). This figure is mainly based on Théron and Moret [28] and Renard [24]

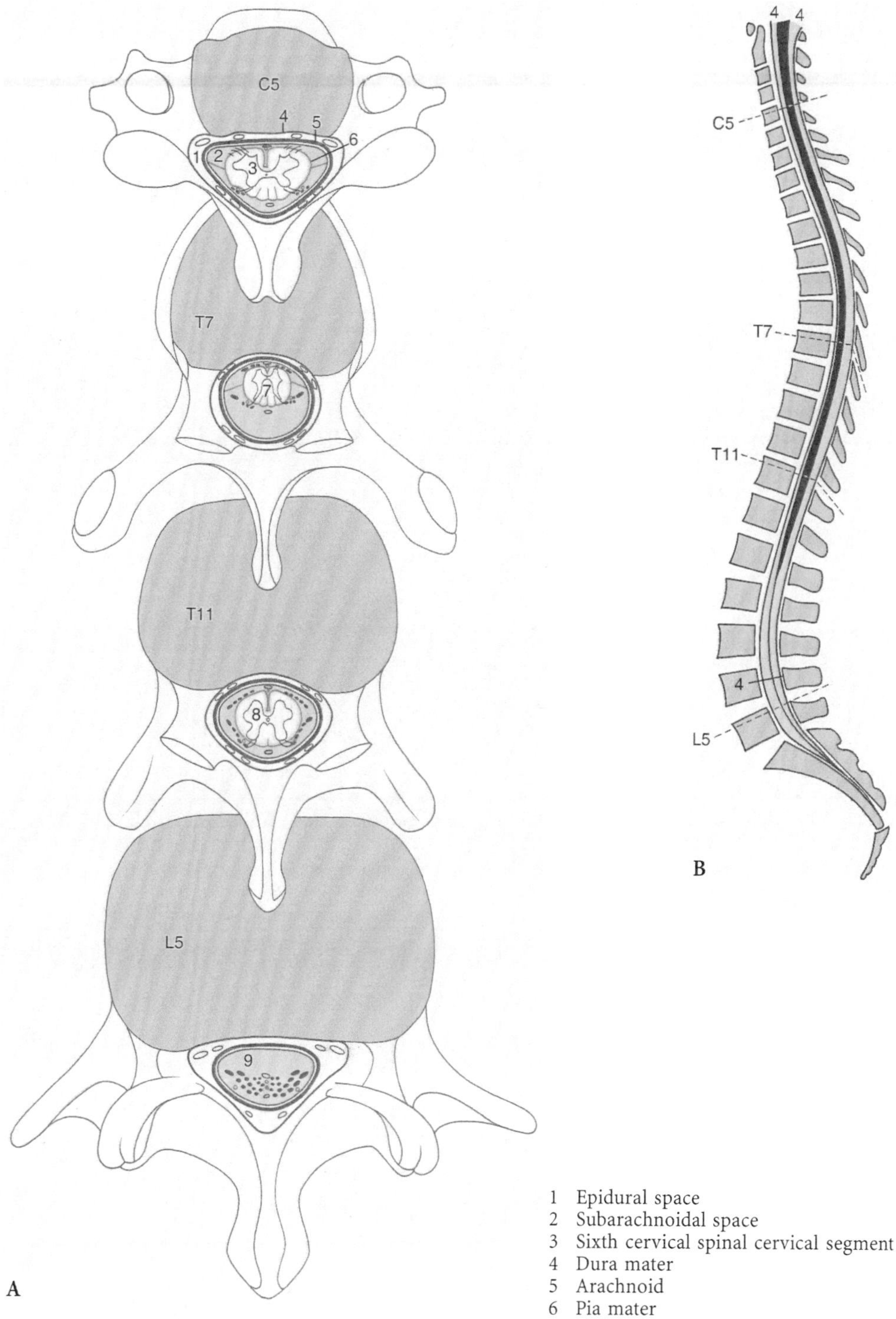

1 Epidural space
2 Subarachnoidal space
3 Sixth cervical spinal cervical segment
4 Dura mater
5 Arachnoid
6 Pia mater
7 Eighth thoracic spinal segment
8 Second and third lumbar spinal segment
9 Cauda equina (fila radicularia)

Fig. 4.32 A, B. Semidiagrammatic presentation of the spinal cord and the meninges within the vertebral canal at four different levels (1/1×). Outlines of the vertebrae as in a superior view; the position of the sections through the spinal cord has been indicated in the median sectioned vertebral column (1/5x), shown in **B**

5 Brain Slices

Introduction

This pictorial chapter presents four series of brain slices cut in different directions, as follows:

13 coronal sections
4 sections perpendicular to the axis of the brain stem
6 sagittal sections
9 horizontal sections

All of the slices shown are taken from brains of adult individuals with no record of neurological diseases. The brains were fixed for at least 2 months in formalin. They were then embedded in gelatin and sliced on a rotary-blade commercial meat slicer. Due to the elasticity of the large gelatin blocks, we initially experienced considerable difficulties in maintaining the intended plane of sectioning throughout a series. However, this problem was ingeniously solved by Mr. A Binnendijk, our laboratory assistant, by embedding the brains in boxes, prepared from styrofoam plates, and by subsequently slicing the brains while still encased in the surrounding box. This procedure yielded perfect, well-oriented, continuous series of slices.

Coronal Sections

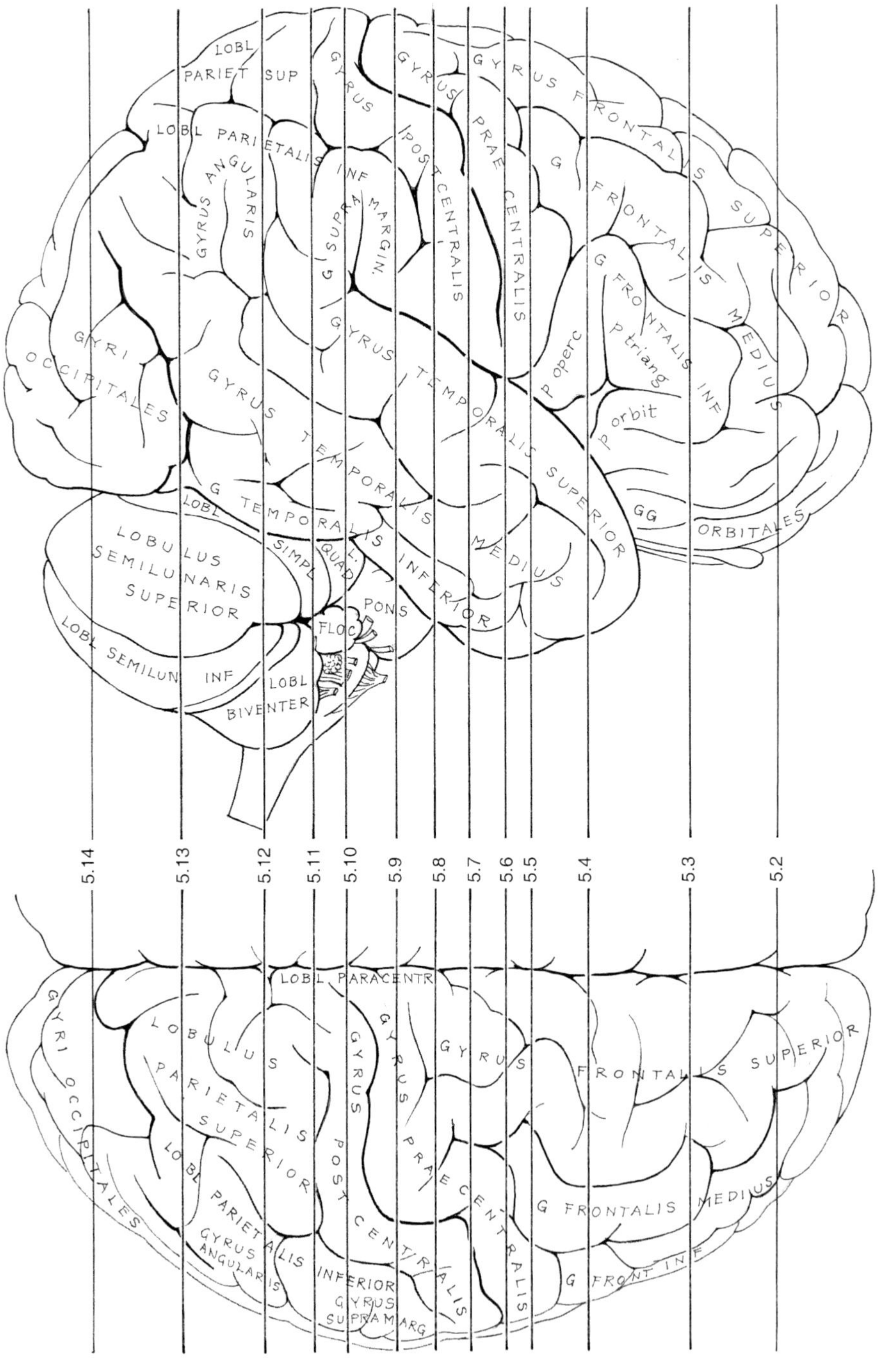

A

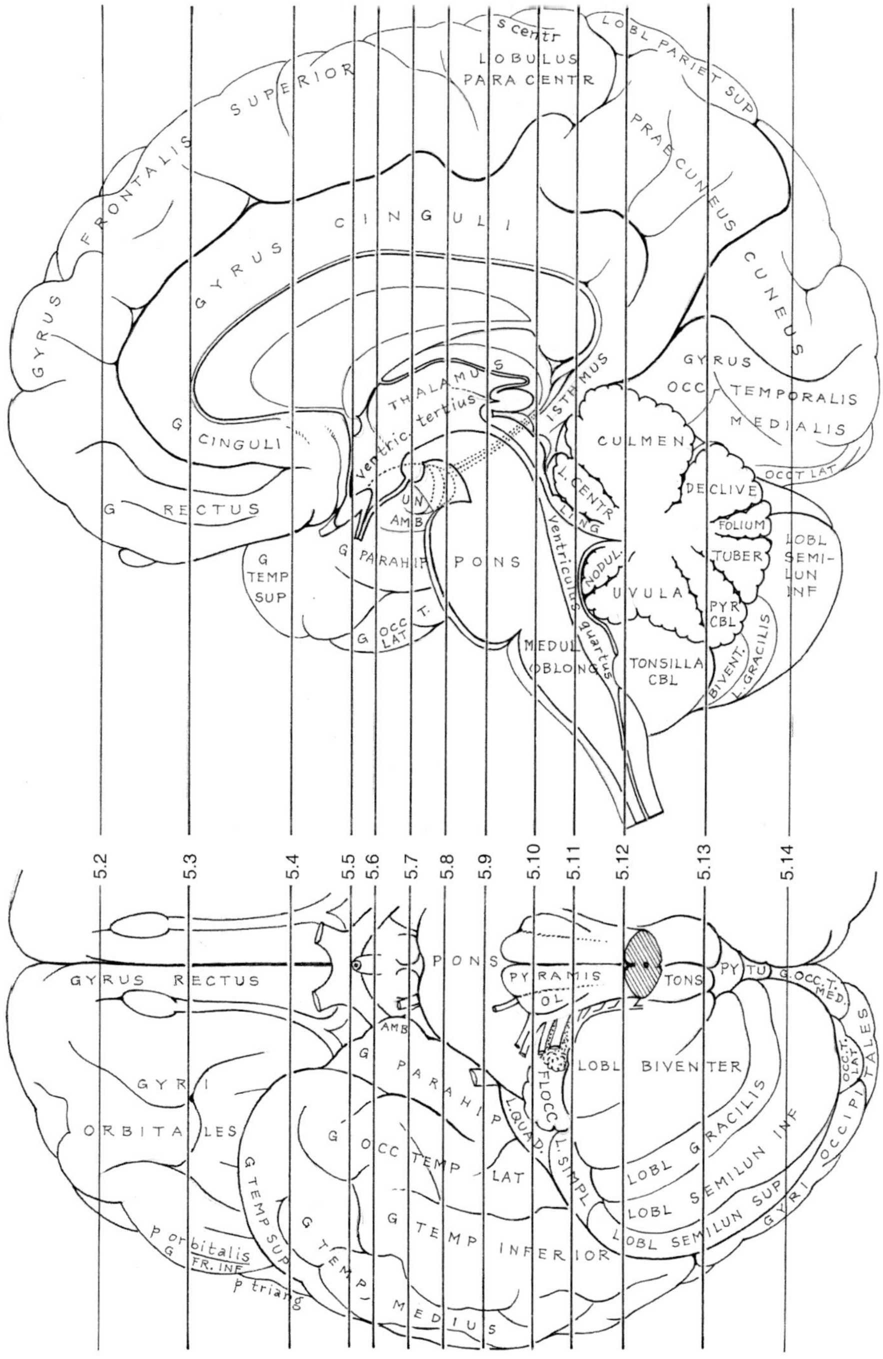

Fig. 5.1 A, B. Key diagrams showing the level and plane of the coronal sections in Figs. 5.2–5.14 (2/3×)

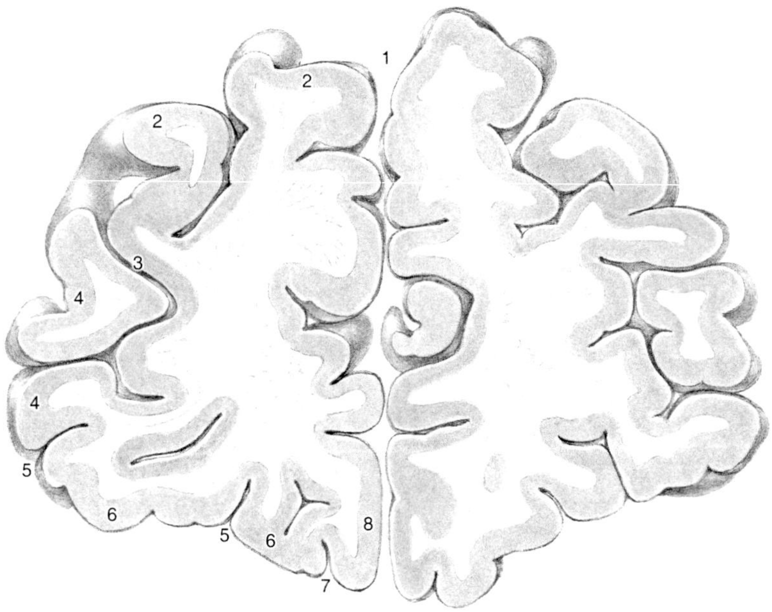

1 Longitudinal cerebral fissure
2 Superior frontal gyrus
3 Superior frontal sulcus
4 Middle frontal gyrus
5 Orbital sulci
6 Orbital gyri
7 Olfactory sulcus
8 Straight gyrus

Fig. 5.2. Section through the anterior part of the cerebral hemispheres (6/5×)

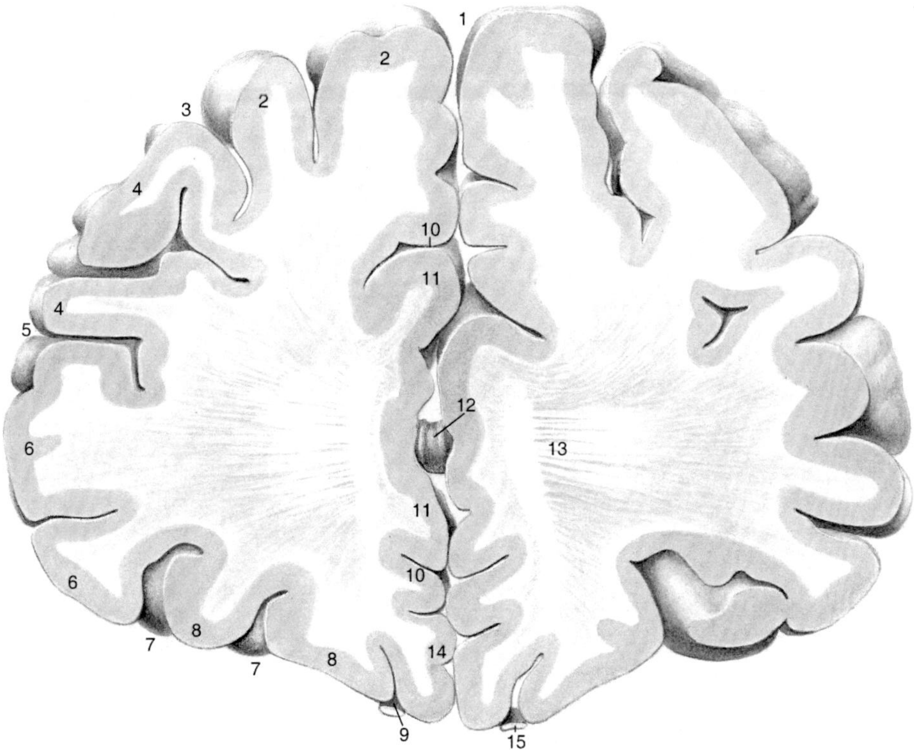

1 Longitudinal cerebral fissure
2 Superior frontal gyrus
3 Superior frontal sulcus
4 Middle frontal gyrus
5 Inferior frontal sulcus
6 Inferior frontal gyrus
7 Orbital sulci
8 Orbital gyri
9 Olfactory sulcus
10 Cingulate sulcus
11 Cingulate gyrus
12 Genu of corpus callosum
13 Radiation of corpus callosum
14 Straight gyrus
15 Olfactory tract

Fig. 5.3. Section through the anterior part of the cerebral hemispheres (6/5×)

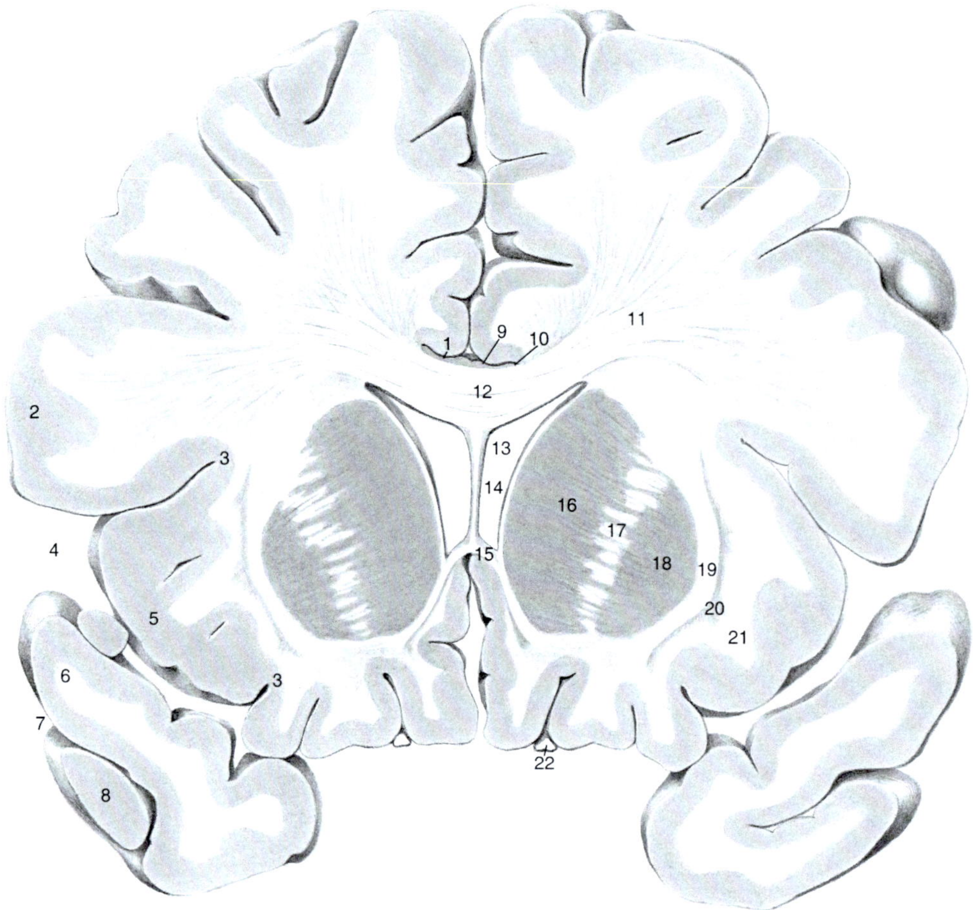

1 Sulcus of corpus callosum
2 Inferior frontal gyrus
3 Circular sulcus of insula
4 Lateral sulcus
5 Short gyri of insula
6 Superior temporal gyrus
7 Superior temporal sulcus
8 Middle temporal gyrus
9 Medial longitudinal stria
10 Lateral longitudinal stria
11 Radiation of corpus callosum
12 Truncus of corpus callosum
13 Septum pallucidum
14 Lateral ventricle, anterior horn
15 Rostrum of corpus callosum
16 Head of caudate nucleus
17 Anterior limb of internal capsule
18 Putamen
19 External capsule
20 Claustrum
21 Extreme capsule
22 Olfactory tract

Fig. 5.4. Section through the head of the caudate nucleus and the putamen (6/5×)

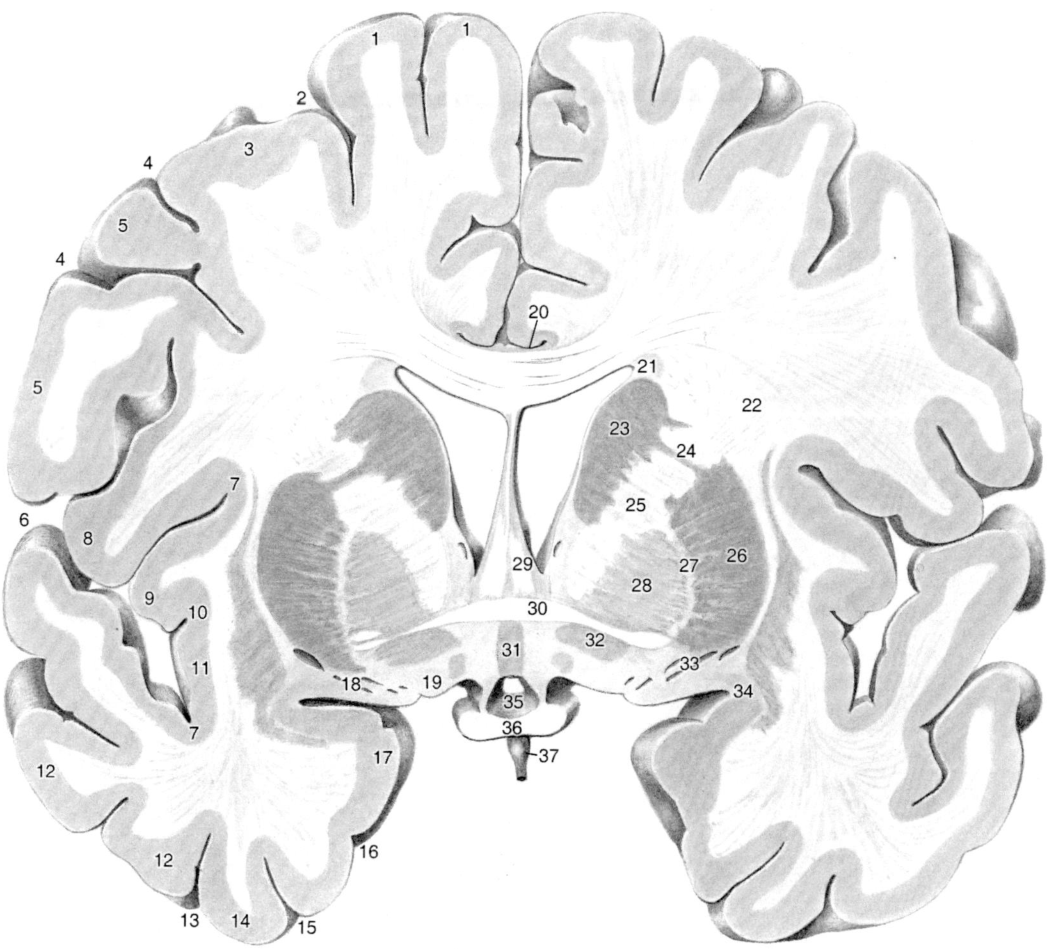

1 Superior frontal gyrus
2 Superior frontal sulcus
3 Middle frontal gyrus
4 Precentral sulcus
5 Precentral gyrus
6 Lateral sulcus, posterior branch
7 Circular sulcus of insula
8 Inferior frontal sulcus, opercular part
9 Short gyri of insula
10 Central sulcus of insula
11 Long gyrus of insula
12 Middle temporal gyrus
13 Inferior temporal sulcus
14 Lateral occipitotemporal gyrus
15 Collateral sulcus
16 Rhinal sulcus
17 Parahippocampal gyrus
18 Anterior perforated substance
19 Diagonal gyrus
20 Indusium griseum
21 Subependymal layer
22 Corona radiata
23 Head of caudate nucleus
24 Caudatolenticular grey bridges
25 Internal capsule, anterior limb
26 Putamen
27 External medullary lamina of globus pallidus
28 Globus pallidus
29 Column of fornix
30 Anterior commissure
31 Lamina terminalis
32 Substantia innominata
33 Middle cerebral artery, striate branches
34 Lateral olfactory stria
35 Supraoptic recess
36 Optic chiasm
37 Infundibulum

Fig. 5.5. Section through the anterior commissure and the optic chiasm (6/5×)

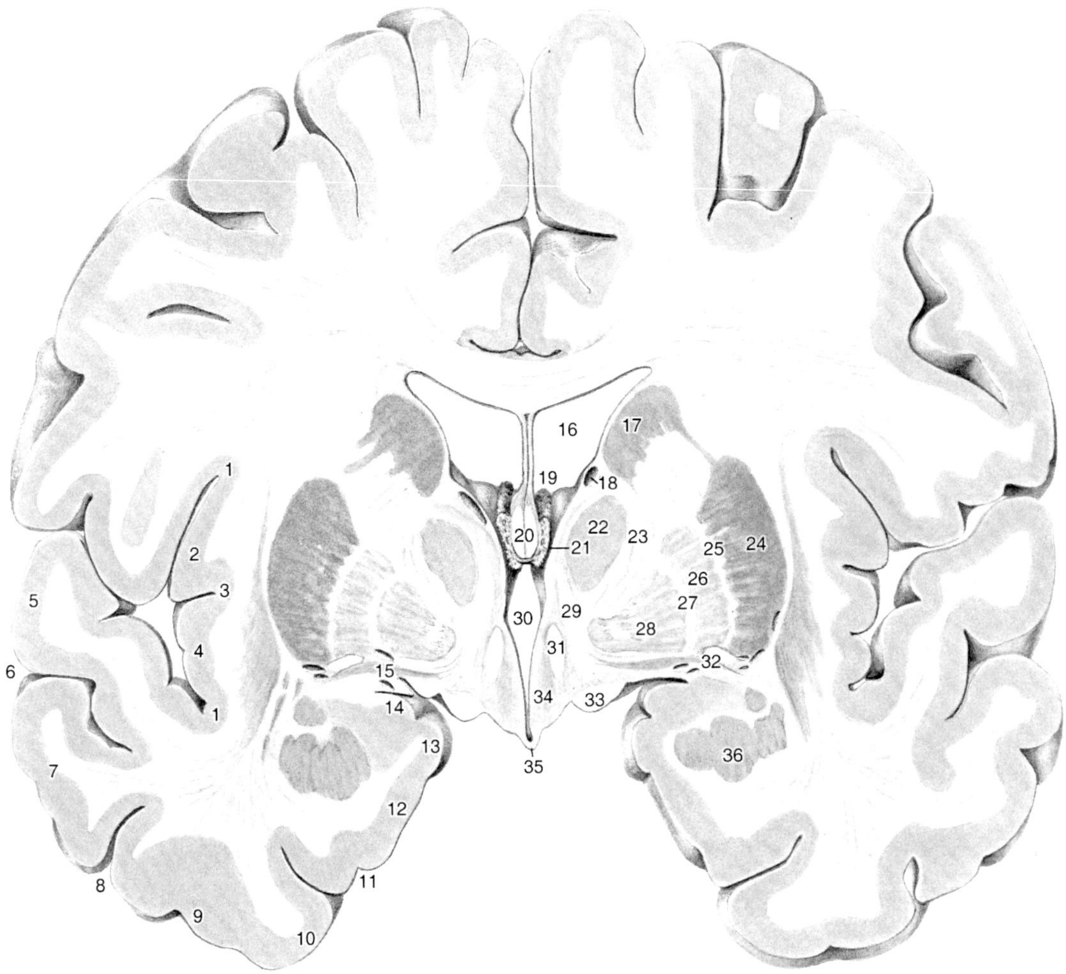

1 Circular sulcus of insula
2 Short gyri of insula
3 Central sulcus of insula
4 Long gyrus of insula
5 Superior temporal gyrus
6 Superior temporal sulcus
7 Middle temporal gyrus
8 Inferior temporal sulcus
9 Inferior temporal gyrus
10 Lateral occipitotemporal gyrus
11 Rhinal sulcus
12 Parahippocampal gyrus
13 Ambient gyrus
14 Semilunar gyrus
15 Middle cerebral artery, striatal branches
16 Lateral ventricle, central part
17 Body of caudate nucleus
18 Thalamostriatal vein
19 Choroid plexus of lateral ventricle
20 Body of fornix
21 Interventricular foramen
22 Anterior thalamic nucleus
23 Genu of internal capsule
24 Putamen
25 External medullary lamina of globus pallidus
26 Globus pallidus, lateral or external part
27 Internal medullary lamina of globus pallidus
28 Globus pallidus, medial or internal part
29 Inferior thalamic peduncle
30 Third ventricle
31 Column of fornix
32 Anterior commissure
33 Optic tract
34 Hypothalamus
35 Infundibulum
36 Amygdaloid body

Fig. 5.6. Section through the interventricular foramen, the infundibulum and the amygdaloid body (6/5×)

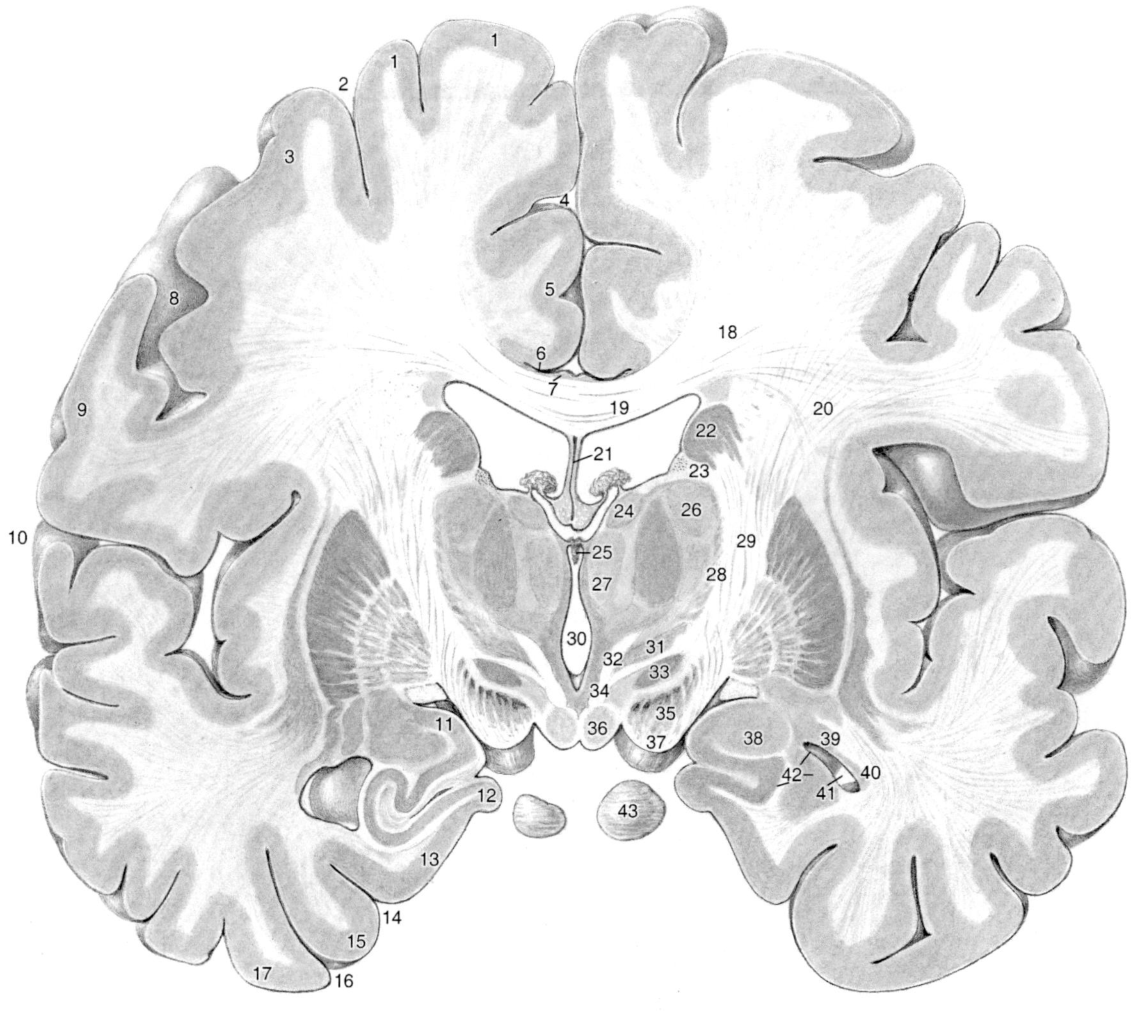

1 Superior frontal gyrus
2 Precentral sulcus
3 Precentral gyrus
4 Cingulate sulcus
5 Cingulate gyrus
6 Sulcus of corpus callosum
7 Indusium griseum
8 Central sulcus
9 Postcentral gyrus
10 Lateral sulcus, posterior branch
11 Uncinate gyrus
12 Ambient gyrus
13 Parahippocampal gyrus
14 Collateral sulcus
15 Lateral occipitotemporal gyrus
16 Occipitotemporal sulcus
17 Inferior temporal gyrus
18 Radiation of corpus callosum
19 Body of corpus callosum
20 Corona radiata
21 Septum pellucidum
22 Body of caudate nucleus
23 Stria terminalis
24 Anterior thalamic nucleus
25 Choroid plexus of third ventricle
26 Ventral lateral thalamic nucleus
27 Mediodorsal thalamic nucleus
28 Reticular thalamic nucleus
29 Internal capsule, posterior limb
30 Third ventricle
31 Zona incerta
32 Mamillothalamic tract
33 Subthalamic nucleus
34 Principal mamillary fasciculus
35 Substantia nigra
36 Mamillary body
37 Cerebral peduncle
38 Amygdaloid body
39 Stria terminalis
40 Tail of caudate nucleus
41 Lateral ventricle, inferior horn
42 Pes of hippocampus
43 Pons

Fig. 5.7. Section through the anterior end of the hippocampus, the mamillary body and the mamillothalamic tract (6/5×)

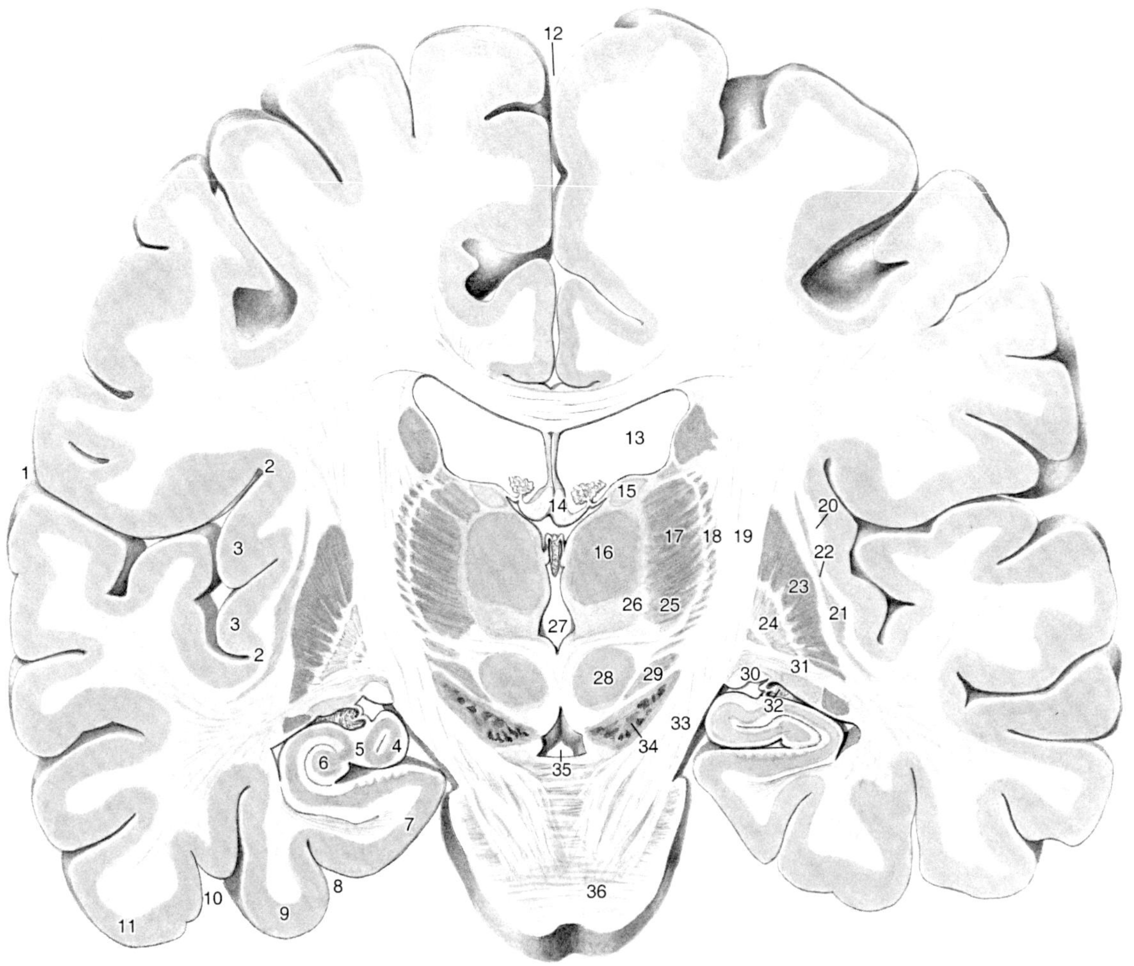

1 Lateral sulcus, posterior branch
2 Circular sulcus of insula
3 Long gyrus of insula
4 Intralimbic gyrus
5 Hippocampal sulcus
6 Dentate gyrus
7 Parahippocampal gyrus
8 Collateral sulcus
9 Lateral occipitotemporal gyrus
10 Occipitotemporal sulcus
11 Inferior temporal gyrus
12 Longitudinal cerebral fissure
13 Lateral ventricle, central part
14 Body of fornix
15 Lateral dorsal thalamic nucleus
16 Mediodorsal thalamic nucleus
17 Ventral lateral thalamic nucleus
18 Reticular thalamic nucleus
19 Internal capsule, posterior limb
20 Extreme capsule
21 Claustrum
22 External capsule
23 Putamen
24 Globus pallidus
25 Ventral posterolateral thalamic nucleus
26 Centromedian thalamic nucleus
27 Third ventricle
28 Red nucleus
29 Subthalamic nucleus
30 Optic tract
31 Internal capsule, sublentiform part
32 Choroid plexus of lateral ventricle
33 Cerebral peduncle
34 Substantia nigra
35 Interpeduncular fossa
36 Pons

Fig. 5.8. Section through the thalamus, the cerebral peduncle and the pons (6/5×)

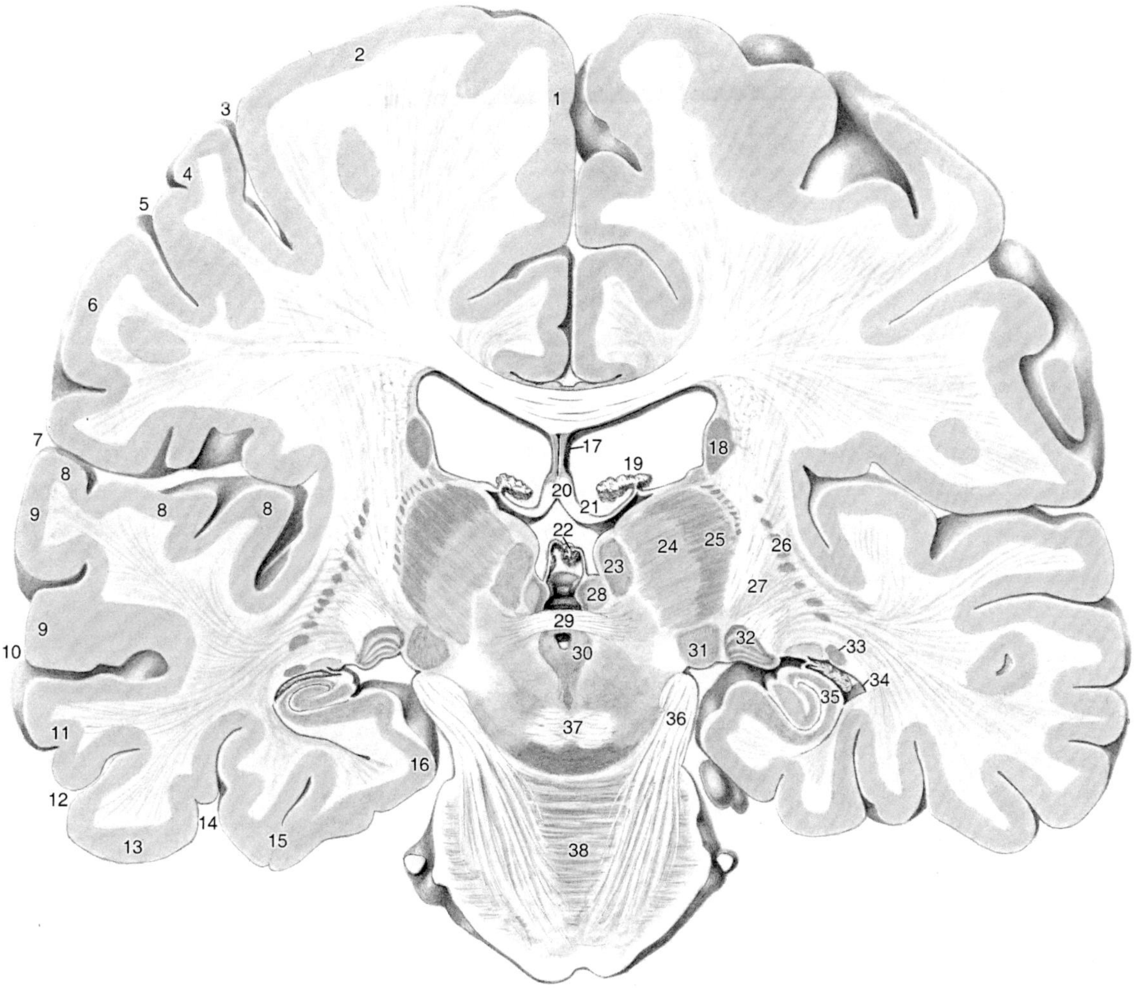

1 Paracentral lobule
2 Precentral gyrus
3 Central sulcus
4 Postcentral gyrus
5 Postcentral sulcus
6 Inferior parietal lobule
7 Temporal plane
8 Transverse temporal gyrus (Heschl)
9 Superior temporal gyrus
10 Superior temporal sulcus
11 Middle temporal gyrus
12 Inferior temporal sulcus
13 Inferior temporal gyrus
14 Occipitotemporal sulcus
15 Lateral occipitotemporal gyrus
16 Parahippocampal gyrus
17 Septum pellucidum
18 Tail of caudate nucleus
19 Choroid plexus of lateral ventricle
20 Body of fornix
21 Crus of fornix
22 Choroid plexus of third ventricle
23 Mediodorsal thalamic nucleus
24 Pulvinar thalamic nuclei
25 Lateral posterior thalamic nucleus
26 Caudolenticular grey bridges
27 Internal capsule, retrolentiform part
28 Habenular nuclei
29 Posterior commissure
30 Cerebral aqueduct
31 Medial geniculate body
32 Lateral geniculate body
33 Tail of caudate nucleus
34 Lateral ventricle, inferior horn
35 Hippocampus
36 Cerebral peduncle
37 Decussation of superior cerebellar peduncles
38 Pons

Fig. 5.9. Section through the posterior part of the thalamus (6/5×)

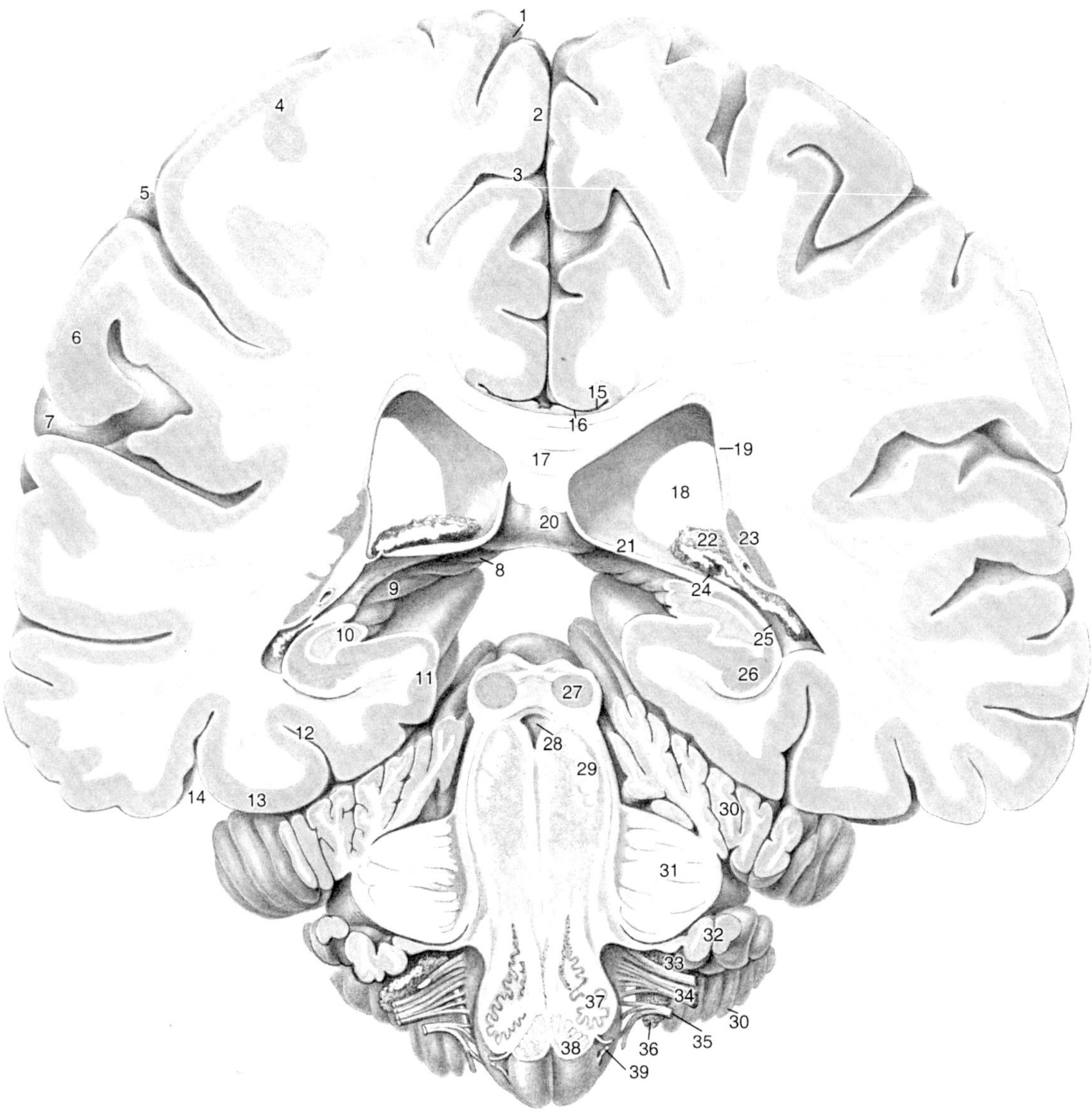

1 Central sulcus
2 Paracentral lobule
3 Cingulate sulcus
4 Postcentral gyrus
5 Postcentral sulcus
6 Inferior parietal lobule
7 Lateral sulcus, posterior branch
8 Fasciolar gyrus
9 Fasciolar cinerea
10 Dentate gyrus
11 Parahippocampal gyrus
12 Collateral sulcus
13 Lateral occipitotemporal gyrus
14 Occipitotemporal sulcus
15 Sulcus of corpus callosum
16 Indusium griseum
17 Body of corpus callosum
18 Lateral ventricle
19 Subependymal layer
20 Splenium of corpus callosum
21 Crus of fornix
22 Choroid plexus of lateral ventricle
23 Tail of caudate nucleus
24 Fimbria of hippocampus
25 Alveus of hippocampus
26 Hippocampus
27 Inferior colliculus
28 Cerebral aqueduct
29 Superior cerebellar peduncle
30 Cerebellar hemisphere
31 Middle cerebellar peduncle
32 Flocculus
33 Glossopharyngeal nerve (IX)
34 Vagus nerve (X)
35 Accessory nerve (XI)
36 Choroid plexus of fourth ventricle
37 Olive
38 Pyramid
39 Hypoglossal nerve (XII)

Fig. 5.10. Section through the inferior colliculus and the inferior olive. The thickness of this slice is three times the standard of 2 mm (6/5×)

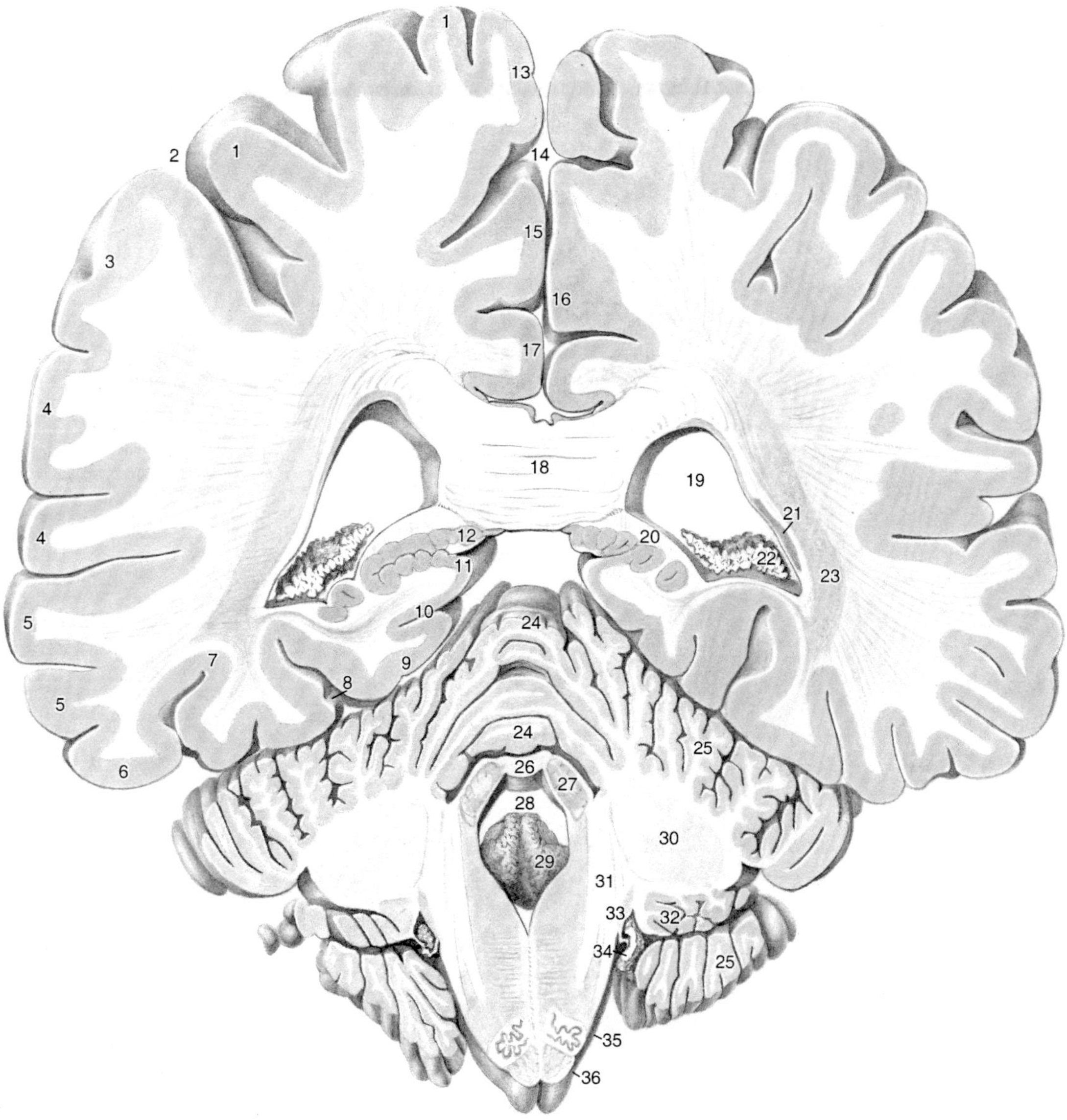

1 Superior parietal lobule
2 Intraparietal sulcus
3 Inferior parietal lobule
4 Superior temporal gyrus
5 Middle temporal gyrus
6 Inferior temporal gyrus
7 Lateral occipitotemporal gyrus
8 Collateral sulcus
9 Medial occipitotemporal gyrus
10 Calcarine sulcus
11 Isthmus of cingulate gyrus
12 Fasciolar gyrus
13 Paracerebral lobule
14 Cingulate sulcus
15 Precuneus
16 Subparietal sulcus
17 Cingulate gyrus
18 Splenium of corpus callosum
19 Lateral ventricle
20 Fimbria of hippocampus
21 Tail of caudate nucleus
22 Choroid plexus of lateral ventricle
23 Optic radiation
24 Vermis of cerebellum
25 Cerebellar hemisphere
26 Superior medullary velum
27 Superior cerebellar peduncle
28 Fourth ventricle
29 Choroid plexus of fourth ventricle
30 Middle cerebellar peduncle
31 Inferior cerebellar peduncle
32 Flocculus
33 Lateral recess of fourth ventricle
34 Choroid plexus of fourth ventricle
35 Olive
36 Pyramid

Fig. 5.11. Section through the splenium of the corpus callosum and the fourth ventricle (6/5×)

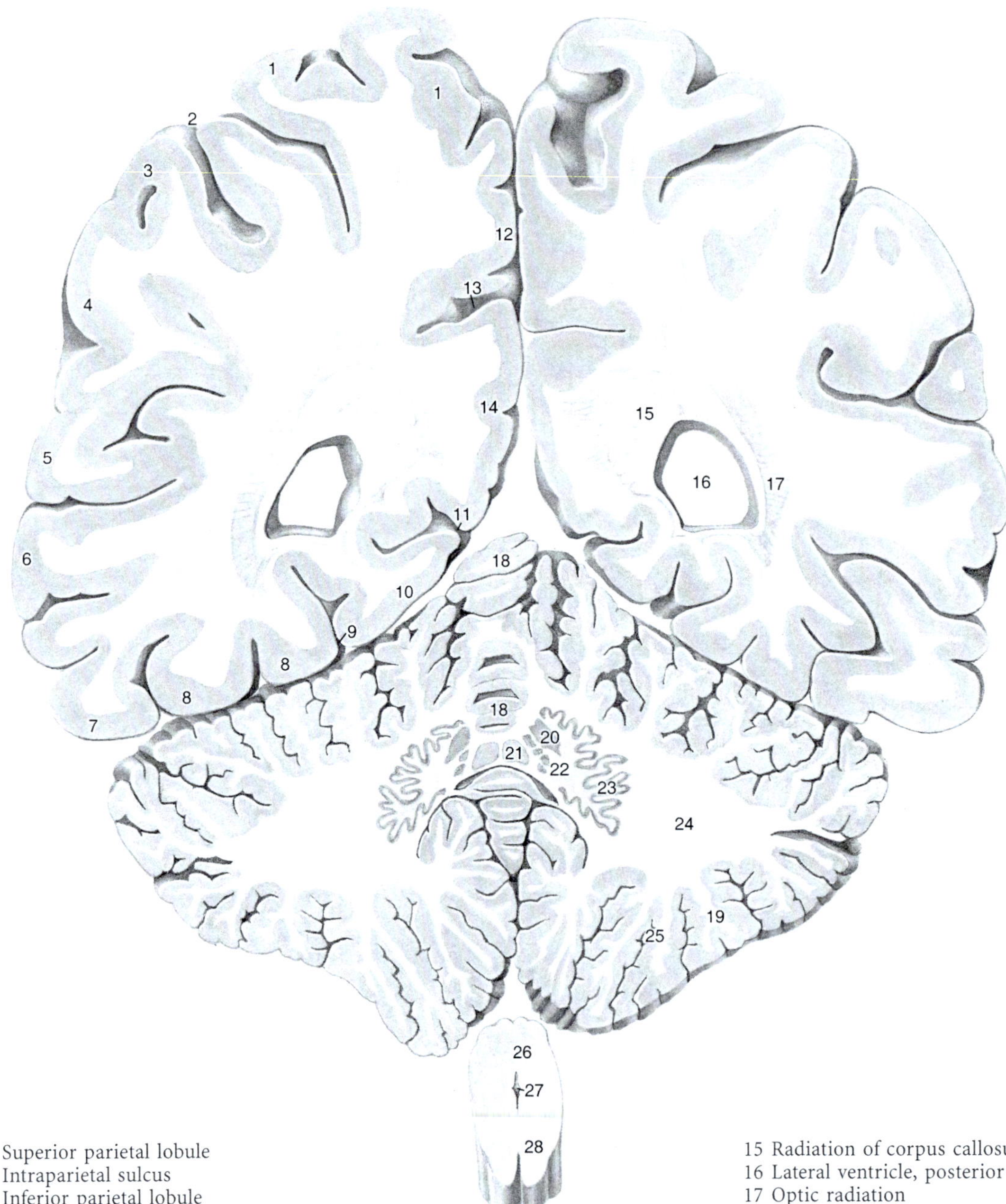

1 Superior parietal lobule
2 Intraparietal sulcus
3 Inferior parietal lobule
4 Angular gyrus
5 Superior temporal gyrus
6 Middle temporal gyrus
7 Inferior temporal gyrus
8 Lateral occipitotemporal gyrus
9 Collateral sulcus
10 Medial occipitotemporal gyrus
11 Calcarine sulcus
12 Precuneus
13 Subparietal sulcus
14 Cingulate gyrus
15 Radiation of corpus callosum
16 Lateral ventricle, posterior horn
17 Optic radiation
18 Vermis of cerebellum
19 Cerebellar hemisphere
20 Emboliform nucleus
21 Fastigial nucleus
22 Globose nucleus
23 Dentate nucleus
24 Corpus medullare of cerebellum
25 Laminae albae of cerebellum
26 Medulla oblongata
27 Central canal
28 Spinal cord

Fig. 5.12. Section through the posterior horns of the lateral ventricles and the central cerebellar nuclei (6/5×)

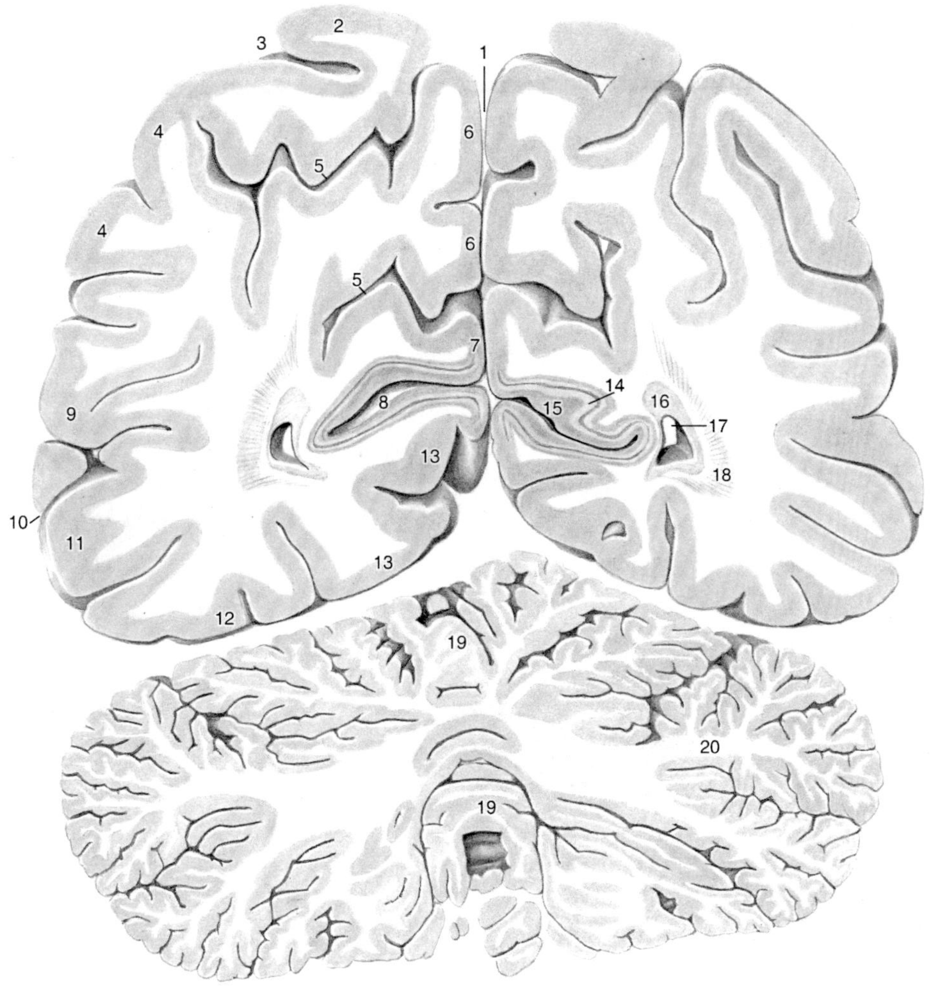

1 Longitudinal cerebral fissure
2 Superior parietal lobule
3 Intraparietal sulcus
4 Inferior parietal lobule
5 Parieto-occipital sulcus
6 Precuneus
7 Cuneus
8 Calcarine sulcus
9 Middle temporal gyrus
10 Anterior occipital sulcus
11 Occipital gyri
12 Lateral occipitotemporal gyrus
13 Medial occipitotemporal gyrus
14 Line of Gennari
15 Striate area
16 Radiation of corpus callosum
17 Lateral ventricle, posterior horn
18 Optic radiation
19 Vermis of cerebellum
20 Cerebellar hemisphere

Fig. 5.13. Section through the deepest part of the calcarine sulcus (6/5×)

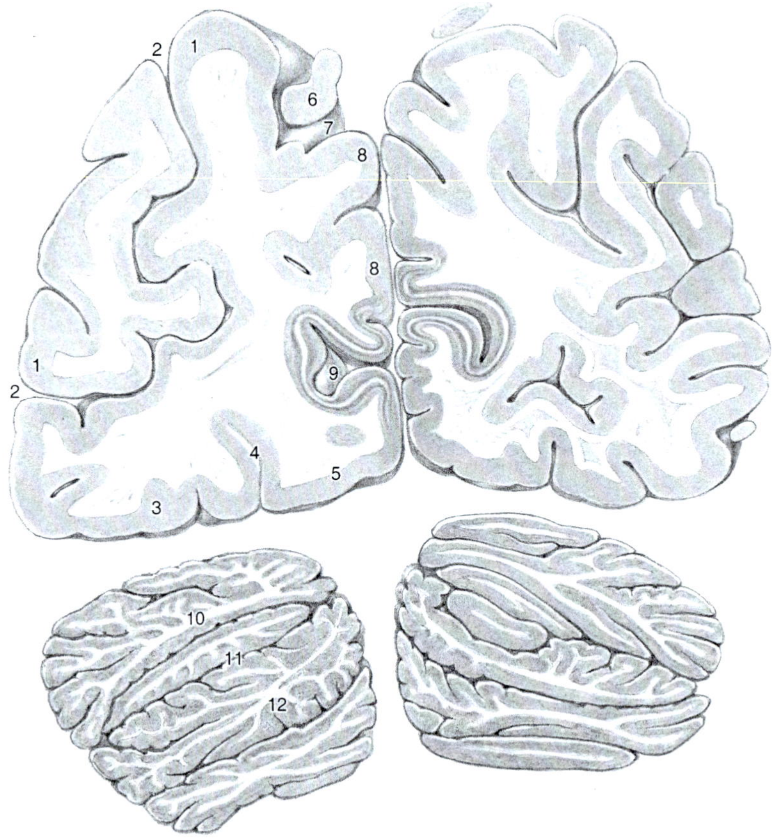

1 Occipital gyri
2 Occipital sulci
3 Lateral occipitotemporal gyrus
4 Collateral sulcus
5 Medial occipitotemporal gyrus
6 Precuneus
7 Parieto-occipital sulcus
8 Cuneus
9 Calcarine sulcus
10 Superior semilunar lobule
11 Horizontal fissure
12 Inferior semilunar lobule

Fig. 5.14. Section through the posterior part of the cerebral hemisphere (6/5×)

Sections Perpendicular to the Axis of the Brain Stem

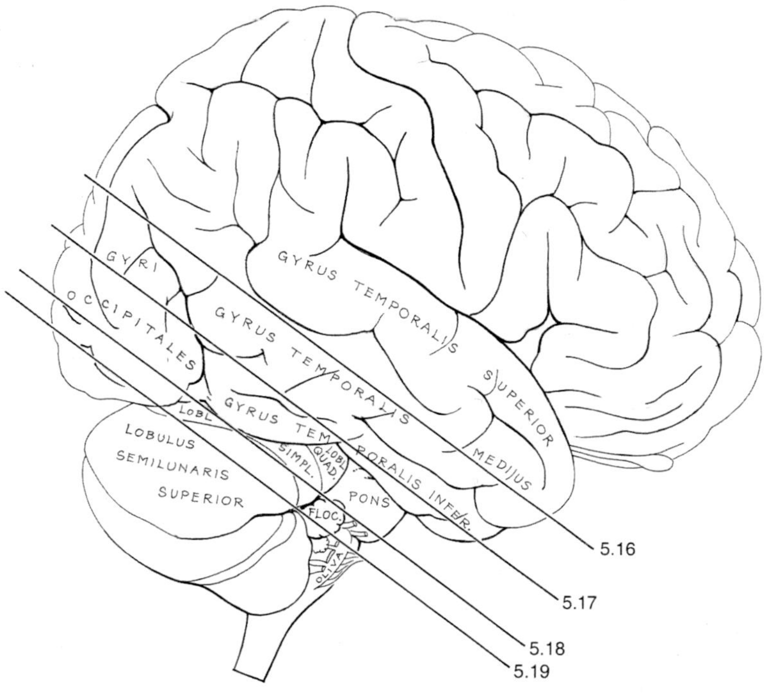

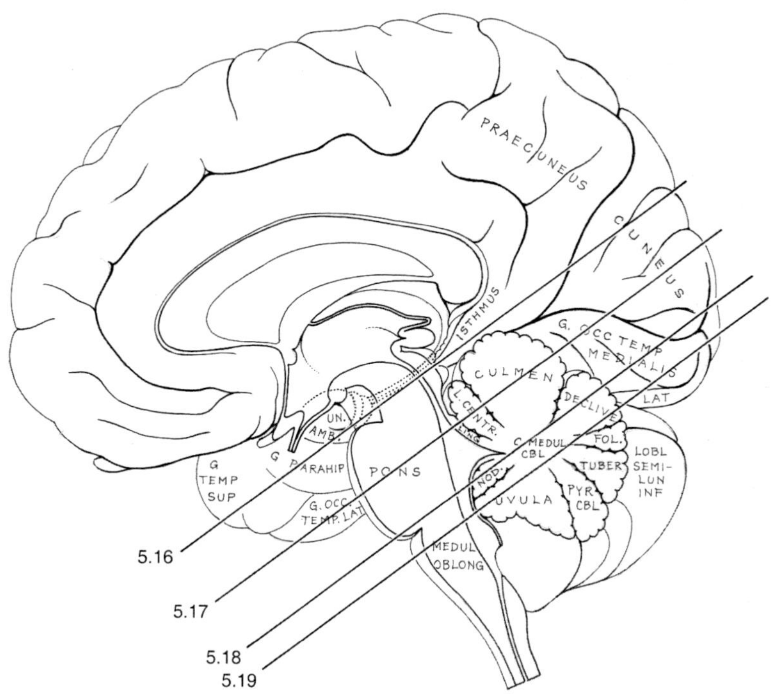

Fig. 5.15. Key diagrams showing the level and plane of the sections perpendicular to the axis of the brain stem in Figs. 5.16–5.19 (2/3×)

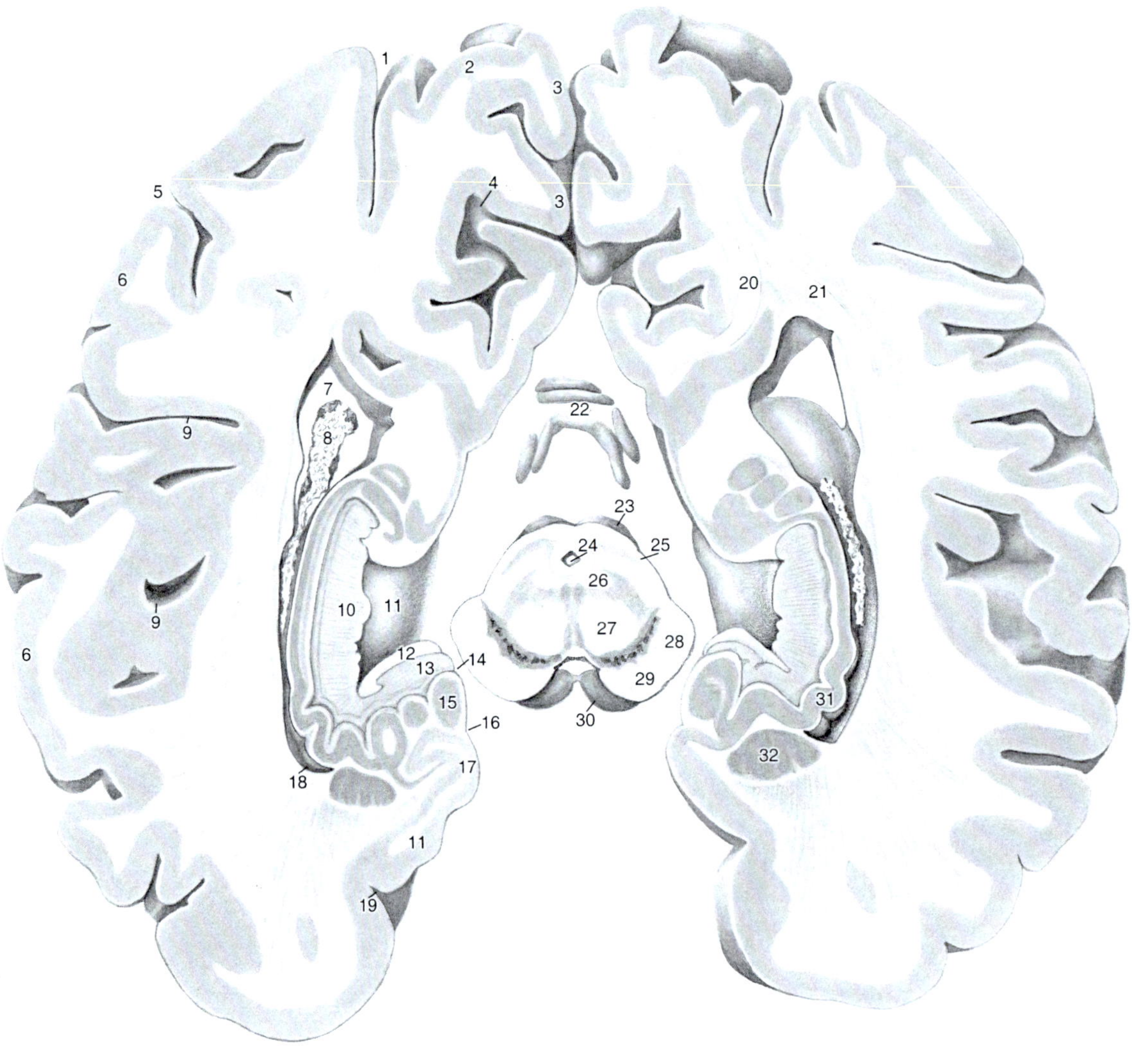

1 Lunate sulcus
2 Occipital gyri
3 Cuneus
4 Parieto-occipital sulcus
5 Anterior occipital sulcus
6 Middle temporal gyrus
7 Lateral ventricle, posterior horn
8 Choroid plexus of lateral ventricle
9 Superior temporal sulcus
10 Dentate gyrus
11 Parahippocampal gyrus
12 Intralimbic gyrus
13 Limbus Giacomini
14 Hippocampal sulcus
15 Uncinate gyrus
16 Incisure of uncus
17 Ambient gyrus
18 Lateral ventricle, inferior horn
19 Rhinal sulcus
20 Radiation of corpus callosum
21 Optic radiation
22 Culmen
23 Inferior colliculus
24 Cerebral aqueduct
25 Brachium of inferior colliculus
26 Tegmentum of midbrain
27 Superior cerebellar peduncle
28 Substantia nigra
29 Cerebral peduncle
30 Pons
31 Ammon's horn
32 Amygdaloid body

Fig. 5.16. Section through the middle of the midbrain (6/5×)

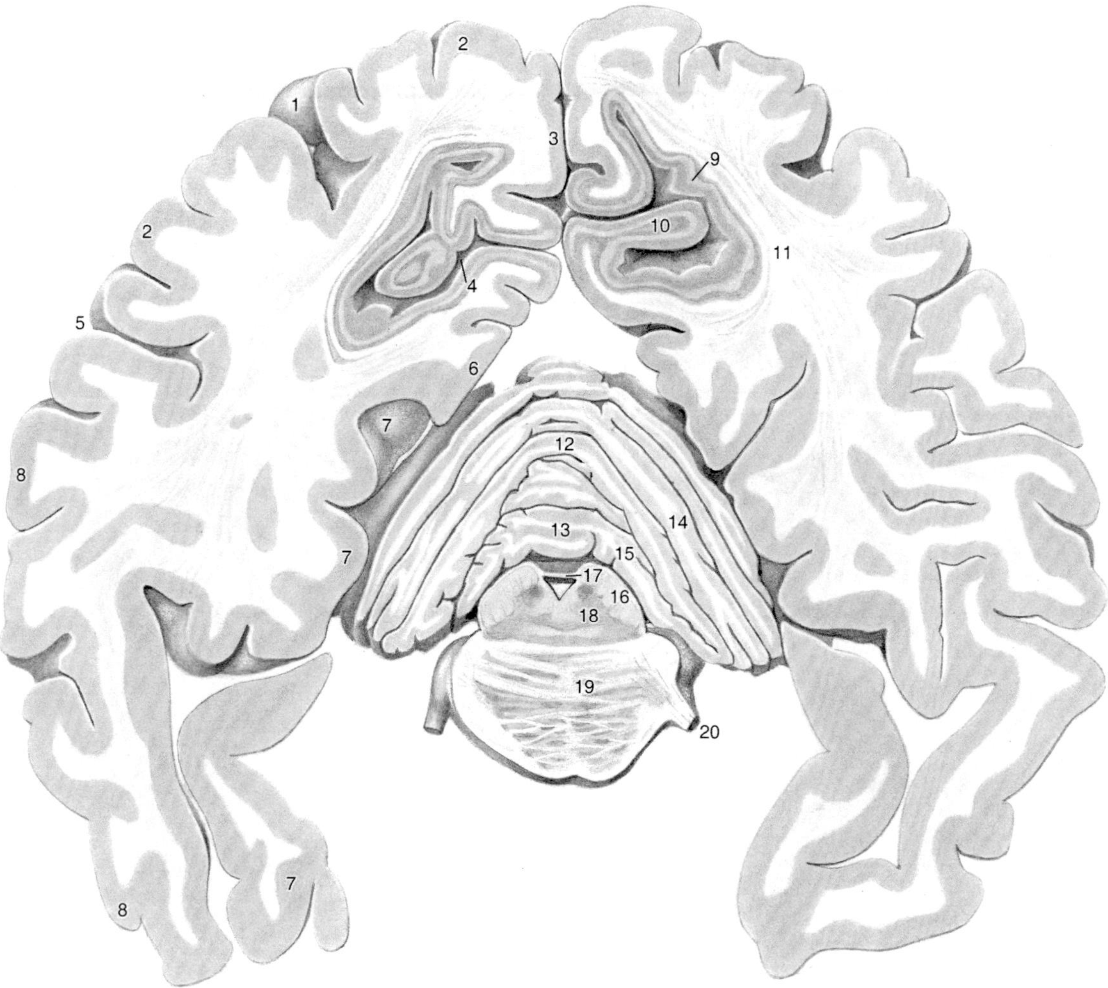

1 Lunate sulcus
2 Occipital gyri
3 Cuneus
4 Calcarine sulcus
5 Anterior occipital sulcus
6 Medial occipitotemporal gyrus
7 Lateral occipitotemporal gyrus
8 Inferior temporal gyrus
9 Line of Gennari
10 Striate area
11 Optic radiation
12 Culmen
13 Central lobule
14 Quadrangular lobule
15 Wing of central lobule
16 Superior cerebellar peduncle
17 Superior medullary velum
18 Pontine tegmentum
19 Pons
20 Trigeminal nerve (V)

Fig. 5.17. Section through the pons at the level of the entrance of the trigeminal nerve (6/5×)

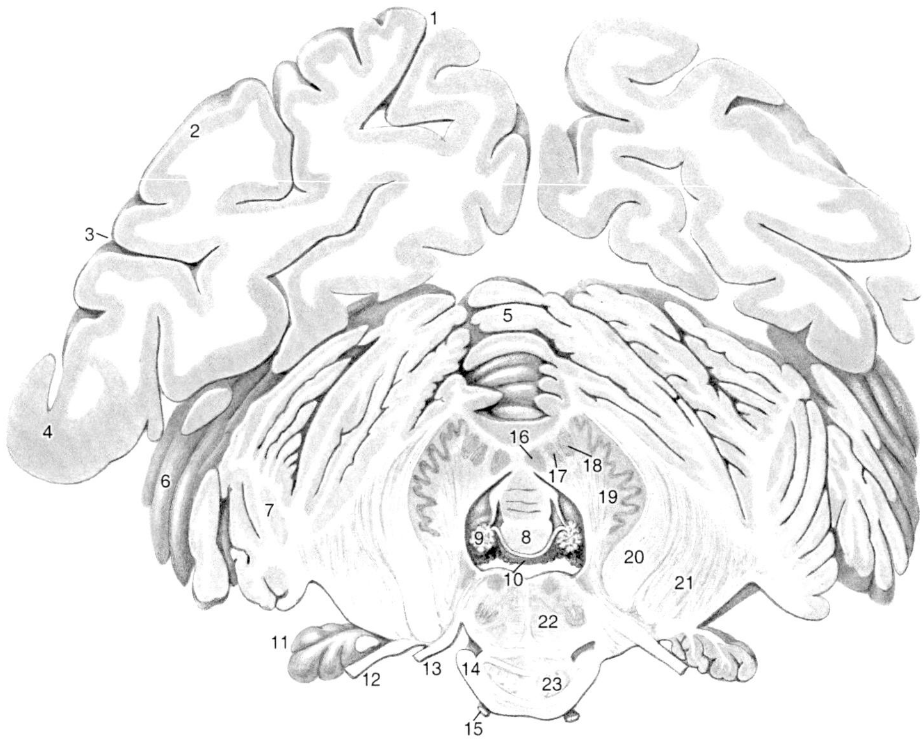

1 Calcarine sulcus
2 Occipital gyri
3 Anterior occipital sulcus
4 Inferior temporal gyrus
5 Declive
6 Superior semilunar lobule
7 Simple lobule
8 Nodule
9 Choroid plexus of fourth ventricle
10 Fourth ventricle
11 Flocculus
12 Vestibulocochlear nerve (VIII)
13 Facial nerve (VII)
14 Pons
15 Abducent nerve (VI)
16 Fastigial nucleus
17 Globose nucleus
18 Emboliform nucleus
19 Dentate nucleus
20 Inferior cerebellar peduncle
21 Middle cerebellar peduncle
22 Pontine tegmentum
23 Corticospinal tract

Fig. 5.18. Section through the transitional area of pons and medulla oblongata (6/5×)

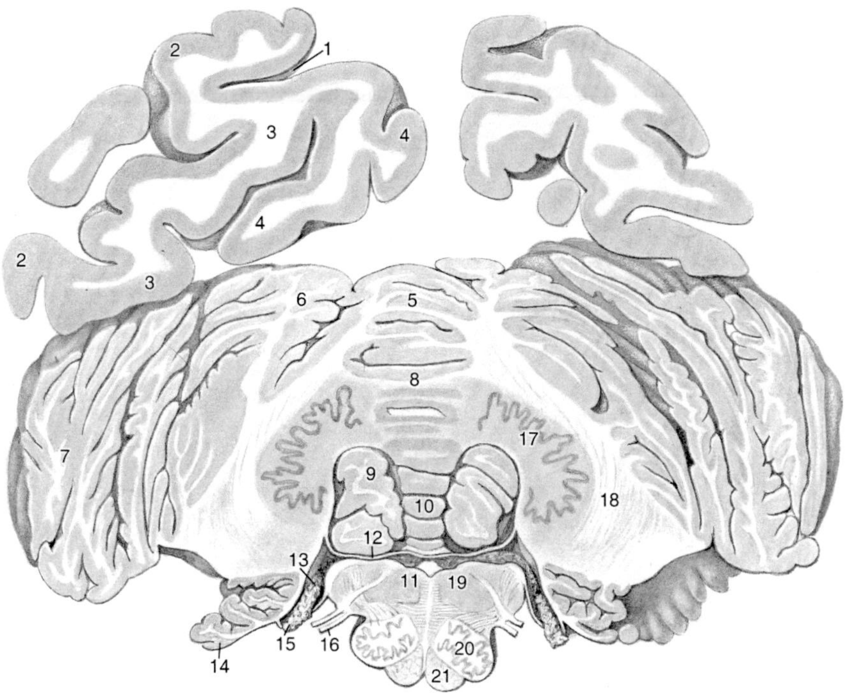

1 Calcarine sulcus
2 Occipital gyri
3 Lateral occipitotemporal gyrus
4 Medial occipitotemporal gyrus
5 Declive
6 Simple lobule
7 Superior semilunar lobule
8 Pyramid of vermis
9 Tonsil of cerebellum
10 Uvula of vermis
11 Fourth ventricle
12 Tela choroidea of fourth ventricle
13 Lateral recess of fourth ventricle
14 Flocculus
15 Choroid plexus of fourth ventricle
16 Glossopharyngeal nerve (IX)
17 Dentate nucleus
18 Middle cerebellar peduncle
19 Myelencephalic tegmentum
20 Inferior olivary nucleus
21 Corticospinal tract

Fig. 5.19. Section through the medulla oblongata (6/5×)

Sagittal Sections

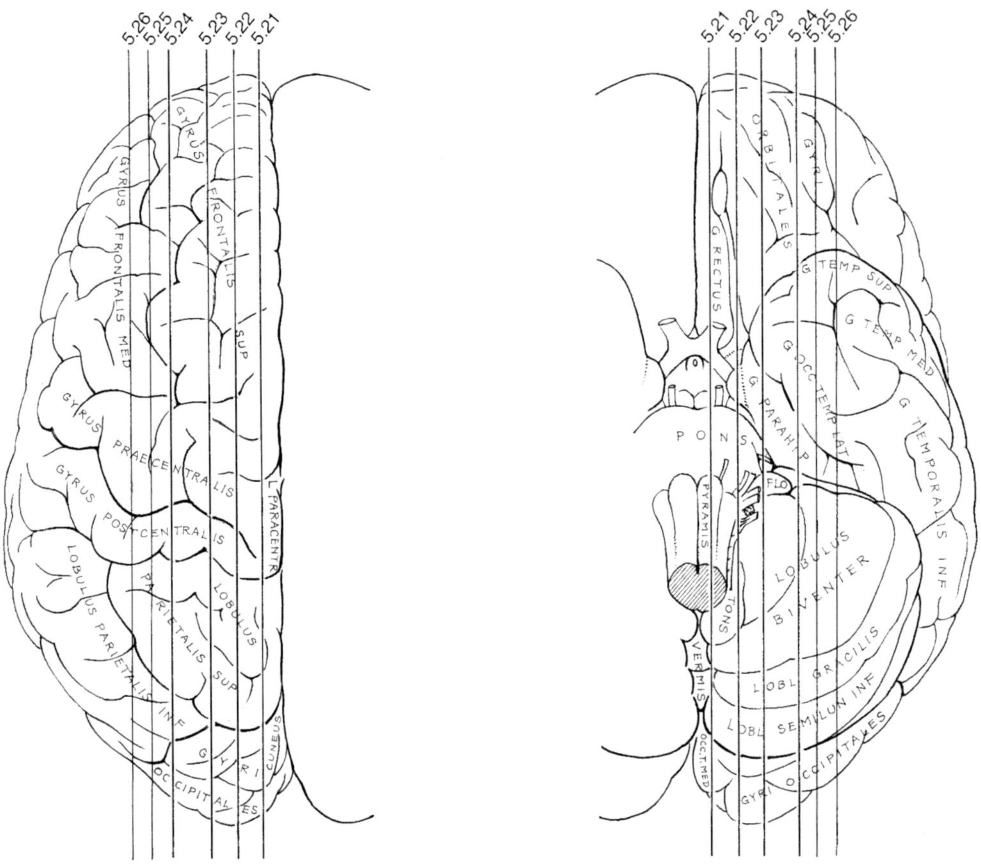

Fig. 5.20. Key diagrams showing the level and plane of the sagittal sections in Figs. 5.21–5.26

1 Postcentral sulcus
2 Parieto-occipital sulcus
3 Cuneus
4 Striatal area
5 Calcarine sulcus
6 Medial occipitotemporal gyrus

Fig. 5.21. Section through the mamillary body, the red nucleus and the fornix (6/5×)

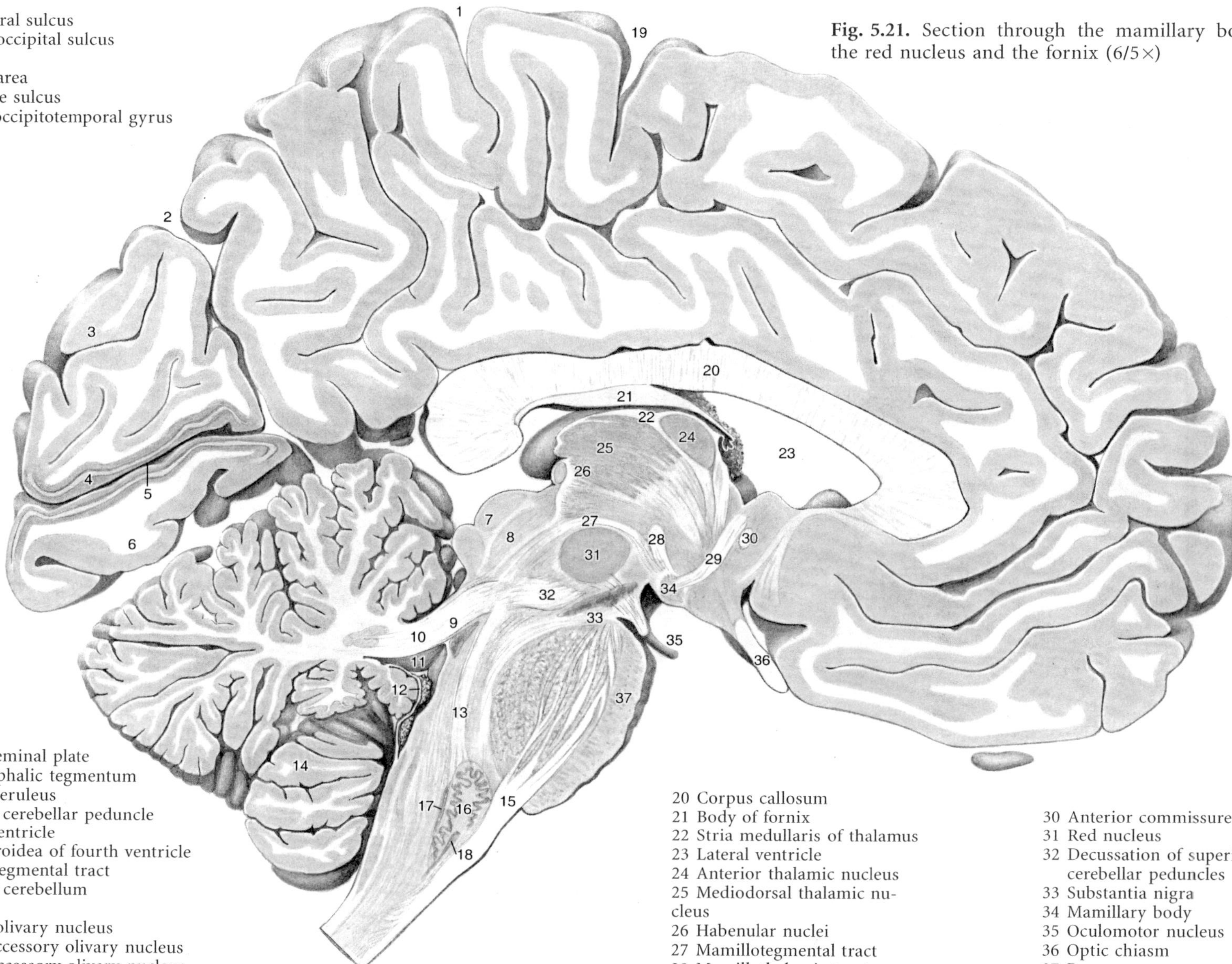

7 Quadrigeminal plate
8 Mesencephalic tegmentum
9 Locus coeruleus
10 Superior cerebellar peduncle
11 Fourth ventricle
12 Tela choroidea of fourth ventricle
13 Central tegmental tract
14 Tonsil of cerebellum
15 Pyramid
16 Inferior olivary nucleus
17 Dorsal accessory olivary nucleus
18 Medial accessory olivary nucleus
19 Paracentral sulcus

20 Corpus callosum
21 Body of fornix
22 Stria medullaris of thalamus
23 Lateral ventricle
24 Anterior thalamic nucleus
25 Mediodorsal thalamic nucleus
26 Habenular nuclei
27 Mamillotegmental tract
28 Mamillothalamic tract
29 Column of fornix

30 Anterior commissure
31 Red nucleus
32 Decussation of superior cerebellar peduncles
33 Substantia nigra
34 Mamillary body
35 Oculomotor nucleus
36 Optic chiasm
37 Pons

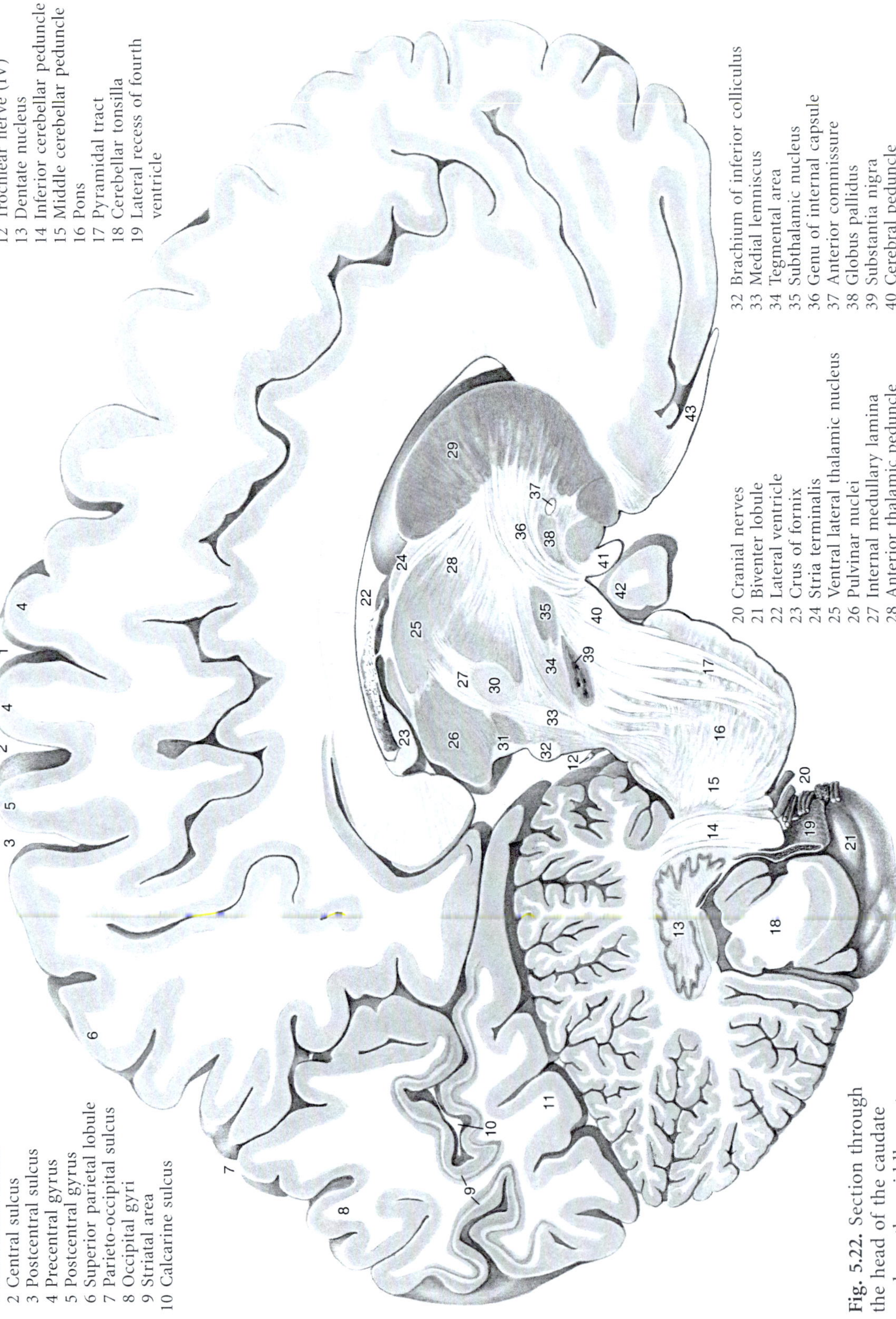

1 Paracentral sulcus
2 Central sulcus
3 Postcentral sulcus
4 Precentral gyrus
5 Postcentral gyrus
6 Superior parietal lobule
7 Parieto-occipital sulcus
8 Occipital gyri
9 Striatal area
10 Calcarine sulcus

11 Medial occipitotemporal gyrus
12 Trochlear nerve (IV)
13 Dentate nucleus
14 Inferior cerebellar peduncle
15 Middle cerebellar peduncle
16 Pons
17 Pyramidal tract
18 Cerebellar tonsilla
19 Lateral recess of fourth ventricle

20 Cranial nerves
21 Biventer lobule
22 Lateral ventricle
23 Crus of fornix
24 Stria terminalis
25 Ventral lateral thalamic nucleus
26 Pulvinar nuclei
27 Internal medullary lamina
28 Anterior thalamic peduncle
29 Head of caudate nucleus
30 Centromedian thalamic nucleus
31 Brachium of superior colliculus

32 Brachium of inferior colliculus
33 Medial lemniscus
34 Tegmental area
35 Subthalamic nucleus
36 Genu of internal capsule
37 Anterior commissure
38 Globus pallidus
39 Substantia nigra
40 Cerebral peduncle
41 Optic tract
42 Uncus
43 Olfactory tract

Fig. 5.22. Section through the head of the caudate nucleus, the middle part of the thalamus and the cerebral peduncle (6/5×)

1 Central sulcus
2 Parieto-occipital sulcus
3 Corpus callosum
4 Lateral ventricle
5 Crus of fornix
6 Fasciolar gyrus
7 Occipital gyri
8 Optic radiation
9 Calcarine sulcus
10 Striate area

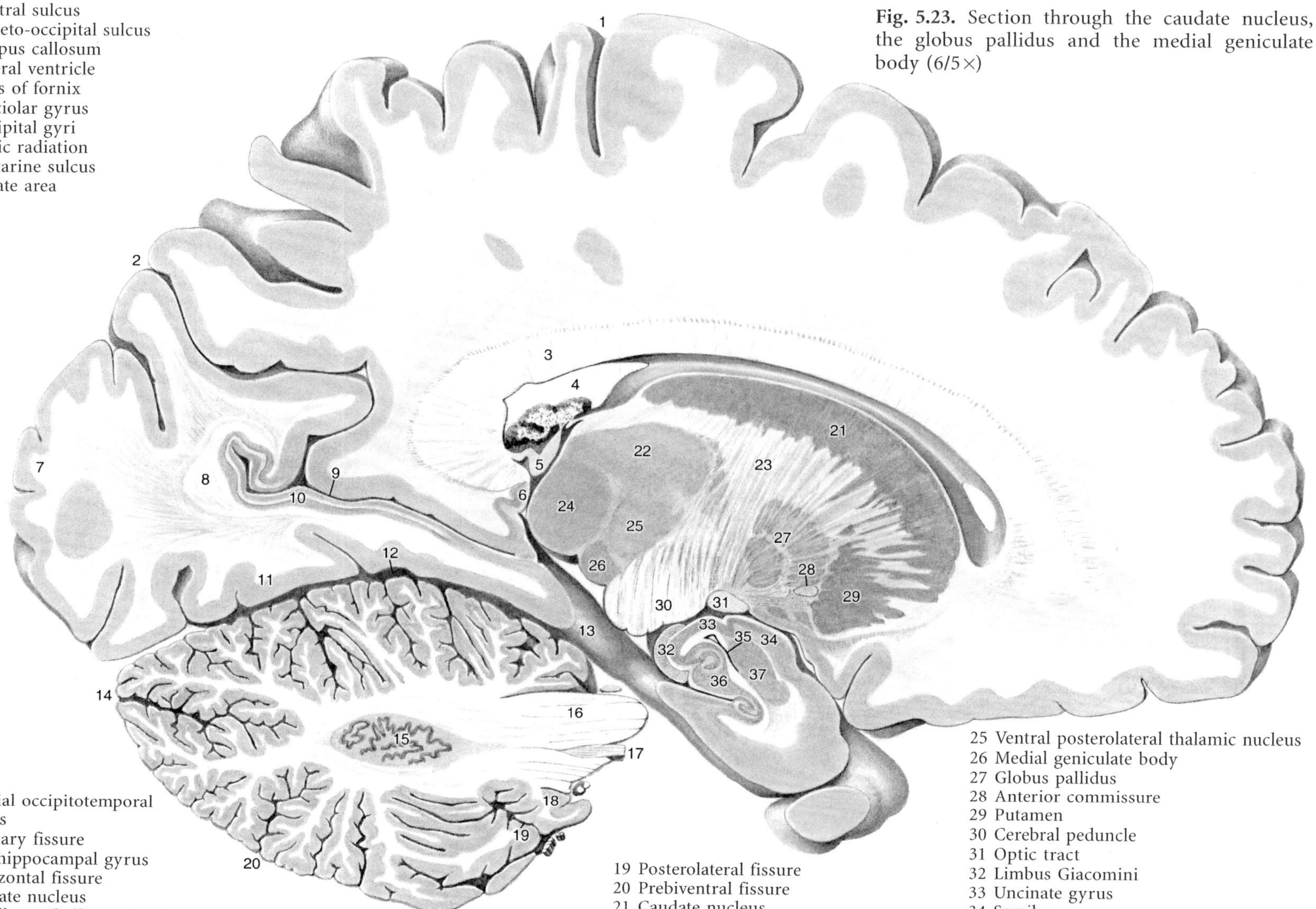

Fig. 5.23. Section through the caudate nucleus, the globus pallidus and the medial geniculate body (6/5×)

11 Medial occipitotemporal gyrus
12 Primary fissure
13 Parahippocampal gyrus
14 Horizontal fissure
15 Dentate nucleus
16 Middle cerebellar peduncle
17 Trigeminal nerve (V)
18 Flocculus

19 Posterolateral fissure
20 Prebiventral fissure
21 Caudate nucleus
22 Lateral posterior thalamic nucleus
23 Internal capsula
24 Pulvinar nuclei

25 Ventral posterolateral thalamic nucleus
26 Medial geniculate body
27 Globus pallidus
28 Anterior commissure
29 Putamen
30 Cerebral peduncle
31 Optic tract
32 Limbus Giacomini
33 Uncinate gyrus
34 Semilunar gyrus
35 Lateral ventricle, inferior horn
36 Ammon's horn
37 Amygdaloid body

Fig. 5.24. Section through the hippocampus, the putamen and the amygdaloid body (6/5×)

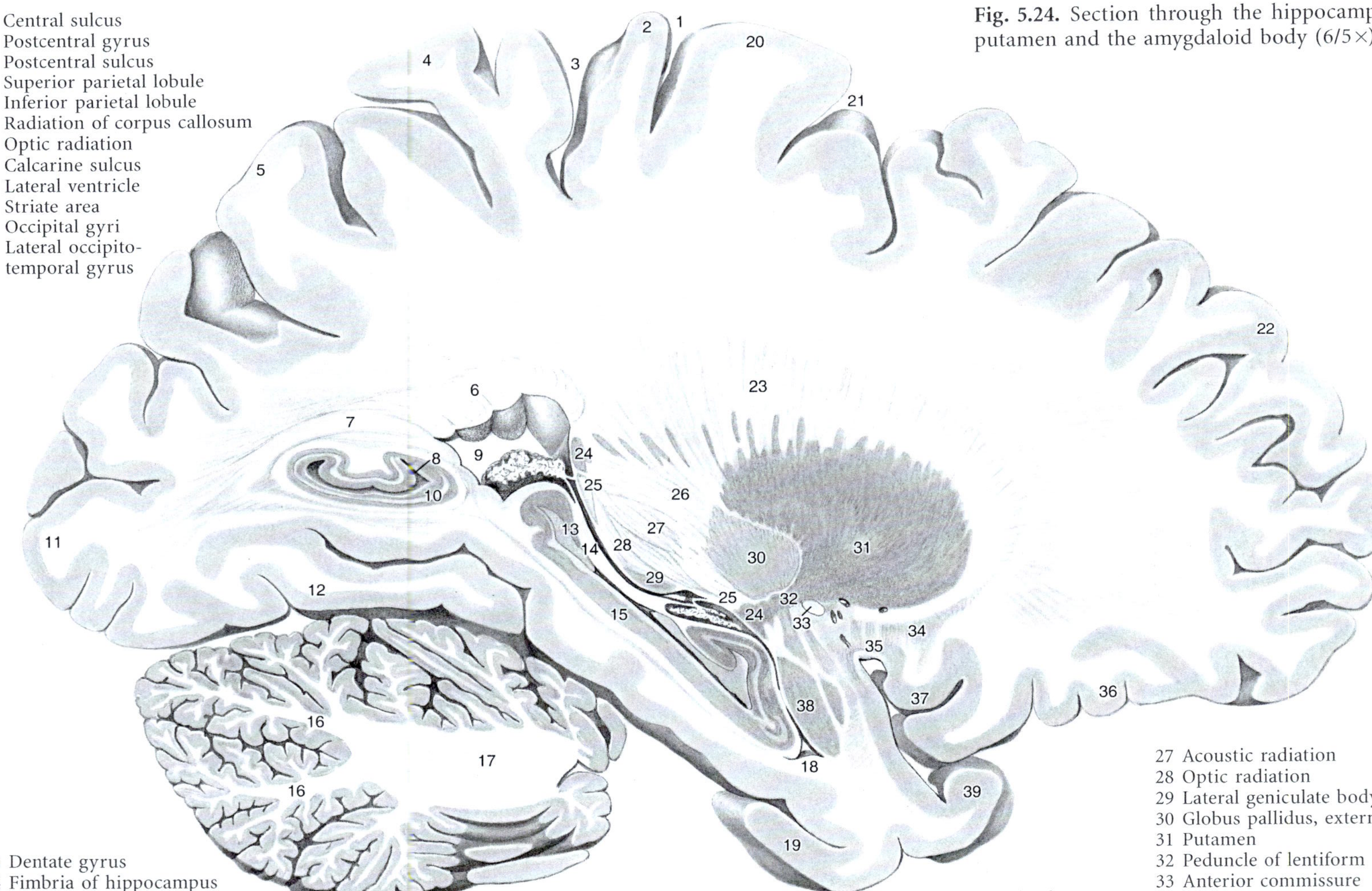

1 Central sulcus
2 Postcentral gyrus
3 Postcentral sulcus
4 Superior parietal lobule
5 Inferior parietal lobule
6 Radiation of corpus callosum
7 Optic radiation
8 Calcarine sulcus
9 Lateral ventricle
10 Striate area
11 Occipital gyri
12 Lateral occipito-temporal gyrus
13 Dentate gyrus
14 Fimbria of hippocampus
15 Subiculum
16 Cerebellar white laminae
17 Cerebellar medullar body
18 Lateral ventricle, inferior horn
19 Lateral occipitotemporal gyrus
20 Precentral gyrus
21 Precentral sulcus
22 Frontal gyri
23 Corona radiata
24 Tail of caudate nucleus
25 Stria terminalis
26 Internal capsule
27 Acoustic radiation
28 Optic radiation
29 Lateral geniculate body
30 Globus pallidus, external part
31 Putamen
32 Peduncle of lentiform nucleus
33 Anterior commissure
34 Claustrum
35 Limen of insula
36 Orbital gyri
37 Pole of insula
38 Amygdaloid body
39 Superior temporal gyrus

Fig. 5.25. Section through the lateral part of the putamen (6/5×)

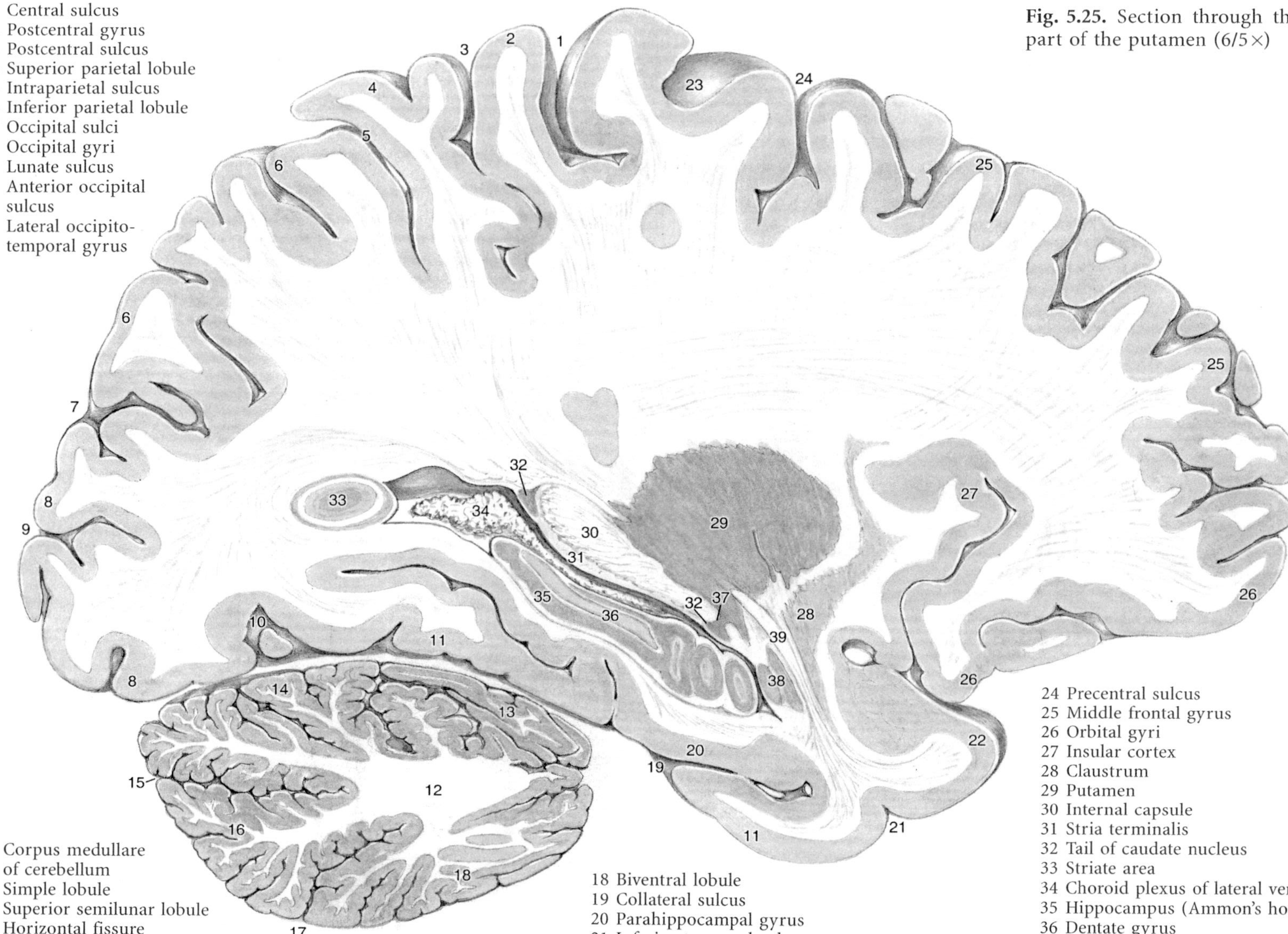

1 Central sulcus
2 Postcentral gyrus
3 Postcentral sulcus
4 Superior parietal lobule
5 Intraparietal sulcus
6 Inferior parietal lobule
7 Occipital sulci
8 Occipital gyri
9 Lunate sulcus
10 Anterior occipital sulcus
11 Lateral occipito-temporal gyrus
12 Corpus medullare of cerebellum
13 Simple lobule
14 Superior semilunar lobule
15 Horizontal fissure
16 Inferior semilunar lobule
17 Prebiventral fissure
18 Biventral lobule
19 Collateral sulcus
20 Parahippocampal gyrus
21 Inferior temporal sulcus
22 Superior temporal gyrus
23 Precentral gyrus
24 Precentral sulcus
25 Middle frontal gyrus
26 Orbital gyri
27 Insular cortex
28 Claustrum
29 Putamen
30 Internal capsule
31 Stria terminalis
32 Tail of caudate nucleus
33 Striate area
34 Choroid plexus of lateral ventricle
35 Hippocampus (Ammon's horn)
36 Dentate gyrus
37 Peduncle of lentiform nucleus
38 Amygdala
39 Anterior commissure, anterior part

Fig. 5.26. Section through the insula (6/5×)

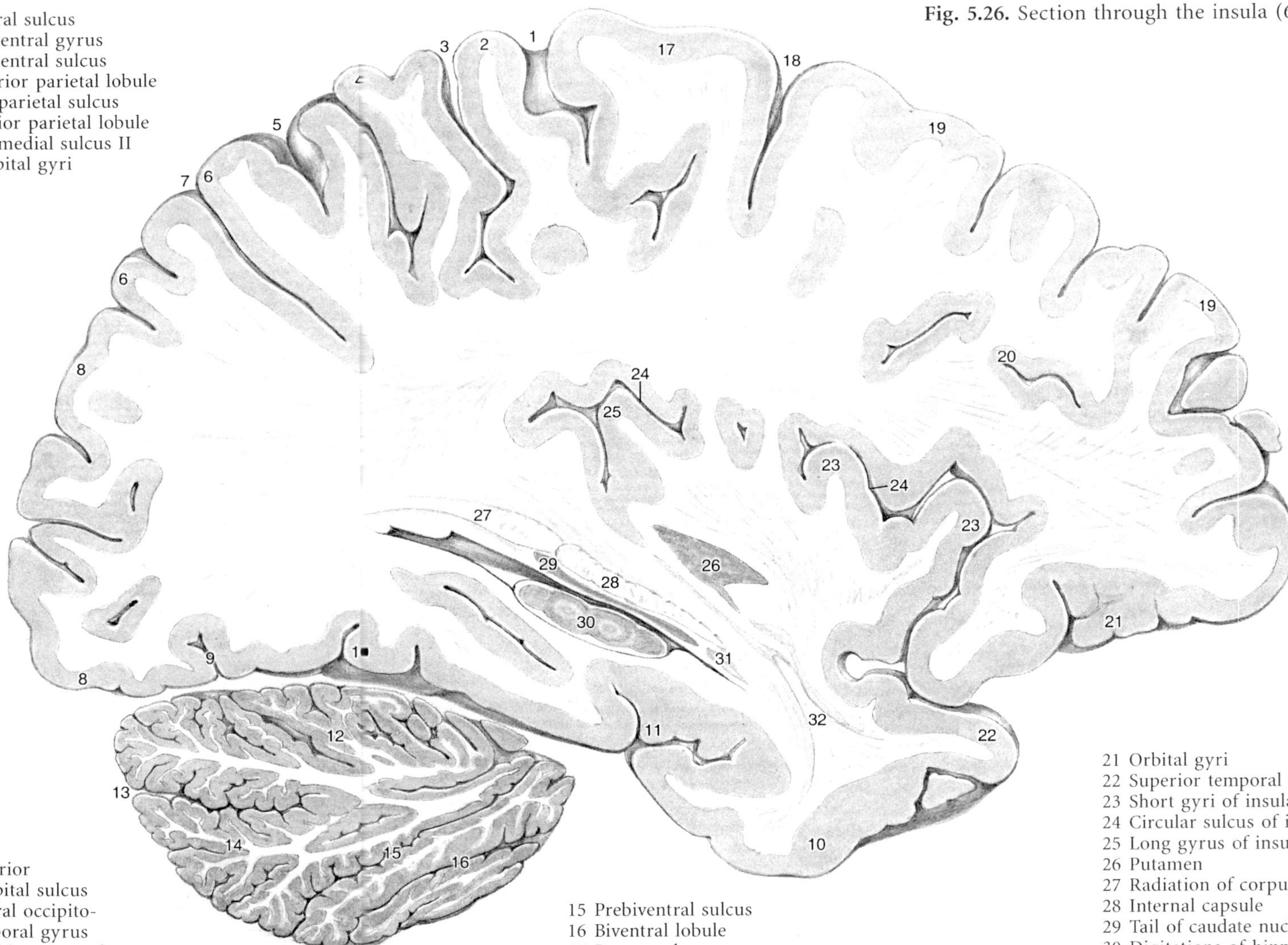

1 Central sulcus
2 Postcentral gyrus
3 Postcentral sulcus
4 Superior parietal lobule
5 Intraparietal sulcus
6 Inferior parietal lobule
7 Intermedial sulcus II
8 Occipital gyri
9 Anterior occipital sulcus
10 Lateral occipito-temporal gyrus
11 Parahippocampal gyrus
12 Superior semilunar lobule
13 Horizontal fissure
14 Inferior semilunar lobule
15 Prebiventral sulcus
16 Biventral lobule
17 Precentral gyrus
18 Precentral sulcus
19 Middle frontal gyrus
20 Inferior frontal sulcus
21 Orbital gyri
22 Superior temporal gyrus
23 Short gyri of insula
24 Circular sulcus of insula
25 Long gyrus of insula
26 Putamen
27 Radiation of corpus callosum
28 Internal capsule
29 Tail of caudate nucleus
30 Digitations of hippocampus
31 Stria terminalis
32 Anterior commissure, anterior part

Horizontal Sections

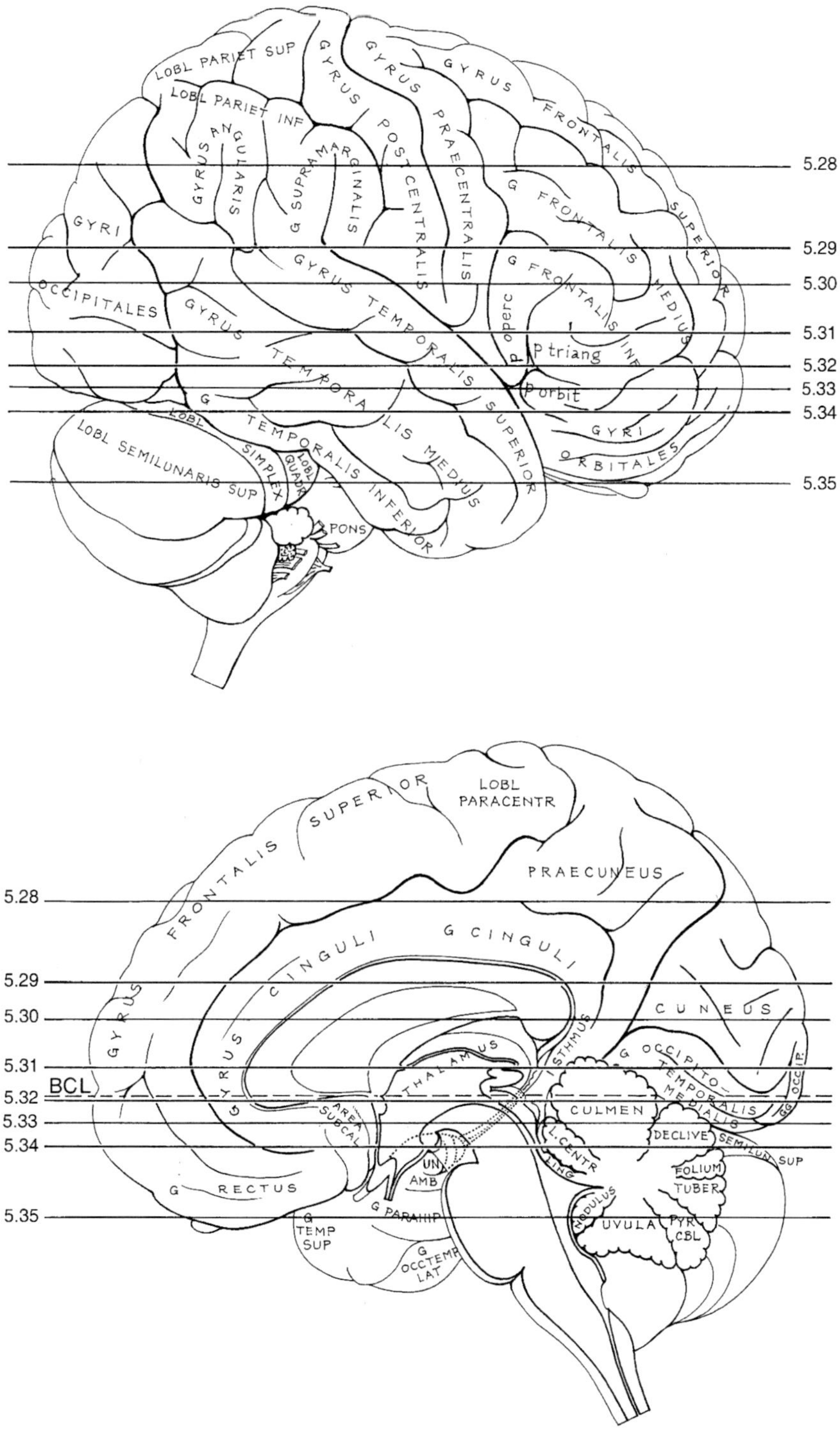

Fig. 5.27. Key diagrams showing the level and plane of the horizontal sections in Figs. 5.28–5.35 (2/3×). The planes of sectioning are about parallel to the bicommissural line of Talairach (*BCL*; *broken line* just above 5.32), corresponding to the CT 0° direction

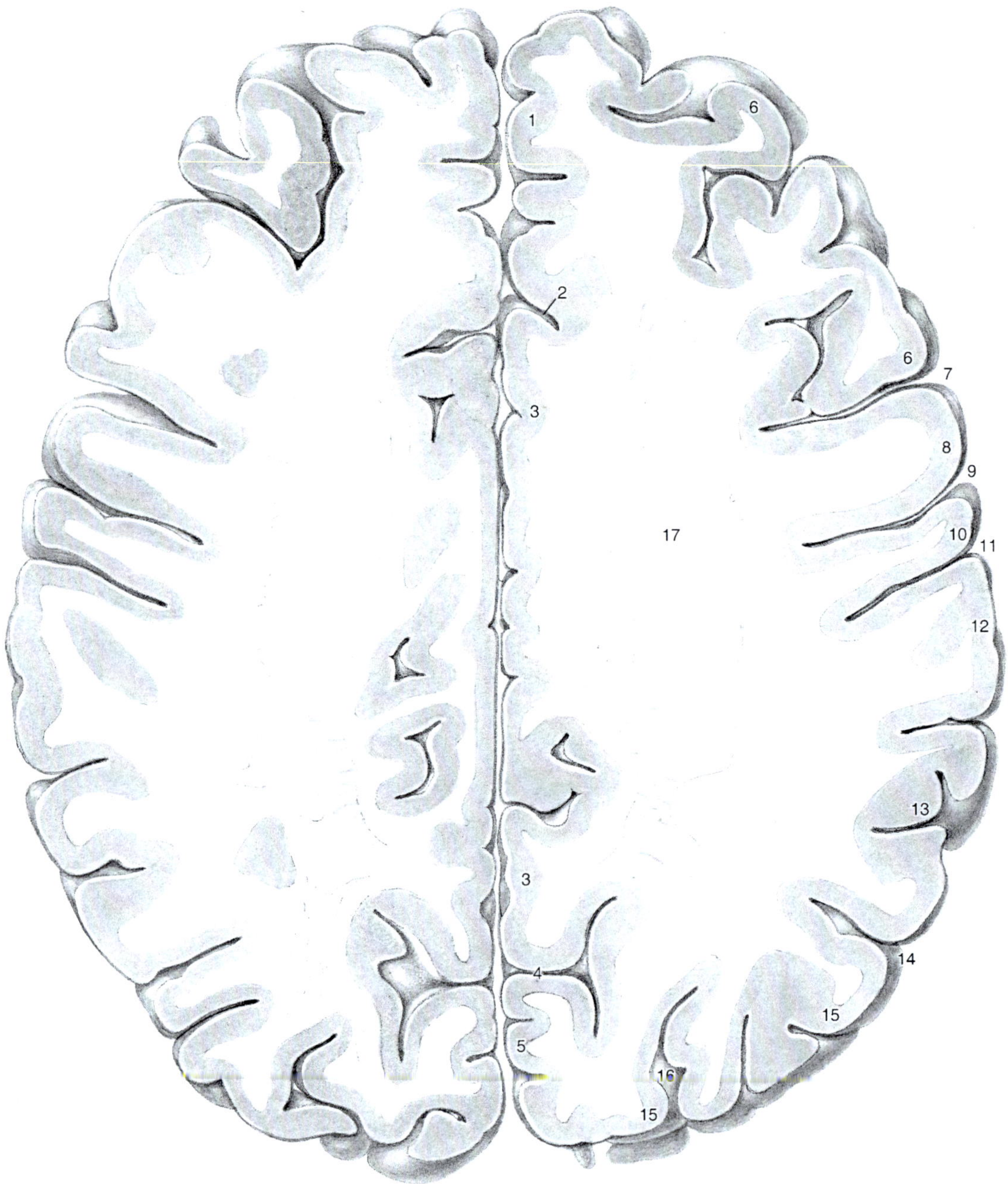

1 Superior frontal gyrus
2 Cingulate sulcus
3 Cingulate gyrus
4 Subparietal sulcus
5 Precuneus
6 Middle frontal gyrus
7 Precentral sulcus
8 Precentral gyrus
9 Central sulcus
10 Postcentral gyrus
11 Postcentral sulcus
12 Supramarginal gyrus
13 Angular gyrus
14 Anterior occipital sulcus
15 Occipital gyri
16 Lunate sulcus
17 Corona radiata

Fig. 5.28. Section through the corona radiata above the corpus callosum (6/5×)

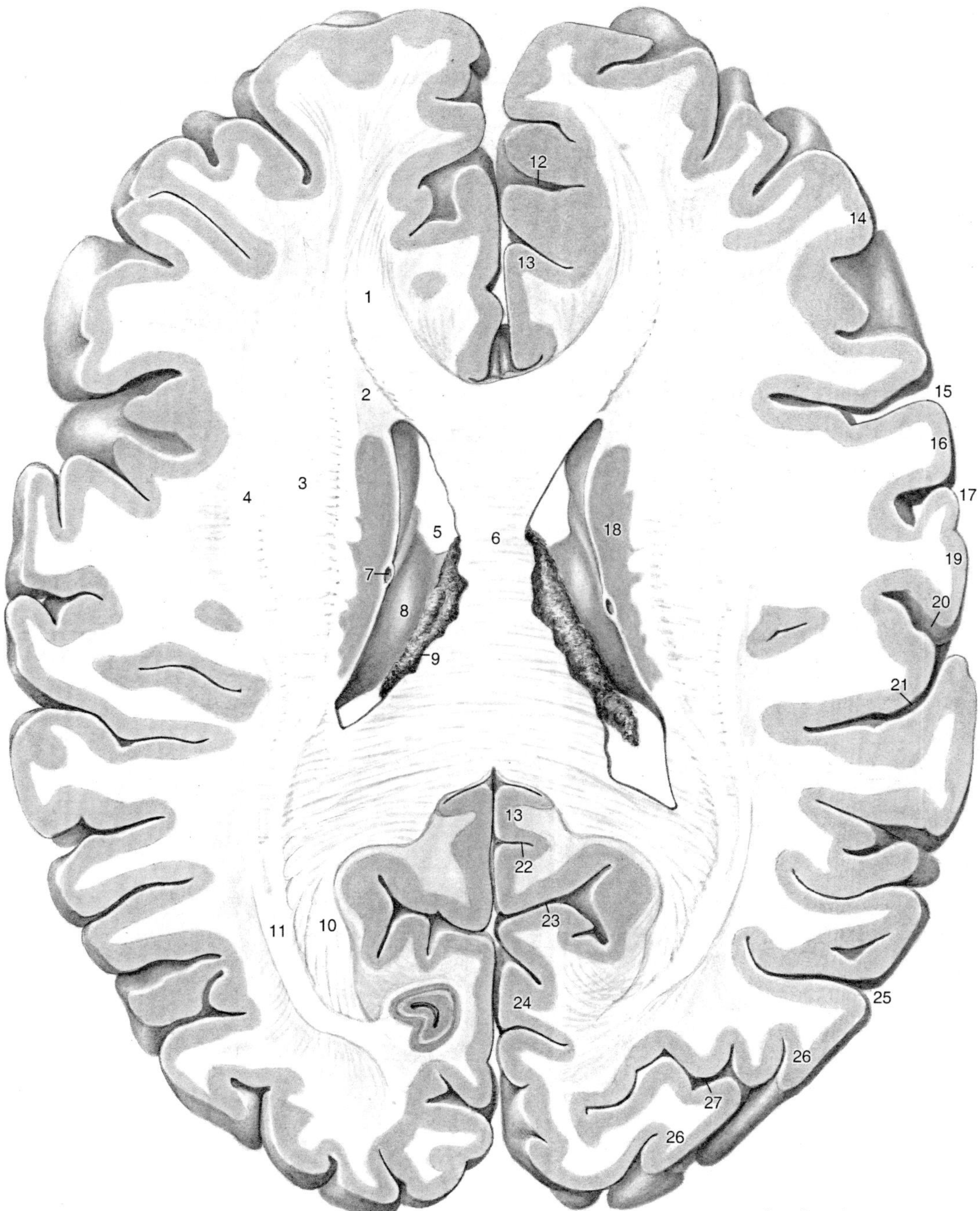

1 Forceps minor of corpus callosum
2 Superior occipitofrontal fasciculus
3 Corona radiata
4 Superior longitudinal fasciculus
5 Lateral ventricle, central part
6 Body of corpus callosum
7 Thalamostriate vein
8 Lamina affixa
9 Choroid plexus of lateral ventricle
10 Forceps major of corpus callosum
11 Stratum sagittale
12 Cingulate sulcus
13 Cingulate gyrus
14 Frontal gyri
15 Precentral sulcus
16 Precentral gyrus
17 Central sulcus
18 Body of caudate nucleus
19 Postcentral gyrus
20 Postcentral sulcus
21 Lateral sulcus, posterior branch
22 Subparietal sulcus
23 Parieto-occipital sulcus
24 Cuneus
25 Anterior occipital sulcus
26 Occipital sulci
27 Lunate sulcus

Fig. 5.29. Section through the corpus callosum and the body of the caudate nucleus. The thickness of this slice is twice the standard of 2 mm (6/5×)

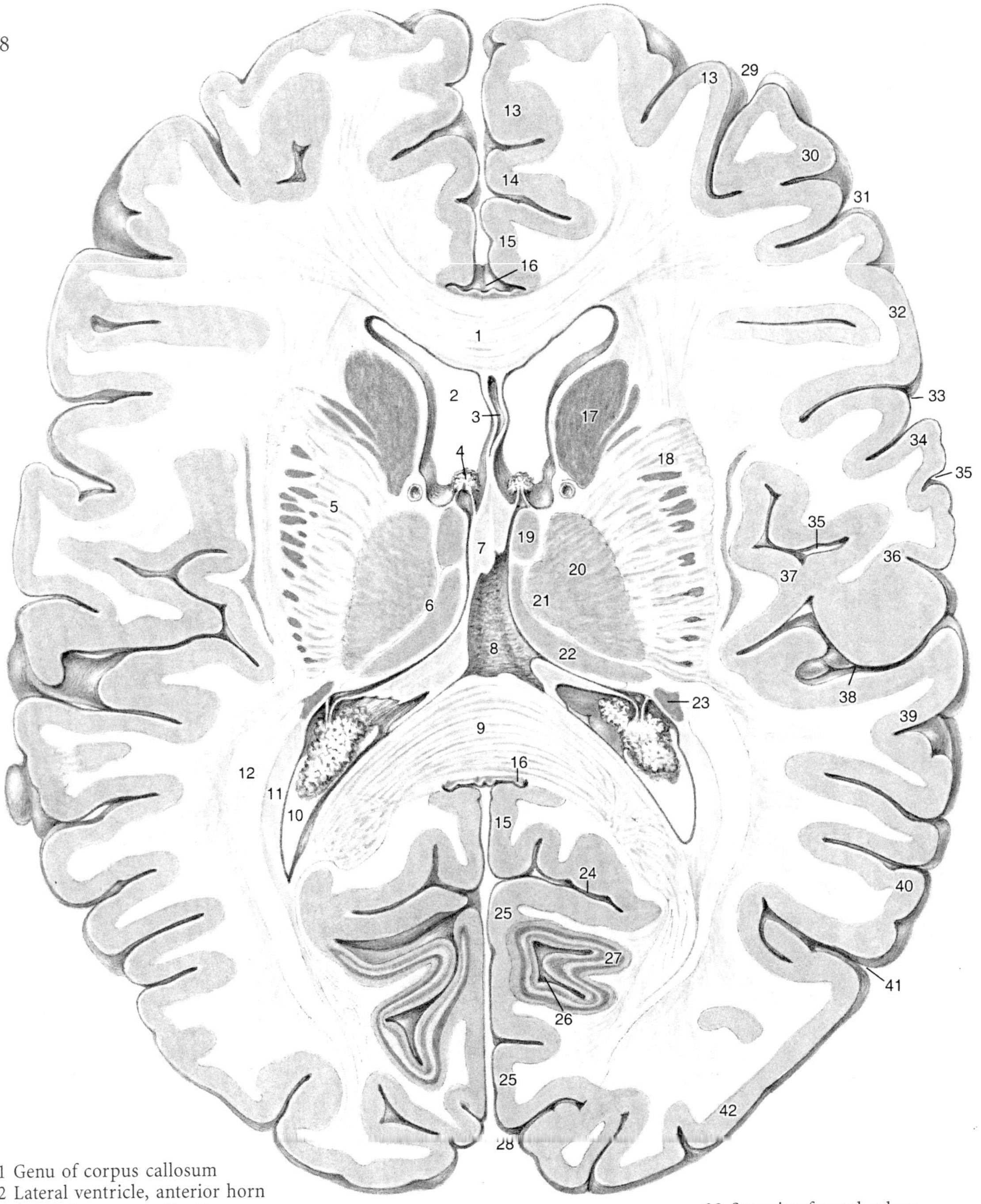

1 Genu of corpus callosum
2 Lateral ventricle, anterior horn
3 Cave of septum pellucidum
4 Choroid plexus of lateral ventricle
5 Internal capsule
6 Internal medullary lamina
7 Fornix
8 Roof of third ventricle
9 Splenium of corpus callosum
10 Lateral ventricle, posterior horn
11 Tapetum
12 Optic radiation
13 Superior frontal gyrus
14 Cingulate sulcus
15 Cingulate gyrus
16 Indusium griseum
17 Head of caudate nucleus
18 Caudatolenticular grey bridges
19 Anterior thalamic nucleus
20 Ventral nuclei of thalamus
21 Mediodorsal thalamic nucleus
22 Lateral dorsal thalamic nucleus
23 Tail of caudate nucleus
24 Parieto-occipital sulcus
25 Cuneus
26 Calcarine sulcus
27 Striate area
28 Lunate sulcus
29 Superior frontal sulcus
30 Middle frontal gyrus
31 Inferior frontal sulcus
32 Inferior frontal gyrus
33 Precentral sulcus
34 Precentral gyrus
35 Central sulcus
36 Postcentral gyrus
37 Long gyrus of insula
38 Lateral sulcus, posterior branch
39 Superior temporal gyrus
40 Middle temporal gyrus
41 Anterior occipital sulcus
42 Occipital gyri

Fig. 5.30. Section through the caudatolenticular grey bridges (6/5×)

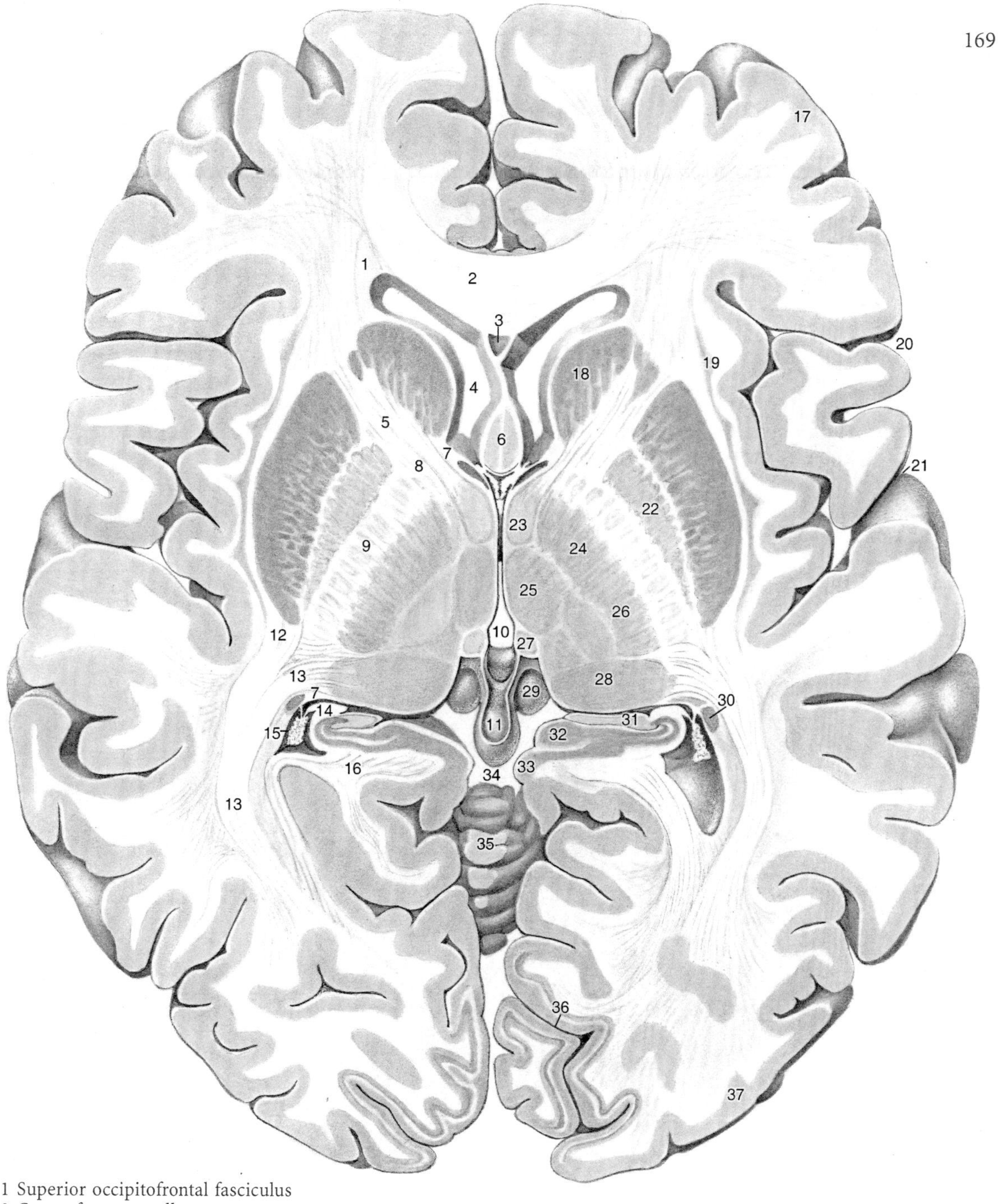

1 Superior occipitofrontal fasciculus
2 Genu of corpus callosum
3 Cavum of septum pallucidum
4 Lateral ventricle, anterior horn
5 Internal capsule, anterior limb
6 Fornix
7 Stria terminalis
8 Internal capsule, genu
9 Internal capsule, posterior limb
10 Third ventricle
11 Suprapineal recess
12 Superior longitudinal fasciculus
13 Optic radiation
14 Fimbria of hippocampus
15 Lateral ventricle, inferior horn
16 Radiation of corpus callosum
17 Frontal gyri
18 Head of caudate nucleus
19 Claustrum
20 Lateral sulcus, ascending branch
21 Lateral sulcus, posterior branch
22 Lentiform nucleus
23 Anterior thalamic nucleus
24 Ventral lateral thalamic nucleus
25 Mediodorsal thalamic nucleus
26 Lateral posterior thalamic nucleus
27 Habenular nuclei
28 Pulvinar nuclei
29 Superior colliculus
30 Tail of caudate nucleus
31 Fasciola cinerea
32 Fasciolar gyrus
33 Gyri of Anders Retzius
34 Pineal body
35 Vermis of cerebellum
36 Calcarine sulcus
37 Occipital gyri

Fig. 5.31. Section through the striate body, the thalamus and the internal capsule. The thickness of this slice is twice the standard of 2 mm (6/5×)

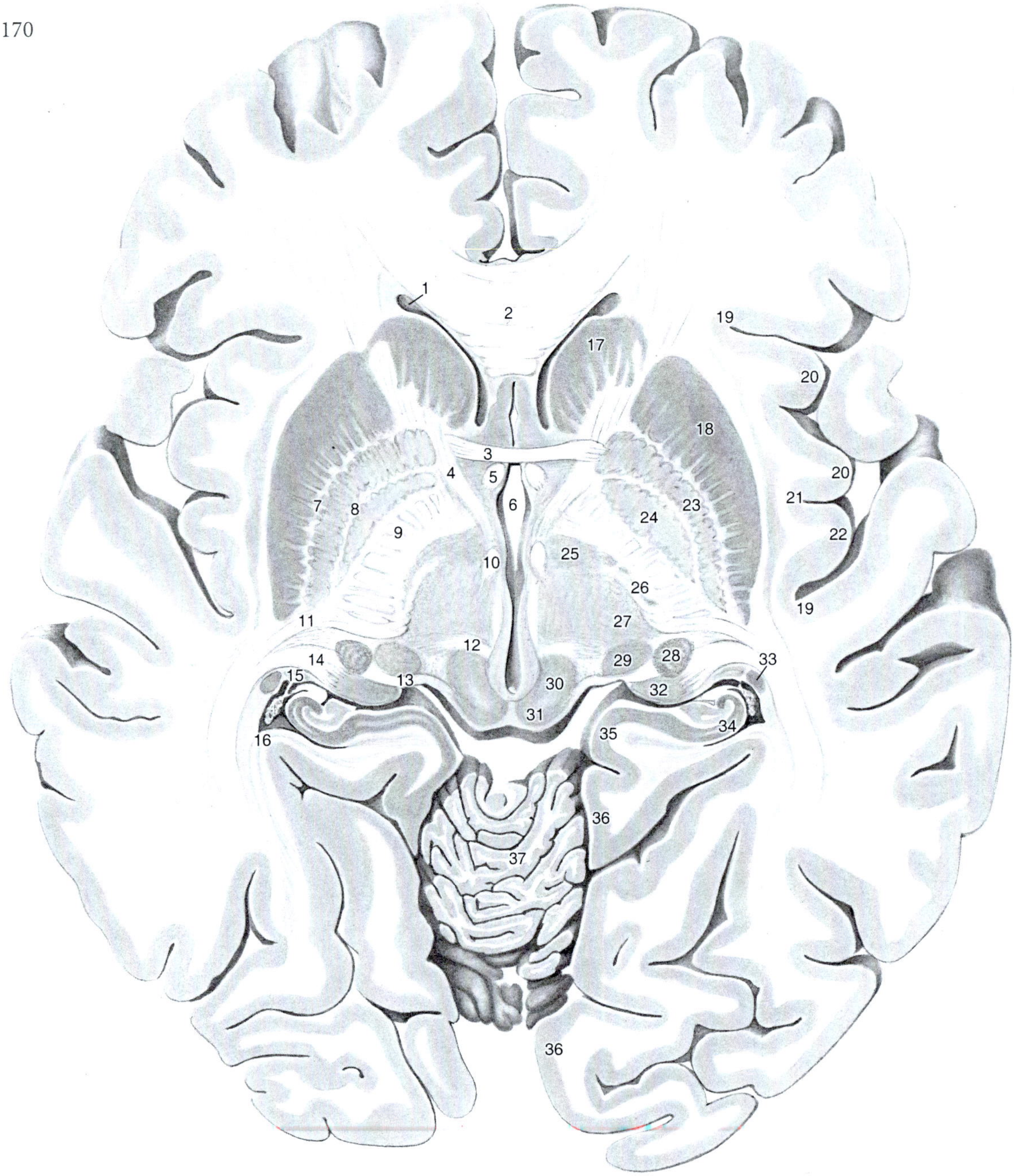

1 Lateral ventricle, anterior horn
2 Genu of corpus callosum
3 Anterior commissure
4 Anterior thalamic peduncle
5 Column of fornix
6 Third ventricle
7 External medullary lamina
8 Internal medullary lamina
9 Internal capsule, posterior limb
10 Mamillothalamic tract
11 Internal capsule, retrolentiform part
12 Posterior commissure
13 Brachium of superior colliculus
14 Optic radiation
15 Stria terminalis
16 Lateral ventricle, inferior horn
17 Head of caudate nucleus
18 Putamen
19 Circular sulcus of insula
20 Short gyri of insula
21 Central sulcus of insula
22 Long gyrus of insula
23 Globus pallidus, external part
24 Globus pallidus, internal part
25 Zona incerta
26 Reticular thalamic nucleus
27 Ventral posterolateral thalamic nucleus
28 Lateral geniculate body
29 Medial geniculate body
30 Pretectal area
31 Superior colliculus
32 Pulvinar nuclei
33 Tail of caudate nucleus
34 Hippocampus
35 Parahippocampal gyrus
36 Medial occipitotemporal gyrus
37 Anterior lobe of cerebellum

Fig. 5.32. Section through the striate body, the anterior commissure and the superior colliculus (6/5×)

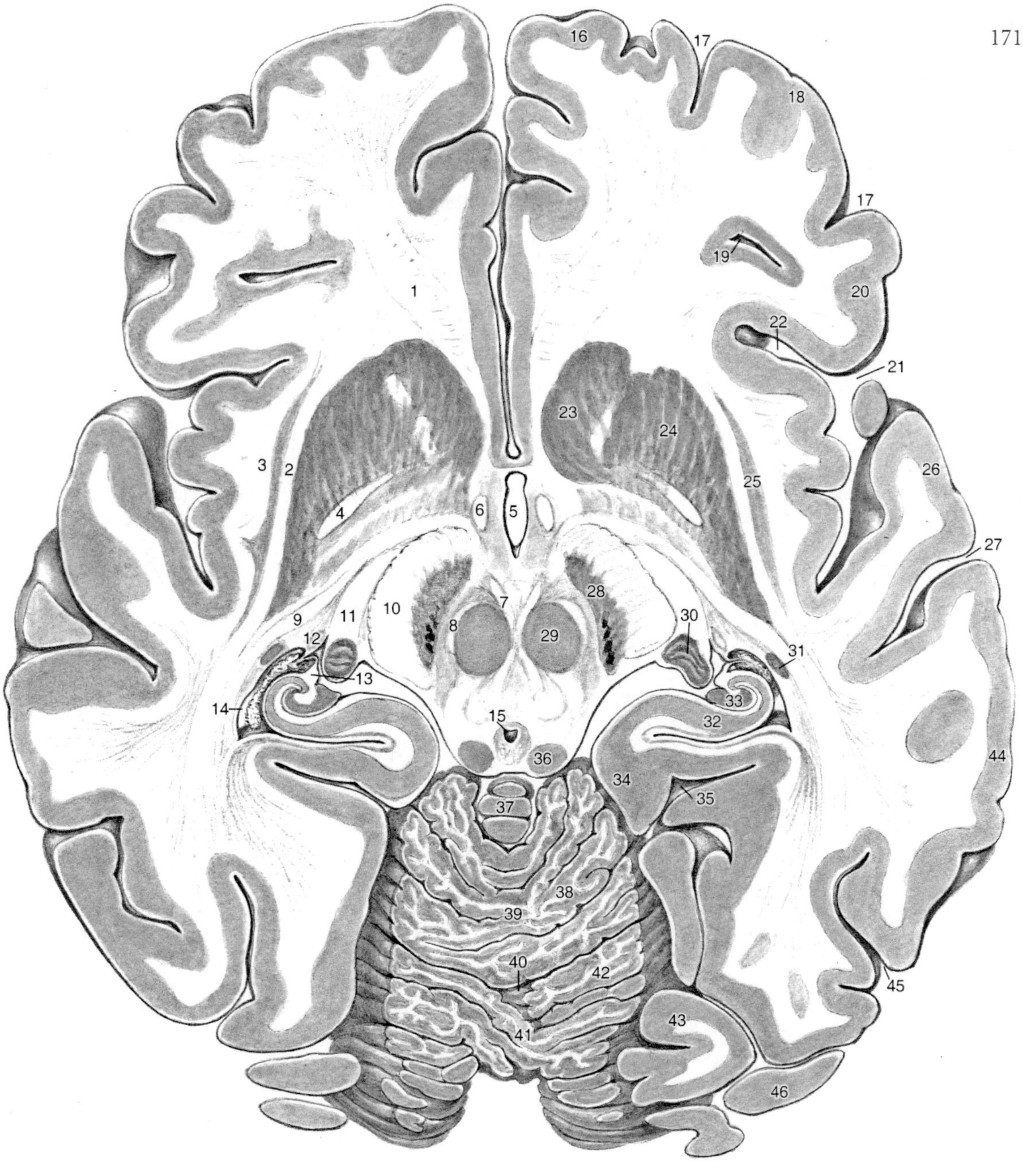

1 Minor forceps of corpus callosum
2 External capsule
3 Extreme capsule
4 Anterior commissure
5 Third ventricle
6 Mamillothalamic tract
7 Habenulo-interpeduncular tract
8 Medial forebrain bundle
9 Ansa lenticularis
10 Cerebral peduncle
11 Optic tract
12 Stria terminalis
13 Fornix
14 Lateral ventricle, inferior horn
15 Cerebral aqueduct
16 Superior frontal gyrus
17 Orbital sulci
18 Orbital gyri
19 Lateral sulcus, anterior branch
20 Middle frontal gyrus, orbital part
21 Lateral sulcus
22 Lateral sulcus, ascending branch
23 Head of caudate nucleus
24 Putamen
25 Claustrum
26 Superior temporal gyrus
27 Superior temporal sulcus
28 Substantia nigra
29 Red nucleus
30 Lateral geniculate body
31 Tail of caudate nucleus
32 Hippocampus (Ammon's horn)
33 Dentate gyrus
34 Parahippocampal gyrus
35 Collateral sulcus
36 Inferior colliculus
37 Central cerebellar lobule
38 Quadrangular lobule
39 Culmen
40 Primary fissure
41 Declive
42 Simple lobule
43 Lateral occipitotemporal gyrus
44 Middle temporal gyrus
45 Anterior occipital sulcus
46 Occipital gyri

Fig. 5.33. Section through the minor forceps and the inferior colliculus (6/5×)

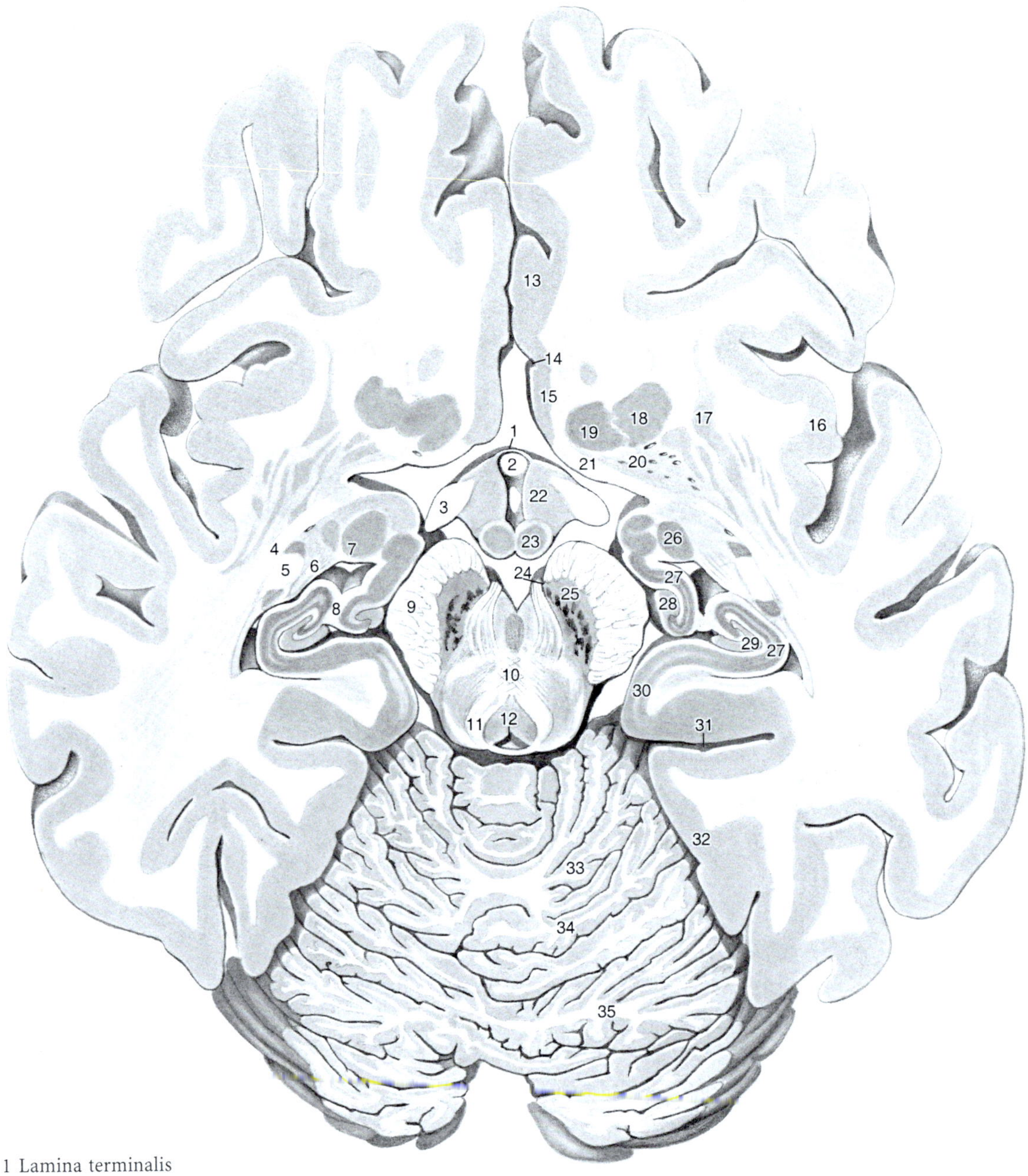

1 Lamina terminalis
2 Third ventricle
3 Optic tract
4 Anterior commissure
5 Internal capsule, sublentiform part
6 Stria terminalis
7 Lateral ventricle, inferior horn
8 Fimbria of hippocampus
9 Cerebral peduncle
10 Decussation of superior cerebellar peduncles
11 Superior cerebellar peduncle
12 Cerebral aqueduct
13 Cingulate gyrus
14 Anterior parolfactory sulcus
15 Subcallosal area
16 Insula
17 Claustrum
18 Putamen
19 Accumbens nucleus
20 Anterior perforate substance
21 Diagonal gyrus
22 Hypothalamus
23 Mamillary body
24 Oculomotor nerve (III)
25 Substantia nigra
26 Amygdaloid body
27 Ammon's horn
28 Limbus Giacomini
29 Dentate gyrus
30 Parahippocampal gyrus
31 Collateral sulcus
32 Lateral occipitotemporal gyrus
33 Anterior lobe of cerebellum
34 Primary fissure
35 Posterior lobe of cerebellum

Fig. 5.34. Section through the mamillary body and the cerebral peduncle (6/5×)

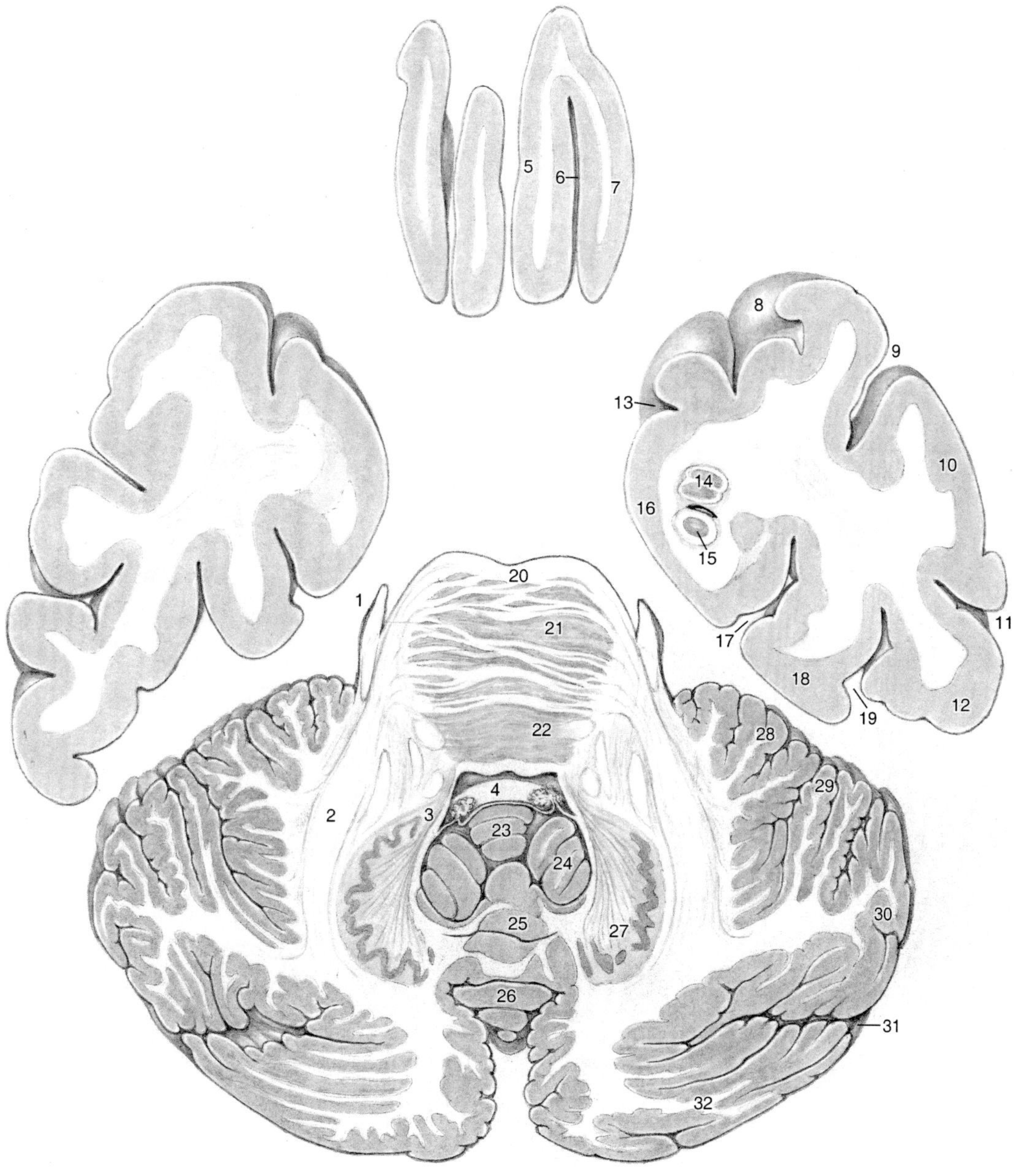

1 Trigeminal nerve (V)
2 Middle cerebellar peduncle
3 Superior cerebellar peduncle
4 Fourth ventricle
5 Straight gyrus
6 Olfactory sulcus
7 Orbital gyri
8 Superior temporal gyrus
9 Superior temporal sulcus
10 Middle temporal gyrus
11 Inferior temporal sulcus
12 Inferior temporal gyrus
13 Rhinal sulcus
14 Amygdala
15 Hippocampus
16 Parahippocampal gyrus
17 Collateral sulcus
18 Lateral occipitotemporal gyrus
19 Occipitotemporal sulcus
20 Pons
21 Pontine nuclei
22 Tegmentum of myelencephalon
23 Nodule
24 Tonsil of cerebellum
25 Uvula
26 Pyramid
27 Dentate nucleus
28 Quadrangular lobule
29 Simple lobule
30 Superior semilunar lobule
31 Horizontal fissure
32 Inferior semilunar lobule

Fig. 5.35. Section through temporal lobe, pons and cerebellum (6/5×)

Section II Structure of Spinal Cord and Brain Parts

6 Topography of Spinal Cord, Brain Stem and Cerebellum

Spinal Cord

Introduction

The spinal cord with its meninges is located within the vertebral canal (Figs. 4.30–4.32). The border of the spinal cord with the brain stem is set rostral to the origin of the first cervical nerve, at the level of the pyramidal decussation. However, the transition of the motor and sensory nuclei of the cord into the corresponding structures of the caudal brain stem (involved in the innervation of structures that have evolved from occipital somites, see Chap. 2) is a gradual one.

The cord occupies the entire length of the vertebral canal during early stages of development. Due to the continued growth of the vertebral column it gradually lags behind, and in the adult only reaches to the upper level of the second lumbar vertebra (Fig. 4.32). This process is known as the *ascensus medullae*. The caudal tip of the spinal cord tapers into the *conus medullaris*, and continues as a thread (*filum terminale*) to the level of the sacrum (Fig. 3.15).

The spinal cord can be subdivided into 31 segments (8 cervical, 12 thoracic, 5 lumbar, 5 sacral and 1 coccygeal). Each segment gives rise to dorsal and ventral rootlets, which unite into a pair of *dorsal and ventral roots*. The dorsal roots contain the *spinal ganglia* (Figs. 3.14, 3.15). The ganglia are located within the intervertebral foramina, where the dorsal and ventral roots combine into a pair of spinal nerves (Fig. 4.30). Due to the ascensus of the cord, the spinal segments are located more rostral than their corresponding vertebrae. As a consequence, the dorsal and ventral roots descend over some distance through the vertebral canal to their exit through the intervertebral foramen, caudal (i.e., rostral for the cervical roots 1–7) to the corresponding vertebra. The bundle of lumbar, sacral and coccygeal roots surrounding the filum terminale, caudal to the cord, is known as the *cauda equina* (horse tail; Fig. 4.32). At cervical and lumbar levels the cord is enlarged (Fig. 3.15). These enlargements (*intumescentiae*) innervate the extremities.

The ventricular system is represented by the *central canal*, which is often obliterated locally. A deep *anterior median fissure* is present at the ventral aspect of the cord (Fig. 3.15). Shallow median and dorsolateral sulci demarcate the *posterior funiculi* (or *dorsal columns*) at its dorsal aspect. The dorsal rootlets enter the cord in the dorsolateral sulcus. The glial *posterior median septum* extends from the central canal dorsally to the median dorsal sulcus (Fig. 6.1).

The grey matter of the cord is butterfly-shaped and can be subdivided into *dorsal and ventral horns* and the *intermediate zone*. The dorsal horn is a laminated structure. It is a derivative of the embryological alar plate. The ventral horn contains groups of large motoneurons. It takes its origin from the basal plate of the neural tube. The intermediate grey is located in between. It was defined more precisely in the cytoarchitectonic studies of Rexed [32] (see below) and serves as an intermediary between the motoneurons and many of their afferent systems.

The dorsal horn is covered by a bundle of unmyelinated (grey) fibres (*dorsolateral tract of Lissauer* [22]), which reaches the meningeal surface at the dorsolateral sulcus. Most of the fibres of the dorsolateral tract are derived from the dorsal roots. The ventral horn is surrounded by the lateral and ventral white matter of the cord. The dorsal white matter contained between the dorsal horn and the dorsal median septum corresponds to the *posterior funiculus*. The ventral rootlets, which emerge from the ventral horn, are taken as the arbitrary border between the *lateral and anterior funiculus* of the spinal white matter. The lateral funiculus can be subdivided into the posterolateral and the anterolateral funiculi (Fig. 6.1). Myelinated fibres that cross the midline ventral to the central canal are known as the *commissura alba* (alba = white). The commissura alba serves as the decussation for short and long pathways of the spinal cord. Unmyelinated fibres cross the midline dorsal and ventral to the central canal in the *anterior and posterior grey commissures*. The shape of the spinal grey matter differs at the level of the cervical and lumbar enlargements, where the ventral horn bulges out laterally and the dorsal horn medially (Figs. 6.7, 6.9). This configuration is related to the reversed somatotopical localization in the dorsal and the ventral horn (Chap. 16 and Fig. 16.5).

The Grey Matter of the Cord

The grey matter of the spinal cord of the cat has been subdivided into 10 laminae on the basis of cytoarchitectonic criteria by Rexed [32]. This lamination has been applied to other species, including humans [35] (Fig. 6.2). Laminae I–IV belong to the dorsal horn. Lamina I is a thin layer of fairly large *marginal cells*. Laminae II and III consist of the small cells of the *substantia gelatinosa*. The marginal cells are located within the dorsolateral funiculus at the border of the substantia gelatinosa. Lamina IV contains larger cells of the *nucleus proprius* of the dorsal horn, located in the centre of this structure. Laminae V–X occupy the base of the dorsal horn and the central region of the ventral horn. In the human spinal cord, lamina V and VI cannot be separated. Many neurons in this region give rise to axons terminating within the cord itself (*propriospinal neurons*). They are collectively known as the *intermediate zone* of the spinal cord. Lamina VIII is located in the ventromedial ventral horn and consists of a group of *commissural neurons*, which send their axons into the ventral white commissure. Groups of large *motoneurons* in the ventral horn, which give rise to the ventral roots, belong to lamina IX. Lamina X consists of small, densely packed neurons surrounding the central canal.

An important question is whether both the dendrites and the somata of the cells are restricted to Rexed's laminae. This appears to be the case for the marginal cells of lamina I and for the cells of the substantia gelatinosa (lamina II). Dendrites of lamina IV cells typically invade the substantia gelatinosa. In the intermediate zone, dendritic trees are flattened in the rostrocaudal direction and generally surpass the borders of the laminae. Dendrites of motoneurons may extend far beyond the motoneuronal columns. The packing of these dendrites in longitudinal bundles is most prominent in the ventral and ventrolateral motoneuronal groups and is typical for the phrenic motor nucleus [34]. Rexed's lamination of the spinal grey matter, therefore, serves a descriptive rather than a functional purpose.

Several additional nuclei are present at certain levels of the cord. The *central cervical nucleus* (not illustrated) is located centrally in the intermediate grey of the upper cervical cord and the caudal brain stem. *Clarke's dorsal column* [4] is a discrete, roundish collection of large neurons, located medially at the base of the dorsal horn at thoracic and high lumbar levels (Fig. 6.8). *Spinal border cells* are located medially in the lumbar ventral horn. The central cervical nucleus,

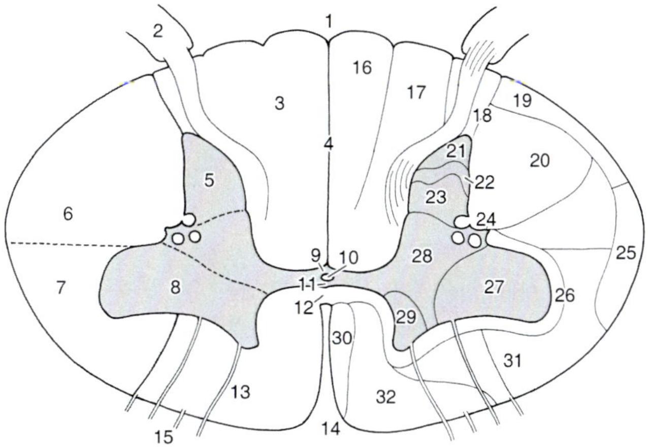

1 Posterior median sulcus
2 Dorsal root
3 Dorsal funiculus (column)
4 Posterior median septum
5 Dorsal horn
6 Posterolateral funiculus
7 Anterolateral funiculus
8 Ventral horn
9 Posterior grey commissure
10 Central canal
11 Anterior grey commissure
12 Ventral (white) commissure
13 Anterior funiculus
14 Anterior median fissure
15 Ventral root
16 Gracile fascicle
17 Cuneate fascicle
18 Dorsolateral fascicle (Lissauer [22])
19 Dorsal spinocerebellar tract
20 Lateral (crossed) pyramidal tract
21 Dorsal horn, marginal cells (lamina I)
22 Dorsal horn, substantia gelatinosa (laminae II and III)
23 Dorsal horn, nucleus proprius (lamina IV)
24 Intermediate zone, processus reticularis
25 Ventral spinocerebellar tract
26 Ground bundles (fasciculi proprii)
27 Lateral motor column (lamina IX)
28 Intermediate zone (laminae V–VIII)
29 Medial motor column (lamina IX)
30 Anterior (uncrossed) pyramidal tract
31 Anterolateral fascicle
32 Medial longitudinal fascicle

Fig. 6.1. The subdivision of the white and grey matter in the spinal cord

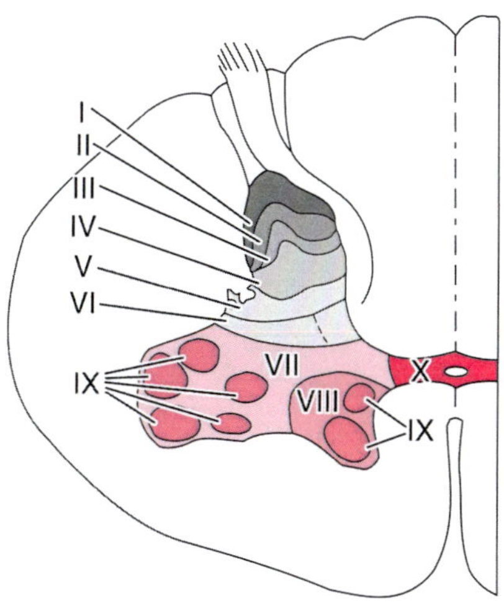

Fig. 6.2. The grey matter of the spinal cord, subdivided according to Rexed [32], at the level of the sixth cervical segment

Clarkes's column and the spinal border cells receive dorsal root afferents and give rise to spinocerebellar pathways (Fig. 6.3, and Chap. 20, Fig. 20.13 B). The grey matter of the thoracic cord between the dorsal and ventral horns protrudes laterally as the *intermediate horn* (Figs. 6.5, 6.8). It contains visceromotor neurons (*nucleus intermediolateralis*) which give rise to preganglionic, orthosympathetic axons (Fig. 6.5). These axons leave the cord via the thoracic ventral roots. At certain intervals, transverse strands consisting of dendrites and unmyelinated fibres connect the intermediolateral nucleus with the *intermediomedial nucleus* (or central autonomic cell column) located adjacent to the central canal. More than one transverse strand may be present at each thoracic segment [1, 37]. Intermediomedial and -lateral nuclei are also represented at the level of the second and third sacral segments, where they give rise to the parasympathetic preganglionic outflow [26] (Fig. 6.5).

Primary Afferents

Axons of spinal ganglion cells divide near the cell body in thicker peripheral and smaller central branches. The central branches enter the spinal cord in the dorsal roots. Like most primary afferents they bifurcate into ascending and descending branches. The thin fibres of the dorsal roots join the dorsolateral tract at the apex of the dorsal horn; the larger fibres enter the dorsal columns. From here, both descending and ascending branches of the root fibres emit collaterals that enter the dorsal horn and terminate in all layers of the spinal grey matter, including the motor neurons (Fig. 6.3). The ascending branches extend in the dorsal columns to terminate in the dorsal column nuclei, located in the caudal brain stem (Figs. 16.3, 16.4).

The primary afferents of the dorsal roots can be subdivided according to myelinization, size and function (Table 16.1). Erlanger and Gasser [12] subdivided the fibres of the peripheral nerves in the groups A–C according to their conduction velocity, which, in myelinated fibres, is proportional to their size. Group A fibres were further subdivided into coarse A-alpha, intermediate A-beta and thinner A-gamma and delta fibres. Unmyelinated (C) and thin, myelinated (A delta) fibres subserve nociceptive transmission. B-fibres are the thin, preganglionic fibres of the autonomic nervous system. A functional subdivision of the primary afferents, mostly used in studies of the motor system (Lloyd [23]), distinguished large, myelinated IA and IB fibres, which carry information from muscle spindles and Golgi tendon receptors, respectively, and the thinner group II–III flexor reflex afferents. Unmyelinated C fibres were designated as group IV.

Primary afferents are the first link in spinal and supraspinal reflexes and in the sensory conduction pathways (Figs. 6.4, 6.5). Propriospinal neurons innervating motoneurons are known as the *interneurons*. Interneurons differ with respect to their location in the intermediate zone, the neurotransmitters and neuropeptides they contain, and their inclusion in particular reflex paths and/or supraspinal motor systems (Fig. 6.4). The first link in the myotatic or stretch reflex is the Ia muscle spindle afferents. Muscle spindles are located in parallel with the regular muscle fibres. Ia afferents, therefore, convey feedback on the length of a particular muscle. Ia afferents terminate directly on motoneurons innervating the same muscle. Collaterals of the Ia fibres terminate on inhibitory (gamma-aminobutyric acid, GABA, -containing) Ia interneurons located in lamina VII, just dorsal to the motor nuclei. These neurons project to motoneurons innervating the antagonist muscles. Ib primary afferents carry information from Golgi tendon organs [17]. These receptors are located in muscle tendons, in series with the muscle fibres, and convey feedback on the force developed in a particular muscle. Ib fibres inhibit the motoneurons innervating this muscle via Ib inhibitory interneurons in the ventral intermediate grey. However, the effect of the Golgi tendon organs is not limited to autogenic inhibition, but also may produce excitatory effects in particular muscle groups [27]. The pathway of the flexor reflex is more complicated and includes two or more spinal interneurons. The first link consists of thin A or C fibres, which

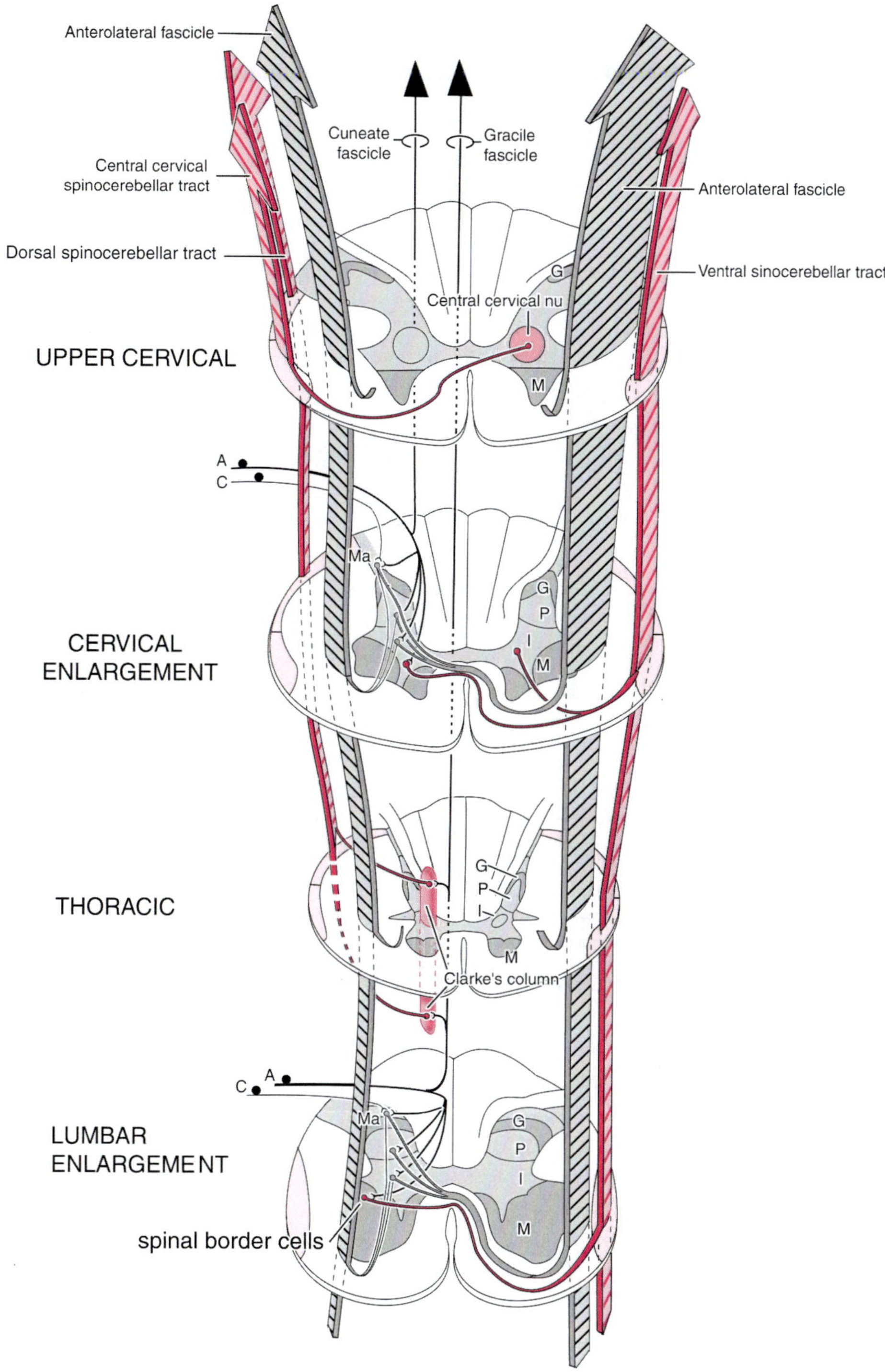

Fig. 6.3. The origin and localization of the ascending fibre tracts of the spinal cord. The anterolateral fascicle, which includes the spinothalamic tract as one of its components, is shown in *grey and black*, the spinocerebellar tracts are shown in *red*. *A*, myelinated primary afferent A fibre; *C*, unmyelinated primary afferent C fibre; *G*, Substantia gelatinosa; *I*, intermediate zone; *P*, nucleus proprius of the dorsal horn; *M*, motor nuclei of the ventral horn; *Ma*, marginal cell layer

terminate on polymodal marginal cells [5]. *Renshaw cells* [31] are GABAergic neurons, located in the medial anterior horn. They receive collaterals of axons of alpha motoneurons and provide these cells with a recurrent inhibition. Recurrent inhibition is primarily concerned with control of proximal muscles and appears to be absent for motoneurons innervating distal muscles of the limbs [6]. *Commissural interneurons* are located in lamina VIII. Their axons cross in the anterior white commissure. Commissural neurons, like other interneurons, can be excitatory or inhibitory. Cervical lamina VIII commissural interneurons use glycine to inhibit contralateral neck motoneurons (Fig. 17.6). Primary afferents are also the first link in long ascending sensory pathways, subserving conscious perception of sensory stimuli and in the spinocerebellar pathways (see Fig. 6.5 and Chaps. 16 and 20).

The White Matter of the Cord

It has already been mentioned that the posterior funiculus consists mainly of primary afferent fibres, which ascend without synaptic interruption to the dorsal column nuclei. The white matter of the anterior and lateral funiculi contains propriospinal fibres, which originate and terminate within the cord, and long ascending and descending pathways that cross the border with the brain stem. The propriospinal fibres surround the grey matter as the *ground bundles* (*fasciculi proprii*) of the spinal cord (Fig. 6.1). Long ascending tracts are located more peripherally (Fig. 6.3). The large, myelinated axons of Clarke's column occupy the dorsolateral periphery of the lateral funiculus as the uncrossed *dorsal spinocerebellar tract* (Figs. 6.6–6.10). Fibres that originate within the contralateral grey matter and decussate in the commissura alba take up a more ventral position in the lateral and anterior funiculus. They include the *anterolateral fasciculus* and the more superficially located *ventral spinocerebellar tract*. The latter stems from contralateral spinal border cells of the lumbar cord. Spinocerebellar fibres from cervical levels are mostly crossed and join the ventral spinocerebellar tract (Chap. 20). The anterolateral fasciculus contains fibres terminating at various levels and in different structures of the brain stem and the thalamus. The latter are known as the *spinothalamic tracts* (Chap. 20).

Long descending fibre systems stem from the cerebral cortex as the *pyramidal* or *corticospinal tract* (Chap. 21), from the contralateral magnocellular red nucleus as the *rubrospinal tract*, from the lateral vestibular nucleus as the *lateral vestibulospinal tract*, and from various regions of the brain stem as the *medial longitudinal fascicle* (Fig. 6.4). After its partial decussation at the border of the brain stem and the cord, the crossed fibres of the pyramidal tract descend into the dorsal half of the lateral funiculus as the *lateral pyramidal tract*. The uncrossed fibres of the *anterior pyramidal tract* are located in the anterior funiculus, next to the anterior median fissure. The medial longitudinal fascicle is located in the anterior funiculus, next to the uncrossed fibres of the pyramidal tract. It is distinguished by the large diameter of the fibres of some of its component tracts. The lateral vestibulospinal tract takes up a more lateral position at the border of the anterior and lateral funiculi. Scattered descending fibres, which cannot be recognized as circumscribed tracts, descend in the anterior and lateral funiculus. They include serotoninergic raphespinal fibres, noradrenergic coerulaeospinal fibres and descending fibres from reticular centres involved in visceromotor control. The *rubrospinal tract*, which is present in most mammals as a distinct bundle of coarse fibres ventral to the lateral pyramidal tract, cannot be recognized in the human spinal cord (Fig. 6.4). From their position in the ventral and lateral funiculus, the descending fibres, or their collaterals, enter the grey matter, where most of them terminate on interneurons of the intermediate zone. Corticofugal fibres, descending in the pyramidal tract, fibres of the rubrospinal tract and vestibulospinal fibres travelling in the medial longitudinal fasciculus and the anterior and lateral funiculi also terminate in the dorsal horn and on medial motoneurons of lamina IX.

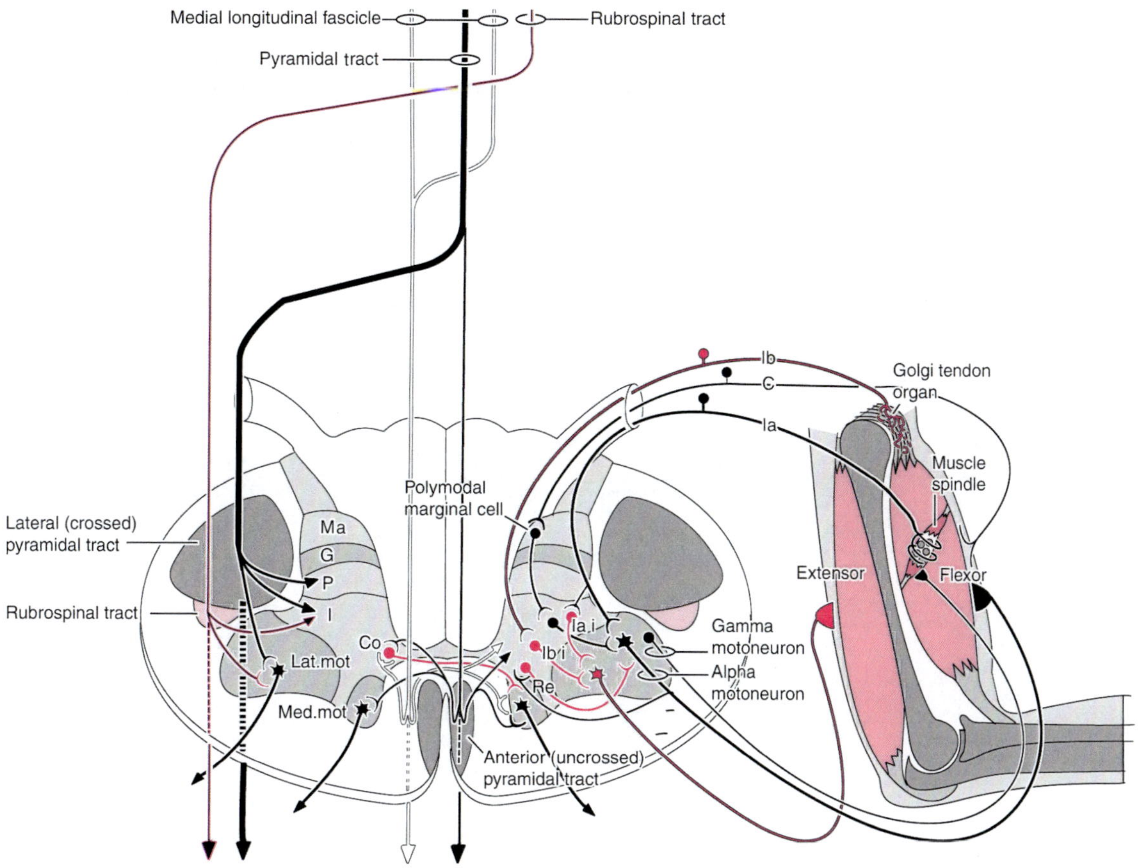

Fig. 6.4. Descending supraspinal paths (left side) and somatic reflex paths (right side) in the spinal cord. Both the rubrospinal and the lateral pyramidal tract terminate on interneurons in the intermediate grey and on motoneurons. Components of the medial longitudinal fascicle and the anterior pyramidal tract terminate bilaterally. The myotatic reflex path consists of Ia primary muscle spindle afferents, which terminate on motoneurons innervating the same muscle, and on Ia inhibitory interneurons (Ia.i) in the intermediate zone, which inhibit motoneurons innervating the antagonist muscles. Muscle spindles are innervated by small gamma motoneurons. Ib Golgi tendon organ afferents inhibit the agonist muscles through Ib inhibitory interneuron (Ib.i). Flexor reflex afferents include small A (*A*) and C (*C*) fibres that terminate on polymodal marginal cells, which are connected with flexor muscles through one or more interneurons. Renshaw cells (Re) inhibit motoneurons, and receive recurrent collaterals of the motoneuron axons. Cornucommissural neurons located in the ventromedial anterior horn can be excitatory or inhibitory. Inhibitory commissural neurons subserve inhibition of motoneurons in vestibulocollic reflex paths (Fig. 17.6). *Ia.i*, Ia inhibitory interneuron; *Ib.i*, Ib inhibitory interneuron; *Co*, cornucommissural neuron; *G*, substantia gelatinosa; *I*, intermediate grey; *Lat.mot*, lateral motor column; *Ma*, marginal cell layer; *Med.mot*, medial motor column; *P*, nucleus proprius of the posterior horn; *Re*, Renshaw cell

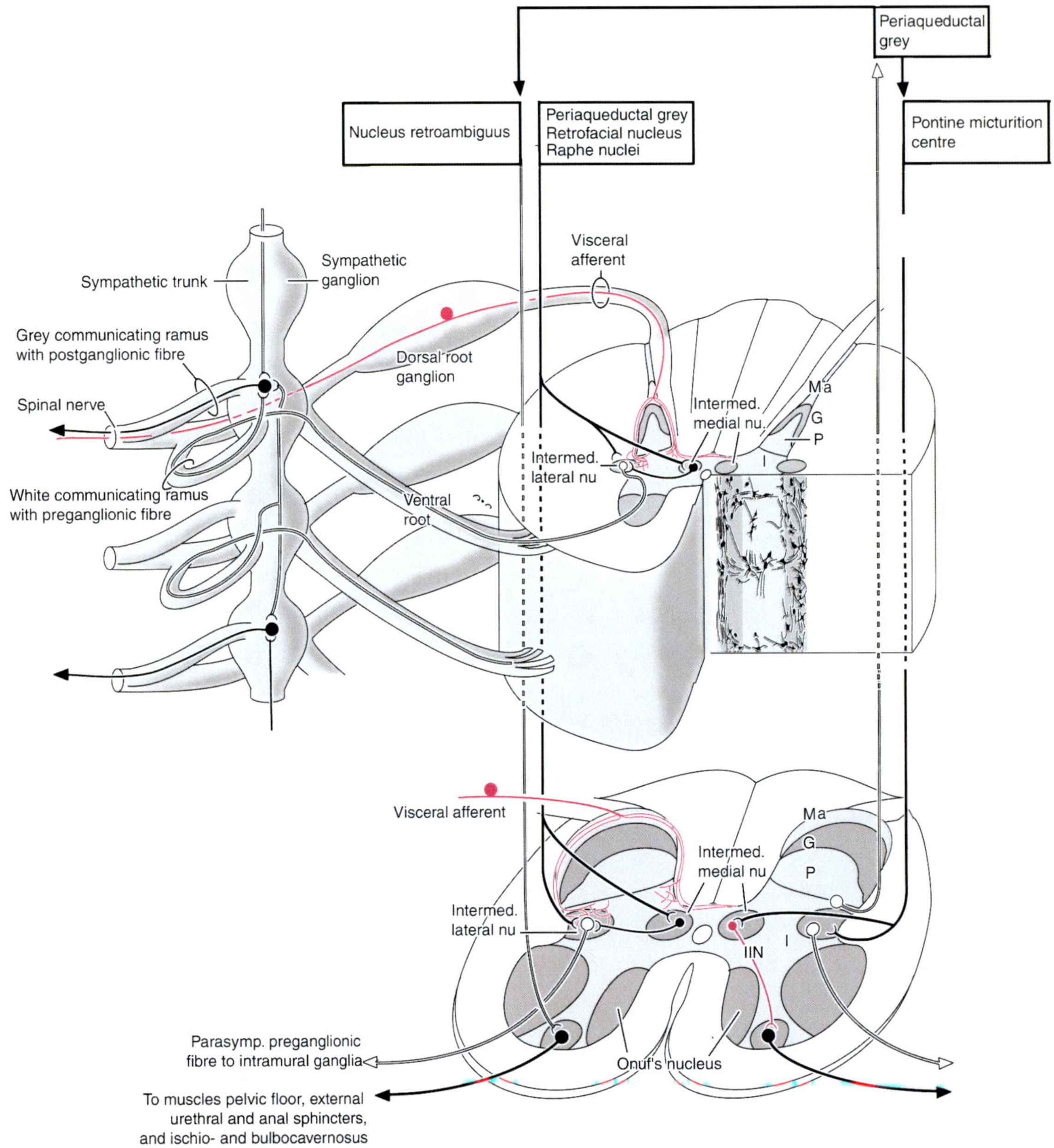

Fig. 6.5. The visceral reflex paths in the spinal cord. Preganglionic neurons are located in and near the intermediolateral nucleus of the intermediate horn of the thoracic cord and the dorsolateral intermediate grey of the sacral cord. The nucleus is connected with the intermediomedial nucleus through periodical strands of fibres and dendrites, illustrated in the frontal section of the thoracic cord (redrawn from [1]). Visceral afferents form a plexus surrounding the substantia gelatinosa and terminate in the medial and lateral intermediate grey. Their termination includes preganglionic neurons, and neurons giving rise to ascending pathways.

Thoracic preganglionic fibres are myelinated. They join the ventral roots and reach the sympathetic trunk through the white communicating rami. They ascend or descend into the trunk and terminate on postganglionic neurons located in its ganglia or in the ganglia of the prevertebral plexus. Postganglionic (B [12]) fibres are unmyelinated. They join the spinal nerves through the grey communicating rami, or pass along blood vessels to reach their destination.

Sacral preganglionic fibres belong to the parasympathetic system. They terminate on juxta- or intramural ganglia or on neurons of the autonomous plexus of the gut. Centres innervating the spinal autonomous nuclei include the hypothalamus, the periaqueductal grey, the retrofacial nucleus and the raphe nuclei.

Connections involved in micturition, indicated in the right half of the sacral cord, are based on a diagram of Holstege [20]. The pontine micturition centre innervates preganglionic neurons responsible for the contraction of the detrusor muscle of the bladder, and interneurons that inhibit the motoneurons of Onuf's nucleus, which innervate the external sphincter of the urethra. The periaqueductal grey receives afferent visceral information from the sacral cord, and controls micturition through the pontine micturition centre, and other vegetative functions through the nucleus retroambiguus, Onuf's nucleus and the muscles it innervates.

G, Substantia gelatinosa; *I*, intermediate grey; *IIN*, inhibitory interneuron; *Intermed.lateral nu*, intermediolateral nucleus; *Intermed.-medial nu*, intermediomedial nucleus; *Ma*, marginal cell layer; *P*, nucleus proprius of the dorsal horn

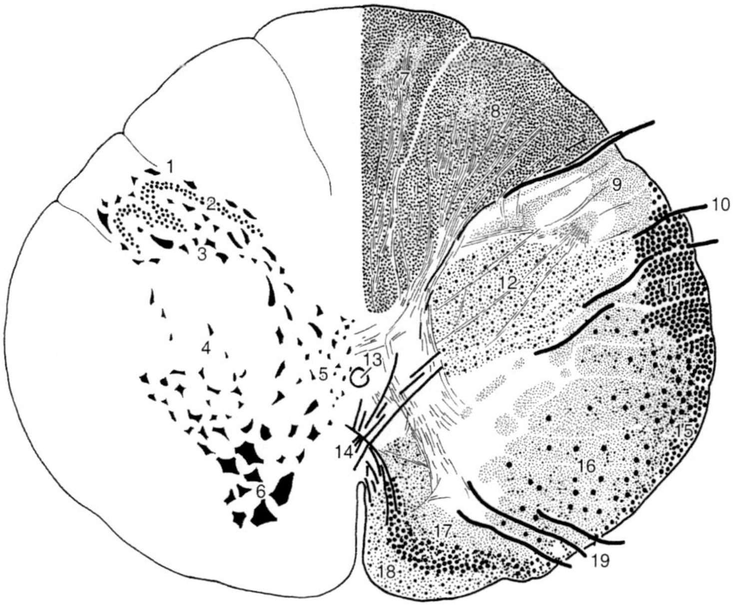

1 Dorsal horn, marginal cells (lamina I)
2 Dorsal horn, substantia gelatinosa (laminae II and III)
3 Dorsal horn, nucleus proprius (lamina IV)
4 Intermediate zone processus reticularis
5 Intermediate zone (laminae V–VIII)
6 Motor column (lamina IX)
7 Gracile fascicle
8 Cuneate fascicle
9 Dorsolateral fascicle (Lissauer [22])
10 Accessory nerve, spinal roots
11 Dorsal spinocerebellar tract
12 Lateral (crossed) pyramidal tract
13 Central canal
14 Ventral (white) commissure
15 Ventral spinocerebellar tract
16 Anterolateral fascicle
17 Medial longitudinal fascicle
18 Anterior (uncrossed) pyramidal tract
19 Ventral root

Fig. 6.6. Section through the first cervical segment (9/1×)

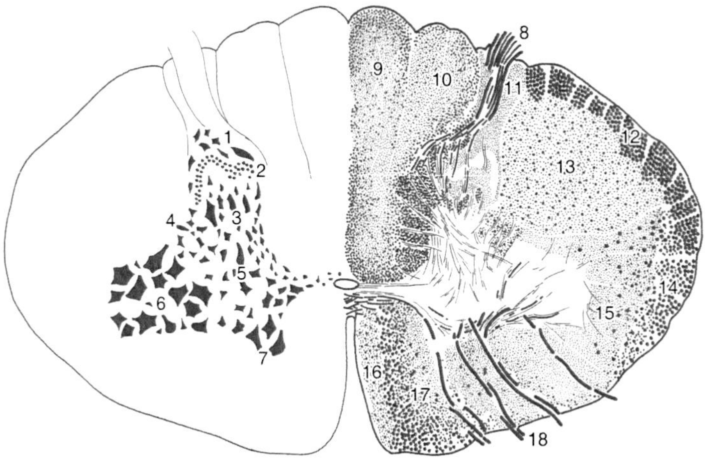

1 Dorsal horn, marginal cells (lamina I)
2 Dorsal horn, substantia gelatinosa (laminae II and III)
3 Dorsal horn, nucleus proprius (lamina IV)
4 Intermediate zone, processus reticularis
5 Intermediate zone (laminae V–VIII)
6 Lateral motor column (lamina IX)
7 Medial motor column (lamina IX)
8 Dorsal root
9 Gracile fascicle
10 Cuneate fascicle
11 Dorsolateral fascicle
12 Dorsal spinocerebellar tract
13 Lateral (crossed) pyramidal tract
14 Ventral spinocerebellar tract
15 Anterolateral fascicle
16 Anterior (uncrossed) pyramidal tract
17 Medial longitudinal fascicle
18 Ventral root

Fig. 6.7. Section through the transition of fifth to sixth cervical segment (9/1×)

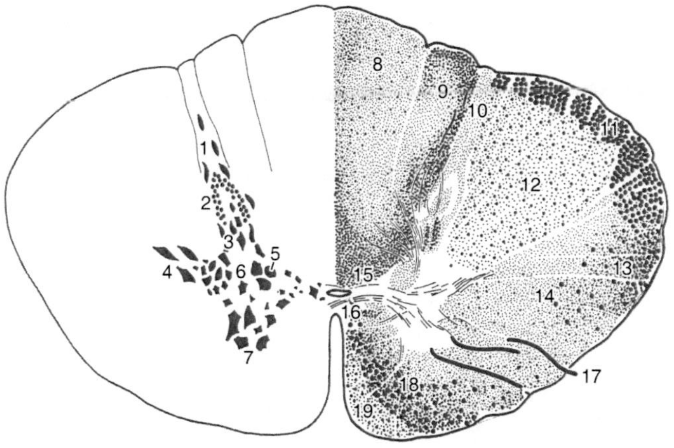

1 Dorsal horn, marginal cells (lamina I)
2 Dorsal horn, substantia gelatinosa (laminae II and III)
3 Dorsal horn, nucleus proprius (lamina IV)
4 Intermediolateral nucleus (lateral horn)
5 Thoracic nucleus (Clarke's column [4])
6 Intermediate zone
7 Motor column (lamina IX)
8 Gracile fascicle
9 Cuneate fascicle
10 Dorsolateral fascicle [22]
11 Dorsal spinocerebellar tract
12 Lateral (crossed) pyramidal tract
13 Ventral spinocerebellar tract
14 Anterolateral fascicle
15 Central canal
16 Ventral (white) commissure
17 Ventral root
18 Medial longitudinal fascicle
19 Anterior (uncrossed) pyramidal tract

Fig. 6.8. Section through the fifth thoracic segment (9/1×)

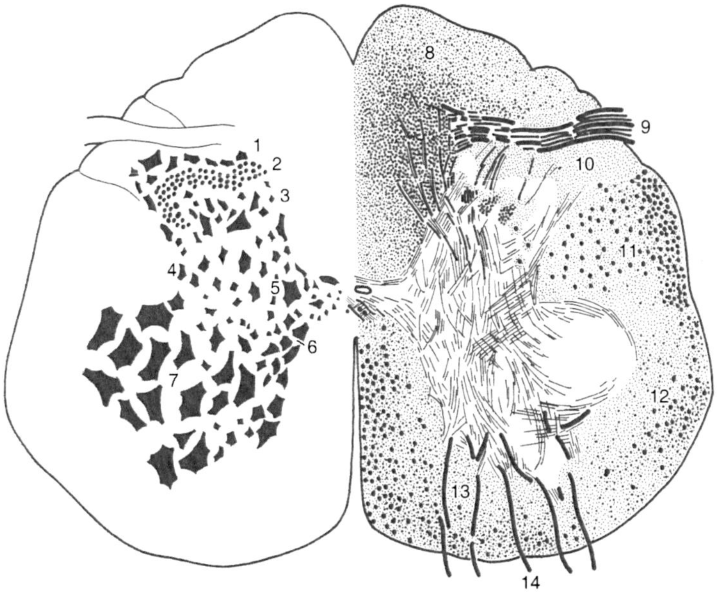

1 Dorsal horn, marginal cells (lamina I)
2 Dorsal horn, substantia gelatinosa (laminae II and III)
3 Dorsal horn, nucleus proprius (lamina IV)
4 Intermediate zone, reticular process
5 Intermediate zone (laminae V–VII)
6 Cornucommissural nucleus (lamina VIII)
7 Lateral motor column (lamina IX)
8 Dorsal funiculus
9 Dorsal root
10 Dorsolateral fasciculus (Lissauer [22])
11 Posterolateral funiculus
12 Anterolateral funiculus
13 Anterior funiculus
14 Ventral root

Fig. 6.9. Section through the fifth lumbar segment (9/1×)

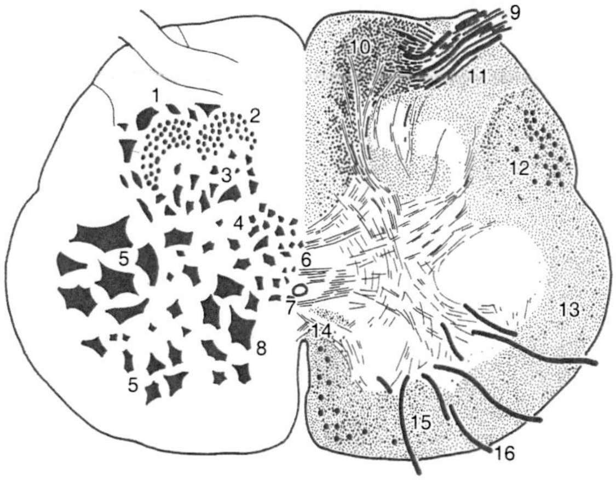

1 Dorsal horn, marginal cells (lamina I)
2 Dorsal horn, substantia gelatinosa (laminae II and III)
3 Dorsal horn, nucleus proprius (lamina IV)
4 Intermediate zone (laminae V–VIII)
5 Lateral motor column (lamina IX)
6 Posterior grey commissure
7 Anterior grey commissure
8 Medial motor column (lamina IX)
9 Dorsal root
10 Dorsal funiculus
11 Dorsolateral fasciculus (Lissauer [22])
12 Posterolateral funiculus
13 Anterolateral funiculus
14 Ventral (white) commissure
15 Anterior funiculus
16 Ventral root

Fig. 6.10. Section through the transition of fourth to fifth sacral segment (9/1×)

Topography of the Brain Stem and the Cerebellum

Introduction

The brain stem forms the central axis of the brain and comprises the medulla oblongata (myelencephalon), the pons (metencephalon) with the cerebellum, and the mesencephalon. The description of the internal structure of these subdivisions of the brain stem will be preceded by an account of the cranial nerve nuclei found therein.

Cranial Nerve Nuclei in the Brain Stem

The brain stem harbours the centres of origin and termination of ten (III–XII) of the twelve cranial nerves (Fig. 6.11). At first sight the arrangement of these cranial nerve nuclei does not show a definite pattern; however, the classical investigations of Gaskell [15, 16], Herrick [18] and many others (for a review, see Nieuwenhuys [28]) have revealed that these centres form part of functional zones, each of which is specifically related to one of the fibre categories of which the cranial nerves are composed. Before elucidating this zonal pattern, the fibre categories or nerve components of peripheral nerves in general deserve some comment.

A single nerve fibre in a spinal or cranial nerve may be afferent or efferent and be concerned with the innervation of either somatic or visceral structures. A combination of these two subdivisions yields the following four categories of peripheral nerve fibres:

1. General somatic afferent fibres, which transmit impulses from the skin, skeletal muscles, joints and ligaments
2. General visceral afferent fibres, which convey impulses from receptors in visceral organs and blood vessels centrally
3. General visceral efferent fibres, which supply the smooth musculature of internal organs, the cardiac muscle and glands
4. General somatic efferent fibres, which innervate skeletal muscles derived from myotomes. The denotation "general" has been added because fibres belonging to these four categories occur in both spinal and cranial nerves. The cranial nerves may, in addition, contain nerve fibres related to special structures that occur in the head region. These fibres can be classified into three additional categories:
5. Special somatic afferent fibres, which are associated with the receptors in the retina and in the cochlea
6. Special visceral afferent fibres, which supply the visceral sense organs, i.e. the organs of taste and smell
7. Special visceral efferent fibres, which innervate muscles derived from the mesenchyme of the visceral, bronchial arches.

Whereas a spinal nerve usually contains fibres of all four "general" categories, there is wide variation among the cranial nerves as regards the types of fibres they carry. Some of them have fibres of only one type, but in others fibres of two or more categories are present.

Returning now to the brain stem, the organizational pattern discovered by Gaskell and Herrick and their followers is that the cranial nerve nuclei are essentially arranged into seven longitudinal zones, and that each of these zones is specifically related to fibres of one of the categories mentioned above. Thus, as a cranial nerve composed of fibres of more than one type enters the brain, the fibres of its constituent types sort themselves out and pass to "their own" specific zone.

The zonal or columnar pattern displayed by the cranial nerve nuclei in the brain stem is diagrammatically represented in Fig. 6.12. As in the spinal cord, the afferent centres are situated in the alar lamina, whereas the efferent centres are located in the basal lamina. The sulcus limitans, which in the embryonic neuraxis marks the boundary between these two fundamental subdivisions, is only recognizable to a small extent in the adult. It should be noted that most of the zones are only partly occupied by cranial nerve nuclei. This may be related to the reduction of some components of some nerves during foetal development. The various zones and their constituent primary afferent or efferent centres will now be briefly reviewed, passing from lateral to medial.

The special somatic afferent (SSA) zone contains the nuclei of termination of the cochlear and vestibular division of the eighth nerve.

The general somatic afferent (GSA) zone includes the three sensory nuclei of the trigeminal nerve, i.e. the mesencephalic, the *princeps* (or principal) and the *spinal nuclei*. The latter nucleus, which also receives some fibres from the seventh, ninth and tenth nerves, is caudally continuous with the apical part of the dorsal horn of the spinal cord. The sensory trigeminal nuclei have shifted ventrolaterally during development; hence, the princeps nucleus and the rostral part of the spinal nucleus lie in the adult ventral rather than medial to the vestibular nuclei.

In the adult brain, the special visceral afferent (SVA) and general visceral afferent (GVA) zones are represented by a single cell mass, which receives the corresponding components of the seventh, ninth and tenth nerves. The latter unite in a well-defined fibre system, the *solitary tract*.

The general visceral efferent (GVE) zone contains four nuclei, the dorsal nucleus of the tenth nerve, the *inferior* and *superior salivatory nuclei* of the ninth and seventh nerves, respectively, and the accessory nucleus of Edinger-Westphal [11, 39] of the third nerve. These nuclei represent together the cranial division of the parasympathetic system. They give rise to preganglionic fibres that terminate in various autonomic ganglia. The dorsal nucleus of the tenth nerve is visible as a caudal ridge in the floor of the fourth ventricle: the *vagal triangle* (Fig. 3.10). The inferior and superior salivatory "nuclei" consist of scattered cells and, therefore, cannot be recognized in the transverse sections of this atlas.

The special visceral efferent (SVE), or branchiomotor zone, contains the motor nuclei of the fifth and seventh nerves as well as the *nucleus ambiguus*, which gives rise to fibres that pass peripherally as components of the ninth, tenth and eleventh (cranial root) nerves. The spinal nucleus of the eleventh nerve, which is situated in the lateral part of the base of the ventral horn of the upper four cervical segments, also belongs to this zone. As with the general somatic afferent nuclei, the cell masses of the branchiomotor zone have migrated away from their original periventricular position. Their efferent fibres, however, have retained their original position. As a consequence, these fibres make a loop, directed at the floor of the fourth ventricle before they leave the brain stem in a lateral direction. In the case of the seventh nerve, this loop is known as the *genu* (knee) of the facial nerve. It can be recognized as an elevation in the floor of the fourth ventricle: the *facial collicle* (Figs. 3.10, 6.25).

The general somatic efferent (GSE) zone may be considered a rostral continuation of the anterior horn of the spinal cord. It comprises the nuclei of origin of the twelfth, sixth, fourth and third nerves. All four of these nuclei are located near the median plane of the brain stem. The nucleus of the twelfth nerve forms an elevation in the caudal floor of the fourth ventricle: the *hypoglossal triangle* (Figs. 3.10, 6.23).

Medulla Oblongata

In the diagrams of Figs. 6.13–6.15 different systems of the medulla oblongata are indicated in different colours and shades. These diagrams refer to the sections depicted in Figs. 6.20 and 6.24, where it has been attempted to depict cell size and shape on the left side, and the typical texture of myelinated fibres of different calibres on the right. In transverse sections the medulla oblongata can be subdivided into dorsolateral "sensory" and ventromedial "motor" areas, which roughly correspond to the dorsal and ventral horns of the spinal cord; derivatives of, respectively, the embryological alar and basal plates. Their approximate border is demarcated by the fibres of the ninth, tenth and eleventh nerves, which pass ventral to or through the spinal tract of the trigeminal nerve on the way to their exit (Fig. 6.14). Apart from the general somatic afferent spinal nucleus of the trigeminal nerve and the dorsomedially located general and special visceral afferent nucleus of the solitary tract, the sensory area of the caudal medulla oblongata harbours the dorsal column nuclei.

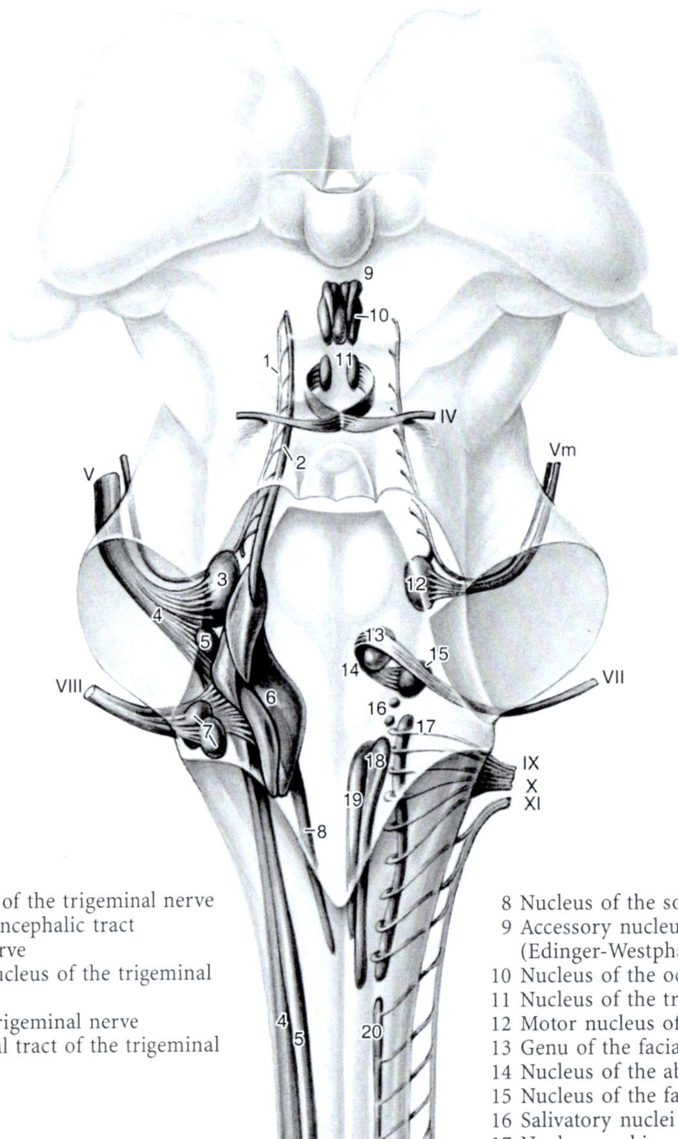

1 Mesencephalic tract of the trigeminal nerve
2 Nucleus of the mesencephalic tract of the trigeminal nerve
3 Principal sensory nucleus of the trigeminal nerve
4 Spinal tract of the trigeminal nerve
5 Nucleus of the spinal tract of the trigeminal nerve
6 Vestibular nuclei
7 Cochlear nuclei

8 Nucleus of the solitary tract
9 Accessory nucleus of the oculomotor nerve (Edinger-Westphal [11, 39])
10 Nucleus of the oculomotor nerve
11 Nucleus of the trochlear nerve
12 Motor nucleus of the trigeminal nerve
13 Genu of the facial nerve
14 Nucleus of the abducens nerve
15 Nucleus of the facial nerve
16 Salivatory nuclei
17 Nucleus ambiguus
18 Dorsal nucleus of the vagal nerve
19 Nucleus of the hypoglossal nerve
20 Nucleus of the spinal root of the accessory nerve
Vm portio minor (motor root) of the trigeminal nerve

Fig. 6.11. The cranial nerve nuclei as viewed from the dorsal side (5/3×). *Left*: Sensory nuclei; *right*: motor nuclei. *Roman numerals* indicate the corresponding cranial nerves

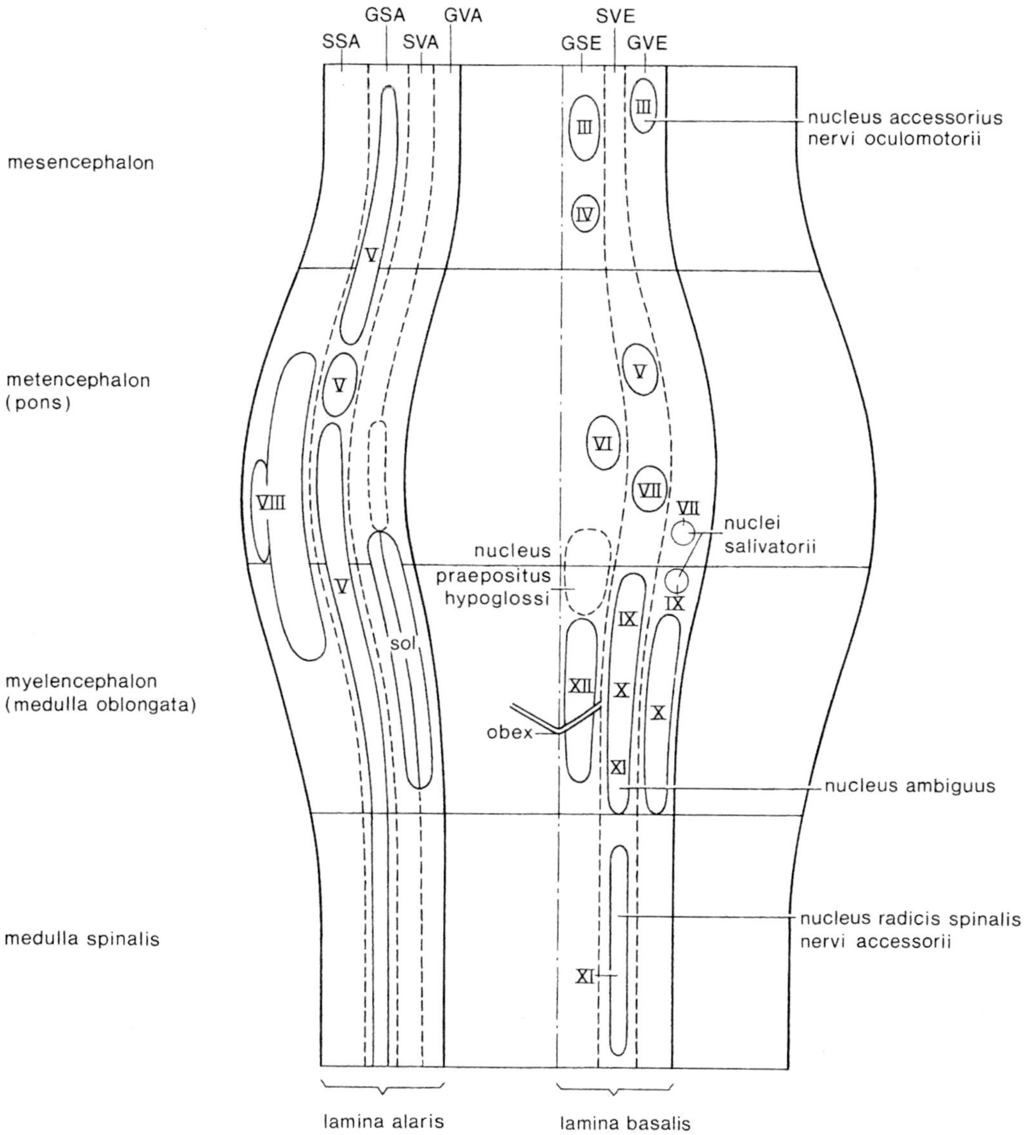

Fig. 6.12. Position of the cranial nerve nuclei in longitudinal columns. The *initials* at the top indicate the functional system to which each column belongs. The subdivisions of the brain stem are indicated at *left*. *III–XII*, nuclei of the cranial nerves; *GSA*, general sensory afferent column; *GSE*, general somatic efferent column; *GVA*, general visceral afferent column; *GVE*, general visceral efferent column; *Sol*, nucleus of the solitary tract; *SSA*, special sensory afferent column; *SVE*, special visceral efferent column; *VA*, special visceral afferent column

The spinal trigeminal nucleus and the nucleus of the solitary tract receive their primary afferents from descending branches of root fibres that bifurcate in ascending and descending branches at their entrance into the brain stem. In the case of the spinal trigeminal nucleus these fibres descend into the *spinal tract of the trigeminal nerve*, from the level of the pons, where the trigeminal nerve enters the brain stem, penetrating the middle cerebellar peduncle. The *solitary tract* consists of descending branches of the facial, vagal and glossopharyngeal nerves. The spinal trigeminal nucleus is directly continuous with the dorsal horn of the cervical cord. Its caudal subnucleus shares its laminar organization with the dorsal horn (Figs. 6.21 and 6.22). The spinal tract of the trigeminal nerves continues into the dorsolateral tract at the apex of the dorsal horn.

The *dorsal column nuclei* receive the ascending branches of spinal root fibres from the dorsal columns. The dorsal columns can be subdivided into the medial *gracile fascicle* and the lateral *cuneate fascicle* (Fig. 6.21), which transmit sensory information from the lower and upper extremities, respectively. A corresponding subdivision can be observed for the dorsal column nuclei. The medial *gracile nucleus* receives root fibres from the gracile fascicle, and the *medial* and *lateral cuneate nuclei* from the cuneate fascicle. The medial and lateral cuneate nuclei differ in their central connections. The medial cuneate nucleus, like the gracile nucleus, gives rise to fibres that arch ventrally as internal arcuate fibres, cross the ventral midline and ascend in the *medial lemniscus*, an important sensory conduction pathway that terminates in the thalamus, conveying the sensory information to the cortex (Fig. 6.22). The main projection of the lateral cuneate nucleus, on the other hand, is to the cerebellum. Most of its fibres join the restiform body; some of its fibres ascend in the contralateral medial lemniscus. The connections of the nuclei of the dorsal column and the trigeminal nerve are considered in more detail in Chap. 16.

In the rostral medulla oblongata the dorsal column, nuclei gradually are replaced by the *vestibular nuclei* (Figs. 6.24–6.27), which are located in the floor of the fourth ventricle. The *cochlear nuclei* are located lateral and dorsal to the restiform body, in the floor of the lateral recess of the fourth ventricle. The vestibulocochlear nerve enters the brain stem immediately caudal to the middle cerebellar peduncle (Fig. 6.15). Its rostral vestibular division passes ventral to the restiform body to enter the vestibular nuclei. Here its fibres split into ascending and descending branches. Bundles of descending branches demarcate the *descending or spinal vestibular nucleus* (Fig. 6.24). They belong to the *juxtarestiform body*, which, in addition, contains Purkinje cell axons from the cerebellar cortex and efferent fibres of the fastigial nuclei. Together with the restiform body it constitutes the inferior cerebellar peduncle. The *lateral vestibular Deiters' nucleus* [10] is located in the cerebellar portion of the juxtarestiform body; the medial vestibular nucleus is located medial to the juxtarestiform body (Figs. 6.24, 6.25). The *superior vestibular nucleus*, which receives the ascending root of the vestibular nerve, is found at a more rostral level (Fig. 6.25).

The cochlear division of the vestibulocochlear nerve enters the *ventral cochlear nucleus*, where it divides into ascending and descending branches (Fig. 6.15). The descending branches ultimately enter the *dorsal cochlear nucleus*, which caps the ventral cochlear nucleus caudally and dorsally. Efferent cochlear systems exit the dorsal cochlear nucleus as the *dorsal acoustic stria* and as the *intermediate acoustic stria* and the *trapezoid body*, issued by the ventral cochlear nucleus. The dorsal acoustic stria are part of the *stria medullaris*, located in the floor of the fourth ventricle (Figs. 6.15, 6.24).

The ventromedial motor area of the caudal medulla oblongata contains the general visceral efferent *dorsal nucleus of the vagal nerve* and the general somatic efferent *nucleus of the hypoglossal nerve*. Both are located dorsally; the special visceral efferent (branchiomotor) *nucleus ambiguus* is located ventromedial to the spinal nucleus of the trigeminal nerve. Prominent structures in the ventromedial medulla oblongata are the pyramidal tract and the inferior olive. Both can be recognized in the surface relief of the ventral medulla oblongata (Fig. 3.12).

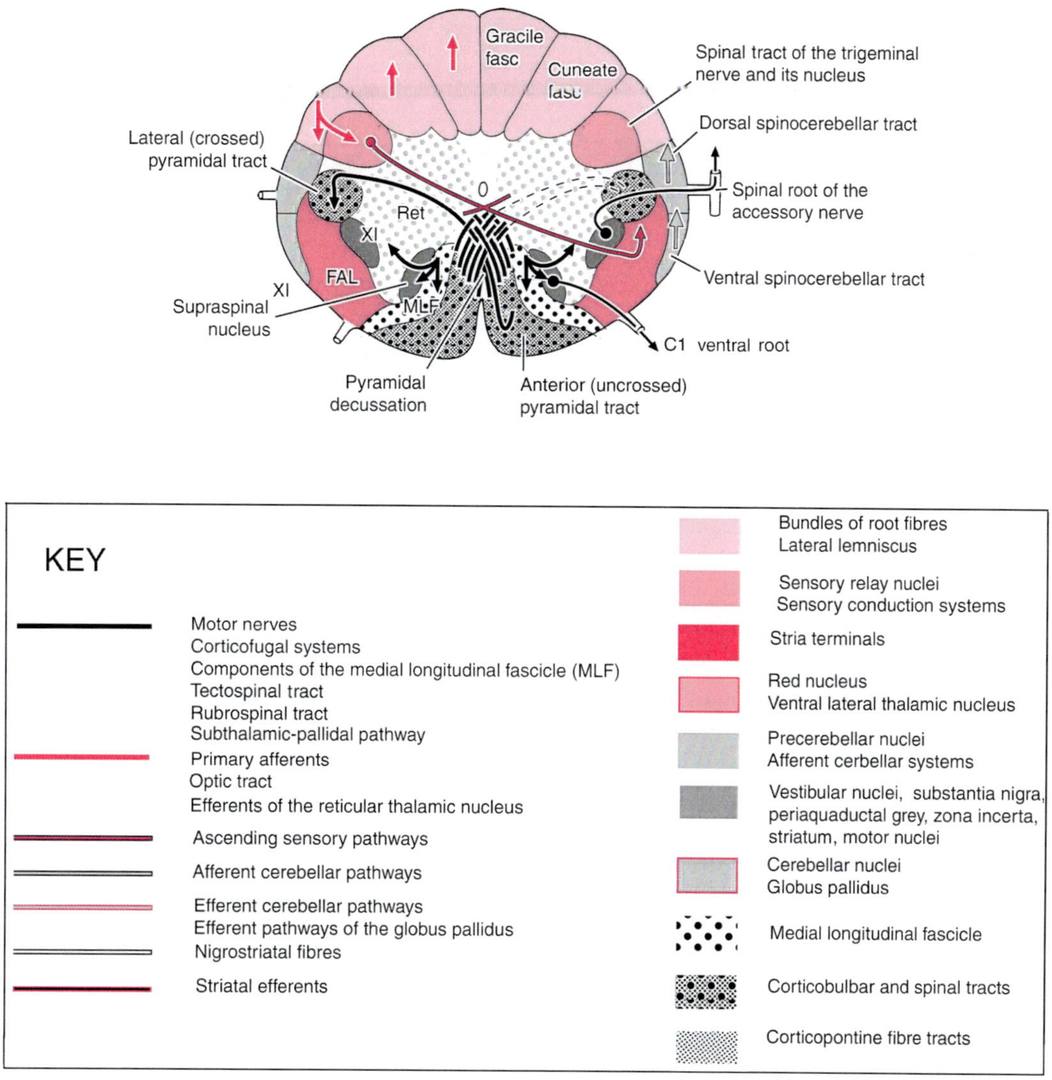

Fig. 6.13. Diagrammatic section through pyramidal decussation, showing the course and direction of fibre tracts, with key for Figs. 6.13–6.19. *XI*, Spinal nucleus of the accessory nerve; *FAL*, anterolateral fascicle; *MLF*, medial longitudinal fascicle; *Ret*, reticular formation

At rostral levels of the medulla oblongata, the motor nuclei of the hypoglossal and vagal nerve are replaced by the *nucleus prepositus hypoglossi*, located medially in the floor of the fourth ventricle. This nucleus is a link in the oculomotor system. The nucleus ambiguus still occupies its position ventral to the spinal nucleus of the trigeminal nerve (Fig. 6.24).

The *pyramidal tract* emerges at the surface at the caudal border of the pons and decussates at the transition of the medulla oblongata with the spinal cord. The dorsal boundary of the pyramidal tract with the medial lemniscus is sharp because the fibre composition of the two tracts is quite different. Fibres of the medial lemniscus are medium-sized, most fibres of the pyramidal tract are small, with an admixture of coarse fibres. A considerable proportion of the fibres of the pyramidal tract leaves it and crosses the midline to terminate on neurons of various brain stem nuclei (Chap. 21). The *inferior olive* consists of the folded cellular lamina of the principal nucleus of the inferior olive and the medial and dorsal accessory olives. Its efferent fibres cross the midline at the level of the olive, pass medial and lateral to the spinal tract of the trigeminal nerve and join the restiform body to proceed towards the cerebellum (Figs. 6.23, 6.24). The inferior olive receives projections from the spinal cord, the dorsal column nuclei and a group of nuclei located at the junction of the mesencephalon with the diencephalon, which descend into the tegmental tracts. One of these tracts, the *central tegmental tract*, forms a fleece of myelinated fibres around the olive, which is known as the *amiculum* (Chap. 20).

Of the fibre systems of the lateral funiculus of the cord, the *dorsal spinocerebellar tract* shifts dorsally, superficial to the spinal tract of the trigeminal nerve. Joined by fibres originating from the reticular formation, the external cuneate nucleus and the contralateral inferior olive, it constitutes the *restiform body*, which belongs to the *inferior cerebellar peduncle*. The restiform body enters the cerebellum at the level of the metencephalon. The fibres of the spino-, cuneo- and reticulocerebellar tracts are coarse to medium-sized and contrast with the small olivocerebellar fibres of uniform calibre. The anterolateral fasciculus, with the ventral spinocerebellar and spinothalamic tracts, keep to their ventrolateral position. The medial longitudinal fasciculus is located in the floor of the fourth ventricle, along the midline. It is characterized by its high content of coarse fibres. It is separated from the pyramidal tract by the medial lemniscus (Figs. 6.22–6.24).

Throughout most of its extent, the central region of the medulla oblongata is occupied by aggregations of loosely arranged cells of different types and sizes, known as the *reticular formation*. The term reticular formation refers to the arrangement of the dendrites of the cells in this area in bundles that together form a net-like pattern [33]. The traversing fibre systems pass through the interstices of this network. Caudally, the reticular formation is continuous with the intermediate grey matter of the spinal cord; rostrally, it continues into the intralaminar nuclei of the thalamus and into certain cellular aggregates of the subthalamus.

The reticular formation can be subdivided into lateral, medial and median zones. The subdivision of the reticular formation, the raphe nuclei and the classification of the indolaminergic cell groups of Dahlström and Fuxe [8] is considered in Chap. 22 (Figs. 22.1, 22.2).

The *lateral zone of the bulbar reticular formation* (or *the lateral tegmental field*), consists mainly of small neurons (which is why it is often designated as the *parvocellular reticular formation*) and is situated directly medial to the spinal trigeminal and solitary tract nuclei (Figs. 6.22–6.24). It contains numerous interneurons that project to the motor nuclei of the brain stem. Its ventrolateral portion, known as the *superficial ventrolateral reticular area*, harbours the adrenergic cell group C1 and the noradrenergic cell group A1. More rostrally, this area continues into the *paragigantocellular nucleus* [29], which contains the noradrenergic cell groups A5 and A7. The most caudal portion of the ventrolateral reticular area corresponds to the *retroambiguus nucleus*. The cells of the noradrenergic cell group A2 and the adrenergic group C2 are located in and around the nucleus of the solitary tract.

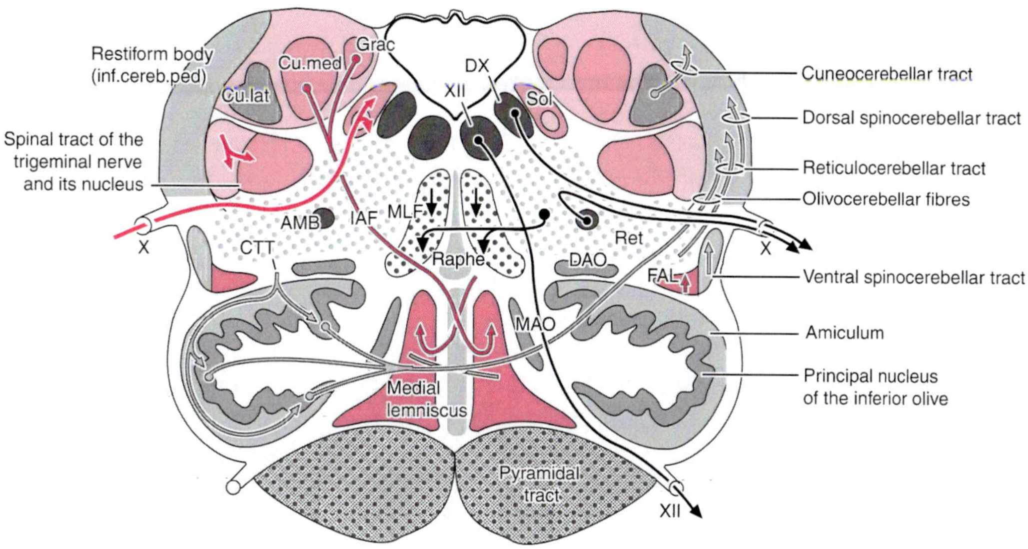

Fig. 6.14. Diagrammatic section through the medulla oblongata, showing the course and direction of fibre tracts. For key see Fig. 6.13. *X*, vagal nerve; *XII*, nucleus of the hypoglossal nerve; *AMB*, nucleus ambiguus; *CTT*, central tegmental tract; *Cu.lat*, lateral cuneate nucleus; *Cu.med*, medial cuneate nucleus; *DAO*, dorsal accessory olive; *DX*, dorsal vagal nucleus; *Grac*, gracile nucleus; *IAF*, internal arcuate fibres; *inf.cer.ped*, inferior cerebellar peduncle (restiform body); *MAO*, medial accessory olive; *MLF*, medial longitudinal fascicle; *Raphe*, nucleus raphes obscurus and raphes pallidus; *Ret*, reticular formation; *Sol*, nucleus of the solitary tract

The *medial zone* of the reticular formation (or the *medial tegmental field*) contains numerous large neurons, with sparsely branching dendrites oriented in the transverse plane, like those of the spinal intermediate grey matter. At the level of the caudal medulla the medial zone is represented by the *central nucleus of the medulla oblongata*. In the rostral medulla the medial zone is known as the *gigantocellular reticular nucleus*. The axons of the neurons of the medial reticular formation generally bifurcate into ascending and descending branches, which emit numerous collaterals along their course. These neurons may serve as links in multisynaptic pathways ascending to the diencephalon and descending to the spinal cord. The medial reticular formation receives afferents from the spinal cord, the cerebellum (mainly from the fastigial nucleus), relay centres of most sensory cranial nerves and from motor and premotor cortical areas. Its fibre connections suggest that the medial reticular formation is integrated into both sensory and motor pathways.

The median zone consists of the raphe nuclei. Many of the neurons of the raphe nuclei are serotoninergic; serotoninergic neurons also are present in the adjoining reticular formation. The raphe nuclei at the level of the medulla oblongata include the *nucleus raphes pallidus*, *raphes obscurus* and *raphes magnus*. They correspond with the serotoninergic cell groups B1, B2 and B3 respectively (Fig. 22.2).

Three more compact reticular nuclei can be distinguished in the reticular formation. Two of them, the nucleus of lateral funiculus (or nucleus reticularis lateralis) and the nucleus of the anterior funiculus (or paramedian reticular nucleus), are located in the medulla oblongata (Figs. 6.22, 6.23). The third one, i.e. the *nucleus reticularis tegmenti pontis* is, as its name implies, located in the tegmentum of the pons (Fig. 6.27). The *nucleus of the lateral funiculus* is located among the ascending spinal fibres of the anterolateral fasciculus and the more dorsally located rubrospinal tract and receives afferents from both. The *nucleus of the anterior funiculus* or *paramedian reticular nucleus* consists of cell groups located among the fibres of the medial longitudinal fascicle and receives afferents from several descending components. The three reticular nuclei project to the cerebellum. The nuclei of the anterior and lateral funiculus give rise to external arcuate fibres, which pass lateral to the spinal tract of the trigeminal nerve to join the restiform body (Fig. 6.23). The nucleus reticularis tegmenti pontis projects to the cerebellum via the rachium pontis.

Metencephalon and the Cerebellum

The metencephalon is known in colloquial neuroanatomy as the pons, after the ventral prominence of the pontine nuclei. The metencephalon is wedge-shaped; the rostrocaudal extent of the pons greatly exceeds its dorsal, ventricular boundary (Fig. 3.7). As a consequence, the pontine nuclei are still contained in sections passing through the tegmentum and tectum of the mesencephalon (Figs. 6.29, 6.30). Sections of the metencephalon (Figs. 6.25–6.28) pass through the broad attachment of the cerebellum to the brain stem, which contains the three cerebellar peduncles, and through the rostral half of the fourth ventricle and the pons. Structures at the level of the metencephalon are summarized in the diagram of Fig. 6.16.

The metencephalon can be subdivided into three zones. The ventral zone contains the pontine nuclei and is known as the pes pontis. The intermediate zone or pontine tegmentum is bordered by the pes pontis, the middle cerebellar peduncles and the fourth ventricle. The dorsal zone corresponds to the cerebellum, located in the roof of the rostral half of the fourth ventricle. Rostrally the cerebellum is represented by the *superior medullary velum*, a plate-like extension of the cerebellum that contains the *lingula*, the ventral-most lobule of the anterior cerebellum (Figs. 3.8, 6.29). The white matter of the velum is an extension of the cerebellar commissure. The decussation and exit of the trochlear nerve from the superior medullary velum marks the border of the pons with the mesencephalon or the pontine tegmentum. The boundary of the metencephalon with the medulla oblongata is an anatomical conven-

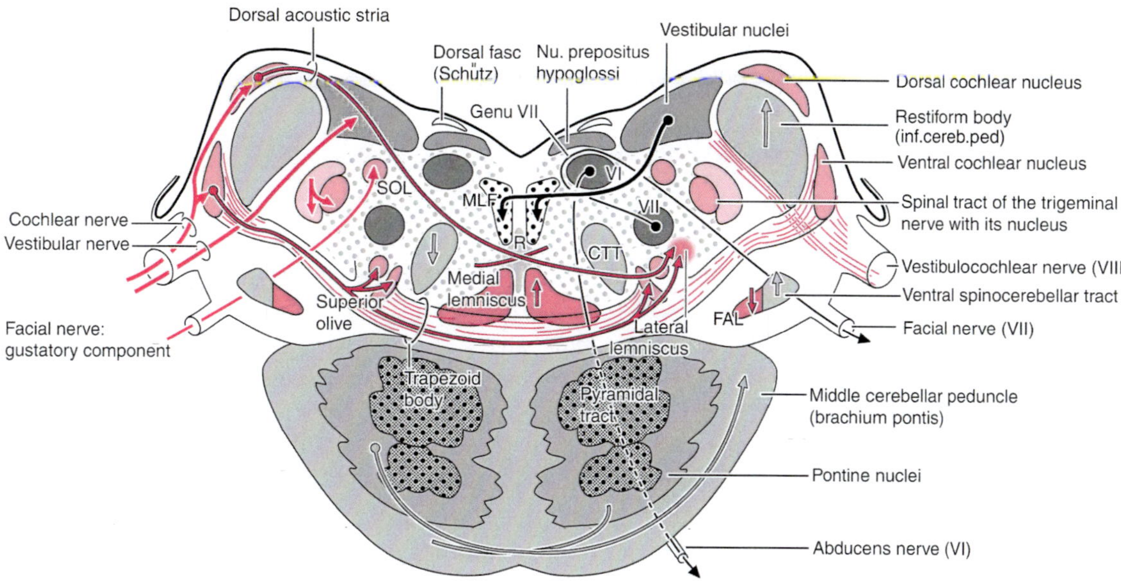

Fig. 6.15. Diagrammatic section through the cochlear nuclei and the trapezoid body, showing the course and direction of fibre tracts. For key see Fig. 6.13. *VI*, Nucleus of the abducens nerve; *CTT*, central tegmental tract; *FAL*, anterolateral fascicle; *inf.cereb.ped*, inferior cerebellar peduncle (restiform body); *MLF*, medial longitudinal fascicle; *R*, nucleus raphes magnus; *SOL*, nucleus of the solitary tract

tion, based on the prominence of the pons and the cerebellum. Structures contained in the medulla oblongata can be traced, without obvious changes, into the tegmentum pontis.

The cerebellum is covered by a uniform, three-layered *cortex*. A monolayer of large Purkinje [30, 38] cells separates the layer of granule cells from the superficial, cell-poor molecular layer. The anatomy of the cerebellum is dealt with in more detail in Chap. 20.

The cerebellar white matter extends as narrow laminae in the lobules and folia of the cerebellum. The branching white matter in a sagittal section is known as the *arbor vitae* (Fig. 3.8). The laminae of the white matter contain afferent mossy and climbing fibres on their way to the cerebellar cortex. Mossy fibres terminate on dendrites of granule cells. These synaptic complexes (glomeruli) are located in islands within the granular layer (Fig. 6.26). Mossy fibres take their origin as the spinocerebellar tracts from the spinal cord, the lateral cuneate nucleus, the vestibular and reticular nuclei and the pons. Climbing fibres exclusively originate from the contralateral inferior olive. They terminate on the Purkinje cell dendrites in the molecular layer. In addition, the white matter laminae contain Purkinje cell axons on their way to the cerebellar and vestibular nuclei. White matter laminae are subdivided into a number of symmetrically disposed, parasagittal compartments. Each compartment contains the Purkinje cell axons from a longitudinal cortical zone and the climbing fibres innervating these Purkinje cells. These longitudinal zones may extend over the entire cerebellum.

At the midline the white matter of both sides of the cerebellum is continuous as the cerebellar commissure. This commissure consists of a ventral portion, containing the decussating fibres of the fastigial nucleus, and a rostral and dorsal portion containing branches of mossy fibres terminating contralaterally (Fig. 6.27). Rostrally, the cerebellar commissure extends into the superior medullary velum. The olivocerebellar fibres do not participate in the cerebellar commissure; instead they decussate in the medulla oblongata and distribute strictly unilaterally.

Four cerebellar nuclei can be distinguished in the central white matter of the cerebellum (Figs. 6.25–6.27). The large *dentate* or *lateral nucleus* consists of a folded cellular sheet, very similar in appearance to the inferior olive. The *fastigial (medial) cerebellar nucleus* is located next to the midline. The *emboliform* (*anterior interposed*) and *globose* (*posterior interposed*) *nuclei* occupy an intermediate position. Efferent fibres of the dentate nucleus collect in its hilus and, together with fibres from the globose and emboliform nuclei, constitute the *superior cerebellar peduncle* or *brachium conjunctivum*. The majority of the efferents from the fastigial nucleus decussate within the cerebellar commissure (Fig. 6.27) and form the *uncinate tract*. This tract hooks over the brachium conjunctivum to join the juxtarestiform body, lateral to the lateral vestibular nucleus. *Uncrossed fastigiobulbar fibres* pass ventrally in the lateral wall of the fourth ventricle, where they join the juxtarestiform body, medial to the lateral vestibular nucleus. The uncinate tract and the direct fastigiobulbar fibres terminate in the medial and descending vestibular nuclei and the medial reticular formation (Fig. 6.16).

Of the three cerebellar peduncles, the middle cerebellar peduncle or brachium pontis is a purely afferent system, conveying mossy fibres from the pontine nuclei. The *inferior cerebellar peduncle* consists of the lateral, afferent restiform body and the medial, partially efferent, juxtarestiform body. In the dorsolateral medulla oblongata, the restiform body is located lateral to the vestibular nuclei with the juxtarestiform body, and ventromedial to the cochlear nuclei (Fig. 6.24). It is composed of the dorsal spinocerebellar tract and of external arcuate fibres from the reticular nuclei of the lateral and anterior funiculus, which join the restiform body by passing dorsally, lateral to the spinal tract of the trigeminal nerve. Olivocerebellar fibres from the contralateral inferior olive, on their way to the restiform body, pass lateral to and through the spinal tract (Fig. 6.23). The restiform body enters the cerebellum rostrally in metencephalon, medial to the middle cerebellar peduncle. In the cerebellum its fibres occupy a position rostral and dorsal to the cerebellar nuclei (Fig. 6.27).

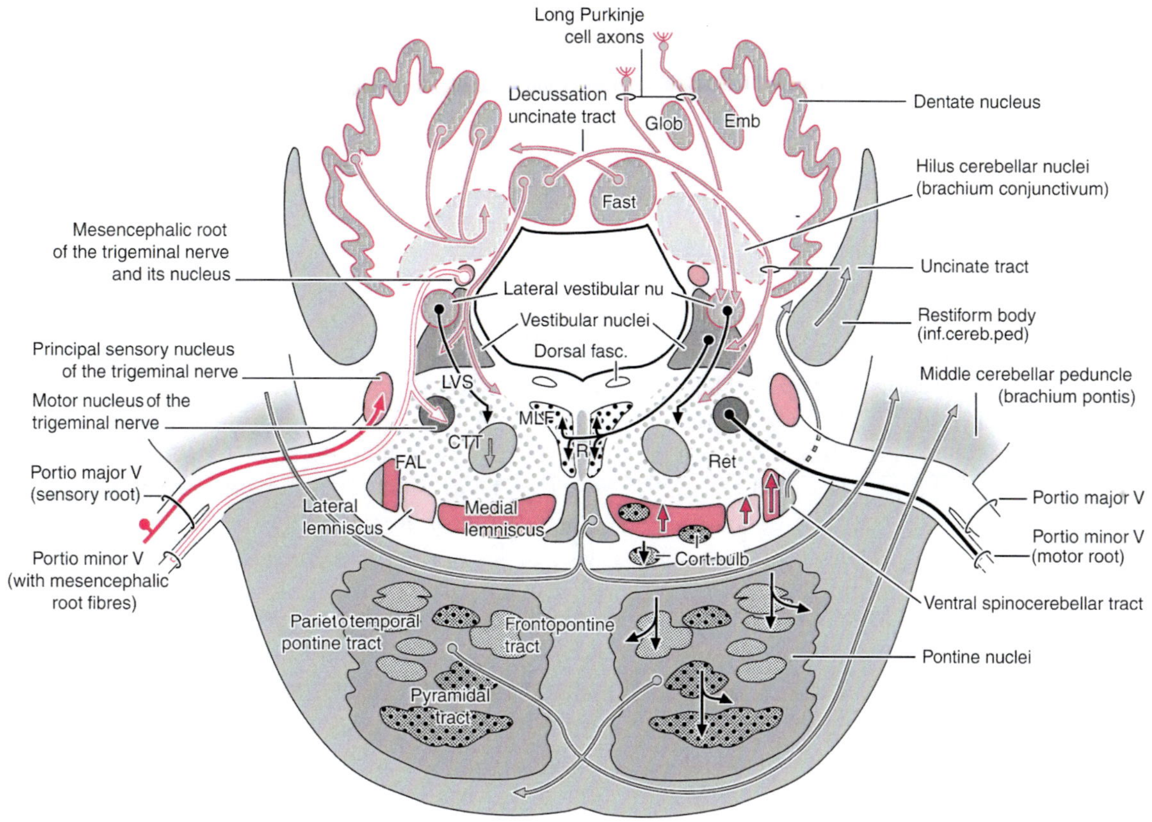

Fig. 6.16. Diagrammatic section through the cerebellar nuclei and the pontine tegmentum, showing the course and direction of fibre tracts. For key see Fig. 6.13. *Cort.bulb*, corticobulbar fibre bundles (areae nebulosae); *CTT*, central tegmental tract; *Emb*, emboliform (anterior interposed) nucleus; *FAL*, anterolateral fasciculus; *Fast*, fastigial nucleus; *Glob*, globose (posterior interposed) nucleus; *inf.cereb.ped*, inferior cerebellar peduncle (restiform body); *LVS*, lateral vestibulospinal tract; *MLF*, medial longitudinal fascicle; *R*, nucleus raphes pontis; *Ret*, reticular formation

The *middle cerebellar peduncle* or *brachium pontis* takes its origin from the pontine nuclei and the nucleus reticularis tegmenti pontis, mainly on the contralateral side (Figs. 6.27, 6.28).

The *pontine nuclei* are contained in the pes pontis. They form the main link between the cerebral cortex and the contralateral cerebellum. The pontine nuclei receive corticopontine fibres via the cerebral peduncle and as collaterals from axons of the pyramidal tract, which passes through the pons on its way to the medulla oblongata (Figs. 6.25–6.30). Corticobulbar fibres detach from the pyramidal tract to enter the tegmentum in small bundles, dispersed in the medial lemniscus, known as the areae nebulosae (Figs. 6.27–6.29). The pontine nuclei give rise to transversely oriented fibres, most of which cross the midline of the pons. Laterally they assemble into the middle cerebellar peduncle.

The *nucleus reticularis tegmenti pontis* is a compact reticular nucleus located in the ventromedial tegmentum (Figs. 6.27, 6.28). It remains separated from the pontine nuclei by fibre bundles of the medial lemniscus. The nucleus is reciprocally connected with the cerebellum. It receives *the crossed descending branch of the superior cerebellar peduncle* (Fig. 6.27). The efferent fibres of this nucleus join the middle cerebellar peduncle and terminate within the cerebellum.

The *pontobulbar body* is a group of cells located ventral to the cochlear nuclei (Figs. 6.24, 6.25). It is a remnant of the pontine migration and is located at its site of origin. It may project to the cerebellum.

The *superior cerebellar peduncle* or *brachium conjunctivum* is the main efferent pathway from the cerebellum. It is located in the lateral wall of the fourth ventricle, medial to the restiform body (Figs. 3.10, 6.27). It decussates more rostrally, at the border of the metencephalon and the mesencephalon (Figs. 6.27, 6.29). The dorsomedial portion of the peduncle and its ventral-most fibres are the first to decussate, the latter crossing immediately dorsal to the interpeduncular nucleus. The middle part of the superior cerebellar peduncle decussates at a more rostral level. The ventral spinocerebellar tract leaves the area of the anterolateral fascicle in the rostral metencephalon, and passes rostral to the entrance of the trigeminal nerve to enter the cerebellum along the superior cerebellar peduncle.

The vestibular and sensory trigeminal nuclei and the nucleus of the solitary tract continue from the medulla oblongata into the dorsolateral portion of the tegmentum pontis. The position of the rostral pole of the vestibular nuclear complex in the transitional region of the brain stem with the cerebellum is worth noting. Cerebellar efferents, such as the uncinate tract and Purkinje cell axons from certain regions of the cerebellar cortex, enter the vestibular nuclei in the juxtarestiform body, and vestibular afferents gain access to the cerebellum at this point (Figs. 6.26, 6.27). The sensory trigeminal nuclei are represented by the rostral pole of the spinal trigeminal nucleus (the "*pars oralis*") and the *principal sensory nucleus*, which is located rostrally in the metencephalon, at the level where the trigeminal nerve enters the brain stem, passing through the middle cerebellar peduncle (Figs. 6.25–6.27). The rostral pole of the nucleus of the solitary tract is represented at this level by the *nucleus ovalis*. This nucleus is located within a fibre bundle, the *area ovalis*, next to the principal sensory nucleus of the trigeminal nerve. This bundle contains ascending gustatory fibres and/or fibres of the uncrossed ascending dorsal trigeminothalamic tract (Figs. 6.25, 6.26).

The medial lemniscus is located at the border of the pes and the tegmentum pontis (Figs. 6.25–6.29). Fibres from the contralateral spinal and principal sensory nuclei of the trigeminal nerve join the medial lemniscus and occupy its medial part as the "*trigeminal lemniscus*". Fibres from the principal nucleus decussate in the rostral metencephalon, immediately caudal to the *interpeduncular nucleus*. The anterolateral fasciculus with the spinothalamic tracts is located in the ventrolateral tegmentum, in between the medial and lateral lemniscus (Fig. 6.29). The ventral spinocerebellar tract separates from the anterolateral fasciculus to join the superior cerebellar peduncle.

The efferent pathways of the vestibular and cochlear nuclei arise at this level of the brain stem (Figs. 6.15, 6.16). The vestibular nuclei give rise to crossed and uncrossed, ascending and descending pathways that are located within or in close proximity to the medial longitudinal fascicle (Figs. 6.26, 6.27). Fibres from the cochlear nuclei leave these nuclei in a ventral direction, pass ventral to the spinal root of the trigeminal nerve and cross the midline of the caudal metencephalon, dorsal to the pes pontis in the *trapezoid body* (Figs. 6.24–6.26). The fibre bundles of the trapezoid body traverse the medial lemniscus on both sides of the midline and turn rostrally, lateral to the medial lemniscus as the *lateral lemniscus*. The dorsal acoustic stria, which pass dorsal to the restiform body, and the intermediate acoustic stria, which pass through the spinal tract of the trigeminal nerve, join the trapezoid body after or prior to its decussation, respectively. In the rostral metencephalon the lateral lemniscus shifts dorsally to terminate in the inferior colliculus of the mesencephalon (Figs. 6.29, 6.30). The nuclei composing the *superior olivary complex* are located within the trapezoid body. They include the *medial* and *lateral superior olive*, the *periolivary nuclei* and the *nucleus of the trapezoid body*. These nuclei are links in different ascending and descending auditory pathways. The *ventral and dorsal nuclei of the lateral lemniscus* are located among the fibres of this bundle. The vestibular and acoustic pathways are considered in more detail in Chaps. 17 and 18.

Both the general somatic and the general and special visceral efferent nuclei are represented in the ventromedial tegmentum pontis. The large branchiomotor nucleus of the facial nerve gives rise to fibres that are directed to the floor of the fourth ventricle, where they make a rostrally directed loop, the *knee* (*genu*) *of the facial nerve*, to leave the brain stem ventral to the spinal tract of the trigeminal nerve, immediately caudal to the middle cerebellar peduncle (Figs. 6.25, 6.26). Preganglionic parasympathetic fibres of the facial nerve, which take their origin from the superior salivatory nucleus (not illustrated), leave the brain stem in a separate nerve, the *intermediate nerve* (Fig. 3.12), located between the facial and vestibulocochlear nerves.

The branchiomotor nucleus of the trigeminal nerve is located more rostrally, medial to the root of the trigeminal nerve (Figs. 6.26, 6.27). It gives rise to the smaller one of the two subdivisions of this nerve (the *portio minor*). The larger *portio major* contains the central axons of the trigeminal ganglion cells, which divide into ascending branches that terminate in the principal sensory nucleus and descending branches that constitute the spinal root of the trigeminal nerve. A certain proportion of the portio minor fibres are afferent. These fibres originate from large, round neurons of the *mesencephalic nucleus of the trigeminal nerve*, located lateral to the central grey matter of the mesencephalon and in the lateral corner of the fourth ventricle in the rostral metencephalon (Figs. 6.27–6.30). The fibres descend as the *mesencephalic root of the trigeminal nerve* and then pass between the motor and the principal sensory nucleus of the trigeminal nerve to join the portio minor. They innervate muscle spindles in the muscles of mastication and mechanorceptors of the teeth. Their cells of origin, although located within the CNS, are the analogues of the peripheral ganglion cells giving rise to primary root fibres.

The motor nucleus of the abducens nerve is located within the genu of the facial nerve. Its fibres leave the brain stem in a ventral direction, immediately caudal to the pons and lateral to the pyramid (Fig. 6.26).

The reticular formation is traversed by the central tegmental tract on its way from the parvocellular red nucleus to the inferior olive. The subdivision of the reticular formation into lateral, medial and median zones is still evident at the level of the metencephalon (Figs. 6.25–6.27). The lateral pontine tegmentum constitutes the enlarged rostral end of the lateral zone. Its medial part consists of scattered small neurons. Its lateral part is formed by a complex of cells that surrounds the superior cerebellar peduncle. This complex can be divided into two aggregations of cells, the medial and lateral parabrachial nuclei, and a ventral extension known as the Kölliker-Fuse nucleus (Figs.

6.28, 22.2 C). The Kölliker-Fuse nucleus [14, 21] corresponds, at least in part, to the pneumotaxic centre. The parabrachial nuclei are links in ascending viscerosensory pathways and in nociceptive projections to the hypothalamus and the amygdaloid nucleus. The lateral pontine tegmentum also contains two separate regions involved in the control of micturition. One of those, the M-region of Holstege [20], is located in the dorsomedial part of the lateral pontine tegmentum (Fig. 6.5). It corresponds in position to the micturition reflex centre first described by Barrington [2]. The other micturition centre is in the ventrolateral pontine tegmentum and was designated by Holstege as the L-region [19] (see Chap. 20).

The medial zone of the reticular formation includes the rostral pole of the gigantocellular reticular nucleus and the *nucleus pontis caudalis and oralis*. The region of the medial reticular formation located rostral and caudal to the nucleus of the abducens nerve is known as the *paramedian pontine reticular formation* and corresponds to the horizontal gaze centre (see Chap. 19).

Serotoninergic nuclei are located in the raphe of the metencephalon. They include the *nucleus raphe pontis* (cell group B5), located between the nucleus reticularis tegmenti pontis of both sides (Fig. 6.27), and cell groups B6 and B8, located within the *central superior nucleus*, which is an oval nucleus situated immediately caudal to the decussation of the superior cerebellar peduncle (Figs. 6.28, 6.29). The noradrenergic nuclei are represented by the conspicuous *nucleus of the locus coeruleus*, located in the lateral corner of the rostral pole of the fourth ventricle (Fig. 6.28). Its name refers to the pigmentation of its neurons (coeruleus = dark blue). It corresponds to the A6 group of Dahlström and Fuxe [7]. Noradrenergic cells are also present more ventrally in a region known as the nucleus subcoeruleus, along the superior cerebellar peduncle (A4), and in the central gray matter of the mesencephalon. The cells of the noradrenergic group A5 occupy the ventrolateral tegmentum (Fig. 22.2 B, C). Two cholinergic cell groups have been identified at the pontomesencephalic junction (Fig. 22 C). One group is located in the pontine central grey (group Ch6 of Mesulam [24]). The other corresponds to the *nucleus pedunculus pontinus* in the dorsolateral tegmentum of the pons and the mesencephalon (group Ch5; Fig. 6.29). The connections of these noradrenergic and cholinergic cell groups are reviewed in more detail in Chap. 22.

Mesencephalon

The decussation of the trochlear nerve in the superior medullary velum and the rostral edge of the pons mark the border of the mesencephalon and the metencephalon on the dorsal and ventral sides of the brain stem, respectively, and the decussation of the superior cerebellar peduncle marks the transition of the tegmentum pontis into the tegmentum mesencephali (Figs. 6.29, 6.30). The border of the mesencephalon with the diencephalon is set caudal to the *posterior commissure* dorsally and caudal to the *mamillary bodies* ventrally (Fig. 6.33). Laterally it follows the caudal border of the optic tracts, where the pes pedunculi emerges on the ventral surface. The geniculate bodies and the pulvinar of the thalamus are located lateral to the rostral mesencephalon (Figs. 6.32, 6.33). Structures at the level of the mesencephalon are summarized in the diagrams of Figs. 6.17 and 6.18. These diagrams refer to the Figs. 6.29–6.33.

The mesencephalon can be subdivided into three dorsoventrally arranged zones, the *tectum* (roof), the *tegmentum mesencephali* and the *pes pedunculi* with the *substantia nigra*. The *periaqueductal grey*, surrounding the cerebral aqueduct, should be indicated as the fourth subdivision of the mesencephalon. Rostrally, the periaqueductal grey is continuous with the grey matter of the hypothalamus; caudally, it extends into the floor of the fourth ventricle (Figs. 6.29–6.33). It contains a diffuse fibre system, the *dorsal longitudinal fascicle* of Schütz [36], which interconnects the hypothalamus with visceral centres in the central grey and the lower brain stem and serves as a focus for ascending catecholaminergic fibres from lower levels.

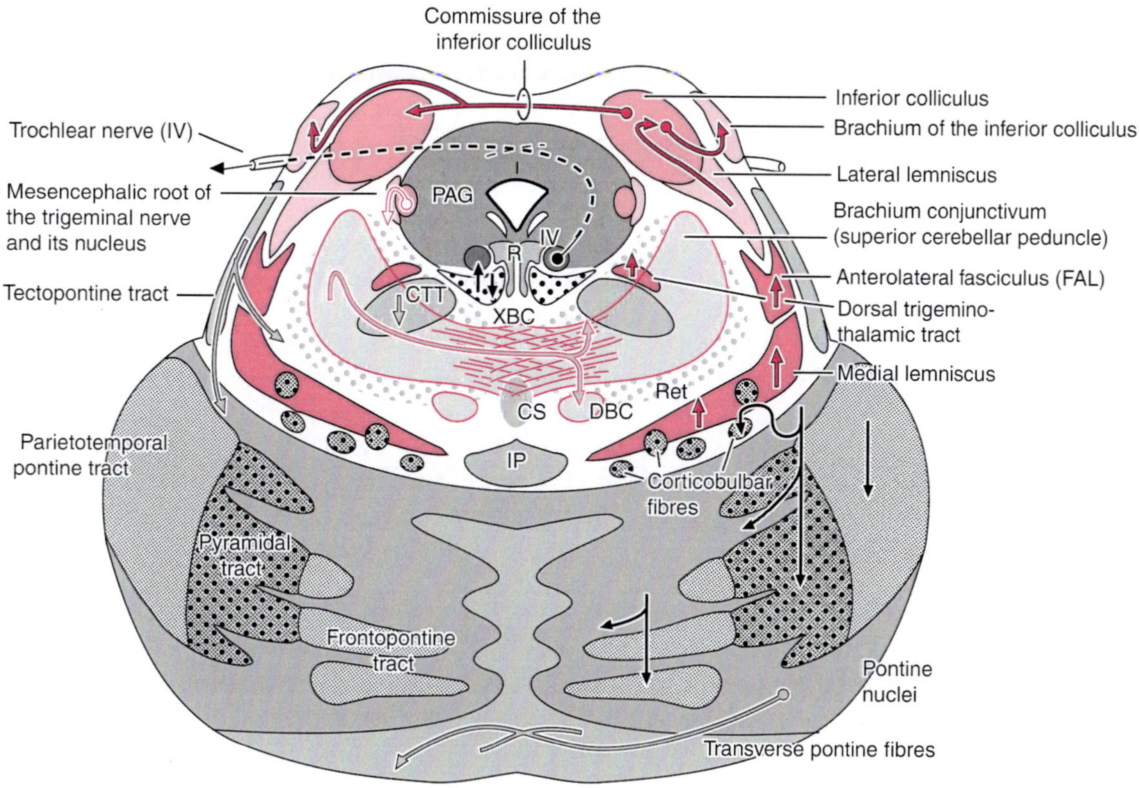

Fig. 6.17. Diagrammatic section through the inferior colliculus, showing the course and direction of fibre tracts. For key see Fig. 6.13. The trochlear nerve, which exits caudal to the inferior colliculus, is indicated with a *broken line*. *IV*, Nucleus of the trochlear nerve; *CS*, central superior nucleus; *CTT*, central tegmental tract; *DBC*, descending branch of the brachium conjunctivum (superior cerebellar peduncle); *IP*, interpeduncular nucleus; *MLF*, medial longitudinal fascicle; *PAG*, periaqueductal grey; *R*, dorsal raphe nucleus; *Ret*, reticular formation; *XBC*, decussation of the brachium conjunctivum (superior cerebellar peduncle)

The tectum mesencephali consists of the inferior and superior colliculi. The *inferior colliculus* is an important link in the acoustic system. On cross section it appears as an oval cellular area, which receives the lateral lemniscus. It gives rise to an ascending pathway, the *brachium of the inferior colliculus*, which extends on the dorsolateral surface of the mesencephalon and terminates in the *medial geniculate body*. The *intercollicular nucleus* forms a bridge between the inferior colliculi of both sides, dorsal to the central grey (Figs. 6.29, 6.30).

The *superior colliculus* is a layered structure. Its superficial layers contain small cells that receive visual afferents from the optic tract and the visual cortex. Larger cells of the intermediate and deep layers of the superior colliculus receive a mixture of somatosensory, acoustic, cerebellar and nonvisual cortical afferents (Figs. 6.30, 6.31). The efferent pathways of the superior colliculus are also segregated into ascending connections of the superficial layers with visual centres, such as the pretectum and the visual thalamic nuclei, and widespread connections of the deep layers with thalamic, reticular and precerebellar nuclei and the spinal cord. Among the efferent pathways the *tectospinal tract* is quite prominent. Its coarse fibres radiate from the intermediate layers of the superior colliculus to the midline, where they cross in the *dorsal tegmental decussation* (Fig. 6.31). They descend as a separate bundle, the *predorsal fascicle*, ventral to the medial longitudinal fascicle (Fig. 6.28). In the pons they become incorporated into this fascicle to descend to cervical levels of the cord. The superior colliculus is one the major centres for the initiation of eye movements and their coordination with associated movements of head and neck (Chap. 19).

The *parabigeminal nucleus* is located superficially, ventrolateral to the tectum (Figs. 6.28–6.30). It is reciprocally connected with the superior colliculus.

The *pretectum* is an ill-defined region, extending between the rostral margin of the superior colliculus and the thalamus. It contains the *nucleus of the optic tract* and the *olivary pretectal nucleus*, two small cell masses that receive retinal afferents from the optic tract and serve as essential links in two different oculomotor pathways. The *olivary pretectal nucleus* is a relay in the pupillary light reflex, and the nucleus of the optic tract constitutes a centre for horizontal optokinetic reactions. The *nucleus of the optic tract* is closely allied to a system surrounding the rostral mesencephalon as a ring-shaped assembly of nuclei (*nuclei of the transverse peduncular tract*), which receives input from the optic tract and supplies the brain stem and the cerebellum with optokinetic information used in compensatory eye movements (Chap. 19). The connections of the anterior, posterior and medial pretectal nuclei bear a close resemblance to those of the superior colliculus (Chap. 19).

The pes pedunculi or *cerebral peduncle* is the main system carrying fibres from the cerebral cortex to the brain stem (Figs. 6.29–6.33). Caudally, it enters the pes pontis. It comprises the large medial *frontopontine* and the lateral *parietotemperopontine tracts*, flanking the centrally located pyramidal tract. The latter is distinguished by its content of coarse myelinated fibres; the corticopontine tracts are populated by smaller fibres. The *substantia nigra* is located dorsal to the peduncle. The dorsal layer of the substantia nigra consists of the *pars compacta*, groups of closely packed pigmented neurons (Figs. 5.16, 5.22, 5.34, 6.29, 6.33). The pigment granules that gradually accumulate in these cells in early life are a by-product of the synthesis of the neurotransmitter of these neurons, dopamine. The cells of the pars compacta are reciprocally connected with the striatum (Figs. 6.18, 6.19). Ventrally, the substantia nigra extends as a reticular neuropil, with scattered, nonpigmented neurons known as the *pars reticulata*. These neurons use GABA as their inhibitory neurotransmitter. They project to the thalamus, the superior colliculus and the reticular formation.

The region located between the tectum and the pes is the *tegmentum mesencephali*. It contains the motor nuclei of the oculomotor and trochlear nerves, the crossed ascending limb of the superior cerebellar peduncle, that surrounds and penetrates the red nucleus, and the somatosensory pathways of the medial lemniscus, the dorsal trigeminothalamic and the spinothalamic tracts on their way from the tegmentum pontis to the thalamus.

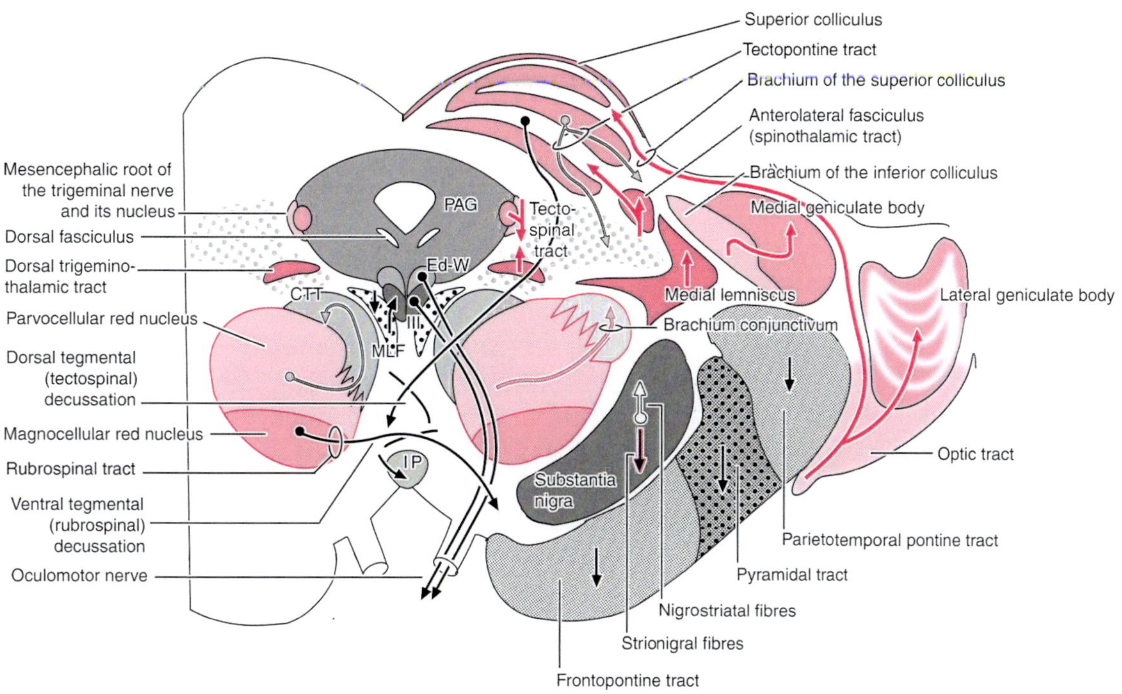

Fig. 6.18. Diagrammatic section through the superior colliculus and the geniculate bodies, showing the course and direction of fibre tracts. For key see Fig. 6.13. *III*, Oculomotor nucleus; *CTT*, central tegmental tract; *Ed-W.*, nucleus of Edinger-Westphal; *IP*, interpeduncular nucleus; *MLF*, medial longitudinal fascicle; *PAG*, periaqueductal grey

The general efferent somatic *nuclei of the trochlear and oculomotor nerves* are located near the midline, ventral to the central grey (Figs. 6.30–6.32). They are enveloped by the fibres of the medial longitudinal fascicle, which carries the preoculomotor afferents from the vestibular nuclei. The roots of the trochlear nerve pass dorsally through the caudal periaqueductal grey to their decussation in the superior medullary velum. The oculomotor nerve leaves the ventral brain stem, medial to the cerebral peduncle in the *interpeduncular fossa* (Fig. 6.32). The general visceral efferent nucleus that provides the oculomotor nerve with preganglionic parasympathetic fibres is the *accessory oculomotor nucleus* of *Edinger-Westphal* [11, 39], located dorsal and rostral to its somatic nucleus (Fig. 6.33).

Fibres of the crossed ascending branch of the superior cerebellar peduncle surround and pass through the *red nucleus* (Figs. 6.31–6.33). Rostrolateral to the red nucleus, they reassemble to enter the thalamus. During their passage through the red nucleus most if not all of the fibres of the superior cerebellar peduncle emit collaterals terminating on the cells of this nucleus. During its course through the mesencephalon the superior peduncle also innervates the oculomotor nuclei, parts of the central grey, the superior colliculus, the pretectum and the reticular formation.

The red nucleus is a conspicuous cell mass. Caudally and medially it is traversed by root fibres of the oculomotor nerve, rostrally by the habenulo-interpeduncular tract. Its main efferent systems are the central tegmental tracts and the *rubrobulbar- and spinal tract*. The red nucleus consists of two subnuclei. The caudal *magnocellular red nucleus*, which mainly, but not exclusively, consists of large neurons, gives rise to the rubrobulbar- and spinal tract, which decussates in the *ventral tegmental decussation* and descends to terminate on motoneurons of the facial nerve nucleus, interneurons in the lateral tegmental field, the nucleus of the lateral funiculus in the caudal medulla oblongata and at all levels of the spinal cord. In humans the magnocellular red nucleus is relatively small (Fig. 6.31). A rubrobulbar tract probably exists in the human brain, but the presence of a rubrospinal tract has never been substantiated. The rostral *parvocellular red nucleus* (Figs. 6.31 – 6.33) gives rise to the *central tegmental tract*, which takes its origin from the dorsomedial aspect of the red nucleus (Fig. 6.32). It is located dorsolateral to the red nucleus and descends on the ipsilateral side, traversing the superior cerebellar peduncle. In the tegmentum of the pons it occupies a central position in the reticular formation (Fig. 6.29). It terminates in the principal nucleus of the inferior olive. The parvocellular red nucleus in lower mammals is small and less well circumscribed than in primates and includes a number of cell groups located at the mesodiencephalic border, which have become known under different names, such as Bechterew's medial nucleus [3].

The *medial tegmental tract* is the smaller of the two tegmental tracts. It takes its origin from the nucleus of Darkschewitsch [9], located in the central grey at the level of the posterior commissure (Fig. 6.33). The tract occupies the ventral part of the medial longitudinal fascicle and descends to the level of the inferior olive, where it terminates in the rostral part of the medial accessory olive.

The *interstitial nucleus of Cajal* is a group of fairly large neurons located dorsal to the red nucleus and ventral to the central grey (Fig. 6.32). It serves as a coordinating centre for eye and head movements, with bilateral connections to the nuclei of the third and fourth nerves, the bulbar reticular formation and the vestibular nuclei. The *interstitiospinal tract* descends in the medial longitudinal fascicle. The *rostral interstitial nucleus of the medial longitudinal fascicle* is located more rostrally (not illustrated). It is the premotor nucleus for conjugated vertical eye movements, also known as the vertical gaze centre (Chap. 19).

Where the lateral lemniscus terminates in the inferior colliculus, the medial lemniscus and the anterolateral fascicle, with the spinothalamic tracts, shift dorsally (Fig. 6.32). The anterolateral fascicle takes up a position ventrolateral to the superior colliculus, where its spinotectal component terminates. At rostral levels of the superior colliculus the anterolateral fascicle consists of a pure collection of spinothalamic fibres (Fig. 6.33). The medial lemniscus and the spinothalamic tracts enter the thalamus more rostrally.

The serotoninergic raphe nuclei of the mesencephalon are represented by the *nucleus raphes dorsalis* (group B7, Fig. 6.17), located in the ventral central grey at the level of the inferior colliculus, and the *nucleus linearis caudalis*, located in the rostral mesencephalon. The mesencephalic dopaminergic cell groups include the pars compacta of the substantia nigra, and a cell group located ventral to the red nucleus in the so-called "*ventral tegmental area*". The dopaminergic cells in the substantia nigra project to the caudate nucleus and the putamen. The principal connections of the dopaminergic elements in the ventral tegmental area pass ventrally through the lateral hypothalamus to terminate in a variety of limbic structures, including the cingulate cortex, the accumbens nucleus and the amygdala.

A more or less circumscribed fibre system, first described by Forel [13], is located within the reticular formation of the mesencephalon, lateral to the central grey and dorsal to the central tegmental tract. It contains the *uncrossed ascending trigeminothalamic pathway* originating from the dorsal part of the principal sensory nucleus of the trigeminal nerve (Chap. 16). This region also is the focus for fibres of the *ascending reticular activating system*, which connect the gigantocellular and lateral pontine reticular formation with the thalamus, and the *dorsal catecholaminergic and cholinergic bundles*, which ascends from noradrenergic, serotoninergic and cholinergic cell groups to join the medial forebrain bundle in the hypothalamus (Figs. 6.30–6.32). The location of the ventral catecholaminergic and cholinergic bundles cannot be identified without histochemical staining techniques.

The *habenulo-interpeduncular tract or fasciculus retroflexus* takes its origin from the habenular nuclei in the diencephalon. It consists of a core of unmyelinated axons, surrounded by myelinated fibres. It passes through the caudal thalamus and the medial part of the red nucleus to terminate in the interpeduncular nucleus, located in the floor of the interpeduncular fossa (Figs. 6.31–6.33).

Addendum: Diencephalon and the Corpus Striatum

The origin and termination in the thalamus and the corpus striatum of pathways mentioned in the previous paragraphs are summarized in the diagram of Fig. 6.19, which refers to the sections depicted in Figs. 6.35–6.41. The topography and the connections of the structures present at these levels are discussed more fully in Chaps. 7–14.

The dorsal thalamus and the hypothalamus are located in the lateral wall of the third ventricle. The roof of the third ventricle with the choroid plexus is attached to a ridge on the dorsolateral thalamus (the *habenulum*), which contains the *medullary stria* and the *habenular nuclei*. The latter give rise to the habenulo-interpeduncular tract (fasciculus retroflexus) illustrated in Fig. 6.33. The corpus striatum is located lateral to the thalamus. It consists of the caudate nucleus, the putamen and the globus pallidus. The *caudate nucleus* and the *putamen* are partially separated by the internal capsule. The thalamus and the caudate nucleus both border on the lateral ventricle. The caudate nucleus is located in its lateral wall. The *stria terminalis*, an efferent system of the amygdaloid body, is located in the corner between the caudate nucleus and the thalamus. The *globus pallidus* is located in between the putamen laterally and the internal capsule medially. It consists of external and internal segments, separated by the *medial medullary lamina* (Figs. 6.36–6.41). *The lateral medullary lamina* separates the globus pallidus from the putamen. The *internal capsule* continues as the cerebral peduncle.

The *internal medullary lamina*, with the *intralaminar nuclei*, divides the dorsal thalamus into the *medial dorsal thalamic nucleus* and the *ventral and lateral nuclear groups*. Rostrally and dorsally the internal medullary lamina splits to include the *anterior nucleus of the thalamus*.

The pulvinar and the geniculate bodies, with their afferent systems, are illustrated in Figs. 6.18, 6.32, and 6.33. The dorsal thalamus is surrounded by the *external medullary lamina*; the *reticular nucleus of the thalamus* is located between the external medullary lamina and the internal capsule. It is perforated by thalamocortical and corticothalamic fibres carried by the internal capsule. Collaterals of these fibres terminate on neurons of the reticular nucleus. The neurons of the reticular nucleus are inhibitory and project back to the thalamic nuclei. The region wedged between the dorsal thalamus and the hypothalamus is known as the *subthalamus*. It contains the *zona incerta* and the *subthalamic nucleus*. In sections like those illustrated in Figs. 6.19 and 6.37, the zona incerta, the subthalamic nucleus and the substantia nigra occupy dorsoventral, partially overlapping positions.

Somatosensory conduction systems such as the spinothalamic and trigeminothalamic tracts, the medial lemniscus and the dorsal trigeminothalamic tract terminate in the nuclei of the ventral posterior complex. This complex includes the medial and lateral ventral posterior nuclei and several subnuclei, which serve as relays in pathways for deep sensation, and in nociceptive and thermoceptive, gustatory and general visceral pathways. The ventral lateral and ventral anterior thalamic nuclei are the more rostrally and dorsally located nuclei of the ventral nuclear group (Chap. 16).

The *ventral lateral nucleus* is the main termination of the superior cerebellar peduncle. After its course through the red nucleus and the so-called *prerubral field* (or the H or Hauben (tegmentum) field of Forel [13]) it passes through the external medullary lamina, dorsal to the zona incerta (also known as the *thalamic fascicle* or field H2 of Forel) to its termination in the ventral lateral nucleus (Chap. 20).

The caudate nucleus and the putamen issue large numbers of small, thinly myelinated fibres, which traverse and terminate in the external and internal segments of the globus pallidus. After traversing the globus pallidus, striatal fibres continue by passing through the internal capsule to terminate in the substantia nigra. The globus pallidus contains large neurons, which give rise to coarse, heavily myelinated fibres, which give this nucleus its pale appearance. Fibres from the external segment loop rostral and ventral to the internal capsule (Figs. 6.19, 6.38, 6.39) as the *ansa lenticularis*. These fibres terminate in the subthalamic nucleus. The internal segment of the globus pallidus gives rise to fibres passing through the internal capsule and assembling ventral to the zona incerta. This fibre bundle is known as the *lenticular fascicle* and, where it becomes located ventral to the zona incerta, as the field H1 of Forel. It curves medially around the zona incerta (through the field H) and reaches its termination in the ventral anterior nucleus of the thalamus, via the fasciculus thalamicus (or field H2 of Forel).

Projection neurons of the striatum and the globus pallidus are GABAergic and inhibitory. An excitatory pathway from the subthalamic nucleus reciprocates the inhibitory projection from the external segment of the globus pallidus, but terminates in its internal segment. Dopaminergic afferents from the pars compacta of the substantia nigra innervate both the caudate nucleus and the putamen. The course of this pathway is not precisely known. The bundles of small striatal and coarse pallidal fibres passing through the internal capsule are known as the "*comb system*".

References

1. Barber RP, Phelps PE, Houser CR, Crawford GD, Salvaterra PM, Vaughn JE (1984) The morphology and distribution of neurons containing choline acetyltransferase in the adult rat spinal cord: an immunocytochemical study. J Comp Neurol 229:329–346
2. Barrington FJF (1925) The effect of lesion of the hind- and midbrain on micturition in the cat. J Exp Physiol 15:81–102
3. Bechterew W von (1899) Leitungsbahnen im Gehirn und Rückenmark., 2nd edn. Arthur Georgi, Leipzig
4. Clarke JAL (1851) Researches into the structure of the spinal cord. Phil Trans Roy Soc London 141:607–621
5. Craig AD, Kniffki KD (1985) Spinothalamic lumbosacral lamina I cells responsive to skin and muscle stimulation in the cat. J Physiol 365:197–221
6. Cullheim S, Kellerth JO (1978) A morphological study of the axons and recurrent axon collaterals of cat α-motoneurones supplying different functional types of muscle unit. J Physiol 281:301–314

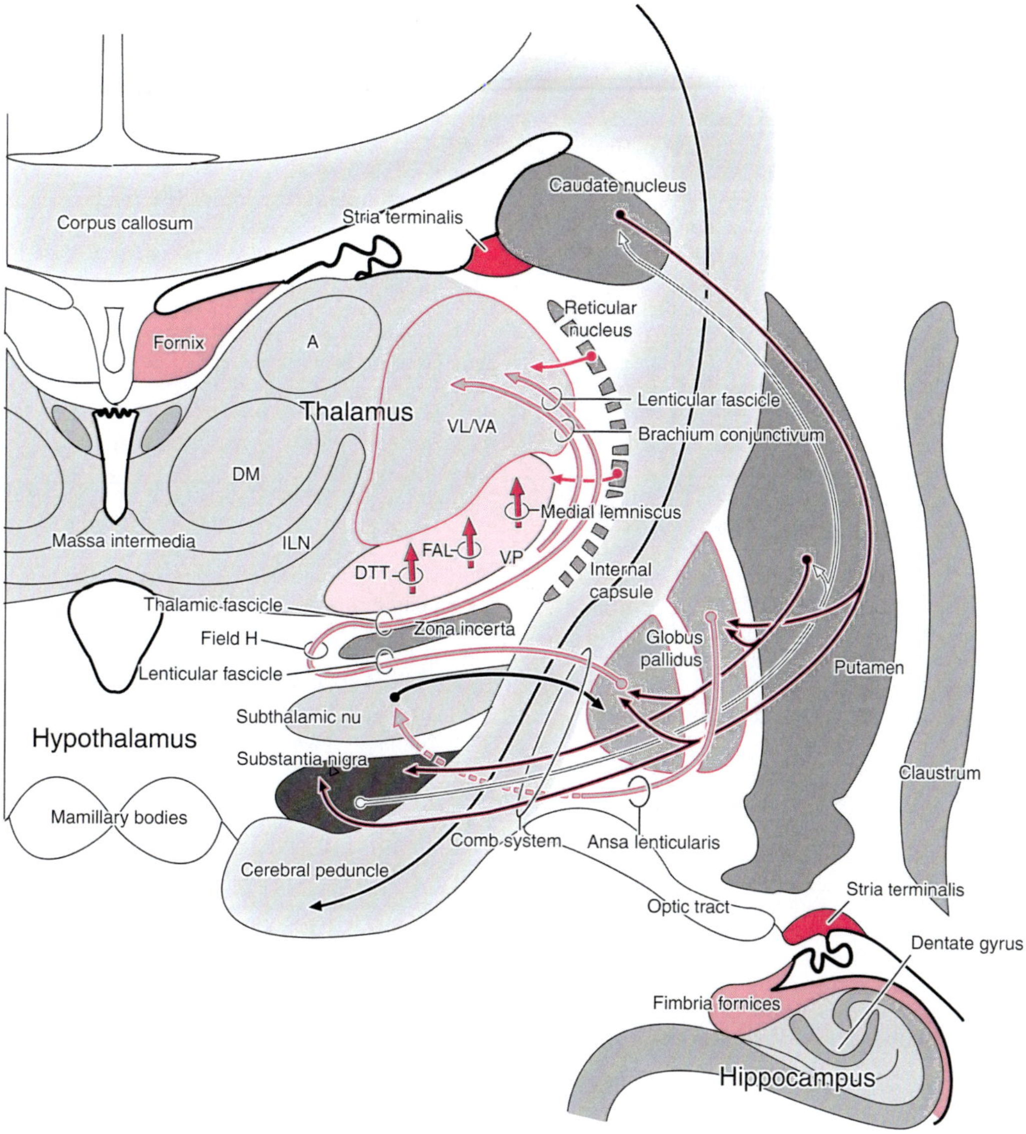

Fig. 6.19. Diagrammatic section through the thalamus, the basal ganglia and the hippocampus, showing the termination of the ascending sensory systems in the ventroposterior complex of the thalamus, the course and the termination of the brachium conjunctivum (superior cerebellar peduncle) in the ventrolateral nucleus of the thalamus and the components of the striatal circuit.

Note passage of the lenticular fasciculus through the internal capsule as part of the comb system. The ansa lenticularis passes rostral to the internal capsule and is indicated as a *broken line*. For key see Fig. 6.13. *A*, Anterior nucleus of the thalamus; *DM*, dorsomedial nucleus of the thalamus; *DTT*, dorsal trigemino-thalamic tract; *FAL*, anterolateral fasciculus (spino- and trigemino-thalamic tracts); *ILN*, intralaminar nuclei; *VL/VA*, ventral lateral and ventral anterior nuclei of the thalamus; *VP*, ventroposterior complex of the thalamus

7. Dahlström A, Fuxe K (1964) Evidence for the exitstance of monoamine-containing neurons in the central nervous system. I. Demonstration of monoamines in the cell bodies of the brain stem neurons. Acta Physiol Scand [Suppl] 62 (247):1–55
8. Dahlström A, Fuxe K (1964) Localization of monoamines in the lower brain stem. Experientia 20: 398–399
9. Darkschewitsch LO (1889) Über den oberen Kern des Oculomotorius. Arch f Anat u Physiol Anat Abth: 107–116
10. Deiters O (1865) Untersuchungen über Gehirn und Rückenmark des Menschen und der Säugethiere. Friedrich Vieweg, Braunschweig
11. Edinger L (1885) Ueber den Verlauf der centralen Hirnnervenbahnen mit Demonstration von Präparaten. Arch f Psychiat u Nervenkrankh Berlin 16: 858–859
12. Erlanger J, Gasser HS (1937) Electrical signs of nervous activity. University of Pennsylvania Press, Philadelphia
13. Forel A (1877) Untersuchungen über die Haubenregion und ihre oberen Verknüpfungen im Gehirne des Menschen und einiger Säugethiere, mit Beitragen zu Methoden der Gehirnuntersuchung. Arch f Psychiat u Nervenkrankh 7:392–495
14. Fuse G (1913) Die Randgebiete des Pons und des Mittelhirns. Arb Hirnant Inst Zürich 7:211–258
15. Gaskell WH (1986) On the structure, distribution and function of the nerves that innervate the visceral and vascular systems. J Physiol 7:1–81
16. Gaskell WH (1989) On the relation between the structure, function, distribution and origin of the cranial nerves; together with a theory of the origin of the nervous system of vertebrata. J Physiol 10: 153–211
17. Golgi C (1880) Dui nervi nei tendini dell'uomo e di altri vertebrati e di un nuovo organo nervoso terminale musculo-tendineo. Mem Roy Acad Sci Tor 32:359–385
18. Herrick CJ (1913) Anatomy of the brain. The Reference Handbook of the Medical Sciences, vol 2. Wood, New York, pp 274–342
19. Holstege G, Griffiths D, de Wall H, Dalm E (1986) Anatomical and physiological observations on supraspinal control of bladder and urethral sphincter muscles in the cat. J Comp Neurol 250:449–461
20. Holstege G, Mouton LJ, Gerrits NM (2004) Emotional motor system. In: Paxinos G, Mai JK (eds) The human nervous system. Elsevier, Amsterdam, pp 1306–1324
21. Koelliker A (1896) Nervensystem des Menschen und der Thiere. Handbuch der Gewebelehre des Menschen, vol 2. Wilhelm Engelmann, Leipzig
22. Lissauer H (1886) Beitrag zum Faserverlauf im Hinterhorn des menschlichen Rückenmarks und zum Verhalten desselben bei Tabes dorsalis. Arch Psych 17:377–438
23. Lloyd DPC (1943) Neuron patterns controlling transmission of ipsilatral hind limb reflexes in cat. J Neurophysiol 6:293–315
24. Mesulam MM, Mufson EJ, Levey AI, Wainer BH (1984) Atlas of cholinergic neurons in the forebrain and upper brainstem of the macaque based on monoclonal choline acetyltransferase immunohistochemistry and acetylcholinesterase histochemistry. Neuroscience 12:669–686
25. Muskens LJJ (1913) De rolbeweging en de opstijgende vestibularisverbreiding (Fasciculus Deiters ascendens). Versl Kon Acad Wet: 1478
26. Nadelhaft I, Roppolo J, Morgan C, de Groat WC (1983) Parasympathetic preganglionic neurons and visceral primary afferents in monkey sacral spinal cord revealed following application of horseradish peroxidase to pelvic nerve. J Comp Neurol 216: 36–52
27. Nichols TR (1996) Force feedback: Update and speculations. http://www.ac.wwu.edu/~chalmers/forcefeedback.html
28. Nieuwenhuys R (1974) Topological analysis of the brain stem: a general introduction. J Comp Neurol 156:255–276
29. Olszewski J, Baxter D (1954) Cytoarchitecture of the human brain stem. Karger, Basel
30. Purkinje JE (1838) Untersuchungen aus der Nerven- und Hirnanatomie: über die scheinbar kanalikulöse Beschaffenheit der elementaren Nervencylinder. Berichte über die Versammlung Deutscher Naturforscher und Ärzte in Prague, September 1837:177–179
31. Renshaw B (1946) Central effects of centripetal impulses in axons of spinal ventral roots. J Neurophysiol 9:191–204
32. Rexed B (1954) A cytoarchitectonic atlas of the spinal cord in the cat. J Comp Neurol 100: 297–379
33. Scheibel ME, Davies TL, Scheibel AB (1973) Maturation of reticular dendrites: loss of spines and development of bundles. Exp Neurol 38:301–310
34. Schoenen J (1982) Dendritic organization of the human spinal cord: the motoneurons. J Comp Neurol 211:226–247
35. Schoenen J, Faull RLM (2004) Spinal cord: cyto- and chemoarchitecture. In: Mai JK (ed) The human nervous system. Elsevier, Amsterdam, pp 190–249
36. Schütz H (1891) Anatomische Untersuchungen über den Faserverlauf im centralen Hohlengrau und den Nervenfaserschwund in denselben bei der progressiven Paralyse der Irren. Arch Psychiat u Nervenkrankh 22:527–587
37. Vera PL, Ellenberger HH, Haselton JR, Haselton CL, Schneiderman N (1986) The intermediolateral nucleus: an 'open' or 'closed' nucleus? Brain Res 386:84–92
38. Voogd J (2000) The Purkinje cell. In: Koehler PJ, Bruyn GW, Pearce JMS (eds) Neurological eponyms. Oxford University Press, Oxford, pp 37–43
39. Westphal CFO (1887) Ueber ein Fall von chronischer progressiver Lähmung der Augenmuskeln (Ophthalmoplegia externa) nebst Beschreibung von Ganglienzellengruppen im Bereiche des Oculomotoriuskerns. Arch. f Psychiat u Nervenkrankh Berlin 18:846–871

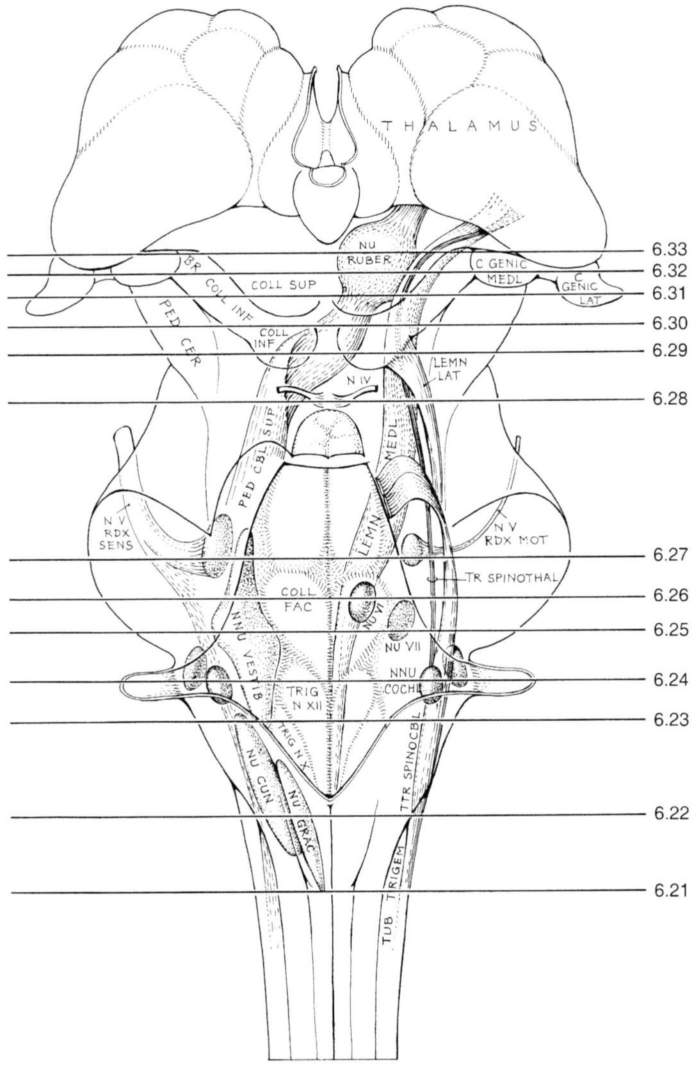
THALAMUS
NU RUBER
BR COLL INF
COLL SUP
C GENIC MEDL
C GENIC LAT
PED CER
COLL INF
LEMN LAT
N IV
PED CBL SUP
MEDL
LEMN
N V RDX SENS
N V RDX MOT
TR SPINOTHAL
COLL FAC
NU VI
NU VII
NNU VESTIB
NNU COCHL
TRIG N XII
TRIG N X
NU CUN
NU GRAC
TTR SPINOCBL
TUB TRIGEM
6.33
6.32
6.31
6.30
6.29
6.28
6.27
6.26
6.25
6.24
6.23
6.22
6.21

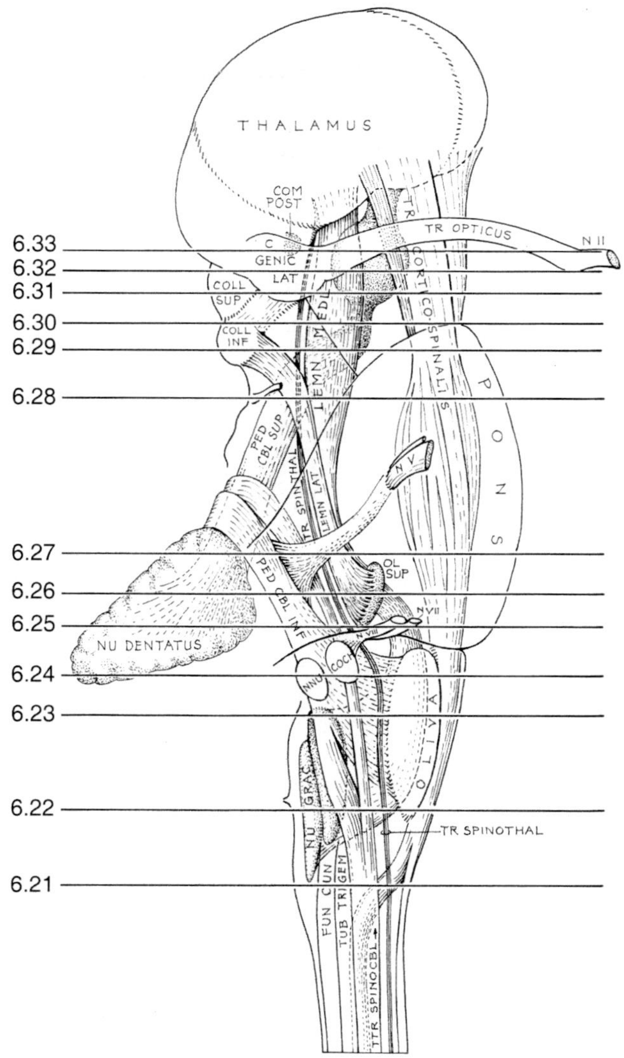

Fig. 6.20 A, B. Key diagrams showing level and plane of the sections illustrated in Figs. 6.21–6.33. *Left,* dorsal view; *right,* lateral view

BR COLL INF, brachium of the inferior colliculus; *C GENIC LAT,* lateral geniculate body; *C GENIC MEDL,* medial geniculate body; *COLL FAC,* facial colliculus; *COLL INF,* inferior colliculus; *COLL SUP,* superior colliculus; *COM POST,* posterior commissure; *FUN CUN,* cuneate fascicle; *LEMN LAT,* lateral lemniscus; *LEMN MED,* medial lemniscus; *N II,* optic nerve; *N IV,* trochlear nerve; *N V RDX MOT,* motor root of the trigeminal nerve (portio minor); *N V RDX SENS,* sensory root of the trigeminal nerve (portio major); *N V,* trigeminal nerve; *N VIII,* vestibulocochlear nerve; *NNU COCHL,* cochlear nuclei; *NNU VESTIB,* vestibular nuclei; *NU CUN,* cuneate nucleus; *NU GRAC,* gracile nucleus; *NU VI,* nucleus of the abducens nerve; *NU VII,* nucleus of the facial nerve; *OL SUP,* superior olive; *PED CBL INF,* inferior cerebellar peduncle (restiform body); *PED CBL SUP,* superior cerebellar peduncle (brachium conjunctivum); *PED CER,* cerebral peduncle; *TR CORTICO SPINALIS,* corticospinal tract; *TR OPTICUS,* optic tract; *TR SPINOTHAL,* spinothalamic tract; *TRIG N X,* vagal triangle; *TRIG N XII,* hypoglossal triangle; *TTR SPINOCB,* spinocerebellar tracts; *TUP TRIGEM,* trigeminal tubercle

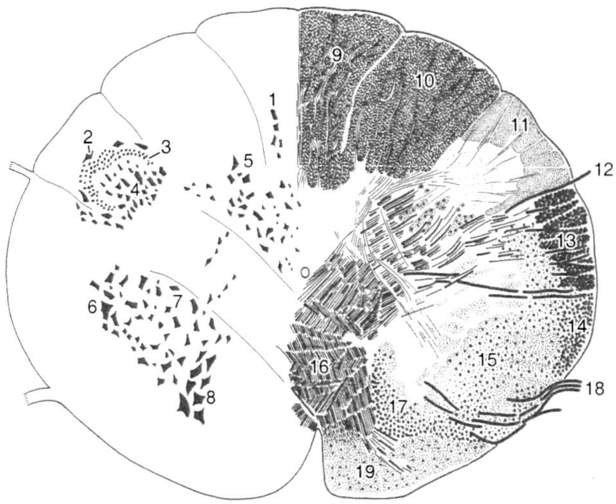

1 Gracile nucleus
2 Nucleus of the spinal tract of the trigeminal nerve, subnucleus caudalis: marginal cells
3 Nucleus of the spinal tract of the trigeminal nerve, subnucleus caudalis: substantia gelatinosa
4 Nucleus of the spinal tract of the trigeminal nerve, subnucleus caudalis: nucleus proprius
5 Medial cuneate nucleus
6 Retroambiguus nucleus
7 Central nucleus of the medulla oblongata
8 Supraspinal nucleus
9 Gracile fascicle
10 Cuneate fascicle
11 Spinal tract of the trigeminal nerve
12 Accessory nerve, bulbar roots
13 Dorsal spinocerebellar tract
14 Ventral spinocerebellar tract
15 Anterolateral fascicle
16 Decussation of the pyramidal tract
17 Medial longitudinal fascicle
18 C1 ventral root
19 Pyramidal tract

Fig. 6.21. Section through the pyramidal decussation (7/1×)

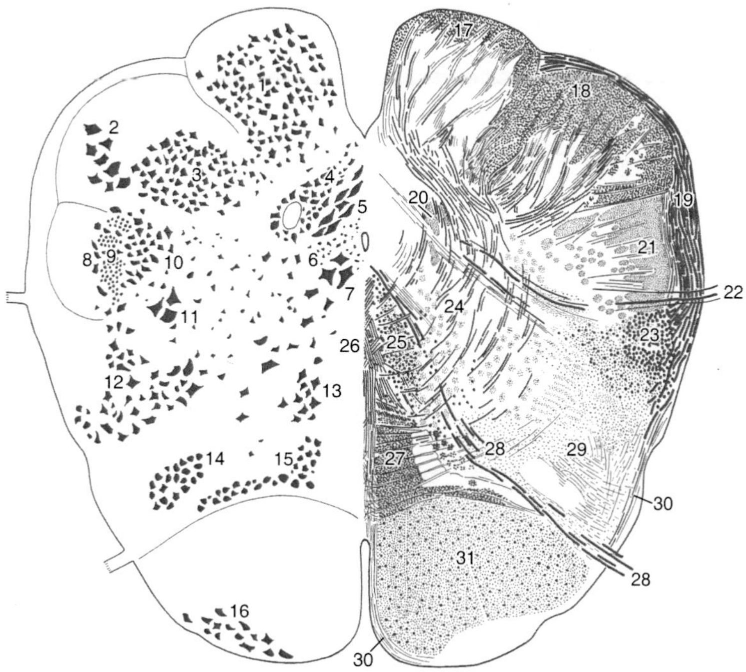

1 Gracile nucleus
2 Lateral cuneate nucleus
3 Medial cuneate nucleus
4 Nucleus of the solitary tract
5 Dorsal nucleus of the vagal nerve
6 Intercalated nucleus
7 Nucleus of the hypoglossal nerve
8 Nucleus of the spinal tract of the trigeminal nerve, subnucleus caudalis: marginal cells
9 Nucleus of the spinal tract of the trigeminal nerve, subnucleus caudalis: substantia gelatinosa
10 Nucleus of the spinal tract of the trigeminal nerve, subnucleus caudalis: nucleus proprius
11 Ambiguus nucleus
12 Nucleus of the lateral funiculus (lateral reticular nucleus)
13 Nucleus of the anterior funiculus (nucleus reticularis paramedianus)
14 Dorsal accessory olive
15 Medial accessory olive
16 Arcuate nuclei
17 Gracile fascicle
18 Cuneate fascicle
19 Inferior cerebellar peduncle (restiform body)
20 Solitary tract
21 Spinal tract of the trigeminal nerve
22 Accessory nerve: bulbar roots
23 Anterolateral fascicle
24 Internal arcuate fibres
25 Medial longitudinal fascicle
26 Decussation of the internal arcuate fibres destined for the medial lemniscus
27 Medial lemniscus
28 Hypoglossal nerve
29 Amiculum (fleece) of the inferior olive
30 External arcuate fibres
31 Pyramidal tract

Fig. 6.22. Section through the dorsal column nuclei (7/1×)

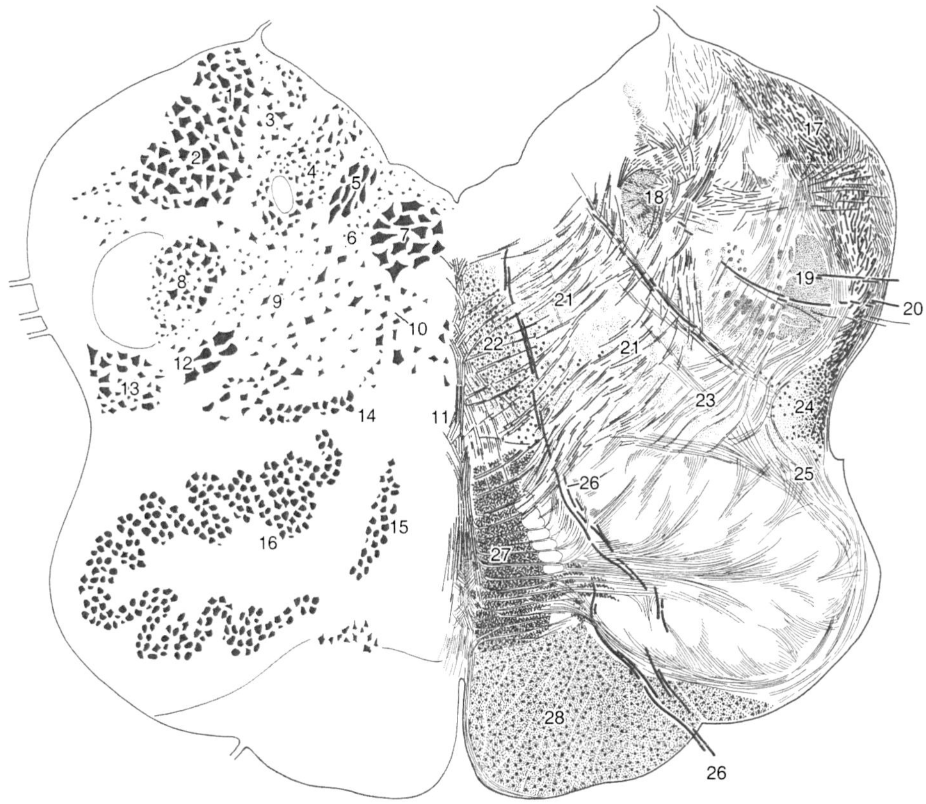

1 Medial cuneate nucleus
2 Lateral cuneate nucleus
3 Spinal vestibular nucleus
4 Nucleus of the solitary tract
5 Dorsal nucleus of the vagal nerve
6 Intercalated nucleus
7 Nucleus of the hypoglossal nerve
8 Nucleus of the spinal tract of the trigeminal nerve, interpolated part
9 Central nucleus of the medulla oblongata
10 Nucleus of the anterior funiculus (nucleus reticularis paramedianus)
11 Nucleus raphes obscurus
12 Ambiguus nucleus
13 Nucleus of the lateral funiculus (nucleus reticularis lateralis)
14 Dorsal accessory olive
15 Medial accessory olive
16 Inferior olive, principal nucleus
17 Inferior cerebellar peduncle (restiform body)
18 Solitary tract
19 Spinal tract of the trigeminal nerve
20 Vagal nerve
21 Internal arcuate fibres
22 Medial longitudinal fascicle
23 Olivocerebellar fibres
24 Anterolateral fascicle
25 Amiculum (fleece) of the inferior olive
26 Hypoglossal nerve
27 Medial lemnicus
28 Pyramidal tract

Fig. 6.23. Section through the middle part of the inferior olive (7/1×)

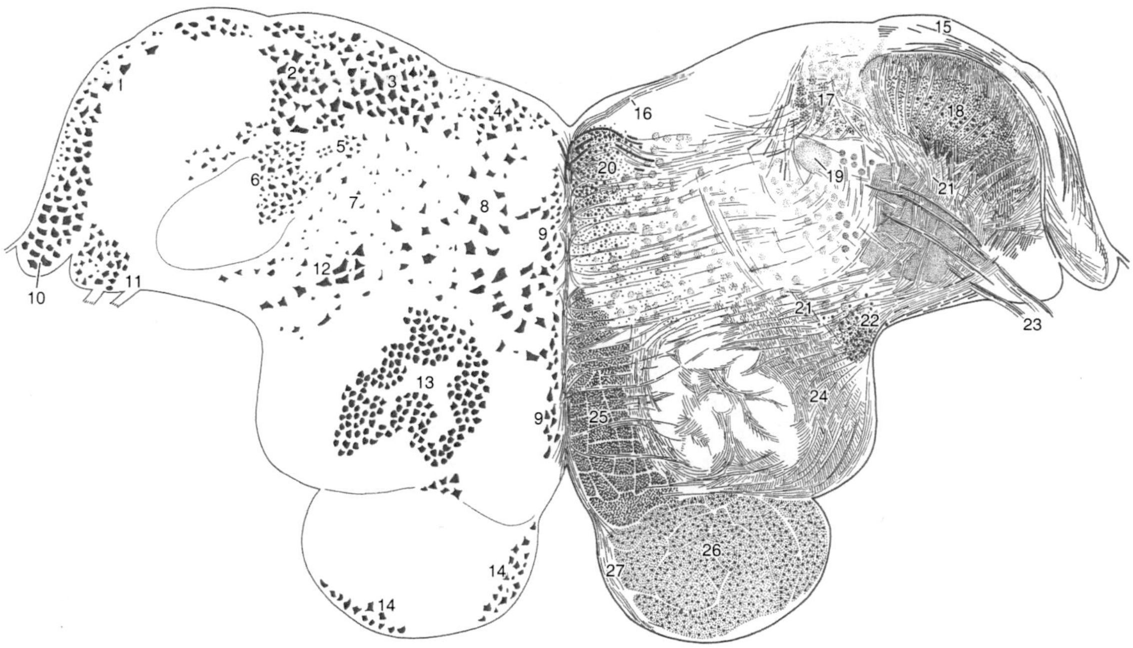

1 Dorsal cochlear nucleus
2 Spinal vestibular nucleus
3 Medial vestibular nucleus
4 Nucleus prepositus hypoglossi
5 Nucleus of the solitary tract
6 Nucleus of the spinal tract of the trigeminal nerve, oral part
7 Lateral reticular formation
8 Gigantocellular reticular nucleus
9 Nucleus raphes magnus
10 Ventral cochlear nucleus
11 Pontobulbar body
12 Ambiguus nucleus
13 Inferior olive, principal nucleus
14 Arcuate nuclei
15 Dorsal acoustic striae
16 Striae medullares
17 Descending branch of the vestibular nerve
18 Inferior cerebellar peduncle (restiform body)
19 Solitary tract
20 Medial longitudinal fascicle
21 Olivocerebellar fibres
22 Anterolateral fascicle
23 Glossopharyngeal nerve
24 Central tegmental tract
25 Medial lemniscus
26 Pyramidal tract
27 External arcuate fibres

Fig. 6.24. Section through the cochlear nuclei (7/1×)

1 Dentate nucleus
2 Globose (posterior interposed) nucleus
3 Lateral vestibular nucleus (Deiters [10])
4 Medial vestibular nucleus
5 Spinal vestibular nucleus
6 Nucleus prepositus hypoglossi
7 Oval nucleus of the nucleus of the solitary tract
8 Lateral reticular formation
9 Nucleus of the spinal tract of the trigeminal nerve, oral part
10 Nucleus of the facial nerve
11 Gigantocellular reticular nucleus
12 Nucleus raphes magnus
13 Superior olive
14 Nucleus of the trapezoid body
15 Pontobulbar body
16 Pontine nuclei

17 Superior cerebellar peduncle (brachium conjunctivum)
18 Floccular peduncle
19 Inferior cerebellar peduncle (restiform body)
20 Fibres of vestibulomesencephalic and lateral vestibulospinal tracts
21 Oval fascicle, ascending branch of the solitary tract
22 Facial nerve
23 Medial longitudinal fascicle
24 Central tegmental tract
25 Trapezoid body
26 Middle cerebellar peduncle (brachium pontis)
27 Vestibulocochlear nerve
28 Facial nerve
29 Medial lemniscus
30 Abducens nerve
31 Pyramidal tract

Fig. 6.25 A, B. Section through the vestibular nuclei and the motor nucleus of the facial nerve (7/1×)

1 Dentate nucleus
2 Emboliform (anterior interposed) nucleus
3 Globose (posterior interposed) nucleus
4 Fastigial nucleus
5 Superior vestibular nucleus
6 Lateral vestibular nucleus
7 Medial vestibular nucleus
8 Nucleus of the abducens nerve
9 Lateral reticular formation
10 Principal sensory nucleus of the trigeminal nerve
11 Nucleus of the facial nerve
12 Gigantocellular reticular nucleus
13 Nucleus raphes magnus
14 Lateral superior olive
15 Medial superior olive
16 Nucleus of the trapezoid body
17 Pontine nuclei

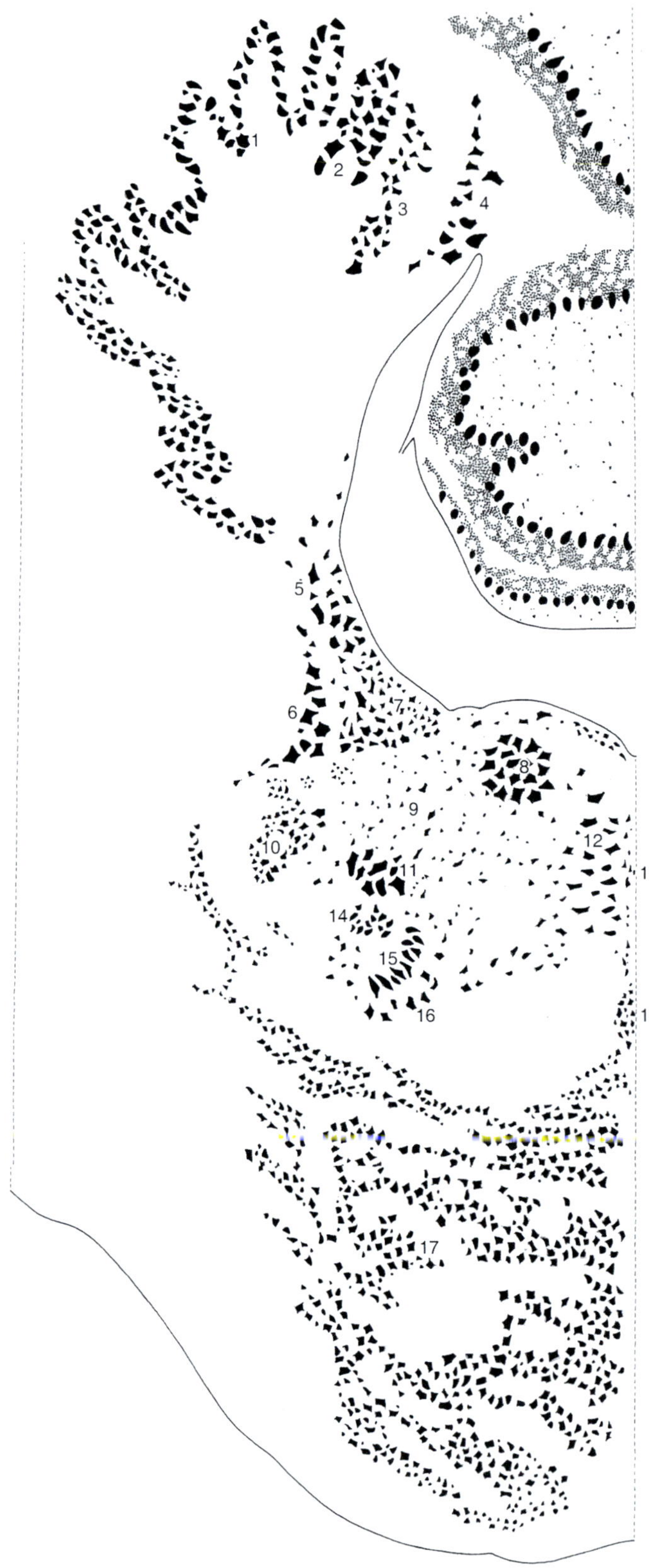

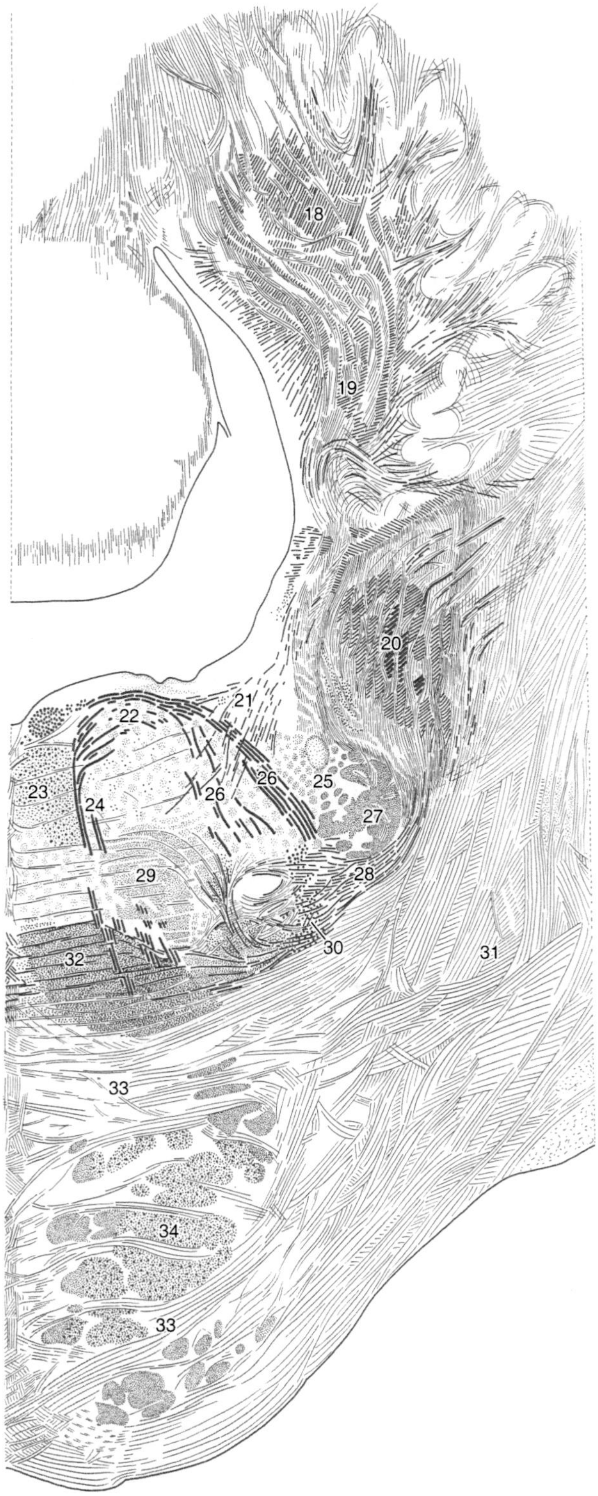

18 Superior cerebellar peduncle (brachium conjunctivum)
19 Inferior cerebellar peduncle (juxtarestiform body)
20 Inferior cerebellar peduncle (restiform body)
21 Fibres of vestibulomesencephalic and lateral vestibulospinal tracts
22 Knee (genu) of the facial nerve
23 Medial longitudinal fascicle
24 Abducens nerve
25 Oval fascicle, ascending branch of the solitary tract
26 Facial nerve
27 Spinal tract of the trigeminal nerve
28 Trapezoid body
29 Central tegmental tract
30 Anterolateral fascicle
31 Middle cerebellar peduncle (brachium pontis)
32 Medial lemniscus
33 Pontocerebellar fibres
34 Pyramidal tract

Fig. 6.26 A, B. Section through the abducens nucleus, the superior olive and the trapezoid body (7/1×)

1 Dentate nucleus
2 Emboliform (anterior interposed) nucleus
3 Globose (posterior interposed) nucleus
4 Fastigial nucleus
5 Superior vestibular nucleus
6 Periaqueductal grey
7 Mesencephalic nucleus of the trigeminal nerve
8 Principal sensory nucleus of the trigeminal nerve
9 Motor nucleus of the trigeminal nerve
10 Nucleus reticularis pontis caudalis
11 Lateral reticular formation
12 Ventral nucleus of the lateral lemniscus
13 Nucleus raphes pontis
14 Nucleus reticularis tegmenti pontis
15 Pontine nuclei

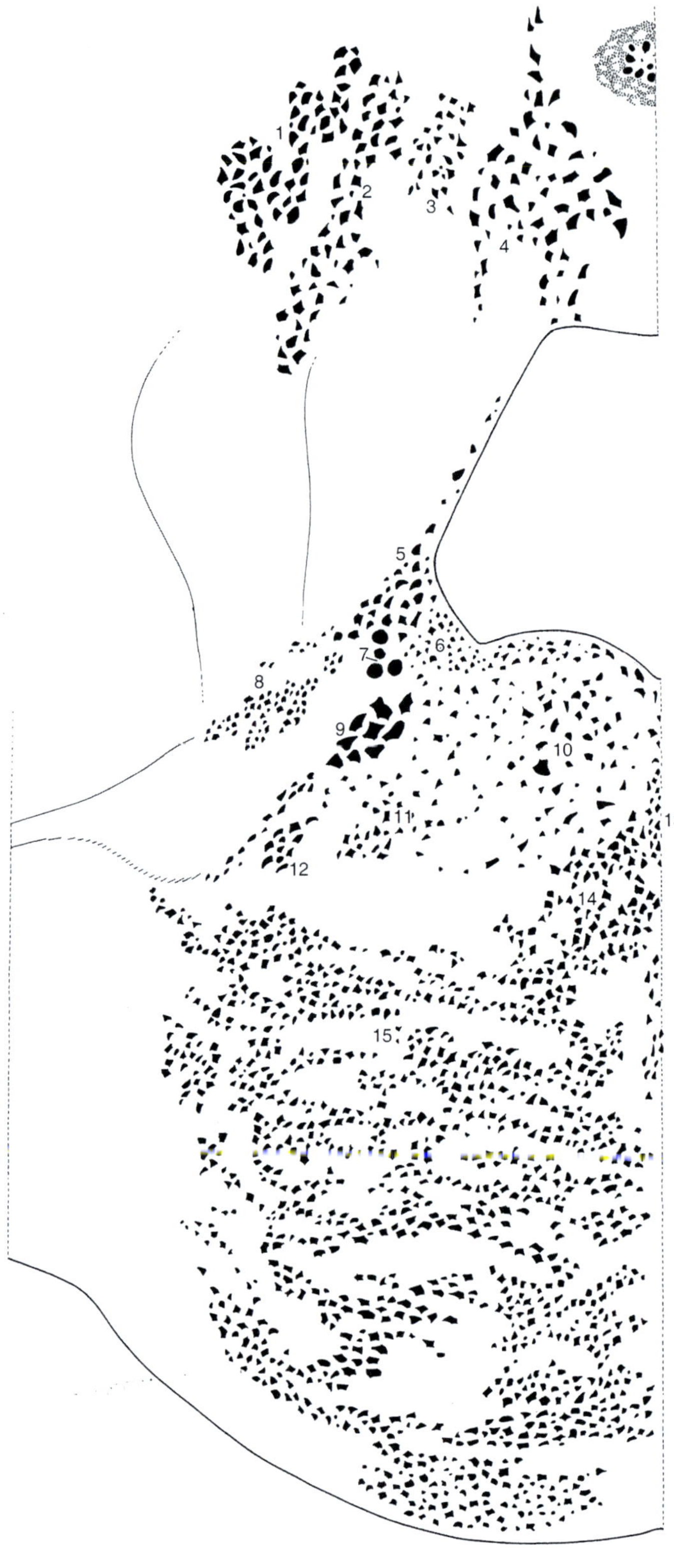

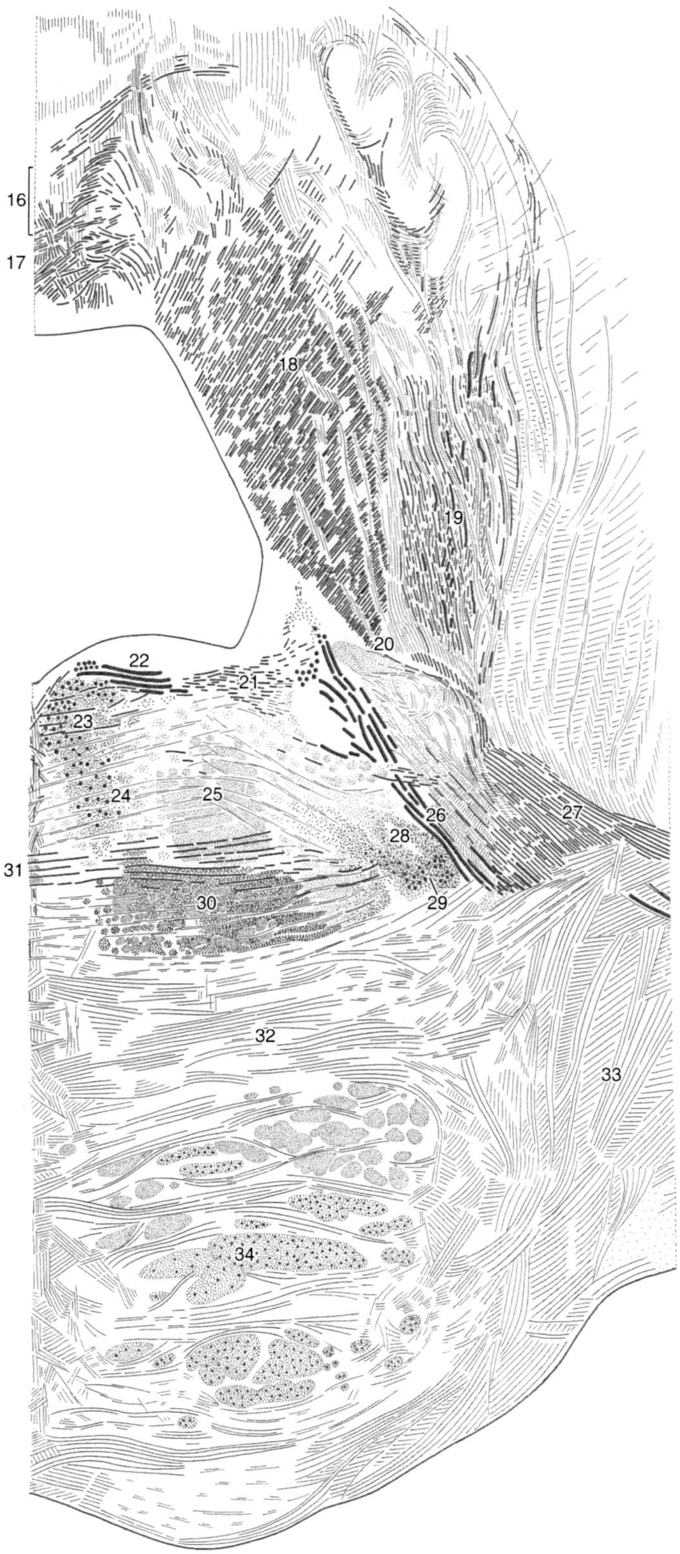

16 Cerebellar commissure: decussation of afferent mossy fibre systems
17 Cerebellar commissure: decussation of the uncinate tract
18 Superior cerebellar peduncle (brachium conjunctivum)
19 Inferior cerebellar peduncle (restiform body)
20 Oval fasciculus, ascending branch of the solitary tract
21 Vestibulomesencephalic tract (ascending tract of Deiters [25])
22 Knee (genu) of the facial nerve
23 Medial longitudinal fascicle
24 Tectospinal tract (predorsal fascicle)
25 Central tegmental tract
26 Trigminal nerve, portio minor
27 Trigeminal nerve, portio major
28 Lateral lemniscus
29 Anterolateral fascicle
30 Medial lemniscus
31 Ventral sensory trigeminal decussation
32 Pontocerebellar fibres
33 Middle cerebellar peduncle (brachium pontis)
34 Pyramidal tract

Fig. 6.27 A, B. Section through the principal sensory and motor nuclei of the trigeminal nerve (7/1×)

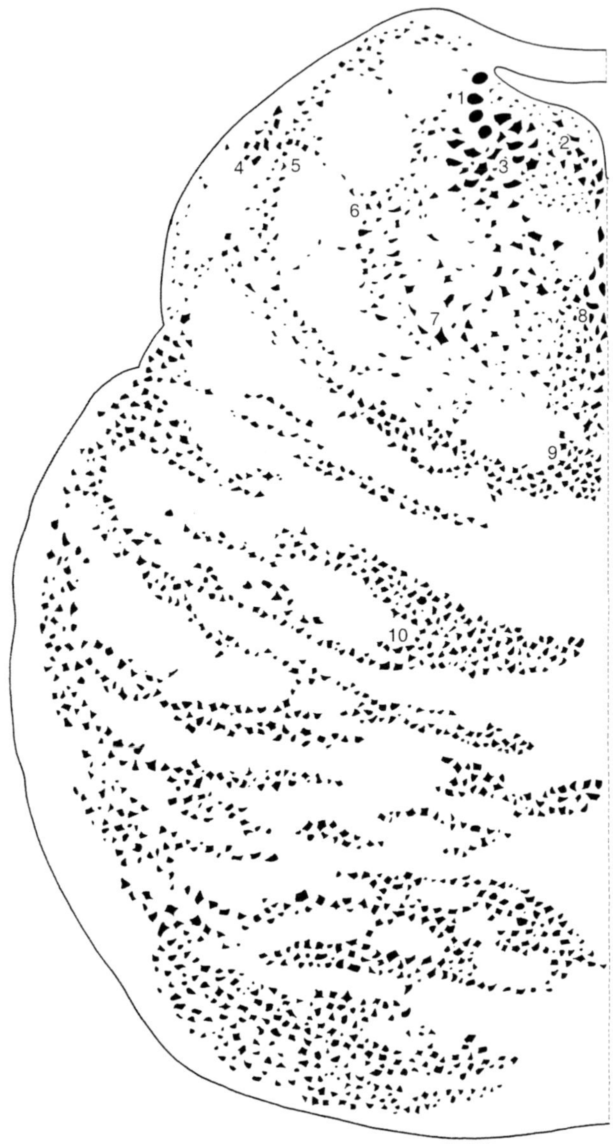

1 Mesencephalic nucleus of the trigeminal nerve
2 Periaqueductal grey
3 Locus coeruleus
4 Dorsal nucleus of the lateral lemniscus
5 Lateral parabrachial nucleus
6 Medial parabrachial nucleus
7 Nucleus reticularis pontis oralis
8 Central superior nucleus
9 Nucleus reticularis tegmenti pontis
10 Pontine nuclei

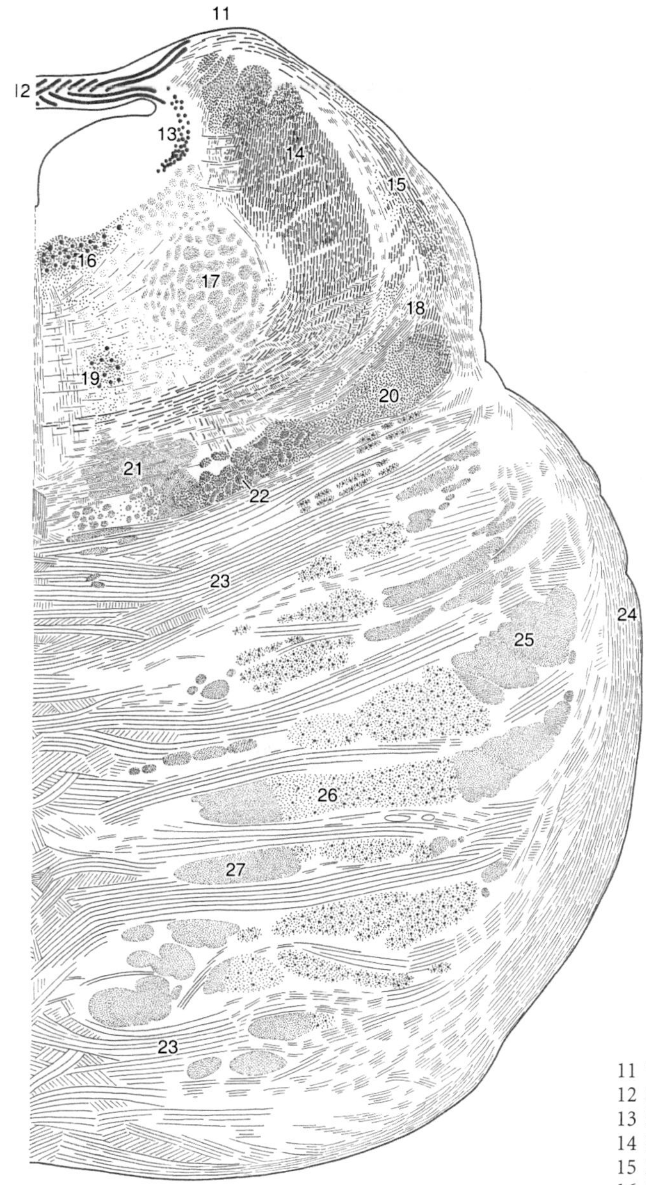

11 Trochlear nerve
12 Decussation of the trochlear nerves
13 Mesencephalic tract of the trigeminal nerve
14 Superior cerebellar peduncle (brachium conjunctivum)
15 Lateral lemniscus
16 Medial longitudinal fascicle
17 Central tegmental tract
18 Anterolateral fascicle
19 Tectospinal tract (predorsal fascicle)
20 Medial lemniscus
21 Descending branch of the superior cerebellar peduncle (brachium conjunctivum)
22 Corticotegmental fibres (areae nebulosae)
23 Pontocerebellar fibres
24 Middle cerebellar peduncle (brachium pontis)
25 Parietotemporopontine tract
26 Pyramidal tract
27 Frontopontine tract

Fig. 6.28 A, B. Section through the decussation of the trochlear nerves (7/1×)

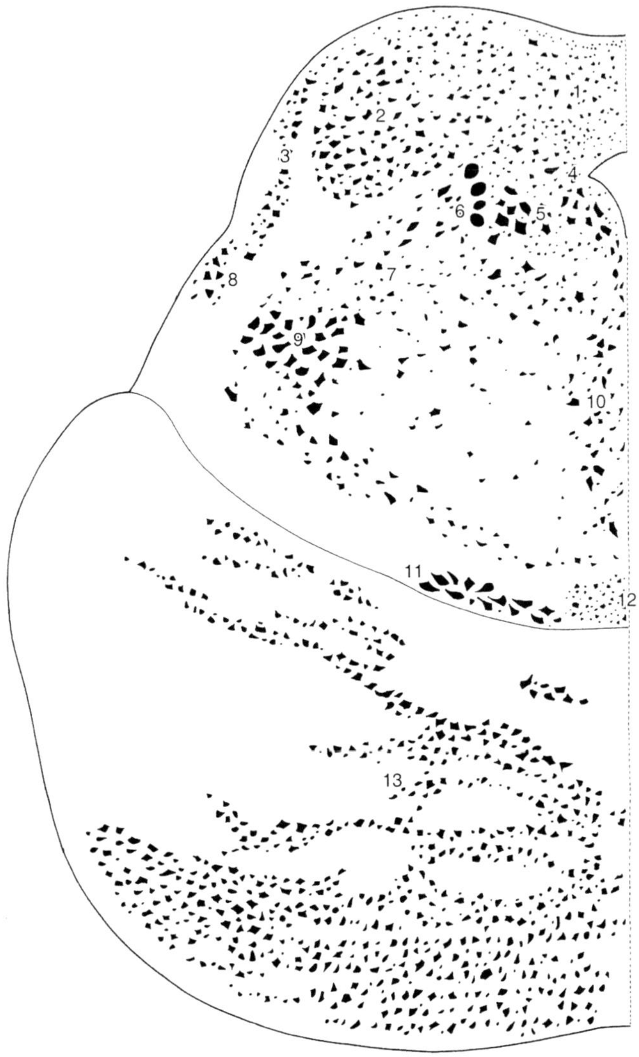

1 Intercollicular nucleus
2 Inferior colliculus, central nucleus
3 Inferior colliculus, external nucleus
4 Periaqueductal grey
5 Locus coeruleus
6 Mesencephalic nucleus of the trigeminal nerve
7 Cuneiform nucleus
8 Parabigeminal nucleus
9 Nucleus pedunculopontinus, pars compacta
10 Superior central nucleus
11 Substantia nigra, pars compacta
12 Interpeduncular nucleus
13 Pontine nuclei

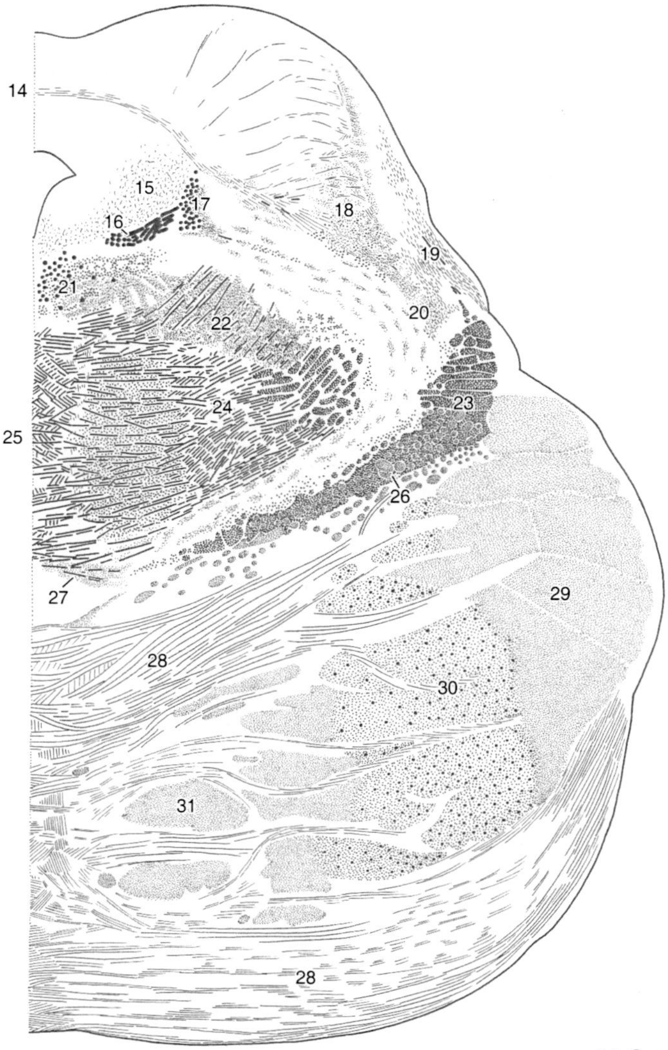

14 Commissure of the inferior colliculus
15 Dorsal longitudinal fascicle
16 Trochlear nerve
17 Mesencephalic tract of the trigeminal nerve
18 Lateral lemniscus
19 Tectopontine tract
20 Anterolateral fascicle
21 Medial longitudinal fascicle
22 Central tegmental tract
23 Medial lemniscus
24 Superior cerebellar peduncle (brachium conjunctivum)
25 Decussation of the superior cerebellar peduncles
26 Corticotegmental fibres (areae nebulosae)
27 Mamillary peduncle
28 Pontocerebellar fibres
29 Parietotemporopontine tract
30 Pyramidal tract
31 Frontopontine tract

Fig. 6.29 A, B. Section through the inferior colliculus and the decussation of the superior cerebellar peduncles (7/1×)

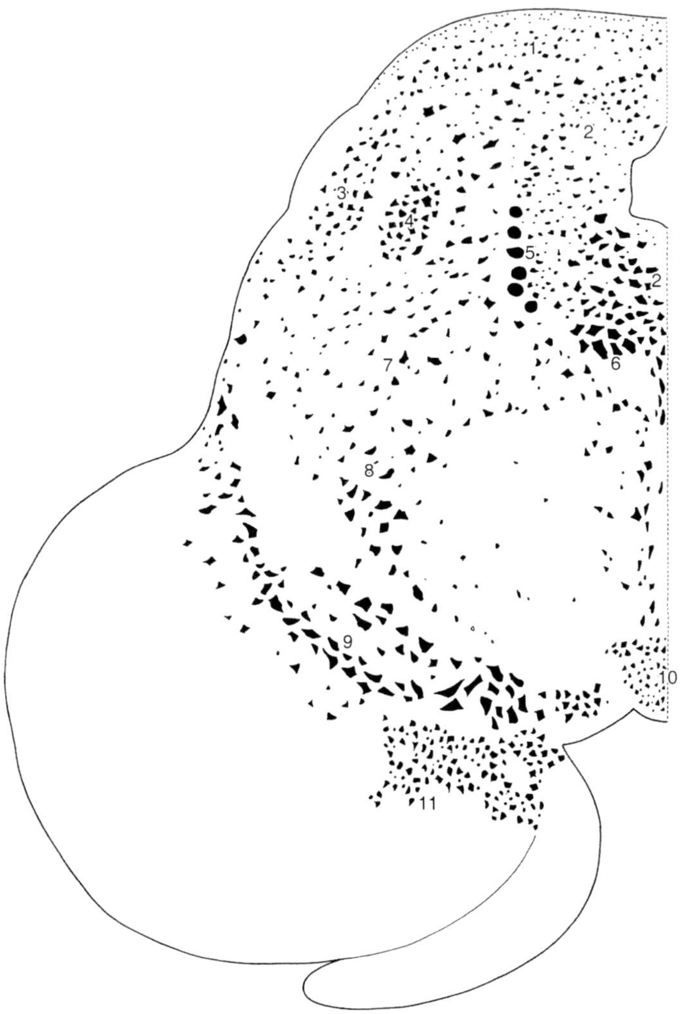

1 Intercollicular nucleus
2 Periaqueductal grey
3 Parabigeminal nucleus
4 Central nucleus of the inferior colliculus
5 Mesencephalic nucleus of the trigeminal nerve
6 Nucleus of the trochlear nerve
7 Cuneiform nucleus
8 Pedunculopontine nucleus, pars dissipata
9 Substantia nigra, pars compacta
10 Interpeduncular nucleus
11 Pontine nuclei

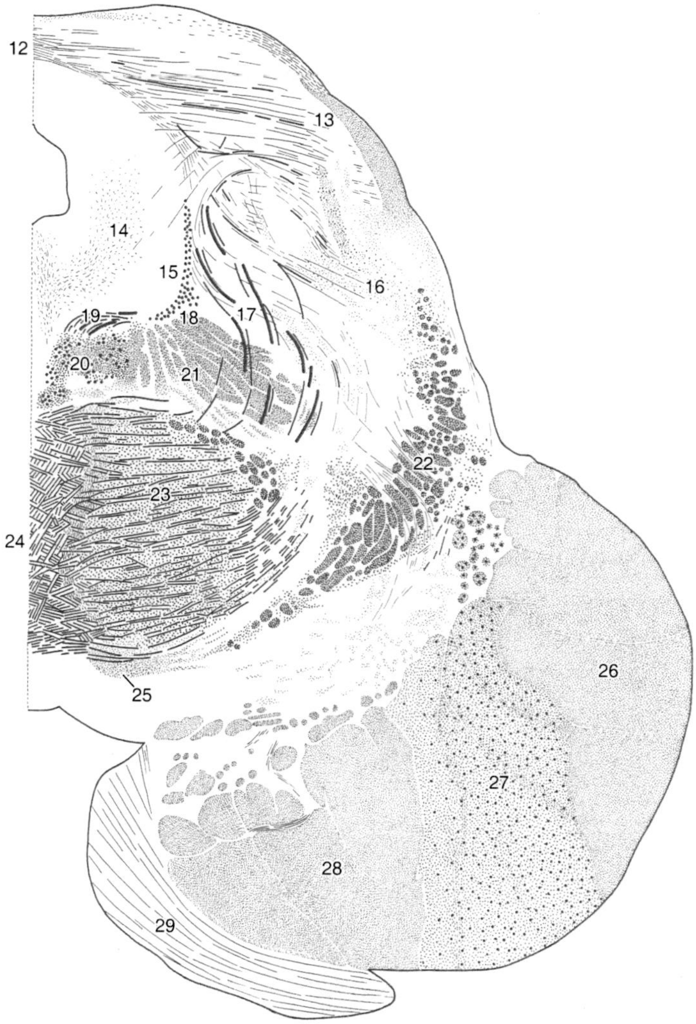

12 Commissure of the inferior colliculus
13 Brachium of the inferior colliculus
14 Dorsal longitudinal fascicle
15 Mesencephalic tract of the trigeminal nerve
16 Anterolateral fascicle
17 Tectopontine tract
18 Dorsal trigeminothalamic tract
19 Trochlear nerve
20 Medial longitudinal fascicle
21 Central tegmental tract
22 Medial lemniscus
23 Superior cerebellar peduncle (brachium conjunctivum)
24 Decussation of the superior cerebellar peduncle
25 Mamillary peduncle
26 Parietotemporopontine tract
27 Pyramidal tract
28 Frontopontine tract
29 Pontocerebellar fibres

Fig. 6.30 A, B. Section through the intercollicular area and the trochlear nucleus (7/1×)

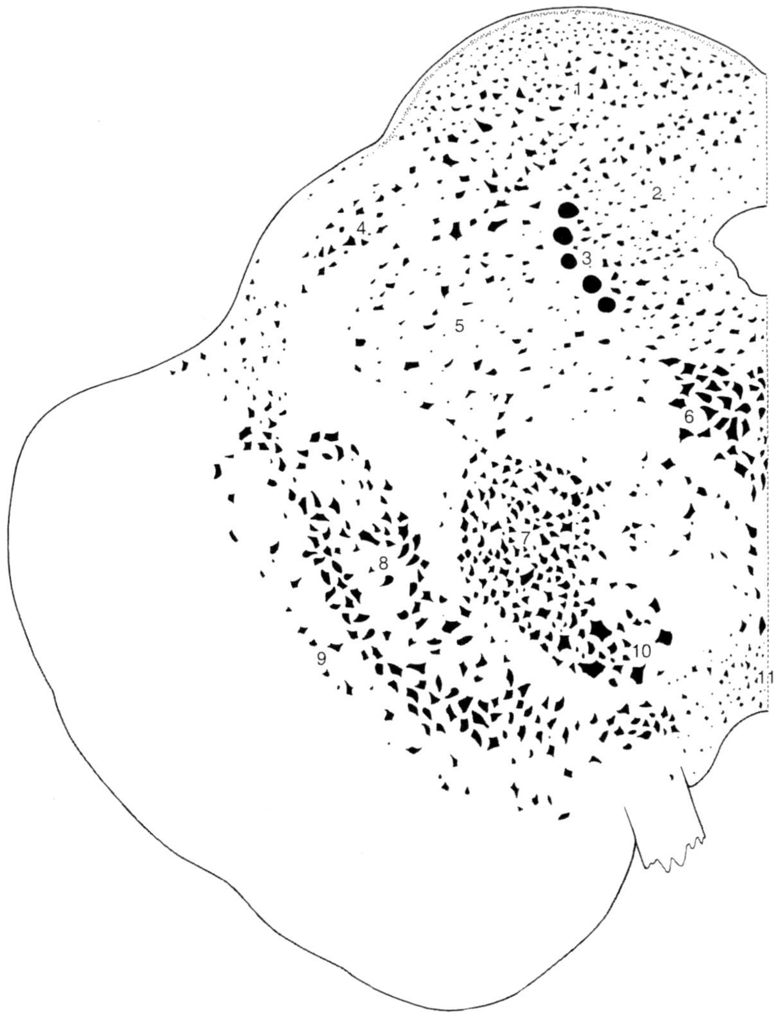

1 Superior colliculus
2 Periaqueductal grey
3 Mesencephalic nucleus of the trigeminal nerve
4 Parabigeminal nucleus
5 Cuneiform nucleus
6 Nucleus of the oculomotor nerve
7 Red nucleus, parvocellular part
8 Substantia nigra, pars compacta
9 Substantia nigra, pars reticulata
10 Red nucleus, magnocellular part
11 Interpeduncular nucleus

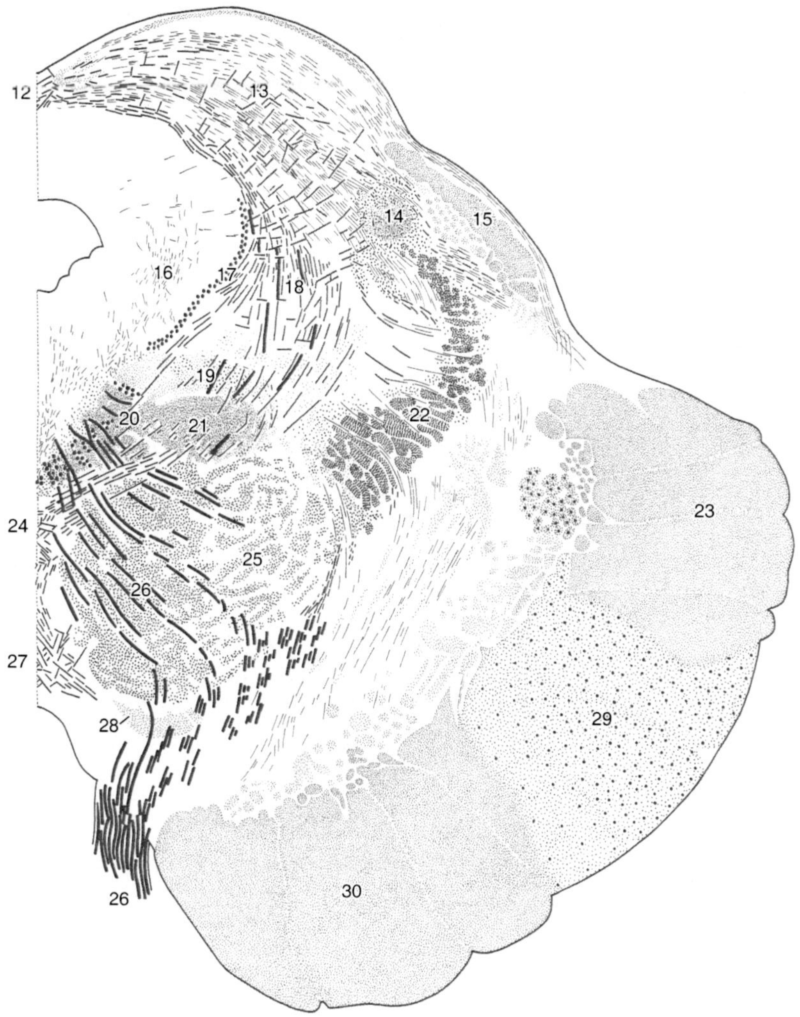

12 Commissure of the superior colliculus
13 Superior colliculus
14 Anterolateral fascicle
15 Brachium of the inferior colliculus
16 Dorsal longitudinal fascicle
17 Mesencephalic tract of the trigeminal nerve
18 Tectospinal tract
19 Dorsal trigeminothalamic tract
20 Medial longitudinal fascicle
21 Central tegmental tract
22 Medial lemniscus
23 Parietotemporopontine tract
24 Dorsal tegmental (tectospinal) decussation
25 Superior cerebellar peduncle (brachium conjunctivum)
26 Oculomotor nerve
27 Ventral tegmental (rubrospinal) decussation
28 Mamillary peduncle
29 Pyramidal tract
30 Frontopontine tract

Fig. 6.31 A, B. Section through the superior colliculus and the oculomotor nuclei (7/1×)

1 Stratum zonale of the superior colliculus
2 Superficial grey layer of the superior colliculus
3 Middle grey layer of the superior colliculus
4 Deep grey layer of the superior colliculus
5 Periaqueductal grey
6 Mesencephalic nucleus of the trigeminal nerve
7 Cuneiform nucleus
8 Medial geniculate body, dorsal subnucleus
9 Medial geniculate body, ventral subnucleus
10 Interstitial nucleus
11 Nucleus of Edinger-Westphal [11, 39]
12 Nucleus of the oculomotor nerve
13 Red nucleus, parvocellular part
14 Substantia nigra, pars compacta
15 Substantia nigra, pars reticulata

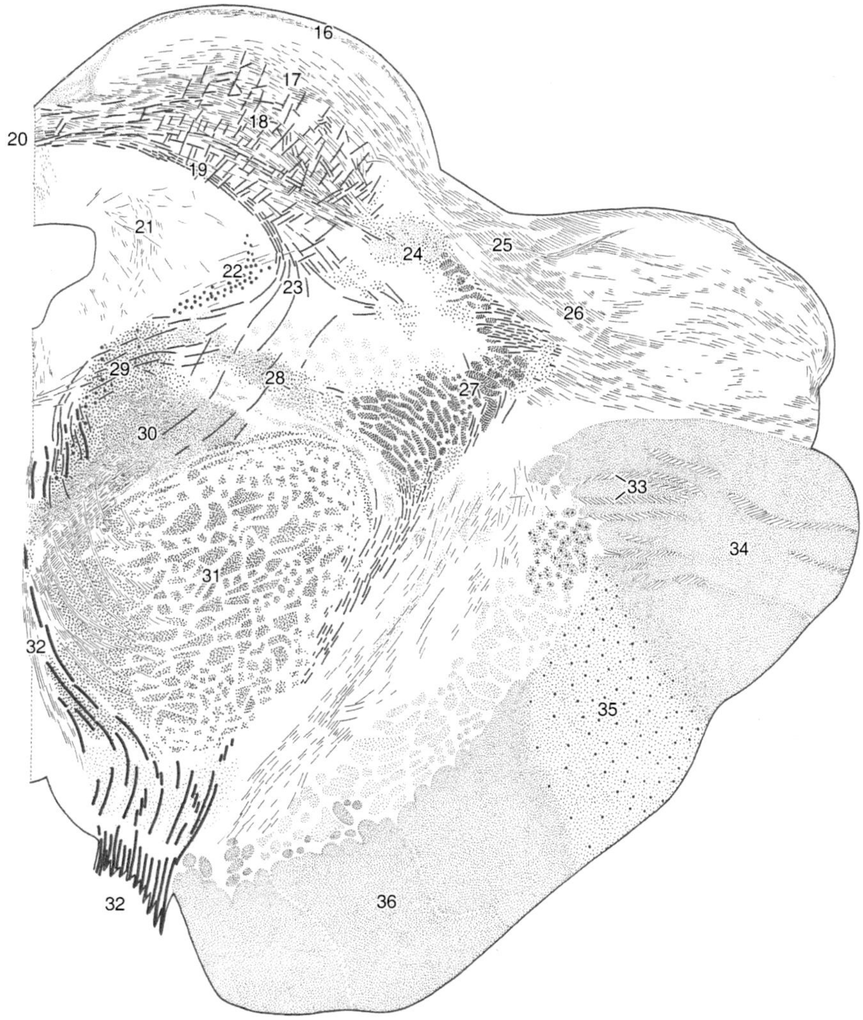

16 Stratum zonale of the superior colliculus
17 Stratum opticum of the superior colliculus
18 Stratum lemnisci of the superior colliculus
19 Deep white layer of the superior colliculus
20 Commissure of the superior colliculus
21 Dorsal longitudinal fascicle
22 Mesencephalic tract of the trigeminal nerve
23 Tectospinal tract
24 Spinothalamic tract
25 Brachium of the superior colliculus
26 Brachium of the inferior colliculus
27 Medial lemniscus
28 Dorsal trigeminothalamic tract
29 Medial longitudinal fascicle
30 Central tegmental tract
31 Superior cerebellar peduncle (brachium conjunctivum)
32 Oculomotor nerve
33 Strionigral fibres
34 Parietotemporopontine tract
35 Pyramidal tract
36 Frontopontine tract

Fig. 6.32 A, B. Section through the red nucleus and the medial geniculate body (7/1×)

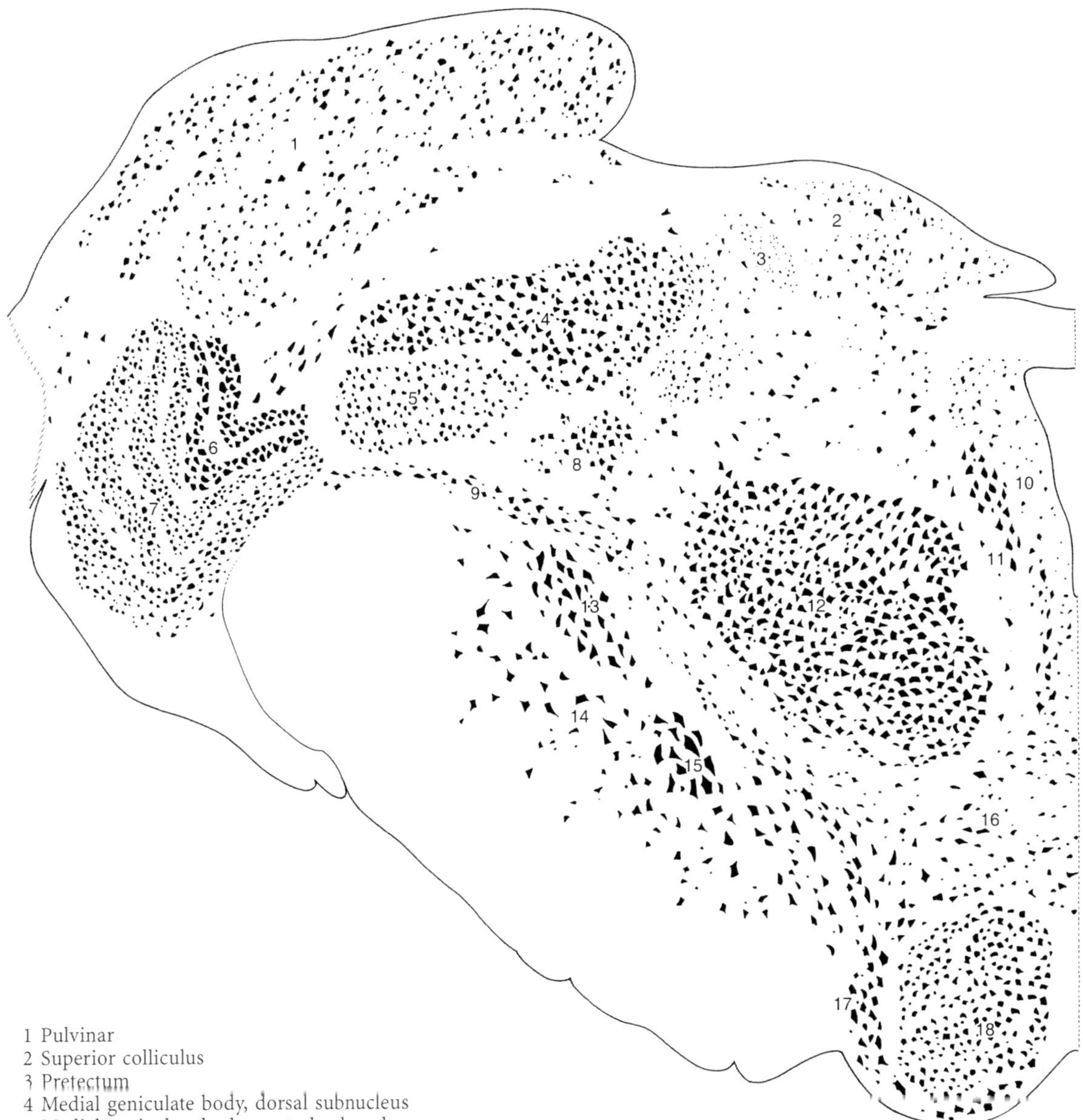

1 Pulvinar
2 Superior colliculus
3 Pretectum
4 Medial geniculate body, dorsal subnucleus
5 Medial geniculate body, ventral subnucleus
6 Lateral geniculate body, magnocellular laminae
7 Lateral geniculate body, parvocellular laminae
8 Ventral posteromedial nucleus
9 Peripeduncular nucleus
10 Periaqueductal grey
11 Darkschewitsch nucleus [9]
12 Red nucleus, parvocellular part
13 Subthalamic nucleus
14 Substantia nigra, pars reticulata
15 Substantia nigra, pars compacta
16 Ventral tegmental area
17 Mamillary body, lateral nucleus
18 Mamillary body, medial nucleus

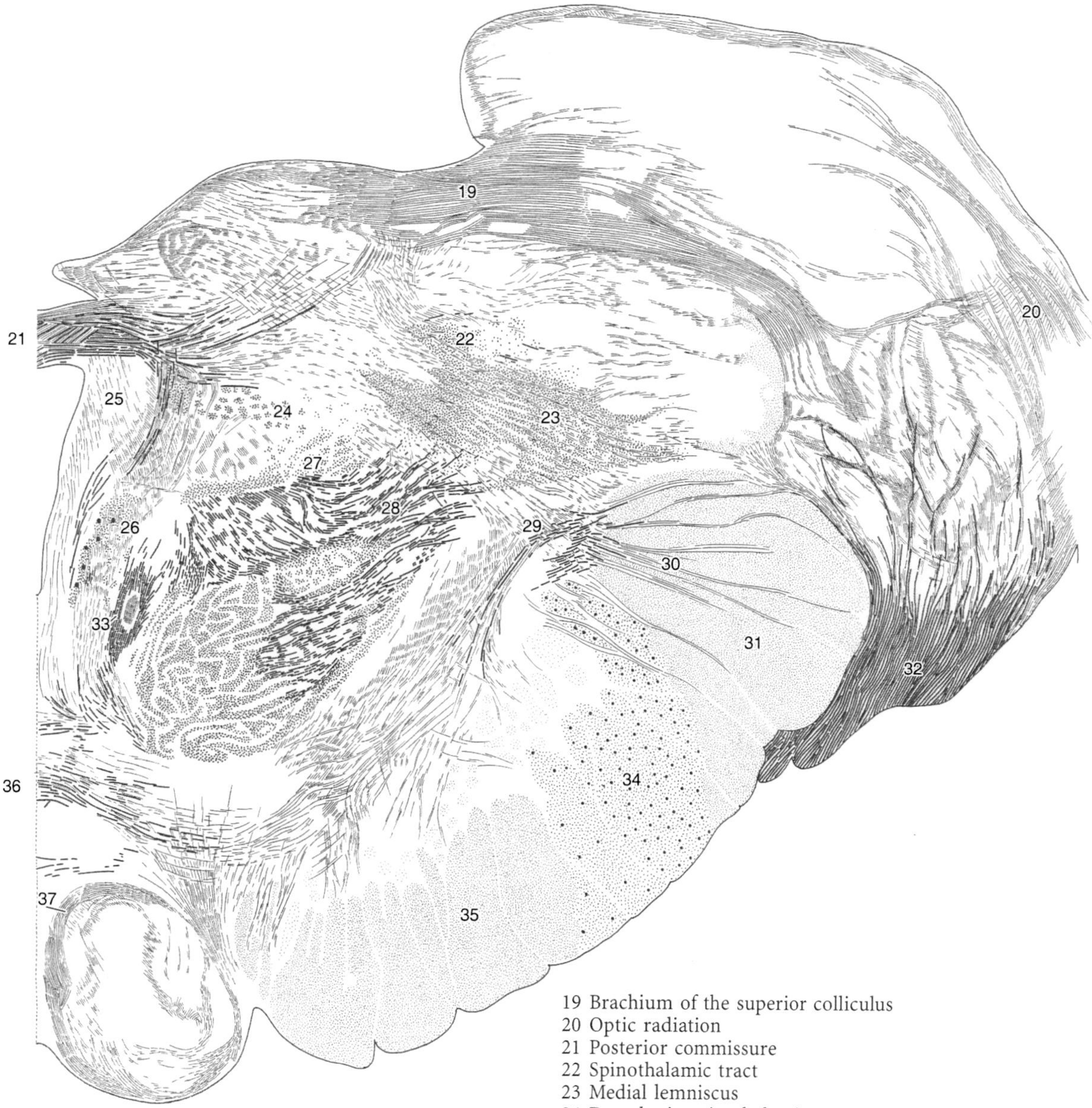

19 Brachium of the superior colliculus
20 Optic radiation
21 Posterior commissure
22 Spinothalamic tract
23 Medial lemniscus
24 Dorsal trigeminothalamic tract
25 Dorsal longitudinal fascicle
26 Medial longitudinal fascicle
27 Central tegmental tract
28 Superior cerebellar peduncle (brachium conjunctivum)
29 Pallidoreticular fibres
30 Strionigral fibres
31 Parietotemporal pontine tract
32 Optic tract
33 Habenulointerpeduncular tract (fasciculus retroflexus)
34 Pyramidal tract
35 Frontopontine tract
36 Supramamillary commissure
37 Main mamillary fasciculus

Fig. 6.33 A, B. Section through the posterior commissure, the medial and lateral geniculate bodies and the mamillary body (7/1×)

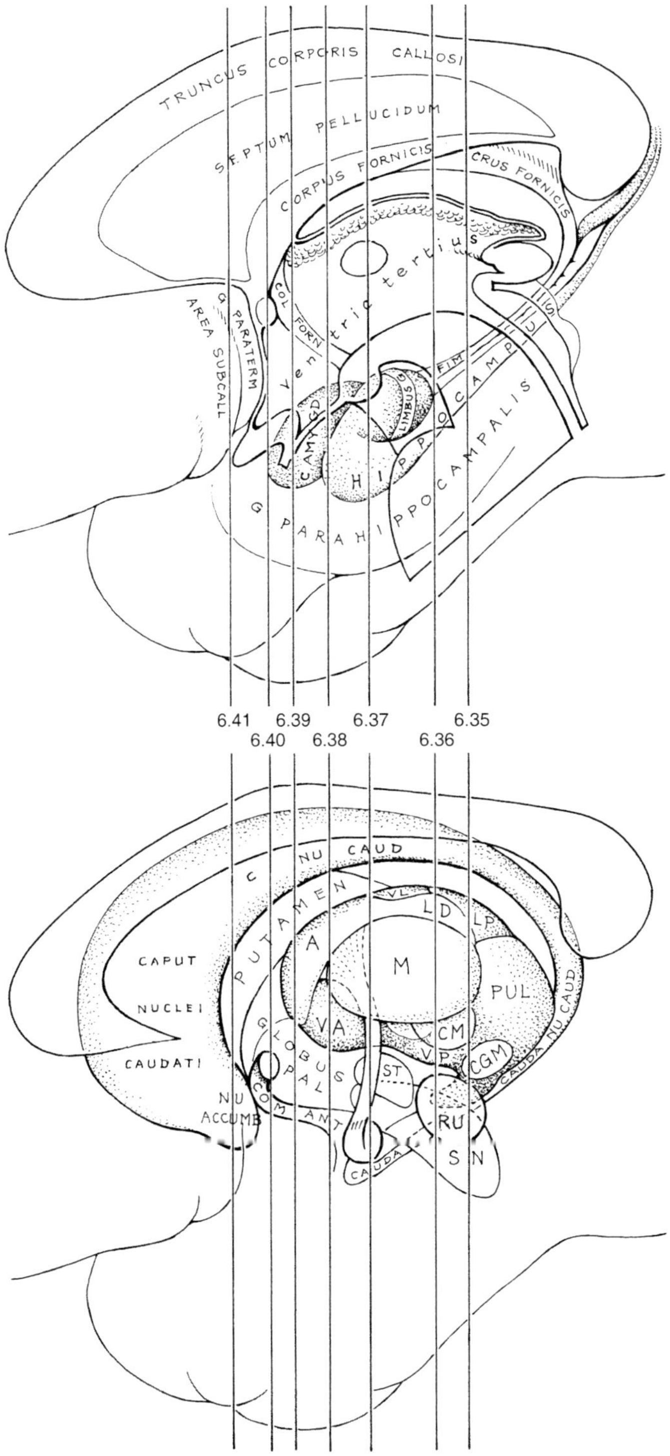

Fig. 6.34. Key diagrams showing level and plane of the sections illustrated in Figs. 6.35–6.41. *Top*, medial view; *bottom*, deep structures exposed from the same side. *A*, anterior thalamic nucleus; *C NU CAUD*, body of the caudate nucleus; *C.AMYGD*, amygdaloid nucleus; *CGM*, medial geniculate body; *CM*, nucleus of the centre médian; *COM ANT*, anterior commissure; *FIM*, fibria; *LD*, lateral dorsal thalamic nucleus; *LP*, lateral posterior thalamic nucleus; *M*, massa intermedia; *NU ACCUMB*, nucleus accumbens; *PUL*, pulvinar of the thalamus; *RU*, red nucleus; *SN*, substantia nigra; *ST*, subthalamic nucleus; *VA*, ventral anterior thalamic nucleus; *VL*, ventral lateral thalamic nucleus; *VP*, ventroposterior thalamic nucleus

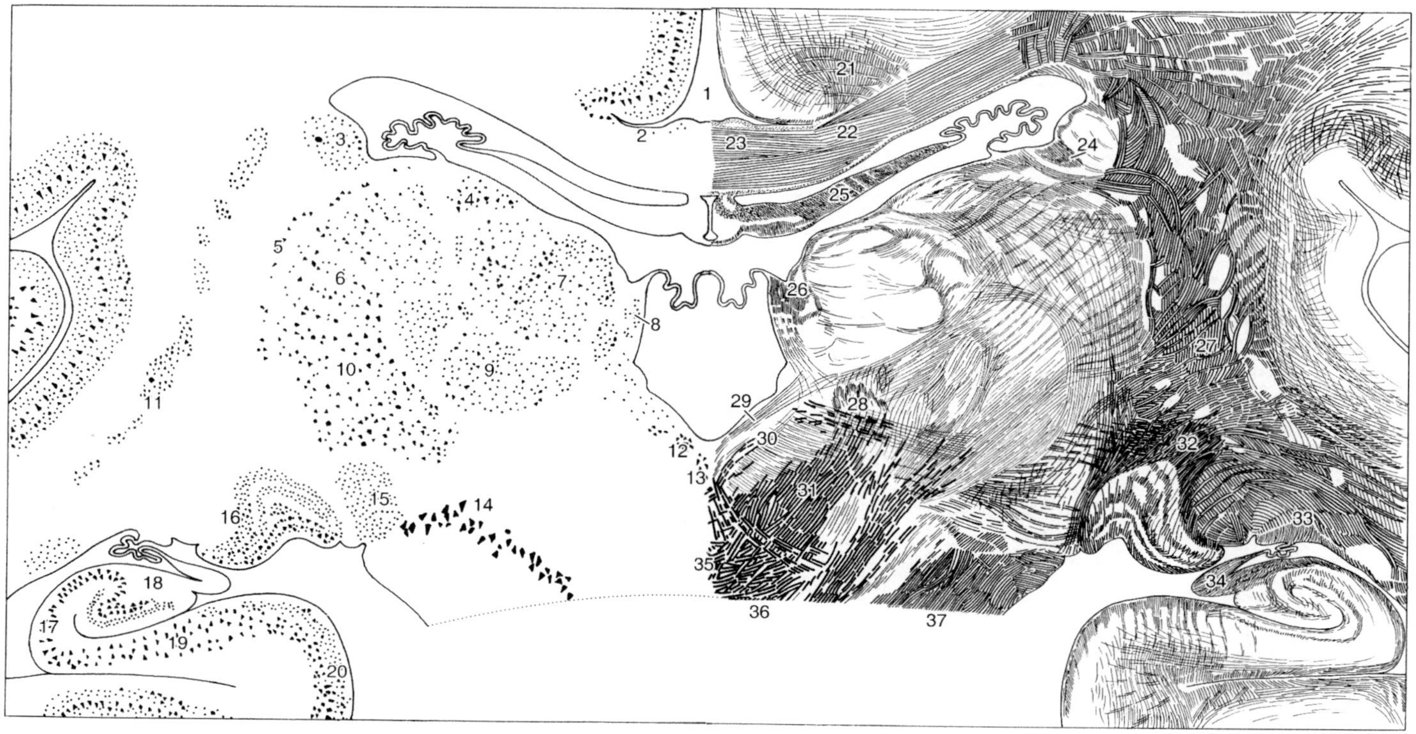

Fig. 6.35. Section through the medial and lateral geniculate bodies (5/2×)

1 Cingulate gyrus
2 Indusium griseum
3 Tail of the caudate nucleus
4 Lateral dorsal thalamic nucleus
5 Reticular nucleus of the thalamus
6 Lateral posterior thalamic nucleus
7 Medial dorsal thalamic nucleus of the thalamus
8 Habenular nuclei
9 Nucleus of the centre médian
10 Ventral posterolateral nucleus
11 Bridges of grey matter connecting the caudate nucleus with the putamen (pontes grisei)
12 Interstitial nucleus
13 Nucleus of Edinger-Westphal [11, 39]
14 Substantia nigra
15 Medial geniculate body
16 Lateral geniculate body
17 Ammon's horn
18 Fascia dentata
19 Subiculum
20 Entorhinal cortex
21 Cingulum
22 Lateral longitudinal stria
23 Medial longitudinal stria
24 Stria terminalis
25 Crus of the fornix
26 Stria medullaris
27 Internal capsule, retrolenticular part
28 Habenulointerpeduncular tract (fasciculus retroflexus)
29 Dorsal longitudinal fascicle
30 Medial longitudinal fascicle
31 Superior cerebellar peduncle (brachium conjunctivum)
32 Optic radiation
33 Stria terminalis
34 Fimbria
35 Decussation of the superior cerebellar peduncles
36 Medial lemniscus
37 Cerebral peduncle

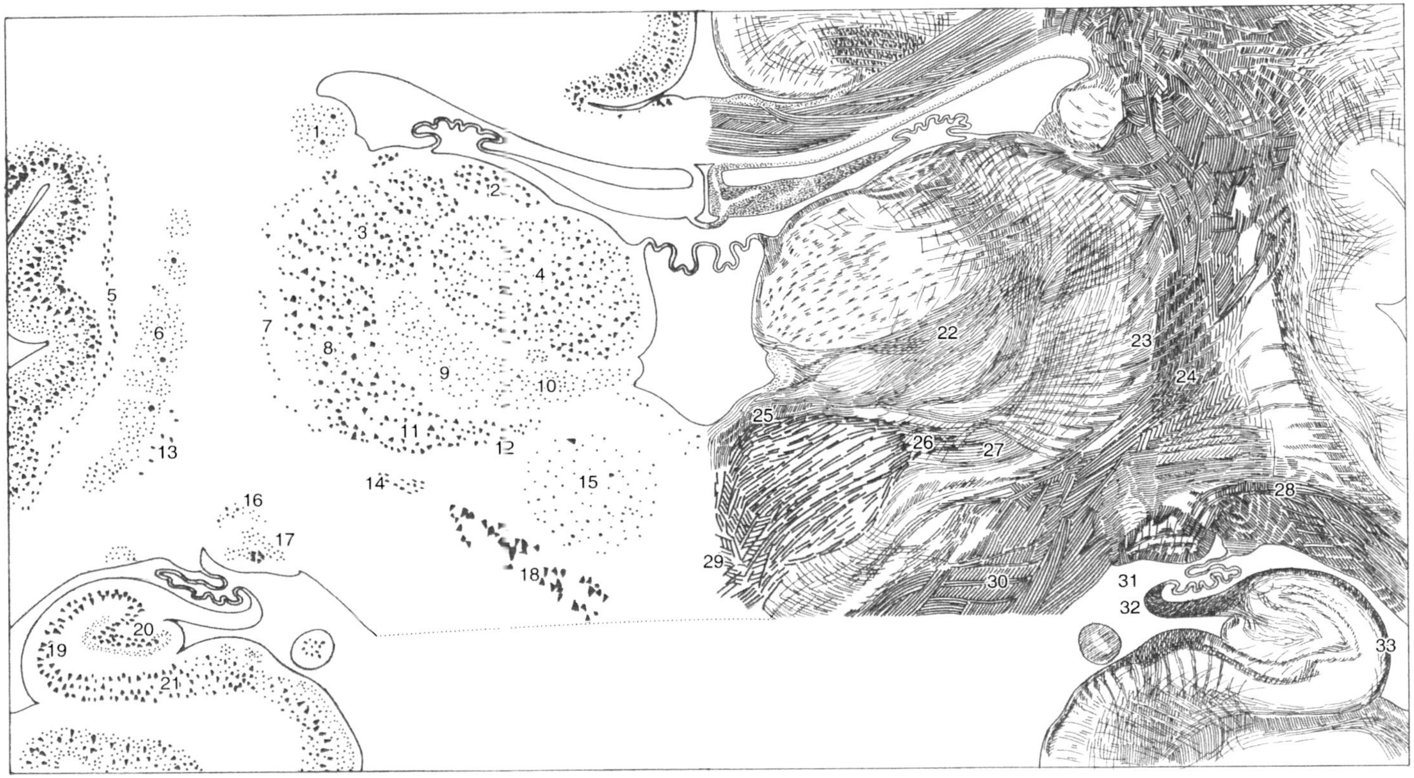

Fig. 6.36. Section through the thalamus and the caudal end of the putamen (5/2×)

1 Tail of the caudate nucleus
2 Lateral dorsal nucleus
3 Lateral posterior thalamic nucleus
4 Medial dorsal thalamic nucleus of the thalamus
5 Claustrum
6 Putamen
7 Reticular nucleus of the thalamus
8 Ventral posterolateral thalamic nucleus
9 Nucleus of the centre médian
10 Parafascicular nucleus
11 Ventral posteromedial thalamic nucleus
12 Ventral posteromedial thalamic nucleus, parvocellular subnucleus
13 Globus pallidus, external segment
14 Zona incerta
15 Red nucleus
16 Pregeniculate thalamic nucleus
17 Lateral geniculate body
18 Substantia nigra
19 Ammon's horn
20 Fascia dentata
21 Subiculum
22 Internal medullary lamina
23 External medullary lamina
24 Internal capsule, posterior limb
25 Medial longitudinal fascicle
26 Superior cerebellar peduncle (brachium conjunctivum)
27 Medial lemniscus
28 Internal capsule, sublenticular part
29 Decussation of the superior cerebellar peduncles
30 Cerebral peduncle
31 Optic tract
32 Fimbria
33 Alveus

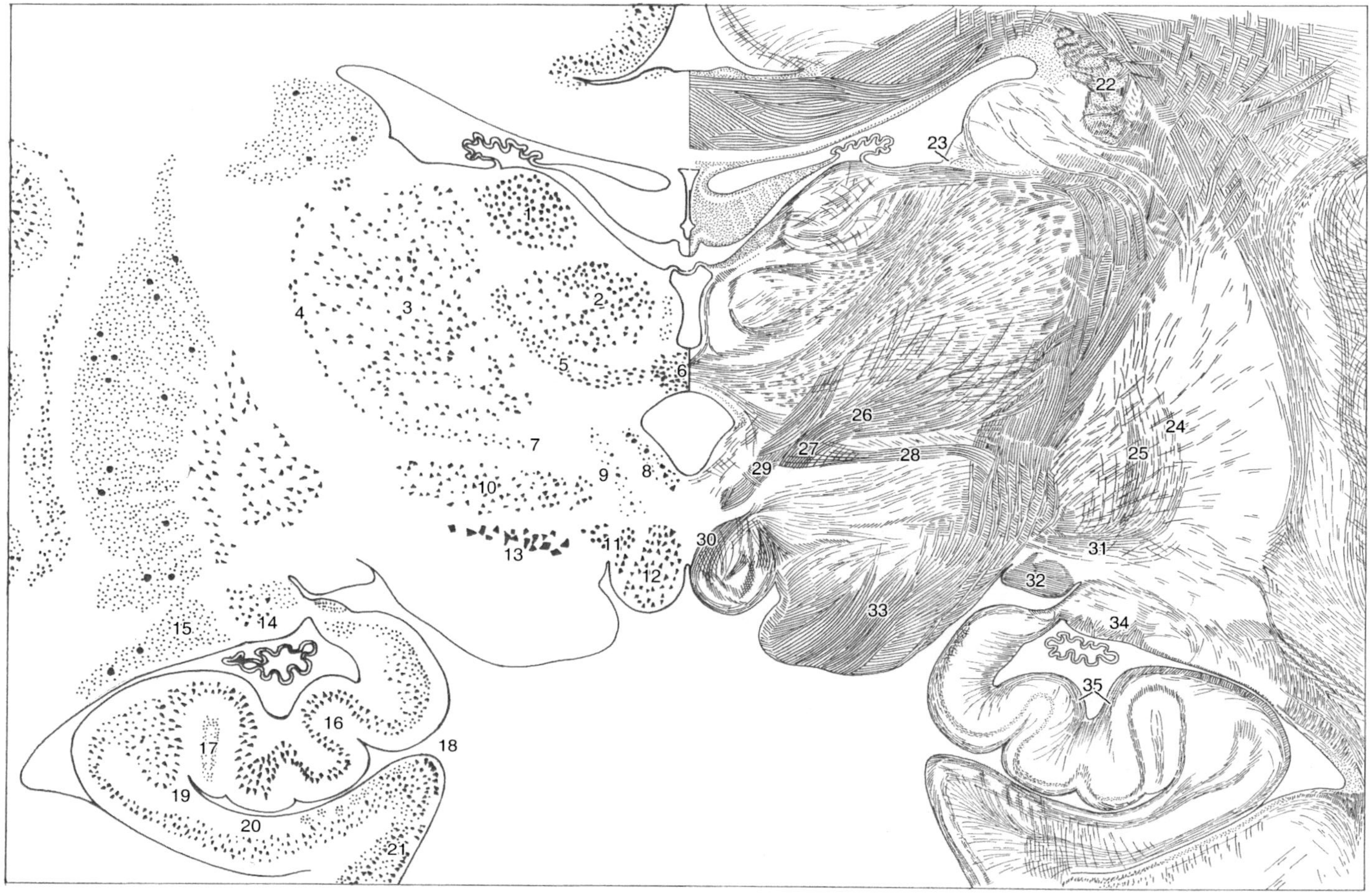

Fig. 6.37. Section through the centre of the third ventricle, the mamillary body and the hippocampus (5/2×)

1 Anterior nucleus of the thalamus
2 Medial nucleus of the thalamus
3 Ventral lateral thalamic nucleus
4 Reticular nucleus of the thalamus
5 Intralaminar thalamic nuclei
6 Median nuclei of the thalamus
7 Zona incerta
8 Posterior nucleus of the hypothalamus
9 Lateral area of the hypothalamus
10 Subthalamic nucleus
11 Mamillary body, lateral nucleus
12 Mamillary body, medial nucleus
13 Substantia nigra
14 Amygdaloid body, accessory basal nucleus
15 Peduncle of the lentiform nucleus
16 Ammon's horn
17 Fascia dentata
18 Incisure of the uncus
19 Hippocampal sulcus
20 Subiculum
21 Entorhinal cortex
22 Superior occipitofrontal fascicle
23 Stria terminalis
24 Lateral medullary lamina
25 Medial medullary lamina
26 Thalamic fascicle
27 Tegmental area H
28 Lenticular fascicle
29 Mamillothalamic tract
30 Main mamillary fascicle
31 Ansa lenticularis
32 Optic tract
33 Cerebral peduncle
34 Stria terminalis
35 Alveus

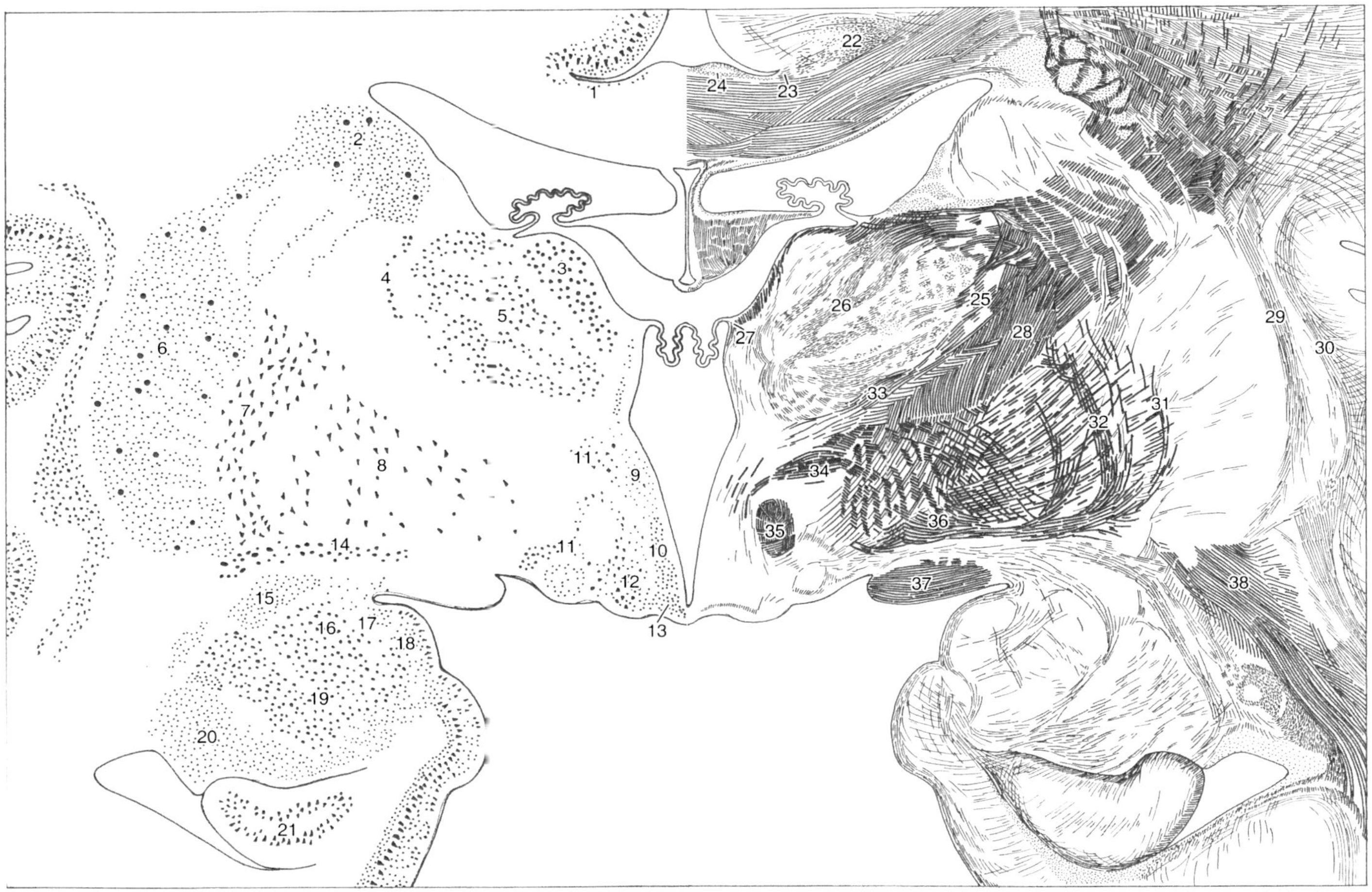

Fig. 6.38. Section through the rostral part of the thalamus, the amydaloid body and the rostral pole of the hippocampus (5/2×)

1 Indusium griseum
2 Body of the caudate nucleus
3 Anterior nucleus of the thalamus
4 Reticular nucleus of the thalamus
5 Ventral anterior thalamic nucleus
6 Putamen
7 Globus pallidus, external segment
8 Globus pallidus, internal segment
9 Posterior nucleus of the hypothalamus
10 Ventromedial nucleus of the hypothalamus
11 Lateral area of the hypothalamus
12 Tuberal nuclei
13 Infundibular nucleus
14 Innominate substance
15 Central nucleus of the amygdaloid body
16 Accessory basal nucleus of the amygdaloid body
17 Medial nucleus of the amygdaloid body
18 Cortical nucleus of the amygdaloid body
19 Basal nucleus of the amygdaloid body
20 Lateral nucleus of the amygdaloid body
21 Ammon's horn
22 Cingulum
23 Lateral longitudinal stria
24 Medial longitudinal stria
25 External medullary lamina
26 Internal medullary lamina
27 Stria medullaris
28 Internal capsule, posterior limb
29 External capsule
30 Extreme capsule
31 Lateral medullary lamina
32 Medial medullary lamina
33 Thalamic fascicle
34 Lenticular fascicle
35 Column of the fornix
36 Ansa lenticularis
37 Optic tract
38 Anterior commissure

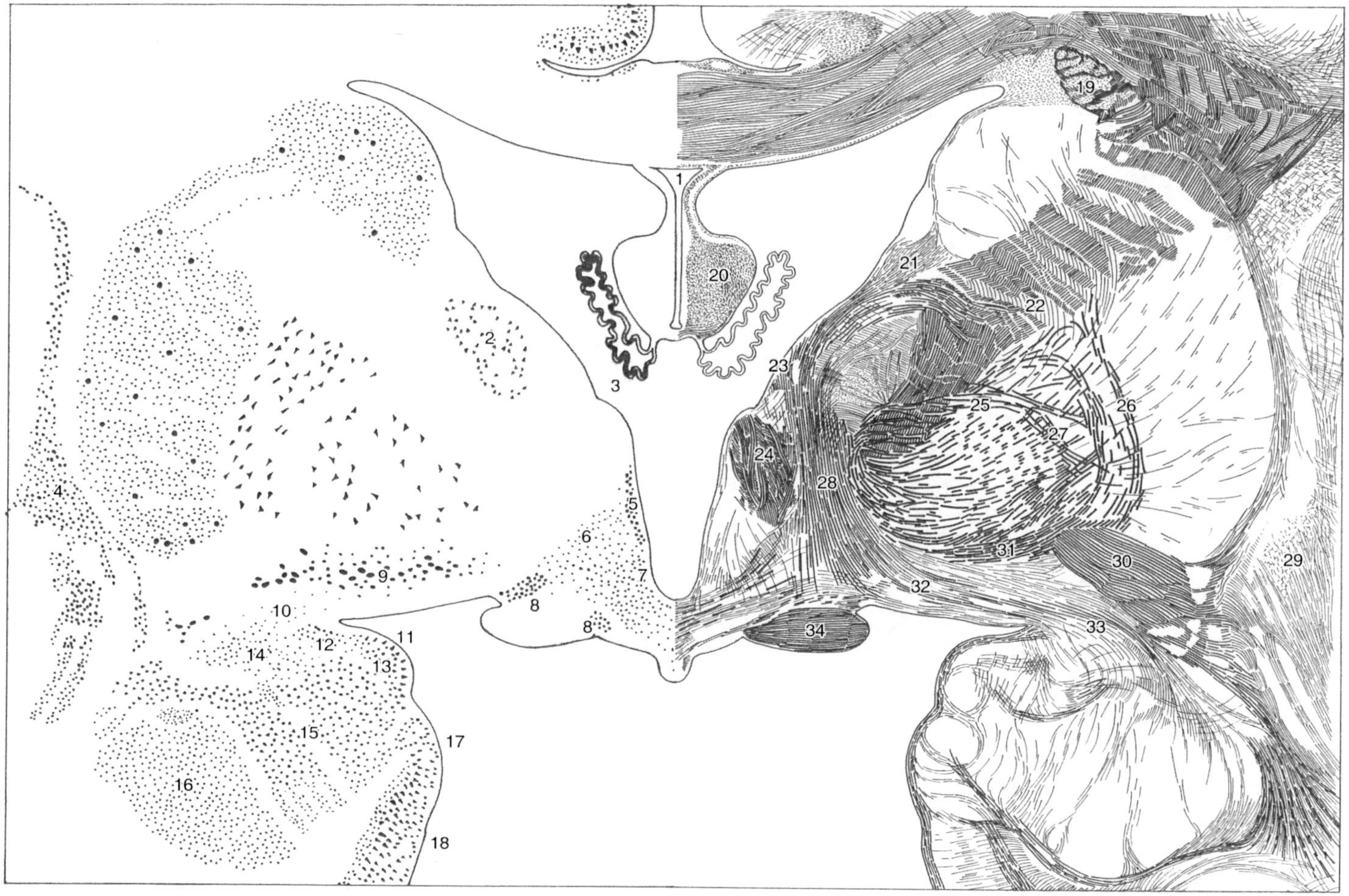

Fig. 6.39. Section through the interventricular foramen, the hypothalamus and the amygdaloid body (5/2×)

1 Cave of the septum pellucidum
2 Ventral anterior thalamic nucleus
3 Interventricular foramen
4 Claustrum
5 Paraventricular nucleus
6 Lateral area of the hypothalamus
7 Anterior nucleus of the hypothalamus
8 Supraoptic nucleus
9 Innominate substance
10 Bed nucleus of the stria terminalis
11 Semilunar gyrus
12 Medial nucleus of the amygdaloid body
13 Cortical nucleus of the amygdaloid body
14 Accessory basal nucleus of the amygdaloid body
15 Basal nucleus of the amygdaloid body
16 Lateral nucleus of the amygdaloid body
17 Ambiens gyrus
18 Parahippocampal gyrus
19 Superior occipitofrontal fascicle
20 Corpus of the fornix
21 Stria terminalis
22 Comb system: caudato-pallidal fibres
23 Stria medullaris
24 Column of the fornix
25 Lenticular fascicle
26 Lateral medullary lamina
27 Medial medullary lamina
28 Inferior thalamic peduncle
29 Inferior occipitofrontal fascicle
30 Anterior commissure
31 Ansa lenticularis
32 Ansa peduncularis
33 Amydalofugal fibers
34 Optic tract

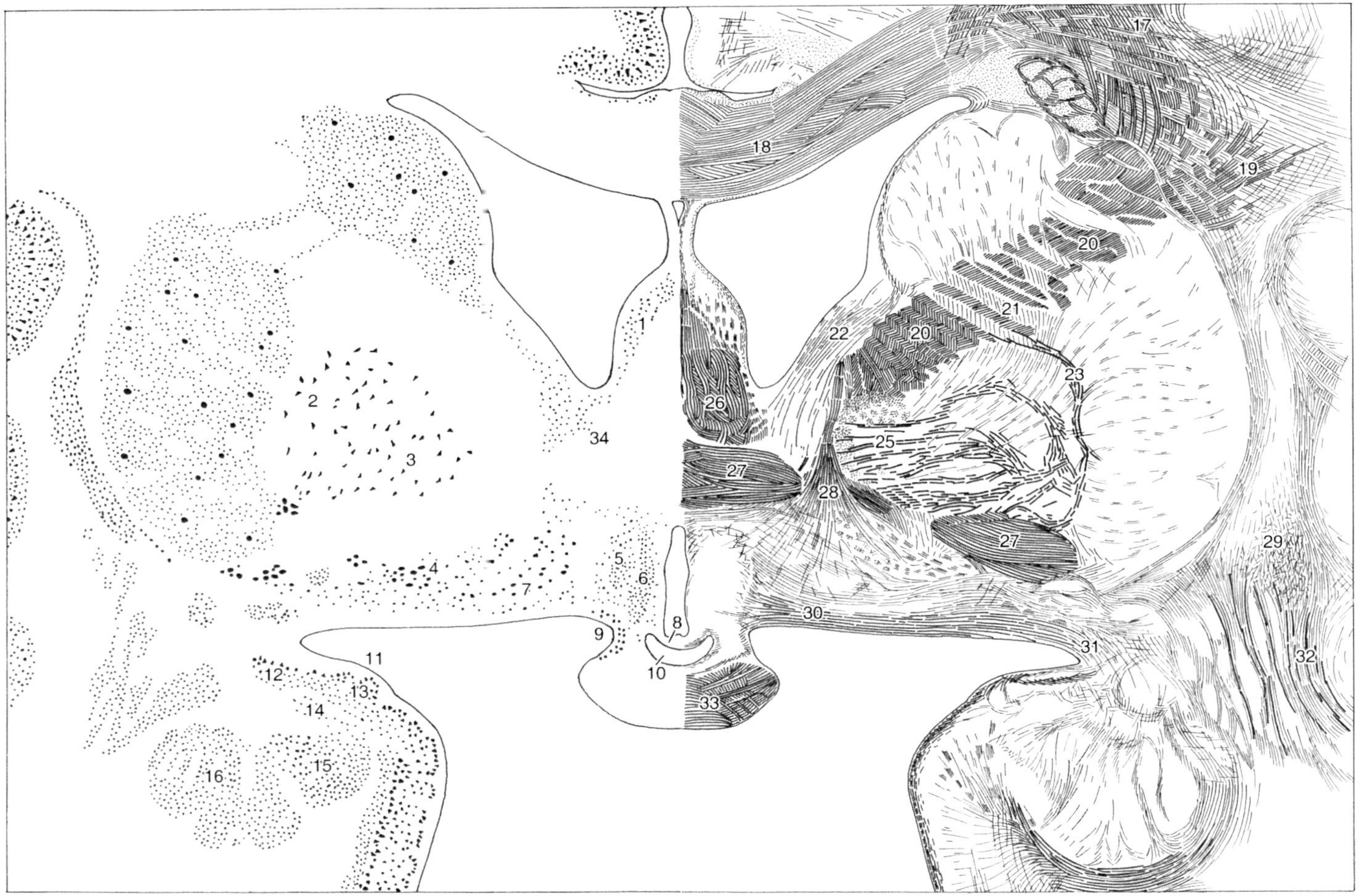

1 Septal nuclei
2 Globus pallidus, external segment
3 Globus pallidus, internal segment
4 Innominate substance
5 Lateral preoptic nucleus
6 Medial preoptic nucleus
7 Nucleus of the diagonal band
8 Lamina terminalis
9 Supraoptic nucleus
10 Optic recess
11 Semilunar gyrus
12 Anterior nucleus of the amygdaloid body
13 Cortical nucleus of the amygdaloid body
14 Accessory basal nucleus of the amygdaloid body
15 Basal nucleus of the amygdaloid body
16 Lateral nucleus of the amygdaloid body
17 Radiation of the corpus callosum
18 Trunk of the corpus callosum
19 Corona radiata
20 Internal capsule, anterior limb
21 Caudatopallidal fibres
22 Anterior thalamic peduncle
23 Lateral medullary lamina
24 Medial medullary lamina
25 Lenticular fascicle
26 Column of the fornix
27 Anterior commissure
28 Stria terminalis
29 Occipitofrontal fascicle
30 Diagonal band
31 Lateral olfactory stria
32 Uncinate fascicle
33 Optic chiasm
34 Bed nucleus of the stria terminalis

Fig. 6.40. Section through the anterior commissure and the optic chiasm (5/2×)

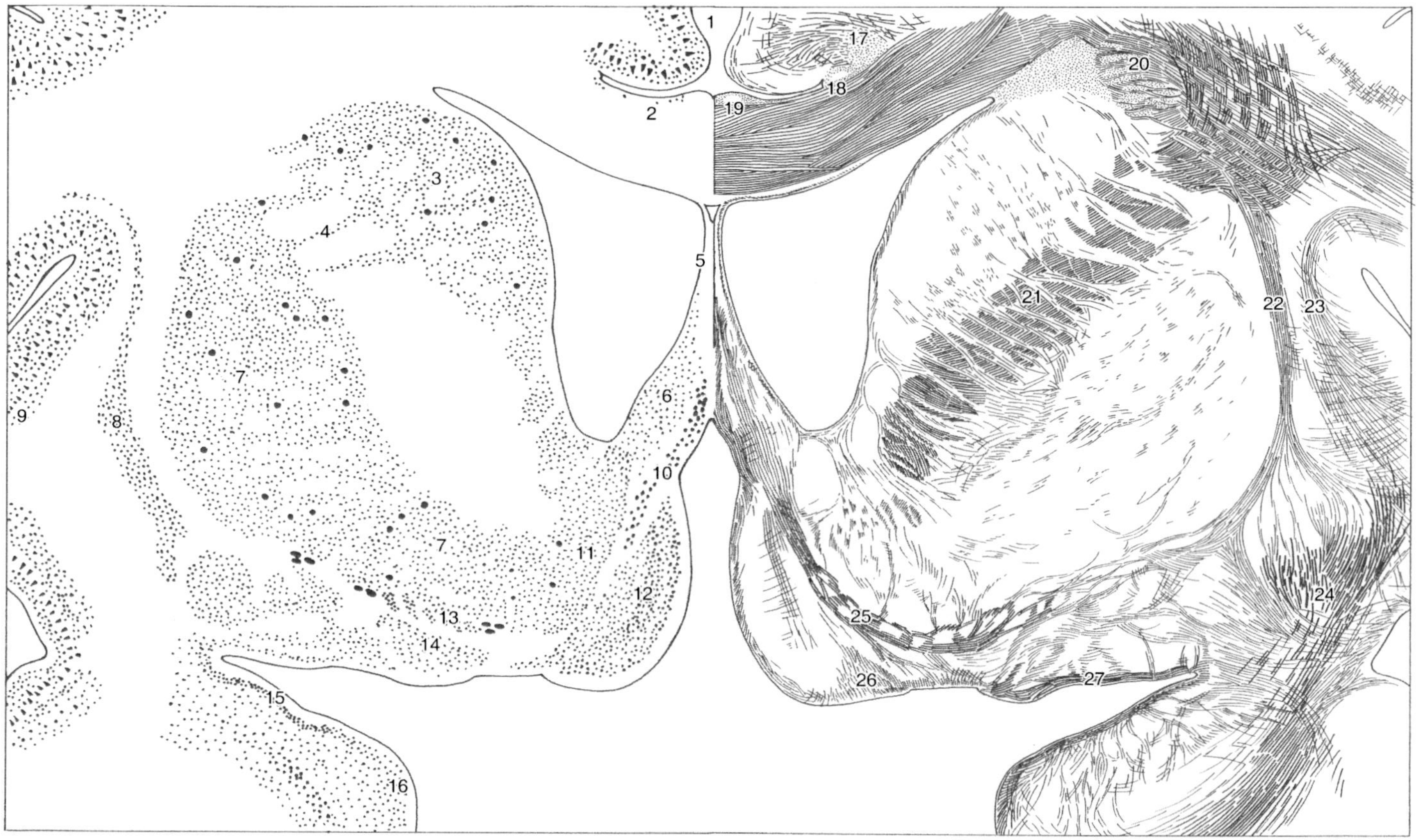

1 Cingulate gyrus
2 Indusium griseum
3 Body of the caudate nucleus
4 Bridges of grey matter connecting the caudate nucleus with the putamen (pontes grisei)
5 Septum pellucidum
6 Septal nuclei
7 Putamen
8 Claustrum
9 Insula
10 Nucleus of the diagonal band
11 Accumbens nucleus
12 Subcallosal area
13 Innominate substance
14 Anterior olfactory nucleus
15 Prepiriform cortex
16 Entorhinal cortex
17 Cingulum
18 Lateral longitudinal stria
19 Medial longitudinal stria
20 Superior occipitofrontal fascicle
21 Internal capsule, anterior limb
22 External capsule
23 Extreme capsule
24 Uncinate fascicle
25 Diagonal band
26 Medial olfactory stria
27 Lateral olfactory stria

Fig. 6.41. Section through the septal area (5/2×)

7 Diencephalon: Introduction and Epithalamus

Introduction

Herrick [9, 10] divided the diencephalon into four longitudinal zones: the epithalamus, dorsal thalamus, ventral thalamus and hypothalamus. This subdivision has found wide acceptance and has been claimed to be applicable to all vertebrate groups, including mammals [3, 27] and humans [5, 11, 12, 26]. Although we presently know (see Chap. 2) that two of Herrick's zones, i.e. the dorsal and ventral thalami, are direct derivatives of neuromeres and that their topographically horizontal orientation is due to the strong curvature of the rostral part of the brain, it cannot be denied that Herrick's subdivision offers a convenient starting point for a discussion of the various diencephalic cell masses and their connections. For this practical reason, the same partitioning will be used in the following chapters.

Epithalamus

The *epithalamus* encompasses the habenular complex, epiphysis cerebri (pineal gland) and a nervous strip of tissue passing over the dorsomedial surface of the dorsal thalamus directly adjacent to the line of attachment of the membranous roof of the diencephalon or *taenia thalami.*

The *habenular complex* is a small accumulation of grey matter medial to the caudal part of the dorsal thalamus directly underneath the diencephalic ventricular surface. It receives afferents via the stria medullaris and gives origin to the habenulo-interpeduncular tract or fasciculus retroflexus (Fig. 7.1). The *stria medullaris* is a complex bundle of fibres originating from the septal region and the lateral preoptico-hypothalamic continuum. These fibres assemble in the region behind the anterior commissure and pass along the taenia thalami to the habenula. Some fibres, ascending with the stria medullaris, decussate in the habenular commissure (Fig. 3.8) to terminate in the contralateral habenular complex. The *habenulo-interpeduncular tract* descends from the habenular complex to the basal region of the midbrain, through which many of its fibres pass to the *interpeduncular nucleus* (Fig. 7.1). The latter is a derivative of the isthmic and first rhombencephalic neuromere (Fig. 2.11). The stria medullaris, habenulo-interpeduncular tract and intercalated habenular complex together constitute a dorsal diencephalic conduction system along which impulses originating from telencephalic and diencephalic limbic regions are conveyed to the rostral part of the brain stem.

The habenular complex includes medial and lateral nuclei (Fig. 6.35). The cells in the medial nucleus are smaller and more densely packed than those in the lateral nucleus.

By way of the stria medullaris, the *medial habenular nucleus* (MH) receives fibres from the medial septal-diagonal band complex [29]. The MH is also in receipt of fibres from the

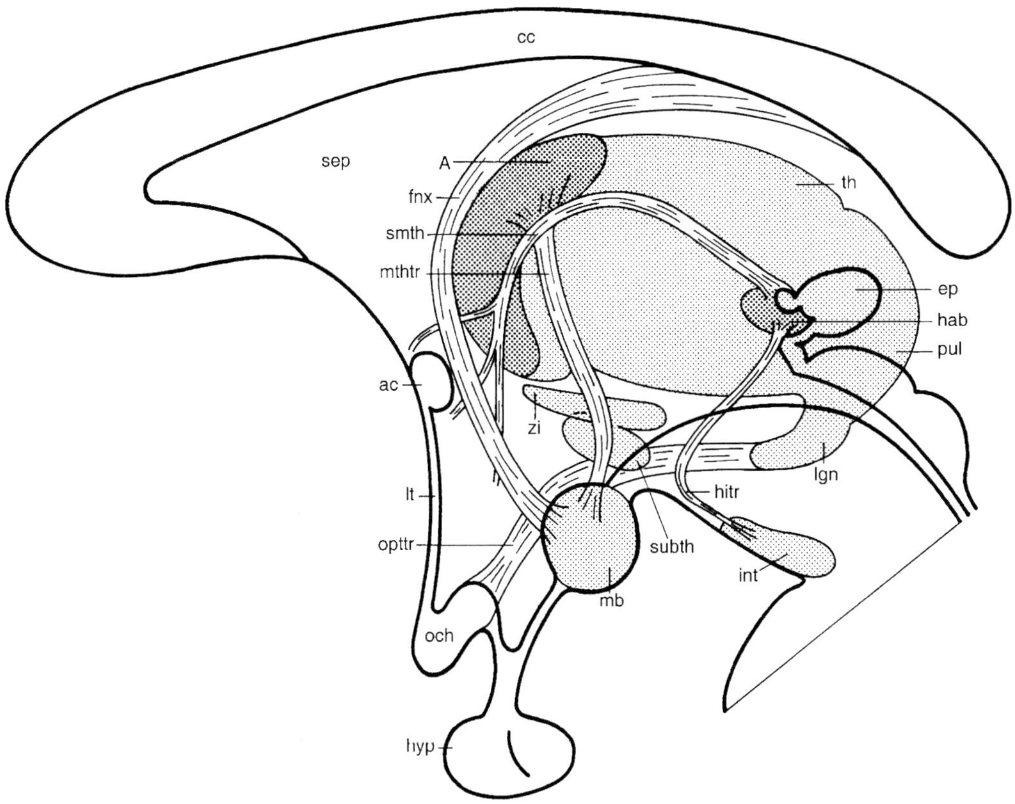

Fig. 7.1. Reconstruction of diencephalic and adjacent structures in medial view. Reproduced from [32]. *A*, nucleus anterior thalami; *ac*, anterior commissure; *cc*, corpus callosum; *ep*, epiphysis; *hab*, habenular ganglion; *hitr*, habenulo-interpeduncular tract; *hyp*, hypophysis; *mthtr*, mamillothalamic tract; *fnx*, fornix; *int*, interpeduncular nucleus; *lgn*, lateral geniculate nucleus; *lt*, lamina terminalis; *mb*, mamillary body; *och*, optic chiasm; *opttr*, optic tract; *pul*, pulvinar; *sep*, septum pellucidum; *smth*, stria medullaris thalami; *subth*, subthalamic nucleus; *th*, thalamus; *zi*, zona incerta

periaqueductal grey matter [28], a serotoninergic projection from the mesencephalic raphe nucleus and an adrenergic innervation from the superior cervical ganglion [4, 14]. The efferents of the MH terminate exclusively in the interpeduncular nucleus [8, 16]. The efferent fibres of the interpeduncular nucleus, which project massively to the dorsal raphe and central superior nuclei, form a further link in the "medial habenular path" [6, 15, 21].

The *lateral habenular nucleus* (LH) receives its afferents principally from the nuclei of the diagonal band, lateral preopticohypothalamic area, substantia innominata and the medial pallidal segment; whereas the ventral tegmental area, mesencephalic raphe nuclei and ventral part of the periaqueductal grey matter give rise to ascending input to this nucleus [7, 8, 17, 22–25, 28, 31]. Fibres originating from the LH descend in the habenulo-interpeduncular tract and terminate in various mesencephalic centres including the pars compacta of the substantia nigra, ventral part of the periaqueductal grey matter, dorsal raphe and central superior nuclei and mesencephalic reticular formation [1, 7, 8, 23].

The functional role of the habenular nuclei is not clear. Judging from their fibre connections, they represent processing stations in strictly separate medial and lateral habenular paths. The medial habenular path, which is synaptically interrupted in the medial habenular and interpeduncular nuclei, conveys information from medial septal and ultimately hippocampal antecedents. In contrast, the lateral preopticohypothalamic area and the globus pallidus provide the LH with the majority of its afferents. The efferents of the LH bypass the interpeduncular nucleus, but the medial and lateral habenular paths converge upon the mesencephalic raphe nuclei. To what extent these two afferent systems interact at the level of the raphe nuclei is presently unknown. The projection from the LH to pars compacta of the substantia nigra presumably closes a nigro-striato-pallido-habenulo-nigral circuit [8].

Andres et al. [2] recently compiled data from the literature suggesting that the habenular complex engages in a variety of biological functions, including pain processing, reproduction, nutrition, stress responses, sleep-waking cycles and even learning. Based on a detailed structural and ultrastructural analysis, they delineated five distinct subnuclei in the MH and ten in the LH. New and refined functional and behavioural studies are required to assess whether discrete functions can be attributed to all of these newly identified subnuclei.

The *pineal gland* or *epiphysis cerebri* develops from a median evagination of the roof of the diencephalon between the habenular and posterior commissures (Figs. 2.5 B, C, 2.6). It is gradually transformed into a solid structure connected with the caudal diencephalon by the pineal stalk. In primates this stalk is short, but it is long and slender in many other mammalian groups, including rodents. In adult humans, the pineal gland is a small conical body between the suprapineal recess (a dorsocaudal extension of the third ventricle) and the superior colliculi (Fig. 3.8).

The pineal parenchyma is composed of four types of cells: pinealocytes, glia-like interstitial cells, perivascular phagocytes and classic neurons [18]. Of these, the pinealocytes are by far the most numerous (more than 90%). The pineal gland is surrounded by a capsule formed by the pia mater. From this capsule, septa of connective tissue extend into the gland, subdividing it into small lobules. Via these septa, blood vessels and nerve fibres reach the interior of the gland.

Pinealocytes are modified neuronal elements. They are provided with one or more long, ramifying processes that terminate with club-shaped endings close to blood vessels or freely in the extracellular space. Pinealocytes produce the hormone *melatonin*, a derivative of serotonin. Melatonin secretion is very high during the night but almost undetectable during the day [33]. The synthesis and secretion of melatonin in the pineal gland are regulated by light received by retinal photoreceptors and transmitted to the brain by the optic nerves. The long, multisynaptic pathway along which this photic information is transferred to the pineal gland is well-established, at least in rodents such as rats [13, 19, 20, 30]. The retina projects by way of retinohypothalamic fibres to the *suprachiasmatic nucleus*, a small cell mass directly dorsal to the optic chiasm. This nucleus, which functions as a circadian pacemaker for the melatonin rhythm, sends a projection to an adjacent hypothalamic centre, the *paraventricular nucleus*. Cells in the autonomic subdivision of this nucleus send their axons to the rostral part of the *intermediolateral column* of the thoracic spinal cord, where they synapse on preganglionic sympathetic motor neurons. These neurons in turn contact postganglionic sympathetic motor neurons in the *superior cervical ganglion* which emit the fibres reaching the pineal gland. The postganglionic fibres enter the apex of the pineal gland from the region of the tentorium cerebelli as single or paired *nervi conarii* in humans and rodents, respectively. They release norepinephrine from their terminals, which form synapses on the surface of the pinealocytes, which in turn convert serotonin into melatonin. The hormone is released into the extracellular space of the pineal gland, from which it enters the cerebrospinal fluid or general circulation. The sympathetic fibres innervating the pineal gland contain neuropeptide Y in addition to norepinephrine.

Sympathetic input represents the principal but by no means only innervation of the pineal gland. Neuroanatomical tracer studies, often combined with immunohistochemical results for neurotransmitters, have shown that pinealopetal nerve fibres also originate from parasympathetic ganglia, the trigeminal ganglion and several diencephalic and mesencephalic cell masses [18]. The latter, which collectively form the central innervation of the pineal gland, reach it directly via the pineal stalk. The parasympathetic fibres originate from the sphenopalatine and otic ganglia, and contain the neuropeptides vasoactive intestinal peptide and histidine isoleucine. The trigeminal pinealopetal neurons contain substance P, calcitonine gene-related peptide and pituitary adenylate cyclase-activating peptide. The final elements forming the central innervation of the pineal gland include vasopressinergic and oxytocinergic neurons in the paraventricular nucleus, histaminergic tuberomamillary neurons, serotonergic neurons in the dorsal raphe nucleus, and GABAergic neurons in the so-called intergeniculate leaflet of the lateral geniculate body. Because receptors for all neurotransmitters and neuropeptides mentioned above have been shown to be present on the pinealocyte cell membrane, it may be concluded that regulation of the activity of these elements is very complex [18].

References

1. Aghajanian GK, Wang RY (1977) Habenular and other midbrain raphe afferents demonstrated by a modified retrograde tracing technique. Brain Res 122:229–242
2. Andres KH, Von Düring M, Veh RW (1999) Subnuclear organization of the rat habenular complexes. J Comp Neurol 407:130–150
3. Ariëns Kappers CU, Huber GC, Crosby EC (1936) The comparative anatomy of the nervous system of vertebrates, including man. MacMillan, New York
4. Björklund A, Lindvall O (1984) Dopamine-containing systems in the CNS. In: Björklund A, Hökfelt T (eds) Handbook of chemical neuroanatomy, vol 2/ Part 1: Classical transmitters in the CNS. Elsevier, Amsterdam, pp 55–122
5. Gilbert MS (1935) The early development of the human diencephalon. J Comp Neurol 62:1–115
6. Groenewegen HJ, Ahlenius S, Haher SN, Kowall NW, Nauta WJH (1986) Cytoarchitecture, fiber connections, and some histochemical aspects of the interpeduncular nucleus in the rat. J Comp Neurol 249:65–102
7. Herkenham M, Nauta WJH (1977) Afferent connections of the habenular nuclei in the rat. A horseradish peroxidase study, with a note on the fiber-of-passage problem. J Comp Neurol 173:123–146
8. Herkenham M, Nauta WJH (1979) Efferent connections of the habenula nuclei in the rat. J Comp Neurol 187:19–48
9. Herrick CJ (1910) The morphology of the forebrain in Amphibia and Reptilia. J Comp Neurol 20:413–547
10. Herrick CJ (1913) Anatomy of the brain. In: The reference handbook of the medical sciences, vol 2. Wood, New York, pp 274–342
11. Kostovic I (1990) Zentralnervensystem. In: Hinrichsen KV (ed) Humane Embryologie. Springer, Berlin, pp 381–448

12. Kuhlenbeck H (1954) The human diencephalon: a summary of development, structure, function and pathology. Confin Neurol 14:1–230
13. Larsen PJ, Enquist LW, Card JP (1998) Characterization of the multisynaptic neuronal control of the rat pineal gland using viral transneuronal tracing. Eur J Neurosci 10:128–145
14. Lavoie B, Parent A (1991) Serotoninergic innervation of the thalamus in the primate: an immunohistochemical study. J Comp Neurol 312:1–18
15. Maciewicz R, Taber-Pierce E, Ronner S, Foote WE (1981) Afferents to the central superior raphe nucleus in the cat. Brain Res 216:414–421
16. Marchand ER, Riley JN, Moore RY (1980) Interpeduncular nucleus afferents in the rat. Brain Res 193:339–352
17. McBride RL (1981) Organization of afferent connections of the feline lateral habenular nucleus. J Comp Neurol 198:89–99
18. Møller M, Baeres FMM (2002) The anatomy and innervation of the mammalian pineal gland. Cell Tissue Res 309:139–150
19. Moore RY (1996) Entrainment pathways and the functional organization of the circadian system. Progr Brain Res 111:103–119
20. Moore RY, Card JP (1986) Visual pathways and the entrainment of circadian rhythms. Ann NY Acad Sci 453:123–133
21. Nauta WJH (1958) Hippocampal projections and related neural pathways to the midbrain in the cat. Brain 81:319–340
22. Nauta WJH (1974) Evidence of a pallidohabenular pathway in the cat. J Comp Neurol 156:19–27
23. Nauta WJH (1979) Expanding borders of the limbic system concept. In: Rasmussen T, Marino R (eds) Functional neurosurgery. Raven, New York, pp 7–24
24. Nauta WJH, Mehler WR (1966) Projections of the lentiform nucleus in the monkey. Brain Res 1:3–42
25. Parent A, Gravel S, Boucher R (1980) The origin of forebrain afferents to the habenula in rat, cat and monkey. Brain Res Bull 6:23–38
26. Reinoso-Suarez F (1966) Development of the human diencephalon. In: Hassler R, Stephan H (eds) Evolution of the forebrain. Thieme, Stuttgart, pp 296–304
27. Rose JE (1942) The ontogenetic development of the rabbit's diencephalon. J Comp Neurol 77:61–129
28. Sutherland FJ (1982) The dorsal diencephalic conduction system: a review of the anatomy and functions of the habenular complex. Neurosci Biobehav Rev 6:1–13
29. Swanson LW, Cowan WM (1979) The connections of the septal region in the rat. J Comp Neurol 186:621–655
30. Teclemariam-Mesbah R, Ter Horst GJ, Postema F, Wortel J, Buijs R (1999) Anatomical demonstration of the suprachiasmatic nucleus-pineal pathway. J Comp Neurol 406:171–182
31. Troiano R, Siegel A (1975) The ascending and descending connections of the hypothalamus in the cat. Exp Neurol 49:161–173
32. Van Dongen PAM, Nieuwenhuys R (1998) Diencephalon. In: Nieuwenhuys R, Ten Donkelaar HJ, Nicholson C (eds) The central nervous system of vertebrates. Springer, Berlin Heidelberg New York, pp 1844–1862
33. Vaughan GM, Pelham RW, Pang SF et al (1976) Nocturnal elevation of plasma melatonin and urinary 5-hydroxyindoleacetic acid in young men: attempts at modification by brief changes in environmental lighting and sleep and by autonomic drugs. J Clin Endocr Metab 42:752–764

8 Diencephalon: Dorsal Thalamus

Introductory Notes

The thalamus is a large, ovoid nuclear complex in the wall of the diencephalon caudal to the interventricular foramen (Figs. 3.19, 3.20). Laterally, a narrow band of myelinated fibres, the *external medullary lamina*, separates the main body of the thalamus from the reticular thalamic nucleus. The main mass of the thalamus is known as the dorsal thalamus. The reticular thalamic nucleus and some adjacent structures belong to the *ventral thalamus* (see below). The terms dorsal and ventral thalamus are derived from their topographical position in the lateral diencephalic wall during development (see Chap. 2).

The *reticular thalamic nucleus* (R), which borders laterally on the internal capsule, covers the lateral and ventrolateral surface of the dorsal thalamus as a thin sheet of grey matter. It is penetrated by bundles of thalamocortical and corticothalamic fibres that detach from the internal capsule and enter the thalamus. Fibres from the superior cerebellar peduncle and globus pallidus pass through the ventral portion of the external medullary lamina to the anterior part of the thalamus.

Corticothalamic and thalamocortical fibres that detach from the corona radiata and internal capsule to enter the thalamus dorsally at its rostral and caudal poles are termed *thalamic peduncles*. These peduncles are medial to the long corticofugal pathways that descend to the brain stem and spinal cord (Fig. 8.1). The *anterior thalamic peduncle* breaks away from the anterior limb of the internal capsule, and its fibres form a reciprocal connection with the prefrontal and orbitofrontal parts of the cortex and cingulate gyrus. The *superior* and *posterior thalamic peduncles* diverge from the posterior limb of the internal capsule, and their fibres form a two-way connection between the thalamus and the central parietal and occipitotemporal areas. The *inferior thalamic peduncle* reaches the thalamus at its ventromedial side medially to the posterior limb of the internal capsule. It contains fibres connecting the thalamus with the orbitofrontal, insular and temporal cortices and the amygdaloid complex. Together with fibres connecting the amygdala with the hypothalamus, the thalamocortical, corticothalamic and amygdalothalamic fibres in the inferior thalamic bundle constitute the *ansa peduncularis*, which passes ventrally to the lentiform nucleus through the basal telencephalon (Fig. 9.1). The amygdalothalamic and amygdalohypothalamic components of the ansa peduncularis are collectively designated the *ventral amygdalofugal pathway* (Fig. 13.2).

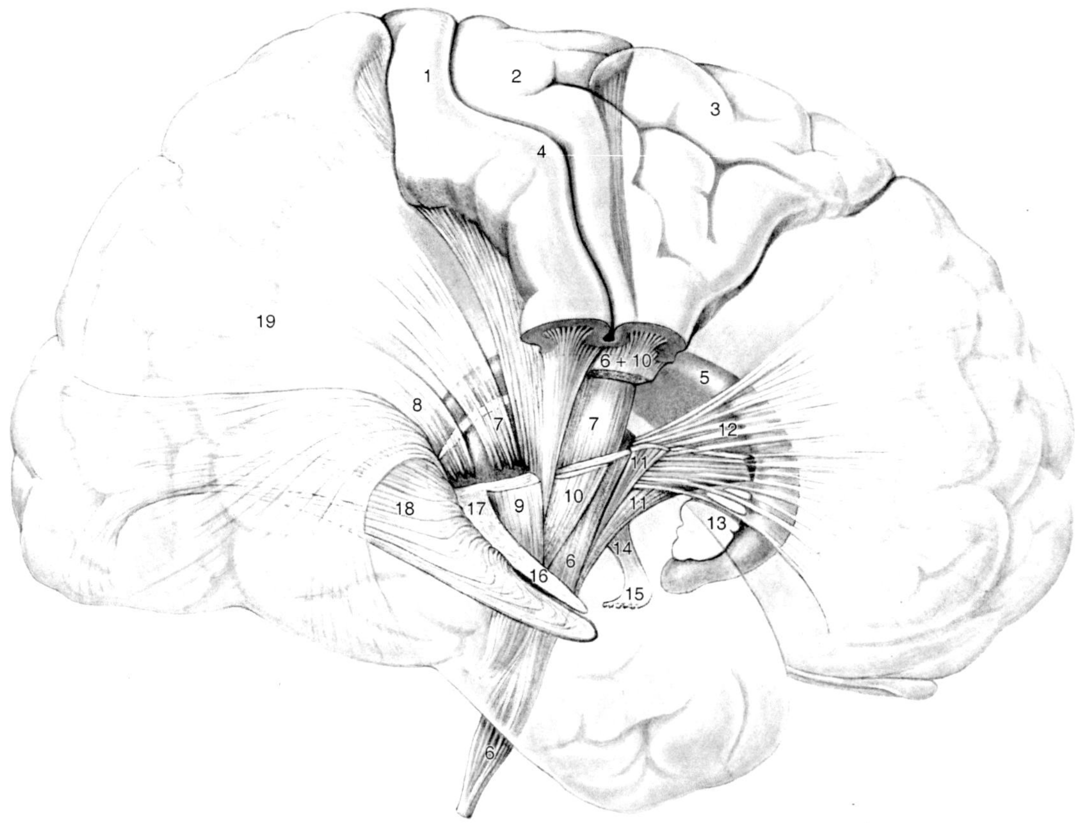

1 Postcentral gyrus (areas 2, 1, 3)
2 Precentral gyrus (area 4)
3 Frontal gyri (areas 6, 8)
4 Central sulcus
5 Caudate nucleus
6 Pyramidal tract
7 Superior thalamic peduncle
8 Posterior thalamic peduncle
9 Parietopontine tract
10 Corticotegmental fibres
} Internal capsule, posterior limb

11 Frontopontine tract
12 Anterior thalamic peduncle
} Internal capsule, anterior limb

13 Putamen
14 Inferior thalamic peduncle
15 Ansa peduncularis
16 Temporopontine tract (internal capsule, sublentiform part)
17 Occipitopontine tract (internal capsule, retrolentiform part)
18 Optic radiation
19 Sagittal stratum

Fig. 8.1. Lateral view of the thalamic peduncles and the internal capsule (1/1×). The proximal part of the corticopontine and corticospinal fibre bundles has been removed to show the thalamic peduncles, which are illustrated as compact bundles for reasons of clarity. The origin of the pyramidal tract in the cerebral cortex has been emphasized

Parcellation of the Thalamus

Cytoarchitectonic and myeloarchitectonic differences allow subdivision of the thalamus into a number of nuclear groups, each of which can be further subdivided into two or more smaller units [43, 75, 121] (Figs. 6.35–6.39, 8.2). Most of these units are also or even more clearly distinguishable in material treated with histochemical or neurochemical markers [82, 93, 147, 152]. The nomenclature of the thalamic nuclei in the present work (Table 8.1, Figs. 8.2, 8.3) is based on the studies of Walker [243], Jones [99, 100], Hirai and Jones [82] and Steriade et al. [222]. Other nomenclatures frequently used for nuclei of the human thalamus are by Hassler [75] and Olszewski [162]. For a detailed concordance of their nomenclatures and the one used here, see [82] and [100].

A curved sheet of white matter, the *internal medullary lamina*, divides the thalamus into the *mediodorsal thalamic nucleus* and *ventral* and *lateral groups of thalamic nuclei* (Figs. 6.36, 8.2). The bifurcated and most rostral part of the internal medullary lamina encloses the *anterior nuclear group*. Caudally, the ventral group is replaced by the *medial geniculate body* (MG) which, together with the *lateral geniculate body* (LG), belongs to the metathalamus. The periventricular zone of the rostral part of the thalamus contains small cell masses: the elongated *parataenial* (Pt) and *paraventricular* (Pv) nuclei and the more compact *reuniens nucleus* (Re). Together, these masses constitute the *midline nuclear group*. The reuniens nucleus bears a close relationship with the interthalamic adhesion, when present. The latter structure is a local fusion of the two thalami behind the interventricular foramen (Fig. 3.8), which is lacking in 25–30% of human brains.

The lateral nuclear group includes the *lateral posterior nucleus* (LP) and constitutes the massive caudal pole of the thalamus known as the *pulvinar* (PUL). Rostrally it tapers into the *lateral dorsal nucleus* (LD) in the split internal medullary lamina, together with the anterior nuclear group. The ventral group of nuclei extends more rostrally. It is divided into the *ventral anterior* (VA), *ventral lateral* (VL) and *ventral posterior* (VP) *nuclei*.

Within the *ventral anterior nucleus*, a magnocellular part (VAmg) can be delineated ventromedially.

The *ventral lateral nucleus* can be clearly subdivided into separate *anterior* (VLa) and *posterior* (VLp) portions [82, 100, 222]. The ventral posterior nucleus has two major divisions, the *ventral posterolateral* (VPL) and *ventral posteromedial* (VPM) *nuclei*. An ill-defined *ventral posterior inferior nucleus* (VPI) lies inferiorly between the VPM and VPL.

The *posterior thalamic nuclear group* lies caudomedially to the ventral posterior nucleus, ventral to the rostral part of the pulvinar and dorsomedial to the medial geniculate body, in a region traversed by the fibres of the medial lemniscus. It comprises the diffuse *posterior* nucleus (Po) and the more compact *limitans* (L) and *suprageniculate* (Sg) *nuclei*.

As their name indicates, the *intralaminar nuclei* are enmeshed by the fibres of the internal medullary lamina (Fig. 8.2). Rostral and caudal groups of intralaminar nuclei can be distinguished. The caudal group includes the large *centromedian nucleus* (CM), the *parafascicular nucleus* (Pf) medial to it, and surrounding the fibre bundles of the habenulo-interpeduncular tract. The rostral group consists of the central medial, paracentral and central lateral nuclei. These three nuclei are not easily separated [99]. The *central medial nucleus* (CeM) lies in the expanded medial part of the internal medullary lamina. The thin *paracentral nucleus* (Pc) covers the rostral part of the mediodorsal nucleus. The *central lateral nucleus* (CL) is located dorsally and caudally to the paracentral nucleus. Its caudal part extends over the caudal surface of the mediodorsal nucleus.

The thalamus parcellation presented above is mainly based on topographical relations and structural properties of the various entities. Almost all thalamic nuclei receive afferents from one or more subcortical sources and are strongly and reciprocally related to some area of the cortex. Evaluation of the connectional

Table 8.1. Classification of thalamic nuclei

EPITHALAMUS	
Habenular complex	
Medial habenular nucleus	MH
Lateral habenular nucleus	LH
DORSAL THALAMUS	
Anterior nuclear group	A
Anterodorsal nucleus	Ad
Anteroventral nucleus	Av
Anteromedial nucleus	Am
Ventral nuclear group	V
Ventral anterior nucleus	VA
Magnocellular division	VAmc
Ventral lateral nucleus	VL
Anterior part	VLa
Posterior part	VLp
Medial part	VLm
Ventral posterior complex	VP
Ventral posteromedial n.	VPM
Ventromedial posterior n.	VMpo
Ventral posterolateral n.	VPL
Ventral posterior inferior n.	VPI
Mediodorsal nucleus	MD
Magnocellular part	MDmc
Parvocellular part	MDpc
Paralaminar part	MDpl
Lateral nuclear group	L
Lateral dorsal nucleus	LD
Lateral posterior nucleus	LP
Pulvinar	PUL
Anterior part	PULa
Lateral part	PULl
Medial part	PULm
Inferior part	PULi
Medial geniculate body	MG
Ventral nucleus	MGv
Dorsal nucleus	MGd
Medial nucleus	MGm
Lateral geniculate body	LG
Dorsal nucleus	LGd
Ventral nucleus	LGv
Posterior nuclear group	P
Suprageniculate nucleus	Sg
Limitans nucleus	Li
Posterior nucleus	Po
Midline nuclear group	Mi
Parataenial nucleus	Pt
Paraventricular nucleus	Pv
Reuniens nucleus	Re
Intralaminar nuclear group	I
Rostral nuclei	
Paracentral nucleus	Pc
Central lateral nucleus	CL
Central medial nucleus	CeM
Caudal nuclei	
Parafascicular nucleus	Pf
Centromedian nucleus	CM
VENTRAL THALAMUS	
Reticular nucleus	R

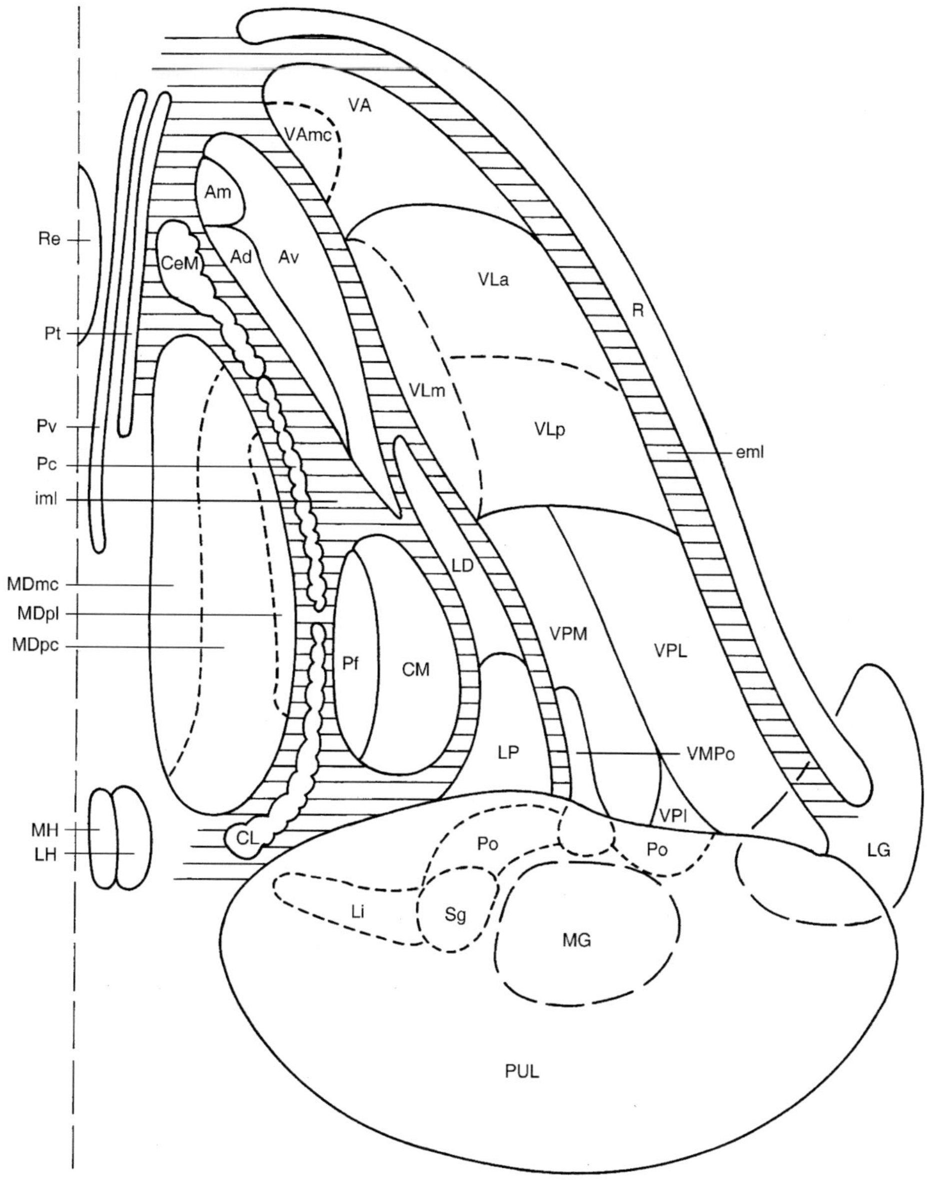

Fig. 8.2. The principal epithalamic and thalamic nuclei, projected upon a horizontal plane. *eml*, external medullary lamina; *iml*, internal medullary lamina. For other abbreviations, see Table 8.1

properties of the various thalamic nuclei has led to the distinction of sensory relay, motor relay, association and limbic nuclei and to the categorization of specific and non-specific nuclei.

Sensory relay nuclei represent the final link in sensory projections ascending to the primary sensory fields of the neocortex. The ventral posterior nucleus, lateral geniculate body and medial geniculate body (i.e. the thalamic processing stations for somatic, hearing and visual perception) belong to this group.

Motor relay nuclei are interposed in projections leading from the cerebellar nuclei and centres belonging to the extrapyramidal motor system, such as the substantia nigra and globus pallidus, to the motor and premotor cortical fields. The ventral anterior and ventral lateral thalamic nuclei (often collectively designated as the VA-VL complex or motor thalamus) belong to this group [137, 199, 222].

Association nuclei are characterized by strong reciprocal relations with the association areas of the neocortex. This group comprises the mediodorsal nucleus, which is specifically related to the prefrontal cortex and the complex formed by the lateral posterior and pulvinar nuclei. The latter is connected with a second vast area of association cortex encompassing large parts of the temporal, parietal and occipital lobes.

The *limbic nuclei* comprise the anterior nuclear complex and lateral dorsal nucleus, which project to the cingulate and entorhinal cortices, presubiculum and parasubiculum, all structures forming part of the limbic system.

The thalamic nuclei of the four classes discussed above are often designated *specific nuclei* because they are reciprocally connected with specific areas or regions of the cerebral cortex. Their efferent fibres terminate preferentially in layers 3 and 4 at the middle depth of the cortex. These projections are discontinuous and arranged in patches or strips. The well-known projections of alternate laminae of the lateral geniculate to alternate strips in layer IV of the visual cortex form the centres of the ocular dominance columns [86, 90, 248]. Another striking example is the 1:1 relationship in the projection from representations of the vibrissae in the ventral posterior nucleus of the thalamus to the somatosensory cortex in rats and mice, where the vibrissae are represented by barrel-like cell condensations in the fourth cortical layer [233, 255].

Until recently it was customary to contrast the *specific thalamic nuclei* from a group of *non-specific thalamic nuclei* including the intralaminar, midline and sometimes also the posterior nuclei [107]. This group was reportedly characterized by diffuse cortical efferents which ignore the boundaries between cortical fields [106]. These efferents were described as terminating preferentially in the outermost, plexiform layer of the cortex and therefore in an optimal position to modulate the excitability in neurons of all layers that extend apical dendrites into this superficial layer.

The non-specific nuclei and their diffuse cortical projections were considered to form the final link in a functional system designated as the *ascending reticular activating system* (ARAS). Stimulation of this system produces general cortical activation associated with the well-known behavioural and cortico-electrical arousal reaction. It is important to note that the differences between specific and non-specific thalamic nuclei are considered less extreme than was previously thought. Detailed studies with modern tracing techniques have shown that each intralaminar or midline nucleus has a restricted field of cortical termination that overlaps only slightly with the projections of adjacent nuclei [13, 15, 62, 234]. Recent studies of Jones [101–103] on the chemical identities of thalamocortical relay neurons in monkeys have shown that "specific" thalamocortical fibres, projecting in an area-specific manner to the middle layer of the cortex, and "non-specific" thalamocortical fibres, projecting to the outermost cortical layer over wide areas, indeed exist. However, the fibres of the latter category do not emanate from a separate group of nuclei but from a matrix of cells extending throughout the whole thalamus. These cells are distinguished by immunoreactivity for the calcium-binding protein calbindin. Jones [101, 102] found that in certain thalamic nuclei such as VPM and VPL, a core of cells distinguished by immunoreactivity for another calcium-binding protein, parvalbumin, is superimposed upon the matrix. These core elements project in a highly ordered fashion to the middle layers (III–IV) of the cortex in area-specific manner. According to Jones, the core cells and nuclei in which a core is present receive subcortical inputs that are highly ordered topographically, whereas the matrix cells and nuclei in which the matrix is enhanced receive more diffuse subcortical inputs whose distribution is not restricted by thalamic nuclear borders.

The thalamus represents a key integrative structure for the processing of pain, and it is known that fibres conveying nociceptive impulses terminate in many thalamic cell masses, including VPM, VPL, VPI, VMP, CeM, Pc, Pf and the posterior complex [250, 251]. Within all of these cell masses, the nociceptive fibres have a strong predilection for terminating in relation to the calbindin-immunoreactive matrix cells [100].

Thalamic and Thalamocortical Circuits

1. The majority of thalamic neurons represent relay cells that project to the cortex or to the striatum. These relay cells receive a direct subcortical input in the sensory relay nuclei (MG, LG, VP); this input is topically organized and consists of thick fibres forming focal terminal fields with large boutons containing round vesicles.

2. The thalamic nuclei contain inhibitory GABAergic interneurons amounting to about 30% in primates [149, 167, 227, 244].

3. The thalamic relay nuclei typically contain triadic synapse complexes [177] composed of (a) a large terminal of a specific thalamic afferent fibre, (b) a primary dendrite of a thalamic relay cell and (c) a dendrite of an inhibitory interneuron. The latter dendrite, which contains flattened, pleiomorphic vesicles, is postsynaptic to the incoming afferent terminal and presynaptic to the dendrite of the relay cell. The afferent terminal makes multiple asymmetric synapses with the dendrite of the relay cell and that of the interneuron. The dendrite of the interneuron finally makes symmetric synapses with the dendrite of the thalamic relay cell. The inhibitory interneurons are supposedly involved in the postsynaptic inhibition which usually follows excitation of thalamic relay cells by afferent volleys. However, this effect could also be mediated by cells of the reticular nucleus [99] (see below). The triads just discussed regularly form part of larger, glomerular structures in which various presynaptic and postsynaptic elements are arranged.

4. The various thalamic nuclei receive, apart from “specific” afferents, “non-specific” cholinergic, noradrenergic and serotoninergic projections from the brain stem. Although distributed to all thalamic nuclei, these three fibre systems concentrate their terminations in certain nuclei [222].

5. The specific nuclei have strictly topically ordered projections to their cortical target regions, with endings predominantly in layer IV. The so-called non-specific nuclei have a more diffuse projection to the cortex, with endings not just in layer IV but also others, especially I and VI.

6. Corticothalamic fibres arise from all parts of the cerebral cortex, and all thalamic nuclei are in receipt of such fibres. The corticothalamic fibres are of two different kinds, thin type-I fibres arising from cells in layer VI and coarse type-II fibres arising from cells in lamina V [68, 212]. The type-I fibres participate in formation of topographically organized corticothalamic projections, which reciprocate the thalamocortical projections arising from particular thalamic nuclei. Their sparsely branching terminal systems are characterized by many small, en passant boutons. These terminals containing round vesicles make asymmetric synapses on the distal dendrites of both relay cells and interneurons within the neuropil outside the glomeruli. The type-II fibres are formed by collateral branches of layer V pyramidal neurons, whose main axons continue to

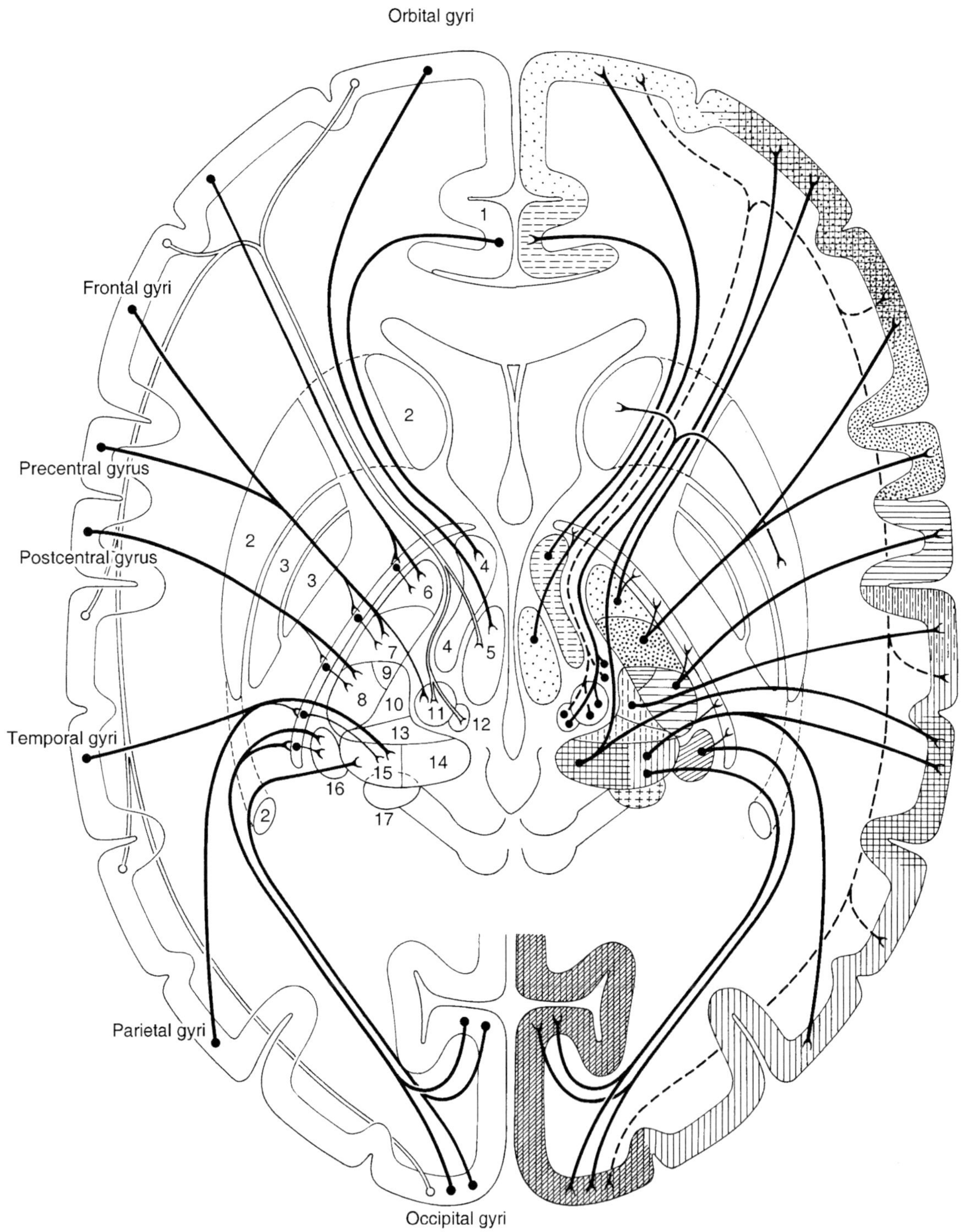

Fig. 8.3. Connections between the thalamic nuclei and the cerebral cortex I: diagrammatic horizontal section. *Left:* corticothalamic projections. *Right:* thalamocortical projections. For explanation of numbers see Fig. 8.4

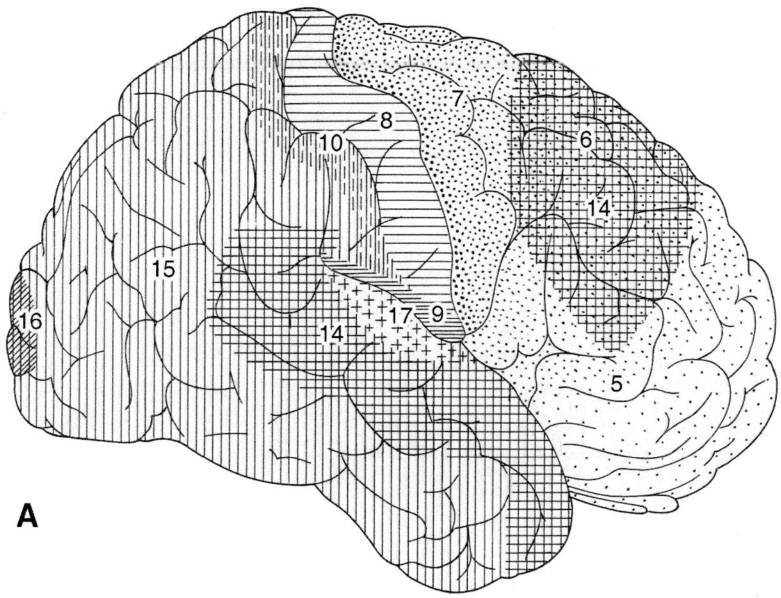

A

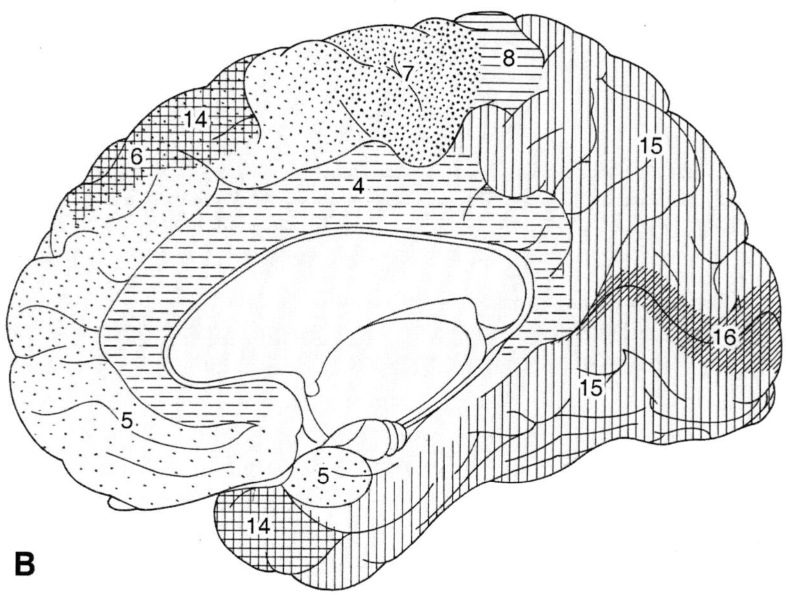

B

1 Cingulate gyrus
2 Corpus striatum
3 Globus pallidus
4 Anterior thalamic nuclei
5 Mediodorsal thalamic nucleus
6 Ventral anterior thalamic nucleus
7 Ventral lateral nucleus
8 Ventral posterior complex
9 Ventromedial posterior nucleus
10 Lateral posterior nucleus
11 Centromedian nucleus
12 Parafascicular nucleus
13 Pulvinar thalami, anterior part
14 Pulvinar thalami, medial part
15 Pulvinar thalami, lateral part
16 Lateral geniculate body
17 Medial geniculate body

Fig. 8.4 A, B. Connections between the thalamic nuclei and the cerebral cortex II: cortical projection areas of the thalamic nuclei. **A** lateral view; **B** medial view. The shading is the same as in Fig. 8.3

other subcortical sites in the brain stem and spinal cord. These coarse cortical efferents do not reciprocate thalamocortical afferent systems but rather connect the site of termination of one thalamocortical projection to another's area of origin. They terminate in grape-like clusters of large boutons that form excitatory synapses on the somata and proximal dendrites of thalamocortical relay cells [222]. Remarkably, these endings are indistinguishable from those of the principal (subcortical) afferents of the sensory and motor thalamic relay nuclei. The marked differences in location and character of their respective terminals suggest that the synaptic effect mediated by type-II fibres, which originate in layer V, is much more powerful than that of type-I fibres, originating in layer VI. The latter are considered to exert a modulatory feedback action on specific sensory, motor, or other information traversing the thalamic relay nuclei. The type II corticothalamic afferents, on the other hand, are thought to form part of feed-forward systems, transmitting information from one cortical area to another via the thalamic relay of the second area [83, 191, 210]. In the mediodorsal and pulvinar nuclei, type-II fibres prevail and form their main driving afferents. On that account, Sherman and Guillery [210–212] placed them in a separate category of *higher order* nuclei as opposed to *first order* nuclei receiving their main driving input from ascending specific afferents. In the present account, mediodorsal and pulvinar nuclei are designated as *association nuclei* because they obviously represent thalamic links in cortico-cortical association systems.

7. As already mentioned, the reticular thalamic nucleus forms a thin neuronal sheet that largely covers the rostral, lateral and ventrolateral surfaces of the dorsal thalamus. Its neurons are all GABAergic and project to the dorsal thalamus, where they synapse on thalamocortical relay cells [84, 195]. The reticular thalamic nucleus is penetrated by bundles of thalamocortical and their reciprocating corticothalamic fibres. Since these projections are highly specific and topographically arranged, it may be said that each thalamic nucleus has its "own" sector in the reticular thalamic nucleus. During their course through the reticular thalamic nucleus, thalamocortical and type-I corticothalamic fibres emit short collaterals which form excitatory synaptic contacts with the reticular cells [23, 31, 45, 98, 150, 205], and these project back to the thalamic nuclei from which they receive their afferents, thus closing negative feedback loops.

Below, the main afferent and efferent connections of the various thalamic nuclei are briefly discussed. Pictorial surveys of the thalamic efferent projections are presented in Figs. 8.3 and 8.4.

Ventral Nuclear Group

The ventral group of thalamic nuclei comprises three major subdivisions, the *ventral anterior nucleus* (VA), *ventral lateral nuclear complex* (VL) and *ventral posterior complex* (VP). The last-named relays impulses of sensory systems to cortical regions, while the more rostral nuclei (VA, VL) relay impulses from the cerebellum, substantia nigra and basal ganglia.

The *ventral anterior nucleus* (VA) receives afferents from the pars reticulata of the substantia nigra and, to a lesser extent, from the internal segment of the globus pallidus [95, 168]. It is reciprocally connected with the frontal eye field (area 8) and prefrontal cortex, including the cingulate cortex [7, 58, 113, 138, 220]. Some of the corticothalamic connections of the prefrontal cortex with the ventral anterior nucleus are bilateral [57].

The *ventral lateral nuclear complex* (VL) has been subdivided into pars anterior (VLa), pars posterior (VLp) and pars medialis (VLm). The VLa is the principal terminus of afferents from the internal segment of the globus pallidus [44, 75, 125, 156, 163, 198, 199]. These fibres give off collateral branches to the centromedian nucleus (Fig. 9.1). The VLp receives a massive, topographically organized projection from the cerebellar nuclei [5, 168, 198, 199] while the

VLm receives fibres from the pars reticulata of the substantia nigra [32, 94]. The VLp projects to the primary motor cortex, the VLa to the premotor cortex, including the supplementary motor area, and the VLm rather diffusely to the premotor and adjacent prefrontal area [99, 198, 206, 223].

The *ventral posterior complex* (VP) comprises two major divisions, the ventral posterolateral (VPL) and ventral posteromedial (VPM) nuclei, as well as the smaller and less distinct ventral posterior inferior nucleus (VPI). The latter is situated inferiorly between VPL and VPM. The recently described [10, 19, 40] and much debated [38, 178, 249, 252] ventromedial posterior nucleus (VMpo) may be considered a satellite of the VP complex.

The VPL and VPM represent the principal somatosensory relay nuclei. The medial cuneate and gracile nuclei project to the contralateral VPL via the medial lemniscus in an organized manner, and the same nucleus also receives spinothalamic afferents from the contralateral side of the spinal cord [22]. Fibres arising from the ventral two thirds of the principal trigeminal nucleus decussate in the rostral pons and ascend via the trigeminal lemniscus to the VPM. A set of fibres comparable to the spinothalamic tract originates from the caudal part of the spinal trigeminal nucleus. After decussation, these fibres form the lateral trigeminothalamic tract, which also terminates in the VPM. A third afferent projection to the VPM is formed by uncrossed fibres originating from the dorsal part of the principal trigeminal nucleus [26]. The medial lemniscal and trigeminal lemniscal fibres terminate in the VP complex in a series of parallel laminae, each representing a specific region of the body but including different sensory modalities along their anteroposterior axis. Within the VP complex, the projections of the spinothalamic and lateral trigeminothalamic tracts are somatotopically organized. They overlap with the termination of the medial lemniscus [16, 17, 28, 37, 99, 105, 135, 142].

The VP complex is reciprocally connected with areas 3b, 2 and 1 of the first somatosensory area in the postcentral gyrus and the second somatosensory area in the parietal operculum [55, 99, 105, 129, 172, 247].

The ventral nuclear group also comprises cell assemblies forming relays in gustatory and vestibular projections. Secondary gustatory fibres, arising from the rostral part of the nucleus solitarius, terminate in the medial part of the VPM, which in turn projects to the insular operculum and orbitofrontal cortex [99, 176]. Fibres ascending from the vestibular nuclear complex have been reported to terminate in the rostral part of VPL, in VLp and in VPI [35, 127, 134, 181].

The *ventromedial posterior thalamic nucleus* (VMpo), first described by Craig et al. [19, 40], is an ellipsoid cell mass posteromedial to VPM, ventral to the anterior pulvinar, and rostrodorsal to the medial geniculate body (Fig. 8.2). It is characterized by a dense plexus of calbindin-immunoreactive fibres and shows ultrastructural features typical for sensory relay nuclei [10]. The VMpo receives a densely clustered, topographically organized input from spinothalamic and trigeminothalamic lamina I neurons. It is known that almost all of these neurons are nociception- or thermoreception-specific [34, 39]. The VMpo projects to the dorsal posterior portion of the insular cortex [40]. Craig et al. [19, 38, 40] concluded that the VMpo is a specific thalamic relay nucleus for nociception and thermoreception. However, it is known that lamina I fibres project to several other thalamic nuclei, including VPL, VPM and the ventrocaudal part of the mediodorsal nucleus; hence it seems likely that the VMpo does not represent the only nucleus subserving the sensory modalities mentioned [60, 178, 249, 250, 252].

Anterior Nuclear Group and Lateral Dorsal Nucleus

The anterior nuclear group (A) of the thalamus consists of the anteromedial (Am), anterodorsal (Ad) and anteroventral nuclei (Av) [82, 222]. It is enclosed within the split internal medullary lamina together with the lateral dorsal nucleus (LD). The connections of the anterior nuclear group and lateral dorsal nucleus with the limbic cortex are rather similar [99].

The anterior nuclear group receives the mamillothalamic tract. The lateral mamillary nucleus projects bilaterally to the Ad and the medial mamillary nucleus to Am and Av [126, 173, 237]. There are no projections from the mamillary body to the lateral dorsal nucleus, which receives a projection from the pretectum [184, 186]. The anterior and lateral dorsal nuclei are reciprocally connected with the limbic cortex of the cingulate gyrus, retrosplenial area and pre- and parasubiculum [111, 185, 235, 239, 242]. Although each of the anterior subnuclei has been related to a special field within the limbic cortex [190], their projections appear to be rather diffuse and overlapping.

Mediodorsal Nucleus

The mediodorsal thalamic nucleus (MD), which is particularly large in humans, is surrounded by the internal medullary lamina with its nuclei. Rostromedially to it are the nuclei of the midline (paraventricular nucleus, nucleus reuniens and parataenial nucleus). The mediodorsal nuclei can be subdivided into medial magnocellular (MDmc) and lateral parvocellular (MDpc) parts [162]. The large cells on the border of the lateral division of the nucleus with the internal medullary lamina (pars paralaminaris) belong to the central lateral nucleus [99]. The medial, magnocellular part of the mediodorsal nucleus is reciprocally connected with olfactory-related areas in the medial prefrontal and orbitofrontal regions of the neocortex [2, 180, 256]. It also receives fibres from the lateral amygdala [1] both directly and indirectly by way of the bed nucleus of the stria terminalis [61, 204], entorhinal and perirhinal cortices and cortex of the temporal pole. Direct projections from the olfactory cortex to the mediodorsal nucleus [115] are few in primates [194]. The lateral, parvocellular part of the mediodorsal nucleus is reciprocally connected with the frontal eye field (area 8) and entire prefrontal cortex [2, 58, 91, 194, 206, 225]. The lateral part of the nucleus receives afferent connections from the superior colliculus, substantia nigra, vestibular nuclei and midbrain tegmental fields [94, 124, 238]. The mediodorsal nucleus receives a projection from the ventral pallidum [71, 72, 76, 194, 258]. Through this pathway, the ventral striatum (nucleus accumbens) is connected with the prefrontal cortex.

Medial Geniculate Body

The medial geniculate body is a complex of three nuclei: ventral (MGv), dorsal (MGd) and medial or magnocellular (MGm).

The MGv forms the principal auditory relay nucleus. It receives afferents from the central nucleus of the inferior colliculus, and its cortical projections are focused on the primary auditory area (area 41). The MGv and its target area in the temporal operculum have a tonotopic representation. The morphological substrate of this representation is the lamellar organization of incoming fibres and relay cells first described by Morest [153].

The MGd receives inputs from the pericentral nucleus of the inferior colliculus and several other mesencephalic structures, including the superior colliculus and dorsal tegmentum [29, 236]. It is connected with the auditory association cortex in the temporal plane and the superior temporal gyrus [27, 99, 145, 217].

The MGm receives fibres from the inferior colliculus and collaterals from the medial lemniscus and spinothalamic tract. It maintains diffuse connections with auditory and surrounding non-auditory areas of the cortex. Its thalamocortical fibres terminate predominantly in layer I but also in layers III and IV [74], similarly to projections of the intralaminar nuclei.

Lateral Geniculate Body

The lateral geniculate body is a small, rostrolaterally directed projection of the posterior thalamus (Figs. 3.19, 3.20, 7.1). In most mammals, this body comprises two separate nuclei, dorsal lateral geniculate (LGd) and ventral lateral geniculate (LGv). The LGd projects to the cortex and represents the principal thalamic relay nucleus for the visual system. The LGv develops from the ventral thalamus (Figs. 2.19 B, 2.20 B) and does not project to the cortex. The human lateral geniculate body comprises only the LGd. The LGv is represented in man by a separate cell mass known as the pregeniculate nucleus.

The LGd is a laminated nucleus (Fig. 6.33). In humans it consists of two ventrally placed magnocellular laminae (1 and 2) and four parvocellular laminae (3–6) [81]. Laminae 1, 4 and 6 receive fibres from the contralateral, laminae 2, 3 and 5 from the ipsilateral eye. The basic primate pattern of the lateral geniculate body consists of two magnocellular and two parvocellular laminae [110]. Each of these laminae contains a full representation of the contralateral visual hemifield. In man and other primates, the parvocellular laminae are split and interdigitate to constitute four parvocellular layers. As a result, representations of the visual hemifield in the parvocellular laminae are incomplete and complementary to the representations in the other parvocellular laminae carrying information from the same eye [36]. Apart from the magnocellular and parvocellular laminae, intercalated (koniocellular) laminae have been distinguished. Their connections are discussed in Chap. 19. As detailed there, parasol ganglion cells of the primate retina, which give rise to a rapidly conducting (Y-like) system, project to the magnocellular laminae whereas midget ganglion cells, which belong to a slowly conducting (X-like) system, project to the parvocellular laminae.

The LGd is reciprocally connected with the primary visual cortex (area 17). Thalamocortical fibres from the parvocellular laminae terminate in deep and superficial sublayers of layer IV. The magnocellular laminae project to an intermediary sublayer within layer IV. Additionally, magnocellular layers project to layer VI [78, 90, 109, 169, 183, 248]. In primates the dorsal part of the lateral geniculate body does not project beyond area 17, but in lower mammals it projects to both area 17 and the visual association cortex of areas 18 and 19 [188].

The projection from one eye through the appropriate laminae to the visual cortex is discontinuous in most primates and organized in ocular dominance strips in layer IV. These strips form the basis of the ocular dominance columns that extend perpendicularly to the surface through all cortical layers. These columns, approximately 0.5 mm wide, branch off roughly parallel from the representation of the horizontal meridian. Their width is constant throughout the representations of the central and peripheral parts of the visual field [85–89].

The connection between LGd and visual cortex is known as the optic radiation. Its ventral part consists of fibres to the portion of the visual cortex below the calcarine sulcus; this portion represents the upper quadrant of the visual field. The fibres curve rostrally in the lateral wall of the inferior horn of the lateral ventricle before turning towards the occipital lobe. Deep lesions of the temporal lobe may therefore lead to upper quadrant anopsia [9]. Within the optic radiation, thalamocortical fibres are separated from the more medial corticothalamic fibres [253].

Lateral Nuclear Group

The lateral group of thalamic nuclei consists of the lateral dorsal (LD) and lateral posterior (LP) nuclei and the pulvinar. The LD connections have been discussed with the anterior nucleus. The pulvinar is large in primates and can be subdivided into anterior, medial, lateral and inferior nuclei [162]. The nuclei of the lateral group are reciprocally connected with the large expanse of association cortex in the posterior parietal, occipital and temporal lobes [7, 11, 27, 30, 41, 42, 59, 70, 92, 108, 112, 136, 146, 166, 168, 207, 219, 229–232, 245, 257]. The lateral posterior nucleus and anterior pulvinar are connected with areas 5 and 7 in the superior and inferior parietal lobules [27]. The projection of the lateral posterior nucleus is rostral to that of the anterior pulvinar. The projections from the lateral and inferior nuclei of the pulvinar include the cortex at the parieto-occipital junction, circumstriate areas 18 and 19 and primary visual area (area 17). Fibres from the pulvinar terminate in layer I of area 17 and layers II and IV of the circumstriate cortex. These connections are reciprocated by projections from pyramidal cells in layer V of area 17 and layer VI of the circumstriate cortex [12, 30, 41, 160, 161, 169, 183, 229, 230, 245].

The projection of the inferior pulvinar extends into the inferotemporal visual association cortex. The medial pulvinar projects to the superior temporal gyrus, dorsal wall of the superior temporal sulcus and temporal pole. In addition, separate cell populations in the medial pulvinar project to the frontal eye field [7, 92] and limbic areas [8]. The LP-pulvinar complex has been consistently shown to receive input from the motor-related cortical fields of the frontal lobe [189].

The principal subcortical input to the LP-pulvinar complex arises in the visuomotor region of the midbrain, i.e. the superior colliculus and pretectum. Projections from the deep, non-visual layers of the superior colliculus terminate in the medial pulvinar [11, 12, 91, 131, 228]. Direct connections from the retina to the inferior pulvinar have been described [77].

Sherman and Guillery [66, 210–212] reported that (a) in the LP-pulvinar complex the subcortical projections are much less massive than those to sensory and motor relay nuclei and do not attain all parts of this complex and (b) the corticothalamic type II afferents to this complex are strongly developed and form its principal input. These authors believe that the LP-pulvinar complex largely concerns transmitting information on the output of one cortical area to another and that it plays a key role in cortico-cortical communication and higher cortical functions. Moreover, evidence suggests that the pulvinar plays an important role in visual salience, that is discriminating relevant from irrelevant visual stimuli [187].

After lesions of the striate cortex, the extrageniculate pathway from the superior colliculus via the pulvinar to the visual association areas is responsible for residual visual discrimination, as demonstrated in both experimental animals and humans [33, 208, 246].

Posterior Nuclear Complex

The posterior complex of thalamic nuclei is caudomedial to the VP, ventral to the medial portion of the rostral pulvinar and rostrodorsal to the medial geniculate body. It comprises the compact limitans (L) and suprageniculate (Sg) nuclei and the more diffuse posterior nucleus (Po) (Fig. 8.2). These nuclei all project to the retroinsular cortex surrounding S2 and the adjacent insular cortex [27, 104, 155]. This complex as a whole receives subcortical afferents from the superior colliculus [73], inferior colliculus [120, 151, 236], medial lemniscus [21] and the spinothalamic and trigeminothalamic tracts [22, 24, 140, 143, 179]. Most nuclei of the posterior complex are multimodal, and none appear to have the organizational specificity of a typical sensory relay nucleus [99]. The medial part of the posterior nucleus receives numerous terminations of spinothalamic fibres [22], and a certain proportion of its neurons respond to noxious stimuli [4, 171].

Intralaminar and Midline Nuclei

The intralaminar thalamic nuclei are made up of rostral and caudal groups. The former includes central medial (CeM), paracentral (Pc) and central lateral (CL) nuclei. The caudal group consists of the centromedian (CM) and parafascicular (Pf) nuclei. The midline nuclei include the parataenial (Pt), paraventricular (Pv) and reuniens (Re) nuclei. In humans and primates generally, the centromedian nucleus is large (Figs. 6.35, 6.36), but the midline nuclei are small and difficult to delimit [99].

The intralaminar and midline nuclei are collectively designated as non-specific thalamic nuclei. Until recently it was generally believed that these nuclei project diffusely to the cerebral cortex without respecting the boundaries between areas. However, studies with modern anatomical tracing techniques [62, 234] have shown that most of the individual nuclei project preferentially to particular cortical areas and there is little overlap in the projection fields of adjacent intralaminar/midline (I/Mi) nuclei. In spite of these new findings, there is still some justification for the still common designation "non-specific nuclei" [130]:

a. Instead of receiving a massive input from one particular source, as with specific relay nuclei, the I/Mi nuclei receive converging inputs from many different sources.
b. Their cortical projections, although less diffuse than previously thought, lack the topographical precision of projections of the specific relay nuclei.
c. Whereas the specific relay nuclei are reciprocally connected to small sectors of the reticular thalamic nuclei, I/Mi nuclei are innervated by wide expanses of that nucleus.

An important feature shared by all I/Mi nuclei is that they project to the cortex as well as the striatum. Moreover, the targets of the thalamocortical and thalamostriatal projections of a given nucleus are interconnected through corticostriatal connections [62].

The intralaminar nuclei receive ascending projections from the spinal grey matter, particularly from laminae VII and VIII [209] and a wide variety of brain stem nuclei, including the spinal trigeminal nucleus, the medial and lateral vestibular nuclei, various parts of the reticular formation, particularly the mesencephalic cuneiform nucleus, the parabrachial nuclear complex, the deep layers of the superior colliculi, several pretectal nuclei and the noradrenergic locus coeruleus, the (mainly) serotoninergic dorsal raphe nucleus, and the cholinergic pedunculopontine and laterodorsal tegmental nuclei [20, 25, 26, 49, 53, 73, 128, 141, 157, 193, 201, 215, 216, 222, 254]. Recent studies in rat [116–119] have shown that the individual intralaminar nuclei receive different combinations of inputs from the various brain stem nuclei. The projections from laminae VII and VIII of the spinal grey matter and corresponding parts of the spinal trigeminal nucleus terminate mainly in the rostral intralaminar nuclei. These projections and their thalamic targets form part of the *medial pain system*, which is implicated in affective-emotional aspects of pain [209, 240]. (The fibres forming the *lateral pain system* originate from laminae I and V of the spinal dorsal horn and from the corresponding zones of the spinal trigeminal nucleus, and they terminate mainly in the ventral posterior thalamic complex. This system, including its cortical projections to SI, is concerned with signaling the sensory-discriminative aspects of pain [251].) Together with the projections from the noradrenergic, serotoninergic and cholinergic brain stem nuclei, the reticulothalamic fibres form a non-specific ascending system involved in the regulation of general brain states such as sleep, wakefulness, attention and arousal.

The rostral intralaminar nuclei are major targets of the cerebellar nuclei [6, 79, 218, 226], whereas the caudal intralaminar nuclei receive a heavy projection from the internal segment of the globus pallidus [114, 125, 158, 163, 164]. This latter input consists mainly of collaterals of pallidal fibres terminating in the ventral lateral thalamic nucleus (Fig. 9.1).

As already mentioned, all intralaminar nuclei project to the striatum as well as the cerebral cortex. Within these nuclei, cells giving rise to branching axons terminating in both cortex and striatum have been observed, but according to most authors these elements are rare, and the various nuclei consist of largely separate populations of striatally and cortically projecting cells [132, 133, 196, 197, 203]. The thalamostriatal projection was first demonstrated in human material [241] and later in experimental animals [47, 48, 159, 174, 175]. In cats both the rostral and caudal groups project to the entire striatum in an overlapping fashion [139, 192]. Studies in primates [50, 51, 54, 122, 133, 165, 196, 197] have shown that the projections of different intralaminar nuclei tend to focus on particular components or sectors of the striatum. Thus, CM provides heavy input to a region of the putamen dominated by corticostriatal fibres from the primary sensory and motor areas, whereas Pf projects specifically to sectors of the caudate nucleus dominated by inputs from association areas [132, 133, 196, 197].

The cortical projections of the intralaminar nuclei are rather diffuse, but each of these nuclei appears to have a regional zone of dominance [13, 62, 80, 106, 147, 148, 234]. The anterior parietal association cortex (area 5) represents a major target of efferents from the rostral intralaminar nuclei, and these nuclei also project to the cingulate, entorhinal and sparsely to the prefrontal cortex. The CM and CL project preferentially to the primary motor cortex [108, 123, 124, 132, 133, 139, 170, 222]. Corticothalamic fibres reciprocate the intralaminar-cortical projections [132].

The *thalamic midline nuclei* receive afferents from numerous subcortical centres, which include the reticular formation of the brain stem and noradrenergic locus coeruleus, the mainly serotoninergic dorsal raphe nucleus, cholinergic laterodorsal tegmental nucleus, parabrachial nuclei, mesencephalic periaqueductal grey matter, hypothalamus and bed nucleus of the stria terminalis [117, 234]. The midline nuclei project to both the striatum and cerebral cortex, but contrary to what is seen in intralaminar nuclei, the efferents of the midline nuclei are directed primarily to the cortex and to a much lesser degree the striatum. The cortical efferents of the midline nuclei are mainly directed to the anterior cingulate cortex, entorhinal cortex and subiculum [3, 96, 241, 242]. These cortical projections are reciprocal. The subcortical efferents of the midline nuclei project to the accumbens nucleus, adjacent parts of the ventral striatum and amygdaloid complex [14, 56, 63, 144, 224].

Classically, the intralaminar and midline nuclei have been considered to form the non-specific thalamus involved in modulating cortical activity during sleep, arousal, selective attention and pain perception. Van der Werf et al. [234] recently reviewed the afferent and efferent connections of individual intralaminar and midline nuclei. Combining these connectional data with results of functional and clinical studies led them to propose that as a whole, i.e. working in concert, these nuclei play an important role in arousal and awareness but certain groups or clusters of them carry out more specific tasks within this overall functional domain. Thus they suggested that the midline nuclei (minus the reuniens nucleus) are involved in viscerolimbic functions, the rostral intralaminar nuclei in cognitive functions, and the caudal intralaminar nuclei in limbic motor functions.

Functional Aspects

1. The dorsal thalamus is often characterized as the gateway to the cortex. This is correct insofar as the principal though not sole route by which subcortical neuronal activity influences the cerebral cortex passes through the thalamus. Impulses carrying information from all sense organs except olfactory terminate in particular thalamic nuclei, where they are projected to specific cortical areas by thalamic relay cells (Fig. 8.5 A). The same holds true for information from other parts of the brain, such as the cerebellum or basal ganglia, that cannot be labeled as sensory. However, it should be emphasized that in the thalamus the flow of information to the cortex is not just passively relayed but also subjected to complex modulatory influences [210, 212–214]. As diagrammatically shown by Fig. 8.5 B, in addition to their driving sensory or intrinsic afferents, the thalamic relay cells receive:

a. Direct excitatory, glutamatergic input from cortical layer VI neurons

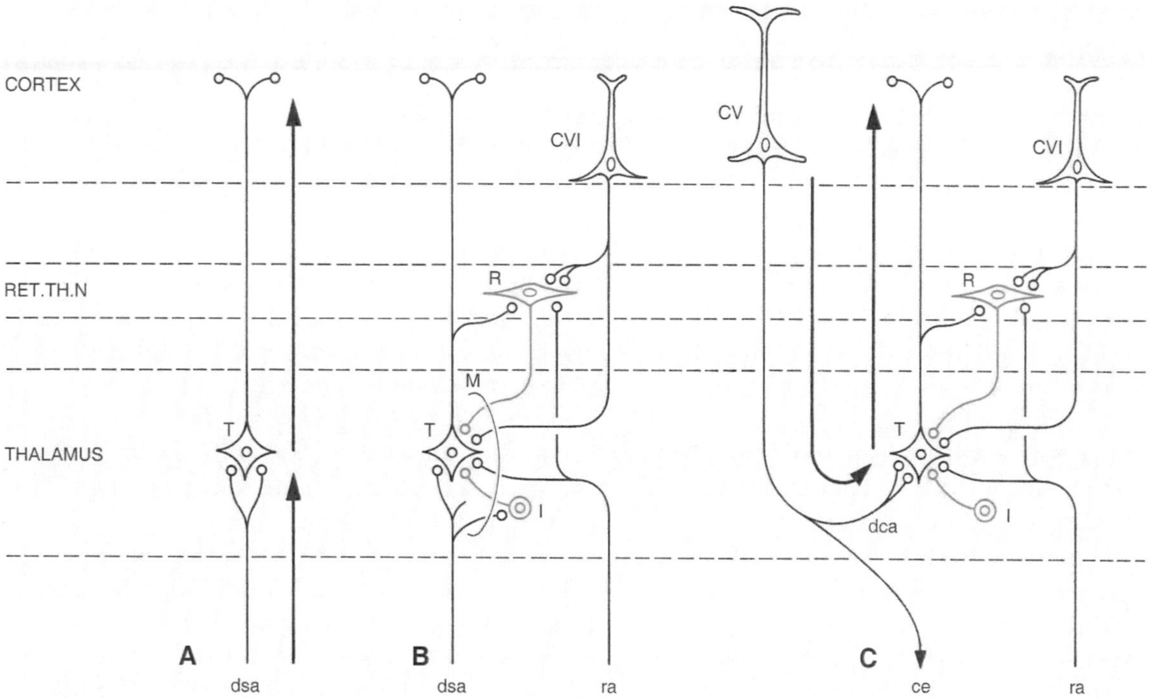

Fig. 8.5 A–C. Thalamic circuitry. **A**, First-order thalamic relay, driven by a sensory afferent fibre. **B**, First-order relay, to which the thalamic modulatory apparatus is added. **C**, Higher-order thalamic relay, driven by an axonal branch of a pyramidal layer V cell of the cortex. Inhibitory elements in red. For details see text. Based on Sherman and Guillery [213]. *CV, CVI*, lamina V and lamina VI cortical pyramidal cells. *ce*, cortical efferent fibre; *dca*, driving cortical afferent fibre; *dsa*, driving sensory afferent fibre; *I*, thalamic interneuron; *M*, thalamic modulatory apparatus; *R*, thalamic reticular neuron; *RET.TH.N.*, reticular thalamic nucleus; *ra*, cholinergic fibre ascending from the mesopontine tegmentum; *T*, thalamic relay neuron; *o*, excitatory terminals

b. Indirect inhibitory, GABAergic input from the same cortical layer via the reticular thalamic nucleus
c. Excitatory input from the upper pontine and lower mesencephalic tegmentum by neurons employing acetylcholine and nitric oxide as transmitters (these fibres originate from the pedunculopontine tegmental and laterodorsal tegmental nuclei and terminate in virtually all thalamic nuclei [222])
d. An inhibitory, GABAergic input from thalamic local circuit neurons

Both ionotropic and metabotropic receptors are postsynaptic to the four modulatory inputs to the relay cells. Activation of ionotropic receptors produces fast postsynaptic potentials with short latency, whereas activation of metabotropic receptors leads to slow, long-lasting responses. The membranes of all thalamic relay cells have special calcium channels, the transient (T)-type Ca^{2+} channels. These enable thalamic neurons to respond to excitatory stimuli in one of two very different response modes known as *tonic* and *burst* [212, 213]. The activity of the T channels is voltage-dependent. At relatively depolarized membrane potentials, they are inactivated and play no role in the firing properties of the neurons. Under these circumstances they show the tonic firing mode. The T channels are activated by hyperpolarization, which initiates the burst firing mode. The shift from burst to tonic mode requires depolarization of sufficient amplitude and duration, and the reverse shift from tonic to burst requires a similarly sustained hyperpolarization. The metabotropic receptors are of great importance for the production of the sustained hyperpolarization-depolarization to activate resp. inactivate the T receptors. By way of these metabotropic receptors, the modulatory inputs shown in Fig. 8.5 B can exert sustained effects on the overall responsiveness of thalamic relay cells and therewith on their firing mode. During the tonic mode of firing, relay cells generate trains of action potentials, the number and frequency of which are closely related to intensity and duration of the underlying depolarization [222]. Under these circumstances the information received via the driving afferents is almost linearly transferred to the cerebral cortex. During the burst mode, the information relay to the cerebral cortex is non-linearly transformed and hence the accuracy of the messages relayed through the thalamus is compromised. However, burst firing provides a much higher signal:noise ratio than tonic firing, implying that it may be superior for detecting stimuli [65]. The differences just discussed have led to the concept that tonic firing permits more faithful and accurate signal analysis, while burst firing is better for signal detection and may be used as a "wake-up call" for novel and potentially interesting or dangerous stimuli [64, 213, 214].

2. We have seen that parts of the mediodorsal thalamic nucleus and pulvinar region do not receive their dominant or "driver" afferents from the periphery or lower centres but rather from the cortex itself. These afferents come from large pyramidal cells in layer V of the cortex; most and possibly all of them are branches of long, descending axons passing to the brain stem or the spinal cord. The thalamocortical relay cells impinged upon by this layer V input are also in receipt of layer VI cortical, subcortical and local inputs, which together form part of the thalamic modulatory apparatus (Fig. 8.5 C). It may be concluded that the thalamus not only processes information from the periphery and lower centres destined for the cortex but also plays a potentially significant role in cortico-cortical communication [66, 212, 213]. Spatial and functional relationships between cortico-thalamo-cortical connections and the abundant direct cortico-cortical routes are unexplored. However, because layer V pyramidal neurons are usually provided with a rich system of intracortical collaterals, it seems likely that both projection systems have "initial common paths".

It has already been mentioned that Guillery and colleagues [66, 68, 212, 213] designated the thalamic nuclei which receive their driving afferents from ascending pathways as first-order relays (Fig. 8.5 B) and those receiving driving afferents from layer V of the cortex as higher-order relays (Fig. 8.5C). Their views on the possible function of these higher-order relays may be summarized as follows:

a. The long, descending axons of the layer V cells pass to premotor and motor centres in the lower brain stem and spinal cord and hence concern movement control
b. The thalamic branches of the layer V axons send copies of motor instructions to the higher-order thalamic relays
c. These motor efference copies, fed into the cortex by series of cortico-thalamo-cortical circuits, contribute substantially to perceptual processing in associative ("higher") cortical areas
d. In brief, the cortex not only processes sensory inputs but also monitors motor outputs.

3. The "status" of thalamic intralaminar and midline nuclei has changed considerably during the last decades. Traditionally [97, 154, 221] these nuclei were considered to give rise to widespread projections terminating in superficial layers of the cortex and to form the final link in an aspecific, ascending reticular system that plays a crucial role in regulating wakefulness and sleep. As for the cortical projections, it has already been mentioned that Groenewegen and collaborators [15, 63, 234] found that the individual intralaminar and midline nuclei have restricted cortical termination fields. As for the ascending reticular system, there can be no doubt that the brain stem reticular formation exerts a powerful activating influence on the entire cerebral cortex. However, the activation state of the cerebral cortex is not regulated exclusively via the intralaminar and midline nuclei but also by multiple output pathways of the reticular formation, including projections to the specific thalamic nuclei (Fig. 8.5B), reticular thalamic nucleus [67, 69] and direct [52] and indirect extrathalamic projections to the cortex. The latter are relayed in the posterior hypothalamus [182, 200] and basal forebrain (Fig. 10.5) [18, 46, 202].

References

1. Aggleton JP, Mishkin M (1984) Projections of the amygdala to the thalamus in the cynomolgus monkey. J Comp Neurol 222:56–68
2. Akert K (1964) Comparative anatomy of frontal cortex and thalamo-frontal connections. In: Warren JM, Akert K (eds) The frontal granular cortex and behavior. McGraw Hill, New York, pp 372–396
3. Amaral DG, Cowan WM (1980) Subcortical afferents to the hippocampal formation in the monkey. J Comp Neurol 189:573–591
4. Apkarian AV, Shi T (1994) Squirrel monkey lateral thalamus. I. Somatic nociresponsive neurons and their relation to spinothalamic terminals. J Neurosci 14:6779–6795
5. Asanuma C, Thach WT, Jones EG (1983) Distribution of cerebellar terminations and their relation to other afferent terminations in the ventral lateral thalamic region of the monkey. Brain Res Rev 5:237–266
6. Asanuma C, Thach WT, Jones EG (1983) Anatomical evidence for segregated focal groupings of efferent cells and their terminal ramifications in the cerebellothalamic pathway of the monkey. Brain Res Rev 5:267–298
7. Asanuma C, Anderson RA, Cowan WM (1985) The thalamic relations of the caudal inferior parietal lobule and the lateral prefrontal cortex in monkeys. Divergent cortical projections from all clusters in the medial pulvinar nucleus. J Comp Neurol 241:357–381
8. Baleydier C, Manguiere F (1985) Anatomical evidence for medial pulvinar connections with the posterior cingulate cortex, the retrosplenial area and the posterior parahippocampal gyrus in monkeys. J Comp Neurol 232:219–229
9. Basle TL, Wilson CL, Crandall PH (1982) Asymmetry and ventral course of the human geniculostriate pathway as determined by hippocampal visual evoked potentials and subsequent visual field defects after temporal lobectomy. Exp Brain Res 47:317–328
10. Beggs J, Jordan S, Ericson AC, Blomqvist A, Craig AD (2003) Synaptology of trigemino- and spinothalamic lamina I terminations in the posterior ventral medial nucleus of the macaque. J Comp Neurol 459:334–354
11. Bender DB (1981) Retinotopic organization of macaque pulvinar. J Neurophysiol 46:672–693
12. Benevento LA, Standage GP (1983) The organization of projections of the retinorecipient and nonretinorecipient nuclei of the pretectal complex and layers of superior colliculus to the lateral pulvinar and medial pulvinar in the macaque monkeys. J Comp Neurol 217:307–336
13. Bentivoglio M, Balercia G, Kruger L (1991) The specificity of the non-specific thalamus: the midline nuclei. In: Holstege G (ed) Progress in brain research, vol 87. Elsevier, Amsterdam, pp 53–80

14. Berendse HW, Groenewegen HJ (1990) Organization of the thalamostriatal projection in the rat, with special emphasis on the ventral striatum. J Comp Neurol 299:187–228
15. Berendse HW, Groenewegen HJ (1991) Restricted cortical termination fields of the midline and intralaminar thalamic nuclei in the rat. Neuroscience 42:73–102
16. Berkley KJ (1980) Spatial relationships between the terminations of somatic sensory and motor pathways in the rostral brain stem of cats and monkeys I. Ascending somatic sensory inputs to the lateral diencephalon. J Comp Neurol 193:283–317
17. Berkley KJ (1983) Spatial relationships between the terminations of sensory motor pathways in the rostral brain stem of cats and monkeys II. Cerebellar projections compared with those of the ascending somatic sensory pathways in lateral diencephalon. J Comp Neurol 220:229–251
18. Berntson GG, Shafi R, Sarter M (2002) Specific contributions of the basal forebrain corticopetal cholinergic system to electroencephalographic activity and sleep/waking behaviour. Eur J Neurosci 16:2453–2461
19. Blomqvist A, Zhang ET, Craig AD (2000) Cytoarchitectonic and immunohistochemical characterization of a specific pain and temperature relay, the posterior portion of the ventral medial nucleus, in the human thalamus. Brain 123:601–619
20. Bobillier R, Seguin S, Petitjean F et al (1976) The raphe nuclei of the cat brain stem: a topographical atlas of their efferent projections as revealed by autoradiography. Brain Res 113:449–486
21. Boivie J (1978) Anatomical observations on the dorsal column nuclei, their thalamic projection and the cytoarchitecture of some somatosensory thalamic nuclei in the monkey. J Comp Neurol 178:17–48
22. Boivie J (1979) An anatomical reinvestigation of the termination of the spinothalamic tract in the monkey. J Comp Neurol 186:343–370
23. Bourassa J, Pinault D, Deschnes M (1995) Corticothalamic projections from the cortical barrel field to the somatosensory thalamus in rats: a single-fibre study using biocytin as an anterograde tracer. Eur J Neurosci 7:19–30
24. Bowsher D (1957) Termination of the central pain pathway: the conscious appreciation of pain. Brain 80:606–622
25. Bowsher D (1975) Diencephalic projections from the midbrain reticular formation. Brain Res 95: 211–220
26. Burton H, Craig AD Jr (1979) Distribution of trigeminothalamic projection cells in cat and monkey. Brain Res 161:515–521
27. Burton H, Jones EG (1976) The posterior thalamic region and its cortical projection in New World and Old World monkeys. J Comp Neurol 168:249–302
28. Burton H, Loewy AD (1977) Projections to the spinal cord from medullary somatosensory relay nuclei. J Comp Neurol 173:773–792
29. Calford MB, Aitken LM (1983) Ascending projections to the medial geniculate body of the cat: evidence for multiple, parallel auditory pathways through thalamus. J Neurosci 3:2365–2380
30. Carcio CA, Harting KJ (1978) Organization of pulvinar afferents to area 18 in the squirrel monkey: evidence for stripes. Brain Res 143:155–161
31. Carman JB, Cowan WM, Powell TPS (1964) Cortical connections of the thalamic reticular nucleus. J Anat 98:587–598
32. Carpenter MB, Nakano K, Kim R (1976) Nigrothalamic projections in the monkey demonstrated by autoradiographic technics. J Comp Neurol 165: 401–416
33. Chalupa LM (1977) A review of cat and monkey studies implicating the pulvinar in visual function. Behav Biol 20:146–167
34. Christensen BN, Perl ER (1970) Spinal neurons specifically excited by noxious or thermal stimuli: marginal zone of the dorsal horn. J Neurophysiol 33:293–307
35. Condé F, Condé H (1978) Thalamic projections of the vestibular nuclei in the cat as revealed by retrograde transport of horseradish peroxidase. Neurosci Lett 9:141–146
36. Connolly M, Van Essen D (1984) The representation of the visual field in parvicellular and magnocellular layers of the lateral geniculate nucleus in the macaque monkey. J Comp Neurol 226:544–565
37. Craig AD Jr, Burton H (1981) Spinal and medullary lamina I projection to nucleus submedius in medial thalamus: a possible pain center. J Neurophysiol 45:443–466
38. Craig AD, Blomqvist A (2002) Is there a specific lamina I spinothalamocortical pathway for pain and temperature sensations in primates? J Pain 3:95–101
39. Craig AD, Kniffki K-D (1985) Spinothalamic lumbosacral lamina I cells responsive to skin and muscle stimulation in the cat. J Physiol (Lond) 365: 197–221
40. Craig AD, Bushnell MC, Zhang E-T, Blomqvist A (1994) A thalamic nucleus specific for pain and temperature sensation. Nature 372:770–773
41. Curcio CA, Harting JK (1978) Organization of pulvinar afferents to area 18 in the squirrel monkey: Evidence for stripes. Brain Res 143:155–161
42. Cusick CG, Scripter JL, Darensbourg JG, Weber JT (1993) Chemoarchitectonic subdivisions of the visual pulvinar in monkeys and their connectional relations with the middle temporal and rostral dorsolateral visual areas, MT and DLr. J Comp Neurol 336:1–30
43. Dekaban A (1954) Human thalamus. An anatomical, developmental and pathological study: development of the human thalamic nuclei. J Comp Neurol 100:63–97

44. DeVito JL, Anderson ME (1982) An autoradiographic study of efferent connections of the globus pallidus in macaca mulatta. Exp Brain Res 46:107–117
45. Domesick VB (1969) Projections from the cingulate cortex in the rat. Brain Res 12:296–320
46. Dringenberg HC, Olmstead MC (2003) Integrated contributions of basal forebrain and thalamus to neocortical activation elicited by pedunculopontine tegmental stimulation in urethane-anesthetized rats. Neuroscience 119:839–853
47. Droogleever-Fortuyn J (1953) Anatomical basis of cortico-subcortical relationships. Electroencephalogr Clin Neurophysiol 4:149–162
48. Droogleever-Fortuyn J, Stefens R (1951) On the anatomical relations of the intralaminar and midline cells of the thalamus. Electroencephalogr Clin Neurophysiol 3:393–400
49. Edwards SB, De Olmos JS (1976) Autoradiographic studies of the projections of the midbrain reticular formation: ascending projections of nucleus cuneiformis. J Comp Neurol 165:417–431
50. Flaherty AW, Graybiel AM (1993) Two input systems for body representations in the primate striatal matrix: evidence in the squirrel monkey. J Neurosci 13:1120–1137
51. Flaherty AW, Graybiel AM (1994) Input-output organization of the sensorimotor striatum in the squirrel monkey. J Neurosci 14:599–610
52. Foote SL (1987) Extrathalamic modulation of cortical function. Annu Rev Neurosci 10:67–95
53. Foster GA, Schultzberg M, Goldstein M, Hökfelt T (1985) Ontogeny of phenylethanolamine N-methyltransferase- and tyrosine hydroxylase-like immunoreactivity in presumptive adrenaline neurones of the foetal rat central nervous system. J Comp Neurol 236:348–381
54. François C, Percheron G, Parent A et al (1991) Topography of the projection from the complex of the thalamus to the sensorimotor striatal territory in monkeys. J Comp Neurol 354:127–149
55. Friedman DP, Jones EG (1981) Thalamic input to areas 3a and 2 in monkeys. J Neurophysiol 45:59–85
56. Gimenez-Amaya JM, McFarland NR, De las Heras S, Haber SN (1995) Organization of thalamic projections to the ventral striatum in the primate. J Comp Neurol 354:127–149
57. Goldman PS (1979) Contralateral projections to the dorsal thalamus from frontal association cortex in the rhesus monkey. Brain Res 166:166–171
58. Goldman-Rakic PS, Porrino LJ (1985) The primate mediodorsal (MD) nucleus and its projection to the frontal lobe. J Comp Neurol 242:535–560
59. Gray D, Gutierrez C, Cusick CG (1999) Neurochemical organization of inferior pulvinar complex in squirrel monkeys and macaques revealed by acetylcholinesterase histochemistry, calbindin and cat-301 immunostaining, and *Wisteria floribunda* agglutinin binding. J Comp Neurol 409:452–468
60. Graziano A, Jones EG (2004) Widespread thalamic terminations of fibers arising in the superficial medullary dorsal horn of monkeys and their relation to calbindin immunoreactivity. J Neurosci 24:248–256
61. Groenewegen HJ (1988) Organization of the afferent connections of the mediodorsal thalamic nucleus in the rat, related to the mediodorsal-prefrontal topography. Neuroscience 24:379–431
62. Groenewegen HJ, Berendse HW (1994) The specificity of the 'non-specific' midline and intralaminar thalamic nuclei. Trends Neurosci 17:52–57
63. Groenewegen HJ, Becker NEK, Lohman AHM (1980) Subcortical afferents of the nucleus accumbens septi in the cat, studied with retrograde axonal transport of horseradish peroxidase and bisbenzimide. Neuroscience 5:1903–1916
64. Guido W, Weyand T (1995) Burst responses in thalamic relay cells of the awake behaving cat. J Neurophysiol 74:1782–1786
65. Guido W, Lu SM, Vaughan GM, Godwin DW, Sherman SM (1995) Receiver operating characteristic (ROC) analysis of neurons in the cat's lateral geniculate nucleus during tonic and burst response mode. Vis Neurosci 12:723–741
66. Guillery RW (1995) Anatomical evidence concerning the role of the thalamus in corticocortical communication: a brief review. J Anat 187:583–592
67. Guillery RW, Harting JK (2003) Structure and connections of the thalamic reticular nucleus: advancing views over half a century. J Comp Neurol 463:360–371
68. Guillery RW, Sherman SM (2002) The thalamus as a monitor of motor outputs. Phil Trans R Soc Lond B 357:1809–1821
69. Guillery RW, Feig SL, Lozsádi DA (1998) Paying attention to the thalamic reticular nucleus. Trends Neurosci 21:28–32
70. Gutierrez C, Cola MG, Seltzer B, Cusick C (2000) Neurochemical and connectional organization of the dorsal pulvinar complex in monkeys. J Comp Neurol 419:61–86
71. Haber SN, Groenewegen HJ, Grove EA, Nauta WJH (1985) Efferent connections of the ventral pallidum: evidence of a dual striato-pallidofugal pathway. J Comp Neurol 235:322–335
72. Haber SN, Lynd-Balta E, Mitchell SJ (1993) The organization of the descending ventral pallidal projections in the monkey. J Comp Neurol 329:111–128
73. Harting JK, Huerta MF, Frankfurter AJ, Strominger NL, Royce GJ (1980) Ascending pathways from the monkey superior colliculus: an autoradiographic analysis. J Comp Neurol 192:853–882
74. Hashikawa T, Molinari M, Rausell E, Jones EG (1995) Patchy and laminar terminations of medial geniculate axons in monkey auditory cortex. J Comp Neurol 362:195–208
75. Hassler R (1959) Anatomy of the thalamus. In: Schaltenbrand G, Bailey P (eds) Introduction to stereotaxis with an atlas of the human brain. Thieme, Stuttgart, pp 230–290

76. Heimer L, Switzer RD, Van Hoesen GW (1982) Ventral striatum and ventral pallidum. Components of the motor system? Trends Neurosci 5:83–87
77. Hendrickson AE, Wilson ME, Toyne MJ (1970) The distribution of optic nerve fibers in Macaca mulatta. Brain Res 23:425–427
78. Hendrickson AE, Wilson JR, Ogren MP (1978) The neuroanatomical organization of pathways between the dorsal lateral geniculate nucleus and visual cortex in Old World and New World primates. J Comp Neurol 182:123–135
79. Hendry SHC, Jones EG, Graham J (1979) Thalamic relay nuclei for cerebellar and certain related fiber systems in the cat. J Comp Neurol 185:679–714
80. Herkenham M (1980) Laminar organization of thalamic projections to rat neocortex. Science 207: 532–534
81. Hickey TL, Guillery RW (1979) Variability of laminar patterns in the human lateral geniculate nucleus. J Comp Neurol 183:221–246
82. Hirai T, Jones EG (1989) A new parcellation of the human thalamus on the basis of histochemical staining. Brain Res Rev 14:1–34
83. Hoogland PV, Wouterlood FG, Welker E, Van der Loos H (1991) Ultrastructure of giant and small thalamic terminals of cortical origin, a study of the projections from the barrel cortex in mice using *Phaseolus vulgaris* leuco-agglutinin (PHA-L). Exp Brain Res 87:159–172
84. Houser CR, Vaughn JE, Barber RP, Roberts E (1980) GABA neurons are the major cell type of the nucleus reticularis thalami. Brain Res 200:341–354
85. Hubel DH, Wiesel TN (1962) Receptive fields, binocular interaction and functional architecture in the cat's visual cortex. J Physiol (Lond) 160:106–154
86. Hubel DH, Wiesel TN (1969) Anatomical demonstration of columns in the monkey striate cortex. Nature (London) 221:747–750
87. Hubel DH, Wiesel TN (1972) Laminar and columnar distribution of geniculo-cortical fibers in the macaque monkey. J Comp Neurol 146:421–443
88. Hubel DH, Wiesel TN (1974) Uniformity of monkey striate cortex: a parallel relationship between field size, scatter and magnification factor. J Comp Neurol 15:295–306
89. Hubel DH, Wiesel TN (1977) Ferrier lecture. Functional architecture of macaque monkey visual cortex. Proc R Soc Lond Biol 198:1–59
90. Hubel DH, Wiesel TN, LeVay S (1977) Plasticity of ocular dominance columns in monkey striate cortex. Philos Trans R Soc Lond Biol 278:377–409
91. Huerta MF, Harting JK (1983) Sublamination within the superficial gray layer of the squirrel monkey: an analysis of the tectopulvinar projection using anterograde and retrograde transport methods. Brain Res 261:119–126
92. Huerta MF, Krubitzer LA, Kaas JH (1986) Frontal eye field as defined by intracortical microstimulation in squirrel monkeys, owl monkeys and macaque monkeys: I. Subcortical connections. J Comp Neurol 253:415–439
93. Hunt CA, Pang DZ, Jones EG (1991) Distribution and density of GABA cells in intralaminar and adjacent nuclei of monkey thalamus. Neuroscience 43:185–196
94. Ilinsky IA, Jouandet ML, Goldman-Rakic PS (1985) Organization of the nigrothalamocortical system in the rhesus-monkey. J Comp Neurol 236:315–330
95. Ilinsky IA, Tourtellotte WG, Kultas-Ilinsky K (1993) Anatomical distinctions between the two basal ganglia afferent territories in the primate motor thalamus. Stereotact Funct Neurosurg 60:62–69
96. Insausti R, Amaral DG, Cowan WM (1987) The entorhinal cortex of the monkey: II. Cortical afferents. J Comp Neurol 264:356–395
97. Jasper HH (1958) Reticular-cortical systems and theories of the integrative action of the brain. In: Harlow HF, Woolsey CN (eds) Biological and biochemical bases of behavior. University of Wisconsin Press, Madison, pp 37–61
98. Jones EG (1975) Some aspects of the organization of the thalamic reticular complex. J Comp Neurol 162:285–308
99. Jones EG (1985) The thalamus. Plenum, New York
100. Jones EG (1997) A description of the human thalamus. In: Steriade M, Jones EG (eds) Thalamus, vol II. Elsevier, Amsterdam, pp 425–499
101. Jones EG (1998) Viewpoint: the core and matrix of thalamic organization. Neuroscience 85:331–345
102. Jones EG (2001) The thalamic matrix and thalamocortical synchrony. Trends Neurosci 24:595–601
103. Jones EG (2002) Thalamic circuitry and thalamocortical synchrony. Phil Trans R Soc Lond B 357: 1659–1673
104. Jones EG, Burton H (1976) Areal differences in the laminar distribution of thalamic afferents in cortical fields of the insular, parietal and temporal regions of primates. J Comp Neurol 168:197–248
105. Jones EG, Friedman DP (1982) Projection patterns of functional components of thalamic ventrobasal complex in monkey somatosensory cortex. J Neurophysiol 48:521–544
106. Jones EG, Leavitt RY (1974) Retrograde axonal transport and the demonstration of non-specific projections to the cerebral cortex and striatum from thalamic intralaminar nuclei in the rat, cat and monkey. J Comp Neurol 154:349–378
107. Jones EG, Powell TPS (1971) An analysis of the posterior group of the thalamic nuclei on the basis of its afferent connections. J Comp Neurol 143:185–216
108. Jones EG, Wise SP, Coulter JD (1979) Differential thalamic relationships of sensory-motor and parietal cortical fields in monkeys. J Comp Neurol 183:833–882
109. Kaas JH, Lin CS, Casagrande VA (1976) The relay of ipsilateral and contralateral retinal input from the lateral geniculate nucleus to striate cortex in the owl monkey: a transneuronal transport study. Brain Res 106:371–378

110. Kaas JH, Huerta MF, Weber JT, Harting JK (1978) Patterns of retinal terminations and laminar organization of the lateral geniculate nucleus of primates. J Comp Neurol 182:517–554
111. Kaitz SS, Robertson RT (1981) Thalamic connections with limbic cortex. II. Corticothalamic projections. J Comp Neurol 195:527–545
112. Kasdon DL, Jacobson S (1978) The thalamic afferents to the inferior parietal lobule of the rhesus monkey. J Comp Neurol 177:685–706
113. Kievit J, Kuypers HGJM (1977) Organization of the thalamo-cortical connexions to the frontal lobe in the rhesus monkey. Exp Brain Res 29: 299–322
114. Kim R, Nakano K, Jayaraman A, Carpenter B (1976) Projections of the globus pallidus and adjacent structures: an autoradiographic study in the monkey. J Comp Neurol 169:263–290
115. Krettek JE, Price JL (1977) The cortical projections of the mediodorsal nucleus and adjacent thalamic nuclei in the rat. J Comp Neurol 171: 157–191
116. Krout KE, Loewy AD (2000) Parabrachial nucleus projections to midline and intralaminar thalamic nuclei of the rat. J Comp Neurol 428:475–494
117. Krout KE, Loewy AD (2000) Periaqueductal gray matter projections to midline and intralaminar thalamic nuclei of the rat. J Comp Neurol 424: 111–141
118. Krout KE, Loewy AD, Westby GW, Redgrave P (2001) Superior colliculus projections to midline and intralaminar thalamic nuclei of the rat. J Comp Neurol 431:198–216
119. Krout KE, Belzer RE, Loewy AD (2002) Brainstem projections to midline and intralaminar thalamic nuclei of the rat. J Comp Neurol 448:53–101
120. Kudo M, Niimi K (1980) Ascending projection of the inferior colliculus in the cat: an autoradiographic study. J Comp Neurol 191:545–556
121. Kuhlenbeck H (1954) The human diencephalon: a summary of development, structure, function and pathology. Confin Neurol 14:1–230
122. Künzle H (1975) Bilateral projections from precentral motor cortex to the putamen and other parts of the basal ganglia. An autoradiographic study in the Macaca fascicularis. Brain Res 88:195–209
123. Künzle H (1978) An autoradiographic analysis of the efferent connections from premotor and adjacent prefrontal regions (areas 6 and 9) in Macaca fascicularis. Brain Behav Evol 15:185–234
124. Künzle H, Akert K (1977) Efferent connections of cortical area 8 (frontal eye field) in Macaca fascicularis. A reinvestigation using the autoradiographic technique. J Comp Neurol 173:147–164
125. Kuo JS, Carpenter MB (1973) Organization of pallidothalamic projections in the rhesus monkey. J Comp Neurol 151:201–236
126. Kuypers HGJM, Bentivoglio M, Catsman-Berrevoets CE, Bharos AT (1980) Double retrograde labeling through divergent axon collaterals, using two fluorescent tracers with the same excitation wavelength which label different features of the cell. Exp Brain Res 40:383–392
127. Lang W, Büttner-Ennever JA, Büttner U (1979) Vestibular projections to the monkey thalamus: an autoradiographic study. Brain Res 177:3–17
128. Lavoie B, Parent A (1991) Serotoninergic innervation of the thalamus in the primate: an immunohistochemical study. J Comp Neurol 312:1–18
129. Lin C-S, Merzenich MM, Sur M, Kaas JH (1979) Connections of areas 3b and 1 of the parietal somatosensory strip with the ventroposterior nucleus in the owl monkey (Aotus trivirgatus). J Comp Neurol 185:355–372
130. Llinás R, Paré D (1997) Coherent oscillations in specific and nonspecific thalamocortical networks and their role in cognition. In: Steriade M, Jones EG, McCormick DA (eds) Thalamus, vol II. Elsevier, Amsterdam, pp 501–516
131. Lysakowski A, Standage GP, Benevento LA (1986) Histochemical and architectonic differentiation of zones of pretectal and collicular inputs to the pulvinar and dorsal lateral geniculate nuclei in the macaque. J Comp Neurol 250:431–448
132. Macchi G, Bentivoglio M (2003) The thalamic intralaminar nuclei and the cerebral cortex. In: Jones EG, Peters A (eds) Cerebral cortex, vol 5: Sensory-motor areas and aspects of cortical connectivity. Plenum Press, New York, pp 355–401
133. Macchi G, Bentivoglio M, Molinari M, Miniacchi D (1984) The thalamo-caudate versus thalamo-cortical projections as studied in the cat with fluorescent retrograde double labeling. Exp Brain Res 54:225–239
134. Maciewicz R, Phipps BS, Bry J, Highstein SM (1982) The vestibulothalamic pathway: contribution of the ascending tract of Deiters. Brain Res 252:1–11
135. Mantyh PW (1983) The spinothalamic tract in the primate: a re-examination using WGA-HRP. Neuroscience 9:847–862
136. Markowitsch HJ, Emmans D, Irle E, Streicher M, Preilowski B (1985) Cortical and subcortical afferent connections of the primate's temporal pole: a study of rhesus monkeys, squirrel monkeys and marmosets. J Comp Neurol 242:425–458
137. Matelli M, Luppino G (1993) Cortical projections of motor thalamus. In: Miniacchi D, Molinari M, Macchi G, Jones EG (eds) Thalamic networks for relay and modulation. Pergamon, Oxford, pp 165–174
138. McFarland NR, Haber SN (2002) Thalamic relay nuclei of the basal ganglia form both reciprocal and nonreciprocal cortical connections, linking multiple frontal cortical areas. J Neurosci 22: 8117–8132
139. McGuinness C, Krauthammer GM (1980) Afferent projections to the centrum-medianum of the cat as demonstrated by retrograde transport of horseradish peroxidase. Brain Res 184:255–269

140. Mehler WR (1962) The anatomy of the so-called "pain tract" in man: an analysis of the course and distribution of the ascending fibers of the fasciculus anterolateralis. In: French JD, Porter RW (eds) Basic research in paraplegia. Thomas, Springfield, pp 26–55
141. Mehler WR (1966) The posterior thalamic region in man. Confin Neurol 27:18–29
142. Mehler WR (1969) Some neurological species differences a posteriori. Ann NY Acad Sci 167:424–468
143. Mehler WR (1974) Central pain and the spinothalamic tract. Adv Neurol 4:127–146
144. Mehler WR (1980) Subcortical afferent connections of the amygdala in the monkey. J Comp Neurol 190:733–762
145. Mesulam MM, Pandya DN (1973) The projections of the medial geniculate complex within the sylvian fissure of the rhesus monkey. Brain Res 60:315–333
146. Mesulam MM, Van Hoesen GW, Pandya DN, Geschwind M (1977) Limbic and sensory connections of the inferior parietal lobule (area PG) in the rhesus monkey: a study with a new method for horseradish peroxidase histochemistry. Brain Res 136:393–414
147. Molinari M, Leggio MG, Dell'Anna ME, Gianetti S, Macchi G (1993) Structural evidence in favour of a relay function for the anterior intralaminar nuclei. In: Miniacchi G, Molinari M, Macchi G, Jones EG (eds) Thalamic networks for relay and modulation. Pergamon, Oxford, pp 197–208
148. Molinari M, Leggio MG, Dell'Anna ME, Gianetti S, Macchi G (1994) Chemical compartmentation and relationships between calcium-binding protein immunoreactivity and layer-specific cortical and caudate-projecting cells in the anterior intralaminar nuclei of the cat. Eur J Neurosci 6:299–312
149. Montero VM (1986) Localization of gamma-aminobutyric acid (GABA) in type 3 cells and demonstration of their source to F2 terminals in cat lateral geniculate nucleus. A Golgi-electron-microscopic GABA-immunocytochemical study. J Comp Neurol 254:228–245
150. Montero VM, Guillery RW, Woolsey CN (1977) Retinotopic organization within the thalamic reticular nucleus demonstrated by a double label autoradiographic technique. Brain Res 138:407–421
151. Moore RY, Goldberg JM (1960) Projections of the inferior colliculus in the monkey. Exp Neurol 14:429–438
152. Morel A, Magnin M, Jeanmonod D (1997) Multiarchitectonic and stereotactic atlas of the human thalamus. J Comp Neurol 387:588–630
153. Morest DK (1965) The laminar structure of the medial geniculate body of the cat. J Anat 99:143–160
154. Moruzzi G, Magoun HW (1949) Brain stem reticular formation and activation of the EEG. Electroencephalogr Clin Neurophysiol 1:455–473
155. Mufson EJ, Mesulam MM (1984) Thalamic connections of the insula in the rhesus monkey and comments on the paralimbic connectivity of the medial pulvinar nucleus. J Comp Neurol 227:109–120
156. Nauta HJW (1979) Projections of the pallidal complex: an autoradiographic study in the cat. Neuroscience 4:1853–1873
157. Nauta WJH, Kuypers HGJM (1958) Some ascending pathways in the brain stem reticular formation. In: Jasper HH et al (eds) Reticular formation of the brain. Little Brown, Toronto, pp 3–31
158. Nauta WJH, Mehler WR (1966) Projections of the lentiform nucleus in the monkey. Brain Res 1:3–42
159. Nauta WJH, Whitlock DG (1954) An anatomical analysis of the non-specific thalamic projection system. In: Delafresnaye JF (ed) Brain mechanisms and consciousness. Thomas, Springfield, pp 81–1160
160. Ogren MP, Hendrickson AE (1976) Pathways between striate cortex and subcortical regions in Macaca mulatta and Saimiri sciureus: evidence for a reciprocal pulvinar connection. Exp Neurol 53:780–800
161. Ogren MP, Hendrickson AE (1977) The distribution of pulvinar terminals in visual areas 17 and 18 of the monkey. Brain Res 137:343–350
162. Olszewski J (1952) The thalamus of the Macaca mulatta. Karger, Basel
163. Parent A, DeBellefeuille L (1982) Organization of efferent projections from the internal segment of globus pallidus in primate as revealed by fluorescence retrograde labeling method. Brain Res 245:201–213
164. Parent A, DeBellefeuille L (1983) The pallidointralaminar and pallidonigral projections in primate as studied by retrograde double-labeling method. Brain Res 278:11–28
165. Parent A, Mackey A, DeBellefeuille L (1983) The subcortical afferents to caudate nucleus and putamen in primate: a fluorescence retrograde double labeling study. Neuroscience 10:1137–1150
166. Pearson RCA, Brodal P, Powell TPS (1978) The projection of the thalamus upon the parietal lobe in the monkey. Brain Res 144:143–148
167. Penny GR, Fitzpatrick D, Schmechel D, Diamond IT (1983) Glutamic acid decarboxylase immunoreactive neurons and horseradish peroxidase-labeled projection neurons in the ventral posterior nucleus of the cat and Galago senegalensis. J Neurosci 3:1868–1887
168. Percheron G, François C, Talbi B et al (1993) The primate motor thalamus analyzed with reference to subcortical afferent territories. Stereotact Funct Neurosurg 60:32–41
169. Perkel DJ, Bullier J, Kennedy H (1986) Topography of the afferent connectivity of area-17 in the macaque monkey. A double-labeling study. J Comp Neurol 253:374–402

170. Petras JM (1969) Some efferent connections of the motor and somatosensory cortex of simian primates and felid, canid and procyonid carnivores. Ann NY Acad Sci 167:469–505
171. Poggio GF, Mountcastle VB (1960) A study of the functional contributions of the lemniscal and spinothalamic systems to somatic sensibility. Bull Johns Hopkins Hosp 106:266–316
172. Pons JP, Kaas JH (1985) Connections of area 2 of somatosensory cortex with the anterior pulvinar and subdivisions of the ventroposterior complex in macaque monkeys. J Comp Neurol 240:16–36
173. Poremba A, Kubota Y, Gabriel M (1994) Afferent connections of the anterior thalamus in rabbits. Brain Res Bull 33:361–365
174. Powell TPS, Cowan WM (1954) The connexions of the midline and intralaminar nuclei of the thalamus of the rat. J Anat 88:307–319
175. Powell TPS, Cowan WM (1956) A study of thalamostriate relations in the monkey. Brain 79:364–390
176. Pritchard TC, Hamilton RB, Morse JR, Morgren R (1986) Projections of thalamic gustatory and lingual areas in the monkey, Macaca fascicularis. J Comp Neurol 244:213–228
177. Ralston HJ III (1971) Evidence for presynaptic dendrites and a proposal for their mechanism of action. Nature 230:585–587
178. Ralston HJ III (2003) Pain, the brain, and the (calbindin) stain. J Comp Neurol 459:329–333
179. Ralston HJ III, Ralston DD (1992) The primate dorsal spinothalamic tract: evidence for a specific termination in the posterior nuclei (Po/SG) of the thalamus. Pain 48:107–118
180. Ray JP, Price JL (1992) The organization of the thalamocortical connections of the mediodorsal thalamic nucleus in the rat, related to the ventral forebrain-prefrontal cortex topography. J Comp Neurol 323:167–197
181. Raymond J, Sans A, Marty R (1974) Projections thalamiques des noyaux vestibulaires: étude histologique chez le chat. Exp Brain Res 20:273–283
182. Rempel-Clower NL, Barbas H (1998) Topographic organization of connections between the hypothalamus and prefrontal cortex in the rhesus monkey. J Comp Neurol 398:393–419
183. Rezak M, Benevento LA (1979) A comparison of the organization of the projections of the dorsal lateral geniculate nucleus, the inferior pulvinar and adjacent lateral pulvinar to primary visual cortex (area 17) in the Macaque monkey. Brain Res 167:19–40
184. Robertson RT (1983) Efferents of the pretectal complex: separate populations of neurons project to lateral thalamus and to inferior olive. Brain Res 258:91–95
185. Robertson RT, Kaitz SS (1981) Thalamic connections with limbic cortex. I. Thalamocortical projections. J Comp Neurol 195:501–525
186. Robertson RT, Thompson SM, Kaitz SS (1983) Projections from the pretectal complex to the thalamic lateral dorsal nucleus of the cat. Exp Brain Res 51:157–171
187. Robinson DL, Petersen SE (1992) The pulvinar and visual salience. Trends Neurosci 15:127–132
188. Rodieck RW (1979) Visual pathways. Annu Rev Neurosci 2:193–225
189. Romansky LM, Giguere M, Bates JF, Goldman-Rakic PS (1997) Topographic organization of medial pulvinar connections with the prefrontal cortex in the rhesus monkey. J Comp Neurol 379:313–332
190. Rose JE, Woolsey C (1948) Structure and relations of limbic cortex and anterior thalamic nuclei in rabbit and cat. J Comp Neurol 89:79–347
191. Rouiller EM, Welker E (1991) Morphology of corticothalamic terminals arising from the auditory cortex of the rat: a *Phaseolus vulgaris*-leucoagglutinin (PHA-L) tracing study. Hearing Res 56:179–190
192. Royce GJ, Mourey RJ (1985) Efferent connections of the centromedian and parafascicular thalamic nuclei: an autoradiographic investigation in the cat. J Comp Neurol 235:277–300
193. Royce GJ, Bromley S, Cracco C (1991) Subcortical projections to the centromedial and parafascicular thalamic nuclei in the cat. J Comp Neurol 306:129–155
194. Russchen FT, Amaral DG, Price JL (1987) The afferent input to the magnocellular division of the mediodorsal thalamic nucleus in the monkey, Macaca fascicularis. J Comp Neurol 256:175–210
195. Rustioni A, Schmechel DE, Spreafico R, Cuenod M (1983) Excitatory and inhibitory aminoacid putative neurotransmitters in the ventralis posterior complex: an autoradiographic and immunocytochemical study in rats and cats. In: Macchi O, Rustioni A, Spreafico R (eds) Somatosensory integration in the thalamus. Elsevier, Amsterdam, pp 365–383
196. Sadikot AF, Parent A, François C (1992) Efferent connections of the centromedian and parafascicular thalamic nuclei in the squirrel monkey: a PHA-L study of subcortical projections. J Comp Neurol 315:137–159
197. Sadikot AF, Parent A, Smith Y, Bolam JP (1992) Efferent connections of the centromedian and parafascicular thalamic nuclei in the squirrel monkey: a light and electron microscopic study of the thalamostriatal projection in relation to striatal heterogeneity. J Comp Neurol 320:228–242
198. Sakai ST, Inase M, Tanji J (1999) Pallidal and cerebellar inputs to thalamocortical neurons projecting in the supplementary motor area in *Macaca fuscata*: a triple-labeling light microscopic study. Anat Embryol 199:9–19
199. Sakai ST, Stepniewska I, Qi HX, Kaas JH (2000) Pallidal and cerebellar afferents to pre-supplementary motor area thalamocortical neurons in the owl monkey: a multiple labeling study. J Comp Neurol 417:164–180

200. Saper CB (1985) Organization of cerebral cortical afferent systems in the rat. II. Hypothalamocortical projections. J Comp Neurol 237:21–46
201. Saper CB, Loewy AD (1980) Efferent connections of the parabrachial nucleus in the rat. Brain Res 197:291–317
202. Sarter M, Bruno JP (2000) Cortical cholinergic inputs mediating arousal, attentional processing and dreaming: differential afferent regulation of the basal forebrain by telencephalic and brainstem afferents. Neuroscience 95:933–952
203. Sato M, Itoh K, Mizuno N (1979) Distribution of thalamo-caudate neurons in the cat as demonstrated by horseradish peroxidase. Exp Brain Res 34:143–153
204. Saunders RC, Rosene DL, Mishkin M (1986) The bed nucleus of the stria terminalis in the rhesus monkey: connections with the amygdala and the medial dorsal nucleus. Soc Neurosci Abstr 12:976
205. Scheibel ME, Scheibel AB (1966) The organization of the nucleus reticularis thalami: a Golgi study. Brain Res 1:43–62
206. Schell GR, Strick PL (1984) The origin of thalamic inputs to the arcuate, premotor and supplementary motor areas. J Neurosci 4:539–560
207. Schmahmann JD, Pandya DN (1990) Anatomical investigation of projections from thalamus to posterior parietal cortex in the rhesus monkey: a WGA-HRP and fluorescent tracer study. J Comp Neurol 295:299–326
208. Schneider GE (1969) Two visual systems. Science 163:795–802
209. Sewards TV, Sewards MA (2002) The medial pain system. Brain Res Bull 59:163–180
210. Sherman SM, Guillery RW (1996) Functional organization of thalamocortical relays. J Neurophysiol 76:1367–1395
211. Sherman SM, Guillery RW (1998) On the actions that one nerve cell can have on another: distinguishing "drivers" from "modulators". Proc Natl Acad Sci USA 95:7121–7126
212. Sherman SM, Guillery RW (2001) Exploring the thalamus. Academic Press, San Diego
213. Sherman SM, Guillery RW (2002) The role of the thalamus in the flow of information to the cortex. Phil Trans R Soc Lond B 357:1695–1708
214. Sherman SM, Koch C (1998) Thalamus. In: Shepherd GM (ed) The synaptic organization of the brain, 4th edn. Oxford University Press, Oxford, pp 289–328
215. Shigenaga Y, Nakatani Z, Nishimori T et al (1983) The cells of origin of cat trigeminothalamic projections: especially in the caudal medulla. Brain Res 277:201–222
216. Shiroyama T, Kayahara T, Yasui Y, Nomura J, Nakano K (1995) The vestibular nuclei of the rat project to the lateral part of the thalamic parafascicular nucleus (centromedian nucleus in primates). Brain Res 704:130–134
217. Sousa-Pinto A (1973) Cortical projections of the medial geniculate body in the cat. Ergeb Anat Entwicklungsgesch 48:1–40
218. Stanton GB (1980) Topographical organization of ascending cerebellar projections from the dentate and interposed nuclei in Macaca mulatta: an antegrade degeneration study. J Comp Neurol 190:699–731
219. Stanton GB, Cruce WLR, Goldberg NE, Robinson DL (1977) Some ipsilateral projections to areas PF and PG of inferior parietal lobule in monkeys. Neurosci Lett 6:243–250
220. Stanton GB, Bruce C, Goldberg ME (1982) Organization of subcortical projections from saccadic eye movement sites in the macaque frontal eye fields. Soc Neurosci Abstr 8:293
221. Steriade M (1981) Mechanisms underlying cortical activation: neuronal organization and properties of the midbrain reticular core and intralaminar thalamic nuclei. In: Pompeiano O, Ajmone-Marsan C (eds) Brain mechanisms of perceptual awareness. Raven Press, New York, pp 327–377
222. Steriade M, Jones EG, McCormick DA (1997) Thalamus, vol I: Organisation and function. Elsevier, Amsterdam
223. Strick PL (1973) Light microscopic analysis of the cortical projection of the thalamic ventrolateral nucleus in the cat. Brain Res 55:1–24
224. Swanson LW, Cowan WM (1975) A note on the connections and development of the nucleus accumbens. Brain Res 92:324–330
225. Tanaka D Jr (1976) Thalamic projections of the dorsomedial prefrontal cortex in the rhesus monkey (Macaca mulatta). Brain Res 110:21–38
226. Thach WT, Jones EG (1979) The cerebellar dentatothalamic connection: terminal field, lamellae, rods and somatotopy. Brain Res 169:168–172
227. Tömböl T (1969) Two types of short axon (Golgi 2nd) interneurons in the specific thalamic nuclei. Acta Morphol Hung 17:285–297
228. Trojanowski JQ, Jacobson S (1975) A combined horseradish peroxidase-autoradiographic investigation of reciprocal connections between superior temporal gyrus and pulvinar in squirrel monkey. Brain Res 85:347–353
229. Trojanowski JQ, Jacobson S (1976) Areal and laminar distribution of some pulvinar cortical efferents in rhesus monkey. J Comp Neurol 169:371–391
230. Trojanowski JQ, Jacobson S (1977) The morphology and laminar distribution of cortico-pulvinar neurons in the rhesus monkey. Exp Brain Res 28:51–62
231. Ungerleider LO, Desimone R, Oalkin TW, Mishkin M (1984) Subcortical projections of area MT in the macaque. J Comp Neurol 223:368–387
232. Updyke BV (1983) A re-evaluation of the functional organization and cytoarchitecture of the feline lateral posterior complex, with observations of adjoining cell groups. J Comp Neurol 219:143–181

233. Van der Loos H (1976) Barreloids in mouse somatosensory thalamus. Neurosci Lett 2:1–6
234. Van der Werf YD, Witter MP, Groenewegen HJ (2002) The intralaminar and midline nuclei of the thalamus. Anatomical and functional evidence for participation in processes of arousal and awareness. Brain Res Rev 39:107–140
235. Van Groen T, Wyss JM (1992) Projections from the laterodorsal nucleus of the thalamus to the limbic and visual cortices in the rat. J Comp Neurol 324:427–448
236. Van Noort J (1969) The structure and connections of the inferior colliculus. Thesis, University of Leiden
237. Veazey RB, Amaral DG, Cowan WM (1982) The morphology and connections of the posterior hypothalamus in the cynomolgus monkey (Macaca fascicularis). II. Efferent connections. J Comp Neurol 207:135–156
238. Velayos JL, Reinoso Suarez F (1982) Topographic organization of the brainstem afferents to the mediodorsal nucleus. J Comp Neurol 206:17–27
239. Vogt BA (1985) Cingulate cortex. In: Peters A, Jones EG (eds) Cerebral cortex, vol 4: association and auditory cortices. Plenum, New York, pp 63–88
240. Vogt BA, Sikes RW (2000) The medial pain system, cingulate cortex, and parallel processing of nociceptive information. Progr Brain Res 122: 223–235
241. Vogt C, Vogt O (1941) Thalamusstudien I–III: I. Zur Einführung. II. Homogenität und Grenzgestaltung der Grisea des Thalamus. III. Das Griseum centrale (Centrum medianum Luys). J Psychol Neurol 50:32–154
242. Vogt BA, Rosene DL, Pandya DN (1979) Thalamic and cortical afferents differentiate anterior from posterior cingulate cortex in the monkey. Science 204:205–207
243. Walker AE (1938) The primate thalamus. University of Chicago Press
244. Weber AJ, Kalil RE (1983) The percentage of interneurons in the dorsal lateral geniculate nucleus of the cat and observations on several variables that affect the sensitivity of horseradish peroxidase as a retrograde marker. J Comp Neurol 220:336–346
245. Weber JT, Yin TCT (1984) Subcortical projections of the inferior parietal cortex (area 7) in the stumptailed monkey. J Comp Neurol 222:206–230
246. Weiskrantz L, Warrington EK, Sanders MD, Marshall J (1974) Visual capacity in the hemianopic field following a restricted cortical ablation. Brain 97:709–728
247. Whitsel BL, Rustioni A, Dreyer DA et al (1978) Thalamic projections to S-I in macaque monkey. J Comp Neurol 178:385–510
248. Wiesel TN, Hubel DH, Lam D (1974) Autoradiographic demonstration of ocular-dominance columns in the monkey striate cortex by means of transneuronal transport. Brain Res 79:273–279
249. Willis WD (1995) Cold, pain and the brain. Nature 373:19–20
250. Willis WD (1997) Nociceptive functions of thalamic neurons. In: Steriade M, Jones EG, McCormick DA (eds) Thalamus, vol II. Elsevier, Amsterdam, pp 373–424
251. Willis WD, Westlund KN (1997) Neuroanatomy of the pain system and of the pathways that modulate pain. J Clin Neurophysiol 14:2–31
252. Willis WD Jr, Zhang X, Honda CN, Giesler GJ Jr (2002) A critical review of the role of the proposed VMpo nucleus in pain. J Pain 3:79–94
253. Woodward WR, Coull BM (1984) Localization and organization of geniculocortical and corticofugal fiber tracts within the subcortical white matter. Neuroscience 12:1089–1099
254. Woolf NJ, Harrison JB, Buchwald JS (1990) Cholinergic neurons of the feline pontomesencephalon. II. Ascending anatomical projections. Brain Res 520:55–72
255. Woolsey TA, Van der Loos H (1970) The structural organization of layer IV in the somatosensory region (SI) of mouse cerebral cortex. Brain Res 17:205–242
256. Yarita L, Iino M, Tanabe T, Kogure S, Takagi SF (1980) A transthalamic olfactory pathway to orbitofrontal cortex in the monkey. J Neurophysiol 43:69–85
257. Yeterian EH, Pandya DP (1985) Corticothalamic connections of the posterior parietal cortex in the rhesus monkey. J Comp Neurol 237:408–427
258. Zahm DS, Zaborszky L, Alheid GF, Heimer L (1987) The ventral striatopallidothalamic projection: II. The ventral pallidothalamic link. J Comp Neurol 255:592–605

9 Diencephalon: Ventral Thalamus or Subthalamus

Introductory Notes

The ventral thalamus occupies a relatively narrow zone of the diencephalon intercalated between the dorsal thalamus and hypothalamus. Its main nuclear groups are the reticular nucleus, zona incerta, pregeniculate nucleus and corpus subthalamicum or subthalamic nucleus (Figs. 6.36, 6.37). Noteworthy is that Kahle [20] and Richter [52, 53] assert that the subthalamic nucleus forms part of a separate – subthalamic – zone. However, most authors do not make this distinction and treat subthalamus and ventral thalamus as synonyms.

The Swiss neurobiologist and psychiatrist August Forel (1848–1931) was the first to analyse the microscopical structure of the subthalamic region (a term introduced by him) [16]. He regarded this region as representing the rostral continuation of the mesencephalic tegmentum. Forel noticed that the caudomedial part of the subthalamic region is occupied by a dense feltwork of fibres which he designated with the letter H and that more rostrally and laterally, the zona incerta splits this fibre mass into dorsal and ventral subfields which he named H1 and H2, respectively. The H stood for "Haubenfeld" (Haube is the German term for tegmentum). The three formations mentioned are now generally designated as *tegmental fields H, H1 and H2 of Forel* (Figs. 6.37, 6.38, 9.1). Numerous fibres arising from the contralateral cerebellar nuclei traverse field H on their way to the thalamus. Moreover, as discussed below, efferents from the globus pallidus contribute to all three tegmental fields.

It is important to note that the term ventral thalamus is purely descriptive and merely indicates the topographical position of this diencephalic subdivision in the adult stage. Ontogenetically, the ventral thalamus derives from the parencephalon anterius, a neuromere rostral to that from which the dorsal thalamus develops (see Chap. 2, Fig. 2.11). This means that all fibres projecting from the dorsal thalamus to the telencephalon and vice versa must pass through the interposed parencephalon anterius. Due to the enormous expansion of the dorsal thalamus, the adjacent part of this neuromere is transformed into a thin neuronal sheet, the thalamic reticular nucleus. As already discussed, this nucleus plays an important role in modulating the transfer of information through the thalamus to the cerebral cortex (Fig. 8.5 B).

Subthalamic Cell Masses

The *thalamic reticular nucleus* is a thin neuronal sheet covering the rostral, ventral and lateral aspects of the dorsal thalamus and separated from it by the external medullary lamina (Figs. 6.35, 6.38, 8.2). The continuity of this neuronal sheet is interrupted at intervals by bundles of thalamocortical and corticothalamic fibres passing through it; hence the name "reticular" nucleus [62]. The reticular nucleus is composed of large inhibitory, GABAergic elements, the axons of which pass exclusively to the thalamus. The large, disc-like dendritic arbors of these cells lie in the same plane as the thin sheet formed by the thalamic reticular nucleus itself. Dendrodendritic and axoaxonic synapses have been observed in the nucleus [48]; moreover there is evidence that the reticular cells may be electrically coupled [32]. Synaptic relations between the reticular nucleus on the one hand and thalamus and cortex on the other have already been discussed. In summary:

a. The thalamocortical and corticothalamic axons passing through the reticular nucleus emit short collaterals making synaptic contact with reticular cells.
b. A given thalamic nucleus and its corresponding cortical area have connections with the same region of the reticular nucleus
c. Reticular cells send axons to the dorsal thalamus terminating in the nuclei that provide their inputs (Fig. 8.5 B).

It follows that the reticular nucleus can be divided into distinct sectors, each specifically related to distinct parts of the thalamocortical pathway. It should be emphasized that the topographical organization just outlined holds only for "specific" thalamic nuclei [34]. Projections from the reticular nucleus to "non-specific" intralaminar and midline nuclei and the corresponding cortical inputs to the reticular nucleus are arranged far more loosely than those related to specific nuclei [26].

In addition to its extensive thalamic and cortical inputs, the reticular nucleus receives afferents from the globus pallidus and basal telencephalon and a large number of nuclei in the brain stem including the substantia nigra, superior colliculus, periaqueductal grey, cuneiform nucleus, pedunculopontine tegmental nucleus and parabrachial nuclei. Each of these nuclei are also known to have direct projections to nuclei of the dorsal thalamus. It has been established that efferents from each brain stem nucleus project to distinct zones in the reticular nucleus and that the efferents of some brain stem nuclei (e.g. from the substantia nigra) project to a particular thalamic nucleus as well as to the corresponding, functionally associated reticular sector [26].

In the thalamus the flow of information to the cortex is controlled by a complex modulatory apparatus (Fig. 8.5 B) in which the reticular thalamic nucleus is prominent. This modulatory apparatus controls the firing mode of thalamocortical projection neurons and therewith the type of information transferred from thalamus to cortex. As discussed in the previous chapter, there are two main patterns of firing in the thalamocortical system, tonic and burst modes and thalamic relay neurons switch between them in response to sustained changes in their membrane potential. During each mode a different type of information is transmitted through the thalamus. The tonic mode is thought to be related to detailed signal analysis, the burst mode to signal detection. Excitatory inputs from the cerebral cortex and brain stem depolarize the thalamic relay neurons and hence cause them to switch the response mode from burst to tonic, whereas inhibitory inputs from the thalamic reticular nucleus have a hyperpolarizing effect that leads to a switch from tonic to burst. The spatial and temporal patterns of these two modes determine the functioning of the thalamocortical system. For example, it is conceivable that a set of relay neurons in a thalamic nucleus in the tonic firing mode creates a focus of attention and detailed analysis in the corresponding area of the cortex, whereas the surrounding relay cells are brought or kept in the burst mode [58]. It seems likely that excitatory layer VI corticothalamic projections, excitatory thalamoreticular projections and inhibitory reticulothalamic projections (Fig. 8.5 B), *which are all topographically organized*, play a critical role in the creation of such focused attentional mechanisms. It is important to note that despite considerable overall reciprocity between the dorsal thalamus and thalamic reticular nucleus, there is no strict reciprocity between individual thalamic relay cells and reticular thalamic neurons. It has been shown that axons of a reticular thalamic neuron which receives input from a particular thalamic relay cell project to an area just adjacent to the dendritic arbor of that thalamic relay cell [47]. Physiological evidence suggests that similar disynaptic thalamoreticulothalamic pathways exist between adjacent, functionally related thalamic nuclei [10, 11]. These local disynaptic pathways are supposedly involved in mechanisms of lateral or surround inhibition [18].

Although afferents from the various brain stem nuclei generally project to particular territories of the thalamic reticular nucleus [26], these inputs are less focused than those from the dorsal thalamus and cerebral cortex [25]. Accordingly, they are expected to have rather widespread actions on the reticular nucleus and, through this, on thalamocortical transmission in general [18]. Because all neurons in the thalamic reticular nucleus are GABAergic, these elements may be expected to exert inhibitory influence on each other through their local chemical contacts. It has been suggested that the electrical coupling of reticular neurons enables the nucleus to generate general synchronized thalamocortical activity, in this way "closing" the thalamocortical gate for instance during sleep [17]. For a comprehensive review on the structure and (putative) function of the reticular thalamic nucleus, see [46].

The *pregeniculate nucleus* of primates is homologous with the ventral lateral geniculate nucleus of non-primate species. The anlage of this nucleus is ventral to that of the dorsal lateral geniculate nucleus (Figs. 2.19 B, 2.20 B); however, due to the latter's rotation, the topographical relations change considerably during human development, and in adults the pregeniculate nucleus lies above the dorsal lateral geniculate nucleus (Fig. 6.36).

The pregeniculate nucleus is a visual nucleus. It receives prominent input from the retina that is retinotopically organized and mainly contralateral. Additional afferents come from the visual cortical areas (17–19), pretectum and superior colliculus [8, 9, 63]. The pregeniculate nucleus is also a major termination site of noradrenergic, serotoninergic and cholinergic afferents from the brain stem [62].

The efferent projections of the pregeniculate nucleus pass to the pretectum, superior colliculi and pontine nuclei [8, 9]. Moreover, this nucleus projects with fibres containing neuropeptide Y to the hypothalamic suprachiasmatic nucleus [19]. The pregeniculate nucleus has been implicated in eye and head movements and in the maintenance of circadian rhythms relative to cycles of light and darkness [62].

The *zona incerta* is usually regarded as a rostral continuation of the brain stem reticular formation [24, 56, 57]. It is situated in the caudal part of the diencephalon, just ventral to the thalamus. Its rostral part occupies a position dorsal to the subthalamic nucleus (corpus subthalamicum), whereas its caudal part is ventrally bounded by the cerebral peduncle. The zona incerta is largely surrounded by pallidal efferents which pass via fields H2, H and H1 of Forel to the thalamus (Fig. 9.1). It is composed of rather loosely arranged cell groups that are continuous laterally with the thalamic reticular nucleus (Fig. 6.37). There are two morphologically distinct cell types in the primate zona incerta: large fusiform or polygonal principal cells and small round interneurons [34].

The zona incerta derives its input from the medial prefontal, cingulate, somatosensory and motor cortices [41, 55, 57], central amygdaloid nucleus [55], substantia innominata [7], ventromedial hypothalamic nucleus [28] and a variety of structures in the brain stem such as superior colliculus, periaqueductal grey matter, parvocellular part of the red nucleus, mesencephalic reticular formation, medial and lateral parabrachial nuclei, cerebellar nuclei, sensory trigeminal nuclear complex and dorsal column nuclei [27, 36, 39, 40, 55, 57, 64].

Efferents from the zona incerta terminate in the thalamus, preferentially the intralaminar and association nuclei, several hypothalamic areas, pretectum, superior colliculus, periaqueductal grey matter, parvocellular part of the red nucleus, mesencephalic and rhombencephalic reticular formation, pedunculopontine nucleus, nucleus raphes magnus, inferior olivary complex and spinal cord [27, 31, 39, 49, 50, 56, 61, 66].

Recent studies mainly in rats led to the distinction of four different sectors within the zona incerta – rostral, dorsal, ventral and caudal – each having a largely distinct cytoarchitecture, immunohistochemical profile and pattern of connections [21, 24, 27, 34, 39, 41]. The *rostral* sector contains numerous dopaminergic and somatostatinergic cells [24]. Dopaminergic cells were originally described as the A13 group by Dahlström and Fuxe [12]. They innervate a variety of structures including the central nucleus of the amygdala, nucleus of the diagonal band of Broca and hypothalamic paraventricular nucleus [6, 14, 65]. The *dorsal* sector receives inputs from the cingulate cortex, ventromedial hypothalamic nucleus, and many brain stem structures, while it has outputs to the intralaminar thalamic nuclei and back to the brain stem nuclei that provide its input [41, 49, 55]. The *ventral* sector receives afferents principally from somatosensory centres such as the somatosensory cortex, superior colliculus, trigeminal complex and dorsal column nuclei. It has outputs mainly to the thalamic association nuclei and superior colliculus [21, 27, 49, 55]. The ventral sector also contains a population of GABAergic cells, and these are reported to project to the somatosensory cortex [33, 34, 44]. A separate cell mass in the mesodiencephalic transitional area also known as the *peripeduncular nucleus* presumably represents the *caudal sector* of the zona incerta (Fig. 6.33). This nucleus has been reported to be bidirectionally connected with a variety of structures including the basal ganglia, ventromedial hypothalamic nucleus, superior and inferior colliculi, nuclei of the lateral lemniscus, periaqueductal grey and cuneiform nucleus. Projections to the lateral hypothalamic area and amygdaloid complex have also been described [1, 2, 22, 23, 37, 59].

The zona incerta in rats has also been shown to be connected with the cerebellar interposed nucleus [40] and parvocellular red nucleus [39]. The former projects heavily to the contralateral zona incerta, mainly to a medial zone involving all four cytoarchitectonically defined sectors of that structure. There is also a small, mainly ipsilateral projection from the zona incerta back to the interposed nucleus. The parvocellular red nucleus is reciprocally connected to the zona incerta. The area of termination of the rubral projection corresponds to that of the interposed nucleus.

The zona incerta has been implicated in a variety of functions including visual, somatosensory and nociceptive processing, arousal and attention, locomotion, onset of saccadic eye movements and feeding, drinking and sexual behaviour [27, 34, 39, 51].

The *subthalamic nucleus* or *corpus Luysi* is in the caudal part of the diencephalon, ventral to the zona incerta and dorsal to the posterior limb of the internal capsule and its transition into the cerebral peduncle (Figs. 5.7, 5.8, 5.22). It is composed of fairly large triangular and polygonal cells (Figs. 6.33, 6.37). The subthalamic nucleus is embedded in the circuitry of the basal ganglia, which will be discussed in Chap. 14. It receives input from the lateral segment of the globus pallidus [3, 4, 5, 13, 43, 45], cortex of the frontal lobe [29, 30, 35] and mesencephalic pedunculopontine tegmental nucleus [15, 38, 54]. It projects to the medial and lateral segments of the globus pallidus, pedunculopontine tegmental nucleus and reticular part of the substantia nigra [4, 5, 42, 45, 60].

Subthalamic Fibre Systems

It was already mentioned that Forel [16] distinguished three fibre fields in the subthalamic region: the medial field H (also known as the prerubral field), dorsolateral field H1 and ventrolateral field H2. Separated by the zona incerta, H1 and H2 merge rostromedially with field H. Coarse, heavily myelinated fibres originating from the medial segment of the globus pallidus (GPm) contribute to all three Forel's fields (Fig. 9.1). These *pallidofugal* fibres assemble in two different fibre systems, the *ansa lenticularis* and the *fasciculus lenticularis*. The former passes medially along the basal surface of GPm and curves around the internal capsule. The fibres of the latter emerge from the dorsomedial aspect of GPm, traverse the posterior limb of the internal capsule, and then pass ventromedially through field H2. The fibres of the ansa lenticularis and fasciculus lenticularis merge in field H, from where they pass dorso-

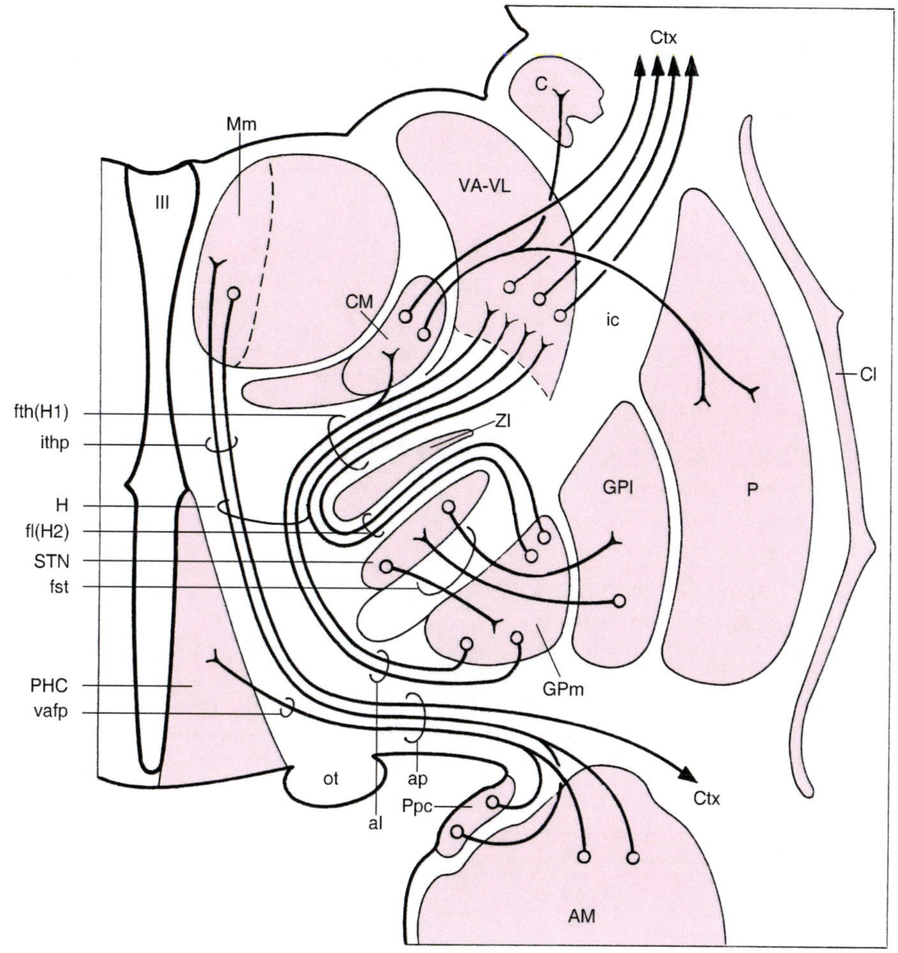

AM	Amygdaloid complex
al	Ansa lenticularis
ap	Ansa peduncularis
C	Caudate nucleus
Cl	Claustrum
CM	Centromedian thalamic nucleus
Ctx	Cerebral cortex
fl	Fasciculus lenticularis
fst	Fasciculus subthalamicus
fth	Fasciculus thalamicus
GPl	Globus pallidus, lateral segment
GPm	Globus pallidus, medial segment
H, H1, H2	Tegmental fields of Forel
ic	Internal capsule
ithp	Inferior thalamic peduncle
Mm	Magnocellular part of mediodorsal thalamic nucleus
ot	Optic tract
P	Putamen
PHC	Preopticohypothalamic continuum
Ppc	Prepiriform cortex
STN	Subthalamic nucleus
VA	Ventral anterior thalamic nucleus
VL	Ventral lateral thalamic nucleus
vafp	Ventral amygdalofugal projection
ZI	Zona incerta
III	Third ventricle

Fig. 9.1. Some fibre bundles interconnecting telencephalic and diencephalic structures, semidiagrammatically represented in a transverse plane

laterally through field H1, forming part of another conspicuous bundle, the *fasciculus thalamicus*. Other components of this bundle include fibres from the brachium conjunctivum and medial lemniscus.

As their name indicates, the fibres of the thalamic fasciculus pass to the dorsal thalamus, where they are distributed to various parts of the ventral nuclear group. The pallidofugal fibres project to the ventral anterior nucleus and particularly to the anterior part of the ventral lateral nucleus, giving off collateral branches to the centromedian nucleus. The brachium conjunctivum, which originates from the contralateral cerebellar nuclei, terminates in the posterior part of the ventral lateral nucleus, whereas the ventral posterolateral nucleus is the target of fibres of the medial lemniscus. The latter arise from the contralateral gracile and medial cuneate nuclei. The pallidosubthalamic and subthalamopallidal fibres, which traverse the internal capsule and interweave with it at right angles, are collectively designated as the *fasciculus subthalamicus* (Fig. 9.1).

References

1. Arnault P, Roger M (1987) The connections of the peripeduncular area studied by retrograde and anterograde transport in the rat. J Comp Neurol 250:163-176
2. Berk ML, Finkelstein JA (1981) Afferent projections to the preoptic area and hypothalamic regions in the rat brain. Neuroscience 6:1601–1624
3. Carpenter MB, Fraser RAR, Shriver JE (1968) The organization of pallidosubthalamic fibers in the monkey. Brain Res 11:522–559
4. Carpenter MB, Batton RR, Carleton SC, Keller JT (1981) Interconnections and organization of pallidal and subthalamic nucleus neurons in the monkey. J Comp Neurol 197:579–603
5. Carpenter MB, Carleton SC, Keller JT, Conte P (1981) Connections of the subthalamic nucleus in the monkey. Brain Res 224:1–29
6. Cheung S, Ballew JR, Moore KE, Lookingland KJ (1998) Contribution of dopamine neurons in the medial zona incerta to the innervation of the central nucleus of the amygdala, horizontal diagonal band of Broca and hypothalamic paraventricular nucleus. Brain Res 808:174–181
7. Ciriello J, Caverson MM (1984) Direct pathway from neurons in the ventrolateral medulla relaying cardiovascular afferent information to the supraoptic nucleus in the cat. Brain Res 292:221–228
8. Conley M, Friederich-Ecsy B (1993) Functional organization of the ventral lateral geniculate complex of the tree shrew (*Tupaia belangeri*). I. Nuclear subdivisions and retinal projections. J Comp Neurol 328:1–20
9. Conley M, Friederich-Ecsy B (1993) Functional organization of the ventral lateral geniculate complex of the tree shrew (*Tupaia belangeri*). II. Connections with the cortex, thalamus, and brainstem. J Comp Neurol 328:21–42
10. Crabtree JW, Isaac JT (2002) New intrathalamic pathways allowing modality-related and cross-modality switching in the dorsal thalamus. J Neurosci 22:8754–8761
11. Crabtree JW, Collingridge GL, Isaac JT (2002) A new intrathalamic pathway linking modality-related nuclei in the dorsal thalamus. Nat Neurosci 1:389–394
12. Dahlström A, Fuxe K (1964) Evidence for the existence of monoamine-containing neurons in the central nervous system. I. Demonstration of monoamines in the cell bodies of brain stem neurons. Acta Physiol Scand [Suppl] 62(232):1–55
13. DeVito JL, Anderson ME (1982) An autoradiographic study of efferent connections of the globus pallidus in macaca mulatta. Exp Brain Res 46:107–117
14. Eaton MJ, Wagner CK, Moore KE, Lookingland KJ (1994) Neurochemical identification of A_{13} dopaminergic neuronal projections from the medial zona incerta to the horizontal limb of the diagonal band of Broca and the central nucleus of the amygdala. Brain Res 659:201–207
15. Edley SM, Graybiel AM (1983) The afferent and efferent connections of the feline nucleus tegmenti pedunculopontinus, pars compacta. J Comp Neurol 217:187–215
16. Forel A (1877) Untersuchungen über die Haubenregion und ihre oberen Verknüpfungen im Gehirn des Menschen und einiger Säugethiere, mit Beiträgen zu den Methoden der Gehirnuntersuchung. Arch F Psychiat 7:393–495
17. Groenewegen HJ, Witter MP (2004) Thalamus. In: Paxinos G (ed) The rat nervous system. Academic Press, San Diego, pp 407–453
18. Guillery RW, Harting JK (2003) Structure and connections of the thalamic reticular nucleus: advancing views over half a century. J Comp Neurol 463:360–371
19. Harrington ME, Nance DM, Rusak B (1987) Double-labeling of neuropeptide Y-immunoreactive neurons which project from the geniculate to the suprachiasmatic nuclei. Brain Res 410:275–282
20. Kahle W (1956) Zur Entwicklung des menschlichen Zwischenhirns. Dtsch Z Nervenheilkd 175:259–318

21. Kim U, Gregory E, Halls WC (1992) Pathway from the zona incerta to the superior colliculus in the rat. J Comp Neurol 321:555–575
22. Kita H, Oomura Y (1982) An HRP study of the afferent connections to rat medial hypothalamic region. Brain Res Bull 8:53–62
23. Kita H, Oomura Y (1982) An HRP study of the afferent connections to rat lateral hypothalamic region. Brain Res Bull 8:63–71
24. Kolmac C, Mitrofanis J (1999) Distribution of various neurochemicals within the zona incerta: an immunohistochemical and histochemical study. Anat Embryol 199:265–280
25. Kolmac C, Mitrofanis J (2001) Induction of Fos-like immunoreactivity in the ventral thalamus after electrical or chemical stimulation of various subcortical centres of rats. Neurosci Lett 301:195–198
26. Kolmac CI, Mitrofanis J (1998) Patterns of brainstem projection to the thalamic reticular nucleus. J Comp Neurol 396:531–543
27. Kolmac CI, Power BD, Mitrofanis J (1998) Patterns of connections between zona incerta and brainstem in rats. J Comp Neurol 396:544–555
28. Krieger MS, Conrad LCA, Pfaff DW (1979) An autoradiographic study of the efferent connections of the ventromedial nucleus of the hypothalamus. J Comp Neurol 183:785–816
29. Künzle H (1978) An autoradiographic analysis of the efferent connections from premotor and adjacent prefrontal regions (areas 6 and 9) in Macaca fascicularis. Brain BehavEvol15:185–234
30. Künzle H, Akert K (1977) Efferent connections of cortical area 8 (frontal eye field) in Macaca fascicularis. A reinvestigation using the autoradiographic technique. J Comp Neurol 173:147–164
31. Kuypers HGJM, Maisky VA (1975) Retrograde axonal transport of horseradish peroxidase from spinal cord to brain stem cell groups in the cat. Neurosci Lett 1:9–14
32. Landisman CE, Long MA, Beierlein M et al (2002) Electrical synapses in the thalamic reticular nucleus. J Neurosci 22:1002–1009
33. Lin CS, Nicolelis MAL, Schneider JS, Chapin JK (1990) A major direct GABAergic pathway from zona incerta to neocortex. Science 248:1553–1556
34. Ma TP, Hu X-J, Anavi Y, Rafols JA (1992) Organization of the zona incerta in the macaque: a Nissl and Golgi study. J Comp Neurol 320:273–290
35. Matsumura M, Kojima J, Gardiner TW, Hikosaka O (1992) Visual and oculomotor functions of monkey subthalamic nucleus. J Neurophysiol 67:1615–1632
36. May PJ, Sun W, Halls WC (1997) Reciprocal connections between the zona incerta and the pretectum and superior colliculus of the cat. Neuroscience 77:1091–1114
37. Mehler WR (1980) Subcortical afferent connections of the amygdala in the monkey. J Comp Neurol 190:733–762
38. Mesulam M-M, Mash D, Hersh L, Bothwell M, Geula C (1992) Cholinergic innervation of the human striatum, globus pallidus, subthalamic nucleus, substantia nigra, and red nucleus. J Comp Neurol 323:252–268
39. Mitrofanis J (2002) Distinctive patterns of connectivity between the zona incerta and the red nucleus of rats. Anat Embryol 205:283–289
40. Mitrofanis J, deFonseka R (2001) Organisation of connections between the zona incerta and the interposed nucleus. Anat Embryol 204:153–159
41. Mitrofanis J, Mikuletic L (1999) Organisation of the cortical projection to the zona incerta of the thalamus. J Comp Neurol 412:173–185
42. Nauta HJW, Cole M (1978) Efferent projections of the subthalamic nucleus: an autoradiographic study in monkey and cat. J Comp Neurol 180:1–16
43. Nauta WJH, Mehler WR (1966) Projections of the lentiform nucleus in the monkey. Brain Res 1:3–42
44. Nicolelis MAL, Chapin JK, Lin RCS (1992) Somatotopic maps within the zona incerta relay parallel GABAergic somatosensory pathway to the neocortex, superior colliculus, and brainstem. Brain Res 577:134–141
45. Parent A, Hazrati L-N (1995) Functional anatomy of the basal ganglia. II. The place of subthalamic nucleus and external pallidum in basal ganglia circuitry. Brain Res Rev 20:128–154
46. Pinault D (2004) The thalamic reticular nucleus: structure, function and concept. Brain Res Rev 46:1–31
47. Pinault D, Deschnes M (1998) Anatomical evidence for a mechanism of lateral inhibition in the rat thalamus. Eur J Neurosci 10:3462–3469
48. Pinault D, Smith Y, Deschnes M (1997) Dendrodendritic and axoaxonic synapses in the thalamic reticular nucleus of the adult rat. J Neurosci 17:3215–3233
49. Power BD, Kolmac CI, Mitrofanis J (1999) Evidence for a large projection from the zona incerta to the dorsal thalamus. J Comp Neurol 404:554-565
50. Ricardo JA (1981) Efferent connections of the subthalamic region in the rat. II. The zona incerta. Brain Res 214:43–60
51. Ricardo JA (1983) Hypothalamic pathways involved in metabolic regulatory functions, as identified by tracktracing methods. Adv Metab Dis 10:1–30
52. Richter C (1966) Über die Entwicklung des Globus pallidus und des Corpus subthalamicum beim Menschen. In: Hassler R, Stephan H (eds) Evolution of the forebrain. Thieme, Stuttgart, pp 285–295
53. Richter E (1965) Die Entwicklung des Globus Pallidus und des Corpus Subthalamicum. Springer, Berlin
54. Rinvik E, Grofova I, Hammond C, Féger J, Deniau JM (1979) A study of the afferent connections of the subthalamic nucleus in the monkey and cat using HRP technique. In: Poirier LJ, Sourkes TL, Bédard PJ (eds) Advances in neurology. Raven Press, New York, pp 53–70
55. Roger M, Cadusseau J (1985) Afferents to the zona incerta in the rat: a combined retrograde and anterograde study. J Comp Neurol 241:480–492

56. Romanowski CA, Mitchell IJ, Crossman AR (1985) The organisation of the efferent projections of the zona incerta. J Anat 143:75–95
57. Shammah-Lagnado SJ, Negrao N, Ricardo JA (1985) Afferent connections of the zona incerta: a horseradish peroxidase study in the rat. Neuroscience 15:109–134
58. Sherman SM, Koch C (1998) Thalamus. In: Shepherd GM (ed) The synaptic organization of the brain. Fourth edition. Oxford University Press, Oxford, pp 289–328
59. Simerly RB, Swanson LW (1986) The organization of neural inputs to the medial preoptic nucleus of the rat. J Comp Neurol 246:312–342
60. Smith Y, Hazrati L-N, Parent A (1990) Efferent projections of the subthalamic nucleus in the squirrel monkey as studied by the PHA-L anterograde tracing method. J Comp Neurol 294:306–323
61. Steriade M, Parent A, Ropert N, Kitsikis A (1982) Zona incerta and lateral hypothalamic afferents to the midbrain reticular core of cat – an HRP and electrophysiological study. Brain Res 238:13–28
62. Steriade M, Jones EG, McCormick DA (1997) Thalamus, Vol I: Organisation and function. Elsevier, Amsterdam
63. Updyke BV (1977) Topographic organization of the projections from cortical areas 17, 18, and 19 onto the thalamus, pretectum and superior colliculus in the cat. J Comp Neurol 173:81–122
64. Veazey RB, Severin CM (1980) Efferent projections of the deep mesencephalic nucleus (pars lateralis) in the rat. J Comp Neurol 190:231–244
65. Wagner CK, Eaton MJ, Moore KE, Lookingland KJ (1995) Efferent projections of the medial zona incerta containing A_{13} dopaminergic neurons: a PHA-L anterograde tract tracing study in the rat. Brain Res 677:229–237
66. Watanabe K, Kawana E (1982) The cells of origin of the incertofugal projections to the tectum, thalamus, tegmentum and spinal cord in the rat: a study using the autoradiographic and horseradish peroxidase methods. Neuroscience 7:2389–2406

10 Diencephalon: Hypothalamus

Introductory Notes

The human hypothalamus comprises only 4 cm^3 of neural tissue, or about 0.3% of the normal adult brain volume [151, 205]. Nevertheless it is critically involved in coordination and integration of autonomic, endocrine, and behavioural responses necessary for the maintenance of circadian and circannual rhythms, wake/sleep cycles, thermoregulation, the balance of fluid, electrolytes, and energy, food ingestion, agonistic behaviour, and reproduction.

Topography and Subdivision

The hypothalamus encompasses the most ventral part of the diencephalon, of which it forms the floor and contributes to the lateral walls of the third ventricle. Its upper boundary is marked on the ventricular side by a shallow groove, the hypothalamic sulcus (Fig. 3.8). Caudally the hypothalamus passes gradually over into the periventricular and tegmental grey matter of the mesencephalon. However, it is customary to define the posterior margin of the hypothalamus as a ventral plane just caudal to the mamillary bodies. The latter are paired small eminences on the basal aspect of the brain. The rostral boundary of the hypothalamus coincides with a vertical plane passing from the foramen of Monro to the middle of the optic chiasm. The preoptic region, which flanks the rostral end of the third ventricle, extends from the lamina terminalis to the rostral boundary of the hypothalamus. Although this region is of telencephalic origin, it is structurally and functionally so closely tied in with the hypothalamus that they are treated here together.

During ontogenesis, the floor of the hypothalamus forms a hollow, finger-shaped process, the infundibulum or hypophysial stalk. Two structures, the median eminence and neurohypophysis, develop from this process. The former is a neurohemal contact zone which forms a functional interface between the hypothalamus and the anterior lobe of the pituitary (see below). It is situated in the anterior wall of the infundibulum. The neurohypophysis develops from the distalmost part of the infundibulum.

Table 10.1. The preopticohypothalamic continuum: principal nuclei (n) and areas (a)

Preoptic region	Anterior region	Tuberal region	Mamillary region
		Periventricular zone	
Periventricular preoptic n. (PePn)	Suprachiasmatic n. (SCH) Paraventricular n. (PVH) ↕	Infundibular n. (INF)	
		Medial zone	
Medial preoptic n. (MPN)	Paraventricular n. (PVH) Anterior hypothalamic n. (AHN)	Ventromedial n. (VMH) Dorsomedial n. (DMH)	posterior hypothalamic a. (PHA) dorsal premamillary n. (PMD) ventral premamillary n. (PMV) medial mamillary n. (MM) lateral mamillary n. (LM)
			Tuberomamillary n. (TM) ↕
		Lateral zone	
			Tuberomamillary n. (TM)
Lateral preoptic a. (LPA)	←	Lateral hypothalamic a. (LHA)	→
	Supraoptic n. (SO)		

Le Gros Clark [204, 205] divided the preoptico-hypothalamic continuum into four rostrocaudal levels or regions, preoptic, supraoptic or anterior, tuberal, and mamillary, while Crosby and Woodburne [72] distinguished three mediolaterally arranged zones, i.e. periventricular, medial, and lateral (Table 10.1) (Figs. 6.37–6.40, 10.1).

The periventricular zone consists of a few layers of small cells among which larger, neurosecretory elements are intermingled. The latter form a single concentration, the medial magnocellular part of the paraventricular nucleus. The cell layers alternate with thin sheets of fine, mostly unmyelinated fibres. In addition to the medial part of the paraventricular nucleus, the periventricular preoptic, compact suprachiasmatic, and infundibular nuclei in the basal hypothalamus represent derivatives of the periventricular zone.

The medial zone, which is relatively cellular, contains a number of variably discrete cell masses: the rather compact medial preoptic nucleus, more diffuse anterior nucleus, discrete posterior magnocellular part of the paraventricular nucleus, the dorsomedial and ventromedial nuclei (which are often poorly delimited from each other), discrete dorsal and ventral premamillary nuclei, medial and lateral mamillary nuclei, and diffuse posterior hypothalamic area.

The lateral zone is partly separated from the medial zone by the postcommissural fornix, a large bundle connecting the hippocampal formation with the mamillary body (Figs. 6.38, 10.1). The lateral zone contains the magnocellular neurosecretory supraoptic nucleus, but most of its territory is occupied by the diffuse lateral preoptic and lateral hypothalamic areas.

The tuberomamillary nucleus is a rather diffusely organized group of large cells in the caudal tuberal and rostral mamillary regions. It forms part of the medial hypothalamic zone but extends laterally well into the lateral zone. The tuberomamillary nucleus contains a large population of histaminergic neurons (see Fig. 10.5).

Major Conduction Channels

The hypothalamus forms part of the greater limbic system, a functional entity discussed in the final chapter of this work. Suffice it to mention here that this entity extends throughout the brain and that the hypothalamus is interposed between its rostral telencephalic and caudal mesencephalic and rhombencephalic parts. The septal region, hippocampus, and amygdaloid complex are major components of the telencephalic part of the greater limbic system, while its caudal part includes the mesencephalic central grey, the parabrachial, solitary, and raphe nuclei, and an array of cytoarchitectonically poorly differentiated cell masses known as the lateral paracore.

Several hypothalamic fibre systems including the hypothalamo-hypophyseal pathways, medial forebrain bundle, dorsal longitudinal fasciculus of Schütz, mamillothalamic bundle, fornix, fasciculus mamillaris princeps, and mamillary peduncle form part of the circuitry of the greater limbic system.

The *hypothalamo-hypophyseal pathways* are shown in Fig. 10.1. The magnocellular nuclei of the anterior hypothalamus, i.e. the supraoptic and (magnocellular) paraventricular nuclei, give rise to axons that descend through the infundibular stalk to the posterior lobe of the pituitary. These axons, which together form the supraoptico-paraventriculo-hypophyseal tract, transport colloid droplets containing the hormones oxytocin and vasopressin to the posterior pituitary lobe or neurohypophysis, where they are released into the blood. The cells of the infundibular nucleus are involved in controlling secretion of the anterior pituitary hormones by regulating hormones that stimulate or inhibit the liberation of hormones produced in the pituitary. Each of the latter has a corresponding regulating hormone. These regulating hormones pass from the infundibular nucleus along the axons of its constituent cells to the median eminence, where they are released from the axon terminals into capillaries of the hypophyseal portal system. This system forms a vascular link between the infundibulum and adenohypophysis. It is noteworthy that the median eminence, a conspicuous neurohemal organ in the anterior wall of the infundibular stalk, like the axons of infundibular nucleus neurons also receives neurosecretory fibres from several other centres, including the parvocellular part of the paraventricular nucleus and the medial septal nucleus.

The *fasciculus medialis telencephali* or *medial forebrain bundle* (MFB) may be considered the central longitudinal pathway of the limbic forebrain-midbrain continuum. It is an assemblage of loosely arranged, mostly thin fibres extending from the septal area to the tegmentum of the midbrain. It traverses the lateral preoptico-hypothalamic area, the scattered neurons of which are collectively designated as the bed nucleus of the medial forebrain bundle. This bundle is highly complex, comprising a variety of short and long ascending and descending links [257, 403, 405, 406]. In the transitional area of the diencephalon and mesencephalon, the MFB fibres are rearranged into a smaller medial and a larger lateral stream [153, 169, 253]. The medial stream roughly maintains the sagittal orientation of the bundle's hypothalamic trajectory. It passes through the medial parts of the mesencephalic and rhombencephalic tegmental areas, just next to the raphe nuclei. The medial stream contains descending fibres, by which several hypothalamic centres project to the raphe nuclei and adjacent part of the medial reticular formation [153, 253], and also fibres which ascend from the raphe nuclei to the lateral hypothalamus, from where they pass to a variety of diencephalic and telencephalic centres (cf. Chap. 22 on reticular and monoaminergic systems). The lateral stream of fibres from MFB to brain stem sweeps laterally and caudally over the dorsal border of the substantia nigra into a ventrolateral tegmental position immediately caudal to the substantia nigra; at this level it curves dorsomedially and then descends through the central tegmental area to the lateral tegmental field of the pons and medulla oblongata. This stream contains fibres descending from the central nucleus of the amygdala [162, 279, 390], the bed nucleus of the stria terminalis [160, 390], and several hypothalamic areas [153, 169, 214, 306]. These descending fibres terminate in a variety of brain stem centres, among which the substantia nigra (pars compacta), parabrachial nuclei, locus coeruleus, nucleus subcoeruleus, the noradrenergic A1, A2, and

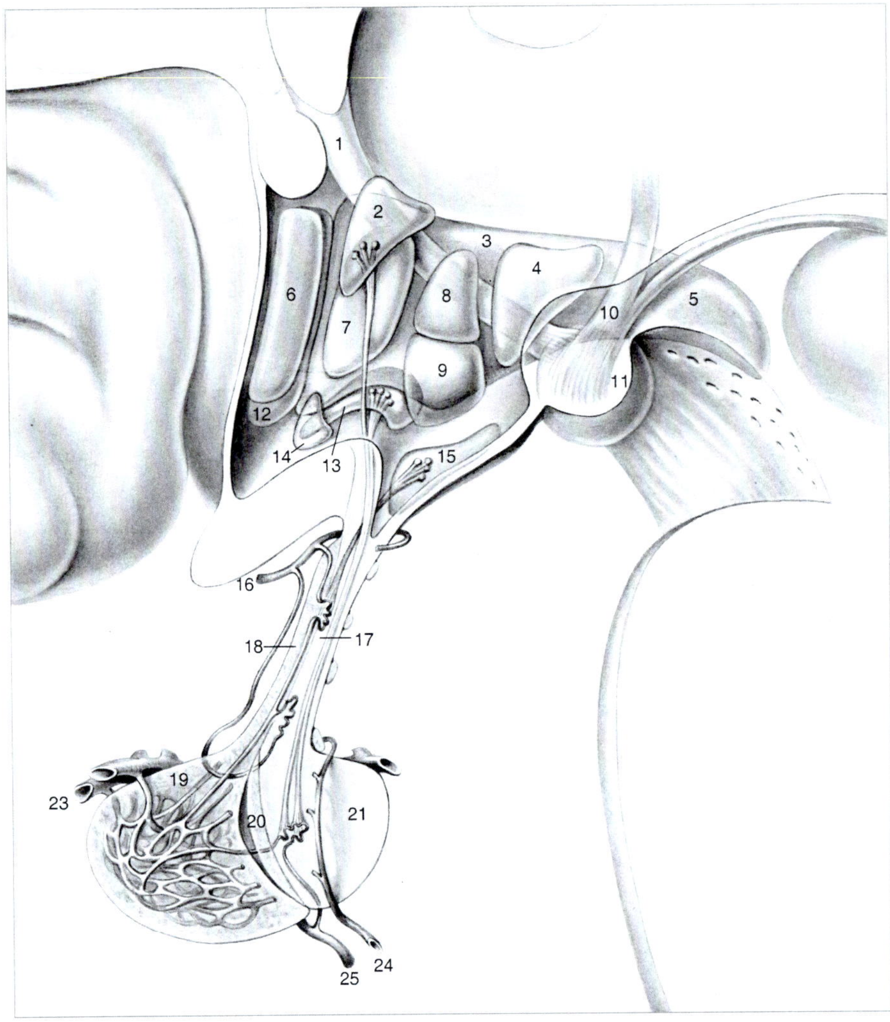

1 Column of fornix
2 Paraventricular nucleus
3 Lateral hypothalamic area
4 Posterior hypothalamic nucleus
5 Ventral tegmental area
6 Medial preoptic nucleus
7 Anterior hypothalamic nucleus
8 Dorsomedial nucleus
9 Ventromedial nucleus
10 Fasciculus mamillaris princeps
11 Mamillary body
12 Lateral preoptic nucleus
13 Supraoptic nucleus
14 Suprachiasmatic nucleus
15 Infundibular nucleus
16 Superior hypophysial artery (right)
17 Infundibulum
18 Pars infundibularis } Anterior lobe of hypophysis
19 Pars distalis } Anterior lobe of hypophysis
20 Pars intermedia } Anterior lobe of hypophysis
21 Posterior lobe of hypophysis
22 Posterior intercavernous sinus
23 Anterior intercavernous sinus
24 Inferior hypophysial artery (left)
25 Inferior hypophysial artery (right)

Fig. 10.1. Hypothalamic nuclei and the relationship between the hypothalamus and pituitary gland (×4)

A5 groups, superficial ventrolateral reticular area, and dorsal vagal complex may be especially mentioned. Most of these descending MFB projections are reciprocated by corresponding ascending projections [405, 406]. Indeed, the MFB is a major descending and ascending link between the forebrain and brain stem.

The *dorsal longitudinal fasciculus of Schütz* or *periventricular fibre system* [309, 349], much like the medial forebrain bundle, is a composite system consisting of thin ascending and descending fibres. This fascicle extends from the posterior part of the hypothalamus to the caudal medulla oblongata and occupies a periventricular position over its entire length. Rostrally its fibres become incorporated into the hypothalamic periventricular system. In the older literature [73, 254] it has been reported that most of the ascending and descending projections contained within the dorsal longitudinal fascicle are synaptically interrupted in either the central grey matter of the midbrain or in the dorsal tegmental nucleus of Gudden. More recently it has, however, been established that substantial fibre contingents passing directly from the forebrain to the autonomic centres of the lower medulla oblongata and back are also present [129, 292, 312].

The *fornix* is a large, compact, arch-shaped bundle that originates from the hippocampal formation and projects to the septal region and hypothalamus. Its final segment passes as the columna fornicis through the wall of the hypothalamus and terminates largely in the mamillary body (Figs. 6.38–6.40) [194, 225, 226, 327, 364].

The *fasciculus mamillaris princeps*, containing the efferents of the mamillary nuclei, constitutes a large compact bundle that passes dorsally for a short distance and then splits into two components, the large mamillothalamic and smaller mamillotegmental tracts (Fig. 10.1). The *mamillothalamic tract*, which passes to the anterior thalamic nucleus, forms part of the so-called Papez circuit. The mamillotegmental tract curves caudally into the tegmentum of the midbrain to terminate in the dorsal tegmental nucleus and nucleus reticularis tegmenti pontis of Bechterew [74, 291].

The *mamillary peduncle* receives fibres from the superior central and dorsal tegmental nuclei and the ventral tegmental area. It passes ventrally and ascends along the ventral surface of the midbrain to the mamillary body, where most of its fibres terminate. Some of them join the medial forebrain bundle and spread to the lateral preoptico-hypothalamic zone and septum [244, 255]. In rats, the projections to the mamillary body from the above three cell masses have been found to be topographically organized [9, 326].

A synopsis of the main afferent and efferent connections of the hypothalamus is presented in Tables 10.2 and 10.3 and the accompanying Figs. 10.2 and 10.3.

Functional Aspects 1: Introductory Notes

The hypothalamus is critically involved in the regulation of endocrine functions, control of autonomic reactions, and generation of basic behavioural patterns. The endocrine functions are regulated along the hypothalamo-hypophyseal pathways discussed above. The cells of origin of these pathways may be considered neuroendocrine motoneurons representing the final common pathway for central neural influences on hormone secretion from the anterior and posterior lobes of the pituitary gland [252, 361].

Several hypothalamic centres exert regulatory influences on preganglionic autonomic neurons in the brain stem and spinal cord (Fig. 10.3). Thus the paraventricular nucleus, referred to as "master controller of the autonomic system" [209], projects directly to the dorsal vagal motor nucleus, nucleus ambiguus, and thoracic (sympathetic) and sacral (parasympathetic) preganglionic cell columns [308, 360]. This projection shows a distinct functional differentiation, in that selective clusters of paraventricular neurons address specific sites of preganglionic neurons [343]. Indirect projections from the paraventricular nucleus to the sympathetic and parasympathetic efferent

Table 10.2. Hypothalamic afferents. For termination abbreviations, see Table 10.1

Origin	Termination	References
Neocortex (N), various areas including prefrontal, cingulate, prelimbic, infralimbic	AHN, VMH, PHA, MM, LHA	[2, 9, 109, 206, 263, 290, 327, 386]
Hippocampal subicular cortex (SUB)	MM, ML, LPA, LHA	[250, 270, 327, 386]
Lateral septal nucleus (LS)	MPN, AHN, VMH, PMV	[2, 270, 294]
Amygdaloid complex (AM)	MPN, AHN, VMH, PMV, LHA	[195, 251, 254]
Bed nucleus of stria terminalis (BST)	LPA, PVH, SO, DMH, LHA	[91, 92, 386]
Nucleus accumbens (ACB)	LHA	[118, 121, 387]
Retina (R)	SCH	[82, 242, 338]
Periaqueductal grey (PAG)	MPN, AHN, VMH, DMH, PHA, LHA	[2, 57, 218, 386]
Parabrachial nucleus (PB)	MPN, PVH, VMH, DMH, PHA, LHA	[2, 112, 200, 280, 311, 386]
Serotoninergic dorsal raphe (DR) and central superior (CS) nuclei	MPN, LPA, AHN, PVH, SCH, LHA	[2, 256]
Noradrenergic cell groups A1, A2, A5, A6 (locus coeruleus), A7	MPN, PVH, SO, PHA, LHA	[256]
Adrenergic cell groups C1, C2	PVH, SO	[319]
Nucleus of solitary tract (NTS)	PVH, DMH, LHA	[63, 64, 292]
Superficial ventrolateral reticular area: "ventrolateral medulla" (VLM)	PVH, SO	[65, 77, 223]
Spinal nucleus of trigeminal, caudal part (SPVc) Laminae X, V, and I of spinal cord	SCH, PVH, DMH, PHA, LHA	[49, 50, 51, 52, 67]

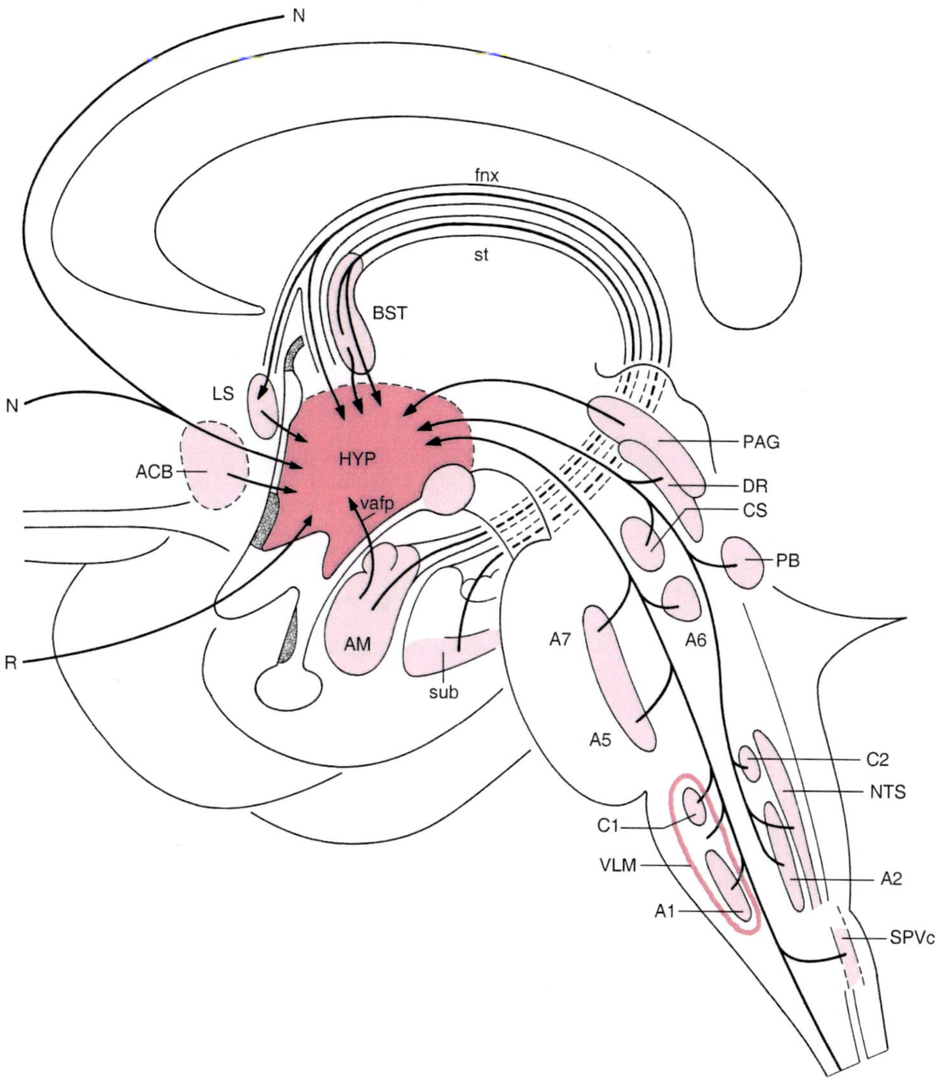

Fig. 10.2. Main afferent connections of the hypothalamus. *fnx*, fornix; *st*, stria terminalis; *vafp*, ventral amygdalofugal pathway. For other abbreviations, see Table 10.2

Table 10.3. Hypothalamic efferents. For origin abbreviations, see Table 10.1

Origin	Termination	References
LHA, PHA	Entire neocortex (N)	[185, 188, 277, 290, 296, 307]
PHA	Hippocampus: dentate gyrus (DG), Ammon's horn (CA)	[122, 300, 307, 421]
MPN, AHN, VMH, PMV	Lateral septal nucleus (LS)	[15, 55, 70, 138, 315, 335]
LHA	Medial septal nucleus (MS)	[286, 358]
VMH, PMV, LHA	Amygdaloid complex (AM)	[12, 55, 307, 315]
VMH, PMV	Bed nucleus of stria terminalis (BST)	[55, 196]
LPA, LHA	Nucleus of the diagonal band (NDB)	[358]
VMH, LHA	Substantia innominata (SI)	[315]
MPN, AHN, VMH, LHA	Thalamus: midline (MI) and mediodorsal (MD) nuclei	[296, 298, 315]
INF	Median eminence (ME)	[135, 373]
SO, PVN	Neurohypophysis (NH)	[166]
MPN, VMH	Zona incerta (ZI)	[196, 371]
MPN, DMH, VMH, PMD, LHA	Periaqueductal grey (PAG)	[30, 297, 298, 384, 416]
PVH, LHA	Parabrachial nuclei (PB)	[107, 298, 393]
PVH, PHA, LHA	Dorsal vagal complex: Nucleus of solitary tract (NTS), area postrema (AP), and dorsal motor nucleus vagus nerve (DMX)	[168, 312, 404]
MPN, DMH	Ventral tegmental area (VTA)	[15, 99, 297, 371]
MPN, PMV, PHA, LPA, LHA	Dorsal raphe (DR) and central superior (CS) nuclei	[55]
VMH	Mesencephalic reticular formation (MRF)	[153, 169, 314, 323]
MPN	Mesencephalic locomotor region (MLR)	[15, 371]
PVH, DMH	Locus coeruleus (A6)	[18, 297]
LHA	Rhombencephalic reticular formation (RRF)	[153, 169, 314, 323]
AHN, PVH, PHA, LHA	Superficial ventrolateral reticular area: ventrolateral modulla (VLM)	[84]
PVH, DMH, LHA	Laminae I and X of spinal cord, Intermediolateral column (IML)	[127, 166, 288, 414]

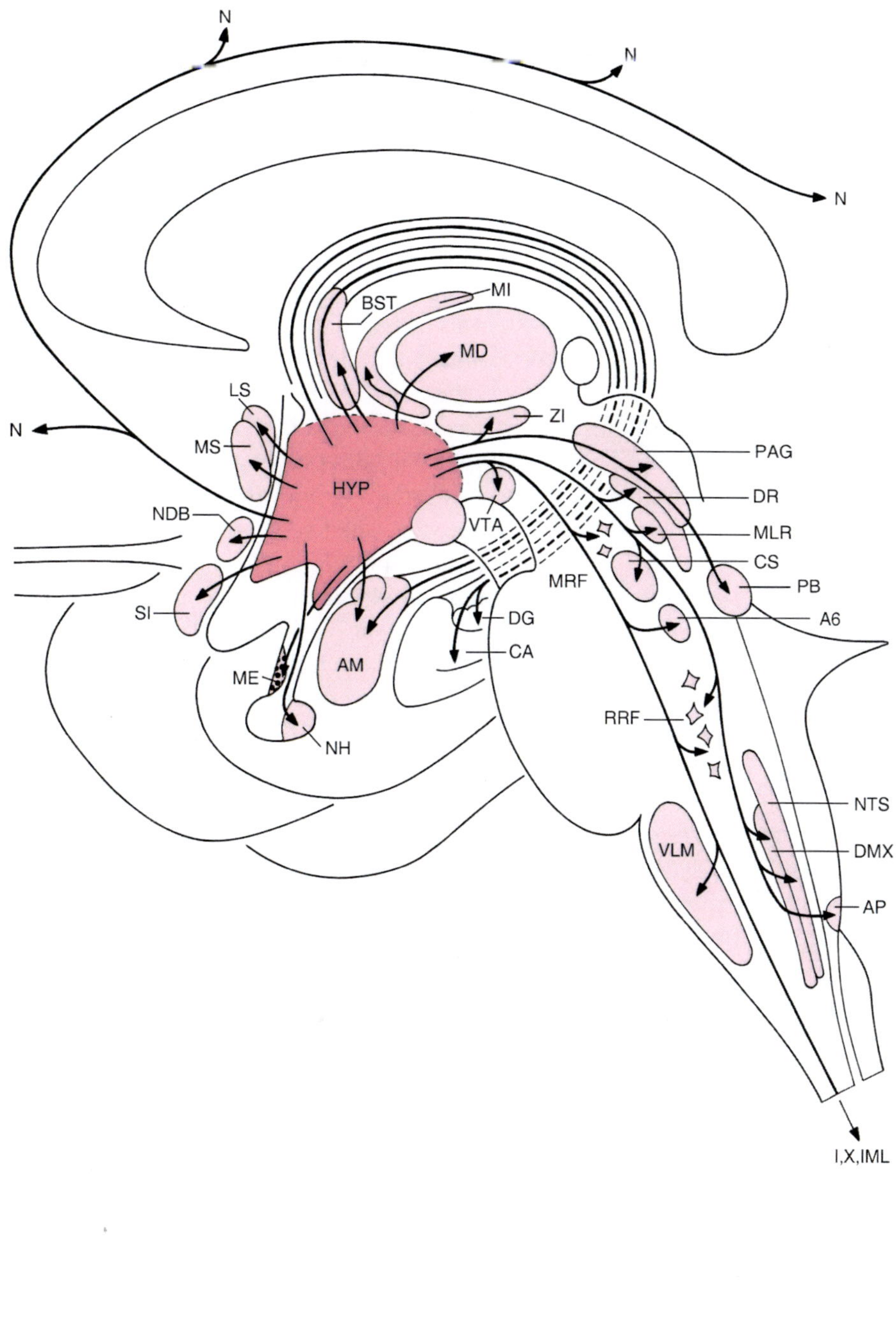

Fig. 10.3. Main efferent connections of the hypothalamus. For abbreviations, see Table 10.3.

centres are synaptically interrupted in autonomic premotor nuclei such as the A5 noradrenergic cell group and ventrolateral medulla. Apart from the paraventricular nucleus, the dorsomedial nucleus and lateral hypothalamic area also provide descending connections to the autonomic nuclei of the lower brain stem and spinal cord [209, 360].

The classic work of Hess and Brügger [145] and others have shown that by electrical stimulation, characteristic behavioural responses can be elicited from different hypothalamic loci. The somatomotor neurons involved in these responses are attained via multisynaptic pathways relayed in, among other structures, the zona incerta, tegmental pedunculopontine nucleus, and reticular formation (Fig. 10.3). In the hypothalamus, these somatomotor responses are integrated with endocrine and autonomic responses into motivated or goal-oriented behaviours that can be divided into two broad classes, one related primarily to survival of the individual and another to survival of the species as a whole [361]. Thermoregulative, ingestive, defensive, and aggressive behaviours belong in the first, sexual and parental behaviours in the second. In the next section, morphological substrates of five motivated behaviours, viz. thermoregulation, eating, drinking, sexual behaviour and defensive behaviour, are discussed. This exposé will be prefaced by an outline of the neural circuitry underlying three important general hypothalamic functions: circadian rhythms, waking and sleep, and the stress response.

For a comprehensive review of the functional histology and neuropathology of the human hypothalamus, we refer to the recent works of Swaab [350–354].

Functional Aspects 2: Organization of Specific Hypothalamic Functional Systems

Circadian Rhythms

The small suprachiasmatic nucleus (SCH), located in the rostroventral part of the hypothalamus (Fig. 10.1), is considered the major pacemaker of the brain and to coordinate a wide spectrum of endocrine, physiological, and behavioural circadian rhythms (Fig. 10.4) [151, 152, 350, 351]. This nucleus is indistinct with Nissl staining but can be clearly recognized with antisera against a number of neuropeptides. Neuron populations are present that are immunoreactive for vasopressin (VP), vasoactive intestinal polypeptide (VIP), neuropeptide-Y (NPY), and neurotensin (NT), all of which show a particular spatial organization. In addition, somatostatin, galanin, and preproenkephalin have been found in the SCH [351]. Individual SCH neurons in isolation have the capacity to generate circadian oscillations in electrical activity and secretion [68]. It has been recently demonstrated that SCH cells are electrically coupled, promoting synchronization of neural activity and also required for the maintenance of circadian rhythms in behaviour [211].

Our knowledge of the fibre connections of the SCH, mainly based on rodents [31, 35, 119, 237, 238, 241, 341, 363, 412, 413], has recently been extended by a series of thorough studies on the human brain [78, 79, 81–83]. Immunohistochemical technique and a newly developed postmortem tracing procedure were used in these studies [80].

The SCH itself generates biological rhythms with periods of approximately 24 h [37, 236]. These endogenous SCH rhythms are adjusted daily (entrained) to the environmental light/dark cycle through a direct projection from the retina to the SCH known as the *retinohypothalamic tract* [82, 176, 207, 240, 241, 338]. In rodents [338] as well as humans [82], this tract terminates in the ventral part of the SCH. It has been shown that the rat retinohypothalamic tract originates from a distinct subset of retinal gan-

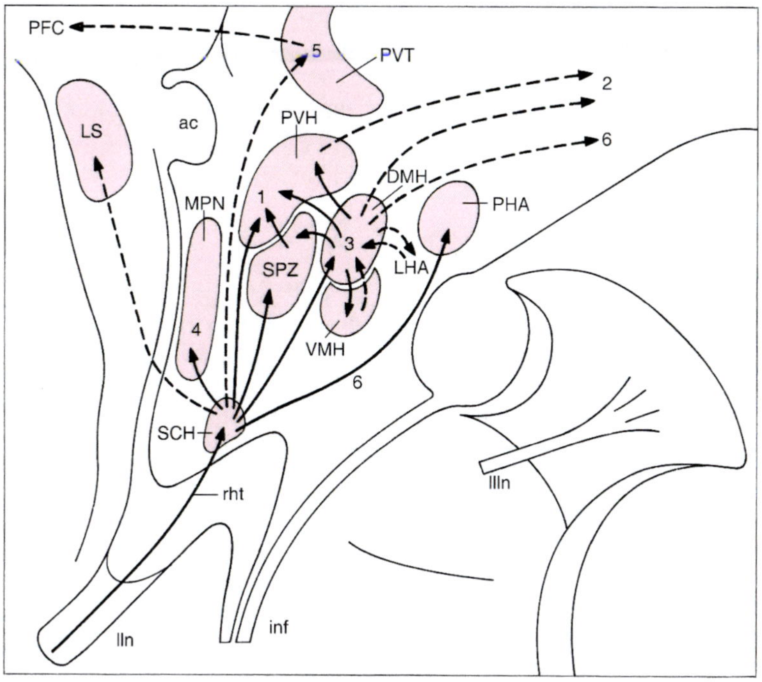

Fig. 10.4. Pathways involved in the mediation of photoperiodic information. *ac*, anterior commissure; *DMH*, dorsomedial hypothalamic nucleus; *inf*, infundibulum; *LHA*, lateral hypothalamic area; *LS*, lateral septal nucleus; *MPN*, medial preoptic nucleus; *PFC*, prefrontal cortex; *PHA*, posterior hypothalamic area; *PVH*, paraventricular hypothalamic nucleus; *PVT*, paraventricular thalamic nucleus; *rht*, retinohypothalamic tract; *SCH*, suprachiasmatic nucleus; *SPZ*, subparaventricular zone; *VMH*, ventromedial hypothalamic nucleus; *IIn*, *IIIn*, cranial nerves; *1*, *2*, etc. pathways discussed in the text

glion cells [242] which costore glutamate and pituitary adenylate-cyclase-activating polypeptide (PACAP) as neurotransmitters [130].

Experimental studies have shown that the SCH in rat, in addition to its retinal input, receives afferents from a number of other sources including the intergeniculate leaflet (the homologue of the primate pregeniculate nucleus), pretectum, midbrain raphe nuclei, ventromedial hypothalamic nucleus, posterior hypothalamic area, lateral septum, ventral subiculum, infralimbic cortex, and paraventricular thalamic nucleus [133, 174, 182, 233, 238, 239, 335, 370]. These SCH afferent connections are topographically organized [238]. The intergeniculate leaflet, midbrain raphe nuclei, and pretectum, which all receive afferents from the retina, like the retinohypothalamic tract project mainly to the ventral part of the SCH. These structures form relays in indirect routes along which photic input from the retina attains the SCH. Except for the paraventricular thalamic nucleus, the remaining cell masses project mainly to the dorsal part of the SCH. The paraventricular thalamic efferents finally appeared to terminate diffusely throughout the SCH. Evidence suggests that the neurons in several of the nuclei projecting to the SCH, including the lateral septal nucleus, ventromedial hypothalamic nucleus, posterior hypothalamic area, and paraventricular thalamic nucleus, use an excitatory amino acid as their principal neurotransmitter [237].

The SCH, i.e. the circadian pacemaker, exerts control over the sleep-wake cycle and is also involved in regulating a number of endocrine and behavioural processes. The circuitry mediating these influences is diagrammatically shown in Fig. 10.4. (Pathways indicated by continuous arrows have been established in humans [1, 81–83] and rodents [46, 62, 161, 180, 245, 380, 384, 409, 413]; pathways indicated by dashed arrows are known only from experimental studies in rodents.) The numbers used in the following synopsis correspond to those in Fig. 10.4.

1. The rostroventral part of the paraventricular nucleus (PVH) receives a direct projection from the SCH which is augmented by indirect projections via the subparaventricular zone (SPZ) and dorsomedial hypothalamus nucleus (DMH). The subparaventricular zone is cytoarchitectonically not clearly delimitable from the surrounding hypothalamic grey; however, it is characterized by a dense network of vasopressin (VP)- and vasoactive intestinal polypeptide (VIP)-containing fibres originating from the SCH [79]. The DMH also receives a major projection from the SCH consisting mainly of VIP fibres [62, 79, 81]. The rostroventral part of the PVN, on which the direct and indirect projections from the SCN converge, contains numerous neurons that synthesize and release VP and corticotropin-releasing hormone. Hence it seems likely that the SCH influences secretion of these hormones by these routes.

2. Many neurons in the PVH and DMH have long, descending axons that pass to autonomic centres in the brain stem and spinal cord [166, 213, 366, 380]. These fibres form part of the projections by which the SCH controls secretion of melatonin from the pineal gland [203, 377] and affects secretion of glucocorticoids from the adrenal cortex [47] and glucagon and insulin from the pancreas [45]. The projection to the pineal includes sympathetic preganglionic neurons in the rostral part of the spinal intermediolateral cell column and postganglionic neurons in the superior cervical ganglion. The projection to the adrenal appeared to be formed by VIP- or VP-containing SCH efferents, oxytocin-containing fibres originating from the autonomic division of the PVH, and neurons in the intermediolateral column of the spinal cord [47]. Paths by which the SCH influences pancreatic hormone secretion include a sympathetic projection corresponding to that of the adrenal and parasympathetic projection via PVH and the dorsal motor nucleus of the vagus [45]. Multisynaptic pathways connecting the SCH with the pineal, adrenal, and pancreas have been experimentally established with the aid of retrograde transneural virus tracers.

The SCH may utilize the large DMH-PVH pathway to exert control over other autonomic centres as well. There is evidence that in the human brain, VP-containing fibres descend from the dorsal part of the PVN to the locus coeruleus, parabrachial nuclei, nucleus of the solitary tract, and area postrema [78, 107, 392, 393].

3. Experimental studies in rodents have shown that the DMH is reciprocally connected with the ventromedial hypothalamic nucleus (VMH) and lateral hypothalamic area (LHA) [212, 313, 314, 378, 379, 384]. It is known that this complex of interconnected centres is involved in the regulation of food intake. It is possible that the SCH influences this function by its projection to DMH.

4. The medial preoptic nucleus (MPN) has been implicated in, among other functions, reproductive behaviour and regulation of body temperature. A direct projection from the SCH to MPN indicates that these functions may also be influenced by the SCH.

5. A recent viral transneural tracing study [372] has shown that in the rat, via a relay in the paraventricular thalamic nucleus (PVT), the SCH projects to the medial prefrontal cortex. It has been suggested [372] that this projection modulates higher-level brain functions such as attention, mood, and working memory.

6. It is well-established that the circadian oscillator exerts strong regulatory influence on sleeping and waking [69, 190]. The morphological basis for this influence may be two pathways connecting the SCH with structures forming part of the neural circuitry underlying the sleep-wake cycle. (The latter circuitry is complex and will be discussed below.) The first of these pathways consists of fibres passing from the SCH to two groups of posterior hypothalamic neurons, one producing hypocretins and the other producing melanin-concentrating hormone. This projection has been demonstrated in both rats and humans [2]. The second, indirect pathway connects the SCH with the noradrenergic locus coeruleus. It is synaptically interrupted in the DMH [18]. Worthy of note is that the DMH also sends a mainly GABAergic projection to the sleep-promoting ventrolateral preoptic nucleus [62]. Physiological evidence has recently shown that the central circadian pacemaker regulating the sleep-wake cycle also gets feedback information about the specific sleep state [88]. However, the anatomical pathways underlying this feedback loop remain to be established.

Wakefulness and Sleep

The circuitry for regulation of sleep, wakefulness, and arousal includes the following key structures (Fig. 10.5) [120, 149, 266, 316, 348]:

a. Two cell masses, the *laterodorsal tegmental* (LDT) and *pedunculopontine tegmental* (PPT) nuclei, both situated in the tegmentum of the caudal midbrain and rostral pons. These nuclei contain numerous neurons that use acetylcholine as a neurotransmitter. The cholinergic neurons in the PPT are collectively designated as CH5, those in the LDT as CH6 [230, 231]. The activity of LDT-PPT neurons varies with different behavioural states. They fire rapidly during wakefulness, but their activity varies markedly during sleep. It is important to note that periodically during the night, humans enter a state of "active" sleep with rapid eye movements (hence named REM sleep) and a loss of muscle tone except for the muscles involved in respiration. In contrast, normal or "inactive" sleep is generally designated as non-REM (NREM) sleep. During this phase the LDT-PPT neurons are inactive, but during REM sleep they are even more active than in wakefulness [316].

b. Noradrenergic neurons clustered in the *locus coeruleus* (LC), in the rostral pontine tegmentum (Fig. 6.28). These neurons have been shown to be most active during waking, less active during NREM sleep, and silent during REM sleep [108].

c. Mesencephalic serotoninergic neurons in the *dorsal raphe* (DR) and *central superior* (CS) nuclei. The activity pattern of these neurons closely resembles that of the LC cells [316].

d. A group of histaminergic cells clustered in the *tuberomamillary nucleus* (TMN) in the posterior hypothalamus [5, 120, 267, 411]. The TMN neurons contain several other neurotransmitters and modulators in addition to histamine, including GABA, galanin, and substance P [4, 192, 208]. Direct recordings of TMN neurons have shown their activity is high during waking and attention, low during NREM sleep, and absent during REM sleep [120, 191, 316]. It is well known that in humans, lesions of the posterior hypothalamus due to viral infection, local infarction, or tumor may lead to prolonged

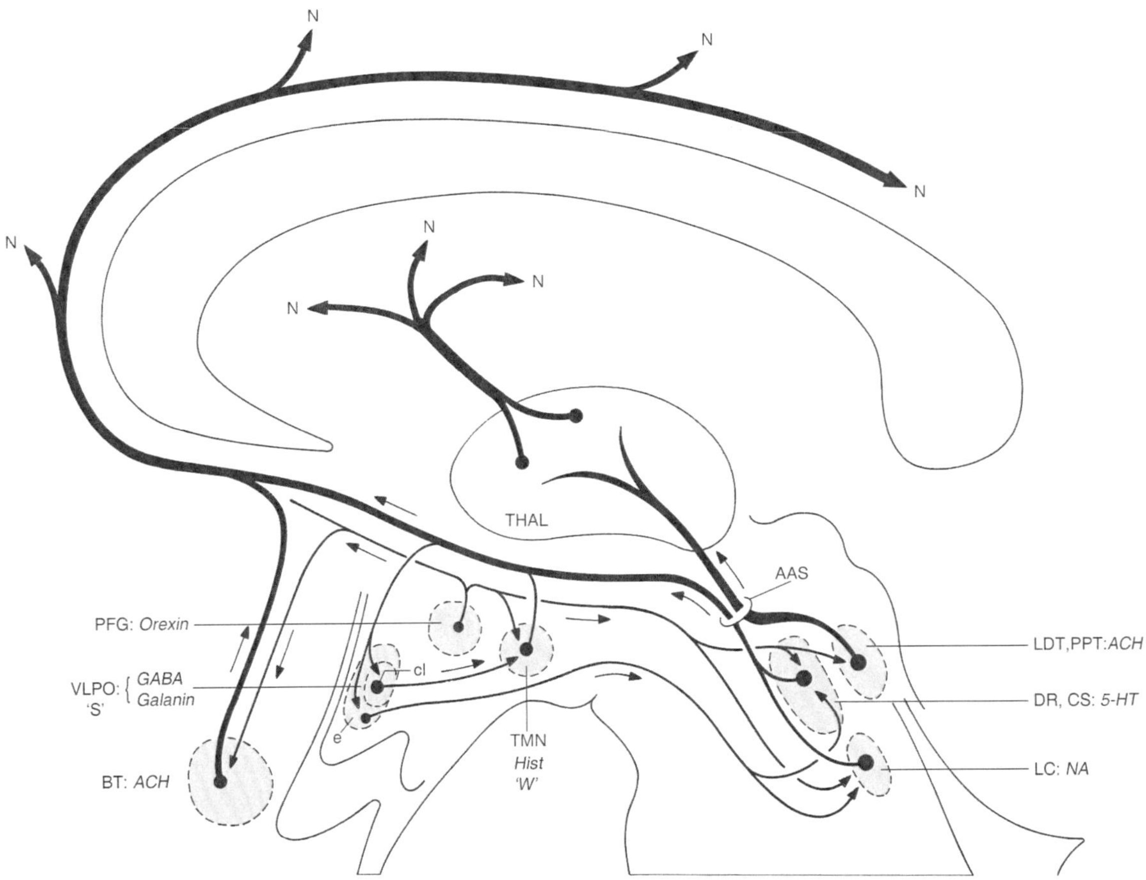

Fig. 10.5. Circuitry subserving the regulation of sleep, arousal, and wakefulness. *AAS*, ascending arousal system; *BT*, basal telencephalon; *cl*, dense cluster of neurons in VLPO; *CS*, central superior nucleus; *DR*, dorsal raphe nucleus; *e*, diffuse extension of VLPO; *LC*, locus coeruleus; *LDT*, laterodorsal tegmental nucleus; *N*, neocortex; *PFG*, perifornical group; *PPT*, pedunculopontine tegmental nucleus; *S*, sleep-promoting centre; *THAL*, thalamus; *TMN*, tuberomamillary nucleus; *VLPO*, ventrolateral preoptic area; *W*, wake-promoting region

states of hypersomnia [134]. Thus clinical as well as experimental evidence indicates that the posterior hypothalamus contains a region (W) promoting wakefulness.

e. A group of diffusely arranged neurons in the intermediate region of the hypothalamus characterized by presence of the neuropeptide *orexin* (also known as *hypocretin*) [87, 304, 348]. Because these cells more or less surround the fornix, they are collectively designated as the *perifornical group* (PFG). There is evidence that degeneration or agenesis of the orexinergic hypothalamic neurons in humans [258, 272, 381] and mice [5] lead to narcolepsy. In this disorder individuals exhibit daytime sleepiness and lapse unexpectedly into bouts of REM sleep. Among the perifornical neurons in the rat, 53% appeared to increase their firing rates during both wakefulness and REM sleep but decreased activities during NREM sleep. A further 38% of the neurons in this area appeared to be activated only during wakefulness [7].

f. Sherin et al. [324] recently described a sleep-promoting region in the *ventrolateral preoptic area* (VLPO) of the rat. During NREM sleep, this area appeared to activate the immediate-early gene c-fos, indicating an increase in metabolic activity in this behavioural state. The neurons in this area produce GABA and the inhibitory peptide galanin [325]. A corresponding cell group containing galanin has been identified in monkeys and humans [316]. The VLPO comprises a dense cluster of neurons (VLPO cluster or cl) surrounded by a more diffuse extension from the nucleus (extended VLPO or VLPO e) [324, 325]. Electrophysiological recordings have identified sleep-active neurons in the VLPO region of rat [374]. A sleep-promoting centre (S) in the human preoptic and adjacent basal telencephalic area was postulated over 70 years ago by von Economo [407, 408] based on clinical and pathological observations in certain patients suffering from virus encephalitis (see, however, [6]).

g. Groups of cholinergic neurons are situated in the *medial septal nucleus* (Ch1), *nucleus of the diagonal band* (Ch2, 3), and *substantia innominata-nucleus basalis complex* (Ch4) (Figs. 6.38–6.40) [137, 229–231, 269, 310]. These neurons are collectively designated as the cholinergic basal telencephalon (BT). They have been shown particularly sensitive to novel and motivational stimuli [417].

Fibres ascending from the mesopontine cholinergic (LDT, PPT), noradrenergic (LC), and serotoninergic (DR, CS) cell groups and from other parts of the reticular formation together form an *ascending arousal system* (AAS). This system divides into two major branches at the junction of the midbrain and diencephalon. The dorsal branch, which includes numerous cholinergic fibres from the LDT and PPT, passes to the thalamus itself and the reticular thalamic nucleus [95, 126, 330]. It has already been pointed out that these cholinergic fibres contribute to the thalamic modulatory apparatus (Fig. 8.5B). Their influence is considered crucial in activating thalamocortical transmission [342]. The ventral branch of the ascending arousal system, which consists mainly of serotoninergic and noradrenergic fibres, enters the hypothalamus. Traveling rostrally through the medial forebrain bundle, it attains the mediobasal telencephalon. During their course through the hypothalamus, noradrenergic and serotoninergic fibres of the ventral branch of the ascending arousal system are joined by histaminergic fibres from the TMN, orexinergic fibres from the PFG, and further rostrally by numerous axons from the cholinergic BT. All of these components innervate widespread regions of the forebrain, including the neocortex (N). The neocortical innervation by these components is commonly designated as "diffuse". However, it is known that they clearly have preferred regions and laminae of termination. Hence it may be expected that each component influences the processing of information in the cortex in its own specific way [108].

A detailed discussion of the known and proposed mechanisms underlying regulation of the behavioural states wakefulness, NREM sleep, and REM sleep is beyond the scope of the present work (for reviews, see [149, 266]). Here we confine ourselves to a few salient and well-established aspects of this regulation.

a. The dorsal ascending arousal system and the noradrenergic, serotoninergic, histaminergic, and BT-cholinergic components of the ventral arousal system conspire to promote the activated brain state of waking, whereas the VLPO plays a crucial role in the initiation and maintenance of NREM sleep.

b. The mesopontine cholinergic system, consisting of the LDT and the PPT, acts alone to promote the activated state of REM sleep. There is evidence that normally the noradrenergic LC and serotoninergic DR and CS are tonically active and inhibit the mesopontine cholinergic system. However, at onset of REM sleep, noradrenergic and serotoninergic neurons are inhibited by a local GABAergic element, which leads to disinhibition of the cholinergic neurons. Thus released from their monoamine-mediated suppression, these neurons fire rapidly during REM sleep [215, 259, 260, 303]. The source of activation of the GABAergic neurons involved is not known.

c. Reciprocal relationships exist between the sleep-promoting VLPO and the wakefulness-promoting posterior hypothalamic (TMN) and brainstem regions (LC, DR, CS). The VLPO is innervated by noradrenergic terminals from the LC, serotoninergic axons from the DR and CS, and histaminergic inputs from the TMN [61]. The wake-promoting neuromodulators noradrenaline and serotonin inhibit VLPO neurons [113]. These neurons do not respond to histamine, but the histaminergic TMN neurons also contain GABA and galanin, which might inhibit the VLPO [5].

The VLPO sends descending projections to the various components of the ascending arousal system [98, 114, 316, 325]. Neurons of the tightly clustered subgroup of the VLPO (cl) intensely innervate the somata of the TMN, whereas those of the diffuse, extended VLPO (e) less intensely innervate the DRN, CS, and LC [325, 339, 422]. Because the VLPO neurons synthesize GABA and galanin, their descending projections are likely to be inhibitory. Thus a reciprocal inhibitory interaction likely exists between the sleep-promoting VLPO and the monoaminergic wake-promoting centres in the caudal hypothalamus and brain stem. Saper et al. [316] propose that this arrangement forms the dynamic basis for a pattern of bistability in sleep-wake regulation: "When VLPO neurons fire rapidly during sleep, they would inhibit the monoaminergic cell groups, thus disinhibiting and reinforcing their own firing. Similarly, when monoamine neurons fire at a high rate during wakefulness, they would inhibit the VLPO, thereby disinhibiting their own firing" ([316], p. 728).

d. It has already been mentioned that degeneration of the orexinergic perifornical neurons leads to narcolepsy, a disturbance exhibiting two principal symptoms, excessive daytime sleepiness and sudden transitions from the waking state to REM sleep. Immunohistochemical studies [165, 243, 271] have shown that the orexin-containing neurons have widespread axonal ramifications which, among many other targets, innervate all components of the ascending arousal system, i.e. the cholinergic BT, histaminergic TMN, serotoninergic DR and CS, and, particularly densely, the noradrenergic LC. Because orexin has been shown to exert an excitatory action on these centres [29, 40, 43, 96], it seems plausible that this neuropeptide helps maintain wakefulness by increasing activity of the ascending arousal system [316], thus preventing untimely transitions from waking to sleep, as in narcolepsy. The second main symptom of narcolepsy, sudden attacks of REM sleep, is more difficult to explain. However, it seems likely that the LC, the noradrenergic output of which favours cortical arousal of waking but opposes REM-associated activation, is prominently involved. This cell group, as mentioned, is densely innervated by orexinergic fibres, and sends inhibitory projections to the REM-generating neurons in the LDT-PPT complex [165]. A complicating aspect is, however, that this complex also receives direct orexinergic projections. For further details on the interaction of orexinergic neurons with other structures involved in regulating wakefulness and sleep, see [186, 266, 316, 348]. Remarkably, the orexinergic neurons not only play a role in the regulation of arousal states but also influence feeding and energy metabolism [304, 318, 348].

Stress Response

The so-called *hypothalamo-pituitary-adrenal* (HPA) *axis* plays a crucial role in the response to stress (Fig. 10.6). Its structural elements are:

a. Parvocellular neurons in the hypothalamic paraventricular nucleus (PVH), which produce corticotropin-releasing hormone (CRH). In rats, CRH-producing cells are located in a distinct subnucleus of the PVH, but in humans they are scattered throughout the nucleus [283].
b. Corticotrope cells in the anterior pituitary gland (APG) that produce adrenocorticotrope hormone (ACTH).
c. Elements in the adrenal cortex (AC) producing the so-called stress hormones, i.e. the glucocorticoids (GCS) *corticosterone* and *cortisol.*

These steroid hormones have a wide range of effects, from increasing blood pressure and glucose levels to suppressing inflammatory and immune responses. Most of the CRH-producing neurons in the PVH project to the neurohemal contact zone in the median eminence (ME). Attaining the anterior pituitary via the hypothalamo-hypophyseal portal circulation, CRH stimulates the secretion of ACTH from the corticotropes and ACTH in turn stimulates the release of adrenal glucocorticoids. Studies mainly in rodents [360] have shown that CRH-producing cells in the PVH receive the following inputs:

a. Afferent fibres of the glossopharyngeal and vagus nerves terminating in the nucleus of the solitary tract (NST). The viscerosensory information carried by these fibres is mediated to the CRH cells by a direct solitario-hypothalamic projection.
b. Afferents from the parabrachial complex (PB), the noradrenergic locus coeruleus (LC) and A1 and A2 groups, and the adrenergic C1 and C2 groups. Since all of these cell groups are in receipt of afferents from the solitary nucleus, their efferent projections may also participate in transmission of viscerosensory information to the CRH cells.
c. Projections from the periaqueductal grey (PAG) which may relay somatosensory, particularly nociceptive information.
d. Fibres from the subfornical organ (SFO) which transmit information from blood-borne chemical signals.
e. Efferents from most other hypothalamic regions, converging on the CRH cells. These fibres form the morphological substrate of the interrelationship of the various hypothalamic functions. Thus it is known that activity of the HPA axis is strongly influenced by the suprachiasmatic "Zeitgeber" and that the various motivated behaviours are often accompanied by stress.
f. Numerous local circuit neurons surrounding the PVH. Glutamatergic excitatory as well as GABAergic inhibitory elements are present, and both cell types have been shown to synapse with CRH cells [21, 38, 75, 144, 234, 299, 376].
g. Three typical limbic structures, viz. the cingulate cortex (CING), hippocampal formation (HF) and amygdaloid complex (AM) projecting to the PVH. By way of these structures, highly processed information concerning the external and internal environment is relayed to the PVH. They are, for example, involved in the decision of which stimuli should be interpreted as stressful based on experience. The projection from the amygdala is synaptically interrupted in the bed nucleus of the stria terminalis (BST), whereas that from the hippocampus is relayed in the septum (S) and BST [76]. The structures and connections just mentioned are collectively designated as the "forebrain arms" of the stress system [222]. It is important to note that no component of these arms (with the possible exception of the BST) is in direct monosynaptic contact with the CRH cells. They all exert influence on these neurosecretory elements via the local pool of interneurons mentioned above [21, 144]. The hippocampus exerts a tonic and predominantly inhibitory effect on CRH neurons [140–142], whereas a stimulatory influence on these elements is ascribed to the amygdala [10].

The stress response is not confined to activity along the HPA axis but also involves auto-

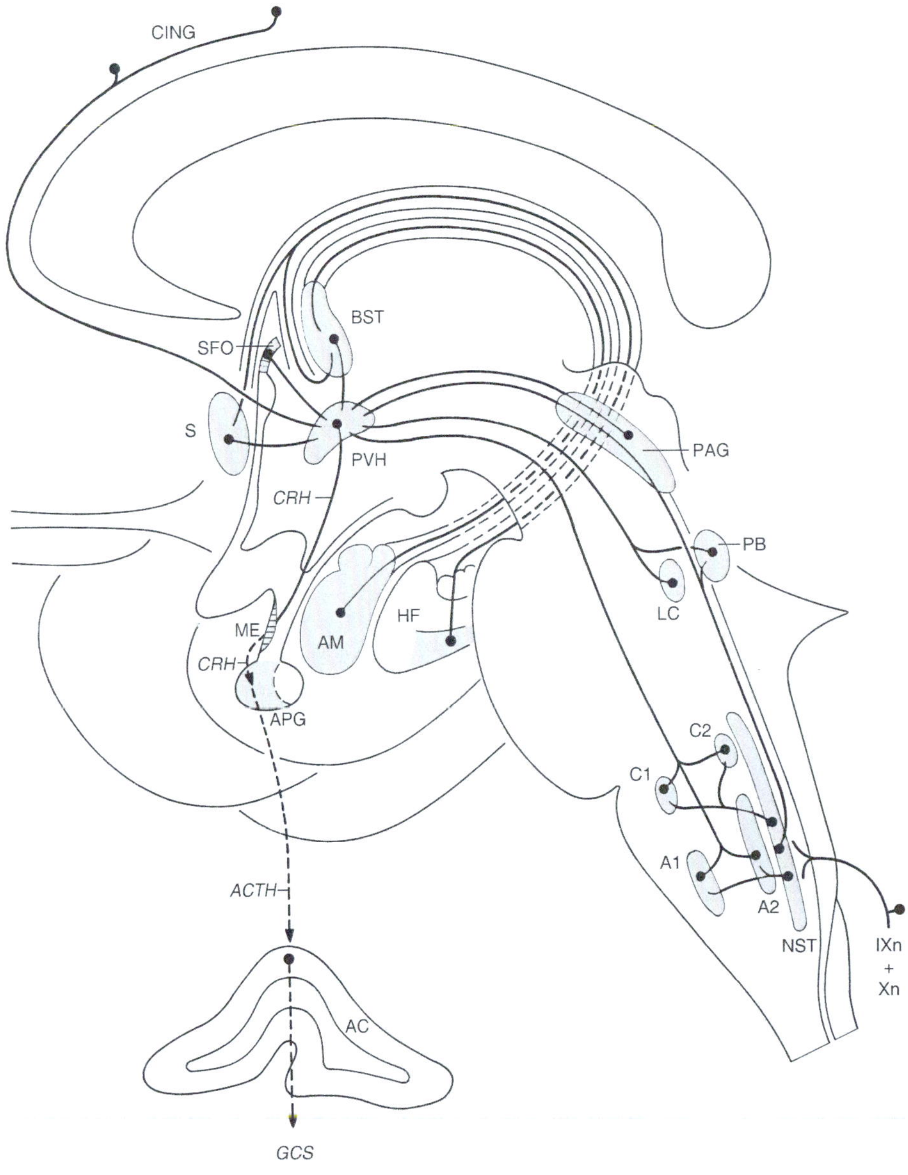

Fig. 10.6. Circuitry subserving the stress response. *A1*, *A2*, noradrenergic cell groups; *AC*, adrenal cortex; *AM*, amygdala; *APG*, anterior pituitary gland; *BST*, bed nucleus of stria terminalis; *CING*, cingulate cortex; *CRH*, corticotropin-releasing hormone; *GCS*, glucocorticoids; *HF*, hippocampal formation; *LC*, locus coeruleus; *ME*, median eminence; *NST*, nucleus of the solitary tract; *PAG*, periaqueductal gray; *PB*, parabrachial nuclei; *PVH*, paraventricular hypothalamic nucleus; *S*, septal nuclei; *SFO*, subfornical organ

nomic and behavioural responses. The CRH is crucially involved in coordinating these responses [265]. Parvocellular neurons synthesizing CRH in the PVH send descending projections to the lower brain stem and spinal cord and may well mediate the autonomic responses.

The brain is an important target of the hormones produced at the various levels of the HPA axis. Adrenal corticosteroids exert a selective negative feedback effect on the synthesis of CRH in the PVH and on the hippocampus [101]. The latter region is particularly rich in corticosteroid receptors. Two different types of CRH receptors, CRH-R1 and CRH-R2, have been identified so far. They show distinct yet overlapping distribution in the brain. In the rhesus monkey both types were found in the neocortex, hippocampal formation, and amygdala [305]. It was recently demonstrated [248] that interaction of CRH with CRH-R1 receptors in limbic forebrain regions modulates anxiety-related behaviour (i.e. a behavioural component of the stress response) and that this effect is independent of HPA system activity.

It is important to note that stress and stress-associated dysfunction of CRH neuronal circuitries has been implicated in the development of specific psychiatric disorders such as major depression and anxiety disorders [10, 222]. There is a strong increase of CRH activity in major depression; remarkably, the total number of CRH neurons of major depressed patients appeared to be four times higher than in controls [284, 285].

The suprachiasmatic nucleus (SCH) modulates activity of the HPA axis through a disynaptic pathway in which SCH neurons synapse on neurons in the dorsomedial hypothalamic nucleus that are presynaptic to CRF neurons in the PVH [48, 62].

Thermoregulation

In mammals, it is essential for functioning of the organism that the body temperature remains constant. The organism regulates temperature by two mechanisms, one concerned with dissipation of heat and the other with its production and conservation. Raised ambient temperature induces heat loss by cutaneous vasodilatation, sweating, and depressed heat production. When the ambient temperature drops, extra heat is produced by shivering, and heat loss is diminished by constriction of cutaneous blood vessels. Long-term exposure to cold enhances the release of thyroxine, which increases body heat by stimulating tissue metabolism.

The preoptico-hypothalamic continuum plays a prominent role in thermoregulation. Physiological and clinical evidence suggests the presence of rostral and caudal control centres.

The *rostral heat control centre* is located in the preoptic anterior hypothalamic area (POAH). This contains numerous neurons sensitive to increased blood temperature. Many of these warmth-sensitive elements also respond to skin temperature [39]. Stimulation of the POAH induces cutaneous vasodilatation and sweating, and the area also shows a high level of c-fos activity following hyperthermia [320]. Hyperthermia, hypothermia, and poikilothermia may occur when the POAH is damaged by infarction, trauma, or surgery [352].

The *caudal heat control centre* is located in the posterior hypothalamic area. This area is temperature-blind, i.e. does not contain neurons that respond to heating or cooling. However, electrical or chemical stimulation there can produce shivering, vasoconstriction, and increased metabolic heat production [385]. Large lesions in the posterior hypothalamus may impair heat production, which results in poikilothermia [352].

Little is known about the specific circuits involved in thermoregulation. It seems likely that information derived from cutaneous thermoreceptors is relayed by fibres belonging to the spinohypothalamic component of the anterolateral system and its trigeminal equivalent. Cold exposure leads to the secretion of thyreotropin-releasing hormone (TRH) to activate the thyroid axis. In humans the TRH-producing cells are concentrated in the dorsocaudal part of the paraventricular nucleus [106]. The afferents involved in temperature-related reactions of these elements are unknown. The same holds true for afferents to the somatomotor neurons involved in shivering. In humans the body temperature, which is maintained close to 37 °C (98.6 °F), varies somewhat diurnally, decreasing to a minimum during sleep. The suprachiasmatic nucleus (SCH) is responsible for this circadian fluctuation in temperature [249] and likely influences the temperature control system via its projection to the anteroventral hypothalamus [82].

As regards the central circuitry involved in relaying autonomic thermoregulatory responses, several direct and indirect projections connect the hypothalamus with the sympathetic and parasympathetic preganglionic cell groups in the lower brain stem and spinal cord. All of these may mediate thermoregulatory impulses. There is evidence that in rats, an indirect pathway connecting the preoptic region with the spinal intermediolateral column, which is synaptically interrupted in the periaqueductal grey and medullary raphe nuclei, plays a prominent role in thermoregulation [58, 143, 210, 246, 426].

Food Intake

Classic studies in the 1940s and 1950s revealed that in rats, large bilateral electrolytic lesions centred in the region of the ventromedial hypothalamic nucleus (VMH) produce hyperphagia and obesity [146], while lesions at the same level in the lateral hypothalamic area (LHA) result in aphagia [14]. Moreover, electrical stimulation appeared to elicit feeding behaviour from the LHA and feeding suppression from the VMH [13, 90]. Collectively, these findings provided the foundation of a *dual-centre hypothesis*, according to which the VMH contains a "satiety centre", whereas the LHA contains a reciprocally interrelated "feeding centre" [14, 136, 340]. More recent observations have thrown considerable doubt on this hypothesis. Thus it was found that lesions restricted entirely to the VMH do not lead to obesity, whereas slightly more dorsal lesions, which included the paraventricular nucleus (PVH), appeared to be effective [115]. Moreover, evidence suggested that aphagia following LHA lesions is due to damage to fibres passing through the medial forebrain bundle rather than to destruction of specific populations of neurons in the LHA [344, 386]. Finally, it was found that connections between the two feeding control centres are only weakly developed [214]. Thus the neuroanatomical basis for lesion- and stimulation-induced changes in ingestive behaviour remained unclear [32, 360]. Another tenacious problem involved in study of central feeding-related circuitry was that the specific stimulus or signal and the specific entrance to this circuitry remained unknown.

Recent morphological, physiological, and molecular studies have provided new insights on the role of the hypothalamus in regulating food intake and body weight. It has been shown that interacting populations of peptidergic neurons in the ventromedial hypothalamus and lateral hypothalamic area play a crucial role in these regulations and that the paraventricular nucleus is also critically involved. However, it appeared that the ventromedial hypothalamic satiety centre corresponds with the infundibular (or arcuate) nucleus (INF) rather than with the VMH. An important step forward was also identification of the gene responsible for obesity in rodents and the characterization of its protein product called *leptin* [425]. Leptin is an adipocyte-derived hormone whose circulating levels vary directly as a function of adipose tissue mass. It has been shown that circulating leptin provides a physiologically important negative feedback signal to control body weight by affecting both ingestive behaviour and metabolic activity [53, 125, 268, 419]. The infundibular nucleus has to be considered the principal monitor of circulating leptin-based signals to the brain. It is a major site of leptin receptor expression, and because the blood-brain barrier is absent locally, its constituent neurons are directly accessible to circulating hormones [100, 123]. Recent knowledge concerning mediators and neural pathways involved in the spread of leptin-induced effects over the brain may be summarized as follows (Fig. 10.7 A) [105, 318, 348].

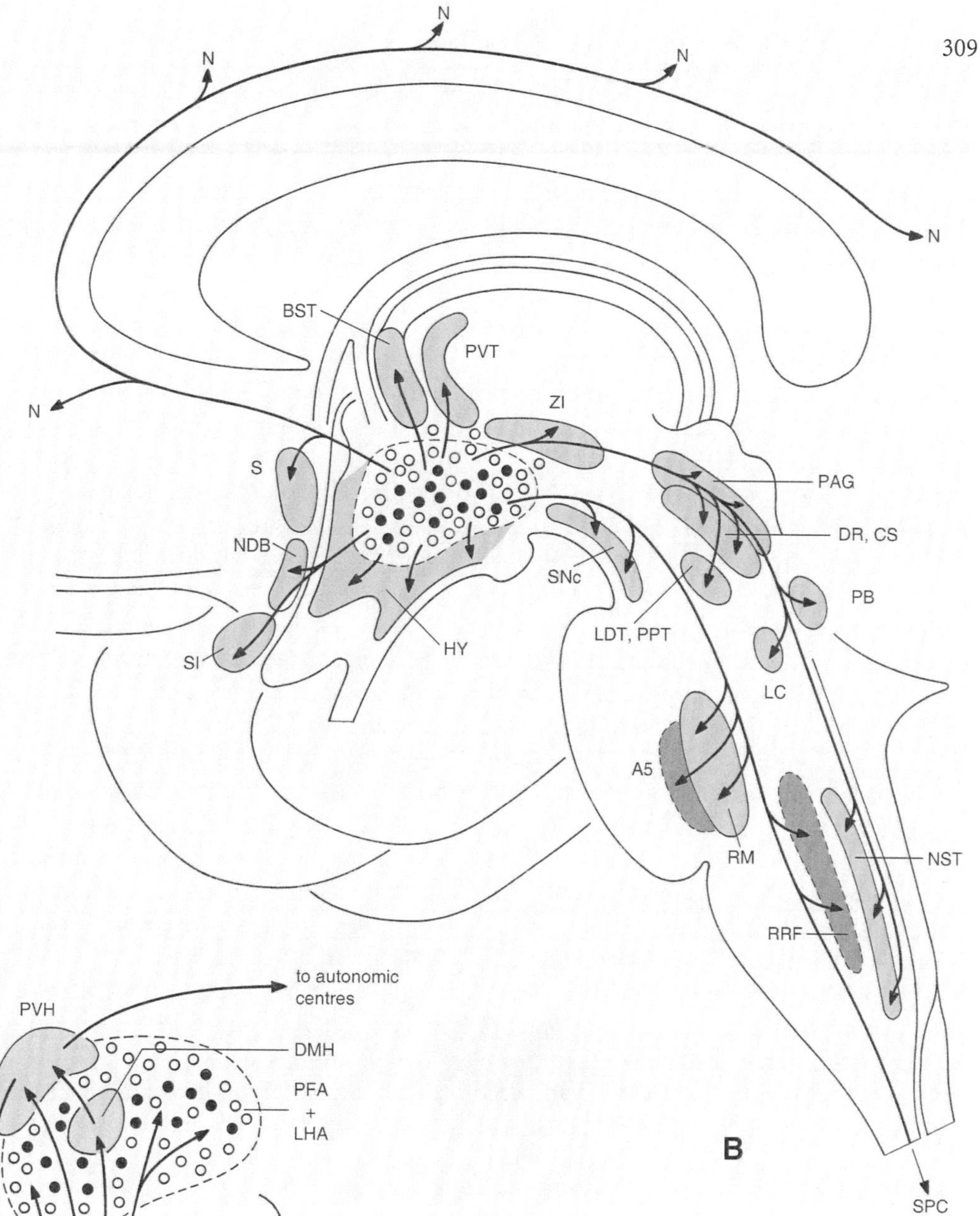

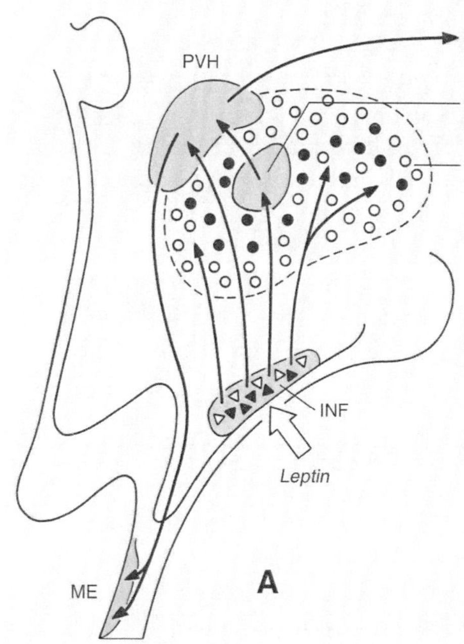

Fig. 10.7 A, B. Pathways and cell groups involved in the regulation of feeding. **A** Efferents of the infundibular nucleus (*INF*). Leptin-sensitive neurons in this nucleus project to the paraventricular hypothalamic nucleus (*PVH*), dorsomedial hypothalamic nucleus (*DMH*), and populations of orexin- and melanin-concentrating hormone-containing neurons in the perifornical (*PFA*) and lateral hypothalamic areas (*LHA*). *ME*, median eminence. **B** Efferents and areas of termination of the population of hypothalamic orexinergic cells. Only the most prominent terminal fields are included (based on [243, 388]). *A5*, pontine noradrenergic cell group; *BST*, bed nucleus of stria terminalis; *CS*, central superior nucleus; *DR*, dorsal raphe nucleus; *HY*, hypothalamus; *LC*, locus coeruleus; *LDT*, laterodorsal tegmental nucleus; *N*, neocortex; *NDB*, nucleus of the diagonal band of Broca; *NST*, nucleus of the solitary tract; *PAG*, periaqueductal gray; *PB*, parabrachial nucleus; *PPT*, pedunculopontine tegmental nucleus; *PVT*, paraventricular thalamic nucleus; *RM*, raphe magnus nucleus; *RRF*, rhombencephalic reticular formation; *S*, septal nuclei; *SI*, substantia innominata; *SNc*, substantia nigra, pars compacta; *SPC*, spinal cord; *ZI*, zona incerta. △, cells containing α-MSH, ▲, cells containing AgRP + NPA; ○, cells containing MCH; •, cells containing orexin/hypocretin

1. The leptin-sensitive neurons in the infundibular nucleus form two populations of peptidergic neurons, one expressing neuropeptide Y (NPY) and agouti-related protein (AgRP) and the other expressing α-melanocyte-stimulating hormone (α-MSH) [3, 59, 228, 321]. These three neuropeptides are all known to participate in the regulation of food intake.

2. Leptin-responsive peptidergic infundibular neurons project to the paraventricular (PVH) and dorsomedial (DMH) hypothalamic nuclei and to the perifornical (PFA) and lateral hypothalamic (LHA) areas [19, 20, 42, 97, 318].

3. Projections to the PVH potentially connect leptin-sensitive infundibular elements with the parvocellular neurosecretory neurons comprising the central limbs of the hypothalamo-pituitary-adrenal and hypothalamo-pituitary-thyroid axes. The activities of these axes are known to be influenced by leptin [3, 171]. The PVH is also a major source of long descending fibres passing to sympathetic and parasympathetic preganglionic neurons in the lower brain stem and spinal cord [369]. These fibres form part of multisynaptic pathways through which the PVN among others influences the secretion of adrenal and pancreatic hormones [45, 47]. The direct projection from the infundibular nucleus to the PVH may be supplemented by an indirect pathway routed through the DMH. Endowed with leptin receptors [100, 123, 321], this nucleus receives afferents from the infundibular nucleus and projects strongly to the PVH [124].

4. Two populations of peptidergic neurons were recently identified in the perifornical and lateral hypothalamic areas, one containing *melanin-concentrating hormone* (MCH), the other containing *hypocretin* (also known as *orexin*) (H/O) [87, 304]. In rodents and humans the MCH and H/O neurons comprise segregated though spatially overlapping populations. The MCH cells extend in a wider area and partly surround the H/O population [42, 97]. It has been established that the two lateral hypothalamic populations receive input from the chemically specified AgRP+NPY and α-MSH neurons in the infundibular nucleus [42, 97].

5. Immunohistochemical studies in rodents [85, 271, 388] and humans [243, 388] have shown that the H/O neurons project widely throughout the neuraxis (Fig. 10.7 B). High densities of H/O-positive fibres have been observed in numerous areas, including the entire hypothalamus (HY), septum (S), nucleus of the diagonal band (NDB), substantia innominata (SI), paraventricular thalamic nucleus (PVT), zona incerta (ZI), periaqueductal grey (PAG), dorsal raphe (DR) and central superior (CS) nuclei, laterodorsal (LDT) and pedunculopontine tegmental (PPT) nuclei, locus coeruleus, gigantocellular reticular nucleus (GR), and nucleus of the solitary tract (NST). It is noteworthy that moderate numbers of H/O fibres appeared to be present throughout the cerebral cortex and that in the spinal cord these fibres are concentrated in the marginal zone, intermediolateral nucleus, and area surrounding the central canal.

6. Projections of the MCH neurons, which so far have only been studied in rat [33], generally show a striking resemblance to those of H/O neurons. The only major difference seems to be that the substantia innominata and locus coeruleus are less densely innervated by MCH fibres than by H/O fibres.

7. It has been shown that the lateral hypothalamic H/O and MCH systems project back to the infundibular nucleus [33, 42, 97] and that the H/O fibres make direct synaptic contacts with NPY cells, i.e. one of the infundibular neuronal groups that express leptin receptors [164]. Hence it seems likely that the H/O and MCH systems, rather than being passive recipients of information from the infundibular nucleus, can actively modify this information [322].

The findings on mediators and pathways involved in the regulation of feeding summarized above imply a revival of the classic dual-centre concept! Chemically specified cell groups in the mediobasal hypothalamic infundibular nucleus, important targets of circulating leptin, are reciprocally connected with chemically specified cell groups in the lateral hypothalamus that have widespread projections throughout the central nervous system. For detailed discussions of the possible functional significance of this newly discovered circuitry, see the review articles [105, 318, 348, 375, 415]. Here we confine ourselves to some notes on the status of the H/O- and MCH-expressing neuronal populations.

a. There is good evidence that these peptidergic populations are involved in feeding behaviour. They not only are in receipt of afferents which mediate leptin-based signals but also express leptin receptors themselves, albeit in much lower concentrations than observed in the infundibular nucleus [124, 164]. Intracerebroventricular injections of H/O and MCH have been shown to affect food intake [170, 278, 282, 304]. Moreover, the expression of H/O mRNA is increased by food deprivation [304], and MCH-knockout mice have reduced food intake [328].

b. However, it is important to note that intracerebroventricular injections of H/O affect several functions other than feeding such as blood pressure [281], heart rate [302], body temperature [261], pain perception [41], and secretion of several hormones [281, 301].

c. The lateral hypothalamic area (LHA) in which the O/H and MCH neuronal populations are embedded receives, besides afferents from the infundibular nucleus, input from a variety of other structures ranging from the medial prefrontal cortex to the nucleus of the solitary tract [31, 36, 167, 189, 214, 254, 293]. Many of these afferent systems reciprocate efferent projections of the LHA. Although it has not been shown that all of the afferent systems specifically address the O/H and MCH neuronal populations, combining connectional data with the functional findings mentioned above, strongly suggests that these neuronal populations are involved in more functions than just regulation of feeding-related activities.

d. As previously discussed, H/O neurons are involved in the regulation of wakefulness and sleep. Without going into details, it may be recalled that (i) the H/O neurons send particularly dense projections to a number of monoaminergic and cholinergic groups, all representing key components of the ascending arousal system, (ii) the H/O neurons exert excitatory influence on the cell groups mentioned, and (iii) deficiency in hypocretin neurotransmission results in humans and mice in narcolepsy, a serious wakefulness disorder. These data have led some authors [348, 375] to conclude that the H/O system, though participating in several functions, is primarily involved in regulating the arousal cycle.

Thirst

Two hypothalamic cell masses (the supraoptic and paraventricular nuclei), two circumventricular organs (the subfornical and vascular organs of the lamina terminalis), and the central projections of the IXth and Xth cranial nerves form essential parts of the neural circuitry associated with regulation of thirst and drinking behaviour (Fig. 10.8).

The *supraoptic* (SO) and *paraventricular* (PVH) *nuclei* are distinct cell masses in the anterior hypothalamus (Figs. 6.39, 10.1). Large neurons in these nuclei, which synthesize the peptide neurohormones *vasopressin* and *oxytocin*, give rise to axons descending through the infundibular stalk to the posterior pituitary, thus forming the supraoptico- and paraventriculo-hypophyseal tracts (Fig. 10.1). The hormones are transported along the axons of these pathways and released into blood vessels of the neurohypophysis. Vasopressin is responsible for antidiuresis, promoting reabsorption of water by the kidney; oxytocin stimulates smooth muscle in the uterus and mammary glands to contract. Vasopressin and oxytocin are produced by separate neurons which collectively constitute the so-called magnocellular neurosecretory system. Apart from the magnocellular elements just discussed, the PVH contains two other populations of peptidergic neurons. The first consists of elements which synthesize corticotropin-releasing factor (CRF). These neurons form a *parvocellular neurosecretory system*. Their axons pass to the external lamina of the median eminence, where their terminals contact the hypophyseal portal capillaries [16, 337]. It has already been discussed that CRF, once liberated in the hypophyseal portal system, activates ACTH secretion by the anterior pituitary, which in turn leads to corticosteroid release from the adrenals. The elements of the other population give rise to descending projections consisting of thin, unmyelinated fibres which innervate primarily a number of autonomic centres in the brain stem and spinal cord [44, 166, 359]. These neurons contain a multitude of neuroactive peptides such as vasopressin, neurotensin, enkephalin, somatostatin, and cholecystokinin. Each of these peptides occurs alone or in various combinations in individual parvocellular neurons [201]. In the PVH of rodents, the neuronal populations just discussed are concentrated in distinct subnuclei, but in human PVH such an arrangement into subnuclei is absent [350]. The vasopressinergic neurons in the PVH and SO as well as the caudally projecting peptidergic PVH elements participate in body water regulation.

Two *circumventricular organs* (CVOs), the *subfornical organ* (SFO) and *vascular organ of the lamina terminalis* (OVLT), are involved in regulating thirst and drinking behaviour. For general notes on CVOs, see Chap. 4.

The SFO contains many neurons, a dense fenestrated capillary plexus and a high density of receptors for the peptide hormone angiotensin II (A II). It is known that hypovolemic thirst is accompanied by increased circulating levels of this hormone which, by way of the SFO neurons and their targets, elicits a spectrum of responses aimed at the maintenance of blood pressure and replacement of lost body water (see below) [361].

The OVLT and immediately adjacent median preoptic nucleus may be designated together as the anteroventral third ventricular area (AV3V) [175]. Like the SFO, the OVLT is rich in A II receptors. The AV3V is known to be especially sensitive to hyperosmolality of the blood, another stimulus to thirst [41].

Apart from humoral stimuli, mechanoceptive information, derived from volume receptors in the walls of the great veins and atria and from baroreceptors in the carotid sinus, participate in the regulation of water balance and arousal of thirst. This viscerosensory information is carried over the glossopharyngeal and vagus nerves (IX, X) to the nucleus of the solitary tract (NST).

The following summary of neural circuitry underlying body water regulation and control of thirst and drinking behaviour is largely based on a series of publications by Swanson and colleagues [360–362, 367, 368] (Fig. 10.8).

1. The SFO and the AV3V are reciprocally connected [383]. Interestingly, the projection from the SFO to the OVLT uses A II as a neurotransmitter.

2. The SFO and AV3V project to the SO and PVH.

3. Viscerosensory information from the vagus and glossopharyngeus nerves is relayed to the PVN by direct projections from the nucleus of the solitary tract and by less direct inputs via the parabrachial nuclear complex (PB) and so-called ventrolateral medulla (VLM). The latter includes the lateral paragigantocellular nucleus, a ventrolateral extension of the medial reticular formation in which the A1 noradrenergic and C1 adrenergic cell groups are embedded. The ventrolateral medulla is known to play a prominent role in the maintenance of blood pressure. It receives excitatory afferents from the nucleus of the solitary tract (NST). Its efferents to the PVH use among others noradrenaline, adrenaline, galanin, and neuropeptide Y as neurotransmitters.

4. Intrahypothalamic afferents to the PVH include fibres from the dorsomedial (DMH), infundibular (INF), and suprachiasmatic nuclei (SCH). The latter may provide the PVH with information about the circadian cycle.

5. Presumed cognitive inputs from the cerebral cortex are relayed to the PVH by the bed nucleus of the stria terminalis (BST) and ventral lateral septal nucleus (VLS), which in turn receive inputs from the prefrontal cortex (PF), amygdaloid complex (AM), and hippocampal formation (HF) [296, 362]. The control of thirst involves neuroendocrine, autonomic, and somatomotor responses.

6. Neuroendocrine response is mediated by the vasopressinergic PVH and SO neurons, the axons of which pass to the posterior pituitary, where they release their product into general circulation. The name vasopressin refers to the

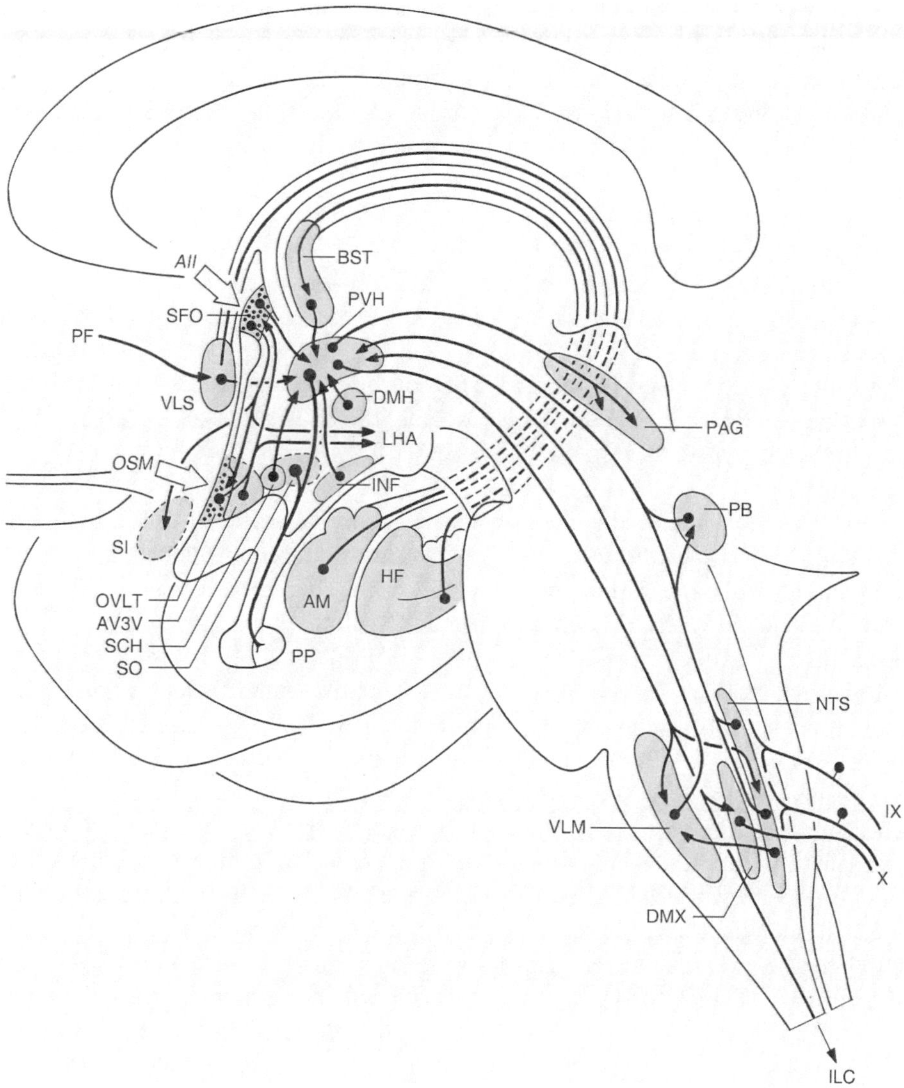

Fig. 10.8. The neural circuitry associated with the regulation of thirst and drinking behaviour. *AII*, angiotensin II; *AM*, amygdaloid complex; *AV3V*, anteroventral third ventricular area; *BST*, bed nucleus of stria terminalis; *DMH*, dorsomedial hypothalamic nucleus; *DMX*, dorsal motor vagus nucleus; *HF*, hippocampal formation; *ILC*, spinal intermediolateral cell column; *INF*, infundibular nucleus; *LHA*, lateral hypothalamic area; *NTS*, nucleus of the solitary tract; *OSM*, osmolality; *OVLT*, vascular organ of the lamina terminalis; *PAG*, periaqueductal grey; *PB*, parabrachial nuclei; *PF*, prefrontal cortex; *PP*, posterior pituitary; *PVH*, paraventricular hypothalamic nucleus; *SCH*, suprachiasmatic nucleus; *SFO*, subfornical organ; *SI*, substantia innominata; *SO*, supraoptic nucleus; *VLM*, ventrolateral medulla; *VLS*, ventral lateral septal nucleus; *IX*, *X*, cranial nerves

fact that this hormone increases blood pressure by vasoconstriction. Its principal function is, however, water-conserving action on the distal nephron; hence its alternative name *antidiuretic hormone*.

7. The neural substrate of the visceromotor response to thirst is formed by axons descending from the PVH to the nucleus of the solitary tract (NST), dorsal motor vagus nucleus (DMX), ventrolateral medulla (VLM), and spinal intermediolateral column (ILC) [127, 288]. All of these centres are involved in regulation of blood pressure.

8. The somatomotor response to thirst includes general arousal to and the behaviour act of drinking. The arousal is believed to be mediated by fibres passing from the SFO to the substantia innominata (SI) and projections from the SFO and AV3V to the lateral hypothalamic area (LHA) [184]. As already discussed, the substantia innominata and lateral hypothalamic area contain populations of neurons that, forming part of the ascending arousal system, innervate the entire neocortex (Fig. 10.5). The presumably polysynaptic pathway which connects the hypothalamus with the lower rhombencephalic motor centres involved in the act of drinking are not known. The fibres descending from the PVH to the periaqueductal grey (PAG) may well form part of this pathway.

Sexual Behaviour

There is ample functional, morphological, and neuroendocrinological evidence that a complex of highly interconnected hypothalamic preoptic and basal telencephalic cell groups are involved in the initiation and coordination of reproductive behaviour (Fig. 10.9 A). This complex, designated here as the telodiencephalic reproductive complex (TDRC), encompasses the hypothalamic ventromedial (VMH), infundibular (INF), ventral premamillary (PMV), and suprachiasmatic (SCH) nuclei, the medial preoptic nucleus (MPN) and, in the basal telencephalon, the bed nucleus of the stria terminalis (BST), lateral septal nucleus (LS), nucleus of the diagonal band of Broca (NDB), and medial amygdaloid nucleus (MEA). The extensive physiological literature on the neural control of reproduction is not reviewed here. Suffice it that lesion and stimulation studies have shown that in rodents the MPN is essential for male sexual behaviour [131, 227], while the VMH plays a comparable role in females [193, 275].

The considerable differences in sexual behaviour between males and females are reflected in differing size, cell number, and structure of several TRDC cell groups, which include the medial preoptic area, BST, VMH, and SCH.

In rats, the preoptic area contains an intensely staining group of cells which is three to eight times larger in males than in females [116, 117]. Therefore it was designated as the sexually dimorphic nucleus of the preoptic area (SDN-POA). A probable homologue is present in the human preoptic area [355]. The human SDN-POA is an ovoid, densely packed collection of large cells between the dorsolateral supraoptic nucleus and the rostral pole of the paraventricular nucleus. Morphometric analysis revealed that the volume of this cell group is more than twice as large in young adult men as in women and contains about twice as many cells in men [150, 356]. Contrary to what might be expected, the VMH is larger in male rats than in females [221].

The BST is a telencephalic structure situated at the junction of the septum, accumbens nucleus, and preoptic region (Fig. 6.40). Like all other components of the TDRC, it forms part of the limbic system and is characterized by an extraordinarily large number of neuromediators, particularly neuropeptides [256]. Walter et al. [410] subdivided the human BST on the basis of distribution patterns of 13 different neuropeptides into three main areas, lateral, central, and medial. The central area has been found to be clearly dimorphic. It is 40% smaller in women than men [427] and has almost twice as many somatostatin-containing neurons in men [197]. It has also been reported that a particular darkly staining posteromedial component of human BST is 2.5 times larger in males than in females [11]. Volumetrical

[89, 148] and neurochemical [216, 391] sex differences have also been reported for the BST in several rodent species.

The SCH contains several groups of peptidergic neurons, each of which is concentrated in a particular subnucleus [350]. One of these, consisting of vasoactive intestinal polypeptide (VIP)-expressing neurons, was found to be twice as large in young men (10 to 30 years of age) than in young women and contain twice as many cells. Interestingly, in the 41–65-year age group, this sex difference was reversed, and it disappeared altogether after the age of 65 [356, 357]. It is known that VIP-containing neurons in the SCH project to the MPN, where they directly innervate neurosecretory, gonadotropin-releasing, and hormone-producing cells [389].

Phenomena of sexual dimorphism are not confined to structures within the TDRC. The sacral spinal cord contains a distinct, ventrally situated group of motoneurons known as the *nucleus of Onuf*. The axons emerging from this nucleus supply, among other structures, the muscles of the pelvic floor and the bulbocavernosus and ischiocavernosus muscles. It has been established that in human and canine spinal cord, Onuf's nucleus contains more motoneurons in males than in females.

The TDRC contains concentrations of neurons producing gonadotropin-releasing hormone and is an important target of circulating gonadal steroid hormones.

Gonadotropin-releasing hormone (GnRH), also known as *luteinizing hormone-releasing hormone* (LHRH), is a decapeptide essential for mammalian reproduction. It is released from nerve terminals in the median eminence and carried via the hypothalamo-hypophyseal portal system to the anterior pituitary, where it stimulates the synthesis and secretion of the gonadotropic *luteinizing hormone* (LH) and *follicle-stimulating hormone* (FSH). The distribution of neurons expressing GnRH in the human brain has been studied with the aid of antibodies against this hormone [27, 28, 93] or utilizing hybridization histochemistry to visualize GnRH mRNA [287]. It appeared that GnRH-producing perikarya are widely scattered over the entire hypothalamus and adjacent basal telencephalic regions, with concentrations in the septum, nuclei of the diagonal band of Broca, dorsal and ventral parts of the medial preoptic area, and infundibular nucleus (Fig. 10.9 B). A retrograde tracer study [333] revealed that in rat approximately 50% of GnRH neurons in the preoptic and basal telencephalic regions project to the median eminence. The remaining GnRH-containing perikarya send their axons presumably to other parts of the brain. It has been observed that from the areas where these perikarya are concentrated, GnRH-containing fibres are distributed to various extrahypothalamic structures including the olfactory bulb, amygdala, cingulate cortex, and periaqueductal grey [332, 418]. There is evidence that GnRH functions not only as a neurohormone but also as a neurotransmitter or neuromodulator in the central regulation of sexual behaviour [247, 274].

The gonadal steroid hormones *estrogen* and *testosterone* are known to play a crucial role in mediating reproductive functions. Experimental evidence indicates that a primary way in which these steroids exert their influence is by altering the biosynthetic and electrical activity of specific groups of neurons known to be involved in the control of sexual behaviour and gonadotropin secretion [335].

The distribution of estrogen-sensitive (ES) and androgen-sensitive (AS) neurons has been studied with different techniques in several groups of mammals including rodents [66, 273, 317, 336, 345–347], carnivores [181, 289], monkeys [139, 183, 232, 276, 420], and man [103, 198, 199, 264]. Considerable numbers of ES and AS neurons have been consistently found in the MEA, BST, LS, NDB, MPN, VMH, and INF. In humans, sex-related differences have been observed in some of these nuclei. Thus, in females ES neurons in the NDB and VMH appeared to be more heavily labeled than in males [420].

Gonadal steroids are known to participate in the feedback regulation of synthesis and release of GnRH. Concerning estrogen, this regulation is apparently indirect. Despite a considerable overlap between the distributions of ES and GnRH cells, the latter do not possess receptors for estrogen [273].

Estrogen-sensitive and AS cells are not confined to the hypothalamus and adjacent regions of the basal telencephalon. Studies in rodents [66, 273, 317, 346, 347] have shown that such cells are widely distributed over the entire neuraxis. Neurons sensitive to circulating androgen and estrogen are found predominantly in (greater) limbic structures with direct connections with the hypothalamus such as the olfactory bulb, olfactory cortex, hippocampal formation, periaqueductal grey, parabrachial nuclei, nucleus of the solitary tract, and spinal laminae I, II, and X. Sex-related differences have been observed in some of these structures. Thus, in the hippocampal formation, AS neurons appeared to be more heavily labeled than ES neurons. Heavily labeled AS neurons have been observed in several structures involved in the central relay of somatosensory information including the vestibular nuclei, cochlear nuclei, nucleus of the lateral lemniscus, and medial geniculate nucleus. Clearly labeled AS neurons have also been found in the motor nuclei associated with the Vth, VIIth, Xth, and XIIth cranial nerves and in spinal lamina IX. Cerebellar Purkinje cells and the large precerebellar nuclei, i.e. inferior olive and pontine grey, appeared sensitive to androgen, although the parvocellular part of the lateral cerebellar nucleus clearly contained ES neurons. The data just discussed are mainly based on [66], to which the reader is referred.

It is noteworthy that, according to some recent studies on the distribution of gonadal steroid receptors in the human brain [181, 199, 420], the medial and lateral mamillary nuclei contain receptors for both androgen and estrogen in a sexually dimorphic pattern.

The circuitry of the telodiencephalic reproductive complex, as diagrammatically depicted in Fig. 10.9 A, is mainly based on experimental studies in the rat [54, 334, 335, 360, 362]. The MPN and VMH form important nodal points in this circuitry. These nuclei receive highly processed somatosensory information from cortical association areas which attains the hypothalamus via the hippocampal formation and amygdaloid complex. The hippocampo-hypothalamic and amygdalo-hypothalamic projections carrying this information are partly synaptically interrupted in the LS and BST, respectively. The amygdaloid complex also relays to the hypothalamus olfactory information derived from the main and accessory olfactory systems. The accessory olfactory or vomeronasal system [128, 329] comprises the vomeronasal organ or Jacobson's organ, vomeronasal nerve, and a special part of the olfactory bulb known as the accessory olfactory bulb. The last-named projects mainly to the medial amygdaloid nucleus. The main and accessory olfactory systems are both chemosensory systems, but the accessory olfactory system is specialized in the transduction of pheromonal information. Pheromones are chemical messengers produced by other members of the same species. In mammals with well-developed accessory olfactory systems such as insectivores and rodents, pheromonal communication plays a prominent role in reproductive behaviour. However, in humans this system is vestigial, and recent reports suggesting that it might be functional are controversial [128].

The central reproductive network is not confined to the telencephalon and diencephalon but extends over all parts of the neuraxis. In Fig. 10.10 some reproduction-related centres and their ascending (A) and descending connections (B) are diagrammatically shown. The centres indicated are: peripeduncular nucleus (PP), ventral tegmental area (VTA), nucleus retroambiguus (NRA), A1 and A2 noradrenergic cell groups, and of course the lumbosacral spinal cord. It should be emphasized that none of these centres is *exclusively* concerned with reproduction and that many entertain reciprocal connections with the MPN and/or VMH.

The *peripeduncular nucleus* (PP), situated in the mesodiencephalic transitional area, presumably represents the caudal sector of the zona incerta (Fig. 6.33). Its connections have already been discussed in the section on the ventral thalamus. The PP is strongly and bidirectionally connected with the VMH [17, 214, 315].

The *ventral tegmental area* (VTA) is a rather diffuse cell mass in the ventromedial tegmentum of the midbrain (Fig. 6.33). This area con-

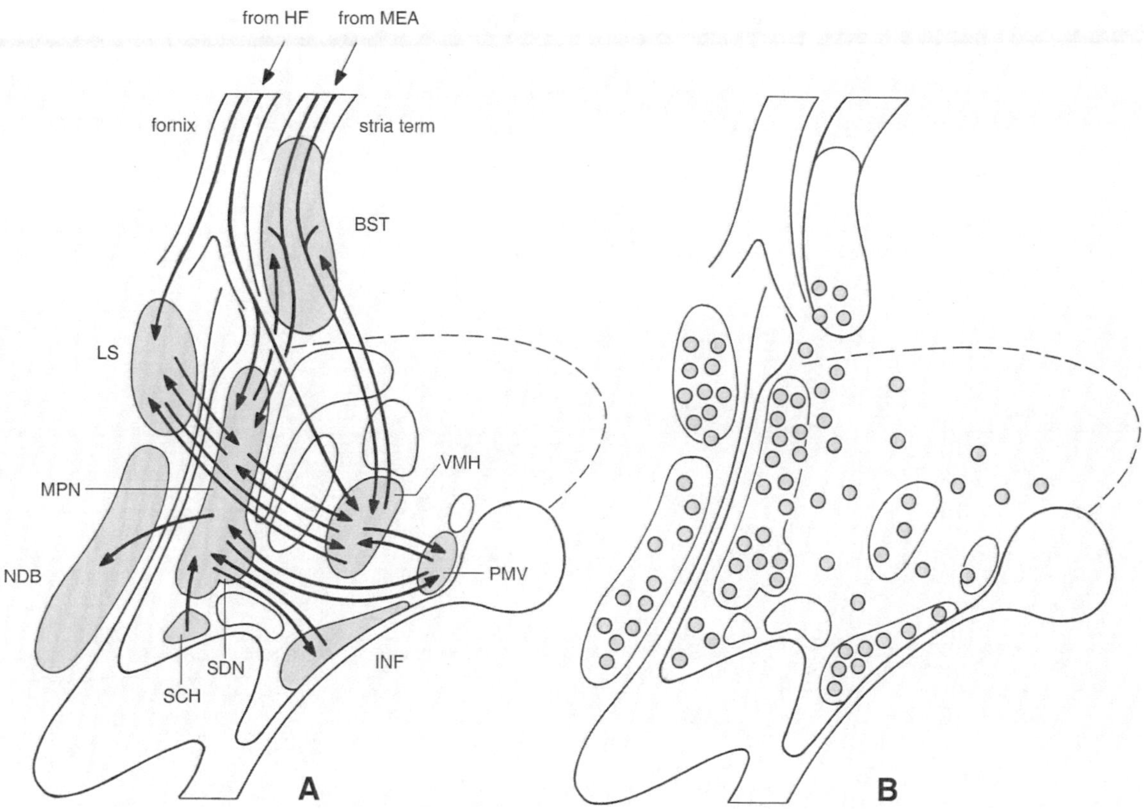

Fig. 10.9 A, B. Organization of structures involved in the regulation of sexual behaviour. **A** Cell groups and fibre connections forming the telodiencephalic reproductive complex. **B** The distribution of GnRH-containing neurons. *BST*, bed nucleus of stria terminalis; *HF*, hippocampal formation; *INF*, infundibular nucleus; *LS*, lateral septal nucleus; *MEA*, medial amygdala; *MPN*, medial preoptic nucleus; *NDB*, nucleus of the diagonal band of Broca; *PMV*, ventral premamillary nucleus; *SCH*, suprachiasmatic nucleus; *SDN*, sexually dimorphic nucleus of the preoptic area; *VMH*, ventromedial hypothalamic nucleus

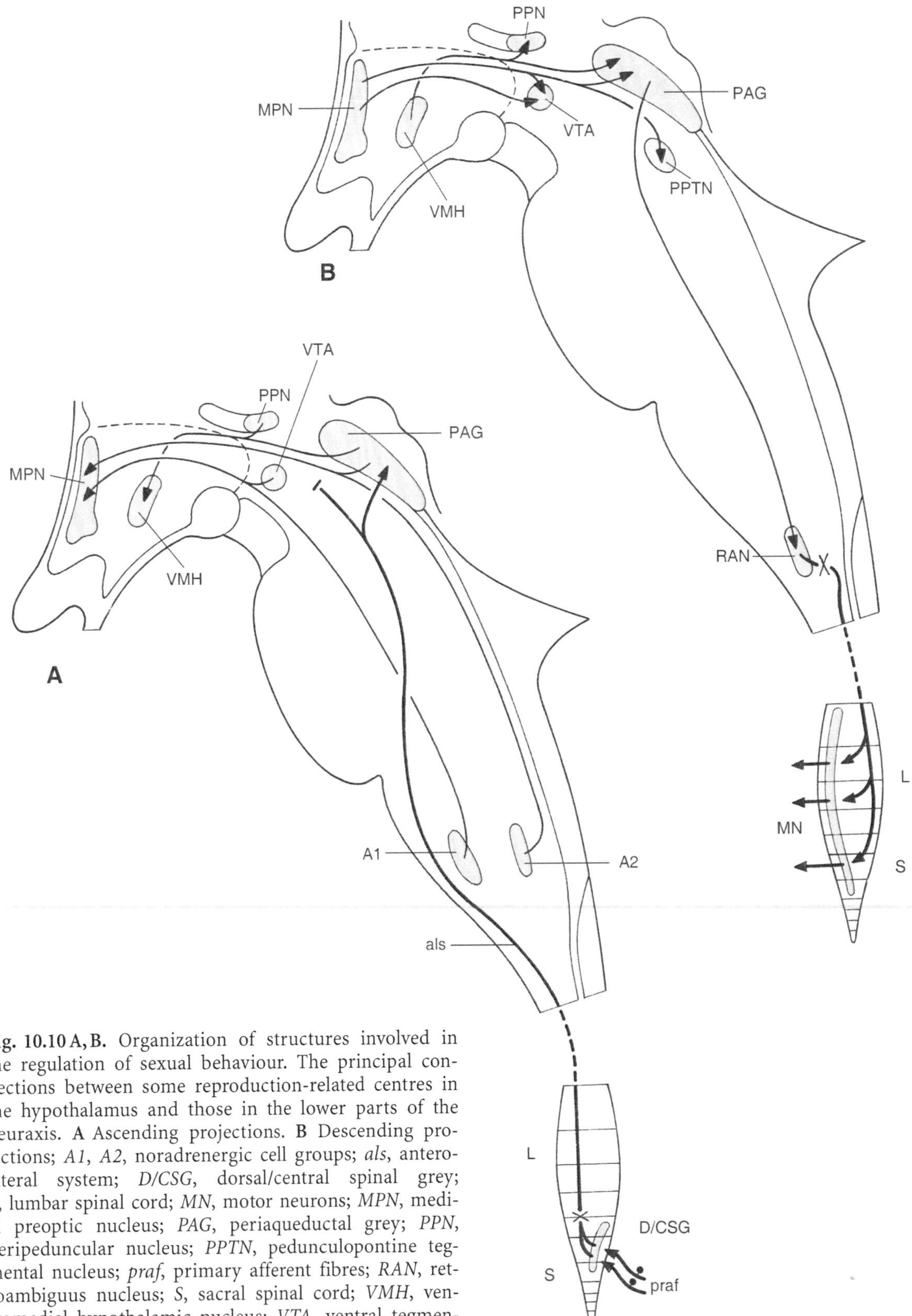

Fig. 10.10 A, B. Organization of structures involved in the regulation of sexual behaviour. The principal connections between some reproduction-related centres in the hypothalamus and those in the lower parts of the neuraxis. **A** Ascending projections. **B** Descending projections; *A1*, *A2*, noradrenergic cell groups; *als*, anterolateral system; *D/CSG*, dorsal/central spinal grey; *L*, lumbar spinal cord; *MN*, motor neurons; *MPN*, medial preoptic nucleus; *PAG*, periaqueductal grey; *PPN*, peripeduncular nucleus; *PPTN*, pedunculopontine tegmental nucleus; *praf*, primary afferent fibres; *RAN*, retroambiguus nucleus; *S*, sacral spinal cord; *VMH*, ventromedial hypothalamic nucleus; *VTA*, ventral tegmental area

tains numerous dopaminergic neurons which project to the nucleus accumbens. The VTA is included here in the group of reproduction-related centres because (a) it receives afferents from almost all telencephalic and diencephalic members of the same group, including the LS, BST, MPN, and VMH [214, 313, 358, 365] and (b) it is strongly activated during human ejaculation (see below) [156, 157, 159].

The *periaqueductal grey* (PAG) or mesencephalic central grey is a conspicuous midline structure which surrounds the cerebral aqueduct (Figs. 6.29–6.32). It consists mainly of small, densely packed cells. The PAG receives input from a variety of structures including several limbic cortical fields (prelimbic, infralimbic, anterior cingulate) [132, 217], the amygdaloid complex [177, 178], MPN [15, 30, 220], VMH [30, 217, 402], zona incerta (ZI) [163, 217, 220], ATV [187], dorsal raphe and central superior nuclei [30, 220], parabrachial nuclei [30, 220], pontine and medullary reticular formation [30, 220], and spinal cord [224, 398]. Many of these regions receive reciprocal projections from the PAG [60, 94, 218, 219]. The PAG plays an important role in a variety of functions and behavioural patterns including defense reactions, reproductive behaviour, vocalization, autonomic regulation, and pain modulation [23, 25, 26].

Physiological studies using microinjection of excitatory amino acids have provided evidence of four longitudinal zones or columns within the PAG: dorsomedial, dorsolateral, lateral, and ventrolateral [24, 25, 329]. Without going into detail, it may be mentioned that stimulation of the lateral zone evokes defensive behaviour accompanied by increased blood pressure and heart rate, whereas stimulation of the ventrolateral zone elicits immobility ("freezing") and decreased blood pressure and heart rate. The caudal part of the lateral zone of the PAG is involved in reproductive behaviour. In female rats, lordosis can be evoked by stimulation of this zone [329]. In the cat, fibres originating from the lateral dorsal horn and lateral intermediate zone of the lower lumbar and upper sacral spinal segments decussate at the level of their parent neurons, ascend in the contralateral lateral funiculus, and terminate in the lateral part of the caudal PAG [398]. The cells of origin of this spinomesencephalic projection are situated in the area of termination of primary afferents (praf) from vagina or penis, pelvic floor, perineal skin, and related structures [398]. In cats [395] as well as monkeys [400], a compact group of neurons in the lateral PAG sends a dense projection to the nucleus retroambiguus.

The *nucleus retroambiguus* (NRA) is a compact, rostrocaudally oriented column of cells found laterally in the caudal rhombencephalon (Fig. 6.21) [262]. It has been shown involved in expiration and expiration-related activities such as vocalization and vomiting. To mediate these activities, the NRA sends projections to motoneuronal cell groups innervating larynx, pharynx, intercostal abdominal wall, and pelvic floor muscles [102, 154, 158, 235]. Recent studies demonstrated that the NRA also projects to a distinct set of motoneuronal cell groups in the lumbosacral cord that participates in producing female (lordosis) and male mating postures [394, 395, 399, 401]. In the cat, the NRA-lumbosacral projection shows a highly remarkable and dynamic dimorphism. The density of this projection in the area of its target motoneuronal cell groups appeared to depend strongly on the estrous cycle: it was almost nine times greater in estrous than non-estrous females [397]. Evidence suggested that the greater strength of the NRA-motoneuronal pathway in estrous cases is caused by outgrowth of the terminal NRA fibres and that this outgrowth is in all likelihood induced by estrogen [397].

The NRA-lumbosacral projection in male cats innervates slightly different motoneuronal cell groups than in female cats, presumably because during mating males adopt a different posture than females. This projection in males is stronger than in non-estrous females but weaker than in estrous females [396]. No signs of plasticity could be detected in the male NRA-lumbosacral projection [396].

It has been suggested [156, 157] that in cats, the ascending projection from the lumbosacral cord to the PAG and descending PAG-NRA-lumbosacral motoneuronal projection form the afferent and efferent limbs of a circuit generating the posture necessary for copulation. Higher limbic structures with access to the PAG can influence this spino-brainstem-spinal circuit [157].

The *noradrenergic A1 and A2 groups* are both situated in the caudal part of the medulla oblongata. The cells of the A1 group surround the nucleus of the lateral funiculus, whereas those of the A2 group are embedded in the nucleus of the solitary tract, dorsal motor nucleus of the vagus, and intervening area (Figs. 6.22, 6.23). Both cell groups are known to project to the hypothalamus. Experiments using c-Fos as a marker of neuronal activity revealed that in female rats, the A1 and A2 groups are strongly activated after mating. Many of the activated cells appeared to contain estrogen receptors [423].

How do sensory impulses from the genitals attain the brain and, ultimately, consciousness? We have seen that primary afferents from these organs enter the spinal cord and terminate in the lateral dorsal horn and lateral intermediate zone of the lower lumbar and upper sacral spinal segments, and that neurons in this area of termination project to the PAG. It is important to note that this projection forms part of a large, composite assembly of fibres known as the *anterolateral system*. This system's cells of origin are mainly located in the dorsal and central parts of the spinal grey. Most of its fibres decussate in the anterior white commissure and ascend in the contralateral anterolateral funiculus. (In mammals whose pyramidal tract is situated in the posterior funiculus, such as rodents, they may well encroach upon the posterolateral funiculus.) Fibres or collaterals of the anterolateral system terminate in the basal telencephalon [50], thalamus, hypothalamus [49, 52, 424], and a large number of structures in the brain stem [224], several of which form part of or are directly connected with the reproduction-related network (Fig. 10.10 A). The spinothalamic component of the anterolateral funiculus, known to form part of the circuitry concerned with perception of temperature, touch, and pain, is most probably also involved in gaudiception.

Using positron emission tomography, Holstege and collaborators [156, 157, 159] recently addressed the question which brain structures are activated during ejaculation in the human male. The strongest activation was found in a number of structures in the mesencephalon and diencephalon including the VTA (i.e. "reward area"), lateral central tegmental field, and ventralis posterior, midline, and intralaminar thalamic nuclei. Telencephalic activity was found in the lateral putamen and certain parts of the prefrontal, temporal, parietal, and insular cortices, exclusively on the right side. Remarkably, strong ejaculation-related activation was also observed in the cerebellum. It is worthy of note that the phrenologist Franz Joseph Gall (1757–1829) hypothesized almost two centuries ago that the cerebellum plays an important role in sexual excitation, albeit on rather trivial grounds. The neurohistorian Stanley Finger [104] relates that Gall's hypothesis was "based on a boyhood observation, in which he associated thick necks, like those found in a bull or a stallion, with large cerebellums and sexual prowess" (l.c.p. 211).

Defensive Behaviour

Threatening stimuli may evoke a characteristic pattern of behavioural and autonomic responses commonly designated the affective defense reaction (Fig. 10.11). In cats, the somatomotor components of this reaction include ear retraction, hissing, growling, and striking with the forepaw. The autonomic response pattern consists of pupillary dilatation, piloerection, increased respiratory rate and amplitude, and a sympathetic cardiovascular reaction comprising accelerated heart rate, elevated blood pressure, and vasoconstriction in cutaneous, renal, and splanchnic beds but vasodilatation in skeletal muscles. The classic work of Hess and Brügger [145], Hunsperger [172, 173], and others has shown that this defense response can be elicited by electrical stimulation of a brain area extending from the hypothalamus into the mesencephalic grey matter. Later it appeared that defensive behaviour in cats can

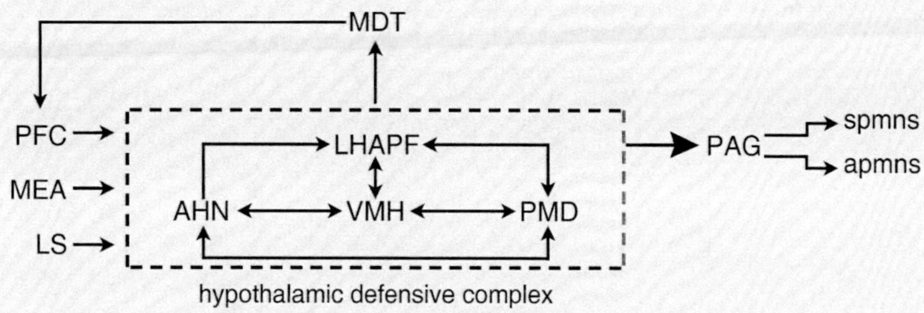

Fig. 10.11. Diagram of the principal circuit mediating defensive behaviour. Largely based on [54, 298]. *AHN*, anterior hypothalamic nucleus; *LHAPF*, lateral hypothalamic area, perifornical part; *LS*, lateral septal nucleus; *MDT*, mediodorsal thalamic nucleus; *MEA*, medial amygdaloid nucleus; *PAG*, periaqueductal grey; *PFC*, prefrontal cortex; *PMD*, dorsal premamillary nucleus; *spmns*, somatic premotor neurons; *apmns*, autonomic premotor neurons; *VMH*, ventromedial hypothalamic nucleus

also be elicited by electrical stimulation of the amygdala, particularly its medial nucleus [25, 71, 147, 179]. Fernandez de Molina and Hunsperger [34] showed that electrical stimulation along a continuous zone that includes the amygdala, stria terminalis, hypothalamus, and periaqueductal grey elicits defensive behaviour. According to Hilton and colleagues [71, 147] this zone continues through the pontomedullary tegmentum caudally to the level of the dorsal motor nucleus of the vagus nerve. Lesion and electrical stimulation experiments have shown that the lateral septal nucleus (LS) is involved in defensive behaviour as well [34, 331]. This nucleus is known to receive its principal input from the hippocampal formation and be strongly and reciprocally connected with the hypothalamus. Tumors in the human septal area are also associated with heightened defensiveness [8].

Within the hypothalamus, several structures have been reported to be directly involved in defensive behaviour. Hess and Brügger [145] found that in the cat this behaviour can readily be elicited from a limited zone centred in the *perifornical part of the lateral hypothalamic area* (LHApf). A comparable area has been identified in the rat [202, 298]. According to Fuchs et al. [110, 111], the hypothalamic zone from which defensive behaviour can be elicited electrically in cats also includes the *ventromedial* (VMH) and *anterior hypothalamic nuclei* (AH). They presented anatomical and physiological evidence that the principal pathway subserving affective defensive behaviour in this species passes from the VMH via the AHN to the PAG. These authors emphasized that the AHN receives input from key structures such as the amygdala and septal area and may well subserve an integrating function with respect to affective defense behaviour.

In the medial hypothalamic zone of the rat, not only the AH and the VM but also the *dorsal premamillary nucleus* (PMD) is involved in defensive behaviour. In this species all three of these nuclei presented abundant c-Fos immunoreactive neurons after the display of defensive behavioural responses [56]. Moreover, these nuclei have been shown to be highly interconnected in a neural system clearly distinct from those mediating ingestive and reproductive behaviours [54, 55, 295, 362].

In the rat an area encompassing the perifornical zone of the lateral hypothalamic area plus the ventrolateral pole of the VMH has been designated the *hypothalamic "attack area"* (HAA) [298]. This area has been shown to project to the lateral septal nucleus, mediodorsal thalamic

Table 10.4. Hypothalamic functions and their principal (●) and other (○) nodal points. For abbreviations, see Table 10.1

	LPA	MPN	AHN	PVH	SO	SCH	DMH	VMH	INF	TM	PMD	PMV	PHA	LHA
Circadian rhythms		○		○		●	○							
Sleep-wake cycle	●									●				●
Stress response				●										
Thermoregulation		●	●										●	
Food intake				●			●		●					●
Thirst				●	●		○		○					
Sexual behaviour		●				○		●				●		
Defensive behaviour			●					●			●			●

nucleus (MDT), PAG, and various structures in the mesencephalic and rhombencephalic tegmentum [298]. The MDT, also associated with aggressive behaviour [22], projects massively to the prefrontal cortex, which in turn exerts an inhibitory influence on hypothalamically elicited attack behaviour [86].

The cell groups together forming the *hypothalamic defensive complex* (Fig. 10.11) project to the PAG via the dorsal longitudinal fasciculus of Schütz [54, 214, 298, 402]. As already mentioned, this mesencephalic structure is subdivisible in four longitudinal zones or columns: dorsal, dorsolateral, lateral, and ventrolateral. Electrical or chemical stimulation of the lateral column evokes defensive behaviour associated with increased heart rate, blood pressure, and respiration and a shift in blood from the viscera to the hind limb muscles [57], whereas stimulation of the ventrolateral column elicits immobility (freezing) and decreased blood pressure and heart rate [24]. Efferents from the PAG which mediate these responses, descend to somatic (spmns) and autonomic premotor neuron pools (apmns) in the lower brain stem and spinal cord [24, 155].

Table 10.5. Hypothalamic functional centres and their overall localization. For abbreviations, see Table 10.1

Circadian pacemaker	SCH
Sleep-promoting region	VLPA
Wake-promoting region	TM
Rostral heat control centre	MPN-AHN
Caudal heat control centre	PHA
Satiety centre	INF
Feeding centre	LHA
Male sex centre	MPN
Female sex centre	VMH
Hypothalamic attack area	LHA-PF

Functional Aspects 3: Summary and Concluding Remarks

1. In the preceding section, a survey of the organization of specific hypothalamic functions is presented.
2. It appears that all of these functions have characteristic neuronal networks and that all of these networks extend well beyond the hypothalamus (Figs. 10.4–10.11).
3. The various networks are not mutually independent; rather they share numerous nodal points (Table 10.4) and fibre paths.
4. Physiological experiments (lesions, electrical or chemical stimulation) led to the functional characterization of a number of hypothalamic loci or "centres" (Table 10.5).

5. Known histological landmarks are often used as a reference system for localizing these functional centres (Tables 10.4, 10.5). However, it should be emphasized that most do not coincide with distinct cytoarchitectonic entities. In fact, many behaviourally determined hypothalamic centres show considerable overlap, as do their efferent connections [298].

References

1. Abrahamson EE, Moore RY (2001) The posterior hypothalamic area: chemoarchitecture and afferent connections. Brain Res 889:1–22
2. Abrahamson EE, Leak RK, Moore RY (2001) The suprachiasmatic nucleus projects to posterior hypothalamic arousal systems. Neuroreport 12:435–440
3. Ahima RS, Prabakaran D, Mantzoros C et al (1996) Role of leptin in the neuroendocrine response to fasting. Nature 382:250–252
4. Airaksinen MS (1992) Multiple neurotransmitters in the tuberomammillary nucleus: comparison of rat, mouse, and guinea pig. J Comp Neurol 323:103–116
5. Airaksinen MS, Paetau A, Paljarvi L et al (1991) Histamine neurons in human hypothalamus: anatomy in normal and Alzheimer diseased brains. Neuroscience 44:465–481
6. Akert TK (1965) The anatomical substrate of sleep. Progr Brain Res 18:9–19
7. Alam MN, Gong H, Alam T et al (2002) Sleep-waking discharge patterns of neurons recorded in the rat perifornical lateral hypothalamic area. J Physiol 538:619–631
8. Albert DJ, Walsh ML, Jonik RH (1993) Aggression in humans: what is its biological foundation? Neurosci Biobehav Rev 17:405–425
9. Allen GV, Hopkins DA (1989) Mamillary body in the rat: topography and synaptology of projections from the subicular complex, prefrontal cortex, and midbrain tegmentum. J Comp Neurol 286:311–336
10. Allen JP, Allen CF (1975) Amygdalar participation in tonic ACTH secretion in the rat. Neuroendocrinology 19:115–125
11. Allen LS, Gorski RA (1990) Sex difference in the bed nucleus of the stria terminalis of the human brain. J Comp Neurol 302:697–706
12. Amaral DG, Veazey RB, Cowan WM (1982) Some observations on hypothalamo-amygdaloid connections in the monkey. Brain Res 252:13–27
13. Anand BK, Brobeck JR (1951) Hypothalamic control of food intake in rats and cats. Yale J Biol Med 24:123–140
14. Anand BK, Brobeck JR (1951) Localization of a "feeding center" in the hypothalamus of the rat. Proc Soc Exp Biol Med 77:323–324
15. Anderson CH, Shen CL (1980) Efferents of the medial preoptic area in the guinea pig: an autoradiographic study. Brain Res Bull 5:257–265
16. Antunes JL, Carmel PW, Zimmerman EA (1977) Projections from the paraventricular nucleus to the zona externa of the median eminence of the rhesus monkey: an immunohistochemical study. Brain Res 137:1–10
17. Arnault P, Roger M (1987) The connections of the peripeduncular area studied by retrograde and anterograde transport in the rat. J Comp Neurol 258:463–476
18. Aston-Jones G, Chen S, Zhu Y, Oshinsky ML (2001) A neural circuit for circadian regulation of arousal. Nat Neurosci 4:732–738
19. Bai FL, Yamano M, Shiotani Y et al (1985) An arcuato-paraventricular and -dorsomedial hypothalamic neuropeptide Y-containing system which lacks noradrenaline in the rat. Brain Res 331:172–175
20. Baker RA, Herkenham M (1995) Arcuate nucleus neurons that project to the hypothalamic paraventricular nucleus: neuropeptidergic identity and consequences of adrenalectomy on mRNA levels in the rat. J Comp Neurol 358:518–530
21. Bali B, Kovacs KJ (2003) GABAergic control of neuropeptide gene expression in parvocellular neurons of the hypothalamic paraventricular nucleus. Eur J Neurosci 18:1518–1526
22. Bandler RJ (1971) Direct chemical stimulation of the thalamus: effects on aggressive behavior in the rat. Brain Res 26:81–93
23. Bandler R, Depaulis A (1991) Midbrain periaqueductal gray control of defensive behavior in the cat and the rat. In: Depaulis A, Bandler R (eds) The midbrain periaqueductal gray matter: functional, anatomical and neurochemical organization. Plenum, New York, pp 175–198
24. Bandler R, Keay KA (1996) Columnar organization in the midbrain periaqueductal gray and the integration of emotional expression. Progr Brain Res 107:285–300
25. Bandler R, Carrive P, Depaulis A (1991) Emerging principles of organization of the midbrain periaqueductal gray matter. In: Depaulis A, Bandler R (eds) The midbrain periaqueductal gray matter: functional, anatomical and neurochemical organization. Plenum, New York, pp 1–8
26. Bandler R, Carrive P, Zhang SP (1991) Integration of somatic and autonomic reactions within the midbrain periaqueductal grey: viscerotopic, somatotopic and functional organization. Progr Brain Res 87:269–305
27. Barry J (1976) Characterization and topography of LH-RH neurons in the human brain. Neurosci Lett 3:287–291
28. Barry J (1977) Immunofluorescence study of LRF neurons in man. Cell Tissue Res 181:1–14

29. Bayer L, Eggermann E, Serafin M et al (2001) Orexins (hypocretins) directly excite tuberomammillary neurons. Eur J Neurosci 14:1571–1575
30. Beitz AJ (1982) The organization of afferent projections to the midbrain periaqueductal gray of the rat. Neuroscience 7:133–159
31. Berk ML, Finkelstein JA (1981) Afferent projections to the preoptic area and hypothalamic regions in the rat brain. Neuroscience 6:1601–1624
32. Bernardis LL, Bellinger LL (1987) The dorsomedial hypothalamic nucleus revisited: 1986 update. Brain Res 434:321–381
33. Bittencourt JC, Presse F, Arias C et al (1992) The melanin-concentrating hormone system of the rat brain: an immuno- and hybridization histochemical characterization. J Comp Neurol 319:218–245
34. Blanchard RJ, Blanchard DC (1968) Limbic lesions and reflexive fighting. J Comp Physiol Psychol 66: 603–605
35. Bons N, Combes A, Szafarczyk A, Assenmacher I (1983) Extrahypothalamic efferent connections from the suprachiasmatic nucleus in the rat. C R Séances Acad Sci [III] 297:347–350
36. Borison HL (1989) Area postrema: chemoreceptor circumventricular organ of the medulla oblongata. Progr Neurobiol 32:351–390
37. Bos NPA, Mirmiran M (1990) Circadian rhythms in spontaneous neuronal discharges of the cultured suprachiasmatic nucleus. Brain Res 511:158–162
38. Boudaba C, Schrader LA, Tasker JG (1997) Physiological evidence for local excitatory synaptic circuits in the rat hypothalamus. J Neurophysiol 77: 3396–3400
39. Boulant JA (1996) Hypothalamic neurons regulating body temperature. In: Fregly MJ, Blatteis CM (eds) APS handbook of physiology. Section 4: Environmental physiology. Oxford University Press, New York, pp 105–126
40. Bourgin P, Huitron-Resendiz S, Spier AD et al (2000) Hypocretin-1 modulates rapid eye movement sleep through activation of locus coeruleus neurons. J Neurosci 20:7760–7765
41. Bourque CW, Oliet SH (1997) Osmoreceptors in the central nervous system. Annu Rev Physiol 59: 601–619
42. Broberger C, De Lecea L, Sutcliffe JG, Hökfelt T (1998) Hypocretin/orexin- and melanin-concentrating hormone-expressing cells form distinct populations in the rodent lateral hypothalamus: relationship to the neuropeptide Y and agouti gene-related protein systems. J Comp Neurol 402:460–474
43. Brown RE, Sergeeva O, Eriksson KS, Haas HL (2001) Orexin A excites serotonergic neurons in the dorsal raphe nucleus of the rat. Neuropharmacology 40:457–459
44. Buijs RM (1978) Intra- and extrahypothalamic vasopressin and oxytocin pathways in the rat. Pathways to the limbic system, medulla oblongata and spinal cord. Cell Tissue Res 192:423–435
45. Buijs RM, Chun SJ, Niijima A, Romijn HJ, Nagai K (2001) Parasympathetic and sympathetic control of the pancreas: a role for the suprachiasmatic nucleus and other hypothalamic centers that are involved in the regulation of food intake. J Comp Neurol 431:405–423
46. Buijs RM, Markman M, Nunes-Cardoso B, Hou YX, Shinn S (1993) Projections of the suprachiasmatic nucleus to stress-related areas in the rat hypothalamus: a light and electron microscopic study. J Comp Neurol 335:42–54
47. Buijs RM, Wortel J, Van Heerikhuize JJ et al (1999) Anatomical and functional demonstration of a multisynaptic suprachiasmatic nucleus adrenal (cortex) pathway. Eur J Neurosci 11:1535–1544
48. Buijs RM, Wortel J, Van Heerikhuize JJ, Kalsbeek A (1997) Novel environment induced inhibition of corticosterone secretion: physiological evidence for a suprachiasmatic nucleus mediated neuronal hypothalamo-adrenal cortex pathway. Brain Res 758:229–236
49. Burstein R (1996) Somatosensory and visceral input to the hypothalamus and limbic system. Progr Brain Res 107:257–267
50. Burstein R, Cliffer KD, Giesler GJ Jr (1987) Direct somatosensory projections from the spinal cord to the hypothalamus and telencephalon. J Neurosci 7:4159–4164
51. Burstein R, Wang JL, Elde RP, Giesler GJ Jr (1990) Neurons in the sacral parasympathetic nucleus that project to the hypothalamus do not also project through the pelvic nerve – a double labeling study combining Fluoro-gold and cholera toxin B in the rat. Brain Res 506:159–165
52. Burstein R, Falkowsky O, Borsook D, Strassman A (1996) Distinct lateral and medial projections of the spinohypothalamic tract of the rat. J Comp Neurol 373:549–574
53. Campfield LA, Smith FJ, Guisez Y, Devos R, Burn P (1995) Recombinant mouse OB protein: evidence for a peripheral signal linking adiposity and central neural networks. Science 269:546–549
54. Canteras NS, Simerly RB, Swanson LW (1992) Projections of the ventral premammillary nucleus. J Comp Neurol 324:195–212
55. Canteras NS, Simerly RB, Swanson LW (1994) Organization of projections from the ventromedial nucleus of the hypothalamus: a Phaseolus vulgaris-leucoagglutinin study in the rat. J Comp Neurol 348:41–79
56. Canteras NS, Chiavegatto S, Valle LE, Swanson LW (1997) Severe reduction of rat defensive behavior to a predator by discrete hypothalamic chemical lesions. Brain Res Bull 44:297–305
57. Carrive P, Morgan MM (2004) Periaqueductal gray. In: Paxinos G, Mai JK (eds) The human nervous system, 2nd edn. Elsevier, Amsterdam, pp 393–423
58. Chen XM, Hosono T, Yoda T, Fukuda Y, Kanosue K (1998) Efferent projection from the preoptic area for the control of non-shivering thermogenesis in rats. J Physiol 512:883–892

59. Cheung CC, Clifton DK, Steiner RA (1997) Proopiomelanocortin neurons are direct targets for leptin in the hypothalamus. Endocrinology 138:4489–4492
60. Chi CC (1970) An experimental silver study of the ascending projections of the central gray substance and adjacent tegmentum in the rat with observations in the cat. J Comp Neurol 139:259–272
61. Chou TC, Bjorkum AA, Gaus SE et al (2002) Afferents to the ventrolateral preoptic nucleus. J Neurosci 22:977–990
62. Chou TC, Scammell TE, Gooley JJ et al (2003) Critical role of dorsomedial hypothalamic nucleus in a wide range of behavioral circadian rhythms. J Neurosci 23:10691–10702
63. Ciriello J, Calaresu FR (1980) Autoradiographic study of ascending projections from cardiovascular sites in the nucleus tractus solitarii in the cat. Brain Res 186:448–453
64. Ciriello J, Calaresu FR (1980) Monosynaptic pathway from cardiovascular neurons in the nucleus tractus solitarii to the paraventricular nucleus in the cat. Brain Res 193:529–533
65. Ciriello J, Caverson MM (1984) Direct pathway from neurons in the ventrolateral medulla relaying cardiovascular afferent information to the supraoptic nucleus in the cat. Brain Res 292:221–228
66. Clancy AN, Whitman C, Michael RP, Albers HE (1994) Distribution of androgen receptor-like immunoreactivity in the brains of intact and castrated male hamsters. Brain Res Bull 33:325–332
67. Cliffer KD, Burstein R, Giesler GJ Jr (1991) Distributions of spinothalamic, spinohypothalamic, and spinotelencephalic fibers revealed by anterograde transport of PHA-L in rats. J Neurosci 11:852–868
68. Colwell CS (2005) Bridging the gap: coupling single-cell oscillators in the suprachiasmatic nucleus. Nat Neurosci 8:10–12
69. Colwell CS, Michel S (2003) Sleep and circadian rhythms: do sleep centers talk back to the clock? Nat Neurosci 6:1005–1006
70. Conrad LCA, Pfaff DW (1976) Efferents from medial basal forebrain and hypothalamus in the rat. I. An autoradiographic study of the medial preoptic area. J Comp Neurol 169:185–220
71. Coote JH, Hilton SM, Zbrozyna AW (1973) The ponto-medullary area integrating the defence reaction in the cat and its influence on muscle blood flow. J Physiol 229:257–274
72. Crosby EC, Woodburne RT (1940) The comparative anatomy of the preoptic area and the hypothalamus. Res Publ Assoc Nerv Ment Dis 20:52–169
73. Crosby EC, Humphrey T, Lauer EW (1962) Correlative anatomy of the nervous system. MacMillan, New York
74. Cruce JAF (1977) An autoradiographic study of the descending connections of the mammillary nuclei of the rat. J Comp Neurol 176:631–644
75. Csaki A, Kocsis K, Kiss J, Halasz B (2002) Localization of putative glutamatergic/aspartatergic neurons projecting to the supraoptic nucleus area of the rat hypothalamus. Eur J Neurosci 16:55–68
76. Cullinan WE, Herman JP, Watson SJ (1993) Ventral subicular interaction with the hypothalamic paraventricular nucleus: evidence for a relay in the bed nucleus of the stria terminalis. J Comp Neurol 332:1–20
77. Cunningham ET Jr, Sawchenko PE (1991) Reflex control of magnocellular vasopressin and oxytocin secretion. Trends Neurosci 14:406–411
78. Dai J (1998) The anatomical organization of the human hypothalamus in relation to circadian rhythmus: immunohistochemical and postmortem tracing studies. Thesis, Graduate School Neurosciences Amsterdam, Netherlands Institute for Brain Research, University of Amsterdam
79. Dai J, Swaab DF, Buijs RM (1997) Distribution of vasopressin and vasoactive intestinal polypeptide (VIP) fibers in the human hypothalamus with special emphasis on suprachiasmatic nucleus efferent projections. J Comp Neurol 383:397–414
80. Dai J, Swaab DF, Buijs RM (1998) Recovery of axonal transport in "dead neurons". Lancet 351:499–500
81. Dai J, Swaab DF, Van der Vliet J, Buijs RM (1998) Postmortem tracing reveals the organization of hypothalamic projections of the suprachiasmatic nucleus in the human brain. J Comp Neurol 400:87–102
82. Dai J, Van der Vliet J, Swaab DF, Buijs RM (1998) Human retinohypothalamic tract as revealed by in vitro postmortem tracing. J Comp Neurol 397:357–370
83. Dai J, Van der Vliet J, Swaab DF, Buijs RM (1998) Postmortem anterograde tracing of intrahypothalamic projections of the human dorsomedial nucleus of the hypothalamus. J Comp Neurol 401:16–33
84. Dampney RA, Czachurski J, Dembowsky K, Goodchild AK, Seller H (1987) Afferent connections and spinal projections of the pressor region in the rostral ventrolateral medulla of the cat. J Auton Nerv Syst 20:73–86
85. Date Y, Ueta Y, Yamashita H et al (1999) Orexins, orexigenic hypothalamic peptides, interact with autonomic, neuroendocrine and neuroregulatory systems. Proc Natl Acad Sci USA 96:748–753
86. De Bruin JPC (2004) Orbital prefrontal cortex, dopamine and social-agonistic behavior of male Long-Evans rats. Aggress Behav 16:231–248
87. De Lecea L, Kilduff TS, Peyron C et al (1998) The hypocretins: hypothalamus-specific peptides with neuroexcitatory activity. Proc Natl Acad Sci USA 95:322–327
88. Deboer T, Vansteensel MJ, Détári L, Meijer JH (2003) Sleep states alter activity of suprachiasmatic nucleus neurons. Nat Neurosci 6:1086–1090
89. Del Abril A, Segovia S, Guillamon A (1987) The bed nucleus of the stria terminalis in the rat: regional sex differences controlled by gonadal steroids early after birth. Brain Res 429:295–300
90. Delgado JM, Anand BK (1953) Increase of food intake induced by electrical stimulation of the lateral hypothalamus. Am J Physiol 172:162–168

91. Dong HW, Swanson LW (2003) Projections from the rhomboid nucleus of the bed nuclei of the stria terminalis: implications for cerebral hemisphere regulation of ingestive behaviors. J Comp Neurol 463:434–472
92. Dong HW, Swanson LW (2004) Organization of axonal projections from the anterolateral area of the bed nuclei of the stria terminalis. J Comp Neurol 468:277–298
93. Dudas B, Mihaly A, Merchenthaler I (2000) Topography and associations of luteinizing hormone-releasing hormone and neuropeptide Y-immunoreactive neuronal systems in the human diencephalon. J Comp Neurol 427:593–603
94. Eberhart JA, Morrell JI, Krieger MS, Pfaff DW (1985) An autoradiographic study of projections ascending from the midbrain central gray, and from the region lateral to it, in the rat. J Comp Neurol 241:285–310
95. Edley SM, Graybiel AM (1983) The afferent and efferent connections of the feline nucleus tegmenti pedunculopontinus, pars compacta. J Comp Neurol 217:187–215
96. Eggermann E, Serafin M, Bayer L et al (2001) Orexins/hypocretins excite basal forebrain cholinergic neurones. Neuroscience 108:177–181
97. Elias CF, Saper CB, Maratos-Flier E et al (1998) Chemically defined projections linking the mediobasal hypothalamus and the lateral hypothalamic area. J Comp Neurol 402:442–459
98. Ericson H, Blomqvist A, Kohler C (1991) Origin of neuronal inputs to the region of the tuberomammillary nucleus of the rat brain. J Comp Neurol 311:45–64
99. Fahrbach SE, Morrell JI, Pfaff DW (1986) Identification of medial preoptic neurons that concentrate estradiol and project to the midbrain in the rat. J Comp Neurol 247:364–382
100. Fei H, Okano HJ, Li C et al (1997) Anatomic localization of alternatively spliced leptin receptors (Ob-R) in mouse brain and other tissues. Proc Natl Acad Sci USA 94:7001–7005
101. Feldman S, Conforti N (1980) Participation of the dorsal hippocampus in the glucocorticoid feedback effect on adrenocortical activity. Neuroendocrinology 30:52–55
102. Feldman JL, Loewy AD, Speck DF (1985) Projections from the ventral respiratory group to phrenic and intercostal motoneurons in cat: an autoradiographic study. J Neurosci 5:1993–2000
103. Fernández-Guasti A, Kruijver FP, Fodor M, Swaab DF (2000) Sex differences in the distribution of androgen receptors in the human hypothalamus. J Comp Neurol 425:422–435
104. Finger S (1994) Origins of neuroscience. Oxford University Press, New York
105. Flier JS, Maratos-Flier E (1998) Obesity and the hypothalamus: novel peptides for new pathways. Cell 20, 92:437–440
106. Fliers E, Noppen NW, Wiersinga WM, Visser TJ, Swaab DF (1994) Distribution of thyrotropin-releasing hormone (TRH)-containing cells and fibers in the human hypothalamus. J Comp Neurol 350:311–323
107. Fodor M, Gorcs TJ, Palkovits M (1992) Immunohistochemical study on the distribution of neuropeptides within the pontine tegmentum-particularly the parabrachial nuclei and the locus coeruleus of the human brain. Neuroscience 46:891–908
108. Foote SL, Morrison JH (1987) Extrathalamic modulation of cortical function. Annu Rev Neurosci 10:67–95
109. Freedman LJ, Insel TR, Smith Y (2000) Subcortical projections of area 25 (subgenual cortex) of the macaque monkey. J Comp Neurol 421:172–188
110. Fuchs SAG, Edinger HM, Siegel A (1985) The organization of the hypothalamic pathways mediating affective defense behavior in the cat. Brain Res 330:77–92
111. Fuchs SAG, Edinger HM, Siegel A (1985) The role of the anterior hypothalamus in affective defense behavior elicited from the ventromedial hypothalamus of the cat. Brain Res 330:93–107
112. Fulwiler CE, Saper CB (1984) Subnuclear organization of the efferent connections of the parabrachial nucleus in the rat. Brain Res Rev 7:229–259
113. Gallopin T, Fort P, Eggermann E et al (2000) Identification of sleep-promoting neurons in vitro. Nature 404:992–995
114. Gervasoni D, Peyron C, Rampon C et al (2000) Role and origin of the GABAergic innervation of dorsal raphe serotonergic neurons. J Neurosci 20:4217–4225
115. Gold RM (1973) Hypothalamic obesity: the myth of the ventromedial nucleus. Science 182:488–490
116. Gorski RA, Gordon JH, Shryne JE, Southam AM (1978) Evidence for a morphological sex difference within the medial preoptic area of the rat brain. Brain Res 148:333–346
117. Gorski RA, Harlan RE, Jacobson CD, Shryne JE, Southam AM (1980) Evidence for the existence of a sexually dimorphic nucleus in the preoptic area of the rat. J Comp Neurol 193:529–539
118. Groenewegen HJ, Russchen FT (1984) Organization of the efferent projections of the nucleus accumbens to pallidal hypothalamic, and mesencephalic structures: a tracing and immunohistochemical study in the cat. J Comp Neurol 223:347–367
119. Groenewegen HJ, Van Dijk CA (1984) Efferent connections of the dorsal tegmental region in the rat, studied by means of anterograde transport of the lectin Phaseolus vulgaris-leucoagglutinin (PHA-L). Brain Res 304:367–371
120. Haas H, Panula P (2003) The role of histamine and the tuberomamillary nucleus in the nervous system. Nat Rev Neurosci 4:121–130
121. Haber SN, Wolfe DP, Groenewegen HJ (1990) The relationship between ventral striatal efferent fibers and the distribution of peptide-positive woolly fibers in the forebrain of the rhesus monkey. Neuroscience 39:323–338

122. Haglund L, Swanson LW, Köhler C (1984) The projection of the supramammillary nucleus to the hippocampal formation: an immunohistochemical and anterograde transport study with the lectin PHA-L in the rat. J Comp Neurol 229:171–185
123. Hakansson ML, Brown H, Ghilardi N, Skoda RC, Meister B (1998) Leptin receptor immunoreactivity in chemically defined target neurons of the hypothalamus. J Neurosci 18:559–572
124. Hakansson M, De Lecea L, Sutcliffe JG, Yanagisawa M, Meister B (1999) Leptin receptor- and STAT3-immunoreactivities in hypocretin/orexin neurons of the lateral hypothalamus. J Neuroendocrinol 11:653–663
125. Halaas JL, Gajiwala KS, Maffei M et al (1995) Weight-reducing effects of the plasma protein encoded by the obese gene. Science 269:543–546
126. Hallanger AE, Wainer BH (1988) Ultrastructure of ChAT-immunoreactive synaptic terminals in the thalamic reticular nucleus of the rat. J Comp Neurol 278:486–497
127. Hallbeck M, Blomqvist A (1999) Spinal cord-projecting vasopressinergic neurons in the rat paraventricular hypothalamus. J Comp Neurol 411: 201–211
128. Halpern M, Martinez-Marcos A (2003) Structure and function of the vomeronasal system: an update. Progr Neurobiol 70:245–318
129. Hancock MB (1976) Cells of origin of hypothalamospinal projections in the rat. Neurosci Lett 3:179–184
130. Hannibal J (2002) Neurotransmitters of the retino-hypothalamic tract. Cell Tissue Res 309:73–88
131. Hansen S, Kohler C, Goldstein M, Steinbusch HV (1982) Effects of biotenic acid-induced neuronal degeneration in the medial preoptic area and the lateral hypothalamic area on sexual behavior in the male rat. Brain Res 239:213–232
132. Hardy SGP, Leichnetz GR (1981) Cortical projections to the periaqueductal gray in the monkey: a retrograde and orthograde horseradish peroxidase study. Neurosci Lett 22:97–101
133. Hay-Schmidt A, Vrang N, Larsen PJ, Mikkelsen JD (2003) Projections from the raphe nuclei to the suprachiasmatic nucleus of the rat. J Chem Neuroanat 25:293–310
134. Haymaker W (1969) Bing's local diagnosis in neurological diseases, 15th edn. Mosby, St. Louis
135. Haymaker W (1969) Hypothalamo-pituitary neural pathways and the circulatory system of the pituitary. In: Haymaker W, Anderson E (eds) The hypothalamus. Charles C Thomas, Springfield, pp 219–250
136. Hayward JN (1977) Functional and morphological aspects of hypothalamic neurons. Physiol Rev 57:574–658
137. Hedreen JC, Struble RG, Whitehouse PJ, Price DL (1984) Topography of the magnocellular basal forebrain system in the human brain. J Neuropathol Exp Neurol 43:1–21
138. Heimer L (2000) Basal forebrain in the context of schizophrenia. Brain Res Brain Res Rev 31:205–235
139. Herbison AE, Horvath TL, Naftolin F, Leranth C (1995) Distribution of estrogen receptor-immunoreactive cells in monkey hypothalamus: relationship to neurones containing luteinizing hormone-releasing hormone and tyrosine hydroxylase. Neuroendocrinology 61:1–10
140. Herman JP, Schafer MK, Young EA et al (1989) Evidence for hippocampal regulation of neuroendocrine neurons of the hypothalamo-pituitary-adrenocortical axis. J Neurosci 9:3072–3082
141. Herman JP, Cullinan WE, Young EA, Akil H, Watson SJ (1992) Selective forebrain fiber tract lesions implicate ventral hippocampal structures in tonic regulation of paraventricular nucleus corticotropin-releasing hormone (CRH) and arginine vasopressin (AVP) mRNA expression. Brain Res 592:228–238
142. Herman JP, Cullinan WE, Morano MI, Akil H, Watson SJ (1995) Contribution of the ventral subiculum to inhibitory regulation of the hypothalamo-pituitary-adrenocortical axis. J Neuroendocrinol 7:475–482
143. Hermann DM, Luppi PH, Peyron C, Hinckel P, Jouvet M (1997) Afferent projections to the rat nuclei raphe magnus, raphe pallidus and reticularis gigantocellularis pars alpha demonstrated by iontophoretic application of choleratoxin (subunit b). J Chem Neuroanat 13:1–21
144. Herman JP, Tasker JG, Ziegler DR, Cullinan WE (2002) Local circuit regulation of paraventricular nucleus stress integration: glutamate-GABA connections. Pharmacol Biochem Behav 71:457–468
145. Hess WR, Brügger M (1943) Das subkortikale Zentrum der affektiven Abwehrreaktion. Helv Physiol Pharmacol Acta 1:33–52
146. Hetherington AW (1940) Hypothalamic lesions and adiposity in the rat. Anat Rec 78:149–172
147. Hilton SM, Zbrozyna AW (1963) Amygdaloid region for defence reactions and its efferent pathway to the brain stem. J Physiol 165:160–173
148. Hines M, Davis FC, Coquelin A, Goy RW, Gorski RA (1985) Sexually dimorphic regions in the medial preoptic area and the bed nucleus of the stria terminalis of the guinea pig brain: a description and an investigation of their relationship to gonadal steroids in adulthood. J Neurosci 5:40–47
149. Hobson JA, Pace-Schott EF (2002) The cognitive neuroscience of sleep: neuronal systems, consciousness and learning. Nat Rev Neurosci 3:679–693
150. Hofman MA, Swaab DF (1989) The sexually dimorphic nucleus of the preoptic area in the human brain: a comparative morphometric study. J Anat 164:55–72
151. Hofman MA, Swaab DF (1992) The human hypothalamus: comparative morphometry and photoperiodic influences. In: Swaab DF, Hofman MA, Mirmiran M, Ravid R, Van Leeuwen FW (eds)

The human hypothalamus in health and disease. Elsevier, Amsterdam, pp 133–149 (Progress in Brain Research Vol 93)
152. Hofman MA, Swaab DF (1993) Diurnal and seasonal rhythms of neuronal activity in the suprachiasmatic nucleus of humans. J Biol Rhythms 8:283–295
153. Holstege G (1987) Some anatomical observations on the projections from the hypothalamus to brainstem and spinal cord: an HRP and autoradiographic tracing study in the cat. J Comp Neurol 260:98–126
154. Holstege G (1989) Anatomical study of the final common pathway for vocalization in the cat. J Comp Neurol 284:242–252
155. Holstege G (1998) The emotional motor system in relation to the supraspinal control of micturition and mating behavior. Behav Brain Res 92:103–109
156. Holstege G, Georgiadis JR (2003) Neurobiology of cat and human sexual behavior. Int Rev Neurobiol 56:213–225
157. Holstege G, Georgiadis JR (2004) The emotional brain: neural correlates of cat sexual behavior and human male ejaculation. Progr Brain Res 143:39–45
158. Holstege G, Kuypers HGJM (1982) The anatomy of brain stem pathways to the spinal cord in cat. A labeled amino acid tracing study. Prog Brain Res 57:145–175
159. Holstege G, Georgiadis JR, Paans AM et al (2003) Brain activation during human male ejaculation. J Neurosci 23:9185–9193
160. Holstege G, Meiners L, Tan K (1985) Projections of the bed nucleus of the stria terminalis to the mesencephalon, pons, and medulla oblongata in the cat. Exp Brain Res 58:379–391
161. Hoorneman EMD, Buijs RM (1982) Vasopressin fiber pathways in the rat brain following suprachiasmatic nucleus lesioning. Brain Res 243:235–241
162. Hopkins DA, Holstege G (1978) Amygdaloid projections to the mesencephalon, pons and medulla oblongata in the cat. Exp Brain Res 32:529–548
163. Hopkins DA, Niessen LW (1976) Substantia nigra projections to the reticular formation, superior colliculus and central gray in the rat, cat and monkey. Neurosci Lett 2:253–259
164. Horvath TL, Diano S, Van den Pol AN (1999) Synaptic interaction between hypocretin (orexin) and neuropeptide Y cells in the rodent and primate hypothalamus: a novel circuit implicated in metabolic and endocrine regulations. J Neurosci 19:1072–1087
165. Horvath TL, Peyron C, Diano S et al (1999) Hypocretin (orexin) activation and synaptic innervation of the locus coeruleus noradrenergic system. J Comp Neurol 415:145–159
166. Hosoya Y, Matsushita M (1979) Identification and distribution of the spinal and hypophyseal projection neurons in the paraventricular nucleus of the rat. A light and electron microscopic study with the horseradish peroxidase method. Exp Brain Res 35:315–331
167. Hosoya Y, Matsushita M (1980) Cells of origin of the descending afferents to the lateral hypothalamic area in the rat, studied with the horseradish peroxidase method. Neurosci Lett 18:231–236
168. Hosoya Y, Matsushita M (1981) A direct projection from the hypothalamus to the area postrema in the rat, as demonstrated by the HRP and autoradiographic methods. Brain Res 214:144–149
169. Hosoya Y, Matsushita M (1981) Brainstem projections from the lateral hypothalamic area in the rat, as studied with autoradiography. Neurosci Lett 24:111–116
170. Houser CR, Vaughn JE, Barber RP, Roberts E (1980) GABA neurons are the major cell type of the nucleus reticularis thalami. Brain Res 200: 341–354
171. Huang Q, Rivest R, Richard D (1998) Effects of leptin on corticotropin-releasing factor (CRF) synthesis and CRF neuron activation in the paraventricular hypothalamic nucleus of obese (ob/ob) mice. Endocrinology 139:1524–1532
172. Hunsperger RW (1956) Affektreaktionen auf elektrische Reizung im Hirnstamm der Katze. Helv Physiol Acta 14:70–92
173. Hunsperger RW (1956) Role of the substantia grisea centralis mesencephalis in electrically induced rage reactions. Prog Neurobiol 1:289–294
174. Hurley KM, Herbert H, Moga MM, Saper CB (1991) Efferent projections of the infralimbic cortex of the rat. J Comp Neurol 308:249–276
175. Johnson AK (1985) The periventricular anteroventral third ventricle (AV3V): its relationship with the subfornical organ and neural systems involved in maintaining body fluid homeostasis. Brain Res Bull 15:595–601
176. Johnson RF, Morin LP, Moore RY (1988) Retinohypothalamic projections in the hamster and rat demonstrated using cholera toxin. Brain Res 462:301–312
177. Jürgens U, Ploog D (1970) Cerebral representation of vocalization in the squirrel monkey. Exp Brain Res 10:532–554
178. Jürgens U, Pratt R (1979) Role of the periaqueductal gray in vocal expression of emotion. Brain Res 167:367–379
179. Kaada BR (1972) Stimulation and regional ablation of the amygdaloid cortex with reference to functional representations. In: Eleftheriou BE (ed) The neurobiology of the amygdala. Plenum, New York, pp 205–282
180. Kalsbeek A, Teclemariam-Mesbah R, Pevet P (1993) Efferent projections of the suprachiasmatic nucleus in the golden hamster (Mesocricetus auratus). J Comp Neurol 332:293–314
181. Kashon ML, Arbogast JA, Sisk CL (1996) Distribution and hormonal regulation of androgen receptor immunoreactivity in the forebrain of the male European ferret. J Comp Neurol 376:567–586

182. Kawano H, Decker K, Reuss S (1996) Is there a direct retina-raphe-suprachiasmatic nucleus pathway in the rat? Neurosci Lett 212:143–146
183. Keefer DA, Stumpf WE (1975) Estrogen localization in the primate brain. In: Stumpf WE, Grant LD (eds) Anatomical neuroendocrinology. Karger, Basel, pp 153–165
184. Kelly AB, Watts AG (1996) Mediation of dehydration-induced peptidergic gene expression in the rat lateral hypothalamic area by forebrain afferent projections. J Comp Neurol 370:231–246
185. Kievit J, Kuypers HGJM (1975) Basal forebrain and hypothalamic connections to frontal and parietal cortex in the rhesus monkey. Science 187:660–662
186. Kilduff TS, Peyron C (2000) The hypocretin/orexin ligand-receptor system: implications for sleep and sleep disorders. Trends Neurosci 23:359–365
187. Kirouac GJ, Li S, Mabrouk G (2004) GABAergic projection from the ventral tegmental area and substantia nigra to the periaqueductal gray region and the dorsal raphe nucleus. J Comp Neurol 469:170–184
188. Kita H, Oomura Y (1981) Reciprocal connections between the lateral hypothalamus and the frontal cortex in the rat: electrophysiological and anatomical observations. Brain Res 213:1–16
189. Kita H, Oomura Y (1982) An HRP study of the afferent connections to rat lateral hypothalamic region. Brain Res Bull 8:63–71
190. Klein DC, Moore RY, Reppert SM (1991) Suprachiasmatic nucleus: the mind's clock. Oxford University Press, New York
191. Ko EM, Estabrooke IV, McCarthy M, Scammell TE (2003) Wake-related activity of tuberomammillary neurons in rats. Brain Res 992:220–226
192. Kohler C, Ericson H, Watanabe T et al (1986) Galanin immunoreactivity in hypothalamic neurons: further evidence for multiple chemical messengers in the tuberomammillary nucleus. J Comp Neurol 250:58–64
193. Kow LM, Pfaff DW (1998) Mapping of neural and signal transduction pathways for lordosis in the search for estrogen actions on the central nervous system. Behav Brain Res 92:169–180
194. Krayniak PF, Siegel A, Meibach RC, Fruchtman D, Scrimenti M (1979) Origin of the fornix system in the squirrel monkey. Brain Res 160:401–411
195. Krettek JE, Price JL (1978) Amygdaloid projections to subcortical structures within the basal forebrain and brainstem in the rat and cat. J Comp Neurol 178:225–254
196. Krieger MS, Conrad LCA, Pfaff DW (1979) An autoradiographic study of the efferent connections of the ventromedial nucleus of the hypothalamus. J Comp Neurol 183:785–816
197. Kruijver FP, Zhou JN, Pool CW et al (2000) Male-to-female transsexuals have female neuron numbers in a limbic nucleus. J Clin Endocr Metab 85:2034–2041
198. Kruijver FP, Balesar R, Espila AM, Unmehopa UA, Swaab DF (2002) Estrogen receptor-alpha distribution in the human hypothalamus in relation to sex and endocrine status. J Comp Neurol 454:115–139
199. Kruijver FP, Balesar R, Espila AM, Unmehopa UA, Swaab DF (2003) Estrogen-receptor-beta distribution in the human hypothalamus: similarities and differences with ER alpha distribution. J Comp Neurol 466:251–277
200. Krukoff TL, Harris KH, Jhamandas JH (1993) Efferent projections from the parabrachial nucleus demonstrated with the anterograde tracer Phaseolus vulgaris leucoagglutinin. Brain Res Bull 30:163–172
201. Kupfermann I (1991) Hypothalamus and limbic system: peptidergic neurons, homeostasis, and emotional behavior. In: Kandel ER, Schwartz JH, Jessel TM (eds) Principles of neural science. Elsevier, New York, pp 735–749
202. Lammers JH, Kruk MR, Meelis W, Van der Poel AM (1988) Hypothalamic substrates for brain stimulation-induced patterns of locomotion and escape jumps in the rat. Brain Res 449:294–310
203. Larsen PJ, Enquist LW, Card JP (1998) Characterization of the multisynaptic neuronal control of the rat pineal gland using viral transneuronal tracing. Eur J Neurosci 10:128–145
204. Le Gros Clark WE (1936) The topography and homologies of the hypothalamic nuclei in man. J Anat 70:203–214
205. Le Gros Clark WE (1938) Morphological aspects of the hypothalamus. In: Le Gros Clark WE, Beattie J, Riddoch G, Dott NM (eds) The hypothalamus. Oliver and Boyd, Edinburgh, pp 1–68
206. Leichnetz GR, Astruc J (1976) The efferent projections of the medial prefrontal cortex in the squirrel monkey (Saimiri sciureus). Brain Res 109:455–472
207. Levine JD, Weiss ML, Rosenwasser AM, Miselis RR (1991) Retinohypothalamic tract in the female albino rat: a study using horseradish peroxidase conjugated to cholera toxin. J Comp Neurol 306:344–360
208. Lin J (2000) Brain structures and mechanisms involved in the control of cortical activation and wakefulness, with emphasis on the posterior hypothalamus and histaminergic neurons. Sleep Med Rev 4:471–503
209. Loewy AD (1991) Forebrain nuclei involved in autonomic control. Progr Brain Res 87:253–268
210. Loewy AD, McKellar S (1981) Serotonergic projections from the ventral medulla to the intermediolateral cell column in the rat. Brain Res 211:146–152
211. Long MA, Jutras MJ, Connors BW, Burwell RD (2005) Electrical synapses coordinate activity in the suprachiasmatic nucleus. Nat Neurosci 8:61–66
212. Luiten PG, Room P (1980) Interrelations between lateral, dorsomedial and ventromedial hypothala-

mic nuclei in the rat. An HRP study. Brain Res 190:321–332
213. Luiten PGM, Ter Horst GJ, Karst H, Steffens AB (1985) The course of paraventricular hypothalamic efferents to autonomic structures in medulla and spinal cord. Brain Res 329:374–378
214. Luiten PGM, Ter Horst GJ, Steffens AB (1987) The hypothalamus, intrinsic connections and outflow pathways to the endocrine system in relation to the control of feeding and metabolism. Prog Neurobiol 28:1–54
215. Maloney KJ, Mainville L, Jones BE (1999) Differential c-Fos expression in cholinergic, monoaminergic, and GABAergic cell groups of the pontomesencephalic tegmentum after paradoxical sleep deprivation and recovery. J Neurosci 19:3057–3072
216. Malsbury CW, McKay K (1987) A sex difference in the pattern of substance P-like immunoreactivity in the bed nucleus of the stria terminalis. Brain Res 420:365–370
217. Mantyh PW (1982) Forebrain projections to the periaqueductal gray in the monkey, with observations in the cat and rat. J Comp Neurol 206:146–158
218. Mantyh PW (1983) Connections of midbrain periaqueductal gray in the monkey. I. Ascending efferent projections. J Neurophysiol 49:567–581
219. Mantyh PW (1983) Connections of midbrain periaqueductal gray in the monkey. II. Descending efferent projections. J Neurophysiol 49:582–594
220. Marchand JE, Hagino N (1983) Afferents to the periaqueductal gray in the rat. A horseradish peroxidase study. Neuroscience 9:95–106
221. Matsumoto A, Arai Y (1983) Sex difference in volume of the ventromedial nucleus of the hypothalamus in the rat. Endocrinol Jpn 30:277–280
222. Mayer EA, Fanselow MS (2003) Dissecting the components of the central response to stress. Nat Neurosci 6:1011–1012
223. McKellar S, Loewy AD (1981) Organization of some brain stem afferents to the paraventricular nucleus of the hypothalamus in the rat. Brain Res 217:351–357
224. Mehler WR (1969) Some neurological species differences a posteriori. Ann NY Acad Sci 167:424–468
225. Meibach RG, Siegel A (1975) The origin of fornix fibers which project to the mammillary bodies in the rat: a horseradish peroxidase study. Brain Res 88:508–512
226. Meibach RC, Siegel A (1977) Efferent connections of the hippocampal formation of the rat. Brain Res 124:197–224
227. Merari A, Ginton A (1975) Characteristics of exaggerated sexual behavior induced by electrical stimulation of the medial preoptic area in male rats. Brain Res 86:97–108
228. Mercer JG, Hoggard N, Williams LM et al (1996) Coexpression of leptin receptor and preproneuropeptide Y mRNA in arcuate nucleus of mouse hypothalamus. J Neuroendocrinol 8:733–735
229. Mesulam MM, Mufson EJ, Levey AI, Wainer BH (1983) Cholinergic innervation of cortex by the basal forebrain: cytochemistry and cortical connections of the septal area, diagonal band nuclei, nucleus basalis (substantia innominata), and hypothalamus in the rhesus monkey. J Comp Neurol 214:170–197
230. Mesulam MM, Mufson EJ, Wainer BH, Levey AI (1983) Central cholinergic pathways in the rat: an overview based on an alternative nomenclature (Ch1–Ch6). Neuroscience 10:1185–1201
231. Mesulam MM, Mufson EJ, Levey AI, Wainer BH (1984) Atlas of cholinergic neurons in the forebrain and upper brainstem of the macaque based on monoclonal choline acetyltransferase immunohistochemistry and acetylcholinesterase histochemistry. Neuroscience 12:669–686
232. Michael RP, Clancy AN, Zumpe D (1995) Distribution of androgen receptor-like immunoreactivity in the brains of cynomolgus monkeys. J Neuroendocrinol 7:713–719
233. Mikkelsen JD, Vrang N (1994) A direct pretecto-suprachiasmatic projection in the rat. Neuroscience 62:497–505
234. Miklos IH, Kovacs KJ (2002) GABAergic innervation of corticotropin-releasing hormone (CRH)-secreting parvocellular neurons and its plasticity as demonstrated by quantitative immunoelectron microscopy. Neuroscience 113:581–592
235. Miller AD, Tan LK, Lakos SF (1989) Brainstem projections to cats' upper lumbar spinal cord: implications for abdominal muscle control. Brain Res 493:348–356
236. Mirmiran M, Koster-Van Hoffen GC, Bos NPA (1995) Circadian rhythm generation in the cultured suprachiasmatic nucleus. Brain Res Bull 38:275–283
237. Moga MM, Moore RY (1996) Putative excitatory amino acid projections to the suprachiasmatic nucleus in the rat. Brain Res 743:171–177
238. Moga MM, Moore RY (1997) Organization of neural inputs to the suprachiasmatic nucleus in the rat. J Comp Neurol 389:508–534
239. Moga MM, Weis RP, Moore RY (1995) Efferent projections of the paraventricular thalamic nucleus in the rat. J Comp Neurol 359:221–238
240. Moore RY (1973) Retinohypothalamic projection in mammals: a comparative study. Brain Res 49:403–409
241. Moore RY, Lenn NJ (1972) A retinohypothalamic projection in the rat. J Comp Neurol 146:1–14
242. Moore RY, Speh JC, Card JP (1995) The retinohypothalamic tract originates from a distinct subset of retinal ganglion cells. J Comp Neurol 352:351–366
243. Moore RY, Abrahamson EA, Van den Pol AN (2001) The hypocretin neuron system: an arousal system in the human brain. Arch Ital Biol 139:195–205

244. Morest DK (1961) Connexions of the dorsal tegmental nucleus in rat and rabbit. J Anat 95:229–246
245. Morin LP, Goodless-Sanchez N, Smale L, Moore RY (1994) Projections of the suprachiasmatic nuclei, subparaventricular zone and retrochiasmatic area in the golden hamster. Neuroscience 61:391–410
246. Morrison SF (1999) RVLM and raphe differentially regulate sympathetic outflows to splanchnic and brown adipose tissue. Am J Physiol 276: R962–R973
247. Moss RL, McCann SM (1973) Induction of mating behavior in rats by luteinizing hormone-releasing factor. Science 181:177–179
248. Müller MB, Zimmermann S, Sillaber I et al (2003) Limbic corticotropin-releasing hormone receptor 1 mediates anxiety-related behavior and hormonal adaptation to stress. Nat Neurosci 6: 1100–1107
249. Nagai K, Nagai N, Shimizu K et al (1996) SCN output drives the autonomic nervous system: with special reference to the autonomic function related to the regulation of glucose metabolism. Progr Brain Res 111:253–272
250. Nauta WJH (1956) An experimental study of the fornix in the rat. J Comp Neurol 104:247–272
251. Nauta WJH (1961) Fibre degeneration following lesions of the amygdaloid complex in the monkey. J Anat 95:515–531
252. Nauta WJH (1963) Central nervous organization and the endocrine motor system. In: Nalbanov AV (ed) Advances in neuroendocrinology. University of Illinois, Urbana, pp 5–21
253. Nauta WJH (1979) Expanding borders of the limbic system concept. In: Rasmussen T, Marino R (eds) Functional neurosurgery. Raven, New York, pp 7–24
254. Nauta WJH, Haymaker W (1969) Hypothalamic nuclei and fiber connections. In: Haymaker W, Anderson E, Nauta WJH (eds) The hypothalamus. Thomas, Springfield, pp 136–209
255. Nauta WJH, Kuypers HGJM (1958) Some ascending pathways in the brain stem reticular formation. In: Jasper HH et al (eds) Reticular formation of the brain. Little Brown, Toronto, pp 3–31
256. Nieuwenhuys R (1985) Chemoarchitecture of the brain. Springer, Berlin
257. Nieuwenhuys R, Geeraedts LMG, Veening JG (1982) The medial forebrain bundle of the rat. I. General introduction. J Comp Neurol 206:49–81
258. Nishino S, Ripley B, Overeem S, Lammers GJ, Mignot E (2000) Hypocretin (orexin) deficiency in human narcolepsy. Lancet 355:39–40
259. Nitz D, Siegel J (1997) GABA release in the dorsal raphe nucleus: role in the control of REM sleep. Am J Physiol 273:R451–R455
260. Nitz D, Siegel JM (1997) GABA release in the locus coeruleus as a function of sleep/wake state. Neuroscience 78:795–801
261. Nowak KW, Mackowiak P, Switonska MM, Fabis M, Malendowicz LK (2000) Acute orexin effects on insulin secretion in the rat: in vivo and in vitro studies. Life Sci 66:449–454
262. Olszewski J, Baxter D (1954) Cytoarchitecture of the human brain stem. Karger, Basel
263. Öngür D, An X, Price JL (1998) Prefrontal cortical projections to the hypothalamus in macaque monkeys. J Comp Neurol 401:480–505
264. Österlund MK, Keller E, Hurd YL (2000) The human forebrain has discrete estrogen receptor alpha messenger RNA expression: high levels in the amygdaloid complex. Neuroscience 95:333–342
265. Owens MJ, Nemeroff CB (1991) Physiology and pharmacology of corticotropin-releasing factor. Pharmacol Rev 43:425–473
266. Pace-Schott EF, Hobson JA (2002) The neurobiology of sleep: genetics, cellular physiology and subcortical networks. Nat Rev Neurosci 3:591–605
267. Panula P, Airaksinen MS, Pirvola U, Kotilainen E (1990) A histamine-containing neuronal system in human brain. Neuroscience 34:127–132
268. Pelleymounter MA, Cullen MJ, Baker MB et al (1995) Effects of the obese gene product on body weight regulation in ob/ob mice. Science 269: 540–543
269. Perry RH, Candy JM, Perry EK, Thompson J, Oakley AE (1984) The substantia innominata and adjacent regions in the human brain: histochemical and biochemical observations. I Anat 138: 713–732
270. Petrovich GD, Canteras NS, Swanson LW (2001) Combinatorial amygdalar inputs to hippocampal domains and hypothalamic behavior systems. Brain Res Brain Res Rev 38:247–289
271. Peyron C, Tighe DK, Van den Pol AN et al (1998) Neurons containing hypocretin (orexin) project to multiple neuronal systems. J Neurosci 18:9996–10015
272. Peyron C, Faraco J, Rogers W et al (2000) A mutation in a case of early onset narcolepsy and a generalized absence of hypocretin peptides in human narcoleptic brains. Nat Med 6:991–997
273. Pfaff D, Keiner M (1973) Atlas of estradiol-concentrating cells in the central nervous system of the female rat. J Comp Neurol 151:121–158
274. Pfaff DW (1973) Luteinizing hormone-releasing factor potentiates lordosis behavior in hypophysectomized ovariectomized female rats. Science 182:1148–1149
275. Pfaff DW, Sakuma Y (1979) Facilitation of the lordosis reflex of female rats from the ventromedial nucleus of the hypothalamus. J Physiol 288: 189–202
276. Pfaff DW, Gerlach JL, McEwen BS et al (1976) Autoradiographic localization of hormone-concentrating cells in the brain of the female rhesus monkey. J Comp Neurol 170:279–293
277. Porrino LF, Goldman-Rakic PS (1982) Brainstem innervation of prefrontal and anterior cingulate

cortex in the rhesus monkey revealed by retrograde transport of HRP. J Comp Neurol 205:63–76

278. Presse F, Sorokovsky I, Max JP, Nicolaidis S, Nahon JL (1996) Melanin-concentrating hormone is a potent anorectic peptide regulated by food deprivation and glucopenia in the rat. Neuroscience 71:735–745
279. Price JL, Amaral DG (1981) An autoradiographic study of the projections of the central nucleus of the monkey amygdala. J Neurosci 1:1242–1259
280. Pritchard TC, Hamilton RB, Norgren R (2000) Projections of the parabrachial nucleus in the old world monkey. Exp Neurol 165:101–117
281. Pu S, Jain MR, Kalra PS, Kalra SP (1998) Orexins, a novel family of hypothalamic neuropeptides, modulate pituitary luteinizing hormone secretion in an ovarian steroid-dependent manner. Regul Pept 78:133–136
282. Qu D, Ludwig DS, Gammeltoft S et al (1996) A role for melanin-concentrating hormone in the central regulation of feeding behaviour. Nature 380:243–247
283. Raadsheer FC, Sluiter AA, Ravid R, Tilders FJ, Swaab DF (1993) Localization of corticotropin-releasing hormone (CRH) neurons in the paraventricular nucleus of the human hypothalamus: age-dependent colocalization with vasopressin. Brain Res 615:50–62
284. Raadsheer FC, Hoogendijk WJ, Stam FC, Tilders FJ, Swaab DF (1994) Increased numbers of corticotropin-releasing hormone expressing neurons in the hypothalamic paraventricular nucleus of depressed patients. Neuroendocrinology 60:436–444
285. Raadsheer FC, Van Heerikhuize JJ, Lucassen PJ et al (1995) Corticotropin-releasing hormone mRNA levels in the paraventricular nucleus of patients with Alzheimer's disease and depression. Am J Psychiatry 152:1372–1376
286. Raisman G (1966) The connexions of the septum. Brain 89:317–348
287. Rance NE, Young WS, III, McMullen NT (1994) Topography of neurons expressing luteinizing hormone-releasing hormone gene transcripts in the human hypothalamus and basal forebrain. J Comp Neurol 339:573–586
288. Ranson RN, Motawei K, Pyner S, Coote JH (1998) The paraventricular nucleus of the hypothalamus sends efferents to the spinal cord of the rat that closely appose sympathetic preganglionic neurones projecting to the stellate ganglion. Exp Brain Res 120:164–172
289. Rees HD, Switz GM, Michael RP (1980) The estrogen-sensitive neural system in the brain of female cats. J Comp Neurol 193:789–804
290. Rempel-Clower NL, Barbas H (1998) Topographic organization of connections between the hypothalamus and prefrontal cortex in the rhesus monkey. J Comp Neurol 398:393–419
291. Ricardo JA (1983) Hypothalamic pathways involved in metabolic regulatory functions, as identified by tracktracing methods. Adv Metab Dis 10:1–30
292. Ricardo JA, Koh ET (1977) Direct projections from the nucleus of the solitary tract to the hypothalamus, amygdala, and other forebrain structures in the rat. Anat Rec 187:693
293. Ricardo JA, Koh ET (1978) Anatomical evidence of direct projections from the nucleus of the solitary tract to the hypothalamus and the amygdala, and other forebrain structures in the rat. Brain Res 153:1–26
294. Risold PY, Swanson LW (1997) Connections of the rat lateral septal complex. Brain Res Brain Res Rev 24:115–195
295. Risold PY, Canteras NS, Swanson LW (1994) Organization of projections from the anterior hypothalamic nucleus: a Phaseolus vulgaris-leucoagglutinin study in the rat. J Comp Neurol 348:1–40
296. Risold PY, Thompson RH, Swanson LW (1997) The structural organization of connections between hypothalamus and cerebral cortex. Brain Res Brain Res Rev 24:197–254
297. Roeling TA, Veening JG, Peters JP, Vermelis ME, Nieuwenhuys R (1993) Efferent connections of the hypothalamic "grooming area" in the rat. Neuroscience 56:199–225
298. Roeling TA, Veening JG, Kruk MR et al (1994) Efferent connections of the hypothalamic "aggression area" in the rat. Neuroscience 59:1001–1024
299. Roland BL, Sawchenko PE (1993) Local origins of some GABAergic projections to the paraventricular and supraoptic nuclei of the hypothalamus in the rat. J Comp Neurol 332:123–143
300. Room P, Groenewegen HJ (1986) Connections of the parahippocampal cortex in the cat. II. Subcortical afferents. J Comp Neurol 251:451–473
301. Russell SH, Small CJ, Dakin CL et al (2001) The central effects of orexin-A in the hypothalamic-pituitary-adrenal axis in vivo and in vitro in male rats. J Neuroendocrinol 13:561–566
302. Russell SH, Small CJ, Kennedy AR et al (2001) Orexin A interactions in the hypothalamo-pituitary gonadal axis. Endocrinology 142:5294–5302
303. Sakai K, Crochet S, Onoe H (2001) Pontine structures and mechanisms involved in the generation of paradoxical (REM) sleep. Arch Ital Biol 139: 93–107
304. Sakurai T, Amemiya A, Ishii M et al (1998) Orexins and orexin receptors: a family of hypothalamic neuropeptides and G protein-coupled receptors that regulate feeding behavior. Cell 92:573–585
305. Sánchez MM, Young LJ, Plotsky PM, Insel TR (1999) Autoradiographic and in situ hybridization localization of corticotropin-releasing factor 1 and 2 receptors in nonhuman primate brain. J Comp Neurol 408:365–377

306. Sandrew BB, Edwards DL, Poletti CE, Foote WE (1986) Amygdalospinal projections in the cat. Brain Res 373:235–239
307. Saper CB (1985) Organization of cerebral cortical afferent systems in the rat. II. Hypothalamocortical projections. J Comp Neurol 237:21–46
308. Saper CB (1995) Central autonomic system. In: Paxinos G (ed) The rat nervous system. Academic Press, San Diego, pp 107–135
309. Saper CB (2004) Hypothalamus. In: Paxinos G, Mai JK (eds) The human nervous system, 2nd edn. Elsevier, Amsterdam, pp 393–423
310. Saper CB, Chelimski TC (1984) A cytoarchitectonic and histochemical study of nucleus basalis and associated cell groups in the normal human brain. Neuroscience 13:1023–1037
311. Saper CB, Loewy AD (1980) Efferent connections of the parabrachial nucleus in the rat. Brain Res 197:291–317
312. Saper CB, Loewy AD, Swanson LW, Cowan WM (1976) Direct hypothalamo-autonomic connections. Brain Res 117:305–312
313. Saper CB, Swanson LW, Cowan WM (1976) The efferent connections of the ventromedial nucleus of the hypothalamus of the rat. J Comp Neurol 169:409–442
314. Saper CB, Swanson LW, Cowan WM (1979) An autoradiographic study of the efferent connections of the lateral hypothalamic area in the rat. J Comp Neurol 183:689–706
315. Saper CB, Swanson LW, Cowan WM (1979) Some efferent connections of the rostral hypothalamus in the squirrel monkey (Saimiri sciureus) and cat. J Comp Neurol 184:205–242
316. Saper CB, Chou TC, Scammell TE (2001) The sleep switch: hypothalamic control of sleep and wakefulness. Trends Neurosci 24:726–731
317. Sar MN, Stumpf WE (1975) Distribution of androgen-concentrating neurons in rat brain. In: Stumpf WE, Grant LP (eds) Anatomical neuroendocrinology. Karger, Basel, pp 120–133
318. Sawchenko PE (1998) Toward a new neurobiology of energy balance, appetite, and obesity: the anatomists weigh in. J Comp Neurol 402:435–441
319. Sawchenko PE, Swanson LW (1982) The organization of noradrenergic pathways from the brainstem to the paraventricular and supraoptic nuclei in the rat. Brain Res Rev 4:275–325
320. Scammell TE, Price KJ, Sagar SM (1993) Hyperthermia induces c-fos expression in the preoptic area. Brain Res 618:303–307
321. Schwartz MW, Seeley RJ, Campfield LA, Burn P, Baskin DG (1996) Identification of targets of leptin action in rat hypothalamus. J Clin Invest 98:1101–1106
322. Schwartz MW, Woods SC, Porte D Jr, Seeley RJ, Baskin DG (2000) Central nervous system control of food intake. Nature 404:661–671
323. Shen CL (1983) Efferent projections from the lateral hypothalamus in the guinea pig: an autoradiographic study. Brain Res Bull 11:335–347
324. Sherin JE, Shiromani PJ, McCarley RW, Saper CB (1996) Activation of ventrolateral preoptic neurons during sleep. Science 271:216–219
325. Sherin JE, Elmquist JK, Torrealba F, Saper CB (1998) Innervation of histaminergic tuberomammillary neurons by GABAergic and galaninergic neurons in the ventrolateral preoptic nucleus of the rat. J Neurosci 18:4705–4721
326. Shibata H (1987) Ascending projections to the mammillary nuclei in the rat: a study using retrograde and anterograde transport of wheat germ agglutinin conjugated to horseradish peroxidase. J Comp Neurol 264:205–215
327. Shibata H (1989) Descending projections to the mammillary nuclei in the rat, as studied by retrograde and anterograde transport of wheat germ agglutinin-horseradish peroxidase. J Comp Neurol 285:436–452
328. Shimada M, Tritos NA, Lowell BB, Flier JS, Maratos-Flier E (1998) Mice lacking melanin-concentrating hormone are hypophagic and lean. Nature 396:670–674
329. Shipley MT, Murphy AZ, Rizvi TA, Ennis M, Behbehani MM (1996) Olfaction and brainstem circuits of reproductive behavior in the rat. Progr Brain Res 107:355–377
330. Shute CCD, Lewis PR (1967) The ascending cholinergic reticular system: Neocortical, olfactory and subcortical projections. Brain 90:497–520
331. Siegel A, Skog E (1970) Effects of electrical stimulation of the septum upon attack behavior elicited from the hypothalamus in the cat. Brain Res 23:371–380
332. Silverman AJ, Krey LC, Zimmerman EA (1979) A comparative study of the luteinizing hormone releasing hormone (LHRH) neuronal networks in mammals. Biol Reprod 20:98–110
333. Silverman AJ, Jhamandas J, Renaud LP (1987) Localization of luteinizing hormone-releasing hormone (LHRH) neurons that project to the median eminence. J Neurosci 7:2312–2319
334. Simerly RB (1998) Organization and regulation of sexually dimorphic neuroendocrine pathways. Behav Brain Res 92:195–203
335. Simerly RB, Swanson LW (1988) Projections of the medial preoptic nucleus: a Phaseolus vulgaris leucoagglutinin anterograde tract-tracing study in the rat. J Comp Neurol 270:209–242
336. Simerly RB, Chang C, Muramatsu M, Swanson LW (1990) Distribution of androgen and estrogen receptor mRNA-containing cells in the rat brain: an in situ hybridization study. J Comp Neurol 294:76–95
337. Sofroniew MY, Weindl A, Schinko I, Wetzstein R (1979) The distribution of vasopressin-, oxytocin-, and neurophysin-producing neurons in the guinea pig brain. Cell Tissue Res 196:367–384
338. Speh JC, Moore RY (1993) Retinohypothalamic tract development in the hamster and rat. Brain Res Dev Brain Res 76:171–181

339. Steininger TL, Gong H, McGinty D, Szymusiak R (2001) Subregional organization of preoptic area/anterior hypothalamic projections to arousal-related monoaminergic cell groups. J Comp Neurol 429:638–653
340. Stellar E (1994) The physiology of motivation. Psychol Rev 101:301–311
341. Stephan FK, Berkley KJ, Moss RL (1981) Efferent connections of the rat suprachiasmatic nucleus. Neuroscience 6:2625–2641
342. Steriade M, McCormick DA, Sejnowski TJ (1993) Thalamocortical oscillations in the sleeping and aroused brain. Science 262:679–685
343. Strack AM, Sawyer WB, Hughes JH, Platt KB, Loewy AD (1989) A general pattern of CNS innervation of the sympathetic outflow demonstrated by transneuronal pseudorabies viral infections. Brain Res 491:156–162
344. Stricker EM (1984) Biological bases of hunger and satiety: therapeutic implications. Nutr Rev 42:333–340
345. Stumpf WE (1970) Estrogen neurons in the periventricular brain. Am J Anat 129:207–218
346. Stumpf WE, Sar M (1975) Hormone-architecture of the mouse brain with 3H-estradiol. In: Stumpf WE, Grant LO (eds) Anatomical neuroendocrinology. Karger, Basel, pp 82–103
347. Stumpf WE, Sar M, Keefer SA (1975) Atlas of estrogen target cells in the rat brain. In: Stumpf WE, Grant LO (eds) Anatomical neuroendocrinology. Karger, Basel, pp 104–119
348. Sutcliffe JG, De Lecea L (2002) The hypocretins: setting the arousal threshold. Nat Rev Neurosci 3:339–349
349. Sutin J (1966) The periventricular stratum of the hypothalamus. Int Rev Neurobiol 9:263–300
350. Swaab DF (1997) Neurobiology and neuropathology of the human hypothalamus. In: Bloom FE, Björklund A, Hökfelt T (eds) The primate nervous system, Part 1. Elsevier, Amsterdam, pp 39–137 (Handbook of Chemical Neuroanatomy vol 13)
351. Swaab DF (1999) Biological rhythms in health and disease: the suprachiasmatic nucleus and the autonomic nervous system. In: Appenzeller (ed) The autonomic nervous system, Part 1. Normal functions. Elsevier, Amsterdam, pp 467–521 (Handbook of Clinical Neurology vol 74)
352. Swaab DF (2004) The human hypothalamus: basic and clinical aspects. Part II: Neuropathology of human hypothalamus and adjacent brain. Elsevier, Amsterdam
353. Swaab DF (2004) The human hypothalamus: basic and clinical aspects. Part I: Nuclei of the human hypothalamus. Elsevier, Amsterdam
354. Swaab DF (2004) Neuropeptides in hypothalamic disorders. Int Rev Cytol 240:305–375
355. Swaab DF, Fliers E (1985) A sexually dimorphic nucleus in the human brain. Science 228:1112–1115
356. Swaab DF, Hofman MA (1995) Sexual differentiation of the human hypothalamus in relation to gender and sexual orientation. Trends Neurosci 18:264–270
357. Swaab DF, Zhou JN, Ehlhart T, Hofman MA (1994) Development of vasoactive intestinal polypeptide neurons in the human suprachiasmatic nucleus in relation to birth and sex. Brain Res Dev Brain Res 79:249–259
358. Swanson LW (1976) An autoradiographic study of the efferent connections of the preoptic region in the rat. J Comp Neurol 167:227–256
359. Swanson LW (1977) Immunohistochemical evidence for a neurophysin-containing autonomic pathway arising in the paraventricular nucleus of the hypothalamus. Brain Res 128:346–353
360. Swanson LW (1987) The hypothalamus. In: Björklund A, Swanson LW (eds) Integrated systems of the CNS, part I. Elsevier, Amsterdam, pp 125–277 (Handbook of Chemical Neuroanatomy vol 5)
361. Swanson LW (1989) The neural basis of motivated behavior. Acta Morphol Neerl-Scand 26:165–176
362. Swanson LW (2000) Cerebral hemisphere regulation of motivated behavior. Brain Res 886:113–164
363. Swanson LW, Cowan WM (1975) The efferent connections of the suprachiasmatic nucleus of the hypothalamus. J Comp Neurol 160:1–12
364. Swanson LW, Cowan WM (1977) An autoradiographic study of the organization of the efferent connections of the hippocampal formation in the rat. J Comp Neurol 172:49–84
365. Swanson LW, Cowan WM (1979) The connections of the septal region in the rat. J Comp Neurol 186:621–655
366. Swanson LW, Kuypers HGJM (1980) The paraventricular nucleus of the hypothalamus: cytoarchitectonic subdivisions and organization of projections to the pituitary, dorsal vagal complex, and spinal cord as demonstrated by retrograde fluorescence double-labeling methods. J Comp Neurol 194:555–570
367. Swanson LW, Lind RW (1986) Neural projections subserving the initiation of a specific motivated behavior in the rat: new projections from the subfornical organ. Brain Res 379:399–403
368. Swanson LW, Mogenson GJ (1981) Neural mechanisms for the functional coupling of autonomic, endocrine and somatomotor responses in adaptive behaviour. Brain Res Rev 3:1–34
369. Swanson LW, Sawchenko PE (1983) Hypothalamic integration: organization of the paraventricular and supraoptic nuclei. Annu Rev Neurosci 6:269–324
370. Swanson LW, Cowan WM, Jones EG (1974) An autoradiographic study of the efferent connections of the ventral lateral geniculate nucleus in the albino rat and the cat. J Comp Neurol 156:143–163
371. Swanson LW, Mogenson GJ, Simerly RB, Wu M (1987) Anatomical and electrophysiological evidence for a projection from the medial preoptic area to the mesencephalic and subthalamic locomotor regions in the rat. Brain Res 405:108–122

372. Sylvester CM, Krout KE, Loewy AD (2002) Suprachiasmatic nucleus projection to the medial prefrontal cortex: a viral transneuronal tracing study. Neuroscience 114:1071–1080
373. Szentagothai J, Flerko B, Mess B (1968) Hypothalamic control of the anterior pituitary: an experimental-morphological study. Akademiai Kiado, Budapest
374. Szymusiak R, Alam N, Steininger TL, McGinty D (1998) Sleep-waking discharge patterns of ventrolateral preoptic/anterior hypothalamic neurons in rats. Brain Res 803:178–188
375. Taheri S, Zeitzer JM, Mignot E (2002) The role of hypocretins (orexins) in sleep regulation and narcolepsy. Annu Rev Neurosci 25:283–313
376. Tasker JG, Boudaba C, Schrader LA (1998) Local glutamatergic and GABAergic synaptic circuits and metabotropic glutamate receptors in the hypothalamic paraventricular and supraoptic nuclei. Adv Exp Med Biol 449:117–121
377. Teclemariam-Mesbah R, Ter Horst GJ, Postema F, Wortel J, Buijs R (1999) Anatomical demonstration of the suprachiasmatic nucleus-pineal pathway. J Comp Neurol 406:171–182
378. Ter Horst GJ, Luiten PG (1986) The projections of the dorsomedial hypothalamic nucleus in the rat. Brain Res Bull 16:231–248
379. Ter Horst GJ, Luiten PG (1987) Phaseolus vulgaris leuco-agglutinin tracing of intrahypothalamic connections of the lateral, ventromedial, dorsomedial and paraventricular hypothalamic nuclei in the rat. Brain Res Bull 18:191–203
380. Ter Horst GJ, Luiten PGM (1986) The projections of the dorsomedial hypothalamic nucleus in the rat. Brain Res Bull 16:231–248
381. Thannickal TC, Moore RY, Nienhuis R et al (2000) Reduced number of hypocretin neurons in human narcolepsy. Neuron 27:469–474
382. Thompson RH, Swanson LW (1998) Organization of inputs to the dorsomedial nucleus of the hypothalamus: a reexamination with Fluorogold and PHAL in the rat. Brain Res Brain Res Rev 27:89–118
383. Thompson RH, Swanson LW (2003) Structural characterization of a hypothalamic visceromotor pattern generator network. Brain Res Brain Res Rev 41:153–202
384. Thompson RH, Canteras NS, Swanson LW (1996) Organization of projections from the dorsomedial nucleus of the hypothalamus: a PHA-L study in the rat. J Comp Neurol 376:143–173
385. Thornhill JA, Halvorson I (1994) Electrical stimulation of the posterior and ventromedial hypothalamic nuclei causes specific activation of shivering and nonshivering thermogenesis. Can J Physiol Pharmacol 72:89–96
386. Ungerstedt U (1971) Stereotaxic mapping of the monoamine pathways in the rat brain. Acta Physiol Scand [Suppl] 367:1–49
387. Usuda I, Tanaka K, Chiba T (1998) Efferent projections of the nucleus accumbens in the rat with special reference to subdivision of the nucleus: biotinylated dextran amine study. Brain Res 797: 73–93
388. Van den Pol AN (1999) Hypothalamic hypocretin (orexin): robust innervation of the spinal cord. J Neurosci 19:3171–3182
389. Van der Beek EM, Horvath TL, Wiegant VM, Van den HR, Buijs RM (1997) Evidence for a direct neuronal pathway from the suprachiasmatic nucleus to the gonadotropin-releasing hormone system: combined tracing and light and electron microscopic immunocytochemical studies. J Comp Neurol 384:569–579
390. Van der Kooy D, Koda LY, McGinty JF, Gerfen CR, Bloom FE (1984) The organization of projections from the cortex, amygdala, and hypothalamus to the nucleus of the solitary tract in rat. J Comp Neurol 224:1–24
391. Van Leeuwen FW, Caffe AR, De Vries GJ (1985) Vasopressin cells in the bed nucleus of the stria terminalis of the rat: sex differences and the influence of androgens. Brain Res 325:391–394
392. Van Zwieten EJ, Ravid R, Hoogendijk WJ, Swaab DF (1994) Stable vasopressin innervation in the degenerating human locus coeruleus in Alzheimer's disease. Brain Res 649:329–333
393. Van Zwieten EJ, Ravid R, Swaab DF (1996) Differential vasopressin and oxytocin innervation of the human parabrachial nucleus: no changes in Alzheimer's disease. Brain Res 711:146–152
394. VanderHorst VGJM, Holstege G (1995) Caudal medullary pathways to lumbosacral motoneuronal cell groups in the cat: evidence for direct projections possibly representing the final common pathway for lordosis. J Comp Neurol 359:457–475
395. VanderHorst VGJM, Holstege G (1996) A concept for the final common pathway of vocalization and lordosis behavior in the cat. Progr Brain Res 107:327–342
396. VanderHorst VGJM, Holstege G (1997) Nucleus retroambiguus projections to lumbosacral motoneuronal cell groups in the male cat. J Comp Neurol 382:77–88
397. VanderHorst VGJM, Holstege G (1997) Estrogen induces axonal outgrowth in the nucleus retroambiguus-lumbosacral motoneuronal pathway in the adult female cat. J Neurosci 17:1122–1136
398. VanderHorst VGJM, Mouton LJ, Blok BF, Holstege G (1996) Distinct cell groups in the lumbosacral cord of the cat project to different areas in the periaqueductal gray. J Comp Neurol 376:361–385
399. VanderHorst VGJM, de Weerd H, Holstege G (1997) Evidence for monosynaptic projections from the nucleus retroambiguus to hindlimb motoneurons in the cat. Neurosci Lett 224:33–36
400. VanderHorst VGJM, Terasawa E, Ralston HJ III, Holstege G (2000) Monosynaptic projections from the lateral periaqueductal gray to the nucleus retroambiguus in the rhesus monkey: implications for vocalization and reproductive behavior. J Comp Neurol 424:251–268

401. VanderHorst VGJM, Terasawa E, Ralston HJ III, Holstege G (2000) Monosynaptic projections from the nucleus retroambiguus to motoneurons supplying the abdominal wall, axial, hindlimb, and pelvic floor muscles in the female rhesus monkey. J Comp Neurol 424:233–250
402. Veening J, Buma P, Ter Horst GJ et al (1991) Hypothalamic projections to the PAG in the rat: topographical, immuno-electronmicroscopical and functional aspects. In: Depaulis A, Bandler R (eds) The midbrain periaqueductal gray matter: functional, anatomical and neurochemical organization. Plenum, New York, pp 387–415
403. Veening JG, Swanson LW, Cowan WM, Nieuwenhuys R, Geeraedts LMG (1982) The medial forebrain bundle of the rat. II. An autoradiographic study of the topography of the major descending and ascending components. J Comp Neurol 206: 82–108
404. Veening JG, Te LS, Posthuma P, Geeraedts LM, Nieuwenhuys R (1987) A topographical analysis of the origin of some efferent projections from the lateral hypothalamic area in the rat. Neuroscience 22:537–551
405. Vertes RP (1984) A lectin horseradish peroxidase study of the origin of ascending fibers in the medial forebrain bundle of the rat. The upper brainstem. Neuroscience 11:669–690
406. Vertes RP (1984) A lectin horseradish peroxidase study of the origin of ascending fibers in the medial forebrain bundle of the rat. The lower brainstem. Neuroscience 11:651–668
407. Von Economo C (1930) Sleep as a problem of localization. J Nerv Ment Dis 71:249–259
408. Von Economo C (1929) Die Encephalitis lethargica, ihre Nachkrankheiten und ihre Behandlung. Urban und Schwarzenberg, Vienna
409. Vrang N, Larsen PJ, Moller M, Mikkelsen JD (1995) Topographical organization of the rat suprachiasmatic-paraventricular projection. J Comp Neurol 353:585–603
410. Walter A, Mai JK, Lanta L, Gorcs T (1991) Differential distribution of immunohistochemical markers in the bed nucleus of the stria terminalis in the human brain. J Chem Neuroanat 4:281–298
411. Watanabe T, Taguchi Y, Shiosaka S et al (1984) Distribution of the histaminergic neuron system in the central nervous system of rats; a fluorescent immunohistochemical analysis with histidine decarboxylase as a marker. Brain Res 295:13–25
412. Watts AG, Swanson LW (1987) Efferent projections of the suprachiasmatic nucleus: II. Studies using retrograde transport of fluorescent dyes and simultaneous peptide immunohistochemistry in the rat. J Comp Neurol 258:230–252
413. Watts AG, Swanson LW, Sanchez-Watts G (1987) Efferent projections of the suprachiasmatic nucleus: I. Studies using anterograde transport of Phaseolus vulgaris leucoagglutinin in the rat. J Comp Neurol 258:204–229
414. Westerhaus MJ, Loewy AD (1999) Sympathetic-related neurons in the preoptic region of the rat identified by viral transneuronal labeling. J Comp Neurol 414:361–378
415. Willie JT, Chemelli RM, Sinton CM, Yanagisawa M (2001) To eat or to sleep? Orexin in the regulation of feeding and wakefulness. Annu Rev Neurosci 24:429–458
416. Willis WD Jr, Zhang X, Honda CN, Giesler GJ Jr (2002) A critical review of the role of the proposed VMpo nucleus in pain. J Pain 3:79–94
417. Wilson FA, Rolls ET (1990) Neuronal responses related to novelty and familiarity of visual stimuli in the substantia innominata, diagonal band of Broca and periventricular region of the primate basal forebrain. Exp Brain Res 80:40–120
418. Witkin JW, Paden CM, Silverman AJ (1982) The luteinizing hormone-releasing hormone (LHRH) systems in the rat brain. Neuroendocrinology 35:429–438
419. Woods SC, Seeley RJ, Porte D Jr, Schwartz MW (1998) Signals that regulate food intake and energy homeostasis. Science 280:1378–1383
420. Wu SS, Nathanielsz PW, McDonald TJ (1995) Immunocytochemical distribution of androgen receptors in the hypothalamus and pituitary of the fetal baboon in late gestation. Brain Res Dev Brain Res 84:278–281
421. Wyss JM, Swanson LW, Cowan WM (1979) A study of subcortical afferents to the hippocampal formation in the rat. Neuroscience 4:463–476
422. Yang QZ, Hatton GI (1997) Electrophysiology of excitatory and inhibitory afferents to rat histaminergic tuberomammillary nucleus neurons from hypothalamic and forebrain sites. Brain Res 773:162–172
423. Yang SP, Voogt JL (2001) Mating-activated brainstem catecholaminergic neurons in the female rat. Brain Res 894:159–166
424. Zhang X, Wenk HN, Gokin AP, Honda CN, Giesler GJ Jr (1999) Physiological studies of spinohypothalamic tract neurons in the lumbar enlargement of monkeys. J Neurophysiol 82:1054–1058
425. Zhang Y, Proenca R, Maffei M et al (1994) Positional cloning of the mouse obese gene and its human homologue. Nature 372:425–432
426. Zhang YH, Hosono T, Yanase-Fujiwara M, Chen XM, Kanosue K (1997) Effect of midbrain stimulations on thermoregulatory vasomotor responses in rats. J Physiol 503:177–186
427. Zhou JN, Hofman MA, Gooren LJ, Swaab DF (1995) A sex difference in the human brain and its relation to transsexuality. Nature 378:68–70

11 Telencephalon: Introduction and Olfactory System

Introduction

During ontogeny, the telencephalic hemispheres evaginate bilaterally from the dorsalmost alar plate of the secondary prosencephalon (Figs. 2.2 C, 2.3 C, 2.4, 2.5 B, C). There are two principal divisions of the telencephalic hemispheres: the *roof or pallium* and the *basis or subpallium*. Classically, the pallium is divided into three longitudinal zones, medial, dorsal and lateral. During development, the neuroblasts move outward from the ventricular matrix to form cortical structures in all of these pallial zones. The *medial pallium* matures into the *hippocampal formation*, in which a compact cell layer is sandwiched between periventricular and subpial plexiform layers. A three-layered structure is also formed within the *lateral pallium*. Here we find a superficial molecular layer, containing relatively few scattered cells, a cell-dense intermediate layer and a deep layer displaying a moderate cell density. The cortical formation, developing from the lateral pallium, receives direct projections from the olfactory bulb. Hence it is designated in the adult brain as the *olfactory cortex*. This olfactory cortex is, particularly in macrosmatic mammals, subdivided in prepiriform and piriform (or periamygdaloid) cortical fields. The mammalian dorsal pallium gives rise to a highly complex, six-layered cortical formation which, due to its late phylogenetic appearance, is designated as the *neocortex*. The differentiation of the neocortex during development is preceded by the formation of a compact *cortical plate* (Figs. 2.10 J, 2.19 A, 2.20). In primitive mammals, such as the opossum or the hedgehog, the extent of the neocortex is still limited (Fig. 11.2 A), but in some mammalian groups, particularly the primates and the whales, this formation attains amazing dimensions (Fig. 11.2 B) and becomes by far the largest centre of the brain.

The recent studies of Puelles and colleagues [48, 67–69] have shown that the lateral pallium also gives rise to the *dorsolateral claustrum* or *claustrum proper* and that a fourth pallial district – the *ventral pallium* – is intercalated between the lateral pallium and the subpallium (Figs. 2.24 B, 2.25, 11.1, 11.2, Table 11.1). This new entity extends, just like the other telencephalic histogenetic units, from the ventricle to the pial surface. It comprises the anlagen of the *olfactory bulb* and the *ventromedial claustrum* or *endopiriform nucleus*, and participates in the formation of the *olfactory cortex* and of the *amygdaloid complex*. It is of particular importance that the investigations of Puelles and colleagues on gene expression patterns have fully confirmed the results of the classical studies of Holmgren [38] and Källén [43], according to which a considerable portion of the amygdaloid complex is of pallial origin.

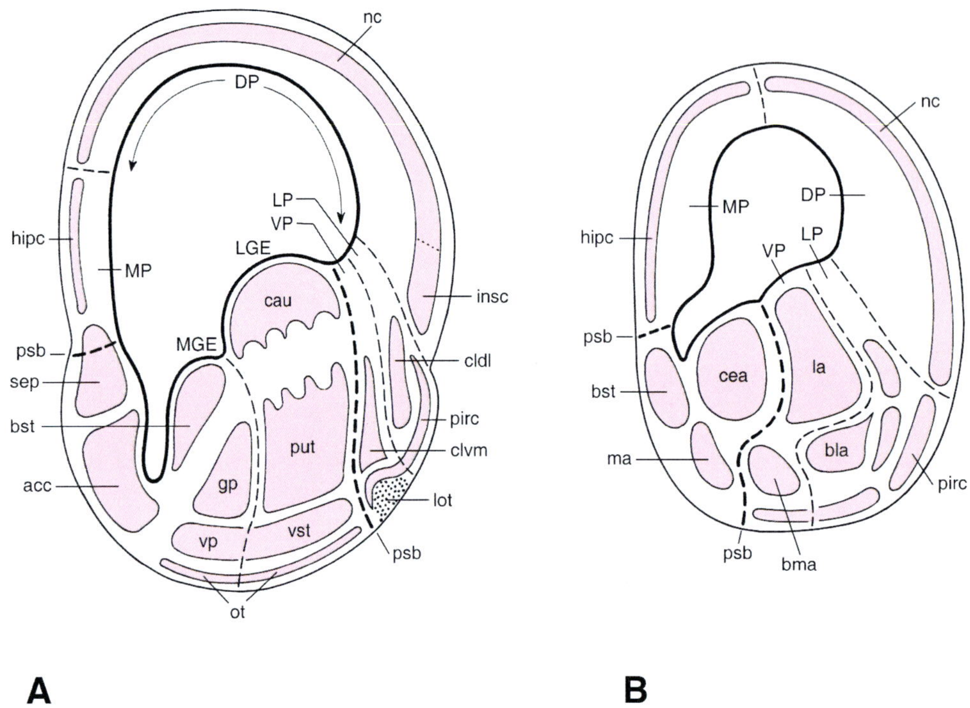

ac Anterior commissure
acc Accumbens nucleus
bla Basolateral amygdaloid nucleus
bma Basomedial amygdaloid nucleus
bst Bed nucleus of stria terminalis
cau Caudate nucleus
cc Corpus callosum
cea Central amygdaloid nucleus
cldl Claustrum, dorsolateral part
clvm Claustrum, ventromedial part
DP Dorsal pallium
gp Globus pallidus
hipc Hippocampal cortex
ic Internal capsule
ins Insula
insc Insular cortex
isCal Island of Calleja
la Lateral amygdaloid nucleus
LGE Lateral ganglionic eminence
lot Lateral olfactory tract
lv Lateral ventricle
LP Lateral pallium
ma Medial amygdaloid nucleus
MGE Medial ganglionic eminence
MP Medial pallium
nc Neocortex
ndb Nucleus of the diagonal band
ot Olfactory tubercle
pirc Piriform cortex
psb Palliosubpallial boundary
put Putamen
sep Septum
VP Ventral pallium
vp Ventral pallidum
vst Ventral striatum

Fig. 11.1 A, B. The main subdivision of the mammalian telencephalon shown in schematic cross sections through the rostral (**A**) and caudal (**B**) sectors of an undeformed cerebral hemisphere. Slightly modified from Puelles and Rubenstein [68]. Within the sections, the *thick, dashed lines* mark the palliosubpallial boundary, whereas the *thin dashed lines* indicate the boundaries between the telencephalic zones, discussed in the text

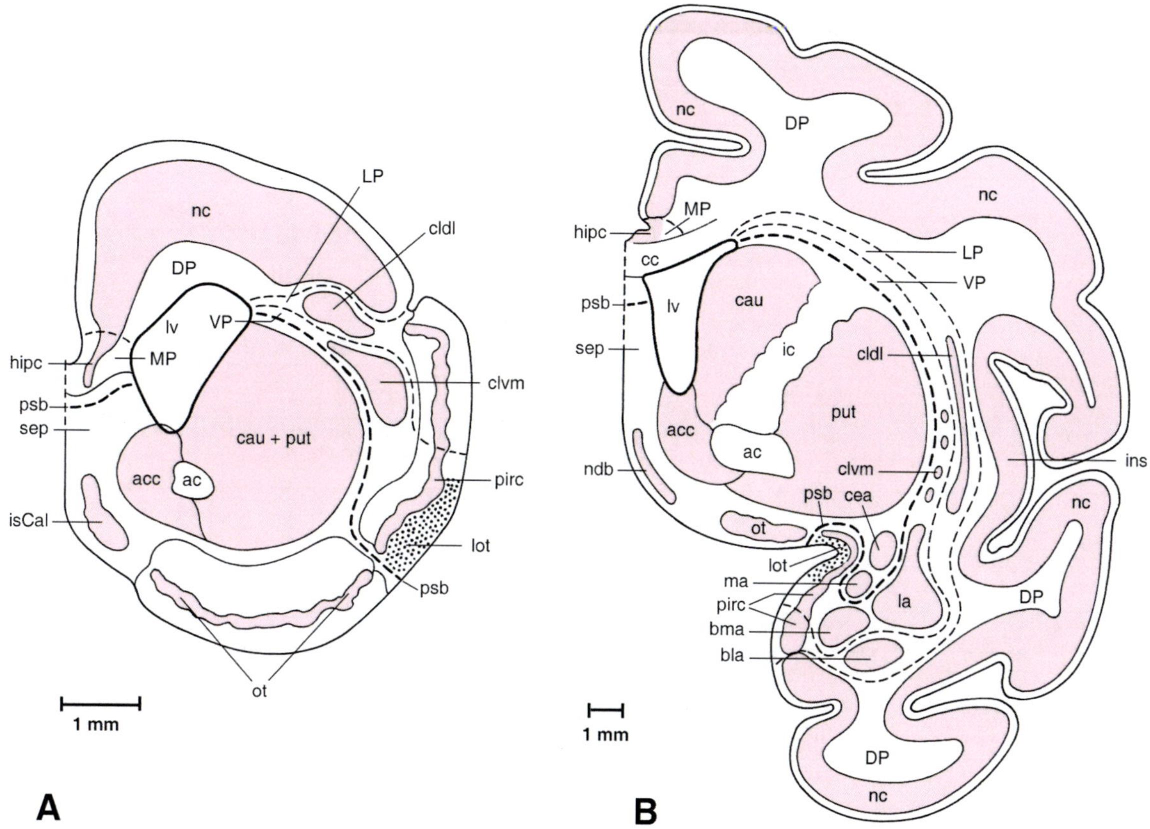

Fig. 11.2 A, B. Transverse hemisections through the rostral part of the telencephalic hemispheres of the hedgehog, *Erinaceus europaeus* (**A**) and the rhesus monkey, *Macaca mulatta* (**B**). The significance of the *dashed lines* and the abbreviations is the same as in Fig. 11.1. In the hedgehog, the central olfactory structures are strongly developed, but the neocortex (which develops within the dorsal pallium) is only small. In the macaque, and in primates in general, these relations are reversed. The enormous expansion of the neocortex leads in this group to several deformations, such as (1) the ventromedial shift of the constituents of the lateral and ventral pallial zones, and (2) the rostroventral displacement of the amygdaloid complex. **A** is partly based on illustrations in Stephan et al. [84] and in Dinopoulos et al. [25]; **B** is partly based on photomicrographs in Lauer [45] and Krieg [44]. Explanations see in Fig. 11.1

Table 11.1. Telencephalic embryonic zones and their principal derivatives in the adult mammalian brain. Based on Puelles et al. [130, 157, 162, 166]

medial pallial zone		hippocampal formation	
dorsal pallial zone		neocortex	
lateral pallial zone		dorsolateral claustrum olfactory cortex	basolateral amygd n cortical amygd n
ventral pallial zone		olf bulb ventromedial claustrum olfactory cortex	lateral amygd n basomedial amygd n cortical amygd n
palliosubpallial boundary			
striatal zone	septum	accumbens n caudate n putamen substantia innominata olfactory tubercle	central amygd n
pallidal zone		bed n stria terminalis external pallidal segment substantia innominata olfactory tubercle	medial amygd n
entopeduncular zone		internal pallidal segment n basalis Meynert n diagonal band	
			amygdaloid complex

Traditionally (see, e.g. [3, 36]), the subpallium is subdivided topographically into a small, medial *septum* and a much larger, lateral *strio-amygdaloid complex* (Table 2.2). However, the studies of Puelles and colleagues [48, 67–69] have shown that this region is composed of three parallel, molecularly characteristic zones: *striatal*, *pallidal* and *entopeduncular* (Figs. 2.24B, 2.25). The derivatives of these three zones are indicated in Table 11.1. The principal differences with the traditional subdivision of the subpallium are that: (1) only the central and medial nuclei of the amygdaloid complex are of subpallial origin and that (2) all of the subpallial zones, and even a restricted part of the ventral pallial zone, participate in the formation of the septum (Fig. 2.24B).

In the discussion of the structure and fibre connections of the telencephalon, which occupies the second part of the present chapter (Chap. 11) and the following four chapters (Chaps. 12–15). This brain part is subdivided into the following functional systems and complexes:

- The *olfactory system*, encompassing the olfactory bulb, the olfactory tubercle, the prepiriform and periamygdaloid cortices, and the cortical nucleus of the amygdala
- The *hippocampus* and some related structures: the *precommissural septum* and the *limbic lobe*
- The *amygdala* and the *claustrum*
- The *basal ganglia*
- The *neocortex*

It is worthy of note that none of these functional entities is the product of a single histogenetic zone. It is true that the hippocampal formation and the neocortex derive principally from the medial and dorsal pallial zones, respectively, but both of these structures import the precursors of their inhibitory interneurons from the subpallium (see Chap. 2 and Figs. 2.9 and 2.25).

Olfactory System

Gross Features

The parts of the brain that are involved in the transmission of olfactory signals are usually designated as the *olfactory brain* or *rhinencephalon*. Its principal parts are the olfactory bulb and the targets of its efferent projections, i.e. the retrobulbar area or anterior olfactory nucleus; the olfactory tubercle, the prepiriform, periamygdaloid and adjoining entorhinal cortices; and certain nuclei of the amygdaloid complex. In macrosmatic mammals, all of these structures are well developed. In rodents they occupy most of the basal surface of the telencephalon (Fig. 11.3) and in macrosmatic forms with a small neopallium, such as the hedgehog, almost the entire basal surface and a considerable part of the lateral telencephalic surface belong to the rhinencephalon (Figs. 11.2 A, 11.3 A).

The olfactory bulb is the primary centre in the olfactory system and serves as a relay station for all olfactory impulses between the olfactory mucosa and the higher olfactory centres. The olfactory bulbs are separate forward extensions of the telencephalon, which develop from evaginations of the hemisphere walls. In marsupials and insectivores, they contain a large olfactory ventricle, but in most other mammals these cavities are either small or entirely obliterated in the adult stage. In most mammals short olfactory peduncles connect the olfactory bulbs with the cerebral hemispheres, but due to a strong rostrally directed expansion of the frontal lobes, these peduncles are drawn out into long stalks in primates.

The macroscopic relations of the olfactory parts of the human brain are shown in Fig. 11.3 C. In the human, the *olfactory bulbs* are small, flattened ovoid bodies that rest on the cribriform plate of the ethmoid bone. Arising from the posterior poles of the bulbs, the *olfactory tracts or peduncles* pass backwards over the basal surface of the frontal lobe to become attached to the hemispheres. At their sites of attachment, the tracts bifurcate into medial and lateral olfactory striae. The initial parts of these striae border, together with the *diagonal band of Broca*, a territory known as the *anterior perforated substance*. The *medial olfactory stria* extends towards the subcallosal area on the medial surface of the frontal lobe; the *lateral olfactory stria* passes laterally and then bends sharply around the limen insulae to enter the rostromedial part of the temporal lobe. A small *intermediate olfactory stria* continues the course of the olfactory tract for a short distance and then fans out in the anterior perforated substance.

During its transition from the frontal to the temporal lobe, the lateral olfactory stria is accompanied by the small *gyrus ambiens* on its lateral side. This gyrus is occupied by the *olfactory cortex*. On topographical grounds, just as in macrosmatic mammals, the olfactory cortex can be subdivided into *prepiriform* and *piriform* (or *periamygdaloid*) areas (Fig. 11.3).

Olfactory Receptors

The receptive elements of the olfactory apparatus are slender, bipolar cells found in specialized areas of the pseudostratified epithelium that lines the nasal cavities. The olfactory epithelium occupies an area of approximately 1 cm^2 on each side of the posterodorsal parts of the nasal cavities.

The receptive elements are scattered among supporting cells in the olfactory epithelium. Each of them has two processes, a coarse apical one, extending to the epithelial surface, and a fine basally directed one. The apical process expands into a knob-like protrusion from which several microvilli extend into the mucus covering the epithelium. The basal processes of the receptive elements continue into extremely thin (0.2–0.4 μm) unmyelinated axons that group together into small bundles or *olfactory fila*, which penetrate the cribriform plate and enter the olfactory bulb. Within the olfactory bulb, the primary olfactory fibres synapse with various types of neurons, among which the *mitral cells* are the most conspicuous (Fig. 11.4). The olfactory fila collectively constitute the *ol-*

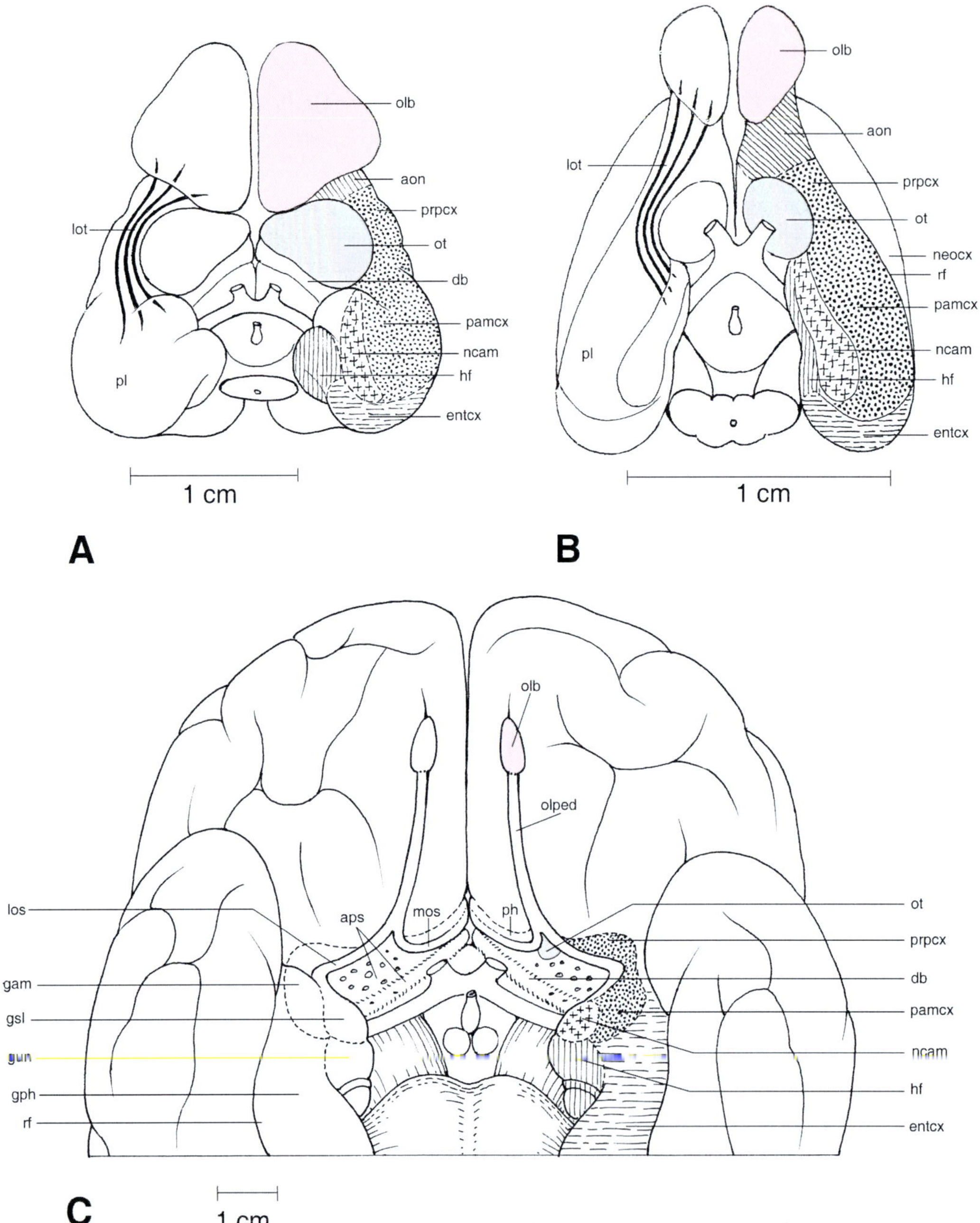

aon Anterior olfactory nucleus
aps Anterior perforated space
db Diagonal band
entcx Entorhinal cortex
gam Gyrus ambiens
gph Gyrus parahippocampalis
gun Gyrus uncinatus
gsl Gyrus semilunaris
hf Hippocampal formation
los Lateral olfactory stria
lot Lateral olfactory tract
mos Medial olfactory stria
ncam Nucleus corticalis amygdalae
neocx Neocortex
olb Olfactory bulb
olped Olfactory peduncle
ot Olfactory tubercle
pamcx Periamygdaloid cortex
ph Precommissural hippocampus
pl Piriform lobe
prpcx Prepiriform cortex
rf Rhinal fissure

Fig. 11.3 A–C. Basal views of the rostral parts of the brains of the hedgehog, *Erinaceus europaeus* (**A**), the rat, *Mus rattus* (**B**) and the human (**C**), showing the size of the olfactory bulb and of other rhinencephalic structures in relation to the whole forebrain. In **C**, the temporal lobes have been displaced laterally in order to expose underlying structures, and the semilunar and uncinate gyri have been rotated towards the plane of the page. **A** is based on illustrations in Stephan [83]; **B** is modified from Heimer [34]; and **C** is reproduced from Nieuwenhuys [61]. Explanations see in Fig. 11.4

factory nerve (nerve I). The transmitter used by the olfactory neurosensory cells is glutamate [4, 6].

There are about 6 million olfactory sensory neurons in the human nasal mucosa [53]. These elements are morphologically uniform, but their molecular phenotype is highly diverse. A large multigene family, first identified in the rat, appears to encode for more than 1,000 different olfactory receptors [9, 29, 100], each of which interacts with a specific odorous ligand. The olfactory sensory neurons expressing a given receptor are restricted to one of four partially overlapping zones in the olfactory epithelium [55, 70]. Remarkably, the axons of neurosensory cells expressing the same odorant receptor all converge on a few target glomeruli [52, 88].

The olfactory neurosensory cells are short-lived, with an average life span of only 30-60 days. They are continuously replaced from basal cells in the epithelium [78]. This means that throughout life, newly formed primary olfactory axons are growing back to the olfactory bulb, making functionally correct synaptic connections in specific glomeruli.

Olfactory Bulb

The olfactory bulb (Fig. 11.4) has a laminar organization which is well-defined in many mammals, including the human foetus [39], but becomes less conspicuous as the brain matures [18, 19]. Passing from superficial to deep, these layers are as follows:

1. The *olfactory nerve layer*, consisting of densely interwoven, extremely fine axons of the olfactory neurosensory cells.
2. The *glomerular layer*, which contains the conspicuous glomeruli, specialized condensations of neuropil in which the terminal arborizations of the olfactory nerve fibres synapse with the dendrites of three types of secondary olfactory neurons, namely mitral, tufted and periglomerular cells. The two cell types first mentioned lie in deeper layers of the olfactory bulb; the latter elements are small and granular and, as their name implies, surround the glomeruli. In young adult humans, approximately 8,000 glomeruli have been counted [49]. Experiments using the 2-deoxyglucose technique [80] or high-resolution fMRI [96] have shown that the intrinsic, conformational properties of odour molecules are mapped into spatial patterns of glomerular activation. Specific odorants consistently activate small groups of adjacent glomeruli. Together, these "odour images" form a spatial, "odotopic" map in the glomerular sheet [81, 95]. The morphological basis of this functional map is formed by the convergent connections between specific sets of receptor neurons and specific glomeruli, already mentioned.
3. The *external plexiform layer*, consisting mainly of interlacing dendritic processes, particularly the primary and secondary dendrites of the mitral and tufted cells, and the peripherally directed *efferent dendrites* of the internal granule cells. However, it also contains the perikarya of superficial interneurons and of the relatively large tufted cells.
4. The *mitral cell layer*, made up of a thin sheet of the somata of large mitral cells. Some cell bodies of granule cells are also found in this layer.
5. The *internal plexiform layer*, composed of ascending dendrites of deep granule cells, axons of mitral and tufted cells and their recurrent collaterals, and axons of centrifugal fibres from other parts of the brain. Moreover, the perikarya of some granule cells and of somewhat larger short-axon cells are found in this layer.
6. The *granule cell layer*, consisting of several zones of densely packed granule cells, separated from each other by bundles of nerve fibres. Some larger neurons are scattered in this layer.
7. A central fibre zone, occupied by bulbar efferents and afferents. In animals with persistent olfactory ventricles, this zone is usually designated as the periventricular zone; however, in the adult human, these ventricles are totally obliterated.

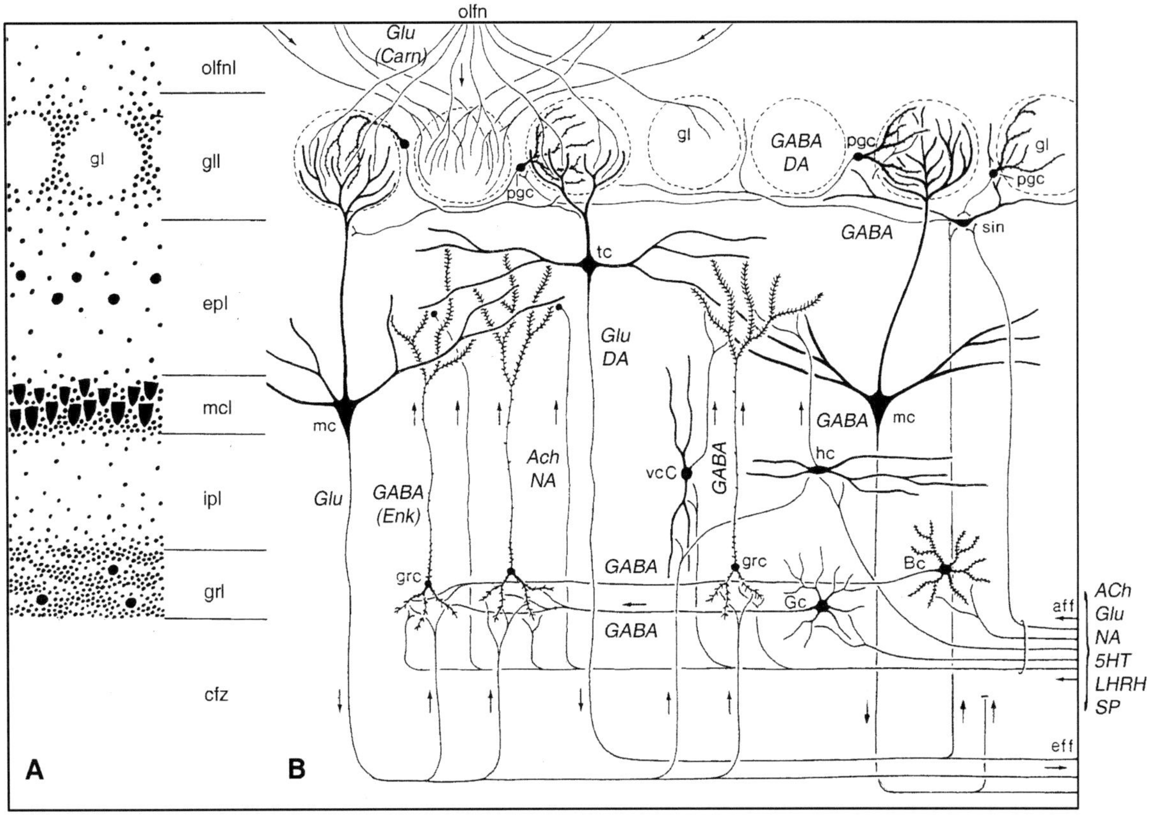

Fig. 11.4 A, B. The olfactory bulb. **A** Cytoarchitecture; **B** neuronal elements and fibres as observed in Golgi preparations (somewhat simplified). Modified from Nieuwenhuys [61]. *Ach*, acetylcholine; *aff*, centrifugal afferents; *Bc*, Blanes cell; *Carn*, carnosine; *cfz*, central fibre zone; *DA*, dopamine; *eff*, efferent fibres; *epl*, external plexiform layer; *Gc*, Golgi cell; *gl*, glomerulus; *gll*, glomerular layer; *Glu*, glutamate; *grc*, granule cell; *grl*, granule cell layer; *hc*, horizontal cell; *ipl*, internal plexiform layer; *mc*, mitral cell; *mcl*, mitral cell layer; *NA*, noradrenaline; *olfn*, olfactory nerve; *olfnl*, olfactory nerve layer; *pgc*, periglomerular cell; *sin*, superficial interneuron; *SP*, substance P; *tc*, tufted cell; *vcC*, vertical cell of Cajal

The microcircuitry of the olfactory bulb has been the subject of numerous studies, the older of which have been ably summarized by Stephan [83]. In the present work, however, only a brief survey of the structural and functional relations of the various bulbar elements can be given. In preparing this survey, we have relied heavily on the review articles by Macrides and Davis [47], Mori [54], Shipley et al. [82] and Shepherd et al. [81]. The following aspects will be discussed consecutively: the olfactory projection, the roles of the periglomerular and granule cells, and the centrifugal fibres and their targets.

The *olfactory projection* is formed by the primary olfactory neurosensory cells and the second-order olfactory projection neurons, i.e. the mitral and tufted cells. The primary olfactory elements have already been discussed. In general the large *mitral cells* and the somewhat smaller *tufted cells* have a single main or primary dendrite, terminating as a tuft of branches within a glomerulus, and several secondary or accessory dendrites branching in the external plexiform layer. The main dendrites enter into synaptic contact with olfactory nerve fibres and with axons and dendrites of periglomerular cells. The synaptic relations of the secondary dendrites will be dealt with below.

The axons of the mitral and tufted cells pass radially through the deeper layers of the olfactory bulb. During their course through the bulb, they emit numerous collaterals, which contact granule cells and deep interneurons. The main axons of the mitral and the middle and deep tufted cells gain myelin sheaths and become grouped together in bundles, which, turning posteriorly, constitute the secondary olfactory projection. The fibres of this projection convey olfactory information to higher-order olfactory structures. The axons of superficially situated tufted cells project mainly to other sites in the same olfactory bulb, thus forming an intrabulbar association system. The middle and deep tufted cells project to sites in the olfactory cortex different from those of mitral cells. Mitral cells use glutamate as a transmitter at their axon terminal and dendritic output sites. The same holds true for most of the tufted cells, but some of these elements appear to be dopaminergic.

It has already been mentioned that all of the sensory axons terminating in a glomerulus express the same olfactory receptor. This means that the glomerulus, including all of the primary sensory axons converging on it and all of the mitral and tufted cells connected to it, forms the basic functional unit in the processing of odour stimuli [5, 81].

The *periglomerular cells* and the *granule cells* have several features in common. Both maintain reciprocal dendrodendritic synaptic contacts with mitral and tufted cells, and both have these two cell types as their main targets. However, the periglomerular and granule cells also show marked differences, the most salient being that the former are regular short-axon cells, whereas the latter are amacrine, i.e. axon-less elements. Because of their position and small size, the periglomerular cells are often designated as superficial granule cells. Their dendrites enter glomeruli, where they receive impulses from olfactory nerve terminals and also form "reciprocal" or bidirectional synapses with the dendritic branches of mitral and tufted cells. The ultrastructural features of these dendrodendritic contacts suggest that mitral and tufted cell dendrites are excitatory to the periglomerular dendrites and that these latter dendrites are inhibitory to mitral and tufted cell dendrites. The axons of the periglomerular cells course along the periphery of two to five glomeruli, creating inhibitory synapses on the somata and dendrites of other periglomerular cells and on the initial parts of the primary dendrites of mitral and tufted cells. Thus, the periglomerular cells exert an inhibitory influence on the mitral and tufted cells in two different places and in two different ways, namely via interglomerular dendrodendritic synapses and via subglomerular axodendritic synapses. The population of periglomerular cells is neurochemically heterogeneous. Immunohistochemical evidence indicates that most of these elements are GABAergic, that many are dopaminergic and that GABA and dopamine colocalize within some of them.

The abundant amacrine (deep) granule cells have several short, basal dendrites and a long, peripherally coursing dendrite which ramifies in the external plexiform layer among the secondary dendrites of mitral and tufted cells. The branching distal portions of these long granule cell dendrites are densely studded with conspicuous spines (gemmules). The granule cells receive axodendritic synapses from recurrent collaterals of mitral and tufted cells, and there is also a considerable input to the granule cells from the telencephalon proper (see below).

The mitral/tufted cell secondary dendrites and the spines of the peripheral granule cell dendrites are richly interconnected by dendrodendritic synapses, organized in reciprocally oriented pairs. There is both ultrastructural and electrophysiological evidence indicating that the mitral/tufted dendrite-to-gemmule synapses are excitatory, whereas the adjacent gemmule-to-mitral/tufted dendrite synapses are inhibitory. These peculiar synapse pairs constitute extremely short inhibitory pathways from mitral cell to mitral cell or from tufted cell to tufted cell. The loops formed by the collaterals of mitral and tufted cells and by granule cells provide for self- and surround inhibition of the former elements. The inhibition delivered by the granule-to-mitral synapses is very powerful and is the principal means of mediating control of the output of the olfactory bulb. Most probably, GABA is the neurotransmitter, released by the dendrodendritic synapses of the granule spines onto the mitral/tufted cell dendrites. It is noteworthy that these dendrodendritic synapses form the sole output of the granule cells.

Like many other sensory regions, the olfactory bulb is under extensive and highly differentiated control by the brain. Experimental neuroanatomical studies [7, 20, 21, 24, 56, 87, 90, 97] have shown that *centrifugal fibres*, originating from many different brain regions, pass directly to the olfactory bulb. Most of these bulbar afferents arise from olfactory-related structures and reciprocate bulbar efferents; others form non-olfactory, general modulatory inputs. The "olfactory-related" afferents originate principally from the anterior olfactory nucleus and the olfactory cortex. The sources of the general modulatory inputs include cholinergic neurons in the basal forebrain (particularly the nucleus of the horizontal limb of the diagonal band), serotoninergic neurons in the dorsal raphe nucleus and noradrenergic elements in the locus coeruleus.

The anterior olfactory nucleus projects massively and bilaterally to the olfactory bulb. These centrifugal afferents terminate mainly in the granular and internal plexiform layers, but some ascend to the subglomerular zone of the external plexiform layer. The bulbar afferents originating from the olfactory cortex consist of axon collaterals of glutamatergic pyramidal neurons, which terminate mostly in the granular layer.

The noradrenergic and serotoninergic bulbar afferents distribute mostly within the granule cell layer and the internal plexiform layer, as well as in the glomerular layer. Serotoninergic fibres have been observed to terminate within glomeruli. The cholinergic afferents distribute relatively evenly through the bulbar laminae.

In the external plexiform layer, cholinergic, noradrenergic and other afferents terminate preferably on dendritic spines of granule cells. At these strategic sites they are in an optimal position to modulate the dendrodendritic inhibition of mitral and tufted cells. Such centrifugal control is also exerted at the levels of the glomerular and the granule cell layers. In the glomeruli, the intraglomerular terminals of the serotoninergic fibres already mentioned are involved. In the granule cell layer, bulbar afferents make synaptic contacts with the basal dendrites of granule cells.

Most layers of the olfactory bulb contain scattered, medium-sized interneurons. A detailed discussion of these elements, some of which are shown in Fig. 11.4, is beyond the scope of the present work. In general, it can be stated that these elements are inhibitory in nature, use GABA as a neurotransmitter and form a link between excitatory bulbar afferents and small interneurons, i.e. the periglomerular and granule cells. Because the latter elements are also inhibitory, it may be assumed that the medium-sized interneurons exert a disinhibitory influence on the mitral and tufted cells which, as discussed, are the targets of the granule and periglomerular cells [47, 57, 58].

In addition to the classical neurotransmitters acetylcholine, glutamate and GABA and the monoamines dopamine, noradrenaline and serotonin, many neuropeptides have been localized in the olfactory bulb. Most of these neuropeptides are colocalized with one of the classical transmitters. The following combinations could be mentioned: glutamate + carnosine in the olfactory neurosensory cells, glutamate + substance P in tufted cells and GABA + enkephalins in periglomerular and granule cells [81]. Additionally, substance P and luteinizing hormone-releasing hormone (LHRH) have been demonstrated in centrifugal afferents to the bulb [31]. The significance of these neuropeptides is not yet understood. We consider it likely that they are involved in non-synaptic or volume transmission.

Secondary Olfactory Projections and Their Sites of Termination

The axons of the projection neurons in the olfactory bulb pass backward in the olfactory tract and are distributed to the various secondary olfactory areas (see Fig. 11.7). Experimental neuroanatomical studies have conclusively shown that the macroscopic elevation known as the medial olfactory stria does not receive any secondary olfactory fibres [35, 74, 89]. After having distributed fibres to the anterior olfactory nucleus and the olfactory tubercle, the entire secondary olfactory projection enters the lateral olfactory stria (Fig. 5.37). Following that stria, the fibres reach the prepiriform, periamygdaloid and entorhinal cortices and the cortical amygdaloid nucleus, where they terminate [15, 34, 66, 89].

In primates, including humans, the *anterior olfactory nucleus* is only poorly developed and consists of some small groups of neurons situated in the caudal part of the olfactory bulb, in the olfactory peduncle and in the area where this peduncle joins the hemisphere [18]. In macrosmatic mammals, the nucleus forms a ring, surrounding the caudal part of the bulbar ventricle or its vestige, separating the olfactory bulb from the remaining olfactory structures (Fig. 11.3 A, B).

The *olfactory tubercle* is an ambiguous and somewhat enigmatic structure. In macrosmatic mammals, it is a conspicuous eminence on the base of the hemispheres (Fig. 11.3 A, B), which microscopically shows a distinct trilaminar pattern. In spite of this cortex-like structure, the olfactory tubercle has much in common with the adjacent regions of the striatum. Both structures show an intense staining for acetylcholinesterase, apparently related to intrinsic cholinergic neurons [33], and the deeper layers of the tubercle contain numerous medium-sized, densely spiny neurons, which resemble the most common striatal cell type. The somata of these elements form numerous cell bridges that directly link the olfactory tubercle with the nucleus accumbens and the caudate-putamen [51]. Moreover, the ventral striatal regions and the olfactory tubercle both receive a strong dopaminergic projection from the ventral tegmental area.

The olfactory tubercle and adjacent parts of the substantia innominata and the nucleus of the diagonal band contain aggregates of small neurons that are known as the *islands of Calleja* [16]. These islands receive terminal and vomeronasal nerve fibres [27, 60]. In rat, the islands of Calleja contain receptors for estrogens and cells that produce LHRH and are therefore supposed to be involved in reproductive functions.

In humans, the olfactory tubercle is poorly developed but the three layers that are so distinct in macrosmatic forms can be recognized nevertheless. These include: a superficial, plexiform fibre layer, a compact middle layer of pyramidal and ellipsoid cells and a deep stratum of multiform cells [18, 59]. The olfactory tubercle is situated immediately behind the bifurcation of the medial and lateral olfactory striae, forming part of the *anterior perforated space* (or *substance* or *area*). This area is characterized by the presence of numerous penetrating, small blood vessels. When these vessels are pulled out, their sites of entrance manifest themselves as a number of perforations, hence the name. The homologue of the olfactory tubercle of macrosmatic mammals is confined to the most rostral part of the anterior perforated substance; its caudal part, which is traversed by numerous transversely oriented fibres, is known as the diagonal band (Fig. 11.3C).

The olfactory tubercle receives sensory fibres from the olfactory bulb and association fibres from the primary olfactory cortex [15].

The *primary olfactory cortex* is the largest of the olfactory areas. As already mentioned, it can be topographically subdivided into prepiriform and piriform (or periamygdaloid) areas (Fig. 11.3). (In some recent publications [60, 82] it was proposed to designate these two subdivisions *together* as piriform cortex.)

The primary olfactory cortex has been studied in great detail, not only because of its salient position in the olfactory system, but also because of its prototypical value. It has many features in common with the neocortex, but is more simply built and comprises only three instead of six layers. Thus, it is well suited to serve as a model system for the analysis of basic structural and functional features of cortical organization. C.U. Ariëns Kappers [1–3] considered the olfactory or piriform cortex as the phylogenetically oldest part of the cortex; hence, he designated it as the *palaeocortex*. He believed that this ancient structure has given rise to the neocortex in the course of evolution.

The following survey of the structure of the primary olfactory cortex is principally based on two recent reviews by Haberly [30] and Neville and Haberly [60], but some previous studies [11, 65, 85, 91] have also been consulted. The primary olfactory cortex comprises three layers, which are usually designated as layers I, II and III (Fig. 11.5). Layer I is a superficial plexiform layer, which contains only sparsely scattered neurons. It is mainly composed of tangentially passing axons and ascending dendritic expansions of pyramidal and semilunar cells. The plexiform layer can be subdivided into layer Ia, a superficial sublamina dominated by afferent fibres from the olfactory bulb, and layer Ib, a deep sublamina dominated by association fibres from other parts of the primary olfactory cortex. Layer II consists of densely packed neuronal somata. It can be divided into a superficial layer IIa, in which semilunar cells are concentrated, and a deep layer IIb, in which pyramidal cell somata prevail. Layer III displays a moderate cell density. It contains pyramidal neurons, which tend to be concentrated in the superficial part, and multipolar neurons, which are most common in the deeper parts.

The neurons in the primary olfactory cortex can be classified as belonging to one of two basic types: pyramidal cells and non-pyramidal cells. Two populations of typical pyramidal cells can be distinguished: superficial pyramidal cells, which have their somata in layer IIb, and deep pyramidal cells, which have their somata in layer III. All pyramidal neurons have apical and basal dendritic tufts, but the lengths of their apical dendritic trunks are determined by the depth of their somata. The axons of the pyramidal cells descend radially through lamina III and issue numerous collaterals. Short collaterals course in all directions, establishing synaptic contacts with both pyramidal and non-pyramidal neurons. Long collaterals ascend to the superficial plexiform layer, where they take a tangential course. It has been experimentally established that long collaterals from pyramidal neurons situated in the anterior part of the primary olfactory cortex pass to the posterior primary olfactory cortex, where they are concentrated in a superficial sublayer of layer Ib, whereas long collaterals from the posterior primary olfactory cortex project predominantly to the deeper parts of layer Ib in the anterior and posterior parts of the primary olfactory cortex (see Fig. 11.6). The distribution of synaptic contacts over the receptive surface of the pyramidal neurons in the primary olfactory cortex closely resembles that observed on pyramidal neurons in other parts of the cortex: asymmetrical synapses with associated round vesicles are concentrated on dendritic spines, but are lacking on the somata; symmetrical synapses with associated

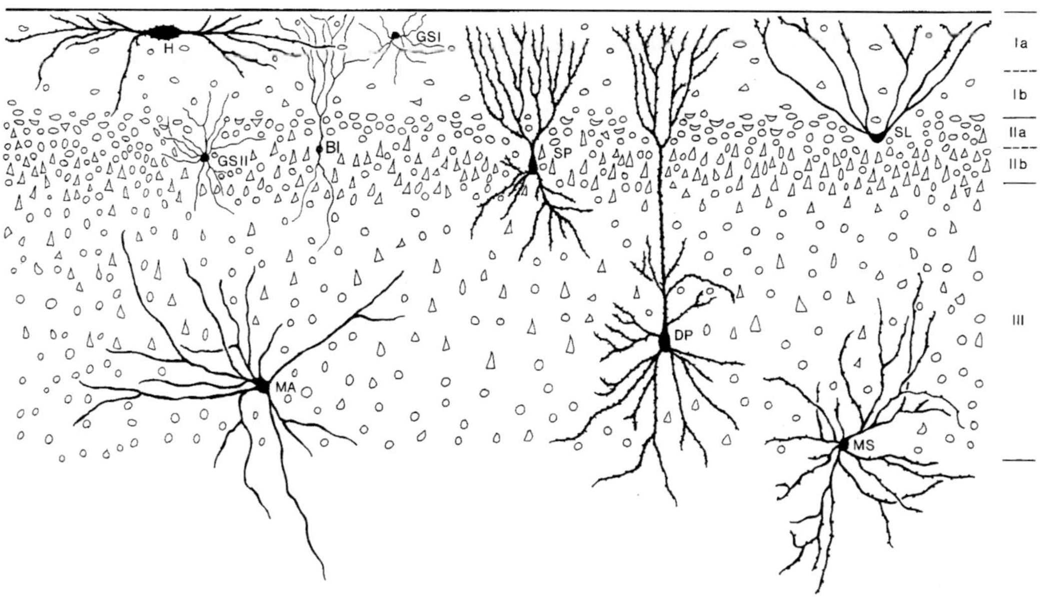

Fig. 11.5. Cytoarchitecture and major cell types of the primary olfactory cortex. *BI*, bipolar cell; *DP*, deep pyramidal cell; *GSI*, glomerular soma cell in layer I; *GSII*, globular soma cell in layer II; *H*, horizontal cell; *M*, multipolar cell; *MA*, multipolar aspiny neuron; *MS*, multipolar spiny neuron; *SP*, superficial pyramidal cell; *SL*, semilunar cell; *Ia–III*, cortical layers. (Modified from Neville and Haberly [60])

pleiomorphic vesicles are found in a very high density on the initial axonal segments and at a moderate density over all other parts, including perikarya and both apical and basal dendrites. The axons of the pyramidal cells in the primary olfactory cortex, like those in other parts of the cortex, create typical asymmetrical synapses with their target neurons.

It has already been mentioned that the somata of semilunar cells are concentrated in the superficial part of layer II. These elements emit dendrites or groups of dendrites from the poles of their crescent-shaped cell bodies (Fig. 11.5). These dendrites extend peripherally and ramify in the plexiform layer. Their axons descend radially through the cortex to enter the underlying white matter. Although at first sight the semilunar elements differ considerably from typical pyramidal cells, they should be considered as aberrant members of the pyramidal cell group.

As in other part of the cortex, the non-pyramidal cells form a heterogeneous group. Most, though not all, of these neurons have smooth dendrites, and Golgi studies suggest that their axons generally do not extend beyond the cortex. This group includes large horizontal cells, which are found exclusively in sublayer Ia; small bipolar or bitufted cells; small stellate cells with globular somata, being very common in layers I and II; and medium-sized and large multipolar cells, which are mainly confined to layer III.

Most non-pyramidal cells in the primary olfactory cortex, as in the hippocampal formation and neocortex, use GABA as neurotransmitter and are inhibitory in nature. A notable exception is a group of large multipolar neurons with spiny dendrites (cell MS in Fig. 11.5), which have been shown to be excitatory in nature with glutamate as neurotransmitter [37]. A synopsis of the circuitry and synaptology of the primary olfactory cortex is presented in Fig. 11.6. This synopsis is largely based on the excellent reviews of Haberly [30] and Neville and Haberly [60], to which the reader is referred for details.

The data presented in Fig. 11.6 can be summarized as follows:

1. Both extrinsic and intracortical association fibres pass tangentially in distinct subzones of layer I.
2. Typical pyramidal cells are present, which have apical dendrites that branch in layer I. The axonal ramifications of these cells spread both within and beyond the primary olfactory cortex (see below); together these ramifications form an extensive associational system, which makes synaptic contacts with pyramidal cells and local circuit neurons.
3. The primary olfactory cortex contains local circuit neurons of diverse types. Most of these elements are inhibitory in nature and use GABA as their neurotransmitter. They form a part of the circuits that mediate both feed-forward and feed-back inhibition onto pyramidal cells.

The primary olfactory cortex receives projections from dopaminergic, serotoninergic and noradrenergic neurons in the brain stem. It has been established that these monoaminergic projections (among possible other functions) exert an excitatory influence on a particular group of inhibitory interneurons, situated in the superficial part of layer III [28].

Many neuropeptides have been demonstrated in the primary olfactory cortex, most of which are colocalized with GABA. Thus, it has been shown that vasoactive intestinal peptide (VIP) and cholecystokinin (CCK) are present in separate subpopulations of bipolar cells [28].

It has already been mentioned that the axons of neurosensory cells expressing a particular odorant receptor converge to a few discrete glomeruli in the olfactory bulb and that the "convergence spots" of the various (>1,000!) odorant receptors together form a sensory, "odotopic" map in the glomerular sheet. Recently, evidence has been presented that a similar sensory map is also present in the primary olfactory cortex. Using a genetic, transneuronal labelling technique, it was demonstrated that pyramidal neurons in this cortex, which receive input from a particular odorant receptor, form discrete clusters [101]. This observation is in line with an earlier finding [10] that la-

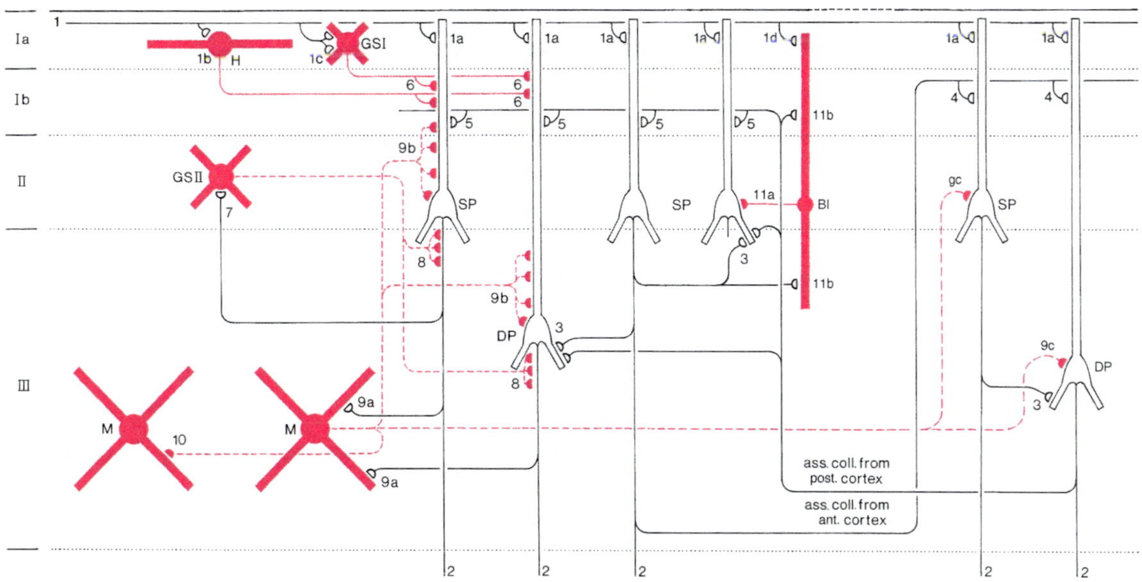

Fig. 11.6. Circuitry and synaptology of the primary olfactory cortex. *Dashed lines* indicate probable, though not proven connections. Excitatory cells: black outlines; (putative) inhibitory elements: in red. The *numbers* used in the following correspond to those in the figure. For abbreviations, see Fig. 11.5.

1. The mitral cells in the olfactory bulb are the principal source of afferent input to the primary olfactory cortex. These afferents pass via the lateral olfactory tract to layer Ia of the cortex, making excitatory synaptic contacts with the apical dendrites of pyramidal cells (1a) and with horizontal (1b) elements.
2. Main axons of the pyramidal cells leave the cortex.
3. Local axon collaterals of pyramidal cells make excitatory synaptic contacts with the basal dendrites of other pyramidal cells.
4. Long associational collaterals of superficial pyramidal cells in the anterior part of the primary olfactory cortex pass through the deepest part of the cortex and then ascend in the posterior part of the cortex to the superficial zone of layer Ib, where they make excitatory synapses with the apical dendrites of superficial and deep pyramidal cells.
5. Long associational collaterals of deep pyramidal cells in the posterior part of the primary olfactory cortex, ascend in the anterior part of the cortex to the deep zone of layer Ib, where they make excitatory synaptic contacts with the apical dendrites of superficial and deep pyramidal cells.
6. The axons of the horizontal cells and the superficial stellate cells contact the apical dendrites of pyramidal cells and mediate forward inhibition.
7. Small stellate cells situated in layer II receive axosomatic contacts from superficial pyramidal cells.
8. The initial segments of the pyramidal cells are densely covered with symmetrical synapses. It seems likely that these contacts are made by the axons of one variant of small stellate cells present in layer II. If this assumption is confirmed, these elements would occupy a position in the circuitry of the primary olfactory cortex comparable to that of the chandelier cells in the hippocampus and the neocortex.
9. Physiological studies have shown that interneurons in the deeper parts of the olfactory cortex form part of feed back loops. Most probably, large multipolar cells receiving an excitatory synaptic input from pyramidal cells (9a) and making inhibitory synapses with elements of the same type (9b) subserve this function. Many of the large multipolar cells are basket cells, i.e. their axons participate in the formation of basket-like tangles of terminal branches around the somata of pyramidal cells. Long axonal branches of these cells provide a potential substrate for surround inhibition (9c).
10. Physiological evidence indicates that feedback inhibitory neurons inhibit each other.
11. It has been recently established that small bipolar or bitufted cells with vertically oriented dendrites are an additional source of basket endings on pyramidal cells (11a) [26, 76]. Because these cells are supposed to receive excitatory input from afferent fibres in layer I (1d), as well as from collaterals of pyramidal cells (11b), they possibly function in both feed forward and feed back capacities

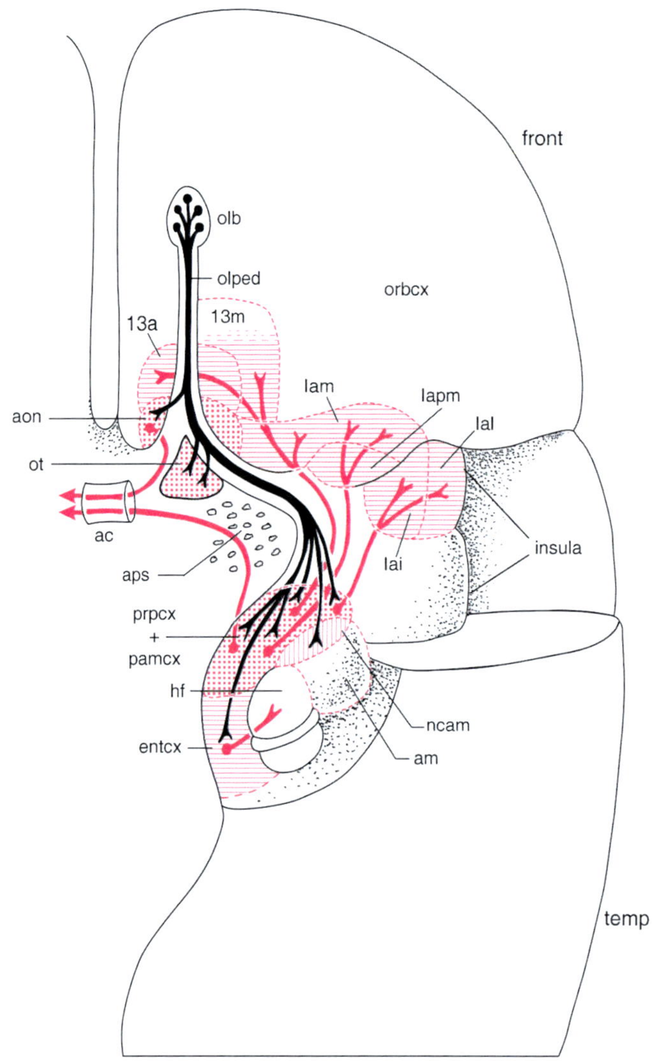

Fig. 11.7. Olfactory centres and pathways, projected upon the basal surface of the human brain. The anterior part of the temporal lobe has been removed. Areas of termination of secondary olfactory projections are *dotted*. Neocortical targets of efferents from the primary olfactory cortex (*prpcx* + *pamcx*) are *dashed*. In this and the following figure, experimental data on fibre connections in the telencephalon of the macaque monkey [12–14, 63, 72] have been extrapolated to the human. *ac*, anterior commissure; *am*, amygdaloid complex; *aon*, anterior olfactory nucleus; *aps*, anterior perforated substance; *entcx*, rostral, olfactory part of entorhinal cortex; *front*, frontal lobe; *hf*, hippocampal formation; *Iai*, intermediate agranular insular area; *Iam*, medial agranular insular area; *Iapm*, posteromedial agranular insular area; *ncam*, cortical nucleus of amygdala; *olb*, olfactory bulb; *olped*, olfactory peduncle; *orbcx*, orbital cortex; *ot*, olfactory tubercle; *pamcx*, periamygdaloid cortex; *prpcx*, prepiriform cortex; *temp*, temporal lobe; *13a,m*, subdivisions of area 13

belling a few mitral cells lying close together in the mitral cell layer with an axonal tracer results in well-delineated patches of labelled fibres in the primary olfactory cortex.

Two structures which border on the primary olfactory cortex, i.e. the *cortical nucleus of the amygdala* and the *entorhinal cortex* (Fig. 11.7), also receive direct projections from the olfactory bulb. In primates, the olfactory projections are confined to the rostral parts of both of these structures. In the entorhinal cortex, the secondary olfactory fibres form a thin, subpial sheet [15, 40, 89].

The primary olfactory cortex projects to several "higher-order" areas (see below), but also gives rise to a substantial system of association fibres, which extends through all centres that receive direct projections from the olfactory bulb [15, 41]. The anterior olfactory nucleus and the primary olfactory cortex are interconnected by fibres passing through the anterior commissure. By the same route, fibres originating from the anterior olfactory nucleus project to the deeper layers of the contralateral olfactory bulb [46].

Further Spread of Olfactory Information

The primary olfactory cortex projects to several other parts of the forebrain, including the amygdala, the hippocampus, the hypothalamus, the dorsal thalamus and the neocortex [15, 66]. The fibres passing to the hippocampus and the amygdala arise principally from the periamygdaloid cortex and from the adjacent rostral, olfactory part of the entorhinal cortex. These fibres form part of polysynaptic routes by which the rhinencephalon is connected with the septum and the hypothalamus. The fibres that connect the primary olfactory cortex with the hypothalamus join the medial forebrain bundle and terminate predominantly in the caudal part of the lateral hypothalamic area. These olfactohypothalamic projections are involved in the control of appetite and feeding behaviour. The projection from the primary olfactory cortex to the dorsal thalamus is relatively small and terminates mainly in the medial part of the mediodorsal nucleus [75]. This part of the mediodorsal thalamic nucleus is known to be reciprocally connected with several areas in the posterior orbitofrontal and rostral insular cortex. As will be discussed below, these cortical areas are also in receipt of more strongly developed, monosynaptic projections directly from the primary olfactory cortex.

The olfactory projections to the neocortex have been thoroughly studied by Carmichael et al. [15]. This study forms part of a comprehensive research programme that includes the architectonic subdivision of the orbital and adjacent insular cortex in the macaque monkey and humans, and a detailed experimental analysis of the connectional networks within these cortical regions in the monkey [12–14, 63, 64, 66, 72]. The following data are derived from these studies.

The projections from the primary olfactory cortex terminate in four insular and two orbital areas (Fig. 11.7). All of these cortical areas are agranular or dysgranular, which means that they have a poorly developed or absent layer IV. The olfactory neocortical projections differ considerably from the pathways for other sensory modalities. In the somatosensory, visual and auditory systems, sensory information reaches the cortex via a thalamic relay, and the thalamocortical fibres terminate mainly in a well-differentiated layer IV. In contrast, the olfactory neocortical areas receive their major sensory input directly from the primary olfactory cortex, without a thalamic relay, and the afferents carrying this input terminate principally in layer I.

The orbital prefrontal cortex can be subdivided into some 20 different architectonic areas, which are interrelated by an elaborate set of connections [14, 63, 64] (Fig. 11.8). The intrinsic network formed by these connections receives, apart from the olfactory input already discussed, extrinsic afferents from several other sensory modalities. Thus, it has been shown that gustatory, general visceral, somatosensory and visual projections terminate in restricted areas of the orbital network, especially in its caudal and lateral parts (Fig. 11.8 B). Anatomical and physiological data [13, 14, 17, 71–73] indicate

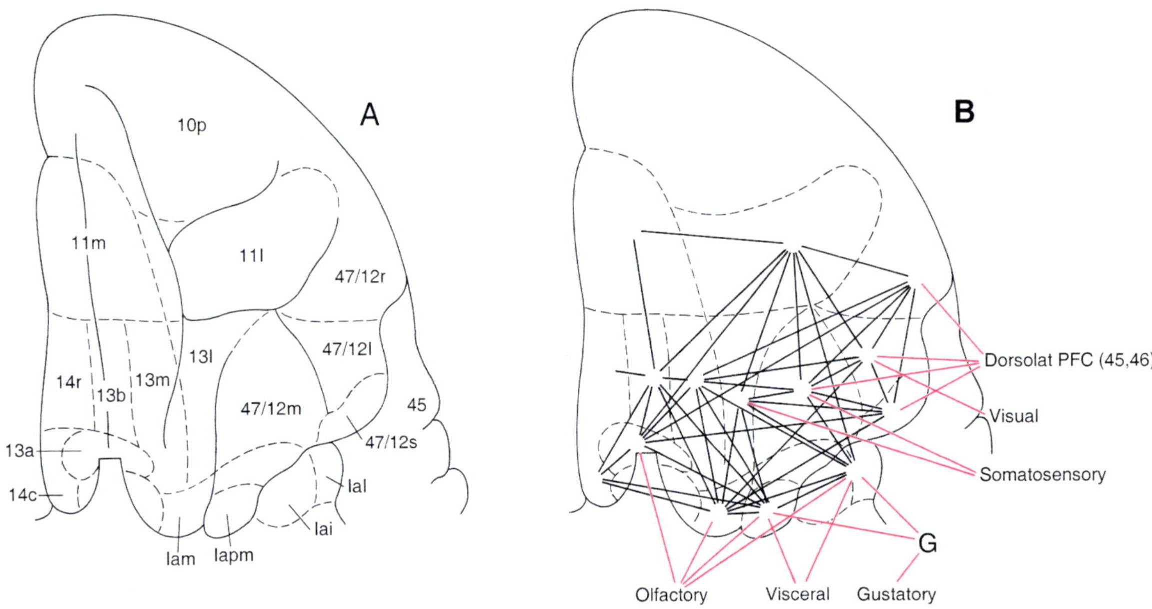

Fig. 11.8 A, B. The orbitofrontal connectional network. **A** Architectonic subdivision of the human orbital prefrontal cortex [64]. The areas 12–14 correspond to similarly numbered areas in the macaque monkey [12, 92]. The areas 10, 11, 45 and 47 were previously identified by Brodmann [8]. It is important to note that all of the (sub)areas shown can also be identified in the macaque monkey [12, 64]. **B** Connections between the architectonic areas shown in **A**, and some of the extrinsic inputs to this orbital prefrontal network [13, 14]. Most, but not all of the connections shown are reciprocal. The olfactory afferents arise, as shown in Fig. 11.7, from the prepiriform cortex. The gustatory afferents originate from the primary gustatory cortex (*G*), situated in the rostral agranular insula, which in its turn receives afferents from the dorsal part of the parvocellular ventral posteromedial thalamic nucleus. The visceral afferents stem from the ventral part of the same thalamic nucleus. The somatosensory input originates from the primary and secondary somatosensory cortical areas (SI, SII), and from the somatosensory association area 7b. The visual afferents arise from the temporal visual association area 21. Based on illustrations in [14, 63, 64]

that the orbital areas in receipt of these sensory inputs and the cortico-cortical connections between these areas are involved in the analysis and integration of food-related sensations and play an important role in the control of feeding. Olfactory and gustatory inputs, which converge in the caudal part of the orbital network, are probably involved in generating the sensation of the flavour associated with different foods [63]. The visceral afferents may provide signals on nutritive intake and satiety, such as stomach distension and blood sugar levels. These internal signals serve to modulate feeding behaviour [17]. It is known that the activity of gustatory neurons in the orbital cortex is modulated by hunger and satiety [17]. The somatosensory signals appear to arise from the head and face regions of the somatosensory cortical area [72]. Oral somesthetic inputs localize flavour in the mouth and provide for an "oral sense," i.e. a unified sensory perception of smell, taste and texture [14]. In the area receiving visual inputs, neurons are present that show selective responses to the sight of food. Interestingly, these neurons, in analogy to the gustatory elements mentioned above, show a sensory-specific reduction of their responses to foods after satiation [17].

Functional imaging techniques have been used to identify the cortical areas activated by olfactory stimuli and to explore the nature of olfactory processing in the human brain. For a detailed discussion of the results of these studies, the reader is referred to the review articles by Zatorre and Jones Gotman [99] and Zald and Pardo [98]; here, we confine ourselves to a few notes.

1. As might be expected from the anatomical data discussed above, olfactory-related activity has been consistently detected in the regions of the primary olfactory and orbitofrontal cortices.
2. Monorhinal odorous stimuli activate the primary olfactory cortex bilaterally, consistent with the commissural connections between these areas via the anterior commissure (Fig. 11.7).
3. Several functional imaging studies (see [99]) have reported that, following bilaterally presented olfactory stimuli, the response is much higher in the right orbitofrontal cortex than in the left. This finding suggests a functional asymmetry in the higher processing of olfactory information, which concurs with the right hemisphere dominance in olfaction, as derived from clinical evidence [42].
4. It is important to note that the orbitofrontal focus of olfactory activity, as detected by functional neuroimaging studies, is situated considerably more rostral than would be expected on the basis of anatomical findings [98] (Fig. 15.69 A).
5. Not only neurophysiological recordings (see above) but also neuroimaging studies have shown that the activity of the orbitofrontal cortex depends on the motivational changes that accompany feeding. For example, fMRI responses to food-related olfactory stimuli decrease considerably and specifically after feeding to satiation with that particular food [62].

Accessory Olfactory System and Terminal Nerve

The *accessory olfactory system* [32, 50, 79], also known as the *vomeronasal system*, comprises the vomeronasal organ or Jacobson's organ; the vomeronasal nerve, a special part of the olfactory bulb, known as the accessory olfactory bulb; the medial nucleus of the amygdaloid complex; and some parts of the bed nucleus of the stria terminalis. Although the main olfactory system and the accessory olfactory system show many parallels in their structural organization, the two are morphologically distinct and subserve different functions. The accessory olfactory system is primarily involved in the regulation of reproductive behaviour elicited by pheromones, i.e. chemical messengers from other members of the same species.

The *vomeronasal organ* is an elongated tube contained in a bony capsule, located bilaterally along the base of the nasal septum. The tube opens anteriorly via a narrow duct into the floor of the nasal cavity. The vomeronasal chemosensory receptor cells, which are confined to the medial side of the organ, are similar to those in the olfactory epithelium, except that they bear microvilli rather than cilia. The fine axons of these receptor cells assemble in the vomeronasal nerve. After passing through the cribriform plate, they terminate in the accessory olfactory bulb.

The *accessory olfactory bulb* commonly appears as a small protrusion on the mediocaudal surface of the olfactory bulb. Its internal structure replicates most of the features of the main olfactory bulb, and the same seven layers can be recognized. The accessory olfactory bulb projects to the amygdaloid complex and to the bed nucleus of the stria terminalis by way of the accessory olfactory tract. The latter initially forms part of the lateral olfactory tract, but becomes increasingly individuated at more caudal levels.

The accessory olfactory system is well developed in several mammalian groups, including marsupials, insectivores and rodents. In the latter group it has been shown to play a prominent role in sexual behaviour.

The question of whether the adult human possess a functioning accessory olfactory system cannot be answered definitively, as is apparent from the following data, which have been mainly derived from two recent reviews [32, 86].

1. There is strong sociological and behavioural evidence for the existence of pheromonal communication between humans.
2. The widely held view that a vomeronasal organ is only present in humans during embryonic and foetal life and that this organ disappears before birth has proven to be untenable. Recent studies have shown that distinct vomeronasal pits are present in adult humans and electron microscopical examination revealed the presence of putative sensory cells in the pseudostratified epithelial lining of these pits. However, it has never been shown that these putative sensory cells possess axons that unite in a vomeronasal nerve.
3. The accessory olfactory bulb is absent in adult humans. Possibly, pheromonal signals in humans reach the CNS by an alternative route, provided by the terminal nerve.

The *terminal nerve*, also referred to as *nerve zero*, is ubiquitous in vertebrates, including humans [22, 23]. It is a ganglionated cranial nerve that extends from the nose to the basal forebrain regions. Intracranially, its fibres form a loose plexus along the medial and ventral sides of the olfactory bulb and olfactory peduncle. Its peripheral branches are associated with olfactory and vomeronasal fila and project to the nasal mucosa, providing free nerve endings in both the olfactory and non-olfactory epithelia and dense plexuses surrounding Bowman's glands. Branches of the terminal nerve have also been observed to form terminations in the walls of blood vessels. Centrally, the terminal nerve establishes connections with various forebrain regions, including the septum, the olfactory tubercle, the preoptic area and the median eminence. A component of this nerve enters the brain in the immediate vicinity of the lamina terminalis (hence, its name).

The neurons which contribute fibres to the terminal nerve are either scattered along its course or aggregated in some ganglia. The largest of these, known as the "ganglion terminale", is located ventral to the caudal border of the olfactory bulb. Bipolar or fusiform and multipolar neurons have been observed in the terminal nerve. The multipolar neurons are believed to represent postganglionic autonomic elements providing for innervation of glands and blood vessels in the nose. The bipolar cells originate in the olfactory placode and migrate inward during ontogeny [77]. They form part of a centripetal conduction system conveying neural (perhaps chemosensory) messages from the nasal epithelium directly to the brain. Immunocytochemical studies have shown that some of these cells and their processes contain the reproductive peptide LHRH. There is experimental evidence suggesting that the terminal nerve is involved in the regulation of reproductive behaviour [93, 94].

References

1. Ariëns Kappers CU (1909) The phylogenesis of the palaeo-cortex compared with the evolution of the visual neo-cortex. Arch Neurol Psychiatry (London) 4:161–173
2. Ariëns Kappers CU (1929) The evolution of the nervous system in invertebrates, vertebrates and man. Bohn, Haarlem
3. Ariëns Kappers CU, Huber GC, Crosby EC (1936) The comparative anatomy of the nervous system of vertebrates, including man. MacMillan, New York
4. Aroniadou-Anderjaska V, Ennis M, Shipley MT (1997) Glomerular synaptic responses to olfactory nerve input in rat olfactory bulb slices. Neuroscience 79:425–434
5. Belluscio L, Lodovichi C, Feinstein P, Mombaerts P, Katz LC (2002) Odorant receptors instruct functional circuitry in the mouse olfactory bulb. Nature 419:296–300
6. Berkowicz DA, Trombley PQ, Shepherd GM (1994) Evidence for glutamate as the olfactory receptor cell neurotransmitter. J Neurophysiol 71:2557–2561
7. Broadwell RD, Jacobowitz DM (1976) Olfactory relationships of the telencephalon and diencephalon in the rabbit. III. The ipsilateral centrifugal fibers to the olfactory bulbar and retrobulbar formations. J Comp Neurol 170:321–346
8. Brodmann K (1909) Vergleichende Lokalisationslehre der Großhirnrinde in ihren Prinzipien dargestellt auf Grund des Zellenbaues. Barth, Leipzig
9. Buck L, Axel R (1991) A novel multigene family may encode odorant receptors: a molecular basis for odor recognition. Cell 65:175–187
10. Buonviso N, Revial MF, Jourdan F (1991) The projections of mitral cells from small local regions of the olfactory bulb: an anterograde tracing study using PHA-L (Phaseolus vulgaris Leucoagglutinin). Eur J Neurosci 3:493–500
11. Cajal SR (1972) Histologie du système nerveux de l'homme et des vertébrés. Consejo Superior de Investigaciones Cientificas. Instituto Ramon y Cajal, Madrid
12. Carmichael ST, Price JL (1994) Architectonic subdivision of the orbital and medial prefrontal cortex in the macaque monkey. J Comp Neurol 346:366–402
13. Carmichael ST, Price JL (1995) Sensory and premotor connections of the orbital and medial prefrontal cortex of macaque monkeys. J Comp Neurol 363:642–664
14. Carmichael ST, Price JL (1996) Connectional networks within the orbital and medial prefrontal cortex of macaque monkeys. J Comp Neurol 371:179–207
15. Carmichael ST, Clugnet MC, Price JL (1994) Central olfactory connections in the macaque monkey. J Comp Neurol 346:403–434
16. Celleja C (1893) La region olfatoria del cerebro. N Moya, Madrid
17. Critchley HD, Rolls ET (1996) Hunger and satiety modify the responses of olfactory and visual neurons in the primate orbitofrontal cortex. J Neurophysiol 75:1673–1686
18. Crosby EC, Humphrey T (1941) Studies of the vertebrate telencephalon. II. The nuclear pattern of the anterior olfactory nucleus, tuberculum olfactorium and the amygdaloid complex in adult man. J Comp Neurol 47:309–352
19. Crosby EC, Humphrey T, Lauer EW (1962) Correlative anatomy of the nervous system. MacMillan, New York
20. Davis BJ, Macrides F (1981) The organization of centrifugal projections from the anterior olfactory nucleus, ventral hippocampal rudiment, and piriform cortex to the main olfactory bulb in the hamster: An autoradiography study. J Comp Neurol 203:475–493
21. Davis BJ, Macrides F, Youngs WM, Schneider SP, Rosene DL (1978) Efferents and centrifugal afferents of the main and accessory olfactory bulbs in the hamster. Brain Res Bull 3:59–72
22. Demski LS (1993) Terminal nerve complex. Acta Anat (Basel) 148:81–95
23. Demski LS, Schwanzel-Fukuda M (1987) The terminal nerve (nervus terminalis): structure, function, and evolution. Ann NY Acad Sci 519:469
24. Dennis BJ, Kerr DIB (1976) Origins of olfactory bulb centrifugal fibres in the cat. Brain Res 110:593–600
25. Dinopoulos A, Papadopoulos GC, Michaloudi H et al (1992) Claustrum in the hedgehog (Erinaceus europaeus) brain: cytoarchitecture and connections with cortical and subcortical structures. J Comp Neurol 316:187–205
26. Ekstrand JJ, Domroese ME, Feig SL, Illig KR, Haberly LB (2001) Immunocytochemical analysis of basket cells in rat piriform cortex. J Comp Neurol 434:308–328
27. Fallon JH, Loughlin SE, Ribak CE (1983) The islands of Calleja complex of rat basal forebrain. III. Histochemical evidence for a striatopallidal system. J Comp Neurol 218:91–120
28. Gellman RL, Aghajanian GK (1993) Pyramidal cells in piriform cortex receive a convergence of inputs from monoamine activated GABAergic interneurons. Brain Res 600:63–73
29. Glusman G, Bahar A, Sharon D et al (2000) The olfactory receptor gene superfamily: data mining, classification, and nomenclature. Mamm Genome 11:1016–1023
30. Haberly LB (1990) Comparative aspects of olfactory cortex. In: Jones EG, Peters A (eds) Comparative structure and evolution of cerebral cortex, part II. Plenum, New York, pp 137–166 (Cerebral cortex, vol 8B)
31. Halasz N (1990) The vertebrate olfactory system. Akademiai Kiado, Budapest

32. Halpern M, Martinez-Marcos A (2003) Structure and function of the vomeronasal system: an update. Progr Neurobiol 70:245–318
33. Hartz-Schütt CG, Mai JK (1991) [Cholinesterase activity in the human striatum with special consideration of the terminal islands]. J Hirnforsch 32:317–342
34. Heimer L (1969) The secondary olfactory connections in mammals, reptiles and sharks. Ann NY Acad Sci 167:129–146
35. Heimer L (1975) Olfactory projections to the diencephalon. In: Stumpf WE, Grant LD (eds) Anatomical neuroendocrinology. Karger, Basel, pp 30–39
36. Herrick CJ (1910) The morphology of the forebrain in Amphibia and Reptilia. J Comp Neurol 20:413–547
37. Hoffman WH, Haberly LB (1993) Role of synaptic excitation in the generation of bursting-induced epileptiform potentials in the endopiriform nucleus and piriform cortex. J Neurophysiol 70:2550–2561
38. Holmgren N (1925) Points of view concerning forebrain morphology in higher vertebrates. Acta Zool 6:413–477
39. Humphrey T (1940) The development of the olfactory and the accessory olfactory formations in human embryos and fetuses. J Comp Neurol 73:431–468
40. Insausti R, Marcos P, Arroyo-Jimenez MM, Blaizot X, Martinez-Marcos A (2002) Comparative aspects of the olfactory portion of the entorhinal cortex and its projection to the hippocampus in rodents, nonhuman primates, and the human brain. Brain Res Bull 57:557–560
41. Johnson DM, Illig KR, Behan M, Haberly LB (2000) New features of connectivity in piriform cortex visualized by intracellular injection of pyramidal cells suggest that "primary" olfactory cortex functions like "association" cortex in other sensory systems. J Neurosci 20:6974–6982
42. Jones-Gotman M, Zatorre RJ (1993) Odor recognition memory in humans: role of right temporal and orbitofrontal regions. Brain Cogn 22:182–198
43. Källén B (1951) The nuclear development in the mammalian forebrain with special regard to the subpallium. Kgl Fysiogr Sällsk Lund Handl N F 61:1–43
44. Krieg WJS (1975) Interpretative atlas of the monkey's brain. Brain Books, Evanston
45. Lauer EW (1945) The nuclear pattern and fiber connections of certain basal telencephalic centers in the macaque. J Comp Neurol 82:215–254
46. Lohman AHM (1963) The anterior olfactory lobe of the guinea-pig: a descriptive and experimental anatomical study. Acta Anat 53 (Suppl 49):1–109
47. Macrides F, Davis BJ (1983) The olfactory bulb. In: Emson PC (ed) Chemical neuroanatomy. Raven, New York, pp 391–426
48. Medina L, Legaz I, Gonzalez G et al (2004) Expression of Dbx1, Neurogenin 2, Semaphorin 5A, Cadherin 8, and Emx1 distinguish ventral and lateral pallial histogenetic divisions in the developing mouse claustroamygdaloid complex. J Comp Neurol 474:504–523
49. Meisami E, Mikhail L, Baim D, Bhatnagar KP (1998) Human olfactory bulb: aging of glomeruli and mitral cells and a search for the accessory olfactory bulb. Ann NY Acad Sci 855:708–715
50. Meredith M (1987) Vomeronasal organ and nervus terminalis. In: Adelman G (ed) Encyclopedia of neuroscience. Birkhäuser, Boston, pp 1303–1305
51. Millhouse OE, Heimer L (1984) Cell configurations in the olfactory tubercle of the rat. J Comp Neurol 228:571–597
52. Mombaerts P, Wang F, Dulac C et al (1996) Visualizing an olfactory sensory map. Cell 87:675–686
53. Moran DT, Rowley JC III, Jafek BW, Lovell MA (1982) The fine structure of the olfactory mucosa in man. J Neurocytol 11:721–746
54. Mori K (1987) Membrane and synaptic properties of identified neurons in the olfactory bulb. Progr Neurobiol 29:275–320
55. Mori K, Nagao H, Yoshihara Y (1999) The olfactory bulb: coding and processing of odor molecule information. Science 286:711–715
56. Mori K, Satou M, Takagi SF (1979) Axonal projection of anterior olfactory nuclear neurons to the olfactory bulb bilaterally. Exp Neurol 64:295–305
57. Mugnaini E, Oertel WH, Wouterlood FF (1984) Immunocytochemical localization of GABA neurons and dopamine neurons in the rat main and accessory olfactory bulbs. Neurosci Lett 47:221–226
58. Mugnaini E, Wouterlood FG, Dahl AL, Oertel WH (1984) Immunocytochemical identification of GABAergic neurons in the main olfactory bulb of the rat. Arch Ital Biol 122:83–113
59. Nauta WJH, Haymaker W (1969) Hypothalamic nuclei and fiber connections. In: Haymaker W, Anderson E, Nauta WJH (eds) The hypothalamus. Thomas, Springfield, pp 136–209
60. Neville KR, Haberly LB (2004) Olfactory cortex. In: Shepherd GM (ed) The synaptic organization of the brain, 5th edn. Oxford University Press, Oxford, pp 415–454
61. Nieuwenhuys R (1998) Telencephalon. In: Nieuwenhuys R, Ten Donkelaar HJ, Nicholson C (eds) The central nervous system of vertebrates, vol III. Springer, Berlin Heidelberg New York, pp 1871–2008
62. O'Doherty J, Rolls ET, Francis S et al (2000) Sensory-specific satiety-related olfactory activation of the human orbitofrontal cortex. Neuroreport 11: 893–897
63. Öngür D, Price JL (2000) The organization of networks within the orbital and medial prefrontal cortex of rats, monkeys and humans. Cereb Cortex 10:206–219
64. Öngür D, Ferry AT, Price JL (2003) Architectonic subdivision of the human orbital and medial prefrontal cortex. J Comp Neurol 460:425–449
65. O'Leary JL (1937) Structure of the primary olfactory cortex of the mouse. J Comp Neurol 67:1–31

66. Price JL (2004) Olfaction. In: Paxinos G (ed) The human nervous system, 2nd edn. Elsevier, Amsterdam, pp 1197–1211
67. Puelles L (2001) Brain segmentation and forebrain development in amniotes. Brain Res Bull 55:695–710
68. Puelles L, Rubenstein JLR (2002) Forebrain. In: Ramachandran VS (ed) Encyclopedia of the human brain, vol 2. Academic, Amsterdam, pp 299–315
69. Puelles L, Kuwana E, Puelles E et al (2000) Pallial and subpallial derivatives in the embryonic chick and mouse telencephalon, traced by the expression of the genes Dlx-2, Emx-1, Nkx-2.1, Pax-6, and Tbr-1. J Comp Neurol 424:409–438
70. Ressler KJ, Sullivan SL, Buck LB (1993) A zonal organization of odorant receptor gene expression in the olfactory epithelium. Cell 73:597–609
71. Rolls ET (2000) The orbitofrontal cortex and reward. Cereb Cortex 10:284–294
72. Rolls ET, Baylis LL (1994) Gustatory, olfactory, and visual convergence within the primate orbitofrontal cortex. J Neurosci 14:5437–5452
73. Rolls ET, Critchley HD, Mason R, Wakeman EA (1996) Orbitofrontal cortex neurons: role in olfactory and visual association learning. J Neurophysiol 75:1970–1981
74. Rosene DL, Heimer L (1977) Olfactory bulb efferents in the rhesus monkey. Anat Rec 187:698
75. Russchen FT, Amaral DG, Price JL (1987) The afferent input to the magnocellular division of the mediodorsal thalamic nucleus in the monkey, *Macaca fascicularis*. J Comp Neurol 256:175–210
76. Sanides-Kohlrausch C, Wahle P (1990) VIP- and PHI-immunoreactivity in olfactory centers of the adult cat. J Comp Neurol 294:325–339
77. Schwanzel-Fukuda M, Pfaff DW (1994) Luteinizing hormone-releasing hormone (LHRH) and neural cell adhesion molecule (NCAM)-immunoreactivity in development of the forebrain and reproductive system. Ann Endocrinol (Paris) 55:235–241
78. Schwob JE (2002) Neural regeneration and the peripheral olfactory system. Anat Rec 269:33–49
79. Segovia S, Guillamon A (1993) Sexual dimorphism in the vomeronasal pathway and sex differences in reproductive behaviors. Brain Res Brain Res Rev 18:51–74
80. Sharp FR, Kauer JS, Shepherd GM (1975) Local sites of activity-related glucose metabolism in rat olfactory bulb during olfactory stimulation. Brain Res 98:596–600
81. Shepherd GM, Chen WR, Greer CA (2004) Olfactory bulb. In: Shepherd GM (ed) The synaptic organization of the brain, 5th edn. Oxford University Press, Oxford, pp 105–216
82. Shipley MT, McLean JH, Ennis M (1995) Olfactory system. In: Paxinos G (ed) The rat nervous system, 2nd edn. Academic, San Diego, pp 899–926
83. Stephan H (1975) Handbuch der mikroskopischen Anatomie des Menschen, Band 4, Teil 9 Allocortex. Springer, Berlin Heidelberg New York
84. Stephan H, Baron G, Frahm HD (1990) Insectivora with a stereotaxic atlas of the hedgehog brain. Springer, Berlin Heidelberg New York (Comparative brain research in mammals, vol 1)
85. Stevens CF (1969) Structure of cat frontal olfactory cortex. J Neurophysiol 32:184–192
86. Swaab DF (2004) The human hypothalamus: basic and clinical aspects. Part II: Neuropathology of human hypothalamus and adjacent brain. Elsevier, Amsterdam
87. Swanson LW (1976) An autoradiographic study of the efferent connections of the preoptic region in the rat. J Comp Neurol 167:227–256
88. Treloar HB, Feinstein P, Mombaerts P, Greer CA (2002) Specificity of glomerular targeting by olfactory sensory axons. J Neurosci 22:2469–2477
89. Turner BH, Gupta KC, Mishkin M (1978) The locus and cytoarchitecture of the projection areas of the oflactory bulb in *Macaca mulatta*. J Comp Neurol 177:381–396
90. Uemura-Sumi M, Osterstock J, Millhouse OE (1985) The zona incerta: another source of centrifugal fibers to the main olfactory bulb. Neurosci Lett 53:241–245
91. Valverde F (1965) Studies on the piriform lobe. Harvard University Press, Cambridge
92. Walker AE (1940) A cytoarchitectural study of the prefrontal area of the macaque monkey. J Comp Neurol 73:59–86
93. Wirsig CR (1987) Effects of lesions of the terminal nerve on mating behavior in the male hamster. Ann NY Acad Sci 519:241–251
94. Wirsig CR, Leonard CM (1987) Terminal nerve damage impairs the mating behavior of the male hamster. Brain Res 417:293–303
95. Xu F, Greer CA, Shepherd GM (2000) Odor maps in the olfactory bulb. J Comp Neurol 422:489–495
96. Xu F, Liu N, Kida I et al (2003) Odor maps of aldehydes and esters revealed by functional MRI in the glomerular layer of the mouse olfactory bulb. Proc Natl Acad Sci USA 100:11029–11034
97. Záborszky L, Carlsen J, Brashear HR, Heimer L (1986) Cholinergic and GABAergic afferents to the olfactory bulb in the rat with special emphasis on the projection neurons in the nucleus of the horizontal limb of the diagonal band. J Comp Neurol 243:488–509
98. Zald DH, Pardo JV (2000) Functional neuroimaging of the olfactory system in humans. Int J Psychophysiol 36:165–181
99. Zatorre RJ, Jones-Gotman M (2000) Functional imaging of the chemical senses. In: Toga AW, Maziotta JC (eds) Brain mapping, the systems. Academic, San Diego, pp 403–424
100. Zhang X, Firestein S (2002) The olfactory receptor gene superfamily of the mouse. Nat Neurosci 5:124–133
101. Zou Z, Horowitz LF, Montmayour JP, Snapper S, Buck LB (2001) Genetic tracing reveals a stereotyped sensory map in the olfactory cortex. Nature 414:173–179

12 Telencephalon: Hippocampus and Related Structures

Introduction

In this chapter the hippocampus and two related telencephalic structures, the limbic lobe and the precommissural septum, will be dealt with. The hippocampus, which develops from the medial pallium (Figs. 2.24 B, 2.25, 11.1), appears in the sixth week of gestation. In late embryonic stages this structure occupies a considerable part of the medial hemisphere wall (Fig. 12.1 A), but during further development it is more and more overshadowed by the expanding neopallium [80]. From the fourth month onward, the rostral parts of the hippocampal primordium undergo retrogressive changes and gradually become reduced to a narrow, band-like rudiment (Fig. 12.1 B, C). The morphologically caudal part of the hippocampal anlage follows the ventral curvature of the developing hemisphere and becomes incorporated in the temporal lobe. As development proceeds, this temporal part of the hippocampus rolls in on itself along a longitudinal groove, the hippocampal fissure [80]. Due to this "internalization," only relatively small parts of the hippocampus maintain their initial, superficial position in the adult stage (Fig. 3.6).

The reduction of the rostral parts of the hippocampus is clearly associated with the development of the corpus callosum. All eutherian mammals possess a corpus callosum, and in all species studied, the thin, reduced portion of the hippocampus coextends with that commissure (Fig. 12.1 C–E) [214, 234]. In monotremes [1, 74, 81] and marsupials [68, 96, 125, 126], a corpus callosum is lacking. In these forms, a well-developed hippocampus is present throughout the length of the hemispheres (Fig. 12.2). The inrolling, which in eutherians is confined to the postcallosal portion of the hippocampus (Fig. 12.1 C–E), also involves the more rostral parts of that structure in monotremes (Fig. 12.2 B, C) and marsupials.

The hippocampal grey matter shows a laminar differentiation, leading to the formation of the *hippocampal cortex* or *archicortex*. This cortex can be divided into three strip-like areas, *dentate gyrus, Ammon's horn* and *subiculum* (Figs. 12.2 C, 12.5 A). The dentate gyrus and Ammon's horn are typical trilaminar structures, in which a cellular layer is situated between two plexiform/fibre layers (Fig. 12.5 B–D). The subiculum is characterized by the presence of a wide cellular zone, within which separate superficial and deep layers can be discerned (Fig. 12.5 E).

Anatomically, the hippocampus forms part of a large arcuate convolution that Broca [130] called the great limbic lobe. In humans and other primates this lobe includes the cingulate and parahippocampal gyri, which via the ret-

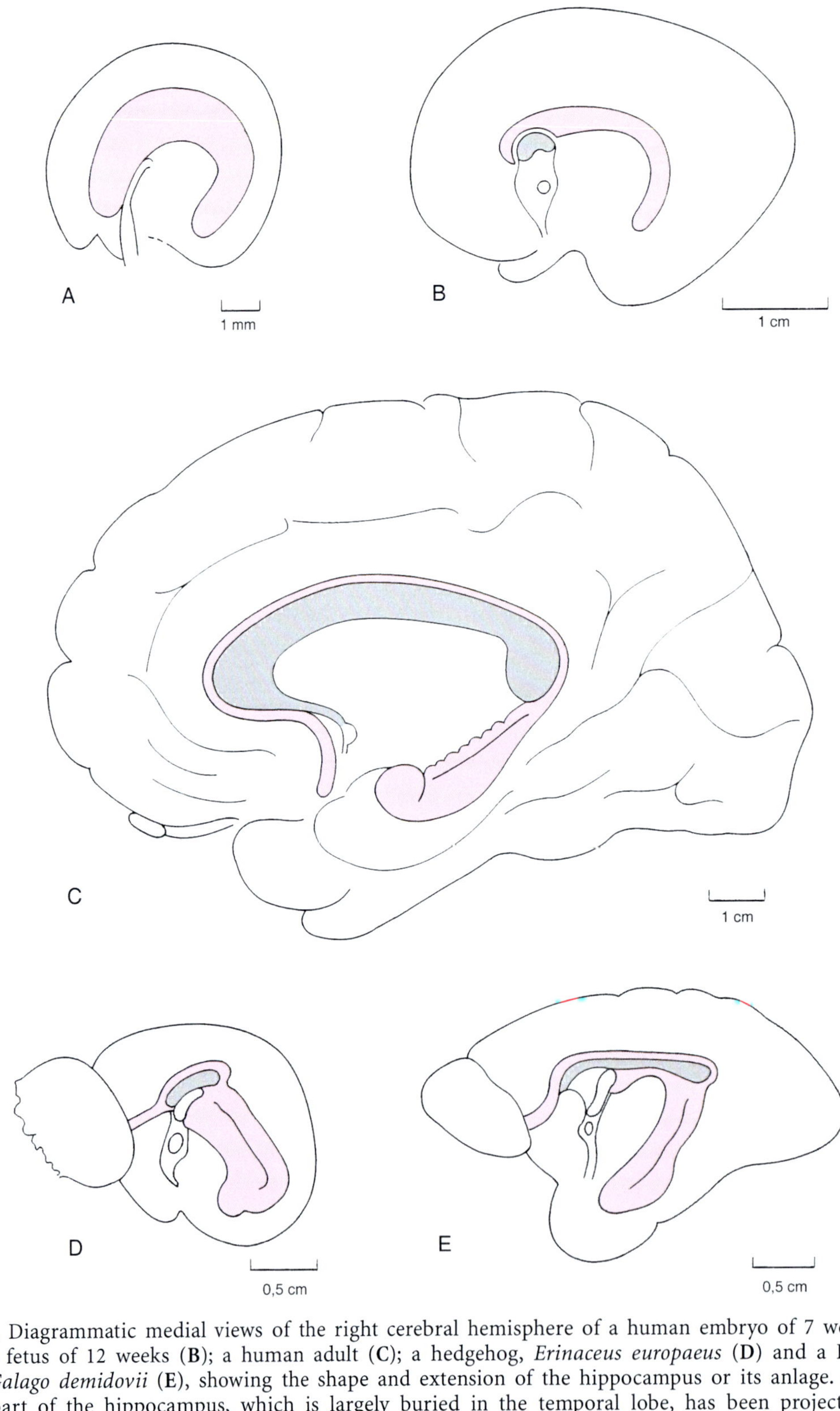

Fig. 12.1. Diagrammatic medial views of the right cerebral hemisphere of a human embryo of 7 weeks (**A**); a human fetus of 12 weeks (**B**); a human adult (**C**); a hedgehog, *Erinaceus europaeus* (**D**) and a Demidoff galago, *Galago demidovii* (**E**), showing the shape and extension of the hippocampus or its anlage. In **C** the ventral part of the hippocampus, which is largely buried in the temporal lobe, has been projected upon the medial surface. The hippocampus is shown in *red*, the corpus callosum in *grey*. **A** and **B** are based on O'Rahilly and Müller [158]; **D** and **E** are simplified from Stephan [214]

rosplenial isthmus of the cingulate gyrus are continuous with each other (Figs. 1.4B, 3.6).

Histologically, the hippocampal cortex is separated from the full-blown, six-layered neocortex throughout its extent by a transitional *mesocortex* [28, 128, 214]. This mesocortex can be divided into an internal *periarchicortical* or *limbic zone* and an external *proneocortical* or *paralimbic zone* [128, 266]. In general, the proneocortex closely resembles the neocortex. The periarchicortex has five cell layers that are organized into an external principal lamina (layers II and III) and an internal principal lamina (layers IV–VI). The internal and external laminae are separated by a cell-poor fibre layer, the lamina dissecans [128, 266]. The periarchicortical and proneocortical zones extend over the limbic lobe. Taking Brodmann's [31, 32] subdivision of the cortex as a basis, it can be said that the *periarchicortex* comprises the infralimbic cortex (area 25), the anterior cingulate cortex (area 24), the retrosplenial areas 29 and 30 and the entorhinal cortex (area 28; Fig. 12.3). The presubiculum and the parasubiculum, two strip-like areas situated between the subiculum and the entorhinal cortex, also form part of the periarchicortical zone. In the present account, these two areas and the subiculum will be designated together as the *subicular complex*. From a comparative point of view, it is highly interesting that in the acallosal monotremes and marsupials, the periarchicortex is directly continuous with a well-differentiated hippocampal formation throughout its extent (Fig. 12.2), whereas in eutherians the cingulate and retrosplenial parts of this cortex are flanked by a poorly developed hippocampal rudiment.

The *proneocortical zone* includes the prelimbic cortex (area 32), the posterior cingulate area 23 and the perirhinal areas 35 and 36 (Fig. 12.3) [159, 160]. It is noteworthy that von Economo and Koskinas [258], in a region of the human medial temporal lobe roughly corresponding to the caudal parts of Brodmann's areas 35 and 36, delineated two separate areas, a medial area TH and a more laterally situated area TF (Fig. 12.3). These two areas, whose existence and distinctness have been confirmed in both the human [16] and the rhesus monkey [219, 220, 257], can be classified as paralimbic or proneocortical [128, 160].

The various areas forming the limbic and paralimbic zones are not only structurally related, but also share some salient connectional features. Thus, the various periarchicortical areas are strongly connected with each other and with the hippocampal formation. A further characteristic of these areas is their connectivity with the anterior thalamic nuclei and/or the thalamic midline nuclei, in particular the reuniens nucleus. The proneocortical or paralimbic areas serve as a connectional interface between widespread neocortical areas and the archicortical and periarchicortical zones.

The *septum* forms part of the medial wall of the cerebral hemispheres. Its ontogenetic development is complex because no less than four of the seven fundamental telencephalic zones participate in its formation (Fig. 2.24B, Table 11.1). The septum can be divided into a dorsal and a ventral part. Due to the expansion of the corpus callosum, the dorsal part is drawn out into a thin plate, composed largely of glial cells and nerve fibres. This membranous structure, which is known as the *septum pellucidum*, is bounded by the basal surface of the corpus callosum and by the rostral convexity of the fornix (Fig. 3.7). The ventral part, or *septum verum*, is situated directly rostral to the lamina terminalis within the paraterminal gyrus (Fig. 3.6). It is bordered dorsally by the rostrum of the corpus callosum, rostrally by the precommissural portion of the hippocampus and caudally by the anterior commissure and the preoptic region. Ventrolaterally it borders on the nucleus accumbens, a large cell mass that occupies a position intermediate between the limbic and striatal or "extra-pyramidal" systems in both location and function. Contrary to prevailing opinion, the septal region is well developed in the human brain [14].

The *septum verum* contains two cell masses, the *lateral septal nucleus* and the *medial septal complex*. The latter comprises the dorsally situated medial septal nucleus and the dorsal or septal limb of the nucleus of the diagonal band

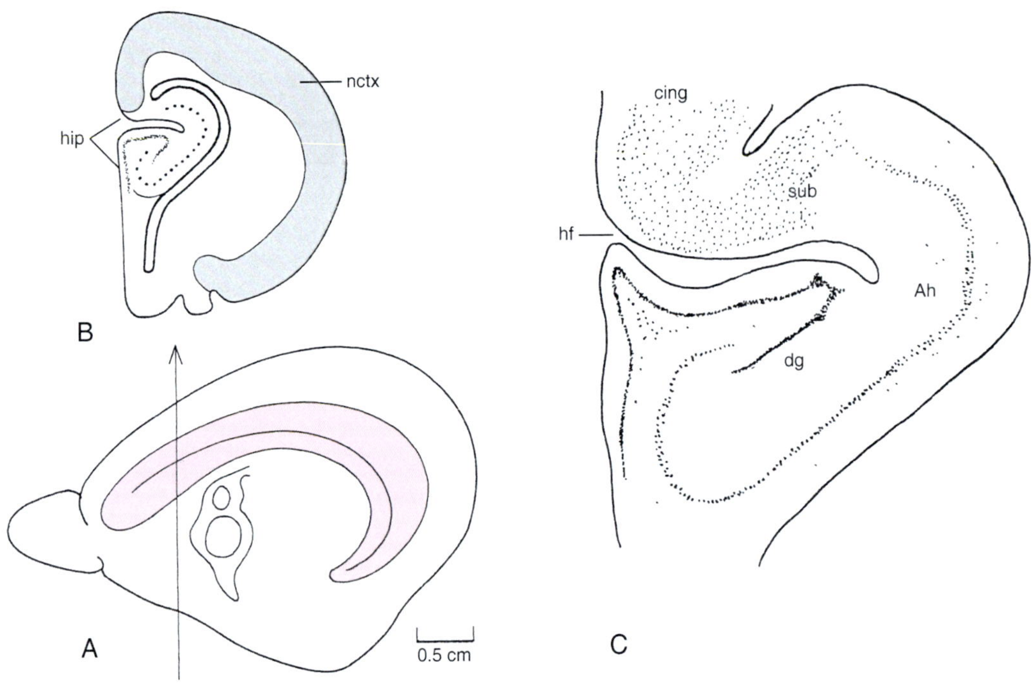

Fig. 12.2 A–C. The hippocampus in the duck-billed platypus, *Ornitorhynchus anatinus*. **A** The hippocampus projected upon the medial surface of the cerebral hemisphere; **B** diagrammatic transverse section at the level indicated; **C** the hippocampal region of the section shown in **B**, at a higher magnification. Based on illustrations in Hines [81]. *Ah*, Ammon's horn; *cing*, cingulate cortex; *dg*, dentate gyrus; *hf*, hippocampal fissure; *hip*, hippocampus; *nctx*, neocortex; *sub*, subiculum

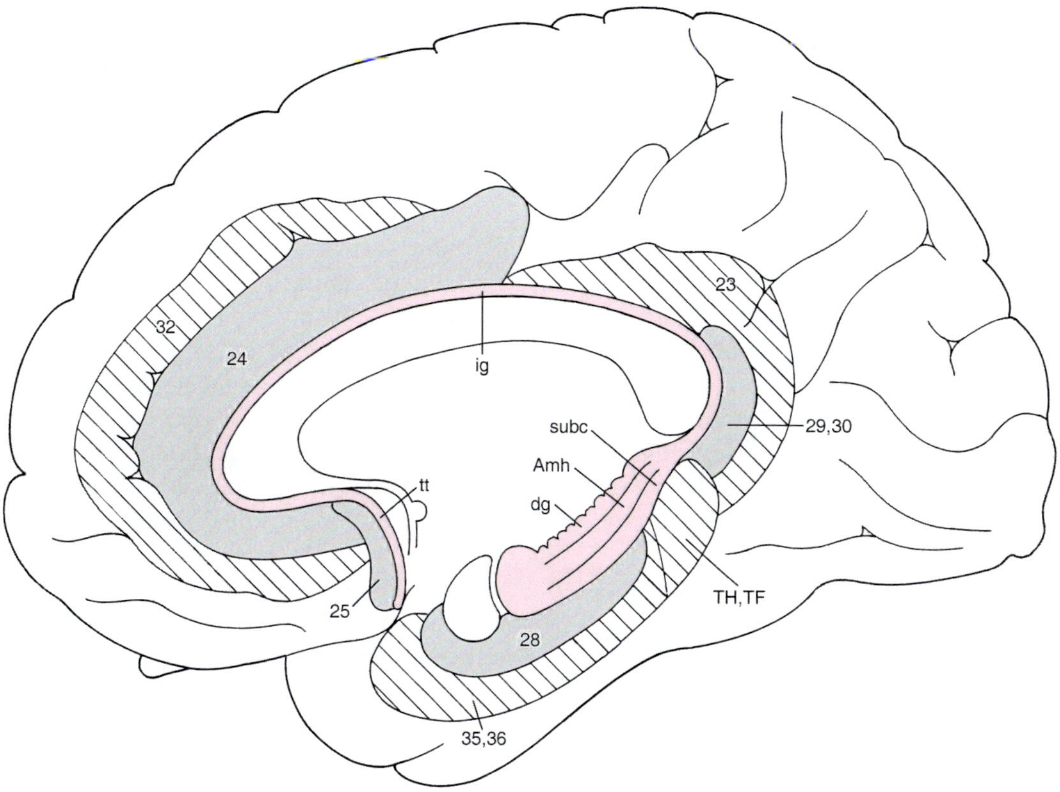

1 Medial longitudinal stria
2 Lateral longitudinal stria
3 Indusium griseum
4 Precommissural fornix
5 Anterior commissure
6 Column of fornix
7 Amygdaloid body
8 Anterior thalamic nucleus
9 Mamillothalamic tract
10 Subiculum
11 Lateral ventricle, inferior horn
12 Ammon's horn (uncinate gyrus)
13 Limbus Giacomini
14 Ammon's horn (intralimbic gyrus)
15 Ammon's horn (hippocampal digitations)
16 Body of fornix
17 Fimbria of hippocampus
18 Dentate gyrus
19 Crus of fornix
20 Commissure of fornix
21 Site of corpus callosum
22 Fasciolar gyrus

Fig. 12.3. Medial view of the human cerebral hemisphere, showing the position of the hippocampal formation (*red*), the limbic or parahippocampal areas (*grey*) and the paralimbic or proneocortical areas (*hatched*). The postcommissural hippocampus is unfolded. *Amh*, Ammon's horn; *dg*, dentate gyrus; *ig*, indusium griseum; *subc*, subicular complex; *TH, TF*, cortical areas according to Von Economo and Koskinas [258]; *tt*, taenia tecta; *numbers*, cortical areas according to Brodmann [31]

of Broca. The ventral or tubercular limb of this nucleus extends caudolaterally along the ventral surface of the hemisphere and marks the caudal boundary of the olfactory tubercle. As its name implies, this nucleus is embedded in a fibre bundle, the diagonal band of Broca (Figs. 6.40, 6.41).

The septal region contains a variety of neurotransmitters and other neuroactive substances. Jakab and Leranth [94] listed 35 different neuropeptides occurring in septal neurons and/or fibres.

The fibre connections of the septum will be discussed on p. 387. In general, these connections show clearly that the septal nuclei serve primarily as way stations between the hippocampal formation and the hypothalamus.

The hippocampus, the perihippocampal mesocortical areas and the septum form, together with the amygdaloid complex, the rostral portion of a large functional complex that extends throughout the brain. This complex, which can be designated as the *greater limbic system*, will be discussed in Chap. 23 of the present work.

Topography and Structure of Hippocampus

The *hippocampus* or *hippocampal formation* is a large C-shaped structure that forms part of the medial wall of the cerebral hemisphere (Figs. 3.18, 12.4). This structure can be subdivided morphologically into three parts, precommissural, supracommissural and retrocommissural. The first two parts are relatively small, vestigial structures; the retrocommissural hippocampus, in contrast, is well developed and represents the main portion of the hippocampal formation. The names of the three parts, it should be noted, refer to their position with respect to the corpus callosum. The *precommissural hippocampus* is a narrow vertical structure situated in the caudal part of the area subcallosa, just rostral to the septum verum. It continues dorsorostrally into the indusium griseum, a strand of hippocampal tissue that extends throughout the length of the corpus callosum. The *indusium griseum* represents the *supracommissural hippocampus*. Two small fibre bundles, the medial and lateral longitudinal striae, are embedded in it. These bundles represent a small supracallosal component of the fornix. The principal, subcallosal fornix will be discussed below. Near the splenium of the corpus callosum, the supracommissural hippocampus is continuous with the *retrocommissural hippocampus*, the expanded, morphologically most caudal end of the hippocampal formation, which is incorporated in the medial temporal lobe. It has already been discussed that during ontogeny this part of the hippocampus rolls in on itself along a longitudinal groove, the hippocampal fissure. This infolding causes the retrocommissural hippocampus to protrude into the inferior horn of the lateral ventricle. The most rostral portion of the retrocommissural hippocampus is recurved dorsally and constitutes a rounded swelling on the medial surface of the hemisphere, known as the uncus. On the ventricular side, this part of the hippocampus is greatly enlarged and differentiated into a series of separate lobules, the hippocampal digitations. As shown by Figs. 5.23, 5.24 and 12.4, the most rostral portion of the hippocampal formation extends for some distance ventral to the amygdaloid complex.

The hippocampal formation constitutes the archipallial part of the cerebral hemisphere; it contains a relatively simple three-layered cortex throughout its extent. The retrocommissural hippocampus is clearly differentiated into three longitudinally arranged structures: the dentate gyrus, Ammon's horn and the subiculum. The dentate gyrus – whose name refers to the toothed or beaded appearance of its surface – is the morphologically most medial strip of the pallium. During early development, the cell layer of the dentate gyrus is directly continuous with that of Ammon's horn, but later this continuity is lost and the dentate layer overlaps and surrounds the free medial edge of the cell layer of Ammon's horn [85]. In this way the characteristic interlocking of these two cell layers, as observed in the adult hippocampus, comes about (Figs. 12.2C, 12.5A). Due to the infolding of the hippocampus, the external

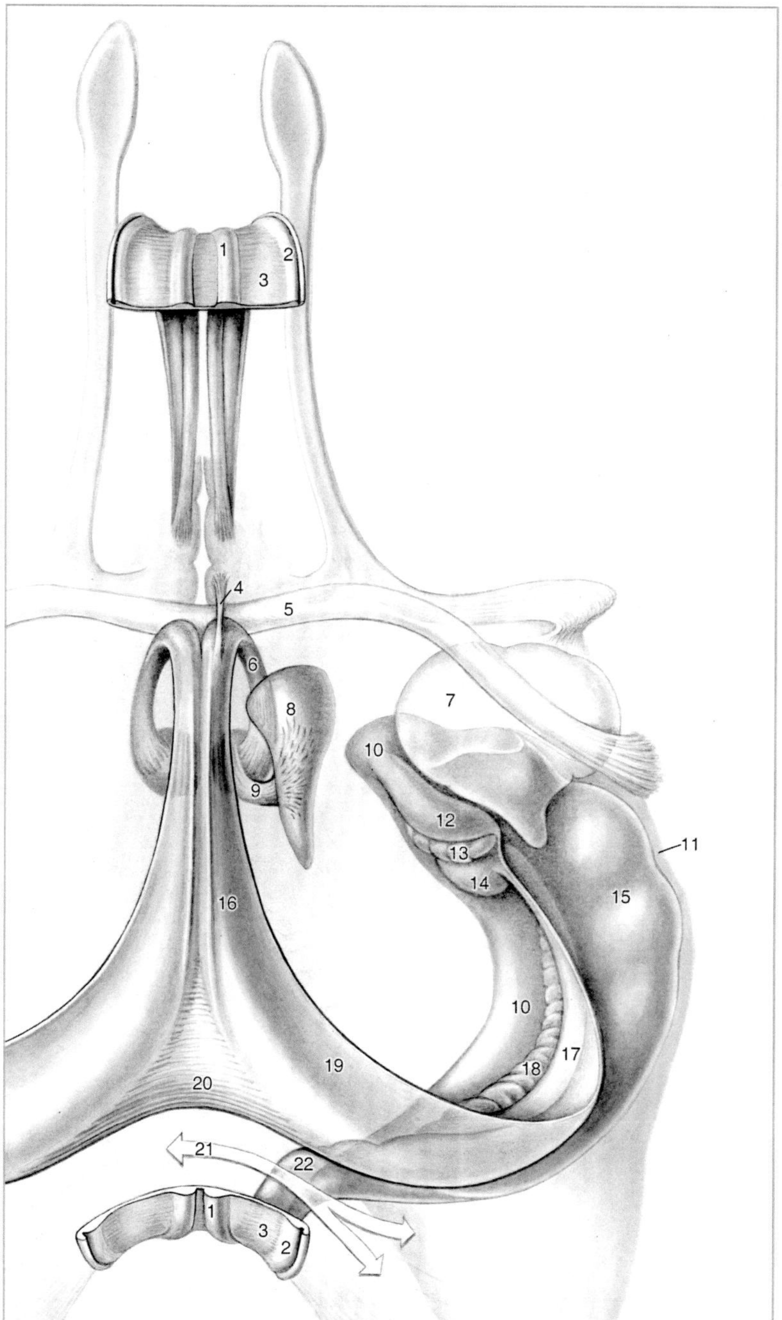

1 Medial longitudinal stria
2 Lateral longitudinal stria
3 Indusium griseum
4 Precommissural fornix
5 Anterior commissure
6 Column of fornix
7 Amygdaloid body
8 Anterior thalamic nucleus
9 Mamillothalamic tract
10 Subiculum
11 Lateral ventricle, inferior horn
12 Ammon's horn (uncinate gyrus)
13 Limbus Giacomini
14 Ammon's horn (intralimbic gyrus)
15 Ammon's horn (hippocampal digitations)
16 Body of fornix
17 Fimbria of hippocampus
18 Dentate gyrus
19 Crus of fornix
20 Commissure of fornix
21 Site of corpus callosum
22 Fasciolar gyrus

Fig. 12.4. The hippocampal formation and other limbic structures isolated from most of their surroundings, seen from above (2/1×)

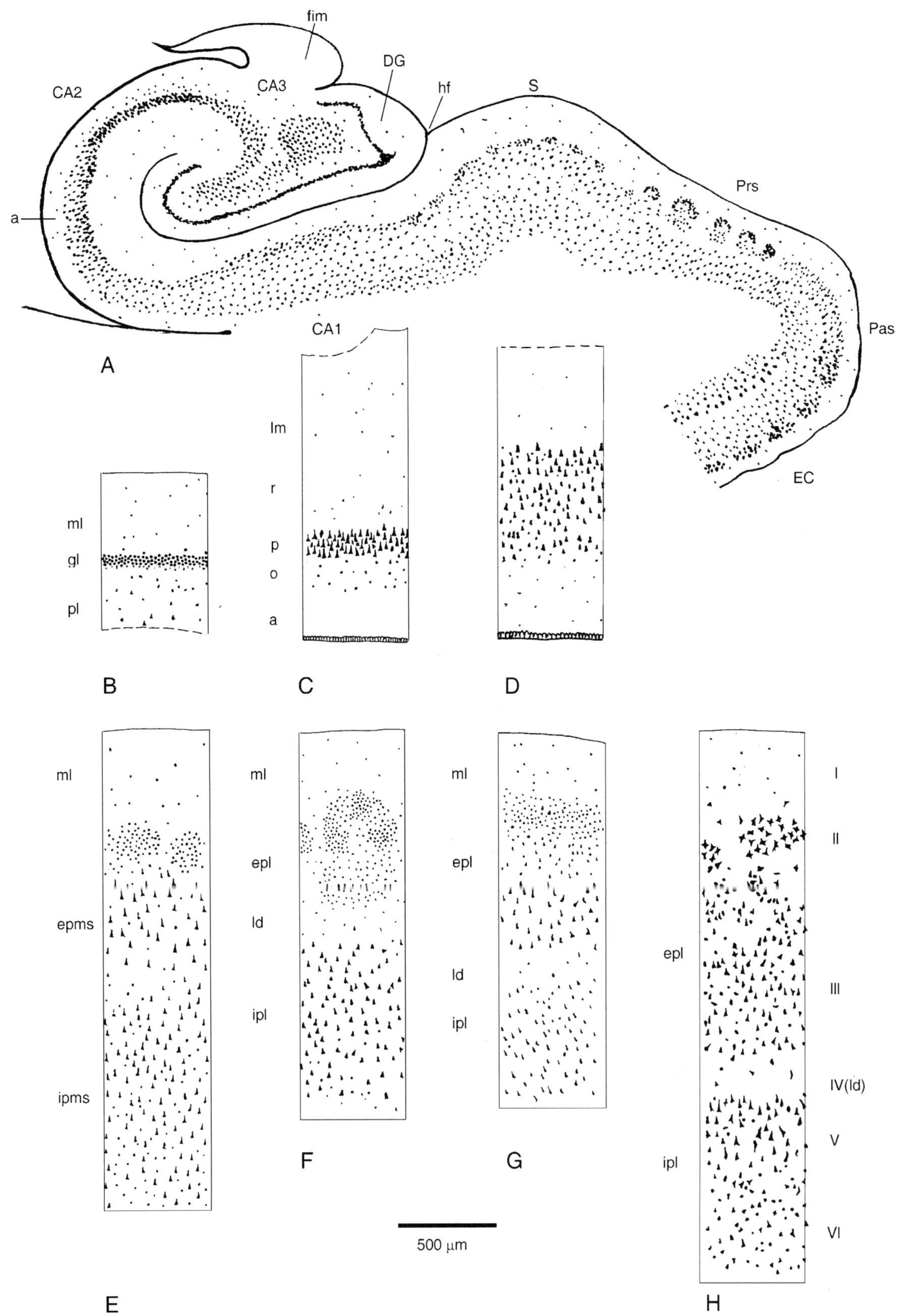
fim
DG
CA2
CA3
hf
S
Prs
a
CA1
Pas
EC
A
lm
r
p
o
a
ml
gl
pl
B
C
D
ml
epms
ipms
E
ml
epl
ld
ipl
F
ml
epl
ld
ipl
G
I
II
epl
III
IV(ld)
V
ipl
VI
H
500 μm

(pial) surfaces of the dentate gyrus and Ammon's horn become closely apposed and finally fuse over a considerable distance. By this fusion, the depth of the hippocampal fissure is greatly reduced. In the adult, the dentate gyrus is situated above the hippocampal fissure, whereas the subiculum is situated below that fissure (Fig. 12.5 A).

The *dentate gyrus* forms the morphologically medial-most part of the cerebral cortex. From external to internal its three layers are: the *molecular layer*, the *granular layer* and the *polymorphic layer* (Fig. 12.5 B). Due to its strong curvature, the dentate gyrus surrounds the, also curved, end of the pyramidal cell layer of Ammon's horn (Fig. 12.5 A).

The granular layer, as its name indicates, is made up of the densely packed somata of granule cells. The spiny dendrites of these elements enter the molecular layer, where they branch extensively (Fig. 12.9). In the molecular layer, the dendrites of the granule cells enter into synaptic contact with afferents from the entorhinal cortex, with commissural/associational afferents and with axons ascending from the polymorph layer. The afferents and terminals from the entorhinal cortex are concentrated in the outer two thirds of the molecular layer, whereas the commissural/associational fibres occupy the inner third of that layer. Human dentate granule cells often give rise to some basal dendrites that extend into the subjacent polymorphic layer [123, 197]. The axons of the granule cells pass through the polymorphic layer and enter Ammon's horn as mossy fibres. In Ammon's horn these mossy fibres make synaptic contact with thorny excrescences of the dendritic shafts of pyramidal cells (Fig. 12.9). The granule cells are excitatory and use glutamate as their primary neurotransmitter substance, but some also contain opioid peptides such as enkephalin and dynorphin [237].

The polymorphic and molecular layers of the dentate gyrus both contain numerous intrinsic neurons. In the polymorphic layer, at least five different types of basket cells have been found [172]. Axons from these neurons innervate the somata of granule cells; axons from other interneurons in the polymorphic layer ascend to the molecular layer, where they arborize in specific superficial, intermediate or deep zones [60]. Prominent among the interneurons in the molecular layer are axo-axonic cells, the axons of which terminate on the axon initial segments of granule cells. The intrinsic neurons in the dentate gyrus are aspiny and GABAergic. However, the superficial zone of the polymorphic layer contains a class of spiny neurons that are excitatory and use glutamate as their neurotransmitter [208]. Amaral [6] called these neurons mossy cells because the proximal parts of their long, sparsely branching dendrites are densely studded with thorny excrescences. The dendrites of the mossy cells are generally confined to the polymorphic layer. Their axons penetrate the granular layer and project for relatively long distances on both sides of the hippocampus. Collectively, they form the commissural and ipsilateral associational projection, which, as already mentioned, occupies the inner third of the molecular layer in the dentate gyrus.

The axons of the dentate granule cells issue several collateral branches within the polymorphic layer before entering the CA3 field of Ammon's horn (Fig. 12.9). These collaterals establish synaptic contacts with local interneurons and also with mossy cells [41].

◀ **Fig. 12.5 A–H.** Cytoarchitecture of the human hippocampal formation. **A** Transverse section at the level of the lateral geniculate nucleus; **B–H** drawings at higher magnification of the various hippocampal fields; **B** dentate gyrus; **C** field CA2; **D** field CA1; **E** subiculum; **F** presubiculum; **G** parasubiculum; **H** entorhinal cortex; *a*, alveus; *DG*, dentate gyrus; *CA1, CA2, CA3*, fields in Ammon's horn; *EC*, entorhinal cortex; *epl*, external principal layer; *epms*, external pyramidal sublayer; *fim*, fimbria of fornix; *gl*, granular layer; *hf*, hippocampal fissure; *ipl*, internal principal layer; *ipms*, internal pyramidal sublayer; *ld*, lamina dissecans; *lm*, stratum lacunosum-moleculare; *ml*, molecular layer; *o*, stratum oriens; *p*, pyramidal layer; *pl*, polymorphic layer; *Pas*, parasubiculum; *Prs*, presubiculum; *r*, stratum radiatum; *S*, subiculum; *I–VI*, cortical layers

Although the dentate granule cells differ considerably from the typical hippocampal pyramids at first sight (Fig. 12.9), it should be emphasized that both evidently belong to the category of pyramid-like cells [156]. It is also worth noting that the local circuitries of the dentate gyrus and Ammon's horn closely resemble each other [194].

Ammon's horn forms the largest part of the hippocampal formation. In addition to a limited number of small elements, its intermediate cellular layer contains the somata of pyramidal cells, which form by far the most numerous class of neurons in Ammon's horn. These elements closely resemble their counterparts in other cortices. Each hippocampal pyramidal cell has a stout apical dendrite, which splits up at a shorter or longer distance from the soma into a number of terminal branches. The apical stem dendrites make up the stratum radiatum, whereas the layer in which the terminal branches of the apical dendrites are situated is known as the stratum lacunosum-moleculare. The basal dendrites of the pyramidal cells ramify in the stratum oriens (Fig. 12.9). The axons of the hippocampal pyramidal cells enter a compact subependymal fibre layer known as the alveus. A certain proportion of these axons descend via the fornix to the lateral septal nucleus; others pass to the entorhinal cortex, but most remain within the hippocampal formation.

Differences in size and packing density of the pyramidal cells allow a subdivision of Ammon's horn into separate cytoarchitectonic fields (Fig. 12.5 A, C, D). Lorente de Nó [130] distinguished in his classic study on the hippocampus four fields, which he designated as CA1–CA4. CA1 is adjacent to the subiculum, whereas CA4 is situated in the so-called hilar region of the dentate gyrus. However, Insausti and Amaral [87] recently proposed to drop the term CA4 and to include the pyramidal cells in the hilar region in CA3.

The afferents to the pyramidal neurons in Ammon's horn show a strict laminar organization. Without going into detail, it may be mentioned that septal afferents terminate on the basal dendrites and on the proximal parts of the apical dendrites, that commissural fibres end on basal dendrites and on intermediate parts of the shafts of the apical dendrites and that afferents from the entorhinal cortex contact the terminal branches of the apical dendrites [214, 222]. Moreover, two sets of intrahippocampal projection fibres, i.e. the mossy fibres already mentioned and the Schaffer collaterals to be discussed below, are arranged in distinct sublayers and impinge upon particular segments of the apical dendrites of the pyramidal neurons.

Local circuit neurons occur in all layers of Ammon's horn. Most of these neurons have smooth dendrites and release GABA as the primary neurotransmitter. Subpopulations of Ammonic interneurons have been shown to contain neuropeptides such as somatostatin, neuropeptide Y and vasoactive intestinal polypeptide or calcium-binding proteins such as calbindin, parvalbumin and calretinin. The classical Golgi studies of Cajal [35] and Lorente de Nó [130] have shown the presence of some 20 different types of interneurons in Ammon's horn. Typical basket cells, synapsing onto the somata of pyramidal neurons, appeared to be present in all Ammonic layers, except for the molecular layer (Fig. 12.6 A, B). Later Golgi studies [203, 204] revealed the presence of axo-axonic cells, making multiple synaptic contacts with the axonal initial segments of pyramidal neurons (Fig. 12.6 C).

Labelling of individual cells with intracellular or neurochemical markers has further increased our insight into the astonishing heterogeneity of the Ammonic interneurons. An important general result of these labelling studies is that the postsynaptic target domains of many interneurons consist of specific segments of the apical and/or basal dendritic arbors of pyramidal cells. The interneuron subtype, designated as oriens-lacumosum moleculare cell, serves as an example. This neuron, which expresses the calcium-binding protein calbindin and the peptide somatostatin, has horizontally oriented dendrites in the stratum oriens. Its axon ascends and ramifies profusely in the stratum lacunosum-moleculare (Fig. 12.6 D). It has been established that these ramifications

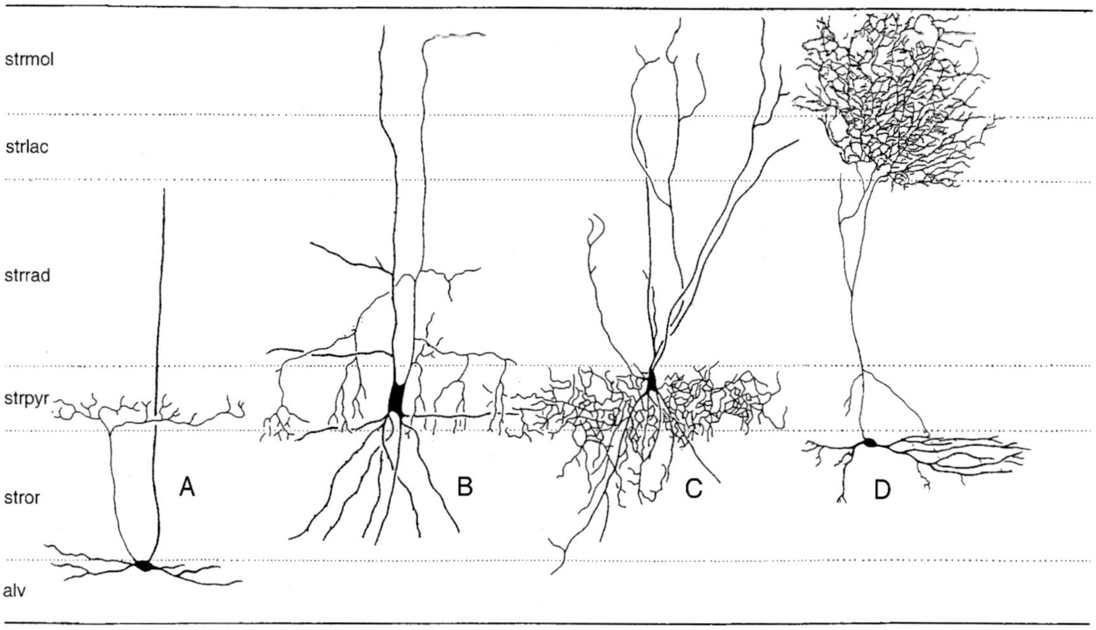

Fig. 12.6 A–D. Some local circuit neurons observed in Ammon's horn, redrawn from the sources indicated. **A** O/A cell [130]; **B** pyramidal basket cell [130]; **C** axo-axonic cell [204]; **D** O/LM cell [60]. *strmol*, stratum moleculare; *strlac*, stratum lacunosum; *strrad*, stratum radiatum; *strpyr*, stratum pyramidale; *stror*, stratum oriens; *alv*, alveus

synapse on the distal dendrites of pyramidal cells, where they exert a GABA-mediated, inhibitory influence. The wealth of literature on the structure and function of hippocampal interneurons has been summarized in some recent reviews [60, 132, 138], to which the reader is referred for further information. However, the synaptic relations of some of these elements are briefly treated in a later section of this chapter, devoted to the microcircuitry of Ammon's horn.

The *subiculum* forms the third part of the hippocampal formation. It is located between Ammon's horn and the presubiculum (Fig. 12.5 A). The subiculum can be divided into a superficial molecular layer and a deep pyramidal cell layer (Fig. 12.5 E). The molecular layer is remarkably wide and harbours the apical tufts of the pyramidal cells. Within the pyramidal layer, separate external and internal sublayers can be distinguished. The pyramidal neurons in the external sublayer are somewhat larger and more widely spaced than those in the internal sublayer. The most superficial zone of the external sublayer contains islands of small, darkly stained cells. Large stellate cells are disseminated throughout the pyramidal layer [28]. The molecular and pyramidal layers both contain numerous GABA-immunoreactive neurons. The density of these elements is considerably higher in the subiculum than in the other parts of the hippocampal formation [98].

The subiculum receives a strong, topographically organized projection from the adjacent CA1 field. The fibres forming this projection terminate in a columnar fashion in the subicular pyramidal layer [12, 230]. Ascending axon collaterals from deep subicular pyramidal cells give rise to a similar columnar pattern [73].

The subiculum constitutes the main output structure of the hippocampus. It projects to the entorhinal cortex and to several subcortical structures, including the septum, the accumbens nucleus, the anterior thalamus, the hypothalamus and the mamillary nuclei [87].

The three periarchicortical areas situated medially to the subiculum, i.e. the presubiculum, the parasubiculum and the entorhinal cortex (Figs. 12.3, 12.5 A), are so closely and intensively connected with the archicortical structures discussed above that they are commonly considered as integral components of the hippocampal formation. All of these structures form part of a single functional complex, designated by Squire and Zola-Morgan [210] as the *medial temporal lobe memory system.*

In the presubiculum, parasubiculum and entorhinal cortex, two principal cell layers, known as the internal and external principal layers, can be distinguished. These two layers are separated by a cell-free lamina dissecans [182]. In the *presubiculum* (Fig. 12.5 F) and *parasubiculum* (Fig. 12.5 G), the superficial zone of the external principal layer is occupied by densely packed small neurons. Throughout most of the presubiculum, this zone is broken up into small islands (Fig. 12.5 A). In the *entorhinal cortex,* the external and internal principal layers are both differentiated into two sublayers, II and III, and V and VI, respectively (Fig. 12.5 H). A distinguishing feature of layer II of the entorhinal cortex is that it is made up of clusters of darkly stained, modified pyramidal cells (Fig. 12.5 A, H).

The human entorhinal cortex has been subdivided into eight [91], nine [112], or even 23 [181] different subareas.

Microcircuitry of Ammon's Horn

The following synopsis of known and probable interneuronal relationships in Ammon's horn (particularly field CA1) of the mammalian hippocampus is based principally on the studies of Schwartzkroin and Lacaille et al. [101, 113, 116–119, 132, 134, 192, 193]. The numbers and other symbols used in the text correspond to those in the accompanying Fig. 12.7.

The highly characteristic *pyramidal cells* (P) with their large somatodendritic surface represent the principal neurons in the circuitry of Ammon's horn (Fig. 12.9). Extrinsic afferents to the pyramidal neurons include fibres from the entorhinal cortex (1), the medial septal nucleus (2) and the contralateral hippocampus (3). These afferent fibres make excitatory synaptic contacts with particular sectors of the dendritic trees of the pyramidal neurons (1a, 2a, 3a). The axons of the CA1 pyramidal cells (4) descend to the alveus, where they bifurcate into a coarse branch, passing to the subiculum, and a smaller branch projecting toward the fimbria. The initial part of these axons issues local collaterals (5), which make excitatory synapses with other pyramidal cells [49, 130, 135] (5a) and with various interneurons (5b, c, d). Ammon's horn contains many different types of inhibitory interneurons, including pyramidal basket cells (B), oriens/alveus neurons (O/A), axo-axonic cells (AA), lacunosum-moleculare cells (L/M) and oriens-lacunosum-moleculare cells (O/LM).

The large *pyramidal basket cells* (Fig. 12.6B) are named as such because they resemble, with regard to the position and size of their somata and the shape of their dendritic trees, the hippocampal pyramidal neurons [35, 130]. Their axonal branches participate in the formation of dense, basket-like plexuses around the somata and proximal dendrites of pyramidal cells (6a). Contacts onto other interneurons were also found (6b, c). The pyramidal basket cells receive excitatory inputs from pyramidal cells (5b) and from afferents in the alveus (2b) and in the stratum radiatum (3b). Other interneurons, among them O/A cells and L/M cells, make inhibitory synapses with the pyramidal basket cells (7b, 9c). The axon collaterals of pyramidal cells terminating on the pyramidal basket cells (5b) and the latter elements with their axonal ramifications impinging upon the pyramidal cells (6a) together form a pathway for recurrent or feedback inhibition, whereas the direct inputs from hippocampal afferents to the pyramidal basket cells (2b, 3b) have been shown to form part of feed-forward inhibitory circuits.

The *O/A cells* (Fig. 12.6A) represent a particular type of basket cell. The axons of these neurons make inhibitory synapses with pyramidal cells (7a) and presumably also with pyramidal basket cells (7b) and L/M interneurons (7c). The O/A cells receive excitatory synapses from pyramidal cell collaterals (5c) and from afferent fibres in the stratum radiatum (3c) and in the alveus (2c). There is ultrastructural evidence suggesting that pyramidal basket cells and L/M interneurons make synaptic contacts with O/A interneurons (6b, 9d). The effect produced by these synapses is unknown. The O/A interneurons mediate, like the pyramidal basket cells, feed-forward and feedback inhibition of CA1 pyramidal cells.

The *axo-axonic cells* (Fig. 12.6C) are specialized interneurons that make multiple inhibitory synaptic contacts, exclusively with the axon initial segments of pyramidal neurons (8). Their axon ramifies in the pyramidal layer and emits numerous terminal segments that "climb" along the axon initial segments of pyramidal neurons. Electron microscopy examination revealed that a single terminal segment provides a series of about eight to 30 symmetrical synapses to a single axonal initial segment of a pyramidal neuron and that terminal segments of several axo-axonic cells converge upon the initial segment of an individual pyramidal cell [203, 204]. Data concerning the input to the axo-axonic cells are lacking; however, the fact that hippocampal axo-axonic cells have roughly the same dendritic distribution as the pyramidal cell may well indicate that they have access to all input available to the latter [204].

The *L/M interneurons,* which occur in the CA1 field, have fusiform or multipolar somata and aspinous, beaded, dendritic processes ramifying in the strata lacunosum, moleculare and radiatum. The varicose axon originates from a primary dendrite and often projects for several millimetres along the stratum lacunosum. It ramifies profusely in the stratum radiatum, some branches reaching as far as the stratum pyramidale and the stratum oriens. Remarkably, the processes of the L/M interneurons are not restricted to the CA1 field, since dendritic and axonal processes of some of these elements were seen ascending in the stratum moleculare, crossing the hippocampal fissure, and entering the adjacent stratum moleculare of the dentate gyrus [113, 116, 117, 119]. The synaptic relationships of the L/M

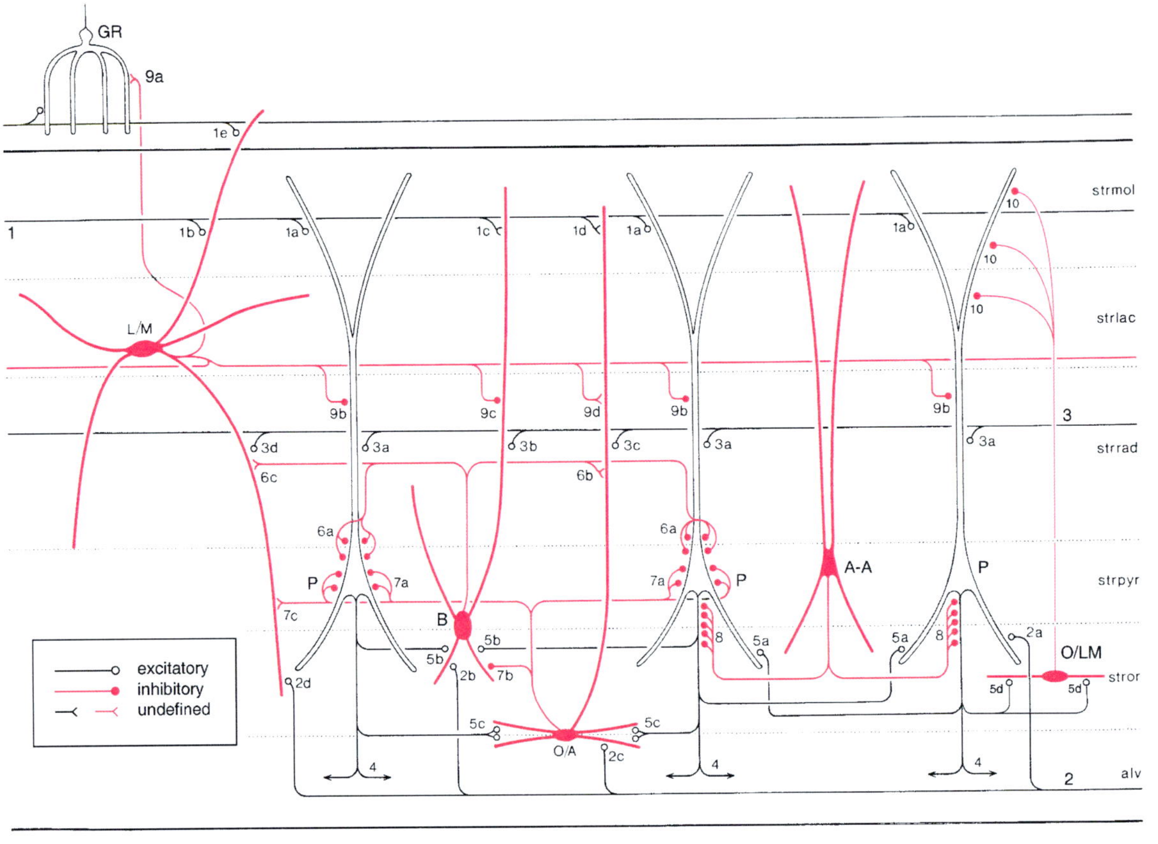

Fig. 12.7. Known and probable interneuronal relationships in Ammon's horn (particularly field CA1) of the mammalian hippocampal formation. For further explanation, see text. *strmol*, stratum moleculare; *strlac*, stratum lacunosum; *strrad*, stratum radiatum; *strpyr*, stratum pyramidale; *stror*, stratum oriens; *alv*, alveus (based principally on studies of Schwartzkroin and Lacaille et al. [101, 113, 116–119, 132, 134, 192, 193])

neurons may be summarized as follows [116, 117, 119]: Afferent fibres in the alveus (2d), the stratum radiatum (3d) and the stratum moleculare of the CA1 region (1b) and the fascia dentata (1e) make excitatory synapses with them. Other interneurons, probably including the pyramidal basket cells and the O/A interneurons, make functionally as yet undetermined synapses with L/M cells (6c, 7c). The axonal ramifications of the L/M interneurons make inhibitory contacts with pyramidal cells (9b), pyramidal basket cells (9c) and probably also with O/A interneurons (9d) in the CA1 region. Some of the axonal branches of the L/M interneurons cross the hippocampal fissure and synapse with granule cells (GR) in the fascia dentata (9a). From these data, it appears that the O/A elements, like the pyramidal basket cells and L/M cells, are inhibitory interneurons. However, whereas the pyramidal basket cells and the O/A cells mediate both feed-forward and feedback inhibition onto CA1 pyramidal cells, L/M interneurons appear to mediate only feed-forward inhibition. Evidence for the presence of excitatory synaptic contacts from the CA1 pyramidal neurons onto L/M interneurons is lacking [117].

The somata of the O/LM interneurons are situated in the stratum oriens and their dendrites spread horizontally in the same layer. Their axons ascend through the stratum pyramidale-radiatum and branch profusely in the stratum lacunosum-moleculare (Fig. 12.6D), where they make GABAergic, inhibitory contacts (10) with the distal dendrites of pyramidal cells [134]. The O/LM cells receive a strong excitatory input via local recurrent collaterals of pyramidal cells (5d) and are, hence, an essential component of feedback circuits [26, 133, 139].

Although the data discussed above and Fig. 12.7 reveal that the local circuitry of the CA1 region is quite complex, it should be emphasized that in reality this circuitry is doubtless far more complex than that. This is immediately apparent from the fact that only five of the 18 or more types of interneurons present in this hippocampal region have been included [35, 130, 166]. Another limitation is that the diagram is only based on qualitative data and that quantitative data, which are available to some extent [11], have been omitted. Finally, it is noteworthy that the hippocampal neurons do not communicate exclusively by means of chemical synapses. There is evidence that many inhibitory interneurons are electrotonically coupled [62] and that the ubiquitous cholinergic, noradrenergic and serotoninergic fibres communicate mainly through non-synaptic or volume transmission with other hippocampal elements [236].

Intrinsic Connections of the Hippocampus

The circuit considered to be the morphological substrate of the main flow of information through the hippocampus begins with a projection from the entorhinal cortex to the dentate gyrus, is relayed to field CA3 of Ammon's horn and then proceeds to field CA1. Projections from field CA1 to the subiculum and from the subiculum to the entorhinal cortex close the circuit [13, 221, 222]. The components of this circuit will be discussed in the following. The numbers used in the text correspond to those in the accompanying Figs. 12.8 and 12.9.

The fibres passing from the entorhinal cortex to the dentate gyrus form a large pathway known as the *temporo-ammonic* or *perforant path* (1). They originate principally from large stellate cells situated in layer II of the entorhinal cortex and perforate through the subiculum and the zone of the obliterated hippocampal fissure. After entering the dentate gyrus, these fibres ramify in the outer two thirds of the molecular layer of that structure, where they form excitatory synapses onto the dendritic spines of granule cells (1a) [33, 84, 225, 263]. It is important to note that the efferents from the entorhinal cortex, which enter the perforant path, do not terminate exclusively in the dentate gyrus, but also supply the fields CA3 and CA1 and the subiculum (1b). The fibres passing to CA3 originate from layer II, whereas those destined for CA1 and the subiculum arise mainly from layer III [263].

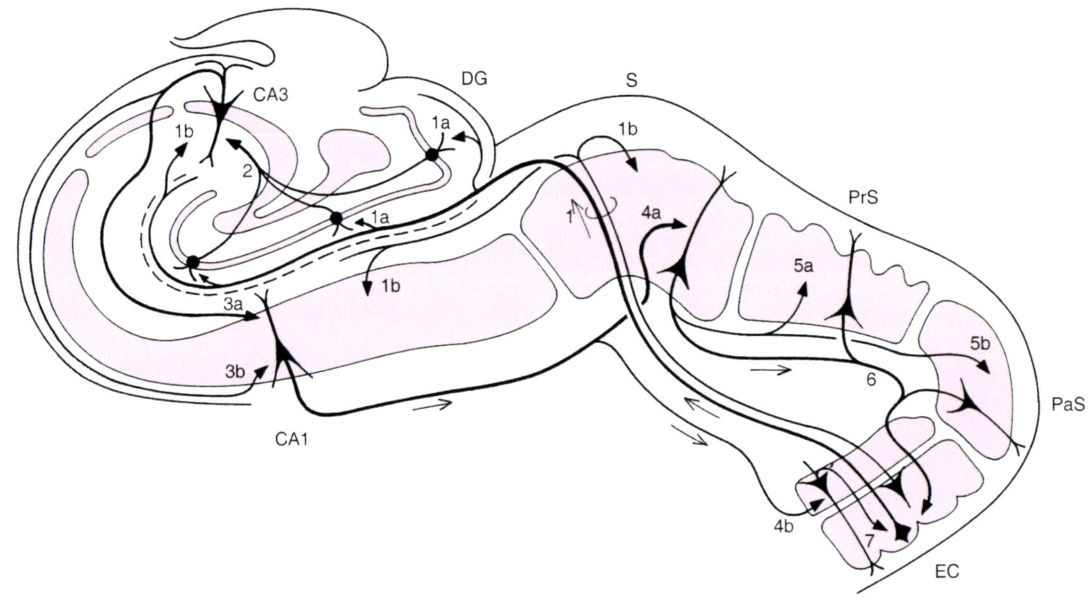

Fig. 12.8. Intrinsic connections of the hippocampal formation. For further explanation, see text. *CA1*, *CA3*, fields in Ammon's horn; *DG*, dentate gyrus; *EC*, entorhinal cortex; *PaS*, parasubiculum; *PrS*, presubiculum; *S*, subiculum

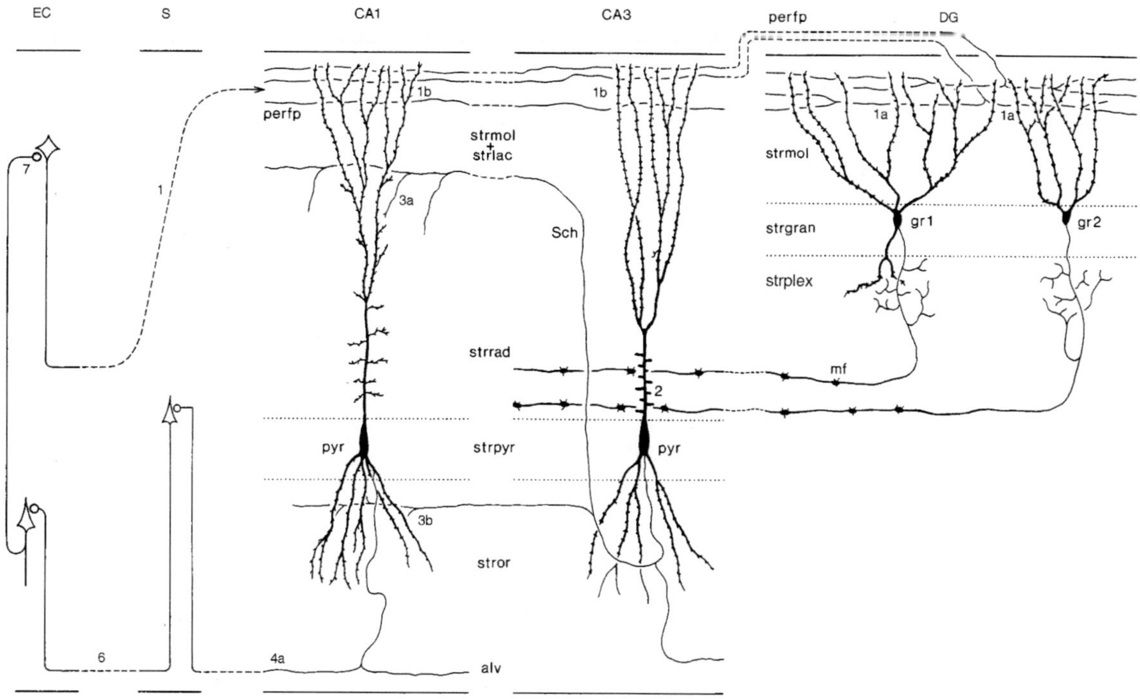

Fig. 12.9. Elements participating in the principal hippocampal circuit. *alv*, alveus; *CA1, CA3*, fields of Ammon's horn; *DG*, dentate gyrus; *EC*, entorhinal cortex; *gr1*, *gr2*, granule cells; *mf*, mossy fibres; *perfp*, perforant path; *pyr*, pyramidal cell; *S*, subiculum; *Sch*, Schaffer collateral; *strgran*, stratum granulare; *strlac*, stratum lacunosum; *strmol*, stratum moleculare; *stror*, stratum oriens; *strplex*, stratum plexiforme; *strpy*, stratum pyramidale; *strrad*, stratum radiatum. The neuronal elements depicted are modified from Cajal [35]

The axons of the granule cells in the dentate gyrus pass as *mossy fibres* to the stratum radiatum of field CA3, in which they form a separate sublayer known as the *stratum lucidum*. The mossy fibres synapse with the large pyramidal cells in CA3 (2). The proximal part of the apical dendrites of these elements are provided with ramified excrescences and these are embraced by the large, irregular varicosities in the mossy fibres to which their name refers. Together, the axonal varicosities and the dendritic excrescences form huge synaptic complexes in which multiple active zones (up to 37!) are present [24, 40, 72]. The mossy fibre varicosities contain large amounts of Zn^{2+} and opioid peptides that are coreleased with the primary neurotransmitter glutamate [15, 213]. Apart from the large pyramidal cells, GABAergic interneurons are major postsynaptic targets of mossy fibres in CA3 [2].

The main axons of the large CA3 pyramidal cells pass to the alveus. However, during their descent through the stratum oriens these axons issue coarse collaterals, many of which penetrate the stratum pyramidale and pass to field CA1, where they form a compact sheet of fibres in the stratum lacunosum. Others, which also pass to CA1, remain within the stratum oriens. These collaterals are collectively known as *Schaffer collaterals*. They issue short branches, which enter into synaptic contact with the apical (3a) and basal dendrites (3b) of the CA1 pyramidal neurons [93, 95, 124]. Physiological experiments have shown that the Schaffer collaterals exert an excitatory action on CA1 pyramidal cells [191]; consistent therewith, the Schaffer collateral synapses, like those of the mossy fibres, are of the type I variety.

The CA1 pyramidal cells project massively to the subiculum (4a) and more lightly to the entorhinal cortex (4b) [12, 57, 58, 230]. The subiculum, in turn, projects to the presubiculum (5a) and parasubiculum (5b) and all three components of the subicular complex project to the entorhinal cortex (6) [83, 103, 104, 183, 188, 200, 201, 268]. All of these projections originate from pyramidal neurons and are considered to be excitatory in nature. The projections from the pre- and parasubiculum terminate in the superficial layers of the entorhinal cortex, where they may innervate the cells that project to the dentate gyrus. The fibres originating from field CA1 and the subiculum pass to the deep layers of the entorhinal cortex [188, 229, 230]. The pyramidal cells in these layers have ascending collaterals [33, 129], and these may form the final link (7) in a second route, along which the hippocampo-entorhinal circuit is closed.

If we survey these data, it appears that the entorhinal cortex, the dentate gyrus, the various fields of Ammon's horn and the subicular complex form part of a closed circuit consisting of a series of unidirectional, excitatory projections (Fig. 12.8).

Major Fibre Systems Connecting the Hippocampal Formation and Its Dependencies with Other Parts of the Brain

Some major fibre systems connect the hippocampal formation and its dependencies with other parts of the brain. These include the fornix, the fasciculus mamillaris princeps and the cingulum.

The *fornix* (Figs. 3.18, 12.4) is a compact fibre bundle connecting the hippocampus with the hypothalamus and various other structures. Its fibres first form the *alveus*, a thin white layer on the ventricular surface of Ammon's horn, and then converge as the *fimbria* along the medial aspect of the hippocampus. Running posterosuperiorly, the fibres of the fimbria enter the *crus of the fornix*, a flattened structure that arches upwards and medially under the splenium of the corpus callosum. In this region a number of fibres decussate to the opposite side, forming a thin sheet of crossing fibres, the *hippocampal commissure, fornical commissure* or *psalterium* (Figs. 3.18, 12.4) [65]. Experimental neuroanatomical studies [10, 48] have shown that in the rhesus monkey, the hippocampal commissure is mainly composed of true commissural fibres, which interconnect the most rostral (or uncal) part of

Ammon's horn and associated dentate gyrus and of decussating fibres, passing from the presubiculum to the contralateral entorhinal cortex.

Proceeding rostrally over the thalamus, the two crura converge and join to form the *corpus of the fornix,* which lies immediately beneath the corpus callosum. However, at the level of the anterior pole of the thalamus the fornical corpus separates again into two bundles, the *columns of the fornix,* which curve ventrally in front of the interventricular foramen and caudal to the anterior commissure to enter the hypothalamus. Immediately behind the interventricular foramen a considerable number of fibres leave the column and pass backwards to the anterior nucleus of the thalamus and to the bed nucleus of the stria terminalis. Other fibres split off from the fornix just above the anterior commissure and constitute a small *precommissural portion of the fornix*. The main bundle of the fornix or *postcommissural fornix* finally traverses the hypothalamus, where most of its fibres terminate in the mamillary body.

The efferent fibres of the mamillary body form a compact bundle, the *fasciculus mamillaris princeps*. This bundle passes dorsally for a short distance and then splits up into two components, the larger mamillothalamic and the smaller mamillotegmental tract. The *mamillothalamic tract,* which passes to the anterior thalamic nuclei, forms part of the so-called Papez circuit (see below). The *mamillotegmental tract* curves caudally into the tegmentum of the midbrain and terminates in the dorsal tegmental nucleus and in the nucleus reticularis tegmenti pontis of Bechterew [46, 173].

The *cingulum* consists of a bundle of short and long association fibres that surround the corpus callosum. Passing through the core of the cingulate and parahippocampal gyri, it extends from the septal area to the uncal region in the temporal lobe (Fig. 12.13).

Afferents to the Hippocampus

Extrinsic inputs, which are presumed to activate and modulate the intrinsic hippocampal circuitry [222], arise from a variety of sources, including (1) various cortical areas, (2) the amygdaloid complex, (3) the medial septal-diagonal band complex, (4) the thalamus, (5) the supramamillary region and (6) monoaminergic cell masses in the brain stem (Figs. 12.10, 12.11).

1. *Cortical afferents.* Pathways leading out of widespread neocortical areas converge upon the hippocampal formation. Most fibres of this large corticohippocampal projection terminate in the entorhinal cortex, i.e. the site of origin of the perforant pathway; others reach the entorhinal cortex after a synaptic interruption in the perirhinal areas 35 and 36 or the posterior parahippocampal areas TH and TF; and yet others bypass the entorhinal cortex to terminate directly in the subicular cortices (Fig. 12.10) [89, 217, 218].

The neocortical areas that project directly to the entorhinal cortex include the ventral or agranular insular cortex, the infralimbic cortex (area 25), the caudal orbitofrontal cortex (area 13), the temporopolar cortex (area 38), several fields in the superior temporal gyrus, the prelimbic cortex (area 32), the anterior and posterior cingulate cortex (areas 24 and 23), the dorsolateral prefrontal cortex (areas 9 and 46) and the retrosplenial cortex (areas 29 and 30) [66, 89, 97, 153, 163, 177, 180, 235, 241, 243, 244].

The projections that reach the entorhinal cortex via a perirhinal or posterior parahippocampal relay originate from the retrosplenial cortex (areas 29 and 30), which in turn receive projections from area 24, 32, 23 and 31 [102], the prefrontal cortex (areas 9 and 46), the superior temporal gyrus (area 22), the parietal cortex (area 7), the occipital cortex (area 19) and the inferior temporal gyrus (area 20) [177, 196, 218, 241, 243, 244].

The neocortical areas that project directly (i.e. bypassing the entorhinal cortex) to the subicular complex include the temporal polar

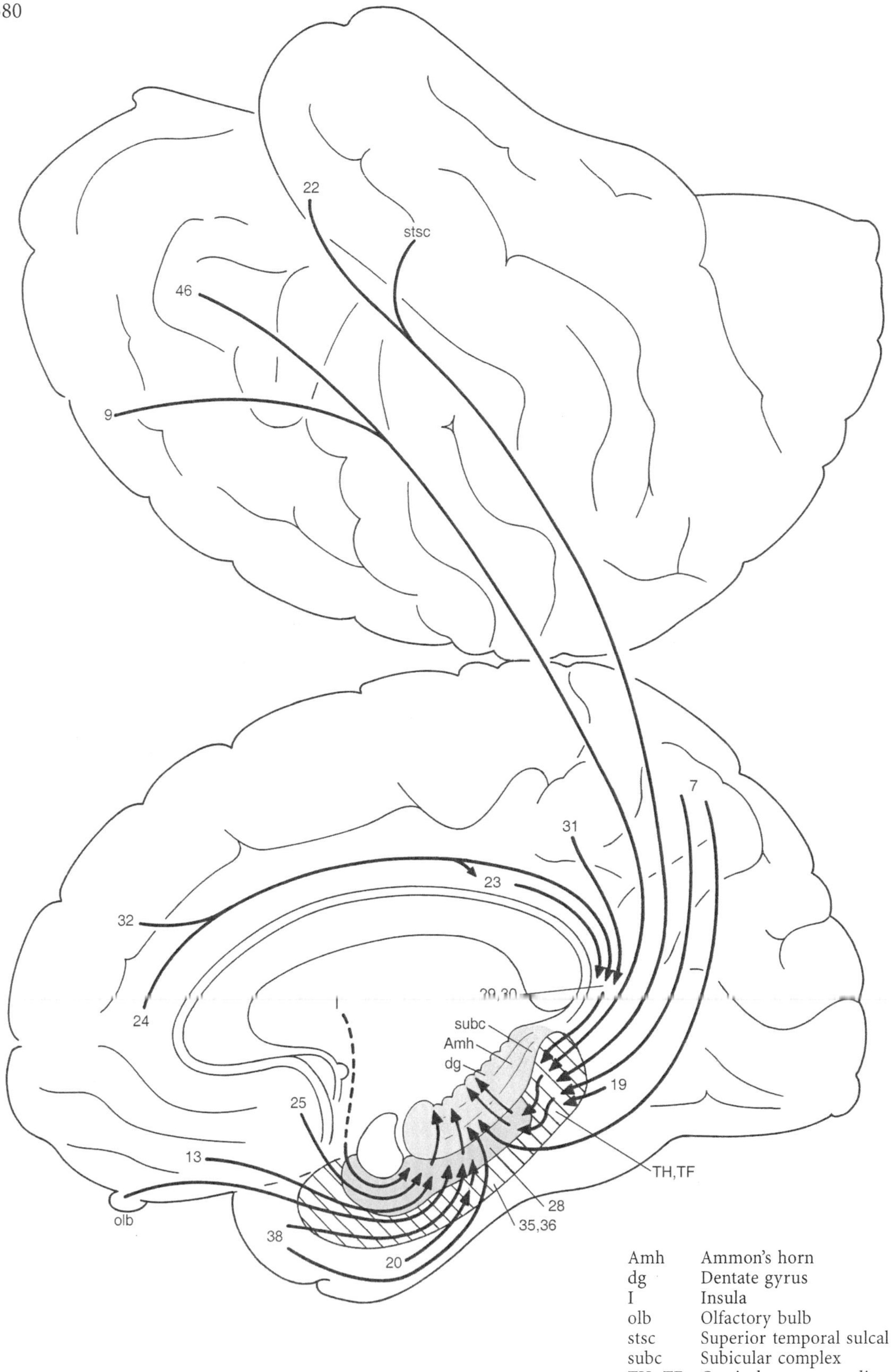

Amh	Ammon's horn
dg	Dentate gyrus
I	Insula
olb	Olfactory bulb
stsc	Superior temporal sulcal cortex
subc	Subicular complex
TH, TF	Cortical areas according to von Economo and Koskinas [258]
7, 9 etc.	Cortical areas according to Brodmann [31]

Fig. 12.10. Direct and indirect cortical afferents of the hippocampus

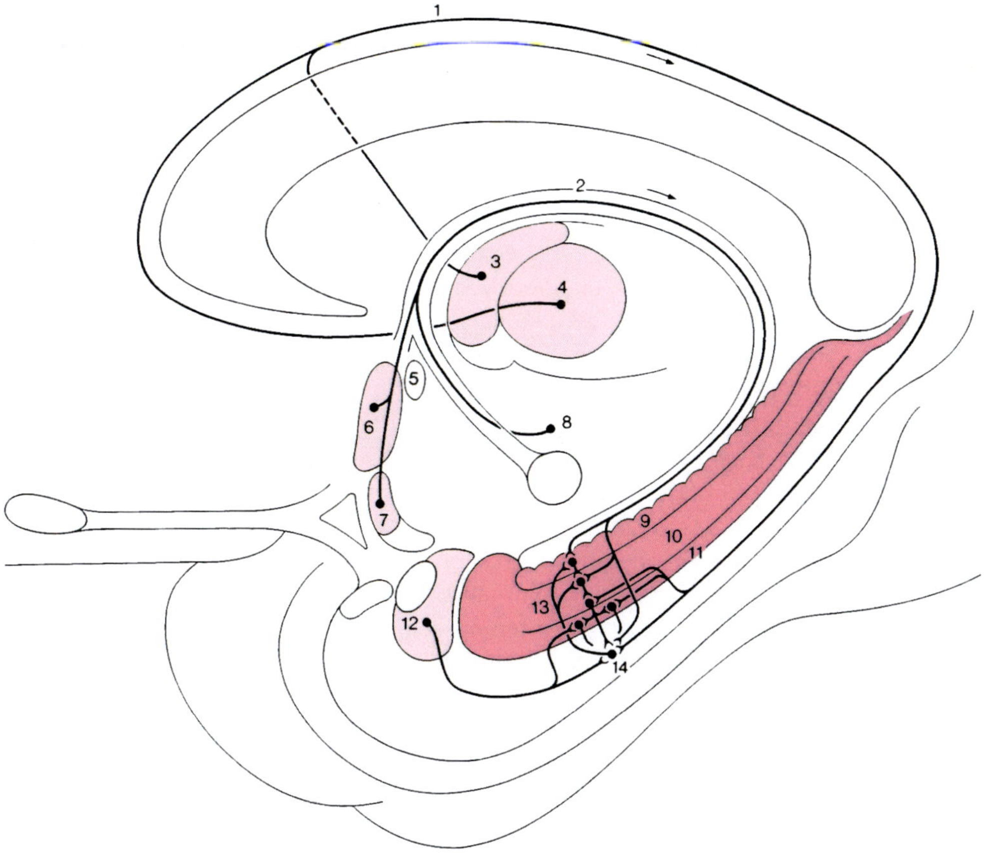

1 Cingulum
2 Fornix
3 Anterior thalamic nucleus
4 Median thalamic nuclei
5 Anterior commissure
6 Medial septal nucleus
7 Nucleus of diagonal band, dorsal part
8 Supramamillary region
9 Dentate gyrus
10 Ammon's horn
11 Subiculum
12 Basal and lateral amygdaloid nuclei
13 Temporo-ammonic tract or "perforant pathway"
14 Entorhinal cortex

Fig. 12.11. Subcortical afferents and intrinsic connections of the hippocampus

area 38, the perirhinal areas 35 and 36, the posterior parahippocampal areas TH and TF, the parietal area 7 and the posterior cingulate area 23 [51, 162, 196, 245, 273].

Substantial contributions to the corticohippocampal projection emanate from areas known to represent sites where sensory information converges. Prominent among these are the superior temporal sulcal cortex, the prefrontal area 46, the orbitofrontal area 13 and the parietal area 7. Pathways leading out of the primary visual, auditory and somatosensory areas of the neocortex, by way of a variable number of intercalated association areas, ultimately converge on these multimodal association areas. Unimodal visual (areas 19, 20) and auditory (area 22) association areas also contribute to the corticohippocampal projection. In the medial temporal lobe, the unimodal and multimodal sensory information conveyed by the various paths is further processed at three hierarchically organized levels, represented by (1) the perirhinal and posterior parahippocampal cortices, (2) the entorhinal cortex and (3) the hippocampal formation itself (Fig. 12.10).

Therefore, it can be concluded that the parahippocampal cortices and ultimately the hippocampus receive a broad spectrum of synthesized sensory-specific as well as multimodal cortical information [241]. Because the entorhinal cortex is also the recipient of olfactory information, either arising from the olfactory bulb [78, 169, 190] or from the prepiriform cortex [70, 110, 111, 179], it seems likely that information derived from all sensory modalities is fed into the hippocampal formation.

Experimental neuroanatomical [177–179, 240, 265, 266] and physiological studies [127, 128, 238, 239] have shown that the hippocampus cannot be considered a homogeneous structure along its spleniotemporal axis. The entorhinal-hippocampal projection is topographically organized so that a lateromedial axis in the entorhinal cortex corresponds to the spleniotemporal axis of the hippocampus. This means that lateral parts of the entorhinal cortex distribute fibres to more splenial portions of the hippocampus, whereas progressively more medial parts of the entorhinal cortex project to increasingly more temporal hippocampal portions. The lateral and medial parts of the entorhinal areas differ considerably with respect to their afferents. The lateral part receives predominantly cortical pathways conveying information from all sensory modalities [177, 265], whereas the medial parts are more influenced by subcortical centres such as the septum, thalamic midline nuclei and amygdala [178]. These differential inputs to the medial and lateral parts of the entorhinal cortex and the topographical organization of the entorhinal-hippocampal perforant path discussed above lead to a convergence of exteroceptive sensory information on the splenial portions of the hippocampus, whereas information derived from limbic-related subcortical centres, presumably reflecting the intrinsic state of the organism, is transmitted in particular to the temporal portions of the hippocampal formation [265].

2. *Amygdaloid afferents.* The hippocampus and the entorhinal cortex receive a substantial input from the amygdaloid complex [7, 19, 20, 90, 109–111, 168, 178, 189]. The lateral, basal and accessory basal amygdaloid nuclei project strongly to the entorhinal cortex and there are additional projections from the basal nucleus to the subicular complex.

3. *Afferents from the medial septal-diagonal band complex.* A major projection to the hippocampus and the entorhinal cortex arises from the medial septal nucleus and the nucleus of the diagonal band [5, 7, 8, 20, 50, 53, 99, 141, 167, 178, 227, 269]. The fibres constituting this projection follow three different routes. The majority of them pass through the fornix-fimbria system, whereas smaller contingents enter the cingulum or take a ventral route via the medial forebrain bundle and then through the amygdaloid complex 24 [178]. The projection to the hippocampus terminates principally in the dentate gyrus and the adjacent region of Ammon's horn [8, 37, 47, 174]. The medial septal nucleus and the nucleus of the diagonal band contain large numbers of cholinergic elements [145–147], and it has been demonstrated that the numerous cholinergic fibres present in the hippocampal formation originate from these neurons [216, 259, 260]. The projection from the septum and

diagonal band complex to the hippocampal and parahippocampal cortices is, however, only partially cholinergic [8], and septal neurons containing several other neurotransmitters, among them GABA [105, 164] and substance P [251], also contribute to this projection. The entorhinal cortex is also in receipt of a – partly cholinergic – projection from the medial septal-diagonal band complex [5, 8, 90].

The hippocampus shows a characteristic, slow, rhythmical electrical activity, usually designated as the hippocampal *theta* rhythm. There is evidence suggesting that both the direct septohippocampal and the indirect septo-entorhinal-hippocampal projections are important in the generation of the hippocampal *theta* rhythm [131, 149].

4. *Afferents from the thalamus.* Two parts of the dorsal thalamus, the anterior nuclei and the midline complex, project to the hippocampal and parahippocampal areas. The fibres originating from the anterior nuclei reach the temporal lobe via the cingulum and terminate in the subicular complex and, more specifically, in the presubiculum [4, 52, 90, 176, 202]. This projection forms part of the so-called circuit of Papez, which will be discussed below. The fibres that originate from the midline nuclei (in rat, cat and monkey most of these fibres arise from the nucleus reuniens thalami [7, 20, 50, 79, 178, 271]) pass to the genu of the corpus callosum, where they join the cingulum [79]. After curving around the callosal splenium, these fibres innervate the entorhinal and subicular cortices and the hippocampal CA1 field [79, 178]. The nucleus reuniens thalami receives input from the parabrachial nuclei, which in turn receive afferents arising from the nucleus of the solitary tract, a primary viscerosensory centre. Hence, the nucleus reuniens may well help transmit information from the visceral periphery to limbic structures [79]. The entorhinal cortex also receives a substantial projection from another midline centre, the paraventricular nucleus [90].

5. *Afferents from the supramamillary region.* A population of large cells, situated directly dorsal to the mamillary body, sends a substantial projection to the hippocampal region via the fornix [7, 71, 90, 167, 174, 269]. Most of the fibres of this supramamillary projection terminate in the dentate gyrus and in the adjacent CA3 field, but the remaining fields of Ammon's horn and the rostral entorhinal cortex are also supplied [71, 90, 178]. A subset of calretinin- and substance P-containing supramamillary cells project specifically to the dentate gyrus [157]. These cells have been proposed to be involved in the regulation of *theta* activity [21].

6. *Afferents from monoaminergic cell groups in the brain stem.* The hippocampal formation receives serotoninergic fibres from the mesencephalic raphe nuclei, noradrenergic fibres from the locus coeruleus and dopaminergic fibres from the ventral tegmental area. Diffusely arranged fibres carrying these three monoamines have been observed in all parts of the hippocampus. However, the deepest zone of the dentate plexiform layer and layer III of the entorhinal cortex are densely innervated by serotoninergic fibres, whereas the entire dentate plexiform layer and the pyramidal layer of field CA3 contain a dense plexus of noradrenergic fibres [227].

Efferents from the Hippocampus

Turning now to the efferent connections of the hippocampus (Fig. 12.12), it should first be stated that investigations using so-called tracer techniques have radically changed our insight into the organization of these connections. Contrary to what was believed for almost a century, the entire postcommissural fornix and considerable part of the precommissural fornix originate from the subiculum rather than from Ammon's horn [142, 224, 225]. The contribution of the latter structure to the fornix has been shown to be only minor and to be confined to the precommissural fornix. In the light of these findings, the hippocampal efferents may be grouped as follows: (1) efferents from Ammon's horn, (2) contributions of the subicular complex to the precommissural fornix, (3) contributions of the subicular complex to the postcommissural fornix and (4) "non-fornical" efferents.

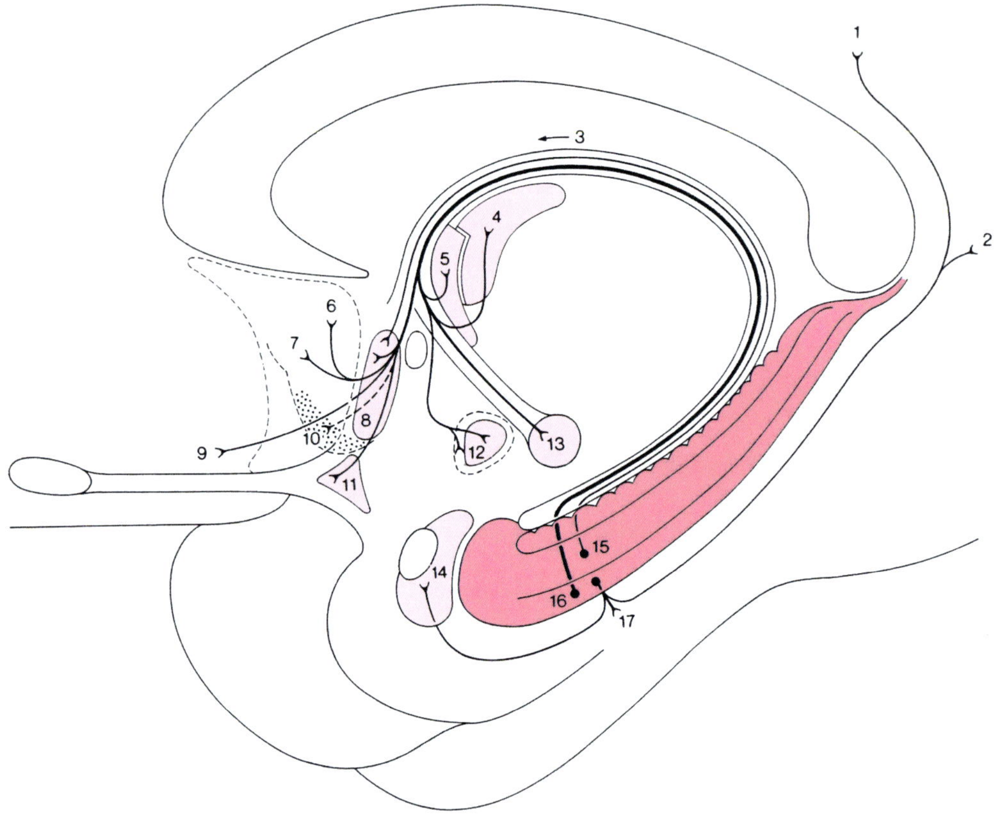

1 Cingulate gyrus (area 23)
2 Retrosplenial cortex (area 29, 20)
3 Fornix
4 Anterior thalamic nucleus
5 Bed nucleus of stria terminalis
6 Precommissural hippocampus
7 Medial prefrontal cortex
8 Septal nuclei
9 Straight gyrus
10 Accumbens nucleus
11 Anterior olfactory nucleus
12 Ventromedial hypothalamic nucleus
13 Mamillary nucleus
14 Amygdaloid body
15 Ammon's horn
16 Subiculum
17 Entorhinal cortex (area 28)

Fig. 12.12. Efferents of the hippocampus

1. *Efferents from Ammon's horn.* The precommissural fornix fibres originating from Ammon's horn terminate exclusively in the lateral septal nucleus [142, 224, 225]. These fibres, which are presumably glutamatergic [274], arise from all cell fields of Ammon's horn [108]. Multiple retrograde labelling studies suggest that most of the pyramidal cells in Ammon's horn that project to the septum also give rise to intrahippocampal collaterals [222]. It is of interest that the lateral septal nucleus projects to the medial septal–diagonal band complex [170], which is known to project back to the hippocampal formation.

2. *Contributions of the subicular complex to the precommissural fornix.* The precommissural fornix fibres originating from the subicular complex are distributed to the lateral septal nucleus, the accumbens nucleus, the caudate nucleus and putamen, the anterior olfactory nucleus, the precommissural hippocampus, the medial part of the frontal cortex and the gyrus rectus [69, 142, 183, 205, 207, 225]. The projection to the accumbens nucleus forms part of the interface of the limbic system with somatomotor mechanisms [150, 151, 272].

3. *Contributions of the subicular complex to the postcommissural fornix.* The postcommissural fornix contains, apart from some hippocampal afferents, only fibres originating from the subicular complex. Most of these fibres terminate in the mamillary body; smaller contingents are distributed to the anterior thalamic nuclei, the bed nucleus of the stria terminalis and the cell-free capsular zone surrounding the ventromedial hypothalamic nucleus [108, 140, 142–144, 183, 224, 225].

The *mamillary body* forms a prominent and characteristic projection on the basal surface of the brain (Figs. 3.4, 3.8). It can be subdivided into a large, small-celled medial nucleus and a much smaller, large-celled lateral nucleus (Fig. 6.33). The medial mamillary nucleus is innervated by the subiculum and the lateral mamillary nucleus is innervated by the subiculum and presubiculum. The mamillary body receives, apart from descending fornical afferents, ascending afferents from the paramedian midbrain zone, by way of a small, compact bundle known as the *mamillary peduncle* (Fig. 6.31). This bundle, which originates mainly from the dorsal tegmental nucleus of Gudden, passes ventrally and then ascends along the ventral surface of the midbrain to the mamillary body, where most of its fibres terminate. Some of its fibres join the medial forebrain bundle and spread to the lateral preopticohypothalamic zone and the medial septal nucleus [152, 154, 155].

Allen and Hopkins [4 a] studied the site and mode of termination of the various afferent systems on the cells in the mamillary body in the rat. They presented ultrastructural evidence suggesting that the subicular afferents (as well as descending afferents from the prefrontal cortex) form excitatory synapses on distal parts of the dendrites of mamillary neurons, whereas the tegmental afferents form inhibitory synapses on the somata, proximal and distal dendrites of these elements. The axons of the mamillary neurons ascend in the principal mamillary fasciculus and bifurcate to form the prominent mamillothalamic and mamillotegmental tracts. The mamillothalamic tract terminates in the anterior thalamic nuclei. The mamillotegmental tract ends in the dorsal tegmental nucleus of Gudden. This descending tract and the reciprocal ascending mamillary peduncle may form part of a "limbic system-midbrain circuit" [30].

4. *Non-fornical efferents.* The non-fornical efferents include fibres projecting to the amygdaloid complex and to the neocortex. The hippocampal projection to the amygdala is substantially weaker than that in the opposite direction [120]. It consists mainly of fibres originating from cells situated along the border of the subiculum and field CA1, which project to the basal amygdaloid nucleus [189].

The hippocampal projection to the neocortex is composed of efferents from the subicular complex, the entorhinal cortex and the perirhinal-posterior parahippocampal complex. Most of the components of the hippocamponeocortical projection have reciprocal fibre systems passing in the opposite direction (Fig. 12.10).

The efferents from the subicular complex terminate in the entorhinal cortex (area 28), the anterior and posterior cingulate cortices (areas 24, 23), the caudal orbitofrontal and medial frontal cortices, the temporal polar cortex (area 38), the inferior temporal cortex (area 20), the parietal cortex (area 7) and the retrosplenial cortex (areas 29, 30) [18, 20, 27, 36, 82, 88, 92, 102, 144, 183, 188, 206, 241, 273].

The entorhinal cortex projects to the perirhinal (areas 35, 36) and posterior parahippocampal (areas TH, TF) cortices, the agranular insular cortex, the prepiriform cortex (area 51), the prelimbic cortex (area 32), the cingulate cortex (areas 23, 24), the retrosplenial cortex (areas 29, 30), the orbitofrontal cortex (area 13), the temporal cortex (areas 20–22) and the parietal cortex (area 7) [66, 102, 107, 205, 217, 218, 241, 246, 264].

The efferents from the perirhinal and posterior parahippocampal cortices pass to the anterior insula, to many parts of the frontal lobe, including areas 11, 12, 13, 9 and 46, to temporal areas 20–22, to cingulate area 24, to posterior parietal area 7, to occipital area 19 and to the retrosplenial areas 29 and 30. The latter project in turn to the cingulate areas 23 and 24 [102, 121, 241, 246, 264].

By means of the diverging projections just enumerated, hippocampal output is disseminated over numerous and widespread neocortical areas, most belonging in the categories of unimodal and multimodal association areas. As will be discussed below, these hippocampo-neocortical projections are supposed to play a crucial role in establishing long-term memories.

The Circuit of Papez

The postcommissural fornix is prominent within a system of centres and connections known as the "circuit of Papez" [165]. This circuit comprises the hippocampus, the postcommissural fornix, the mamillary body, the mamillothalamic tract, the anterior thalamic nucleus, the thalamocingulate projection and the cingulate gyrus. Fibres passing via the cingulum bundle from the cingulate gyrus back to the hippocampus close the circuit. Papez [165] believed that this circuit forms an essential part of the structural basis of emotions. However, there is no sound biological basis for this view [30]. As will be discussed below, most structures in the Papez circuit play a role in memory.

Synopsis of Anatomical Data: Functional and Clinical Aspects

The functional and clinical aspects of the hippocampus are summarized here and related to the anatomical data as illustrated in Fig. 12.13; the numbers used in this figure correspond to those in this section.

The *medial temporal lobe* is a morphologically heterogeneous structure, which encompasses the well-differentiated temporal portion of the hippocampal formation, the entorhinal cortex, the areas TH and TF and the perirhinal areas 35 and 36 (Fig. 12.3). The entorhinal cortex and the various parts of the hippocampal formation are interconnected by a unidirectional series of connections, which together form the so-called *intrinsic hippocampal circuit* (Figs. 12.8, 12.13: 1). Projections from widespread, mostly associational cortical areas reach the medial temporal lobe and converge either directly or via one or two synaptic interruptions on the hippocampal formation (Fig. 12.13: 2). These neocortico-hippocampal projections are reciprocated by extensive hippocampo-neocortical projections. The amygdaloid complex also forms part of the medial temporal lobe. This complex is reciprocally connected with the hippocampal formation (Fig. 12.13: 3).

The medial temporal lobe, which roughly corresponds to the parahippocampal gyrus, plays a prominent role in learning and memory. This was first demonstrated dramatically by a now famous clinical case, patient HM, who underwent bilateral removal of the medial temporal lobe to ameliorate intractable epilepsy. After this operation, patient HM showed a severe amnestic syndrome. He suffered a profound loss in his ability to form short-term memories and appeared to be unable to encode new information into long-term memory. The severe and lasting amnesia was in sharp contrast to his preserved early memories and overall intelligence. Schizophrenic patients who underwent similar operations showed a comparable symptomatology [45, 195]. Later studies in both humans [171, 275] and non-human primates [276] have shown that bilateral damage limited to the hippocampal formation is sufficient to produce enduring memory impairment. This is exemplified by another classical clinical case, patient RB, described by Zola-Morgan et al. in 1986 [275]. This patient developed a substantial, long-lasting memory impairment as a result of a hypotensive period during coronary bypass surgery. No loss of memory had formed before the operation. RB's brain was subjected to postmortem analysis. The only pathology that could be associated with his memory disturbance was complete loss of neurons in the hippocampal CA1 field. Because the amnesia in RB was less severe than that in HM, it has been suggested that the severity of the memory impairment may depend on how much of the hippocampal formation and adjacent cortex is affected by the lesion.

The inability to form new memories, which characterizes the onset of Alzheimer's disease, can also be attributed to damage of the hippocampal formation and adjacent cortical areas [86, 242, 247]. In this disease there is an almost total loss of pyramidal neurons in the CA1 field and in the subiculum. These neurons form important links in the intrinsic hippocampal circuit (Fig. 12.8) and also give rise to hippocampal efferents that course via the fornix to several subcortical telencephalic and diencephalic targets (Figs. 12.12, 12.13: 4). In Alzheimer's disease, severe pathological changes are also observed in layer II of the entorhinal cortex and in the molecular layer of the dentate gyrus, i.e. the sites of origin and termination, respectively, of the perforant path. As shown in Fig. 12.8, this path represents another salient link in the intrinsic hippocampal circuit.

The finding that patients who have suffered damage to the medial temporal lobe have an enduring impairment in their ability to form new long-term memories, whereas older memories remain largely intact, and comparable observations in animal models have led to the concept that this brain part is involved in the formation of new memories, but is not their final repository [209, 211]. It is believed that the medial temporal lobe forms new memories on the basis of the convergent inflow of sensory information from the neocortex (Fig. 12.10), but that the new memories, once formed, are transferred to and consolidated in neocortical areas. The process of forming and consolidating new memories requires that the anatomical pathways between the medial temporal lobe and the neocortex and those of the associated intrinsic hippocampal circuit are intact (Figs. 12.13: 1, 2).

The hippocampal formation is strongly and reciprocally connected with the *precommissural septum*. The medial septal complex projects massively to all parts of the hippocampus, including the subiculum [226] and the entorhinal cortex [5]. Significant portions of the septohippocampal projection arise from populations of cholinergic and GABAergic neurons. The medial septal complex receives projections from virtually all parts of the hypothalamus [23, 43, 44, 187, 233] and from monoaminergic cell groups in the brain stem [227]. It seems likely that the impulses travelling along these ascending projections modulate the flow of information through the intrahippocampal circuit in relation to the prevailing behavioural state [227].

The lateral septal nucleus receives a topographically organized glutamatergic input from Ammon's horn and the subiculum via the *precommissural fornix* (Fig. 12.13: 5) [142, 224, 225] and projects, again in a topographically organized way, to the medial preopticohypothalamic zone (Fig. 12.13: 6). Different parts of the lateral septal nucleus project to specific nuclei within this zone [22, 175, 226]. Many of these nuclei form part of the systems involved

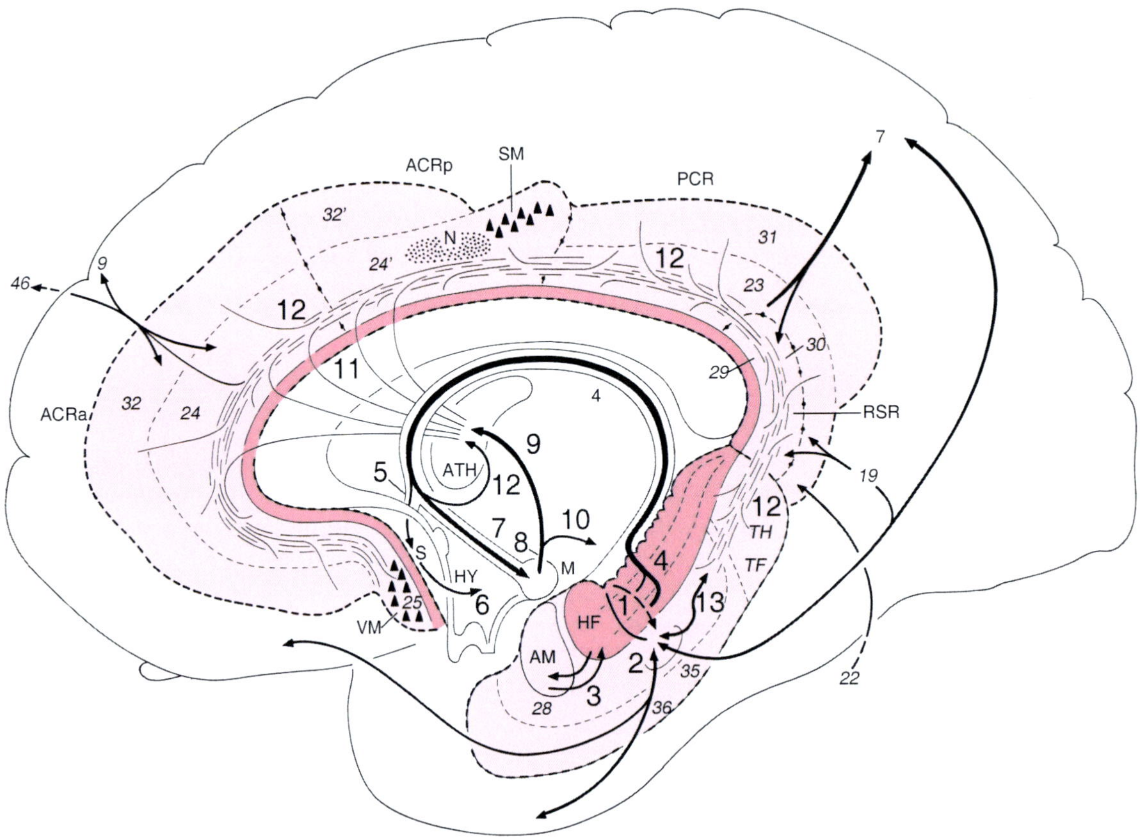

ACRa Anterior part of anterior cingulate region
ACRp Posterior part of anterior cingulate region
AM Amygdala
ATH Anterior thalamic nucleus
HF Hippocampal formation
HY Hypothalamus
M Mamillary body
N Nociceptive area
PCR Posterior cingulate region
RSR Retrosplenial region
S Septum
SM Somatomotor area
VM Visceromotor area

Fig. 12.13. The principal connections of the hippocampal formation (*dark red*), the medial temporal lobe and the cingulate gyrus (both in *light red*). The *heavy dashed lines* delineate the cingulate and parahippocampal gyri, including their parts of the hippocampal formation. The *heavy dotted lines* show the borders of the four principal cingulate regions. Cytoarchitectonic areas are indicated with *small italicized numbers* or *letters*; *thin dashed lines* delineate some of these areas. The *large numbers* are explained in the text

in regulating behaviours essential for the survival of the individual or species. Thus, the paraventricular hypothalamic nucleus plays a prominent role in ingestive behaviour (Figs. 10.7, 10.8) and is also critically involved in the stress response (Fig. 10.6), the medial preoptic and ventromedial hypothalamic nuclei occupy central positions in the circuits related to reproductive behaviour (Fig. 10.9), whereas the anterior hypothalamic nucleus and the dorsal premamillary nucleus occupy corresponding positions in the hypothalamic defense complex (Fig. 10.11). Given the topographical organization of the hippocampo-septal and septo-hypothalamic projections, it seems likely that specific hippocampal domains, by way of these projections, selectively influence particular motivated behaviours [175, 223]. In light of this, it is not surprising that lesions of the septal region may lead to marked changes in ingestive, sexual and agonistic behaviour [3, 30].

The *postcommissural fornix* (Fig. 12.13: 7), which originates from the subiculum, projects to the *mamillary body* in the most posterior part of the hypothalamus. Other afferents to the mamillary body come from the lateral septal nucleus, the medial preoptic nucleus and several hypothalamic nuclei [199]. A direct frontomamillary pathway, originating mainly from the orbital areas 10, 11 and 47, has recently been described [115]. The mamillary body gives rise to a large rostrodorsally directed efferent system, the *principal mamillary fasciculus* (Fig. 12.13: 8), which bifurcates into the ascending *mamillothalamic* and the descending *mamillotegmental tracts* (Fig. 12.13: 9, 10). The mamillothalamic tract distributes its fibres to all three components of the *anterior thalamic nucleus*, where, in turn, massive projections to the cingulate gyrus originate (Fig. 12.13: 11). Studies, mainly in rats, have shown that the area of termination of this projection includes the retrosplenial cortex and extends as far posteriorly as the subicular complex and the entorhinal cortex [215]. Vogt et al. [254] reported that, in the macaque monkey, the thalamocingulate system projects primarily to the retrosplenial cortex. The indirect hippocampomamillothalamic projection (Fig. 12.13: 7–9) is supplemented by fibres which branch off from the postcommissural fornix and pass directly to the anterior thalamic nucleus (Fig. 12.13: 12). As has already been mentioned, the postcommissural fornix, mamillary body, mamillothalamic tract, anterior thalamic nucleus and thalamocingulate projection form part of the so-called *Papez circuit*.

The knowledge concerning the functional role of the circuitry just summarized is still fragmentary and in part controversial. As regards the fornix, some authors [63, 75, 228] reported that bilateral damage of this bundle leads to impaired memory in humans, but patients with bilateral fornix lesions in whom no memory deficits were evident have also been described [267]. In monkeys, fornix transection has been found to produce only transient memory impairment [276].

The mamillary bodies have also been linked to memory function because they are consistently damaged in Korsakoff's syndrome. This syndrome, which includes selective anterograde and retrograde amnesia, confabulations and severe learning disabilities, appears as a sequel to Wernicke's encephalopathy, an alcohol-induced disease caused by thiamine deficiency [38, 231]. However, in Wernicke's encephalopathy, the mamillary nuclei are not the only site of damage. It has been suggested that the critical lesion site responsible for the memory disturbance includes the mediodorsal (not the anterior!) nuclei of the thalamus, with involvement of the mamillary bodies playing only a secondary role [106, 137, 249, 250]. In monkeys, mamillary lesions have been found to produce, just like fornix transactions, only transient memory impairment [276]. Experiments, mainly in rats, suggest a specific role of the mamillary bodies in spatial orientation [25, 212] and in spatial working memory tasks [42, 185, 186, 248]. In monkeys, working memory tasks elevate glucose metabolism in the anterior thalamic nuclei [61].

The mamillotegmental tract (Fig. 12.13: 10) terminates mainly in the dorsal tegmental nucleus of Gudden. By way of this tract, the mamillary body (and ultimately the hippocampus) may influence the activity in the upper brain stem.

The *cingulate gyrus*, which dorsally surrounds the corpus callosum, occupies a large fraction of the medial surface of the human telencephalon (Fig. 3.6). This structure is traditionally divided into an anterior cingulate region (ACR), comprised of areas 25, 24 and 32; a posterior cingulate region (PCR), with areas 23 and 31; and a retrosplenial region (RSR), composed of areas 29 and 30. The anterior cingulate region can be further subdivided, on both structural and functional grounds, into separate anterior and posterior subregions. These subregions are designated here as ACRa and ACRp (Fig. 12.13). Vogt [252, 256], who proposed this subdivision, emphasized that ACRa receives massive amygdala input, while ACRp receives significant afference from the posterior parietal cortex. The posterior parts of areas 24 and 32, which comprise ACRp, are designated as areas 24′ and 32′ (Fig. 12.13). The various cingulate regions are strongly and reciprocally connected by association fibres [161, 163, 246, 254], many of which travel with the *cingulum bundle* (Fig. 12.13: 12).

ACR is reciprocally connected with the prefrontal areas 9 and 46. It receives afferents from somatosensory (area 7) and auditory (area 22) association areas [114, 160, 161] and projects to several subcortical structures, including the septal nuclei, the amygdala and the hypothalamus [160]. ACR is associated with affective experiences and plays a role in cognitive processes, such as the selection of behaviour in response to conflicting stimuli and memory [34]. Moreover, it contains a sensory area concerned with nociception, as well as somatomotor and autonomic effector areas. Electrical stimulation of the ACR in humans produces pleasure, fear, agitation, or anger [148]. In monkeys, bilateral damage of ACR results in increased tameness and social indifference [64]. Patients with lesions in ACR are apathetic and show a marked decrease in spontaneous and emotional motor responses [56]. Intentional lesions in ACR (cingulotomy) or cingulum bundle (cingulumotomy) can reduce chronic pain, anxiety and depressive and obsessive-compulsive behaviours [17, 148, 261].

ACR and RSR, which are strongly interconnected and share an input from the auditory association cortex, are jointly activated in auditory verbal memory tasks [67]. It has been demonstrated that in the macaque monkey, ACR and the ventromedial prefrontal and orbitofrontal cortices are involved in long-term recognition memory processes [270]. In rats, the ACR, together with the prefrontal and temporal cortices, shows robust increases in activity specifically following remote memory retrieval. Damage to these areas appeared to produce selective remote memory deficiencies [262].

ACRp harbours a nociceptive and a somatomotor area (Fig. 12.13). The nociceptive area represents the cortical end station of the medial pain system, which is particularly concerned with the affective (suffering) component of pain [198, 253]. It is noteworthy that ACR not only plays a role in the perception of pain, but also in its control. Fibres passing from this region to the periaqueductal grey form an essential link in the descending inhibitory pain system.

The dorsal part of ACRp contains an area characterized by the presence of very large pyramidal neurons in its fifth layer (Fig. 12.13). This area, which plays an important role in the cognitive control and execution of voluntary movements [232], is reciprocally connected with the primary motor and premotor cortices and gives rise to a corticospinal projection that terminates in the intermediate zone of the spinal cord [54, 55, 76, 246].

ACRa is engaged in autonomic regulation. Stimulation of a subgenual subregion, roughly corresponding to area 25, produces changes in the heart rate, blood pressure and respiration pattern. Area 25 projects to several subcortical centres known to be involved in autonomic control, among them the amygdaloid complex, the perifornical region of the hypothalamus, the periaqueductal grey and the lateral parabrachial nucleus [39, 59, 184]. Stimulation of ACRa may also lead to behavioural responses, such as vocalization and defense and escape reactions. Most likely, the projections descending from this region to the hypothalamus and the periaqueductal grey also have a role in these responses.

PCR is in receipt of massive projection from the posterior parietal cortex [122]. Other cortical afferents to PCR include fibres from the prefrontal areas 9 and 46 and from the visual and auditory association areas [102]. The afferents to RSR closely resemble those to PCR, with the reservation that the projection from the posterior parietal cortex to RSR is much weaker than that to PCR [102].

The known functions of *PCR* and *RSR* are all related to memory. PCR and RSR are both involved in working memory and visuospatial memory [255]. PCR has been shown to play an important role in autobiographical memory retrieval [136] and it has already been mentioned that RSR and ACR are jointly activated in auditory verbal memory tasks [67]. Patients with damage to RSR generally show an enduring anterograde amnesic syndrome with variable retrograde amnesia [102].

The *cingulate gyrus* is closely associated with the *medial temporal lobe*, the latter roughly corresponding to the *parahippocampal gyrus*. As already mentioned, these two structures together form a large arcuate convolution, Broca's [29] *great limbic lobe*. The cingulate gyrus and the parahippocampal gyrus are both accompanied by a part of the hippocampal formation. In humans and other eutherians, the hippocampal component of the cingulate gyrus (or its equivalent) are poorly developed (Fig. 12.1 C–E), but we have seen that in the acallosal monotremes (Fig. 12.2) and marsupials, a well-differentiated hippocampal formation extends throughout the length of the great limbic lobe. ACRa, PCR and RSR are not only reciprocally connected with each other, but also with the medial temporal lobe. The large *cingulum bundle*, which extends from the cingulate gyrus into the medial temporal lobe, provides for these connections (although this bundle also harbours numerous fibres that neither originate nor end in the great limbic lobe). The cingulate neurons, which send their axons via the cingulum bundle to the medial temporal lobe (Fig. 12.13: 13), are thought to receive afferents from the various association cortices and, hence, to form part of the large neocortico-temporo-hippocampal projection already discussed (Fig. 12.13: 2). The remarkable convergence and concentration of this system on the temporal hippocampal formation is related to the equally remarkable reduction of the remaining parts of this formation.

References

1. Abbie AA (1940) Cortical lamination in the monotremata. J Comp Neurol 72:429–467
2. Acsady L, Kamondi A, Sik A, Freund T, Buzsaki G (1998) GABAergic cells are the major postsynaptic targets of mossy fibers in the rat hippocampus. J Neurosci 18:3386–3403
3. Albert DJ, Walsh ML, Jonik RH (1993) Aggression in humans: what is its biological foundation? Neurosci Biobehav Rev 17:405–425
4. Alksen JF, Blackstad TW, Walberg F, White Jr LE (1966) Electron microscopy of axons degeneration: a valuable tool in experimental neuroanatomy. Ergebn Anat Entwicklgesch 39:1–32

4a. Allen GV, Hopkins DA (1989) Mamillary body in the rat: topography and synaptology of projections from the subicular complex, profrontal cortex, midbrain tegmentum. J Comp Neurol 286:311–336

5. Alonso A, Köhler C (1984) A study of the reciprocal connections between the septum and the entorhinal area using anterograde and retrograde axonal transport methods in the rat brain. J Comp Neurol 225:327–343
6. Amaral DG (1978) A Golgi study of cell types in the hilar region of the hippocampus in the rat. J Comp Neurol 182:851–914
7. Amaral DG, Cowan WM (1980) Subcortical afferents to the hippocampal formation in the monkey. J Comp Neurol 189:573–591
8. Amaral DG, Kurz J (1985) An analysis of the origins of the cholinergic and noncholinergic septal projections to the hippocampal formation of the rat. J Comp Neurol 240:37–59
9. Amaral DG, Insausti R, Cowan WM (1983) Evidence for a direct projection from the superior temporal gyrus to the entorhinal cortex in the monkey. Brain Res 275:263–277
10. Amaral DG, Insausti R, Cowan WM (1984) The commissural connections of the monkey hippocampal formation. J Comp Neurol 224:307–336
11. Amaral DG, Ishizuka N, Claiborne B (1990) Neurons, numbers and the hippocampal network. Progr Brain Res 83:1–11
12. Amaral DG, Dolorfo C, Alvarez-Royo P (1991) Organization of CA1 projections to the subiculum: a PHA-L analysis in the rat. Hippocampus 1:415–435
13. Andersen P, Bliss TVP, Skrede KK (1971) Lamellar organization of hippocampal pathways. Exp Brain Res 13:222–238
14. Andy OJ, Stephan H (1968) The septum in the human brain. J Comp Neurol 133:383–410

15. Aniksztejn L, Charton G, Ben Ari Y (1987) Selective release of endogenous zinc from the hippocampal mossy fibers in situ. Brain Res 404:58–64
16. Bailey P, Von Bonin G (1951) The isocortex of man. University of Illinois Press, Urbana
17. Ballantine HT, Levy BS, Dagi T, Giriunas IB (1975) Cingulotomi for psychiatric illness: report of 13 years experience. In: Sweet WH, Obrador S, Martin-Rodrigues JG (eds) Neurosurgical treatment in psychiatry, pain and epilepsy. University Park Press, Baltimore, pp 333–353
18. Barbas H, Blatt GJ (1995) Topographically specific hippocampal projections target functionally distinct prefrontal areas in the rhesus monkey. Hippocampus 5:511–533
19. Barbas H, De Olmos J (1990) Projections from the amygdala to basoventral and mediodorsal prefrontal regions in the rhesus monkey. J Comp Neurol 300:549–571
20. Beckstead RM (1978) Afferent connections of the entorhinal area in the rat as demonstrated by retrograde cell-labeling with horseradish peroxidase. Brain Res 152:249–264
21. Berger B, Esclapez M, Alvarez C, Meyer G, Catala M (2001) Human and monkey fetal brain development of the supramammillary-hippocampal projections: a system involved in the regulation of theta activity. J Comp Neurol 429:515–529
22. Berk ML, Finkelstein JA (1981) Afferent projections to the preoptic area and hypothalamic regions in the rat brain. Neuroscience 6:1601–1624
23. Berk ML, Finkelstein JA (1982) Efferent connections of the lateral hypothalamic area of the rat: an autoradiographic investigation. Brain Res Bull 8:511–526
24. Blackstad TW, Kjaerheim A (1961) Special axodendritic synapses in the hippocampal cortex: electron and light microscopic studies on the layer of mossy fibers. J Comp Neurol 117:133–159
25. Blair HT, Cho J, Sharp PE (1998) Role of the lateral mammillary nucleus in the rat head direction circuit: a combined single unit recording and lesion study. Neuron 21:1387–1397
26. Blasco-Ibanez JM, Freund TF (1995) Synaptic input of horizontal interneurons in stratum oriens of the hippocampal CA1 subfield: structural basis of feed-back activation. Eur J Neurosci 7:2170–2180
27. Blatt GJ, Rosene DL (1998) Organization of direct hippocampal efferent projections to the cerebral cortex of the rhesus monkey: projections from CA1, prosubiculum, and subiculum to the temporal lobe. J Comp Neurol 392:92–114
28. Braak H (1980) Architectomics of the human telencephalic cortex. Springer, Berlin Heidelberg New York
29. Broca P (1878) Anatomie comparée des circonvolutions cerebrales: le grand lobe limbique et la scissure dans la serie des mammiferes. Rev Anthropol (Paris) 2:285–498
30. Brodal A (1981) Neurological anatomy in relation to clinical medicine, 3rd edn. Oxford University Press, New York
31. Brodmann K (1909) Vergleichende Lokalisationslehre der Großhirnrinde in ihren Prinzipien dargestellt auf Grund des Zellenbaues. Barth, Leipzig
32. Brodmann K (1914) Physiologie des Gehirns. In: Von Bruns P (ed) Neue Deutsche Chirurgie, vol 11. Enke, Stuttgart, pp 85–426
33. Buckmaster PS, Alonso A, Canfield DR, Amaral DG (2004) Dendritic morphology, local circuitry, and intrinsic electrophysiology of principal neurons in the entorhinal cortex of macaque monkeys. J Comp Neurol 470:317–329
34. Bush G, Luu P, Posner MI (2000) Cognitive and emotional influences in anterior cingulate cortex. Trends Cogn Sci 4:215–222
35. Cajal SR (1972) Histologie du système nerveus de l'homme et des vertébrés. Consejo Superior de Investigaciones Cientificas. Instituto Ramon y Cajal, Madrid
36. Carmichael ST, Price JL (1995) Limbic connections of the orbital and medial prefrontal cortex in macaque monkeys. J Comp Neurol 363:615–641
37. Chandler JP, Crutcher HA (1983) The septohippocampal projection in the rat: an electron microscopic horseradish peroxidase study. Neuroscience 10:685–696
38. Charness ME, De la Paz RL (1987) Mamillary body atrophy in Wernicke's encephalopathy: antemortem identification using magnetic resonance imaging. Ann Neurol 22:595–600
39. Chiba T, Kayahara T, Nakano K (2001) Efferent projections of infralimbic and prelimbic areas of the medial prefrontal cortex in the Japanese monkey, *Macaca fuscata*. Brain Res 888:83–101
40. Chicurel ME, Harris KM (1992) Three-dimensional analysis of the structure and composition of CA3 branched dendritic spines and their synaptic relationships with mossy fiber boutons in the rat hippocampus. J Comp Neurol 325:169–182
41. Claiborne BJ, Amaral DG, Cowan WM (1986) A light and electron microscopic analysis of the mossy fibers of the rat dentate gyrus. J Comp Neurol 246:435–458
42. Conejo NM, Gonzalez-Pardo H, Vallejo G, Arias JL (2004) Involvement of the mammillary bodies in spatial working memory revealed by cytochrome oxidase activity. Brain Res 1011:107–114
43. Conrad LCA, Pfaff DW (1976) Efferents from medial basal forebrain and hypothalamus in the rat. I. An autoradiographic study of the medial preoptic area. J Comp Neurol 169:185–220
44. Conrad LCA, Pfaff DW (1976) Efferents from medial basal forebrain and hypothalamus in the rat. II. An autoradiographic study of the anterior hypothalamus. J Comp Neurol 169:221–262
45. Corkin S, Amaral DG, Gonzalez RG, Johnson KA, Hyman BT (1997) H.M.'s medial temporal lobe lesion: findings from magnetic resonance imaging. J Neurosci 17:3964–3979
46. Cruce JAF (1977) An autoradiographic study of the descending connections of the mammillary nuclei of the rat. J Comp Neurol 176:631–644

47. Crutcher HA, Madison R, Davis JN (1981) A study of the rat septohippocampal pathway using anterograde transport of horseradish peroxidase. Neuroscience 6:1961–1973
48. Demeter S, Rosene DL, Van Hoesen GW (1985) Interhemispheric pathways of the hippocampal formation, presubiculum, and entorhinal and posterior parahippocampal cortices in the rhesus monkey: the structure and organization of the hippocampal commissures. J Comp Neurol 233:30–47
49. Deuchars J, Thomson AM (1996) CA1 pyramid-pyramid connections in rat hippocampus in vitro: dual intracellular recordings with biocytin filling. Neuroscience 74:1009–1018
50. DeVito JL (1980) Subcortical projections to the hippocampal formation in squirrel monkey (Saimiri sciureus). Brain Res Bull 5:285–289
51. Ding SL, Van Hoesen G, Rockland KS (2000) Inferior parietal lobule projections to the presubiculum and neighboring ventromedial temporal cortical areas. J Comp Neurol 425:510–530
52. Domesick VB (1970) The fasciculus cinguli in the rat. Brain Res 20:19–32
53. Domesick VB (1976) Projections of the nucleus of the diagonal band of Broca in the rat. Anat Rec 184:391–392
54. Dum RP, Strick PL (1993) Cingulate motor areas. In: Vogt BA, Gabriel M, Leichnetz GR (eds) Neurobiology of cingulate cortex and limbic thalamus. Birkhäuser, Boston, pp 415–441
55. Dum RP, Strick PL (1996) Spinal cord terminations of the medial wall motor areas in macaque monkeys. J Neurosci 16:6513–6525
56. Eslinger PJ, Damasio AR (1985) Severe disturbance of higher cognition after bilateral frontal lobe ablation: patient EVR. Neurology 35:1731–1741
57. Finch DM, Babb TL (1980) Neurophysiology of the caudally directed hippocampal efferent system in the rat: projections to the subicular complex. Brain Res 197:11–26
58. Finch DM, Nowlin NL, Babb TL (1983) Demonstration of axonal projections of neurons in the rat hippocampus and subiculum by intracellular injection of HRP. Brain Res 271:201–216
59. Freedman LJ, Insel TR, Smith Y (2000) Subcortical projections of area 25 (subgenual cortex) of the macaque monkey. J Comp Neurol 421:172–188
60. Freund TF, Buzsaki G (1996) Interneurons of the hippocampus. Hippocampus 6:347–470
61. Friedman HR, Janas JD, Goldman-Rakic PS (2005) Enhancement of metabolic activity in the diencephalon of monkeys performing working memory tasks: a 2-deoxyglucose study in behaving rhesus monkeys. J Cogn Neurosci 2:18–31
62. Fukuda T, Kosaka T (2000) Gap junctions linking the dendritic network of GABAergic interneurons in the hippocampus. J Neurosci 20:1519–1528
63. Gaffan D, Gaffan EA (1991) Amnesia in man following transection of the fornix. A review. Brain 114:2611–2618
64. Glees P, Cole J, Whitty CWM, Cairns H (1950) The effects of lesions in the cingular gyrus and adjacent areas in monkeys. J Neurol Neurosurg Psychiatry 13:178–190
65. Gloor P, Salanova V, Olivier A, Quesney LF (1993) The human dorsal hippocampal commissure. An anatomically identifiable and functional pathway. Brain 116:1249–1273
66. Goldman-Rakic PS, Selemon LD, Schwartz ML (1984) Dual pathways connecting the dorsolateral prefrontal cortex with the hippocampal formation and parahippocampal cortex in the rhesus monkey. Neuroscience 12:719–743
67. Grasby PM, Frith CD, Friston KJ et al (1993) Functional mapping of brain areas implicated in auditory–verbal memory function. Brain 116:1–20
68. Gray PA (1924) The cortical lamination pattern of the opossum, *Didelphis virginiana*. J Comp Neurol 37:221–263
69. Groenewegen HJ, Room P, Witter MP, Lohman AHM (1982) Cortical afferents of the nucleus accumbens in the cat, studied with anterograde and retrograde transport techniques. Neuroscience 7: 977–995
70. Habets AMMC, Lopes da Silva FH, De Quartel FW (1980) Autoradiography of the olfactory-hippocampal pathway in the cat with special reference to the perforant path. Exp Brain Res 38:257–265
71. Haglund L, Swanson LW, Köhler C (1984) The projection of the supramammillary nucleus to the hippocampal formation: an immunohistochemical and anterograde transport study with the lectin PHA-L in the rat. J Comp Neurol 229:171–185
72. Hamlyn LH (1962) The fine structure of the mossy fibre endings in the hippocampus of the rabbit. J Anat 96:112–120
73. Harris E, Witter MP, Weinstein G, Stewart M (2001) Intrinsic connectivity of the rat subiculum: I. Dendritic morphology and patterns of axonal arborization by pyramidal neurons. J Comp Neurol 435:490–505
74. Hassiotis M, Paxinos G, Ashwell KWS (2004) Cyto- and chemoarchitecture of the cerebral cortex of the Australian echidna (*Tachyglossus aculeatus*). I. Areal organization. J Comp Neurol 475:493–517
75. Hassler R, Riechert T (1957) Ueber einen Fall von doppelseitiger Forniocotomie bei sogenannter temporaler Epilepsie. Acta Neurochir (Wien) 5:330–340
76. Hatanaka N, Tokuno H, Hamada I et al (2003) Thalamocortical and intracortical connections of monkey cingulate motor areas. J Comp Neurol 462: 121–138
77. Hayakawa T, Zyo K (1985) Afferent connections of Gudden's tegmental nuclei in the rabbit. J Comp Neurol 235:169–181
78. Heimer L (1968) Synaptic distribution of centripetal and centrifugal nerve fibers in the olfactory system of the rat. An experimental anatomical study. J Anat 103:413–432

79. Herkenham M (1978) The connections of the nucleus reuniens thalami: evidence for a direct thalamo-hippocampal pathway in the rat. J Comp Neurol 177:589–610
80. Hines M (1922) Studies in the growth and differentiation of the telencephalon in man. The Fissura hippocampi. J Comp Neurol 34:79–171
81. Hines M (1929) The brain of Ornithorhynchus anatinus (Monotremata). Phil Trans R Soc Lond B 217:155–187
82. Hirose S, Ino T, Takada M et al (1992) Topographic projections from the subiculum to the limbic regions of the medial frontal cortex in the cat. Neurosci Lett 139:61–64
83. Hjorth-Simonsen A (1971) Hippocampal efferents to the ipsilateral entorhinal area: an experimental study in the rat. J Comp Neurol 142:417–438
84. Hjorth-Simonsen A (1972) Projection of the lateral part of the entorhinal area to the hippocampus and fascia dentata. J Comp Neurol 146:219–232
85. Humphrey T (1965) The development of the human hippocampal formation correlated with some aspects of its phylogenetic history. In: Hassler R, Stephan H (eds) Evolution of the forebrain. Thieme, Stuttgart, pp 104–116
86. Hyman BT, Van Hoesen GW, Kromer LJ, Damasio AR (1986) Perforant pathway changes and the memory impairment of Alzheimer's disease. Ann Neurol 20:472–481
87. Insausti R, Amaral DG (2004) Hippocampal formation. In: Paxinos G, Mai JK (eds) The human nervous system, 2nd edn. Elsevier, Amsterdam, pp 871–914
88. Insausti R, Munoz M (2001) Cortical projections of the non-entorhinal hippocampal formation in the cynomolgus monkey (Macaca fascicularis). Eur J Neurosci 14:435–451
89. Insausti R, Amaral DG, Cowan WM (1987) The entorhinal cortex of the monkey: II. Cortical afferents. J Comp Neurol 264:356–395
90. Insausti R, Amaral DG, Cowan WM (1987) The entorhinal cortex of the monkey: III. Subcortical afferents. J Comp Neurol 264:396–408
91. Insausti R, Tunon T, Sobreviela T, Insausti AM, Gonzalo LM (1995) The human entorhinal cortex: a cytoarchitectonic analysis. J Comp Neurol 355:171–198
92. Irle E, Markowitsch HJ (1982) Widespread cortical projections of the hippocampal formation in the cat. Neuroscience 7:2637–2647
93. Ishizuka N, Weber J, Amaral DG (1990) Organization of intrahippocampal projections originating from CA3 pyramidal cells in the rat. J Comp Neurol 295:580–623
94. Jakab RL, Leranth C (1995) Septum. In: Paxinos G (ed) The rat nervous system. Academic, San Diego, pp 405–422
95. Johnston D, Amaral DG (2004) Hippocampus. In: Shepherd GM (ed) The synaptic organization of the brain, 5th edn. Oxford University Press, Oxford, pp 455–498
96. Johnston JB (1913) The morphology of the septum, hippocampus, and pallial commissures in reptiles and mammals. J Comp Neurol 23:371–478
97. Jones EG, Powell TPS (1970) An anatomical study of converging sensory pathways within the cerebral cortex of the monkey. Brain 93:793–824
98. Jongen-Rêlo AL, Pitkänen A, Amaral DG (1999) Distribution of GABAergic cells and fibers in the hippocampal formation of the macaque monkey: an immunohistochemical and in situ hybridization study. J Comp Neurol 408:237–271
99. Kemper TL (1976) The organization and connections of the human septum and septal area. Anat Rec 184:444
100. Kievit J, Kuypers HGJM (1975) Basal forebrain and hypothalamic connections to frontal and parietal cortex in the rhesus monkey. Science 187:660–662
101. Knowles WD, Schwartzkroin PA (1981) Axonal ramifications of hippocampal Ca1 pyramidal cells. J Neurosci 1:1236–1241
102. Kobayashi Y, Amaral DG (2003) Macaque monkey retrosplenial cortex: II. Cortical afferents. J Comp Neurol 466:48–79
103. Köhler C (1985) Intrinsic projections of the retrohippocampal region in the rat brain. I. The subicular complex. J Comp Neurol 236:504–522
104. Köhler C, Shipley MT, Srebro B, Harkmark W (1978) Some retrohippocampal afferents to the entorhinal cortex. Cells of origin as studied by the HRP method in the rat and mouse. Neurosci Lett 10:115–120
105. Köhler C, Chan-Palay V, Wu JY (1984) Septal neurons containing glutamic acid decarboxylase immunoreactivity project to the hippocampal region in the rat brain. Anat Embryol (Berl) 169:41–44
106. Kopelman MD (1995) The Korsakoff syndrome. Br J Psychiatry 166:154–173
107. Kosel KC, Van Hoesen GW, Rosene DL (1982) Nonhippocampal cortical projections from the entorhinal cortex in the rat and rhesus monkey. Brain Res 244:201–213
108. Krayniak PF, Siegel A, Meibach RC, Fruchtman D, Scrimenti M (1979) Origin of the fornix system in the squirrel monkey. Brain Res 160:401–411
109. Krettek JE, Price JL (1974) Projections from the amygdala to the perirhinal and entorhinal cortices and the subiculum. Brain Res 71:150–154
110. Krettek JE, Price JL (1977) Projections from the amygdaloid complex to the cerebral cortex and thalamus in the rat and cat. J Comp Neurol 172:687–722
111. Krettek JE, Price JL (1977) Projections from the amygdaloid complex and adjacent olfactory structures to the entorhinal cortex and to the subiculum in the rat and cat. J Comp Neurol 172:723–752

112. Krimer LS, Hyde TM, Herman MM, Saunders RC (1997) The entorhinal cortex: an examination of cyto- and myeloarchitectonic organization in humans. Cereb Cortex 7:722–731
113. Kunkel DD, Lacaille JC, Schwartzkroin PA (1988) Ultrastructure of stratum lacunosum-moleculare interneurons of hippocampal CA1 region. Synapse 2:382–394
114. Künzle H (1978) An autoradiographic analysis of the efferent connections from premotor and adjacent prefrontal regions (areas 6 and 9) in Macaca fascicularis. Brain Behav Evol 15:185–234
115. L'vovich AI (2001) Descending pathways of the frontal lobe cortex to nuclei of the hypothalamic mamillary bodies in craniocerebral trauma in humans. Neurosci Behav Physiol 31:371–374
116. Lacaille JC, Schwartzkroin PA (1988) Stratum lacunosum-moleculare interneurons of hippocampal CA1 region. I. Intracellular response characteristics, synaptic responses, and morphology. J Neurosci 8:1400–1410
117. Lacaille JC, Schwartzkroin PA (1988) Stratum lacunosum-moleculare interneurons of hippocampal CA1 region. II. Intrasomatic and intradendritic recordings of local circuit synaptic interactions. J Neurosci 8:1411–1424
118. Lacaille JC, Mueller AL, Kunkel DD, Schwartzkroin PA (1987) Local circuit interactions between oriens/alveus interneurons and CA1 pyramidal cells in hippocampal slices: electrophysiology and morphology. J Neurosci 7:1979–1993
119. Lacaille JC, Kunkel DD, Schwartzkroin PA (1989) Electrophysiological and morphological characterization of hippocampal interneurons. In: Chan-Palay V, Köhler C (eds) The hippocampus: new vistas. Liss, New York, pp 287–305 (Neurology and neurobiology, vol 52)
120. Lammers HJ (1972) The neural connections of the amygdaloid complex in mammals. In: Eleftheriou BE (ed) The neurobiology of the amygdala. Plenum, New York, pp 123–144
121. Lavenex P, Suzuki WA, Amaral DG (2002) Perirhinal and parahippocampal cortices of the macaque monkey: projections to the neocortex. J Comp Neurol 447:394–420
122. Leichnetz GR (2001) Connections of the medial posterior parietal cortex (area 7m) in the monkey. Anat Rec 263:215–236
123. Lim C, Blume HW, Madsen JR, Saper CB (1997) Connections of the hippocampal formation in humans: I. The mossy fiber pathway. J Comp Neurol 385:325–351
124. Lim C, Mufson EJ, Kordower JH et al (1997) Connections of the hippocampal formation in humans: II. The endfolial fiber pathway. J Comp Neurol 385:352–371
125. Loo YT (1930) The forebrain of the opossum, *Didelphis virginiana*. I. Gross anatomy. J Comp Neurol 51:1–64
126. Loo YT (1931) The forebrain of the opossum, *Didelphis virginiana*. J Comp Neurol 52:1–48
127. Lopes da Silva FH, Groenewegen HJ, Holsheimer J et al (1985) The hippocampus as a set of partially overlapping segments with a topographically organized system of inputs and outputs: the entorhinal cortex as a sensory gate, the medial septum as a gain-setting system and the ventral striatum as a motor interface. In: Buzsaki G, VanderWolf CH (eds) Electrical activity of archicortex. Academiai Kiado, Budapest, pp 83–106
128. Lopes da Silva FH, Witter MP, Boeijinga PH, Lohman AH (1990) Anatomic organization and physiology of the limbic cortex. Physiol Rev 70: 453–511
129. Lorente de Nó R (1934) Studies on the structure of the cerebral cortex. I. The area entorhinalis. J Psychol Neurol 45:381–439
130. Lorente de Nó R (1934) Studies on the structure of the cerebral cortex. II. Continuation of the study of the ammonic system. J Psychol Neurol 46:113–177
131. Mabuchi M (1967) Corticofugal projections to the subthalamic nucleus, the red nucleus and the adjacent areas in the monkey. Proc Jpn Acad 43:818
132. Maccaferri G, Lacaille JC (2003) Interneuron diversity series: hippocampal interneuron classifications – making things as simple as possible, not simpler. Trends Neurosci 26:564–571
133. Maccaferri G, McBain CJ (1995) Passive propagation of LTD to stratum oriens-alveus inhibitory neurons modulates the temporoammonic input to the hippocampal CA1 region. Neuron 15:137–145
134. Maccaferri G, Roberts JD, Szucs P, Cottingham CA, Somogyi P (2000) Cell surface domain specific postsynaptic currents evoked by identified GABAergic neurones in rat hippocampus in vitro. J Physiol 524:91–116
135. MacVicar BA, Dudek FE (1980) Local synaptic circuits in rat hippocampus: interactions between pyramidal cells. Brain Res 184:220–223
136. Maddock RJ, Garrett AS, Buonocore MH (2001) Remembering familiar people: the posterior cingulate cortex and autobiographical memory retrieval. Neuroscience 104:667–676
137. Mair WG, Warrington EK, Weiskrantz L (1979) Memory disorder in Korsakoff's psychosis: a neuropathological and neuropsychological investigation of two cases. Brain 102:749–783
138. McBain CJ, Fisahn A (2001) Interneurons unbound. Nat Rev Neurosci 2:11–23
139. McBain CJ, DiChiara TJ, Kauer JA (1994) Activation of metabotropic glutamate receptors differentially affects two classes of hippocampal interneurons and potentiates excitatory synaptic transmission. J Neurosci 14:4433–4445
140. Meibach RG, Siegel A (1975) The origin of fornix fibers which project to the mammillary bodies in the rat: a horseradish peroxidase study. Brain Res 88:508–512

141. Meibach RC, Siegel A (1977) Efferent connections of the septal area in the rat: an analysis utilizing retrograde and anterograde transport methods. Brain Res 119:1–20
142. Meibach RC, Siegel A (1977) Efferent connections of the hippocampal formation of the rat. Brain Res 124:197–224
143. Meibach RC, Siegel A (1977) Thalamic projections of the hippocampal formation: evidence for an alternate pathway involving the internal capsule. Brain Res 134:1–12
144. Meibach RC, Siegel A (1977) Subicular projections to the posterior cingulate cortex in rats. Exp Neurol 57:264–274
145. Mesulam MM, Mufson EJ, Wainer BH, Levey AI (1983) Central cholinergic pathways in the rat: an overview based on an alternative nomenclature . Neuroscience 10:1185–1201
146. Mesulam MM, Mufson EJ, Levey AI, Wainer BH (1983) Cholinergic innervation of cortex by the basal forebrain: Cytochemistry and cortical connections of the septal area, diagonal band nuclei, nucleus basalis (substantia innominata), and hypothalamus in the rhesus monkey. J Comp Neurol 214:170–197
147. Mesulam MM, Mufson EJ, Levey AI, Wainer BH (1984) Atlas of cholinergic neurons in the forebrain and upper brainstem of the macaque based on monoclonal choline acetyltransferase immunohistochemistry and acetylcholinesterase histochemistry. Neuroscience 12:669–686
148. Meyer G, MacElhaney M, Martin W, MacGraw CP (1973) Stereotaxic cingulotomy with results of acute stimulation and serial psychological testing. In: Laitinen LV, Livingston KE (eds) Surgical approaches in psychiatry. Medical and Technical Publications, Lancaster, pp 39–58
149. Mitchell SJ, Rawlins JNP, Steward O, Olton DS (1982) Medial septal area lesions disrupt theta rhythm and cholinergic staining in medial entorhinal cortex and produce impaired radial arm maze behavior in rats. J Neurosci 2:292–302
150. Mogenson GJ (1984) Limbic-motor integration – with emphasis on initiation of exploratory and goal-directed locomotion. In: Bandler R (ed) Modulation of sensorimotor activity during alterations in behavioral states. Liss, New York, pp 121–137
151. Mogenson GJ, Swanson LW, Wu M (1984) Evidence that projections from substantia innominata at zona incerta mesencephalic locomotor region contribute to locomotor activity. Brain Res 334:65–76
152. Morest DK (1961) Connexions of the dorsal tegmental nucleus in rat and rabbit. J Anat 95:229–246
153. Nauta WJH (1964) Some efferent connections of the prefrontal region in the monkey. In: Warren JM, Akert K (eds) The frontal granular cortex and behavior. McGraw-Hill, New York, pp 397–409
154. Nauta WJH, Haymaker W (1969) Hypothalamic nuclei and fiber connections. In: Haymaker W, Anderson E, Nauta WJH (eds) The hypothalamus. Thomas, Springfield, pp 136–209
155. Nauta WJH, Kuypers HGJM (1958) Some ascending pathways in the brain stem reticular formation. In: Jasper HH et al (eds) Reticular formation of the brain. Little Brown, Toronto, pp 3–31
156. Nieuwenhuys R (1998) Telencephalon. In: Nieuwenhuys R, Ten Donkelaar HJ, Nicholson C (eds) The central nervous system of vertebrates, vol III. Springer, Berlin Heidelberg New York, pp 1871–2008
157. Nitsch R, Leranth C (1993) Calretinin immunoreactivity in the monkey hippocampal formation. II. Intrinsic GABAergic and hypothalamic non-GABAergic systems: an experimental tracing and co-existence study. Neuroscience 55:797–812
158. O'Rahilly R, Müller F (1999) The embryonic human brain. An atlas of developmental stages, 2nd edn. Wiley, New York
159. Pandya DN, Seltzer B (1982) Association areas of the cerebral cortex. Trends Neurosci 5:386–390
160. Pandya DN, Yeterian EH (1985) Architecture and connections of cortical association areas. In: Peters A, Jones EG (eds) Cerebral cortex, vol 4. Association and auditory cortices. Plenum, New York, pp 3–61
161. Pandya DN, Yeterian EH (1990) Prefrontal cortex in relation to other cortical areas in rhesus monkey: architecture and connections. Prog Brain Res 85:63–94
162. Pandya DN, Van Hoesen GW, Domesick VB (1973) A cingulo-amygdaloid projection in the rhesus monkey. Brain Res 61:369–373
163. Pandya DN, Van Hoesen GW, Mesulam M-M (1981) Efferent connections of the cingulate gyrus in the rhesus monkey. Exp Brain Res 42:319–330
164. Panula P, Revuelta AV, Cheney PL, Wu J-Y, Costa E (1984) An immunohistochemical study on the location of GABAergic neurons in rat septum. J Comp Neurol 222:69–80
165. Papez JW (1937) A proposed mechanism of emotion. Arch Neurol Psychiatry 38:725–743
166. Parra P, Gulyás AI, Miles R (1998) How many subtypes of inhibitory cells in the hippocampus? Neuron 20:983–993
167. Pasquier DA, Reinoso-Suarez F (1978) The topographic organization of hypothalamic and brain stem projections to the hippocampus. Brain Res Bull 3:373–389
168. Pitkänen A, Kelly JL, Amaral DG (2002) Projections from the lateral, basal, and accessory basal nuclei of the amygdala to the entorhinal cortex in the macaque monkey. Hippocampus 12:186–205
169. Pritchard TC, Hamilton RB, Morse JR, Morgren R (1986) Projections of thalamic gustatory and lingual areas in the monkey, *Macaca fascicularis*. J Comp Neurol 244:213–228

170. Raisman G (1966) An experimental analysis of the efferent projection of thc hippocampus. Brain 88:963–996
171. Rempel-Clower NL, Zola SM, Squire LR, Amaral DG (1996) Three cases of enduring memory impairment after bilateral damage limited to the hippocampal formation. J Neurosci 16:5233–5255
172. Ribak CE, Seress L (1983) Five types of basket cell in the hippocampal dentate gyrus: a combined Golgi and electron microscopic study. J Neurocytol 12:577–597
173. Ricardo JA (1983) Hypothalamic pathways involved in metabolic regulatory functions, as identified by tracktracing methods. Adv Metab Dis 10:1–30
174. Riley JN, Moore RY (1981) Diencephalic and brainstem afferents to the hippocampal formation of the rat. Brain Res Bull 6:437–444
175. Risold PY, Swanson LW (1997) Connections of the rat lateral septal complex. Brain Res Brain Res Rev 24:115–195
176. Robertson RT, Kaitz SS (1981) Thalamic connections with limbic cortex. I. Thalamocortical projections. J Comp Neurol 195:501–525
177. Room P, Groenewegen HJ (1986) Connections of the parahippocampal cortex. I. Cortical afferents. J Comp Neurol 251:415–450
178. Room P, Groenewegen HJ (1986) Connections of the parahippocampal cortex in the cat. II. Subcortical afferents. J Comp Neurol 251:451–473
179. Room P, Groenewegen HJ, Lohman AHM (1984) Inputs from the olfactory bulb and olfactory cortex to the entorhinal cortex in the cat. Exp Brain Res 56:488–496
180. Room P, Russchen FT, Groenewegen HJ, Lohman AHM (1985) Efferent connections of the prelimbic (area 32) and the infralimbic (area 25) cortices: an anterograde tracing study in the cat. J Comp Neurol 242:40–55
181. Rose M (1926) Der Allocortex bei Tier und Mensch. J Psychol Neurol 34:1–99
182. Rose M (1935) Cytoarchitektonik und Myeloarchitektonik der Grosshirnrinde. In: Bumke O, Foerster O (eds) Allgemeine Neurologie I, Anatomie. Springer, Berlin, pp 588–778 (Handbuch der Neurologie, vol I)
183. Rosene DL, Van Hoesen GW (1977) Hippocampal efferents reach widespread areas of the cerebral cortex and amygdala in the rhesus monkey. Science 198:315–317
184. Russell SH, Small CJ, Dakin CL et al (2001) The central effects of orexin-A in the hypothalamic-pituitary-adrenal axis in vivo and in vitro in male rats. J Neuroendocrinol 13:561–566
185. Santin LJ, Rubio S, Begega A, Arias JL (1999) Effects of mammillary body lesions on spatial reference and working memory tasks. Behav Brain Res 102:137–150
186. Santin LJ, Aguirre JA, Rubio S et al (2003) c-Fos expression in supramammillary and medial mammillary nuclei following spatial reference and working memory tasks. Physiol Behav 78:733–739
187. Saper CB, Swanson LW, Cowan WM (1979) An autoradiographic study of the efferent connections of the lateral hypothalamic area in the rat. J Comp Neurol 183:689–706
188. Saunders RC, Rosene DL (1988) A comparison of the efferents of the amygdala and the hippocampal formation in the rhesus monkey: I. Convergence in the entorhinal, prorhinal, and perirhinal cortices. J Comp Neurol 271:153–184
189. Saunders RC, Rosene DL, Van Hoesen GW (1988) Comparison of the efferents of the amygdala and the hippocampal formation in the rhesus monkey: II. Reciprocal and non-reciprocal connections. J Comp Neurol 271:185–207
190. Scalia F (1966) Some olfactory pathways in the rabbit brain. J Comp Neurol 126:285–310
191. Schwartzkroin PA (1986) Regulation of excitability in hippocampal neurons. In: Isaacson RL, Pribram KH (eds) The hippocampus. Plenum, New York, pp 113–136
192. Schwartzkroin PA, Kunkel DD (1985) Morphology of identified interneurons in the CA1 regions of guinea pig hippocampus. J Comp Neurol 232:205–218
193. Schwartzkroin PA, Mathers LH (1978) Physiological and morphological identification of a nonpyramidal hippocampal cell type. Brain Res 157:1–10
194. Schwartzkroin PA, Scharfman HE, Sloviter RS (1990) Similarities in circuitry between Ammon's horn and dentate gyrus: local interactions and parallel processing. Progr Brain Res 83/86:269–286
195. Scoville WB, Milner B (1957) Loss of recent memory after bilateral hippocampal lesions. J Neurol Neurosurg Psychiat 20:11–21
196. Seltzer B, Van Hoesen GW (1979) A direct inferior parietal lobule projection to the presubiculum in the rhesus monkey. Brain Res 179:157–161
197. Seress L, Mrzljak L (1987) Basal dendrites of granule cells are normal features of the fetal and adult dentate gyrus of both monkey and human hippocampal formations. Brain Res 405:169–174
198. Sewards TV, Sewards MA (2002) The medial pain system. Brain Res Bull 59:163–180
199. Shibata H (1989) Descending projections to the mammillary nuclei in the rat, as studied by retrograde and anterograde transport of wheat germ agglutinin-horseradish peroxidase. J Comp Neurol 285:436–452
200. Shipley MT (1974) Presubiculum afferents to the entorhinal area and the Papez circuit. Brain Res 67:162–168
201. Shipley MT (1975) The topographical and laminar organization of the presubiculum's projection to the ipsi- and contralateral entorhinal cortex in the guinea pig. J Comp Neurol 160:127–145
202. Shipley MT, Sörensen KE (1975) On the laminar organization of the anterior thalamus projections to the presubiculum in the guinea pig. Brain Res 86:473–477

203. Somogyi P, Nunzi MG, Gorio A, Smith AD (1983) A new type of specific interneuron in the monkey hippocampus forming synapses exclusively with the axon initial segments of pyramidal cells. Brain Res 259:137–142
204. Somogyi P, Freund TF, Hodgson AJ et al (1985) Identified axo-axonic cells are immunoreactive for GABA in the hippocampus and visual cortex of the cat. Brain Res 332:143–149
205. Sørensen KE (1985) Projections of the entorhinal area to the striatum, nucleus accumbens, and cerebral cortex in the guinea pig. J Comp Neurol 238:308–322
206. Sørensen KE, Shipley MT (1979) Projections from the subiculum to the deep layers of the ipsilateral presubicular and entorhinal cortices in the guinea pig. J Comp Neurol 188:313–334
207. Sørensen KE, Witter MP (1983) Entorhinal efferents reach the caudato-putamen. Neurosci Lett 35:259–264
208. Soriano E, Frotscher M (1993) Spiny nonpyramidal neurons in the CA3 region of the rat hippocampus are glutamate-like immunoreactive and receive convergent mossy fiber input. J Comp Neurol 333:435–448
209. Squire LR (1992) Memory and the hippocampus: a synthesis from findings with rats, monkeys, and humans. Psychol Rev 99:195–231
210. Squire LR, Zola-Morgan S (1991) The medial temporal lobe memory system. Science 253:1380–1386
211. Squire LR, Stark CE, Clark RE (2004) The medial temporal lobe. Annu Rev Neurosci 27:279–306
212. Stackman RW, Taube JS (1998) Firing properties of rat lateral mammillary single units: head direction, head pitch, and angular head velocity. J Neurosci 18:9020–9037
213. Stengaard-Pedersen K, Fredens K, Larsson LI (1981) Enkephalin and zinc in the hippocampal mossy fiber system. Brain Res 212:230–233
214. Stephan H (1975) Allocortex. Springer, Berlin Heidelberg New York (Handbuch der mikroskopischen Anatomie des Menschen, Band 4, Teil 9)
215. Steriade M, Jones EG, McCormick DA (1997) Organisation and function. Elsevier, Amsterdam (Thalamus, vol I)
216. Storm-Mathisen J, Opsahl MW (1978) Aspartate and/or glutamate may be transmitters in hippocampal efferents to septum and hypothalamus. Neurosci Lett 9:65–70
217. Suzuki WA (1996) The anatomy, physiology and functions of the perirhinal cortex. Curr Opin Neurobiol 6:179–186
218. Suzuki WA, Amaral DG (1994) Topographic organization of the reciprocal connections between the monkey entorhinal cortex and the perirhinal and parahippocampal cortices. J Neurosci 14: 1856–1877
219. Suzuki WA, Amaral DG (2003) Where are the perirhinal and parahippocampal cortices? A historical overview of the nomenclature and boundaries applied to the primate medial temporal lobe. Neuroscience 120:893–906
220. Suzuki WA, Amaral DG (2003) Perirhinal and parahippocampal cortices of the macaque monkey: cytoarchitectonic and chemoarchitectonic organization. J Comp Neurol 463:67–91
221. Swanson LW (1979) The hippocampus – new anatomical insights. Trends Neurosci 2:9–12
222. Swanson LW (1982) Normal hippocampal circuitry. Neurosci Res Prog Bull 20:624–637
223. Swanson LW (2000) Cerebral hemisphere regulation of motivated behavior. Brain Res 886:113–164
224. Swanson LW, Cowan WM (1975) Hippocampo-hypothalamic connections: origin in subicular cortex, not Ammon's horn. Science 189:303–304
225. Swanson LW, Cowan WM (1977) An autoradiographic study of the organization of the efferent connections of the hippocampal formation in the rat. J Comp Neurol 172:49–84
226. Swanson LW, Cowan WM (1979) The connections of the septal region in the rat. J Comp Neurol 186:621–655
227. Swanson LW, Köhler C, Björklund A (1987) The limbic region. I. The septohippocampal system. In: Björklund A, Hökfelt T, Swanson LW (eds) Integrated systems of the CNS, part I. Elsevier, Amsterdam, pp 125–277 (Handbook of chemical neuroanatomy, vol 5)
228. Sweet WH, Talland GA, Ervin FR (1959) Loss of recent memory following section of fornix. Trans Am Neurol Assoc 84:76–82
229. Tamamaki N, Nojyo Y (1995) Preservation of topography in the connections between the subiculum, field CA1, and the entorhinal cortex in rats. J Comp Neurol 353:379–390
230. Tamamaki N, Abe K, Nojyo Y (1987) Columnar organization in the subiculum formed by axon branches originating from single CA1 pyramidal neurons in the rat hippocampus. Brain Res 412:156–160
231. Tanaka Y, Miyazawa Y, Akaoka F, Yamada T (1997) Amnesia following damage to the mammillary bodies. Neurology 48:160–165
232. Tanji J, Shima K, Matsuzaka Y (2002) Reward-based planning of motor selection in the rostral cingulate motor area. Adv Exp Med Biol 508:417–423
233. Ter Horst GJ, Luiten PGM (1986) The projections of the dorsomedial hypothalamic nucleus in the rat. Brain Res Bull 16:231–248
234. Tilney F (1939) The hippocampus and its relations to the corpus callosum. J Nerv Ment Dis 89:433–513
235. Turner BH, Mishkin M, Knapp M (1980) Organization of the amygdalopetal projections from modality-specific cortical association areas in the monkey. J Comp Neurol 191:515–543
236. Umbriaco D, Garcia S, Beaulieu C, Descarries L (1995) Relational features of acetylcholine, noradrenaline, serotonin and GABA axon terminals in the stratum radiatum of adult rat hippocampus (CA1). Hippocampus 5:605–620

237. Van Daal JH, Zanderink HE, Jenks BG, Van Abeelen JH (1989) Distribution of dynorphin B and methionine-enkephalin in the mouse hippocampus: influence of genotype. Neurosci Lett 97:241–244
238. Van Groen T, Lopes da Silva FH (1985) Septotemporal distribution of entorhinal projections to the hippocampus in the cat: electrophysiological evidence. J Comp Neurol 238:1–10
239. Van Groen T, Lopes da Silva FH (1986) The organization of the reciprocal connections between the subiculum and the entorhinal cortex in the cat. II. An electrophysiological study. J Comp Neurol 251:111–120
240. Van Groen T, Van Haren FJ, Witter MP, Groenewegen HJ (1986) The organization of the reciprocal connections between the subiculum and the entorhinal cortex in the cat. I. A neuroanatomical tracing study. J Comp Neurol 250:485–497
241. Van Hoesen GW (1982) The parahippocampal gyrus. New observations regarding its cortical connections in the monkey. Trends Neurosci 5:345–350
242. Van Hoesen GW, Hyman BT (1990) Hippocampal formation: anatomy and the patterns of pathology in Alzheimer's disease. Prog Brain Res 83:445–457
243. Van Hoesen GW, Pandya DN (1975) Some connections of the entorhinal area (area 28) and perirhinal area (area 35) cortices of the rhesus monkey. I. Temporal lobe afferents. Brain Res 95:1–24
244. Van Hoesen GW, Pandya DN, Butters M (1975) Some connections of the entorhinal area (area 28) and perirhinal area (area 35) cortices of the rhesus monkey. II. Frontal afferents. Brain Res 95:25–38
245. Van Hoesen GW, Rosene DL, Mesulam M-M (1979) Subicular input from temporal cortex in the rhesus monkey. Science 205:608–610
246. Van Hoesen GW, Morecraft RJ, Vogt BA (1993) Connections of the monkey cingulate cortex. In: Vogt BA, Gabriel M (eds) Neurobiology of cingulate cortex and limbic thalamus. Birkhäuser, Boston, pp 249–284
247. Van Hoesen GW, Augustinack JC, Dierking J, Redman SJ, Thangavel R (2000) The parahippocampal gyrus in Alzheimer's disease. Clinical and preclinical neuroanatomical correlates. Ann NY Acad Sci 911:254–274
248. Vann SD, Aggleton JP (2003) Evidence of a spatial encoding deficit in rats with lesions of the mammillary bodies or mammillothalamic tract. J Neurosci 23:3506–3514
249. Victor M, Adams RD, Collins GH (1971) The Wernicke-Korsakoff syndrome. A clinical and pathological study of 245 patients, 82 with postmortem examinations. Contemp Neurol Ser 7:1–206
250. Victor M, Adams RD, Collins GH (1989) The Wernicke-Korsakoff syndrome. Davis, Philadelphia
251. Vincent SR, McGeer EG (1982) A substance-P projection to the hippocampus. Brain Res 215:349–351
252. Vogt BA (1993) Structural organization of cingulate cortex: areas, neurons, and somatodendritic transmitter receptors. In: Vogt BA, Gabriel M (eds) Neurobiology of cingulate cortex and limbic thalamus. Birkhäuser, Boston, pp 19–70
253. Vogt BA, Sikes RW (2000) The medial pain system, cingulate cortex, and parallel processing of nociceptive information. Progr Brain Res 122: 223–235
254. Vogt BA, Pandya DN, Rosene DL (1987) Cingulate cortex of the rhesus monkey. I. Cytoarchitecture and thalamic afferents. J Comp Neurol 262:256–270
255. Vogt BA, Vogt LJ, Perl DP, Hof PR (2001) Cytology of human caudomedial cingulate, retrosplenial, and caudal parahippocampal cortices. J Comp Neurol 438:353–376
256. Vogt BA, Hof PR, Vogt LJ (2004) Cingulate gyrus. In: Paxinos G (ed) The human nervous system, 2nd edn. Elsevier, Amsterdam, pp 915–949
257. Von Bonin G, Bailey P (1947) The neocortex of Macaca mulatta. University of Illinois Press, Urbana
258. Von Economo C, Koskinas GN (1925) Die Cytoarchitektonik der Hirnrinde des erwachsenen Menschen. Springer, Berlin
259. Wainer BH, Levey AI, Rye DB, Mesulam MM, Mufson EJ (1985) Cholinergic and non-cholinergic septohippocampal pathways. Neurosci Lett 54:45–52
260. Walaas I (1983) The hippocampus. In: Emson PC (ed) Chemical neuroanatomy. Raven, New York, pp 337–358
261. Wilson DH, Chang AE (1974) Bilateral anterior cingulectomy for the relief of intractable pain. Report of 23 patients. Confin Neurol 36:61–68
262. Wiltgen BJ, Brown RAM, Talton LE, Silva AJ (2004) New circuits for old memories: the role of the neocortex in consolidation. Neuron 44:101–108
263. Witter MP, Amaral DG (1991) Entorhinal cortex of the monkey: V. Projections to the dentate gyrus, hippocampus, and subicular complex. J Comp Neurol 307:437–459
264. Witter MP, Groenewegen HJ (1986) Connections of the parahippocampal cortex in the cat. III. Cortical and thalamic efferents. J Comp Neurol 252:1–31
265. Witter MP, Room P, Groenewegen HJ, Lohman AHM (1986) Connections of the parahippocampal cortex in the cat. V. Intrinsic connections; Comments on input/output connections with the hippocampus. J Comp Neurol 252:78–94
266. Witter MP, Groenewegen HJ, Lopes da Silva FH, Lohman AH (1989) Functional organization of the extrinsic and intrinsic circuitry of the parahippocampal region. Progr Neurobiol 33:161–253

267. Woolsey RM, Nelson JS (1975) Asymptomatic destruction of the fornix in man. Arch Neurol 32:566–568
268. Wouterlood FG, Van Haeften T, Eijkhoudt M et al (2004) Input from the presubiculum to dendrites of layer-V neurons of the medial entorhinal cortex of the rat. Brain Res 1013:1–12
269. Wyss JM, Swanson LW, Cowan WM (1979) A study of subcortical afferents to the hippocampal formation in the rat. Neuroscience 4:463–476
270. Xiang JZ, Brown MW (2004) Neuronal responses related to long-term recognition memory processes in prefrontal cortex. Neuron 42:817–829
271. Yanagihara M, Ono K, Niimi K (1985) Thalamic projections to the hippocampal formation in the cat. Neurosci Lett 61:31–35
272. Yang CR, Mogenson GJ (1985) An electrophysiological study of the neural projections from the hippocampus to the ventral pallidum and the subpallidal areas by way of the nucleus accumbens. Neuroscience 15:1015–1024
273. Yukie M (2000) Connections between the medial temporal cortex and the CA1 subfield of the hippocampal formation in the Japanese monkey (*Macaca fuscata*). J Comp Neurol 423:282–298
274. Zaczek R, Hedreen JC, Coyle JT (1979) Evidence for a hippocampal-septal glutamatergic pathway in the rat. Exp Neurol 65:145–156
275. Zola-Morgan S, Squire LR, Amaral DG (1986) Human amnesia and the medial temporal region: enduring memory impairment following a bilateral lesion limited to field CA1 of the hippocampus. J Neurosci 6:2950–2967
276. Zola-Morgan S, Squire LR, Amaral DG (1989) Lesions of the hippocampal formation but not lesions of the fornix or the mammillary nuclei produce long-lasting memory impairment in monkeys. J Neurosci 9:898–913

13 Telencephalon: Amygdala and Claustrum

Amygdala

Topography, Subdivision and Structure

The *amygdala,* or *corpus amygdaloideum,* is a large nuclear complex situated in the dorsomedial portion of the temporal lobe, where it forms part of the rostromedial and rostrodorsal walls of the inferior horn of the lateral ventricle (Figs. 5.6, 5.7, 5.23, 5.24, 6.39–6.41). Its name is derived from the Greek (amygdalon) or Latin (amygdalum) word for almond, to which Burdach [28] compared its shape in the early nineteenth century. The volume of the human amygdala, as determined by MRI studies in normal subjects, is about 1700 mm^3 [22].

The amygdaloid complex develops from the caudal portion of the ventral telencephalic wall. During the formation of the temporal lobe, it is displaced ventrally and rostrally, and within that lobe it rotates dorsomedially, due to the expansion of the neocortex [83, 100].

The cytoarchitectonic organization of the human amygdala has been studied by Brockhaus [23] and Crosby and Humphrey [41]. Brockhaus, being in the German cytoarchitectonic tradition, parcelled the amygdala into more than 30 grisea, but Crosby and Humphrey distinguished a more restricted number of nuclei, among them the superficially situated cortical and medial nuclei and the more deeply located lateral, basal, accessory basal and central nuclei (Fig. 13.1). The amygdaloid nuclei have been grouped in different ways. Crosby et al. [41, 42] united the medial, cortical and central nuclei in a corticomedial group and the basal, accessory basal and lateral nuclei in a basolateral group. In contrast, Stephan and coworkers [202, 203], as a result of extensive quantitative comparative studies, separated the cortical amygdaloid nucleus from the corticomedial group and assigned it to the basolateral group. This subdivision corresponds to the one Heimer and colleagues [8–10, 69] put forward in their "extended amygdala" concept. According to this concept, which is based on the classical comparative neuroanatomical and neuroembryological studies of Johnston [88] and on extensive tract-tracing and immunohistochemical data, the central and medial amygdaloid nuclei, the bed nucleus of the stria terminalis (BST) and certain sectors of the substantia innominata (SI) constitute a functional complex, which is characterized by a highly organized system of pathways to many hypothalamic and brain stem areas (see section The BST and the "Extended Amygdala" Concept).

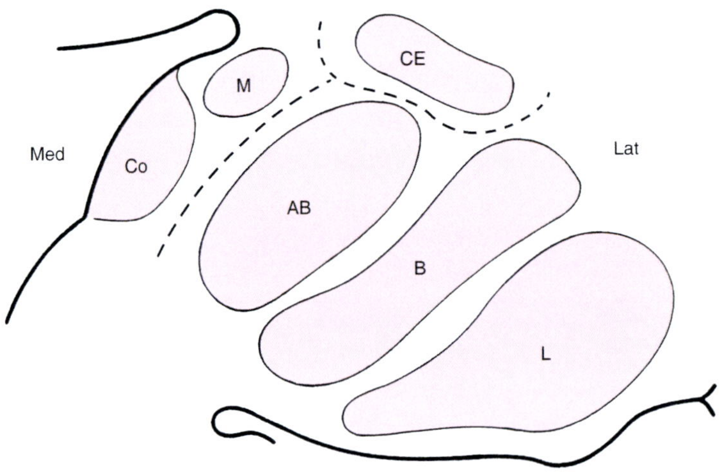

Fig. 13.1. A prototypical transverse section through the human amygdala, showing the position of the various nuclei. The *dashed lines* indicate the boundaries between the three nuclear groups. Dorsomedial group (*Med*): cortical nucleus (*CO*); medial nucleus (*M*). Basolateral group (*Lat*): accessory basal nucleus (*AB*), basal nucleus (*B*) and lateral nucleus (*L*). *CE*, central nucleus

The structural organization of the amygdaloid complex has been discussed in several comprehensive reviews [14, 46, 130, 164], from which the following data are derived.

In the superficial amygdaloid grisea, the neurons tend to form layers. The *cortical nucleus* has three layers: a cell-poor superficial or molecular layer, a middle cell-dense layer and a deep layer consisting of less densely distributed neurons. These three layers can also be recognized in the *medial nucleus* [14, 46].

The neuronal somata in the various nuclei forming the *basolateral complex* vary in size. In the macaque monkey [14] as well as in humans [23], separate magnocellular and parvocellular parts have been delineated within these nuclei. Golgi studies [20, 66, 67, 130] have shown that the neurons in the basolateral nuclei, just like those in the neocortex, can be grouped into two main categories: spiny pyramidal-like neurons and spine-sparse non-pyramidal neurons. The pyramidal-like neurons constitute the predominant cell type in the basolateral nuclei. Many of these elements have three to five "basal-like" stem dendrites of moderate calibre and one "apical-like" dendrite of larger calibre and greater length. The dendrites of the pyramidal-like neurons are, except for their most proximal parts, densely covered with spines. There is evidence indicating that the pyramidal-like elements are projection neurons [130]. The non-pyramidal cells of the basolateral amygdala are characterized by spine-sparse dendrites. These elements are sparsely scattered among the pyramidal-like neurons. Most of them are small and have thin, sparsely branching dendrites, but some attain the size of the pyramidal-like elements. The dense local axonal arborizations exhibited by the non-pyramidal cells suggest that most are local circuit neurons, like their cortical counterparts [130]. Apart from the morphological appearance of their neurons, the basolateral nuclear complex shares several other characteristics with the neocortex. Like the neocortical non-pyramidal cells, amygdaloid non-pyramidal cells contain GABA, calcium-binding proteins and a variety of peptides, including neuropeptide Y, vasoactive intestinal peptide (VIP) and cholecystokinin (CCK) [130]. The basolateral nuclei also maintain strong, direct reciprocal connections with the adjacent temporal neocortex. For these and other reasons as discussed below, the basolateral part of the amygdaloid complex is, in spite of its non-laminar structure, often considered as a neocortex-like structure [30, 207]. It is important to note that the *lateral amygdaloid nucleus*, which is the principal recipient of sensory information from the neocortex, is by far the largest cell mass in the human amygdala.

The *central amygdaloid nucleus* can be divided into medial and lateral sections [14, 46, 164]. Both of these are mainly composed of medium-sized cells, but the elements in the lateral section are more closely packed than those in the medial one. Golgi studies [66, 67, 130] have shown that the principal cell type in the lateral section of the central nucleus closely resembles the medium-sized spiny neurons in the adjacent striatum.

The amygdaloid complex contains a large number of neurotransmitters [14, 46, 164]. The central nucleus is characterized by the presence of an extraordinarily vast array of different neuropeptides.

Development

During development, the basal part of the telencephalic wall thickens and then protrudes into the ventricular cavity. Thus, a longitudinally oriented intraventricular protrusion is formed. The rostral portion of this protrusion gives rise to the striatum (caudate nucleus + putamen + globus pallidus), whereas the amygdala develops from its caudal portion. Because of this common origin, the striatum and the amygdala are classically, collectively designated as the *basal ganglia* or the *strio-amygdaloid complex* [111]. In many mammals, including humans, the basal intraventricular protrusion splits up into separate medial and lateral ganglionic eminences (Figs. 2.19 A, 2.20 A, 11.1 A). The question as to how the developing amygdala relates to these separate eminences has been addressed by several authors. Most of them [75, 82, 83, 111] have come to the conclusion that both of the eminences contribute to the amygdaloid complex, but that the lateral one provides the major source of its cells.

The Swedish investigators Holmgren [77] and Källén [91] thoroughly analysed the histogenesis of the various cell masses in the vertebrate telencephalon. They found that the striatum in mammals is of an entirely subpallial origin, but that the subpallium and the pallium both contribute to the formation of the amygdaloid complex. According to these authors, the cortical, medial and central amygdaloid nuclei develop from the subpallium, whereas the basolateral amygdaloid nuclei along with the claustrum stem from a common pallial primordium. The findings of Holmgren [77] and Källén [91] have been extended and largely confirmed by the recent investigations of Puelles and Rubenstein and their colleagues [134, 167, 168]. These authors studied the development of the cell masses in the embryonic telencephalon by analysing the expression patterns of a large number of developmental regulatory genes. They established that the claustro-amygdaloid primordium encompasses two separate pallial divisions, which they designated as the lateral and ventral pallium, and that the cortical and medial amygdaloid nuclei are of subpallial origin (Fig. 11.1 B). The only major difference between the results of Holmgren and Källén and those of Puelles and Rubenstein is that, according to the former, the cortical amygdaloid nucleus is of subpallial, but according to the latter it is of pallial origin. The studies just discussed have also shown that an overall subdivision of the embryonic mammalian telencephalon into a thick-walled subpallium and a thin-walled pallium, as presented in many textbooks, cannot be made. In the caudal part of the developing telencephalic hemispheres the pallial primordium clearly encroaches upon the thickened portion of the wall (Fig. 11.1 B).

Fibre System Related to the Amygdala

Three large fibre bundles, the lateral olfactory stria, the stria terminalis and the ventral amygdalofugal pathway, connect the amygdala with other part of the brain (Fig. 13.2). The *lateral olfactory stria* carries secondary olfactory fibres to the cortical and medial amygdaloid nuclei. Its course will be considered below. The *stria terminalis*, i.e. the dorsal amygdalofugal pathway, emerges from the caudomedial aspect of the amygdala. From there it runs a remarkably long, curved course along the medial border of the caudate nucleus to the anterior commissure. Immediately dorsocaudal to that commissure, it splits up into precommissural, commissural and postcommissural components. The stria terminalis is composed of both amygdalofugal and amygdalopetal fibres.

The *ventral amygdalofugal pathway* is a large assemblage of rather loosely arranged fibres, which extends from the amygdaloid complex to the rostral part of the diencephalon. It arises from the dorsomedial part of the amygdala, then passes medially and somewhat rostrally through the sublenticular regions of the substantia innominata and the substantia perforata anterior. Some of its fibres pass rostrally to the mediofrontal cortex; others spread in the lateral preoptico-hypothalamic zone; and still others enter the inferior thalamic peduncle to terminate in the mediodorsal thalamic nucleus (Fig. 9.1). Contrary to what its name suggests, the ventral amygdalofugal pathway conducts impulses in both directions.

Afferents to the Amygdala

The afferents of the amygdaloid complex can be grouped into the following six categories: (1) fibres originating from the olfactory bulb and from the olfactory part of the cerebral cortex; (2) fibres arising from the basomedial telencephalon and from the hypothalamus; (3) thalamic afferents; (4) direct afferents from the brain stem; (5) hippocampal afferents and (6) projections from various areas of the neocortex.

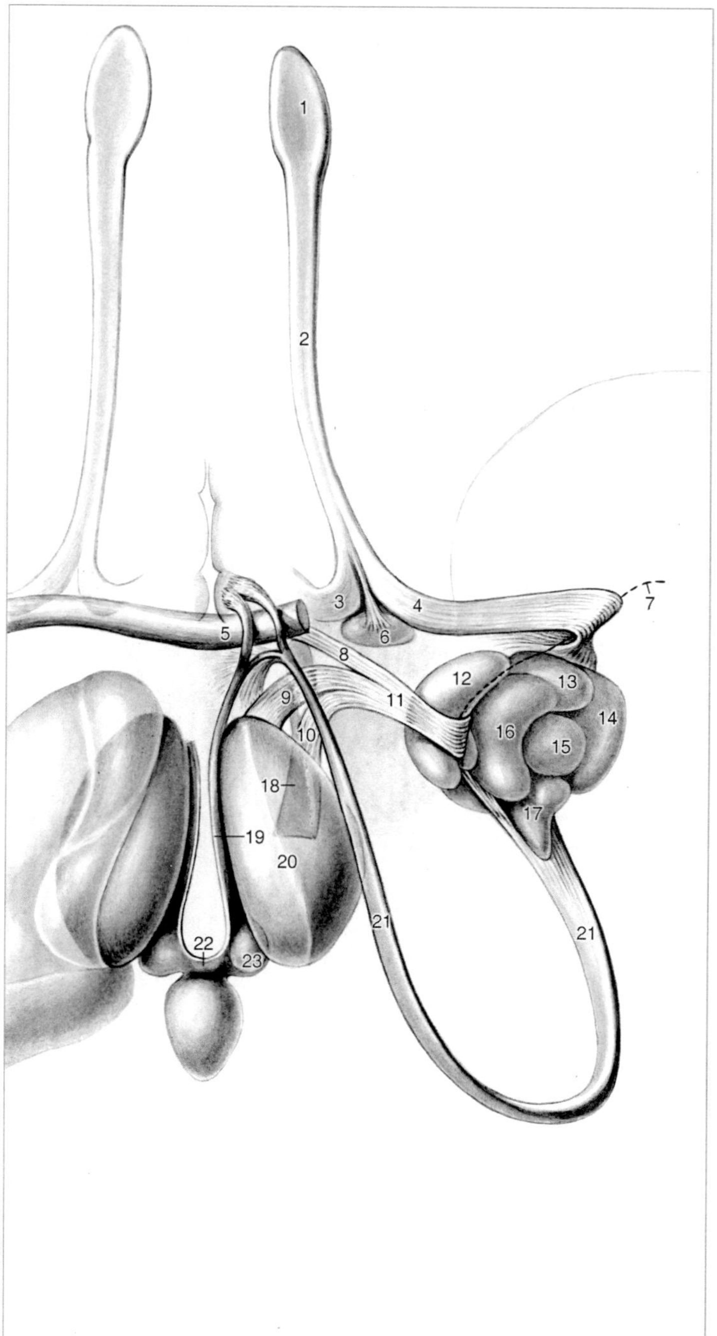

1 Olfactory bulb
2 Olfactory tract
3 Medial olfactory stria
4 Lateral olfactory stria
5 Anterior commissure
6 Olfactory tubercle
7 Limen of insula
8 Diagonal band
9 Inferior thalamic peduncle
10 Medial forebrain bundle
11 Ventral amygdalofugal fibres
12 Cortical nucleus
13 Anterior nucleus
14 Lateral nucleus
15 Central nucleus
16 Medial nucleus
17 Basal nucleus
(12–17: Amygdaloid complex)
18 Lateral hypothalamic area
19 Stria medullaris of thalamus
20 Mediodorsal thalamic nucleus
21 Stria terminalis
22 Habenular commissure
23 Habenular nuclei

Fig. 13.2. The rhinencephalic structures and the amygdaloid complex isolated from most of their surroundings, seen from above (2/1×). The *interrupted line* indicates the limen of the insula and its continuation around the "hilus" of the temporal lobe

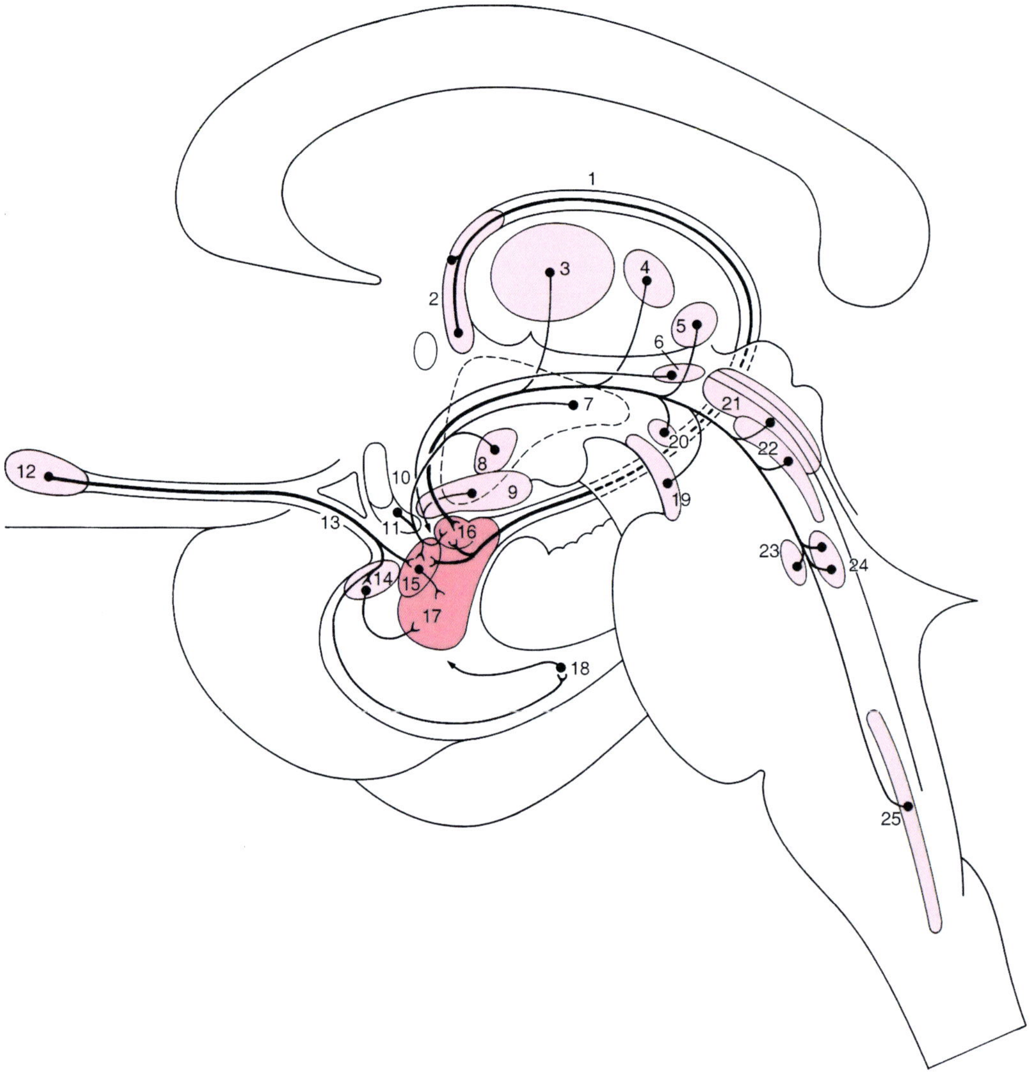

1 Stria terminalis
2 Bed nucleus of stria terminalis
3 Midline thalamic nuclei
4 Parafascicular nucleus
5 Medial geniculate body
6 Peripeduncular nucleus
7 Lateral hypothalamic area
8 Ventromedial hypothalamic nucleus
9 Substantia innominata
10 Ventral amygdalofugal fibres
11 Nucleus of the diagonal band, ventral part
12 Olfactory bulb
13 Lateral olfactory stria
14 Prepiriform cortex
15 Medial and cortical amygdaloid nuclei
16 Central amygdaloid nucleus
17 Basal and lateral amygdaloid nuclei
18 Entorhinal cortex
19 Substantia nigra, pars compacta
20 Ventral tegmental area
21 Mesencephalic central grey
22 Dorsal raphe nucleus
23 Locus coeruleus
24 Lateral parabrachial nucleus
25 Solitary nucleus

Fig. 13.3. Olfactory and subcortical afferent connections of the amygdaloid body

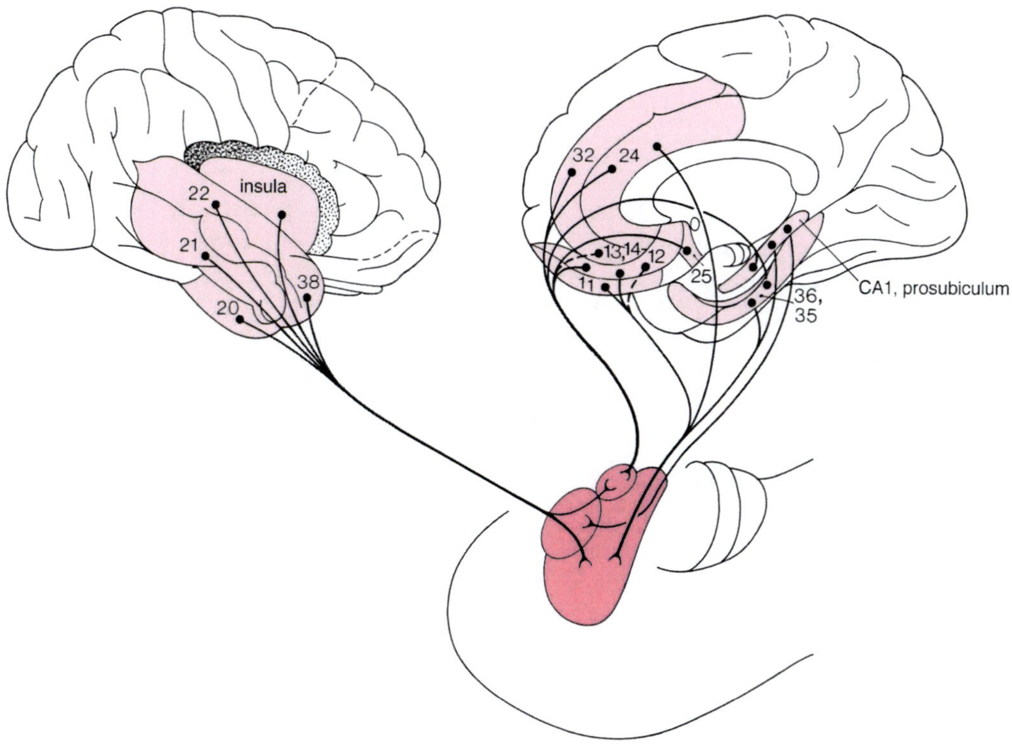

Fig. 13.4. Cortical afferents to the amygdaloid body. *Numbers* indicate fields of Brodmann

These fibre categories, which are represented diagrammatically in Figs. 13.3 and 13.4, will now be discussed briefly.

1. Secondary olfactory fibres originating from the olfactory bulb pass by way of the lateral olfactory tract to the amygdala, where they terminate mainly in the cortical nucleus [32, 179, 213]. Apart from this direct olfactory projection, the amygdala may be indirectly influenced by olfactory impulses along several routes. Thus it has been established that the prepiriform cortex distributes fibres to the basolateral complex and that the entorhinal cortex projects to the central, basolateral and cortical nuclei [152, 162, 183, 216, 219]. Moreover, it has been suggested that the deep amygdaloid nuclei receive indirect olfactory input via intrinsic connections from the cortical part of the amygdala [152].

2. Three cell masses situated in the basomedial telencephalon, the bed nucleus of the stria terminalis (BST), the nucleus of the horizontal limb of the diagonal band (HDB) and the substantia innominata (SI), send fibres to the amygdala. The BST and the HDB project mainly to the medial and central amygdaloid nuclei. The fibres arising from the SI terminate principally in the basal and accessory basal nuclei. The large cholinergic neurons embedded in the SI contribute substantially to this projection [7, 11, 15, 184].

Several hypothalamic structures, including the paraventricular, ventromedial and infundibular nuclei, as well as the lateral hypothalamic area (LHA) project to the medial part the amygdala and to the central nucleus [37, 107, 149, 173, 184, 189, 204, 220]. In the monkey, this hypothalamo-amygdalar connection originates mainly from the ventromedial nucleus and from the caudal portion of the LHA [13, 135]. It is important to note that these hypothalamic projections to the amygdala are much less substantial than the corresponding amygdalohypothalamic projections [14].

3. The amygdala receives input from both dorsal and ventral thalamic regions. The dorsal thalamic centres that have been found to distribute fibres to the amygdala include the midline nuclear complex, the parafascicular nucleus, the parvocellular part of the ventral posteromedial nucleus (VPMpc) and the medial geniculate nucleus (MGN). The midline nuclei and the parafascicular nucleus project predominantly to the central amygdaloid nucleus [7, 135, 149, 153, 154]. The VPMpc, which is a relay for visceral/gustatory information from the brain stem, projects to a portion of the lateral amygdaloid nucleus that also receives fibres from the visceral/gustatory cortex [14, 227]. The magnocellular portion of the MGN sends heavy projections to the lateral, accessory basal and central amygdaloid nuclei [113, 212].

The ventral thalamic peripeduncular nucleus projects to the medial cortical, central and lateral amygdaloid nuclei [7, 149, 151, 184].

4. Brain stem neurons located in the periaqueductal grey matter (PAG), the substantia nigra pars compacta, the ventral tegmental area, the dorsal raphe nucleus, the locus coeruleus, the lateral parabrachial nucleus, the nucleus of the solitary tract and the ventrolateral medulla project to the central amygdaloid nucleus [7, 35, 135, 148, 149, 151, 184, 220]. Most of the brain stem centres mentioned project to the amygdala via the ventral amygdalofugal pathway. However, the serotoninergic fibres from the dorsal raphe nucleus and the noradrenergic fibres from the locus coeruleus reach the amygdala via both the stria terminalis and the ventral pathway. It is noteworthy that the lateral parabrachial nucleus is by far the most important subdiencephalic source of amygdalar afferent connections [135, 151].

5. Hippocampal afferents to the amygdaloid complex arise from field CA1 and from the presubiculum [180, 183, 191]. (The presubiculum is a narrow zone of cortex, interposed between CA1 and the subiculum proper.) The hippocampal afferents, which terminate mainly in the lateral, basal and central nuclei, provide a route along which memory information that is processed in the inferomedial portion of the temporal lobe reaches the amygdala [131].

6. The following parts of the neocortex have been shown to project to the amygdala:

a. The auditory association cortex situated in the anterior half of the superior temporal gyrus (i.e. the rostral part of area 22) [200, 214, 217].

b. The visual association cortex, which occupies most of the inferior part of the temporal lobe (areas 20 and 21) [74, 86, 200, 214, 222, 225].
c. Polysensory association areas, among them the perirhinal areas 35 and 36, the temporopolar area 38 and the cortex forming the dorsal bank of the superior temporal sulcus [7, 74, 86, 214, 217].
d. Areas in the anterior part of the insula, involved in visceral and gustatory functions. These cortical areas receive their principal thalamic afferents from the parvocellular part of the ventral posteromedial nucleus [7, 92, 137, 144, 152, 219].
e. The posterior part of the insula, which contains a somatosensory association area [193, 200]. The latter is reciprocally connected with the second somatosensory area, SII [59]. SII is known to be involved in the processing of nociceptive information.
f. A large number of areas situated in the frontal lobe, including the medial prefrontal areas 11 and 12, the prelimbic area 32, the infralimbic area 25, the anterior cingulate area 24 and the caudal orbitofrontal areas 13 and 14 [7, 31, 63, 99, 114, 155, 199, 200, 215, 217, 218].

All of these neocortical areas project to the deep or basolateral part of the amygdala. The fibres originating from the temporal regions and from the insula terminate mainly in the lateral nucleus, whereas the basal and accessory basal nuclei are the principal targets of the orbitofrontal, medial prefrontal and anterior cingulate regions [200]. The posterior orbitofrontal cortical areas project to the medial, olfactory-related amygdaloid nuclei as well. It is known that these cortical areas receive higher-order olfactory projections (Fig. 11.7). Several cortical areas, including the visual and auditory association areas, the insula and the orbitofrontal cortex, have been shown to project to the central amygdaloid nucleus, too [65, 144, 214]. Most cortico-amygdaloid projections are topographically organized [74, 218]. Interestingly, the various projections from sensory (association) cortices (visual auditory, somatosensory and gustatory/visceral) to the lateral nucleus all target separate domains within that nucleus.

Intra-Amygdaloid Connections

Extensive intrinsic connections exist within the amygdaloid complex. There is general agreement that the major intra-amygdaloid connections arise in the lateral and basal nuclei and terminate in the more medial nuclei [14, 106, 152, 158, 160]. The principal intrinsic connections are summarized in Fig. 13.5. The lateral nucleus projects to the basal, accessory basal, cortical and medial nuclei [158, 159, 201]. The basal nucleus projects to the accessory basal, central (especially the medial part) and medial nuclei [5, 171]. Intrinsic afferents arising from the accessory basal nucleus have been traced to the central nucleus, especially its medial division, as well as to the medial and cortical nuclei [5, 163]. The medial nucleus projects to the central and cortical nuclei, whereas the central nucleus sends its intrinsic efferents mainly to the cortical nucleus [5, 163]. Projections from the medial, central and cortical nuclei to the basal and lateral nuclei have been shown to be present, but these are much more weekly developed than those in the reverse direction [14]. Thus, the major flow of information through the amygdala is unidirectional and follows a lateral to medial direction.

Efferents from the Amygdala

The efferent projections of the amygdaloid complex are distributed to: (1) the septo-preoptico-hypothalamic continuum; (2) the dorsal thalamus; (3) a variety of cell groups located in the brain stem; (4) the corpus striatum and (5) a considerable number of cortical areas (Figs. 13.6, 13.7). These connections will now be discussed.

1a. A large number of fibres originating mainly, but not exclusively from the corticomedial nuclear group passes with the stria terminalis to the area (directly dorsocaudal to the anterior commissure) in which the fibres, as has already been mentioned, are split into three components, precommissural, commissural and postcommissural. The precommissural fibres curve caudally around the anterior commissure and terminate in the medial preoptic and anterior hypothalamic areas as well as in the para-

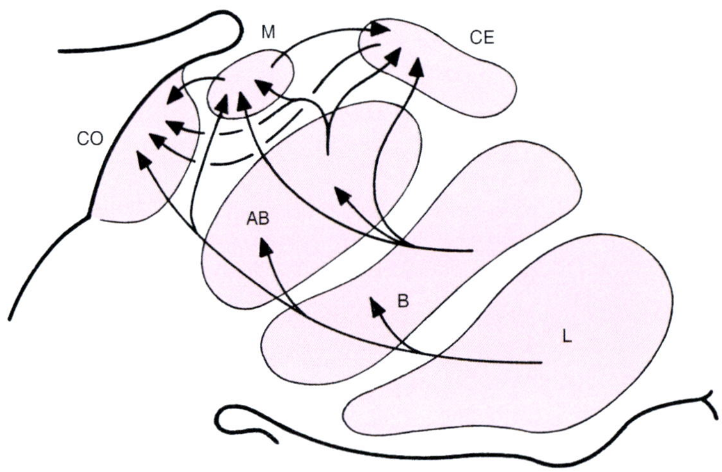

Fig. 13.5. Diagrammatic transverse section through the amygdala to illustrate the organization of the major intrinsic connections. *CO*, cortical nucleus; *M*, medial nucleus; *AB*, accessory basal nucleus; *B*, basal nucleus; *L*, lateral nucleus; *CE*, central nucleus

ventricular, ventromedial and premamillary nuclei [19, 45, 70, 105, 128, 129, 165, 209]. The commissural fibres connect the two cortical amygdaloid nuclei [110]. The postcommissural fibres spread to the BST, an elongated cell mass that accompanies the stria terminalis throughout most of its extent (see below), and to the region of the anterior hypothalamic nucleus. These postcommissural fibres originate from the medial, basal cortical and central amygdaloid nuclei [116]. The projection to the BST is strongly developed and topographically organized. The basal and central nuclei project to the lateral part of the BST, whereas the medial and cortical nuclei project to its medial part [105, 224].

1b. Fibres originating from the basal, accessory basal and central nuclei have been shown to pass via the ventral amygdalofugal pathway to a variety of structures, including the basal septal region, the substantia innominata, the magnocellular basal forebrain nuclei (particularly the nucleus basalis of Meynert and the nucleus of the horizontal limb of the diagonal band), the BST and the lateral hypothalamic area [105, 146, 147]. The projection to the latter area arises mainly from the central amygdaloid nucleus.

2. A certain proportion of the ventral amygdalofugal fibres join the inferior thalamic peduncle and pass to the mediodorsal nucleus of the thalamus (Fig. 9.1) [102, 103, 146]. The fibres of this projection originate principally from the basolateral nuclear group and terminate in the medial, magnocellular portion of the medial nucleus [6]. The central and mediodorsal amygdaloid nuclei project to the thalamic midline nuclei [6, 163].

3. The central amygdaloid nucleus gives rise to a large projection, whose constituent fibres initially travel medially via the ventral amygdalofugal pathway and then turn caudally, to descend through the most lateral part of the hypothalamus and the tegmental regions of the brain stem [79, 163, 195, 216]. The most caudal fibres of this so-called amygdalotegmental projection reach the cervical cord [139, 187]. In the brain stem, fibres are distributed to numerous centres, including the peripeduncular nucleus, the substantia nigra pars compacta, the ventral tegmental area, the cuneiform nucleus, the periaqueductal grey matter, the raphe nuclei, the parabrachial nuclei, the locus coeruleus and the area subcoerulea, the rhombencephalic reticular formation, the dorsal motor nucleus of the vagus and the nucleus of the solitary tract [60, 79, 80, 140, 163, 192, 208, 216]. The medial and lateral parabrachial nuclei and the dorsal vagal complex, all areas known to be implicated in autonomic control, are richly supplied. A sizable proportion of the neurons in the central nucleus that project to the brain stem are GABAergic [89].

4. Fibres originating mainly from the basal and accessory basal amygdaloid nuclei pass via the stria terminalis to the ventral as well as the dorsal striatum. The former includes the accumbens nucleus and certain striatal-like portions of the olfactory tubercle; the latter is composed of the caudate nucleus and the putamen [68, 71, 72]. The fibres entering the dorsal striatum are topographically organized and terminate predominantly in the ventral portions of the putamen and ventral and caudal part of the caudate nucleus [93, 185, 186]. In the caudate nucleus, the amygdaloid fibres preferentially terminate in the acetylcholinesterase-poor striosomes [170]. The amygdalostriate projection overlaps the striatal projections from the cingulate cortex, the ventral tegmental area and the mesencephalic raphe nuclei [93]. Like the amygdalostriatal system, all of these striatal afferents avoid the anterodorsolateral striatal sector, which has been shown to be the main region to which corticostriatal projections from the sensorimotor cortex are distributed. As mentioned before, these observations suggest a division of the striatum into "limbic" and "non-limbic" compartments [93].

5. Projections arising in specific amygdaloid nuclei terminate in distinct areas of the cerebral cortex. Thus, the lateral and accessory basal nuclei project strongly to the entorhinal cortex [85, 104, 190], which represents the principal source of cortical input to the hippocampal formation (Figs. 12.11, 13.6, 13.7); the accessory basal and cortical nuclei project to the hippocampal fields CA1–CA3 [191] and the basal nucleus projects to the subiculum, which represents the principal source of hippocampal output [104, 218].

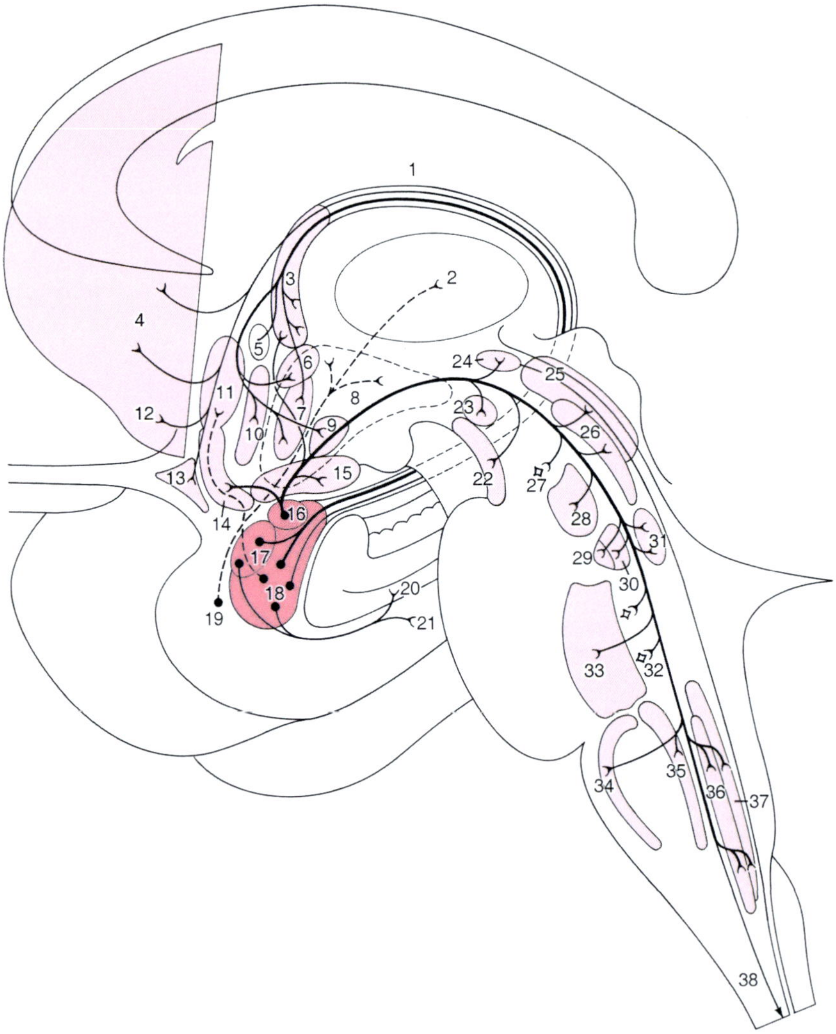

1 Stria terminalis
2 Mediodorsal thalamic nucleus, medial part
3 Bed nucleus of stria terminalis
4 Caudate nucleus + putamen
5 Anterior commissure
6 Paraventricular nucleus
7 Anterior hypothalamic nucleus
8 Lateral hypothalamic area
9 Ventromedial hypothalamic nucleus
10 Preoptic nucleus
11 Septal nuclei + nucleus of diagonal band, dorsal part
12 Accumbens nucleus
13 Olfactory tubercle
14 Nucleus of diagonal band, ventral part
15 Substantia innominata + basal nucleus of Meynert
16 Central amygdaloid nucleus
17 Cortical + medial amygdaloid nuclei
18 Basal + lateral amygdaloid nuclei
19 Periamygdaloid cortex
20 Subiculum
21 Entorhinal cortex
22 Substantia nigra, pars compacta
23 Ventral tegmental area
24 Peripeduncular nucleus
25 Mesencephalic central grey
26 Dorsal raphe nucleus
27 Cuneiform nucleus
28 Central superior nucleus
29 Nucleus subcoeruleus
30 Locus coeruleus
31 Parabrachial nuclei
32 Rhombencephalic reticular formation
33 Raphe magnus nucleus
34 Raphe pallidus nucleus
35 Raphe obscurus nucleus
36 Dorsal motor nucleus of vagus nerve
37 Solitary nucleus
38 Amygdalospinal fibres

Fig. 13.6. Subcortical and hippocampal efferents of the corpus amygdaloideum

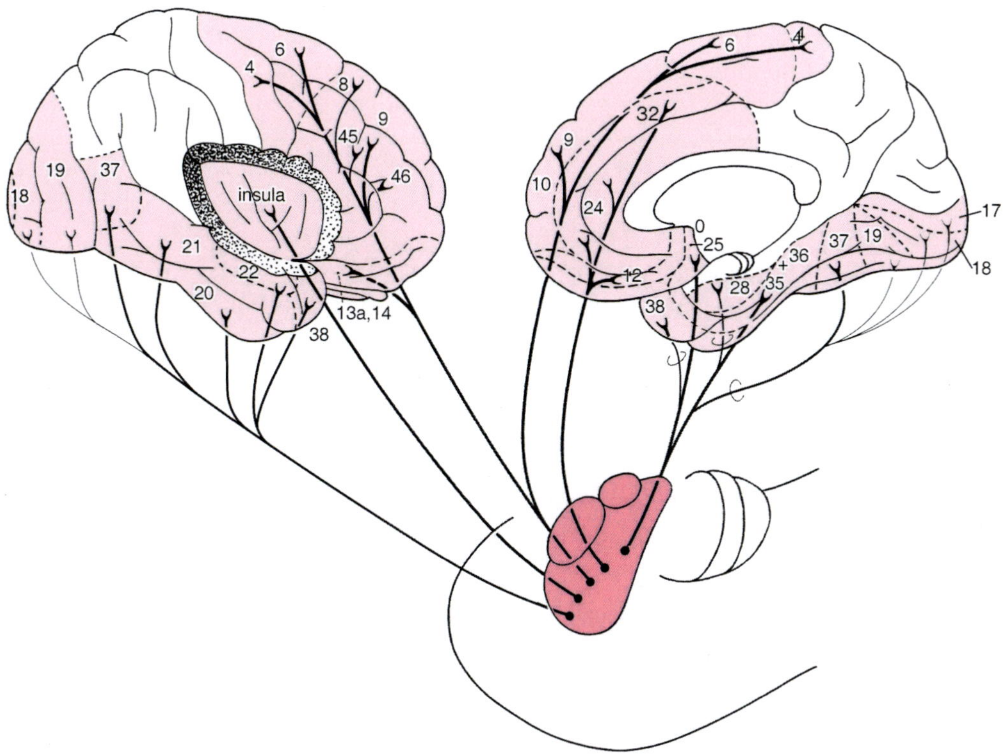

Fig. 13.7. Efferents of the amygdaloid body to the cerebral cortex. *Numbers* indicate fields of Brodmann

The neocortical projection fields of the amygdala include vast areas of the temporal occipital, insular and frontal cortices [12, 14, 103, 125]. The basal and accessory basal nuclei project to all portions of the temporal and occipital cortex related to vision. This projection is organized in a gradient manner, with the heaviest projections being directed to the visual association cortex in the inferior temporal lobe (areas 20, 21, 37) and progressively lighter projections going to the more caudal peristriate (18, 19) and striate (17) areas [12, 14, 115]. The amygdala also projects to the auditory association cortex in the rostral half of the superior temporal gyrus (area 22) and to the polysensory temporal polar cortex (area 38) [12, 166]. The latter projection arises mainly from the lateral and accessory basal nuclei [142]. The perirhinal cortical areas 35 and 36 are innervated by the basal and accessory basal nuclei [31]. These two nuclei also contribute projections to the insula, but this area receives fibres originating from the medial and cortical nuclei as well [144]. The basal and accessory basal nuclei project to a large number of areas in the frontal lobe, including the motor and premotor areas, the dorsolateral areas 8, 9, 45 and 46, the medial areas 24, 25 and 32 and the orbital areas 12, 13a and 14. The medial and orbital areas mentioned are heavily innervated [12, 14, 16, 17, 87, 122, 123, 161].

The Bed Nucleus of the Stria Terminalis and the "Extended Amygdala" Concept

The *bed nucleus of the stria terminalis* (BST) is composed mostly of small and medium-sized neurons interspersed with sporadic large neuronal elements [147]. This arch-shaped cell mass comprises a small caudolateral part, a very elongated intermediate part and an enlarged rostromedial part. The caudolateral part is situated directly dorsal to the amygdaloid complex (Fig. 6.39). The intermediate part forms a matrix of cells scattered along and among the fibre bundles of the stria terminalis. The enlarged rostromedial part envelops the anterior commissure close to its decussation, wedged between the caudate nucleus dorsolaterally, the internal capsule laterally and the anterior horn of the lateral ventricle medially (Fig. 6.40). Rostrally it is bordered by the accumbens nucleus and ventrally by the preoptic and the anterior hypothalamic areas. Detailed studies on the distribution of neurotransmitters, neuropeptides and other marker substances in the BST [118, 221] have shown that several (3–5) chemoarchitectonically distinct sectors can be delineated within this nucleus.

The BST represents the core of the "extended amygdala" concept as put forward by Heimer and colleagues [8, 10, 34, 69]. These authors confirmed the observation of Johnston [88] that the BST is caudoventrally directly continuous with the central and medial amygdaloid nuclei, and they also noticed that subventricular strands of cells connect the rostrolateral part of the BST with the ventromedial amygdala. These cell strands, which form part of the *substantia innominata* (SI), accompany the fibre bundles of the ventral amygdalofugal pathway. Furthermore, Heimer and colleagues designated the ring-shaped continuum formed by the ventromedial amygdala, the cell strands in the SI and the BST, as "the extended amygdala". It is noteworthy that in terms of this concept, the amygdala is not only extended but also reduced because the cortical nucleus and the basolateral complex are excluded from it [69].

The extended amygdala has been subdivided into a central and a medial division on the basis of histochemical and connectional differences [9, 46, 69]. The central division includes the central amygdaloid nucleus, the lateral part of the BST and an associated sublenticular cell column. The medial division includes the medial amygdaloid nucleus, the medial part of the BST and another, somewhat more ventrally located sublenticular cell column. Both subdivisions form part of largely separate neuronal circuits. Without going into detail, let it be said that the central division of the extended amygdala projects to the lateral hypothalamus and to a variety of brain stem areas, including the periaqueductal grey, the parabrachial nuclei, the reticular formation and the nucleus of the solitary tract, whereas the output of the medial division of the extended amygdala is confined mainly to the medial hypothalamus [9].

Fibre Connections of the BST

The BST receives afferents from the cerebral cortex, the amygdala, the hypothalamus and several centres in the brain stem. The cortical afferents originate from the insula [92, 131, 228], the prelimbic and infralimbic areas [84, 194], the lateral entorhinal area [132] and the subiculum [205]. The BST is one of the major terminal fields of the amygdala. Most of the amygdaloid afferents stem from the medial and central nuclei [29, 38, 52, 105, 157, 224]. Hypothalamic afferents of the BST include fibres originating from the paraventricular [37] and ventromedial nuclei [107, 189]. In the brain stem, the periaqueductal grey matter, the ventral tegmental area, the parabrachial nuclei, the dorsal vagal complex and the noradrenergic cell groups A1 and A2 send fibres to the BST [188, 198, 211]. Ricardo and Koh [174, 175] have emphasized that the BST receives the most dense pathways ascending from the interoceptive portion of the nucleus of the solitary tract to the forebrain.

The BST projects substantially to the amygdala, the mediobasal telencephalon, the hypothalamus, the thalamus and to a considerable number of grisea located in the brain stem [36, 37, 49–51, 78, 206]. The fibres passing to the amygdala travel largely along the ventral "amygdalofugal" pathway and terminate principally in the central and medial nuclei [38, 176]. A considerable number of fibres, which also follow the ventral pathway, branch off earlier to innervate the SI. Other fibres originating from the BST take a rostral course and terminate mainly in the accumbens nucleus. The fibres that descend from the BST to lower levels of the brain form three pathways, periventricular, ventral and lateral [49, 50]. The periventricular and ventral pathways are small and confined to the hypothalamus. The former innervates the parvocellular part of the paraventricular hypothalamic nucleus, whereas the latter distributes its fibres to the medial preoptic nucleus and to the anterior and ventromedial hypothalamic nuclei. The lateral pathway is large and forms part of the medial forebrain bundle. It sends fibres to numerous hypothalamic centres, among them the dorsomedial, ventromedial, paraventricular and supramamillary nuclei and the lateral hypothalamic area, and also to the thalamic midline nuclei. However, most of the fibres of the lateral pathway descend to the brain stem by way of the medial forebrain bundle; there they pass caudally through the mesencephalic central tegmental field and through the dorsolateral parts of the pons and the medulla oblongata [78]. These fibres have been observed to project substantially to the dopaminergic neurons in the substantia nigra, pars compacta and the ventral tegmental area [61], the periaqueductal grey and the adjoining cuneiform nucleus, the mesencephalic raphe nuclei, the parabrachial nuclei [141], the locus coeruleus, the lateral part of the rhombencephalic reticular formation, the nucleus of the solitary tract and the dorsal vagal nucleus [49, 192, 198, 216].

Functional and Clinical Aspects

The human amygdala is a large nuclear complex which, due to the enormous expansion of the neocortex, is ventrally and rostrally displaced and has come to lie in the rostromedial part of the temporal lobe. The basolateral part of the amygdaloid complex is directly connected with many cortical structures and receives visual, auditory, somatosensory and gustatory input. Olfactory input is mainly restricted to the cortical nucleus. Via the extensive intrinsic connections within the amygdaloid complex, these sensory stimuli from the environment reach the centromedial amygdala. The extended amygdala (the centromedial amygdala, the BST and related structures) gives rise to a highly organized system of pathways to many hypothalamic and brain stem areas and forms the main output channel of the amygdaloid complex. The extended amygdala is ideally suited to generate endocrine, autonomic and somatomotor aspects of emotional and motivational states [73].

Electrical stimulation of a variety of loci in the amygdaloid complex of the cat, the central nucleus in particular, initially leads to an orienting response, which is followed either by flight or defense behaviours [90]. The following motor and autonomic concomitants of this stimulation were noted: arrest of spontaneous behaviour followed by searching movements, retraction of the nictitating membrane and pupillary dilation, pilo-erection, micturition, growling, hissing, posturing for attack, elevation of blood pressure with bradycardia, respiratory alterations, changes of gastric motility and secretion and masticatory movements with sniffing. The defence reaction is elicited most easily by stimulating the basomedial amygdaloid nuclei, but the basolateral and central nuclei are also positive sites [76].

Electrical stimulation of the amygdala has also been shown to increase plasma levels of corticosterone, thus indicating an excitatory effect of the amygdala on the hypothalamo-pituitary-adrenal axis (Fig. 10.6) [55]. Many of the effects of electrical stimulation of the amygdala can be correlated with activation of hypothalamic and brain stem centres receiving direct projections from the extended amygdala.

Monkeys with bilateral lesions of the amygdala show a number of dramatic behavioural changes and deficits, which are collectively designated as the Klüver-Bucy syndrome [97, 98, 223]. Animals with such lesions appeared to be insensitive to stimuli that normally arouse intense fear. They were remarkably tame and showed no evidence of aggression or rage. The animals, although not blind, failed to recognize objects by sight and picked them up indiscriminately and repeatedly, usually without realizing that they had seen them before. They were unable to visually distinguish normal food from non-food items and showed bizarre sexual behaviour, seeking indiscriminate partnership with both male and female animals, animals of other species and even inanimate objects. Their behavioural changes were ascribed to a "psychic blindness", i.e. the inability to evaluate the significance of environmental events. The behavioural impairments of amygdalectomized monkeys are most severe in the wild, resulting in the death of the animal if left unattended in its troop [96]. Rodents and primates with bilateral lesions of the amygdala cannot be conditioned to associate stimuli with fear [43, 44, 112].

Electrical stimulation of the amygdala in awake human patients (during surgery for severe epilepsy) elicits feelings of fear or anxiety, as well as autonomic reactions of fear [33, 64].

Behavioural changes closely resembling the Klüver-Bucy syndrome have repeatedly been reported in human pathology. The most common causes included bilateral temporal lobe resections for treatment of epilepsy, traumatic brain injury and encephalitis, especially of herpes simplex etiology [62, 121].

In light of the effects of amygdalectomy in monkeys, attempts have been made to attenuate aggressive or virulent behaviour in humans by neurosurgically inducing lesions in the amygdala. Some authors reported marked improvements, but others failed to find a decrease in aggressive behaviour in their patient population as a result of this (for review, see [56]).

Facial expressions of emotion play an important communicative role in primates, and fearful expressions, in particular, represent potent danger signals to conspecifics [96]. Electrophysiological studies in monkeys have identified neurons in the amygdala that show selective responses for faces [117, 177]. Remarkable deficits in the perception of the emotional meaning of faces, especially fearful faces, have been found in patients with selective bilateral amygdaloid damage due to a rare inheritable disorder, Urbach-Wiethe disease [1–4, 197]. It is worthy of note that Urbach-Wiethe patients do not exhibit any of the symptoms of the Klüver-Bucy syndrome, which may have to do with the slowly progressing character of the illness.

The amygdala receives a direct projection from the olfactory bulb, which terminates mainly in the cortical nucleus (Fig. 11.7). Olfactory performance impairments have not been found in patients with a lesion specifically confined to the amygdala. However, direct recordings from the amygdala in epileptic patients prior to surgical treatment revealed that stimulation with odorants can induce large and reproducible evoked potentials [81]. Possibly, these potentials reflect the emotional component of the stimuli. Imaging studies [181, 182, 229] have shown that the human amygdala is involved in the hedonic judgement of odorants. Highly unpleasant odours elicited strong activity in the amygdala bilaterally as well as in the left orbitofrontal region [229].

The experimental and clinical data discussed indicate that the amygdala plays a crucial role in determining the emotional, motivational and social significance of complex sensory inputs and in initiating appropriate integrated neuroendocrine, autonomic and behavioural responses.

It is well-known that emotionally arousing experiences tend to be remembered well. Evidence from both animal and human studies, recently reviewed by McGaugh [133], indicates that the amygdala, particularly the basolateral complex, is critically involved in the consolidation of memories of emotional experiences.

Claustrum

Topography, Subdivision and Structure

The claustrum is a thin sheet of grey matter, embedded in the white matter of the cerebral hemispheres and largely situated between the putamen and the insular cortex. It is separated from these structures by the *extreme capsule* laterally and the *external capsule* medially (Figs. 5.4–5.8, 5.31–5.34, 6.35–6.41). The claustrum can be divided into a compact dorsal or insular part and a more diffuse temporal or ventral part [24, 39, 101, 124, 169, 196]. The latter, which is fragmented by fibre bundles related to the anterior commissure and the uncinate fasciculus, extends laterally to the amygdaloid complex.

The ontogeny of the claustrum has been the subject of much debate in the literature [26, 47, 57, 58, 77, 167, 168, 178]. Brodmann [26] and Rose [178] considered it to be derived from the insular cortex. Brodmann thought that the claustrum is the product of a doubling of the deepest (i.e. the sixth) layer of the insular cortex, but according to Rose it develops directly from the insular cortical plate. Most authors held, however, that the claustrum arises independently from both the insular cortex and the striatum [47, 57, 58, 77, 91, 167, 168]. According to Holmgren [77] and Källén [91] the claustrum is clearly of pallial derivation. Puelles and colleagues [167, 168] recently reported that the pallium is composed of four fundamental units, medial, dorsal, lateral and ventral, and that the latter two contribute to formation of the claustrum (Figs. 2.25, 11.1, 11.2).

In the claustrum, two common cell types can be distinguished: medium to large, spiny stellate or fusiform cells, and small, granular, spine-free elements [21, 48, 119]. The spiny cells have long, coarse axons, which project out of the nucleus, whereas the spine-free neurons have fine axons with dense local arborizations. These elements are GABAergic and can be subdivided into different subsets on both morphological and histochemical grounds [172, 226]. Histochemically, these subsets are characterized by the presence of different neuropeptides or calcium-binding proteins. Like those of the neocortex, the GABAergic neurons of the claustrum originate in the subpallium and migrate tangentially to this structure during development [134].

Fibre Connections

Experimental neuroanatomical studies in the cat [53, 119, 120, 126, 145, 150] and in the macaque monkey [54, 94, 108, 109, 136, 138, 143, 156, 210] have shown that many neocortical areas project in a topically ordered fashion to the claustrum, and that all of these corticoclaustral connections are reciprocated by claustrocortical projections. The neocortical regions that have been found to maintain reciprocal connections with the neocortex include the primary somato-

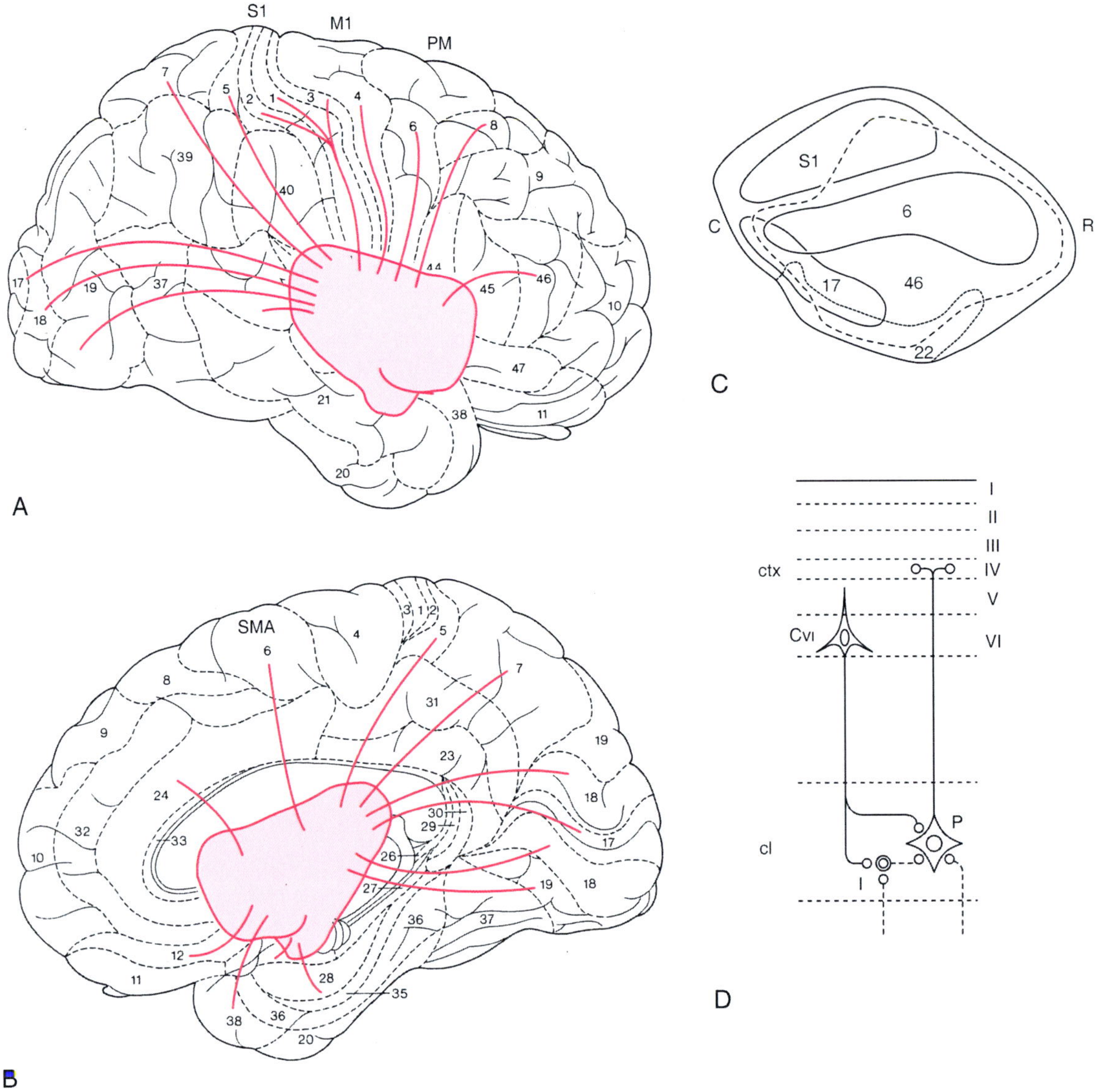

Fig. 13.8 A–D. The claustrum. The outline of the claustrum has been projected on lateral (**A**) and medial views (**B**) of the right cerebral hemisphere, on which the position of the fields of Brodmann [25, 27] are indicated by *numbers*. Fibre connections between the claustrum and the cortex are diagrammatically shown in *red*. **C** Reconstruction of the claustrum of the macaque monkey, in which the zones of origin of the projections to some cortical fields are indicated. **D** The cortico-claustro-cortical circuitry. *C*, caudal; *cl*, claustrum; *ctx*, cerebral cortex; *CVI*, lamina VI cortical pyramidal cell; *I*, claustral interneuron; *M1*, primary motor cortex; *P*, claustral projection neuron; *PM*, premotor cortex; *R*, rostral; *S1*, primary somatosensory cortex; *SMA*, supplementary motor area; **C** is based on [156]

sensory cortex S1 (areas 3, 1, 2), the somatosensory association areas 5 and 7, the visual areas 17, 18 and 19, the auditory association area 22, the primary motor cortex (area 4), the premotor cortex (area 6, lateral part), the supplementary motor cortex (area 6, medial part), the prefrontal area 46, the orbitofrontal area 12, the temperopolar area 38 and the cingulate area 24 (Fig. 13.8 A, B). The entorhinal area 28 and the paleocortical, olfactory area 51 are connected with the ventral part of the claustrum (Fig. 13.8 B). It is remarkable that the primary auditory areas 41 and 42, contrary to the primary somatosensory (3, 1, 2) and visual (17) areas, are not connected with the claustrum.

In general, specific cortical areas are related to specific zones of the claustrum, but these zones may differ considerably in size and overlap with each other in many places [156, 210] (Fig. 13.8C). Anatomical and electrophysiological studies have shown that the visual claustral zone is retinotopically organized [18, 120] and that the somatosensory zone contains one or possibly multiple orderly representations of the contralateral body surface [138, 150].

The data on the claustrocortical and corticoclaustrum connections, reviewed above, are based principally on experiments in which anterograde or retrograde tracers have been injected in particular cortical areas. Because of its inaccessibility, the claustrum's subcortical connections have been analysed much less intensively. Studies in the cat [48, 119] have shown the presence of sparse afferent pathways to the claustrum, originating from the intralaminar thalamic central lateral nucleus, the lateral geniculate complex and the lateral hypothalamic area. In the rat, the basolateral amygdaloid nuclei are reciprocally connected with the claustrum [95, 127]. In the rhesus macaque, a cholinergic projection, originating from the nucleus basalis, has been found to terminate in the most ventral part of the claustrum [48].

Circuitry and Function

Until the existence of substantial subcortical claustral afferents has been demonstrated, the neocortical projections must be regarded as the dominant or 'driver' afferents of the claustrum. In the visual cortex of the cat, the corticoclaustral projections arise from pyramidal neurons in layer VI. The same layer also contains pyramidal cells projecting to the lateral geniculate nucleus, but it has been shown that the claustrum- and the geniculate-projecting elements form separate populations [119, 150]. The axons of the claustrum-projecting cells form asymmetric synapses with both spiny projection neurons and aspiny interneurons in the nucleus [119]. The synaptic relations of the interneurons are unknown. The axons of the claustral projection neurons terminate mainly in layer IV, at least in the visual cortex [119, 150]. The connections of the claustrum, as diagrammatically shown in Fig. 13.8 D, are reminiscent of those of the pulvinar. This thalamic structure receives, just like the claustrum, its driving afferents from the cortex, and both structures project back to the cortex (see Fig. 8.5 C). However, the cortical afferents to the pulvinar arise from pyramidal neurons in layer V, and the thalamic modulatory apparatus is presumably much more complex than that in the claustrum. It is remarkable that the nuclei of the basolateral amygdaloid group, which develop from the same pallial sectors as the claustrum [167, 168] (Fig. 11.1), are also strongly and reciprocally connected with the neocortex. Swanson and Petrovich [207] went so far as to consider these amygdaloid nuclei as a ventromedial extension of the claustrum.

According to the present state of our knowledge, the claustrum must be regarded as a cortical satellite. The fact that many cortical areas representing highly different functions converge on the claustrum strongly suggests that some kind of integration occurs within this nucleus, all the more so since many of the cortical projection zones show a considerable overlap. However, the nature of this integration and its morphological substrate are unknown at present. Crick and Koch [40] recently conjectured that the claustrum is involved in (visual) consciousness.

References

1. Adolphs R, Tranel D (2000) Emotion recognition and the human amygdala. In: Aggleton JP (ed) The amygdala, a functional analysis. Oxford University Press, Oxford, pp 587–630
2. Adolphs R, Tranel D, Damasio H, Damasio A (1994) Impaired recognition of emotion in facial expressions following bilateral damage to the human amygdala. Nature 372:669–672
3. Adolphs R, Tranel D, Damasio H, Damasio AR (1995) Fear and the human amygdala. J Neurosci 15:5879–5891
4. Adolphs R, Tranel D, Damasio AR (1998) The human amygdala in social judgment. Nature 393:470–474
5. Aggleton JP (1985) A description of intra-amygdaloid connections in old world monkeys. Exp Brain Res 57:390–399
6. Aggleton JP, Mishkin M (1984) Projections of the amygdala to the thalamus in the cynomolgus monkey. J Comp Neurol 222:56–68
7. Aggleton JP, Burton MJ, Passingham RE (1980) Cortical and subcortical afferents to the amygdala of the rhesus monkey *(macaca mulatta)*. Brain Res 190:347–368
8. Alheid GF, Heimer L (1988) New perspectives in basal forebrain organization of special relevance for neuropsychiatric disorders: the striatopallidal, amygdaloid, and corticopetal components of substantia innominata. Neuroscience 27:1–39
9. Alheid GF, Heimer L (1996) Theories of basal forebrain organization and the "emotional motor system". Progr Brain Res 107:461–484
10. Alheid GF, De Olmos JS, Beltramino CA (1995) Amygdala and extended amygdala. In: Paxinos G (ed) The rat nervous system, 2nd edn. Academic Press, San Diego, pp 495–578
11. Amaral DG, Bassett JL (1989) Cholinergic innervation of the monkey amygdala: an immunohistochemical analysis with antisera to choline acetyltransferase. J Comp Neurol 281:337–361
12. Amaral DG, Price JL (1984) Amygdalo-cortical projections in the monkey (Macaca fascicularis). J Comp Neurol 230:465–496
13. Amaral DG, Veazey RB, Cowan WM (1982) Some observations on hypothalamo-amygdaloid connections in the monkey. Brain Res 252:13–27
14. Amaral DG, Price JL, Pitkänen A, Carmichael ST (1992) Anatomical organization of the primate amygdaloid complex. In: Aggleton JP (ed) The amygdala: neurobiological aspects of emotion, memory, and mental dysfunction. Wiley-Liss, New York, pp 1–66
15. Asanuma C, Thach WT, Jones EG (1983) Distribution of cerebellar terminations and their relation to other afferent terminations in the ventral lateral thalamic region of the monkey. Brain Res Rev 5:237–266
16. Avendano C, Price JL, Amaral DG (1983) Evidence for an amygdaloid projection to premotor cortex but not to motor cortex in the monkey. Brain Res 264:111–117
17. Barbas H, De Olmos J (1990) Projections from the amygdala to basoventral and mediodorsal prefrontal regions in the rhesus monkey. J Comp Neurol 300:549–571
18. Barbas H, Mesulam MM (1981) Organization of afferent input to subdivisions of area 8 in the rhesus monkey. J Comp Neurol 200:407–431
19. Berk ML, Finkelstein JA (1981) Afferent projections to the preoptic area and hypothalamic regions in the rat brain. Neuroscience 6:1601–1624
20. Braak H, Braak E (1983) Neuronal types in the basolateral amygdaloid nuclei of man. Brain Res Bull 11:349–365
21. Brand S (1981) A serial section Golgi analysis of the primate claustrum. Anat Embryol (Berl) 162: 475–488
22. Brierley B, Shaw P, David AS (2002) The human amygdala: a systematic review and meta-analysis of volumetric magnetic resonance imaging. Brain Res Brain Res Rev 39:84–105
23. Brockhaus H (1938) Zur normalen und pathologischen Anatomie der Mandelkerngebiete. J Psychol Neurol 49:1–136
24. Brockhaus H (1940) Die Cyto- und Myeloarchitektonik des Cortex claustralis und des Claustrum beim Menschen. J Psychol Neurol 49:249–348
25. Brodmann K (1905) Beiträge zur histologischen Lokalisation der Großhirnrinde. Dritte Mitteilung. Die Rindenfelder der niederen Affen. J Psychol Neurol 4:176–226
26. Brodmann K (1909) Vergleichende Lokalisationslehre der Großhirnrinde in ihren Prinzipien dargestellt auf Grund des Zellenbaues. Barth, Leipzig
27. Brodmann K (1914) Physiologie des Gehirns. In: Von Bruns P (ed) Neue Deutsche Chirurgie, vol 11. Enke, Stuttgart, pp 85–426
28. Burdach K (1819) Vom Baue und Leben des Gehirns. Dyk, Leipzig
29. Canteras NS, Simerly RB, Swanson LW (1995) Organization of projections from the medial nucleus of the amygdala: a PHAL study in the rat. J Comp Neurol 360:213–245
30. Carlsen J, Heimer L (1988) The basolateral amygdaloid complex as a cortical-like structure. Brain Res 441:377–380
31. Carmichael ST, Price JL (1995) Sensory and premotor connections of the orbital and medial prefrontal cortex of macaque monkeys. J Comp Neurol 363:642–664
32. Carmichael ST, Clugnet MC, Price JL (1994) Central olfactory connections in the macaque monkey. J Comp Neurol 346:403–434
33. Chapman WP, Schroeder HR, Geyer G et al (1954) Physiological evidence concerning importance of the amygdaloid nuclear region in the integration of circulatory function and emotion in man. Science 120:949–950

34. Ciriello J, Calaresu FR (1980) Autoradiographic study of ascending projections from cardiovascular sites in the nucleus tractus solitarii in the cat. Brain Res 186:448–453
35. Ciriello J, Schultz CG, Roder S (1994) Collateral axonal projections from ventrolateral medullary non-catecholaminergic neurons to central nucleus of the amygdala. Brain Res 663:346–351
36. Conrad LCA, Pfaff DW (1976) Efferents from medial basal forebrain and hypothalamus in the rat. I. An autoradiographic study of the medial preoptic area. J Comp Neurol 169:185–220
37. Conrad LCA, Pfaff DW (1976) Efferents from medial basal forebrain and hypothalamus in the rat. II. An autoradiographic study of the anterior hypothalamus. J Comp Neurol 169:221–262
38. Coolen LM, Wood RI (1998) Bidirectional connections of the medial amygdaloid nucleus in the Syrian hamster brain: simultaneous anterograde and retrograde tract tracing. J Comp Neurol 399:189–209
39. Cowan WM, Raisman G, Powell TS (1965) The connexions of the amygdala. J Neurol Neurosurg Psychiatry 28:137–151
40. Crick FC, Koch C (2005) What is the function of the claustrum? Philos Trans R Soc Lond B Biol Sci 360:1271–1279
41. Crosby EC, Humphrey T (1941) Studies of the vertebrate telencephalon. II. The nuclear pattern of the anterior olfactory nucleus, tuberculum olfactorium and the amygdaloid complex in adult man. J Comp Neurol 47:309–352
42. Crosby EC, Humphrey T, Lauer EW (1962) Correlative anatomy of the nervous system. MacMillan, New York
43. Davis M (1992) The role of the amygdala in fear and anxiety. Annu Rev Neurosci 15:353–375
44. Davis M (1992) The role of the amygdala in conditioned fear. In: Aggleton JP (ed) The amygdala: neurobiological aspects of emotion, memory and mental dysfunction. Wiley-Liss, New York, pp 255–306
45. De Olmos JS (1972) The amygdaloid projection field in the rat as studied with the cupric-silver method. In: Eleftheriou BE (ed) The neurobiology of the amygdala. Plenum, New York, pp 145–204
46. De Olmos JS (2004) Amygdala. In: Paxinos G, Mai JK (eds) The human nervous system. Elsevier, Amsterdam, pp 739–868
47. De Vries E (1910) Bemerkungen zur Ontogenie und vergleichenden Anatomie des Claustrums. Folia Neurobiol 4:481–513
48. DeFelipe J, Jones EG (1988) Cajal on the cerebral cortex. An anatomical translation of the complete writings. Oxford University Press, New York
49. Dong HW, Swanson LW (2003) Projections from the rhomboid nucleus of the bed nuclei of the stria terminalis: implications for cerebral hemisphere regulation of ingestive behaviors. J Comp Neurol 463:434–472
50. Dong HW, Swanson LW (2004) Organization of axonal projections from the anterolateral area of the bed nuclei of the stria terminalis. J Comp Neurol 468:277–298
51. Dong HW, Swanson LW (2004) Projections from bed nuclei of the stria terminalis, posterior division: implications for cerebral hemisphere regulation of defensive and reproductive behaviors. J Comp Neurol 471:396–433
52. Dong HW, Petrovich GD, Swanson LW (2001) Topography of projections from amygdala to bed nuclei of the stria terminalis. Brain Res Brain Res Rev 38:192–246
53. Druga R (1982) Claustro-neocortical connections in the cat and rat demonstrated by HRP tracing technique. J Hirnforsch 23:191–202
54. Druga R, Rokyta R, Benes V Jr (1990) Claustro-neocortical projections in the rhesus monkey (projections to area 6). J Hirnforsch 31:487–494
55. Dunn JD, Whitener J (1986) Plasma corticosterone responses to electrical stimulation of the amygdaloid complex: cytoarchitectural specificity. Neuroendocrinology 42:211–217
56. Eichelman B (1983) The limbic system and aggression in humans. Neurosci Biobehav Rev 7:391–394
57. Faul J (1926) The ontogenetic development of the claustrum in mammals. Proc Kon Ned Akad Wet (Amst) 29:642–647
58. Filimonoff IN (1966) The claustrum, its origin and development. J Hirnforsch 8:503–528
59. Friedman DP, Murray EA, O'Neill JB, Mishkin M (1986) Cortical connections of the somatosensory fields of the lateral sulcus of macaques: evidence for a corticolimbic pathway for touch. J Comp Neurol 252:323–347
60. Fudge JL, Haber SN (2000) The central nucleus of the amygdala projection to dopamine subpopulations in primates. Neuroscience 97:479–494
61. Fudge JL, Haber SN (2001) Bed nucleus of the stria terminalis and extended amygdala inputs to dopamine subpopulations in primates. Neuroscience 104:807–827
62. Gerstenbrand F, Poewe W, Aichner F, Saltuari L (1983) Klüver-Bucy syndrome in man: experiences with posttraumatic cases. Neurosci Biobehav Rev 7:413–417
63. Ghashghaei HT, Barbas H (2002) Pathways for emotion: interactions of prefrontal and anterior temporal pathways in the amygdala of the rhesus monkey. Neuroscience 115:1261–1279
64. Gloor P, Olivier A, Quesney LF (1981) The role of the amygdala in the expression of psychic phenomena in temporal lobe seizures. In: Ben-Ari Y (ed) The amygdaloid complex. Elsevier North-Holland, New York, pp 489–507
65. Goldman-Rakic PS (1984) Modular organization of prefrontal cortex. Trends Neurosci 7:419–424
66. Hall E (1972) Some aspects of the structural organization of the amygdala. In: Eleftheriou BE (ed) The neurobiology of the amygdala. Plenum, New York, pp 95–122
67. Hall E (1972) The amygdala of the cat: a Golgi study. Z Zellforsch Mikrosk Anat 134:439–458

68. Heimer L (1978) The olfactory cortex and the ventral striatum. In: Livingston KE, Hornykiewicz O (eds) Limbic mechanisms: the continuing evolution of the limbic system concept. Plenum, New York, pp 95–187
69. Heimer L (2000) Basal forebrain in the context of schizophrenia. Brain Res Brain Res Rev 31:205–235
70. Heimer L, Nauta WJH (1969) The hypothalamic distribution of the stria terminalis in the rat. Brain Res 13:284–297
71. Heimer L, Wilson RD (1975) The subcortical projections of the allocortex: similarities in the neural associations of the hippocampus, the piriform cortex, and the neocortex. In: Santini M (ed) Golgi Centennial Symposium proceedings. Raven, New York, pp 177–193
72. Heimer L, Switzer RD, Van Hoesen GW (1982) Ventral striatum and ventral pallidum. Components of the motor system? Trends Neurosci 5:83–87
73. Heimer L, Harlan RE, Alheid GF, Garcia MM, De Olmos J (1997) Substantia innominata: a notion which impedes clinical-anatomical correlations in neuropsychiatric disorders. Neuroscience 76:957–1006
74. Herzog AG, Van Hoesen GW (1976) Temporal neocortical afferent connections to the amygdala in the rhesus monkey. Brain Res 15:57–69
75. Hewitt W (1958) The development of the human caudate and amygdaloid nuclei. J Anat 92:377–382
76. Hilton SM, Zbrozyna AW (1963) Amygdaloid region for defence reactions and its efferent pathway to the brain stem. J Physiol 165:160–173
77. Holmgren N (1925) Points of view concerning forebrain morphology in higher vertebrates. Acta Zool 6:413–477
78. Holstege G, Meiners L, Tan K (1985) Projections of the bed nucleus of the stria terminalis to the mesencephalon, pons, and medulla oblongata in the cat. Exp Brain Res 58:379–391
79. Hopkins DA (1975) Amygdalotegmental projections in the rat, cat and rhesus monkey. Neurosci Lett 1:263–270
80. Hopkins DA, Holstege G (1978) Amygdaloid projections to the mesencephalon, pons and medulla oblongata in the cat. Exp Brain Res 32:529–548
81. Hudry J, Ryvlin P, Royet JP, Mauguiere F (2001) Odorants elicit evoked potentials in the human amygdala. Cereb Cortex 11:619–627
82. Humphrey T (1968) The development of the human amygdala during early embryonic life. J Comp Neurol 132:135–165
83. Humphrey T (1968) The development of the human amygdaloid complex. In: Eleftheriou BE (ed) The neurobiology of the amygdala. Plenum, New York, pp 21–80
84. Hurley KM, Herbert H, Moga MM, Saper CB (1991) Efferent projections of the infralimbic cortex of the rat. J Comp Neurol 308:249–276
85. Insausti R, Amaral DG, Cowan WM (1987) The entorhinal cortex of the monkey: III. Subcortical afferents. J Comp Neurol 264:396–408
86. Iwai E, Yukie M (1987) Amygdalofugal and amygdalopetal connections with modality-specific visual cortical areas in macaques (Macaca fuscata, M. mulatta, and M. fascicularis). J Comp Neurol 261:362–387
87. Jacobson S, Trojanowski JQ (1975) Amygdaloid projections to prefrontal granular cortex in rhesus monkey demonstrated with horseradish peroxidase. Brain Res 100:132–139
88. Johnston JB (1923) Further contributions to the study of the evolution of the forebrain. J Comp Neurol 35:337–481
89. Jongen-Rêlo AL, Amaral DG (1998) Evidence for a GABAergic projection from the central nucleus of the amygdala to the brainstem of the macaque monkey: a combined retrograde tracing and in situ hybridization study. Eur J Neurosci 10:2924–2933
90. Kaada BR (1972) Stimulation and regional ablation of the amygdaloid cortex with reference to functional representations. In: Eleftheriou BE (ed) The neurobiology of the amygdala. Plenum, New York, pp 205–282
91. Källén B (1951) The nuclear development in the mammalian forebrain with special regard to the subpallium. Kgl Fysiogr Sällsk Lund Handl N F 61:1–43
92. Kapp BS, Schwaber JS, Driscoll PA (1985) Frontal cortex projections to the amygdaloid central nucleus in the rabbit. Neuroscience 15:327–346
93. Kelley AE, Domesick VB, Nauta WJH (1982) The amygdalostriatal projection in the rat – an anatomical study by anterograde and retrograde tracing methods. Neuroscience 7:615–630
94. Kemp JM, Powell TPS (1970) The cortico-striate projection in the monkey. Brain 93:525–547
95. Kita H, Kitai ST (1990) Amygdaloid projections to the frontal cortex and the striatum in the rat. J Comp Neurol 298:40–49
96. Kling AS, Brothers LA (1992) The amygdala and social behavior. In: Aggleton JP (ed) The amygdala: neurobiological aspects of emotion, memory, and mental dysfunction. Wiley-Liss, New York, pp 353–378
97. Klüver H, Bucy PC (1937) Psychic blindness and other symptoms following bilateral temporal lobectomy in rhesus monkeys. Am J Physiol 119: 352–353
98. Klüver H, Bucy PC (1939) Preliminary analysis of functions of the temporal lobes in monkeys. Arch Neurol Psychiatry (London) 42:979–1000
99. Koikegami H (1963) Amygdala and other related limbic structures; experimental studies on the anatomy and function. I. Anatomical researches with some neurophysiological observations. Acta Med Biol 10:161–277
100. Kordower JH, Piecinski P, Rakic P (1992) Neurogenesis of the amygdaloid nuclear complex in the rhesus monkey. Brain Res Dev Brain Res 68:9–15
101. Kowianski P, Dziewiatkowski J, Kowianska J, Morys J (1999) Comparative anatomy of the

claustrum in selected species: A morphometric analysis. Brain Behav Evol 53:44–54
102. Krettek JE, Price JL (1974) A direct input from the amygdala to the thalamus and the cerebral cortex. Brain Res 67:169–174
103. Krettek JE, Price JL (1977) Projections from the amygdaloid complex to the cerebral cortex and thalamus in the rat and cat. J Comp Neurol 172: 687–722
104. Krettek JE, Price JL (1977) Projections from the amygdaloid complex and adjacent olfactory structures to the entorhinal cortex and to the subiculum in the rat and cat. J Comp Neurol 172:723–752
105. Krettek JE, Price JL (1978) Amygdaloid projections to subcortical structures within the basal forebrain and brainstem in the rat and cat. J Comp Neurol 178:225–254
106. Krettek JE, Price JL (1978) A description of the amygdaloid complex in the rat and cat with observations on intra-amygdaloid axonal connections. J Comp Neurol 178:255–280
107. Krieger MS, Conrad LCA, Pfaff DW (1979) An autoradiographic study of the efferent connections of the ventromedial nucleus of the hypothalamus. J Comp Neurol 183:785–816
108. Künzle H (1975) Bilateral projections from precentral motor cortex to the putamen and other parts of the basal ganglia. An autoradiographic study in the macaca fascicularis. Brain Res 88:195–209
109. Künzle H, Akert K (1977) Efferent connections of cortical area 8 (frontal eye field) in Macaca fascicularis. A reinvestigation using the autoradiographic technique. J Comp Neurol 173:147–164
110. Lammers HJ (1972) The neural connections of the amygdaloid complex in mammals. In: Eleftheriou BE (ed) The neurobiology of the amygdala. Plenum, New York, pp 123–144
111. Lammers GJ (1976) On the development of the strio-amygdaloid complex in the chinese hamster. Thesis, University of Nijmegen
112. LeDoux JE (2000) Emotion circuits in the brain. Annu Rev Neurosci 23:155–184
113. LeDoux JE, Farb C, Ruggiero DA (1990) Topographic organization of neurons in the acoustic thalamus that project to the amygdala. J Neurosci 10:1043–1054
114. Leichnetz GR, Astruc J (1977) The course of some prefrontal corticofugals to the pallidum, substantia innominata, and amygdaloid complex in monkeys. Exp Neurol 54:104–109
115. Leichnetz GR, Spencer RF, Smith DJ (1984) Cortical projections to nuclei adjacent to the oculomotor complex in the medial dien-mesencephalic tegmentum in the monkey. J Comp Neurol 228: 359–387
116. Leonard CM, Scott JW (1971) Origin and distribution of the amygdalofugal pathways in the rat: an experimental neuroanatomical study. J Comp Neurol 141:313–330
117. Leonard CM, Rolls ET, Wilson FAW, Baylis GC (1985) Neurons in the amygdala of the monkey with responses selective for faces. Behav Brain Res 15:159–176
118. Lesur A, Gaspar P, Alvarez C, Berger B (1989) Chemoanatomic compartments in the human bed nucleus of the stria terminalis. Neuroscience 32:181–194
119. LeVay S, Sherk H (1981) The visual claustrum of the cat. I. Structure and connections. J Neurosci 1:956–980
120. LeVay S, Sherk H (1981) The visual claustrum of the cat. II. The visual field map. J Neurosci 1:981–992
121. Lilly R, Cummings JL, Benson DF, Frankel M (1983) The human Klüver-Bucy syndrome. Neurology 33:1141–1145
122. Llamas A, Avendano C, Reinoso-Suarez F (1977) Amygdaloid projections to prefrontal and motor cortex. Science 195:794–796
123. Llamas A, Avendano C, Reinoso-Suarez F (1985) Amygdaloid projections to the motor premotor and prefrontal areas of the cat's cerebral cortex: a topographical study using retrograde transport of horseradish peroxidase. Neuroscience 15:651–657
124. Macchi G (1941) Considerazioni sull'origine e il significato del claustre nell'uomo. Rass Biol Umana 2:134–143
125. Macchi G, Bentivoglio M, Rossini P, Tempesta E (1978) The basolateral amygdaloid projections to the neocortex in the cat. Neurosci Lett 9:347–351
126. Macchi G, Bentivoglio M, Minciacchi D, Molinari M (1983) Claustroneocortical projections studied in the cat by means of multiple retrograde fluorescent tracing. J Comp Neurol 215:121–134
127. Majak K, Pikkarainen M, Kemppainen S, Jolkkonen E, Pitkänen A (2002) Projections from the amygdaloid complex to the claustrum and the endopiriform nucleus: a *Phaseolus vulgaris* leucoagglutinin study in the rat. J Comp Neurol 451:236–249
128. Maragos WF, Newman SW, Lehman MN, Powers JB (1989) Neurons of origin and fiber trajectory of amygdalofugal projections to the medial preoptic area in Syrian hamsters. J Comp Neurol 280:59–71
129. McBride RL, Sutin J (1977) Amygdaloid and pontine projections to the ventromedial nucleus of the hypothalamus. J Comp Neurol 174:377–396
130. McDonald AJ (1992) Cell types and intrinsic connections of amygdala. In: Aggleton JP (ed) The amygdala: neurobiological aspects of emotion, memory and mental dysfunction. Wiley-Liss, New York, pp 67–96
131. McDonald AJ (1998) Cortical pathways to the mammalian amygdala. Progr Neurobiol 55:257–332
132. McDonald AJ, Shammah-Lagnado SJ, Shi C, Davis M (1999) Cortical afferents to the extended amygdala. Ann NY Acad Sci 877:309–338
133. McGaugh JL (2004) The amygdala modulates the consolidation of memories of emotionally arousing experiences. Annu Rev Neurosci 27:1–28

134. Medina L, Legaz I, Gonzalez G et al (2004) Expression of Dbx1, Neurogenin 2, Semaphorin 5A, Cadherin 8, and Emx1 distinguish ventral and lateral pallial histogenetic divisions in the developing mouse claustroamygdaloid complex. J Comp Neurol 474:504–523
135. Mehler WR (1980) Subcortical afferent connections of the amygdala in the monkey. J Comp Neurol 190:733–762
136. Mesulam MM, Mufson EJ (1982) Insula of the old world monkey. III. Efferent cortical output and comments on function. J Comp Neurol 212:38–52
137. Mesulam MM, Mufson EJ (1985) The insula of Reil in man and monkey: architectonics, connectivity and function. In: Peters A, Jones EO (eds) Association and auditory cortices. Plenum, New York, pp 179–226 (Cerebral cortex, vol 4)
138. Minciacchi D, Granato A, Barbaresi P (1991) Organization of claustro-cortical projections to the primary somatosensory area of primates. Brain Res 553:309–312
139. Mizuno N, Takahashi O, Satoda T, Matsushima R (1985) Amygdalospinal projections in the macaque monkey. Neurosci Lett 53:327–330
140. Moga MM, Gray TS (1985) Evidence for corticotropin-releasing factor, neurotensin, and somatostatin in the neural pathway from the central nucleus of the amygdala to the parabrachial nucleus. J Comp Neurol 241:275–284
141. Moga MM, Saper CB, Gray TS (1989) Bed nucleus of the stria terminalis: cytoarchitecture, immunohistochemistry, and projection to the parabrachial nucleus in the rat. J Comp Neurol 283: 315–332
142. Moran MA, Mufson EJ, Mesulam MM (1987) Neural inputs into the temporopolar cortex of the rhesus monkey. J Comp Neurol 256:88–103
143. Morecraft RJ, Geula C, Mesulam MM (1992) Cytoarchitecture and neural afferents of orbitofrontal cortex in the brain of the monkey. J Comp Neurol 323:341–358
144. Mufson EJ, Mesulam MM, Pandya DN (1981) Insular interconnections with the amygdala in the rhesus monkey. Neuroscience 7:1231–1248
145. Narkiewicz O (1964) Degenerations in the claustrum after regional neocortical ablations in the cat. J Comp Neurol 123:335–355
146. Nauta WJH (1961) Fibre degeneration following lesions of the amygdaloid complex in the monkey. J Anat 95:515–531
147. Nauta WJH, Haymaker W (1969) Hypothalamic nuclei and fiber connections. In: Haymaker W, Anderson E, Nauta WJH (eds) The hypothalamus. Thomas, Springfield, pp 136–209
148. Norgren R (1976) Taste pathways to hypothalamus and amygdala. J Comp Neurol 166:17–30
149. Norita M, Kawamura K (1980) Subcortical afferents to the monkey amygdala: an HRP study. Brain Res 190:225–230
150. Olson CR, Graybiel AM (1980) Sensory maps in the claustrum of the cat. Nature 288:479–481
151. Ottersen OP (1981) Afferent connections to the amygdaloid complex of the rat with some observations in the cat. III. Afferents from the lower brain stem. J Comp Neurol 202:335–356
152. Ottersen OP (1982) Connections of the amygdala of the rat. IV: Corticoamygdaloid and intraamygdaloid connections as studied with axonal transport of horseradish peroxidase. J Comp Neurol 205:30–48
153. Ottersen OP, Ben-Ari Y (1979) Afferent connections to the amygdaloid complex of the rat and cat. J Comp Neurol 187:401–424
154. Palay SL, Chan-Palay V (1974) Cerebellar cortex. Cytology and organization. Springer, Berlin Heidelberg New York, p 348
155. Pandya DN, Van Hoesen GW, Domesick VB (1973) A cingulo-amygdaloid projection in the. rhesus monkey. Brain Res 61:369–373
156. Pearson RCA, Brodal P, Gatter KC, Powell TPS (1982) The organization of the connections between the cortex and the claustrum in the monkey. Brain Res 234:435–441
157. Petrovich GD, Swanson LW (1997) Projections from the lateral part of the central amygdalar nucleus to the postulated fear conditioning circuit. Brain Res 763:247–254
158. Pitkänen A, Amaral DG (1991) Demonstration of projections from the lateral nucleus to the basal nucleus of the amygdala: a PHA-L study in the monkey. Exp Brain Res 83:465–470
159. Pitkänen A, Amaral DG (1998) Organization of the intrinsic connections of the monkey amygdaloid complex: projections originating in the lateral nucleus. J Comp Neurol 398:431–458
160. Pitkänen A, Savander V, LeDoux JE (1997) Organization of intra-amygdaloid circuitries in the rat: an emerging framework for understanding functions of the amygdala. Trends Neurosci 20: 517–523
161. Porrino LJ, Crane AM, Goldman-Rakic PS (1981) Direct and indirect pathways from the amygdala to the frontal lobe in rhesus monkeys. J Comp Neurol 198:121–136
162. Powell TPS, Cowan WM, Raisman G (1965) The central olfactory connexions. J Anat 99:791–813
163. Price JL, Amaral DG (1981) An autoradiographic study of the projections of the central nucleus of the monkey amygdala. J Neurosci 1:1242–1259
164. Price JL, Russchen FT, Amaral DG (1987) The limbic region. II. The amygdaloid complex. In: Bjorklund A, Hökfelt T, Swanson LW (eds) Integrated systems of the CNS, part I. Elsevier, Amsterdam, pp 279–388 (Handbook of chemical neuroanatomy, vol 5)
165. Price JL, Slotnick BM, Revial MF (1991) Olfactory projections to the hypothalamus. J Comp Neurol 306:447–461
166. Puelles L (2001) Brain segmentation and forebrain development in amniotes. Brain Res Bull 55:695–710

167. Puelles L, Rubenstein JLR (2002) Forebrain. In: Ramachandran VS (ed) Encyclopedia of the human brain, vol 2. Academic Press, Amsterdam, pp 299–315
168. Puelles L, Kuwana E, Puelles E et al (2000) Pallial and subpallial derivatives in the embryonic chick and mouse telencephalon, traced by the expression of the genes Dlx-2, Emx-1, Nkx-2.1, Pax-6, and Tbr-1. J Comp Neurol 424:409–438
169. Rae ASL (1954) The form and structure of the human claustrum. J Comp Neurol 100:15–40
170. Ragsdale CW Jr, Graybiel AM (1988) Fibers from the basolateral nucleus of the amygdala selectively innervate striosomes in the caudate nucleus of the cat. J Comp Neurol 269:506–522
171. Raisman G (1966) An experimental analysis of the efferent projection of the hippocampus. Brain 88:963–996
172. Real MA, Dávila JC, Guirado S (2003) Expression of calcium-binding proteins in the mouse claustrum. J Chem Neuroanat 25:151–160
173. Renaud LP, Hopkins DA (1977) Amygdala afferents from the mediobasal hypothalamus: an electrophysiological and neuroanatomical study in the rat. Brain Res 121:201–213
174. Ricardo JA (1983) Hypothalamic pathways involved in metabolic regulatory functions, as identified by tracktracing methods. Adv Metab Dis 10:1–30
175. Ricardo JA, Koh ET (1977) Direct projections from the nucleus of the solitary tract to the hypothalamus, amygdala, and other forebrain structures in the rat. Anat Rec 187:693
176. Robertson RT (1983) Efferents of the pretectal complex: separate populations of neurons project to lateral thalamus and to inferior olive. Brain Res 258:91–95
177. Rolls ET (1984) Neurons in the cortex of the temporal lobe and in the amygdala of the monkey with responses selective for faces. Hum Neurobiol 3:209–222
178. Rose M (1928) Die Ontogenie der Inselrinde. J Psychol Neurol 36:182–209
179. Rosene DL, Heimer L (1977) Olfactory bulb efferents in the rhesus monkey. Anat Rec 187:698
180. Rosene DL, Van Hoesen GW (1977) Hippocampal efferents reach widespread areas of the cerebral cortex and amygdala in the rhesus monkey. Science 198:315–317
181. Royet JP, Hudry J, Vigouroux M (2000) Application de l'imagérie cérébrale à l'étude de l'olfaction. In: Christen Y, Collet L (eds) Rencontres IPSEN en ORL. Editions Irvinn, Neuilly-sur-Seine, pp 85–99
182. Royet JP, Zald D, Versace R et al (2000) Emotional responses to pleasant and unpleasant olfactory, visual, and auditory stimuli: a positron emission tomography study. J Neurosci 20:7752–7759
183. Russchen FT (1982) Amygdalopetal projections in the cat. I. Cortical afferent connections. A study with retrograde and anterograde tracing techniques. J Comp Neurol 206:159–179
184. Russchen FT (1982) Amygdalopetal projections in the cat. II. Subcortical afferent connections. A study with retrograde tracing techniques. J Comp Neurol 207:157–176
185. Russchen FT, Price JL (1984) Amygdalostriatal projections in the rat, topographical organization and fiber morphology shown using the lectin PHA-L as an anterograde tracer. Neurosci Lett 47:15–22
186. Russchen FT, Bakst I, Amaral DG, Price JL (1985) The amygdalostriatal projections in the monkey. An anterograde tracing study. Brain Res 329:241–257
187. Sandrew BB, Edwards DL, Poletti CE, Foote WE (1986) Amygdalospinal projections in the cat. Brain Res 373:235–239
188. Saper CB, Loewy AD (1980) Efferent connections of the parabrachial nucleus in the rat. Brain Res 197:291–317
189. Saper CB, Swanson LW, Cowan WM (1976) The efferent connections of the ventromedial nucleus of the hypothalamus of the rat. J Comp Neurol 169:409–442
190. Saunders RC, Rosene DL (1988) A comparison of the efferents of the amygdala and the hippocampal formation in the rhesus monkey: I. Convergence in the entorhinal, prorhinal, and perirhinal cortices. J Comp Neurol 271:153–184
191. Saunders RC, Rosene DL, Van Hoesen GW (1988) Comparison of the efferents of the amygdala and the hippocampal formation in the rhesus monkey: II. Reciprocal and non-reciprocal connections. J Comp Neurol 271:185–207
192. Schwaber JS, Kapp BS, Higgins G (1980) The origin and extent of direct amygdala projections to the region of the dorsal motor nucleus of the vagus and the nucleus of the solitary tract. Neurosci Lett 20:15–20
193. Seltzer B, Pandya DN (1983) The distribution of posterior parietal fibers in the corpus callosum of the Rhesus monkey. Exp Brain Res 49:147–150
194. Sesack SR, Deutch AY, Roth RH, Bunney BS (1989) Topographical organization of the efferent projections of the medial prefrontal cortex in the rat: an anterograde tract-tracing study with Phaseolus vulgaris leucoagglutinin. J Comp Neurol 290:213–242
195. Shaikh MB, Brutus M, Siegel HE, Siegel A (1985) Forebrain structures regulating flight behavior in the cat. Brain Res Bull 14:217–221
196. Sherk H (1986) The claustrum and the cerebral cortex. In: Jones EG, Peters A (eds) Cerebral cortex, vol 5. Plenum, New York, pp 467–499
197. Siebert M, Markowitsch HJ, Bartel P (2003) Amygdala, affect and cognition: evidence from 10 patients with Urbach-Wiethe disease. Brain 126:2627–2637
198. Sofroniew MY (1983) Direct reciprocal connections between the bed nucleus of the stria terminalis and dorsomedial medulla oblongata: evi-

dence from immunohistochemical detection of tracer proteins. J Comp Neurol 213:399–405
199. Stefanacci L, Amaral DG (2000) Topographic organization of cortical inputs to the lateral nucleus of the macaque monkey amygdala: a retrograde tracing study. J Comp Neurol 421:52–79
200. Stefanacci L, Amaral DG (2002) Some observations on cortical inputs to the macaque monkey amygdala: an anterograde tracing study. J Comp Neurol 451:301–323
201. Stefanacci L, Farb CR, Pitkänen A et al (1992) Projections from the lateral nucleus to the basal nucleus of the amygdala: a light and electron microscopic PHA-L study in the rat. J Comp Neurol 323:586–601
202. Stephan H, Andy OJ (1977) Quantitative comparison of the amygdala in insectivores and primates. Acta Anat (Basel) 98:130–153
203. Stephan H, Frahm HD, Baron G (1987) Comparison of brain structure volumes in Insectivora and primates. VII. Amygdaloid components. J Hirnforsch 28:571–584
204. Swanson LW (1976) An autoradiographic study of the efferent connections of the preoptic region in the rat. J Comp Neurol 167:227–256
205. Swanson LW, Cowan WM (1977) An autoradiographic study of the organization of the efferent connections of the hippocampal formation in the rat. J Comp Neurol 172:49–84
206. Swanson LW, Cowan WM (1979) The connections of the septal region in the rat. J Comp Neurol 186:621–655
207. Swanson LW, Petrovich GD (1998) What is the amygdala? Trends Neurosci 21:323–331
208. Takeuchi Y, Matsushima R, Hopkins DA (1983) Direct amygdaloid projections to the dorsal motor nucleus of the vagus nerve: a light and electron microscopic study in the rat. Brain Res 280:143–147
209. Tanaka M, Ikeda T, Hayashi S et al (1997) Nitrergic neurons in the medial amygdala project to the hypothalamic paraventricular nucleus of the rat. Brain Res 777:13–21
210. Tanne-Gariépy J, Boussaoud D, Rouiller EM (2002) Projections of the claustrum to the primary motor, premotor, and prefrontal cortices in the macaque monkey. J Comp Neurol 454:140–157
211. Terenzi MG, Ingram CD (1995) A combined immunocytochemical and retrograde tracing study of noradrenergic connections between the caudal medulla and bed nuclei of the stria terminalis. Brain Res 672:289–297
212. Turner BH, Herkenham M (1991) Thalamoamygdaloid projections in the rat: a test of the amygdala's role in sensory processing. J Comp Neurol 313:295–325
213. Turner BH, Gupta KC, Mishkin M (1978) The locus and cytoarchitecture of the projection areas of the olfactory bulb in Macaca mulatta. J Comp Neurol 177:381–396
214. Turner BH, Mishkin M, Knapp M (1980) Organization of the amygdalopetal projections from modality-specific cortical association areas in the monkey. J Comp Neurol 191:515–543
215. Valverde F (1965) Studies on the piriform lobe. Harvard University Press, Cambridge
216. Van der Kooy D, Koda LY, McGinty JF, Gerfen CR, Bloom FE (1984) The organization of projections from the cortex, amygdala, and hypothalamus to the nucleus of the solitary tract in rat. J Comp Neurol 224:1–24
217. Van Hoesen GW (1981) The differential distribution, diversity and sprouting of cortical projections to the amygdala in the rhesus monkey. In: Ben-Ari Y (ed) The amygdaloid complex. Elsevier North-Holland, Amsterdam, pp 77–90
218. Van Hoesen GW (1985) Neural systems of the nonhuman primate forebrain implicated in memory. Ann NY Acad Sci 444:97–112
219. Veening JG (1978) Cortical afferents of the amygdaloid complex in the rat: an HRP study. Neurosci Lett 8:191–195
220. Veening JG (1978) Subcortical afferents of the amygdaloid complex in the rat: an HRP study. Neurosci Lett 8:197–202
221. Walter A, Mai JK, Lanta L, Gorcs T (1991) Differential distribution of immunohistochemical markers in the bed nucleus of the stria terminalis in the human brain. J Chem Neuroanat 4:281–298
222. Webster MJ, Ungerleider LG, Bachevalier J (1991) Connections of inferior temporal areas TE and TEO with medial temporal-lobe structures in infant and adult monkeys. J Neurosci 11:1095–1116
223. Weiskrantz L (1956) Behavioral changes associated with ablation of the amygdaloid complex in monkeys. J Comp Physiol Psychol 49:381–391
224. Weller KL, Smith DA (1982) Afferent connections to the bed nucleus of the stria terminalis. Brain Res 232:255–270
225. Whitlock DG, Nauta WJH (1956) Subcortical projections from the temporal neocortex in Macaca mulatta. J Comp Neurol 106:183–212
226. Wójcik S, Dziewiatkowski J, Spodnik E et al (2004) Analysis of calcium binding protein immunoreactivity in the claustrum and the endopiriform nucleus of the rabbit. Acta Neurobiol Exp 64:449–460
227. Yasui Y, Itoh K, Sugimoto T, Kaneko T, Mizuno N (1987) Thalamocortical and thalamo-amygdaloid projections from the parvicellular division of the posteromedial ventral nucleus in the cat. J Comp Neurol 257:253–268
228. Yasui Y, Breder CD, Saper CB, Cechetto DF (1991) Autonomic responses and efferent pathways from the insular cortex in the rat. J Comp Neurol 303:355–374
229. Zald DH, Pardo JV (1997) Emotion, olfaction, and the human amygdala: amygdala activation during aversive olfactory stimulation. Proc Natl Acad Sci USA 94:4119–4124

14 Telencephalon: Basal Ganglia

Introduction

The term basal ganglia refers to a group of closely connected cell masses forming a continuum extending from the basis of the telencephalon, via the central part of the diencephalon, into the tegmentum of the mesencephalon. This complex classically encompasses the striatum (the nucleus caudatus and the putamen), the globus pallidus, the subthalamic nucleus and the substantia nigra.

During the first decades of the previous century, the concept was developed that two independent systems, the pyramidal and the extrapyramidal, converge on the bulbar and spinal motor apparatus. In contrast to the direct corticospinal, pyramidal system, the extrapyramidal system was thought to be an array of centres which, together with their emerging fibres, constitute a multisynaptic descending system. Striopallidal, pallidoreticular and reticulospinal pathways were considered to be the principal links in this system. Its highest centre, the striatum, was believed to receive its main input from the thalamus. Experimental hodological studies have shown that, although these pathways do indeed exist, the striatum and the other 'extrapyramidal' centres are not interconnected in a unidirectional, chain-like fashion. Rather, they and their emerging fibre systems constitute a number of interrelated loops or circuits, from which output systems emerge at several levels. However, it appears that the idea of independently operating cortical pyramidal and subcortical extrapyramidal motor systems has to be abandoned, as the cortex, including its sensory and motor fields, projects strongly to the caudate-putamen complex and the principal output channel of the latter, via the globus pallidus and the thalamus, converges on the motor, premotor and prefrontal areas of the cortex. The term extrapyramidal motor system is convenient, however, and still frequently used because it denotes a highly interconnected array of subcortical centres and, in humans, these centres are implicated in the pathogenesis of a number of distinct motor disturbances, including involuntary movements, muscular rigidity, tremor and akinesia.

The concept of 'basal ganglia' has been considerably widened by Heimer and his associates [179, 181, 182]. These authors advanced a view, the essence of which may be summarized as follows: The caudate nucleus, putamen and globus pallidus, as usually delineated, represent only the dorsal part of the striatal complex. The nucleus accumbens, which closely resembles the caudate nucleus and the putamen both cytoarchitectonically and histochemically, and the greater part of the olfactory tubercle should be considered as a ventral portion of the striatum. The rostral part of the substantia innominata, another basal forebrain structure, represents a ventral extension of the globus pallidus. The ventral striatum and the ventral pallidum are nodal points in a loop system which forms a striking pendant to the aforementioned principal striatal circuit. This loop includes the allocortex, the nucleus accumbens (i.e. the principal part of the ventral striatum), the ventral pallidum, the mediodorsal thalamic nucleus, the prefrontal, prelimbic and cingulate cortical regions, and premotor cortical area 6. The well-known fact that the nucleus accumbens receives a strong dopaminergic input from the ventral tegmental area, whereas the caudate-putamen complex receives a similar projection from the substantia nigra, further substantiates the interpretation of the nucleus accumbens as a part of the ventral striatum.

Given that the allocortex forms part of the limbic system, Heimer et al. [182] conjectured that, whereas the dorsal striatopallidal system plays a pre-eminent role in initiating motor activities stemming from cognitive activities, the ventral striatopallidum has a role in initiating movements in response to emotionally or motivationally powerful stimuli.

A concept related to that of Heimer and his associates was put forward by Kelley et al. [218]. These authors found that, in the rat, a robust amygdalostriatal projection is present, which is distributed to all parts of the caudate-putamen complex except its antero-dorsolateral quadrant. They pointed out that this projection widely overlaps the striatal projections from the hippocampus, the cingulate cortex, the ventral tegmental area and the mesencephalic raphe nuclei. Like the amygdalostriatal system, all of these striatal afferents avoid the antero-dorsolateral striatal sector. Kelley et al. [218] further established that this striatal sector represents the main region to which the cortico-striatal projection is distributed from the sensorimotor cortex. In view of these findings, they interpreted the large striatal region receiving a direct projection from the amygdala as 'limbic' and the remainder as the 'non-limbic' striatal compartment. Kelley et al. [218] suggested that the amygdala, which is known to play a role in the neural mechanism underlying motivation and adaptive behaviour, has access to the initiation and patterning of somatomotor behaviour via its massive projection to the striatum. It is important to note that the nucleus accumbens is included in both the "ventral" striatum and the "limbic" striatum, but that the latter entity extends further dorsally than the former.

Experimental neuroanatomical studies [113, 114, 354] have shown that a substantial amygdalostriatal projection is also present in the monkey. In these animals the target area of the amygdala includes the accumbens nucleus, a ventromedial zone of the caudate nucleus extending caudally into the tail of that structure and the ventral part of the putamen.

Although the basal ganglia were considered to be exclusively concerned with motor control for a very long time, during the past few decades it has become increasingly clear that these structures are also involved in cognitive functions such as procedural learning and working memory tasks.

Cell Masses

The basal ganglia, as defined here, form a functional rather than a morphological entity. As extensively discussed in Chap. 2, the substantia nigra stems from two different mesencephalic matrix zones (Fig. 2.18), the subthalamic nucleus develops from the caudal hypothalamic area of the secondary prosencephalon (Fig. 2.11), whereas the basal ganglia in the stricter sense, i.e. the caudate nucleus, the putamen and the globus pallidus, are derivatives of the telencephalic subpallium. The subpallium comprises three longitudinal zones, lateral, intermediate and medial, all of which participate in the formation of the basal ganglia. The lateral or striatal zone gives rise to the caudate nucleus and the putamen; the external and internal segments of the globus pallidus stem from the intermediate or pallidal and the medial or entopeduncular zone, respectively (Figs. 2.24, 2.25, Table 11.1).

In the adult human brain, the *caudate nucleus* is an elongated, curved cell mass, which is in direct contact with the lateral ventricle throughout its extent (Figs. 3.21–3.24). It consists of an enlarged rostral part known as the *head* of the caudate nucleus, which protrudes into the anterior horn of the lateral ventricle; a short intermediate *body,* which lies dorsolateral to the thalamus; and a long *tail,* which enters the temporal lobe and terminates in the vicinity of the amygdaloid complex. The *accumbens nucleus* is a medial expansion of the head of the caudate nucleus, which extends around the ventral aspect of the lateral ventricle into the medial or septal wall of the hemisphere (Figs. 3.24, 5.34).

During development, increasing numbers of fibres passing to and from the dorsal, pallial parts of the hemispheres penetrate the striatal zone. In primates including humans, these fibres unite to form a more or less continuous *internal capsule,* dividing the striatal zone into a dorsomedial *caudate nucleus* and a ventrolateral *putamen* (Figs. 3.24, 11.2 B, 14.1 A, 5.3–5.8). It is important to note that the separation of these structures is always incomplete. Rostroventrally and caudally, the caudate nucleus and the putamen remain connected and, at intermediate levels, striatal cell bridges are interspersed between the bundles of the internal capsule (Fig. 3.24). In some mammalian groups, including the insectivores and rodents, the corticofugal and corticopetal fibres passing through the striatum assemble in a large number of small bundles that do not unite to form an internal capsule. In these groups, the undivided striatum is usually designated as the caudate-putamen complex (Fig. 11.2 A).

The caudate-putamen complex, divided or undivided, is by far the largest subcortical cell mass in the mammalian brain. In humans, its fresh volume is, according to Schröder et al. [369], approximately 10 cm^3.

The putamen is situated medial to the insula and constitutes, with the pallidum or globus pallidus, the lentiform nucleus, a cone-shaped complex with the apex directed inwards (Figs. 5.31, 5.32). The putamen constitutes the outer part and the globus pallidus the inner part of the lentiform nucleus. The globus pallidus can be subdivided into an external and an internal segment. The anterior limb of the internal capsule largely separates the caudate nucleus from the putamen, whereas the posterior limb of the internal capsule occupies the space between the lentiform nucleus and the thalamus (Fig. 5.31).

Structurally, the *caudate nucleus* and the *putamen* are identical. They have a homogeneous structure throughout and contain numerous medium-sized neurons, among which conspicuous large cells are sparsely scattered (Figs. 6.35–6.41). A quantitative analysis has revealed that about 100 million medium-sized and some 600,000 large cells are present in the human striatum.

The majority ($\approx 75\%$) of the medium-sized neurons in the striatum belong in the category of medium spiny projection neurons. Afferents from the cerebral cortex, the thalamus and the pars compacta of the substantia nigra converge upon these elements, which are all inhibitory and use GABA as their principal neurotransmitter. In primates, two general types of striatal medium spiny neurons can be distinguished

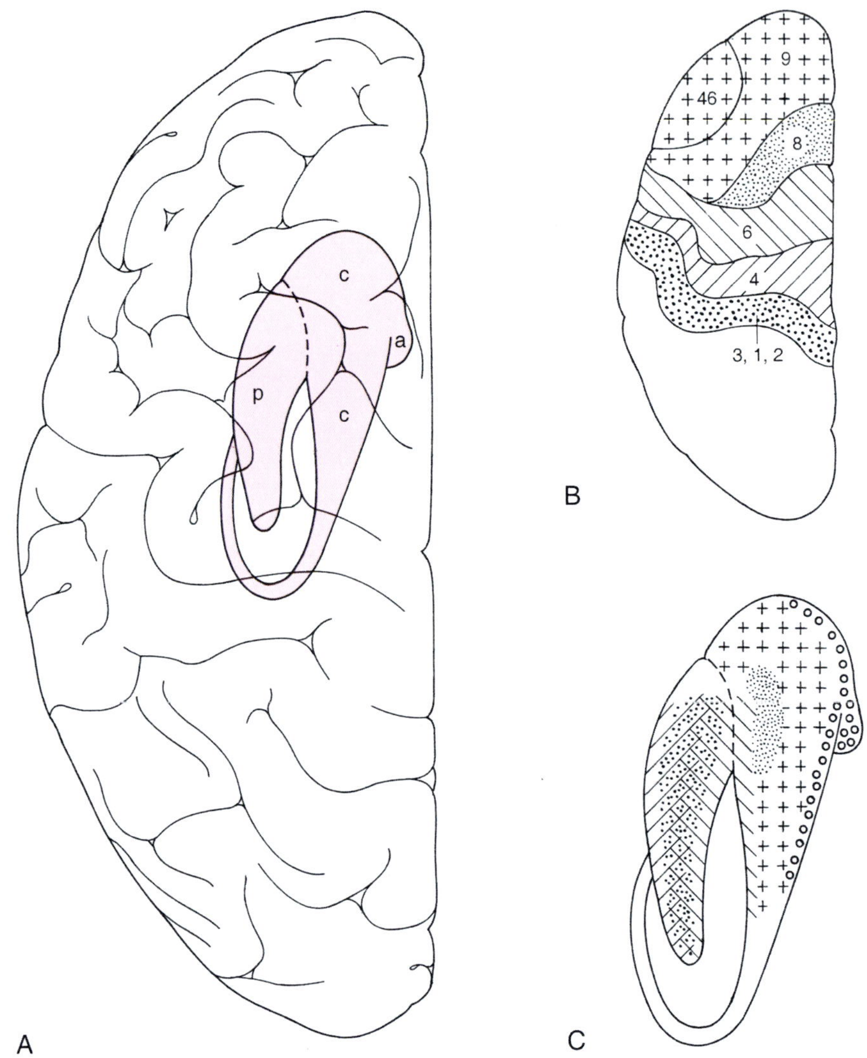

Fig. 14.1. **A** View from above of the left cerebral hemisphere showing the location of the striatum: *a*, accumbens nucleus; *c*, caudate nucleus; *p*, putamen. **B** The position of some cytoarchitectural areas, numbered according to Brodmann, on the upper surface of the right cerebral hemisphere. **C** The striatal projection zones of the cortical areas shown in **B**. The orbital and medial prefrontal cortices project to the ventral striatum. Only a small rim of the projection area of these cortices (indicated by *small, open circles*) is visible from above

according to their neuroactive peptide content and their site of termination. GABAergic striatal neurons containing enkephalin project principally to the external pallidal segment, whereas those containing substance P project to the internal pallidal segment or to the substantia nigra pars reticulata [32, 33, 157, 341].

The caudate nucleus and the putamen also contain, apart from the abundant medium spiny cells, a fair amount of interneurons, which are all aspiny. These include conspicuous large cholinergic cells and several types of small to medium-sized GABAergic elements.

The medium spiny cells occupy a central position in the circuitry of the caudate nucleus and the putamen. The name of these neurons refers to the fact that their dendrites, which radiate and ramify in all directions, are densely covered with spines (Fig. 14.2) [37, 82, 132, 220]. The principal afferent systems to the striatum are glutamatergic fibres from the cerebral cortex [98, 221, 391] and from the intralaminar thalamic nucleus [351], dopaminergic fibres from the substantia nigra [233] and serotoninergic fibres from the dorsal raphe nucleus [243, 319]. All of these extrinsic afferents synapse with the medium spiny cells, the glutamatergic and dopaminergic fibres mainly contacting the distal portion of the dendritic arbour of these neurons. As regards the synaptology of these afferents, Parent and Hazrati [313] pointed out that the terminals of the corticostriatal fibres make asymmetrical contacts, primarily with the heads of dendritic spines, whereas the thalamostriatal fibres make asymmetrical contacts with dendritic shafts (Fig. 14.3). The nigrostrial dopaminergic fibres make symmetrical synapses, and many of them selectively contact the necks of those dendritic spines that receive cortical input (Fig. 14.3). Hence, they are optimally positioned to specifically modulate cortical influences. There is evidence suggesting that the cortical and thalamic afferents to the striatum may not converge upon the same medium spiny cells [86]. The axons of the medium spiny cells, which pass to the globus pallidus and the pars reticulata of the substantia nigra, issue numerous collateral branches within the striatum, which contact other neurons of the same type and provide for lateral or surround inhibition [130, 156].

The large cholinergic interneurons have extensive dendritic trees and elaborate axonal fields (Fig. 14.2). Their excitatory input is derived principally from the thalamostriate system [242]. The axonal branches of the large cholinergic interneurons terminate on other neurons of the same type and on medium spiny cells; however, there is no unanimity in the literature regarding the exact site of termination on the receptive surface of the latter cells. Thus, Parent and Hazrati [313] indicate that the axonal branches of the cholinergic interneurons terminate primarily on the proximal somatodendritic domain of the medium spiny cells (Fig. 14.3), while according to Izzo and Bolam [202] these axonal branches target dendritic shafts and spines, without showing preference for the proximal dendritic sectors. The remaining striatal interneurons, which are all medium to small, aspiny, GABAergic elements, can be subdivided into three main types: I, II and III [159, 216, 429].

The type I GABAergic striatal interneurons are medium-sized and co-contain the calcium-binding protein parvalbumin. Their varicose dendrites as well as their axons branch profusely, while remaining relatively close to the soma (Fig. 14.2). Their axonal terminals form baskets around the somata of the medium spiny neurons. Remarkably, the dendrites of the type I neurons are connected by gap junctions, indicating that these elements are coupled in a continuous network. Similar GABA- and parvalbumin-containing, electrotonically coupled basket cells are known to be present in both the hippocampus and the neocortex [429]. The type I neurons receive a powerful input from the cortex.

The second type of aspiny, GABA-containing interneuron stains positively for the neuropeptides somatostatin and neuropeptide Y and co-contains the diffusible second messenger nitric oxide. Compared to the type I cells, they have larger somata, longer dendrites and more extended axonal fields (Fig. 14.2). The type II cells also receive their principal input from the cerebral cortex. Their axonal terminals synapse

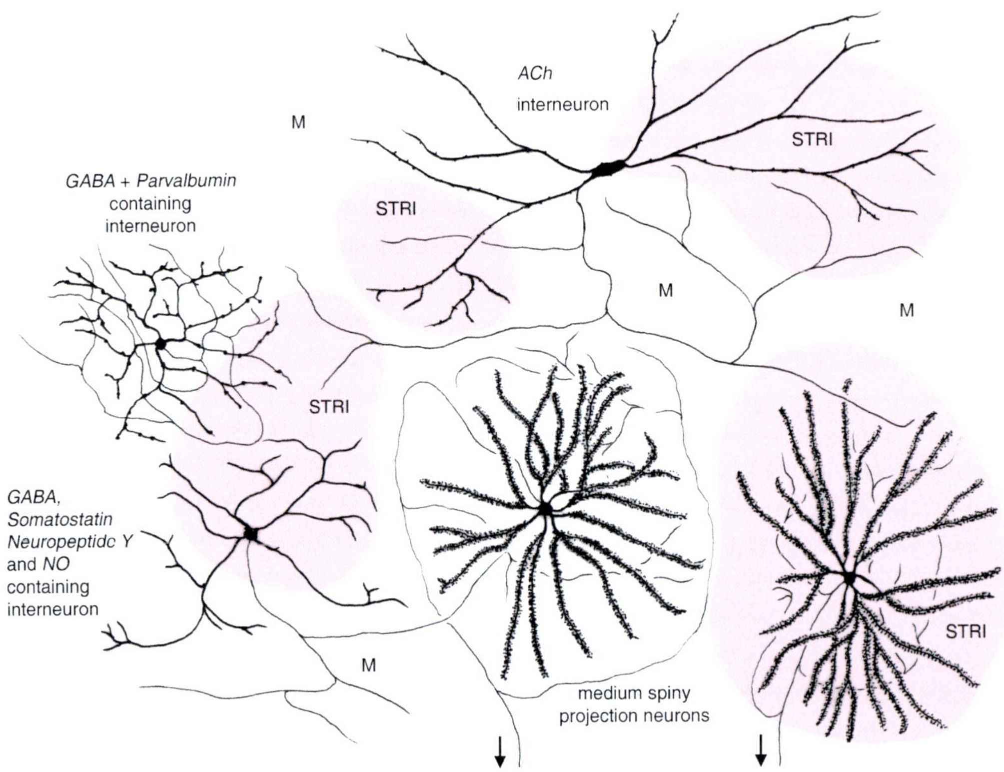

Fig. 14.2. The mosaic internal organization of the striatum (caudate nucleus and putamen) in relation to the dendritic fields and axonal ramifications of the major cell types. The striosomal compartment (*STRI*) is shown in *red*, while the matrix (*M*) is *white*. The cell types shown are discussed in the text. Modified from [428]

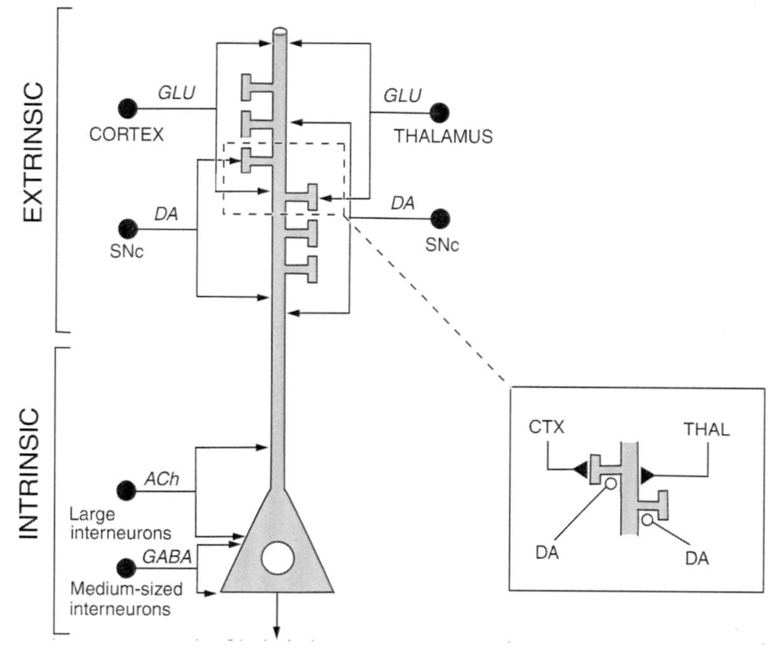

Fig. 14.3. Various extrinsic and intrinsic inputs to the medium spiny projection neurons of the striatum. The *insert* shows the modes of termination of corticostriatal (*CTX*), thalamostriatal (*THAL*) and dopaminergic (*DA*) nigrostriatal afferents. For further explanation, see text. (Reproduced from [313])

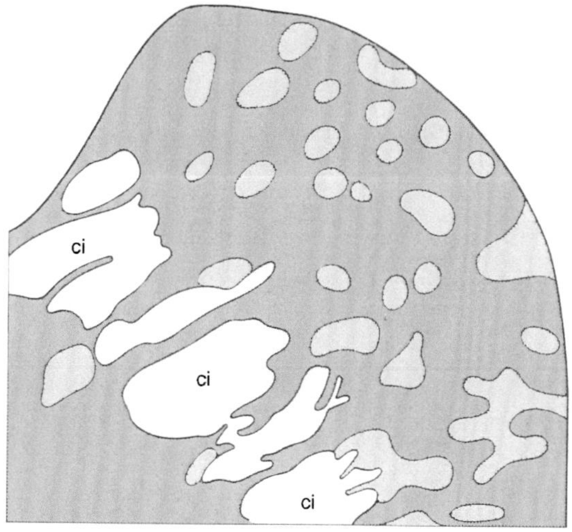

Fig. 14.4. Transverse section through the caudate nucleus of the rhesus monkey, stained for acetylcholinesterase (AChE); striosomes poor in AChE are embedded in an AChE-rich matrix. *ci*, capsula interna

on the somata and proximal dendrites of medium spiny cells, but do not form pericellular baskets. Johannes et al. [208] recently investigated the human putamen by means of the NADPH-diaphorase staining method. This method is selective for nitric oxide synthase, the enzyme which produces nitric oxide. They reported the presence of no less than 12 different subtypes of NOS-containing neurons showing clear morphological differences.

The third type of aspiny, GABAergic interneuron is generally medium-sized and has the distinguishing feature that it contains the calcium-binding protein calretinin [20]. Although these elements are reportedly by far the most abundant in the human striatum [251, 310], little is known as yet about their morphology and connections.

In the striatum, the medium spiny projection neurons are far more abundant than interneurons. In the human putamen the cells of the latter type account for less than 15% of the total number of neurons [131].

A remarkable recent finding is that the striatum harbours a population of dopaminergic neurons [26, 63, 251]. In humans most of these neurons have medium-sized perikarya and three to five varicose dendrites. They are scattered throughout the striatum, but abound particularly in its ventral portion. In monkeys, these elements have been found to increase markedly in response to experimental nigrostriatal dopaminergic deafferentiation [26]. There is evidence suggesting that the intrinsic dopaminergic neurons in the deafferented striata are generated de novo and may stem either from in situ progenitor cells in the striatal parenchyma and/or migration of precursors from the nearby subventricular zone [63, 251].

Although the caudate-putamen complex is cytoarchitectonically a homogeneous structure, histochemical and particularly immunohistochemical studies have revealed a remarkable heterogeneity within the complex. The first evidence for this chemoarchitectural heterogeneity came from studies in which a staining technique for the enzyme acetyl cholinesterase (AChE) was applied. These studies showed that, in the caudate-putamen complex, 300- to 600-μm-wide zones of low AChE activity stand out against an otherwise AChE-rich background. Graybiel and Ragsdale [136–138], who first identified these zones, designated them as striatal bodies or striosomes (Figs. 14.2, 14.4). In recent years, it has gradually become clear that throughout most of the caudate-putamen complex the striosomes and the 'matrix' in which they are embedded represent chemoarchitectonically distinct tissue compartments. These compartments are related to the intrinsic structure of the complex and to the organization of its afferent and efferent connections.

An extensive discussion of the organization of the striatal matrix and striosomal compartments is beyond the scope of the present work. Some aspects of their chemodifferentiation, microcircuitry and connectivity should, however, be mentioned.

1. In addition to a low AChE concentration, the striosomes also show high enkephalin, substance P, GABA and neurotensin immunoreactivity [120, 140].
2. The complementary matrix compartment, in addition to a high AChE concentration, also shows a dense plexus of somatostatin-containing fibres [120].
3. In the striosomes μ-opiate receptor binding is remarkably dense [76, 332] and staining for parvalbumin scarce [337].
4. In the human striatum, within the striosomes, two chemically distinct domains can be distinguished: a central core and a peripheral, ring-like zone [337].
5. In the caudate nucleus of the rhesus monkey, cell clusters, some with cell-poor capsules, appear to match the striosomes [129].
6. In the rat, medium spiny neurons, including their dendritic trees and local axon arborizations, strictly adhere to the striatal compartment boundaries, lying in their entirety either in a striosome or in the matrix [327]. The same holds true for about three quarters of the medium spiny cells in the squirrel monkey. The remaining medium spiny neurons in this species, however, appear to have dendrites that cross from one compartment into the other in both striosomes and matrix [420].

7. The majority of the striatal interneurons are located in the matrix [55, 337]. However, the dendrites of these elements readily cross the borders between matrix and striosomes (Fig. 14.2) and, hence, are considered to form association connections between the two compartments [55, 120, 224, 232].
8. Ultrastructural analysis has shown that in the human striatum, the matrix and striosomal compartments show marked differences in the frequency of occurrence of various types of synapses (symmetric, asymmetric, axospinous, axodendritic etc.) [345].
9. Studies with anterograde tracers have shown that the connectivity of the various striatal afferents to the matrix and striosome compartments is heterogeneous. For example, several cortical areas project to either the striosomes or the matrix, but not to both. Motor and somatosensory cortices project exclusively to the matrix compartment, whereas the limbic cortex preferably innervates the striosomes [84, 223]. In cortical regions that project to both the striosomes and the matrix, particular layers within a given cortical region project to one compartment or the other, but not to both [9, 64, 121, 223]. Afferents from the basolateral amygdaloid nucleus project selectively to striosomes [339], but the terminals of the thalamostriatal fibres avoid these structures [127, 134, 356, 357]. The dopaminergic innervation to the striatum is also heterogeneous, with cells in the ventral zone of the compact part of the substantia nigra projecting to the striosomes, whereas the remaining mesencephalic dopaminergic cell groups prefer the matrix [122, 135, 138].
10. The striatal projection neurons are differentially distributed over the two striatal compartments. After injecting a retrograde tracer into the internal pallidal segment, the external pallidal segment and the substantia nigra, pars reticulata, labelled neurons appear to be most densely distributed in the matrix compartment, whereas after injections centred in the compact part of the substantia nigra, labelled neurons are located preferentially in the striosomes [34, 120, 125].
11. In cats and primates, the projection neurons lying in the extrastriosomal matrix are arranged in clusters [76, 125, 373], which have been called matrisomes [141]. Flaherty and Graybiel [97] tested the hypothesis that these matrisomes consist of neurons projecting to a single extrastriatal target by small paired injections of distinguishable retrograde tracers in the external and internal pallidal segments of squirrel monkeys. It appeared that individual neurons project to only one of these targets, but that, within individual matrisomes, GPe- and GPi-projecting neurons are extensively intermixed. Desban et al. [77] studied the distribution of striatonigral and striatopallidal neurons in the matrix compartment of the caudate nucleus in the cat. They identified two types of striatonigral neurons: poorly collateralized and well-collateralized cells. The former were distributed in clusters and the latter outside these clusters. Neurons innervating the entopeduncular nucleus (i.e. the feline homologue of the primate internal pallidal segment) were distributed outside the matrix clusters of aggregated striatonigral neurons and appeared to be intermingled with the non-aggregated, well-collateralized striatonigral neurons.
12. Holt et al. [191] compared the distribution of a large number of chemical markers in different parts of the human striatum. Only in the dorsal parts of the caudate nucleus and the putamen did the distribution patterns of these chemicals adhere to the classical striosome-matrix organization. The findings in the ventral parts of these structures were indicative of a more complicated pattern of compartmentalization.

From the foregoing synopsis it can be concluded that the caudate-putamen complex displays a complex mosaic-like heterogeneity and that many structural and connectional features fit into mosaic. The functional significance of this compartmentalization remains to be elucidated.

The *accumbens nucleus*. As has already been pointed out, Heimer and colleagues [179, 181, 182] proposed on structural, histochemical and connectional grounds that together the nucleus accumbens and medium-sized cell territories of the olfactory tubercle represent the ventral section of the striatum. This concept is corroborated by the fact that the dorsal and ventral sectors of the striatum both receive a robust dopaminergic projection from the ventral midbrain [252, 288] and by the results of some experimental hodological studies. Thus, it appears that each of the major cortical regions projects to a defined, but partially overlapping longitudinal zone of the striatum: the sensorimotor area projects onto the dorsolateral part of the caudate-putamen; the associational areas of the neocortex project to a more medial sector of the caudate-putamen; the mesocortex projects mainly to the medial and ventral regions of the caudate-putamen, but also to the ventral striatum; and, finally, the allocortex projects mainly to the ventral striatum, but also to the medial and ventral parts of the caudate-putamen [268, 329]. Moreover, Haber et al. [163] concluded that, with regard to its efferent connections, the nucleus accumbens cannot be distinguished from other parts of the striatum.

The dorsal and ventral striatal sectors show a structural and connectional continuity; this does not necessarily mean, however, that all parts of the striatum are equivalent. Groenewegen and Russchen [146] demonstrated that, in the cat, the efferents of the lateral and medial parts of the accumbens, though very similar, are not identical. Using the patterns of Timm's sulfide-silver technique for zinc, AChE activity and cholecystokinin immunoreactivity, Záborszky et al. [436] showed that the accumbens nucleus can be divided into a peripheral and a central subterritory, which they referred to as the 'shell' and 'core', respectively. Subsequent studies have revealed that such a distinction can also be made with antibodies against substance P and calbindin-28, a calcium-binding protein (for a review see Zahm and Brog [438]). The core region, which merges dorsally with the caudate-putamen, shows a complex striosome/matrix-like compartmentalization. The shell region also shows chemoarchitectural inhomogeneities, but these bear little resemblance to those in the core (for details, see Groenewegen et al. [149], Jongen-Rêlo et al. [211, 212]). Experimental hodological studies (summarized in Jongen-Rêlo et al. [212]) indicate that the afferent and efferent connections are also different in the core and shell regions.

The *olfactory tubercle* forms part of the basal olfactory area, i.e. the telencephalic territory that receives direct projections from the olfactory bulb (see Chap. 11 "Olfactory System"). It is situated directly behind the anterior olfactory nucleus and superficially covers the head of the caudate nucleus and the nucleus accumbens. Its size is directly proportional to that of the olfactory bulb [399]. In macrosmatic mammals, it forms a prominent mass on the ventral surface of the hemisphere (hence its name).

In most mammals, three layers can be distinguished in the olfactory tubercle: an external plexiform layer, an intermediate densocellular layer of pyramidal and ellipsoid cells and a deeper layer of loosely arranged polymorphic cells. At many places, bridges of cells extend between the polymorphic cell layer and the ventral striatum-accumbens region. A characteristic feature of the olfactory tubercle is the presence of numerous islets of closely crowded, mostly small cells. These so-called islands of Calleja can occur in all three layers.

The olfactory tubercle receives a direct projection from the olfactory bulb, which, however, is mainly confined to its rostral and lateral parts. Heimer and Wilson [41] incorporated the polymorphic layer of the olfactory tubercle into the striatal domain because (a) this layer is structurally continuous with the head of the caudate nucleus and the nucleus accumbens; (b) the ventral parts of the striatum, the nucleus accumbens and the polymorphic cell layer form a continuous projection field for allocortical fibres originating from the hippocampus and the prepiriform cortex and for dopaminergic fibres from the ventral tegmental area; and (c) the projection areas of the accumbens and the olfactory tubercle are contiguous and outline a discrete, unitary region underneath the anterior commissure.

The *globus pallidus* or *pallidum* is a triangular mass of cells lying along the medial aspect of the putamen. It is separated from the latter by a fibre sheet, the external medullary lamina. The globus pallidus differs considerably in structure from the striatal nuclei. It is chiefly composed of large, widely spaced, fusiform cells (Figs. 6.37–6.40). The total number of pallidal cells in humans is about the same as that of the large striatal neurons, i.e. 600,000 [409]. Many bundles of myelinated fibres traverse the globus pallidus and in fresh preparations give it a paler color than the putamen or the caudate nucleus. An internal medullary lamina divides the globus pallidus into internal and external segments.

The pallidal neurons have long, sparsely spiny, poorly branching dendrites. Together, the dendrites of individual cells form flattened, disc-like territories oriented parallel to the external border of the pallidum and perpendicular to the incoming striatal fibres. These striatofugal axons, which are thin and thinly myelinated, issue numerous thin, unmyelinated collaterals parallel to the pallidal dendrites, with which they repeatedly synapse [81]. It has already been mentioned that the striatal afferents projecting to the external pallidal segment express enkephalin, whereas those projecting to the internal pallidal segment express substance P.

All large neurons within both pallidal segments are GABAergic, inhibitory projection neurons. The elements in the external segment project to the subthalamic nucleus, whereas those in the internal segment project to the thalamus, the lateral habenular nucleus and the pedunculopontine nucleus in the caudal mesencephalic tegmentum.

Substantia Innominata, Basal Nucleus of Meynert, Magnocellular Basal Forebrain and Ventral Pallidum. The *substantia innominata* is a rather ill-defined, flattened cell mass situated directly ventral to the putamen and the globus pallidus; it is partly separated from the latter by the anterior commissure. Its rostral part covers the olfactory tubercle; medially, it borders on the lateral pre-optic and lateral hypothalamic areas, and its caudal part is situated between the globus pallidus and the nucleus centralis amygdalae (Figs. 6.38–6.41). The substantia innominata is mainly composed of rather loosely arranged, medium-sized cells, but also contains large, deeply stained, fusiform and multipolar neurons. The population of large cells is collectively designated as the *basal nucleus of Meynert*. It becomes progressively larger and more conspicuous with increasing cerebralization, attaining its greatest development in cetaceans and primates. The basal nucleus is known to project mainly to the neocortex [83, 222, 323].

The mediobasal telencephalon contains groups of large, multipolar neurons, which are collectively denoted as the *magnocellular basal forebrain* (MBF). This MBF can be topographically subdivided into a rostromedial, an intermediate and a caudolateral part. The rostromedial part is situated in the medial septal nucleus; the intermediate part co-extends roughly with the vertical and horizontal limbs of the nucleus of the diagonal band of Broca (DBBv, DBBh), whereas the large caudolateral part is embedded in the substantia innominata and corresponds to the basal nucleus of Meynert. The MBF is neurochemically heterogeneous and comprises cholinergic, GABAergic and nitric oxide (NO)-containing cells. The cholinergic MBF neurons, which project in a topographically ordered fashion to the neocortex and several other telencephalic centres, were subdivided by Mesulam et al. [274] into four groups: Ch1–Ch4, a parcellation that can also be applied to the human brain [178, 360]. The Ch1 group is found in the medial septal nucleus, the Ch2 group corresponds to the DBBv, the small Ch3 group is embedded in the DBBh and the Ch4 group forms part of the basal nucleus of Meynert. According to Mesulam et al. [274] the cholinergic neurons that constitute Ch4 can be separated into five subgroups: anteroventral (Ch4 av), anterolateral (Ch4 al), intermedioventral (Ch4 iv), interomediodorsal (Ch4 id) and posterior (Ch4 p).

Large GABAergic neurons are distributed over all parts of the MBF, without forming clearly distinct subpopulations [361]. A roughly equal number of cholinergic and GABAergic MBF neurons project to the cerebral cortex [361]. The NO-containing neurons in the basal forebrain project to microvessels in the neocortex and are supposed to play a role in the regulation of cortical blood flow [410].

Heimer and Wilson [181] found that, in the rat, fibres originating from the nucleus accumbens and the olfactory tubercle converge on the rostral, subcommissural part of the substantia innominata. Similar connections have been found in the cat [146] and monkey [163]. Since this ventral part of the substantia innominata appears to be related to the ventral striatum in the same way as the main or dorsal part of the globus pallidus is related to the caudate-putamen complex, it has been referred to as the *ventral pallidum*. Within the ventral pallidum, separate subcommissural and ventral parts have been distinguished by the differential distribution of enkephalin and substance P immunoreactivity and by differences with respect to input and output characteristics [150, 163, 164, 396].

The relatively poorly characterized part of the substantia innominata caudal to the ventral pallidum, according to Alheid and Heimer [4], shares many characteristics with the medial and central part of the amygdala; for this reason, they included this caudal part of the substantia innominata with the bed nucleus of the stria terminalis in a new entity, the extended amygdala. The interpretation by Altheid and Heimer [4] was corroborated by studies performed by Grove [152, 153], which showed that the afferent and efferent connections of the caudal part of the substantia innominata of the rat strongly resemble those of neighbouring components of the amygdala.

The *subthalamic nucleus* is a conspicuous cell mass that is situated in the caudal part of the diencephalon, ventral to the zona incerta and dorsal to the internal capsule and its transition into the cerebral peduncle. Its medial part overlies the rostral portion of the substantia nigra (Figs. 6.33, 6.37). The subthalamic nucleus is composed of fairly large, triangular and polygonal cells. The dense dendritic trees of these cells form large ellipsoidal domains parallel to the rostrocaudal axis of the nucleus [338]. The subthalamic neurons receive an excitatory, glutamatergic input from the cortex and an inhibitory, GABAergic input from the external pallidal segment. The subthalamic neurons are excitatory and use glutamate as their neurotransmitter [387]. Their branching axons project to the external and internal segments of the globus pallidus, the ventral pallidum and the substantia nigra part reticulata.

The *substantia nigra* is the largest cell mass of the mesencephalon. Lying between the tegmentum and the cerebral peduncle, its most rostral part extends into the diencephalon and closely approaches the globus pallidus.

According to cytoarchitectonic criteria, the substantia nigra can be subdivided into a dorsal, cell-rich *pars compacta* and a ventral, less cellular *pars reticulata* (Figs. 6.29–6.33). This subdivision into two components is supported by the chemoarchitecture of this nuclear complex. The pars compacta is mainly composed of large, darkly pigmented cells, which synthesize dopamine. (The pigment, neuromelanin, is a by-product of catecholamine metabolism.) The cells in the pars reticulata are somewhat smaller than those in the pars compacta; most of these elements are GABAergic. It is noteworthy that the pars reticulata of the substantia nigra has many features in common with the internal segment of the globus pallidus.

The dopaminergic cells in the pars compacta of the substantia nigra, designated as the A9 group in the terminology of Dahlström and Fuxe [68], form a continuum with the other dopaminergic cell groups, A10 and A8. The former is embedded in the ventral tegmental area and the latter in the nucleus parabrachialis pigmentosus.

The pars compacta of the substantia nigra has been further divided into dorsal and ventral tiers. The cells in the dorsal tier are loosely arranged and have mediolaterally oriented dendritic trees. The dorsal cells merge medially with the adjacent dopamine cell group of the ventral tegmental area. All of the dorsal tier cells, including those in the ventral tegmental area, contain calbindin, a calcium-binding protein. In contrast, the neurons in the densocellular ventral tier are calbindin-negative [167]. The dendrites of these elements are dorsoventrally oriented and penetrate into the underlying pars reticulata [432]. The ventral tier encompasses a number of cell columns that extend from the densocellular zone into the pars reticulata. Dendrites of pars compacta cells are interconnected by dopaminergic synapses as well as by electrotonic gap junctions [413].

The quantitatively most important input to the substantia nigra comes from the striatum and consists of axons of GABAergic, inhibitory medium spiny cells, which co-contain substance P. These axons terminate in the pars reticulata, where they contact both the local neurons and the ventrally extending dendrites of the pars compacta cells [343]. Excitatory afferents to the substantia nigra originate from glutamatergic neurons in the subthalamic nucleus and from glutamatergic and cholinergic neurons in the pedunculopontine tegmental nucleus [35, 245, 277].

The dopaminergic neurons in the pars compacta give rise to a large nigrostriatal system, which reciprocates the striatonigral projection. The efferents from the GABAergic pars reticulata cells include nigrothalamic fibres to certain parts of the ventral anterior and mediodorsal nuclei [49, 198], nigrotectal fibres to the superior colliculus and nigrotegmental fibres to the pedunculopontine tegmental nucleus [194, 344].

The *ventral tegmental area* lies in the ventromedial part of the tegmentum of the midbrain (Fig. 6.33). Rostrally, it is continuous with the lateral hypothalamic area; caudally it extends to the level of the caudal pole of the interpeduncular nucleus.

The *nucleus parabrachialis pigmentosus* forms a thin sheet between the nucleus ruber and the pars compacta of the substantia nigra [170, 227]. The dopaminergic neurons contained within this nucleus are calbindin-positive and form part of the dorsal tier [103].

The *pedunculopontine tegmental nucleus* is the caudalmost cell mass that is clearly integrated in the circuitry of the basal ganglia. It is situated in the caudal part of the tegmentum of the midbrain and extends into the rostralmost part of the pontine tegmentum (Fig. 6.29). This nucleus contains numerous cholinergic neurons, which are collectively designated as cell group Ch5 [272–274] and roughly corresponds to the so-called mesencephalic locomotor region. The latter has been given this name because electrical stimulation of this region in postmamillary decerebrate mammals induces coordinated locomotion on a treadmill [142, 384]. It has been claimed that, at least in the rat, not the pedunculopontine tegmental nucleus itself, but rather a medially adjacent 'midbrain extrapyramidal area' is specifically included in basal ganglia circuitry [246, 355, 398].

Fibre Connections

Introductory Notes

In the course of the last two decades, an extraordinarily large number of experimental studies have appeared on the connections of the striatum and related 'extrapyramidal' centres. Within the limits set for the present work, only a brief and simplified overview of these connections can be presented. For details, the reader is referred to the papers cited and to the reviews by Parent and Hazrati [313, 314], Haber [158] and Haber and Johnson Gdowski [159], in which much of the pertinent literature is compiled.

The Direct Circuit: Cerebral Cortex – Striatum – Globus Pallidus – Thalamus – Cerebral Cortex; or: Cerebral Cortex – Striatum – Substantia Nigra, pars reticulata – Thalamus – Cortex

The circuitry as described in this paragraph is illustrated by Figs. 14.1 B,C, 14.5, 14.6 and 14.7. Contrary to the views of previous workers, it has been established that the whole of the neocortex sends fibres to both the caudate nucleus and the putamen and that all parts of these two cell masses receive fibres from the

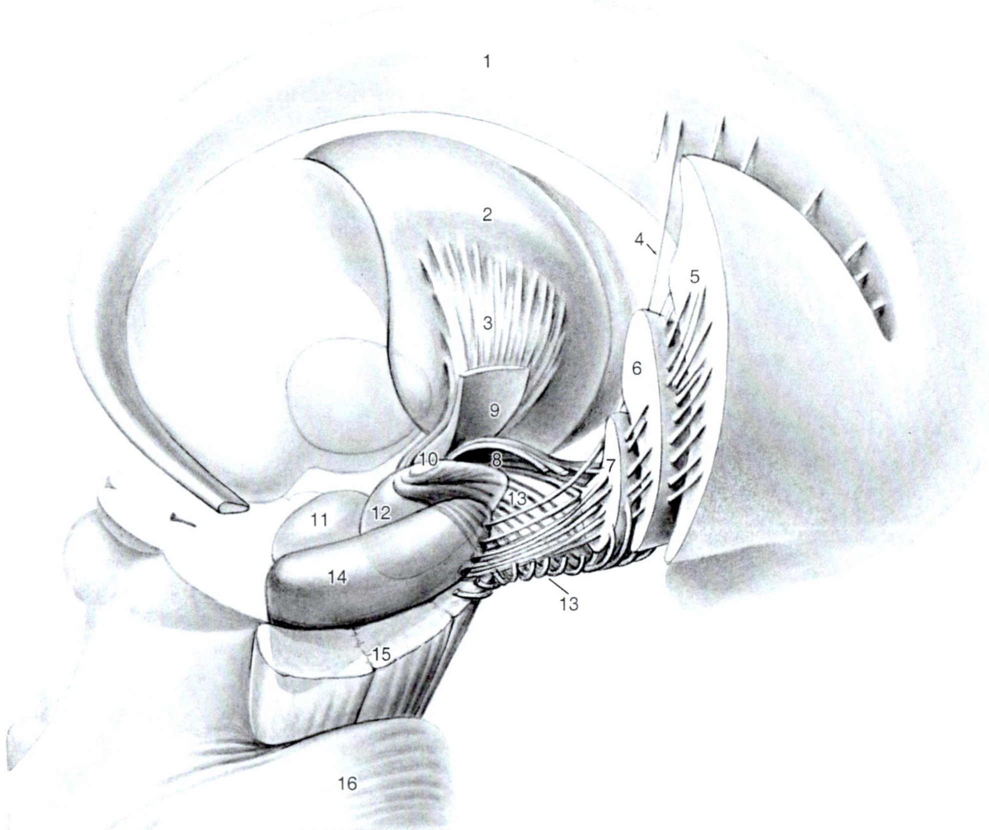

1 Caudate nucleus
2 Ventral lateral thalamic nucleus
3 Thalamic fasciculus
4 Strionigral fibres
5 Putamen
6 Globus pallidus, external part
7 Globus pallidus, internal part
8 Lenticular fasciculus
9 Reticular thalamic nucleus
10 Zona incerta
11 Red nucleus
12 Subthalamic nucleus
13 Ansa lenticularis
14 Substantia nigra
15 Cerebral peduncle
16 Pons

Fig. 14.5. Some nuclei and fibre bundles of the so-called extrapyramidal system in a lateral view (2.5×). Of the fibres originating from the occipital, removed part of the lentiform nucleus, only the ansa lenticularis is represented

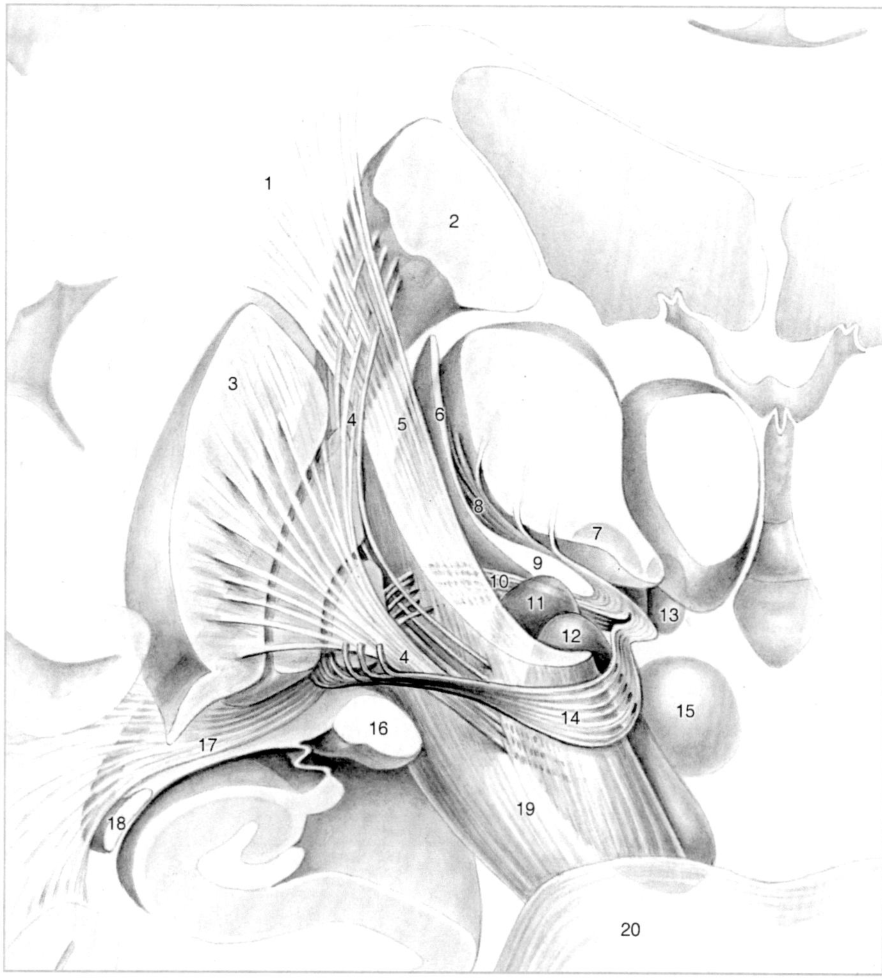

1 Corona radiata
2 Body of caudate nucleus
3 Putamen
4 Strionigral fibres
5 Internal capsule, posterior limb
6 Reticular thalamic nucleus
7 Centromedian thalamic nucleus
8 Thalamic fasciculus
9 Zona incerta
10 Lenticular fasciculus
11 Subthalamic nucleus
12 Substantia nigra
13 Parafascicular thalamic nucleus
14 Ansa lenticularis
15 Red nucleus
16 Optic tract
17 Internal capsule, sublentiform part
18 Tail of caudate nucleus
19 Cerebral peduncle
20 Pons

Fig. 14.6. Some nuclei and fibre bundles of the so-called extrapyramidal system in a frontal view (2.5×). The fibres originating from the removed anterior part of the lentiform nucleus are not illustrated, with the exception of the ansa lenticularis

cortex. It was originally held that this cortico-striate projection is arranged on a simple topographical basis, in that particular cortical areas project to proximal portions of the striatum [44, 219]. However, more recent studies [128, 268, 372] have shown that the terminal fields of the corticostriate fibres originating from circumscribed cortical regions are organized in longitudinally arranged strips or bands, which often span much of the length of the nucleus (Fig. 14.1 B, C). It was formerly held that the sensorimotor cortex projects bilaterally to the striatum, but that the corticostriate projections from the remaining cortical areas are strictly ipsilateral [45, 221]. However, it has been established in the rat that all major regions of the cerebral cortex project to the striatum on both sides of the brain, albeit with strong ipsilateral preponderance [268]. The corticostriate fibres, which use the excitatory amino acid glutamate as their neurotransmitter [98], establish direct synaptic contacts with the striatal efferent neurons, i.e. the medium spiny cells [391].

Efferent fibres from the striatum converge towards the globus pallidus, where they constitute a massive fibre system that passes radially through the external and internal segments of the globus pallidus (GPe, GPi). During its transit through the globus pallidus, this system emits numerous collateral and terminal arborizations at right angles to its stem fibres. These disc-like terminal fields synapse with pallidal neurons, which are likewise oriented perpendicular to the main stream of the striatofugal bundle. The latter then leaves the GPi and passes to the substantia nigra, where its fibres terminate mainly in the pars reticulata (SNr). The striopallidonigral bundle is topographically organized [175, 329]. It was previously believed that the striatal projections terminating in the GPe and GPi are collateral branches of striatonigral fibres [100, 101], but studies using double retrograde tracing [16, 302] have shown that these three projections come mainly from separate cell populations. The vast majority of the striatopallidal and striatonigral fibres are known to contain GABA and to exert an inhibitory influence. In many of these fibres, GABA co-exists with one or several neuropeptides. Thus, neurons projecting to the GPe contain enkephalin, those arborizing in the GPi are enriched with substance P and dynorphin and striatonigral neurons contain either enkephalin, substance P and/or dynorphin [313].

GPi and SNr constitute the principal output structures of the (dorsal) striatum. Both structures project massively to the thalamus. In primates, the fibres emanating from GPi initially constitute two separate bundles, the lenticular fasciculus and the ansa lenticularis. These two bundles merge in Forel's field H, after which they ascend as a single bundle, the thalamic fasciculus, to the rostral part of the thalamus (Figs. 9.1, 14.5, 14.6) [300]. The fibres of this pallidothalamic projection, which is topographically organized [80], terminate principally in the anterior or oral part of the ventral lateral thalamic nucleus and give off collaterals to the centromedian nucleus (Figs. 9.1, 14.7 A) [94, 328]. The nigrothalamic fibres, which, as already mentioned, arise from cells in the pars reticulata, pass rostrally and terminate in the ventral anterior (VA) and in the lateral (paralaminar) part of the mediodorsal thalamic nucleus (MDpl) (Fig. 14.7 B) [49, 198]. Both the pallidothalamic and nigrothalamic projections are known to be GABAergic [102, 326].

The final link in the direct striatal circuits is formed by fibres passing from the target nuclei of the pallidothalamic and nigrothalamic projections to the cerebral cortex (Fig. 14.7). The ventral anterior and ventral lateral thalamic nuclei send their efferent fibres to the motor, premotor and supplementary motor areas, whereas the mediodorsal thalamic nucleus is known to project to the prefrontal cortex, including the frontal eye field. These thalamocortical projections innervate cortical layers I and III, and to a lesser extent layer V [414]. They are glutamatergic and activate cortical neurons.

As a result of these data the following concept was developed to explain the basic function of the synaptic relations between the various elements forming the direct striatal circuit [74, 325, 389, 429]: The fibres of the cortico-striatal projection exert an excitatory influence

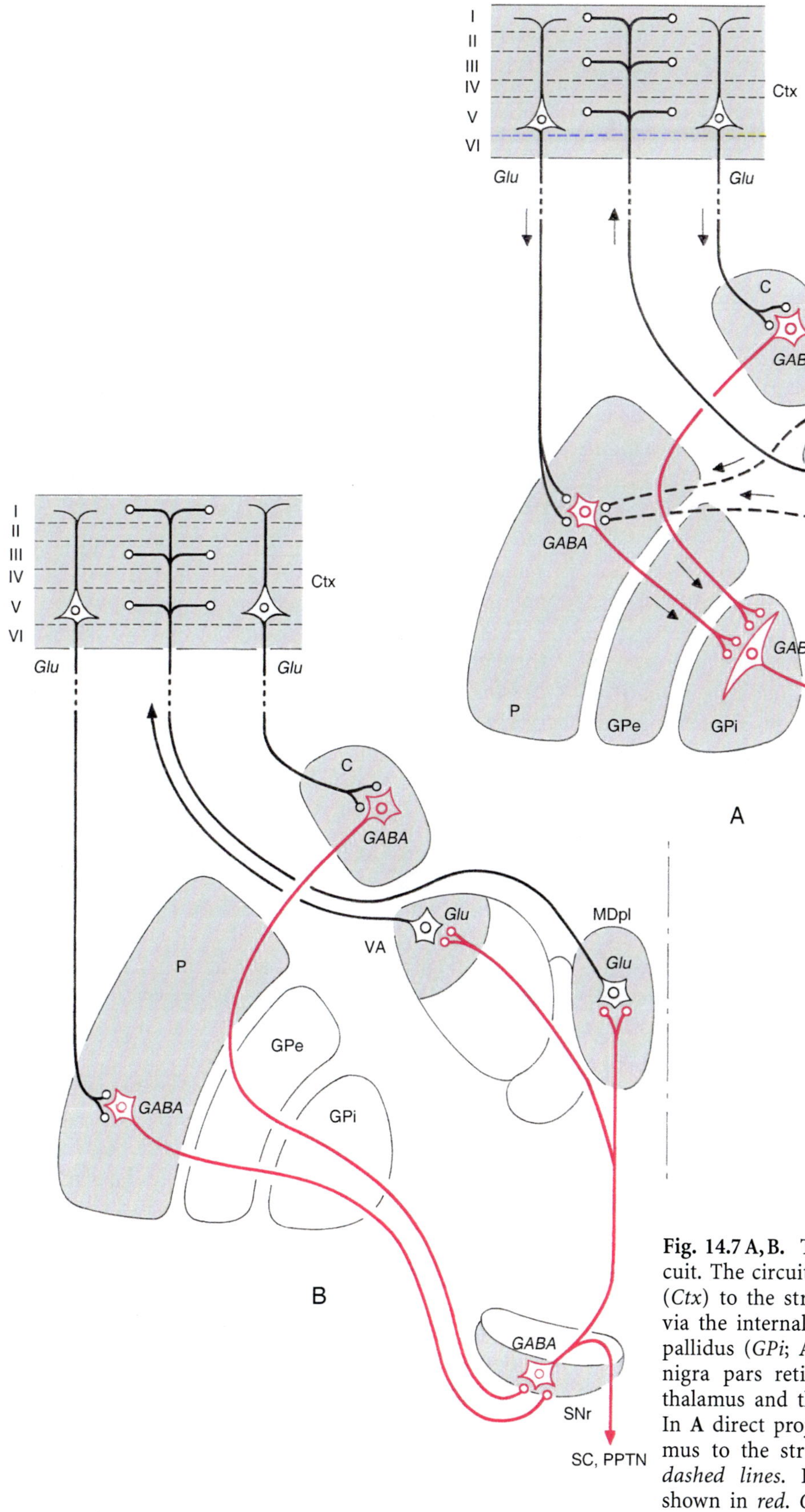

Fig. 14.7 A, B. The direct striatal circuit. The circuit passes from the cortex (*Ctx*) to the striatum, and next, either via the internal segment of the globus pallidus (*GPi*; **A**) or via the substantia nigra pars reticulata (*SNr*; **B**) to the thalamus and then back to the cortex. In **A** direct projections from the thalamus to the striatum are indicated by *dashed lines*. Inhibitory neurons are shown in *red*. *C*, caudate nucleus; *CM*, centromedian thalamic nucleus; *MDpl*, paralaminar part of mediodorsal thalamic nucleus; *P*, putamen; *Pf*, parafascicular thalamic nucleus; *PPTN*, pedunculopontine tegmental nucleus; *SC*, superior colliculus; *VA*, ventral anterior thalamic nucleus; *VL*, ventral lateral thalamic nucleus. For further explanation see text

on the medium spiny neurons. The increased activity in these GABAergic elements leads to a decreased activity of pallidal and SNr neurons. Because these neurons are also GABAergic and inhibitory, their inhibition results in an increase in activity of the thalamocortical neurons on which they impinge. Therefore, excitation of striatal neurons by cortical efferents ultimately produces a *disinhibition* (and hence an activation) of thalamocortical neurons. In other words, the direct circuit supports thalamocortical interactions by *positive feedback* [429].

Recordings of single cell activity in the basal ganglia of awake monkeys have shown that the striatal medium spiny cells fire very rarely and in episodes that only last for about 0.1–3 s, while the pallidal and SNr neurons, on the other hand, fire tonically at very high rates. In doing so, they keep the thalamic neurons in a permanent state of inhibition. However, when phasic excitatory inputs from the cortex transiently activate the direct pathway, the tonically active pallidal and nigral neurons are briefly suppressed, thus permitting the thalamus and cortex to become activated [72, 429].

It should be emphasized that in the foregoing as well as in Fig. 14.7, only a very simplified picture of the neuronal relations in the principal striatal circuit has been presented. For example, the thalamocortical projection is treated as a simple, unidirectional relay back to the cortex. However, it is known that the thalamocortical projections are reciprocated by strongly developed corticothalamic projections and that the flow of information to the cortex is not just passively relayed but, through the intermediary of interneurons, is subjected to complex modulatory influences (Fig. 8.5) [265, 376–378, 401]. Another simplification concerns the neglect of striatal interneurons. It is believed that these elements help to regulate the duration, strength and spatial pattern of the disinhibition resulting from the excitation by the corticostriatal afferents [429].

Subdivision of Direct Striatal Circuit

Initially, the direct striatal circuit described above was taken as the morphological substrate of the motor functions of the basal ganglia (see e.g. Evarts and Thach [93], Kemp and Powell [221]). Information derived from the entire neocortex is processed in the stratum, GPi and the thalamus, respectively, and then fed back into the motor and premotor cortices, i.e. the principal sites of origin of the motor part of the pyramidal tract. Subsequent anatomical and physiological findings promoted the concept that influences from the sensorimotor and association cortices remain segregated in the basal ganglia-thalamocortical pathways. Thus, DeLong and Georgopoulos [73] suggested that two separate loops pass through the basal ganglia: (1) a 'motor' loop passing largely through the putamen, which receives input from the sensorimotor cortex, converging upon certain premotor cortical areas, and (2) a 'complex' loop passing through the caudate nucleus, which receives input from the association areas and whose influences are ultimately returned to certain portions of the prefrontal cortex. Further studies, summarized in references [2, 3], advanced the concept that the basal ganglia participate in five functionally segregated subcircuits or loops: motor, premotor, oculomotor, dorsolateral prefrontal and limbic. Each of these subcircuits is considered to originate in a specific area of the cerebral cortex, to be funnelled through separate sectors of the stratum, the globus pallidus (or the SNr) and the thalamus, and to be ultimately directed back to the same cortical area from which the subcircuit originated. In Figs. 14.1B,C and 14.8, the topographical organization of the initial, corticostriatal and the final, pallido(nigro)-thalamocortical parts of the various subcircuits is presented diagrammatically.

The *motor loop* originates from the primary motor cortex, M1, which corresponds to area 4 of Brodmann. This area is known to send strong projections to somatomotor cell groups in the brain stem and spinal cord via the pyramidal tract. The striatal projections from M1 terminate in the lateral part of the putamen,

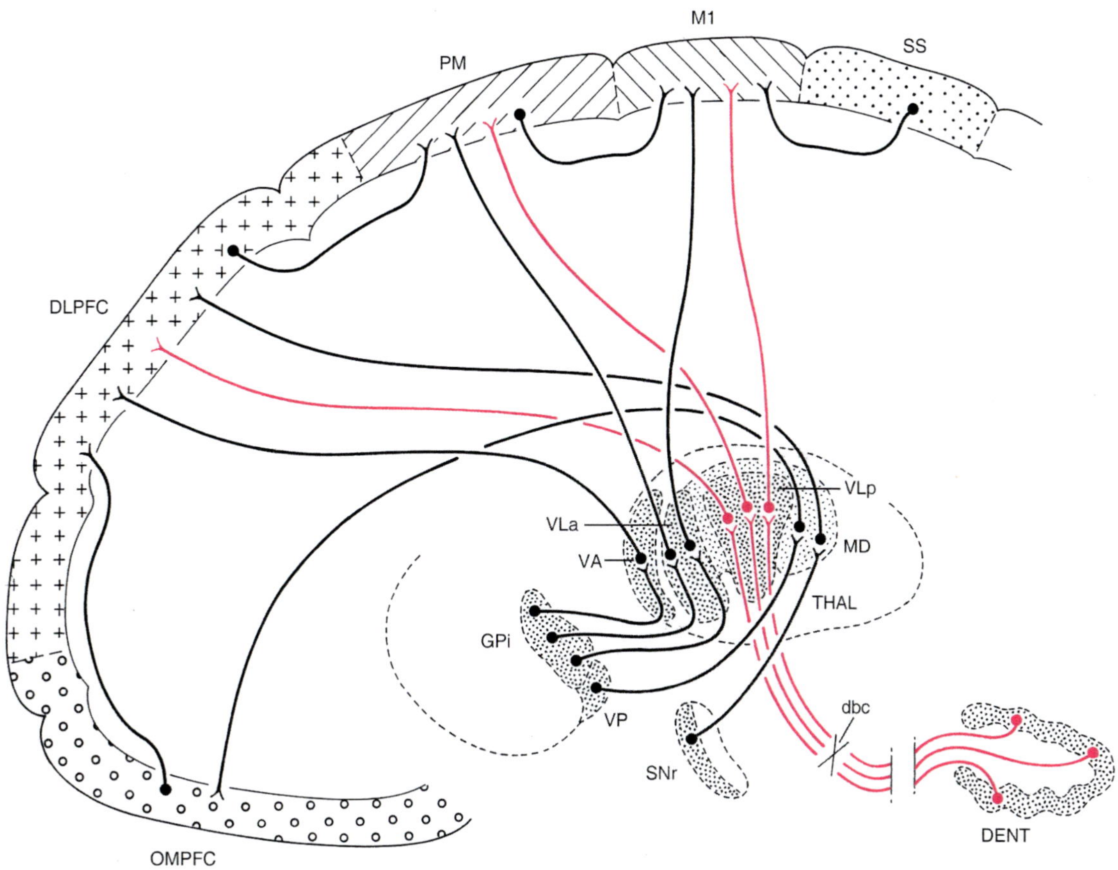

Fig. 14.8. Sagittally oriented scheme, showing that the efferent projections from the basal ganglia (in *black*) and the cerebellar dentate nucleus (in *red*) are synaptically interrupted in separate thalamic nuclei but terminate in overlapping cortical fields in the frontal lobe. *DENT*, dentate nucleus; *DLPFC*, dorsolateral prefrontal cortex; *dbc*, decussation of brachium conjunctivum; *GPi*, internal part of globus pallidus; *MD*, mediodorsal thalamic nucleus; *M1*, primary motor cortex; *OMPFC*, orbital and medial prefrontal cortex; *PM*, premotor cortex; *SNr*, pars reticulata of substantia nigra; *SS*, somatosensory cortex; *VA*, ventral anterior thalamic nucleus; *VLa*, *VLp*, anterior and posterior parts of ventral lateral thalamic nucleus; *VP*, ventral pallidum

but not in its most rostral part [263, 406]. The lateral putamen projects to the posterolateral part of the GPi, which in turn projects to the anterior part of the ventral lateral thalamic nucleus (VLa) [217, 359]. Fibres passing from the VLa to M1 complete the motor loop. M1 and its dependent subcortical areas are concerned with the voluntary control of body movements.

The *premotor loop* originates from area 6 of Brodmann, which includes the dorsolateral premotor area (PM) and the supplementary motor area (SMA) and presupplementary motor area (PreSMA), both of which are situated on the medial aspect of the hemisphere. Area 6 projects to somatomotor centres in the brain stem and spinal cord, but these projections are much sparser than those from M1 [159]. The striatal projections from area 6 terminate in the dorsomedial zone of the putamen, the most lateral zone of the caudate nucleus and in the cell bridges that interconnect these two zones (Figs. 3.23, 14.1 C) [199, 406]. The PM and the SMA are primarily concerned with the sequence generation of movements and motor learning [159]. The PreSMA participates in the control of complex arm movements [199].

Experimental neuroanatomical studies [235, 406] have shown that, in the monkey, the projections from the motor and premotor cortices to the striatum are somatotopically organized. The corticostriatal input zones from hindlimb, forelimb and orofacial representations are, in this order, arranged from dorsal to ventral in the putamen. Neuroimaging studies [119, 256] have suggested a similar dorsoventral representation of leg, arm and face in the human putamen. The somatosensory areas 3, 1 and 2 project in an overlapping fashion to the putamen (Fig. 14.1 C) and these parietal projections follow the same somatotopic organization as those from the motor and premotor areas. These connectional data suggest that the putamen represents an important centre for integration of movement-related and sensory feedback information [72]. Evidence recently compiled by Romanelli et al. [350] indicates that this somatotopical organization is not confined to the putamen, but rather is maintained throughout the motor and premotor loops. It is noteworthy that the somatosensory input to the putamen is not confined to proprioceptive and tactile impulses but also includes somatotopically organized nociceptive information [28].

The *oculomotor loop* originates from the frontal eye field (FEF), which corresponds to area 8 of Brodmann, and from the supplementary eye field (SEF), situated in the rostral dorsomedial part of area 6 [364]. These two functional areas, which represent different aspects of oculomotor processing, are strongly and reciprocally interconnected [187, 196]. The corticostriatal projections from FEF and SEF converge on the oculomotor striatal area, which is situated in the central and lateral parts of the head and body of the caudate nucleus (Fig. 14.1) [187, 196, 318, 382]. These projections terminate exclusively in the striatal matrix and avoid striosomes [382]. Efferents from the oculomotor striatum pass to the SNr, which, via the VA and MD thalamic nuclei, project back to FEF and SEF [382]. The nigrothalamic fibres emit collaterals that descend to pre- and paraoculomotor networks in the brain stem. FEF and SEF also project directly to these networks [381]. The oculomotor loop is involved in the control of saccadic eye movements in complex behavioural contexts [188].

The *dorsolateral prefrontal loop* originates principally in Brodmann's areas 9 and 46. The striatal projection field of these cortical areas occupies a large part of the head of the caudate nucleus and extends caudally over the body and the tail of the same structure (Fig. 14.1 C) [9, 433]. These parts of the striatum project to the dorsomedial portion of GPi and the rostral SNr, which, via the VA and MD thalamic nuclei, project back upon the dorsolateral prefrontal cortex [105, 282]. The various centres forming part of the dorsolateral prefrontal loop have been implicated in cognitive functions, such as planning the order and timing of future behaviour, procedural learning and working memory tasks [115, 306]. It is assumed that, in general, the neural activity in the striatum is subordinate to that in the cortex [158]. However, it has recently been shown that in certain learning-related processes the striatum takes precedence over the prefrontal cortex [320].

The *limbic loop*, finally, originates from areas forming part of the medial and orbitofrontal cortical regions. The cytoarchitectonic organization of these regions has recently been analysed by Öngür et al. [304, 305] and the areal subdivision used here is largely based on their work. The medial areas participating in the limbic loop include areas 14, 24, 25 and 32, whereas the orbital areas involved are areas 11, 13, 47/12 and some adjacent agranular insular areas (Fig. 11.8). The various orbital and medial prefrontal cortical areas (OMPFC) are strongly interconnected [47]. They project to the ventral striatum, composed of the accumbens nucleus, a ventromedial strip of the caudate nucleus and part of the ventral putamen [57, 88, 166, 433]. The ventral striatal regions that receive fibres from the OMPFC in turn project to the ventral pallidum, formed by rostral and ventral extensions of the GP. The ventral pallidum then projects to medial and central parts of the MD thalamic nucleus, which are reciprocally interconnected with the OMPFC [304]. Experimental neuroanatomical studies have shown that, in monkeys, the projections from the orbital and medial prefrontal cortices remain segregated in all processing stations of the limbic loop [95, 340]. Physiological and imaging studies have shown that the OMPFC plays an important role in mood and emotions [91, 322], the analysis of food-related sensations [46, 349], the reward-guided choice of behaviours [348, 370] and hedonic experience [231].

Although these anatomical and functional data suggest that the basal ganglia are prominent components of a family of subcortico-cortical loops, it should be emphasized that parallel processing is not the only principle of organization in forebrain circuits associated with the basal ganglia. Rather, it has been demonstrated that the various loops are interconnected at several levels. Thus, the axons and dendrites of the striatal interneurons often cross functional boundaries and have been suggested to form the morphological substrate for exchange of information across regions [159]. Lynd-Balta and Haber [255] reported that fibres originating from the ventral striatum synapse with dopaminergic neurons in the substantia nigra that innervate the dorsal striatum and suggested that, via this striato-nigro-striatal loop, the limbic-related striatum communicates with the somatomotor-related striatum. Joel and Weiner [206] emphasized that the thalamocortical connections closing the various basal ganglia-thalamocortical loops are always supplemented by thalamocortical projections terminating in one or several cortical regions that are not the source of the cortico-striate input of the loop. They held that the various loops may well interact via these 'asymmetrical' or 'open' thalamocortical pathways. Finally, it should be mentioned that the various cortical regions forming the sites of origin and termination of these five cortico-striatal-thalamocortical loops are tightly interconnected [47, 87, 283, 334].

A Brief Excursion to the Cerebellum

The development of ideas concerning the function of the cerebellum shows a remarkable parallel to that of the basal ganglia. Until recently, it was generally believed that the cerebellum is devoted entirely to motor control. A complex cerebrocerebellar circuit, comprising a cerebrocerebellar feed-forward or afferent limb and a cerebellocerebral feed-back or efferent limb, was considered to represent an essential component of the cerebellar motor control system. Topographically organized corticopontine and pontocerebellar projections form the feed-forward limb, whereas the feed-back limb is comprised of the cerebellar corticonuclear projection, the efferents from the deep cerebellar nuclei (particularly the dentate nucleus) terminating in the posterior part of the ventral lateral thalamic nucleus (VLp) and the thalamocortical projection arising from that nucleus. It was held that along the consecutive neuronal links in this circuit, information derived from widespread regions of the cerebral cortex is funnelled through the thalamus to the primary motor cortex [10, 40]. In the 1990s, the idea that the cerebellum is exclusively involved in motor control was challenged by neuroanatomical, neurophysiological and clinical evidence, as well as by functional neuroimaging results.

As regards neuroanatomy, the transsynaptic retrograde tracer studies of Middleton and Strick [193, 278–281] revealed that distinct regions within the deep cerebellar nuclei form the origin of projections to different cortical areas, among which are not only motor, premotor and oculomotor areas, but also the prefrontal areas 9 and 46 (Fig. 14.8). The cerebellar neurons, labelled after injecting tracer into the prefrontal areas, were confined to the most ventral and caudal portions of the contralateral dentate nucleus. This region of the dentate nucleus appeared to differ clearly from the more rostral and dorsal regions of the nucleus, which were labelled by retrograde transneural transport from the primary motor and premotor cortex. On the basis of these findings, Middleton and Strick concluded that the cerebrocerebellar circuit, just like the direct cortico-striato-thalamocortical circuit, *does not converge on the primary motor cortex, but rather is composed of a number of parallel loops*. They suggested that the loop involving the prefrontal cortex operates in parallel with that serving the motor areas of the cerebral cortex, but has a cognitive rather than a motor function. It is important to note that the cerebellofugal and pallidofugal projections remain separated at the thalamic level, but overlap in the cortex (Fig. 14.8).

The hypothesis that the cerebellum is involved in cognitive processes has received support from physiological, neuroimaging and clinical studies. As regards physiology, Middleton and Strick [279], who recorded the activity of single neurons in the dentate nucleus in awake trained monkeys, found that units located in ventral regions of the nucleus are involved in working memory. Neuroimaging studies [79, 250, 365] indicate that the cerebellum (particularly the lateral cortex and the dentate nucleus) is involved in activities as diverse as working memory, implicit and explicit learning, linguistic processing and mental imagery.

Finally, evidence has been presented that patients with cerebellar diseases have cognitive dysfunctions related to the cerebellar disorder itself. Drepper et al. [85] reported that patients with isolated degenerative cerebellar disease show deficits in cognitive associative learning. Schmahmann and Sherman [368], who examined a cohort of patients with diseases confined to the cerebellum, found that behavioural changes were clinically prominent in patients with lesions involving the posterior lobe of the cerebellum and the vermis. According to Schmahmann and Sherman [367, 368] these deficits conformed to an identifiable syndrome, which they termed the *cerebellar cognitive affective syndrome*. Its defining features are disturbances of executive planning, impaired spatial cognition, linguistic difficulties and personality change characterized by flattening or blunting of affect. Schmahmann [366] pointed out that these disturbances are similar to the functional affiliations of the cerebral cortical regions with which the cerebellum has reciprocal interconnections. He proposed that disruption of the neural circuitry linking the cerebellum with the associative and paralimbic neocortical regions produces the observed cognitive and affective deficits. Although it is quite possible that the cerebrocerebellar circuitry involves an equivalent of the striatal limbic loop, sufficient experimental neuroanatomical evidence for the presence of such a cerebellar loop is lacking as yet.

Thalamostriatal Projections

The thalamo-cortico-striatal segment of the direct striatal circuit is supplemented by fibres that pass directly from the thalamus to the striatum (Fig. 14.7). These thalamostriatal projections originate principally from the centromedian and parafascicular (CM-Pf) thalamic nucleus and from the VA-VL complex. The constituent fibres of these projections are glutamatergic and synapse with the medium-sized striatal projection neurons [221]. Collectively, they provide the second-largest source of excitatory input to the striatum [159]. It has already been mentioned that considerable numbers of fine, presumably collateral fibres leave the fasciculus thalamicus before it reaches the VA and VL thalamic nuclei. These fibres enter the internal medullary lamina and terminate in the CM-Pf complex (Figs. 9.1, 14.7A) [80, 94, 300]. The fibres that project from this complex to the striatum are topographically organized. The lateral part of CM projects principally to the dorsal putamen (the sensorimotor striatal territory) and the medial part of CM projects to the head of the caudate nucleus (the associative striatal territory), whereas Pf projects to the ventral or limbic striatum [293, 315, 356, 357, 386]. Interestingly, the cortical efferents from these various parts of the CM-Pf complex project specifically to the cortical areas that have the same striatal targets as the direct thalamostriatal projections from these parts [184, 209]. In other words, a particular part of the CM-Pf complex has two projections to a specific zone of the striatum, a direct one and an indirect one, the latter of which is synaptically interrupted in the cerebral cortex. Single-axon tracing studies have shown that virtually all CM-Pf neurons project to both the striatum and the cerebral cortex in the rat [78], but that more than half of the CM-Pf neurons project solely to the striatum or to the cerebral cortex in the squirrel monkey [315].

The VA-VL complex also contributes significantly to striatal innervation [263, 264, 293, 356, 357, 386]. We have seen that the various parts of the VA-VL complex have reciprocal connections with specific frontal cortical areas (Fig. 14.8) and that these cortical areas project to particular sectors of the striatum (Fig. 14.1B, C). It has been shown that, just as in the CM-Pf complex, the direct thalamostriatal projections from the various parts of the VA-VL complex target the same striatal zones as do their indirect thalamo-cortico-striatal projections [263]. Thus, we can conclude that the thalamostriatal projections maintain the functional distinction of different striatal regions and of the related subdivisions or loops of the direct striatal circuit [159].

The Indirect Circuit: Cerebral Cortex – Striatum – External Segment of Globus Pallidus – Subthalamic Nucleus – Internal Segment of Globus Pallidus and Substantia Nigra, pars reticulata – Thalamus – Cerebral Cortex

Indirect circuitry as described in this paragraph is illustrated by Fig. 14.9. The external and internal segments of the globus pallidus (GPe and GPi) are engaged in different circuits in the basal ganglia. The striatal efferents passing to the GPi arise from medium-sized spiny neurons (MSNs) that co-contain GABA and substance P, whereas the striatal projections to the GPe arise from MSNs that co-contain GABA and enkephalin. These two populations of MSNs also express different kinds of dopamine receptors on their surfaces. Although there are a number of different dopamine receptor subtypes present in the striatum and elsewhere, they can be generally grouped into two classes: the D1 class and the D2 class. The D1 class receptors are preferentially expressed in the substance P-containing MSNs forming part of the direct circuit, whereas the D2 class receptors are specifically expressed on the enkephalin-containing MSNs [122, 403]. The latter, which form part of the indirect circuit, project to the GPe, where they exert an inhibitory influence on the constituent neurons. The efferents of the GPe, which are also GABAergic and inhibitory, project in a topographically organized fashion to the subthalamic nucleus (STN) [50, 260] and the latter has been shown to project back to both parts of the GP and to the substantia nigra, pars reticulata (SNr) [50, 51, 299, 379, 388]. The STN also receives a glutamatergic excitatory input from the cerebral cortex. These cortical afferents arise mainly, but not exclusively from the primary motor, premotor and supplementary motor cortices [294–296]. The subthalamic efferents have an excitatory effect on pallidal and nigra neurons and use glutamate as their neurotransmitter [1]. Inhibitory fibres from the striatum and excitatory fibres from the subthalamic nucleus converge on the same neurons in the GPi (Fig. 14.9). Therefore, the sequential actions of the various neurons involved in the indirect

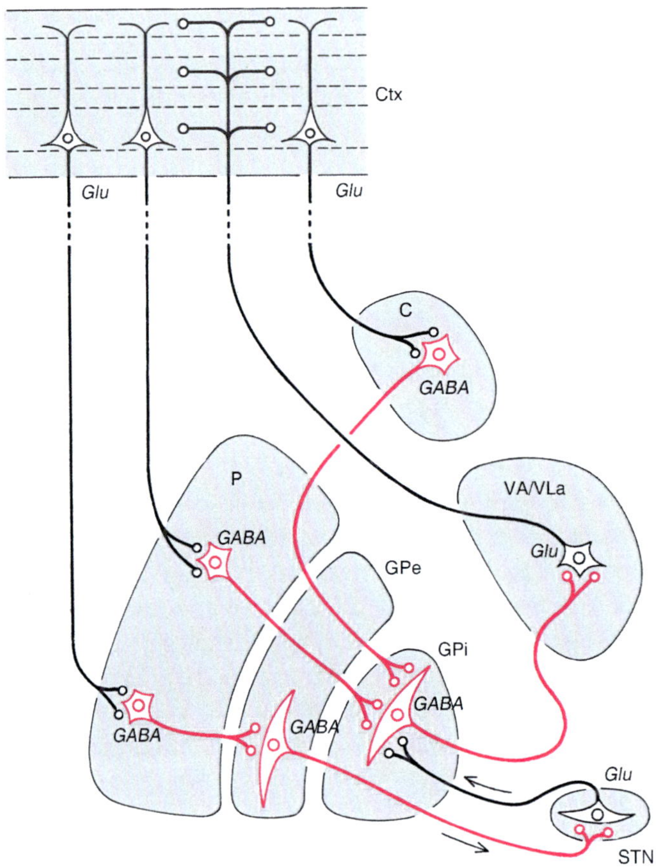

Fig. 14.9. The indirect striatal circuit, which includes a loop, passing via the subthalamic nucleus from the external to the internal part of the globus pallidus (*GPe*, *GPi*). *C*, caudate nucleus; *Ctx*, cortex; *P*, putamen; *STN*, subthalamic nucleus; *VA*, ventral anterior thalamic nucleus; *VLa*, anterior part of ventral lateral thalamic nucleus

circuit are as follows: cortical activation of the inhibitory, GABA and enkephalin-containing MSNs reduces the output of GPe activity, resulting in a decreased inhibition or disinhibition of STN neurons, which causes an increase in the glutamatergic activation of GPe/SNr. The activation of the inhibitory output of these centres reduces the thalamic activation of cortical neurons. Thus, the effect of activating the indirect circuit on its cortical target neurons is opposite to that of activating the direct circuit: *negative feed-back rather than positive feed-back* [73, 123, 389]. As will be discussed below, this scheme of two opposing circuits plays an important role when attempting to explain certain aspects of the pathophysiology of the basal ganglia.

Two questions concerning the organization of the direct and indirect circuits should be briefly addressed: (1) are these circuits completely separate, and (2) can the indirect circuit, just like the direct one, be subdivided into a number of separate functional loops. The answer to the first question is: no. Haber and Johnson Gdowski [159] cite evidence showing that there are links between the two circuits at several levels. For example, MSNs of the direct and indirect circuits are interconnected by collaterals within the striatum [434] and the GPe neurons forming part of the indirect circuit project directly to the GPi and the SNr, which are embedded in the direct circuit [316, 362, 363]. As regards the second question, Joel and Weiner [207] reported that the indirect circuit can indeed be subdivided into a number of functional loops. They distinguished motor, associative and limbic 'closed' loops, which fit into the pattern of the direct loops. However, they emphasized that, in addition to these 'closed' indirect loops, 'open' indirect loops are present that interconnect the closed indirect loops at the level of the striatum.

Caveat: Given the presence of so many links and shunts, the scheme of two opposing circuits, which will be used in the following sections of this chapter, is an evident oversimplification of the circuitry of the basal ganglia.

The Striato-Nigro-Striatal Circuit

The striato-nigro-striatal circuitry is depicted in Fig. 14.10. It has already been mentioned that fibres originating from the caudate nucleus and the putamen traverse the globus pallidus and subsequently descend to the substantia nigra and the adjacent ventral tegmental area (VTA). Projections from the striatum to the midbrain show an inversed dorsoventral topography as do the ascending midbrain projections. Thus, the dorsal parts of the striatum project to the ventral regions of the midbrain, while the ventral striatal areas project mainly dorsally [159]. More specifically, the lateral, sensorimotor part of the striatum projects to the lateral part of the SNr, including some of the dopaminergic cell columns that extend from the SNc into this region. The projection from the central, associative part of the striatum, which overlaps widely with that of the sensorimotor part [313], terminates principally in the ventral densocellular region of the SNc and the associated SNr. Finally, the ventromedial striatum innervates a large dorsomedial region, including a large part of the SNc, the medial part of the SNr and the VTA [255]. The striatonigral fibres contain GABA and substance P and inhibit the neurons in the SNc.

As already discussed, the dopaminergic neurons in the midbrain can be grouped into a dorsal tier and a ventral tier on cytoarchitectonic and chemoarchitectonic grounds [170]. The dorsal tier encompasses the loosely arranged cells, which form the most dorsal part of the SNc, the nucleus parabrachialis pigmentosus and the VTA. The ventral tier consists of the densocellular SNc, including the cell columns that extend from it into the SNr. The mesencephalic dopaminergic neurons give rise to a major, topographically organized projection to the striatum, which is composed of extremely fine axons [48, 58, 408, 411]. The dorsal tier and the most medial part of the densocellular ventral tier project to the ventromedial striatum; the remainder of the densocellular tier projects primarily to the central, associative striatal zone; whereas the ventrally extending cell columns of the ventral tier project to

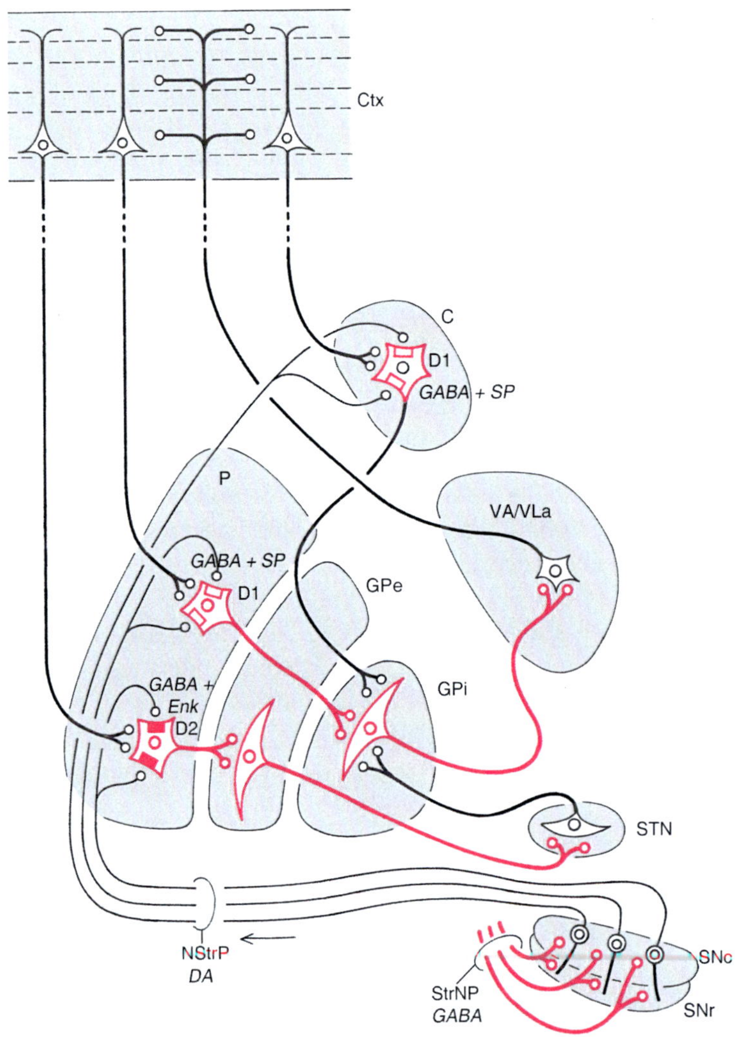

Fig. 14.10. Contacts made by nigrostriatal dopaminergic fibres with striatal neurons forming part of the direct and indirect striatal circuits. *C*, caudate nucleus; *Ctx*, cortex; *GPe*, *GPi*, external and internal part of the globus pallidus; *NStrP*, nigrostriatal projection; *P*, putamen; *SNc*; substantia nigra, pars compacta; *SNr*, substantia nigra, pars reticulata; *STN*, subthalamic nucleus; *StrNP*, striatonigral projection; dopamine (*D1*, *D2*) receptors are indicated by *small rectangles*

the dorsolateral, sensorimotor striatal zone [168, 254].

The mesencephalic dopaminergic neurons establish direct synaptic contacts with medium spiny striatal projection neurons [110], and it is known that striatonigral fibres in turn synapse with the somata and dendrites of SNc neurons [177, 421]. The axodendritic contacts are mainly located in the SNr, into which many SNc cells extend their long, ventrally directed dendrites. As discussed above, the various striatal territories are reciprocally connected with particular groups of dopaminergic cells. However, the nigrostriatal fibres do not exclusively form part of the closed striato-nigro-striatal loops. It has been established that ventral tier neurons receiving input from the ventromedial striatum project to a more dorsolaterally situated striatal region [255, 391]. This feed-forward projection is considered to be the first link of a 'spiraling' system: via this system the limbic striatum influences the more dorsally and laterally situated associative and sensorimotor striatal sectors [159, 168]. François et al. [103] reported that each mesencephalic dopaminergic cell group projects primarily to one specific striatal region but also participates, albeit to a lesser extent, in the innervation of all of the remaining striatal parts.

The functions of the midbrain dopamine neurons are very complex and cannot be adequately discussed here (for reviews, see [426, 429]). Let it suffice to mention that these neurons play a key role in recognizing and predicting rewards and, hence, in selecting and guiding behaviours to obtain rewards, as well as in reward-based or reinforcement learning.

It has already been mentioned that (a) the ascending axons of midbrain dopamine cells terminate directly on striatal MSNSs; (b) the MSNs can be divided into two populations, one co-containing GABA and substance P and one co-containing GABA and enkephalin; (c) the GABA + substance P elements project to the GPi and the SNr and form part of the direct striatal pathway (Fig. 14.7), whereas the GABA + enkephalin elements project to GPe and form part of the indirect circuit (Fig. 14.9); and (d) the GABA + substance P neurons of the direct circuit express dopamine D1-class receptors and the GABA + enkephalin indirect circuit neurons express D2-class receptors (Fig. 14.10).

There is physiological evidence that the principal effect of D1-class receptor activation is excitatory and that of D2-class receptor activation is inhibitory. This means that the release of dopamine in the striatum would excite MSNs projecting to GPi and inhibit MSNs projecting to GPe. The latter would lead to increased inhibition of STN neurons and, because of reduced firing, the excitatory effect of the GPi neurons would be diminished. Therefore, although their synaptic actions are different, the dopaminergic inputs to the direct and indirect circuits produce the same effect: they reduce the inhibition of thalamocortical neurons and, ultimately, facilitate movements initiated in the motor cortex [2, 73, 389, 426].

The SNc neurons receive, in addition to their input from the striatum, afferents from the bed nucleus of the stria terminalis, the mesencephalic raphe nuclei, the pedunculopontine tegmental nucleus and the locus coeruleus [243–245]. Most of these nuclei change their activity in relation to arousal, motivation and emotionally driven behaviour [159].

It is important to note that the action of dopamine in the striatum is not confined to a synaptic influence on MSNs. Thus, it has been reported that dopamine can diffuse away from its sites of release in the striatum [118] and that in this structure a high proportion of dopamine receptors are on neuronal membranes at non-synaptic sites [435]. These findings indicate that, apart from its synaptic activity, dopamine is also involved in non-synaptic or volume transmission in the striatum [301]. Moreover, pharmacological evidence suggests that the nigrostriatal dopaminergic fibres exert an additional presynaptic action on glutamatergic corticostriatal projections [12, 226], where they dampen striatal excitation. There is also converging evidence indicating that tonic release of dopamine attenuates MSN's neuronal response, while phasic release potentiates striatal response [42, 53]. Finally, it should be mentioned that the axons of mesencephalic dopaminergic neurons do not exclusively pass to the caudate nucleus and the putamen, but also innervate the GPe, GPi and the subthalamic nucleus [104, 176].

Additional Striatal Connections

With the aid of the modern anterograde and retrograde tracer techniques, a considerable number of additional 'extrapyramidal' fibre connections have been established. A few of these will be mentioned here (see Fig. 14.11 for illustration).

As we know, the nucleus subthalamicus not only sends numerous fibres to the globus pallidus, but also projects to the caudate nucleus and the putamen [15], the pars reticulata of the substantia nigra [342, 388] and the pedunculopontine tegmental nucleus [89, 203]. The latter cell mass is known to receive massive afferents from the two output structures of the basal ganglia, i.e. the internal segment of the globus pallidus and the pars reticulata of the substantia nigra. The neurons constituting the latter projection utilize GABA as their neurotransmitter [17, 18, 54, 89, 244, 311].

The pedunculopontine tegmental nucleus has been shown to give rise to a small descending and a much larger ascending efferent projection. The former will be considered below; the latter is distributed mainly to the pars compacta of the substantia nigra, the subthalamic nucleus and the internal segment of the globus pallidus [17, 18, 54, 89, 244, 311]. The pedunculopontine tegmental nucleus is known to exert an excitatory action upon neurons in the various target structures [313]. This excitatory effect is principally mediated by acetylcholine. Clearly, these connections close a considerable number of additional 'extra-pyramidal' loop systems.

Input Systems

The reticular formation of the brain stem has a quantitatively very important access to the 'extra-pyramidal' circuitry. The reticular formation, particularly its mesencephalic portion, is one of the principal sources of afferent fibres to the intralaminar nuclei of the thalamus and, as pointed out earlier, the latter project to both the caudate nucleus and the putamen. A second important input system to be mentioned here is the mesostriatal serotoninergic projection, which originates mainly from the dorsal raphe nucleus and terminates throughout the striatum, but more significantly in its ventral and medial regions [6, 30, 124, 297, 386, 404].

Output Channels

Because the main outflow from the striatum and the nuclei related to it converges via the globus pallidus and the thalamus upon the motor, premotor and supplementary motor cortical areas, the fibres originating from these cortical areas constitute the principal output channel of the basal ganglia (for output channels see Fig. 14.11). Prominent among the systems emanating from the cortical areas mentioned is, of course, the motor part of the pyramidal tract, but these areas also give rise to a number of other projections, some of which may well consist of collaterals of pyramidal axons. Such projections terminate in the ventral lateral thalamic nucleus [237, 269], the subthalamic nucleus [43, 173, 333], the compact part of the substantia nigra [235, 358, 412] or both parts of the structure [289] and the pedunculopontine tegmental nucleus [89]. All of these pathways feed back into centres included in one of the striatal circuits, thus closing additional subsidiary loops. The remaining output pathways of the basal ganglia include pallidohabenular, nigrotectal, nigrotegmental and pedunculopontino-reticular projections.

The pallidohabenular fibres originate in primates mainly in a peripheral zone of the medial pallidal segment, which encroaches upon the lateral hypothalamus [80, 185, 298, 309, 312, 317] and terminate in the lateral habenular nucleus. Fibres originating from this nucleus descend in the habenulo-interpeduncular tract, bypass the interpeduncular nucleus and terminate in various mesencephalic centres, including the periaqueductal grey matter, the compact part of the substantia nigra and the dorsal

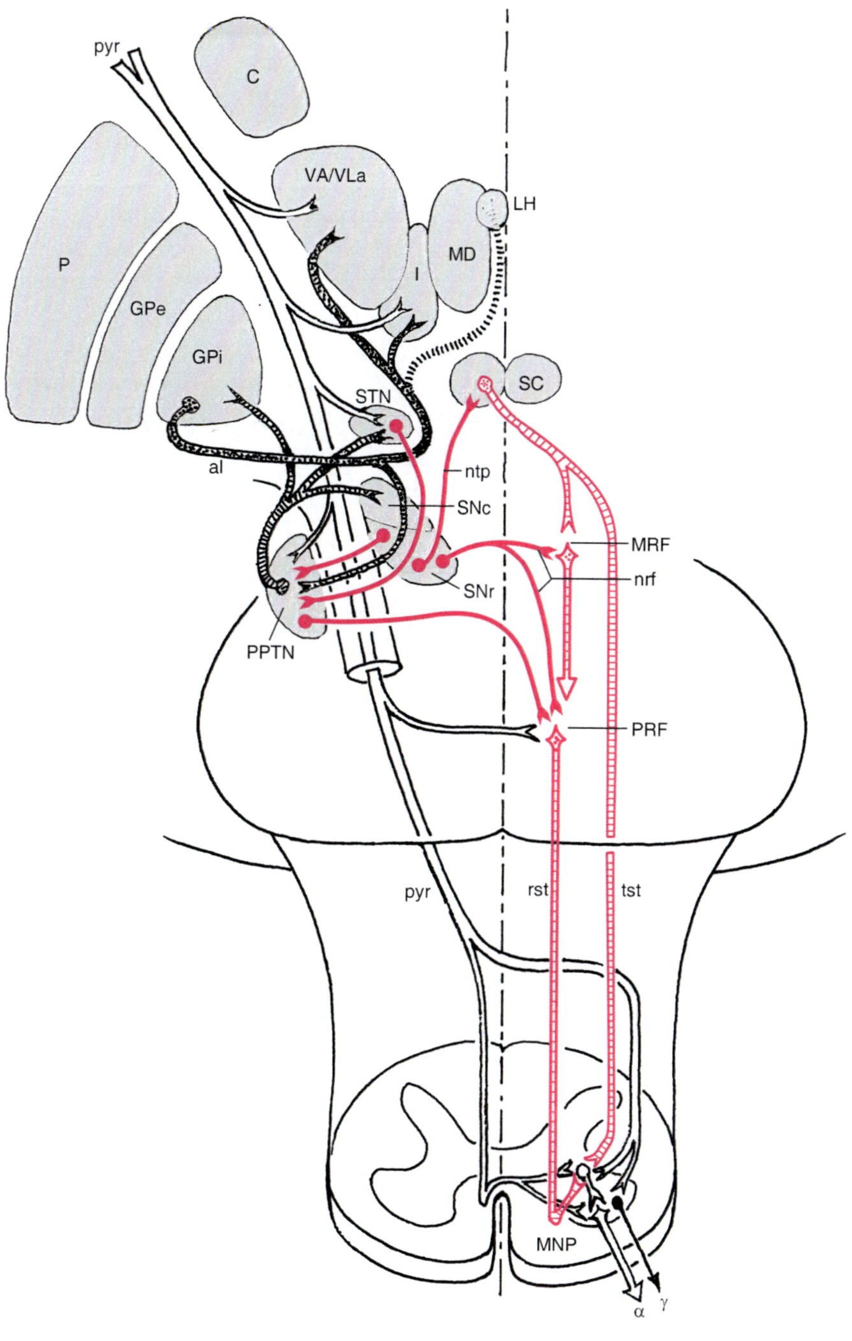

Fig. 14.11. Ascending efferents from the pedunculopontine tegmental nucleus and the principal descending output channels of the basal ganglia. The pyramidal output of these ganglia is added in *open contours*. *al*, ansa lenticularis; *C*, caudate nucleus; *GPe*, external part of the globus pallidus; *GPi*, internal part of the globus pallidus; *I*, intralaminar thalamic nuclei; *LH*, lateral habenular nucleus; *MD*, mediodorsal thalamic nucleus; *MNP*, motor neuron pool; *MRF*, mesencephalic reticular formation; *nrf*, nigroreticular fibres; *ntp*, nigrotectal projection; *P*, putamen; *PPTN*, pedunculopontine tegmental nucleus; *PRF*, pontine (and medullary) reticular formation; *pyr*, pyramidal tract; *rst*, reticulospinal tract; *SC*, superior colliculus; *SNc*, substantia nigra, pars compacta; *SNr*, substantia nigra, pars reticulata; *STN*, subthalamic nucleus; *tst*, tectospinal tract; *VA*, ventral anterior thalamic nucleus; *VLa*, anterior part of ventral lateral thalamic nucleus; *α*, *α* motoneurons; *γ*, *γ* motoneurons

raphe nucleus [186]. The projections to the two latter centres may well add further loops to the circuitry of the basal ganglia.

The nigrotectal and nigroreticular projections originate both from the pars reticulata of the substantia nigra [17, 133, 139, 194, 197, 344]. Both are thought to be inhibitory and to use GABA as their neurotransmitter [56, 58].

The massive nigrotectal projection terminates, mainly ipsilaterally, in the middle grey layer of the superior colliculus, where its fibres enter into synaptic contact with cells of origin of the predorsal bundle [27]. The latter represents the principal pathway from the superior colliculus to those centres in brain stem and spinal cord involved in initiating orienting movements of the head and eyes [188]. Olazabal and Moore [303] reported that, in rats, bats and cats, dopaminergic neurons situated in the lateral part of the substantia nigra and in the adjacent lateral tegmentum project to the ipsilateral inferior colliculus. They conjectured that impulses travelling along these fibres may influence orienting movements of the pinnae, head and eyes to auditory stimuli mediated by the inferior colliculus.

The nigrotegmental fibres terminate mainly in the pedunculopontine tegmental nucleus, but some descend to the medullary reticular formation. The pedunculopontine tegmental nucleus, in addition to a projection from the substantia nigra, pars reticulata, also receives afferents from the motor cortex [238], globus pallidus [89, 203, 299] and the subthalamic nucleus [80]. The pedunculopontine tegmental nucleus overlaps with the so-called mesencephalic locomotor region. (It has already been mentioned that electrical stimulation of this region elicits walking movements.) The ascending efferents from the pedunculopontine tegmental nucleus have already been discussed; its descending efferents, in contrast, are principally non-cholinergic and terminate in the pontine and medullary parts of the medial reticular formation [89, 200, 204, 290] and in the cervical spinal cord [394]. Together with the reticulospinal neurons on which they impinge, these descending connections constitute a truly *extrapyramidal motor system*, by which the basal ganglia influence postural muscle tone and locomotion (Fig. 14.11) [117, 407].

Connections of the Ventral Striatum

The ventral striatum is considered to be that portion of the striatum associated with the limbic system by virtue of its afferent connections from allocortical and mesocortical areas as well as from the amygdala [163] (Fig. 14.12). In addition to the accumbens nucleus, the ventral striatum comprises the ventromedial and caudoventral parts of both the caudate nucleus and the putamen [112, 392, 395].

Apart from a strong input from limbic cortical regions and the amygdala, the nucleus accumbens and adjacent sectors of the caudate nucleus and the putamen receive afferents from the midline nuclear complex of the thalamus and from the ventral midbrain.

The cortical afferents, which originate from the limbic, infralimbic, medial orbital, agranular insular, perirhinal and entorhinal cortices and from the hippocampal subiculum [106, 111], project bilaterally with an ipsilateral predominance to the ventral striatum. Each of these cortical areas has a longitudinally oriented striatal terminal field that overlaps slightly with those of adjacent cortical area [23, 24, 148, 268].

With regard to the thalamostriate projections, Berendse and Groenewegen [21] established that, in the rat, the various midline and intralaminar thalamic nuclei project to longitudinally oriented striatal sectors and that the terminal sectors of the midline nuclei (i.e. the paraventricular, paratenial, intermediodorsal and rhomboid nuclei) are mainly situated in the ventral striatum, whereas those of the intralaminar nuclei occupy most of the dorsal striatum. Similar midline thalamic projections to the ventral striatum have been found in the primate [126]. Correlation of these results with those of previous studies on the organization of the projections of the midline and intralaminar thalamic nucleus on the prefrontal cortex [22] and those of the prefrontal cortex to the striatum [23, 24, 145] revealed that the projection zones of individual prefrontal cortical areas converge in the

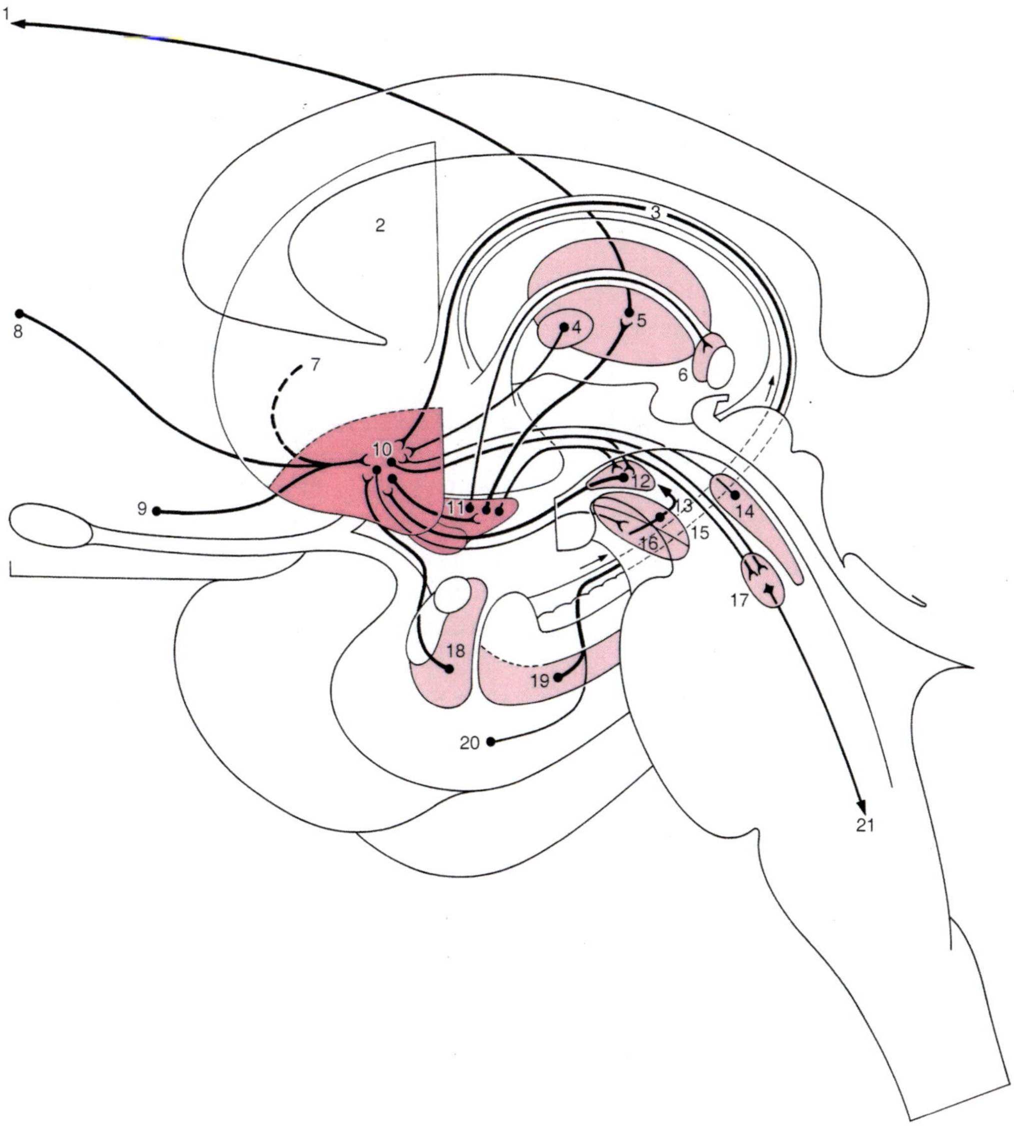

1 Prefrontal cortex
2 Dorsal striatum
3 Fornix
4 Median thalamic nuclei
5 Mediodorsal thalamic nucleus
6 Lateral habenular nucleus
7 Insular cortex
8 Prelimbic cortex
9 Medial prefrontal cortex
10 Ventral striatum
11 Ventral pallidum
12 Ventral tegmental area
13 Nigrostriatal projection
14 Dorsal raphe nucleus
15 Substantia nigra, pars compacta
16 Substantia nigra, pars reticulata
17 Pedunculopontine tegmental nucleus
18 Basal and accessory basal amygdaloid nuclei
19 Subiculum
20 Entorhinal and perirhinal cortices
21 Pedunculopontine-reticular fibres

Fig. 14.12. Principal connections of the ventral striatum and the ventral pallidum

striatum with those of 'their own' midline and intralaminar afferent nuclei [145]. The amygdalostrial projection arises from the basal and accessory basal nuclei and terminates mainly in the ventral striatum [113, 114]. Its fibres are topographically organized [147, 352, 354]. In addition to the projections from the ventral striatum, the prefrontal cortex receives massive input from the amygdaloid nuclei mentioned [230, 336]. Double-labelling experiments have shown that, in rats, numerous neurons in the basolateral amygdaloid nucleus project to both the prefrontal cortex and the nucleus accumbens [380]. The topographical organization and the projection patterns of these neurons, and of the amygdaloprefrontal and amygdalostriatal connections in general, correspond to those of the prefrontostriatal projection.

The ventral striatum receives a strong dopaminergic projection from the ventral midbrain. It was formerly held that this projection originates principally from the ventral tegmental area [18, 404, 405], but the studies of Lynd-Balta and Haber [253] have shown that sizeable proportions of this projection arise from the dorsal tier and from the medial, densocellular zone of the ventral tier of the substantia nigra. The neurons projecting to the ventral striatum are not topographically organized in the ventral mesencephalon [253].

The dopaminergic innervation of the ventral striatum is unevenly distributed. This uneven distribution of dopaminergic terminals is paralleled by the uneven distribution of the neuropeptides enkephalin, substance P and dynorphin and the neurotransmitter GABA [418]. It is noteworthy that many neurons in the ventral tegmental area, which project to the ventral striatum, contain cholecystokinin [436] and that, in some of these elements, cholecystokinin is co-localized with dopamine [189, 190].

The efferents of the ventral striatum pass primarily to extrapyramidal centres, such as the ventral pallidum, the substantia nigra and the pedunculopontine tegmental nucleus, but the nucleus basalis magnocellularis and a number of limbic-related structures, among them the bed nucleus of the stria terminalis, the lateral hypothalamic area and the medial part of the mediodorsal thalamic nucleus, are also innervated (Fig. 14.12) [146, 163, 183].

The projections from the nucleus accumbens to the ventral pallidum contain a major GABA-ergic component [285] and substance P-, enkephalin- and dynorphin-positive fibres have also been shown to be present within this projection [160, 161]. The neurons containing these neuropeptides most probably use GABA as their principal neurotransmitter. The enkephalin- and substance P-positive fibres form dense and highly characteristic terminal plexuses within the ventral pallidum [13, 160, 161]. It has been demonstrated that the accumbofugal efferents to the ventral pallidum project monosynaptically to output cells of that structure [431]. The efferents from the ventral striatum, which are topographically organized, also project to limited portions of the external and internal pallidal segments. However, the ventral striatal output remains segregated from the dorsal striatal efferent projections to these pallidal structures [163].

Fibres from the ventral striatum projecting to the substantia nigra extend throughout the rostral-caudal extent of that structure. Although the densest terminal fields occur in the medial portion, numerous fibres also extend laterally to innervate the dorsal tier of dopaminergic neurons in the substantia nigra [163, 255]. The ventral striatal fibres passing to the substantia nigra enter into direct synaptic contact with dopaminergic neurons that innervate the dorsal striatum [391]. It has already been mentioned that, via this striato-nigro-striatal circuit, the ventral striatum may help integrate the limbic system with other output systems of the striatum [163].

The projections from the ventral striatum to the pedunculopontine tegmental nucleus have been shown to arise from only a restricted part of the ventral striatum in the monkey, namely the medial nucleus accumbens and the ventromedial caudate nucleus [163]. We have seen that the pedunculopontine tegmental nucleus maintains reciprocal connections with the internal segment of the globus pallidus, the substantia nigra and the subthalamic nucleus and also receives a strong input from the motor cortex (Fig. 14.11).

The ventral striatum projects heavily to the basal nucleus of Meynert, which is embedded in the substantia innominata (Figs. 6.38-6.41). As will be discussed in a later section of the present chapter, this basal magnocellular nucleus is the main source of cholinergic fibres to the cerebral cortex and the amygdala.

The outputs from the ventral striatum to the limbic-related bed nucleus of the stria terminalis, lateral hypothalamic area and medial part of the mediodorsal thalamic nucleus are much smaller than the massive ventral striatal projections to the ventral pallidum and the substantia nigra [163].

The efferent projections from the core region of the nucleus accumbens differ in some respect from those of the shell region of the same structure. Without going into detail, the following may be mentioned: (a) core and shell project to different parts of the ventral pallidum; (b) both regions project to the midbrain (but with a bias for the core projection to innervate the substantia nigra and the shell projection to reach primarily the ventral tegmental area); and (c) the shell, in contrast to the core, projects diffusely throughout the rostrocaudal extents of the lateral hypothalamus [183, 438]. Deniau et al. [75] presented anatomical and physiological evidence indicating that, in the rat, the core of the nucleus accumbens innervates the dorsal region of the substantia nigra, pars reticulata, which, via certain parts of the mediodorsal and ventral medial thalamic nuclei, projects mainly to the prelimbic and to a lesser degree to the orbital areas of the prefrontal cortex.

The efferent projections of the ventral pallidum terminate primarily in the lateral habenular nucleus, the mediodorsal thalamic nucleus, the subthalamic nucleus, the lateral hypothalamus, the substantia nigra and the dorsocaudal part of the mesencephalic tegmentum [150, 162, 163, 438]. In the rat, efferent fibres from the ventral pallidum have in addition been traced to numerous other areas, including the prefrontal cortex, the ventral striatum, the basolateral, lateral and central amygdaloid nuclei, the lateral septum, the area tegmentalis ventralis, the rostral raphe nuclei, the periaqueductal grey and the locus coeruleus [150]. The projections from the ventral pallidum to the mediodorsal thalamic nucleus, the subthalamic nucleus and the lateral hypothalamus are distinctly topographically organized [150, 162]. As discussed in a previous section, the fibres passing from the ventral pallidum to the mediodorsal thalamic nucleus are a link in the *limbic loop*, which forms part of the *direct striatal circuit*.

It is noteworthy that, whereas the ventral pallidothalamic projection is strongly developed in the rat, surprisingly few ventral pallidal terminations were found in the mediodorsal nucleus of the thalamus in the cat and in the monkey [165, 397]. This may mean that the loops from the various prefrontal subareas through the ventral striatopallidothalamic system back to the frontal lobe are much less prominent in these species than in the rat. Like the dorsal pallidothalamic fibres, the ventral pallidal efferents presumably carry GABA as their neurotransmitter [236].

Following this discussion of the ventral striatopallidal fibres and the circuits they form part of, brief consideration will be given to some of the remaining efferent projections of the ventral pallidum.

The strongly developed projection from the ventral pallidum to the subthalamic nucleus is reciprocated by an equally strong efferent projection from the latter [144]. The circuit, thus closed, presumably corresponds to the dorsal pallido-subthalamo-pallidal circuit [150].

With regard to the ventral pallidonigral projections, Haber et al. [165] reported that, in the monkey, these efferents from the ventral pallidum overlap widely, suggesting convergence of terminals from different pallidal regions. In the rat, the ventral pallidonigral fibres terminate predominantly in the medial part of the pars reticulata, an area which also receives input from the ventral striatum [150]. As has already been discussed, the efferents from the substantia nigra, pars reticulata terminate in several thalamic nuclei, in the superior colliculus and in the caudolateral mesencephalic tegmentum.

The ventral pallidohabenular projection terminates in the lateral habenula, which sends its efferents to several targets, including the mesencephalic serotoninergic cell groups [185]. The latter also receive a direct input from the ventral pallidum.

Input from the nucleus accumbens, an integral part of the ventral striatum, may reach the caudolateral mesencephalic tegmentum along four different routes: (1) accumbo-tegmental, (2) accumbo-nigro-tegmental, (3) accumbo-ventral pallido-nigrotegmental and (4) accumbo-ventral pallido-tegmental. The caudolateral mesencephalic tegmentum, which includes the mesencephalic locomotor area, gives rise to a small descending projection. The fibres from this projection terminate in the pontine and medullary reticular formation and in the spinal cord [116, 393].

Locomotion is an essential component of various adaptive behaviours, such as fight and flight reactions and food and water procurement. Pharmacological [335] and physiological experiments [285–287] have provided evidence suggesting that the nucleus accumbens influences locomotor behaviour and that some of the pathways enumerated above (accumbens → ventral pallidum → caudal mesencephalic tegmentum) may well form part of the extrapyramidal projection system along which these influences reach the spinal motor mechanism. The ethopharmacological experiments of Hooks and Kalivas [192] revealed that a circuit containing the ventral pallidum is critical for the manifestation of novelty-induced locomotor activity.

The limbic system is known to play a key role in initiating the various adaptive or goal-oriented behaviours of which locomotion is a component. Because the nucleus accumbens receives massive projections from the hippocampal formation and the amygdala, both essential components of the limbic system, this cell mass has been hypothesized to constitute the functional interface between the limbic and the motor system [192, 284, 286]. Pennartz et al. [324] critically reviewed the morphological, electrophysiological and ethological literature on the nucleus accumbens. They arrived at the conclusion that, although the nucleus accumbens most probably influences locomotion, it is not correct to associate this particular behavioural state with the activity of this centre as a whole. They pointed out that the accumbens nucleus is a collection of neuronal ensembles, or groups, with specific input-output relationships and with different functional and behavioural connotations. This view has been corroborated by the results of tracing experiments carried out by Groenewegen et al. [151], which showed that the neurons projecting directly or indirectly (i.e. via the ventral pallidum) to the caudolateral mesencephalic tegmentum (including the mesencephalic locomotor region) are located in specific subregions of the nucleus accumbens, preferentially innervated by fibres originating from specific limbic territories. For example, the ventral striatal projections to the caudolateral mesencephalic tegmentum originate mainly from clusters of cells in the border region between the 'shell' and the 'core' regions of the accumbens nucleus, and these clusters are preferentially innervated by fibres stemming from the prelimbic cortex and from the basolateral amygdaloid nuclei.

Connections of the Substantia Innominata, Basal Nucleus of Meynert and Related Cell Groups

Before presenting a survey of the afferent and efferent connections of the substantia innominata and related cell groups, it would be appropriate to recall a few data concerning these grisea.

1. The substantia innominata (SI) is an ill-defined cell mass, situated in the substriatal part of the telencephalon (Figs. 6.38–6.41).
2. The studies of Heimer and associates [180–182] have shown that the rostral third of the SI is occupied by a ventral extension of the globus pallidus; the term SI has since been reserved for the area caudal to this ventral pallidum.
3. The SI is rostromedially continuous with the bed nucleus of the stria terminalis and merges caudolaterally with the amygdaloid complex.

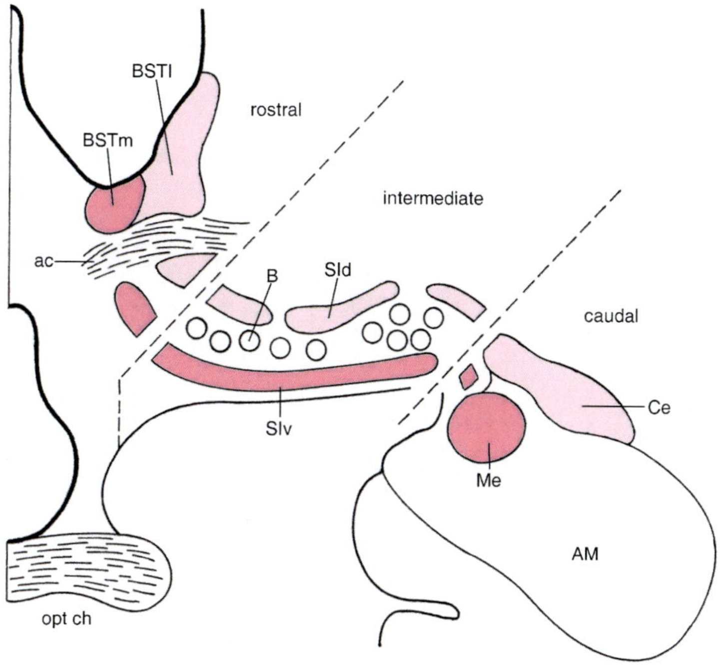

Fig. 14.13. Three rostrocaudally arranged sections through the basal telencephalon, showing the continuities of the dorsal and ventral zones of the substantia innominata (*SId*, *SIv*) with parts of the bed nucleus of the stria terminalis (*BST*) and the amygdaloid complex (*AM*). *ac*, anterior commissure; *B*, basal nucleus of Meynert; *BSTl,m*, lateral and medial parts of BST; *Ce*, central amygdaloid nuclei; *Me*, medial amygdaloid nucleus; *opt ch*, optic chiasm

Table 14.1. Connections of the substantia innominata of the rat. Based on Groves [152, 153]

Afferents	**Efferents**
Dorsal division	
bed nucleus stria terminalis lateralis	bed nucleus stria terminalis
central & basolateral amygdaloid nucleus*	central & basolateral amygdaloid nucleus
	lateral preoptic area
	paraventricular hypothalamic nucleus
lateral hypothalamus*	lateral hypothalamus
dorsal raphe nucleus*	
parabrachial nucleus*	
nucleus of solitary tract*	
Ventral division	
bed nucleus stria terminalis medialis	bed nucleus stria terminalis medialis
medial amygdaloid nucleus*	medial amygdaloid nucleus
medial preoptic area*	medial & lateral preoptic area
anterior hypothalamus	anterior hypothalamus
ventromedial hypothalamus	
Both divisions	
	mediodorsal thalamic nucleus
	reticular thalamic nucleus
dorsomedial & posterior hypothalamus	dorsomedial & posterior hypothalamus
supramamillary region	supramamillary region
ventral tegmental area*	ventral tegmental area
Selectivity not determined	
	main & accessory olfactory bulb
	piriform cortex
entorhinal cortex*	entorhinal cortex
perirhinal cortex*	perirhinal cortex
	infralimbic cortex
	prelimbic cortex
insular cortex*	insular cortex
medial prefrontal cortex	
midline thalamic nucleus	

Russchen et al. [353] have shown that the afferents marked with an *asterisk* (*) are also present in the macaque monkey. In this species, additional afferents to the SI appeared to originate from the orbitofrontal cortex, the periaqueductal grey and the reticular formation

4. The SI can be divided into separate dorsal and ventral divisions (SId, SIv), each of which is affiliated to specific parts of the bed nucleus of the stria terminalis (BST) and of the amygdala. Thus, two continua, one consisting of the lateral part of the BST, the dorsal division of the SI and the central amygdaloid nucleus and the other comprising the medial part of the BST, the ventral division of the SI and the medial amygdaloid nucleus, can be distinguished (Fig. 14.13) [4, 152, 153, 180].
5. Large multipolar and fusiform neurons are embedded in the SI. In primates, these neurons are concentrated in the basal nucleus of Meynert. These SI neurons represent the lateral part of a continuum of large neurons, known as the magnocellular basal forebrain (MBF). This continuum encompasses, apart from the large SI neurons, similar elements situated in the horizontal and ventral limbs of the nucleus of the diagonal band of Broca (DBBh, DBBv), and in the medial septal nucleus (MS).
6. The MBF contains two large intermingled, chemically specified populations of corticopetally projecting neurons, one cholinergic [178, 272–274, 331, 360], the other GABAergic [107, 108, 361, 419]. The cholinergic MBF neurons have been subdivided into four groups, Ch1–Ch4, and the last of these groups, which is co-extensive with the basal nucleus of Meynert, is further subdivided into five subgroups, Ch4av, Ch4al, Ch4iv, Ch4id and Ch4p [178, 272, 275]. Each of these (sub)groups projects to a different part of the telencephalon [273, 275]. Apart from numerous cholinergic and GABAergic elements, the MBF also contains a sizeable proportion of neurons whose neurotransmitters have not yet been identified [374].

The connections of the SI have been thoroughly studied in the rat by Grove [152, 153]. Her principal results are summarized in Table 14.1. Grove pointed out that (a) the efferent projections of the SI largely reciprocate its afferent connections; (b) the dorsal and ventral divisions receive afferent input from largely different sets of structures and maintain similarly distinct efferent connections and (c) the connections of the dorsal and ventral divisions closely resemble those of their respective adjoining parts of the BST and the amygdala (Fig. 14.13). Combining the results of her own studies on the connections of the substantia innominata with those of others on the connections and functions of the various parts of the amygdala, she conjectured that the functional complex to which the ventral substantia innominata belongs might play a role *inter alia* in the initiation of sexual and investigatory behaviour in response to complex (chemo)sensory stimuli, whereas the dorsal substantia innominata, with its associated forebrain and brain stem structures, forms a functional assembly concerned, at least in part, with monitoring and influencing autonomical and visceral functions, such as blood pressure, heart rate and respiration.

Turning now to the *magnocellular basal forebrain* (MBF), it should be mentioned at the outset that our knowledge of this complex has enormously increased during the last 20 years, and that in the present work only a cursory survey of its connections and (possible) functions can be presented. For a series of comprehensive reviews on the MBF, the reader is referred to "Behavioral Brain Research" (2000), vol. 115, issue 2.

The MBF receives afferents from the brain stem, the hypothalamus, the amygdala and a number of cortical areas (Fig. 14.14) [201, 210, 271, 353, 375]. Most of the data on these afferents pertain specifically to the basal nucleus of Meynert (BNM), which is by far the largest cell mass in the MBF. The quantitatively most important brain stem afferents to the MBF arise from the dorsal raphe nucleus, the locus coeruleus and the ventral tegmental area. Less substantial projections originate from the medial parabrachial nucleus and from cells in the mesopontine reticular formation [375, 401]. The hypothalamic afferents to the MBF arise principally from the lateral hypothalamic area [66]. An extensive projection system connects the amygdaloid complex with the MBF. This system originates from the basal and accessory basal nuclei and terminates principally in the BNM and in the horizontal limb of the nucleus of the diagonal band of Broca [307]. The corti-

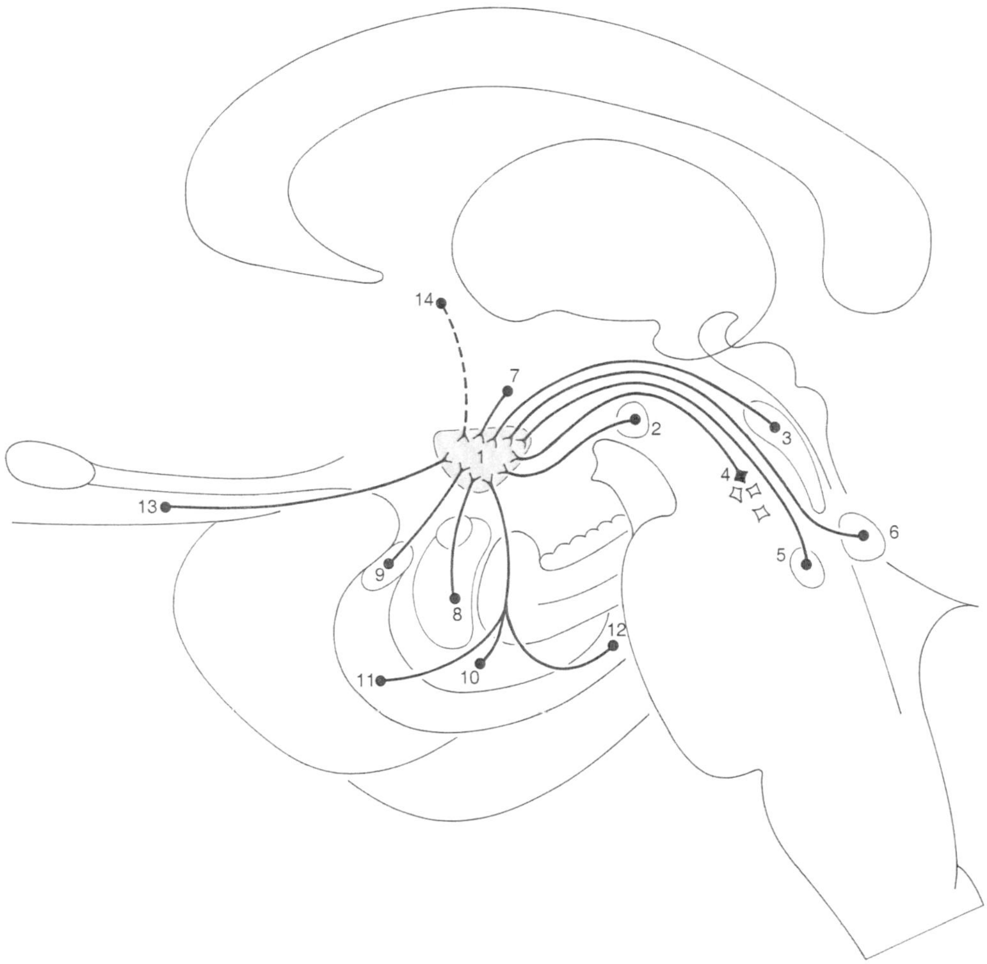

1 Basal nucleus of Meynert
2 Ventral tegmental area
3 Dorsal raphe nucleus
4 Mesopontine reticular formation
5 Locus coeruleus
6 Medial parabrachial nucleus
7 Hypothalamus
8 Basolateral amygdala
9 Prepiriform cortex
10 Entorhinal cortex (area 28)
11 Perirhinal cortex (areas 35, 36)
12 Areas TH, TF
13 Orbitofrontal cortex
14 Anterior insular cortex

Fig. 14.14. Afferents of the basal nucleus of Meynert

Table 14.2. Principal efferents from the magnocellular basal forebrain. Based on Semba [374]

Neurotransmitter	ACh	GABA	Unidentified
Telencephalon			
Neocortex	+	+	+
Hippocampus	+	+	
Amygdala	+		
Thalamus			
Reticular thalamic nucleus	+	+	
Hypothalamus			
Preoptic/anterior hypothalamus	+	+	+
Posterior hypothalamus	+	+	+
Brain stem			
Interpeduncular nucleus	+		
Pedunculopontine tegmental nucleus	–		+
Laterodorsal tegmental nucleus	–		+
Dorsal raphe nucleus	–		+
Locus coeruleus	–		+
Periaqueductal grey	–		+
Ventral tegmental area	–		+
Parabrachial nuclei	–		+
Dorsal motor vagus nucleus	–		+
Nucleus of solitary tract	–		+

cal afferents to the MBF arise from the prepiriform, entorhinal, perirhinal, orbitofrontal and anterior insular cortices [249, 271, 353]. There is evidence that, at least in rats, the afferents from the orbitofrontal cortex exclusively terminate on GABAergic MBF neurons [437].

The efferents from the MBF pass to a variety of brain structures, including the neocortex, the hippocampus, the amygdaloid complex, the thalamus, the hypothalamus and the brain stem (Table 14.2) [374]. It has already been mentioned that the large MBF projection neurons can be subdivided on account of their neurotransmitter content into: (1) a population of cholinergic neurons, (2) a population of GABAergic neurons and (3) a population that can be negatively characterized as non-cholinergic and non-GABAergic.

The basal forebrain cholinergic projection system encompasses the following components [5, 225, 248, 270, 273, 275, 371]:

1. Neurons within the medial septal nucleus (Ch1) and the vertical limb of the nucleus of the diagonal band (Ch2), which provide a substantial cholinergic projection to the hippocampus. The fibres pass through the fornix and terminate in Ammon's horn and the fascia dentata.
2. Neurons within the horizontal limb of the nucleus of the diagonal band (Ch3) and the magnocellular preoptic nucleus, which project to the olfactory bulb and the piriform and entorhinal cortices.
3. Neurons situated within the substantia innominata (Ch4), which innervate the basolateral amygdala and provide a major cholinergic projection to the entire neocortex. There is a specific, but overlapping topography in the organization of this projection. In primates, the corticopetal cholinergic cells are subdivided according to the topography of their projections. Thus, Ch4am is the major source of fibres passing to medial cortical areas, including the cingulate cortex; Ch4al projects to the frontoparietal cortex; Ch4i to the pre-

frontal, lateral peristriate, middle temporal and inferotemporal regions; and Ch4p projects to the superior temporal gyrus and the temporal pole region [270, 273, 275, 371].

In the human cerebral cortex, all areas and all cell layers contain cholinergic axons, albeit with considerable regional variations [276]. Within the cortex, cholinergic fibres have been observed to contact pyramidal neurons, (excitatory) spiny stellate cells, as well as GABAergic interneurons [14, 261, 262].

Fibres belonging to the substantial GABAergic projection to the cortex originate from all parts of the MBF and have been traced to many cortical areas, including frontal, premotor, motor, somatosensory, visual, cingulate and retrosplenial fields, as well as to the hippocampus [96, 143, 154, 385]. The cortically projecting GABAergic MBF neurons have been shown to make multiple synaptic contacts with GABAergic interneurons [107, 108]. The latter are extensively collateralized, each contacting hundreds of pyramidal neurons [109]. These morphological data suggest that activation of the GABAergic MBF projection to the cortex may lead to widespread disinhibition of pyramidal neurons. Thus, it seems likely that the MBF cholinergic and GABAergic corticopetal projections exert an activating (excitatory and disinhibitory) influence on cortical output cells [361, 374, 401].

The MBF forms part of the network involved in the regulation of sleep, wakefulness, arousal and alertness. This network has been discussed in Chap. 10. Apart from the MBF, the following neuronal aggregations are important nodal points in this network (Fig. 14.15): (1) a sleep-promoting region, 'S' in the preoptic and anterior hypothalamic areas; (2) a wake-promoting region, 'W' in the posterior hypothalamus, which includes the histaminergic tuberomamillary nucleus; (3) the largely cholinergic laterodorsal tegmental and pedunculopontine tegmental nuclei; (4) the serotoninergic dorsal raphe and central superior nuclei and (5) the noradrenergic locus coeruleus. The cell masses mentioned under 3–5 are all located in the *mesopontine tegmentum* and are sometimes collectively denoted with that term. Figure 14.15 shows that the MBF projects to the 'S' and 'W' regions and is reciprocally connected with all of the five mesopontine tegmental nuclei mentioned. Figure 14.15 also shows that many of the centres of the sleep-wakefulness network project to the cerebral cortex. These projections, which include noradrenergic, serotoninergic, histaminergic, orexinergic (not shown in Fig. 14.15, but see Figs. 10.5 and 10.7), cholinergic and GABAergic fibres, collectively form the corticopetal output of the ventral or extrathalamic branch of the *ascending arousal system* (AAS). The dorsal, thalamic branch of this system originates from cholinergic cells in the laterodorsal tegmental and pedunculopontine tegmental nuclei and from non-cholinergic, presumably glutamatergic cells in the same nuclei, as well as in the more caudal parts of the reticular formation and involves synaptic interruptions in the specific and aspecific thalamic nuclei as well as in the reticular thalamic nucleus (Fig. 14.15) [401]. Remarkably, the MBF not only contributes to the ventral branch of the AAS, but also has access to the circuitry related to the dorsal branch of the same system by way of a substantial projection to the reticular thalamic nucleus [169, 308]. The neurons in this nucleus provide a topographically organized, GABAergic innervation of all thalamic nuclei and, through this connection, the reticular neurons influence the firing mode of the thalamocortical projection neurons in these nuclei (Fig. 14.15) [401].

Impulses travelling along the various corticopetal output pathways of the AAS are supposed to bring about collectively a general cortical activation, which manifests itself in the well-known behavioural and cortico-electrical arousal reaction. However, given the fact that each of the contributing pathways has its own transmitter profile and its own characteristic mode of termination in the cortex, it might be expected that each of these pathways influences the processing of cortical information in its own specific way [99].

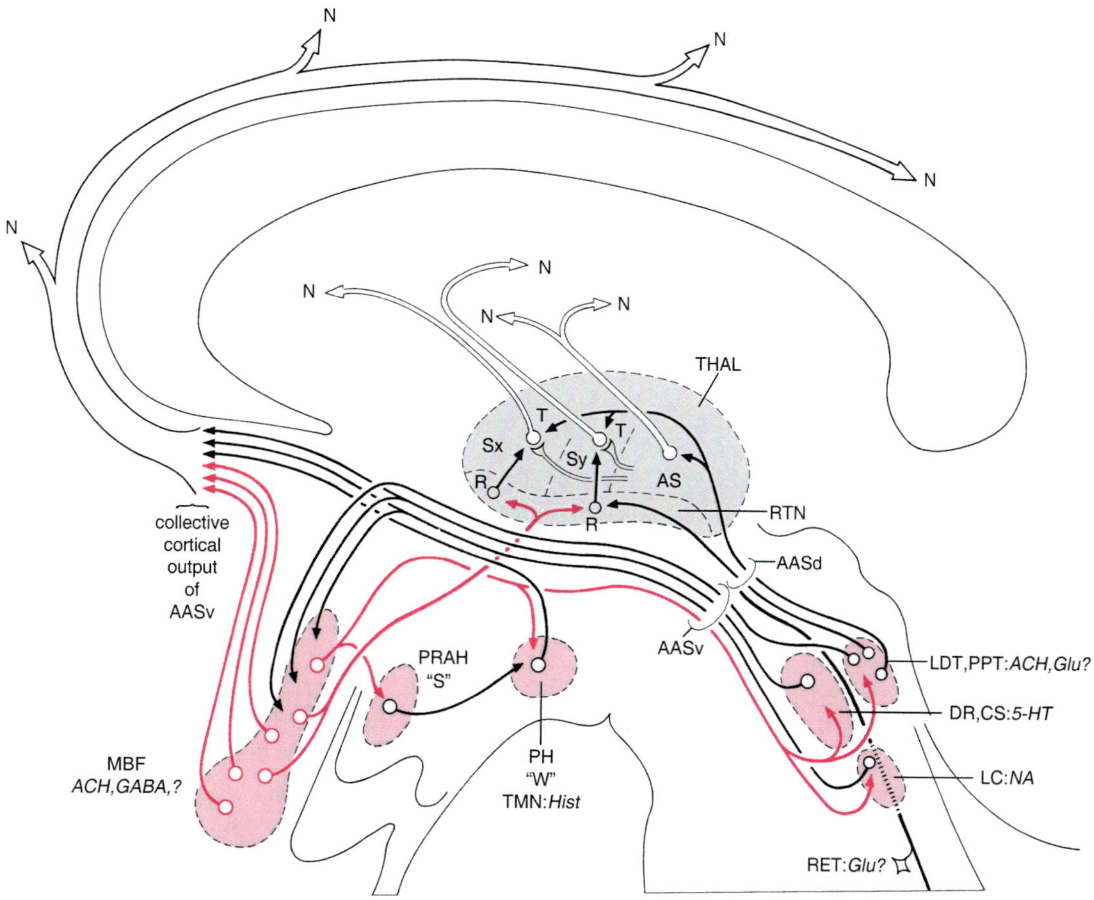

Fig. 14.15. The magnocellular basal forebrain (*MBF*), the network involved in the regulation of sleep, wakefulness, arousal and alertness, and the ascending arousal system (*AAS*). Efferents from the MBF are shown in *red*. Other abbreviations: *AASd,v*, dorsal and ventral branches of AAS; *AS*, aspecific thalamic nucleus; *CS*, central superior raphe nucleus; *DR*, dorsal raphe nucleus; *LC*, locus coeruleus; *LDT*, laterodorsal tegmental nucleus; *N*, neocortex; *PH*, posterior hypothalamic area; *PPT*, pedunculopontine tegmental nucleus; *PRAH*, preoptic-anterior hypothalamic area; *R*, GABAergic neurons in RTN; *RET*, pontine and medullary reticular formation; *RTN*, reticular thalamic nucleus; *'S'*, Sleep-promoting hypothalamic region; *Sx,y*, specific thalamic nuclei; *T*, thalamocortical relay neurons; *THAL*, thalamus; *TMN*, tuberomamillary nucleus; *'W'*, wake-promoting hypothalamic region

Clinical Aspects

Introductory Note

The basal ganglia have been implicated in a variety of neurological and psychiatric disorders. Before commenting on the morphological substrates and (possible) pathophysiological mechanisms involved in some of these disorders, it would be appropriate to present a brief, summarizing overview of the circuitry of the basal ganglia (see also Figs. 14.16 and 14.17).

Résumé of the Circuitry of the Basal Ganglia

1. The basal ganglia receive their principal afferents from the cerebral cortex and project back to the cortex via the thalamus. Within this overall sequence of cortico-subcortico-cortical connections, two different circuits, a direct one and an indirect one, can be distinguished, both of which being composed of consecutive sets of excitatory glutamatergic and inhibitory GABAergic neurons (Fig. 14.16).

The *direct circuit* is composed of excitatory corticostriatal, inhibitory striatopallidal, inhibitory pallidothalamic and excitatory thalamocortical projections. The fibres of the corticostriatal projection terminate on medium spiny projection neurons that co-contain GABA and substance P along with dopamine D1-class receptors. The striatopallidal projection terminates in the internal segments of the globus pallidus (GPi) through the direct circuit, cortical activation of the striatum results in reinforcing thalamic output back to the cortex ('positive feed-back').

The *indirect circuit* includes an excitatory corticostriatal projection that terminates onto striatal medium spiny projection neurons that co-contain GABA and enkephalin along with dopamine D2-class receptors, an inhibitory striatopallidal projection terminating in the external segment of the globus pallidus (GPe), an inhibitory projection from the GPe to the subthalamic nucleus (STN) and an excitatory projection from STN to the GPi. Cortical activation of the indirect circuit produces disinhibition of subthalamic neurons, resulting in an increase of the subthalamic activation of the GPi. Finally, this increase in GPi activation reduces thalamic activation of cortical neurons. Thus, the result of cortical activation of the indirect circuit is opposite to that of the direct circuit: reduction rather than reinforcement of cortical activity ('negative feed-back'). The pars reticulata of the substantia nigra forms part of both the direct and the indirect striatal circuit. In both circuits this structure occupies a position comparable to that of the GPi.

2. The projections passing from the cerebral cortex to the basal ganglia, which return to the cortex via the thalamus, form several parallel and functionally segregated loop systems. Prominent among these are a *motor loop,* which involves the motor and premotor cortices, an *associative or cognitive loop,* involving the dorsolateral prefrontal cortex, and a *limbic loop,* involving the orbital and medial prefrontal cortex (Fig. 14.17).

3. The flow of information from the cerebral cortex through the basal ganglia is at the level of the striatum modulated by dopaminergic fibres originating from the ventral mesencephalon. These dopaminergic fibres, which reach all functional domains of the striatum (Fig. 14.17), form part of a system that is critically involved in signalling the importance of events (including rewarding, aversive, novel and unexpected stimuli) and in driving motivation. The information provided by this system plays a crucial role in one of the principal functions of the basal ganglia, i.e. the selection and implementation of purposeful actions in response to environmental and internal cues.

4. The direct and indirect circuits (Fig. 14.16) are influenced differently by the dopaminergic projection from the ventral midbrain. This is so because the striatal projection neurons forming part of the direct circuit have D1-type dopamine receptors that facilitate transmission, whereas the striatal projection neurons forming part of the indirect circuit have D2-type receptors that reduce transmission. Although their synaptic actions are different, the dopamine inputs to the two circuits

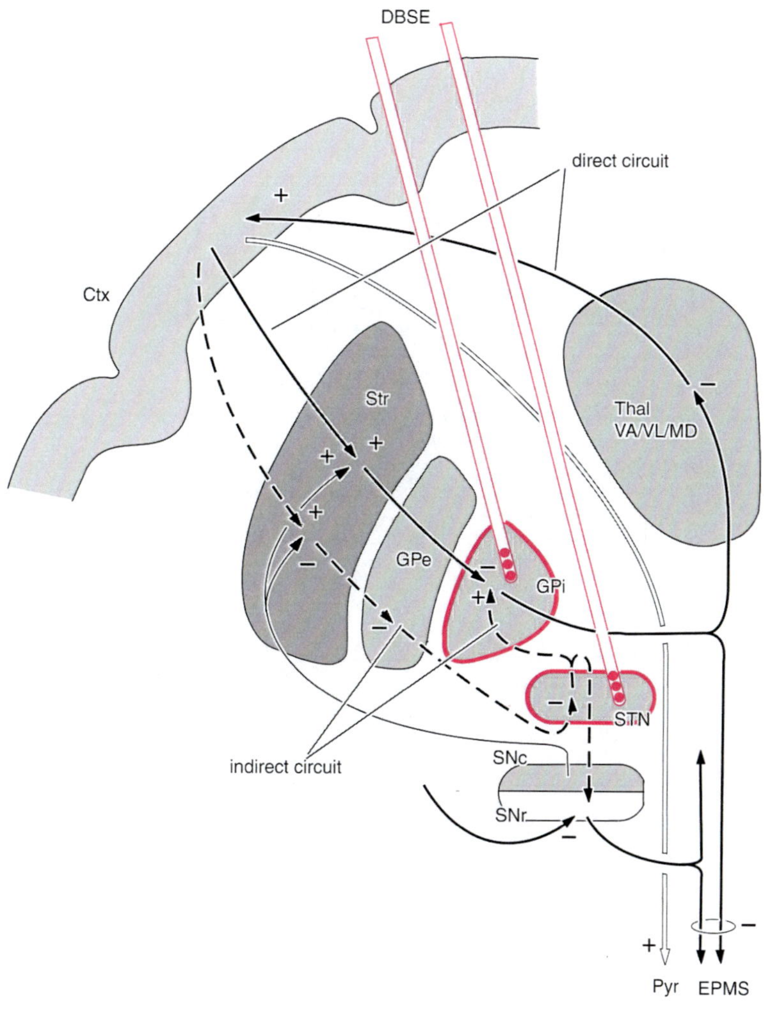

Fig. 14.16. Summary diagram of the circuitry of the basal ganglia. The cell masses which in Parkinson's disease are the targets of surgical interventions are surrounded by *bold red lines*. *Ctx*, cortex; *DBSE*, deep brain stimulation electrodes; *EPMS*, extrapyramidal motor systems; *GPe,i*, external and internal segments of globus pallidus; *MD*, mediodorsal thalamic nucleus; *SNc,r*, compact and reticular part of substantia nigra; *STN*, subthalamic nucleus; *Str*, striatum (caudate nucleus + putamen); *VA*, ventral anterior thalamic nucleus; *VL*, ventral lateral thalamic nucleus; *Pyr*, pyramidal tract; *Thal*, thalamus

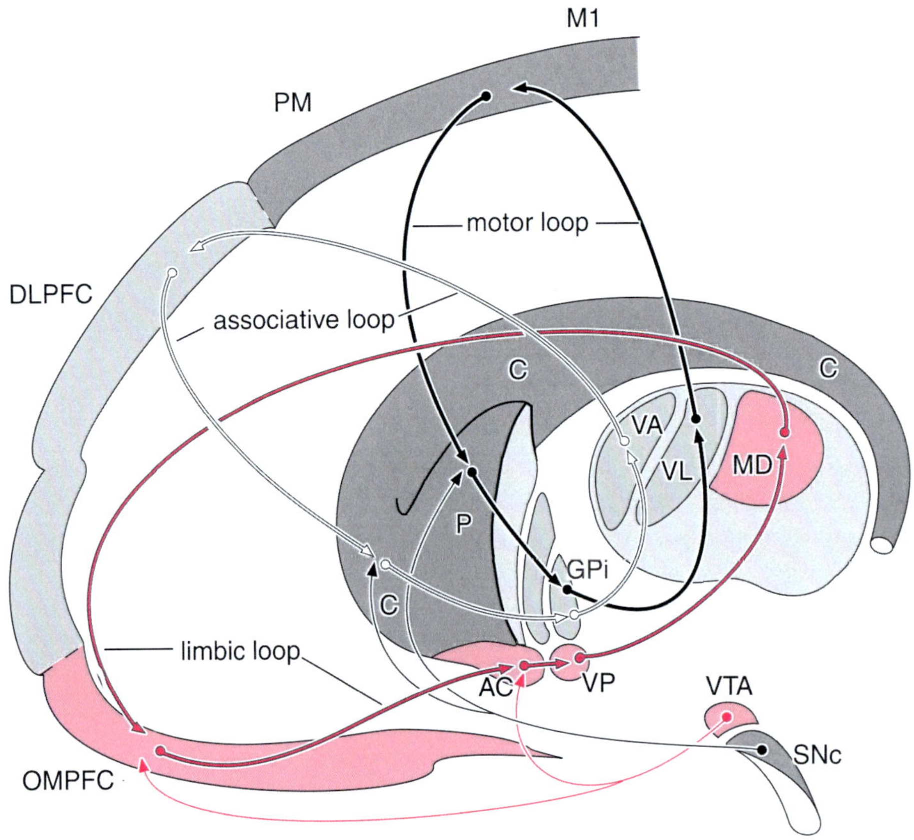

Fig. 14.17. The three principal functional loops by which the basal ganglia are connected with the cerebral cortex (*thick arrows*), and the dopaminergic pathways along which the flow of information through these loops is modulated (*thin arrows*). *AC*, accumbens nucleus; *C*, caudate nucleus; *DLPFC*, dorsolateral prefrontal cortex; *M1*, primary motor cortex; *OMPFC*, orbital and medial prefrontal cortex; *P*, putamen; *PM*, premotor cortex; *VP*, ventral pallidum; *VTA*, ventral tegmental area. For other abbreviations, see Fig. 14.16

ultimately lead to synergizing actions: enhancement of the positive feed-back to the cortex (through the direct circuit) and inhibition of the negative feed-back to the cortex (through the indirect circuit).

5. The parallel processing systems discussed under points 2 and 3 and depicted in Fig. 14.17 do not act in isolation. Anatomically, links between these systems have been demonstrated at cortical, striatal, pallidal and subthalamic levels. The preparation, implementation and selection of complex goal-oriented behaviours require that information be channeled from the limbic to the cognitive and motor loops.

6. The ascending outputs of the GPi and the SNr reach the cerebral cortex via the thalamus, to contribute to the various circuits and loops discussed above. However, within the motor and premotor cortices, these outputs also reach the cells of origin of the pyramidal tract, and the motor actions of the basal ganglia are largely mediated via this fibre system (Fig. 14.16).

7. The GPi and the SNr also give rise to descending projections which, after one or several synaptic interruptions, reach the motor neuron pools in the brain stem and spinal cord.

8. The basal nucleus of Meynert forms by far the largest and the most conspicuous cell mass in the so-called magnocellular basal forebrain. This nucleus receives afferents from many, predominantly limbic, territories and projects in a topographically ordered fashion to the entire cerebral cortex.

Disorders of the Basal Ganglia

Parkinson's disease (PD) or *paralysis agitans* is a common neurodegenerative disorder that usually begins during the fifth or sixth decade of life. The disease is characterized by progressive slowness of movement (bradykinesia), muscular rigidity, tremor and disturbances in posture and gait. The patients have difficulty in initiating and executing movements, and the movements are slower and smaller than normal. The muscular rigidity predominates in the flexor muscles of the limbs. The facial muscles exhibit an unnatural immobility (mask face). Spontaneous ocular movements are infrequent. Tremor, with a frequency of 3 to 6 per second, is the characteristic involuntary movement in PD. The tremor involves the hands more than other parts of the body. It is most prominent at rest and is suppressed during voluntary movements. Walking is difficult to initiate. The gait is slow, shuffling and composed of small steps. The normal pendulum of the arms during walking are lacking. Apart from these motor disturbances, PD is also characterized by the frequent occurrence of cognitive deficits, decreased drive and motivation and depression. As regards cognitive deficits, neuropsychological tests have shown that PD patients often have problems in adequately handling the available exteroreceptive sensory information [61, 62, 205] and in the planning and execution of constructional tasks [247].

The neuropathological hallmark of PD is the massive loss of the pigmented, dopaminergic neurons in the pars compacta of the substantia nigra, which leads to a reduction in the dopamine content in the striatum, most severely in the putamen [195]. A significant loss of neurons has also been found in the dopaminergic area tegmentalis ventralis [31], the noradrenergic locus coeruleus, the serotoninergic raphe nuclei and the gigantocellular reticular nucleus [38, 39]. Moreover, evidence has recently been presented suggesting that in PD the striatal medium spiny cells, i.e. the targets of the nigrostriatal dopaminergic fibres, are also affected. On the basis of a Golgi analysis, Stephens et al. [400] reported that the size of the dendritic trees, the number of dendritic branches and the density of spines of these neurons are all significantly reduced in PD.

In attempt to compensate for the loss of dopaminergic neurons and for the consequent reduction of striatal dopamine, it was discovered that administration of levodopa, a precursor of dopamine which – contrary to dopamine itself – is able to cross the blood-brain barrier, has a dramatic, although temporary, beneficial effect on the motor symptoms of PD.

Surgical interventions play an important role in the treatment of PD, particularly in patients with decreased response to levodopa. Current targets of these interventions are the GPi and the STN. In both of these structures, inducing stereotactic lesions [321, 383] and chronic high-frequency stimulation, delivered through stereotactically implanted electrodes [52, 70], may lead to marked improvements in the motor symptoms in PD (Fig. 14.16).

By serendipity, it was discovered that the meperidine derivative 1-methyl-4-phenyl-1,2,3,6-tetrahydroxypyridine (MPTP) is highly toxic to the dopaminergic neurons in the substantia nigra and induces a typical Parkinsonian syndrome in both humans and non-human primates [241]. As a result an animal model was developed for the experimental study of PD [71].

A working model of the pathophysiology of PD has also been developed based on knowledge of the functional circuitry of the basal ganglia, observations in PD patients and the results of studies in MPTP-treated primates [29, 71, 427]. According to this model, loss of dopaminergic input from the substantia nigra to the striatum decreases activity in the direct circuit and increases activity in the indirect circuit, due to the different actions of dopamine on the two circuits, via the D1-type and the D2-type receptors, respectively. The two circuits converge on the inhibitory striatal output neurons in the GPi and the SNr (Figs. 14.7, 14.9, 14.10, 14.16). Because the lack of stimulation of the D1-type receptors on the striatal projection neurons that form part of the direct circuit leads to a *decrease in the inhibition* of the GPi and SNr, while lack of stimulation of the D2-type receptors on the striatal projection neurons forming part of the indirect circuit ultimately leads to an *increase in the excitation* of the GPi and STn neurons, the dopamine deficiency increases the activity of these output neurons and increases the inhibition of the thalamocortical neurons on which they impinge. Thus, the resultant reduced cortical motor output explains the hypokinetic features of PD. The increase in excitation of the GPi and SNr neurons via the indirect circuit is the result of an increased excitatory drive from the STN.

The scheme just outlined is consistent with the findings of increased electrical activity in the GPi and STN in MPTP-treated monkeys [25] and of increased metabolic activity in the GPi of PD patients [90]. The surgical interventions mentioned earlier, which have been shown to be effective in both MPTP-treated monkeys [19, 70] and parkinsonian patients [228, 229, 234, 346], are supposed to interfere with the excessive inhibition of GPi neurons on thalamocortical neurons or with the excessive excitation of the STN neurons or the GPi and SNr elements (Fig. 14.16). Because in both the GPi and the STN, high-frequency stimulations appear to have the same clinical effect as lesions, stimulation presumably 'blocks' the neurons in these centres.

It should be emphasized that this model, although helpful, does not reflect the true complexity of the basal ganglia. Many fibre connections have not been taken into consideration (see Figs. 14.7 and 14.11) and, as discussed in the section dealing with the striatonigro-striatal circuit of the present chapter, the action of dopamine is not properly described by the dichotomy of its action on D1-type and D2-type receptors. Moreover, several recent findings are not in harmony with the predictions and propositions of the model. Thus, it has been shown that in patients with parkinsonism, high-frequency stimulation of the STN does not 'block' the neurons in that centre [11] and that in non-human MPTP primates high-frequency stimulation of the STN at therapeutic levels increases rather than decreases the activity in the GPi [174].

We have seen that dopamine exerts its influence on all loops through the basal ganglia (Fig. 14.17). However, the fact that the dopaminergic projections to the putamen are most severely affected in PD explains why the motor symptoms are dominant in this disease [195].

Huntington's chorea is a rare autosomal dominant disease of middle age, characterized by rapid, jerky, involuntary movements of the face, arms and legs and by progressive mental deterioration. The most striking neuropathological feature of this disease is severe atrophy of the striatum subsequent to a massive loss of medium spiny projection neurons. The striatal interneurons are largely spared, and this holds in particular for the elements expressing the calcium-binding protein calretinin [251]. It has been reported that in the earlier stages of the disease the striatal projection neurons participating in the indirect circuit are preferentially lost [72]. Consequently, inhibition of neurons in the GPe is reduced, which leads to excessive discharge of these neurons and to inhibition of the neurons in the STN. This inhibition of the STN neurons in turn reduces inhibition or disinhibits thalamocortical neurons in VA and VL and, ultimately, increases the excitatory input to the motor cortical areas, which could explain the choreatic movements (Figs. 14.9, 14.16). In later stages of Huntington's chorea, rigidity and akinesia develop, while the choreatic movements continue. The development of these hypokinetic symptoms is thought to be associated with the loss of striatal projection neurons that form part of the direct circuit [72].

Hemibalism is a disorder characterized by vigorous involuntary movements of the extremities with a repetitive, stereotypical character on one side of the body. The hyperkinesia is known to be caused by focal (commonly vascular) lesions of the contralateral STN [155, 171, 172, 425]. It is assumed that the STN lesions reduce the excitation of inhibitory GPi neurons, which in turn causes a hyperactivity of disinhibited thalamocortical projection neurons [72, 155].

We have seen that the STN is currently considered to play a key role in the pathophysiological origin of the parkinsonian state and is therefore an important target for surgical treatment of PD. It is important to note that stereotactic lesions in the STN in parkinsonian patients only very rarely lead to hemibalism. Gurudi and Obeso [155] have offered a possible explanation for this important clinical observation, which can be summarized as follows: (1) Dopamine depletion leads to a reduced activity in the direct GABAergic projection to the GPi. (2) STN lesions reduce excitation of the GPi, and theoretically this should induce hemiballism. (3) However, an STN lesion also simultaneously further reduces the hypoactivity in the GPe that is a feature of PD and hence may compensate for GPi hypoactivity, thus self-stabilizing basal ganglia output activity and reducing the risk of hemiballism.

The basal ganglia have been associated with the pathogenesis of several mental disorders, in particular schizophrenia, obsessive-compulsive disorder (OCD) and drug addiction. However, in general this association is less compelling than in PD or Huntington's disease and does not involve massive death of neurons. In these mental disorders, functional disturbances in the circuitry of the basal ganglia, particularly in the limbic loop and its dopaminergic innervation, are more apparent (Fig. 14.17).

Schizophrenia is a very common mental disease, which usually begins during young adulthood, and is characterized by a range of symptoms that typically include disordered thought, delusions of persecution, hallucinations, blunting of affect, social withdrawal and bizarre behaviour. The notion that dysfunctioning of the basal ganglia might be involved in the pathophysiology of schizophrenia evolved in the early 1950s as a result of the serendipitous finding that phenothiazines (such as chlorpromazine) can be effectively used for the treatment of this psychosis. The fact that phenothiazines act as dopamine receptor antagonists led to the so-called dopamine theory of schizophrenia, which posits that the mesencephalic dopaminergic projections to the nucleus accumbens and orbital and medial prefrontal cortices are hyperactive in this disease, which leads to disturbances in the flow of information through the limbic loop (Fig. 14.17) [259, 402]. In the 1990s this theory received a new impulse from the discovery of a new type of dopamine receptor (D3) and from the finding that, in schizophrenia, the density of the receptors of this type is considerably elevated in the accumbens nucleus and adjacent parts of the ventral striatum [180, 213, 390].

Obsessive compulsive disorder (OCD) is a chronic and disabling condition, which severely impairs personal, social and professional life. Patients with OCD suffer from intrusive, unwanted thoughts or ideas that are distressing (obsessive) and urge them to perform repetitive ritualistic behaviours or mental acts to conjure this distress.

There is some evidence that the orbitofrontal cortex may be implicated in the pathophysiology of OCD. Functional neuroimaging studies have shown that the metabolic activity in the orbitofrontal cortex is higher in OCD patients than in normal control subjects [239, 240] and that this activity is decreased after successful pharmacotherapy [59, 215]. Moreover, it has been found that the volume of the left anterior orbitofrontal cortex is in OCD patients significantly smaller than in normal controls [130].

Drug addiction is characterized by poorly managed, motivated behaviour, exemplified by an uncontrollable drive to seek drugs and decreased incentive to seek non-drug rewards [214, 417]. Experiments with animal models of addiction and neuroimaging studies in addicted subjects suggest that dysfunctioning of the glutamatergic projection from the prefrontal cortex to the accumbens nucleus may account for the cardinal features of addiction, and that this dysfunctioning in turn is probably due to abnormalities in the synaptic transmission from mesocortical dopaminergic fibres to the pyramidal neurons giving origin to the prefronto-accumbens projection.

We will conclude this section with some notes on the pathological changes observed in *Alzheimer's disease* (AD) and related disorders. AD and its variant in the elderly, *senile dementia of the Alzheimer type* (SDAT), are chronic neurodegenerative diseases that lead to mental deterioration and ultimately to death. Macroscopically, these diseases are characterized by a severe atrophy of the cerebral cortex, with thinning of the gyri and widening of the sulci. Cortical neuritic plaques and neurofibrillary tangles are their classical histopathological hallmarks. In the 1980s it was found that the amounts of choline acetyl transferase and acetylcholinesterase, i.e. the enzymes which are responsible for the synthesis and the inactivation of acetylcholine, are considerably reduced in the cerebral cortex in these diseases [69] and that the cholinergic neurons in the basal nucleus undergo a profound and selective degeneration [92, 266, 267, 422, 423]. As a result of these findings the so-called cholinergic hypothesis was developed, which posits that the role of the nucleus basalis of Meynert in AD/SDAT is comparable to that of the substantia nigra in PD [60, 65, 330]. However, later studies [415, 416] have convincingly shown that in AD/SDAT neuronal shrinkage rather than neuronal loss prevails in the basal nucleus, and that this degenerative shrinkage is secondary to the pathological changes in the cerebral cortex.

It is worthy of note that in AD/SDAT some other centres that project directly to the cortex, namely the noradrenergic locus coeruleus [36, 257, 258] and the largely serotonergic dorsal raphe nucleus [67, 424, 430], are also affected. Finally, it should be mentioned that degenerative changes in the basal nucleus of Meynert, accompanied by a deficiency of the cholinergic cortical projection system, is seen not only in AD and SDAT, but also in other disorders with similar deterioration of memory and cognitive functions, such as Creutzfeldt-Jakob's disease [8], Korsakoff's disease [7], many cases of PD [7, 292, 347, 424] and the remarkable Parkinsonism-dementia complex of Guam [291].

References

1. Albin RL, Aldridge JW, Young AB, Gilman S (1989) Feline subthalamic nucleus neurons contain glutamate-like but not GABA-like or glycine-like immunoreactivity. Brain Res 491:185–188
2. Alexander GE, Crutcher MD (1990) Functional architecture of basal ganglia circuits: neural substrates of parallel processing. Trends Neurosci 13:266–271
3. Alexander GE, DeLong MR, Strick PL (1986) Parallel organization of functionally segregated circuits linking basal ganglia and cortex. Annu Rev Neurosci 9:357–381
4. Alheid GF, Heimer L (1988) New perspectives in basal forebrain organization of special relevance for neuropsychiatric disorders: the striatopallidal, amygdaloid, and corticopetal components of substantia innominata. Neuroscience 27:1–39

5. Amaral DG, Kurz J (1985) An analysis of the origins of the cholinergic and noncholinergic septal projections to the hippocampal formation of the rat. J Comp Neurol 240:37–59
6. Andén NE, Dahlström A, Fuxe K et al (1966) Ascending monoamine neurons to the telencephalon and diencephalon. Acta Physiol Scand 67:313–326
7. Arendt T, Arendt A, Tennstedt A (1983) Loss of neurons in the nucleus basalis of Meynert in Alzheimer's disease, paralysis agitans and Korsakoffs disease. Acta Neuropathol (Berl) 61:101–108
8. Arendt T, Bigl V, Arendt A (1984) Neurone loss in the nucleus basalis of Meynert in Creutzfeldt-Jakob Disease. Acta Neuropathol (Berl) 65:85–88
9. Arikuni T, Kubota K (1986) The organization of prefrontocaudate projections and their laminar origin in the macaque monkey: a retrograde study using HRP-gel. J Comp Neurol 244:492–510
10. Asanuma C, Thach WT, Jones EG (1983) Distribution of cerebellar terminations and their relation to other afferent terminations in the ventral lateral thalamic region of the monkey. Brain Res 286:237–265
11. Ashby P, Kim YJ, Kumar R, Lang AE, Lozano AM (1999) Neurophysiological effects of stimulation through electrodes in the human subthalamic nucleus. Brain 122 (Pt 10):1919–1931
12. Bamford NS, Robinson S, Palmiter RD et al (2004) Dopamine modulates release from corticostriatal terminals. J Neurosci 24:9541–9552
13. Beach TG, McGeer EG (1984) The distribution of substance P in the primate basal ganglia: an immunohistochemical study of baboon and human brain. Neuroscience 13:29–52
14. Beaulieu C, Somogyi P (1991) Enrichment of cholinergic synaptic terminals on GABAergic neurons and coexistence of immunoreactive GABA and choline acetyltransferase in the same synaptic terminals in the striate cortex of the cat. J Comp Neurol 304:666–680
15. Beckstead RM (1983) A reciprocal axonal connection between the subthalamic nucleus and the neostriatum in the cat. Brain Res 275:137–142
16. Beckstead RM, Cruz CJ (1986) Striatal axons to the globus pallidus entopeduncular nucleus and substantia nigra come mainly from separate cell populations in the cat. Neuroscience 19:147–158
17. Beckstead RM, Frankfurter A (1982) The distribution and some morphological features of substantia nigra neurons that project to the thalamus, superior colliculus and pedunculopontine nucleus in the monkey. Neuroscience 7:2377–2388
18. Beckstead RM, Domesick VB, Nauta WJH (1979) Efferent connections of the substantia nigra and ventral tegmental area in the rat. Brain Res 175:191–217
19. Benazzouz A, Boraud T, Feger J et al (1996) Alleviation of experimental hemiparkinsonism by high-frequency stimulation of the subthalamic nucleus in primates: a comparison with L-Dopa treatment. Mov Disord 11:627–632
20. Bennett BD, Bolam JP (1993) Characterization of calretinin-immunoreactive structures in the striatum of the rat. Brain Res 609:137–148
21. Berendse HW, Groenewegen HJ (1990) Organization of the thalamostriatal projections in the rat, with special emphasis on the ventral striatum. J Comp Neurol 299:187–228
22. Berendse HW, Groenewegen HJ (1991) Restricted cortical termination fields of the midline and intralaminar thalamic nuclei in the rat. Neuroscience 42:73–102
23. Berendse HW, Galis-de Graaf Y, Groenewegen HJ (1992) Topographical organization and relationship with ventral striatal compartments of prefrontal corticostriatal projections in the rat. J Comp Neurol 316:314–347
24. Berendse HW, Groenewegen HJ, Lohman AHM (1992) Compartmental distribution of ventral striatal neurons projecting to the mesencephalon in the rat. J Neurosci 12:2079–2103
25. Bergman H, Wichmann T, Karmon B, DeLong MR (1994) The primate subthalamic nucleus. II. Neuronal activity in the MPTP model of parkinsonism. J Neurophysiol 72:507–520
26. Betarbet R, Turner R, Chockkan V et al (1997) Dopaminergic neurons intrinsic to the primate striatum. J Neurosci 17:6761–6768
27. Bickford ME, Hall WC (1992) The nigral projection to predorsal bundle cells in the superior colliculus of the rat. J Comp Neurol 319:11–33
28. Bingel U, Glascher J, Weiller C, Buchel C (2004) Somatotopic representation of nociceptive information in the putamen: an event-related fMRI study. Cereb Cortex 14:1340–1345
29. Blandini F, Nappi G, Tassorelli C, Martignoni E (2000) Functional changes of the basal ganglia circuitry in Parkinson's disease. Prog Neurobiol 62:63–88
30. Bobillier R, Seguin S, Petitjean F et al (1976) The raphe nuclei of the cat brain stem: a topographical atlas of their efferent projections as revealed by autoradiography. Brain Res 113:449–486
31. Bogerts B, Hantsch J, Herzer M (1983) A morphometric study of the dopamine-containing cell groups in the mesencephalon of normals, Parkinson patients, and schizophrenics. Biol Psychiatry 18:951–969
32. Bolam JP, Izzo PN (1988) The postsynaptic targets of substance P-immunoreactive terminals in the rat neostriatum with particular reference to identified spiny striatonigral neurons. Exp Brain Res 70:361–377
33. Bolam JP, Smith Y (1990) The GABA and substance P input to dopaminergic neurones in the substantia nigra of the rat. Brain Res 529:57–78
34. Bolam JP, Izzo PN, Graybiel AM (1988) Cellular substrate of the histochemically defined striosome/matrix system of the caudate nucleus: a combined Golgi and immunocytochemical study in cat and ferret. Neuroscience 24:853–875

35. Bolam JP, Francis CM, Henderson Z (1991) Cholinergic input to dopaminergic neurons in the substantia nigra: a double immunocytochemical study. Neuroscience 41:483–494
36. Bondareff W, Mountjoy CQ, Roth M (1982) Loss of neurons of origin of the adrenergic projections to cerebral cortex (nucleus locus coeruleus) in senile dementia. Neurology 32:164–168
37. Braak H, Braak E (1982) Neuronal types in the striatum of man. Cell Tissue Res 227:319–342
38. Braak H, Braak E (2000) Pathoanatomy of Parkinson's disease. J Neurol 247 Suppl 2:II3–10
39. Braak H, Rub U, Sandmann-Keil D et al (2000) Parkinson's disease: affection of brain stem nuclei controlling premotor and motor neurons of the somatomotor system. Acta Neuropathol (Berl) 99:489–495
40. Brooks VB, Thach WT (1981) Cerebellar control of posture and movement. In: Brooks VB (ed) Handbook of physiology. Section I: The nervous system. Vol 2: Motor control. Part II. American Physiological Society, Bethesda, pp 877–946
41. Büttner-Ennever JA, Büttner U, Cohen B, Baumgartner O (1982) Vertical gaze paralysis and the rostral interstitial nucleus of the medial longitudinal fasciculus. Brain 105:125–149
42. Calabresi P, Centonze D, Gubellini P et al (2000) Synaptic transmission in the striatum: from plasticity to neurodegeneration. Prog Neurobiol 61: 231–265
43. Canteras NS, Shammah-Lagnado SJ, Silva BA, Ricardo JA (1990) Afferent connections of the subthalamic nucleus: a combined retrograde and anterograde horseradish peroxidase study in the rat. Brain Res 513:43–59
44. Carman JB, Cowan WM, Powell TPS (1963) The organization of the cortico-striate connexions in the rabbit. Brain 86:525–562
45. Carman JB, Cowan WM, Powell TP, Webster KE (1965) A bilateral cortico-striate projection. J Neurol Neurosurg Psychiatry 28:71–77
46. Carmichael ST, Price JL (1995) Sensory and premotor connections of the orbital and medial prefrontal cortex of macaque monkeys. J Comp Neurol 363:642–664
47. Carmichael ST, Price JL (1996) Connectional networks within the orbital and medial prefrontal cortex of macaque monkeys. J Comp Neurol 371:179–207
48. Carpenter MB, Peter Ph (1972) Nigrostriatal and nigrothalamic fibers in the rhesus monkey. J Comp Neurol 144:93–117
49. Carpenter MB, Nakano K, Kim R (1976) Nigrothalamic projections in the monkey demonstrated by autoradiographic technics. J Comp Neurol 165: 401–416
50. Carpenter MB, Batton RR, Carleton SC, Keller JT (1981) Interconnections and organization of pallidal and subthalamic nucleus neurons in the monkey. J Comp Neurol 197:579–603
51. Carpenter MB, Carleton SC, Keller JT, Conte P (1981) Connections of the subthalamic nucleus in the monkey. Brain Res 224:1–29
52. Ceballos-Baumann AO, Boecker H, Bartenstein P et al (1999) A positron emission tomographic study of subthalamic nucleus stimulation in Parkinson disease: enhanced movement-related activity of motor-association cortex and decreased motor cortex resting activity. Arch Neurol 56:997–1003
53. Cepeda C, Levine MS (1998) Dopamine and N-methyl-D-aspartate receptor interactions in the neostriatum. Dev Neurosci 20:1–18
54. Charara A, Parent A (1994) Brainstem dopaminergic, cholinergic and serotoninergic afferents to the pallidum in the squirrel monkey. Brain Res 640:155–170
55. Chesselet MF, Graybiel AM (1986) Striatal neurons expressing somatostatin-like immunoreactivity: evidence for a peptidergic interneuronal system in the cat. Neuroscience 17:547–571
56. Chevalier G, Thierry AM, Shibazaki T, Feger J (1981) Evidence for a gabaergic inhibitory nigrotectal pathway in the rat. Neurosci Lett 21:67–70
57. Chiba T, Kayahara T, Nakano K (2001) Efferent projections of infralimbic and prelimbic areas of the medial prefrontal cortex in the Japanese monkey, Macaca fuscata. Brain Res 888:83–101
58. Childs JA, Gale K (1983) Neurochemical evidence for a nigrotegmental GABAergic projection. Brain Res 258:109–114
59. Choi JS, Kang DH, Kim JJ et al (2004) Left anterior subregion of orbitofrontal cortex volume reduction and impaired organizational strategies in obsessive-compulsive disorder. J Psychiatr Res 38:193–199
60. Collerton D (1986) Cholinergic function and intellectual decline in Alzheimer's disease. Neuroscience 19:1–28
61. Cools AR (1980) Role of the neostriatal dopaminergic activity in sequencing and selecting behavioral strategies: facilitation of processes involved in selecting the best strategy in an stressful situation. Behav Brain Res 1:361–378
62. Cools AR, Jaspers R, Schwartz M et al (1984) Basal ganglia and switching motor programs. In: McKenzie JS, Kemm RE, Wilcock LN (eds) Basal ganglia structure and function. Plenum, New York, pp 513–544
63. Cossette M, Lecomte F, Parent A (2005) Morphology and distribution of dopaminergic neurons intrinsic to the human striatum. J Chem Neuroanat 29:1–11
64. Cowan RL, Wilson CJ (1994) Spontaneous firing patterns and axonal projections of single corticostriatal neurons in the rat medial agranular cortex. J Neurophysiol 71:17–32
65. Coyle JT, Price DL, DeLong MR (1983) Alzheimer's disease: a disorder of cortical cholinergic innervation. Science 219:1184–1190
66. Cullinan WE, Zaborszky L (1991) Organization of ascending hypothalamic projections to the rostral forebrain with special reference to the innervation

of cholinergic projection neurons. J Comp Neurol 306:631–667
67. Curcio CA, Kemper T (1984) Nucleus raphe dorsalis in dementia of the Alzheimer type neurofibrillary changes and neuronal packing density. J Neuropathol Exp Neurol 43:359–368
68. Dahlström A, Fuxe K (1964) Evidence for the existence of monoamine-containing neurons in the central nervous system. I. Demonstration of monoamines in the cell bodies of brain stem neurons. Acta Physiol Scand [Suppl] 62(232):1–55
69. Davies P (1979) Neurotransmitter related enzymes in senile dementia of the Alzheimer type. Brain Res 171:310–327
70. Davis KD, Taub E, Houle S et al (1997) Globus pallidus stimulation activates the cortical motor system during alleviation of parkinsonian symptoms. Nat Med 3:671–674
71. DeLong MR (1990) Primate models of movement disorders of basal ganglia origin. Trends Neurosci 13:281–285
72. DeLong MR (2000) The basal ganglia. In: Kandel ER, Schwartz JH, Jessel TM (eds) Principles of neural science, 4th edn. McGraw-Hill, New York, pp 853–867
73. DeLong MR, Georgopoulos AP (1981) Motor function of the basal ganglia. In: Brookhart JM, Mountcastle VB, Brooks VB (eds) Handbook of physiology, sect 1: The nervous system, vol 2: Motor control, part 2. American Physiological Society, Bethesda, pp 1017–1061
74. Deniau JM, Chevalier G (1985) Disinhibition as a basic process in the expression of striatal functions. II. The striato-nigral influence on thalamocortical cells of the ventromedial thalamic nucleus. Brain Res 334:227–233
75. Deniau JM, Menetrey A, Thierry AM (1994) Indirect nucleus accumbens input to the prefrontal cortex via the substantia nigra pars reticulata: a combined anatomical and electrophysiological study in the rat. Neuroscience 61:533–545
76. Desban M, Kemel ML, Glowinski J, Gauchy C (1993) Spatial organization of patch and matrix compartments in the rat striatum. Neuroscience 57:661–671
77. Desban M, Gauchy C, Glowinski J, Kemel ML (1995) Heterogeneous topographical distribution of the striatonigral and striatopallidal neurons in the matrix compartment of the cat caudate nucleus. J Comp Neurol 352:117–133
78. Deschênes M, Bourassa J, Doan VD, Parent A (1996) A single-cell study of the axonal projections arising from the posterior intralaminar thalamic nuclei in the rat. Eur J Neurosci 8:329–343
79. Desmond JE, Fiez JA (1998) Neuroimaging studies of the cerebellum: language, learning and memory. Trends Cogn Sci 2:355–362
80. DeVito JL, Anderson ME (1982) An autoradiographic study of efferent connections of the globus pallidus in macaca mulatta. Exp Brain Res 46:107–117
81. DiFiglia M, Rafols JA (1988) Synaptic organization of the globus pallidus. J Electron Microsc Tech 10:247–263
82. DiFiglia M, Pasik P, Pasik T (1976) A Golgi study of neuronal types in the neostriatum of monkeys. Brain Res 114:245–256
83. Divac I (1975) Magnocellular nuclei of the basal forebrain project to neocortex, brain stem, and olfactory bulb: review of some functional correlates. Brain Res 93:385–398
84. Donoghue JP, Herkenham M (1986) Neostriatal projections from individual cortical fields conform to histochemically distinct striatal compartments in the rat. Brain Res 365:397–403
85. Drepper J, Timmann D, Kolb FP, Diener HC (1999) Non-motor associative learning in patients with isolated degenerative cerebellar disease. Brain 122 (Pt 1):87–97
86. Dubé L, Smith AD, Bolam JP (1988) Identification of synaptic terminals of thalamic or cortical origin in contact with distinct medium-size spiny neurons in the rat neostriatum. J Comp Neurol 267:455–471
87. Dum RP, Strick PL (2005) Frontal lobe inputs to the digit representations of the motor areas on the lateral surface of the hemisphere. J Neurosci 25:1375–1386
88. Eblen F, Graybiel AM (1995) Highly restricted origin of prefrontal cortical inputs to striosomes in the macaque monkey. J Neurosci 15:5999–6013
89. Edley SM, Graybiel AM (1983) The afferent and efferent connections of the feline nucleus tegmenti pedunculopontinus, pars compacta. J Comp Neurol 217:187–215
90. Eidelberg D, Moeller JR, Dhawan V et al (1994) The metabolic topography of parkinsonism. J Cereb Blood Flow Metab 14:783–801
91. Elliott R, Rubinsztein JS, Sahakian BJ, Dolan RJ (2002) The neural basis of mood-congruent processing biases in depression. Arch Gen Psychiatry 59:597–604
92. Etienne P, Robitaille Y, Wood P et al (1986) Nucleus basalis neuronal loss, neuritic plaques and choline acetyltransferase activity in advanced Alzheimer's disease. Neuroscience 19:1279–1291
93. Evarts EV, Thach WT (1969) Motor mechanisms of the CNS: cerebrocerebellar interrelations. Annu Rev Physiol 31:451–498
94. Fénelon G, Francois C, Percheron G, Yelnik J (1990) Topographic distribution of pallidal neurons projecting to the thalamus in macaques. Brain Res 520:27–35
95. Ferry AT, Öngür D, An X, Price JL (2000) Prefrontal cortical projections to the striatum in macaque monkeys: evidence for an organization related to prefrontal networks. J Comp Neurol 425:447–470
96. Fisher RS, Buchwald NA, Hull CD, Levine MS (1988) GABAergic basal forebrain neurons project to the neocortex: the localization of glutamic acid decarboxylase and choline acetyltransferase in feline corticopetal neurons. J Comp Neurol 272:489–502

97. Flaherty AW, Graybiel AM (1993) Output architecture of the primate putamen. J Neurosci 13:3222–3237
98. Fonnum F, Storm-Mathisen J, Divac I (1981) Biochemical evidence for glutamate as neurotransmitter in corticostriatal and corticothalamic fibres in rat brain. Neuroscience 6:863–873
99. Foote SL, Morrison JH (1987) Extrathalamic modulation of cortical function. Annu Rev Neurosci 10:67–95
100. Fox CA, Rafols JA (1975) The radial fibers in the globus pallidus. J Comp Neurol 159:177–200
101. Fox CA, Rafols JA, Cowan WM (1975) Computer measurements of axis cylinder diameters of radial fibers and "comb" bundle fibers. J Comp Neurol 159:201–224
102. Francois C, Percheron G, Yelnik J (1984) Localization of nigrostriatal, nigrothalamic and nigrotectal neurons in ventricular coordinates in macaques. Neuroscience 13:61–76
103. Francois C, Yelnik J, Tande D, Agid Y, Hirsch EC (1999) Dopaminergic cell group A8 in the monkey: anatomical organization and projections to the striatum. J Comp Neurol 414:334–347
104. Francois C, Savy C, Jan C et al (2000) Dopaminergic innervation of the subthalamic nucleus in the normal state, in MPTP-treated monkeys, and in Parkinson's disease patients. J Comp Neurol 425:121–129
105. Francois C, Tande D, Yelnik J, Hirsch EC (2002) Distribution and morphology of nigral axons projecting to the thalamus in primates. J Comp Neurol 447:249–260
106. French SJ, Totterdell S (2002) Hippocampal and prefrontal cortical inputs monosynaptically converge with individual projection neurons of the nucleus accumbens. J Comp Neurol 446:151–165
107. Freund TF, Gulyas AI (1991) GABAergic interneurons containing calbindin D28K or somatostatin are major targets of GABAergic basal forebrain afferents in the rat neocortex. J Comp Neurol 314:187–199
108. Freund TF, Meskenaite V (1992) gamma-Aminobutyric acid-containing basal forebrain neurons innervate inhibitory interneurons in the neocortex. Proc Natl Acad Sci USA 89:738–742
109. Freund TF, Martin KA, Smith AD, Somogyi P (1983) Glutamate decarboxylase-immunoreactive terminals of Golgi-impregnated axoaxonic cells and of presumed basket cells in synaptic contact with pyramidal neurons of the cat's visual cortex. J Comp Neurol 221:263–278
110. Freund TF, Powell JF, Smith AD (1984) Tyrosine hydroxylase-immunoreactive boutons in synaptic contact with identified striatonigral neurons, with particular reference to dentritic spines. Neuroscience 13:1189–1215
111. Friedman DP, Aggleton JP, Saunders RC (2002) Comparison of hippocampal, amygdala, and perirhinal projections to the nucleus accumbens: combined anterograde and retrograde tracing study in the Macaque brain. J Comp Neurol 450:345–365
112. Fudge JL, Haber SN (2002) Defining the caudal ventral striatum in primates: cellular and histochemical features. J Neurosci 22:10078–10082
113. Fudge JL, Kunishio K, Walsh P, Richard C, Haber SN (2002) Amygdaloid projections to ventromedial striatal subterritories in the primate. Neuroscience 110:257–275
114. Fudge JL, Breitbart MA, McClain C (2004) Amygdaloid inputs define a caudal component of the ventral striatum in primates. J Comp Neurol 476: 330–347
115. Fuster JM (1997) The prefrontal cortex: anatomy, physiology and neuropsychology of the frontal lobe. Lippincott-Raven, Philadelphia
116. Garcia-Rill E, Skinner RD (1987) The mesencephalic locomotor region. II. Projections to reticulospinal neurons. Brain Res 411:13–20
117. Garcia-Rill E, Skinner RD, Miyazato H, Homma Y (2001) Pedunculopontine stimulation induces prolonged activation of pontine reticular neurons. Neuroscience 104:455–465
118. Garris PA, Ciolkowski EL, Pastore P, Wightman RM (1994) Efflux of dopamine from the synaptic cleft in the nucleus accumbens of the rat brain. J Neurosci 14:6084–6093
119. Gerardin E, Lehericy S, Pochon JB et al (2003) Foot, hand, face and eye representation in the human striatum. Cereb Cortex 13:162–169
120. Gerfen CR (1984) The neostriatal mosaic: compartmentalization of corticostriatal input and striatonigral output systems. Nature 311:461–464
121. Gerfen CR (1989) The neostriatal mosaic: striatal patch-matrix organization is related to cortical lamination. Science 246:385–388
122. Gerfen CR (1992) The neostriatal mosaic: multiple levels of compartmental organization in the basal ganglia. Annu Rev Neurosci 15:285–320
123. Gerfen CR, Wilson CJ (1996) The basal ganglia. In: Swanson LW, Björklund A, Hökfelt T (eds) Integrated systems of the CNS, Part III. Elsevier, Amsterdam, pp 371–468
124. Geyer MA, Puerto A, Dawsey WI et al (1976) Histologic and enzymatic studies of the mesolimbic and mesostriatal serotonergic pathways. Brain Res 106:241–256
125. Giménez-Amaya J-M, Graybiel AM (1990) Compartmental origins of the striatopallidal projection in the primate. Neuroscience 34:111–126
126. Giménez-Amaya J-M, McFarland NR, De las Heras S, Haber SN (1995) Organization of thalamic projections to the ventral striatum in the primate. J Comp Neurol 354:127–149
127. Goldberg JM, Fernández C (1980) Efferent Vestibular System in the Squirrel Monkey: anatomical Location and Influence on Afferent Activity. J Neurophysiol 43:986–1025
128. Goldman PS, Nauta WJH (1977) Columnar distribution of cortico-cortical fibers in the frontal as-

sociation, limbic and motor cortex of the developing rhesus monkey. Brain Res 122:393–413
129. Goldman-Rakic PS (1982) Cytoarchitectonic heterogeneity of the primate neostriatum: subdivision into Island and Matrix cellular compartments. J Comp Neurol 205:398–413
130. Goldstein RZ, Volkow ND (2002) Drug addiction and its underlying neurobiological basis: neuroimaging evidence for the involvement of the frontal cortex. Am J Psychiatry 159:1642–1652
131. Graveland GA, DiFiglia M (1985) The frequency and distribution of medium-sized neurons with indented nuclei in the primate and rodent neostriatum. Brain Res 327:307–311
132. Graveland GA, Williams RS, DiFiglia M (1985) A golgi study in the human neostriatum: neurons and afferent fibers. J Comp Neurol 234:317–333
133. Graybiel AM (1978) Organization of the nigrotectal connection: an experimental tracer study in the cat. Brain Res 143:339–348
134. Graybiel AM (1984) Correspondence between the dopamine islands and striosomes of the mammalian striatum. Neuroscience 13:1157–1187
135. Graybiel AM (1990) Neurotransmitters and neuromodulators in the basal ganglia. Trends Neurosci 13:244–254
136. Graybiel AM, Ragsdale CW (1978) Histochemically distinct compartments in the striatum of human, monkey and cat demonstrated by acetylcholinesterase staining. Proc Natl Acad Sci USA 75:5723–5726
137. Graybiel AM, Ragsdale Jr CW (1979) Fiber connections of the basal ganglia. In: Bloom RE, Kreutzberg GW, Cuénod M (eds) Development and chemical specificity of neurons. Elsevier, Amsterdam, pp 239–283
138. Graybiel AM, Ragsdale Jr CW (1983) Biochemical anatomy of the striatum. In: Emson PC (ed) Chemical neuroanatomy. Raven Press, New York, pp 427–504
139. Graybiel AM, Sciascia TR (1975) Origin and distribution of nigrotectal fibers in the cat. Neurosci Abstr 1:174
140. Graybiel AM, Ragsdale CW, Yoneoka ES, Elde RH (1981) An immunohistochemical study of enkephalins and other neuropeptide in the striatum of the cat with evidence that the opiate peptides are arranged to form mosaic patterns in register with the striosomal compartments visible by acetylcholinesterase staining. Neuroscience 6: 377–397
141. Graybiel AM, Flaherty AW, Giménez-Amaya J-M (1991) Striosomes and matrisomes. In: Bernardi G, Carpenter MB, di Chiara G, Morelli M, Stanzione P (eds) The basal ganglia, vol 3. Plenum, New York, pp 3–12
142. Grillner S, Shik MI (1973) On the descending control of the lumbosacral spinal cord from the "mesencephalic locomotor region". Acta Physiol Scand 87:320–333
143. Gritti I, Mainville L, Mancia M, Jones BE (1997) GABAergic and other noncholinergic basal forebrain neurons, together with cholinergic neurons, project to the mesocortex and isocortex in the rat. J Comp Neurol 383:163–177
144. Groenewegen HJ, Berendse HW (1990) Connections of the subthalamic nucleus with ventral striatopallidal parts of the basal ganglia in the rat. J Comp Neurol 294:607–622
145. Groenewegen HJ, Berendse HW (1994) The specificity of the 'nonspecific' midline and intralaminar thalamic nuclei. Trends Neurosci 17:52–57
146. Groenewegen HJ, Russchen FT (1984) Organization of the efferent projections of the nucleus accumbens to pallidal hypothalamic, and mesencephalic structures: a tracing and immunohistochemical study in the cat. J Comp Neurol 223: 347–367
147. Groenewegen HJ, Becker NEK, Lohman AHM (1980) Subcortical afferents of the nucleus accumbens septi in the cat, studied with retrograde axonal transport of horseradish peroxidase and bisbenzimide. Neuroscience 5:1903–1916
148. Groenewegen HJ, Room P, Witter MP, Lohman AHM (1982) Cortical afferents of the nucleus accumbens in the cat, studied with anterograde and retrograde transport techniques. Neuroscience 7:977–995
149. Groenewegen HJ, Meredith GE, Berendse HW, Voorn P, Wolters JG (1989) The compartmental organization of the ventral striatum in the rat. In: Crossman AR, Sambrook AM (eds) Neural mechanisms in disorders of movement. Libbey, London, pp 45–52
150. Groenewegen HJ, Berendse HW, Haber SN (1993) Organization of the output of the ventral striatopallidal system in the rat: ventral pallidal efferents. Neuroscience 57:113–142
151. Groenewegen HJ, Wright CI, Beijer AV (1996) The nucleus accumbens: gateway for limbic structures to reach the motor system? Prog Brain Res 107:485–511
152. Grove EA (1988) Neural associations of the substantia innominata in the rat: afferent connections. J Comp Neurol 277:315–346
153. Grove EA (1988) Efferent connections of the substantia innominata in the rat. J Comp Neurol 277:347–364
154. Gulyas AI, Gorcs TJ, Freund TF (1990) Innervation of different peptide-containing neurons in the hippocampus by GABAergic septal afferents. Neuroscience 37:31–44
155. Guridi J, Obeso JA (2001) The subthalamic nucleus, hemiballismus and Parkinson's disease: reappraisal of a neurosurgical dogma. Brain 124: 5–19
156. Gúzman JN, Hernandez A, Galarraga E et al (2003) Dopaminergic modulation of axon collaterals interconnecting spiny neurons of the rat striatum. J Neurosci 23:8931–8940

157. Haber S, Elde R (1982) The distribution of enkephalin immunoreactive fibers and terminals in the monkey central nervous system: an immunohistochemical study. Neuroscience 7:1049–1095
158. Haber SN (2003) The primate basal ganglia: parallel and integrative networks. J Chem Neuroanat 26:317–330
159. Haber SN, Johnson Gdowski M (2004) The basal ganglia. In: Paxinos G, Mai JK (eds) The human nervous system, 2nd edn. Elsevier, Amsterdam, pp 676–738
160. Haber SN, Nauta WJH (1983) Ramifications of the globus pallidus in the rat as indicated by patterns of immunohistochemistry. Neuroscience 9:245–260
161. Haber SN, Watson SJ (1985) The comparative distribution of enkephalin dynorphin and substance P in the human globus pallidus and basal forebrain. Neuroscience 14:1011–1024
162. Haber SN, Groenewegen HJ, Grove EA, Nauta WJH (1985) Efferent connections of the ventral pallidum: evidence of a dual striato-pallidofugal pathway. J Comp Neurol 235:322–335
163. Haber SN, Lynd E, Klein C, Groenewegen HJ (1990) Topographic organization of the ventral striatal efferent projections in the rhesus monkey: a anterograde tracing study. J Comp Neurol 293:282–298
164. Haber SN, Wolfe DP, Groenewegen HJ (1990) The relationship between ventral striatal efferent fibers and the distribution of peptide-positive woolly fibers in the forebrain of the rhesus monkey. Neuroscience 39:323–338
165. Haber SN, Lynd-Balta E, Mitchell SJ (1993) The organization of the descending ventral pallidal projections in the monkey. J Comp Neurol 329:111–128
166. Haber SN, Kunishio K, Mizobuchi M, Lynd-Balta E (1995) The orbital and medial prefrontal circuit through the primate basal ganglia. J Neurosci 15: 4851–4867
167. Haber SN, Ryoo H, Cox C, Lu W (1995) Subsets of midbrain dopaminergic neurons in monkeys are distinguished by different levels of mRNA for the dopamine transporter: comparison with the mRNA for the D2 receptor, tyrosine hydroxylase and calbindin immunoreactivity. J Comp Neurol 362:400–410
168. Haber SN, Fudge JL, McFarland NR (2000) Striatonigrostriatal pathways in primates form an ascending spiral from the shell to the dorsolateral striatum. J Neurosci 20:2369–2382
169. Hallanger AE, Levey AI, Lee HJ, Rye DB, Wainer BH (1987) The origins of cholinergic and other subcortical afferents to the thalamus in the rat. J Comp Neurol 262:105–124
170. Halliday G (2004) Substantia nigra and locus coeruleus. In: Paxinos G, Mai JK (eds) The human nervous system, 2nd edn. Elsevier, Amsterdam, pp 449–463
171. Hamani C, Saint-Cyr JA, Fraser J, Kaplitt M, Lozano AM (2004) The subthalamic nucleus in the context of movement disorders. Brain 127:4–20
172. Hammond C, Feger J, Bioulac B, Souteyrand JP (1979) Experimental hemiballism in the monkey produced by unilateral kainic acid lesion in corpus Luysii. Brain Res 171:577–580
173. Hartmann von Monakow K, Akert K, Künzle H (1978) Projections of the precentral motor cortex and other cortical areas of the frontal lobe to the subthalamic nucleus in the monkey. Exp Brain Res 33:395–403
174. Hashimoto T, Elder CM, Okun MS, Patrick SK, Vitek JL (2003) Stimulation of the subthalamic nucleus changes the firing pattern of pallidal neurons. J Neurosci 23:1916–1923
175. Hazrati LN, Parent A (1992) The striatopallidal projection displays a high degree of anatomical specificity in the primate. Brain Res 592:213–227
176. Hedreen JC (1999) Tyrosine hydroxylase-immunoreactive elements in the human globus pallidus and subthalamic nucleus. J Comp Neurol 409:400–410
177. Hedreen JC, DeLong MR (1991) Organization of striatopallidal, striatonigral, and nigrostriatal projections in the macaque. J Comp Neurol 304:569–595
178. Hedreen JC, Struble RG, Whitehouse PJ, Price DL (1984) Topography of the magnocellular basal forebrain system in the human brain. J Neuropathol Exp Neurol 43:1–21
179. Heimer L (1978) The olfactory cortex and the ventral striatum. In: Livingston KE, Hornykiewicz O (eds) Limbic mechanisms: the continuing evolution of the limbic system concept. Plenum, New York, pp 95–187
180. Heimer L (2000) Basal forebrain in the context of schizophrenia. Brain Res Brain Res Rev 31:205–235
181. Heimer L, Wilson RD (1975) The subcortical projections of the allocortex: similarities in the neural associations of the hippocampus, the piriform cortex, and the neocortex. In: Santini M (ed) Golgi Centennial Symposium proceedings. Raven, New York, pp 177–193
182. Heimer L, Switzer RD, Van Hoesen GW (1982) Ventral striatum and ventral pallidum. Components of the motor system? Trends Neurosci 5:83–87
183. Heimer L, Zahm DS, Churchill L, Kalivas PW, Wohltmann C (1991) Specificity in the projection patterns of accumbal core and shell in the rat. Neuroscience 41:89–125
184. Herkenham M (1986) New perspectives on the organization and evolution of nonspecific thalamocortical projections. In: Jones EG, Peters A (eds) Cerebral cortex, vol 5: Sensory-motor areas and aspects of cortical connectivity. Plenum Press, New York, pp 403–445
185. Herkenham M, Nauta WJH (1977) Afferent connections of the habenular nuclei in the rat. A

horseradish peroxidase study, with a note on the fiber-of-passage problem. J Comp Neurol 173: 123–146
186. Herkenham M, Nauta WJH (1979) Efferent connections of the habenula nuclei in the rat. J Comp Neurol 187:19–48
187. Hikosaka O, Sakamoto M, Usui S (1989) Functional properties of monkey caudate neurons. I. Activities related to saccadic eye movements. J Neurophysiol 61:780–798
188. Hikosaka O, Takikawa Y, Kawagoe R (2000) Role of the basal ganglia in the control of purposive saccadic eye movements. Physiol Rev 80:953–978
189. Hökfelt T, Goldstein M, Fuxe K et al (1980) Histochemical identification of adrenaline containing cells with special reference to neurons. In: Fuxe K, Goldstein M, Hökfelt B, Hökfelt T (eds) Central adrenaline neurons. Basic aspects and their role in cardio-vascular functions. Pergamon, Oxford, pp 19–47
190. Hökfelt T, Skirboll L, Rehfeld JF et al (1980) A subpopulation of mesencephalic dopamine neurons projecting to limbic areas contains a cholecystokinin-like peptide: evidence from immunohistochemistry combined with retrograde tracing. Neuroscience 5:2093–2124
191. Holt DJ, Graybiel AM, Saper CB (1997) Neurochemical architecture of the human striatum. J Comp Neurol 384:1–25
192. Hooks MS, Kalivas PW (1995) The role of mesoaccumbens–pallidal circuitry in novelty-induced behavioral activation. Neuroscience 64:587–597
193. Hoover JE, Strick PL (1999) The organization of cerebellar and basal ganglia outputs to primary motor cortex as revealed by retrograde transneuronal transport of herpes simplex virus type 1. J Neurosci 19:1446–1463
194. Hopkins DA, Niessen LW (1976) Substantia nigra projections to the reticular formation, superior colliculus and central gray in the rat, cat and monkey. Neurosci Lett 2:253–259
195. Hornykiewicz O (2001) Chemical neuroanatomy of the basal ganglia – normal and in Parkinson's disease. J Chem Neuroanat 22:3–12
196. Huerta MF, Kaas JH (1990) Supplementary eye fields as defined by intracortical microstimulation: connections in macaques. J Comp Neurol 293:299–330
197. Huerta MF, Van Lieshout DP, Harting JK (1991) Nigrotectal projections in the primate Galago crassicaudatus. Exp Brain Res 87:389–401
198. Ilinsky IA, Jouandet ML, Goldman-Rakic PS (1985) Organization of the nigrothalamocortical system in the rhesus-monkey. J Comp Neurol 236:315–330
199. Inase M, Tokuno H, Nambu A, Akazawa T, Takada M (1999) Corticostriatal and corticosubthalamic input zones from the presupplementary motor area in the macaque monkey: comparison with the input zones from the supplementary motor area. Brain Res 833:191–201
200. Inglis WL, Winn P (1995) The pedunculopontine tegmental nucleus: where the striatum meets the reticular formation. Prog Neurobiol 47:1–29
201. Irle E, Markowitsch HJ (1985) Afferent connections of the substantia innominate/basal nucleus of Meynert in carnivores and primates. J Hirnforsch 27:343–367
202. Izzo PN, Bolam JP (1988) Cholinergic synaptic input to different parts of spiny striatonigral neurons in the rat. J Comp Neurol 269:219–234
203. Jackson A, Crossman AR (1981) Subthalamic projections to nucleus tegmenti pedunculopontinus in the rat. Neurosci Lett 22:17–22
204. Jackson A, Crossman AR (1983) Nucleus tegmenti pedunculopontinus efferent connections with special reference to the basal ganglia, studied in the rat by anterograde and retrograde transport of horseradish peroxidases. Neuroscience 10:725–765
205. Jaspers R, Schwartz M, Sontag KH, Cools AR (1984) Caudate nucleus and programming behavior in cats: role of dopamine in switching motor patterns. Behav Brain Res 14:17–28
206. Joel D, Weiner I (1994) The organization of the basal ganglia-thalamocortical circuits: open interconnected rather than closed segregated. Neuroscience 63:363–379
207. Joel D, Weiner I (1997) The connections of the primate subthalamic nucleus: indirect pathways and the open-interconnected scheme of basal ganglia-thalamocortical circuitry. Brain Res Brain Res Rev 23:62–78
208. Johannes S, Reif A, Senitz D, Riederer P, Lauer M (2003) NADPH-diaphorase staining reveals new types of interneurons in human putamen. Brain Res 980:92–99
209. Jones EG (1998) The thalamus of primates. In: Bloom FE, Björklund A, Hökfelt T (eds) The primate nervous system, Part II. Elsevier, Amsterdam, pp 1–298
210. Jones EG, Burton H, Saper CB, Swanson LW (1976) Midbrain, diencephalic and cortical relationships of the basal nucleus of Meynert and associated structures in primates. J Comp Neurol 167:385–418
211. Jongen-Relo AL, Groenewegen HJ, Voorn P (1993) Evidence for a multi-compartmental histochemical organization of the nucleus accumbens in the rat. J Comp Neurol 337:267–276
212. Jongen-Relo AL, Voorn P, Groenewegen HJ (1994) Immunohistochemical characterization of the shell and core territories of the nucleus accumbens in the rat. Eur J Neurosci 6:1255–1264
213. Joyce JN, Gurevich EV (1999) Dopamine D3 receptor: from anatomy to neuropsychiatry. Neurosci News 2:11–21
214. Kalivas PW, Volkow N, Seamans J (2005) Unmanageable motivation in addiction: a pathology in prefrontal-accumbens glutamate transmission. Neuron 45:647–650

215. Kang DH, Kim JJ, Choi JS et al (2004) Volumetric investigation of the frontal-subcortical circuitry in patients with obsessive-compulsive disorder. J Neuropsychiatry Clin Neurosci 16:342–349
216. Kawaguchi Y, Wilson CJ, Augood SJ, Emson PC (1995) Striatal interneurones: chemical, physiological and morphological characterization. Trends Neurosci 18:527–535
217. Kayahara T, Nakano K (1996) Pallido-thalamo-motor cortical connections: an electron microscopic study in the macaque monkey. Brain Res 706:337–342
218. Kelley AE, Domesick VB, Nauta WJH (1982) The amygdalostriatal projection in the rat – an anatomical study by anterograde and retrograde tracing methods. Neuroscience 7:615–630
219. Kemp JM, Powell TPS (1970) The cortico-striate projection in the monkey. Brain 93:525–547
220. Kemp JM, Powell TPS (1971) The structure of the caudate nucleus of the cat: light and electron microscopy. Philos Trans R Soc Lond [Biol] 262:383–401
221. Kemp JM, Powell TPS (1971) The site of the termination of afferent fibers in the caudate nucleus. Philos Trans R Soc Lond [Biol] 262:413–427
222. Kievit J, Kuypers HGJM (1975) Basal forebrain and hypothalamic connections to frontal and parietal cortex in the rhesus monkey. Science 187:660–662
223. Kincaid AE, Wilson CJ (1996) Corticostriatal innervation of the patch and matrix in the rat neostriatum. J Comp Neurol 374:578–592
224. Kita H, Kosaka T, Heizmann CW (1990) Parvalbumin-immunoreactive neurons in the rat neostriatum: a light and electron microscopic study. Brain Res 536:1–15
225. Kitt CA, Mitchell SJ, DeLong MR, Wainer BH, Price DL (1987) Fiber pathways of basal forebrain cholinergic neurons in monkeys. Brain Res 406:192–206
226. Kornhuber J, Kornhuber ME (1986) Presynaptic dopaminergic modulation of cortical input to the striatum. Life Sci 39:669–674
227. Koutcherov Y, Huang X-F, Halliday G, Paxinos G (2004) Organization of human brain stem nuclei. In: Paxinos G, Mai JK (eds) The human nervous system, 2nd edn. Elsevier, Amsterdam, pp 267–320
228. Krack P, Benazzouz A, Pollak P et al (1998) Treatment of tremor in Parkinson's disease by subthalamic nucleus stimulation. Mov Disord 13:907–914
229. Krack P, Pollak P, Limousin P et al (1998) Subthalamic nucleus or internal pallidal stimulation in young onset Parkinson's disease. Brain 121 (Pt 3):451–457
230. Krettek JE, Price JL (1977) Projections from the amygdaloid complex to the cerebral cortex and thalamus in the rat and cat. J Comp Neurol 172:687–722
231. Kringelbach ML (2004) Food for thought: hedonic experience beyond homeostasis in the human brain. Neuroscience 126:807–819
232. Kubota Y, Kawaguchi Y (1993) Spatial distributions of chemically identified intrinsic neurons in relation to patch and matrix compartments of rat neostriatum. J Comp Neurol 332:499–513
233. Kubota Y, Inagaki S, Kito S (1986) Innervation of substance P neurons by catecholaminergic terminals in the neostriatum. Brain Res 375:163–167
234. Kumar R, Lozano AM, Kim YJ et al (1998) Double-blind evaluation of subthalamic nucleus deep brain stimulation in advanced Parkinson's disease. Neurology 51:850–855
235. Künzle H (1975) Bilateral projections from precentral motor cortex to the putamen and other parts of the basal ganglia. An autoradiographic study in the macace fascilcularis. Brain Res 88:195–209
236. Kuroda M, Price JL (1991) Synaptic organization of projections from basal forebrain structures to the mediodorsal thalamic nucleus of the rat. J Comp Neurol 303:513–533
237. Kuypers HGJM (1966) Discussion. In: Purpura DP, Yahr MD (eds) The thalamus. Columbia University Press, New York, pp 122–126
238. Kuypers HGJM, Lawrence DG (1967) Cortical projections to the red nucleus and the brain stem in the rhesus monkey. Brain Res 4:151–188
239. Kwon JS, Kim JJ, Lee DW et al (2003) Neural correlates of clinical symptoms and cognitive dysfunctions in obsessive-compulsive disorder. Psychiatry Res 122:37–47
240. Lacerda AL, Dalgalarrondo P, Caetano D et al (2003) Elevated thalamic and prefrontal regional cerebral blood flow in obsessive-compulsive disorder: a SPECT study. Psychiatry Res 123:125–134
241. Langston JW, Ballard P, Tetrud JW, Irwin I (1983) Chronic Parkinsonism in humans due to a product of meperidine-analog synthesis. Science 219:979–980
242. Lapper SR, Bolam JP (1992) Input from the frontal cortex and the parafascicular nucleus to cholinergic interneurons in the dorsal striatum of the rat. Neuroscience 51:533–545
243. Lavoie B, Parent A (1990) Immunohistochemical study of the serotoninergic innervation of the basal ganglia in the squirrel monkey. J Comp Neurol 299:1–16
244. Lavoie B, Parent A (1994) Pedunculopontine nucleus in the squirrel monkey: projections to the basal ganglia as revealed by anterograde tract-tracing methods. J Comp Neurol 344:210–231
245. Lavoie B, Parent A (1994) Pedunculopontine nucleus in the squirrel monkey: cholinergic and glutamatergic projections to the substantia nigra. J Comp Neurol 344:232–241
246. Lee HJ, Rye DB, Hallanger AE, Levey AI, Wainer BH (1988) Cholinergic vs. noncholinergic efferents from the mesopontine tegmentum to the ex-

trapyramidal motor system nuclei. J Comp Neurol 275:469–492
247. Lees AJ, Smith E (1983) Cognitive deficits in the early stages of Parkinson's disease. Brain 106: 257–270
248. Lehman J, Nagy IJ, Atmadja S, Fibiger HC (1980) The nucleus basalis magnocellularis: the origin of a cholinergic projection to the neocortex to the rat. Neuroscience 5:1161–1174
249. Leichnetz GR, Astruc J (1977) The course of some prefrontal corticofugals to the pallidum, substantia innominata, and amygdaloid complex in monkeys. Exp Neurol 54:104–109
250. Leiner HC, Leiner AL, Dow SR (1995) The underestimated cerebellum. Hum Brain Map 2:244–254
251. Lévesque M, Bedard A, Cossette M, Parent A (2003) Novel aspects of the chemical anatomy of the striatum and its efferents projections. J Chem Neuroanat 26:271–281
252. Lindvall O, Björklund A (1983) Dopamine- and norepinephrine-containing neuron systems: their anatomy in the rat brain. In: Emson PC (ed) Clinical neuroanatomy. Raven, New York, pp 229–255
253. Lynd-Balta E, Haber SN (1994) The organization of midbrain projections to the ventral striatum in the primate. Neuroscience 59:609–623
254. Lynd-Balta E, Haber SN (1994) The organization of midbrain projections to the striatum in the primate: sensorimotor-related striatum versus ventral striatum. Neuroscience 59:625–640
255. Lynd-Balta E, Haber SN (1994) Primate striatonigral projections: a comparison of the sensorimotor-related striatum and the ventral striatum. J Comp Neurol 345:562–578
256. Maillard L, Ishii K, Bushara K et al (2000) Mapping the basal ganglia: fMRI evidence for somatotopic representation of face, hand, and foot. Neurology 55:377–383
257. Mann DMA, Yates PO, Marcyniuk B (1984) A comparison of changes in the nucleus basalis and locus coeruleus in Alzheimer's disease. J Neurol Neurosurg Psychiatry 47:201–203
258. Marcyniuk B, Mann DMA, Yates PO (1986) Loss of nerve cells from locus coeruleus in Alzheimer's disease is topographically arranged. Neurosci Lett 64:247–252
259. Matthysse S (1973) Antipsychotic drug actions: a clue to the neuropathology of schizophrenia? Fed Proc 32:200–205
260. McBride RL, Larsen KD (1980) Projections of the feline globus pallidus. Brain Res 189:3–14
261. McCormick DA, Prince DA (1985) Two types of muscarinic response to acetylcholine in mammalian cortical neurons. Proc Natl Acad Sci USA 82:6344–6348
262. McCormick DA, Prince DA (1986) Mechanisms of action of acetylcholine in the guinea-pig cerebral cortex in vitro. J Physiol 375:169–194
263. McFarland NR, Haber SN (2000) Convergent inputs from thalamic motor nuclei and frontal cortical areas to the dorsal striatum in the primate. J Neurosci 20:3798–3813
264. McFarland NR, Haber SN (2001) Organization of thalamostriatal terminals from the ventral motor nuclei in the macaque. J Comp Neurol 429:321–336
265. McFarland NR, Haber SN (2002) Thalamic relay nuclei of the basal ganglia form both reciprocal and nonreciprocal cortical connections, linking multiple frontal cortical areas. J Neurosci 22: 8117–8132
266. McGeer PL (1984) The 12th JAF Stevenson memorial lecture. Aging, Alzheimer's disease, and the cholinergic system. Can J Physiol Pharmacol 62: 741–754
267. McGeer PL, McGeer EG, Suzuki J, Dolman CE, Nagai T (1984) Aging, Alzheimer's disease and the cholinergic system of the basal forebrain. Neurol 34:741–745
268. McGeorge AJ, Faull RLM (1989) The organization of the projection from the cerebral cortex to the striatum in the rat. Neuroscience 29:503–537
269. Mehler WR (1981) The basal ganglia – circa 1982: a review and commentary. Appl Neurophysiol 44:261–290
270. Mesulam MM, Geula C (1988) Nucleus basalis (Ch4) and cortical cholinergic innervation in the human brain: observations based on the distribution of acetylcholinesterase and choline acetyltransferase. J Comp Neurol 275:216–240
271. Mesulam MM, Mufson EJ (1984) Neural inputs into the nucleus basalis of the substantia innominata (CH4) in the rhesus monkey. Brain 107:253–274
272. Mesulam MM, Mufson EJ, Wainer BH, Levey AI (1983) Central cholinergic pathways in the rat: an overview based on an alternative nomenclature. Neuroscience 10:1185–1201
273. Mesulam MM, Mufson EJ, Levey AI, Wainer BH (1983) Cholinergic innervation of cortex by the basal forebrain: Cytochemistry and cortical connections of the septal area, diagonal band nuclei, nucleus basalis (substantia innominata), and hypothalamus in the rhesus monkey. J Comp Neurol 214:170–197
274. Mesulam MM, Mufson EJ, Levey AI, Wainer BH (1984) Atlas of cholinergic neurons in the forebrain and upper brainstem of the macaque based on monoclonal choline acetyltransferase immunohistochemistry and acetylcholinesterase histochemistry. Neuroscience 12:669–686
275. Mesulam MM, Mufson EJ, Wainer BH (1986) Threedimensional representation and cortical projection topography of the nucleus basalis (Ch4) in the macaque: concurrent demonstration of choline acetyltransferase and retrograde transport with a stabilized tetramethylbenzidine method for horseradish peroxidase. Brain Res 367:301–308
276. Mesulam MM, Hersh LB, Mash DC, Geula C (1992) Differential cholinergic innervation within functional subdivisions of the human cerebral

cortex: a choline acetyltransferase study. J Comp Neurol 318:316–328
277. Mesulam MM, Mash D, Hersh L, Bothwell M, Geula C (1992) Cholinergic innervation of the human striatum, globus pallidus, subthalamic nucleus, substantia nigra, and red nucleus. J Comp Neurol 323:252–268
278. Middleton FA, Strick PL (1994) Anatomical evidence for cerebellar and basal ganglia involvement in higher cognitive function. Science 266: 458–461
279. Middleton FA, Strick PL (1998) Cerebellar output: motor and cognitive channels. Trends Cogn Sci 2:348–354
280. Middleton FA, Strick PL (2000) Basal ganglia and cerebellar loops: motor and cognitive circuits. Brain Res Brain Res Rev 31:236–250
281. Middleton FA, Strick PL (2001) Cerebellar projections to the prefrontal cortex of the primate. J Neurosci 21:700–712
282. Middleton FA, Strick PL (2002) Basal-ganglia 'projections' to the prefrontal cortex of the primate. Cereb Cortex 12:926–935
283. Miyachi S, Lu X, Inoue S et al (2005) Organization of multisynaptic inputs from prefrontal cortex to primary motor cortex as revealed by retrograde transneuronal transport of rabies virus. J Neurosci 25:2547–2556
284. Mogenson GJ (1984) Limbic-motor integration – with emphasis on initiation of exploratory and goal-directed locomotion. In: Bandler R (ed) Modulation of sensorimotor activity during alterations in behavioral states. Liss, New York, pp 121–137
285. Mogenson GJ, Nielsen MA (1983) Evidence that an accumbens to subpallidal GABAergic projection contributes to Locomotor activity. Brain Res Bull 11:309–314
286. Mogenson GJ, Jones DL, Yim CY (1980) From motivation to action: functional interface between the limbic system and the motor system. Prog Neurobiol 14:69–97
287. Mogenson GJ, Swanson LW, Wu M (1984) Evidence that projections from substantia innominata to zona incerta mesencephalic locomotor region contribute to locomotor activity. Brain Res 334:65–76
288. Moore RY, Bloom FE (1978) Central catecholamine neuron systems: anatomy and physiology of the dopamine systems. Annu Rev Neurosci 1:129–169
289. Naito A, Kita H (1994) The cortico-nigral projection in the rat: an anterograde tracing study with biotinylated dextran amine. Brain Res 637:317–322
290. Nakamura Y, Tokuno H, Moriizumi T, Kitao Y, Kudo M (1989) Monosynaptic nigral inputs to the pedunculopontine tegmental nucleus neurons which send their axons to the medial reticular formation in the medulla oblongata. An electron microscopic study in the cat. Neurosci Lett 103:145–150
291. Nakano I, Hirano A (1983) Neuron loss in the nucleus basalis of Meynert in Parkinsonism – dementia complex of Guam. Ann Neurol 13:87–91
292. Nakano I, Hirano A (1984) Parkinson's disease: neuron loss in the nucleus basalis without concomitant Alzheimer's disease. Ann Neurol 15:415–418
293. Nakano K, Hasegawa Y, Tokushige A et al (1990) Topographical projections from the thalamus, subthalamic nucleus and pedunculopontine tegmental nucleus to the striatum in the Japanese monkey, Macaca fuscata. Brain Res 537:54–68
294. Nambu A, Tokuno H, Inase M, Takada M (1997) Corticosubthalamic input zones from forelimb representations of the dorsal and ventral divisions of the premotor cortex in the macaque monkey: comparison with the input zones from the primary motor cortex and the supplementary motor area. Neurosci Lett 239:13–16
295. Nambu A, Tokuno H, Hamada I et al (2000) Excitatory cortical inputs to pallidal neurons via the subthalamic nucleus in the monkey. J Neurophysiol 84:289–300
296. Nambu A, Tokuno H, Takada M (2002) Functional significance of the cortico-subthalamo-pallidal 'hyperdirect' pathway. Neurosci Res 43:111–117
297. Nauta HJW (1972) Neural associations of the frontal cortex. Acta Neurobiol Expl 32:125–140
298. Nauta HJW (1974) Evidence of a pallidohabenular pathway in the cat. J Comp Neurol 156:19–27
299. Nauta HJW, Cole M (1978) Efferent projections of the subthalamic nucleus: an autoradiographic study in monkey and cat. J Comp Neurol 180:1–16
300. Nauta HJW, Mehler WR (1966) Projections of the lentiform nucleus in the monkey. Brain Res 1:3–42
301. Nieuwenhuys R (2000) Comparative aspects of volume transmission, with sidelight on other forms of intercellular communication. Prog Brain Res 125:49–126
302. Oertel WH, Mugnaini E (1984) Immunocytochemical studies of gaba-ergic neurons in the rat basal ganglia and their relations to other neuronal systems. Neurosci Lett 15:159–164
303. Olazabal UE, Moore JK (1989) Nigrotectal projection to the inferior colliculus: horseradish peroxidase transport and tyrosine hydroxylase immunohistochemical studies in rats, cats, and bats. J Comp Neurol 282:98–118
304. Öngür D, Price JL (2000) The organization of networks within the orbital and medial prefrontal cortex of rats, monkeys and humans. Cereb Cortex 10:206–219
305. Öngür D, Ferry AT, Price JL (2003) Architectonic subdivision of the human orbital and medial prefrontal cortex. J Comp Neurol 460:425–449

306. Packard MG, Knowlton BJ (2002) Learning and memory functions of the Basal Ganglia. Annu Rev Neurosci 25:563–593
307. Paré D, Smith Y (1994) GABAergic projection from the intercalated cell masses of the amygdala to the basal forebrain in cats. J Comp Neurol 344:33–49
308. Paré D, Smith Y, Parent A, Steriade M (1988) Projections of brainstem core cholinergic and non-cholinergic neurons of cat to intralaminar and reticular thalamic nuclei. Neuroscience 25: 69–86
309. Parent A (1979) Identification of the pallidal and peri pallidal cells projecting to the habenula in monkey. Neurosci Lett 15:159–164
310. Parent A (2002) Chemical anatomy of the human striatum in health and disease. In: The human brain 2002. Abstracts of an ICRCCS meeting; Rome, Oct 5–10, 2002, p 45
311. Parent A, DeBellefeuille L (1982) Organization of efferent projections from the internal segment of globus pallidus in primate as revealed by fluorescence retrograde labeling method. Brain Res 245:201–213
312. Parent A, DeBellefeuille L (1983) The pallidointralaminar and pallidonigral projections in primate as studied by retrograde double-labeling method. Brain Res 278:11–28
313. Parent A, Hazrati L-N (1995) Functional anatomy of the basal ganglia. I. The cortico-basal ganglia-thalamo-cortical loop. Brain Res Rev 20:91–127
314. Parent A, Hazrati L-N (1995) Functional anatomy of the basal ganglia. II. The place of subthalamic nucleus and external pallidum in basal ganglia circuitry. Brain Res Rev 20:128–154
315. Parent M, Parent A (2005) Single-axon tracing and three-dimensional reconstruction of centre median-parafascicular thalamic neurons in primates. J Comp Neurol 481:127–144
316. Parent A, Sato F, Wu Y et al (2000) Organization of the basal ganglia: the importance of axonal collateralization. Trends Neurosci 23:S20–S27
317. Parent M, Lévesque M, Parent A (2001) Two types of projection neurons in the internal pallidum of primates: single-axon tracing and three-dimensional reconstruction. J Comp Neurol 439: 162–175
318. Parthasarathy HB, Schall JD, Graybiel AM (1992) Distributed but convergent ordering of corticostriatal projections: analysis of the frontal eye field and the supplementary eye field in the macaque monkey. J Neurosci 12:4468–4488
319. Pasik P, Pasik T, Pecci Saavedra J, Holstein GR (1981) Light and electron microscopic immunocytochemical localization of serotonin in the basal ganglia of cats and monkeys. Anat Rec 199:194
320. Pasupathy A, Miller EK (2005) Different time courses of learning-related activity in the prefrontal cortex and striatum. Nature 433:873–876
321. Patel NK, Heywood P, O'Sullivan K et al (2003) Unilateral subthalamotomy in the treatment of Parkinson's disease. Brain 126:1136–1145
322. Patterson JC, Ungerleider LG, Bandettini PA (2002) Task-independent functional brain activity correlation with skin conductance changes: an fMRI study. Neuroimage 17:1797–1806
323. Pearson RCA, Gatter KC, Brodal P, Powell TSP (1982) The projection of the basal nucleus of Meynert upon the neocortex in the monkey. Brain Res 259:132–136
324. Pennartz CMA, Groenewegen HJ, Lopes da Silva FH (1994) The nucleus accumbens as a complex of functionally distinct neuronal ensembles: an integration of behavioural, electrophysiological and anatomical data. Prog Neurobiol 42:719–761
325. Penney JB Jr, Young AB (1983) Speculations on the functional anatomy of basal ganglia disorders. Annu Rev Neurosci 6:73–94
326. Penney JB Jr, Young AB (1986) Striatal inhomogeneities and basal ganglia function. Mov Disord 1:3–15
327. Penny GR, Wilson CJ, Kitai ST (1988) Relationship of the axonal and dendritic geometry of spiny projection neurons to the compartmental organization of the neostriatum. J Comp Neurol 269:275–289
328. Percheron G (2004) Thalamus. In: Paxinos G, Mai JK (eds) The human nervous system, 2nd edn. Elsevier, Amsterdam, pp 592–675
329. Percheron G, Yelnik J, Francois C (1984) A Golgi analysis of the primate globus pallidus. III. Spatial organization of the striato-pallidal complex. J Comp Neurol 227:214–227
330. Perry EK (1986) The cholinergic hypothesis–ten years on. Br Med Bull 42:63–69
331. Perry RH, Candy JM, Perry EK, Thompson J, Oakley AE (1984) The substantia innominata and adjacent regions in the human brain: histochemical and biochemical observations. I Anat 138: 713–732
332. Pert CB, Kuhar MI, Snyder SH (1976) Opiate receptor: Autoradiographic localization in rat brain. Proc Natl Acad Sci USA 73:3729–3733
333. Petras JM (1969) Some efferent connections of the motor and somatosensory cortex of simian primates and felid, canid and procyonid carnivores. Ann NY Acad Sci 167:469–505
334. Petrides M, Pandya DN (1999) Dorsolateral prefrontal cortex: comparative cytoarchitectonic analysis in the human and the macaque brain and corticocortical connection patterns. Eur J Neurosci 11:1011–1036
335. Pijnenburg AJJ, Honig WMM, Van der Heyden JAM, Van Rossum JM (1976) Effects of chemical stimulation of the mesolimbic dopamine system upon locomotor activity. Eur J Pharmacol 35:45–58
336. Porrino LJ, Crane AM, Goldman-Rakic PS (1981) Direct and indirect pathways from the amygdala to the frontal lobe in rhesus monkeys. J Comp Neurol 198:121–136

337. Prensa L, Giménez-Amaya J-M, Parent A (1999) Chemical heterogeneity of the striosomal compartment in the human striatum. J Comp Neurol 413:603–618
338. Rafols JA, Fox CA (1976) The neurons in the primate subthalamic nucleus: a Golgi and electron microscopic study. J Comp Neurol 168:75–111
339. Ragsdale CW Jr, Graybiel AM (1988) Fibers from the basolateral nucleus of the amygdala selectively innervate striosomes in the caudate nucleus of the cat. J Comp Neurol 269:506–522
340. Ray JP, Price JL (1993) The organization of projections from the mediodorsal nucleus of the thalamus to orbital and medial prefrontal cortex in macaque monkeys. J Comp Neurol 337:1–31
341. Reiner A, Medina L, Haber SN (1999) The distribution of dynorphinergic terminals in striatal target regions in comparison to the distribution of substance P-containing and enkephalinergic terminals in monkeys and humans. Neuroscience 88:775–793
342. Ricardo JA (1980) Efferent connections of the subthalamic region in the rat. I. The subthalamic nucleus of Luys. Brain Res 202:257–271
343. Rinvik E, Ottersen OP (1993) Terminals of subthalamonigral fibres are enriched with glutamate-like immunoreactivity: an electron microscopic, immunogold analysis in the cat. J Chem Neuroanat 6:19–30
344. Rinvik E, Grofova I, Ottersen OP (1976) Demonstration of nigrotectal and nigroreticular projections in the cat by axonal transport of proteins. Brain Res 112:388–394
345. Roberts RC, Knickman JK (2002) The ultrastructural organization of the patch matrix compartments in the human striatum. J Comp Neurol 452:128–138
346. Rodriguez MC, Guridi OJ, Alvarez L et al (1998) The subthalamic nucleus and tremor in Parkinson's disease. Mov Disord 13 Suppl 3:111–118
347. Rogers JD, Brogan D, Mirra SS (1985) The nucleus basalis of Meynert in neurological disease: a quantitative morphological study. Ann Neurol 17:162–170
348. Rolls ET (2000) The orbitofrontal cortex and reward. Cereb Cortex 10:284–294
349. Rolls ET, Baylis LL (1994) Gustatory, olfactory, and visual convergence within the primate orbitofrontal cortex. J Neurosci 14:5437–5452
350. Romanelli P, Esposito V, Schaal DW, Heit G (2005) Somatotopy in the basal ganglia: experimental and clinical evidence for segregated sensorimotor channels. Brain Res Brain Res Rev 48: 112–128
351. Royce GJ (1987) Recent research on the centromedian and parafascicular nuclei. In: The basal ganglia II: Structure and function – current concepts. Plenum, New York, pp 293–319 (Advances in behavioral biology, vol 32)
352. Russchen FT, Price JL (1984) Amygdalostriatal projections in the rat, topographical organization and fiber morphology shown using the lectin PHA-L as an anterograde tracer. Neurosci Lett 47:15–22
353. Russchen FT, Amaral DG, Price JL (1985) The afferent connections of the substantia innominata in the monkey, Macaca fascicularis. J Comp Neurol 242:1–27
354. Russchen FT, Bakst I, Amaral DG, Price JL (1985) The amygdalostriatal projections in the monkey. An anterograde tracing study. Brain Res 329:241–257
355. Rye DB, Saper CB, Lee HJ, Wainer BH (1987) Pedunculopontine tegmental nucleus of the rat: cytoarchitecture, cytochemistry, and some extrapyramidal connections of the mesopontine tegmentum. J Comp Neurol 259:483–528
356. Sadikot AF, Parent A, Francois C (1992) Efferent connections of the centromedian and parafascicular thalamic nuclei in the squirrel monkey: a PHA-L study of subcortical projections. J Comp Neurol 315:137–159
357. Sadikot AF, Parent A, Smith Y, Bolam JP (1992) Efferent connections of the centromedian and parafascicular thalamic nuclei in the squirrel monkey: a light and electron microscopic study of the thalamostriatal projection in relation to striatal heterogeneity. J Comp Neurol 320:228–242
358. Sakai ST (1988) Corticonigral projections from area 6 in the raccoon. Exp Brain Res 73:498–504
359. Sakai ST, Inase M, Tanji J (1999) Pallidal and cerebellar inputs to thalamocortical neurons projecting in the supplementary motor area in *Macaca fuscata*: a triple-labeling light microscopic study. Anat Embryol 199:9–19
360. Saper CB, Chelimski TC (1984) A cytoarchitectonic and histochemical study of nucleus basalis and associated cell groups in the normal human brain. Neuroscience 13:1023–1037
361. Sarter M, Bruno JP (2002) The neglected constituent of the basal forebrain corticopetal projection system: GABAergic projections. Eur J Neurosci 15:1867–1873
362. Sato F, Lavallee P, Lévesque M, Parent A (2000) Single-axon tracing study of neurons of the external segment of the globus pallidus in primate. J Comp Neurol 417:17–31
363. Sato F, Parent M, Lévesque M, Parent A (2000) Axonal branching pattern of neurons of the subthalamic nucleus in primates. J Comp Neurol 424:142–152
364. Schlag J, Schlag-Rey M (1987) Evidence for a supplementary eye field. J Neurophysiol 57:179–200
365. Schmahmann JD (1996) From movement to thought: anatomic substrates of the cerebellar contribution to cognitive processing. Hum Brain Map 4:174–198
366. Schmahmann JD (1998) Dysmetria of thought: clinical consequences of cerebellar dysfunction on cognition and affect. Trends Cogn Sci 2:362–371

367. Schmahmann JD (2004) Disorders of the cerebellum: ataxia, dysmetria of thought, and the cerebellar cognitive affective syndrome. J Neuropsychiatry Clin Neurosci 16:367–378
368. Schmahmann JD, Sherman JC (1998) The cerebellar cognitive affective syndrome. Brain 121 (Pt 4):561–579
369. Schröder KF, Hopf A, Lange H, Thörner G (1975) Morphometrisch-statistische Strukturanalysen des Striatum, Pallidum und Nucleus subthalamicus beim Menschen. I. Striatum. J Hirnforschung 16:333–350
370. Schultz W, Tremblay L, Hollerman JR (2000) Reward processing in primate orbitofrontal cortex and basal ganglia. Cereb Cortex 10:272–284
371. Selden NR, Gitelman DR, Salamon-Murayama N, Parrish TB, Mesulam MM (1998) Trajectories of cholinergic pathways within the cerebral hemispheres of the human brain. Brain 121 (Pt 12): 2249–2257
372. Selemon LD, Goldman-Rakic PS (1985) Longitudinal topography and interdigitation of corticostriatal projections in the rhesus monkey. J Neurosci 5:776–794
373. Selemon LD, Goldman-Rakic PS (1990) Topographic intermingling of striatonigral and striatopallidal neurons in the rhesus monkey. J Comp Neurol 297:359–376
374. Semba K (2000) Multiple output pathways of the basal forebrain: organization, chemical heterogeneity, and roles in vigilance. Behav Brain Res 115:117–141
375. Semba K, Reiner PB, McGeer EG, Fibiger HC (1988) Brainstem afferents to the magnocellular basal forebrain studied by axonal transport, immunohistochemistry, and electrophysiology in the rat. J Comp Neurol 267:433–453
376. Sherman SM, Guillery RW (2001) Exploring the thalamus. Academic Press, San Diego
377. Sherman SM, Guillery RW (2002) The role of the thalamus in the flow of information to the cortex. Phil Trans R Soc Lond B 357:1695–1708
378. Sherman SM, Koch C (1998) Thalamus. In: Shepherd GM (ed) The synaptic organization of the brain, 4th edn. Oxford University Press, Oxford, pp 289–328
379. Shink E, Bevan MD, Bolam JP, Smith Y (1996) The subthalamic nucleus and the external pallidum: two tightly interconnected structures that control the output of the basal ganglia in the monkey. Neuroscience 73:335–357
380. Shinonaga Y, Takada M, Mizuno N (1994) Topographic organization of collateral projections from the basolateral amygdaloid nucleus to both the prefrontal cortex and nucleus accumbens in the rat. Neuroscience 58:389–397
381. Shook BL, Schlag-Rey M, Schlag J (1990) Primate supplementary eye field: I. Comparative aspects of mesencephalic and pontine connections. J Comp Neurol 301:618–642
382. Shook BL, Schlag-Rey M, Schlag J (1991) Primate supplementary eye field. II. Comparative aspects of connections with the thalamus, corpus striatum, and related forebrain nuclei. J Comp Neurol 307:562–583
383. Siegel KL, Metman LV (2000) Effects of bilateral posteroventral pallidotomy on gait of subjects with Parkinson disease. Arch Neurol 57:198–204
384. Skinner RD, Garcia-Rill E (1984) The mesencephalic locomotor region (MLR) in the rat. Brain Res 323:385–389
385. Smith ML, Hale BD, Booze RM (1994) Calbindin-D28k immunoreactivity within the cholinergic and GABAergic projection neurons of the basal forebrain. Exp Neurol 130:230–236
386. Smith Y, Parent A (1986) Differential connections of caudate nucleus and putamen in the squirrel monkey (Saimiri sciureus). Neuroscience 18:347–371
387. Smith Y, Parent A (1988) Neurons of the subthalamic nucleus in primates display glutamate but not GABA immunoreactivity. Brain Res 453: 353–356
388. Smith Y, Hazrati LN, Parent A (1990) Efferent projections of the subthalamic nucleus in the squirrel monkey as studied by the PHA-L anterograde tracing method. J Comp Neurol 294:306–323
389. Smith Y, Bevan MD, Shink E, Bolam JP (1998) Microcircuitry of the direct and indirect pathways of the basal ganglia. Neuroscience 86:353–387
390. Sokoloff P, Giros B, Martres MP, Bouthenet ML, Schwartz JC (1990) Molecular cloning and characterization of a novel dopamine receptor (D3) as a target for neuroleptics. Nature 347:146–151
391. Somogyi P, Bolam JP, Smith AD (1981) Monosynaptic cortical input and local axon collaterals of identified striatonigral neurons. A light and electron microscopic study in using the Golgi-peroxidase transport degeneration procedure. J Comp Neurol 195:567–584
392. Spangler EM, Henkel CK, Miller IJ Jr (1982) Localization of the motor neurons to the tensor tympani muscle. Neurosci Lett 32:23–27
393. Spann BM, Grofova I (1989) Origin of ascending and spinal pathways from the nucleus tegmenti pedunculopontinus in the rat. J Comp Neurol 283:13–27
394. Spann BM, Grofova I (1991) Nigropedunculopontine projection in the rat: an anterograde tracing study with phaseolus vulgaris-leucoagglutinin (PHA-L). J Comp Neurol 311:375–388
395. Spatz WB (1975) An efferent connection of the solitary cells of Meynert. A study with horseradish peroxidase in the marmoset Callithrix. Brain Res 92:450–455
396. Spooren WPJM, Veening JG, Groenewegen HJ, Cools AR (1991) Efferent connections of the striatopallidal and amygdaloid components of the substantia innominata in the cat: projections to

the nucleus accumbens and caudate nucleus. Neuroscience 44:431–447
397. Spooren WPJM, Veening JG, Cools AR (1993) Descending efferent connections of the sub-pallidal areas in the cat: projections to the subthalamic nucleus, the hypothalamus, and the midbrain. Synapse 15:104–123
398. Steininger TL, Rye DB, Wainer BH (1992) Afferent projections to the cholinergic pedunculopontine tegmental nucleus and adjacent midbrain extrapyramidal area in the albino rat. I. Retrograde tracing studies. J Comp Neurol 321:515–543
399. Stephan H (1975) Handbuch der mikroskopischen Anatomie des Menschen. Band 4, Teil 9 Allocortex. Springer, Berlin Heidelberg New York
400. Stephens B, Mueller AJ, Shering AF et al (2005) Evidence of a breakdown of corticostriatal connections in Parkinson's disease. Neuroscience 132:741–754
401. Steriade M, Jones EG, McCormick DA (1997) Thalamus, vol I: Organisation and function. Elsevier, Amsterdam
402. Stevens JR (1973) An anatomy of schizophrenia? Arch Gen Psychiatry 29:177–189
403. Surmeier DJ, Song WJ, Yan Z (1996) Coordinated expression of dopamine receptors in neostriatal medium spiny neurons. J Neurosci 16:6579–6591
404. Szabo J (1980) Distribution of striatal afferents from the mesencephalon in the cat. Brain Res 188:3–21
405. Szabo J (1980) Organization of the ascending striatal afferents in monkeys. J Comp Neurol 189:307–321
406. Takada M, Tokuno H, Nambu A, Inase M (1998) Corticostriatal projections from the somatic motor areas of the frontal cortex in the macaque monkey: segregation versus overlap of input zones from the primary motor cortex, the supplementary motor area, and the premotor cortex. Exp Brain Res 120:114–128
407. Takakusaki K, Habaguchi T, Ohtinata-Sugimoto J, Saitoh K, Sakamoto T (2003) Basal ganglia efferents to the brainstem centers controlling postural muscle tone and locomotion: a new concept for understanding motor disorders in basal ganglia dysfunction. Neuroscience 119:293–308
408. Tanaka C, Ishikawa M, Shimada S (1982) Histochemical mapping of catecholaminergic neurons and their ascending fiber pathways in the rhesus monkey brain. Brain Res Bull 9:255–270
409. Thörner G, Lange H, Hopf A (1975) Morphometrisch-statistische Strukturanalysen des Striatum, Pallidum und Nucleus subthalamicus beim Menschen. II. Pallidum. J Hirnforschung 16:404–413
410. Tong CK, Hamel E (2000) Basal forebrain nitric oxide synthase (NOS)-containing neurons project to microvessels and NOS neurons in the rat neocortex: cellular basis for cortical blood flow regulation. Eur J Neurosci 12:2769–2780
411. Ungerstedt U (1971) Stereotaxic mapping of the monoamine pathways in the rat brain. Acta Physiol Scand [Suppl] 367:1–49
412. Usunoff KO, Romansky KY, Malinov OB et al (1982) Electron microscopic evidence for the existence of a cortical tract in the rat. J Hirnforschung 23:23–29
413. Vandecasteele M, Glowinski J, Venance L (2005) Electrical synapses between dopaminergic neurons of the substantia nigra pars compacta. J Neurosci 25:291–298
414. Vaughan GM, Pelham RW, Pang SF et al (1976) Nocturnal elevation of plasma melatonin and urinary 5-hydroxyindoleacetic acid in young men: attempts at modification by brief changes in environmental lighting and sleep and by autonomic drugs. J Clin Endocr Metab 42:752–764
415. Vogels OJM (1990) The nucleus basalis of Meynert complex and adjacent structures in normal aging and Alzheimer's disease. Thesis, University of Nijmegen, The Netherlands
416. Vogels OJM, Broere CA, ter Laak HJ et al (1990) Cell loss and shrinkage in the nucleus basalis Meynert complex in Alzheimer's disease. Neurobiol Aging 11:3–13
417. Volkow ND, Fowler JS, Wang GJ (2004) The addicted human brain viewed in the light of imaging studies: brain circuits and treatment strategies. Neuropharmacology 47 Suppl 1:3–13
418. Voorn P, Jorritsma-Byham B, Van Dijk C, Buijs RM (1986) The dopaminergic innervation of the ventral striatum in the rat: a light- and electron-microscopical study with antibodies against dopamine. J Comp Neurol 251:84–99
419. Walker LC, Price DL, Young WS III (1989) GABAergic neurons in the primate basal forebrain magnocellular complex. Brain Res 499:188–192
420. Walker RH, Arbuthnott GW, Baughman RW, Graybiel AM (1993) Dendritic domains of medium spiny neurons in the primate striatum: relationships to striosomal borders. J Comp Neurol 337:614–628
421. Wassef M, Berod A, Sotelo C (1981) Dopaminergic dendrites in the pars reticulata of the rat substantia nigra and their striatal input: combined immunocytochemical localization of tyrosine hydroxylase and anterograde degeneration. Neuroscience 6:2125–2139
422. Whitehouse PJ, Price DL, Clark WA, Coyle TJ, De Long MR (1981) Alzheimer's disease: evidence for selective loss of cholinergic neurons in the nucleus basalis. Ann Neurol 10:122–126
423. Whitehouse PJ, Price DL, Struble RG, Aark AW, Coyle JT (1982) Alzheimer's disease and senile dementia: loss of neurons in the basal forebrain. Science 215:1237–1239
424. Whitehouse PJ, Hedreen JC, White CL (1983) Basal forebrain neurons in the dementia of Parkinson disease. Ann Neurol 13:243–248

425. Whittier JR (1947) Ballism and the subthalamic nucleus (nucleus hypothalamicus; corpus Luysii). Arch Neurol Psychiatr 58:672–692
426. Wichmann T, DeLong MR (1996) Functional and pathophysiological models of the basal ganglia. Curr Opin Neurobiol 6:751–758
427. Wichmann T, Vitek J, DeLong MR (1995) Parkinson's disease and the basal ganglia: lessons from the laboratory and from neurosurgery. Neuroscientist 1:236–244
428. Wilson CJ (1998) Basal ganglia. In: Shepherd GM (ed) The synaptic organization of the brain, 4th edn. Oxford University Press, Oxford, pp 329–375
429. Wilson CJ (2004) Basal ganglia. In: Shepherd GM (ed) The synaptic organization of the brain. Oxford University Press, Oxford, pp 361–413
430. Yamamoto T, Hirano A (1985) Nucleus raphe dorsalis in Alzheimer's disease: neurofibrillary tangles and loss of large neurons. Ann Neurol 17:573–577
431. Yang CR, Mogenson GJ (1985) An electrophysiological study of the neural projections from the hippocampus to the ventral pallidum and the subpallidal areas by way of the nucleus accumbens. Neuroscience 15:1015–1024
432. Yelnik J, Francois C, Percheron G, Heyner S (1987) Golgi study of the primate substantia nigra. I. Quantitative morphology and typology of nigral neurons. J Comp Neurol 265:455–472
433. Yeterian EH, Pandya DN (1991) Prefrontostriatal connections in relation to cortical architectonic organization in rhesus monkeys. J Comp Neurol 312:43–67
434. Yung KK, Smith AD, Levey AI, Bolam JP (1996) Synaptic connections between spiny neurons of the direct and indirect pathways in the neostriatum of the rat: evidence from dopamine receptor and neuropeptide immunostaining. Eur J Neurosci 8:861–869
435. Yung KKL, Bolam JP, Smith AD et al (1995) Immunocytochemical localization of D1 and D2 dopamine receptors in the basal ganglia of the rat: light and electron microscopy. Neuroscience 65:709–730
436. Záborszky L, Alheid GF, Beinfeld MC et al (1985) Cholecystokinin innervation of the ventral striatum: a morphological and radioimmunological study. Neuroscience 14:427–453
437. Zaborszky L, Gaykema RP, Swanson DJ, Cullinan WE (1997) Cortical input to the basal forebrain. Neuroscience 79:1051–1078
438. Zahm DS, Brog JS (1992) On the significance of subterritories in the "accumbens" part of the rat ventral striatum. Neuroscience 50:751–767

15 Telencephalon: Neocortex

Introduction

The neocortex is an ultracomplex, six-layered structure that develops from the dorsal pallial sector of the telencephalic hemispheres (Figs. 2.24, 2.25, 11.1). All mammals, including monotremes and marsupials, possess a neocortex, but in reptiles, i.e. the ancestors of mammals, only a three-layered neocortical primordium is present [509, 511]. The term neocortex refers to its late phylogenetic appearance, in comparison to the "palaeocortical" olfactory cortex and the "archicortical" hippocampal cortex, both of which are present in all amniotes [509].

The size of the neocortex varies greatly among the various mammalian groups. In some insectivores, such as the hedgehog, its

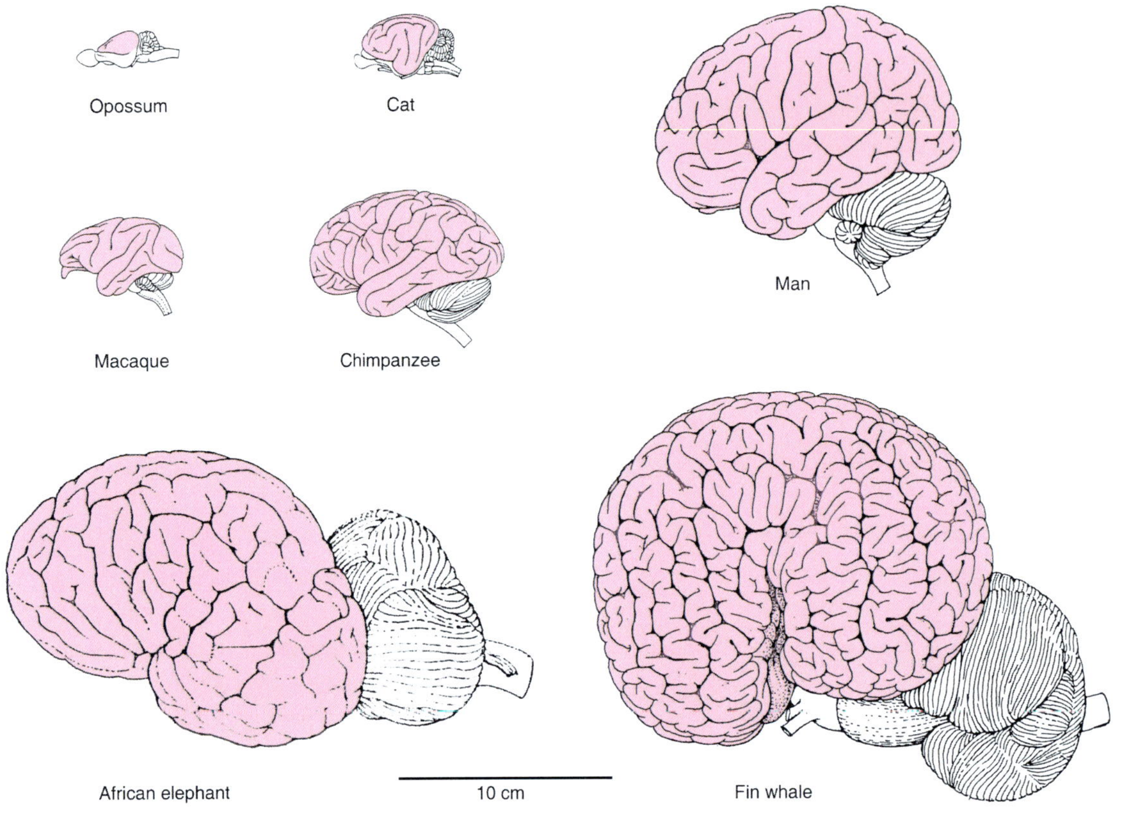

Fig. 15.1. Brains of some mammals drawn on the same scale. The neocortex is shown in *red*

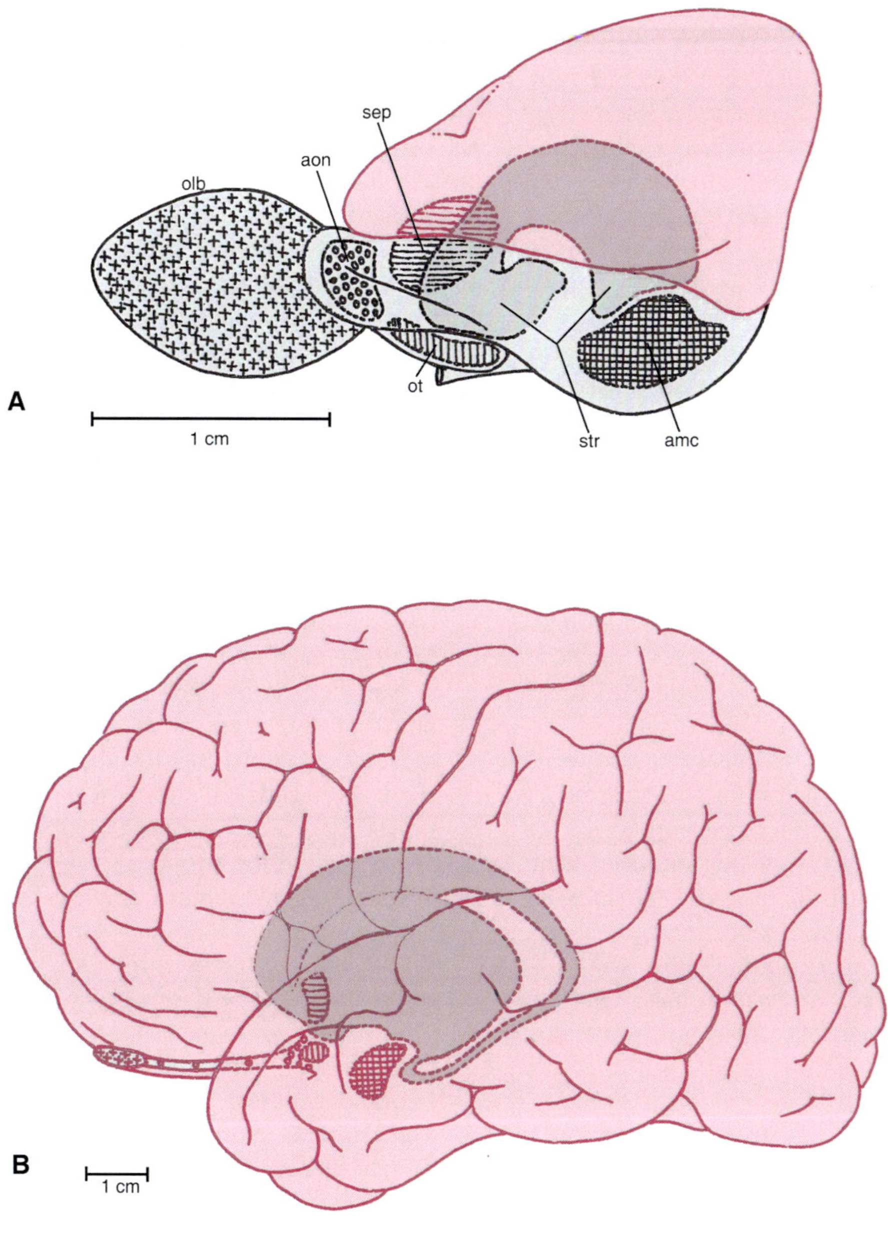

Fig. 15.2. The telencephalon of the opossum (**A**) and human (**B**). The position and extent of the neocortex (shown in *red*) and of a number of subcortical centres (marked in **A** and **B** by the same *hatchings/symbols*) are indicated. In the opossum, the neocortex covers the remaining parts of the telencephalic hemispheres like a cap. In humans, the enormously expanded neocortex almost completely surrounds all subcortical centres. *amc*, amygdaloid complex; *aon*, anterior olfactory nucleus; *olb*, olfactory bulb; *ot*, olfactory tubercle; *sep*, precommissural septum; *str*, corpus striatum

size does not exceed that of the "older" parts of the cortex, but in primates and cetaceans it attains remarkable proportions, becoming by far the largest centre in the brain (Figs. 15.1, 15.2). Stephan and Andy [709] determined the average index of progression for the neocortex (and for many other brain structures) in a number of insectivores, prosimians and simians, including humans. These indices express how many times larger the neocortex is in a particular group of species than in that of a typical basal insectivore of the same size. It was found that in prosimians the neocortex is on average 14.5 times, and in the simians 45.5 times larger than in basal insectivores. In humans, the neocortex appeared to be 156 times larger than that of basal insectivores.

As regards the functional differentiation of the neocortex, in many "primitive" mammals (marsupials, insectivores and rodents) much of the neocortex is occupied by projection areas which either receive, via the thalamus, impulses directly related to the various special senses, or are concerned with the steering of motor activity. The sensory fields comprise the somatosensory cortex, receiving impulses from sense organs situated in the skin, the muscles and the joints, then the primary visual cortex and finally the primary auditory cortex.

There is evidence suggesting that already in primitive mammals the various projection areas just mentioned are separated from each other by narrow strips of non-projection cortex (see Cajal [84], p. 830 and figure 531). What we now see is that these strips expand enormously to form a large area known as the parietotemporal association cortex. Another area of association cortex develops in front of the motor cortical areas (Fig. 15.3). This region is known as the prefrontal association cortex. All cortical areas maintain afferent and efferent connections with subcortical centres. However, these various association cortical areas are primarily connected with other cortical fields.

The neocortex shows a laminated structure throughout its extent. Whereas in the piriform and hippocampal parts of the mammalian cortex three layers can be distinguished, in the neocortex six layers are usually recognized (Fig. 15.4B). The characterization and determination of these layers is generally based on Nissl-stained material. Although the study of such material, in which the dendritic and axonal processes of the neurons remain largely unstained, yields in itself very little insight into the structural and functional organization of the cortex, the results of such *cytoarchitectonic analyses* are important because they provide a very useful general framework for studies employing other, more critical techniques.

Beginning at the surface, the six neocortical layers are as follows: (I) lamina molecularis, (II) lamina granularis externa, (III) lamina pyramidalis externa, (IV) lamina granularis interna, (V) lamina pyramidalis interna and (VI) lamina multiformis.

I. The *lamina molecularis* or *lamina zonalis* contains only very few cell bodies.
II. The *lamina granularis externa* is composed of small, densely packed cell bodies. The name of this layer is misleading because most of its constituent somata belong to small pyramidal neurons which, like all typical cortical pyramids, direct their apices toward the surface.
III. The *lamina pyramidalis externa* is a thick layer in which pyramidal somata prevail. These somata increase progressively in size from superficial to deep.
IV. The *lamina granularis interna* consists of small, densely packed, pyramidal and non-pyramidal somata.
V. The *lamina pyramidalis interna* consists mainly of medium-sized and large, loosely arranged pyramidal somata.
VI. The *lamina multiformis* is composed of relatively tightly packed, spindle-shaped somata. Golgi material reveals that most of these somata belong to modified pyramidal cells.

The neocortex is frequently also designated as *isocortex* [784, 791] or *homogenetic cortex*. Brodmann [70, 71] used the latter term to indicate that, even though some cortical areas in the adult brain may have more or less than six layers, the entire neocortex has passed through a basic, six-layered stage during ontogeny (Fig. 15.4A).

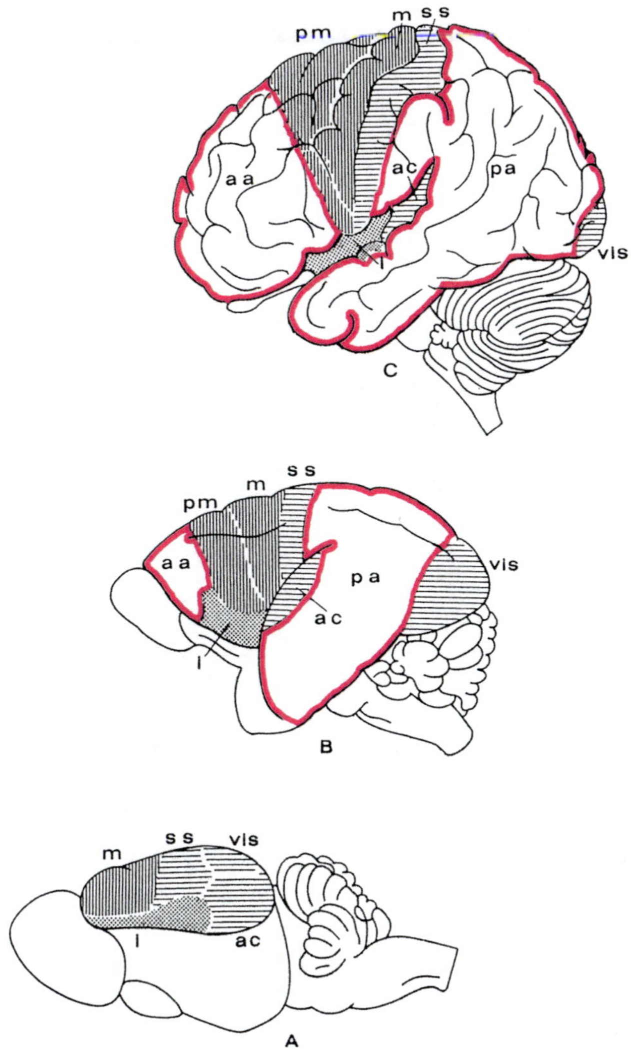

Fig. 15.3. Lateral views of the brains of the hedgehog (**A**), the prosimian *Galago* (**B**) and human (**C**) to show the evolutionary differentiation of the neocortex. In the hedgehog almost the entire neocortex is occupied by sensory (ss, vis, ac) and motor (m) areas. In the prosimian *Galago* the sensory cortical areas are separated by an area occupied by association cortex (pa). A second area of association cortex (aa) is found in front of the premotor cortex (pm). In the human the association areas are strongly developed. *aa*, anterior association area; *ac*, auditory cortex; *i*, insular cortex; *m*, motor cortex; *pa*, posterior association area; *pm*, premotor cortex; *ss*, somatosensory cortex; *vis*, visual cortex

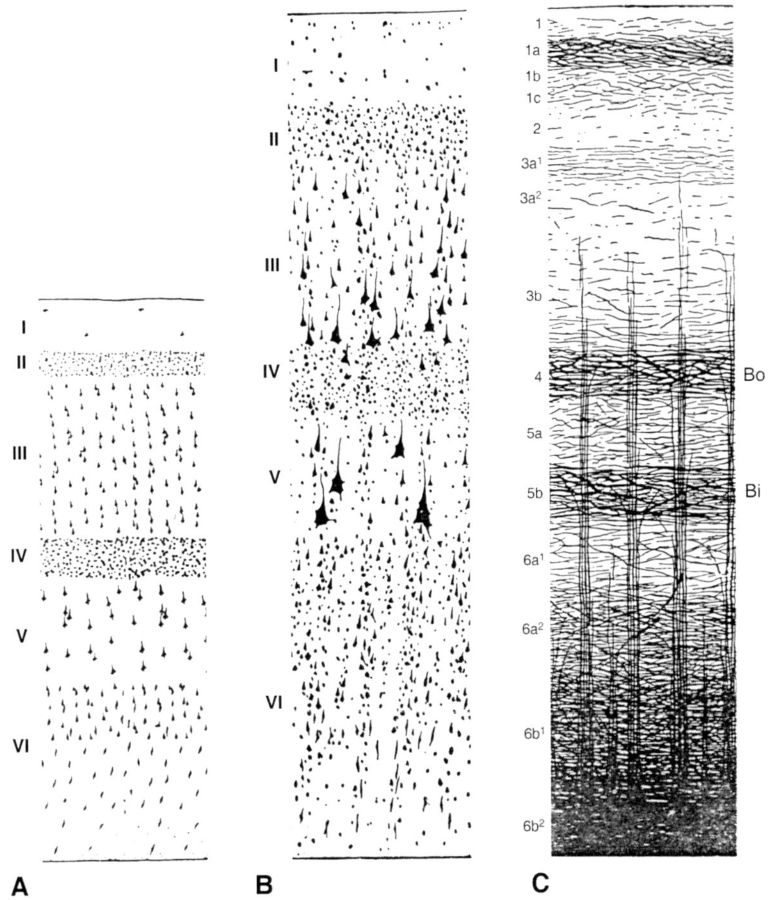

Fig. 15.4. The laminar pattern of the human neocortex: **A** in a newborn, diagrammatic; **B** Nissl preparation, adult; **C** myelin sheath preparation, adult. After Vogt and Vogt [786] and Rose [637]. *Bi, Bo,* inner and outer stripes of Baillarger

Apart from the neuronal perikarya, several other structural elements in the cortex show a more or less distinct laminar arrangement. Myelin-stained and reduced silver preparations reveal the presence of tangentially oriented fibre concentrations at several levels, and Golgi material shows that these fibres and their terminal ramifications may contribute to plexiform zones, which are likewise tangentially arranged. Thus, lamina I contains a dense plexus of horizontally running extrinsic and intrinsic fibres that contact the apical dendritic bouquets of pyramidal neurons situated in the deeper layers. Tangential plexuses of fibres in layer IV and deep in layer V are known as the outer and inner striae of Baillarger. These striae are probably formed primarily by intrinsic cortical axons, i.e. axons of local circuit neurons and collateral branches of pyramidal cell axons [340]. The outer stria of Baillarger is particularly well developed in the visual cortex, where it is referred to as Gennari's (or Vicq d'Azyr's) line (Fig. 5.13 C).

Although tangential lamination is a prominent feature of the neocortex, many of its constituent elements show an evident radial orientation. Prominent among these are the pyramidal cells. The apical dendritic shafts of these ubiquitous elements extend peripherally, and many of them reach the most superficial layers before forming their terminal tufts. The axons of the pyramidal cells are also radially oriented. They emerge from the basis of the cell bodies and descend toward the white matter.

The pyramidal cell axons acquire a myelin sheath at a short distance from the soma and assemble in bundles that increase in size as they descend and as more axons are added. These bundles are known as radial fasciculi. The apical dendrites of the pyramidal cells are also arranged in bundles [194, 196, 565, 568]. These axonal and dendritic bundles impose upon the neuronal cell bodies an arrangement in slender radially oriented columns that extend throughout the thickness of the cortex. Not only the main axons of the pyramidal cells, but also their collateral branches often show a radial orientation, and the same holds true for the axonal systems of many intrinsic cortical neurons. Many extrinsic afferents also take a radial course after having entered the cortex from the deep white matter.

The systematic study of the disposition of myelinated fibres in brain structures is known as *myeloarchitectonics*. Detailed myeloarchitectonic analyses of the human cerebral cortex have been carried out by C. and O. Vogt [784, 786, 789–791]. So far as the neocortex is concerned, the layers distinguished in these analyses correspond to the cytoarchitectonic layers of Brodmann and others (Fig. 15.4 B); however, they are designated with arabic, rather than with roman numerals (Fig. 15.4 C).

Lorente de Nó [416] concluded on the basis of the study of extensive Golgi material that the main pattern of connections between cortical neurons is in a vertical direction. He advanced the idea that the cerebral cortex essentially consists of small, radially arranged sets of neurons having a thalamic afferent fibre as their axis. According to Lorente de Nó, these column-like elementary units contain all types of cortical cells and within their confines the whole process of the transmission of impulses from the afferent fibre to the efferent axon is accomplished. Electrophysiological studies of the somaesthetic [494] and the visual cortex [308] have provided powerful evidence for the presence of functionally discrete radial columns or modules in these sensory areas. Later, morphological studies [242] showed that in other areas the cortical modules are organized around cortico-cortical afferents rather than thalamic inputs. The cortical columns will be further discussed in a later section of the present chapter.

Some quantitative data on the cerebral cortex are presented in Table 15.1.

Table 15.1. Quantitative data on the human cerebral cortex

Volume (both hemispheres)	517 cm^3 (males) 440 cm^3 (females)	Pakkenberg and Gundersen [525]
Surface (both hemispheres)	1470–2275 cm^2	Blinkov and Glezer [56] Elias and Schwartz [173] Pakkenberg and Gundersen [525]
Depth of neocortex	1.5–5 mm	von Economo and Koskinas [796]
Total number of neurons (both hemispheres)	22.8×10^9	Pakkenberg and Gundersen [525]

Sulcal Pattern

The human cerebral cortex, as that of other large mammals, is strongly folded (Fig. 15.1) [31]. Convolutions or gyri are separated by fissures or sulci. In humans, almost two thirds of the neocortex is hidden away in the depth of the sulci [64]. The sulcal pattern of the human cerebral hemispheres has been the subject of numerous studies, as for instance those of Retzius [615], Bailey and von Bonin [30] and Ono et al. [519]. These studies have shown that the overall sulcal pattern on the telencephalic surface is highly characteristic for humans (Fig. 15.5 A–C) and that the same holds true for other mammalian species. However, the individual sulci show a considerable intersubject variability. They may vary in position and course (Fig. 15.5 G). Many sulci may show one or several interruptions (Fig. 15.5 A, F) and some may be doubled over a certain part of their trajectory (Fig. 15.5 E). The sulcal pattern not only varies among individuals but also varies between the two hemispheres in the same individual. Some sulci, including the lateral and central sulci and the parieto/occipital and cingulate sulci, mark the boundaries between cerebral lobes (Fig. 1.4).

Structural and Functional Subdivision of Neocortex

Structural Subdivision 1: Cytoarchitecture

Although the laminar basic pattern is recognizable throughout the neocortex, this structure is not homogeneous. Differences in the relative thickness and cell density of the various layers, and in the size, shape and arrangement of the neuronal perikarya, are present (Figs. 15.6, 15.7) and have been used to divide the neocortex into cytoarchitectural areas. Similarly, differences in the pattern of myelinated fibres (local development and distinctness of the striae of Baillarger, length of the radial fasciculi) have been used in myeloarchitectonic parcellations of the cortex (see below).

During the first half of the twentieth century, several (groups of) investigators, namely Campbell [85], Brodmann [70, 71], von Economo and Koskinas [796], Bailey and von Bonin [30] and Sarkissov et al. [649], have produced cytoarchitectonic maps of the human cerebral cortex (Table 15.2). In all of these maps, the cortex is parcellated into a number of juxtaposed areas or fields. Brodmann's map, which is by far the most widely used, is shown in Fig. 15.8. Brodmann distinguished 44 sharply delineated areas, each of which he designated with a different figure. He emphasized that the boundaries between these areas generally do not coincide with the sulci on the cerebral surface. Von Economo and Koskinas [796] and Sarkissov et al. [649] subdivided several of Brodmann's areas into smaller units; hence, their total number of fields is somewhat larger

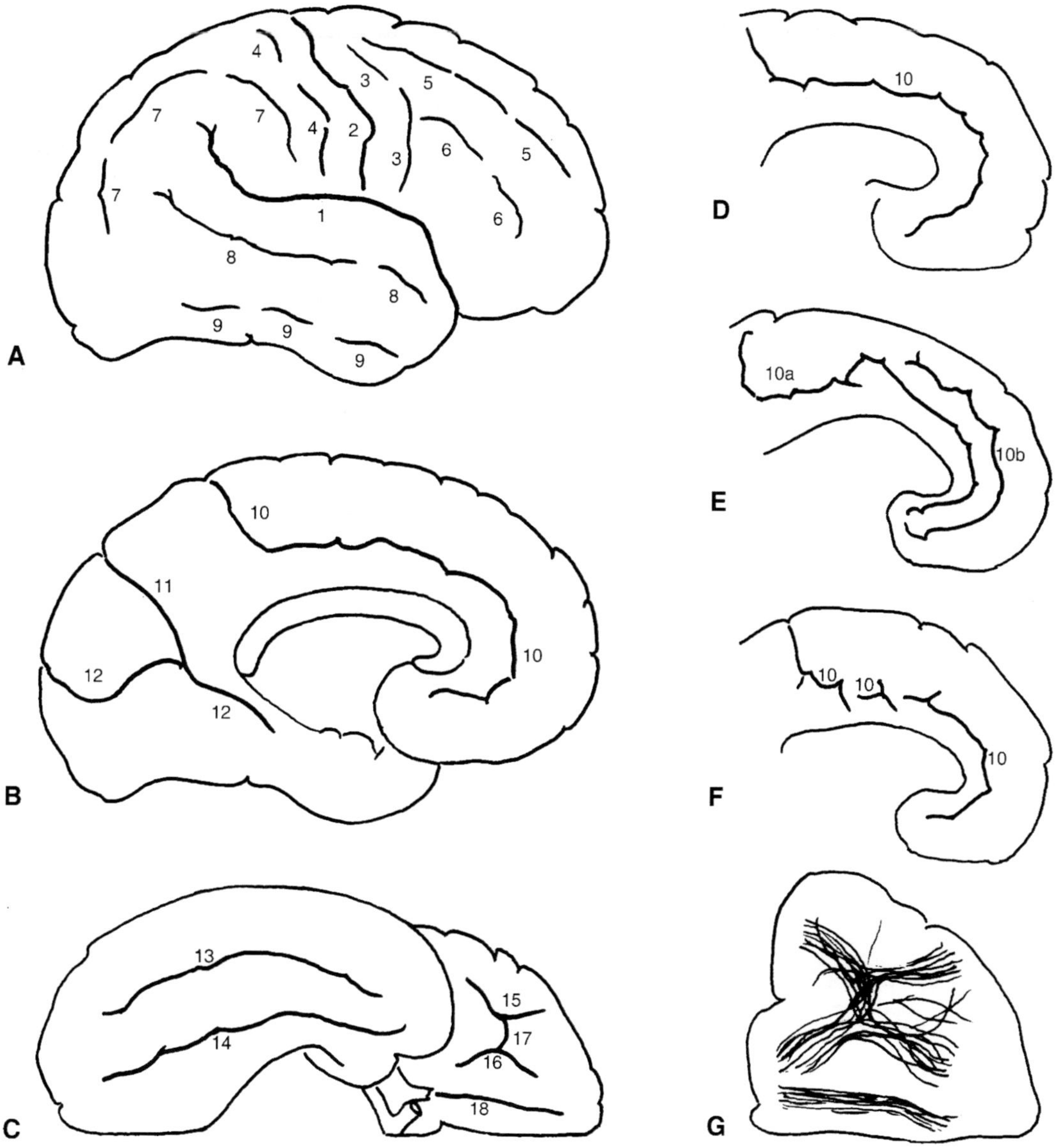

1 Lateral sulcus (Sylvian fissure)
2 Central sulcus (of Rolando)
3 Precentral sulcus
4 Postcentral sulcus
5 Superior frontal sulcus
6 Inferior frontal sulcus
7 Intraparietal sulcus
8 Superior temporal sulcus
9 Inferior temporal sulcus
10 Cingulate sulcus
11 Parieto-occipital sulcus
12 Calcarine sulcus
13 Occipitotemporal sulcus
14 Collateral sulcus
15 Lateral orbital sulcus
16 Medial orbital sulcus
17 Arcuate orbital sulcus
18 Olfactory sulcus

Fig. 15.5 A–G. Synopsis of cerebral sulci. **A–C** Lateral, medial and ventral aspect of a cerebral hemisphere, showing the most important sulci. **D–F** Three variations in the course of the cingulate sulcus, as depicted by Retzius [615]. In **D** there is a single, continuous cingulate sulcus; in **E** the sulcus is doubled (*10a,b*) and in **F** it is interrupted twice. **G** Ten variants in the course of the sulci on the orbitofrontal cortex, derived from Kanai [358], have been transferred to a standard outline of this region

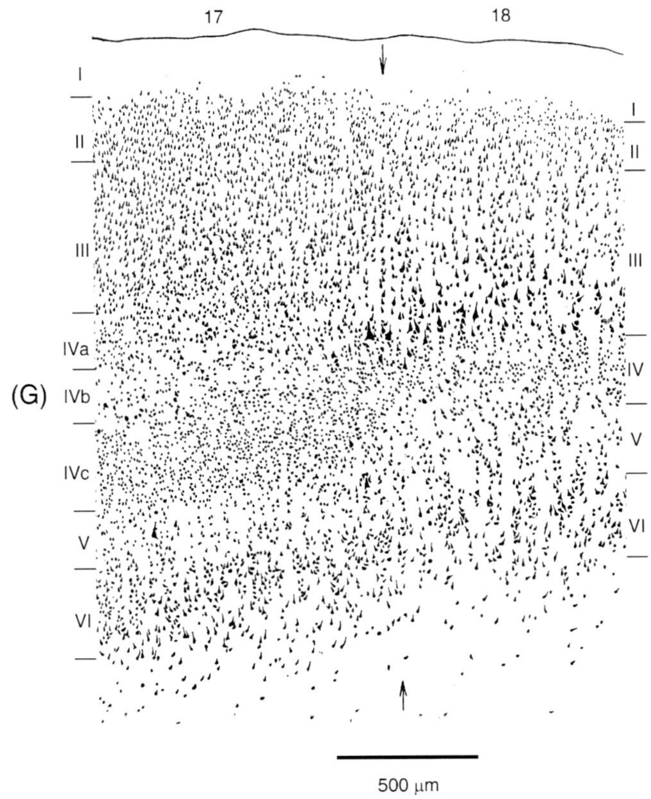

Fig. 15.6. Areas 17 and 18 and their border zone in the human brain. Reproduced from Vogt and Vogt [786]. The position of the line of Gennari (*G*) corresponds with that of the cell-poor layer IVb

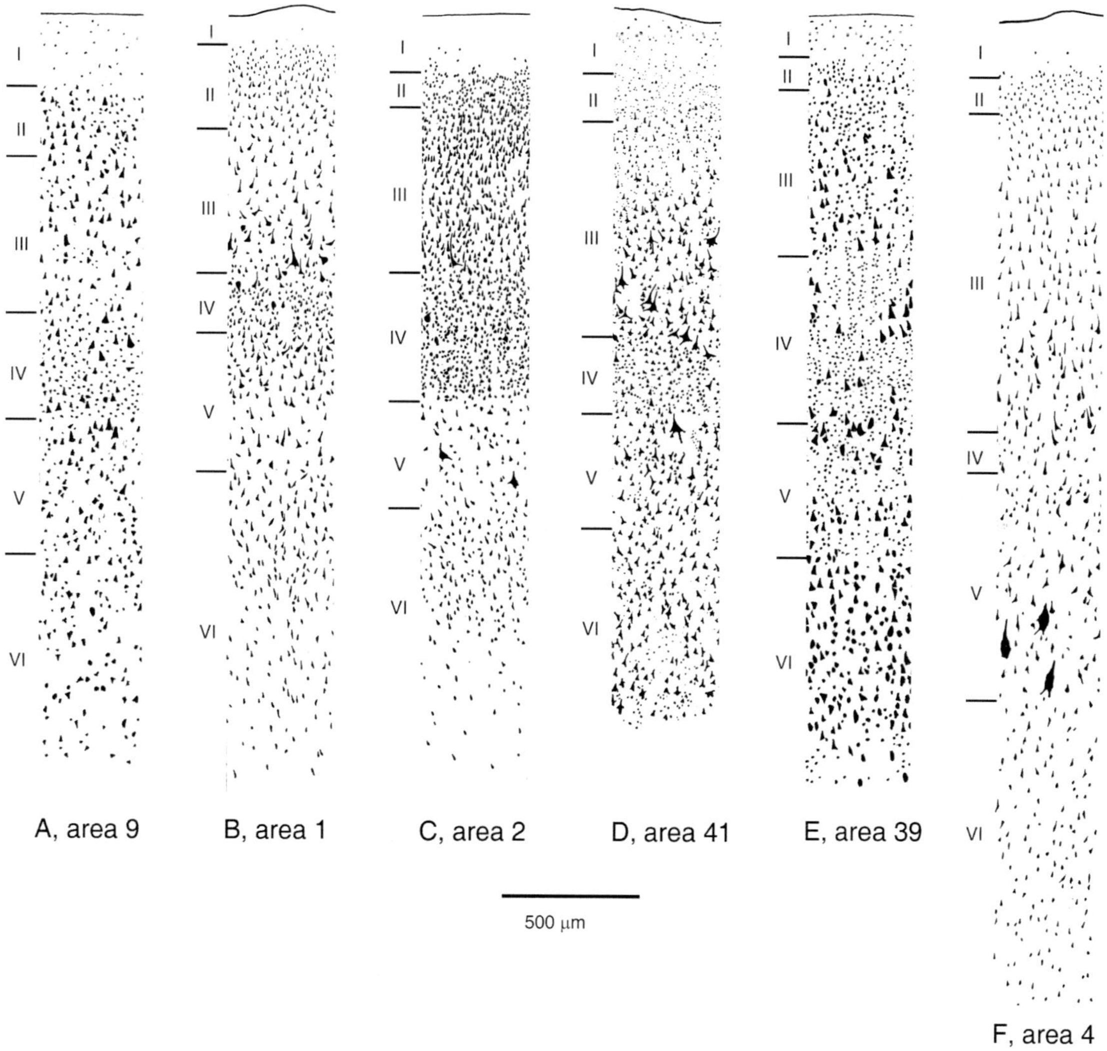

Fig. 15.7. Cytoarchitecture of several neocortical areas, designated with Brodmann's [70, 71] numbers

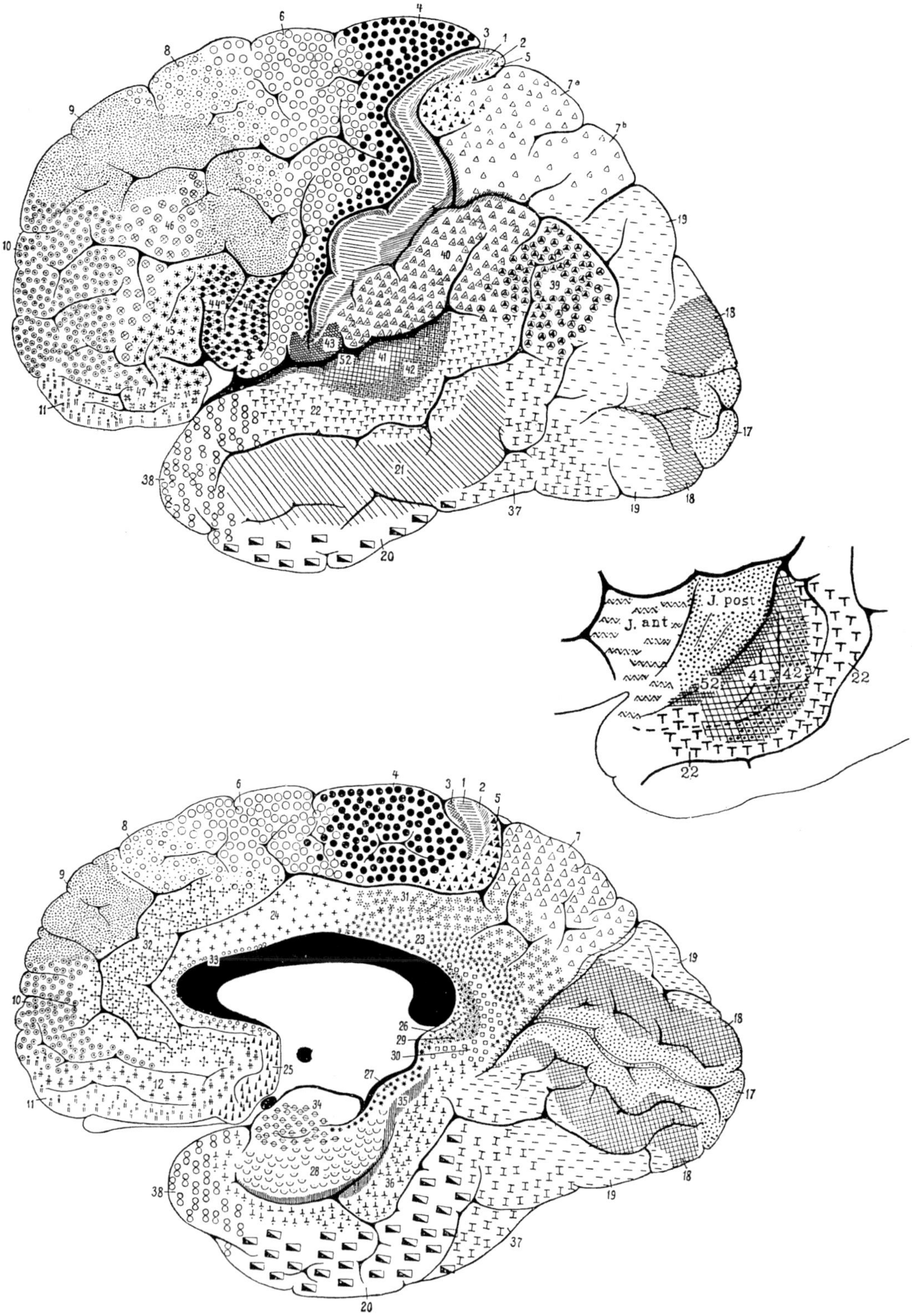

Fig. 15.8. Brodmann's famous map of the cytoarchitecture of the human cerebral cortex. The latest (1914) version is shown here [71]. In the pericentral and retrosplenial regions some (parts of) cytoarchitectonic areas have been transferred to the free surface of the hemisphere. The small auxiliary chart shows the cytoarchitecture of the insula and the upper surface of the superior temporal gyrus

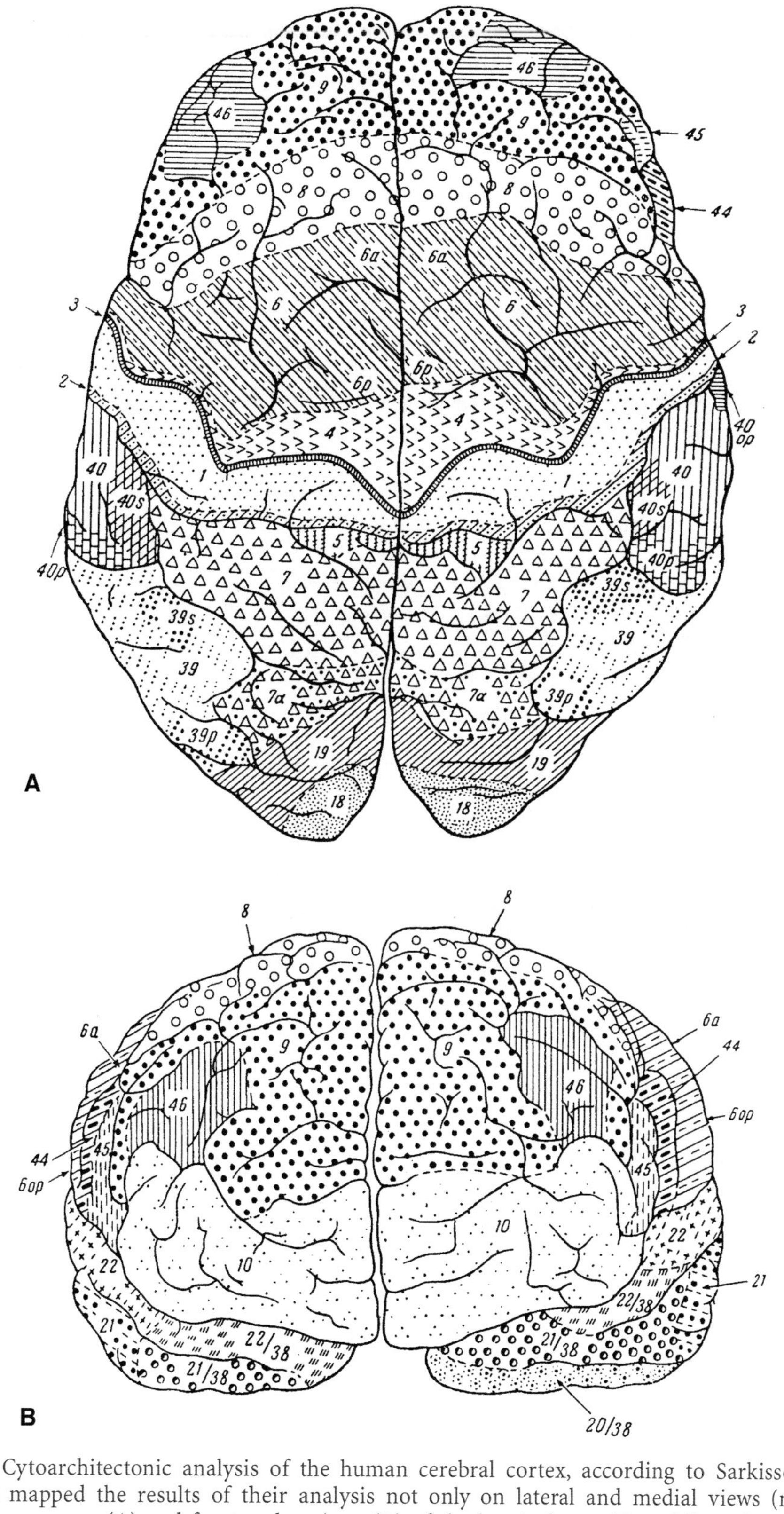

Fig. 15.9 A, B. Cytoarchitectonic analysis of the human cerebral cortex, according to Sarkissov et al. [649]. These authors mapped the results of their analysis not only on lateral and medial views (not reproduced here), but also on upper (**A**) and frontopolar views (**B**) of the hemispheres. They followed Brodmann's numbering scheme and introduced self-explanatory symbols for subareas and transitional zones

Table 15.2. Architectonic subdivisions of the human neocortex

Author(s)	Type of analysis	No. of fields	Judgement of borders
Campbell (1905) [85]	Cyto- (myelo-)	14	Sharp
Brodmann (1909, 1914) [70, 71]	Cyto-	44 [a]	Sharp
von Economo and Koskinas (1925) [796]	Cyto-	54	Generally not sharp
Vogt and Vogt (1919, 1926) [784, 791]	Myelo-	>200	Very sharp
Bailey and von Bonin (1951) [30]	Cyto-	8	Generally not sharp
Sarkissov et al. (1955) [649]	Cyto-	52	Many transition zones

[a] The highest numbered area in Brodmann's map is area 52; however, because areas numbered 13–16 and 48–51 are lacking the total number of areas in his map amounts to 44

than that of Brodmann. According to von Economo and Koskinas, who designated the various cytoarchitectonic areas with combinations of letters and figures (PA1, TE2 etc.), the interareal boundaries are generally not sharp. The map of Sarkissov and colleagues (Fig. 15.9) is patterned after that of Brodmann. Their nomenclature followed Brodmann's numbering scheme. However, they intercalated transition zones between several of the areas that Brodmann had distinguished.

Several authors have attempted to allocate the various cytoarchitectonic areas to a smaller number of variant types. It is worthy of note in this context that as early as 1874, Betz [51] pointed out that the human cerebral cortex is divided by the central sulcus of Rolando into an anterior part in which pyramidal cells predominate and a posterior part where granular cells prevail. In the anterior part he described the giant pyramidal cells that bear his name. Von Economo [795], who also focussed on the distribution of pyramidal and granular cells, recognized five fundamental cytoarchitectural types in the human neocortex, which he labelled 1–5 (Fig. 15.10). The cortical types labelled 2, 3 and 4 contain the six typical neocortical layers described above, although not all of these are developed to the same degree in the different types. Von Economo designated these areas as *homotypical,* thus contrasting them to the *heterotypical* cortical types 1 and 5, in which, at least in the fully developed cortex, the six layers cannot be clearly discerned.

The cortex of fundamental type 1 is distinguished by its lack of distinct granular layers II and IV, whereas the pyramidal layers III and V are strongly developed. Reference to Fig. 15.10 B, C shows that this *agranular cortex* occupies a region situated directly in front of the central sulcus, which corresponds to the areas 4 and 6 of Brodmann, and that this type of cortex is also found in the paracentral lobule on the medial surface of the hemisphere and in the rostral part of the cingulate gyrus. Within this region, area 4 is characterized by the presence of the giant pyramidal cells of Betz in its fifth layer (Fig. 15.7 F). Because the agranular cortex gives rise to prominent corticobulbar and corticospinal projections, it may be considered the prototype of *"motor" cortex.*

The other type of heterotypical cortex (labelled 5) is characterized by its richness in small, granular cells and the strongly developed layers II and IV. Layers III and V, on the other hand, are poorly developed. Areas belonging to this type of cortex are found in the anterior part of the postcentral gyrus, along the calcarine fissure and in a limited part of the upper surface of the superior temporal gyrus. The type 5 cortex is called *granular cortex* or *koniocortex* (in Greek *konios* = dust). It is characteristic of those cortical areas that receive the great afferent systems, i.e. the somatosensory projection, the acoustic projection and the optic radiation. A particularly well-differentiated granular cortex is found in the area striata (area 17 of Brodmann). As shown in

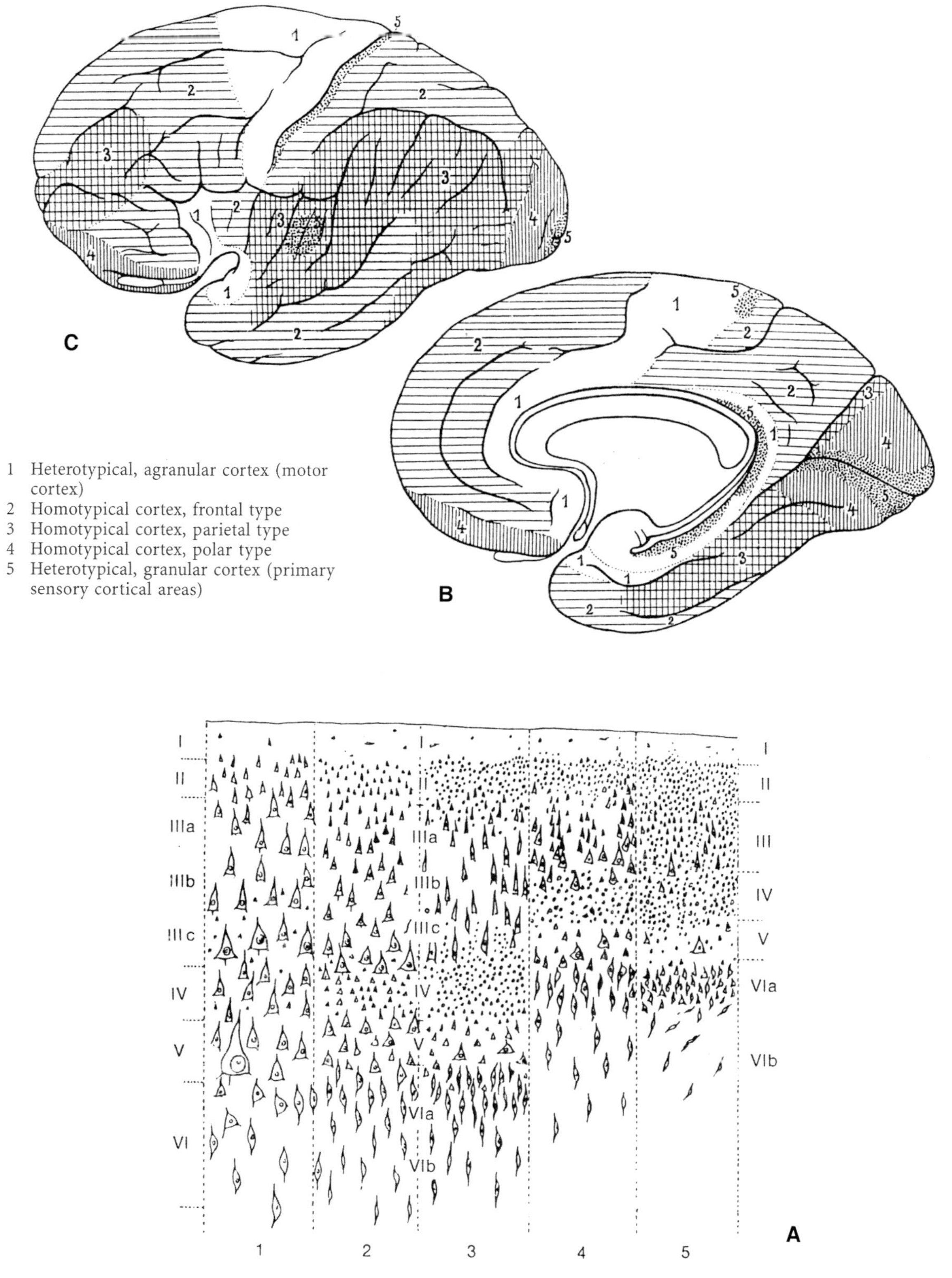

Fig. 15.10. The five principal types of neocortex (**A**), as distinguished by von Economo [795] and their distribution over the medial (**B**) and lateral surfaces (**C**) of the cerebral hemispheres

Fig. 15.6, the number of granular cells is exceedingly high here and the inner granular layer is clearly subdivided into three sublayers, IVa,b and c. The middle of these, which is relatively poor in cells, coincides with the layer of myelinated fibres known as the line of Gennari or Vic d'Azyr.

The three types of homotypical cortex (2, 3 and 4; Fig. 15.10) occupy an intermediate position between the agranular (1) and granular (5) types. They are named according to their main, but not exclusive localization: frontal type (2), parietal type (3) and polar type (4). Types 2 and 3 both contain numerous small and medium-sized pyramidal cells, but in type 3, granular cells are more prominent than in type 2 (Fig. 15.7 A, E). The polar type (4) occupies small areas near the frontal and occipital poles, hence its name. It is characterized by its thinness and high content of granular cells. In both the parietal (3) and polar (4) types of cortex, the multiform layer (VI) is particularly well developed (Fig. 15.7 E). The homotypical cortical types (2, 3 and 4) occupy much of what Flechsig [191, 192] and many later investigators have provisionally designated as *association cortex.*

If we consider the cytoarchitectonic analyses of Campbell [85], Brodmann [70, 71], von Economo and Koskinas [796], Bailey and von Bonin [30] and Sarkissov et al. [649] in light of the classification of cortical types just discussed, the following conclusions can be drawn:

1. The heterotypical agranular and granular areas, i.e. the primary motor and primary sensory fields, have been recognized and delineated in all of these analyses.
2. The homotypical cortex was left largely undivided by Campbell and by Bailey and von Bonin; hence, the number of areas distinguished by these authors is considerably smaller than that recognized by the others (Table 15.2).
3. With regard to the location and subdivision of the three types of homotypical cortex, the results of Brodmann, von Economo and Koskinas and Sarkissov et al. are in general closely comparable. Exceptions should be made, however, for the medial temporal lobe and the orbitofrontal cortex. In both of these regions the analyses of the "classical" cytoarchitectonists, as well as those of later investigators, have led to widely diverging results (Fig. 15.11) [518, 718].

In his notable monograph of 1909, Brodmann [70] did not confine himself to the human brain, but also presented cytoarchitectonic analyses of the cortex of a considerable number of other mammalian species, among them two primates, a prosimian, a carnivore and an insectivore. The number of areas identified in these animals was smaller than that in the human. Thus, he delineated 24 areas in the macaque *Cercopithecus,* 26 in the marmoset *Callithrix,* 14 in the prosimian *Lemur* and 12 in the insectivore *Erinaceus.* In the monkeys studied, homologues of the human primary motor areas 4 and 6 and the primary sensory areas 3, 1, 2 and 17 could be identified, and the same holds true for several of the human "association" areas, such as the frontal areas 9 and 12, the parietal areas 5 and 7 and the temporal areas 20–22 (Fig. 15.12). Brodmann's comparative cytoarchitectonic explorations laid the foundation for later electrical stimulation studies as well as for experimental analyses of cortical fibre connections (see below).

Structural Subdivision 2: Myeloarchitecture

It has already been mentioned that C. and O. Vogt [784, 786, 789–791] subjected the human cerebral cortex to a detailed myeloarchitectonic analysis. The myelinated fibres in the cortex show two principal orientations, tangential and radial. The tangential fibres tend to form laminae, which in general can be readily identified in conjunction with the corresponding layers observed in Nissl preparations (Fig. 15.4 B, C). The radially oriented fibres are arranged in bundles, termed radii, which ascend from and descend to the subcortical white matter. The Vogts noticed that the number and distinctness of the tangential fibre layers show considerable local differences in the cortex, and that the same holds true for the extent to which the radii penetrate into the cortex. Focussing on

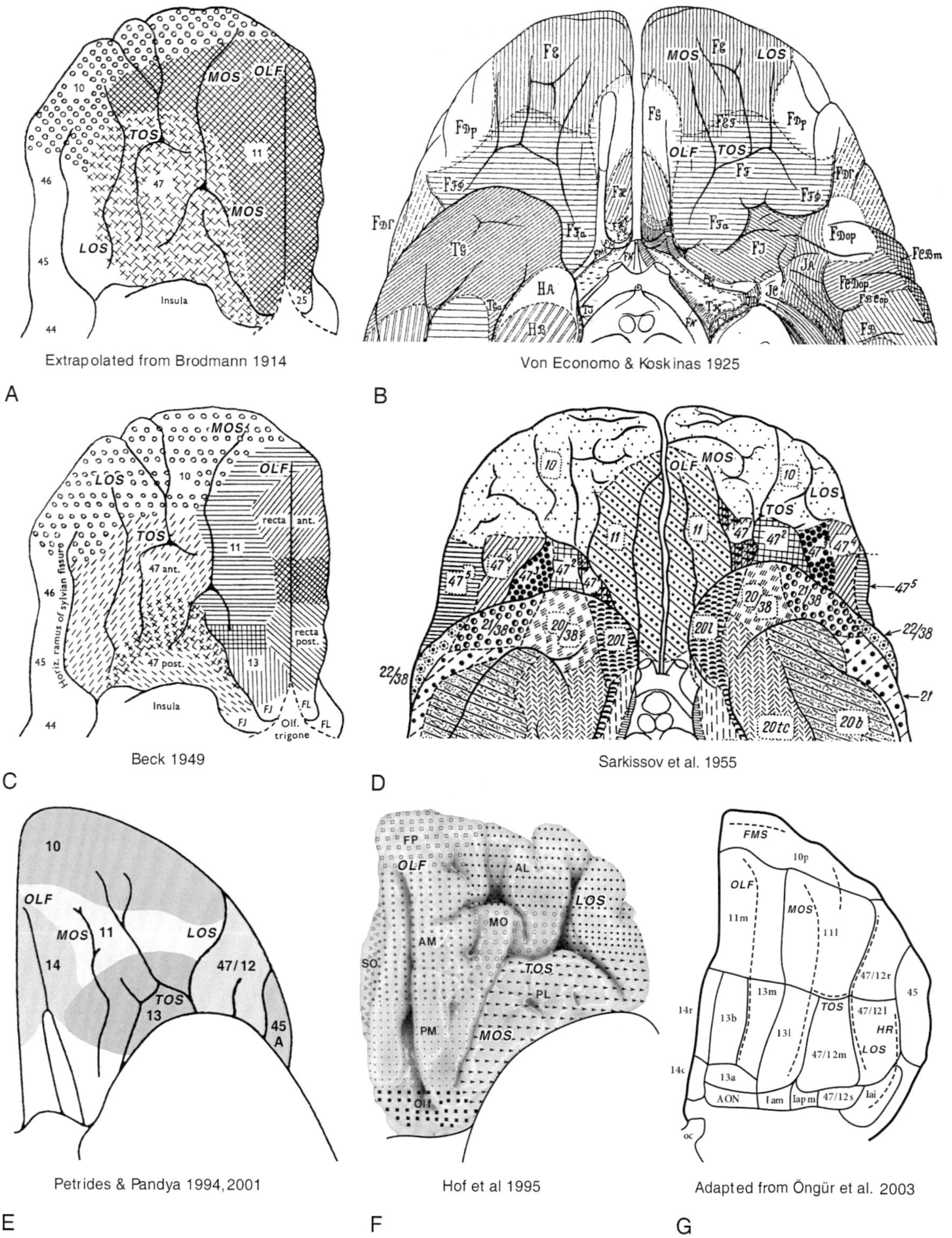

Fig. 15.11 A–G. Cytoarchitectonic parcellations of the human orbitofrontal cortex. Note that the analyses of the seven (groups of) authors have led to widely diverging results. Courtesy of Drs. H. Uylings and G. Rajkowska. *LOS*, lateral orbital sulcus; *MOS*, medial orbital sulcus; *OLF*, olfactory sulcus; *TOS*, transverse orbital sulcus. References cited: Brodmann [71], von Economo and Koskinas [796], Beck [44], Sarkissov et al. [649], Petrides and Pandya [577], Hof et al. [301], Öngür et al. [518]

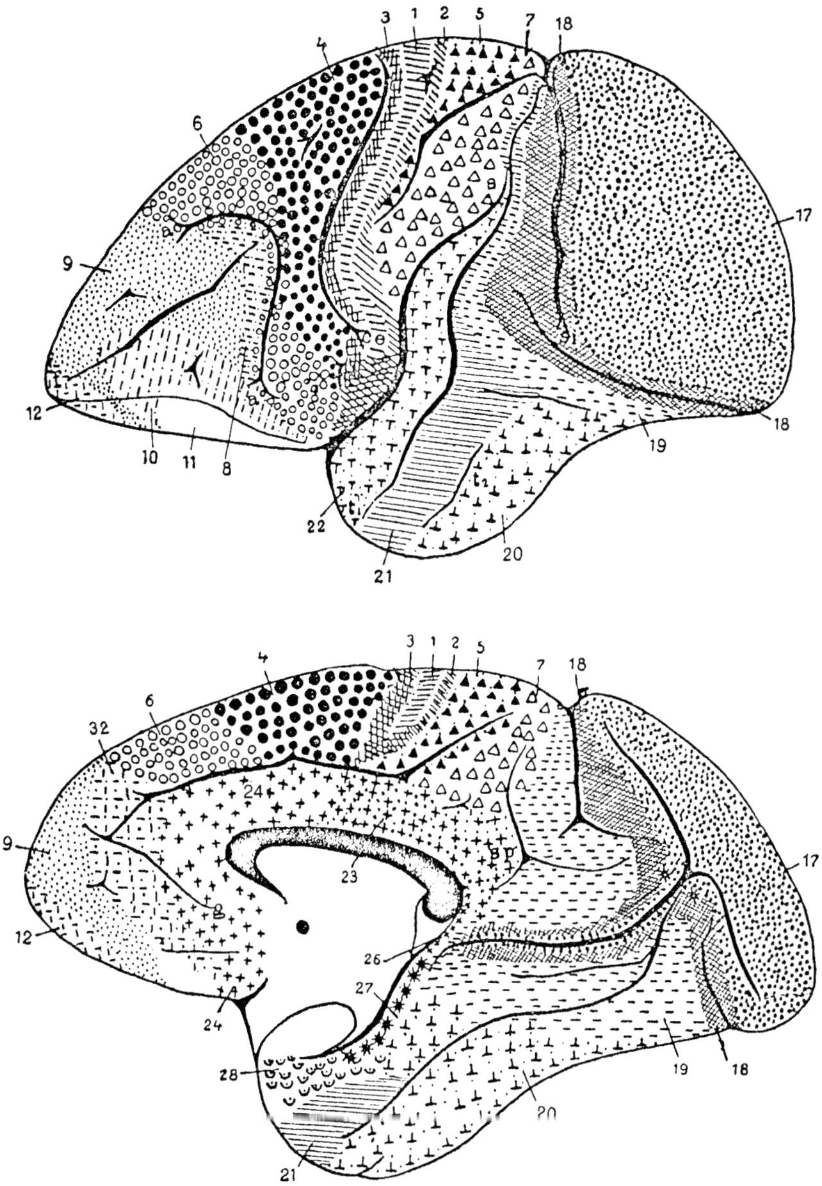

Fig. 15.12. Cytoarchitectonic map of the cerebral cortex of the macaque *Cercopithecus*, according to Brodmann [70]

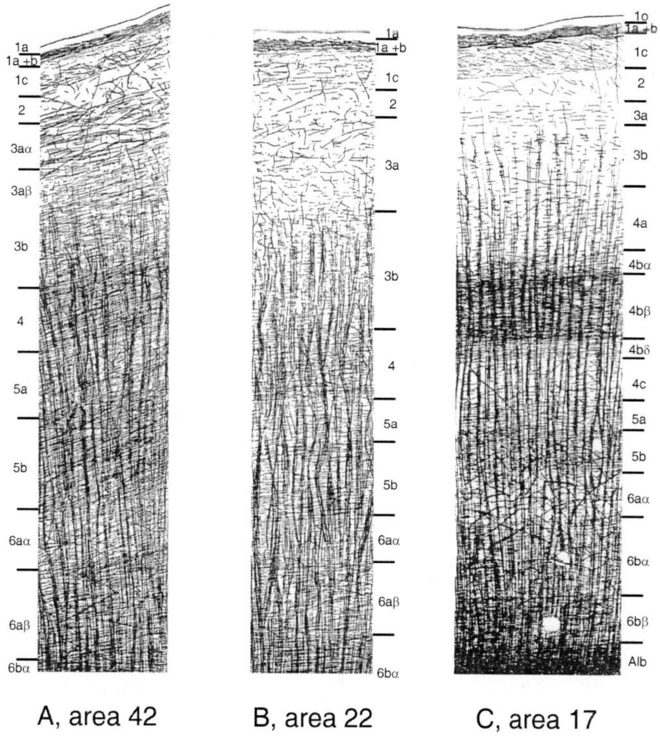

Fig. 15.13 A–C. Myeloarchitecture of some cortical areas, according to the Vogts. The *numbers* have been added according to the Brodmann scheme. Reproduced from Rose [637]

these differences, they were able to delineate a large number of areas whose boundaries were found to coincide with those of Brodmann's cytoarchitectonic fields (Fig. 15.13) [787]. They emphasized that the areal boundaries are very sharp and omnilaminar, i.e. characterized by changes in all layers. In their parcellation of the cortex, the Vogts went much further than Brodmann. Many of the areas distinguished by the latter were subdivided into several subareas, which in turn were claimed to be subdivisible into still smaller units, termed campuli [787].

The Vogts did not investigate the temporal and occipital lobes; hence, their myeloarchitectonic map of the human cortex remained incomplete [64].

Structural Subdivision 3: Myelogenesis

Myelinated fibres also took a central position in the investigations of Flechsig [191–193]. However, this author did not study the disposition of the myelinated fibres in the mature cortex, but rather the temporal order in which the fibres of the white matter immediately underneath the different parts of the cortex become myelinated during development. Flechsig distinguished 45 myelogenetic areas, which he numbered according to their sequences of myelination during foetal and early postnatal life (Fig. 15.14). He classed these areas in three categories, *primordial areas* (1–16), which show signs of myelination before birth, *intermediate areas* (17–36), which become myelinated between birth and the second postnatal month and *terminal areas* (37–45), which myelinate after the second postnatal month.

Flechsig observed that the prenatally myelinating fibres belong to the large sensory and motor projections. For this reason he characterized the primordial areas as *projection areas*. According to Flechsig, the postnatally maturing fibres form associative links between different parts of the cortex; hence, he designated the intermediate and terminal areas as *association areas*. He speculated that these areas subserve higher cognitive and mental functions. In his view, clusters of late-maturing areas form three association centres in the human brain: a large *posterior association centre,* occupying much of the parietal, occipital and temporal lobes; an *anterior association centre,* expanding in front of the motor projection areas; and a small *insular association centre* (Fig. 15.14).

Structural Subdivision 4: Connectivity

Cytoarchitectonic studies form an indispensable first step in the analysis of any part of the central nervous system. They should logically be followed by exploration of the fibre connections of the delineated grisea. The fibre connections may be used as auxiliary criteria in the detection and homologization of morphological entities, but secondly and more importantly, they may give salient clues as to the functional significance of their targets. Experimental studies on two sets of connections, thalamocortical and cortico-cortical, have led to very useful functional subdivisions of the neocortex. It is important to note that, because humans, for obvious reasons, cannot be subjected to interventional techniques, our knowledge of the fibre connections mentioned is almost entirely based on studies in non-human primates, particularly the rhesus macaque.

The thalamic nuclei and their connections have been extensively discussed in Chap. 8. Let it suffice to recall here that the four groups of specific thalamic nuclei, sensory relay nuclei, motor relay nuclei, limbic nuclei and association nuclei, project to specific areas or regions of the cortex and together innervate the entire cortex (Fig. 8.4).

The *sensory relay nuclei,* i.e. the lateral posterior nucleus, the lateral geniculate body and the medial geniculate body, project to and functionally define the primary sensory cortices. The ventral posterior nucleus receives afferents from the somatosensory pathways and projects to the areas 3, 1 and 2, which collectively form the primary somatosensory area, S1. The lateral geniculate body receives afferents from the retina and projects to the primary visual cortex, V1, which corresponds to

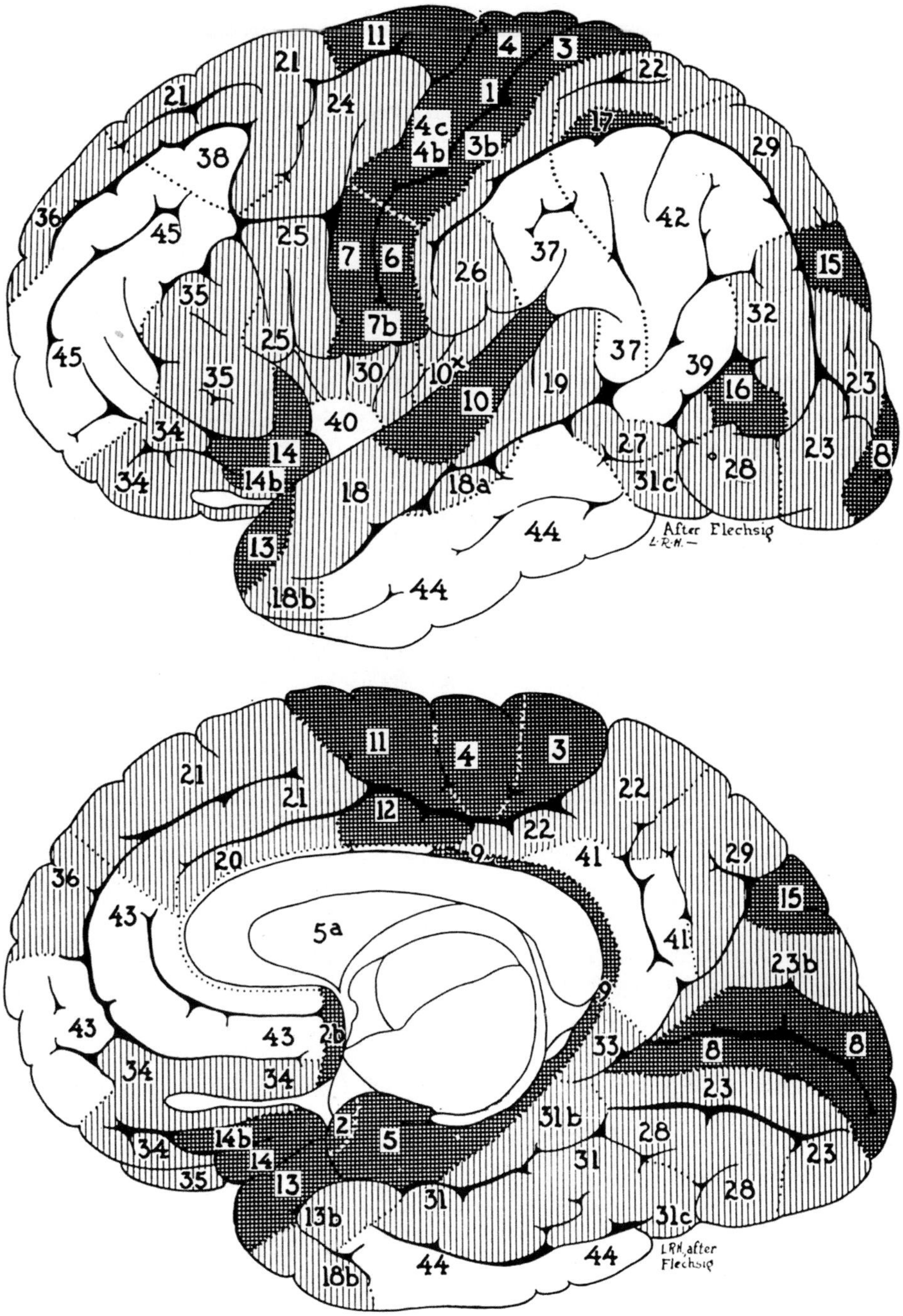

Fig. 15.14. Myelogenetic subdivision of the human cerebral cortex, according to Flechsig [193]. The *numbers* of the areas indicate the sequence of myelination during fetal and early postnatal life: *1–16* (*cross-hatched*), early myelinating, primordial areas; *17–36* (*vertical lines*), intermediate areas; *37–45* (*white*), terminal or "association" areas. Reproduced from von Bonin [793]

area 17. This area is also known as the striate area, because of the presence of the line of Gennari, or the calcarine cortex, because it surrounds the calcarine sulcus. The medial geniculate body, which represents the thalamic processing station in the auditory pathway, sends its efferents to the primary auditory cortex, A1, which consists of the areas 41 and 42.

The *motor relay nuclei* comprise the ventral anterior nucleus (VA) and the anterior and posterior divisions of the ventral lateral nucleus (VLa, VLp). VA is the principal target of fibres ascending from the pars reticulata of the substantia nigra and projects to the frontal eye field (area 8) and adjacent parts of the prefrontal cortex. VLa is the terminus of afferents from the internal segment of the globus pallidus and projects to the premotor cortex (area 6). VLp, finally, receives a massive input from the cerebellar nuclei and sends its efferents to the primary motor cortex, M1 (area 4). It should be mentioned that the efferent projections from the three motor relay nuclei, VA, VLa and VLp, are focussed on the cortical areas mentioned but that their terminal fields overlap to some extent.

The *limbic nuclei* comprise the anterior nuclear complex and the lateral dorsal nucleus. These structures are linked by being the targets of hippocampal fibres arising from the subicular complex, reaching the thalamus via the fornix, the mamillary nuclei and the mamillothalamic tract (Fig. 12.13). They are the source of thalamic fibres to the cingulate gyrus (areas 23, 24, 32), including the retrosplenial areas 29 and 30, this projection extending as far posteriorly and inferiorly as the presubiculum and parasubiculum and the entorhinal cortex (area 28) [712]. The sources of the afferents to the anterior nuclear complex and the lateral dorsal nucleus, as well as the targets of their efferents, all form part of the limbic system.

The *association nuclei* include the mediodorsal nucleus and the pulvinar, two cell masses that are particularly well developed in primates. The pulvinar involves more than one half of the whole thalamic volume in humans [550]. Both cell masses actually represent cell complexes. They can be subdivided into several subsidiary nuclei, each with receiving afferents from one or more specific subcortical source and sending efferents to a specific cortical area or region. However, the distinguishing feature of these two cell masses as a whole is that their subcortical afferents are rather weakly developed and that they receive their principal, *driving* afferents from the cortex. They function primarily as links in cortico-thalamo-cortical association paths, hence their name [680, 681]. The mediodorsal nucleus projects to the prefrontal cortex, which corresponds to Flechsig's anterior association centre (Fig. 15.14). The cortical output of the pulvinar is to the areas of parieto-temporo-occipital cortex, intercalated between the primary somatosensory, auditory and visual areas, i.e. to Flechsig's posterior association centre. However, these projections are less specific than was previously thought. Thus, the mediodorsal nuclei send axons to several cortical regions other than the prefrontal, among them the cingulate, insular and parietal regions, as well as the premotor and motor accessory areas, while the pulvinar has additional connections with the frontal eye field and with the anterior and posterior cingulate, retrosplenial, insular and parahippocampal areas.

The question of whether a cortical area that receives projection fibres from a specific thalamic element possesses structural characteristics that permit its morphological delimitation was specifically addressed by Rose and Woolsey [634] in a study on the structure and thalamic connections of the cingulate gyrus in rabbit and cat. They found that the cingulate cortex in these animals can be subdivided into three cytoarchitectonic areas, and that each of these areas co-extends the distribution field of a particular nucleus within the anterior nuclear group. Such close correlations between cytoarchitecture and the projection fields of individual thalamic nuclei were later also found in many other regions of the cerebral cortex [342, 401, 636, 670].

Experimental studies on the cortico-cortical connections in the rhesus monkey have considerably increased our insights into the functional organization of the association cortices. The following summary of the results of these studies is principally based on review articles by Pandya and colleagues [526, 529, 532, 579], Van Hoesen [773] and Mesulam [463–465]. The latter made a constructive effort to integrate cytoarchitectural and clinical findings in humans with experimental evidence obtained from the non-human primate to form a working model of the human cerebral cortex (Fig. 15.15).

1. The association areas can be subdivided into two main types: modality-specific (unimodal) and high-order (heteromodal). Each primary sensory area is adjoined by a modality-specific sensory association area. These unimodal sensory association areas can be further subdivided into proximal or upstream and distal or downstream components. Proximal areas are only one synapse away from the corresponding primary sensory area, whereas distal areas are at a distance of two or more synapses from the relevant primary area.
2. The unimodal sensory association areas together occupy most of the post-Rolandic neocortex.

   The *somatosensory unimodal association cortex* (SA) is situated in the parietal lobe, directly behind the primary somatosensory areas 3, 1 and 2, which are collectively designated as S1. It is supposed to occupy parts of areas 5 and 7 in the superior parietal lobule and may also include parts of area 40 in the anterior portion of the inferior parietal lobule. The subdivision of the somatosensory association cortex into proximal and distal areas in the human remains to be elucidated.

   The *visual unimodal association cortex* (VA) occupies much of the occipital lobe and extends far anteriorly into the lower parts of the temporal lobe. Its proximal zone consists of the areas 18 and 19 which, together forming the circumstriate belt, surround the primary visual of striate cortex (V1). The distal zone of the unimodal visual association cortex includes areas 20, 21 and 37.

   The *unimodal auditory association cortex* (AA) covers the superior temporal gyrus (area 22). It flanks the primary auditory cortex, A1 (areas 41 and 42), which is located in Heschl's gyrus on the posterior aspect of the temporal plane. The connectivity found in the monkey brain suggests that the posterior part of the superior temporal cortex (area 22) displays the projections of the proximal auditory association cortex, whereas the more anterior part of this gyrus may represent the distal auditory association cortex.

   Unimodal association areas for taste and vestibular sensation have not been identified so far. Some areas in the posterior orbitolateral cortex and anterior insula (13a, I am) may represent unimodal olfactory association areas (see Figs. 11.7, 11.8).
3. Mesulam [463, 464] considers the premotor cortex anterior to the primary motor area, M1, as a motor analogue of the modality-specific sensory association areas, chiefly because it provides the principal cortical input to M1. According to Mesulam, this *motor association area* (MA) is composed of area 6, including the supplementary motor region (M2), posterior area 8 and area 44. It has reciprocal connections with the unimodal sensory association areas [526, 532].
4. The high-order heteromodal association areas, which are sometimes also designated as polymodal or supramodal areas, receive their principal cortical afferents from (1) unimodal areas, particularly their distal zones, (2) other heteromodal areas and (3) paralimbic areas. In the primate brain, three heteromodal association areas, parietotemporal, medial temporal and prefrontal, have been identified.

The *parietotemporal heteromodal association area* is situated at the juncture of the unimodal sensory association cortices, from which it receives convergent bimodal and trimodal afferents. This area includes the caudal part of the superior parietal lobule (area 7), most of the inferior parietal lobule (areas 39 and 40) and a forwardly extending strip, formed by lateral temporal cortex within the banks of the superior temporal sulcus at the juncture of areas 21 and 22 (Fig. 15.15).

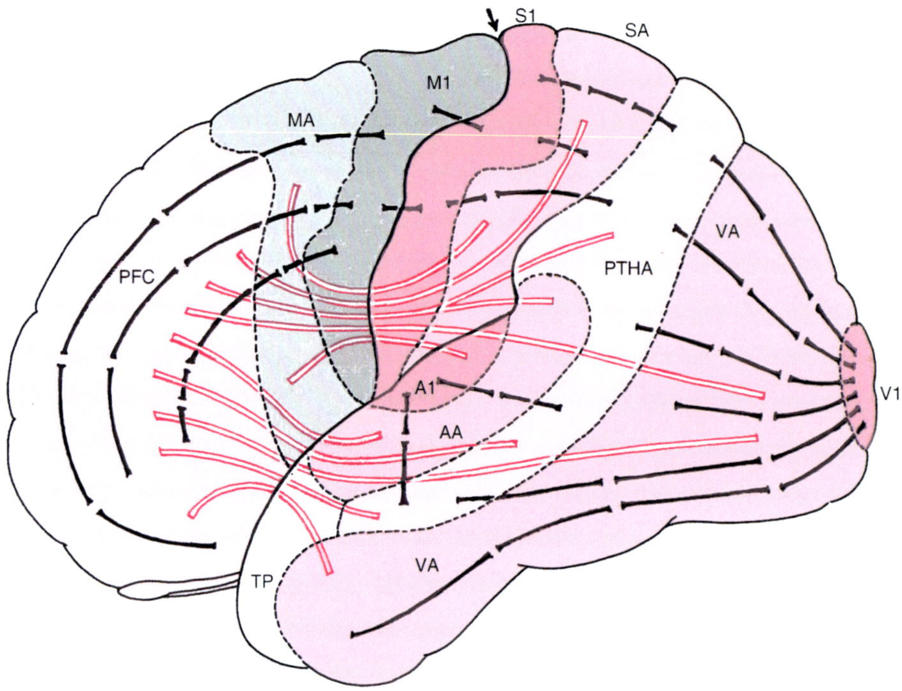

Fig. 15.15. Subdivision of the human neocortex into functional zones. This subdivision is largely based on experimental studies of cortico-cortical connections in monkeys. The primary somatosensory (*S1*), visual (*V1*) and auditory (*A1*) cortical areas (in *dark red*) project via short connections to the adjacent unimodal somatosensory (*SA*), visual (*VA*) and auditory (*AA*) association areas (in *medium red*). These unimodal sensory association areas in turn project to an elongated parietotemporal heteromodal association area (*PTHA*, in *light red*). The unimodal and heteromodal sensory association areas project massively to the prefrontal cortex (*PFC*, in *light red*). These long associational projections are diagrammatically indicated by *red-outlined*, white fibres. Sequences of short connections successively link the various parts of the PFC with the motor association area (*MA*, in *light grey*) and the primary motor area (*M1*, in *dark gray*) (based on Mesulam [463, 465]). *TP*, temporopolar cortex

The *medial temporal heteromodal association area* is formed by the perirhinal areas 35 and 36, which are interposed between the entorhinal area 28 and the visual association areas 19, 37 and 20.

The *prefrontal heteromodal association area* is situated in front of the motor association cortex and includes areas 9 and 10, 45–47 and the anterior parts of areas 8, 11 and 12. It receives afferents from the unimodal sensory association areas, particularly from their distal parts and from the parietotemporal and medial temporal heteromodal association areas. An important output system of the prefrontal heteromodal association area is formed by sequences of short association fibres that successively link the anterior orbitofrontal cortex, the polar and lateral prefrontal areas 9, 10 and 46, the motor association cortex and the primary motor cortex.

If we survey the data just discussed, it may be concluded that most of the human neocortex is occupied by association areas of various kinds and that the boundaries between these areas do not closely correspond to those of the cytoarchitectonic fields, as delineated by Brodmann and others.

Combining connectional data with the results of functional studies has revealed some basic features of the flow of information through the cortex. These features, which may be designated as hierarchical processing, feedback and parallel processing, will now be briefly discussed.

Hierarchical Processing. We have seen that multisynaptic feed-forward systems can be traced from the various primary sensory cortices via successive association areas to the premotor and motor cortices. In light of the results of physiological experiments on the visual system [308, 309], it is assumed that the sequential processing of information within these feed-forward systems becomes progressively more complex and may be characterized on this account as hierarchical [180, 770]. These feed-forward systems arise largely from pyramidal neurons in layer III and terminate in and around layer IV of the cortical area they send axons to [623].

Feed-back. The great majority (75% or more) of the cortical feed-forward connections are reciprocated by descending systems projecting back to areas from which an input was received. The fibres forming these feed-back systems originate nearly always from the infragranular layers V and VI and terminate largely in layers I and VI [623].

Parallel Processing. Extensive anatomical and physiological studies on the visual system of the rhesus monkey have shown that different features of the visual image remain segregated in the striate cortex and in their further projection to the extrastriate visual areas (see Chap. 19 for references and details). These extrastriate visual areas are organized into two hierarchically organized and functionally specialized processing pathways: an occipitoparietal pathway or "dorsal stream", concerned with spatial vision and movement, and an occipitotemporal pathway or "ventral stream" for object identification (Fig. 19.6). Areas along both pathways are organized hierarchically, such that the initial, low-level inputs are transformed into more complex and specific representations through successive stages of processing. Both pathways project separately to different parts of the prefrontal cortex. There is evidence suggesting that parallel processing channels are also present in the central somatosensory and auditory systems.

It should be emphasized that the various processing streams do not operate in isolation. There is substantial intermixing and cross-talk between them at successive levels of processing. Association fibres forming lateral connections between areas at the same processing level differ from ascending and descending fibres in that they terminate in a columnar pattern involving all cortical layers. It is assumed that via these interlinking fibres the information processed within the individual channels is combined into behaviourally relevant percepts. Together, the processing streams and their interconnections form *distributed hierarchical networks* subserving a *distributed hierarchical processing of information* [180, 770].

Functional Subdivision

The first systematic attempt to localize different functions in different regions of the cerebral cortex was made by Franz Joseph Gall (1758–1828) and his collaborator Johann Spurzheim (1758–1832), the founders of *phrenology* [187, 256]. Gall and Spurzheim maintained that the cerebral cortex is composed of discrete organs or regions that represent different mental faculties ("*seelische Kräfte*"), and that there are as many such organs as there are mental faculties. They distinguished 35 such faculties, including arithmetic, hope, speech, causality, destructiveness and parental love (Fig. 15.16). Gall and Spurzheim, moreover, suggested that the organs subserving these faculties correspond to prominences of the overlying skull. On that account, they asserted to be able to disclose a person's intellectual and moral abilities by palpating the bumps of his/her skull.

Gall and Spurzheim's doctrine was strongly contested by Marie-Jean-Pierre Flourens (1794–1867), who stated on the basis of extensive ablation experiments that the crucial factor determining how an animal organism is affected by brain damage is the amount of tissue removed and not its location. In short: only the *size* and not the *site* of the lesion matters.

In spite of its empirical weakness, phrenology has had a positive and lasting influence on modern neuroscience because it stimulated the discussion of whether specific functions and qualities of the mind could be localized within the convolution of the brain. During the second half of the nineteenth century a number of clinical, anatomical and experimental landmark studies appeared that seemed to favour localization.

In 1865, Paul Broca (1824–1880) described several patients in whom circumscript lesions at the basis of the third convolution of the left frontal lobe had led to a *motor aphasia*, i.e. the inability to produce articulated speech [67]. A few years later, Carl Wernicke (1848–1905) reported on a second type of language disorder, namely the inability to comprehend speech while having speech production relatively unaffected. He associated this *sensory aphasia* with damage to the posterior part of the left hemisphere, in the region where the occipital, parietal and temporal regions meet [811]. It is important to note that Wernicke in his remarkable 1874 monograph, and in a later textbook [812], did not confine himself to attributing the different forms of aphasia to particular areas or centres. He designed circuit diagrams for reading, writing and speech and postulated that certain disorders of communication could result from damage to the connections between the various processing centres included in these diagrams.

During the last decade of the nineteenth century, and the first decade of the twentieth century, several authors, including Hugo Liepmann (1863–1925) and Joseph Jules Déjerine (1849–1917), extended and substantiated Wernicke's idea that the interruption of cortico-cortical fibre connections ("disconnection") may cause various kinds of behavioural impairment. Liepmann [410, 411] published several studies on apraxias, the higher-order disorders in the execution of skilled movements. He showed that several sites of damage, including the corpus callosum and the white matter deep to the supramarginal gyrus, could produce apraxia by disconnecting pathways joining selected sensory, motor and language areas. Déjerine [143, 144] analysed the various types of acquired reading disorders (alexias). He found, inter alia, that pure alexia, the inability to read with a preserved ability to write, may be caused by a lesion in the left occipital lobe, which disconnects the calcarine cortex from the angular gyrus.

In 1868, John Harlow (1819–1907) reported on the case of Phineas Gage, who exhibited profound changes in personality, social conduct and emotional stability, following an accident in which an explosion drove an iron bar – more than a meter in length and weighing about 6 kg – through the anterior part of his brain, destroying the prefrontal cortex bilaterally [278]. Harlow's case study was followed by several other publications, reporting profound character changes following lesions of the frontal lobes [187].

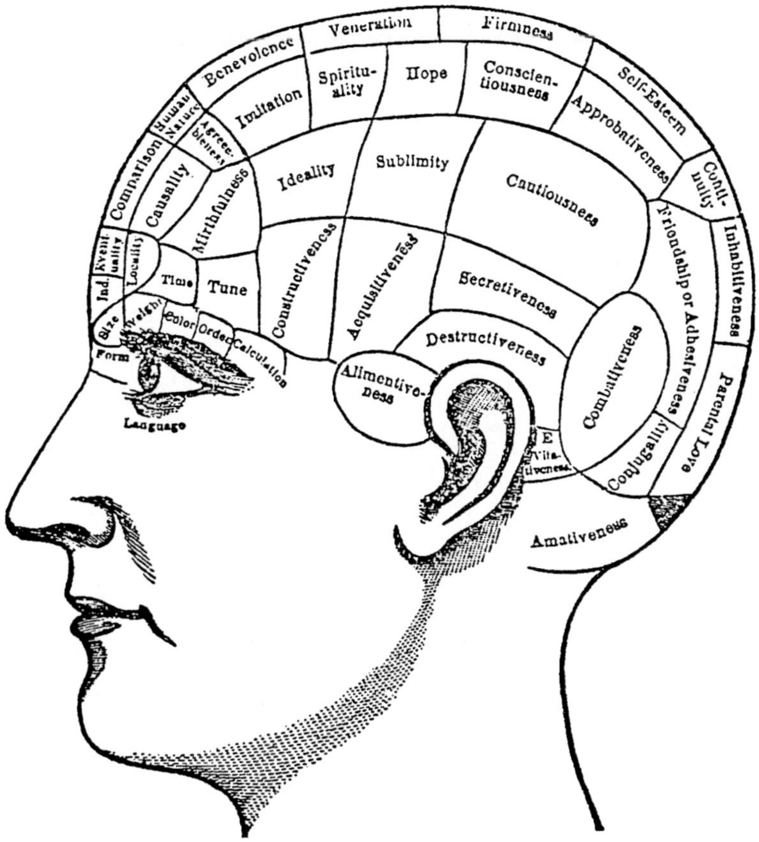

Fig. 15.16. Gall and Spurzheim's "phrenological head", showing 35 mental faculties, residing in specific organs of the cerebral cortex, projected upon the surface of the skull

The landmark experiments in 1870 of Gustav Theodor Fritsch (1838–1927) and Eduard Hitzig (1838–1907) led to the discovery of a motor zone in the cortex. They found that electrical stimulation of the anterior part of the cerebral cortex in dogs produces distinct motor reactions on the contralateral side and presented evidence suggesting a topographical arrangement of muscle groups in the area that was stimulated [212]. Their experiments were confirmed by David Ferrier (1843–1928) for various mammals, including monkeys [182]. Ferrier also induced lesions in the precentral cortex of monkeys. He noted that the lesions were associated with paralysis and weakness of muscles on the opposite side. The experiments of Fritsch and Hitzig [212] and Ferrier [182] confirmed the inferences on the existence of somatotopically arranged motor centres in the brain made by John Hughlings Jackson (1835–1911) from observations on patients suffering from a certain type of epileptic seizure, now commonly designated as Jacksonian fits. This kind of convulsion starts locally in a group of muscles and then spreads in an orderly fashion to other muscles [326]. For example, the contractions may begin in the lower part of the face and spread to the ipsilateral arm and, in turn, to the corresponding leg or vice versa.

During the last quarter of the nineteenth century, there was an extensive search for the localization of sensory centres in the cortex (see Finger [187] for a detailed account). As regards audition, Ferrier [182] and Ferrier and Yeo [183] reported on the effects of lesions of the superior temporal gyrus in monkeys. It was found that animals with bilateral lesions were totally irresponsive to hearing and that animals with unilateral lesions were deaf in the contralateral ear. In 1891, Charles K. Mills (1845–1931) presented two cases of totally deaf patients, one with a marked atrophy of both superior temporal convolutions, the other with severely damaged superior temporal gyri, due to successive strokes [471]. These findings were considered to be highly supportive of Ferrier's experimental findings with monkeys.

In 1881, Hermann Munk (1839–1912) reported that lesions of the occipital lobes in dogs and monkeys lead to blindness [499] and during the 1880s, several clinicians identified blindness with damage to the occipital cortex [256]. Salomon Henschen (1847–1930) collected over 160 clinical cases of blindness and hemianopsia after cortical lesions from the literature [288]. This vast amount of material led him to identify the "centre of vision" or "cortical retina" with the striate cortex [256].

The fact that lesions in the region of the central sulcus can cause somatosensory disturbances in human patients was recognized in 1884 by Moses Allen Starr (1854–1932). He noticed that in the majority of cases of such disturbances, the lesion was posterior to the central sulcus in the postcentral gyrus or the adjacent parietal region [705]. In the 1890s, Frederick Mott (1853–1926) [493] and Hermann Munk [500] reported that, in monkeys, lesions in the region of the central sulcus cause loss of touch and pressure sensations as well as paralysis. Neither Mott nor Munk produced or analysed lesions confined to the postcentral gyrus. Due to their studies, the prevailing view at the end of the nineteenth century was that the somatosensory and motor areas overlap [187].

During the period just discussed, several investigators rejected the idea of cortical localization. Prominent among them were Charles-Edouard Brown-Séquard (1817–1894) and Friedrich Goltz (1834–1902). Brown-Séquard [73] rejected Broca's idea that speech can be localized in the cortex. He presented evidence showing that lesions outside Broca's region can affect speech and he cited patients with damage to Broca's area who did not exhibit speech defects [187]. Brown-Séquard proposed that cells subserving speech and other functions are distributed over many parts of the brain. Goltz [248, 249] mistrusted both the results obtained from electrical stimulation of the cortex and the observations made on animals with local cortical ablations after short survival times. Goltz performed large and bilateral ablations of the cerebral cortex in dogs and kept these animals alive for a long time. He observed that in these animals permanent deficits were limited to higher psychic functions such as intelligence and memory. Like Flourens,

Goltz stated that it was the size and not the location of the lesion that determined the severity of its effects on such higher functions [257].

In spite of the objections just briefly outlined, it may be stated that at the beginning of the twentieth century, i.e. at the time that the architectonists began their mapping studies, the idea that at least certain functions can be localized to particular areas of the cerebral cortex prevailed. The monograph in which Campbell [85] presented the first cytoarchitectonic map of the human cerebral cortex was entitled: "Histological studies on the localization of cerebral function". Brodmann [70, 71] and von Economo [795] also emphasized that the cytoarchitectonic fields shown in their maps actually represent functional entities. Brodmann stated that the cerebral cortex should not be considered as a single organ, but rather a complex of organs, and that each of the different organs (i.e. the separate cytoarchitectonic fields) subserves a particular set of functions. He pointed out that many of these fields coincide with well-known physiological centres, specifying that his area 17 co-extends with the "clinical visual area of Henschen" and that his precentral region (area 4) corresponds to the "electromotor zone". However, Brodmann took issue with the idea of the phrenologists that complex mental functions such as memory, will or fantasy could be localized in circumscribed cortical areas. Rather, he believed that these functions result from the conjoint activities of a large number of areas distributed more or less widely over the cortical surface.

During the twentieth century, the study of the effects of brain lesions in human patients continued to play a prominent role in the localization-anti-localization debate. That the outcome of such studies could differ considerably is illustrated by the work of Constantin von Monakow (1853–1930) and Karl Kleist (1879–1960), two clinical neurologists who both had a vast amount of clinical material at their disposal.

Von Monakow [329, 797, 798] admitted that there was some sort of localization of elementary sensory and motor functions, but he emphasized that the answer to the question of what is actually localized remains obscure. He believed that "*Leistungen*", i.e. purposeful accomplishments, are not the product of single cortical areas, but rather involve the whole brain, which functions as a dynamic organizing entity. Von Monakow introduced the concept of *diaschizis of collaboration,* which implies that a breakdown in the widely correlated structures of the brain would result in a gross corresponding defect syndrome instead of the loss of function in the circumscribed area actually hit by the lesion. As regards central visual disturbances, he maintained that many of these are due to interruptions of pathways rather than to the lesion as such. We will see that many of the ideas of von Monakow reemerged and became dominant in the second half of the twentieth century.

The material of Kleist [373, 374] included, apart from numerous regular clinical cases, 300 persons who had sustained local brain injuries during World War I. He summarized his findings concerning the localization of functions in the cerebral cortex in a famous map, shown here in Fig. 15.17. It will be seen that Kleist subdivided the cortex according to Brodmann (see Fig. 15.8), and that he provided almost all of the cytoarchitectonic areas distinguished by the latter with a functional label. The overall functions of the primary sensory and motor areas, which were well known by the end of World War I, were correctly indicated. However, Kleist went far beyond that by attributing all sorts of higher cognitive and mental functions and faculties to many other areas. Thus, he associated temporal area 21 with acoustic awareness ("*akustische Aufmerksamkeit*"), prefrontal area 10 with motor skill ("*motorische Handlungsfolgen*") and orbitofrontal area 11 with personal and social ego ("*Selbst- und Gemeinschafts-Ich*"). It was particularly because of this detailed localization of psychic functions that many of his colleagues disposed of Kleist's map as "brain mythology" [119]. Uttal [744] recently characterized this map as a modern manifestation of phrenological thinking.

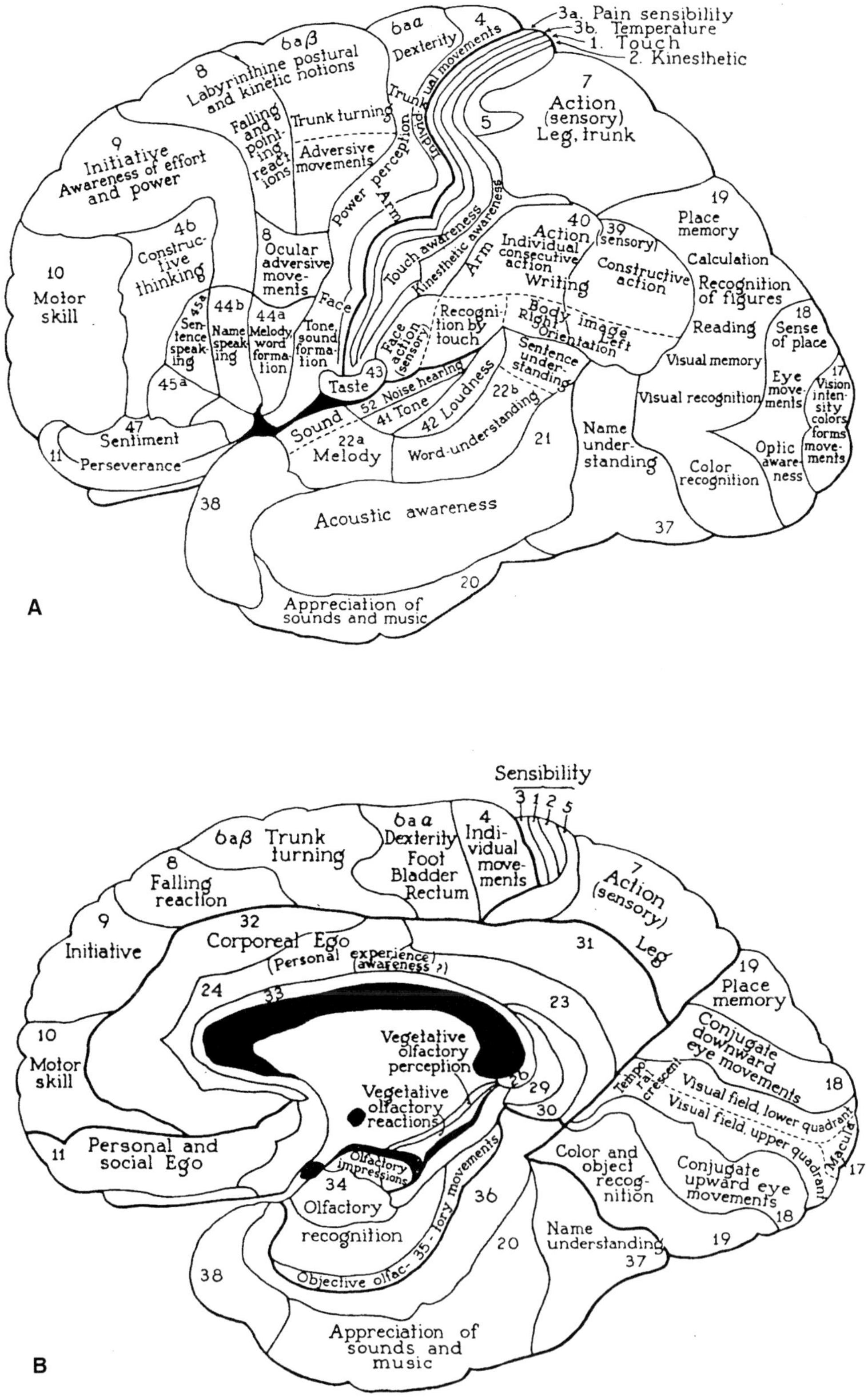

Fig. 15.17 A, B. Localization of functions in the cerebral cortex, according to Kleist [373]. The *numbers* indicate Brodmann's cytoarchitectonic areas

The most ardent anti-localizationist of the twentieth century was, beyond any doubt, Karl Lashley (1890–1954). He subjected animals, mostly rats, to complex maze-learning tests and analysed their performance before and after damaging various parts of the cerebral cortex [392, 393]. He found that their performance in the maze after operation was entirely independent of the location of the cortical lesion and only depended on its size. On the basis of his experimental results, Lashley formulated two general principles of brain organization, equipotentiality and mass action. With the term *equipotentiality* he designated the apparent capacity of any intact part of a functional area to carry out the functions lost by destruction of the whole. He considered it likely that this capacity holds only for associative parts of the cortex and for functions more complex than simple sensory or motor co-ordination. Lashley pointed out that the equipotentiality is not absolute but subject to the principle of *mass action,* whereby the efficiency of performance of a complex function may be reduced in proportion to the extent of brain injury within an equipotential area. Overall, Lashleys ideas about functional localization had much in common with those of Flourens and Goltz, but contrary to these authors, he accepted the general concepts of specialized sensory and motor areas in the cortex [187].

Lashley not only challenged the functional but also the structural subdivision of the cerebral cortex. He and his collaborator Clark [394] independently studied the brains of two spider monkeys *(Ateles geoffroyi),* both producing a cytoarchitectonic map. The two maps showed little agreement. They then compared their two specimens and found that most of the differences were not observer-dependent. They noticed that there was considerable variation in size and appearance of corresponding areas from brain to brain and that some areas could be seen in the one brain, but not in the other. Furthermore, they remained unable to locate the majority of the neocortical areas recognized by other investigators in this monkey and in the rhesus macaque. Lashley and Clark concluded that standard architectonic maps are of little value for the planning of experimental work, because the areal subdivisions are in large part anatomically meaningless, and because individual variation is too great to make the map significant for a single specimen. They emphasized that the preparation of maps representing functional units required that objective criteria and standards be developed and that other methods of confirmation be used, particularly the study of fibre connections.

Difficulties with the cytoarchitectonic method, somewhat similar to those of Lashley and Clark, were experienced by Bailey and von Bonin in their cytoarchitectonic studies of the cortex of the rhesus macaque [794], chimpanzee [29] and human [30]. With regard to the human cortex, they stated that vast areas are so closely similar in structure that any attempt at subdividing them would be unprofitable, if not impossible. It has already been mentioned that the total number of areas distinguished by Bailey and von Bonin [30] is much lower than that of other cytoarchitectonists (Table 15.2). Braak [64] criticized the results of these authors. He stated that they missed or neglected obvious cytoarchitectonic boundaries and suspected that they were content with superficial observations.

Several authors have attempted to localize functional areas in the cortex by electrical stimulation. Vogt and Vogt [784, 785, 787, 788] stimulated the cortex in monkeys *(Cercopithecus)* under general anaesthesia, and Foerster [197] did the same in patients who underwent brain operations under local anaesthesia. They found that, although the classical motor cortex on stimulation at low strength typically reacted with isolated contractions of small muscle groups, quite characteristic, more complex motor effects could be evoked from many other cortical areas. Thus, eye and head movements to the contralateral side could be elicited from areas 8 and 19, and stimulation of the parietal cortex produced synergic flexion of the contralateral arm and leg (Fig. 15.18C). Their procedure of identifying functionally equivalent areas involved three steps: (1) They plotted the results of the stimulation experiments in the *Cercopithecus* monkeys on a cytoarchitectonic

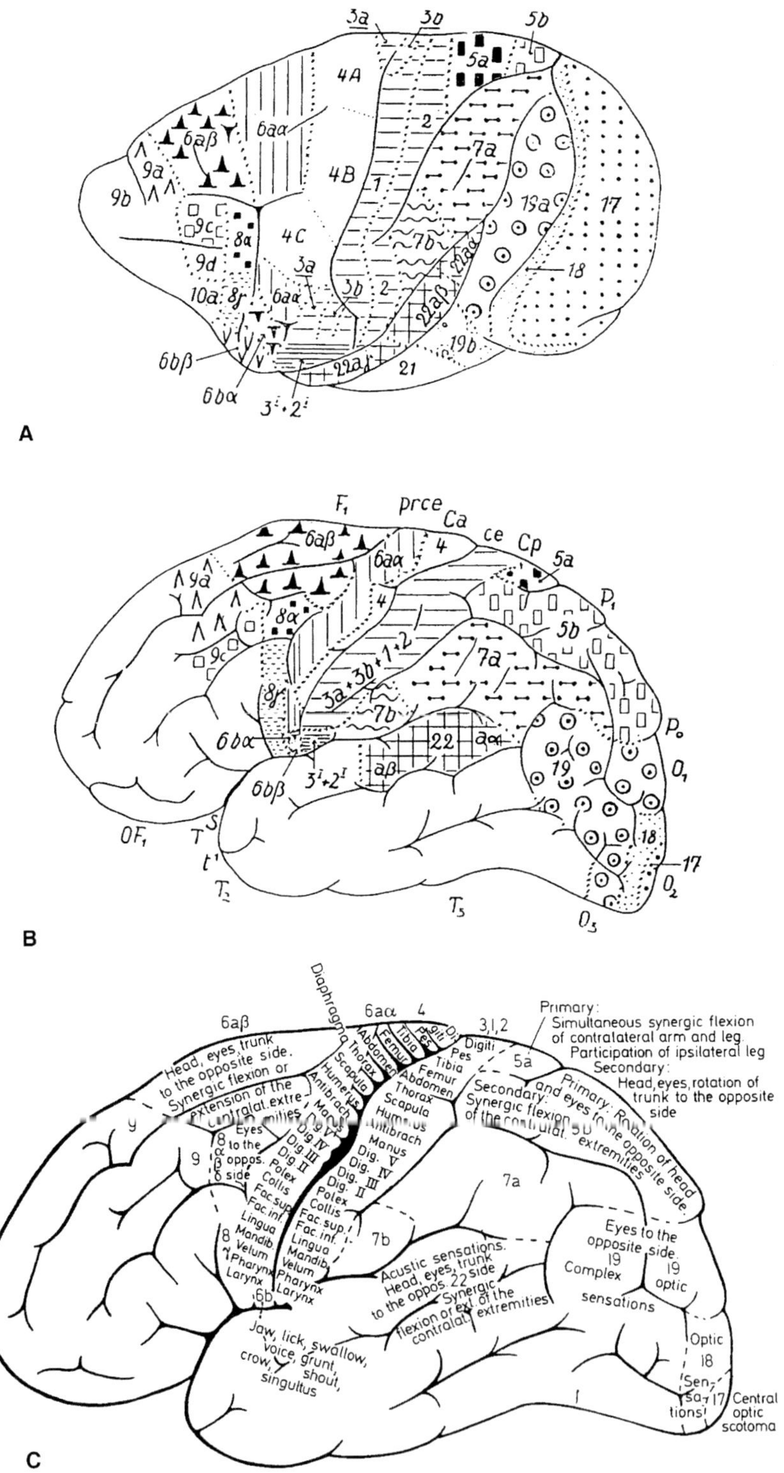

Fig. 15.18 A–C. Functional mapping by electrical stimulation. **A** The various effects of electrical stimulation of the cerebral cortex of the monkey *Cercopithecus*, plotted with *particular symbols* on a cytoarchitectonic map of the same species. **B** Comparable stimulation effects, elicited in humans, plotted with the *same symbols* on a cytoarchitectonic map of the human cortex. **C** Chart of motor effects elicited by electrical stimulation of the human cerebral cortex. **A** and **B** are reproduced from Vogt and Vogt [787]; **C** is taken from Foerster [197]

map of the same species, using different symbols for the various reaction patterns (Fig. 15.18 A). It appeared that almost all of the areas characterized by particular reactions exactly matched with a particular cytoarchitectonic area. (2) Next, they compared the cytoarchitecture of *Cercopithecus* monkeys with that of humans and were able to identify and to delineate in the latter equivalents of all the structurally and functionally characterized areas in the monkey brain. (3) Finally, they transferred the results of the stimulation experiments in humans (Fig. 15.18 C) to their map of the human brain, using the same symbols for the various reaction patterns as had been used in the monkey (Fig. 15.18 B). They concluded that many cortical areas in the monkey area are structurally and functionally equivalent (i.e. homologous and analogous) to particular areas in the human cortex. It is noteworthy that Vogt and Vogt [787] proposed to reserve the term *cortical field* ("*Rindenfeld*") for areas representing both structural and functional units.

Electrical stimulation experiments on the human cerebral cortex in awake patients undergoing brain surgery have also been performed by Penfield and his associates [544, 545, 548]. These experiments have extended our knowledge concerning the somatotopic organization of the primary somatosensory and primary motor areas considerably. They showed that, in both of these cortical areas, the different parts of the body are represented in territories of very different sizes. Thus, in the somatosensory cortex, the face and hand areas are very large, whereas the trunk and the proximal parts of limbs occupy relatively small territories (Fig. 15.19 A). In the motor cortex those parts of the body capable of performing the most differentiated and delicate movements have the largest representation (Fig. 15.19 B).

Woolsey and collaborators [830–833] have carefully mapped the somatosensory and motor cortical areas in a variety of mammals, using electrical stimulation and also recording evoked cortical potentials. These studies have given results concordant with those in humans.

Studies using single-cell recordings and electrical stimulation of cells have generated important findings with regard to the functional organization of the neocortex. Thus, it has been shown that most of the post-Rolandic cortex is devoted to processing specific aspects of a single sensory modality and that numerous separate areas for processing unimodal sensory information are present in this vast region. Monkeys have over 30 cortical areas for processing visual information, at least 15 for somatosensory information, and some 20 for auditory information [353]. Moreover, it has been shown that many of these areas are occupied by orderly representations or maps of receptor surfaces.

Recording the activity from single neurons under particular experimental or behavioural conditions has yielded important clues about the functional significance of the areas in which the elements studied are embedded. Thus, in the visual cortex, there are neurons that respond selectively to various directions of motion, specific colours, the orientation of line segments, the density of visual texture and many other visual features [313, 536]. In the extrastriate infratemporal visual area IT, many cells respond only or best to highly complex stimuli and some are selective for faces [258]. Cells in the anterior premotor cortex play a role in the preparation of movements [828], whereas the activity of neurons in certain parts of the prefrontal cortex is correlated with delayed-response performance and hence with working memory [215].

Returning to the systems level of functional localization in the neocortex, it should be mentioned that in 1965 Norman Geschwind (1926–1984) published a series of seminal papers on disconnection syndromes in animals and humans [227]. Referring to the studies of Wernicke, Liepmann and Déjerine (see above), he posited that many disorders of the higher functions of the nervous system, such as the aphasias, apraxias and agnosias, may be considered disturbances produced by anatomical disconnection of primary receptive and motor areas from one another. He emphasized the importance of long cortico-cortical connections emanating from the unimodal sensory association areas and pointed out that disconnection of cortical areas can be achieved by le-

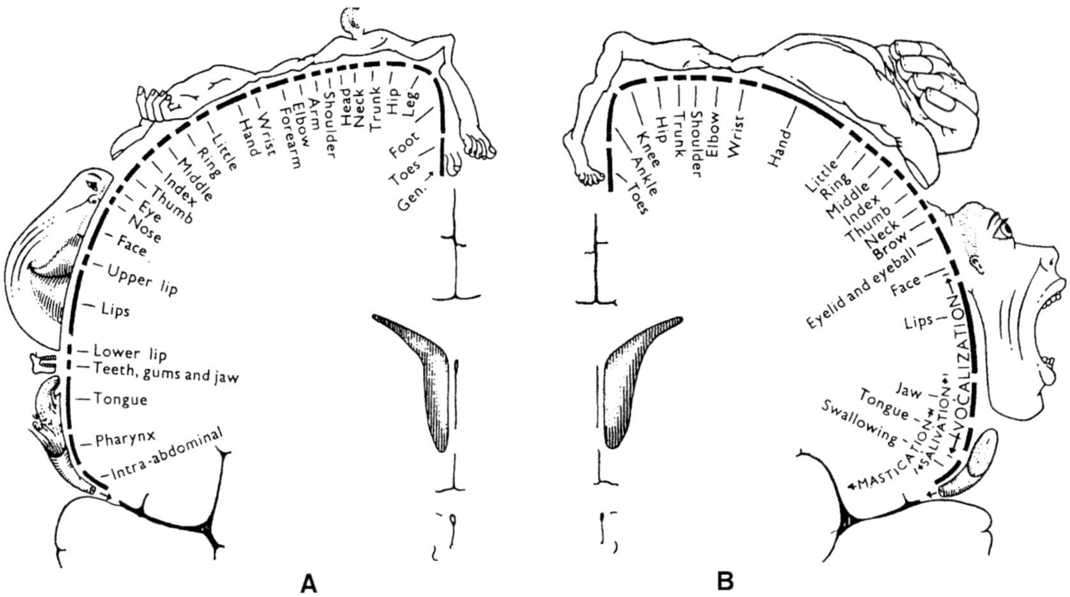

Fig. 15.19. Diagrams showing the relative proportion of the representation of the different parts of the body in (A) the postcentral primary somatosensory cortex, and (B) the precentral primary motor cortex. From Penfield and Rasmussen [548]

sions involving either white matter connections or by damage to association areas that constitute obligatory way stations between the primary sensory, motor, and limbic regions of the brain. Herewith Geschwind provided a general conceptual framework for higher brain disturbances, based on the neocortical wiring pattern (Fig. 15.20). Geschwind's papers became the manifesto of behavioural neurology and played an important role in the development of the current concept that networks of interconnected multiple specialized cortical territories form the morphological substrate of higher brain functions [3, 92, 464, 465].

It is noteworthy that, according to Catani and ffytche [92], not only disconnection, but also local hyperconnectivity or combinations of these two abnormalities may play a role in the pathogenesis of neurological and psychiatric disorders. Thus, it has been suggested that, in autism, a combination of fronto-frontal hyperconnectivity with frontal disconnection from other brain regions may be present [115].

Functional neuroimaging techniques such as positron emission tomography (PET) and functional magnetic resonance imaging (fMRI) play a prominent role in localizing functions and functional complexes in the neocortex. Maps constructed by these techniques depict neural activity as coloured spots or blobs, either on the surface of the brain or on brain sections in various directions. PET and fMRI monitor changes in blood flow or metabolism in specific regions of the brain while human subjects perform various sensory, motor or cognitive tasks. The blood flow signals and metabolic signals reflect changes in neuronal, particularly synaptic, activity [602].

Neuroimaging studies have not only confirmed and extended previous knowledge about localization of function in the primary sensory and motor cortical areas, but have also demonstrated a pervasive form of localization in a wide variety of cognitive tasks [83]. The activity of identifying and localizing cortical areas related to specific sensorimotor or behavioural events, using imaging techniques, is generally denoted as *brain mapping*. The resultant maps or charts (Fig. 15.21) show a striking resemblance to the early-nineteenth-century, phrenology-based maps (Fig. 15.16), as well as to the early-twentieth-century, clinical evidence-based map of Kleist (Fig. 15.17). All of these maps show detailed functions, allocated to distinct territories of the cortex. However, an important difference is that the phrenology maps and the Kleist map, contrary to neuroimaging maps, are typical "one-function–one-brain-area maps".

It stands to reason that, in order to become scientifically meaningful, the raw products of brain imaging, i.e. the activity-related blobs and patches, have to be placed in the context of the structural and functional architecture of the brain. As regards morphological architecture, many authors confine themselves to indicating the location of the spots of activity detected in terms of the gross anatomy of the cortex, for instance: activity in "the inferior parietal lobule" or in "the posterior part of the superior frontal sulcus". Many others use Brodmann's map (Fig. 15.8), as extrapolated by Talairach and Tournoux [727] onto a standardized brain atlas, as a frame of reference. Given the considerable interindividual variability of the human brain at both the gross anatomical and the structural levels, neither of these two approaches is entirely satisfactory [66]. We will come back to this issue in a later section.

Insight into the functional architecture of the neocortex may be gained by considering the following features of neuroimaging data.

1. In most cognitive tasks, two or more cortical areas are activated [83]. The activated areas may be considered as nodal points in the networks underlying the different cognitive processes being investigated.
2. Reviews of functional neuroimaging results across different cognitive domains clearly show that cortical regions, such as the prefrontal cortex and the parietal area 7, are engaged in a wide variety of cognitive demands. The most parsimonious explanation of this kind of activation is that they reflect cognitive processes that are tapped by tasks in different domains [83]. However, it is also conceivable that more refined analyses will lead to a further functional parcellation of the areas involved.

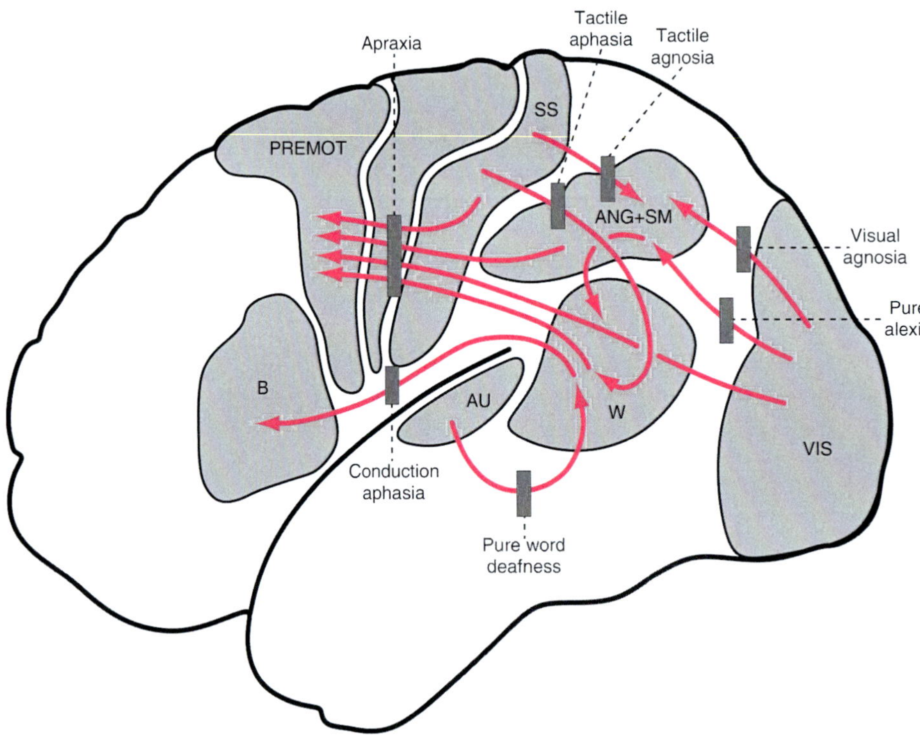

Fig. 15.20. Geschwinds' disconnection syndromes. Cortico-cortical pathways in the human brain and the disconnection syndromes to which their interruption may lead. Geschwind and his predecessors Liepmann and Déjerine also indicated that damage to sectors of the corpus callosum may also lead to typical disconnection syndromes. These have been omitted from this scheme (based on [92]). *ANG+ SM*, angular and supramarginal gyri; *AU*, auditory cortex; *B*, Broca's speech area; *PREMOT*, premotor cortex; *SS*, somatosensory cortex; *VIS*, visual cortex; *W*, Wernicke's speech area

---

a Attention [684]
b Calculation [684]
c Activation arising from a working memory task [66]
d Processing of incorrect arithmetic equations [461]
e Maternal love [39]
f Performance of a verbal working memory task [583]
g Reading aloud irregular words [289]
h Meditative state [418]
j Finger opposition task [713]
k Visual-spatial orienting [450]
l Coherence processes in language comprehension [184]
m Executive function in working memory [523]
n Embarrassment [724]
o Perception of visual motion: V5 [808]
p Written sentence comprehension [109]
q Processing of color patterns [38]
r Verb comprehension [259]
s Judgement of the grammaticality of sentences [807]
t Selective auditory attention [413]
u Evaluation of unpleasant, arousing words [435]
v Cross-modal sensory processing [396]
w Error processing [460]
x Categorization of written object descriptions [260]
y Pursuit eye movements [50]
z Painful visceral sensation [662]
a[1] Saccades [684]
b[1] Visual stimulation [506]

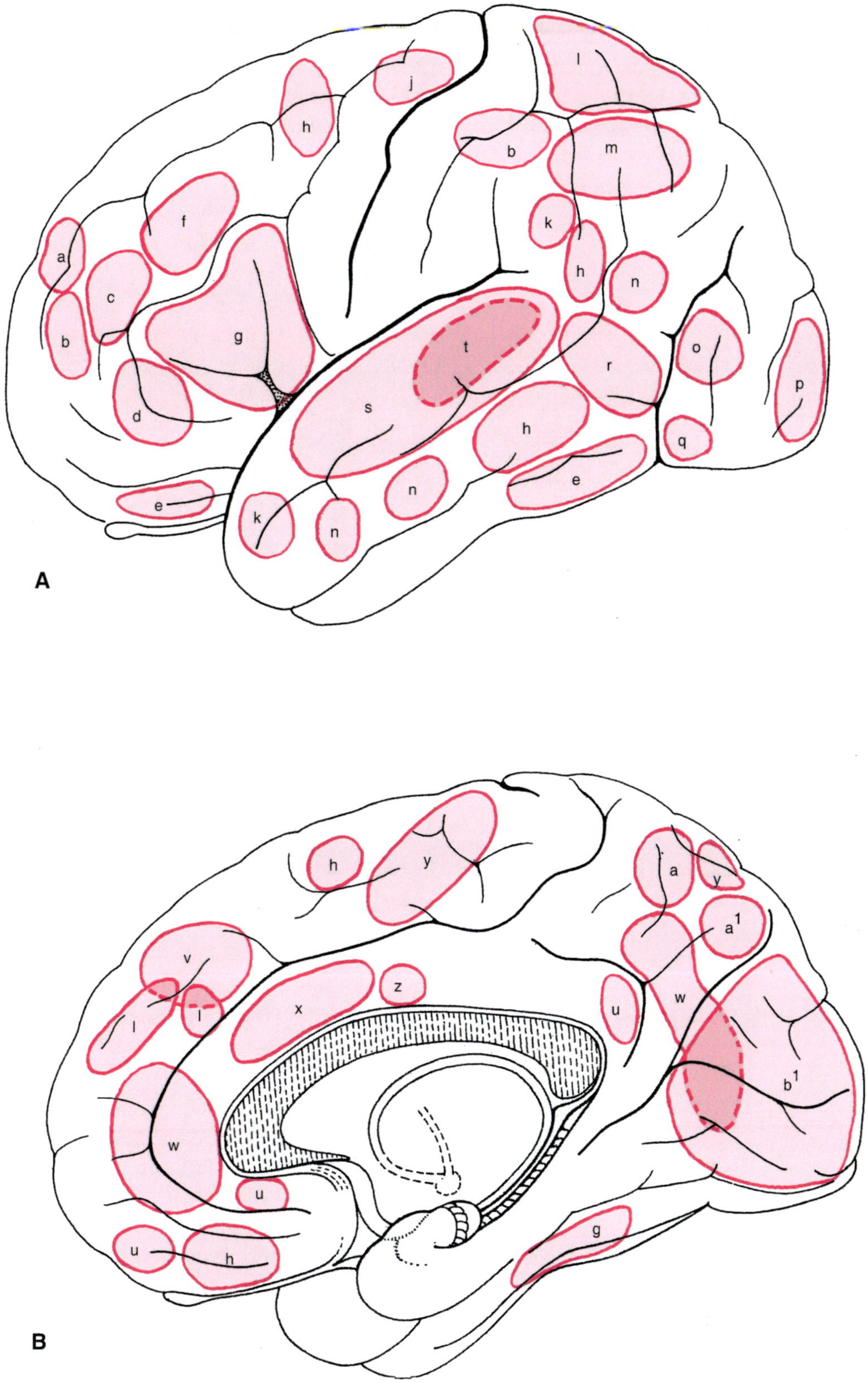

Fig. 15.21. Activation loci, found in a number of functional imaging experiments, projected upon the lateral (**A**) and medial (**B**) surface of the cerebral hemisphere. For significance of letters, see opposite pape

3. Combining neuroimaging data with the results of single-cell recordings in non-human primates may help determine the functional specialization of a given area in more detail.
4. The functional role played by any cortical area is defined largely by its connections. Thus, it is important to complement the results of neuroimaging experiments with studies aimed at determining the afferent and efferent connections by which the areas (spots of activity) detected are embedded in the cortical network.

Although techniques have recently become available to globally trace the tracts of the living human brain [40, 108, 335, 488, 588], information concerning the connectivity of particular areas in the human cortex still must be derived and extrapolated from experimental studies in monkeys. Nevertheless, we have at present a fairly accurate picture of the connectional frameworks involved in a large number of specific cognitive and behavioural operations, such as spatial orientation, object recognition, language [464, 465], attention [553] and decision forming [522]. The neural circuitry related to many of these operations is not confined to the neocortex, but also includes one or several subcortical centres. The cerebellum is consistently activated in a variety of cognitive processes, but the nature of its role in these processes is still obscure [83].

Structural and Functional Subdivision: Overview

1. The structure of the neocortex is not homogeneous throughout.
2. Many authors have attempted to subdivide the neocortex or parts thereof into structural entities, using cytoarchitectonic and/or myeloarchitectonic criteria. Most of these authors considered the resultant architectonic areas or fields not merely as structural entities, but also as functional units. We have seen that, on the basis of clinical evidence, Kleist provided nearly all of the areas delineated by Brodmann (Fig. 15.8) with a detailed functional label (Fig. 15.17), and that the Vogts sought to identify structural-functional units in the neocortex of monkey and humans, by combining the results of cytoarchitectonic analyses with those of electrical stimulation experiments (Fig. 15.18).
3. The primary sensory and motor areas are structurally and functionally distinct. They can be easily identified as heterotypical agranular and granular areas (types 1 and 5 of von Economo: Fig. 15.10), and as early maturing myelogenetic projection areas (Fig. 15.14). Hodologically, the primary sensory areas are defined by receiving the thalamocortical endsegments of the great sensory projections, whereas the primary motor cortex is characterized by its massive corticobulbar and corticospinal projections.
4. The remainder of the neocortex is formed by Flechsig's late-maturing association cortex. In the vast region occupied by this cortex, the structural differences between the different areas are rather subtle, which explains why the results of the various cytoarchitectonists are less concordant here than in the primary areas. All of the areas forming part of the association cortex occupy, as homotypical areas, an intermediate position in von Economo's classification system (Fig. 15.10).
5. Experimental studies on the cortico-cortical connections are and have been of paramount importance for both the morphological and functional interpretation of the association cortex. However, because such studies cannot be performed on humans, they have to be carried out on laboratory animals, preferably monkeys. Hence, in practice, the acquisition of data on the fibre connections of a given area in the human cortex involves the following four steps: (1) identifying and locating an area in the human cortex; (2) identifying and locating the homologue of that human area in the cortex of the monkey; (3) establishing the fibre connections of that area in the monkey cortex by using experimental tract-tracing techniques, and (4) extrapolating the results of these experiments to the human brain. A

serious limitation of this indirect method is that many areas in the human neocortex have no obvious homologue in the monkey (Figs. 15.8, 15.12) [120].

Experimental studies in the monkey have shown that (a) the primary somatosensory, visual and auditory cortical areas project via short connections to adjacent unimodal association areas; (b) these unimodal sensory association areas project in turn to two strips of heteromodal sensory association cortex, situated in the lateral and medial parts of the parietal and temporal lobes; (c) the unimodal and heteromodal sensory association areas project massively to the prefrontal cortex which, on this account, should be considered as a higher association cortex, and (e) sequences of short connections successively link the various prefrontal areas, the premotor area and the primary motor area (Fig. 15.15). It is important to note that the boundaries between the various association areas thus defined do not show a close correspondence to those of the cytoarchitectonic fields, as delineated by Brodmann and others.

6. A combination of connectional data with the results of microphysiological studies has shown that the wiring of the neocortex may be characterized as a distributed hierarchical network that contains numerous intertwined processing streams.
7. Unit activity recordings, the effect of lesions and particularly the results of neuroimaging studies have firmly established functional segregation as a principle of neocortical organization, not only in the primary cortices, but also in the association regions (Fig. 15.21). This having been said, it should immediately be added that the precise morphological identification of the functionally segregated domains is a difficult problem and much current research is devoted to this (see below).
8. As regards the functional organization of the neocortex, there are three different concepts or paradigms: localizationism, anti-localizationism or holism and (dis)connectionism.

Localizationism claims that each region of the neocortex represents an independent organ, dedicated to a complete and distinct function. This idea was first put forward by the phrenologists Gall and Spurzheim (Fig. 15.16), culminated in the work of Kleist (Fig. 15.17) and is still tangible in many current mapping studies aimed exclusively at localizing functions. Such functional cartography with the aid of neuroimaging techniques has been characterized by some [508, 601, 744, 745] as "neophrenology".

Anti-localizationism or holism holds that higher cognitive and mental functions are distributed homogeneously throughout the cortex and require the integrated activity of the entire cortex. The degree of cognitive or behavioural impairment following a cortical lesion is simply proportional to the amount of tissue destroyed, with localization being entirely irrelevant. The most prominent anti-localizationists were Flourens, in the early nineteenth century, Goltz, in the late nineteenth century, and Lashley, in the twentieth century.

The presently prevailing concept of *(dis)connectionism* has its roots in the clinical work of Wernicke, Liepmann, Déjerine and Geschwind and has been further developed by Mesulam [463–465], Van Essen [180, 770], Friston [201, 211] and many others. It states that, although functional localization and functional specialization are important principles, they do not offer a complete or sufficient explanation of cortical organization. The physiological changes elicited in the cortex by particular cognitive processes or behaviours should be explained in terms of distributed patterns of changing neural activity in networks of interconnected, functionally specialized areas. Cognitive and mental abilities are not the product of single and separate cortical regions, but rather result from the functional integration of the elementary processing operations occurring in a smaller or larger number of functional areas. The neural mechanism of many neurological disorders can be understood as dysfunction within specific neural circuits.

Structural and Functional Localization in the Neocortex: Current Research and Perspectives

Current research on the subdivision of the neocortex concentrates on the following two interrelated questions: (1) what are the true and fundamental neocortical units, and (2) how do we determine and visualize the location and extent of these units.

As regards the first question, we have seen that Campbell, Brodmann, the Vogts and von Economo held that the cytoarchitectonic and myeloarchitectonic fields they delineated represent fundamental structural as well as functional units. Although there is much respect and appreciation for the work of the classical architectonists, it is generally felt that their analyses show two serious methodological shortcomings and, hence, do not provide a reliable baseline for further studies on the organization of the cortex. These shortcomings are: (1) the classical parcellations are based on the study of microscopic sections stained according to a single technique, either the Nissl technique for neuronal perikarya or the Weigert technique for myelinated fibres, and (2) the areal boundaries were established using purely visual inspection of histological sections and were, hence, influenced by varying observer-dependent conditions, such as individual abilities in pattern recognition [660]. In current architectonic studies these methodological shortcomings are avoided by using several complementary histological techniques [747] and/or applying observer-independent procedures [659, 660]. Thus, Carmichael and Price [86] presented an architectonic division of the orbital and medial prefrontal cortex (OMPFC) of the macaque monkey based on nine different cytoarchitectonic, myeloarchitectonic and immunohistochemical stains. These stains included Nissl, myelin and acetylcholinesterase, and immunohistochemical stains for parvalbumin (PV), calbindin (CB) and a membrane-bound glycoprotein. A cortical area was defined as distinct if it was differentiated in at least three different stains. This analysis subdivided many of the areas originally described by Brodmann [70] and Walker [802] and resulted in the identification of 22 different areas in the OMPFC of the macaque monkey. A similar architectonic analysis of the human OMPFC, carried out by Öngür et al. [518], revealed that all of the areas recognized in the macaque monkey have counterparts in humans. On the basis of these results and of studies on the fibre connections of the various areas in the monkey [88, 89], Öngür et al. concluded that each of the cortical areas distinguished is a module with specific input-output relations and a unique role in information processing. They considered it likely that much of the cortex consists of such *discrete structural and functional modules*.

Observer-independent procedures in the structural parcellation of the cortex include computer-assisted quantification of cortical morphology in terms of (a) cell density, neuron size and laminar thickness [605, 659, 660]; (b) packing density and laminar distribution of myelinated fibres [12]; and (c) receptor architecture, i.e. the density and laminar distribution of different receptor types as visualized by autoradiography [230, 851]. Areal boundaries are fixed at locations where these features change significantly.

Roland and Zilles [626] expect that combining the results of quantitative morphological studies that apply observer-independent procedures with data derived from functional neuroimaging studies will lead to the detection of *functional cortical fields*. They advance the hypothesis that the organization of the cortex is based on such functional fields, each occupying a certain, relatively large territory of the cortex, and postulate that all neurons and synapses within these fields perform a co-operative computation.

A preliminary reconnaissance of the current literature on the neocortex, aimed at detecting entities comparable to, or at least foreshadowing the *modules* of Öngür et al. [518] or the *functional fields* of Roland and Zilles [626], yields the following results:

1. Just as the classical architectonists indicated, the human neocortex can be subdivided into a number of juxtaposed structural fields.
2. Recent investigations on the primary sensory and motor areas generally confirm the results of the classical architectonists [7,

231, 598–600], although two subfields [230] have been detected within the primary motor cortex and three subfields [489] within the primary auditory cortex.

3. Detailed quantitative analyses have confirmed the presence of Brodmann's frontal areas 9 and 46 [604, 605] and 44 and 45 [6, 8] as distinct architectonic entities.
4. In many regions of the neocortex, including the orbitofrontal cortex [518, 751], the parietal cortex [849], the cingulate region [783], the retrosplenial region [376] and the extrastriate visual region [352, 408, 850], the number of architectural areas detected is larger, and sometimes considerably larger than in the classical studies.
5. Taking these data together, we estimate that about 150 juxtaposed structural and potentially functional entities are present in the human neocortex.

Comparative architectonic studies [516, 518, 577, 580, 581, 605] have shown that many of the delineable structural entities in the human neocortex have distinct counterparts (homologues) in the neocortex of the macaque monkey. The results of experimental studies on the afferent and efferent connections of these comparable areas in the cortex of the monkey can be transferred to the human brain and may provide important clues as to the functional significance of the entities involved. Using data collected in the macaque connectivity database CoCoMac [710], Passingham et al. [541] demonstrated that each cortical area has a unique pattern of cortico-cortical connections, a defining *"connectional fingerprint"*. They suggested that the connectional fingerprints underlie the cell-firing differences observed between areas during different tasks or task events. Passingham et al. [541] designate this pattern as a *"functional fingerprint"*. They indicate that functional brain imaging will be a useful tool for detecting such functional fingerprints because it enables us to compare activations across many cortical areas and across a wide range of tasks.

The second central questions in current research on the neocortex – how do we determine and visualize the location and extent of structural and functional units – has to do with two further shortcomings, or rather limitations, of the work of the classical architectonists. Firstly, these studies were all based on a single or a few brains and, hence, interindividual variations in the position and extent of the various architectonic areas were completely neglected. Secondly, the classical architectonists only analysed the superficially exposed parts of the cortex, leaving the sulcal cortices largely unexplored. This is a serious limitation indeed because, as already mentioned, in humans almost two thirds of the cortex are hidden away in the depths of the sulci.

As regards the first limitation, the pioneering studies of Filimonov [186] on the visual areas 17, 18 and 19, and of Kononova [383] on the frontal areas 44 and 45 have brought to light that these areas show a considerable intersubject variability (Fig. 15.22). Their observations have been fully confirmed by later studies on the same areas [6–8]. Similar variability has also been reported for many other cortical areas, including the primary motor [230, 598, 599] and somatosensory cortices [231, 232], the primary auditory cortex [489, 600], the prefrontal areas 9 and 46 [604, 605] and the orbitofrontal cortex (Fig. 15.23) [751].

Interindividual variability appears to be a general feature of neocortical architectonic areas and it should be added that this microstructural variation is superimposed upon the also considerable macrostructural variation pertaining to the overall size and shape of the hemispheres as well as to the sulcal and gyral pattern. It will be clear that this variability seriously hampers the establishment of spatial relationships between structural and functional domains.

In human brain imaging studies, it is common practice to determine the location of the activation foci detected by transferring these loci to the three-dimensional version of Brodmann's chart, incorporated in the stereotaxic reference system of Talairach and Tournoux [727]. Because the considerable structural variability of the cortex is completely neglected in this procedure, it is apt to lead to erroneous conclusions [748].

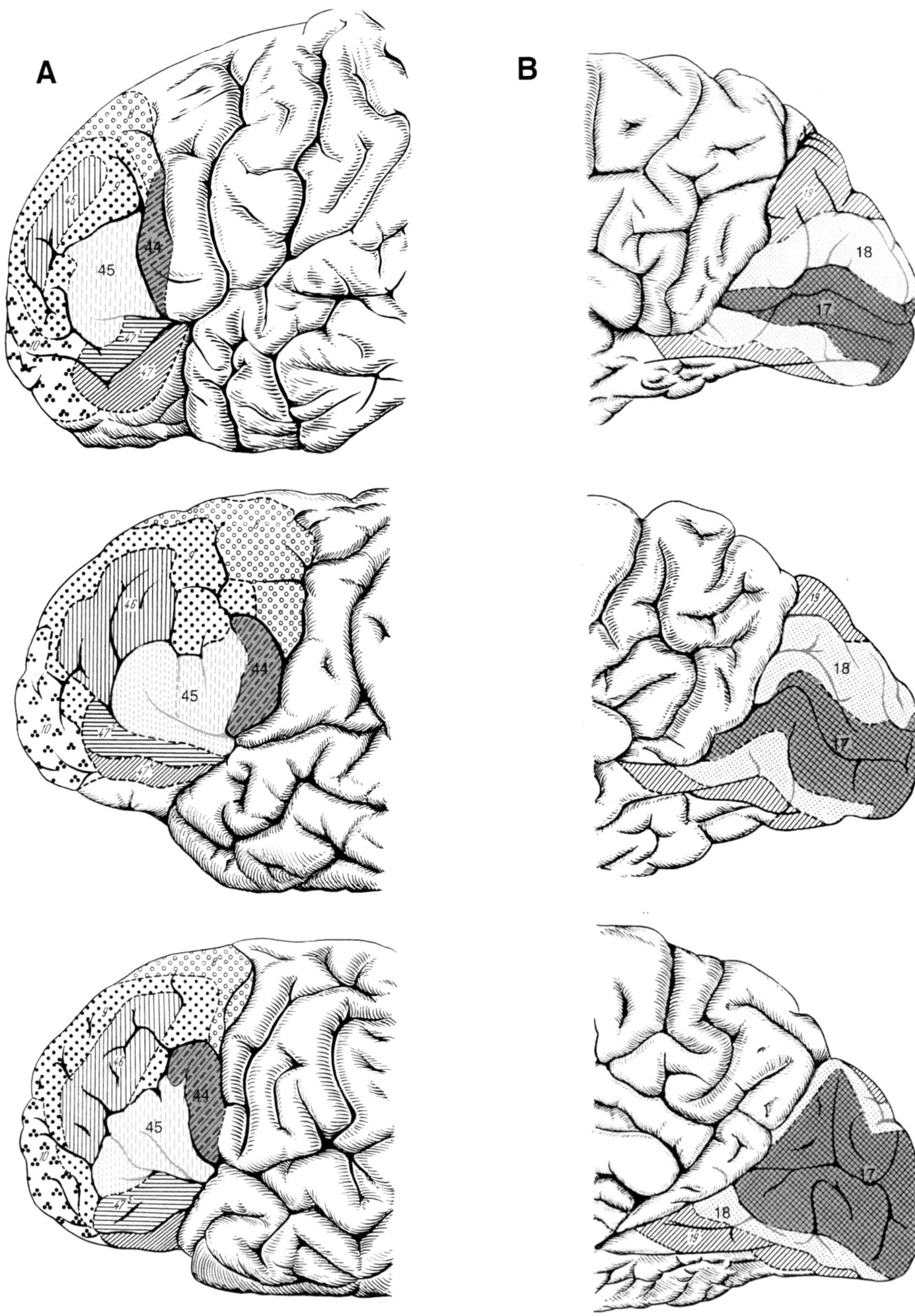

Fig. 15.22 A, B. Interindividual variability of cytoarchitectonic areas in normal human adults. **A** Prefrontal cortical areas, among them areas *44* and *45*, which together form Broca's motor speech region; **B** the visual cortical areas *17*, *18* and *19*. Reproduced from [748]. References cited: Kononova [383], Filimonov [186]

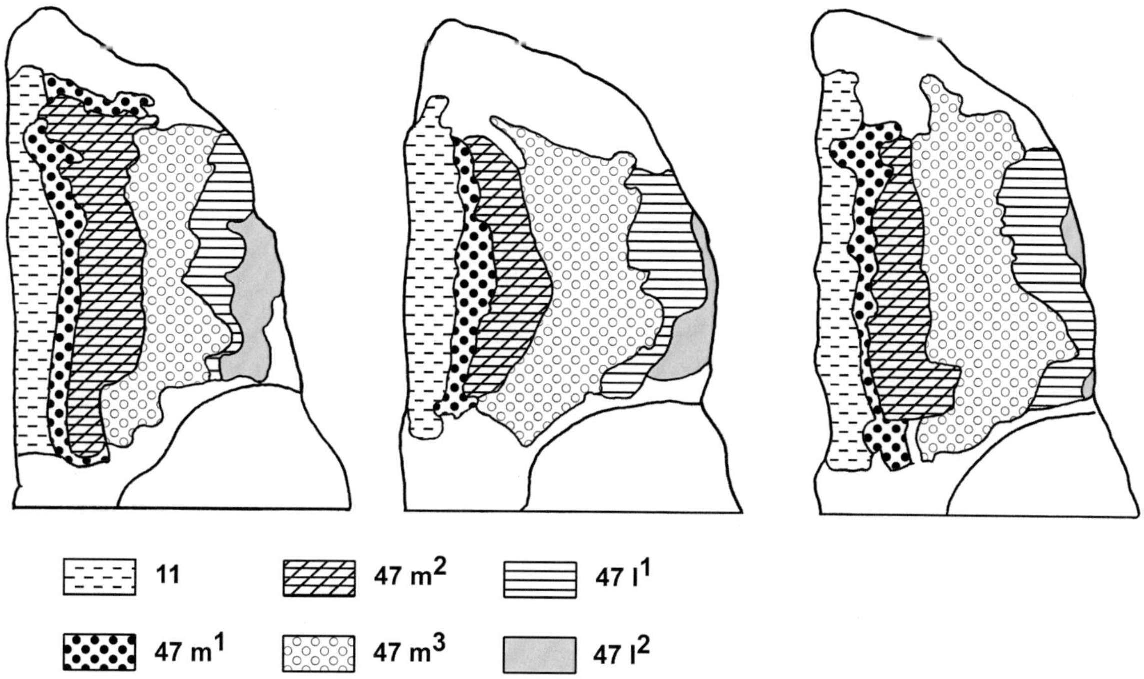

Fig. 15.23. Cytoarchitectonic (sub)areas covering the orbital surface of the frontal lobe in three different brains. Note the considerable interindividual variability (cf. Fig. 15.11) (based on Uylings et al. [751])

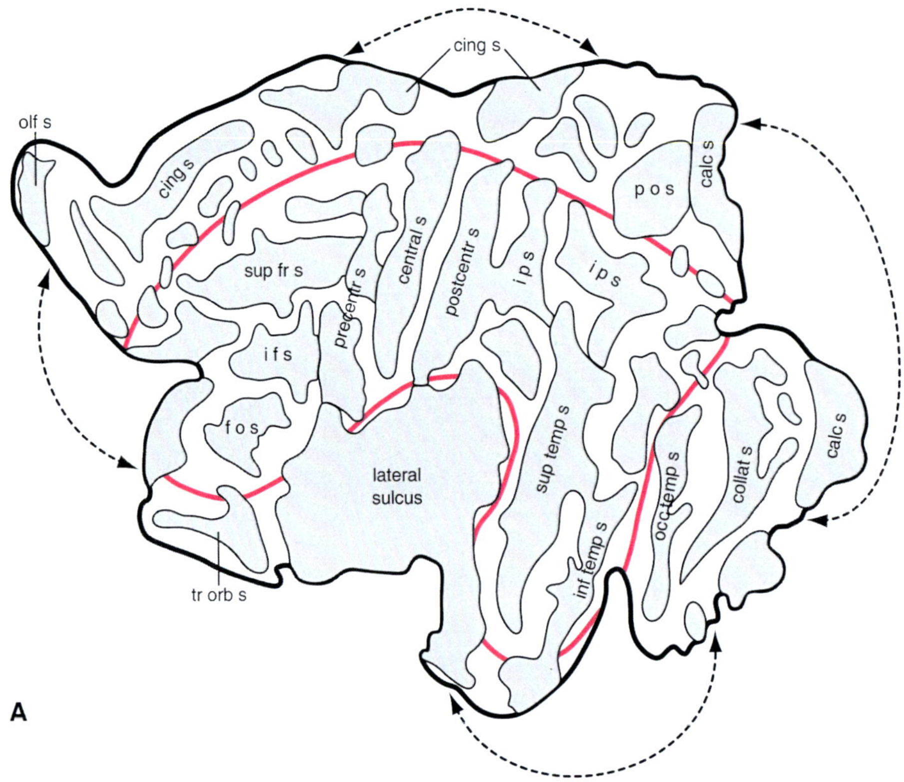
cing s
olf s
cing s
calc s
p o s
sup fr s
precentr s
central s
postcentr s
i p s
i p s
i f s
f o s
lateral sulcus
sup temp s
inf temp s
occ-temp s
collat s
calc s
tr orb s

A

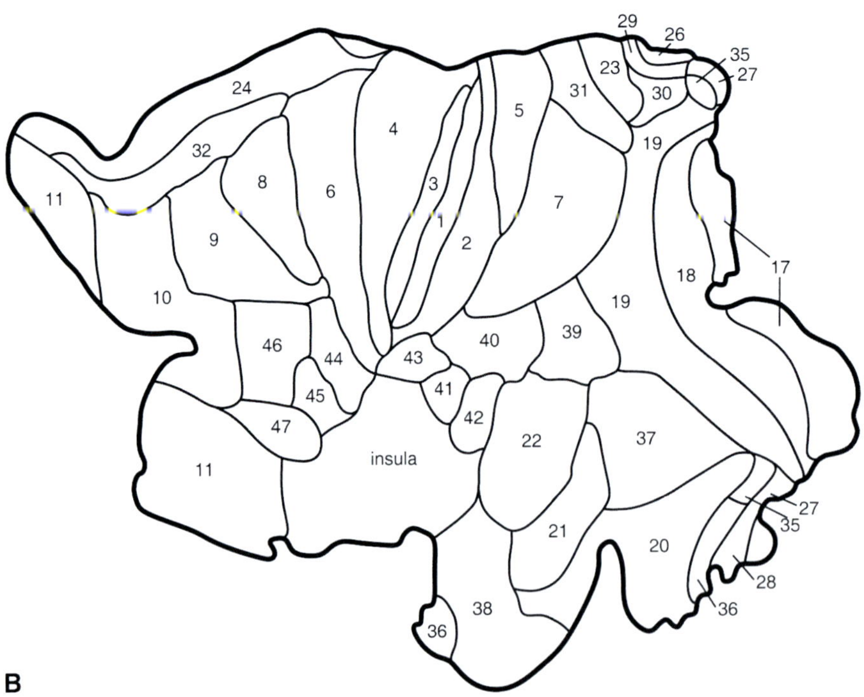
29
26
35
27
23
30
31
24
5
4
19
32
3
11
8
6
1
7
9
2
17
18
10
19
46
44
43
40
39
45
41
42
47
22
37
insula
11
27
35
21
20
28
38
36
36

B

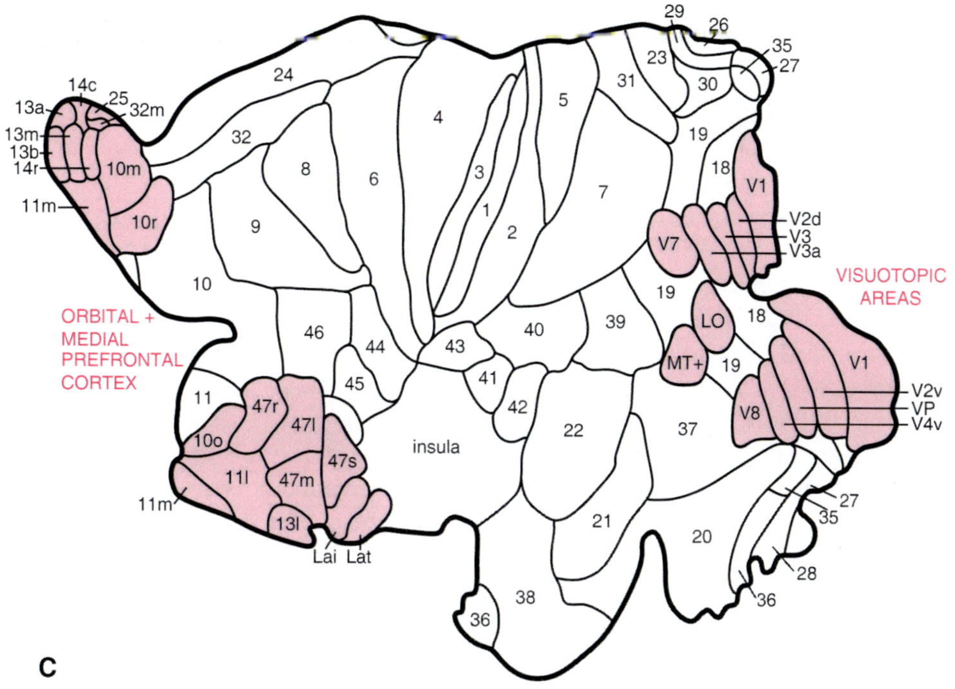

Fig. 15.24 A–C. Flat maps of the surface of the left human hemisphere, according to Van Essen [765]. In order to avoid gross distortions, several cuts have been made in the regions corresponding to the medial surface of the hemisphere. The *dashed, arrow-headed curves* in (**A**) connect corresponding points. In (**A**) the sulcal surfaces are *shaded* and the principal sulci are *labelled.* In order to facilitate orientation, outlines of the structures visible in a lateral view of the hemisphere are shown in *red.* In (**B**) The cytoarchitectonic fields of Brodmann (see Fig. 15.8) have been extended over the sulcal parts of the neocortex. In (**C**) the results of a detailed architectonic study of the orbital and medial prefrontal cortex [518] and of a number of functional imaging studies on the localizations of visuotopic areas [275, 590, 736, 764] have been superimposed on Brodmann's subdivision. It is noteworthy that the 30 (sub)areas delimited, together occupy 20% of the neocortical surface. Extrapolating from these data, it may be estimated that the neocortex is composed of some 150 (sub)areas or units. *calc s*, calcarine sulcus; *central s*, central sulcus; *cing s*, cingulate sulcus; *collat s*, collateral sulcus; *fos*, fronto-orbital sulcus; *ifs*, inferior frontal sulcus; *inf temp s*, inferior temporal sulcus; *ips*, intraparietal sulcus; *occ temp s*, occipitotemporal sulcus; *olf s*, olfactory sulcus; *pos*, parieto-occipital sulcus; *postcentr s*, postcentral sulcus; *precentr s*, precentral sulcus; *sup fr s*, superior frontal sulcus; *sup temp s*, superior temporal sulcus; *tr orb s*, transverse orbital sulcus

In order to cope with the problem of intersubject variability of brain structures, computer-based, three-dimensional probability reference systems have been developed that incorporate data on the gross anatomy of a group of brains and provide an adequate probabilistic framework for both microstructural and functional studies [452–454, 624, 625, 627]. Spatial normalization procedures render it possible to fit structural data derived from individual brains (or groups of brains) into this reference system and the same holds true for the results of neuroimaging studies [66]. It is necessary to superimpose structural and functional data within the reference system so that structural-functional relationships can be rigorously tested. This procedure provides the basis for making sound statistical statements concerning the probability that a particular activity focus lies within a certain architectonic unit.

It will be appreciated that the generation of probability maps of all of the, say 150, structural units of the human neocortex is a formidable task. Two groups of neuroanatomists, that of Zilles and Amunts [6–8, 97, 169, 230–232, 489, 490, 599, 600], and that of Uylings [747, 748, 750, 751] are currently working on this mega-project, which will take at least 12 years to complete. Within the framework of this project, the neocortex of 10–15 normal human subjects will be analysed, using various staining techniques and observer-independent procedures. If the final results of this project matched with the results of large series of carefully carried out functional imaging studies would yield a (probabilistic) functional profile for all of the architectural units delineated, this neophrenological achievement would still represent no more than a first step in the analysis of the structural-functional relationships in the neocortex. The next step would be to *systematically* analyse how these various units or centres are integrated in the circuits and processing streams subserving higher cognitive and mental functions.

The final problem left by the classical architectonists, i.e. the representation and charting of the vast neocortical areas hidden in the sulci, has been successfully tackled by Van Essen and colleagues [160, 180, 762–764, 766, 767]. These authors have worked out computerized procedures by which the entire surface of a cerebral hemisphere can be unfolded and flattened into a two-dimensional reconstruction or flat map. Three such flat maps are shown in Fig. 15.24. In the first (Fig. 15.24 A), the sulcal surfaces are shaded and selected sulci are labelled. In the second (Fig. 15.24 B), the cytoarchitectonic fields, which in Brodmann's chart (Fig. 15.8) cover the superficially exposed parts of the hemisphere, have been extended over the sulcal domains. This map should be considered as provisional and tentative because: (1) the borders of the sulcal parts of the fields have not been determined, but rather estimated, and (2) data concerning the interindividual variability of the fields have not been included. In the third flat map (Fig. 15.24 C), the results of a detailed architectonic study of the orbital and medial prefrontal cortex [268, 518] and of a number of fMRI studies on the localization of visuotopic areas [275, 590, 736] have been superimposed on Brodmann's parcellation. Flat maps allow the entire neocortical surface to be seen in a single view; in the present authors' opinion, therefore, they are very well suited for visualizing the results of such studies. If the data concerning the localization and variability of the approximately 150 neocortical architectonic units, once established (see above), would be transferred to a surface-based atlas, this mega-operation would yield a very useful probabilistic framework for the topical specification of fMRI results and other functional data. A probabilistic map of the human visual cortex is already available [771].

Neocortical Afferents

In the neocortex, the fibres of all extrinsic afferent systems follow a radial course. Golgi studies and experimental investigations using anterograde degeneration and tracer techniques have shown that most of these afferent systems, after having entered the cortex from the deep white matter, distribute their terminal branches preferably in one or more layers.

Lorente de Nó [414, 416] provided a detailed description of thalamic afferent fibres in the rodent neocortex. His Golgi material displayed two different laminar distributions of terminal arborizations: the "specific" distribution, which is densely aggregated in lamina IV, with or without extension to lamina III (a, b in figure 15.25 A), and the "unspecific" distribution, which is sparsely distributed throughout all cortical layers, but appears predominantly in lamina VI (c in Fig. 15.25 A). The "specific" afferents were considered to originate from "specific" thalamic sensory relay nuclei, such as the medial and lateral geniculate bodies and the ventral posterior nucleus, and to terminate in specific sensory cortical fields. The "unspecific" afferents, which were observed to extend over multiple cortical fields, were believed to originate from other, as yet undetermined thalamic sources. Frost and Caviness [213] studied the intracortical distributions of the projections of a number of different thalamic loci to the neocortex in the mouse using an axon degeneration technique. They found that in this animal virtually the entire tangential extent of the neocortex receives a projection from the thalamus. The areas of termination of thalamofugal axons appeared to be segregated into three tiers: an outer tier in layer I, a middle tier in layer IV and/or III and an inner tier in layer VI. In most fields, terminating axons belonging to the middle or the inner tier or both were found to extend over some distance into layer V. Herkenham [291–293] examined the cortical projections of individual thalamic nuclei in the rat using tritiated amino acids as anterograde tracers. He found that the thalamic nuclei can be grouped into three classes according to the laminar patterns of their cortical projections. The first class includes the thalamic relay nuclei for vision, audition and somatic sensibility. Their cortical projections terminate mainly in lamina IV, lamina III or both. The fibres of this class clearly correspond to the "specific" afferents of Lorente de Nó [416]. The second class includes the intralaminar thalamic nuclei, which issue sparse but widespread projections to deep cortical layers (laminae V, VI or both). The third class encompasses a number of nuclei that share a pattern of dense, widespread projections to lamina I, though terminations in other laminae may or may not be present. Because these nuclei projecting to lamina I generally occupy a position adjacent to the intralaminar nuclei, they were collectively designated by Herkenham [293] as the paralaminar nuclei. The finding that different (groups of) thalamic nuclei project in a particular laminar fashion to smaller or larger parts of the neocortex has been confirmed by a large number of studies on different animals, although with regard to the projections of certain nuclei different results have been obtained by different authors. Thus, Berendse and Groenewegen [49], who studied the cortical projections of the midline and intralaminar thalamic nuclei in the rat using PHA-L as an anterograde tracer, reported that the fibres originating from the intralaminar nuclei do not terminate exclusively in the deep cortical layers (as reported by Herkenham), but also project to lamina I. Corresponding findings have been reported for the cat and the monkey [209, 364, 640, 641].

The cortical projections of some thalamic centres have been analysed in detail by placing multiple small focal injections of anterograde tracers within their confines or by introducing horseradish peroxidase (HRP) into the cortex using a micropipette after recording from individual thalamocortical fibres. Such an analysis of the ventroanterior-ventrolateral complex of the thalamus of the cat revealed, for instance, that fibres originating from the ventrolateral or the caudal part of this complex are distributed in laminae I, III and IV of the parietal cortex, whereas fibres arising from rostral or dorsomedial portions of the complex are almost always confined to lamina I [356]. However, the most detailed information of this type presently available concerns the projection from the lateral geniculate body to the primary visual cortex in primates. In this group, the lateral geniculate nucleus is a laminar structure in which separate magnocellular, parvocellular and intercalated zones can be distinguished. In

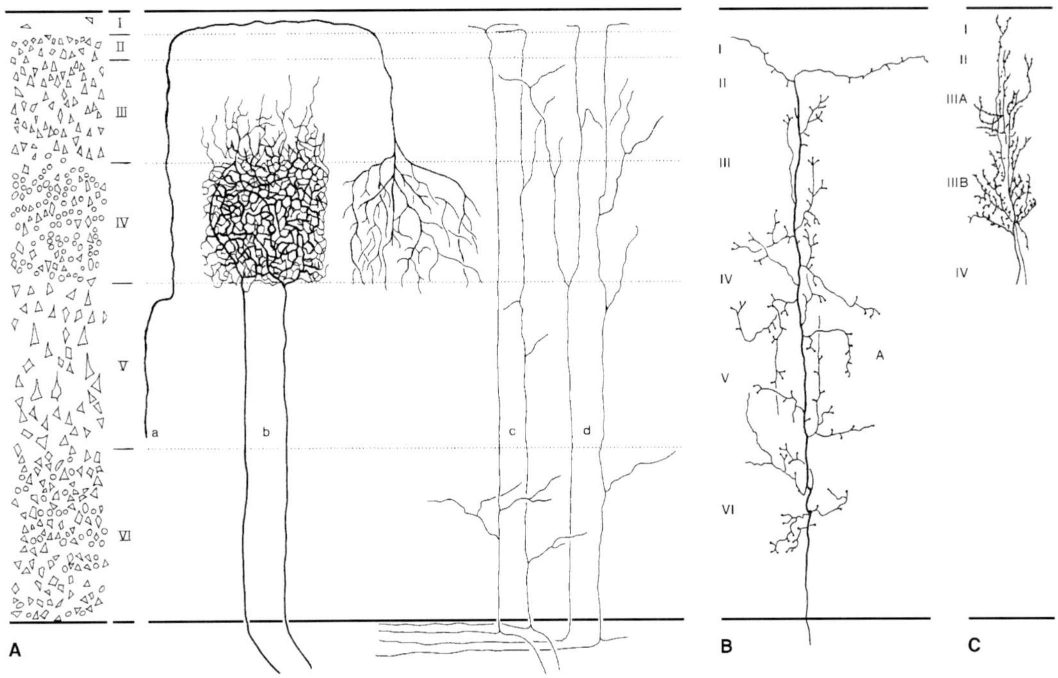

Fig. 15.25 A–C. Neocortical afferents. A "Specific" (*a,b*) and "unspecific" afferents (*c,d*) as observed in Golgi material of the somatosensory cortex of the mouse. *Left,* same cortical area as observed in Nissl preparations (based on Lorente de Nó [414]). B Cortico-cortical afferent fibre arborization in the auditory cortex of the macaque monkey (redrawn from Szentágothai [721]). C Callosal axons in the somatosensory cortex of the same species, anterogradely labelled by horseradish peroxidase injected in the corpus callosum. Redrawn from Jones [339]

the primate primary visual cortex, the laminar pattern is very distinct, and lamina IV can be divided into four subzones, designated as IVA, IVB, IVCα and IVCβ (Fig. 15.26). Investigations of many different authors, summarized by Fitzpatrick et al. [189, 190] and Lund [425], revealed that there are at least six discrete populations of geniculocortical axons, differing markedly from one another in laminar distribution and tangential spread (Fig. 15.26):

1. Coarse fibres originating from the magnocellular layers of the lateral geniculate nucleus projecting to lamina IVCα with widespreading terminal fields that span the entire depth of the lamina.
2. Fibres resembling those mentioned above, the terminal fields of which are, however, confined to the upper half of lamina IVCα.
3. Axons originating presumably from both the parvocellular and the intercalated layers of the lateral geniculate nucleus forming small, dense clusters of terminal branches in lamina IVA and contributing some rising collaterals to the adjacent zone of lamina III.
4. Fibres originating from the intercalated layers of the lateral geniculate nucleus, the terminal arborizations of which participate in the formation of small, dome-shaped formations in lamina III.
5. Fibres from the parvocellular layers forming small, dense terminal fields in lamina IVCβ.
6. Fine fibres likewise originating from the parvocellular layers of the lateral geniculate nucleus, extending over large distances horizontally, that contribute terminals to upper lamina VI and to lamina I (for a detailed discussion of the visual system the reader is referred to Chap. 19).

Although the thalamus is a major source of subcortical input, it is not the only one. In fact, more than ten different extrathalamic subcortical structures projecting to the neocortex have been identified [734]. In most of these, the nature of the neurotransmitter utilized by their constituent neurons has been established.

These structures include: (1) the claustrum (Fig. 13.8); (2) the basolateral amygdaloid nuclei (Fig. 13.7); (3) the basal forebrain, including the nucleus basalis of Meynert, which sends numerous cholinergic as well as GABAergic fibres to the cortex (Fig. 14.15); (4) the hypothalamus, comprising orexinergic, histaminergic and melanin-concentrating hormone (MCH)-containing neocortically projecting neuron groups (Figs. 10.5, 10.7 B, 14.15); (5) the mesencephalic dorsal raphe and central superior nuclei, which provide the entire neocortex with richly ramifying serotoninergic fibres; (6) the ventral tegmental area and the substantia nigra, sending dopaminergic axons mainly, but not exclusively, to the prefrontal cortex; and (7) the locus coeruleus in the rostral pontine tegmentum, which is the source of noradrenergic fibres spreading over the entire neocortex.

The claustral and amygdaloid efferents have been discussed in Chap. 13. The neocortical projections from the basal forebrain, the hypothalamus and the upper brain stem together constitute the ventral branch of the ascending arousal system (Figs. 10.5, 14.15). A detailed discussion of these various projections is beyond the scope of the present work. As far as the cholinergic, GABAergic and monoaminergic projections are concerned, it may be stated that, although all of these projections innervate the entire cortex, they are not to be considered as "diffuse and non-specific" [538]. Rather, they show a high degree of anatomical specificity, both for particular cortical areas and for particular laminae within a single cortical area (Fig. 15.27 A–C) [198, 264, 331, 467, 492, 651]. In some regions, for instance the primary visual cortex, different monoamines may show clearly complementary laminar patterns of innervation (Fig. 15.27 C, D). These data strongly suggest that the effects of the cholinergic, GABAergic and monoaminergic systems are not generalized excitation or inhibition but rather region-specific enhancement or diminution of activity in limited neuronal ensembles during certain stages of information processing [198].

Immuno-electron microscopic studies have shown that cholinergic [13, 43, 307], dopaminergic [677], noradrenergic [534] and serotoninergic axons [141, 535] in the neocortex form synaptic contacts with pyramidal as well as

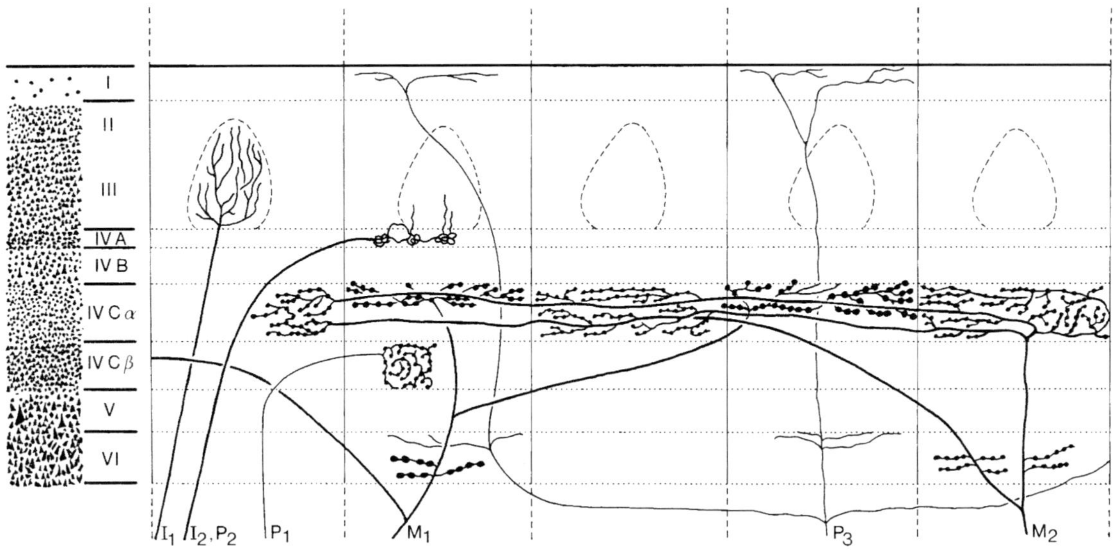

Fig. 15.26. Distribution of thalamic inputs from the lateral geniculate nucleus to the primate primary visual cortex of the macaque monkey. M_1, M_2, input from magnocellular layers; P_1–P_3, input from parvocellular layers; I_1, I_2, input from intercalated layers. The ocular dominance bands are indicated by *dashed lines*, 400–500 μm apart. *Left,* same cortical area as observed in Nissl preparations. Redrawn from Lund [425]

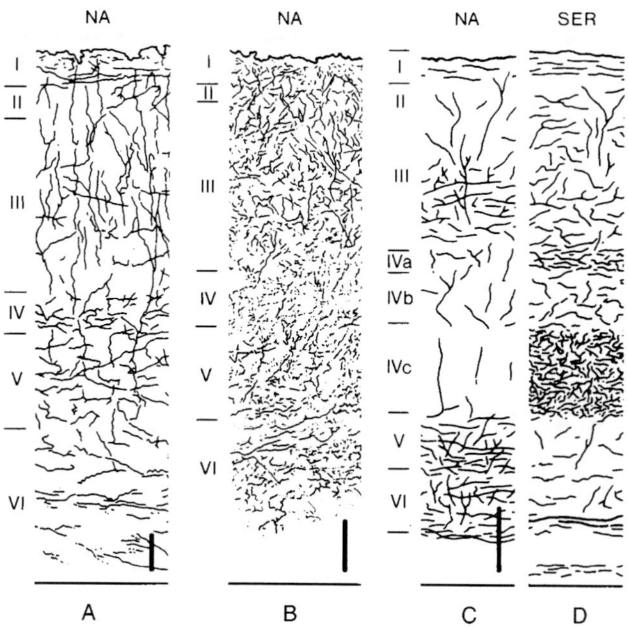

Fig. 15.27 A–D. Monoaminergic innervation of three different regions of the neocortex of the squirrel monkey. **A–C** Noradrenergic (*NA*) innervation of the dorsolateral prefrontal cortex (areas 9, 10), the primary somatosensory cortex (areas 3, 1, 2) and the primary visual cortex (area 17), respectively. **D** Serotoninergic (*SER*) innervation of the primary visual cortex. *Bars* represent 200 μm. Slightly modified from Morrison and Magistretti [491]

with (GABAergic) non-pyramidal cells. Descarries and colleagues have devoted a long series of publications, summarized in [148], to the synaptology of cholinergic and monoaminergic axons. They found that, in the neocortex, these axons are provided with varicosities and terminals which, although filled with typical "synaptic" vesicles, quite often lack the membrane functional complexes that are typical of synapses. They consider it likely that these non-junctional structures are involved in non-synaptic or volume transmission. Evidence in support of the occurrence of this mode of neurotransmission in the neocortex has been presented by Smiley and Goldman-Rakic for both the serotoninergic [692] and for the dopaminergic projections [691].

It is known that the various ascending extrathalamic pathways modulate the responsiveness of cortical neurons that process sensory input, co-ordinate motor output and perform higher brain functions, such as mood, attention, motivation, cognition, learning and memory [198, 264].

The dopaminergic projection is a key modulator of motivational cognitive and motor functions [244, 652] and the serotoninergic projection helps regulate wake-sleep cycles and modulates the sensory gating of behaviourally relevant cues in the environment [465].

The cholinergic projection to the neocortex is implicated in arousal, sleep-wake cycles, visual information processing learning, memory and selective attentional functions [239, 650, 835].

The noradrenergic projection to the neocortex arises, as mentioned, from the locus coeruleus in the rostrodorsal pons. Traditionally, this nucleus has been thought to play a role in vigilance and in responsiveness to novel stimuli [21, 199].

Recently, Aston-Jones and Cohen [19, 20] and Bouret and Sara [59] have developed new concepts on the function of the locus coeruleus. Aston-Jones and Cohen point out that during the waking state, the locus coeruleus neurons exhibit two modes of activity, phasic and tonic. Phasic activity is driven by the outcome of task-related decision processes and is proposed to optimize task performance (exploitation). When utility in the task declines, locus coeruleus neurons exhibit a tonic activity mode, associated with disengagement from the current task and a search for alternative behaviours (exploration). The locus coeruleus is supposed to produce these patterns of activity under the control of the cingulate and orbitofrontal cortices, both of which project strongly to it.

From studies of the cognitive context governing the activity of locus coeruleus neurons in rats and monkeys, Bouret and Sara concluded that these neurons are activated in situations requiring interruption of on-going behaviour and rapid adaptation. This locus coeruleus activation appeared to occur whenever there is a change in the environmental imperative, such as the appearance of a novel, unexpected sensory event. They suppose that the noradrenaline signals from the locus coeruleus facilitate changes in the forebrain networks that are mediating specific cognitive functions.

In addition to thalamic and extrathalamic subcortical afferents, each particular neocortical area also receives a strong input from other neocortical areas. These association fibres come from either the same or the opposite hemisphere; in the latter case they are called callosal fibres. According to Lorente de Nó [414, 416], the association fibres give off collaterals in the deep laminae, especially VI, but their main territory of distribution is in laminae I–IV, and especially II and III (d in Fig. 15.25 A). Goldman and Nauta [242] made small focal injections with tritiated amino acids into various areas of the association cortex of monkeys. They found that the anterogradely transported label accumulated in narrow, 200- to 300-μm-wide columns in relatively distant regions. On the basis of a study of Golgi material, Szentágothai [721, 723] concluded that the arborization space and pattern of individual cortico-cortical axons correspond in size with the columns revealed by isotopic labelling. He observed that these fibres pass radially through the cortex and issue relatively short branches at all levels (Fig. 15.25 B). Cortico-cortical fibres defining a narrow radial zone of termination have also been observed in the primate sensorimotor cortex [286, 339]. Such a fibre, labelled by HRP injected in the corpus callosum is depicted in Fig. 15.25C. The major layers of termination of callosal fibres in the primate somatosensory areas are laminae I–IV. In the motor cortex, they terminate in a comparable pattern in laminae I–III. Studies on cortico-cortical projections based on the tracing of individual fibres, like the one cited above, are scant. However, an overall picture of the mode of termination of these fibres can be deduced from axon degeneration studies of cortico-cortical projections and from tracer studies in which such fibres are labelled en masse. A detailed survey of this literature, presented by Innocenti [322], revealed that most cortico-cortical fibres terminate in layers III and IV in primates.

Rockland and Pandya [623] reported that, in the visual cortex of the rhesus monkey, the pyramidal neurons of the superficial layers project to the middle layer (principally layer IV) of their cortical target areas, whereas the deep layer pyramids project outside the middle layers, mainly to layers I and VI. It has already been mentioned that Felleman and Van Essen [180] classified the projections to layer IV as feed-forward pathways and those projecting outside layer IV as feed-back projections.

From the foregoing, it appears that most of the extrinsic cortical afferent systems distribute themselves in layered arrays, and it may be added that, in the primary sensory cortices, an elaborate laminar segregation of thalamic inputs appears to be reflected in the stratification visible in the cellular architecture. Morphological evidence suggests that all extrinsic neocortical afferents are excitatory, raising the question of how the intrinsic machinery of this cortex is driven by these afferents.

Lamina I contains the apical dendritic bouquets of pyramidal cells of laminae II, III and V. Because in this layer only sparsely scattered intrinsic neurons are present, it is reasonable to assume that the extrinsic thalamic afferents spreading in it synapse principally with pyramidal neurons and thus have a direct access to the long-axonal outflow of the cortex. Superficial pyramids situated in lamina II extend not only their apical dendrites in lamina I; these elements often have basal dendrites which pass laterally from the soma and ascend to layer I [779]. These elements are most probably strongly excited by the lamina I extrinsic afferents. It is interesting to note that the afferents from different thalamic nuclei, which, after having traversed the cortex, spread in lamina I, terminate in different subzones of that layer [290] and the apical dendritic tufts of the pyramids thus receive stratified input from different sources.

Until recently, it was widely held that the specific sensory thalamic afferents mainly distributed to lamina IV (or its sublayers) of the primary sensory cortices contact only a single category of neurons, i.e. the spiny stellate cells, and that the input is then processed sequentially by hierarchically organized chains of neurons (see e.g. Hubel and Wiesel [308, 310] and Eccles [166]). Golgi-electron microscopy studies combined with degeneration and tracer experiments [814, 815] conclusively showed that the specific sensory thalamic afferents to the cortex synapse with pyramidal cells and with non-pyramidal intrinsic cortical elements. According to White [814, 815], this finding challenges the concept that thalamic input is processed by hierarchically organized chains of neurons and lends support to the notion that the function of the cerebral cortex depends heavily on parallel processing mechanisms. However, it should be noted that in highly differentiated sensory cortices different groups of local circuit neurons may well be involved in the parallel processing of sensory information. We have seen that, in the primary visual cortex of primates, the magnocellular zones of the lateral geniculate nucleus project to sublamina IVCα, whereas the parvocellular zones project to IVCβ (Fig. 15.26). These two sublaminae are characterized by the presence of different classes of spiny stellate cells and by an almost complete lack of pyramidal neurons [422–424, 446]. Moreover, it has been established that the principal projection of the IVCα spiny stellate neurons is to sublamina IVB, whereas the majority of IVCβ spiny stellate neurons project to sublaminae IIIC and IVA [190, 422]. These findings suggest a continued separation of magnocellularly and parvocellularly derived information within the primate primary visual cortex [425].

We have seen that the deeper layers of the cortex (V and VI) also receive direct thalamic inputs. It is known that these layers contain many pyramidal and pyramid-like cells and only a relatively small number of intrinsic elements. The axons of many of these deep pyramidal cells possess recurrent, ascending collaterals and the axons of a certain proportion of the local circuit neurons in laminae V and VI are also directed to more superficial cortical layers. There is experimental evidence indicating that, in the deeper layers of the primary somatosensory cortex of the mouse, thalamocortical afferents contact both pyramidal and non-pyramidal cells [294, 295, 813]. The deep pyramidal cells may well receive an "aspecific" thalamic input via collaterals spreading in lamina VI (Fig. 15.25 A (c) and a "specific" thalamic input via their apical dendrites in lamina IV.

Finally, the callosal and ipsilateral association fibres form nearly all their synapses with the spiny dendrites of pyramidal cells [322, 815]. Because these fibres terminate principally in laminae III and IV, it seems likely that the superficial pyramidal neurons form their main target.

Neocortical Neurons and Their Synaptic Relationships

Introductory Note

The neurons of the neocortex can be classified as belonging to two basic types: pyramidal cells and non-pyramidal cells. The former can be subdivided into typical and atypical pyramidal cells (Figs. 15.28–15.30). The elements belonging in these categories will now be discussed.

Typical Pyramidal Cells

The typical pyramidal cells constitute the largest and most characteristic category of neocortical neurons. In the visual cortex of rat, cat and monkey, according to Winfield et al. [824], these elements account for about 65% of the total neuronal population, and in other cortical areas this percentage may be even higher. A typical pyramidal neuron is provided with a radially oriented apical dendrite that forms a terminal tuft in laminae I and II, with basal dendrites that radiate out from the base of the soma and with an axon descending downward to leave the cortex (Fig. 15.29). The apical dendrites may give rise to one or more horizontal or obliquely ascending side branches. All of the dendrites of typical pyramidal cells are more or less densely covered with spines (Fig. 15.30 A) [561]. Positionally, the apical dendritic tuft with its ramifications in laminae I and II is the only feature shared by all typical pyramidal cells. The somata and their basal dendritic systems may vary in position from lamina II to lamina VI, and the length of the apical dendrites varies accordingly (Fig. 15.29). In the following discussion of the typical pyramidal cells, the afferents impinging upon the various parts of their receptive surface will be considered first; the relation between the position of their somata and the destination of their axons will then be dealt with. Finally, attention will be paid to the patterns of distribution of their collaterals.

The somata of the pyramidal cells are not under the direct influence of any extrinsic afferent system. Rather, they are specifically addressed by one type of local circuit neuron, i.e. the basket cell (Fig. 15.31 A), although several other types of non-pyramidal cells may contribute to some extent to the somatic innervation of pyramidal cells [137]. The somata of the basket cells, which, like those of the pyramidal cells, vary in distance from the pia and are concentrated in laminae III and V [177, 436]. Their poorly ramifying dendrites, which bear few or no spines, radiate in all directions, with a tendency toward vertical and horizontal trajectories. The axons are either ascending or

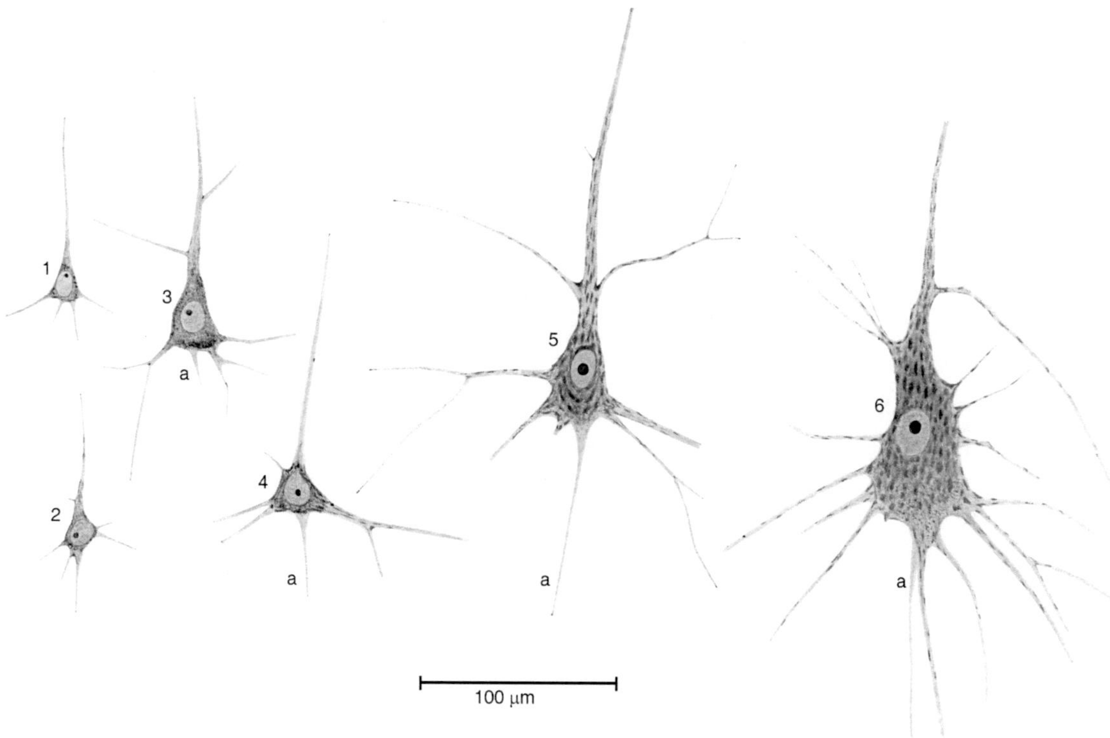

Fig. 15.28. Pyramidal neurons, as observed in Nissl preparations of the human neocortex (reproduced from [786]). *1*, Largest pyramidal cell from the IIIrd layer of area 17; *2*, stellate cell from the VIth layer of area 18; *3*, large pyramidal cell from the IIIrd layer of area 18; *4*, Meynert cell from the Vth layer of area 17; *5*, largest pyramidal cell from the IIIrd layer of area 4; *6*, Betz cell from the Vth layer of area 4. Note that the Meynert cell and the Betz cell, both of which represent by far the largest cell type in their respective cytoarchitectonic areas, and which, hence, are both designated as "giant cells" [84], differ considerably in absolute size. *a*, axon

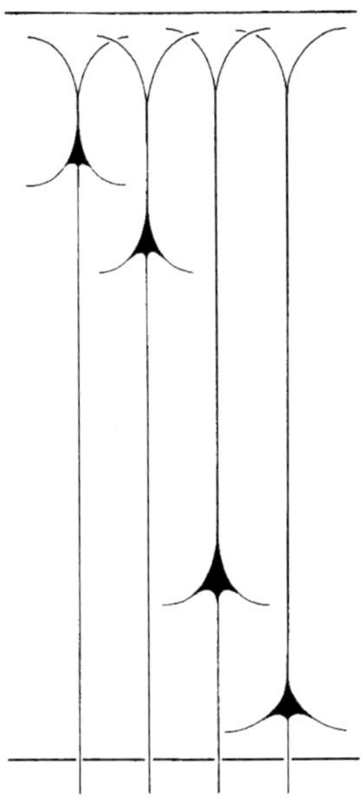

Fig. 15.29. Typical pyramidal neurons in the mammalian neocortex

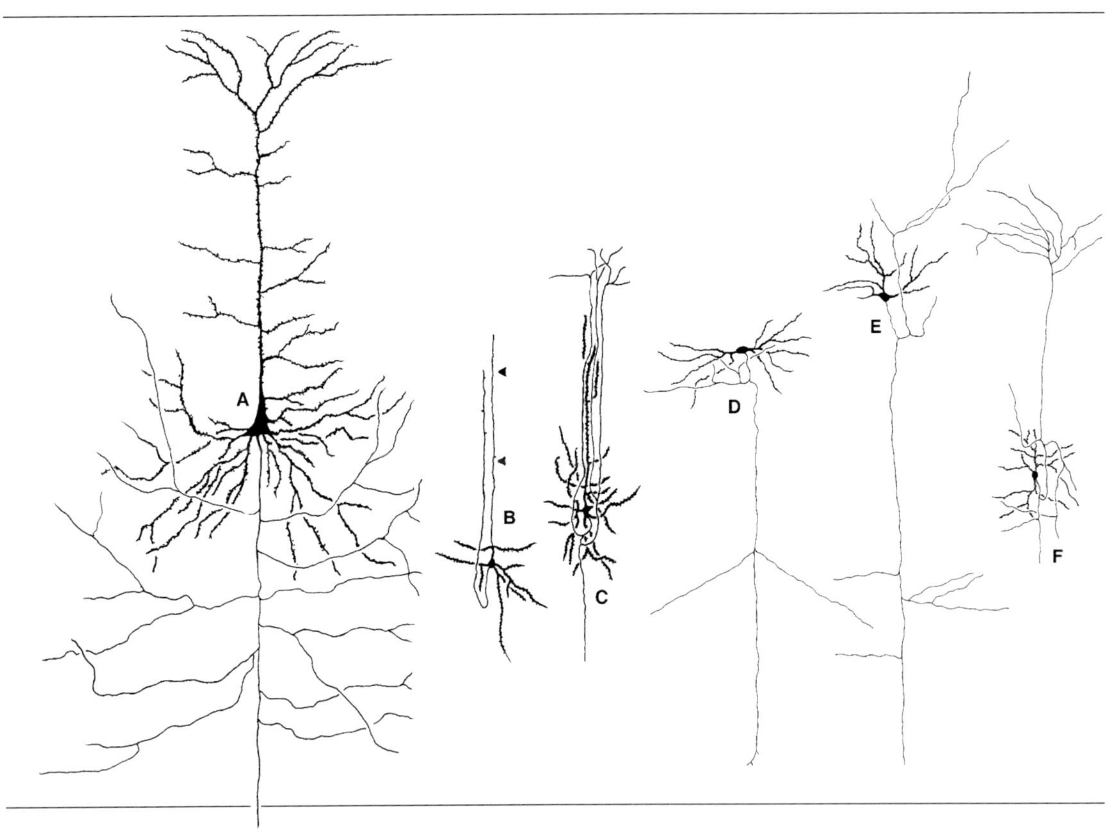

Fig. 15.30 A–F. Typical and atypical pyramidal neurons in the primate neocortex. **A** Typical pyramidal cell in the human neocortex [84]; **B** small pyramidal neuron with poorly developed apical dendrite (*arrowheads*) and a recurving axon from the striate cortex of the macaque monkey [422]; **C** a "star pyramid" in layer IV of the somatosensory cortex of the squirrel monkey [336]; **D–F**, spiny stellate cells in layers IVB, IVA and IVCβ, respectively, of the primary visual cortex of the macaque monkey [422, 426]

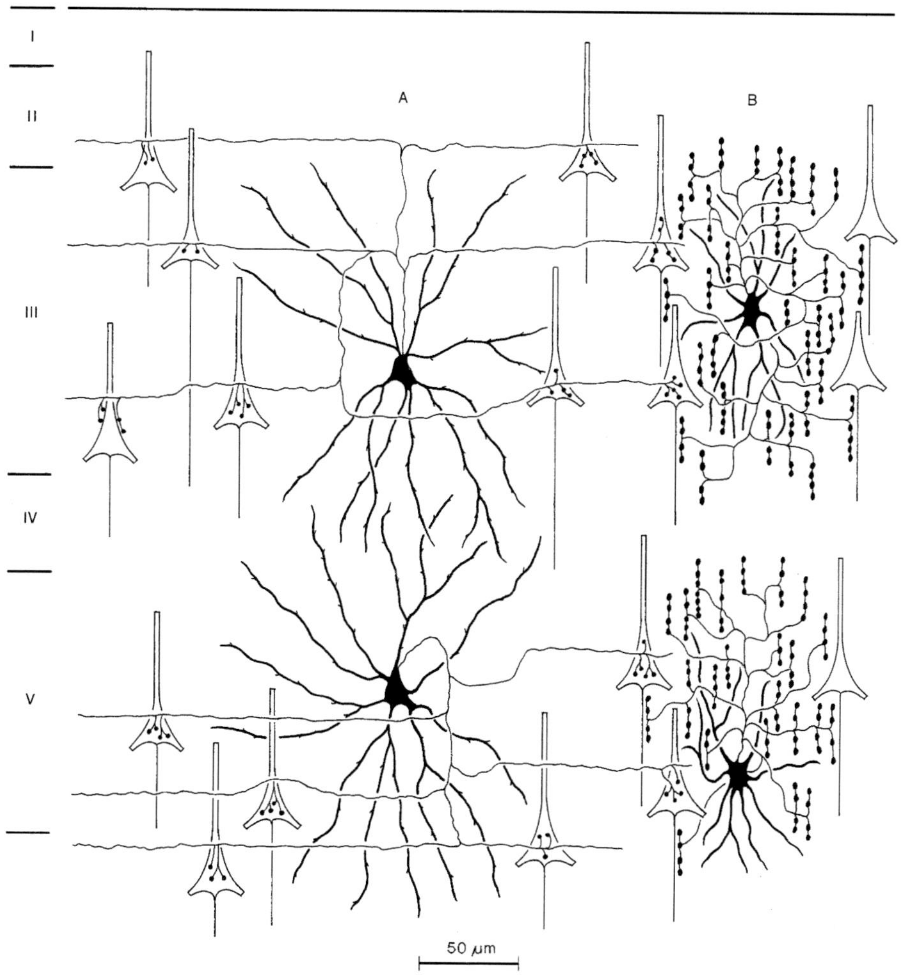

Fig. 15.31 A, B. Intrinsic cortical neurons contacting specific parts of the receptive surface of the pyramidal cells. **A** Basket cells, whose axonal terminals participate in the formation of pericellular nests surrounding the somata and proximal dendrites of pyramidal cells (partly based on Jones [339]). **B** Chandelier cells, whose numerous vertical axonal terminal portions specifically contact the axon initial segments of pyramidal cells. Based on Peters [555], Fairén et al. [177], Somogyi et al. [701] and Jones [339]

descending and provide four or more horizontal collaterals at various levels. The horizontal collaterals, which can extend for 1 mm or more in either direction, issue at intervals short side branches. These side branches terminate in pericellular baskets surrounding the somata and proximal dendrites of pyramidal cells. The basket cell terminals contain flattened vesicles, and the synapses made with the pyramidal cells are of the symmetrical variety. They use GABA as a neurotransmitter and exert a powerful inhibitory action on their target elements [343]. There is evidence suggesting that the basket cells are major recipients of thalamic and commissural inputs [555, 690, 700]. Moreover, the somata of the basket cells are contacted by numerous GABAergic terminals, which may well arise from the axons of other basket cells [287].

Like the somata, the axon hillocks and initial axonal segments of the pyramidal cells are also the target of a special category of local circuit neurons, i.e. the axo-axonic or chandelier cells (Fig. 15.31 B) [176, 555, 699]. These elements occur in laminae II–V, but are most prominent in lamina III. Their dendrites may be grouped in an upper and a lower tuft or spread in all directions. The axon ramifies many times and produces a dense plexus in the vicinity of the parent soma. From this axonal plexus, numerous (up to 300) vertically oriented arrays of terminals arise that contact the initial axonal segments of pyramidal cells. The axo-axonic synapses between the chandelier cells and the pyramidal neurons are symmetrical, and the vesicles in the presynaptic axon terminals are flattened [178, 345, 690].

It has been shown that the axon terminals of the chandelier cells contain the inhibitory neurotransmitter GABA; hence, it may be assumed that these elements inhibit the pyramidal cells [570]. Because the chandelier cells exert this action at a strategic site, i.e. the trigger zone for the initiation of action potentials, they may be considered to have a decisive influence on the output from the pyramidal neurons. It has been found that the number of axo-axonal synapses along the initial segments of pyramidal cells is much larger in the supragranular layers than in the infragranular layers [570, 690]. Because the pyramidal neurons projecting via the corpus callosum to the contralateral hemisphere are situated mainly in laminae II and III, it has been suggested that the chandelier cells principally influence cortico-cortical circuitry [570, 698].

In the hippocampal cortex, the perikarya of the pyramidal neurons are concentrated in a single layer, and all of these elements extend their basal dendrites in one and the same plexiform zone, the stratum oriens (Fig. 12.9). However, in the neocortex, the pyramidal somata are situated at different levels and, given the laminar organization of many extrinsic and intrinsic cortical fibre systems, it is to be expected that the basal dendritic systems of different pyramidal neurons are involved in different synaptic relationships, depending on the laminae or sublaminae within which these dendritic systems occur. Thus, it has been established that the basal dendrites of pyramidal cells situated in laminae II and III receive callosal afferents [100] and that lamina III border pyramids extend their basal dendrites into lamina IV, where they are contacted by terminals of thalamocortical fibres [303, 643]. Intrinsic cortical axons, i.e. axons of local circuit neurons and collateral branches of pyramidal cells, are concentrated in lamina IV and deep in lamina V, where they form the external and internal striae of Baillarger [340, 756]. The basal dendrites of certain groups of pyramidal neurons exhibit a specific affinity to the terminal ramifications of the axons within these striae. In the primate primary visual cortex, the external stria of Baillarger (also known there as the line of Gennari) is strongly developed and occupies a discrete sublayer IVB, in which no thalamic afferents are present (Fig. 15.26). According to Valverde [755], its axonal components include horizontal collaterals of descending axons of pyramidal and stellate cells of sublayer IVA, horizontal branches of axons of neurons located within sublayer IVB and ascending ramifications of spiny stellate cells residing in sublayer IVC. Lund [422] reported that the basal dendrites of pyramidal neurons situated in sublayer IVA turn sharply downward from the soma to enter sublayer IVB, where they fan out horizontally, markedly avoiding any ramifications within lamina IVA (Fig. 15.32E). To the author's knowledge, the question of which neurons contribute axons or axonal branches to the internal stria of Baillarger has never been specifically addressed; however, the Golgi studies by Cajal [84] and Valverde [755, 756] suggest that the axons of both superficial and deep pyramids issue numerous horizontally oriented collaterals into this stria and that certain local circuit neurons with ascending axons, situated in lamina VI, also contribute collaterals to it. In the primate motor and visual cortices, basal dendrites of pyramidal cells form a conspicuous plexus within the domain of the internal stria. In both of these cortices, large pyramidal neurons occur, known in the motor cortex as Betz cells and in the visual cortex as Meynert cells (Fig. 15.28 (4, 6). Beyond their extraordinary size, these neurons have several other features in common. Both cells are characterized by the presence of long horizontally oriented basal dendrites that extend into the internal stria, and both further contribute to this stria with additional horizontal dendrites emanating from the lateral surface of the soma and even from the proximal stem of the apical dendrite [63, 64, 84, 654, 657]. The long, coarse collaterals of the Meynert cells may also contribute to the internal stria (Fig. 15.35) [622].

Whereas the horizontal axonal systems discussed so far contact the basal dendrites of pyramidal neurons situated at one particular level, the neocortex also contains vast numbers of vertically oriented axonal elements, which potentially contact the basal dendritic systems of pyramidal elements situated at different levels. These vertical axonal elements, which play a prominent role in the so-called radial coupling of neuronal elements and hence in the columnar functional organization of the cortex, may be categorized as follows:

1. *Thalamocortical and cortico-cortical association fibres* (Fig. 15.25).
2. *Axons and recurrent collaterals of pyramidal neurons* (Figs. 15.30 B, C, 15.32 D–F). Ultrastructural analyses of pyramidal neurons labelled by intracellular injections with HRP have conclusively shown that the axon collaterals of these elements make contact predominantly with the dendrites of other pyramidal cells [218, 370, 457].
3. *The vertically elongate axonal systems of some types of cortical local circuit neurons.* These elements, which will be discussed below, include spiny stellate cells (Fig. 15.30 D–F) and various types of smooth or sparsely spiny non-pyramidal neurons (Fig. 15.36 (M–Q).

The recurrent ascending axons or axonal branches of pyramidal neurons, as well as the ascending and descending axons of spiny stellate cells, have been observed to assemble in highly characteristic, radially oriented bundles [177, 423, 755, 756].

Turning now to the apical dendritic shafts of the typical pyramidal cells, it is important to note that these conspicuous processes are well placed to receive input from a variety of axonal pathways known to terminate within specific cortical layers. There is considerable variation in apical shaft length, ranging from essentially no shaft at all (pyramidal cells in the superficial zone of lamina II) to apical dendrites of 2 mm or more (pyramidal cells in lamina VI). Moreover, particularly in primates, the apical dendrites of many deep pyramidal cells do not extend into the subpial cortical zone, but rather terminate in a variably developed terminal tuft deeper within the cortex. Given these variations in length and in position of their apical dendrites, it is evident that different pyramidal cells may receive different samples of lamina-specific extracortical and intracortical afferents, and these differences are still further accentuated by the fact that the apical dendrites of different pyramidal neurons may exhibit different specific affinities to particular afferent systems. The evidence for such specific affinities includes (1) the presence of distinct, lamina-specific differences in the density of spines along the apical dendrites, (2) the presence of lamina-specific side branches on the apical dendrites and (3) direct proof that apical dendritic segments of different pyramidal cells passing through a particular layer

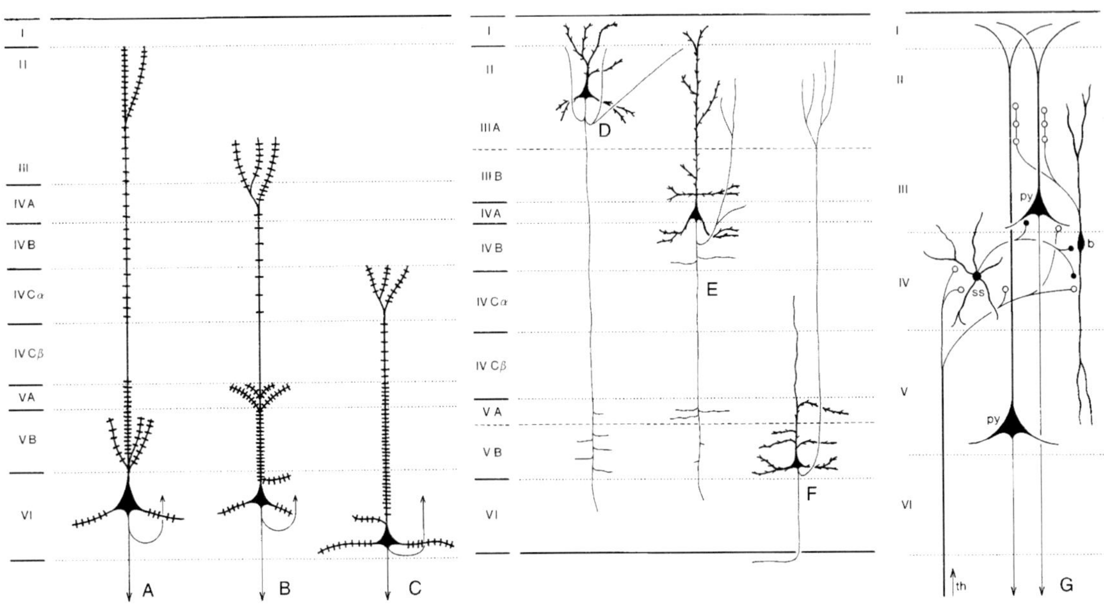

Fig. 15.32 A–G. Features of pyramidal neurons. **A-F** Semidiagrammatic representations of pyramidal neurons in the primary visual cortex of the rhesus monkey. The apical dendrites of the elements **A–C** issue their side branches and terminal tufts in particular (sub)layers; the density of spines along these processes show distinct lamina-specific differences. The elements **D–F**, which are situated at different levels, contribute axon collaterals to particular levels. **G** shows two pyramidal neurons, whose apical dendrites receive multiple asymmetrical synapses (*open circles*) from the axons of a bipolar cell. **G** also shows that smooth or sparsely spiny, non-pyramidal cells (*ss*) form symmetrical synapses (*filled circles*) with pyramidal (*py*) and bipolar cells (*b*) and that thalamic afferents (*th*) contact pyramidal, bipolar and smooth or sparsely spiny, non-pyramidal cells forming asymmetrical synapses (**A–C** are redrawn from Lund [423]; **D–F** are redrawn from Lund and Boothe [426]; **G** is based on Peters [554])

may receive highly different numbers of synapses from the afferents concentrated in that layer. These three aspects will now be briefly discussed.

1. *The presence of distinct, lamina-specific differences in the density of spines along the apical dendrites.* Spines are present in abundance on the dendrites of pyramidal neurons, where they function as the primary postsynaptic structures of the cell [178]. Quantitative analyses have shown that these spines are not evenly distributed along the dendrites. Characteristically, the most proximal portions of the dendrites emanating from the pyramidal somata are devoid of spines. From these initial segments onward, the concentration of spines increases gradually, attains a maximum some 80 μm away from the soma and then declines again distally [240, 561, 753]. Superimposed over this general pattern of spine distribution, the apical dendrites of pyramidal neurons may display distinct, lamina-specific changes in spine density. For example, Lund [422] found that in the primary visual cortex of the monkey, the apical dendrites of pyramidal cells lying in laminae V and VI show markedly fewer spines in lamina IVCβ than in the densely spiny portion of the same shafts in laminae V and VI. The number of spines on these apical dendrites in general increases again in laminae above IVCα, but not to the level shown on the proximal portion of the shaft (Fig. 15.32A,B). According to Lund [422], this reduction in spines on the apical dendritic shaft as it passes through laminae IVCβ is true of the great majority of the pyramidal cells of laminae V and VI, including the giant pyramidal cells of Meynert. It is known that lamina IVCβ receives afferents specifically from the parvocellular layers of the lateral geniculate nucleus. Thus, the findings of Lund [422] might suggest that, in the monkey, the deep pyramids in the visual cortex are not the primary targets of these particular afferents.

2. *The presence of lamina-specific side branches on the apical dendrites.* The apical dendrites of many pyramidal neurons issue small numbers (some three to eight) of oblique or horizontally oriented side branches (Fig. 15.30 A). At first sight, these side branches seem to be randomly distributed along the apical shafts. However, systematic studies of large numbers of apical dendrites have shown that in several cortical areas these processes are given off at selective levels, i.e. as they pass through particular layers. Thus, Lorente de Nó [415] observed that, in the entorhinal cortex, the ascending shafts of the pyramids of the deeper laminae always issue some side branches in lamina III, but give off none at all during their passage through lamina II. Lund and Boothe [426] observed that the apical dendrites of a particular population of lamina VI pyramidal neurons in the primate visual cortex issue side branches selectively in lamina IVCα, whereas the apical dendrites of another lamina VI pyramidal population in this cortex give off side branches at the border of lamina V and IVCβ and split up into their terminal dendritic tufts in lamina IVA (Fig. 15.32 B,C). According to Lund [425], these various dendritic patterns suggest that deeper pyramidal neurons may receive thalamic input within laminae IVCα and IVA.

3. *Direct proof that apical dendritic segments of different pyramidal cells passing through a particular layer may receive highly different numbers of synapses from the afferents concentrated in that layer.* The thalamocortical connectivity of pyramidal neurons has been the subject of a series of quantitative studies by White and collaborators [294–297, 817, 818, 820]. Hersch and White [294, 295] reported that, in layer IV of the primary somatosensory cortex of the mouse, the proportion of thalamocortical synapses received by apical dendrites belonging to various sizes of Golgi-impregnated lamina V and lamina VI pyramidal neurons ranged from 1.3% to 14.6% of all the asymmetrical synapses formed onto these dendrites. In subsequent work in the same cortex, the number of thalamocortical synapses in lamina IV on the apical dendrites belonging to deep pyramids that project from the primary somatosensory cortex to either the ventrobasal nucleus of the thalamus [296, 818], the primary motor cortex [820] or the ipsilateral striatum [297] was determined. It appeared that these three populations of pyramidal neurons have very different thalamocortical connectivity patterns. Thus only 0.3–0.9% of the total number

of apical dendritic synapses in lamina IV of corticostriatal projection neurons were made with thalamic afferents. For the cortico-cortical and corticothalamic elements examined, these values were 1–7% and 7–20%, respectively.

So far, I have dealt with tangentially oriented afferent systems establishing synaptic contacts with particular segments of the apical dendrites of pyramidal neurons. However, there is also evidence for the presence of radially oriented afferents that make repeated contacts with these processes. Thus, Scheibel and Scheibel [656] mentioned that "non-specific" cortical afferent fibres originating from the brain stem and the medial thalamus break up into a series of branches that ascend radially through the cortex, establishing sequences of axodendritic contacts with the spines of apical shafts and terminal branches of pyramidal neurons. It is also known that the axons of a certain category of cortical local circuit neurons, known as bipolar cells, typically give rise to vertically oriented branches which parallel the trajectories of clustered pyramidal apical dendrites (Fig. 15.32 G). These branches form multiple asymmetrical synapses with the spines of the apical dendrites [179, 554]. Apart from "non-specific" cortical afferents and the axons of bipolar cells, collateral branches of pyramidal cell axons may also make repeated, climbing fibre-like contacts with the apical dendrites of (other) pyramidal cells [218].

The final part of the dendritic system of typical pyramidal neurons to be considered is the set of dendritic branches emanating from the tip of the apical dendrite. These so-called apical dendritic tufts or terminal dendritic bouquets extend into lamina I, in which, together with numerous tangentially running axons, they constitute a typical plexiform zone. The axonal components observed within this zone include:

1. Fibres originating from the intralaminar and midline thalamic nuclei [49, 640].
2. Monoaminergic fibres arising from the brain stem [438].
3. Recurrent collaterals of the axons of pyramidal neurons, principally situated in laminae II and III [414, 656, 755].
4. Ascending axons from multipolar or bitufted neurons situated in deeper cortical layers. These elements, which have smooth or sparsely spinous dendrites, are known as Martinotti cells (Fig. 15.36 Q). Large elements of this type are found in laminae V and VI, but they are also present in the more superficial cortical layers [177]. Marin-Padilla [438] reported that the morphology of the axonal termination of Martinotti neurons resembles quite closely the arborization of the apical dendritic tufts of pyramidal cells and considered it likely that Martinotti cells form dual sets with pyramidal neurons of similar cortical depth. With regard to the function of the Martinotti cells, Marin-Padilla [438] suggested that these elements are inhibitory and that the inhibition takes place specifically between the axonal terminals of a given Martinotti cell and the dendritic tufts of the pyramidal neurons with which it forms a dual set.
5. Axons of so-called horizontal neurons, which are situated within lamina I itself [84]. The dendrites and axonal branches of these elements are entirely confined to lamina I and extend parallel to the surface of the cortex (Fig. 15.36 D). Some of their axonal branches may attain a considerable length. There is immunohistochemical evidence (summarized by Vogt [779]) indicating that the horizontal neurons are all GABAergic and also contain the neuropeptide cholecystokinin.

In summary, it may be stated that the apical dendritic branches of neocortical pyramidal neurons receive input from various sources. Thalamic afferents and recurrent collaterals of pyramidal cells presumably exert an excitatory influence on these dendrites, whereas the axonal endings of the Martinotti cells and the horizontal cells may well exert an inhibitory influence on them. The horizontal cells have been suggested to receive a specific input from monoaminergic fibres originating from neurons situated in the brain stem [438], but thalamic afferents may also impinge on these intrinsic lamina I elements and the monoaminergic fibres may also directly contact the pyramidal apical dendrites. However, it should be emphasized that so far none of the synaptic contacts suggested by light microscopy material has been verified with the aid of experimental ultrastructural techniques.

Having discussed the afferents making contact with the various parts of the receptive surface of the typical pyramidal neurons, I will now turn to the axons of these elements. It has already been mentioned that these processes all leave the cortex and pass either to other ipsilateral or contralateral cortical regions or to one or several subcortical centres. The latter may include the striatum, i.e. the nucleus caudatus and the putamen, the various "specific" and "non-specific" thalamic nuclei, the nucleus ruber, the superior colliculus or tectum mesencephali, the pontine nuclei, the medulla oblongata and the spinal cord.

Retrograde tracing studies have shown that the cell bodies of pyramidal neurons projecting to particular cortical or subcortical targets are preferentially located in particular cortical layers or sublayers (Fig. 15.33). The following summary of the laminar relationships of cortical efferent cells is based on reviews by Jones [337] and White [815], to which the reader is referred for the primary sources.

Cortico-cortical and callosally projecting fibres arise predominantly from pyramidal neurons in lamina II and III; however, in rodents and primates, significant numbers of these fibres have been found to originate from elements situated in the infragranular layers. As regards the superficial layers, it has been established that the smaller, more superficially situated pyramids tend to project to ipsilateral cortical areas situated nearby, whereas the larger, more deeply placed cells tend to project to contralateral and to more remote ipsilateral cortical areas.

Pyramidal neurons situated in lamina V have been shown to project subcortically to the intralaminar and other "aspecific" thalamic nuclei, the striatum, the red nucleus, the tectum, the medulla oblongata and the spinal cord. The smallest and most superficially situated elements in this layer project to the striatum, while the largest and most deeply situated cells project to the spinal cord. The elements projecting to the remaining subcortical sites tend to occupy an intermediate position.

The corticothalamic projections to the "specific" thalamic relay nuclei arise exclusively from large pyramids in lamina VI.

Although most cortical neuronal populations projecting to a particular cortical or subcortical target show a distinct laminar specificity, it is not uncommon to find some degree of overlap in the boundaries demarcating different populations of projection neurons. This raises the question of the extent to which projections to particular targets of cortical efferent neurons are made up by collaterals of axons projecting to other centres. In this context, it may be recalled that Cajal [84] noted in his Golgi studies of the brains of rodents that numerous subcortical centres are supplied by collaterals of corticofugal fibres. Thus, he observed that, during their descent through the internal capsule, such corticofugal fibres issued numerous collaterals to the striatum or thalamus and that pyramidal tract neurons in the brain stem gave off collateral branches to several centres, including the red nucleus, the pontine nuclei, and the dorsal column nuclei. Double-labelling experiments, i.e. experiments in which two different and distinctive retrogradely transported labels are injected into two different known terminal fields, have revealed that double-projecting neurons do occur in the neocortex. Catsman-Berrevoets and Kuypers [93] reported the presence of double-labelled cells in the motor cortex of the monkey after injecting markers into the magnocellular part of the red nucleus and spinal cord, and Rustioni and Hayes [644] found double-labelled cells in the sensory cortex of the cat after injecting markers into the dorsal column nuclei and spinal cord. However, in these and other comparable experiments, the number of double-labelled cells appeared to be very small, implying that the degree of subcortical collateralization of corticofugal fibres is limited, too [337]. The abundance of such collaterals observed by Cajal [84] might well have to do with the fact that the Golgi material studied by that author was exclusively derived from the brains of very young animals.

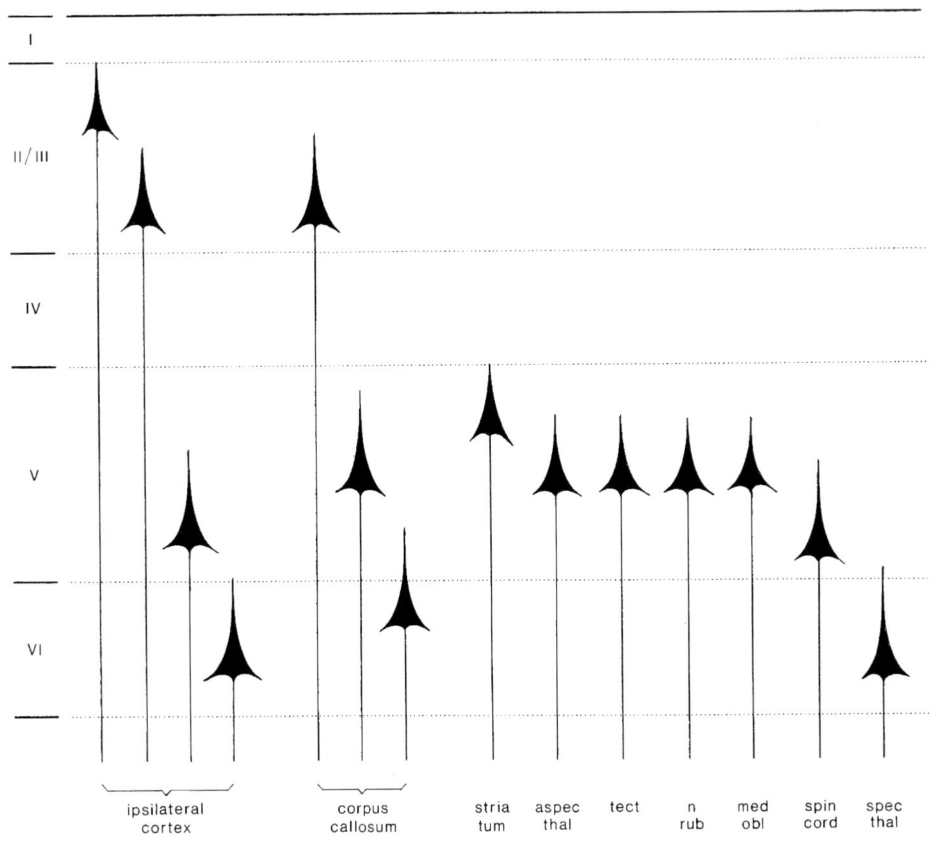

Fig. 15.33. Laminar location of the perikarya of pyramidal cells projecting to other parts of the cerebral cortex and to subcortical centres. *aspec thal*, aspecific thalamic nuclei; *tect*, tectum; *n rub*, nucleus ruber; *med obl*, medulla oblongata; *spec thal*, specific thalamic nuclei; *spin cord*, spinal cord. Based on White [815]

The axons of all typical pyramidal neurons release a number of collaterals before entering the subgriseal white matter (Fig. 15.30 A). These collaterals may ramify within close proximity in the parent cell body or may descend, ascend or travel for shorter or longer horizontal distances within the cortex. Early studies of the intracortical distribution of pyramidal cell axon collaterals were based exclusively on the study of Golgi material, but the real extent of these processes has only recently been revealed by experimental studies in which single pyramidal cells were intracellularly injected with HRP, or in which small extracellular injections of tracer substances, such as biocytin or biotinylated dextran amine (BDA), are placed in the trajectory of fibres or collaterals. Given the numerical preponderance of pyramidal neurons, there can be no doubt that the intracortical collateral branches of these neurons together constitute the largest single category of axons in the neocortex. The endings of these collateral branches, like those of the main axons, all make synapses of the asymmetrical/round vesicle variety and use the excitatory amino acids glutamate and aspartate as neurotransmitters.

The local axon collaterals of pyramidal neurons may show quite characteristic distribution patterns. Thus, Lund and Boothe [426] found that the axons of pyramidal cells situated in different layers of the primary visual cortex of the macaque monkey selectively issue short, horizontal, collateral branches in different layers (Fig. 15.32 D, E).

Apart from local collaterals, the axons of pyramidal neurons may also give rise to long, horizontally disposed branches (Figs. 15.34, 15.35). Pyramidal neurons emitting such long-range collaterals have been the subject of numerous studies, of which the following may be mentioned (for the sake of brevity, the animals studied, the cortical areas and laminae in which the parent somata were located and an indication of the length of the collaterals observed are included in the references): DeFelipe et al. ([138]: monkey sensory motor cortex, lamina III, up to 6 mm); Kisvárday et al. ([370]: cat visual cortex, lamina III, 1500 μm); Gabbott et al. ([218]: cat visual cortex, lamina V, up to 2.64 mm); Ojima et al. ([515]: cat auditory cortex, laminae II, III, 0.7–2.5 mm); McGuire et al. ([457]: macaque visual cortex, lamina III, 2 mm); Kisvárday and Eysel ([368]: visual cortex, lamina III, up to 2.8 mm); Melchitzky et al. ([459]: macaque prefrontal cortex, lamina III, 6 mm); and Rockland and Knutson ([622]: macaque primary visual cortex, lamina VI, 8 mm).

As regards the extent of the long-range collaterals, it has been observed that processes of this type do not remain within the cytoarchitectonic area in which their parent soma lies, but may project to adjacent cortical areas [138, 218].

The number of major collaterals with long horizontal trajectories issued by the main axons of pyramidal neurons varies; most of the cells examined by DeFelipe et al. [138] had one to three long collaterals, whereas the elements studied by Ojima et al. [515] issued two to five such processes.

The long-range primary collaterals are usually coarse and well myelinated.[1] They give off thin, unmyelinated, bouton-laden secondary branches that are predominantly oriented perpendicular to the cortical surface. Remarkably, these secondary branches are emitted in clusters at more or less regular intervals (Fig. 15.35). Because of the overall vertical orientation of the individual secondary branches, these clusters often have a column-like appearance. Secondary branches arising from different major collaterals of the same parent axon in different layers often converge upon the same cluster [138, 370, 515]; the same holds true for the focussed terminal branches of major collaterals of different pyramidal cells [138]. Rockland and Knutson [622] observed that the axons of Meynert cells, i.e. large pyramidal elements in the infragranular layers of the primary visual cortex, issue a number of coarse, remarkably long (up to 8.0 mm) collateral branches, which pass horizontally through the VIth layer of that cortical area. Each of

[1] Myelinated fibres fail to impregnate with the Golgi stain; this may explain why these long-range collaterals remained unnoticed in the classical Golgi studies by Cajal [84] and Lorente de Nó [414, 416].

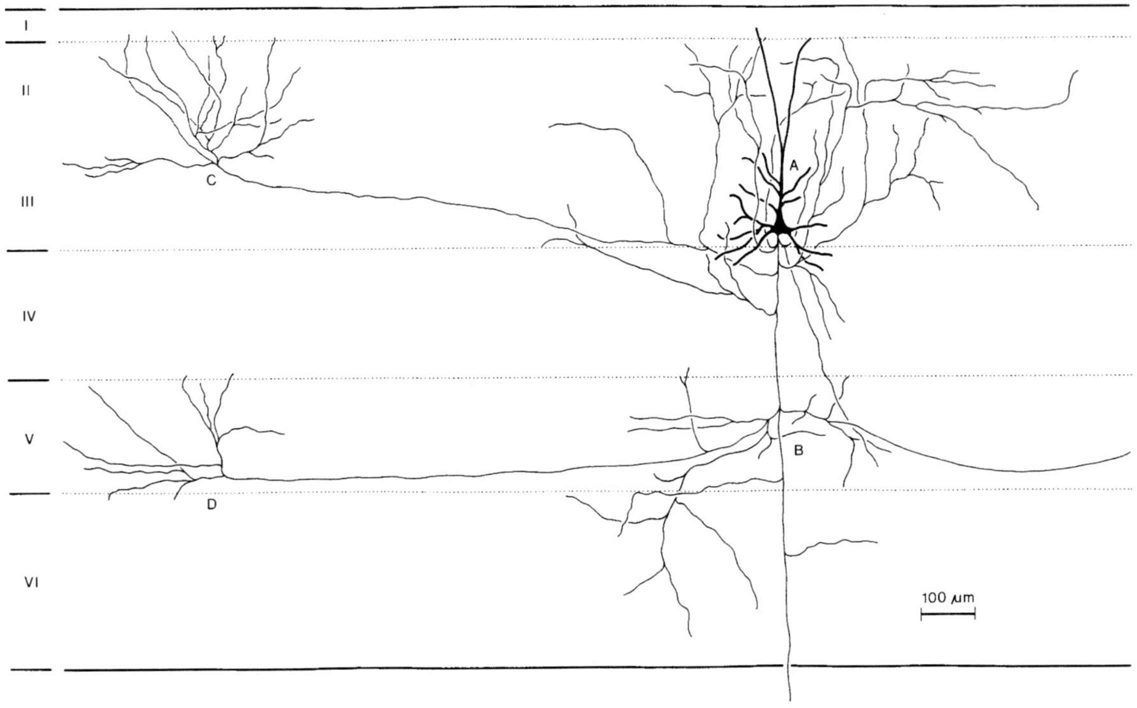

Fig. 15.34. A pyramidal neuron in layer III of the primary visual cortex of the cat. The element was intracellularly filled with horseradish peroxidase and reconstructed from 80-µm-thick serial sections. The intracortical system of collaterals forms distinct clusters, one near the cell's dendritic field in layer III (**A**), another just below the cell in layer V (**B**), and two at a distance of some 1000 µm from the soma in layers III and V (**C, D**). Modified from Kisvárday et al. [370]

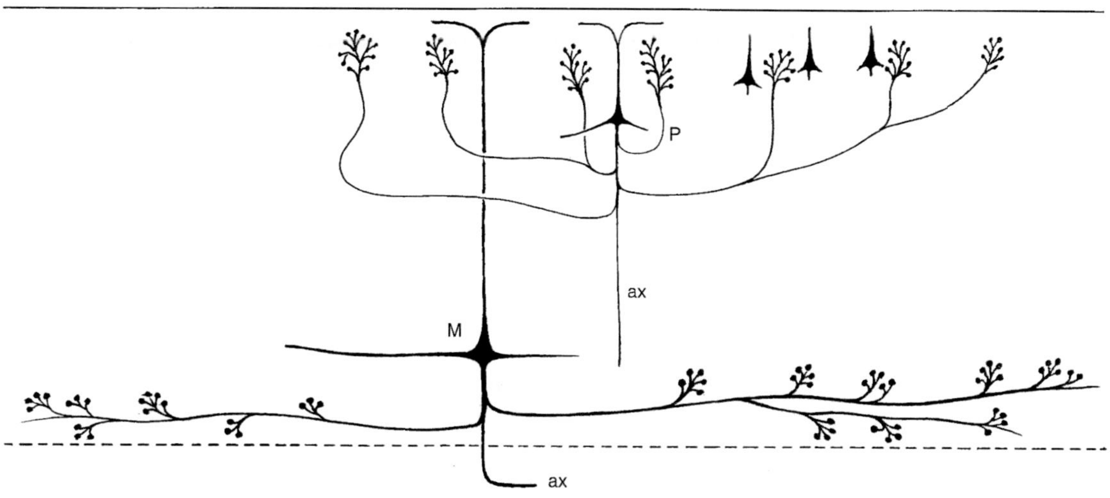

Fig. 15.35. Diagrammatic representation of a Meynert cell (*M*) and a lamina III pyramidal cell (*P*) in the primary visual cortex of the macaque monkey. The Meynert collaterals have (1) relatively small terminal clusters concentrated in lamina VI; (2) large terminal specializations; (3) a small number of terminations per cluster, and (4) a termination-free zone in the vicinity of the soma. On all of these points, these collaterals differ from those of the smaller supragranular pyramidal cells. Modified from Rockland and Knutson [622]. *ax*, axon

these collaterals forms a number of small terminal clusters, containing a small number of large terminal boutons (up to 3.0 μm), and each of these collaterals has an initial termination-free zone. All of these features appeared to be markedly different from those of horizontal intrinsic connections in the supragranular layers of the primary visual cortex (Fig. 15.35). It is noteworthy that the basal dendrites of Meynert cells show a remarkable asymmetrical spread (Fig. 15.35) [84]. It has been proposed [515] that their asymmetrical dendrites make Meynert cells sensitive to visual motion, and in particular to motion in one direction. Comparable asymmetry does not characterize the collateral axonal fields of supragranular pyramidal neurons [622]. Li et al. [409] recently investigated how Meynert cell collaterals are mapped in relation to the functional architecture of the primary visual area in macaque monkeys. They found that such collaterals may cross over several pairs of ocular dominance columns and contact both left- and right-eye ocular dominance columns. According to Li et al. [409], these findings suggest hat the system of Meynert intrinsic collaterals is involved with binocular interactions over wide sectors of the visual field. The main axons of the Meynert cells project to the middle temporal visual area (MT) and to subcortical targets including the superior colliculus [778].

In order to gain insight into the way in which pyramidal neurons participate in the intrinsic circuitry of the neocortex, the synaptic connections of axon collaterals belonging to neurons of this type have been examined with the electron microscope in several areas and in different species [128, 172, 218, 365, 370, 456, 457, 459, 816, 817, 819, 825]. The results of these studies may be summarized as follows:

1. Collaterals of pyramidal cells only form synapses of the asymmetrical/round vesicle variety and, hence, may be considered to excite their targets.
2. Practically all of the synaptic contacts made by these collaterals are on dendritic spines or dendritic shafts.
3. Different types of pyramidal neurons may show striking differences in their local output relationships. Most pyramidal cells examined appeared to synapse predominantly with other pyramidal elements. However, the collaterals of groups of corticothalamic projection neurons in the primary sensory cortex of the mouse were observed to form over 90% of their synapses with the smooth dendrites of non-pyramidal neurons [819]. Pyramidal neurons whose axon collaterals contact the dendrites of pyramidal and non-pyramidal cells in about equal numbers have also been found [456, 825].
4. The number of synapses made by the collaterals of a given pyramidal neuron with one other individual pyramidal neuron is presumably generally very limited [218, 370, 456, 457, 720, 721]. However, the collaterals of one pyramidal cell contact numerous other pyramidal cells and, conversely, one pyramidal cell receives the converging input of numerous other pyramidal cells.
5. Neurons contacted by the collaterals of pyramidal neurons have been identified as non-spiny bipolar cells [456], non-spiny multipolar cells [819] and basket cells [457], and it has also been established that some of the dendrites postsynaptic to pyramidal collaterals are immunoreactive to GABA [370]. It is known that most types of non-spiny or sparsely spiny non-pyramidal cells, including basket cells, chandelier cells and double bouquet cells, use GABA as a neurotransmitter and that these elements are the principal source of the GABAergic, symmetrical synapses that impinge upon the somata, proximal dendrites and axon initial segments of pyramidal neurons [306]. All in all, it seems reasonable to assume that the GABAergic interneurons in the neocortex receive input directly from the axon collaterals of pyramidal neurons and in turn synapse with pyramidal neurons. These circuits probably provide the morphological substrate for both feed-forward and feed-back inhibition of pyramidal neurons [306, 815].

If we survey the data concerning the typical pyramidal neurons discussed above, it appears that these elements show a quite remarkable structural diversity. This diversity may concern their size, their laminar position, the branching pattern of their dendrites, the density of spines along their apical dendrites, their affinity to particular afferent systems, the cortical or subcortical target regions to which their main axons project, the distribution of their axonal collaterals and their patterns of intracortical synaptic output. Certain structural properties are clearly correlated. It has been demonstrated that the somata of pyramidal neurons projecting to a particular target are not only located in one and the same layer or sublayer, but also show striking similarities with regard to dendritic morphology, thalamocortical connectivity and distribution of their axon collaterals. It seems likely that all pyramidal neurons projecting to a particular target are in receipt of similar extracortical and intracortical inputs and that they participate in a similar way in the intrinsic circuitry of the cerebral cortex.

Atypical Pyramidal Cells

In the mammalian neocortex, neurons occur that lack one or several of the features characterizing typical pyramidal cells, but that are nevertheless considered to belong to the pyramidal cell group. These atypical pyramids include: (a) elements in which the apical dendrites are shortened (Fig. 15.32 B, C), reduced (Fig. 15.30 B) or absent [171, 756, 775]; (b) "star pyramids", in which the dendrites radiate out from the soma in all directions, rather than forming a basal skirt (Fig. 15.30 C); (c) sparsely spiny pyramidal cells [149, 277, 315, 836, 837]; (d) "pure" projection pyramidal neurons, whose axons do not emit any intracortical collaterals [234]; and (e) intrinsic pyramidal neurons, whose axons remain within the cortex and often ascend (Fig. 15.30 B) or form ascending and descending branches (Fig. 15.30 C) [363, 756]. However, the most remarkable modified pyramidal neurons are doubtless the so-called *spiny stellate cells* (Fig. 15.30 D–F). These elements occur exclusively in lamina IV of primary sensory areas of the neocortex, where they may be abundant [336, 422, 823]. Their small, spherical or ellipsoid somata in Nissl preparations often form one or several conspicuous granular zones, which is why laminae I–III and V–VI in the sensory cortices are often designated as the supragranular and infragranular layers. The dendrites of these spiny stellate cells are emitted at several points from their soma and are generally confined to the fourth layer or to the sublayer in which their soma is situated. They may be strongly stratified horizontally or may show a radiating or even vertically elongated distribution [423]. The spiny stellate cells are most probably the principal, but not the exclusive targets of the thalamocortical afferents terminating in lamina IV of the sensory cortical areas [208]. Ahmed et al. [4] studied the connectivity of spiny stellate cells in the visual cortex of the cat. They presented evidence suggesting that, of the asymmetrical synapses on these elements, 45% comes from lamina VI pyramidal cells, 28% from other spiny stellate cells and 6% from thalamic afferents. The source of the remaining 21% of asymmetrical synapses remained obscure. Spiny stellate cells presumably receive inhibitory afferents from neurogliaform cells, basket cells and chandelier cells [338, 424, 815].

The axon of the spiny stellate cells may ascend to more superficial layers or descend to deeper layers. Many of these elements are provided with both ascending and descending axonal branches. Short collateral branches establishing synaptic contacts within the sublayer of origin are invariably present (Fig. 15.30 D–F). The axon terminals of spiny stellate cells form asymmetrical synapses mainly with dendritic spines [446, 646, 779]. The investigations of Lund and collaborators [422–424, 426] have shown that the spiny stellate cells, present in the various sublayers of layer IV of the monkey primary visual cortex, differ markedly with respect to their projections to other layers. It seems likely that the short local collaterals of the spiny stellate cells mainly contact other spiny stellate cells [646] and that their ascending and descending axonal branches establish numerous synaptic contacts with superficial and deep pyramidal neurons, but direct evidence for the existence of such outputs is lacking [137]. Nevertheless, it may be safely assumed that spiny stellate cells play a crucial role in the radial propagation of the activity fed by thalamocortical afferents into layer IV of primary sensory cortical areas.

Although spiny stellate cells represent typical local circuit neurons and lack an apical dendrite, they nevertheless have to be considered as modified pyramidal neurons [423, 756]. The reasons for this interpretation are as follows:

1. The dendrites of pyramidal neurons as well as those of spiny stellate cells are densely covered with spines and the axon terminals of both cell types form synapses of the asymmetrical type, mainly with dendritic spines [403, 646].
2. In lamina IV of the primary sensory areas containing spiny stellate cells, neurons provided with a more or less developed apical dendrite are frequently observed that in all other respects cannot be distinguished from typical spiny stellate neurons. The "star pyramids" in the somatosensory cortex of the squirrel monkey [336] are a good example of such an intermediate form (compare Fig. 15.30 C with F).
3. Peinado and Katz [543] have presented evidence that during ontogeny lamina IV stellate cells initially extend an apical dendrite to lamina I and only later lose this process and develop their mature stellate morphology.

Local Circuit Neurons

In the preceding sections, the neocortical pyramidal neurons have been discussed. It was pointed out that this category not only encompasses true pyramidal cells, but also atypical elements that have lost one or several of the typical pyramidal characteristics. Pyramidal neurons account for 60–85% of the total neuronal population of the neocortex [240, 556, 589, 824]. The remaining 15–40% of neocortical neurons include a variety of morphological types that have the following features in common:

1. They are evidently non-pyramidal, i.e. they have no conical soma and lack a dominant apical dendrite. On that account, the group is often referred to as non-pyramidal, but this designation is not entirely satisfactory because many neurons belonging to the pyramidal category do not show a pyramidal morphology (Fig. 15.30 C–F).
2. Their dendrites bear only few spines or are entirely spine-free. This is a very important distinguishing feature, even though the dendrites of some otherwise typical pyramidal neurons are also aspinous [315].
3. Their somata have both symmetrical and asymmetrical axosomatic synapses, whereas pyramidal cell bodies possess only symmetrical axosomatic synapses [557].
4. Their axons do not leave the cortex, which is why the cells under consideration are often referred to as local circuit neurons. However, it should be recalled that many pyramidal cells also possess exclusively intracortical axons (Fig. 15.30 B–F) and, hence, also belong to the category of cortical local circuit neurons.
5. With a single exception, their axon terminals contain flattened vesicles and form symmetrical synapses with their postsynaptic targets, both features suggesting an inhibitory function [177, 403, 539].
6. An inhibitory function is also suggested by the fact that most of the cells under consideration use GABA as their primary neurotransmitter [285, 306, 371, 558, 616].
7. A certain proportion (25–30%) of the GABAergic cortical neurons also express one or several neuropeptides. The neuropeptides detected in cortical neurons include substance P, vasoactive intestinal polypeptide (VIP), cholecystokinin (CCK), neuropeptide Y (NPY), somatotropin-release-inhibiting factor (SRIF), corticotropin-releasing factor (CRF) and tachykinin (TK) [304, 350, 440]. Several subpopulations of GABAergic cortical neurons appeared to be definable on the basis of their immunoreactivity for certain neuropeptides. To give a few examples: in the visual cortex of the rat a subpopulation of bipolar cells can be labelled with antibod-

ies to VIP [559], whereas in the neocortex of the monkey, subpopulations of chandelier cells and of double bouquet cells have been shown to be immunoreactive for CRF [406] and TK [140], respectively. Another subpopulation of double bouquet cells has been shown to be immunoreactive for SRIF [133].

8. It has been shown that differential immunoreactivity for the calcium-binding proteins parvalbumin (PV), calbindin (CB) and calretinin (CR) can be used as a selective marker for different subpopulations of neocortical local circuit neurons [105, 134, 174]. In the primate neocortex, among the most characteristic types of neurons immunoreactive for PV are chandelier cells and large basket cells, whereas for CB they are double bouquet cells, and for CR, double bouquet and bipolar cells [134].

In brief, the mammalian neocortex contains a large population of non-pyramidal, inhibitory interneurons with smooth or sparsely spinous dendrites. These elements use GABA as their primary neurotransmitter and some also produce one or several neuropeptides. The population of neocortical neurons thus outlined is morphologically heterogeneous, and numerous authors, including Lorente de Nó [416], Jones [336], Feldman and Peters [179], Peters and Jones [561], Fairén et al. [177], Lund [424], Lund and Yoshioka [428], Lund and Lewis [427], Lund et al. [430] and DeFelipe [136] have attempted to subdivide this population of neurons into different groups or types, using the size and shape of the somata, the shape of the dendritic field and the number, distribution, length and branching patterns of individual dendrites, the preferred direction of axons and axonal branches and configuration of axonal terminals as criteria. The following discussing of the non-pyramidal, smooth or sparsely spinous local circuit neurons in the mammalian neocortex is based on the publications cited above, on the excellent characterizations of particular cell types [338, 343, 554, 555, 563, 695], published in the first volume of the *Cerebral Cortex* [560], and on other sources to be quoted below. As has already been indicated, many of the morphological types of neocortical interneurons can be subdivided into two or more biochemically definable subgroups. The synaptic relationships of the various types of neocortical interneurons to be discussed are semidiagrammatically indicated in Fig. 15.37 B.

Stellate neurons are found in all cortical layers (Fig. 15.36 A, C, D) [416]. The dendrites of these elements radiate from the soma in all directions and branch infrequently. Their axonal arborization forms a local plexus occupying approximately the same territory as that covered by the dendrites. Distinctive axonal terminals are lacking [563] (Fig. 15.36 B). Because of the limited spread of their axonal system, the stellate cells are also referred to as local plexus neurons.

Neurogliaform or spiderweb cells form a special class of stellate cells (Fig. 14.36 B). These elements have a small, spherical soma and short sinuous dendrites. Their axons arborize profusely around the soma, forming a dense feltwork. Characteristically, small empty spaces appear in this feltwork, representing the positions occupied by the unstained somata of other neurons [338]. Neurogliaform or spiderweb cells have been observed in all layers of the cortex, but they are particularly concentrated in the primate somatosensory and primary visual cortex. The function of the spiderweb cells is unknown. Jones [338] considered it likely that the elements concentrated in lamina IV of sensory cortices receive thalamic afferents and synapse mainly with the spiny stellate cells of that layer. Kisvárday et al. [371] observed synaptic contacts between the axons of neurogliaform cells and the distal dendrites of pyramidal cells.

There are numerous neurons in the neocortex which, judging from the disposition of their dendritic tree, would fall in the category of stellate cells, but which are distinguished from the elements in that category by the course and/or mode of termination of their axons. The chandelier cells and basket cells, which have already been discussed in a previous section, represent two distinct types of such specialized stellate cells.

Chandelier cells have been thus named because their profuse axonal plexuses give rise to

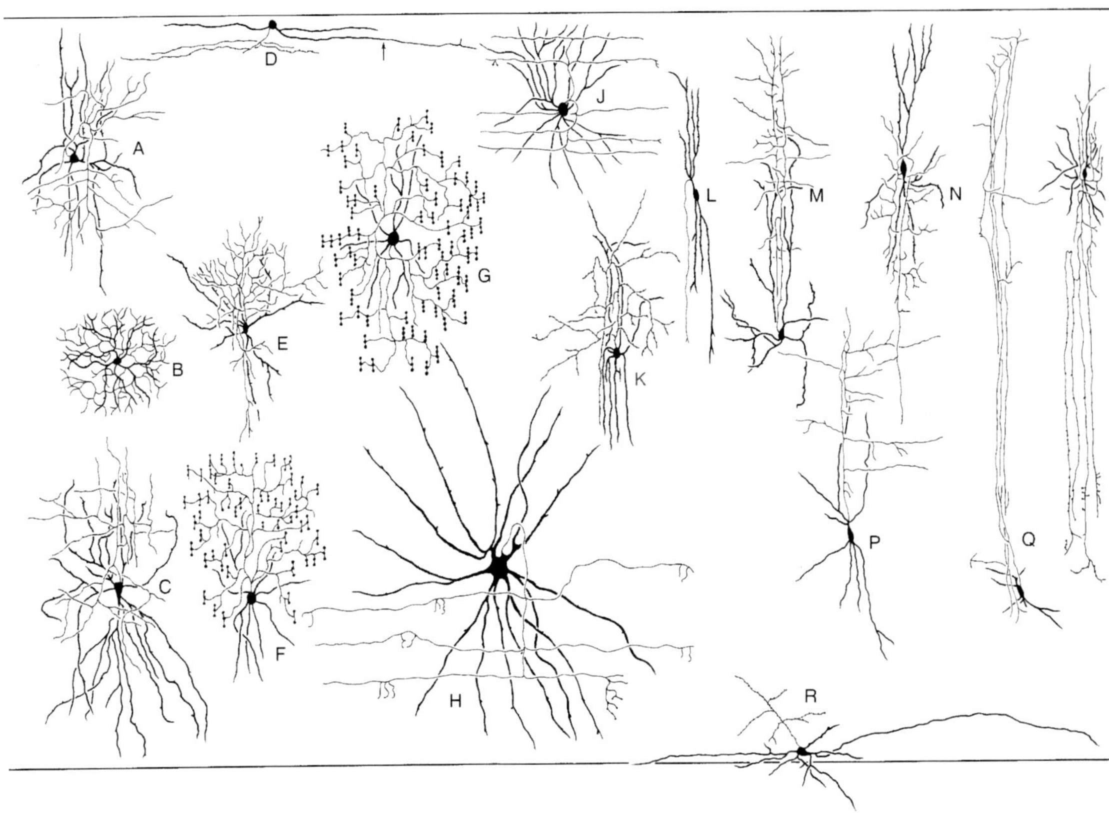

Fig. 15.36. Local circuit neurons with smooth or sparsely spiny dendrites in the neocortex. *A*, Multipolar neuron with axonal arcades in lamina III of the somatosensory cortex of the squirrel monkey (Jones [336]); *B*, neurogliaform or spiderweb cell from layer IV of the somatosensory cortex of the squirrel monkey (Jones [338]); *C*, stellate neuron with sparsely spiny dendrites extending from lamina IVC to lamina VI in the primary visual cortex of the macaque (Lund [422]); *D*, horizontal cell in lamina I of the neocortex of the hedgehog (Valverde and Facal-Valverde [757]); *E*, stellate cell from lamina IV in rat visual cortex (Peters and Saint Marie [563]); *F*, *G*, chandelier cells in laminae III and V of the neocortex of the macaque (Jones [340]); *H*, *J*, basket cells in laminae II and V of the neocortex of the macaque (Jones [340]); *K*, neuron with local, beaded axon in lamina IV of the prefrontal cortex of the macaque (Lund and Lewis [427]); *L*, bipolar or vertical cascade neuron in lamina III of the prefrontal cortex of the macaque (Lund and Lewis [427]); *M*, neuron with rising axon in lamina IV of the prefrontal cortex of the macaque (Lund and Lewis [427]); *N*, neuron with axon connecting laminae II and III in the prefrontal cortex of the macaque (Lund and Lewis [427]); *O*, double bouquet cell in lamina II of the primary visual cortex of the macaque (Werner et al. [810]; *P*, neuron with rising axon in upper zone of lamina V of the primary auditory cortex of the cat (Fairén et al. [177]); *Q*, Martinotti cell in lamina V of the visual cortex of the cat (Wahle [801]); *R*, horizontally oriented neuron in lamina VI of the prefrontal cortex of the macaque (Lund and Lewis [427])

a large number (up to 300) of highly characteristics, vertically oriented "candles", each consisting of a preterminal axonal branch that forms a short row of terminal boutons. These vertical arrays, which are designated as axon terminal cartridges, synapse with the axon initial segments of pyramidal cells (Figs. 15.31B, 15.36F, G) [177, 555]. The axon initial segments of spiny stellate cells may also receive synapses from chandelier cells [424].

Chandelier cells occur in layers II–V, but are most common in layer II. Because the pyramidal neurons projecting to the ipsilateral and contralateral neocortex are mainly situated in layers II and III and because the axon initial segments of these supragranular pyramids receive a much richer synaptic supply from the chandelier cells than the infragranular elements, it has been suggested that chandelier cells principally influence cortico-cortical circuitry [570, 698]. Chandelier cells typically express one or both of the calcium-binding proteins PV and CB [134].

It is noteworthy that the axon initial segments of pyramidal neurons are not exclusively contacted by chandelier cells. Gonchar et al. [250] detected in the visual cortex of rat and monkey a population of somatostatin-expressing, GABAergic neurons, whose axons innervate somata, dendritic spines and initial segments of pyramidal neurons. It is also noteworthy that, according to DeFelipe [135], loss of chandelier cells may represent a key component in the aetiology of epilepsy.

Basket cells are among the largest non-pyramidal cells in the neocortex (Figs. 15.31A, 15.36H, J). Their poorly ramifying dendrites radiate in all directions, but vertically oriented dendrites prevail in some and give these cells a bitufted appearance. The axons of the basket cells are either ascending or descending and give rise to four or more horizontal branches at various levels. These collateral branches are myelinated and may reach a length of 1 mm or more. At intervals, they issue short ascending or descending terminal branches that contribute to the formation of pericellular baskets around pyramidal cell bodies. Each terminal branch forms a series of synapses with its target soma. One basket cell contributes to numerous baskets, and terminal axonal branches of several basket cells contribute to a single, pericellular plexus. The somata of basket cells are concentrated in laminae II and V.

These cells – with their radiating short dendrites and their long horizontal axonal branches – form a distinct group of cortical local circuit neurons, which have been designated as *large basket cells* [177]. Cells of this type have so far only been observed in the cat and monkey [343]. They can express PV, CB, NPY, CCK and occasionally CR and SRIF. They never express VIP [440].

In the neocortex of different mammals, both "higher" and "lower" small, smooth or sparsely spinous intrinsic neurons have been observed, whose axons consistently produce multiple synaptic contacts on the somata of pyramidal cells. Fairén et al. [177] named these elements *small basket cells*. According to their observations, these small basket cells are multipolar elements with rather wide dendritic fields, sometimes showing a predominant dendritic tuft oriented towards the pial surface. The axon is primarily descending and typically produces frequently ramifying and "curvy" collateral branches. They differ from other basket cells in that they express VIP [440].

A third class of basket cells is formed by the so-called *nest basket cells*. These elements, which were only recently shown to represent a distinct class of soma-targeting cell, occupy an intermediate position between large and small basket cells [805]. Their primary axonal branches are long and ramify sparsely, much like those of the large basket cells, but their clusters of secondary axonal branches more closely resemble those of the small basket cells. The nest basket cells do not typically express CR and never express VIP [440].

Approximately 50% of all inhibitory neocortical interneurons are basket cells [440].

The classical descriptions of basket cells presented by Cajal [84] created the impression that these elements synapse exclusively with the somata and proximal dendrites of pyramidal neurons. However, in several later studies [207, 369, 697, 700], evidence was presented

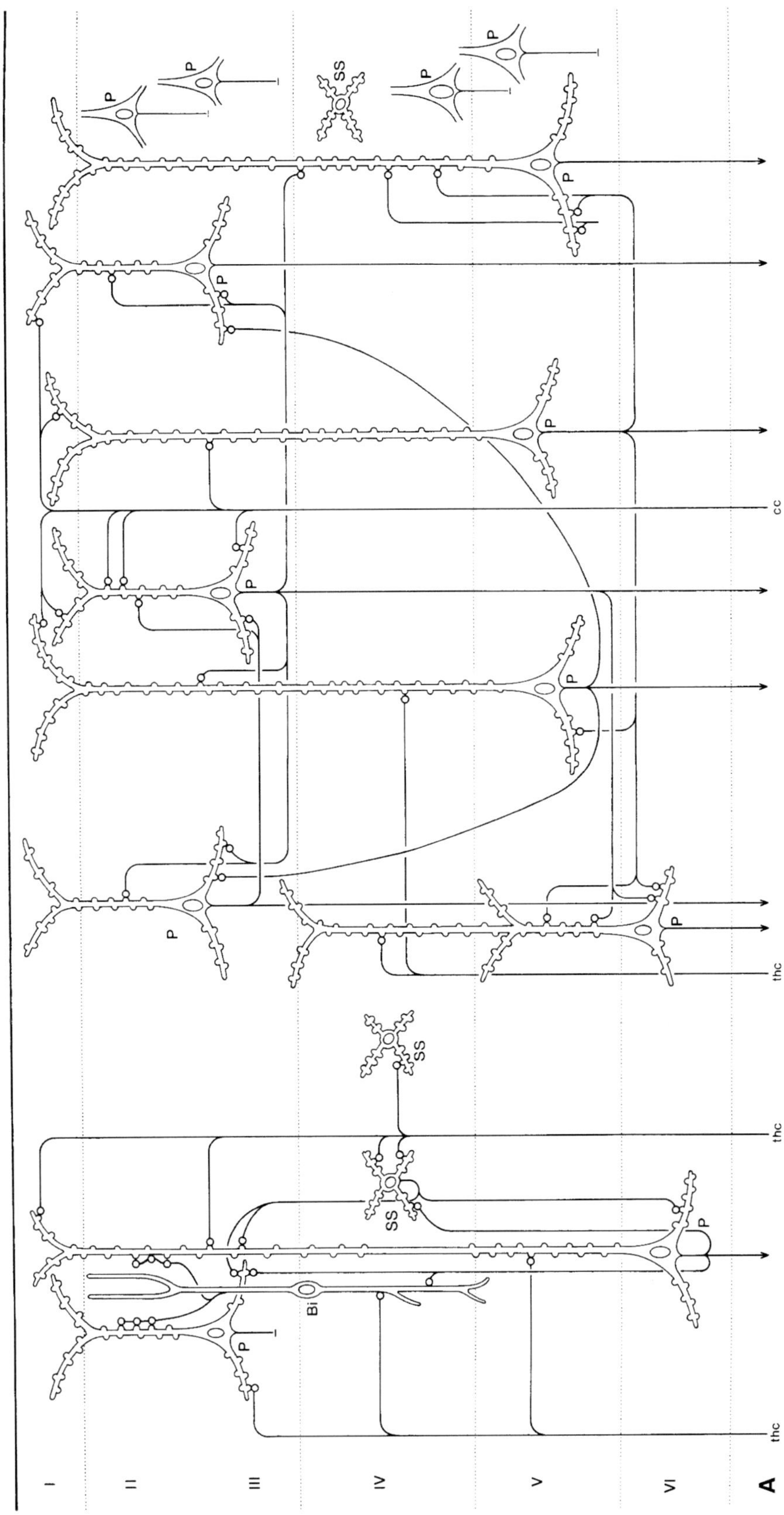

I
II
III
IV
V
VI
A
P
SS
Bi
thc
cc

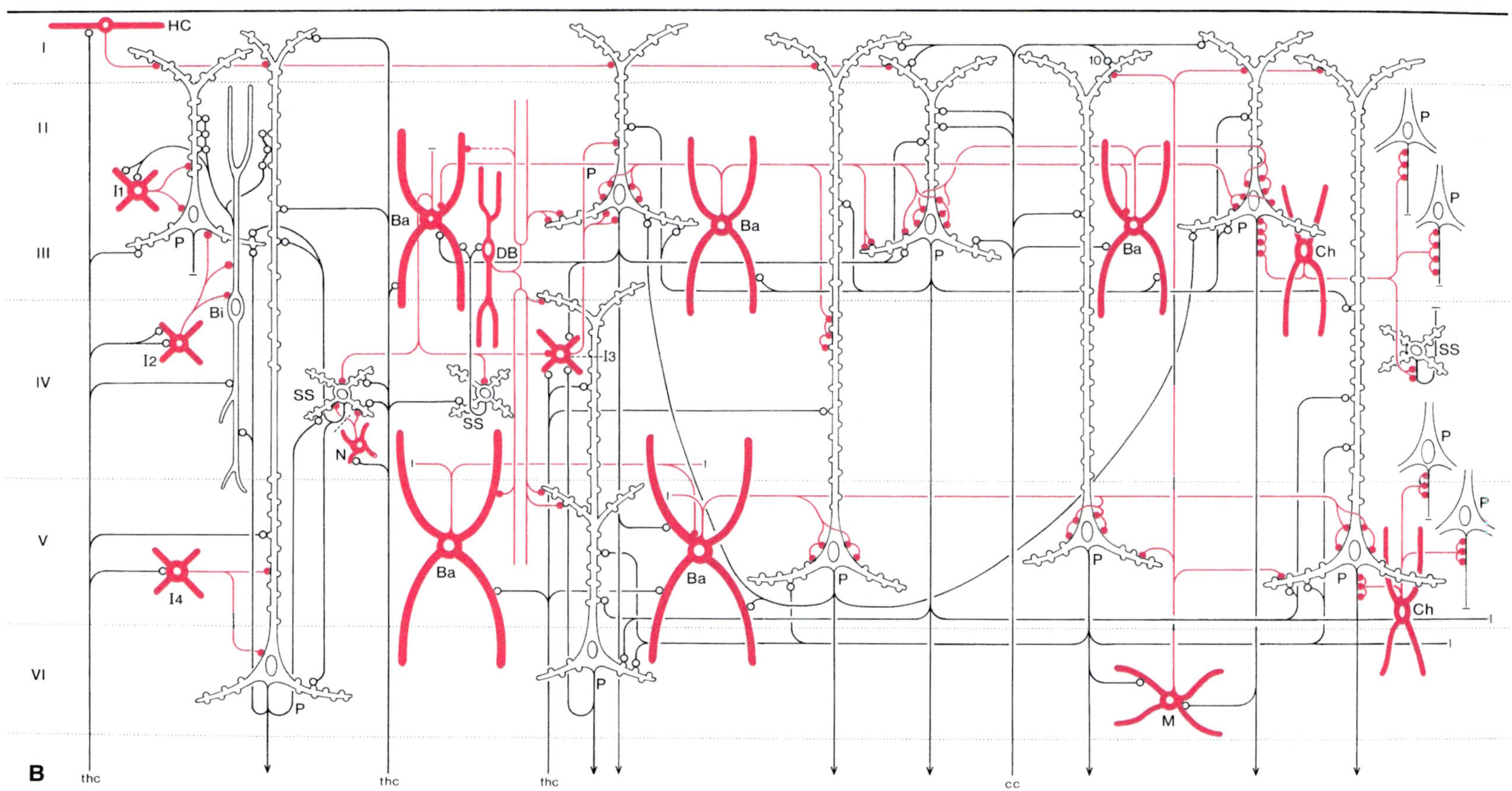

Fig. 15.37. Neocortical circuits showing excitatory elements in **A** and excitatory plus inhibitory elements in **B**. *Ba*, basket cells; *Bi*, bipolar cell; *Ch*, chandelier cell; *cc*, cortico-cortical fibre; *DB*, double bouquet cell; *HC*, horizontal cell of Cajal; *I1*, *I2*, etc., different types of interneurons mentioned in the text; *M*, Martinotti cell; *N*, neurogliaform or spiderweb cell; *P*, pyramidal neurons; *SS*, spiny stellate cells; *thc*, thalamocortical fibres; *I*, *II*, etc., cortical layers. Excitatory neurons and their synaptic terminals are shown by open profiles in *black*; inhibitory neurons and their terminals by filled profiles in *red*. Modified from Nieuwenhuys [509]

suggesting that axon terminals of basket cells also synapse with the somata of spiny stellate cells, the distal dendrites and axons of pyramidal cells and the somata and dendrites of non-pyramidal neurons. Kisvárday et al. [372], using biocytin as a label and taking advantage of the fact that large basket cells are GABAergic and contain PV, demonstrated that in the visual cortex (area 18) of the cat these cells not only synapse with pyramidal neurons, but also establish an average of four to six perisomatic contacts onto other large basket cells. A large basket cell in lamina II was found to synapse with the somata of 58 other large basket cells, whereas a large lamina V basket cell appeared to contact 33 of its fellow cells. From these observations, Kisvárday et al. [372] concluded that large basket cells form an interconnected network in area 18 of the visual cortex. Assuming that the GABAergic large basket cells are inhibitory, they proposed that: (a) a large basket cell provided direct perisomatic inhibition onto a number of pyramidal cells (at least 200–300) and in the range of about 50 other basket cells, (b) the target cells of the directly inhibited basket cells become facilitated via a disinhibitory effect and (c) the number of neurons disinhibited through this process may well greatly exceed the number of elements that are directly inhibited by the large basket cells.

Vertically Oriented Neurons. If we consult the extensive inventories of neocortical neurons provided by Cajal [84], Lorente de Nó [414], Jones [336], Feldman and Peters [179], Fairén et al. [177], Lund [422, 424], Lund and Yoshioka [428], Lund and Lewis [427], Lund et al. [429, 430], DeFelipe [136] and others, it appears that numerous smooth or sparsely spinous local circuit neurons in this structure show an overall vertical orientation. This orientation may concern their dendritic trees (Fig. 15.36 K), their axonal systems (Fig. 15.36 M–Q) or both (Fig. 15.36 L). Among the neurons with vertically oriented axons, elements with rising axons (Fig. 15.36 M, P, Q), descending axons (Fig. 15.36 L, N) and both descending and rising axons (Fig. 15.36 O) may be distinguished. A discussion of all of the types of vertically oriented cortical interneurons described in the literature is not possible here. However, some, i.e. the bipolar cells, the bitufted cells, the double bouquet cells and the elements with ascending axons known as Martinotti cells, may be briefly commented upon.

Bipolar cells have a small, spindle-shaped, vertically oriented cell body, from which one primary ascending and one primary descending dendrite arises (Fig. 15.32 G (b)) [554]. These two processes and their ramifications, which bear few or no spines, extend for long distances through the depth of the cortex, producing a narrow and very elongated dendritic tree. The axons of the bipolar cells usually arise from one of the primary dendrites and form a plexus that is also narrow in spread and vertical in orientation [554]. Bipolar cells occur in layers II–VI and typically express VIP and CR [440].

Ultrastructural investigations have shown that there are two different populations of bipolar neurons, one forming symmetrical synapses and the other forming asymmetrical synapses [106, 177, 559, 562]. The bipolar cells with axons forming symmetrical synapses release principally GABA, but also express VIP [440, 458, 571]. The axons of these inhibitory elements preferentially synapse with dendritic shafts.

A certain proportion of the neocortical bipolar neurons that form symmetrical synapses can be labelled with antibodies to ChAT, a specific marker for cholinergic neurons. The axon terminals of these cholinergic bipolar cells most commonly synapse with small- to medium-sized dendritic shafts and less frequently with apical dendrites and with the somata of neurons [307, 540].

The bipolar cells that form asymmetrical synapses are excitatory by releasing only VIP [440]. The axons of these elements give rise to vertically oriented branches which parallel the trajectories of clustered apical dendrites of pyramidal neurons, forming multiple synapses with spines on these processes (Fig. 15.32 G) [562]. However, axonal branches of the excitatory bipolar cells have also been observed to contact the shafts of apical dendrites and the somata and dendrites of non-pyramidal cells.

Bitufted cells have dendrites arising mainly from the upper and lower poles of the soma, forming, as the name implies, two dendritic tufts [557]. The dendrites diverge initially, but at some distance from the soma often assume a radial orientation. Many neurons of this type have local axonal plexuses which partly overlap with the field occupied by their dendritic trees [557]. From these local axonal plexuses, radially oriented branches may extend into the adjoining cortical layers. Bitufted cells are dendritic-targeting cells that occur in layers II–VI [702]. They can express NPY, SRIF, VIP, CCK, CB and CR, but not PV [440].

Among the bitufted cells, elements are found whose axons produce tight, radially oriented long plexuses of thin, parallel axonal branches. These plexuses are generally designated as "bundles" or "horse-tails". Smooth or sparsely spinous cortical neurons provided with these highly characteristic long, fascicular axonal systems have been commonly referred to as double bouquet cells [177, 554, 561, 695]. Cajal [84] used the name *cellules à double bouquet dendritiques* to refer to a number of cell types with highly different axonal ramification patterns, including elements with the long radially oriented arrays of axonal branches discussed above. However, it has become customary to designate only smooth or sparsely spinous cells showing this particular axonal pattern as double bouquet cells. In the current literature, even elements with these long vertical axonal branches whose dendritic trees are not of a bitufted appearance are still referred to by this name (Fig. 15.36 O) [336, 695].

Typical double bouquet cells have only been observed in laminae II and III of the neocortex of cats and primates [33, 695]. It has been shown that double bouquet cells form symmetrical synapses containing flat or pleiomorphic vesicles with their target structures and use GABA as their principal neurotransmitter [139, 695]. It was formerly believed that the vertically oriented axons of double bouquet cells mainly synapse with the apical dendrites of pyramidal neurons [84, 103, 721]. However, the studies by Somogyi and Cowey [694] and by DeFelipe et al. [139, 140, 146] showed that the axon terminals of these cells do not form synapses with apical dendrites, but rather with basal dendrites and with oblique branches of the apical dendrites of pyramidal neurons and with postsynaptic structures belonging to non-pyramidal neurons. Double bouquet cells express CB [139, 146] and can also express VIP or CCK [440].

Martinotti cells are multipolar or bitufted neurons with smooth or sparsely spinous dendrites. Their distinguishing feature is a long ascending axon which reaches layer I, where it forms a terminal arborization. The axon emerges either from the upper surface of the soma or from an ascending dendrite. The initial part of the axon gives rise to a number of descending collaterals, which form a local terminal plexus [84, 177, 442, 643, 754]. "Classical" Martinotti cells have a single ascending axon, but Jones et al. [351] and Wahle [801] described "double bouquet type" Martinotti cells whose axon branched into a bundle of two to eight long, ascending collaterals (Fig. 15.36 Q).

Martinotti cells occur in all cortical layers except layer I, but they have been most frequently found in layers V and VI. They use GABA as a neurotransmitter [696] and may additionally contain a neuropeptide, e.g. TK [351] or somatostatin [801]. Little is known with certainty about the afferent and efferent connections of the Martinotti cells. On the basis of a Golgi study of the visual cortex of the mouse, Ruiz-Marcos and Valverde [643] suggested that Martinotti cells situated in layer V of the cortex receive terminals from both superficial and deep pyramidal cells, that their local axonal plexuses contact deep pyramidal cells and that their terminal plexuses in layer I impinge upon the apical dendritic tufts of pyramidal cells. It has been observed that the axons of certain Martinotti cells, after having reached lamina I, give rise to long, horizontal collaterals [414]. According to Szentágothai [721], these branches may run for several millimetres through lamina I, making synapses with the apical dendrites of numerous pyramidal cells.

Horizontal Cells. Local circuit neurons showing an overall horizontal orientation are almost exclusively found in layers I and VI. It is important to note that these layers, although far apart in the adult cortex, are both derivatives of a single embryonic pallial zone, i.e. the primordial plexiform layer. During the formation of the cortex, immature bipolar cells migrate peripherally and together form a compact cortical plate, which splits up the primordial plexiform layer into a superficial and deep zone. The former becomes layer I, and the latter gives rise to the deep zone of layer VI in the mature cortex. The intervening layers are all derivatives of the cortical plate (Fig. 2.10 H–K) [437].

Horizontal Cells of Cajal. Bipolar neurons occurring in layer I of the cortex are known as the *horizontal cells of Cajal.* These elements are provided with one or a few long, smooth dendrites that pursue a course parallel to the cortical surface. Their axons, which pass horizontally like the dendrites, may attain a considerable length (Fig. 15.36 D). Curiously enough, horizontal cells in layer I may have more than one axon [84, 757]. The elements under discussion are GABAergic and additionally contain the neuropeptide CCK [779]. The axons of the horizontal cells probably enter into synaptic contacts with the apical dendritic branches of pyramidal neurons [416, 757]. Several authors [84, 167, 438, 537, 757] have expressed the opinion that the conspicuous Cajal-Retzius cells found in layer I of the immature cortex undergo morphological changes and transform to horizontal cells.

Horizontal Cells in Layer VI. The *deep zone of layer VI* contains numerous medium-sized horizontal cells (Fig. 15.36 R) [427, 497, 498]. One or a few long dendrites arise from the ends of their fusiform cell bodies, but some shorter dendrites may also arise from the upper and lower surface of the soma. Their axons, which like the principal dendrites often pursue a horizontal course, give off varicose side branches. In Golgi material, the axons of the horizontal cells under discussion can rarely be traced to their final destination, and very little is known about the afferent and efferent connections of these elements. The fact that the horizontal cells in layer VI most probably synthesize GABA [760] and that their cell bodies have both symmetrical and asymmetrical axosomatic synapses [556] indicates that they represent inhibitory local circuit neurons.

Remaining Inhibitory Interneurons. There are many other smooth or sparsely spinous, inhibitory neurons in the neocortex that are difficult to classify or that have not yet been recognized as belonging to a particular morphological type [136]. In Fig. 15.37 B some of these elements, designated provisionally as I1–I4, are included.

Element I1, which is situated in layer II, receives excitatory afferents from bipolar cells, whereas its axon forms symmetrical, presumably inhibitory synapses with the apical dendrites and the somata of superficial pyramidal neurons [562].

Element I2, which is situated in layer IV, is contacted by thalamic afferents and its axonal endings impinge upon the somata of superficial pyramidal neurons, as well as on the somata and proximal dendrites of bipolar cells [562].

Element I3 is a local plexus neuron situated in the upper part of layer IV. It receives afferents from thalamocortical fibres, from axon collaterals of superficial and deep pyramidal neurons, and from basket cells. Its ascending axon forms synaptic contacts with the apical dendritic shafts, somata, and axon initial segments of superficial pyramidal neurons [815].

Element I4, finally, is a local plexus neuron situated in the upper part of layer V. It receives terminals from thalamocortical fibres, and its axonal endings establish synaptic contacts with the somata and apical dendritic shafts of deep pyramidal neurons [815].

It is noteworthy that defects in the neurotransmission of GABAergic neocortical interneurons may well play a role in the pathophysiology of schizophrenia. A detailed discussion of the voluminous literature on this subject is beyond the scope of the present work (see [46] for a review). However, the results of two recent studies, those of Lewis et al. [407] and of Konopaske [384], may be briefly mentioned.

Lewis et al. [407] found that a deficiency in signalling by brain-derived neurotropic factor through its receptor, the tyrosine kinase Trk (tropomyosin-related kinase) B, leads to reduced GABA synthesis in the PV-containing subpopulation of inhibitory GABAergic neurons in the dorsolateral prefrontal cortex of individuals with schizophrenia. They indicate that the resulting alteration in perisomatic inhibition of pyramidal neurons contributes to a diminished capacity for the synchronized neuronal activity that is required for working memory function.

Konopaske et al. [384] presented evidence suggesting that in schizophrenia the neurotransmission in another type of neocortical interneuron, the chandelier cell, might also be compromised. As has been discussed, the axon terminals of these GABAergic elements form linear arrays, termed cartridges, that synapse on the axon initial segments of pyramidal neurons (Fig. 15.31 B). These cartridges are immunoreactive for the GABA membrane transporter-1, which regulates the duration and efficiency of GABAergic neurotransmission. Konopaske et al. [384] reported that the density of GABA membrane transporter-1-immunoreactive cartridges is reduced in schizophrenia, particularly, but not exclusively, in the prefrontal cortex.

Microcircuitry of the Neocortex

Introduction

The data concerning the synaptic connections of neocortical neurons and principal neocortical afferents, discussed in the previous section, were employed to compose two summary diagrams of the microcircuitry of the neocortex (Fig. 15.37 A, B). Moreover, the principal results of some recent studies on the connectivity of a certain type of interneuron have been summarized in an additional diagram (Fig. 15.38). (The symbols used in the text – Ba, Bi etc. – correspond to those in the figures and are explained in the legend of Fig. 15.37.) Before commenting on these diagrams, a few cautionary remarks should be made.

1. The assembled data are derived from many different species and from many different cortical areas; hence, it can not be expected a priori that the diagrams present a reliable picture of *the* microcircuitry of *the* mammalian cortex. However, it is worthy of note that the extensive ontogenetic and cytoarchitectonic studies of Brodmann [70] have shown that all neocortical areas in all of the many species he investigated represent variations on a common basic plan, which he designated as "der sechsschichtige tektogenetische Grundtypus". Rockel et al. [620] counted the number of neuronal cell bodies in a narrow strip (30 μm) through the depth of the neocortex in several functional areas (motor, somatosensory, primary visual, frontal, parietal and temporal) and in many species (mouse, rat, cat, monkey and human). With the exception of the primary visual cortex in primates the same absolute number (about 110) of neurons was found in all areas and in all species. In the primate primary visual cortex there appeared to be approximately 2.5 times more neurons. According to Rockel et al. [620] these findings suggest that the intrinsic structure of the neocortex is basically uniform and that differences in cytoarchitecture and function reflect differences in extrinsic connections. On the basis of a study of a vast array of Golgi material, Szentágothai [721] concluded that the various neocortical cell types show only little variation from mouse to human. Moreover, the ratio of different cell types has been shown to be essentially the same in the motor and visual cortices of rats and cats [824]. It may be concluded that the neocortex of different mammalian species is built according to a common basic plan, and that, hence, the structural features assembled in Fig. 15.37 A, B may well present an image of that plan. However, the reservation should be made that the spiny stellate cells depicted (SS) occur only in specialized primary sensory cortices and that excitatory bipolar cells (Bi) have been demonstrated so far only in the primary visual cortex of the rat [559, 562] and cat [177].

2. If our diagram presents an image of the microcircuitry of the neocortex, it does so

merely in a qualitative sense. The total number of neocortical synapses in the neocortex has been estimated to be 3×10^{14} [104]. Our diagram contains 157 synapses. Because excitatory (grossly: pyramidal) neurons are by far more numerous than inhibitory (grossly: non-pyramidal) neurons and because the intracortical axon terminal- and collateral systems of most of the former are vastly more extensive than those of the latter, the excitatory synapses may be expected to outnumber the inhibitory synapses. The validity of these estimates has been confirmed by the quantitative synaptological studies of DeFelipe et al. [142]. These authors found that in neocortical areas as different as the hindlimb area of the rat somatosensory cortex and the human anterolateral temporal cortex, symmetrical synapses formed 10.7% and 11.5% of the total synapse population, respectively. However, in our diagram there are 76 excitatory and 82 inhibitory synapses. More specifically, about 10% of the excitatory synapses (7 out of 76) is formed by terminals of cortico-cortical fibres. Because cortico-cortical projections constitute by far the largest neocortical input system, this percentage is most probably much higher.

3. Several functionally important structural features have been neglected. For example, it is known that different thalamic nuclei project to different sets of neocortical (sub)layers and that the same holds true for particular cell types within some thalamic nuclei, for instance the lateral geniculate nucleus. In our diagrams no distinction has been made between these various projections. Rather, they are taken together as if representing one single thalamocortical system.

Networks of Pyramidal Neurons

The pyramidal neurons (P) are doubtless the principal neurons in the neocortex (Fig. 15.37 A). They are not only by far the most numerous cellular elements in that structure, but also constitute its sole output system and its largest input system. Separate sets of deep pyramidal neurons project to different subcortical targets, whereas cortico-cortical fibres arise mainly from superficial pyramidal neurons.

The very extensive axon collateral systems of pyramidal neurons primarily contact other pyramidal neurons. There is evidence suggesting that superficial pyramidal neurons contact other superficial and deep pyramidal neurons and that deep pyramidal neurons impinge on other deep and superficial pyramidal neurons. It is known that, in the primary visual cortex, the axon collateral systems of pyramidal neurons constitute reciprocal patchy networks, which can be traced over distances of up to 7 mm [368]. It has been suggested that these networks link sites with similar physiological characteristics, such as orientation preference. This would imply that, within the primary visual cortex, different interwoven networks of pyramidal axon collaterals are present and that the extent of these networks would be confined to that cortical area.

Another remarkable local network of interconnected pyramidal neurons is found in the prefrontal cortex. The prefrontal cortex (PFC) has been identified as the key neocortical region supporting working memory [244]. Extracellular recordings during delayed response tasks have shown that a considerable fraction of prefrontal cortical neurons remain active after the cue (e.g. a particular sensory stimulus or event) and until the task is completed. Such activity, which can persist for several seconds, has been proposed as the neural correlate of working memory [804]. It has been recently established [806] that the PFC of the ferret contains a network of heavily interconnected cells with characteristic dual apical dendrites and a particularly wide and dense system of basal dendrites. These “hyper-reciprocally” connected cells appeared to have synaptic and functional properties essential for supporting persistent activity in the PFC.

A continuous network of excitatory elements potentially involving the entire neocortex and even extending into the hippocampal region is constituted by the ipsilaterally and contralaterally projecting cortico-cortical pyramidal neurons. The continuity of this network is emphasized by the fact that the cortico-cortical fibres terminate throughout the neocortex mainly in the superficial layers, where the cortically projecting pyramidal neurons are concentrated. The presence of this strongly developed, ubiquitous network warrants the conclusion that the neocortex communicates first and foremost with itself. However, the fact that the cortico-cortical fibres most probably impinge not only on superficial pyramidal neurons, but also on the apical dendrites and terminal dendritic bouquets of deep pyramidal neurons indicates that the various subcortical centres to which these elements project are continuously kept informed about the successive transformation of data occurring along the cortico-cortical processing streams. This applies in particular to the caudate-putamen complex and to the pontine nuclei, centres which are known to receive projections from almost the entire neocortex.

Thalamocortical fibres (thc) synapse directly with pyramidal neurons, although the number of contacts made by such fibres with particular types of pyramidal neurons is subject to considerable variation [815].

Interneuronal Systems

All types of neocortical local circuit neurons have been reported to establish synaptic contacts with pyramidal neurons. Most of these elements, namely spiny stellate cells (SS), bipolar cells (Bi), neurogliaform cells (N), basket cells (Ba), horizontal cells of Cajal (HC) and cells of the provisional types I3 and I4, have been suggested to receive a direct input from the thalamus, and some of these (SS, Bi, Ba, I3) are also contacted by axon collaterals of pyramidal neurons. Martinotti cells (M) are primarily intercalated between different pyramidal neurons. The nature of the input to one type of local circuit neuron, the chandelier cell (Ch), is entirely unknown.

Several types of neocortical interneurons are in receipt of afferent contacts from other interneurons and/or establish efferent contacts with such elements. Prominent among them are the spiny stellate cells and the basket cells. Spiny stellate cells are contacted by basket cells and most probably by neurogliaform cells, chandelier cells and other spiny stellate cells and probably impinge on basket cells and double bouquet cells (DB). Basket cells receive afferents from other basket cells and probably from spiny stellate cells and double bouquet cells and make efferent contacts with spiny stellate cells and probably with smooth stellate elements (e.g. I3).

Excitatory local circuit neurons, i.e. spiny stellate cells and some bipolar cells, are, as far as it is known, confined to primary sensory areas. Inhibitory local circuit neurons are of many different types and occur throughout the neocortex. Some types of local circuit neurons are lamina-specific, i.e. their somata are situated mainly in one or in two adjacent layers (SS, DB, HC, M).

In the highly differentiated primary visual cortex of the rhesus monkey, thalamocortical fibres carrying particular kinds of visual information terminate in sharply defined sublayers of lamina IV (Fig. 15.26) [189, 190, 425]. The detailed Golgi studies carried out by Lund and her associates [424, 428, 430] showed that each of these thalamic recipient sublayers, and some other layers as well, contain several types of local circuit neurons. The dendritic trees of all of these elements are strictly confined to the layer or zone in which their soma is situated. Their axons project to more superficial layers, to deeper layers or to both. These interlaminar projections are highly specific, targeting from one to four laminar divisions, depending on the type of neuron. The strictly radial orientation of all of these projections is presumably connected with the preservation of the detailed topographical map existing within one of the thalamic recipient sublayers. Some of these types of local circuit neurons may well be involved in the selective processing and transfer of one particular type of visual information.

Important clues to the specific functional roles of the various types of inhibitory interneurons derive from their differential axonal arborizations (Fig. 15.36) and from the location of their synaptic connections with pyramidal neurons [25]. Most of these inhibitory cell types distribute their synaptic contacts preferentially to selected membrane domains of pyramidal elements [440, 702, 815]. Thus, chandelier cells (Ch) terminate on the axon initial segments of pyramidal neurons, whereas basket cells (Ba) target the somata and proximal dendrites of these elements (Fig. 15.31). The contacts of the chandelier and basket cells are optimally localized for controlling the output and oscillatory synchronization of groups of pyramidal neurons [206]. Small stellate cells such as those belonging to the provisional types I1, I2 and I4 may also participate in these specific functions. Neurogliaform cells (N) preferentially innervate the proximal and mid-range dendritic domains. (This feature of the neurogliaform cells is not shown in Fig. 15.37B.) The neurogliaform neurons are optimally positioned to influence dendritic processing and integration of synaptic inputs [440]. Finally, the preferential innervation of distal dendritic branches and apical tuft regions of pyramidal neurons by double bouquet cells (DB), horizontal cells of Cajal (HC) and Martinotti cells (M), allows these elements to affect local dendritic integration.

Double bouquet cells are abundant in the primate neocortex [146]. It has been frequently observed that the dendritic spines postsynaptic to the double bouquet cells receive, in addition, an asymmetrical synapse from a putative excitatory terminal [137]. In the human neocortex at least 47% of the spines postsynaptic to double bouquet cells were found to be in receipt of such a dual innervation [146].

Dendritic spines are the principal targets of the thalamocortical and cortico-cortical afferents and of the local axonal arborizations of pyramidal and spiny stellate cells, which are known to form asymmetrical, excitatory synapses [128, 137, 815]. However, because the majority of dendritic spines of pyramidal neurons are outside layer IV, i.e. the principal area of termination of thalamocortical afferents and spiny stellate cell axons, it may be assumed that the main source of asymmetrical (excitatory) synapses on spines with dual innervation is from intracortical axons (local and cortico-cortical). Hence, it is reasonable to assume that double bouquet cells control the level of excitation in circuits linking pyramidal neurons with other pyramidal neurons [146].

During the last decade, important new findings concerning the connectivity and the regulation of the activity of neocortical inhibitory interneurons have been reported. Thus, it has been shown that (1) groups of inhibitory interneurons are coupled through gap junctions, (2) the axons of certain types of interneurons form autapses at their own soma or dendrites, and (3) certain types of interneurons are specifically or preferentially contacted by particular extrathalamic neuromodulatory systems. These phenomena, and their (putative) functional significance, will now be briefly discussed (Fig. 15.38).

1. *Electrical coupling of neocortical inhibitory interneurons.* Gap junctions or electrical synapses are specialized loci where channels bridge the plasma membranes of two adjacent nerve cells. By providing a low-resistance reciprocal pathway for ions and small organic molecules, such electrical connections permit the direct transmission of electrical signals between neurons [107, 223, 693]. Gap junctions have been observed between smooth neurons in electronmicroscopic studies of the primate neocortex [688, 689]. Electrophysiological studies on slices of rat neocortex established the existence of a network of electrically coupled inhibitory interneurons [222]. These interneurons, which were functionally characterized as fast-spiking cells, could be identified as PV-expressing basket cells [729]. Fukuda et al. [214] found that large PV-containing cells in layer II/III of the cat visual cortex form about 60 gap junctions with other cells. Most, though not all of these junctions were between proximal dendrite sites. Further studies revealed that other types of neocortical inhibitory interneurons, designated as low-threshold-spiking [235], multipolar bursting [54] and late-spiking cells [99] also form type-specific, electrically coupled networks. The same holds true for neurogliaform cells [683]. However, the network formed by the latter cells appeared to be not strictly type-specific. In slices of rat somatosensory cortex, gap junctions connected approximately 50% of neurogliaform cells, but 20% of these elements also formed gap junctions with other interneurons, including fast-spiking basket cells, regular-spiking non-pyramidal cells

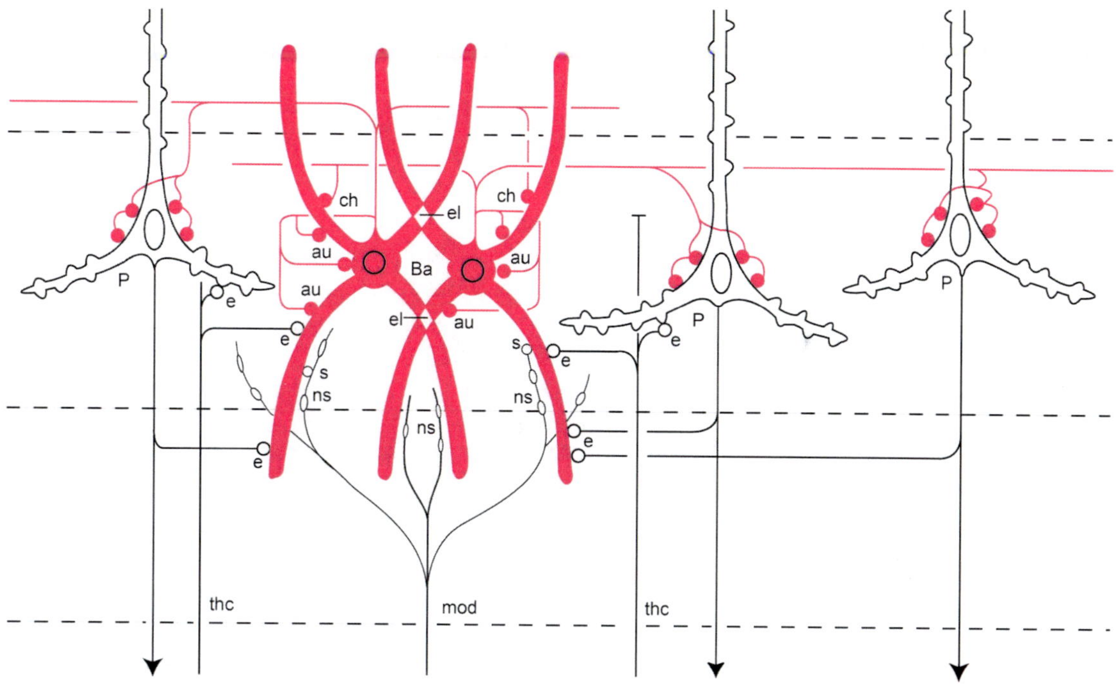

Fig. 15.38. Neocortical micronetwork. This micronetwork includes some specific afferent fibres from the thalamus (*thc*), a fibre representing a neuromodulatory system (*mod*), originating from an extrathalamic subcortical centre, a number of pyramidal neurons (*P*) and, finally, a set of GABAergic, inhibitory interneurons, represented by two basket cells (*Ba*). The thalamocortical fibres form excitatory synapses (*e*) with pyramidal and basket cells. The axons of the basket cells make inhibitory contacts with the somata of the pyramidal cells. When the basket cells are triggered by their thalamocortical afferents, they will exert a strong feed-forward inhibition on the pyramidal cells. The axons of the pyramidal neurons leave the cortex. These processes are provided with collateral branches, which establish excitatory synaptic contacts with other, related pyramidal neurons (not shown) and with the inhibitory interneurons (i.e. the basket cells) from which they receive input. The pyramidal axon collaterals and the basket cells form pathways for feed-back inhibition. The neuromodulatory system selectively addresses the set of inhibitory interneurons (Ba) with which they make some synaptic (*s*) and numerous non-synaptic (*ns*) contacts. This system may exert an excitatory or an inhibitory influence, depending on the neurotransmitter used by its afferent fibres and/or by the receptors expressed by their target neurons. The interneurons are interconnected by excitatory, reciprocal electrical synapses (*el*) as well as by inhibitory chemical synapses (*ch*). Moreover, the axons of the interneurons (Ba) emit collateral branches that form numerous autaptic contacts (*au*) with their own soma and proximal dendrites

and bitufted cells. It is not known whether the networks of inhibitory interneurons extend indefinitely across the neocortex or whether they have distinct boundaries [107].

The cerebral cortex displays synchronized, rhythmic activity, and a variety of oscillations accompany states of sensory perception, motor performance, arousal and sleep [45, 82]. The diverse networks of inhibitory interneurons most probably operate as precision clockworks for the entrainment of the various types of cortical oscillations.

2. *Autaptic innervation of neocortical inhibitory interneurons.* Autapses are transmitter release sites made by the axon of a neuron on its own soma or dendrites. Tamás et al. [728] studied the synaptic relations of various types of neurons in the visual cortex of the cat during intracellular biocytin labelling and correlated light and electron microscopy. They found that the axons of two types of inhibitory interneurons, namely basket cells and dendrite-targeting cells form significant numbers (10–20) of synaptic contacts with their own somatodendritic surfaces. These autapses appeared to be domain-specific, i.e. those formed by basket cells concentrated on the perisomatic region, whereas those formed by dendrite-targeting cells were located on more distal dendrites. Bacci et al. [23] recorded autaptic activity in neocortical fast-spiking GABAergic interneurons. It appeared that in these neurons the autaptic activity has significant inhibitory effects on repetitive firing and increased the current threshold for evoking action potentials.

As noted above, during cortical oscillations, fast-spiking elements (as well as other types of inhibitory neurons) probably synchronize the activity of groups of pyramidal neurons [45, 82]. It is believed that inhibitory autaptic transmission enables fast-spiking cells "to sense their own firing and regulate it in phase with that of other fast-spiking cells, resulting in synchronous inhibitory neurotransmission onto pyramidal neurons" [25].

3. *Several extrathalamic neurotransmitter systems ascend to the neocortex, where they specifically or preferentially target particular types of inhibitory interneurons.* In a previous section of the present chapter, it was shown that several neurotransmitter systems, including those containing acetylcholine, serotonin, dopamine and noradrenaline, ascend to the neocortex, and that interneurons are a major target of these systems. It was also demonstrated that, although a certain proportion of the terminals of these systems make classical synapses, many others are not associated with a specialized postsynaptic structure. The neurotransmitter molecules released by these non-synaptic terminals diffuse over some distance through the extracellular space to act on extrasynaptic receptors. This mode of intercellular communication – known as volume transmission – enables the various extrathalamic neurotransmitter systems to exert a simultaneous modulatory influence on large numbers of interneurons.

Xiang et al. [835] studied the influence of acetylcholine on the excitability of two types of inhibitory interneurons in layer V of the rat visual cortex, namely fast-spiking cells and low-threshold spike cells. They found that acetylcholine elicits hyperpolarization in fast-spiking cells through activation of muscarine receptors, resulting in disinhibition of their pyramidal cell targets. The low-threshold spike cells, on the other hand, appeared to be excited by acetylcholine through the activation of nicotine receptors. Bacci et al. [25] cited evidence suggesting that the dopaminergic, noradrenergic and serotoninergic systems exert similar selective and differential modulations on particular subgroups of inhibitory interneurons, thus providing a substrate for fine control of information flow through cortical networks.

Yoshimura and Callaway [841, 842] recently provided some further examples of "fine-scale specificity" in the organization of cortical networks. Studying the mutual functional relationships between pyramidal neurons and those between pyramidal neuron and fast-spiking interneurons in layer II/III of the rat visual cortex, they found that (1) layer II/III pyramidal neurons only share common input from layer IV and from within layer II/III in the minority of cases in which they are directly interconnected to each other; (2) fast-spiking inhibitory interneurons connect preferentially to neighbouring pyramidal cells that provide them (via their axon collaterals) with recurrent excitation; and (3) these pairs of reciprocally connected fast-spiking interneurons and pyramidal cells share common specific excitatory input. Yoshimura and Callaway [841] considered it likely that these sets of reciprocally connected neurons contribute to the synchronization of activity within particular neuronal subpopulations.

Each neocortical area contains at least several types of pyramidal neurons. In a previous section of this chapter, it was shown that these types of pyramidal neurons may differ from each other with regard to (1) the laminar position of their perikarya; (2) the laminar spread of their dendritic branches (Fig. 15.32); (3) the laminar density of their dendritic spines, and, hence, their laminar afference (Fig. 15.32); (4) the destination of their axons (Fig. 15.33); and (5) the spread and extent of their cortical axon collaterals (Figs. 15.34, 15.35). These data indicate that the various types of pyramidal cells are differently embedded in the microcircuitry of the neocortex. Given the fact that pyramidal neurons of the same type are strongly and reciprocally connected by axon collaterals [806, 841], it may be concluded that each particular neocortical area contains a number of networks of interconnected, type-specific pyramidal neurons. The number of pyramidal networks present within a given neocortical area is unknown and the same holds true for the actual extent of these networks.

In light of the recent literature reviewed above, it seems likely that the various pyramidal elements belonging to a particular network are in receipt of afferents from cohorts of inhibitory interneurons, each cohort contacting a specific domain of the receptive surface of the pyramidal neurons involved. The inhibitory interneurons forming these cohorts are all of the same type and are generally reciprocally connected by inhibitory chemical synapses and by excitatory electrical synapses [729].

Thalamic inputs selectively contact and strongly excite the interneurons belonging to particular cohorts, while others receive weaker or no thalamic inputs [235].

Finally, each of the various cohorts of inhibitory interneurons impinging on a particular pyramidal network is specifically addressed by one or more of the extrathalamic modulatory systems. The influence exerted by these systems depends on the nature of the neurotransmitter used by their afferent fibres and on the receptor types expressed by their target neurons.

It is important to note that not only the inhibitory input, but also the excitatory input to pyramidal neurons belonging to the same network may be specific. Yoshimura et al. [842] studied the connections to pyramidal neurons in layer II/III of the visual cortex of the rat. They found that reciprocally connected pyramidal neurons share common excitatory input from layer IV and within layer II/III. However, adjacent layer II/III neurons that were not connected to each other appeared to share very little (if any) common excitatory input from layers IV and II/III.

The question arises as to how strict the separation is between the various pyramidal networks, including their satellite cohorts of inhibitory interneurons? The answer to this question is largely unknown. However, it seems likely that the abundant double bouquet cells with their vertically oriented axonal systems contact pyramidal neurons belonging to different networks, and we have seen that the neurogliaform cells form gap junctions with several other types of inhibitory interneurons.

Neocortical Columns and Modules

Introduction

During the last 50 years, physiological and morphological evidence has accumulated indicating that several parts of the neocortex of many different mammalian species are composed of radially oriented, column-like units or modules. The concept that column-like modules represent fundamental units of the mammalian neocortex has gained wide acceptance in the literature. In the following, a survey will be presented of the findings and considerations that led to this concept. This survey will be followed by a brief critical and cautionary commentary.

The Investigations of Lorente de Nó: Elementary Units and Glomérulos

The first to propose a modular structure of the neocortex was Lorente de Nó [416]. He claimed that, in small radially oriented cylinders having a specific thalamocortical fibre as their axis, all elements of the cortex are represented and that in these *elementary units* the whole process of transmission of impulses from the afferent fibre to the efferent axon may theoretically be accomplished. Lorente de Nó's concept was based on the observations that (a) the terminal branches of thalamocortical fibres may form discrete cylindrical patches in lamina IV of the cortex, (b) the efferent system of the cortex is formed by radially oriented axons of pyramidal neurons and (c) the axons of many cortical local circuit neurons are likewise radially oriented.

The results of earlier investigations of Lorente de Nó have played an additional role in the development of the ideas concerning columnar organization of the neocortex. In 1922 Lorente de Nó [414] described discrete cylindroid aggregations of cells in lamina IV of what he believed to be the auditory cortex of the mouse, which he termed *"glomérulos"* (Fig. 15.39 A). Golgi material revealed that the specific thalamic afferents to this cortex constitute dense patches of terminal ramifications which coincide with the glomérulos (Fig. 15.39 B) and that the dendrites of the neurons in lamina IV are largely confined to the glomérulo in which their soma is located (Fig. 15.39 C). The glomérulos appeared to contain numerous spinous stellate cells. The axons of these local circuit neurons were found to descend to the deeper cortical layers, where they issue numerous collaterals. Some of these collaterals were observed to ascend to lamina III.

Some 50 years later, it became clear that Lorente de Nó [414] had actually analysed the portion of the somatosensory cortex to which the large mystacial vibrissae on the snout project via relays in the main trigeminal nucleus and in the thalamus. It appeared that each "glomérulo" or *"barrel"*, as they came to be called, is the primary cortical representation of one vibrissal follicle. The vibrissae are arranged in a stereotypical grid-like pattern in which each vibrissa has a unique position. Within the somatosensory cortex, each barrel also occupies a unique position and the topographical arrangement of these structures closely corresponds to that of the vibrissae. The largest barrels in mice are elliptical in cross-section, with a greatest diameter of 300 μm (Woolsey and van der Loos [834]; van der Loos and Woolsey [759]). According to van der Loos [758], barrels are the visible lamina IV counterparts of cortical columns.

Staiger et al. [704] recently studied the spiny neurons in layer IV of the somatosensory cortex of the rat. It appeared that this layer, in addition to numerous spiny stellate cells (58%), also contains star pyramidal cells (25%) as well as pyramidal elements (18%). As regards the spread of the axonal ramifications of these cell types, the authors reported that the axonal arbors of the spiny stellate cells are largely confined to their related columns, that the pyramids show extensive transcolumnar axonal branches and that the star pyramids occupy an intermediate position (Fig. 15.39 D). It is noteworthy that the columns denoted by the authors have no structural boundaries above and below layer IV.

The Columnar Organization of the Somatosensory Cortex

Mountcastle [494–496] analysed the functional organization of the somatosensory cortex in the cat, using microelectrodes to record the activity of single cells. During radial penetration, he encountered cells with similar receptive fields throughout the depth of the cortex that responded to stimulation of the same type of cutaneous receptors located at a particular site. Penetrations parallel to the pial surface and

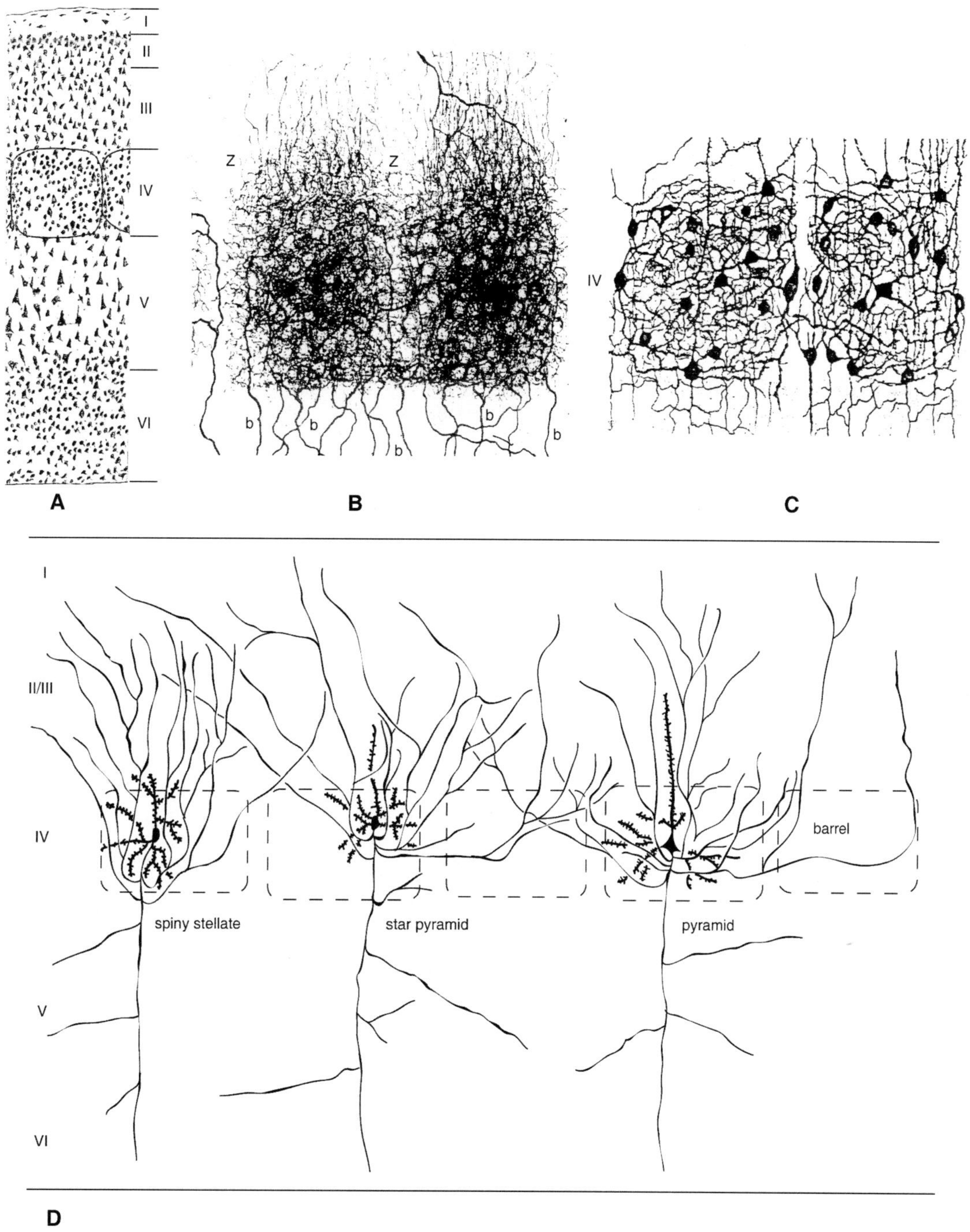

Fig. 15.39 A–D. The somatosensory cortex of rodents. **A** Cytoarchitecture; the fourth layer contains specialized structures consisting of a cylinder of densely packed cells surrounding a region of lower cell density. These structures were designated as glomérulos by Lorente de Nó [414] and as barrels by Woolsey and van der Loos [834]. **B** Specific somatosensory efferents (*b*) from the thalamus form patches of a dense terminal feltwork, which correspond to the glomérulos. These patches are separated by zones (*Z*), in which the axonal feltwork is clearly less dense. The fibres of the feltwork surround the neuronal somata in the fourth cortical layer. **C** Two glomérulos as observed in Golgi preparations. **D** The glomérulos or barrels contain three different types of spiny neurons: spiny stellates, star pyramids and pyramids. These cell types differ with regard to the spread of their axonal arbors. **A**, **B** and **C** are reproduced from Lorente de Nó [414]; **D** is based on Staiger et al. [704]

crossing the radial axis of the cortex appeared to pass through blocks of tissue 300–500 μm in size, and neurons with identical properties were encountered in each of them. Sharp transitions were observed from a block with one set of properties to the adjacent block with different properties. These findings led Mountcastle ([494], p. 430) to hypothesize that "there is an elementary unit of organization in the somatic cortex made up of a vertical group of cells extending through all the cellular layers." He termed this unit a *"column"*. On the basis of experiments in which multiple, closely spaced penetrations were made, Mountcastle concluded that the individual columns have a width of maximally 500 μm. He believed that functionally active columns are able to isolate themselves from their surroundings by exerting an inhibitory action on neurons in their inactive neighbours, a process which he designated as "pericolumnar inhibition".

The Columnar Organization of the Visual Cortex

The detailed physiological and morphological studies performed by Hubel and Wiesel and collaborators [308, 310–314, 404, 412] on the primary visual cortex of the monkey have led to the identification of three different kinds of column-like structure in that area: orientation columns, ocular dominance columns and blobs. Sets of these three column-like structures are conjectured to be united in larger entities, termed hypercolumns.

The *orientation columns* were detected by electrophysiological recordings. During radial penetrations with microelectrodes, cells having identical axes of orientation (i.e. cells that responded strongest to a bar of light in one particular orientation) were encountered throughout the thickness of the cortex. During tangential penetrations, the electrode encountered a shift in the axis of orientation of the light bar of about 10° every 300–100 μm. The morphological substrate of the functional columns thus detected could be visualized with the aid of the 2-deoxyglucose technique of mapping relative changes in metabolic activity following peripheral stimulation.

Microelectrode recordings also revealed the presence of parallel stripes, about 500 μm wide, alternately receiving their input from the ipsilateral and the contralateral eye. The morphological correlate of these *ocular dominance columns* could be strikingly visualized using silver impregnation techniques and autoradiographically by means of transneural transport of radiolabelled amino acids injected into one eye.

In the monkey striate cortex, staining for the mitochondrial enzyme cytochrome oxidase revealed an array of densely staining, peg-like structures in laminae II and III, which can be seen lying in register with less densely staining formations in the cortical layers below lamina IV. These *blobs*, as they were termed, are ellipsoid in tangential sections, measuring roughly 150×200 μm. Their most conspicuous superficial (i.e. supragranular) parts receive a separate projection from the intercalated layers of the lateral geniculate body (Fig. 15.26). Physiological studies have shown that the cells within the blobs respond selectively to the colour of a stimulus, without regard to its orientation.

A *hypercolumn* is considered to represent the circuitry necessary for the analysis of a given, discrete region in the visual field. It is conceived to contain a set of nine orientation columns, together encompassing a complete cycle of orientation through 180°, an adjoining pair of right and left ocular dominance columns and several blobs.

In a previous section, it has been mentioned that the axons of many neocortical pyramidal neurons give rise to long, horizontally running collaterals, which issue clusters of terminal branches at regular intervals (Fig. 15.35). Interestingly, experimental studies using different techniques, i.e. retrograde tracing, 2-deoxyglucose autoradiography and cross-correlation analysis, have produced evidence suggesting that, in the primary visual cortex, these horizontal collaterals mediate communication between functionally related columns, e.g. blobs of similar colour opponency and columns of similar orientation preference (for a review, see Gilbert [237]).

The Auditory Cortex

Evidence of a columnar organization of the primary auditory cortex has also been presented. During microelectrode penetrations normal to the cortical surface, cells with the same characteristic frequency were encountered throughout the entire depth of the cortex [1]. Binaural interaction bands have also been described, exhibiting either binaural summation or inhibition. Experiments in which single-unit mapping was combined with autoradiographic tract tracing revealed that, in the primary auditory cortex of the cat, clusters of summation responses coincide with bands receiving a heavy innervation from the contralateral primary auditory cortex, whereas suppression responses were recorded in regions of sparse contralateral innervation [75].

The Motor Cortex

Within the motor area of the neocortex, there are small loci where stimulation with weak currents produces movements executed by a single muscle. It has been demonstrated that the loci for producing contraction of a given muscle extend perpendicularly from the surface to the depth of the cortex. These radial arrays of stimulation points have been called *cortical motor columns*. The diameter of these columns is approximately 1 mm in cross-section [16]. Using two penetrating microelectrodes, one for stimulation and the other for recording, it was demonstrated that stimulation of a given column produces a pericolumnar zone of inhibition [18]. Whereas the afferent input arriving at the somatosensory columns is modality specific, the cortical motor columns receive polymodal input from sensors in the target muscle itself, from the joint to which the muscle inserts and from the skin overlaying the muscle, i.e. sensors that are stimulated when the target muscle contracts [17].

Columnar Patterns Shown by the Cells of Origin and the Terminal Ramifications of Cortico-cortical Connections

Anatomical evidence for the existence of vertical columns or bands in the cerebral cortex has been provided by studies on the origin and termination of cortico-cortical connections. Thus, Jones et al. [347], using an anterograde tracer, demonstrated that both commissural and ipsilateral cortico-cortical fibres arising and terminating in the somatic sensory cortex of monkeys terminate in distinct vertically oriented columns with a variable width of 500–800 μm. Each of these columns appeared to be separated from its neighbours by a gap of approximately 500 μm. Each column was considered to represent the terminals of a bundle of cortico-cortical axons emanating from cells at the centre of each of the loci in which the tracer was injected. Using a retrograde tracer, Jones et al. [347] also demonstrated that the cells of origin of the cortico-cortical projections investigated, found predominantly in lamina III, are also aggregated in vertically oriented clusters (Fig. 15.40). Goldman and Nauta [242] injected an anterograde tracer into three cytoarchitectonically distinct regions (areas 4, 9 and 12) of the frontal lobe of rhesus monkeys. Neurons in these various regions are known to project both contralaterally and ipsilaterally to cytoarchitectonically diverse areas within the frontal, temporal and parietal lobes. They found that all of these cortico-cortical projections terminate in vertically oriented columns, 300–700 μm wide, which alternate in regular sequence with zones of comparable width that are free of such terminals. Interestingly, the pattern and dimensions of cortical columns in the prefrontal cortex, defined by cortical afferent inputs, appeared to be strikingly similar in the rat, squirrel monkey and rhesus monkey, although the brains of these animals differ considerably in size [76, 325]. In order to determine the pattern of termination of two converging cortico-cortical systems in the same animal, Goldman-Rakic and Schwartz [246] followed a double anterograde labelling strategy. Using rhesus monkeys, they im-

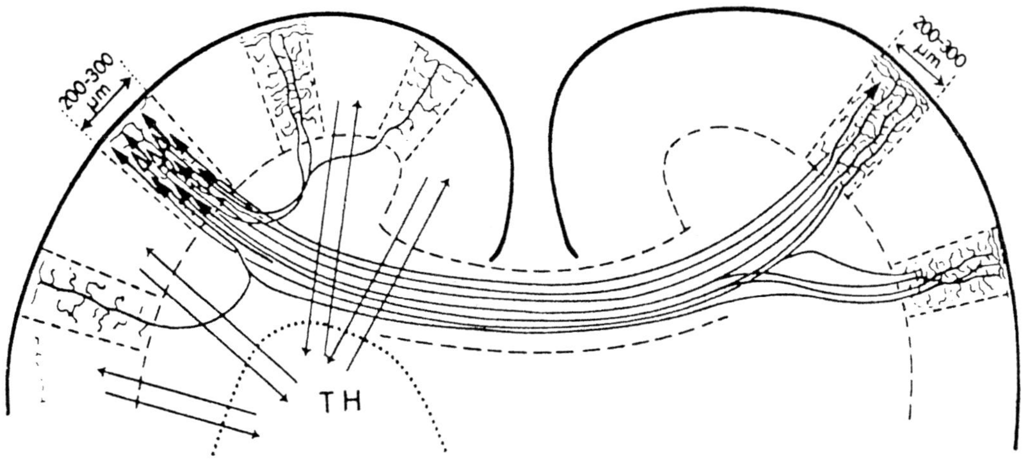

Fig. 15.40. The general arrangement of cortico-cortical fibres, shown diagrammatically in a lisencephalic brain. The fibres interconnect radial columns with a diameter of 200–300 μm. Ipsilateral connections originate mainly from pyramidal neurons in layer III (cells shown at *left* in *outlines*), while commissural connections (shown in *solid black*) derive from layers III, V and VI. Reproduced from Szentágothai [721]. *TH*, thalamus

planted HRP pellets in area 7 of the parietal lobe in one hemisphere and injected a mixture of tritiated amino acids in area 9 of the frontal lobe of the other hemisphere. It appeared that, in the prefrontal cortex, contralateral callosal fibre columns interdigitate with ipsilateral associational fibre columns. Goldman-Rakic and Schwartz [246] also investigated the spatial organization of the populations of cells in the prefrontal cortex projecting to the parietal associational cortex and to the contralateral prefrontal cortex using HRP or fluorescent dyes as retrograde tracers. They demonstrated that these two populations are inversely related in their relative densities over portions of the prefrontal cortex examined. Moreover, in the HRP material, the high-density patches of retrogradely labelled neurons appeared to coincide with the afferent fibre columns revealed by anterograde transport. In a later study (Selemon and Goldman-Rakic [669]), terminal labelling originating from prefrontal and parietal injections in the same hemisphere was investigated in a large number of areas of convergence in both the ipsilateral and the contralateral hemisphere. In most of the target areas, prefrontal and parietal terminal fields formed an array of interdigitating columns, but in some other areas prefrontal and parietal projections converged on the same column or cluster of adjacent columns but terminated within different laminae. For instance, in the depths of the superior temporal sulcus, prefrontal terminals occupied laminae I, III and V, with parietal terminals filling complementary laminae IV and VI.

From the experiments of Goldman-Rakic and Schwartz [246] and Selemon and Goldman-Rakic [669], it appears that projections from two different (heterotopic) cortical regions (in the same or opposite hemispheres) remain segregated within common cortical targets, either by terminating in separate alternating columns or by concentrating in complementary laminae within the same column. The relationship between the termination zones of projections from paired *homotopic* areas was studied by McGuire et al. [457] in the cerebral cortex of the rhesus monkey. Injections with HRP and tritiated amino acids were made in two topographically matched regions of the frontal lobe in each hemisphere. The cortical projections from both of these homotopic pairs of areas appeared to converge in common columnar territories.

Minicolumns and the Radial Unit Hypothesis of Cortical Development

In 1979, Mountcastle [495] introduced what he believed to be "the basic modular unit of the neocortex" under the name *minicolumn*. He characterized this unit as a narrow chain of neurons extending radially across the cellular layers II–VI. Referring to the ontogenetic studies of Rakic (see below), he indicated that the minicolumn is produced by the iterative division of a small cluster of progenitor cells, and that the resultant neuroblasts migrate to their destination in the cortex along radial glial cells. According to Mountcastle, a minicolumn occupies a radial cylinder of cortical space with a diameter of about 30 μm. On the basis of cell counts in the neocortex of various mammals, carried out by Rockel et al. [619], Mountcastle estimated that each minicolumn contains about 110 neurons, except for the striate cortex where the number is about 2.5 times larger. He posited that minicolumns contain all the major cortical neural cell phenotypes, interconnected in the vertical dimension [496]. According to Mountcastle, the cortical columns, which he sometimes also designated as *macrocolumns* [495], are to be considered as aggregations of several hundred minicolumns bound together by short-range horizontal connections [496].

The studies of Rakic [606–610], on the ontogeny of the neocortex in the rhesus monkey, which strongly influenced the development of Mountcastle's minicolumn concept, may be summarized as follows: The matrix zone of the embryonic pallium is divided by glial septa into columns of precursor or stem cells termed "proliferative units". The postmitotic cells produced by these proliferative units find their way to the primordial cortex by following the shafts of elongated, radially oriented glial cells, which stretch across the embryonic pallial wall. Eventually, all postmitotic cells generated in a single proliferative unit form a morphologically identifiable stack of neurons in the cortex. These entities, which may be designated as "ontogenetic" or "embryonic" columns, form, according to Rakic, the fundamental building blocks in the developing neocortex [608]. Each proliferative unit is supposed to produce multiple neuronal phenotypes. Rakic [608] advanced the idea that the proliferative units in the matrix zone of the embryonic pallium constitute a proto-map of prospective cytoarchitectonic areas. Each of these areas is composed of a large number of ontogenetic columns, which become the basic processing units in the cerebral cortex. According to Rakic [608], the relatively constant size of these units in different mammalian species suggests that, during evolution, the cortex has expanded by the addition of such radial units rather than by their enlargement. This postulate was designated as the *"radial unit hypothesis"* [608, 610].

Mountcastle's estimate that a minicolumn contains about 110 cells was, as he indicates, based on the work of Rockel et al. [619]. In this publication and in a later, more extensive study [620], the results of a quantitative comparative analysis of the neocortex were reported. The number of neuronal cell bodies were counted in a narrow strip, 30 µm wide, through the depth of the neocortex in several different functional areas (somatic sensory, visual, motor, frontal, parietal and temporal) in a number of different species (mouse, rat, cat, monkey and human). These counts gave remarkably constant values of 110 ± 10 cells in all areas and in all species studied. The only exception to this similarity appeared to be the binocular part of the visual cortex in a number of primate brains, in which approximately 2.5 times as many neurons were found. Rockel et al. [620] remarked that they chose a width of 30 µm for counting the cortical perikarya because, according to Hubel and Wiesel [312] and others, this is approximately the width of the simplest functional column in the neocortex. However, they pointed out that this is not necessarily the size of the simplest anatomical unit. In fact, their findings imply that, regardless of their areal size, modules will contain constant cell numbers (Swindale [719]).

Dendritic Clusters, Axonal Bundles and Radial Cell Cords As (Possible) Constituents of Neocortical Minicolumns

A number of examples of repeating microarrays of cortical elements could be interpreted as conforming to a minicolumnar radial pattern. These include the clusters of apical dendrites of pyramidal neurons, the bundles of myelinated axons and the column-like cell soma arrays running orthogonal to the horizontal laminae.

The existence of *clusters or bundles of dendrites* in the neocortex was discovered independently and almost synchronously by Peters and Walsh [568] in rat somatosensory cortex and by Fleischhauer et al. [196] in the rabbit and cat sensorimotor cortex. Since then, similar bundles of dendrites have been revealed in a number of other neocortical areas in mice, rats, rabbits and cats [140, 194, 195, 569, 826]. In each of these areas there appeared to be a similar basic organization: the apical dendrites of groups of layer V pyramidal neurons form clusters, which during their ascent are joined by the apical dendrites of pyramidal cells situated in the more superficial layers.

Radially oriented *bundles of myelinated fibres* occur in all parts of the neocortex (Figs. 15.4C, 15.13). These bundles are known as radii or as the radiations of Meynert [85]. As already mentioned in a previous section of this chapter, local differences in the size and peripheral extent of these bundles play a prominent role in the myeloarchitectonic parcellation of the cortex.

Column-like *arrays of cell bodies* can be observed in many parts of the neocortex. They are particularly conspicuous in the human foetal cortex and in the mature temporal cortex of humans and other primates [341]. Brodmann [70], who beautifully depicted these cell arrays (Fig. 15.41), indicated that there is a correla-

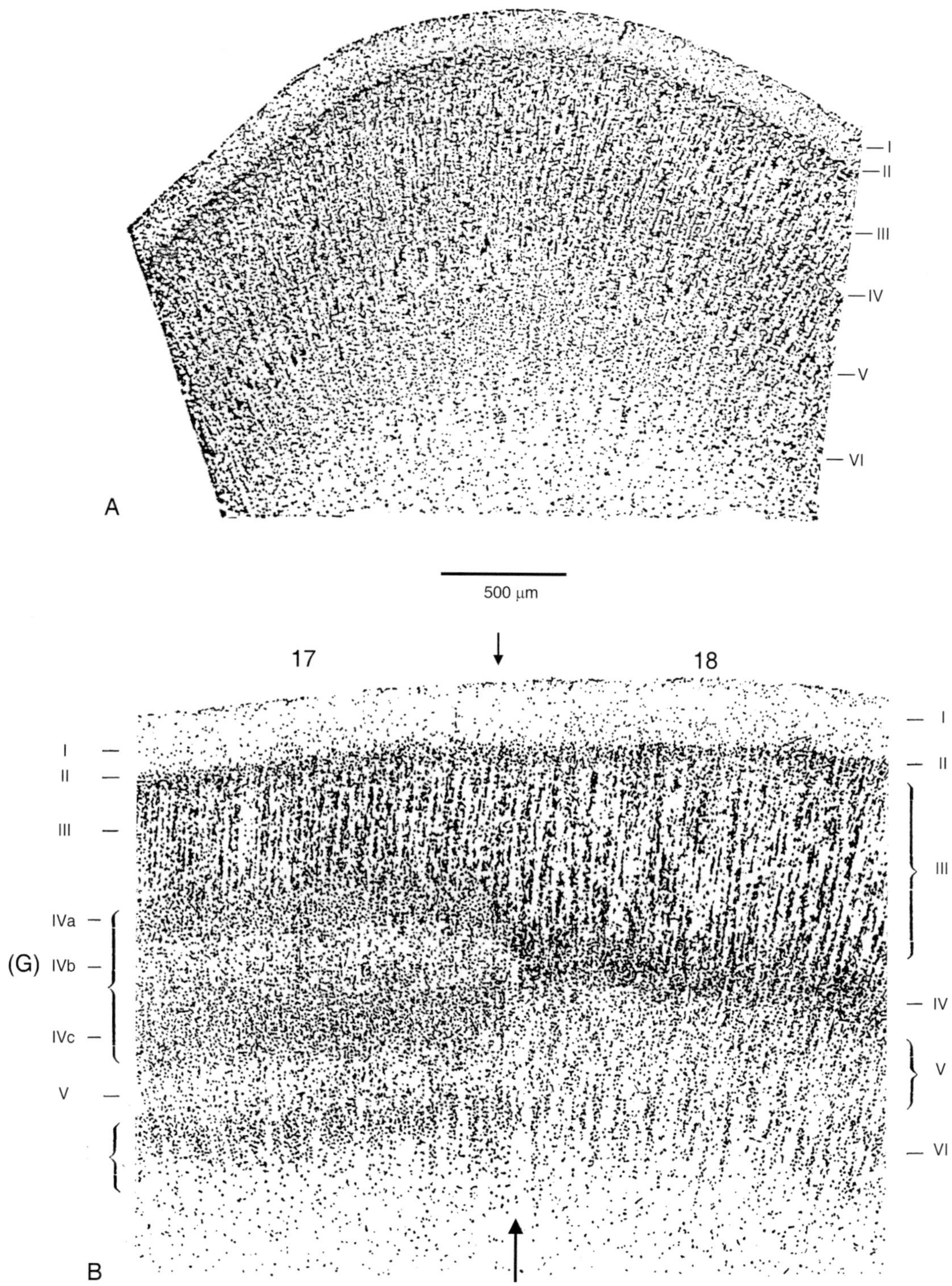

Fig. 15.41. Radial arrays of cell bodies in the primary motor area 4 (**A**) and the visual areas 17 and 18 (**B**) of 8-month-old human fetuses. Reproduced from Brodmann [70]. G, location of line of Gennari

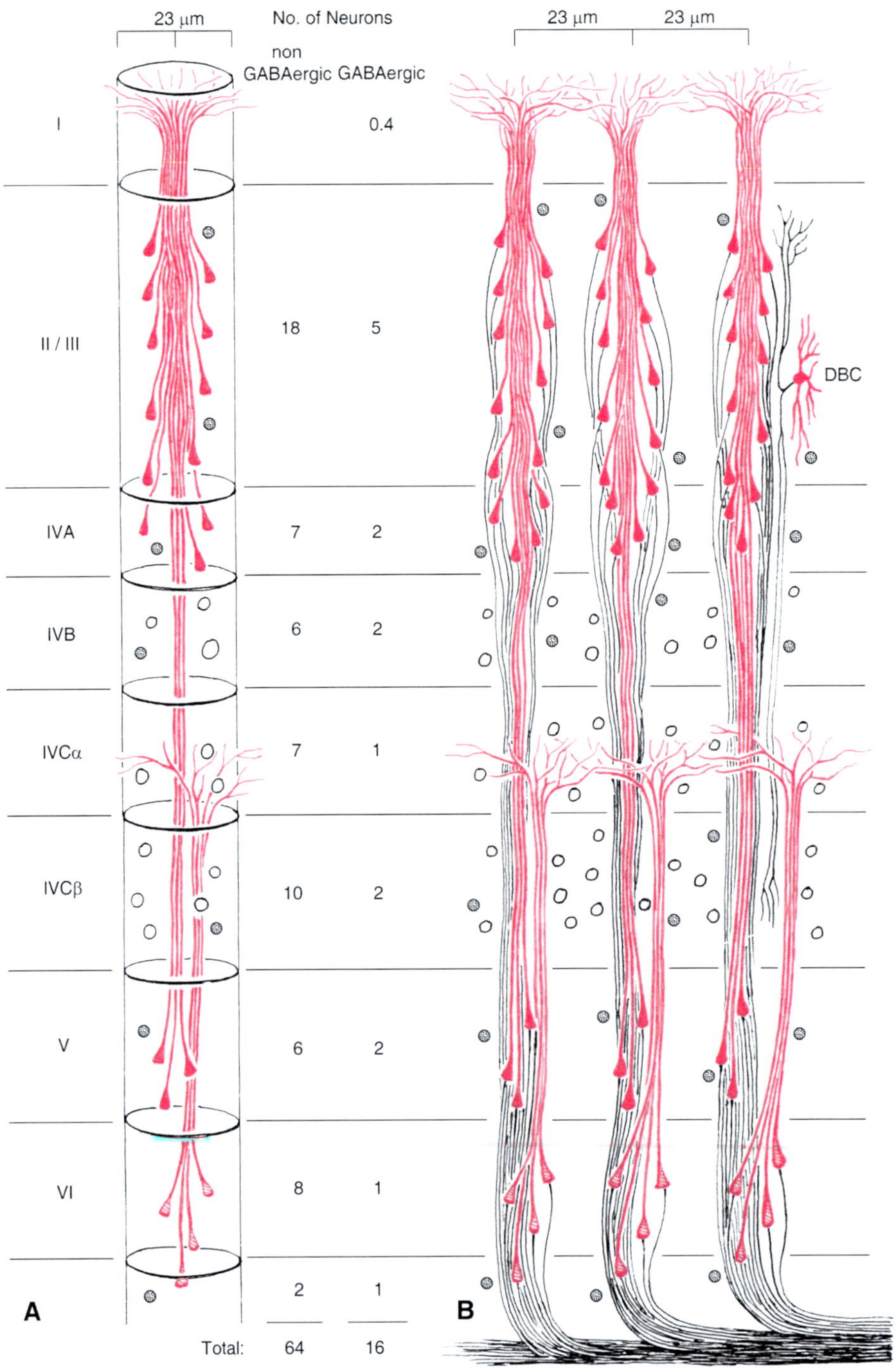

Fig. 15.42 A, B. Arrangement of pyramidal neurons and double bouquet cells in 23-μm-wide radial modules of the striate cortex of the macaque monkey. Neuronal somata and dendrites are shown in *red*, axons in *black*. **A** The arrangement of the apical dendrites of pyramidal cells in a module; for clarity, only one half of the neurons present are shown. The somata of spiny stellate cells are represented by *open circles*, while those of inhibitory, GABAergic neurons are represented by *circles filled with small dots*. The total *number* of spiny (excitatory) and GABAergic (inhibitory) neurons within each layer are given on the *right*. **B** A representation of three modules to show the arrangement of the dendrites and axons of the pyramidal cells. Each module contains one double bouquet cell (*DBC*). The axonal system of the double bouquet cell runs alongside the clustered apical dendrites of pyramidal neurons in layers IV and II/III and then alongside the bundle of myelinated axons. Modified from Peters and Sethares [566, 567]

tion between them and the bundles of myelinated fibres, and several later authors, among them Fleischhauer [195] and Buxhoeveden and Casanova [81], arrived at a similar conclusion. However, von Economo and Koskinas [796], who thoroughly investigated this relation, stated that in certain cortical regions, in which the neuronal perikarya were clearly arranged in vertical columns, a corresponding pattern of radial fibre bundles was absent. It is important to note that neither in the foetal nor in the adult stage do the arrays of cell bodies extend uninterruptedly through all cortical layers.

The phenomenon of apical dendrite bundling led Peters and Sethares [564–566] to the concept that the neocortex is composed of *pyramidal cell modules*. In accordance with earlier studies on rodents and cats, these authors found that the apical dendrites of the pyramidal cells in layer V of the primary visual cortex of the rhesus monkey cluster together in distinct and regularly spaced groups as they ascend through the various sublayers of lamina IV and pass into superficial layers II/III. The apical dendrites of the pyramidal neurons in those layers are then added to the clusters of layer V apical dendrites so that the clusters of apical dendrites gradually increase in size as they ascend. Upon reaching layer I, the apical dendrites split up into their terminal tufts (Fig. 15.42 A). When fully formed, the clusters comprise the dendrites of some 50 pyramidal cells. The apical dendrites of the pyramidal neurons in layer VI do not add to the clusters, but rather ascend independently to form their terminal rufts in layer IV (Fig. 15.42 A).

On the basis of these findings, Peters and Sethares [566] proposed that the primary visual cortex is composed of pyramidal cell modules. This concept is shown diagrammatically in Fig. 15.42 B, where three such modules are depicted. The average diameters of the modules are equal to the mean centre-to-centre spacing of the clusters of apical dendrites, namely about 23 μm.

The primary visual cortex of the rhesus monkey also contains prominent vertical bundles of myelinated fibres. These bundles, which are largely composed of pyramidal cell axons, are regularly arranged, and their average centre-to-centre spacing is similar to that of the dendritic clusters. In the deeper layers of the cortex, each myelinated axon bundle can be paired with a nearby cluster of apical dendrites (Fig. 15.42 B) [566]. Peters and Sethares envisioned that the pyramidal cell modules represent the basic functional units in the cortex, and that in essence they are equivalent to Mountcastle's minicolumns.

Column-like domains of spontaneously coactive pyramid-like neurons have been observed in slices of neonatal rat cortex [844]. The neurons within each domain appeared to be coupled by gap junctions. It has been suggested [341] that these domains are comparable to the pyramidal cell modules. However, their diameter (50–120 μm) is much larger than that of the modules, whereas their number of constituent cells (35 ± 17) is smaller.

The investigations of Vercelli et al. [776] have shown that pyramidal cells that share an apical dendritic bundle may project to specific targets. Studying the rat visual cortex, they found that neurons projecting to the ipsilateral or contralateral cortex form bundles together and with neurons projecting to the striatum, but not with those projecting to the lateral geniculate body or the superior colliculus or through the cerebral peduncle. Bundles composed of the apical dendrites of neurons projecting to the striatum and through the cerebral peduncle were also observed. The layer VI neurons projecting to the lateral geniculate body formed separate dendritic bundles. Vercelli et al. [776] concluded that the neocortex is organized in minicolumns of output neurons. They considered it likely that the neurons in an output minicolumn are closely related. Before leaving the pyramidal cell modules, it may be mentioned that Peters and Sethares [567] related one type of interneuron, namely the double bouquet cells, to these entities.

Double bouquet cells are abundant throughout the primate neocortex [140, 146]. The cell bodies of these elements are situated in layer II and upper layer III. Their axons form tightly packed bundles of collaterals that descend to layer V, where they disperse in horizontal branches. These vertically oriented double bouquet axonal systems or horse tails are equally spaced [140, 146]. Peters and Sethares [567] found that the average centre-to-centre spacing of the horse tails in the primary visual cortex of the rhesus monkey is about 23 μm, a value similar to that of the centre-to-centre spacing of the clusters of apical dendrites forming the axes of the pyramidal cell modules and that of myelinated axon bundles that contain the efferent axons of these neurons. It appeared that each pyramidal cell module is associated with one horse tail axon bundle (Fig. 15.42 B). Within the superficial layers II/III, the double bouquet cell axons run alongside the apical dendritic clusters, while in layer IVC they are closely associated with the bundles of myelinated axons. It is important to note that the double bouquet axonal branches do not synapse with the apical dendrites but rather with basal dendrites and with side branches of the apical dendrites of pyramidal neurons and with postsynaptic structures belonging to local circuit neurons [1, 136]. The branches of apical and basal dendrites of pyramidal neurons extend horizontally for hundred of micra; hence, it is clear that the influence exerted by the double bouquet cells does not remain confined to narrow cylindroid modules, but rather is dispersed over pyramidal neurons with apical dendrites located in many different apical dendritic bundles [341].

The third major radial configurations in the neocortex, i.e. the arrays of neuronal perikarya as observed in Nissl preparations, deserve little comment. On the basis of a study of the development of the human auditory cortex, Krmpotic-Nemanic [387] suggested that there is a direct continuum between the ontogenetic radial columns as described by Rakic and adult cell columns. These cellular columns are the focus of two recent reviews by Buxhoeveden and Casanova [80, 81]. The authors indicate that the visibility of these structures in Nissl preparations depends on the linear arrangement of pyramidal cells and the existence of cell-poor, neuropil zones surrounding them. In layers II, IV and VI of the adult neocortex, the radial arrays are usually one cell wide. Buxhoeveden and Casanova equate these radial arrays with Mountcastle's minicolumns.

Microcircuitry of Neocortical Columns

Two noted neuroscientists, Szentágothai and Eccles, attempted to present the microcircuitry of the neocortical columns. Szentágothai's [721–723] opinion was that "the basic unit of cortical architecture – the true cortical columns – is anatomically based not on the mode of termination of the specific sensory afferents, as one might logically assume, but on that of cortico-cortical afferents". The size of these columnar units of cortico-cortical afferent terminations, according to Szentágothai, is remarkably constant, measuring around 200–300 μm in diameter (Fig. 15.40). On the basis of light and electron microscopy data, Szentágothai indicated how the various types of presumed excitatory and inhibitory interneurons are connected with the cortical afferent system and with the cortical output neurons, i.e. the pyramidal cells. He emphasized that the internal connectivity of each columnar unit is predominantly vertical, in a way that would favour a subdivision of each column into narrower microcolumns, and that the space of many of these microcolumns is occupied by the dendritic and/or axonal ramifications of individual interneurons (Fig. 15.43). Eccles' [166] sketch of the neuronal structure of the cortical module conforms to that of Szentágothai and is largely based on the work of the latter.

Neocortical Columns and Modules: A Critical Commentary

In the preceding pages, the evidence leading to the concept that the mammalian neocortex is composed of repetitive radially oriented columnar units has been reviewed. Although this concept has dominated the neurobiological literature on the cortex for the last 50 years, it is by no means generally accepted. Numerous authors, including Towe [738], Valverde [756], Swindale [719], Purves et al. [597], Jones [341]

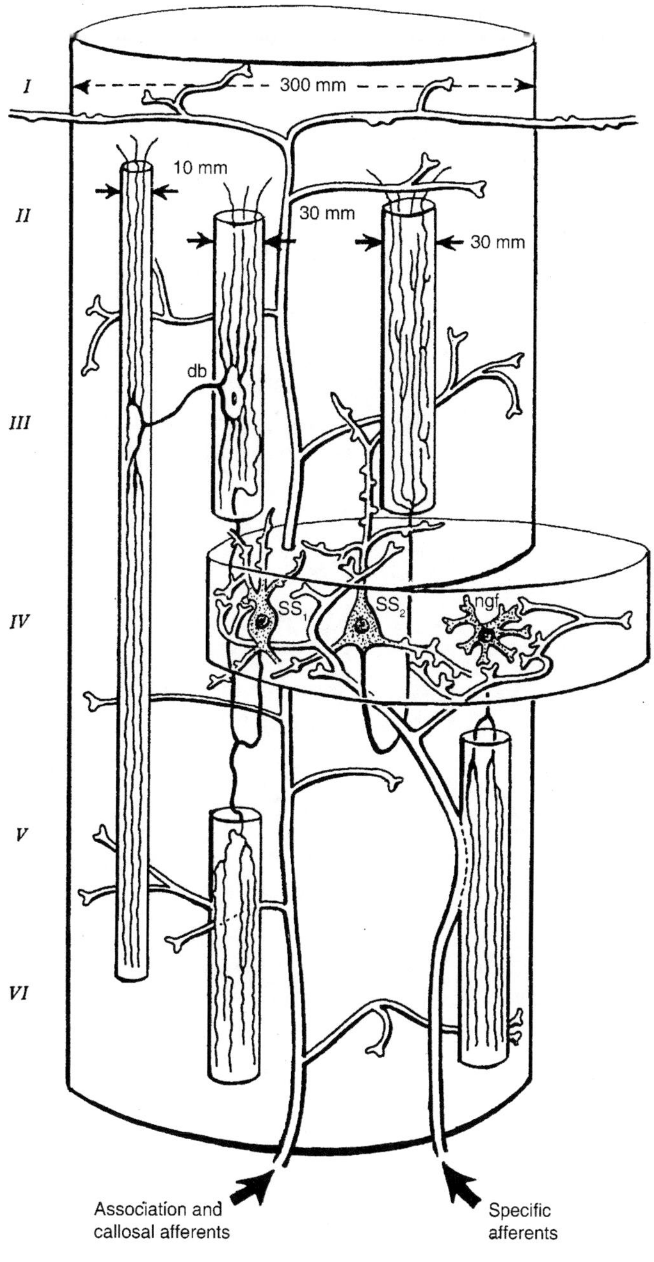

Fig. 15.43. Some features of the organization of a neocortical module or column, "defined" by a cortico-cortical (either association or callosal) afferent, ascending in its centre, as drawn by Szentágothai [721]. The module consists of a tall cylinder, 300 μm in diameter, extending through the full thickness of the cortex. The flat cylinder of the same diameter corresponds to the termination space of a specific thalamocortical afferent. This afferent fibre forms synaptic contacts with two different types of spiny stellate cells, *SS1* and *SS2*, as well as with a neurogliaform cell: ngf; SS1 has both an ascending and a descending arborization; SS2 has only an ascending arborization, whereas ngf has only a descending arborization. The ascending arborization of SS1 synapses with a typical double bouquet cell (*db*), whose bifurcating axon forms a long radial slender arborization. It should be mentioned that Szentágothai considered all of the neuronal elements depicted to be excitatory in nature. However, it is now known that neurogliaform cells and double bouquet cells are GABAergic, inhibitory interneurons. Reproduced from Szentágothai [721]

and Horton and Adams [305], have challenged the validity of this concept. Some of the main objections may be summarized as follows:

1. The original concept of Rakic, according to which all neurons forming a radial unit, both pyramidal and non-pyramidal, stem from a small cluster of progenitor cells in the pallial neuroepilium, is untenable. As has been discussed in Chap. 2, many, if not most, of the neocortical non-pyramidal cells are produced in the subpallial matrix and migrate tangentially into the developing pallium (Fig. 2.25).

2. The thesis that columnar structures form the basic structural and functional units of the neocortex has so far not been substantiated by direct evidence concerning the specific neuronal composition and the specific mode of operation of these units. The statement that the columns contain all cell phenotypes [76, 474] has not been proven and is even unprovable because an indeterminate number of cortical neurons have not been characterized in detail as yet. Moreover, recent histochemical and physiological studies have shown that the diversity of neocortical interneuron types is much larger than was previously thought [136, 507].

It is important to note that the models presented by Szentágothai [721, 723] and Eccles [166] are not based on analyses of the local connectivity within particular morphological units; rather, findings on cortical neurons and their connections derived from different cortical areas in different species were assembled and placed within the spatial framework of a cylinder with a diameter of 300 µm. Hence, these models should be considered as theoretical constructs, showing at best how the local circuitry within columns of cortical tissue, defined by streams of cortico-cortical afferents, could be organized. Mountcastle's [495] definition, stating that "a cortical column is a complex processing and distributing unit that links a number of inputs to several outputs", does not contain any information concerning the specific functional design of these columns.

3. According to the original concept of Lorente de Nó [416], the neocortex is built up of cylindrically shaped units, composed of vertical arrays of interconnected neurons that include all cortical layers. He designated these units as "elementary units of cortical operation", because he believed that within them "the whole process of the transmission of impulses from the afferent fibre to the efferent axon may be accomplished". Lorente de Nó did not specify the pattern of connectivity within these units, but during the last decades, several attempts have been made to characterize the basic neocortical microcircuit (see Fig. 15.44) [136, 157, 679]. However, in none of these attempts has the question been specifically addressed as to whether the elements involved in these microcircuits are confined to small, cylindrically shaped spaces. The pyramidal cell modules described by Peters and Sethares (Fig. 15.42) [565–567] could be the starting point for this type of research. Such a module would comply with Lorente de Nó's elementary unit concept if it could be shown that: (a) all axons of the constituent pyramidal neurons pass to the same target(s); (b) all constituent pyramid neurons are strongly and reciprocally connected by axon collaterals; (c) all sets of interneurons impinging on the various somatodendritic compartments of the pyramidal cells are situated within the confines of the module; and (d) all of its afferents terminate within the 23-µm-wide cylindrical space of the module. As regards (c), we have seen that the

Fig. 15.44 A–C. Neocortical microcircuits; excitatory elements in *black*, inhibitory elements in *red*. **A** *Basic neocortical circuit for sensory areas*, modified from Shepherd [679]. The diagram includes elements representing superficial and deep pyramidal cells (*P1*, *P2*), spiny stellate cells (*SS*) and inhibitory interneurons (*I1*, *I2*). Afferent inputs (*AFFs*) excite pyramidal cells (*1*, *2*, *5*, *6*) as well as inhibitory cells (*1*, *4*). In addition, there is feed-forward excitation (*AFF3*) through spiny stellate cells. The axons of the P cells have collaterals (*ac*) that recurrently and laterally excite the P cells (*ac1,2*) as well as inhibitory cells (*ac3*) that provide for feed-back and lateral inhibition. **B** *Canonical neocortical circuit*, based on the studies of Douglas et al. [157], as represented by Shepherd [679]. *P1* and *P2* represent the superficial and deep populations of (excitatory) pyramidal cells. Element *I* represents the population of inhibitory, GABAergic neurons. Neurons within each population form connections with other members of

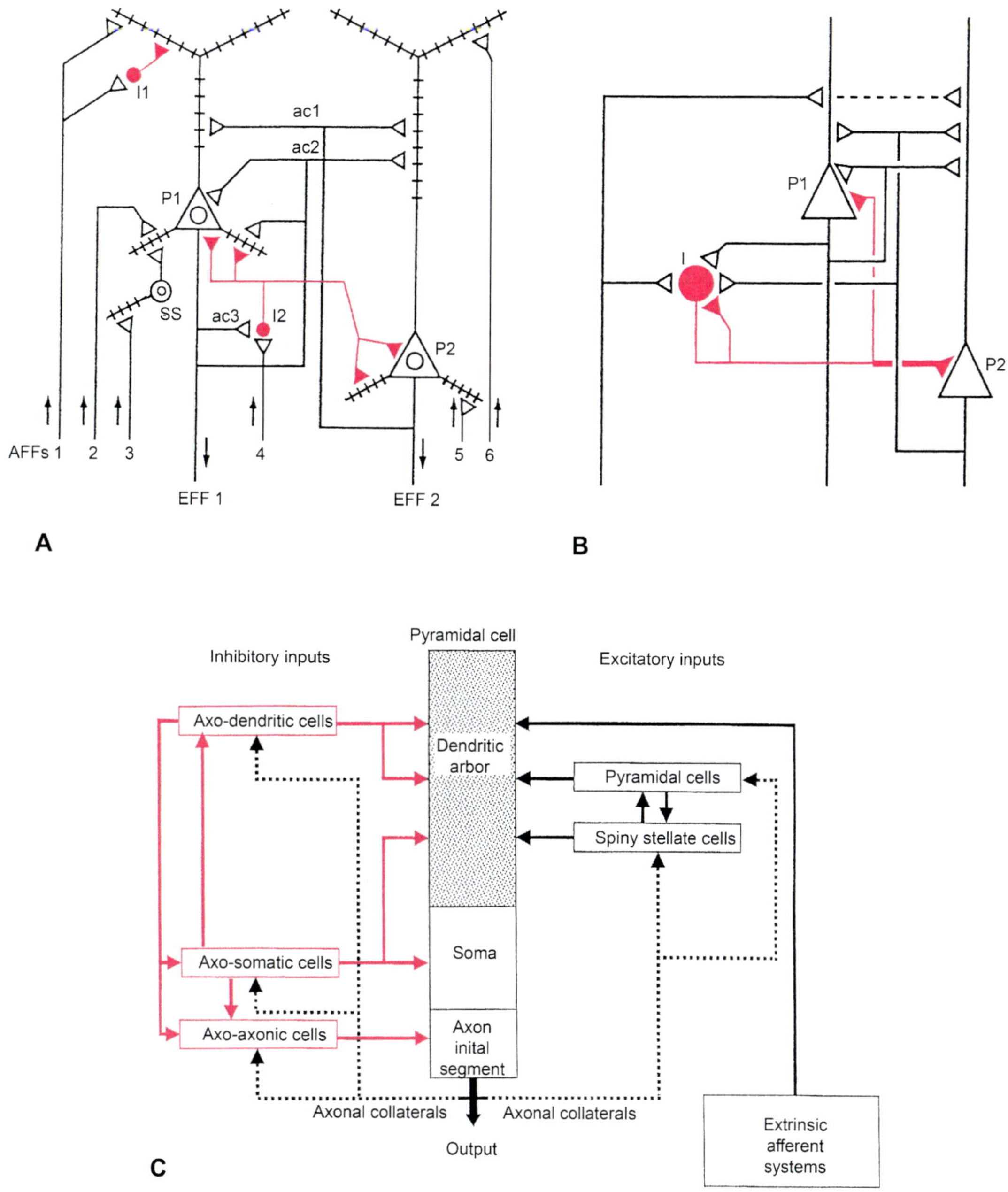

that population. All three populations receive direct activation from the thalamus, but because the thalamic input provides only about 10% of the excitatory input, 90% of the excitation stems from intracortical connections between P cells. The excitation of nearby P cells is provided by axon collaterals of these elements. These axon collaterals also excite the interneuron-pool to provide feed-back and lateral inhibition. Modified from Shepherd [679]. **C** *Skeleton of the basic neocortical microcircuit,* according to DeFelipe [136]. The diagram shows the synaptic inputs and main patterns of local connections of the pyramidal cell. Excitatory inputs to this element arrive exclusively to its dendritic arbor and originate from extrinsic afferent systems and spiny cells (which include other pyramidal neurons and spiny stellate cells). Inhibitory inputs, which originate from GABAergic interneurons, terminate on the proximal dendrites, soma and axon initial segment. These interneurons are interconnected between one another, with the exception of the axo-axonic chandelier cells. Reproduced from DeFelipe [136]

double bouquet cells conform to the modules (Fig. 15.42 B) but that their axons most probably do not exclusively contact the dendrites belonging to the pyramidal cells situated within "their own" module. As regards (d), the cortical afferents are not of an appropriate size to match an individual pyramidal cell module. The terminal arbors of thalamocortical and cortico-cortical afferents are generally 200 µm or more in diameter (Figs. 15.40, 15.43). This does not exclude, of course, that these afferents participate in the formation of larger processing units (see below). An important general question to be addressed in the near future is: what is the relationship (if any) between the cylindrically shaped units proposed by Lorente de Nó, Mountcastle and others and the neocortical functional networks, discussed in the previous section of the present chapter.

The various column-like cortical entities have been distinguished on the basis of highly different morphological and/or physiological properties, e.g. similarity in physiological response properties, similarity in motor responses following stimulation, spatial segregation of sets of thalamocortical afferents, high concentrations of the enzyme cytochrome oxidase, concentrations of spiny stellate cells in lamina IV and the mode of termination of bundles of cortico-cortical fibres. It is unlikely that all of these entities are derivatives of one and the same basic cortical module, especially as they may differ considerably in size. It cannot be excluded that, as Szentágothai [722] surmised, the units defined by cortico-cortical afferents represent the "true cortical columns". With regard to these structures, however, it should be emphasized that (a) their presence has so far not been established in monotremes, marsupials or insectivores, (b) their intrinsic structure is unknown and (c) their functional significance remains to be elucidated [305].

It is also important to note that the vertically aligned rows of cell bodies observed in Nissl preparations of many cortical areas, which are often equated to cortical minicolumns, never extend uninterruptedly from layers II to VI and that most of the column-like formations described have no definite boundaries [621].

5. Although column-like entities of a given type are readily apparent in the neocortex of some species, they are often not detectable in other, sometimes closely related animals. The following examples of such inconsistencies are quoted from Purves et al. [597] and Horton and Adams [305], to whom the reader is referred for primary sources and details:

(a) Ocular dominance columns are present in Old World monkeys, but reportedly, absent in some New World monkeys.

(b) Remarkably, it was found in squirrel monkeys that some members of this species have ocular dominance columns in only part of the visual cortex, leaving other regions of binocular cortex bereft of columns.

(c) Blobs in the primary visual cortex of the rhesus monkey are reportedly concerned with processing information about colour. However, these structures are also present in nocturnal primates with cone-poor retinas (and, therefore, poor colour vision). Moreover, blobs are absent in some non-primate species with cone-rich retinas and excellent colour vision, such as the tree shrew and the ground squirrel.

(d) Barrels are well developed in the mouse, rat, squirrel, porcupine and walrus, but are absent in other species that have prominent facial whiskers such as the dog, cat, raccoon and tree shrew. Furthermore, barrels occur in the guinea pig, which hardly uses its whiskers, and in the chinchilla, another cavimorph that has no whisking behaviour at all.

6. The barrel cortex of rodents and the primate visual cortex are frequently cited as archetypical examples of columnar structures. However, Jones ([341], p. 5021) has pointed out that these two structures "represent end-points in the evolution of the two orders, one of which noses and whisks its way around its environment and the other of which extracts an extraordinary richness of visual detail from its environment".

If we survey the data discussed above, the following conclusions seem to be warranted:

1. During the last 50 years, numerous different types of classes of column-like entities have been detected in different parts of the neo-

cortex of many different mammalian species with the aid of a variety of physiological and anatomical techniques. There is little evidence of a systematic relation between these classes of "columns".

2. There is very little evidence in favour of the concepts that (a) the entire mammalian neocortex is composed of column-like entities, (b) all of these entities represent variations on one and the same theme, (c) all of these entities essentially have the same structure and (d) they all essentially subserve the same function.

Comparative Aspects

If we compare the neocortex with the "older" parts of the mammalian cortex, i.e. the prepiriform or olfactory cortex (Figs. 11.5, 11.6) and the hippocampal cortex (Figs. 12.6–12.9), the following similarities and differences can be observed:

1. Pyramidal neurons occupy a central position in the circuitry of all of these cortices. These elements are provided with two, spatially separated, dendritic systems; an apical dendritic system, expanding in the most superficial cortical layer; and a basal dendritic system, the laminar position of which may vary considerably.
2. In all cortices, the pyramidal neurons are excitatory in nature and potentially constitute a continuous network extending throughout the cortex. Nearby connections in this network are provided by axon collaterals of pyramidal neurons, whereas longer intracortical connections consist of the main axons or axonal branches of more remote pyramidal neurons. In the neocortex, each pyramidal neuron receives by far its largest input from other pyramidal neurons.
3. Pyramidal neurons constitute the output system of all of these cortices.
4. In all cortices, the most superficial layer contains numerous tangentially running afferent fibres, making excitatory synaptic contacts with the apical dendritic extensions of pyramidal neurons. The sources of these fibres include the olfactory bulb, the thalamus and other cortical regions. The fibre composition of the superficial layer varies from region to region. Thus, in the prepiriform cortex, secondary olfactory fibres occupy a superficial position, whereas cortico-cortical fibres form a deeper zone. In the sensory part of the neocortex of primitive mammals, numerous thalamocortical fibres reach lamina I, but in the hippocampus and in neocortical association area, cortico-cortical fibres prevail in this layer. In lamina I of all cortices, fibres from different sources tend to be arranged in different sublaminae.
5. In neocortical sensory regions, thalamic afferents terminate massively in lamina IV. In the primary visual cortex of prosimians and primates, different laminae of the lateral geniculate nucleus project to different sublayers of lamina IV.
6. Inhibitory interneurons, using GABA as their neurotransmitter, occur in all cortices. All inhibitory interneurons in all cortices impinge directly with some or all of their terminals on pyramidal neurons. In all cortices, separate sets of inhibitory interneurons terminate mainly or exclusively on one of the three receptive domains of the pyramidal cells, i.e. the dendritic compartment, the soma and the axon initial segment.
7. Some inhibitory interneurons receive their principal input from extrinsic afferents and mediate feed-forward inhibition of pyramidal neurons, e.g. the horizontal cells in lamina I of the piriform cortex and in the neocortex. Other inhibitory interneurons receive their main input via axon collaterals of pyramidal neurons, thus forming part of feed-back loops (e.g. the large multipolar cells in lamina III of the piriform cortex) and still others, receiving excitatory inputs from extrinsic afferents and from axon collaterals of pyramidal cells, are involved in both feed-forward and feed-back inhibition of pyramidal neurons, e.g. basket cells in the hippocampus and the neocortex.
8. Inhibitory interneurons, exerting their influence on other inhibitory interneurons, and

thus a disinhibition of pyramidal neurons, are present in all cortices. The elements involved in such inter-interneuronal inhibitory circuits may be of the same type, e.g. the large multipolar cells in the piriform cortex and the basket cells in the neocortex, or of different types, e.g. the so-called L/M cells and the basket cells in the hippocampus and the basket cells and double bouquet cells in the neocortex.

9. In the neocortex as well as in the hippocampus groups of inhibitory interneurons have been observed that are interconnected via chemical synapses and gap junctions and also innervate themselves via autapses [24, 101, 362].
10. Excitatory interneurons have so far only been demonstrated in the neocortex. These elements, the spiny stellate cells, represent transformed pyramidal neurons. They are abundant in primary sensory cortices, where they play a prominent role in the radial propagation of the activity fed by thalamocortical afferents into lamina IV of these cortices.

Synopsis of Main Neocortical Regions

Introduction

In this section an overview of the structure and principal connections of the main neocortical regions will be presented. We will confine ourselves here mainly to cortico-cortical projections. Neocortical afferents have been discussed in a previous section of the present chapter and efferents of the motor cortex will be dealt with in Chap. 21. For discussion of the relations of the neocortex with the basal ganglia and cerebellum, we refer the reader to Chaps. 14 and 20, respectively. In this overview the classical subdivision of the hemisphere into the occipital, parietal, temporal, limbic, frontal and insular lobes will be followed (Fig. 1.4). It is important to note that this subdivision was based originally on the bones of the skull and, hence, has little to do with brain function. However, it appears to be possible to subdivide all of them into a number of functional domains. Throughout the description, the still widely used area parcellation of Brodmann (Fig. 15.8) will be used as a reference system. The discussion of the various lobes is prefaced with some remarks on the neocortical association and commissural connection and on the functional and structural asymmetry of the two cerebral hemispheres.

Association and Commissural Connections

Association and commissural fibres, which interconnect different neocortical areas, make up most of the white matter of the cerebral hemisphere. Commissural fibres cross in the corpus callosum and in the caudal part of the anterior commissure. Association systems can be rather arbitrarily subdivided into short and long association fibres. Short association fibres may remain within the grey matter of the cortex or pass through the superficial white matter between neighbouring cortical areas as U fibres. Long association systems are located in deeper parts of the white matter, lateral and medial to the corona radiata and the internal capsule. In the human brain, long association systems are mainly known from gross dissection [145, 421], although imaging techniques have recently become available, rendering it possible to visualize these bundles in the living human brain [40, 108, 335, 488, 588]. The most important long association systems are named and depicted in Figs. 15.45 and 15.46. It can be seen that these systems interconnect cortical regions in different lobes within the same hemisphere.

The *superior occipitofrontal (or subcallosal) fasciculus* is situated dorsolateral to the caudate nucleus, immediately underneath the most medial part of the radiation of the corpus callosum. Its fibres connect the occipital and temporal regions with the frontal lobe.

The *superior longitudinal fasciculus* is located along the laterosuperior border of the putamen. It is separated from the superior occipitofrontal fasciculus by the fibres of the most proximal part of the corona radiata. It forms a large arcuate bundle that interconnects the frontal lobe with the three post-Rolandic lobes. Its posterior part splits up into a brachium posterius, which fans out into the parietal, occipital and posterior temporal lobes, and a strongly curved brachium anterius, which is connected with the anterior part of the temporal lobe.

The *inferior occipitofrontal fasciculus* and the *uncinate fasciculus* form part of a single fibre complex. The compact, intermediate parts of these bundles are situated directly underneath the claustrum and pass via the limen insulae from the frontal to the temporal lobe. The inferior occipitofrontal fasciculus consists of fibres that connect the lateral parts of the frontal lobe with the inferior temporal and medial and lateral occipitotemporal gyri and with the occipital lobe. The uncinate fasciculus connects the inferior frontal gyrus and the orbital surface of the frontal lobe with anterior portions of the temporal lobe.

The *inferior longitudinal fasciculus* follows the lateral wall of the posterior and inferior horns of the lateral ventricle. It connects the occipital lobe with most parts of the temporal lobe. Tusa and Ungerleider [739] provided experimental evidence indicating that the inferior longitudinal fasciculus in the rhesus monkey consists of a series of U fibres that sequentially connect adjacent regions in the striate, prestriate and inferior temporal cortex. On that account they proposed replacing the term "inferior longitudinal fasciculus" by the term "occipitotemporal projection system".

The *cingulum,* finally, is a large system of shorter and longer association fibres, which is situated in the white matter of the cingulate gyrus, following this convolution throughout its entire length (Fig. 12.13). Anteriorly, it extends around the genu of the corpus callosum to the subcallosal area. Posteriorly, it arches around the splenium of the corpus callosum and continues downward and forward within the parahippocampal gyrus. From there, its constituent fibres fan out in the adjoining parts of the medial temporal lobe.

Although the neocortical association systems have already been dealt with in a previous section of the present chapter (Structural Subdivision 4: Connectivity), it may be good to repeat some general principles applying to the organization of the connections (Figs. 15.15, 15.46, 15.47). It should be emphasized that our knowledge of the cortico-cortical connections is almost entirely based on experimental studies in non-human primates [463–465, 526, 529, 532, 579, 773].

1. Cascades of short association fibres interconnect the primary, modality-specific areas of the cortex, which receive their sensory input from the thalamus, with the modality-specific parasensory association areas. The somatosensory association cortex is found in the parietal lobe, the visual association areas in the occipital lobe and the temporal lobe below the superior temporal sulcus, and the auditory association areas are in the superior temporal gyrus and the temporal operculum. These short connections will be reviewed according to their modality.
2. Modality-specific areas of parasensory association cortex are connected with multimodal sensory areas located at their borders. These areas constitute a belt extending from the junction of the occipital and parietal lobes, through the caudal part of the superior temporal gyrus, into the superior temporal sulcus. Long association systems connect the modality-specific parasensory association cortex and the multimodal areas in the occipital, temporal and parietal lobes with the premotor and prefrontal cortex of the frontal lobe. Short association fibres interconnect the prefrontal cortex, the premotor area and the motor cortex. Short association fibres interconnect the motor cortex and the primary somatosensory cortex.
3. Connections from the parasensory and multimodal association cortices and the prefrontal cortex to limbic structures pass by way of the cingulum to the medial part of the temporal lobe. Other fibres originating from parasensory association cortices reach limbic structures via the insula.
4. Most association connections are reciprocal.
5. Connections from the primary sensory areas to their neighbouring association areas usual-

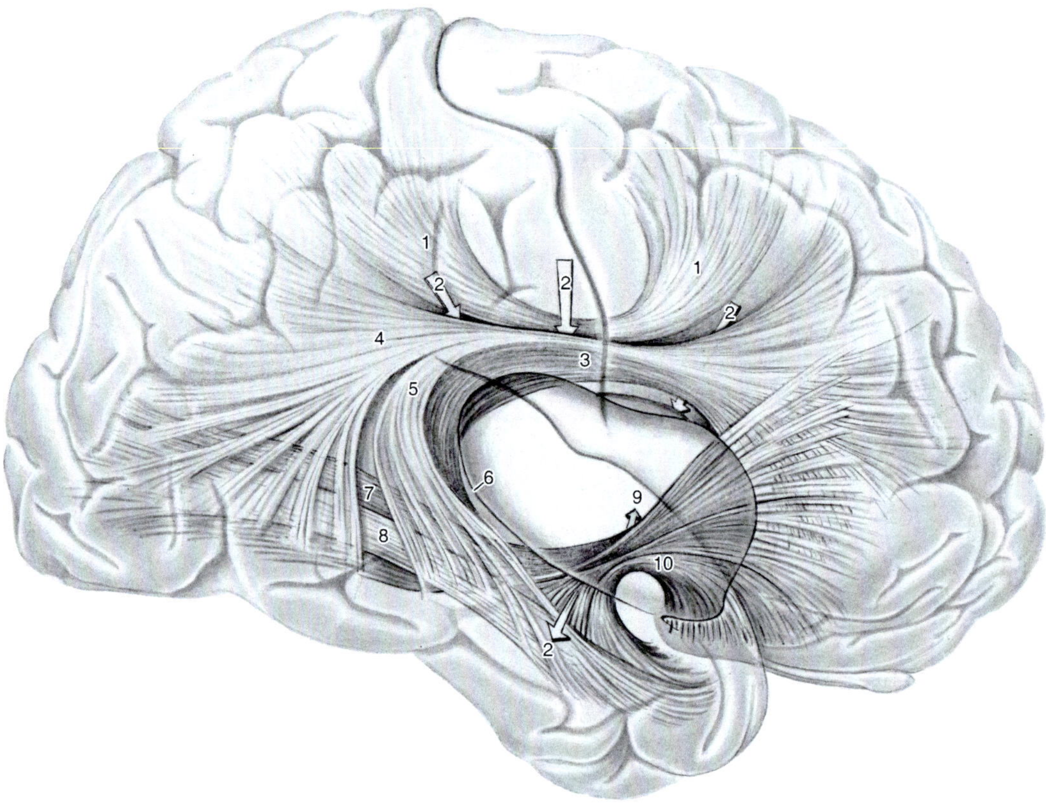

1 Superior occipitofrontal fasciculus
2 Site of corona radiata
3 Superior longitudinal fasciculus
4 Superior longitudinal fasciculus, brachium posterius
5 Superior longitudinal fasciculus, brachium anterius
6 Outline of insula
7 Inferior occipitofrontal fasciculus
8 Inferior longitudinal fasciculus
9 Site of anterior commissure
10 Uncinate fasciculus

Fig. 15.45. Long association bundles of the right cerebral hemisphere in a lateral view (1/1×). Part of the superior longitudinal fascicle has been removed to show the superior occipitofrontal fascicle

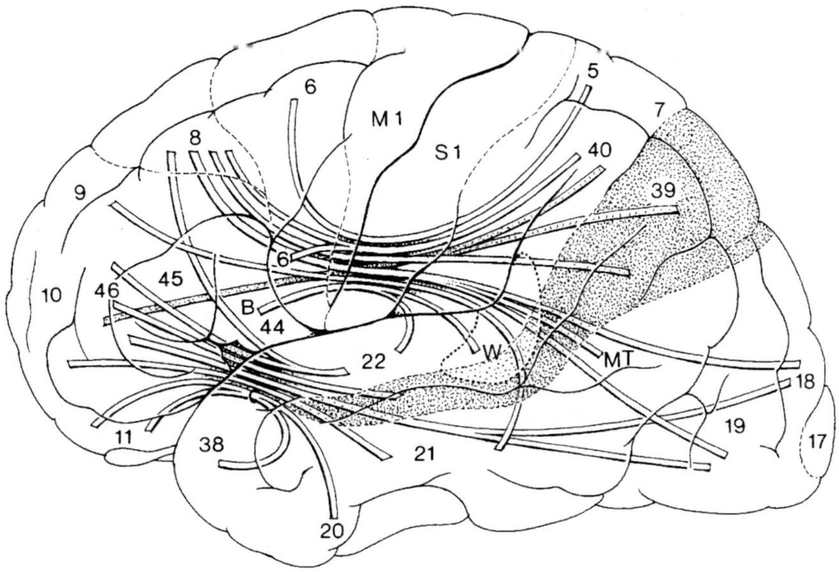

B Broca's speech region
M1 primary motor area
MT Middle temporal visual association area
S1 primary sensory area
W Wernicke's speech region

Fig. 15.46. Long association connections of the neocortex of the left hemisphere. The multimodal association cortex in the superior temporal sulcus and the parieto-occipital sulcus is *dotted*

ly originate from the supragranular layers and terminate in and around layer IV. This type of connection is therefore considered to transfer information in a forward direction. The reverse (feed-back) connections originate in the infragranular layers and terminate in layers I and VI [623]. The laminar analysis of association connections may therefore reveal the direction of information transfer.

Fibres interconnecting neocortical regions in the two hemispheres pass through two commissural systems, the anterior commissure and the corpus callosum (Fig. 3.7). Both of these commissures develop in the commissural plate, i.e. the thickened, most rostral part of the lamina terminalis (Fig. 2.6).

The *anterior commissure* is throughout most of its extent a compact, transversely oriented bundle (Fig. 12.4). It comprises a small anterior limb and a much larger posterior limb (Fig. 15.48). The anterior limb interconnects olfactory structures, such as the anterior olfactory nuclei and the primary olfactory cortices, on the two sides (Fig. 11.7). The posterior limb, which passes laterally and somewhat backward through the most inferior parts of the lentiform nucleus (Figs. 5.5, 5.6, 5.21–5.24), interconnects the anterior portions of the middle and inferior temporal gyri.

A *corpus callosum* is lacking in monotremes and marsupials [511]. In eutherians, the size of the corpus callosum parallels that of the neocortex. Thus, it is small in primitive mammals such as the hedgehog (Fig. 12.1 D), larger in prosimians (Fig. 12.1 E), and attains its maximal size in humans (Fig. 12.1 C). During ontogeny, the human corpus callosum manifests itself initially as a small, compact bundle in the most dorsal part of the commissural plate (Fig. 2.6 A). During further development, it thickens, expands backwardly (Fig. 2.6 B–D) and ultimately covers the entire diencephalon (Fig. 7.1).

In the human adult, the corpus callosum is a broad plate of fibres that interconnects the neocortical portions of the two hemispheres. It forms the floor of the longitudinal fissure and the roof of the lateral ventricles (Figs. 5.4–5.11).

The corpus callosum is divided into a curved rostral part, the *genu*, a large middle part, the *body or truncus*, and a thickened caudal part, the *splenium*. A tapering *rostrum* connects the posteriorly recurved part of the genu with the lamina terminalis (Figs. 3.7, 15.48). The fibres of the corpus callosum, which fan out into the white matter of both hemispheres, form the *radiation of the corpus callosum* (Figs. 5.4–5.10). The extension of these fibres in the most frontal parts of the hemisphere is called *forceps minor*, whilst the similar but wider sweep of fibres toward the occipital poles is called *forceps major* (Figs. 5.29, 15.48). The fibres destined for the basal parts of the temporal and occipital lobes form a thin fibre plate, the *tapetum*, which accompanies the dorsolateral wall of the inferior horn of the lateral ventricle (Fig. 15.48).

Commissural connections are homotopic or heterotopic. *Homotopic fibres* interconnect corresponding cortical areas in both hemispheres. The fibres interconnecting the frontal lobes pass in the anterior half of the corpus callosum, the others pass in the posterior half, with the parietal fibres anterior to those from the temporal lobe, and the occipital lobe fibres most posteriorly, in the splenium [156, 533].

Experimental studies in rhesus monkeys [126, 361] have shown that there are considerable regional variations as regards the distribution of callosal afferents. Thus, commissural fibres are absent from most of the primary visual cortex (area 17) and the same holds true for the areas representing the distal parts of the limbs in both the primary somatosensory and the primary motor cortices. Most of the association areas, on the other hand, are strongly interconnected by callosal fibres.

Heterotopic commissural connections connect a cortical area with non-corresponding areas in the contralateral hemisphere. The pattern of heterotopic connections of a particular cortical area often mimics its association connections with other areas in the ipsilateral hemisphere. It is sometimes possible to distinguish between forward and backward types of heterotopic connections on the basis of their laminar origin and termination [126].

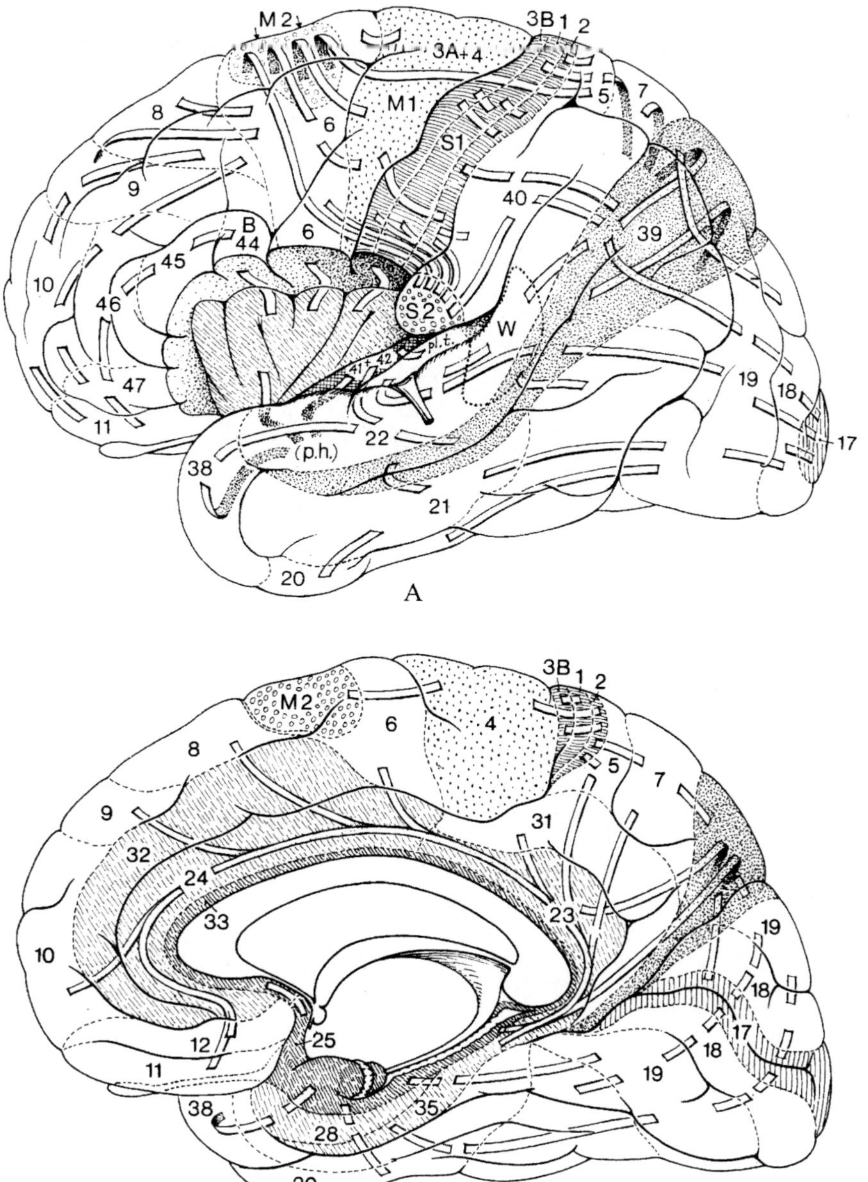

B Broca's speech region
p.h. parahippocampal gyrus
pl.t. planum temporale
W Wernicke's speech region

Fig. 15.47 A, B. Short association connections of the cerebral cortex. **A** Lateral view. The frontal and part of the parietal operculum has been removed; the temporal operculum is retracted. **B** Medial view. The primary and secondary sensory areas (*S1* and *S2*), the primary and secondary motor areas (*M1* and *M2*), the primary visual and auditory cortices, the paralimbic association cortex and the insula, and the limbic cortex are indicated with *different shadings*. The multimodal cortex in the superior temporal sulcus and the parieto-occipital sulcus is *dotted*

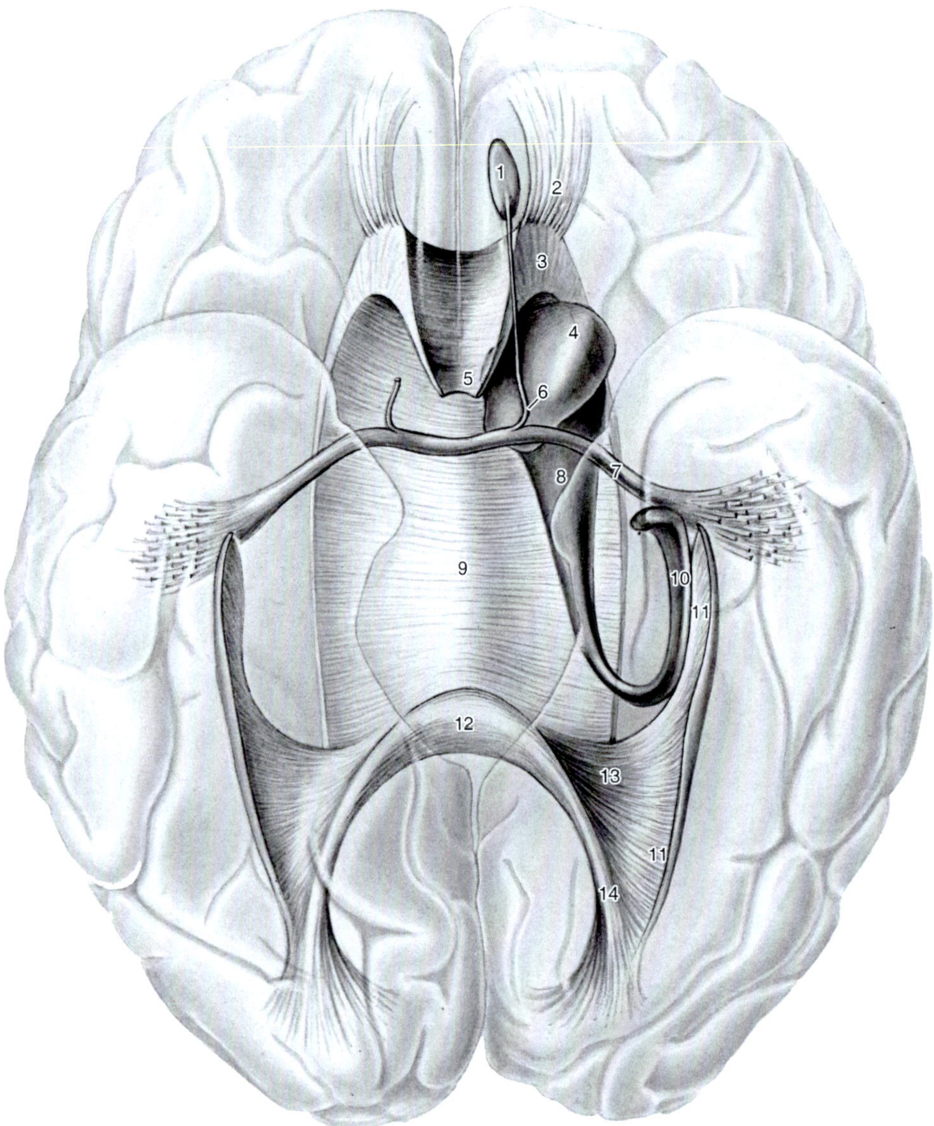

1 Olfactory bulb
2 Forceps minor
3 Genu of corpus callosum
4 Head of caudate nucleus
5 Rostrum of corpus callosum
6 Anterior commissure, anterior limb
7 Anterior commissure, posterior limb
8 Body of caudate nucleus
9 Body of corpus callosum
10 Tail of caudate nucleus
11 Tapetum
12 Splenium of corpus callosum
13 Radiation of corpus callosum
14 Forceps major

Fig. 15.48. Commissural connections of the telencephalon as seen from the basal side of the brain (1/1×)

Association and commissural connections often originate from and terminate in strips, which in turn are separated from each other by strips lacking these particular connections. This pattern of alternating positive and negative strips has a periodicity of 200–1000 μm [243, 319, 332, 349]. In the somatosensory and auditory cortices, the strips extend perpendicular to the borders of the somatotopic or the isofrequency maps, but a periodicity in the origin and termination of cortico-cortical connections has also been observed for visual association areas and for the multimodal, frontal and paralimbic association cortices. Cells of origin and their homotopic terminations are located in the same strips. It is not certain whether the ipsilateral association and commissural pathways are complementary with respect to the strips [76, 246, 337, 344, 349, 667].

Association and commissural connections are involved in many higher functions of the nervous system. Damage to the primary sensory or parasensory association area may result in perceptual deficits. In contrast, the interruption of commissural connections or the association connections between the unimodal area and multimodal or paralimbic association areas may lead to disconnection syndromes. These syndromes have been extensively studied in experimental animals and in humans [227, 462, 703].

Functional and Structural Asymmetry of the Two Hemispheres

Functional asymmetry of the two hemispheres is a salient feature of human brain organization and cognition. This phenomenon is also termed "hemispheric specialization", "functional lateralization" and "cerebral dominance". These terms all refer to the fact that the right and left cerebral hemispheres have different roles in mediating behaviour and higher mental processes [226].

Evidence for functional asymmetry of the cerebral hemispheres is mainly derived from studies of the following types.

1. Clinicopathological studies. The modern era of neuroscientific investigation into cerebral asymmetry began in the 1860s and 1870s with the discovery of Broca [67] and Wernicke [811] that language functions are lateralized to the left hemisphere. The notion of a dominant hemisphere was first presented by Hughlings Jackson on the basis of the frequent occurrence of motor (and other kinds of) aphasia in right-handed persons when they suffer a hemiplegia of the right limbs, while a left-sided hemiplegia in such persons usually is not accompanied by aphasia [69].
2. Studies on the laterality of motor control. Such studies have shown that in all races more than 90% of the population is naturally more skilled with the right hand than with the left [112, 114]. As is well known, the right hand is controlled by the left hemisphere.
3. Studies in which one of the cerebral hemispheres is temporarily inactivated. In the so-called Wada test [799], a barbiturate is injected into the right or left internal carotid artery, by which the ipsilateral hemisphere is inactivated for a short while. The test forms part of the presurgical assessment of hemispheric language dominance in patients suffering from intractable epilepsy. In the dominant hemisphere, such an injection transiently blocks speech. Results with this test have shown that 92–99% of dextral individuals are left hemisphere-dominant for language, while the pattern in non-dextral individuals includes leftward but also bilateral or even rightward dominance for language [417].
4. Studies in patients in whom the two cerebral hemispheres are surgically separated by transection of the corpus callosum and the anterior and hippocampal commissures. Operations of this type are carried out to control intractable seizures [799]. Such "split-brain" individuals offer a unique opportunity to test the functions of each hemisphere independently of the other [224, 225].
5. Neuroimaging studies, which render it possible to identify and localize changes in metabolic activity in the brain, correlated with cognitive and mental tasks, have been

widely applied to the study of functional asymmetries. Moreover, MRI scans can be used for in vivo morphometry of brain structures [417, 708].

Overall, the studies listed above have shown that in right-handed persons the left hemisphere is mainly concerned with verbal and linguistic functions, mathematical skills and analytical thinking, whereas the right hemisphere is primarily involved in spatial relationships, musical and artistic functions and in the recognition and expression of emotions.

Left-handed persons form a heterogeneous group. Only 15% of left-handers show the expected right-hemisphere dominance; a full 70% of them is left-hemisphere dominant, while the remaining 15% has bilateral language abilities [226].

Crow and Mitchell [123, 473] concluded from lesion studies in stroke patients and functional imaging studies of healthy people that some language functions, including discourse planning/comprehension, understanding humour, sarcasm and metaphors, are mediated by the right hemisphere rather than the left. It may be added that, according to these authors, the language deficits in patients with schizophrenia can be interpreted as abnormalities of lateralization.

Possible morphological correlates of the functional asymmetries just discussed have been described by numerous authors. Geschwind and Levitsky [229] reported left-right asymmetries in the temporal speech region. They measured the size of the planum temporale, i.e. the posterior region of the superior surface of the temporal lobe, in 100 brains and found it to be larger on the left than on the right in 65, with no asymmetry in 24 and with a reversed pattern in 11 brains. Their observations have been confirmed by several other investigators, among them Witelson and Pallie [829], Wada et al. [800], Rubens [642] and Steinmetz [708]. Habib et al. [271] demonstrated that the leftward volume asymmetry of the planum temporale is related to the degree of right-handedness. Asymmetries in the planum temporale have also been demonstrated in the volume of cytoarchitectonically defined areas. Thus, Galaburda et al. [221] measured an auditory association area, designated as Tpt in four brains. Distinct leftward asymmetries were seen in this area, whereby in one case the left area Tpt was 620% larger than the right. Moreover, a perfect rank-order correlation was found between the magnitude of asymmetry of area Tpt and the degree of planum temporale asymmetry.

It is generally accepted that areas 44 and 45 of Brodmann in the inferior part of the frontal lobe constitute the cytoarchitectonic correlates of Broca's speech region. Amunts et al. [6] delineated these two areas in 10 human brains. They found a left-larger-than-right asymmetry of area 44 in all brains studied.

Histological asymmetries have also been found in areas related to language. Scheibel et al. [655] reported differences in the complexity of the dendritic trees of pyramidal cells between the speech areas in the left hemisphere and their counterparts on the right side, and Hayes and Lewis [281] found that the pyramidal neurons in layer III of Broca's area 45 are consistently larger on the left than on the right side. Finally, it may be mentioned that evidence, recently summarized by Hutsler and Galuske [318], indicates that the width of individual columns in the posterior language-oriented areas is greater in the left hemisphere than in the right.

Occipital Lobe

The occipital lobe occupies the most posterior portion of the hemisphere on its lateral, medial and basal surfaces (Figs. 1.4, 15.49). On the lateral or convex surface of the hemisphere, the anterior boundary is formed by a somewhat irregular parieto-occipital line which, starting from the superior end of the parieto-occipital sulcus, partly follows the anterior occipital sulcus (Fig. 3.2). On the medial surface, the anterior boundary of the occipital lobe is marked by the parieto-occipital sulcus (Fig. 3.6). An arbitrary, transversely oriented occipitotemporal line, cutting through the medial and lateral

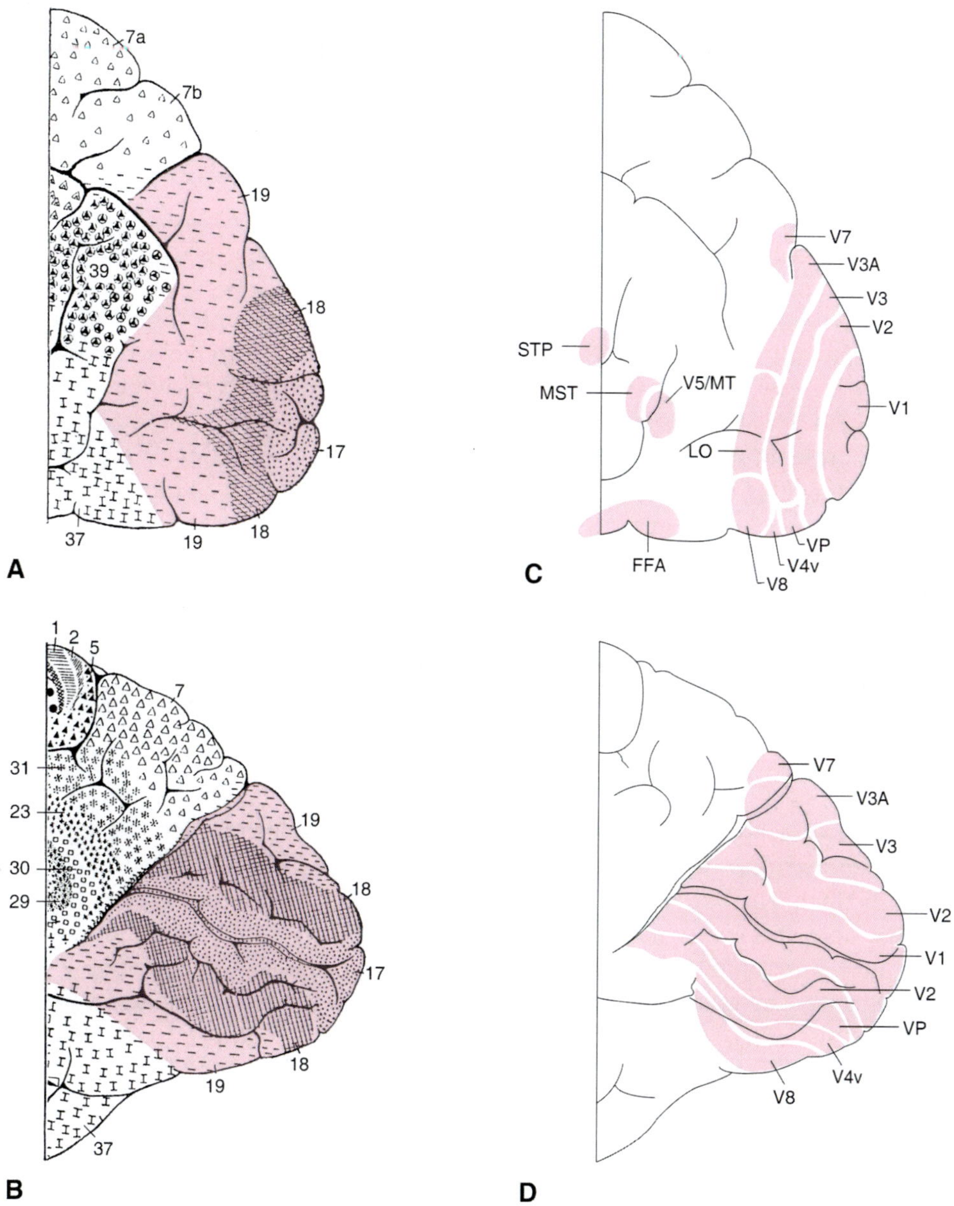

Fig. 15.49 A–D. The occipital lobe. Subdivision (*numbers*) according to Brodmann, lateral (**A**) and medial (**B**) views. The occipital lobe is shown in *red*. Functional areas adopted from Tootell et al. [737], lateral (**C**) and medial (**D**) views. *FFA*, fusiform face area; *LO*, lateral occipital area; *MST*, middle superior temporal visual area; *MT*, middle temporal visual area; *STP*, superior temporal polysensory area; *V1*, *V2*, *VP*, etc., visual areas

occipitotemporal gyri, forms the anterior boundary of the occipital lobe on the basal side (Fig. 3.5). On the medial aspect of the hemisphere, the deep calcarine sulcus separates the cuneus from the medial occipitotemporal gyrus. Anteriorly, the calcarine sulcus joins the parieto-occipital sulcus (Fig. 3.6).

The occipital lobe is occupied by three concentrically arranged cytoarchitectonic areas, the areas 17, 18 and 19 of Brodmann (Figs. 15.8, 15.49 A, B).

Area 17 surrounds the calcarine sulcus. It is also known as the area striata because it is characterized by the presence of a dense layer of myelinated fibres, known as the line of Gennari. This line is clearly visible in unstained, macroscopical sections (Figs. 5.13, 5.14, 5.30, 5.31). Histologically, the area striata is typified by its richness of small, granular cells ("koniocortex"). The internal granular layer is subdivided into three sublayers, IVa–c (Figs. 15.6, 15.41 B). The transition zone between layers V and VI contains the large cells of Meynert (Figs. 15.28 (4), 15.35).

Areas 18 and 19 show a homotypical lamination pattern. The borderline between areas 17 and 18 is very distinct. The line of Gennari stops abruptly and the various sublayers of layer IV merge into a single internal granular layer (Figs. 15.6, 15.41 B). The anterior borders of areas 18 and 19, on the other hand, are hard to establish. The borders of areas 17 and 18 show a considerable intersubject variability (Fig. 15.22 B) [7, 186].

Area 17, which receives its principal afferent projection from the lateral geniculate body, represents the *primary visual cortex (V1)*. The areas 18 and 19, which surround area 17, are collectively designated as the *parastriate belt*. The areas 18 and 19, which are strongly visually responsive, receive their principal afferents either directly or indirectly from the striate area and are, hence, also characterized as *extrastriate visual areas*. However, the total extrastriate visual cortex is not confined to the occipital lobe, but rather extends anteriorly into the parietal and temporal lobes. The overall extent of extrastriate visual cortex (or *visual association cortex*), as estimated from a combination of fMRI, PET and lesion studies, includes Brodmann's areas 7, 18–21, 37 and 39 (Fig. 15.50) [161]. By this estimate, roughly a third of human neocortex is associated with visual processing [771].

Our present-day knowledge of the central visual system is largely based on experimental studies in non-human primates, particularly the rhesus monkey. These studies have shown that there is a far-reaching "division of labour" in the analysis and processing of visual information. Many cortical areas outside the primary visual cortex appear to be specialized for the processing of specific aspects of vision, such as form, motion or colour.

The extrastriate cortex of the macaque monkey has been shown to contain a large number of visual areas defined by three or more of the following criteria: (i) presence of retinotopic representation of the contralateral visual hemifield or of part of it; (ii) functional specialization of neurons; (iii) specific effects of lesions; (iv) specific connectivity pattern and (v) specific cyto-, myelo- and/or chemoarchitecture [850]. More than 30 different areas have been identified in the extrastriate visual cortex of the rhesus monkey [180, 408, 761, 770, 771]. Some of these are labelled with a capital "V", standing for visual area, followed by a number: V2, V3 etc. Others are designated with combinations of letters, indicating their topographical position. Mainly on the basis of functional imaging studies, putative equivalents of 8–10 of the functional areas detected in the extrastriate cortex of the rhesus monkey have been identified in the human brain (Figs. 15.24 C, 15.49 C, D).

As regards V2, Zilles [849] indicates that this area corresponds to Brodmann's area 18, although Kaas [352] claims that V2 is much narrower than area 18. V2 is strongly activated during shape discrimination [267].

Area 19 of Brodmann has been shown to harbour a considerable number of functional areas within its confines, including V3, VP, V3A, V4v, V8 and V5/MT. V3, which adjoins the anterior border of the upper part of V2, contributes to dynamic form perception. Many of its neurons are both orientation- and mo-

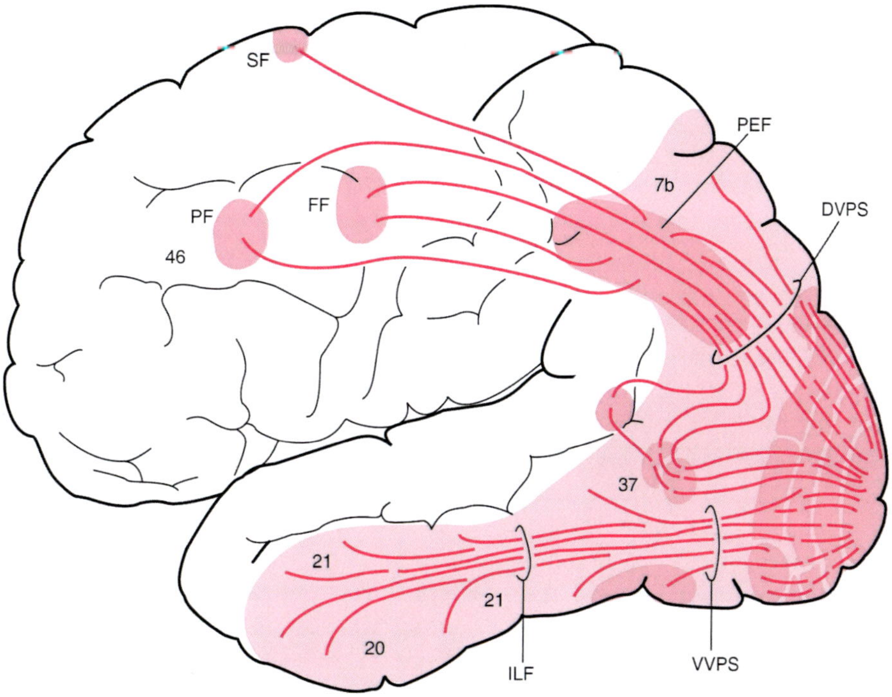

Fig. 15.50. Principal visual projections. The total brain area dominated by visual projections is shown in *light red*; visual centres in *darker red*. *DVPS*, dorsal visual processing stream; *FF*, frontal eye field; *ILF*, inferior longitudinal fasciculus; *PEF*, parietal eye field; *PF*, prefrontal eye field; *SF*, supplementary eye field; *VVPS*, ventral visual processing stream; *7b*, *20*, etc., areas according to Brodmann

tion selective [849]. VP, the ventroposterior visual area, is situated anteriorly to the basal part of V2. It has been argued that this area is not a separate functional entity, but rather represents the ventral part of V3 [847]. V3A occupies the most superior part of the occipital lobe, extending over both the lateral and the medial surface of the hemisphere. V3A, V7 and adjoining areas in the parietal lobe form a complex that is activated by saccades and attention [124]. LO, the lateral occipital area, is situated anteriorly to V3. Human imaging studies have shown that this area is involved in shape processing. It responds strongly to images of objects, but not to scrambled control stimuli [241]. V8, which is sometimes designated as V4 [737], borders laterally to LO and extends basally over the posterior parts of the lateral and medial occipitotemporal gyri. Functional imaging studies have shown that this area is involved in colour perception [60, 266, 275, 846]. Lesions in this region cause *central colour blindness (achromatopsia).* As its name indicates, the fusiform face area (FFA) is located in the fusiform (or lateral occipitotemporal) gyrus, directly in front of V8. Functional imaging studies and single unit electrophysiology from nonhuman primates have shown that this area is selectively responsive to faces [241, 658, 737]. Lesions in FFA lead to *prosopagnosia, a selective difficulty identifying faces.* V5 or MT, the middle temporal visual area, is located in the area where the inferior temporal sulcus meets the anterior occipital sulcus. This area, which is common to all primates studied so far [352], is morphologically characterized by its dense myelination [761]. Functional imaging studies have shown that this area is strongly activated by moving visual stimuli [266, 317, 846]. Bilateral lesions that include this area cause a severe *impairment in detecting the movements of objects,* known as *cortical akinetopsia.* MT is surrounded by several other specialized areas which process higher order aspects of motion perception and motor planning [653]. One of these, the middle superior temporal visual area, MST, is situated anteriorly to MT [241].

For a detailed discussion of the visual system, the reader is referred to Chap. 19. Here we confine the discussion to a brief survey of the principal connections of the various components of the occipital lobe (Fig. 15.50).

Efferent fibres from the lateral geniculate body pass to the primary visual cortex (V1) and to the middle temporal visual area (MT) [687]. In V1, the fibres from the magnocellular, parvocellular and intercalated layers of the lateral geniculate body terminate in different layers, particularly in the various sublayers of layer IV (Fig. 15.26). The remaining visual centres in the occipital lobe receive their thalamic afferents principally from the pulvinar. Additional afferents to the occipital lobe arise from the amygdala (Fig. 13.7) [205] and the claustrum (Fig. 13.8). The afferents from the thalamus and the claustrum are reciprocated by efferent fibres. Several visual cortical centres, including V1–V3A, MT and MST, project to the pontine nuclei [210].

Short association fibres interconnect area 17(V1) with many extrastriate visual areas, including V2, V3, V3A, V4, V5/MT and MST [5, 161, 742, 768, 769].

Analysis of the connections of the various extrastriate visual areas has led to the concept that these areas are organized into two pathways: an occipitoparietal pathway or "dorsal stream" and an occipitotemporal pathway or "ventral stream" [32, 180, 181, 623, 686, 706, 735, 741, 743]. As discussed in a previous section of the present chapter (Structural Subdivision 4: Connectivity), the sequential processing of information in the areas along both pathways is progressively more complex and may, hence, be characterized as hierarchical.

The *dorsal, occipitoparietal processing stream* is engaged in the perception of relative spatial location of objects as well as visual guidance of movements towards objects *("where" system).* It includes V1, V2, V3, V3A, V4, V5/MT, MST and STP (superior temporal polymodal) (i.e. a polysensory area situated in the dorsal bank of the superior temporal sulcus, and the so-called parietal eye field [32, 241, 686]. STP responds to visual, auditory and somatosensory stimuli. It interacts with the inferior temporal cortex. In the parietal lobe of the rhesus monkey, the territory surrounding the intraparietal sulcus contains a series of at least five different areas, all of which are concerned with highly specialized visual functions (see next section). These areas are collectively designated as the parietal eye field. Neuroimaging studies suggest that all of the areas included in this field have a functional equivalent (analogue) in the human brain [124]. The parietal eye field functions as a visual-motor interface. It is reciprocally connected with V1, V2 and V3 through direct and indirect pathways passing through MT, MST and V3A. The frontal lobe contains three fields which give rise to eye movements when stimulated, namely the frontal eyefield, situated in the most posterior part of the middle frontal gyrus, the prefrontal eye field, located more anteriorly in area 46, and the supplementary eye field, situated dorsally, in the most rostral part of area 6. These frontal eye fields are heavily and reciprocally connected with the parietal eye field. Moreover, most of the occipital extrastriate areas, intercalated in the dorsal processing stream, also project directly to the frontal and prefrontal eye fields [434, 579, 653].

The *ventral occipitotemporal processing stream* is concerned with pattern discrimination and the visual identification of objects (*"what" system*). Areas V1, V2, LO, V4/V8 and the entire inferotemporal cortex, composed of areas 37, 20 and 21, are included in the ventral stream. Most of the fibres, interconnecting the various areas participating in this stream, pass forwardly in the inferior longitudinal fasciculus (Fig. 15.45) [579]. The separation of the dorsal and ventral streams is not absolute. Thus, it is known that in the macaque monkey, the posterior part of the inferotemporal cortex is reciprocally connected with the parietal and frontal eye fields [155].

Interruption of the dorsal and ventral processing streams leads to distinct disconnection syndromes. Thus, judgements of spatial relations are impaired by lesions in the posterior parietal region, whereas object recognition is selectively impaired by lesions in inferior occipital and temporal areas [743].

Parietal Lobe

The parietal lobe is bounded on the lateral surface anteriorly by the central sulcus, posteriorly by the parieto-occipital line, and inferiorly by the posterior ramus of the lateral sulcus and an arbitrary parietotemporal line. The latter passes horizontally through the border area of the supramarginal and angular gyri, on the one hand, and the superior and middle temporal gyri on the other (Figs. 3.2, 15.51). On the medial surface, the boundaries are: the parieto-occipital sulcus, the subparietal sulcus and a vertical line connecting the superior limit of the central sulcus with the cingulate sulcus (Fig. 3.6).

The parietal lobe can be subdivided into four parts, the postcentral gyrus, the superior parietal lobule, the inferior parietal lobule and the parietal operculum. The postcentral gyrus runs parallel to the precentral gyrus and is situated between the central and postcentral sulci. The horizontally oriented intraparietal sulcus divides the region posterior to the postcentral sulcus into superior and inferior parietal lobules. The latter consists of two gyri, the supramarginal gyrus, arching around the upturned end of the lateral sulcus, and the angular gyrus, which surrounds the ascending, terminal part of the superior temporal sulcus (Fig. 3.2). The parietal operculum is not visible on the free surface of the hemisphere. It is situated deep to the posterior part of the lateral sulcus and connects the postcentral gyrus and the anterior part of the supramarginal gyrus with the insula. In Fig. 15.51 A, C, the parietal operculum is brought to the external surface of the hemisphere by rotation.

Brodmann [70, 71] divided the parietal lobe into nine cytoarchitectonic areas: 1, 2, 3, 5, 7, 31, 39, 40 and 43 (Figs. 15.8, 15.51 A, B). Areas 3, 1 and 2 occupy the postcentral gyrus as three narrow strips. They extend for some distance over the medial surface of the hemisphere into the anterior part of the superior parietal lobule. Area 3 was later subdivided into a small transitional zone 3a, situated in or near the fundus of the central sulcus [388], and a larger zone 3b. The areas 5 and 7 occu-

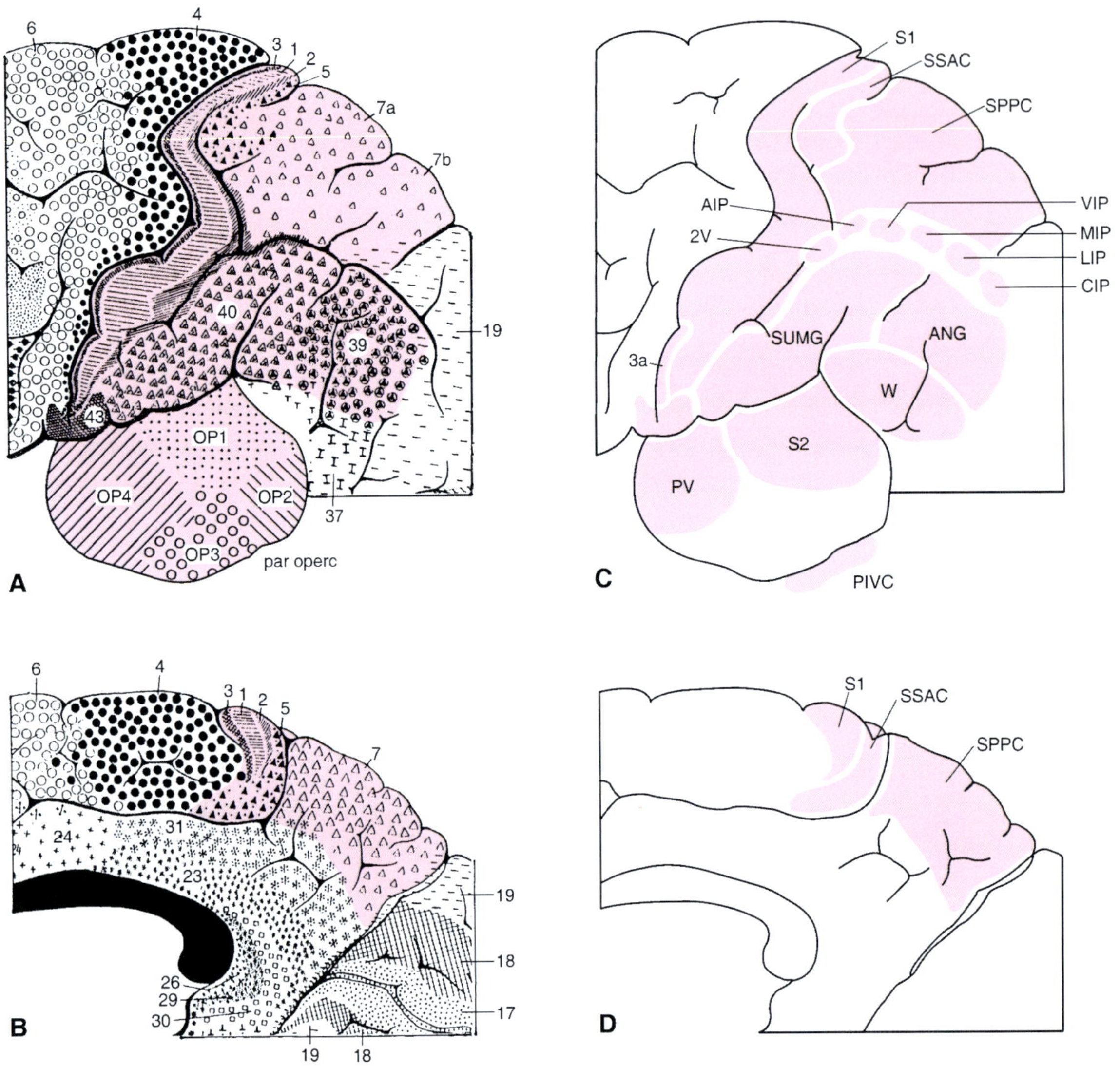

Fig. 15.51 A–D. The parietal lobe. Subdivision according to Brodmann, lateral (**A**) and medial views (**B**). The parietal lobe is shown in *red*. Functional areas, lateral (**C**) and medial views (**D**). In **A** and **C**, the parietal operculum is brought to view by rotation. *AIP*, anterior intraparietal area; *ANG*, angular gyrus; *CIP*, caudal intraparietal area; *LIP*, lateral intraparietal area; *MIP*, medial intraparietal area; *OP1–4*, parietal opercular areas, according to Eickhoff et al. [169, 170]; *par operc*, parietal operculum; *PIVC*, parieto-insular vestibular cortex; *PV*, parietal ventral area; *SPPC*, superior polymodal parietal cortex; *SSAC*, somatosensory association cortex; *SUMG*, supramarginal gyrus; *S1*, primary somatosensory cortex; *S2*, second somatosensory area; *VIP*, ventral intraparietal area; *W*, Wernicke's region; *1*, *2*, *3*, etc., areas according to Brodmann; *7a*, *b*, subdivisions of area 7; *2V*, *3a*, vestibular cortical areas

py most of the superior temporal lobule. On the medial side of the hemisphere, area 7 extends over the precuneus. Areas 39 and 40 roughly coincide with the angular and supramarginal gyri, respectively. Area 31 is situated on the medial side of the hemisphere, where it adjoins the posterior part of the cingulate gyrus. It forms part of the limbic system and will not be further considered here. The cytoarchitecture of the parietal operculum was recently analysed by Eickhoff et al. [169, 170]. Four cytoarchitectonic areas, termed OP1–4, were identified, of which OP4 and possibly OP3 correspond approximately with Brodmann's area 43.

Cytoarchitectonically, four of the five fundamental cortex types, as distinguished by von Economo [795], are represented in the parietal lobe. Area 3 a, with its strongly developed pyramidal layers and inconspicuous granular layers, closely resembles the adjacent primary motor cortex (type 1). Area 3 b, which has a strongly developed internal granular layer IV and poorly developed pyramidal layers III and V, must be classified as a typical koniocortex (type 5). The remaining parietal areas are all homotypical, belonging either to types 2 or 3 (Fig. 15.10).

In the following discussion of the connectional and functional relations of the parietal lobe, this structure will be provisionally divided into four parts, the somatosensory and vestibular cortices, and the superior, intermediate and inferior polymodal parietal cortices (Fig. 15.51 C, D).

The *somatosensory cortex* occupies the postcentral gyrus and part of the parietal operculum. The cytoarchitectonic areas 3 b, 1 and 2, which cover the postcentral gyrus, are collectively known as the primary somatosensory cortex, S1. (Area 3 a is usually included in the motor cortex.) The contralateral half of the body surface is somatotopically mapped onto the S1 region of each hemisphere (Figs. 15.18 C, 15.19 A). The lower limb is represented in the upper part, the face in the lower part, and the trunk and upper limb in the intermediate part of S1. Each of the three primary sensory areas contain a complete body map [354] (Fig. 16.5 F).

The parietal operculum contains two somatosensory areas, the parietal ventral area and the second somatosensory area. The parietal ventral area (PV) is located immediately adjacent to the S1 cortex. It is posteriorly followed by the second somatosensory area (S2) (Fig. 15.51 C) [153]. PV and S2 roughly correspond to the cytoarchitectonic areas OP4 and OP1, respectively [169]. Areas PV and S2 both contain a somatotopic representation of the contralateral body surface. These two representations form mirror reversals of each other along their common border. The somatosensory system will be extensively discussed in Chap. 16. Here we confine the discussion to mentioning that the three parts of S1 are reciprocally connected with each other and with the various parts of the superior parietal lobule [346, 348, 389, 586, 781]. S2 is densely interconnected with the primary sensory area 3 b, PV and the inferior parietal lobule, whereas PV is interconnected with area 3 b, the superior and inferior parietal lobules, the premotor cortex, the frontal eye field and the medial auditory belt areas [154]. Some of these connections are diagrammatically indicated in Fig. 15.52 A.

The *vestibular cortex* includes an elongated area located in the inferior part of somatosensory area 3 a, an area 2 v situated around the anterior tip of the intraparietal sulcus and the parieto-insular vestibular cortex (Figs. 15.51 C, 15.72). The different vestibular cortical areas are interconnected and the PIVC occupies a central position in this network [265]. The vestibular system will be discussed in Chap. 17.

The *superior polymodal parietal cortex* occupies the superior parietal lobule, except for its most anterior part (area 5), which forms the (unimodal) somatosensory association cortex. There is evidence suggesting that this polymodal region is important for hand–eye coordination [333, 334, 733]. It is reciprocally connected with the prestriate cortex, via the dorsal processing stream, somatosensory area 5 and the dorsal part of area 8, which contains the smooth eye movement-related frontal eye field [400]. The superior parietal lobule also projects to the dorsal part of the premotor cortex (area 6) and to the supplementary motor area on the medial surface of the frontal lobe [444, 579].

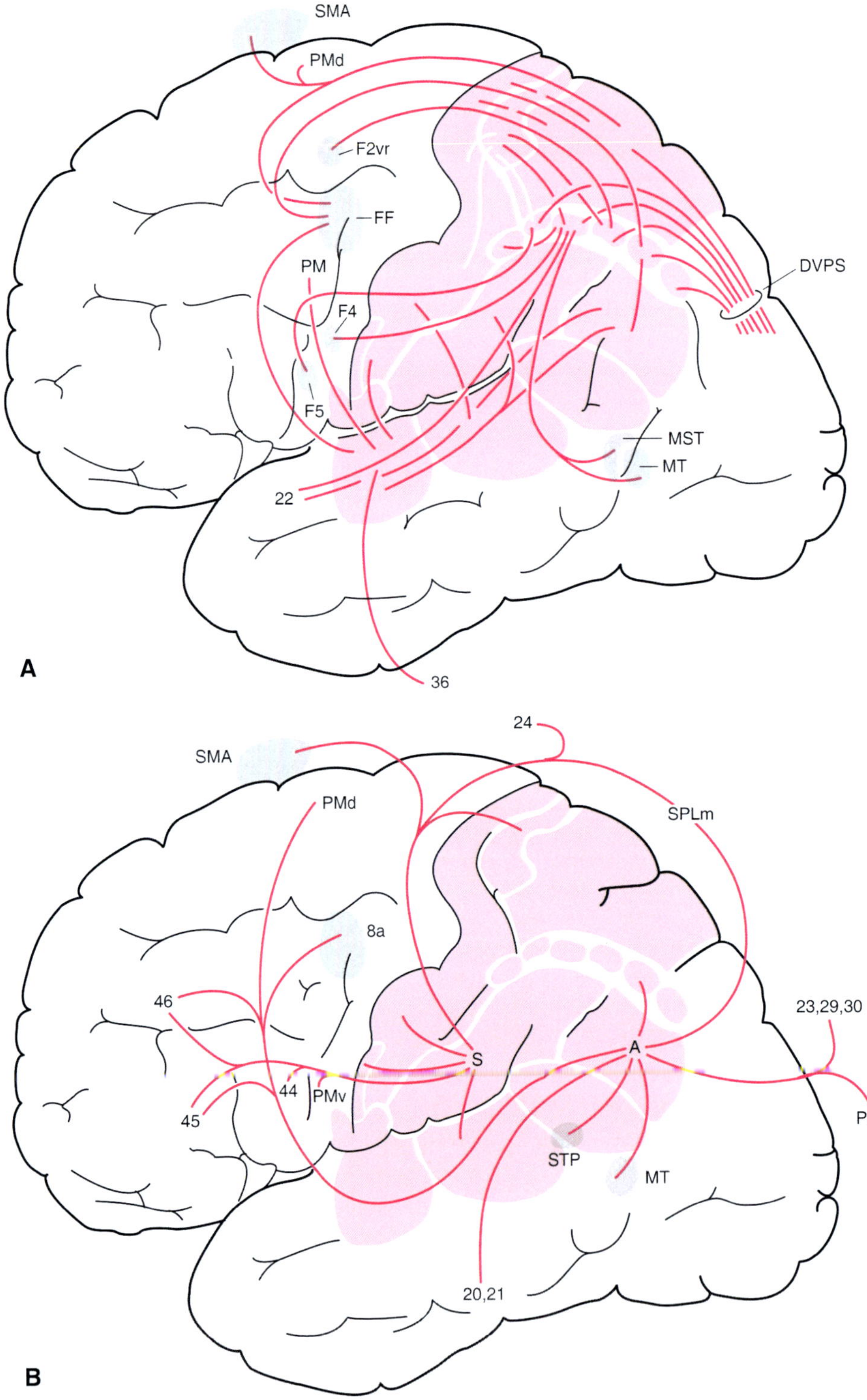

Fig. 15.52 A, B. Principal connections of the parietal lobe. Connections of the superior and opercular (**A**) and inferior parts (**B**) of the parietal lobe. *A*, angular gyrus; *DVPS*, dorsal visual processing stream; *FF*, frontal eye field; *F2vr*, *F4*, *F5*, specialized areas within the premotor cortex; *MST*, middle superior temporal visual area; *MT*, middle temporal visual area; *PHC*, parahippocampal cortex; *PM*, premotor cortex; *PMd,v*, dorsal and ventral subdivisions of premotor cortex; *S*, supramarginal gyrus; *SMA*, supplementary motor area; *SPL,m*, medial part of superior parietal lobule; *STP*, superior temporal polysensory area; *20*, *21*, etc., areas according to Brodmann

The *intermediate polymodal parietal cortex* is formed by a number of functional areas, which occupy the banks of the intraparietal sulcus. Recent morphological and physiological studies in monkeys and neuroimaging studies in humans have shown that the cortex bordering the intraparietal sulcus harbours multiple, highly specialized functional areas, some of which can also be morphologically characterized. In general, these areas serve as interfaces between perceptive and motor systems. They receive varying combinations of inputs from the surrounding visual, somatosensory, vestibular and auditory cortices and send strongly developed feed-forward projections to particular parts of the premotor cortex, which are reciprocated by feed-back projections [444]. The various intraparietal sulcal areas and their premotor targets form functional units, each of which is dedicated to a particular aspect of sensorimotor transformation [445]. The following survey of the various intraparietal sulcal areas is principally based on two review articles, by Culham and Kanwisher [124] and Grefkes and Fink [253], to which the reader is referred for references and details. The approximate positions of these areas are indicated in Fig. 15.51 C.

The *anterior intraparietal area (AIP)*, which is located on the lateral bank of the anterior intraparietal sulcus (ips), is concerned with tactile and visual object processing. It plays a crucial role in the crossmodal transfer of object information between the sensorimotor and visual systems. The AIP is connected to the ventral premotor area, especially to a subunit known as F5. The neurons in this subunit discharge during specific object-related hand movements. It seems likely that AIP in combination with F5 transforms visual and somatosensory object data into finger movements for object grasping and manipulation.

The *ventral intraparietal area (VIP)* is located in the fundus of the ips. It receives projections from several visual areas, especially from MT and MST, from somatosensory, auditory and vestibular areas, and from other polymodal cortices. It is strongly connected with F4, another subunit of the ventral premotor cortex. There is evidence suggesting that VIP is involved in the perception of self-movements and object movements in near extrapersonal space. Premotor subunit F4 is known to be concerned with the transformation of object locations into appropriate movements towards them [445].

The *medial intraparietal area (MIP)* is located in the intermediate part of the medial bank of the ips. It receives somatosensory and visual afferents and is strongly connected with subunit F2vr, forming part of the dorsal premotor area. MIP and F2vr are known to be involved in the planning, execution and monitoring of reaching movements.

The *lateral intraparietal area (LIP)* forms part of a network of areas mediating saccades. It receives input from several visual areas and is interconnected with the frontal eye field (FF) and the superior colliculus. In the macaque monkey, LIP is found in the lateral wall of the ips, hence its name. However, comparative functional studies have shown convincingly that the human LIP equivalent is located in the medial ips rather than in the lateral ips.

The *caudal intraparietal area (CIP)*, finally, is situated in the medial bank of the posterior ips. It receives afferents from several visual areas, including V3, V3A and V4. Experiments in macaque monkeys have shown that neurons in CIP are involved in the analysis of three-dimensional object features and are especially responsive to axis and surface orientations of objects in space. Neuroimaging studies have demonstrated that CIP is also activated during the analysis of surface and pattern orientation in humans.

Some of the connections of the intraparietal cortical areas just discussed are indicated in Fig. 15.52 A.

Choi et al. [97] recently studied the cytoarchitecture of the cortex within the anterior ventral bank of the human ips. They delineated two areas, which were termed the human intraparietal area 1 (hIP1) and the human intraparietal area 2 (hIP2). The areas hIP1 and hIP2 appeared to correspond with the functionally defined areas VIP and AIP, respectively.

The *inferior polymodal parietal cortex* occupies the inferior parietal lobule, composed of the angular and supramarginal gyri. As mentioned before, the extension of areas 39 and 40 of Brodmann correspond roughly to those of the angular and supramarginal gyri, respectively.

There are some differences of opinion concerning the interpretation of the structures under discussion. Geschwind [227] considered the human inferior parietal lobule as a "new anatomical structure". Crosby et al. [122] maintained that areas 39 and 40 have not been recognized in the macaque, and Zilles [849] designated these areas as "human specific", which means that they have no homologues in non-human primates. One of the reasons for the interpretation of these authors may be that Brodmann [70], in his map of the cortex of *Cercopithecus* (Fig. 15.12), labelled the entire region posteroinferior to the intraparietal sulcus as area 7 and did not delineate areas 39 and 40 in this species. However, it should be noted that Brodmann in his book repeatedly emphasizes that identical numbers do not indicate absolute homologies on the brain maps. In discussing the map of *Cercopithecus*, he indicates that area 7 of this species should be considered as an undifferentiated primordium of all human parietal areas, except for area 5, which is clearly distinguishable in both species. In line with this, Petrides and Pandya [579] interpreted the anterior and posterior parts of the inferior parietal lobule of the rhesus monkey (their areas PF and PG) as homologous to the human supramarginal and angular gyri, respectively. We fully agree with this interpretation because the areas mentioned occupy identical relative or topological positions in the monkey and the human. Hence, we feel justified in extrapolating the results of experimental studies on the fibre connections of the inferior parietal areas PF and PG in the macaque to the human situation.

Wernicke's speech region extends for an undetermined distance over the left inferior parietal lobule [463, 549]. The core of this region occupies the left temporal plane behind the transverse gyrus of Heschl and the posterior part of the left superior temporal gyrus. Damage to this area gives rise to receptive or Wernicke's aphasia, involving deficits in the comprehension of both spoken and written language. Long association fibres connect Wernicke's region, via the anterior limb of the superior longitudinal fasciculus, with the premotor region of the frontal lobe, including Broca's speech region in the inferior frontal gyrus [219, 346, 398, 532].

The following survey of the fibre connections of the angular (area 39, PG) and supramarginal gyri (area 40, PF) is principally based on the experimental studies of Cavada and Goldman-Rakic [94, 95], Neal et al. [503–505], Andersen et al. [11], Petrides and Pandya [579] and Gregoriou et al. [254], which were all carried out in the macaque monkey.

The *angular gyrus* is connected with visual areas of both the dorsal (MST) and the ventral (infratemporal cortex) processing streams, the medial part of the superior parietal lobule, the intraparietal oculomotor area LIP, the superior temporal polysensory area STP, and with limbic areas of the posterior cingulate, retrosplenial and parahippocampal regions. Frontal connections of the angular gyrus are principally directed to the anterior part of the dorsal premotor cortex, area 8a and prefrontal areas 45 and 46 (Fig. 15.52B). There is evidence suggesting that the angular gyrus may play a role in the visual guidance of arm movements [254].

The most prominent connections of the *supramarginal gyrus* include those to the somatosensory areas S1, S2 and area 5, the medial part of the superior parietal lobule, the vestibular cortex, the anterior cingulate area 24, the supplementary motor area, the ventral premotor cortex, and the adjacent area 44, and prefrontal areas 45 and 46 (Fig. 15.52B). Unit activity studies in monkeys indicate that the supramarginal gyrus is involved in the organization of co-ordinated hand and face movements [840]. Petrides and Pandya [579] suggested that the interactions between rostral inferior parietal cortex and the ventral frontal region in the monkey may be necessary for gestural communication, which may have preceded the evolution of linguistic communication. An alternative view on the development of brain mechanisms that gave rise to language has been recently presented by Gil-da-Costa et al. [236]. These authors pointed out that monkeys possess a repertoire of species-specific vocalizations that – like human speech – seem to encode meaning in arbitrary sound patterns. Interestingly, they found that conspecific calls activate putative homologues of Wernicke's and Broca's speech regions in the rhesus monkey.

In the foregoing, we have divided the posterior parietal cortex provisionally into nine functional subregions (Fig. 15.51 C, D). If we survey the data collected on the connections and functions of these subregions, it appears that what has been said about the areas located in the banks of the intraparietal sulcus hold true for the remaining subregions as well. They are all specifically connected with one or some sections of premotor cortex, and all are dedicated to the selection and preparation of specific motor patterns, including eye movements, reaching movement, grasping movements and articulated speech [10, 445, 828].

It has already been mentioned that lesions of the inferior parietal lobule in the left hemisphere may lead to Wernicke's aphasia. When damage spares Wernicke's speech region, but involves the adjacent, more superior parts of the left inferior parietal lobule, complex combinations of anomia (naming disorders), alexia (reading impairment), construction deficits, alcalculia (difficulty in performing simple arithmetic calculations), agraphia (loss of writing skills), finger agnosia (inability to recognize, distinguish and name one's fingers or those of other persons) and right-left disorientation (inability to name or point to the right and left sides of objects or body parts) develop. The last four symptoms are collectively referred to as *Gerstmann's syndrome*.

Lesions in the inferior parietal lobule of the right hemisphere lead to characteristic deficits, including dressing difficulties, constructional difficulties and multimodal neglect of the left hemispace [463].

Bilateral lesions of the dorsal parts of posterior parietal lobules may give rise to the rare but dramatic *syndrome of Balint*, which is characterized by (1) simultanagnosia, (2) oculomotor apraxia and (3) optic ataxia. Simultanagnosia is the inability to see all the components of the visual scene in an integrated way. Oculomotor apraxia or psychic paralysis of gaze is the inability to voluntarily direct gaze toward a specific part of the visual field. Optic, or visuomotor, ataxia is the inability to direct movement of an extremity using visual guidance [151].

Temporal Lobe

The temporal lobe presents large lateral and basal surfaces and a more limited medial surface. The lateral surface is dorsally separated from the frontal and anterior parietal lobes by the lateral sulcus, and more caudally from the posterior parietal lobe by an arbitrary parietotemporal line (Fig. 3.2). The caudal border of the temporal lobe is formed on the lateral surface by the anterior occipital sulcus and on the basal side by the occipitotemporal sulcus and an arbitrary, transversely oriented occipitotemporal line (Figs. 1.4, 3.5, 15.53). Ventromedially, the temporal lobe is separated from the limbic lobe by the collateral and rhinal sulci (Fig. 3.6).

Two grooves, the superior and inferior temporal sulci, which roughly parallel the lateral sulcus, divide the lateral surface of the temporal lobe into superior, middle and inferior temporal gyri. The basal surface of the temporal lobe is formed by the basal part of the inferior temporal gyrus and most of the lateral and medial occipitotemporal gyri.

The superior temporal gyrus includes the temporal operculum because the gyrus extends medially to meet the inferior part of circular sulcus around the insula. On the opercular surface of the temporal lobe, which forms the floor of the lateral sulcus, there are one or two, more or less transversely running convolutions, the transverse temporal gyri of Heschl. The areas anterior and posterior to these gyri are known as the planum polare and planum temporale, respectively (Fig. 18.1 B).

Brodmann [70, 71] divided the cortex covering the temporal lobe into eight cytoarchitec-

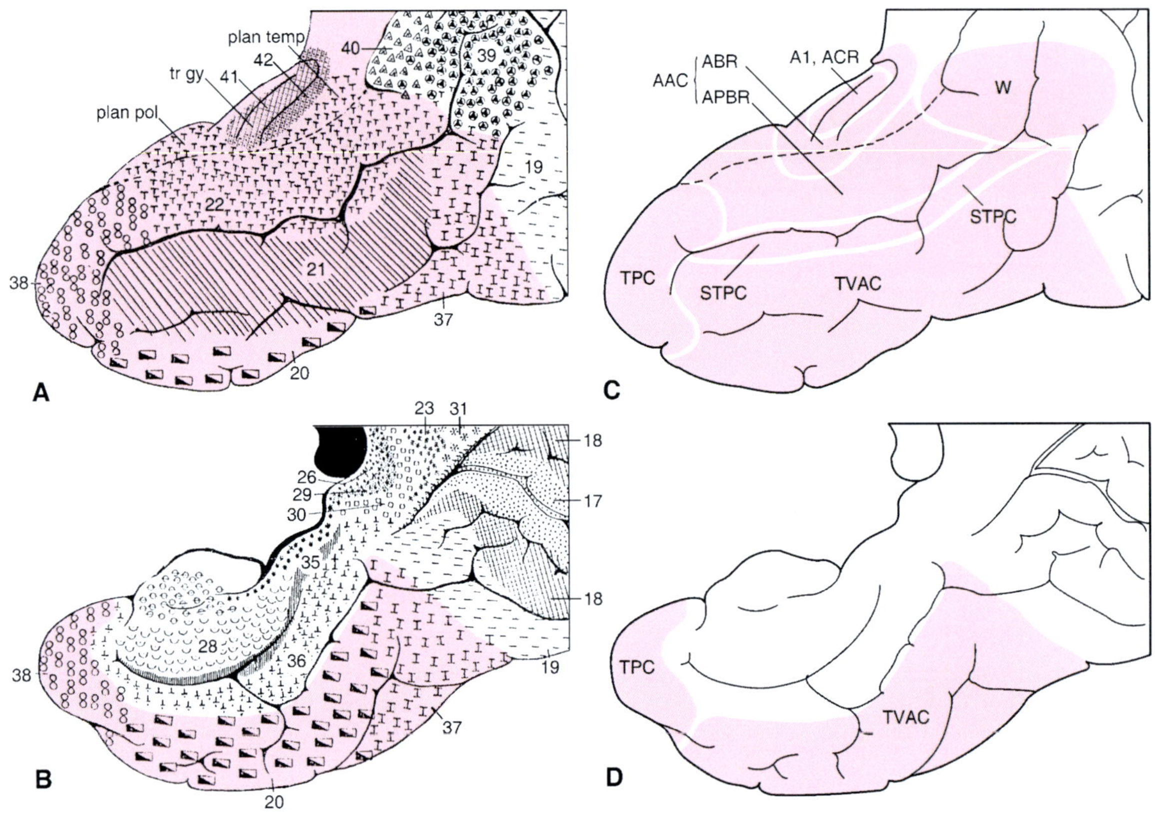

Fig. 15.53 A–D. The temporal lobe. Subdivision according to Brodmann, lateral (**A**) and medial (**B**) views. The temporal lobe is shown in *red*. Functional areas, lateral (*C*) and medial (*D*) views. In **A** and **C**, the upper surface of the superior temporal gyrus or temporal operculum is brought to view by rotation. The approximate boundary between the temporal operculum and the lateral surface of the superior temporal gyrus is indicated by a *dashed line*. *A1*, primary auditory cortex; *AAC*, auditory association cortex; *ABR*, auditory belt region; *ACR*, auditory core region; *APBR*, auditory parabelt region; *plan pol*, planum polare; *plan temp*, planum temporale, *STPC*, superior temporal polymodal cortex; *TPC*, temporopolar cortex; *TVAC*, temporal visual association cortex; *tr gy*, transverse gyrus of Heschl; *W*, Wernicke's region; *19*, *20*, etc., areas according to Brodmann

tonic areas: 41, 42, 22, 21, 20, 36, 37 and 38 (Figs. 15.8, 15.53 A, B). Area 41, the primary auditory cortex, is situated on the opercular surface of the temporal lobe. It forms a rather narrow strip which, extending from anterolateral to posteromedial, corresponds roughly with the (anterior) transverse temporal gyrus of Heschl. Area 42 is located directly posterior to area 41 and partly surrounds the latter (Fig. 15.8 insert). Most of this area is opercular, but its lateral part extends for some distance over the free lateral surface of the superior temporal gyrus. Areas 22, 21 and 20 were designated by Brodmann as area temporalis superior, -media and -inferior, respectively, indicating that they roughly correspond to the gyri of the same name. Area 20 extends basally over the anterior part of the lateral occipitotemporal gyrus. Area 36 is situated on the medial surface of the temporal lobe. It is bounded superiorly by the rhinal and collateral sulci and inferiorly by area 20. Area 37 occupies the posterior part of the temporal lobe, extending over its lateral and basomedial surface. Area 38, finally, covers the temporal pole.

In what follows, the temporal lobe will be divided into the following five functional regions: (1) the primary auditory cortex, (2) the auditory association cortex, including Wernicke's region, (3) the temporal visual association cortex, (4) the superior temporal polymodal cortex and (5) the temporopolar cortex (Fig. 15.53 C, D). The auditory system, including the auditory cortices, will be extensively discussed in Chap. 18; hence we confine the discussion here on these subjects to a brief overview.

The *primary auditory cortex*, or auditory core region, corresponds with area 41 of Brodmann. It has the typical koniocortical structure of primary sensory cortices, with a well-developed layer IV, consisting of densely packed granule cells. Morosan et al. [489] recently subdivided the primary auditory cortex into posteromedial, central and anterolateral subareas, which they designated Te1.1, Te1.0 and Te1.2, respectively. A similar tripartitioning of the primary auditory cortex was proposed for the monkey (Fig. 18.3 g). The functional significance of this parcellation remains to be elucidated.

The *auditory association cortex* can be divided into a belt and a parabelt region. The belt region borders on the primary auditory core, surrounding it anteriorly, laterally and posteriorly. Most of it is confined to the temporal operculum, but laterally it extends a short distance onto the superficial aspect of the superior temporal gyrus. The belt region consists of several cytoarchitectonic fields, among them Brodmann's area 42 [220]. The structure of the belt areas is intermediate between the typical koniocortex of area 41 and the homotypical structure of area 22 and the remainder of the temporal lobe.

The belt region is surrounded in turn by an extensive parabelt region that covers the remaining anterior and posterior parts of the temporal operculum and the lateral surface of the superior temporal gyrus with the exception of its anterior pole. It corresponds largely with Brodmann's area 22.

Short association projections connect the auditory core with the belt and the latter with the parabelt cortex [273, 485, 486, 528, 670]. Neuroimaging studies on the response to auditory stimuli show a spread of activity, beginning with activation by simple stimuli of a limited region of the temporal operculum, and expanding with more complex stimuli to the surface of the superior temporal gyrus [478].

Commissural connections of the auditory cortex are predominantly homotopical. Lesions in the core, belt or parabelt result in degenerating fibres in corresponding areas in the contralateral hemisphere [272, 486, 670]. The auditory commissural fibres are located in the posterior part of the truncus of the corpus callosum, together with the commissural fibres of the parietal lobe [531].

The auditory association cortex in the posterior parabelt region grades on the left side into the heteromodal cortex of Wernicke's speech region. The precise architectonic borders of this region are not known, but it is generally assumed that it includes the posterior portions of the planum temporale and superior temporal gyrus and the most basal parts of the angular and supramarginal gyri [26, 398, 465]. Interruption of the auditory, visual or tactile projections to Wernicke's region lead to pure word deafness, pure alexia and tactile aphasia, respectively [227] (Fig. 15.20). As already mentioned, damage to Wernicke's region itself causes *receptive or Wernicke's aphasia*.

Experimental studies in rhesus macaques [274, 576, 579, 631, 632] have shown that the auditory association cortex is strongly and reciprocally connected with the prefrontal cortex. Romanski et al. [631, 632] found that these audito-prefrontal connections form two streams, anterior and posterior, connecting different sectors of the auditory association cortex with largely different prefrontal regions.

The anterior stream connects the anterior belt and parabelt cortex reciprocally with the frontal pole (area 10), the anterior part of area 46 and a ventral prefrontal region, including areas 12 and 45. In contrast, the posterior stream mainly interconnects the posterior belt and parabelt cortex with the frontal eye field (area 8) and the posterior part of area 46 (Fig. 15.54 A). Romanski et al. [632] pointed out that the regions receiving input from the posterior belt and parabelt are implicated in spatial processing, whereas those receiving input from the anterior belt and parabelt are implicated in non-spatial, higher functions. They drew a parallel with the central visual system, where separate processing streams, the dorsal "where" stream and the ventral "what" stream, follow different courses and terminate ultimately in roughly the same dorsolateral spatial and ventrolateral object-processing regions of the frontal lobe [822].

The *temporal visual association cortex* represents a rostral extension of the circumstriate belt, below the superior temporal sulcus. It includes areas 20, 21 and 37 of Brodmann, which may be collectively designated as the inferotemporal visual cortex. The posterior part of this region houses several functionally defined areas, including the middle temporal visual area, MT, the middle superior temporal area, MST, and the fusiform face area, FFA (Fig. 15.49 C). In rhesus macaques, neurons in the inferotemporal cortex respond almost exclusively to visual stimulation and not to stimulation in other sensory modalities [150]. Many of these neurons respond selectively to various complex visual features of objects [255, 730]. Recordings of field potentials from the surface of the brain in patients have shown that an area in the anterobasal temporal lobe is specifically involved in the processing of written words [512].

The inferotemporal cortex, including FFA but excluding MT and MIT, forms part of the ventral, occipitotemporal visual processing stream, which is concerned with pattern discrimination and visual identification of objects. The inferior longitudinal fascicle, which forms the axis of this processing stream, contains, apart from long occipitotemporal projections, numerous short fibres that sequentially connect adjacent regions in the striate, peristriate and inferotemporal cortices [155, 579, 715, 739, 809] (Fig. 15.45). Efferents from the basal amygdaloid nucleus pass caudally in or close to the inferior longitudinal fascicle, to terminate in the various cortical regions forming part of the ventral visual processing stream. These amygdalar efferents terminate mainly in the most superficial and deep layers of their cortical targets and hence resemble cortico-cortical feed-back fibres [204].

Efferents from the inferotemporal cortical region, particularly its rostral parts, relay extensively processed visual information to: (1) the adjacent superior temporal polymodal region [127], (2) limbic cortical regions, including the temporopolar cortex, the perirhinal area 36 and the subiculum [647, 774]; (3) the lateral amygdaloid nucleus [204, 298]; and (4) the prefrontal cortex. The projection to the latter passes via the uncinate fasciculus and terminates mainly in areas 11 and 47/12 [579] (Fig. 15.54 B). Commissural fibres from the inferotemporal cortex cross in the posterior part of the body of the corpus callosum and in the anterior commissure [531].

Experimental neuroanatomical studies have shown that the cortex of the upper bank of the superior temporal sulcus in the rhesus monkey contains an elongated region that receives converging input from surrounding auditory, vis-

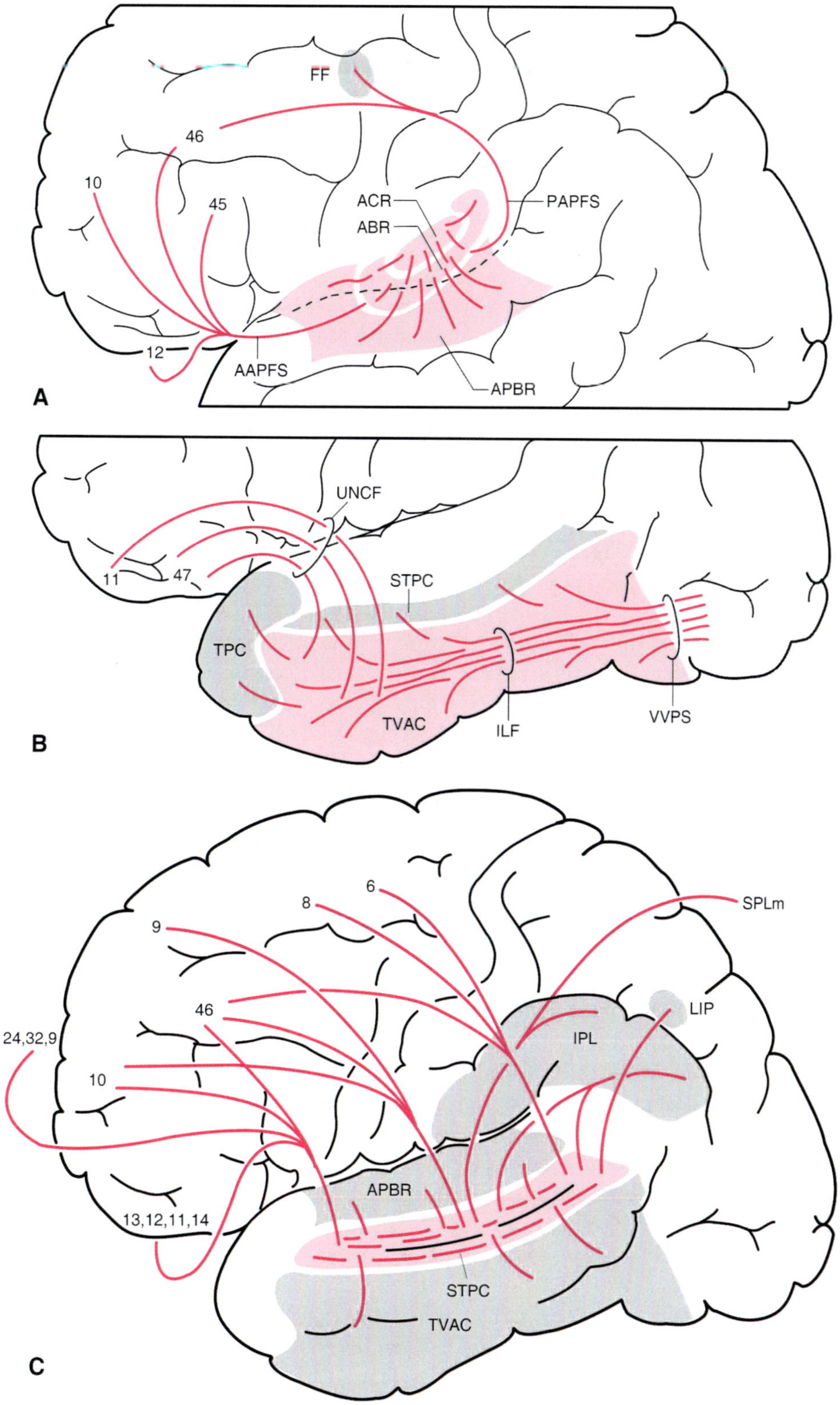

Fig. 15.54 A–C. Principal connections of the temporal lobe. Cortical auditory pathways (**A**); connections of the temporal visual association cortex (**B**) and connections of the superior temporal polymodal cortex (**C**). *AAPFS*, anterior audito-prefrontal processing stream; *ABR*, auditory belt region; *ACR*, auditory core region; *APBR*, auditory parabelt region; *FF*, frontal eye field; *ILF*, inferior longitudinal fasciculus; *IPL*, inferior parietal lobule; *LIP*, lateral intraparietal area; *PAPFS*, posterior audito-prefrontal processing stream; *SPLm*, medial part of superior parietal lobule; *STPC*, superior temporal polymodal cortex; *TPC*, temporopolar cortex; *TVAC*, temporal visual association cortex; *UNCF*, uncinate fasciculus; *VVPS*, ventral visual processing stream; *6, 8, 9*, etc., areas according to Brodmann

ual and somatosensory association areas [346, 670, 675]. Electrophysiological studies have revealed that neurons in this region respond to auditory, visual or somatosensory stimulation, including some whose response is bi- or trimodal [42, 47, 74, 150, 299, 472, 551, 707]. These findings have led to the conclusion that the sulcal cortex in question is polymodal in nature and should therefore be designated as the *superior temporal polymodal or STP cortex* [150, 346, 670].

The elongated strip of STP cortex has been divided on cytoarchitectonic and chemoarchitectonic grounds into four rostrocaudally arranged units, termed TPO1–TPO4 [125, 127, 670, 673, 676]. These units are tied together in a sequence of reciprocal connections, comprising shorter and longer, rostrally directed feed-forward and caudally directed feed-back projections [673].

The extrinsic connections of the STP cortex can be divided into a post-Rolandic and a pre-Rolandic group. The post-Rolandic connections relate the STP cortex to auditory, visual and somatosensory association cortices and to several limbic areas.

Auditory afferents to the STP cortex originate from the superior temporal cortex. They show a topical organization in that the caudal, middle and rostral zones of the superior temporal cortex project to areas TPO4, TPO2–3 and TPO1, respectively [675].

Visual projections to the STP cortex include afferents from (1) the LIP area in the intraparietal sulcus, which terminate in the TPO4, (2) the caudal inferior parietal lobule, terminating in areas TPO2–4, (3) the caudal inferotemporal cortex, targeting areas TPO3–4, and (4) the rostral inferotemporal cortex terminating in area TPO1. The inferotemporal cortex also projects indirectly to area TPO1 via some visual association areas, known as IPa, TEa and TEm, situated in the lateral bank of the superior temporal sulcus [155, 638, 647, 675, 772]. It is noteworthy that the visual projections converging on the STP cortex originate from areas forming part of both the dorsal and the ventral visual streams. Thus, the LIP area and the inferior parietal lobule belong to the dorsal stream, whereas the inferotemporal cortex belongs to the ventral stream.

Somatosensory afferents to the STP cortex emanate reportedly from mid-portions of the inferior parietal lobule and the medial parietal lobe [530, 671].

In addition to the sensory inputs just discussed, the STP cortex has extensive afferent connections with several limbic cortical structures. As with cortical sensory input, these projections have a rostral-to-caudal topographical organization. Projections from parahippocampal area 35, for example, target area TPO1; those from entorhinal area 28 target mid-area TPO, whereas area TPO4 receives input from the insular limbic cortex. Furthermore, areas TPO2–4 have afferent connections with anterior cingulate area 24, posterior cingulate area 23, and retrosplenial areas 29 and 30. Seltzer and Pandya [675], who described the limbic connections just mentioned, pointed out that certain sets of cortical areas projecting to the same rostrocaudal segment of the STP cortex are themselves interconnected. They mention as examples that the parahippocampal area 35 and rostral superior temporal cortex, which target area TPO1, have strong reciprocal connections and that the same holds true for the entorhinal area 28 and middle superior temporal cortex, which target areas TPO2–3.

Almost all of the sensory and limbic afferent systems to the various sectors of the STP cortex are reciprocated by efferent systems [37, 674].

The pre-Rolandic connections of the STP cortex are organized according to the rostral-to-caudal differentiation of this multimodal region. Area TPO1 projects to basal (areas 13, 12, 11 and 14), medial (areas 24, 32 and 9) and lateral (areas 10 and 46) sectors of the frontal lobe. Areas TPO2–3 project to rostral subdivisions of the lateral prefrontal cortex, namely areas 46, 9 and 10, whereas area TPO4 projects to caudal subdivisions (areas 46, 8 and 6) of the lateral frontal lobe (Fig. 15.54C). All of these efferent projections, except for those to the basal (orbitofrontal) cortex, are reciprocated by afferent systems [672]. Luppino et al. [432] reported that not only the caudal, but also the rostral part of the STP cortex projects to the premotor cortex (area 6).

The areal organization of the cortex surrounding the human superior temporal sulcus has not been explored so far. However, it seems reasonable to assume that this cortex, which forms the border region of areas 21 and 22 of Brodmann, contains a polymodal zone comparable to that in non-human primates [463]. Recent brain-imaging studies, summarized by Zilbovicius et al. [848], have shown that in humans, the superior temporal sulcal (STS) cortex is involved in the processing and integration of complex visual and auditory information, conducive to understanding of the mental states and intentions of other individuals. The information involved in this social perception includes eye gazes, gestures, facial displays of emotions and voice perception. Zilbovicius et al. [848] also cite brain-imaging results, suggesting that in autism, a mental disorder in which communication deficits are most prominent, anatomical and functional abnormalities in the STS cortex are implicated.

The *temporopolar cortex,* which corresponds to area 38 of Brodmann, is commonly designated as a paralimbic structure. Morphologically, it lies beyond the limbic lobe, but structurally it forms part of the so-called paralimbic belt. As discussed in Chap. 11 (and in the next section of the present chapter), the limbic lobe forms a large arcuate convolution on the medial aspect of the hemisphere, which extends from the frontal, via the parietal, into the temporal lobe. As shown by Fig. 12.3, its frontal and temporal ends are formed by infralimbic area 25 and perirhinal areas 35 and 36, respectively. According to Mesulam [463], the gap between the ends of the limbic lobe is bridged by three structures, the caudal orbitofrontal cortex (the posterior parts of areas 12, 13, 14), the anterior insula, and the temporopolar cortex (Fig. 15.55). Together with the cortices covering the various parts of the limbic lobe, these structures form the paralimbic belt. Architectonically, the belt areas provide a continuous transition zone between the simple, three-layered olfactory and hippocampal cortices, on the one hand, and the surrounding neocortical regions, on the other [463, 479].

The temporopolar cortex receives afferents from the adjacent auditory and visual association areas and has reciprocal connections with numerous paralimbic and limbic areas, including the caudal orbitofrontal cortex, anterior insula, entorhinal cortex, and the subicular complex [439, 479] (Fig. 15.56).

It has been demonstrated that single neurons in the temporopolar cortex of macaque monkeys exhibit oscillatory activities in response to the representation of complex visual stimuli such as photographs of familiar human faces, familiar foods, and familiar non-food objects [501, 502]. Bilateral lesions of the human temporopolar cortex cause marked retrograde amnesia, which includes the loss of all remote memories, but spares anterograde learning and recent memory [359, 360]. Typically, patients with these lesions not only lose episodic, autobiographical memories, but also the ability to recognize faces and names of famous persons learned in the remote past [359, 731]. Sewards and Sewards [678] inferred from these lesion data that the neuronal activities providing the awareness of remote (consolidated) object recognition occur in the temporopolar cortex. The fact that patients with retrograde recognition memory loss due to lesions of the temporopolar cortex do not lose the ability to recognize objects learned in the recent past, nor the ability to learn to recognize new objects, indicated, according to Sewards and Sewards, that there must be at least one other cortical area in which neuronal activities produce recognition awareness. They adduce clinical and experimental evidence indicating that such a centre, providing recognition awareness of objects learned in the recent past, does exist and is located in another paralimbic region, the medial orbitofrontal cortex.

Limbic Lobe and Paralimbic Belt

The limbic lobe, as defined by Broca [68], consists of a large, arciform convolution on the medial aspect of the cerebral hemisphere, which surrounds the interhemispheric commissures and the upper brain stem (Fig. 1.4B).

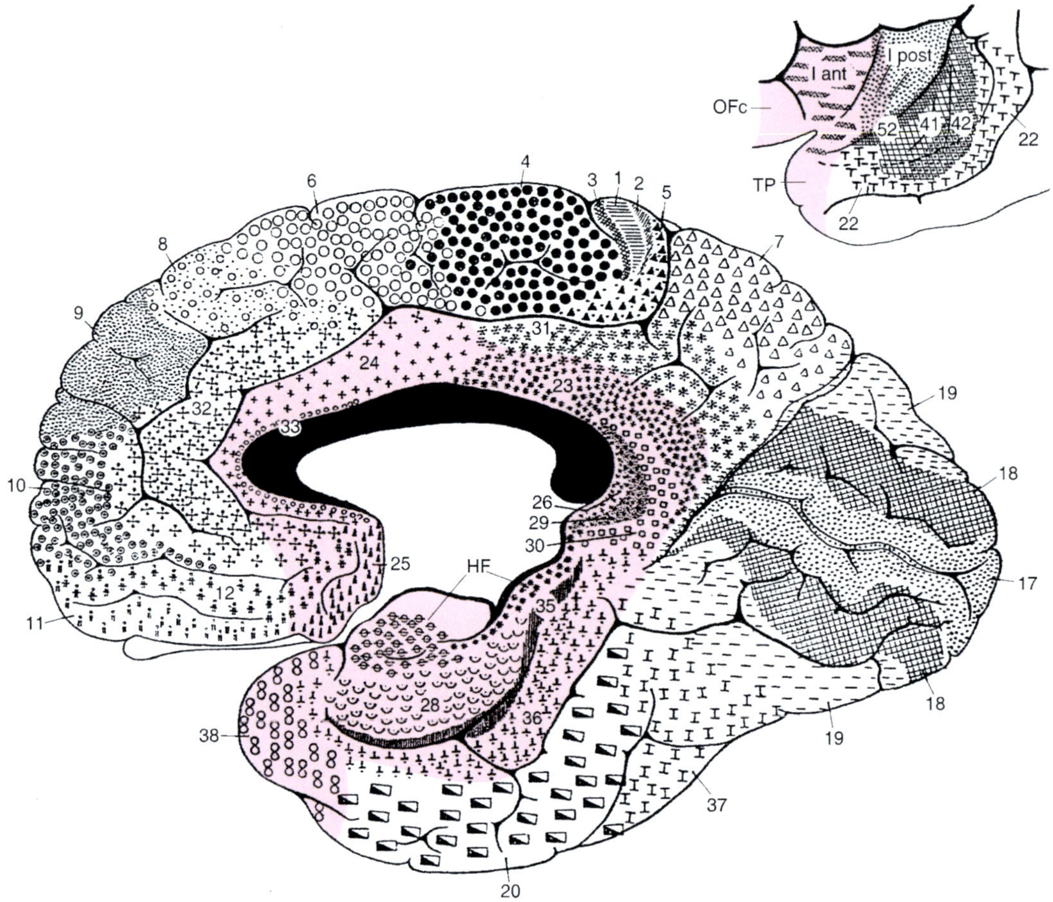

Fig. 15.55. Brodmann's cytoarchitectonic map of the medial hemisphere wall. Paralimbic areas are shown in *red*. The *insert* represents a small part of the lateral surface of the hemisphere. The lateral sulcus is opened to expose the insula. In combination with the main figure, it illustrates that the caudal orbitofrontal cortex, anterior insula and temporopolar cortex bridge the gap between the prelimbic area 25 and perirhinal areas 35 and 36, thus closing the paralimbic belt. Based on Mesulam [463]. *HF*, hippocampal formation; *I ant*, *post*, anterior and posterior insula; *OFc*, caudal orbitofrontal cortex; *TP*, temporopolar cortex

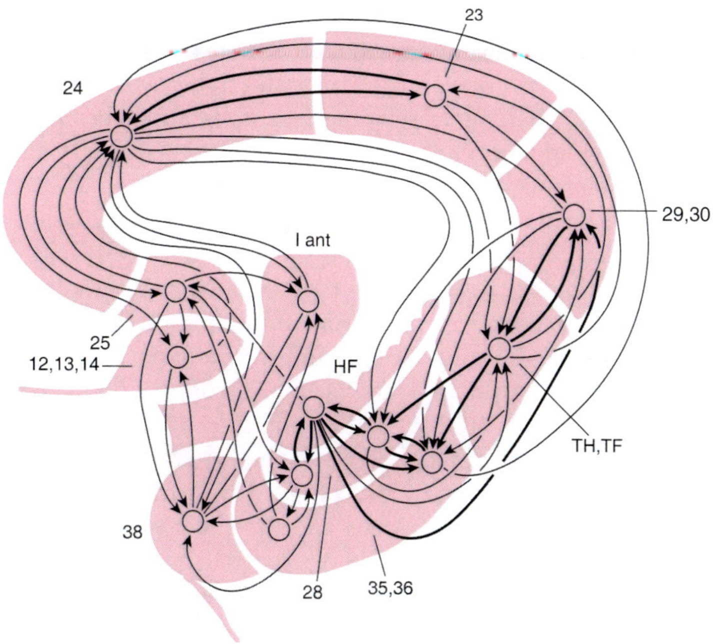

Fig. 15.56. Intrinsic cortical connections of structures forming the paralimbic belt. Particularly strongly developed projections are indicated by *heavy lines*. *HF*, hippocampal formation; *I ant*, anterior insula; *TH*, *TF*, medial temporal cortical areas according to von Economo and Koskinas [796]. *12, 13* etc., areas according to Brodmann

The limbic lobe includes the cingulate and parahippocampal gyri and also the hippocampal formation. Cytoarchitectonically, the limbic lobe encompasses the infralimbic cortex (area 25), the anterior and posterior cingulate cortices (areas 24 and 23), the retrosplenial cortex (areas 26, 29 and 30), the perirhinal cortex (areas 35 and 36), the entorhinal cortex (area 28) and the various parts of the hippocampal formation (dentate gyrus, Ammon's horn and subicular complex). Von Economo and Koskinas [796] demonstrated that a region roughly corresponding to the posterior parts of Brodmann's areas 35 and 36 contains two distinct separate entities, a superior area TH and an inferior area TF. As discussed in the previous section, the cytoarchitectonic areas just enumerated constitute, together with the caudal orbitofrontal cortex (areas 12, 13, 14 and caudal part of area 11), the anterior insula and the temporopolar cortex (area 38), the paralimbic belt (Fig. 15.55).

The intrinsic [55, 96, 152, 324, 377, 397, 439, 479, 482, 717, 772, 782, 843] and extrinsic connections of the various (para)limbic cortical structures [96, 152, 323, 324, 346, 377, 380, 397, 441, 480, 527, 579, 639, 716, 772, 782, 843] have been extensively studied in the rhesus monkey. The principal results of these studies are summarized and extrapolated to the human brain in Figs. 15.56 and 15.57. It will be seen that the various belt structures (a) are strongly and reciprocally interconnected (Fig. 15.56); (b) receive extrinsic cortical afferents from numerous unimodal (7, 19, 20, 21, 22) and polymodal (STSC, 9, 10, 11, 46) sensory areas (Fig. 15.57A); and (c) send their extrinsic cortical efferents mainly to the same association areas (Fig. 15.57B). For a further discussion of the limbic lobe, the reader is referred to Chap. 12. The limbic lobe and the more extensive paralimbic belt form part of a large functional entity, designated as the greater limbic system [510]. This functional system will be treated in the final chapter (Chap. 23) of the present work.

Frontal Lobe

The frontal lobe is strongly developed and comprises approximately one third of the entire hemisphere surface. On the lateral, convex surface it extends from the central sulcus to the frontal pole. Basolaterally, it is separated from the temporal lobe by the lateral sulcus. On the medial surface, the frontal lobe is separated from the limbic lobe by the cingulate sulcus, and from the parietal lobe by an arbitrary vertical line connecting the superior limit of the central sulcus with the cingulate sulcus.

The superolateral surface of the frontal lobe is traversed by three sulci, the precentral sulcus and the superior and inferior frontal sulci. The precentral sulcus runs parallel to the central sulcus. These two sulci border the precentral gyrus. The arciform superior and inferior frontal sulci divide the convex surface in front of the precentral gyrus into three convolutions, the superior, middle and inferior frontal gyri (Fig. 3.2). Short anterior and ascending branches of the lateral sulcus divide the inferior frontal gyrus into three parts: pars opercularis, pars triangularis and pars orbitalis (Fig. 3.2). These areas in the dominant hemisphere (usually the left in right-handed individuals) correspond to the region of Broca [67] associated with the motor aspects of speech.

The basal surface of the frontal lobe overlies the bony orbit; hence, the cortex in this region is designated as the orbitofrontal cortex. The olfactory bulb and tract lie in a longitudinal sulcus near the medial margin of the frontal lobe, known as the olfactory sulcus (Figs. 3.4, 3.5). The concave area lateral to the olfactory bulb and tract bears a number of sulci, which together form an H-shaped configuration. These sulci divide this area into four orbital gyri, anterior, lateral, posterior and medial. The latter is separated from a fifth orbital gyrus, the gyrus rectus, by the olfactory sulcus (Figs. 3.4, 3.5, 15.5C, G).

The cytoarchitectonic subdivision of the frontal lobe, according to Brodmann [70, 71], is shown in Fig. 15.58A, B. Among the 14 frontal areas distinguished by that author, there are agranular (4, 6, 24, 25, 32), dysgranular (8,

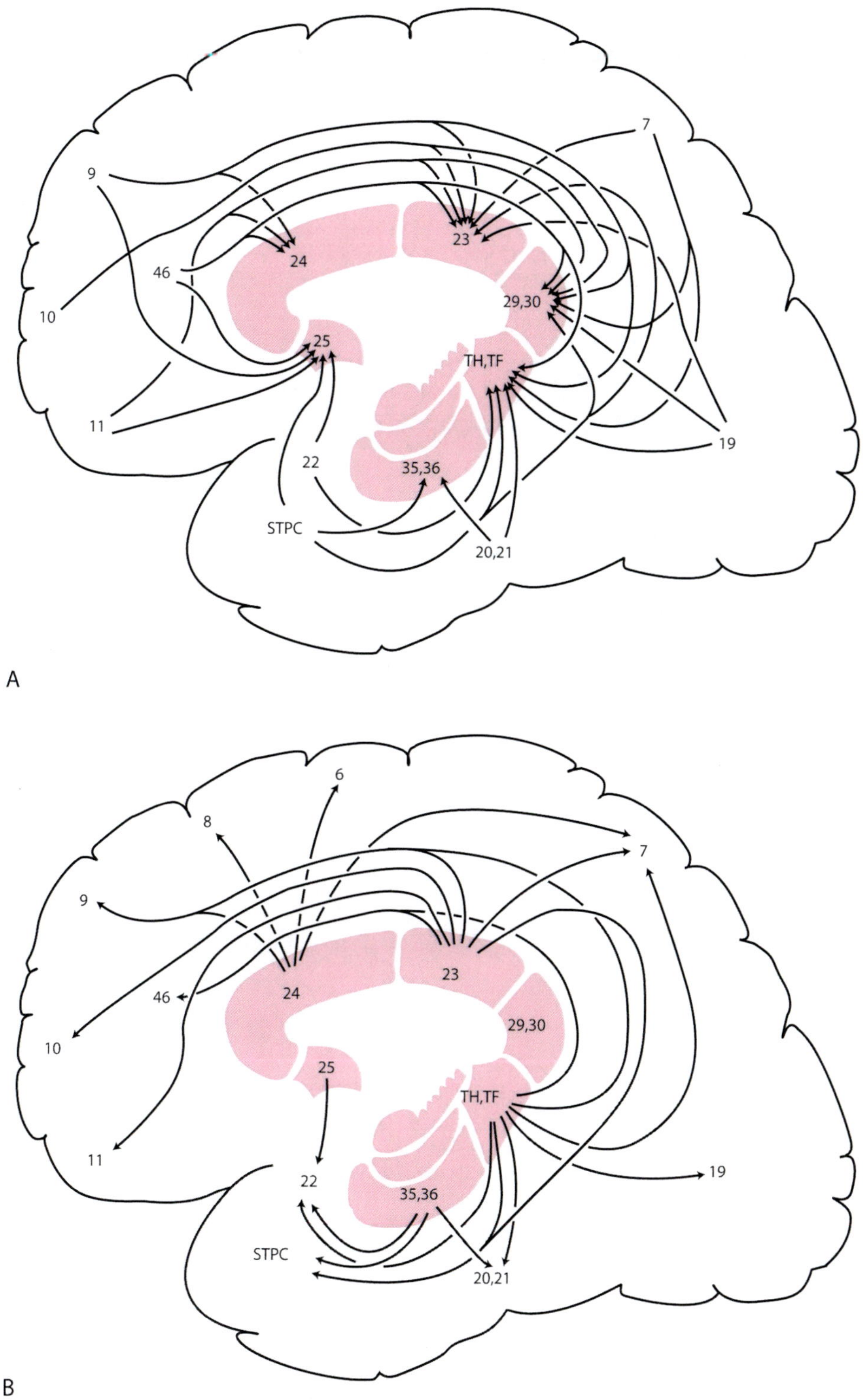

Fig. 15.57 A, B. Extrinsic cortical connections of paralimbic structures. **A** Afferents; **B** efferents. *STPC*, superior temporal polymodal cortex. Numbers and other abbreviations as in Fig. 15.56

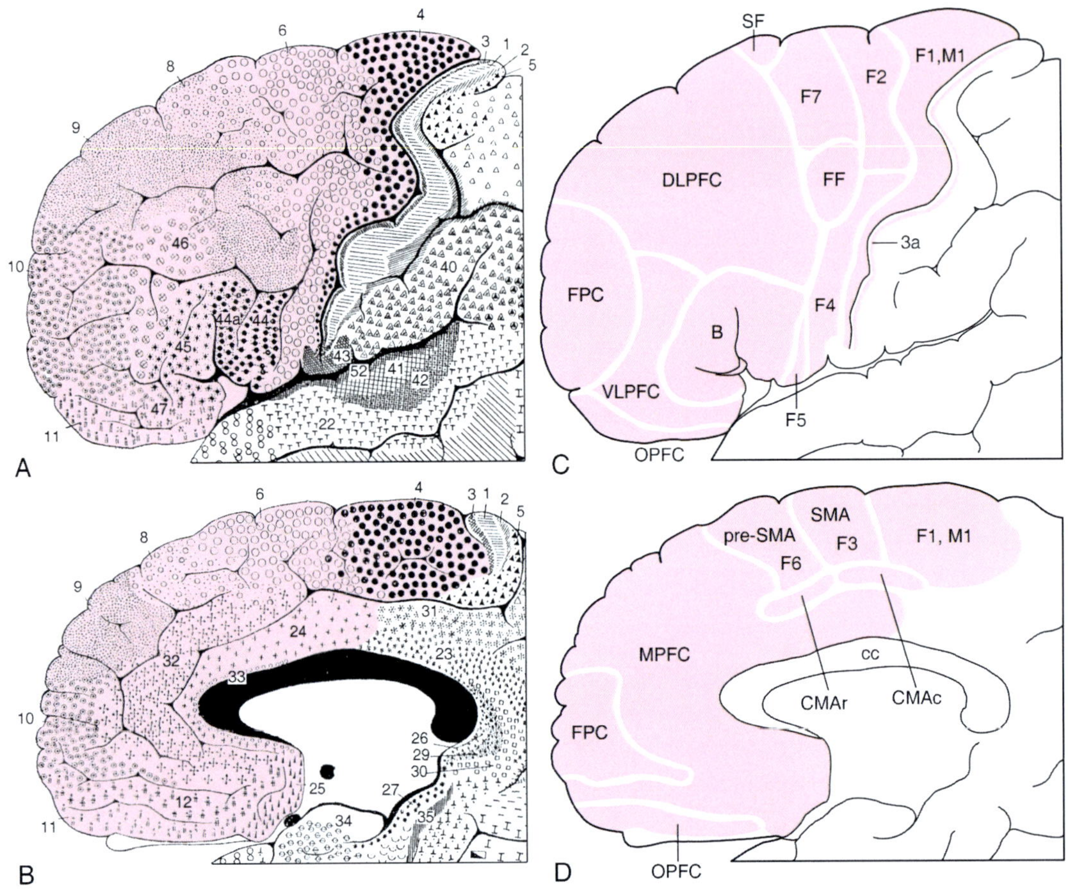

Fig. 15.58 A–D. The frontal lobe. Subdivision according to Brodmann, lateral (**A**) and medial (**B**) views. The frontal lobe is shown in *red*. Functional areas, lateral (**C**) and medial (**D**) views. *B*, Broca's speech region; *cc*, corpus callosum; *CAMc,r*, caudal and rostral cingulate motor area; *DLPFC*, dorsolateral prefrontal cortex; *F1–F7*, subdivisions of motor cortex; *FF*, frontal eye field; *FPC*, frontopolar cortex; *M1*, primary motor cortex; *MPFC*, medial prefrontal cortex; *OPFC*, orbital prefrontal cortex; *pre-SMA*, pre-supplementary motor cortex; *SF*, supplementary eye field; *SMA*, supplementary motor area; *VLPFC*, ventrolateral prefrontal cortex

44, 45), as well as granular areas (9, 10, 11, 12, 46, 47) (Fig. 15.7 A, F). According to the cytoarchitectonic typology of von Economo [795], areas 4, 6, 24 and 25 are heterotypical and belong to type 1, whereas all of the remaining frontal areas are homotypical. Of these, area 46 and part of area 10 belong to type 3, the orbitofrontal areas 11, 12 and 47 to type 4 and the remaining homotypical frontal areas to type 2 (Fig. 15.10).

The frontal cortex can be divided into two large functional domains, the motor cortex and the (associative) prefrontal cortex. The *motor cortex* is situated in front of the central sulcus and extends over the medial surface of the hemisphere. The *prefrontal cortex* occupies the large region that lies rostral to the precentral motor cortex [581] (Fig. 15.58 C, D).

The motor cortex comprises the primary motor cortex (M1) and the non-primary motor cortex. Matelli et al. [443, 445] have subdivided the somatic motor cortex of the rhesus monkey into seven areas designated F1–F7. This parcellation was based on cytoarchitectonic, histochemical, neurochemical, hodological and functional data. Modern architectonic analyses and functional evidence, mostly from neuroimaging studies, indicate that the organization of the human motor cortex closely resembles that of the rhesus monkey. Indeed, almost all of the structural/functional entities distinguished in the former can also be identified in the latter [188, 445, 585]. Hence, the following parcellation of the motor cortex applies to both species (Fig. 21.6).

Area F1 (or M1) represents the *primary motor cortex*. It corresponds to area 4 of Brodmann, which is characterized by the presence of the giant pyramidal cells of Betz (Figs. 15.7 F, 15.28 (6). Lassek [395] counted approximately 34,000 Betz cells in area 4 of the human brain.

Electrical stimulation studies of the primary motor cortex in both man [197, 544, 545, 548] and experimental animals [261, 784, 785, 830, 831] have revealed the presence of a topographical map within that cortex of the contralateral body half, comparable to that found in the adjacent primary sensory area (Figs. 15.18 C, 15.19).

Area 3 a is represented by a narrow strip of cortex, located in the fundus of the central sulcus. Its cytoarchitecture is similar to that of area 4, but it receives information from muscle receptors through the thalamus. Its cortical connections closely resemble those of the primary motor cortex.

The primary motor cortex receives afferents from both subcortical and cortical sources. The subcortical afferents arise principally from the contralateral dentate nucleus, via the posterior part of the ventral lateral thalamic nucleus and from the ipsilateral globus pallidus, via the anterior part of the same thalamic nucleus (Fig. 14.8). The cortical afferents to the primary motor cortex originate from several non-primary motor areas [164, 475], including the dorsal and ventral premotor areas [481, 711], the supplementary motor area [481] and the cingulate motor area [480], and from the primary somatosensory cortex (S1) and the somatosensory association cortex (area 5) [316, 542, 587]. The projection from S1, which arises from areas 1, 2 and 3b, is topographically organized. The somatosensory association area is primarily concerned with the analysis of proprioceptive information. It seems likely that the efferents from this association area to the primary motor cortex provide the latter with information on the localization of body parts, necessary for the control of limb movements.

The primary motor cortex contributes substantially to the pyramidal tract. For a detailed discussion of this fibre system, the reader is referred to Chap. 21 and Fig. 21.9. Here, we confine the discussion to mentioning that this tract arises principally from the motor and somatosensory cortical areas surrounding the central sulcus. Its fibres originate from pyramidal neurons situated in cortical layer V and via the internal capsule descend to the brain stem and spinal cord. The pyramidal fibres originating from the precentral gyrus (areas 3 a and 4) pass to premotor interneurons and also directly to motoneurons. The direct corticomotoneuronal connections are established by coarse fibres, derived from the giant cells of Betz. These direct pyramidal tract connections to motoneurons, especially to those innervating distal extremity muscles, appear to provide the capacity to execute highly fractionated movements, as typified by independent movements of the digits [390].

It has long been thought that the relation between the primary motor cortex and the skeletal muscular system is fixed and that this cortex is exclusively concerned with the execution of movements and is not responsible for the design of movement patterns. Both concepts have appeared to be incorrect. Recent evidence, summarized by Graziano [252], has shown that the connectivity between primary motor cortex and muscles is not fixed but plastic, changing constantly on the basis of feedback from the periphery. Moreover, it has been shown that with electrical stimulation trains of longer duration than used in the classical mapping studies cited above (500 ms instead of 50 ms), complex movements resembling meaningful actions, such as putting the hand to the mouth, defensive gestures, reaching motions and shaping the hand as if to grasp an object, can be elicited from the precentral cortex (Fig. 21.11 C).

Area 6 of Brodmann lies immediately anterior to the primary motor cortex. Just like the latter, it extends on the medial wall of the hemisphere (Fig. 15.58 A, B). The portion of area 6 situated on the lateral surface of the hemisphere corresponds to the *premotor cortex (PM), which can be subdivided into a dorsal (PMD) and a ventral zone (PMV).* Both of these zones can be further subdivided into caudal and rostral parts. The *caudal part of the ventral premotor zone, PMVc or area F4,* is located directly in front of the representation of orofacial movements in the primary motor cortex. The *rostral part of the ventral premotor zone, PMVr or area F5*, in the rhesus monkey contains two different functional areas, a caudal area F5ab and a more rostrally situated area F5c. It seems likely that the most rostroventral part of area 6 of non-human primates is homologous to Broca's speech region [445] in humans. The *dorsal premotor zone (PMD)* can be structurally and functionally subdivided into a *caudal PMDc* or *area F2* and a *rostral PMDr* or *area F7* (Fig. 15.58 C).

The medial portion of Brodmann's area 6 is occupied by two functional areas, the caudal *supplementary motor area (SMA) or area F7*, which is also known as M2, and the rostral *presupplementary motor area (pre-SMA) or area F6* [447] (Fig. 15.58 D). The medial wall of the hemisphere contains two further areas related to motor control. These areas, which may be designated as the *caudal and rostral cingulate motor areas, CMAc and CMAr*, are buried in the cingulate sulcus [188, 585]. The CMAc, which borders on the SMA and the rostral part of the primary motor cortex, forms a subfield of Brodmann's area 24. It contains a population of gigantopyramidal neurons, similar to those in the adjacent primary motor cortex [62]. The CMAr is located roughly at the same rostrocaudal level as the pre-SMA (Fig. 15.58 D). It also forms part of Brodmann's area 24 but probably encroaches upon area 32. CMAr and CMAc are also denoted as M3 and M4, respectively [482, 483].

Classically, the central skeletomotor system was looked upon as strictly hierarchical. The efferent pyramidal neurons in the primary motor cortex were collectively designated as the "upper motoneuron" or "final common pathway" for the central control of movements, impinging either directly or indirectly (i.e. via interneurons) on the "lower motoneurons" in the brain stem and spinal cord. Projections from other motor-related areas, converging upon the primary motor cortex, were thought to represent a hierarchical level superimposed on that of the "upper motoneuronal system". Recent experimental neuroanatomical studies, mainly in rhesus monkeys, have shown that the organization of the central motor system is much more complex than indicated by the hierarchical model just sketched. It has been shown that the motor part of the pyramidal tract, i.e. the "upper motoneuronal system", does not originate exclusively from the primary motor cortex. Rather, this fibre system has appeared to originate from many of the non-primary motor areas discussed above, including the PMVc and PMDc [162, 282], the SMA [163, 283] and the CMAr and CMAc [162, 163, 283]. Consequently, all of these areas have the potential to influence the generation and control of movements independently of the primary motor cortex. All of the areas giving rise to pyramidal tract fibres also project to the primary motor cortex [131, 431, 475, 480, 682, 711]. Thus, they influence motor activity through at least two pathways, corticospinal projections and corticocortical projections to the primary motor cortex. The various non-primary motor areas mentioned, which are all somatotopically organized [282, 283, 480], do not project only to M1, but they are also strongly and reciprocally interconnected [131, 431, 480, 711]. Dum and Strick [164] recently studied the digit representations in M1, PMD, PMV and SMA of cebus monkeys. They concluded that the subareas of the various clusters concerned with the generation and control of hand movements form a densely interconnected network, within which a clear hierarchical organization is lacking.

Matelli et al. [445] have pointed out that the various motor areas can be grouped into two major classes. The caudal motor areas F1, F2, F3, F4 and F5 form the first class. They are typified by their projections to the spinal cord, to F1 and to each other. The rostral areas F6 and F7 form the second class. These areas do not project to the spinal cord, but their descending output terminates in various part of the brain stem. Furthermore, these areas are not directly connected with F1. According to Matelli et al. [445], the two classes also differ with respect to the sources of their principal "extrinsic" cortical inputs. The areas forming the caudal group receive a primary cortical input from the parietal lobe; therefore, they may be characterized as parieto-dependent motor areas. The rostral areas, on the other hand, receive their primary cortical input from the prefrontal cortex and may, hence, be characterized as prefronto-dependent motor areas. The areas forming the caudal group collectively correspond to Mesulam's [463] motor association area (Fig. 15.15). As regards the relationship with the parietal lobe, we have seen that this lobe contains a number of centres in which sensory information from various sources is processed, and that each of these centres is specifically and reciprocally connected with a particular frontal motor area. The various parietal centres and their frontal targets form functional circuits, each of which is dedicated to a particular aspect of sensory-motor transformation [445, 828, 847]. These parietofrontal circuits have been briefly dealt with in the section on the parietal lobe of the present chapter and will be further discussed in Chap. 21.

The areas forming the rostral group of Matelli et al. [445], i.e. the pre-SMA or F6 and the PMDr or F7, are tightly connected to the prefrontal cortex, particularly area 46, and to the caudal motor areas [419, 431, 433, 475]. They are believed to play a role in cognitive aspects of motor control such as temporal planning of actions and motivation [445].

The cingulate motor areas do not fit into the categorization of Matelli et al. [445]. Just like the premotor areas of the caudal group, CMAr and CMAc are reciprocally connected with the primary motor cortex [480, 481] and project directly to the spinal cord [163, 283]. However, unlike the caudal group and conforming with the rostral group, CMAr and CMAc receive a substantial projection from the prefrontal cortex [481]. The two cingulate motor areas are strongly interconnected [482], and neither projects only to the primary motor cortex (M1), but also to the supplementary motor area (M2) [480]. All of these connections are somatotopically organized.

The prefrontal afferents to the cingulate motor areas comprise a strong projection from the dorsolateral prefrontal cortex and less substantial projections from the ventrolateral prefrontal and caudal orbitofrontal cortices [481]. It is important to note that the cingulate motor areas, in addition to the prefrontal and motor-related cortical afferents discussed above, also receive inputs from diverse and widespread limbic cortical areas, including cingulate areas 24, 23 and 32, retrosplenial areas 29 and 30 and temporal areas 35, TF and TH [482].

The functional roles of CMAr and CMAc are not well understood. Neuroimaging studies during motor tasks, involving the execution of various arm movements, have shown that CMAr is activated in relation to complex tasks, whereas CMAc is activated during simpler tasks [585]. Morecraft and Van Hoesen [482] suggested that the cingulate motor areas may play an important role in advancing emotional, motivational and memory-related information generated in the domain of the limbic lobe directly to the neocortical motor areas and, hence, to the voluntary motor system.

The non-primary motor cortex also comprises the frontal eye field and the supplementary eye field which, as their names indicate, are both involved in the control of eye movements, and Broca's speech region (Fig. 15.58 C).

The *frontal eye field (FF)* is situated at the caudal end of the middle frontal gyrus, in the vicinity of the precentral sulcus [653]. In Brodmann's map, this region is situated within the confines of area 6. However, Foerster [197] identified FF with a subfield of area 8, designated by Vogt and Vogt [784] as area $8\alpha\beta\delta$ (Fig. 15.18 C). According to Rosano et al. [633], FF is predominantly situated just anterior to area 6, in a transition region between area 6 and area 8 and extending into area 8 proper. FF is functionally divisible into a rostral subregion concerned with the generation of saccades (FF sac) and a caudal subregion concerned with the generation of smooth eye movements (FF sem) [434] (Fig. 19.14 A).

The *supplementary eye field (SF)* is located in the rostrodorsal part of a subfield of Brodmann's area 6, designated by Vogt and Vogt [784] as area $6\alpha\beta$ (Fig. 15.58 C) and by Matelli et al. [445] as a part of PMDr or F7 (Fig. 15.58 C).

The neocortex contains a number of highly interconnected regions that make direct contributions to the initiation and control of voluntary eye movements. In addition to FF and SF, these regions include the medial part of area 7, the lateral intraparietal area (LIP), the middle superior temporal visual area (MST) and the prefrontal eye field (PF), forming part of area 46 of Brodmann (Figs. 15.50, 15.51, 15.52 A). In a recent review, Lynch and Tian [434] summarized our current knowledge of the connections and functions of the cortical eye fields mentioned above. They point out that: (1) each of these eye fields is reciprocally connected to most or all of the other eye fields; (2) most of the eye fields receive direct input from several regions of visual association cortex; (3) in each of these fields electrical stimulation evokes eye movements; (4) surgical lesions or chemical inactivation of each field produces transient impairments of eye movements; and (5) each field demonstrates increased activity during eye movement tasks in functional imaging experiments in humans. Lynch and Tian also report that the saccade subregion of the frontal eye field (FFsac) and the pursuit region of the same field (FFsem) are selectively connected with distinct subregions in each of the other eye fields. On that account they propose that there are two parallel cortical oculomotor networks, one devoted to the control of saccadic eye movements and a second devoted to the control of pursuit eye movements.

The network mediating gaze control encompasses, apart from the cortical eye fields just discussed, numerous subcortical structures, including the thalamus, the basal ganglia, the cerebellum, the superior colliculus, the paramedian pontine reticular formation and, as a matter of course, the various oculomotor nuclei. The structural and functional organization of this network will be discussed in Chap. 19.

Broca's speech region is located in the opercular and triangular parts of the inferior frontal gyrus of the dominant (generally the left) hemisphere (Fig. 3.2). It is widely accepted that areas 44 and 45 of Brodmann constitute the cytoarchitectonic correlates of Broca's region [2, 6, 8, 70, 71, 746, 849] (Fig. 15.58 A, C). Both areas contain a thin, inconspicuous inner granular layer and can, hence, be classified as dysgranular [6]. The location and extent of areas 44 and 45 vary among subjects [383] (Fig. 15.22 A). A quantitative study of ten different brains [6] revealed that, although the volume of area 44 varied considerably among subjects, the volume of this area was larger on the left than on the right side in all ten brains. Area 45 did not show such interhemispheric differences. It seems likely that during evolution, areas 44 and 45 have developed from a primordium situated in the ventral premotor region in non-human primates [2, 591].

Damage to Broca's region generally leads to a *motor aphasia*. This aphasia, which is also known as *expressive or Broca's aphasia,* consists of a cluster of linguistic features including non-fluent, effortful speech, impaired repetition, and relatively preserved comprehension [26]. Nonfluency itself, apart from aspontaneous, sparse and slow speech, includes reduced or absence of grammaticality and reduced number of words per utterance (generally one or a few, expressed in a "telegraphic" style) [26, 151].

Broca's anterior speech region is strongly connected with Wernicke's posterior speech region and projects to the primary motor cortex. The fibres interconnecting Broca's and Wernicke's regions follow two different routes, dorsal and ventral. The fibres following the dorsal route pass dorsally on leaving Wernicke's region, arch around the posterior part of the lateral sulcus, and then travel rostrally beneath the supramarginal and somatosensory areas in the parietal operculum, to reach Broca's region. These fibres form part of the so-called arcuate fasciculus, i.e. the anterior limb of the superior longitudinal fascicle [145, 398, 532] (Figs. 15.45, 15.46, 15.59). The fibres following the ventral route pass directly forward from Wernicke's region and reach Broca's region by way of the extreme capsule, passing directly beneath the insular cortex [130], in which some of them may be synaptically interrupted. The efferent fibres from Broca's region pass to the lower part of area 4, in which the muscles of larynx, tongue and lips are represented.

It has already been mentioned that damage to Wernicke's region leads to a *sensory or receptive aphasia,* in which the comprehension of written and spoken language is severely affected. Interrupting the connections between Wernicke's and Broca's regions and between the fibres passing from Broca's region to the inferior primary motor cortex results in two other types of aphasia, known as conduction aphasia and subcortical motor aphasia. *Conduction aphasia* is characterized by a marked impairment in repeating spoken language, accompanied by preserved comprehension [26, 48, 151]. *Subcortical motor aphasia* clinically closely resembles *cortical motor aphasia* due to damage of Broca's region [165].

The anatomical structures and connections just discussed (and the clinical syndromes attached to their damage or interruption) together form the classical or Wernicke-Geschwind model of the neural architecture of language [48, 227, 228, 811, 812] (Fig. 15.59). During the past decades an enormous body of literature dealing with cerebral language organization and its disorders has accumulated. A detailed discussion of this literature is beyond the scope of the present work. Here, we only present a brief survey of some of the main results.

1. The classical model of the neural organization of language was exclusively based on postmortem studies on the brains of patients who had suffered from disturbances in comprehending and/or producing articulated speech. Numerous later studies, based on diverse approaches, such as intra-operative electrical stimulation of the brain [514, 547, 549], application of radioisotopes for the localization of brain infarcts [366], structural neuroimaging in patients with aphasic disorders (e.g. [385]) and functional neuroimaging in normal subjects during the execution of linguistic tasks (e.g. [200, 451, 593]), have all confirmed that the posterior part of the left inferior frontal gyrus and the left parietotemporal junction are critically important for the processing of language. However, several other authors [73, 821] failed to demonstrate any consistent association between aphasic syndromes and lesion locations.

2. Pure forms of the various aphasias do occur, but are very rare because the lesions involved (mostly infarctions in the left middle cerebral artery territory) seldom conform to a single functional site [65].

3. It has become clear that, apart from Broca's and Wernicke's regions, numerous other brain areas participate in speech and language (Fig. 15.60).

(a) Functional neuroimaging studies [668, 740] have shown that *the intermediate and posterior parts of the left superior temporal gyrus* play a crucial role in storage and retrieval of linguistic information and that the *left middle and inferior temporal gyri* are critically involved in lexical and semantic processing [320, 321, 645].

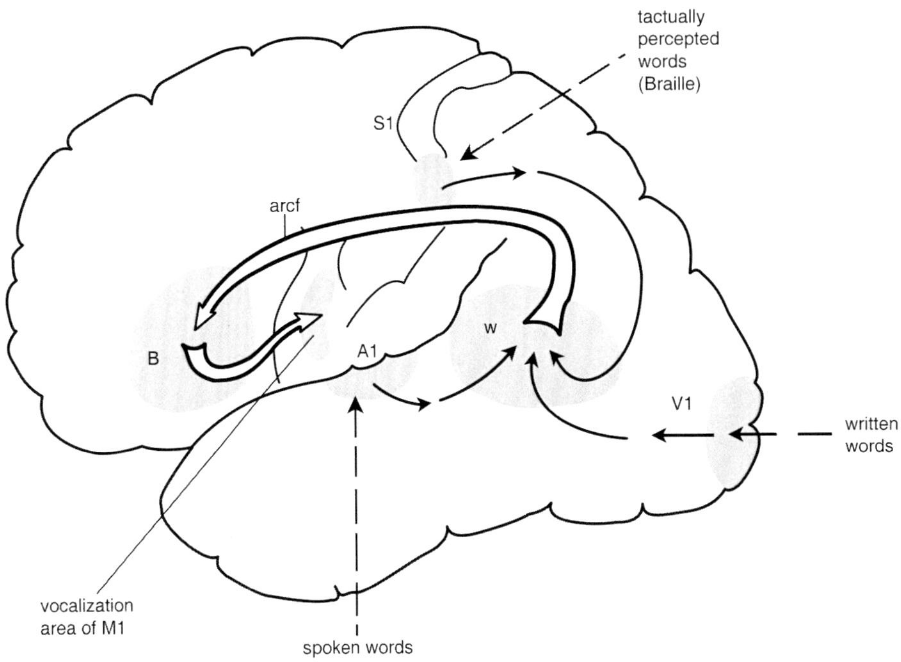

Fig. 15.59. The classical or Wernicke-Geschwind model of the circuitry related to the comprehension and production of language. *A1*, primary auditory area; *arcf*, arcuate fasciculus; *B*, Broca's speech region; *M1*, primary motor cortex; *S1*, primary somatosensory cortex; *V1*, primary visual cortex; *W*, Wernicke's speech region

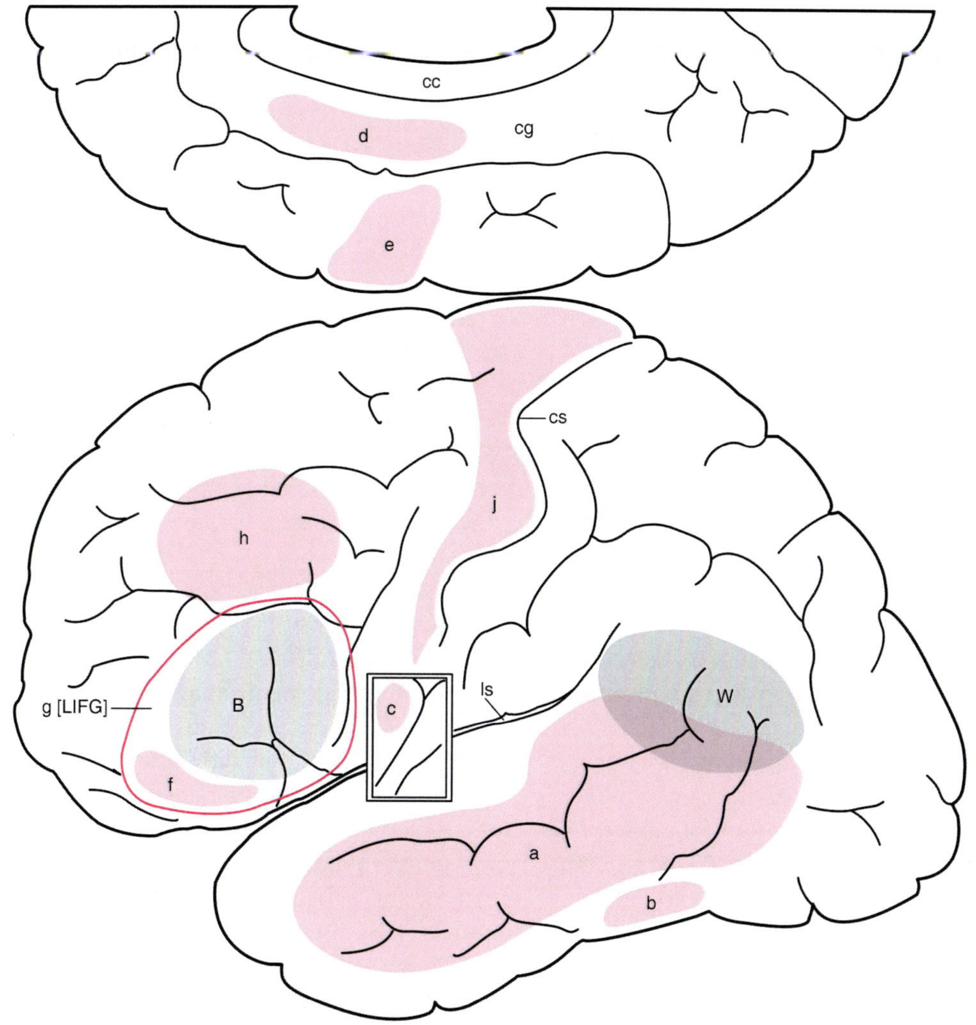

Fig. 15.60. Cortical areas involved in the comprehension and production of language. The classic regions of Wernicke (*W*) and Broca (*B*) are shown in *grey*; more recently discovered language-related areas are shown in *red*. The *small window* shows part of the insula. *a–g*, language-related areas discussed in the text; *cc*, corpus callosum; *cg*, cingulate gyrus; *cs*, central sulcus; *LIFG*, left inferior frontal cortex; *ls*, lateral sulcus

(b) Electrical stimulation in epileptic patients [77, 420] and fMRI studies [289, 593] have shown that an area situated in *the intermediate part of the left lateral occipitotemporal gyrus* (designated as the *"basal temporal language area"*) is involved in word retrieval.

(c) A study involving computerized lesion reconstruction in a series of aphasia patients [158] revealed that in patients with a disorder in co-ordinating the movements for speech articulation, an area situated in the *anterosuperior insula* was specifically damaged. The authors concluded that this insular area is specialized for the motor planning of speech. It has been suggested [466] that fibres passing from Wernicke's to Broca's region through the insula are synaptically interrupted in the posterior insular cortex.

(d) Functional neuroimaging studies have shown that the *anterior cingulate cortex* is involved in linguistic activities such as selecting verbs to lists of nouns [603] and translation [594].

(e) The *supplementary motor area* is consistently activated in functional imaging studies involving speech [159, 593].

(f) Neuroimaging studies [58, 185, 572, 573] have identified a small region in the left inferior frontal gyrus that becomes active during specific linguistic activities, such as processing semantic relationships between words or phrases and retrieving semantic information. This region corresponds roughly to Brodmann's area 47.

(g) Hagoort [276] has recently pointed out that the language-relevant part of the frontal cortex, apart from the classical Broca region, also includes the inferior prefrontal region corresponding to area 47 and presumably also the antero-inferior part of area 6. He denoted this new entity as *left inferior frontal cortex (LIFG)*. Hagoort adduces evidence indicating that LIFG is crucial in unification operations required for binding single word information received from memory into larger semantic, syntactic and phonological structures.

(h) Based on clinical evidence, Damasio [129] reported that certain higher-level aspects of language formation require an intact prefrontal cortex surrounding Broca's region. Hagoort [276] concluded from fMRI studies on subjects performing verbal control tasks that a network of areas consisting of the anterior cingulate cortex and *a dorsolateral prefrontal region encompassing parts of Brodmann's areas 9 and 46* is involved in verbal action planning and attentional control.

(i) Remarkably, the *primary motor cortex* is also involved in semantic processing. Hauk et al. [280] showed action words referring to face, arm and leg actions (e.g. to lick, pick or kick) in a passive reading test to normal volunteers under fMRI monitoring. The mere reading of these words appeared to activate loci of the motor cortex involved in the actual movement of the tongue, fingers or feet. These findings suggest that the brain areas that are used to perform a particular action are also involved in the comprehension of the words related to that action [132].

(k) Functional neuroimaging studies have shown that the right (non-dominant) hemisphere contributes substantially to many aspects of language comprehension and production, including the detection of syntactic errors, comprehending contextual and figurative meaning and prosody (melody, timing and intonation) [58, 391, 473].

4. There is evidence that the cortical domains dedicated to language are compartmentalized into separate systems for processing different aspects of language. Ojemann [514] cites lesion studies indicating the presence of separate areas for handling different languages, for handling different grammatical classes of words or for the naming of specific semantic categories, such as "animals" or "tools". Functional neuroimaging research, summarized by Bookheimer [58], has shown that within the left inferior frontal lobe separate subsystems are responsible for different aspects of language. It appeared that a rostroventral area, corresponding to Brodmann's areas (BAs) 47 and 45, contributes to semantic processing; an intermediate area, corresponding to BAs 45 and 44, has a role in syntactic processing; whereas a dorsocaudal area including parts of BAs 44 and 6 are involved in phonological processing. Similar subsystems, subserving partic-

ular specialized language-related functions have also been reported for Wernicke's region [827] and for both Broca's and Wernicke's regions [300].

5. The data reviewed indicate that the neural network involved in the comprehension and production of language is much more extensive than was originally envisioned. However, the precise wiring of this network remains to be established. The marked functional specialization detected within language-processing areas suggests that several parallel, though interrelated subnetworks are present. The language-related (sub)networks do not function in isolation. Neuroimaging studies, cited by Bookheimer [58] and Hagoort [276], have reported increased activity in Broca's region during various nonlinguistic tasks. Mesulam [464] indicates that many cortical nodes participate in the function of more than one cognitive network. Finally, it should be mentioned that the circuitry related to language functions is presumably not confined to the neocortex. Clinical studies, reviewed in [26, 48, 391], have shown that local lesions in the basal ganglia, thalamus and even cerebellum may lead to language deficits.

The *prefrontal cortex* is distinguished from the motor and premotor areas by (1) the presence of a pronounced internal granular layer, and (2) its strong reciprocal connections with the mediodorsal thalamic nucleus (MD) [402, 635]. On account of these features, the prefrontal cortex is sometimes denoted as the *granular frontal cortex* or as the *MD-projection cortex*. However, neither of these two criteria is absolute. Thus, the caudobasal part of the prefrontal cortex is dysgranular rather than granular, and studies with modern tracing techniques in the rhesus monkey have shown that most prefrontal areas that receive MD input are also connected to other thalamic nuclei, and that, conversely, the MD projections do not pass exclusively to the prefrontal cortex [592, 749]. Another important feature of the prefrontal cortex is that it receives inputs from all unimodal and heteromodal sensory association areas. On the basis of these afferents, the prefrontal cortex may be qualified as a *high-order heteromodal association area* [463] (Fig. 15.15).

Two groups of authors, Petrides and Pandya [580, 581] and Price and collaborators [86, 516, 518], modified Brodmann's original parcellation of the human prefrontal cortex on the basis of comparative studies on the brain of the rhesus monkey and the human. Both groups took Walker's [802] cytoarchitectonic analysis of the frontal lobe of the rhesus monkey as their starting point. In this analysis (Figs. 15.61 A, 15.62 A, 15.63 A), Walker used Brodmann's numbering scheme. With regard to their position, his areas 6, 8, 9, 10, 24, 45 and 46 are directly comparable to the similarly numbered areas of Brodmann's map (Fig. 15.58 A, B). However, Walker's area 46 is very extensive and occupies a considerable part of the lateral surface of the frontal lobe (Fig. 15.61 A). Contrary to its human counterpart, Walker's area 24 extends to the basal surface of the brain, and his area 25 largely corresponds to Brodmann's area 32 in the human brain (Figs. 15.58 B, 15.62 A). Walker designated an area occupying the ventrolateral part of the frontal lobe, corresponding positionally with parts of Brodmann's areas 11 and 47, as area 12 (Figs. 15.59 A, 15.63 A). However, Brodmann used number 12 in his latest map of the human cortex [71] to specify quite another area, occupying a ventromedial position in the frontal lobe (Fig. 15.58 B). Finally, Walker delineated two "new" areas on the basal surface of the frontal lobe, which he labelled as areas 13 and 14 (Fig. 15.63), neglecting the fact that Brodmann [70] used these numbers to specify other (insular) areas in the cortex of non-primate species.

The modifications of Brodmann's parcellation of the frontal cortex introduced by Walker [802] are clearly reflected in the comparative architectonic studies of Petrides and Pandya [578, 580, 581] and Price and collaborators [86, 516, 518]. As shown by Figs. 15.61 B, C, 15.62 B–E and 15.63 B, C, both groups use a mixed Brodmann-Walker nomenclature in their parcellations of the monkey and human cortex [748]. The most important deviations from Brodmann's subdivision of the frontal cortex, introduced by the two groups of investigators mentioned, are the following:

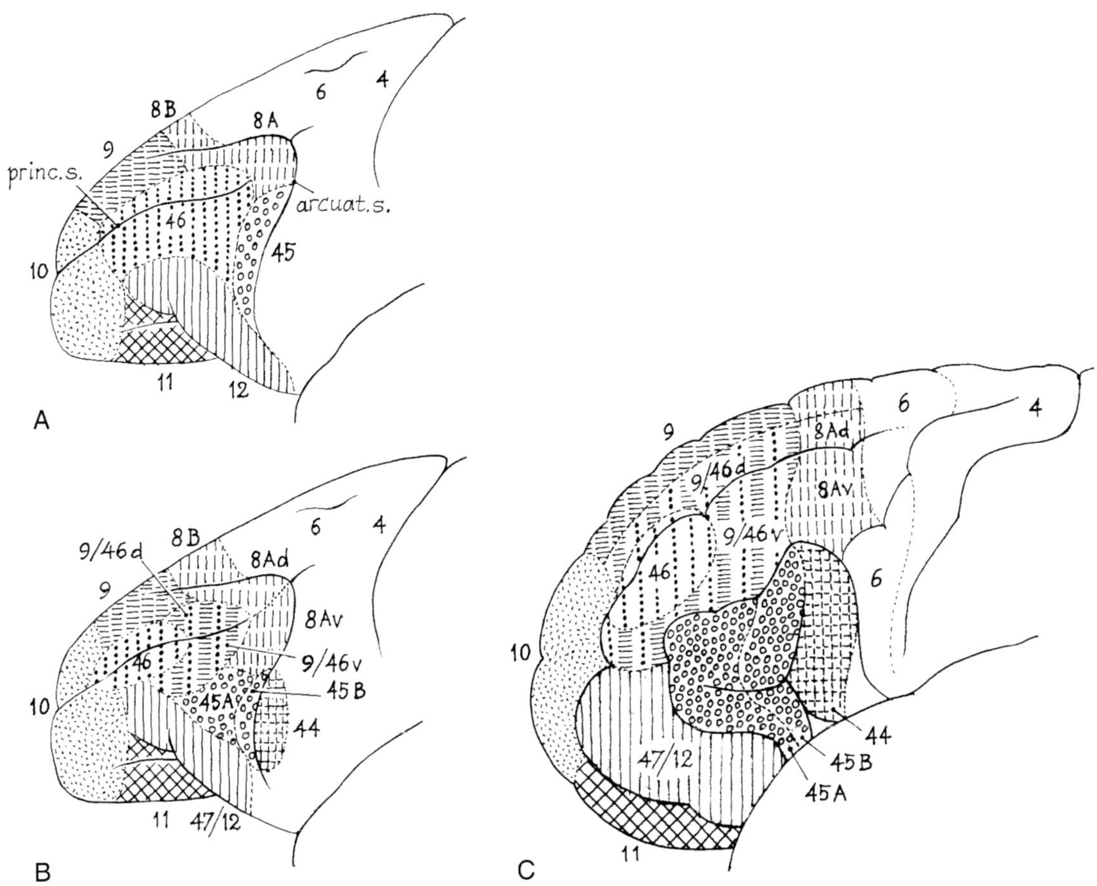

Fig. 15.61 A–C. Cytoarchitectonic maps of the lateral frontal cortex. A Rhesus monkey, according to Walker [802]. B Rhesus monkey, according to Petrides and Pandya [578]. C Human, according to Petrides and Pandya [578]. *arcuat s*, arcuate sulcus; *princ s*, principal sulcus

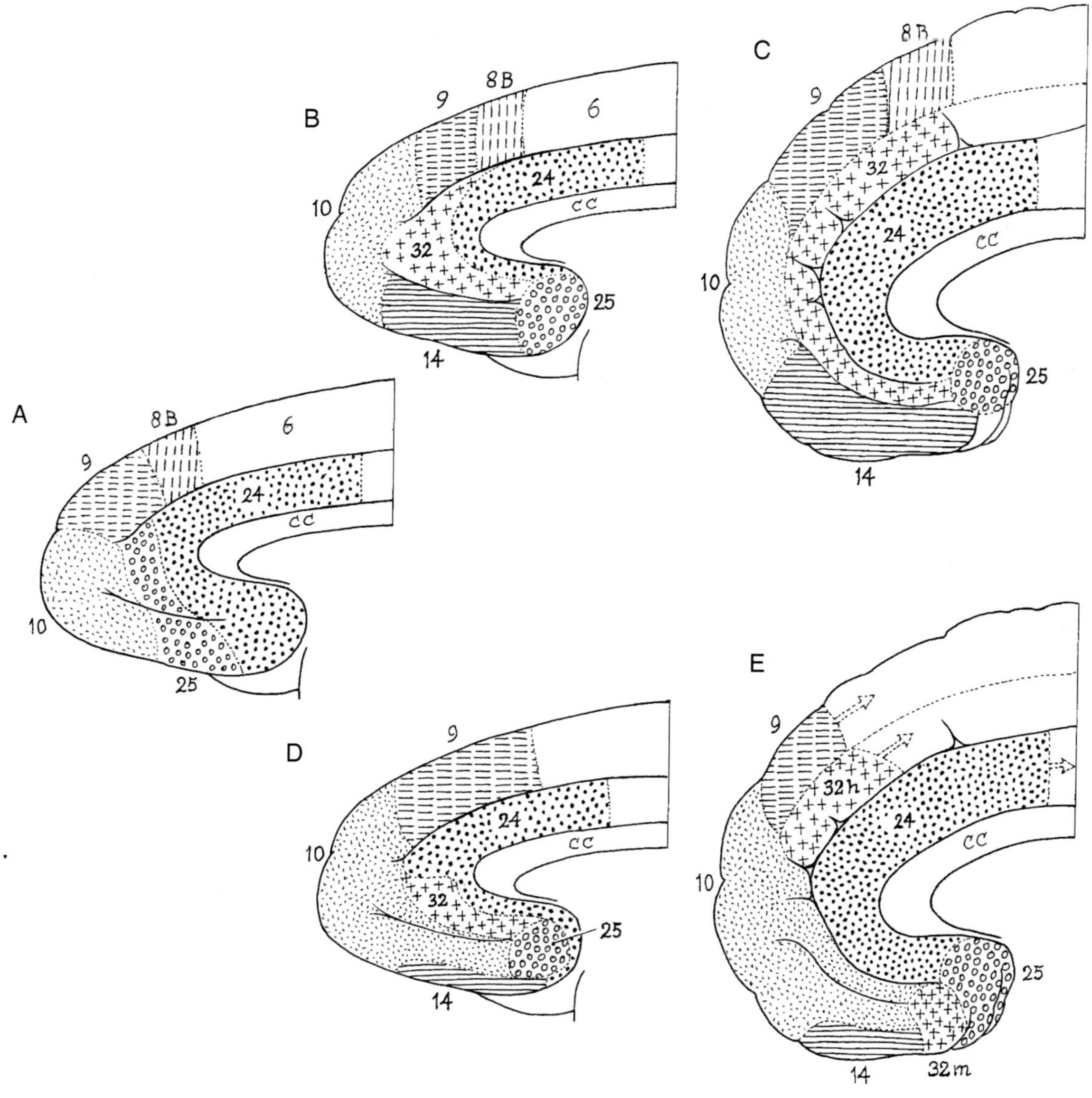

Fig. 15.62 A–E. Cytoarchitectonic maps of the medial frontal cortex. **A** Rhesus monkey, according to Walker [802]. **B, C** Rhesus monkey and human, respectively, according to Petrides and Pandya [578]. **D, E** Rhesus monkey and human, respectively, according to Öngür and Price [516]. *cc*, corpus callosum

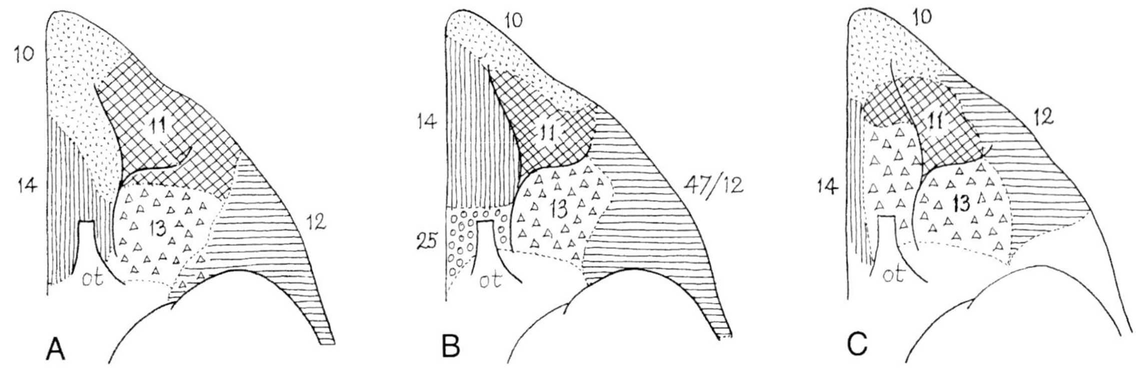

Fig. 15.63. Cytoarchitectonic maps of the orbitofrontal cortex of the rhesus monkey, according to Walker [802] (**A**), Petrides and Pandya [578] (**B**) and Öngür and Price [516] (**C**). *ot*, caudal part of the olfactory tract

1. In Brodmann's map, area 46, which occupies a central position on the lateral surface of the frontal lobe, is separated from area 8 by the ventrocaudal part of area 9 (Fig. 15.58 A). However, Petrides and Pandya [578] observed that the architecture of the portion of the middle frontal gyrus, labelled area 9 in the map of Brodmann, is more akin to area 46. They also observed that this cortex corresponds in architecture to the cortex found in the caudal part of the sulcus principalis in the rhesus monkey, and that has been included in area 46 by Walker (Fig. 15.61 A). Petrides and Pandya therefore labelled the portion of area 9 that lies on the middle frontal gyrus as area 9/46 (Fig. 15.61 C), to acknowledge its architectonic similarity to area 46 and its inclusion with area 9 in Brodmann's map.

2. In Brodmann's map, the region rostroventral to area 45 has been labelled area 47 (Fig. 15.58 A). Petrides and Pandya [580] observed that this region has architectonic features comparable to those of Walker's area 12 in the rhesus monkey (Fig. 15.61 A). They therefore labelled this region area 47/12 (Fig. 15.61 C).

3. On the medial surface of the human frontal lobe, Petrides and Pandya [581] (Fig. 15.62 C) as well as Price and collaborators [516, 518] have replaced area 12 and part of area 11 of Brodmann (Fig. 15.58 B) by area 14 of Walker (Fig. 15.63 A).

4. It is remarkable that in the map of Price and collaborators, the medial part of Brodmann's area 10 is extraordinarily large and abuts directly on area 24, splitting area 32 into a dorsal, human (h) and a ventral monkey (m) part (Fig. 15.62 E).

5. We have seen that Walker, in his architectonic analysis of the frontal lobe of the rhesus monkey, has delineated two "new" areas on the orbital surface of that lobe, which he labelled areas 13 and 14 (Fig. 15.63 A). Petrides and Pandya [581] and Price and collaborators [516, 518] confirmed the presence of these areas in the rhesus monkey (Fig. 15.63 B, C) and also identified them in the human (Fig. 15.11 E, G).

It is important to note that Price and collaborators divided most of the areas they delineated into two to four subareas (Figs. 11.8, 15.11 G).

We discussed the results of the architectonic analysis of Petrides and Pandya and Price and collaborators in some detail because they form the basis of extensive hodological studies on the frontal lobe of the rhesus monkey and that of a sound extrapolation of the results of these studies to the human brain. In what follows, wherever necessary, Brodmann's areas will be denoted with the letters "BA" and areas derived from the mixed Brodmann-Walker parcellation as "BWA".

The various cytoarchitectonic areas forming the prefrontal cortex are strongly and reciprocally interconnected [89, 581]. An important output channel from the prefrontal cortex is formed by sequences of short association fibres which, via the premotor cortex, converge upon the primary motor cortex [419, 711].

The fibre connections of the prefrontal cortex are multifarious. The afferents from the various association areas are reciprocated by efferent projections. It receives cholinergic and GABAergic afferents from the basal nucleus of Meynert, histaminergic, orexinergic and melanin-concentrating hormone-containing fibres from the hypothalamus, serotoninergic fibres from the mesencephalic raphe nuclei, dopaminergic fibres from the ventral tegmental area and noradrenergic fibres from the locus coeruleus. It has reciprocal connections with various parts of the limbic lobe (Fig. 15.57), the amygdala (Figs. 13.6, 13.7), septum and hypothalamus, and it projects to the caudate nucleus (Fig. 14.1), the periaqueductal grey (PAG) and the pons. Several of these connections will be further considered below.

The prefrontal cortex is critically involved in complex brain functions such as orientation and attention, decision making on the basis of current exteroceptive and interoceptive information and past experience, planning and sequencing of actions, emotionality and personality. Large bilateral lesions of the prefrontal cortex lead to a disorder known as *frontal lobe syndrome*. The essential features of this syndrome are [151]: (1) diminished capacity to sustain attention and concentration; (2) lack of spontaneity and initiative; (3) inhibition and impulsivity of affect, thought and action; (4)

inability to plan, organize and execute complex behaviour; and (5) loss of social decorum.

Comparable changes have been observed in psychiatric patients who underwent *prefrontal leucotomy*. The principle of this operation is to cut the fibre connections to and from the prefrontal region. This was done by inserting a "leucotome" through a burr hole and moving the instrument in a coronal plane (Fig. 15.64). As a rule these operations were performed bilaterally. Prefrontal leucotomy was introduced by the Portuguese neurologist Egas Moniz in 1936 [476] and was thenceforward strongly advocated by the Americans Walter J. Freeman and James W. Watts [202, 203]. In the two decades following its introduction, tens of thousands of severely disturbed psychiatric patients underwent leucotomies all over the world [330, 752]. For his discovery, Moniz was awarded the Nobel Price in 1949. The principal target group of the intervention consisted of chronic, hospitalized patients, suffering from depression, anxiety states and obsessive-compulsive disorders. Although Moniz [477] qualified prefrontal leucotomy euphorically as "highly effective and always safe", it soon became clear that the intervention produced serious changes in personality, including apathy, slowness, lack of initiative, carelessness, poor judgement and inhibited behaviour in social situations [803]. From a neuroanatomical point of view, the operations were very crude, with poor control of the actual place of the section, due to the considerable interindividual variations of brain size and shape, and skull-brain relationships. Furthermore, there were unintended side effects, such as haemorrhages, sometimes found far from the site of the section [468, 469] (Fig. 15.65). Fortunately, the advent of effective psychopharmaceuticals put an end to the era of prefrontal leucotomy.

The prefrontal cortex can be divided into three major regions: lateral, medial and orbital.

The *lateral prefrontal cortex* encompasses BAs 8, 9, 46, 10 and 47 (Fig. 15.58 A). BAs 9 and 46 together correspond to BWAs 9, 9/46d, 9/46v and 46. BA 47 roughly corresponds to BWAs 47/12 (Fig. 61 C). This cortical region is known to participate in numerous higher-order brain functions, such as integrating sensory analysis with motor activity, selective attention, working memory, planning and reasoning. In what follows, a brief survey of the principal connections of the lateral prefrontal cortex (LPFC) will be presented, to which some functional notes will be attached.

Experimental studies in rhesus monkeys [578, 580, 581] have shown that the various prefrontal areas are tightly and reciprocally interconnected and that this connectional network extends uninterruptedly over the medial prefrontal and orbitofrontal regions [89, 516].

The LPFC represents the highest level of sensorimotor integration. It receives highly processed information from multiple sensory modalities and plays a prominent role in the planning and organization of voluntary goal-directed behaviour. The sensory information is conveyed to the LPFC by massive projections emanating from post-Rolandic unimodal and polymodal sensory association area (Figs. 15.15, 15.46).

A combination of electrophysiological recordings and anatomical tract-tracing in rhesus monkeys [245, 513, 631, 632, 822], as well as neuroimaging studies in humans [113, 116, 147, 455], have shown that the LPFC is organized in separate dorsal spatial cognition- and ventral object cognition- and pattern cognition domains. For a discussion of the relevant electrophysiological and neuroimaging data, the reader is referred to the publications cited; here we confine ourselves to mentioning that the visual and auditory projection streams involved in the processing of spatial information converge upon the dorsal part of the LPFC, whereas visual and auditory streams, concerned with pattern discrimination, mainly target the ventral part of the LPFC (Fig. 15.66). However, this functional segregation is not absolute. In both domains units processing both object identity and location have been found [613].

All the pathways that convey sensory information from posterior cortical areas to the LPFC contain reciprocal fibres, projecting back to areas from which input was received [581, 582]. These feed-back fibres form part of the

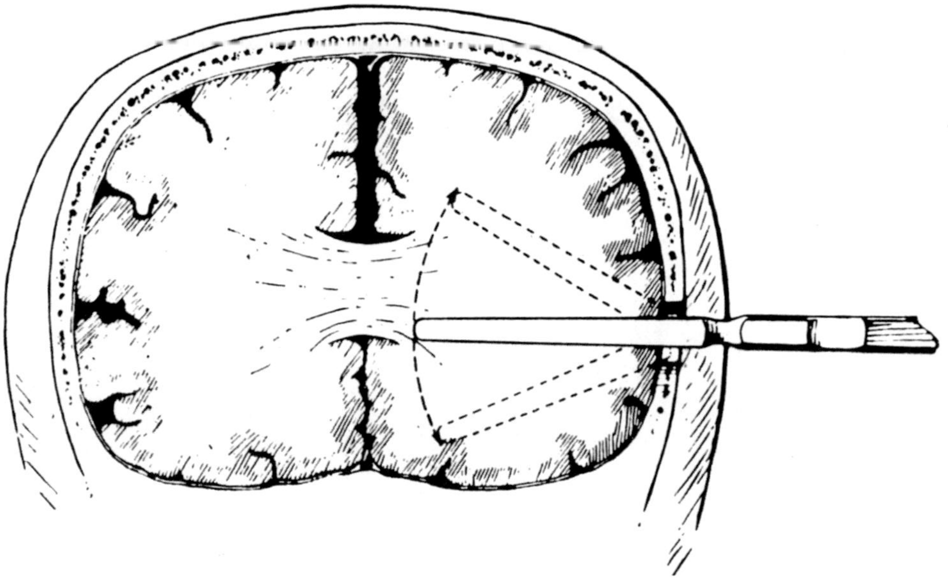

Fig. 15.64. Drawing showing how the brain is incised during a prefrontal leucotomy. An identical incision is made on the opposite side. Reproduced from Freeman and Watts [202]

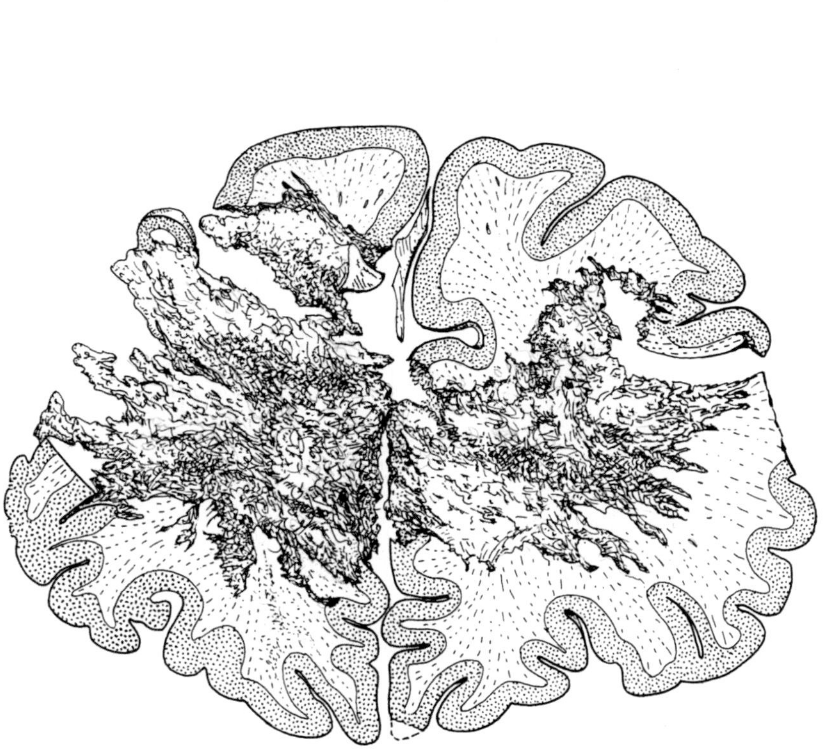

Fig. 15.65. Section of frontal lobes in plane of leucotomy incision, 4 months after operation in which the left anterior cerebral artery was injured. Based on a photograph in Freeman and Watts [202]

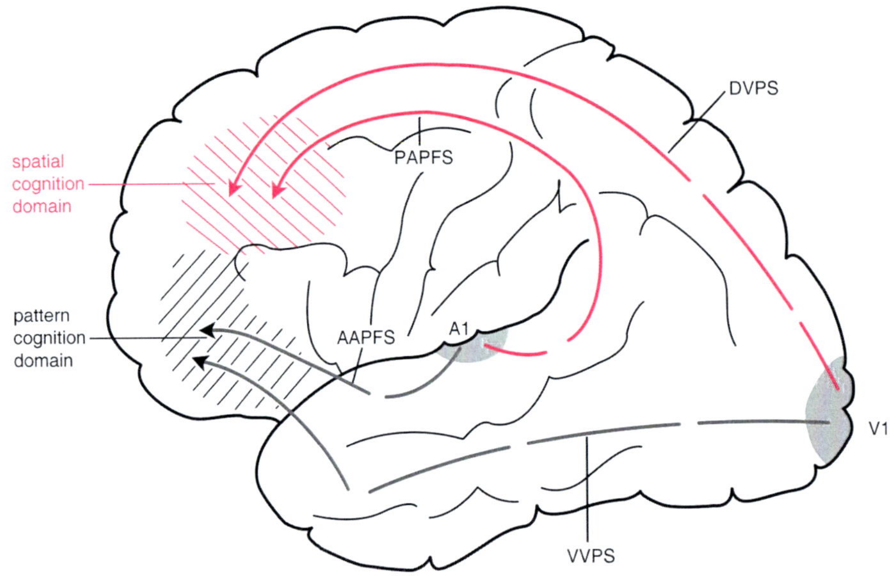

Fig. 15.66. Schematic diagram showing that visual and auditory projection streams involved in the processing of spatial information, target the dorsal part of the lateral prefrontal cortex (shown in *red*), whereas projection streams subserving pattern discrimination target the ventral part of the lateral prefrontal cortex (in *black*). *A1*, primary auditory cortex; *AAPFS*, anterior audito-frontal stream; *DVPS*, dorsal visual processing stream; *PAPFS*, posterior audito-prefrontal stream; *V1*, primary visual cortex; *VVPS*, ventral visual processing stream

morphological substrate of *attention*, i.e. the mechanism that enables us to direct our processing resources to a subset of the available information [124]. This mechanism allows us to selectively process the information relevant to current goals. A combination of the results of neuroimaging studies in humans with data derived from tract-tracing studies in monkeys suggests that, apart from the LPFC, the visual attentional control system involves several way stations in the dorsal and ventral visual processing streams (Fig. 15.67) [552, 553]. Egner and Hirsch [168] recently presented neuroimaging evidence suggesting that the LPFC exerts attentional control on post-Rolandic cortical areas by amplifying task-relevant information rather than by inhibiting distracting stimuli.

The organization of mental processes, such as the sequencing of complex goal-directed behaviours, not only depends on the selective attentional control of incoming sensory information, but also on the ability to keep the information selected available for some time. This ability to transiently maintain and manipulate a limited amount of information to guide thought or behaviour is known as *working memory* [405]. The LPFC plays a key role in working memory. Classic experiments on monkeys and chimpanzees in the 1930s [327, 328] have shown that, following bilateral lesions of the LPFC, the animals were still fully able to perform auditory or visual discrimination tasks or recall spatial orientation of objects, providing tests were made in immediate memory. However, the animals appeared to be unable to perform such tasks if a delay of more than a few seconds was introduced between the stimulus and the response. The relation between this temporal integrative function and the LPFC has been substantiated electrophysiologically [215, 217]. Cells in the LPFC of the monkey were found to fire persistently at high rates while the animals retained an item of visual information in short-term memory. Lesion studies in patients [575] and neuroimaging studies in normal human subjects [574] have also shown that the LPFC is critically involved in working memory.

Apart from the projections from the various sensory association areas, the LPFC receives afferents from the limbic lobe as well as from the mediodorsal, ventral anterior, and the anterior and posterior parts of the ventral lateral thalamic nucleus, and from a large number of subcortical, extrathalamic structures.

The afferents from the limbic lobe reach the LPFC via two fibre systems: the dorsal and ventral limbic pathways [579]. The *dorsal limbic pathway* originates from the rostral and caudal cingulate cortex (areas 24 and 23) and the retrosplenial cortex (area 30) and passes forward in the cingulate bundles to the LPFC. The *ventral limbic pathway* originates from the posterior parahippocampal cortex, particularly area TF, and reaches the LPFC via the extreme capsule [247, 397]. Both limbic pathways are bidirectional, i.e. they also contain fibres originating from the LPC and terminating in the cingulate, retrosplenial and parahippocampal cortices [247, 377, 397]. Neuroimaging studies have shown that various subregions of the cingulate cortex can be activated by stimuli related to simple emotions, such as happiness, sadness, anger and fear [780]. It seems likely that the cinguloprefrontal projections may subserve the assigning of emotional valence to the sensory information processed in the LPFC. It is well known that the hippocampal formation and adjacent medial temporal structures represent essential components of the memory system (Chap. 12). Because these structures are tightly connected to all other parts of the limbic lobe (Figs. 12.13, 15.56), it has been suggested that the limbico-prefrontal pathways may also form part of the network involved in the encoding and retrieval of long-term memory [579, 685].

The LPFC is reciprocally connected with the lateral, parvocellular division of the mediodorsal thalamic nucleus [712]. As discussed in Chap. 8, this nucleus belongs to the so-called higher-order thalamic relays, which means that it receives its principal or "driver" afferents mainly from the cortex and plays a potentially significant role in cortico-cortical communication [681].

Fibres passing from the ventral anterior (VA) and anterior ventral lateral (VLa) thalamic nuclei to the LPFC represent the final stage

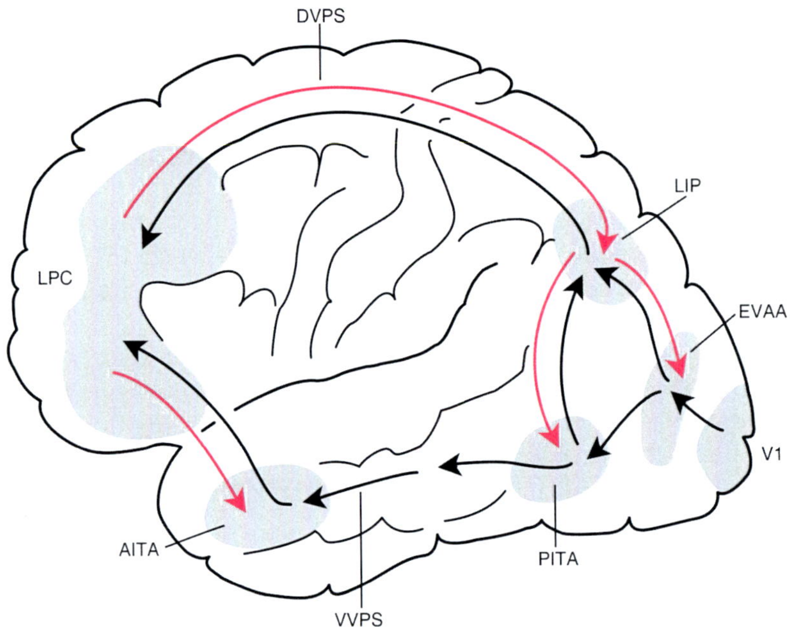

Fig. 15.67. Some centres involved in the processing of visual information. Their feed-forward connections (shown in *black*) originate from the primary visual cortex and proceed to the lateral prefrontal cortex, via the dorsal and ventral visual processing streams. *Red arrows* indicate potential feedback connections, forming part of the visual attentional control system. Based on Pessoa et al. [552, 553]. *AITA*, anterior inferior temporal area; *DVPS*, dorsal visual processing stream; *EVAA*, extrastriate visual association area; *LIP*, lateral intraparietal area; *LPC*, lateral prefrontal cortex; *PITA*, posterior inferior temporal area; *V1*, primary visual cortex; *VVPS*, ventral visual processing stream

of the *associative or cognitive loop*, forming part of the *direct striatal circuit* (for details, see Chap. 14). This loop is composed of: (1) corticostriate fibres which project to the caudate nucleus, (2) striopallidal fibres passing to the internal segment of the globus pallidus, (3) pallidothalamic fibres terminating in VA and VLa, and (4) the thalamocortical fibres already mentioned, which pass back to the same cortical region from which the loop originated (Figs. 14.1, 14.8, 14.17).

Classically, the striatum has been regarded as a part of the motor system, particularly because disturbances in its functions can lead to severe movement disorders. However, it has gradually become clear that the striatum also participates in cognitive functions. Lesions or electrical stimulation of the caudate nucleus result in cognitive deficits, as revealed by delayed response and delayed alternation tasks [839]. Neuroimaging studies indicate that the striatum is involved in procedural learning and memory processes [524], and patients with basal ganglia disorders exhibit impairments in the planning and execution of constructional tasks [399] and in the switching from one behavioural strategy to another [110, 111].

Just like the striatum, the cerebellum is not only concerned with motor control, but also participates in cognitive functions. The prefrontal cortex and the contralateral cerebellar cortex are interconnected by a complex corticocerebellar circuit, comprising a feed-forward or afferent limb and a feed-back or efferent limb. Topographically organized corticopontine and pontocerebellar projections form the feed-forward limb, whereas the feed-back limb is composed of cerebellar corticonuclear projections, efferents from the dentate nucleus to the posterior part of the ventral lateral thalamic nucleus (VLp), and efferents from the latter to the prefrontal cortex (Fig. 14.8). These connections are considered to represent the morphological substrate for the cerebellar control of cognitive processing. For a survey of the evidence indicating the participation of the cerebellum in cognitive functions, the reader is referred to the section "A brief excursion to the cerebellum" in Chap. 14. It has recently been suggested [611] that the prefronto-cerebellar system may facilitate the skilled execution of routine cognitive operations.

The afferents to the LPFC from extrathalamic subcortical centres do not differ from those to other neocortical regions. The structure and overall functions of these afferent systems have been discussed in the section "Neocortical afferents" of the present chapter. Let it suffice to recall that the neurons in several of these subcortical centres are particularly responsive to novel and motivationally relevant sensory stimuli and that their efferents modulate the excitability of the cortical neurons on which they impinge.

The LPFC exerts its control on motor behaviour by cascades of short projections, which reach the caudal premotor areas (F2–F5) and the primary motor cortex (M1, F1), after synaptic interruption in the rostral premotor areas (F6, F7) [419, 433, 445, 475, 578, 580]. The caudal premotor and primary motor areas, which send direct projections to the spinal cord, are closely involved in movement execution, whereas the rostral premotor areas process more "cognitive" aspects of motor control, such as sequence generation and motor learning [269]. It has already been mentioned that the LPFC has a substantial direct projection to the cingulate motor areas.

Based on fMRI evidence, several recent studies [238, 378, 379, 612] suggest that the human *frontal polar cortex* is specifically concerned with complex cognitive functions. Koechlin et al. [378] found that this cortical region is selectively activated when subjects held in mind a main goal while performing concurrent (sub)goals. Ramnani and Owen [612] concluded that the frontal polar cortex has a specific role in integrating the outcomes of two or more separate cognitive operations in the pursuit of a higher behavioural goal.

The complex of symptoms shown by patients with damage to the LPFC has been termed the *dysexecutive syndrome* [27]. Patients with this syndrome are characterized by diminished insight and judgement, poor planning and decision making and impairments of attention, working memory and the temporal organization of recent events [216, 483].

The *orbital prefrontal cortex (OPFC)* has been subdivided in different ways. Brodmann [70, 71], who did not study this region in detail, subdivided it into two areas: 11 and 47 (Fig. 15.11 A). Uylings et al. [751], who recently subjected the OPFC to a thorough cytoarchitectonic analysis, took Brodmann's parcellation as their point of departure. They confirmed the presence of Brodmann's areas 11 and 47, but subdivided the latter into three medial and two lateral subareas (Fig. 15.23). At present, the subdivisions of the LPFC, presented by Petrides and Pandya [580, 581] and Price and collaborators [86, 516, 518], which, as discussed before, are both derived from Walker's parcellation of the frontal cortex of the rhesus monkey [802], are most widely used. According to these subdivisions, the OPFC encompasses four areas, BWAs 11, 13, 14 and 47/12 (Fig. 15.11 E, G). It has already been mentioned that Price and colleagues [516, 518] subdivided each of these four areas into several subareas. There is an anterior-posterior trend in the cytoarchitecture of the OPFC [751], which manifests itself as a gradual reduction in the granularity of layer IV. Hence, the posterior portions of BWAs 13, 14 and 47/12 are either dysgranular or agranular. Mesulam [463] included this dysgranular/agranular region of the OPFC in what he called the paralimbic belt (Figs. 15.55, 15.56).

Apart from medially extending portions of BAs 9 and 10 and BWA 14, the *medial prefrontal cortex (MPFC)* includes Brodmann's anterior limbic area 24, prelimbic area 32 and infralimbic area 25 (Figs. 15.58 B, 15.62 C). BAs 24 and 25 are situated within the confines of the limbic lobe and form part of the paralimbic belt (Figs. 15.55–15.57).

Because the OPFC and MPFC share many connectional and functional features, they are commonly considered to form a single complex, the *orbital and medial prefrontal cortex (OMPFC)*.

The detailed experimental hodological studies of Price and collaborators [89, 516, 595] in the rhesus monkey have shown that the cytoarchitectonic areas of the OMPFC can be allocated to two groups or networks, one restricted to the OPFC and the other confined to the MPFC. In Fig. 15.68 A–D, these findings are extrapolated to the human brain.

As already discussed in Chap. 11, the OPFC receives extrinsic sensory afferents from five different sources. Olfactory, gustatory, autonomic, visual and somatosensory projections have been traced to restricted and different areas in the caudal and lateral parts of this region [89, 90] (Fig. 15.68 D). The olfactory projections originate from the primary olfactory cortex (Fig. 11.7). The gustatory and autonomic afferents stem from different parts of the parvocellular section of the ventral posteromedial thalamic nucleus and reach the caudal OPFC via synaptic relays in the insular cortex (see the next section of the present chapter). The visual afferents originate from the inferior temporal cortex and the somatosensory projections arise from the first and second somatosensory cortex and from the parietal association cortex. Anatomical, physiological and neuroimaging data [88, 89, 121, 355, 386, 628–630] indicate that the orbital areas in receipt of these sensory inputs and the network in which they are embedded are involved in the analysis and integration of food-related sensations and play an important role in the control of feeding. The caudal sectors of the orbital network provide a basis for convergence from the various unimodal sensory areas onto multimodal areas in the centromedial region of the OPFC (Fig. 15.68 D).

Fig. 15.68 A–E. The orbital and medial prefrontal cortex (OMPFC). Cytoarchitectonic parcellation of the medial (**A**) and orbital prefrontal cortex (**B**), according to Öngür et al. [518]. **C** Intrinsic connections comprising the medial cortico-cortical network (in *black*), and four prominent output channels of the medial prefrontal cortex (in *red*), originating from: (*1*) the infralimbic cortex; (*2*) the prelimbic cortex, (*3*) the anterior cingulate cortex, and (*4*) the cingulate motor areas. Based on Carmichael and Price [89], Price et al. [595], An et al. [9], Öngür et al. [517] and Morecraft and Van Hoesen [482]. **D** Intrinsic connections comprising the orbital cortico-cortical network (in *black*) and unimodal sensory inputs to the orbital prefrontal cortex (in *red*). Based on Carmichael and Price [89] and Price et al. [595]. **E** Connections between the orbital and medial cortico-cortical networks. Based on Carmichael and Price [89]. *cc*, corpus callosum; *CMA*, cingulate motor area; *facial n*, facial nucleus; *hypoth*, hypothalamus; *PAG*, periaqueductal grey

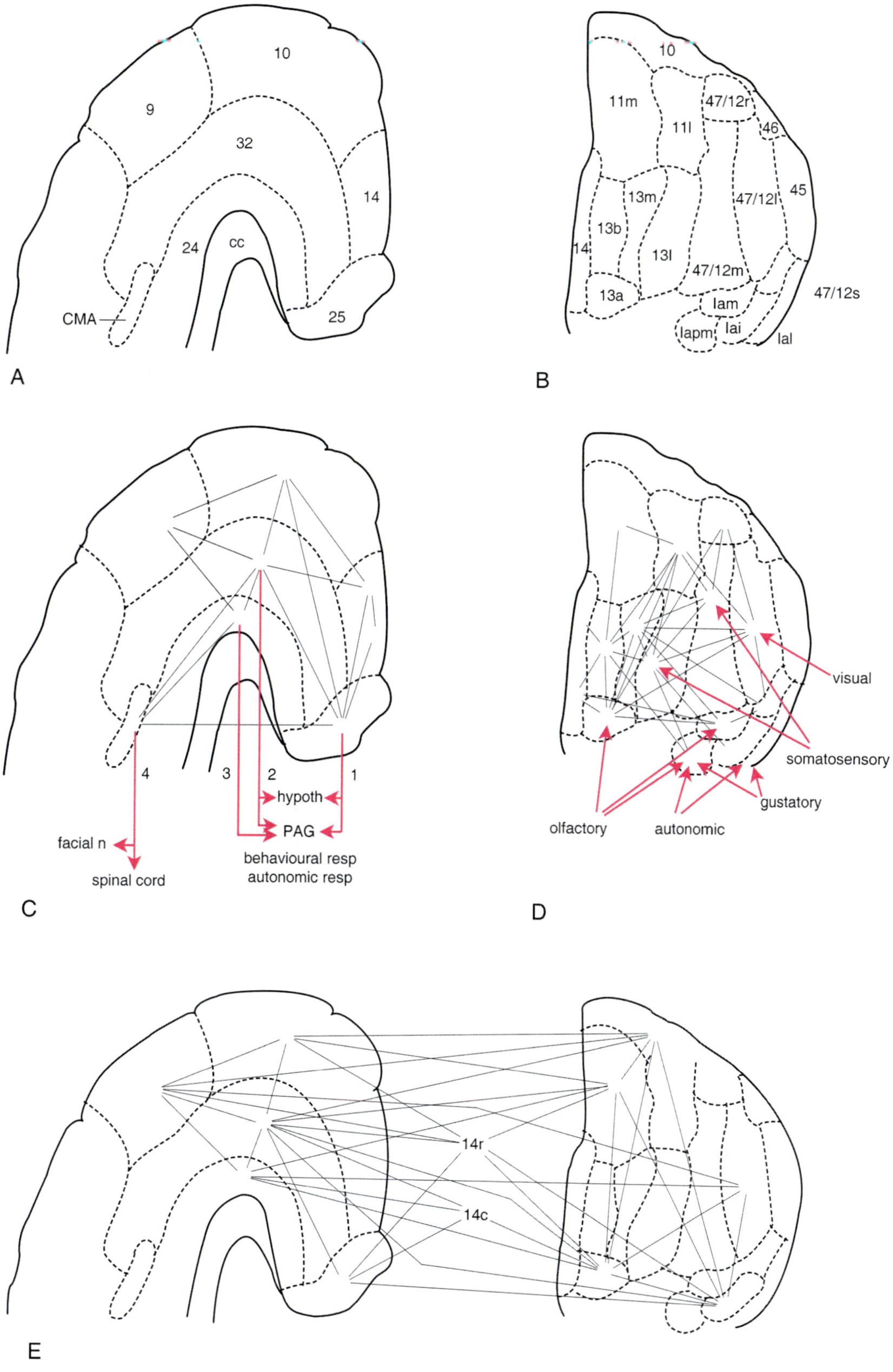

10
9
32
14
24
cc
CMA
25
A
10
11m
47/12r
11l
46
13m
47/12l
45
13b
14
13l
47/12m
47/12s
13a
Iam
Iapm
Iai
Ial
B
4
3
2
1
hypoth
PAG
facial n
behavioural resp
autonomic resp
spinal cord
C
visual
somatosensory
gustatory
olfactory
autonomic
D
14r
14c
E

Mention should be made here of a remarkable problem with regard to the exact localization of the olfactory zone in the OPFC. The orbitofrontal focus of olfactory activity in humans, as assessed by neuroimaging studies [251, 845], appeared to be situated considerably more anteriorly (Fig. 15.69 A) than that derived from an extrapolation of the experimental results of Carmichael et al. [90] in the rhesus monkey (Fig. 15.68 D).

It is important to note that the processing of sensory information in the OPFC is not confined to the physical and chemical properties of the stimuli but also involves their affective or emotional significance. Emotions can be classed as positive or negative. Positive emotions are elicited by rewards or positive reinforcers; negative emotions are evoked by punishments or negative reinforcers. A large meta-analysis of neuroimaging studies in humans [386] has shown that different subregions of the OPFC play different roles in the processing of emotional information. It appeared that activity in the medial OPFC is related to the monitoring, learning and memory of the reward value of positive reinforcers, whereas lateral OPFC activity is related to the evaluation of negative reinforcers ("punishers") that can lead to a change in behaviour (Fig. 15.69 B). This meta-analysis [386] also yielded evidence indicating that in the OPFC more complex or abstract reinforcers (such as monetary gain and loss) are represented more anteriorly than less complex reinforcers (such as taste).

The OMPFC projects densely to the hypothalamus and the mesencephalic periaqueductal grey (PAG) and moderately to lightly to many brainstem centres, including the ventral tegmental area, the mesencephalic raphe nuclei, the pedunculopontine nucleus, the locus coeruleus and the parabrachial nuclei. Almost all of these centres form part of the greater limbic system, discussed in Chap. 23. The fibres projecting to the hypothalamus arise from prelimbic area 32 and infralimbic area 25 on the medial surface of the frontal lobe, and from the orbital areas 13 a, 12/47 l and Iai [86, 517]. These three orbital areas are strongly connected to the medial network (Fig. 15.68 E). The medial wall areas 25 and 32 project to the anterior and ventromedial hypothalamic nuclei (AHN, VMH) in the medial hypothalamus. The orbital areas 13 a, 12/47 l and Iai selectively innervate the lateral hypothalamic area (LHA), particularly its posterior part [517]. The AHN and VMH are nodal points in networks subserving motivated or goal-oriented behaviours, i.e. behaviours essential in maintaining the individual and species, such as thermoregulation, agonistic behaviour and reproduction (see Chap. 10). The posterior LHA is involved in autonomic control.

The PAG can be subdivided into four longitudinal zones or columns: dorsal dorsolateral, lateral and ventrolateral [34, 91]. This structure is known to play a prominent role in co-ordinating behavioural and autonomic responses to escapable and unescapable stressful situations. Electrical or chemical stimulation of its lateral column evokes co-ordinated response strategies, such as threat display, fight or flight, accompanied by hypertension, tachycardia and a shift in blood from the viscera to the limb muscles. In contrast, stimulation of the ventrolateral PAG column elicits quiescence ("freezing") and decreasing of blood pressure and heart rate. Retrograde and anterograde tracing experiments in rhesus monkeys [9] have shown that the medial prefrontal areas 24, 25 and 32 and the orbital areas 13 a, 27/12 l and Iai project to the PAG and that projections from distinct cortical areas terminate primarily in individual PAG columns. The projections from areas 25 and 32 terminate primarily in the dorsolateral columns, bilaterally. Fibres from area 24 predominantly end in the lateral column, whereas fibres from orbital areas 13 a, 47/12 l and Iai terminate mainly in the ventrolateral columns. It seems likely that the OMPFC, by way of these projections, influences the integrated behavioural and autonomic responses mediated by the lateral and ventrolateral PAG columns. The functional significance of the cortical input to the dorsolateral column is not known. It is, however, important to note that this column receives a strong projection from the ventromedial hypothalamic nucleus [517], and that this nucleus is involved in both sexual and defensive behaviour (see Chap. 10).

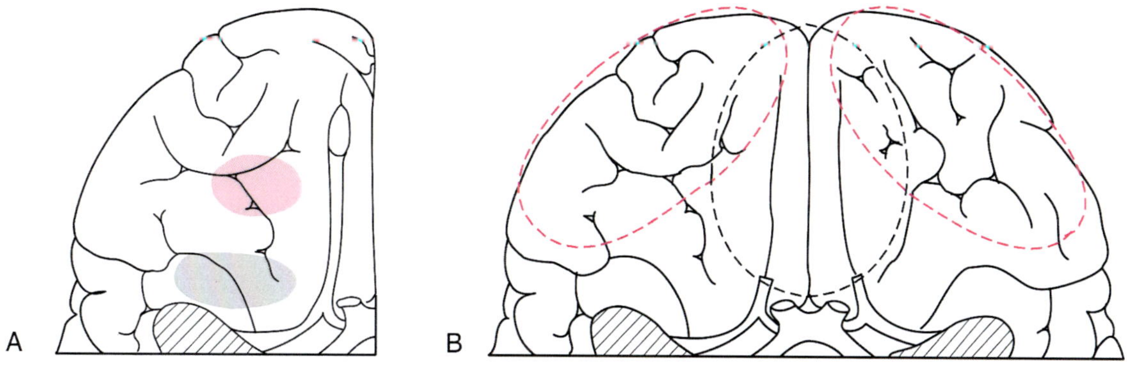

Fig. 15.69 A, B. The functions of the human orbitofrontal cortex. **A** The putative olfactory projection area (in *red*), as based on a meta-analysis of neuroimaging data, appears to be situated substantially rostrally to that extrapolated from experimental hodological studies in the rhesus monkey (in *grey*). Based on Gottfried and Zald [251]. **B** A large meta-analysis of neuroimaging data showed that there is a medial-lateral functional distinction in the orbitofrontal cortex, such that activity in the medial orbitofrontal cortex (*black dashed ellipse*) is related to the learning, monitoring and memory of the reward value of positive reinforcers, whereas lateral orbitofrontal cortex activity (*red dashed ellipses*) is related to the evaluation of negative reinforcers (punishers). Based on Kringelbach [386]. In the figures, temporopolar regions have been removed to allow visualization of the insula and caudal orbitofrontal cortex. The cut surfaces of the "temporal peduncles" are *hatched*

The projections from the OMPFC to the hypothalamus and PAG form part of the *limbic motor system*, which is also designated as the *emotional motor system* [302]. Fibres descending from the central nucleus of the amygdala and the bed nucleus of the stria terminalis also contribute to this system.

The complex formed by the strongly interconnected rostral and caudal cingulate motor areas M3 and M4, which is also embedded in the medial prefrontal network, projects via the corticobulbospinal tract to the facial nucleus and spinal cord. The complex is interconnected with other cortical motor areas, such as the primary motor cortex, the supplementary and presupplementary motor areas and the lateral premotor cortex [279, 481] and receives afferents from the lateral prefrontal areas 9 and 46 and from many limbic cortical areas, including posterior cingulate area 23, retrosplenial areas 29 and 30, temporal areas 35, TH and TF, temporopolar area 38 [482], as well as from the amygdala [483]. In light of these connectional data, Morecraft and Van Hoesen [482] suggested that the cingulate motor areas form a strategic cortical entry point for limbic influence on the voluntary motor system. They pointed out that patients with damage to the anterior cingulate cortex are often characterized as akinetic or docile, attention impaired, mute and lack emotional tone.

As indicated in Fig. 15.68 E, all medial prefrontal areas are connected to several orbital areas [89].

The data concerning the wiring of the OMPFC discussed above may be summarized as follows:

1. The cytoarchitectonic (sub)areas forming the medial and orbital prefrontal cortices are both embedded into a network of cortico-cortical connections.
2. The orbital network contains a "sensory pole" involved in the analysis and integration of food-related sensations.
3. The medial network contains a "motor pole", which potentially exerts influence on (a) subcortical networks subserving the execution of integrated motivated behaviours, (b) centres involved in autonomic control and (c) the voluntary motor system, via the cingulate motor areas.
4. The two networks are strongly interconnected. By way of these connections, sensory information screened for emotionally and motivationally relevant features is transferred to the medial network.

There is a certain resemblance between the limbic or emotional-motivational sensory-motor transfer system just outlined and the cognito-motor transfer system through the lateral prefrontal cortex (LPFC). In the latter system, highly processed information from sensory association areas in the parietal and temporal cortex converges onto the LPFC. The LPFC then produces voluntary motor actions based on this sensory input via sequences of short connections to premotor and motor areas (Fig. 15.15).

The OMPFC is connected to a large number of other brain structures, including the anterior temporal lobe, the amygdala and other limbic structure, and the basal ganglia. These connections will now be briefly discussed.

Experimental hodological studies in rhesus monkeys [36, 274, 381, 484, 576, 579, 581, 631, 672] have shown that the OMPFC is strongly interconnected with various parts of the anterior temporal lobe. Without going into details, it may be mentioned that many orbital areas receive afferents from the auditory association cortex in the superior temporal gyrus, the inferotemporal visual association cortex, and the multimodal sensory association cortex in the superior temporal sulcus (Fig. 15.54).

The amygdaloid complex projects to various parts of the OMPFC [90, 233, 484] (Figs. 15.70 A, 15.71). These amygdalocortical projections originate from the basal nucleus, and to a lesser extent from the accessory basal and lateral nuclei, and terminate principally in (a) the caudolateral entry zone for food-related information, (b) a medial orbital zone, which forms an interface between the orbital and medial networks, and (c) the medial prefrontal output areas 24 and 32 (Fig. 15.70 A). Most of these projections are reciprocated by projections from the cortex back to the amygdala.

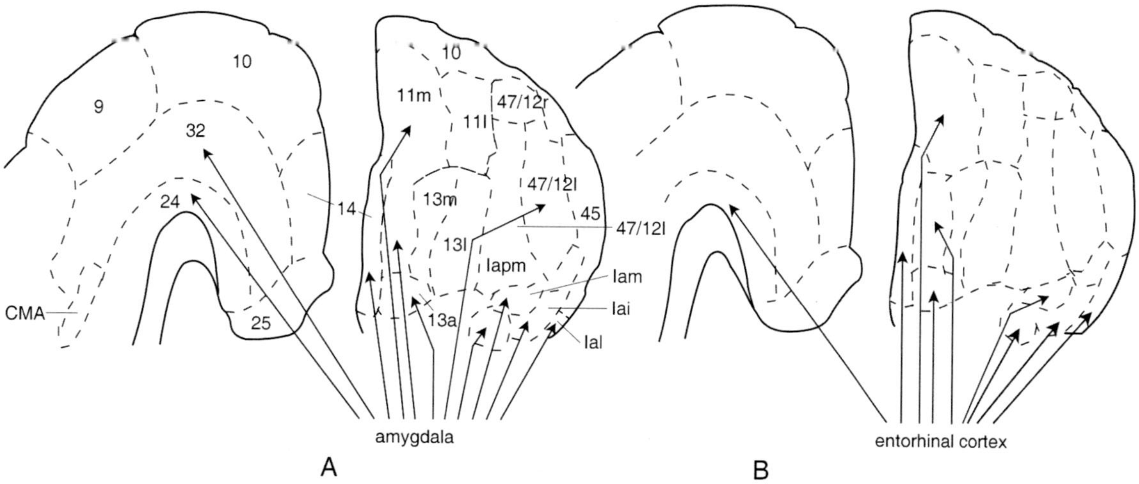

Fig. 15.70 A, B. Limbic inputs to the orbital and medial prefrontal cortex. A Afferents from the amygdaloid complex. B Afferents from the entorhinal cortex. Based on Carmichael and Price [87]

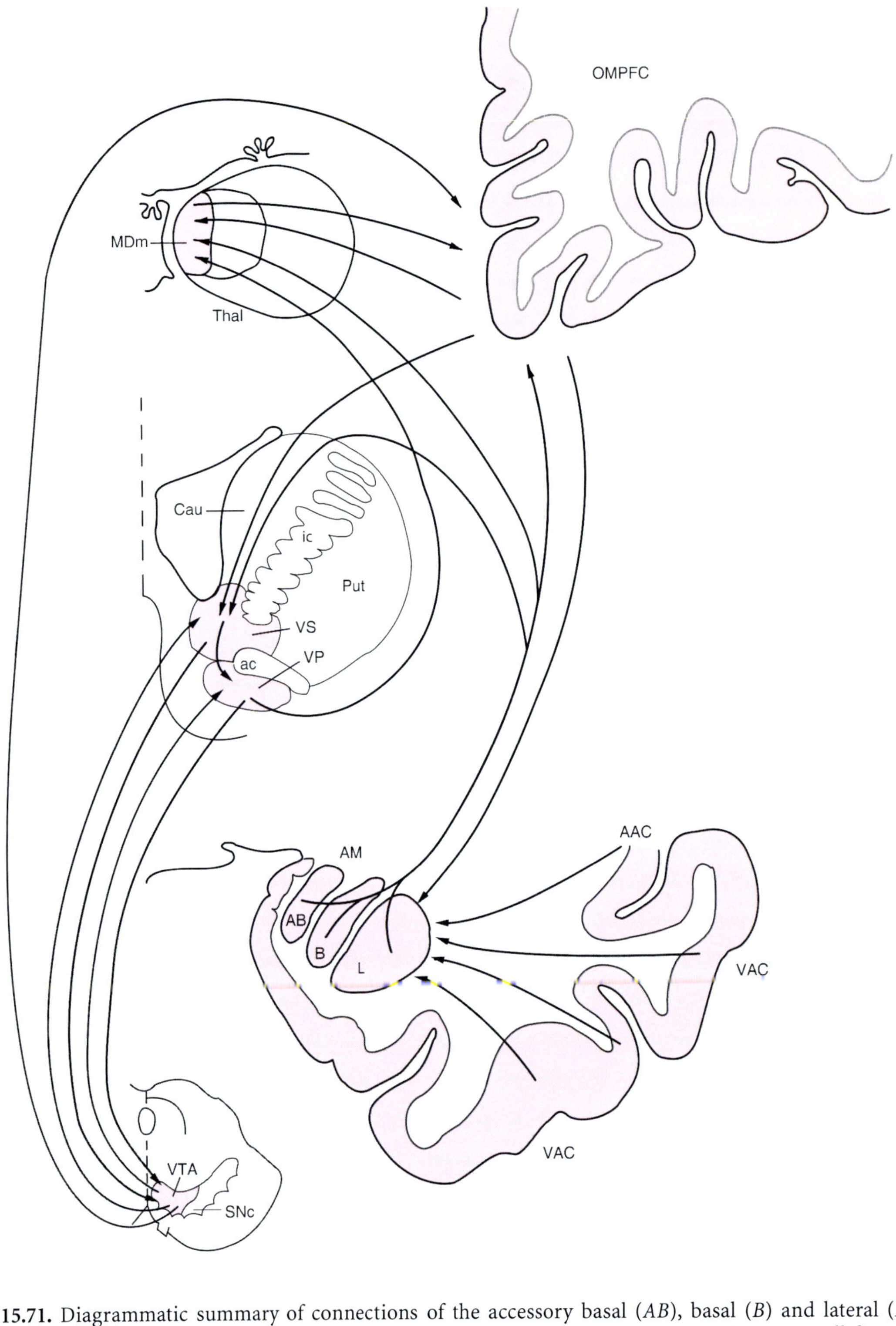

Fig. 15.71. Diagrammatic summary of connections of the accessory basal (*AB*), basal (*B*) and lateral (*L*) nuclei of the amygdaloid complex (*AM*), orbital and medial prefrontal cortex (*OMPFC*), magnocellular part of mediodorsal thalamic nucleus (*MDm*), ventral striatum (*VS*), ventral pallidum (*VP*) and mesencephalic ventral tegmental area (*VTA*) and adjacent pars compacta of substantia nigra (*SNc*). Partly based on Price et al. [595]. *AAC*, auditory association cortex; *ac*, anterior commissure; *Cau*, caudate nucleus; *ic*, internal capsule; *Put*, putamen; *Thal*, thalamus; *VAC*, visual association cortex

The basolateral nuclei of the amygdala are known to receive highly processed visual, auditory and somatosensory information and to be involved in evaluating the emotional and social significance of this information [41, 584] (see Chap. 13).

In addition to receiving inputs and sending outputs to the amygdala, the OMPFC is also connected with many other limbic structures, including the subiculum, entorhinal area 28, perirhinal areas 35 and 36, and parahippocampal areas TF and TH [87, 382, 484]. The subiculum, entorhinal and perirhinal areas are almost exclusively connected with the medial and caudal parts of the orbitofrontal cortex (Fig. 15.70 B). In contrast, the parahippocampal cortex is primarily connected with the medial prefrontal cortex, especially BAs 24, 25 and 32.

Apart from the direct projections from limbic structures to the OMPFC, there are also indirect projections, which are synaptically interrupted in the thalamus. The amygdala, entorhinal cortex and subiculum all project to the medial, magnocellular part of the mediodorsal thalamic nucleus (MDm), which in turn is reciprocally connected with all parts of the OMPFC [614].

As discussed in Chap. 14, the neocortex, basal ganglia and thalamus are interconnected by a number of parallel circuits or loops. One of these, designated as the *limbic loop*, involves the OMPFC and MDm. It originates from the medial BAs 24, 25 and 32 and the orbital BWAs 11, 13, 14 and 47/12. All of these areas project to the ventral striatum, composed of the accumbens nucleus and adjacent sectors of the caudate nucleus and putamen [270, 839]. The ventral striatum, in turn, projects to the ventral pallidum formed by rostral and ventral extensions of the globus pallidus. The ventral pallidum, then, projects to MDm, and connections between this thalamic nucleus and the OMPFC form the final link of the limbic loop (Figs. 14.17, 15.71).

Dopaminergic neurons in the mesencephalic ventral tegmental area (VTA) and adjacent parts of the substantia nigra, pars compacta (SNc) project to the OMPFC and ventral striatum. The ventral striatum projects back to the VTA and the latter is also reciprocally connected with the ventral pallidum [269] (Fig. 15.71).

There is physiological [14, 15, 61, 628, 666] and neuroimaging evidence [375, 664, 665] indicating that the limbic circuit, including its mesencephalic extension, is involved in the expectation and evaluation of reward and in the reward-guided selection of goal-oriented behaviours. Dopaminergic neurons in the VTA and adjacent SNc are known to play a key role in recognizing and predicting rewards [664, 665]. Dysregulation in the circuits interconnecting the orbitofrontal cortex, ventral striatum and dopaminergic midbrain are associated with several mental health disorders, including depression [448, 449], drug addiction [175, 357, 663, 792] and schizophrenia [284].

Lesions of the orbitofrontal cortex often induce dramatic changes in personality, as exemplified by the famous case of Phineas Gage [278], discussed in a previous section of the present chapter. Patients with such lesions are irritable, impulsive and disinhibited, with a characteristic tendency to tactlessness, vulgarity and disregard for social and moral principles. In addition, orbitofrontal patients exhibit a severe disorder of attention, and there are marked abnormalities in the realms of reasoning and decision-making [216, 483].

Medial prefrontal lesions commonly lead to apathy, manifesting as a severe reduction in spontaneity, motivation and general motility. Patients with such lesions also show a flattened affect and reduced interest in their environment. Large bilateral lesions may lead to akinetic mutism [216, 483].

Insula

The insula or island of Reil is buried in the depths of the lateral sulcus. It is covered by adjacent parts of the frontal, parietal and temporal lobe, known as the frontal, frontoparietal and temporal opercula (Figs. 2.23, 3.3). The insula is shaped like a triangle, the apex of which is directed anterobasally. The insular cortex is separated from surrounding opercular cortices by a limiting or circular sulcus. A central sulcus divides the insular surface into a larger anterosuperior and a smaller posteroinferior portion. The anterior part is covered by some short gyri (gyri breves); the posterior part is incompletely separated into two long gyri (gyri longi) (Fig. 3.3).

As regards the structure of the insula, Brodmann [70] indicated that its cortex can be divided into an agranular anterior part and a granular posterior part, and that the boundary between these areas is formed by a vertical line which partly coincides with the central insular sulcus (Fig. 15.8). In the insular cortex of the rhesus macaque and human, Mesulam and Mufson [466] recognized three concentrically arranged zones, a rostroventral agranular zone, an intermediate dysgranular zone and a caudodorsal granular zone (Fig. 15.72 A). The intermediate zone is termed dysgranular because the granule cells in layers II and IV are rather scarce and do not display complete laminar differentiation.

The fibre connections and functions of the insula have been extensively reviewed by Mesulam and Mufson [466] and Augustine [22]. Much of what follows has been derived from these studies.

The insula receives afferents from the dorsal thalamus and is connected with the amygdala and with a large number of cortical areas (Table 15.3). The thalamic nuclei which project to the insula include the ventromedial posterior nucleus (VMPo), the ventral posterior superior and ventral posterior inferior nuclei, as well as the parvocellular part of the ventral posteromedial nucleus.

The ventromedial posterior nucleus (VMPo), which is situated in the posterobasal part of the thalamus (Fig. 8.2) [57, 118], is in receipt of nociceptive and thermoreceptive spinothalamic and trigeminothalamic lamina I neurons [98, 117]. This nucleus sends a somatotopically organized projection to the posterosuperior part of the insular cortex [72, 118]. This projection area may be designated as the *insular nociceptive and thermoreceptive cortex INTC.*

The ventral posterior superior (VPS) and ventral posterior inferior (VPI) nuclei form part of a shell surrounding the ventral posterior thalamic complex. These nuclei receive afferents from the vestibular nuclear complex and project to several cortical areas, including an area in the posterosuperior insula and adjacent operculum known as the *parietoinsular vestibular cortex, PIVC* [79]. This vestibular cortical area overlaps with the INTC.

The most medial sector of the parvocellular part of the ventral posteromedial thalamic nucleus (VMpc,m) is in receipt of gustatory projections from the most rostral part of the nu-

Table 15.3. Connections of the insula

	Ia	Idg	Ig	aff	eff
Somatosensory					
SI (areas 3, 1, 2)			++	+	
Area 5			++	+	
Area 7b			++	+	+
VMPo (thal.)			++	+	
Vestibular					
VPS, VPI (thal.)			++	+	
Auditory					
Audit. assoc. ctx (STP, PA)		+	++	+	
Motor					
Medial premotor (area 6)		++	++		+
High-order association					
Ant. orbfr. ctx (area 11)		++	++	+	+
Prefr. ctx (areas 45, 46)			++	+	+
Banks sup. temp. sulcus		+	++	+	+
Olfactory					
Olfactory cortex	++	+		+	+
Gustatory					
VPMpc, med (thal.)			++	+	
General viscerosensory					
VPMpc, lat (thal.)		++	+	+	
Limbic					
Entorhinal ctx (area 28)	++	+		+	+
Perirhinal ctx (areas 35, 36)		+			+
Temporopolar ctx (area 38)	++	++		+	+
Post orbfr. ctx (areas 12, 13)	++	+		+	+
Cingulate gyrus (areas 23, 24)		++	++	+	+
Amygdala, corticomedial	++	++		+	+
Amygdala, basolateral	++	++		+	+

Based on [22, 79, 466, 596, 648]

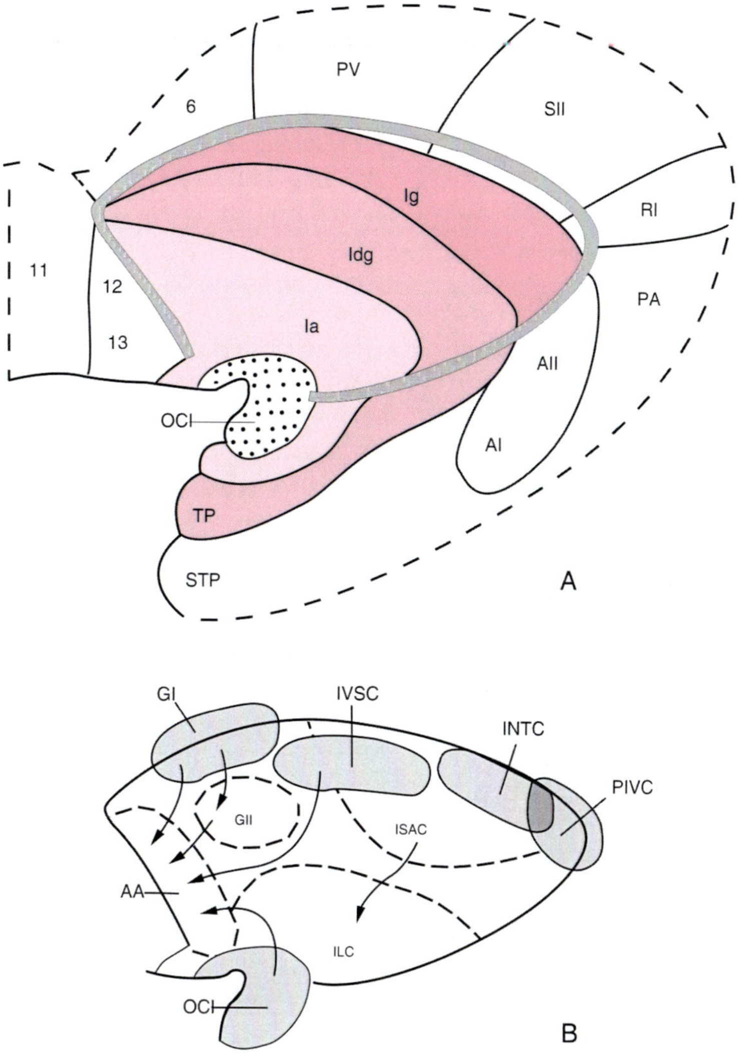

Fig. 15.72 A, B. The insula of the rhesus monkey. **A** Reconstruction of the insula and surrounding regions. The three types of insular cortex, agranular (*Ia*), dysgranular (*Idg*) and granular (*Ig*) are indicated with different hues of *red*. The *heavy grey line* represents the circular insular sulcus. Modified from Mesulam and Mufson [466]. **B** The approximate positions of functional areas in the insula and the course of some fibre connections. Primary sensory areas are *shaded*. *AA*, agranular anterior zone; *AI, AII*, primary and secondary auditory areas; *GI, II*, primary and secondary gustatory areas; *ILC*, insular limbic cortex; *INTC*, insular nociceptive and thermoreceptive cortex; *ISAC*, insular somatic association cortex; *IVSC*, insular viscerosensory cortex; *OCI*, primary olfactory cortex; *PA*, postauditory cortex; *PV*, parietal ventral area; *PIVC*, parietoinsular vestibular cortex; *RI*, retroinsular cortex; *SII*, secondary somatosensory area; *STP*, superior temporal plane; *TP*, temporopolar cortex

cleus of the solitary tract. This medial sector of the VPMpc projects to the granular anterosuperior part of the insula and the adjacent portion of the frontal operculum. This area, which represents the *primary gustatory cortex, GI,* projects to a more basally situated dysgranular and agranular insular area which, hence, may be designated as the *secondary gustatory cortex, GII* [596].

Via a synaptic relay in the external medial parabrachial nucleus, the intermediate and lateral sectors of VPMpc receive viscerosensory information from the intermediate and caudal parts of the nucleus of the solitary tract and project to an insular area located directly posterior to the gustatory areas. This area, which is known as the *insular viscerosensory cortex, IVSC,* shows an organotopic ordering. Physiological mapping experiments revealed that neurons responding to gastrointestinal sensations are located in its anterior part, whereas neurons responding to cardiovascular and respiratory afferents are located more posteriorly [648]. A similar topographical organization is observed in the postgustatory part of the nucleus of the solitary tract, where axons from the gastrointestinal tract end more rostrally than those from cardiovascular and respiratory structures. Hence, it seems likely that this topographical and viscerotopical ordering is maintained throughout the central viscerosensory pathway [648].

If we survey that data concerning the thalamic projections just discussed, it appears that the superior tier of the insula contains four anteroposteriorly arranged primary sensory areas: gustatory, viscerosensory, somatosensory (pain and temperature) and vestibular (Fig. 15.72 B).

The insula is also in receipt of primary and higher-order sensory projections from other cortical areas. Fibres from the primary olfactory (prepiriform and periamygdaloid) cortex project to an *agranular anterior zone (AA)* of the insula. This zone also receives afferents from the primary and secondary gustatory cortices and from the primary viscerosensory cortex (Fig. 15.72 B). As discussed in Chap. 11, it participates with the adjacent caudal orbitofrontal cortex in the formation of an "orbital network", functioning in the analysis and integration of food-related sensory information (Figs. 11.7, 11.8) [89].

The insula receives afferents from the primary somatosensory cortex SI, the somatosensory association areas 5 and 7b, the primary vestibular areas 3a and 2v (located in the basal part of the central sulcus and the anterior tip of the intraparietal sulcus, respectively) and auditory association areas situated in the anterior and posterior parts of the temporal plane. All of these sensory cortical areas project to the posterosuperior part of the insula, which may hence be characterized as the *insular somatic association cortex (ISAC).* Some high-order association areas, including the anterior orbitofrontal cortex (area 11), the prefrontal cortex (areas 45, 46) and the polymodal sensory association cortex occupying the banks of the superior temporal sulcus, are also known to be connected with the insula.

Finally, it may be mentioned that a considerable number of limbic cortical areas, including the entorhinal, perirhinal, temporopolar, posterior orbitofrontal and cingulate cortices, as well as the amygdaloid complex, are reciprocally connected with agranular and dysgranular sectors in the anterior and anterobasal parts of the insula. We designate these areas collectively as the *insular limbic cortex, ILC.* It has been suggested [466] that the insular somatic association cortex and the insular limbic cortex represent way stations in a somatolimbic projection (Fig. 15.72 B) and that this projection may provide a means for interrelating events in the extrapersonal world with relevant motivational states.

The insular cortex has been implicated in olfactory, gustatory, viscerosensory, visceromotor, somatosensory and vestibular functions. All of these functions, except for the visceromotor ones, can be readily associated with structures and connections discussed above, which are diagrammatically summarized in Fig. 15.72 B.

Olfaction. Electrical stimulation of the anterior insula in humans may lead to olfactory sensations [546].

Taste. The results of both stimulation and ablation experiments in monkeys point to the presence of a gustatory centre in the anterior insula [28, 714]. The detection of taste-sensitive neurons in the same area further documents the presence of this centre [838]. Verhagen et al. [777] recently reported that, apart from numerous taste-sensitive neurons, the primary gustatory cortex in the rhesus monkey also contains elements responding to non-taste properties of oral stimuli, related to the texture (viscosity, grittiness) or temperature of food. Neurons responding to combinations of these inputs also appeared to be present. These observations further substantiate the concept that the anterior insula and the adjacent caudal orbitofrontal cortex are involved in the analysis and integration of food-related information.

Viscerosensory Functions. Electrical stimulation of the insula in humans produces nausea and a variety of gastric and abdominal sensations [546, 547]. Neuroimaging studies in humans demonstrated activation of an insular area, just caudal to the gustatory cortex, with visceral stimuli such as air hunger, maximal inspiration and Valsalva manoeuvre [35, 367].

Visceromotor Control. Electrical stimulation of the insular region in humans may elicit a variety of visceromotor phenomena, including vomiting and other alterations of gastrointestinal motility, respiratory arrest, as well as changes in heart rate and blood pressure [521, 546]. Ischaemic strokes involving the insula are frequently accompanied by atrial fibrillation and electrocardiographic abnormalities [102, 470, 520, 732]. These findings indicate that the insula plays a role in autonomic regulation. It is known that the subgenual part of the cingulate gyrus (BA25), which is involved in similar functions, projects to several subcortical autonomic control centres, including the perifornical part of the hypothalamus, the periaqueductal grey and the parabrachial nuclei [9, 96, 517]. However, such connections have so far not been described for the insular cortex. It is remarkable that the cardiac dysfunctions mentioned occur much more frequently after right-sided strokes than after left-sided ones [102, 470]. This finding suggests that the autonomic control of cardiac activity is lateralized and is mediated by the right-sided insular cortex.

Somatosensory Functions. Physiological studies in rhesus monkeys [617, 618, 661] have shown that neurons in the granular part of the insula respond to innocuous cutaneous stimuli. A PET study [78] revealed that the human insula can be activated by vibrotactile stimulation. The involvement of the insula in protopathic sensibility is documented by clinical evidence. Thus, Biemond [52] described a patient in whom a lesion involving the insula and SII was associated with a dramatic loss of pain perception, and Birklein et al. [53] recently reported that isolated insular infarction may lead to contralateral elimination of cold, cold pain and pinprick perception.

Vestibular Functions. Neurons in the parieto-insular vestibular cortex (PIVC) of the rhesus monkey respond to vestibular stimuli [262]. However, most of these elements also responded to somatosensory and visual stimulation and were, hence, classified as polymodal vestibular units [263].

Finally, it may be mentioned that some recent imaging studies [725, 726] have shown that the volume of the insula is significantly reduced in schizophrenia.

References

1. Abeles M, Goldstein MH Jr (1970) Functional architecture in cat primary auditory cortex: columnar organization and organization according to depth. J Neurophysiol 33:172–187
2. Aboitiz F, Garcia VR (1997) The evolutionary origin of the language areas in the human brain. A neuroanatomical perspective. Brain Res Rev 25:381–396
3. Absher JR, Benson DF (1993) Disconnection syndromes: an overview of Geschwind's contributions. Neurology 43:862–867
4. Ahmed B, Anderson JC, Douglas RJ, Martin KAC, Nelson JC (1994) Polyneuronal innervation of spiny stellate neurons in cat visual cortex. J Comp Neurol 341:39–49
5. Allman JM, Kaas JH (1974) The organization of the second visual area (VII) in the owl monkey: a second order transformation of the visual hemifield. Brain Res 76:247–265
6. Amunts K, Schleicher A, Burgel U et al (1999) Broca's region revisited: cytoarchitecture and intersubject variability. J Comp Neurol 412:319–341

7. Amunts K, Malikovic A, Mohlberg H, Schormann T, Zilles K (2000) Brodmann's areas 17 and 18 brought into stereotaxic space - where and how variable? Neuroimage 11:66–84
8. Amunts K, Schleicher A, Ditterich A, Zilles K (2003) Broca's region: cytoarchitectonic asymmetry and developmental changes. J Comp Neurol 465:72–89
9. An X, Bandler R, Öngür D, Price JL (1998) Prefrontal cortical projections to longitudinal columns in the midbrain periaqueductal gray in macaque monkeys. J Comp Neurol 401:455–479
10. Andersen RA, Buneo CA (2002) Intentional maps in posterior parietal cortex. Annu Rev Neurosci 25:189–220
11. Andersen RA, Asanuma C, Essick G, Siegel RM (1990) Corticocortical connections of anatomically and physiologically defined subdivisions within the inferior parietal lobule. J Comp Neurol 296:65–113
12. Annese J, Pitiot A, Dinov ID, Toga AW (2004) A myelo-architectonic method for the structural classification of cortical areas. Neuroimage 21:15–26
13. Aoki C, Kabak S (1992) Cholinergic terminals in the cat visual cortex: ultrastructural basis for interaction with glutamate-immunoreactive neurons and other cells. Vis Neurosci 8:177–191
14. Apicella P, Ljungberg T, Scarnati E, Schultz W (1991) Responses to reward in monkey dorsal and ventral striatum. Exp Brain Res 85:491–500
15. Apicella P, Scarnati E, Ljungberg T, Schultz W (1992) Neuronal activity in monkey striatum related to the expectation of predictable environmental events. J Neurophysiol 68:945–960
16. Asanuma H (1975) Recent developments in the study of the columnar arrangement of neurons within the motor cortex. Physiol Rev 55:143–156
17. Asanuma H (1987) Cortical motor columns. In: Adelman G (ed) Encyclopedia of neuroscience, vol 1. Birkhäuser, Boston, pp 281–282
18. Asanuma H, Rosen I (1973) Spread of mono- and polysynaptic connections within cat's motor cortex. Exp Brain Res 16:507–520
19. Aston-Jones G, Cohen JD (2005) Adaptive gain and the role of the locus coeruleus-norepinephrine system in optimal performance. J Comp Neurol 493: 99–110
20. Aston-Jones G, Cohen JD (2005) An integrative theory of locus coeruleus-norepinephrine function: adaptive gain and optimal performance. Annu Rev Neurosci 28:403–450
21. Aston-Jones G, Chiang C, Alexinsky T (1991) Discharge of noradrenergic locus coeruleus neurons in behaving rats and monkeys suggests a role in vigilance. Prog Brain Res 88:501–520
22. Augustine JR (1996) Circuitry and functional aspects of the insular lobe in primates including humans. Brain Res Rev 22:229–244
23. Bacci A, Huguenard JR, Prince DA (2003) Functional autaptic neurotransmission in fast-spiking interneurons: a novel form of feedback inhibition in the neocortex. J Neurosci 23:859–866
24. Bacci A, Rudolph U, Huguenard JR, Prince DA (2003) Major differences in inhibitory synaptic transmission onto two neocortical interneuron subclasses. J Neurosci 23:9664–9674
25. Bacci A, Huguenard JR, Prince DA (2005) Modulation of neocortical interneurons: extrinsic influences and exercises in self-control. Trends Neurosci 28:602–610
26. Bachman DL, Albert ML (1991) The cerebral organization of language. In: Peters A, Jones EG (eds) Normal and altered states of function. Plenum, New York, pp 213–262 (Cerebral cortex, vol 9)
27. Baddeley AD (1986) Working memory. Oxford University Press, Oxford
28. Bagshaw MH, Pribram KH (1953) Cortical organization in gustation (Macaca mulatta). J Neurophysiol 16:499–508
29. Bailey P, von Bonin G (1950) The isocortex of the chimpanzee. University of Illinois Press, Urbana
30. Bailey P, von Bonin G (1951) The isocortex of man. University of Illinois Press, Urbana
31. Baillarger JGF (1853) De l'étendue de la surface du cerveau et de ses rapports avec le developpement de l'intelligence. Ann Med Psychol 5:1–9
32. Baizer JS, Ungerleider LG, Desimone R (1991) Organization of visual inputs to the inferior temporal and posterior parietal cortex in macaques. J Neurosci 11:168–190
33. Ballesteros Yáñez I, Muñoz A, Contreras J et al (2005) Double bouquet cell in the human cerebral cortex and a comparison with other mammals. J Comp Neurol 486:344–360
34. Bandler R, Keay KA (1996) Columnar organization in the midbrain periaqueductal gray and the integration of emotional expression. Progr Brain Res 107:285–300
35. Banzett RB, Mulnier HE, Murphy K et al (2000) Breathlessness in humans activates insular cortex. Neuroreport 11:2117–2120
36. Barbas H, Ghashghaei H, Dombrowski SM, Rempel-Clower NL (1999) Medial prefrontal cortices are unified by common connections with superior temporal cortices and distinguished by input from memory-related areas in the rhesus monkey. J Comp Neurol 410:343–367
37. Barnes CL, Pandya DN (1992) Efferent cortical connections of multimodal cortex of the superior temporal sulcus in the rhesus monkey. J Comp Neurol 318:222–244
38. Barrett NA, Large MM, Smith GL et al (2001) Human cortical processing of colour and pattern. Hum Brain Mapp 13:213–225
39. Bartels A, Zeki S (2004) The neural correlates of maternal and romantic love. Neuroimage 21:1155–1166
40. Basser PJ, Pajevic S, Pierpaoli C, Duda J, Aldroubi A (2000) In vivo fiber tractography using DT-MRI data. Magn Reson Med 44:625–632
41. Baxter MG, Murray EA (2002) The amygdala and reward. Nat Rev Neurosci 3:563–573

42. Baylis GC, Rolls ET, Leonard CM (1987) Functional subdivisions of the temporal lobe neocortex. J Neurosci 7:330–342
43. Beaulieu C, Somogyi P (1991) Enrichment of cholinergic synaptic terminals on GABAergic neurons and coexistence of immunoreactive GABA and choline acetyltransferase in the same synaptic terminals in the striate cortex of the cat. J Comp Neurol 304:666–680
44. Beck E (1949) A cytoarchitectural investigation into the boundaries of cortical areas 13 and 14 in the human brain. J Anat 83:147–157
45. Beierlein M, Gibson JR, Connors BW (2000) A network of electrically coupled interneurons drives synchronized inhibition in neocortex. Nat Neurosci 3:904–910
46. Benes FM, Berretta S (2001) GABAergic interneurons: implications for understanding schizophrenia and bipolar disorder. Neuropsychopharmacology 25:1–27
47. Benevento LA, Fallon J, Davis BJ, Rezak M (1977) Auditory-visual interaction in single cells in the cortex of the superior temporal sulcus and the orbital frontal cortex of the macaque monkey. Exp Neurol 57:849–872
48. Benson DF (1985) Aphasia and related disorders: a clinical approach. In: Mesulam M-M (ed) Principles of behavioral neurology. FA Davis Company, Philadelphia, pp 193–238
49. Berendse HW, Groenewegen HJ (1991) Restricted cortical termination fields of the midline and intralaminar thalamic nuclei in the rat. Neuroscience 42:73–102
50. Berman RA, Colby CL, Genovese CR et al (1999) Cortical networks subserving pursuit and saccadic eye movements in humans: an FMRI study. Hum Brain Mapp 8:209–225
51. Betz W (1874) Anatomischer Nachweis zweier Gehirnzentralen. Zentralbl Med Wiss 12:578–580
52. Biemond A (1956) The conduction of pain above the level of the thalamus opticus. Arch Neurol Psychiatry 75:221–231
53. Birklein F, Rolke R, Muller-Forell W (2005) Isolated insular infarction eliminates contralateral cold, cold pain, and pinprick perception. Neurology 65:1381
54. Blatow M, Rozov A, Katona I et al (2003) A novel network of multipolar bursting interneurons generates theta frequency oscillations in neocortex. Neuron 38:805–817
55. Blatt GJ, Rosene DL (1998) Organization of direct hippocampal efferent projections to the cerebral cortex of the rhesus monkey: projections from CA1, prosubiculum, and subiculum to the temporal lobe. J Comp Neurol 392:92–114
56. Blinkov SM, Glezer I (1968) Das Zentralnervensystem in Zahlen und Tabellen. Fischer, Jena
57. Blomqvist A, Zhang ET, Craig AD (2000) Cytoarchitectonic and immunohistochemical characterization of a specific pain and temperature relay, the posterior portion of the ventral medial nucleus, in the human thalamus. Brain 123:601–619
58. Bookheimer S (2002) Functional MRI of language: new approaches to understanding the cortical organization of semantic processing. Annu Rev Neurosci 25:151–188
59. Bouret S, Sara SJ (2005) Network reset: a simplified overarching theory of locus coeruleus noradrenaline function. Trends Neurosci 28:574–582
60. Bouvier SE, Engel SA (2006) Behavioral deficits and cortical damage loci in cerebral achromatopsia. Cereb Cortex 16:183–191
61. Bowman EM, Aigner TG, Richmond BJ (1996) Neural signals in the monkey ventral striatum related to motivation for juice and cocaine rewards. J Neurophysiol 75:1061–1073
62. Braak H (1976) A primitive gigantopyramidal field buried in the depth of the cingulate sulcus of the human brain. Brain Res 109:219–223
63. Braak H (1976) On the striate area of the human isocortex. A Golgi- and pigmentarchitectonic study. J Comp Neurol 166:341–364
64. Braak H (1980) Architectomics of the human telencephalic cortex. Springer, Berlin Heidelberg New York
65. Brain R (1961) Speech disorders. Aphasia, apraxia and agnosia. Butterworth, London
66. Brett M, Johnsrude IS, Owen AM (2002) The problem of functional localization in the human brain. Nat Rev Neurosci 3:243–249
67. Broca P (1865) Sur le siège de la faculté du langage articulé. Bull Soc d'Anthropol Paris 6:377–393
68. Broca P (1878) Anatomie comparée des circonvolutions cerebrales: le grand lobe limbique et la scissure dans la serie des mammiferes. Rev Anthropol (Paris) 2:285–498
69. Brodal A (1981) Neurological anatomy in relation to clinical medicine, 3rd edn. Oxford University Press, New York, IX, 1053 p
70. Brodmann K (1909) Vergleichende Lokalisationslehre der Großhirnrinde in ihren Prinzipien dargestellt auf Grund des Zellenbaues. Barth, Leipzig
71. Brodmann K (1914) Physiologie des Gehirns. In: Von Bruns P (ed) Neue Deutsche Chirurgie, Vol 11. Enke, Stuttgart, pp 85–426
72. Brooks JC, Zambreanu L, Godinez A, Craig AD, Tracey I (2005) Somatotopic organisation of the human insula to painful heat studied with high resolution functional imaging. Neuroimage 27:201–209
73. Brown-Séquard C-E (1877) Aphasia as an effect of brain-disease. Dublin J Med Sci 63:209–225
74. Bruce C, Desimone R, Gross CG (1981) Visual properties of neurons in a polysensory area in superior temporal sulcus of the macaque. J Neurophysiol 46:369–384
75. Brugge JF, Reale RA (1985) Auditory cortex. In: Peters A, Jones EG (eds) Association and auditory cortices. Plenum, New York, pp 229–271 (Cerebral cortex, vol 4)

76. Bugbee NM, Goldman-Rakic PS (1983) Columnar organization of corticocortical projections in squirrel and rhesus monkeys: similarity of column width in species differing in cortical volume. J Comp Neurol 220:355–364
77. Burnstine TH, Lesser RP, Hart J Jr et al (1990) Characterization of the basal temporal language area in patients with left temporal lobe epilepsy. Neurology 40:966–970
78. Burton H, Videen TO, Raichle ME (1993) Tactile-vibration-activated foci in insular and parietal-opercular cortex studied with positron emission tomography: mapping the second somatosensory area in humans. Somatosens Mot Res 10:297–308
79. Büttner-Ennever JA, Gerrits NM (2004) Vestibular system. In: Paxinos G, Mai JK (eds) The human nervous system, 2nd edn. Elsevier, Amsterdam, pp 1212–1240
80. Buxhoeveden DP, Casanova MF (2002) The minicolumn and evolution of the brain. Brain Behav Evol 60:125–151
81. Buxhoeveden DP, Casanova MF (2002) The minicolumn hypothesis in neuroscience. Brain 125:935–951
82. Buzsaki G, Draguhn A (2004) Neuronal oscillations in cortical networks. Science 304:1926–1929
83. Cabeza R, Nyberg L (2000) Imaging cognition II: An empirical review of 275 PET and fMRI studies. J Cogn Neurosci 12:1–47
84. Cajal SR (1972) Histologie du système nerveux de l'homme et des vertébrés. Consejo Superior de Investigaciones Cientificas. Tome II. Instituto Ramon y Cajal, Madrid
85. Campbell AW (1905) Histological studies on the localization of cerebral function. Cambridge University Press, Cambridge, MA
86. Carmichael ST, Price JL (1994) Architectonic subdivision of the orbital and medial prefrontal cortex in the macaque monkey. J Comp Neurol 346:366–402
87. Carmichael ST, Price JL (1995) Limbic connections of the orbital and medial prefrontal cortex in macaque monkeys. J Comp Neurol 363:615–641
88. Carmichael ST, Price JL (1995) Sensory and premotor connections of the orbital and medial prefrontal cortex of macaque monkeys. J Comp Neurol 363: 642–664
89. Carmichael ST, Price JL (1996) Connectional networks within the orbital and medial prefrontal cortex of macaque monkeys. J Comp Neurol 371:179–207
90. Carmichael ST, Clugnet MC, Price JL (1994) Central olfactory connections in the macaque monkey. J Comp Neurol 346:403–434
91. Carrive P, Morgan MM (2004) Periaqueductal gray. In: Paxinos G, Mai JK (eds) The human nervous system, 2nd edn. Elsevier, Amsterdam, pp 393–423
92. Catani M, ffytche DH (2005) The rises and falls of disconnection syndromes. Brain 128:2224–2239
93. Catsman-Berrevoets CE, Kuypers HGJM (1981) A search for corticospinal collaterals to thalamus and mesencephalon by means of multiple retrograde fluorescent tracers in cat and rat. Brain Res 218:15–33
94. Cavada C, Goldman-Rakic PS (1989) Posterior parietal cortex in rhesus monkey: I. Parcellation of areas based on distinctive limbic and sensory corticocortical connections. J Comp Neurol 287:393–421
95. Cavada C, Goldman-Rakic PS (1989) Posterior parietal cortex in rhesus monkey: II. Evidence for segregated corticocortical networks linking sensory and limbic areas with the frontal lobe. J Comp Neurol 287:422–445
96. Chiba T, Kayahara T, Nakano K (2001) Efferent projections of infralimbic and prelimbic areas of the medial prefrontal cortex in the Japanese monkey, Macaca fuscata. Brain Res 888:83–101
97. Choi HJ, Zilles K, Mohlberg H et al (2006) Cytoarchitectonic identification and probabilistic mapping of two distinct areas within the anterior ventral bank of the human intraparietal sulcus. J Comp Neurol 495:53–69
98. Christensen BN, Perl ER (1970) Spinal neurons specifically excited by noxious or thermal stimuli: marginal zone of the dorsal horn. J Neurophysiol 33:293–307
99. Chu Z, Galarreta M, Hestrin S (2003) Synaptic interactions of late-spiking neocortical neurons in layer 1. J Neurosci 23:96–102
100. Cipolloni PB, Peters A (1983) The termination of callosal fibres in the auditory cortex of the rat. A combined Golgi-electron microscope and degeneration study. J Neurocytol 12:713–726
101. Cobb SR, Halasy K, Vida I et al (1997) Synaptic effects of identified interneurons innervating both interneurons and pyramidal cells in the rat hippocampus. Neuroscience 79:629–648
102. Colivicchi F, Bassi A, Santini M, Caltagirone C (2004) Cardiac autonomic derangement and arrhythmias in right-sided stroke with insular involvement. Stroke 35:2094–2098
103. Colonnier ML (1966) The structural design of the neocortex. In: Eccles JC (ed) Brain and conscious experience. Springer, Berlin Heidelberg New York, pp 1–23
104. Colonnier ML (1981) The electron-microscopic analysis of the neuronal organization of the cerebral cortex. In: Adelman G, Dennis SG, Schmitt FO, Worden FG (eds) The organization of the cerebral cortex. MIT Press, Cambridge, pp 125–153
105. Condé F, Lund JS, Jacobowitz DM, Baimbridge KG, Lewis DA (1994) Local circuit neurons immunoreactive for calretinin, calbindin D-28k or parvalbumin in monkey prefrontal cortex: distribution and morphology. J Comp Neurol 341:95–116
106. Connor JR, Peters A (1984) Vasoactive intestinal polypeptide-immunoreactive neurons in rat visual cortex. Neuroscience 12:1027–1044
107. Connors BW, Long MA (2004) Electrical synapses in the mammalian brain. Annu Rev Neurosci 27:393–418

108. Conturo TE, Lori NF, Cull TS et al (1999) Tracking neuronal fiber pathways in the living human brain. Proc Natl Acad Sci USA 96:10422–10427
109. Cooke A, Zurif EB, DeVita C et al (2002) Neural basis for sentence comprehension: grammatical and short-term memory components. Hum Brain Mapp 15:80–94
110. Cools AR, Jaspers R, Schwartz M et al (1984) Basal ganglia and switching motor programs. In: McKenzie JS, Kemm RE, Wilcock LN (eds) Basal ganglia structure and function. Plenum, New York, pp 513–544
111. Cools AR, Van den Bercken JHL, Horstink MWI, Van Spaendonck KMP, Berger HJC (1984) Cognitive and motor shifting aptitude in Parkinson's disease. J Neurol Neurosurg Psychiatry 47:443–453
112. Corballis MC (2003) From mouth to hand: gesture, speech, and the evolution of right-handedness. Behav Brain Sci 26:199–208
113. Corbetta M, Miezin FM, Dobmeyer S, Shulman GL, Petersen SE (1991) Selective and divided attention during visual discriminations of shape, color, and speed: functional anatomy by positron emission tomography. J Neurosci 11:2383–2402
114. Coren S, Porac C (1977) Fifty centuries of right-handedness: the historical record. Science 198: 631–632
115. Courchesne E, Pierce K (2005) Why the frontal cortex in autism might be talking only to itself: local over-connectivity but long-distance disconnection. Curr Opin Neurobiol 15:225–230
116. Courtney SM, Ungerleider LG, Keil K, Haxby JV (1996) Object and spatial visual working memory activate separate neural systems in human cortex. Cereb Cortex 6:39–49
117. Craig AD, Kniffki K-D (1985) Spinothalamic lumbosacral lamina I cells responsive to skin and muscle stimulation in the cat. J Physiol (Lond) 365:197–221
118. Craig AD, Bushnell MC, Zhang E-T, Blomqvist A (1994) A thalamic nucleus specific for pain and temperature sensation. Nature 372:770–773
119. Creutzfeldt OD (1983) Cortex cerebri. Springer, Berlin Heidelberg New York
120. Crick F, Jones E (1993) Backwardness of human neuroanatomy. Nature 361:109–110
121. Critchley HD, Rolls ET (1996) Hunger and satiety modify the responses of olfactory and visual neurons in the primate orbitofrontal cortex. J Neurophysiol 75:1673–1686
122. Crosby EC, Humphrey T, Lauer EW (1962) Correlative anatomy of the nervous system. MacMillan, New York
123. Crow TJ (1997) Schizophrenia as failure of hemispheric dominance for language. Trends Neurosci 20:339–343
124. Culham JC, Kanwisher NG (2001) Neuroimaging of cognitive functions in human parietal cortex. Curr Opin Neurobiol 11:157–163
125. Cusick CG (1997) The superior temporal polysensory region in monkeys. In: Rockland KS, Kaas JH, Peters A (eds) Extrastriate cortex in primates. Plenum, New York, pp 435–468 (Cerebral cortex, vol 12)
126. Cusick CG, Kaas JH (1986) Interhemispheric connections of cortical sensory and motor representations in primates. In: Lepore F, Ptito M, Jasper HH (eds) Two hemispheres - one brain. Functions of the corpus callosum. Liss, New York, pp 83–102
127. Cusick CG, Seltzer B, Cola M, Griggs E (1995) Chemoarchitectonics and corticocortical terminations within the superior temporal sulcus of the rhesus monkey: evidence for subdivisions of superior temporal polysensory cortex. J Comp Neurol 360:513–535
128. Czeiger D, White EL (1993) Synapses of extrinsic and intrinsic origin made by callosal projection neurons in mouse visual cortex. J Comp Neurol 330:502–513
129. Damasio AR (1992) Aphasia. N Engl J Med 326:531–539
130. Damasio H, Damasio AR (1980) The anatomical basis of conduction aphasia. Brain 103:337–350
131. Dancause N, Barbay S, Frost SB et al (2006) Ipsilateral connections of the ventral premotor cortex in a new world primate. J Comp Neurol 495:374–390
132. de Lafuente V, Romo R (2004) Language abilities of motor cortex. Neuron 41:178–180
133. de Lima AD, Morrison JH (1989) Ultrastructural analysis of somatostatin-immunoreactive neurons and synapses in the temporal and occipital cortex of the macaque monkey. J Comp Neurol 283:212–227
134. DeFelipe J (1997) Types of neurons, synaptic connections and chemical characteristics of cells immunoreactive for calbindin-D28K, parvalbumin and calretinin in the neocortex. J Chem Neuroanat 14:1–19
135. DeFelipe J (1999) Chandelier cells and epilepsy. Brain 122:1807–1822
136. DeFelipe J (2002) Cortical interneurons: from Cajal to 2001. Prog Brain Res 136:215–238
137. DeFelipe J, Farinas I (1992) The pyramidal neuron of the cerebral cortex: morphological and chemical characteristics of the synaptic inputs. Prog Neurobiol 39:563–607
138. DeFelipe J, Conley M, Jones EG (1986) Long-range focal collateralization of axons arising from corticocortical cells in monkey sensory-motor cortex. J Neurosci 6:3749–3766
139. DeFelipe J, Hendry SHC, Jones EG (1989) Synapses of double bouquet cells in monkey cerebral cortex visualized by calbindin immunoreactivity. Brain Res 503:49–54
140. DeFelipe J, Hendry SHC, Hashikawa T, Molinari M, Jones EG (1990) A microcolumnar structure of monkey cerebral cortex revealed by immunocytochemical studies of double bouquet cell axons. Neuroscience 37:655–673

141. DeFelipe J, Hendry SHC, Hashikawa T, Jones EG (1991) Synaptic relationships of serotonin-immunoreactive terminal baskets on GABA neurons in the cat auditory cortex. Cereb Cortex 1:117–133
142. DeFelipe J, Alonso-Nanclares L, Arellano JI (2002) Microstructure of the neocortex: comparative aspects. J Neurocytol 31:299–316
143. Déjerine JJ (1891) Sur un cas de cécité verbale avec agraphie, suivi d'autopsie. CR Hebdomadaire Séances Mém Soc Biol 43:187–201
144. Déjerine JJ (1892) Contribution à l'étude anatomopathologique et clinique des différentes variétés de cécité verbale. C R Hebdomadaire Séances Mém Soc Biol 44:61–90
145. Déjerine JJ (1895) Anatomie des centres nerveux. Rueff et Cie, Paris
146. del Rio MR, DeFelipe J (1995) A light and electron microscopic study of calbindin D-28k immunoreactive double bouquet cells in the human temporal cortex. Brain Res 690:133–140
147. Demb JB, Desmond JE, Wagner AD et al (1995) Semantic encoding and retrieval in the left inferior prefrontal cortex: a functional MRI study of task difficulty and process specificity. J Neurosci 15:5870–5878
148. Descarries L, Mechawar N (2000) Ultrastructural evidence for diffuse transmission by monoamine and acetylcholine neurons of the central nervous system. Prog Brain Res 125:27–47
149. Deschênes M, Labelle A, Landry P (1979) Morphological characterization of slow and fast pyramidal tract cells in the cat. Brain Res 178:251–274
150. Desimone R, Gross CG (1979) Visual areas in the temporal cortex of the macaque. Brain Res 178: 363–380
151. Devinsky O (1992) Behavioral neurology. Hodder and Staughton, London
152. Ding SL, Van Hoesen G, Rockland KS (2000) Inferior parietal lobule projections to the presubiculum and neighboring ventromedial temporal cortical areas. J Comp Neurol 425:510–530
153. Disbrow E, Roberts T, Krubitzer L (2000) Somatotopic organization of cortical fields in the lateral sulcus of Homo sapiens: evidence for SII and PV. J Comp Neurol 418:1–21
154. Disbrow E, Litinas E, Recanzone GH, Padberg J, Krubitzer L (2003) Cortical connections of the second somatosensory area and the parietal ventral area in macaque monkeys. J Comp Neurol 462:382–399
155. Distler C, Boussaoud D, Desimone R, Ungerleider LG (1993) Cortical connections of inferior temporal area TEO in macaque monkeys. J Comp Neurol 334:125–150
156. Dougherty RF, Ben Shachar M, Bammer R, Brewer AA, Wandell BA (2005) Functional organization of human occipital-callosal fiber tracts. Proc Natl Acad Sci USA 102:7350–7355
157. Douglas R, Martin KAC, Whitteridge D (1989) A canonical microcircuit for neocortex. Neural Comput 1:480–488
158. Dronkers NF (1996) A new brain region for coordinating speech articulation. Nature 384:159–161
159. Dronkers NF (2001) Language, neural basis of. In: Wilson RA, Keil FC (eds) The MIT encyclopedia of the cognitive sciences. MIT Press, Cambridge, pp 448–451
160. Drury HA, Van Essen DC, Anderson CH et al (1996) Computerized mappings of the cerebral cortex: a multiresolution flattening method and a surface-based coordinate system. J Cogn Neurosci 8:1–28
161. Drury HA, Van Essen DC, Corbetta M, Snyder AZ (1999) Surface-based analyses of the human cerebral cortex. In: Toga AW (ed) Brain warping. Academic Press, New York, pp 337–363
162. Dum RP, Strick PL (1991) The origin of corticospinal projections from the premotor areas in the frontal lobe. J Neurosci 11:667–689
163. Dum RP, Strick PL (1996) Spinal cord terminations of the medial wall motor areas in macaque monkeys. J Neurosci 16:6513–6525
164. Dum RP, Strick PL (2005) Frontal lobe inputs to the digit representations of the motor areas on the lateral surface of the hemisphere. J Neurosci 25:1375–1386
165. Duus P (1976) Neurologisch-topische Diagnostik. Thieme, Stuttgart
166. Eccles JC (1984) The cerebral neocortex: a theory of its operation. In: Jones EG, Peters A (eds) Functional properties of cortical cells. Plenum, New York, pp 1–36 (Cerebral cortex, vol 2)
167. Edmunds SM, Parnavelas JG (1982) Retzius-Cajal cells: an ultrastructural study in the developing visual cortex of the rat. J Neurocytol 11:427–446
168. Egner T, Hirsch J (2005) Cognitive control mechanisms resolve conflict through cortical amplification of task-relevant information. Nat Neurosci 8:1784–1790
169. Eickhoff SB, Schleicher A, Zilles K, Amunts K (2006) The human parietal operculum. I. Cytoarchitectonic mapping of subdivisions. Cereb Cortex 16:254–267
170. Eickhoff SB, Amunts K, Mohlberg H, Zilles K (2006) The human parietal operculum. II. Stereotaxic maps and correlation with functional imaging results. Cereb Cortex 16:268–279
171. Einstein G, Fitzpatrick D (1991) Distribution and morphology of area 17 neurons that project to the cat's extrastriate cortex. J Comp Neurol 303:132–149
172. Elhanany E, White EL (1990) Intrinsic circuitry: synapses involving the local axon collaterals of corticocortical projection neurons in the mouse primary somatosensory cortex. J Comp Neurol 291:43–54
173. Elias H, Schwartz D (1969) Surface areas of the cerebral cortex of mammals determined by stereological methods. Science 166:111–113
174. Elston GN, Gonzalez-Albo MC (2003) Parvalbumin-, calbindin-, and calretinin-immunoreactive neurons in the prefrontal cortex of the owl mon-

key (Aotus trivirgatus): a standardized quantitative comparison with sensory and motor areas. Brain Behav Evol 62:19–30
175. Everitt BJ, Robbins TW (2005) Neural systems of reinforcement for drug addiction: from actions to habits to compulsion. Nat Neurosci 8:1481–1489
176. Fairén A, Valverde F (1980) A specialized type of neuron in the visual cortex of cat: a Golgi and electron microscope study of chandelier cells. J Comp Neurol 194:761–779
177. Fairén A, DeFelipe J, Regidor J (1984) Nonpyramidal neurons: general account. In: Peters A, Jones EG (eds) Cellular components of the cerebral cortex. Plenum, New York, pp 201–254 (Cerebral cortex, vol 1)
178. Feldman ML (1984) Morphology of the neocortical pyramidal neuron. In: Peters A, Jones EG (eds) Cellular components of the cerebral cortex. Plenum, New York, pp 123–200 (Cerebral cortex, vol 1)
179. Feldman ML, Peters A (1978) The forms of nonpyramidal neurons in the visual cortex of the rat. J Comp Neurol 179:761–793
180. Felleman DJ, Van Essen DC (1991) Distributed hierarchical processing in the primate cerebral cortex. Cereb Cortex 1:1–47
181. Felleman DJ, Burkhalter A, Van Essen DC (1997) Cortical connections of areas V3 and VP of macaque monkey extrastriate visual cortex. J Comp Neurol 379:21–47
182. Ferrier D (1876) The functions of the brain. Smith, Elder & Comp., London
183. Ferrier D, Yeo G (1885) A record of experiments on the effects of lesions of different regions of the cerebral hemispheres. Phil Trans R Soc Lond B 175:479–564
184. Ferstl EC, von Cramon DY (2002) What does the frontomedian cortex contribute to language processing: coherence or theory of mind? Neuroimage 17:1599–1612
185. Fiez JA (1997) Phonology, semantics, and the role of the left inferior prefrontal cortex. Hum Brain Mapp 5:79–83
186. Filimonov IN (1932) Über die Variabilität der Grosshirn Rindenstruktur. Mitt. 2: Regio occipitalis beim erwachsenen Menschen. J Psychol Neurol 44:1–96
187. Finger S (1994) Origins of neuroscience. Oxford University Press, New York
188. Fink GR, Frackowiak RS, Pietrzyk U, Passingham RE (1997) Multiple nonprimary motor areas in the human cortex. J Neurophysiol 77:2164–2174
189. Fitzpatrick D, Itoh K, Diamond IT (1983) The laminar organization of the lateral geniculate body and the striate cortex in the squirrel monkey (Saimiri sciureus). J Neurosci 3:673–702
190. Fitzpatrick D, Lund JS, Blasdel GG (1985) Intrinsic connections of macaque striate cortex: afferent and efferent connections of lamina 4C. J Neurosci 5:3329–3349
191. Flechsig P (1896) Gehirn und Seele. Veit & Co., Leipzig
192. Flechsig P (1901) Developmental (myelogenetic) localisation of the cerebral cortex in the human subject. Lancet 2:1027–1029
193. Flechsig P (1920) Anatomie des menschlichen Gehirns und Rückenmarks auf myelogenetischer Grundlage. Thieme, Leipzig
194. Fleischhauer K (1974) On different patterns of dendritic bundling in the cerebral cortex of the cat. Z Anat Entwicklungsgesch 143:115–126
195. Fleischhauer K (1978) Cortical architectonics: the last 50 years and some problems of today. In: Brazier MAB, Petsche H (eds) Architectonics of the cerebral cortex. Raven, New York, pp 99–117
196. Fleischhauer K, Petsche H, Wittkowski W (1972) Vertical bundles of dendrites in the neocortex. Z Anat Entwicklungsgesch 136:213–223
197. Foerster O (1936) Motorische Felder und Bahnen. In: Bumke O, Foerster O (eds) Handbuch der Neurologie, Bd 6. Springer, Berlin, Heidelberg, pp 1–357
198. Foote SL (1987) Extrathalamic modulation of cortical function. Annu Rev Neurosci 10:67–95
199. Foote SL, Aston-Jones G, Bloom FE (1980) Impulse activity of locus coeruleus neurons in awake rats and monkeys is a function of sensory stimulation and arousal. Proc Natl Acad Sci USA 77:3033–3037
200. Frackowiak RS (1994) Functional mapping of verbal memory and language. Trends Neurosci 17:109–115
201. Frackowiak RS, Friston KJ (1994) Functional neuroanatomy of the human brain: positron emission tomography – a new neuroanatomical technique. J Anat 184:211–225
202. Freeman W, Watts JW (1942) Psychosurgery: intelligence, emotion and social behavior following prefrontal lobotomy for mental disorders. Thomas, Springfield
203. Freeman W, Watts JW (1950) Psychosurgery in the treatment of mental disorders and intractable pain, 2nd edn. Thomas, Springfield
204. Freese JL, Amaral DG (2005) The organization of projections from the amygdala to visual cortical areas TE and V1 in the macaque monkey. J Comp Neurol 486:295–317
205. Freese JL, Amaral DG (2006) Synaptic organization of projections from the amygdala to visual cortical areas TE and V1 in the macaque monkey. J Comp Neurol 496:655–667
206. Freund TF (2003) Interneuron Diversity series: rhythm and mood in perisomatic inhibition. Trends Neurosci 26:489–495
207. Freund TF, Martin KAC, Smith AD, Somogyi P (1983) Glutamate decarboxylase-immunoreactive terminals of Golgi-impregnated axoaxonic cells and of presumed basket cells in synaptic contact with pyramidal neurons of the cat's visual cortex. J Comp Neurol 221:263–278

208. Freund TF, Martin KAC, Soltesz I, Somogyi P, Whitteridge D (1989) Arborisation pattern and postsynaptic targets of physiologically identified thalamocortical afferents in striate cortex of the macaque monkey. J Comp Neurol 289:315–336
209. Friedman DP, Bachevalier J, Ungerleider LG, Mishkin M (1987) Widespread projections to layer 1 of primate cortex. Soc Neurosci Abstr 13:251
210. Fries W (1990) Pontine projection from striate and prestriate visual cortex in the macaque monkey: an anterograde study. Vis Neurosci 4:205–216
211. Friston K (2002) Beyond phrenology: what can neuroimaging tell us about distributed circuitry? Annu Rev Neurosci 25:221–250
212. Fritsch G, Hitzig E (1870) Über die elektrische Erregbarkeit des Grosshirns. Arch Anat Physiol Wiss Med 37:300–332
213. Frost DO, Caviness VS Jr (1980) Radial organization of thalamic projections to the neocortex in the mouse. J Comp Neurol 194:369–393
214. Fukuda T, Kosaka T, Singer W, Galuske RA (2006) Gap junctions among dendrites of cortical GABAergic neurons establish a dense and widespread intercolumnar network. J Neurosci 26:3434–3443
215. Fuster JM (1973) Unit activity in prefrontal cortex during delayed-response performance: neuronal correlates of transient memory. J Neurophysiol 36:61–78
216. Fuster JM (2001) The prefrontal cortex - an update: time is of the essence. Neuron 30:319–333
217. Fuster JM, Alexander GE (1971) Neuron activity related to short-term memory. Science 173:652–654
218. Gabbott PLA, Martin KAC, Whitteridge D (1987) Connections between pyramidal neurons in layer 5 of cat visual cortex (area 17). J Comp Neurol 259:364–381
219. Galaburda AM, Pandya DH (1983) The intrinsic architectonic and connectional organization of the superior temporal region of the rhesus monkey. J Comp Neurol 221:169–184
220. Galaburda AM, Sanides F (1980) Cytoarchitectonic organisation of the human auditory cortex. J Comp Neurol 190:597–610
221. Galaburda AM, LeMay M, Kemper TL, Geschwind N (1978) Right-left asymmetries in the brain. Science 199:852–856
222. Galarreta M, Hestrin S (1999) A network of fast-spiking cells in the neocortex connected by electrical synapses. Nature 402:72–75
223. Galarreta M, Hestrin S (2001) Electrical synapses between GABA-releasing interneurons. Nat Rev Neurosci 2:425–433
224. Gazzaniga MS (1989) Organization of the human brain. Science 245:947–952
225. Gazzaniga MS (2005) Forty-five years of split-brain research and still going strong. Nat Rev Neurosci 6:653–659
226. Gazzaniga MS, Hutsler JJ, Keil FC (2001) Hemispheric specialization. In: Wilson RA, Keil FC (eds) The MIT encyclopedia of the cognitive sciences. MIT Press, Cambridge, MA, pp 369–372
227. Geschwind N (1965) Disconnexion syndromes in animals and man. Brain 88:237–294, 585–644
228. Geschwind N (1970) The organization of language and the brain. Science 170:940–944
229. Geschwind N, Levitsky W (1968) Human brain: left-right asymmetries in temporal speech region. Science 161:186–187
230. Geyer S, Ledberg A, Schleicher A et al (1996) Two different areas within the primary motor cortex of man. Nature 382:805–807
231. Geyer S, Schleicher A, Zilles K (1999) Areas 3a, 3b, and 1 of human primary somatosensory cortex. Neuroimage 10:63–83
232. Geyer S, Schormann T, Mohlberg H, Zilles K (2000) Areas 3a, 3b, and 1 of human primary somatosensory cortex. Part 2. Spatial normalization to standard anatomical space. Neuroimage 11: 684–696
233. Ghashghaei HT, Barbas H (2002) Pathways for emotion: interactions of prefrontal and anterior temporal pathways in the amygdala of the rhesus monkey. Neuroscience 115:1261–1279
234. Ghosh S, Fyffe RE, Porter R (1988) Morphology of neurons in area 4 gamma of the cat's cortex studied with intracellular injection of HRP [corrected and issued with original paging in J Comp Neurol 1988 Nov 8; 277(2)]. J Comp Neurol 277:290–312
235. Gibson JR, Beierlein M, Connors BW (1999) Two networks of electrically coupled inhibitory neurons in neocortex. Nature 402:75–79
236. Gil-da-Costa R, Martin A, Lopes MA et al (2006) Species-specific calls activate homologs of Broca's and Wernicke's areas in the macaque. Nat Neurosci 9:1064–1070
237. Gilbert CD (1992) Horizontal integration and cortical dynamics. Neuron 9:1–13
238. Gilbert SJ, Spengler S, Simons JS, Frith CD, Burgess PW (2006) Differential functions of lateral and medial rostral prefrontal cortex (area 10) revealed by brain-behavior associations. Cereb Cortex 16:1783–1789
239. Giovannini MG, Bartolini L, Kopf SR, Pepeu G (1998) Acetylcholine release from the frontal cortex during exploratory activity. Brain Res 784: 218–227
240. Globus A, Scheibel AB (1967) Pattern and field in cortical structure: the rabbit. J Comp Neurol 131:155–172
241. Goebel R, Muckli L, Kim D-S (2004) Visual system. In: Paxinos G, Mai JK (eds) The human nervous system, 2nd edn. Elsevier, Amsterdam, pp 1280–1305
242. Goldman PS, Nauta WJH (1977) Columnar distribution of cortico-cortical fibers in the frontal association, limbic and motor cortex of the developing rhesus monkey. Brain Res 122:393–413

243. Goldman-Rakic PS (1984) Modular organization of prefrontal cortex. Trends Neurosci 7:419–424
244. Goldman-Rakic PS (1995) Cellular basis of working memory. Neuron 14:477–485
245. Goldman-Rakic PS (1996) The prefrontal landscape: implications of functional architecture for understanding human mentation and the central executive. Philos Trans R Soc Lond B Biol Sci 351:1445–1453
246. Goldman-Rakic PS, Schwartz ML (1982) Interdigitation of contralateral and ipsilateral columnar projections to frontal association cortex in primates. Science 216:755–757
247. Goldman-Rakic PS, Selemon LD, Schwartz ML (1984) Dual pathways connecting the dorsolateral prefrontal cortex with the hippocampal formation and parahippocampal cortex in the rhesus monkey. Neuroscience 12:719–743
248. Goltz F (1881) Über die Verrichtungen des Grosshirns. Archiv für die gesamte Neurologie und Psychiatrie 26:1–49
249. Goltz F (1888) Über die Verrichtungen des Grosshirns. Pflügers Archiv für die gesamte Physiologie 42:419–467. In: Von Bonin G (1960) (ed) Some papers on the cerebral cortex. Translated as, "On the functions of the hemispheres". Thomas, Springfield, pp 118–158
250. Gonchar Y, Turney S, Price JL, Burkhalter A (2002) Axo-axonic synapses formed by somatostatin-expressing GABAergic neurons in rat and monkey visual cortex. J Comp Neurol 443:1–14
251. Gottfried JA, Zald DH (2005) On the scent of human olfactory orbitofrontal cortex: meta-analysis and comparison to non-human primates. Brain Res Rev 50:287–304
252. Graziano M (2006) The organization of behavioral repertoire in motor cortex. Annu Rev Neurosci 29:105–134
253. Grefkes C, Fink GR (2005) The functional organization of the intraparietal sulcus in humans and monkeys. J Anat 207:3–17
254. Gregoriou GG, Borra E, Matelli M, Luppino G (2006) Architectonic organization of the inferior parietal convexity of the macaque monkey. J Comp Neurol 496:422–451
255. Gross CG (1994) How inferior temporal cortex became a visual area. Cereb Cortex 4:455–469
256. Gross CG (1997) From Imhotep to Hubel and Wiesel; the story of visual cortex. In: Jones EG, Peters A (eds) Cerebral cortex, vol 12. Plenum, New York, pp 1–58
257. Gross CG (1999) History of localization of function in the cerebral cortex. In: Wilson RA, Keil FC (eds) The MIT encyclopedia of the cognitive sciences. MIT Press, Cambridge, pp 203–204
258. Gross CG, Bender DB, Rocha-Miranda CE (1969) Visual receptive fields of neurons in inferotemporal cortex of the monkey. Science 166:1303–1306
259. Grossman M, Koenig P, DeVita C et al (2002) Neural representation of verb meaning: an fMRI study. Hum Brain Mapp 15:124–134
260. Grossman M, Smith EE, Koenig P et al (2002) The neural basis for categorization in semantic memory. Neuroimage 17:1549–1561
261. Grünbaum ASF, Sherrington CS (1903) Observations on the physiology of the cerebral cortex of the higher apes. Proc R Soc Lond B Biol Sci 69:206–209
262. Grüsser OJ, Pause M, Schreiter U (1990) Localization and responses of neurones in the parieto-insular vestibular cortex of awake monkeys (Macaca fascicularis). J Physiol 430:537–557
263. Grüsser OJ, Pause M, Schreiter U (1990) Vestibular neurones in the parieto-insular cortex of monkeys (Macaca fascicularis): visual and neck receptor responses. J Physiol 430:559–583
264. Gu Q (2002) Neuromodulatory transmitter systems in the cortex and their role in cortical plasticity. Neuroscience 111:815–835
265. Guldin WO, Grüsser OJ (1998) Is there a vestibular cortex? Trends Neurosci 21:254–259
266. Gulyás B (1997) Functional organization of human visual cortical areas. In: Jones EG, Peters A (eds) Extrastriate cortex in primates. Plenum, New York, pp 743–775 (Cerebral cortex, vol 12)
267. Gulyás B, Roland PE (1994) Processing and analysis of form, colour and binocular disparity in the human brain: functional anatomy by positron emission tomography. Eur J Neurosci 6:1811–1828
268. Gusnard DA, Ollinger JM, Schulman GL et al (2003) Persistence and brain circuitry. Proc Natl Acad Sci USA 100:3479–3484
269. Haber SN, Johnson Gdowski M (2004) The basal ganglia. In: Paxinos G, Mai JK (eds) The human nervous system, 2nd edn. Elsevier, Amsterdam, pp 676–738
270. Haber SN, Kunishio K, Mizobuchi M, Lynd-Balta E (1995) The orbital and medial prefrontal circuit through the primate basal ganglia. J Neurosci 15:4851–4867
271. Habib M, Robichon F, Levrier O, Khalil R, Salamon G (1995) Diverging asymmetries of temporo-parietal cortical areas: a reappraisal of Geschwind/Galaburda theory. Brain Lang 48:238–258
272. Hackett TA, Stepniewska I, Kaas JH (1998) Subdivisions of auditory cortex and ipsilateral cortical connections of the parabelt auditory cortex in macaque monkeys. J Comp Neurol 394:475–495
273. Hackett TA, Stepniewska I, Kaas JH (1999) Callosal connections of the parabelt auditory cortex in macaque monkeys. Eur J Neurosci 11:856–866
274. Hackett TA, Stepniewska I, Kaas JH (1999) Prefrontal connections of the parabelt auditory cortex in macaque monkeys. Brain Res 817:45–58
275. Hadjikhani N, Liu AK, Dale AM, Cavanagh P, Tootell RB (1998) Retinotopy and color sensitivity in human visual cortical area V8. Nat Neurosci 1:235–241
276. Hagoort P (2005) On Broca, brain, and binding: a new framework. Trends Cogn Sci 9:416–423

277. Hamada I, Sakai M, Kubota K (1981) Morphological differences between fast and slow pyramidal tract neurons in the monkey motor cortex as revealed by intracellular injection of horseradish peroxidase by pressure. Neurosci Lett 22:233–238
278. Harlow JM (1868) Recovery from the passage of an iron bar through the head. Bull Massachusetts Med Soc 2:3–20
279. Hatanaka N, Tokuno H, Hamada I et al (2003) Thalamocortical and intracortical connections of monkey cingulate motor areas. J Comp Neurol 462:121–138
280. Hauk O, Johnsrude I, Pulvermuller F (2004) Somatotopic representation of action words in human motor and premotor cortex. Neuron 41:301–307
281. Hayes TL, Lewis DA (1993) Hemispheric differences in layer III pyramidal neurons of the anterior language area. Arch Neurol 50:501–505
282. He SQ, Dum RP, Strick PL (1993) Topographic organization of corticospinal projections from the frontal lobe: motor areas on the lateral surface of the hemisphere. J Neurosci 13:952–980
283. He SQ, Dum RP, Strick PL (1995) Topographic organization of corticospinal projections from the frontal lobe: motor areas on the medial surface of the hemisphere. J Neurosci 15:3284–3306
284. Heimer L (2000) Basal forebrain in the context of schizophrenia. Brain Res Rev 31:205–235
285. Hendry SHC, Jones EG (1981) Sizes and distributions of intrinsic neurons incorporating tritiated GABA in monkey sensory-motor cortex. J Neurosci 1:390–408
286. Hendry SHC, Jones EG (1983) The organization of pyramidal and non-pyramidal cell dendrites in relation to thalamic afferent terminations in the monkey somatic sensory cortex. J Neurocytol 12:277–298
287. Hendry SHC, Houser CR, Jones EG, Vaughn JE (1983) Synaptic organization of immunocytochemically identified GABA neurons in the monkey sensory-motor cortex. J Neurocytol 12:639–660
288. Henschen SE (1893) On the visual path and centre. Brain 16:170–180
289. Herbster AN, Mintun MA, Nebes RD, Becker JT (1997) Regional cerebral blood flow during word and nonword reading. Hum Brain Mapp 5:84–92
290. Herkenham M (1979) The afferent and efferent connections of the ventromedial thalamic nucleus in the rat. J Comp Neurol 183:487–517
291. Herkenham M (1980) Laminar organization of thalamic projections to the rat neocortex. Science 207:532–535
292. Herkenham M (1980) The laminar organization of thalamic projections to neocortex. Trends Neurosci 3:17–18
293. Herkenham M (1986) New perspectives on the organization and evolution of nonspecific thalamocortical projections. In: Jones EG, Peters A (eds) Sensory-motor areas and aspects of cortical connectivity. Plenum, New York, pp 403–446 (Cerebral cortex, vol 5)
294. Hersch SM, White EL (1981) Quantification of synapses formed with apical dendrites of Golgi-impregnated pyramidal cells: variability in thalamocortical inputs, but consistency in the ratios of asymmetrical to symmetrical synapses. Neuroscience 6:1043–1051
295. Hersch SM, White EL (1981) Thalamocortical synapses involving identified neurons in mouse primary somatosensory cortex: a terminal degeneration and golgi/EM study. J Comp Neurol 195:253–263
296. Hersch SM, White EL (1981) Thalamocortical synapses with corticothalamic projection neurons in mouse SmI cortex: electron microscopic demonstration of a monosynaptic feedback loop. Neurosci Lett 24:207–210
297. Hersch SM, White EL (1982) A quantitative study of the thalamocortical and other synapses in layer IV of pyramidal cells projecting from mouse SmI cortex to the caudate-putamen nucleus. J Comp Neurol 211:217–225
298. Herzog AG, Van Hoesen GW (1976) Temporal neocortical afferent connections to the amygdala in the rhesus monkey. Brain Res 15:57–69
299. Hikosaka K, Iwai E, Saito H, Tanaka K (1988) Polysensory properties of neurons in the anterior bank of the caudal superior temporal sulcus of the macaque monkey. J Neurophysiol 60:1615–1637
300. Hoen M, Pachot-Clouard M, Segebarth C, Dominey PF (2006) When Broca experiences the Janus syndrome: an ER-fMRI study comparing sentence comprehension and cognitive sequence processing. Cortex 42:605–623
301. Hof PR, Mufson EJ, Morrison JH (1995) Human orbitofrontal cortex: cytoarchitecture and quantitative immunohistochemical parcellation. J Comp Neurol 359:48–68
302. Holstege G, Mouton LJ, Gerrits NM (2004) Emotional motor system. In: Paxinos G, Mai JK (eds) The human nervous system, 2nd edn. Elsevier, Amsterdam, pp 1306–1324
303. Hornung JP, Garey LJ (1981) The thalamic projection to cat visual cortex: ultrastructure of neurons identified by Golgi impregnation or retrograde horseradish peroxidase transport. Neuroscience 6:1053–1068
304. Hornung JP, De Tribolet N, Tork I (1992) Morphology and distribution of neuropeptide-containing neurons in human cerebral cortex. Neuroscience 51:363–375
305. Horton JC, Adams DL (2005) The cortical column: a structure without a function. Philos Trans R Soc Lond B Biol Sci 360:837–862
306. Houser CR, Vaughn JE, Hendry SHC, Jones EG, Peters A (1984) GABA neurons in the cerebral cortex. In: Jones EG, Peters A (eds) Functional properties of cortical cells. Plenum, New York, pp 63–89 (Cerebral cortex, vol 2)

307. Houser CR, Crawford GD, Salvaterra PM, Vaughn JE (1985) Immunocytochemical localization of choline acetyltransferase in rat cerebral cortex: a study of cholinergic neurons and synapses. J Comp Neurol 234:17–34
308. Hubel DH, Wiesel TN (1962) Receptive fields, binocular interaction and functional architecture in the cat's visual cortex. J Physiol 160:106–154
309. Hubel DH, Wiesel TN (1965) Receptive fields and functional architecture in two non-striate visual areas (18 and 19) of the cat. J Neurophysiol 18:229–289
310. Hubel DH, Wiesel TN (1968) Receptive fields and functional architecture of monkey striate cortex. J Physiol 195:215–243
311. Hubel DH, Wiesel TN (1969) Anatomical demonstration of columns in the monkey striate cortex. Nature (London) 221:747–750
312. Hubel DH, Wiesel TN (1972) Laminar and columnar distribution of geniculo-cortical fibers in the macaque monkey. J Comp Neurol 146:421–443
313. Hubel DH, Wiesel TN (1977) Functional architecture of macaque monkey visual cortex. Proc R Soc Lond Biol 198:1–59
314. Hubel DH, Wiesel TN, Stryker MP (1978) Anatomical demonstration of orientation columns in macaque monkey. J Comp Neurol 177:361–380
315. Hübener M, Schwarz C, Bolz J (1990) Morphological types of projection neurons in layer 5 of cat visual cortex. J Comp Neurol 301:655–674
316. Huerta MF, Pons TP (1990) Primary motor cortex receives input from area 3a in macaques. Brain Res 537:367–371
317. Huk AC, Heeger DJ (2002) Pattern-motion responses in human visual cortex. Nat Neurosci 5:72–75
318. Hutsler J, Galuske RA (2003) Hemispheric asymmetries in cerebral cortical networks. Trends Neurosci 26:429–435
319. Imig TJ, Brugge JF (1978) Sources of termination of callosal axons related to binaural and frequency maps in primary auditory cortex of the cat. J Comp Neurol 183:637–660
320. Indefrey P, Levelt WJM (2000) The neural correlates of language production. In: Gazzaniga MS (ed) The new cognitive neurosciences. MIT Press, Cambridge, pp 845–865
321. Indefrey P, Levelt WJM (2004) The spatial and temporal signatures of word production components. Cognition 92:101–144
322. Innocenti GM (1986) General organization of callosal connections in the cerebral cortex. In: Jones EG, Peters A (eds) Sensory-motor areas and aspects of cortical connectivity. Plenum, New York, pp 291–354 (Cerebral cortex, vol 5)
323. Insausti R, Muñoz M (2001) Cortical projections of the non-entorhinal hippocampal formation in the cynomolgus monkey (Macaca fascicularis). Eur J Neurosci 14:435–451
324. Insausti R, Amaral DG, Cowan WM (1987) The entorhinal cortex of the monkey: II. Cortical afferents. J Comp Neurol 264:356–395
325. Isseroff A, Schwartz ML, Dekker JJ, Goldman-Rakic PS (1984) Columnar organization of callosal and associational projections from rat frontal cortex. Brain Res 293:213–223
326. Jackson JH (1931) Selected writings of John Hughlings Jackson, 2 vols. Hodder & Stoughton, London
327. Jacobsen CF (1935) Function of frontal association areas in primates. Arch Neurol Psychiatry 33:549–569
328. Jacobsen CF, Wolfe JB, Jackson TA (1935) An experimental analysis of the function of the frontal association areas in primates. J Nerv Ment Dis 52:1–14
329. Janzen R (1972) Die letzten 100 Jahre Lokalisationsforschung an der Grosshirnrinde. Z Neurol 202:75–79
330. Jasper HH (1995) A historical perspective: the rise and fall of prefrontal lobotomy. In: Jasper HH, Riggio S, Goldman-Rakic PS (eds) Epilepsy and the functional anatomy of the frontal lobe. Raven, New York, pp 97–114
331. Javoy-Agid F, Scatton B, Ruberg M et al (1989) Distribution of monoaminergic, cholinergic, and GABAergic markers in the human cerebral cortex. Neuroscience 29:251–259
332. Jenny AB (1979) Commissural projections of the cortical hand motor area in monkey. J Comp Neurol 188:137–146
333. Johnson PB, Ferraina S, Caminiti R (1993) Cortical networks for visual reaching. Exp Brain Res 97:361–365
334. Johnson PB, Ferraina S, Bianchi L, Caminiti R (1996) Cortical networks for visual reaching: physiological and anatomical organization of frontal and parietal lobe arm regions. Cereb Cortex 6:102–119
335. Jones DK, Simmons A, Williams SC, Horsfield MA (1999) Non-invasive assessment of axonal fiber connectivity in the human brain via diffusion tensor MRI. Magn Reson Med 42:37–41
336. Jones EG (1975) Varieties and distribution of non-pyramidal cells in the somatic sensory cortex of the squirrel monkey. J Comp Neurol 160:205–267
337. Jones EG (1984) Laminar distribution of cortical efferent cells. In: Peters A, Jones EG (eds) Cellular components of the cerebral cortex. Plenum, New York, pp 521–553 (Cerebral cortex, vol 1)
338. Jones EG (1984) Neurogliaform or spiderweb cells. In: Peters A, Jones EG (eds) Cellular components of the cerebral cortex. Plenum, New York, pp 409–418 (Cerebral cortex, vol 1)
339. Jones EG (1986) Connectivity of the primate sensory-motor cortex. In: Jones EG, Peters A (eds) Sensory-motor areas and aspects of cortical connectivity. Plenum, New York, pp 113–184 (Cerebral cortex, vol 5)

340. Jones EG (1987) Cerebral cortex. In: Adelman G (ed) Encyclopedia of neuroscience, vol 1. Birkhäuser, Boston, pp 209–211
341. Jones EG (2000) Microcolumns in the cerebral cortex. Proc Natl Acad Sci USA 97:5019–5021
342. Jones EG, Burton H (1976) Areal differences in the laminar distribution of thalamic afferents in cortical fields of the insular, parietal and temporal regions of primates. J Comp Neurol 168:197–248
343. Jones EG, Hendry SHC (1984) Basket cells. In: Peters A, Jones EG (eds) Cellular components of the cerebral cortex. Plenum, New York, pp 309–336 (Cerebral cortex, vol 1)
344. Jones EG, Leavitt RY (1974) Retrograde axonal transport and the demonstration of non-specific projections to the cerebral cortex and striatum from thalamic intralaminar nuclei in the rat, cat and monkey. J Comp Neurol 154:349–378
345. Jones EG, Powell TP (1969) Synapses on the axon hillocks and initial segments of pyramidal cell axons in the cerebral cortex. J Cell Sci 5:495–507
346. Jones EG, Powell TPS (1970) An anatomical study of converging sensory pathways within the cerebral cortex of the monkey. Brain 93:793–824
347. Jones EG, Burton H, Porter R (1975) Commissural and cortico-cortical "columns" in the somatic sensory cortex of primates. Science 190:572–574
348. Jones EG, Coulter JD, Hendry SHC (1978) Intracortical connectivity of architectonic fields in the somatic sensory, motor and parietal cortex of monkeys. J Comp Neurol 181:291–348
349. Jones EG, Coulter JD, Wise SP (1979) Commissural columns in the sensory motor cortex of monkeys. J Comp Neurol 188:113–136
350. Jones EG, Hendry SHC, DeFelipe J (1987) GABA-peptide neurons of the primate cerebral cortex: a limited cell class. In: Jones EG, Peters A (eds) Further aspects of cortical function. Raven, New York, pp 237–266 (Cerebral cortex, vol 6)
351. Jones EG, DeFelipe J, Hendry SHC, Maggio JE (1988) A study of tachykinin-immunoreactive neurons in monkey cerebral cortex. J Neurosci 8:1206–1224
352. Kaas JH (1997) Theories of visual cortex organization in primates. In: Jones EG, Peters A (eds) Extrastriate cortex in primates. Plenum, New York, pp 91–125 (Cerebral cortex, vol 12)
353. Kaas JH (2002) Neocortex. In: Ramachandran VS (ed) Encyclopedia of the human brain, vol 3. Academic Press, Amsterdam, pp 291–303
354. Kaas JH (2004) Somatosensory system. In: Paxinos G, Mai JK (eds) The human nervous system, 2nd edn. Elsevier, Amsterdam, pp 1059–1092
355. Kadohisa M, Rolls ET, Verhagen JV (2004) Orbitofrontal cortex: neuronal representation of oral temperature and capsaicin in addition to taste and texture. Neuroscience 127:207–221
356. Kakei S, Shinoda Y (1990) Parietal projection of thalamocortical fibers from the ventroanterior-ventrolateral complex of the cat thalamus. Neurosci Lett 117:280–284
357. Kalivas PW, Volkow N, Seamans J (2005) Unmanageable motivation in addiction: a pathology in prefrontal-accumbens glutamate transmission. Neuron 45:647–650
358. Kanai T (1938) Über die Furchen und Windungen der Orbitalfläche des Stirnhirns bei Japanern. Okajima's Fol anat japon 18:229–306
359. Kapur N, Ellison D, Smith MP, McLellan DL, Burrows EH (1992) Focal retrograde amnesia following bilateral temporal lobe pathology. A neuropsychological and magnetic resonance study. Brain 115:73–85
360. Kapur N, Ellison D, Parkin AJ et al (1994) Bilateral temporal lobe pathology with sparing of medial temporal lobe structures: lesion profile and pattern of memory disorder. Neuropsychologia 32:23–38
361. Karol EA, Pandya DN (1971) The distribution of the corpus callosum in the Rhesus monkey. Brain 94:471–486
362. Katsumaru H, Kosaka T, Heizmann CW, Hama K (1988) Gap junctions on GABAergic neurons containing the calcium-binding protein parvalbumin in the rat hippocampus (CA1 region). Exp Brain Res 72:363–370
363. Katz LC (1987) Local circuitry of identified projection neurons in cat visual cortex brain slices. J Neurosci 7:1223–1249
364. Kaufman EFS, Rosenquist AC (1985) Efferent projections of the thalamic intralaminar nuclei in the cat. Brain Res 335:257–279
365. Keller A, Asanuma H (1993) Synaptic relationships involving local axon collaterals of pyramidal neurons in the cat motor cortex. J Comp Neurol 336:229–242
366. Kertesz A, Lesk D, McCabe P (1977) Isotope localization of infarcts in aphasia. Arch Neurol 34:590–601
367. King AB, Menon RS, Hachinski V, Cechetto DF (1999) Human forebrain activation by visceral stimuli. J Comp Neurol 413:572–582
368. Kisvárday ZF, Eysel UT (1992) Cellular organization of reciprocal patchy networks in layer III of cat visual cortex (area 17). Neuroscience 46:275–286
369. Kisvárday ZF, Martin KA, Whitteridge D, Somogyi P (1985) Synaptic connections of intracellularly filled clutch cells: a type of small basket cell in the visual cortex of the cat. J Comp Neurol 241:111–137
370. Kisvárday ZF, Martin KA, Freund TF et al (1986) Synaptic targets of HRP-filled layer III pyramidal cells in the cat striate cortex. Exp Brain Res 64:541–552
371. Kisvárday ZF, Gulyas A, Beroukas D et al (1990) Synapses, axonal and dendritic patterns of GABA-immunoreactive neurons in human cerebral cortex. Brain 113:793–812

372. Kisvárday ZF, Beaulieu C, Eysel UT (1993) Network of GABAergic large basket cells in cat visual cortex (area 18): implication for lateral disinhibition. J Comp Neurol 327:398–415
373. Kleist K (1934) Gehirnpathologie. Johann Ambrosius Barth, Leipzig
374. Kleist K (1959) Die Lokalisation im Grosshirn und ihre Entwicklung. Psychiatr Neurol 137:289–309
375. Knutson B, Adams CM, Fong GW, Hommer D (2001) Anticipation of increasing monetary reward selectively recruits nucleus accumbens. J Neurosci 21:RC159(1–5)
376. Kobayashi Y, Amaral DG (2000) Macaque monkey retrosplenial cortex: I. three-dimensional and cytoarchitectonic organization. J Comp Neurol 426:339–365
377. Kobayashi Y, Amaral DG (2003) Macaque monkey retrosplenial cortex: II. Cortical afferents. J Comp Neurol 466:48–79
378. Koechlin E, Basso G, Pietrini P, Panzer S, Grafman J (1999) The role of the anterior prefrontal cortex in human cognition. Nature 399:148–151
379. Koechlin E, Ody C, Kouneiher F (2003) The architecture of cognitive control in the human prefrontal cortex. Science 302:1181–1185
380. Köhler C, Shipley MT, Srebro B, Harkmark W (1978) Some retrohippocampal afferents to the entorhinal cortex. Cells of origin as studied by the HRP method in the rat and mouse. Neurosci Lett 10:115–120
381. Kondo H, Saleem KS, Price JL (2003) Differential connections of the temporal pole with the orbital and medial prefrontal networks in macaque monkeys. J Comp Neurol 465:499–523
382. Kondo H, Saleem KS, Price JL (2005) Differential connections of the perirhinal and parahippocampal cortex with the orbital and medial prefrontal networks in macaque monkeys. J Comp Neurol 493:479–509
383. Kononova EP (1935) [The variability of structure of the cortex of the brain: inferior frontal gyrus of adult man] in Russian. Brain Res Inst Publ 1:49–118
384. Konopaske GT, Sweet RA, Wu Q, Sampson A, Lewis DA (2006) Regional specificity of chandelier neuron axon terminal alterations in schizophrenia. Neuroscience 138:189–196
385. Kreisler A, Godefroy O, Delmaire C et al (2000) The anatomy of aphasia revisited. Neurology 54: 1117–1123
386. Kringelbach ML (2005) The human orbitofrontal cortex: linking reward to hedonic experience. Nat Rev Neurosci 6:691–702
387. Krmpotic-Nemanic J, Kostovic I, Nemanic D (1984) Prenatal and perinatal development of radial cell columns in the human auditory cortex. Acta Otolaryngol 97:489–495
388. Krubitzer L, Huffman KJ, Disbrow E, Recanzone G (2004) Organization of area 3a in macaque monkeys: contributions to the cortical phenotype. J Comp Neurol 471:97–111
389. Künzle H (1978) Cortico-cortical efferents of primary motor and somatosensory regions of the cerebral cortex in Macaca fascicularis. Neuroscience 3:25–39
390. Kuypers HGJM (1987) Pyramidal tract. In: Adelman G (ed) Encyclopedia of neuroscience, vol 2. Birkhäuser, Boston, pp 1018–1020
391. Lalouschek W (1997) Brain and language: beyond the left cerebral hemisphere. Contributions of subcortical structures, cerebellum, and right hemisphere to language. Evolution and Cognition 3:133–147
392. Lashley KS (1929) Brain mechanisms and intelligence. University of Chicago Press, Chicago
393. Lashley KS (1933) Integrative functions of the cerebral cortex. Physiol Rev 13:1–42
394. Lashley KS, Clark G (1946) The cytoarchitecture of the cerebral cortex of Ateles: A critical examination of architectonic studies. J Comp Neurol 85:223–305
395. Lassek AM (1940) The human pyramidal tract. II. A numerical investigation of the Betz cells of the motor area. Arch Neurol Psychiatry 44:718–724
396. Laurienti PJ, Wallace MT, Maldjian JA et al (2003) Cross-modal sensory processing in the anterior cingulate and medial prefrontal cortices. Hum Brain Mapp 19:213–223
397. Lavenex P, Suzuki WA, Amaral DG (2002) Perirhinal and parahippocampal cortices of the macaque monkey: projections to the neocortex. J Comp Neurol 447:394–420
398. Lee Seldon H (1985) The anatomy of speech perception. Human auditory cortex. In: Peters A, Jones EO (eds) Cerebral cortex, vol 4. Association and auditory cortices. Plenum, New York, pp 273–327
399. Lees AJ, Smith E (1983) Cognitive deficits in the early stages of Parkinson's disease. Brain 106:257–270
400. Leichnetz GR (2001) Connections of the medial posterior parietal cortex (area 7m) in the monkey. Anat Rec 263:215–236
401. Leonard CM (1969) The prefrontal cortex of the rat. I. Cortical projections of the mediodorsal nucleus. II. Efferent connections. Brain Res 12:321–343
402. Leonard CM (1972) The connections of the dorsomedial nuclei. Brain Behav Evol 6:524–541
403. LeVay S (1973) Synaptic patterns in the visual cortex of the cat and monkey. Electron microscopy of Golgi preparations. J Comp Neurol 150:53–85
404. LeVay S, Hubel DH, Wiesel TN (1975) The pattern of ocular dominance columns in macaque visual cortex revealed by a reduced silver stain. J Comp Neurol 159:559–576
405. Lewis DA, Gonzalez-Burgos G (2006) Pathophysiologically based treatment interventions in schizophrenia. Nat Med 12:1016–1022

406. Lewis DA, Lund JS (1990) Heterogeneity of chandelier neurons in monkey neocortex: corticotropin-releasing factor- and parvalbumin-immunoreactive populations. J Comp Neurol 293:599–615
407. Lewis DA, Hashimoto T, Volk DW (2005) Cortical inhibitory neurons and schizophrenia. Nat Rev Neurosci 6:312–324
408. Lewis JW, Van Essen DC (2000) Mapping of architectonic subdivisions in the macaque monkey, with emphasis on parieto-occipital cortex. J Comp Neurol 428:79–111
409. Li H, Fukuda M, Tanifuji M, Rockland KS (2003) Intrinsic collaterals of layer 6 Meynert cells and functional columns in primate V1. Neuroscience 120:1061–1069
410. Liepmann H (1900) Das Krankheitsbild der Apraxia (motorischen Asymbolie) auf Grund eines Falles von einseitiger Apraxie. Monatsschr Psychiatr Neurol 8:15–44, 102–132, 182–197
411. Liepmann H, Maas O (1907) Fall von linksseitiger Agraphie und Apraxie bei rechtsseitiger Lähmung. Z Psychiatr Neurol 10:214–227
412. Livingstone MS, Hubel DH (1984) Anatomy and physiology of a color system in the primate visual cortex. J Neurosci 4:309–356
413. Loose R, Kaufmann C, Auer DP, Lange KW (2003) Human prefrontal and sensory cortical activity during divided attention tasks. Hum Brain Mapp 18:249–259
414. Lorente de Nó R (1922) La corteza cereral del ratón. Trabajos Cajal Madrid 20:41–80
415. Lorente de Nó R (1934) Studies on the structure of the cerebral cortex. I. The area entorhinalis. J Psychol Neurol 45:381–439
416. Lorente de Nó R (1938) Architectonics and structure of the cerebral cortex. In: Fulton JF (ed) Physiology of the nervous system. Oxford University Press, London, pp 291–330
417. Loring DW, Meador KJ, Lee GP et al (1990) Cerebral language lateralization: evidence from intracarotid amobarbital testing. Neuropsychologia 28:831–838
418. Lou HC, Kjaer TW, Friberg L et al (1999) A ^{15}O-H_2O PET study of meditation and the resting state of normal consciousness. Hum Brain Mapp 7:98–105
419. Lu MT, Preston JB, Strick PL (1994) Interconnections between the prefrontal cortex and the premotor areas in the frontal lobe. J Comp Neurol 341:375–392
420. Luders H, Lesser RP, Hahn J et al (1991) Basal temporal language area. Brain 114:743–754
421. Ludwig E, Klingler J (1956) Atlas cerebri humani. Karger, Basel
422. Lund JS (1973) Organization of neurons in the visual cortex, area 17, of the monkey (Macaca mulatta). J Comp Neurol 147:455–496
423. Lund JS (1984) Spiny stellate neurons. In: Peters A, Jones EG (eds) Cellular components of the cerebral cortex. Plenum, New York, pp 255–308 (Cerebral cortex, vol 1)
424. Lund JS (1987) Local circuit neurons of macaque monkey striate cortex: I. Neurons of laminae 4C and 5A. J Comp Neurol 257:60–92
425. Lund JS (1988) Anatomical organization of macaque monkey striate visual cortex. Annu Rev Neurosci 11:253–288
426. Lund JS, Boothe RG (1975) Interlaminar connections and pyramidal neuron organisation in the visual cortex. J Comp Neurol 159:305–334
427. Lund JS, Lewis DA (1993) Local circuit neurons of developing and mature macaque prefrontal cortex: Golgi and immunocytochemical characteristics. J Comp Neurol 328:282–312
428. Lund JS, Yoshioka T (1991) Local circuit neurons of macaque monkey striate cortex: III. Neurons of laminae 4B, 4A, and 3B. J Comp Neurol 311:234–258
429. Lund JS, Henry GH, MacQueen CL, Harvey AR (1979) Anatomical organization of the primary visual cortex (area 17) of the cat. A comparison with area 17 of the macaque monkey. J Comp Neurol 184:599–618
430. Lund JS, Hawken MJ, Parker AJ (1988) Local circuit neurons of macaque monkey striate cortex: II. Neurons of laminae 5B and 6. J Comp Neurol 276:1–29
431. Luppino G, Matelli M, Camarda R, Rizzolatti G (1993) Corticocortical connections of area F3 (SMA-proper) and area F6 (pre-SMA) in the macaque monkey. J Comp Neurol 338:114–140
432. Luppino G, Calzavara R, Rozzi S, Matelli M (2001) Projections from the superior temporal sulcus to the agranular frontal cortex in the macaque. Eur J Neurosci 14:1035–1040
433. Luppino G, Rozzi S, Calzavara R, Matelli M (2003) Prefrontal and agranular cingulate projections to the dorsal premotor areas F2 and F7 in the macaque monkey. Eur J Neurosci 17:559–578
434. Lynch JC, Tian JR (2005) Cortico-cortical networks and cortico-subcortical loops for the higher control of eye movements. Prog Brain Res 151: 461–501
435. Maddock RJ, Garrett AS, Buonocore MH (2003) Posterior cingulate cortex activation by emotional words: fMRI evidence from a valence decision task. Hum Brain Mapp 18:30–41
436. Marin-Padilla M (1969) Origin of the pericellular baskets of the pyramidal cells of the human motor cortex: a Golgi study. Brain Res 14:633–646
437. Marin-Padilla M (1978) Dual origin of the mammalian neocortex and evolution of the cortical plate. Anat Embryol (Berl) 152:109–126
438. Marin-Padilla M (1984) Neurons of layer I. A development analysis. In: Peters A, Jones EG (eds) Cellular components of the cerebral cortex. Plenum, New York, pp 447–478 (Cerebral cortex, vol 1)
439. Markowitsch HJ, Emmans D, Irle E, Streicher M, Preilowski B (1985) Cortical and subcortical afferent connections of the primate's temporal pole: a study of rhesus monkeys, squirrel monkeys and marmosets. J Comp Neurol 242:425–458

440. Markram H, Toledo-Rodriguez M, Wang Y et al (2004) Interneurons of the neocortical inhibitory system. Nat Rev Neurosci 5:793–807
441. Martin-Elkins CL, Horel JA (1992) Cortical afferents to behaviorally defined regions of the inferior temporal and parahippocampal gyri as demonstrated by WGA-HRP. J Comp Neurol 321:177–192
442. Martinotti C (1890) Beitrag zum Studium der Hirnrinde und dem Centralursprung der Nerven. Int Monatsschr Anat Physiol 7:69–90
443. Matelli M, Luppino G, Rizzolatti G (1991) Architecture of superior and mesial area 6 and the adjacent cingulate cortex in the macaque monkey. J Comp Neurol 311:445–462
444. Matelli M, Govoni P, Galletti C, Kutz DF, Luppino G (1998) Superior area 6 afferents from the superior parietal lobule in the macaque monkey. J Comp Neurol 402:327–352
445. Matelli M, Luppino G, Geyer S, Zilles K (2004) Motor Cortex. In: Paxinos G, Mai JK (eds) The human nervous system, 2nd edn. Elsevier, Amsterdam, pp 973–996
446. Mates SL, Lund JS (1983) Neuronal composition and development in lamina 4C of monkey striate cortex. J Comp Neurol 221:60–90
447. Matsuzaka Y, Aizawa H, Tanji J (1992) A motor area rostral to the supplementary motor area (presupplementary motor area) in the monkey: neuronal activity during a learned motor task. J Neurophysiol 68:653–662
448. Mayberg HS, Liotti M, Brannan SK et al (1999) Reciprocal limbic-cortical function and negative mood: converging PET findings in depression and normal sadness. Am J Psychiatry 156:675–682
449. Mayberg HS, Brannan SK, Tekell JL et al (2000) Regional metabolic effects of fluoxetine in major depression: serial changes and relationship to clinical response. Biol Psychiatry 48:830–843
450. Mayer AR, Dorflinger JM, Rao SM, Seidenberg M (2004) Neural networks underlying endogenous and exogenous visual-spatial orienting. Neuroimage 23:534–541
451. Mazoyer BM, Tzourio N, Frak V et al (1993) The cortical representation of speech. J Cogn Neurosci 5:467–479
452. Mazziotta J, Toga A, Evans A et al (2000) A probabilistic approach for mapping the human brain: the International Consortium for Brain Mapping (ICBM). In: Toga AW, Mazziotta J (eds) Brain mapping: The systems. Academic Press, San Diego, pp 141–156
453. Mazziotta J, Toga A, Evans A et al (2001) A probabilistic atlas and reference system for the human brain: International Consortium for Brain Mapping (ICBM). Philos Trans R Soc Lond B Biol Sci 356:1293–1322
454. Mazziotta J, Toga A, Evans A et al (2001) A four-dimensional probabilistic atlas of the human brain. J Am Med Inform Assoc 8:401–430
455. McCarthy G, Puce A, Constable RT et al (1996) Activation of human prefrontal cortex during spatial and nonspatial working memory tasks measured by functional MRI. Cereb Cortex 6:600–611
456. McGuire BA, Hornung JP, Gilbert CD, Wiesel TN (1984) Patterns of synaptic input to layer 4 of cat striate cortex. J Neurosci 4:3021–3033
457. McGuire PK, Bates JF, Goldman-Rakic PS (1991) Interhemispheric integration: I. Symmetry and convergence of the corticocortical connections of the left and the right principal sulcus (PS) and the left and the right supplementary motor area (SMA) in the rhesus monkey. Cereb Cortex 1:390–407
458. Meinecke DL, Peters A (1987) GABA immunoreactive neurons in rat visual cortex. J Comp Neurol 261:388–404
459. Melchitzky DS, Sesack SR, Pucak ML, Lewis DA (1998) Synaptic targets of pyramidal neurons providing intrinsic horizontal connections in monkey prefrontal cortex. J Comp Neurol 390:211–224
460. Menon V, Adleman NE, White CD, Glover GH, Reiss AL (2001) Error-related brain activation during a Go/NoGo response inhibition task. Hum Brain Mapp 12:131–143
461. Menon V, Mackenzie K, Rivera SM, Reiss AL (2002) Prefrontal cortex involvement in processing incorrect arithmetic equations: evidence from event-related fMRI. Hum Brain Mapp 16:119–130
462. Mesulam MM (1983) The functional anatomy and hemispheric specialization for directed attention. The role of the parietal lobe and its connectivity. Trends Neurosci 6:384–387
463. Mesulam MM (1985) Principles of behavioral neurology. Davis, Philadelphia
464. Mesulam MM (1998) From sensation to cognition. Brain 121:1013–1052
465. Mesulam MM (2002) Brain anatomy and networks. In: Ramachandran VS (ed) Encyclopedia of the human brain, vol 1. Academic Press, Amsterdam, pp 469–480
466. Mesulam MM, Mufson EJ (1985) The insula of Reil in man and monkey. In: Peters A, Jones EG (eds) Association and auditory cortices. Plenum, New York, pp 179–226 (Cerebral cortex, vol 4)
467. Mesulam MM, Hersh LB, Mash DC, Geula C (1992) Differential cholinergic innervation within functional subdivisions of the human cerebral cortex: a choline acetyltransferase study. J Comp Neurol 318:316–328
468. Meyer A, Beck E (1945) Neuropathological problems arising from prefrontal leucotomy. J Ment Sci 91:411–422
469. Meyer A, Beck E (1954) Prefrontal leucotomy and related operations: anatomical aspects of success and failure. Oliver and Boyd, Edinburgh/London
470. Meyer S, Strittmatter M, Fischer C, Georg T, Schmitz B (2004) Lateralization in autonomic dysfunction in ischemic stroke involving the insular cortex. Neuroreport 15:357–361

471. Mills CK (1891) On the localization of the auditory centre. Brain 14:465–472
472. Mistlin AJ, Perrett DI (1990) Visual and somatosensory processing in the macaque temporal cortex: the role of 'expectation'. Exp Brain Res 82: 437–450
473. Mitchell RLC, Crow TJ (2005) Right hemisphere language functions and schizophrenia: the forgotten hemisphere? Brain 128:963–978
474. Mitrofanis J, deFonseka R (2001) Organisation of connections between the zona incerta and the interposed nucleus. Anat Embryol 204:153–159
475. Miyachi S, Lu X, Inoue S et al (2005) Organization of multisynaptic inputs from prefrontal cortex to primary motor cortex as revealed by retrograde transneuronal transport of rabies virus. J Neurosci 25:2547–2556
476. Moniz E (1936) Essai d'un traitement chirurgical de certaine psychoses. Bull Acad Med 115:385–392
477. Moniz E (1937) Prefrontal leucotomy in the treatment of mental disorders. Am J Psychiatry 93:1379–1385
478. Moore JK, Linthicum FH (2004) Auditory system. In: Paxinos G, Mai JK (eds) The human nervous system, 2nd edn. Elsevier, Amsterdam, pp 1241–1279
479. Moran MA, Mufson EJ, Mesulam MM (1987) Neural inputs into the temporopolar cortex of the rhesus monkey. J Comp Neurol 256:88–103
480. Morecraft RJ, Van Hoesen GW (1992) Cingulate input to the primary and supplementary motor cortices in the rhesus monkey: evidence for somatotopy in areas 24c and 23c. J Comp Neurol 322:471–489
481. Morecraft RJ, Van Hoesen GW (1993) Frontal granular cortex input to the cingulate (M3), supplementary (M2) and primary (M1) motor cortices in the rhesus monkey. J Comp Neurol 337:669–689
482. Morecraft RJ, Van Hoesen GW (1998) Convergence of limbic input to the cingulate motor cortex in the rhesus monkey. Brain Res Bull 45:209–232
483. Morecraft RJ, Yeterian EH (2002) Prefrontal cortex. In: Ramachandran VS (ed) Encyclopedia of the human brain, vol 4. Academic Press, Amsterdam, pp 11–26
484. Morecraft RJ, Geula C, Mesulam MM (1992) Cytoarchitecture and neural afferents of orbitofrontal cortex in the brain of the monkey. J Comp Neurol 323:341–358
485. Morel A, Kaas JH (1992) Subdivisions and connections of auditory cortex in owl monkeys. J Comp Neurol 318:27–63
486. Morel A, Garraghty PE, Kaas JH (1993) Tonotopic organization, architectonic fields, and connections of auditory cortex in macaque monkeys. J Comp Neurol 335:437–459
487. Mori K, Satou M, Takagi SF (1979) Axonal projection of anterior olfactory nuclear neurons to the olfactory bulb bilaterally. Exp Neurol 64:295–305
488. Mori S, Crain BJ, Chacko VP, van Zijl PC (1999) Three-dimensional tracking of axonal projections in the brain by magnetic resonance imaging. Ann Neurol 45:265–269
489. Morosan P, Rademacher J, Schleicher A et al (2001) Human primary auditory cortex: cytoarchitectonic subdivisions and mapping into a spatial reference system. Neuroimage 13:684–701
490. Morosan P, Schleicher A, Amunts K, Zilles K (2005) Multimodal architectonic mapping of human superior temporal gyrus. Anat Embryol 210: 401–406
491. Morrison JH, Magistretti PJ (1983) Monoamines and peptides in cerebral cortex. Trends Neurosci 6:146–151
492. Morrison JH, Foote SL, O'Connor D, Bloom FE (1982) Laminar, tangential and regional organization of the noradrenergic innervation of monkey cortex: dopamine-beta-hydroxylase immunohistochemistry. Brain Res Bull 9:309–319
493. Mott FW (1894) The sensory functions of the central convolutions of the cerebral cortex. J Physiol 15:464–487
494. Mountcastle VB (1957) Modality and topographic properties of single neurons of cat's somatic sensory cortex. J Neurophysiol 20:408–434
495. Mountcastle VB (1979) An organizing principle for cerebral function: the unit module and distributed system. In: Schmitt FO, Worden FG (eds) The neurosciences fourth study program. MIT Press, Cambridge, pp 21–42
496. Mountcastle VB (1997) The columnar organization of the neocortex. Brain 120:701–722
497. Mrzljak L, Uylings HBM, Kostovic I, Van Eden CG (1988) Prenatal development of neurons in the human prefrontal cortex: I. A qualitative Golgi study. J Comp Neurol 271:355–386
498. Mrzljak L, Uylings HBM, Van Eden CG, Judás M (1990) Neuronal development in human prefrontal cortex in prenatal and postnatal stages. Prog Brain Res 85:185–222
499. Munk H (1881) Über die Funktionen der Grosshirnrinde. 3te Mitteilung, pp 28–53. A. Hirschwald, Berlin. In: Von Bonin G (1960) (ed) Some papers on the cerebral cortex. Translated as "On the functions of the cortex". Charles C Thomas, Springfield, IL, pp 97–117
500. Munk H (1892) Über die Fühlsphären des Grosshirns. Sitzungsberichte der Kaiserlichen Akademie der Wissenschaft 101:679–723
501. Nakamura K, Mikami A, Kubota K (1991) Unique oscillatory activity related to visual processing in the temporal pole of monkeys. Neurosci Res 12:293–299
502. Nakamura K, Mikami A, Kubota K (1992) Oscillatory neuronal activity related to visual short-term memory in monkey temporal pole. Neuroreport 3:117–120

503. Neal JW, Pearson RCA, Powell TPS (1986) The organization of the corticocortical projection of area 5 upon area 7 in the parietal lobe of the monkey. Brain Res 381:164–167
504. Neal JW, Pearson RCA, Powell TPS (1990) The ipsilateral cortico-cortical connections of area 7b, PF, in the parietal and temporal lobes of the monkey. Brain Res 524:119–132
505. Neal JW, Pearson RCA, Powell TPS (1990) The ipsilateral corticocortical connections of area 7 with the frontal lobe in the monkey. Brain Res 509:31–40
506. Nebel K, Stude P, Wiese H et al (2005) Sparse imaging and continuous event-related fMRI in the visual domain: a systematic 2190N comparison. Hum Brain Mapp 24:130–143
507. Nelson S (2002) Cortical microcircuits: diverse or canonical? Neuron 36:19–27
508. Nichols MJ, Newsome WT (1999) The neurobiology of cognition. Nature 402:C35–C38
509. Nieuwenhuys R (1994) The neocortex. An overview of its evolutionary development, structural organization and synaptology. Anat Embryol (Berl) 190:307–337
510. Nieuwenhuys R (1996) The greater limbic system, the emotional motor system and the brain. Prog Brain Res 107:551–580
511. Nieuwenhuys R (1998) Telencephalon. In: Nieuwenhuys R, Ten Donkelaar HJ, Nicholson C (eds) The central nervous system of vertebrates, vol III. Springer, Berlin Heidelberg New York, pp 1871–2008
512. Nobre AC, Allison T, McCarthy G (1994) Word recognition in the human inferior temporal lobe. Nature 372:260–263
513. O'Scalaidhe SP, Wilson FA, Goldman-Rakic PS (1997) Areal segregation of face-processing neurons in prefrontal cortex. Science 278:1135–1138
514. Ojemann GA (1991) Cortical organization of language. J Neurosci 11:2281–2287
515. Ojima H, Honda CN, Jones EG (1991) Patterns of axon collateralization of identified supragranular pyramidal neurons in the cat auditory cortex. Cereb Cortex 1:80–94
516. Öngür D, Price JL (2000) The organization of networks within the orbital and medial prefrontal cortex of rats, monkeys and humans. Cereb Cortex 10:206–219
517. Öngür D, An X, Price JL (1998) Prefrontal cortical projections to the hypothalamus in macaque monkeys. J Comp Neurol 401:480–505
518. Öngür D, Ferry AT, Price JL (2003) Architectonic subdivision of the human orbital and medial prefrontal cortex. J Comp Neurol 460:425–449
519. Ono M, Kubik S, Abernathy CD (1990) Atlas of the cerebral sulci. Thieme, Stuttgart
520. Oppenheimer SM (1994) Neurogenic cardiac effects of cerebrovascular disease. Curr Opin Neurol 7:20–24
521. Oppenheimer SM, Gelb A, Girvin JP, Hachinski VC (1992) Cardiovascular effects of human insular cortex stimulation. Neurology 42:1727–1732
522. Opris I, Bruce CJ (2005) Neural circuitry of judgment and decision mechanisms. Brain Res Rev 48:509–526
523. Osaka N, Osaka M, Kondo H et al (2004) The neural basis of executive function in working memory: an fMRI study based on individual differences. Neuroimage 21:623–631
524. Packard MG, Knowlton BJ (2002) Learning and memory functions of the Basal Ganglia. Annu Rev Neurosci 25:563–593
525. Pakkenberg B, Gundersen HJG (1997) Neocortical neuron number in humans: effect of sex and age. J Comp Neurol 384:312–320
526. Pandya DN (1987) Association cortex. In: Adelman G (ed) Encyclopedia of neuroscience, vol I. Birkhäuser, Boston, pp 80–83
527. Pandya DN, Kuypers HGJM (1969) Cortico-cortical connections in the rhesus monkey. Brain Res 13:13–36
528. Pandya DN, Sanides F (1973) Architectonic parcellation of the temporal operculum in rhesus monkey, and its projection pattern. Z Anat Entwicklungsgesch 139:127–161
529. Pandya DN, Seltzer B (1982) Association areas of the cerebral cortex. Trends Neurosci 5:386–390
530. Pandya DN, Seltzer B (1982) Intrinsic connections and architectonics of posterior parietal cortex in the rhesus monkey. J Comp Neurol 204:196–210
531. Pandya DN, Seltzer B (1986) The topography of commissural fibers. In: Lepore F, Ptito M, Jasper HH (eds) Two hemispheres - one brain. Functions of the corpus callosum. Liss, New York, pp 47–73
532. Pandya DN, Yeterian EH (1985) Architecture and connections of cortical association areas. In: Peters A, Jones EG (eds) Cerebral cortex, vol 4. Association and auditory cortices. Plenum, New York, pp 3–61
533. Pandya DN, Karol EA, Heilbronn D (1971) The topographical distribution of interhemispheric projections in the corpus callosum of the rhesus monkey. Brain Res 32:31–43
534. Papadopoulos GC, Parnavelas JG, Buijs RM (1987) Monoaminergic fibers form conventional synapses in the cerebral cortex. Neurosci Lett 76:275–279
535. Papadopoulos GC, Parnavelas JG, Buijs RM (1987) Light and electron microscopic immunocytochemical analysis of the serotonin innervation of the rat visual cortex. J Neurocytol 16:883–892
536. Parker AJ, Newsome WT (1998) Sense and the single neuron: probing the physiology of perception. Annu Rev Neurosci 21:227–277
537. Parnavelas JG, Edmunds SM (1983) Further evidence that Retzius-Cajal cells transform to nonpyramidal neurons in the developing rat visual cortex. J Neurocytol 12:863–871

538. Parnavelas JG, Papadopoulos GC (1989) The monoaminergic innervation of the cerebral cortex is not diffuse and nonspecific. Trends Neurosci 12:315–319
539. Parnavelas JG, Sullivan K, Lieberman AR, Webster KE (1977) Neurons and their synaptic organization in the visual cortex of the rat. Electron microscopy of Golgi preparations. Cell Tissue Res 183:499–517
540. Parnavelas JG, Kelly W, Franke E, Eckenstein F (1986) Cholinergic neurons and fibres in the rat visual cortex. J Neurocytol 15:329–336
541. Passingham RE, Stephan KE, Kotter R (2002) The anatomical basis of functional localization in the cortex. Nat Rev Neurosci 3:606–616
542. Pavlides C, Miyashita E, Asanuma H (1993) Projection from the sensory to the motor cortex is important in learning motor skills in the monkey. J Neurophysiol 70:733–741
543. Peinado A, Katz LC (1990) Development of cortical spiny stellate cells: retraction of a transient apical dendrite. Soc Neurosci Abstr 16:1127
544. Penfield W (1958) The excitable cortex in conscious man. Liverpool Press, Liverpool
545. Penfield W, Boldrey E (1937) Somatic motor and sensory representation in the cerebral cortex of man as studied by electrical stimulation. Brain 60:389–443
546. Penfield W, Faulk ME Jr (1955) The insula; further observations on its function. Brain 78:445–470
547. Penfield W, Jasper H (1954) Epilepsy and the functional anatomy of the human brain. Little, Brown, Boston
548. Penfield W, Rasmussen AJ (1950) Cerebral cortex in man. A clinical study of localization of function. MacMillan, New York
549. Penfield W, Roberts L (1959) Speech and brain mechanisms. Princeton University Press, Princeton
550. Percheron G (2004) Thalamus. In: Paxinos G, Mai JK (eds) The human nervous system, 2nd edn. Elsevier, Amsterdam, pp 592–675
551. Perrett DI, Harries MH, Bevan R et al (1989) Frameworks of analysis for the neural representation of animate objects and actions. J Exp Biol 146:87–113
552. Pessoa L, Kastner S, Ungerleider LG (2002) Attentional control of the processing of neural and emotional stimuli. Cogn Brain Res 15:31–45
553. Pessoa L, Kastner S, Ungerleider LG (2003) Neuroimaging studies of attention: from modulation of sensory processing to top-down control. J Neurosci 23:3990–3998
554. Peters A (1984) Bipolar cells. In: Peters A, Jones EG (eds) Cellular components of the cerebral cortex. Plenum, New York, pp 381–408 (Cerebral cortex, vol 1)
555. Peters A (1984) Chandelier cells. In: Peters A, Jones EG (eds) Cellular components of the cerebral cortex. Plenum, New York, pp 361–380 (Cerebral cortex, vol 1)
556. Peters A (1985) The visual cortex of the rat. In: Peters A, Jones EG (eds) Visual cortex. Plenum, New York, pp 19–80 (Cerebral cortex, vol 3)
557. Peters A (1987) Cortical neurons. In: Adelman G (ed) Encyclopedia of neuroscience, vol 1. Birkhäuser, Boston, pp 282–284
558. Peters A, Fairen A (1978) Smooth and sparsely-spined stellate cells in the visual cortex of the rat: a study using a combined Golgi-electron microscopic technique. J Comp Neurol 181:129–171
559. Peters A, Harriman KM (1988) Enigmatic bipolar cell of rat visual cortex. J Comp Neurol 267:409–432
560. Peters A, Jones EG (1984) Cellular components of the cerebral cortex. Plenum, New York
561. Peters A, Jones EG (1984) Classification of cortical neurons. In: Peters A, Jones EG (eds) Cellular components of the cerebral cortex. Plenum, New York, pp 107–122 (Cerebral cortex, vol 1)
562. Peters A, Kimerer LM (1981) Bipolar neurons in rat visual cortex: a combined Golgi-electron microscope study. J Neurocytol 10:921–946
563. Peters A, Saint Marie PL (1984) Smooth and sparsely spinous nonpyramidal cells forming local axonal plexures. In: Peters A, Jones EG (eds) Cellular components of the cerebral cortex. Plenum, New York, pp 419–445 (Cerebral cortex, vol 1)
564. Peters A, Sethares C (1991) Layer IVA of rhesus monkey primary visual cortex. Cereb Cortex 1:445–462
565. Peters A, Sethares C (1991) Organization of pyramidal neurons in area 17 of monkey visual cortex. J Comp Neurol 306:1–23
566. Peters A, Sethares C (1996) Myelinated axons and the pyramidal cell modules in monkey primary visual cortex. J Comp Neurol 365:232–255
567. Peters A, Sethares C (1997) The organization of double bouquet cells in monkey striate cortex. J Neurocytol 26:779–797
568. Peters A, Walsh TM (1972) A study of the organization of apical dendrites in the somatic sensory cortex of the rat. J Comp Neurol 144:253–268
569. Peters A, Yilmaz E (1993) Neuronal organization in area 17 of cat visual cortex. Cereb Cortex 3:49–68
570. Peters A, Proskauer CC, Ribak CE (1982) Chandelier cells in rat visual cortex. J Comp Neurol 206: 397–416
571. Peters A, Meinecke DL, Karamanlidis AN (1987) Vasoactive intestinal polypeptide immunoreactive neurons in the primary visual cortex of the cat. J Neurocytol 16:23–38
572. Petersen SE, Fox PT, Posner MI, Mintun M, Raichle ME (1988) Positron emission tomographic studies of the cortical anatomy of single-word processing. Nature 331:585–589
573. Petersen SE, Fox PT, Posner MI, Mintun M, Raichle ME (1989) Positron emission tomographic studies of the cortical anatomy of single-word processing. J Cogn Neurosci 1:153–170

574. Petrides M (2000) Mapping prefrontal cortical systems for the control of cognition. In: Toga AW, Mazziotta JC (eds) Brain mapping: the systems. Academic Press, San Diego, pp 159–176
575. Petrides M, Milner B (1982) Deficits on subject-ordered tasks after frontal- and temporal-lobe lesions in man. Neuropsychologia 20:249–262
576. Petrides M, Pandya DN (1988) Association fiber pathways to the frontal cortex from the superior temporal region in the rhesus monkey. J Comp Neurol 273:52–66
577. Petrides M, Pandya DN (1994) Comparative architectonic analysis of the human and the macaque frontal cortex. In: Boller F, Grafman J (eds) Handbook of neuropsychology, vol 9. Elsevier, Amsterdam, pp 17–58
578. Petrides M, Pandya DN (1999) Dorsolateral prefrontal cortex: comparative cytoarchitectonic analysis in the human and the macaque brain and corticocortical connection patterns. Eur J Neurosci 11:1011–1036
579. Petrides M, Pandya DN (2002) Association pathways of the prefrontal cortex and functional observations. In: Stuss DT, Knight RT (eds) Principles of frontal lobe function. Oxford University Press, pp 31–50
580. Petrides M, Pandya DN (2002) Comparative cytoarchitectonic analysis of the human and the macaque ventrolateral prefrontal cortex and corticocortical connection patterns in the monkey. Eur J Neurosci 16:291–310
581. Petrides M, Pandya DN (2004) The frontal cortex. In: Paxinos G, Mai JK (eds) The human nervous system, 2nd edn. Elsevier, Amsterdam, pp 950–972
582. Petrides M, Pandya DN (2006) Efferent association pathways originating in the caudal prefrontal cortex in the macaque monkey. J Comp Neurol 498:227–251
583. Petrides M, Alivisatos B, Meyer E, Evans AC (1993) Functional activation of the human frontal cortex during the performance of verbal working memory tasks. Proc Natl Acad Sci USA 90:878–882
584. Phelps EA, LeDoux JE (2005) Contributions of the amygdala to emotion processing: from animal models to human behavior. Neuron 48:175–187
585. Picard N, Strick PL (1996) Motor areas of the medial wall: a review of their location and functional activation. Cereb Cortex 6:342–353
586. Pons TP, Kaas JH (1986) Corticocortical connections of area 2 of somatosensory cortex in macaque monkeys: a correlative anatomical and electrophysiological study. J Comp Neurol 248:313–336
587. Porter LL, Sakamoto K (1988) Organization and synaptic relationships of the projection from the primary sensory to the primary motor cortex in the cat. J Comp Neurol 271:387–396
588. Poupon C, Clark CA, Frouin V et al (2000) Regularization of diffusion-based direction maps for the tracking of brain white matter fascicles. Neuroimage 12:184–195
589. Powell TPS (1981) Certain aspects of the intrinsic organization of the cerebral cortex. In: Pompeiano O, Ajmone Marsan C (eds) Brain mechanisms and perceptual awareness. Raven, New York, pp 1–19
590. Press WA, Brewer AA, Dougherty RF, Wade AR, Wandell BA (2001) Visual areas and spatial summation in human visual cortex. Vision Res 41:1321–1332
591. Preuss TM (1995) The argument from humans to animals in cognitive neuroscience. In: Gazzaniga MS (ed) The cognitive neurosciences. MIT Press, Boston, pp 1227–1241
592. Preuss TM, Goldman-Rakic PS (1991) Myelo- and cytoarchitecture of the granular frontal cortex and surrounding regions in the strepsirhine primate Galago and the anthropoid primate Macaca. J Comp Neurol 310:429–474
593. Price CJ (2000) The anatomy of language: contributions from functional neuroimaging. J Anat 197:335–359
594. Price CJ, Green DW, von Studnitz R (1999) A functional imaging study of translation and language switching. Brain 122:2221–2235
595. Price JL, Carmichael ST, Drevets WC (1996) Networks related to the orbital and medial prefrontal cortex; a substrate for emotional behavior? Prog Brain Res 107:523–536
596. Pritchard TC, Norgren R (2004) Gustatory system. In: Paxinos G, Mai JK (eds) The human nervous system, 2nd end. Elsevier, Amsterdam, pp 1171–1196
597. Purves D, Riddle DR, LaMantia AS (1992) Iterated patterns of brain circuitry (or how the cortex gets its spots). Trends Neurosci 15:362–368
598. Rademacher J, Caviness VS Jr, Steinmetz H, Galaburda AM (1993) Topographical variation of the human primary cortices: implications for neuroimaging, brain mapping, and neurobiology. Cereb Cortex 3:313–329
599. Rademacher J, Burgel U, Geyer S et al (2001) Variability and asymmetry in the human precentral motor system. A cytoarchitectonic and myeloarchitectonic brain mapping study. Brain 124: 2232–2258
600. Rademacher J, Morosan P, Schormann T et al (2001) Probabilistic mapping and volume measurement of human primary auditory cortex. Neuroimage 13:669–683
601. Raichle ME (1999) Modern phrenology: maps of human cortical function. Ann N Y Acad Sci 882: 107–118
602. Raichle ME (2003) Functional brain imaging and human brain function. J Neurosci 23:3959–3962
603. Raichle ME, Fiez JA, Videen TO et al (1994) Practice-related changes in human brain functional anatomy during nonmotor learning. Cereb Cortex 4:8–26

604. Rajkowska G, Goldman-Rakic PS (1995) Cytoarchitectonic definition of prefrontal areas in the normal human cortex: I. Remapping of areas 9 and 46 using quantitative criteria. Cereb Cortex 5:307–322
605. Rajkowska G, Goldman-Rakic PS (1995) Cytoarchitectonic definition of prefrontal areas in the normal human cortex: II. Variability in locations of areas 9 and 46 and relationship to the Talairach Coordinate System. Cereb Cortex 5:323–337
606. Rakic P (1971) Neuron-glia relationship during granule cell migration in developing cerebellar cortex. A Golgi and electronmicroscopic study in *Macacus rhesus*. J Comp Neurol 141:283–312
607. Rakic P (1972) Mode of cell migration to the superficial layers of fetal monkey neocortex. J Comp Neurol 145:61–84
608. Rakic P (1988) Specification of cerebral cortical areas. Science 241:170–176
609. Rakic P (1990) Principles of neural cell migration. Experientia 46:883–891
610. Rakic P (1995) A small step for the cell, a giant leap for mankind: a hypothesis of neocortical expansion during evolution. Trends Neurosci 18: 383–388
611. Ramnani N (2006) The primate cortico-cerebellar system: anatomy and function. Nat Rev Neurosci 7:511–522
612. Ramnani N, Owen AM (2004) Anterior prefrontal cortex: insights into function from anatomy and neuroimaging. Nat Rev Neurosci 5:184–194
613. Rao SC, Rainer G, Miller EK (1997) Integration of what and where in the primate prefrontal cortex. Science 276:821–824
614. Ray JP, Price JL (1993) The organization of projections from the mediodorsal nucleus of the thalamus to orbital and medial prefrontal cortex in macaque monkeys. J Comp Neurol 337:1–31
615. Retzius G (1896) Das Menschenhirn. Studien in der makroskopischen Morphologie. Norstedt and Söner, Stockholm
616. Ribak CE (1978) Aspinous and sparsely-spinous stellate neurons in the visual cortex of rats contain glutamic acid decarboxylase. J Neurocytol 7:461–478
617. Robinson CJ, Burton H (1980) Organization of somatosensory receptive fields in cortical areas 7b, retroinsula, postauditory and granular insula of M. fascicularis. J Comp Neurol 192:69–92
618. Robinson CJ, Burton H (1980) Somatic submodality distribution within the second somatosensory (SII), 7b, retroinsular, postauditory, and granular insular cortical areas of M. fascicularis. J Comp Neurol 192:93–108
619. Rockel AJ, Hiorns RW, Powell TP (1974) Numbers of neurons through full depth of neocortex. J Anat 118:371
620. Rockel AJ, Hiorns RW, Powell TP (1980) The basic uniformity in structure of the neocortex. Brain 103:221–244
621. Rockland KS, Ichinohe N (2004) Some thoughts on cortical minicolumns. Exp Brain Res 158:265–277
622. Rockland KS, Knutson T (2001) Axon collaterals of Meynert cells diverge over large portions of area V1 in the macaque monkey. J Comp Neurol 441:134–147
623. Rockland KS, Pandya DN (1979) Laminar origins and terminations of cortical connections of the occipital lobe in the rhesus monkey. Brain Res 179:3–20
624. Roland PE, Zilles K (1994) Brain atlases – a new research tool. Trends Neurosci 17:458–467
625. Roland PE, Zilles K (1996) The developing European computerized human brain database for all imaging modalities. Neuroimage 4:S39–S47
626. Roland PE, Zilles K (1998) Structural divisions and functional fields in the human cerebral cortex. Brain Res Rev 26:87–105
627. Roland PE, Geyer S, Amunts K et al (1997) Cytoarchitectural maps of the human brain in standard anatomical space. Hum Brain Mapp 5:222–227
628. Rolls ET (2000) The orbitofrontal cortex and reward. Cereb Cortex 10:284–294
629. Rolls ET, Baylis LL (1994) Gustatory, olfactory, and visual convergence within the primate orbitofrontal cortex. J Neurosci 14:5437–5452
630. Rolls ET, Critchley HD, Mason R, Wakeman EA (1996) Orbitofrontal cortex neurons: role in olfactory and visual association learning. J Neurophysiol 75:1970–1981
631. Romanski LM, Bates JF, Goldman-Rakic PS (1999) Auditory belt and parabelt projections to the prefrontal cortex in the rhesus monkey. J Comp Neurol 403:141–157
632. Romanski LM, Tian B, Fritz J et al (1999) Dual streams of auditory afferents target multiple domains in the primate prefrontal cortex. Nat Neurosci 2:1131–1136
633. Rosano C, Sweeney JA, Melchitzky DS, Lewis DA (2003) The human precentral sulcus: chemoarchitecture of a region corresponding to the frontal eye fields. Brain Res 972:16–30
634. Rose JE, Woolsey C (1948) Structure and relations of limbic cortex and anterior thalamic nuclei in rabbit and cat. J Comp Neurol 89:79–347
635. Rose JE, Woolsey CN (1948) The orbitofrontal cortex and its connections with the mediodorsal nucleus in rabbit, sheep and cat. Res Publ Assoc Nerv Ment Dis 27:210–232
636. Rose JE, Woolsey CN (1949) The relations of thalamic connections, cellular structure and evocable electrical activity in the auditory region of the cat. J Comp Neurol 91:441–466
637. Rose M (1935) Cytoarchitektonik und Myeloarchitektonik der Groszhirnrinde. In: Bumke O, Foerster O (eds) Handbuch der Neurologie Band I: Anatomie. Springer, Berlin, pp 589–778
638. Rosene DL, Pandya DN (1983) Architectonics and connections of the posterior parahippocampal gyrus in the rhesus monkey. Soc Neurosci Abstr 9:222

639. Rosene DL, Van Hoesen GW (1977) Hippocampal efferents reach widespread areas of the cerebral cortex and amygdala in the rhesus monkey. Science 198:315–317
640. Royce GJ (1987) Recent research on the centromedian and parafascicular nuclei. In: Carpenter MB, Jayaraman A (eds) The basal ganglia II: Structure and function – current concepts. Plenum, New York, pp 293–319 (Advances in behavioral biology, vol 32)
641. Royce GJ, Mourey RJ (1985) Efferent connections of the centromedian and parafascicular thalamic nuclei: an autoradiographic investigation in the cat. J Comp Neurol 235:277–300
642. Rubens AB (1977) Anatomical asymmetries of human cerebral cortex. In: Harnad S, Doty RW, Goldstein L, Jaynes J, Krauthamer G (eds) Lateralization in the nervous system. Academic Press, New York, pp 503–517
643. Ruiz-Marcos A, Valverde F (1970) Dynamic architecture of the visual cortex. Brain Res 19:25–39
644. Rustioni A, Hayes NL (1981) Corticospinal tract collaterals to the dorsal column nuclei of cats. An anatomical single and double retrograde tracer study. Exp Brain Res 43:237–245
645. Saffran E, Sholl A (1999) Clues to the functional and neural architecture of word meaning. In: Brown CM, Hagoort P (eds) The neurocognition of language. Oxford University Press, Oxford, pp 241–273
646. Saint Marie RL, Peters A (1985) The morphology and synaptic connections of spiny stellate neurons in monkey visual cortex (area 17): a Golgi-electron microscopic study. J Comp Neurol 233:213–235
647. Saleem KS, Suzuki W, Tanaka K, Hashikawa T (2000) Connections between anterior inferotemporal cortex and superior temporal sulcus regions in the macaque monkey. J Neurosci 20:5083–5101
648. Saper CB (2002) The central autonomic nervous system: conscious visceral perception and autonomic pattern generation. Annu Rev Neurosci 25:433–469
649. Sarkissov SA, Filimonoff IN, Kononowa EP, Preobraschenskaja IS, Kukuew LA (1955) Atlas of the cytoarchitectonics of the human cerebral cortex. Medgiz, Moscow
650. Sarter M, Bruno JP (2000) Cortical cholinergic inputs mediating arousal, attentional processing and dreaming: differential afferent regulation of the basal forebrain by telencephalic and brainstem afferents. Neuroscience 95:933–952
651. Sarter M, Bruno JP (2002) The neglected constituent of the basal forebrain corticopetal projection system: GABAergic projections. Eur J Neurosci 15:1867–1873
652. Sawaguchi T, Goldman-Rakic PS (1991) D1 dopamine receptors in prefrontal cortex: involvement in working memory. Science 251:947–950
653. Schall JD (1997) Visuomotor areas of the frontal lobe. In: Jones EG, Peters A (eds) Extrastriate cortex in primates. Plenum, New York, pp 527–638 (Cerebral cortex, vol 12)
654. Scheibel AB (1979) Development of axonal and dendritic neuropil as a function of evolving behavior. In: Schmidt FO, Worden FG (eds) The neurosciences fourth study program. MIT Press, Cambridge, p VII+1185
655. Scheibel AB, Paul LA, Fried I et al (1985) Dendritic organization of the anterior speech area. Exp Neurol 87:109–117
656. Scheibel ME, Scheibel AB (1970) Elementary processes in selected thalamic and cortical subsystems: the structural substrates. In: Schmitt FO (ed) The neurosciences second study program. Rockefeller University Press, New York, pp 443–457
657. Scheibel ME, Scheibel AB (1978) The dendritic structure of the human Betz cells. In: Brazier MAB, Petsche H (eds) Architectonics of the cerebral cortex. Raven, New York, pp 43–57
658. Schiltz C, Sorger B, Caldara R et al (2006) Impaired face discrimination in acquired prosopagnosia is associated with abnormal response to individual faces in the right middle fusiform gyrus. Cereb Cortex 16:574–586
659. Schleicher A, Amunts K, Geyer S, Morosan P, Zilles K (1999) Observer-independent method for microstructural parcellation of cerebral cortex: A quantitative approach to cytoarchitectonics. Neuroimage 9:165–177
660. Schleicher A, Amunts K, Geyer S et al (2000) A stereological approach to human cortical architecture: identification and delineation of cortical areas. J Chem Neuroanat 20:31–47
661. Schneider RJ, Friedman DP, Mishkin M (1993) A modality-specific somatosensory area within the insula of the rhesus monkey. Brain Res 621:116–120
662. Schnitzler A, Seitz RJ, Freund HJ (2000) The somatosensory system. In: Toga AW, Mazziotta JC (eds) Brain mapping, the systems. Academic Press, San Diego, pp 291–329
663. Schoenbaum G, Roesch MR, Stalnaker TA (2006) Orbitofrontal cortex, decision-making and drug addiction. Trends Neurosci 29:116–124
664. Schultz W (1998) Predictive reward signal of dopamine neurons. J Neurophysiol 80:1–27
665. Schultz W (2002) Getting formal with dopamine and reward. Neuron 36:241–263
666. Schultz W, Tremblay L, Hollerman JR (2000) Reward processing in primate orbitofrontal cortex and basal ganglia. Cereb Cortex 10:272–284
667. Schwarz ML, Goldman-Rakic PS (1984) Callosal and intrahemispheric connectivity of the prefrontal association cortex in Rhesus monkey: relation between intraparietal and principal sulcus cortex. J Comp Neurol 226:403–420

668. Scott SK, Johnsrude IS (2003) The neuroanatomical and functional organization of speech perception. Trends Neurosci 26:100–107
669. Selemon LD, Goldman-Rakic PS (1988) Common cortical and subcortical targets of the dorsolateral prefrontal and posterior parietal cortices in the rhesus monkey: evidence for a distributed neural network subserving spatially guided behavior. J Neurosci 8:4049–4068
670. Seltzer B, Pandya DN (1978) Afferent cortical connections and architectonics of the superior temporal sulcus and surrounding cortex in the rhesus monkey. Brain Res 149:1–24
671. Seltzer B, Pandya DN (1984) Further observations on parieto-temporal connections in the rhesus monkey. Exp Brain Res 55:301–312
672. Seltzer B, Pandya DN (1989) Frontal lobe connections of the superior temporal sulcus in the rhesus monkey. J Comp Neurol 281:97–113
673. Seltzer B, Pandya DN (1989) Intrinsic connections and architectonics of the superior temporal sulcus in the rhesus monkey. J Comp Neurol 290:451–471
674. Seltzer B, Pandya DN (1991) Post-rolandic cortical projections of the superior temporal sulcus in the rhesus monkey. J Comp Neurol 312:625–640
675. Seltzer B, Pandya DN (1994) Parietal, temporal, and occipital projections to cortex of the superior temporal sulcus in the rhesus monkey: a retrograde tracer study. J Comp Neurol 343:445–463
676. Seltzer B, Cola MG, Gutierrez C et al (1996) Overlapping and nonoverlapping cortical projections to cortex of the superior temporal sulcus in the rhesus monkey: double anterograde tracer studies. J Comp Neurol 370:173–190
677. Sesack SR, Snyder CL, Lewis DA (1995) Axon terminals immunolabeled for dopamine or tyrosine hydroxylase synapse on GABA-immunoreactive dendrites in rat and monkey cortex. J Comp Neurol 363:264–280
678. Sewards TV, Sewards MA (2002) On the neural correlates of object recognition awareness: relationship to computational activities and activities mediating perceptual awareness. Conscious Cogn 11:51–77
679. Shepherd GM (2004) Introduction to synaptic circuits. In: Shepherd GM (ed) The synaptic organization of the brain, 5th edition. Oxford University Press, Oxford, pp 1–38
680. Sherman SM, Guillery RW (2001) Exploring the thalamus. Academic Press, San Diego
681. Sherman SM, Guillery RW (2002) The role of the thalamus in the flow of information to the cortex. Phil Trans R Soc Lond B 357:1695–1708
682. Shimazu H, Maier MA, Cerri G, Kirkwood PA, Lemon RN (2004) Macaque ventral premotor cortex exerts powerful facilitation of motor cortex outputs to upper limb motoneurons. J Neurosci 24:1200–1211
683. Simon A, Oláh S, Molnar G, Szabadics J, Tamás G (2005) Gap-junctional coupling between neurogliaform cells and various interneuron types in the neocortex. J Neurosci 25:6278–6285
684. Simon O, Kherif F, Flandin G et al (2004) Automatized clustering and functional geometry of human parietofrontal networks for language, space, and number. Neuroimage 23:1192–1202
685. Simons JS, Spiers HJ (2003) Prefrontal and medial temporal lobe interactions in long-term memory. Nat Rev Neurosci 4:637–648
686. Sincich LC, Horton JC (2003) Independent projection streams from macaque striate cortex to the second visual area and middle temporal area. J Neurosci 23:5684–5692
687. Sincich LC, Park KF, Wohlgemuth MJ, Horton JC (2004) Bypassing V1: a direct geniculate input to area MT. Nat Neurosci 7:1123–1128
688. Sloper JJ (1972) Gap junctions between dendrites in the primate neocortex. Brain Res 44:641–646
689. Sloper JJ, Powell TP (1978) Gap junctions between dendrites and somata of neurons in the primate sensori-motor cortex. Proc R Soc Lond B Biol Sci 203:39–47
690. Sloper JJ, Powell TPS (1979) An experimental electron microscopic study of afferent connections to the primate motor and somatic sensory cortices. Philos Trans R Soc Lond B Biol Sci 285:199–226
691. Smiley JF, Goldman-Rakic PS (1996) Serotonergic axons in monkey prefrontal cerebral cortex synapse predominantly on interneurons as demonstrated by serial section electron microscopy. J Comp Neurol 367:431–443
692. Smiley JF, Levey AI, Ciliax BJ, Goldman-Rakic PS (1994) D1 dopamine receptor immunoreactivity in human and monkey cerebral cortex: predominant and extrasynaptic localization in dendritic spines. Proc Natl Acad Sci USA 91:5720–5724
693. Söhl G, Maxeiner S, Willecke K (2005) Expression and functions of neuronal gap junctions. Nat Rev Neurosci 6:191–200
694. Somogyi P, Cowey A (1981) Combined Golgi and electron microscopic study on the synapses formed by double bouquet cells in the visual cortex of the cat and monkey. J Comp Neurol 195: 547–566
695. Somogyi P, Cowey A (1984) Double bouquet cells. In: Peters A, Jones EG (eds) Cellular components of the cerebral cortex. Plenum, New York, pp 337–360 (Cerebral cortex, vol 1)
696. Somogyi P, Hodgson AJ (1985) Antisera to gamma-aminobutyric acid. III. Demonstration of GABA in Golgi-impregnated neurons and in conventional electron microscopic sections of cat striate cortex. J Histochem Cytochem 33:249–257
697. Somogyi P, Soltesz I (1986) Immunogold demonstration of GABA in synaptic terminals of intracellularly recorded, horseradish peroxidase-filled basket cells and clutch cells in the cat's visual cortex. Neuroscience 19:1051–1065
698. Somogyi P, Hodgson AJ, Smith AD (1979) An approach to tracing neuron networks in the cere-

bral cortex and basal ganglia. Combination of Golgi staining, retrograde transport of horseradish peroxidase and anterograde degeneration of synaptic boutons in the same material. Neuroscience 4:1805–1852
699. Somogyi P, Freund TF, Cowey A (1982) The axo-axonic interneuron in the cerebral cortex of the rat, cat and monkey. Neuroscience 7:2577–2607
700. Somogyi P, Kisvárday ZF, Martin KA, Whitteridge D (1983) Synaptic connections of morphologically identified and physiologically characterized large basket cells in the striate cortex of cat. Neuroscience 10:261–294
701. Somogyi P, Freund TF, Hodgson AJ et al (1985) Identified axo-axonic cells are immunoreactive for GABA in the hippocampus and visual cortex of the cat. Brain Res 332:143–149
702. Somogyi P, Tamás G, Lujan R, Buhl EH (1998) Salient features of synaptic organisation in the cerebral cortex. Brain Res Rev 26:113–135
703. Sperry R (1986) Consciousness, personal identity, and the divided brain. In: Lepore F, Ptito M, Jasper HH (eds) Two hemispheres - one brain: functions of the corpus callosum. Liss, New York, pp 3–20
704. Staiger JF, Flagmeyer I, Schubert D et al (2004) Functional diversity of layer IV spiny neurons in rat somatosensory cortex: quantitative morphology of electrophysiologically characterized and biocytin labeled cells. Cereb Cortex 14:690–701
705. Starr MA (1884) Cortical lesions of the brain: A collection and analysis of the American cases of localized cerebral disease. Am J Med Sci 88:114–141
706. Steele GE, Weller RE, Cusick CG (1991) Cortical connections of the caudal subdivision of the dorsolateral area (V4) in monkeys. J Comp Neurol 306:495–520
707. Stein BE, Meredith MA, Wallace MT (1993) The visually responsive neuron and beyond: multisensory integration in cat and monkey. Prog Brain Res 95:79–90
708. Steinmetz H (1996) Structure, functional and cerebral asymmetry: in vivo morphometry of the planum temporale. Neurosci Biobehav Rev 20: 587–591
709. Stephan H, Andy OJ (1964) Quantitative comparisons of brain structures from insectivores to primates. Am Zool 136:59–74
710. Stephan KE, Kamper L, Bozkurt A et al (2001) Advanced database methodology for the Collation of Connectivity data on the Macaque brain (CoCoMac). Philos Trans R Soc Lond B Biol Sci 356:1159–1186
711. Stepniewska I, Preuss TM, Kaas JH (2006) Ipsilateral cortical connections of dorsal and ventral premotor areas in New World owl monkeys. J Comp Neurol 495:691–708
712. Steriade M, Jones EG, McCormick DA (1997) Thalamus, vol I: Organisation and function. Elsevier, Amsterdam
713. Stokking R, Zuiderveld KJ, Viergever MA (2001) Integrated volume visualization of functional image data and anatomical surfaces using normal fusion. Hum Brain Mapp 12:203–218
714. Sudakov K, MacLean PD, Reeves A, Marino R (1971) Unit study of exteroceptive inputs to claustrocortex in awake, sitting, squirrel monkey. Brain Res 28:19–34
715. Suzuki W, Saleem KS, Tanaka K (2000) Divergent backward projections from the anterior part of the inferotemporal cortex (area TE) in the macaque. J Comp Neurol 422:206–228
716. Suzuki WA, Amaral DG (1994) Perirhinal and parahippocampal cortices of the macaque monkey: cortical afferents. J Comp Neurol 350:497–533
717. Suzuki WA, Amaral DG (1994) Topographic organization of the reciprocal connections between the monkey entorhinal cortex and the perirhinal and parahippocampal cortices. J Neurosci 14: 1856–1877
718. Suzuki WA, Amaral DG (2003) Where are the perirhinal and parahippocampal cortices? A historical overview of the nomenclature and boundaries applied to the primate medial temporal lobe. Neuroscience 120:893–906
719. Swindale NV (1990) Is the cerebral cortex modular? Trends Neurosci 13:487–492
720. Szentágothai J (1975) The 'module-concept' in cerebral cortex architecture. Brain Res 95:475–496
721. Szentágothai J (1978) The Ferrier Lecture, 1977. The neuron network of the cerebral cortex: a functional interpretation. Proc R Soc Lond B Biol Sci 201:219–248
722. Szentágothai J (1979) Local neuron circuits of the neocortex. In: Schmitt FO, Worden FG (eds) The neurosciences fourth study program. MIT Press, Cambridge, pp 399–415
723. Szentágothai J (1983) The modular architectonic principle of neural centers. Rev Physiol Biochem Pharmacol 98:11–61
724. Takahashi H, Yahata N, Koeda M et al (2004) Brain activation associated with evaluative processes of guilt and embarrassment: an fMRI study. Neuroimage 23:967–974
725. Takahashi T, Suzuki M, Hagino H et al (2004) Bilateral volume reduction of the insular cortex in patients with schizophrenia: a volumetric MRI study. Psychiatry Res 131:185–194
726. Takahashi T, Suzuki M, Zhou SY et al (2005) Volumetric MRI study of the short and long insular cortices in schizophrenia spectrum disorders. Psychiatry Res 138:209–220
727. Talairach J, Tournoux P (1988) Co-Planar Stereotaxic Atlas of the Human Brain. 3-D proportioned system: an approach to cerebral imaging. Thieme, Stuttgart
728. Tamás G, Buhl EH, Somogyi P (1997) Massive autaptic self-innervation of GABAergic neurons in cat visual cortex. J Neurosci 17:6352–6364

729. Tamás G, Buhl EH, Lorincz A, Somogyi P (2000) Proximally targeted GABAergic synapses and gap junctions synchronize cortical interneurons. Nat Neurosci 3:366–371
730. Tanaka K (1996) Inferotemporal cortex and object vision. Annu Rev Neurosci 19:109–139
731. Tanaka Y, Miyazawa Y, Hashimoto R, Nakano I, Obayashi T (1999) Postencephalitic focal retrograde amnesia after bilateral anterior temporal lobe damage. Neurology 53:344–350
732. Tatschl C, Stollberger C, Matz K et al (2006) Insular involvement is associated with QT prolongation: ECG abnormalities in patients with acute stroke. Cerebrovasc Dis 21:47–53
733. Thier P, Andersen RA (1998) Electrical microstimulation distinguishes distinct saccade-related areas in the posterior parietal cortex. J Neurophysiol 80:1713–1735
734. Tigges J, Tigges M (1985) Subcortical sources of direct projections to visual cortex. In: Peters A, Jones EG (eds) Visual cortex. Plenum, New York, pp 351–378 (Cerebral cortex, vol 3)
735. Tigges J, Tigges M, Anschel S et al (1981) Areal and laminar distribution of neurons interconnecting the central visual cortical areas 17, 18, 19 and MT in squirrel monkey (Saimiri). J Comp Neurol 202:539–560
736. Tootell RB, Hadjikhani N (2001) Where is 'dorsal V4' in human visual cortex? Retinotopic, topographic and functional evidence. Cereb Cortex 11:298–311
737. Tootell RB, Hadjikhani NK, Mendola JD, Marrett S, Dale AM (1998) From retinotopy to recognition: fMRI in human visual cortex. Trends Cogn Sci 2:174–183
738. Towe AL (1975) Notes on the hypothesis of columnar organization in somatosensory cerebral cortex. Brain Behav Evol 11:16–47
739. Tusa RJ, Ungerleider LG (1985) The inferior longitudinal fasciculus: a reexamination in humans and monkeys. Ann Neurol 18:583–591
740. Ullman MT (2001) A neurocognitive perspective on language: the declarative/procedural model. Nat Rev Neurosci 2:717–726
741. Ungerleider LG, Desimone R (1986) Cortical connections of visual area MT in the macaque. J Comp Neurol 248:190–223
742. Ungerleider LG, Mishkin M (1979) The striate projection zone in the superior temporal sulcus of Macaca mulatta: location and topographic organization. J Comp Neurol 188:347–367
743. Ungerleider LG, Mishkin M (1982) Two cortical visual systems. In: Ingle et al (eds) Analysis of visual behavior. MIT Press, Cambridge, pp 549–586
744. Uttal WR (2001) The new phrenology: the limits of localizing cognitive processes in the brain. MIT Press, Cambridge
745. Uttal WR (2002) Précis of the new phrenology: the limits of localizing cognitive processes in the brain. Brain and Mind 3:221–228
746. Uylings HBM, Malofeeva LI, Bogolepova IN, Amunts K, Zilles K (1999) Broca's language area from a neuroanatomical and developmental perspectives. In: Brown CM, Hagoort P (eds) The neurocognition of language. Oxford University Press, Oxford, pp 319–336
747. Uylings HBM, Zilles K, Rajkowska G (1999) Optimal staining methods for delineation of cortical areas and neuron counts in human brains. Neuroimage 9:439–445
748. Uylings HBM, Sanz-Arigita E, de Vos K et al (2000) The importance of a human 3D database and atlas for studies of prefrontal and thalamic functions. Prog Brain Res 126:357–368
749. Uylings HBM, Groenewegen HJ, Kolb B (2003) Do rats have a prefrontal cortex? Behav Brain Res 146:3–17
750. Uylings HBM, Rajkowska G, Sanz-Arigita E, Amunts K, Zilles K (2005) Consequences of large interindividual variability for human brain atlases: converging macroscopical imaging and microscopical neuroanatomy. Anat Embryol (Berl) 210:423–431
751. Uylings HBM, Sanz-Arigita EJ, de Vos K et al (2007) 3-D cytoarchitectonic parcellation of human orbitofrontal cortex. Correlation with postmortem MRI. [submitted]
752. Valenstein ES (1990) The prefrontal area and psychosurgery. Prog Brain Res 85:539–553
753. Valverde F (1967) Apical dendritic spines of the visual cortex and light deprivation in the mouse. Exp Brain Res 3:337–352
754. Valverde F (1976) Aspects of cortical organization related to the geometry of neurons with intracortical axons. J Neurocytol 5:509–529
755. Valverde F (1985) The organizing principles of the primary visual cortex in the monkey. In: Peters A, Jones EG (eds) Visual cortex. Plenum, New York, pp 207–257 (Cerebral cortex, vol 3)
756. Valverde F (1986) Intrinsic neocortical organization: some comparative aspects. Neuroscience 18:1–23
757. Valverde F, Facal-Valverde MV (1986) Neocortical layers I and II of the hedgehog (Erinaceus europaeus). I. Intrinsic organization. Anat Embryol (Berl) 173:413–430
758. Van der Loos H (1979) The development of topological equivalencies in the brain. In: Meisami A, Brazier MAB (eds) Neuronal growth and differentiation. Raven, New York, pp 331–336
759. Van der Loos H, Woolsey TA (1973) Somatosensory cortex: structural alterations following early injury to sense organs. Science 179:395–398
760. Van Eden CG, Mrzljak L, Voorn P, Uylings HB (1989) Prenatal development of GABA-ergic neurons in the neocortex of the rat. J Comp Neurol 289:213–227
761. Van Essen DC (1985) Functional organization of primate visual cortex. In: Peters A, Jones EG (eds) Visual cortex. Plenum, New York, pp 259–329 (Cerebral cortex, vol 3)

762. Van Essen DC (2002) Windows on the brain: the emerging role of atlases and databases in neuroscience. Curr Opin Neurobiol 12:574–579
763. Van Essen DC (2004) Surface-based approaches to spatial localization and registration in primate cerebral cortex. Neuroimage 23 [Suppl 1]:S97–107
764. Van Essen DC (2005) A Population-Average, Landmark- and Surface-based (PALS) atlas of human cerebral cortex. Neuroimage 28:635–662
765. Van Essen DC (2006) SumsDB website: *http://sumsdb.wustl.edu:8081/sums/index.jsp*
766. Van Essen DC, Drury HA (1997) Structural and functional analyses of human cerebral cortex using a surface-based atlas. J Neurosci 17:7079–7102
767. Van Essen DC, Maunsell JHR (1980) Two-dimensional maps of the cerebral cortex. J Comp Neurol 191:255–281
768. Van Essen DC, Maunsell JHR, Bixby JL (1981) The middle temporal visual area in the macaque: myelo-architecture, connections, functional properties and topographic organization. J Comp Neurol 199:293–326
769. Van Essen DC, Newsome WT, Maunsell JHR, Bixby JL (1986) The projections from striate cortex (V1) to areas V2 and V3 in the macaque monkey: asymmetries, areal boundaries, and patchy connections. J Comp Neurol 244:451–481
770. Van Essen DC, Anderson CH, Felleman DJ (1992) Information processing in the primate visual system: an integrated systems perspective. Science 255:419–423
771. Van Essen DC, Lewis JW, Drury HA et al (2001) Mapping visual cortex in monkeys and humans using surface-based atlases. Vision Res 41:1359–1378
772. Van Hoesen GW (1982) The parahippocampal gyrus. New observations regarding its cortical connections in the monkey. Trends Neurosci 5:345–350
773. Van Hoesen GW (1993) The modern concept of association cortex. Curr Opin Neurobiol 3:150–154
774. Van Hoesen GW, Rosene DL, Mesulam M-M (1979) Subicular input from temporal cortex in the rhesus monkey. Science 205:608–610
775. Vercelli AE, Assal F, Innocenti GM (1992) Emergence of callosally projecting neurons with stellate morphology in the visual cortex of the kitten. Exp Brain Res 90:346–358
776. Vercelli AE, Garbossa D, Curtetti R, Innocenti GM (2004) Somatodendritic minicolumns of output neurons in the rat visual cortex. Eur J Neurosci 20:495–502
777. Verhagen JV, Kadohisa M, Rolls ET (2004) Primate insular/opercular taste cortex: neuronal representations of the viscosity, fat texture, grittiness, temperature, and taste of foods. J Neurophysiol 92:1685–1699
778. Vogt Weisenhorn DM, Illing RB, Spatz WB (1995) Morphology and connections of neurons in area 17 projecting to the extrastriate areas MT and 19DM and to the superior colliculus in the monkey Callithrix jacchus. J Comp Neurol 362:233–255
779. Vogt BA (1991) The role of layer I in cortical function. In: Peters A, Jones EG (eds) Normal and altered states of function. Plenum, New York, pp 49–80 (Cerebral cortex, vol 9)
780. Vogt BA (2005) Pain and emotion interactions in subregions of the cingulate gyrus. Nat Rev Neurosci 6:533–544
781. Vogt BA, Pandya DH (1978) Cortico-cortical connections of somatic sensory cortex (areas 3, 1 and 2) in the rhesus monkey. J Comp Neurol 177:179–192
782. Vogt BA, Pandya DN (1987) Cingulate cortex of the rhesus monkey: II. Cortical afferents. J Comp Neurol 262:271–289
783. Vogt BA, Hof PR, Vogt LJ (2004) Cingulate gyrus. In: Paxinos G (ed) The human nervous system, 2nd edn. Elsevier, Amsterdam, pp 915–949
784. Vogt C, Vogt O (1919) Allgemeinere Ergebnisse unserer Hirnforschung. J Psychol Neurol 25:279–461
785. Vogt C, Vogt O (1926) Die vergleichend architektonische und die vergleichend physiologische Felderung der Grosshirnrinde. Naturwissenschaften 14:1191–1194
786. Vogt C, Vogt O (1942) Morphologische Gestaltungen unter normalen und pathogenen Bedingungen. J Psychol Neurol 50:162–524
787. Vogt C, Vogt O (1954) Gestaltung der topistischen Hirnforschung und ihre Förderung durch den Hirnbau und seine Anomalien. J Hirnforsch 1:1–46
788. Vogt C, Vogt O (1956) Weitere Ausführungen zum Arbeitsprogramm des Hirnforschungsinstitutes in Neustadt/Schwarzwald. J Hirnforsch 2:404–427
789. Vogt O (1910) Die myeloarchitektonische Felderung des menschlichen Stirnhirns. J Psychol Neurol 15:221–232
790. Vogt O (1911) Die Myeloarchitektonik des Isocortex parietalis. J Psychol Neurol 18:379–390
791. Vogt O (1927) Architektonik der menschlichen Hirnrinde. Allg Z Psychiatr 86:247–274
792. Volkow ND, Fowler JS, Wang GJ (2004) The addicted human brain viewed in the light of imaging studies: brain circuits and treatment strategies. Neuropharmacology 47 Suppl 1:3–13
793. Von Bonin G (1950) Essay on the cerebral cortex. Charles C Thomas, Springfield, IL
794. Von Bonin G, Bailey P (1947) The neocortex of Macaca mulatta. University of Illinois Press, Urbana
795. Von Economo C (1927) Zellaufbau der Groszhirnrinde des Menschen. Springer, Berlin
796. Von Economo C, Koskinas GN (1925) Die Cytoarchitektonik der Hirnrinde des erwachsenen Menschen. Springer, Berlin
797. Von Monakow C (1911) Lokalisation der Hirnfunktionen. J Psychol Neurol 17:185–200

798. Von Monakow C (1914) Die Lokalisation im Grosshirn und der Abbau der Funktion durch Kortikale Herde. J.F. Bergmann, Wiesbaden
799. Wada J, Rasmussen T (1960) Intracarotid injection of sodium amytal for the lateralisation of cerebral speech dominance. J Neurosurg 17:226–282
800. Wada JA, Clarke R, Hamm A (1975) Cerebral hemispheric asymmetry in humans. Cortical speech zones in 100 adults and 100 infant brains. Arch Neurol 32:239–246
801. Wahle P (1993) Differential regulation of substance P and somatostatin in Martinotti cells of the developing cat visual cortex. J Comp Neurol 329:519–538
802. Walker AE (1940) A cytoarchitectural study of the prefrontal area of the macaque monkey. J Comp Neurol 73:59–86
803. Walker AE (1944) Psychosurgery: collective review. Int Abstr Surg 78:1–11
804. Wang XJ (2001) Synaptic reverberation underlying mnemonic persistent activity. Trends Neurosci 24:455–463
805. Wang Y, Gupta A, Toledo-Rodriguez M, Wu CZ, Markram H (2002) Anatomical, physiological, molecular and circuit properties of nest basket cells in the developing somatosensory cortex. Cereb Cortex 12:395–410
806. Wang Y, Markram H, Goodman PH et al (2006) Heterogeneity in the pyramidal network of the medial prefrontal cortex. Nat Neurosci 9:534–542
807. Wartenburger I, Heekeren HR, Burchert F et al (2004) Neural correlates of syntactic transformations. Hum Brain Mapp 22:72–81
808. Watson JD, Myers R, Frackowiak RS et al (1993) Area V5 of the human brain: evidence from a combined study using positron emission tomography and magnetic resonance imaging. Cereb Cortex 3:79–94
809. Webster MJ, Ungerleider LG, Bachevalier J (1991) Connections of inferior temporal areas TE and TEO with medial temporal-lobe structures in infant and adult monkeys. J Neurosci 11:1095–1116
810. Werner L, Winkelmann E, Koglin A, Neser J, Rodewohl H (1989) A Golgi deimpregnation study of neurons in the rhesus monkey visual cortex (areas 17 and 18). Anat Embryol (Berl) 180:583–597
811. Wernicke C (1874) Der aphasische Symptomenkomplex. Cohen und Weigert, Breslau
812. Wernicke C (1881) Lehrbuch der Gehirnkrankheiten. Fischer, Kassel
813. White EL (1978) Identified neurons in mouse Sml cortex which are postsynaptic to thalamocortical axon terminals: a combined Golgi-electron microscopic and degeneration study. J Comp Neurol 181:627–661
814. White EL (1986) Termination of thalamic afferents in the cerebral cortex. In: Jones EG, Peters A (eds) Sensory-motor areas and aspects of cortical connectivity. Plenum, New York, pp 271–289 (Cerebral cortex, vol 5)
815. White EL (1989) Cortical circuits. Synaptic organization of the cerebral cortex: Structure, function and theory. Birkhäuser, Boston
816. White EL, Czeiger D (1991) Synapses made by axons of callosal projection neurons in mouse somatosensory cortex: emphasis on intrinsic connections. J Comp Neurol 303:233–244
817. White EL, Hersch SM (1981) Thalamocortical synapses of pyramidal cells which project from SmI to MsI cortex in the mouse. J Comp Neurol 198: 167–181
818. White EL, Hersch SM (1982) A quantitative study of thalamocortical and other synapses involving the apical dendrites of corticothalamic projection cells in mouse SmI cortex. J Neurocytol 11:137–157
819. White EL, Keller A (1987) Intrinsic circuitry involving the local axon collaterals of corticothalamic projection cells in mouse SmI cortex. J Comp Neurol 262:13–26
820. White EL, Hersch SM, Belford GR (1982) Quantitative studies of thalamocortical synapses with labeled pyramidal cells in mouse SmI cortex. Soc Neurosci Abstr 8:853
821. Willmes K, Poeck K (1993) To what extent can aphasic syndromes be localized? Brain 116:1527–1540
822. Wilson FA, Scalaidhe SP, Goldman-Rakic PS (1993) Dissociation of object and spatial processing domains in primate prefrontal cortex. Science 260:1955–1958
823. Winer JA (1982) The stellate neurons in layer IV of primary auditory cortex (AI) of the cat: a study of columnar organization. Soc Neurosci Abstr 8:1020
824. Winfield DA, Gatter KC, Powell TPS (1980) An electron microscopic study of the types and proportions of neurons in the cortex of the motor and visual areas of the cat and rat. Brain 103:245–258
825. Winfield DA, Brooke RN, Sloper JJ, Powell TP (1981) A combined Golgi-electron microscopic study of the synapses made by the proximal axon and recurrent collaterals of a pyramidal cell in the somatic sensory cortex of the monkey. Neuroscience 6:1217–1230
826. Winkelmann E, Brauer K, Berger U (1975) Zur columnaren Organisation von Pyramidenzellen im visuellen Cortex der Albinoratte. Z Mikrosk Anat Forsch 89:239–256
827. Wise RJ, Scott SK, Blank SC et al (2001) Separate neural subsystems within 'Wernicke's area'. Brain 124:83–95
828. Wise SP, Boussaoud D, Johnson PB, Caminiti R (1997) Premotor and parietal cortex: corticocortical connectivity and combinatorial computations. Annu Rev Neurosci 20:25–42
829. Witelson SF, Pallie W (1973) Left hemisphere specialization for language in the newborn. Neuroanatomical evidence of asymmetry. Brain 96:641–646

830. Woolsey CN (1958) Organization of somatic sensory and motor areas of the cerebral cortex. In: Harlow HF, Woolsey CN (eds) Biological and biochemical bases of behavior. University of Wisconsin Press, Madison, pp 63–81
831. Woolsey CN (1964) Cortical localization as defined by evoked potential and electrical stimulation studies. In: Schaltenbrand G, Woolsey CN (eds) Cerebral localization and organization. University of Wisconsin Press, Madison, pp 17–32
832. Woolsey CN, Marshall WH, Bard P (1942) Representation of cutaneous tactile sensibility in the cerebral cortex of the monkey as indicated by evoked potentials. Bull Johns Hopkins Hosp 70:399–441
833. Woolsey CN, Marshall WH, Bard P (1943) Note on the organization of the tactile sensory area of the cerebral cortex of the chimpanzee. J Neurophysiol 6:287–291
834. Woolsey TA, van der Loos H (1970) The structural organization of layer IV in the somatosensory region (SI) of mouse cerebral cortex. Brain Res 17:205–242
835. Xiang Z, Huguenard JR, Prince DA (1998) Cholinergic switching within neocortical inhibitory networks. Science 281:985–988
836. Yamamoto T, Samejima A, Oka H (1987) Morphological features of layer V pyramidal neurons in the cat parietal cortex: an intracellular HRP study. J Comp Neurol 265:380–390
837. Yamamoto T, Samejima A, Oka H (1987) Morphology of layer V pyramidal neurons in the cat somatosensory cortex: an intracellular HRP study. Brain Res 437:369–374
838. Yaxley S, Rolls ET, Sienkiewicz ZJ (1990) Gustatory responses of single neurons in the insula of the macaque monkey. J Neurophysiol 63:689–700
839. Yeterian EH, Pandya DN (1991) Prefrontostriatal connections in relation to cortical architectonic organization in rhesus monkeys. J Comp Neurol 312:43–67
840. Yokochi H, Tanaka M, Kumashiro M, Iriki A (2003) Inferior parietal somatosensory neurons coding face-hand coordination in Japanese macaques. Somatosens Mot Res 20:115–125
841. Yoshimura Y, Callaway EM (2005) Fine-scale specificity of cortical networks depends on inhibitory cell type and connectivity. Nat Neurosci 8:1552–1559
842. Yoshimura Y, Dantzker JL, Callaway EM (2005) Excitatory cortical neurons form fine-scale functional networks. Nature 433:868–873
843. Yukie M (2000) Connections between the medial temporal cortex and the CA1 subfield of the hippocampal formation in the Japanese monkey (Macaca fuscata). J Comp Neurol 423:282–298
844. Yuste R, Peinado A, Katz LC (1992) Neuronal domains in developing neocortex. Science 257:665–669
845. Zald DH, Pardo JV (2000) Functional neuroimaging of the olfactory system in humans. Int J Psychophysiol 36:165–181
846. Zeki S (1997) The color and motion systems as guides to conscious visual perception. In: Jones EG, Peters A (eds) Extrastriate cortex in primates. Plenum, New York, pp 777–809 (Cerebral cortex, vol 12)
847. Zeki S (2003) Improbable areas in the visual brain. Trends Neurosci 26:23–26
848. Zilbovicius M, Meresse I, Chabane N et al (2006) Autism, the superior temporal sulcus and social perception. Trends Neurosci 29:359–366
849. Zilles K (2004) Architecture of the human cerebral cortex. In: Paxinos G, Mai JK (eds) The human nervous system, 2nd edn. Elsevier, Amsterdam, pp 997–1055
850. Zilles K, Clarke S (1997) Architecture, connectivity, and transmitter receptors of human extrastriate visual cortex. In: Jones EG, Peters A (eds) Extrastriate cortex in primates. Plenum, New York, pp 673–742 (Cerebral cortex, vol 12)
851. Zilles K, Schleicher A, Palomero-Gallagher N, Amunts K (2002) Quantitative analysis of cyto- and receptorarchitecture of the human brain. In: Toga AW, Mazziotta JC (eds) Brain mapping: The methods, 2nd edn. Academic Press, San Diego, pp 573–602

Section III Functional Systems

16 General Sensory Systems and Taste

Introduction

Conscious perception of cutaneous and proprioceptive sensation and taste depends on conduction pathways that connect peripheral receptors with centres in the diencephalon and the telencephalon. Most of these connections, with the exception of the pathway for taste, are crossed. Initial processing of sensory information occurs in the spinal cord and the dorsal column nuclei for the trunk and the extremities, and in the sensory nuclei of the trigeminal nerve for the face. The first relay for taste is located in the nucleus of the solitary tract in the lower brain stem. Spinal conduction pathways are the anterolateral fasciculus and the dorsal column-medial lemniscus pathway. Both of these systems are crossed. Corresponding paths take their origin from the sensory nuclei of the trigeminal nerve.

Ascending fibres from the spinal cord cross in the ventral white commissure and join the anterolateral fasciculus, a composite pathway, which terminates at different levels of the brain stem, the thalamus, the hypothalamus and in restricted regions of the telencephalon (Figs. 16.1, 16.2). The medial lemniscus takes its origin from the dorsal column nuclei in the caudal medulla. Its fibres cross at this level and proceed towards the thalamus. The dorsal column nuclei receive primary afferents from the dorsal roots that ascend in the dorsal columns and the postsynaptic dorsal column pathway, which takes its origin from the dorsal horn (Figs. 16.3, 16.4).

Tactile sensibility requires a precise localization of the stimulus and active exploration of shapes and surfaces. Receptors in deep tissues, muscle, tendons and joints subserve the sense of position and movement. Discriminative qualities of tactile sensation and position sense are primarily conducted by the dorsal column-medial lemniscus pathway and the corresponding pathway from the sensory nuclei of the trigeminal nerve, both of which are relayed in the ventrobasal nuclear complex of the thalamus. In patients with lesions of the dorsal columns the localization of tactile stimuli and position sense is mostly intact, therefore, other ascending spinal systems must be involved.

The pain-related circuitry extends over both the cognitive and the emotional-motivational domains of the brain. It includes the spinothalamic component of the anterolateral fasciculus, the trigeminothalamic tract and pathways that connect the spinal cord and the sensory nuclei of the trigeminal nerve with the brain stem, the hypothalamus and the basal forebrain.

Conscious perception of sensory stimuli always involves a projection of the thalamus to one or more sensory areas of the cerebral cortex, and the processing of the sensory input in adjacent cortical fields. Some systems subserving motivational and emotional aspects of sensation, circumvent the thalamus and terminate in the amygdaloid nucleus and/or the basal telencephalon. Descending systems from the cerebral cortex and brain stem centres, which generally are in receipt of sensory information, control sensory transmission at lower levels in a reciprocal manner.

Peripheral Receptors and Sensory Afferents

Peripheral receptors are the endings of the peripheral branches of the axons of the cells of the spinal and cranial ganglia. Their central branches enter the spinal cord or the brain stem. Peripheral receptors can be subdivided into encapsulated endings and free nerve endings (Table 16.1). Peripheral receptors recently were reviewed by Kaas [91] and Willis and Westlund in 2004 [164]. In encapsulated endings the distal axon ramifies among a core of modified epithelial or Schwann cells, which is surrounded by cellular lamellae and a connective tissue capsule of modified perineuronal cells. In Merkel disc endings a large axon expands into a disc just below the epidermis. The disc is closely applied to the base of a Merkel cell, which is inserted into the epidermis with spike-like protrusions. Merkel cells contain large dense-cored vesicles filled with various peptides. These cells have the potential to release the peptides, together with glutamate [82, 134]. In free endings the distal axon is only surrounded by an incomplete sheath of Schwann cells. Transduction may take place at the gaps between the Schwann cells, where the axon is freely exposed to its surroundings.

Different encapsulated endings have been distinguished in the skin. Ruffini endings, like the Merkel discs, are slowly adapting mechanoreceptors that detect local pressure on the skin, as in touch. Merkel disks occur in the epidermis of glabrous skin and in aggregates (touch domes, *Haarscheibe*) in hairy skin. Ruffini endings are present in the dermis. In the hair follicles they occur as aggregates of lanceolar endings. They also occur in deep tissues, joints and the periodontal ligament. Meissner corpuscles are rapidly adapting receptors that are mainly found in dermal papillae of glabrous skin. They are sensitive to moving stimuli and may be involved in sensing shape and texture. Pacinian corpuscles are large, encapsulated, rapidly adapting receptors. In glabrous skin they are located in the subcutis, but they occur in all deep tissues. High frequency vibrations are perceived by Pacinian corpuscles. Encapsulated endings in fasciae, joints and muscles are responsible for the perception of position and movement. Apart from the Ruffini endings and Pacinian corpuscles, they occur as Golgi tendon endings and muscle spindles.

Fibres from peripheral nerves can be classified according to their conduction velocity, a property that is directly related to their calibre. Two classifications are in use. The classification of Erlanger and Gasser [66] in A alpha-delta, B and C fibres was introduced for sensory nerves, and the roman numeral classification of Lloyd [105] for muscle nerves (Table 16.1).

Sensory afferents from the Merkel discs and encapsulated endings generally are medium-sized myelinated fibres in the A-beta (group II) or A-gamma (group III) range. Muscle spindle primary endings and Golgi spindle afferents are among the largest myelinated fibres (A-alpha or group I) in peripheral nerves. The secondary endings of muscle spindles are derived from smaller (A-beta or group II) fibres.

Free nerve endings are related to sensations of pain, itch, warmth and cold. Specific nociceptors are thin, myelinated fibres (A-delta or group III) that are activated by tissue damage. Stimulation of these fibres leads to sharp pricking pain. Polymodal afferents are thin

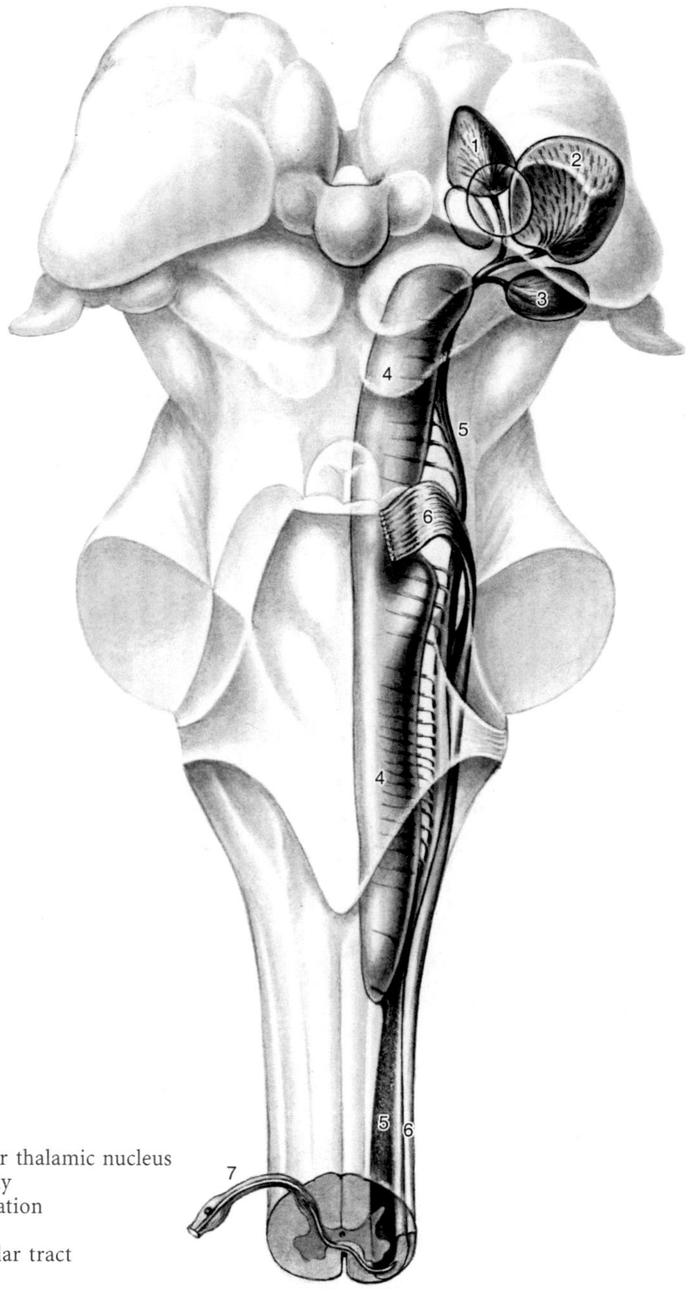

1 Intralaminar nuclei
2 Lateral ventroposterior thalamic nucleus
3 Medial geniculate body
4 Medial reticular formation
5 Spinothalamic tract
6 Anterior spinocerebellar tract
7 Spinal dorsal root

Fig. 16.1. The anterolateral fasciculus. The position of the ascending spinal tracts and related nuclei in a dorsal view (5/3×)

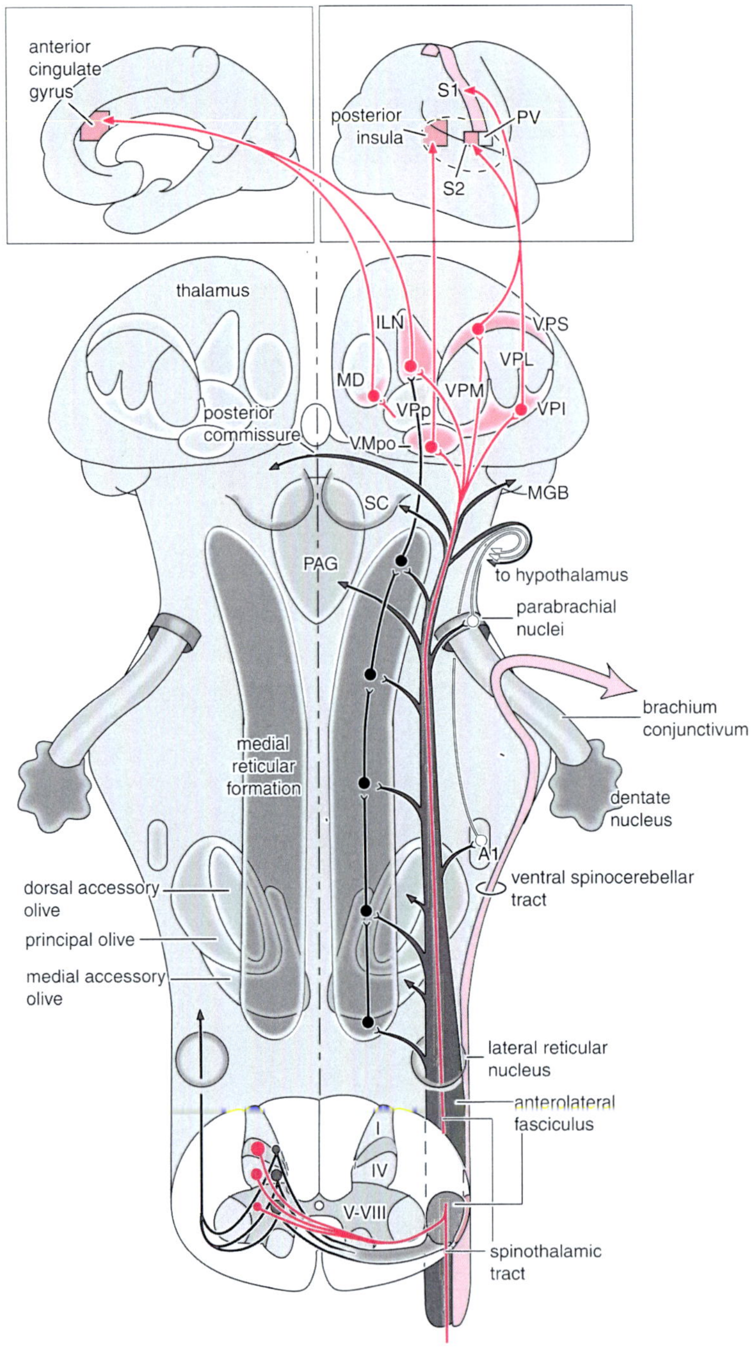

Fig. 16.2. The anterolateral fasciculus is a composite system arising from different spinal cord laminae. Its terminations include the medial reticular formation, lateral reticular nucleus, medial and dorsal accessory olives, parabrachial nuclei, periaqueductal grey, the amygdaloid nucleus and the hypothalamus. A small proportion of its fibres, known as the spinothalamic tract, terminates in the thalamus. It is shown in *red*. Its terminations include the intralaminar nuclei, the caudal medial dorsal nucleus, and the ventromedial posterior nucleus. In the ventral posterior complex, spinothalamic fibres terminate mainly, but not exclusively along its rostral and caudal borders and along the laminae separating the medial and lateral ventral posterior nuclei, here indicated as the inferior and superior ventral posterior nuclei. These regions can be distinguished as

and unmyelinated fibres (C or group IV) that respond to mechanical, thermal and/or chemical stimuli. They give rise to a sensation of dull, burning pain or itch.

Two major classes of C fibres have been distinguished in rodents. Peptidergic C fibres express the peptides calcitonin gene-related peptide (CGRP) and substance P. Peptidergic C fibres depend for their integrity on nerve growth factor (NGF) and express the NGF receptor tyrosine kinase (TRK). In non-peptidergic C fibres TRK is downregulated but another receptor tyrosine kinase (RET) is upregulated. These neurons bind the lectin IB4 [150]. A subset of non-peptidergic C fibres that express Mrgprd, a member of a large family of sensory neuron-specific G protein coupled receptors, exclusively innervate the epidermis. Their terminals are located under the stratum corneum, superficial to the terminals of the peptidergic C fibres which are located in the stratum spinosum [170]. Peptidergic and non-peptidergic C fibres also differ in their central connections and their functional properties.

Electrical stimulation of visceral nerves at low intensities results in vague sensations of fullness and nausea, but higher intensities cause a sensation of pain. Pain elicited by the activation of specific nociceptors is called nociceptive pain. However, pain may also result from injury to sensory fibres or from damage to the central nervous system. These types of pain are designated as non-nociceptive or neuropathic pain.

The cell bodies of the primary afferents are located in the ganglia of the dorsal roots and the cranial nerves. Muscle spindle afferents from facial and masticatory muscles are an exception: their cell bodies are located in the mesencephalic nucleus of the trigeminal nerve in the upper brain stem. Large pale cells of the ganglia give rise to myelinated afferents; the axons of the small and dark ganglion cells are unmyelinated.

Architecture of the Dorsal Horn of the Spinal Cord

The spinal dorsal horn (and its equivalent, the pars caudalis of the spinal tract nucleus of the trigeminal nerves) constitutes one of the main relay stations for primary afferents of the dorsal roots. The spinal grey matter can be subdivided into ten dorsoventrally arranged laminae of Rexed [135], of which I–VI fall within the confines of the spinal dorsal horn (Fig. 6.2). Lamina I contains fairly large neurons, with dendrites extending tangentially over the surface of the dorsal horn. Different morphological classes of lamina I cells can be distinguished and have been identified as the relay cells for different sensory submodalities [169]. Laminae II and III are mainly composed of densely packed small local circuit neurons. The absence of myelin in layer II is responsible

(continued text of Fig. 16.2) the "shell" or "matrix" from the "core" of the ventroposterior complex, represented by the medial and lateral ventroposterior nuclei and the specialized neuronal aggregates they contain, which are innervated by the medial lemniscus. Thalamic nuclei of the spino- and trigeminothalamic system project to the primary and secondary somatosensory cortex, the posterior insula and the anterior gyrus cinguli. Spinothalamic fibres originate from different laminae of the spinal cord. The distribution of the specific nociceptive and thermoceptive neurons of lamina I is shown in Fig. 16.6. The multisynaptic projection of the spinal cord to the intralaminar nuclei, which passes through the medial reticular formation, is known as the "medial spinothalamic tract". *I–VIII*, cytoarchitectonic laminae I–VIII of the spinal grey matter; *A1*, noradrenergic cell group A1; *ILN*, intralaminar nuclei; *MD*, medial dorsal thalamic nucleus; *MGB*, medial geniculate body; *PAG*, periaqueductal grey; *PV*, ventral parietal area; *SC*, superior colliculus; *S1(2)*, primary (secondary) somatosensory cortex; *SC*, superior colliculus; *VMpo*, posterior ventromedial thalamic nucleus; *VPI*, inferior ventroposterior thalamic nucleus; *VPL*, lateral ventroposterior thalamic nucleus; *VPM*, medial ventroposterior thalamic nucleus; *VPp*, parvocellular medial ventroposterior thalamic nucleus; *VPS*, superior ventroposterior thalamic nucleus

Table 16.1. Peripheral receptors and their afferent fibres. Fibres in peripheral nerves can be subdivided into myelinated (A) fibres and unmyelinated (C) fibres. Myelinated fibres are further subdivided into four classes (A-alpha – A-delta) according to their conduction velocity, which correspond with their calibre [66]. For afferents innervating the musculoskeletal system, another classification into four groups indicated with roman numerals (I–IV) is often used [105]

Encapsulated nerve endings

Type	Adaptation	Localization	Sensation	Type afferent	Calibre
Merkel discs	Slowly adapting	Glabrous skin: epidermis; hairy skin: aggregates	Touch	A-beta	7–12 ηm
Ruffini endings	Slowly adapting	Skin: dermis deep tissues, joints		A-beta (II)	7–12 ηm
Meissner corpuscles	Rapidly adapting	Skin: dermis lanceolate endings in hair follicles	Shape and texture	A-beta	7–12 ηm
Pacinian corpuscles	Rapidly adapting	Deep tissues	Vibration	A-beta/gamma II/(III)	3–12 ηm
Golgi tendon organs	Slowly adapting	Tendons	Position and movement	A-alpha (Ib)	12–22 ηm
Muscle spindles, annulospiral or primary endings	Slowly adapting	Striated muscles		A-alpha (Ia)	12–22 ηm
Muscle spindles, secondary endings	Slowly adapting			A-beta (II)	6–12 ηm

Free nerve endings

Type	Localization	Sensation	Type afferent	Calibre
Wide dynamic range: specific nociceptors	Skin, teeth, joints, muscles, viscera	Initial sharp "primary" pain	A-delta (III)	2–5 ηm
Polymodal afferents	Skin, teeth, joints, muscles, viscera	Burning, "secondary" pain	C (IV)	Unmyelinated
Innocuous cooling and heat	Skin	Warmth and cold	C	Unmyelinated
Itch	Skin	Itch	C	Unmyelinated

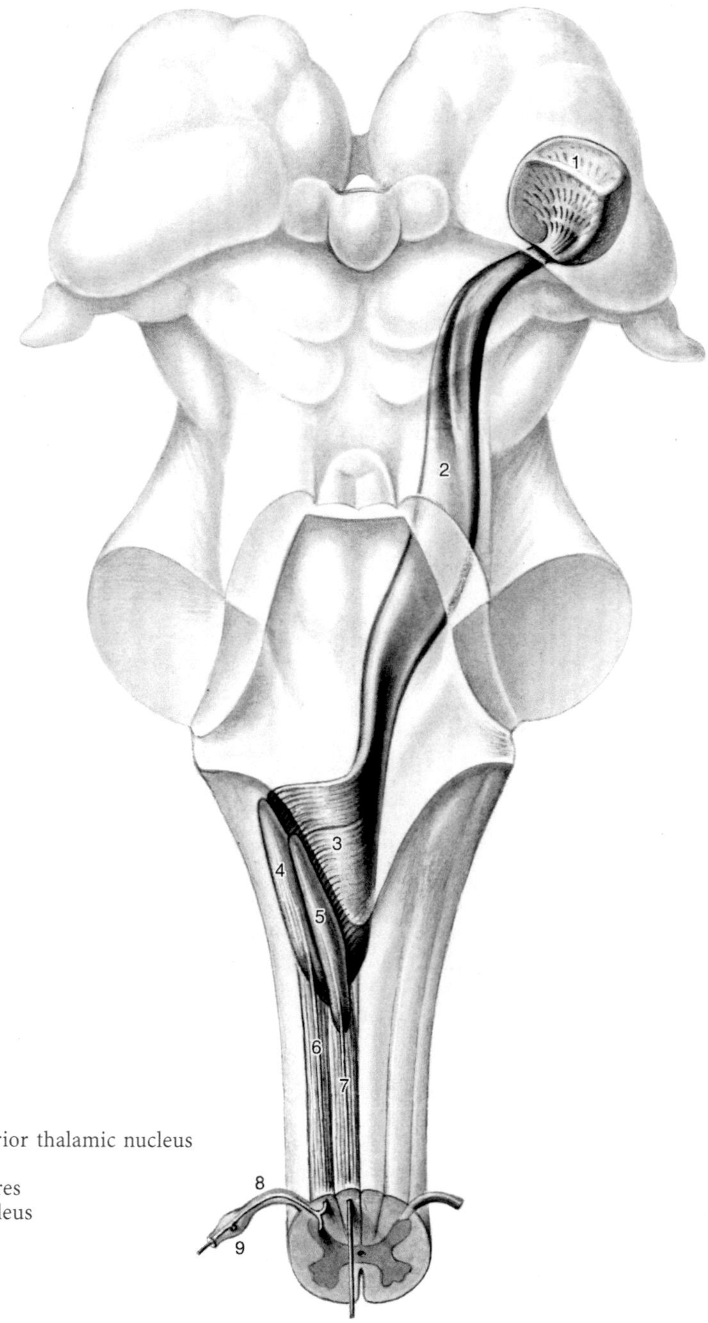

1 Lateral ventroposterior thalamic nucleus
2 Medial lemniscus
3 Internal arcuate fibres
4 Medial cuneate nucleus
5 Gracile nucleus
6 Cuneate fascicle
7 Gracile fascicle
8 Spinal dorsal root
9 Spinal ganglion

Fig. 16.3. The medial lemniscus. Position of tracts and nuclei in a dorsal view (5/3×)

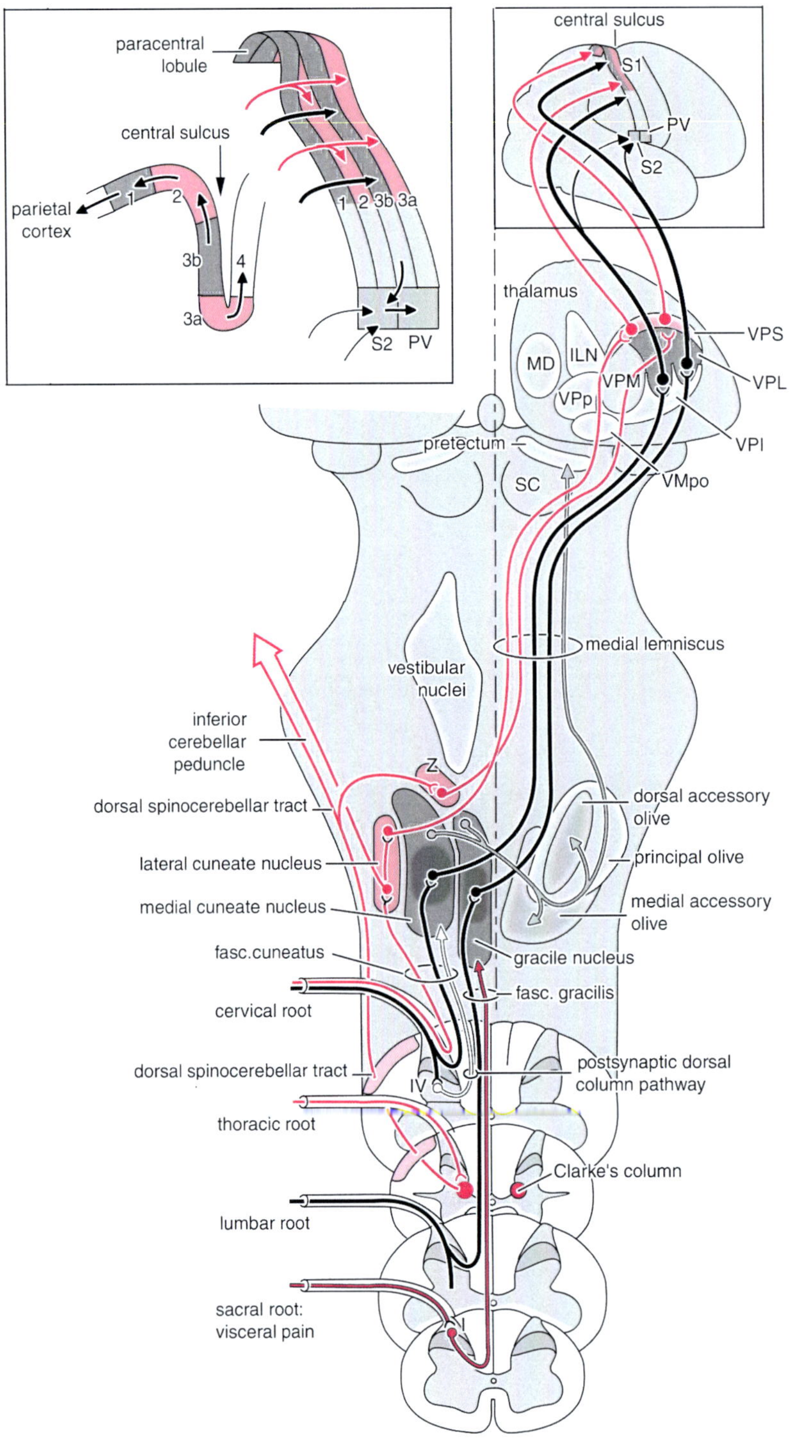

Fig. 16.4. The dorsal column-medial lemniscus pathway. Root fibres from sacral and lumbar dorsal roots ascend into the fasciculus gracilis; root fibres from cervical roots ascend into the fasciculus cuneatus, to terminate in the corresponding dorsal column nuclei. Fibres from the dorsal column nuclei decussate in the caudal medulla oblongata and ascend into the medial lemniscus to the lateral ventroposterior nucleus of the thalamus.

for its translucent appearance and the name of the substantia gelatinosa. Layer IV, or the nucleus proprius of the dorsal horn, contains large neurons with dendrites spreading equally in a longitudinal and a mediolateral direction. The dorsal dendrites enter laminae I and II. Multipolar neurons of different size occupy the base of the dorsal horn (laminae V and VI) and the intermediate grey matter (lamina VII). The dendrites of these disk-shaped neurons mainly extend in the transverse plane.

At their entrance into the cord, the unmyelinated C fibres and the thin, myelinated A-delta fibres separate from the dorsal root to join Lissauer's dorsolateral fasciculus on top of the dorsal horn. The larger, myelinated root fibres proceed to the dorsal columns, where many of them bifurcate into ascending and descending branches. These branches emit collaterals, which enter the spinal grey matter.

A-delta and C fibres conduct pain, temperature and itch. A-delta fibres terminate mainly in laminae I and V, whereas C fibres innervate principally lamina I and the outer layer of the substantia gelatinosa. In rodents peptidergic C fibres terminate on relay cells of the spinothalamic tract in layer I. Non-peptidergic C fibres terminate on interneurons of layer II [31, 170].

Collaterals from the myelinated fibres of the dorsal columns, chiefly cutaneous mechanoreceptors, terminate in the inner half of the substantia gelatinosa (lamina III), and in layers III–VI of the dorsal horn. Collaterals of hair follicle afferent fibres turn back in lamina V to terminate in the nucleus proprius and as the "flame-shaped arborizations" of Scheibel and Scheibel [140] in laminae III and II. Collaterals of slowly adapting cutaneous mechanoreceptors terminate in laminae III, IV and dorsal V. Projections from rapidly adapting mechanoreceptors, including the Pacinian corpuscles, are mainly restricted to lamina III, and laminae V and VI of the base of the dorsal horn. Collaterals from a single cutaneous afferent form narrow, radially oriented longitudinal sheets in laminae III–V of the dorsal horn. A-delta and C fibres terminate as longitudinal sheets in laminae I and II. These superficial sheets are in a somatotopical register with the sheets in the deeper laminae. Together they form the basis for the somatotopical localization in the dorsal horn [36, 37, 77] (Fig. 16.5 A). A wide variety of cutaneous and deep afferents, including those from muscle spindle and tendon organs, terminate in laminae VII through X (see also the topographical review, the Spinal Cord).

The cutaneous projections to the layers of the dorsal horn are somatotopically organized. At the level of the lumbar and cervical enlargements the digits are represented medially in

(continued text of Fig. 16.4) These fibres take their origin mainly, but not exclusively, from a central region of the gracile and medial cuneate nuclei that contains the representation of the hand and digits (Fig. 16.5 C). A component of the medial lemniscus arising from the lateral cuneate nucleus and group Z (shown in *red*), which transmits proprioceptive information from the upper extremity (lateral cuneate nucleus) and lower extremity (group Z), terminates in the superior ventroposterior nucleus. The lateral ventroposterior nucleus innervates areas 3 b and 1 of the primary somatosensory cortex and the secondary somatosensory cortex (S2). The thalamocortical projection of the superior ventroposterior nucleus terminates in areas 3 a and 2 (*inset*). A cascade of association fibres interconnects the somatosensory areas and the ventral parietal somatosensory area. Area 3 a is connected with the primary motor cortex. In addition, rostral and ventral parts of the gracile and medial cuneate nucleus project to brain stem nuclei, including the accessory olives, tectum and pretectum. The dorsal columns contain the postsynaptic dorsal column pathway from neurons in lamina IV of the dorsal horn. Sacral lamina III neurons transmit visceral pain through a pathway located in the fasciculus gracilis. *I, IV,* cytoarchitectonic laminae I and IV of the spinal grey matter; *1, 2, 3 a, 3 b, 4, 5,* Brodmann areas 1, 2, 3 a, 3 b, 4, 5; *ILN,* intralaminar nuclei; *MD,* medial dorsal thalamic nucleus; *PV,* ventral parietal somatosensory area; *S1(2),* primary (secondary) somatosensory cortex; *SC,* superior colliculus; *VMpo,* posterior ventromedial thalamic nucleus; *VPI,* inferior ventroposterior thalamic nucleus; *VPL,* lateral ventroposterior thalamic nucleus; *VPM,* medial ventroposterior thalamic nucleus; *VPp,* parvocellular medial ventroposterior thalamic nucleus; *VPS,* superior ventroposterior nucleus; *Z,* group Z

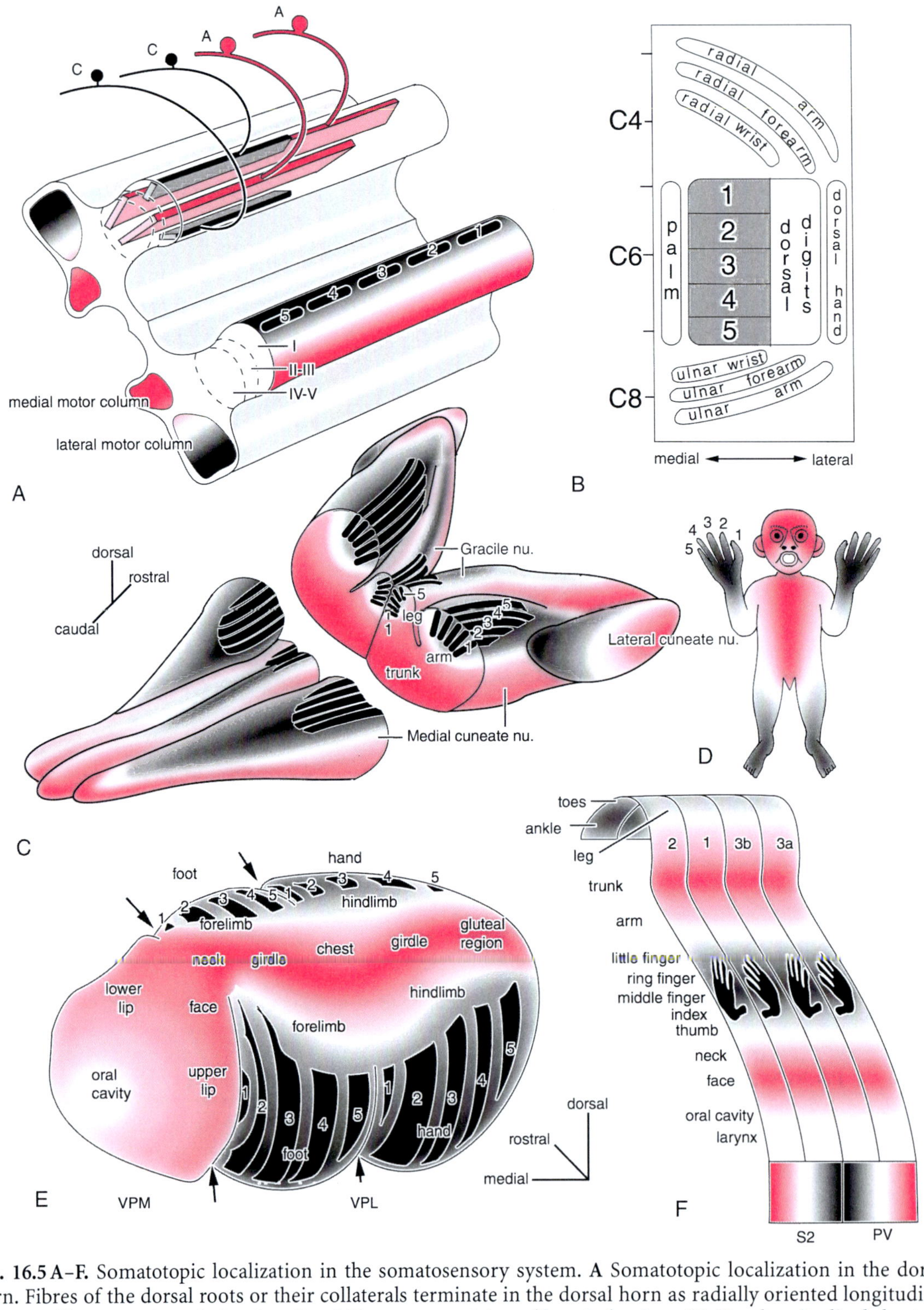

Fig. 16.5 A–F. Somatotopic localization in the somatosensory system. **A** Somatotopic localization in the dorsal horn. Fibres of the dorsal roots or their collaterals terminate in the dorsal horn as radially oriented longitudinal sheets. C fibres terminate in laminae I and II, cutaneous A-beta fibres in laminae III–V as longitudinal sheets in superficial and deep parts of the dorsal horn. Superficial and deep sheets, representing the same body part, are in topographically aligned register. This organization is the basis of the somatotopic localization in the dorsal horn, as shown in the right dorsal horn. Notice that the somatotopical localization in the dorsal horn, and the topical localization for proximal (*red*) and distal (*black*) muscles in the medial and lateral motor columns of the

the dorsal horn and the skin of the more proximal extremity and the girdle region in successively more lateral and rostral parts of the dorsal horn (Fig. 16.5 A and B). Afferents from the fingers terminate at cervical levels C5–7 in a rostrocaudal sequence in separate, elongated columns, with the thumb most rostrally [37, 68, 77, 98].

The main ascending pathways from the spinal cord are the spinocerebellar tracts (see Chap. 20, Cerebellum) and the anterolateral fasciculus. The latter includes the spinothalamic tract, which arises from lamina I and IV–IX. The nucleus proprius (lamina IV) of the dorsal horn gives rise to the spinocervical and the postsynaptic dorsal column pathways. Spinocervical fibres terminate on the lateral cervical nucleus, located in the dorsolateral funiculus at upper cervical levels. In humans the lateral cervical nucleus is vestigious [154] and its connections will not be considered here. The postsynaptic dorsal column pathway will be discussed in the next section of this chapter.

Dorsal Column-Medial Lemniscus System

The myelinated dorsal root fibres of the dorsal columns, which carry tactile and proprioceptive information, are somatotopically organized. Fibres from sacral and lumbar roots ascend medially, in the fasciculus gracilis; those from cervical roots ascend laterally, in the fasciculus cuneatus. A small contingent of thoracic fibres takes an intermediate position (Figs. 16.3, 16.4). The dorsal columns are not the exclusive pathways for the conduction of tactile and proprioceptive information because the primary afferents in the dorsal columns emit collaterals to the spinal grey matter rostral, but also caudal, to their level of entry. An important contingent of sacral and lumbar root fibres, carrying proprioceptive information from the lower limbs, leaves the dorsal columns to terminate in the spinal grey matter, particularly in Clarke's dorsal column, which gives rise to the dorsal spinocerebellar tract (Fig. 16.4). The dorsal columns continue into the medulla oblongata, where their fibres terminate on the dorsal column nuclei.

The dorsal column nuclei consist of the medial nucleus gracilis and the more laterally located medial and lateral cuneate nuclei. The dorsal column fibres terminate in a somatotopical manner over the whole length of the gracile and cuneate nuclei. Clusters of round cells with bushy dendrites are present at mid-rostrocaudal levels. They occupy the medial part of the gracile and the central region of the medial cuneate nucleus. In the medial cuneate nucleus this region is known as the pars rotunda. In other parts of the nuclei, triangular, fusiform and multipolar cells prevail. A large proportion of these cells is inhibitory [12]. The

(continued text of Fig. 16.5) anterior horn, are reversed. **B** Detailed somatotopical localization in the C4–C8 dorsal horn of the macaque monkey (redrawn with permission from Florence et al. [68]). **C** Diagram of the somatotopical localization in the dorsal column nuclei. The digits of the hand are represented in the central portion of the medial cuneate nucleus, the toes in the medial gracile nucleus. The order of the toes is reversed with respect to the fingers [129]. **D** Figurine showing the key for the somatotopic localization in the other panels. **E** Somatotopical localization in the right medial and lateral ventroposterior thalamic nuclei of the squirrel monkey. Fibrous laminae (*arrows*) separate the ventral parts of the medial and lateral ventral posterior nuclei and the representations of the digits from the hand and foot (redrawn with permision from Kaas et al. [92]). **F** Diagram of the somatotopical localization in the right somatosensory cortices. Notice mirror-shaped organization in adjacent somatosensory areas, exemplified by the position of palm and digits. *1, 2, 3a, 3b*, Brodmann areas 1, 2, 3a and 3b ; *1–5*, digits 1–5; *I–V*, cytoarchitectonic laminae I–V of the dorsal horn; *A*, A-beta root fibres; *C*, C fibres of the dorsal roots; *PV*, ventral parietal somatosensory area; *S2*, secondary somatosensory area; *VPL*, lateral ventroposterior thalamic nucleus; *VPM*, medial ventroposterior thalamic nucleus

pars rotunda contains the representation of the glabrous and hairy skin of the hand, with fingers 1–5 represented laterally to medially in discrete columns. The arm and the rump are represented as concentric sheets in the lateral, ventral and medial periphery of the medial cuneate nucleus (Fig. 16.5C). A similar localization of the lower extremity is present in the gracile nucleus, with the foot being represented in the dorsomedial cluster region. The sequence of the toes is reversed with respect to the fingers [129]. In the rostral and caudal poles of the rostral parts of the nuclei the somatotopic organization is much less precise [43, 68, 69, 146, 167]. Rapidly and slowly adapting cutaneous afferents activate the cells of the cluster regions. Afferents from Pacinian corpuscles are restricted to the caudal poles of the nuclei; slowly adapting muscle and joint afferents terminate in the ventral and rostral parts of the nuclei [63].

The lateral cuneate nucleus and group Z, a collection of large neurons rostral to the gracile nucleus, both receive information from muscle spindle afferents. Muscle spindle afferents from the forelimb ascend in the dorsal columns and terminate in a somatotopical pattern in the lateral cuneate nucleus. Group Z receives a bilateral input from the hindlimb as collaterals of the dorsal spinocerebellar tract [2, 75, 89, 139] (Fig. 16.4). No inhibitory interneurons appear to be present in the lateral cuneate nucleus.

In addition to primary afferents, the medial cuneate and gracile nuclei receive the postsynaptic dorsal column pathway, which takes its origin from the nucleus proprius of the dorsal horn and terminates in all regions of the nuclei (Fig. 16.4) [47]. The postsynaptic dorsal column pathway is considered to provide nociceptive and viscerosomatic input to the dorsal column nuclei and to be involved in the adaptation of receptive field size and excitability of its neurons [38, 64].

A specific component of the postsynaptic dorsal column pathway has been identified in the medial fasciculus gracilis. This pathway arises mainly from cells located in laminae III and X in the vicinity of the central canal of the sacral dorsal horn, which are in receipt of pelvic nerve afferents [112]. It terminates on neurons in the nucleus gracilis (Fig. 16.4) and has been shown to play an important role in viscerosensory transmission and visceral pain [3, 4]. Restricted lesions in the medial part of the fasciculus gracilis have been reported to alleviate or even eliminate intractable pelvic visceral pain [114].

Efferents from the diffusely organized ventral and rostral parts of the gracile and internal cuneate nuclei are more widely distributed than the efferents from the cell clusters in the dorsal parts of these nuclei (Fig. 16.4). Both regions project to the contralateral ventroposterior complex of the thalamus through the medial lemniscus, but their ventral and rostral parts also project to the cerebellum, the inferior olive, the tectum and pretectum, the red nucleus and back to the dorsal horn [18, 20, 22, 44, 102]. Fibres from the lateral cuneate nucleus and group Z join the medial lemniscus to terminate in the contralateral ventroposterior complex of the thalamus [30]. The main projection of the lateral cuneate nucleus is to the cerebellum.

Sensory Relay Nuclei of the Thalamus and the Somatosensory Cortex

The main sensory relay nuclei of the thalamus belong to the ventroposterior complex. This complex consists of the medial (VPM), lateral (VPL) and inferior (VPI) ventroposterior nuclei (Chap. 8, Fig. 8.2). VPM and VPL are separated by a fibrous lamina; a similar lamina partially bisects the VPL. VPI is located ventral to VPM and VPL. Its neurons invade the fibrous laminae between and within VPM and VPL. A superior ventroposterior nucleus (VPS) was distinguished, located dorsal and rostral to the VPL and VPM, on the basis of the afferents proceeding from deep receptors and projections to the somatosensory cortex [91, 99]. VPM and VPL have been indicated as the "core" of the ventroposterior complex. VPI and VPS surround VPM and VPL as its "shell".

The region posterior to the VPM and the VPI, located between the nucleus limitans medially, the pulvinar dorsally and the medial geniculate body ventrally, is traversed by the fibres of the medial lemniscus and the spinothalamic tract. It is known as the posterior nucleus or the regio basalis [123]. It contains the posterior division of the ventromedial nucleus (VMpo), a nucleus receiving a dense projection from nociceptive and thermoreceptive neurons in lamina I of the dorsal horn and the caudal spinal tract nucleus of the trigeminal nerve, recently identified by Craig and Blomqvist in this region [28, 51, 59].

The nuclei of the ventroposterior complex project to the primary sensory cortex (S1) in the postcentral gyrus and to adjacent areas in the parietal operculum and the insula. The connections of VMpo in sensory transmission shall be discussed in the section on the anterolateral fasciculus with the termination of the spinothalamic tract.

The dorsal column nuclei project to the contralateral VPL through the medial lemniscus. The principal sensory nucleus of the trigeminal nerve projects bilaterally to the VPM as the trigeminal component of the medial lemniscus and the uncrossed dorsal trigeminothalamic tract (see the section on the Trigeminal System). The gustatory parvocellular medial ventroposterior nucleus (VPp) is located medial to the VPM.

Fibres of the medial lemniscus terminate in a series of parallel laminae or rods that are present both in the VPM and the VPL and extend anteroposteriorly throughout these nuclei. Rostrally and dorsally, they continue into the VPS. The rods stain intensely for cytochrome oxidase, and the intervening matrix is only lightly stained. Large, parvalbumin-reactive neurons prevail in the rods. Small, 28-kDa calbindin-reactive neurons populate the matrix and the shell of the ventroposterior complex [131, 132]. Each rod represents a specific region of the body, but includes different sensory modalities along its anteroposterior axis. Tactile input reaches the VPM and the VPL, and input from deep receptors is segregated in the VPS [71, 90, 91]. The neurons located within the rods, which receive an input from the medial lemniscus, project to middle layers of the primary somatosensory cortex. The small cells of the matrix receive spinothalamic and trigeminothalamic afferents and project to its superficial layers [71, 90, 131, 133]. The main relay nucleus for the conduction of pain and temperature is the VMpo [55, 166].

VPL, VPM and VPS are somatotopically organized (Fig. 16.5 E). The intraoral cavity is represented medially in the VPM, along the border of the parvocellular VPM medial to the face. The hand and foot occupy large ventral regions, separated by a fibrous lamella between the medial and lateral VPL, with a mediolateral sequential representation of the digits. Hindlimb, trunk and forelimb are located from lateral to medial in the dorsal, undivided part of the nucleus [91, 92, 103].

VPL, VPM and VPS project somatotopically to four cytoarchitectonic areas in the postcentral gyrus, collectively known as the primary somatosensory cortex (S1). The four areas of Brodmann [33] are numbered 3 a, 3 b, 1 and 2 from anterior to posterior. They extend from the paracentral lobule on the medial wall of the hemisphere to the lateral sulcus. Two additional somatosensory areas are located adjacent to S1 in the parietal operculum. S2 is located caudally, and the rostral ventral parietal area PV more rostrally [49, 100]. These areas also have been delineated in the human brain [61, 65].

Each of the primary sensory areas contains a complete body map. The representation of the body surface is illustrated in Fig. 16.5 F. It follows the order as established by Penfield and Bodrey [122] by stimulation of the cortex in human patients. In the primary sensory areas the foot is located on the medial surface of the hemisphere, followed by the leg and trunk. The neck and the back of the head are included in the representation of the trunk. Shoulder, arm, hand and fingers are located more laterally, with the little finger medially and the thumb laterally. The face is next, with a bilateral representation of the intraoral cavity, lips, tongue [62], pharynx and larynx. The representation of the body image in areas 3 a, 3 b, 1, 2, S2 and PV reverses at the areal borders. This is most distinct for the palm of the hand with the digits and for the face (Fig. 16.5 F). Because detailed maps depend on multiple microelectrode recordings, the reversal of the representation at the areal borders has not been confirmed for the human brain, with the exception of the reversal at the S2/PV border [61].

The following account of the subdivision, the functional connectivity and the results of imaging studies of the somatosensory cortex is largely based on the recent review of the somatosensory cortex in human and nonhuman primates by Kaas [91]. Results from human brain mapping with PET and fMRI were discussed by Schnitzler [141].

Area 3 a is located at the bottom of the central sulcus. With its well-defined pyramidal layers and less distinct granular layers it resembles the primary motor cortex. It shares a projection from the VPS with area 2 and can be activated by muscle spindle and deep receptors (Fig. 16.4, inset). Area 3 b occupies the caudal bank of the central sulcus. With its well-developed layer IV and small pyramidal neurons, it is the prototype of a koniocortex. It shares projections from VPM and VPL with areas 1 and 2. Rapidly and slowly adapting cutaneous inputs are segregated in rostrocaudally arranged columns. Area 1 occupies the rostral half of the crown of the postcentral gyrus. It is dominated by rapidly adapting cutaneous afferents. Area 2 is located still more caudally. It contains a complex representation of both cutaneous and noncutaneous receptors. Unlike areas 3 a, 3 b and 1, area 2 does not extend on to the mesial hemisphere [78].

Receptive field size increases, and the organization becomes more complex, when one advances from area 3 b to area 1. Area 3 a projects to the motor cortex, to area 2 and S2. Each of areas 3 b, 1 and 2 is reciprocally connected with the more caudally located anterior and posterior parietal sensory areas. The callosal connections of S1 follow a similar pattern. They are sparse, or less dense, in the representations of the glabrous skin of the hand.

Several additional sensory areas have been distinguished in the cortex of the superior and inferior parietal lobules, the gyrus cinguli, the parietal operculum and the adjoining insula. These areas receive projections from primary sensory areas or from each other, but lack connections from the ventroposterior complex of the thalamus. They receive thalamocortical projections from the pulvinar, lateral posterior and/or ventral lateral nuclei instead. Somatosensory areas S2 and PV were already mentioned. S2 receives a minor projection from the ventroposterior complex (Fig. 16.4). In monkeys, two somatotopically organized areas are located more ventrally in the dorsal insula, collectively known as the ventral somatosensory area (VS) [49]. They may be links in corticolimbic pathways for tactile memory. The caudal insula contains the projection of Craig's nociceptive and thermoceptive VMpo nucleus [51, 59].

Posterior parietal areas generally maintain reciprocal connections with more rostral sensory areas and project to the motor, premotor and prefrontal cortex, the superior temporal cortex and the limbic cortex. The human superior parietal lobule was subdivided by Brodmann [33] into the small rostral area 5 and the larger area 7. The inferior parietal lobule consists of the angular area 39 and the supramarginal area 40. In monkeys, areas 5 and 7 occupy the superior and inferior parietal lobule, respectively. Pandya and Seltzer [120] in their studies of the monkey parietal lobe used a modified nomenclature of von Economo and Koskinas [159] (Fig. 21.6). The primary somatosensory cortex (more specifically area 2) provides direct inputs from both proprioceptive and cutaneous receptors to the adjoining area PE of the superior parietal lobule and rostral area PF of the inferior parietal lobule. PE projects to area PEc in the caudal superior parietal lobule and PGm on its medial surface; PF is connected with PG in the caudal inferior parietal lobule. A number of polymodal areas have been distinguished in the monkey intraparietal sulcus. The rostral part of the dorsal bank of the intraparietal sulcus belongs to area PE. It is distinguished as area PEip and, like area PE, receives somatosensory input. The rostral part of the ventral bank is known as the anterior intraparietal area (AIP) and receives somatosensory input from neighbouring fields. The ventral, lateral and medial intraparietal areas (VIP, LIP and MIP) are visuomotor fields which receive input from visual association areas (see Chap. 19, Visual Areas, Dorsal and Ventral Processing Streams, and Fig. 19.6). The posterior parietal cortex and the areas of the intraparietal sulcus are connected with the primary motor cortex and premotor areas in the frontal lobe. They are involved in the transformation of sensory input into goal-directed movements (see Chap. 21, Connections and Properties of Motor and Premotor Areas, and Fig. 21.6). These parietal areas also have been studied in the human brain [32, 79].

Large lesions involving S1 cause loss of fine, tactile discrimination and the perception of shapes and texture, but leave a crude localization of touch intact. After dorsal column lesions in monkeys, the 3b hand area can no longer be activated, and other spinal pathways are unable to substitute for this loss. Posterior parietal sensory areas are involved in higher-order processing of sensory information and allow convergence with visual and cerebellar afferent input. Functions attributed to the posterior parietal areas include the creation of a mental image of the body, the guidance of hand and eye movements and the recognition and memory of objects by touch. In humans, lesions of these areas lead to neglect of information from the contralateral visual space or body half.

Descending Control by the Somatosensory Cortex

Pyramidal tract fibres from layer Vb pyramidal cells in SI and S2 project topically to the contralateral sensory nuclei of the trigeminal nerve, the dorsal column nuclei and the dorsal horn. The main projections to the principal trigeminal sensory nucleus and the dorsal column nuclei originate from areas 3b, 1 and 2. In the cuneate and gracile nuclei, area 3b projects to their core, which contains the representation of the hand or foot and the digits; areas 1 and 2 project to the rostral, proprioceptive region of the nuclei. In the caudal subnucleus of the trigeminal spinal tract nucleus and the dorsal horn, these cortical areas project preferentially to the superficial layers I, II and III, and more sparsely to the deeper layers [17, 45, 46].

Ascending Pathways from the Spinal Cord: The Anterolateral Fasciculus and the Spinothalamic Tract, Conduction of Pain, Itch and Temperature

The anterolateral fasciculus and the spinocerebellar tracts are the main ascending pathways from the spinal cord to the brain stem. The spinocerebellar tracts are discussed in Chap. 20 (Localization in the Cerebellum). Most of the fibres joining the anterolateral fasciculus cross in the ventral white commissure, but a distinct ipsilateral component exists. The anterolateral fasciculus of Gowers [76] is a composite pathway arising from different laminae of the spinal grey matter and terminating at different levels of the brain stem, the diencephalon and the basal telencephalon. Although some of its components constitute distinct pathways with a circumscribed laminar origin and a particular area of termination, the origin of the majority of its fibres is more diffuse and the distribution of these fibres is characterized by collateralization to multiple targets. There is a rough localization within the anterolateral fasciculus, with fibres from lower levels of the cord occupying its medial portion and fibres from higher levels being added ventrolaterally.

Nowhere along its ascending course does the anterolateral fasciculus constitute a circumscribed bundle. In the caudal medulla it still occupies its superficial, anterolateral position. At the level of the pontomesencephalic junction it moves dorsally, medial to and intermingling with the lateral lemniscus to become located medial to the brachium of the inferior colliculus. At their entrance into the thalamus the remaining fibres occupy a position on top of the dorsolateral wing of the medial lemniscus. Some of the fibres of the anterolateral fasciculus (re)cross in the rostral medulla and in the tectal, posterior and supraoptic commissures [48, 109, 115]. The terminations of the anterolateral fasciculus are summarized in Figs. 16.1 and 16.2.

One of the components of the anterolateral fasciculus, the spinothalamic tract, is an essential element in the circuitry involved in the perception of pain, itch and temperature from the contralateral half of the body. The focussing on pain conduction by the anterolateral fasciculus is justified by its clinical importance. However, clinical and experimental evidence also supports a role in the transmission of tactile and proprioceptive information. Complete lesions of the dorsal columns or the dorsal column nuclei do not affect the localization of tactile stimuli and the awareness of body position, but seriously impair motor performance by the forelimb. The localization of touch and position sense were preserved in a human case with transection of the entire spinal cord, with the exception of the anterolateral fasciculus [14, 156, 157, 162]. These observations should be kept in mind when the anatomy of spinothalamic and other components of the anterolateral fasciculus is discussed in the context of pain conduction in the following paragraphs.

Pain-related projection neurons in the spinal cord transmit information to a large number of regions of the brain stem, the diencephalon and the basal telencephalon. Schematizing somewhat, the following ascending systems can be distinguished:

1. the ascending connections of specific nociceptive and thermoreceptive neurons of lamina I of the dorsal horn;
2. the projections of polymodal nociceptive neurons of other laminae of the spinal grey matter (with a few exceptions, it is not possible to distinguish between the connections from these neurons and neurons with a different functional signature); and
3. the multisynaptic medial pain system.

The presence of a pathway for the conduction of visceral pain in the fasciculus gracilis, and the more general role of the postsynaptic dorsal column pathway in pain conduction, has been discussed in the section on the dorsal column-medial lemniscus pathway.

The spinothalamic tract takes its origin from multiple laminae and from all levels of the spinal cord. In monkeys the largest contribution is from the upper cervical cord, followed by the cervical and lumbar enlargements [7]. Most spinothalamic tract cells occur in lamina I (50% [51]), in the lamina IV/V border zone and lamina VI mostly laterally, and scattered in laminae VII and VIII.

Lamina I cells receive both A-delta and C fibres. The C fibres belong to the peptidergic class [150]. Three morphological and functional groups of lamina I neurons, each with its own distinct pattern of termination, can be distinguished: fusiform, pyramidal and multipolar cells. In the cat the dominant pattern of lamina I is formed by the nociceptive-specific, fusiform cells. These cells can be activated by pinch or noxious heat. The pyramidal cells are thermoreceptive and are activated by innocuous cooling; multipolar neurons are a mixture of polymodal (heat, pinch and cold-sensitive) and nociceptive-specific neurons [6, 83, 169]. An additional population of lamina I spinothalamic cells, sensitive to histamine and involved in the perception of itch, was recently identified [5]. The polymodal neurons of lamina I do not project to the thalamus, but are involved in spinal motor or sympathetic reflex pathways [53, 142]. The specific nociceptive fusiform and thermoreceptive pyramidal cells contribute to the spinothalamic tract. However, the projections of these neurons are not restricted to the thalamus, but also include the medullary and pontine medial reticular formation, the A1 noradrenergic group and other cellgroups in the ventrolateral medulla oblongata, the subcoerulean and parabrachial nuclei, the intercollicular nucleus and the central grey matter [50, 104] (Fig. 16.6).

In rodents, non-peptidergic C fibres terminate on interneurons in lamina II, which innervate relay cells of lamina V. These neurons, identified with transneuronal transport of a genetically expressed lectin tracer, wheat germ agglutinin, project to the hypothalamus, the amygdala, the bed nucleus of the stria terminalis and the globus pallidus [31].

Spinothalamic lamina IV–V cells respond to noxious and innocuous stimuli in small receptive fields. Some of these neurons function as second-order cells in the non-peptidergic nociceptive C fibre pathway. Lamina VII and VIII have large and complex receptive fields and receive convergent inputs from muscles and skin. This type of neuron prevails among the spinothalamic tract cells of the upper cervical cord [7]. Lamina VII and VIII spinothalamic neurons collateralize to the bulbar and pontine reticular formation) [73]. Specific nociceptors, therefore, seem to be restricted to laminae I and IV–VI; specific thermoreceptors only occur in lamina I. Laminae V–VII are involved in the transmission of tactile and proprioceptive information to the thalamus [7, 9, 151].

The monkey lamina I projections of the spinothalamic tract are found in the VMpo and the ventrocaudal medial dorsal thalamic nucleus. The projections of lamina I neurons of the pars caudalis of the spinal trigeminal nucleus are very similar. The projection to VMpo is somatotopically organized, with the hindlimb located in the caudal and the face in the rostral portions of the nucleus. Pain, itch and temperature remain segregated within the VMpo. Scattered terminations are present in VPM, VPL and VPI [28, 51, 52, 55, 56, 59]. In anterograde tracing experiments where the injections include other lamina (i.e. laminae IV/V and VII/VIII) apart from lamina I, the distribution of spinothalamic fibres is more extensive: They extend more rostrally in the ventrobasal complex where they concentrate along its shell and its fibrous laminae and form patches in the matrix of the nucleus. In these cases they also occur in the VPI and the intralaminar nuclei [8, 19, 29, 51, 74, 109]. The preferential projection of lamina IV/V and VII/VIII neurons to the VPL was confirmed in a recent retrograde labeling study in the monkey [52]. In Figs. 16.2 and 16.8, the spino- and trigemino-thalamic projections to VPM and VPL are shown as extensions of the VPI.

With respect to projections of the spinothalamic tract to the intralaminar nuclei, it has been established that their origin is restricted to the laminae of the spinal grey below lamina I. They terminate diffusely in the centrolateral intralaminar nucleus, in the adjoining lateral paralaminar region of the mediodorsal nucleus, and more sparsely, in other intralaminar and midline nuclei of the thalamus [8]. Observations of projections of lamina I neurons to the medial thalamus can be explained by their termination in the ventrocaudal medial dorsal nucleus [1, 8, 57, 67, 73, 165].

The role of the cerebral cortex in pain perception is still being debated. The thalamocortical projection of the VMpo targets the dorsal posterior insular cortex, with the face located rostral and lateral to the extremities [35]. The ventrocaudal medial dorsal nucleus and the centrolateral nucleus project to the anterior gyrus cinguli [35, 51, 80] (Fig. 16.6). The small calbindin immunoreactive neurons of the VPM and VPL matrix, which receive spinothalamic input, project to superficial layers of S1 [132, 133]. The projection of VPI includes the somatotopically organized area S2 in the parietal operculum [60, 74, 100].

In early studies in humans it was reported that pain is processed simultaneously in the contralateral primary somatosensory cortex and bilaterally in S2. Tactile stimuli are processed in a serial manner, with early activation of the primary somatosensory cortex, followed by S2 [124]. However, more recent studies show that pain is also serially processed, with an early activation of S1 followed by a bilateral second activation of S1 together with S2 and the insula, and late activation of the cingulate cortex, the region of the amygdala and the hippocampus [93]. Cingulate areas are involved in attention to and escape from pain, and its affective-motivational aspects [87, 137, 158]. S1 and S2 subserve the sensory-discriminative component of pain. The relative importance of projections of the ventrobasal complex to the primary somatosensory cortex, of the projection of VPI to S2 and of the projection of the VMpo to the insula in pain transmission is still being disputed. The role of the primary somatosensory cortex was emphasized by Willis et al. [164, 165]. A-delta and C fibre activation of S2 and the insula, without clear signs of involvement of S1, was observed by others [70,

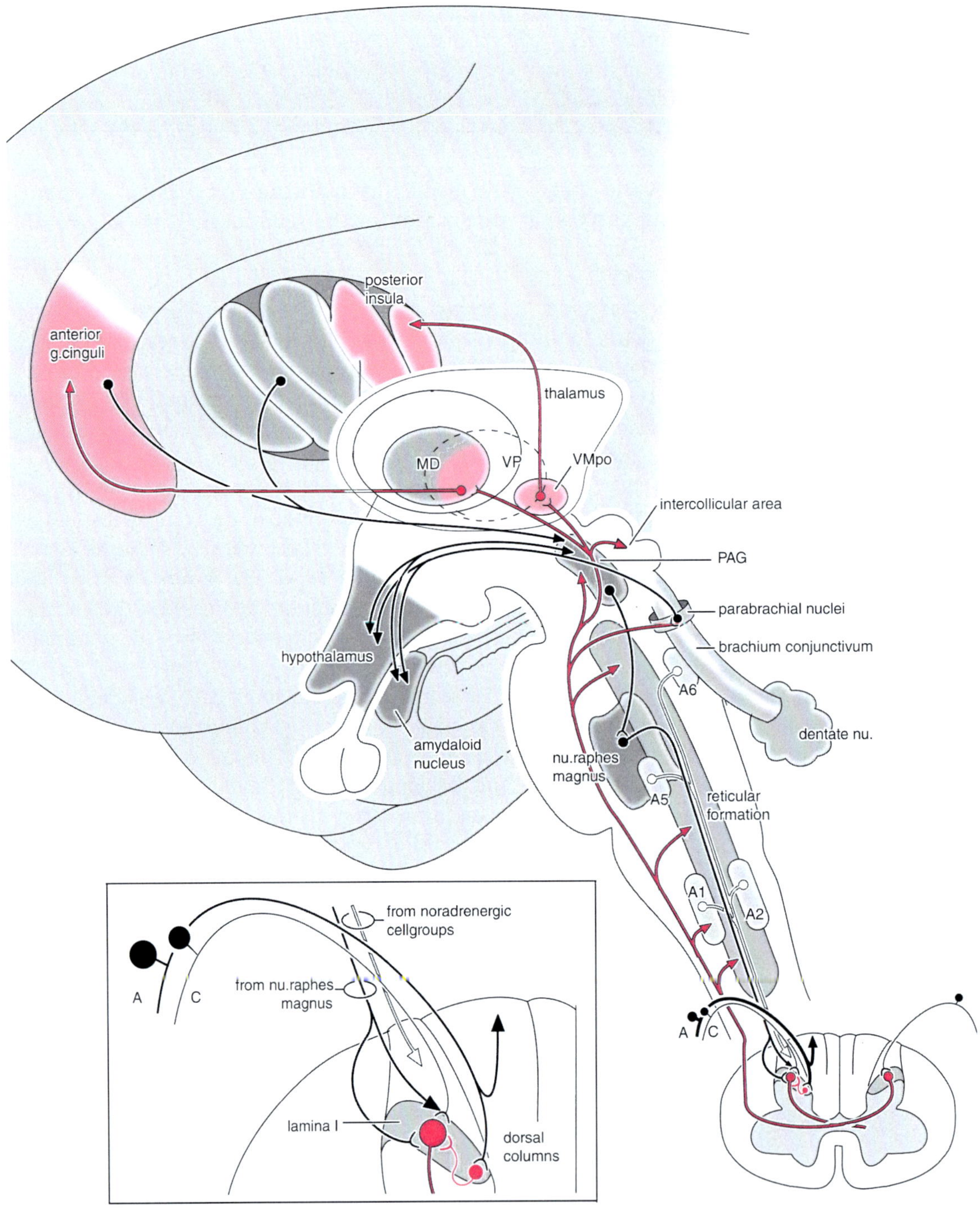

Fig. 16.6. The projection of specific nociceptive and thermoceptive neurons of lamina I of the dorsal horn (shown in *red*) and the descending control systems for pain. Lamina I neurons give rise to projections to the brain stem (reticular formation, A1 noradrenergic group, parabrachial nuclei, intercollicular area and periaqueductal grey) and terminate in the posterior ventromedial and caudal medial dorsal nuclei of the thalamus. These nuclei project to the caudal insula and the anterior cingulate gyrus, respectively. The parabrachial nuclei project to the hypothalamus and the amygdaloid nucleus. The direct C fibre pathway to the amygdaloid nucleus, synapsing in lamina II and V of the dorsal horn, described in the text, is not indicated. Control systems for pain descend from the periaqueductal grey via the nucleus raphe magnus and from noradrenergic cell groups 1, 2, 5 and 6. Inset shows the dorsal horn with presumed inhibitory pathway from A-beta root fibres to the relay cells of the C fibres. *A*, A-beta root fibres; *A1–6*, noradrenergic cell groups 1–6; *C*, C root fibres; *MD*, medial dorsal thalamic nucleus; *PAG*, periaqueductal grey; *VMpo*, posterior ventromedial thalamic nucleus; *VP*, ventral posterior thalamic nucleus

130]. The role of VMpo-insular projection was discussed by Craig [55].

The hypothalamus and the central nucleus of the amygdala are targeted by ascending spinal pathways contained in the anterolateral fasciculus [39, 40]. One of these pathways, identified with transneuronal antegrade transport in rodents, takes its origin from lamina V neurons, which receive a non-peptidergic C fibre input via lamina II interneurons. This pathway terminates in the amygdaloid nucleus, the bed nucleus of the stria terminalis and the globus pallidus [31]. An indirect pathway connecting lamina I of the dorsal horn, the spinal cord and the pars caudalis of the spinal trigeminal nucleus caudalis to these centres is interrupted in the parabrachial nuclei (Fig. 16.6). The spinal projections to the parabrachial nuclei are bilateral and diffuse [27]. In the lateral part of the parabrachial nuclear complex they synapse with separate neuronal populations projecting to the central nucleus of the amygdala [23, 88, 126], the intralaminar nuclei of the thalamus and the hypothalamus [24, 106]. Different populations of lamina I neurons have been held responsible for this projection to the parabrachial nuclei, among them the large cells of Waldeyer [128, 161]. An alternative nociceptive spinal pathway to the paraventricular nucleus of the hypothalamus includes the catecholaminergic cells situated in the A1 region of the ventrolateral medulla oblongata [119] (Fig. 16.2). Such pathways are believed to be involved in behavioural (e.g. vocalization, defence), emotional-affective (e.g. fear), autonomic (e.g. cardiorespiratory) and endocrine (e.g. adrenocortical) reactions to noxious events.

The mesencephalic periaqueductal grey receives afferents from spinal laminae I, VI, VII and VIII, and projects to the intralaminar thalamic nuclei [6, 50, 113, 163]. The periaqueductal grey also represents an essential component of the descending pain control system (see next section Control Mechanisms of Nociceptive Transmission).

Vogt and Sikes [158] and Sewards and Sewards [143] proposed the presence of a separate medial, multisynaptic nociceptive pathway, mediating the motivational aspects of pain, with relays in laminae VII and VIII of the spinal grey matter, the periaqueductal grey, the hypothalamus and the intralaminar nuclei of the thalamus, projecting to the anterior cingulate cortex (Fig. 16.2). The concept of medial, multisynaptic and lateral, direct spino-thalamo-cortical pathways dates from Bishop [25] and Mehler [109]. It remains difficult, however, to disentangle the role of such a multisynaptic system in pain transmission from the multiple functions of the brain stem and the hypothalamus in premotor and homeostatic state control, functions that also critically depend on the input from primary afferents.

Control Mechanisms of Nociceptive Transmission

Two control mechanisms influence the traffic of nociceptive impulses at the level of the spinal dorsal horn. The first of these is the "gate control mechanism", which was first described by Melzack and Wall [110]. Collaterals of coarse mechanoreceptor (touch and pressure) primary afferents, which ascend into the spinal dorsal columns, were believed to activate inhibitory interneurons at segmental entry level, which in turn act on secondary nociceptive neurons (Fig. 16.6, inset). The latter are, as we have seen, driven by the small-diameter A-delta and C primary afferents. The notion that the balance of activity in small (nociceptive) and large (nonnociceptive) primary afferent fibres is important in pain transmission has led to the use of transcutaneous electrical stimulation of peripheral nerves and of stimulation of the dorsal columns of the spinal cord by implanted electrodes for certain types of peripheral pain.

The second mechanism that can effectively modulate the experience of pain (and, hence, produce analgesia) is the descending pain control system. This system was first noticed after discovering that electrical stimulation of the mesencephalic periaqueductal grey produces profound analgesia [13]. This effect of central grey stimulation can be explained by the activation of descending projections into the spinal dorsal horn that are synaptically interrupted in a complex known as the rostral ventromedial medulla oblongata, which consists of the nucleus raphe magnus and the adjacent medial reticular formation (Fig 16.6). The nucleus raphe magnus contains numerous serotoninergic neurons, and it has been demonstrated that its cells project massively to the superficial dorsal horn, i.e. the region where the nociceptive primary afferents contact the cells of origin of the various pain-related ascending systems. Both serotoninergic and non-serotoninergic cells of this complex influence nociceptive transmission in a bidirectional manner through a direct effect onto nociceptive primary afferents, as well as through an indirect effect mediated by local interneurons [107]. The same serotoninergic neurons may mediate cutaneous vasoconstriction in response to painful stimuli [26]. Noradrenergic cells situated in rhombencephalic catecholaminergic cell groups A5, A6 and A7, which also project to the superficial dorsal horn, contribute to the descending pain-inhibiting system (Fig. 16.6). These cell groups are in receipt of strong afferent projections from the periaqueductal grey, the rostral ventromedial medulla oblongata and the superficial laminae of the dorsal horn [11, 107]. Opiates exert their influence on nociceptive transmission both at the level of the periaqueductal grey and at the termination of the primary afferents. Specific nociceptive neurons in lamina I are selectively inhibited by morphine [54].

Most probably, centres above the mesencephalic level also form part of the descending pain control system. The periaqueductal grey has strong reciprocal connections with the hypothalamus and receives afferents from prefrontal, cingulate and insular cortical areas. By means of the descending system just discussed, pain sensations are under the control of the highest levels of the brain.

The Trigeminal System

The trigeminus is a mixed nerve (Figs. 16.7, 16.8). Its sensory (major) portion carries touch, nociception, temperature and proprioception from the face, the facial and masticatory muscles, the temporomandibular joint and the intraoral cavity. The anatomy of the trigeminal nerve and nuclei was reviewed by Usunoff et al. [155] and Waite and Ashwell [160]. Somatosensory fibres of the trigeminal nerve enter the pons in the sensory root (portio major) and are distributed to the principal sensory nucleus and the spinal trigeminal nucleus. Many of the A-beta and gamma fibres contained in the portio major bifurcate and then terminate in the principal sensory nucleus and in different subdivisions of the spinal trigeminal nucleus [118]. Thin A-delta and C fibres descend into the spinal tract, which continues as Lissauer's dorsolateral fasciculus on top of the dorsal horn. They terminate in the caudal subnucleus of the spinal trigeminal nucleus and in the upper cervical dorsal horn. Fibres of the mandibular division of the trigeminal nerve descend into the dorsal part of the tract, whereas the ophthalmic division occupies the ventral part. Fibres of the maxillary nerve occupy an intermediate position and descend less far caudally than the other two divisions [101, 137]. Small somatosensory components of the facial, glossopharyngeal and vagal nerves join the spinal trigeminal tract (Fig. 16.8).

The caudal subnucleus of the spinal trigeminal nucleus is continuous with the dorsal horn. The laminar architecture of the dorsal horn and the caudal subnucleus and the synaptic relations of the afferent A-delta and C fibres terminating in both structures is very similar. The caudal subnucleus gives rise to ascending fibres, which cross in the caudal medulla oblongata and join the spinothalamic tract [147]. Like the spinothalamic tract, they terminate in the VMpo, the medial dorsal nucleus and the shell of the ventroposterior complex and in the intralaminar nuclei [19, 41, 51, 72]. Widespread projections to the reticular formation,

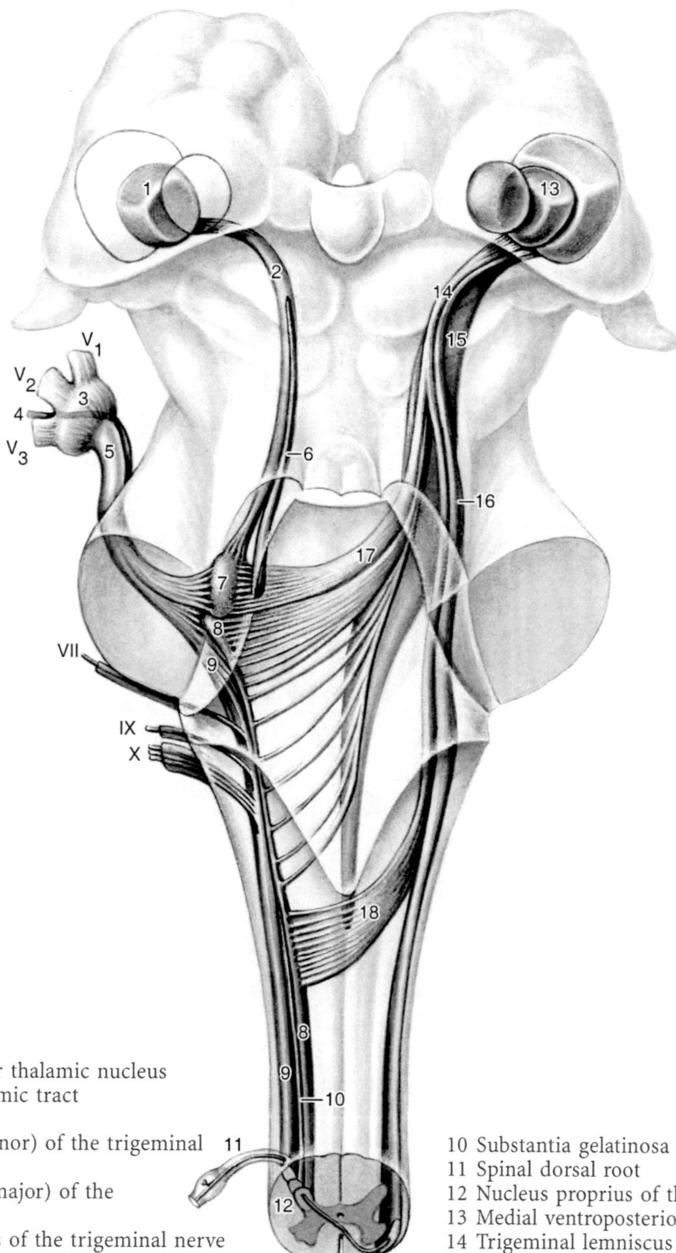

1 Medial ventroposterior thalamic nucleus
2 Dorsal trigeminothalamic tract
3 Trigeminal ganglion
4 Motor root (portio minor) of the trigeminal nerve
5 Sensory root (portio major) of the trigeminal nerve
6 Mesencephalic nucleus of the trigeminal nerve
7 Principal sensory nucleus of the trigeminal nerve
8 Spinal nucleus of the trigeminal nerve
9 Spinal tract of the trigeminal nerve
10 Substantia gelatinosa
11 Spinal dorsal root
12 Nucleus proprius of the dorsal horn
13 Medial ventroposterior thalamic nucleus
14 Trigeminal lemniscus
15 Medial lemniscus
16 Spinothalamic tract
17 Decussation of the trigeminal lemniscus
18 Decussation of the spinothalamic tract

Fig. 16.7. The central connections of the trigeminal nerve. Position of nerves, tracts and nuclei in a dorsal view (5/3×). *Roman numerals* indicate the corresponding cranial nerves
V_1, Ophthalmic nerve; V_2, Maxillary nerve; V_3, Mandibular nerve

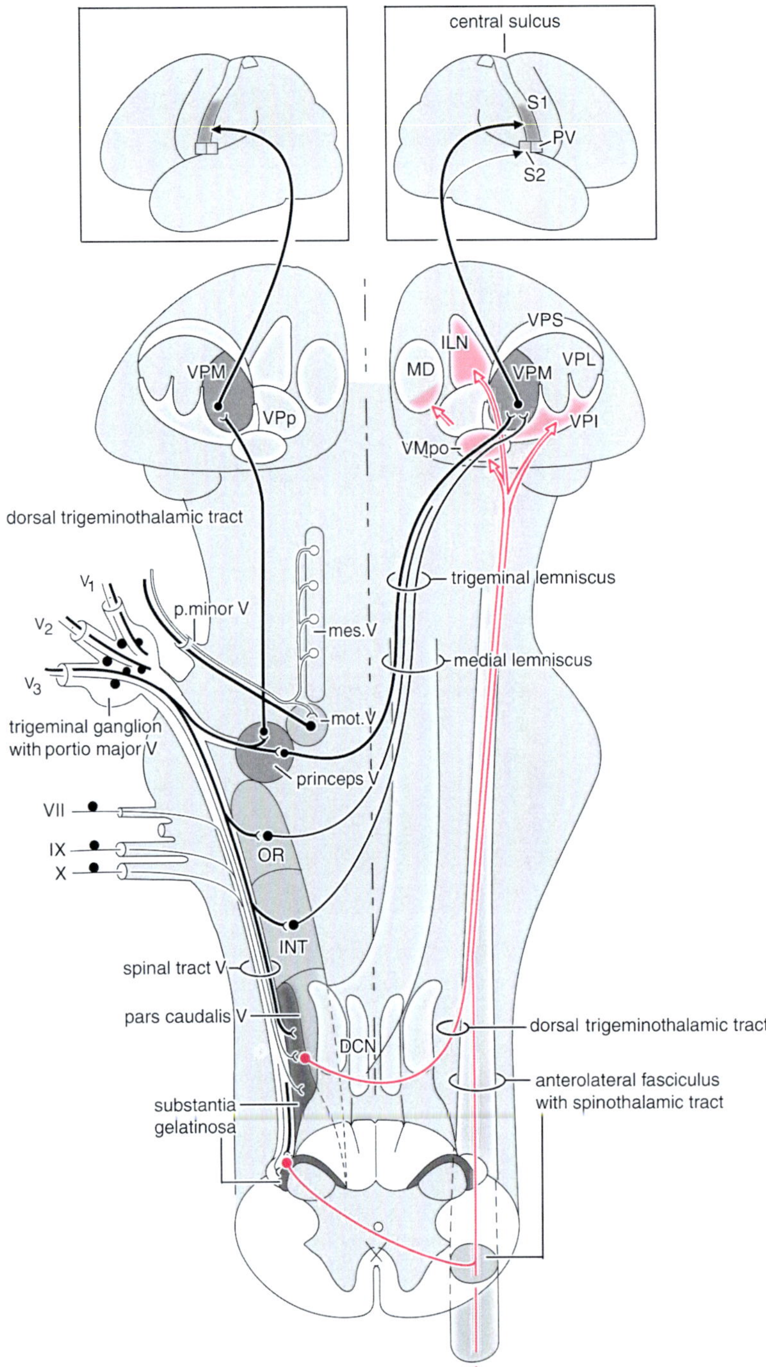

Fig. 16.8. Connections of the trigeminal nerve. Fibres from the principal sensory nucleus and the pars oralis and interpolaris of the spinal nucleus of the trigeminal nerve decussate at the level of the pons and join the medial lemniscus as the trigeminal lemniscus. The trigeminal lemniscus terminates in the medial ventroposterior (VPM) nucleus of the thalamus. The uncrossed dorsal trigeminothalamic tract from the principal sensory nucleus also terminates in the VPM. VPM projects laterally in the primary somatosensory cortex (S1). Perioral and intraoral structures are represented bilaterally in the VPM and lateral S1. Trigeminothalamic fibres from all layers of the subnucleus caudalis of the spinal nucleus of the trigeminal nerve join the spinothalamic tract to terminate in the intralaminar nuclei, the posterior ventromedial nucleus (VMpo), the caudal medial dorsal nucleus (MD) and the shell of the ventroposterior complex. The termination of trigeminothalamic fibres from specific nociceptive and thermoreceptive neurons of lamina I of the pars caudalis is limited to the MD and VMpo nuclei. The axons of neurons located in the mesencephalic nucleus of the trigeminal nerve join the motor root (portio minor) of the trigeminal nerve and send branches to the motoneurons innervating the jaw muscles, located in the motor nucleus of the trigeminal nerve. *VII–X*, cranial nerves VII–X; *DCN*, dorsal column nuclei; *ILN*, intralaminar nuclei; *INT*, pars interpolaris of the spinal nucleus of the trigeminal nerve; *MD*, medial dorsal thalamic nucleus; *mes.V*, mesencephalic nucleus of the trigeminal nerve; *mot.V*, motor nucleus of the trigeminal nerve; *OR*, pars oralis of the spinal nucleus of the trigeminal nerve; *PV*, ventral parietal somatosensory area; *S1(2)*, primary (secondary) somatosensory cortex; V_1, ophthalmic nerve; V_2, maxillary nerve; V_3, mandibular nerve; *VMpo*, posterior ventromedial thalamic nucleus; *VPI*, inferior ventroposterior thalamic nucleus; *VPL*, lateral ventroposterior thalamic nucleus; *VPM*, medial ventroposterior thalamic nucleus; *VPp*, parvocellular medial ventroposterior thalamic nucleus; *VPS*, superior ventroposterior thalamic nucleus

the parabrachial nuclei, the periaqueductal grey and the tectum mesencephali resemble the connections of the spinal cord through the anterolateral fasciculus.

Projections from specific nociceptive and thermoreceptive neurons in lamina I of the pars caudalis are similar to their spinal counterparts. They are located in the rostral VMpo and rostrally in the posterior insula [34, 51, 58]. The pars caudalis also gives rise to an intranuclear ascending system, which terminates in the principal sensory nucleus [42, 72, 86, 121]. The essential role of the caudal subnucleus in pain and temperature transmission from the face is exemplified by Sjöqvist's tractotomy of the spinal tract at the level of the rostral border of the pars caudalis [101, 148], which effectively eliminates pain and temperature sensitivity in the ipsilateral face, leaving the tactile sense intact.

The rostral portion of the spinal trigeminal nucleus can be subdivided into the pars oralis and the interpolaris. The borders between these subnuclei and with the principal sensory nucleus are fairly arbitrary. The subnucleus oralis may be included with the principal nucleus. The nuclei are connected with the facial motor nucleus, the brain stem, the thalamus and the cerebellum.

The structure and function of the principal sensory nucleus of the trigeminal nerve in the discriminative aspects of touch are very similar to the gracile and medial cuneate nuclei. The projection of the trigeminal nerve to the principal sensory nucleus is somatotopically organized, with the face ventrally, the jaw dorsally and the intraoral cavity dorsally and caudally extending into the pars oralis of the spinal trigeminal nuclei. This dorsoventral reverted representation of the face is repeated in more caudal subdivisions of the spinal subnucleus [144, 145].

The entire principal sensory nucleus gives rise to a crossed pathway joining the medial lemniscus as the trigeminal lemniscus (Fig. 16.8). The uncrossed dorsal trigeminothalamic tract arises from the dorsal part of the principal sensory nucleus and conducts information from the intraoral cavity [41, 108, 149, 153]. The trigeminal lemniscus terminates in the VPM. The termination of the uncrossed trigeminothalamic is found in the medial VPM, lateral to the gustatory parvocellular VPM. The VPM, therefore, contains a bilateral representation of the intraoral cavity. The cortical projections of the sensory relay nuclei of the thalamus and how the anatomy affects the perception of touch, pain and temperature have been discussed in the section of this chapter on the Ascending Pathways from the Spinal Cord.

Unlike other primary afferents, the Ia muscle-spindle afferents from the jaw muscles, which enter the brain stem in the trigeminal nerve, arise from cells located within the central nervous system. These cells constitute the mesencephalic nucleus of the trigeminal nerve, located alongside the central grey matter of the mesencephalon. The axons of these cells descend into the mesencephalic tract of the trigeminal nerve and then bifurcate. One branch terminates on the motoneurons of the trigeminal motor nucleus. The main branches of these neurons are the Ia afferents from the masticatory muscles. They exit the brain stem, together with the axons of the motoneurons innervating the masticatory muscles, into the portio minor of the trigeminal nerve, which joins the mandibular division of the trigeminal nerve (Figs. 16.8, 21.12).

The Visceral Afferent System

The visceral afferent system was recently reviewed by Pritchard and Norgren [125]. Visceral afferent fibres and small contingents of somatosensory fibres enter the medulla oblongata in the facial, glossopharyngeal and vagus nerves. The somatosensory fibres join the spinal tract of the trigeminal nerve. Special visceral afferents that innervate taste buds in the fungiform papillae of the anterior half of the tongue and foliate papillae along its lateral edges course in the chorda tympani and the intermediate nerve, usually included with the facial nerve. Axons innervating taste buds in the circumvallate papillae enter the brain stem in the glossopharyngeal nerve. Taste buds of the palate and the pharynx are innervated by the vagus nerve. General visceral afferents, which innervate the baroreceptors of the carotid sinus and the arch of the aorta, chemoreceptors of the glomus caroticum, the heart, the bronchi and the abdominal viscera, are contained in the glossopharyngeal and vagus nerves (Figs. 16.9, 16.10).

The visceral afferents join the solitary tract and terminate on its nucleus. Special visceral afferents terminate in its rostral, gustatory part (Figs. 16.9, 16.10). Ascending gustatory fibres of the facial nerve terminate in the nucleus ovalis, a rostral prolongation of the solitary nucleus surrounded by a capsule (fasciculus ovalis, Fig. 6.27) located dorsal to the principal sensory nucleus of the trigeminal nerve [10, 15, 136]. General visceral afferents descend more caudally to terminate in the caudal and commissural parts of the solitary nucleus and the area postrema. Some of them terminate in the ventral part of the lateral cuneate nucleus and in the dorsal nucleus of the vagus nerve [15, 94, 95, 116]. Sacral and lumbar fibres arising from neurons in the base of the dorsal horn, which can be activated by distension of the colon and the uterine cervix, also terminate in the caudal solitary nucleus [21, 81, 85] (Figs. 16.9, 16.10).

The rostral, gustatory portion of the solitary nucleus gives rise to an uncrossed ascending pathway terminating in the caudal half of the parvocellular VPM [16]. This gustatory pathway is located in the same region, ventrolateral to the periaqueductal grey, as the dorsal trigeminothalamic tract, which conveys somatosensory information from the ipsilateral intraoral cavity to the VPM. The connections of the nucleus ovalis, a rostral extension of the nucleus of the solitary tract located within the confines of the dorsal principal sensory nucleus, are not clear. It may have been included in many experimental studies in the principal sensory nucleus. The caudal, general viscerosensory portion of the nucleus on the solitary tract is connected with the rostral part of the parvocellular VPM through a bilateral, indirect pathway that contains an extra synapse in the parabrachial nuclei [24, 126]. As discussed in the section on pain transmission, different populations of parabrachial neurons also access the amygdala, the lateral hypothalamus and the intralaminar nuclei.

The parvocellular VPM projects to the gustatory cortex, located in the rostral, granular cortex of the insula and the inner frontal operculum [127]. Evidence from clinical, imaging and experimental studies supporting this localization was reviewed by Pritchard [125] and Zatore and Jones-Gotman [168]. In recent magnetoencephalographic and fMRI studies, the frontal operculum, the ventral part of the central sulcus and the transition of the parietal operculum and insula were activated by gustatory stimulation [97, 117]. Secondary projections from the gustatory cortex were traced to the dysgranular insular cortex and the lateral orbitofrontal cortex (Fig. 16.10). Convergence of olfactory and gustatory input was found in the latter (Kobayashi et al. [96]). Separate cortical targets of the caudal gustatory, and rostral general viscerosensory, portions of the parvocellular VPM have not yet been considered.

The caudal portion of the solitary nucleus constitutes an important centre for the regulation of visceral, cardiovascular and respiratory functions (reviewed by Blessing [26]). It projects to vasomotor interneurons in the caudal ventrolateral medulla, involved in the maintenance of blood pressure, to vagal parasympathetic premotor neurons controlling heart rate and to the motoneurons of the nucleus ambiguus innervating the striated muscle of the oesophagus involved in swallowing [26, 84]. The dorsal longitudinal fascicle, located medially in the floor of the fourth ventricle, connects the caudal part of the solitary nucleus with the central grey matter [111]. Crossed solitariospinal fibres descend from the caudal part of the nucleus to terminate in the autonomic centres of the spinal cord [152].

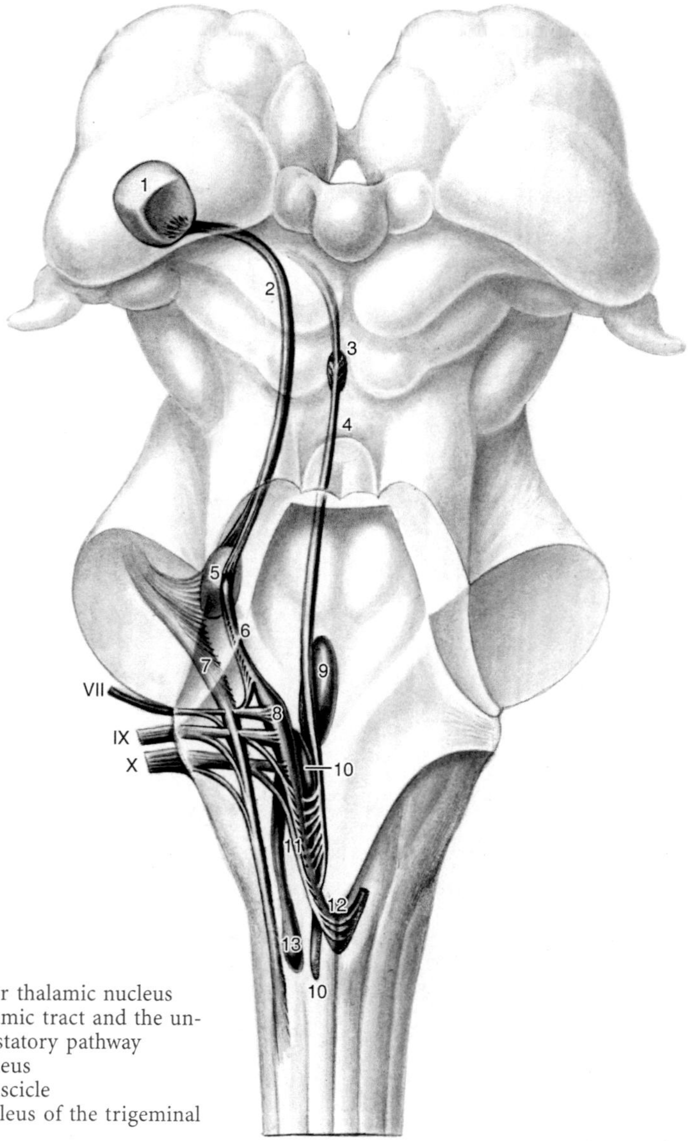

1 Medial ventroposterior thalamic nucleus
2 Dorsal trigeminothalamic tract and the uncrossed ascending gustatory pathway
3 Dorsal tegmental nucleus
4 Dorsal longitudinal fascicle
5 Principal sensory nucleus of the trigeminal nerve
6 Nucleus ovalis
7 Spinal tract of the trigeminal nerve
8 Nucleus of the solitary tract
9 Nucleus prepositus hypoglossi
10 Dorsal nucleus of the vagal nerve
11 Solitary tract
12 Obex
13 Nucleus ambiguus

Fig. 16.9. The solitary tract. Position of nerves, tracts and nuclei in a dorsal view (5/3×). *Roman numerals* indicate the corresponding cranial nerves

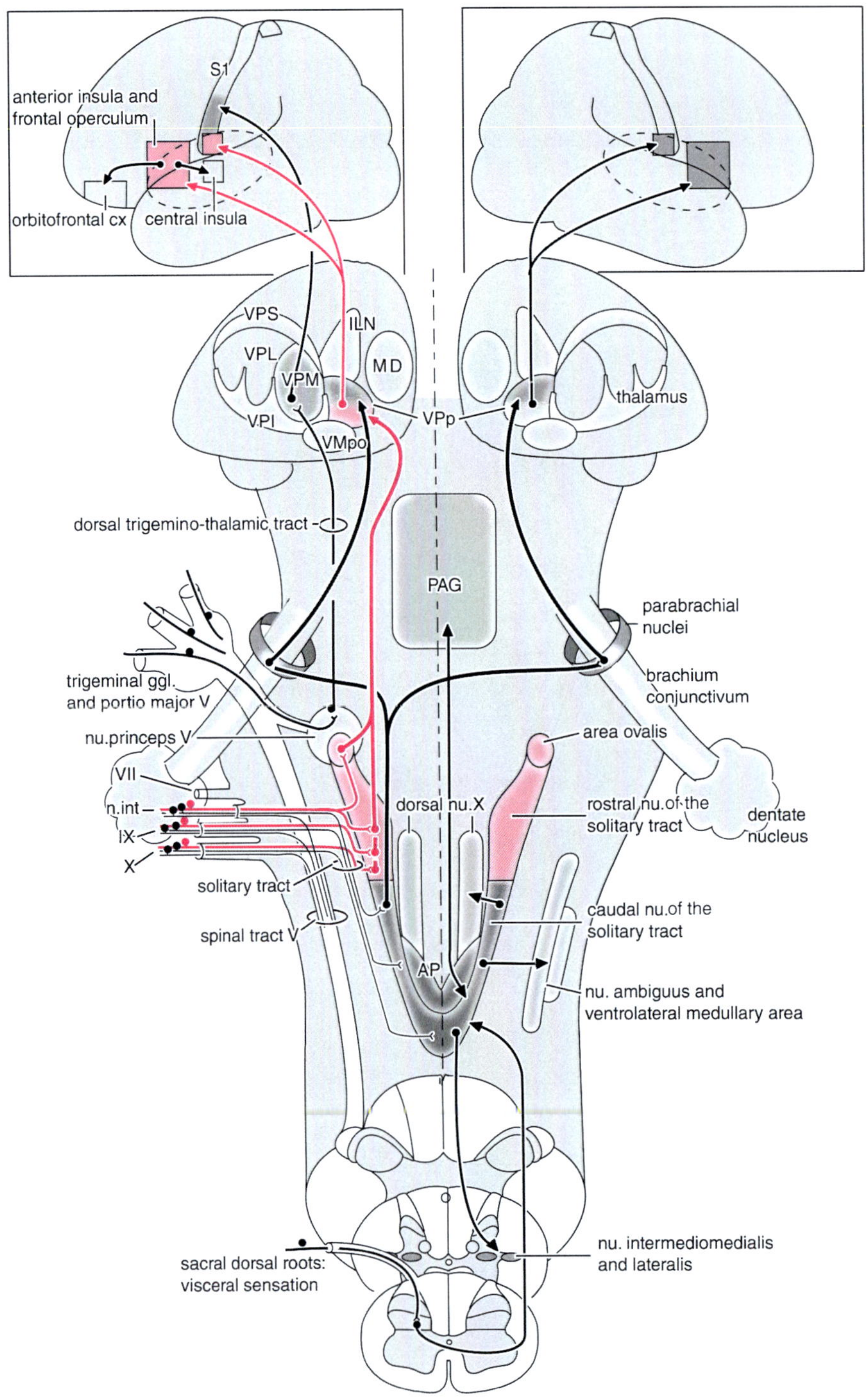

Fig. 16.10. The connections of the nucleus of the solitary tract. The nucleus of the solitary tract can be subdivided into a rostral gustatory portion, which receives gustatory afferents from the facial, glossopharyngeal and vagal nerves, and a caudal portion, which receives general visceral afferents from these nerves. The uncrossed gustatory pathway, from the rostral solitary nucleus to the caudal half of the parvocellular medial ventroposterior nucleus of the thalamus (VPp), ascends near the dorsal trigeminothalamic tract. In Fig. 16.9 the two tracts were not distinguished. VPp projects to the rostral insula and the frontal operculum, and to the base of the postcentral gyrus. Here the gustatory and the trigeminal projections appear to overlap. The caudal part of the solitary nucleus projects bilaterally to rostral VPp. This projection is interrupted in the parabrachial nuclei. The

References

1. Albe-Fessard D, Boivie J, Grant G, Levante A (1975) Labeling of cells in the medulla oblongata and the spinal cord of the monkey after injections of horseradish peroxidase in the thalamus. Neurosci Lett 1:75–80
2. Albright BC, Friedenbach DJ (1982) The distribution of lateral funicular and cortical fibers to the dorsal column, Z and X nuclei in the prosimian Galago. Neuroscience 7:1175–1185
3. Al-Chaer ED, Feng Y, Willis WD (1999) Comparative study of viscerosomatic input onto postsynaptic dorsal column and spinothalamic tract neurons in the primate. J Neurophysiol 82:1876–1882
4. Al-Chaer ED, Lawand NB, Westlund KN, Willis WD (1996) Visceral nociceptive input into the ventral posterolateral nucleus of the thalamus: a new function for the dorsal column pathway. J Neurophysiol 76:2661–2674
5. Andrew D, Craig AD (2001) Spinothalamic lamina I neurons selectively sensitive to histamine: a central neural pathway for itch. Nat Neurosci 4:72–77
6. Andrew D, Krout KE, Craig AD (2003) Differentiation of lamina I spinomedullary and spinothalamic neurons in the cat. J Comp Neurol 458:257–271
7. Apkarian AV, Hodge CJ (1989) Primate spinothalamic pathways: I. A quantitative study of the cells of origin of the spinothalamic pathway. J Comp Neurol 288:447–473
8. Apkarian AV, Hodge CJ, (1989) Primate spinothalamic pathways: II. The cells of origin of the dorsolateral and ventral spinothalamic pathways. J Comp Neurol 288:474–492
9. Apkarian AV, Hodge CJ (1989) Primate spinothalamic pathways: III. Thalamic terminations of the dorsolateral and ventral spinothalamic pathways. J Comp Neurol 288:493–511
10. Arvidsson J, Thomander L (1984) An HRP study of the central course of sensory intermediate and vagal fibers in peripheral facial nerve branches in the cat. J Comp Neurol 223:35–45
11. Bajic D, Proudfit HK (1999) Projections of neurons in the periaqueductal gray to pontine and medullary catecholamine cell groups involved in the modulation of nociception. J Comp Neurol 405:359–379
12. Barbaresi P, Spreafico R, Frassoni G, Rustioni A (1986) GABAergic neurons are present in the dorsal column nuclei but not in the ventroposterior complex of rats. Brain Res 382:305–326
13. Basbaum AI, Clanton CH, Fields HL (1976) Opiate and stimulus-produced analgesia: functional anatomy of a medullospinal pathway. Proc Natl Acad Sci USA 73:4685–4688
14. Beck C (1976) Forlimb performance by squirrel monkeys (Saimiri sciureus) before and after dorsal column lesions. J Comp Physiol Psychol 90:353–362
15. Beckstead RM, Norgren R (1979) An autoradiographic examination of the central distribution of the trigeminal, facial, glossopharyngeal, and vagal nerves in the monkey. J Comp Neurol 184:455–472
16. Beckstead RM, Morse JR, Norgren R (1980) The nucleus of the solitary tract in the monkey: projections to the thalamus and brain stem nuclei. J Comp Neurol 190:259–282
17. Bentivoglio M, Rustioni A (1986) Corticospinal neurons with branching axons to the dorsal column nuclei in the monkey. J Comp Neurol 253:260–276
18. Berkley KJ (1975) Different targets of different neurons in nucleus gracilis of the cat. J Comp Neurol 163:285–303
19. Berkley KJ (1980) Spatial relationships between the terminations of somatic sensory and motor pathways in the rostral brainstem of cats and monkeys. I. Ascending somatic sensory inputs to lateral diencephalon. J Comp Neurol 193:283–317
20. Berkley KJ, Hand PJ (1978) Projections to the inferior olive of the cat. II. Comparisons of input from the gracile, cuneate and the spinal trigeminal nuclei. J Comp Neurol 180:253–264
21. Berkley KJ, Hubscher CH, Wall PD (1993) Neuronal responses to stimulation of the cervix, uterus, colon, and skin in the rat spinal cord. J Neurophysiol 69:545–556
22. Berkley KJ, Blomqvist A, Pelt A, Flink R (1980) Differences in the collateralization of neuronal projections from the dorsal column nuclei and lateral cervical nucleus to the thalamus and tectum in the cat: an anatomical study using two different double-labeling techniques. Brain Res 202:273–290
23. Bernard JF, Huang GF, Besson JM (1992) Nucleus centralis of the amygdala and the globus pallidus ventralis: electrophysiological evidence for an involvement in pain processes. J Neurophysiol 68:551–569

(continued text of Fig. 16.10) caudal part of the solitary nucleus also projects to the periaqueductal grey, the dorsal motor nucleus of the vagal nerve, the nucleus ambiguus and the adjacent ventrolateral medullary area and to the visceral nuclei in the spinal cord. It receives connections from the base of the sacral dorsal horn, mediating visceral sensation. *VII–X*, cranial nerves VII–X; *AP*, area postrema; *cx*, cortex; *dorsal nu.X*, dorsal nucleus of the vagal nerve; *ggl*, ganglion; *ILN*, intralaminar nuclei; *MD*, medial dorsal thalamic nucleus; *n.int*, intermediate nerve; *S1*, primary somatosensory cortex; *VMpo*, posterior ventromedial thalamic nucleus; *VPI*, inferior ventroposterior nucleus; *VPL*, lateral ventroposterior thalamic nucleus; *VPM*, medial ventroposterior thalamic nucleus; *VPp*, parvocellular medial ventroposterior thalamic nucleus; *VPS*, superior ventroposterior thalamic nucleus

24. Bester H, Menendez L, Besson JM, Bernard JF (1995) Spino(trigemino) parabrachiohypothalamic pathway: electrophysiological evidence for an involvement in pain processes. J Neurophysiol 73:568–585
25. Bishop GH (1959) The relation between nerve fiber size and sensory modality: phylogenetic implications of the afferent innervation of the cortex. J Nerv Mental Dis 128:89–114
26. Blessing W (2004) Lower brain stem regulation of visceral, cardiovascular, and respiratory function. In: Paxinos G, Mai JK (eds) The human nervous system. Elsevier, Amsterdam, pp 465–478
27. Blomqvist A, Ma W, Berkley KJ (1989) Spinal input to the parabrachial nucleus in the cat. Brain Res 480:29–36
28. Blomqvist A, Zhang ET, Craig AD (2000) Cytoarchitectonic and immunohistochemical characterization of a specific pain and temperature relay, the posterior portion of the ventral medial nucleus, in the human thalamus. Brain 123:601–619
29. Boivie J (1979) An anatomical reinvestigation of the termination of the spinothalamic tract in the monkey. J Comp Neurol 186:343–369
30. Boivie J, Grant G, Albe-Fessard DL, Llevante A (1975) Evidence for a projection to the thalamus from the external cuneate nucleus of the monkey. Neurosci Lett 1:3–8
31. Braz JM, Nassar MA, Wood JN, Basbaum AI (2005) Parallel "pain" pathways arise from subpopulations of primary afferent nociceptor. Neuron 47:787–793
32. Bremmer F, Schlack A, Shah NJ, Zafiris O, Kubischek M, Hoffmann K, Zilles K, Fink GR (2001) Polymodal motion processing in posterior parietal and premotor cortex: a human fMRI study strongly implies equivalencies between humans and monkeys. Neuron 29:287–296
33. Brodmann K (1909) Vergleichende Lokalisationslehre der Grosshirnrinde in ihren Prinzipien dargestellt auf Grund des Zellenbaues. Barth, Leipzig
34. Brooks JC, Nurmikko TJ, Bimson EE, Singh KD, Roberts N (2002) fMRI of thermal pain: effects of stimulus laterality and attention. Neuroimage 15:293–301
35. Brooks JCW, Zambreanu L, Godinez A, Craig AD, Tracey I (2005) Somatotopic organisation of the human insula to painful heat studied with high resolution functional imaging. Neuroimage 27:201–209
36. Brown AG (1981) Organization in the spinal cord. The anatomy and physiology of identified neurones. Springer Verlag, Berlin Heidelberg New York
37. Brown PB, Fuchs JL (1975) Somatotopic representation of hindlimb skin in cat dorsal horn. J Neurophysiol 38:1–9
38. Brown AG, Brown PB, Fyffe RE, Pubols LM (1983) Receptive field organization and response properties of spinal neurones with axons ascending the dorsal columns in the cat. J Physiol 337:575–588
39. Burstein R, Wang JL, Elde RP, Giesler GJ (1990) Neurons in the sacral parasympathetic nucleus that project to the hypothalamus do not also project through the pelvic nerve. A double labeling study combining Fluoro-gold and cholera toxin B in the rat. Brain Res 506:159–165
40. Burstein R, Falkowsky O, Borsook D, Strassman A (1996) Distinct lateral and medial projections of the spinohypothalamic tract of the rat. J Comp Neurol 373:549–574
41. Burton H, Craig AD (1979) Distribution of trigeminothalamic projection cells in cat and monkey. Brain Res 161:515–521
42. Burton H, Craig AD, Poulos DA, Molt JT (1979) Efferent projections from temperature sensitive recording loci within the marginal zone of the nucleus caudalis of the spinal trigeminal complex in the cat. J Comp Neurol 183:753–777
43. Carpenter MB, Stein BM, Shriver JE (1968) Central projections of spinal dorsal roots in the monkey. II. Lower thoracic, lumbosarcral and coccygeal dorsal roots. Am J Anat 123:75–118
44. Cheek MD, Rustioni A, Trevino DL (1975) Dorsal column nuclei projections to the cerebellar cortex in cats as revealed by the use of the retrograde transport of horseradish peroxidase. J Comp Neurol 164:31–46
45. Cheema SS, Rustioni A, Whitsel BL (1984) Light and electron microscopic evidence for a direct corticospinal projection to superficial laminae of the dorsal horn in cats and monkeys. J Comp Neurol 225:276–290
46. Cheema S, Rustioni A, Whitsel BL (1985) Sensorimotor cortical projections to the primate cuneate nucleus. J Comp Neurol 240:196–211
47. Cliffer KD, Willis WD (1994) Distribution of the postsynaptic dorsal column projection in the cuneate nucleus of monkeys. J Comp Neurol 345:84–93
48. Cliffer KD, Burstein R, Giesler GJ (1991) Distributions of spinothalamic, spinohypothalamic, and spinoteleccephalic fibers revealed by antegrade transport of PHA-L in rats. J Neurosci 11:852–868
49. Coq JO, Qi H, Collins CE, Kaas JH (2004) Anatomical and functional organization of somatosensory areas of the lateral fissure of the New World titi monkey (Callicebus moloch). J Comp Neurol 476:363–387
50. Craig AD (1995) Distribution of brainstem projections from spinal lamina I neurons in the cat and the monkey. J Comp Neurol 361:225–248
51. Craig AD (2004) Distribution of trigeminothalamic and spinothalamic lamina I terminations in the macaque monkey. J Comp Neurol 477:119–148
52. Craig AD (2006) Retrograde analyses of spinothalamic projections in the macaque monkey: Input to ventral posterior nuclei. J Comp Neurol 499:965–978
53. Craig AD, Kniffki KD (1985) Spinothalamic lumbosacral lamina I cells responsive to skin and muscle stimulation in the cat. J Physiol 365:197–221

54. Craig AD, Serrano LP (1994) Effects of systemic morphine on lamina I spinothalamic tract neurons in the cat. Brain Res 636:233–244
55. Craig AD, Blomqvist A (2002) Is there a specific lamina I spinothalamocortical pathway for pain and temperature sensations in primates? J Pain 3:95–101
56. Craig AD, Zhang ET (2006) Retrograde analyses of spinothalamic projections in the macaque monkey: Input to posterolateral thalamus. J Comp Neurol 499:953–964
57. Craig AD Jr, Linington AJ, Kniffki KD (1989) Cells of origin of spinothalamic tract projections to the medial and lateral thalamus in the cat. J Comp Neurol 289:568–585
58. Craig AD, Zhang ET, Blomqvist A (1999) A distinct thermoreceptive subregion of lamina I in nucleus caudalis of owl monkey. J Comp Neurol 404:221–234
59. Craig AD, Bushnell MC, Zhang ET, Blomqvist A (1994) A thalamic nucleus specific for pain and temperature sensation. Nature 372:770–773
60. Cusick CG, Gould HJ, 3rd (1990) Connections between area 3b of the somatosensory cortex and subdivisions of the ventroposterior nuclear complex and the anterior pulvinar nucleus in squirrel monkeys. J Comp Neurol 292:83–102
61. Disbrow E, Roberts T, Krubitzer L (2000) Somatotopic organization of cortical fields in the lateral sulcus of Homo sapiens: evidence for SII and PV. J Comp Neurol 418:1–21
62. Disbrow EA, Hinkley LB, Roberts TP (2003) Ipsilateral representation of oral structures in human anterior parietal somatosensory cortex and integration of inputs across the midline. J Comp Neurol 467:487–495
63. Dykes RW (1983) Parallel processing of somatosensory information: a theory. Brain Res Rev 6:47–115
64. Dykes RW, Craig AD (1998) Control of size and excitability of mechanosensory receptive fields in dorsal column nuclei by homolateral dorsal horn neurons. J Neurophysiol 80:120–129
65. Eickhoff SB, Amunts K, Mohlberg H, Zilles K (2006) The human parietal operculum. II. Stereotaxic maps and correlation with functional imaging results. Cereb Cortex 16:268–279
66. Erlanger J, Gasser HS (1937) Electrical signs of nervous activity. University of Pennsylvania Press, Philadelphia
67. Fields HL, Basbaum AI (1978) Brainstem control of spinal pain-transmission neurons. Annu Rev Physiol 40:217–248
68. Florence SL, Wall JT, Kaas JH (1989) Somatotopic organization of inputs from the hand to the spinal gray and cuneate nucleus of monkeys with observations on the cuneate nucleus of humans. J Comp Neurol 286:48–70
69. Florence SL, Wall JT, Kaas JH (1991) Central projections from the skin of the hand in squirrel monkeys. J Comp Neurol 311:563–578
70. Forss N, Raij TT, Seppa M, Hari R (2005) Common cortical network for first and second pain. Neuroimage 24:132–142
71. Friedman DP, Jones EG (1981) Thalamic input to areas 3a and 2 in monkeys. J Neurophysiol 45:59–85
72. Ganchrow D (1978) Intratrigeminal and thalamic projections of nucleus caudalis in the squirrel monkey (Saimiri sciureus): a degeneration and autoradiographic study. J Comp Neurol 178:281–312
73. Giesler GJ, Yezierski RP, Gerhart KD, Willis WD (1981) Spinothalamic tract neurons that project to medial and/or thalamic nuclei; evidence for a physiologically novel population of spinal neurons. J Neurophysiol 46:1285–1308
74. Gingold SI, Greenspan JD, Apkarian AV (1991) Anatomic evidence of nociceptive inputs to primary somatosensory cortex: relationship between spinothalamic terminals and thalamocortical cells in squirrel monkeys. J Comp Neurol 308:467–490
75. Gordon G, Grant G (1982) Dorsolateral spinal afferents to some medullary sensory nuclei. An anatomical study in the cat. Wxp Brain Res 46:12–23
76. Gowers R (1886) Bemerkungen über die anterolateral aufsteigende Degeneration im Rückenmark. Neurol Centralbl 5:97–99
77. Grant G (1993) Projection patterns of primary sensory neurons studied by transganglionic methods: somatotopy and target-related organization. Brain Res Bull 30:199–208
78. Grefkes C, Geyer S, Schormann T, Roland P, Zilles K (2001) Human somatosensory area 2: observer-independent cytoarchitectonic mapping, interindividual variability, and population map. Neuroimage 14:617–631
79. Grefkes C, Ritzl A, Zilles K, Fink GR (2004) Human medial intraparietal cortex subserves visuomotor coordinate transformation. Neuroimage 23:1494–1506
80. Groenewegen HJ, Witter MP (2004) Thalamus. In: Paxinos G (ed) The rat nervous system. Elsevier, Amsterdam, pp 407–453
81. Guilbaud G, Berkley KJ, Benoist JM, Gautron M (1993) Responses of neurons in thalamic ventrobasal complex of rats to graded distension of uterus and vagina and to uterine suprafusion with bradykinin and prostaglandin F2 alpha. Brain Res 614:285–290
82. Haeberle H, Fujiwara M, Chuang J, Medina MM, Panditrao MV, Bechstedt S, Howard J, Lumpkin EA (2004) Molecular profiling reveals synaptic release machinery in Merkel cells. Proc Natl Acad Sci USA 101:14503–14508
83. Han ZS, Zhang ET, Craig AD (1998) Nociceptive and thermoreceptive lamina I neurons are anatomically distinct. Nat Neurosci 1:218–225
84. Hayakawa T, Zheng JQ, Seki M, Yajima Y (1998) Synaptology of the direct projections from the nucleus of the solitary tract to pharyngeal motoneu-

rons in the nucleus ambiguus of the rat. J Comp Neurol 393:391–401
85. Hubscher CH, Berkley KJ (1994) Responses of neurons in caudal solitary nucleus of female rats to stimulation of vagina, cervix, uterine horn and colon. Brain Res 664:1–8
86. Ikeda M, Matsushita M, Tanami T (1982) Termination and cells of origin of the ascending intra-nuclear fibers in the spinal trigeminal nucleus of the cat. A study with the horseradish peroxidase technique. Neurosci Lett 31:215–220
87. Iwata K, Kamo H, Ogawa A, Tsuboi Y, Noma N, Mitsuhashi Y, Taira M, Koshikawa N, Kitagawa J (2005) Anterior cingulate cortical neuronal activity during perception of noxious thermal stimuli in monkeys. J Neurophysiol 94:1980–1991
88. Jasmin L, Burkey AR, Card JP, Basbaum AI (1997) Transneuronal labeling of a nociceptive pathway, the spino-(trigemino-)parabrachio-amygdaloid, in the rat. J Neurosci 17:3751–3765
89. Johansson H, Silfvenius H (1977) Axon-collateral activation by dorsal spinocerebellar tract fibres of group I relay cells of nucleus Z in the cat medulla oblongata. J Physiol 265:341–369
90. Jones EG, Friedman DP (1982) Projection pattern of functional components of thalamic ventrobasal complex on monkey somatosensory cortex. J Neurophysiol 48:521–544
91. Kaas JH (2004) Somatosensory system. In: Paxinos G, Mai JK (eds) The human nervous system. Elsevier, Amsterdam, pp 1059–1092
92. Kaas JH, Nelson RJ, Sur M, Dykes RW, Merzenich MM (1984) The somatotopic organization of the ventroposterior thalamus of the squirrel monkey, Saimiri sciureus. J Comp Neurol 226:111–140
93. Kakigi R, Inui K, Tran DT, Qiu Y, Wang X, Watanabe S, Hoshiyama M (2004) Human brain processing and central mechanisms of pain as observed by electro- and magneto-encephalography. J Clin Med Assoc 67:377–386
94. Kalia M, Mesulam MM (1980) Brain stem projections of sensory and motor components of the vagus complex in the cat: I. The cervical vagus and nodose ganglion. J Comp Neurol 193:435–465
95. Kalia M, Mesulam MM (1980) Brain stem projections of sensory and motor components of the vagus complex in the cat: II. Laryngeal, tracheobronchial, pulmonary, cardiac, and gastrointestinal branches. J Comp Neurol 193:467–508
96. Kobayashi M, Sasabe T, Takeda M, Kondo Y, Yoshikubo S, Imamura K, Onoe H, Sawada T, Watanabe Y (2002) Functional anatomy of chemical senses in the alert monkey revealed by positron emission tomography. Eur J Neurosci 16:975–980
97. Kobayakawa T, Wakita M, Saito S, Gotow N, Sakai N, Ogawa H (2005) Location of the primary gustatory area in humans and its properties, studied by magnetoencephalography. Chem Senses 30:1226–1227
98. Koerber HR, Brown PB (1982) Somatotopic organization of hindlimb cutaneous nerve projections to cat dorsal horn. J Neurophysiol 48:481–489
99. Krubitzer LA, Kaas JH (1992) The somatosensory thalamus of monkeys: cortical connections and a redefinition of nuclei in marmosets. J Comp Neurol 319:123–140
100. Krubitzer L, Clarey J, Tweedale R, Elston G, Calford M (1995) A redefinition of somatosensory areas in the lateral sulcus of macaque monkeys. J Neurosci 15:3821–3839
101. Kunc Z (1964) Tractus spinalis nervi trigemini; fresh anatomic data and their significance for surgery. Publishing House of the Czechoslovak Academy of Sciences., Prague
102. Kuypers HJGM, Maisky VA (1975) Retrograde axonal transport of horseradish peroxidase from spinal cord to brain stem cell groups in the cat. Neurosci Lett 1:9–14
103. Lenz FA, Dostrovsky JO, Tasker RR, Yamashiro K, Kwan HC, Murphy JT (1988) Single-unit analysis of the human ventral thalamic nuclear group: somatosensory responses. J Neurophysiol 59:299–316
104. Lima D, Mendes-Ribeiro JA, Coimbra A (1991) The spino-latero-reticular system of the rat: projections from the superficial dorsal horn and structural characterization of marginal neurons involved. Neuroscience 45:137–152
105. Lloyd DPC (1943) Neuron patterns controlling transmission of ipsilatral hind limb reflexes in cat. J Neurophysiol 6:293–315
106. Ma W, Blomqvist A, Berkley KJ (1989) Spino-diencephalic relays through the parabrachial nucleus in the cat. Brain Res 480:27–50
107. Mason P (1999) Central mechanisms of pain modulation. Curr Opin Neurobiol 9:436–441
108. Matsushita M, Ikeda M, Okado N (1982) The cells of origin of the trigeminothalamic, trigeminospinal and trigeminocerebellar projections in the cat. Neuroscience 7:1439–1454
109. Mehler WR (1962) The anatomy of the so-called "pain tract" in man: an analysis of the course and distributionof the ascending fibers of the fasciculus anterolateralis. In: French JD, Porter RW (eds) Basis research in paraplegia. Thomas, Springfield, pp 26–55
110. Melzack R, Wall PD (1965) Pain mechanisms: a new theory. Science 150:971–979
111. Morest DK (1967) Experimental study of the projections of the nucleus of the tractus solitarius and the area postrema in the cat. J Comp Neurol 130:277–300
112. Morgan C, Nadelhaft I, de Groat WC (1981) The distribution of visceral primary afferents from the pelvic nerve to Lissauer's tract and the spinal gray matter and its relationship to the sacral parasympathetic nucleus. J Comp Neurol 201: 415–440
113. Mouton LJ, Klop E, Holstege G (2001) Lamina I-periaqueductal gray (PAG) projections represent only a limited part of the total spinal and caudal medullary input to the PAG in the cat. Brain Res Bull 54:167–174

114. Nauta HJ, Hewitt E, Westlund KN, Willis WD (1997) Surgical interruption of a midline dorsal column visceral pain pathway. Case report and review of the literature. J Neurosurg 86:538–542
115. Newman HM, Stevens RT, Apkarian AV (1996) Direct spinal projections to limbic and striatal areas: anterograde transport studies from the upper cervical spinal cord and the cervical enlargement in squirrel monkey and rat. J Comp Neurol 365:640–658
116. Nomura S, Mizuno N (1982) Central distribution of afferent and efferent components of the glossopharyngeal nerve: an HRP study in the cat. Brain Res 236:1–13
117. Ogawa H, Wakita M, Hasegawa K, Kobayakawa T, Sakai N, Hirai T, Yamashita Y, Saito S (2005) Functional MRI detection of activation in the primary gustatory cortices in humans. Chem Senses 30:583–592
118. Olszewski J (1950) On the anatomical and functional organization of the spinal trigeminal nucleus. J Comp Neurol 92:401–413
119. Pan B, Castro-Lopes JM, Coimbra A (1999) Central afferent pathways conveying nociceptive input to the hypothalamic paraventricular nucleus as revealed by a combination of retrograde labeling and c-fos activation. J Comp Neurol 413:129–145
120. Pandya DN, Seltzer B (1982) Intrinsic connections and architectonics of posterior parietal cortex in the rhesus monkey. J Comp Neurol 204:196–210
121. Panneton WM, Burton H (1982) Origin of ascending intratrigeminal pathways in the cat. Brain Res 236:463–470
122. Penfield W, Bodrey E (1937) Somatic motor and sensory representation in the cerebral cortex of man as studied by electrical stimulation. Brain 60:389–443
123. Percheron G (2004) Thalamus. In: Paxinos G, May JK (eds) The human nervous system. Elsevier, Ammsterdam, pp 592–675
124. Ploner M, Schmitz F, Freund HJ, Schnitzler A (1999) Parallel activation of primary and secondary somatosensory cortices in human pain processing. J Neurophysiol 81:3100–3104
125. Pritchard TC, Norgren R (2004) Gustatory system. In: Paxinos G, Mai JK (eds) The human nervous system. Elsevier, Amsterdam, pp 1171–1196
126. Pritchard TC, Hamilton RB, Norgren R (2000) Projections of the parabrachial nucleus in the old world monkey. Exp Neurol 165:101–117
127. Pritchard TC, Hamilton RB, Morse JR, Norgren R (1986) Projections of thalamic gustatory and lingual areas in the monkey, Macaca fascicularis. J Comp Neurol 244:213–228
128. Puskar Z, Polgar E, Todd AJ (2001) A population of large lamina I projection neurons with selective inhibitory input in rat spinal cord. Neuroscience 102:167–176
129. Qi H-X, Kaas JH (2006) Organization of primary afferent projections to the gracile nucleus of the dorsal column system of primates. J Comp Neurol 499:183–217
130. Qiu Y, Noguchi Y, Honda M, Nakata H, Tamura Y, Tanaka S, Sadato N, Wang X, Inui K, Kakigi R (2005) Brain Processing of the Signals Ascending Through Unmyelinated C Fibers in Humans: An Event-Related Functional Magnetic Resonance Imaging Study. Cereb Cortex 16:1289–1295
131. Rausell E, Jones EG (1991) Histochemical and immunocytochemical compartments of the thalamic VPM nucleus in monkeys and their relationship to the representational map. J Neurosci 11:210–225
132. Rausell E, Jones EG (1991) Chemically distinct compartments of the thalamic VPM nucleus in monkeys relay principal and spinal trigeminal pathways to different layers of the somatosensory cortex. J Neurosci 11:226–237
133. Rausell E, Bae CS, Vinuela A, Huntley GW, Jones EG (1992) Calbindin and parvalbumin cells in monkey VPL thalamic nucleus: distribution, laminar cortical projections, and relations to spinothalamic terminations. J Neurosci 12:4088–4111
134. Reinisch CM, Tschachler E (2005) The touch dome in human skin is supplied by different types of nerve fibers. Ann Neurol 58:88–95
135. Rexed B (1954) A cytoarchitectonic atlas of the spinal cord in the cat. J Comp Neurol 100:297–351
136. Rhoton AL (1968) Afferent connections of the facial nerve. J Comp Neurol 133:89–100
137. Rolls ET, O'Doherty J, Kringelbach ML, Francis S, Bowtell R, McGlone F (2003) Representations of pleasant and painful touch in the human orbitofrontal and cingulate cortices. Cereb Cortex 13:308–317
138. Rustioni A, Sanyal S, Kuypers HG (1971) A histochemical study of the distribution of the trigeminal divisions in the substantia gelatinosa of the rat. Brain Res 32:45–52
139. Rustioni A, Hayes NL, O'Neill S (1979) Dorsal column nuclei and ascending spinal afferents in macaques. Brain 102:95–125
140. Scheibel ME, Scheibel AB (1968) Terminal axonal patterns in cat spinal cord. II. The dorsal horn. Brain Res 9:32–58
141. Schnitzler A, Rüdiger JS, Freund HJ (2000) The somatosensory system. In: Toga AW, Mazziotta JC (eds) Brain mapping. The systems. Academic Press, San Diego, pp 291–329
142. Schouenborg J, Sjölund BH (1983) Activity evoked by A- and C-afferent fibers in rat dorsal horn neurons and its relation to a flexion reflex. J Neurophysiol 50:1108–1121
143. Sewards TV, Sewards MA (2002) The medial pain system: neural representations of the motivational aspect of pain. Brain Res Bull 59:163–180

144. Shigenaga Y, Chen IC, Suemune S, Nishimori T, Nasution ID, Yoshida A, Sato H, Okamoto T, Sera M, Hosoi M (1986) Oral and facial representation within the medullary and upper cervical dorsal horns in the cat. J Comp Neurol 243:388–408
145. Shigenaga Y, Okamoto T, Nishimori T, Suemune S, Nasution ID, Chen IC, Tsuru K, Yoshida A, Tabuchi K, Hosoi M, et al. (1986) Oral and facial representation in the trigeminal principal and rostral spinal nuclei of the cat. J Comp Neurol 244:1–18
146. Shriver JE, Stein BM, Carpenter MB (1968) Central projections of spinal dorsal roots in the monkey. I. Cervical and upper thoracic dorsal roots. Am J Anat 123:27–74
147. Sie PG (1956) Localization of fibre systems within the white matter of the medulla oblongata and the cervical cord in man. Thesis Leiden Eduard Ijdo
148. Sjöqvist O (1938) Eine neue Operationsmethode bei Trigminusneuralgie: Durchschneidung des Tractus spinalis trigemini. Zentralbl Neurochir 2:274–281
149. Smith RL (1975) Axonal projections and connections of the principal sensory trigeminal nucleus in the monkey. J Comp Neurol 163:347–376
150. Snider WD, McMahon SB (1998) Tackling pain at the source: new ideas about nociceptors. Neuron 20:629–632
151. Stevens RT, Apkarian AV, Hodge CJ (1991) The location of spinothalamic axons within spinal cord white matter in cat and squirrel monkey. Somatosens Mot Res 8:97–102
152. Torvik A (1957) The spinal projection of the nucleus of the solitary tract. An experimental study in the cat. J Anat 91:314–322
153. Torvik A (1979) The ascending fibers from the main trigeminal nucleus. An experimntal study in the cat. Am J Anat 100:1–16
154. Truex RC, Taylor MJ, Smythe MQ, Gildenberg PL (1970) The lateral cervical nucleus of cat, dog and man. J Comp Neurol 139:93–104
155. Usunoff KG, Marani E, Schoen JH (1997) The trigeminal system in man. Adv Anat Embryol Cell Biol 136:1–126
156. Vierck CJ (1977) Absolute and differential sensitivities to touch stimuli after spinal cord lesions in monkeys. Brain Res 134:529–539
157. Vierck CJ, Cooper BY (1998) Cutaneous texture discrimination following transection of the dorsal spinal column in monkeys. Somatosens Mot Res 15:309–315
158. Vogt BA, Sikes RW (2000) The medial pain system, cingulate cortex, and parallel processing of nociceptive information. Prog Brain Res 122:223–235
159. von Economo C, Koskinas GN (1925) Die Cytoarchitektonik der Hirnrinde des ewachsenen Menschen. Springer, Berlin
160. Waite PME, Ashwell KWS (2004) Trigminal sensory system. In: Paxinos G, Mai JK (eds) The human nervous system. Elsevier, Amsterdam, pp 1093–1124
161. Waldeyer W (1888) Das Gorilla-Rückenmark. Abh Kön Akad Wiss Berlin
162. Wall PD, Noordenbos W (1977) Sensory functions which remain in man after complete transection of dorsal columns. Brain 100:641–653
163. Wiberg M, Westman J, Blomqvist A (1986) The projection to the mesencephalon from the sensory trigeminal nuclei. An anatomical study in the cat. Brain Res 399:51–68
164. Willis WD, Westlund KN (2004) Pain system. In: Paxinos G, May JK (eds) The human nervous system. Elsevier, Amsterdam, pp 1125–1170
165. Willis WD Jr, Zhang X, Honda CN, Giesler GJ Jr (2001) Projections from the marginal zone and deep dorsal horn to the ventrobasal nuclei of the primate thalamus. Pain 92:267–276
166. Willis WD, Xijing Z, Honda CN, Giesler GJ (2002) A crtitical review of the role of the proposed VMpo nucleus in pain. J Pain 2:79–94
167. Xu J, Wall JT (1996) Cutaneous representations of the hand and other body parts in the cuneate nucleus of a primate, and some relationships in previously described cortical representations. Somatosens Mot Res 13:187–197
168. Zatore RJ, Jones-Gotman M (2000) Functional imaging of the chemical senses. In: Toga AW, Mazziotta JC (eds) Brain mapping: the systems. Academic Press, San Diego, pp 403–424
169. Zhang ET, Craig AD (1997) Morphology and distribution of spinothalamic lamina I neurons in the monkey. J Neurosci 17:3274–3284
170. Zylka MJ, Rice FL, Anderson DJ (2005) Topographically distinct epidermal nociceptive circuits revealed by axonal tracers targeted to Mrgprd. Neuron 45:17–25

17 Vestibular System

Introduction

The vestibular system provides information about the position and motion of the head in space. The receptive elements are the hair cells of the membranous labyrinth. These cells are located in the cristae ampullares of the ampullae of the semicircular canals and in the maculae of the utricle and the saccule. The first-order elements are bipolar neurons, the somata of which constitute the vestibular ganglion, located in the internal auditory meatus. The peripheral processes of these cells terminate on the hair cells. The central processes of the bipolar cells constitute the vestibular part of the eighth cranial nerve, which enters the brain stem at the level of the pontomedullary junction.

The Vestibular Labyrinth, the Vestibular Nerve and the Vestibular Nuclei

The three semicircular canals are oriented in three approximately perpendicular planes. Displacement of the endolymphe in the horizontal canal in the direction of the ampulla increases the efferent discharge in the ganglion cell axons, and displacement away from the ampulla reduces their resting discharge. The effects of displacement of the endolymphe in the anterior and posterior vertical canals are reversed: displacement away from the ampullae excites the ganglion cells, and displacement toward the ampullae inhibits them. The semicircular canals provide information about the magnitude and the direction of angular accelerations in the plane of each canal. The two maculae of the otoliths are situated at right angles, with the macula utriculi in a horizontal position and the macula sacculi in a vertical position. The otolith organs detect the position of the head with respect to gravity and respond to linear accelerations, the utriculus in the fore-aft and intraaural directions and the sacculus in the dorso-ventral and fore-aft directions.

Two types of hair cells, which connect with different types of vestibular ganglion cells, can be distinguished [26]. The flask-shaped type I hair cells, which are surrounded by an afferent nerve calyx, occur in the central parts of the maculae (the striolar area) and the cristae ampullaris. Type I hair cells are innervated by ganglion cells that emit large to medium-sized axons in the vestibular nerve. These axons are characterized by an irregular resting discharge. The cylindrical type II hair cells, which are contacted by multiple afferent endings, occur

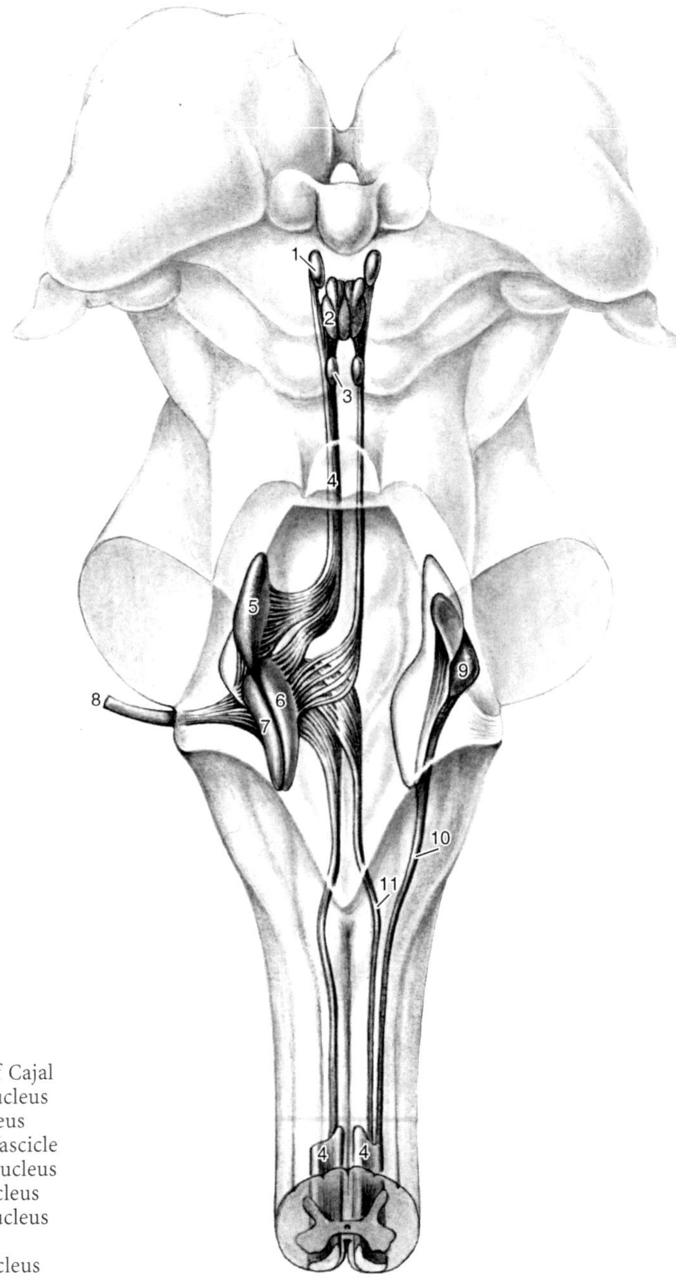

1 Interstitial nucleus of Cajal
2 Oculomotor nerve nucleus
3 Trochlear nerve nucleus
4 Medial longitudinal fascicle
5 Superior vestibular nucleus
6 Medial vestibular nucleus
7 Inferior vestibular nucleus
8 Vestibular nerve
9 Lateral vestibular nucleus
10 Lateral vestibulospinal tract
11 Medial vestibulospinal tract

Fig. 17.1. The vestibular system. Position of nerve, tracts and nuclei in a dorsal view (5/3×). The lateral vestibulospinal tract is shown on the right.

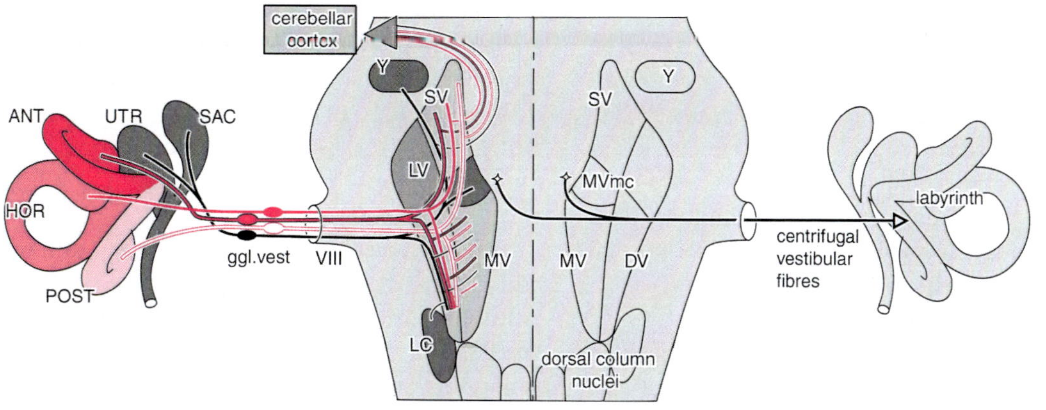

Fig. 17.2. The connections of the vestibular system. Rootfibres of the vestibular nerve bifurcate in ascending and descending branches at their entrance in the brain stem. The approximate position of the components of the vestibular nerve in the ascending and descending bundles is indicated. Fibres innervating the maculae of the sacculus and the utriculus are located most laterally (black). Fibres innervating the semicircular canals occupy successively more medial positions, with fibres innervating the anterior canal (red) laterally and the posterior canal (light red) medially. Projections of the sacculus predominate in the magnocellular medial vestibular nucleus, the group Y and the lateral cuneate nucleus. The vestibular nerve does not terminate in the lateral vestibular nucleus. Vestibular rootfibres reach the cerebellum via the superior vestibular nucleus. They terminate as mossy fibres on the ipsilateral side. Their distribution corresponds to the secondary vestibulocerebellar fibres illustrated in Fig. 17.3. Centrifugal fibres innervating the hair cells and the primary afferents of the maculae and the cristae ampullaris take their origin from the reticular formation adjoining the vestibular nuclei. *ANT*, anterior semicircular canal; *DV*, descending (inferior) vestibular nucleus; ggl.vest, vestibular ganglion; *HOR*, horizontal semicircular canal; *LC*, lateral cuneate nucleus; *LV*, lateral vestibular nucleus; *MV*, medial vestibular nucleus; *MVmc*, magnocellular medial vestibular nucleus; *POST*, posterior semicircular canal; *SAC*, sacculus; *SV*, superior vestibular nucleus; *UTR*, utriculus; *VIII*, vestibular nerve; *Y*, group Y

in the periphery of the maculae and the cristae ampullaris. They are innervated by ganglion cells with small to medium-sized, slowly conducting axons, with a regular resting discharge. The sensitivity of the coarse, irregular fibres of the vestibular nerve to linear or angular head displacements is much higher than that of the small, regular afferents. The central distribution of the two types of primary vestibular afferents is very similar

The vestibular nuclear complex is situated beneath the floor of the lateral part of the fourth ventricle. It comprises four cell masses, the superior, lateral, medial and descending (or inferior) vestibular nuclei (Figs. 6.23–6.26). The lateral vestibular nucleus can be subdivided into dorsal and ventral portions. The dorsal part contains the giant cells of Deiters' nucleus and is located within the juxtarestiform body with its efferent cerebellar components. In this chapter the name "lateral vestibular nucleus" will be reserved for this dorsal portion. The ventral part of the lateral vestibular nucleus contains a mixture of cells of different sizes and is located medial to the juxtarestiform body. It will be indicated as the magnocellular part of the medial vestibular nucleus (MVmc). Several accessory nuclei have been distinguished in the vestibular complex [10]. Cellgroup X is located lateral to the inferior vestibular nucleus, rostral to the lateral cuneate nucleus; group Y caps the restiform body and is located within the floccular peduncle. The parasolitary nucleus is a caudal extension of the vestibular nuclei, located lateral to the nucleus of the solitary tract (Figs. 6.23, 17.7). The nucleus prepositus hypoglossi is located in the floor of the fourth ventricle, medial to the medial vestibular nucleus.

At its entrance in the brain stem the fibres of the vestibular nerve divide into ascending and descending branches (Figs 17.1, 17.2). Most ascending branches terminate in the superior vestibular nucleus, but some ascend to the cerebellum in the juxtarestiform body. The descending branches of the vestibular nerve constitute a conspicuous bundle, situated among the fibres of the juxtarestiform body in the descending vestibular nucleus. The fibres of this bundle issue numerous transversely orientated collaterals, which terminate in the descending and medial vestibular nuclei. The lateral vestibular nucleus receives no primary vestibular fibres [48]. It is innervated by Purkinje cell axons of the B zone of the anterior vermis (Fig. 17.3).

The bifurcating axons innervating different end-organs distribute to different parts of the vestibular complex with a great degree of overlap [14, 40, 41, 49, 51, 71]. Still, direct, monosynaptic convergence of sensory information from vestibular nerve afferents innervating different end-organs on individual neurons in the vestibular nuclei is relatively rare [45]. Fibres innervating the sacculus are found most laterally in the ascending and descending branches of vestibular root fibres, followed by the fibres innervating the utriculus (Fig. 17.2). Anterior, horizontal and posterior canal fibres are distributed latero-medially in this order.

Macular afferents mainly terminate in the lateral periphery of the superior vestibular nucleus, semicircular canal afferents in its central portion and medial periphery. The subdivision of the superior vestibular nucleus in central and peripheral portions is also an important feature in the distribution of vestibulo-cerebellar neurons and Purkinje cell afferents from the vestibulocerebellum (see section on Efferents from the Vestibular Nuclei). Afferents from the maculae terminate in the MVmc and the adjacent spinal and medial vestibular nuclei. Distinct projections of the sacculus are present in the groups Y and X, and the caudal ventral cochlear nucleus. Fibres innervating the semicircular canals terminate in the medial and spinal vestibular nuclei. Saccular afferents are rare in the medial vestibular nucleus. Fibres innervating the maculae terminate in the rostrolateral lateral cuneate nucleus. Group Y is not directly innervated by canal afferents, but receives a projection from the anterior canal, through the superior vestibular nucleus (Highstein, personal communication). Macular and canal afferents also terminate in the gigantocellular and parvocellular reticular formation, adjacent to the vestibular nuclear complex, particularly in a cell group lying immediately cau-

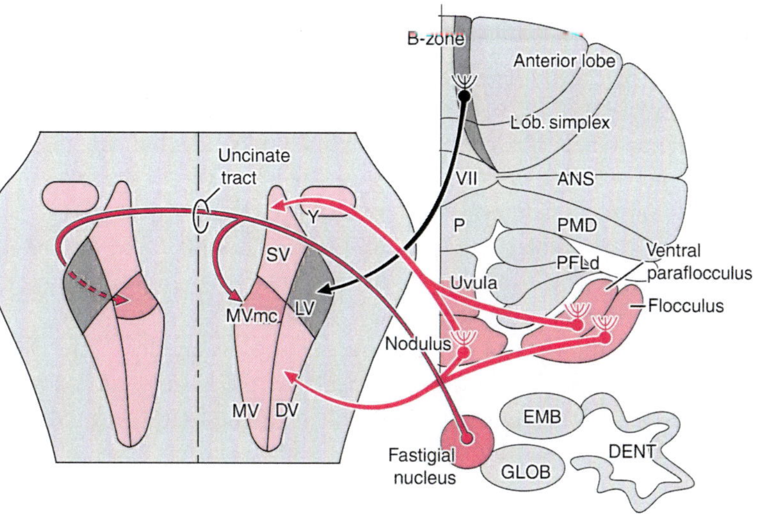

Fig. 17.3. The connections of the vestibular system. Cerebellar afferents take their origin from the Purkinje cells of the cerebellar cortex. The B zone of the anterior vermis projects to the lateral vestibular nucleus. Purkinje cells of the vestibulocerebellum (nodulus and adjoining uvula, flocculus and ventral paraflocculus) project to the other vestibular nuclei and group Y. The fastigial nucleus innervates the magnocellular medial vestibular nucleus through the crossed uncinate tract and an ipsilateral pathway. In the classical nomenclature of the cerebellum the lobulus simplex is known as the posterior quadrangular lobule with the declive, and the ventral paraflocculus as the accessory paraflocculus. *ANS*, ansiform (semilunar) lobule; *DENT*, dentate nucleus; *DV*, descending vestibular nucleus; *EMB*, emboliform (anterior interposed) nucleus; *GLOB*, globose (posterior interposed) nucleus; *LV*, lateral vestibular nucleus; *MV*, medial vestibular nucleus; *MVmc*, magnocellular medial vestibular nucleus; *P*, pyramid; *PFLd*, dorsal parafocculus (tonsil); *PMD*, paramedian (gracile) lobule; *SV*, superior vestibular nucleus; *VII*, lobule VII of Larsell (folium and tuber vermis); *Y*, group Y

doventral to the abducens nucleus [14]. There is experimental evidence suggesting that this reticular cell group projects primarily to the contralateral abducens nucleus [39].

Primary vestibular afferents reach the cerebellum through the juxtarestiform body. They terminate as mossy fibres, mainly ipsilaterally in the nodulus and the ventral uvula, with smaller contributions to the cortex in the bottom of the deep fissures, the lingula and the ventral paraflocculus. The flocculus lacks an innervation by the vestibular nerve [25]. Saccular and utricular afferents prevail in the uvula, where they terminate as longitudinal aggregates of mossy fibre rosettes. Canal afferents terminate in the nodulus [41, 51].

It should be noted that the vestibular nerve, in addition to afferent fibres, also carries a number of efferent fibres. The latter arise from a group of neurons that is interposed between the abducens and the superior vestibular nuclei embedded in the reticular formation (Fig. 17.2). The projection to one labyrinth has a bilaterally symmetrical origin from the cell groups on each side of the brain stem [27]. In the cat about 20% of the efferent vestibular neurons give off axons that terminate on hair cells of both labyrinths [17]. These neurons both express acetyltransferase and calcitonin gene-related peptide (CGRP) [53]. They terminate on the calyx of type I hair cells and directly on type II hair cells. These fibres have a predominantly excitatory influence on the afferent activity in the vestibular nerve and extend the dynamic range of the afferents during the large accelerations accompanying voluntary head movements [27].

Non-vestibular Afferents, Intrinsic and Commissural Connections of the Vestibular Nuclei

Apart from afferents from the vestibular ganglion the vestibular nuclei receive input from various other sources, among which are the cerebellum, the C2 and C3 spinal roots and the grey matter of the spinal cord, the nucleus prepositus hypoglossi, the interstitial nucleus of Cajal and the cerebral cortex.

The cerebellovestibular fibres arise from Purkinje cells located in the flocculonodular lobe, the uvula and the anterior vermis, and from the fastigial nucleus (Fig. 17.3). The caudal vermis (nodulus and uvula) and the flocculus project in a complementary manner [30, 68] to the ipsilateral superior, medial and inferior vestibular nuclei. The flocculus mainly projects to the central and dorsolateral superior and rostrally in the medial vestibular nucleus, and to cell group Y located in the floccular peduncle; regions that contain the vestibulo-ocular relay cells. Terminals of Purkinje cells from the uvula and the nodulus are found in the periphery of the superior vestibular nucleus and in caudal parts of the medial and descending vestibular nuclei, which mainly contain neurons projecting to the cerebellum and the spinal cord.

The principal afferents to the lateral vestibular nucleus are the axons of Purkinje cells which are located in the paramedian B zone of the anterior vermis (Fig. 17.3). Because of this, the lateral vestibular nucleus should be included in the cerebellar nuclei. This allocation is supported by the observation that climbing fibre collaterals innervate the cerebellar nuclei and the lateral vestibular nucleus (Chap. 20 and Fig. 20.6), but are absent from other parts of the vestibular complex [55, 64, 68]. The cerebellovestibular fibres originating from the fastigial nucleus project bilaterally to the MVmc, the spinal vestibular nucleus and adjacent parts of the medial vestibular nucleus [8]. The crossed fastgiobulbar pathway corresponds to the uncinate tract, which decussates in the cerebellar commissure (Figs. 6.27, 17.3).

Root fibres of the upper cervical roots terminate in the medial and spinal vestibular nuclei and the nucleus prepositus [50] (Fig. 17.4). Other relays for proprioceptive input from upper cervical roots to the vestibular nuclei are the central cervical nucleus and certain regions of the upper cervical grey matter. They project to the rostromedial MVmc and adjacent parts of the medial and spinal vestibular nuclei, where they contact vestibulospinal and vestibulo-oculomotor neurons involved in vestibulocollic and cervico-ocular reflexes [42, 43]. Spinovestibular fibres, arising from caudal levels of the spinal cord, ascend ipsilaterally with the dorsal spinocerebellar tract and terminate, possibly as collaterals, in the lateral vestibular nucleus and in the caudal parts of the medial and descending vestibular nuclei [52].

Vestibular nuclear afferents descending from higher levels of the neuraxis arise from the interstitial nucleus of Cajal and from the nucleus of Edinger-Westphal. The fibres emanating from the interstitial nucleus of Cajal descend in the medial longitudinal fascicle and terminate in the ipsilateral superior and medial vestibular nuclei and the nucleus prepositus hypoglossi whereas those from the nucleus of Edinger-Westphal project to the medial and descending vestibular nuclei [13, 14, 37]. The vestibular areas of the cerebral cortex, which provide a bilateral projection to the vestibular nuclei (Fig. 17.4), will be discussed below. It is still unclear how optokinetic information from the accessory optic system reaches the vestibular nuclei. Possible pathway pass through the nucleus prepositus hypoglossi or the nucleus reticularis tegmenti pontis [15, 23]. These pathways are considered in more detail in Chap. 19.

Intrinsic connections between the superior, medial and spinal vestibular nuclei are reciprocally organized (Fig. 17.5). The MVmc occupies a unique position, because it receives an input from all ipsilateral nuclei, but does not reciprocate these connections. Commissural pathways interconnecting the vestibular nuclei of both sides are widely divergent. They originate mainly from the medial vestibular nucleus, the adjacent spinal nucleus and from the periphery of the superior nucleus, and terminate in all nuclei, with relative sparing of the central part of the superior vestibular nucleus, the MVmc and the descending nucleus [19]. The commissural system contains inhibitory and excitatory components, responsible for the convergence and synergistic combination of information of the semicircular canals and the otholiths of both sides on the vestibular neurons [56].

The nucleus prepositus hypoglossi receives input from the vestibular nuclei, the cerebellum, from motor pathways involved in controlling eye and head movements, and from pathways that are involved in processing visual and proprioceptive signals. This nucleus and the surrounding perihypoglossal nuclei, in turn, project to nearly every region in the vestibular nuclei [9, 44].

Efferents from the Vestibular Nuclei

The vestibular nuclei distribute their efferents more widely in the neuraxis than any other special sensory system. These efferents can be grouped as follows:

1. Projections to the cerebellum (Fig. 17.7).
2. Projections to the spinal cord (Fig. 17.6) and to oculomotor centres (Fig. 17.8).
3. Connections with the thalamus and the cerebral cortex (Fig. 17.6).

Vestibular Projections to the Cerebellum

Secondary vestibulocerebellar fibres originate from the inferior, medial and the periphery of the superior nuclei and project bilaterally to the ipsilateral nodulus, uvula, the bottom of the deep transverse fissures and the flocculus (Fig. 17.7). Like the primary vestibulocerebellar fibres they terminate as mossy fibres in the granular layer of the cerebellar cortex [20, 66]. Unlike the primary afferents, they include the flocculus in their projection. Some of the vestibulo-cerebellar projections use acetylcholine as a neurotransmitter [5, 6]. Nodulus and uvula, the main targets of the vestibulocerebellar projections, are characterized by a complicated array of intrinsic cholinergic systems and receptors [35].

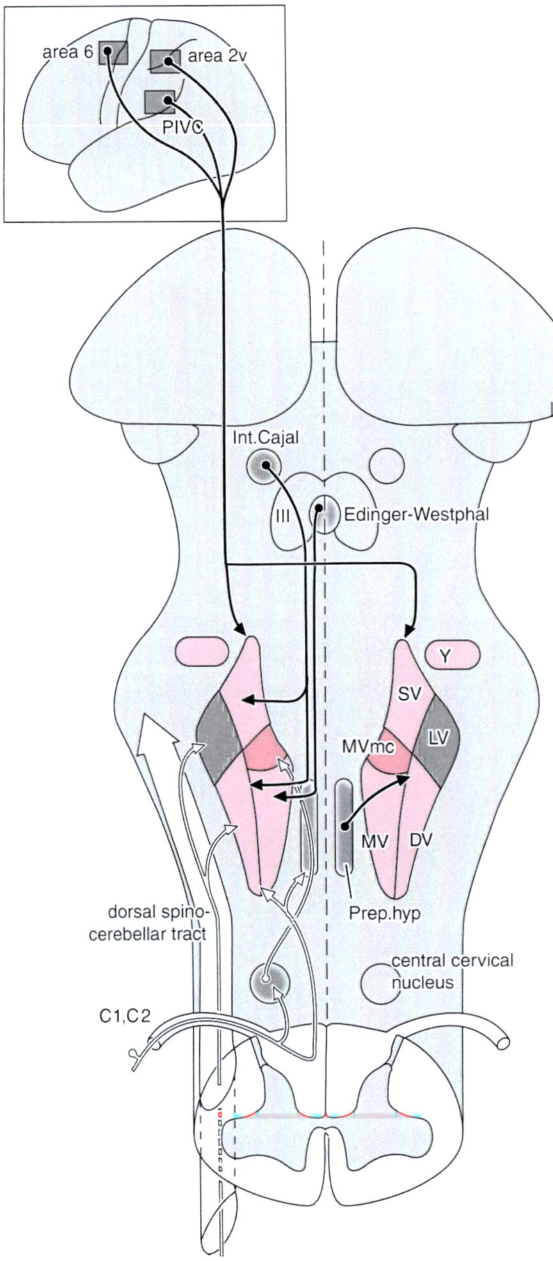

Fig. 17.4. The connections of the vestibular system. Non-vestibular afferent connections of the vestibular nuclei. Spinovestibular afferents include upper cervical root fibres, projections of the central cervical nucleus and collaterals of the dorsal spinocerebellar tract. Other afferent systems descend from the interstitial nucleus of Cajal and the nucleus of Edinger-Westphal. Cortical afferents from vestibular cortical areas terminate bilaterally. Of these areas, a region in premotor area 6, area 2v in the rostral tip of the intraparietal sulcus, and the parieto-insular vestibular cortical area PIVC are indicated. The nucleus prepositus hypoglossi projects to all vestibular nuclei, with the exception of the lateral nucleus; *III*, nucleus of the oculomotor nerve; *DV*, descending vestibular nucleus; *Int.Cajal*, interstitial nucleus of Cajal; *LV*, lateral vestibular nucleus; *MV*, medial vestibular nucleus; *MVmc*, magnocellular medial vestibular nucleus; *PIVC*, parieto-insular vestibular cortex; *Prep.hyp*, nucleus prepositus hypoglossi; *SV*, superior vestibular nucleus; *Y*, group Y

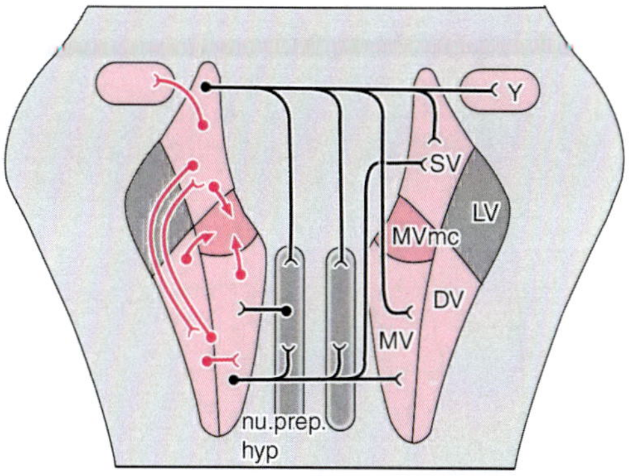

Fig. 17.5. The connections of the vestibular system. Intrinsic and commissural connections of the vestibular nuclei. Intrinsic connections, indicated in red, interconnect the superior (SV), descending (DV) and medial (MV) vestibular nuclei. Intrinsic pathways focus on the magnocellular medial nucleus (MVmc) and connect the superior nucleus with group Y. Commissural pathways interconnect all nuclei with the exception of the lateral (LV) and the magnocellular medial nuclei. The nucleus prepositus hypoglossi (nu.prep.hyp) is reciprocally connected with all vestibular nuclei

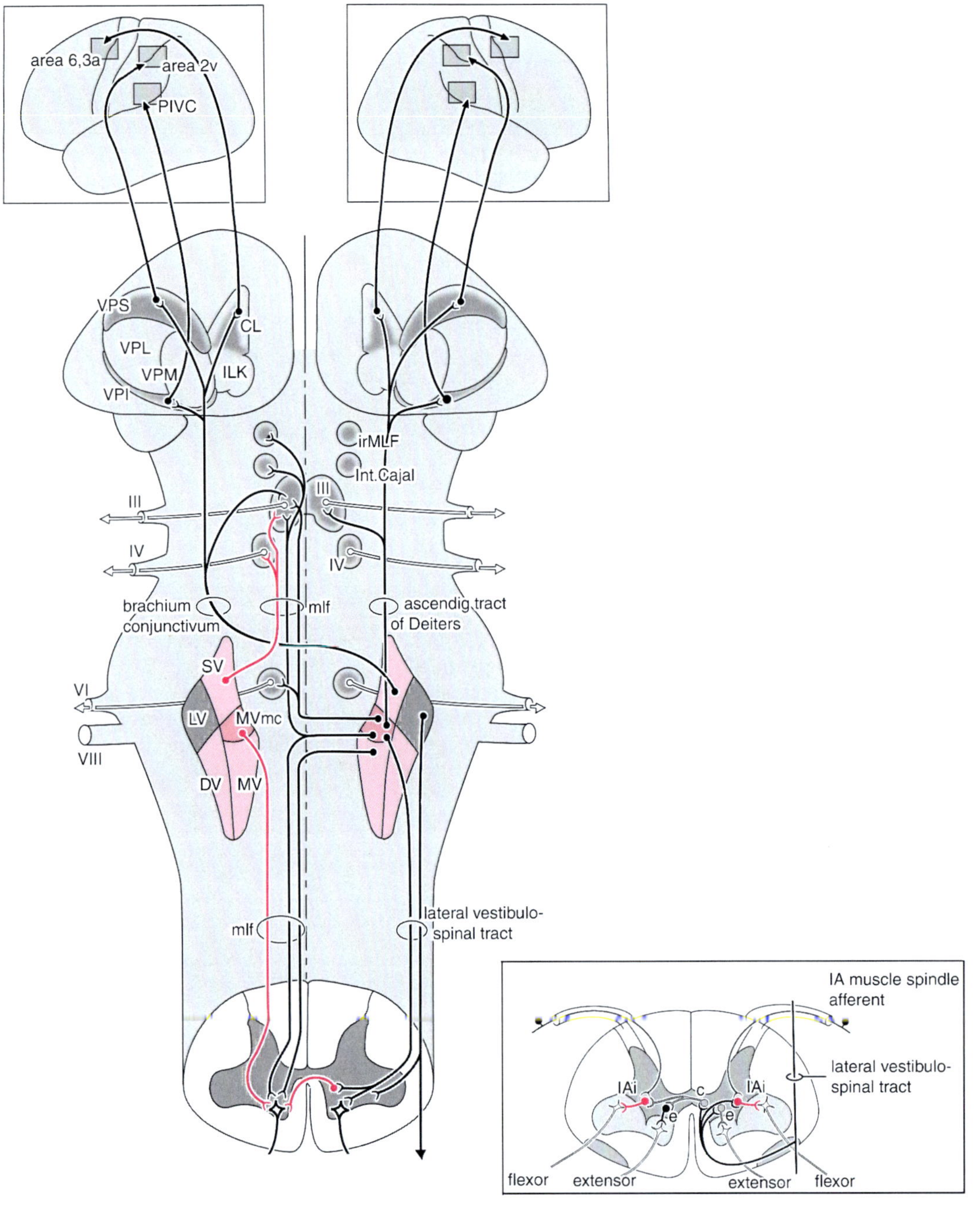

Fig. 17.6. The connections of the vestibular system. Efferent connections of the vestibular nuclei. Inset: spinal connections of the lateral vestibulospinal tract. Excitatory connections of the vestibular nuclei (black) are crossed, inhibitory connections (red) are always uncrossed. Exceptions to this rule are the ascending tract of Deiters and the lateral vestibulospinal tract, both of which are excitatory but course on the ipsilateral side. Vestibulo-oculomotor pathways terminate in the oculomotor nuclei, the interstitial nucleus of Cajal and the rostral interstitial nucleus of the medial longitudinal fascicle. The connections of the semicircular canals with the external eye muscles are detailed in Fig. 17.8. Vestibulospinal tracts descend in the medial longitudinal fascicle and as the lateral vestibulospinal tract. The medial longitudinal fascicle contains the direct (inhibitory) and the crossed (excitatory) vestibulospinal tracts. The lateral vestibulospinal tract takes its origin from the lateral vestibular nucleus. The vestibulo-cervical component of the lateral vestibulospinal tract excites ipsilateral neck

Vestibular climbing fibre pathways recently were reviewed by Barmack [4]. Vestibular climbing fibre pathways relay in two subnuclei of the medial accessory olive: the group beta and the dorsomedial cell column (Fig. 17.7). Excitatory projections to these subnuclei originate from the contralateral superior vestibular nucleus and group Y. Inhibitory projections, like all inhibitory projections from the vestibular nuclei, are ipsilateral and take their origin from a caudal cell group of the vestibular complex, known as the nucleus parasolitarius [7]. Signals from the anterior and posterior semicircular canals are relayed to the group beta. Neurons of the dorsomedial cell column respond preferentially to otolithic stimulation [4]. Climbing fibres from the group beta and the dorsomedial cell column terminate in a series of parallel zones in the nodulus, the uvula and the pyramis. In the nodulus and the uvula these zones interdigitate with climbing fibre zones innervated by the dorsal cap and the ventrolateral outgrowth (Fig. 19.13). In addition, projections of the medial and lateral vestibular nuclei terminate in the caudal medial accessory olive, which innervates the A zone of the anterior vermis [24]. The nucleus prepositus hypoglossi inhibits olivary subnuclei relaying optokinetic information to the flocculus through a pathway co-localizing GABA and acetylcholine [16].

Projections to the Spinal Cord and the Oculomotor Centres

Bilateral ascending and descending projections of the vestibular nuclei are among the main components of the medial longitudinal fascicle (MLF). Ipsilateral components of the MLF generally are inhibitory, and crossed systems are excitatory. Crossed ascending fibres also use the brachium conjunctivum as a vehicle. The ipsilaterally ascending tract of Deiters [47] and the ipsilaterally descending lateral vestibulospinal tract are exceptions to the rule that excitatory vestibular pathways are always crossed (Fig. 17.6). These pathways serve as the final common route for reflexes stabilizing the eye, head and body in space, and for the execution of smooth pursuit movements (see also Chap. 19). An additional vestibulospinal pathway originates from the caudal vestibular nuclei to terminate in the cervical dorsal horn and intermediate grey [3].

The lateral vestibulospinal tract is an uncrossed pathway from the lateral vestibular nucleus located lateral to the MLF in the centre of the bulbar reticular formation, that descends along the entire spinal cord at the junction of the ventral and lateral funiculi (Fig. 17.5). Its main component consists of coarse fibres arising from the non-vestibular, dorsally located lateral vestibular nucleus. These fibres exert a facilitatory influence on spinal reflex activity and on extensor muscle tone [11]. Inhibition of flexor muscles is mediated by Ia inhibitory interneurons [36] (Fig. 17.6).

◀ motoneurons and inhibits contralateral motoneurons through an inhibitory spinal interneuron. The main component of the lateral vestibulospinal tract descends to lumbar levels. It exerts a facilitatory influence on extensor motoneurons through excitatory interneurons (inset: e). Flexor motoneurons are inhibited by Ia inhibitory interneurons (Iai). Vestibulo-cortical pathways ascend in the brachium conjunctivum and the ascending tract of Deiters. They are relayed by the superior and inferior thalamic nuclei that surround the medial and lateral ventral posterior nuclei and the central lateral intralaminar nucleus, to polysensory cortical areas. These include portions of areas 6 and areas 3 a, and 2 v, located in the rostral intraparietal sulcus and the parieto-insular vestibular cortex. *c*, commissural interneuron; *CL*, central lateral intralaminar nucleus; *DV*, descending vestibular nucleus; *e*, excitatory interneuron; *III, IV, VI* (nuclei of the) oculomotor, trochlear and abducens nerves; *IAi*; Ia inhibitory interneuron; *ILK*, intralaminar nuclei of the thalamus; *Int.Cajal*, interstitial nucleus of Cajal; *irMLF*, interstitial nucleus of the medial longitudinal fascicle; *LV*, lateral vestibular nucleus; *mlf*, medial longitudinal fascicle; *MV*, medial vestibular nucleus; *MVmc*, magnocellular medial vestibular nucleus; *PIVC*, parieto-insular vestibular cortex; *SV*, superior vestibular nucleus; *VIII*, vestibular nerve; *VPI*, inferior ventral posterior thalamic nucleus; *VPL*, lateral ventral posterior thalamic nucleus; *VPM*, medial ventral posterior thalamic nucleus; *VPS*, superior ventral posterior thalamic nucleus

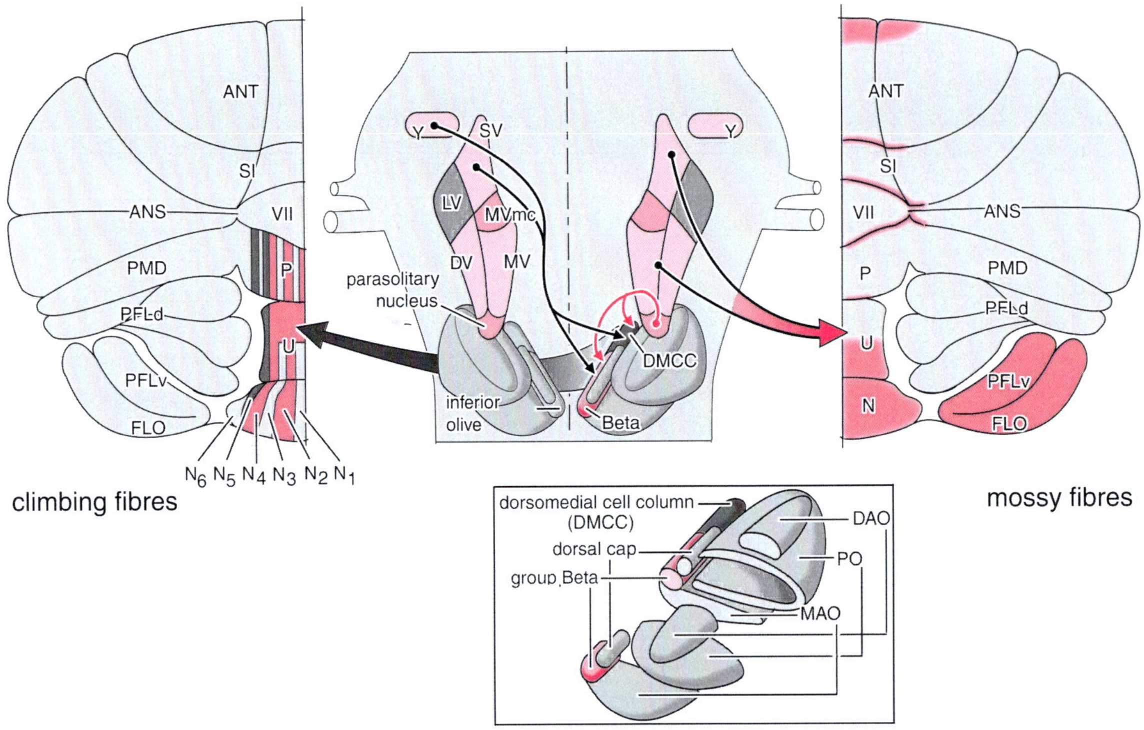

Fig. 17.7. The connections of the vestibular system. Efferent connections with the cerebellum. Climbing fibre paths are relayed by two subnuclei of the medial accessory olive, the group beta and the dorsomedial cell column. Their position is shown in a diagram of the transected inferior olive (inset). They receive an inhibitory projection from the ipsilateral vestibular nuclei (from the parasolitary nucleus) and an excitatory input from the contralateral superior vestibular nucleus and group Y. They project to longitudinal zones in the contralateral nodulus (N2, 4 and 5), the uvula and the pyramis. These zones interdigitate with zones receiving climbing fibres from the dorsal cap (Fig. 19.18). Mossy fibres from the superior, the medial, and the descending vestibular nucleus terminate bilaterally in the nodulus, the flocculus and adjacent lobules, in the rostral anterior lobe and in the depth of the transverse fissures. *VII*, lobule VII of Larsell; *ANS*, ansiform (semilunar) lobule; *ANT*, anterior lobe; *DAO*, dorsal accessory olive; *DMCC*, dorsomedial cell column; *DV*, descending vestibular nucleus; *FLO*, flocculus; *LV*, lateral vestibular nucleus; *MAO*, medial accessory olive; *MV*, medial vestibular nucleus; *MVmc*, magnocellular medial vestibular nucleus; *N*, nodulus; N_{1-6}, zones 1–6 of the nodulus; *P*, pyramis; *PFLd*, dorsal paraflocculus (tonsil); *PFLv*, ventral paraflocculus (accessory paraflocculus); *PMD*, paramedian (gracile) lobule; *PO*, principal olive; *SI*, simplex lobule (declive and lobulus quadrangularis posterior); *SV*, superior vestibular nucleus; *U*, uvula; *Y*, group Y

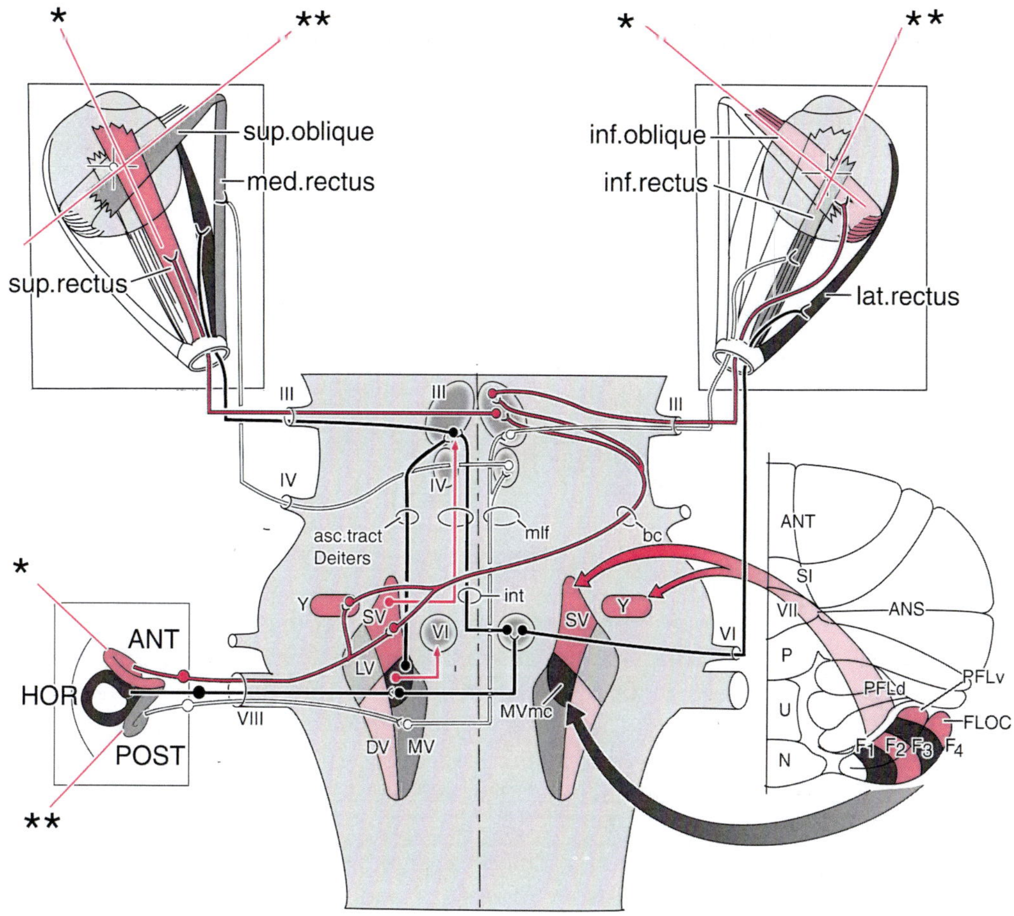

Fig. 17.8. The connections of the vestibular system. The three-neuron arc of the vestibulo-ocular reflex connecting the semicircular canals with the extraocular muscles. The system connecting the anterior semicircular canal with the ipsilateral superior rectus and contralateral inferior oblique muscles is indicated in red. It passes through the superior vestibular nucleus, the group Y and the brachium conjunctivum. The connections of the horizontal canal with the ipsilateral medial and the contralateral lateral rectus muscles are indicated in *dark grey*. They pass through the magnocellular medial vestibular nucleus and the uncrossed ascending tract of Deiters. The latter connection is duplicated by the internuclear pathway from the abducens nerve nucleus. This pathway recrosses on its way to the medial rectus motoneurons in the oculomotor nucleus. The connections of the posterior canal with the ipsilateral superior oblique and contralateral inferior rectus muscles (*open contours*) run in the contralateral medial longitudinal fascicle. Inhibitory pathways (indicated in *red*) connect the vestibular nuclei with the ipsilateral oculomotor nuclei. Purkinje cells located in zones F1 and F3 of the flocculus and the ventral (accessory) paraflocculus inhibit relay cells of the horizontal canal in the magnocellular medial vestibular nucleus. Zones F2 and F4 project to relay cells of the anterior canal in the superior vestibular nucleus and group Y. The climbing fibre innervation of these zones by the accessory optic system and the nucleus of the optic tract is illustrated in Fig. 19.13. Note approximate colinearity between the planes of the anterior (*asterisk*) and posterior (*double asterisk*) semicircular canals and the pulling directions of the extraocular muscles they innervate. *III, IV, VI*, nuclei of the oculomotor, trochlear and abducens nerves; *VI–VIII*, lobules VI–VIII of Larsell; *ANS*, ansiform (semilunar) lobule; *ANT*, anterior lobe; *ANT*, anterior semicircular canal; *bc*, brachium conjunctivum; *DV*, descending vestibular nucleus; F_1–F_4, Purkinje cell zones 1–4 of the flocculus; *FLOC*, flocculus; *HOR*, horizontal semicircular canal; *int*, internuclear pathway; *LV*, lateral vestibular nucleus; *mlf*, medial longitudinal fascicle; *MV*, medial vestibular nucleus; *MVmc*, magnocellular medial vestibular nucleus; *N*, nodulus; *P*, pyramis; *PFLd*, dorsal paraflocculus (tonsil); *PFLv*, ventral (accessory) paraflocculus; *POST*, posterior semicircular canal; *SI*, simplex lobule (declive and lobulus quadrangularis posterior); *SV*, superior vestibular nucleus; *U*, uvula; *Y*, group Y

Vestibulo-ocular Reflex Pathways

The fibres that connect the vestibular nuclei with the oculomotor nuclei are links in elementary three-neuron reflex arcs of the vestibulo-ocular reflex. This reflex stabilizes the retinal image by producing compensatory conjugate eye movements for head movements perceived by the labyrinths. These pathways are also used by the optokinetic reflex (see Chap. 19). The final common path of these reflexes is formed by three excitatory and three inhibitory pathways that connect the vestibular nuclei with the motoneurons innervating pairs of synergistic external eye muscles.

The six external eye muscles are located in three perpendicular planes, which are approximately co-linear with the planes of the three semicircular canals [62] (Fig. 17.8). The vestibulo-oculomotor system is composed of three pairs, each consisting of an excitatory and an inhibitory subsystem. Excitatory subsystems connect a semicircular canal with the eye muscles producing a compensatory eye movement in the plane of the semicircular canal. The inhibitory systems innervate their antagonists. [31–33, 54, 70]. The alignment of the planes of the external eye muscles and the semicircular canals is not precise, additional connections between the semicircular canals and the oculomotor nuclei compensate for these aberrations [21, 22]. The direction of the compensatory eye movements is always the opposite of the direction of the angular acceleration of the head, as perceived by the semicircular canals. Similar reflex pathways exist for the otoliths [12].

A lateral rotation of the head excites the ipsilateral horizontal canal by a displacement of the endolymphe in the direction of the ampulla. Horizontal canal fibres terminate on two groups of excitatory neurons located in the MVmc, projecting to the contralateral abducens nucleus and the ipsilateral medial rectus pool of motoneurons in the nucleus of the oculomotor nerve and thus cause a compensatory conjugate eye movement consisting of adduction of the ipsilateral, and abduction of the contralateral eye. The excitatory pathway to the ipsilateral medial rectus motoneurons ascends in the ascending tract of Deiters, located in the lateral wing of the MLF (Fig. 17.8). It is duplicated by an internuclear pathway, which arises from the contralateral abducens nucleus, and then recrosses to ascend in the MLF. The corresponding inhibitory pathway takes its origin from the MVmc and terminates in the ipsilateral abducens nucleus. This inhibitory pathway, like some of the inhibitory vestibulospinal pathways, uses glycine as a neurotransmitter; the ascending inhibitory pathways to the fourth and third nuclei are GABAergic [63].

The connections for eye movements produced by stimulation of the anterior and posterior vertical canals are illustrated in Fig. 17.8. It should be noted that excitation of the vertical canals results from a movement of the endolymphe away from the ampullae. The excitatory system from the anterior canal innervates the ipsilateral superior rectus and the contralateral inferior oblique muscles, and the excitatory pathway from the posterior canal terminates on motoneurons of the ipsilateral superior oblique and the contralateral inferior rectus muscles. Motoneurons innervating these four muscles are located on the contralateral side, but the axons innervating the superior rectus through the oculomotor nerve, and the superior oblique through the trochlear nerve, cross within the brain stem. Excitatory connections from the anterior canal take their origin from the dorsal group Y and the dorsal superior vestibular nucleus and use the superior cerebellar peduncle to terminate in the contralateral oculomotor nucleus. A second group of excitatory anterior canal neurons is located in the rostral medial vestibular nucleus. It gives rise to branching axons, which terminate in the oculomotor nuclei and the spinal cord [28, 46] (Fig. 17.6) Similar neurons with branching axons exist for the posterior canal [34]. Excitatory posterior canal fibres. arise from the rostral medial vestibular nucleus and ascend in the contralateral MLF to the fourth and third nucleus. The inhibitory pathways for the vertical canals from the central portion of the superior vestibular nucleus ascend in the ipsilateral MLF. The vertical canal pathways also connect with the interstitial nucleus of Cajal.

Vestibulo-collic Reflex Pathways

The vestibulo-collic reflex [57–60, 65] stabilizes head position in response to head perturbations: it produces head movements in the planes of the stimulated semicircular canals. Co-activation of the anterior and posterior canals produces movements around a bitemporal axis, co-activation of all ipsilateral or contralateral canals causes tilts around an occipitonasal axis, and the activation of the horizontal canals, which excite the contralateral and inhibit the ipsilateral neck muscles, causes rotation of the head. The more than 30 neck muscles constitute a complicated system, innervated from the upper cervical cord. Motoneurons of each muscle receive excitatory and/or inhibitory input from all six semicircular canals. Groups of vestibulo-spinal axons branch to innervate motoneurons of muscles that participate in co-contractions producing a purposeful movement. This motor pool includes the spinal accessory motor nucleus, which innervates the sterno-cleido-mastoid and the trapezius muscles. This system recently was reviewed by Shinoda et al. [60].

The organization of vestibulospinal systems subserving the vestibulo-collic reflex is relatively simple (Fig. 17.6). The cell bodies of the excitatory and inhibitory canal-activated vestibulospinal neurons are located in the MVmc and the adjoining medial and inferior vestibular nuclei. Ipsilaterally descending fibres, activated by the ipsilateral anterior or posterior canal, join the lateral vestibulospinal tract. The canal-activated components of the lateral vestibulospinal tract terminate in the cervical cord, where they make excitatory connections with motoneurons of neck muscles and inhibitory commissural interneurons, which inhibit contralateral motoneurons. Two other fibre systems, corresponding to the medial and the crossed vestibulospinal tracts, descend in or adjacent to the MLF [11]. The medial vestibulospinal tract is located ipsilaterally in the dorsolateral bulbar MLF and consists of medium-sized and small fibres. It contains the inhibitory connections from each semicircular canal to the cervical motoneurons. Fibres of the crossed vestibulospinal tract decussate at their level of origin and become located next to the midline in the MLF. Coarse fibres account for a large proportion of the fibres of this tract. It contains the excitatory connections from all ipsilateral canals to the neck motoneurons. A contribution of reticulospinal systems to the vestibulocollic reflex was discussed by Wilson and Schor [69].

Projections from the Vestibular Nuclei to the Thalamus and the Cerebral Cortex

Vestibular pathways to the cerebral cortex are still imperfectly known. However, there is evidence that cells located within the superior, medial and inferior vestibular nuclei give rise to fibres that terminate in the thalamus. Most of these fibres ascend with the superior cerebellar peduncle. They decussate in the isthmus region and pass diffusely through the midbrain tegmentum and the fields of Forel to terminate in the thalamus. A smaller contingent of vestibulothalamic fibres remains on the ipsilateral side and follows the ascending tract of Deiters (Fig. 17.6). In primates the vestibulothalamic fibres terminate bilaterally in the shell surrounding the ventral posterior nuclei (ventral posterior superior nucleus, VPS, and ventral posterior inferior nucleus, VPI) [1] and in the central lateral intralaminar nucleus [18, 61]. In primates the thalamic projections of the vestibular complex largely overlap with projections of proprioceptive origin (see Chap. 16, The Sensory Relay Nuclei of the Thalamus).

Polymodal cortical areas responding to vestibular stimulation are located in the same regions or adjacent to regions receiving the propioceptive and nociceptive input. They include the parietoinsular vestibular cortex (PIVC), areas located in the neck, trunk, and hand divisions of area 3a in the central sulcus, area 2v located in the anterior tip of the intraparietal sulcus and part of area 6 (area 6pa), bordering on the smooth pursuit subfield of the frontal eye field. In monkeys, PIVC receives thalamic afferents from the VPI and areas 2v and 3a from the VPS [1, 18]. The different vestibular cortical areas are interconnected; the PIVC occupies a central position in this network [29]. Similar vestibular cortical areas also were localized in the human brain using fMRI after bilateral galvanic stimulation of the labyrinth [38]. In the human brain most of these areas are lateralized. The PIVC was located in the left parieto-temporal junction, superficial to Heschl's gyrus and caudal to the parietal speech areas. Foci in the armfield in the right central sulcus and in a region in the left intraparietal sulcus may correspond to similar foci in the monkey area 3 and 2. Bilateral activity was present in the premotor area, in areas 6a and 8a. The latter were not found in earlier studies, using caloric stimulation and, therefore, may represent input from the otoliths.

Most of the vestibular cortical fields project bilaterally to the vestibular nuclei. Vestibular nuclei which are connected with skeletomotor centres (mainly the medial vestibular nucleus) receive their cortical afferents mainly from areas 3a and 6. Nuclei with an output to oculomotor centres (rostral-medial and superior vestibular nuclei, group y) receive projections from the parietoinsular cortex [2, 29, 67] (Fig. 17.4).

References

1. Akbarian S, Grüsser OJ, Guldin WO (1992) Thalamic connections of the vestibular cortical fields in the squirrel monkey (Saimiri sciureus). J Comp Neurol 326:423–441
2. Akbarian S, Grüsser OJ, Guldin WO (1994) Corticofugal connections between the cerebral cortex and brainstem vestibular nuclei in the macaque monkey. J Comp Neurol 339:421–437
3. Bankoul S, Neuhuber WL (1990) A cervical primary afferent input to vestibular nuclei as demonstrated by retrograde transport of wheat germ agglutinin-horseradish peroxidase in the rat. Exp Brain Res 79:405–411
4. Barmack NH (2005) Inferior olive and oculomotor system. Prog Brain Res 151:269–291
5. Barmack NH, Baughman RW, Eckenstein FP (1992) Cholinergic innervation of the cerebellum of rat, rabbit, cat, and monkey as revealed by choline acetyltransferase activity and immunohistochemistry. J Comp Neurol 17:233–249
6. Barmack NH, Baughman RW, Eckenstein F.P, Shojaku H (1992) Secondary vestibular cholinergic projection to the cerebellum of rabbit and rat as revealed by choline acetyltransferase immunohistochemistry, retrograde and orthograde tracers. J Comp Neurol 317:250–270
7. Barmack NH, Fredette BJ, Mugnaini E (1998) Parasolitary nucleus: a source of GABAergic vestibular information to the inferior olive of rat and rabbit. J Comp Neurol 392:352–372
8. Batton RR 3rd, Jayaraman A, Ruggiero D, Carpenter MB (1977) Fastigial efferent projections in the monkey: an autoradiographic study. J Comp Neurol 174:281–305
9. Belknap DB, McCrea RA (1988) Anatomical connections of the prepositus and abducens nuclei in the squirrel monkey. J Comp Neurol 268:13–28
10. Brodal A, Pompeiano O (1957) The vestibular nuclei in cat. J Anat 91:438–54
11. Busch HFM (1961) An anatomical analysis of the white matter in the brain stem of the cat. Van Gorcum, Assen
12. Büttner-Ennever JA (1999) A review of otolith pathways to brainstem and cerebellum. Ann NY Acad Sci 871:51–64
13. Carleton SC, Carpenter MB (1983) Afferent and efferent connections of the medial, inferior and lateral vestibular nuclei in the cat and monkey. Brain Res 278:29–51
14. Carleton SC, Carpenter MB (1984) Distribution of primary vestibular fibers in the brainstem and cerebellum of the monkey. Brain Res 294:281–298
15. Cazin L, Precht W, Lannou J (1980) Pathways mediating optokinetic responses of vestibular nucleus neurons in the rat. Pflügers Arch 384:19–29
16. De Zeeuw CI, Wentzel P, Mugnaini E (1993) Fine structure of the dorsal cap of the inferior olive and its GABAergic and non-GABAergic input from the nucleus prepositus hypoglossi in rat and rabbit. J Comp Neurol 327:63–82
17. Dechesne C, Raymond J, Sans A (1984) The efferent vestibular system in the cat: a horseradish peroxidase and fluorescent retrograde tracers study. Neuroscience 11:893–901
18. Ebata S, Sugiuchi Y, Izawa Y, Shinomiya K, Shinoda Y (2004) Vestibular projection to the periarcuate cortex in the monkey. Neurosci Res 49:55–68
19. Epema AH, Gerrits NM, Voogd J (1988) Commissural and intrinsic connections of the vestibular nuclei in the rabbit: a retrograde labeling study. Exp Brain Res 71:129–146

20. Epema AH, Gerrits NM, Voogd J (1990) Secondary vestibulocerebellar projections to the flocculus and uvulo-nodular lobule of the rabbit: A study using HRP and double fluorescent tracer techniques. Exp Brain Res 80:72–82
21. Ezure K, Graf W (1984) A quantitative analysis of the spatial organization of the vestibulo-ocular reflexes in lateral- and frontal-eyed animals. I. Orientation of semicircular canals and extraocular muscles. Neuroscience 12:85–93
22. Ezure K, Graf W (1984) A quantitative analysis of the spatial organization of the vestibulo-ocular reflexes in lateral- and frontal-eyed animals. II. Neuronal networks underlying vestibulo-oculomotor coordination. Neuroscience 12:95–109
23. Gerrits NM, Voogd J (1986) The nucleus reticularis tegmenti pontis and the adjacent rostral paramedian reticular formation: differential projections to the cerebellum and the caudal brain stem. Exp Brain Res 62:29–45
24. Gerrits NM, Voogd J, Magras IN (1985) Vestibular afferents of the inferior olive and the vestibulo-olivocerebellar climbing fiber pathway to the flocculus in the cat. Brain Res 322:355–363
25. Gerrits NM, Epema AH, van Linge A, Dalm E (1989) The primary vestibulocerebellar projection in the rabbit: absence of primary afferents in the flocculus. Neurosci Lett 105:27–33
26. Goldberg JM (1991) The vestibular end organs: Morphological and physiological diversity of afferents. Curr Opin Neurobiol 1:229–235
27. Goldberg JM, Fernandez C (1980) Efferent vestibular system in the squirrel monkey: anatomical location and influence on afferent activity. J Neurophysiol 43:986–1025
28. Graf W, Ezure K (1986) Morphology of vertical canal related second order vestibular neurons in the cat. Exp Brain Res 63:35–48
29. Guldin WO, Grüsser OJ (1998) Is there a vestibular cortex? Trends Neurosci pp 254–259
30. Haines DE (1976) Cerebellar corticonuclear and corticovestibular fibers of the anterior lobe vermis in a prosimian primate (Galago senegalensis). J Comp Neurol 170:67–95
31. Highstein SM (1973) Synaptic linkage in the vestibulo-ocular and cerebello-vestibular pathways to the VIth nucleus in the rabbit. Exp Brain Res 17:301–314
32. Highstein SM (1973) The organization of the vestibulo-oculomotor and trochlear reflex pathways in the rabbit. Exp Brain Res 17:285–300
33. Highstein SM, Ito M (1971) Differential localization within the vestibular nuclear complex of the inhibitory and excitatory cells innervating 3d nucleus oculomotor neurons in rabbit. Brain Res 29:358–362
34. Isu N, Uchino Y, Nakashima H, Satoh S, Ichikawa T, Watanabe S (1988) Axonal trajectories of posterior canal-activated secondary vestibular neurons and their coactivation of extraocular and neck flexor motoneurons in the cat. Exp Brain Res 70:181–191
35. Jaarsma D, Ruigrok TJ, Caffe R, Cozzari C, Levey AI, Mugnaini E, Voogd J (1997) Cholinergic innervation and receptors in the cerebellum. Prog Brain Res 11:467–496
36. Jankovska E, Lindström S (1972) Morphology of interneurons mediating Ia reciprocal inhibition of motoneurones in the spinal cord of the cat. J Physiol London 266:805–832
37. Kokkoroyannis T, Scudder CA, Balaban CD, Highstein SM, Moschovakis AK (1996) Anatomy and physiology of the primate interstitial nucleus of Cajal I. efferent projections. J Neurophysiol 75:725–739
38. Lobel E, Kleine JF, Leroy-Willig A, Van de Moortele PF, Le Bihan D, Grüsser OJ, Berthoz A (1999) Cortical areas activated by bilateral galvanic vestibular stimulation. Ann NY Acad Sci 871:313–323
39. Maciewicz RJ, Kaneko CR, Highstein SM, Eagen K (1977) Vestibular and medullary brain stem afferents to the abducens nucleus in the cat. Brain Res 123:229–240
40. Maklad A, Fritzsch B (2002) The developmental segregation of posterior crista and saccular vestibular fibers in mice: a carbocyanine tracer study using confocal microscopy. Brain Res Dev Brain Res 140:223–236
41. Maklad A, Fritzsch B (2003) Partial segregation of posterior crista and saccular fibers to the nodulus and uvula of the cerebellum in mice, and its development. Brain Res Dev Brain Res 140:223–236
42. Matsushita M, Xiong G (2001) Uncrossed and crossed projections from the upper cervical spinal cord to the cerebellar nuclei in the rat, studied by anterograde axonal tracing. J Comp Neurol 432:101–118
43. Matsushita M, Gao X, Yaginuma H (1995) Spinovestibular projections in the rat, with particular reference to projections from the central cervical nucleus to the lateral vestibular nucleus. J Comp Neurol 361:334–344
44. McCrea RA, Horn AK (2005) Nucleus prepositus. Prog Brain Res 151:205–230
45. McCrea RA, Gdowski G, Luan H (2001) Current concepts of vestibular nucleus function: transformation of vestibular signals in the vestibular nuclei. Ann NY Acad Sci 942:328–344
46. Minor LB, McCrea RA, Goldberg JM (1990) Dual projections of secondary vestibular axons in the medial longitudinal fasciculus to extraocular motor nuclei and the spinal cord of the squirrel monkey. Exp Brain Res 83:9–21
47. Muskens LJJ (1913) De rolbeweging en de opstijgende vestibularisverbreiding (Fasciculus Deiters ascendens). Versl Kon Acad Wet: 1478
48. Naito Y, Newman A, Lee WS, Beykirch K, Honrubia V (1995) Projections of the individual vestibular end-organs in the brain stem of the squirrel monkey. Hear Res 87:141–155

49. Naito Y, Tateya I, Hirano S, Inoue M, Funabiki K, Toyoda H, Ueno M, Ishizu K, Nagahama Y, Fukuyama H, Ito J (2003) Cortical correlates of vestibulo-ocular reflex modulation: a PET study. Brain 12:1562–1578
50. Neuhuber WL, Zenker W (1989) Central distribution of cervical primary afferents in the rat, with emphasis on proprioceptive projections to vestibular, perihypoglossal, and upper thoracic spinal nuclei. J Comp Neurol 280:231–253
51. Newlands SD, Vrabec JT, Purcell IM, Stewart CM, Zimmerman BE, Perachio AA (2003) Central projections of the saccular and utricular nerves in macaques. J Comp Neurol 466:31–47
52. Pompeiano O, Brodal A (1957) Spinovestibular fibers in the cat; an experimental study. J Comp Neurol 108:353–381
53. Popper P, Ishiyama A, Lopez I, Wackym PA (2002) Calcitonin gene-related peptide and choline acetyltransferase colocalization in the human vestibular periphery. Audiol Neurootol 7:298–302
54. Reisine H, Highstein SM (1979) The ascending tract of Deiters' conveys a head velocity signal to medial rectus motoneurons. Brain Res 170:172–176
55. Ruigrok TJ, Voogd J (1990) Cerebellar nucleo-olivary projections in the rat: an anterograde tracing study with Phaseolus vulgaris-leucoagglutinin (PHA-L). J Comp Neurol 298:315–333
56. Shimazu H, Precht W (1966) Inhibition of central vestibular neurons from the contralateral labyrinth and its mediating pathway. J Neurophysiol 29:467–492
57. Shinoda Y, Sugiuchi Y, Futami T, Ando N, Kawasaki T (1994) Input patterns and pathways from the six semicircular canals to motoneurons of neck muscles. I. The multifidus muscle group. J Neurophysiol 72:2691–2702
58. Shinoda Y, Sugiuchi Y, Futami T, Kakei S, Izawa Y, Na J (1996) Four convergent patterns of input from the six semicircular canals to motoneurons of different neck muscles in the upper cervical cord. Ann NY Acad Sci 781:261–275
59. Shinoda Y, Sugiuchi Y, Futam T, Ando N, Yagi J (1997) Input patterns and pathways from the six semicircular canals to motoneurons of neck muscles. II. The longissimus and semispinalis muscle groups. J Neurophysiol 77:1234–1258
60. Shinoda Y, Sugiuchi Y, Izawa Y, Hata Y (2005) Long descending motor tract axons and their control of neck and axial muscles. Prog Brain Res 151:527–563
61. Shiroyama T, Kayahara T, Yasu Y, Nomura J, Nakano K (1999) Projections of the vestibular nuclei to the thalamus in the rat: a Phaseolus vulgaris leucoagglutinin study. J Comp Neurol 407:318–332
62. Simpson JI, Graf W (1985) The selection of reference frames by nature and its investigators. Rev Oculomot Res 1:3–16
63. Spencer RF, Wenthold RJ, Baker R (1989) Evidence for glycine as an inhibitory neurotransmitter of vestibular, reticular, and prepositus hypoglossi neurons that project to the cat abducens nucleus. J Neurosci. 9:2718–2736
64. Sugihara I, Ebata S, Shinoda Y (2004) Functional compartmentalization in the flocculus and the ventral dentate and dorsal group y nuclei: an analysis of single olivocerebellar axonal morphology. J Comp Neurol 470:113–133
65. Sugiuchi Y, Kakei S, Izawa Y, Shinoda Y (2004) Functional synergies among neck muscles revealed by branching patterns of single long descending motor-tract axons. Prog Brain Res 143:411–421
66. Thunnissen IE, Epema AH, Gerrits NM (1989) Secondary vestibulocerebellar mossy fiber projection to the caudal vermis in the rabbit. J Comp Neurol 262–277
67. Ventre J, Faugier-Grimaud S (1988) Projections of the temporo-parietal cortex on vestibular complex in the macaque monkey (Macaca fascicularis). Exp Brain Res 72:653–658
68. Voogd J (1964) The cerebellum of the cat. Van Gorcum, Assen
69. Wilson VJ, Schor RH (1999) The neural substrate of the vestibulocollic reflex. What needs to be learned. Exp Brain Res 129:483–493
70. Yamamoto M, Shimoyama I, Highstein SM (1978) Vestibular nucleus neurons relaying excitation from the anterior canal to the oculomotor nucleus. Brain Res 140:31–32
71. Yingcharoen K, Siegborn J, Grant G (2003) Brainstem projections of different branches of the vestibular nerve: an experimental study by transganglionic transport of horseradish peroxidase in the cat. III. The saccular nerve. Exp Brain Res 151:190–196

18 Auditory System

Auditory Centres

The central auditory system processes information from the organ of Corti in the cochlea. The perilymphatic space of the cochlea consists of the scala vestibuli and the scala tympani, which are continuous at the apex (the helicotrema) of the cochlea. The membranous ductus cochlearis is situated between the two scalae and contains the endolymph. The organ of Corti is located on the basilar membrane, which separates the ductus cochlearis from the scala tympani. Reissner's membrane forms the separation of the ductus from the scala vestibuli. The piston-like footplate of the stapes, the final link in the chain of auditory ossicles, sets up a wave in the scala vestibuli and tympani, and introduces a travelling wave along the basilar membrane. The basilar membrane acts as a frequency analyzer. Short waves, created by the higher frequencies, maximally displace the the narrow basal part of the membrane; long waves, created by the lower frequencies, maximally displace the broader apical segments (Fig. 17.3). Hair cells of the organ of Corti transduce these vibrations into receptor potentials. Two groups of hair cells can be distinguished according to their position. Inner hair cells are the principal receptor cells. Outer hair cells undergo rapid length changes when stimulated by a tone stimulus, causing localized amplification of the motion of the basilar membrane, which increases the frequency selectivity and the sensitivity of the inner hair cells [16, 70]. The cochlear microphonic, a low-amplitude potential recorded from the external auditory meatus, reflects the activity of the hair cells, principally or exclusively that of the outer hair cells, during the transduction process. Otoacoustic emissions, audible sounds emitted by the ear, are produced by the contractions of the outer hair cells and can be used to check the function of these cells [48].

Each inner hair cell is innervated by multiple type 1 bipolar ganglion cells, located in the spiral ganglion. The thick myelinated axons of the type 1 ganglion cells are the major component of the cochlear nerve. Type 2 ganglion cells each innervate multiple outer hair cells. Their thin, unmyelinated axons account for 5% of the fibres of the cochlear nerve. In addition, the cochlear nerve contains an efferent cochlear system, arising from the brain stem and terminating on the hair cells and the cells of the spiral ganglion [99] (Fig. 18.4).

In addition to several rhombencephalic cell masses, the central auditory system includes a mesencephalic, a diencephalic and a telencephalic centre (Figs. 18.1, 18.2). The rhombencephalic cell masses are the cochlear nuclei, the superior olivary complex and the nuclei of the lateral lemniscus.

The human cochlear nuclei are composed of a ventral and a dorsal nucleus, which are comparable in their cytoarchitecture to those of

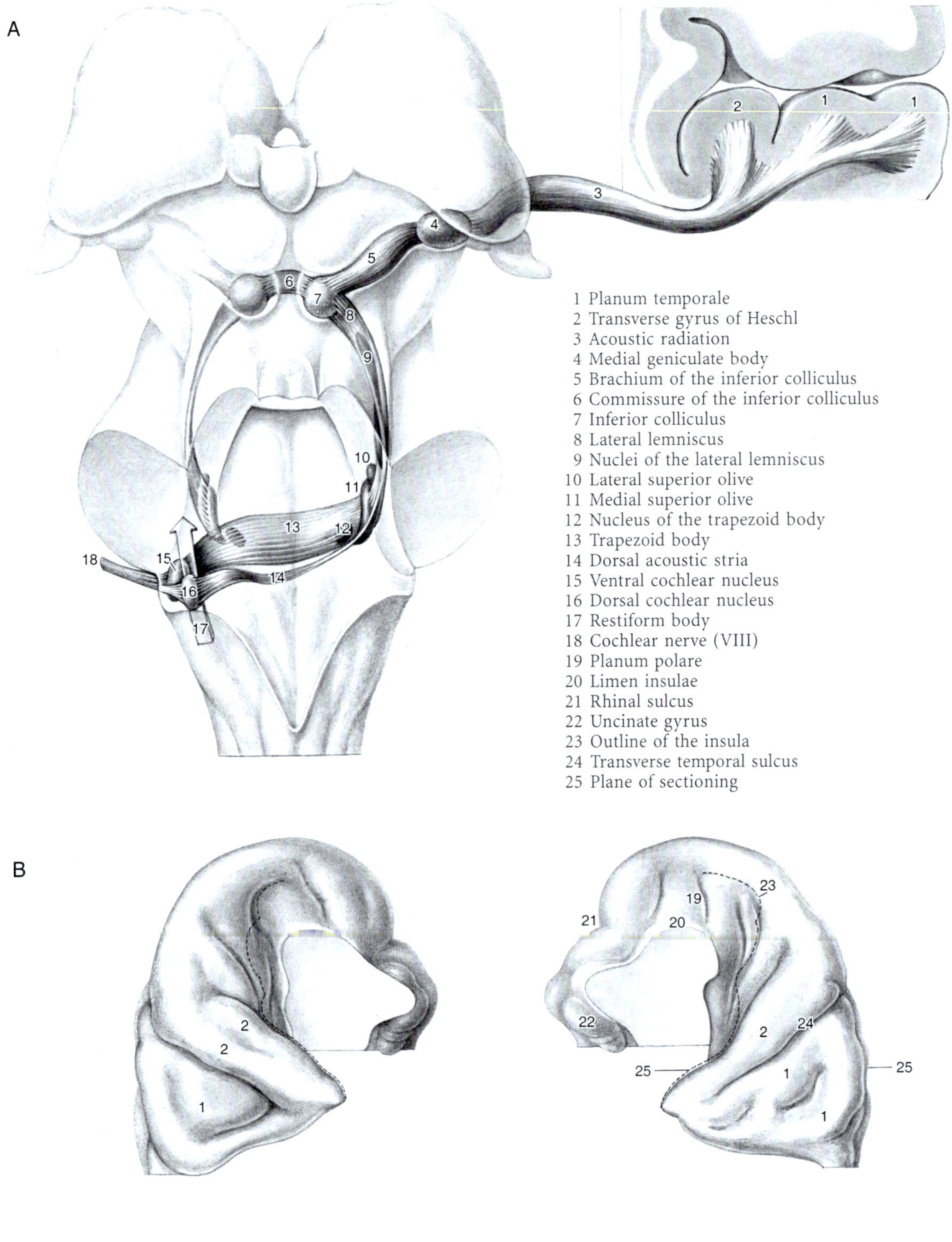

Fig. 18.1 A, B. The auditory system. **A** Position of nerve, nuclei and tracts in a dorsal view (5/3). The transverse gyrus of Heschl and the planum temporale have been drawn in their true position relative to the brain stem and the thalamus. **B** Dorsal view of the left and right temporal lobes, to show the position of the transverse gyri of Heschl and the temporal plane (1/1×)

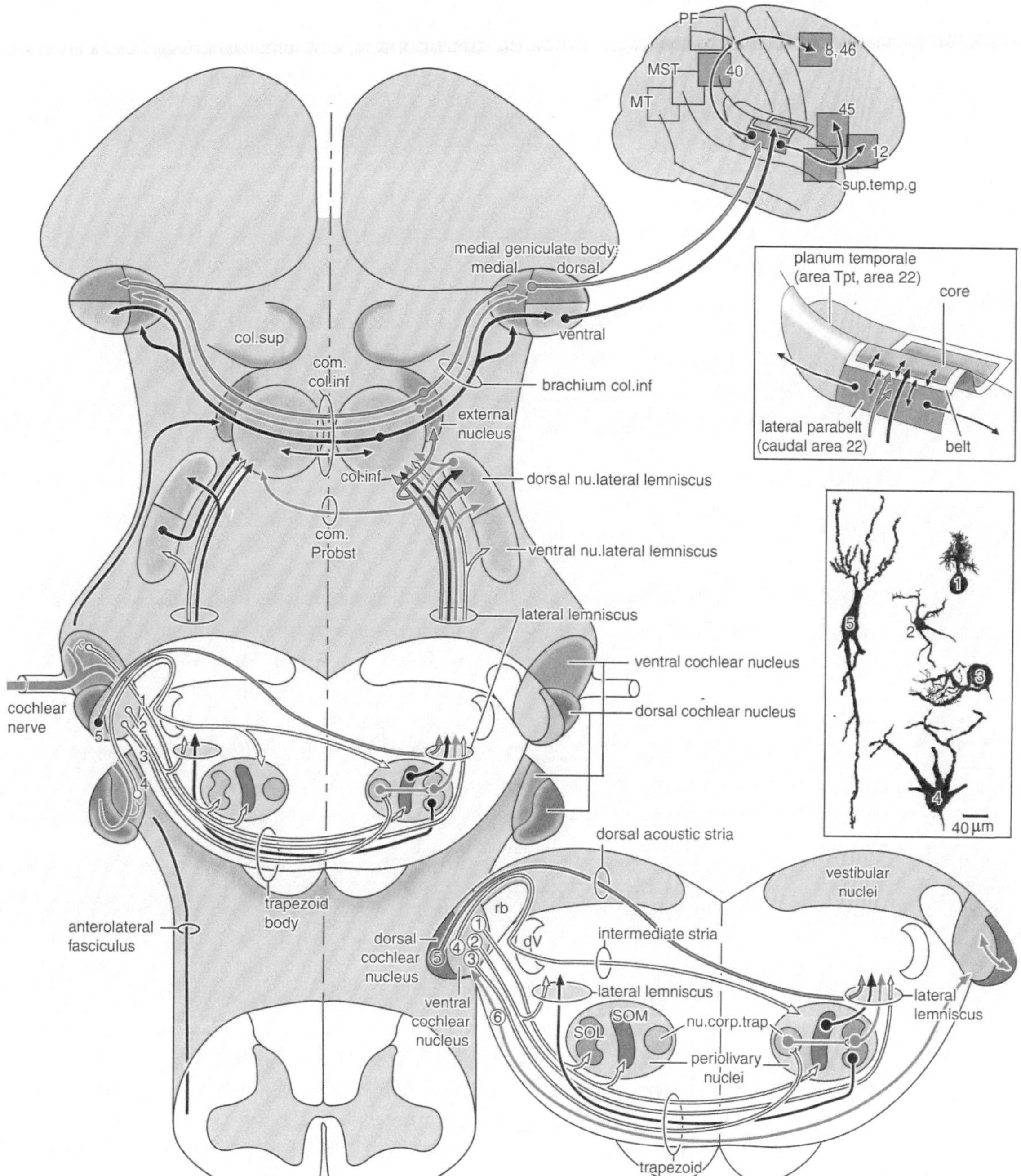

Fig. 18.2. The connections of the auditory system: ascending projections. The brain stem is transected at the level of the cochlear nuclei. This section is also depicted in the *lower right panel*. The six efferent systems of the cochlear nuclei, enumerated in the text, are indicated by their respective *numbers*. The cochlear nerve and the inhibitory connections in the brain stem are shown in *red*. Connections of the dorsal cochlear nucleus, the external nucleus of the inferior colliculus and the medial and dorsal subdivisions of the medial geniculate body are shown as contours filled with *grey*. The positions of the auditory cortex and the dorsal and ventral auditory streams are indicated in the diagram of the right cerebral hemisphere. The areas MT, MST and PF of the dorsal visual stream occupy a more peripheral position (compare to Fig. 19.6 D). The *upper inset* shows a diagram of the core and belt areas of the auditory cortex and their connections. The *lower inset* depicts five of the cochlear nuclei cell types that give rise to the corresponding efferent systems (redrawn from Brawer et al. [8]). *1*, spherical bushy cell; *2*, stellate cell; *3*, bushy multipolar cell; *4*, octopus ▶

other mammals [61, 65] (Fig. 6.24). The ventral cochlear nucleus comprises a rostral area of spherical cells. a central area of multipolar, globular cells and stellate cells, a caudal area of so-called octopus cells and a laterodorsal cap of small neurons (Fig. 18.2). The dorsal cochlear nucleus is well developed but does not show the characteristic laminated pattern that is prominent in mammals that are able to move their pinnas.

The superior olivary complex is embedded in the trapezoid body, a large, transversely oriented fibre stream. It comprises the medial and lateral superior olivary nuclei and the medial nucleus of the trapezoid body (Figs. 6.25, 6.26, 17.1). The medial olivary nucleus is an elongated cell mass. In most mammals, but not in humans, the lateral superior olive is S-shaped. These three cell masses are surrounded by a zone containing small groups of cells that vary in size and shape. The latter cell groups are collectively designated as the periolivary nuclei.

The nuclei of the lateral lemniscus constitute an elongated strand of cells that lies embedded in the fibre bundles of the lateral lemniscus as it ascends through the pontine tegmentum to the midbrain. Two nuclei, a ventral and a dorsal, can be distinguished (Figs. 6.27, 6.28, 18.1). The lateral superior olive, the nucleus of the trapezoid body and the ventral nucleus of the lateral lemniscus are relatively small in the human brain. These structures, which are typically enlarged in echolocating species, are related to the high-frequency domain [62–65]. At the mesencephalic level the auditory system is represented by the inferior colliculus (Figs. 6.29, 6.30). This structure, which is a most important relay station in both the ascending and descending auditory projections, consists of a large, laminated central nucleus and a dorsal and lateral peripheral zone or cortex. The specific diencephalic nucleus for hearing is the medial geniculate body, which forms part of the dorsal thalamus. This nucleus comprises three main divisions: medial, dorsal and ventral (Fig. 6.32). The ventral division has a laminar organization. The telencephalic auditory centres occupy the cortex covering the upper surface of the temporal lobe. Recent reviews of the auditory system, including the organ of Corti, were published by Malmierca and Merchán [53] and Moore and Linthicum [66].

Auditory Pathways

Figures 18.1 and 18.2 show that the auditory centres are interconnected by distinct fibre streams. The cochlear nuclei give rise to three more or less transversely orientated bundles, the trapezoid body and the dorsal and intermediate acoustic stria. The trapezoid body crosses in the ventral part of the tegmentum. After reaching the lateral part of the pons, the bundle makes an abrupt rostral turn and continues as the lateral lemniscus.

The nuclei of the superior olivary complex lie embedded in the trapezoid body. The dorsal acoustic stria passes over the inferior cerebellar peduncle. Arching ventrally, this bundle traverses the reticular formation to reach the region of the contralateral superior olivary complex, where it continues into the lateral lemniscus. The intermediate acoustic striae detach from the dorsal acoustic striae after passing over the restiform body, traverse the spinal nucleus of the trigeminal nerve and join the trapezoid body before its decussation. The lateral lemniscus is situated in the dorsolateral

(continued text to Fig. 18.2) cell; *5*, pyramidal or fusiform cell of the dorsal cochlear nucleus. 8, 12, 40, 45, 46, areas 8, 12, 40, 45, 46 of Brodmann; *col.sup*, superior colliculus; *com.col.inf*, commissure of the inferior colliculus; *com.Probst*, Probst's commissure; *dors.nu.ll*, dorsal nucleus of the lateral lemniscus; *dV*, spinal tract of the trigeminal nerve; *MST*, middle superior temporal visual area; *MT*, medial temporal visual area; *Nu.-corp.trap*, nucleus of the trapezoid body; *PF*, parietal visual areas; *rb*, restiform body; *SOM*, medial superior olive; *SOL*, lateral superior olive; *sup.temp.g*, superior temporal gyrus; *ventr.nu.ll*, ventral nucleus of the lateral lemniscus

part of the pons and the caudal midbrain (Fig. 6.28). It ascends to the inferior colliculus, where its fibres terminate. The brachium of the inferior colliculus, a prominent bundle visible on the surface of the midbrain, connects the inferior colliculus with the medial geniculate body (Figs. 6.29–6.32). In addition, the inferior colliculi of both sides are interconnected by commissural fibres. The final link in the auditory system is formed by the auditory radiation through which the medial geniculate body is connected with the superior temporal cortex. Passing laterally, this radiation traverses the sublenticular portion of the internal capsule.

The primary auditory cortex (known as its core area) corresponds roughly to the transverse gyrus of Heschl. It is surrounded by a belt area that extends on to the planum temporale, i.e. the region on the superior temporal plane lying posterior to the transverse gyrus and extending back to the end of the Sylvian fossa. There are considerable variations in the size and convolution of the supratemporal plane in the two hemispheres (Fig. 18.1B). Thus, Heschl's gyrus is usually solitary on the left, but double on the right side, and the left planum temporale is commonly much larger than the right [23–26]. These gross anatomical asymmetries may well represent the morphological substrate for language lateralization.

Although several commissures interconnect the auditory centres at various levels of the brain stem, the main multisynaptic excitatory connection between the organ of Corti and the auditory cortex is a crossed one.

Ascending Auditory Projection

The cochlear division of cranial nerve VIII enters the central nervous system just caudal to its vestibular division. The axons of the type 1 cells of the spiral ganglion bifurcate in the centre of the ventral cochlear nucleus, dividing it into the anterior and posterior ventral nucleus, and distribute in a tonotopical manner to both the dorsal and the ventral cochlear nuclei (Figs. 18.2, 18.3b). The thin, unmyelinated axons of type 2 spiral ganglion cells contact clusters of small interneurons scattered throughout the cochlear nuclei [42]. Within the ventral cochlear nucleus the bifurcating type I fibres form dorsoventrally arranged sheets, with fibres from the apical spiral ganglion representing low frequencies and terminating ventrally, and fibres from the basal turn of the spiral ganglion representing high frequencies and terminating dorsally [103] (Fig. 18.3b). Posterior branches converge upon the caudal pole of the ventral nucleus. From here the root fibres enter the dorsal cochlear nucleus. Root fibres show profuse collateral branching and establish specific types of synaptic contacts with various populations of neurons present in the cochlear nuclear complex. These neuronal populations give rise to several parallel pathways. These pathways terminate in the superior olivary complex, in the nuclei of the lateral lemniscus and in the inferior colliculus (Fig. 18.2 (1–6)).

1. The spherical bushy cells located in the anterior portion of the ventral cochlear nucleus project to the ipsilateral lateral superior olivary nucleus and to the medial superior olivary nucleus of both sides (Fig. 18.2 (1)). Both superior olivary nuclei project in turn to the dorsal nucleus of the lateral lemniscus and the inferior colliculus; the projection of the medial superior olive is an ipsilateral one, and the projections of the lateral superior olive are bilateral [36, 77]. The projection of the medial superior olive and the contralateral projection of the lateral superior olive are excitatory; the ipsilateral projection of the lateral superior olive is glycinergic and inhibitory [75].

The cells in the medial superior olivary nucleus are bipolar and orientated horizontally. From both the medial and lateral poles of these cells a dendritic tuft arises. The laterally extending dendrites receive their input from fibres of the ipsilateral side, whereas the medially extending dendrites receive input from the contralateral side [107]. The projections of the cochlear nuclei to the superior olives are tonotopically organized, with the lower frequencies located in the dorsal part and higher frequen-

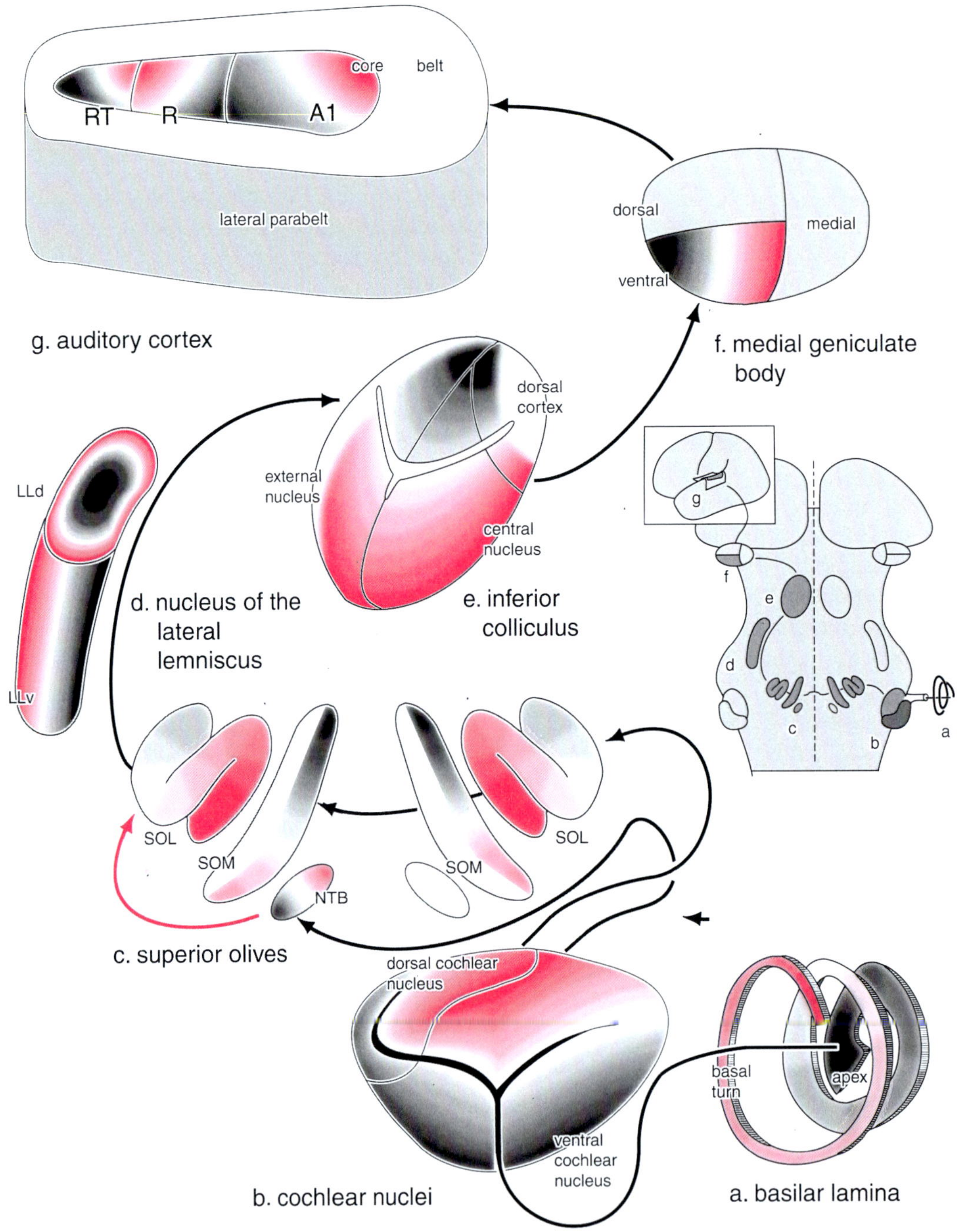

Fig. 18.3. Tonotopic organization of the auditory system. High frequency representations are indicated in *red*, low frequency representations in *black*. Diagram of the basilar lamina (*a*) Sagittal section through the cochlear nuclei showing laminar tonotopic profiles extending from the ventral into the dorsal cochlear nucleus (*b*, cat [103]). The superior olives of the cat (*c*, [103]). Tonotopic localization in the lateral superior olive is biased towards higher frequencies, and biased in the medial superior olive towards lower frequencies. The medial superior olive receives a projection to its lateral half from the ipsilateral cochlear nuclei and to its medial half from the contralateral side. The lateral superior olive receives tonotopically arranged projections from the ipsilateral and contralateral cochlear nuclei. The ipsilateral projection is excitatory; the contralateral projection is mediated by the nucleus of the trapezoid body and is inhibitory (*red*). Laminar tonotopical profiles are present in the ventral nucleus of the lateral lemniscus and a concentric lamination in the dorsal nu-

cies in the ventral part of the medial superior olive (Fig. 18.3c). In the lateral superior olive, the frequency scale runs from lower frequencies in the lateral part of the nucleus to higher frequencies in its medial portion [103]. The medial superior olive is biased towards the lower frequency range, and the lateral superior olive towards the higher frequency range. The highly specific spatial organization of afferents to cells in the medial superior olivary nucleus has to do with its role in interpreting interaural time differences, which are maximal at low frequencies. A map of interaural time differences is present across the anteroposterior axis of the medial superior olive, with the lowest values at the anterior pole and increasing values in more posterior positions. For the localization of sound, interaural level differences, which are maximal at high frequencies, are encoded in the lateral superior olive [123].

2. Bilateral projections to the inferior colliculus are mainly derived from stellate cells in the central portion of the ventral cochlear nucleus [74, 124] (Fig. 18.2 (2)). This region also provides a (possibly collateral) projection to the ventral nucleus of the lateral lemniscus.

3. The terminations of the axons of the globular bushy cells in the central portion of the ventral cochlear nucleus envelop the cells of the contralateral nucleus of the trapezoid body with their calyces (of Held [35]). These cells may correspond to the multipolar bushy cells illustrated by Brawer et al. [8] (Fig. 18.2 (3)). The neurons of the nucleus of the trapezoid body use glycine as their inhibitory neurotransmitter. Their axons terminate in the ipsilateral lateral superior olivary nucleus. As a consequence, the cells of the lateral superior olive are excited by input from the ipsilateral ear (see 1) and inhibited from the contralateral ear (Fig. 18.3c). The lateral superior olive is involved in the analysis of interaural differences in frequency and intensity. Its cells do not display the mediolateral polarity of the medial superior olive's cells, but are primarily orientated in a rostrocaudal direction (Voogd, unpublished observations).

4. The most extensive projections to the periolivary nuclei are derived from the multipolar and stellate cells of the posterior division of the ventral cochlear nucleus and the octopus cells of the posterior pole [101] (Fig. 18.2 (4)). Octopus cells are so named because they are provided with multiple, straight, dorsally directed dendrites [79]. Their axons join the intermediate acoustic striae, which terminate bilaterally in the periolivary nuclei and the contralateral ventral nucleus of the lateral lemniscus. Their terminations may include the neurons of the olivocochlear bundle (see The Descending Auditory Projection). The periolivary nuclei project to the inferior colliculus. In addition, they are involved in multisynaptic systems, descending from higher auditory centres and terminating in the cochlear nuclei [102] (Fig. 18.4). A direct projection to the suprageniculate nucleus, one of the thalamic nuclei of the posterior group, bypassing the inferior colliculus, has been traced from a cell group located between the superior olive and the ventral nucleus of the lateral lemniscus in a bat [14].

5. Pyramidal (fusiform) and giant cells are the relay cells of the dorsal cochlear nucleus. They project to the contralateral dorsal nucleus of the lateral lemniscus, the inferior colliculus and, in the rat at least, to the medial geniculate body [55, 73, 124] (Fig. 18.2 (5)). The structure of the dorsal cochlear nucleus resembles the cerebellar cortex. Granule cells receive mossy fibres from the dorsal column and pon-

◀ cleus (*d*, rat [57, 58]). Intrinsic system of tonotopically arranged lamellae in the central nucleus, and the dorsal and external cortex of the inferior colliculus of the rat (*e*, [92]). Laminar organization in the ventral nucleus of the medial geniculate body of the rat (*f*, [121]). Diagram of the core, belt and parabelt areas of the auditory cortex of the monkey (*g*). Three tonotopically organized areas can be distinguished in the core. Redrawn with permission from Kaas [45]. *A1*, primary auditory area 1; *LLd*, dorsal nucleus of the lateral lemniscus; *LLv*, ventral nucleus of the lateral lemniscus; *NTB*, nucleus of the trapezoid body; *R*, rostral primary auditory area; *RT*, rostrotemporal primary auditory area; *SOL*, lateral superior olive; *SOM*, medial superior olive

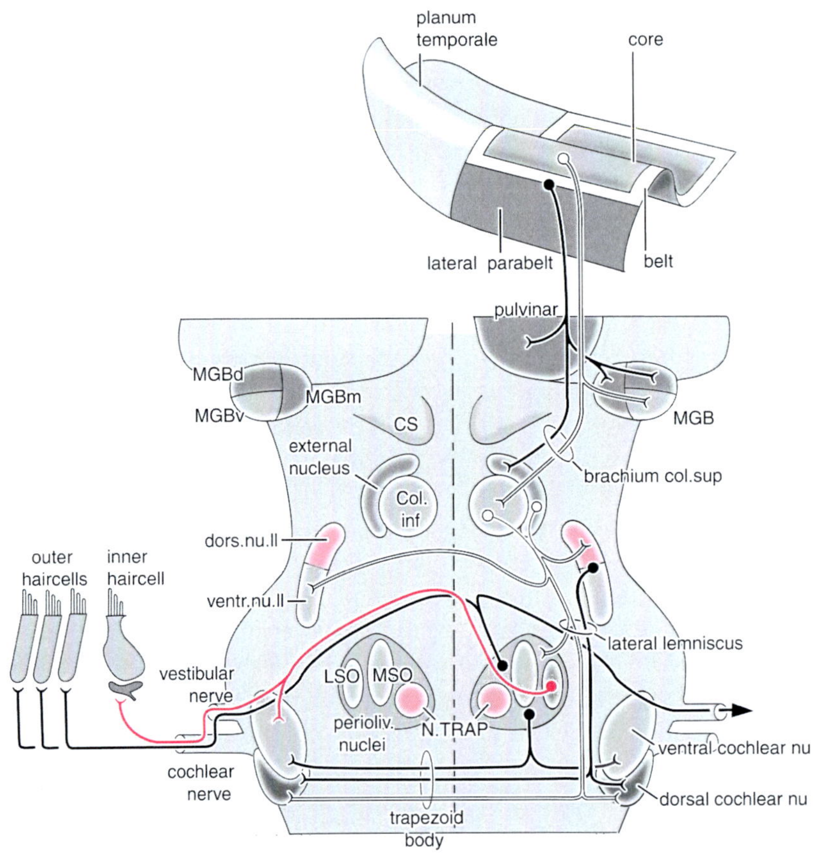

Fig. 18.4. The descending connections of the auditory system. Inhibitory systems are indicated in *red*. The origin of systems descending from the auditory cortex is indicated in the diagram of the core and belt areas. *Col.inf*, inferior colliculus; *CS*, superior colliculus; *LSO*, lateral superior olive; *MGBd*, dorsal nucleus of the medial geniculate body; *MGBv*, ventral nucleus of the medial geniculate body; *MGBm*, medial nucleus of the medial geniculate body; *MSO*, medial superior olive; *N.TRAP*, nucleus of the trapezoid body; *Perioliv.nuclei*, periolivary nuclei

tine nuclei and also from different auditory centres. Their axons run in a superficial molecular layer, where they contact the apical dendrites of the pyramidal cells and different kinds of interneurons, which provide feed-forward inhibition to the pyramidal cells. Auditory root fibres terminate on the basal dendrites and the giant cells in the deep layers of the dorsal cochlear nucleus [71]. The human dorsal cochlear nucleus lacks granule cells and a molecular layer. Granule cells, which cover the outer surface of the cochlear nuclei in lower mammals, are virtually absent in the human cochlear nuclei [61, 62].

6. Glycinergic neurons connect the cochlear nuclei of the two sides and interconnect the dorsal and ventral cochlear nuclei [20, 50, 81, 91, 116] (Fig. 18.2 (6)).

The Nuclei of the Lateral Lemniscus and the Inferior Colliculus

The dorsal and ventral nuclei of the lateral lemniscus are links in pathways ascending from the cochlear nuclei and the superior olives to the inferior colliculus, and descending from the colliculus to lower auditory centres. The ascending connections of the dorsal nucleus of the lateral lemniscus are GABAergic. Both nuclei are tonotopically organized. In the ventral nucleus, the tonotopic localization consists of interdigitating laminae; in the dorsal nucleus, the organization is concentric, with high frequencies represented in the periphery and low frequencies represented in the centre of the nucleus [57, 59] (Fig. 18.3 D).

The inferior colliculus can be subdivided into a central nucleus, a dorsal cortex and a lateral cortex (or external nucleus) [28, 41, 68, 103]. Dorsal to the aqueduct, the commissure of the inferior colliculus crosses the midline. This is a true commissure, interconnecting the inferior colliculi. Additionally, it contains the fibres of the crossed ascending pathway from the colliculus (Fig. 18.2). The inferior colliculus of the rat possesses an intrinsic, bilaterally symmetrical laminar organization. Interneurons of the colliculus distribute their axons in diverging planes that extend from the caudal colliculus into the central nucleus and dorsal cortex and lateral cortex. In the central nucleus and the dorsal cortex they are stacked from ventromedial to dorsolateral and in the lateral cortex from ventrolateral to dorsomedial [92] (Fig. 18.3 E). Within the central nucleus these lamina correspond with the tonotopic distribution of the fibres of the lateral lemniscus, with low frequencies represented dorsally and high frequencies represented ventrally in the colliculus. Certain types of neurons of the central nucleus are confined to one lamina; other cells cut across them [56, 76, 78, 82].

The central nucleus receives tonotopically ordered projections from the contralateral cochlear nuclei, both lateral superior olives, the ipsilateral medial superior olive and the nuclei of the lateral lemniscus (Fig. 18.2). The ventral nucleus of the lateral lemniscus is an important relay between the cochlear nuclei and the central nucleus. The dorsal nucleus of the lateral lemniscus receives its main afferents from the superior olives. Its projection to the inferior colliculus is bilateral and GABAergic. Its crossed efferents decussate in Probst's commissure [85], dorsal to the decussation of the brachium conjunctivum. The external nucleus of the inferior colliculus receives inputs from the contralateral dorsal cochlear nucleus, the ipsilateral and contralateral central nucleus and the dorsal nucleus of the lateral lemniscus [15, 29, 73, 80, 103]. Additionally, the external nucleus receives somatosensory input from the spinal cord, the dorsal column nuclei, the spinal trigeminal nucleus, the parabrachial area and the deep layers of the superior colliculus [2, 15, 103]. The dorsal cortex receives its fibres from the nucleus sagulum, a group of small cells lateral to the dorsal nucleus of the lateral lemniscus [37], and from the auditory cortex [4, 38, 103]. Auditory cortical projections terminate in all three subdivisions of the inferior colliculus and generally adhere to its tonotopical organization [93] (Fig. 18.2).

Efferent fibres of the inferior colliculus collect in the brachium of the inferior colliculus, which terminates in the medial geniculate body. This projection is mainly ipsilateral, with a contralateral component. In the cat large GABAergic axons of the brachium account for 20% of its fibres [90, 119]. The central nucleus and the cortex of the inferior colliculus differ in their efferent projections. The central nucleus projects bilaterally to the external nucleus, the parabrachial region of the lateral tegmentum, the interstitial nucleus of the brachium of the inferior colliculus, the ventral and magnocellular divisions of the medial geniculate body and the posterior group of the thalamus. The external nucleus projects to the dorsal division, and the dorsal cortex projects to the medial division of the medial geniculate body [41, 64, 67, 103] (Fig. 18.2).

Medial Geniculate Body and Auditory Cortex

The medial geniculate body protrudes on the lateral side of the mesencephalon and is situated medial to the lateral geniculate and ventral to the pulvinar (Figs. 3.10, 3.20, 6.32, 6.33). It can be subdivided into ventral, medial and dorsal nuclei. The ventral nucleus is laminated: the splayed dendrites of its principal cells are arranged in sheets that correspond with the isofrequency planes, with higher frequencies represented medially and lower frequencies represented laterally [67, 118, 120] (Fig. 18.3F). The cellular organization of the dorsal nucleus is less distinct. The medial nucleus is characterized by the presence of large neurons [41, 118].

It is noteworthy that from the level of the inferior colliculus onwards, the ascending auditory projection can be subdivided into a "core" projection and a "belt" projection [9]. The ventral nucleus of the medial geniculate body receives a tonotopically ordered projection from the central nucleus of the inferior colliculus, with lower frequencies located laterally (Fig. 3F) [121]. Its predominant projection is to the primary auditory cortical areas, which constitute the core of the auditory cortex. The medial and dorsal nuclei of the medial geniculate body receive their afferents from the external and dorsal cortex of the inferior colliculus, respectively. They also project to the core, but their main targets are in the belt surrounding the core [13, 44] (Fig. 18.2).

Core and belt areas of the auditory cortex are situated in the temporal operculum and the adjacent superior temporal gyrus. Three tonotopically arranged primary areas have been distinguished in nonhuman primates (Fig. 18.3g). They are known as area A1, the rostral (R) and the rostrotemporal (RT) areas [10, 33, 34, 44, 86] (Fig. 18.3g). The callosal connections of the core are tonotopically organized. Cortical association fibres interconnect both the primary areas and the core with the belt. The lateral parabelt forms the third stage in the processing of auditory information by the cerebral cortex. It is interconnected with the belt but not with the core (Fig. 18.2). It receives its main afferents from the belt areas and its thalamic afferents from the medial and dorsal nuclei of the medial geniculate, the suprageniculate nucleus and the medial pulvinar, but not from the ventral nucleus of the medial geniculate body. Callosal connections are made with homotopic portions of the contralateral parabelt and belt regions. The neurons of the parabelt, therefore, likely depend on belt rather than thalamic or core inputs for auditory activation [45].

In the human brain the koniocortex of the core area occupies Heschl's transverse gyrus or straddles the separating sulcus when two transverse gyri are present [23, 34]. It corresponds to area 41 of Brodmann. The belt areas include the cortex medial to the transverse gyrus bordering on the circular sulcus, adjacent regions of the planum temporale and the more lateral area 42 of Brodmann. The structure of the belt is intermediate between the koniocortex and the homotypical cortex of area 22 on the surface of the superior temporal gyrus. The lateral parabelt occupies the rostral half of area 22; the planum temporale (area Tpt) forms its caudal half. The lateral parabelt can be subdivided according to cytoarchitectonic criteria into internal and external subfields. Area Tpt is larger on the left side in right-handed subjects (Fig. 18.1). This asymmetry contributes substantially to the asymmetry of the planum temporale [23–26, 34, 100].

Different "streams" have been distinguished in the connections of the auditory belt and parabelt, which are comparable to and may share functions with similar streams in the visual system (compare Figs. 18.2 and 19.6D). One stream, originating from the anterior belt and parabelt, is directed laterally and rostrally at the superior temporal gyrus and sulcus and the inferior and polar prefrontal cortex of areas 10, 12, 13, 45 and, especially, 45. The connections of area 44, which, together with area 45 constitute Broca's area for explicit speech, were not studied. The anterior stream subserves identification and recognition of auditory stimuli. A caudal stream, originating from the caudal parabelt, is directed at the adjacent superior temporal gyrus and sulcus and the caudal inferior parietal lobule, but spares the medial superior temporal (MST), the middle temporal (MT) and the parietal (PF) visual areas. The parietal target corresponds to area 40 in the supramarginal gyrus. The prefrontal connections of the caudal stream include middle prefrontal areas 8a, 12 and 46 [33, 45, 84, 89]. These areas are involved in spatial localization of auditory stimuli.

The asymmetry of the planum temporale in the human brain suggests that the left planum may belong to the "language substrate" of the brain [22, 23]. However, no such specific activation was found in fMRI studies that compared listening to tones and words [6, 7]. The classical studies of Wernicke [17, 117] identified the posterior part of the left superior temporal gyrus as the area responsible for speech perception. Its destruction causes sensory aphasia. Experimental studies in nonhuman primates and functional imaging in humans showed that more extensive areas are involved in these processes. Neurons in the lateral parabelt, which occupies the more anterior portions of the superior temporal gyrus, are more broadly tuned to the particular bandwidth of a noise pulse than the pure tones that activate the primary auditory areas. Other neurons in this area are tuned for direction and rate of frequency modulation as a first stage in the localization of auditory stimuli or are selective for different types of monkey calls and the right temporal order of the syllables therein [86–88]. Similar neurons responding to complex sounds, including animal and human vocalizations, were located in the ventrolateral prefrontal cortex of areas 45 and 12, regions that receive input from the anterior parabelt [89]. Lateralization for phonetic information was found in the same region of the left superior temporal gyrus, located anterior to Heschl's gyrus in humans [98]. A more posterior region located in the superior temporal sulcus is involved in the comprehension of speech. Lesions of this region result in conduction aphasia, where a patient can understand speech but not repeat it. The left posterior superior temporal gyrus, which belongs to Wernicke's area, is activated when perceiving and retrieving words from long-term memory and the transient representation of phonetic sequences. A region at its junction with the planum temporale is linked to speech production rather than perception [122].

Functional imaging in humans showed that the right parietal cortex is involved in sound localization. The region concerned is area 40 in the supramarginal gyrus [30], but others consider the superior parietal lobule as the target for the projection of the dorsal visual and the caudal auditory streams and the inferior parietal lobule as the site of supramodal spatial processing [11]. Lesions of this area result in apraxia, where the patients are able to name an object, but unable to handle it [39]. Similarities in the perception of the evolution of sound in time, in sound localization and in speech perception suggested a role for the dorsal stream in the extraction of the verbal message in a spoken sentence, while the ventral stream is involved in the identification of the speaker's voice [5].

More areas than enumerated above are involved in the recognition of sounds and speech. Prosody and pitch, like musical perception, generally are lateralized to secondary auditory areas on the right [3, 98]. Experienced musicians, however, showed a dominance of the left side. The degree of activation of the left auditory and posterior dorsolateral prefrontal cortex correlates with the absolute pitch ability and the early commencement of training in experienced musicians [72, 95, 96]. An early start of musical training also results in an increased cortical representation for musical stimuli [83]. Areas in the middle frontal gyrus, which are among the targets of the posterior parabelt and are involved in the spatial processing of sound, may include dorsal area 8, directing gaze to a region of interest and area 46, which is involved in short-term memory storage and the control of smooth pursuit [11, 52]. Anterior and posterior parahippocampal areas are activated by the processing of novel and familiar words, respectively [94]. The orbital frontal cortex, which receives projections from the anterior parabelt, belongs to the emotional and motivational reward system. Projections to the lateral amygdala have been traced from the auditory receptive suprageniculate and posterior intralaminar nuclei of the thalamus [19].

Descending Auditory Projection

Parallel to the pathway from the organ of Corti to the auditory cortex, there is an uninterrupted chain of neurons conducting impulses in the opposite direction (Fig. 18.4). The final link in this descending system is formed by the olivocochlear bundle, which originates from the superior olivary complex. Olivocochlear neurons receive convergent projections from the cochlear nuclei and the descending auditory pathway [69]. The olivocochlear bundle enters the vestibular nerve and joins the cochlear nerve via Oord's vestibulocochlear anastomosis. After entering the cochlea, they terminate on hair cells of the organ of Corti. The studies of Warr et al. [32, 107–111] indicate that the olivocochlear bundle consists of two separate efferent systems, which differentially innervate the two types of hair cells situated in the organ of Corti. The medial system originates from large periolivary neurons located medial to the superior olives (medial nucleus of the trapezoid body). The medial system is cholinergic. Its coarse axons project bilaterally to the cochlea, where they form large synaptic terminals at the base of the outer hair cells and emit collaterals to the ventral cochlear nuclei [40, 105, 106, 110, 112]. Most fibres of the lateral olivocochlear system take their origin from small neurons located within the lateral superior olive; a second population consists of large shell cells located in the periphery of this nucleus. Two neurochemically distinct types of neurons can be distinguished. One is cholinergic, but colocalizes CGRP; the other uses GABA as a neurotransmitter [104, 105]. Both groups project to inner hair cells, where they make "en passage" synapses with the primary afferent fibres just beneath the inner hair cells. The crossed projection of the small neurons is topically organized; shell cells project diffusely over the entire length of the ipsilateral cochlea, and provide collaterals to the ventral cochlear nucleus [40, 113]. The olivocochlear neurons thus form a system by which the brain can influence its own auditory input. The medial system can suppress cochlear gain by its action on the outer hair cells. It is part of the olivocochlear reflex (see Auditory Reflex Pathways). The function of the lateral system is less well understood [31]. It may control the contractile properties of the outer hair cells.

The higher links in the corticocochlear projection are constituted by projections (1) of the auditory cortex to the medial geniculate, the inferior colliculus, the superior olive and the cochlear nuclei; (2) from the inferior colliculus to the periolivary and the cochlear nuclei; and (3) from the periolivary nuclei to the cochlear nuclei (Fig. 18.4).

The corticothalamic and the thalamocortical connections of the medial geniculate body are reciprocally organized. The core areas project to the ventral and medial nuclei of the medial geniculate, and belt areas project to the dorsal and medial nuclei, the suprageniculate nucleus and the nucleus limitans. The parabelt is connected with the medial pulvinar [33]. The core area also projects to the central nucleus of the inferior colliculus; this projection is tonotopically organized [21]. Projections from the belt and parabelt prevail in the external nucleus and the dorsal cortex [41, 120]. In the rat, the primary auditory cortex projects both to the central nucleus and the dorsal and external cortex. This projection is organized according to the same laminated, tonotopical pattern as the intrinsic and commissural system of the inferior colliculus [93]. Direct projections of the auditory cortex to the superior olives and the cochlear nuclei have been documented only for the rat [114, 115].

Descending connections from the inferior colliculus terminate in the ipsilateral dorsal and the contralateral ventral nucleus of the lateral lemniscus. The central and external nuclei of the inferior colliculus and the nuclei of the lateral lemniscus give rise to descending pathways, which terminate in the ipsilateral periolivary nuclei and bilaterally in the dorsal cochlear nuclei [12, 41, 54, 97]. The nuclei of the lateral lemniscus project to the ipsilateral periolivary nuclei [12, 41, 51, 97].

The periolivary cochleonuclear projection originates, as its name implies, from the cell groups situated in the periolivary area. These cell groups receive impulses from the ventral cochlear nuclei and from fibres descending from the inferior colliculus and the nuclei of the lateral lemnius. Adams [1] mentioned two groups of periolivary cells that impinge upon the cochlear nuclear complex: a lateral group of multipolar cells whose members are located around the ipsilateral lateral superior olivary nucleus, and a medial group of small neurons situated ventral to the nucleus of the trapezoid body. This medial group projects bilaterally to the cochlear nuclei.

Auditory Reflex Pathways

The auditory cell masses in the brain stem serve not only as relay nuclei in ascending and descending projections, but also as centres for the olivocochlear reflex, the startle response and the protective reflexes of the inner ear muscles.

The olivocochlear reflex (Fig. 18.5 A) suppresses cochlear gain by its action on inner hair cells. Interneurons in the posterior ventral cochlear nucleus, activated by a tone stimulus, connect with neurons of the medial olivocochlear system, which reduce the outer hair cell's contribution to cochlear amplification in a frequency-specific manner. The reflex may reduce the response of the cochlea to naturally occurring environmental noise, thus "unmasking" biologically significant stimuli, or may function as a protective devise, reducing receptor damage during intense acoustic exposure [49]. Separate populations of medial olivocochlear neurons subserve the ipsilateral and contralateral olivocochlear reflexes. A minority of the medial olivocochlear neurons project bilaterally [18].

Efferents from the cochlear nuclei enter the reticular formation, where they synapse with neurons of the ascending reticular activating system. Impulses entering the reticular formation along this path give rise to the auditory evoked startle response [46].

Fibres passing from the lateral zone of the inferior colliculus to the superior colliculus [2, 27, 51] may be considered as important links between eyes and head in response to auditory stimuli. Tectopontine fibres from the inferior colliculus also terminate in the pontine nuclei and the nucleus reticularis tegmenti pontis-nuclei, which also receive direct projections from the cochlear nuclei [46]. Auditory tecto-ponto-cerebellar pathways converge upon lobule VII of the caudal vermis. Similar visual tecto-ponto-cerebellar pathways are discussed in Chap. 20 (Fig. 19.17).

Pathways passing from the ventral cochlear nuclei to the motor nuclei of the trigeminal and facial nerves constitute reflex arcs that link

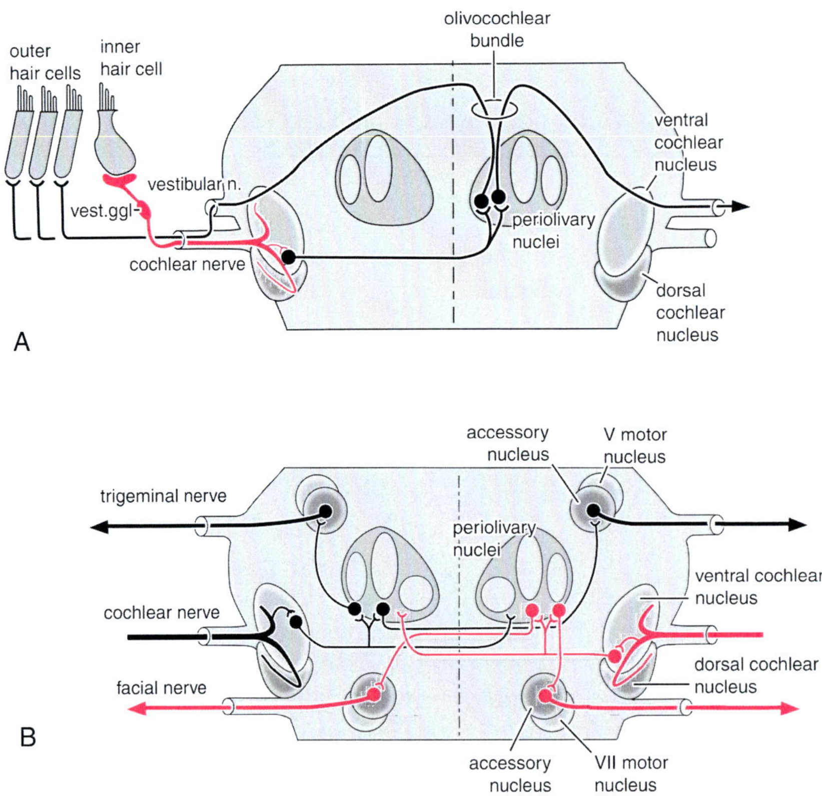

Fig. 18.5. The auditory reflex paths. **A** The olivocochlear reflex. **B** The stapedius reflex paths are indicated in *red*, and the tensor tympani pathways in *black*. *Vest ggl*, vestibular ganglion

the organ of Corti with the tensor tympani and the stapedius muscles. In response to sounds of high intensity, these muscles contract reflexively and dampen the vibration of the ear ossicles (Fig. 18.5B). In this way the organ of Corti is protected from damage by excessive stimulation. Studies using the retrograde tracer technique [43, 47, 60] have shown that the stapedius and the tensor tympani motoneurons both form separate, accessory cell groups situated close to but clearly beyond the confines of the facial and motor trigeminal nuclei. The first link in both reflex pathways is the cochlear nerve, they both contain neurons of the ventral cochlear nucleus and cells situated ventrolateral to the lateral superior olive. Projections of neurons in this position to the stapedius or the tensor tympani motoneurons are bilateral [43, 47, 60].

References

1. Adams JC (1983) Cytology of periolivary cells and the organization of their projection in the cat. J Comp Neurol 215:275–289
2. Aitkin LM, Kenyon CE, Philpott P (1981) The representation of the auditory and somatosensory systems in the external nucleus of the cat inferior colliculus. J Comp Neurol 196:25–40
3. Alain C, Arnott SR, Hevenor S, Graham S, Grady CL (2001) "What" and "where" in the human auditory system. Proc Natl Acad Sci USA 98:12301–12306
4. Andersen RA, Roth GL, Aitkin LM, Merzenich MM (1980) The efferent projections of the central nucleus and the pericentral nucleus of the inferior colliculus in the cat. J Comp Neurol 194:649–662
5. Belin P, Zatorre RJ, Lafaille P, Ahad P, Pike B (2000) Voice-selective areas in human auditory cortex. Nature 403:309–312
6. Binder JR, Frost JA, Hammeke TA, Rao SM, Cox RW (1996) Function of the left planum temporale in auditory and linguistic processing. Brain 119 (Pt 4):1239–1247
7. Binder JR, Swanson SJ, Hammeke TA, Morris GL, Mueller WM, Fischer M, Benbadis S, Frost JA, Rao SM, Haughton VM (1996) Determination of language dominance using functional MRI: a comparison with the Wada test. Neurology 46:978–984
8. Brawer JA, Morest DK, Cohen-Kane E (1974) The neuronal architecture of the cochlear nucleus of the cat. J Comp Neurol 153:251–300
9. Brodal A (1981) Neurological anatomy in relation to clinical medicine. Oxford Univ Press, New York
10. Brodmann K (1909) Vergleichende Lokalisationslehre der Grosshirnrinde in ihren Prinzipien dargestellt auf Grund des Zellenbaues. Barth, Leipzig
11. Bushara KO, Weeks RA, Ishii K, Catalan MJ, Tian B, Rauschecker JP, Hallett M (1999) Modality-specific frontal and parietal areas for auditory and visual spatial localization in humans. Nat Neurosci 2:759–766
12. Caicedo A, Herbert H (1993) Topography of descending projections from the inferior colliculus to auditory brainstem nuclei in the rat. J Comp Neurol 328:377–392
13. Calford MB, Aitkin LM (1983) Ascending projections to the medial geniculate body of the cat: evidence for multiple, parallel auditory pathways through thalamus. J Neurosci 3:2365–2380
14. Casseday JH, Kobler JB, Isbey SF, Covey E (1989) Central acoustic tract in an echolocating bat: an extralemniscal auditory pathway to the thalamus. J Comp Neurol 287:247–259
15. Coleman JR, Clerici WJ (1987) Sources of projections to subdivisions of the inferior colliculus in the rat. J Comp Neurol 262:215–226
16. Dallos P (1997) Outer hair cells: the inside story. Ann Otol Rhinol Laryngol Suppl 168:16–22
17. Dalman AJE, Eling P (2000) Wernicke's aphasia. In: Koehler PJ, Bruyn GW, Pearce JMS (eds) Neurological eponyms. Oxford Univ Press: 244–249
18. De Venecia RL, Liberman MC, Guinan JJ Jr, Brown MC (2005) Medial olivocochler reflex interneurons are located in the posteroventral cochlear nucleus: a kainic acid lesion study in guinea pigs. J Comp Neurol 487:345–360
19. Doron NN, Ledoux JE (1999) Organization of projections to the lateral amygdala from auditory and visual areas of the thalamus in the rat. J Comp Neurol 412:383–409
20. Doucet JR, Ross AT, Gillespie MB, Ryugo DK (1999) Glycine immunoreactivity of multipolar neurons in the ventral cochlear nucleus which project to the dorsal cochlear nucleus. J Comp Neurol 408:515–531
21. FitzPatrick KA, Imig TJ (1978) Projections of auditory cortex upon the thalamus and midbrain in the owl monkey. J Comp Neurol 177:537–555
22. Galaburda AM (1993) The planum temporale. Arch Neurol 50:457
23. Galaburda A, Sanides F (1980) Cytoarchitectonic organization of the human auditory cortex. J Comp Neurol 190:597–610
24. Galaburda AM, Pandya DN (1983) The intrinsic architectonic and connectional organization of the superior temporal region of the rhesus monkey. J Comp Neurol 221:169–184
25. Galaburda AM, Sanides F, Geschwind N (1978) Human brain. Cytoarchitectonic left-right asymmetries in the temporal speech region. Arch Neurol 35:812–817

26. Galaburda AM, LeMay M, Kemper TL, Geschwind N (1978) Right-left asymmetrics in the brain. Science 199:852–862
27. Garcia Del Cano G, Gerrikagoitia I, Alonso-Cabria A, Martinez-Millan L (2006) Organization and origin of the connection from the inferior to the superior colliculi in the rat. J Comp Neurol 499: 716–731
28. Geniec P, Morest DK (1971) The neuronal architecture of the human posterior colliculus. A study with the Golgi method. Acta Otolaryngol Suppl 295:1–33
29. Glendenning KK, Masterton RB (1983) Acoustic chiasm: efferent projections of the lateral superior olive. J Neurosci 3:1521–1537
30. Griffiths TD, Rees G, Rees A, Green GG, Witton C, Rowe D, Buchel C, Turner R, Frackowiak RS (1998) Right parietal cortex is involved in the perception of sound movement in humans. Nat Neurosci 1:74–79
31. Groff A, Liberman MC (2003) Modulation of cochlear afferent response by the lateral olivocochlear system: activation via electrical stimulation of the inferior colliculus. J Neurophysiol 90:3178–3200
32. Guinan JJ Jr, Warr WB, Norris BE (1983) Differential olivocochlear projections from lateral versus medial zones of the superior olivary complex. J Comp Neurol 221:358–370
33. Hackett TA, Stepniewska I, Kaas JH (1998) Subdivisions of auditory cortex and ipsilateral cortical connections of the parabelt auditory cortex in macaque monkeys. J Comp Neurol 394:475–495
34. Hackett TA, Preuss TM, Kaas JH (2001) Architectonic identification of the core region in auditory cortex of macaques, chimpanzees, and humans. J Comp Neurol 441:197–222
35. Held H (1893) Die centrale Gehörgeleitung. Arch Anat Physiol Anat Abth 201–248
36. Henkel CK, Spangler KM (1983) Organization of the efferent projections of the medial superior olivary nucleus in the cat as revealed by HRP and autoradiographic tracing methods. J Comp Neurol 221:416–428
37. Henkel CK, Shneiderman A (1988) Nucleus sagulum: projections of a lateral tegmental area to the inferior colliculus in the cat. J Comp Neurol 271: 577–588
38. Herbert H, Aschoff AOJ (1991) Topography of projections from the auditory cortex to the inferior colliculus in the rat. J Comp Neurol 304:103–122
39. Hodges JR, Spatt J, Patterson K (1999) "What" and "how": evidence for the dissociation of object knowledge and mechanical problem-solving skills in the human brain. Proc Natl Acad Sci USA 96: 9444–9448
40. Horvath M, Kraus KS, Illing RB (2000) Olivocochlear neurons sending axon collaterals into the ventral cochlear nucleus of the rat. J Comp Neurol 422:95–105
41. Huffman RF, Henson OW (1990) The descending auditory pathway and acousticomotor systems: connections with the inferior colliculus. Brain Res Brain Res Rev 15:295–323
42. Hurd LB, Hutson KA, Morest DK (1999) Cochlear nerve projections to the small cell shell of the cochlear nucleus: the neuroanatomy of extremely thin sensory axons. Synapse 33:83–117
43. Joseph MP, Guinan JJ, Fullerton BC, Norris BE, Kiang NY (1985) Number and distribution of stapedius motoneurons in cats. J Comp Neurol 232:43–54
44. Kaas JH, Hackett TA (1998) Subdivisions of auditory cortex and levels of processing in primates. Audiol Neurootol 3:73–85
45. Kaas JH, Hackett TA (2000) Subdivisions of auditory cortex and processing streams in primates. Proc Nat Acad Sci USA 97:11793–11799
46. Kandler K, Herbert H (1991) Auditory projections from the cochlear nucleus to pontine and mesencephalic reticular nuclei in the rat. Brain Res 562:230–242
47. Keller JT, Saunders MC, Ongkiko CM, Johnson J, Frank E, Van Loveren H, Tew JM (1983) Identification of motoneurons innervating the tensor tympani and tensor veli palatini muscles in the cat. Brain Res 270:209–215
48. Kemp DT, Ryan S (1991) Otoacoustic emission tests in neonatal screening programmes. Acta Otolaryngol Suppl 482:73–84
49. Kirk EC, Smith DW (2003) Protection from acoustic trauma is not a primary function of the medial olivocochlear efferent system. J Assoc Res Otolaryngol 4:445–465
50. Kolston J, Osen KK, Hackney CM, Ottersen OP, Storm-Mathisen J (1992) An atlas of glycine- and GABA-like immunoreactivity and colocalization in the cochlear nuclear complex of the guinea pig. Anat Embryol (Berl) 186:443–465
51. Kudo M (1981) Projections of the nuclei of the lateral lemniscus in the cat: an autoradiographic study. Brain Res 221:57–69
52. Maeder PP, Meuli RA, Adriani M, Bellmann A, Fornari E, Thiran JP, Pittet A, Clarke S (2001) Distinct pathways involved in sound recognition and localization: a human fMRI study. Neuroimage 14:802–816
53. Malmierca MM, Merchán MA (2004) Auditory system. In: Paxinos G (ed) The rat nervous system. Elsevier, Amsterdam, pp 997–1082
54. Malmierca MS, Le Beau FE, Rees A (1996) The topographical organization of descending projections from the central nucleus of the inferior colliculus in guinea pig. Hear Res 93:167–180
55. Malmierca MS, Merchán MA, Henkel CK, Oliver DL (2002) Direct projections from cochlear nuclear complex to auditory thalamus in the rat. J Neurosci 22:10891–10897
56. Malmierca MS, Blackstad TW, Osen KK, Karagulle T, Molowny RL (1993) The central nucleus of the inferior colliculus in rat: a Golgi and computer reconstruction study of neuronal and laminar structure. J Comp Neurol 332:1–27

57. Malmierca MS, Leergaard TB, Bajo VM, Bjaalie JG, Merchán MA (1998) Anatomic evidence of a three-dimensional mosaic pattern of tonotopic organization in the ventral complex of the lateral lemniscus in cat. J Neurosci 18:10603–10618
58. Merchán MA, Berbel P (1996) Anatomy of the ventral nucleus of the lateral lemniscus in rats: a nucleus with a concentric laminar organization. J Comp Neurol 372:245–263
59. Merchán MA, Saldana E, Plaza I (1994) Dorsal nucleus of the lateral lemniscus in the rat: concentric organization and tonotopic projection to the inferior colliculus. J Comp Neurol 342:259–278
60. Mizuno N, Nomura S, Konishi A, Uemura-Sumi M, Takahashi O, Yasui Y, Takada M, Matsushima R (1982) Localization of motoneurons innervating the tensor tympani muscles: an horseradish peroxidase study in the guinea pig and cat. Neurosci Lett 31:205–208
61. Moore JK (1980) The primate cochlear nuclei: loss of lamination as a phylogenetic process. J Comp Neurol 193:609–629
62. Moore JK (1987) The human auditory brain stem: a comparative view. Hear Res 29:1–32
63. Moore JK (2000) Organization of the human superior olivary complex. Microsc Res Tech 51:403–412
64. Moore RY, Goldberg JM (1966) Projections of the inferior colliculus in the monkey. Exp Neurol 14:429–438
65. Moore JK, Osen KK (1979) The cochlear nuclei in man. Am J Anat 154:393–418
66. Moore JK, Linthicum JR (2004) Auditory system. In: Paxinos G, Mai J (eds) The human nervous system, 2nd edn. Elsevier, Amsterdam, pp 1241–1279
67. Morest DK (1965) The laminar structure of the medial geniculate body in the cat. J Anat 99:143–160
68. Morest DK, Oliver DL (1984) The neuronal architecture of the inferior colliculus in the cat: defining the functional anatomy of the auditory midbrain. J Comp Neurol 222:209–236
69. Mulders WH, Robertson D (2002) Inputs from the cochlea and the inferior colliculus converge on olivocochlear neurons. Hear Res 167:206–213
70. Nobili R, Mammano F, Ashmore J (1998) How well do we understand the cochlea? Trends Neurosci 21:159–167
71. Oertel D, Young ED (2004) What's a cerebellar circuit doing in the auditory system? Trends Neurosci 27:104–110
72. Ohnishi TM, Matsuda H, Asada T, Aruga M, Hirakata M, Nishikawa M, Katoh A, Imabayashi E (2001) Functional anatomy of musical perception in musicians. Cereb Cortex 11:754–760
73. Oliver DL (1984) Neuron types in the central nucleus of the inferior colliculus that project to the medial geniculate body. Neuroscience 11:409–424
74. Oliver DL (1987) Projections to the inferior colliculus from the anteroventral cochlear nucleus in the cat: possible substrates for binaural interaction. J Comp Neurol 264:24–46
75. Oliver DL (2000) Ascending efferent projections of the superior olivary complex. Microsc Res Tech 51:355–363
76. Oliver DL, Morest DK (1984) The central nucleus of the inferior colliculus in the cat. J Comp Neurol 222:237–264
77. Oliver DL, Shneiderman A (1989) An EM study of the dorsal nucleus of the lateral lemniscus: inhibitory, commissural, synaptic connections between ascending auditory pathways. J Neurosci 9:967–982
78. Oliver DL, Kuwada S, Yin TC, Haberly LB, Henkel CK (1991) Dendritic and axonal morphology of HRP-injected neurons in the inferior colliculus of the cat. J Comp Neurol 303:75–100
79. Osen KK (1969) Cytoarchitecture of the cochlear nuclei in the cat. J Comp Neurol 136:453–484
80. Osen KK (1972) Projection of the cochlear nuclei on the inferior colliculus in the cat. J Comp Neurol 144:355–372
81. Ostapoff EM, Morest DK, Parham K (1999) Spatial organization of the reciprocal connections between the cat dorsal and anteroventral cochlear nuclei. Hear Res 130:75–93
82. Paloff AM, Usonoff KG, Hinova-Palova DV (1982) Ultrastructure of Golgi-impregnated and gold-toned neurons in the central nucleus of the inferior colliculus of the cat. J Hirnforsch 33:361–407
83. Pantev C, Oostenveld R, Engelien A, Ross B, Roberts LE, Hoke M (1998) Increased auditory cortical representation in musicians. Nature 392:811–814
84. Petrides M, Pandya DN (2002) Comparative cytoarchitectonic analysis of the human and the macaque ventrolateral prefrontal cortex and corticocortical connection patterns in the monkey. Eur J Neurosci 16:291–310
85. Probst M (1902) Experimentelle Untersuchungen über die Anatomie und Physiologie der Leitungsbahnen des Gehirnstammes. Arch Anat Physiol Anat Abth Suppl 147–254
86. Rauschecker JP (1998) Parallel processing in the auditory cortex of primates. Audiol Neurootol 3:86–103
87. Rauschecker JP (1998) Cortical processing of complex sounds. Curr Opin Neurobiol 8:516–521
88. Rauschecker JP, Tian B (2000) Mechanisms and streams for processing of "what" and "where" in auditory cortex. Proc Natl Acad Sci USA 97:11800–11806
89. Romanski LM, Goldman-Rakic PS (2002) An auditory domain in primate prefrontal cortex. Nat Neurosci 5:15–16
90. Saint Marie RL, Shneiderman A, Stanforth DA (1997) Patterns of gamma-aminobutyric acid and glycine immunoreactivities reflect structural and functional differences of the cat lateral lemniscal nuclei. J Comp Neurol 389:264–276
91. Saint Marie RL, Benson CG, Ostapoff EM, Morest DK (1991) Glycine immunoreactive projections from the dorsal to the anteroventral cochlear nucleus. Hear Res 51:11–28

92. Saldāna E, Merchán MA (1992) Intrinsic and commissural connections of the rat inferior colliculus. J Comp Neurol 319:417–437
93. Saldāna E, Feliciano M, Mugnaini E (1996) Distribution of descending projections from primary auditory neocortex to inferior colliculus mimics the topography of intracollicular projections. J Comp Neurol 371:15–40
94. Saykin AJ, Johnson SC, Flashman LA, McAllister TW, Sparling M, Darcey TM, Moritz CH, Guerin SJ, Weaver J, Mamourian A (1999) Functional differentiation of medial temporal and frontal regions involved in processing novel and familiar words: an fMRI study. Brain 122:1963–1971
95. Schlaug G, Jancke L, Huang Y, Steinmetz H (1995) In vivo evidence of structural brain asymmetry in musicians. Science 267:699–701
96. Schlaug G, Jancke L, Huang Y, Staiger JF, Steinmetz H (1995) Increased corpus callosum size in musicians. Neuropsychologia 33:1047–1055
97. Schofield BR, Cant NB (1999) Descending auditory pathways: projections from the inferior colliculus contact superior olivary cells that project bilaterally to the cochlear nuclei. J Comp Neurol 409:210–223
98. Scott KS, Blank CC, Rosen S, Wise RJS (2000) Identification of a pathway for intelligible speech in the left temporal lobe. Brain 123:2400–2406
99. Spoendlin H, Schrott A (1989) Analysis of the human auditory nerve. Hear Res 143:25–38
100. Sweet RA, Dorph-Petersen K-A, Lewis DA (2005) Mapping auditory core, latral belt and parabelt cortices in human superior temporal gyrus. J Comp Neurol 491:270–289
101. Thompson AM, Thompson GC (1991) Posteroventral cochlear nucleus projections to olivocochlear neurons. J Comp Neurol 303:267–285
102. Thompson AM, Schofield BR (2000) Afferent projections of the superior olivary complex. Microsc Res Tech 51:330–354
103. Van Noort J (1969) The structure and connections of the inferior colliculus. Van Gorcum, Assen
104. Vetter DE, Mugnaini E (1992) Distribution and dendritic features of three groups of rat olivocochlear neurons. A study with two retrograde cholera toxin tracers. Anat Embryol (Berl) 185:1–16
105. Vetter DE, Adams JC, Mugnaini E (1991) Chemically distinct rat olivocochlear neurons. Synapse 7:21–43
106. Vetter DE, Saldāna E, Mugnaini E (1993) Input from the inferior colliculus to medial olivocochlear neurons in the rat: a double label study with PHA-L and cholera toxin. Hear Res 70:173–186
107. Warr WB (1966) Fiber degeneration following lesions in the anterior ventral cochlear nucleus of the cat. Exp Neurol 14:453–474
108. Warr WB (1975) Olivocochlear and vestibular efferent neurons of the feline brain stem: their location, morphology and number determined by retrograde axonal transport and acetylcholinesterase histochemistry. J Comp Neurol 161:159–181
109. Warr WB (1978) The olivocochlear bundle: its origins and terminations in the cat. In: Naunton RF, Fernandez C (eds) Evoked electrical activity in the auditory nervous system. Academic Press, New York, pp 43–63
110. Warr WB (1980) Efferent components of the auditory system. Ann Otol Rhinol Laryngol Suppl 89:114–120
111. Warr WB, Guinan JJ Jr (1979) Efferent innervation of the organ of corti: two separate systems. Brain Res 173:152–155
112. Warr WB, Beck JE (1996) Multiple projections from the ventral nucleus of the trapezoid body in the rat. Hear Res 93:83–101
113. Warr WB, Boche JB, Neely ST (1997) Efferent innervation of the inner hair cell region: origins and terminations of two lateral olivocochlear systems. Hear Res 108:89–111
114. Weedman DL, Ryugo DK (1996) Pyramidal cells in primary auditory cortex project to cochlear nucleus in rat. Brain Res 706:97–102
115. Weedman DL, Ryugo DK (1996) Projections from auditory cortex to the cochlear nucleus in rats: synapses on granule cell dendrites. J Comp Neurol 371:311–324
116. Wenthold RJ (1987) Evidence for a glycinergic pathway connecting the two cochlear nuclei: an immunocytochemical and retrograde transport study. Brain Res 415:183–187
117. Wernicke C (1874) Der aphasische Symptomencomplex; eine psychologische Srudie auf Anatomischer Basis. Cohn & Weigert, Breslau
118. Winer JA, Morest DK (1983) The medial division of the medial geniculate body of the cat: implications for thalamic organization. J Neurosci 3:2629–2651
119. Winer JA, Saint Marie RL, Larue DT, Oliver DL (1996) GABAergic feedforward projections from the inferior colliculus to the medial geniculate body. Proc Natl Acad Sci USA 93:8005–8010
120. Winer JA, Larue DT, Diehl JJ, Hefti BJ (1998) Auditory cortical projections to the cat inferior colliculus. J Comp Neurol 400:147–174
121. Winer JA, Sally SL, Larue DT, Kelly JB (1999) Origins of medial geniculate body projections to physiologically defined zones of rat primary auditory cortex. Hear Res 130:42–61
122. Wise RJ, Scott SK, Blank SC, Mummery CJ, Murphy K, Warburton EA (2001) Separate neural subsystems within 'Wernicke's area'. Brain 124:83–95
123. Yin TC, Chan JC (1990) Interaural time sensitivity in medial superior olive of cat. J Neurophysiol 64:465–488
124. Zook JM, Casseday JH (1985) Projections from the cochlear nuclei in the mustache bat, Pteronotus parnellii. J Comp Neurol 237:307–324

19 Visual System

Retina

The visual system consists of a number of parallel pathways, each serving a specific function. Parallel processing of visual information already is present at the level of the retina (Figs. 19.1, 19.2). The structure and function of the retina have been reviewed extensively [54, 114, 155, 188, 189, 309].

The retina is a laminated structure. Its layers are confined between the external and internal limiting membranes. The external limiting membrane consists of extensions of the radial glial, the Müller cells of the retina (Fig. 19.1), which are connected through adherent junctions with the inner segments of the photoreceptors. The endfeet of the Müller cells and the associated basal membrane constitute the internal limiting membrane.

The following nine layers are generally distinguished:

1. Layer of rods and cones (Fig. 19.1: LCR)
2. Outer limiting membrane (OLM)
3. Outer nuclear layer (ONL), containing the nuclei of the photoreceptors
4. Outer plexiform layer (OPL) containing the synaptic connections of photoreceptors and horizontal cells
5. Inner nuclear layer (INL) containing the nuclei of bipolar, horizontal and many amacrine cells
6. Inner plexiform layer, subdivided into a vitreal OFF sublayer and an inner ON sublayer
7. Ganglion cell layer (GCL)
8. Layer of optic nerve fibres (LON)
9. Internal limiting membrane (ILM)

Photoreceptors are composed of an outer and inner segment, a cell body with the nucleus and a synaptic terminal. The outer and inner segments protrude through the external limiting membrane into the cleft between the retina and the pigment epithelium. The outer segment contains the membrane-bound photopigments; the inner segment is filled with mitochondria. Outer and inner segments are interconnected by a modified cilium. The cell body with the nucleus is located in the outer nuclear layer. Synaptic contacts are established in the outer plexiform layer.

The inner segment of the cones is cone-shaped; rods are more elongated structures.

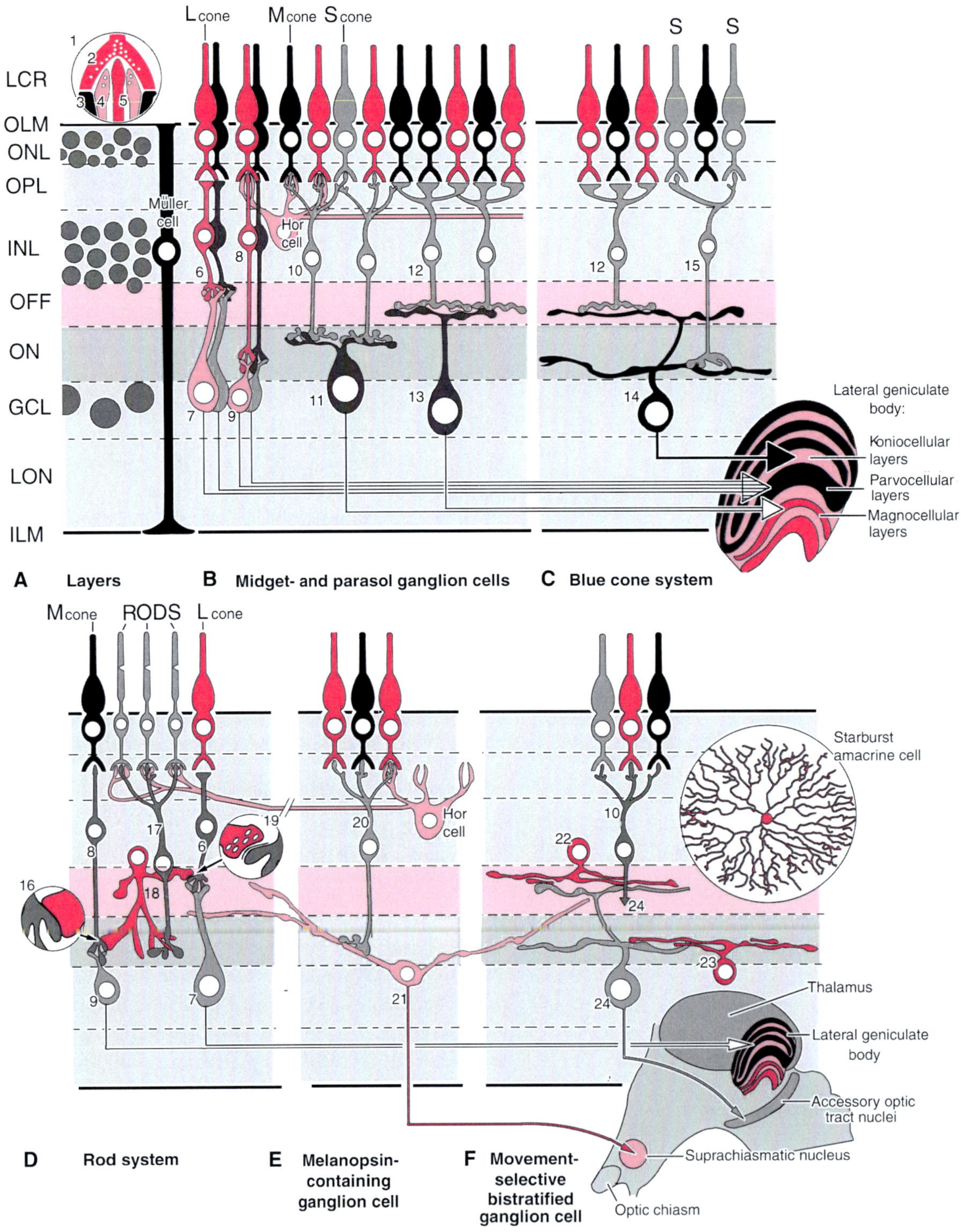

Fig. 19.1 A–F. The primate retina and its connections with the lateral geniculate body, the nuclei of the accessory optic system and the suprachiasmatic nucleus. **A Layers.** *LCR*, layer of cones and rods; *OLM*, outer limiting membrane; *ONL*, outer nuclear layer; *OPL*, outer plexiform layer; *INL*, inner nuclear layer; *OFF*, OFF sublayer of inner plexiform layer; *ON*, ON sublayer of inner plexiform layer; *GCL*, ganglion cell layer: *LON*, layer of optic nerve fibres; *ILM*, inner limiting membrane. **B Midget and parasol ganglion cell pathways.** Long wavelength (*L cone*, red-sensitive, shown in *red*) and medium wavelength (*M cone*, green-sensitive,

Three types of cones can be distinguished in the retinas of Old World monkeys and humans [143]. They differ in their sensitivity to light of different wavelengths. Long wavelength (L) cones are maximally sensitive to red light, medium wavelength (M) cones are maximally sensitive to green light, and short wavelength (S) cones are maximally sensitive to blue light. Cones perceive light of higher intensity (photopic vision); rods are sensitive to low intensity illumination (scotopic vision) and are able to reliably transduce the absorption of single photons. In darkness, the photoreceptors are depolarized and produce glutamate from their synaptic endings. When light falls upon the photoreceptors they hyperpolarize and the neurotransmitter release to their second-order neurons ceases. The synaptic endings of rods and cones are located in the outer plexiform layer and are known as the rod spherules and the cone pedicles.

The main cell types and the basic circuitry of the retina were recognized by Cajal [33] and Polyak [225]. Bipolar neurons, with cell bodies located in the inner nuclear layer, connect cones with the ganglion cells, which are located in an inner, vitreal layer of the retina. They give rise to the axons of the optic nerve [237]. These simple connections are exemplified by the midget and parasol ganglion cell systems (Fig. 19.1 B). Horizontal cells, located in the inner nuclear layer, synapse with photoreceptors in the outer plexiform layer; amacrine cells, located in the vitreal part of the inner nuclear layer, establish connections between bipolar, ganglion and other amacrine cells in the inner plexiform layer. Horizontal and amacrine cells are nonspiking, inhibitory neurons that conduct impulses in all directions [161, 184]. Many different types of horizontal and amacrine cells exist; a few are illustrated in Fig. 19.1.

Horizontal cells form an electrotonically coupled network. Their dendrites contact cone pedicles in triad synapses (Fig. 19.1 B, inset). The L, M, and the S cones connect with two different types of horizontal cell. Axons of axon-bearing horizontal cells terminate on rod spherules (Fig. 19.1 D). The long axons of these cells are electrotonically isolated from the dendritic cone feedback system. Horizontal cells are believed important for contrast enhancement and the formation of the antagonistic surround of the ganglion cell receptive field [14, 161, 190, 299].

Essential features of the retina are the existence of ON pathways (activated by light) and OFF pathways (activated during darkness), as

shown in *black*) cones each have their private line with the parvocellular layers of the lateral geniculate body (shown in *black*). Each cone is connected through an OFF (flat) bipolar (*6*) and an ON (invaginating) bipolar (*8*) with a midget ON (*7*) and a midget OFF ganglion cell (*9*), respectively. ON-diffuse bipolar cells (*10*) and OFF-diffuse bipolar cells (*12*) receive information from randomly organized cones and are connected with ON and OFF parasol ganglion cells (*11, 13*), respectively. Parasol cells project to the magnocellular layers of the lateral geniculate body. *Inset (1)*: triad synapse between a cone pedicle (*2*) and the basal, flat dendrites of OFF bipolar cells (*3*), the dendrites of horizontal cells (Hor cell) (*4*), and the invaginating dendrite of an ON bipolar (*5*). **C Blue cone system.** Short wavelength cones (*S cone*, blue-sensitive, shown in *grey*) connect with bistratified ganglion cells (*14*), which project to the interlaminar (koniocellular) layers of the lateral geniculate body. S cone-ON bipolars (*15*) connect with ganglion cell dendrites in the ON layer of the inner plexiform layer. L and M cones, responsible for the sensation of yellow, connect with diffuse bipolar cells (*12*) in the OFF layer. **D Rod pathways.** Rod bipolars (*17*) connect multiple rods with AII amacrine cells (*18*). These amacrine cells establish gap junctions (*16*) with the axon of ON cone bipolar cells (*8*) and conventional inhibitory synapses (*19*) with OFF cone bipolar cells (*9*). The rod signal, therefore, is carried by cone pathways to the lateral geniculate nucleus. **E Light-sensitive, melanopsin-containing ganglion cells** project to the suprachiasmatic nucleus. These ganglion cells also receive input from photoreceptors (*20*). **F** One type of direction-sensitive ganglion cell (*24*) is a bistratified neuron, which receives input from starburst amacrine cells (*22*). The cell body of displaced starburst cells is located in the ganglion cell layer (*23*). The precise connections with the photoreceptors are unknown (*24*). Direction-sensitive ganglion cells project to the accessory optic system. *Inset* shows tangential view of a starburst amacrine cell

well as the centre-surround structure of the receptive fields of many ganglion cells, with an antagonism between the effects of light falling on the central and surrounding portions of the receptive fields of these cells [156].

ON bipolar cells are depolarized by increase of light; OFF bipolar cells hyperpolarize in light [147, 314]. ON and OFF bipolar cells differ in the morphology of their synaptic contacts with the photoreceptors, in the glutamate receptors present at these synapses and in the sublayer of the inner plexiform layer where their connections with ON- and OFF-type ganglion cells and amacrine cells are established (Fig. 19.1 B). Dendrites of ON bipolar cells invaginate into the synaptic ending of the photoreceptors and are located opposite a synaptic ribbon (Fig. 19.1 (1), inset). OFF bipolar cells synapse with the flattened base of the synaptic ending. Horizontal cells make synaptic connections on both sides of the invaginating dendrite. The composite of an invaginating dendrite and the horizontal cell terminals is known as a triad [41].

Metabotropic glutamate receptors in ON bipolar cells reverse the sign of a photoreceptor, hyperpolarized by light, into depolarization of the ON bipolar cell. Ionotropic glutamate receptors conserve the sign of the hyperpolarized photoreceptor in the flattened synapses with the OFF bipolar cells [59, 265]. OFF bipolar cells synapse in the outer OFF-sublayer of the inner plexiform layer; ON bipolar cells synapse in the inner, vitreal ON-sublayer [66, 320].

The connections of ON and OFF bipolar cells with ganglion cell dendrites in the ON and OFF inner plexiform sublayers explain the presence of ganglion cells with ON and OFF centre receptive fields. The circuitry underlying the antagonistic surround of the receptive fields of these ganglion cells, however, is incompletely known.

The microcircuitry of rods and cones differs for the central and peripheral retina (Fig. 19.2). The highest concentration of cones is found in the fovea, and rods are absent from this region. Rods are most numerous in the peripheral retina, where the concentration of the cones falls off rapidly. Photoreceptors are absent from the optic disk, the origin of the optic nerve.

The following parallel pathways with different central connections take their origin from the retina.

1. Densely packed L and M cones in the fovea each have their private connections through "midget" ON and OFF cone-bipolar cells, with "midget" ON and OFF ganglion cells, respectively [238, 311] (Fig. 19.1 B). These ganglion cells project to the upper four parvocellular layers of the lateral geniculate body [170, 220]. They are responsible for the optimal resolution in the visual system, which is determined by the packing density of the L and M cones in the fovea. Moreover, they subserve the perception of red and green. The receptive fields of ON and OFF midget ganglion cells are colour-opponent, with a red or green centre and an antagonistic green or red surround [54, 56, 57].

2. Large ON and OFF "parasol" ganglion cells [55, 238, 311] receive input from multiple ON and OFF diffuse bipolar cells (Fig. 19.1B: 10, 12). Diffuse bipolar cells are contacted by a random collection of L, M and S cones and, therefore, are not colour-sensitive [17]. Parasol ganglion cells project to the two magnocellular layers of the lateral geniculate body [170, 220] and provide them with a luminance-contrast signal.

3. A separate retinal system is responsible for blue-yellow opponent perception [34, 56]. ON-type S-bipolar cells (Fig. 19.1C: 15) synapse selectively with several S cones and terminate in the ON inner plexiform sublayer on dendrites of small, S-type bistratified ganglion cells (Fig. 19.1 C: 14). L and M cones project, through OFF-type diffuse bipolar cells, to a set of dendrites of these cells in the OFF sublayer. The ON-type small stratified ganglion cells, therefore, are excited by blue and inhibited by a combination of red and green, i.e. yellow. They project to the interlaminar (koniocellular) layers of the lateral geniculate nucleus. The spatial resolution of this system is relatively low and does not contribute to this attribute of the visual system, which is determined by the L and M midget ganglion cells and their pro-

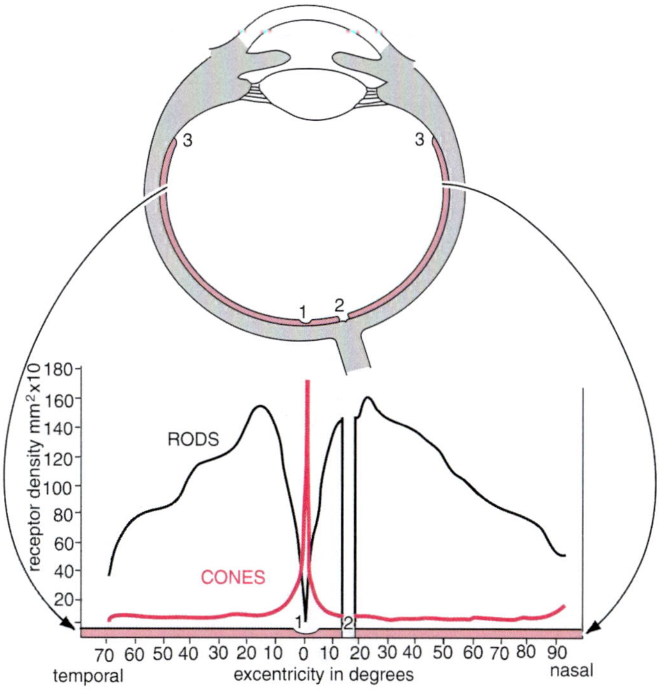

Fig. 19.2. Distribution of rods and cones in the human retina. Redrawn from Kolb et al. [155]. *1,* Fovea centralis; *2,* Optic disk; *3,* Ora serrata

jection to the parvocellular lateral geniculate. OFF-type S ganglion cells are either rare or absent.

4. The central connections of the rod photoreceptors include an extra synapse with an amacrine cell (the AII amacrine cell). Rods use cone bipolar and ganglion cells for their central projections [14, 154, 299, 310] (Fig. 19.1 D). Numerous rods synapse with an ON rod bipolar cell; OFF rod bipolar cells do not exist. Rod bipolar cells contact the basal dendrites of AII amacrine cells in the ON internal plexiform sublayer (Fig. 19.1: 17, 18). AII amacrine cells are electrotonically coupled. Dopamine, released by other amacrine cells, regulates the coupling of AII amacrine cells and adjusts their responsiveness during different states of light adaptation. Gap junctions link the dendrites of AII amacrine cells with ON cone-bipolar cells in the ON inner plexiform sublayer (Fig. 19.1 D: 16, inset). Inhibitory (glycinergic) synapses connect the lobular appendages of the AII amacrine cells with OFF cone bipolar cells in the OFF sublayer (Fig. 19.1 D: 19, inset). The rod system, therefore, uses the cone system as its output. As a consequence, the target ganglion cells of the rod system .project to the parvocellular and/or magnocellular layers of the lateral geniculate nucleus.

5. Large, light-sensitive ganglion cells contain the photopigment melanopsin (Fig. 19.1 E: 21). Their dendrites form a bilayered, photoreceptive net in the OFF and ON inner plexiform sublayers [11, 109, 227]. In addition, these cells are driven by bipolar cells [7, 242]. Axons of melanopsin-containing ganglion cells terminate in the suprachiasmatic nucleus, the intergeniculate leaflet and the pretectal olivary nucleus. Melanopsin-containing ganglion cells are responsible for the retained pupillary light reflex and circadian responses in transgenic mice, lacking both rod and cone photoreceptors [175, 200].

6. Several types of ganglion cells are direction-selective. One of them is a bistratified ganglion cell, which receives input from starburst amacrine cells (Fig. 19.1F, inset) [63–65, 190] The cell bodies of starburst amacrine cells are located in the inner nuclear layer and, as displaced amacrine cells, in the ganglion cell layer (Fig. 19.1F: 22, 23). By feed-forward excitation and inhibition they release both acetylcholine and GABA. The starburst amacrine cells are important for direction sensitivity. The microcircuitry of these direction-sensitive neurons is not completely known (for a hypothesis, see [327]). Direction-sensitive ganglion cells project to the accessory optic system and the nucleus of the optic tract.

Retinal ganglion cells can be subdivided on both structural and functional grounds into several categories. The main populations in primates are the midget and parasol cells, mentioned previously. Midget ganglion cells are red/green-opponent, with small receptive fields and low luminance contrast gain. Midget ganglion cells react to light steps with a tonic, sustained response. Their axons are small and slowly conducting. Parasol ganglion cells are broadband, i.e. not selective for a particular wavelength, with large receptive fields and a high luminance contrast gain. Their axons are large, with a high conduction velocity. They react to light steps with a phasic, transient response [148]. Midget cells project to the upper four parvocellular laminae of the lateral geniculate body and parasol cells to the lower two magnocellular layers. In some respects, the midget and parasol cells are similar to the X and Y cells of the cat's retina [167, 170, 272]. However, it has been proposed that the midget cell population is a new acquisition in primates, serving higher resolution and colour vision. In this case, the X and Y cells of the cat retina would be homologous to the parasol cells of the primate [254]. A rest group consists of ganglion cells of different shapes and functional properties, with projections to multiple visual centres. The small, bistratified blue/yellow-opponent ganglion cells, with their projections to the intercalated layers of the lateral geniculate body; the melanotropic cells, with their suprachiasmatic connections and the direction-sensitive ganglion cells (which provide input to the accessory optic system) form distinct groups. In primates, only a small proportion of the ganglion cells projects to the lateral geniculate body and to one or more of these other targets [84, 219, 220, 272].

Optic Nerve, Chiasm and Tract

The axons of the ganglion cells pass over the inner surface of the retina and converge towards the posterior pole of the eye, where they pierce through foramina in the sclera and then constitute the optic nerve. Only after leaving the retina do the axons of the ganglion cells acquire a myelin sheath. The optic nerves pass to the optic chiasm, which is situated at the base of the brain in the most rostral part of the hypothalamus (Figs. 19.3, 19.4). Within the optic chiasm a partial decussation takes place: fibres from the nasal halves of the retinae, including the projection of the monocular visual field, cross to the opposite side; those from the temporal halves of the retinas remain uncrossed (Fig. 19.4). After this partial decussation, the axons of the retinal ganglion cells continue without interruption behind the chiasm as the two diverging optic tracts. These arch around the lateral sides of the diencephalon until they reach the lateral geniculate bodies, where most of their fibres terminate. However, some fibres continue in a mediocaudal direction and pass via the brachium of the superior colliculus to the superior colliculus, the pretectal region and the terminal nuclei of the accessory optic system (Fig. 19.4). Within the optic tract, retinal fibres with different destinations are sorted. Small, presumably midget cell fibres occupy a deep position; coarse fibres, including the parasol cell axons and fibres destined for the superior colliculus, are located superficially [230, 231].

It is worthy to note that some retinofugal fibres or their collaterals leave the optic chiasm and enter the anterior part of the hypothalamus, terminating in the suprachiasmatic nucleus [58]. This is a small periventricular cell mass that lies, as its name implies, directly above the chiasm (see Chap. 10). Via this retinohypothalamic projection the visual system participates in the regulation of various behavioural rhythms. These fibres belong to the system of melanotropic ganglion cells, which, in rodents, also terminates in the intergeniculate leaflet, the olivary pretectal nucleus, in different hypothalamic nuclei, the amygdala, the superior collicus and the periaqueductal grey [110, 200]. In primates, retinohypothalamic fibres terminate more extensively in the hypothalamus [197].

Lateral Geniculate Body and the Visual Cortex: Retinotopical Projections to the Lateral Geniculate Body and the Primary Visual Cortex (V1)

As mentioned previously, most fibres of the optic tract end in the lateral geniculate body. This thalamic nucleus has a distinct laminated cellular structure consisting of six concentric layers. These layers are usually numbered from 1 to 6, beginning from the ventral, concave surface. The magnocellular (M) layers 1 and 2 occupy the hilar region. The parvocellular (P) layers 3–6 consist of relatively small cells. Still smaller cells are found in the intercalated or koniocellular (K) layers. Layers K1–K6 are located below the six main layers of the nucleus (Fig. 19.4).

The retinal fibres terminate in a highly orderly fashion in layers of the lateral geniculate body. Layer M1 and layers P4 and P6 receive fibres from the contralateral eye, and layer M2 and layers P3 and P5 receive fibres from the ipsilateral eye (Fig. 19.4). Moreover, parasol cells project to the M layers, and midget ganglion cells to the P layers. K layers are innervated by the same retina as the overlying main layer. Small bistratified blue/yellow-opponent ganglion cells project to layers K3 and K4; other types of retinal ganglion cells innervate other K layers [115]. Ventral K layers receive an extraretinal projection from the superior colliculus and cholinergic input from the parabigeminal nucleus [105, 106, 115, 158].

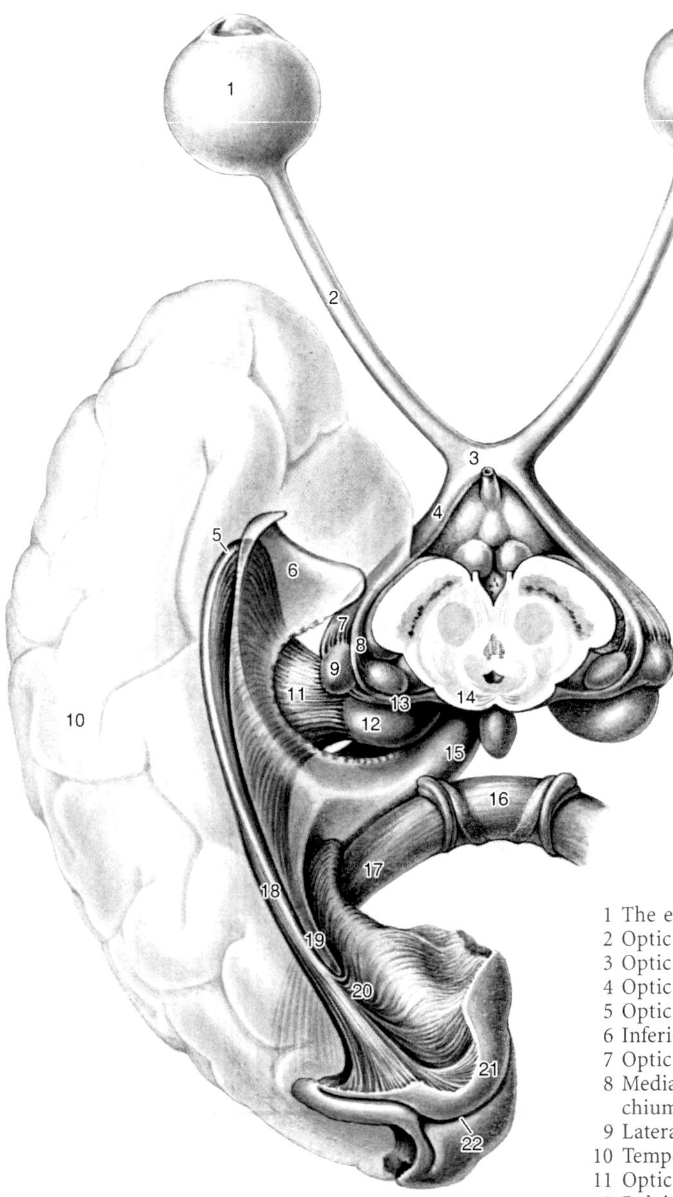

1 The eye
2 Optic nerve
3 Optic chiasm
4 Optic tract
5 Optic radiation, temporal knee
6 Inferior horn of the lateral ventricle
7 Optic tract, geniculate root
8 Medial root of the optic tract joining brachium of the superior colliculus
9 Lateral geniculate body
10 Temporal lobe
11 Optic radiation
12 Pulvinar of the thalamus
13 Brachium of the superior colliculus
14 Superior colliculus
15 Central part of the lateral ventricle
16 Splenium of the corpus callosum
17 Radiation of the corpus callosum
18 Sagittal stratum
19 Posterior horn of the lateral ventricle
20 Optic radiation, occipital knee
21 Area striata: V1
22 Calcarine sulcus

Fig. 19.3. The retinogeniculocortical projection in a ventral view (1/1×). Of the cerebral hemispheres, only the right temporal lobe and the inferior portion of the occipital lobe have been depicted

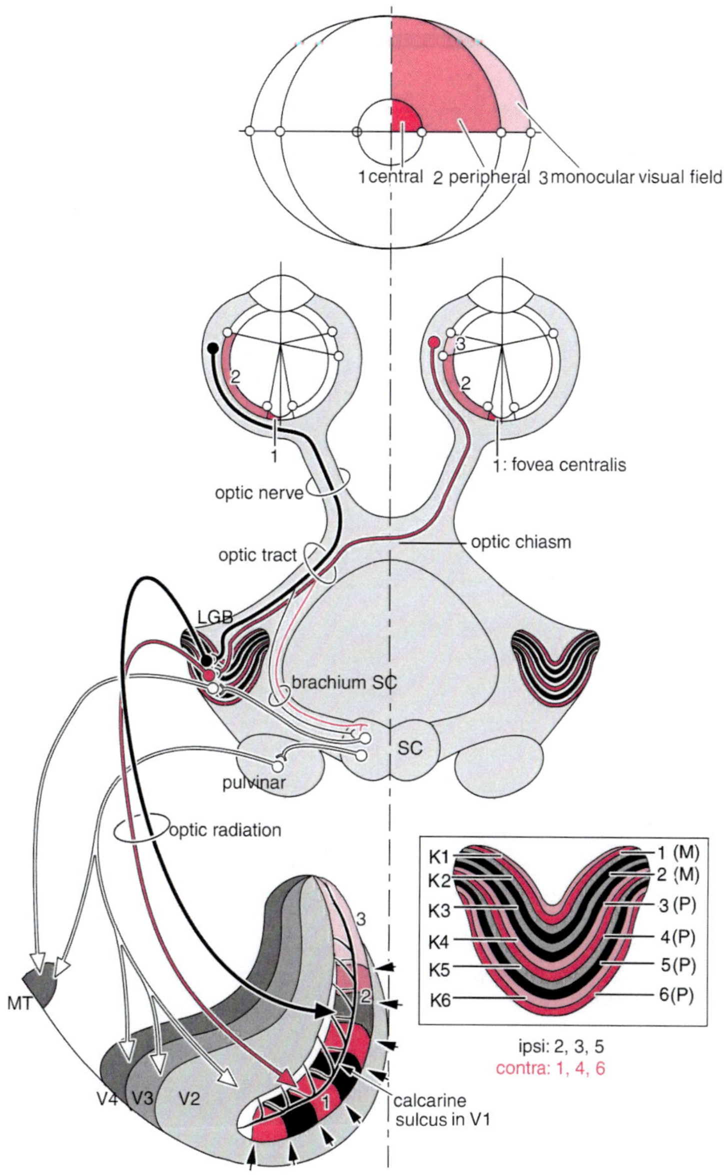

Fig. 19.4. The retinogeniculocortical projection and the extrageniculate and extrastriate visual pathways. The position of tracts and nuclei corresponds to that in Fig. 19.3. The projections of the upper right quadrant of the visual field are indicated; for the right eye this includes the projection of the monocular visual field (3). The contralateral upper quadrant is represented in the striate visual area (V1) in the lower bank of the calcarine sulcus. Projections of the contralateral (*red*) and ipsilateral retina (*black*) remain segregated in different laminae of the lateral geniculate body (*LGB*) and in the ocular dominance columns of the striate cortex illustrated in the lower bank of the calcarine sulcus (*arrows*). The representation of the central visual field (*1*) is located posteriorly; the monocular visual field (*3*) projects in the anterior striate area adjoining the splenium. Several retinocortical pathways to nonprimary visual areas avoid the lateral geniculate body and synapse in the superior colliculus and the pulvinar. The projection of the koniocellular layers of the lateral geniculate body to the middle temporal visual area (*MT*) avoids the primary visual cortex (Fig. 19.7). *Inset*: Layers of the lateral geniculate body. Parvocellular (*P*), magnocellular (*M*) and koniocellular (*K*) layers 1, 4 and 6 (*red*) receive retinal afferents from the contralateral eye, layers 2, 3 and 5 (*black*) from the ipsilateral eye. *SC*, superior colliculus; *V1–V4*, visual cortical areas V1–4

The efferent fibres of the lateral geniculate body constitute the optic radiation or geniculocalcarine tract, which terminates in the primary visual cortex. The primary visual area (V1) corresponds to Brodmann's area 17. It is also designated as the striate cortex (area striata), because it contains the highly characteristic line of Gennari, a macroscopically visible layer of myelinated fibres bisecting the internal granular layer (sublayer 4B; Figs. 5.13, 5.21, 5.24, 5.31). The striate cortex surrounds the calcarine sulcus on the medial side of the occipital lobe (Fig. 19.1). The fibres of the optic radiation first traverse the retrolenticular part of the internal capsule, then arch around the lateral ventricle and finally pass posteriorly towards the occipital cortex. Figure 19.3 shows that many fibres of the optic radiation do not reach their destination by the shortest route. Those arising from the lateral part of the lateral geniculate body and terminating ventral to the calcarine fissure sweep forward into the temporal lobe and pass laterally over the inferior horn of the lateral ventricle before turning backwards.

There is a precise point-to-point projection from the retina to the lateral geniculate body, and from the latter to the visual cortex. In the lateral geniculate body the contralateral hemifield is represented in each successive layer, with the fovea caudally [296]. Figures 19.4 and 19.5 illustrate the following features of the projection of the retina to V1 [136, 137, 295].

1. As a consequence of the partial decussation of the retinofugal fibres in the chiasm, the right half of the visual field, which projects on the nasal half of the right retina and the temporal half of the left retina, is represented in the left occipital lobe, and vice versa.

2. The upper quadrant of the visual field is represented in the contralateral V1 cortex of the lower bank of the calcarine sulcus, while the lower quadrant is represented in the upper bank of this sulcus.

3. The fovea, i.e. the retinal areas concerned with central vision, projects to a relatively large area in the posterior part of the primary visual cortex. Portions of the retina concerned with the periphery of the binocular field project to a smaller, intermediate part. The nasal periphery of the retina, concerned with the temporal, monocular periphery of the visual field, projects to the extreme anterior part of the visual cortex, next to the splenium of the corpus callosum.

4. The segregation of the projections of the right and left retinae is maintained in the projection of the lateral geniculate body to the V1 cortex. The projections of geniculate layers innervated by the contralateral retina to layer IV of the V1 cortex are organized in columns, separated by columns receiving afferents from geniculate layers innervated by the ipsilateral retina (Figs. 19.4, 19.5). The borders between these ocular dominance columns [125, 126, 317] are quite distinct at the level of cortical layer IV receiving the primary input, where most of the cells are monocularly driven, though less so in more superficial cortical layers where binocularly driven cells prevail. In humans the ocular dominance columns are 0.5 mm wide and are orientated at right angles to the V1/V2 boundary. They are most regular in the representation of the periphery of the visual field and absent from the representation of the monocular crescent. Their width does not vary much in central and peripheral representations of the visual field (Fig. 19.5) [39, 169].

Visual Areas: Localization

Many visual areas have been distinguished in the cerebral cortex of primates, including humans, on the basis of architecture, connectivity, visual topography and/or functional characteristics (Fig. 19.6) [295]. The striate or primary visual area V1 corresponds to Brodmann's area 17. Layer 4 can be subdivided into four sublayers (4A, 4B, 4Cα and 4Cβ). Gennari's line of myelinated fibres is located in sublayer 4B. Spiny stellate neurons, the recipients of the afferents of the lateral geniculate body, are present in sublayers 4A and 4C alpha and beta. Pyramidal cells are absent from both 4C sublayers, but present in 4A and B. Dendrites from pyramidal neurons of the lower

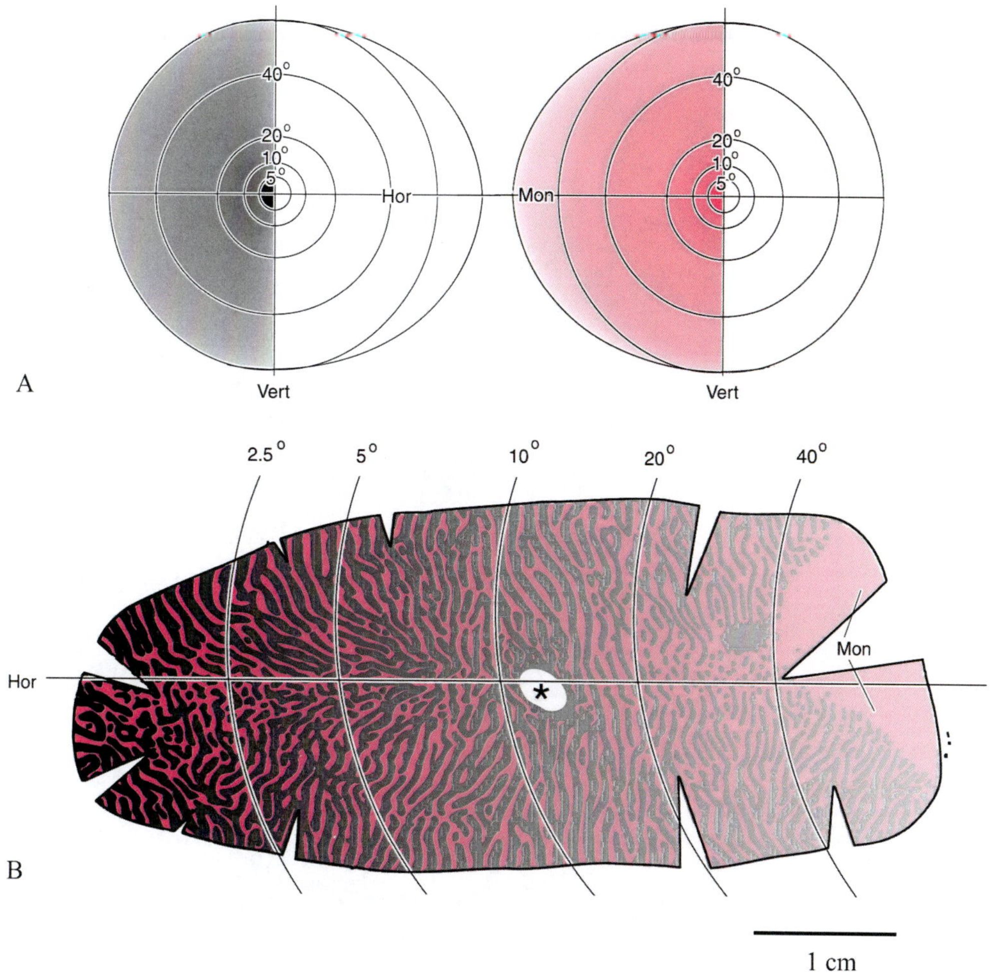

Fig. 19.5. **A** Left and right retina with the projection of the right half of the visual field. **B** Computer reconstruction of the ocular dominance columns in the left primary visual cortex of a macaque monkey. Ocular dominance columns receiving projections from the contralateral retina are indicated in *red*, those receiving projections from the ipsilateral retina in shades of *black*. *Asterisk*, projection of the blind spot. Ocular dominance columns are absent from the representation of the monocular visual field. Modified with permission from LeVay et al. [169]. Approximate eccentricity coordinates are derived from Van Essen et al. [296]. *Hor*, horizontal meridian; *Mon*, monocular field of the retina and its projection on the visual cortex; *Vert*, vertical meridian

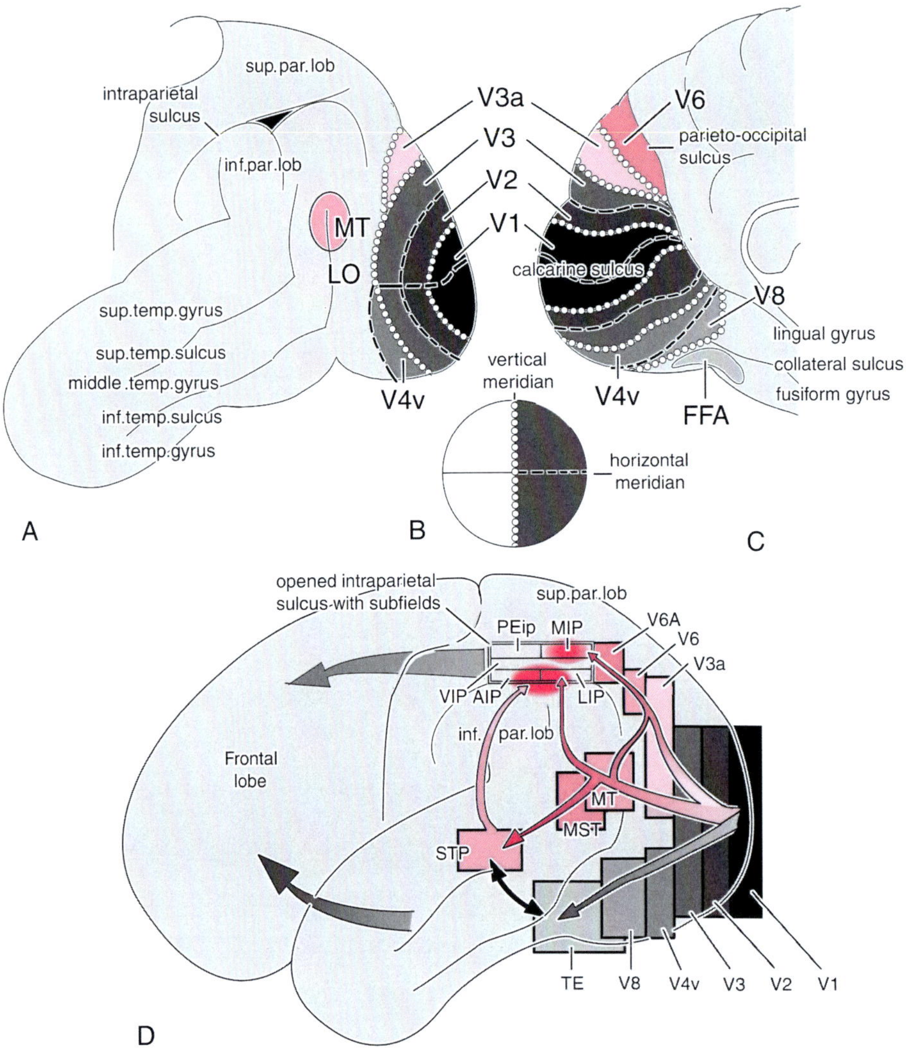

Fig. 19.6 A–D. The visual system. Visual areas (*V1–V8*; **A–C**) and visual streams (**D**). The boundaries of visual areas that have been identified in the human brain are indicated in lateral (**A**) and medial (**C**) views of the occipital and temporal lobes. These boundaries correspond to the projections of the vertical and horizontal meridians of the contralateral retina (**B**). The retinotopical organization of each successive visual hemifield is the mirror image of its neighbour's. The representation of the horizontal meridian of V1 (*broken lines*) corresponds to the bottom of the calcarine fissure, which separates the projections of the upper and lower quadrants of the contralateral retina in its lower and upper banks. It continues on the convexity of the hemisphere as the border between upper and lower quadrant representations of V2 and V3. Of V4, its ventral division V4v is illustrated, V4d is part of the ill-defined lateral occipital area (LO), located between V3 and the middle temporal visual area (MT). V1 and V2, and V3 with V3a and V4v are bordered by the common representation of the vertical meridian (*open circles*). The border of V2 with V3 and of V4v with V8 corresponds to the projection of the horizontal meridian (*broken lines*). V6 is located in the apex of the cuneus and in the wall of the parieto-occipital sulcus. (**D**) The dorsal visual stream and its visual areas is shown in *red*, the ventral visual stream in *shades of grey*. A dorsal component of the dorsal stream, focussing on the medial bank of the intraparietal sulcus and a ventral component, focussing on its lateral bank and the inferior parietal lobule can be distinguished [233]. *AIP*, anterior intraparietal area; *FFA*, fusiform face area; *inf.par.lob*, inferior parietal lobule; *LIP*, lateral intraparietal area; *LO*, lateral occipital area; *MST*, medial superior temporal visual area; *MT*, middle temporal visual area; *MIP*, medial intraparietal area; *PEip*, intraparietal part of area PE; *STP*, superior temporal polysensory area; *sup.par.lob*, superior parietal lobule; *TE*, inferior temporal area; *V1–V6A*, visual areas V1–V6A; *VIP*, ventral intraparietal area

layers traverse layer 4 [178]. Large pyramidal cells (of Meynert) are present at the border of layer 5/6. V1 is bordered by the second visual area V2. V2 corresponds to Brodmann's area 18. At the area 17/18 border Gennari's line ends and the sublayers of V1 layer 4 coalesce into a single internal granular layer (Fig. 19.7).

In V1 layer 2/3 populations of neurons expressing cytochrome oxidase (CO) are segregated in clusters or "blobs", with CO-negative cells residing in the "interblobs" (Fig. 19.7). The blobs continue, as indistinct columns, through the deeper layers of the striate cortex. Series of blobs occupy the centres of the ocular dominance columns. Sublayers 4A and 4C, which receive direct projections from the lateral geniculate body, display strong CO activity [37]. A specific localization of CO-immunoreactive cells in alternating thin and thick stripes, separated by nonimmunoreactive interstripes, is also present in V2. These stripes are orientated perpendicular to the V1/V2 border (Fig. 19.7).

Brodmann's areas 18 and 19 are difficult to separate. Most higher-order visual areas in area 19 and beyond, therefore, have been delineated according to non-cytoarchitectonic criteria. The presence of a visuotopic organization in many of these visual areas has proved to be a valuable tool to recognize their boundaries. This criterion could also be applied in humans, using functional magnetic resonance imaging [283, 286]. The visual areas illustrated in Fig. 19.6 A–C were identified in the human brain; many more visual areas have been distinguished in nonhuman primates.

The contiguous V1, V2 and V3 visual areas contain mirror-symmetrical visuotopic maps of the contralateral hemifield. As a consequence, the vertical meridian forms the border between the representations of the visual field in V1 and V2. The horizontal meridian separates areas V2 and V3 (Fig. 19.6). V2 and V3, like V1, consist of ventral and dorsal portions, which contain paired representations of the contralateral, upper and lower quadrants. Interestingly, some of the borders of visual areas also could be established in the human brain by tracing callosal afferents from large lesions of the opposite occipital cortex. This is possible because the termination of callosal fibres in V1–V4 shows a strong preference for the representations of the vertical meridian, i.e. for the V1/V2 and the outer V3 borders [43].

Visual areas beyond V3 have been arranged in a dorsal stream, directed at the parietal lobe and a ventral stream directed temporally. The areas V3a, V6, MT and MST and the areas of the intraparietal sulcus belong to the dorsal stream. The accessory area V3a is located dorsally, near the parieto-occipital sulcus. It shares the representation of the vertical meridian with dorsal area V3. It contains a complete representation of the contralateral hemifield, bisected by the horizontal meridian [281]. V6 is located at the dorsal apex of the cuneus and in the posterior occipito-parietal sulcus. In monkeys another area, V6A is located on the anterior bank of the parieto-occipital fissure and V6 occupies its bottom [76, 223]. The middle temporal area (MT or area V5) is located in the ascending limb of the inferior temporal sulcus. MT is characterized by its heavy myelinization and, therefore, can be localized in the human brain [43]. The medial superior temporal area (MST) is located anterior to MT. Multiple visual areas have been mapped in the intraparietal and inferior parietal cortex of monkeys. Their localization in the human brain is still tentative [52].

Ventral stream areas V4v and V4d [281], the ventral and dorsal divisions of area V4, probably are independent cortical areas. V4d is included in the lateral occipital area (LO) located between V3 and MT on the convexity of the occipital lobe [187, 284]. V4v shares its vertical meridian with ventral V3. Area V8 is located in the collateral sulcus and the lingual (medial occipitotemporal) gyrus, rostral to V4d, with which it shares its horizontal meridian. V8 contains a complete representation of the contralateral visual field [100].

Visual areas have been arrranged in different functional streams by tracing their fibre connections and ordering their electrophysiological and psychophysical properties in hierarchies of increasing complexity and specificity. The complexity of the cortico-cortical connec-

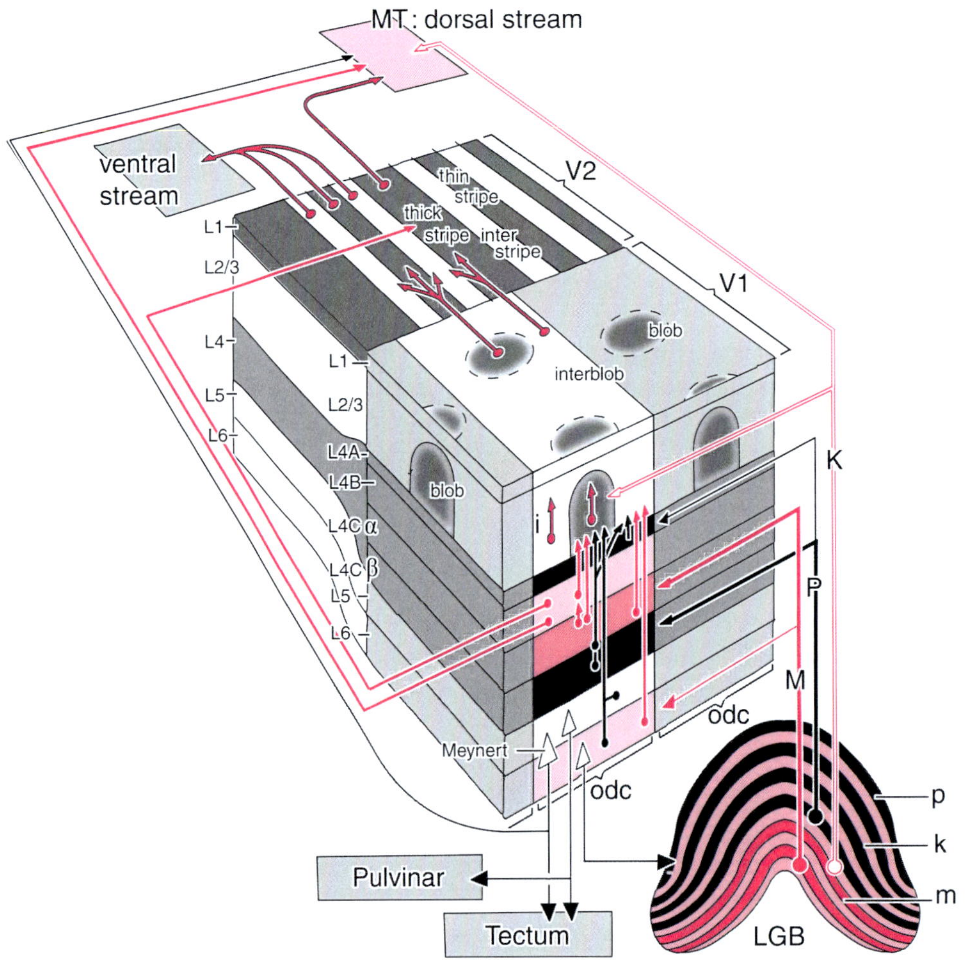

Fig. 19.7. The dorsal and ventral visual streams. Both streams take their origin from visual areas *V1* and *V2*. Magnocellular (*M, m, red*), parvocellular (*P, p, black*) and koniocellular (*K, k, red open contour*) pathways take their origin from the corresponding layers of the lateral geniculate body (*LGB*). Intrinsic connections are shown for one ocular dominance column (*odc*). The main direct and indirect projections of the M pathway are located in the V1 layers 4Cα and 4B, respectively. The M pathway can be traced from layer 4B into the thick stripes of V2 and the middle temporal visual area (*MT*) of the dorsal visual stream. The three pathways intermingle in the blobs and interblobs of V1 layer 2/3. Intrinsic connections (*i*) of this layer, and its projections into the ventral and dorsal visual streams via the thin and thick stripes and the interstripes of V2 (*filled arrows*), therefore, carry signals from all three types of layers of the lateral geniculate body. Projections from Meynert pyramidal cells at the layer 5/6 border, and of layer 6 pyramidal cells, are indicated in the lower part of the figure. Both Meynert cells and the koniocellular pathway project to area MT. *L1–L6*, cortical layers 1–6

tions between visual areas makes them less useful for this purpose. Of the theoretical number of possible connections between the 32 visual areas distinguished by Felleman and Van Essen [68] in the macaque monkey, 31% have been verified, and more must be present. Distinguishing feed-forward and feed-backward connections by their origin and termination from different layers does not help much because there are many exceptions to this rule [24].

Visual Areas: Thalamocortical Projections of the Lateral Geniculate Body

Parallel processing in the visual cortex begins with the differential projections of the lateral geniculate body to the primary visual cortex (Fig. 19.7). The projection of the magnocellular layers is focussed on sublayer 4Cβ, with a minor contribution to layer 6. Parvocellular layers innervate 4Cα and 4A [126, 178]. Projections to these cortical laminae from layers receiving their retinal afferents from the contralateral or the ipsilateral eye terminate in alternating strips and are responsible for the organization of V1 in ocular dominance columns. The K pathway, originating from the koniocellular layers, projects to the cytochrome oxidase (CO)-immunoreactive blobs in layer 2/3 and to layer 1 [69, 158]. Layers K1, K4 and K6 innervate blobs located in the contralateral ocular dominance columns; layers K2 and K3 innervate blobs in the ipsilateral columns [115].

The segregation in the projection of different populations of ganglion cells in different layers of the lateral geniculate body and in the projection of these layers to the cortex makes it possible to distinguish different functional streams in the processing of visual information. The parvocellular (P) stream, named after the corresponding layers of the lateral geniculate body, is a slow system, monopolized by the input of the midget ganglion cells. This system carries both chromatic and achromatic information. Chromatic information is provided by the red/green colour opponency of the M and L midget ganglion cells and allows red-green discrimination in colour vision. The L and M channels are used synergistically for achromatic contrast detection, i.e. for distinguishing light from dark as used in object vision [93]. The magnocellular (M) stream is a fast system that receives its retinal input from parasol cells. These cells are also involved in achromatic contrast detection and may perceive motion.

The koniocellular stream is driven by the small blue/yellow-opponent bistratified ganglion cells. These cells are totally committed to colour, and carry no achromatic contrast information [93]. The koniocellular stream is not exclusively concerned with colour vision. Projections of other ganglion cell types with widely spanning dendrites prevail in certain of the koniocellular layers. Moreover, the koniocellular stream exists in primates with a rod-dominated retina and only one type of cone, which lacks colour vision, such as the owl monkey and the bushbaby [18, 213, 324]. Finally, cells that share the immunocytochemical characteristics of the koniocellular population, and their projection to the CO blobs in layer 2/3, are present in the parvocellular layers of the lateral geniculate body as displaced koniocellular elements [115].

Gouras [93] suggested that chromatic and achromatic information carried by the midget ganglion cells is split at the level of the primary visual cortex. Chromatic information on red/green discrimination would be carried by the displaced koniocellular elements in the parvocellular layers. The koniocellular projection to the blobs in layer 2/3, therefore, would include information on the entire colour spectrum. Information on achromatic contrast would be fed into other channels, i.e. into the parvocellular projection to layers 4Cβ and 4A.

As a useful simplification, the P stream in the projection of the lateral geniculate body to V1 and beyond may be considered to represent object form and the K stream object colour. Information from the M stream is used to generate the physiological properties of orientation specificity, monocular fusion and detection of motion, but how this is accomplished is still far from clear.

Visual Areas: Functional Streams. Intrinsic and Interareal Connections of V1 and V2

Figure 19.7 illustrates the intrinsic and interareal connections of V1 and V2 [70, 158, 159, 178, 325, 328]. The magnocellular pathway (shown in red in Fig. 19.7) can be followed from sublayer 4Cα into layer 4B. Layer 3 blobs receive magnocellular input directly from 4Cα, and indirectly via 4B. Magnocellular input reaches layer 3 interblobs from stellate cells in the bottom of 4Cα and from layer 6. The parvocellular input (shown in black in Fig. 19.7) is distributed from sublayer 4Cβ to the blobs and interblobs of layer 3 and to the interblobs via layer 4A, which also receives a direct projection from the parvocellular layers of the lateral geniculate body.

The interareal projection of layer 4B pyramidal cells to V2, V6 and area MT [261] can be considered as an almost pure prolongation of the magnocellular pathway. Magnocellular and parvocellular streams become mixed at the level of the interblobs of lamina 3; and all three of the M, P and K streams innervate the blobs in this layer. These intrinsic projections to the blobs and the interblobs terminate deep in layer 3 (sometimes indicated as sublayer 3B). Interareal connections of the blobs and interblobs with area V2 take their origin from a more superficially located population of pyramidal cells of layers 2 and 3B.

Blobs were found to project to CO-immunoreactive thin stripes and the CO-negative interstripes of V2; interblobs connect with the CO-immunoreactive thick stripes and the interstripes of this visual area [260, 262, 321]. The thick stripes, moreover, receive an input from layer 4B. V2 thick stripes next project to visual area MT, which also receives M input from V1 layer 4B. Moreover, the Meynert pyramidal cells of the layer 5/6 border project to MT, in addition to the superior colliculus [74, 75]. MT is a key area in the dorsal, occipitoparietal processing stream. Thin stripes and interstripes connect with visual areas belonging to the inferior temporal ventral processing stream, to be considered in more detail in the following paragraphs.

The 4Cα – 4B – V2 thick stripes – MT pathway is clearly dominated by magnocellular input, with some admixture of the parvocellular stream at the level of the interblobs. Cells in 4Cα were found to be orientation-selective. Cells in layer 4B are both orientation- and direction-selective, and direction selectivity is maintained in MT and subsequent stages in the dorsal processing stream. The pathways converging upon the ventral, occipitotemporal processing stream, which include the V1 blobs and the V2 thin stripes, rely on input from all layers of the lateral geniculate body and, therefore, are more difficult to characterize. Recent studies using desoxyglucose uptake by pools of active neurons may have solved this problem.

In the original studies of Livingstone and Hubel [172, 173], blobs were found to contain higher-order, red/green or blue/yellow colour-opponent cells. Cells in the interblobs were orientation-selective and arranged in long, linear sequences (the orientation columns), which were walled off from the blobs. This preferential location of colour and orientation-selective cells has been contested [171, 232, 324]. However, it was confirmed in desoxyglucose studies of the orientation columns [298] and of colour processing in the macaque visual cortex [285].

A repeating system of orientation columns, containing neurons preferring the same orientation of a bar-shaped stimulus presented to the eye, was first described in the primary visual area [127]. These orientation columns intersect with the ocular dominance columns at right angles. Cells labelled in the desoxyglucose uptake studies on the presentation of two orthogonal stimuli were located in two sets of maximally separated orientation columns. In V1 these columns were most distinct in sublayer 4Cα, absent from 4Cβ and quite robust in the more superficial layers, where they avoided the blobs. In V2 they were centred on the CO-immunoreactive thick stripes extending into the interstripes, but absent from the thin stripes [298]. On the other hand, desoxyglucose uptake on the presentation of coloured stimuli was stimulated in the blobs and, in V2, was concentrated in the thin CO-immunoreactive stripes [285].

The preferential localization of the orientation columns in the interblobs and the V2 thick stripes only holds for the periphery of the visual field. In the V1 and V2 representations of the fovea, the orientation columns extend into the blobs, and the preference for the V2 thick stripes decreases [298]. The processing of chromatic information, which depends on input from parvocellular and koniocellular layers of the lateral geniculate body, therefore appears to be segregated in the V1 blobs and V2 thin stripes and is continued in the ventral processing stream. Similarly, orientation selectivity, which depends on the magnocellular input, is segregated in the V1 interblobs and the V2 thick stripes in the periphery of the visual field and is transmitted to the dorsal stream, which depends on the periphery rather than on central regions of the visual field (see next section on Visual Areas: Dorsal and Ventral Processing Streams). Information used in the perception of object shape by the ventral processing stream may depend on input from all layers of the lateral geniculate. This information may be transmitted by the blobs in the central representation of the visual field, where orientation-selective cells are present within these structures. A function of the blobs in the processing of colour, contrast and orientation would explain their presence in primates without colour vision.

Visual Areas: Dorsal and Ventral Processing Streams

The observations discussed in the previous paragraphs give some idea how visual information in V1 and V2 is distributed over the two major processing streams, each including multiple visual areas (Fig. 19.6 D). The dorsal, occipitoparietal stream carries information on spatial relations to the parietal lobe. The ventral, occipitotemporal stream brings information on object properties to the inferotemporal cortex. This concept, as formulated by Ungerleider and Mishkin [290], originally was based on differential effects of parietal and inferior temporal lesions. It received support from anatomical and electrophysiological evidence and imaging studies in humans. Generally, receptive field size and latency increase in subsequent stages of a stream, and a retinotopic localization makes place for a representation, independent of position in the visual field. Areas V4, V8 and inferior temporal area TE are included in the ventral stream (indicated in grey in Fig. 19.6 D). Areas V3a (accessory V3 area), V6, V6A MT, MST (medial superior temporal area), STP (superior temporal polysensory area) and a series of visual areas in the intraparietal sulcus and the inferior parietal lobule belong to the dorsal stream (indicated in red in Fig. 19.6 D). This list is not exhaustive; other areas not mentioned in this chapter have been distinguished in either stream [292].

The dorsal and the ventral stream also are distinguished by their preference for peripheral or central vision, and their involvement in scotopic or photopic vision, respectively. Injections of retrograde tracers in the parietal target areas of the dorsal stream in monkeys resulted in retrograde labelling of cells in V6, MT, V3a, MST and STP and, predominantly, in peripheral field representations of V2, V3 and V4. Injections of the ventral stream in inferior temporal areas labelled cells prevailing in the foveal representations of these areas, but mainly in V4 [3]. The consequences of this distribution were discussed in the previous paragraph. Scotopic, rod-dependent vision activates visual areas in the dorsal stream, but does not affect the foveal representations of V1, V2 and V3 and the colour-sensitive area V8 [99].

As noted earlier, the V1 and V2 input of MT, V3a and V6, the key areas in the dorsal stream, is dominated by the magnocellular layers of the geniculate body. MT, V3a and V6 receive their input from V1 layer 4B via the strongly myelinated fibres of the line of Gennari [286]. Retinotopic localization in MT is not well developed. Its cells have small receptive fields and are directionally and orientation selective. Illusory movements also activate its cells. Cells with a similar direction of motion preference are segregated in columns, alternating with columns with an opposite direction preference [1]. Estimations of target speed, used in smooth pursuit of objects, are made from pooled populations of many neurons [226, 284]. MT is connected with visuomotor areas in the lateral and medial banks of the intraparietal fissure [77, 241, 256]. Area V3a is involved in the distinction of real motion from self motion. Motion and direction-selective cells also occur in area V3a. Apart from linking the dorsal visual stream to visuomotor areas in the dorsal bank of the intraparietal sulcus, little is known about the function of area V6.

MT is connected with the more rostrally located area MST and area V6. MST lacks a retinotopic organization, and receptive fields of its cells are large. Its cells are sensitive to disparity, or the expansion or contraction of an image. The area is thought to be involved in the perception of depth, optic flow and, like MT, has been implicated in smooth pursuit [133, 179, 291]. Cells with preferences for contraction or expansion of an image are clustered in alternating columns [80]. In monkeys, area MTS is connected with the superior temporal polysensory area (STP) in the dorsal bank of the superior temporal sulcus. This area responds to visual, auditory and somatosensory stimuli, and interacts with inferotemporal area TE, the final stage in the ventral visual stream, and the inferior parietal lobule, one of the targets of the dorsal stream [233]. Different visuomotor areas, innervated by the dorsal visual stream, have been recognized in the parieto-occipital sulcus in monkeys and, tentatively, in humans (Figs. 19.6 D, 21.6 A) [46, 47, 52]. These areas function as a visual-motor interface. They maintain strong connections with the superior colliculus and the pontine nuclei and are connected through several reciprocal parallel pathways with premotor and eye fields in the frontal lobe [180]. Cells in the lateral intraparietal area (LIP or the "parietal eye field") are modulated by eye position. LIP, in the lateral bank of the intraparietal sulcus, encodes attended positions relative to the fovea and guides eye movements. The anterior intraparietal area AIP plays an essential role in sensory transformation for grasping. The medial intraparietal area MIP, located in the medial bank, encodes location relative to the arm and is involved in reaching. Area V6A in the caudal intraparietal sulcus and the ventral intraparietal area (VIP), located in the bottom of the intraparietal sulcus, are head-centred areas encoding stimulus position relative to the head and are concerned with movements towards the mouth and head.

Nomenclature and connections are also complex in the ventral visual processing stream (Fig. 19.6 D). Area V4 was often considered as the gate keeper of the ventral stream. Originally it was subdivided into dorsal (V4d) and ventral (V4v) parts, with representations of the ventral and dorsal quadrants of the contralateral hemifield, respectively. A high percentage of its cells were found to be colour-selective in monkeys and humans [330]. However, in subsequent fMRI studies dorsal V4 could not be distinguished [253]. Instead, an area activated to a similar degree by scanning objects, faces, common objects and abstract 3-D objects was localized in the same lateral occipital (LO) region, between V3 and MT [187, 284]. Retinotopy in LO was crude or nonexistent, and colour-selective activity was found to be absent [282]. A colour-selective area called V8, with a complete representation of the contralateral hemifield, was located rostral to V4v, in the collateral sulcus [100]. Higher-order object areas, selective for specific objects such as faces or buildings, were found lateral to the collateral sulcus in the fusiform (lateral occipitotemporal) gyrus or in more rostral and medial inferior temporal regions. Best known is the fusiform face area in the posterior fusiform gyrus, activated during matching or recognition of faces, and the locus of lesions causing prosopagnosia, the inability to recognize faces in human patients [111, 112]. It was emphasized that recognition of faces, like numbers and words, is associated with a central visual field bias, whereas the recognition of buildings, in a more medial region bordering on the parahippocampal gyrus, is dependent on peripheral vision [186]. Attributes like colour and form apparently are carried by different subdivisions of the ventral stream, to be used for semantic purposes in higher-order areas, which may be dependent on the central or peripheral visual field for their information. Representation of objects in the ventral temporal cortex is not restricted to small selective patches, but rather should be considered as a distributed system extending from LO to rostral TE [138, 139, 289]. Functions in this region may be lateralized: responses on perception of objects are stronger on the right side, responses to imagination of the same objects are stronger on the left [138, 139]. Reciprocal connections of the rostral inferior temporal areas with the inferior orbital and medial prefrontal cortex are a prominent feature of the ventral visual stream [245].

The original assumption of Ungerleider and Mishkin [290] that the dorsal and the ventral stream are distinguished by the contents of the information they carry, spatial information by the dorsal stream, object properties by the ventral stream, has been criticized by Goodale and Milner [91, 92] and Rizzolatti and Matelli [233]. According to these authors the main feature distinguishing the two streams is the use of this information carried by them rather than its contents. Parietal areas use visual information in the control of actions, the inferotemporal cortex uses it in perception of objects. This difference is well illustrated by one of Goodale's patients, with a lesion interrupting the ventral stream in the lateral occipital area [144]. This patient was unable to recognize drawings of objects presented to her, but retained her visuomotor skills in grasping similar objects, with concomitant activation of the expected regions in the dorsal stream in fMRI.

The dorsal stream was further subdivided by Rizzolatti and Matelli into a dorsal component, formed by area V6, V6a and visuomotor area MIP, and focussing on the superior parietal lobule, and a ventral component formed by area MT, LIP AIP, VIP and other visuomotor areas of the inferior parietal lobule (Fig. 19.6D). The dorsal component provides spatial information to the superior parietal lobule used for direct action, information that is used in the visuomotor transformations by area AIP for grasping objects. The ventral component, in addition, receives information on object properties from the ventral stream, through its connections with STP, used by inferior parietal area PF for action imitation by the premotor cortex. The localization of the parietal syndromes of optic ataxia, a disorder of visually guided arm movements towards a goal, in the superior parietal lobule, and of ideomotor apraxia, the inability to replicate a gesture following a verbal command, in the inferior parietal lobule, is in accordance with this dorsoventral subdivision.

The projection of the lateral geniculate body to the striate cortex is thought to be essential for conscious vision. However, pathways to higher visual areas, which exclude either the striate cortex or the lateral geniculate body, exist and are supposed to function in "blindsight," i.e. unconscious perception of movement and in conscious perception of movement in patients with damage to the striate area. These pathways include:

1. Projections of the superficial layers of the superior colliculus, which receive a substantial contribution from the optic tract, to the pulvinar. The pulvinar innervates the extrastriatal visual areas in the occipital, temporal and parietal lobes (Fig. 19.4).
2. Projections of the superficial layers of the superior colliculus to area MT, via the koniocellular layers of the lateral geniculate body (Figs. 19.4, 19.7) [263].

Visuomotor Pathways

Visuomotor pathways include reflex paths for accommodation, vergence and stabilization of the eye and pathways involved in voluntary saccades and smooth pursuit of moving objects. Two main systems driving conjugate eye movements exist. One is the three-neuron reflex arc of the vestibulo-ocular reflex (VOR), connecting each semicircular canal, through the vestibular nuclei, with the muscle pairs that move the eye in the plane of this semicircular canal. This system was developed in lower vertebrates to stabilize the position of the eye in space and remained present, scarcely altered, in the mammalian brain. It was discussed and illustrated (Fig. 17.7) in Chap. 17. Pathways used by the VOR are shared by the optokinetic reflex, another reflex that stabilizes the eye and, presumably, by smooth pursuit movements, although the latter involve different areas of the vestibular nuclei than those subtending the basic VOR. The VOR and the optokinetic reflex receive their information about "retinal slip" from the accessory optic system and the nucleus of the optic tract. Smooth pursuit involves the processing of information on direction and speed of an object to be traced by visual cortical areas and the concomitant suppression of the VOR. The second system generates saccades. It is controlled by the superior colliculus, by several visual cortical areas and by the cerebellum.

The Saccade Generating System

Saccades are rapid eye movements that redirect the line of sight to an area of interest, ensuring that the target falls on the very small fovea. The brain stem burst generator for saccadic eye movements was reviewed by Scudder et al. [250] and Büttner and Büttner-Ennever [32]. Its structure is summarized in Fig. 19.8.

Motor neurons innervating the external eye muscles generate a burst-tonic discharge. The burst serves to move the eye, the tonic discharge keeps it in position. The burst is generated by short-lead burst cells, which establish excitatory connections with motoneurons. For horizontal eye movements these cells are located in the medial part of the nucleus reticularis pontis caudalis, in the so-called pontine paramedian reticular formation (PPRF), rostral to the abducens nucleus. Excitatory burst cells (EBC's) are glutaminergic. Inhibitory burst cells (IBC's) are located in the medial part of the medullary paragiganto reticular nucleus, caudal to the abducens nucleus. IBC's use glycine as their neurotransmitter [269]. EBC's project to the ipsilateral abducens nucleus; the projection of the IBS's is crossed. The projection of the IBC's also may include the bilateral oculomotor and trochlear nerve nuclei [36]. The abducens nerve innervates the lateral rectus muscle. The medial rectus muscle is innervated by the oculomotor nerve. Its motoneurons receive their excitatory input through an internuclear pathway, originating from interneurons in the abducens nucleus (Figs. 17.8, 19.8).

Short lead burst neurons for motoneurons of the oculomotor and trochlear nerves lie in the rostral interstitial nucleus of the medial longitudinal fascicle (riMLF). The riMLF is a wing-shaped nucleus located dorsal to the rostral pole of the red nucleus and rostral to the interstitial nucleus of Cajal. In monkeys and humans the cells of the riMLF can be recognized by their high content of parvalbumin [121]. They share this property with the excitatory and inhibitory burst neurons in the PPRF [122]. The projection of the riMLF to the motoneurons of the oculomotor and trochlear nerve nuclei is mainly ipsilateral. In the monkey this projection is excitatory [268, 307, 308]. Recently, inhibitory neurons with projections to the contralateral inferior rectus and inferior oblique motor pools have been located to the interstitial nucleus of Cajal [123, 205]. The riMLF projects to the ipsilateral interstitial nucleus. Clinicopathological evidence indicates that Parinaud's syndrome (vertical gaze paralysis) results from damage to the riMLF or its efferent fibres [222].

The tonic discharge that maintains the position of the eye after the completion of the saccade is generated by two neural integrators (a position signal is derived by integration of the velocity signal): the nucleus prepositus hypoglossi for horizontal eye movements, and the interstitial nucleus of Cajal for vertical and rotatory eye movements. The projections of the nucleus prepositus to the abducens nucleus are bilateral; the ipsilateral projection is excitatory, the contralateral projection is crossed and, again, uses glycine as its neurotransmitter [259]. Both the nucleus prepositus hypoglossi and Cajal's interstitial nucleus are dependent for their function on reciprocal connections with the vestibular nuclei and, for the nucleus prepositus with the cerebellum (Figs. 17.3, 17.4, 19.17; see also Chap. 17). Coarse fibres originating from the interstitial nucleus des-

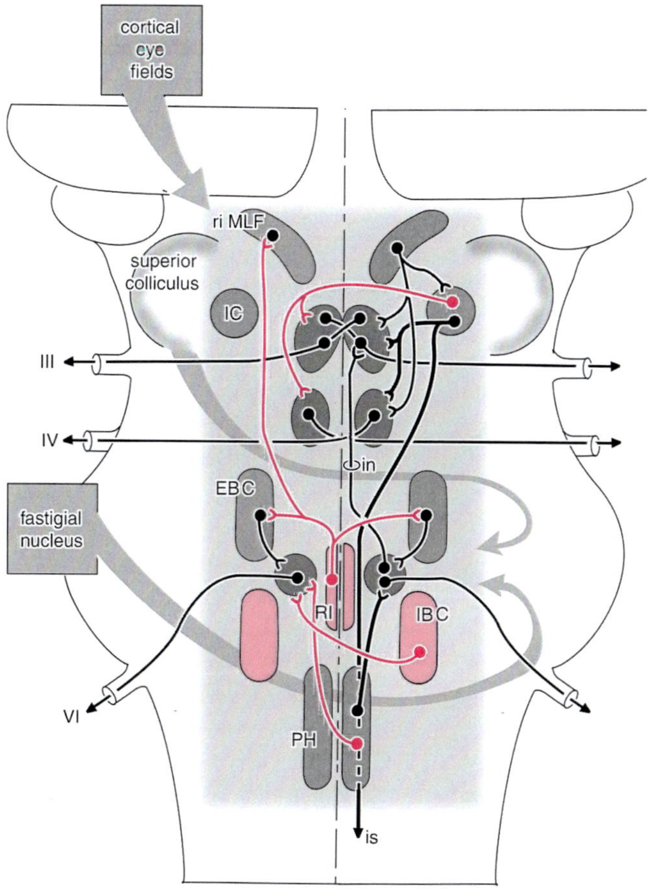

Fig. 19.8. The saccade-generating system. Horizontal saccades are generated in the horizontal gaze centre, located in the pontine and medullary reticular formation; the vertical gaze centre is located in the rostral mesencephalon. The horizontal gaze centre consists of groups of excitatory (*EBC*) and inhibitory (*IBC*) burst cells with connections to the abducens nucleus. Gaze holding is achieved by integration of the velocity signal issued by the excitatory burst cells and by neurons of the nucleus prepositus hypoglossi (*PH*). The excitatory and inhibitory burst cells for vertical and rotatory saccades both are located in the rostral interstitial nucleus of the medial longitudinal fascicle (*riMLF*) and the interstitial nucleus of Cajal (*IC*), respectively. The excitatory connections of the EBN's and the riMLF with the oculomotor nuclei are ipsilateral pathways; the inhibitory projection of the IBNs and the IC are crossed. The neural integrator for these eye movements is located in the IC. Omnipause neurons in the pontine nucleus raphe interpositus (*RI*) inhibit the excitatory burst cells at both locations. The cortical eye fields, the superior colliculus and the cerebellar fastigial nucleus modulate the saccade-generating system. Inhibitory connections are indicated in *red*. *is*, interstitiospinal tract; *III, IV, VI*, oculomotor (III), trochlear (IV) and abducens (VI) nerves

cend as the interstitospinal tract in the medial longitudinal fascicle.

Omnipause neurons, located in the paired pontine nucleus raphe interpositus, are glycinergic elements that fire when the eye is fixating and pause before and during a saccade, inhibiting the EBC's in the PPRF and the riMLF [28, 124]. Afferents of the omnipause neurons include the superior colliculus and the frontal and supplementary eye fields.

The saccade-generating system is used by the superior colliculus and the frontal eye fields to generate goal-directed saccades. Projections from the contralateral fastigial nucleus mediate saccade size by determining the acceleration and the deceleration of saccades and their long term adaptation (Fig. 19.8).

Superior Colliculus

The superior colliculus plays a major role in orienting behaviour, directing the line of sight towards objects of interest by movements of the eyes, head and body (Figs. 19.9–19.12). A retinotopic map is present in superficial layers; auditory, somatosensory and visual signals initiating orienting movements converge in the deeper layers upon premotor neurons with connections to saccade and head movement networks. The anatomy of the superior colliculus was recently reviewed by May [192]. The superior colliculus contains alternating grey and white layers (Figs. 6.31, 6.32, and 19.9). From superficial to deep, these layers are:

1. The mainly fibrous stratum zonale
2. The stratum griseum superficiale
3. The stratum opticum
4. The stratum griseum medium
5. The stratum album medium or stratum lemnisci
6. The stratum griseum profundum
7. The stratum album profundum

On the basis of connections and functions, the superior colliculus can be partitioned into a superficial zone and a deep zone. The superficial zone, which consists of layers 1–3, receives primarily visual afferents and gives rise to ascending fibres. The deep zone, which includes layers 4–7, receives heterogeneous multimodal inputs and sends ascending and descending fibres to centres in the brain stem, the thalamus and the spinal cord. Dendritic arbors of neurons in superficial layers generally are restricted to these layers. Large multipolar neurons in deeper layers extend their dendrites into the superficial layers. The connections between the superficial and deep layers are formed by axon collaterals of small L neurons of the superficial layers (see Efferent Connections of the Superior Colliculus).

Terminations of retinotectal fibres in superficial layers, and several afferent systems terminating in intermediate layers of the colliculus, display a mediolateral periodicity. Cholinergic markers (choline acetyltransferase, CHAT and acetylcholinesterase, AChE) are evenly distributed in the neuropil of the superficial layers. Some of this activity is of intrinsic origin, but it is mainly related to the bilateral, cholinergic projection of the parabigeminal nucleus to these layers [101]. In the intermediate grey layer, CHAT and AChE are distributed in patches [95, 96] related to cholinergic afferents originating from the pedunculopontine nucleus (Fig. 19.9). This patchy distribution stands for a similar distribution of several substances in this layer, such as enkephalin [97], and for discontinuities in the distribution of most afferent systems and the origin of some efferent systems of the intermediate layers of the colliculus. This periodicity has been observed in most mammals, including humans [95, 305].

Afferent Connections of the Superficial Layers of the Superior Colliculus

Afferents of the superficial layers of the colliculus originate from both retinas and the ipsilateral cerebral cortex. The retinal fibres pass from the optic tract through the brachium of the superior colliculus (Figs. 19.3, 19.4) and the superficial stratum opticum to terminate in the stratum griseum superficiale and zonale. The retinocollicular projection in primates is bilateral with a contralateral dominance.

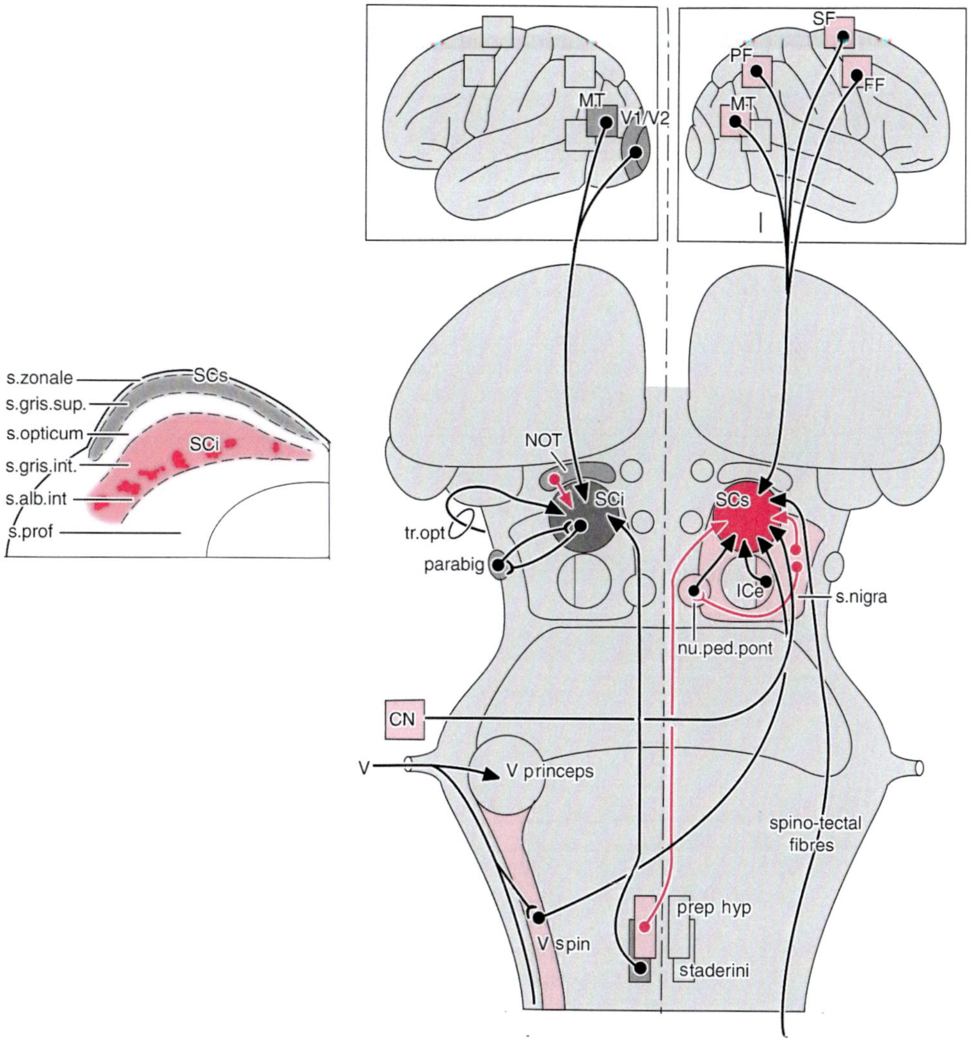

Fig. 19.9. The afferent connections of the superior colliculus. Layers of the superior colliculus are indicated in the small figure on the *left*. Acetylcholinesterase in the intermediate layers (*red*) is distributed in longitudinal columns, which appear as irregular patches in this transverse section, in the superficial layers (*grey*) it is more evenly distributed. The cholinergic projection to the intermediate layers originates from the nucleus pedunculopontinus, and for the superficial layers from the parabigeminal nucleus. In the diagram of the brain stem the afferent connections of the superficial layers of the colliculus are indicated on the *left*, afferent connections of the intermediate layers on the *right*. Inhibitory connections are indicated in *red*. *CN*, cerebellar nuclei; *FF*, frontal visual area; *Ice*, external nucleus of the inferior colliculus; *MT*, middle temporal visual area; *nu.ped.pont*, nucleus pedunculopontinus; *NOT*, nucleus of the optic tract; *parabig*, parabigeminal nucleus; *PF*, parietal visual area; *LIP*, lateral intraparietal area; *prep.hyp*, nucleus prepositus hypoglossi; *s.alb.int*, intermediate white layer; *s.gris.int*, intermediate grey layer; *s.gris.sup*, superficial grey layer; *s.nigra*, substantia nigra, pars reticulata; *s.prof*, deep layers of the superior colliculus; *s.zonale*, stratum zonale; *Sci*, intermediate layers of the superior colliculus; *SCs*, superficial layers of the superior colliculus; *SF*, supplementary visual area; *staderini*, nucleus intercalatus of Staderini; *tr.opt*, optic tract; *V princeps*, principal sensory nucleus of the trigeminal nerve; *V spin*, spinal nucleus of the trigeminal nerve; *V*, trigeminal nerve; *V1/V2*, visual areas V1/V2

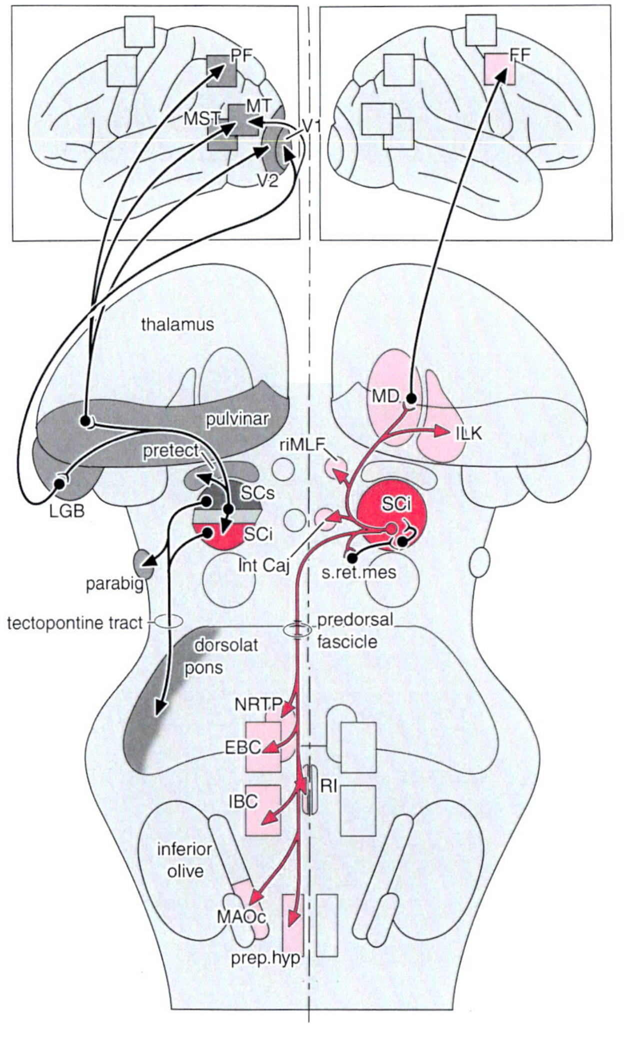

Fig. 19.10. The efferent connections of the superior colliculus. Ascending connections originating from the superficial layers of the superior colliculus are shown in *black* on the *left*. Collaterals of these neurons descend into intermediate layers of the colliculus. The tectopontine tract takes its origin both from superficial and deep layers. Efferent connections of intermediate layers of the colliculus, contained in the descending limb of the predorsal fascicle and its uncrossed, ventrally ascending collateral projection, are shown in *red* on the right. Note termination of dorsally ascending fibres from the superficial layers in the pulvinar, and the projection of this nucleus to post-rolandic visual areas. The ventrally ascending collateral projection of the intermediate layers terminates in the mediodorsal nucleus of the thalamus, which projects to the frontal eye field. The mesencephalic reticular formation provides feedback from the predorsal output back to the colliculus. *EBC*, excitatory burst cells; *FF*, frontal eye field; *IBC*, inhibitory burst cells; *ILK*, intralaminar thalamic nuclei; *Int Caj*, interstitial nucleus of Cajal; *LGB*, lateral geniculate body; *MAOc*, caudal part of medial accessory olive; *MD*, mediodorsal thalamic nucleus; *MST*, medial superior temporal visual area; *MT*, middle temporal visual area; *NRTP*, nucleus reticularis tegmenti pontis; *parabig*, parabigeminal nucleus; *PF*, parietal visual area; *prep.hyp*, nucleus prepositus hypoglossi; *pretect*, pretectum; *RI*, nucleus raphe interpositus; *riMLF*, rostral interstitial nucleus of the medial longitudinal fascicle; *s.ret.mes*, reticular formation of the mesencephalon; *SCi*, intermediate layers of the superiorcolliculus; *SCs*, superficial layers of the superior colliculus

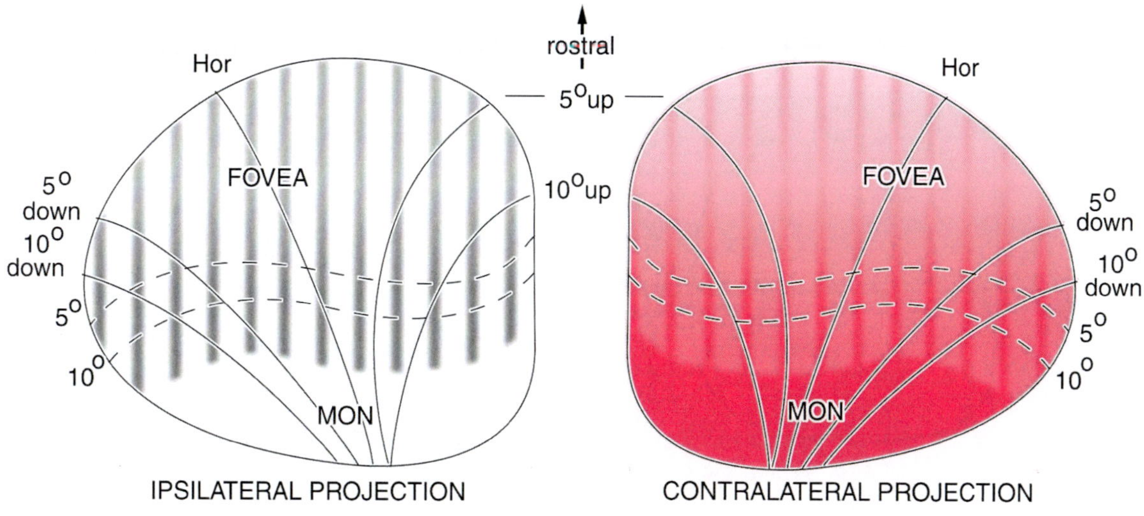

Fig. 19.11. Projections of the left hemifield on to the ipsilateral (*grey*) and contralateral (*red*) superior colliculus in the monkey. The contralateral projection predominates and includes the monocular visual field (MON). The projections of the fovea and the peripheral retina of both sides interdigitate. Visual field coordinates were taken from Cynader and Berman [53]. *Hor*, horizontal meridian

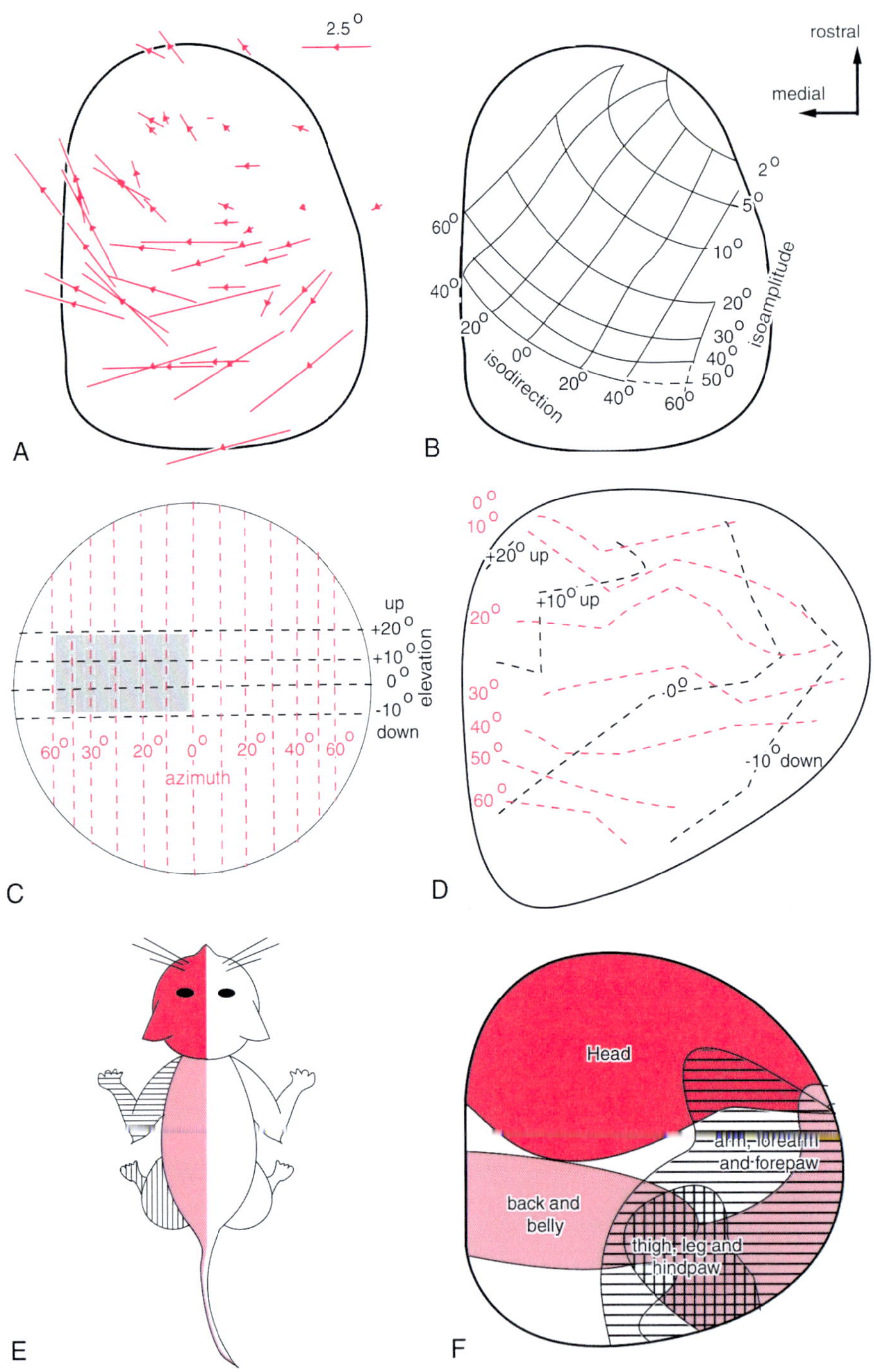

Fig. 19.12 A–F. Maps of the superior colliculus. **A** Saccadic motor map of the superior colliculus. *Arrows* show orientation and amplitude of saccades generated from the intermediate layers of the monkey superior colliculus. Redrawn from Sparks [266]. **B** Plot of the isodirection and isoamplitude lines of saccades illustrated in **A**. Redrawn from Sparks [266]. **C** Map of the auditory field. The area of the contralateral auditory field depicted in **D** is *shaded*. **D** Map of the contralateral auditory field in the intermediate layers of the superior colliculus of the cat. Note corresponding orientation of coordinates of the auditory and visual fields (Fig. 19.11). Redrawn with permission from Middlebrooks and Knudsen [198]. **E, F** Somatosensory map of the contralateral body in the intermediate layers of the superior colliculus of the cat. Redrawn with permission from Meredith et al. [196]

Retinal fibres distribute in a retinotopical pattern. The map of the contralateral hemifield is oriented somewhat obliquely. The vertical meridian coincides with the rostral border of the superior colliculus; the horizontal meridian bisects the colliculus in lateral and medial halves (Fig. 19.11). The fovea is represented rostrolaterally, peripheral regions caudally, the upper visual field medially and the lower field laterally. The monocular field is represented only as a contralateral projection to the caudal colliculus. The projections to the representation of the fovea are relatively light. The fovea is expanded and contains gaps in the contralateral projection, which are filled in by patches on the ipsilateral side. In the more posteromedial representation of the periphery, the input from the contralateral eye becomes more continuous, but gaps filled by the ipsilateral projections remain present in deeper portions of the superficial grey [53, 128, 224, 279, 318] (Fig. 19.11). This distribution is reminiscent of the ocular dominance columns in the primary visual cortex (Fig. 19.5), but is less orderly.

Diverse types of ganglion cells, accounting for less than 10% of the total population, contribute to the retinotectal projection. Only a small minority of the parasol cells shares projections with the lateral geniculate body; midget ganglion cells do not project to the superior colliculus [170, 219].

Corticotectal fibres take their origin from layer V pyramidal cells, with the exception of the primary visual area, where they also originate from Meynert's pyramidal cells at layer V/VI border (Fig. 19.7) [38, 74, 75]. A large proportion, if not all, of the corticotectal fibres, at least in the cat, are branches of neurons that also project to the pontine nuclei [5, 150]. There are distinct differences in their laminar terminations [16, 104, 174, 176, 185, 279]. Most projections are derived from visual and visuomotor areas. In primates, contributions of motor, somatosensory, auditory and limbic areas and large portions of the frontal lobe are sparse or absent [48, 72, 174, 185]. In Galago, higher-order somatosensory areas S2 and PV in the parietal operculum were found to project to dense patches in the intermediate grey. Primary somatosensory areas do not project to the colliculus [319]. Areas 17 (V1) and 18 (V2) project to the stratum zonale and griseum superficiale and constitute the major corticotectal input (Fig. 19.9). The V2 corticotectal projection is derived from band-like clusters of layer V pyramidal cells that are aligned with the direction-selective, cytochrome oxidase-thick stripes in the superficial layers, i.e. from a direction- and orientation-specific region of V2 [2]. Retinotopy is maintained in the projections of V1 and V2. Area 19 (which includes several visual areas) projects to a deeper portion of the stratum griseum superficiale. Visual area MT, located in the ascending limb of the inferior temporal sulcus, projects both to superficial and intermediate layers. Neighbouring area MST does not project to the superior colliculus. Tectal projections from the inferior temporal visual areas are disputed (Fig. 19.9) [174, 271]. Parietal and frontal visual areas project to superficial, intermediate and deep layers of the superior colliculus. The frontal eye field innervates the AchE-positive patches in this region, which contain the premotor neurons issuing the commands for saccades in the contralateral visual field [157, 163]. Parietal afferents terminate more deeply and arise from the lateral intraparietal visual area (LIP) in the lateral bank of the intraparietal sulcus, an area implicated in the control of saccades [181]. The connections of these visuomotor areas will be discussed in the section of this chapter on The Frontal and Parieto-Temporal Eye Fields.

Subcortical afferents to superficial layers of the colliculus include the pretectum, the parabigeminal nucleus, the nucleus of the optic tract, the nucleus intercalatus and the substantia nigra [304].

Afferent Connections of the Intermediate and Deep Layers of the Superior Colliculus

Intermediate and deep layers receive somatosensory, auditory, striatal and cerebellar afferents. Several of these systems terminate in patches in the intermediate grey layer. Cerebellar afferents will be discussed together with other cerebellar connections of the visual system in the section of this chapter on The Visual System and the Cerebellum.

Somatosensory afferents are derived from all levels of the spinal cord, the dorsal column nuclei as well as the nucleus interpolaris of the spinal trigeminal nucleus. All these projections are crossed and terminate in patches in the intermediate grey layer. The trigeminal projection is restricted to the rostral portion of the superior colliculus [316]. The rostral position of the trigeminal afferents is in accordance with the somatosensory map of the contralateral body in the intermediate layers of the superior colliculus (Fig. 19.12 E, F). Head and arm occupy most of the space in the rostral and caudolateral colliculus, respectively. The smaller representation of the leg overlaps with the arm. Receptive fields of neurons in these layers are large [196].

Auditory projections originate from the dorsal nucleus of the periolivary nuclei, lateral lemniscus, the external nucleus of the inferior colliculus and the nucleus of the brachium of the inferior colliculus [61, 306] (see Chap. 18). A map of the contralateral auditory field in the intermediate layers of the superior colliculus has the same orientation as the map of the visual field (compare Figs. 19.11, 19.12 B, C). The expansion of the central auditory field is less prominent than the expansion of the fovea in the visual field map, but this is probably due to the fact that the auditory map is based on experiments in the cat. Receptive fields of auditory-responsive neurons are large; an auditory stimulus may activate the entire population of auditory-responsive neurons, with a peak response in a more restricted region [198].

Neurons of the rostrolateral pars reticulata of the substantia nigra provide the superior colliculus with a major GABAergic projection [6, 132, 280, 300]. The main projection is to the ipsilateral intermediate grey stratum, where it distributes in longitudinal bands, which overlap with the projection of the pedunculopontine nucleus, the frontal eye fields and the AChE-positive patches and alternate with somatosensory projections. Within these patches nigral afferents contact premotor cells [103, 107, 146]. It should be noted that the pars reticulata both innervates the pedunculopontine nucleus, which is the major source of the cholinergic projection to the AChE-positive patches in the intermediate grey layer, and the patches themselves [101]. The basal ganglia control saccadic eye movements by disinhibition of the sustained inhibitory input from the substantia nigra, thus selecting appropriate input from the frontal eye fields [117]. A similar, discontinuous GABAergic projection to the intermediate grey in the rostral superior colliculus stems from the nucleus prepositus hypoglossi [49, 116, 194] and the entopeduncular nucleus in the rat [275]. Another source of an inhibitory projection to the intermediate layers of the colliculus is the zona incerta [151]. Cerebellar afferents, some of which are also compartmentalized in the intermediate layers, will be discussed in the section on The Visual System and the Cerebellum.

Efferent Connections of the Superior Colliculus

Different types of projection neurons have been distinguished in the superior colliculus on the basis of their laminar distribution, their participation in one or more efferent pathways and their functional characteristics. Small neurons located in superficial layers give rise to a dorsal ascending pathway. In primates these neurons have been designated as the L neurons [202]. Collaterals of L neurons terminate in deeper layers of the superior colliculus, and constitute the main connection between the superficial and deep layers of the colliculus. The dorsal ascending pathway passes through and terminates in the pretectum. It proceeds towards the ventral geniculate nucleus, the inferior and lateral pulvinar and the lateral geniculate body, where it terminates in the intercalated (koniocellular) layers (Fig. 19.10) [105, 106].

Axons of neurons participating in the predorsal bundle decussate in the dorsal tegmental decussation (Figs. 6.30, 19.10) and descend ventrolaterally to the medial longitudinal fascicle, lateral to the central superior nucleus. In primates, few, if any, predorsal fibres reach the spinal cord; in other mammalian species the system terminates in the cervical cord as the tectospinal tract. Predorsal neurons are a heterogeneous population, mainly residing in the intermediate layers of the superior colliculus. In cats, rats and primates, axons destined for the predorsal bundle have been shown to emit an ascending collateral before their decussation [12, 94, 229]. These ascending collaterals form the ventral ascending bundle, which terminates in the interstitial nucleus of Cajal, the riMLF, the mesencephalic reticular formation lateral to the interstitial nucleus, the supraoculomotor central grey, the nucleus of Edinger-Westphal, the zona incerta, several thalamic intralaminar and midline nuclei, and in the paralaminar portion of the dorsomedial thalamic nucleus (Fig. 19.10) [94, 105, 106, 120].

The predorsal bundle terminates in the nucleus reticularis pontis oralis and caudalis and the paragigantocellular nucleus, regions that contain the excitatory and inhibitory short lead burst cells of the horizontal saccadic system, located rostral and caudal to the abducens nucleus. Terminations are present in two precerebellar nuclei, the dorsomedial nucleus reticularis tegmenti pontis and the caudal medial accessory olive. In addition, projections to the nucleus prepositus hypoglossi and the nucleus raphe interpositus have been observed [94, 102, 105, 120, 149]. A projection to the olive was absent in studies using individually injected predorsal axons [94, 251]. Tecto-olivary fibres, therefore, may originate from a separate population of tectal neurons.

In lower mammals tectospinal neurons project to the intermediate zone of the cervical spinal cord; direct projections to motoneurons have not been observed. Bilateral connections with motoneurons innervating neck muscles are effectuated by interneurons. Other disynaptic connections of the superior colliculus with spinal motoneurons use Cajal's interstitial nucleus or the medullary reticular formation, both of which receive collaterals from the predorsal fascicle, and project to the spinal cord via the interstitiospinal and medial reticulospinal tracts [255].

An uncrossed descending pathway leaves the colliculus laterally, coursing superficially along the parabigeminal nucleus. It terminates in this nucleus, in the lateral pontine grey and in the pontine reticular formation. It is known as the tectopontine tract, first described by Münzer and Wiener [207] (Fig. 19.10). It takes its origin from a population of small neurons in all layers of the colliculus, including L neurons [12, 201, 202, 206, 229].

Three types of projection neurons were distinguished in the primate tectum: the superficial L cells and the X and T cells [202]. For the X and T cells, functional profiles were described [203]. T cells are small- to medium-sized neurons located in the stratum opticum and the intermediate grey that participate in the ipsilateral and crossed descending and/or the dorsal ascending bundles. They correspond to the vectorial long-lead burst neurons that burst 20 ms before a saccade in their movement field. X cells are large, multipolar predorsal bundle cells, preferentially located in the intermediate grey. Their axons emit commissural and ventrally ascending collaterals. The activity in X cells is not related to spontaneous saccades. This categorization in primates does not correspond with observations made in retrograde, double-labelling studies in the rat and data on the cat [12, 94, 229]. Predorsal bundle cells, in these studies, always are located in the intermediate grey, emit ipsilaterally ascending collaterals that terminate in the intralaminar nuclei and do not participate in the tectal commissure. A more widespread population of neurons, projecting to the ipsilateral tegmentum, does not participate in the tectal commissure either. The conclusion that, in primates, the long-lead burst neurons (T cells) form a separate population to be distinguished from the predorsal cells with their ipsilaterally ascending collaterals and crossed descending axons, distributing to the contralateral saccade-generating system, therefore, should be viewed with some circumspection.

Another group of cells, located in the rostral colliculus at the level of the representation of the fovea, are tectoreticular cells, which are tonically active during fixation of a (remembered) target [98]. They may correspond to predorsal bundle cells projecting to omnipause neurons in the nucleus raphe interpositus, which inhibit the excitatory short-lead burst neurons, and are preferentially located in the rostral colliculus superior [31].

Long-lead burst neurons in the superior colliculus with spike burst tightly coupled to the onset of a saccade have been located in the intermediate grey layer [183, 203, 266, 267]. These cells are arranged topographically in a map of motor (saccadic) space that specifies the direction and amplitude of a saccade, preceded by the discharge of a cell in a particular location (Fig. 19.12 A,B). Cells discharging before contralaterally directed small saccades are located rostrally, before large saccades caudally, before saccades with upwards components medially and with downwards components laterally in the colliculus. The localization of increased metabolic activity during saccades, with the desoxyglucose method, showed that this map remains stationary [204].

According to the foveation hypothesis, redirecting the line of sight towards objects of interest is achieved by excitation spreading from a target defined in the superficial retinotopical map to the underlying pre-saccadic cells. Since the superficial retinotopic map and the deep saccadic maps are aligned, this would lead to an appropriate saccade, bringing the target on the fovea. This simple scheme has been elaborated on the basis of many experimental studies of this system.

1. The connections between the superficial and intermediate layers of the colliculus, which contain the retinotopic and saccadic maps, have remained enigmatic for a long time. However, the descending collaterals of the L cells offer a likely morphological substrate for this connection (Fig. 19.10) [202].

2. Saccades to a particular target, specified by the premotor neurons of the motor map, may be interrupted by the appearance of another target in another position. Correction for the direction and amplitude of the intended saccade could be achieved by the interposition of another neuron in the pathway from the superficial layers to the premotor neurons. These visual-responsive neurons receive feedback about the intended saccade from cells in the mesencephalic reticular formation, innervated by collaterals from the premotor cells (Fig. 19.10). They compute direction and amplitude of the corrected saccade from information on the new target and the metrics of the initial saccade [203]. Visual-responsive cells have not yet been anatomically verified.

3. In models of saccade generation by the superior colliculus, long-lead burst neurons are interposed between the premotor cells of the superior colliculus and the short-lead burst neurons of the gaze centres [250]. A few of these cells have been located in the pontine reticular formation [251, 252].

4. Saccades can be initiated by visual, auditory and somatosensory stimuli. Visual targets can be located in the superficial retinotopic map of the contralateral visual field, but a crude visual map is also present in intermediate layers of the colliculus [195]. Auditory and somatosensory maps in the intermediate layers are based on systematic variations in the receptive fields of neurons, presumably the premotor neurons, of this layer. The map of the contralateral auditory field localizes potential targets for a saccade in head-centred coordinates, i.e. their localization is independent from the position of the eyes in orbit. Amplitude and direction of saccades in the saccadic motor map, however, are coded in eye-centred coordinates, i.e. the initial position of the eye in the orbit is taken into account. The calculation of the correct direction and amplitude of a saccade to an auditory target, therefore, should reckon with the initial position of the eye. This is achieved by a shift in the location of the auditory map, relative to the stationary saccadic map, coupled to changes of the position of the eyes in orbit. Systematic changes in auditory receptive field organization in premotor neurons, accompanying changes in position of the eyes in orbit, have been observed in awake animals [145, 152, 218]. Similar transformations may occur in the tectal somatosensory map. The responsible circuits for these transformations are not known.

5. One distinguishing feature of the intermediate layers of the superior colliculus, usual-

ly disregarded by the map makers for this region, is the discontinuous projection of most afferent systems in most species and their relationship to the distribution of AChE and other substances in this layer. In early reviews of this topic [103, 129] it was suggested that nigral, pedunculopontine and corticotectal afferents from the frontal eye field contact predorsal bundle cells in the AChE-positive patches, whereas somatosensory afferents terminate in intermediate regions. In rodents, predorsal bundle cells are distinctly clustered. In primates, a relationship between the localization of long-lead burst cells and the AChE patches could not be found [183]. The significance of the parcellation of the intermediate layers, therefore, remains unknown.

6. The superior colliculus also is supposed to initiate orientating movements of head and body. Generally the tectospinal pathways and/or projections to reticular areas with spinal projections are held responsible [255]. The latter include paramedian pontine neurons in the nucleus pontis oralis and caudalis and the nucleus reticularis gigantocellularis, with their crossed ponto- and bulbospinal and medial bulbospinal tracts, and regions in the mesencephalic reticular formation [26, 50, 51, 236, 255]. Probably these systems display the same amount of complexity as the vestibulospinal pathways, which mediate the spinal labyrinthine reflexes, discussed in Chap. 17.

Accessory Optic System and the Nucleus of the Optic Tract

The accessory optic system (AOS) consists of the dorsal, lateral and medial terminal nuclei located in a side branch of the optic tract, in the periphery of the rostral mesencephalon. These nuclei register speed and direction of global movements of the visual surround. This information is relayed to the cerebellum by the inferior olive and is used in the long-term adaptation of the VOR. Low speed preference of AOS neurons, along with their large receptive fields, suggests that they are suited to complement the vestibular system in detecting self-motion.

The nucleus of the optic tract, located in the pretectum, also receives afferents from the optic tract and projects to the inferior olive. It is not a terminal nucleus of the accessory optic system, but lies adjacent to the dorsal terminal nucleus. The dorsal terminal nucleus and the nucleus of the optic tract are difficult to separate. The nucleus of the optic tract is an important station in the subcortical pathway mediating the horizontal optokinetic reflex. This reflex is especially effective at low speeds, and thus complements the VOR in stabilizing the image on the retina. The anatomy of the AOS was recently reviewed by Giolli et al. [86].

The primate AOS receives optic tract fibres that detach from their collicular branch in the brachium of the superior colliculus (Fig. 19.13). These fibres turn ventrally, along the surface of the mesencephalon, and form the superior fasciculus of the AOS, which crosses the cerebral peduncle as the transpeduncular tract of Von Gudden [301]. This tract terminates in the medial terminal nucleus of the AOS, located immediately medial to the peduncle. The dorsal and the lateral terminal nucleus lie embedded among the fibres of the superior fasciculus. The inferior fasciculus of the AOS, which detaches from the optic tract close to the chiasm to terminate in the medial terminal nucleus, is rudimentary in primates. The AOS is present in humans.

The terminal nuclei of the AOS and the nucleus of the optic tract are bilaterally interconnected by, mainly, GABAergic pathways. Commissural projections decussate in the posterior commissure [88, 294]. The medial and lateral terminal nuclei of the AOS are connected with a cell group, located in the ventral tegmentum, dorsal to the medial terminal nucleus, via a pathway crossing in the posterior commissure [13, 85, 87]. The connections of this area, known as the *ventral tegmental reflex zone*, are very similar to those of the medial terminal nucleus.

The projections of the retina to the nucleus of the optic tract and the terminal nuclei of the AOS are mainly crossed and originate from direction-selective ganglion cells, discussed in the section on the Retina of this chapter (Fig. 19.1 F).

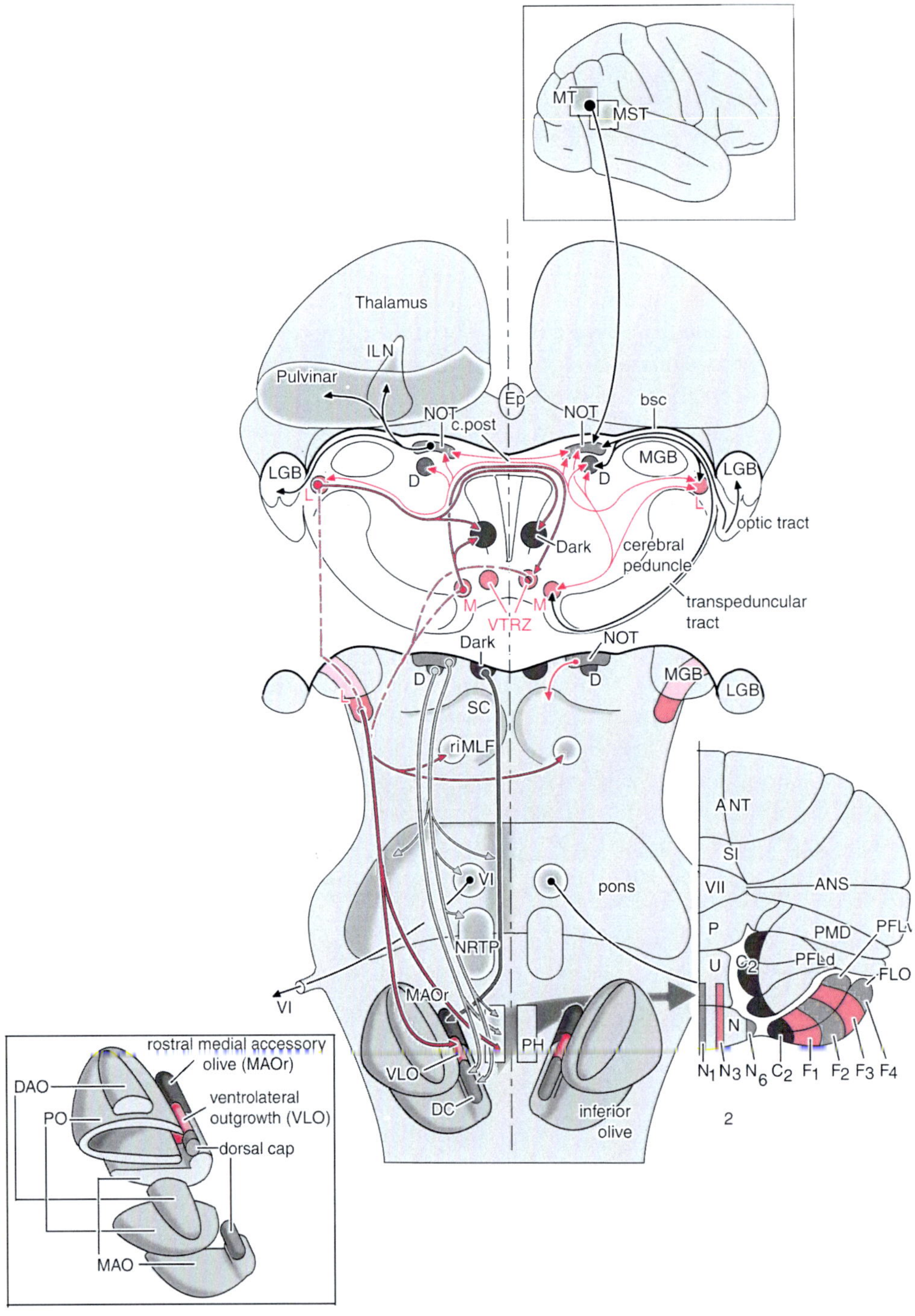

Fig. 19.13. The accessory optic system and the nucleus of the optic tract. The rostral brain stem is transected at the level of this system. The cerebellar target-zones of the accessory optic system are illustrated in a diagram of the flattened cerebellum. The inset shows the subnuclei of the inferior olive. The superior fascicle of the accessory optic system detaches from the optic tract where it runs in the brachium of the superior colliculus. It continues as the transpeduncular tract on the surface of the cerebral peduncle. It projects to the dorsal (*D*), lateral (*L*) and medial (*M*) terminal nuclei. The nucleus of the optic tract (NOT) is located dorsal to the terminal nucleus in the pretectum. The terminal nuclei and the NOT are interconnected by inhibitory pathways (*red*). Commissural connections decussate in the posterior commissure. The terminal nuclei and the NOT give rise to descending pathways, terminating in the inferior olive. The lateral and medial terminal nuclei, with the

Neurons of the medial and lateral terminal nuclei detect near-vertical movements across the retina of the contralateral eye, preferentially in the plane of one of the vertical semicircular canals [258, 259]. Similar, but binocular, neurons are found in the ventral tegmental relay zone. These nuclei project to precerebellar, accessory optic, vestibular and reticular nuclei [13, 85, 87]. Their main precerebellar target is a small subnucleus of the inferior olive: the ipsilateral ventrolateral outgrowth (VLO). The VLO gives rise to a climbing fibre projection to the contralateral nodulus, flocculus and ventral (accessory) paraflocculus. Other projections of the medial terminal nucleus are located in the contralateral dorsolateral pontine nuclei; the lateral terminal nucleus innervates the same structure on the ipsilateral side (not illustrated). A projection to the lateral pole of the nucleus reticularis tegmenti pontis has only been observed for the lateral terminal nucleus. Most optokinetic centres project to the medial and dorsomedial aspects of this nucleus, that also receives many other visuomotor afferents.

The accessory optic nuclei project to the interstitial nucleus of Cajal and to the nuclei of Darkschewitsch and Bechterew. Some of these projections branch off from commissural systems in the posterior commissure. Projections to the vestibular nuclei are moderate or weak. All three the terminal nuclei project to the nucleus prepositus [29, 30].

The dorsal terminal nucleus of the AOS and the nucleus of the optic tract both register horizontal movements of the visual surround. Both nuclei project to the dorsal cap of the inferior olive. However, the two nuclei should be distinguished because the response properties of neurons of the optic tract nucleus differ from those of the dorsal terminal nucleus [78]. Moreover, the efferent connections of the optic tract nucleus are believed to be more extensive than those of the AOS terminal nuclei [29, 120, 209]. Ascending projections to the magnocellular layers of the lateral geniculate body, intralaminar nuclei, pulvinar and zona incerta and Darkschewitsch nucleus have been reported. Descending pathways to the superficial layers of the superior colliculus are GABAergic. Other projections target, the ipsilateral nucleus prepositus hypoglossi, the abducens nucleus and the caudally adjoining reticular formation and the same precerebellar nuclei that received projections from the AOS terminal nuclei, i.e. the ipsilateral, dorsolateral and medial pontine nuclei and the dorsal cap of the inferior olive. A termination in the medial nucleus reticularis tegmenti pontis was present in primates [29, 209] and rats [199, 287], but not in rabbits [120]. Through the dorsal cap of the inferior olive, the nucleus of the optic tract and the dorsal terminal nucleus of the AOS project to the cerebellar nodulus, flocculus and ventral paraflocculus (Fig. 19.13).

ventral tegmental reflex zone (VTRZ) as an intermediary, project to the ventrolateral outgrowth of the inferior olive (VLO). This subnucleus projects to a central zone in the cerebellar nodulus (N_3) and to paired zones (F_1 and F_4) in the flocculus and the ventral paraflocculus. This system is shown in *red*. The dorsal terminal nucleus and the NOT project to the dorsal cap of the inferior olive (DC), which provides climbing fibres to paired zones in the nodulus and the flocculus (N_1 and N_6, F_2 and F_4). This system is shown in *grey*. The lateral and medial terminal nuclei and the NOT (not indicated) also project to the C_2 zone in the flocculus and the paraflocculus, via Darkschewitsch (*dark*) nucleus and the rostral medial accessory olive (MAOr). The inhibitory projection of the NOT to the superior colliculus is indicated in *red*. *ANS*, ansiform lobule; *ANT*, anterior lobe; *Bsc*, brachium of the superior colliculus; *c.post*, posterior commissure; *D*, dorsal terminal nucleus; *DAO*, dorsal accessory olive; *Dark*, nucleus of Darkschewitsch; *DC*, dorsal cap; *Ep*, epiphysis; F_{1-4}, Climbing fibre zones 1–4 of the flocculus and the ventral paraflocculus; *FLO*, flocculus; *ILN*, intralaminar thalamic nuclei; *L*, lateral terminal nucleus; *LGB*, lateral geniculate body; *M*, medial terminal nucleus; *MAOr*, rostral medial accessory olive; *MGB*, medial geniculate body; *MST*, medial superior temporal visual area; *MT*, middle temporal visual area; N_{1-3}, climbing fibre zones 1–3 of the nodulus; *NOT*, nucleus of the optic tract; *NRTP*, nucleus reticularis tegmenti pontis; *P*, pyramid; *PO*, principal olive; *PFLd*, dorsal paraflocculus; *PFLv*, ventral paraflocculus; *PH*, nucleus prepositus hypoglossi; *PMD*, paramedian lobule; *riMLF*, rostral interstitial nucleus of the medial longitudinal fascicle; *SC*, superior colliculus; *SI*, simplex lobule; *U*, uvula; *VII*, lobule VII: visual vermis; *VLO*, ventrolateral outgrowth; *VTRZ*, ventral tegmental reflex zone

In primates, cortical regions with projections to the nucleus of the optic tract include the areas MT and MST in the inferior temporal sulcus [119, 209], the frontal and supplementary eye fields [257] and the caudal banks of the superior temporal sulcus [176]. The latter region also connects with the terminal nuclei of the AOS. A projection from the primary visual cortex [118] appears unlikely.

The optokinetic reflex presumably uses the vestibulo-oculomotor pathway as its final link (Fig. 17.8). However, it is not immediately clear how the nucleus of the optic tract gains access to the vestibular nuclei. Some direct connections have been reported, but the nucleus prepositus hypoglossi and the pontine and olivary subnuclei targeted by this nucleus have been suggested as possible links [29]. The role of the cerebellum will be discussed in the section on The Visual System and the Cerebellum of this chapter.

The Frontal and Parietal Eye Fields

Three eye fields, which give rise to eye movements when stimulated, have been distinguished in the frontal lobe, and one has been distinguished in the parietal lobe (Fig. 19.14). Their properties and connections were reviewed by Lynch and Tian [180]. In humans the frontal eye field is located in the precentral sulcus, in the transitional of the areas 6 and 8, extending into area 8. The supplementary eye field occupies a more dorsomedial position in the interhemispheric fissure. A prefrontal eye field was distinguished in area 46 [177, 180, 239, 240]. In human patients, lesions including the frontal eye field produce paralysis of conjugate gaze to the opposite side, with deviation of eyes and head to the side of the lesion. Spontaneous recovery usually occurs in a few days. Experimental inactivation of the frontal eye field profoundly impairs saccadic eye movements and smooth pursuit and causes a contralateral visual neglect, which resolves within days or weeks. Lesion studies of the supplementary eye field have reported only mild oculomotor impairment. The influence of the prefrontal eye field on voluntary eye movements appears to be inhibitory [180].

The parietal eye field is located in the lateral bank of the intraparietal sulcus. It is known as the lateral intraparietal area (LIP). Presaccadic activity is observed in the parietal eye field. It encodes attended locations relative to the fovea and guides eye movements [46, 47, 52]. The intraparietal sulcus contains several subregions with different connections and functional properties (see section on Visual Areas: Functional Streams). Damage to this region causes visuospatial neglect, constructional apraxia, gaze apraxia and disorders of spatial cognition.

The three frontal eye fields, the parietal and the occipitotemporal visual fields MT and MST, are heavily interconnected. The specificity of these connections is illustrated by the observation that the parietal eye field (LIP) is interconnected with frontal eye fields, but the neighbouring areas MIP and VIP are not (Fig. 19.14 A) [45]. Apart from its reciprocal connections with the frontal eye fields, the parietal eye field is connected with the visual areas V2, V3, V4, MT and MST. It should be noted that most of these connections are reciprocal. Moreover, most of the visual areas that project to the parietal eye field also project directly to the frontal eye field [180]. Separate subdivisions of the frontal eye field subserve saccades and smooth pursuit, and activity in the other frontal and parietal fields is associated with both types of eye movements [221]. In the frontal eye field, neural signals related to saccades are located more dorsally than those related to smooth pursuit (Fig. 19.14).

The frontal eye fields are connected with striatum, claustrum, thalamus and centres such as the superior colliculus, the vertical and horizontal gaze centres, the nucleus prepositus hypoglossi and the precerebellar nucleus reticularis tegmenti pontis and the pontine nuclei (Fig. 19.14). Thalamic connections of the parietal eye fields, like those of areas MT and MST, are with the inferior, lateral and medial pulvinar [89]. The frontal and supplementary eye fields connect with the mediodorsal, ventral anterior and intralaminar nuclei. The supple-

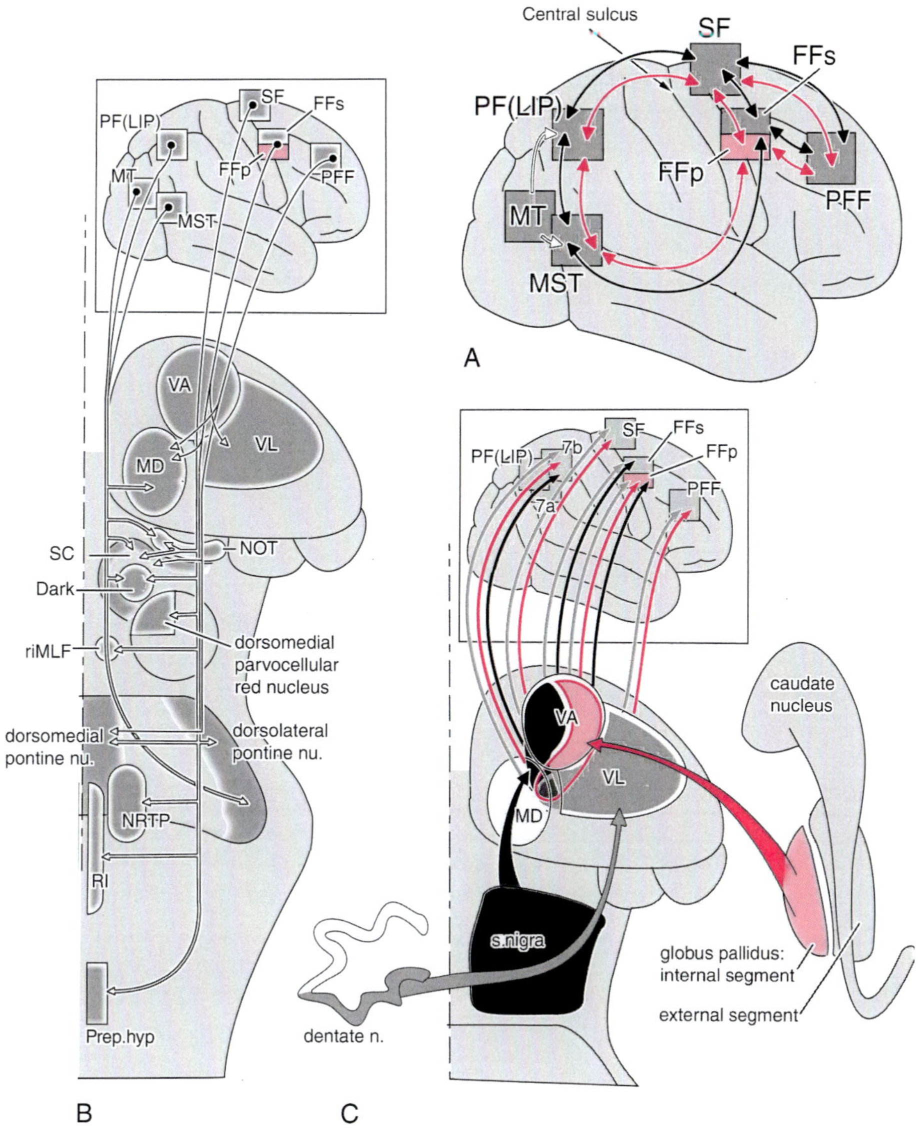

Fig. 19.14 A–C. The cortical eye fields. **A** The cortical eye fields are interconnected by association fibres. Each eye field contains separate subdivisions for saccades and smooth pursuit. These subfields have been localized in the frontal eye field (saccades: FFs, pursuit: FFp). Interconnections for pursuit are shown in *red*, for saccades in *black*. The parietal eye field corresponds to the lateral intraparietal area (LIP). It belongs to a series of intraparietal areas, illustrated in Figs. 19.6 and 21.6. Redrawn from Lynch and Tian [180]. **B** Efferent connections of the cortical eye fields. **C** Afferent connections of the cortical eye fields from the substantia nigra (*black*), the dentate nucleus (*red*) and the internal segment of the globus pallidus (*grey*). The projections of these grisea to the thalamus remain largely segregated in the mediodorsal, ventrolateral and ventral anterior nuclei of the thalamus, but they overlap or interdigitate in the junctional region of these nuclei. This region gives rise to differential projections to the different visual areas. The parietal eye field does not receive afferents from the cerebellum or the striatum, but these systems do innervate neighbouring parietal areas (based on data from Lynch and Tian [180]. *7b*, medial intraparietal area MIP; *dark*, nucleus of Darkschewitsch; *FFp*, pursuit subfield of frontal visual area; *FFs*, saccade subfield of frontal visual area; *MD*, mediodorsal thalamic nucleus; *MST*, medial superior temporal visual area; *MT*, middle temporal visual area; *NOT*, nucleus of the optic tract; *NRTP*, nucleus reticularis tegmenti pontis; *PF*, parietal visual field; *PFF*, prefrontal visual field; *Prep.hyp*, nucleus prepositus hypoglossi; *RI*, nucleus raphes interpositus; *riMLF*, rostral interstitial nucleus of the medial longitudinal fascicle; *SC*, superior colliculus; *SF*, supplementary visual field; *s.nigra*, substantia nigra; *VA*, Ventral anterior thalamic nucleus; *VL*, ventrolateral thalamic nucleus

mentary frontal eye field receives more projections from the medial portion of the ventral lateral nucleus and its medially adjoining subnucleus X. The saccade and smooth pursuit regions of the frontal eye field receive nonoverlapping thalamocortical projections [130, 131, 162, 270, 278]. These and other studies allow a prediction of the putative cerebellar, striatal and nigral input to the frontal and parietal eye fields. It was concluded that both the saccade and the pursuit regions of the frontal eye field receive input from the substantia nigra and the cerebellar nuclei [182]. The pursuit region and the supplementary eye field, in addition, are the target of the inner segment of the globus pallidus. In the supplementary eye field this is combined with a cerebellar input [278]. The parietal eye field lacks afferent connections from the cerebellum or the striatum. Neighbouring area MIP (7b) does receive thalamic projections mediating cerebellar nuclear, pallidal and nigral input to this lobule (Fig. 19.14C) [44, 45].

Subcortical connections of the eye fields include the superior colliculus and other saccadic premotor centres and several precerebellar nuclei. Parietal afferents from LIP terminate in intermediate layers of the superior colliculus; area MT projects both to superficial and deeper layers; area MST lacks a tectal projection. The frontal and supplementary eye fields project to the stratum zonale; their termination in deeper layers has been variously reported to occupy patches in the intermediate grey layer or to be restricted to the stratum opticum.

Of the gaze centres, the riMLF is innervated by both subdivisions of the frontal eye field and the supplementary eye field. The innervation of the PPRF is weak and does not target the saccade generator. Omnipause neurons in the nucleus raphe interpositus receive a projection from the supplementary eye field either alone [130] or in combination with the frontal eye field [270]. The nucleus of the optic tract and the nucleus prepositus hypoglossi have been mentioned occasionally as targets for the frontal eye fields; the subthalamic nucleus and the zona incerta are targets shared by all frontal eye fields.

Targets of the visual eye fields among the precerebellar nuclei include the pontine nuclei and the nucleus reticularis tegmenti pontis, as well as the nucleus of Darkschewitsch and the dorsomedial parvocellular red nucleus, both of which project to the inferior olive. The primate nucleus of Darkschewitch receives a projection from the frontal eye field and posterior parietal area 7 [67]. Both frontal and supplementary eye fields and parietal area 5 [25] project to the dorsomedial parvocellular red nucleus.

Corticopontine fibres from the frontal eye field have been reported to predominate in the dorsolateral pontine nucleus [162, 257] or in the medial pons [131], where they overlap with a projection from the supplementary eye field [130, 257]. In the lateral and dorsolateral pontine nuclei, projections of the frontal eye field may overlap with projections from parietal areas 5, 7a, 7b and LIP, areas MT and MST and the peripheral representations of V1, V2, V3 and V3a [4, 16, 20, 73, 247–249], the pretectum [312], the medial and lateral terminal nuclei and the ventral tegmental relay zone of the AOS, the nucleus of the optic tract, and, caudally at least, with the projection of the uncrossed tectopontine tract (Figs. 19.10, 19.14, 19.17). Both frontal eye fields project, rather diffusely, over the entire rostrocaudal extent of the nucleus reticularis tegmenti pontis. Convergence is possible with the crossed tectopontine projection of the predorsal fascicle (Figs. 19.10, 19.14). The nucleus reticularis tegmenti pontis does not receive a projection from postrolandic visual areas.

Several eye fields are reciprocally connected with the striatum, through the globus pallidus or the substantia nigra, and with the superior colliculus and the cerebellum. Connections of the striatum are illustrated in Fig. 19.14 and discussed in Chap. 14; cerebellar pathways are reviewed in Chap. 20 and in the section on The Visual System and the Cerebellum of this chapter. Reciprocal pathways from the superior colliculus to the frontal eye field and the parietal eye field were found to differ (Fig. 19.10). Premotor cells, located in the intermediate layers of the superior colliculus, receive a direct input from the frontal eye field. Their axons bifurcate before their entrance in the dorsal tegmental decussation and give rise to an ipsilateral ascending collateral pathway, terminating in the paralaminar portion of the mediodorsal thalamic nucleus, which, in turn, projects to the saccadic region of the frontal eye field [278]. This pathway provides the frontal eye field with a corollary discharge, encoding upcoming eye movements [264]. A recurrent system destined for the parietal visual field takes its origin from superficial layers, whereas parietal afferents terminate in intermediate layers of the superior colliculus. These fibres ascend in the ipsilateral dorsal pathway and relay in the pulvinar. This pathway, supposedly, transmits visual information [45, 181].

Can the questions on the pathways used for smooth pursuit be answered by this account of the connections of the cortical eye fields? There is no doubt that these pathways pass through the early visual areas 1 and 2 and areas MT and MST, with MT as the area responsible for direction discrimination of the pursuit behaviour [217]. Parietal visual area LIP may be involved by encoding the location of an event relative to the fovea and guiding eye movements [46, 47]. The frontal pursuit area of the frontal eye field processes information in terms of image and eye velocity and plays a role in modulating the gain of the visual-motor responses, using feedback information from the cerebellum [40, 276].

Different alternatives for the connections of the frontal eye field with the oculomotor neurons exist. These connections have the appearence of a network, with the nucleus of the optic tract, the superior colliculus, the nucleus prepositus hypoglossi and the cerebellum as its nodal points. Several of these pathways are shared with the vestibulo-ocular and the optokinetic reflexes. Access to the saccade-generating system as a final link can be gained through the superior colliculus or through the cerebellum. Access to the vestibulo-oculomotor pathways is possible through the nucleus prepositus hypoglossi, a target of the frontal eye field, the nucleus of the optic tract and the superior colliculus, or via the cerebellum. Connections of cerebellar visuomotor areas will be discussed in the section on The Visual System and the Cerebellum of this chapter.

Pretectum and the Visual Reflexes

The pretectal region is situated immediately rostral to the superior colliculus at the level of the posterior commissure (Figs. 19.15, 19.16). According to an analysis in the monkey [134], this complex can be subdivided into five different cell groups: the nucleus of the optic tract, the pretectal olivary nucleus and the medial, anterior and posterior pretectal nulei. A similar parcellation is possible in the human brain [15]. Fibres from the retina pass to the pretectal region, predominantly to the contralateral side, where they terminate in the nucleus of the optic tract. The pretectal olivary nucleus receives a bilateral projection. The posterior and medial pretectal nuclei are more sparsely innervated by the retina [10, 35, 134, 313]. The nucleus of the posterior commissure is sometimes included in the pretectum. The posterior commissure itself contains decussating and commissural fibres from the nucleus of the optic tract, the terminal nuclei of the AOS and from various long ascending brain stem systems.

The connections of the optic tract nucleus were considered in the section on The Accessory Optic System and the Nucleus of the Optic Tract. The olivary pretectal nucleus is considered together with the pupillary light reflex. The anterior and posterior pretectal nuclei receive afferents from contralateral somatosensory nuclei of the brain stem and projections from the primary and secondary somatosensory cortices [9, 315, 316]. Descending pathways terminate in the superior colliculus, the ipsilateral pontine nuclei and the inferior olive [23, 234, 312]. The uncrossed projection of the anterior and posterior pretectal nuclei to the inferior olive is distinguished by its termination in the dorsal accessory olive, an olivary subnucleus serving as the main relay in somatosensory spinocerebellar climbing fibre paths [142, 153]. The rodent equivalent of the medial pretectal nucleus contains cells that represent

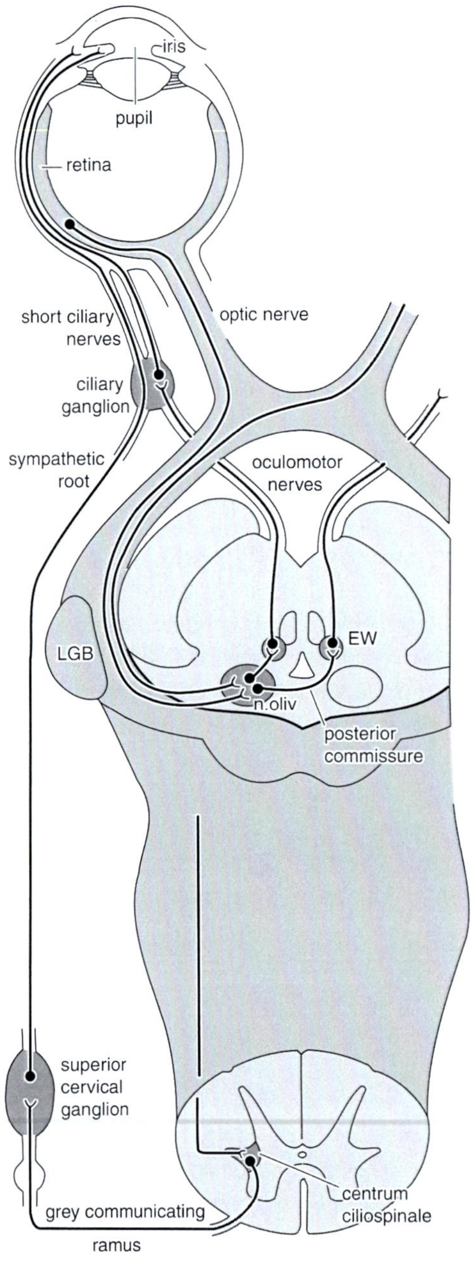

Fig. 19.15. The neural reflex arcs of the visual system: the pupillary light reflex. *EW*, nucleus of Edinger-Westphal; *LGB*, lateral geniculate body; *n.oliv*, pretectal olivary nucleus

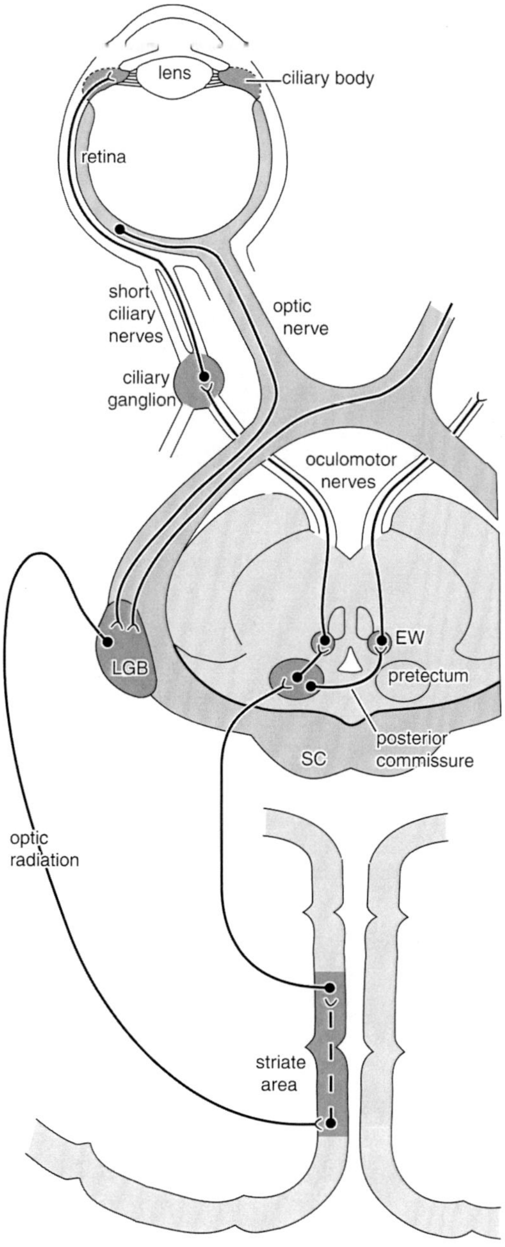

Fig. 19.16. The neural reflex arcs of the visual system: the accommodation reflex. *EW*, nucleus of Edinger-Westphal; *LGB*, lateral geniculate body

the source of a (mainly contralateral) centrifugal projection to the retina [140, 141].

The light reflex, i.e. constriction of the pupil on illumination of the eye, is mediated by a pathway that involves the following links (Fig. 19.15):

1. Axons of retinal ganglion cells, which pass via the optic nerve and tract to the pretectal region, particularly the olivary pretectal nucleus [19, 288].
2. Axons of neurons of the olivary pretectal nucleus project to the Edinger-Westphal nuclei of both sides. Their fibres decussate in the posterior commissure [329]. Additional pathways, relaying in the lateral terminal nucleus of the AOS, have been reported in primates [29, 42].
3. Axons of the parasympathetic preganglionic neurons of the accessory oculomotor nuclei, which pass with the oculomotor nerves to the ciliary ganglia. Here they synapse with postganglionic neurons.
4. Postganglionic neurons whose axons innervate the sphincter pupillae muscle of the iris.

The pretectal efferents to the ipsilateral accessory oculomotor nucleus provide for the direct pupillary light reflex, those to the contralateral nucleus for the consensual pupillary light reflex.

In addition to the response to light, pupillary constriction also occurs following the initiation of ocular convergence. The neural pathway for this convergence-induced pupillary constriction is believed to be independent of that for light-induced pupillary constriction, since the two reflexes are dissociated in the clinical phenomenon, known as the Argyll Robertson sign. This pathway terminates on a group of small neurons outside the classical oculomotor nucleus [29].

The centre for pupillary dilation is located in the intermediolateral cell column of the upper thoracic cord. The preganglionic sympathetic fibres originating from this so-called centrum ciliospinale ascend through the sympathetic trunk and synapse in the superior cervical ganglion with postganglionic elements. The axons of the latter accompany the branches of the internal carotid artery and traverse the ciliary ganglion before they innervate the dilatator pupillae muscle of the iris. Fibres descending through the lateral part of the medulla oblongata and of the spinal cord synapse in the sympathetic ciliospinal centre. The origin of these fibres is unknown.

The pathway for the accommodation reflex, i.e. increase in the curvature of the lens for near vision, is (as far as its mesencephalic centres and efferent limbs are concerned) closely comparable to the pathway for the light reflex. In both pathways the pretectal region, the accessory oculomotor nucleus and the ciliary ganglion represent successive relay stations. However, the afferent limb of the accommodation reflex includes the visual cortex and is thus much longer and much more complex than that of the light reflex (Fig. 19.16). The final link on the efferent side of the accommodation pathway is formed by the postganglionic fibres to the ciliary muscle.

Visual System and the Cerebellum

In non-human primates the oculomotor cerebellum [303] includes the following regions: the flocculus, the ventral and dorsal paraflocculus, the visual vermis and the nodulus/uvula (Figs. 19.17, 19.18). Parts of the ansiform lobule, probably, should be added to this list. In the human cerebellum the dorsal paraflocculus is represented by the tonsilla, the ventral paraflocculus by the accessory paraflocculus, and the ansiform lobule by the superior and inferior semilunar lobules. It is not known whether the connections and functional properties of these lobules in subhuman primates are shared by their human counterparts. Visual projections to precerebellar nuclei have been reviewed in the previous paragraphs. Here these connections are summarized and related to the oculomotor cerebellum and its efferent pathways. To determine which part of the cerebellum is involved in a particular visual subsystem, it is important to remember that the function of the cerebellum depends on the in-

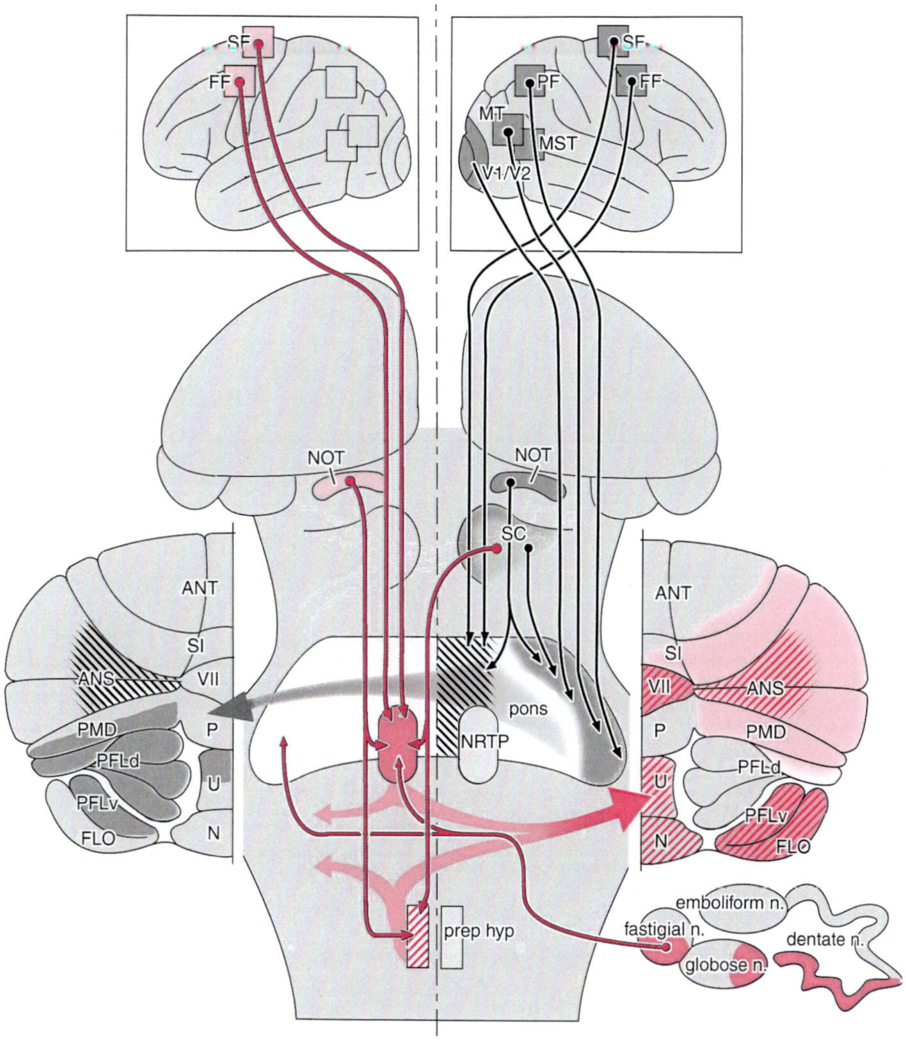

Fig. 19.17. The visual system and the cerebellum: mossy fibre afferents. Cerebellar projections are illustrated in flattened maps of the cerebellum. Of the bilaterally distributed mossy fibre afferents, only the contralateral projection has been illustrated. The projections of the left nucleus reticularis tegmenti pontis (NRTP) to the right half of the cerebellum and the cerebellar nuclei are shown in *red shading*, and of the left nucleus prepositus hypoglossi in *red hatching*. Projections of the right medial and lateral pontine nuclei to the left cerebellum are shown in *black hatching* and *grey shading*, respectively. Note convergence of afferent projections to the NRTP and the lateral pontine nuclei. Frontal and post-rolandic visual areas project to medial and lateral pontine nuclei, respectively. The caudal oculomotor region of the fastigial nucleus receives a collateral projection from the NRTP and projects back to this nucleus and to the pontine nuclei. *ANS*, ansiform lobule; *ANT*, anterior lobe; *FF*, frontal eye field; *FLO*, flocculus; *MST*, medial superior temporal visual area; *MT*, middle temporal visual area; *N*, nodulus; *NOT*, nucleus of the optic tract; *NRTP*, nucleus reticularis tegmenti pontis; *P*, pyramid; *PF*, parietal visual area; *PFLd*, dorsal paraflocculus; *PFLv*, ventral paraflocculus; *PMD*, paramedian lobule; *prep.hyp*, nucleus prepositus hypoglossi; *SC*, superior colliculus; *SF*, supplementary eye field; *SI*, simplex lobule; *U*, uvula; *V1/V2*, visual areas V1/V2; *VII*, lobule VII: visual vermis

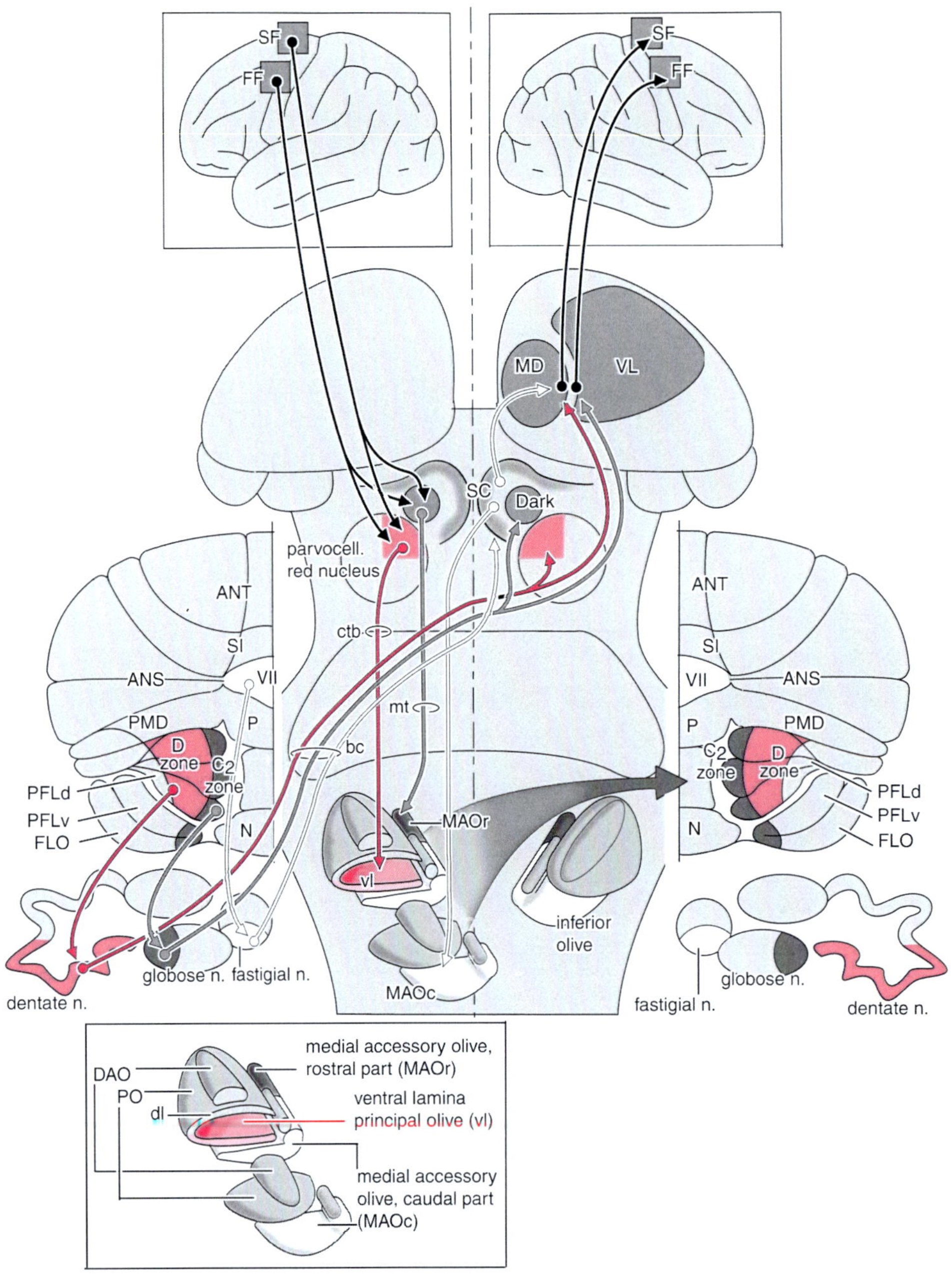

Fig. 19.18. The visual system and the cerebellum: climbing fibre afferents and the efferent projections of oculomotor regions of the cerebellar nuclei. Cerebellar target zones are illustrated on the right, corticonuclear projections of these zones in a flattened map of the cerebellum and the cerebellar nuclei on the left. Inset shows subdivision of the inferior olive. Subnuclei receiving afferents from visual centers include the ventral lamina of the principal olive (from the dorsomedial parvocellular red nucleus; shown in *red*), the rostral medial accessory olive MAOr from Darkschewitsch nucleus (shown in *dark grey*) and the caudal medial accessory olive MAOc from the superior colliculus (shown in *light grey*). The ventral lamina of the principal nucleus of the left inferior olive projects to the D_1 zone of the dorsal paraflocculus and the caudal dentate nucleus. The rostral medial accessory olive provides climbing fibres to the C_2 zone of this lobule and the lateral globose (posterior interposed) nucleus. The caudal medial accessory olive projects to the visual vermis (lobule VII) and the caudal pole of the fastigial nucleus. The corticonuclear projection to these "oculomotor" regions of the cerebellar nuclei are illustrated for the left cerebellar nuclei. Note reciprocity in the organization of these corticonuclear and efferent connections with the connections of the rostromedial red nucleus,

nervation of the Purkinje cells by two afferent systems: the mossy fibre/parallel fibre system and the climbing fibres. Climbing fibres always innervate a sagittal strip of Purkinje cells and their target cells in the cerebellar nuclei. They are closely allied to the output systems of the cerebellar cortex, both anatomically and functionally. Mossy fibres are more widely distributed. Mossy fibre/parallel fibre systems generally innervate multiple sagittal strips of Purkinje cells (see Chap. 20).

The main mossy fibre sources of the visual system are the pontine nuclei, the nucleus reticularis tegmenti pontis, the nucleus prepositus hypoglossi and, possibly, the cell groups in the pontine and medullary reticular formation collectively known as the cell groups of the paramedian tracts [27].

The pontine nuclei provide the majority of the mossy fibres innervating the paraflocculus. They take their origin from lateral and medial cell columns, each of which may include several of the ill-defined nuclei distinguished in the pons. Neurons innervating successively more distal folia of the paraflocculus are located in more peripheral regions of the pontine grey [79, 212]. The lateral column receives a corticopontine projection from post-rolandic visual areas 7a and 7b, and LIP (the parietal visual area), occipitoparietal areas MT and MST and from regions of V1–V3a representing the periphery of the visual field [73]. These cortical afferents overlap with projections from the ipsilateral superior colliculus, the nucleus of the optic tract and the medial and lateral terminal nuclei of the AOS (Fig. 19.17). These lateral column projections terminate on neurons providing mossy fibres to the dorsal paraflocculus [79, 90, 211, 322]. Units in the dorsolateral pons discharge during smooth pursuit or in response to movement of large textured patterns. Most of these units are direction-selective. Lesions of the dorsolateral pons cause deficits in smooth pursuit and optokinetic-induced eye movements, but these deficits recover within a few days [191, 208].

The frontal eye fields project to the medial pontine nuclei. Here the frontopontine projection overlaps with neurons projecting to the ansiform lobule, but not with neurons projecting to the paraflocculus [323]. The pontine grey only provides a minor contribution to mossy innervation of the primate ventral paraflocculus [191, 208].

The nucleus reticularis tegmenti pontis consists of different afferent territories. Projections from premotor cortical areas, including the frontal and supplementary eye fields, terminate predominantly in medial and dorsomedial regions of the nucleus [20, 108]. The same region of the nucleus receives a contralateral projection from the superior colliculus, which may arise as a collateral system from the predorsal fascicle, vestibular afferents, a projection from the nucleus of the optic tract and a projection from the caudal, oculomotor portion of the fastigial nucleus [273]. Projections from the AOS have not been verified in anterograde tracing studies. The central portion of the nucleus receives fibres from the primary motor cortex and the dentate and anterior interposed (emboliform) cerebellar nuclei. Parietal afferents terminate most laterally, but this projection does not include the parietal and parieto-occipital visual areas. Physiological studies favour a rostrocaudal division of the nucleus,

Darkschewitsch nucleus, the superior colliculus and the frontal eye fields. The dorsal paraflocculus, which receives its main mossy fibre afferent projection from post-rolandic visual fields through the lateral pontine nuclei (Fig. 19.17), receives its climbing fibres from olivary subnuclei innervated by the frontal eye fields. *ANS*, ansiform lobule; *ANT*, anterior lobe; *bc*, brachium conjunctivum; *ctb*, central tegmental tract; *DAO*, dorsal accessory olive; *Dark*, Darkschewitsch nucleus; *dl*, dorsal lamina of the principal olive; *FF*, frontal eye field; *FLO*, flocculus; *MAOc*, caudal medial accessory olive; *MAOr*, rostral medial accessory olive; *MD*, mediodorsal thalamic nucleus; *mt*, medial tegmental tract; *N*, nodulus; *P*, pyramid; *PFLd*, dorsal paraflocculus; *PFLv*, ventral paraflocculus; *PMD*, paramedian lobule; *PO*, pontine nuclei; *SC*, superior colliculus; *SF*, supplementary eye field; *SI*, simplex lobule; *VII*, lobule VII: visual vermis; *vl*, ventral lamina of the principal olive; *VL*, ventrolateral thalamic nucleus

with the rostral nucleus being involved in smooth pursuit. Stimulation in its caudal half evokes saccade-like eye movements [274, 326]. However, there is little anatomical support for such a rostrocaudal division.

The nucleus reticularis tegmenti pontis projects bilaterally to widespread regions of the cerebellar cortex [81, 83]. It provides the cerebellar nuclei, including the caudal oculomotor region of the fastigial nucleus, but not the vestibular nuclei [209], with a collateral projection (Fig. 19.17). This projection is complementary to the collateral mossy fibre projection of the lateral reticular nucleus [82, 244]. Projections to the oculomotor cerebellum take their origin from the visually dominated medial and dorsal periphery of the nucleus, and adjoining regions of the dorsomedial pontine grey. In subhuman primates these projections include the flocculus, the ventral paraflocculus, and the visual vermis (lobule VII) [21, 22, 71, 83, 90, 160, 211, 216, 277].

The nucleus prepositus hypoglossi receives afferents from the superior colliculus, via the predorsal fascicle, the terminal nuclei of the accessory optic system, the nucleus of the optic tract and the frontal (but not the post-rolandic) eye fields. With its mossy fibre projection to flocculus, ventral paraflocculus, lobule VII and the adjoining ansiform lobule, and the nodulus uvula, it outlines the oculomotor cerebellum, with the exception of the dorsal paraflocculus (Fig. 19.17) [8].

The nucleus reticularis tegmenti pontis only accounts for a minority of the afferents to the flocculus (cat: 10%, [246] and, presumably, to the ventral paraflocculus). The majority of the mossy fibres terminating in these lobules are derived from the reticular formation (50%), the vestibular nuclei (20%) and the nucleus prepositus hypoglossi (20%). The afferent connections of the cell groups of the paramedian tracts [27], which provide most of these reticular mossy fibre afferents, however, are largely unknown. The role of the vestibular nuclei as a link in optokinetic afferent pathways is difficult to evaluate: the interstitial nucleus of Cajal and the nucleus prepositus hypoglossi serve as a visual/vestibular interface, but direct projections from the nucleus of the optic tract, the AOS, the superior colliculus and the eye fields to the vestibular nuclei are either weak or absent. Mossy fibre projections to the ventral uvula/nodulus are almost exclusively derived from the vestibular nerve and nuclei.

Climbing fibre input to the oculomotor cerebellum is derived from subnuclei of the medial accessory and the principal olive. The dorsal cap and the VLO are often considered as two caudal subnuclei of the principal olive. These two subnuclei are part of a system that provides the cerebellum with information on retinal slip in the planes of the horizontal and the vertical anterior semicircular canals, used in the calibration of the vestibulo-ocular and optokinetic reflexes. Each subnucleus innervates paired Purkinje cell zones extending from the flocculus into the ventral paraflocculus and issues a collateral projection to Purkinje cell zones of the nodulus (Fig. 19.13) [243]. The lateral and medial terminal nuclei of the AOS provide the VLO and its climbing fibres with information about retinal slip in a vertical plane, colinear with the plane of the ipsilateral semicircular canal [168]. The nucleus of the optic tract and the dorsal terminal nucleus of the AOS supply the dorsal cap with information on similar movements of the visual surround in the horizontal plane [259] (Fig. 19.13). Purkinje cells of the VLO-innervated zones inhibit the anterior canal vestibulo-ocular relay cells in the superior vestibular nucleus. Purkinje cells of the dorsal cap-innervated zones inhibit the horizontal canal cells in the medial vestibular nuclei [60] (Fig. 17.8). The climbing fibre input to the flocculus and the ventral paraflocculus is used in the gating of the mossy fibre input to these lobules, as discussed in the previous paragraphs [293]. The molecular basis of this process is the long-term depression (LTD) of the parallel fibre-Purkinje cell synapses resulting from simultaneous activation of climbing and parallel fibres terminating on the same Purkinje cell (see Chap. 20).

Flocculus and ventral paraflocculus and their connections with the vestibulo-oculomotor pathways (Chap. 17, Figs. 17.3, 17.8) thus may be considered as links in a parallel, modifiable cerebellar loop, superimposed upon the three neuronal reflex arcs of the VOR, but these lobules also constitute a link in the pathway for the optokinetic reflex and smooth pursuit. A specific role has been proposed for the flocculus in long-term adaptation of the VOR and for the ventral paraflocculus in smooth pursuit [210, 228]. However, the anatomical evidence for such a distinction is still rather meagre.

A circumscribed climbing fibre path taking its origin from the superior colliculus synapses in the contralateral caudal medial accessory olive and projects to the ipsilateral visual vermal lobule VII, with a collateral projection to its cerebellar target nucleus: the caudal portion of the fastigial nucleus (Fig. 19.18) [135]. The information transmitted by this pathway is not known. It may carry an efference copy of motor commands generated by the intermediate layers of the superior colliculus. In lobule VII, it coincides with mossy fibre projections from the nucleus reticularis tegmenti pontis and the adjoining medial pontine grey and the nucleus prepositus hypoglossi, which relay information from the superior colliculus, the nucleus of the optic tract and the frontal eye fields. Like the flocculus, lobule VII receives a strong reticular input of unknown signature. Its corticonuclear target, the caudal fastigial nucleus, innervates the saccade-generating system in the brain stem and supplies reciprocal projections to the superior colliculus, the nucleus reticularis tegmenti pontis and the pontine nuclei (Figs. 19.17, 19.18) [193, 273]. Coincident mossy and climbing fibre projections to lobule VII subserve long-time adaptation of saccades [164–166, 215]. Involvement of the visual vermis in smooth pursuit was extensively discussed by Robinson and Fuchs [235].

The visual climbing fibre projections of the principal and rostral medial accessory olive are rarely mentioned. The frontal and parietal eye fields access the ventral lamina of the principal olive via the dorsomedial parvocellular red nucleus and the central tegmental tract [302]. Afferents of the nucleus of Darkschewitsch include the same eye fields, the nucleus of the optic tract and the medial and lateral terminal nuclei of the AOS (Figs. 19.13, 19.14 and 19.18) [214]. The Darkschewitsch nucleus innervates the rostral medial accessory olive through the medial tegmental tract. Climbing fibres from the ventral lamina of the principal olive innervate a Purkinje cell zone of the dorsal parafloculus (D_1 zone) and its target nucleus, the caudal and ventral dentate nucleus. The Purkinje cell zone innervated by the rostral medial accessory olive (C_2 zone) extends over the flocculus, the ventral and dorsal paraflocculus and beyond (Fig. 19.18) [302]. Its nuclear target, the lateral pole of the posterior interposed nucleus, receives a collateral climbing fibre projection. The third and most lateral D_2 zone of the paraflocculus receives its climbing fibres from the dorsal lamina of the principal olive. The input of this system is derived from non-visual motor and premotor areas (see Chap. 21, Fig. 21.12). In the dorsal paraflocculus the climbing fibres coincide with mossy fibre projections from the dorsolateral pontine nuclei, i.e. with post-rolandic and tectal afferents. Contrary to the climbing fibres innervating the C_2 and D_1 zones of the dorsal paraflocculus, these mossy fibres do not carry a signal from the frontal eye fields. The visual signature of the ventrocaudal dentate and the lateral posterior interposed nuclei, the target nuclei of the D_1 and C_2 zones of the dorsal paraflocculus, has been recognized [297, 322]. These cerebellar nuclei provide reciprocal connections to the superior colliculus, the parvocellular red nucleus, the Darkschewitsch nucleus and, via the thalamus, to the frontal eye fields and parietal area 7b [44, 45, 193, 302].

The climbing fibre input of the uvula/nodulus is mainly vestibular. It receives optokinetic input from collaterals of climbing fibres of the dorsal cap and the VLO, innervating the flocculus [243], and labyrinthine input from group β and the dorsomedial cell column, subnuclei of the medial accessory olive, which also receive a projection from the nucleus of the optic tract [120]. Its vestibular mossy fibre input does not seem to encode specific parameters of eye movement. It influences slow-phase eye velocity contingent on head position and serves as a mediator for the velocity storage mechanism of the optokinetic reflex in the vestibular nuclei.

It can be concluded that four cerebellar areas are involved in visuomotor control. The flocculus and the ventral paraflocculus subserve adaptation of the vestibulo-ocular and the optokinetic reflex and, possibly, smooth pursuit. The dorsal paraflocculus provides feedback to the frontal eye fields from post-rolandic visual areas. Lobule VII adapts eye movements through its connections with the brain stem saccade-generating system. The nodulus influences postural reflex mechanisms through its connections with the vestibular nuclei. A role of the ansiform lobule (corresponding to the semilunar lobules of the human cerebellum) in the regulation of eye movements, as originally proposed by Fadiga and Pupilli [62], is supported by the presence of mossy fibre input of the frontal eye fields to these lobules [113, 323].

References

1. Albright TD, Desimone R, Gross CG (1984) Columnar organization of directionally selective cells in visual area MT of the macaque. J Neurophysiol 51:16-31
2. Apel PL, O'Brien BJ, Olavaria JF (1997) Distribution of neurons projecting to the superior colliculus correlates with thick stripes in macaque visual area V2. J Comp Neurol 377:313-323
3. Baizer JS, Ungerleider LG, Desimone R (1991) Organization of visual inputs to the inferior temporal and posterior parietal cortex in macaques. J Neurosci 11:168-190
4. Baizer JS, Desimone R, Ungerleider LG (1993) Comparison of subcortical connections of inferior temporal and posterior parietal cortex in monkeys. Vis Neurosci 10:59-72
5. Baker J, Gibson A, Mower G, Robinson F, Glickstein M (1983) Cat visual corticopontine cells project to the superior colliculus. Brain Res 265:227-232
6. Beckstead RM, Frankfurter A (1982) The distribution and some morphological features of substantia nigra neurons that project to the thalamus, superior colliculus and pedunculopontine nucleus in the monkey. Neuroscience 7:2377-2388
7. Belenky MA, Smeraski CA, Provencio I, Sollars PJ, Pickard GE (2003) Melanopsin retinal ganglion cells receive bipolar and amacrine cell synapses. J Comp Neurol 460:380-393
8. Belknap DB, McCrea RA (1988) Anatomical connections of the prepositus and abducens nuclei in the squirrel monkey. J Comp Neurol 268:13-28
9. Berkley KJ, Mash DC (1978) Somatic sensory projections to the pretectum in the cat. Brain Res 158:445-449
10. Berman N (1977) Connections of the pretectum. J Comp Neurol 174:227-254
11. Berson DM, Dunn FA, Takao M (2002) Phototransduction by retinal ganglion cells that set the circadian clock. Science 295:1070-1073
12. Bickford ME, Hall WC (1989) Collateral projections of predorsal bundle cells of the superior colliculus in the rat. J Comp Neurol 283:86-106
13. Blanks RH, Clarke RJ, Lui F, Giolli RA, Van Pham S, Torigoe Y (1995) Projections of the lateral terminal accessory optic nucleus of the common marmoset (Callithrix jacchus). J Comp Neurol 354:511-532
14. Bloomfield SA, Dacheux RF (2001) Rod vision: pathways and processing in the mammalian retina. Prog Retin Eye Res 20:351-384
15. Borostyankoi-Baldauf Z, Herczeg L (2002) Parcellation of the human pretectal complex: a chemoarchitectonic reappraisal. Neurosci Res 110:527-540
16. Boussaoud D, Desimone R, Ungerleider LG (1992) Subcortical connections of visual areas MST and FST in macaques. Vis Neurosci 9:291-302
17. Boycott BB, Wässle H (1991) Morphological classification of bipolar cells of the primate retina. Eur J Neurosci 3:1069-1088
18. Boyd JD, Mavity-Hudson JA, Casagrande VA (2000) The connections of layer 4 subdivisions in the primary visual cortex (V1) of the owl monkey. Cereb Cortex 10:644-662
19. Breen LA, Burde RM, Loewy AD (1983) Brainstem connections to the Edinger-Westphal nucleus of the cat: a retrograde tracer study. Brain Res 261:303-306
20. Brodal P (1978) The corticopontine projection in the rhesus monkey. Origin and principles of organization. Brain 101:251-283
21. Brodal P (1980) The projection from the nucleus reticularis tegmenti pontis to the cerebellum in the rhesus monkey. Exp Brain Res 38:29-36
22. Brodal P (1982) Further observations on the cerebellar projections from the pontine nuclei and the nucleus reticularis tegmenti pontis in the rhesus monkey. J Comp Neurol 204:44-55
23. Bull MS, Mitchell SK, Berkley KJ (1990) Convergent inputs to the inferior olive from the dorsal column nuclei and pretectum in the cat. Brain Res 525:1-10
24. Bullier J (2004) Communications between cortical areas of the visual system. In: Chalupa LM, Werner JS (eds) The Visual Neurosciences. MIT Press, Cambridge MA, pp 522-540

25. Burman K, Darian-Smith C, Darian-Smith I (2000) Macaque red nucleus: origins of spinal and olivary projections and terminations of cortical inputs. J Comp Neurol 423:179–196
26. Busch HFM (1961) An anatomical analysis of the white matter in the brain stem of the cat. Van Gorcum, Assen
27. Büttner-Ennever JA, Horn AK (1996) Pathways from cell groups of the paramedian tracts to the floccular region. Ann NY Acad Sci 781:532–540
28. Büttner-Ennever JA, Cohen B, Pause M, Fries W (1988) Raphe nucleus of the pons containing omnipause neurons of the oculomotor system in the monkey, and its homologue in man. J Comp Neurol 267:307–321
29. Büttner-Ennever JA, Cohen B, Horn AK, Reisine H (1996) Efferent pathways of the nucleus of the optic tract in monkey and their role in eye movements. J Comp Neurol 373:90–107
30. Büttner-Ennever JA, Cohen B, Horn AK, Reisine H (1996) Pretectal projections to the oculomotor complex of the monkey and their role in eye movements. J Comp Neurol 366:348–359
31. Büttner-Ennever JA, Horn AK, Henn V, Cohen B (1999) Projections from the superior colliculus motor map to omnipause neurons in monkey. J Comp Neurol 413:55–67
32. Büttner U, Büttner-Ennever JA (1988) Present concepts of oculomotor organization. Rev Oculomot Res 2:3–32
33. Cajal RY (1892) The structure of the retina. Thomas, Springfield
34. Calkins DJ, Tsukamoto Y, Sterling P (1998) Microcircuitry and mosaic of a blue-yellow ganglion cell in the primate retina. J Neurosci 18:3373–3385
35. Campbell G, Lieberman AR (1985) The olivary pretectal nucleus: experimental anatomical studies in the rat. Philos Trans R Soc Lond B Biol Sci 310:573–609
36. Carpenter MB, Pereira AB, Guha M (1992) Immunocytochemistry of oculomotor afferents in the squirrel monkey (Saimiri sciureus). J Hirnforsch 33:151–167
37. Carroll EW, Wong-Riley MT (1984) Quantitative light and electron microscopic analysis of cytochrome oxidase-rich zones in the striate cortex of the squirrel monkey. J Comp Neurol 222:1–17
38. Catsman-Berrevoets CE, Kuypers HGJM (1981) A search for corticospinal collaterals to thalamus and mesencephalon by means of multiple retrograde fluorescent tracers in cat and rat. Brain Res 218:15–33
39. Cheng K, Waggoner RA, Tanaka K (2001) Human ocular dominance columns as revealed by high-field functional magnetic resonance imaging. Neuron 25:359–374
40. Chou IH, Lisberger SG (2004) The role of the frontal pursuit area in learning in smooth pursuit eye movements. J Neurosci 24:4124–4133
41. Chun MH, Grunert U, Martin PR, Wässle H (1996) The synaptic complex of cones in the fovea and in the periphery of the macaque monkey retina. Vision Res 36:3383–3395
42. Clarke RJ, Blanks RH, Giolli RA (2003) Midbrain connections of the olivary pretectal nucleus in the marmoset (Callithrix jacchus): implications for the pupil light reflex pathway. Anat Embryol (Berl) 207:149–155
43. Clarke S, Miklossy J (1990) Occipital cortex in man: organization of callosal connections, related myelo- and cytoarchitecture, and putative boundaries of functional visual areas. J Comp Neurol 298:188–214
44. Clower DM, Dum RP, Strick PL (2005) Basal ganglia and cerebellar inputs to 'AIP'. Cereb Cortex 15:913–920
45. Clower DM, West RA, Lynch JC, Strick PL (2001) The inferior parietal lobule is the target of output from the superior colliculus, hippocampus, and cerebellum. J Neurosci 21:6283–6291
46. Colby CL, Duhamel JR (1996) Spatial representation for action in parietal cortex. Brain Res Cogn Brain Res 5:105–115
47. Colby CL, Duhamel JR, Goldberg ME (1993) Ventral intraparietal area of the macaque: anatomic location and visual response properties. J Neurophysiol 69:902–914
48. Collins CE, Lyon DC, Kaas JH (2005) Distribution across cortical areas of neurons projecting to the superior colliculus in new world monkeys. Anat Rec 285:619–627
49. Corvisier JH, Hardy O (1991) Possible excitatory and inhibitory feedback to the superior colliculus: a combined retrograde and immunocytochemical study in the prepositus hypoglossi nucleus of the guinea pig. Neurosci Res 12:486–502
50. Cowie RJ, Robinson DL (1994) Subcortical contributions to head movements in macaques. I. Contrasting effects of electrical stimulation of a medial pontomedullary region and the superior colliculus. J Neurophysiol 72:2648–2664
51. Cowie RJ, Smith MK, Robinson DL (1994) Subcortical contributions to head movements in macaques. II. Connections of a medial pontomedullary head-movement region. J Neurophysiol 72:2665–2682
52. Culham JC, Kanwisher NG (2001) Neuroimaging of cognitive functions in human parietal cortex. Curr Opin Neurobiol 11:157–163
53. Cynader M, Berman N (1972) Receptive-field organization of monkey superior colliculus. J Neurophysiol 35:187–201
54. Dacey DM (1996) Circuitry for color coding in the primate retina. Proc Natl Acad Sci USA 93:582–588
55. Dacey DM, Petersen MR (1992) Dendritic field size and morphology of midget and parasol ganglion cells of the human retina. Proc Natl Acad Sci USA 89:9666–9670
56. Dacey DM, Lee BB (1994) The 'blue-on' opponent pathway in primate retina originates from a distinct bistratified ganglion cell type. Nature 367:731–735

57. Dacey DM, Packer OS (2003) Colour coding in the primate retina: diverse cell types and cone-specific circuitry. Curr Opin Neurobiol 13:421–427
58. Dai J, Van der Vliet J, Swaab DF, Buijs RM (1998) Human retinohypothalamic tract as revealed by in vitro postmortem tracing. J Comp Neurol 397:357–370
59. De Vries SH (2000) Bipolar cells use kainate and AMPA receptors to filter visual information into separate channels. Neuron 28:628–629
60. De Zeeuw CI, Wylie DR, DiGiorgi PL, Simpson JI (1994) Projections of individual Purkinje cells of identified zones in the flocculus to the vestibular and cerebellar nuclei in the rabbit. J Comp Neurol 349:428–447
61. Edwards SB, Ginsburgh CL, Henkel CK, Stein BE (1979) Sources of subcortical projections to the superior colliculus in the cat. J Comp Neurol 184:309–329
62. Fadiga E, Pupilli GC (1964) Teleceptive Components of the Cerebellar Function. Physiol Rev 44:432–486
63. Famiglietti EV (1985) Starburst amacrine cells: morphological constancy and systematic variation in the anisotropic field of rabbit retinal neurons. J Neurosci 5:562–577
64. Famiglietti EV (1991) Synaptic organization of starburst amacrine cells in rabbit retina: analysis of serial thin sections by electron microscopy and graphic reconstruction. J Comp Neurol 309:40–70
65. Famiglietti EV (1992) Dendritic co-stratification of ON and OFF directionally selective ganglion cells with starburst amacrine cells in rabbit retina. J Comp Neurol 324:322–335
66. Famiglietti EV, Kolb H (1976) Structural basis for ON- and OFF-center responses in retinal ganglion cells. Science 194:193–195
67. Faugier-Grimaud S, Ventre J (1989) Anatomic connections of inferior parietal cortex (area 7) with subcortical structures related to vestibulo-ocular function in a monkey (Macaca fascicularis). J Comp Neurol 280:1–14
68. Felleman DJ, Van Essen DC (1991) Distributed hierarchical processing in the primate cerebral cortex. Cereb Cortex 1:1–47
69. Fitzpatrick D, Itoh K, Diamond IT (1983) The laminar organization of the lateral geniculate body and the striate cortex in the squirrel monkey (Saimiri sciureus). J Neurosci 3:673–702
70. Fitzpatrick D, Lund JS, Blasdel GG (1985) Intrinsic connections of macaque striate cortex: afferent and efferent connections of lamina 4C. J Neurosci 5:3329–3349
71. Frankfurter A, Weber JT, Harting JK (1977) Brain stem projections to lobule VII of the posterior vermis in the squirrel monkey: as demonstrated by the retrograde axonal transport of tritiated horseradish peroxidase. Brain Res 124:135–139
72. Fries W (1984) Cortical projections to the superior colliculus in the macaque monkey: a retrograde study using horseradish peroxidase. J Comp Neurol 230:55–76
73. Fries W (1990) Pontine projection from striate and prestriate visual cortex in the macaque monkey: an anterograde study. Vis Neurosci 4:205–216
74. Fries W, Distel H (1982) Large layer VI neurons of monkey striate cortex (Meynert cells) project to superior colliculus. Proc R Soc Lond B Biol Sci 219:53–59
75. Fries W, Keizer K, Kuypers HJGM (1985) Large layer VI cells in macaque striate cortex (Meynert cells) project to both superior colliculus and prestriate area MT. Exp Brain Res 58:613–616
76. Galletti C, Fattori P, Gamberini M, Kutz DF (1999) The cortical visual area V6: brain location and visual topography. Eur J Neurosci 11:3922–3936
77. Galletti C, Gamberini M, Kutz DF, Fattori P, Luppino G, Matelli M (2001) The cortical connections of area V6: an occipito-parietal network processing visual information. Eur J Neurosci 13:1572–1588
78. Gamlin PD (2005) The pretectum: connections and oculomotor-related roles. Prog Brain Res 151:379–405
79. Gayer NS, Faull RL (1988) Connections of the paraflocculus of the cerebellum with the superior colliculus in the rat brain. Brain Res 449:253–270
80. Geesaman BJ, Born RT, Andersen RA, Tootell RB (1997) Maps of complex motion selectivity in the superior temporal cortex of the alert macaque monkey: a double-label 2-desoxyglucose study. Cereb Cortex 7:749–757
81. Gerrits NM, Voogd J (1986) The nucleus reticularis tegmenti pontis and the adjacent rostral paramedian reticular formation: differential projections to the cerebellum and the caudal brain stem. Exp Brain Res 62:29–45
82. Gerrits NM, Voogd J (1987) The projection of the nucleus reticularis tegmenti pontis and adjacent regions of the pontine nuclei to the central cerebellar nuclei in the cat. J Comp Neurol 258:52–69
83. Gerrits NM, Epema AH, Voogd J (1984) The mossy fiber projection of the nucleus reticularis tegmenti pontis to the flocculus and adjacent ventral paraflocculus in the cat. Neuroscience 11:627–644
84. Giolli RA, Towns LC (1980) A review of axon collateralization in the mammalian visual system. Brain Behav Evol 17:364–390
85. Giolli RA, Blanks RH, Torigoe Y (1984) Pretectal and brain stem projections of the medial terminal nucleus of the accessory optic system of the rabbit and rat as studied by anterograde and retrograde neuronal tracing methods. J Comp Neurol 227:228–251
86. Giolli RA, Blanks RH, Lui F (2005) The accessory optic system: basic organization with an update on connectivity, neurochemistry, and function. Prog Brain Res 151:407–440
87. Giolli RA, Peterson GM, Ribak CE, McDonald HM, Blanks RH, Fallon JH (1985) GABAergic neurons comprise a major cell type in rodent visual relay nuclei: an immunocytochemical study of pretectal and accessory optic nuclei. Exp Brain Res 61:194–203

88. Giolli RA, Torigoe Y, Clarke RJ, Blanks RH, Fallon JH (1992) GABAergic and non GABAergic projections of accessory optic nuclei, including the visual tegmental relay zone, to the nucleus of the optic tract and dorsal terminal accessory optic nucleus in rat. J Comp Neurol 319:349–358
89. Giolli RA, Gregory KM, Suzuki DA, Blanks RH, Lui F, Betelak KF (2001) Cortical and subcortical afferents to the nucleus reticularis tegmenti pontis and basal pontine nuclei in the macaque monkey. Vis Neurosci 18:725–740
90. Glickstein M, Gerrits N, Kralj-Hans I, Mercier B, Stein J, Voogd J (1994) Visual pontocerebellar projections in the macaque. J Comp Neurol 349:51–72
91. Goodale MA, Milner AD (1992) Separate visual pathways for perception and action. Trends Neurosci 15:20–25
92. Goodale MA, Meenan JP, Bulthoff HH, Nicolle DA, Murphy KJ, Racicot CI (1994) Separate neural pathways for the visual analysis of object shape in perception and prehension. Curr Biol 4:604–610
93. Gouras P (2005) Color vision. In: Kolb H, Fernandez E, Nelson R (eds) Webvision. The organization of the retina and the visual system. http://webvision.med.utah.edu
94. Grantyn R, Shapovalov AI, Shiriaev BI (1982) Combined morphological and electrophysiological description of connections between single primary afferent fibres and individual motoneurons in the frog spinal cord. Exp Brain Res 48:459–462
95. Graybiel AM (1979) Periodic-compartmental distribution of acetylcholinesterase in the superior colliculus of the human brain. Neuroscience 4:643–650
96. Graybiel AM, Illing RB (1994) Enkephalin-positive and acetylcholinesterase-positive patch systems in the superior colliculus have matching distributions but distinct developmental histories. J Comp Neurol 340:297–310
97. Graybiel AM, Brecha N, Karten HJ (1984) Cluster-and-sheet pattern of enkephalin-like immunoreactivity in the superior colliculus of the cat. Neuroscience 12:191–214
98. Guitton D (1999) Gaze shifts in three-dimensional space: A closer look at the superior colliculus. J Comp Neurol 413:77–82
99. Hadjikhani N, Tootell RB (2000) Projection of rods and cones within human visual cortex. Hum Brain Mapp 9:55–63
100. Hadjikhani N, Liu AK, Dale AM, Cavanagh P, Tootell RB (1998) Retinotopy and color sensitivity in human visual cortical area V8. Nat Neurosci 1:235–241
101. Hall WC, Fitzpatrick D, Klatt LL, Raczkowski D (1989) Cholinergic innervation of the superior colliculus in the cat. J Comp Neurol 287:495–514
102. Harting JK (1977) Descending pathways from the superior collicullus: an autoradiographic analysis in the rhesus monkey (Macaca mulatta). J Comp Neurol 173:583–612
103. Harting JK, Van Lieshout DP (1991) Spatial relationships of axons arising from the substantia nigra, spinal trigeminal nucleus, and pedunculopontine tegmental nucleus within the intermediate gray of the cat superior colliculus. J Comp Neurol 305:543–558
104. Harting JK, Updyke BV, Van Lieshout DP (1992) Corticotectal projections in the cat: anterograde transport studies of twenty-five cortical areas. J Comp Neurol 324:379–414
105. Harting JK, Hall WC, Diamond IT, Martin GF (1973) Anterograde degeneration study of the superior colliculus in Tupaia glis: evidence for a subdivision between superficial and deep layers. J Comp Neurol 148:361–386
106. Harting JK, Huerta MF, Frankfurter AJ, Strominger NL, Royce GJ (1980) Ascending pathways from the monkey superior colliculus: an autoradiographic analysis. J Comp Neurol 192:853–882
107. Harting JK, Huerta MF, Hashikawa T, Weber JT, Van Lieshout DP (1988) Neuroanatomical studies of the nigrotectal projection in the cat. J Comp Neurol 278:615–631
108. Hartmann-von Monakow K, Akert K, Künzle H (1981) Projection of precentral, premotor and prefrontal cortex to the basilar pontine grey and to nucleus reticularis tegmenti pontis in the monkey (Macaca fascicularis). Schweiz Arch Neurol Neurochir Psychiatr 129:189–208
109. Hattar S, Liao HW, Takao M, Berson DM, Yau KW (2002) Melanopsin-containing retinal ganglion cells: architecture, projections, and intrinsic photosensitivity. Science 295:1065–1070
110. Hattar S, Kumar M, Park A, Tong P, Tung J, Yau KW, Berson DM (2006) Central projections of melanopsin-expressing retinal ganglion cells in the mouse. J Comp Neurol 497:326–349
111. Haxby JV, Horwitz B, Ungerleider LG, Maisog JM, Pietrini P, Grady CL (1994) The functional organization of human extrastriate cortex: a PET-rCBF study of selective attention to faces and locations. J Neurosci 14:6336–6353
112. Haxby JV, Ungerleider LG, Horwitz B, Maisog JM, Rapoport SI, Grady CL (1996) Face encoding and recognition in the human brain. Proc Natl Acad Sci USA 93:922–927
113. Hayakawa Y, Nakajima T, Takagi M, Fukuhara N, Abe H (2002) Human cerebellar activation in relation to saccadic eye movements: a functional magnetic resonance imaging study. Ophthalmologica 216:399–405
114. Hendry SH, Calkins DJ (1998) Neuronal chemistry and functional organization in the primate visual system. Trends Neurosci 21:344–349
115. Hendry SH, Reid RC (2000) The koniocellular pathway in primate vision. Annu Rev Neurosci 23:127–153
116. Higo S, Kawano J, Matsuyama T, Kawamura S (1992) Differential projections to the superior collicular layers from the perihypoglossal nuclei in the cat. Brain Res 599:19–28

117. Hikosaka O, Takikawa Y, Kawagoe R (2000) Role of the basal ganglia in the control of purposive saccadic eye movements. Physiol Rev 80:953–978
118. Hoffmann KP, Distler C, Erickson R (1991) Functional projections from striate cortex and superior temporal sulcus to the nucleus of the optic tract (NOT) and dorsal terminal nucleus of the accessory optic tract (DTN) of macaque monkeys. J Comp Neurol 313:707–724
119. Hoffmann KP, Distler C, Ilg U (1992) Callosal and superior temporal sulcus contributions to receptive field properties in the macaque monkey's nucleus of the optic tract and dorsal terminal nucleus of the accessory optic tract. J Comp Neurol 321:150–162
120. Holstege G, Collewijn H (1982) The efferent connections of the nucleus of the optic tract and the superior colliculus in the rabbit. J Comp Neurol 209:139–175
121. Horn AK, Büttner-Ennever JA (1998) Premotor neurons for vertical eye movements in the rostral mesencephalon of monkey and human: histologic identification by parvalbumin immunostaining. J Comp Neurol 392:413–427
122. Horn AK, Büttner-Ennever JA, Henn V (1995) Histological identification of premotor neurons for horizontal saccades in monkey and man by parvalbumin immunostaining. J Comp Neurol 359:350–363
123. Horn AK, Helmchen C, Wahle P (2003) GABAergic neurons in the rostral mesencephalon of the macaque monkey that control vertical movements. Ann NY Acad Sci 1004:19–28
124. Horn AK, Büttner-Ennever JA, Wahle P, Reichenberger I (1994) Neurotransmitter profile of saccadic omnipause neurons in nucleus raphe interpositus. J Neurosci 14:2032–2046
125. Hubel DH, Wiesel TN (1963) Shape and arrangement of columns in cat's striate cortex. J Physiol 165:559–568
126. Hubel DH, Wiesel TN (1972) Laminar and columnar distribution of geniculo-cortical fibers in the macaque monkey. J Comp Neurol 146:421–450
127. Hubel DH, Wiesel TN (1974) Sequence regularity and geometry of orientation columns in the monkey striate cortex. J Comp Neurol 158:267–293
128. Hubel DH, LeVay S, Wiesel TN (1975) Mode of termination of retinotectal fibers in macaque monkey: an autoradiographic study. Brain Res 96:25–40
129. Huerta MF, Harting JK (1984) Connectional organization of the superior colliculus. Trends Neurosci 7:286–289
130. Huerta MF, Kaas H (1990) Supplementary eye field as defined by intracortical microstimulation: connections in macaques. J Comp Neurol 293:299–330
131. Huerta MF, Krubitzer LA, Kaas JH (1986) Frontal eye field as defined by intracortical microstimulation in squirrel monkeys, owl monkeys, and macaque monkeys: I. Subcortical connections. J Comp Neurol 253:415–439
132. Huerta MF, Van Lieshout DP, Harting JK (1991) Nigrotectal projections in the primate Galago crassicaudatus. Exp Brain Res 87:389–401
133. Huk ACD, Dougherty RF, Heeger DJ (2002) Retinotopy and functional subdivision of human areas MT and MST. J Neurosci 22:7195–7205
134. Hutchins B, Weber JT (1985) The pretectal complex of the monkey: a reinvestigation of the morphology and retinal terminations. J Comp Neurol 232:425–442
135. Ikeda Y, Noda H, Sugita S (1989) Olivocerebellar and cerebelloolivary connections of the oculomotor region of the fastigial nucleus in the macaque monkey. J Comp Neurol 284:463–488
136. Inouye T (1909) Die Sehstörungen bei Schussverletzungen der kortikalen Sehsphäre nach Beobachtungen an Verwundeten der letzten Japanischen Kriege. W. Engelmann, Leipzig
137. Inouye T (2000) Visual disturbances following gunshot wounds of the cortical visual area. Translated by Glickstein M and Fahle M. Oxford University Press, Oxford
138. Ishai A, Ungerleider LG, Haxby JV (2000) Distributed neural systems for the generation of visual images. Neuron 28:979–990
139. Ishai A, Ungerleider LG, Martin A, Haxby JV (2000) The representation of objects in the human occipital and temporal cortex. J Cogn Neurosci 12 Suppl 2:35–51
140. Itaya SK (1980) Retinal efferents from the pretectal area in the rat. Brain Res 201:436–441
141. Itaya SK (1985) Centrifugal fibers to the cat retina from periaquaductal grey matter. Brain Res 326:362–365
142. Itoh K, Takada M, Yasui Y, Kudo M, Mizuno N (1983) Direct projections from the anterior pretectal nucleus to the dorsal accessory olive in the cat: an anterograde and retrograde WGA-HRP study. Brain Res 272:350–353
143. Jacobs GH, Deegan JF, 2nd (1999) Uniformity of colour vision in Old World monkeys. Proc Biol Sci 266:2023–2028
144. James TW, Culham J, Humphrey GK, Milner AD, Goodale MA (2003) Ventral occipital lesions impair object recognition but not object-directed grasping: an fMRI study. Brain 126:2463–2475
145. Jay MF, Sparks DL (1987) Sensorimotor integration in the primate superior colliculus. II. Coordinates of auditory signals. J Neurophysiol 57:35–55
146. Jayaraman A, Batton RR, 3rd, Carpenter MB (1977) Nigrotectal projections in the monkey: an autoradiographic study. Brain Res 135:147–152
147. Kaneko A (1970) Physiological and morphological identification of horizontal, bipolar and amacrine cells in goldfish retina. J Physiol 207:623–633
148. Kaplan E (2004) The M, P, and K pathways of the primate visual system. In: Chalupa LM, Werner JS (eds) The Visual Neurosciences. MIT Press, Cambridge Mass, pp 481–493

149. Kawamura K, Brodal A, Hoddevik G (1974) The projection of the superior colliculus onto the reticular formation of the brain stem. An experimental anatomical study in the cat. Exp Brain Res 19:1–19
150. Keizer K, Kuypers HGJM, Ronday HK (1987) Branching cortical neurons in cat which project to the colliculi and to the pons: a retrograde fluorescent double-labeling study. Exp Brain Res 67:1–15
151. Kim U, Gregory E, Hall WC (1992) Pathway from the zona incerta to the superior colliculus in the rat. J Comp Neurol 321:555–575
152. King AJ (2004) The superior colliculus. Curr Biol 14:R335–338
153. Kitao Y, Nakamura Y, Kudo M, Moriizumi T, Tokuno H (1989) The cerebral and cerebellar connections of pretecto-thalamic and pretecto-olivary neurons in the anterior pretectal nucleus of the cat. Brain Res 484:304–313
154. Kolb H, Famiglietti EV (1974) Rod and cone pathways in the inner plexiform layer of cat retina. Science 186:47–49
155. Kolb H, Fernandez E, Nelson R (2005) Webvision. The organization of the retina and visual system. http://webvision.med.utah.edu
156. Kuffler SW (1953) Discharge patterns and functional organization of mammalian retina. J Neurophysiol 16:37–68
157. Künzle H (1976) Thalamic projections from the precentral motor cortex in Macaca fascicularis. Brain Res 105:253–267
158. Lachica EA, Casagrande VA (1992) Direct W-like geniculate projections to the cytochrome oxidase (CO) blobs in primate visual cortex: axon morphology. J Comp Neurol 319:141–158
159. Lachica EA, Beck PD, Casagrande VA (1992) Parallel pathways in macaque monkey striate cortex: anatomically defined columns in layer III. Proc Natl Acad Sci USA 89:3566–3570
160. Langer T, Fuchs AF, Scudder CA, Chubb MC (1985) Afferents to the flocculus of the cerebellum in the rhesus macaque as revealed by retrograde transport of horseradish peroxidase. J Comp Neurol 235:1–25
161. Lee BB, Dacey DM, Smith VC, Pokorny J (1999) Horizontal cells reveal cone type-specific adaptation in primate retina. Proc Natl Acad Sci USA 96:14611–14616
162. Leichnetz GR, Gonzalo-Ruiz A (1996) Prearcuate cortex in the Cebus monkey has cortical and subcortical connections like the macaque frontal eye field and projects to fastigial-recipient oculomotor-related brainstem nuclei. Brain Res Bull 41:1–29
163. Leichnetz GR, Spencer RF, Hardy SG, Astruc J (1981) The prefrontal corticotectal projection in the monkey; an anterograde and retrograde horseradish peroxidase study. Neuroscience 6:1023–1041
164. Lemij HG, Collewijn H (1991) Short-term nonconjugate adaptation of human saccades to anisometropic spectacles. Vision Res 31:1955–1966
165. Lemij HG, Collewijn H (1991) Long-term nonconjugate adaptation of human saccades to anisometropic spectacles. Vision Res 31:1939–1954
166. Lemij HG, Collewijn H (1992) Nonconjugate adaptation of human saccades to anisometropic spectacles: meridian-specificity. Vision Res 32: 453–164
167. Lennie P (1980) Parallel visual pathways: a review. Vision Res 20:561–594
168. Leonard CS, Simpson JI, Graf W (1988) Spatial organization of visual messages of the rabbit's cerebellar flocculus. I. Typology of inferior olive neurons of the dorsal cap of Kooy. J Neurophysiol 60:2073–2090
169. LeVay S, Connolly M, Houde J, Van Essen DC (1985) The complete pattern of ocular dominance stripes in the striate cortex and visual field of the macaque monkey. J Neurosci 5:486–501
170. Leventhal AG, Rodieck RW, Dreher B (1981) Retinal ganglion cell classes in the Old World monkey: morphology and central projections. Science 213:1139–1142
171. Leventhal AG, Thompson KG, Liu D, Zhou Y, Ault SJ (1995) Concomitant sensitivity to orientation, direction, and color of cells in layers 2, 3, and 4 of monkey striate cortex. J Neurosci 15:1808–1818
172. Livingstone MS, Hubel DH (1984) Anatomy and physiology of a color system in the primate visual cortex. J Neurosci 4:309–356
173. Livingstone MS, Hubel DH (1988) Segregation of form, color, movement, and depth: anatomy, physiology, and perception. Science 240:740–749
174. Lock TM, Baizer JS, Bender DB (2003) Distribution of corticotectal cells in macaque. Exp Brain Res 151:455–470
175. Lucas RJ, Douglas RH, Foster RG (2001) Characterization of an ocular photopigment capable of driving pupillary constriction in mice. Nature Neuroscience 4:621–626
176. Lui F, Gregory KM, Blanks RH, Giolli RA (1995) Projections from visual areas of the cerebral cortex to pretectal nuclear complex, terminal accessory optic nuclei, and superior colliculus in macaque monkey. J Comp Neurol 363:439–460
177. Luna B, Thulborn KR, Strojwas MH, McCurtain BJ, Berman RA, Genovese CR, Sweeney JA (1998) Dorsal cortical regions subserving visually guided saccades in humans: an fMRI study. Cereb Cortex 8:40–47
178. Lund JS (1988) Anatomical organization of macaque monkey striate visual cortex. Annu Rev Neurosci 11:253–288
179. Luppino G, Calzavara R, Rozzi S, Matelli M (2001) Projections from the superior temporal sulcus to the agranular frontal cortex in the macaque. Eur J Neurosci 14:1035–1040
180. Lynch JC, Tian JR (2005) Cortico-cortical networks and cortico-subcortical loops for the higher control of eye movements. Prog Brain Res 151:461–501

181. Lynch JC, Graybiel AM, Lobeck LJ (1985) The differential projection of two cytoarchitectonic subregions of the inferior parietal lobule of macaque upon the deep layers of the superior colliculus. J Comp Neurol 235:241–254
182. Lynch JC, Hoover JE, Strick PL (1994) Input to the primate frontal eye field from the substantia nigra, superior colliculus, and dentate nucleus demonstrated by transneuronal transport. Exp Brain Res 100:181–186
183. Ma TP, Graybiel AM, Wurtz RH (1991) Location of saccade-related neurons in the macaque superior colliculus. Exp Brain Res 85:21–35
184. MacNeil MA, Heussy JK, Dacheux RF, Raviola E, Masland RH (1999) The shapes and numbers of amacrine cells: matching of photofilled with Golgi-stained cells in the rabbit retina and comparison with other mammalian species. J Comp Neurol 413:303–326
185. Maioli MG, Domeniconi R, Squatrito S, Riva Sanseverino E (1992) Projections from cortical visual areas of the superior temporal sulcus to the superior colliculus, in macaque monkeys. Arch Ital Biol 130:157–166
186. Malach R, Levy I, Hasson U (2002) The topography of high-order human object areas. Trends Cogn Sci 6:176–184
187. Malach R, Reppas JB, Benson RR, Kwong KK, Jiang H, Kennedy WA, Ledden PJ, Brady TJ, Rosen BR, Tootell RB (1995) Object-related activity revealed by functional magnetic resonance imaging in human occipital cortex. Proc Natl Acad Sci USA 92:8135–8139
188. Masland RH (2001) The fundamental plan of the retina. Nat Neurosci 4:877–886
189. Masland RH (2001) Neuronal diversity in the retina. Curr Opin Neurobiol 11:431–436
190. Masland RH, Tauchi M (1985) A possible amacrine cell substrate for the detection of stimulus motion. Neurosci Res Suppl 2:S185–199
191. May JG, Keller EL, Suzuki DA (1988) Smooth-pursuit eye movement deficits with chemical lesions in the dorsolateral pontine nucleus of the monkey. J Neurophysiol 59:952–977
192. May PJ (2005) The mammalian superior colliculus: laminar structure and connections. Progr Brain Res 151:321–378
193. May PJ, Hartwich-Young R, Nelson J, Sparks DL, Porter JD (1990) Cerebellotectal pathways in the macaque: implications for collicular generation of saccades. Neuroscience 36:305–324
194. McCrea RA, Baker R (1985) Anatomical connections of the nucleus prepositus of the cat. J Comp Neurol 237:377–407
195. Meredith MA, Stein BE (1985) Descending efferents from the superior colliculus relay integrated multisensory information. Science 227:657–659
196. Meredith MA, Clemo HR, Stein BE (1991) Somatotopic component of the multisensory map in the deep laminae of the cat superior colliculus. J Comp Neurol 312:353–370
197. Mick G, Cooper H, Magnin M (1993) Retinal projection to the olfactory tubercle and basal telencephalon in primates. J Comp Neurol 327:205–219
198. Middlebrooks JC, Knudsen EI (1984) A neural code for auditory space in the cat's superior colliculus. J Neurosci 4:2621–2634
199. Mihailoff GA, Kosinski RJ, Azizi SA, Border BG (1989) Survey of noncortical afferent projections to the basilar pontine nuclei: a retrograde tracing study in the rat. J Comp Neurol 282:617–643
200. Morin LP, Blanchard JH, Provencio I (2003) Retinal ganglion cell projections to the hamster suprachiasmatic nucleus, intergeniculate leaflet, and visual midbrain: bifurcation and melanopsin immunoreactivity. J Comp Neurol 465:401–416
201. Moschovakis AK, Karabelas AB (1985) Observations on the somatodendritic morphology and axonal trajectory of intracellularly HRP-labeled efferent neurons located in the deeper layers of the superior colliculus of the cat. J Comp Neurol 239:276–308
202. Moschovakis AK, Karabelas AB, Highstein SM (1988) Structure-function relationships in the primate superior colliculus. I. Morphological classification of efferent neurons. J Neurophysiol 60:232–262
203. Moschovakis AK, Karabelas AB, Highstein SM (1988) Structure-function relationships in the primate superior colliculus. II. Morphological identity of presaccadic neurons. J Neurophysiol 60:263–302
204. Moschovakis AK, Gregoriou GG, Savaki HE (2001) Functional imaging of the primate superior colliculus during saccades to visual targets. Nat Neurosci 4:1026–1031
205. Moschovakis AK, Scudder CA, Highstein SM, Warren JD (1991) Structure of the primate oculomotor burst generator. II. Medium-lead burst neurons with downward on-directions. J Neurophysiol 65:218–229
206. Mower G, Gibson A, Glickstein M (1979) Tectopontine pathway in the cat: laminar distribution of cells of origin and visual properties of target cells in dorsolateral pontine nucleus. J Neurophysiol 42:1–15
207. Münzer E, Wiener H (1902) Das Zwischen- und Mittelhirn des Kaninchen und die Beziehungen dieser Teile zum übrigen Centralnervensystem, mit besonderer Berücksichtigung des Pyramidenbahn und Schleife. Mschr f Psychiat u Neurol 12:241–279
208. Mustari MJ, Fuchs AF, Wallman J (1988) Response properties of dorsolateral pontine units during smooth pursuit in the rhesus macaque. J Neurophysiol 60:664–686
209. Mustari MJ, Fuchs AF, Kaneko CR, Robinson FR (1994) Anatomical connections of the primate pretectal nucleus of the optic tract. J Comp Neurol 349:111–128

210. Nagao S (1992) Different roles of flocculus and ventral paraflocculus for oculomotor control in the primate. Neuroreport 3:13–16
211. Nagao S, Kitamura T, Nakamura N, Hiramatsu T, Yamada J (1997) Differences of the primate flocculus and ventral paraflocculus in the mossy and climbing fiber input organization. J Comp Neurol 382:480–498
212. Nikundiwe AM, Bjaalie JG, Brodal P (1994) Lamellar organization of pontocerebellar neuronal populations. A multi-tracer and 3-D computer reconstruction study in the cat. Eur J Neurosci 6:173–186
213. O'Keefe LP, Levitt JB, Kiper DC, Shapley RM, Movshon JA (1998) Functional organization of owl monkey lateral geniculate nucleus and visual cortex. J Neurophysiol 80:594–609
214. Onodera S (1984) Olivary projections from the mesodiencephalic structures in the cat studied by means of axonal transport of horseradish peroxidase and tritiated amino acids. J Comp Neurol 227:37–49
215. Optican LM, Robinson DA (1980) Cerebellar-dependent adaptive control of primate saccadic system. J Neurophysiol 44:1058–1076
216. Osanai R, Nagao S, Kitamura T, Kawabata I, Yamada J (1999) Differences in mossy and climbing afferent sources between flocculus and ventral and dorsal paraflocculus in the rat. Exp Brain Res 124:248–264
217. Osborne LC, Bialek W, Lisberger SG (2004) Time course of information about motion direction in visual area MT of macaque monkeys. J Neurosci 24:3210–3222
218. Peck CK, Baro JA, Warder SM (1993) Sensory integration in the deep layers of superior colliculus. Prog Brain Res 95:91–102
219. Perry VH, Cowey A (1984) Retinal ganglion cells that project to the superior colliculus and pretectum in the macaque monkey. Neuroscience 12:1125–1137
220. Perry VH, Oehler R, Cowey A (1984) Retinal ganglion cells that project to the dorsal lateral geniculate nucleus in the macaque monkey. Neuroscience 12:1101–1123
221. Petit L, Haxby JV (1999) Functional anatomy of pursuit eye movements in humans as revealed by fMRI. J Neurophysiol 82:463–471
222. Pierrot-Deseilligny CH, Chain F, Gray F, Serdaru M, Escourolle R, Lhermitte F (1982) Parinaud'd syndrome: electro-oculographic and anatomical analyss of six vascular cases with deductions about vertical gaze organization in the premotor structures. Brain 105:667–696
223. Pitzalis S, Galletti C, Huang RS, Patria F, Committeri G, Galati G, Fattori P, Sereno MI (2006) Wide-field retinotopy defines human cortical visual area V6. J Neurosci 26:7962–7973
224. Pollack JG, Hickey TL (1979) The distribution of retino-collicular axon terminals in rhesus monkey. J Comp Neurol 185:587–602
225. Polyak SL (1941) The retina. University of Chicago Press, Chicago IL
226. Priebe NJ, Lisberger SG (2004) Estimating target speed from the population response in visual area MT. J Neurosci 24:1907–1916
227. Provencio I, Rollag MD, Castrucci AM (2002) Photoreceptive net in the mammalian retina. This mesh of cells may explain how some blind mice can still tell day from night. Nature 415:493
228. Rambold H, Churchland A, Selig Y, Jasmin L, Lisberger SG (2002) Partial ablations of the flocculus and ventral paraflocculus in monkeys cause linked deficits in smooth pursuit eye movements and adaptive modification of the VOR. J Neurophysiol 87:912–924
229. Redgrave P, Odekunle A, Dean P (1986) Tectal cells of origin of predorsal bundle in rat: location and segregation from ipsilateral descending pathway. Exp Brain Res 63:279–293
230. Reese BE, Guillery RW (1987) Distribution of axons according to diameter in the monkey's optic tract. J Comp Neurol 260:453–460
231. Reese BE, Cowey A (1990) Fibre organization of the monkey's optic tract: I. Segregation of functionally distinct optic axons. J Comp Neurol 295:385–400
232. Ringach DL, Shapley RM, Hawken MJ (2002) Orientation selectivity in macaque V1: diversity and laminar dependence. J Neurosci 22:5639–5651
233. Rizzolatti G, Matelli M (2003) Two different streams form the dorsal visual system: anatomy and functions. Exp Brain Res 153:146–157
234. Robertson RT (1983) Efferents of the pretectal complex: separate populations of neurons project to lateral thalamus and to inferior olive. Brain Res 258:91–95
235. Robinson FR, Fuchs AF (2001) The role of the cerebellum in voluntary eye movements. Annu Rev Neurosci 24:981–1004
236. Robinson FR, Phillips JO, Fuchs AF (1994) Coordination of gaze shifts in primates: brainstem inputs to neck and extraocular motoneuron pools. J Comp Neurol 346:43–62
237. Rockhill RL, Daly FJ, MacNeil MA, Brown SP, Masland RH (2002) The diversity of ganglion cells in a mammalian retina. J Neurosci 22:3831–3843
238. Rodieck RW, Watanabe M (1993) Survey of the morphology of macaque retinal ganglion cells that project to the pretectum, superior colliculus, and parvocellular laminae of the lateral geniculate nucleus. J Comp Neurol 338:289–303
239. Rosano C, Sweeney JA, Melchitzky DS, Lewis DA (2003) The human precentral sulcus: chemoarchitecture of a region corresponding to the frontal eye fields. Brain Res 972:16–30
240. Rosano C, Krisky CM, Welling JS, Eddy WF, Luna B, Thulborn KR, Sweeney JA (2002) Pursuit and saccadic eye movement subregions in human frontal eye field: a high-resolution fMRI investigation. Cereb Cortex 12:107–115

241. Rozzi S, Calzavara R, Belmalih A, Borra E, Gregoriou GG, Matelli M, Luppino G (2006) Cortical connections of the inferior parietal cortical convexity of the macaque monkey. Cereb Cortex 16:1389–1417
242. Ruby NF, Brennan TJ, Xie X, Cao V, Franken P, Heller HC, O'Hara BF (2002) Role of melanopsin in circadian responses to light. Science 298:2211–2213
243. Ruigrok TJ (2003) Collateralization of climbing and mossy fibers projecting to the nodulus and flocculus of the rat cerebellum. J Comp Neurol 466:278–298
244. Ruigrok TJ, Cella F, Voogd J (1995) Connections of the lateral reticular nucleus to the lateral vestibular nucleus in the rat. An anterograde tracing study with Phaseolus vulgaris leucoagglutinin. Eur J Neurosci 7:1410–1413
245. Saleem KS, Suzuki W, Tanaka K, Hashikawa T (2000) Connections between anterior inferotemporal cortex and superior temporal sulcus regions in the macaque monkey. J Neurosci 20:5083–5101
246. Sato Y, Kawasaki T, Ikarashi K (1983) Afferent projections from the brain stem to the three floccular zones in cats. II Mossy fiber projections. Brain Res 272:37–48
247. Schmahmann JD, Pandya DN (1989) Anatomical investigation of projections to the basis pontis from posterior parietal association cortices in rhesus monkey. J Comp Neurol 289:53–73
248. Schmahmann JD, Pandya DN (1991) Projections to the basis pontis from the superior temporal sulcus and superior temporal region in the rhesus monkey. J Comp Neurol 308:224–248
249. Schmahmann JD, Pandya DN (1993) Prelunate, occipitotemporal, and parahippocampal projections to the basis pontis in rhesus monkey. J Comp Neurol 337:94–112
250. Scudder CA, Kaneko CS, Fuchs AF (2002) The brainstem burst generator for saccadic eye movements: a modern synthesis. Exp Brain Res 142:439–462
251. Scudder CA, Moschovakis AK, Karabelas AB, Highstein SM (1996) Anatomy and physiology of saccadic long-lead burst neurons recorded in the alert squirrel monkey. I. Descending projections from the mesencephalon. J Neurophysiol 76:332–352
252. Scudder CA, Moschovakis AK, Karabelas AB, Highstein SM (1996) Anatomy and physiology of saccadic long-lead burst neurons recorded in the alert squirrel monkey. II. Pontine neurons. J Neurophysiol 76:353–370
253. Sereno MI, Dale AM, Reppas JB, Kwong KK, Belliveau JW, Brady TJ, Rosen BR, Tootell RB (1995) Borders of multiple visual areas in humans revealed by functional magnetic resonance imaging. Science 268:889–893
254. Shapley R, Perry VH (1986) Cat and monkey retinal ganglion cells and their visual functional roles. Trends Neurosci 9:229–235
255. Shinoda Y, Sugiuchi Y, Izawa Y, Hata Y (2005) Long descending motor tract axons and their control of neck and axial muscles. Prog Brain Res 151:527–563
256. Shipp S, Blanton M, Zeki S (1998) A visuo-somatomotor pathway through superior parietal cortex in the macaque monkey: cortical connections of areas V6 and V6A. Eur J Neurosci 10:3171–3193
257. Shook BL, Schlag-Rey M, Schlag J (1990) Primate supplementary eye field: I. Comparative aspects of mesencephalic and pontine connections. J Comp Neurol 301:618–642
258. Simpson JI (1984) The accessory optic system. Annu Rev Neurosci 7:13–41
259. Simpson JI, Leonard CS, Soodak RE (1988) The accessory optic system of rabbit. II. Spatial organization of direction selectivity. J Neurophysiol 60:2055–2072
260. Sincich LC, Horton JC (2002) Divided by cytochrome oxidase: a map of the projections from V1 to V2 in macaques. Science 295:1734–1737
261. Sincich LC, Horton JC (2003) Independent projection streams from macaque striate cortex to the second visual area and middle temporal area. J Neurosci 23:5684–5692
262. Sincich LC, Horton JC (2005) Input to V2 thin stripes arises from V1 cytochrome oxidase patches. J Neurosci 25:10087–10093
263. Sincich LC, Park KF, Wohlgemuth MJ, Horton JC (2004) Bypassing V1: a direct geniculate input to area MT. Nat Neurosci 7:1123–1128
264. Sommer MA, Wurtz RH (2000) Composition and topographic organization of signals sent from the frontal eye field to the superior colliculus. J Neurophysiol 83:1979–2001
265. Soucy E, Wang Y, Nirenberg S, Nathans J, Meister M (1998) A novel signaling pathway from rod photoreceptors to ganglion cells in mammalian retina. Neuron 21:481–493
266. Sparks DL (1988) Neural cartography: sensory and motor maps in the superior colliculus. Brain Behav Evol 31:49–56
267. Sparks DL, Holland R, Guthrie BL (1976) Size and distribution of movement fields in the monkey superior colliculus. Brain Res 113:21–34
268. Spencer RF, Wang SF (1996) Immunohistochemical localization of neurotransmitters utilized by neurons in the rostral interstitial nucleus of the medial longitudinal fasciculus (riMLF) that project to the oculomotor and trochlear nuclei in the cat. J Comp Neurol 366:134–148
269. Spencer RF, Wenthold RJ, Baker R (1989) Evidence for glycine as an inhibitory neurotransmitter of vestibular, reticular, and prepositus hypoglossi neurons that project to the cat abducens nucleus. J Neurosci 9:2718–2736
270. Stanton GB, Goldberg ME, Bruce CJ (1988) Frontal eye field efferents in the macaque monkey: I. Subcortical pathways and topography of striatal and thalamic terminal fields. J Comp Neurol 271:473–492

271. Steele GE, Weller RE (1993) Subcortical connections of subdivisions of inferior temporal cortex in squirrel monkeys. Vis Neurosci 10:563–583
272. Stone J, Dreher B (1982) Parallel processing of information in the visual pathways. A general principle of sensory coding. Trends Neurosci 5:441–446
273. Sugita S, Noda H (1991) Pathways and terminations of axons arising in the fastigial oculomotor region of macaque monkeys. Neurosci Res 10:118–136
274. Suzuki DA, Yamada T, Hoedema R, Yee RD (1999) Smooth-pursuit eye-movement deficits with chemical lesions in macaque nucleus reticularis tegmenti pontis. J Neurophysiol 82:1178–1186
275. Takada M, Tokuno H, Ikai Y, Mizuno N (1994) Direct projections from the entopeduncular nucleus to the lower brainstem in the rat. J Comp Neurol 342:409–429
276. Tanaka M, Lisberger SG (2002) Role of arcuate frontal cortex of monkeys in smooth pursuit eye movements. I. Basic response properties to retinal image motion and position. J Neurophysiol 87:2684–2699
277. Thielert CD, Thier P (1993) Patterns of projections from the pontine nuclei and the nucleus reticularis tegmenti pontis to the posterior vermis in the rhesus monkey: a study using retrograde tracers. J Comp Neurol 337:113–126
278. Tian JR, Lynch JC (1997) Subcortical input to the smooth and saccadic eye movement subregions of the frontal eye field in Cebus monkey. J Neurosci 17:9233–9247
279. Tigges J, Tigges M (1981) Distribution of retinofugal and corticofugal axon terminals in the superior colliculus of squirrel monkey. Invest Ophthalmol Vis Sci 20:149–158
280. Tokuno H, Takada M, Kondo Y, Mizuno N (1993) Laminar organization of the substantia nigra pars reticulata in the macaque monkey, with special reference to the caudato-nigrotectal link. Exp Brain Res 92:545–548
281. Tootell RB, Hamilton SL (1989) Functional anatomy of the second visual area (V2) in the macaque. J Neurosci 9:2620–2644
282. Tootell RB, Hadjikhani N (2001) Where is 'dorsal V4' in human visual cortex? Retinotopic, topographic and functional evidence. Cereb Cortex 11:298–311
283. Tootell RB, Tsao D, Vanduffel W (2003) Neuroimaging weighs in: humans meet macaques in "primate" visual cortex. J Neurosci 23:3981–3989
284. Tootell RB, Dale AM, Sereno MI, Malach R (1996) New images from human visual cortex. Trends Neurosci 19:481–489
285. Tootell RB, Nelissen K, Vanduffel W, Orban GA (2004) Search for color 'center(s)' in macaque visual cortex. Cereb Cortex 14:353–363
286. Tootell RB, Mendola JD, Hadjikhani NK, Ledden PJ, Liu AK, Reppas JB, Sereno MI, Dale AM (1997) Functional analysis of V3A and related areas in human visual cortex. J Neurosci 17:7060–7078
287. Torigoe Y, Blanks RH, Precht W (1986) Anatomical studies on the nucleus reticularis tegmenti pontis in the pigmented rat. II. Subcortical afferents demonstrated by the retrograde transport of horseradish peroxidase. J Comp Neurol 243:88–105
288. Trejo LJ, Cicerone CM (1984) Cells in the pretectal olivary nucleus are in the pathway for the direct light reflex of the pupil in the rat. Brain Res 300:49–62
289. Tsao DY, Freiwald WA, Knutsen TA, Mandeville JB, Tootell RB (2003) Faces and objects in macaque cerebral cortex. Nat Neurosci 6:989–995
290. Ungerleider LG, Mishkin M (1982) Two cortical visual systems. In: Ingle DJ, Goodale MA, Mansfield RJW (eds) Analysis of visual behaviour. MIT Press, Cambridge MA, pp 549–586
291. Ungerleider LG, Haxby JV (1994) 'What' and 'where' in the human brain. Curr Opin Neurobiol 4:157–165
292. Ungerleider LG, Pasternak T (2004) Ventral and dorsal cortical processing streams. In: Chalupa LM, Werner JS (eds) The visual neurosciences. MIT Press, Cambridge MA, pp 541–561
293. Van der Steen J, Simpson JI, Tan J (1994) Functional and anatomic organization of three-dimensional eye movements in rabbit cerebellar flocculus. J Neurophysiol 72, 31–46
294. Van der Want JJ, Nunes Cardozo JJ, Van der Togt C (1992) GABAergic neurons and circuits in the pretectal nuclei and the accessory optic system of mammals. Prog Brain Res 90:283–305
295. Van Essen DC (2004) Organization of visual areas in macaque and human cerebral cortex. In: Chalupa LC, Werner JS (eds) The visual Neurosciences. MIT Press, Cambridge MA, pp 507–521
296. Van Essen DC, Newsome WT, Maunsell JH (1984) The visual field representation in striate cortex of the macaque monkey: asymmetries, anisotropies, and individual variability. Vision Res 24:429–448
297. van Kan PL, Houk JC, Gibson AR (1993) Output organization of intermediate cerebellum of the monkey. J Neurophysiol 69:57–73
298. Vanduffel W, Tootell RB, Schoups AA, Orban GA (2002) The organization of orientation selectivity throughout macaque visual cortex. Cereb Cortex 12:647–662
299. Vaney DI (1997) Neuronal coupling in rod-signal pathways of the retina. Invest Ophthalmol Vis Sci 38:267–273
300. Vincent SR, Hattori T, McGeer EG (1978) The nigrotectal projection: a biochemical and ultrastructural characterization. Brain Res 151:159–164
301. Von Gudden B (1870) Über einen bisher nicht beschriebenen Nervenfaserstrang im Gehirne der Saugerthiere und des Menschen. Arch Psychiat u Nerv Krankh 2:364–366
302. Voogd J (2004) Cerebellum and precerebellar nuclei. In: Paxinos G, Mai JK (eds) The human nervous system. Elsevier, Amsterdam, pp 322–392
303. Voogd J, Barmack NH (2005) Oculomotor cerebellum. Prog Brain Res 151:231–268

304. Voogd J, Nieuwenhuys R, Van Dongen PAM, Ten Donkelaar HJ (1998) Mammals. In: Nieuwenhuys R, Ten Donkelaar HJ, Nicholson V (eds) The central nervous system of vertebrates, vol 3. Springer, Heidelberg, pp 1637–2097
305. Wallace MN (1988) Lattices of high histochemical activity occur in the human, monkey, and cat superior colliculus. Neuroscience 25:569–583
306. Wallace MN, Fredens K (1989) Relationship of afferent inputs to the lattice of high NADPH-diaphorase activity in the mouse superior colliculus. Exp Brain Res 78:435–445
307. Wang SF, Spencer RF (1996) Spatial organization of premotor neurons related to vertical upward and downward saccadic eye movements in the rostral interstitial nucleus of the medial longitudinal fasciculus (riMLF) in the cat. J Comp Neurol 366:163–180
308. Wang SF, Spencer RF (1996) Morphology and soma-dendritic distribution of synaptic endings from the rostral interstitial nucleus of the medial longitudinal fasciculus (riMLF) on motoneurons in the oculomotor and trochlear nuclei in the cat. J Comp Neurol 366:149–162
309. Wässle H (2004) Parallel processing in the mammalian retina. Nat Rev Neurosci 5:747–757
310. Wässle H, Grunert U, Chun MH, Boycott BB (1995) The rod pathway of the macaque monkey retina: identification of AII-amacrine cells with antibodies against calretinin. J Comp Neurol 361:537–551
311. Watanabe M, Rodieck RW (1989) Parasol and midget ganglion cells of the primate retina. J Comp Neurol 289:434–454
312. Weber JT, Harting JK (1980) The efferent projections of the pretectal complex: an autoradiographic and horseradish peroxidase analysis. Brain Res 194:1–28
313. Weber JT, Young R, Hutchins B (1981) Morphologic and autoradiographic evidence for a laminated pretectal olivary nucleus in the squirrel monkey. Brain Res 224:153–159
314. Werblin FS, Dowling JE (1969) Organization of the retina of the mudpuppy, Necturus maculosus. II. Intracellular recording. J Neurophysiol 32:339–355
315. Wiberg M, Blomqvist A (1984) The projection to the mesencephalon from the dorsal column nuclei. An anatomical study in the cat. Brain Res 311:225–244
316. Wiberg M, Westman J, Blomqvist A (1987) Somatosensory projection to the mesencephalon: an anatomical study in the monkey. J Comp Neurol 264:92–117
317. Wiesel TN, Hubel DH, Lam DM (1974) Autoradiographic demonstration of ocular-dominance columns in the monkey striate cortex by means of transneuronal transport. Brain Res 79:273–279
318. Wilson ME, Toyne MJ (1970) Retino-tectal and cortico-tectal projections in Macaca mulatta. Brain Res 24:395–406
319. Wu CW, Bichot NP, Kaas JH (2005) Somatosensory areas S2 and PV project to the superior colliculus of a prosimian primate, Galago garnetti. Somatosens Mot Res 22:221–231
320. Wu SM, Gao F, Maple BR (2000) Functional architecture of synapses in the inner retina: segregation of visual signals by stratification of bipolar cell axon terminals. J Neurosci 20:4462–4470
321. Xiao Y, Felleman DJ (2004) Projections from primary visual cortex to cytochrome oxidase thin stripes and interstripes of macaque visual area 2. Proc Natl Acad Sci USA 101:7147–7151
322. Xiong G, Nagao S (2002) The lobulus petrosus of the paraflocculus relays cortical visual inputs to the posterior interposed and lateral cerebellar nuclei: an anterograde and retrograde tracing study in the monkey. Exp Brain Res 147:252–263
323. Xiong G, Hiramatsu T, Nagao S (2002) Cortico-pontocerebellar pathway from the prearcuate region to hemispheric lobule VII of the cerebellum: an anterograde and retrograde tracing study in the monkey. Neurosci Lett 322:173–176
324. Xu X, Bosking WH, White LE, Fitzpatrick D, Casagrande VA (2005) Functional organization of visual cortex in the prosimian bush baby revealed by optical imaging of intrinsic signals. J Neurophysiol 94:2748–2762
325. Yabuta NH, Callaway EM (1998) Functional streams and local connections of layer 4C neurons in primary visual cortex of the macaque monkey. J Neurosci 18:9489–9499
326. Yamada T, Suzuki DA, Yee RD (1996) Smooth pursuit-like eye movements evoked by microstimulation in macaque nucleus reticularis tegmenti pontis. J Neurophysiol 76:3313–3324
327. Yoshida KW, Watanabe D, Ishikane H, Tachibana M, Pastan I, Nakanishi S (2001) A key role of starburst amacrine cells in originating retinal directional selectivity and optokinetic eye movements. Neuron 30:771–780
328. Yoshioka T, Levitt JB, Lund JS (1994) Independence and merger of thalamocortical channels within macaque monkey primary visual cortex: anatomy of interlaminar projections. Vis Neurosci 11:467–489
329. Young MJ, Lund RD (1994) The anatomical substrates subserving the pupillary light reflex in rats: origin of the consensual pupillary response. Neurosci Res 62:481–496
330. Zeki S, Watson JD, Lueck CJ, Friston KJ, Kennard C, Frackowiak RS (1991) A direct demonstration of functional specialization in human visual cortex. J Neurosci 11:641–649

20 Cerebellum

Introduction

The cerebellum (the "small brain") is located in the posterior fossa of the skull. Transverse fissures divide its surface into narrow folds called folia. The cerebellum processes information from many different sources, including the spinal cord, the brain stem and the cerebral cortex, and projects to many different centres in the brain involved in postural adaptations and the generation of movements. The cerebellum is covered by a cortex; the cerebellar nuclei occupy a central position in the cerebellar white matter. Three layers are distinguished in the cortex. A monolayer formed by the perikarya of the large Purkinje cells separates the superficial, cell-poor molecular layer from the deep layer of the granule cells. The Purkinje cells form the sole output element of the cerebellar cortex. The granular and molecular layers both contain several types of interneurons.

The cerebellum is characterized by three main features. The first is that the flattened dendritic trees of the Purkinje cells, which extend into the molecular layer, are orientated perpendicularly to the parallel fibres, the axons of the granule cells (Fig. 20.4). This configuration was discovered more than a century ago by Cajal in the avian cerebellum ([27] cited from [105]).

The second feature is the rectilinear arrangement of Purkinje cell aggregates and their second afferent system: the climbing fibres, with respect to the transverse fissures. The Purkinje cells and the climbing fibres are organized in parallel longitudinal zones that cross the transverse fissures at right angles. Purkinje cells of each zone project to a particular cerebellar or vestibular nucleus. Consequently, the output of the cerebellum is organized modularly (Fig. 20.7).

The third feature of the cerebellum is the remarkable plasticity of the synaptic connections between the parallel fibres and the dendrites of the Purkinje cells. Each Purkinje cell receives more than 100,000 synapses from the parallel fibres that traverse its dendritic tree. At any given moment, however, only a small fraction of these synaptic connections is available [54]. Most of these synaptic connections are totally or partially closed by a process known as *long-term depression* (LTD), a process that affects the ionotropic glutamate receptors responsible for the transmission at the parallel fibre/Purkinje cell synapse. LTD depends on the simultaneous activation of climbing fibres and the parallel fibre afferents of a Purkinje cell. LTD sculpts the cerebellar cortical network, according to previous experience, and, most likely, forms the basis of the main function of the cerebellum, i.e. the optimization of movements. The flattening of the Purkinje cell dendritic trees and, at a higher level of organization, the parasagittal organization of the Purkinje cell aggregates guarantees maximal access of parallel fibres to the Purkinje cells.

Gross Anatomy

The gross anatomy of the cerebellum is characterized by transverse fissures of different depths, which subdivide the cerebellum into lobes, lobules and folia (leaflets). Two longitudinal sulci, which are shallow on the anterior (superior) surface of the cerebellum but deep on its inferior aspect, demarcate the folial chain of the vermis from the hemispheres (Figs. 3.8, 3.11, 3.13). The classical nomenclature of the cerebellum is used in this work to describe the subdivision of the folial chains in lobes and lobules. Bolk's [15] comparative anatomical nomenclature is based on the relative constancy of this lobular pattern among mammalian species. Larsell [99, 100] introduced a simple system of roman numerals I–X for the lobules of the vermis and indicated their counterparts in the hemisphere with the prefix H. These nomenclatures can be compared in Fig. 20.2. Schmahmann [145, 146] adapted the cerebellar nomenclature for use in imaging studies in a recent MRI atlas of the human cerebellum.

The folial chain of the vermis is straight. In the anterior lobe and the lobulus simplex (declive and posterior quadrangular lobule; lobule VI of Larsell), located behind the primary fissure, the border between vermis and hemispheres is indistinct; more caudally the paramedian sulcus forms a clear boundary (Fig. 20.2). The folial chain of the hemisphere is curved. Two loops are present, one in the region of the semilunar lobules (Bolk's ansiform lobule) and the other in the region of the tonsil (the paraflocculus of comparative anatomical nomenclature, Fig. 20.2). In nonhuman primates the folial loop of the paraflocculus is directed laterally. Its distal segment is located as the petrosal lobule in the subarcuate fossa of the petrosal bone. In the human cerebellum the folial loop of the tonsil is directed medially (Fig. 3.13) and a petrosal lobule is absent. The most caudal lobules of vermis and hemisphere are the nodulus and the flocculus, both located on the ventral surface of the cerebellum.

The cerebellum is covered with a cortex; its central white matter contains the cerebellar nuclei (Figs. 5.16, 5.18, 5.19). When unfolded and stretched out on a single plane, the cortex of vermis and hemispheres would be shaped as three long (more than 100 cm) and fairly narrow parallel strips (Fig. 20.3) [17, 164]. In the anterior portion of the cerebellum, the cortex of vermis and hemispheres is broadly continuous; in the posterior cerebellum vermis and hemisphere are isolated from each other by interruptions of the cortex in the bottom of the paramedian sulcus. Within the folial chains of vermis and hemispheres the cortex is always continuous, with the exception of the flocculus, which is separated from the adjoining tonsil by an interruption of the cortex in the posterolateral fissure.

Three peduncles connect the cerebellum with the brain stem (Figs. 3.10, 3.13, 20.1). The inferior cerebellar peduncle consists of the restiform body and the juxtarestiform body. The restiform body is an afferent system, containing the spinocerebellar and olivocerebellar tracts and other afferents from the medulla oblongata. The juxarestiform body is located medial to the restiform body within the vestibular nuclei and carries efferent pathways from the cortex and the fastigial nucleus and the descending branch of the vestibular nerve. The middle cerebellar peduncle (brachium pontis) is the largest afferent system of the cerebellum, arising from the pontine nuclei. The superior cerebellar peduncle (brachium conjunctivum) is the main efferent system of the cerebellum, arising from the dentate, emboliform (anterior interposed) and globose (posterior interposed) nuclei.

The cerebellum receives its blood through the posterior inferior cerebellar artery, a branch from the vertebral artery, and through the anterior inferior and superior cerebellar arteries, branches of the basilar artery (Figs. 4.11, 4.17). These arteries also vascularize parts of the brain stem (Fig. 4.20). Blockage of

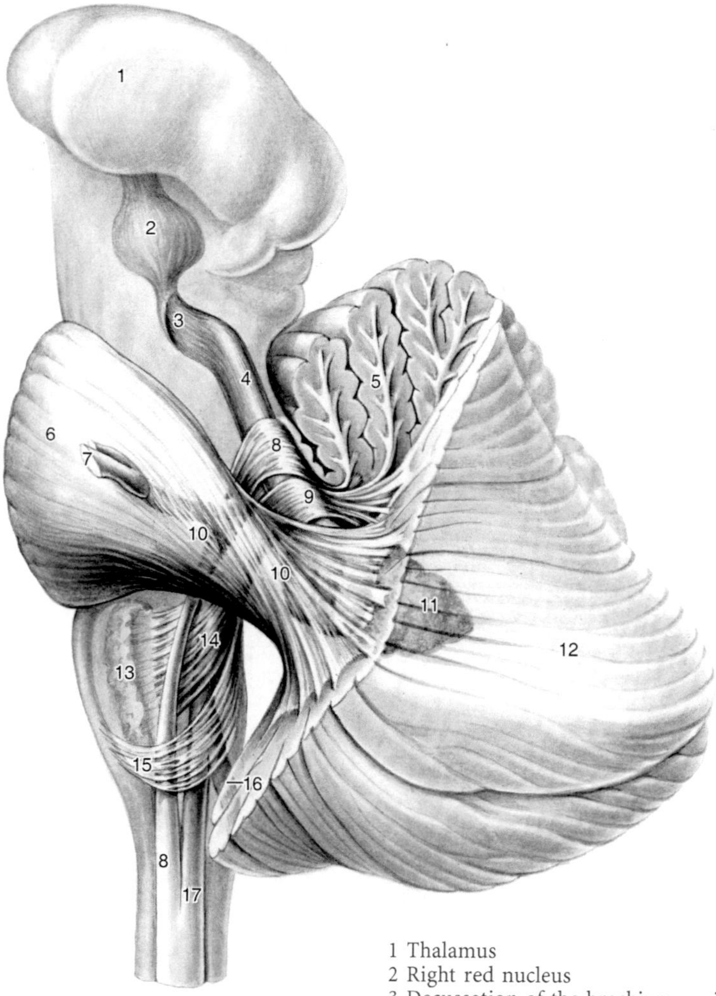

1 Thalamus
2 Right red nucleus
3 Decussation of the brachium conjunctivum
4 Brachium conjunctivum (superior cerebellar peduncle)
5 Cerebellar vermis
6 Pons
7 Trigeminal nerve
8 Ventral spinocerebellar tract
9 Restiform body (inferior cerebellar peduncle)
10 Brachium pontis (middle cerebellar peduncle)
11 Dentate nucleus
12 Cerebellar hemisphere
13 Right inferior olive
14 Olivocerebellar fibres
15 External arcuate fibres
16 Cuneocerebellar tract
17 Dorsal spinocerebellar tract

Fig. 20.1. The peduncles of the cerebellum in a lateral view (3/2×)

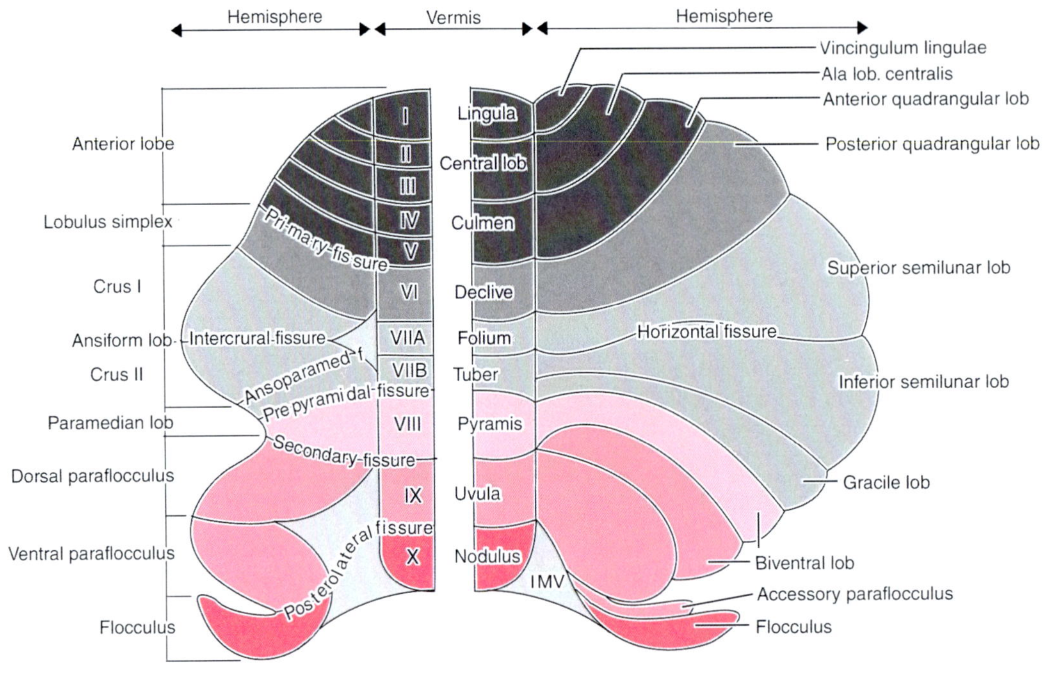

Fig. 20.2. The comparative anatomical nomenclature (*left panel*) and classical nomenclature (*right panel*) of the lobules and fissures of the cerebellum, shown in diagrams of the flattened cerebellar surface. Many of the terms used in the comparative anatomy of the mammalian cerebellum were introduced by Bolk [15]. The *roman numerals*, introduced by Larsell [99] for the cerebellar lobules, are indicated in the left vermis. *iMV*, inferior medullary velum

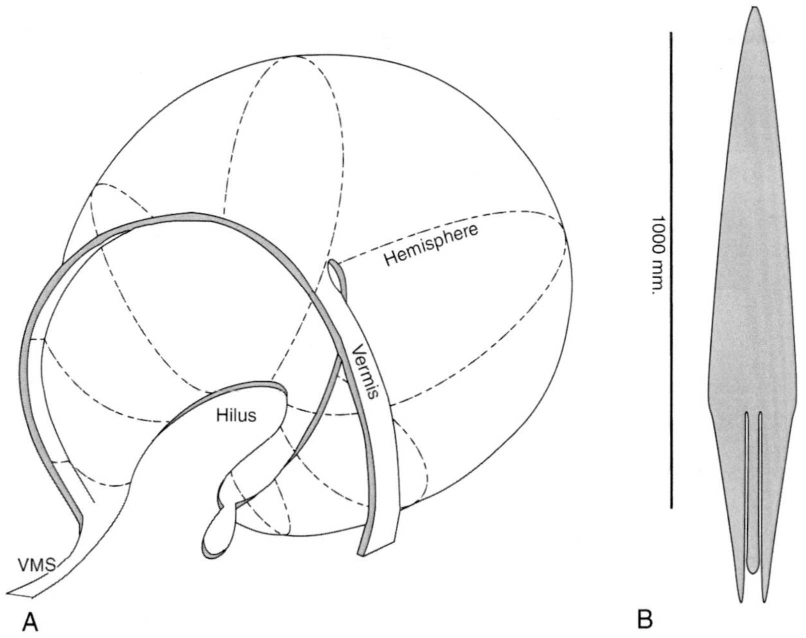

Fig. 20.3. A The cerebellar cortical sheet, showing the discontinuity of the cortex between the caudal vermis and the hemisphere. **B** Length and width of the unfolded cortical sheet of the human cerebellum. Its length exceeds 1 m. Redrawn from Braitenberg and Atwood [17]. *VMS*, superior medullary velum

these vessels, therefore, leads to a severe brain stem syndrome, combined with cerebellar symptoms. The cerebellum is located within the posterior cranial fossa, below the tentorium. Increased pressure in this compartment may cause herniation of the cerebellar tonsils, compression of the brain stem in the foramen magnum and compression of the trochlear nerve by the tentorium.

Lattice Structure of the Cerebellar Cortex and the Microcircuitry of the Cerebellum

The combination of transverse and longitudinal principles in the gross anatomy of the cerebellum can also be recognized in the arrangement of the neurons of the cerebellar cortex and the arrangement of the circuitry of the cerebellum in two perpendicular planes (Figs. 20.4–20.6). The essential features of this circuitry will be presented, together with a description of the main neurons of the cerebellar cortex, the Purkinje cells and the granule cells, and the two types of extracerebellar afferents: the mossy and climbing fibres. The structure of the cortical interneurons complies with the stereotypic architecture of the cerebellum, but their contribution to the processing of information by the cerebellum remains largely unknown. Knowledge of the microcircuitry is essential to evaluate our knowledge of the functional localization within the cerebellum, which is reviewed in the last part of this chapter.

The structure of the cerebellar cortex appears to be uniform throughout the cerebellum. As already mentioned, its elements are arranged in a lattice, i.e. their dendrites and/or axons are orientated in one of two planes, one of which is parallel to the transverse fissures and the other perpendicular to it [17, 28, 45, 129]. The output of the cortex is conducted through the Purkinje cells (Fig. 20.6). These large cells are arranged in a monolayer. Purkinje cells are inhibitory and use gamma-aminobutyric acid (GABA) as their neurotransmitter [86]. Purkinje cell axons traverse the white matter and terminate on neurons of the cerebellar and certain vestibular nuclei. The dendritic trees of the Purkinje cells are flattened on a plane perpendicular to the transverse fissures. They occupy the superficial, cell-poor molecular layer. Proximal dendrites are smooth; distal dendrites are covered with narrow-necked spines (spiny branchlets).

With regard to Purkinje cells, the cerebellar cortex is not homogeneous, but rather a structure subdivided into multiple paired, narrow longitudinal zones (Fig. 20.7). These Purkinje cell zones are characterized by their cortico-nuclear projection to a single cerebellar or vestibular nucleus. Three parallel zones are present in each half of the vermis, and eight or more are present in the hemispheres. Some zones may extend for the entire length of the folial chain of the vermis or hemisphere [174–176, 179]. Multiple zones are present in the nodulus and the flocculus. The longitudinal subdivision of the Purkinje cell layer stems from its origin as a series of discrete Purkinje cell clusters early in development [95, 97, 154, 155] (Fig. 2.17). It can also be recognized in the zonally distributed differences in immunohistochemical properties of the Purkinje cells [78, 131, 161, 177, 181, 182] and in the arrangement of the climbing fibres.

The other main cell type of the cortex are the granule cells (Figs. 21.4, 21.6). These cells are small, extremely abundant and located in the granular layer, underneath the Purkinje cells. Granule cells emit four or five thin dendrites that terminate in digit-like protrusions. Granule cell axons ascend toward the molecular layer, where they split into two branches that run parallel to the transverse fissures and, therefore, are called parallel fibres. Parallel fibres may reach a length of 10 mm or more and terminate with excitatory (glutamatergic) synapses on the spiny branchlets of Purkinje cell dendrites and on the dendrites of cortical interneurons they meet on the way (Fig. 20.6). The parallel fibres are the main determinant of the firing rate of the Purkinje cells. The fast, parallel fibre-evoked spikes of Purkinje cells are known as "simple spikes".

Mossy fibres are cortical afferents that mostly originate outside the cerebellum, in the

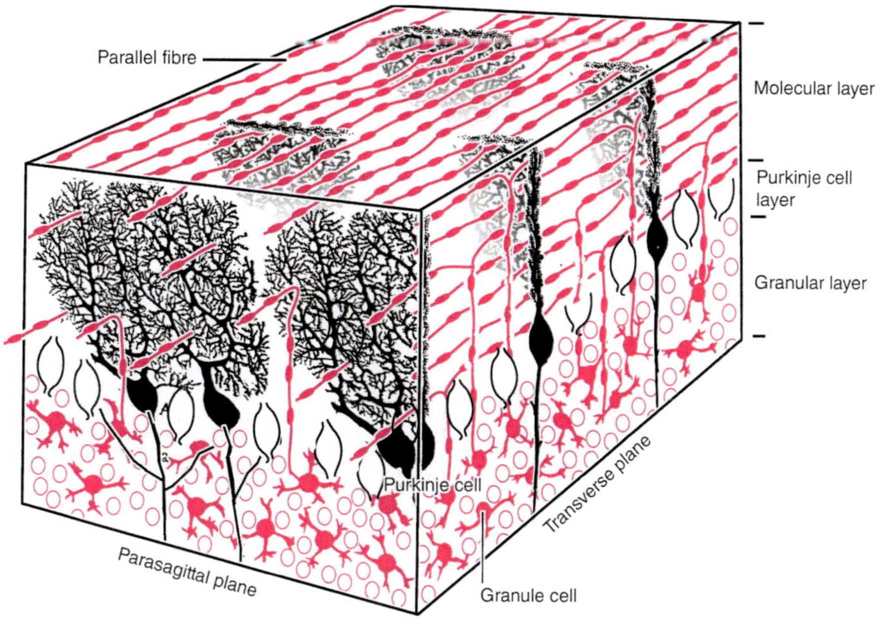

Fig. 20.4. Perpendicular organization of the dendritic trees of the Purkinje cells and the axons of the granule cells, also known as the parallel fibres

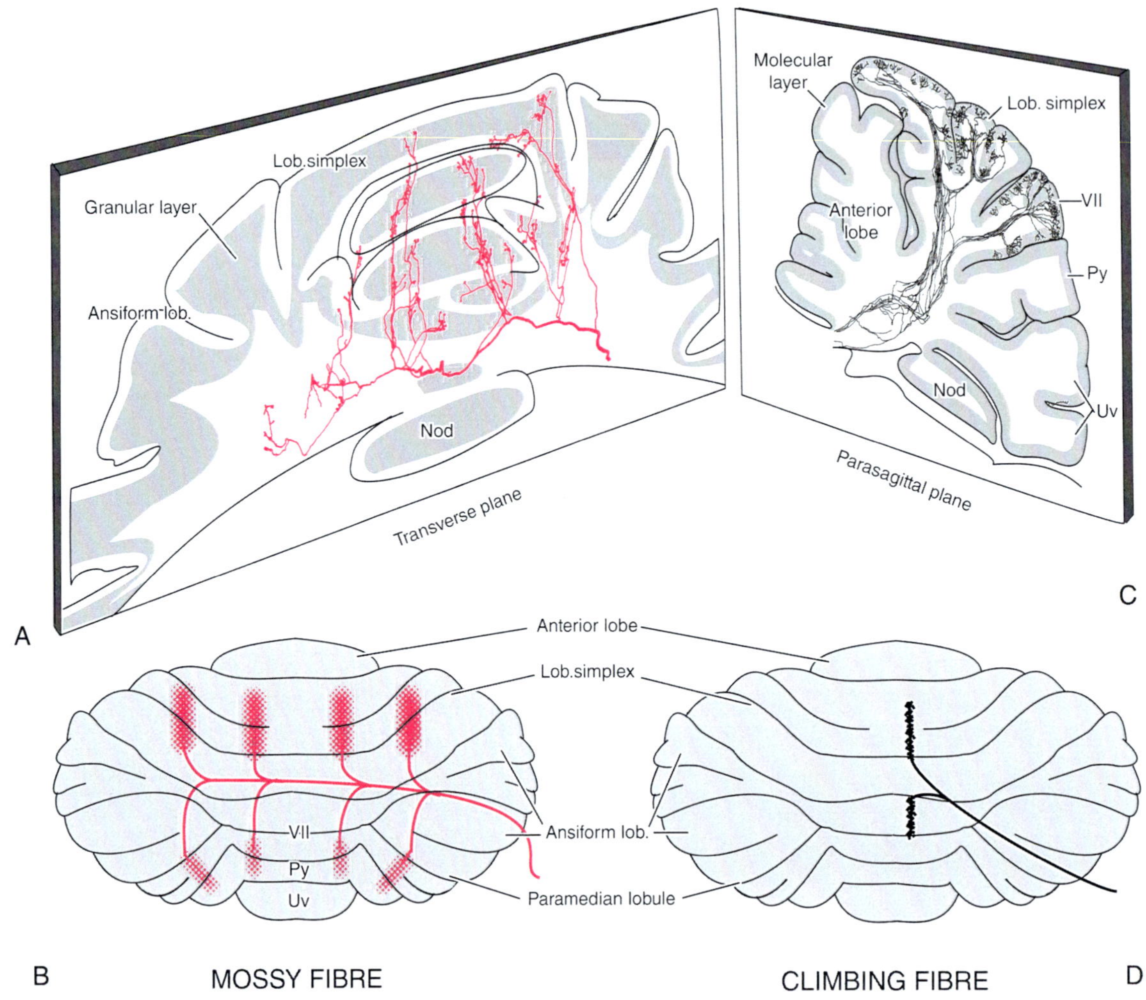

Fig. 20.5 A–D. Mossy and climbing fibres. **A** Mossy fibres (*red*) take their origin from different nuclei in the spinal cord and brain stem. In the cerebellum they are orientated transversely; many mossy fibres cross the midline in the cerebellar commissure (redrawn from a transverse section of the rat cerebellum from Wu et al. [184]). **B** Mossy fibres emit collaterals at certain preferred positions in the anterior and the posterior cerebellum, which terminate in parasagittal aggregates of mossy fibre terminals in the granular layer. **C** Climbing fibres innervate the Purkinje cell dendritic trees in the molecular layer. Climbing fibres originate from the contralateral inferior olive. In the cerebellum they emit 5–10 collaterals that terminate in a one or two Purkinje cell-wide parasagittal strip. The distribution of a small group of climbing fibres originating from a few neighbouring olivary neurons is illustrated in a parasagittal section of the rat cerebellum (redrawn from Sugihara et al. [162]). **D** Climbing fibres terminate in narrow parasagittal strips in anterior and posterior lobules of the cerebellum. The parasagittal mossy fibre terminal aggregates and the climbing fibre strips are in topographical register. They differ because the parasagittal mossy fibre terminal aggregates are multiple and bilaterally distributed. Climbing fibres are distributed unilaterally and terminate in one or two parasagittal strips in the molecular layer (**B**, **D**). *VII*, lobule VII of Larsell; *Nod*, nodulus; *Py*, pyramis; *Uv*, uvula

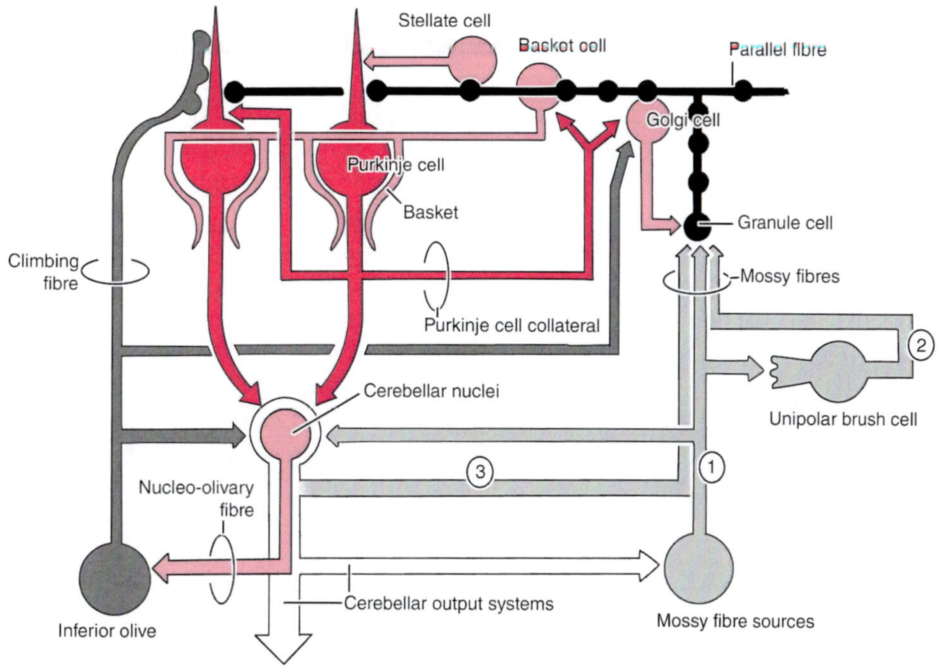

Fig. 20.6. The microcircuitry of the cerebellum. Inhibitory neurons are shown in *red*. They include the Purkinje cells and three types of cortical interneurons. The basket and stellate cells of the molecular layer provide forward inhibition to the Purkinje cells. The Golgi cells inhibit the granule cells in the granular layer. Small GABAergic neurons of the cerebellar nuclei give rise to nucleo-olivary fibres, which inhibit the neurons of the contralateral inferior olive. Both climbing fibres and (some) mossy fibres provide the cerebellar nuclei with excitatory collaterals. Three types of mossy fibres can be distinguished. The most common type originates from nuclei in the spinal cord and the brain stem (*1*). These mossy fibres also terminate with extensive synapses on unipolar brush cells in certain regions of the granular layer. The unipolar brush cells are a second source of the mossy fibres (*2*). Some mossy fibres arise as collaterals from relay cells of the cerebellar nuclei (*3*). Mossy fibre recurrent loops are formed by cerebellar nuclear efferent systems terminating on neurons that give rise to mossy fibres. Climbing fibres terminate on Purkinje cells, Golgi cells and the cerebellar nuclei. The possible innervation of interneurons by mossy fibres, and of stellate and basket cells by climbing fibres is not shown

spinal cord, medulla oblongata and pons, where they arise from the basal pontine nuclei. Mossy fibres are excitatory; most use glutamate as their neurotransmitter. Some secondary vestibulocerebellar mossy fibres are cholinergic [8, 10, 181]. Mossy fibres terminate with large, lobulated terminals (rosettes) in islands of neuropil in the granular layer, where they contact the dendrites of several granule cells. This synaptic complex, which also includes the inhibitory terminals from Golgi cells (Fig. 21.6), is known as the cerebellar glomerulus. Mossy fibres of certain systems produce collaterals that innervate the cerebellar nuclei.

Mossy fibres enter the cerebellum through the restiform body and the middle cerebellar peduncle. Within the white matter of the cerebellum they take a mainly transverse course (Fig. 20.5). At the midline many mossy fibres cross in the cerebellar commissure, to be distributed bilaterally. Mossy fibres emit thin collaterals during their course, which enter the white matter of the lobules. The terminal branches of these collaterals distribute on a plane perpendicular to the transverse fissures and form an elongated plexus located within the granular layer [141, 184]. The collaterals of mossy fibres are not randomly produced, but rather are given off at specific locations. Mossy fibres, therefore, terminate in typical arrays of longitudinal strips in the granular layer, which are specific with respect to their location in certain lobules, their depth profile and their laterality for the different mossy fibre systems. Secondary vestibular mossy fibres, for instance, terminate preferentially in the nodulus, but also in a pattern of bilateral, longitudinal strips in the adjoining uvula and in the granular layer at the bottom of the deep fissures (Fig. 17.7) [107, 114, 167]. Spinocerebellar fibres distribute to the anterior lobe and certain lobules of the posterior lobe (lobulus simplex, pyramis, lobulus gracilis and the lateral belly of the biventral lobule, the paramedian lobule of comparative anatomy; Fig. 20.13). Different spinocerebellar systems have different distributions. Those from the central cervical nucleus overlap completely with the secondary vestibulocerebellar mossy fibres in the bottom of the fissures; others terminate in more apical parts of the granular layer. All spinocerebellar systems display a clear mediolateral periodicity in their termination [115, 116, 172, 185]. Mossy fibres from the pontine nuclei terminate in the apical portions of the lobules and, to a large degree, in the lobules of the hemisphere and the central lobules of the vermis (folium and tuber vermis; Fig. 20.13). The apparent, diffuse termination of these fibres hides a system of alternating stripes of different, but presently unknown, provenance [148–150].

Unipolar brush cells are neurons located in the granular layer of certain lobules, that emit a single dendrite (Fig. 21.6). This dendrite ends in a brush, that forms a large synapse (the synapse “en marron”) with an afferent mossy fibre terminal. The axon of the unipolar brush cell terminates as a mossy fibre on granule cell dendrites. Unipolar brush cells are excitatory, and may serve as a booster system for certain mossy fibre inputs [44, 57].

The climbing fibres constitute the second, main afferent system of the cerebellar cortex (Fig. 20.5). Climbing fibres originate in the medulla oblongata from the contralateral inferior olive, ascend in the restiform body and distribute to the Purkinje cells, with collaterals to the cerebellar nuclei in one half of the cerebellum. Each adult Purkinje cell is innervated by a single climbing fibre, which terminates with multiple, excitatory (glutamatergic) synapses on the smooth, proximal branches of its dendritic tree. The climbing fibre evokes a strong depolarization of the Purkinje cell, resulting in a short spike train (the “complex spike”). Olivocerebellar fibres, however, only conduct impulses at a very slow rate (less than 10 Hz) and the overall contribution of the climbing fibres to the firing rate of the Purkinje cells is small. Other functions of the climbing fibres, therefore, have been proposed [65].

Olivocerebellar axons branch, and each gives rise to seven to ten climbing fibres [162, 163]. Climbing fibres originating from different subnuclei of the inferior olive terminate as discrete longitudinal bands. This zonal pattern is identical to the longitudinal pattern in the projection of the Purkinje cells to the cerebellar nuclei (Fig. 20.7). Each subnucleus of the inferior olive innervates one to three Purkinje cell zones and provides their cerebellar or vestibular target nucleus with a collateral projection [20, 60, 131, 161, 172, 175–177, 179, 182].

An important feature of the inferior olive is the electrotonic coupling of its neurons [37, 38, 106, 156]. Gap junctions are present between their distal dendrites. Afferent systems of the olive regulate the amount of coupling between cells. Excitatory afferents excite olivary neurons, but also increase coupling. Inhibitory afferents inhibit the neurons of the olive and reduce coupling [136]. Excitatory afferents convey somatosensory, vestibular, visual and optokinetic information (see Ruigrok [138] for a review) and include a massive descending pathway known as the central and medial tegmental tract, which serves as the final link in an excitatory feedback loop from the cerebellar nuclei to the olive and which is controlled by the cerebral cortex [173]. Inhibitory afferents of the olive arise from the cerebellar and some vestibular nuclei. The distribution of these nucleo-olivary fibres reciprocates the collateral projection of the subnuclei of the inferior olive to the individual cerebellar and vestibular nuclei [39, 122, 135, 137].

The cerebellar cortex contains two types of intrinsic, inhibitory interneurons (Fig. 20.6). The dendrites of these cells are contained within the molecular layer and receive their main input from the parallel fibres. One set of these neurons, known as basket and stellate cells, resides in the molecular layer. Dendrites and axons of these cells distribute on a plane perpendicular to the transverse fissures. These cells are GABAergic and exert feed-forward inhibition on Purkinje cells. Basket cell axons terminate with terminal baskets around the proximal dendrite and the cell body and with a terminal brush around the initial segment of the axon of the Purkinje cell. Stellate cells project to more distal portions of the Purkinje cell dendritic tree. Golgi cells represent the second type of interneuron. Golgi cell bodies are located in the granular layer, and their dendrites extend into the molecular layer. They colocalize the inhibitory neurotransmitters GABA and glycine [181] and provide feedback inhibition to the granule cells. Their axonal plexus terminates on granule cell dendrites within the glomeruli.

The smallest efferent unit of the cerebellar cortex is the *microzone*. It consists of a narrow, elongated strip of Purkinje cells innervated by a few climbing fibres, which share the same receptive field in the periphery [3, 4] and terminate with axon collaterals on the target neurons of the Purkinje cells in a cerebellar or vestibular nucleus. Microzones have been exclusively studied in the cat, in climbing fibre zones innervated by the dorsal accessory olive [55], which receives somatosensory afferents from the spinal cord and the dorsal column nuclei, and which contains a detailed map of the body surface [61]. Within these zones the microzones and their associated Purkinje cells are arranged in a clear somatotopical pattern. This pattern is repeated in each of the multiple dorsal accessory olive-innervated zones, both in the anterior and posterior regions of the cerebellum. In the anterior lobe these multiple somatotopical representations are due to the transverse branching of two sets of climbing fibres, innervating the C_1 and medial C_3 zones, and the lateral C_3 zone and the Y zone, respectively [50, 51, 53] (Fig. 20.7 C). These zones alternate with zones innervated from other subnuclei of the inferior olive, which lack such a somatotopical arrangement. There are indications for the presence of a microzonal arrangement in other zones as well, but their size, organization and degree of electrotonic coupling between their olivary neurons may be different.

Climbing fibre-defined microzones are of the same magnitude as the terminal plexus of the mossy fibre collaterals (Fig. 20.5). More importantly, microzones and the mossy fibres terminating underneath them share the same receptive field [23, 52, 94]. This correspondence in receptive field organization was first noticed by Eccles [47, 48], who also found a similar relationship for mossy and climbing fibres activated from the motor and sensory cortex [132, 133]. The detailed convergence of sig-

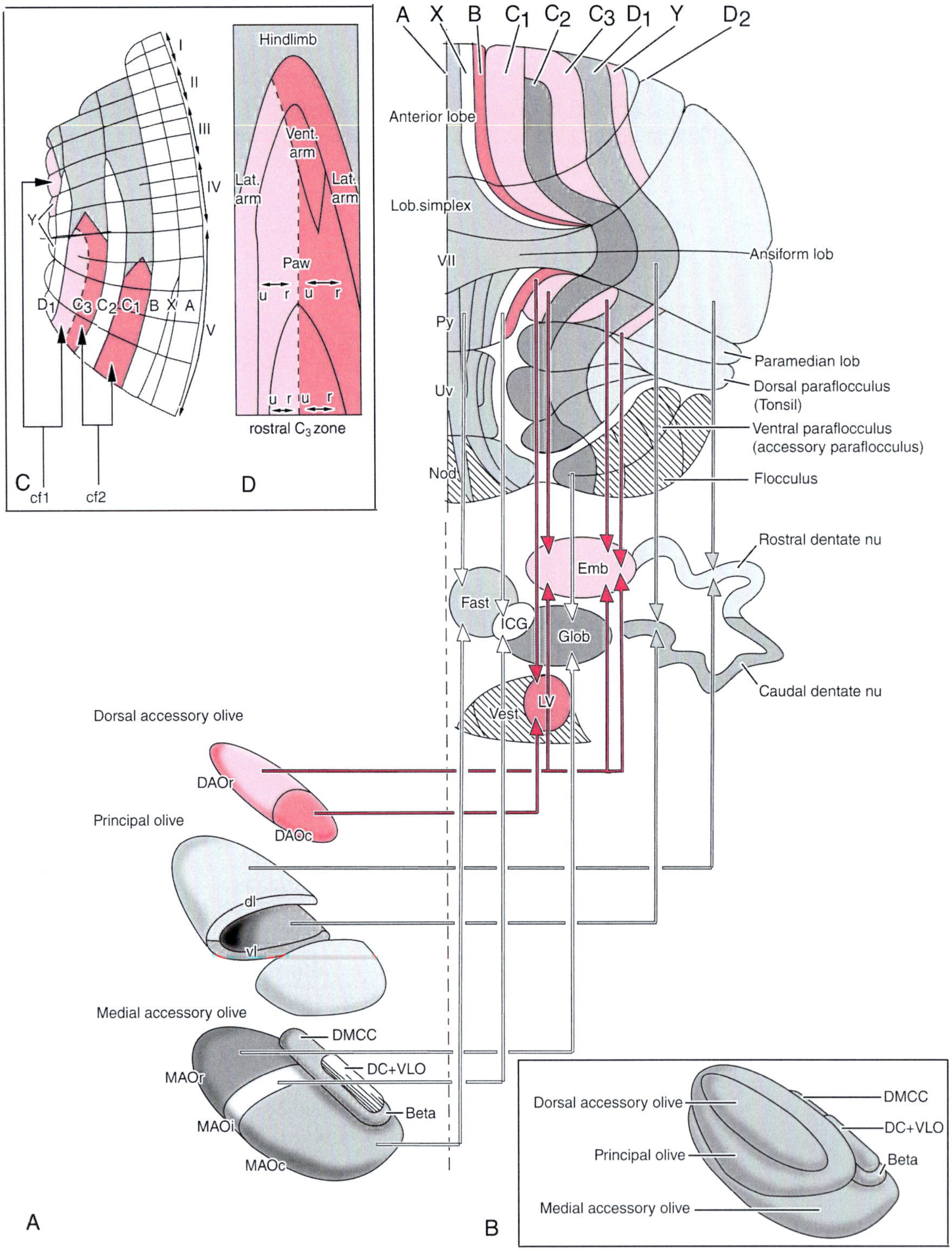

Fig. 20.7. A The modular organization of the cerebellum. Purkinje cells are distributed in longitudinal zones that follow the long axis of the folial chains of vermis and hemispheres. Some of these zones extend over the entire rostrocaudal length of the cerebellum. Each Purkinje cell zone projects to a particular cerebellar or vestibular target nucleus. The C_1, C_3 and Y zones share the emboliform (anterior interposed) nucleus as their target nucleus. The Purkinje cells of each zone and the neurons of their target nucleus are innervated by climbing fibres from a particular subnucleus of the inferior olive. The *arrows* representing the olivocere-

nals of peripheral and cortical provenance, conveyed by different mossy and climbing fibre paths, is remarkable. Mossy fibres differ from climbing fibres because they emit collaterals at multiple locations along their transverse, and often bilateral course through the white matter. Supposedly, the multiple terminal plexuses of these collaterals are located in close proximity to climbing fibres belonging to different olivocerebellar projection zones, but bearing the same signal. An important difference between mossy and climbing fibres is that the former exert their action on Purkinje cells through granule cells and parallel fibres. It has been established that Purkinje cells are preferentially activated by the synapses on the ascending branch of the granule cell axon, i.e. by the granule cells located underneath these Purkinje cells [16, 72]. The synaptic coupling of the parallel fibres with Purkinje cells located along the course of the parallel fibres is much weaker, but this weakness is compensated for by their large numbers. Mossy fibres, therefore, preferentially activate Purkinje cells located above their terminal plexuses, but it should be remembered that their influence is spread along transverse beams of parallel fibres to Purkinje cells with quite different anatomical and functional characteristics.

The mossy fibre–parallel fibre input is orientated at right angles to the output of the cortex through the Purkinje cells of the microzones and the corticonuclear projection zones. This lattice structure guarantees optimal access of the mossy fibre–parallel fibre input to the output systems of the cortex. Within the projection of the zones and microzones to the cerebellar nuclei there is a strong convergence of a large number of Purkinje cells to a much smaller number of nuclear cells. Input through different sets of parallel fibres spaced along a longitudinal zone, therefore, may activate the same group of nuclear neurons.

How are choices made in a network where every input is connected to every output? The existence of hot spots in the cortex, microzones that receive closely related input from mossy and climbing fibres, was already mentioned. Coupling of olivary cells and inhibitory interneurons will also contribute to the regulation of the flow of information in the cortex. The main factor, however, appears to be synaptic plasticity, i.e. changes in the amount of neurotransmitter or the efficacy of the receptor and transducer systems of the synapse. Different types of synaptic plasticity, which may play a role in the cerebellar cortex, were reviewed by Hansel [73]. One kind of plasticity, LTD in the parallel fibre-Purkinje cell synapse, has been studied most extensively [36, 71, 85]. In LTD the efficacy of these synapses is permanently decreased when they are activated si-

bellar projection end at the cerebellar nuclei; their subsequent course to the Purkinje cell zones is not shown, but follows the same trajectory as the corticonuclear projection. The modular organization of the nodulus, the flocculus and the accessory paraflocculus is illustrated in Figs. 17.3, 17.7, 17.8 and 19.13. A cerebellar module consists of a Purkinje cell zone and its target nucleus, its climbing fibre afferent system and a recurrent nucleo-olivary pathway (not illustrated, see Fig. 20.6). **B** The inferior olive in a dorsolateral view. **C** The olivocerebellar projection of the dorsal accessory olive to the C_1, C_3 and Y zones is somatotopically organized. This is illustrated in a diagram of the anterior lobe of the cat. The hind limb is located in the rostral part of the anterior C_1, C_3 and Y zones (*grey*), the fore limb more caudally (*red*). The C_1 and Y zones both contain a single representation of the body surface; the C_3 zone contains a double, mirrored representation. Two different sets of transversely branching climbing fibres innervate the Y and the lateral C_3 zones (*cf1*) and the lateral C_3 and the C_1 zones (*cf2*). **D** Double representation of the forelimb in the rostral C_3 zone. **C** and **D** were redrawn from Ekerot and Larson [51]. *I–VII*, lobules I–VII of Larsell; *A*, A zone *B*, B zone; C_{1-3}, C_{1-3} zones; $D_{1,2}$, $D_{1,2}$ zone; *DAOc*, dorsal accessory olive, caudal part; *DAOr*, dorsal accessory olive, rostral part; *DC*, dorsal cap; *dl*, dorsal lamina of the principal olive; *DMCC*, dorsomedial cell column; *Emb*, emboliform (anterior interposed) cerebellar nucleus; *Fast*, fastigial (medial) cerebellar nucleus; *Glob*, globose (posterior interposed) cerebellar nucleus; *ICG*, interstitial cell groups; *Lat arm*, lateral arm; *LV*, lateral vestibular nucleus; *MAOc*, medial accessory olive, caudal part; *MAOi*, medial accessory olive, intermediate part; *MAOr*, medial accessory olive, rostral part; *Nod*, nodulus; *Py*, pyramis; *r*, radial; *u*, ulnar; *vent arm*, ventral side arm; *vest*, vestibular nuclei; *vl*, ventral lamina of the principal olive; *VLO*, ventrolateral outgrowth; *X*, X zone; *Y*, Y zone

multaneously with the climbing fibre terminating on the same Purkinje cell. LTD is considered the basis for motor learning of the cerebellum, a process that has been extensively studied in long-term adaptation of the vestibulo-ocular reflex that is dependent on the integrity of the cerebellum [33, 83, 84, 104, 120, 151].

In long-term adaptation of the vestibulo-ocular reflex, climbing fibres serve as an error detector. Climbing fibres are activated by movement of the image over the retina, which occurs when the reflex fails to stabilize the image on the retina. The mossy fibre-parallel fibre system conveys information on head and eye movement to the same Purkinje cells. Repeated pairing with the error signal carried by the climbing fibres leads to LTD in the active synapses of the parallel fibres with the Purkinje cells and, ultimately, to disinhibition of the vestibulo-oculomotor neurons in the vestibular nuclei and correction of the error by a change in the gain of the vestibulo-ocular reflex.

Pairing of activity in climbing and parallel fibres also plays a role in the production of conditioned eyeblink responses [58, 87, 165, 186]. When the unconditional stimulus, i.e. touching the cornea, is paired with sound or light as a conditional stimulus, these previously neutral stimuli will acquire the ability to elicit a conditioned eyeblink response. Large lesions of the cerebellar cortex or the nuclei abolish this form of associative learning. It has been found that climbing fibres are activated by the unconditional stimulus and mossy fibres by the conditional stimulus. LTD may develop at the site where the parallel fibres, activated by the conditional stimulus, cross the microzones activated by the unconditional stimulus and thus may represent the sites for memory formation in this form of conditioning. The convergence of the Purkinje cells of these microzones on the same group of target cells in the cerebellar nuclei may explain why different conditional stimuli are effective in eliciting the same response.

These examples illustrate how synaptic plasticity can modify the flow of information in the cortex and the significance of properties of the cortical lattice, such as the focussing of climbing and parallel fibres on particular groups of Purkinje cells, the role of long and narrow Purkinje cell aggregates in the collection of information and the convergence of these Purkinje cells on to the same neurons of the cerebellar nuclei. In LTD, climbing fibres that, as postulated by Marr [110], are closely allied to the output systems of the cerebellum, limit access to the Purkinje cells by mossy fibre-parallel fibre systems, based on previous experience. As a result of LTD only a small fraction of the parallel fibre-Purkinje cell synapses is available at any given moment.

The Cerebellar Nuclei, the Modular Organization of the Cerebellar Output and the Efferent Connections of the Cerebellum

Four cerebellar nuclei can be distinguished, which are arranged into two groups of interconnected nuclei: a medial and caudal group, consisting of the fastigial (medial) and the globose (posterior interposed) nuclei, and a rostral and lateral group consisting of the emboliform (anterior interposed) and the dentate nuclei [124, 159, 183]. A cell group located at the border of the fastigial and posterior interposed nucleus (interstitial cell groups) has been distinguished as a separate nucleus in rat, cat and nonhuman primates [24, 170, 174, 177]. The human dentate nucleus can be subdivided into a rostromedial microgyric portion containing large cells, and a caudal and ventrolateral macrogyric portion containing smaller neurons [41, 42, 173, 183] (Figs. 6.25, 6.26). The lateral vestibular nucleus of Deiters, which traditionally is included in the vestibular nuclei, but does not receive root fibres of the vestibular nerve, should be included in the cerebellar nuclei (Figs. 17.2, 17.3, 20.7). The other "true" vestibular nuclei serve as the target for Purkinje cells of the vestibulocerebellum.

The cerebellar nuclei contain three types of neurons. Excitatory (glutamatergic) neurons of different size give rise to axons that branch profusely and terminate in widely different regions, extending from the spinal cord to the thalamus. Some of their collaterals feed back into the cortex as mossy fibres. Small GABAergic neurons give rise to the nucleo-olivary pathway [9, 122, 135, 181]. Interneurons have been identified as mainly small GABAergic and glycinergic neurons [30].

Among the connections made by excitatory efferents of the nuclei, the projections to the reticular nuclei of the pons and the lower medulla oblongata, and to the parvocellular red nucleus and related nuclei located at the mesodiencephalic border, deserve special attention because these nuclei are links in important, recurrent mossy and climbing fibre pathways [62, 139, 173, 184]. The nucleus reticularis tegmenti pontis and the lateral reticular nucleus give rise to mossy fibres with strong collateral projections to complementary regions in the cerebellar nuclei. Descending pathways to the inferior olive take their origin from the parvocellular red nucleus (the central tegmental tract) and the nuclei of the mesodiencephalic border (the medial tegmental tract) (Figs. 6.23–6.32, 20.11, 20.12).

The corticonuclear projection is organized in a zonal pattern (Fig. 20.7). The zones follow the loops in the folial chains of vermis and hemispheres and thus maintain their orientation perpendicular to the transverse fissures. Corticonuclear projection zones may extend over the entire length of a folial chain or be limited to certain lobules. Each of the cerebellar nuclei serves as the target for one or more of the Purkinje cell zones. The Purkinje cells of each corticonuclear projection zone receives climbing fibres from a subnucleus of the inferior olive, which provides the target nucleus of the zone with a collateral projection. The Purkinje cells of one or more corticonuclear projection zones, their climbing fibre innervation from the inferior olive and their target nucleus function as a unit, an efferent cerebellar module. Different efferent modules operate largely independently of one another [24, 131, 173, 175–177, 182].

Purkinje cells of different corticonuclear projection zones differ in their chemical identity. The differential distribution of 5′-nucleotidase in alternating longitudinal zones in the molecular layer of the cerebellum of the mouse was first observed by Scott [147] and attributed to the Purkinje cells by Marani [108]. The best studied example of the chemical heterogeneity of the Purkinje cells is their immunoreactivity with antibodies against "zebrin," which was extensively studied by Hawkes and his collaborators in rodents, insectivores, carnivores, ungulates and primates [5, 76, 152, 181]. The distribution of the zebrins is identical to that of 5′-nucleotidase [49]. This distribution derives its importance from the observation that it stands for the distribution of a variety of substances in the rodent cerebellum [2, 18, 31, 40, 46, 59, 77, 78, 92, 101, 111, 112, 140, 157, 169, 181] and because zebrin-positive and zebrin-negative Purkinje cell zones differ in their afferent and efferent connections [131, 161, 177, 182]. Unfortunately, the chemical heterogeneity of the cerebellar cortex is rarely taken into account in physiological or molecular studies of the cerebellum, and a unifying hypothesis linking the substances with a similar distribution is still lacking.

Early studies on the modular organization of the cerebellum were reviewed by Voogd and Bigaré [175], Voogd et al. [180] and Brodal and Kawamura [20]. Recent data mainly concern the situation in the rat [24, 25, 131, 161, 177, 182]. Three modules were distinguished in each half of the vermis in carnivores, rodents and nonhuman primates (Fig. 20.7). The medial A zone extends over the entire length of the vermis and projects to the fastigial nucleus. The intermediate X zone (target nucleus: interstitial cell groups) and the lateral B zone (lateral vestibular nucleus of Deiters) are limited to the vermis of the anterior lobe and the declive, and are only represented to a small degree in the pyramis. The A and X zones receive climbing fibres from different cell columns within the caudal medial accessory olive; the B zone is innervated by the caudal half of the dorsal accessory olive.

Three modules, C_1, C_2 and C_3, are present in the paravermal region, also indicated as the "intermediate zone" of the cerebellum from the classical studies of Brodal and Jansen [19, 89, 90]. The C_1 and C_3 zones project to the emboliform (anterior interposed) nucleus. They are limited to the same anterior and posterior regions of the cerebellum as the vermal X and B zones and are found in the anterior lobe, the lobulus simplex, the lobulus gracilis and the biventer (the paramedian lobule of comparative anatomy). The C_1 and C_3 zones receive a somatotopically organized climbing fibre projection from the rostral half of the dorsal accessory olive. The C_2 zone (globose, posterior interposed nucleus) is located between the C_1 and C_3 zones, and extends over the entire cerebellar hemisphere.

The lateral D_1 and D_2 zones project to the dentate nucleus. In rat, cat and monkey, the medial D_1 zone projects to the caudal and ventral dentate, and the lateral D_2 zone to the rostral and dorsal dentate nucleus. D_1 receives a climbing fibre projection from the ventral lamina of the principal olive, D_2 from its dorsal lamina. The D_1 zone is relatively narrow; the D_2 zone probably accounts for much of the width of the cerebellar hemisphere in primates. Although the human dentate nucleus can be subdivided into rostromedial microgyric and caudolateral macrogyric portions, it is not known whether a similar modular organization exists within the lateral hemisphere of the human cerebellum. In rodents and carnivores the Y zone is located between the D_1 and D_2 zones in anterior and posterior portions of the cerebellum. Its connections are similar to the C_1 and C_3 zones of the pars intermedia.

In the rat, Purkinje cells of the B zone and the C_1, C_3 and Y zones, innervated by the dorsal accessory olive, are zebrin-negative. Purkinje cells of the X zone, C_2 zone and the D zones, innervated by the intermediate and rostral medial accessory olive and the principal olive, respectively, are zebrin-positive. Zones innervated by the caudal medial accessory olive, collectively known as the A zone, can be either zebrin-postive or zebrin-negative [161, 177, 182].

The output of the vermis is directed to centres controlling axial and girdle muscles in the brain stem and the spinal cord and to oculomotor centres (Figs. 20.8, 20.9). The most direct connections of the vermis with the spinal cord are effectuated through the lateral vestibulospinal tract, arising from the lateral vestibular nucleus, the target of the B zone, and cerebellospinal fibres from the interstitial cell groups, the target nucleus of the X zone. The latter also send collaterals to the thalamus [13, 14, 113].

The A zone is connected bilaterally through the fastigial nucleus with the vestibular nuclei and the bulbar and pontine medial reticular formation [11]. The crossed fastigiobulbar pathway (uncinate tract) decussates in the cerebellar commissure. The direct fastigiobulbar tract runs in the lateral wall of the fourth ventricle. In the vestibular nuclei they terminate mainly in the rostral, magnocellular medial vestibular nucleus, but circumvent the lateral vestibular nucleus.

Reticulospinal pathways, the medial vestibulospinal tracts and the cervical component of the lateral vestibulospinal tract connect the reticular formation and the vestibular nuclei with the cervical cord (see Chap. 17 and Fig. 17.6 for more details on the vestibulospinal system). The axons of certain medial vestibulospinal tract neurons bifurcate to innervate the oculomotor nuclei. The crossed ascending limb of the uncinate tract connects the fastigial nucleus with the periaqueductal grey and the adjoining reticular formation in the mesencephalon, the superior colliculus, saccade-generating centres in the mesencephalon and the ventral lateral and intralaminar nuclei of the thalamus.

The caudal pole of the fastigial nucleus, which receives a projection from the folium and tuber vermis (lobule VII of Larsell, also known as the oculomotor vermis), projects contralaterally to the excitatory and inhibitory burst cells of the horizontal gaze centre in the paramedian pontine reticular formation, to the rostral interstital nucleus of the medial longitudinal fascicle and the Cajal's interstitial nucleus of the vertical gaze centre and to medial and lateral portions of the pontine nuclei and the nucleus reticularis tegmenti pontis [123] (Figs. 19.17, 20.9).

The output of the emboliform, globose and dentate nuclei is directed through the superior cerebellar peduncle (brachium conjunctivum). This pathway decussates at the pontomesencephalic border and splits into ascending and descending branches (Figs. 20.8. 20.10, 20.12). The ascending branch traverses and surrounds the red nucleus, courses through the subthalamus (prerubral field, zona incerta) and terminates in the thalamus. The descending branch terminates in the nucleus reticularis tegmenti pontis and in the adjoining pontine nuclei. The nucleus reticularis tegmenti pontis gives rise to a bilaterally distributed system of mossy fibres, with strong collateral projections to the cerebellar nuclei (Fig. 19.17). Nucleo-olivary fibres from the emboliform, globose and dentate nuclei constitute a discrete pathway, to be distinguished from the brachium conjunctivum, which decussates at the same level and descends towards the olive, where it terminates in a topical manner.

The emboliform nucleus (anterior interposed nucleus), which receives Purkinje cell axons of the C_1, C_3 and Y zones, projects to the contralateral magnocellular red nucleus and to the thalamus (Fig. 20.10). The magnocellular red nucleus gives rise to the rubrobulbar and spinal tracts, which cross in the mesencephalon. By way of the caudolateral part of the ventral lateral nucleus of the thalamus, the emboliform nucleus projects to the primary motor cortex, the main origin of the pyramidal tract [93]. Descending fibres terminate in the nucleus reticularis pontis. The Purkinje cells of the C_1 and C_3 zones, the emboliform nucleus and its efferent pathways are somatotopically organized.

The output of the C_2 zone through the globose nucleus (posterior interposed nucleus) is mainly directed at the periaqueductal grey, the contralateral superior colliculus and the nucleus of Darkschewitsch [1], located at the mesodiencephalic junction [96] (Figs. 20.1, 20.12). Darkschewitsch nucleus gives rise to the medial tegmental tract, which terminates in the rostral medial accessory olive [35]. This olivary subnucleus projects back to the globose nucleus and the C_2 zone. In the ventral thalamus its fibres cover a fairly broad region, which largely overlaps with the projection of the dentate nucleus. The globose nucleus does not project to the nucleus reticularis pontis. The C_2 zone and the globose nucleus lack a clear somatotopical organization.

The rostrodorsal and the caudolateral portions of the dentate nucleus, which in nonhuman primates are the target nuclei of the D_2 and the D_1 zone, respectively, differ in their efferent connections (Fig. 20.12). The rostrodorsal dentate nucleus is somatotopically organized, projects to the contralateral lateral and rostral parvocellular red nucleus and to lateral regions of the ventral thalamic nucleus. The latter are connected with the motor areas of the frontal lobe. The caudolateral dentate nucleus projects to the dorsomedial subnucleus of the parvocellular red nucleus, adjacent to the fasciculus retroflexus, the superior colliculus and, through medial portions of the ventral nucleus and the dorsomedial nucleus of the thalamus, to the prefrontal cortex (areas 9 and 46), the frontal and supplementary eye fields and the medial intraparietal and rostral inferior parietal areas [32, 80, 96, 118, 119, 128, 171,

[1] Darkschewitsch [34] described the nucleus that bears his name in human fetuses as the rostral, parvocellular nucleus of the oculomotor nerve. The connections of this nucleus with the oculomotor nerve and the posterior commissure have not been confirmed. The origin of the medial tegmental tract from the Darkschewitsch nucleus and its termination in the rostral medial accessory olive was established by Ogawa [126, 127]. The same author observed the great increase in size of this nucleus (renamed as the nucleus ellipticus), in whales together with the magnification of the C_2 zone, the posterior interposed nucleus and the rostral medial accessory olive [125].

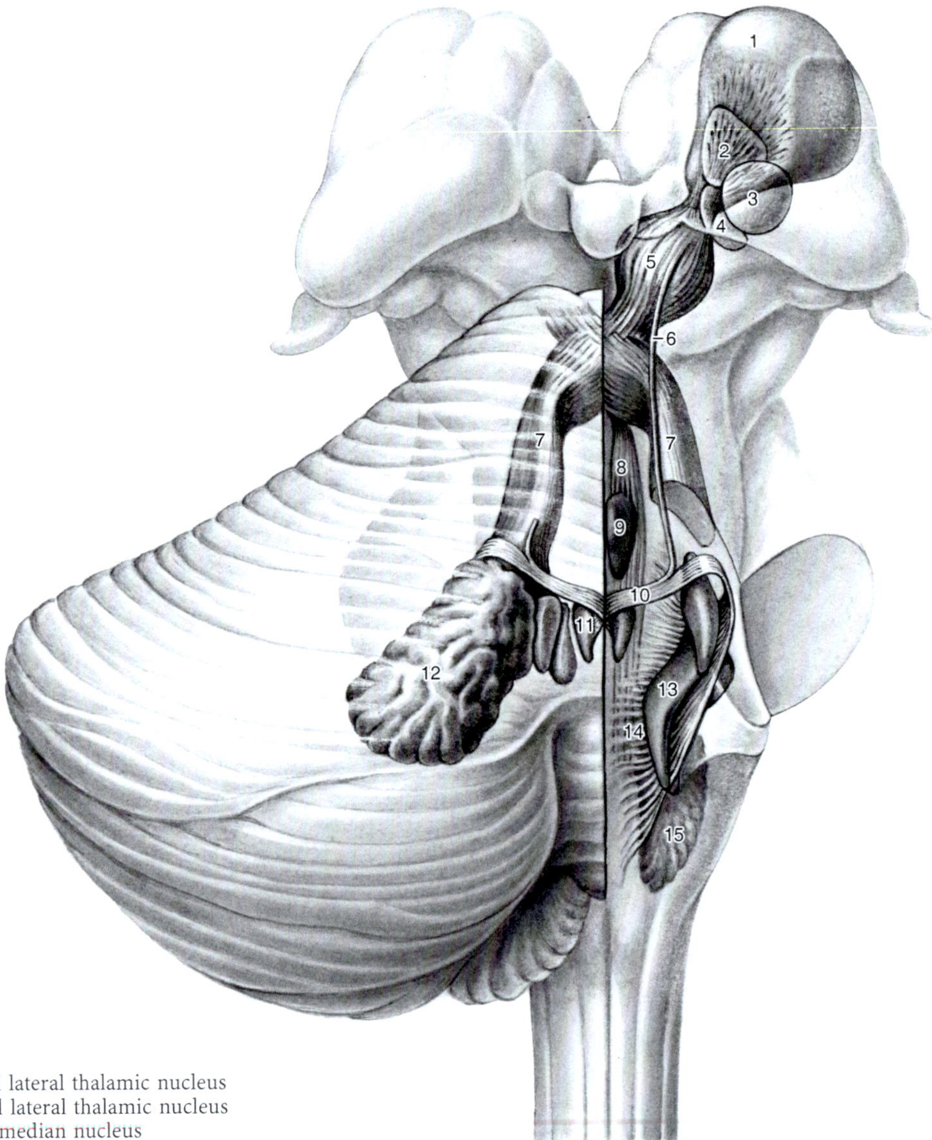

1 Ventral lateral thalamic nucleus
2 Central lateral thalamic nucleus
3 Centromedian nucleus
4 Parafascicular nucleus
5 Red nucleus
6 Uncinate tract, ascending limb
7 Brachium conjunctivum (superior cerebellar peduncle)
8 Brachium conjunctivum, descending branch
9 Nucleus reticularis tegmenti pontis
10 Uncinate tract
11 Fastigial (medial) cerebellar nucleus
12 Dentate nucleus
13 Vestibular nuclei
14 Medullary and pontine reticular formation
15 Inferior olive

Fig. 20.8. The efferent connections of the cerebellum. Position of tracts and nuclei in a dorsal view (5/3×). The cerebellum was split in the midline and the right half removed

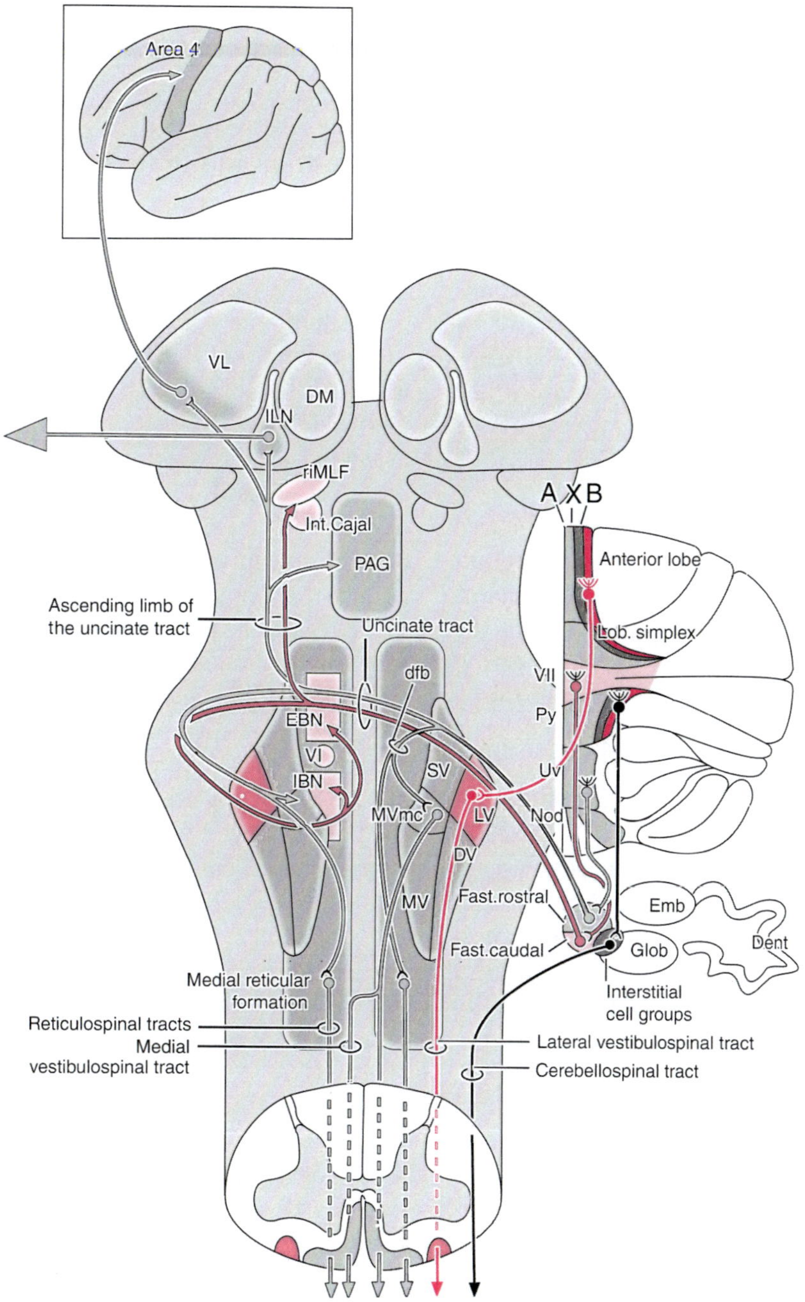

Fig. 20.9. The efferent connections of the vermis and fastigial nucleus. The medial A zone of the vermis projects to the fastigial nucleus. The projection of lobule VII (folium, tuber, the visual vermis) to the caudal pole of the fastigial nucleus is shown in *light red*. The rostral fastigial (*light grey*) nucleus projects bilaterally through the uncinate tract and the direct fastigiobulbar pathway to the vestibular nuclei and the medial reticular formation, which give rise to the medial vestibulospinal and reticulospinal tracts (see Fig. 17.6 for more details). The ascending limb of the uncinate tract terminates in the periaqueductal grey, the intralaminar nuclei and the ventral lateral nucleus of the thalamus. The projection of the caudal pole of the fastigial nucleus is completely crossed and targets the excitatory and inhibitory burst cells in the paramedian pontine reticular formation, the rostral interstitial nucleus of the medial longitudinal fascicle, and Cajal's interstitial nucleus in the mesencephalon (Fig. 19.8). The vermal X zone (*dark grey*) projects to the interstitial cell groups, which give rise to the cerebellospinal tract. The lateral B zone of the vermis (*red*) projects to the lateral vestibular nucleus, the origin of the lateral vestibulospinal tract. *VI*, motor nucleus of the abducens nerve; *VII*, lobule VII of Larsell (folium, tuber), the visual vermis; *A*, A zone; *B*, B zone; *dfb*, direct fastigiobulbar ▶

174]. Through the intralaminar nuclei both parts of the dentate are connected with the striatum [81]. Both parts of the dentate give rise to descending projections to the nucleus reticularis tegmenti pontis and the medial reticular formation.

The central tegmental tract connects the parvocellular red nucleus with the principal olive (Figs. 6.23–6.31). The dorsomedial subnucleus of the parvocellular red nucleus, and its central and lateral portion, project differentially to the ventral and dorsal lamina of the principal olive [160]. A dorsolateral subnucleus of the parvocellular red nucleus projects to the contralateral cervical cord [26].

Apart from its cerebellar afferents, the red nucleus receives a corticorubral projection from the frontal lobe. The corticorubral projection to the magnocellular red nucleus from the primary motor cortex is a collateral projection from the pyramidal tract. The corticorubral projection to the parvocellular red nucleus is an independent pathway from superficial lamina V pyramidal cells of the motor, the premotor and the supplementary and the cingular motor cortex and the frontal and supplementary eye fields. There are only few or no connections from parietal fields; prefrontal projections have not been studied. The projections of the motor and supplementary motor cortex to the rostral and lateral parvocellular red nucleus are somatotopically organized. The dorsomedial subnucleus of the parvocellular red nucleus receives its corticorubral afferents from the rostral premotor area and the frontal and supplementary eye fields. Darkschewitsch nucleus receives projections from the premotor area, the frontal and supplementary eye fields and area 7 [26, 29, 56, 75, 82, 98, 168, 174]. The cerebellar thalamocortical projections of the globose (posterior interposed) and the subdivisions of dentate nucleus, therefore, are reciprocated by the cortico-rubro-olivary pathways.

Localization in the Cerebellum

Different, functional regions have been recognized in the cerebellum, such as the vestibulocerebellum, the spinocerebellum and the pontocerebellum, i.e. the large expanse of the cortex of the hemispheres dominated by corticopontocerebellar connections (Fig. 20.13). However, the designation of such regions is a rather crude way to subdivide the cerebellum. It does not take into account the anatomical overlap and interdigitation of vestibular, spinal and corticopontine connections and neglects the interaction of these divisions in different functional systems. Vestibular, spinal and pontine mossy fibre afferents actually are distributed in a concentric and overlapping manner. Vestibular mossy fibres terminate in anterior and posterior regions of the cerebellum and in the bottom of the deep fissures; spinocerebellar systems terminate more superficially in anterior and posterior regions of the cerebellum. Pontocerebellar fibres terminate in the apex of the lobules and more heavily in folium and tuber vermis (lobule VII of Larsell), the uvula and in more lateral parts of the cerebellum (Figs. 17.7, 20.13). Moreover, both climbing and mossy fibre systems do not terminate diffusely, but rather in discrete, longitudinal strips or aggregates. Mediolateral interdigitation is the hallmark of the climbing fibre zones; it is likely to occur, and has been demonstrated in some instances for different mossy fibre systems [91].

Rather than the term vestibulocerebellum, we prefer to use the term *oculomotor cerebel-*

(continued text of Fig. 20.9) tract; *Dent*, dentate nucleus; *DM*, dorsomedial thalamic nucleus; *DV*, descending vestibular nucleus; *EBC*, excitatory burst cells; *Emb*, emboliform (anterior interposed) nucleus; *Fast*, fastigial (medial) cerebellar nucleus; *Glob*, globose (posterior interposed) cerebellar nucleus; *IBC*, inhibitory burst cells; *ICG*, interstitial cell groups; *ILN*, intralaminar thalamic nuclei; *Int.Cajal*, interstitial nucleus of Cajal; *LV*, lateral vestibular nucleus; *MV*, medial vestibular nucleus; *MVmc*, magnocellular medial vestibular nucleus; *Nod*, nodulus; *PAG*, periaqueductal grey; *Py*, pyramid; *riMLF*, rostral interstitial nucleus of the medial longitudinal fascicle; *SV*, superior vestibular nucleus; *Uv*, uvula; *VL*, ventral lateral thalamic nucleus; *X*, X zone

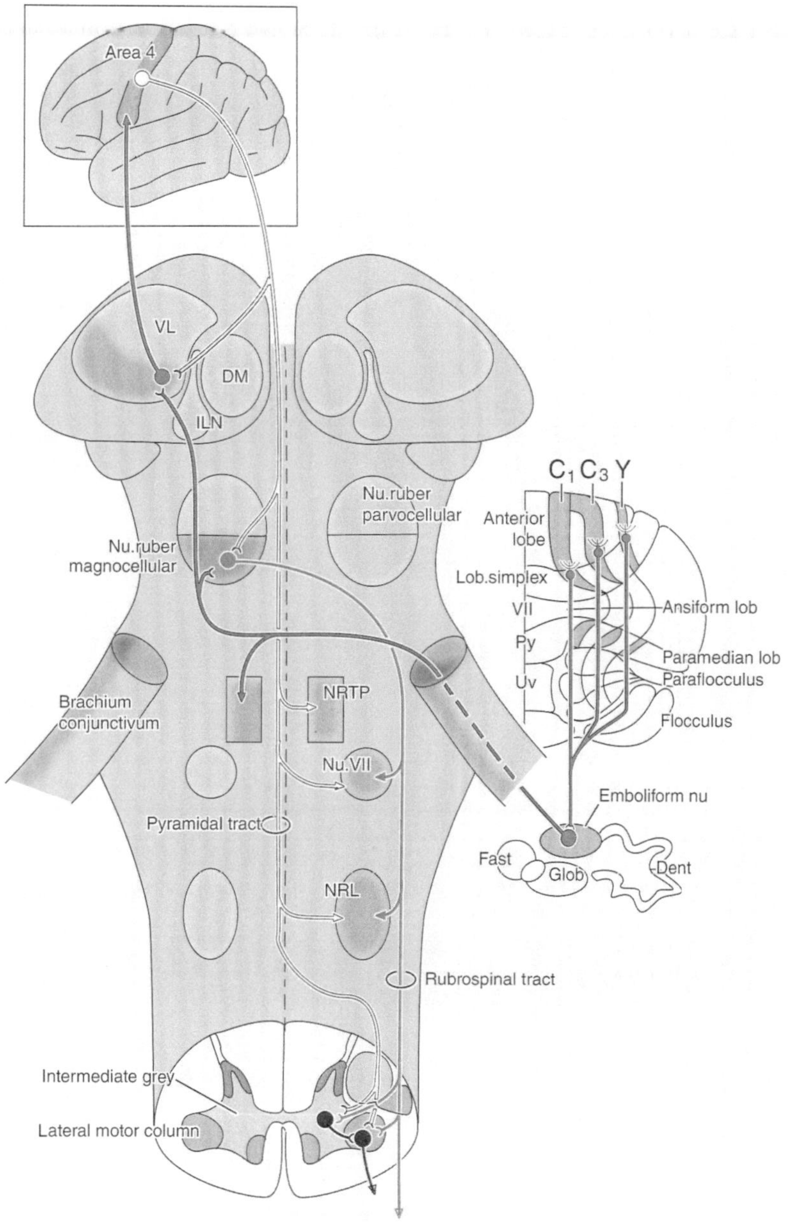

Fig. 20.10. The efferent connections of the C_1, C_3 and Y zones and the emboliform (anterior interposed) cerebellar nucleus. The three zones project to the emboliform nucleus. The efferent pathway of this nucleus joins the brachium conjunctivum, which decussates at the pontomesencephalic border and splits into ascending and descending branches. The descending branch terminates in the nucleus reticularis tegmenti pontis (NRTP). The NRTP gives rise to a recurrent mossy fibre pathway, which provides the cerebellar nuclei with a collateral projection (Fig. 19.17). The ascending branch terminates in the magnocellular red nucleus and the caudal and lateral portion of the ventral lateral nucleus of the thalamus (VL). This subdivision of VL innervates the primary motor cortex (area 4). The cortico- and rubrobulbar and spinal tracts innervate interneurons and motoneurons in the brain stem and the cord. Both tracts also project to the lateral reticular nucleus, which gives rise to a similar recurrent mossy fibre pathway as the NRTP. The primary motor cortex innervates the magnocellular red nucleus with a collateral projection. *VII*, lobule VII of Larsell; $C_{1,3}$, $C_{1,3}$ zones; *Dent*, dentate cerebellar nucleus; *DM*, medial dorsal thalamic nucleus; *Fast*, fastigial (medial) cerebellar nucleus; *Glob*, globose (posterior interposed) cerebellar nucleus; *ILN*, intralaminar thalamic nuclei; *NRL*, lateral reticular nucleus; *NRTP*, nucleus reticularis tegmenti pontis; *Nu.VII*, motor nucleus of the facial nerve; *Py*, pyramid; *Uv*, uvula; *VL*, ventral lateral thalamic nucleus; *Y*, Y zone

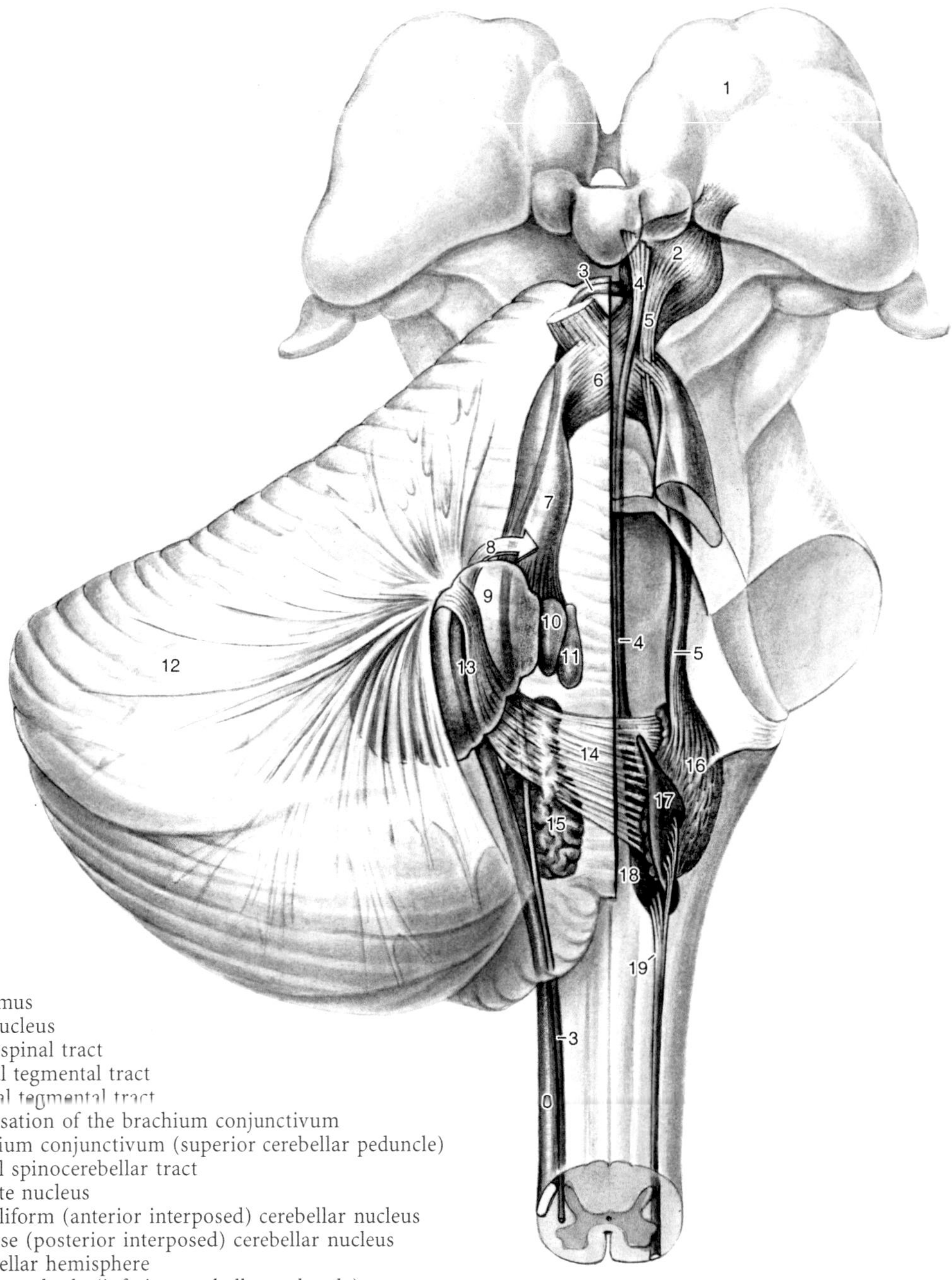

1 Thalamus
2 Red nucleus
3 Rubrospinal tract
4 Medial tegmental tract
5 Central tegmental tract
6 Decussation of the brachium conjunctivum
7 Brachium conjunctivum (superior cerebellar peduncle)
8 Dorsal spinocerebellar tract
9 Dentate nucleus
10 Emboliform (anterior interposed) cerebellar nucleus
11 Globose (posterior interposed) cerebellar nucleus
12 Cerebellar hemisphere
13 Restiform body (inferior cerebellar peduncle)
14 Olivocerebellar fibres
15 Inferior olive
16 Amiculum (fleece) of the inferior olive
17 Dorsal accessory olive
18 Medial accessory olive
19 Spino-olivary tract

Fig. 20.11. The olivocerebellar circuits. Position of tracts and nuclei in a dorsal view (5/3×). The cerebellum was split at the midline and the right half removed

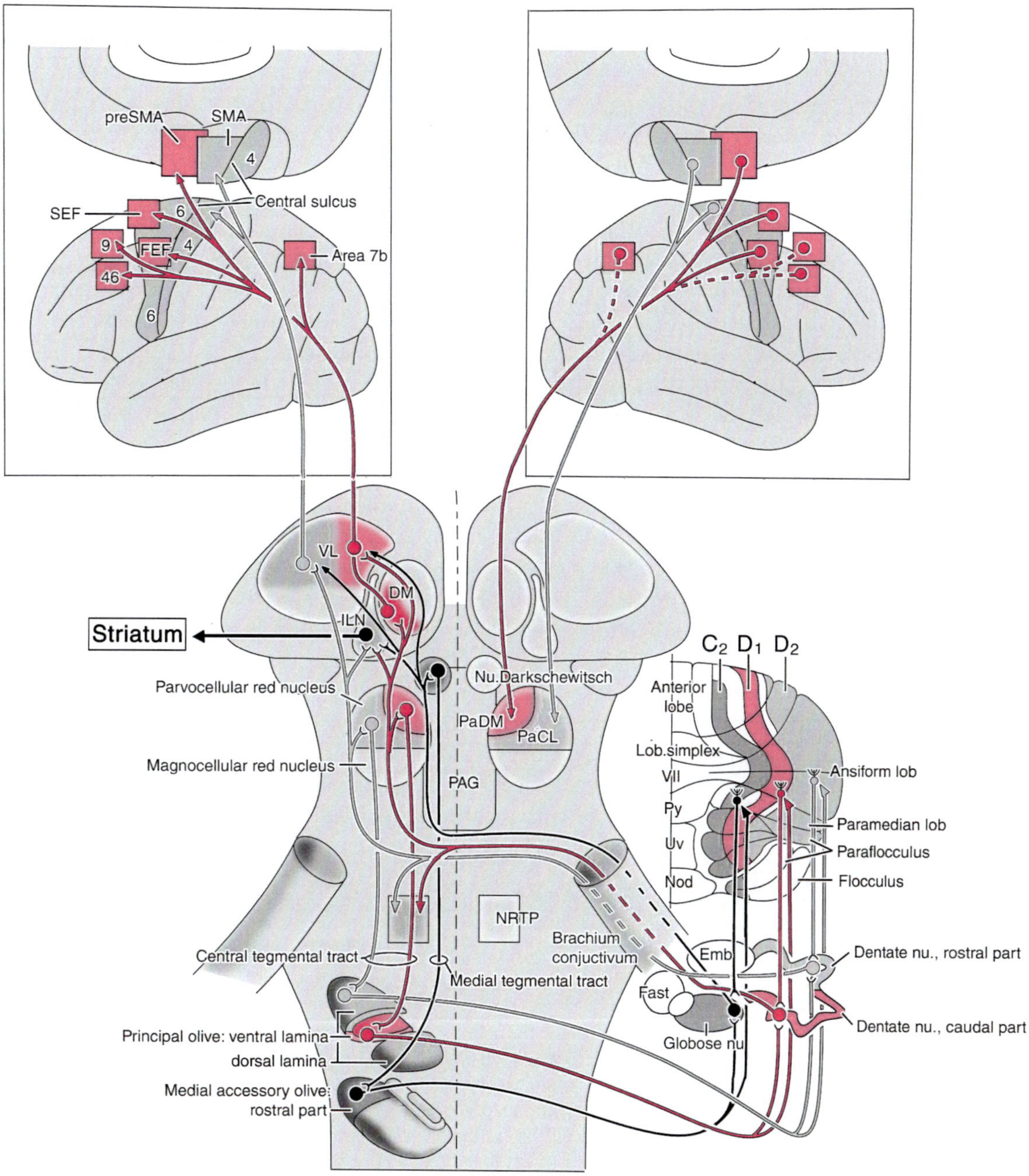

Fig. 20.12. The efferent connections of the C_2, D_1 and D_2 zones and the globose (posterior interposed) and dentate cerebellar nuclei. These Purkinje cell zones and their target nuclei are characterized by strong, recurrent olivocerebellar circuits and extensive connections with the cerebral cortex. The lateral D_2 zone projects to the rostral dentate nucleus (*light grey*). Efferents of the rostral dentate join the brachium conjunctivum, which decussates at the pontomesencephalic border and splits into ascending and descending limbs. The descending limb terminates in the nucleus reticularis tegmenti pontis. The ascending limb terminates in the laterocaudal part of the parvocellular red nucleus, the intralaminar nuclei and the rostrolateral ventral lateral thalamic nucleus. The intralaminar nuclei project to the striatum, the rostrolateral VL projects to the primary motor cortex (area 4) and the premotor cortex (area 6), which includes the supplementary motor cortex (SMA) on the mesial surface of the hemisphere. The laterocaudal parvocellular red nucleus gives rise to an ipsilaterally descending component of the central tegmental tract that terminates in the dorsal lamina of the principal olive. This olivary subnucleus provides the rostral dentate and the D_2 zone with a recurrent climbing fibre projection. The D_1 zone projects to the caudal dentate nucleus (shown in *red*). This nucleus ▶

llum (reviewed by Voogd and Barmack [178]). It includes the flocculus, the paraflocculus (the accessory paraflocculus and the tonsil), the nodulus and adjacent regions of the vermis (ventral uvula) and the visual vermis (lobule VII of Larsell). It is reviewed in more detail in Chaps. 17 and 19 on the vestibular and visual systems (Figs. 17.7, 17.8, 19.13, 19.14, 19.17, 19.18). The cortex of the nodulus, the adjacent uvula and the flocculus are characterized by the presence of unipolar brush cells [45] and a cholinergic microcircuitry [88]. Nodulus and uvula receive primary root fibres from the vestibular nerve, which are largely lacking in the flocculus [64]. Secondary vestibulocerebellar mossy fibres from the vestibular nuclei, some of which are cholinergic [8, 10], terminate in the flocculus and the nodulus-uvula and in the bottom of the deep transverse fissures (Fig. 17.7). Mossy fibre projections to the nodulus-uvula, therefore, are dominated by the labyrinth. In the flocculus, however, most mossy fibres are of a reticular and pontine reticular nuclear origin [178], and relay optokinetic, eye-position and velocity information (Fig. 19.17). Climbing fibre projections from optokinetic subnuclei of the inferior olive (dorsal cap of Kooy, ventrolateral outgrowth) terminate in the flocculus on narrow longitudinal strips of Purkinje cells, which mainly project to groups of vestibulo-oculomotor neurons in the vestibular nuclei, which, in turn, innervate motoneurons of the external eye muscles [166] (Figs. 17.8, 19.13). The flocculus is involved in long-term adaptation of the vestibulo-ocular reflex. A similar modular organization is present in the nodulus-uvula, but here climbing fibre input from the maculae, relayed by specific subnuclei of the olive (group beta, dorsomedial cell column), which is absent from the flocculus, dominates the scene. The output of the nodulus-uvula is mainly to the vestibular nuclei. The uvula-nodulus monitors optokinetic and labyrinthine reflexes and is involved in the pathogenesis of motion sickness.

The oculomotor cerebellum is bordered by an area that includes the tonsil and the lateral belly of the biventral lobule (corresponding to the paraflocculus of lower mammals) and the dorsal uvula. This region receives an exclusive mossy fibre projection from the rostral and lateral pontine nuclei. These regions of the pontine nuclei are a link in the cortico-ponto-cerebellar projection of the striate and extrastriatal visual areas of the cerebral cortex, located in the occipital, parietal and temporal lobes [69, 70, 121, 158] (Figs. 19.17, 19.18). Only the C_2, D_1 and D_2 zones are represented in the para-

(continued text to Fig. 20.12) projects to the dorsomedial part of the contralateral red nucleus, the shell of the medial dorsal thalamic nucleus (DM), the intralaminar nuclei and the rostromedial VL nucleus. The shell of DM and the rostromedial VL project more anteriorly than the lateral and caudal VL. Projections in non-human primates include frontal visuomotor areas (frontal eye field, FEF, supplementary eye field, SEF and area 46 (see Fig. 19.18), area 9, the pre-SMA on the mesial aspect of the hemisphere, the medial intraparietal area and inferior parietal area 7b). The intralaminar nuclei project to the striatum. The component of the central tegmental tract from the dorsomedial parvocellular red nucleus terminates in the ventral lamina of the principal olive, which gives rise to a recurrent climbing fibre path to the D_1 zone and the caudal dentate. The recurrent circuit of the C_2 zone and the globose (posterior interposed) nucleus is shown in *dark grey*. It includes the Darschewitsch nucleus at the mesodiencephalic junction, the medial tegmental tract and the rostral medial accessory olive. The thalamic and cortical projections of the globose nuclei are probably as extensive as those from the dentate nucleus. The globose nucleus lacks a projection to the nucleus reticularis tegmenti pontis. The cortical areas, which receive connections from the dentate nucleus, provide the parvocellular red nucleus with an independent corticorubral projection. Brodmann's areas are indicated with their *numbers*. *VII*, lobule VII of Larsell; *C_2*, C_2 zone; *D_1*, D_1 zone; *D_2*, D_2 zone; *DM*, dorsomedial thalamic nucleus; *Emb*, emboliform (anterior interposed) cerebellar nucleus; *Fast*, fastigial (medial) cerebellar nucleus; *FEF*, frontal eye field; *ILN*, intralaminar thalamic nuclei; *Nod*, nodulus; *NRTP*, nucleus reticularis tegmenti pontis; *PaCL*, caudolateral part of the parvocellular red nucleus; *PaDM*, dorsomedial part of the parvocellular red nucleus; *PAG*, periaqueductal grey; *preSMA*, cortical area located rostral to the supplementary motor cortex; *Py*, pyramid, *SEF*, supplementary eye field; *SMA*, supplementary motor area; *Uv*, uvula; *VL*, ventral lateral thalamic nucleus

flocculus. They project to "visual" regions of the globose and caudal dentate nucleus that are connected with the contralateral superior colliculus and the frontal eye fields. This region of the cerebellar hemisphere, probably, is involved in voluntary eye movements such as smooth pursuit.

The vermal visual area is located in the central lobules of the vermis, folium and tuber vermis. This region is part of the A zone and has been shown to receive visual information from the superior colliculus and visuomotor areas of the cerebral cortex, relayed by mossy fibres from the nucleus reticularis tegmenti pontis and the pontine nuclei [74, 102, 121] (Fig. 19.17) and a tecto-olivary climbing fibre pathway, relayed by the caudal medial accessory olive [79, 123] (Fig. 19.18). The visual vermis projects to the caudal fastigial nucleus, which projects to the horizontal and vertical gaze centres and the superior colliculus [117, 123] (Fig. 20.9). The visual vermis is involved in long-term adaptation of saccades and smooth pursuit [43, 134].

The "spinocerebellum" (Fig. 20.13 B) includes the anterior lobe with the lobulus simplex (declive and lobulus quadrangularis posterior) and the lobulus gracilis with the medial belly of the lobulus biventer (the paramedian lobule of comparative anatomy). It receives mossy fibres from the following sources: (a) primary and secondary vestibulocerebellar pathways [114]; (b) the spinocerebellar tracts, including the external and internal cuneate nuclei [63, 115, 116, 185]; (c) the lateral reticular nucleus [184], which serves as an important relay from the spinal cord, transmitting information from spinal locomotion centres [6, 7] (see Chap. 21, "Locomotion", Fig. 21.17); (d) the trigeminal nuclei; (e) those parts of the pontine nuclei relaying corticopontine projections from the motor and sensory cortex, via collaterals of the pyramidal tract [144]; and (f) the nucleus reticularis tegmenti pontis. The main modules of the spinal cerebellum are the A, X and B zones, with their direct and indirect spinal projections, the C_1 and C_3 zones, which connect with the magnocellular red nucleus and the primary motor cortex, and the D_2 zone, with its somatotopically arranged projection to the rostral and dorsal dentate and its connections with the motor and premotor cortex. The D_1 and C_2 zones are clearly represented in the spinal cerebellum, but their functional significance for this region is not well understood.

The spinocerebellum is somatotopically organized (Fig. 20.13 B). A rough localization in the paravermal region of the hind limb in the ventral lobules of the anterior lobe, the hind limb in its dorsal and caudal parts and the face in the lobulus simplex and of a similar, but reversed sequence in the spinocerebellum of the posterior lobe has been known since it was reported by Snider and Stowell [153] and Adrian [1]. It is the expression of the much finer grain of the somatotopical localization in the microzones of the dorsal accessory olive, innervated B, C_1 and C_3 zones and their corresponding mossy fibre plexus. Much of what we know about the involvement of the cerebellum in movement is based on studies of the spinal cerebellum, its mossy and climbing fibre afferents proceeding from the spinal cord and its effects on the vestibular nuclei, the reticular formation, red nucleus and the motor cortex, transmitted by the classical descending pathways to the cord [66].

Pontocerebellar mossy fibres terminate in all lobules of the cerebellum, with the exception of the nodulus, the flocculus and the accessory paraflocculus (the ventral paraflocculus of the comparative anatomical nomenclature). In medial and intermediate regions of the anterior lobe and the lobulus simplex, the pyramis and the gracile (paramedian) lobule pontocerebellar mossy fibres are restricted to the apex of the lobules (Figs. 19.17, 20.13 C). The pontocerebellar projection is mainly crossed; most of its fibres decussate in the pes pontis. The pontine nuclei relay information from frontal, parietal and occipital areas of the cerebral cortex. Visual and visuomotor areas contribute substantially to the corticopontine projection [21, 22, 69, 143]. In the frontal lobe the origin of the corticopontine projection in humans and monkeys is restricted to the dorsal prefrontal cortex [12]. A contribution of the temporal lobe is small or absent. However, distinct projections from the auditory cortex were mapped

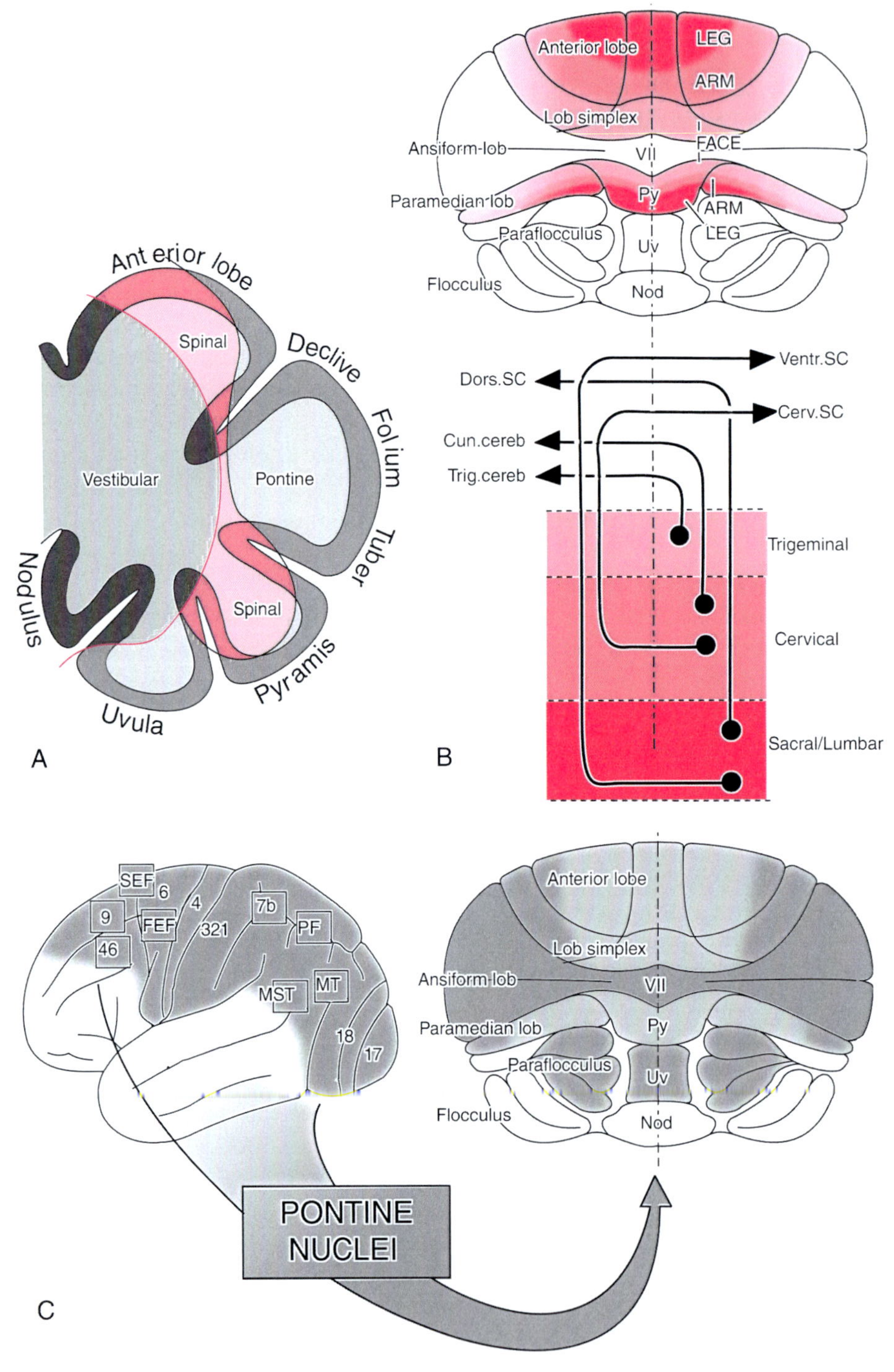

Fig. 20.13 A–C. The distribution of spino- and pontocerebellar mossy fibres. **A** Distribution of the main mossy fibre afferent systems in a midsagittal section of the cerebellum. The main mossy fibre afferent systems are distributed in concentric layers. For the primary and secondary vestibulocerebellar systems these include the nodulus, the ventral uvula, the granular layer in the bottom of the deep fissures and the rostral lobules of the anterior lobe. Spinocerebellar fibres terminate in more apical portions of the anterior lobe, the base of the lobulus simplex and in the pyramid. Pontocerebellar fibres terminate in apical portions of the anterior lobe and the pyramid and in extensive parts of the folium and tuber vermis (lobule VII of Larsell). They are absent from the nodulus. As a consequence, the rostrocaudal distribution of different mossy fibre systems, when followed over the cerebellar surface, is discontinuous. **B** Spinocerebellar fibres take their ori-

in the cat [130]. Their localization is very similar to the more extensive projections from visual cortical areas. Frontal areas that give rise to corticopontine projections, therefore, receive projections from the contralateral cerebellar nuclei (compare Figs. 20.12 and 20.13). Extensive areas in the parietal and occipital lobes, which project to the pontine nuclei, seem to lack cerebellar nuclear afferents.

Function and Malfunction of the Cerebellum

Lesions of the cerebellum result in a loss of motor coordination, known as cerebellar ataxia [67]. There is loss of precision in muscle contraction; movements overshoot or undershoot (dysmetria), and an intention tremor is often present. Loss of coordination and dysmetria also affect speech (cerebellar dysarthria) and ocular movements. Changes in muscle tone, usually hypotonia, are frequently seen. An explanation of the cerebellum's role in motor coordination is still lacking. The laterality and localization of symptoms are usually in accordance with the known anatomy of the input and output channels of the cerebellum. Bilateral, axial ataxia dominates if the vermis is injured. The hind limbs are more affected when the anterior lobe is damaged (l'atrophie cérébelleuse tardive à prédominance corticale of Marie and Alajouanine [109]), and vestibular and oculomotor symptoms result from diseases that affect the vestibulocerebellum. Symptoms from damage to the hemisphere and the interposed and dentate nuclei are lateralized to the ipsilateral side due to the double decussation of the superior cerebellar peduncle and the pyramidal tract. The correlation between symptoms and the topography of lesions remains rather crude, and clinical manifestations of lesions that damage a single cerebellar zone or a single deep cerebellar nucleus have rarely been described.

It was already emphasized in this chapter that the great majority of the experimental, anatomical and physiological studies of the cerebellum have been conducted in the spinal cerebellum, the vestibulocerebellum and the vermal visual area. Large expanses of the cerebellar cortex, including the semilunar lobules (the ansiform lobule) and the biventral lobule and the tonsilla (the paraflocculus) have received very little attention. Damage to these lobules generally results in minor cerebellar symptoms, and complete recovery is the rule. Leiner, Leiner and Dow [103], considering these areas and the great size of the dentate nucleus in humans, especially of its ventrolateral macrogyric portion, suggested that these regions of the cerebellum are connected to the frontal association cortices and subserve cognitive functions. This suggestion has resulted

gin from lumbar, sacral and thoracic levels of the cord as the dorsal and ventral spinocerebellar tracts, and from the cervical cord as the cervicocerebellar and cuneocerebellar tracts. Each of these systems distributes bilaterally in the cerebellum. Lumbosacral afferents terminate in the rostral anterior lobe and the caudal pyramid with the paramedian (gracile) lobule. Cervical and trigeminal afferents terminate more caudally in the anterior lobe and more rostrally in the posterior cerebellum. This distribution is responsible for the double somatotopical localization in the anterior and the posterior cerebellum. **C** Pontocerebellar mossy fibres are the final link in the main cerebrocerebellar mossy fibre system. Corticopontine fibres originate from dorsal prefrontal, premotor and primary motor areas from the parietal and occipital lobes. Projections from ventral prefrontal and temporal areas are scarce. Pontocerebellar mossy fibres terminate bilaterally over the entire cerebellum, with the exception of the nodulus, the flocculus and the ventral (accessory) paraflocculus. In medial portions of the anterior lobe, the pyramid and the paramedian (gracile) lobule the pontocerebellar mossy fibres are restricted to apical portions of the lobules. The cerebrocerebellar projections of visual and visuomotor areas are detailed in Fig. 19.17. The spino- and pontocerebellar projections are illustrated as continuous sheets; however, they actually consist of discrete longitudinal aggregates of mossy fibre terminals (Fig. 20.5). Brodmann's cortical areas are indicated with their *numbers*. *VII*, lobule VII of Larsell (folium/tuber); *FEF*, frontal eye field; *MST*, medial superior temporal visual area; *MT*, middle temporal visual area; *Nod*, nodulus; *PF*, parietal eye field; *Py*, pyramid; *SC*, superior colliculus; *SEF*, supplementary eye field; *Uv*, uvula

in a large number of publications on possible cognitive functions of the cerebellum and the definition of a cognitive cerebellar syndrome (Schmahmann [142]). However, arguments used to define a cognitive function of the cerebellum have been criticized [68]. Anatomical data on the afferent and efferent connections of the primate cerebellar hemisphere, reviewed in this chapter and in Chap. 19 on the visual system, emphasize the visuomotor functions of this region.

References

1. Adrian AD (1943) Afferent areas in the cerebellum connected with the limbs. Brain Res 66:289–315
2. Ahn AH, Dziennis S, Hawkes R, Herrup K (1994) The cloning of zebrin II reveals its identity with aldolase C. Development 120:2081–2090
3. Andersson G, Oscarsson O (1978) Projections to lateral vestibular nucleus from cerebellar climbing fiber zones. Exp Brain Res 32:549–564
4. Andersson G, Oscarsson O (1978) Climbing fiber microzones in cerebellar vermis and their projection to different groups of cells in the lateral vestibular nucleus. Exp Brain Res 32:565–579
5. Armstrong CL, Hawkes R (2000) Pattern formation in the cerebellar cortex. Biochem Cell Biol 78:551–562
6. Armstrong DM (1989) The supraspinal control of mammalian locomotion. J Physiol 405:1–37
7. Arshafsky YI, Gelfand IM, Orlovsky GN (1986) Cerebellum and rythmical movements. Springer, Berlin Heidelberg New York
8. Barmack NH, Baughman RW, Eckenstein FP (1992) Cholinergic innervation of the cerebellum of rat, rabbit, cat, and monkey as revealed by choline acetyltransferase activity and immunohistochemistry. J Comp Neurol 317:233–249
9. Barmack NH, Fredette BJ, Mugnaini E (1998) Parasolitary nucleus: a source of GABAergic vestibular information to the inferior olive of rat and rabbit. J Comp Neurol 392:352–372
10. Barmack NH, Baughman RW, Eckenstein FP, Shojaku H (1992) Secondary vestibular cholinergic projection to the cerebellum of rabbit and rat as revealed by choline acetyltransferase immunohistochemistry, retrograde and orthograde tracers. J Comp Neurol 317:250–270
11. Batton RR, Jayaraman A, Ruggiero D, Carpenter MB (1977) Fastigial efferent projections in the monkey: an autoradiographic study. J Comp Neurol 174:281–305
12. Beck E (1950) The origin, course and termination of the prefronto-pontine tract in the human brain. Brain 73:368–391
13. Bentivoglio M, Kuypers HG (1982) Divergent axon collaterals from rat cerebellar nuclei to diencephalon, mesencephalon, medulla oblongata and cervical cord. A fluorescent double retrograde labeling study. Exp Brain Res 46:339–356
14. Bentivoglio M, Molinari M (1986) Crossed divergent axon collaterals from cerebellar nuclei to thalamus and lateral medulla oblongata in the rat. Brain Res 362:180–184
15. Bolk L (1906) Das Cerebellum der Säugetiere. Fischer, Haarlem
16. Bower JM, Woolston DC (1983) Congruence of spatial organization of tactile projections to granule cell and Purkinje cell layers of cerebellar hemispheres of the albino rat: vertical organization of cerebellar cortex. J Neurophysiol 49:745–766
17. Braitenberg V, Atwood RP (1958) Morphological observations on the cerebellar cortex. J Comp Neurol 109:1–33
18. Brochu G, Maler L, Hawkes R (1990) Zebrin II: a polypeptide antigen expressed selectively by Purkinje cells reveals compartments in rat and fish cerebellum. J Comp Neurol 291:538–552
19. Brodal A (1940) Untersuchungen über die Olivocerebellare Lokalisation. Z Neurol 169:1053
20. Brodal A, Kawamura K (1980) Olivocerebellar projection: a review. Adv Anat Embryol Cell Biol 64:1–137
21. Brodal P (1978) The corticopontine projection in the rhesus monkey. Origin and principles of organization. Brain 101:251–283
22. Brodal P (1979) The pontocerebellar projection in the rhesus monkey: an experimental study with retrograde axonal transport of horseradish peroxidase. Neuroscience 4:193–208
23. Brown IE, Bower JM (2001) Congruence of mossy fiber and climbing fiber tactile projections in the lateral hemispheres of the rat cerebellum. J Comp Neurol 429:59–70
24. Buisseret-Delmas C, Angaut P (1993) The cerebellar olivo-corticonuclear connections in the rat. Prog Neurobiol 40:63–87
25. Buisseret-Delmas C, Yatim N, Buisseret P, Angaut P (1993) The X zone and CX subzone of the cerebellum in the rat. Neurosci Res 16:195–207
26. Burman K, Darian-Smith C, Darian-Smith I (2000) Macaque red nucleus: origins of spinal and olivary projections and terminations of cortical inputs. J Comp Neurol 423:179–196
27. Cajal SR y (1888) Estructura de los centros nerviosos de las aves. I. Cerebelo. Revist trimestr de Histol norm y patol 1
28. Cajal SR y (1972) Histologie du système nerveux de l'homme et des vertebrés. Consejo Superior de Investigaciones Cientificas, Madrid
29. Catsman-Berrevoets CE, Kuypers HJGM, Lemon RN (1979) Cells of origin of the cortical projections to magnocellular and parvocellular red nucleus and

superior colliculus in cynomolgus monkey. An HRP study. Neurosci Lett 12:41–46
30. Chen S, Hillman DE (1993) Colocalization of neurotransmitters in the deep cerebellar nuclei. J Neurocytol 22:81–91
31. Chen S, Hillman DE (1994) Immunohistochemical localization of protein kinase C delta during postnatal development of the cerebellum. Brain Res Dev Brain Res 80:19–25
32. Clower DM, West RA, Lynch JC, Strick PL (2001) The inferior parietal lobule is the target of output from the superior colliculus, hippocampus, and cerebellum. J Neurosci 21:6283–6291
33. Collewijn H, Van der Steen J (1987) Visual control of the vestibulo-ocular reflex in the rabbit: a multilevel interaction. In: Glickstein M, Yeo C, Stein J (eds) Cerebellum and neuronal plasticity. Plenum, London, pp 277–291
34. Darkschewitsch LO (1889) Über den oberen Kern des Oculomotorius. Arch f Anat u Physiol Anat Abth:107–116
35. De Zeeuw CI, Ruigrok TJ (1994) Olivary projecting neurons in the nucleus of Darkschewitsch in the cat receive excitatory monosynaptic input from the cerebellar nuclei. Brain Res 653:345–350
36. De Zeeuw CI, Yeo CH (2005) Time and tide in cerebellar memory formation. Curr Opin Neurobiol 15:667–674
37. De Zeeuw CI, Holstege JC, Ruigrok TJ, Voogd J (1989) Ultrastructural study of the GABAergic, cerebellar, and mesodiencephalic innervation of the cat medial accessory olive: anterograde tracing combined with immunocytochemistry. J Comp Neurol 284:12–35
38. De Zeeuw CI, Holstege JC, Ruigrok TJ, Voogd J (1990) Mesodiencephalic and cerebellar terminals terminate upon the same dendritic spines in the glomeruli of the cat and rat inferior olive: an ultrastructural study using a combination of [^{3}H]leucine and wheat germ agglutinin coupled horseradish peroxidase anterograde tracing. Neuroscience 34:645–655
39. De Zeeuw CI, Gerrits NM, Voogd J, Leonard CS, Simpson JI (1994) The rostral dorsal cap and ventrolateral outgrowth of the rabbit inferior olive receive a GABAergic input from dorsal group Y and the ventral dentate nucleus. J Comp Neurol 341:420–432
40. Dehnes Y, Chaudhry FA, Ullensvang K, Lehre KP, Storm-Mathisen J, Danbolt NC (1998) The glutamate transporter EAAT4 in rat cerebellar Purkinje cells: a glutamate-gated chloride channel concentrated near the synapse in parts of the dendritic membrane facing astroglia. J Neurosci 18:3606–3619
41. Demolé V (1927) Structure et connexions des noyeaux dentelés du cervelet. I. Schweiz Arch Psychiat Neurol 20:271–294
42. Demolé V (1927) Structure et connexions des noyeaux dentelés du cervelet. II. Schweiz Arch Psychiat Neurol 21:73–110
43. Desmurget M, Pelisson D, Grethe JS, Alexander GE, Urquizar C, Prablanc C, Grafton ST (2000) Functional adaptation of reactive saccades in humans: a PET study. Exp Brain Res 132:243–259
44. Dino MR, Schuerger RJ, Liu Y, Slater NT, Mugnaini E (2000) Unipolar brush cell: a potential feedforward excitatory interneuron of the cerebellum. Neuroscience 98:625–636
45. Diño MR, Nunzi MG, Anelli R, Mugnaini E (2000) Unipolar brush cells of the vestibulocerebellum: afferents and targets. Prog Brain Res 124:123–137
46. Dusart I, Morel MP, Sotelo C (1994) Parasagittal compartmentation of adult rat Purkinje cells expressing the low-affinity nerve growth factor receptor: changes of pattern expression after a traumatic lesion. Neuroscience 63:351–356
47. Eccles JC, Provini L, Strata P, Taborikova H (1968) Analysis of electrical potentials evoked in the cerebellar anterior lobe by stimulation of hindlimb and forelimb nerves. Exp Brain Res 6:171–194
48. Eccles JC, Provini L, Strata P, Taborikova H (1968) Topographical investigations on the climbing fiber inputs from forelimb and hindlimb afferents to the cerebellar anterior lobe. Exp Brain Res 6:195–215
49. Eisenman LM, Hawkes R (1989) 5′-Nucleotidase and the mabQ113 antigen share a common distribution in the cerebellar cortex of the mouse. Neuroscience 31:231–235
50. Ekerot CF, Larson B (1979) The dorsal spino-olivocerebellar system in the cat. II. Somatotopical organization. Exp Brain Res 36:219–232
51. Ekerot CF, Larson B (1979) The dorsal spino-olivocerebellar system in the cat. I. Functional organization and termination in the anterior lobe. Exp Brain Res 36:201–217
52. Ekerot CF, Larson B (1980) Termination in overlapping sagittal zones in cerebellar anterior lobe of mossy and climbing fiber paths activated from dorsal funiculus. Exp Brain Res 38:163–172
53. Ekerot CF, Larson B (1982) Branching of olivary axons to innervate pairs of sagittal zones in the cerebellar anterior lobe of the cat. Exp Brain Res 48:185–198
54. Ekerot CF, Jörntell H (2003) Parallel fiber receptive fields: a key to understanding cerebellar operation and learning. Cerebellum 2:101–109
55. Ekerot CF, Garwicz M, Schouenborg J (1991) Topography and nociceptive receptive fields of climbing fibres projecting to the cerebellar anterior lobe in the cat. J Physiol 441:257–274
56. Faugier-Grimaud S, Ventre J (1989) Anatomic connections of the inferior parietal cortex (area 7) with subcortical structures related to vestibulo-ocular function in a monkey (Macaca fascicularis). J Comp Neurol 280:1–14
57. Floris A, Dino M, Jacobowitz DM, Mugnaini E (1994) The unipolar brush cells of the rat cerebellar cortex and cochlear nucleus are calretinin-positive: a study by light and electron microscopic immunocytochemistry. Anat Embryol (Berl) 189:495–520
58. Freeman JH, Rabinak CA (2004) Eyeblink conditioning in rats using pontine stimulation as a con-

ditioned stimulus. Integr Physiol Behav Sci 39:180–191
59. Fusco FR, Viscomi MT, Bernardi G, Molinari M (2001) Localization of ataxin-2 within the cerebellar cortex of the rat. Brain Res Bull 56:343–347
60. Garwicz M (1997) Sagittal zonal organization of climbing fibre input to the cerebellar anterior lobe of the ferret. Exp Brain Res 117:389–398
61. Gellman R, Houk JC, Gibson AR (1983) Somatosensory properties of the inferior olive of the cat. J Comp Neurol 215:228–243
62. Gerrits NM, Voogd J (1986) The nucleus reticularis tegmenti pontis and the adjacent rostral paramedian reticular formation: differential projections to the cerebellum and the caudal brain stem. Exp Brain Res 62:29–45
63. Gerrits NM, Voogd J, Nas WS (1985) Cerebellar and olivary projections of the external and rostral internal cuneate nuclei in the cat. Exp Brain Res 57:239–255
64. Gerrits NM, Epema AH, van Linge A, Dalm E (1989) The primary vestibulocerebellar projection in the rabbit: absence of primary afferents in the flocculus. Neurosci Lett 105:27–33
65. Gibson AR, Horn KM, Pong M (2004) Activation of climbing fibers. Cerebellum 3:212–221
66. Gibson AR, Horn KM, Pong M, Van Kan PL (1998) Construction of a reach-to-grasp. Novartis Found Symp 218:233–245
67. Gilman S, Bloedel JR, Lechtenberg R (1981) Disorders of the cerebellum. Davis, Philadelphia
68. Glickstein M (1993) Motor skills but not cognitive tasks. Trends Neurosci 16:450–451
69. Glickstein M, May JG, 3rd, Mercier BE (1985) Corticopontine projection in the macaque: the distribution of labelled cortical cells after large injections of horseradish peroxidase in the pontine nuclei. J Comp Neurol 235:343–359
70. Glickstein M, Gerrits N, Kralj-Hans I, Mercier B, Stein J, Voogd J (1994) Visual pontocerebellar projections in the macaque. J Comp Neurol 349:51–72
71. Goossens J, Daniel H, Rancillac A, van der Steen J, Oberdick J, Crepel F, De Zeeuw CI, Frens MA (2001) Expression of protein kinase C inhibitor blocks cerebellar long-term depression without affecting Purkinje cell excitability in alert mice. J Neurosci 21:5813–5823
72. Gundappa-Sulur G, De Schutter E, Bower JM (1999) Ascending granule cell axon: an important component of cerebellar cortical circuitry. J Comp Neurol 408:580–596
73. Hansel C, Linden DJ, D'Angelo E (2001) Beyond parallel fiber LTD: the diversity of synaptic and non-synaptic plasticity in the cerebellum. Nat Neurosci 4:467–475
74. Harting JK (1977) Descending pathways from the superior collicullus: an autoradiographic analysis in the rhesus monkey (Macaca mulatta). J Comp Neurol 173:583–612
75. Hartmann- von Monakow KH, Akert K, Künzle H (1979) Projections of precentral and premotor cortex to the red nucleus and other midbrain areas in Macaca fascicularis. Exp Brain Res 34:91–105
76. Hawkes R, Leclerc N (1987) Antigenic map of the rat cerebellar cortex: the distribution of parasagittal bands as revealed by monoclonal anti-Purkinje cell antibody mapQ113. J Comp Neurol 256:29–41
77. Hawkes R, Turner RW (1994) Compartmentation of NADPH-diaphorase activity in the mouse cerebellar cortex. J Comp Neurol 346:499–516
78. Hawkes R, Herrup K (1995) Aldolase C/zebrin II and the regionalization of the cerebellum. J Mol Neurosci 6:147–158
79. Hess DT (1982) The tecto-olivo-cerebellar pathway in the rat. Brain Res 250:143–148
80. Hoover JE, Strick PL (1999) The organization of cerebellar and basal ganglia outputs to primary motor cortex as revealed by retrograde transneuronal transport of herpes simplex virus type 1. J Neurosci 19:1446–1463
81. Hoshi E, Tremblay L, Feger J, Carras PL, Strick PL (2005) The cerebellum communicates with the basal ganglia. Nat Neurosci 8:1491–1493
82. Humphrey DR, Gold R, Reed DJ (1984) Sites, laminar and topographical origins of cortical projections to the major divisions of the red nucleus in the monkey. J Comp Neurol 225:75–94
83. Ito M (1972) Neural design of the cerebellar motor control system. Brain Res 40:81–84
84. Ito M (1982) Cerebellar control of the cerebellar vestibulo-ocular reflex – around the floccular hypothesis. Ann Rev Neurosci 5:275–296
85. Ito M (2001) Cerebellar long term depression: characterization, signal transduction, and functional roles. Physiol Rev 81:1143–1195
86. Ito M, Yoshida M, Obata K (1964) Monosynaptic inhibition of the cerebellar nuclei induced from the cerebellar cortex. Experientia 20:575–576
87. Ivarsson M, Svensson P, Hesslow G (1997) Bilateral disruption of conditioned responses after unilateral blockade of cerebellar output in the decerebrate ferret. J Physiol 502:189–201
88. Jaarsma D, Ruigrok TJ, Caffé R, Cozzari C, Levey AI, Mugnaini E, Voogd J (1997) Cholinergic innervation and receptors in the cerebellum. Prog Brain Res 114:67–96
89. Jansen J, Brodal A (1940) Experimental studies on the intrinsic fibers of the cerebellum. II. The corticonuclear projection. J Comp Neurol 73:267–321
90. Jansen J, Brodal A (1942) Experimental studies on the intrinsic fibers of the cerebellum. III. Corticonuclear projection in the rabbit and the monkey. Norsk Vid Akad Avh I Math Nat Kl 3:1–50
91. Ji Z, Hawkes R (1994) Topography of Purkinje cell compartments and mossy fiber terminal fields in lobules II and III of the rat cerebellar cortex: spinocerebellar and cuneocerebellar projections. Neuroscience 61:935–954
92. Jinno S, Jeromin A, Roder J, Kosaka T (2003) Compartmentation of the mouse cerebellar cortex by neuronal calcium sensor-1. J Comp Neurol 458: 412–424

93. Jörntell H, Ekerot CF (1999) Topographical organization of projections to cat motor cortex from nucleus interpositus anterior and forelimb skin. J Physiol 514:551–566
94. Jörntell H, Garwicz M, Ekerot CF (1996) Relation between cutaneous receptive fields and muscle afferent input to climbing fibres projecting to the cerebellar C3 zone in the cat. Eur J Neurosci 8:1769–1779
95. Kappel RM (1981) The development of the cerebellum in Macaca mulatta. A study of regional differences during corticogenesis. Thesis, Leiden
96. Kievit J (1979) Cerebello-thalamische projecties en de afferente verbindingen naar de frontaalschors in de rhesus aap. Thesis, Rotterdam
97. Korneliussen HK (1968) Comments on the cerebellum and its division. Brain Res 8:229–236
98. Kuypers HG, Lawrence DG (1967) Cortical projections to the red nucleus and the brain stem in the Rhesus monkey. Brain Res 4:151–188
99. Larsell O (1952) The morphogenesis and adult pattern of the lobules and tissues of the cerebellum of the white rat. J Comp Neurol 97:281–356
100. Larsell O, Jansen J (1972) The comparative anatomy and histology of the cerebellum. III. The human cerebellum, cerebellar connections, and cerebellar cortex. University of Minnesota Press, Minneapolis
101. Leclerc N, Schwarting GA, Herrup K, Hawkes R, Yamamoto M (1992) Compartmentation in mammalian cerebellum: Zebrin II and P-path antibodies define three classes of sagittally organized bands of Purkinje cells. Proc Natl Acad Sci USA 89:5006–5010
102. Legg CR, Mercier B, Glickstein M (1989) Corticopontine projection in the rat: the distribution of labelled cortical cells after large injections of horseradish peroxidase in the pontine nuclei. J Comp Neurol 286:427–441
103. Leiner HC, Leiner AL, Dow RS (1986) Does the cerebellum contribute to mental skills? Behav Neurosci 100:443–454
104. Lisberger SG (1998) Physiologic basis for motor learning in the vestibulo-ocular reflex. Otolaryngol Head Neck Surg 119:43–48
105. Llinas R, Sotelo C (1992) The cerebellum revisited. Springer, Heidelberg
106. Llinas R, Baker R, Sotelo C (1974) Electrotonic coupling between neurons in cat inferior olive. J Neurophysiol 37:560–571
107. Maklad A, Fritzsch B (2003) Partial segregation of posterior crista and saccular fibers to the nodulus and uvula of the cerebellum in mice, and its development. Brain Res Dev Brain Res 140:223–236
108. Marani E (1986) Topographic histochemistry of the cerebellum. Prog Histol Cytochem 16:1–169
109. Marie P, Alajouanine A (1922) L'atrophie cérébelleuse tardive à prédominance corticale. Rev Neurol 38:849–885, 1082–1111
110. Marr D (1969) A theory of the cerebellar cortex. J Physiol 202:437–470
111. Martinez-Murillo R, Bentura ML, Fernandez AM, Nieto-Sampedro M, Rodrigo J (1995) Chemical heterogeneity in adult rat cerebellar Purkinje cells as revealed by zebrin I and low-affinity nerve growth factor receptor immunocytochemical expression following injury. J Neurocytol 24:807–817
112. Marzban H, Khanzada U, Shabir S, Hawkes R, Langnaese K, Smalla KH, Bockers TM, Gundelfinger ED, Gordon-Weeks PR, Beesley PW (2003) Expression of the immunoglobulin superfamily neuroplastin adhesion molecules in adult and developing mouse cerebellum and their localisation to parasagittal stripes. J Comp Neurol 462:286–301
113. Matsushita M, Hosoya Y (1978) The location of spinal projection neurons in the cerebellar nuclei (cerebellospinal tract neurons) of the cat. A study with the horseradish peroxidase technique. Brain Res 142:237–248
114. Matsushita M, Wang CL (1987) Projection pattern of vestibulocerebellar fibers in the anterior vermis of the cat: an anterograde wheat germ agglutinin-horseradish peroxidase study. Neurosci Lett 74:25–30
115. Matsushita M, Tanami T (1987) Spinocerebellar projections from the central cervical nucleus in the cat, as studied by anterograde transport of wheat germ agglutinin-horseradish peroxidase. J Comp Neurol 266:376–397
116. Matsushita M, Yaginuma H (1989) Spinocerebellar projections from spinal border cells in the cat as studied by anterograde transport of wheat germ agglutinin-horseradish peroxidase. J Comp Neurol 288:19–38
117. May PJ, Hartwich-Young R, Nelson J, Sparks DL, Porter JD (1990) Cerebellotectal pathways in the macaque: implications for collicular generation of saccades. Neuroscience 36:305–324
118. Middleton FA, Strick PL (1997) Cerebellar output channels. Int Rev Neurobiol 41:61–82
119. Middleton FA, Strick PL (2001) Cerebellar projections to the prefrontal cortex of the primate. J Neurosci 21:700–712
120. Miles FA (1987) The role of the cerebellum in adaptive regulation of the vestibulo-ocular reflex. In: Glickstein M, Yeo C, Stein J (eds) Cerebellum and neuronal plasticity. Plenum, London
121. Mower G, Gibson A, Robinson F, Stein J, Glickstein M (1980) Visual pontocerebellar projections in the cat. J Neurophysiol 43:355–366
122. Nelson BJ, Mugnaini E (1989) Origin of GABAergic inputs to the inferior olive. Exp Brain Res Series 17:86–107
123. Noda H, Sugita S, Ikeda Y (1990) Afferent and efferent connections of the oculomotor region of the fastigial nucleus in the macaque monkey. J Comp Neurol 302:330–348
124. Ogawa T (1935) Beiträge zur vergleichenden Anatomie des Zentralnervensystems der Wassersäugetiere: Über die Kleinhirnkerne der Pinnepedien und Cetaceen. Arb Anat Inst Sendai 17:63–136
125. Ogawa T (1935) Über den Nucleus ellipticus und den Nucleus ruber beim Delphin. Arb Anat Inst Sendai 17:55–61

126. Ogawa T (1939) The tractus tegmenti medialis and its connections with the inferior olive in the cat. J Comp Neurol 70:181–191
127. Ogawa T (1939) Experimentelle Untersuchungen über die mediale und zentrale Haubenbahnen bei der Katze. Arch Psychiatr Nervenkrankh 110:365–444
128. Orioli PJ, Strick PL (1989) Cerebellar connections with the motor cortex and the arcuate premotor area: an analysis employing retrograde transneuronal transport of WGA-HRP. J Comp Neurol 288:612–626
129. Palay SL, Chan-Palay V (1974) Cerebellar cortex. Cytology and organization. Springer, Berlin Heidelberg New York
130. Perales M, Winer JA, Prieto JJ (2006) Focal projections of cat auditory cortex to the pontine nuclei. J Comp Neurol 497:959–980
131. Pijpers A, Voogd J, Ruigrok TJ (2005) Topography of olivo-cortico-nuclear modules in the intermediate cerebellum of the rat. J Comp Neurol 492:193–213
132. Provini L, Redman S, Strata P (1967) Somatotopic organization of mossy and climbing fibres to the anterior lobe of cerebellum activated by the sensorimotor cortex. Brain Res 6:378–381
133. Provini L, Redman S, Strata P (1968) Mossy and climbing fibre organization of the anterior lobe of the cerebellum activated by forelimb and hindlimb areas of the sensorimotor cortex. Exp Brain Res 6:216–233
134. Robinson FR, Fuchs AF (2001) The role of the cerebellum in voluntary eye movements. Annu Rev Neurosci 24:981–1004
135. Ruigrok TJ, Voogd J (1990) Cerebellar nucleo-olivary projections in the rat: an anterograde tracing study with Phaseolus vulgaris-leucoagglutinin (PHA-L). J Comp Neurol 298:315–333
136. Ruigrok TJ, Voogd J (1995) Cerebellar influence on olivary excitability in the cat. Eur J Neurosci 7:679–693
137. Ruigrok TJ, Voogd J (2000) Organization of projections from the inferior olive to the cerebellar nuclei in the rat. J Comp Neurol 426:209–228
138. Ruigrok TJ, Cella F (2004) Precerebellar nuclei and red nucleus. In: Paxinos G (ed) The rat nervous system. Elsevier, Amsterdam, pp 167–204
139. Ruigrok TJ, Cella F, Voogd J (1995) Connections of the lateral reticular nucleus to the lateral vestibular nucleus in the rat. An anterograde tracing study with Phaseolus vulgaris leucoagglutinin. Eur J Neurosci 7:1410–1413
140. Sarna JR, Marzban H, Watanabe M, Hawkes R (2006) Complementary stripes of phospholipase Cbeta3 and Cbeta4 expression by Purkinje cell subsets in the mouse cerebellum. J Comp Neurol 496:303–313
141. Scheibel AB (1977) Sagittal organization of mossy fiber terminal systems in the cerebellum of the rat. Exp Neurol 57:1067–1070
142. Schmahmann JD (1997) The cerebellum and cognition. Academic Press, San Diego
143. Schmahmann JD, Pandya DN (1997) Anatomic organization of the basilar pontine projections from prefrontal cortices in rhesus monkey. J Neurosci 17:438–458
144. Schmahmann JD, Doyon J, Toga AW, Petrides M, Evans AC (2000) MRI atlas of the human cerebellum. Academic Press, San Diego
145. Schmahmann JD, Rosene DL, Pandya DN (2004) Motor projections to the basis pontis in rhesus monkey. J Comp Neurol 478:248–268
146. Schmahmann JD, Doyon J, McDonald D, Holmes C, Lavoie K, Hurwitz AS, Kabani N, Toga A, Evans A, Petrides M (1999) Three-dimensional MRI atlas of the human cerebellum in proportional stereotaxic space. Neuroimage 10:233–260
147. Scott TG (1964) A unique pattern of localization within the cerebellum of the mouse. J Comp Neurol 22:1–8
148. Serapide MF, Cicirata F, Sotelo C, Panto MR, Parenti R (1994) The pontocerebellar projection: longitudinal zonal distribution of fibers from discrete regions of the pontine nuclei to vermal and parafloccular cortices in the rat. Brain Res 644:175–180
149. Serapide MF, Panto MR, Parenti R, Zappala A, Cicirata F (2001) Multiple zonal projections of the basilar pontine nuclei to the cerebellar cortex of the rat. J Comp Neurol 430:471–484
150. Serapide MF, Parenti R, Panto MR, Zappala A, Cicirata F (2002) Multiple zonal projections of the nucleus reticularis tegmenti pontis to the cerebellar cortex of the rat. Eur J Neurosci 15:1854–1848
151. Sillevis Smitt P, Kinoshita A, De Leeuw B, Moll W, Coesmans M, Jaarsma D, Henzen-Logmans S, Vecht C, De Zeeuw C, Sekiyama N, Nakanishi S, Shigemoto R (2000) Paraneoplastic cerebellar ataxia due to autoantibodies against a glutamate receptor. N Engl J Med 342:21–27
152. Sillitoe RV, Marzban H, Larouche M, Zahedi S, Affanni J, Hawkes R (2005) Conservation of the architecture of the anterior lobe vermis of the cerebellum across mammalian species. Prog Brain Res 148:283–297
153. Snider RS, Stowell A (1944) Receiving areas of the tactile, auditory and visual systems in the cerebellum. J Neurophysiol 7:331–357
154. Sotelo C (2004) Cellular and genetic regulation of the development of the cerebellar system. Prog Neurobiol 72:295–339
155. Sotelo C, Wassef M (1991) Cerebellar development: afferent organization and Purkinje cell heterogeneity. Philos Trans R Soc Lond B Biol Sci 331:307–313
156. Sotelo C, Llinas R, Baker R (1974) Structural study of inferior olivary nucleus of the cat: morphological correlates of electrotonic coupling. J Neurophysiol 37:541–569
157. Stainier DY, Gilbert W (1989) The monoclonal antibody B30 recognizes a specific neuronal cell surface antigen in the developing mesencephalic trigeminal nucleus of the mouse. J Neurosci 9:2468–2485

158. Stein JF, Glickstein M (1992) Role of the cerebellum in visual guidance of movement. Physiol Rev 72.967–1017
159. Stilling B (1864) Untersuchungen über den Bau des kleinen Gehirns des Menschen. Fischer, Cassel
160. Strominger NL, Truscott TC, Miller RA, Royce GJ (1979) An autoradiographic study of the rubroolivary tract in the rhesus monkey. J Comp Neurol 183:33–45
161. Sugihara I, Shinoda Y (2004) Molecular, topographic, and functional organization of the cerebellar cortex: a study with combined aldolase C and olivocerebellar labeling. J Neurosci 24:8771–8785
162. Sugihara I, Wu HS, Shinoda Y (1999) Morphology of single olivocerebellar axons labeled with biotinylated dextran amine in the rat. J Comp Neurol 414:131–148
163. Sugihara I, Wu HS, Shinoda Y (2001) The entire trajectories of single olivocerebellar axons in the cerebellar cortex and their contribution to cerebellar compartmentalization. J Neurosci 21:7715–7723
164. Sultan F, Braitenberg V (1993) Shapes and sizes of different mammalian cerebella. A study in quantitative comparative neuroanatomy. J Hirnforsch 34:79–92
165. Svensson P, Ivarsson M, Hesslow G (2000) Involvement of the cerebellum in a new temporal property of the conditioned eyeblink response. Prog Brain Res 124:317–323
166. Tan J, Gerrits NM, Nanhoe R, Simpson JI, Voogd J (1995) Zonal organization of the climbing fiber projection to the flocculus and nodulus of the rabbit: a combined axonal tracing and acetylcholinesterase histochemical study. J Comp Neurol 356:23–50
167. Thunnissen IE, Epema AH, Gerrits NM (1989) Secondary vestibulocerebellar mossy fiber projection to the caudal vermis in the rabbit. J Comp Neurol 290:262–277
168. Tokuno H, Takada M, Nambu A, Inase M (1995) Somatotopical projections from the supplementary motor area to the red nucleus in the macaque monkey. Exp Brain Res 106:351–355
169. Touri F, Hawkes R, Riederer BM (1996) Differential distribution of MAP1a and aldolase c in adult mouse cerebellum. Eur J Neurosci 8:61–68
170. Trott JR, Armstrong DM (1987) The cerebellar corticonuclear projection from lobule Vb/c of the cat anterior lobe: a combined electrophysiological and autoradiographic study. II. Projections from the vermis. Exp Brain Res 68:339–354
171. Van Kan PL, Houk JC, Gibson AR (1993) Output organization of intermediate cerebellum of the monkey. J Neurophysiol 69:57–73
172. Voogd J (1969) The importance of fiber connections in the comparative anatomy of the mammalian cerebellum. In: Llinas R (ed) Neurobiology of cerebellar evolution and development. AMA, Chicago, pp 493–514
173. Voogd J (2003) The human cerebellum. J Chem Neuroanat 26:243–252
174. Voogd J (2004) Cerebellum and precerebellar nuclei. In: Paxinos G, Mai JK (eds) The Human Nervous System. Elsevier, Amsterdam, pp 321–392
175. Voogd J, Bigaré F (1980) Topographical distribution of olivary and cortico-nuclear fibres in the cerebellum: a review. In: Courville J (ed) The olivary nucleus. Anatomy and physiology. Raven, New York, pp 207–234
176. Voogd J, Glickstein M (1998) The anatomy of the cerebellum. Trends Neurosci 21:370–375
177. Voogd J, Ruigrok TJ (2004) The organization of the corticonuclear and olivocerebellar climbing fiber projections to the rat cerebellar vermis: the congruence of projection zones and the zebrin pattern. J Neurocytol 33:5–21
178. Voogd J, Barmack NH (2005) Oculomotor cerebellum. Prog Brain Res 151:231–268
179. Voogd J, Hess DT, Marani E (1987) The parasagittal zonation of the cerebellar cortex in cat and monkey. Topography, distribution of acetylcholinesterase and development. In: King JS (ed) New concepts in cerebellar neurobiology. Liss, New York, pp 183–220
180. Voogd J, Gerrits NM, Hess DT (1987) Parasagittal zonation of the cerebellum in macaques: an analysis based on acetylcholinesterase histochemistry. In: Glickstein M, Yeo C, Stein J (eds) Cerebellum and neuronal plasticity. Plenum, London, pp 15–39
181. Voogd J, Jaarsma D, Marani E (1996) The cerebellum. Chemoarchitecture and anatomy. In: Swanson LW, Björklund A, Hökfelt T (eds) Handbook of chemical neuroanatomy, vol 12 III. Elsevier, Amsterdam, pp 1–369
182. Voogd J, Pardoe J, Ruigrok TJ, Apps R (2003) The distribution of climbing and mossy fiber collateral branches from the copula pyramidis and the paramedian lobule: congruence of climbing fiber cortical zones and the pattern of zebrin banding within the rat cerebellum. J Neurosci 23:4645–4656
183. Weidenreich F (1899) Zur Anatomie der zentralen Kleinhirnkerne der Säuger. Z Morphol Anthropol 1:259–312
184. Wu HS, Sugihara I, Shinoda Y (1999) Projection patterns of single mossy fibers originating from the lateral reticular nucleus in the rat cerebellar cortex and nuclei. J Comp Neurol 411:97–118
185. Yaginuma H, Matsushita M (1989) Spinocerebellar projections from the upper lumbar segments in the cat, as studied by anterograde transport of wheat germ agglutinin-horseradish peroxidase. J Comp Neurol 281:298–319
186. Yeo CH (1991) Cerebellum and classical conditioning of motor responses. Ann NY Acad Sci 627:292–304

21 Motor Systems

Introduction

"The neuronal networks of the brain create behaviour by ordering the execution of different movements and movement complexes which allow the organism to achieve certain goals. The descending connections from the motor cortex" and the brain stem "constitute the channels through which the messages that ultimately will be translated into movements and movement complexes are transmitted to the effector organs, i.e. the assembly of motor neurons and muscles" (Kuypers [119]). Kuypers' anatomical and functional studies in humans and non-human primates in the second half of the last century laid the groundwork for modern concepts of motor control. His ideas have been modified and extended with respect to the organization of the motor and premotor cortex (reviewed by Matelli et al. [139]), the functional anatomy of subcortical motor systems (see also Chaps. 17 and 22), and the contribution of the limbic system and the hypothalamus to motor control (reviewed by Holstege et al. [103]).

Kuypers' Concept of the Motor System: Medial and Lateral Brain Stem Systems and the Projections of the Motor Cortex

Motoneurons of the nuclei of the cranial nerves and the ventral horn are innervated by fibre systems descending from the brain stem and the cerebral cortex, either directly or through intercalated interneurons. Interneurons for the motoneurons of the ventral horn are located in the intermediate zone of the spinal cord. Interneurons innervating the motor nuclei of the trigeminal, facial and hypoglossal nerves and the ambiguus and supraspinal nuclei are located in the lateral tegmental field, which includes the parvocellular reticular formation. The lateral tegmental field extends rostrally as the Koelliker-Fuse area in the lateral pontine tegmentum and the parabrachial nuclei (Fig. 22.3).

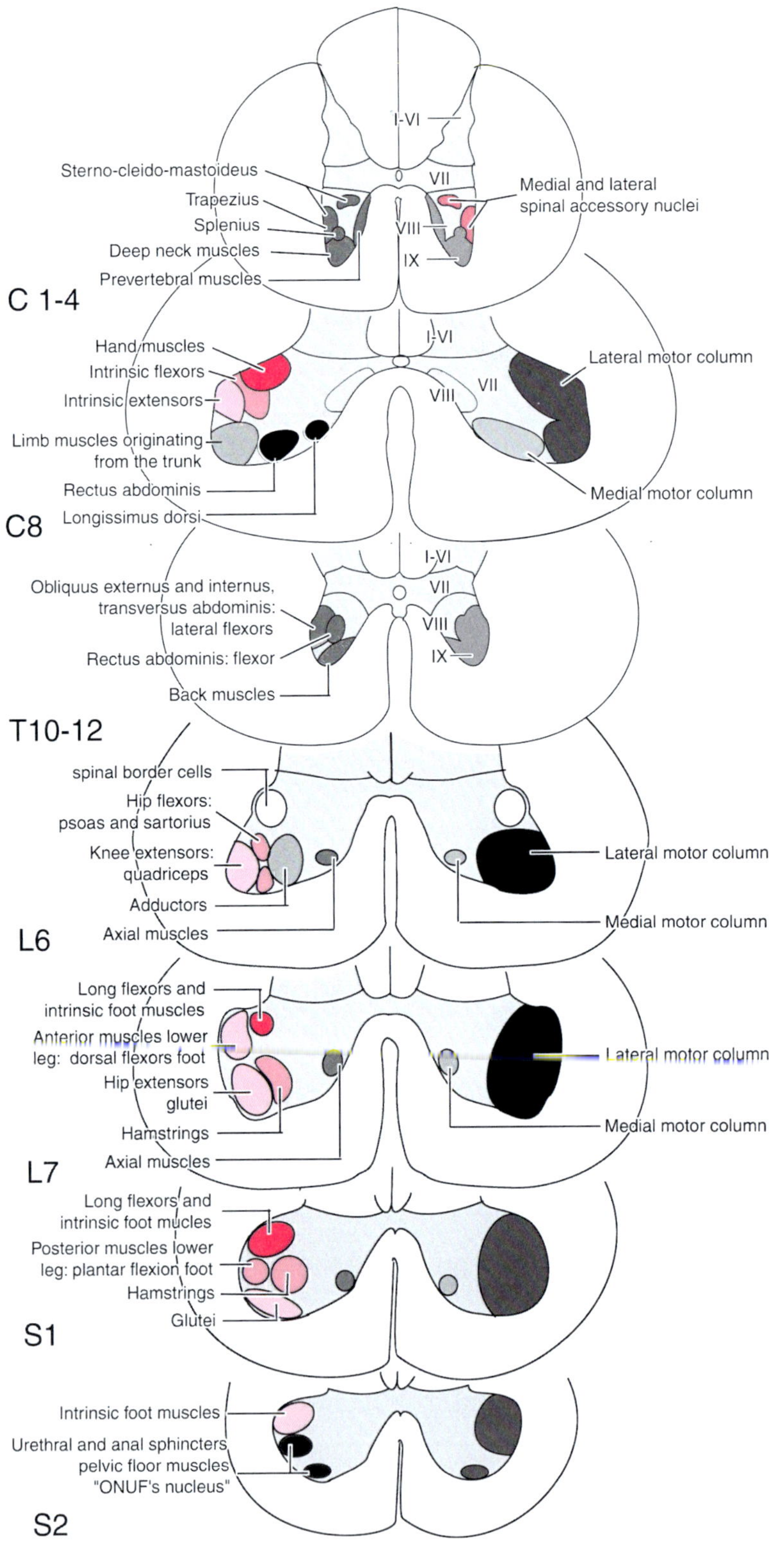

Fig. 21.1. Localization of motoneurons in the ventral horn of the cat. Separate medial and lateral motor columns can be distinguished in the cervical and lumbar enlargements, as shown on the *right side* of the diagrams. The nuclei of the spinal root of the accessory nerve (medial and lateral spinal accessory nuclei) are present at high cervical levels (*light red*). At the level of the enlargements, short hand and foot muscles are innervated by motoneurons located dorsolaterally in the lateral motor column (*left side, red*). Intrinsic flexors (all flexor muscles are shown in *medium red*) and extensors (*light red*) of the limbs occupy a more ventral position. Girdle and trunk muscles are innervated by motoneurons located ventrally and medially. Motoneurons of the medial motor column innervate axial muscles. I–IX, laminae I–X of the spinal grey matter. Redrawn from Holstege [92] data on lumbosacral levels from Van der Horst and Holstege [216]

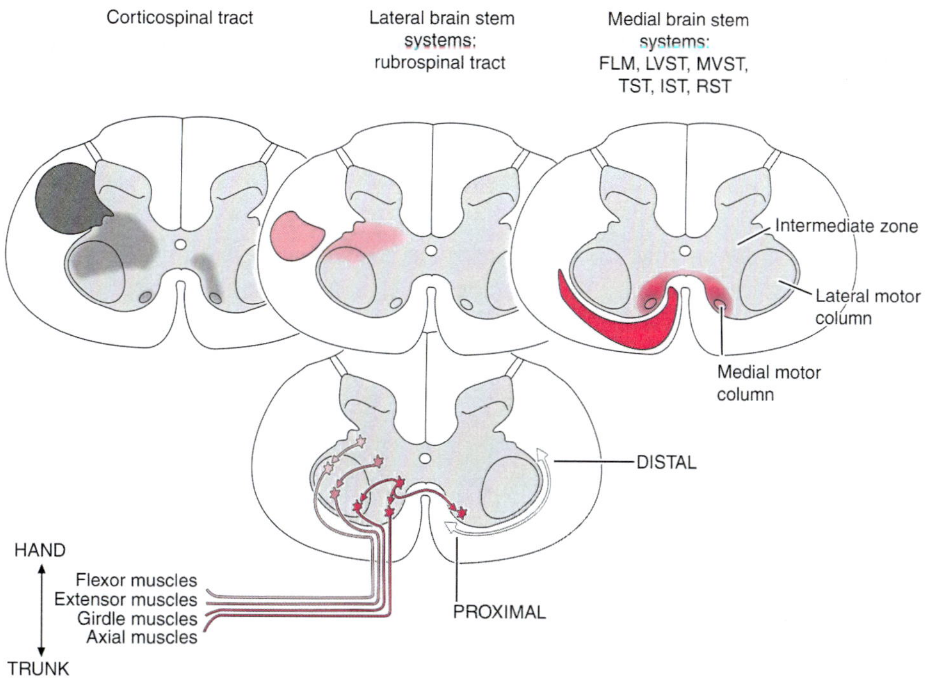

Fig. 21.2. The medial and lateral brain stem systems with projections to spinal interneurons and motoneurons, according to Kuypers [120]. The localization of motoneurons, innervating different muscle groups is shown in the *bottom of the diagram* (see also Fig. 21.1). Intrinsic hand and foot muscles are innervated by motoneurons located in the dorsolateral part of the lateral motor column; intrinsic flexors and extensors of the limbs and girdle muscles occupy a more ventral and medial position; and axial muscles are innervated by the medial motor column. A similar lamination is present in the projection of interneurons, located in the dorsolateral and central intermediate zone, to motoneurons of the lateral motor column. Interneurons in the ventromedial intermediate zone project bilaterally to motoneurons of the medial motor column. The prototype of a lateral brain stem system is the rubrospinal tract. It terminates on interneurons in the dorsolateral intermediate zone and motoneurons in the dorsolateral part of the lateral motor column. Medial brain stem systems descend in the anterior and anterolateral funiculi. They terminate, often bilaterally, on interneurons in the ventromedial intermediate zone and on motoneurons of the medial motor column. The terminations of the corticospinal tract largely overlap with the projections of the medial and lateral brain stem systems (Figs. 21.3, 21.4). Lateral brain stem systems have a preferential influence on distal movements of the limbs; medial brain stem systems target axial muscles. *FLM*, medial longitudinal fascicle; *IST*, interstitiospinal tract; *LVST*, lateral vestibulospinal tract; *MVST*, medial vestibulospinal tract; *RST*, reticulospinal tracts; *TST*, rectospinal tract. Redrawn from Kuypers [118]

Motoneurons in the ventral horn are arranged in longitudinal arrays, which may extend over several segments. The dendrites of motoneurons may extend beyond these longitudinal aggregates, but most of them remain confined within them. The most detailed data on the innervation of skeletal muscles by neurons of the ventral horn are available for the cat [177, 206, 216]; information on the situation in monkeys and humans is less complete [173, 197].

Lateral and medial motor columns can be distinguished at the level of the enlargements; in the upper cervical, thoracic and sacral cord they fuse into a single mass (Figs. 21.1, 21.2). Motoneurons innervating intrinsic muscles of the hand are located caudally and dorsolaterally in the lateral motor column of the cervical enlargement (Figs. 21.1, 21.2). Motoneurons, innervating intrinsic flexors and extensors of the limbs, are located in successively more rostral and ventral parts of the lateral motor column and motoneurons innervating girdle and trunk muscles, are located still more rostrally and medially. Motoneurons of the medial motor column, located ventromedially in the ventral horn, innervate axial muscles [92, 173, 216].

The topical localization of motoneurons in the lumbosacral enlargement differs in some respects from that in the cervical cord. The long toe flexors and the intrinsic footmuscles are located dorsally and caudally in the lateral motor column. Adductors and abductors occupy a medial position and the quadriceps muscle a lateral position in the caudal lumbar enlargement. Anterior muscles of the lower leg are represented lateral to posterior muscles in the caudal lateral motor column. Axial muscles are innervated by motoneurons of the medial motor column.

Motoneurons innervating respiratory muscles and those involved in micturition, defecation and sexual functions deserve special attention. In the cat the phrenic motor column, innervating the diaphragm, is located at C4–6. Abdominal muscles receive their innervation from motoneurons located in the central ventral horn at low cervical and thoracic levels. Intercostal muscles are innervated by the thoracic cord (Fig. 21.1). A column of small motoneurons extending from S1 caudally, which contains Onuf's nucleus [161, 216], innervate the urethral and anal sphincters, the pelvic floor and the ischio- and bulbocavernosus muscles (Figs. 6.5, 21.1).

Interneurons innervating spinal motoneurons are located in the intermediate zone. Those terminating on dorsolateral motoneurons innervating distal muscles of the upper limb occupy a dorsolateral position in the cervical intermediate zone, those innervating motoneurons of more distal, flexor and extensor muscles of the limb occupy a more ventral position. Interneurons terminating on motoneurons of the medial motor column are located in the ventromedial intermediate zone and project bilaterally (Fig. 21.2). A similar organization is present in the lumbosacral enlargement.

Premotor interneurons with monosynaptic connections to motoneurons innervating respiratory muscles and Onuf's nucleus are located in the lateral tegmental field of the brain stem (see Chap. "Brain Stem Control of Micturition and Ejaculation", and Fig. 6.5, and Chap. 22 "Connections of the Reticular Formation and Associated Cell Groups" and Fig. 22.4C).

At cervical levels C3/C4, a population of interneurons was identified in cat and monkey with projections to forelimb motoneurons and interneurons in laminae VI–VIII at C6–Th1, that control grasping movements [2, 4, 109]. These interneurons give rise to an ascending collateral projection to the lateral reticular nucleus. A similar population of interneurons located at C3–C5 gives rise to long descending axons, terminating on lumbar interneurons [3].

Kuypers distinguished between a group of lateral brain stem systems descending in the lateral funiculus and a medial group descending in the ventral lateral and anterior funiculi [118] (Fig. 21.2). The prototype of a lateral brain stem system is the rubrospinal tract. It terminates on dorsolateral interneurons in the intermediate zone and (as shown later [96, 145]), on motoneurons innervating distal muscles of the upper limb. The projection of the rubrospinal tract is focussed at the cervical or the lumbosacral enlargement, although some collateralization to both enlargements occurs [105–107, 203]. Lateral brain stem systems steer movements of the extremities, especially of their distal parts. Medial brain stem systems include the vestibulospinal tracts, the interstitiospinal tract, the tectospinal tract and several reticulospinal systems. They descend in the anterior funiculus and the ventral part of the lateral funiculus and terminate, often bilaterally, in the ventromedial intermediate zone and, for some systems, on motoneurons of the medial motor column. Medial brain stem systems collateralize widely to different levels of the cord. Medial brain stem systems steer postural and orienting movements of head and body and synergistic movements of body and limbs. A high degree of precision in the organization of certain medial brain stem systems was shown by Shinoda et al. [203] in their analysis of vestibular projections to the upper cervical cord subserving the vestibulocollic reflex (Chap. 17, "Vestibulocollic Reflex Pathways"; Fig. 17.6). Lateral brain stem systems, therefore, monitor distal movements; medial brain stem systems exert their actions bilaterally on axial and proximal muscles.

In primates the termination of the corticospinal system overlaps with both the lateral and the medial brain stem systems. Kuypers defined the motor cortex as the region giving origin to the corticospinal (pyramidal) tract. It corresponds to the agranular cortex of Brodmann's areas 4 and caudal area 6 [119]. Area 4 is located in the precentral gyrus and corresponds to the primary motor cortex, characterized by the presence of the giant pyramidal cells of Betz. Betz [19] localized clumps of these cells in layer 5 (his layer 4) of the precentral gyrus and noticed that they gave rise to large-calibre myelinated axons. In the brain of the dog they occupy a position along the sulcus cruciatus, corresponding to the electrically excitable region where Fritsch and Hitzig [62] had obtained muscular contractions on the opposite side of the body. Betz concluded that "These cells clearly possess all attributes of so-called 'motor' cells" [20]. Caudal area 6 corresponds to the premotor cortex [63]; on the medial aspect of the hemisphere it contains the supplementary motor cortex [167, 225]. Both the primary and the supplementary motor cortex are somatotopically organized. In the primary motor cortex the representation of the foot and the leg is located medially and dorsally, the face ventrolaterally and the arm and the hand in an intermediate position. In the supplementary motor cortex the leg is represented rostrally and the face caudally (Figs. 21.9, 21.11 A).

Neurons in the representations of hand and foot in area 4 project to the contralateral dorsolateral intermediate zone in the cervical and lumbosacral enlargement, respectively, and to motoneurons in the lateral motor column, preferentially to those innervating distal limb movements [119, 120, 192] (Figs. 21.2, 21.3). Kuypers emphasized the direct corticomotoneuronal connections, which presumably originate as the large-calibre, myelinated axons from the Betz cells. These direct connections develop postnatally [117]. Intermediate regions of area 4 and the more rostral premotor cortex project bilaterally to the ventromedial intermediate zone and the medial motor column (Fig. 21.4). Neurons in the basolateral part of area 4, containing the representation of face and larynx, project bilaterally, but mainly contralaterally, to the cranial motor nuclei of the facial, trigeminal and hypoglossal nerves and to the ambiguus and supraspinal nuclei. In addition they connect with interneurons innervating motoneurons of the cranial motor nuclei in the lateral tegmental field (Fig. 21.3). The projection of the supplementary motor cortex is restricted to the intermediate zone of the spinal grey. The distribution of the corticospinal tract in the human brain is very similar to that in nonhuman primates [116, 192, 193] (Fig. 21.5).

The motor cortex also projects to centres which give origin to descending brain stem systems [115, 116, 119, 121, 193]. Corticospinal fibres from regions in area 4 representing foot, hand and face, emit collaterals to the magnocellular red nucleus, where they terminate in a somatotopical manner (Fig. 21.3). The magnocellular red nucleus gives rise to the corticobulbar

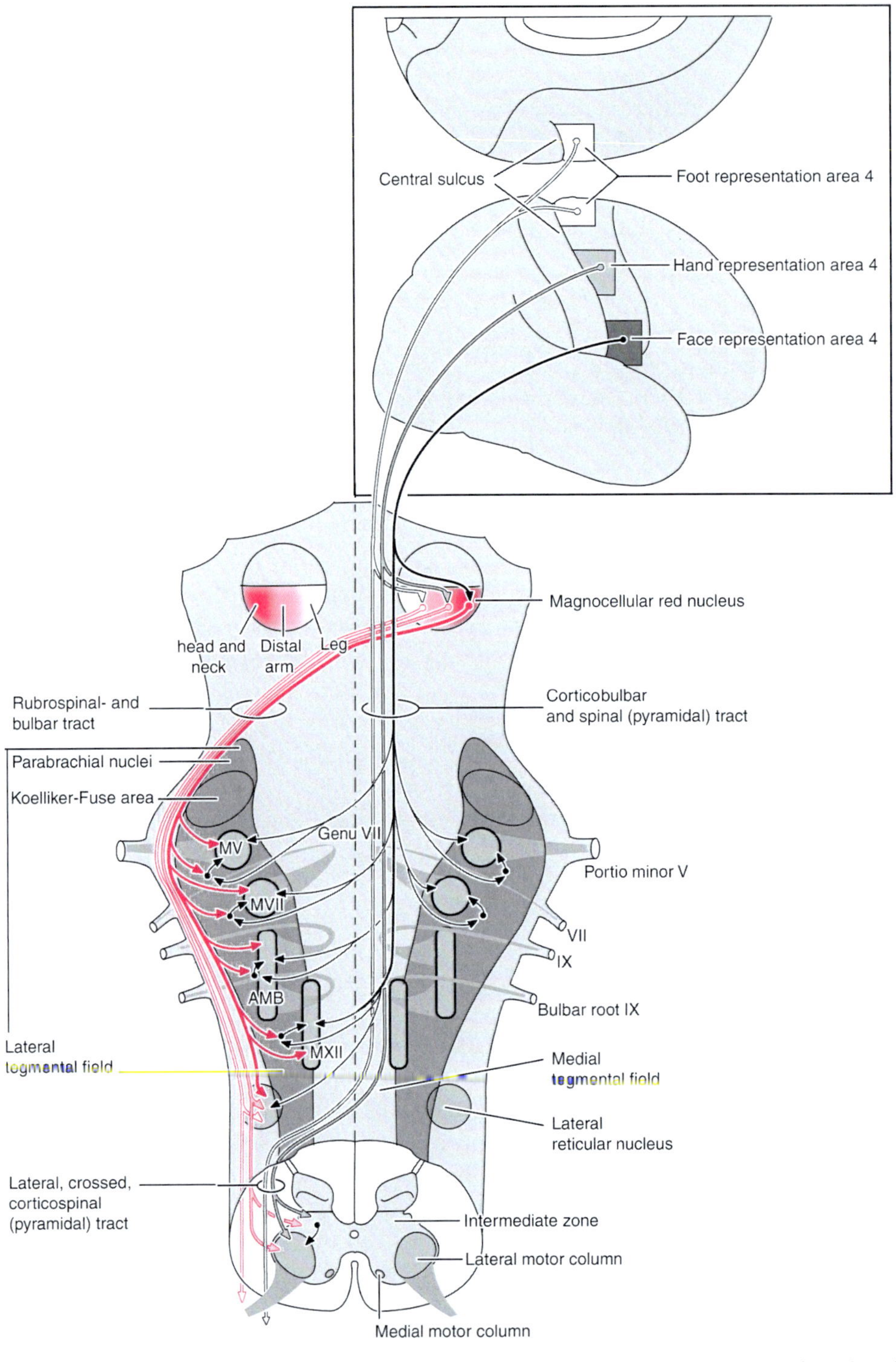

Fig. 21.3. The organization of the somatic motor system according to Kuypers [120]. The foot, hand and face representations in the primary motor cortex (area 4) project, mainly contralaterally, to motoneurons of the brain stem nuclei and the lateral motor column of the ventral horn and to premotor interneurons in the lateral tegmental field and the dorsolateral intermediate zone of the spinal grey matter. Collateral projections terminate, in a somatotopical pattern, in the magnocellular red nucleus. The projections of the rubrobulbar and spinal tract to motoneurons and interneurons are very similar to those of the primary motor cortex. Interneurons in the lateral tegmental field with bilateral projections have not been indicated. *AMB*, ambiguus nucleus; *MV*, motor nucleus of the trigeminal nerve; *MVII*, motor nucleus of the facial nerve; *MXII*, motor nucleus of the hypoglossal nerve

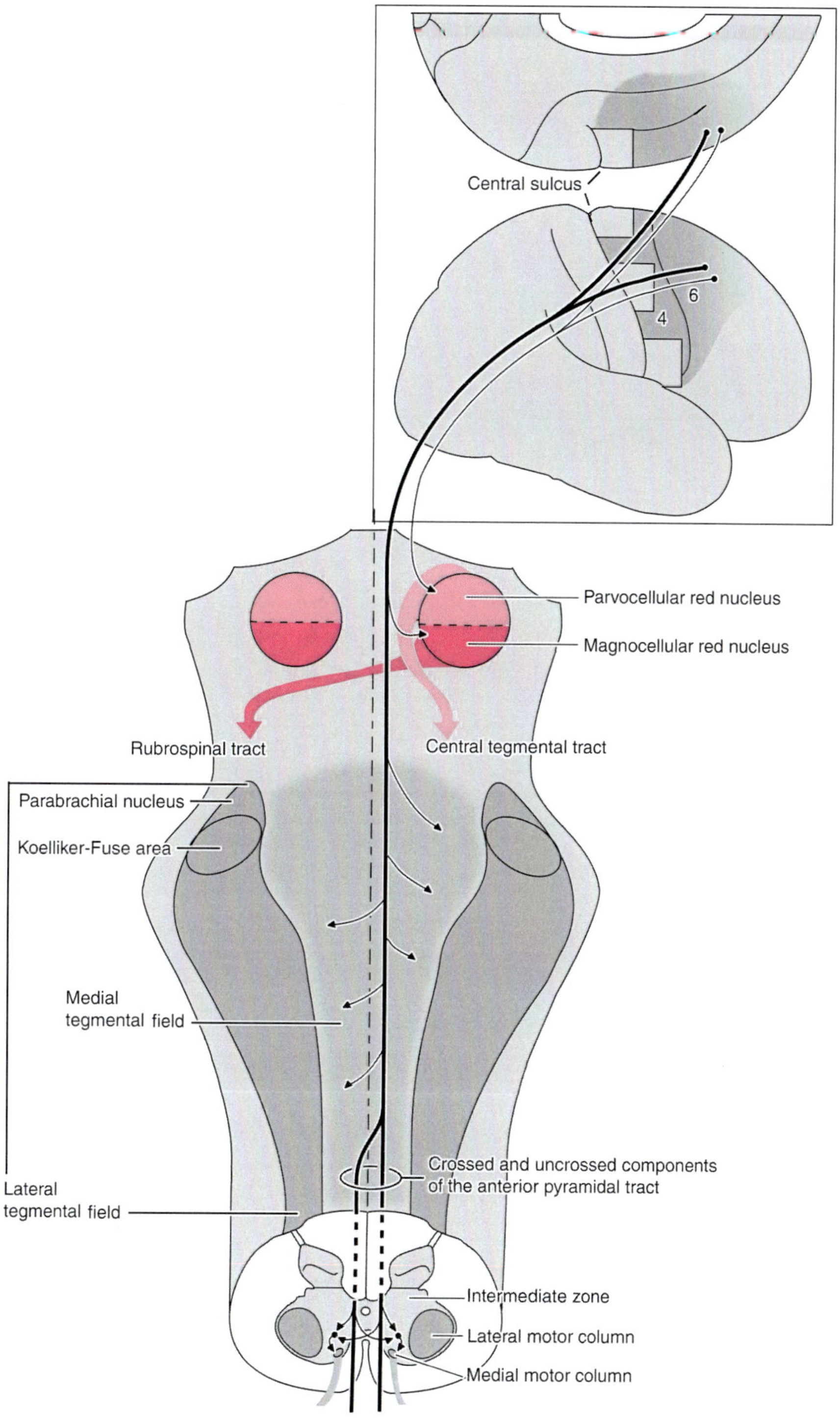

Fig. 21.4. The organization of the somatic motor system according to Kuypers [120]. Regions of the primary motor cortex, interpolated between the foot, hand and face representations, and more rostral premotor areas (area 6) project, bilaterally, to the medial tegmental field, the ventromedial intermediate zone of the spinal cord and to motoneurons of the medial motor column. A separate population of cortical neurons projects to the parvocellular red nucleus. *4*, Area 4; *6*, area 6

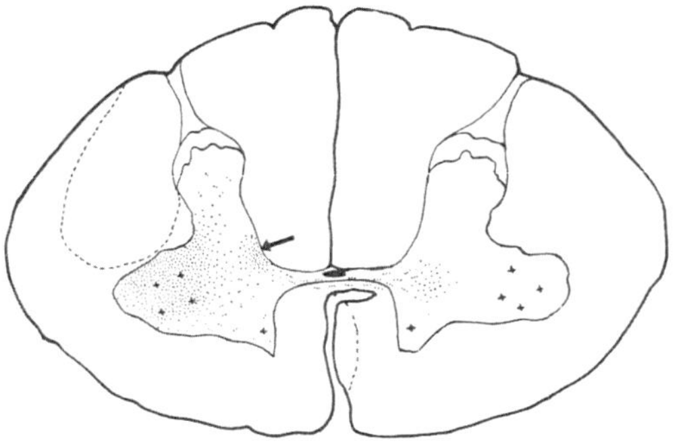

Fig. 21.5. Termination of the corticospinal tract in the human cervical spinal cord (Schoen [192]). Note termination in central dorsal horn, the dorsolateral intermediate zone, the lateral motor column, the ventromedial intermediate zone mainly contralaterally, and the presence of a cluster of terminals medially at the base of the dorsal horn (*arrow*)

and spinal tract. This tract decussates at the level of the red nucleus and terminates in cranial motor nuclei, on interneurons in the lateral tegmental field, the lateral reticular nucleus and focusses at the dorsolateral intermediate zone of the enlargements and on motoneurons innervating distal muscles of the forelimb [96, 145]. In these regions the rubral and cortical projections largely overlap. Convergence of corticospinal and rubrospinal systems was also found on the C3/C4 interneurons with projections to forelimb motoneurons [5]. Intermediate regions of area 4 and the more rostral premotor cortex project to the medial bulbar tegmental field, ipsilaterally at the level of the pons and bilaterally in the medulla oblongata. In the spinal cord, the terminations of reticulospinal systems from the medial pontobulbar reticular formation and other medial brain stem systems tracts in the ventromedial intermediate zone overlap with the cortical projection of the premotor cortex (Figs. 21.2, 21.4).

Lawrence and Kuypers [125, 126] studied the motor behaviour of monkeys after inducing lesions to the pyramidal tract and the descending brain stem systems. With respect to the overlap in the projections of brain stem and the motor cortex they remark: "Descending cortical pathways are capable of guiding a range of activities which includes independent limb movements in addition to total body activity. The corticospinal pathways superimpose speed and agility upon these subcortical mechanisms, and, in addition, provide the capacity for a high degree of fractionation of movements, as exemplified by individual finger movements."

Kuyper's concept of the motor system has been modified in the last decades, by (1) the identification of multiple premotor areas in the motor cortex, (2) the afferent connections and the corticospinal projection of these areas, (3) the specification of the cortical and subcortical innervation of the cranial motor nuclei and (4) the emphasis on limbic and hypothalamic structures in motor control.

The Motor Cortex

The Primary Motor Cortex and the Premotor Areas

The subdivision of the motor cortex in monkeys in the primary motor cortex, the supplementary motor cortex and a premotor area was modified and extended in the last decades after multiple premotor areas were identified. The premotor areas can be defined by their projection to the primary motor cortex [46]. They were delineated using cytoarchitectonic and immunocytochemical criteria. Those located on the convexity of the hemisphere maintain reciprocal connections with areas in the parietal lobe. Specific parietofrontal circuits clarified the function of many of these premotor areas. Together with the primary and the supplementary motor area, they comprise the entire region of the frontal lobe giving origin to the corticobulbar and spinal tract. Premotor areas, like the primary and supplementary motor cortex, project differentially to the intermediate zone and, in some cases, to motoneurons at different levels of the cord. The motor cortex was recently reviewed by Wiesendanger and Wise [224], Geyer et al. [73], Matelli et al. [140], and Strick and colleagues [47, 88, 89].

In monkeys seven premotor areas have been distinguished in area 6, including the supplementary motor area, and two or three premotor areas are present on the medial aspect of the hemisphere in the wall of the cingulate sulcus (Fig. 21.6B). The number of premotor areas is not fixed, however, a further parcellation in the near future appears likely. Matelli et al. [136] originally used cytochrome oxidase histochemistry to delineate five areas (F1–F5) in areas 4 and 6 of the macaque monkey. F1 corresponds to area 4 and is known as the primary motor cortex (*M1*). F2 is located rostral to M1 in caudal and dorsal area 6 and is also known as the caudal part of the dorsal premotor cortex (*PMDc*). F3 corresponds to the supplementary motor cortex (*SMA*), rostrally adjoining M1 on the medial hemisphere. F4 and

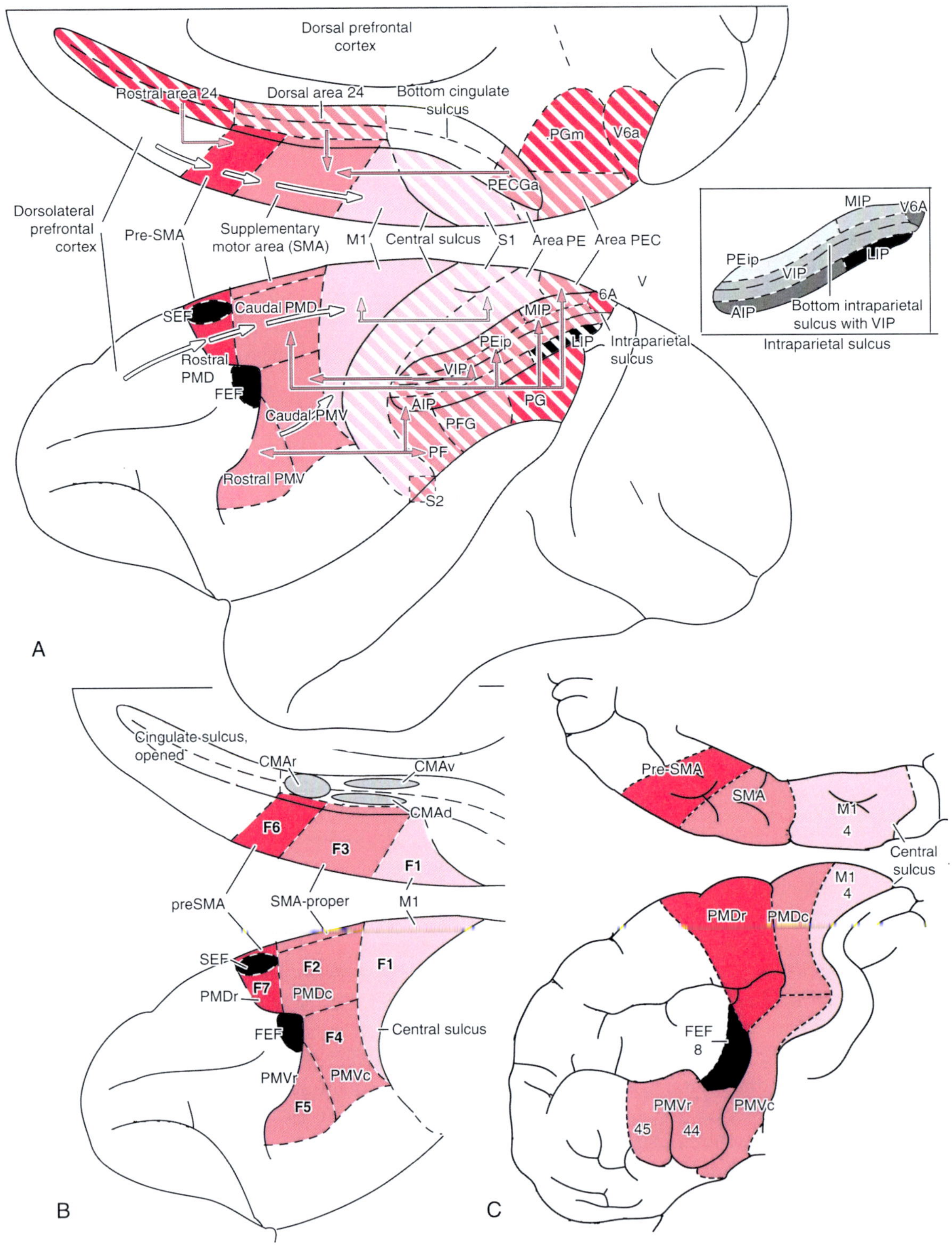

Fig. 21.6. A The primary motor and the premotor areas of the monkey and their connections with the prefrontal, cingulate and parietal cortices. Redrawn from Matelli et al. [140]. The primary motor area (M1) is shown in *light red*. Caudal premotor areas (SMA, caudal PMD and rostral and caudal PMV) project to the primary motor cortex and are reciprocally connected with areas in the parietal lobe (parietal-dependent premotor areas: *medium red*). Rostral premotor areas (pre-SMA and rostral PMD) receive their main afferents

F5 are located rostral to M1 in ventral area 6 and are also known as the caudal and rostral parts of the ventral premotor area, respectively (*PMVc and PMVr*). F2–5 are reciprocally connected with M1 and with certain parietal areas and are known as the *parietal-dependent premotor areas.*

Two additional areas, F6 and F7, are located in rostral area 6. They are connected with M1 through more caudal premotor areas and receive corticocortical connections from prefrontal areas. They are known as the *prefrontal-dependent premotor areas.* F6 is located rostral to SMA [138] and is indicated as the presupplementary motor area (*pre-SMA*). F7 corresponds to the rostral part of the dorsal premotor area (*PMDr*).

Rostral and caudal *cingulate motor areas* are present in the cingulate sulcus. The rostral cingulate motor area (*CMAr*) is part of area 24. The caudal cingulate motor area is divided into the ventral cingulate motor area (*CMAv*), located on the ventral bank of the cingulate sulcus in area 23, and the dorsal cingulate motor area (*CMAd*) in area 6 on its dorsal bank [89] (Figs. 21.6 B, 21.9 C).

The two *visual fields* located in the monkey premotor area, the frontal eye field (*FEF*) located in area 8 in the arcuate sulcus, and the supplementary eye field (*SEF*), which is part of PMDr (Fig. 21.6 B), are considered in Chap. 19 (Fig. 19.14). The descriptive abbreviations (indicated in italic) of the premotor areas will be used in the present chapter, instead of Motelli's F1–F7 nomenclature. Homologues of the monkey premotor areas have been proposed for the human brain [140]. Human PMVr supposedly includes Brodmann's areas 44 and 45 (Fig. 21.6 C).

The Origin of the Corticospinal and Corticobulbar Tracts

Corticospinal and corticobulbar fibres from the pericentral cortex descend in the internal capsule and the middle portion of the cerebral peduncle (Figs. 21.7, 21.8). As a bundle they are characterized by the presence of the large-calibre myelinated axons of the Betz cells among a majority of smaller fibres (Fig. 6.31). After traversing the pontine nuclei they form the pyramidal tract on the ventral surface of the medulla oblongata. Corticospinal fibres decussate at the border of the medulla oblongata with the spinal cord. Fibres descending in the

from the prefrontal cortex (prefrontal-dependent premotor areas: *dark red*) and project to the primary motor cortex via caudal premotor areas. (Pre)motor cortical areas are reciprocally connected with parietal areas and receive projections from the cingulate areas, indicated with *hatching in the same color.* The parietofrontal association system is concentrically arranged: areas next to the central sulcus and successively more distal areas are interconnected. The predominant connections, as tabulated by Matelli et al. [140] are shown as *arrows.* Connections between the premotor areas are not shown. The connections of the frontal and supplementary eye field (FEF and SEF) with the parietal eye field (LIP, shown in *black*) are illustrated in more detail in Fig. 19.14. *Inset* shows nomenclature of areas in the banks of the intraparietal sulcus. There is a good correspondence between the premotor areas shown in this figure, and the premotor areas, as delineated by Strick and collaborators on the basis of their corticospinal projection, with the exception of the subdivision of PMV and the demarcation of the cingulate premotor areas (Fig. 21.9). **B** The macaque primary motor cortex (F1) and the premotor areas F2–F7 as defined by Matelli [73, 136, 140] and their descriptive nomenclature as used in this chapter. Color code as in **A**. Modified from Matelli et al. [140]. **C** Proposed homologies between the monkey and human primary motor and premotor areas. Redrawn from Matelli et al. [140]. *AIP,* anterior intraparietal area; *CMA(r,v,d),* (rostral, ventral, dorsal) cingulate motor area; *FEF,* frontal eye field; *LIP,* lateral intraparietal area (parietal visual field); *M1,* primary motor cortex; *MIP,* medial intraparietal area; *PE,* area PE of Pandya and Seltzer; *PEC,* area PEC of Pandya and Seltzer [164]; *PECGa,* portion of area PEC in cingulate sulcus; *PEip,* intraparietal part of area PE; *PF,* area 40 of Brodman; *PFG,* intermediate area between PF and PG; *PG,* area 39 of Brodmann; *PGm,* medial area PG; *PMD(c,r),* (caudal, rostral) dorsal premotor areas; *PMV(c,d),* (caudal, rostral) ventral premotor areas; *S1,* primary somatosensory area; *S2,* secondary somatosensory area; *SEF,* supplementary eye field; *SMA,* supplementary motor cortex; *V6a,* area V6a; *VIP,* ventral intraparietal area

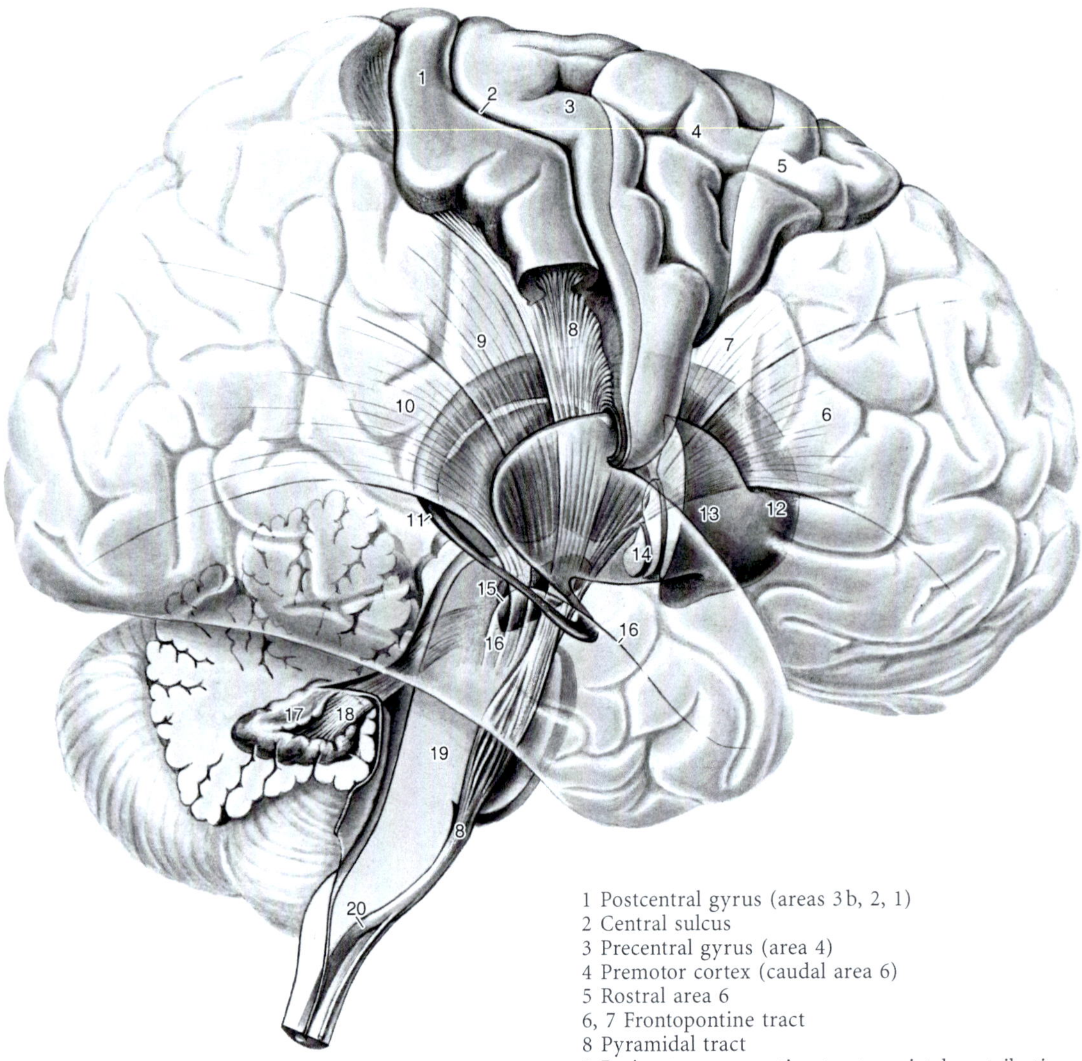

1 Postcentral gyrus (areas 3b, 2, 1)
2 Central sulcus
3 Precentral gyrus (area 4)
4 Premotor cortex (caudal area 6)
5 Rostral area 6
6, 7 Frontopontine tract
8 Pyramidal tract
9 Parietotemporopontine tract, parietal contribution
10 Parietotemporopontine tract, occipital contribution
11 Tail of the caudate nucleus
12 Head of the caudate nucleus
13 Putamen
14 Globus pallidus
15 Substantia nigra
16 Parietotemporopontine tract, temporal contribution
17 Left dentate nucleus
18 Superior cerebellar peduncle
19 Pons
20 Pyramidal decussation

Fig. 21.7. Pictorial survey of the origin of the pyramidal tract in the precentral cerebral cortex, and of the corticopontine systems in a lateral view (1/1×). The brain stem and the cerebellum have been cut in the median plane and the right half has been removed, with the exception of the pyramidal tract

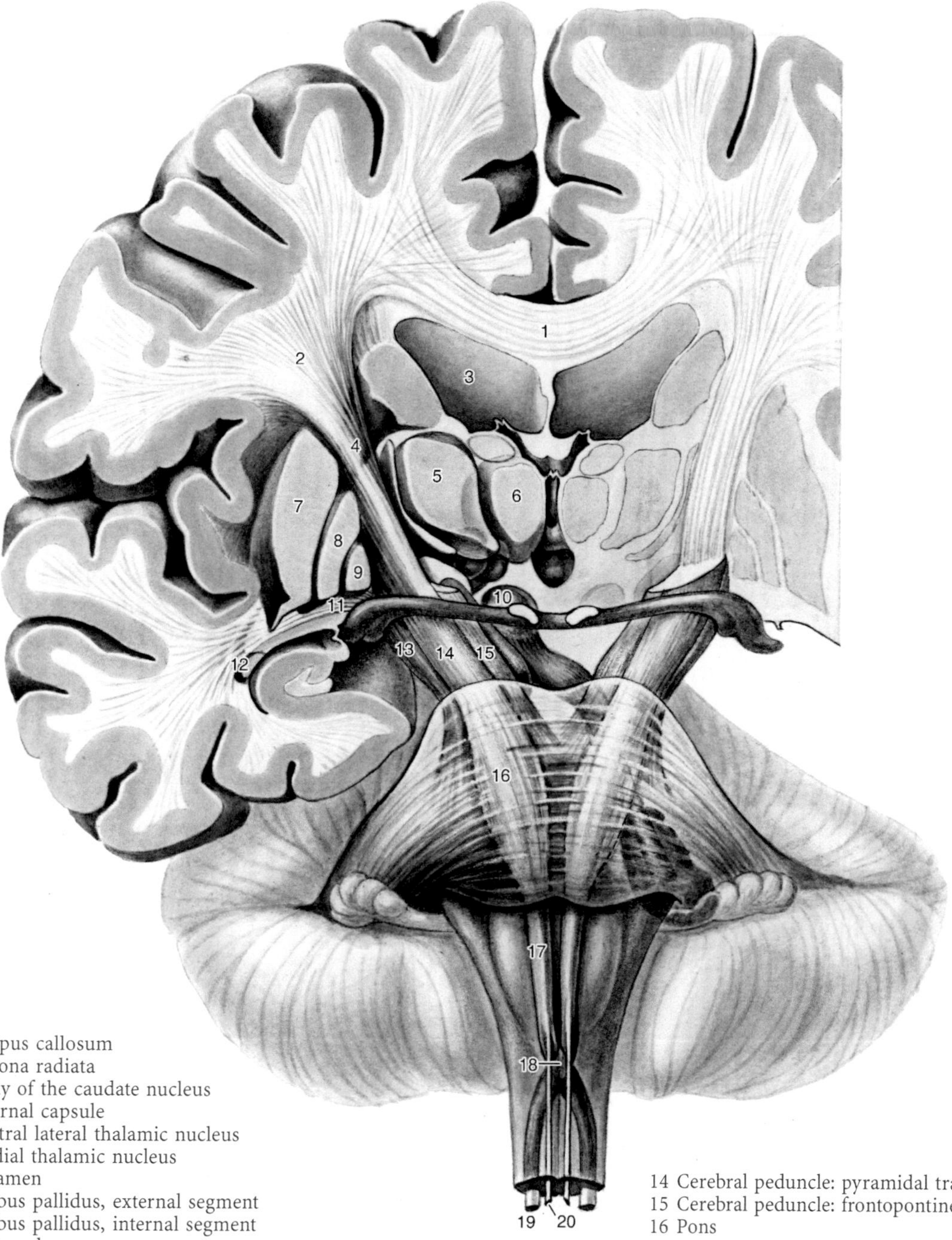

1 Corpus callosum
2 Corona radiata
3 Body of the caudate nucleus
4 Internal capsule
5 Ventral lateral thalamic nucleus
6 Medial thalamic nucleus
7 Putamen
8 Globus pallidus, external segment
9 Globus pallidus, internal segment
10 Red nucleus
11 Internal capsule, retrolentiform part
12 Tail of the caudate nucleus
13 Cerebral peduncle: parietotemporopontine tract
14 Cerebral peduncle: pyramidal tract
15 Cerebral peduncle: frontopontine tract
16 Pons
17 Pyramid
18 Pyramidal decussation
19 Lateral pyramidal tract
20 Anterior pyramidal tract

Fig. 21.8. The long corticofugal fibre system in a frontal view (6/5×). The plane of the section shown in this figure coincides with the long axis of the brain stem

dorsolateral funiculus are mostly crossed; uncrossed fibres travel in the anterior funiculus of the cord. In monkeys 10% of the dorsolateral pyramidal tract is uncrossed, and a similar percentage of the corticospinal fibres terminates ipsilaterally [124]. Corticobulbar fibres separate from the main tract and enter the tegmentum of the pons in small bundles, scattered among the fibres of the medial lemniscus (areae nebulosae, Fig. 6.28). At the level of the medulla oblongata they traverse the region of the medial lemniscus and the inferior olive. Some corticobulbar fibres participate in the pyramidal decussation and subsequently ascend into the medulla oblongata.

Corticospinal and corticobulbar fibres take their origin both from pre- and postcentral areas. The corticospinal projection of the motor cortex was studied by Keizer and Kuypers [113] and Dum and Strick [47] in macaque monkeys, using retrograde neuronal labeling from injections of appropriate retrograde tracers at upper cervical levels of the cord (Figs. 21.9, 21.10). Corticospinal neurons are present in layer 5 of area 4 in the caudal precentral gyrus and the rostral bank of the central sulcus and its extension on the medial surface of the hemisphere. On the convexity of the hemisphere, they extend rostrally into area 6, the inferior limb and the arcuate sulcus and the precentral sulcus (called the precentral dimple by some authors). On the medial aspect of the hemisphere corticospinal neurons are present more rostrally, in the medial aspect of the superior frontal gyrus and the banks of the cingulate sulcus. The corticospinal maps of Keizer and Kuypers and Dum and Strick (Figs. 21.9, 21.10) are very similar, with the exception that Keizer and Kuypers [113] also found corticospinal neurons in the caudal parietal operculum and the insula.

Keizer and Kuypers also mapped the location of corticobulbar neurons, projecting to the contralateral medial bulbar reticular formation (Fig. 21.10). These neurons are located in the superficial part of layer 5. Their distribution partially overlaps with the corticospinal neurons. In regions of overlap some neurons project both to the bulbar reticular formation and the spinal cord. The bulbospinal projection field extends rostrally into the arcuate sulcus, rostral area 6 and, on the medial aspect of the hemisphere, into the rostral cingulate sulcus and the dorsal cingulate gyrus. Ventrally it extends into the orbitofrontal cortex and the rostral insula.

Corticospinal and corticobulbar neurons in the primary and secondary somatosensory areas and the adjacent portions of the superior and inferior parietal lobules have been mapped by several authors [64]. The data in Fig. 21.9 B on the distribution of corticospinal neurons are from a paper by Matelli et al. [138]. Here, corticospinal neurons are present in the primary somatosensory cortex, and rostrally in area PE [164] of the superior parietal lobule. They are absent from its caudal area PEc. Corticospinal neurons are concentrated in areas 2 and 1 of the primary somatosensory cortex. They project to the spinal dorsal horn. Corticobulbar neurons in the primary sensory cortex project to the somatosensory relay nuclei of the trigeminal nerve and the dorsal column nuclei [34]. The projection of the motor cortex to the dorsal column nuclei is restricted to their ventral periphery. Corticospinal neurons extend into the medial bank of the rostral intraparietal sulcus. This area is known as the intraparietal part of area PE (PEip). It differs from other intraparietal areas, such as the medial intraparietal area (MIP), that is found more caudally in the medial bank of the intraparietal sulcus, which lack a corticospinal projection.

When the map of the motor and premotor areas is compared to the corticospinal and corticobulbar projection fields (Figs. 21.6, 21.9, 21.10) it is clear that all these areas give rise to both corticospinal and corticobulbar projections, with the exception of the preSMA and rostral PMD, which are populated by corticobulbar neurons only.

The corticospinal projections of M1 and the different premotor areas were studied in more detail by He et al. [88, 89]. They mapped the distribution of neurons projecting to cervical and to lumbosacral levels of the cord, which control arm and leg movements, respectively.

They concluded that discrete representations of arm and leg are present in M1, PMDc, PMV, SMA proper and in the three cingulate premotor areas. A double representation of arm and leg appears to be present in PMDc and CMAd (Fig. 21.9 C). When the corticobulbar projection field (Fig. 21.10) is compared with the map of topical organization of the corticospinal projections of the different premotor areas (Fig. 21.9 C), it is clear that the locations of corticobulbar neurons overlap and adjoin the arm representations of all premotor areas. A prominent feature of primary and higher somatosensory (Fig. 16.5 F), visual (Fig. 19.6) and acoustic (Fig. 18.3 G) areas, namely that the topical maps in neighbouring areas are reversed, with similar representations bordering each other at the areal borders, can also be observed in the motor cortex (Fig. 21.9 C).

Corticocortical, Cerebellar and Pallidal Connections of the Primary Motor Cortex and the Premotor Areas

The primary motor area and the premotor areas differ in the relative size and the overlap of their representations of arm and leg. More importantly, their projections to upper and lower cervical levels, which control proximal movements of the arm and movements of the hand, respectively, are different [88, 89]. In M1, neurons projecting to the cervical and lumbosacral levels of the cord are completely separated. Two separate regions projecting most densely to the lower cervical cord are present, medially and laterally in the arm area. Proximal and distal arm representations are present in M1, SMA, PMDc and in the three cingulate premotor areas. PMV only projects to the upper cervical and thoracic cord; neurons projecting to low cervical and lumbosacral levels, innervating hand and foot, are scarce.

Direct corticomotoneuronal projections are not the exclusive property of M1. Direct projections to hand muscle motoneurons have been shown for the SMA and the three cingulate motor areas [48]. When the direct projections of M1 and SMA are compared, the former appear to be more extensive. Corticospinal volleys from M1 had larger amplitudes and a faster conduction velocity than those of the SMA [133]. Projections from the SMA to the spinal intermediate zone resemble those of M1 and include dorsolateral laminae V–VII, dorsomedial lamina VI and ventromedial lamina VII and VIII (Fig. 21.5). Projections from CMAd and CMAv are restricted to the dorsolateral and dorsomedial intermediate zone, respectively [48]. It should be remembered that interneurons at high cervical (C3/C4) levels mediate corticospinal and rubrospinal excitation and inhibition of forelimb motoneurons and interneurons at C6/C7 controlling grasping. PMV, which mainly projects to the upper cervical cord, therefore, still has access to motoneurons controlling distal limb movements.

The primary motor area and the various premotor areas differ in their corticocortical connections. M1 is reciprocally connected with the primary sensory area S1 (mainly with proprioceptive area 2), and the rostral superior parietal lobule (area PE, Fig. 21.6 A). M1 receives projections from PMV and PMDc on the lateral convexity and from the SMA and the three cingulate motor areas on the medial surface of the hemisphere [75]. These connections are somatotopically arranged. Strong connections exist between the areas representing the digits of PMV and PMDc and M1; the projections from the SMA and the cingulate motor areas to M1 are less extensive. SMA projects more strongly to PMDc and PMV. The laminar pattern of these corticocortical connections, with terminations in superficial and deep cortical laminae, favours a parallel rather than a hierarchical relationship between M1 and the premotor areas [49].

Reciprocal connections between the parietal lobe and the premotor areas are symmetrically arranged, with respect to the central sulcus (Fig. 21.6 A). The parietal-dependent areas PMV, PMDc, FEF and SMA are connected with the caudal superior parietal lobule (area PEC), most of the cortex in the intraparietal sulcus and the rostral inferior parietal lobule (areas PF and PFG [180]). The connections of the "parietal-dependent" areas will be detailed in

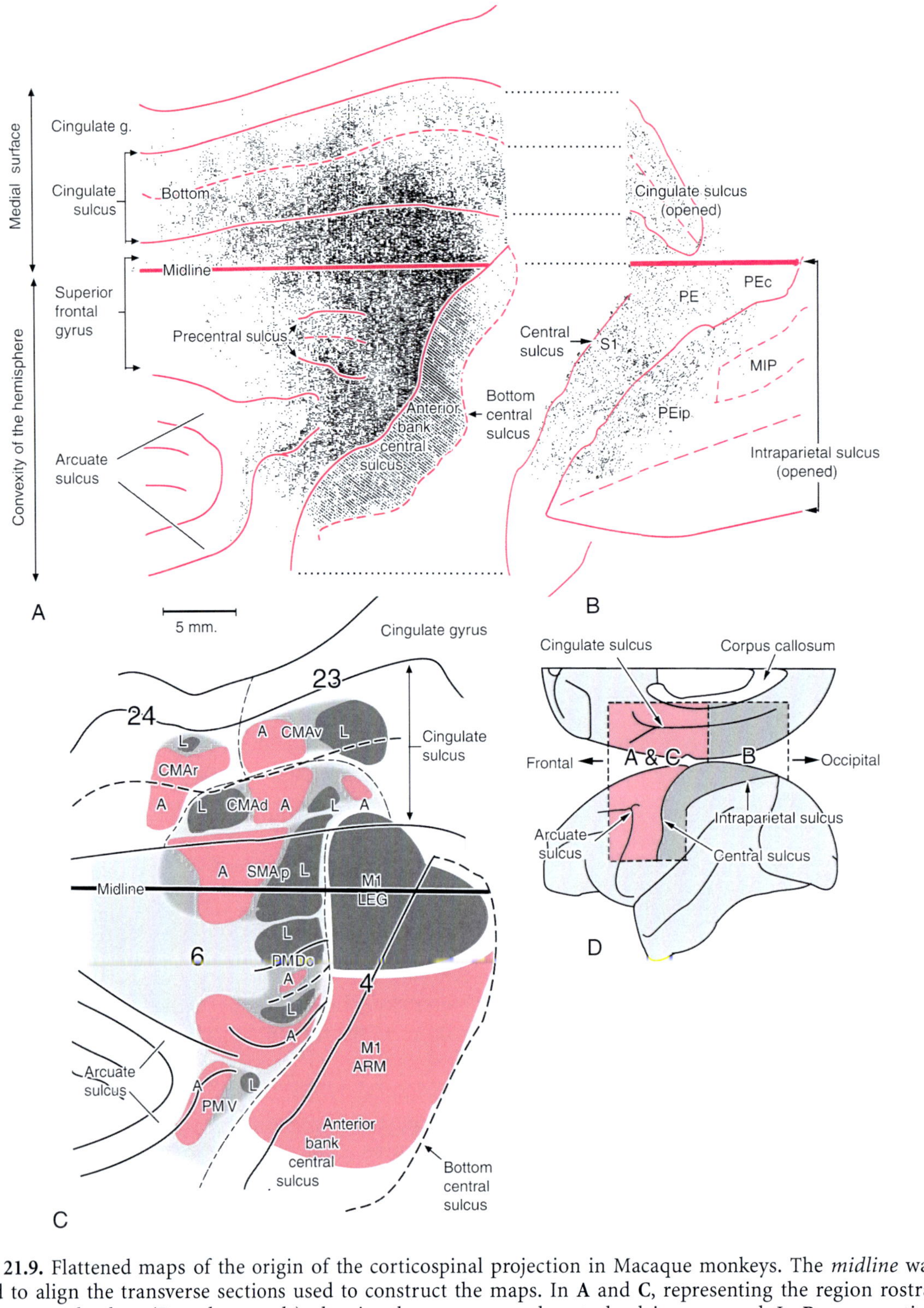

Fig. 21.9. Flattened maps of the origin of the corticospinal projection in Macaque monkeys. The *midline* was used to align the transverse sections used to construct the maps. In **A** and **C**, representing the region rostral to the central sulcus (**D**, *red rectangle*), the cingulate, arcuate and central sulci are opened. In **B**, representing the region caudal to the central sulcus (**D**, *grey rectangle*), the intraparietal sulcus is opened. The bottom of the opened sulci is indicated with a *broken line*. **A** Origin of the corticospinal tract in *Macaca nemestrina* in a flattened map of the region rostral to the bottom of the central sulcus on the convexity and the medial aspect of the cerebral hemisphere. Slightly modified with permission from Dum and Strick [47]. **B** Origin of

the next section "Properties of Primary Motor and Premotor Areas". The more rostral "prefrontal-dependent" premotor areas PMDr, SEF and preSMA are connected with the more caudal parietal cortex (PGm, V6A and PG). Dorsoventral relations are more or less maintained. The superior parietal lobule and the areas in the medial bank of the intraparietal sulcus connect with the dorsal premotor areas (PMDr, PMDc, and SMA); the cortex in the ventral bank of the intraparietal lobule and the inferior parietal lobule are connected with PMV.

Prefrontal connections of the "prefrontal-dependent" premotor areas (preSMA and PMDr), the cingulate motor areas and the frontal and supplementary eyefields take their origin from areas 46, 9 and 11 [18, 73]. In addition, the rostral and caudal cingulate motor areas receive a projection from the amygdala [154] and from many other structures [152].

Cerebellar projections to M1 and the premotor areas take their origin from the dentate and the interposed nuclei and are relayed by the caudal part of the ventral lateral nucleus (VL) of the thalamus (which includes the VPLo, the rostral part of the lateral ventroposterior nucleus, and the medially located area X of Olszewski [160]; see Chap. 21 and Figs. 20.10, 20.12). Striatopallidal projections are transmitted by rostral VL (VLo) and the ventral anterior nucleus (VA). Some overlap of the cerebellar and pallidal innervation of the thalamus occurs at the border regions of these nuclei [185]. Somatotopically arranged cerebellar input to M1 from rostral and dorsal dentate and the anterior interposed nucleus is relayed by the caudal VL and terminates in cortical layers III and V. Pallidal input of M1 is relayed by the VLo and terminates in superficial layer I, mainly rostrally [104, 137, 158, 179]. The parietal-dependent areas SMA and PMDc receive a mainly pallidal input from the VLo. Thalamocortical projections to the prefrontal-dependent preSMA, PMDr, SEF and FEF stem from the VA and area X. These nuclei relay pallidal and cerebellar input to these more rostral premotor areas [135, 146, 179, 186]. The cerebellar projections take their origin from the caudal dentate nucleus. PMV shares afferent connections with both parietal- and prefrontal-dependent premotor areas and receives cerebellar projections from the rostrolateral dentate through caudal VL and area X and a pallidal input through the VLo. The cerebellar input from area X is mainly directed at its rostral subdivision (F5), the pallidal input from VLo to its caudal part (F4) [137, 162].

◀

the corticospinal tract in *Macaca fascicularis* in a flattened map of the region caudal to the central sulcus on the convexity and the medial aspect of the cerebral hemisphere. Modified with permission from Matelli et al. [138]. **C** Somatotopical localization in the primary motor cortex and in premotor areas in *Macaca nemestrina*. Based on the localization of retrogradely filled neurons after injections of tracers at low cervical (arm: *red*) and lumbar (leg: *dark grey*) levels. The approximate extend of the premotor areas is shown in *light grey*. The dorsal premotor areas (PMDc) corresponds to the caudal portion of the dorsal premotor area (Matelli's F2), as illustrated in Fig. 21.6. The ventral premotor area (PMV), however, includes both the caudal and rostral portions of the ventral premotor area (Matelli's F4 and F5) illustrated in the same figure. The pre-SMA is not illustrated because it does not give origin to a corticospinal projection. Rostral (CMAr), dorsal (CMAd) and ventral (CMAv) cingulate premotor areas can be distinguished. Redrawn with permission from He et al. [88, 89]. **D** Figurine of the cerebral hemisphere, showing position of flattened maps A and C in *red*, and B in *grey*. *A*, representation of the arm; *CMAd*, dorsal cingulate premotor area; *CMAr*, rostral cingulate premotor area; *CMAv*, ventral cingulate premotor area; *L*, representation of the leg; *M1*, primary motor cortex; *MIP*, medial intraparietal area; *PE, PEc*, areas PE and PEc of Pandya and Seltzer [164]; *PEip*, intraparietal part of area PE of Pandya and Seltzer [164]; *PMDc*, caudal part of dorsal premotor area; *PMV*, ventral premotor area; *S1*, primary sensory cortex; *SMAp*, supplementary motor area (SMAproper); *6, 23, 24*, areas 6, 23, 24 of Brodmann

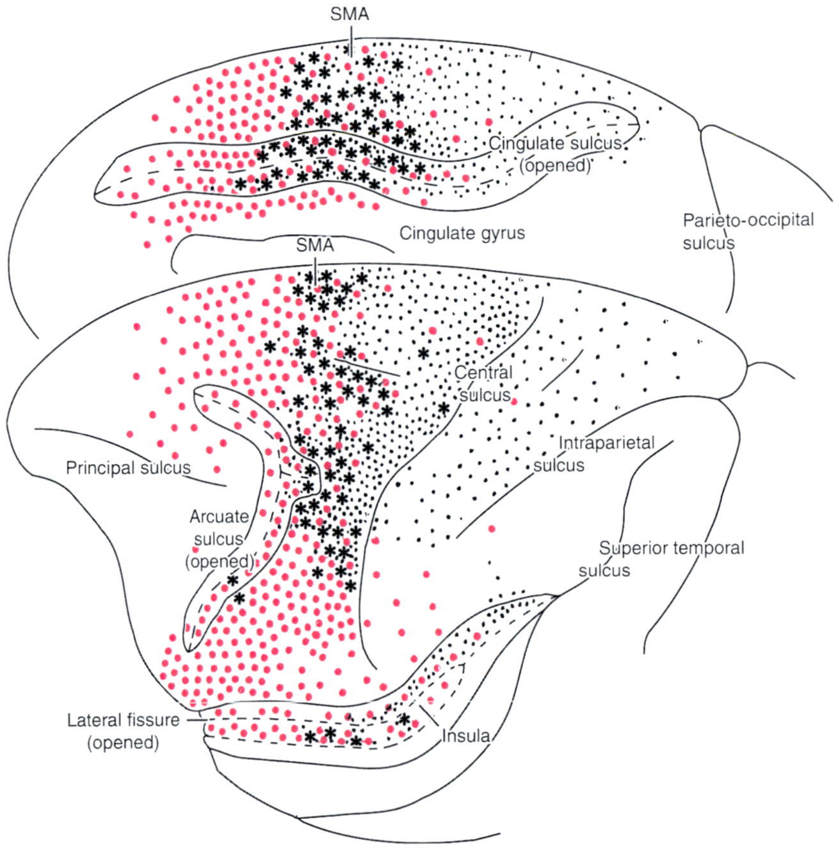

Fig. 21.10. Diagrammatic representation of the distribution of corticospinal neurons (*black dots*), neurons projecting to the bulbar medial tegmental field (*red circles*) and neurons projecting to both targets (*asterisks*) in *Macaca fascicularis*. The cingulate and arcuate sulci and the lateral fissure have been opened. The bottom of these sulci and the insula, located in the bottom of the lateral fissure, are indicated with a *broken line*. Corticospinal and corticobulbar neurons are distributed in partially overlapping regions; corticobulbar neurons occupy a more rostral and ventral position. Corticospinal and corticobulbar neurons are also present in the insula. Compare Fig. 21.9 A. Redrawn with permission from Keizer and Kuypers [113]. *SMA*, supplementary motor area

Properties of the Primary Motor and the Premotor Areas

The Primary Motor Area

M1 contains a somatotopically organized motor map of the body's periphery. Penfield and collaborators [166] showed this arrangement by surface electrical stimulation in human patients under local anaesthesia. The representation of foot, hand and fingers and the orofacial region are most extensive and arranged from medially to laterally (Fig. 21.11 A). Although illustrated as an orderly motor sequence, Penfield's data show much scatter and overlap. The principle of the motor sequence in M1 was confirmed by Fink et al. [60] by positron emission tomography (PET).

When the primary motor cortex of the monkey was studied with intracortical microstimulation, more complicated arrangements invading area 6 were observed. With this technique Kwan et al. [123] found a "nested" organization of the evoked movements of the upper extremity, with finger movements in the centre, surrounded by and overlapping with crescents where movements around the wrist, the elbow and the shoulder could be evoked (Fig. 21.11 B). Such a nested organization is probably present in the foot and the orofacial region of M1. This type of organization is reminiscent of Kuypers concept of the motor cortex, with the digits located caudally in the anterior bank of the central sulcus and proximal body parts more rostrally (Figs. 21.3, 21.4). The nested organization of neurons in M1 with projections to motoneurons innervating distal and proximal forelimb muscles also can be recognized in the figures of He et al. [88].

More recently Graziano [79, 80] observed that, when long, half-second stimulus trains were used instead of the short pulses employed by previous authors, complex, behaviourally relevant movements could be evoked from the primary motor cortex and the rostrally adjacent premotor cortex. Stimulation of a cortical site always elicits movements to the same final posture, regardless of the initial limb configuration. These complex movements are reproducible and fit into several categories, such as reaching, defensive postures, hand-to-mouth movements, manipulation and climbing and leaping with all four limbs (Fig. 21.11 C). Graziano stressed that neurons in the motor cortex may be typically tuned to behaviourally useful patterns of motor output that belong to the monkey's repertoire, and that the motor cortex is subject to constant remapping on the basis of feedback from the periphery.

Although the idea of a rough motor sequence can be maintained, localization in M1 appears to be more complex. Neural activity in M1 reflects many different sensorimotor attributes such as spatial goals, hand motion and rhythmic movements, processed in various premotor areas [196]. Anatomically there is no point-to-point connection of a single M1 cortical neuron with a single muscle; axons of the pyramidal tract collateralize widely in the spinal cord and terminate on multiple motoneurons innervating different muscles [79, 80, 202]. M1 neurons control force, speed and direction of a movement; however, they do not do this on their own, but rather as a population. Georgopoulos et al. [71, 72] studied hundreds of M1 neurons, which fired during a movement in a particular direction. Each individual neuron discharged most intensely with movements in a preferred direction and gradually less when movements were made in directions farther away from the preferred one. The directional tuning curves of the different cells partially overlap. A movement trajectory in a desired direction could be generated by neurons with overlapping tuning curves as a population code for movement direction. The existence of population codes in M1 for the attributes of more complicated movements, as shown by Graziano, are difficult to imagine. As yet the anatomical substrate of these movements is unknown.

Premotor Cortical Areas

Caudal PMD is electrically excitable. Stimulation evokes slowly evolving movements of the upper or lower extremity [76]. It is reciprocally

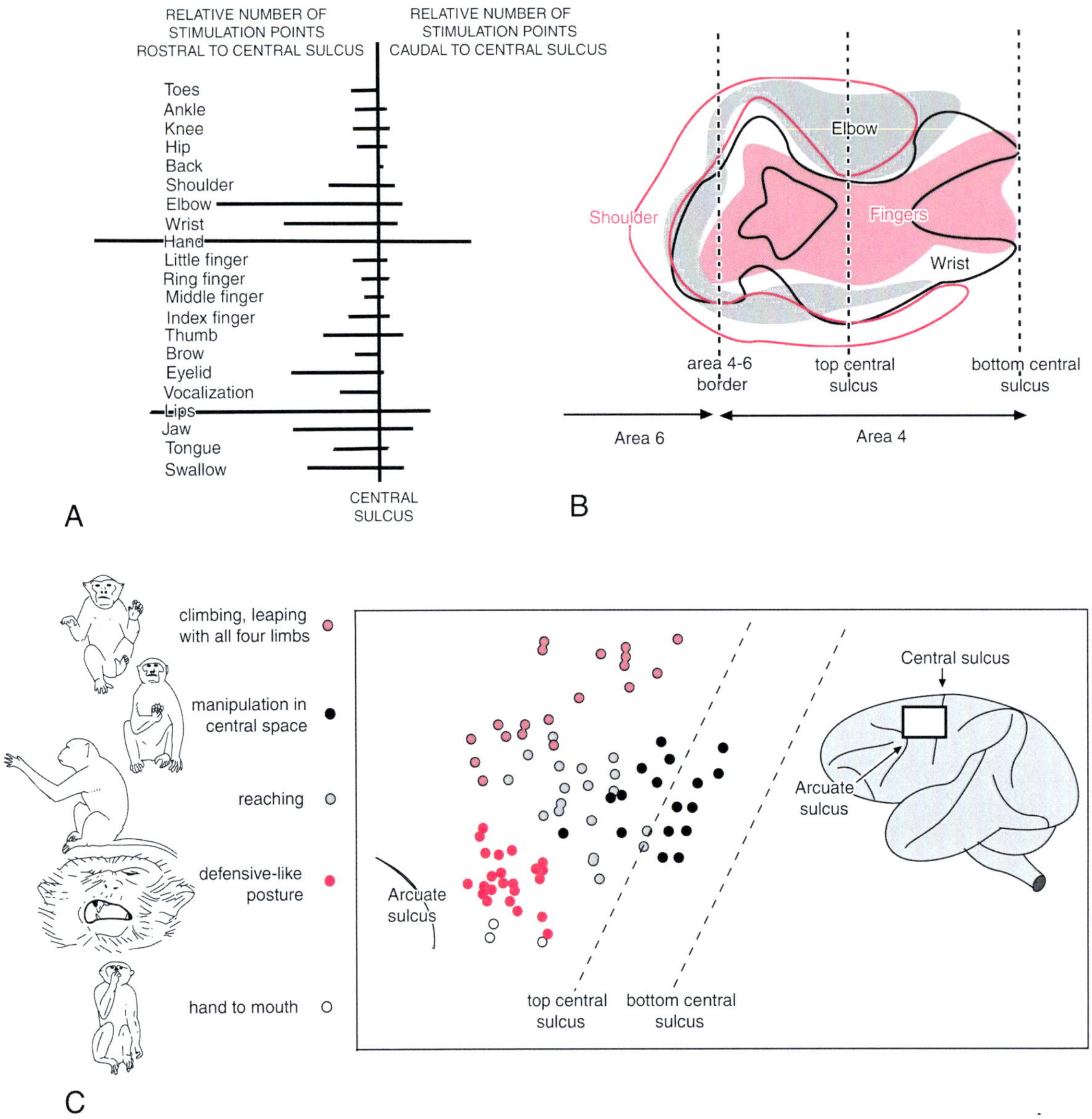

Fig. 21.11. Localization in the motor cortex. **A** Motor sequence in the motor cortex of the human brain. Based on electrical stimulation under local anesthesia. After Penfield and Boldrey [166]. **B** Concentric (nested) organization of movements produced by low-strength intracortical microstimulation in the medial motor cortex of the monkey. Redrawn from Kwan [123]. **C** Complex (purposeful) movement produced by microstimulation at higher strength in the motor cortex of the monkey. Redrawn with permission from Graziano [80]

connected with area PEip, area MIP located more laterally in the medial bank of the intra parietal sulcus, and area PEC in the caudal superior parietal lobule (Fig. 21.6 A). Both PEip and MIP receive proprioceptive information from the superior parietal lobule; in addition, MIP receives visual input from the dorsal component of the dorsal visual stream, through visual area V6A (Fig. 19.6). Neurons in both these areas and in PMD are activated by arm reaching or pointing in specific directions. Reciprocal connections of these intraparietal areas with caudal PMD are used for monitoring and visual guidance of the arm position during the transport phase of hand to target [73, 139, 140, 175].

Rostral PMD is not electrically excitable. It is involved in movement selection and coding object locations for orienting and co-ordinating arm-body movements. This area has been implicated in conditional stimulus-response association tasks [165]. Its actions are exerted through PMDc.

PMV can be subdivided into rostral and caudal parts, corresponding with Matelli's [136] areas F5 and F4, respectively. *PMVr* is reciprocally connected with area AIP, in the lateral bank of the intraparietal sulcus and area PF and PFG, in the rostral inferior parietal lobule (Fig. 21.6 A). AIP and the rostral inferior parietal lobule receive somatosensory input and visual information about 3D features of objects through the ventral and dorsal components of the dorsal visual stream (Fig. 19.6). AIP targets the F5 cortex in the wall of the arcuate sulcus; the inferior parietal lobule is connected with the cortex covering the F5 convexity. Hand movements can be evoked by electrical stimulation of the wall of the arcuate sulcus and the adjacent convexity of PMVr. Orofacial movements are produced by stimulation of the convexity of PMVc [76]. PMVr may exert its actions through its strong connections with the M1 hand and digit representation or through C3/C4 interneurons with projections to forelimb motor neurons, or by activating both pathways. PMV is not connected with the lower cervical cord.

The properties of *PMVr* and its reciprocal connections with AIP have been studied in great detail. Different types of AIP neurons can be distinguished. Motor-dominant neurons and visual-and-motor neurons are preferentially activated during manipulation and grasping of objects. Visual-and-motor and visual-dominant neurons are sensitive to shape, size and orientation of sighted objects. The AIP-PMVr circuit processes incoming binocular visual signals of 3D features of objects and matches these signals with the motor signals from PMVr that will be required for preshaping of the hand to manipulate an object. Types of neurons in PMVr are very similar to those in AIP. Motor neurons in the arcuate sulcus are selective for precision grip, finger prehension or whole-hand prehension or different phases of these movements [111, 187, 188]. They constitute a true vocabulary of motor acts. Visuomotor neurons of this area also discharge during visual presentation of an object, without grasping [73, 156, 175].

The motor properties of neurons located in the cortex of the convexity of PMVr are very similar. These neurons discharge when a monkey observes another individual performing an action that is similar to that encoded by this neuron. They are known as "mirror neurons". Similar cells occur in rostral inferior parietal area PF, which projects to PMVr [65, 175].

PMVc is connected with the ventral intraparietal area (VIP). VIP receives afferents from the dorsal visual stream, with a major input from area MT, and from the primary somatosensory cortex, via the superior parietal lobule. VIP contains bimodal neurons with tactile and visual response properties and unimodal tactile neurons. Their receptive fields emphasize head and face. Tactile and visual receptive fields are aligned. Small fields on the muzzle are aligned with small foveal receptive fields, peripheral fields with fields on the same side of the head or body. This suggests that VIP is involved in the construction of a multisensory, head-centred representation of near extrapersonal space. A subset of VIP neurons is sensitive to optic flow patterns [45]. Bimodal and unimodal neurons also occur in rostral PMVc. This area is concerned with the transformation of object locations into appropriate movements towards the face and the upper body [175]. The extensive corticobulbar projections of PMVc are probably related to this function. Involvement of PMV in mastication, facial expression, swallowing and vocalization will be considered in the next section Cranial Motor Nuclei.

The *supplementary motor area* was the first premotor area to be distinguished as an electrically excitable area with a complete somatotopic map [167, 225]. It is subdivided into the SMAproper (*SMA)* and the more rostral *pre-SMA* [138]. SMA receives afferent connections from preSMA, area 24 and the cortex of the medial aspect of the superior parietal lobule; preSMA belongs to the prefrontal-dependent premotor areas. A specific function of these areas has not been defined. SMA was supposed to be mainly involved in proximal movements, but its projections to motoneurons innervating hand muscles defy this proposition [76, 89]. Its activation by ideation of movements is not an exclusive property of this area but is shared with many other premotor areas [210]. Pre-SMA is distinguished by its activation by higher order movements, requiring learning, whereas the SMA is only activated during their performance [90, 168]. PreSMA controls parietopremotor circuits [175], but a similar function has been attributed to other prefrontal and prefrontal-dependent areas.

Similar premotor areas appear to be present in the human brain [140] (Fig. 21.6B,C). With respect to their role in grasping and pre-shaping of the hand using visual clues on the 3D aspects of the object, and their activation by imitating hand movements, many similarities were noticed in experimental studies on non-human primates and in human functional magnetic resonance (fMRI) studies of AIP and PMVr (human area 44). Apart from its involvement in speech, area 44 is activated by imagining, observing and imitating hand actions [22, 32, 176, 211]. Cattaneo et al. [33], using paired pulse transcranial magnetic stimulation, studied motor evoked potentials in hand muscles used to shape the hand when reaching out to grasp a specific object. They showed that 600 ms before movement, there is an enhancement in the excitability of corticocortical inputs to corticospinal neurons projecting to the specific hand muscles that will be used for the grasp. Presentation of the object alone or movements of the same muscles but without an object did not evoke such a change in excitability.

SMA and pre-SMA also can be distinguished in the human brain. The vertical commissura posterior (VCP) line, passing through the posterior commissure perpendicular to a line passing through the posterior and anterior commissures, separates human SMA and pre-SMA [230]. A Bereitschaftspotential (readiness potential) develops over SMA, just prior to movement initiation; human SMA, therefore, appears to be important in motor planning and timing [43, 78]. Rostral to the VCP line activation was found for complex movements with a cognitive component or selection between motor tasks. Basic or highly practiced movements activated an area caudal to the VCP line [73].

Little is known about the function of the cingulate motor areas. Their connections with the amygdala and other limbic structures suggest that they are involved in emotional and motivational aspects of the generation of movements. Support for this notion can be found in the innervation of the nucleus of the facial nerve (see the next section "The Cranial Motor Nuclei" and Chap. 15).

The Cranial Motor Nuclei

Holstege and Kuypers [93, 98], in a study of propriobulbar connections of the cranial motor nuclei in the cat, concluded that interneurons with ipsilateral projections were located laterally in the lateral tegmental field, which includes the lateral parvocellular reticular formation (Figs. 21.3, 22.1), and neurons with bilateral connections more medially, at the border of the lateral tegmental field and the medial reticular formation (usually indicated as the medial tegmental field). Premotor connections of the nuclei of the cranial nerves were reviewed by Holstege [92] and in several more recent publications on the cranial nerve nuclei and their premotor neurons. The connections of the motor trigeminal nucleus, the facial and hypoglossal nuclei and the ambiguus nucleus, and their involvement in mastication, swallowing and vocalization will be summarized in this section.

The Motor Nucleus of the Trigeminal Nerve

The motor nucleus of the trigeminal nerve can be divided into a lateral group, containing motoneurons innervating jaw closing muscles, and a medial group with the motoneurons of the jaw openers [150] (Fig. 21.12). It is embedded in the lateral tegmental field. Most interneurons projecting to the trigeminal motor neurons are located in the supratrigeminal and intertrigeminal nuclei, which border the motor nucleus on its dorsal and lateral side, and in the reticular formation along its ventromedial aspect.

Early studies of the cortical innervation of the motoneurons of the trigeminal nerve already stressed the paucity of the direct corticomotoneuronal connections in primates [115, 116, 193]. In a recent study in *Macaca fuscata* only few direct corticomotoneuronal projections from the primary and supplementary motor areas and the ventral premotor area were found. This paucity of direct corticomotoneuronal connections may be related to the bilateral co-ordination of movements in mastication. The problem of bilateral co-ordination is probably solved by the interposition of a central pattern generator (CPG) for jaw movements, but such a structure is still largely hypothetical.

Most premotor connections of trigeminal motor neurons take their origin from the lateral tegmental field. Jaw closers of the lateral group receive excitatory projections from the supratrigeminal and intertrigeminal nuclei, the parabrachial nuclei and the area of Koelliker-Fuse and from more caudal regions of the lateral tegmental field [92] (Fig. 21.12). Neurons in the supratrigeminal nucleus are both excitatory and inhibitory [29, 128, 148]. Inhibitory neurons, co-localizing GABA and glycine, project to the contralateral motor nucleus [129]. In the mouse the inhibitory interneurons of the supratrigeminal nucleus receive input from marginal cells of the caudal spinal trigeminal nucleus [128].

Motoneurons of the jaw-opening muscles in the medial group of the motor nucleus receive excitatory input from the caudal lateral tegmental field, especially from the retroambiguus nucleus, and from the oral part of the spinal nucleus of the trigeminal nerve [92, 200].

Neurons of the rostral mesencephalic nucleus of the trigeminal nerve, which innervate the muscle spindles of the jaw muscles, and its caudal part, which innervate pressure mechanoreceptors in the periodontal ligaments, give rise to collateral projections, descending as Probst's tract. These collaterals innervate motoneurons of jaw closers in the lateral group, both directly and by synapsing in the supratrigeminal nucleus (Fig. 21.12). Probst's tract continues into the dorsolateral medulla oblongata to the level of the dorsal vagal nucleus where it terminates in the lateral tegmental field [40, 132, 199]. It traverses the large neurons of Probst's nucleus, located ventral to the nucleus of the solitary tract [41] (Fig. 21.16A). In the rat this nucleus gives rise to a crossed pathway to the lateral trigeminal subnucleus (Fig. 21.12). Excitatory pathways from the mesencephalic nucleus thus have access to motoneurons innervating jaw closers from both sides.

Projections of the motor cortex to trigeminal premotor neurons were studied by Hatanaka et al. in monkeys [87]. Sites in lateral M1 and rostral SMA, where twitches in jaw muscles were evoked on electrical stimulation, project to the supra- and intertrigeminal nuclei, the principal sensory nucleus of the trigeminal nerve and the lateral tegmental field. Projections from M1 distribute mainly contralaterally, from the SMA bilaterally (Fig. 21.12). Stimulation of the ventral premotor area (principal cortical masticatory area, CMaAp [87]) and the medial frontal operculum (deep cortical masti-

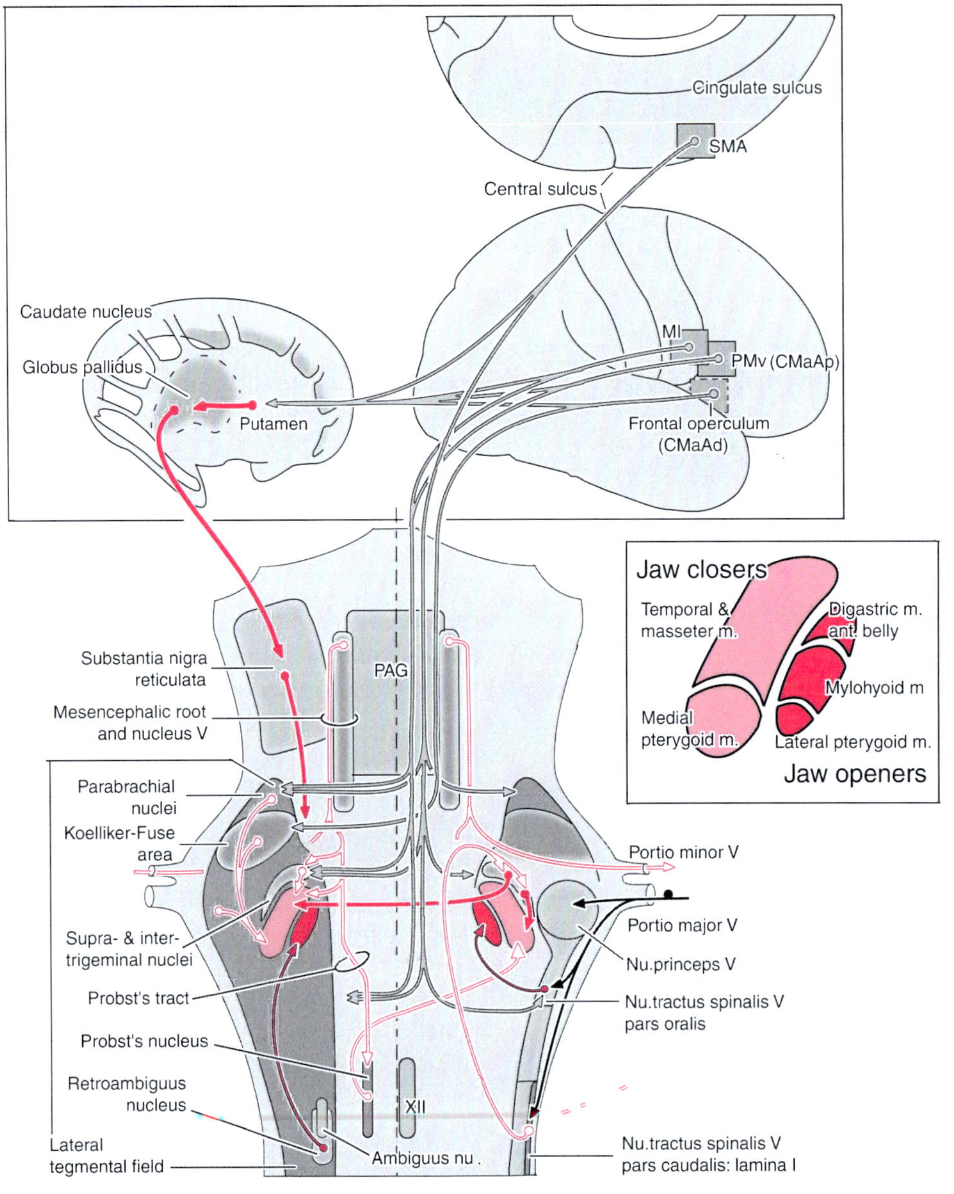

Fig. 21.12. Afferent connections of the motor trigeminal nucleus in the monkey. The localization of motoneurons innervating jaw closers and openers is illustrated in a diagram of a transverse section through the nucleus of the monkey. Redrawn with permission from Mizuno et al. [150] (*inset*). Collaterals from the mesencephalic root of the trigeminal nerve descend in Probst's tract and terminate on motoneurons innervating jaw closers and their excitatory premotor interneurons (*red open contours*) in the supra- and intertrigeminal nuclei and on neurons of Probst's nucleus. This nucleus innervates motoneurons of the contralateral jaw closers. Interneurons innervating jaw openers are located in the retroambiguus nucleus. Primary afferents in the trigeminal nerve, illustrated on the *right side* of the figure, inhibit jaw-opening and jaw-closing motoneurons through inhibitory interneurons in the spinal trigeminal nucleus and the supra- and intertrigeminal nuclei. In monkeys, chewing movements can be elicited from the motor and premotor areas illustrated in this figure. These areas project to interneurons located in various subdivisions of the lateral tegmental field and influence the trigeminal motor system via an inhibitory (*bright red*) striatal loop passing through the globus pallidus and the substantia nigra reticulata. Direct corticomotoneuronal connections to the trigeminal motor nucleus appear to be scarce or absent. A putative central pattern generator for chewing is located in the pontine reticular formation (Fig. 21.15). *CmaAd*, deep cortical masticatory area; *M1*, primary motor cortex; *PAG*, periaqueductal grey; *PMv*, (CmaAp) ventral premotor area (principal cortical masticatory area); *SMA*, supplementary motor cortex; *XII*, hypoglossal nucleus

catory area, CMaAd) evokes rhythmic jaw movements. Projections from CMaAp are similar to those from M1. CMaAd projects to the contralateral parabrachial nuclei and the Koelliker-Fuse area. CMaAd is a gustatory area, receiving projections from the solitary nucleus via the parvocellular ventral posteromedial thalamic nucleus [171, 172]. Trigeminal connections of the cingulate motor areas have not yet been investigated.

Hatanaka et al. [87] also drew attention to the corticostriatal pathways from jaw movement-associated cortical areas. Corticostriatal fibres were traced from all these areas to the putamen. They cite evidence, mainly obtained in the rat, on the connections of the striatum with the lateral tegmental field through the internal segment of the globus pallidus and/or the dorsolateral aspect of the substantia nigra reticulata. These inhibitory pathways may be of a more general importance in controlling the cranial motor nuclei.

A cell group MA, located dorsal to the superior olive, and which may correspond to the pattern generator for mastication or a part thereof, was identified by Hage and Jürgens [82] in the squirrel monkey in a region of the reticular formation, dorsal to the superior olive (Fig. 21.15). This region projects bilaterally to motoneurons of the trigeminal, facial, hypoglossal and ambiguus nuclei [86]. In addition, connections of this region with the lateral bulbar reticular formation, the nucleus of the solitary tract, the periaqueductal grey and several vestibular, cochlear, auditory and precerebellar nuclei were identified. This region receives a strong projection from the periaqueductal grey. Neurons with mastication-correlated activity were restricted to a cell group MA, located in this region, and to the ipsilateral motor nucleus of the trigeminal nerve and in the facial nucleus. They were absent from the nucleus ambiguus. No recordings were made in the hypoglossal nucleus, which contains the motoneurons innervating the jaw-opening geniohyoid muscle.

The Nucleus of the Facial Nerve

The nucleus of the facial nerve consists of four or five subnuclei (Fig. 21.13). Motoneurons of the medial and ventral subnuclei innervate the muscles of the auricle and the platysma, respectively. The lateral subnucleus innervates the perioral muscles. The dorsal subnucleus innervates the frontalis and orbicularis oculi muscles. The orbicularis oculi motor pool is sometimes called the intermediate nucleus [153, 189].

Premotor neurons of the facial nucleus are mainly located in the lateral tegmental field. Motoneurons innervating the frontalis and orbicularis oculi muscles of the dorsal subnucleus receive projections from the parabrachial nuclei, the area of Koelliker-Fuse and the principal and spinal nuclei of the trigeminal nerve (Fig. 21.13). In addition they receive a projection from the rubrobulbar and spinal tract [100]. The lateral subnucleus, innervating the perioral muscles, also receives projections from the parabrachial nuclei and the area of Koelliker-Fuse. Its strongest input is from the more caudal lateral tegmental field and the contralateral retroambiguus nucleus. This nucleus is located next and caudal to the ambiguus nucleus and represents the caudalmost extension of the lateral tegmental field (Figs. 6.21, 22.1). The medial and ventral motoneurons innervating the ear muscles and the platysma receive a projection from the retroambiguus nucleus in the caudal lateral tegmental field. Moreover, fibres from the cervical cord, travelling in the anterolateral funiculus, terminate in these subnuclei.

The blink reflex protects the eyeball by closing the eyelids through contraction of the orbicularis oculi muscle. Several relays have been identified in the pathway of the blink reflex. Its afferent pathway consists of sensory fibres innervating the cornea, that terminate in the nucleus of the spinal tract of the trigeminal nerve (Fig. 21.13). The ipsilateral, short latency orbicularis oculi reflex (R1 component) synapses in the nucleus of the spinal tract of the trigeminal nerve. The bilateral, long-latency R2 component is relayed by the medullary and pontine

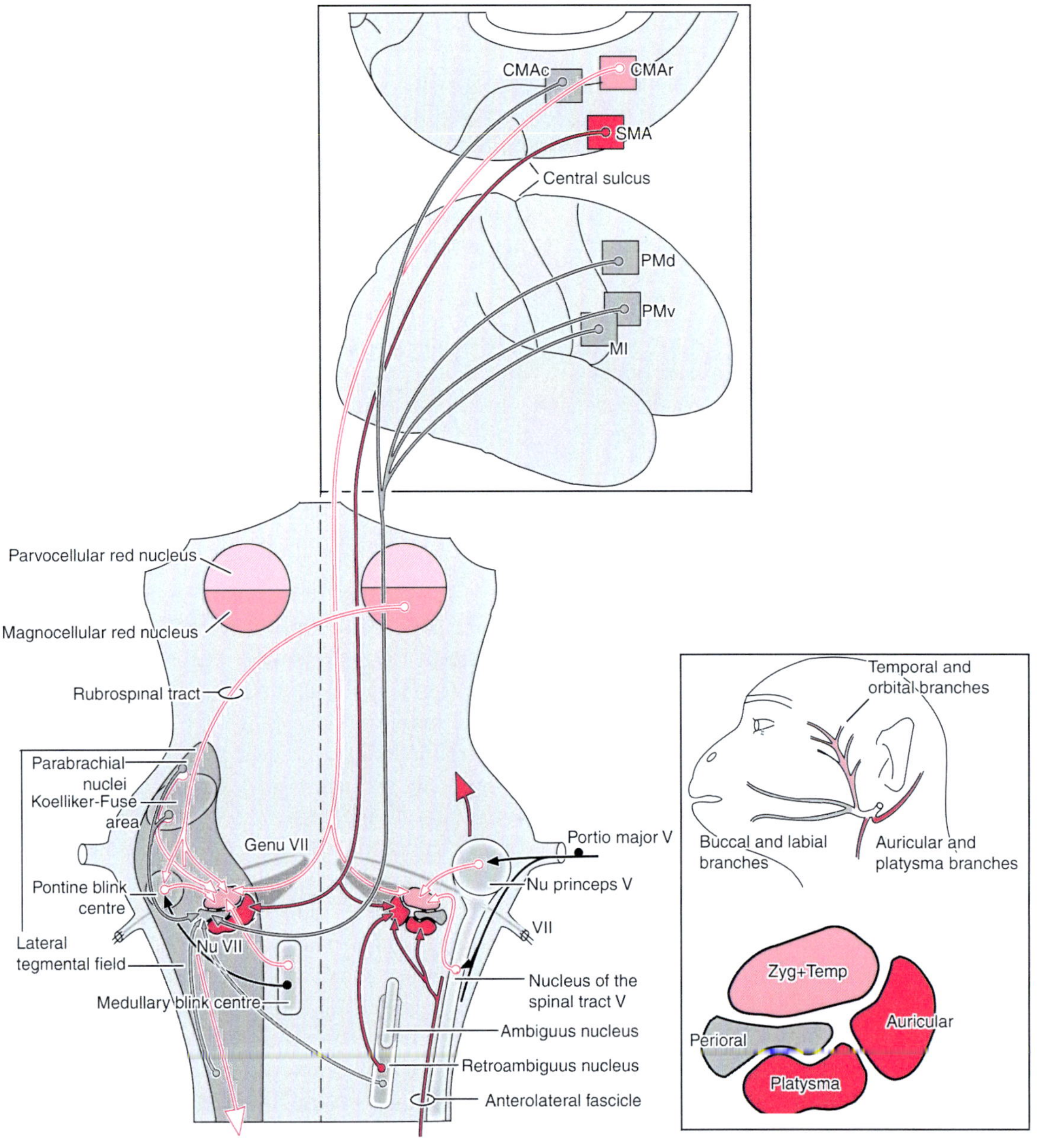

Fig. 21.13. Afferent connections of the motor nucleus of the facial nerve. The localization of motoneurons innervating the facial muscles is illustrated in a diagram of a transverse section through the nucleus of the monkey. Redrawn with permission from Satoda et al. [189] (*inset*). Interneurons for motoneurons innervating ear muscles and the platysma (shown in *red*) are located caudally in the lateral tegmental field in the retroambiguus nucleus. In addition, these motoneurons receive a collateral projection from the spinal cord. Motoneurons innervating perioral muscles receive projections from the lateral tegmental field adjoining the spinal trigeminal nucleus and a strong projection from the parabrachial nuclei and the nucleus of Koelliker-Fuse (shown as *contours filled with grey*). Motoneurons innervating the orbicularis oculi muscle are involved in the blink reflex. The early, ipsilateral (R1) component of the blink reflex is relayed by interneurons located in the principal and spinal trigeminal nuclei (shown as *open red contours*). The late, bilateral (R2) component of the blink reflex is relayed by interneurons in the pontine and medullary blink centres. The magnocellular red nucleus targets both this group of motoneurons and the pontine blink centre. The red nucleus serves as the main link in conditioning the blink reflex by the cerebellum. In primates, different motor and premotor areas project to facial motoneurons and interneurons. The projection of M1, the ventral and dorsal premotor areas and the SMA to motoneurons innervating perioral muscles is crossed; projections from the rostral cingulate motor area to motoneurons innervating auricular muscles, platysma and periorbital and temporal muscles are bilateral. *CMAc*, caudal cingulate premotor area; *CMAr*, rostral cingulate premotor area; *MI*, primary motor cortex; *PMd*, dorsal premotor area; *PMv*, ventral premotor area; *SMA*, supplementary motor area; *VII*, facial nerve; *Zyg+Temp*, subnucleus innervating zygomatic and temporal branches of the facial nerve

blink centres (Fig. 21.13). The medullary blink centre receives trigeminal input via multi-synaptic pathways relaying in the tectum and the reticular formation. The pontine blink centre receives a projection from the rubrobulbar and spinal tract. This connection is used by the cerebellum in conditioning the blink reflex [92]. Transneuronal retrograde rabies virus transport from orbicularis oculi motoneurons in the rat revealed additional projections from vestibular, cochlear and tectal cell groups, which may be involved in the blink reflex [151].

Contrary to the motor nucleus of the trigeminal nerve, the motoneurons of the facial nucleus receive a strong, bilateral projection from the motor cortex (Fig. 21.13). The studies of Morecraft et al. [153] in monkeys showed that the face area of M1, the dorsal and ventral premotor cortex and the caudal cingulate motor area project to the contralateral lateral subnucleus that innervates the perioral muscles. The projection of the SMA is bilateral and targets motoneurons of the medial subnucleus, innervating the muscles of the ear. The rostral cingulate motor area provides the motoneurons of the orbicularis oculi with a bilateral innervation [153]. A more extensive cortical projection to the dorsal subnucleus was reported by Gong et al. [77]. However, their use of retrograde axonal transport methods in a densely innervated region such as the facial motor nucleus makes these observations less reliable.

Two classical clinical observations on the sparing of the upper facial muscles in supranuclear palsy of the facial nerve and on the preservation of emotional facial expressions, such as smiling and laughing, in certain cases of facial paralysis were discussed by Morecraft et al. [153], on the basis of their experimental studies in subhuman primates. Sparing of the upper facial muscles is not due to a bilateral M1 projection to the motoneurons supplying these muscles, as previously assumed, but rather may be the consequence of the bilateral projection of the rostral cingulate motor area. Preserved cingulate motor areas, with projections to upper facial and perioral muscles, in patients who have suffered a stroke, represent a possible substrate for the presence of emotional facial expressions.

The Nucleus Ambiguus and the Nucleus of the Hypoglossal Nerve: Swallowing and Vocalization

The nucleus ambiguus can be subdivided into a dorsal, branchiomotor division, corresponding to the nucleus ambiguus in a narrow sense, and an external formation containing preganglionic parasympathetic neurons. Figure 21.14 illustrates the nucleus ambiguus of the rat [21]; its morphology in primates is very similar [190, 191, 226, 227]. The dorsal division consists of a rostral segment with a compact arrangement of the motoneurons and of intermediate and caudal diffuse segments, where the packing of the cells displays a rostrocaudal periodicity. The stylopharyngeus motor pool of the glossopharyngeal nerve is located dorsal to the facial nucleus in the retrofacial nucleus, together with neurons innervating the cricothyroid muscle via the superior laryngeal nerve. In primates the cricothyroid muscle, one of the laryngeal muscles involved in phonation, surrounds the external formation. The upper oesophagus with the soft palate and the lower oesophagus with the pharynx are innervated by motoneurons of the rostral and intermediate part of the compact nucleus ambiguus, respectively. The caudal part of the nucleus contains a second pool of laryngeal motoneurons, which gives rise to the recurrent laryngeal nerve.

Motoneurons innervating the geniohyoid muscle are located in the ventralmost region of the rostral hypoglossal nucleus of the cat. The other extrinsic tongue muscles are located laterally; the intrinsic muscles of the tongue are innervated by motoneurons of its medial and dorsal parts [212]. Premotor neurons innervating hypoglossal motoneurons are located in the dorsal and parvocellular lateral tegmental field. This region receives input from the cortical tongue area [1]. Direct corticomotoneuronal projections to the hypoglossal nucleus are rare in non-human primates, but strengthened in humans [112].

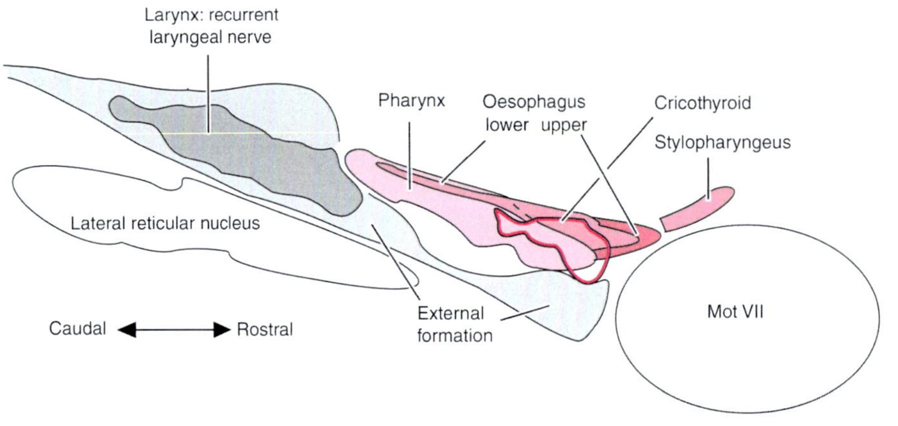

Fig. 21.14. The localization of motoneurons innervating different muscle groups in the nucleus ambiguus of the rat. The external formation contains preganglionic, parasympathetic neurons. The glossopharyngeal nerve innervates the stylopharyngeus muscle. Together with the vagus nerve it gives rise to the nervous plexus of the pharynx and the oesophagus. The cricothyroid muscle is innervated by a branch of the vagus nerve, the superior laryngeal nerve. Redrawn with permission from Bieger and Hopkins [21]. In primates the configuration is very similar [226, 227]. *MotVII*, motor nucleus of the facial nerve

A region in the caudal periaqueductal grey plays a central role in vocalization. Stimulation of this region results in vocalization in many different species [91, 229]. It gives rise to excitatory projections to the ambiguus nucleus and a strong projection to the retroambiguus nucleus [91, 134, 218] and the reticular formation dorsal to the superior olive [86]. The retroambiguus nucleus is connected with motoneurons innervating laryngeal, abdominal and other muscles involved in vocalization [198, 219].

More recently a cellgroup (cellgroup VOC), located dorsal to the superior olive, was identified as a vocal pattern generator, or a part thereof, in the squirrel monkey [83] (Fig. 21.15). Neurons in this group showed an increase in activity before and during frequency-modulated, self-produced vocalizations, such as trill or cackle. Neurons displaying a similar increase in activity were present in the ipsilateral rostral nucleus ambiguus (corresponding to the retrofacial nucleus, containing motoneurons innervating the cricothyroid muscle), the facial nucleus and in the motor nucleus of the trigeminal nerve. The general area containing the VOC group receives a strong projection from the periaqueductal grey. It projects bilaterally to these motor nuclei, the nucleus of the solitary tract, the bulbar lateral tegmental field (the dorsal medullary reticular column, DMRC, see next paragraph) and to multiple auditory, vestibular and precerebellar nuclei [86]. The nucleus of the hypoglossal nerve, which is also implicated in vocalization, also receives a projection from this area, but was not included in the recordings. This area also contains a cell group MA, considered as the pattern generator in mastication (see "The Motor Nucleus of the Trigeminal Nerve").

Three stages can be distinguished in swallowing. The first or oral phase includes mastication. During the second, pharyngeal phase the oral and nasal cavities are closed off from the pharynx and the bolus is displaced by pharyngeal constriction. During the third or oesophageal phase, which completes the swallowing act, the larynx is elevated and sealed. Premotor neurons for the first phase are located in the caudal lateral tegmental field, in an area distinguished as the dorsal medullary reticular column [41, 92, 99] (Fig. 21.16: DMRC). In the rat these premotor neurons project bilaterally to the hypoglossal, facial and trigeminal motor nuclei. Motoneurons of the ambiguus nucleus innervating pharyngeal muscles, which are involved in the second phase, receive input from the nucleus of the solitary tract. Oesophageal motoneurons of the nucleus ambiguus are innervated by neurons of the central nucleus of the nucleus solitary tract [16, 41].

A region controlling movements of the soft palate, the pharynx and the oesophagus, and the supra- and infrahyoid and tongue muscles innervated by the hypoglossal and trigeminal nerves was identified in the cat in the caudal pontine tegmentum, dorsomedial to the superior olive. This cellgroup projects to the dorsal group of the nucleus ambiguus, which innervates the pharynx and the soft palate, to trigeminal motoneurons innervating the jaw-opening muscles and to the geniohyoid motoneurons in the hypoglossal nucleus. It has been suggested that this system is responsible for inhibiting these muscles during the early phase of swallowing [92, 99]. Its location and its connections with the cranial motor nuclei are very similar to that of the supraolivary region containing vocal and masticatory pattern generators in the squirrel monkey, as discussed in the previous paragraphs. The projection of this region in the squirrel monkey to the nucleus of the solitary tract and its equivalent of the rat DMCR in the bulbar lateral tegmental field [86] emphasizes the importance of these regions in the control of orofacial movements.

In a PET study of swallowing in humans, increased blood flow was found bilaterally in ventral M1 (compare Fig. 21.11A), and in the ventral premotor area, the inferior parietal lobule and the anterior insula, the latter all on the right side. Other foci were found in the simplex lobule of the cerebellum and in the striatum [228]. The inferior parietal-ventral premotor circuit is apparently involved in the control of swallowing. Detailed studies of the connections of motor cortical areas involved in swallowing, with premotor neurons of the ambi-

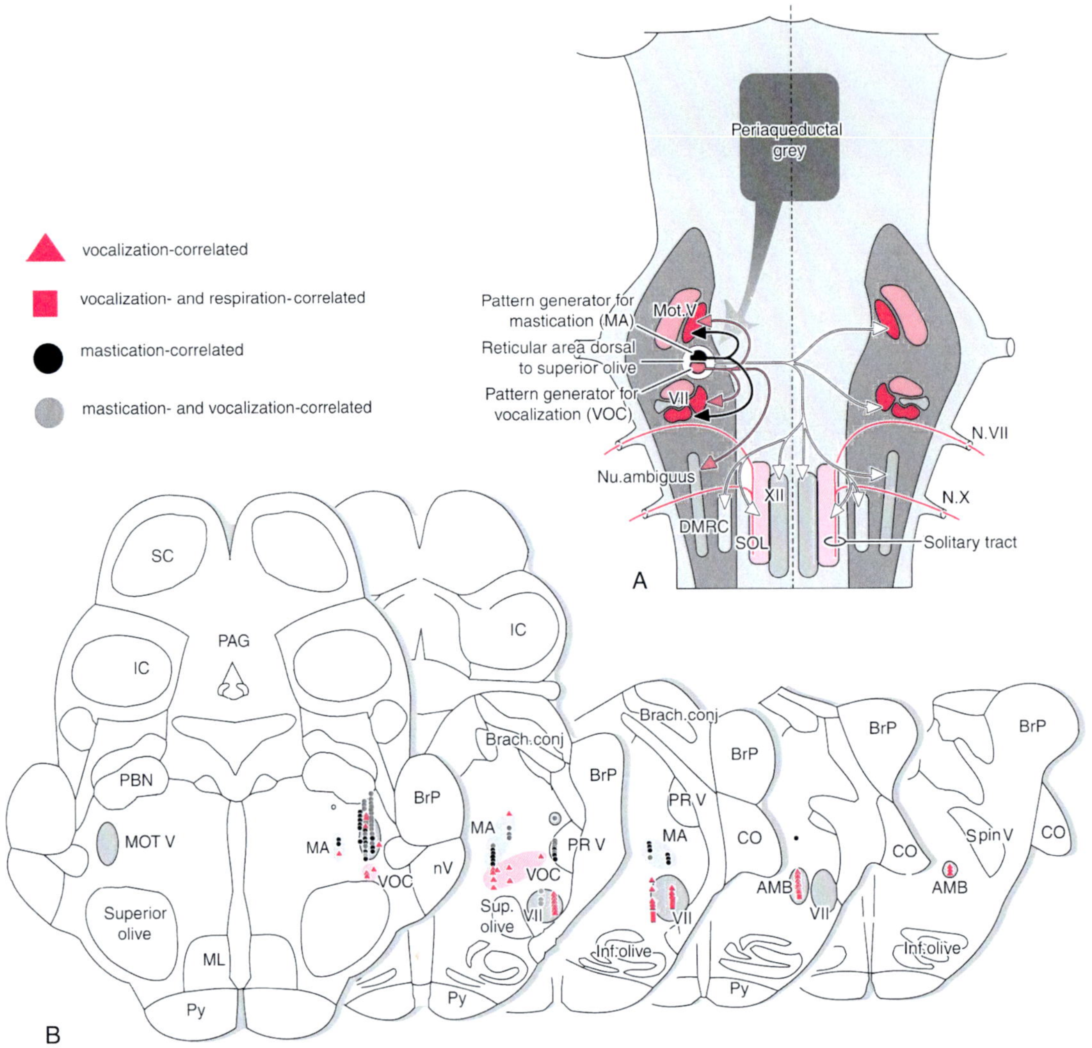

Fig. 21.15. A region of the reticular formation, located dorsal to the superior olive, with widespread connections to cranial motor nuclei and premotor-centres involved in mastication, vocalization and swallowing, has been identified in the squirrel monkey [86]. **A** Projections of this region to the motor nuclei of the trigeminal, facial and hypoglossal nerve, to the ambiguus nucleus and to the dorsomedial reticular column (DMCR) and the nucleus of the solitary tract, two regions involved in swallowing (Fig. 21.16). This region receives a strong projection from the periaqueductal grey. Two neuronal groups, identified in this region, represent the pattern generators for mastication (MA: *black*) and vocalization (VOC: *red*). Projections of both groups to the ipsilateral motor nuclei of the trigeminal and facial nerves and an additional projection of the VOC to the rostral nucleus ambiguus (retrofacial nucleus: innervation of cricothyreoid muscle) were identified by recording from these cells during spontaneous mastication or vocalization. **B** Distribution of vocalization-correlated and mastication-correlated activity in the MA, VOC and the cranial motor nuclei. Mastication-correlated activity (*black and grey units*) is present in MA and in the facial and trigeminal motor nuclei; vocalization-correlated activity (*red units*) was found in neurons of the VOC, the facial and trigeminal nuclei and in the rostral ambiguus nucleus. Redrawn with permission from Hage et al. [84]. Data on respiration-correlated activity were omitted. *AMB*, ambiguus nucleus; *Brach.conj*, brachium conjunctivum; *BrP*, brachium pontis; *CO*, cochlear nuclei; *DMRC*, dorsomedial reticular column; *IC*, inferior colliculus; *MA*, pattern generator for mastication; *ML*, medial lemniscus; *Mot V*, motor nucleus of the trigeminal nerve; *N VII*, facial nerve; *N X*, vagus nerve; *NV*, trigeminal nerve; *PAG*, periaqueductal grey; *PBN*, parabrachial nuclei; *PR V*, principal sensory nucleus of the trigeminal nerve; *Py*, pyramidal tract; *SC*, superior colliculus; *SOL*, nucleus of the solitary tract; *Spin V*, nucleus of the spinal tract of the trigeminal nerve; *Sup.olive*, superior olive; *VII*, nucleus of the facial nerve; *VOC*, pattern generator for vocalization; *XII*, nucleus of the hypoglossal nerve

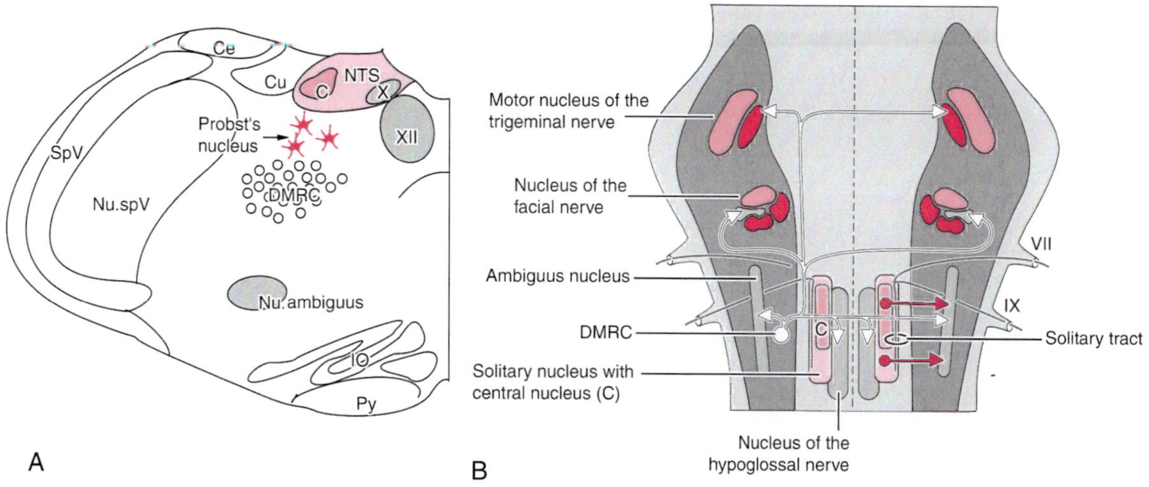

Fig. 21.16. Brain stem connections involved in swallowing. **A** Nuclei involved in the control of swallowing in the rat include the dorsal medullary reticular column (DMRC) in the caudal lateral tegmental field and the nucleus and subnuclei of the solitary tract. Redrawn from Cunningham et al. [41]. **B** Muscles innervated by the hypoglossal, facial and trigeminal motor nuclei are involved in the first, oral phase of swallowing. Premotor neurons of these nuclei are located in the DMRC. Motoneurons of the ambiguus nucleus innervating pharyngeal muscles, which are involved in the second phase of swallowing, receive input from the nucleus of the solitary tract. Oesophageal motoneurons of the nucleus ambiguus are innervated by neurons of the central nucleus of the nucleus solitary tract [41]. *C*, central subnucleus of the nucleus of the solitary tract; *Ce*, lateral cuneate nucleus; *Cu*, medial cuneate nucleus; *DMRC*, dorsomedial reticular column; *NTS*, nucleus of the solitary tract; *Nu.SpV*, nucleus of the spinal tract of the trigeminal nerve; *Py*, pyramidal tract; *SpV*, spinal tract of the trigeminal nerve; *IX*, glossopharyngeal nerve; *VII*, facial nerve; *X*, dorsal nucleus of the vagus nerve; *XII*, nucleus of the hypoglossal nerve

guus nucleus, are not yet available. The presence of different cortical areas involved in swallowing may explain the occurrence of dysphagia in many neurological disorders.

Brain Stem Control of Spinal Motor Systems

Brain stem control of spinal motor systems is effectuated by long descending brain stem pathways. Some pathways involved in specific aspects of motor control are topically organized and terminate on particular cell groups in the spinal cord. Other pathways that subserve more general level-setting functions terminate more diffusely in the spinal cord and in the brain stem. Holstege [92] distinguished a group of descending brain stem pathways of somatic motor control systems from pathways involved in limbic motor control. The first group includes the rubrobulbar and spinal tract (discussed in the section "Kuypers' Concept of the Motor System", this Chapter, Figs. 21.2, 21.3), the vestibulospinal tracts (Chap. 17, "Projections to the Spinal Cord", Fig. 17.6), the projection of the superior colliculus through the tectospinal, interstitiospinal and reticulospinal tracts (Chap. 19, "The Superior Colliculus"). Limbic motor control is effectuated through pathways descending from the hypothalamus, the basal telencephalon, the periaqueductal grey and their relays in the reticular formation of the brain stem.

Two systems, locomotion and micturition and ejaculation, will be reviewed in this section. They illustrate the interactions of the spinal cord with the brain stem, the limbic system, the cerebral cortex and the cerebellum in shaping movement.

Locomotion

Locomotion is dependent on a central pattern generator (CPG) in the spinal cord that initiates and generates the rhythm and shapes the basic pattern of motor bursts of motoneurons in locomotion [13, 52, 61, 108]. Spinal reflexes and cortical and subcortical motor centres are able to adapt locomotion to external and internal circumstances. Their actions on motoneurons are often phase-dependent, i.e. differ in sign or strength in the various phases of the step cycle. Phase dependency can be achieved through a direct action of these systems on the CPG, through the modulation of their spinal interneurons and/or motoneurons by the CPG or through recurrent, modifiable spinal cortical or brain stem loops. The most prominent of these loops passes through the cerebellum. The circuitry of the mammalian CPG is incompletely known. It may or may not include the modifiable interneurons in the spinal reflex pathways and the descending systems. In its simplest form the CPG is modeled as two half centres innervating extensor and flexor muscles, which are connected by inhibitory interneurons [30, 31]. A single CPG, or a hierarchy of CPGs, with one CPG for each extremity, may be present. A recent model of the mammalian CPG was published by Rybak et al. [50, 184].

Locomotion has been mainly studied in cats. Both normal animals and decerebrated and spinal cats have been used. Walking in premamillary decerebrated cats can be spontaneous or induced by stimulating certain brain stem areas, and is known as "controlled locomotion". Locomotion in spinal cats can be induced by stimulating the spinal roots or injecting certain neurotransmitters. Locomotion-related activity in the ventral roots, in spinal cats deprived of their peripheral input, is known as "fictive locomotion".

Phase-dependent modulation of the excitability of spinal interneurons may affect both excitatory and inhibitory interneurons in spinal reflex arcs and long descending corticospinal and brain stem systems, and may suppress or facilitate extensor or flexor bursts (Fig. 21.17 A). Furthermore, direct actions of these systems on the CPG may cause changes in the timing of the step cycle by shortening or lengthening the stance phase, and re-setting the locomotion cycle. It has been shown as the reversal in Ib Golgi tendon reflexes from extensor muscles. Autogenic inhibition by Ib inhibitory interneurons is blocked during fictive locomotion and a disynaptic, excitatory pathway to extensor motoneurons is opened during stance [178]. As a consequence, loading of ankle extensors prolongs the extensor burst during the stance phase and delays the transition to the swing phase [51]. Ia muscle spindle afferents from hip flexors initiate the swing phase when the angle of the hip exceeds a certain value at the end of the stance phase. Cutaneous reflexes, similarly, are phasically modulated. Increasing the load on the extremity during stance activates the group-II-low-threshold cutaneous input from the footpads prolonging the extensor burst during stance. Withdrawing this type of stimulation triggers the onset of flexion [214]. A direct influence of flexor reflex afferents on the half-centre-organization of its interneurons was shown by Jankowska et al. [110].

Phase dependency is also evident from the so-called stumbling response in spinal cats walking on a treadmill. Contact with an obstacle during swing removes the foot from the obstacle and places the foot in front of it. A continuous stimulus applied to the dorsum of the foot during swing thus results in knee, ankle and hip flexion. When applied during stance, the same stimulus increases the latency of reflex amplitude of already active extensors [178]. Phase-dependent modulation of cutaneous reflexes has also been demonstrated in humans [215]. It is not known which interneurons are involved in the adaptation of these reflexes.

Locomotion is also dependent on the activity in descending corticospinal and brain stem systems. Because the circuitry of the CPG and of most of the spinal reflex pathways is not known, the anatomy of locomotion can only be inferred from the behaviour of the experimental preparations. Three groups of descending systems should be distinguished:

1. The mesencephalic locomotion centre with its descending reticulospinal connections;
2. the lateral vestibulospinal tract, with its preferential influence on extensor muscles of the ipsilateral limbs;
3. and the corticospinal and rubrospinal tracts, which preferentially influence the motility of a single, contralateral extremity.

The mesencephalic locomotor centre (MLR; Shik et al. [201]) is located ventral to the inferior colliculus, in the nucleus cuneiformis and/or the adjacent cholinergic and glutamatergic neurons of the nucleus pedunculopontinus, pars compacta (Fig. 6.29). Electrical stimulation or the injection of substance P or GABA antagonists induces locomotion on a treadmill in cats with a premamillary decerebration. Increasing the strength of the stimulus speeds up the pace from walking to trott and ultimately switches the gait to a gallop [67, 69]. A MLR has been identified in nonhuman primates [55, 220]. The MLR may be part of a widespread reticular system regulating vigilance and the state of the motor system. These aspects are considered in Chap. 22 "Connections of the Reticular Formation and Associated Cell Groups".

The MLR receives afferent projections from the bed nucleus of the stria terminalis, the amygdala and the lateral hypothalamus. Prominent, inhibitory connections are present from the reticular part of the substantia nigra and the entopeduncular nucleus (the internal segment of the globus pallidus in primates) and/or the ventral pallidum [67, 92, 205, 208]. There appear to be few cortical and cerebellar afferents. Projections from a dorsal premotor cortical area to the cuneiform nucleus, but not to the nucleus pedunculopontinus, have been reported for the Rhesus monkey by Kuypers and Lawrence [121].

Two descending pathways take their origin from the MLR. One pathways follows Probst's tract in the dorsomedial lateral tegmental field ("pontomedullary locomotor strip"). Caudally it merges with the spinal intermediate zone. The main projection is to the ventromedial tegmental field in the caudal pons and the medulla oblongata (Fig. 21.17 B). This projection is bilateral with an ipsilateral predominance [68, 81, 92, 204]. Injection of cholinergic agonists or *N*-methyl-S-aspartic acid (NMDA) in this region induces stepping [66]. The uncrossed

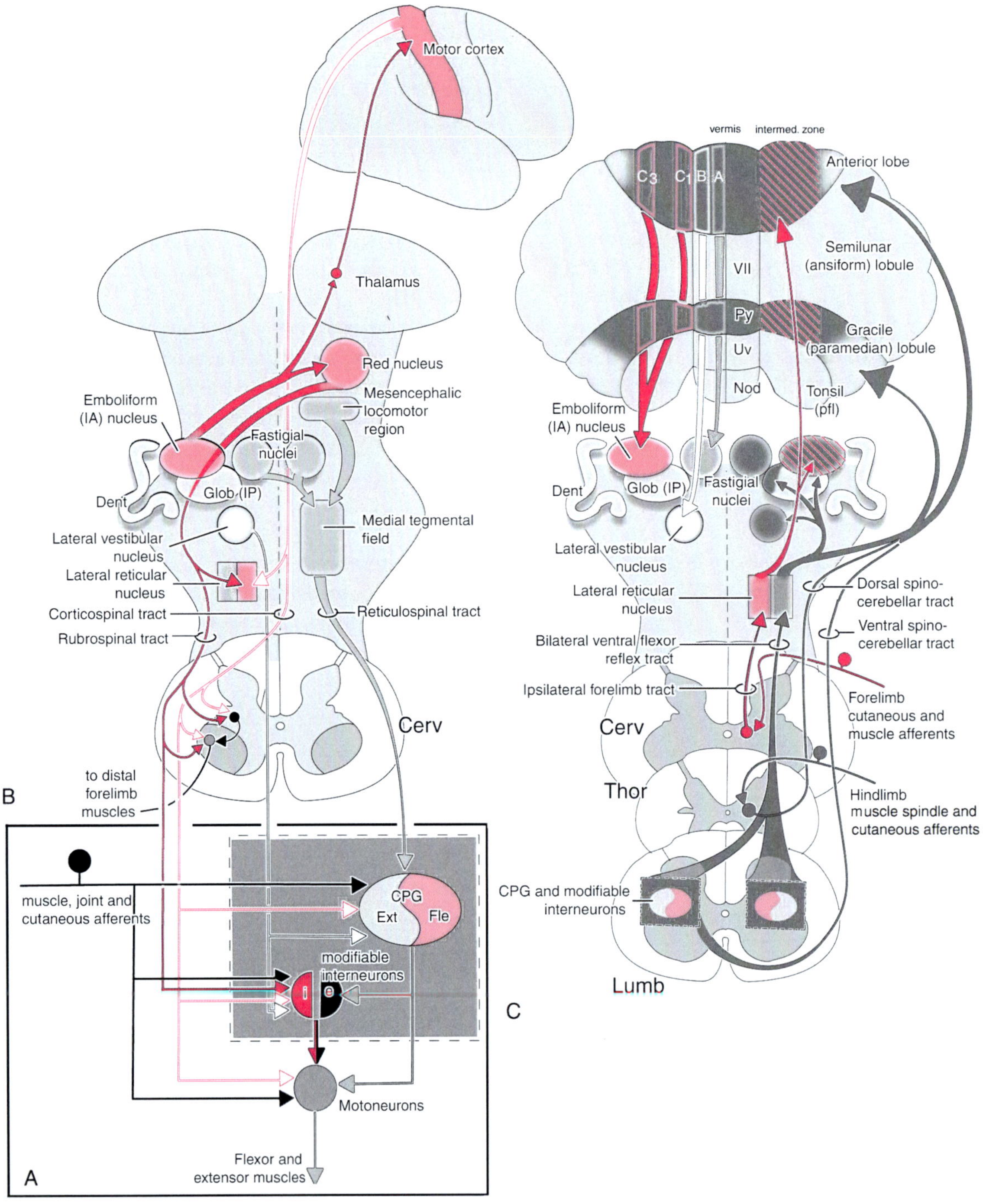

Fig. 21.17. Locomotion. **A** A central pattern generator (CPG) in the spinal cord initiates and generates the rhythm and shapes the basic pattern of motor bursts of flexor and extensor motoneurons in locomotion. Primary afferents and cortical and subcortical motor centres adapt locomotion to external and internal circumstances through direct actions of these systems on the CPG or through the modulation of their excitatory and inhibitory interneurons and/or motoneurons by the CPG. Although primary afferents include Ia and Ib fibres, and cutaneous and joint afferents, and different supraspinal systems influence locomotor activity, their individual connections have not been indicated in the diagram. The *grey rectangle* includes the CPG and the modifiable interneurons included in the different reflex and supraspinal pathways. **B** Descending systems that influence locomotion take their origin from different supraspinal centres. The mesencephalic locomotion region (MLR) projects to the SPG via the pontomedullary medial tegmental field and the medial reticulospinal

medial reticulospinal tract, which contains a large proportion of coarse fibres, descends from this region in the medial longitudinal fascicle, through the anterior funiculus, to terminate bilaterally in laminae VIII and VII of the intermediate zone over the entire length of the spinal cord. These fibres do not terminate, or do so only very sparingly, in the motoneuronal cell groups [94, 142, 143]. The medial reticulospinal tract, supposedly, terminates on the CPG and on modifiable interneurons in flexor and extensor pathways. In its termination, the medial reticulospinal tract resembles the lateral vestibulospinal tract. Both tracts may converge upon the same cornucommissural interneurons and neurons with long ascending collaterals that receive group II and III flexor reflex afferents, but different sets of interneurons are used by the two tracts in their facilitatory effect on extensor motoneurons [44, 122, 178].

The lateral vestibulospinal tract takes its origin from the lateral vestibular nucleus, a nucleus which belongs to the cerebellar nuclei (Chaps. 17, 20). It receives a strong inhibitory input from Purkinje cells in the lateral B zone of the anterior vermis (Fig. 21.17 C) and an excitatory collateral projection from the dorsal spinocerebellar and reticulospinal tracts. The lateral vestibular nucleus influences the level of activity in extensor muscles of the ipsilateral limbs via excitatory interneurons and terminates on cornucommissural interneurons and on interneurons in Ia and flexor reflex afferent pathways. Effects of stimulation of the lateral vestibular nucleus of the cat on extensor activity are maximal during stance and without effect during swing. Stimulation resets the locomotion rhythm; it prolongs the extensor burst and delays the swing phase. Moreover, there is a phasing of impulse activity in many neurons of the nucleus, with peak rates in early stance, timelocked to extensor muscles in the ipsilateral limbs, and a dip during swing. This cyclic modulation disappears after ablation of the cerebellum [8, 13, 122, 141, 183].

The emboliform (anterior interposed) nucleus projects to the contralateral magnocellular red nucleus and, via a relay in the ventral lateral thalamic nucleus, to the primary motor cortex (Fig. 21.17 A). In cat and rat neurons of the magnocellular red nucleus and motor cortical areas are rhythmically modulated during

tract. The lateral vestibular nucleus exerts direct actions on the CPG and connects with extensor motoneurons through modifiable spinal interneurons. The magnocellular red nucleus and the motor cortex are connected with motoneurons innervating distal muscles of the extremities (illustrated for the upper extremity) through the rubrospinal and corticospinal tracts. Spinal interneurons in these pathways can be modulated by the CPG; a direct action of the corticospinal system on the CPG is possible. Different cerebellar nuclei are connected with different supraspinal motor centres. The fastigial nucleus monitors the mesencephalic locomotor centre at the level of the medial tegmental field. The emboliform (anterior interposed) nucleus is connected with the red nucleus and the motor cortex. **C** Ascending spinocerebellar pathways close the loops that influence locomotion. Two pathways convey information to the cerebellum on internal activity in spinal motor centres influenced by segmental afferents. These systems, the ventral spinocerebellar tract and the bilateral ventral flexor reflex tract (BFRT), are shown as originating from a *grey rectangle*, including the CPG and the modifiable spinal interneurons. The cerebellum is shown as a flattened map. The dorsal spinocerebellar tract carries information on external events. A fourth system, the ipsilateral forelimb tract (IFT: shown in *red*), is monosynaptically activated by cutaneous and muscle afferents of the ipsilateral forelimb. The BFRT and the IFT relay in different portions of the lateral reticular nucleus. Mossy fibres of the dorsal and ventral spinocerebellar tract and the BFRT terminate bilaterally in the spinal cerebellum (anterior lobe, gracile lobule and pyramis: *dark grey*). The lateral reticular nucleus, moreover, provides several cerebellar nuclei with a collateral projection of the BFRT. The IFT projects to the ipsilateral intermediate zone and the emboliform nucleus (*red hatching*). The relevant corticonuclear projections are indicated in the left half of the cerebellum. Purkinje cell zones C_1 and C_3 in the intermediate cerebellum project to the emboliform nucleus, the lateral B zone of the vermis projects to the lateral vestibular nucleus and the medial A zone of the vermis to the fastigial nucleus. *Cerv*, cervical cord; *CPG*, central pattern generator; *Dent*, dentate nucleus; *Ext., extensor muscles; Fle*, flexor muscles; *Glob (IP)*, globose (posterior interposed) nucleus; *Lumb*, lumbar cord; *Nod*, nodulus; *pfl*, (paraflocculus) tensil; *i*, inhibitory modifiable interneuron; *IA*, anterior interposed (emboliform) nucleus; *Py*, pyramis; *Thor*, thoracic cord; *Uv*, uvula; *VII*, lobule VII

locomotion. During treadmill locomotion, stimulation of rubrospinal neurons results in phase-dependent facilitation of both flexor and extensor muscles. In both motor cortex and red nucleus, the average discharge of these neurons peaks at the end of stance and during the swing phase of the contralateral forelimb. However, the timing and the number of the peaks varies greatly among individual cells. Stimulation of the red nucleus never re-sets the step cycle, but cortical stimulation may do so. Modulation of the discharge of rubral neurons is dependent on the cerebellum, but some of the rhythmic activity is still present after cerebelectomy. This activity may depend on ascending spinorubral or collateral corticorubral pathways [8, 13, 14, 144, 174, 221]. Phase-dependent facilitation of motor responses on transcranial magnetic stimulation of the motor cortex in humans during gait was higher in leg flexors than in leg extensors with maxima during late swing [194].

In the cat, the red nucleus is mainly concerned with distal movements of the contralateral upper extremity. Movements of the digits in co-ordination with other limb parts are the most important factor for eliciting high discharge rates from the magnocellular red nucleus [147]. Focussing on distal movements of the contralateral limbs also characterizes the corticospinal system. The balance between the rubrospinal and corticospinal systems has shifted from lower mammals and nonhuman primates to humans, the rubrospinal spinal tract having visually disappeared and a large, multifacetted corticospinal system provided with strong, direct corticomotoneural connections having evolved. The consequences of this evolution for the control of locomotion are difficult to evaluate. Spinalized monkeys never recover locomotion [55]. In humans, a complete or nearly complete paralysis is the rule after suffering motorcortical lesions. Spontaneous recovery of rhythmic activity in patients with spinal injury is rare, although some activity may be recovered after training on a treadmill [108, 159, 214]. Experimental evidence on the anatomy and physiology of locomotion in primates is scarce. The presence of a spinal pattern generator and of a mesencephalic locomotor region is disputed, and little is known of how primate reticular formation contributes to posture and locomotion.

The contribution of the primary motor cortex and the premotor areas to locomotion has received only little attention. In PET studies of human subjects during bicycling increased blood flow was found bilaterally in S1, the M1 leg area, the SMA and the anterior cerebellum. During imagined bicycling the only activation was in the SMA [35]. Gait imaginary, similarly, was associated with increased activity in the SMA [149]. In an fMRI study of isolated movements of hand or foot S1, SMA (SMA proper), the caudal cingulate motor area and the ventral premotor area were activated on the contralateral side. Activation of the anterior cerebellar vermis was found ipsilaterally. Movements with both limbs resulted in increased S1, M1 and premotor activity. With movements of both limbs in the same direction the SMA and the cingulate motor cortex were activated bilaterally [42]. These results support a role for the SMA in imagining movement and bilateral co-ordination of actual movement. They do not cast light on functions of the premotor cortex in locomotion. A promising approach to these problems is the study of optic flow, known to be processed in visual parietoprefrontal circuits, in the control of locomotion [195].

Lack of cerebellar control of locomotion is experienced as ataxia, the drunken gait in alcohol intoxication and the main symptom in cerebellar diseases. Coupling between step-cycles in different limbs becomes abnormal and the equilibrium is disturbed. Cerebellar control of locomotion is exerted through the lateral vestibular nucleus, the pontomedullary medial reticular formation, the red nucleus and the motor cortex. The anatomy of cerebellar control was considered in Chap. 20. Purkinje cells of the medial cerebellar (fastigial) nuclei provide the medial vermal A zones with bilateral connections to the medial pontobulbar reticular formation, the link in the descending pathway from the mesencephalic locomotor centre (Fig. 21.17B). The lateral vermal B zone monitor activity in the lateral vestibular nucleus and the lateral vestibulospinal tract (Fig. 21.17C). The C_1 and C_3 zones of the intermediate cerebellum project to the contralateral magnocellular red nucleus via the anterior interposed (emboliform) nucleus, and the primary motor cortex via a thalamic relay (Fig. 21.17C). The lateral D_2 zone of the hemisphere monitors the primary motor cortex and the premotor and parietal areas that give rise to the pyramidal tract (Fig. 20.12).

Phasic activity in the Purkinje cells in the B zone during locomotion is very similar to that seen in the neurons of the lateral vestibular nucleus. However, Purkinje cells are inhibitory and their influence, therefore, would be to dampen, rather than to increase the modulation of neurons in the lateral vestibular nucleus [8]. A similar situation exists for the C_1 and C_3 zones with respect to the modulation of activity in the anterior interposed nucleus and its crossed pathway to the red nucleus and the primary motor cortex [9–12, 54]. The similarity in the topographical organization of receptive fields of spinal withdrawal reflexes and of the mossy and climbing fibres projecting to microzones of the C_1 and C_3 zones in the anterior cerebellum of the cat, suggest a role of the C zones in the acute and adaptive control of these reflexes [59]. Partial reversal of the step-cycle-related modulation of the amplitude of flexor reflexes was obtained after selective lesions of the C_1 and C_3 zones in the rat [169].

Control of the MLR at the level of the pontomedullary medial tegmental field by the A zone and the bilateral projections of the fastigial nucleus was shown by Mori et al. [155]. The possible role of the D_2 zone of the cerebellar hemisphere in locomotion depends on its influence on parietomotor and premotor circuits, a subject which has not yet been studied.

Modulation of cerebellar Purkinje cells is dependent on their mossy and climbing fibre input. Spinocerebellar mossy fibre inputs are conveyed by the spinocerebellar tracts and the lateral reticular nucleus (Fig. 21.17C). Locomotor phase-related activity is present in these systems [13]. The dorsal spinocerebellar and the cuneocerebellar tracts are activated by proprioceptive and exteroceptive stimuli and carry information about external events and the state of the locomotor apparatus [7, 131, 170]. The ventral and rostral spinocerebellar tracts carry information about the internal events, i.e. the excitability of spinal interneurons and, presumably, the state of the spinal pattern generator [15, 53, 130, 163]. The spinocerebellar tracts terminate bilaterally, in vermis and hemisphere (Chap. 20). Interestingly, the lateral reticular nucleus transmits information from two different systems. It receives ascending collaterals from excitatory and inhibitory interneurons through the so-called *bilateral ventral flexor reflex tract* (bVFRT), a component of the anterolateral fasciculus. This tract carries information about activity in spinal motor centres influenced by segmental afferents, which may include the CPG [110]. Like the spinocerebellar tracts, it terminates bilaterally in vermis and hemispheres. bVFRT neurons are under monosynaptic control of the lateral vestibulospinal tract and, possibly, the reticulospinal tract in the ventral funiculus [37, 38, 56, 58]. The second pathway that is synaptically interrupted in the lateral reticular nucleus is the *ipsilateral forelimb tract.* This system is activated by cutaneous and muscle afferents from the ipsilateral forelimb. Like the bVFRT, its neurons are both excitatory and inhibitory. It is relayed by the dorsomedial part of the lateral reticular nucleus, where it converges with afferents from the red nucleus and the motor cortex. Its mossy fibre projection is limited to the ipsilateral intermediate part of the anterior lobe and the paramedian lobule, which contain the C_1 and C_3 zones [36, 57].

Apart from its projection to the cerebellar cortex of the "spinal" cerebellum, the lateral reticular nucleus provides the cerebellar nuclei with a collateral projection. This projection is limited to the anterior and posterior interposed nuclei, the fastigial nucleus and the lateral vestibular nucleus, i.e. to the target nuclei of the Purkinje cells of the spinal cerebellum [181]. The importance of these nuclei in locomotion was confirmed by the expression of the activity-related immediately-early gene c-Fos in just these cerebellar nuclei of the rat during locomotion [182].

Although Purkinje cells, cerebellar nuclear cells and various mossy fibre inputs modulate discharge rate in phase with the step cycle, cerebellar input from the inferior olive, which generates complex spikes, shows little or no relation to the step cycle [12]. In cats, during locomotion, the amplitudes of climbing fibre-evoked potentials are typically depressed when compared with potentials recorded at rest [11], although the degree of depression varies with the phase of locomotion. The lack of a climbing fibre-evoked discharge related to locomotion or other movements (see Gellman et al. [70]) is surprising. Olivary neurons respond reliably to light touch during rest, and foot contact with the floor during locomotion would seem to generate appropriate stimulation for many olivary cells. Armstrong [8] suggested that the olivary neurons are subject to central inhibition during locomotion, and several studies support this suggestion. Electrical stimulation of descending motor pathways such as the corticospinal [127] or rubrospinal tract [223] strongly inhibits olivary responses to peripheral stimulation. Moreover, the inhibitory nucleo-olivary cells of the cerebellar nuclei increase their activity during locomotion (or other movements) and inferior olivary responses to peripheral stimulation are blocked [19, 74]. Clearly, inhibitory mechanisms working on the inferior olive prevent a reliable response to self-produced movement or to stimuli produced by active movement by the animal. Exceptions to this are stimuli produced during exploratory movements or when the paw unexpectedly makes contact with an object during active movement [6, 70]. The inferior olive, apparently, functions as a somatic event detector, responding in particular to unexpected stimuli.

The cerebellum, therefore, is part of at least two different recurrent loops that may control and adapt locomotion. One loop consists of the spinocerebellar and bVFRT afferent pathways, the bilateral A, B and C_1/C_3 zones and their spinal projections through the reticulospinal, lateral vestibulospinal rubrospinal and corticospinal tracts. The other loop is focussed on the ipsilateral forelimb and includes the ipsilateral C_1 and C_2 zones, the red nucleus and the motor cortex. Both loops may correct locomotion by plastic changes in the Purkinje cells of the zones, induced by error, or other signals carried by the climbing fibres. This adaptation is brought about by the suppressing activity in the cerebellar nuclei, induced by the excitatory collaterals of the mossy fibres of the lateral reticular nucleus.

The focussing of the second loop on the ipsilateral forelimb may be species-dependent. Similar loops focussing on the lower extremities may have developed in primates. Moreover cerebellar control of movement in primates has shifted to the cerebellar hemisphere and the motor and premotor cortices at the expense of the lower motor centres.

Brain Stem Control of Micturition and Ejaculation

The lower urinary tract has two functions: continence (referring to the storage phase in which the bladder holds urine) and micturition (denoting the process of urine evacuation by contraction of the detrusor muscle of the bladder and concomitant relaxation of the external urethral sphincter). The detrusor muscle is innervated by an intra- and juxtamural cholinergic plexus, which is postganglionic to parasympathetic preganglionic fibres from the sacral intermediolateral nucleus (Fig. 6.5). The external urethral sphincter is innervated by a subset of motoneurons of Onuf's nucleus. This nucleus is located in the sacral anterior horn (Fig. 21.1). Apart from the urethral sphincter, it innervates the muscles of the pelvic floor, including the anal sphincter, the bulbocavernosus and ischiocavernosus muscles and the penis [213] and thus is also involved in defecation, ejaculation and other sexual functions. Sympathetic preganglionic neurons, located in the upper lumbar intermediolateral nucleus, access the bladder via the parasympathetic juxtamural ganglia [114]. Sympathetic fibres have an inhibitory effect on smooth muscles of the base of the bladder and the urethra. They allow filling of the bladder during the storage phase.

The traditional view of co-ordinated reflex micturition holds that micturition is mediated by a spinal cord-brain stem-spinal cord reflex loop, initiated by input from pelvic nerve afferents signaling bladder distension [101]. In spinalized cats the synergy in micturition is lost, i.e. when the bladder contracts, the sphincter contracts too. A similar situation exists in humans with complete lesions of the spinal cord [103].

The presence of a spinal circuit mediating co-ordinated bladder-sphincter responses, activated by stimulation of the pelvic nerves, was demonstrated by Boggs et al. [28]. The complete anatomy of this spinal circuit, however, is not known. Elements of this circuit include the pelvic nerve afferents terminating as a fibre plexus located in lamina I of the sacral dorsal horn. This plexus extends medially into the region of the dorsal grey commissure (the commissural nucleus) and laterally includes the intermediolateral nucleus [222] (Fig. 6.5). Relaxation of the urethral sphincter is produced by inhibitory interneurons located in the commissural nucleus that terminate on the motoneurons of Onuf's nucleus [25]. In addition, Onuf's nucleus receives supraspinal afferents from the hypothalamus, the lateral pontine reticular formation (L-region, see below) and the nucleus retroambiguus [95] (Fig. 6.5). In subhuman primates Onuf's nucleus may receive a direct projection from the primary motor cortex [157].

A similar spinal circuit exists for ejaculation. Secretion of seminal fluid during the emission phase is under parasympathetic and sympathetic control. Expulsion of the semen is accompanied by contraction of the bulbocavernosus and ischiocavernosus muscles innervated by Onuf's nucleus [39].

Structures mediating the supraspinal reflex loop for micturition were identified in the pontine tegmentum of the cat [101]. The pontine micturition region (M-region of Holstege, Barrington's nucleus [17]) is located in the dorsolateral pontine tegmentum. It projects bilaterally to the sacral intermediolateral nucleus and the commissural nucleus, which contains the interneurons inhibiting Onuf's nucleus (Fig. 6.5). Stimulation of the pontine micturition region results in micturition. The reflex loop is closed by an ascending pathway from spinal neurons located within the terminal plexus of pelvic and pudendal nerve afferents in sacral lamina I that terminates in the periaqueductal grey [217] and a projection of the periaqueductal grey to the pontine micturition region [23]. Stimulation of a more laterally located region in the pontine tegmentum of the cat (L-region [101]), which projects to Onuf's nucleus in the sacral cord, causes contraction of the external urethral sphincter. Connections between the L-region and the pontine micturition region have not been identified [24]. As yet, the function of the L-region is not known.

The presence of a pontine micturition region has been confirmed in PET-studies carried out by Blok et al. [26, 27] in humans. Apart from activation in the pontine tegmentum, they found increased blood flow during micturition in the periaqueductal grey, the preoptic area and the inferior frontal cortex. Withholding micturition decreases activity in the anterior cingulate gyrus. Activity in the pontine micturition region is lateralized to the right side.

Supraspinal structures controlling ejaculation have been identified in PET scans of human subjects [102, 103] and c-Fos expression during ejaculation in rats [39]. They include the periaqueductal grey, the ventral tegmental area, the midbrain lateral central tegmental field, and the parafascicular, posterior and intralaminar thalamic nuclei. In humans certain parts of the prefrontal, temporal, parietal and insular cortices and the cerebellum are activated during ejaculation. Deactivation was found in the hypothalamus and the preoptic area. The activity in the medial amygdala was downregulated. An ascending pathway from the sacral dorsal horn to the periaqueductal grey and the parafascicular nucleus may transmit information from the sexual organs to supraspinal centres. Descending control systems for ejaculation take their origin from the periaqueductal grey, the nucleus retroambiguus and the bed nucleus of the stria terminalis [39].

Ejaculation is the supreme event during copulation. Other components of male and female sexual behaviour and their neural regulation were discussed by Holstege and Georgiadis [97]. The role of the hypothalamus and related structures in reproductive behaviour is considered in Chap. 10 ("Sexual Behaviour").

References

1. Alipour M, Chen Y, Jürgens U (2002) Anterograde projections of the motorcortical tongue area in the saddle-back tamarin (Saguinus fuscicollis). Brain Behav Evol 60:101–116
2. Alstermark B, Lindstrom S, Lundberg A, Sybirska E (1981) Integration in descending motor pathways controlling the forelimb in the cat. 8. Ascending projection to the lateral reticular nucleus from C3-C4 propriospinal also projecting to forelimb motoneurones. Exp Brain Res 42:282-298
3. Alstermark B, Lundberg A, Pinter M, Sasaki S (1987) Subpopulations and functions of long C3-C5 propriospinal neurones. Brain Res 404:395-400
4. Alstermark B, Kummel H, Pinter MJ, Tantisira B (1990) Integration in descending motor pathways controlling the forelimb in the cat. 17. Axonal projection and termination of C3-C4 propriospinal neurones in the C6-Th1 segments. Exp Brain Res 81:447–461
5. Alstermark B, Isa T, Kummel H, Tantisira B (1990) Projection from excitatory C3-C4 propriospinal neurones to lamina VII and VIII neurones in the C6-Th1 segments of the cat. Neurosci Res 8:131–137
6. Andersson G, Armstrong DM (1987) Complex spikes in Purkinje cells in the lateral vermis (b zone) of the cat cerebellum during locomotion. J Physiol 385:107–134
7. Aoyama M, Hongo T, Kudo N (1988) Sensory input to cells of origin of uncrossed spinocerebellar tract located below Clarke's column in the cat. J Physiol 398:233–257
8. Armstrong DM (1986) Supraspinal contributions to the initiation and control of locomotion in the cat. Prog Neurobiol 26:273–361
9. Armstrong DM, Edgley SA (1984) Discharges of nucleus interpositus neurones during locomotion in the cat. J Physiol 351:411–432
10. Armstrong DM, Edgley SA (1984) Discharges of Purkinje cells in the paravermal part of the cerebellar anterior lobe during locomotion in the cat. J Physiol 352:403–424
11. Armstrong DM, Edgley SA (1988) Discharges of interpositus and Purkinje cells of the cat cerebellum during locomotion under different conditions. J Physiol 400:425–445
12. Armstrong DM, Edgley SA, Lidierth M (1988) Complex spikes in Purkinje cells of the paravermal part of the anterior lobe of the cat cerebellum during locomotion. J Physiol 400:405–414
13. Arshafsky YI, Gelfand IM, Orlovsky GN (1986) Cerebellum and rhythmical movements. Springer, Berlin Heidelberg New York
14. Arshavsky YI, Orlovsky GN, Perret C (1988) Activity of rubrospinal neurons during locomotion and scratching in the cat. Behav Brain Res 28:193–199
15. Baldissera F, Hultborn H, Ilert M (1981) Integration in spinal neuronal systems. In: Brooks VB (ed) Handbook of physiology. The nervous system Motor control. American Physiological Society, Bethesda, pp 509–595
16. Barrett RT, Bao X, Miselis RR, Altschuler SM (1994) Brain stem localization of rodent esophageal premotor neurons revealed by transneuronal passage of pseudorabies virus. Gastroenterology 107:728–737
17. Barrington FJF (1925) The effect of lesions of the hind- and mid-brain on micturition in the cat. Quart J Exp Physiol Cogn Med 15:81–102
18. Bates JF, Goldman-Rakic PS (1993) Prefrontal connections of medial motor areas in the rhesus monkey. J Comp Neurol 336:211–228
19. Bengtsson F, Svensson P, Hesslow G (2004) Feedback control of Purkinje cell activity by the cerebello-olivary pathway. Eur J Neurosci 20:2999–3005
20. Betz W (1874) Anatomischer Nachweis zweier Gehirncentra. Centralbl Med Wiss 12:578–580, 595–599
21. Bieger D, Hopkins DA (1987) Viscerotopic representation of the upper alimentary tract in the medulla oblongata in the rat: the nucleus ambiguus. J Comp Neurol 262:546–562
22. Binkofski F, Buccino G, Stephan KM, Rizzolatti G, Seitz RJ, Freund HJ (1999) A parieto-premotor network for object manipulation: evidence from neuroimaging. Exp Brain Res 128:210–213
23. Blok BF, Holstege G (1994) Direct projections from the periaqueductal gray to the pontine micturition center (M region). An anterograde and retrograde tracing study in the cat. Neurosci Lett 166:93–96
24. Blok BF, Holstege G (1999) Two pontine micturition centers in the cat are not interconnected directly: implications for the central organization of micturition. J Comp Neurol 403:209–218
25. Blok BF, de Weerd H, Holstege G (1997) The pontine micturition center projects to sacral cord GABA immunoreactive neurons in the cat. Neurosci Lett 233:109–112
26. Blok BF, Willemsen AT, Holstege G (1997) A PET study on brain control of micturition in humans. Brain 120 (Pt 1):111–121
27. Blok BF, Sturms LM, Holstege G (1998) Brain activation during micturition in women. Brain 121 (Pt 11):2033–2042
28. Boggs JW, Wenzel BJ, Gustafson KJ, Grill WM (2005) Spinal micturition reflex mediated by afferents in the deep perineal nerve. J Neurophysiol 93:2688–2697

29. Bourque MJ, Kolta A (2001) Properties and interconnections of trigeminal interneurons of the lateral pontine reticular formation in the rat. J Neurophysiol 86:2583–2596
30. Brown TG (1911) The intrinsic factors in the act of progression in the mammal. Proc R Soc Lond B84:308–319
31. Brown TG (1912) The factors in rhythmic ativity of the nervous system. Proc R Soc Lond B85:278–289
32. Buccino G, Vogt S, Ritzl A, Fink GR, Zilles K, Freund HJ, Rizzolatti G (2004) Neural circuits underlying imitation learning of hand actions: an event-related fMRI study. Neuron 42:323–334
33. Cattaneo L, Voss M, Brochier T, Prabhu G, Wolpert DM, Lemon RN (2005) A cortico-cortical mechanism mediating object-driven grasp in humans. PNAS 102:898–903
34. Cheema S, Rustioni A, Whitsel BL (1985) Sensorimotor cortical projections to the primate cuneate nucleus. J Comp Neurol 240:196–211
35. Christensen LO, Johannsen P, Sinkjaer T, Petersen N, Pyndt HS, Nielsen JB (2000) Cerebral activation during bicycle movements in man. Exp Brain Res 135:66–72
36. Clendenin M, Ekerot CF, Oscarsson O (1974) The lateral reticular nucleus in the cat. III. Organization of component activated from ipsilateral forelimb tract. Exp Brain Res 21:501–513
37. Clendenin M, Ekerot CF, Oscarsson O, Rosen I (1974) The lateral reticular nucleus in the cat. I. Mossy fibre distribution in cerebellar cortex. Exp Brain Res 21:473–486
38. Clendenin M, Ekerot CF, Oscarsson O, Rosen I (1974) The lateral reticular nucleus in the cat. II. Organization of component activated from bilateral ventral flexor reflex tract (bVFRT). Exp Brain Res 21:487–500
39. Coolen LM, Allard J, Truitt WA, McKenna KE (2004) Central regulation of ejaculation. Physiol Behav 83:203–215
40. Cowie RJ, Holstege G (1992) Dorsal mesencephalic projections to pons, medulla, and spinal cord in the cat: limbic and non-limbic components. J Comp Neurol 319:536–559
41. Cunningham ET Jr, Sawchenko PE (2000) Dorsal medullary pathways subserving oromotor reflexes in the rat: implications for the central neural control of swallowing. J Comp Neurol 417:448–466
42. Debaere F, Wenderoth N, Sunaert S, Van Hecke P, Swinnen SP (2004) Cerebellar and premotor function in bimanual coordination: parametric neural responses to spatiotemporal complexity and cycling frequency. Neuroimage 21:1416–1427
43. Deecke L, Kornhuber HH, Lang W, Lang M, Schreiber H (1985) Timing function of the frontal cortex in sequential motor and learning tasks. Hum Neurobiol 4:143–154
44. Drew T, Prentice S, Schepens B (2004) Cortical and brainstem control of locomotion. Prog Brain Res 143:251–261
45. Duhamel JR, Colby CL, Goldberg ME (1998) Ventral intraparietal area of the macaque: congruent visual and somatic response properties. J Neurophysiol 79:126–136
46. Dum RP, Strick PL (1990) Premotor areas: nodal points for parallel efferent systems involved in the central control of movement. In: Humphrey DR, Freund H-J (eds) Motor control: concepts and issues. Wiley, London
47. Dum RP, Strick PL (1991) The origin of corticospinal projections from the premotor areas in the frontal lobe. J Neurosci 11:667–689
48. Dum RP, Strick PL (1996) Spinal cord terminations of the medial wall motor areas in macaque monkeys. J Neurosci 16:6513–6525
49. Dum RP, Strick PL (2005) Frontal lobe inputs to the digit representations of the motor areas on the lateral surface of the hemisphere. J Neurosci 25:1375–1386
50. Duysens J (2006) How deletions in a model could help explain deletions in the laboratory. J Neurophysiol 95:562–563
51. Duysens J, Pearson KG (1980) Inhibition of flexor burst generation by loading ankle extensor muscles in walking cats. Brain Res 187:321–332
52. Duysens J, Van de Crommert HW (1998) Neural control of locomotion. The central pattern generator from cats to humans. Gait Posture 7:131–141
53. Eccles JC, Hubbard JI, Oscarsson O (1961) Intracellular recording from cells of the ventral spinocerebellar tract. J Physiol 158:486–516
54. Edgley SA, Lidierth M (1988) Step-related discharges of Purkinje cells in the paravermal cortex of the cerebellar anterior lobe in the cat. J Physiol 401:399–415
55. Eidelberg E, Walden JG, Nguyen LH (1981) Locomotor control in macaque monkeys. Brain 104:647–663
56. Ekerot CF (1990) The lateral reticular nucleus in the cat. VI. Excitatory and inhibitory afferent paths. Exp Brain Res 79:109–119
57. Ekerot CF (1990) The lateral reticular nucleus in the cat. VII. Excitatory and inhibitory projection from the ipsilateral forelimb tract (iF tract). Exp Brain Res 79:120–128
58. Ekerot CF (1990) The lateral reticular nucleus in the cat. VIII. Excitatory and inhibitory projection from the bilateral ventral flexor reflex tract (bVFRT). Exp Brain Res 79:129–137
59. Ekerot CF, Garwicz M, Jörntell H (1997) The control of forelimb movements by intermediate cerebellum. Prog Brain Res 114:423–429
60. Fink GR, Frackowiak RS, Pietrzyk U, Passingham RE (1997) Multiple nonprimary motor areas in the human cortex. J Neurophysiol 77:2164–2174
61. Forssberg H, Grillner S (1973) The locomotion of the acute spinal cat injected with clonidine i.v. Brain Res 50:184–186
62. Fritsch GT, Hitzig E (1870) Über die electrische Erregbarkeit des Grosshirns. Arch Anat Physiol: 300–332

63. Fulton JF (1935) A note on the definition of the 'motor' and 'premotor' areas. Brain 58:311–316
64. Galea MP, Darian-Smith I (1994) Multiple corticospinal neuron populations in the macaque monkey are specified by their unique cortical origins, spinal terminations, and connections. Cereb Cortex 4:166–194
65. Gallese V, Fadiga L, Fogassi L, Rizzolatti G (1996) Action recognition in the premotor cortex. Brain 119 (Pt 2):593–609
66. Garcia-Rill E (1991) The pedunculopontine nucleus. Prog Neurobiol 36:363–389
67. Garcia-Rill E, Skinner RD, Jackson MB, Smith MM (1983) Connections of the mesencephalic locomotor region (MLR). I. Substantia nigra afferents. Brain Res Bull 10:57–62
68. Garcia-Rill E, Skinner RD, Gilmore SA, Owings R (1983) Connections of the mesencephalic locomotor region (MLR). II. Afferents and efferents. Brain Res Bull 10:63–71
69. Garcia-Rill E, Kinjo N, Atsuta Y, Ishikawa M, Webber M, Skinner D (1990) Posterior midbrain-induced locomotion. Brain Res Bull 24:499–508
70. Gellman R, Gibson AR, Houk JC (1985) Inferior olivary neurons in the awake cat: detection of contact and passive body displacement. J Neurophysiol 54:40–60
71. Georgopoulos AP, Kalaska JF, Caminiti R, Massey JT (1982) On the relations between the direction of two-dimensional arm movements and cell discharge in primate motor cortex. J Neurosci 2:1527–1537
72. Georgopoulos AP, Caminiti R, Kalaska JF, Massey JT (1983) Spatial coding of movement direction by motor cortical populations. Exp Brain Res Suppl 7:327–335
73. Geyer S, Matelli M, Luppino G, Zilles K (2000) Functional neuroanatomy of the primate isocortical motor system. Anat Embryol (Berl) 202:443–474
74. Gibson AR, Horn KM, Pong M (2002) Inhibitory control of olivary discharge. Ann N Y Acad Sci 978:219–231
75. Godschalk M, Lemon RN, Kuypers HG, Ronday HK (1984) Cortical afferents and efferents of monkey postarcuate area: an anatomical and electrophysiological study. Exp Brain Res 56:410–424
76. Godschalk M, Mitz AR, van Duin B, van der Burg H (1995) Somatotopy of monkey premotor cortex examined with microstimulation. Neurosci Res 23:269–279
77. Gong S, DeCuypere M, Zhao Y, LeDoux MS (2005) Cerebral cortical control of orbicularis oculi motoneurons. Brain Res 1047:177–193
78. Grafton ST (1994) Cortical control of movement. Amnn Neurol 36:3–4
79. Graziano MS (2006) The organization of behavioral repertoire in motor cortex. Annu Rev Neurosci 29:105–134
80. Graziano MS, Aflalo TN, Cooke DF (2005) Arm movements evoked by electrical stimulation in the motor cortex of monkeys. J Neurophysiol 94:4209–4223
81. Grofova I, Keane S (1991) Descending brainstem projections of the pedunculopontine tegmental nucleus in the rat. Anat Embryol (Berl) 184:275–290
82. Hage SR, Jürgens U (2006) On the role of the pontine brainstem in vocal pattern generation: a telemetric single-unit recording study in the squirrel monkey. J Neurosci 26:7105–7115
83. Hage SR, Jürgens U (2006) Localization of a vocal pattern generator in the pontine brainstem of the squirrel monkey. Eur J Neurosci 23:840–844
84. Hage SR, Jürgens U (2006) On the role of the pontine brainstem in vocal pattern generation: a telemetric single-unit recording study in the squirrel monkey. J Neurosci 26:7105–7115
85. Hajnik T, Lai YY, Siegel JM (2000) Atonia-related regions in the rodent pons and medulla. J Neurophysiol 84:1942–1948
86. Hannig S, Jürgens U (2006) Projections of the ventrolateral pontine vocalization area in the squirrel monkey. Exp Brain Res 169:92–105
87. Hatanaka N, Tokuno H, Nambu A, Inoue T, Takada M (2005) Input-output organization of jaw movement-related areas in monkey frontal cortex. J Comp Neurol 492:401–425
88. He SQ, Dum RP, Strick PL (1993) Topographic organization of corticospinal projections from the frontal lobe: motor areas on the lateral surface of the hemisphere. J Neurosci 13:952–980
89. He SQ, Dum RP, Strick PL (1995) Topographic organization of corticospinal projections from the frontal lobe: motor areas on the medial surface of the hemisphere. J Neurosci 15:3284–3306
90. Hikosaka O, Sakai K, Miyauchi S, Takino R, Sasaki Y, Putz B (1996) Brain activation during learning of sequential procedures. Electroencephalogr Clin Neurophysiol Suppl 47:245–252
91. Holstege G (1989) Anatomical study of the final common pathway for vocalization in the cat. J Comp Neurol 284:242–252
92. Holstege G (1991) Descending motor pathways and the spinal motor system: limbic and non-limbic components. Prog Brain Res 87:307–421
93. Holstege G, Kuypers HG (1977) Propriobulbar fibre connections to the trigeminal, facial and hypoglossal motor nuclei. I. An anterograde degeneration study in the cat. Brain 100:239–264
94. Holstege G, Kuypers HG (1982) The anatomy of brain stem pathways to the spinal cord in cat. A labeled amino acid tracing study. Prog Brain Res 57:145–175
95. Holstege G, Tan J (1987) Supraspinal control of motoneurons innervating the striated muscles of the pelvic floor including urethral and anal sphincters in the cat. Brain 110 (Pt 5):1323–1344
96. Holstege G, Tan J (1988) Projections from the red nucleus and surrounding areas to the brainstem and spinal cord in the cat. An HRP and autoradiographical tracing study. Behav Brain Res 28:33–57
97. Holstege G, Georgiadis JR (2003) Neurobiology of cat and human sexual behavior. Int Rev Neurobiol 56:213–225

98. Holstege G, Kuypers HG, Dekker JJ (1977) The organization of the bulbar fibre connections to the trigeminal, facial and hypoglossal motor nuclei. II. An autoradiographic tracing study in cat. Brain 100:264–286
99. Holstege G, Graveland G, Bijker-Biemond C, Schuddeboom I (1983) Location of motoneurons innervating soft palate, pharynx and upper esophagus. Anatomical evidence for a possible swallowing center in the pontine reticular formation. An HRP and autoradiographical tracing study. Brain Behav Evol 23:47–62
100. Holstege G, Tan J, van Ham J, Bos A (1984) Mesencephalic projections to the facial nucleus in the cat. An autoradiographical tracing study. Brain Res 311:7–22
101. Holstege G, Griffiths D, de Wall H, Dalm E (1986) Anatomical and physiological observations on supraspinal control of bladder and urethral sphincter muscles in the cat. J Comp Neurol 250:449–461
102. Holstege G, Georgiadis JR, Paans AM, Meiners LC, van der Graaf FH, Reinders AA (2003) Brain activation during human male ejaculation. J Neurosci 23:9185–9193
103. Holstege G, Mouton LAJ, Gerrits NM (2004) Emotional motor system. In: Paxinos G, Mai JK (eds) The human nervous system. Elsevier, Amsterdam, pp 1306–1324
104. Hoover JE, Strick PL (1999) The organization of cerebellar and basal ganglia outputs to primary motor cortex as revealed by retrograde transneuronal transport of herpes simplex virus type 1. J Neurosci 19:1446–1463
105. Huisman AM, Kuypers HG, Verburgh CA (1981) Quantitative differences in collateralization of the descending spinal pathways from red nucleus and other brain stem cell groups in rat as demonstrated with the multiple fluorescent retrograde tracer technique. Brain Res 209:271–286
106. Huisman AM, Kuypers HG, Verburgh CA (1982) Differences in collateralization of the descending spinal pathways from red nucleus and other brain stem cell groups in cat and monkey. Prog Brain Res 57:185–217
107. Huisman AM, Ververs B, Cavada C, Kuypers HG (1984) Collateralization of brainstem pathways in the spinal ventral horn in rat as demonstrated with the retrograde fluorescent double-labeling technique. Brain Res 300:362–367
108. Hultborn H, Nielsen JB (2007) Spinal control of locomotion – from cat to man. Acta Physiol (Oxf) 189:111–121
109. Isa T, Ohki Y, Seki K, Alstermark B (2006) Properties of propriospinal neurons in the C3–C4 segments mediating disynaptic pyramidal excitation to forelimb motoneurons in the macaque monkey. J Neurophysiol 95:3674–3685
110. Jankowska E, Jukes MG, Lund S, Lundberg A (1967) The effect of DOPA on the spinal cord. 6. Half-centre organization of interneurones transmitting effects from the flexor reflex afferents. Acta Physiol Scand 70:389–402
111. Jeannerod M, Arbib MA, Rizzolatti G, Sakata H (1995) Grasping objects: the cortical mechanisms of visuomotor transformation. Trends Neurosci 18:314–320
112. Jürgens U, Alipour M (2002) A comparative study on the cortico-hypoglossal connections in primates, using biotin dextranamine. Neurosci Lett 328:245–248
113. Keizer K, Kuypers HG (1984) Distribution of corticospinal neurons with collaterals to lower brain stem reticular formation in cat. Exp Brain Res 54:107–120
114. Kluck P (1980) The autonomic innervation of the human urinary bladder, bladder neck and urethra: a histochemical study. Anat Rec 198:439–447
115. Kuypers HGJM (1958) Some projections from the peri-central cortex to the pons and the lower brain stem in monkey and chimpanzee. J Comp Neurol 110:221–255
116. Kuypers HGJM (1958) Corticobulbar connections to the pons and lower brain-stem in man. Brain 81:365–388
117. Kuypers HGJM (1962) Corticospinal connections: postnatal development in the Rhesus monkey. Science 138:678–680
118. Kuypers HGJM (1973) The anatomical organization of the descending pathways and their contribution to motor control especially in primates. In: Desmedt J (ed) New developments in EEG and clinical neurophysilogy. Karger, Basel, pp 38–68
119. Kuypers HGJM (1987) Some aspects of the organization of the output of the motor cortex. In: Ciba foundation symposium: Motor areas of the cerebral cortex. Wiley, Chichester, pp 63–68
120. Kuypers HGJM, Brinkman J (1970) Precentral projections to different parts of the spinal intermediate zone in the rhesus monkey. Brain Res 24:29–48
121. Kuypers HGJM, Lawrence DG (1967) Cortical projections to the red nucleus and the brain stem in the Rhesus monkey. Brain Res 4:151–188
122. Kuze B, Matsuyama K, Matsui T, Miyata H, Mori S (1999) Segment-specific branching patterns of single vestibulospinal tract axons arising from the lateral vestibular nucleus in the cat: A PHA-L tracing study. J Comp Neurol 414:80–96
123. Kwan HC, MacKay WA, Murphy JT, Wong YC (1978) Spatial organization of precentral cortex in awake primates. II. Motor outputs. J Neurophysiol 41:1120–1131
124. Lacroix S, Havton LA, McKay H, Yang H, Brant A, Roberts J, Tuszynski MH (2004) Bilateral corticospinal projections arise from each motor cortex in the macaque monkey: a quantitative study. J Comp Neurol 473:147–161
125. Lawrence DG, Kuypers HG (1968) The functional organization of the motor system in the monkey. I. The effects of bilateral pyramidal lesions. Brain 91:1–14

126. Lawrence DG, Kuypers HG (1968) The functional organization of the motor system in the monkey. II. The effects of lesions of the descending brainstem pathways. Brain 91:15–36
127. Leicht R, Rowe MJ, Schmidt RF (1973) Cortical and peripheral modification of cerebellar climbing fibre activity arising from cutaneous mechanoreceptors. J Physiol 228:619–635
128. Li JL, Wu SX, Tomioka R, Okamoto K, Nakamura K, Kaneko T, Mizuno N (2005) Efferent and afferent connections of GABAergic neurons in the supratrigeminal and the intertrigeminal regions. An immunohistochemical tract-tracing study in the GAD67-GFP knock-in mouse. Neurosci Res 51:81–91
129. Li YQ, Tao FS, Okamoto K, Nomura S, Kaneko T, Mizuno N (2002) The supratrigeminal region of the rat sends GABA/glycine-cocontaining axon terminals to the motor trigeminal nucleus on the contralateral side. Neurosci Lett 330:13–16
130. Lundberg A (1971) Function of the ventral spinocerebellar tract. Exp Brain Res 12:317–330
131. Lundberg A, Oscarsson O (1960) Functional organization of the dorsal spino-cerebellar tract in the cat. VII. Identification of units by antidromic activation from the cerebellar cortex with recognition of five functional subdivisions. Acta Physiol Scand 50:356–374
132. Luschei ES (1987) Central projections of the mesencephalic nucleus of the fifth nerve: an autoradiographic study. J Comp Neurol 263:137–145
133. Maier MA, Armand J, Kirkwood PA, Yang HW, Davis JN, Lemon RN (2002) Differences in the corticospinal projection from primary motor cortex and supplementary motor area to macaque upper limb motoneurons: an anatmical and electrophysiological study. Cereb Cortex 12:381–396
134. Mantyh PW (1983) Connections of midbrain periaqueductal gray in the monkey. I. Ascending efferent projections. J Neurophysiol 49:567–581
135. Matelli M, Luppino G (1996) Thalamic input to mesial and superior area 6 in the macaque monkey. J Comp Neurol 372:59–87
136. Matelli M, Luppino G, Rizzolatti G (1985) Patterns of cytochrome oxidase activity in the frontal agranular cortex of the macaque monkey. Behav Brain Res 18:125–136
137. Matelli M, Luppino G, Fogassi L, Rizzolatti G (1989) Thalamic input to inferior area 6 and area 4 in the macaque monkey. J Comp Neurol 280:468–488
138. Matelli M, Luppino G, Rizzolatti G (1991) Architecture of superior and mesial area 6 and the adjacent cingulate cortex in the macaque monkey. J Comp Neurol 311:445–462
139. Matelli M, Govoni P, Galletti C, Kutz DF, Luppino G (1998) Superior area 6 afferents from the superior parietal lobule in the macaque monkey. J Comp Neurol 402:327–352
140. Matelli M, Luppino G, Geyer S, Zilles K (2004) Motor cortex. In: Paxinos G, Mai JK (eds) The human nervous system. Elsevier, Amsterdam, pp 973–996
141. Matsuyama K, Drew T (2000) Vestibulospinal and reticulospinal neuronal activity during locomotion in the intact cat. II. Walking on an inclined plane. J Neurophysiol 84:2257–2276
142. Matsuyama K, Takakusaki K, Nakajima K, Mori S (1997) Multi-segmental innervation of single pontine reticulospinal axons in the cervico-thoracic region of the cat: anterograde PHA-L tracing study. J Comp Neurol 377:234–250
143. Matsuyama K, Kobayashi Y, Takakusaki K, Mori S, Kimura H (1993) Termination mode and branching patterns of reticuloreticular and reticulospinal fibers of the nucleus reticularis pontis oralis in the cat: an anterograde PHA-L tracing study. Neurosci Res 17:9–21
144. Matsuyama K, Mori F, Nakajima K, Drew T, Aoki M, Mori S (2004) Locomotor role of the corticoreticular-reticulospinal-spinal interneuronal system. Prog Brain Res 143:239–249
145. McCurdy ML, Hansma DI, Houk JC, Gibson AR (1987) Selective projections from the cat red nucleus to digit motor neurons. J Comp Neurol 265:367–379
146. Middleton FA, Strick PL (1997) Cerebellar output channels. Int Rev Neurobiol 41:61–82
147. Miller LE, Gibson AR (2007) The red nucleus. In: The New Encyclopedia of Neural Science. Elsevier, Amsterdam
148. Minkels RF, Juch PJ, van Willigen JD (1995) Interneurones of the supratrigeminal area mediating reflex inhibition of trigeminal and facial motorneurones in the rat. Arch Oral Biol 40:275–284
149. Miyai I, Tanabe HC, Sase I, Eda H, Oda I, Konishi I, Tsunazawa Y, Suzuki T, Yanagida T, Kubota K (2001) Cortical mapping of gait in humans: a near-infrared spectroscopic topography study. Neuroimage 14:1186–1192
150. Mizuno N, Matsuda K, Iwahori N, Uemura-Sumi M, Kume M, Matsushima R (1981) Representation of the masticatory muscles in the motor trigeminal nucleus of the macaque monkey. Neurosci Lett 21:19–22
151. Morcuende S, Delgado-Garcia JM, Ugolini G (2002) Neuronal premotor networks involved in eyelid responses: retrograde transneuronal tracing with rabies virus from the orbicularis oculi muscle in the rat. J Neurosci 22:8808–8818
152. Morecraft RJ, Van Hoesen GW (1998) Convergence of limbic input to the cingulate motor cortex in the rhesus monkey. Brain Res Bull 45:209–232
153. Morecraft RJ, Louie JL, Herrick JL, Stilwell-Morecraft KS (2001) Cortical innervation of the facial nucleus in the non-human primate: a new interpretation of the effects of stroke and related subtotal brain trauma on the muscles of facial expression. Brain 124:176–208
154. Morecraft RJ, McNeal DW, Stilwell-Morecraft KS, Gedney M, Ge J, Schroeder CM, van Hoesen GW (2007) Amygdala interconnections with the cingulate motor cortex in the rhesus monkey. J Comp Neurol 500:134–165

155. Mori S, Matsui T, Kuze B, Asanome M, Nakajima K, Matsuyama K (1998) Cerebellar-induced locomotion: reticulospinal control of spinal rhythm generating mechanism in cats. Ann NY Acad Sci 860:94–105
156. Murata A, Fadiga L, Fogassi L, Gallese V, Raos V, Rizzolatti G (1997) Object representation in the ventral premotor cortex (area F5) of the monkey. J Neurophysiol 78:2226–2230
157. Nakagawa S (1980) Onuf's nucleus of the sacral cord in a South American monkey (Saimiri): its location and bilateral cortical input. Brain Res 191:337–344
158. Nakano K, Hasegawa Y, Kayahara T, Tokushige A, Kuga Y (1993) Cortical connections of the motor thalamic nuclei in the Japanese monkey, Macaca fuscata. Stereotact Funct Neurosurg 60:42–61
159. Nielsen JB (2002) Motoneuronal drive during human walking. Brain Res Rev 40:192–201
160. Olszewski J (1952) The thalamus of Macaca mulatta. Kruger, Basel
161. Onuf(rowicz) B (1900) On the arrangement and function of cells groups of the sacral region of the spinal cord in man. Arch Neurol Psych 3: 387–411
162. Oriolo PJ, Strick PL (1989) Cerebellar connections with the motor cortex and the arcuate premotor area: an analysis employing retrograde transneuronal transport of WGA-HRP. J Comp Neurol 268:612–626
163. Oscarsson O (1965) Functional organization of the spino- and cuneocerebellar tracts. Physiol Rev 43:495–522
164. Pandya DN, Seltzer B (1982) Intrinsic connections and architectonics of posterior parietal cortex in the rhesus monkey. J Comp Neurol 204:196–210
165. Passingham EE (1993) The frontal lobe and voluntary action. Oxford University Press, Oxford
166. Penfield W, Boldrey E (1937) Somatic motor and sensory representation in the cerebral cortex of man as studied by electrical stimulation. Brain 60:389–443
167. Penfield W, Welch K (1951) The supplementary motor area of the cerebral cortex; a clinical and experimental study. AMA Arch Neurol Psychiatry 66:289–317
168. Picard N, Strick PL (1996) Motor areas of the medial wall: a review of their location and functional activation. Cereb Cortex 6:342–353
169. Pijpers A (2007) Functional anatomy of the intermediate cerebellum in the rat. Thesis Erasmus University, Rotterdam
170. Poppele RE, Rankin A, Eian J (2003) Dorsal spinocerebellar tract neurons respond to contralateral limb stepping. Exp Brain Res 149:361–370
171. Pritchard TC, Hamilton RB, Morse JR, Norgren R (1986) Projections of thalamic gustatory and lingual areas in the monkey, Macaca fascicularis. J Comp Neurol 244:213–228
172. Pritchard TC, Hamilton RB, Norgren R (2000) Projections of the parabrachial nucleus in the old world monkey. Exp Neurol 165:101–117
173. Reed AF (1940) The nuclear masses in the cervical spinal cord of Macaca mulatta. J Comp Neurol 72:187–205
174. Rho MJ, Lavoie S, Drew T (1999) Effects of red nucleus microstimulation on the locomotor pattern and timing in the intact cat: a comparison with the motor cortex. J Neurophysiol 81:2297–2315
175. Rizzolatti G, Luppino G, Matelli M (1998) The organization of the cortical motor system: new concepts. Electroencephalogr Clin Neurophysiol 106: 283–296
176. Rizzolatti G, Fogassi L, Gallese V (2002) Motor and cognitive functions of the ventral premotor cortex. Curr Opin Neurobiol 12:149–154
177. Romanes GJ (1951) The motor cell columns of the lumbo-sacral spinal cord of the cat. J Comp Neurol 94:313–363
178. Rossignol S, Dubuc R, Gossard JP (2006) Dynamic sensorimotor interactions in locomotion. Physiol Rev 86:89–154
179. Rouiller EM, Liang F, Babalian A, Moret V, Wiesendanger M (1994) Cerebellothalamocortical and pallidothalamocortical projections to the primary and supplementary motor cortical areas: a multiple tracing study in macaque monkeys. J Comp Neurol 345:185–213
180. Rozzi S, Calzavara R, Belmalih A, Borra E, Gregoriou GG, Matelli M, Luppino G (2006) Cortical connections of the inferior parietal cortical convexity of the macaque monkey. Cereb Cortex 16:1389–1417
181. Ruigrok TJ, Cella F, Voogd J (1995) Connections of the lateral reticular nucleus to the lateral vestibular nucleus in the rat. An anterograde tracing study with Phaseolus vulgaris leucoagglutinin. Eur J Neurosci 7:1410–1413
182. Ruigrok TJ, van der Burg H, Sabel-Goedknegt E (1996) Locomotion coincides with c-Fos expression in related areas of inferior olive and cerebellar nuclei in the rat. Neurosci Lett 214:119–122
183. Russell DF, Zajac FE (1979) Effects of stimulating Deiters' nucleus and medial longitudinal fasciculus on the timing of the fictive locomotor rhythm induced in cats by DOPA. Brain Res 177:588–592
184. Rybak IA, Shevtsova NA, Lafreniere-Roula M, McCrea DA (2006) Modelling spinal circuitry involved in locomotor pattern generation: insights from deletions during fictive locomotion. J Physiol 577:617–639
185. Sakai ST, Inase M, Tanji J (1996) Comparison of cerebellothalamic and pallidothalamic projections in the monkey (Macaca fuscata): a double anterograde labeling study. J Comp Neurol 368:215–228
186. Sakai ST, Inase M, Tanji J (1999) Pallidal and cerebellar inputs to thalamocortical neurons projecting to the supplementary motor area in Macaca fuscata: a triple-labeling light microscopic study. Anat Embryol (Berl) 199:9–19
187. Sakata H, Taira M (1994) Parietal control of hand action. Curr Opin Neurobiol 4:847–856

188. Sakata H, Taira M, Kusunoki M, Murata A, Tanaka Y, Tsutsui K (1998) Neural coding of 3D features of objects for hand action in the parietal cortex of the monkey. Philos Trans R Soc Lond B Biol Sci 353:1363–1373
189. Satoda T, Takahashi O, Tashiro T, Matsushima R, Uemura-Sumi M, Mizuno N (1987) Representation of the main branches of the facial nerve within the facial nucleus of the Japanese monkey (Macaca fuscata). Neurosci Lett 78:283–287
190. Satoda T, Uemura-Sumi M, Tashird T, Takahashi O, Matsushima R, Mizuno N (1990) Localization of motoneurons innervating the stylohyoid muscle in the monkey, cat, rabbit, rat and shrew. J Hirnforsch 31:731–737
191. Satoda T, Takahashi O, Murakami C, Uchida T, Mizuno N (1996) The sites of origin and termination of afferent and efferent components in the lingual and pharyngeal branches of the glossopharyngeal nerve in the Japanese monkey (Macaca fuscata). Neurosci Res 24:385–392
192. Schoen HJR (1964) Comparative aspects of the descending fibre systems in the spinal cord. Progress in Brain Research 11:203–222
193. Schoen JHR (1969) The corticofugal projection on the brain stem and spinal cord in man. Psychiat Neurol Neurochir 72:121–128
194. Schubert M, Curt A, Jensen L, Dietz V (1997) Corticospinal input in human gait: modulation of magnetically evoked motor responses. Exp Brain Res 115:234–246
195. Schubert M, Bohner C, Berger W, Sprundel M, Duysens JE (2003) The role of vision in maintaining heading direction: effects of changing gaze and optic flow on human gait. Exp Brain Res 150:163–173
196. Scott SH (2003) The role of primary motor cortex in goal-directed movements: insights from neurophysiological studies on non-human primates. Curr Opin Neurobiol 13:671–677
197. Sharrard WJW (1955) The distribution of the permanent paralysis in the lower limb in poliomyelitis. J Bone Jt Surg 37:540–558
198. Shiba K, Umezaki T, Zheng Y, Miller AD (1997) The nucleus retroambigualis controls laryngeal muscle activity during vocalization in the cat. Exp Brain Res 115:513–519
199. Shigenaga Y, Yoshida A, Mitsuhiro Y, Doe K, Suemune S (1988) Morphology of single mesencephalic trigeminal neurons innervating periodontal ligament of the cat. Brain Res 448:331–338
200. Shigenaga Y, Yoshida A, Mitsuhiro Y, Tsuru K, Doe K (1988) Morphological and functional properties of trigeminal nucleus oralis neurons projecting to the trigeminal motor nucleus of the cat. Brain Res 461:143–149
201. Shik ML, Severin FV, Orlovsky GN (1966) Control of walking and running by means of electrical stimulation of the midbrain. Biophysics 11:756–765
202. Shinoda Y, Yokota J, Futami T (1981) Divergent projection of individual corticospinal axons to motoneurons of multiple muscles in the monkey. Neurosci Lett 23:7–12
203. Shinoda Y, Sugiuchi Y, Izawa Y, Hata Y (2005) Long descending motor tract axons and their control of neck and axial muscles. Prog Brain Res 151:527–563
204. Skinner RD, Kinjo N, Ishikawa Y, Biedermann JA, Garcia-Rill E (1990) Locomotor projections from the pedunculopontine nucleus to the medioventral medulla. Neuroreport 1:207–210
205. Spann BM, Grofova I (1991) Nigropedunculopontine projection in the rat: an anterograde tracing study with phaseolus vulgaris-leucoagglutinin (PHA-L). J Comp Neurol 311:375–388
206. Sterling P, Kuypers HG (1967) Anatomical organization of the brachial spinal cord of the cat. II. The motoneuron plexus. Brain Res 4:16–32
207. Sukhotinsky I, Reiner K, Govrin-Lippmann R, Belenky M, Lu J, Hopkins DA, Saper CB, Devor M (2006) Projections from the mesopontine tegmental anesthesia area to regions involved in pain modulation. J Chem Neuroanat 32:159–178
208. Takakusaki K, Saitoh K, Harada H, Kashiwayanagi M (2004) Role of basal ganglia-brainstem pathways in the control of motor behaviors. Neurosci Res 50:137–151
209. Takakusaki K, Habaguchi T, Saitoh K, Kohyama J (2004) Changes in the excitability of hindlimb motoneurons during muscular atonia induced by stimulating the pedunculopontine tegmental nucleus in cats. Neuroscience 124:467–480
210. Tanji J (1996) New concepts of the supplementary motor area. Cur Opin Neurobiol 6:782–787
211. Tunik E, Frey SH, Grafton ST (2005) Virtual lesions of the anterior intraparietal area disrupt goal-dependent on-line adjustments of grasp. Nat Neurosci 8:505–511
212. Uemura M, Matsuda K, Kume M, Takeuchi Y, Matsushima R, Mizuno N (1979) Topographical arrangement of hypoglossal motoneurons: an HRP study in the cat. Neurosci Lett 13:99–104
213. Ueyama T, Arakawa H, Mizuno N (1987) Central distribution of efferent and afferent components of the pudendal nerve in rat. Anat Embryol (Berl) 177:37–49
214. Van de Crommert HW, Mulder T, Duysens J (1998) Neural control of locomotion: sensory control of the central pattern generator and its relation to treadmill training. Gait Posture 7:251–263
215. Van Wezel BM, Ottenhoff FA, Duysens J (1997) Dynamic control of location-specific information in tactile cutaneous reflexes from the foot during human walking. J Neurosci 17:3804–3814
216. Vanderhorst VG, Holstege G (1997) Organization of lumbosacral motoneuronal cell groups innervating hindlimb, pelvic floor, and axial muscles in the cat. J Comp Neurol 382:46–76
217. Van der Horst VG, Mouton LJ, Blok BF, Holstege G (1996) Distinct cell groups in the lumbosacral cord of the cat project to different areas in the periaqueductal gray. J Comp Neurol 376:361–385

218. Van der Horst VG, Terasawa E, Ralston HJ, 3rd, Holstege G (2000) Monosynaptic projections from the lateral periaqueductal gray to the nucleus retroambiguus in the rhesus monkey: implications for vocalization and reproductive behavior. J Comp Neurol 424:251–268
219. Van der Horst VG, Terasawa E, Ralston HJ, 3rd (2001) Monosynaptic projections from the nucleus retroambiguus region to laryngeal motoneurons in the rhesus monkey. Neuroscience 107:117–125
220. Vilensky JA, O'Connor BL (1998) Stepping in nonhuman primates with a complete spinal cord transection: old and new data, and implications for humans. Ann NY Acad Sci 860:528–530
221. Vinay L, Padel Y, Bourbonnais D, Steffens H (1993) An ascending spinal pathway transmitting a central rhythmic pattern to the magnocellular red nucleus in the cat. Exp Brain Res 97:61–70
222. Wang HF, Shortland P, Park MJ, Grant G (1998) Retrograde and transganglionic transport of horseradish-conjugated cholera toxin B unit, wheatgerm agglutinin and isolectin B4 from Griffonia simplicifolia I in pimary afferent neurons innervating the rat urinary bladder. Neuroscience 87:275–278
223. Weiss C, Houk JC, Gibson AR (1990) Inhibition of sensory responses of cat inferior olive neurons produced by stimulation of red nucleus. J Neurophysiol 64:1170–1185
224. Wiesendanger M, Wise SP (1992) Current issues concerning the functional organization of motor cortical areas in nonhuman primates. Adv Neurol 57:117–134
225. Woolsey CN, Settlage PH, Meyer DR, Sencer W, Pinto Hamuy T, Travis AM (1952) Patterns of localization in precentral and "supplementary" motor areas and their relation to the concept of a premotor area. Res Publ Assoc Res Nerv Ment Dis 30:238–264
226. Yoshida Y, Mitsumasu T, Miyazaki T, Hirano M, Kanaseki T (1984) Distribution of motoneurons in the brain stem of monkeys, innervating the larynx. Brain Res Bull 13:413–419
227. Yoshida Y, Mitsumasu T, Hirano M, Kanaseki T (1985) Somatotopic representation of the laryngeal motoneurons in the medulla of monkeys. Acta Otolaryngol 100:299–303
228. Zald DH, Pardo JV (1999) The functional neuroanatomy of voluntary swallowing. Ann Neurol 46:281–286
229. Zhang SP, Davis PJ, Bandler R, Carrive P (1994) Brain stem integration of vocalization: role of the midbrain periaqueductal gray. J Neurophysiol 72:1337–1356
230. Zilles K, Schlaug G, Geyer S, Luppino G, Matelli M, Qu M, Schleicher A, Schormann T (1996) Anatomy and transmitter receptors of the supplementary motor areas in the human and nonhuman primate brain. Adv Neurol 70:29–43

22 The Reticular Formation and the Monoaminergic and Cholinergic Cell Groups

Introduction

The area that occupies the central portion of the brain stem is known as the reticular formation. Throughout most of its extent, this area is occupied by aggregations of loosely arranged cells of different types and sizes, and the fibre systems that pass through its territory are likewise mostly diffusely organized. The term 'reticular formation' refers to the fact that the dendrites of the cells in this area are arranged in bundles that together form a net-like pattern [161]. The traversing fibre systems pass through the interstices of this network. The reticular formation is surrounded by cranial nerve and sensory relay nuclei and the long ascending and descending fibre systems of the brain stem. For cytoarchitectonic, chemoarchitectonic and functional reasons, the reticular formation of the medulla oblongata and the pons has been divided into the following three longitudinal zones [29] (Figs. 22.1, 22.2):

1. A median and paramedian zone, which consists of the raphe nuclei. Serotonergic neurons of the groups B1–B9 of Dahlström and Fuxe [43, 44] are located in the raphe nuclei and the adjacent reticular formation.
2. A medial zone, which contains many large cells, known as the *medial tegmental field*.
3. A lateral, largely parvocellular zone, known as the *lateral tegmental field*.

A fourth, intermediate field of the reticular formation, located between the medial magnocellular and the lateral parvocellular fields of the pontobulbar reticular formation, was distinguished by Paxinos and colleagues [41, 95, 109, 140]. The adrenergic cell groups C1 and C2, the noradrenergic groups A1–A6 [75] and the cholinergic cell groups Ch5 and 6 [134] are confined to this intermediate field. It also contains the ambiguus and the facial nucleus and the preganglionic parasympathetic neurons of cranial nerves VII, IX and X [95, 109]. In this chapter it is included into the lateral tegmental field.

Subdivisions of the Reticular Formation

The long, sparsely branching dendritic trees of the neurons of the medial tegmental field are rostrocaudally compressed. Dendrites preferentially extend into the transverse plane [161, 190]. The medial tegmental field contains the *nucleus gigantocellularis*, which extends from midolivary levels into the caudal pons (Fig. 22.1). The giant cells of the nucleus gigantocellularis proper are located fairly dorsally. The more ventrally located cells are slightly smaller and are oriented mediolaterally in cell strands that are continuous across the midline. At pon-

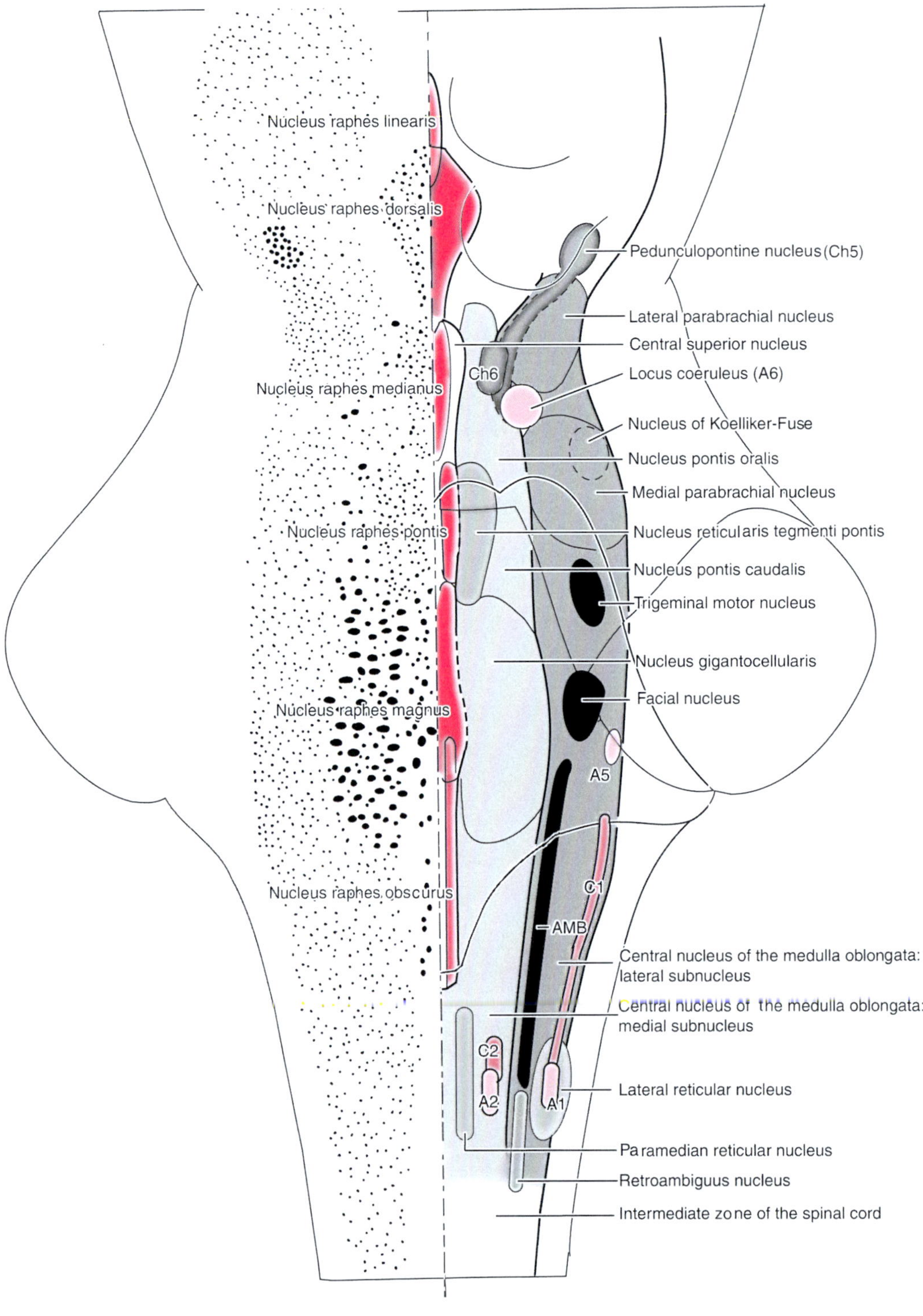

Fig. 22.1. The reticular formation. Semi-diagrammatic representation in a dorsal view of the brain stem. *Left*, cytoarchitecture; *right*, subdivisions of the reticular formation (medial tegmental field, *light grey*; lateral tegmental field, *dark grey*), the raphe nuclei (*red*), the noradrenergic cell groups A1–A6 (*light red*), the adrenergic groups C1 and C2 (*medium red*) and the cholinergic cell groups Ch5 and 6 (*grey*). *AMB*, ambiguus nucleus

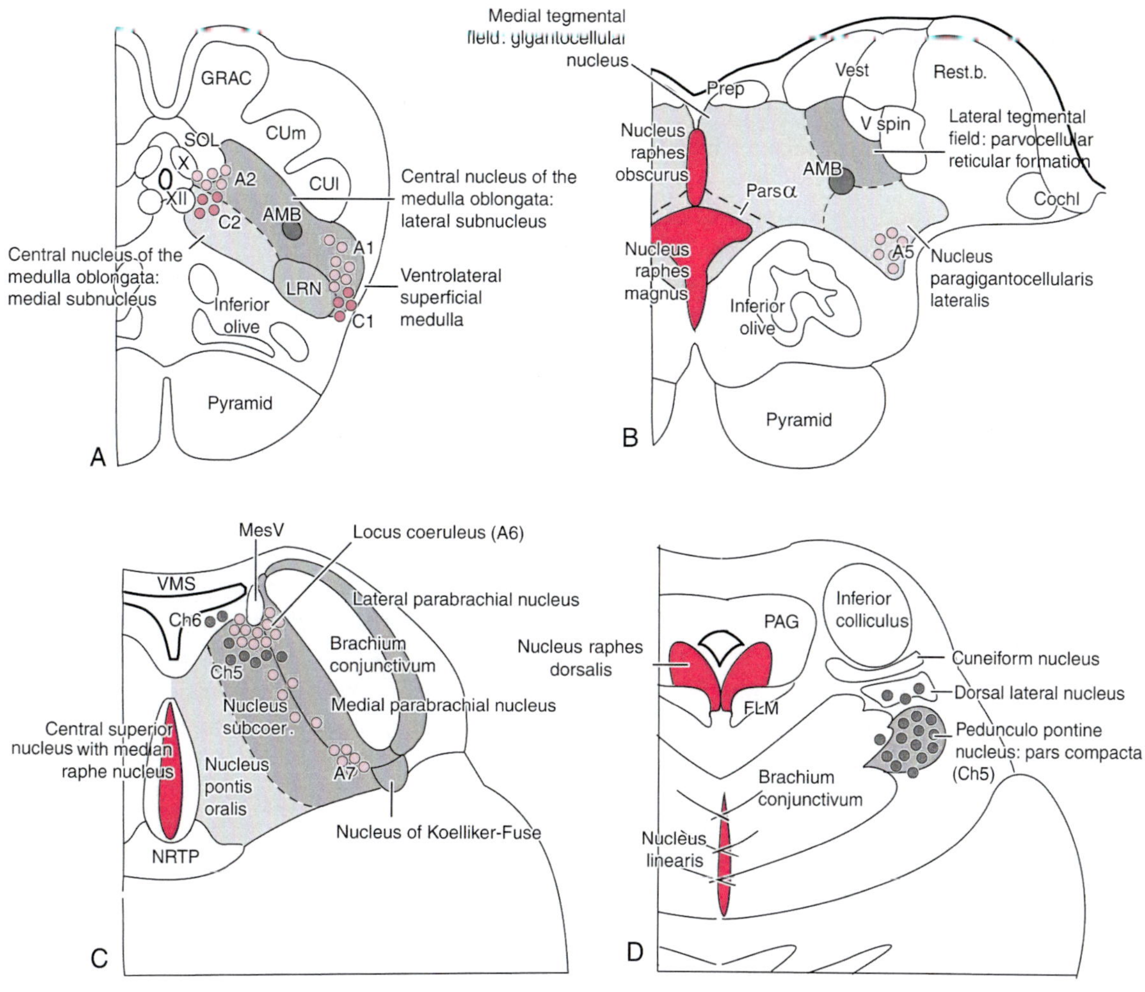

Fig. 22.2 A–D. The reticular formation. Diagrammatic transverse sections through the caudal (**A**) and rostral (**B**) medulla oblongata, the rostral pontine tegmentum (**C**) and the mesencephalon (**D**), to show the position of the medial tegmental field (*light grey*) and the lateral tegmental field (*dark grey*), the raphe nuclei (*red*), the noradrenergic cell groups A1–A7 (*light red*), the adrenergic cell group C1 and S2 (*medium red*) and the cholinergic cell groups Ch5 and Ch6 (*grey*). *A1–A7*, noradrenergic cell groups A1–A7; *AMB*, ambiguus nucleus; *C1*, *C2*, adrenergic cell groups C1, C2; *Cochl*, cochlear nuclei; *CUl*, lateral cuneate nucleus; *CUm*, medial cuneate nucleus; *FLM*, medial longitudinal fascicle; *GRAC*, gracile nucleus; *LRN*, lateral reticular nucleus; *MesV*, mesencephalic tract of the trigeminal nerve; *Nucleus subcoer.*, nucleus subcoeruleus; *NRTP*, nucleus reticularis tegmenti pontis; *PAG*, periaqueductal grey; *Prep*, nucleus prepositus hypoglossi; *Rest.b*, restiform body; *SOL*, nucleus of the solitary tract; *V spin*, spinal tract of the trigeminal nerve; *Vest*, vestibular nuclei; *VMS*, superior medullary velum; *X*, dorsal vagal nucleus; *XII*, nucleus of the hypoglossal nerve

tine and rostral bulbar levels, this ventral subdivision of the gigantocellular nucleus is represented by the *pars alpha of the gigantocellular nucleus* of Meessen and Olszewski [132] (Figs. 6.25, 22.2 B). The pars alpha is located lateral to the nucleus raphe magnus (B3) and contains the lateral wings of this serotoninergic cell group. The *ventral gigantocellular nucleus* occupies the same position at a more caudal bulbar level. It medially adjoins the nucleus raphe obscurus. Both the pars alpha and the ventral subdivision of the gigantocellular nucleus contain a large proportion of GABAergic neurons [95]. A ventrolateral extension of the gigantocellular nucleus contains the noradrenergic cell group A5. Caudally, the medial tegmental field is continuous with the *ventral subdivision of the central nucleus of the medulla oblongata* (Fig. 6.23).

At the level of the motor nucleus of the trigeminal nerve the nucleus gigantocellularis is replaced by the *nucleus reticularis pontis caudalis and oralis* (Figs. 6.26, 6.27). Rostrally, these nuclei extend to the decussation of the superior cerebellar peduncle. The small, medium and scattered giant cells of the nucleus reticularis pontis caudalis are less densely packed. The dorsal part of the nucleus reticularis pontis caudalis and oralis, located rostral and caudal to the nucleus of the abducens nerve, corresponds to the horizontal gaze centre. This region, also known as the *pontine paramedian reticular formation* (PPRF), contains excitatory and inhibitory burst cells with projections to the ipsilateral and contralateral abducens nucleus, respectively, and omnipause neurons, located next to the nucleus raphe pontis (see Chap. 19 "The Saccade Generating System").

The lateral tegmental field contains the *parvocellular reticular formation*, which borders on the nucleus of the spinal tract of the trigeminal nerve. Caudally, it continues as the *dorsal subdivision of the central nucleus of the medulla oblongata*. Its ventrolateral superficial area contains the noradrenergic and adrenergic cell groups A1 and C1 and the retroambiguus nucleus (Figs. 6.21, 22.2 A). In the rostral pons, the lateral tegmental field surrounds the motor trigeminal nucleus as the *supra- and intertrigeminal nuclei* and extends into the *lateral and medial parabrachial nuclei*, the *area of Koelliker-Fuse* and the *subcoeruleus nucleus* (Fig. 6.30, 22.2 C). The noradrenergic cells of the locus coeruleus (group A6) and the ventrolateral cell group A7 are found at this level. Cholinergic neurons of the groups Ch5 and 6 occupy the subcoerulean area and the pontine central grey, respectively. Rostrally, the cholinergic neurons of cell group Ch5 occupy the *pars compacta of the pedunculopontine nucleus* and the *dorsal lateral tegmental nucleus* (Fig. 22.2 D). The small cells of the *cuneiform nucleus* occupy the oblong region between the inferior colliculus and the nucleus pedunculopontinus. The reticular formation of the rostral mesencephalon can be subdivided into a dorsal deep reticular field and the ventral retrorubral area.

Three precerebellar reticular nuclei are contained within the reticular formation (Figs. 22.1, 22.2). Their neurons display a glomerular structure, with dendrites confined to the nuclei. The *lateral reticular nucleus* occupies a ventrolateral and superficial position in the central nucleus of the medulla oblongata. It can be subdivided into ventrolateral parvocellular, ventromedial magnocellular and dorsal subtrigeminal portions (Fig. 6.22). The subtrigeminal portion of the nucleus is located within the rubrospinal tract, and its ventral part within the anterolateral fascicle. Medially, the lateral reticular nucleus extends into the region of the lateral vestibulospinal tract. The *paramedian reticular nucleus* is located next to the midline, within the medial longitudinal fascicle. Its lateral border is formed by the large cells of the *nucleus interfascicularis hypoglossi.* Dorsally, it borders on the cells of the *nucleus of Roller* [156], located lateral to the medial longitudinal fascicle and ventral to the nucleus prepositus hypoglossi. The nucleus of Roller is the caudalmost representative of the nuclei of the paramedian tracts, precerebellar nuclei in the dorsomedial reticular formation with a predominant projection to the vestibulocerebellum. The *nucleus reticularis tegmenti pontis* is located next to the nucleus raphe pontis, dorsal to the medial lemniscus (Figs. 6.27, 22.2 C). As a precerebellar nucleus it transmits information from visuomotor centres (Chap. 19 "The Visual System and the Cerebellum").

Monoaminergic Cell Groups and Pathways

Serotonergic Cell Groups

The description by Dahlström and Fuxe (1964) in the rat of the serotonergic cell groups BI–B9, which are mainly located in the raphe nuclei, has been confirmed in humans [86, 87]. The general organization of the nuclei raphe and the distribution of serotonergic neurons appear to be very similar among mammalian species. The serotonergic neurons can be subdivided into two groups. One is located in the rostral pons and the mesencephalon, and the other in the caudal pons and the medulla oblongata. In the pontine tegmentum the two groups are separated by a gap (Fig. 22.3). The rostral group gives rise to local, cerebellar and ascending projections; the caudal group is connected with the brain stem, the cerebellum and the spinal cord. Serotonergic cells in the medulla are located in the small, ventrally located *nucleus raphe pallidus* (B1) and the more dorsally located *nucleus raphe obscurus* (B2). In the nucleus raphe obscurus, serotonergic neurons are confined to two sheets on both sides of the midline. Rostrally, these serotonergic groups are continuous with the B3 group corresponding to the *nucleus raphe magnus*. The lateral wings of B3 are located in the ventral gigantocellular nucleus pars alpha. Serotonergic neurons in the floor of the fourth ventricle of group B4 are absent in primates. The *nucleus raphe pontis* (B5) contains only a few serotonergic neurons. The largest aggregates of serotonergic neurons are found in the rostral group, in the *nucleus raphe dorsalis* (B8), located in and around the ventromedial part of the periaqueductal grey, and as the *median raphe nucleus* [109] in the central superior nucleus. The serotonergic neurons of the B9 group are a lateral extension of B8, located dorsomedial to the lateral lemniscus. The rostralmost serotonergic neurons occur in the *nucleus linearis* and in the retrorubral field. The nucleus linearis extends from the decussation of the superior cerebellar peduncle to the interpeduncular nucleus, where the serotonergic neurons intermingle with dopaminergic cells of the ventral tegmental area.

A sizeable proportion of the serotonergic neurons is found outside the raphe nuclei in the reticular formation lateral to the inferior olive, surrounding the lateral reticular nucleus, in the ventral gigantocellular nucleus, scattered in the dorsal tegmentum pontis and among the noradrenaline neurons of the locus coeruleus, the nucleus subcoeruleus and the parabrachial nuclei. Scattered neurons in the nucleus pontis oralis are prominent in the human brain stem [12].

Raphe nuclei differ in the proportion of serotonergic neurons they contain. Proportions differ from 80% for the dorsal raphe nucleus to 10–20% for the medullary raphe nuclei [87]. Substance P is often co-localized with serotonin, and several other neuropeptides occur in neurons of the raphe nuclei.

Data on the connections of the raphe nuclei and other monoaminergic cell groups were mostly collected in rodents and carnivores; observations in humans and non-human primates are rare. The raphe nuclei receive their input mainly from limbic structures. The dorsal and median raphe nuclei share afferents from the lateral habenular and the interpeduncular nuclei, hypothalamus and preoptic area, the nucleus of the solitary tract, the substantia nigra, the basal telencephalon and the prefrontal cortex. Afferents from the lateral tegmental field, the locus coeruleus, the tuber cinereum, the amygdala and the entopeduncular nucleus (medial segment of the globus pallidus) are restricted to the dorsal raphe nucleus; septal afferents terminate in the median raphe nucleus (Table 22.1). The nucleus raphe magnus shares many of these sources with the rostral group of raphe nuclei. Afferents to the nucleus raphe pallidus are restricted to the periaqueductal grey, reticular formation, hypothalamus and preoptic area and the prefrontal cortex.

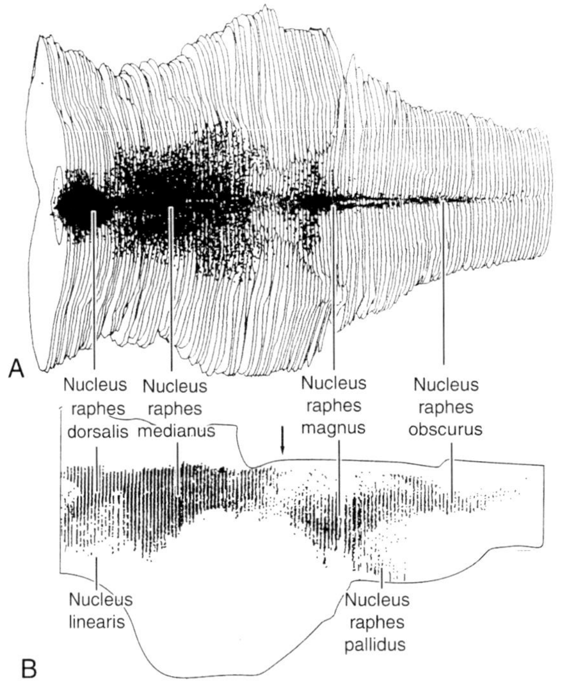

Fig. 22.3. Ventral (**A**) and lateral (**B**) projections of the serotonergic neurons of the raphe nuclei of the human brain stem. *Arrow*, gap between the rostral and caudal raphe nuclei. Redrawn with permission from Törk and Hornung [188]

Table 22.1. Afferent connections of the raphe nuclei

Source	N.r. dorsalis	N.r. medianus	N.r. magnus	N.r. pallidus
Dorsal column nucleus; spinal nucleus V			33	
Nucleus of the solitary tract	1, 70	1		
Periaqueductal grey		15	33, 56, 174	72
Laterodorsal tegmental nucleus	157	15		
Cuneiform nucleus			33	
Substantia nigra	157	15		
Nucleus raphe medianus	146, 157			
Locus coeruleus Subcoeruleus nucleus	157		72	
Medial tegmental field			72	
Lateral gigantocellular nucleus			157	72, 157
Medial parabrachial nucleus Area of Koelliker-Fuse	57, 157			72
Interpeduncular nucleus		15		
Ventral tegmental area	146			
Ventrolateral reticular area	157			
Tectum, pretectum			33	
Parafascicular nucleus			33	
Hypothalamus	146, 157, 159	15, 159	33, 72	89
Tuber cinereum and nuclei tuberis	146			
Preoptic area	41, 146, 157	15, 41	72	72
Lateral habenular nucleus	1, 71, 146, 157	1, 71	72	
Central nucleus of the amygdala	146		72	
Entopeduncular nucleus	146			
Ventral pallidum	151	151		
Zona incerta	146		33	
Bed nucleus, stria terminalis	146	15	72	
Medial septum		15		
Nucleus of the diagonal band	98	15, 98		
Prefrontal cortex	6, 66, 146	6, 15, 66	33, 72	72

Table 22.2. Efferent connections of the raphe nuclei

Destination	N.r. dorsalis	N.r. medianus	N.r. pontis	N.r. magnus	N.r. pallidus, obsc.
Spinal cord	23, 97, 169	23, 169			4, 13, 24, 77, 78, 81–83, 88, 113, 119, 126, 169
Sensory nuclei brain stem	67–69, 186	186		67–69, 126	67–69, 126
Cranial motor nuclei	117	195	117	117	
Cerebellum		21	16, 21, 167		
Pedunculopontine nucleus	176				
Locus coeruleus	93, 135	93, 135	135	135	
Nucleus raphe magnus and pallidus	218	218			
Periaqueductal grey		195			
Substantia nigra	93	93			
Subcommissural organ					
Retina	196				
Superior colliculus, pretectum	197	195		21	21
Interpeduncular nucleus		21, 22			
Thalamus, relay nuclei	21, 59, 131, 197, 208	142, 195			
Intralaminar nuclei	142			21	21
Mamillary body		21, 22			
Hypothalamus			21		
Preoptic area				113	113
Bed nu. stria terminalis	143				
Amygdala	21, 69, 93				
Septum	105	93, 105	105	105	
Striatum, accumbens nu.	21, 93, 193	93			
Olfactory bulb	21				
Cerebral cortex	101, 136, 193, 200, 201	103, 136, 147, 187			
Hippocampus	93, 104	21, 22, 93, 104	104		
Entorhinal cortex	21, 46, 104	104	104		

Serotonergic axons distribute throughout the central nervous system. Two types of axon can be distinguished: thin axons with small varicosities, and coarser axons with large varicosities. Small varicosities lack membrane junctional complexes, the hallmark of the synapse [47]. These fibres release their transmitter diffusely (volume transmission). They are also distinguished by their sensitivity to amphetamine-derived neurotoxins [125]. Large varicosities make synaptic contact with their targets. The two types of axon take their origin from different raphe nuclei: fibres with small varicosities arise from the nucleus raphe dorsalis and magnus. Fibres with coarse varicosities take their origin from the nucleus raphe medianus and pallidus.

Different efferent ascending systems take their origin from the dorsal and the median raphe nucleus. Thin, mainly unmyelinated fibres from the dorsal raphe nucleus collect in a dorsal bundle. This system was distinguished by Forel [50] as a collection of fascicles, containing fibres of different size, located ventrolateral to the periaqueductal grey. It contains serotonergic, noradrenergic and cholinergic elements and ascending fibres from the nucleus pontis oralis and caudalis. The area containing Forel's fascicles coincides approximately with the dorsal trigeminothalamic tract (Figs. 6.31, 6.32). Coarser axons ascend from the median raphe nucleus, in a ventral bundle, through the tegmental area. At the mid-forebrain border, the two systems intermingle. They distribute to their targets through the medial forebrain bundle, the subthalamus, the thalamus and the internal capsule. The dorsal bundle is the more prominent one in primates.

The main projections of the dorsal raphe nucleus are to the lateral geniculate body, the striatum, the entorhinal cortex, the olfactory bulb and the amygdala, and from the median raphe nucleus to the septum, basal forebrain and hippocampus [21]; however, many of these and other destinations are shared by these nuclei (Table 21.2, Chaps. 7–15). Local targets include the periaqueductal grey, the lateral pontine tegmental field with the locus coeruleus, the pedunculopontine nucleus and the substantia nigra. The median raphe nucleus connects with the interpeduncular nucleus, the substantia nigra and the mamillary body. Its cortical projection is rather sparse. Rostral and caudal raphe nuclei are interconnected.

Projections to hippocampus, entorhinal cortex, striatum and amygdala take their origin from different subdivisions of the dorsal raphe nucleus [104, 175]. Neurons of this nucleus projecting to motor, sensorimotor and visual cortex are arranged in a rostral to caudal array [200]. No such topographical arrangement is present in the cortical projection of the median raphe nucleus [136]. Individual serotonergic axons may branch to terminate widely different regions, such as the spinal cord and the cerebral cortex, amygdala and vestibular nuclei [69] and hippocampus and substantia nigra [93].

The caudal raphe nuclei innervate the caudal brain stem and the entire spinal cord. The contribution of the rostral raphe nuclei to the innervation of the spinal cord is limited to cervical levels [23]. The lateral tegmental field and the sensory relay nuclei are innervated by the nucleus raphe magnus and pallidus, cranial motor nuclei by the nucleus raphe pallidus and obscurus. Projections of these nuclei to the rostral brain stem and forebrain are very limited (Table 22.2). The cerebellum receives a sizeable serotonergic projection. Randomly oriented fibres are present in the molecular layer and with a loose plexus in the granular layer. The cerebellar nuclei are densely innervated. These cerebellar afferents originate from scattered neurons in the medullary and pontine reticular formation. The cerebellar projection is topically organized [16, 102, 199]. The serotonergic innervation of the inferior olive is mainly directed at the accessory olives [198].

The nucleus raphe magnus and the adjoining reticular formation densely innervate the marginal layer of the spinal trigeminal nucleus and laminae I and II of the dorsal horn in rodents and primates [24–28]. These fibres descend in the dorsolateral funiculus. The nucleus raphe magnus and pallidus also project to the intermediate grey and innervate the intermediolateral nucleus [97, 119]. The motoneuronal groups receive a serotonergic input from the nucleus raphe pallidus and obscurus and the neighbouring reticular formation. This system travels in the lateral and ventral funiculus [106, 107]. A partial spinal origin of the serotonergic innervation of Onuf's nucleus in the baboon was proposed by Rajaofetra et al. [150].

Only a few of the varicosities in the dorsal horn contribute to classical synapses; synaptic contacts of serotonergic varicosities are more common in the motor columns [127, 155]. Some of the serotonergic terminals on motoneurons co-localize GABA [80]. Serotonergic terminals on somatic motoneurons co-localize substance P and thyreotropin-releasing hormone; terminals on preganglionic parasympathetic neurons do not contain substance P [182, 217].

The dorsal raphe nucleus, the locus coeruleus, the cholinergic pedunculopontine and lateral dorsal tegmental nuclei are interconnected. These nuclei are involved in sleep/waking control. The activity of the dorsal raphe nucleus is high during waking, low during slow wave sleep and abolished during rapid eye movement (REM) sleep (Chap. 10 "Wakefulness and Sleep"). Serotonergic neurons of the nucleus raphe magnus inhibit nociceptive transmission in the superficial laminae of the caudal spinal nucleus of the trigeminal nerve and the spinal dorsal horn (Chap. 16 "Control Mechanisms of Nociceptive Transmission"). The nucleus raphe pallidus and obscurus participate in a system determining the excitability of cranial and spinal motoneurons (this Chap., "Connections of the Reticular Formation and Associated Cell Groups"). Mood disorders are managed with different drugs, modulating serotonin metabolism, serotonin transporters and reuptake systems. A discussion of the relationship between the distribution of serotonergic systems and their neuropharmacology cannot be carried out without a thorough analysis of the different classes of serotonin receptors, however, and is beyond the scope of this chapter.

Adrenergic and Noradrenergic Cell Groups

In rodents the adrenergic and noradrenergic neurons in the medulla oblongata are located in a dorsolateral and a ventrolateral group. The ventrolateral group consists of the noradrenergic group A1 of Dahlström and Fuxe [43, 44] and the adrenergic group C1 of Hökfelt et al. [75]. These neurons are located in the ventrolateral superficial reticular area, near the lateral reticular nucleus, among the fibres of the anterolateral fascicle and the rubrospinal tract (Fig. 22.2A). The C1 group extends most rostrally to the level of the motor nucleus of the facial nerve. The A2 and C2 groups are located within the nucleus of the solitary tract and along the borders of the dorsal vagal nucleus (Fig. 22.2A). Scattered neurons interconnect the two groups. Neurons of the adrenergic C3 group occupy the dorsal midline and the floor of the fourth ventricle at rostral medullary levels. This group is absent in primates. More rostrally, noradrenergic cells are present in the locus coeruleus in the periaqueductal grey (A6), as scattered cells in the nucleus subcoeruleus, in the subependymal extension of the locus coeruleus (A4), ventrolaterally, in the paragigantocellular nucleus (A5) and dispersed within and around the ventral nucleus of the lateral lemniscus (A7) (Fig. 22.2B,C). Noradrenergic cell strands interconnect the groups A5, A6 and A7. Similar localizations of adrenergic and noradrenergic cells have been described for primates, including humans [58, 91, 141]. In the rat and in primates, the locus coeruleus is comprised entirely of noradrenergic neurons.

For the structure of neurotransmitters and connections of the locus coeruleus the present chapter is based on the recent review by Aston-Jones [7], which also provides a complete bibliography on the subject. Afferent connections of the locus coeruleus differ for the locus itself, containing the noradrenergic cell bodies, and the surrounding halo, consisting of their radiating dendrites (the "peri-LC"). The main afferent connections of the locus coeruleus take their origin from the ventrolateral paragigantocellular nucleus, which contains the A5 group, and from a cell group at the medial edge of the nucleus prepositus hypoglossi in the floor of the fourth ventricle (group EF, the epifascicular nucleus [109]). A projection of the A5 group to the locus coeruleus appears to be absent; however, many other transmitter-systems take their origin from the ventrolateral paragigantocellular nucleus, where neurons stained for adrenalin, GABA, met-enkephalin, corticotropin-releasing factor were found. This region also provides a potent excitatory amino acid input to the locus coeruleus [9]. Met-enkephalin and GABA co-exist in certain neurons [191]. Afferents to the ventrolateral paragigantocellular nucleus arise from a wide variety of nuclei with autonomic, visceral and sensory-related functions [192]. Afferents of the noradrenergic A5 group, located within this nucleus, are considered in the next paragraphs.

The EF group inhibits neurons of the locus coeruleus via a GABAergic pathway [8, 9, 48]. The afferent connections of the EF are not known. A projection of lamina I neurons of the dorsal horn was traced to the locus coeruleus by Westlund and Craig in the monkey [203]. Numerous regions project to the peri-LC. Among them, the central nucleus of the amygdala, the preoptic area, different hypothalamic nuclei, the periaqueductal grey, the nucleus of the solitary tract, the prefrontal cortex, the adrenergic cell group C1 [32] and Barrington's micturition area may be mentioned. The locus coeruleus is densely innervated by hypocretin/orexin fibres, originating from the perifornical region of the hypothalamus [49] (see also Chap. 10 "Wakefulness and Sleep"). An indirect projection of the suprachiasmatic nucleus to the locus coeruleus was established by transneuronal retrograde tracing with pseudorabies virus in the rat [10, 35].

Much less is known about the afferent connections of the other noradrenergic cell groups. Neurons of A7 and A5, like the locus coeruleus, receive a projection from the periaqueductal grey. These A7 neurons project to the spinal cord [11, 12]. Both groups receive a projection from the medullary dorsal reticular formation, a supraspinal, pain-modulating area [114]. Group A5, in addition, receives afferents from the paraventricular nucleus of the hypothalamus, the area of Koelliker-Fuse, the parabrachial nuclei and the nucleus of the solitary tract [31, 90]. Group A1 receives afferents from lamina I neurons of the spinal dorsal horn, and, with its projection to the parabrachial nuclei, may serve as a relay in an alternative system mediating emotional and motivational aspects of pain [42, 118, 138]. The location and the connections of the A2 are closely related to the nucleus of the solitary tract.

The projections of the locus coeruleus cover the entire central nervous system. Its fibres ascend in the dorsal and ventral catecholaminergic bundles [96], following a similar course as the serotonergic system. The main projections in the forebrain include olfactory structures, neocortex, hippocampus, thalamus, basal forebrain, preoptic area and hypothalamus. The distribution of noradrenergic fibres in these regions is considered in more detail in the Chaps. 7–15 and 23.

In the brain stem of the rat, the locus coeruleus densely innervates the sensory relay nuclei (spinal nucleus of the trigeminal nerve [61], nucleus of the solitary tract [179], and cochlear nuclei [110, 111]). The tectum, the periaqueductal grey, the pontine nuclei, the interpeduncular nucleus, and the inferior olive also receive a sizeable projection from the locus coeruleus. Large regions of the brain stem, such as the cranial somatic and autonomic motor nuclei and the reticular formation, are sparsely innervated by the locus coeruleus, but receive their afferents from the cell groups A5 and 7 [2, 3, 54, 61, 116]. The nucleus of the solitary tract also is innervated by the A1 and A5 groups [19, 31, 121]. The scarce innervation of the cranial motor nuclei by the locus coeruleus has also been observed in the monkey. The dorsal nucleus of the vagus nerve is an exception to this rule and receives a noradrenergic innervation from the locus coeruleus in rat and monkey [185, 202]. The adrenergic C1 group also projects to the dorsal vagal nucleus [32]. Ascending fibres from the locus coeruleus and the cell groups A5 and A7 innervate a region in the dorsolateral pontine reticular formation, containing the cholinergic neurons of the groups Ch5 and 6, which is involved in the induction of REM sleep [166].

Apart from their projections to local structures, the A1, A2, C1 and C2 groups give rise to ascending fibres that course in a dorsal noradrenergic bundle [96]. A periventricular system from the A1 and A2 cell groups ascends towards the medial nuclei of the thalamus. Both A1 and A2 project to the region of the A5 group in the ventrolateral paragigantocellular nucleus. The projection of the A2 group to this region consists of branching axons that also terminate in the central nucleus of the amygdala [153]. Other neurons of the A1 (and C1) and A2 groups project with branching axons to the central nucleus of the amygdala, the bed nucleus of the stria terminalis, and to the paraventricular hypothalamic nucleus [144, 215, 216]. These connections are part of a more extensive system of projections of the A1 and the A5 groups to the parabrachial nuclei, the Koelliker-Fuse area, the periaqueductal grey, the hypothalamus and the vascular organ of the lamina terminalis [31, 36, 216]. The projections of A1 and A2 (and from the adrenergic C1 group [32]) to the paraventricular nucleus of the hypothalamus mainly target the parvocellular part of this nucleus [160], cells that secrete vasopressin and corticotropin-releasing hormone. The connection of a population of oestrogen-sensitive A1 neurons to gonadotropin-releasing hormone (GnRh) neurons in the preoptic area may be involved in the production of the GnRg/luteinizing hormone surge [152].

The locus coeruleus and the A5 and A7 groups give rise to descending systems to the spinal cord. In the monkey Westlund and colleagues found a projection of the locus coeruleus, through the ipsilateral ventrolateral funiculus. Some of its fibres extending in the dorsal funiculus terminate in the dorsal horn. Fibres from A7 descend in the ipsilateral ventrolateral funiculus and the contralateral dorsolateral funiculus. Fibres from both nuclei cross the midline at all levels of the cord. Terminations of fibres from the locus coeruleus and A7 are present in laminae IV–VIII of the dorsal horn and the intermediate zone, and are heaviest in the motoneuronal cell groups of lamina IX. Lighter terminations are present around the central canal and in superficial laminae of the dorsal horn. With respect to their terminations in autonomic cell groups the two nuclei differ: the locus coeruleus projects bilaterally to the sacral intermediomedial and -lateral nuclei, A7 only projects to the thoracic intermediolateral column. The spinal projection from A5 is a relatively modest one [74, 202, 205–207]. According to Fritschy and Lyons et al. projections of the locus coeruleus of the rat to the ventral horn and the thoracic intermediolateral nucleus are scarce; these projections mainly arise from cell groups A5 and A7 [55, 123, 124]. The rat locus coeruleus innervates the dorsal horn and the intermediate zone. In laminae I and II its fibres descend over the entire length of the spinal cord. A5 and A7 fibres also travel in the ventral and dorsolateral funiculi to terminate in the intermediate zone, the ventral horn and the intermediolateral cell column [53]. A dense projection of the locus coeruleus to the motoneuronal cell columns, with a moderate termination in medial laminae VII and VIII, in lamina X and the base of the dorsal horn and a sparse innervation of the superficial dorsal horn, was found in similar experiments in the rat by Proudfit and Clark [38, 148]. According to these authors, the projection of A7 is restricted to laminae I–IV of the dorsal horn [37]. The A5 projection to the spinal cord targets laminae IV–VI of the base of the dorsal horn, lamina VII of the intermediate zone and the thoracic intermediolateral nucleus [40, 120]. The differences in the spinal projections of the pontine adrenergic cell groups between the experiments of Fritschy and Grzanna and Clark and Proudfit are caused by using two different strains of rats [39, 172]. Similar genetic differences may be responsible for differences in the projections of these cell groups in monkeys, as reported by Westlund and Bowker [202, 204–207]. In the motoneuronal cell columns and the intermediolateral nucleus most synapses are axodendritic. In the dorsal horn the noradrenergic innervation is mainly non-synaptic [149, 155].

Direct projections of the A1 and A2 groups to the spinal cord appear to be absent. These spinal projections and/or those from other neurons in the ventrolateral medulla are shunted through the A5 group in the ventrolateral paragigantocellular nucleus (Fig. 22.4 A). A mainly ipsilateral projection of neurons of the adrenergic C1 group, via the lateral funiculus to the thoracic intermediomedial nucleus, was described by Card et al. in the rat [32].

The noradrenergic innervation of the cerebellum is mostly directed at its cortex. It is present as a randomly oriented plexus in all three cortical layers [52, 62]. Synaptic contacts with Purkinje cell and granule cell dendrites are made by more than 50% of the varicosities [137]. The cerebellar noradrenergic innervation of the cerebellum and the inferior olive stems from the locus coeruleus and the groups A5 and A7 [198].

The noradrenergic system pairs topographic specificity to the branching of individual noradrenergic axons between widely separated targets. Its varicose axons make synaptic contacts, but are also engaged in volume transmission. In some regions, such as the spinal cord, these two modes of transmission are spatially segregated, with thin fibres without synapses prevailing in the dorsal horn, and synaptic transmission in the motoneuronal cell columns. The medullary adrenergic and noradrenergic cell groups, with their spinal output through A5 to the intermediolateral nucleus, their connections with autonomic centres in the brain stem and their ascending connections to the paraventricular nucleus of the hypothalamus and the preoptic area, are involved in a wide range of visceral functions, including cardiovascular control and respiration, and in the control of secretion of different hormones by the hypothalamus. The locus coeruleus is implicated in arousal. Connections of the locus coeruleus and/or cell group A7 with sensory relay nuclei engage these centres in gating mechanisms for sensory input and pain control. The role of the noradrenergic cell groups in the production of anaesthesia was discussed by Hirota and Kushikata [74]. The projections of the rostral noradrenergic cell groups with cranial and spinal motor nuclei subserve level setting in motor control (see also "Connections of the Reticular Formation and Associated Cell Groups").

Dopaminergic Cell Groups

Dopamine-containing neurons are confined to the mesencephalon and the diencephalon [43, 44, 76]. They are located in the olfactory bulb (group A16), in the preoptic area and the hypothalamus (A12–A15), in the periventricular region and surrounding the fasciculus retroflexus (A11), in the ventral tegmental area of Tsai [189] and the nucleus linearis (A11), in the substantia nigra (A9) and in the retrorubral field (A8). The dopaminergic neurons of the substantia nigra are mainly found in the pars compacta and lateralis and are scattered in the pars reticulata. Their projections to the forebrain are considered in the Chaps. 7–15. The dopaminergic projections to the spinal cord and most projections to the brain stem take their origin from the A11 group [18, 169–171].

In the brain stem of the rat, the dopaminergic and the noradrenergic innervation generally overlap. The distribution of the dopaminergic fibres is less ubiquitous and is lacking in some regions (area postrema; lateral cuneate nucleus), which do receive a noradrenergic innervation. The strongest dopamine labelling is found of the nucleus of the solitary tract, the dorsal motor nucleus of the vagus, the nucleus of the solitary tract and the nucleus raphe pallidus and obscurus [100, 194]. The dorsal raphe nucleus, the ventral periaqueductal grey, the parabrachial nuclei, and the locus coeruleus receives projections both from the A10 group in the ventral tegmental area and from the A11 [14, 99, 145]. Fewer dopaminergic fibres are distributed to the reticular formation of the mesencephalon and the pons and the tectum. The dorsal column nuclei, the sensory nuclei of the trigeminal nerve and the cochlear nuclei receive dopaminergic afferents. The dopaminergic innervation differs for different cranial motor nuclei. It forms a dense plexus in the oculomotor, facial, trigeminal and caudal hypoglossal nuclei, and is sparse in the abducens and ambiguus nuclei.

The evidence on dopaminergic projections to the cerebellum is controversial. According to Panagopoulos et al. [139] the entire cerebellum of the rat receives a projection that is very similar to its innervation by the noradrenergic system. According to van Dijken [194] the dopaminergic projection in the rat is scarce or absent. Dopaminergic projections to the cerebellar nuclei and non-dopaminergic projections from this area to the cerebellar cortex were traced in the rat by Ikai et al. [92] from the ventral tegmental area.

Dopaminergic fibres descend, mainly ipsilaterally, in the dorsolateral funiculus and the superficial dorsal horn. Strong dopamine labelling of varicose fibres was found in all laminae of the dorsal horn, including the central canal area (lamina X), sparing the substantia gelatinosa, which only contains few dopaminergic fibres in rat and monkey. In the motoneuronal cell groups, dense dopamine labelling with a fine granular appearance is present. These thin fibres may originate from a longitudinal plexus in laminae I and II. The density of the innervation of Onuf's nucleus exceeds that of neighbouring somatic motoneurons. The sympathetic intermediolateral and -medial cell column are heavily innervated. The dopaminergic fibre plexus in these nuclei is distributed in a ladder-like fashion (compare Fig. 6.5). The innervation of the parasympathetic nuclei is moderate. The remaining areas, such as the intermediate zone (laminae VI–VIII), are sparsely innervated. The precerebellar nuclei (Clarke's column and central cervical nucleus) do not receive a dopaminergic innervation [84, 154]. Dopaminergic fibres generally make axosomatic contacts. The dopaminergic innervation of the cervical dorsal horn is non-synaptic [154].

Evidence for the involvement of dopamine in pain control, excitability control of somatic and visceromotor neurons and other visceral functions was discussed by van Dijken [194] and Holstege et al. [84].

Cholinergic Cell Groups

Cholinergic neurons and pathways have been identified by AChE histochemistry [168] and by the more sensitive immunocytochemical methods, using antibodies against CHAT [115]. In the forebrain, cholinergic neurons are found as large interneurons in the striatum, in the nucleus basalis, and as smaller cells in the nucleus accumbens and around the islands of Calleja of the olfactory tubercle. A collection of small cholinergic neurons in the medial habenular nucleus gives rise to a cholinergic component of the fasciculus habenulo-interpeduncularis [213]. A nomenclature for the non-striatal cholinergic neurons was introduced by Mesulam et al. [133] for the rat and by Mesulam et al. [134] for macaque monkeys. Cholinergic neurons in the medial septal nucleus (group Chl) and the vertical limb of the nucleus of the diagonal band (Ch2) principally innervate the cortex of the hippocampal formation (Chap. 12). The horizontal limb of the nucleus of the diagonal band provides cholinergic fibres to the olfactory bulb (Chap. 23). The nucleus basalis (including the cholinergic cells of the magnocellular preoptic nucleus and scattered cells within the globus pallidus) innervate the cerebral cortex and the amygdaloid complex (Chap. 15). The cholinergic groups Ch5 and Ch6 are located at the junction of the pons with the mesencephalon and correspond to the pedunculopontine nucleus and the lateral dorsal tegmental nucleus, situated in the periaqueductal grey at the level of the rostral fourth ventricle, respectively (Figs. 21.1, 21.2). The connections of these cholinergic groups were reviewed by Saper [158] and Butcher and Woolf [30].

In the rat, the group Ch5 neurons extend dorsolaterally to the caudal substantia nigra [5]. Axons from the group Ch5 neurons ascend in the central reticular fasciculus, dorsolateral to the medial longitudinal fascicle, and provide a cholinergic innervation to the thalamus, the tectum, the substantia nigra, the lateral hypothalamus and the cerebral cortex [133, 214]. These ascending cholinergic pathways are an important constituent of the reticular activating system (Chap. 10 "Wakefulness and Sleep"). Projections from the Ch5 group and from local cholinergic neurons of the intermediate reticular field terminate in the raphe nuclei, the pontine nuclei and in the pontine and medial medullary reticular formation [94, 214]. The diffuse cholinergic projections to the cerebellar cortex and nuclei are similarly derived from the Ch5 group (for a review, see Voogd et al. [198]). Other cholinergic projection systems not included in the classification of Mesulam are the tectal projections from the parabigeminal nucleus (Chap. 19 "The Superior Colliculus"), the mossy fibre projections to the cerebellum from the medial and descending vestibular nuclei (Chap. 17 "Vestibular Projections to the Cerebellum"), the cholinergic projection from the nucleus prepositus hypoglossi to the inferior olive, that co-localizes GABA (Chap. 20 "The Cerebellar Nuclei, the Modular Organization of the Cerebellar Output") and the olivocochlear bundle that arises from cholinergic neurons in and around the lateral superior olive (Chap. 18 "The Ascending Auditory Projection").

Connections of the Reticular Formation and Associated Cell Groups

Connections of the reticular formation, thus far, have appeared in this book as relays in a great variety of neuronal systems, suggesting that the reticular formation is not a nervous centre per se, but rather a mosaic of numerous different functional areas. Many premotor centres for different combinations of cranial nerve nuclei are encountered in the lateral tegmental field. Putative pattern generators for vocalization, mastication and swallowing were located in the lateral reticular formation dorsal to the superior olive and in the medullary lateral tegmental field; the saccade-generating machinery occupies the pontine paramedian reticular formation; and micturition centres were localized in the pontine lateral tegmental field and a locomotor area in its rostral extension, the pedunculopontine nucleus. Circumscribed reticular nuclei function as sources of mossy fibres, relaying information from visuomotor and spinal motor systems to the cerebellum.

Traditionally, more diffusely distributed functions have been attributed to the reticular formation, such as a role in pain transmission and pain control, as a region where different sensory inputs converge, serving as an activating system for the state of vigilance of the subject, as a level-setting system for the excitability of motor systems, regulating muscle tone, and as a system controlling cardiovascular functions and respiration. These ideas are by no means outdated, but rather have been replaced by more sophisticated systems that include parts of the reticular formation, usually in collaboration with the circuitry of one or more of the monoaminergic or cholinergic cell groups, considered in the previous paragraphs.

The idea that some aspects of pain sensation are derived from a flow of information from the spinal cord through the reticular formation to the intralaminar nuclei and certain forebrain structures was studied by Blomqvist and Berkley [20]. They confirmed that the ascending fibres of the anterolateral fasciculus terminate in the medullary reticular formation and are mainly confined to the ipsilateral lateral tegmental field, the ventrolateral paragigantocellular nucleus, the nucleus reticularis pontis caudalis and the area of Koelliker-Fuse. Neurons projecting to the thalamus are located in the central reticular nucleus of the mesencephalon, the parabrachial nuclei and the locus coeruleus and subcoeruleus and scattered in the nucleus reticularis pontis oralis and caudalis and the nucleus gigantocellularis. Very few of these neurons are located within the regions innervated by the spinal cord. Apart from the well-known nociceptive pathways connecting lamina I of the spinal dorsal horn and the spinal nucleus of the trigeminal nerve with the parabrachial nuclei (Chap. 16 "Spinothalamic Tract"), no convincing anatomical evidence for a role of the reticular formation in diffuse sensory conduction pathways is available.

Blomqvist and Berkley [20] pointed out that, rather than participating in ascending conduc-

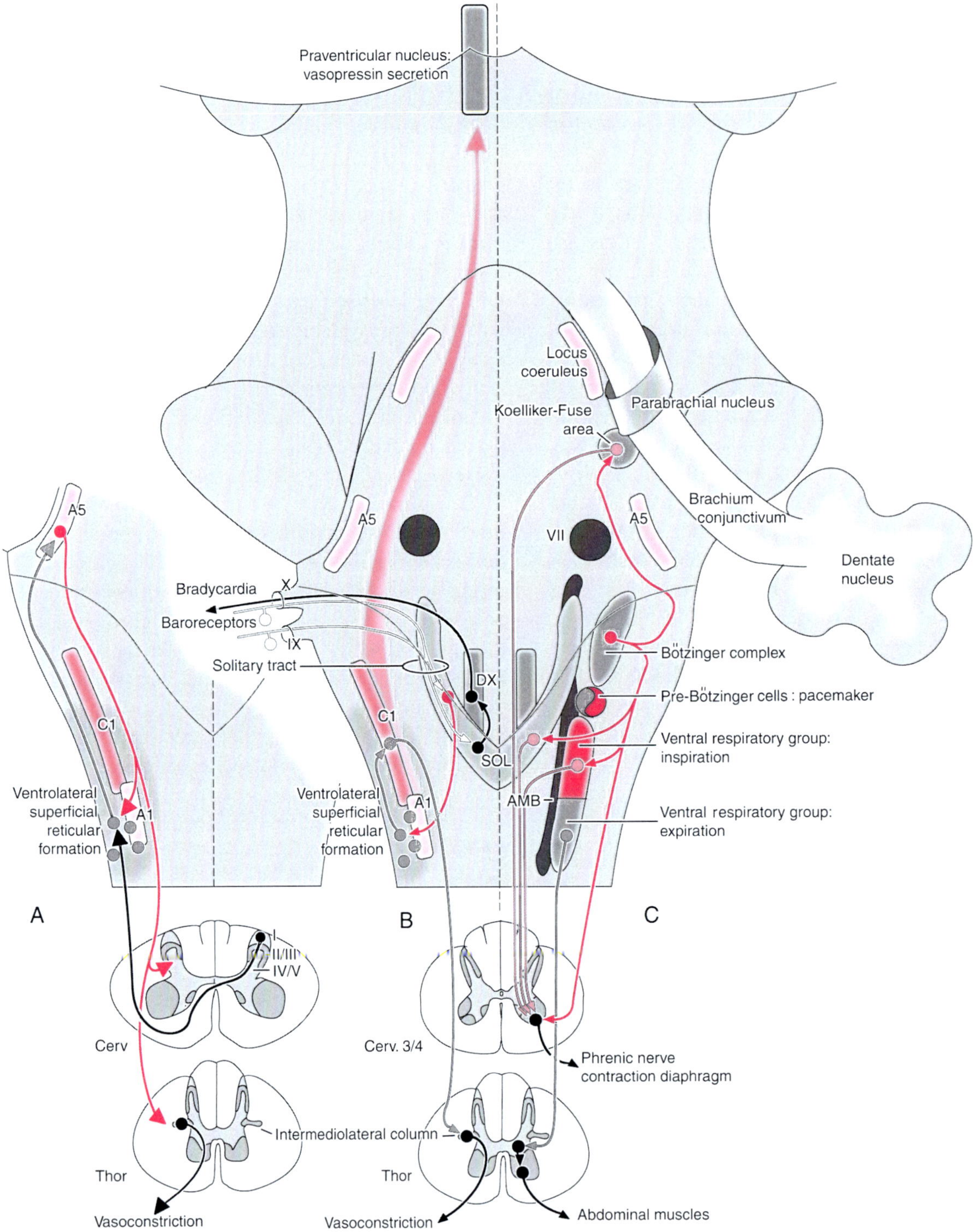

Fig. 22.4. A Spinal connections of neurons located in the ventrolateral superficial reticular formation (VLRF) that receive a projection from laminae I and II neurons of the dorsal horn. These connections with the spinal cord are shunted through the A5 noradrenergic cell group. A5 neurons give rise to an inhibitory projection, terminating on the neurons of the VLRF, the dorsal horn and the intermediolateral cell column [183]. **B** The pathway of the baroreceptor reflex. A descending pathway from the C1 adrenergic cell group terminates on the thoracic (*Thor*) intermediolateral cell column. Preganglionic fibres that originate from this column cause vasoconstriction. The innervation of the heart by motoneurons of the dorsal vagal nucleus is responsible for

tion pathways, regions of the reticular formation that receive a spinal input may exert a reciprocally organized influence on the spinal cord. The most prominent of these regions is the ventrolateral superficial medulla oblongata (VLSM), which contains the noradrenergic and adrenergic groups A1 and C1. Rostrally this region extends into the ventrolateral paragigantocellular nucleus, which harbours the A5 group. This region and the adjoining cell groups in and around the ambiguus nucleus are involved in cardiovascular and respiratory control and may influence sensory transmission in the spinal dorsal horn.

Control of sensory transmission in the dorsal horn has been suggested to be a function of the noradrenergic cell groups of the brain stem. As summarized in the previous paragraphs, the evidence on projections of these cell groups in rat and monkey to the dorsal horn and their termination in the superficial layers involved in nociceptive transmission is incomplete and conflicting. An interesting subsystem of the VLSM was identified by Tavares et al. in the rat [183, 184] (Fig. 22.4 A). A cell group of unknown neurotransmitter-identity located ventral to the spinal tract of the trigeminal nerve, which receives input from superficial laminae of the dorsal horn, projects back to laminae I, II and IV. It gives rise to an ascending, excitatory projection to the A5 group in the paragigantocellular nucleus. A descending system from the A5 group terminates in the spinal dorsal horn and provides the spinally projecting neurons in the VLSM with a collateral projection. According to Tavares et al. [184], the A5 projection is an inhibitory one. The function of this system remains unknown: projections of A5 to the dorsal horn favour its deeper layers or are described as modest. A distinct projection of A5 to the thoracic autonomic nuclei is acknowledged by most authors. The system, therefore, may link nociceptive input to visceromotor functions. Whether the A5 projection to the thoracic visceromotor columns is inhibitory is not known. The presence of a lamina I projection to the VLSM in monkeys [42] suggests that a similar recurrent system may be present in primates.

The ventromedial pars alpha of the gigantocellular reticular nucleus has been identified as a region controlling nociceptive transmission in the caudal spinal nucleus of the trigeminal nerve and the dorsal horn (Chap. 16 "Control Mechanisms of Nociceptive Transmission", Fig. 16.6). Both serotonergic neurons of the nucleus raphe magnus and non-serotonergic neurons are involved in this function. These neurons do not receive a spinal input, but rather are under control of the periaqueductal grey. Their projections focus on superficial and deep layers of the dorsal horn. The nucleus raphe magnus also projects to the visceromotor columns in the spinal cord. Again, somatosensory and visceromotor control functions appear to

◄

slowing down the heart rate. Viscerosensory fibres of the vagal and glossopharyngeal nerves, which innervate baroreceptors, terminate in the caudal nucleus of the solitary tract. Short excitatory connections link the dorsal vagal nucleus with the dorsal vagal nucleus. An inhibitory pathway from the nucleus of the solitary tract (*red*) terminates on a group of excitatory neurons in the ventrolateral superficial reticular formation, at the level of the noradrenergic A1 group. Ascending axons from these cells terminate in C1. The C1 and A1 groups give rise to ascending projections that terminate, among many other regions, in the paraventricular nucleus. Neurons of this nucleus secrete vasopressin. **C** Neurons of the ventral respiratory group are located near the ambiguus nucleus. Inspiratory neurons, projecting to the cervical motor nucleus of the phrenic nerve, are located in its rostral part (*red*), expiratory neurons that innervate abdominal muscles in the thoracic cord are located more caudally (*grey*). Bötzinger's complex, located more rostrally, consists of expiratory neurons that inhibit the inspiratory neurons of the ventral respiratory group and other nuclei with connections to the phrenic motonucleus (area of Koelliker-Fuse, nucleus of the solitary tract) during expiration. This inhibitory connection is shown in *red*. Pre-Bötzinger cells, located between Bötzinger's complex and the ventral respiratory group, are the pace maker for respiration. *AMB*, ambiguus nucleus; *DX*, dorsal vagal nucleus; *SOL*, nucleus of the solitary tract; *I, II–III, IV–V,* laminae I, II–III, IV–V of the spinal grey matter; *VII*, nucleus of the facial nerve; *X, IX*, vagal (X) and glossopharyngeal (IX) roots

be represented in a similar reticular region, containing a mixture of monoaminergic and non-monoaminergic cells. According to Mason [128] the involvement of the periaqueductal grey–ventromedial medulla oblongata system in nociceptive transmission is only one aspect of its more general function in maintaining homeostasis. In particular it was found to be related to the timing of micturition. In this context, it should be remembered that micturition, like nociceptive processes, are dependent on primary afferent input to lamina I neurons of the dorsal horn, which are both under control of this ventromedial area.

The VLSM is one of the best-studied regions in cardiovascular control (Dampney [45]). Both monoaminergic and non-monoaminergic neurons with projections to the thoracic visceromotor columns and the ambiguus and dorsal vagus nuclei are involved in these systems. Figure 22.4 B, illustrating the pathway of the baroceptor reflex, is based on Blessing's review [17]. Baroreceptors in the carotid sinus and the ascending aorta connect with the rostral part of the general visceral portion of the nucleus of the solitary tract; chemoreceptors sensitive to O_2 and CO_2 concentrations terminate more caudally. Short pathways connect the nucleus of the solitary tract with cardiomotor neurons in the dorsal vagal nucleus, subserving slowing down the heart rate. Presympathetic neurons in the VLSM, 50% of which correspond to the adrenergic C1 group, project to the thoracic intermediolateral column [32, 162, 165], and subserve vasoconstriction. The C1 group is inhibited by more caudally located GABAergic neurons in the VLSM [163]. These inhibitory neurons receive an input from baroreceptor-activated cells of the nucleus of the solitary tract, thus causing inhibition of vasomotor tone. Ascending projections from the A1 and A2 groups control vasopressin release by the paraventricular hypothalamic nucleus in a complementing manner (Chap. 10). Other noradrenergic and raphespinal systems are involved in cardiovascular control, but their pathways have been less well characterized.

The respiratory control system resides in the same ventrolateral reticular area (Fig. 22.4C). Inspiratory and expiratory neurons occupy a column near the nucleus ambiguus (ventral respiratory group). Inspiratory neurons are located in its rostral part; they project to motoneurons of the phrenic nerve at C3/C4 that innervate the diaphragm, the main inspiratory muscle. The caudally located expiratory neurons project to spinal interneurons for motoneurons innervating abdominal muscles. Other nuclei projecting to the motoneurons of the phrenic nucleus are the nucleus of the solitary tract, the Bötzinger complex and the area of Koelliker-Fuse (the so-called pneumotactic centre) [77]. The Bötzinger complex is located more rostrally in the ventrolateral medulla oblongata, at a level immediately caudal to the facial nucleus. Its neurons inhibit the inspiratory neurons of the nucleus of the solitary tract and the ventral respiratory group during expiration and project bilaterally to the phrenic motor nucleus. Bötzinger neurons probably use glycine as their neurotransmitter [164]. Pre-Bötzinger cells located rostral to the ventral respiratory group, ventral to the ambiguus nucleus and caudal to Bötzinger's complex serve as a pacemaker for respiration and project to the ventral respiratory group and Bötzinger's complex [173]. Noradrenergic cell groups are involved in the control of respiration. Stimulation of carotid chemoreceptors activates the locus coeruleus and group A5 [63, 108]. The locus coeruleus and the A1 group facilitate, the A5 group inhibits the respiratory rhythm generator [73, 219].

The role of the reticular formation in determining the excitability of spinal and cranial motoneurons and their premotor neurons is well known. It has been studied extensively in relation to the general atonia occurring during REM or paradoxical sleep. Neurons of the dorsal raphe nucleus and the locus coeruleus are active during waking and cease firing during sleep [177]. Mesopontine cholinergic neurons of the pedunculopontine and dorsal lateral tegmental nuclei are active during waking and REM sleep. Cholinergic cells of these nuclei receive a (presumably inhibitory) serotonergic and a catecholaminergic input [85, 112] and receive afferents from the pontine reticular formation. Ascending serotonergic, noradrenergic, cholinergic and reticular fibres constitute the complex fibre system situated in and around Forel's fascicles, ventrolateral to the periaqueductal grey, also indicated as the dorsal catecholaminergic, serotonergic or cholinergic bundle. Together these components of this system interact and combine with other forebrain structures to regulate thalamic and cortical activity and their behavioural state as the so-called reticular activating system (see also Chap. 10 "Wakefulness and Sleep").

General atonia can be caused by systems projecting diffusely to interneurons and/or motoneurons. Inhibition could be produced by inhibitory components of these systems or by local inhibitory interneurons. GABA would be a likely candidate for this kind of inhibition, but it was shown for the atonia during REM sleep that it is transmitted by glycine [34, 51, 122]. Atonia could be evoked from stimulation of almost the entire pontomedullary medial tegmental field in the rat [65]. A more restricted region in the dorsal gigantocellular reticular nucleus of the pontomedullary medial tegmental field was delineated by Takakusaki and colleagues [64, 180, 181]. Stimulation of this region in decerebrate cats caused predominantly inhibitory effects on spinal motoneurons, transmitted by two reticulospinal systems, a rapidly conducting system in the anterior and ventrolateral funiculus and a slower system in the lateral funiculus. The rapid reticulospinal system probably corresponds to the medial reticulospinal tract. This uncrossed pathway descends from the medial pontine and medullary tegmental field, through the anterior funiculus to terminate bilaterally in ventral laminae IX and VII. Only few of its fibres contact motoneurons in lamina IX [79, 129, 130]. The reticulospinal transmission is presumably excitatory. Inhibition, therefore, was produced by local interneurons. The inhibitory neurotransmitter of these interneurons is not known.

Earlier, diffuse projections to the spinal intermediate zone and the spinal and cranial motor nuclei and the visceromotor nuclei of the spinal cord were traced from the region of the locus coeruleus and the nucleus subcoeruleus and from the ventromedial medial tegmental field, caudal to the facial nucleus. The latter includes serotonergic neurons of the nucleus raphe pallidus [4, 79, 220]. Projections of the locus coeruleus and the subcoerulean area were found to terminate with boutons containing spherical vesicles and asymmetrical synapses and with a population of boutons containing microvesicles and symmetrical synapses on spinal motoneurons. The latter were tentatively identified as noradrenergic terminals. Terminals with spherical vesicles presumably contain an excitatory neurotransmitter, but which one is not known. Terminals on spinal motoneurons originating from the ventromedial medullary tegmental field consisted of boutons containing dark core vesicles and asymmetrical synapses, identified as serotonergic terminals, and of a population of boutons containing flattened vesicles and symmetrical synapses. The latter population is GABAergic; GABA and serotonin are co-localized in some terminals [80–83]. The effect of serotonin and noradrenaline on motoneurons, generally, has been found to be facilitatory [209–212]. The presence of GABAergic terminals or terminals with symmetrical synapses in these projections suggests that they are mixed excitatory/inhibitory systems that may shift the excitability of motoneurons in either direction. It is not known whether glycine also acts as a transmitter in these systems. Their involvement in REM sleep, therefore, is still unclear.

Another region influencing muscle tone was delineated in the pontomesencephalic reticular formation as the mesopontine tegmental anaesthesia area in the rat. Injecting GABA agonists into this area produces an irreversible anaesthetic-like state, with loss of spontaneous movement and profound flaccid atonia [178]. Descending connections from this area include the lateral pontine tegmental field and, more caudally, the ventromedial medullary tegmental field. A few fibres reach spinal levels. Apparently, inactivation of the mesopontine anaesthesia area releases a more caudal atonia circuitry from a tonic inhibition, but the circuitry concerned is not known. The descending and ascending connections of this mesopontine area are very similar to those of the cholinergic pedunculopontine nucleus that resides in the same region and contains the mesencephalic locomotor area (Chap. 21 "Locomotion"). The descending cholinergic projections of the pedunculopontine nucleus, similarly, are focussed upon the ventromedial pontomedullary tegmental field. As yet, the relationship between the mesopontine tegmental anaesthesia area, the centres involved in the production of atonia during REM sleep, and the mesencephalic locomotor area are far from clear. Takakusaki et al. [181], who also noticed the close spatial association of the circuitry involved in the inhibition of muscle tone and the execution of locomotion, suggested that activity in these circuits would be balanced by a net cortical excitatory output and a net basal ganglia-reticular nigral inhibitory input to these systems, which could be part of a widespread, reticular assembly regulating vigilance and the state of the motor system.

References

1. Aghajanian GK, Wang RY (1977) Habenular and other midbrain raphe afferents demonstrated by a modified retrograde tracing technique. Brain Res 122:229–242
2. Aldes LD, Shaw B, Chronister RB, Haycock JW (1990) Catecholamine-containing axon terminals in the hypoglossal nucleus of the rat: an immuno-electronmicroscopic study. Brain Res 81:167–178
3. Aldes LD, Chapman ME, Chronister RB, Haycock JW (1992) Sources of noradrenergic afferents to the hypoglossal nucleus in the rat. Brain Res Bull 29:931–942
4. Alstermark B, Kummel H, Tantisira B (1987) Monosynaptic raphespinal and reticulospinal projection to forelimb motoneurones in cats. Neurosci Lett 74:286–290
5. Armstrong DM, Saper CB, Levey AI, Wainer BH, Terry RD (1983) Distribution of cholinergic neurons in rat brain: demonstrated by the immunocytochemical localization of choline acetyltransferase. J Comp Neurol 216:53–68
6. Arnsten AF, Goldman-Rakic PS (1984) Selective prefrontal cortical projections to the region of the locus coeruleus and raphe nuclei in the rhesus monkey. Brain Res 306:9–18
7. Aston-Jones G (2004) Locus coerulaeus, A5 and A7 noradrenergic cell groups. In: Paxinos G (ed) The rat nervous system. Elsevier, Amsterdam, pp 259–294
8. Aston-Jones G, Ennis M, Pieribone VA, Nickell WT, Shipley MT (1986) The brain nucleus locus coeruleus: restricted afferent control of a broad efferent network. Science 234:734–737
9. Aston-Jones G, Shipley MT, Chouvet G, Ennis M, Van Bockstaele E, Pieribone V, Shiekhattar R, Akaoka H, Drolet G, Astier B et al (1991) Afferent regulation of locus coeruleus neurons: anatomy, physiology and pharmacology. Prog Brain Res 88:47–75
10. Aston-Jones G, Chen S, Zhu Y, Oshinsky ML (2001) A neural circuit for circadian regulation of arousal. Nat Neurosci 4:732–738
11. Bajic D, Van Bockstaele EJ, Proudfit HK (2001) Ultrastructural analysis of ventrolateral periaqueductal gray projections to the A7 catecholamine cell group. Neuroscience 104:181–197
12. Baker KG, Halliday GM, Halasz P, Hornung JP, Geffen LB, Cotton RG, Tork I (1991) Cytoarchitecture of serotonin-synthesizing neurons in the pontine tegmentum of the human brain. Synapse 7:301–320
13. Basbaum AI, Clanton CH, Fields HL (1978) Three bulbospinal pathways from the rostral medulla of the cat: an autoradiographic study of pain modulating systems. J Comp Neurol 178:209–224
14. Beckstead RM, Domesick VB, Nauta WJ (1979) Efferent connections of the substantia nigra and ventral tegmental area in the rat. Brain Res 175:191–217
15. Behzadi G, Kalen P, Parvopassu F, Wiklund L (1990) Afferents to the median raphe nucleus of the rat: retrograde cholera toxin and wheat germ conjugated horseradish peroxidase tracing, and selective D-[^{3}H]aspartate labelling of possible excitatory amino acid inputs. Neuroscience 37:77–100
16. Bishop GA, Ho RH (1985) The distribution and origin of serotonin immunoreactivity in the rat cerebellum. Brain Res 331:195–207
17. Blessing W (2001) Lower brain stem regulation of visceral, cardiovascular, and respiratory function. In: Paxinos G, Mai JK (eds) The human nervous system. Elsevier, Amsterdam, pp 465–478
18. Blessing WW, Chalmers JP (1979) Direct projection of catecholamine (presumably dopamine)-containing neurons from hypothalamus to spinal cord. Neurosci Lett 11:35–40
19. Blessing WW, Furness JB, Costa M, West MJ, Chalmers JP (1981) Projection of ventrolateral medullary (A1) catecholamine neurons toward nucleus tractus solitarii. Cell Tissue Res 220:27–40
20. Blomqvist A, Berkley KJ (1992) A re-examination of the spino-reticulo-diencephalic pathway in the cat. Brain Res 579:17–31
21. Bobillier P, Seguin S, Petitjean F, Salvert D, Touret M, Jouvet M (1976) The raphe nuclei of the cat brain stem: a topographical atlas of their efferent projections as revealed by autoradiography. Brain Res 113:449–486

22. Bobillier P, Seguin S, Degueurce A, Lewis BD, Pujol JF (1979) The efferent connections of the nucleus raphe centralis superior in the rat as revealed by radioautography. Brain Res 166:1–8
23. Bowker RM, Westlund KN, Coulter JD (1981) Serotonergic projections to the spinal cord from the midbrain in the rat: an immunocytochemical and retrograde transport study. Neurosci Lett 24:221–216
24. Bowker RM, Westlund KN, Coulter JD (1981) Origins of serotonergic projections to the spinal cord in rat: an immunocytochemical-retrograde transport study. Brain Res 226:187–199
25. Bowker RM, Westlund KN, Coulter JD (1982) Origins of serotonergic projections to the lumbar spinal cord in the monkey using a combined retrograde transport and immunocytochemical technique. Brain Res Bull 9:271–278
26. Bowker RM, Westlund KN, Sullivan MC, Coulter JD (1982) Organization of descending serotonergic projections to the spinal cord. Prog Brain Res 57:239–265
27. Bowker RM, Westlund KN, Sullivan MC, Coulter JD (1982) A combined retrograde transport and immunocytochemical staining method for demonstrating the origins of serotonergic projections. J Histochem Cytochem 30:805–810
28. Bowker RM, Westlund KN, Sullivan MC, Wilber JF, Coulter JD (1983) Descending serotonergic, peptidergic and cholinergic pathways from the raphe nuclei: a multiple transmitter complex. Brain Res 288:33–48
29. Brodal A (1957) The reticular formation of the brain stem. Anatomical aspects and functional correlations. In: The Henderson Trust Lectures. Oliver and Boyd, Edinburgh
30. Butcher LL, Woolf NJ (2004) Cholinergic neurons and networks revisited. In: Paxinos G, Mai JK (eds) The rat nervous system. Elsevier, Amsterdam, pp 1263–1268
31. Byrum CE, Guyenet PG (1987) Afferent and efferent connections of the A5 noradrenergic cell group in the rat. J Comp Neurol 261:529–542
32. Card JP, Sved JC, Craig B, Raizada M, Vazquez J, Sved AF (2006) Efferent projections of rat rostroventrolateral medulla C1 catecholamine neurons: Implications for the central control of cardiovascular regulation. J Comp Neurol 499:840–859
33. Carlton SM, Leichnetz GR, Young EG, Mayer DJ (1983) Supramedullary afferents of the nucleus raphe magnus in the rat: a study using the transcannula HRP gel and autoradiographic techniques. J Comp Neurol 214:43–58
34. Chase MH, Soja PJ, Morales FR (1989) Evidence that glycine mediates the postsynaptic potentials that inhibit lumbar motoneurons during the atonia of active sleep. J Neurosci 9:743–751
35. Chen S, Yang M, Miselis RR, Aston-Jones G (1999) Characterization of transsynaptic tracing with central application of pseudorabies virus. Brain Res 838:171–183
36. Chou TC, Björkum AA, Gaus SE, Lu J, Scammell TE, Saper CB (2002) Afferents to the ventrolateral preoptic nucleus. J Neurosci 22:977–990
37. Clark FM, Proudfit HK (1991) The projection of noradrenergic neurons in the A7 catecholamine cell group to the spinal cord in the rat demonstrated by anterograde tracing combined with immunocytochemistry. Brain Res 547:279–288
38. Clark FM, Proudfit HK (1991) The projection of locus coeruleus neurons to the spinal cord in the rat determined by anterograde tracing combined with immunocytochemistry. Brain Res 538:231–245
39. Clark FM, Proudfit HK (1992) Anatomical evidence for genetic differences in the innervation of the rat spinal cord by noradrenergic locus coeruleus neurons. Brain Res 591:44–53
40. Clark FM, Proudfit HK (1993) The projections of noradrenergic neurons in the A5 catecholamine cell group to the spinal cord in the rat: anatomical evidence that A5 neurons modulate nociception. Brain Res 616:200–210
41. Conrad LC, Pfaff DW (1976) Efferents from medial basal forebrain and hypothalamus in the rat. II. An autoradiographic study of the anterior hypothalamus. J Comp Neurol 169:221–261
42. Craig AD (1995) Distribution of brainstem projections from spinal lamina I neurons in the cat and the monkey. J Comp Neurol 361:225–248
43. Dahlström A, Fuxe K (1964) Localization of monoamines in the lower brain stem. Experientia 20:398–399
44. Dahlström A, Fuxe K (1964) Evidence for the existence of monoamine-containing neurons in the central nervous system. I. Demonstration of monoamines in the cell bodies of the brain stem neurons. Acta Physiol Scand [Suppl] 62 (247):1–55
45. Dampney RA (1994) Functional organization of central pathways regulating the cardiovascular system. Physiol Rev 74:323–364
46. Datiche F, Luppi PH, Cattarelli M (1995) Serotonergic and non-serotonergic projections from the raphe nuclei to the piriform cortex in the rat: a cholera toxin B subunit (CTb) and 5-HT immunohistochemical study. Brain Res 671:27–37
47. Descarries L, Mechawar N (2000) Ultrastructural evidence for diffuse transmission by monoamine and acetylcholine neurons of the central nervous system. Prog Brain Res 125:27–47
48. Ennis M, Aston-Jones G (1989) GABA-mediated inhibition of locus coeruleus from the dorsomedial rostral medulla. J Neurosci 9:2973–2981
49. Espana RA, Reis KM, Valentino RJ, Berridge CW (2005) Organization of hypocretin/orexin efferents to locus coeruleus and basal forebrain arousal-related structures. J Comp Neurol 481:160–178
50. Forel A (1877) Untersuchungen über die Haubenregion und ihre oberen Verknüpfungen im Gehirne des Menschen und einiger Säugethiere, mit Beiträgen zu Methoden der Gehirnuntersuchung. Arch f Psychiat u Nervenkrankh 7:392–495

51. Fort P, Luppi PH, Wenthold R, Jouvet M (1990) Glycine immunoreactive neurons in the medulla oblongata in cats. C R Acad Sci III 311:205–212
52. Fritschy JM, Grzanna R (1989) Immunohistochemical analysis of the neurotoxic effects of DSP-4 identifies two populations of noradrenergic axon terminals. Neuroscience 30:181–197
53. Fritschy JM, Grzanna R (1990) Demonstration of two separate descending noradrenergic pathways to the rat spinal cord: evidence for an intragriseal trajectory of locus coeruleus axons in the superficial layers of the dorsal horn. J Comp Neurol 291:553–582
54. Fritschy JM, Grzanna R (1990) Distribution of locus coeruleus axons within the rat brainstem demonstrated by Phaseolus vulgaris leucoagglutinin anterograde tracing in combination with dopamine-beta-hydroxylase immunofluorescence. J Comp Neurol 293:616–631
55. Fritschy JM, Lyons WE, Mullen CA, Kosofsky BE, Molliver ME, Grzanna R (1987) Distribution of locus coeruleus axons in the rat spinal cord: a combined anterograde transport and immunohistochemical study. Brain Res 437:176–180
56. Gallager DW, Pert A (1978) Afferents to brain stem nuclei (brain stem raphe, nucleus reticularis pontis caudalis and nucleus gigantocellularis) in the rat as demonstrated by microiontophoretically applied horseradish peroxidase. Brain Res 144:257–275
57. Gang S, Nakazono Y, Aoki M, Aoki N (1993) Differential projections to the raphe nuclei from the medial parabrachial-Kolliker-Fuse (NPBM-KF) nuclear complex and the retrofacial nucleus in cats: retrograde WGA-HRP tracing. J Auton Nerv Syst 45:241–244
58. Garver DL, Sladek JR Jr (1975) Monoamine distribution in primate brain. I. Catecholamine-containing perikarya in the brain stem of Macaca speciosa. J Comp Neurol 159:289–304
59. Gonzalo-Ruiz A, Lieberman AR, Sanchez-Anquela JM (1995) Organization of serotoninergic projections from the raphe nuclei to the anterior thalamic nuclei in the rat: a combined retrograde tracing and 5-HT immunohistochemical study. J Chem Neuroanat 8:103–115
60. Groenewegen HJ, Ahlenius S, Naber SN, Kowali NW, Nauta WJ (1986) Cytoarchitecture, fiber connections, and some histochemical aspects of the interpeduncular nucleus of the cat. J Comp Neurol 249:65–102
61. Grzanna R, Chee WK, Akeyson EW (1987) Noradrenergic projections to brainstem nuclei: evidence for differential projections from noradrenergic subgroups. J Comp Neurol 263:76–91
62. Grzanna R, Berger U, Fritschy JM, Geffard M (1989) Acute action of DSP-4 on central norepinephrine axons: biochemical and immunohistochemical evidence for differential effects. J Histochem Cytochem 37:1435–1442
63. Guyenet PG, Koshiya N, Huangfu D, Verberne AJ, Riley TA (1993) Central respiratory control of A5 and A6 pontine noradrenergic neurons. Am J Physiol 264:R1035–1044
64. Habaguchi T, Takakusaki K, Saitoh K, Sugimoto J, Sakamoto T (2002) Medullary reticulospinal tract mediating the generalized motor inhibition in cats: II. Functional organization within the medullary reticular formation with respect to postsynaptic inhibition of forelimb and hindlimb motoneurons. Neuroscience 113:65–77
65. Hajnik T, Lai YY, Siegel JM (2000) Atonia-related regions in the rodent pons and medulla. J Neurophysiol 84:1942–1948
66. Hajos M, Richards CD, Szekely AD, Sharp T (1998) An electrophysiological and neuroanatomical study of the medial prefrontal cortical projection to the midbrain raphe nuclei in the rat. Neuroscience 87:95–108
67. Halberstadt AL, Balaban CD (2003) Organization of projections from the raphe nuclei to the vestibular nuclei in rats. Neuroscience 120:573–594
68. Halberstadt AL, Balaban CD (2006) Anterograde tracing of projections from the dorsal raphe nucleus to the vestibular nuclei. Neuroscience 143: 641–654
69. Halberstadt AL, Balaban CD (2006) Serotonergic and nonserotonergic neurons in the dorsal raphe nucleus send collateralized projections to both the vestibular nuclei and the central amygdaloid nucleus. Neuroscience 140:1067–1077
70. Herbert H (1992) Evidence for projections from medullary nuclei onto serotonergic and dopaminergic neurons in the midbrain dorsal raphe nucleus of the rat. Cell Tissue Res 270:149–156
71. Herkenham M, Nauta WJ (1979) Efferent connections of the habenular nuclei in the rat. J Comp Neurol 187:19–47
72. Hermann DM, Luppi PH, Peyron C, Hinckel P, Jouvet M (1997) Afferent projections to the rat nuclei raphe magnus, raphe pallidus and reticularis gigantocellularis pars alpha demonstrated by iontophoretic application of choleratoxin (subunit b). J Chem Neuroanat 13:1–21
73. Hilaire G, Viemari JC, Coulon P, Simonneau M, Bevengut M (2004) Modulation of the respiratory rhythm generator by the pontine noradrenergic A5 and A6 groups in rodents. Respir Physiol Neurobiol 143:187–197
74. Hirota K, Kushikata T (2001) Central noradrenergic neurones and the mechanism of general anaesthesia. Br J Anaesth 87:811–813
75. Hökfelt T, Fuxe K, Johansson O (1974) Immunochemical evidence for the existence of adrenaline neurons in the rat brain. Brain Res 66:235–251
76. Hökfelt T, Matensson R, Björklund A, Kleinau S, Goldstein M (1984) Distributional maps of tyrosine-hydroxylase-immunoreactive neurons in the rat brain. In: Björklund A, Hökfelt T (eds) Classical transmitters in the CNS. Elsevier, Amsterdam, pp 277–379 (Handbook of chemical neuroanatomy)
77. Holstege G (1991) Descending motor pathways and the spinal motor system: limbic and non-limbic components. Prog Brain Res 87:307–421

78. Holstege G, Kuypers HG (1982) The anatomy of brain stem pathways to the spinal cord in cat. A labeled amino acid tracing study. Prog Brain Res 57:145–175
79. Holstege G, Griffiths D, de Wall H, Dalm E (1986) Anatomical and physiological observations on supraspinal control of bladder and urethral sphincter muscles in the cat. J Comp Neurol 250:449–461
80. Holstege JC (1996) The ventro-medial medullary projections to spinal motoneurons. Ultrastructure, transmitters and functional aspects. Prog Brain Res 107:159–181
81. Holstege JC, Kuypers HG (1982) Brain stem projections to spinal motoneuronal cell groups in rat studied by means of electron microscopy autoradiography. Prog Brain Res 57:177–183
82. Holstege JC, Kuypers HG (1987) Brainstem projections to lumbar motoneurons in rat. I. An ultrastructural study using autoradiography and the combination of autoradiography and horseradish peroxidase histochemistry. Neuroscience 21:345–367
83. Holstege JC, Kuypers HG (1987) Brainstem projections to spinal motoneurons: an update. Neuroscience 23:809–821
84. Holstege JC, van Dijken H, Buijs RM, Godeknegt H, Gosens T, Bongers CM (1996) Distribution of dopamine immunoreactivity in the rat, cat and monkey spinal cord. J Comp Neurol 376:631–652
85. Honda T, Semba K (1994) Serotonergic synaptic input to cholinergic neurons in the rat mesopontine tegmentum. Brain Res 647:299–306
86. Hornung JP (2003) The human raphe nuclei and the serotonergic system. J Chem Neuroanat 26:331–343
87. Hornung JP (2004) Raphe nuclei. In: Paxinos G, Mai JK (eds) The human nervous system. Elsevier, Amsterdam, pp 425–450
88. Hosogai M, Matsuo S, Sibahara T, Kawai Y (1998) Projection of respiratory neurons in rat medullary raphe nuclei to the phrenic nucleus. Respir Physiol 112:37–50
89. Hosoya Y (1980) Hypothalamic projections to the ventral medulla oblongata in the rat, with special reference to the nucleus raphe pallidus: a study using autoradiographic and HRP techniques. Brain Res 344:338–350
90. Hosoya Y, Sugiura Y, Ito R, Kohno K (1990) Descending projections from the hypothalamic paraventricular nucleus to the A5 area, including the superior salivatory nucleus, in the rat. Exp Brain Res 82:513–518
91. Hubbard JE, Di Carlo V (1974) Fluorescence histochemistry of monoamine-containing cell bodies in the brain stem of the squirrel monkey (Saimiri sciureus). II. Catecholamine-containing groups. J Comp Neurol 153:369–384
92. Ikai Y, Takada M, Shinonaga Y, Mizuno N (1992) Dopaminergic and non-dopaminergic neurons in the ventral tegmental area of the rat project, respectively, to the cerebellar cortex and deep cerebellar nuclei. Neuroscience 51:719–728
93. Imai H, Steindler DA, Kitai ST (1986) The organization of divergent axonal projections from the midbrain raphe nuclei in the rat. J Comp Neurol 243:363–380
94. Jones BE (1990) Immunhistochemical study of choline acetyltransferase-immunoreactive processes and cells innervating the pontomedullary reticular formation in the rat. J Comp Neurol 295:485–514
95. Jones BE (1995) Reticular formation: cytoarchitecture, transmitters and projections. In: Paxinos G (ed) The rat nervous system. Academic, San Diego, pp 155–171
96. Jones BE, Friedman L (1983) Atlas of catecholamine perikarya, varicosities and pathways in the brainstem of the cat. J Comp Neurol 215:382–396
97. Jones SL, Light AR (1992) Serotoninergic medullary raphespinal projection to the lumbar spinal cord in the rat: a retrograde immunohistochemical study. J Comp Neurol 322:599–610
98. Kalen P, Wiklund L (1989) Projections from the medial septum and diagonal band of Broca to the dorsal and central superior raphe nuclei: a non-cholinergic pathway. Exp Brain Res 75:401–416
99. Kalen P, Skagerberg G, Lindvall O (1988) Projections from the ventral tegmental area and mesencephalic raphe to the dorsal raphe nucleus in the rat. Evidence for a minor dopaminergic component. Exp Brain Res 73:69–77
100. Kalia M, Fuxe K, Goldstein M, Harfstrand A, Agnati LF, Coyle JT (1984) Evidence for the existence of putative dopamine-, adrenaline- and noradrenaline-containing vagal motor neurons in the brainstem of the rat. Neurosci Lett 50:57–62
101. Kazakov VN, Kravtsov P, Krakhotkina ED, Maisky VA (1993) Sources of cortical, hypothalamic and spinal serotonergic projections: topical organization within the nucleus raphe dorsalis. Neuroscience 56:157–164
102. Kerr CW, Bishop GA (1991) Topographical organization in the origin of serotonergic projections to different regions of the cat cerebellar cortex. J Comp Neurol 308:502–515
103. Kievit J, Kuypers HG (1975) Subcortical afferents to the frontal lobe in the rhesus monkey studied by means of retrograde horseradish peroxidase transport. Brain Res 85:261–266
104. Kohler C, Steinbusch H (1982) Identification of serotonin and non-serotonin-containing neurons of the mid-brain raphe projecting to the entorhinal area and the hippocampal formation. A combined immunohistochemical and fluorescent retrograde tracing study in the rat brain. Neuroscience 7:951–975
105. Kohler C, Chan-Palay V, Steinbusch H (1982) The distribution and origin of serotonin-containing fibers in the septal area: a combined immunohistochemical and fluorescent retrograde tracing study in the rat. J Comp Neurol 209:91–111
106. Kojima M, Sano Y (1983) The organization of serotonin fibers in the anterior column of the mam-

malian spinal cord. An immunohistochemical study. Anat Embryol (Berl) 167:1–11
107. Kojima M, Matsuura T, Tanaka A, Amagai T, Imanishi J, Sano Y (1985) Characteristic distribution of noradrenergic terminals on the anterior horn motoneurons innervating the perineal striated muscles in the rat. An immuno-electromicroscopic study. Anat Embryol (Berl) 171:267–273
108. Koshiya N, Guyenet PG (1994) A5 noradrenergic neurons and the carotid sympathetic chemoreflex. Am J Physiol 267:R519–526
109. Koutcherov Y, Huang X-F, Halliday G, Paxinos G (2004) Organization of human brain stem nuclei. In: Paxinos G, Mai JK (eds) The human nervous system. Elsevier, Amsterdam, pp 267–319
110. Kromer LF, Moore RY (1976) Cochlear nucleus innervation by central norepinephrine neurons in the rat. Brain Res 118:531–537
111. Kromer LF, Moore RY (1980) Norepinephrine innervation of the cochlear nuclei by locus coeruleus neurons in the rat. Anat Embryol (Berl) 158:227–244
112. Kubota Y, Leung E, Vincent SR (1992) Ultrastructure of cholinergic neurons in the laterodorsal tegmental nucleus of the rat: interaction with catecholamine fibers. Brain Res Bull 29:479–491
113. Leanza G, Perez S, Pellitteri R, Russo A, Stanzani S (1995) Branching serotonergic and non-serotonergic projections from caudal brainstem to the medial preoptic area and the lumbar spinal cord in the rat. Neurosci Lett 200:5–8
114. Leite-Almeida H, Valle-Fernandes A, Almeida A (2006) Brain projections from the medullary dorsal reticular nucleus: an anterograde and retrograde tracing study in the rat. Neuroscience 140:577–595
115. Levey AI (1983) Central cholinergic pathways in the rat: an overview based on an alternative nomenclature (Ch1–Ch6). Neuroscience 10:1185–1201
116. Levitt P, Moore RY (1979) Origin and organization of brainstem catecholamine innervation in the rat. J Comp Neurol 186:505–528
117. Li YQ, Takada M, Mizuno N (1993) The sites of origin of serotoninergic afferent fibers in the trigeminal motor, facial, and hypoglossal nuclei in the rat. Neurosci Res 17:307–313
118. Lima D, Mendes-Ribeiro JA, Coimbra A (1991) The spino-latero-reticular system of the rat: projections from the superficial dorsal horn and structural characterization of marginal neurons involved. Neuroscience 45:137–152
119. Loewy AD, McKellar S (1981) Serotonergic projections from the ventral medulla to the intermediolateral cell column in the rat. Brain Res 211:146–152
120. Loewy AD, McKellar S, Saper CB (1979) Direct projections from the A5 catecholamine cell group to the intermediolateral cell column. Brain Res 174:309–314
121. Loewy AD, Wallach JH, McKellar S (1981) Efferent connections of the ventral medulla oblongata in the rat. Brain Res 228:63–80
122. Luppi PH, Gervasoni D, Boissard R, Verret L, Goutagny R, Peyron C, Salvert D, Leger L, Barbagli B, Fort P (2004) Brainstem structures responsible for paradoxical sleep onset and maintenance. Arch Ital Biol 142:397–411
123. Lyons WE, Grzanna R (1988) Noradrenergic neurons with divergent projections to the motor trigeminal nucleus and the spinal cord: a double retrograde neuronal labeling study. Neuroscience 26:681–693
124. Lyons WE, Fritschy JM, Grzanna R (1989) The noradrenergic neurotoxin DSP-4 eliminates the coeruleospinal projection but spares projections of the A5 and A7 groups to the ventral horn of the rat spinal cord. J Neurosci 9:1481–1489
125. Mamounas LA, Mullen CA, O'Hearn E, Molliver ME (1991) Dual serotoninergic projections to forebrain in the rat: morphologically distinct 5-HT axon terminals exhibit differential vulnerability to neurotoxic amphetamine derivatives. J Comp Neurol 314:558–586
126. Manaker S, Fogarty PF (1995) Raphespinal and reticulospinal neurons project to the dorsal vagal complex in the rat. Exp Brain Res 106:79–92
127. Marlier L, Sandillon F, Poulat P, Rajaofetra N, Geffard M, Privat A (1991) Serotonergic innervation of the dorsal horn of rat spinal cord: light and electron microscopic immunocytochemical study. J Neurocytol 20:310–322
128. Mason P (2005) Ventromedial medulla: pain modulation and beyond. J Comp Neurol 493:2–8
129. Matsuyama K, Kobayashi Y, Takakusaki K, Mori S, Kimura H (1993) Termination mode and branching patterns of reticuloreticular and reticulospinal fibers of the nucleus reticularis pontis oralis in the cat: an anterograde PHA-L tracing study. Neurosci Res 17:9–21
130. Matsuyama K, Takakusaki K, Nakajima K, Mori S (1997) Multi-segmental innervation of single pontine reticulospinal axons in the cervico-thoracic region of the cat: anterograde PHA-L tracing study. J Comp Neurol 377:234–250
131. Matsuzaki S, Takada M, Li YQ, Tokuno H, Mizuno N (1993) Serotoninergic projections from the dorsal raphe nucleus to the nucleus submedius in the rat and cat. Neuroscience 55:403–416
132. Meessen H, Olszewski J (1949) A cytoarchitectonic atlas of the rhombencephalon of the rabbit. Karger, Basel
133. Mesulam MM, Mufson EJ, Wainer BH, Levey AI (1983) Central cholinergic pathways in the rat: an overview based on an alternative nomenclature (Ch1–Ch6). Neuroscience 10:1185–1201
134. Mesulam MM, Mufson EJ, Levey AI, Wainer BH (1984) Atlas of cholinergic neurons in the forebrain and upper brainstem of the macaque based on monoclonal choline acetyltransferase immuno-

histochemistry and acetylcholinesterase histochemistry. Neuroscience 12:669–686
135. Morgane PJ, Jacobs MS (1979) Raphe projections to the locus coeruleus in the rat. Brain Res Bull 4:519–534
136. O'Hearn E, Molliver ME (1984) Organization of raphe-cortical projections in rat: a quantitative retrograde study. Brain Res Bull 13:709–726
137. Olschowka JA, Molliver ME, Grzanna R, Rice FL, Coyle JT (1981) Ultrastructural demonstration of noradrenergic synapses in the rat central nervous system by dopamine-beta-hydroxylase immunocytochemistry. J Histochem Cytochem 29:271–280
138. Pan B, Castro-Lopes JM, Coimbra A (1999) Central afferent pathways conveying nociceptive input to the hypothalamic paraventricular nucleus as revealed by a combination of retrograde labeling and c-fos activation. J Comp Neurol 413:129–145
139. Panagopoulos NT, Papadopoulos GC, Matsokis NA (1991) Dopaminergic innervation and binding in the rat cerebellum. Neurosci Lett 130:208–212
140. Paxinos G, Watson C (1986) The rat brain in stereotactic coordinates, 2nd edn. Academic, San Diego
141. Pearson J, Halliday G, Sakamoto N, Michel P (1990) Catecholaminergic neurons. In: Paxinos G (ed) The human nervous system. Academic, San Diego, pp 1023–1049
142. Peschanski M, Besson JM (1984) Coerulear projections to the lateral diencephalon in the rat. An anatomical study using wheat-germ agglutinin conjugated to HRP. Neurosci Lett 46:329–334
143. Petit JM, Luppi PH, Peyron C, Rampon C, Jouvet M (1995) VIP-like immunoreactive projections from the dorsal raphe and caudal linear raphe nuclei to the bed nucleus of the stria terminalis demonstrated by a double immunohistochemical method in the rat. Neurosci Lett 193:77–80
144. Petrov T, Krukoff TL, Jhamandas JH (1993) Branching projections of catecholaminergic brainstem neurons to the paraventricular hypothalamic nucleus and the central nucleus of the amygdala in the rat. Brain Res 609:81–92
145. Peyron C, Luppi PH, Kitahama K, Fort P, Hermann DM, Jouvet M (1995) Origin of the dopaminergic innervation of the rat dorsal raphe nucleus. Neuroreport 6:2527–2531
146. Peyron C, Petit JM, Rampon C, Jouvet M, Luppi PH (1998) Forebrain afferents to the rat dorsal raphe nucleus demonstrated by retrograde and anterograde tracing methods. Neuroscience 82:443–468
147. Porrino LJ, Goldman-Rakic PS (1982) Brainstem innervation of prefrontal and anterior cingulate cortex in the rhesus monkey revealed by retrograde transport of HRP. J Comp Neurol 205:63–76
148. Proudfit HK, Clark FM (1991) The projections of locus coeruleus neurons to the spinal cord. Prog Brain Res 88:123–141
149. Rajaofetra N, Ridet JL, Poulat P, Marlier L, Sandillon F, Geffard M, Privat A (1992) Immunocytochemical mapping of noradrenergic projections to the rat spinal cord with an antiserum against noradrenaline. J Neurocytol 21:481–491
150. Rajaofetra N, Passagia JG, Marlier L, Poulat P, Pellas F, Sandillon F, Verschuere B, Gouy D, Geffard M, Privat A (1992) Serotoninergic, noradrenergic, and peptidergic innervation of Onuf's nucleus of normal and transected spinal cords of baboons (Papio papio). J Comp Neurol 318:1–17
151. Rampon C, Peyron C, Gervasoni D, Pow DV, Luppi PH, Fort P (1999) Origins of the glycinergic inputs to the rat locus coeruleus and dorsal raphe nuclei: a study combining retrograde tracing with glycine immunohistochemistry. Eur J Neurosci 11:1058–1066
152. Rawson JA, Scott CJ, Pereira A, Jakubowska A, Clarke IJ (2001) Noradrenergic projections from the A1 field to the preoptic area in the brain of the ewe and Fos responses to oestrogen in the A1 cells. J Neuroendocrinol 13:129–138
153. Reyes BA, Van Bockstaele EJ (2006) Divergent projections of catecholaminergic neurons in the nucleus of the solitary tract to limbic forebrain and medullary autonomic brain regions. Brain Res 1117:69–79
154. Ridet JL, Sandillon F, Rajaofetra N, Geffard M, Privat A (1992) Spinal dopaminergic system of the rat: light and electron microscopic study using an antiserum against dopamine, with particular emphasis on synaptic incidence. Brain Res 598:233–241
155. Ridet JL, Rajaofetra N, Teilhac JR, Geffard M, Privat A (1993) Evidence for nonsynaptic serotonergic and noradrenergic innervation of the rat dorsal horn and possible involvement of neuron-glia interactions. Neuroscience 52:143–157
156. Roller FCW (1881) Ein kleinzelliger Hypoglossuskern. Arch f Mikr Anat 19:383–395
157. Sakai K, Salvert D, Touret M, Jouvet M (1977) Afferent connections of the nucleus raphe dorsalis in the cat as visualized by the horseradish peroxidase technique. Brain Res 137:11–35
158. Saper CB (1990) Cholinergic system. In: Paxinos G (ed) The human nervous system. Academic, San Diego, pp 1095–1103
159. Saper CB, Swanson LW, Cowan WM (1979) An autoradiographic study of the efferent connections of the lateral hypothalamic area in the rat. J Comp Neurol 183:689–706
160. Sawchenko PE, Swanson LW (1982) The organization of noradrenergic pathways from the brainstem to the paraventricular and supraoptic nuclei in the rat. Brain Res 257:275–325
161. Scheibel ME, Schebel AB (1958) Structural substrates for integrative patterns in the brain stem reticular core. In: Jasper HH, Procter LD (eds) Reticular formation of the brain. Little Brown, Boston, pp 31–55
162. Schreihofer AM, Guyenet PG (1997) Identification of C1 presympathetic neurons in rat rostral ventrolateral medulla by juxtacellular labeling in vivo. J Comp Neurol 387:524–536

163. Schreihofer AM, Guyenet PG (2002) The baroreflex and beyond: control of sympathetic vasomotor tone by GABAergic neurons in the ventrolateral medulla. Clin Exp Pharmacol Physiol 29: 514–521
164. Schreihofer AM, Stornetta RL, Guyenet PG (1999) Evidence for glycinergic respiratory neurons: Botzinger neurons express mRNA for glycinergic transporter 2. J Comp Neurol 407:583–597
165. Schreihofer AM, Stornetta RL, Guyenet PG (2000) Regulation of sympathetic tone and arterial pressure by rostral ventrolateral medulla after depletion of C1 cells in rat. J Physiol 529 Pt 1:221–236
166. Semba K (2004) Aminergic and cholinergic afferents to REM sleep induction regions of the pontine reticular formation in the rat. J Comp Neurol 330:543–556
167. Shinnar S, Maciewicz RJ, Shofer RJ (1975) A raphe projection to cat cerebellar cortex. Brain Res 97:139–143
168. Shute CC, Lewis PR (1967) The ascending cholinergic reticular system: neocortical, olfactory and subcortical projections. Brain 90:497–520
169. Skagerberg G, Bjorklund A (1985) Topographic principles in the spinal projections of serotonergic and non-serotonergic brainstem neurons in the rat. Neuroscience 15:445–480
170. Skagerberg G, Lindvall O (1985) Organization of diencephalic dopamine neurones projecting to the spinal cord in the rat. Brain Res 342:340–351
171. Skagerberg G, Bjorklund A, Lindvall O, Schmidt RH (1982) Origin and termination of the diencephalo-spinal dopamine system in the rat. Brain Res Bull 9:237–244
172. Sluka KA, Westlund KN (1992) Spinal projections of the locus coeruleus and the nucleus subcoeruleus in the Harlan and the Sasco Sprague-Dawley rat. Brain Res 579:67–73
173. Smith JC, Ellenberger HH, Ballanyi K, Richter DW, Feldman JL (1991) Pre-Botzinger complex: a brainstem region that may generate respiratory rhythm in mammals. Science 254:726–729
174. Snowball RK, Dampney RA, Lumb BM (1997) Responses of neurones in the medullary raphe nuclei to inputs from visceral nociceptors and the ventrolateral periaqueductal grey in the rat. Exp Physiol 82:485–500
175. Steinbusch HW, van der Kooy D, Verhofstad AA, Pellegrino A (1980) Serotonergic and non-serotonergic projections from the nucleus raphe dorsalis to the caudate-putamen complex in the rat, studied by a combined immunofluorescence and fluorescent retrograde axonal labeling technique. Neurosci Lett 19:137–142
176. Steininger TL, Wainer BH, Blakely RD, Rye DB (1997) Serotonergic dorsal raphe nucleus projections to the cholinergic and noncholinergic neurons of the pedunculopontine tegmental region: a light and electron microscopic anterograde tracing and immunohistochemical study. J Comp Neurol 382:302–322
177. Steriade M, McVarley RW (1990) Brainstem control of wakefullness and sleep. Plenum, New York
178. Sukhotinsky I, Reiner K, Govrin-Lippmann R, Belenky M, Lu J, Hopkins DA, Saper CB, Devor M (2006) Projections from the mesopontine tegmental anesthesia area to regions involved in pain modulation. J Chem Neuroanat 32:159–178
179. Takahashi Y, Satoh K, Sakumoto T, Toyama M, Shimazu M (1979) A major source of catecholaminergic terminals in the nucleus tractus solitarii. Brain Res: 372–377
180. Takakusaki K, Kohyama J, Matsuyama K, Mori S (2001) Medullary reticulospinal tract mediating the generalized motor inhibition in cats: parallel inhibitory mechanisms acting on motoneurons and on interneuronal transmission in reflex pathways. Neuroscience 103:511–527
181. Takakusaki K, Habaguchi T, Saitoh K, Kohyama J (2004) Changes in the excitability of hindlimb motoneurons during muscular atonia induced by stimulating the pedunculopontine tegmental nucleus in cats. Neuroscience 124:467–480
182. Tallaksen-Greene SJ, Elde R, Wessendorf MW (1993) Regional distribution of serotonin and substance P co-existing in nerve fibers and terminals in the brainstem of the rat. Neuroscience 53:1127–1142
183. Tavares I, Lima D, Coimbra A (1996) The ventrolateral medulla of the rat is connected with the spinal cord dorsal horn by an indirect descending pathway relayed in the A5 noradrenergic cell group. J Comp Neurol 374:84–95
184. Tavares I, Lima D, Coimbra A (1997) The pontine A5 noradrenergic cells which project to the spinal cord dorsal horn are reciprocally connected with the caudal ventrolateral medulla in the rat. Eur J Neurosci 9:2452–2461
185. Ter Horst GJ, Toes GJ, Van Willigen JD (1991) Locus coeruleus projection to the dorsal motor vagus nucleus in the rat. Neuroscience 45:153–160
186. Thompson AM, Moore KR, Thompson GC (1995) Distribution and origin of serotoninergic afferents to guinea pig cochlear nucleus. J Comp Neurol 351:104–116
187. Tigges J, Tigges M, Cross NA, McBride RL, Letbetter WD, Anschel S (1982) Subcortical structures projecting to visual cortical areas in squirrel monkey. J Comp Neurol 209:29–40
188. Törk J, Hornung JP (1990) Raphe nuclei and the serotoninergic system. In: Paxinos G (ed) The human nervous system. Academic, San Diego, pp 1001–1022
189. Tsai C (1925) The optic tracts and centers of the opossum Didelphis viginiana. J Comp Neurol 39:173–216
190. Valverde F (1961) Reticular formation of the pons and the medulla oblongata. A Golgi study. J Comp Neurol 116:71–99
191. Van Bockstaele EJ, Chan J (1996) Electronmicroscopic evidence for coexistence of leucine5-enkephalin and gamma-aminobutyric acid in a sub-

population of axon terminals in the rat locus coeruleus region. Brain Res 746:171–182
192. Van Bockstaele EJ, Pieribone VA, Aston-Jones G (1989) Diverse afferents converge on the nucleus paragigantocellularis in the rat ventrolateral medulla: retrograde and anterograde tracing studies. J Comp Neurol 290:561–584
193. Van Bockstaele EJ, Biswas A, Pickel VM (1993) Topography of serotonin neurons in the dorsal raphe nucleus that send axon collaterals to the rat prefrontal cortex and nucleus accumbens. Brain Res 624:188–198
194. Van Dijken H (2000) The dopaminergic innervation of the brain stem and the spinal cord. Thesis, University of Rotterdam
195. Vertes RP, Martin GF (1988) Autoradiographic analysis of ascending projections from the pontine and mesencephalic reticular formation and the median raphe nucleus in the rat. J Comp Neurol 275:511–541
196. Villar MJ, Vitale ML, Parisi MN (1987) Dorsal raphe serotonergic projection to the retina. A combined peroxidase tracing-neurochemical/high-performance liquid chromatography study in the rat. Neuroscience 22:681–686
197. Villar MJ, Vitale ML, Hokfelt T, Verhofstad AA (1988) Dorsal raphe serotoninergic branching neurons projecting both to the lateral geniculate body and superior colliculus: a combined retrograde tracing-immunohistochemical study in the rat. J Comp Neurol 277:126–140
198. Voogd J, Jaarsma D, Marani E (1996) The cerebellum. In: Hökfelt T (ed) Chemoarchitecture and anatomy. Elsevier, Amsterdam, pp 1–369 (Handbook of chemical neuroanatomy)
199. Walker JJ, Bishop GA, Ho RH, King JS (1988) Brainstem origin of serotonin- and enkephalin-immunoreactive afferents to the opossum's cerebellum. J Comp Neurol 276:481–497
200. Waterhouse BD, Mihailoff GA, Baack JC, Woodward DJ (1986) Topographical distribution of dorsal and median raphe neurons projecting to motor, sensorimotor, and visual cortical areas in the rat. J Comp Neurol 249:460–476, 478–481
201. Waterhouse BD, Border B, Wahl L, Mihailoff GA (1993) Topographic organization of rat locus coeruleus and dorsal raphe nuclei: distribution of cells projecting to visual system structures. J Comp Neurol 336:345–361
202. Westlund KN, Coulter JD (1980) Descending projections of the locus coeruleus and subcoeruleus/medial parabrachial nuclei in monkey: axonal transport studies and dopamine-beta-hydroxylase immunocytochemistry. Brain Res 2:235–264
203. Westlund KN, Craig AD (1996) Association of spinal lamina I projections with brainstem catecholamine neurons in the monkey. Exp Brain Res 110:151–162
204. Westlund KN, Bowker RM, Ziegler MG, Coulter JD (1981) Origins of spinal noradrenergic pathways demonstrated by retrograde transport of antibody to dopamine-beta-hydroxylase. Neurosci Lett 25:243–249
205. Westlund KN, Bowker RM, Ziegler MG, Coulter JD (1982) Descending noradrenergic projections and their spinal terminations. Prog Brain Res 57:219–238
206. Westlund KN, Bowker RM, Ziegler MG, Coulter JD (1983) Noradrenergic projections to the spinal cord of the rat. Brain Res 263:15–31
207. Westlund KN, Bowker RM, Ziegler MG, Coulter JD (1984) Origins and terminations of descending noradrenergic projections to the spinal cord of monkey. Brain Res 292:1–16
208. Westlund KN, Sorkin LS, Ferrington DG, Carlton SM, Willcockson HH, Willis WD (1990) Serotoninergic and noradrenergic projections to the ventral posterolateral nucleus of the monkey thalamus. J Comp Neurol 295:197–207
209. White SR (1985) A comparison of the effects of serotonin, substance P and thyrotropin-rel hormone on excitability of rat spinal motoneuron in vivo. Brain Res 335:63–70
210. White SR (1985) Serotonin and co-localized peptides: effects on spinal motoneuron excitability. Peptides 6 [Suppl]:123–127
211. White SR, Neuman RS (1980) Facilitation of spinal motoneurone excitability by 5-hydroxytryptamine and noradrenaline. Brain Res 21:119–127
212. White SR, Neuman RS (1983) Pharmacological antagonism of facilitatory but not inhibitory effects of seronin and norepinephrine on excitability of spinal motoneurons. Neuropharmacology 22:489–494
213. Woolf NJ, Butcher LL (1985) Cholinergic systems in the rat brain: II. Projections to the interpeduncular nucleus. Brain Res Bull 14:63–83
214. Woolf NJ, Butcher LL (1986) Cholinergic systems in the rat brain: III. Projections from the pontomesencephalic tegmentum to the thalamus, tectum, basal ganglia, and basal forebrain. Brain Res Bull 16:603–637
215. Woulfe JM, Hrycyshyn AW, Flumerfelt BA (1988) Collateral axonal projections from the A1 noradrenergic cell group to the paraventricular nucleus and bed nucleus of the stria terminalis in the rat. Exp Neurol 102:121–124
216. Woulfe JM, Flumerfelt BA, Hrycyshyn AW (1990) Efferent connections of the A1 noradrenergic cell group: a DBH immunohistochemical and PHA-L anterograde tracing study. Exp Neurol 109:308–322
217. Wu W, Elde R, Wessendorf MW (1993) Organization of the serotonergic innervation of spinal neurons in rats – III. Differential serotonergic innervation of somatic and parasympathetic preganglionic motoneurons as determined by patterns of co-existing peptides. Neuroscience 55:223–233
218. Yezierski RP, Bowker RM, Kevetter GA, Westlund KN, Coulter JD, Willis WD (1982) Serotonergic projections to the caudal brain stem: a double la-

bel study using horseradish peroxidase and serotonin immunocytochemistry. Brain Res 239:258–264

219. Zanella S, Roux JC, Viemari JC, Hilaire G (2006) Possible modulation of the mouse respiratory rhythm generator by A1/C1 neurones. Respir Physiol Neurobiol 153:126–138

220. Zemlan FP, Behbehani MM, Beckstead RM (1984) Ascending and descending projections from nucleus reticularis magnocellularis and nucleus reticularis gigantocellularis: an autoradiographic and horseradish peroxidase study in the rat. Brain Res 292:207–220

23 Greater Limbic System

Historical Introduction: From Limbic Lobe to Limbic System

Most of the medial surface of the human cerebral hemispheres is occupied by the greatly expanded neocortex. However, on this side of the brain, a number of non-neocortical structures also exist. Two of these, the central olfactory system and the septum, occupy a superficial position, whereas two others, the amygdaloid complex and the hippocampal formation, lie largely hidden within the hemispheres. Due to the strong development of the neocortex and the related expansion of the temporal lobe, the main efferent systems of the amygdaloid complex and the hippocampal formation have been drawn out into long, arch-shaped bundles, the stria terminalis and the fornix, respectively (Figs. 23.1–23.4). The septum, the amygdala, the hippocampal formation and their efferent bundles constitute together the ring-shaped, medialmost zone of the hemisphere (Figs. 23.3, 23.5). This zone is largely flanked by a second zone, which in primates encompasses two convolutions, the cingulate and the parahippocampal gyri (Figs. 23.1, 23.3, 23.5). In 1878, Broca [17] designated these two gyri together as *le grand lobe limbique* ('the great limbic lobe'). He emphasized that he had introduced this term to denote an anatomical structure and not a functional unit. During the last few decades of the nineteenth and the first few decades of the twentieth century, it was generally believed that most, if not all of the structures included in Broca's limbic lobe are dominated by olfactory projections and thus form part of the rhinencephalon.

In 1937, Papez [99] published a notable paper in which he claimed on theoretical grounds that a circuit, of which the hippocampal formation and the cingulate gyrus form important components, constitutes the neural substrate of emotional behaviour. This theory received some substantiation from the work carried out by Klüver and Bucy [67, 68], who demonstrated that, in monkeys, resections of the anterior portions of the temporal lobes (which included the hippocampal formation and the amygdaloid complex) have, among other effects, a profound influence on affective responses. Somewhat schematically, it may be said that the impact of the publication by Papez and those by Klüver and Bucy was threefold: (1) the idea that the rhinencephalon encompasses almost the entire limbic lobe fell into the background, (2) a direct linkage between emotion and Broca's limbic lobe became established and (3) the amygdaloid complex, a subcortical structure, became incorporated into the limbic lobe.

MacLean [77–80] drew attention to the fact that the various components of Broca's great limbic lobe are strongly and reciprocally connected with a number of subcortical structures,

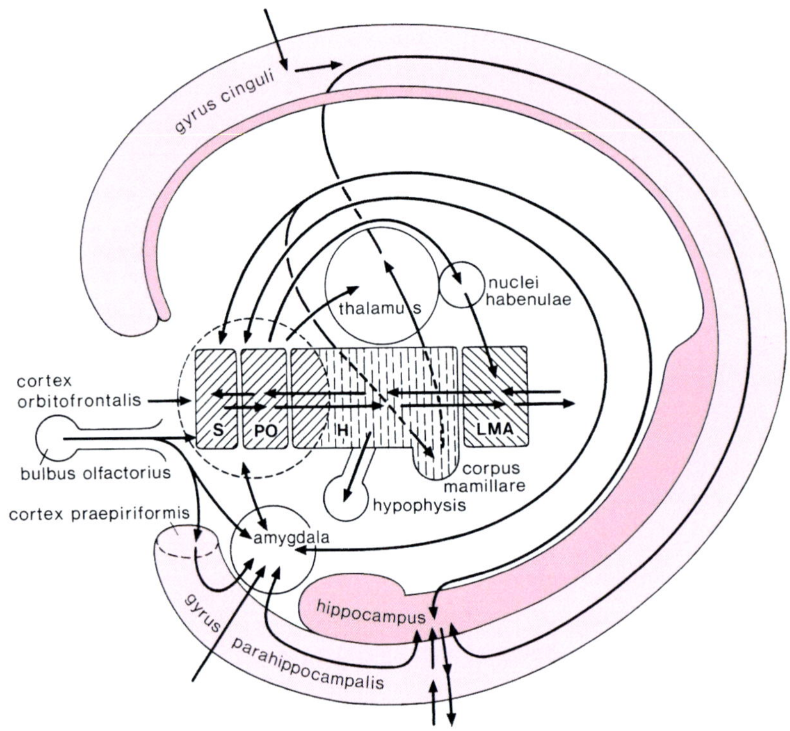

Fig. 23.1. Summary of the limbicohypothalamic complex. Subdivision of the area into central units and rings. *H*, hypothalamus; *LMA*, limbic midbrain area; *PO*, preoptic region; *S*, septum; inner (limbic) ring in *dark red*; outer (paralimbic) ring in *light red*

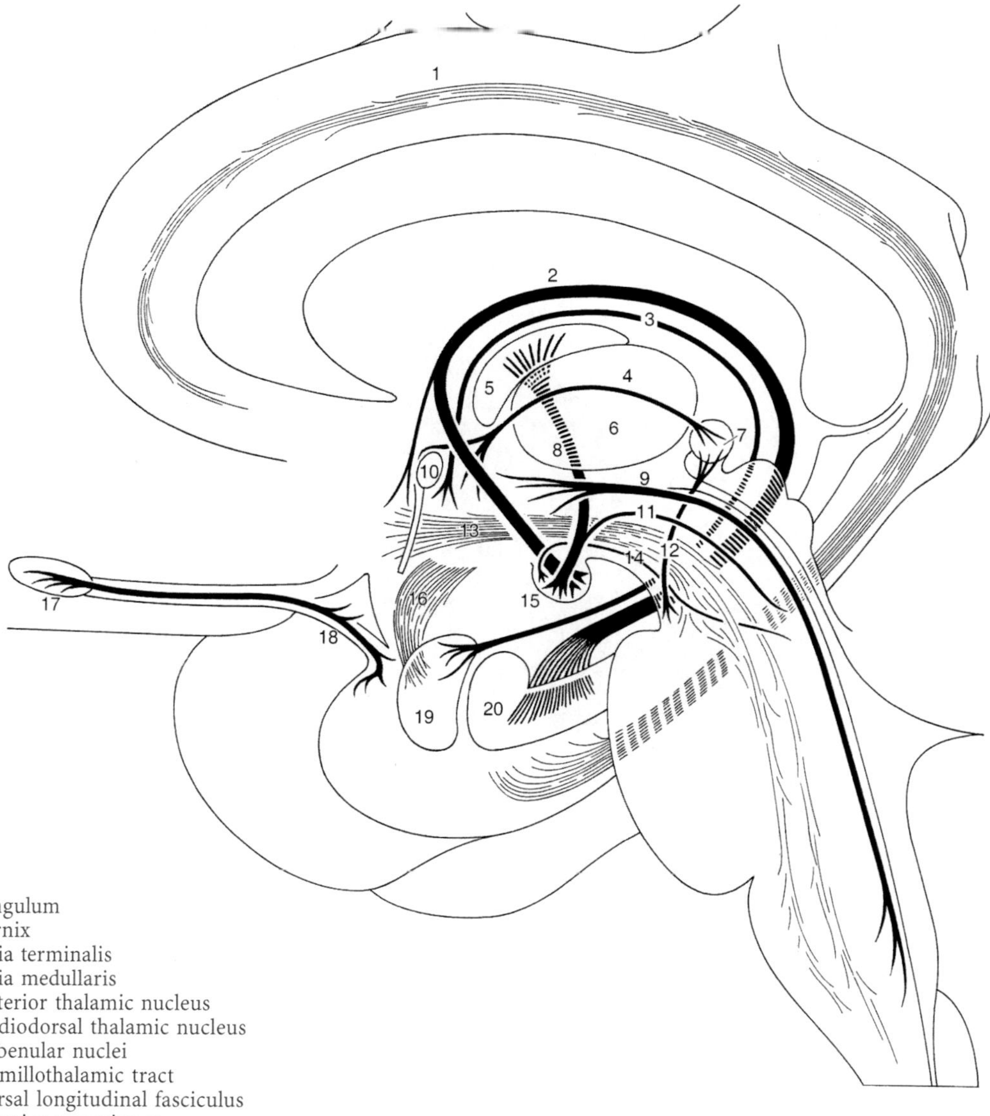

1 Cingulum
2 Fornix
3 Stria terminalis
4 Stria medullaris
5 Anterior thalamic nucleus
6 Mediodorsal thalamic nucleus
7 Habenular nuclei
8 Mamillothalamic tract
9 Dorsal longitudinal fasciculus
10 Anterior commissure
11 Mamillotegmental tract
12 Habenulo-interpeduncular tract
13 Medial forebrain bundle
14 Mamillary peduncle
15 Mamillary body
16 Ansa peduncularis
17 Olfactory bulb
18 Lateral olfactory stria
19 Amygdaloid complex
20 Hippocampus

Fig. 23.2. The major pathways of the limbic system and the rhinencephalon

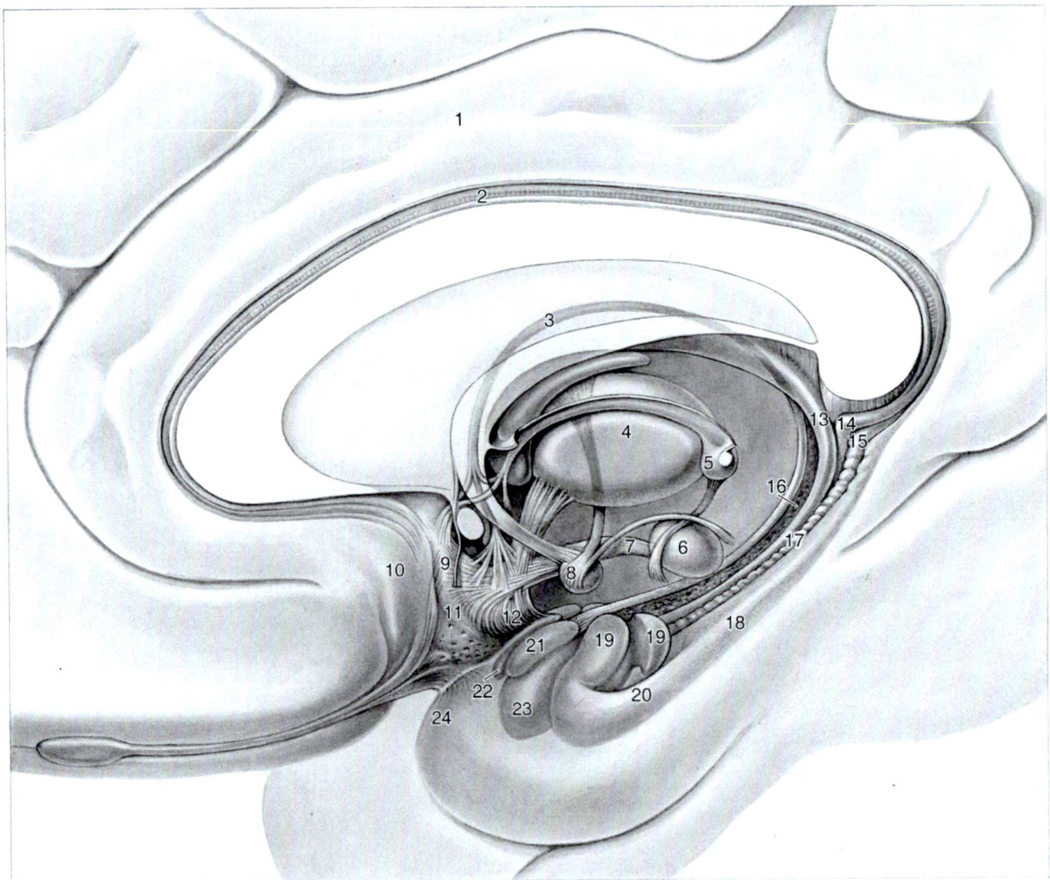

1 Cingulate gyrus
2 Indusium griseum
3 Stria terminalis
4 Mediodorsal thalamic nucleus
5 Habenular nuclei
6 Red nucleus
7 Medial forebrain bundle
8 Mamillary body
9 Septal region
10 Subcallosal area
11 Diagonal gyrus
12 Ventral amygdalofugal pathway
13 Crus of fornix
14 Fasciolar gyrus
15 Fasciola cinerea
16 Choriod fissure
17 Dentate gyrus
18 Subiculum
19 Ammon's horn
20 Site of limbus Giacomini
21 Cortical amygdaloid nucleus
22 Anterior amygdaloid nucleus
23 Basal + lateral amygdaloid nuclei
24 Prepiriform cortex

Fig. 23.3. The structure of the limbic and olfactory systems and some input-output pathways as seen in a medial view (3/2×). Some displacement of structures serves to bring other structures in view. The walls of the third ventricle and the brainstem have been omitted almost completely; of the thalamus only the anterior, mediodorsal and habenular nuclei are illustrated

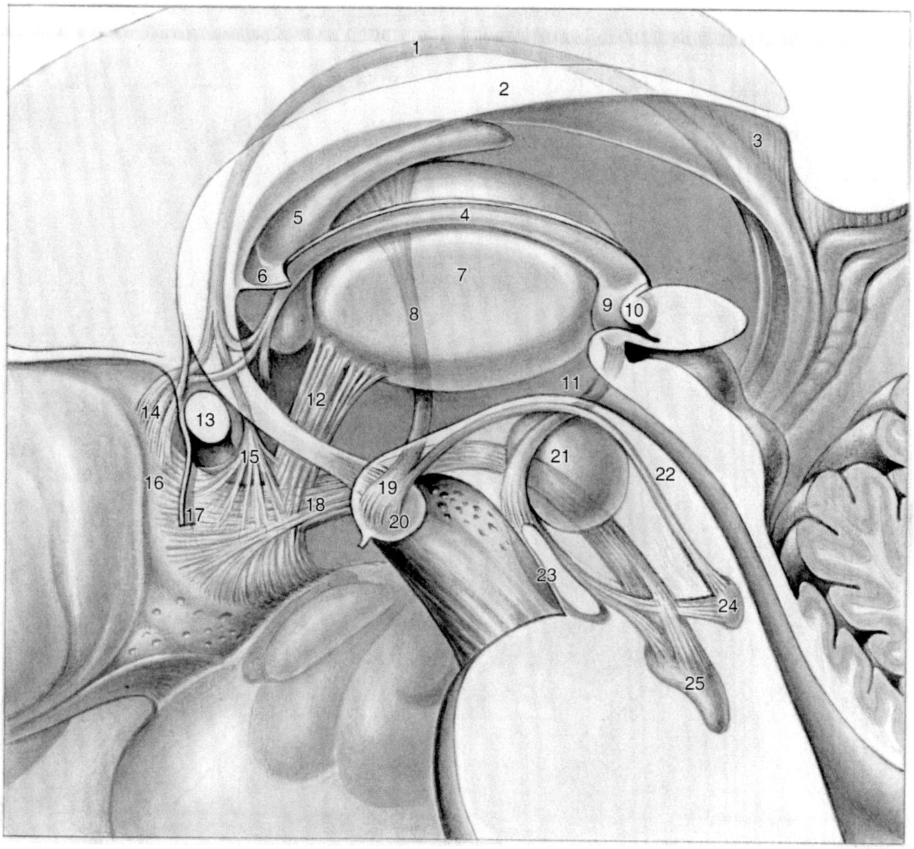

1 Stria terminalis
2 Fornix
3 Commissure of fornix
4 Stria medullaris
5 Anterior thalamic nucleus
6 Tela choroidea of third ventricle
7 Mediodorsal thalamic nucleus
8 Mamillothalamic tract
9 Habenular nuclei
10 Habenular commissure
11 Habenulo-interpeduncular tract
12 Inferior thalamic peduncle
13 Anterior commissure
14 Precommissural components of stria terminalis – stria medullaris fornix
15 Stria terminalis, postcommissural part
16 Septal region
17 Lamina terminalis
18 Medial forebrain bundle
19 Principal mamillary fasciculus
20 Mamillary body
21 Red nucleus
22 Mamillotegmental tract
23 Interpeduncular nucleus
24 Dorsal tegmental nucleus
25 Central superior nucleus

Fig. 23.4. The central part of the limbic area; medial view of nuclei and tracts (5/2×)

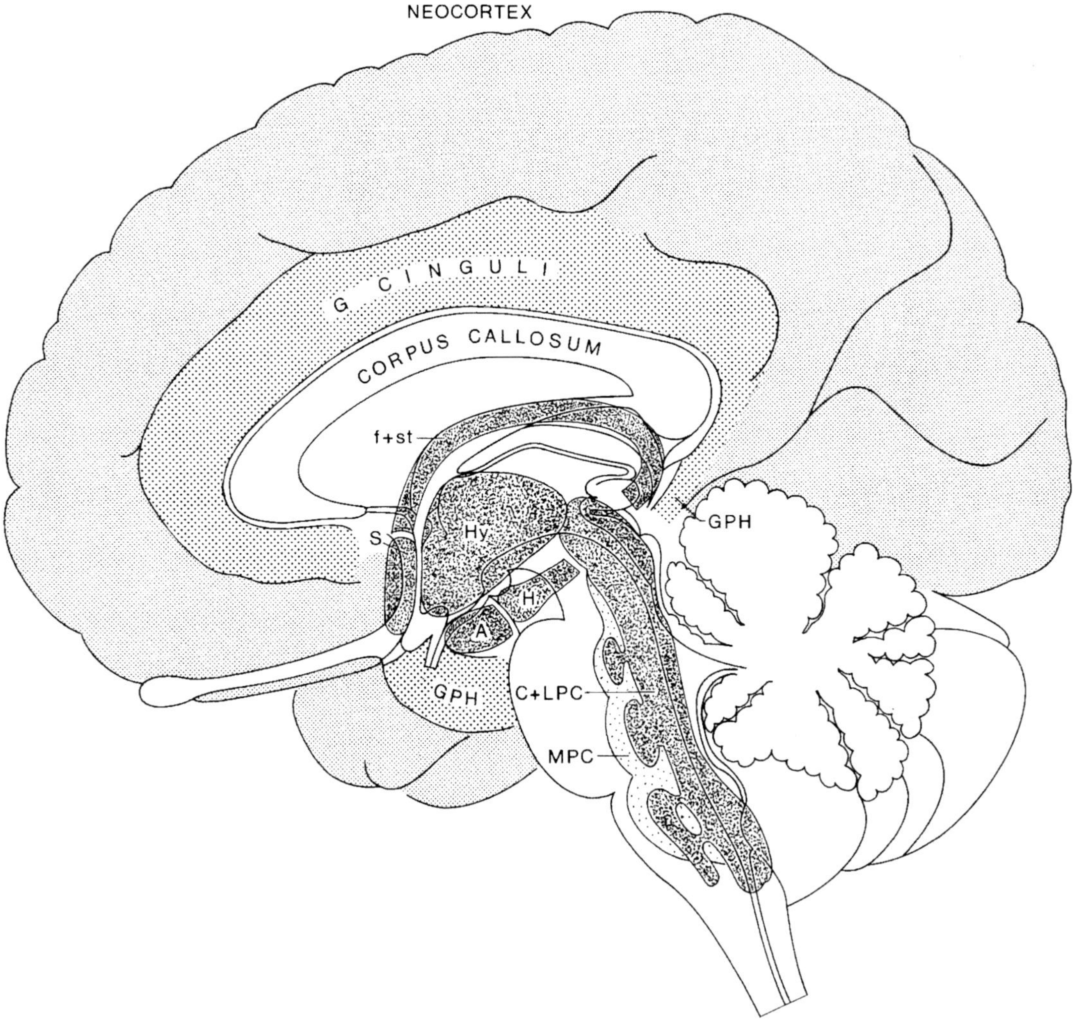

Fig. 23.5. Medial view of the human brain. The positions of the cingulate gyrus (*G CINGULI*), parahippocampal gyrus (*GPH*), septum (*S*), fornix + stria terminalis (*f+st*), hypothalamus (*Hy*), hippocampus (*H*), amygdala (*A*), core + lateral paracore (*C+LPC*) and median paracore (*MPC*) are diagrammatically indicated

particularly the septum, the amygdala, the midline thalamic nuclei, the habenula and the hypothalamus. He suggested that the cortical limbic ring is rostrally closed by two subcortical nodal points, i.e. the more dorsally situated septum and the more ventrolaterally located amygdaloid complex. MacLean cited clinical and experimental evidence suggesting that the lower part of the ring, fed by the amygdaloid complex, is primarily concerned with emotional feelings and with behaviour that ensures self-preservation. As he put it, the circuits of this lower part of the ring are "so to speak, kept busy with the selfish demands of feeding, fighting and self-protection" ([78], p. 340). The structures associated with the septum in the upper part of the ring, on the other hand, would be involved in "expressive and feeling states that are conducive to sociability and the procreation and preservation of the species" ([78], p. 340). These data and aspects led MacLean to the conclusion that the limbic cortex, together with the subcortical structures to which it is directly connected, comprises a functionally integrated system which he designated (in keeping with Broca's terminology) as the limbic system (for the first time in MacLean [77]).

A notable extension to the limbic system concept was made by Nauta [88–91]. Nauta added to the telencephalic limbic 'arch' (in which he included the hippocampal formation and the amygdaloid complex, but not the cingulate and parahippocampal gyri) a neural continuum that may be designated as the 'limbic axis'. This continuum includes, from rostral to caudal, the septal and preoptic regions, the hypothalamus and a number of paramedian mesencephalic structures, including the mesencephalic central grey and the dorsal raphe nucleus (Nauta's 'limbic midbrain area') (Figs. 23.1, 23.4). Nauta pointed out that these various entities are structurally heterogeneous, but that all of them are strongly interconnected by shorter and longer ascending and descending fibres. Taken together, these connections constitute, in Nauta's opinion, one large functional system which he designated as the 'limbic system–midbrain circuit'. He emphasized that the large telencephalic limbic structures, i.e. the hippocampus and the amygdaloid complex, are both reciprocally connected with the rostral pole of the limbic axis and thereby with the limbic system-midbrain circuit.

Nauta's views on the functioning of the limbic system can be summarized as follows: The structures constituting the limbic axis are centrally involved in the regulation of endocrine and visceral effector mechanisms. The functional state of these structures is not affected solely by neural afferents from the hippocampus and the amygdala, but is also modulated by impulses travelling along visceral-sensory pathways ascending from the spinal cord and the medulla oblongata as well as by various blood-borne agents. Moreover, the limbic system as a whole is implicated in the regulation of affective and motivated behaviours.

The medial forebrain bundle may be considered the central longitudinal pathway of the limbic forebrain–midbrain continuum. It is an assemblage of loosely arranged, mostly thin fibres, which traverses the lateral hypothalamic area. The bundle is highly complex, comprising a variety of short and long ascending and descending links [94, 137]. Moreover, the central limbic axis contains a continuous network of thin, unmyelinated fibres. Although this network harbours numerous, quite specific projections interlinking the various septal, preoptic and hypothalamic centres (see e.g. Larsen et al. [72]), these projections only manifest themselves as discrete fibre systems in a few places.

The Concept of a Greater Limbic System (GLS)

The central limbic continuum and its circuitry does not end at the caudal diencephalic or mesencephalic levels, but rather extends throughout the brain stem. The hypothalamic grey matter is caudally directly continuous with a periventricular mesencephalic and rhombencephalic zone. The mesencephalic part of this zone is constituted by the periaqueductal grey, whereas its rhombencephalic extension com-

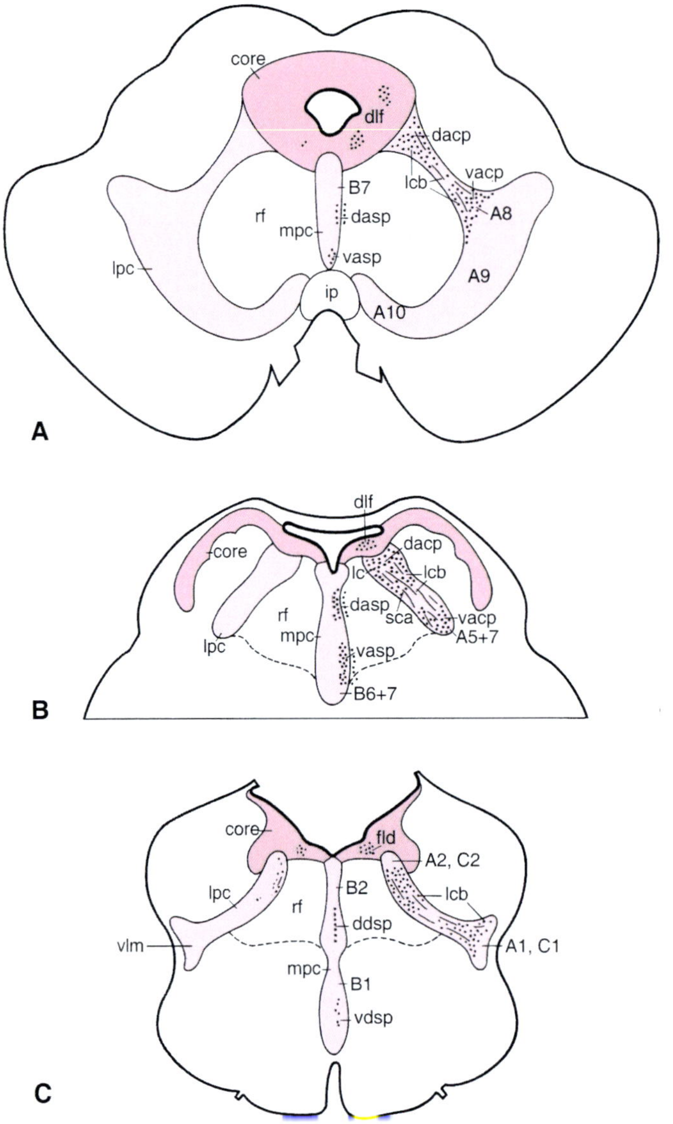

A1–A7	Noradrenergic cell groups
A8–A10	Dopaminergic cell groups
B1–B7	Serotoninergic cell groups
C1, C2	Adrenergic cell groups
dacp	Dorsal ascending catecholamine pathway
dasp	Dorsal ascending serotonin pathway
ddsp	Dorsal descending serotonin pathway
dlf	Dorsal longitudinal fasciculus
ip	Interpeduncular nucleus
lc	Locus coeruleus (A6)
lcb	Longitudinal catecholamine bundle
lpc	Lateral paracore
mpc	Median paracore
rf	Reticular formation
sca	Subcoeruleus area (A6)
vacp	Ventral ascending catecholamine pathway
vasp	Ventral ascending serotonin pathway
vdsp	Ventral descending serotonin pathway
vlm	Ventrolateral medulla

Fig. 23.6. Diagrammatic frontal sections through the brainstem at mesencephalic (**A**), rostral metencephalic (**B**) and myelencephalic levels (**C**) to show the positions of the core (shown in *dark red*), median paracore and (bilateral) lateral paracores (shown in *light red*)

prises the pontine central grey (as defined by Olszewski and Baxter [97]), the parabrachial nuclei and the dorsal vagal complex. In some previous publications [93, 95, 96], Nieuwenhuys and colleagues brought the aforementioned periventricular brain stem structures and the more rostrally situated components of the classical limbic system together under the name 'core of the neuraxis'. In the same publications, it was pointed out that, at the level of the brain stem, this core has two adjuncts, the *median paracore* and the (bilateral) *lateral paracore* (Figs. 23.5, 23.6). The median paracore is constituted by the series of raphe nuclei which extends throughout the brain stem. In most places, the raphe nuclei are directly adjacent to the core region and in some they even penetrate into it. Moreover, fibres of the core region project heavily towards most raphe nuclei.

The lateral paracore consists of a series of grisea which extend from the core region ventrolaterally into the tegmentum. At the mesencephalic level, this series includes the lateral part of the tegmental grey. In the rhombencephalon, the locus coeruleus or A6 group, the nucleus subcoeruleus (A6sc), the Kölliker-Fuse nucleus, the M and L regions of Holstege et al. [56], the nucleus reticularis parvocellularis, the ventrolateral medulla and the cytoarchitectonically ill-defined cell groups A1, A2, A5, A7, C1, C2 and Ch5 constitute this series. All of the rhombencephalic centres mentioned form part of, or are embedded in, the lateral reticular zone or lateral tegmental field.

The median and lateral paracores have the following features in common:

1. As already mentioned, they are both directly continuous with the core region.
2. Both contain large numbers of monoaminergic cells. In the median paracore, numerous serotoninergic cells are found, while in the lateral paracore catecholaminergic neurons prevail; adrenergic neurons are found in the cell groups C1 and C2, whereas the cell groups A1–A7 contain numerous noradrenergic elements.
3. Both paracores lie clearly beyond the trajectories of the large, compact, well-myelinated sensory and motor pathways.
4. Both paracores contain assemblies of thin, longitudinally arranged fibres.

Ascending and descending serotoninergic axons contribute substantially to the median paracore fibre assembly, whereas the large longitudinal catecholamine bundle described by Jones and Friedman [63] occupies a central position in the lateral paracore. The grisea discussed above, i.e. the brain stem portion of the core and the median and lateral paracores, not only form a structural continuum with the limbic forebrain, but also share a number of salient functional and connectional features with the latter. For this reason, we propose uniting the classical limbic system and its caudal extensions into a new entity, the *greater limbic system (GLS)*. The functional, structural and chemical features of this system will now be discussed.

Functional, Structural and Chemical Features of GLS

1. *Generation of integrated somatomotor, visceromotor and endocrine responses.* The GLS generates integrated somatomotor, visceromotor and endocrine responses directly aimed at the survival of the individual and of the species. The classical experiments performed by Bard [7], Hess [48] and others (for reviews, see Jürgens [64] and Swanson [123]) showed that numerous loci are present in the hypothalamus, from which quite characteristic integrated behavioural patterns can be elicited on electrostimulation, including eating, drinking, grooming, fear, attack, rage and reproductive behaviour. Later studies showed that similar responses can be elicited from higher (e.g. septum and amygdala) and lower parts of the brain. In the brain stem, the lateral paracore bundle (see below) is surrounded by a continuous array of loci involved in integrated behavioural and/or autonomic responses (Klemm and Vertes [66]). This array, which extends from the mesencephalic periaqueductal grey to the obex, includes the following (groups of) centres (Fig. 23.7):

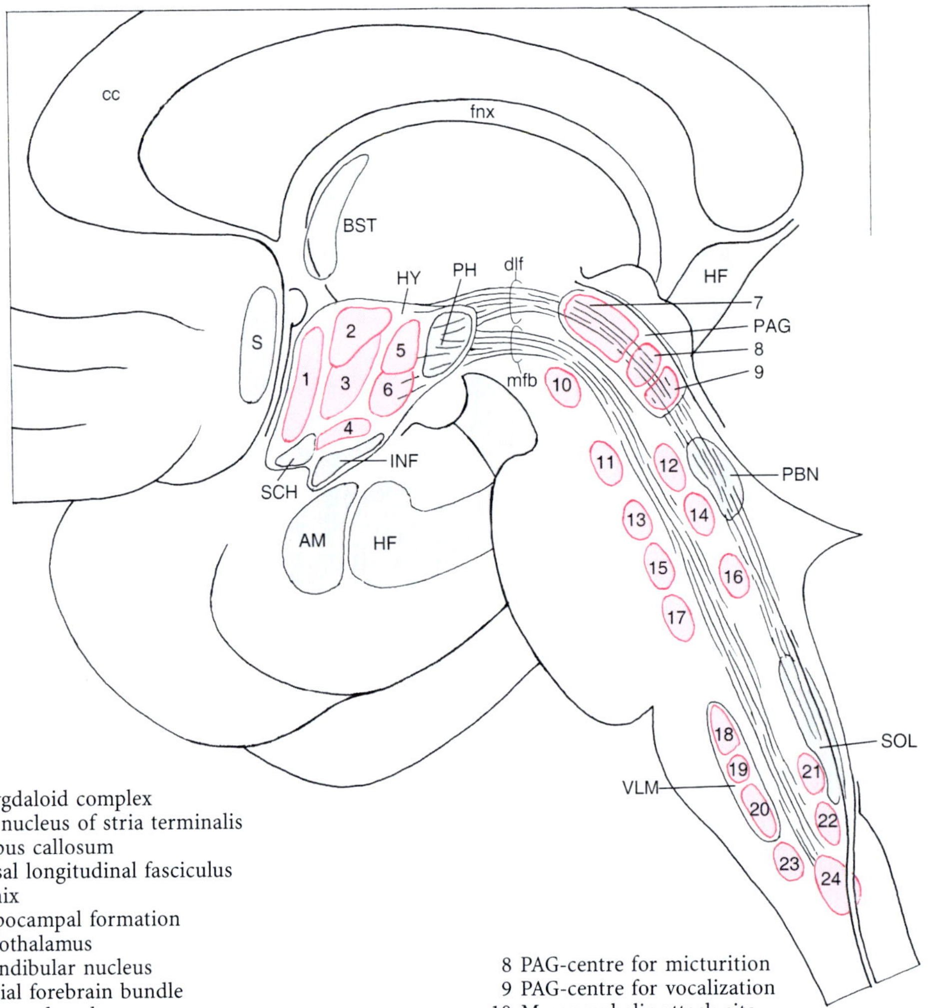

AM Amygdaloid complex
BST Bed nucleus of stria terminalis
cc Corpus callosum
dlf Dorsal longitudinal fasciculus
fnx Fornix
HF Hippocampal formation
HY Hypothalamus
INF Infundibular nucleus
mfb Medial forebrain bundle
PAG Periaqueductal gray
PBN Parabrachial nuclei
PH Posterior hypothalamic nucleus
S Septal region
SOL Solitary nucleus
SCH Suprachiasmatic nucleus
VLM Ventrolateral medulla

1 Medial preoptic nucleus (reproductive behaviour)
2 Paraventricular nuclei (ingestive behaviour)
3 Anterior hypothalamic nucleus (defensive behaviour)
4 Supraoptic nucleus (ingestive behaviour)
5 Dorsomedial hypothalamic n. (ingestive behaviour)
6 Ventromedial hypothalamic n. (defensive and reproductive behaviour)
7 Lateral and ventrolateral PAG (defensive behaviour)
8 PAG-centre for micturition
9 PAG-centre for vocalization
10 Mesencephalic attack site
11 Mesencephalic locomotor region
12 Barrington's (or medial) micturition nucleus
13 Lateral micturition nucleus
14 Nucleus of Kölliker-Fuse (control of respiration)
15 Pontine attack site
16 Vocalization cell group VOC
17 Pontine swallowing centre
18 Bötzinger complex (respiration control)
19 Pre-Bötzinger complex (respiration control)
20 Caudal VLM (cardiovascular control)
21 Medullary swallowing centre
22 Dorsal respiratory group
23 Ventral respiratory group
24 Retroambiguus nucleus (vocalization, reproduction)

Fig. 23.7. Sagittal view of the brain showing the position of centres or loci involved in integrated behavioural and/or autonomic responses (*red*). In the rhombencephalon practically all of these centres are located in the lateral tegmental field

(a) Electrical stimulation studies have shown that the mesencephalic periaqueductal grey (PAG) plays a prominent role in coordinating behavioural and autonomic responses to threatening situations [5, 22]. Such situations may lead to attack, defence or flight reactions, accompanied by hypertension, tachycardia and other autonomic responses or, conversely, to immobility ('freezing') and decreasing of heart rate and blood pressure. The PAG has also been implicated in the vocal expressions of emotions [65, 141], the control of micturition [15, 16] and mating behaviour [26, 132, 134].

(b) Attack behaviour without signs of autonomic arousal, hence designated as 'quiet attack', can be elicited from loci situated in the mesencephalic and pontine tegmentum [113, 117].

(c) Locomotion is an essential component of the procurement phase of various goal-oriented behaviours, such as fight reactions and food and water procurement. The lateral part of the caudal mesencephalic tegmentum contains a locus from which locomotion can be elicited in intact and decerebrate rats and cats [41, 114]. This locus, which is known as the mesencephalic locomotor region, corresponds roughly to the compact part of the pedunculopontine nucleus.

(d) In cats, the supraspinal part of the network regulating micturition includes, apart from the caudolateral part of the PAG, two centres in the tegmentum of the pons, namely Barrington's nucleus [8] or the M-(medial) pontine micturition centre and the L-(lateral) pontine micturition centre [56]. The presence of micturition-related regions in the caudal midbrain and rostral pontine tegmentum has been confirmed by positron emission tomographic (PET) studies in humans [15, 16]. For a detailed discussion of the centres and pathways involved in the regulation of micturition the reader is referred to Chap. 21.

(e) Swallowing is centrally programmed by two centres situated in the rhombencephalic lateral reticular formation, namely the pontine and the medullary swallowing areas. The pontine swallowing area is situated dorsomedial to the superior olivary complex [44, 55]. The medullary swallowing area is located halfway between the solitary and ambiguus nuclei [29, 84]. The circuitry subserving the swallowing act is discussed in Chap. 21.

(f) Cells belonging to the reticular formation form a zone along the lateral surface of the medulla oblongata. This superficial cell zone, which is commonly designated as the 'ventrolateral medulla' (Fig. 23.6), is implicated in cardiovascular and respiratory regulation. The rostral part of this area contains loci for respiratory rhythm generation, known as the Bötzinger and pre-Bötzinger complexes. Three additional lateral tegmental centres form part of the respiratory regulatory network, namely the dorsal and ventral respiratory groups and the Kölliker-Fuse nucleus. The dorsal and ventral respiratory groups are located ventrolateral to the solitary nucleus and close to the ambiguus nucleus, respectively [109]. The Kölliker-Fuse nucleus, which represents the pneumotaxic centre, is situated at pontine levels, just ventral to the parabrachial nuclei.

(g) The circuitry involved in vocalization includes, apart from the caudal portion of the PAG (see above), a group of cells situated dorsal to the superior olivary complex, designated as VOC [43], and the retroambiguus nucleus [50, 134]. The latter is situated directly caudal to the obex, in the transition zone of the reticular formation and the spinal intermediate grey (Fig. 6.21). The retroambiguus nucleus is connected with motoneurons innervating pharyngeal, laryngeal, intercostal and abdominal muscles involved in vocalization [58, 135]; the retroambiguus nucleus is also involved in mating behaviour [54].

2. *Predominance of thin, unmyelinated fibres.* A network of thin and ultrathin, mostly varicose fibres extends from the hypothalamus into the limbic brain stem regions. Within this network, longitudinally running axons tend to aggregate into two loose-textured fibre systems, the dorsal longitudinal fasciculus of Schütz and the medial forebrain bundle (Figs. 23.2, 23.7).

The dorsal longitudinal fasciculus of Schütz extends from the posterior part of the hypothalamus to the caudal medulla oblongata and occupies a periventricular position over its entire length. It is a composite fibre system that, in addition to long ascending and descending fibres directly connecting the hypothalamus with the nucleus solitarius and related autonomic centres, also contains numerous shorter links. Many of its fibres are synaptically interrupted in the periaqueductal grey and in the parabrachial nuclei [39, 105, 136].

The medial forebrain bundle, much like the dorsal longitudinal fasciculus, is a composite fibre system. It has already been mentioned that the rostral part of this bundle occupies a central position in the lateral hypothalamic area. In the transitional area of the diencephalon and mesencephalon, the medial forebrain bundle fibres are rearranged into a smaller medial and a larger lateral stream [49, 60]. The medial fibre stream roughly maintains the sagittal orientation of the hypothalamic trajectory of the bundle. It passes through the medial parts of the mesencephalic and rhombencephalic tegmental areas, just next to the raphe nuclei. The medial stream (or medial paracore bundle) is composed of descending fibres, by which several hypothalamic centres project to the raphe nuclei and to the adjacent parts of the medial reticular formation, and also comprises numerous ascending and descending fibres originating from the raphe nuclei themselves.

The lateral stream of fibres extending from the medial forebrain bundle to the brain stem, or lateral paracore bundle, sweeps laterally and caudally over the dorsal border of the substantia nigra into a ventrolateral tegmental position, from where it descends through the lateral tegmented field of the pons and medulla oblongata. This stream passes through a band-like array of grisea which are all involved in the regulation of specific, motivated behaviours and of related autonomic responses (Fig. 23.7). It contains the innumerable shorter and longer ascending and descending pathways by which these grisea are interconnected.

It will be appreciated that the organization of the behavioural system touched upon above differs considerably from that observed in the various 'classical' sensory and motor systems. In these 'classical' systems, generally discrete centres are interconnected by well-myelinated, through-conducting fibres assembled in discrete bundles. However, in the system concerned with the generation of behaviours directly related to homeostasis and reproduction, 'open-line' polysynaptic connections as described by Ricardo and Koh [105] prevail. The consideration of these authors is worth quoting in full; after indicating that such connections are accessible at each synaptic interruption to side inputs of related ('re-entrant' circuits) or of unrelated origins, they continue as follows:

> The 'open-line' componentry of such systems appears to reflect the need of many homeostatic functions to be guided by several rather than by a single modality of afferent signals; simultaneously, it might serve as a device allowing selective and finely graded modulation of the impulse flow by re-entering circuits (Ricardo and Koh [105], p. 20).

The open-line character of the communication within the system under consideration is further accentuated by the fact that most of its connections are mainly composed of unmyelinated, thin and ultrathin, varicose fibres. Assuming that the varicosities are concerned with neural transmission – synaptic, non-synaptic or both (see below) – this means that these fibres are not only involved in inter-neuronal communication at their end, but rather throughout their extent, or at least at several different levels along their course.

Combined physiological and experimental neuroanatomical studies have elucidated the circuitry underlying several specific behaviours, e.g. grooming, drinking, micturition, aggression and sexual behaviour (see Chaps. 10, 21). It is important to note that the efferents of behaviourally defined centres do not form well-defined bundles at this 'level of resolution' either, but rather spread out, forming diffuse fibre streams [107]. However, the diffuseness of these fibre streams does not exclude connectional specificity. For instance, injections of the anterograde tracer *Phaseolus* in two behaviourally defined hypothalamic areas, i.e. the 'grooming area' and the 'attack area', revealed that fibres of both of these areas project (*inter alia*) to the periaqueductal grey, but terminate in different sectors of that centre [107].

It has already been mentioned that the median and lateral paracores contain groups of monoaminergic neurons and that these neuron groups contribute substantially to the caudal extensions of the medial forebrain bundle, which occupy a central position in both domains. The axons of the noradrenergic (and adrenergic) cells in the lateral paracore as well as those of the serotoninergic cells in the median paracore are thin and varicose and ramify profusely. Collectively, they constitute extensive noradrenergic and serotoninergic networks, which spread far beyond the limits of the GLS over virtually all parts of the central nervous system. It is intriguing, however, that all monoaminergic cell groups receive their principal afferents from limbic centres and that the initial parts of their open-line axonal systems pass through limbic domains.

3. *Presence of Circumventricular Organs.* Circumventricular organs are small, highly special-

ized brain structures which, as their name implies, are situated in the immediate vicinity of the cerebral ventricular system. Most of these organs contain neurons, and, with regard to both their afferent and efferent connections, these neurons are entirely embedded in limbic circuitry. Their most important distinguishing feature is the lack of a blood–brain barrier in their vasculature. The circumventricular organs include the subfornical organ and the organum vasculosum of the lamina terminalis, which are both situated in the telencephalic preoptic region, and the area postrema, which is located in the caudalmost part of the brain in direct contact with the nucleus of the solitary tract. These organs can be considered as chemosensitive zones; they monitor the changing levels of circulating hormones and other substances, forming part of the circuitry underlying, *inter alia*, homeostatic functions of water-electrolytic balance and cardiovascular regulation.

The ventricular system provides an important conduit for informational substances [76, 92]. It is well known that the terminals of monoaminergic fibres penetrate into the ventricular cavities in many places and release their contents into the cerebrospinal fluid. It is also known that peptidergic neurons are frequently located in periventricular areas, and that the peptides released by these neurons readily reach the ventricular cavities. It has been experimentally shown that peptides, migrating via the ventricular system, may reach the extracellular spaces of the periventricular parenchyme throughout the entire brain within a few minutes [104].

4. *Presence of Gonadal Steroid-Receptive Neurons.* Experimental studies using autoradiographic methods for detecting labelled steroid binding [120, 121, 139] or the more recent immunohistochemical techniques for detecting the receptor protein itself [13, 69] have revealed that both the forebrain and the brain stem components of the GLS contain aggregations of gonadal steroid-receptive neurons. In the forebrain component, such aggregations occur *inter alia* in the septal nuclei, the bed nucleus of the stria terminalis (BST), the amygdala, the medial preoptic nucleus, the ventromedial hypothalamic nucleus and the infundibular nucleus, whereas in the brain stem component the periaqueductal grey, the parabrachial nuclei and the solitary nucleus contain numerous steroid-receptive neurons. In many of these centres, the neurons concentrating oestrogen and androgen hormones show a distinct sexual dimorphism with regard to number and distribution (for a review, see Madeira and Lieberman [81]).

5. *Extraordinary Richness and Density of Neuromediators (particularly neuropeptides).* Some time ago, we gathered data from the literature on the localization [93] and density [95] in the central nervous system of 25 different neuromediators, including 16 neuropeptides. Numerous 'classical' limbic forebrain centres appeared to contain an extraordinary diversity of neuropeptides and an extraordinary density of peptidergic neurons and/or fibres and terminals. Interestingly, however, the same appeared to hold true for a number of brain stem centres, including the periaqueductal grey, the lateral parabrachial nucleus, the locus coeruleus and the solitary nucleus.

Neuropeptides may influence behaviour. Characteristic of their action on behavioural processes is their slow onset and long duration [31]. It has been frequently suggested that neuropeptides mainly exert a neuromodulatory (facilitating or attenuating) influence on neural transmission by 'classical' non-peptidergic neuromediators (see e.g. [73]). However, there can be no doubt that many neuropeptides play a key role in the regulation of specific behavioural responses. Thus, leptin, neuropeptide Y and cholecystokinin control food intake, whereas angiotensin II evokes concerted hormonal, vegetative and behavioural responses aimed at the maintenance of fluid homeostasis (see Chap. 10).

The GLS encompasses many different neuronal networks subserving specific, motivated behaviours, such as eating, drinking, aggression and sexual behaviour. Interestingly, high concentrations of the same peptide are often found in different terminal areas of neuronal circuits related to specific, motivated behaviours, which suggests that a single peptide is used at multiple anatomical levels to regulate a particular behavioural process [46]. This feature may be interpreted as a 'peptidergic' specification of the open-line componentry of neuronal systems subserving behavioural functions discussed above.

6. *Prevalence of non-synaptic neurotransmission.* In 1985, Nieuwenhuys [93] expressed the opinion that non-synaptic or volume neurotransmission may well play an important role in the GLS. For a discussion of this hypothesis, the original publication should be consulted. Let it suffice to mention here that, meanwhile, ultrastructural findings strongly suggesting the occurrence of non-synaptic inter-neuronal communication have been reported for numerous limbic structures, including the supraoptic nucleus [87, 101], suprachiasmatic nucleus [23], ventral tegmental area [30], periaqueductal grey [19, 136], locus coeruleus [130], ventrolateral medulla [1] and dorsal vagal complex [9].

A Brief Guide to Discussions of Limbic Structures in Previous Chapters

In the systematic account of the various parts of the neuraxis, presented in Section II of the present work, several components of the GLS were discussed. The following synopsis is intended to facilitate the localization of the pertinent sections and illustrations.

The *hypothalamus* occupies, according to current views, a key position in the limbic continuum. The structure and function of this brain part is discussed in Chap. 10 (pp. 289–336). Its principal afferent and efferent connections are summarized in Tables 10.2 and 10.3 and in Figs. 10.2 and 10.3.

The principal function of the GLS is, as mentioned, the organization of behaviours directly aimed at the survival of the individual (organism) and of the species. In Chap. 10, the circuitry related to several of these motivated or goal-oriented behaviours, including food intake, thirst and drinking behaviour, sexual behaviour and defensive behaviour are discussed and depicted.

The *central olfactory system* is not a component of the GLS as such; yet its efferents pass to the amygdala and hippocampus, as well as to the anterior insula and the caudal orbitofrontal cortex (Figs. 11.7, 11.8), which both form part of the paralimbic belt, as defined by Mesulam [82].

The hippocampal formation and the amygdaloid complex represent the principal telencephalic components of the GLS. The *hippocampal formation* is discussed in Chap. 12 (pp. 361–400), and in the same chapter attention has also been paid to the *septum*, another telencephalic limbic structure, and the *limbic lobe*. The latter forms an interface between the GLS and the neocortex. The *amygdaloid complex* and the related *BST* are discussed in Chap. 13 (pp. 401–426).

Structure and connections of the *ventral or limbic striatum* are treated in a section of Chap. 14 (pp. 456–460).

A brief overview of the intrinsic and extrinsic connections of the *limbic lobe* and the *paralimbic belt* is presented in Chap. 15 (pp. 617–620) and Figs. 15.56 and 15.57.

The *insula*, finally, the anterior part of which forms part of the paralimbic belt, is discussed in the final section of Chap. 15 (pp. 649–653).

Inputs to Limbic Domains

The inputs to limbic regions of the brain can be categorized as follows: (a) humoral, (b) interoceptive, (c) somatosensory, (d) visual, (e) olfactory, (f) cerebellar, (g) extrapyramidal and (h) cortical. These inputs will now be briefly discussed. (For details, please consult the "limbic" sections in the various chapters of Section II of the present work, as specified above.)

(a) *Humoral Input.* The sodium concentration of the blood and the concentration of glucose, free fatty acids and other nutrients are monitored by specialized hypothalamic and extrahypothalamic (e.g. nucleus of the solitary tract) neurons. Moreover, numerous hormones, such as angiotensin II, neuropeptide Y and the gonadal and adrenal steroids, act directly on endocrine-neural transducer cells, which are abundant in limbic domains. Finally, the circumventricular organs, which topographically and functionally lie entirely embedded in the GLS, represent specialized trigger zones for blood and cerebrospinal fluid-borne substances.

(b) *Interoceptive Input.* Cells in lamina I of the spinal dorsal horn give rise to fibres that ascend to the limbic brain stem zone. These fibres, which most probably carry information concerning the physiological status of the various tissues of the body, follow the trajectory of the lateral paracore bundle and terminate *inter alia* in the nucleus of the solitary tract, the ventrolateral medulla, the parabrachial nuclei, the lateral pontine tegmentum, the cuneiform nucleus and the mesencephalic periaqueductal grey, and in the catecholaminergic cell groups A1–A7 and C1 [20, 27, 28] (Fig. 23.8). Other information concerning the internal body state converges via the vagus nerve on the nucleus of the solitary tract. This information is conveyed partly directly and partly indirectly via relays in the parabrachial nuclei or in the periaqueductal grey to prosencephalic limbic regions, including the amygdala [62]. It is important to note that the nucleus of the solitary tract, the parabrachial complex and the periaqueductal grey all maintain strong, reciprocal connections with the hypothalamus, the central nucleus of the amygdala and the BST, all of which represent sites of origin of substantial descending limbic projections [11, 39, 45, 57, 85, 111, 118, 131]. A third route by which interoceptive (particularly cardiovascular) information is conveyed to the limbic prosencephalon passes via the ventrolateral medulla and the catecholaminergic cell groups A1 and C1 embedded therein. The principal afferents of this complex arise from the general viscerosensory zone of the nucleus of the solitary tract [108]. It projects primarily to autonomic centres in the spinal cord, but a certain proportion of its efferents ascend to the forebrain, where they terminate in the hypothalamic supraoptic and paraventricular nuclei [25, 112] and in a number of telencephalic grisea, including the septum, the nucleus accumbens, the hippocampus and the medial prefrontal cortex [140].

(c) *Somatosensory Input.* Protopathic fibres project to two limbic cell masses in the brain stem: the periaqueductal grey and the parabrachial nuclei. The fibres projecting to the periaqueductal grey originate mainly from laminae I and IV/V of the spinal cord and from the most superficial part of the spinal trigeminal nucleus. They carry primarily nociceptive stimuli to the periaqueductal grey and terminate in the lateral zone of that structure [6]. The fibres projecting to the parabrachial complex terminate mainly in the lateral portion of that complex. Some of these fibres have been demonstrated to arise from nociceptive and thermoreceptive lamina I neurons [116]. It is noteworthy that the parabrachial complex contains numerous neurons that are exclusively activated by noxious stimuli applied to several areas of the body and that these 'nociceptive-specific' neurons have been demonstrated to project to the nucleus centralis amygdalae and the retrochiasmatic area of the hypothalamus [11, 12, 61].

Remarkably, some neurons situated in lamina I of the spinal and medullary dorsal horn and in the grey matter surrounding the spinal central canal project directly to many parts of the hypothalamus, including the posterior, ventromedial and paraventricular nuclei and the lateral hypothalamic area and, less densely, to a number of limbic telencephalic regions, including the amygdala, septum, nucleus accumbens, ventral pallidum and orbital cortex [20].

(d) *Visual Input.* A direct visual input reaches the hypothalamus via the retinohypothalamic tract, which terminates in the suprachiasmatic nucleus. This cell group, which projects to various intrahypothalamic and extrahypothalamic limbic centres, may be considered as the endogenous clock of the brain (Fig. 10.4). It is critically involved in the generation and entrainment of circadian rhythms, including patterns of general activity (sleeping/waking), feeding and drinking behaviour and hormonal secretion (see Chap. 10).

(e) *Olfactory Input.* The olfactory system, which in most mammals exerts a strong influence on feeding, mating and several related goal-oriented behaviours, has 'privileged access' (Nauta and Haymaker [91], p. 194) to limbic circuitry. Olfactory projections reach the amygdala, the hippocampus and the lateral hypothalamic area [102].

(f) *Cerebellar Input.* A small cerebellohypothalamic projection originates from the cere-

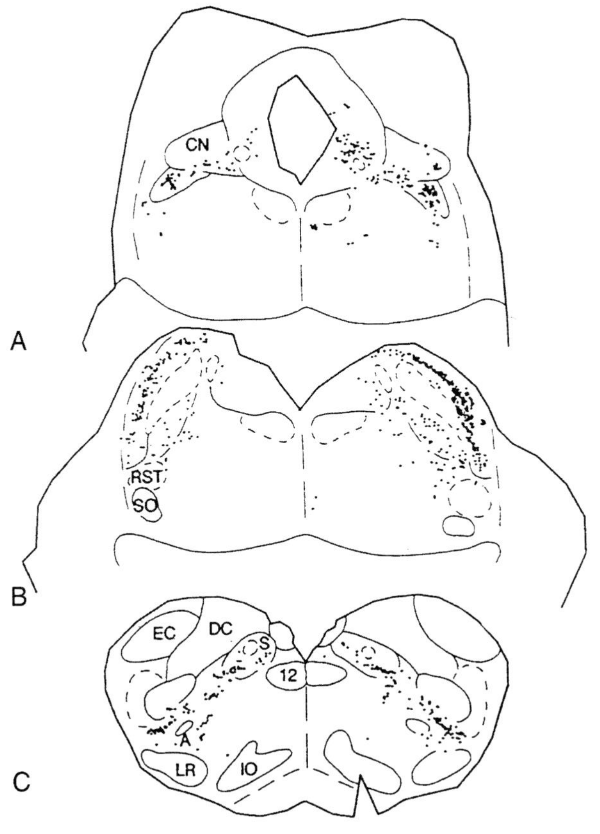

Fig. 23.8. Plots of labelled terminations on transverse sections through the lower mesencephalon (**A**), the pontine region (**B**) and the lower medulla oblongata of a Cynomolgus monkey (**C**), following some small, iontophoretic injections of *Phaseolus vulgaris*-leucoagglutinin in lamina I of the cervical spinal cord (C6–C8); the *left side* is ipsilateral to the injection site. *CN*, cuneiform nucleus; *DC*, dorsal column nuclei; *EC*, external cuneate nucleus; *IO*, inferior olive; *LR*, lateral reticular nucleus; *RST*, rubrospinal tract; *S*, solitary nuclear complex; *SO*, superior olive; *12*, hypoglossal nucleus (reproduced from Craig [27])

bellar nuclei and terminates in the posterior hypothalamus. This projection is reciprocated by hypothalamocerebellar fibres, which reach all parts of the cerebellar cortex and nuclei. Indirect hypothalamocerebellar pathways, relayed through various brain stem nuclei, including the pons and the inferior olive, have also been described (for a review, see Dietrichs and Haines [32]). The functional significance of the hypothalamocerebellar circuits is unknown.

(g) *Extrapyramidal Input.* The structure, connections and function of the basal ganglia have been discussed in Chap. 14. It was pointed out there that the corpus striatum can be subdivided on connectional grounds into a ventral part or ventral striatum and a dorsal part or dorsal striatum. The ventral striatum, which includes the nucleus accumbens and the ventral parts of the caudate nucleus and the putamen, receives its main input from typical limbic structures, such as the hippocampal formation and the amygdala, which is why it is often designated as the limbic striatum. The ventral striatum sends it efferents mainly to a ventral extension of the globus pallidus, situated ventrally, beneath the anterior commissure. This ventral pallidum projects to several typical limbic structures, including the lateral hypothalamic area, the ventrolateral periaqueductal grey and the mesencephalic locomotor region. The ventral pallidal projection to these grisea is reinforced by fibres arising from the nucleus accumbens and the prefrontal cortex. Thus it appears that the ventral striatum, within which the nucleus accumbens occupies a prominent position, receives strong projections from the hippocampus and the amygdala and that this ventral striatum projects either directly or via the ventral pallidum to certain diencephalic and mesencephalic limbic centres, including the mesencephalic locomotor region. Locomotion is an essential component of the procurement phase of many motivated behaviours, and there is experimental evidence [86, 100] suggesting that the ventral striatum, which is situated in a transitional zone between the limbic and the extrapyramidal systems, is critically implicated in initiating this activity. However, Groenewegen et al. [42] cite evidence indicating that, apart from locomotor activity, oral behaviours such as licking, chewing and swallowing can be elicited or influenced by manipulating the ventral striatum, including the accumbens nucleus. Garcia-Rill [40] has expressed the opinion that the caudolateral mesencephalic area, innervated by the ventral striopallidal system, forms part of a much larger complex, extending from the substantia nigra rostroventrally to the laterodorsal tegmental nucleus and locus coeruleus caudodorsally. According to Garcia-Rill, this complex may play a role in regulating and co-ordinating a number of functions that have rhythmicity in common, such as locomotion, respiration and sleep/wake mechanisms.

(h) *Cortical Input.* Together the hippocampal formation and the amygdaloid complex constitute, morphologically and functionally, the rostral pole of the limbic domain. Via the septal region, the hippocampus is strongly and reciprocally connected with the preoptico-hypothalamic continuum (see Chap. 12), and comparable reciprocal connections are present between the hypothalamus and the amygdaloid complex, including its dorsomedial extension, the BST (see Chap. 13). It is of paramount importance that the hippocampus and the amygdala, and with them the entire GLS, are under the control of the neocortex. Most of these cortico-limbic projections are synaptically interrupted in the cingulate and/or parahippocampal gyri, which may thus be designated as paralimbic cortical areas (see Chap. 12 and Figs. 12.13, 15.57 A).

The neocortex not only receives and processes information related to the external world; the viscera, and hence the internal world of the organism, also have a cortical representation. Sensory information from the thoracic and abdominal viscera enters the brain via the vagus nerve and reaches an insular area known as the insular viscerosensory cortex, after synaptic relays in the nucleus of the solitary tract, the parabrachial complex and the parvocellular part of the ventral posteromedial thalamic nucleus [110]. The insular area in which the insular viscerosensory cortex is embedded forms part of the paralimbic belt and is strongly interconnected with other (para)-limbic structures, including the amygdala and the hippocampus (Fig. 15.56, Table 15.3). It may be concluded that, via these various conduction routes, highly processed information concerning both the external and internal environment is fed into the circuitry of the GLS.

The amygdaloid complex and the hippocampal formation form an interface between the 'cognitive brain' and the lower domains of the GLS. It is known that hippocampal formation and the adjacent entorhinal cortex contribute significantly to memory function [138, 142] and that the amygdala plays a key role in evaluating incoming stimuli in relation to the motivational state of the organism (e.g. hunger) or of their direct vital relevance (e.g. appearance of a predator), so as to prepare the organism for an appropriate response. This crucial role of the amygdala is often denoted as the detection of the affective or emotional significance of sensory stimuli. In this context, emotions are classed as positive or negative. As mentioned earlier, positive emotions are elicited by rewards or positive reinforcers, whereas negative emotions are evoked by 'punishers' or negative reinforcers.

At first sight, the functions of the hippocampus and the amygdala seem to be quite different. However, the emotional significance of newly arriving sensory stimuli depends on past experience and the evaluation of incoming signals by comparing them with information stored in memory is thus an essential step in the processing sequence leading from the perception of environmental events to motivated behaviours (see LeDoux [74]). The amygdala and the hippocampal formation, which maintain direct and indirect reciprocal connections, may well be jointly involved in this evaluation process. The descending cortico-limbic projections discussed above are reciprocated by substantial ascending limbico-cortical paths. These ascending and descending conduction routes are both largely funnelled through, and synaptically interrupted in the amygdalo-hippocampal complex. The ascending system terminates mainly in cortical association areas. Impulses travelling along the hippocampo- and amygdalo-cortical projections may be instrumental in adapting the activities of the cognitive brain to the prevailing motivational state of the organism and may lead to conscious emotional experiences.

The Limbic Motor System

The limbic motor system comprises a large number of pathways, descending from subcortical and cortical telencephalic regions to the hypothalamus and to numerous centres contained within the core and paracores of the brain stem. Its most important sites of origin include the septal region, the amygdaloid complex, the BST, and the anterior cingulate and caudal orbitofrontal cortices. The fibre systems descending from these grisea are diagrammatically indicated in Fig. 23.9 and their sites of termination are summarized in Table 23.1.

The concept of a limbic motor system as an independent structural and functional entity was presented by Holstege in 1991 [51]. However, in subsequent publications [52, 53, 58] he replaced the original term limbic motor system by emotional motor system. In the present authors' opinion, this new name narrows the functional significance of the system too far; hence, we maintain the original term.

The principal components of the limbic motor system will now be briefly discussed.

1. The *lateral septal nucleus*, which receives its principal input from the hippocampal formation, projects strongly to the nuclei forming the medial preoptic-hypothalamic zone [10, 106, 125].
2. Several nuclei in the *amygdaloid complex*, including the medial, central, cortical, medial basal and accessory basal nuclei, project via the stria terminalis to the hypothalamus, particularly the medial preoptic, anterior, paraventricular, ventromedial and premamillary nuclei [2]. A strong projection, originating mainly from the central amygdaloid nucleus, passes medially via the ventral amygdalo-fugal pathway and then turns

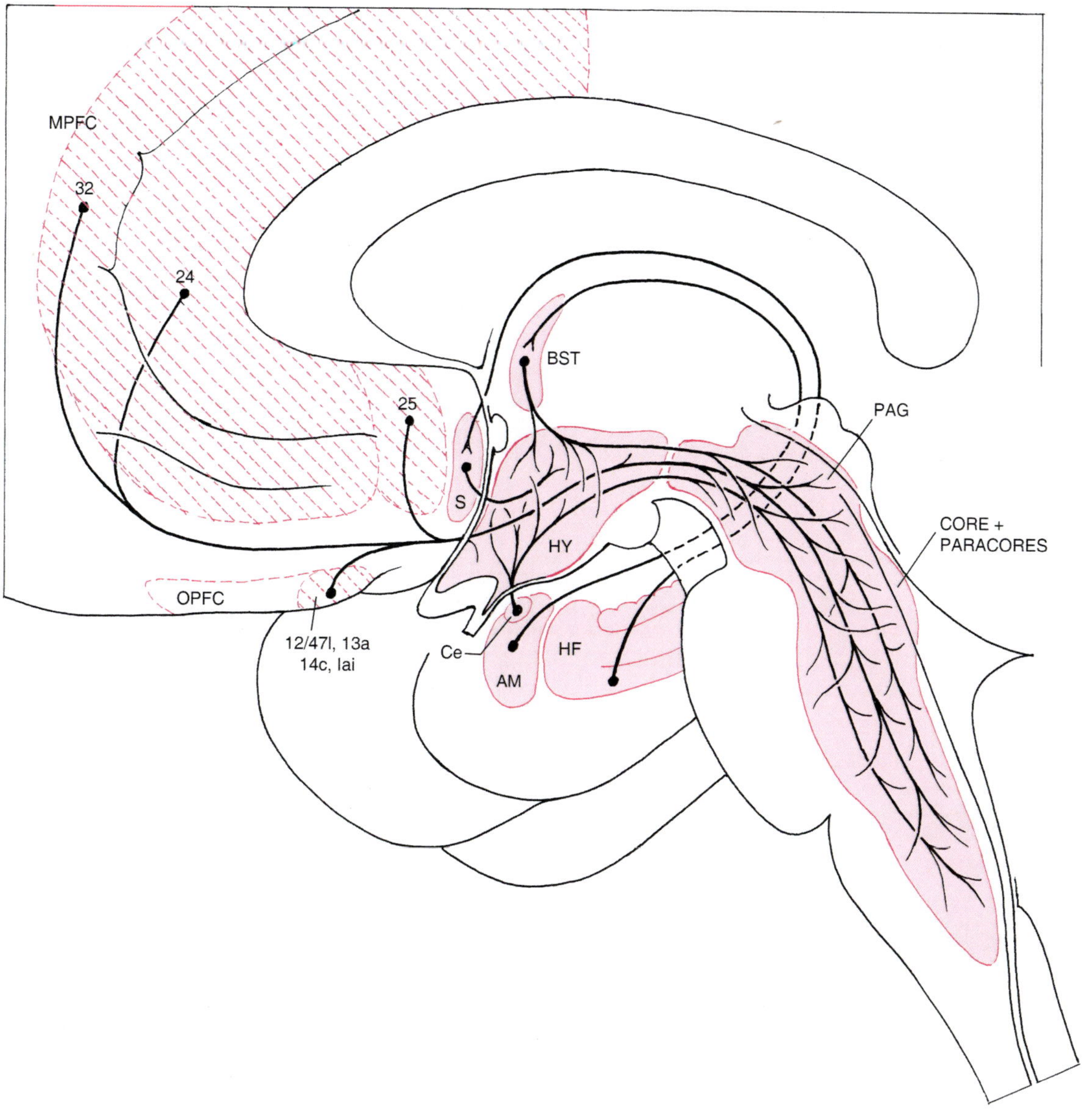

Fig. 23.9. Descending projections from telencephalic limbic structures. For details on the sites of termination of these projections, see Table 23.1. *AM*, amygdala; *BST*, bed nucleus of stria terminalis; *Ce*, central nucleus of amygdala; *HF*, hippocampal formation; *HY*, hypothalamus; *MPFC*, medial prefrontal cortex; *OPFC*, orbital prefrontal cortex; *PAG*, periaqueductal grey; *S*, septal complex; *24, 25, 32*, cortical fields, according to Brodmann; *12/47l, 13a, 14c, Iai*, cortical fields, according to Brodmann-Walker [3, 98]

Table 23.1. Origins and destinations of some descending limbic projections

Destinations	Origins			
	LSN	AM	BST	CING
Hypothalamus				
Medial preoptic n.	+	+	+	+
Anterior hypoth.n.	+	+	+	+
Supraoptic n.				
Paraventricular n.	+	+	+	
Dorsomedial hypoth.n.			+	+
Ventromedial hypoth.n.	+	+	+	+
Tuberomamillary n.				+
Dorsal premamillary n.		+	+	
Ventral premamillary n.		+	+	
Posterior hypoth. n.				+
Lateral hypoth. n.		+	+	+
Mesencephalon				
Periaqueductal grey		+	+	+
Cuneiform n.		+	+	+
Pedunculopontine n.				+
Dopaminergic nn.		+	+	+
Serotoninergic nn.		+	+	+
Pons				
Parabrachial nn.		+	+	+
Lateral tegmental field		+	+	+
Serotoninergic nuclei			+	
Locus coeruleus, subcoeruleus		+	+	+
Medulla oblongata				
Solitary nucleus		+	+	+
Dorsal motor X n.		+	+	+
Lateral tegmental field		+	+	+
Serotoninergic nn.				+

LSN, lateral septal n. [106, 125]; AM, amygdala [2, 59, 103]; BST, bed nucleus of stria terminalis [52, 57]; CING, anterior cingulate cortex [24, 71]

caudally, to descend through the most lateral part of the hypothalamus and the tegmental regions of the brain stem [59, 103]. The fibres of this projection innervate the lateral hypothalamic area and numerous centres in the midbrain, pons and medulla oblongata (Fig. 13.6, Table 23.1).

3. The arch-shaped *bed nucleus of the stria terminalis (BST)* is caudally directly continuous with the medial and central amygdaloid nuclei. It receives afferents from the insula and the prelimbic and infralimbic cortices, the amygdala, the hypothalamus and a number of brain stem centres, among them the periaqueductal grey and the dorsal vagal complex (see Chap. 14). The BST projects substantially to the amygdaloid complex, the mediobasal telencephalon and the hypothalamus and to a large number of grisea located in the brain stem. Its descending fibres enter the hypothalamus and pass, via the medial forebrain bundle, to lateral tegmental parts of the mesencephalon, pons and medulla oblongata. The termination pattern of these descending BST fibres is virtually identical to that of the amygdala [52, 57] (Table 23.1).
4. Fibres, originating from the *anterior cingulate cortex*, comprising Brodmann's areas 24, 25 and 32, pass caudally through the medial wall of the hemisphere and enter the hypothalamus, where several nuclei are supplied. A considerable proportion of these fibres proceeds further caudally to the brain stem. In the brain stem, most of these fibres follow, just like those emanating from the amygdala and BST, the trajectories of the medial and lateral paracores. The patterns of termination of the cingulate projections have been experimentally studied by Chiba et al. [24] and Kuipers et al. [71]. Their main results are summarized in Table 23.1.
5. The studies of An et al. [3] and Öngür et al. [98] have shown that the *caudal orbitofrontal cortex* (areas 13a, 14c, 47/12l) and adjacent *insula* (area Iai, cf. Fig. 15.68B) give rise to fibres that descend to the hypothalamus and midbrain. In the hypothalamus, these fibres terminate mainly in the posteri-

or part of the lateral hypothalamic area. In the midbrain, they supply the periaqueductal grey and, less densely, the substantia nigra and the cuneiform, pedunculo-pontine and dorsal raphe nuclei. Possible pontine and medullary terminations of the fibres originating from the orbitofrontal cortex have not been included in the studies mentioned.

If we survey the data just discussed (and summarized in Table 23.1), it appears that the fibres of the limbic motor system target structures belong to three different functional categories: (1) viscerosensory relay nuclei, (2) monoaminergic cell groups, and (3) components of the limbic motor network.

As regards the first category, the nucleus of the solitary tract and the parabrachial nuclei are major relays for ascending visceral inputs [110]. It seems likely that the limbic motor system influences the transfer of interoceptive information to higher levels of the brain by way of the projections to these nuclei.

As for the second category, the monoaminergic cell groups in the brainstem receive, in addition to the projections from the amygdala, BST and the orbitofrontal and anterior cingulate cortices, afferents from a variety of other limbic structures, and the same holds true for the cholinergic and GABAergic cell groups in the basal forebrain (see Chap. 22). An important overall function of these cell groups is to optimize the activity of the CNS, or parts thereof, during the execution of motivated or goal-oriented behaviours [4, 83]. For example, the pontine and medullary raphe nuclei give rise to diffusely organized descending pathways that terminate among almost all parts of the spinal grey matter, including the dorsal horn and the autonomic and somatic motoneural cell groups. According to Holstege [51, 58], these descending fibres do not produce specific movements, but act as a "level-setting" system. He adduces physiological evidence, indicating that they enhance the excitability of the motoneurons for inputs from other sources, but suppress the transmission of nociceptive impulses through the dorsal horn. Conceivably, these influences enable an animal in a life threatening situation to combine maximal motor activity with neglect of painful stimuli.

Concerning the third category, the limbic motor network, which may be denoted as the executive part of the limbic motor system, is hierarchically organized (Fig. 23.10). Our present-day knowledge of this network is largely based on the seminal studies of Holstege and colleagues [51–53, 58, 133, 134] and Swanson and colleagues [21, 115, 123, 124, 127–129]. Much of what follows is derived from these studies.

The highest level within the hierarchy of the limbic motor network is represented by the so-called *behavioural control column* (BCC) [124, 128]. This column consists of the series of cell masses which together form the medial preoptico-hypothalamic zone (Table 10.1). As discussed in Chap. 10, these cell masses are nodes in circuit systems that regulate the three classes of basic motivated or goal-oriented behaviours required for survival of the individual and of the species: ingestive, agonistic and reproductive. All of these behaviours require somatic, autonomic and neuroendocrine responses, and all usually pass through three sequential phases: initiation, procurement and consummatory [126, 128].

The lowest level in the hierarchy of the limbic motor network is represented by the somatic, autonomic and neuroendocrine motor systems. The *somatic motor system* regulates the contractions of skeletal or voluntary muscle. It consists of pools of motoneurons that extend from the midbrain rostrally to the caudal end of the spinal cord. The *autonomic visceromotor system* is formed by pools of preganglionic sympathetic and parasympathetic neurons, which together also stretch, with some interruptions, from the midbrain to the caudal end of the spinal cord. These neurons innervate, via elements situated in peripheral ganglia and plexuses, cardiac muscle, smooth muscle and glands. The *neuroendocrine motor system* is centred in and around the periventricular zone of the hypothalamus. It consists of pools of secretomotor neurons, which fall into two broad classes, magnocellular neurosecretory and parvocellular neurosecre-

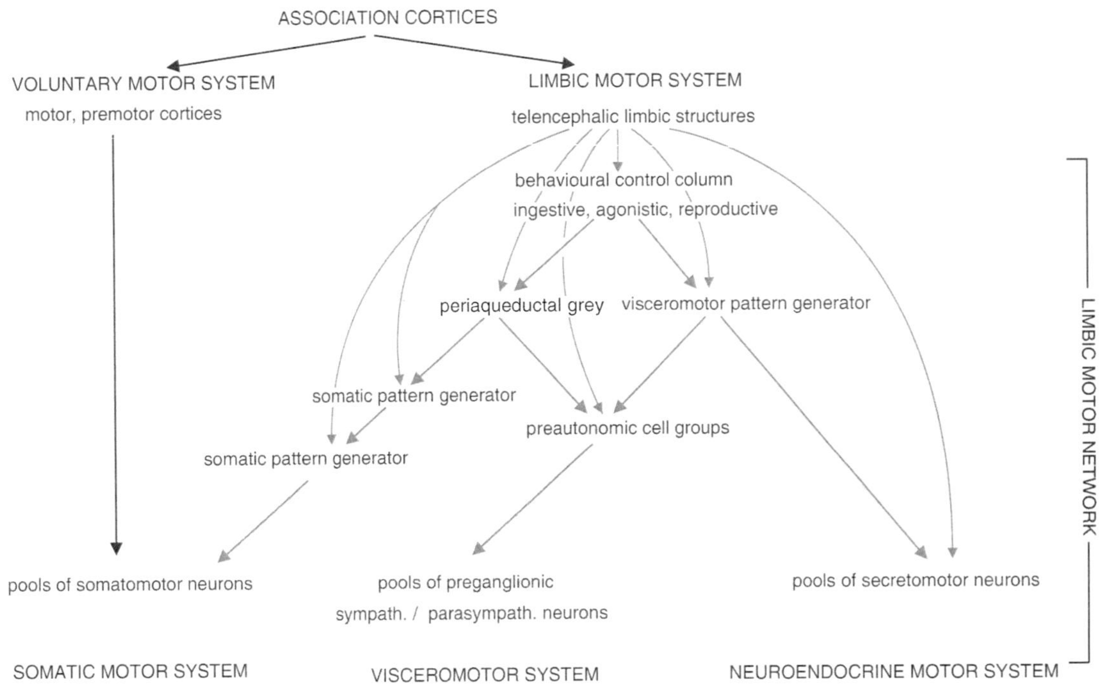

Fig. 23.10. Schematic overview of the organization of the limbic motor network and its principal afferents from telencephalic limbic structures, and of the relationships between the limbic and voluntary motor systems

tory. As discussed in Chap. 10, the magnocellular neurosecretory elements synthesize either the nonapeptide vasopressin (anti-diuretic hormone) or the nonapeptide oxytocin. They project to the posterior pituitary, where their products are released into the bloodstream. There are six different types of parvocellular neurosecretory elements, each of which produces a particular hypophysiotropic hormone. These parvocellular elements regulate the secretion of the various anterior-pituitary hormones, via the hypothalamo-hypophysial portal system.

The pathways descending from the hypothalamic BCC to the pools of motoneurons just described pass via numerous intercalated centres which, somewhat schematically, can be classed as higher intermediate and lower intermediate. The lower intermediate centres encompass several pre-autonomic cell groups and a large number of somatomotor pattern generators.

The *pre-autonomic cell groups* include the paraventricular hypothalamic nucleus, a group of cells in the intermediate part of the lateral hypothalamic zone, Barrington's pontine micturition centre and the caudal, cardiovascular portion of the ventrolateral medulla. As the name of the group indicates, all of these cell masses project to groups of preganglionic sympathetic or parasympathetic neurons in the lower brain stem and/or spinal cord.

The group of *somatomotor pattern generators* consists of a large number of diffusely arranged cell groups, most of which are embedded in the lateral tegmental field (Fig. 23.7). Most of these pattern generators, among which are those subserving locomotion, vocalization, swallowing and respiration, are represented by two or three hierarchically arranged loci, rather than by a single cell group (Chap. 21). The somatic pattern generators project, either directly or indirectly, to pools of somatic motoneurons which innervate specific groups of muscles.

The group of higher intermediate centres encompasses the hypothalamic, visceral pattern generation and the periaqueductal grey matter.

The *hypothalamic visceromotor pattern generator (HVPG)* was recently detected by Swanson and Thompson [124, 128] in the brain of the rat. These authors found that in this species, the hypothalamic zone between the array of discrete nuclei forming the BCC and the periventricular neuroendocrine motor zone is occupied by a network of highly interconnected cell groups which includes, apart from the larger dorsomedial hypothalamic nucleus (DMH), five small nuclei in the preoptic region. The BCC projects in a highly topographic fashion to the adjacent HVPG network, which in turn issues a complex series of projections to various combinations of neuroendocrine motoneuron pools and to particular pre-autonomic cell groups in the paraventricular hypothalamic nucleus. According to Thompson and Swanson ([128], p. 198), "this arrangement is obviously in a position to allow for the coordinate expression of somatomotor, autonomic and neuroendocrine responses that are appropriate for the (three) classes of motivated behavior ...." It is noteworthy that Thompson and Swanson [128] place the DMH, with its incorporation in the HVPG, in an entirely new functional context. In previous publications, cited in Chap. 10, this nucleus has been implicated in the mediation of circadian rhythms (Fig. 10.4) and ingestive behaviours (Figs. 10.7B, 10.8). It is also noteworthy that the various components of the HVPG do not only receive topographically organized projections from the adjacent BCC, but are also differentially innervated by numerous other limbic grisea, including the infralimbic cortex, subiculum, bed nucleus of the stria terminalis, lateral septal nucleus, suprachiasmatic nucleus and parabrachial nucleus [128].

The *mesencephalic periaqueductal grey (PAG)*, as already mentioned, plays a pivotal role in coordinating visceral and behavioural responses to escapable and inescapable threatening situations and is also involved in the control of vocalization, micturition and mating behaviour. It receives afferents from the BCC and projects to numerous lower intermediate centres, including the pontine micturition nucleus of Barrington, the caudal ventrolateral medulla and the retroambiguus nucleus. The latter acts as a somatic pattern generator for both vocalization and mating behaviour.

In concluding this section devoted to the organization of the limbic motor system and its executive part, denoted as the limbic motor network, four general remarks should be made.

1. The picture presented here of the limbic motor network is greatly simplified. In light of the detailed studies of Swanson and collaborators, it could have been extended with countless additional connections, shunts and bypasses. The complexity of the network under consideration may be illustrated by the following two observations: (1) a single hypothalamic nucleus, the DMH, appeared to receive direct projections from at least 40 different brain sites, the vast majority of which were located within the hypothalamus itself [127], and (2) the projection pattern of the small magnocellular part of the bed nucleus of the stria terminalis appeared to include more than 80 distinct terminal fields [38].
2. The connections forming the limbic motor network exhibit, apart from the complexity just alluded to, a remarkable specificity and precision. Thus, Thompson and Swanson ([128], p. 113) reported that each of the six nuclei embedded in the HVPG "generates a pattern of terminal fields that differentially targets a unique set of hypothalamic neuroendocrine motor pools, and of preautonomic parts of the paraventricular nucleus". As regards the organization of the latter nucleus, Strack et al. [119] found that selective clusters of paraventricular neurons address specific sets of preganglionic neurons.
3. A similar specificity and precision is also manifest in the organization of the fibre systems that descend from telencephalic limbic structures to the limbic motor network. One of these is the BST, and we have seen already that this nucleus as a whole innervates numerous centres in the hypothalamus and brain stem (Table 23.1). However, this nucleus is structurally not homogeneous but rather it is composed of a considerable number of separate cell groups. Dong and Swanson [33–38] thoroughly analysed the connections of many of these BST units in the rat. A detailed discussion of the results of these studies is beyond the scope of the present work. However, an important general outcome was that most of these units preferentially project to parts or sectors of the BCC, HVPG network, neuroendocrine motor zone, preautonomic cell groups and somatomotor pattern generators involved in the control of particular basic, vital functions or behaviours. Thus, it was found that one unit (N.B.: the principal nucleus of the posterior division of the BST) preferentially innervates structures that control reproductive behaviour [35], whereas another unit (the anterodorsal area of the BST), targets structures involved in maintaining energy balance and homeostasis [36]. Comparable functional specializations have been observed in the organization of the projections from the lateral septal complex to the medial preoptico-hypothalamic zone [106] and in those from the medial prefrontal and caudal orbitofrontal cortices to the periaqueductal grey [3].
4. Finally, it should be emphasized that the limbic and voluntary motor systems do not operate independently of each other. The sites of origins of these two systems receive afferents from the highest level of the brain, represented by the association cortices, and at the lowest level, they share a final common pathway, the somatomotor system (Fig. 23.10).

Concluding Remarks

Although the term limbic system is widely used in many areas of the neurosciences, the use and the validity of the term and of the concept underlying it have been endlessly debated in the literature. Essentially, criticism of the concept boils down to the following three points: (1) An adequate empirical definition of the limbic system is lacking (e.g. LeDoux [75], Kötter and Meyer [70], Groenewegen et al. [42], Blessing [14]); (2) it is on its way to becoming a synonym for the brain and its workings as a whole (e.g. Brodal [18]) and (3) its superior and inferior ends cannot be sharply defined (e.g. Swanson [122]).

As regards definition, we believe that with the aid of the structural and functional criteria presented above the GLS can be delineated and recognized from the level of telencephalic structures such as the amygdala, the septum and the BST to the level of the obex (see also Herbert [47]).

The criticism of Brodal ([18], p. 690) that as our knowledges increases, the limbic system "appears to be on its way to including all brain regions and functions" is untenable in light of the data discussed in the present chapter. Particularly, the recent studies with modern tract-tracing and immunohistochemical techniques have shown that the GLS, with its loose-textured bundles of thin, mostly unmyelinated fibres and its accumulation of peptidergic neurons, axons and terminals stands in sharp contrast to the "classical" sensory and motor systems.

As regards the third point of criticism, it is true that the superior and inferior ends of the limbic system cannot be sharply defined and have shifted in the course of time. Broca's [17] limbic lobe, encompassing the cingulate and the parahippocampal gyri, which formed an essential part of Papez's [99] "emotional" circuit, and of the limbic system as defined by MacLean [77], were excluded by Nauta [89] from limbic circuitry. According to current views the hypothalamus forms the central and principal part of the limbic system, whereas the cingulate and parahippocampal gyri are considered to form an interface between limbic and neocortical circuitry. The shift of the caudal limit of the limbic system has been even more dramatic. MacLean [77–79] considered the amygdala, the septum and the anterior thalamic nucleus (which project to the cingulate gyrus) as the principal nodal points in the limbic circuitry. Nauta [88, 89, 91] included the hypothalamus and the paramedian midbrain in this circuitry whereas Nieuwenhuys and colleagues [93, 95, 96] added a number of rhombencephalic structures.

In our view the 'mobility' and the 'vagueness' of the upper and lower limits of limbic circuitry do not militate against the GLS concept. As for mobility, it is in the very nature of living concepts to evolve. As for vagueness, not a single functional system in the brain shows sharp and distinct boundaries throughout. The fact that we are unable to indicate where sensibility ends and motricity begins in the neocortical circuitry does not invalidate the concepts of somatosensory, auditory, visual and voluntary motor systems.

References

1. Aicher SA, Schreihofer AM, Kraus JA et al (2001) Muopioid receptors are present in functionally identified sympathoexcitatory neurons in the rat rostral ventrolateral medulla. J Comp Neurol 433:34–47
2. Amaral DG, Price JL, Pitkänen A, Carmichael ST (1992) Anatomical organization of the primate amygdaloid complex. In: Aggleton JP (ed) The amygdala: neurobiological aspects of emotion, memory, and mental dysfunction. Wiley-Liss, New York, pp 1–66
3. An X, Bandler R, Öngür D, Price JL (1998) Prefrontal cortical projections to longitudinal columns in the midbrain periaqueductal gray in macaque monkeys. J Comp Neurol 401:455–479
4. Aston-Jones G, Cohen JD (2005) An integrative theory of locus coeruleus-norepinephrine function: adaptive gain and optimal performance. Annu Rev Neurosci 28:403–450
5. Bandler R, Keay KA (1996) Columnar organization in the midbrain periaqueductal gray and the integration of emotional expression. Progr Brain Res 107:285–300
6. Bandler R, Carrive P, Zhang SP (1991) Integration of somatic and autonomic reactions within the midbrain periaqueductal grey: viscerotopic, somatotopic and functional organization. Prog Brain Res 87:269–305
7. Bard P (1929) The central representation of the sympathetic nervous system as indicated by certain physiologic observations. Arch Neurol Psychiatry 22:230–246
8. Barrington FJF (1925) The effect of lesions of the hind- and mid-brain on micturition in the cat. Q J Exp Physiol Cogn Med 15:81–102
9. Baude A, Shigemoto R (1998) Cellular and subcellular distribution of substance P receptor immunoreactivity in the dorsal vagal complex of the rat and cat: a light and electron microscope study. J Comp Neurol 402:181–196
10. Berk ML, Finkelstein JA (1981) Afferent projections to the preoptic area and hypothalamic regions in the rat brain. Neuroscience 6:1601–1624
11. Bernard JF, Alden M, Besson JM (1993) The organization of the efferent projections from the pontine parabrachial area to the amygdaloid complex: a Phaseolus vulgaris leucoagglutinin (PHA-L) study in the rat. J Comp Neurol 329:201–229
12. Bernard JF, Bester H, Besson JM (1996) Involvement of the spino-parabrachio -amygdaloid and -hypothalamic pathways in the autonomic and affective emotional aspects of pain. Prog Brain Res 107:243–255
13. Blaustein JD (1992) Cytoplasmic estrogen receptors in rat brain: immunocytochemical evidence using three antibodies with distinct epitopes. Endocrinology 131:1336–1342
14. Blessing WW (1997) Inadequate frameworks for understanding bodily homeostasis. Trends Neurosci 20:235–239
15. Blok BF, Willemsen AT, Holstege G (1997) A PET study on brain control of micturition in humans. Brain 120 (Pt 1):111–121
16. Blok BF, Sturms LM, Holstege G (1998) Brain activation during micturition in women. Brain 121 (Pt 11):2033–2042
17. Broca P (1878) Anatomie comparée des circonvolutions cerebrales: le grand lobe limbique et la scissure dans la serie des mammiferes. Rev Anthropol (Paris) 2:285–498
18. Brodal A (1981) Neurological anatomy in relation to clinical medicine, 3rd edn. Oxford University Press, New York, p 1053
19. Buma P, Veening J, Hafmans T, Joosten H, Nieuwenhuys R (1992) Ultrastructure of the periaqueductal grey matter of the rat: an electron microscopical and horseradish peroxidase study. J Comp Neurol 319:519–535
20. Burstein R (1996) Somatosensory and visceral input to the hypothalamus and limbic system. Prog Brain Res 107:257–267
21. Canteras NS, Simerly RB, Swanson LW (1992) Projections of the ventral premamillary nucleus. J Comp Neurol 324:195–212
22. Carrive P, Morgan MM (2004) Periaqueductal gray. In: Paxinos G, Mai JK (eds) The human nervous system, 2nd edn. Elsevier, Amsterdam, pp 393–423
23. Castel M, Feinstein N, Cohen S, Harari N (1990) Vasopressinergic innervation of the mouse suprachiasmatic nucleus: an immuno-electron microscopic analysis. J Comp Neurol 298:172–187
24. Chiba T, Kayahara T, Nakano K (2001) Efferent projections of infralimbic and prelimbic areas of the medial prefrontal cortex in the Japanese monkey, Macaca fuscata. Brain Res 888:83–101
25. Ciriello J, Caverson MM (1984) Direct pathway from neurons in the ventrolateral medulla relaying cardiovascular afferent information to the supraoptic nucleus in the cat. Brain Res 292:221–228
26. Coolen LM, Allard J, Truitt WA, McKenna KE (2004) Central regulation of ejaculation. Physiol Behav 83:203–215
27. Craig AD (1995) Distribution of brainstem projections from spinal lamina I neurons in the cat and the monkey. J Comp Neurol 361:225–248
28. Craig AD (1996) An ascending general homeostatic afferent pathway originating in lamina I. Prog Brain Res 107:225–242
29. Cunningham ET Jr, Sawchenko PE (2000) Dorsal medullary pathways subserving oromotor reflexes in the rat: implications for the central neural control of swallowing. J Comp Neurol 417:448–466
30. Dana C, Vial M, Leonard K et al (1989) Electron microscopic localization of neurotensin binding sites in the midbrain tegmentum of the rat. I. Ventral tegmental area and the interfascicular nucleus. J Neurosci 9:2247–2257

31. De Wied D (1987) Neuropeptides and behavior. In: Adelman G (ed) Encyclopedia of neuroscience, vol II. Birkhäuser, Boston, pp 839–841
32. Dietrichs E, Haines DE (1989) Interconnections between hypothalamus and cerebellum. Anat Embryol (Berl) 179:207–220
33. Dong HW, Swanson LW (2003) Projections from the rhomboid nucleus of the bed nuclei of the stria terminalis: implications for cerebral hemisphere regulation of ingestive behaviors. J Comp Neurol 463:434–472
34. Dong HW, Swanson LW (2004) Organization of axonal projections from the anterolateral area of the bed nuclei of the stria terminalis. J Comp Neurol 468:277–298
35. Dong HW, Swanson LW (2004) Projections from bed nuclei of the stria terminalis, posterior division: implications for cerebral hemisphere regulation of defensive and reproductive behaviors. J Comp Neurol 471:396–433
36. Dong HW, Swanson LW (2006) Projections from bed nuclei of the stria terminalis, anteromedial area: cerebral hemisphere integration of neuroendocrine, autonomic, and behavioral aspects of energy balance. J Comp Neurol 494:142–178
37. Dong HW, Swanson LW (2006) Projections from bed nuclei of the stria terminalis, dorsomedial nucleus: implications for cerebral hemisphere integration of neuroendocrine, autonomic, and drinking responses. J Comp Neurol 494:75–107
38. Dong HW, Swanson LW (2006) Projections from bed nuclei of the stria terminalis, magnocellular nucleus: implications for cerebral hemisphere regulation of micturition, defecation, and penile erection. J Comp Neurol 494:108–141
39. Fulwiler CE, Saper CB (1984) Subnuclear organization of the efferent connections of the parabrachial nucleus in the rat. Brain Res Rev 7:229–259
40. Garcia-Rill E (1991) The pedunculopontine nucleus. Prog Neurobiol 36:363–389
41. Garcia-Rill E, Kinjo N, Atsuta Y et al (1990) Posterior midbrain-induced locomotion. Brain Res Bull 24:499–508
42. Groenewegen HJ, Wright CI, Beijer AV (1996) The nucleus accumbens: gateway for limbic structures to reach the motor system? Prog Brain Res 107: 485–511
43. Hage SR, Jürgens U (2006) Localization of a vocal pattern generator in the pontine brainstem of the squirrel monkey. Eur J Neurosci 23:840–844
44. Hage SR, Jürgens U (2006) On the role of the pontine brainstem in vocal pattern generation: a telemetric single-unit recording study in the squirrel monkey. J Neurosci 26:7105–7115
45. Herbert H, Moga MM, Saper CB (1990) Connections of the parabrachial nucleus with the nucleus of the solitary tract and the medullary reticular formation in the rat. J Comp Neurol 293:540–580
46. Herbert J (1993) Peptides in the limbic system: neurochemical codes for co-ordinated adaptive responses to behavioural and physiological demand. Prog Neurobiol 41:723–791
47. Herbert J (1997) Do we need a limbic system? Trends Neurosci 20:508–509
48. Hess WR (1954) Diencephalon: autonomic and extrapyramidal functions. Grune and Stratton, New York
49. Holstege G (1987) Some anatomical observations on the projections from the hypothalamus to brainstem and spinal cord: an HRP and autoradiographic tracing study in the cat. J Comp Neurol 260:98–126
50. Holstege G (1989) Anatomical study of the final common pathway for vocalization in the cat. J Comp Neurol 284:242–252
51. Holstege G (1991) Descending motor pathways and the spinal motor system: limbic and non-limbic components. Prog Brain Res 87:307–421
52. Holstege G (1992) The emotional motor system. Eur J Morphol 30:67–79
53. Holstege G (1998) The emotional motor system in relation to the supraspinal control of micturition and mating behavior. Behav Brain Res 92:103–109
54. Holstege G, Georgiadis JR (2003) Neurobiology of cat and human sexual behavior. Int Rev Neurobiol 56:213–225
55. Holstege G, Graveland G, Bijker-Biemond C, Schuddeboom I (1983) Location of motoneurons innervating soft palate, pharynx and upper esophagus. Anatomical evidence for a possible swallowing center in the pontine reticular formation. An HRP and autoradiographical tracing study. Brain Behav Evol 23:47–62
56. Holstege G, Meiners L, Tan K (1985) Projections of the bed nucleus of the stria terminalis to the mesencephalon, pons, and medulla oblongata in the cat. Exp Brain Res 58:379–391
57. Holstege G, Griffiths D, De Wall H, Dalm E (1986) Anatomical and physiological observations on supraspinal control of bladder and urethral sphincter muscles in the cat. J Comp Neurol 250:449–461
58. Holstege G, Mouton LJ, Gerrits NM (2004) Emotional motor system. In: Paxinos G, Mai JK (eds) The human nervous system, 2nd edn. Elsevier, Amsterdam, pp 1306–1324
59. Hopkins DA, Holstege G (1978) Amygdaloid projections to the mesencephalon, pons and medulla oblongata in the cat. Exp Brain Res 32:529–548
60. Hosoya Y, Matsushita M (1981) Brainstem projections from the lateral hypothalamic area in the rat, as studied with autoradiography. Neurosci Lett 24:111–116
61. Jhamandas JH, Petrov T, Harris KH, Vu T, Krukoff TL (1996) Parabrachial nucleus projection to the amygdala in the rat: electrophysiological and anatomical observations. Brain Res Bull 39:115–126
62. Jia HG, Rao ZR, Shi JW (1994) An indirect projection from the nucleus of the solitary tract to the central nucleus of the amygdala via the parabrachial nucleus in the rat: a light and electron microscopic study. Brain Res 663:181–190

63. Jones BE, Friedman L (1983) Atlas of catecholamine perikarya, varicosities and pathways in the brainstem of the cat. J Comp Neurol 215:382–396
64. Jürgens U (1974) The hypothalamus and behavioral patterns. Prog Brain Res 41:445–463
65. Jürgens U, Ploog D (1970) Cerebral representation of vocalization in the squirrel monkey. Exp Brain Res 10:532–554
66. Klemm WR, Vertes RP (1990) Brainstem mechanisms of behavior. Wiley, New York
67. Klüver H, Bucy PC (1937) Psychic blindness and other symptoms following bilateral temporal lobectomy in rhesus monkeys. Am J Physiol 119:352–353
68. Klüver H, Bucy PC (1939) Preliminary analysis of functions of the temporal lobes in monkeys. Arch Neurol Psychiatry 42:979–1000
69. Koch M, Ehret G (1989) Immunocytochemical localization and quantitation of estrogen-binding cells in the male and female (virgin, pregnant, lactating) mouse brain. Brain Res 489:101–112
70. Kötter R, Meyer N (1992) The limbic system: a review of its empirical foundation. Behav Brain Res 52:105–127
71. Kuipers R, Mensinga GM, Boers J, Klop EM, Holstege G (2006) Infralimbic cortex projects to all parts of the pontine and medullary lateral tegmental field in cat. Eur J Neurosci 23:3014–3024
72. Larsen PJ, Hay-Schmidt A, Mikkelsen JD (1994) Efferent connections from the lateral hypothalamic region and the lateral preoptic area to the hypothalamic paraventricular nucleus of the rat. J Comp Neurol 342:299–319
73. Leah J, Menetrey D, De Pommery J (1988) Neuropeptides in long ascending spinal tract cells in the rat: evidence for parallel processing of ascending information. Neuroscience 24:195–207
74. LeDoux JE (1987) Emotion. In: Plum F (ed) Higher functions of the brain. Handbook of physiology 1, The nervous system, vol V. American Physiological Society, Bethesda, MD, pp 419–460
75. LeDoux JE (1991) Emotion and the limbic system concept. Concepts Neurosci 2:169–199
76. Lehman M, Silver R (2000) CSF signaling in physiology and behavior. Prog Brain Res 125:415–433
77. MacLean PD (1952) Some psychiatric implications of physiological studies on frontotemporal portion of limbic system (visceral brain). Electroencephalogr Clin Neurophysiol [Suppl 4]:407–418
78. MacLean PD (1970) The triune brain, emotion and scientific bias. In: Schmitt FO (ed) The neurosciences second study program. Rockefeller University Press, New York, pp 336–349
79. MacLean PD (1990) The triune brain in evolution: Role in paleocerebral function. Plenum, New York
80. MacLean PD (1992) The limbic system concept. In: Trimble MR, Bolwig TG (eds) The temporal lobes and the limbic system. Wrightson, Petersfield, pp 1–14
81. Madeira MD, Lieberman AR (1995) Sexual dimorphism in the mammalian limbic system. Prog Neurobiol 45:275–333
82. Mesulam MM (1985) Principles of behavioral neurology. Davis, Philadelphia
83. Mesulam MM (1998) From sensation to cognition. Brain 121:1013–1052
84. Miller AJ (1982) Deglutition. Physiol Rev 62:129–184
85. Moga MM, Herbert H, Hurley KM et al (1990) Organization of cortical, basal forebrain, and hypothalamic afferents to the parabrachial nucleus in the rat. J Comp Neurol 295:624–661
86. Mogenson GJ, Jones DL, Yim CY (1980) From motivation to action: functional interface between the limbic system and the motor system. Prog Neurobiol 14:69–97
87. Morris JF, Pow DV (1991) Widespread release of peptides in the central nervous system: quantitation of tannic acid-captured exocytoses. Anat Rec 231: 437–445
88. Nauta WJH (1958) Hippocampal projections and related neural pathways to the midbrain in the cat. Brain 81:319–340
89. Nauta WJH (1973) Connections of the frontal lobe with the limbic system. In: Laitinen L V, Livingston KE (eds) Surgical approaches in psychiatry. Medical and Technical Publishing, Lancester, pp 303–314
90. Nauta WJH (1979) Expanding borders of the limbic system concept. In: Rasmussen T, Marino R (eds) Functional neurosurgery. Raven, New York, pp 7–24
91. Nauta WJH, Haymaker W (1969) Hypothalamic nuclei and fiber connections. In: Haymaker W, Anderson E, Nauta WJH (eds) The hypothalamus. Thomas, Springfield, pp 136–209
92. Nicholson C (1999) Signals that go with the flow. Trends Neurosci 22:143–145
93. Nieuwenhuys R (1985) Chemoarchitecture of the brain. Springer, Berlin Heidelberg New York
94. Nieuwenhuys R, Geeraedts LMG, Veening JG (1982) The medial forebrain bundle of the rat. I. General introduction. J Comp Neurol 206:49–81
95. Nieuwenhuys R, Veening JG, Van Domburg P (1988) Core and paracores; some new chemoarchitectural entities in the mammalian neuraxis. Acta Morphol Neerl Scand 26:131–163
96. Nieuwenhuys R, Voogd J, Van Huijzen C (1988) The human central nervous system: a synopsis and atlas, 3rd edn. Springer, Berlin Heidelberg New York
97. Olszewski J, Baxter D (1954) Cytoarchitecture of the human brain stem. Karger, Basel
98. Öngür D, An X, Price JL (1998) Prefrontal cortical projections to the hypothalamus in macaque monkeys. J Comp Neurol 401:480–505
99. Papez JW (1937) A proposed mechanism of emotion. Arch Neurol Psychiatry 38:725–743
100. Pijnenburg AJJ, Honig WMM, Van der Heyden JAM, Van Rossum JM (1976) Effects of chemical stimulation of the mesolimbic dopamine system upon locomotor activity. Eur J Pharmacol 35:45–58

101. Pow DV, Morris JF (1989) Dendrites of hypothalamic magnocellular neurons release neurohypophysial peptides by exocytosis. Neuroscience 32:435–439
102. Price JL (2004) Olfaction. In: Paxinos G, Mai JK (ed) The human nervous system, 2nd edn. Elsevier, Amsterdam, pp 1197–1211
103. Price JL, Amaral DG (1981) An autoradiographic study of the projections of the central nucleus of the monkey amygdala. J Neurosci 1:1242–1259
104. Rennels ML, Gregory TF, Blaumanis OR, Fujimoto K, Grady PA (1985) Evidence for a 'paravascular' fluid circulation in the mammalian central nervous system, provided by the rapid distribution of tracer protein throughout the brain from the subarachnoid space. Brain Res 326:47–63
105. Ricardo JA, Koh ET (1978) Anatomical evidence of direct projections from the nucleus of the solitary tract to the hypothalamus and the amygdala, and other forebrain structures in the rat. Brain Res 153:1–26
106. Risold PY, Swanson LW (1997) Connections of the rat lateral septal complex. Brain Res Rev 24: 115–195
107. Roeling TA, Veening JG, Kruk MR et al (1994) Efferent connections of the hypothalamic "aggression area" in the rat. Neuroscience 59:1001–1024
108. Ross CA, Ruggiero DA, Reis DJ (1985) Projections from the nucleus tractus solitarii to the rostral ventrolateral medulla. J Comp Neurol 242: 511–534
109. Saether K, Hilaire G, Monteau R (1987) Dorsal and ventral respiratory groups of neurons in the medulla of the rat. Brain Res 419:87–96
110. Saper CB (2002) The central autonomic nervous system: conscious visceral perception and autonomic pattern generation. Annu Rev Neurosci 25:433–469
111. Saper CB, Loewy AD (1980) Efferent connections of the parabrachial nucleus in the rat. Brain Res 197:291–317
112. Sawchenko PE, Swanson LW (1982) The organization of noradrenergic pathways from the brainstem to the paraventricular and supraoptic nuclei in the rat. Brain Res Rev 4:275–325
113. Shaikh MB, Brutus M, Siegel HE, Siegel A (1985) Topographically organized midbrain modulation of predatory and defensive aggression in the cat. Brain Res 336:308–312
114. Shik ML, Severin FV, Orlovskii GN (1966) Upravlenie khod'boi i begom posredstvom elektricheskoi stimulatsii srednego mozga (Control of walking and running by means of electric stimulation of the midbrain). Biofizika 11:659–666
115. Simerly RB, Swanson LW (1988) Projections of the medial preoptic nucleus: a Phaseolus vulgaris leucoagglutinin anterograde tract-tracing study in the rat. J Comp Neurol 270:209–242
116. Slugg RM, Light AR (1994) Spinal cord and trigeminal projections to the pontine parabrachial region in the rat as demonstrated with Phaseolus vulgaris leucoagglutinin. J Comp Neurol 339:49–61
117. Smith DA, Flynn JP (1979) Afferent projections related to attack sites in the pontine tegmentum. Brain Res 164:103–119
118. Sofroniew MY (1983) Direct reciprocal connections between the bed nucleus of the stria terminalis and dorsomedial medulla oblongata: evidence from immunohistochemical detection of tracer proteins. J Comp Neurol 213:399–405
119. Strack AM, Sawyer WB, Hughes JH, Platt KB, Loewy AD (1989) A general pattern of CNS innervation of the sympathetic outflow demonstrated by transneuronal pseudorabies viral infections. Brain Res 491:156–162
120. Stumpf WE (1975) The brain: an endocrine gland and hormone target. In: Stumpf WE, Grant LO (eds) Anatomical neuroendocrinology. Karger, Basel, pp 2–8
121. Stumpf WE, Sar M (1978) Anatomical distribution of estrogen, androgen, progestin, corticosteroid and thyroid hormone target sites in the brain of mammals: phylogeny and ontogeny. Amer Zool 18:435–445
122. Swanson LW (1987) Limbic system. In: Adelman G (ed) Encyclopedia of neuroscience, vol I. Birkhäuser, Boston, pp 589–591
123. Swanson LW (1987) The hypothalamus. In: Björklund A, Swanson LW (eds) Integrated systems of the CNS, part I. Elsevier, Amsterdam, pp 125–277 (Handbook of Chemical Neuroanatomy, vol 5)
124. Swanson LW (2000) Cerebral hemisphere regulation of motivated behavior. Brain Res 886:113–164
125. Swanson LW, Cowan WM (1979) The connections of the septal region in the rat. J Comp Neurol 186:621–655
126. Swanson LW, Mogenson GJ (1981) Neural mechanisms for the functional coupling of autonomic, endocrine and somatomotor responses in adaptive behavior. Brain Res Rev 3:1–34
127. Thompson RH, Swanson LW (1998) Organization of inputs to the dorsomedial nucleus of the hypothalamus: a reexamination with Fluorogold and PHAL in the rat. Brain Res Rev 27:89–118
128. Thompson RH, Swanson LW (2003) Structural characterization of a hypothalamic visceromotor pattern generator network. Brain Res Rev 41:153–202
129. Thompson RH, Canteras NS, Swanson LW (1996) Organization of projections from the dorsomedial nucleus of the hypothalamus: a PHA-L study in the rat. J Comp Neurol 376:143–173
130. Van Bockstaele EJ, Colago EE, Cheng P et al (1996) Ultrastructural evidence for prominent distribution of the mu-opioid receptor at extrasynaptic sites on noradrenergic dendrites in the rat nucleus locus coeruleus. J Neurosci 16:5037–5048
131. Van der Kooy D, Koda LY, McGinty JF, Gerfen CR, Bloom FE (1984) The organization of projections from the cortex, amygdala, and hypothalamus to the nucleus of the solitary tract in rat. J Comp Neurol 224:1–24

132. Van der Horst VGJM, Mouton LJ, Blok BF, Holstege G (1996) Distinct cell groups in the lumbosacral cord of the cat project to different areas in the periaqueductal gray. J Comp Neurol 376:361–385
133. Van der Horst VGJM, Terasawa E, Ralston HJ 3rd, Holstege G (2000) Monosynaptic projections from the nucleus retroambiguus to motoneurons supplying the abdominal wall, axial, hindlimb, and pelvic floor muscles in the female rhesus monkey. J Comp Neurol 424:233–250
134. Van der Horst VGJM, Terasawa E, Ralston HJ 3rd, Holstege G (2000) Monosynaptic projections from the lateral periaqueductal gray to the nucleus retroambiguus in the rhesus monkey: implications for vocalization and reproductive behavior. J Comp Neurol 424:251–268
135. Van der Horst VGJM, Terasawa E, Ralston HJ 3rd (2001) Monosynaptic projections from the nucleus retroambiguus region to laryngeal motoneurons in the rhesus monkey. Neuroscience 107:117–125
136. Veening J, Buma P, Ter Horst GJ et al (1991) Hypothalamic projections to the PAG in the rat: topographical, immuno-electronmicroscopical and functional aspects. In: Depaulis A, Bandler R (eds) The midbrain periaqueductal gray matter. Plenum, New York, pp 387–415
137. Veening JG, Swanson LW, Cowan WM, Nieuwenhuys R, Geeraedts LMG (1982) The medial forebrain bundle of the rat. II. An autoradiographic study of the topography of the major descending and ascending components. J Comp Neurol 206: 82–108
138. Victor M, Agamanolis D (1990) Amnesia due to lesions confined to the hippocampus: a clinical-pathologic study. J Cognit Neurosci 2:246–257
139. Wood RI, Brabec RK, Swann JM, Newman SW (1992) Androgen and estrogen concentrating neurons in chemosensory pathways of the male Syrian hamster brain. Brain Res 596:89–98
140. Zagon A, Totterdell S, Jones RS (1994) Direct projections from the ventrolateral medulla oblongata to the limbic forebrain: anterograde and retrograde tract-tracing studies in the rat. J Comp Neurol 340:445–468
141. Zhang SP, Davis PJ, Bandler R, Carrive P (1994) Brain stem integration of vocalization: role of the midbrain periaqueductal gray. J Neurophysiol 72:1337–1356
142. Zola-Morgan S, Squire LR (1986) Memory impairment in monkeys following lesions limited to the hippocampus. Behav Neurosci 100:155–160

Subject Index

B

C

D

E

F

G

H

I

J

K

L

N

O

P

R

S

T

U

V

W

X

Y

Z

Kolibri 360 GmbH
Ferdinand-Jühlke-Straße 7
99095 Erfurt, Deutschland
produktsicherheit@kolibri360.de